D1519121

Handbook of Conducting Polymers

Handbook of Conducting Polymers

Second Edition, Revised and Expanded

Edited by

TERJE A. SKOTHEIM
Moltech Corporation
Tucson, Arizona

RONALD L. ELSENBAUMER
The University of Texas at Arlington
Arlington, Texas

JOHN R. REYNOLDS
University of Florida
Gainesville, Florida

MARCEL DEKKER, INC.

NEW YORK · BASEL · HONG KONG

ISBN: 0-8247-0050-3

The publisher offers discounts on this book when ordered in bulk quantities. For more information, write to Special Sales/Professional Marketing at the address below.

This book is printed on acid-free paper.

Copyright © 1998 by MARCEL DEKKER, INC. All Rights Reserved.

Neither this book nor any part may be reproduced or transmitted in any form or by any means, electronic or mechanical, including photocopying, microfilming, and recording, or by any information storage and retrieval system, without permission in writing from the publisher.

MARCEL DEKKER, INC.
270 Madison Avenue, New York, New York 10016
http://www.dekker.com

Current printing (last digit):
10 9 8 7 6 5 4 3 2 1

PRINTED IN THE UNITED STATES OF AMERICA

Preface

The field of electrically conducting and electroactive polymers has developed and expanded at a rapid pace since the publication of the first edition of the *Handbook of Conducting Polymers* in 1986. Building on the exciting discovery of high conductivity in doped polyacetylene in 1977, researchers have continued to develop important structure–property relationships, prepared more well-characterized and ordered materials, unfolded novel electrical and optical properties of these materials, and developed a much deeper understanding of the fundamental processes that lead to electron transport and photon interactions in conjugated polymers. Significant strides have also been made in advancing the technological applications of these materials and their introduction into the marketplace.

The large number of scientific disciplines represented in conducting polymer reseach include many subfields in the areas of chemistry, physics, materials science, and engineering. In fact, extensions into biochemistry, biology, and medicine have now occurred. This has led to a significant amount of interdisciplinary collaboration as researchers have learned to communicate methods and processes across the scientific disciplines.

This second edition of the *Handbook of Conducting Polymers* has been written to provide an update on the fundamental progress and applications made in this field since the publication of the first edition, with contributions from world-renowned authors. The Handbook has been subdivided into the general areas of theory and transport, synthesis, processing, properties, and applications of conducting polymers. The various chapters demonstrate the importance of continued synthesis of new, well-characterized polymers with enhanced properties, along with studies of fundamental physical properties and processes. Moreover, the ability to prepare stable and processable conducting polymers has led to a dramatic increase in proposed applications. Continued applications development will provide the driving force for further research, which in turn will lead to the appearance in the market of additional products based on conducting polymers. Indeed, the next ten years promise to be as fascinating as the previous decade, as the expanding role of these multifunctional materials unfolds.

We wish to express our gratitude to the contributing authors for their efforts in producing a handbook of the finest quality, and to Carol Lowe, Arlene Kyle, and the staff of Marcel Dekker, Inc., for organizational support. We thank everyone for their patience in seeing the project through to the end.

Terje A. Skotheim
Ronald L. Elsenbaumer
John R. Reynolds

Contents

Preface .... *iii*
Contributors .... *ix*

I. Theory and Transport in Conducting Polymers

1. Electronic Structure and Optical Response of Highly Conducting and Semiconducting Conjugated Polymers and Oligomers 1
   J. L. Brédas, K. Cornil, F. Meyers, and D. Beljonne

2. Metal-Insulator Transition in Doped Conducting Polymers 27
   Reghu Menon, C. O. Yoon, D. Moses, and A. J. Heeger

3. Insulator–Metal Transition and Inhomogeneous Metallic State in Conducting Polymers 85
   R. S. Kohlman and Arthur J. Epstein

4. The Dynamics of Solitons in *trans*-Polyacetylene 123
   Christoph Kuhn

5. Electron Spin Dynamics 141
   Maxime Nechtschein

6. π-Electron Models of Conjugated Polymers: Vibrational and Nonlinear Optical Spectra 165
   Zoltán G. Soos, Anna Painelli, Alberto Girlando, and Debasis Mukhopadhyay

II. Synthesis of Conducting Polymers

7. Synthesis of Polyacetylene 197
   Hideki Shirakawa

8. Synthesis of Poly(*para*-phenylene)s 209
   A.-Dieter Schlüter

9. Regioregular, Head-to-Tail Coupled Poly(3-alkythiophene) and Its Derivatives 225
   Richard D. McCullough and Paul C. Ewbank

10. Recent Advances in Heteroaromatic Copolymers 259
    John P. Ferraris and Douglas J. Guerrero

11. Low Band Gap Conducting Polymers 277
    Martin Pomerantz

12. Advances in the Molecular Design of Functional Conjugated Polymers 311
    Jean Roncali

13. The Chemistry and Uses of Polyphenylenevinylenes 343
    Stephen C. Moratti

14. Conjugated Ladder-Type Structures 363
    Ullrich Scherf

15. Synthesis and Properties of Conducting Bridged Macrocyclic Metal Complexes 381
    Michael Hanack, Michael Hees, Patrick Stihler, Götz Winter, and L. R. Subramanian

16. Template Polymerization of Conductive Polymer Nanostructures 409
    Charles R. Martin

III. Processing of Conducting Polymers

17. Colloidal Dispersions of Conducting Polymers 423
    Steven P. Armes

18. Solution Processing of Conductive Polymers: Fibers and Gels from Emeraldine Base Polyaniline ... 437
    Richard V. Gregory

19. Dispersion as the Key to Processing Conductive Polymers 467
    Bernhard Wessling

IV. Properties of Conducting Polymers

20. Electrochemistry of Conducting Polymers 531
    Karl Doblhofer and Krishnan Rajeshwar

21. Ion Implantation Doping of Electroactive Polymers and Device Fabrication 589
    André Moliton

22. Optical Probes of Photoexcitations in Conducting Polymers 639
    Zeev Valy Vardeny and X. Wei

23. Electronic and Chemical Structure of Conjugated Polymers and Interfaces as Studied by
    Photoelectron Spectroscopy 667
    M. Lögdlund, Per Dannetun, and W. R. Salaneck

Contents

24. Conformation-Induced Chromism in Conjugated Polymers 695
    Mario Leclerc and Karim Faïd

25. Structural Studies of Conducting Polymers 707
    Michael J. Winokur

26. Second-Order Nonlinear Optical Materials 727
    J. I. Chen, S. Marturunkakul, L. Li, J. Kumar, and Sukant K. Tripathy

27. The Influence of Charge-State Incorporation on the Nonlinear Optical Properties of Conjugated Polyenes 743
    Charles W. Spangler

28. Molecular and Electronic Structure and Nonlinear Optics of Polyconjugated Materials from Their Vibrational Spectra 765
    M. Del Zoppo, C. Castiglioni, P. Zuliani, and G. Zerbi

V. Applications of Conducting Polymers

29. Electroluminescence in Conjugated Polymers 823
    Richard H. Friend and Neil C. Greenham

30. Fundamentals of Electroluminescence in Paraphenylene-Type Conjugated Polymers and Oligomers 847
    Günther Leising, Stefan Tasch, and Willhelm Graupner

31. Corrosion Inhibition of Metals by Conductive Polymers 881
    Wei-Kang Lu, Sanjay Basak, and Ronald L. Elsenbaumer

32. Conducting Polymers in Microelectronics 921
    Marie Angelopoulos

33. Gas and Liquid Separation Applications of Polyaniline Membranes 945
    Jeanine Conklin, Timothy M. Su, Shu-Chuan Huang, and Richard B. Kaner

34. Chemical and Biological Sensors Based on Electrically Conducting Polymers 963
    Anthony Guiseppi-Elie, Gordon G. Wallace, and Tomakazu Matsue

35. Electrically Conducting Textiles 993
    Hans H. Kuhn and Andrew D. Child

36. Electrochemomechanical Devices: Artificial Muscles Based on Conducting Polymers 1015
    Toribio Fernández Otero and Hans-Jürgen Grande

37. Conductive Polymer/High Temperature Superconductor Assemblies and Devices 1029
    John T. McDevitt, Marvin B. Clevenger, and Steven G. Haupt

38. Transparent Conductive Coatings 1059
    Vaman G. Kulkarni

    Index *1075*

Contributors

Marie Angelopoulos   IBM Research Division, T.J. Watson Research Center, Yorktown Heights, New York

Steven P. Armes   School of Chemistry, Physics, and Environmental Science, University of Sussex, Brighton, East Sussex, England

Sanjay Basak   Department of Chemistry and Biochemistry, The University of Texas at Arlington, Arlington, Texas

D. Beljonne   Service de Chimie des Matériaux Nouveaux, Centre de Recherche en Electronique et Photonique Moléculaires, Université de Mons-Hainaut, Mons, Belgium

J.-L. Brédas   Service de Chimie des Matériaux Nouveaux, Centre de Recherche en Electronique et Photonique Moléculaires, Université de Mons-Hainaut, Mons, Belgium

C. Castiglioni   Department of Industrial Chemistry, Politecnico, Milan, Italy

J. I. Chen   Department of Chemistry, University of Massachusetts—Lowell, Lowell, Massachusetts

Andrew D. Child   Milliken Research Corporation, Spartanburg, South Carolina

Marvin B. Clevenger*   Department of Chemistry and Biochemistry, The University of Texas at Austin, Austin, Texas

Jeanine A. Conklin   PPG Industries, Monroeville, Pennsylvania

J. Cornil   Aspirant FNRS, Service de Chimie des Matériaux Nouveaux, Centre de Recherche en Electronique et Photonique Moléculaires, Université de Mons-Hainaut, Mons, Belgium

---

* *Current affiliation:* Westinghouse Electric Corporation, West Mifflin, Pennsylvania

Per Dannetun* Department of Physics and Measurement Technology, Linköping University, Linköping, Sweden

M. Del Xoppo Department of Industrial Chemistry, Politecnico, Milan, Italy

Karl Doblhofer Fritz-Haber-Institut der Max-Planck-Gesellschaft, Berlin, Germany

Ronald L. Elsenbaumer Materials Science and Engineering, and Department of Chemistry and Biochemistry, The University of Texas at Arlington, Arlington, Texas

Arthur J. Epstein Departments of Physics and Chemistry, The Ohio State University, Columbus, Ohio

Paul C. Ewbank Department of Chemistry, Carnegie Mellon University, Pittsburgh, Pennsylvania

Karim Faïd Department of Chemistry, University of Montreal, Montreal, Quebec, Canada

John P. Ferraris The Department of Chemistry, The University of Texas at Dallas, Richardson, Texas

Richard H. Friend Cavendish Laboratory, University of Cambridge, Cambridge, England

Alberto Girlando Dipartimento di Chimica Generale ed Inorganica, Chimica Analitica e Chimica Fisica, Parma University, Parma, Italy

Hans-Jürgen Grande Laboratory of Electrochemistry, Department of Physical Chemistry, Universidad del País Vasco, San Sebastián, Spain

Willhelm Graupner Institut fur Festkörper Physik, Technischen Universitat Graz, Graz, Austria

Neil C. Greenham Cavendish Laboratory, Cambridge University, Cambridge, England

Richard V. Gregory School of Textiles and Polymer Science, Clemson University, Clemson, South Carolina

Douglas J. Guerrero† The Department of Chemistry, The University of Texas at Dallas, Richardson, Texas

Anthony Guiseppe-Elie ABTECH Scientific, Yardley, Pennsylvania, and Department of Biomedical Engineering, Johns Hopkins University School of Medicine, Baltimore, Maryland

Michael Hanack Institut für Organische Chemie, Lehrstuhl für Organische Chemie II, Universität Tübingen, Tübingen Germany

Steven G. Haupt‡ Department of Chemistry and Biochemistry, The University of Texas at Austin, Austin, Texas

A. J. Heeger Institute for Polymers and Organic Solids, University of California at Santa Barbara, Santa Barbara, California

Michael Hees Institut für Organische Chemie, Lehrstuhl für Organische Chemie II, Universität Tübingen, Germany

---

* *Current affiliation:* Swedish Research Council for Engineering Sciences, Stockholm, Sweden
† *Current affiliation:* Brewer Science, Inc., Rolla, Missouri
‡ *Current affiliation:* Quantum Magnetics, San Diego, California

Contributors

Shu-Chuan Huang Department of Chemistry, University of California at Los Angeles, Los Angeles, California

Richard B. Kaner Department of Chemistry, University of California at Los Angeles, Los Angeles, California

R. S. Kohlman* Department of Physics, The Ohio State University, Columbus, Ohio

Christoph Kuhn Department of Chemistry, University of Pittsburgh, Pittsburgh, Pennsylvania

Hans H. Kuhn Milliken Research Corporation, Spartanburg, South Carolina

Vaman G. Kulkarni Americhem Inc., Concord, North Carolina

J. Kumar Department of Physics, University of Massachusetts—Lowell, Lowell, Massachusetts

Mario Leclerc Department of Chemistry, University of Montreal, Montreal, Quebec, Canada

Günther Leising Institut fur Festkörper Physik, Technischen Universität Graz, Graz, Austria

L. Li Department of Physics, University of Massachusetts—Lowell, Lowell, Massachusetts

M. Lögdlund Department of Physics and Measurement Technology, Linköping University, Linköping, Sweden

Wei-Kang Lu Department of Materials Science and Engineering, The University of Texas at Arlington, Arlington, Texas

Charles R. Martin Department of Chemistry, Colorado State University, Fort Collins, Colorado

S. Marturunkakul Department of Chemistry, University of Massachusetts—Lowell, Lowell, Massachusetts

Tomakazu Matsue Department of Molecular Chemistry and Engineering, Faculty of Engineering, Tohoku University, Sendai, Japan

Richard D. McCullough Department of Chemistry, Carnegie Mellon University, Pittsburgh, Pennsylvania

John T. McDevitt Department of Chemistry and Biochemistry, The University of Texas at Austin, Austin, Texas

Reghu Menon Institute for Polymers and Organic Solids, University of California at Santa Barbara, Santa Barbara, California

F. Meyers Service de Chimie des Matériaux Nouveaux, Centre de Recherche en Electronique et Photonique Moléculaires, Université de Mons-Hainaut, Mons, Belgium

André Moliton Faculty of Sciences, Department of Physics, Université de Limoges, Limoges, France

Stephen C. Moratti Department of Chemistry, University of Cambridge, Cambridge, England

D. Moses Institute for Polymers and Organic Solids, University of California at Santa Barbara, Santa Barbara, California

Debasis Mukhopadhyay Department of Chemistry, Princeton University, Princeton, New Jersey

---

Current affiliation: Los Alamos National Laboratory, Los Alamos, New Mexico

Maxime Nechtschein* Département de Recherche Fondamentale sur la Matière Condensée, CEA—Grenoble, France

Toribio Fernández Otero Department of Physical Chemistry, Universidad del País Vasco, San Sebastián, Spain

Anna Painelli Departimento di Chimica Generale ed Inorganica, Chimica Analytica e Chimica Fisica, Parma University, Parma, Italy

Martin Pomerantz Center for Advanced Polymer Research, Department of Chemistry and Biochemistry, The University of Texas at Arlington, Arlington, Texas

Krishnan Rajeshwar Department of Chemistry and Biochemistry, The University of Texas at Arlington, Arlington, Texas

Jean Roncali Ingénierie Moléculaire et Matériaux Organiques CNRS, Université d'Angers, Angers, France

W. R. Salaneck Department of Physics and Measurement Technology, Linköping University, Linköping, Sweden

Ullrich Scherf Max-Planck-Institut für Polymerforschung, Mainz, Germany

A.-Dieter Schlüter Institut für Organische Chemie, Freie Universität Berlin, Berlin, Germany

Hideki Shirakawa Institute of Materials Science, University of Tsukuba, Tsukuba, Ibaraki, Japan

Zoltán G. Soos Department of Chemistry, Princeton University, Princeton, New Jersey

Charles W. Spangler[†] Department of Chemistry, Northern Illinois University, DeKalb, Illinois

Patrick Stihler Institut für Organische Chemie, Lehrstuhl für Organische Chemie II, Universität Tübingen, Tübingen, Germany

Timothy M. Su City College of San Francisco, San Francisco, California

L. R. Subramanian Institut für Organische Chemie, Lehrstuhl für Organische Chemie II, Universität Tübingen, Tübingen, Germany

Stefan Tasch Institut fur Festkörper Physik, Technischen Universitat Graz, Graz, Austria

Sukant K. Tripathy Department of Chemistry, University of Massachusetts—Lowell, Lowell, Massachusetts

Zeev Valy Vardeny Department of Physics, University of Utah, Salt Lake City, Utah

Gordon G. Wallace Department of Chemistry, Intelligent Polymer Research Institute, University of Wollongong, Wollongong, Australia

X. Wei[‡] Department of Physics, University of Utah, Salt Lake City, Utah

Bernhard Wessling Zipperling Kessler & Company, Ormecon Chemie, Ahrensburg, Germany

---

* Member of CNRS
[†] *Current affiliation:* Department of Chemistry and Biochemistry, Montana State University, Bozeman, Montana
[‡] *Current affiliation:* Los Alamos National Laboratories, Los Alamos, New Mexico

Contributors

Michael J. Winokur Department of Physics, University of Wisconsin, Madison, Wisconsin

Götz Winter Institut für Organische Chemie, Lehrstuhl für Organische Chemie II, Universität Tübingen, Tübingen, Germany

C. O. Yoon Institute for Polymers and Organic Solids, University of California at Santa Barbara, Santa Barbara, California

G. Zerbi Department of Industrial Chemistry, Politecnico, Milan, Italy

P. Zuliani Department of Industrial Chemistry, Politecnico, Milan, Italy

Handbook of Conducting Polymers

1
Electronic Structure and Optical Response of Highly Conducting and Semiconducting Conjugated Polymers and Oligomers

J. L. Brédas, J. Cornil, F. Meyers, and D. Beljonne
Centre de Recherche en Electronique et Photonique Moléculaires, Université de Mons-Hainaut, Mons, Belgium

I. INTRODUCTION

The impact of quantum-chemical calculations in the field of conducting polymers has tremendously increased since the chapters for the first edition of this handbook [1] were written just over a decade ago. At that time, most of the quantum-chemical calculations were restricted to the ground-state properties of the conjugated polymers and were mainly aimed at (1) determining the polymers' intrinsic electronic properties via the evaluation of important parameters such as ionization potentials, electron affinities, bandwidths, and band gaps [2] and (2) describing the geometric and electronic structure modifications taking place upon reaction with reducing or oxidizing agents, thereby characterizing the nature of the charge-storage species formed upon doping, such as solitons, polarons, and bipolarons [3–5]. Furthermore, these early quantum-chemical calculations were performed mostly at the ab initio or semiempirical Hartree–Fock level or at the simple Hückel level, the by now famous Su–Schrieffer–Heeger Hamiltonian [4,6] belonging to the latter class.

Since then, the evolution has been impressive. The explosion in computer power and, consequently, the possibilities of application to larger systems and exploitation of more sophisticated quantum-chemical techniques including electron correlation effects (e.g., the use of post-Hartree–Fock methods in quantum-chemical terminology) have allowed one to venture into the excited-state properties of conjugated polymers or their large oligomers. This has proven essential in providing a deeper understanding of, for instance, (1) the linear and nonlinear optical response of these systems, (2) the characteristics of both singlet and triplet polaron-excitons; and (3) the interactions of conjugated compounds with metals. The latter topic is, however, not covered in this chapter, because conjugated polymer/metal interfaces are discussed thoroughly in the chapter by Lögdlund et al. in this volume as well as in the recent monograph by Salaneck et al. [7].

In Section II we focus on one of the most promising applications of semiconducting conjugated polymers: their use as active layers in light-emitting diodes. We review the results of theoretical studies we performed on oligomers representative of polyparaphenylenevinylene and polythiophene, which deal with (1) the modeling of the linear optical properties of the conjugated polymers involved in the emitting layer, taking into account the vibronic structure; (2) the investigation of the relative locations of, and relaxation phenomena taking place in, the lowest singlet and triplet excited states; and (3) the analysis of the effects of derivatization of the conjugated backbone. We also provide a general discussion of the various terms contributing to the polaron-exciton binding energy.

Section III is devoted to characterizing the changes occurring in the properties of oligothiophenes upon doping in order to rationalize the discrepancies found in the literature regarding the interpretation of the optical absorption spectra. We demonstrate the importance of considering the selection rules imposed by the symmetry of the systems. We also revisit, on that basis, the assignment of optical transitions in conjugated polymers supporting charged defects (i.e., polarons and/or bipolarons).

In the final section, we take a look at the nonlinear

optical response of conjugated oligomers and polymers. In particular, we analyze the dependence of the nonlinear optical properties on the nature of the medium surrounding the chromophore. Most important, this allows us to describe a unified picture of linear and nonlinear polarization in these compounds that comprises geometric structure, electronic structure, and optical response aspects.

It should be emphasized that a quantum-chemical approach, *by taking explicitly into account the fine details of the whole chemical structure*, is able to depict the strong dependence of geometry and electronic properties on subtle effects such as those induced by variation in molecular architecture or chemical substitution. One can then build on one of the major characteristics of organic compounds: the remarkable flexibility in the fine-tuning of their properties. In this context, quantum-chemical calculations allow not only for in-depth interpretations of experimental data but also for the theoretical design of materials with enhanced characteristics.

Our choice of topics offers the opportunity of scanning a wide range of quantum-chemical techniques; note, however, that we will not provide any detailed description of the quantum-chemical methods, as applied to polymers and long oligomers, since such a description can be found, for instance, in Ref. 8. At this stage, however, there is one major caveat that needs to be stressed. The wide availability of quantum-chemical packages has turned them into almost a household item. To cite an example, a software company that delivers a package to draw chemical structures on a computer screen has added for a token price the possibility of running semiempirical Hartree–Fock calculations on the basis of the constructed structures; that the results of such calculations can be totally misconceived is a feature that is, of course, not advertised. As a consequence, the literature is now plagued with results of poorly run calculations in which the methodology that has been followed is not adapted to the system under investigation. In other words, one has to be alert to specific deficiencies and exert caution. Moreover, as will be apparent from this chapter and the references we cite, one actually very often needs a mix of various techniques in order to obtain a reliable answer; the all-purpose quantum-chemical method for conjugated polymers and oligomers has still to materialize.

II. TOWARD A BETTER UNDERSTANDING OF CONJUGATED POLYMER-BASED LIGHT-EMITTING DIODES

Since the first report of electroluminescence from organic conjugated polymers [9], many efforts have been devoted to the design of light-emitting diodes (LEDs) with enhanced performance. Prospects in the fabrication of low-cost, large-area flat panel displays have stimulated the achievement of fast and significant progress in this recent field of research. As a result, organic LED devices [10] are expected in the near future to be in a position to compete with their inorganic counterparts, with the additional advantages of flexibility [11], wide spectral range, and ease of patterning upon metal deposition.

A simple organic LED device is fabricated by sandwiching a polymer or oligomer film between two metallic layers, i.e., a high work function metal (typically indium tin oxide or emeraldine salt) to inject holes into the polymer valence band and a low work function metal (such as aluminum or calcium) to inject electrons into the polymer conduction band upon application of an external electric field. The charge carriers, which relax under the form of positive and negative polarons, then drift in opposite directions within the polymer matrix until they meet to give rise to the formation of singlet and triplet polaron-excitons in a 1:3 statistical ratio that sets the intrinsic upper limit of quantum efficiency at 25% [10]. (Note that if a band process were operative, the efficiency would not be limited by this ratio—we will come back to a discussion of these aspects in the final part of this section.) The radiative decay of singlet excitons leads to light emission, at a wavelength directly dependent on the band gap of the polymer as well as the relaxation processes taking place in the excited state. The bandwidth of the emission peak can be significantly narrowed when the polymer is incorporated into a microcavity [12].

So far, the emphasis has been on polyparaphenylene-vinylene (PPV) and its derivatives as active layers in LEDs, owing to their high yield of luminescence [13] and their ease of processing via soluble precursors in the form of air-stable thin films [14,15]. The first PPV-based LED device was characterized by poor efficiency, on the order of 0.01% (i.e., one emitted photon per 10,000 injected electrons) [9]. Several strategies have been followed to increase the performance of such devices. Among them, attempts to trap the excitons in quantum wells, and thus to prevent nonradiative decay at quenching sites, were found to have a significant impact and have led to improvements of up to two orders of magnitude in the electroluminescence efficiency; such a confinement can be obtained by incorporating, during the thermal conversion, nonconjugated fractions along the polymer backbone [16,17]. Another successful route has been to create preferential sites for the charge carriers to recombine, either at the interface between different polymer layers included in the device [18] or, in PPV-based copolymers, at the interface between homopolymer sequences [19,20].

The achievement of high efficiencies in the LEDs also requires a proper balance of electron and hole injection rates in order to prevent at best quenching of the excitons by charged defects [21]. One has thus to try to minimize the energy barriers between the Fermi energy of the metallic electrodes and the energy of the frontier levels of the polymer layer; one efficient way to do so is to modulate the positions of the band edges of the polymer through derivatization of its conjugated backbone, for instance with π-electron donor and/or acceptor groups [22–24]. It is interesting to note that the grafting of cyano side groups on to the PPV backbone has led to a significant increase in electron affinity of the polymer and has therefore favored the use of metals with high environmental stability such as aluminum as electron-injecting electrodes. In this context, the fabrication of double-layer devices based on PPV and a cyano-substituted derivative was found to be very efficient because both the minimized energy barriers at the polymer/metal interfaces and the strong confinement of the charge carriers at the polymer/polymer interface contribute to the gain of a high quantum efficiency [25]; the best value reported to date stands around 12% [26]. These considerations have also opened the way to the development of new electron and/or hole transporting layers to be incorporated in the devices [27].

Interest has also been devoted in recent years to the use of polythiophene (PT) and its substituted derivatives in LEDs. The color of emitted light in polythiophene-based devices can sweep the entire visible range by controlling the band gap of the systems through the steric hindrance induced by β-substituents on the thiophene rings [28,29]. Furthermore, it was shown that the emission color of organic diodes involving blends of polythiophene chains with varying band gaps could be tuned, depending on the strength of the applied driving voltage [30]. An additional striking feature reported for PT-based diodes is the possibility of making polarized light sources by stretch orientation of the polymers [31]. Note that the corresponding finite-size compounds, oligothiophenes, can also serve as active materials in such devices [32,33].

In the search for new routes toward significant improvements in the performances of organic LEDs, one important task is to provide a good understanding of the intrinsic geometric, electronic, and optical properties of the conjugated polymers that are involved in the devices. To do so, extensive theoretical and experimental investigations have addressed the properties of the corresponding conjugated oligomers. One main advantage of oligomer analogs over the parent polymers is that their well-defined chemical structure together with their improved solubility and processibility give access to detailed interpretations of the experimental measurements; moreover, they are particularly convenient when sophisticated theoretical modelings are to be performed. Quantum-chemical calculations carried out on model molecules thus present a twofold interest as they allow for a direct comparison with experimental data available for the conjugated oligomers and for a precise description of the intrinsic properties of the polymers following extrapolation of the oligomer results at the scale of long chains. In this section, we illustrate that such calculations can prove useful in the context of light-emitting diodes by providing a deep insight into the basic phenomena taking place in the devices. To do so, we have chosen to focus our attention on oligophenylenevinylenes (hereafter denoted PPVn, where n represents the number of phenylene rings) and oligothiophenes (PTn) as well as on their derivatives, due to the tremendous impact of the corresponding polymers on the development of electro-optical devices, as exemplified above.

After having outlined the general features of our theoretical approach, we show that the simulated absorption spectra of PPV oligomers compare very well with experimental data and allow us to describe the nature of the four absorption features observed in the spectrum of the corresponding polymer. We also pay particular attention to the evolution with chain length of the lowest energy feature corresponding to the $1B_u$ excited state, from which singlet polaron-excitons radioactively decay to give rise to luminescence; on the basis of well-resolved optical absorption spectra of PPV oligomers, we have set up a theoretical strategy to model the vibronic couplings and provide, on the basis of the fitting procedure, direct estimates of the extent of relaxation occurring in the $1B_u$ state. Then, dealing with both PPV and PT oligomers, we investigate the evolution with chain length of the relative locations of the lowest singlet excited state (S_1) and lowest triplet excited state (T_1); the latter plays an important role in the context of LEDs because this state can be reached either by intersystem crossing processes or by triplet recombination of the electron–hole pairs. We also discuss in the case of oligothiophenes the way the singlet-to-triplet intersystem crossing is affected by chain length; such an analysis provides valuable information for the design of new molecular architectures, for which such nonradiative decay routes, competing with light emission, would be avoided. The following subsection is related to a theoretical characterization of the geometry relaxation phenomena occurring in both the S_1 and T_1 states of PPV oligomers; our motivation is to gain a refined picture of the singlet and triplet polaron-excitons responsible of the fluorescence and phosphorescence [34] processes, respectively. We then analyze the way the positions of the frontier levels in the five-ring PPV oligomer are affected upon derivatization, either by attaching π-donor and/or acceptor groups to the side of the chain or by introducing nitrogen atoms within the backbone; the changes occurring in the one-electron structure and in the energy of the lowest optical transition are highlighted. We conclude this section by discussing the various terms that contribute to the polaron-exciton binding energy in electroluminescent conjugated polymers.

A. Theoretical Approach

We examine here PPV oligomers that contain two to five phenylene rings and PT oligomers ranging in size from two to six thiophene units; the chemical structures of these compounds as well as those of the derivatized PPV oligomers we have considered are presented in Fig. 1.1. The geometry of the oligomers in their ground state (S_0) has been optimized by means of semiempirical Hartree–Fock techniques derived from the NDDO (neglect of differential diatomic overlap) formalism, namely the AM1 (Austin Model 1) [35] and MNDO (modified neglect of differential overlap) [36] methods; planar conformations are consistently assumed for all the systems. The C–C bond lengths of the PPV oligomers as obtained from AM1 calculations as well as the MNDO-optimized bond lengths in oligothiophenes are found to be in very good agreement with X-ray diffraction data [37,38]. The choice of two different theoretical approaches to optimize the geometries is related to the fact that we require a good description of the bond length alternation along the backbone since this parameter is found to be crucial when estimating the $S_0 \rightarrow S_1$ transition energy [39].

On the basis of the optimized geometries of PPV oligomers and derivatives, transition energies in these systems and related transition intensities are determined with the semiempirical Hartree–Fock INDO (intermediate neglect of differential overlap) Hamiltonian [40] combined with a single configuration interaction (SCI) technique [41]. The CI development is built on the basis of singly excited configurations that are generated by promoting an electron from one of the 20 upper occupied levels to one of the 20 lower unoccupied levels; the electron–electron repulsion term is parameterized with the Mataga–Nishimoto potential, which provides the best reproduction of experimental absorption spectra within the INDO/SCI formalism. The simulation of the absorption spectra is carried out by injecting the transition moments and transition energies calculated at the INDO/SCI level into a frequency-dependent sum-over-states (SOS) expression of the polarizability [42] and then plotting the dispersion of its imaginary part multiplied by the incident frequency. To explicitly incorporate the effects of vibronic couplings in the absorption spectra, we have modified the SOS expression by introducing additional terms relative to the overlap between the vibrational wave functions that we treat within the Franck–Condon approximation. This is shown in the equation

$$\alpha_{ij}(-\omega; \omega) = \frac{1}{\hbar} P(i,j; -\omega, \omega)$$

$$\sum_m \sum_{\nu^x} \frac{\langle g|\mu_i|m\rangle \langle m|\mu_j|g\rangle \prod_{x=1}^{3N-6} \langle 0|\nu^x\rangle^2}{\omega_{mg} + \nu^x E_{ph}^x - \omega - i\Gamma}$$

where $P(i,j; \omega, -\omega)$ is a permutation operator that imposes for any permutation of the i and j indices an equivalent permutation of the ω and $-\omega$ factors; ω is the incident frequency of light while i and j correspond to the Cartesian axes; g and m denote the ground state and an excited state, respectively, ω_{mg} being the transition energy between these two states; μ_i is the ith component of the dipole operator; ν_x and E_{ph}^x are the vibrational quantum number and quantum energy of the x normal mode, respectively; Γ is the damping factor. The Franck–Condon integrals are given within the harmonic approximation by

$$\langle 0|\nu^x\rangle^2 = \frac{e^{-S_x} S_x^\nu}{\nu!}$$

Here, S is the Huang–Rhys factor; it describes the extent to which geometric deformations take place in the excited state and corresponds to the average number of phonons involved in the relaxation process. In the present case, we consider the two most dominant effective modes observed in the fluorescence spectra of PPV and its oligomers at 0.16 and 0.21 eV, in a 1:2 intensity ratio [43]. It should be emphasized that we derive the amplitude of the Huang–Rhys factor from a fitting procedure between the experimental and calculated absorption spectra.

The relative positions of the lowest singlet (S_1) and triplet (T_1) excited states in PPV and PT oligomers of varying chain lengths have been investigated by coupling the INDO Hamiltonian to a multi-reference double-configuration interaction (MRD-CI) [44] scheme that allows us to include singly to fourfold excited configurations in the CI expansion. To do so, configurations are generated following single and double excitations with respect to several reference determinants, among which it was shown that the most relevant two are (1) the ground-state configuration and (2) the determinant resulting from the excitation of one electron from the HOMO (highest occupied molecular orbital) level to the LUMO (lowest unoccupied molecular orbital) level [45]. Such calculations typically involve excitations among the five or six highest occupied and lowest unoccupied molecular orbitals [45,46].

The geometry relaxation phenomena taking place in the S_1 and T_1 states of PPV oligomers are evaluated via a direct optimization of the geometry in these states with the help of the AM1 or MNDO methods. Calculations are then performed on the basis of a level occupancy characterized by one electron promoted from HOMO to LUMO levels (one spin being flipped for the triplet). The total energy of the excited state, as obtained from a limited CI calculation, is then used as the criterion of convergence of the optimization process. Note that the trends provided by this approach are consistent with those obtained by allowing the geometry to relax under the form imposing of a confined soliton–antisoliton pair [47] or by making use of empirical bond order bond length (BOBL) relationships [48], as shown in Ref. 46.

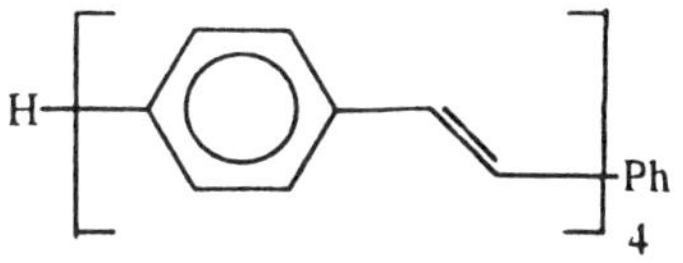

I

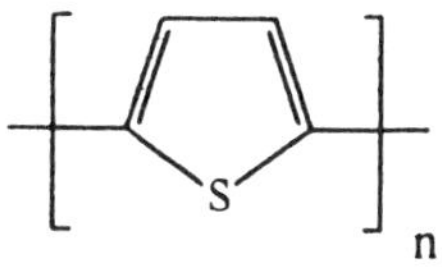

II

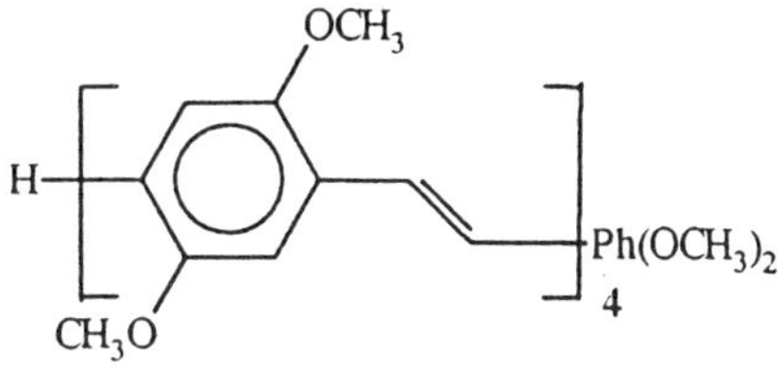

III

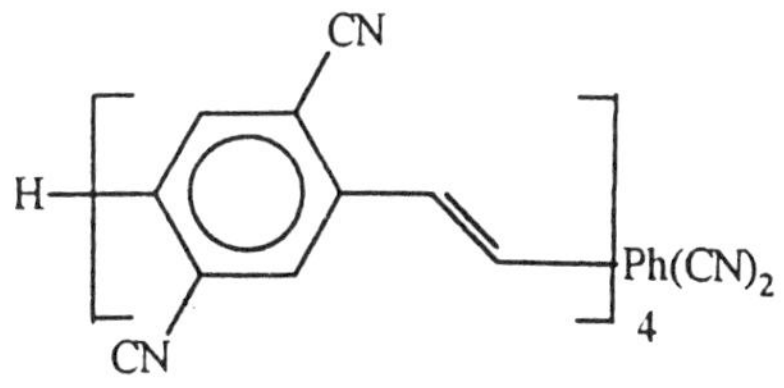

IV

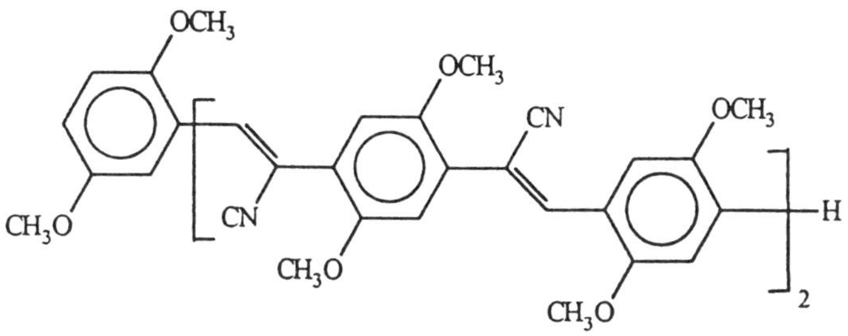

V

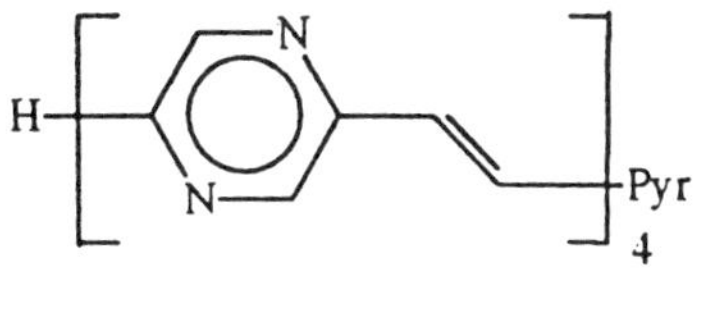

VI

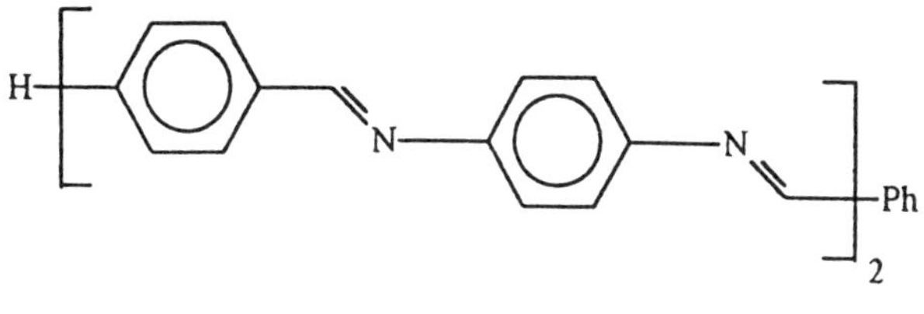

VII

Fig. 1.1 Molecular structures of the conjugated oligomers reviewed in this chapter.

B. Optical Absorption Spectra of PPV Oligomers

We present in Fig. 1.2 the simulated absorption spectrum of the five-ring PPV oligomer together with the experimental spectrum of the polymer [49]. In the latter, three distinct absorption features appear around 3.0, 4.8, and 6.1 eV as well as a weak shoulder at 3.7 eV. Similarly, we observe that four absorption bands dominate the spectrum of the oligomer. It is therefore straightforward to establish a one-to-one correspondence between the two spectra in order to uncover the nature of the absorption peaks in the polymer; furthermore, we aim to distinguish the roles of the delocalized and localized molecular orbitals (the latter are characterized by the existence of very weak LCAO coefficients on the para-carbon atoms of the phenylene rings).

Analysis of the CI expansion coefficients indicates that the lowest energy feature primarily originates from an electron transition between the HOMO and LUMO levels. The second peak results from the mixing of several configurations, each of them involving exclusively delocalized levels. The third band is depicted by $l \rightarrow d^*$ and $d \rightarrow l^*$ transitions (l and d denoting localized and delocalized levels, respectively), which are characterized by similar weight due to the existence of a quasi electron-hole symmetry in such systems. Finally, the highest absorption peak is described by the interaction of configurations associated to transitions between localized levels (within the phenylene rings). The discrepancy in the relative intensities of the absorption peaks as well as the existence of shifts in the transition energies when comparing the two spectra have to be mainly attributed to finite-size effects prevailing in our calculations. However, we stress that the theoretical evolution with chain length (going from three to five phenylene rings) of the simulated spectra follows remarkably well the available experimental measurements [49] and is fully in agreement with the results of the model calculations of Rice and Gartstein [50].

We now focus our attention on the lowest energy singlet excited state (S_1) from which radiative decay takes place in light-emitting diodes. We display in Fig. 1.3 well-resolved optical absorption spectra of phenylenevinylene oligomers containing from two to five rings together with the INDO/SCI simulated spectra for which we fit the amplitude of both the damping factor (which defines the bandwidths of the absorption features) and the Huang-Rhys factor (which governs the relative intensities within the vibronic progression). From the comparison between theory and experiment, we can highlight the following aspects:

1. The similarity in shape of the measured and simulated spectra validates the theoretical approach we have adopted (we have further demonstrated that we could not reach the same level of agreement with the use of a single effective mode at 0.18 eV [51]).
2. As for most conjugated oligomers [52,53], both the experimental and theoretical data show a linear relationship between the 0-0 transition energy and chain length.
3. The fitted values obtained for the Huang-Rhys factor are found to diminish as the chain grows (going from 1.95 for stilbene to 1.60 for the longer chains). This evolution is consistent with the experimental absorption spectrum of improved PPV showing a dominant intensity for the 0-0 transition [53]; such behavior actually illustrates a progressive softening of the relaxation process occurring upon photoexcitation of electron-hole pairs.
4. Direct estimates of the strength of relaxation in the $1B_u$ state can actually be obtained by expanding the relaxation energy into a sum of two contributions corresponding to the energy of an effective mode weighted by the Huang-Rhys factor related to this mode; this leads to values on the order of 0.39 eV for stilbene and 0.34 eV for the other oligomers.

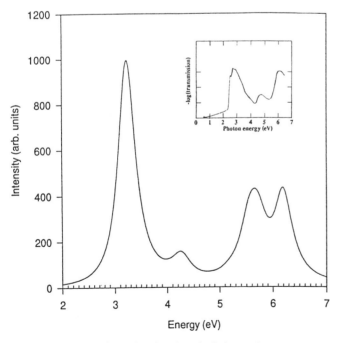

Fig. 1.2 INDO/SCI simulated optical absorption spectrum of the five-ring PPV oligomer. Shown in the insert is the experimental spectrum of the polymer (From Ref. 49.)

It is worth stressing that the trends in PPV oligomers markedly differ from the situation prevailing in the case of polyenes where, due to the degenerate nature of the

C. Relative Locations of the S_1 and T_1 States—Intersystem Crossing Process

We display in Fig. 1.4 the evolution with chain length in oligothiophenes of the $S_0 \to S_1$ transition energy as well as the $S_0 \to T_1$ energy difference, as calculated at the INDO/MRD-CI level. Strikingly, we observe the $S_0 \to T_1$ energy difference to hardly evolve with chain size whereas there occurs a significant red shift of the lowest one-photon allowed excitation (shifts on the order of 0.2 eV and 1.4 eV, respectively, when going from the dimer to the hexamer). The weak evolution of the lowest triplet excitation energy actually reflects the strong confinement of the triplet exciton; this is further supported by optically detected magnetic resonance (ODMR) measurements on polythiophene indicating that only a single thiophene unit is needed for a proper accommodation of the triplet defect [55].

The calculated $S_0 \to T_1$ energy differences compare well with available experimental data. Indeed, a weak and broad peak associated to this transition has been observed in the absorption spectrum of PT3 due to spin–orbit coupling induced by the existence of heavy atoms in the solvent; the experimental value reported at ~ 1.71 eV [56] is in excellent agreement with the 1.68 eV calculated result. Further experiments dealing with charge transfer between C_{60} and oligothiophenes ranging in size from 6 to 11 units have located the position of the T_1 state in these oligomers between 1.57 and 1.71

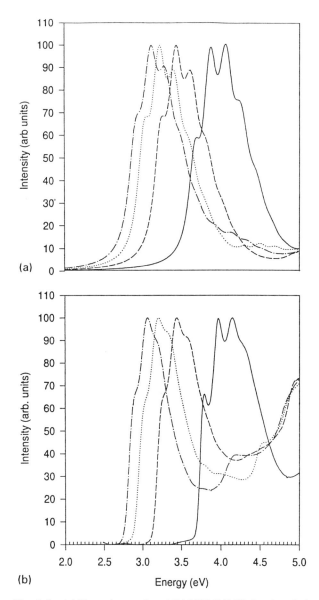

Fig. 1.3 (a) Experimental and (b) INDO/SCI simulated absorption spectra of the (——) two-, (– – –) three-, (· · ·) four-, and (–·–) five-ring PPV oligomers. (From Ref. 51.)

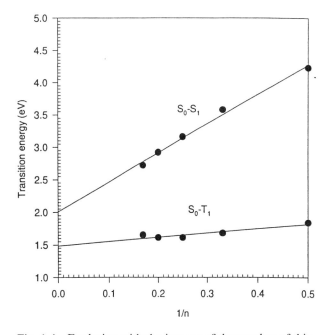

Fig. 1.4 Evolution with the inverse of the number of thiophene units ($1/n$) of the $S_0 \to S_1$ and $S_0 \to T_1$ excitation energies, as calculated at the INDO/MRD-CI level.

ground state of the polymer, the Huang–Rhys factor associated to the $1B_u$ state is found first to decrease, to pass by a minimum, and then to increase as the chain elongates. A combination of theoretical and experimental data [54] has established a coherent picture of the nature of the photogenerated species, going from polaron-excitons in short polyenes to the formation of soliton–antisoliton pairs in long oligomers and *trans*-polyacetylene.

eV [57], i.e., within an energy range similar to that obtained with the MRD-CI calculations. Finally, we mention that a phosphorescence signal has been detected around 1.5 eV in polythiophene [34]; the same value is obtained by extrapolating the theoretical data at the scale of an infinite polymer chain.

Fully similar conclusions can be drawn from the evolution with chain length of the $S_0 \to S_1$ and $S_0 \to T_1$ excitations in oligophenylenevinylenes [45]. Here also, the strong confinement of the triplet exciton is confirmed by ODMR measurements [55].

Recent time-resolved fluorescence measurements have shown a sharp increase in the fluorescence quantum yield ϕ_F of unsubstituted oligothiophenes when going from two to seven units [58–60]. Such an evolution has been closely related to a decrease in the nonradiative decay rate k_{NR} since almost no change is observed for the radiative decay rate k_R. Furthermore, the main nonradiative process was found to originate from singlet-to-triplet intersystem crossing [59]. We have therefore tried to rationalize the evolution with chain size of the nonradiative decay rate in oligothiophenes; this rate has been expressed in PT3 as a sum of two separate contributions [58], as shown below:

$$k_{NR} = k_1 + k_2(T) = k_1 + A_2 \exp[-\Delta E_{ISC}/kT]$$

where k_1 corresponds to nonactivated phenomena while k_2 is the thermally activated intersystem crossing process. Note that this analysis was done under the reasonable assumption that the changes in k_{NR} are driven mainly by the energy difference between the singlet and triplet states involved in the crossing.

Although the S_1–T_1 energy differences are much too large to give rise to efficient crossing, our calculations indicate that a higher lying triplet excited state, the T_4 state, is located within the same energy range as S_1 and is thus in all likelihood the essential triplet state to consider to better comprehend the intersystem crossing processes. Starting from bithiophene, where the position of T_4 is below that of S_1, there occurs a progressive reversal in the ordering of these two states as the chain grows. The crossing between the two states takes place at a chain length corresponding to the trimer, as sketched in Fig. 1.5 (where we observe the position of the T_4 state in the trimer to be largely overestimated due to not taking account of spin–orbit interactions). The experimental trends can therefore be understood on the basis of these considerations. Indeed, the location of T_4 below S_1 in bithiophene makes the intersystem crossing a nonactivated and very efficient process that strongly inhibits the fluorescence. In contrast, ϕ_F is expected to be substantially raised as the chain elongates due to the appearance of an increasing activation energy.

In Table 1.1, we present our calculated estimates of k_{NR}, which are in very good agreement with the experimental data despite the simplicity of our model [45]. Another aspect worth mentioning from the table is that, in

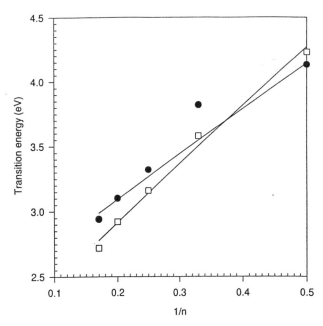

Fig. 1.5 Evolution of the INDO/MRD-CI calculated (□) $S_0 \to S_1$ and (●) $S_0 \to T_4$ transition energies as a function of the inverse of the number of thiophene rings.

terthiophene, substitution of the chain ends by electron acceptors such as formyl groups also results in a strong enhancement of the fluorescence yield [45]. This thus constitutes a nice illustration that simple chemical substitution provides a handle to fine-tune a desired electronic property.

D. Geometry Relaxation Phenomena in the S_1 and T_1 States of Oligophenylenevinylenes

Analysis of the AM1 optimized geometry of the lowest singlet excited state in stilbene shows relatively weak deformations with respect to the ground state. The C–C bond lengths in the phenylene rings are found to be almost unaffected (changes on the order of 0.01 Å) while a more significant change in the bond length alternation takes place in the vinylene moieties (going from 0.11 Å in S_0 to 0.02 Å in S_1); these trends are in excellent agreement with the results of earlier Pariser–Parr–Pople calculations [61]. In contrast, a much more pronounced geometry relaxation occurs for the triplet polaron-exciton; indeed, the phenylene rings adopt a semiquinoid character while the single–double C–C bond pattern is reversed within the vinylene linkages. As a result, the relaxation energy associated to the triplet is calculated to be about twice as much as for the singlet (0.60 and 0.28 eV, respectively).

As the chain elongates, in the S_1 state, only the central portion of the molecule is affected by the relaxation

Electronic Structure and Optical Response

Table 1.1 INDO/MRD-CI S_1–T_4 Energy Difference (in eV) Calculated for the Unsubstituted Oligothiophenes and (in parentheses) Extrapolated from a Linear Relationship Between the Excitation Energies and the Inverse Number of Rings[a]

	$\Delta E(S_1$–$T_4)$	k_{NR}^{th} (10^9 s^{-1})	k_{NR}^{exp} (10^9 s^{-1})			ϕ_F^{exp}		
			Ref 58	Ref 59	Ref 60	Ref 58	Ref 59	Ref 60
Th2	−0.10 (−0.10)	20.1		19.7			0.018	
Th3	0.24 (0.05)	3.70	3.88	5.3	3.7	0.07	0.07	0.05
Th4	0.16 (0.13)	1.23	1.54	1.50		0.20	0.18	
Th5	0.18 (0.18)	1.14	0.82	0.74		0.28	0.36	
Th6	0.22 (0.21)	1.13	0.70	0.73		0.42		

[a] We also include the fluorescence yields, ϕ_F, and nonradiative decay rates, k_{NR}, measured in solution (k_{NR}^{exp}) and calculated on the basis of simple assumptions (k_{NR}^{th}, see text).

process and the geometry of the external units is unchanged with respect to the ground state; the weak deformations extend over a spatial domain corresponding typically to three or four aromatic rings, i.e., some 25–30 Å. As a result, a weak and similar relaxation energy on the order of 0.23 eV is calculated for the three- and four-ring oligomers. Note that these trends are consistent with the estimates provided on the basis of the fitting procedure of the experimental absorption spectra (on the order of 0.34 eV). The slight discrepancy between the two sets of values can be partly attributed to the fact that the experimental data do not obviously originate from fully planar ground-state conformations, as assumed in the calculations.

The triplet exciton in the longer oligomers is characterized by local and strong deformations; the defect is calculated to extend over a single phenylene ring, in agreement with ODMR measurements [55]. The AM1 optimized geometries of the four-ring PPV oligomer in the S_0, S_1, and T_1 states are illustrated in Fig. 1.6.

Similar trends prevail for the oligothiophenes. However, stronger lattice distortions are calculated in the triplet state of the PPV oligomers, especially on the vinylene linkages.

E. Effects of Derivatization on the Frontier Molecular Orbitals

Analysis of the one-electron structure provided by the INDO calculations indicates the way the positions of the frontier levels in the five-ring PPV oligomer evolve upon derivatization. Substitution with π-donor groups (derivative III in Fig. 1.1) leads to an overall destabilization of the HOMO and LUMO levels, the first being the most affected; in contrast, these levels are stabilized when π-acceptor substituents are attached on the backbone (derivative IV in Fig. 1.1), the stronger shift being observed for the LUMO level.

We report in Table 1.2 the shift in the positions of the HOMO and LUMO levels in the various derivatives reviewed here, where we deal with methoxy and cyano groups as the donor and acceptor entities, respectively. The trends we calculate can be rationalized by a simple three-level model involving the frontier levels of the unsubstituted repeat unit and the occupied level of the donor group or the unoccupied level of the acceptor

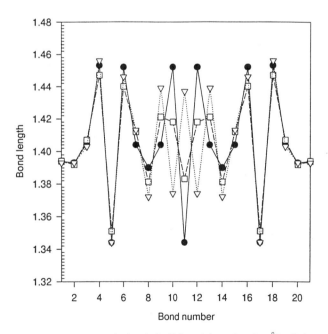

Fig. 1.6 AM1 optimized C–C bond lengths (in Å) of the four-ring PPV oligomer in the ground state (S_0, ——), the lowest singlet excited state (S_1, - - -), and the lowest triplet excited state (T_1, · · ·). We take into account the C–C bonds located on the shortest pathway between the two ends of the chain; bond 11 lies at the center of the molecule.

Table 1.2 INDO Shifts (in eV) of the HOMO and LUMO Levels for Derivatives III–VII with Respect to the Unsubstituted PPV Pentamer

Derivative	HOMO	LUMO	E_g, (eV)[a]
PPV			3.24
III	+0.27	+0.08	3.06
IV	−0.99	−1.17	3.11
V	−0.17	−0.55	2.92
VI	−0.28	−0.52	2.92
VII	−0.32	−0.19	3.32

[a] Energy of the lowest optical transition, as calculated at the INDO/SCI level.

group [62]. The system presenting the two kinds of substituents (derivative V) is characterized by an overall stabilization of the frontier levels (which are located in the energy range delimited by the "donor only" and "acceptor only" cases); this feature reflects the stronger coupling to the backbone of the cyano group with respect to the methoxy substituents. Cyclic voltammetry measurements indicate that the cyano functionalization on the vinylene moieties lowers the HOMO and LUMO levels by 0.45 and 0.62 eV, respectively [63]; these shifts compare very well with the corresponding INDO calculations leading to values of 0.44 and 0.63 eV. These considerations suggest that an appropriate choice in the nature of the side groups allows one to modulate the positions of the frontier levels and hence the color of the emitted light in LED devices [62]. We note that derivatives where nitrogen atoms are incorporated within the phenylene rings or the vinylene linkages behave as chains substituted by π-acceptor groups [62].

Calculations conducted at the correlated level provide valuable information on the evolution upon derivatization of the lowest optical transition and hence of the wavelength of the emitted light. As is the case in the PPV oligomers, the lowest absorption in all the derivatives can be mainly described by an electron transition between the HOMO and LUMO levels; there is, however, one exception in the case of compound VII, poly(1,4-phenylenemethylidynenitrilo-1,4-phenylenenitrilomethylidyne) (PPI), (see Fig. 1.1) for which a mixing of several configurations is needed to properly describe the absorption peak. Note that such behavior is reminiscent of the typical optical properties of the emeraldine base form of the polyaniline family [64], whose molecular structure presents similarities to that of derivative VII.

Except for PPI, the evolution in transition energy when going from one derivative to the next closely follows the trends derived from the examination of the one-electron structure. The asymmetry in stabilization or destabilization of the frontier levels systematically leads to a red shift of the optical transition; the maximum shift is on the order of 0.3 eV in the case of derivatives V and VI (see Table 1.2). These results rationalize the color modulation observed in LED devices when going from PPV to a derivatized chain as active layer. The lowest transition energy in derivative VII is calculated to be weakly blue-shifted, by ~0.08 eV, with respect to the PPV pentamer; this result is consistent with the 0.07 eV blue shift reported from experimental measurements on the corresponding polymer [65].

F. On the Nature of the Exciton Binding Energy

We now turn to a discussion of the issue related to the nature of the emitting species in the lowest excited state of luminescent conjugated polymers, with the help of theoretical calculations including both electron–phonon and electron–electron interactions. Specifically, we address the following propositions that have been reported in the literature:

1. Free charge carriers are generated in the excited state [66–68], and emission is an interband process.
2. Emission originates from a tightly bound electron–hole pair with a binding energy larger than 1 eV [69,70].
3. Emission is from the radiative decay of weakly bound polaron-exciton with a binding energy of a few tenths of an electronvolt [71–73].

Note that the polaron-exciton terminology implies that lattice relaxations are associated with the photogenerated electron–hole pair.

In our opinion, any Hamiltonian used to characterize the nature of the photogenerated species has to incorporate electron–phonon contributions because these correspond to a basic feature of π-conjugated compounds. As mentioned above, a typical manifestation of lattice relaxations taking place in the excited states is the appearance of vibronic progressions in the experimental optical absorption spectra. As a matter of fact, vibronic features could not be observed if the equilibrium positions in the ground and excited states were identical. This is sketched in Fig. 1.7a (where we assume the existence of a single active vibrational mode coupled to the electronic excitation); the orthonormality of the vibrational wave functions would then exclusively allow transitions between vibrational levels of the two states with the same quantum number. In contrast, a displacement of the equilibrium geometry in the excited state (Figs. 1.7b and 1.7c) leads to the appearance of several vibronic features, usually between the zeroth vibrational level of the ground state and various levels of the excited state. When treated within the Franck–Condon approximation, the intensities are weighted by the overlap of

Electronic Structure and Optical Response

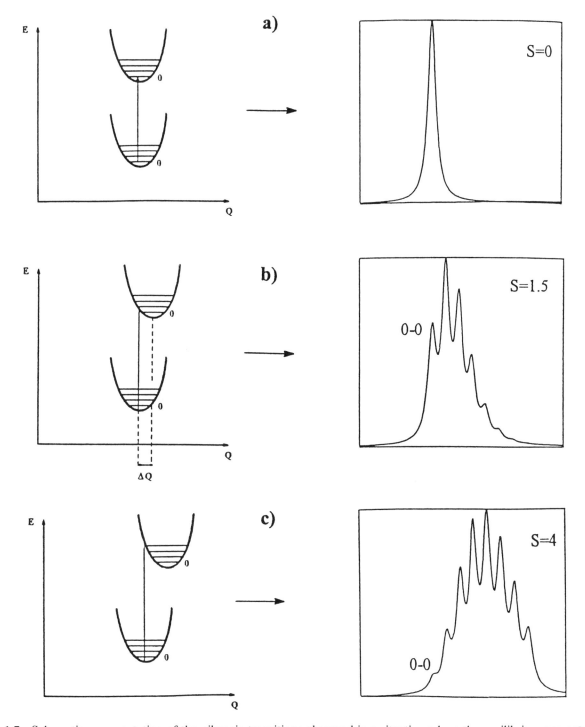

Fig. 1.7 Schematic representation of the vibronic transitions observed in a situation where the equilibrium geometries of the ground and excited electronic states are (a) identical ($S = 0$), (b) weakly displaced ($S = 1.5$), and (c) strongly displaced ($S = 4$). S is the Huang–Rhys factor as defined in the text.

the vibrational wave functions. It is worth stressing that the lowest energy transition (i.e., the 0–0 transition) is to the relaxed geometry of the excited state.

In the framework of models considering rigid and fully delocalized bands, the excited-state relaxations (and thus the vibronic effects) are expected to decrease linearly with the inverse of the number of atoms in the chain; hence, they would be thought to be insignificant at the scale of long conjugated chains. However, the existence of a vibronic progression in the absorption spectra of conjugated macromolecules indicates that self-localization phenomena occur in the excited states.

On the other hand, given the ease of delocalization and polarization of π electrons, electron correlation is another major ingredient to be incorporated in a theoretical modeling. We have thus to consider the influence on the binding of photogenerated electron–hole pairs, which is due to electron–lattice coupling and to electron–electron interactions [74].

We have described above that an analysis of the *absorption spectra* of PPV oligomers indicates that the relaxation energy in the $1B_u$ excited state is on the order of 0.30 eV for the three-, four-, five-ring oligomers; it thus hardly evolves with increasing chain length. In contrast, we have established on the basis of experimental *photoluminescence spectra* reported for the same oligomers [75] that, in the case of emission, the total Huang–Rhys factor of the $1B_u$ excited state decreases as the chain grows; the evolution is linear as a function of the inverse of chain length, tending to 0.2 eV at the limit of long chains. These contradictory results for absorption and emission suggest that the absorption process is actually affected by conformational disorder; the latter is strongly reduced in the emission spectra due to migration of the polaron-excitons toward the most ordered conjugated segments. We emphasize that the relaxation energy estimates provided by the analysis of the photoluminescence spectra are in excellent agreement with direct geometric optimizations of the $1B_u$ excited state performed on the two-, three-, and four-ring oligomers [46] within the AM1/CI formalism (semiempirical Hartree–Fock Austin Model 1 method coupled to a configuration interaction scheme).

We now refer to recent experimental measurements carried out on luminescent conjugated polymers that have dealt with estimating the polaron-exciton binding energy. Note that the binding energy is defined as the difference between the creation energy of two noninteracting polarons of opposite charge and the formation energy of a neutral polaron-exciton [74].

Internal photoemission experiments have been performed on polymers in an LED architecture; the measurements give access to the energy difference between the electron and hole injections and thus the energy gap for creation of two polarons of opposite charge. The data provide a value of 2.45 eV in the case of poly(2-methoxy-5-(2′-ethyl-hexyloxy)paraphenylenevinylene), MEH-PPV, taking into account image charge effects and extrapolating to zero photon energy [76]. Since the 0–0 transition of the polymer peaks at 2.25 eV [77], the binding of the electron–hole pair is estimated to be 0.2 eV ($\pm$0.1 eV). Similar experiments using internal field emission yielded binding energies of 0.2 $\pm$ 0.2 eV in the case of MEH-PPV and poly-2-decyloxyparaphenylene, DO-PPP [77]. The recent fabrication of light electrochemical cells (LECs) has also enabled independent measurements of the energy gap to be done [78]; the emission process in such devices is indeed expected to occur for an applied voltage equal to the energy gap. The results collected for various luminescent polymers indicate that the turn-on voltage is always approximately identical to the measured optical gap; the LEC data can thus be consistent with the semiconductor model in which the exciton binding energy is at most a few times k_BT at room temperature. Such binding energy values might, however, constitute a lower limit, since thermally assisted mechanisms could tend to slightly lower the emission threshold.

According to these important experimental measurements, we conclude that the most reasonable values of the binding energy of polaron-excitons in conjugated polymers lie in an energy range between a few kiloteslas (0.1 eV) and at most 0.4 eV, the latter value corresponding to a number of earlier experimental estimates [71,72]. These results further demonstrate that binding energies as high as 1 eV are not reasonable.

It is informative to note that ultraviolet photoelectron spectroscopy (UPS) measurements performed on a PPV sample indicate that the valence band edge is located at 1.55 $\pm$ 0.10 eV below the Fermi energy [79]; assuming that the Fermi level is located in the middle of the gap, the energy gap corresponds to twice this value. Since UPS incorporates neither relaxation effects nor interactions between the departed electron and the remaining hole, we then subtract from this value twice the polaron relaxation energy (2 $\times$ 0.15 eV from our AM1 calculations [80]) and obtain a value of 2.80 eV for the creation energy of two polarons of opposite signs. Since the 0–0 transition of the same PPV sample, i.e., the formation energy of a neutral polaron-exciton, is measured at 2.45 eV, the binding energy of the polaron-excitons is estimated to be on the order of 0.35 eV. Our estimate lies within the range given above; its uncertainty is, however, very difficult to assess, as is also the case in the other experimental measurements we refer to.

It is important to distinguish the contributions to the polaron-exciton binding energy arising from electron correlation effects and electron–lattice coupling. The results derived from both the analysis of the vibronic structures observed in photoluminescence spectra of long PPV oligomers and direct AM1/CI geometry optimizations [46] indicate that the relaxation energy in the lowest neutral excited state amounts to ~0.20 eV. The amplitude of the polaron relaxation energy cannot be evaluated in a simple way by experimental means; as mentioned above, AM1 calculations provide an estimate on the order of 0.15 eV for each polaron. The comparison of the relaxation energy of two polarons to that of a neutral polaron-exciton leads to the conclusion that the lattice contribution to the binding energy is very weak and could actually even be negative. It is worth stressing that such a conclusion would not be expected in the framework of one-electron models where the relaxation energy of a neutral polaron-exciton is found to be equivalent to that of a doubly charged bipolaron and is thus much larger than in a single polaron (up to some 0.5 eV [81]). On the other hand, very recent highly correlated density matrix renormalization group (DMRG) calculations performed at the extended Hubbard level by Shuai and coworkers [82] conclusively demonstrate that the electron–electron contribution to the polaron-exciton binding energy is in the range from 0.1 eV up to at most 0.3 eV.

A most important result we have obtained is thus that *the small value of the polaron-exciton binding energy results from a cancelation of the electron–electron and electron–lattice contributions*; this occurs indirectly via a cancelation of the electron–electron and electron–lattice contributions. Such behavior clearly demonstrates the need to take correlation effects into account when describing the excited state wave functions. (Note that the fact that polarons and polaron-excitons are found to spread over three to four rings [46,80], i.e., over about 25 Å, does not prevent the first optical transition from evolving linearly with the inverse of the number of rings ($1/n$); the binding energy must indeed be considered relative to the single-particle gap energy, which generally evolves as $1/n$.)

We conclude this section by discussing the implications of the presence of a Stokes shift. It should be borne in mind that, by its general definition, the Stokes shift corresponds to the energy separation between the 0–0 peaks in absorption and emission. There is actually much confusion in the literature; indeed, some authors associate the Stokes shift with the separation between the most intense peaks, even if they differ from the 0–0 bands; others have incorrectly claimed that the absence of a Stokes shift means an absence of relaxation in the excited state, even when clear vibronic effects are observed.

The appearance of a Stokes shift can actually be observed in situations where (1) the emission and absorption processes involve different electronic states due to, for instance, intersystem crossing to a lower triplet state or to the existence of a lower energy A_g excited state; (2) conformational disorder induces the photogenerated electron–hole pairs to travel via a random walk process to lowest energy chain segments before light is emitted [83]; or (3) emission takes place in the same segment after time has allowed for an optimal conformation to be adopted, thus implying that the absorption was not to the absolute minimum. In the case of PPV chains, the latter two aspects can be operational; for instance, for twisted chains, point 3 might be due to ring rotations occurring after excitation in order to reach the optimal excited-state coplanar geometry [46].

III. OPTICAL ABSORPTION SPECTRA OF DOPED OLIGOMERS

In this section, we focus on unsubstituted oligothiophenes ranging in size from two to 11 rings. Note that systems with an even number of thiophene units are characterized by C_{2h} symmetry whereas oligomers with an odd number of rings present C_{2v} symmetry. The geometries of the singly and doubly oxidized oligomers are optimized by means of the AM1 method, the radical cations being treated within the restricted open-shell Hartree–Fock (ROHF) formalism. On that basis, the singlet transition energies and relative intensities are estimated with the help of the nonempirical valence effective Hamiltonian (VEH) method [84,85], which is known for its ability to provide reasonable locations of the defect levels inside the gap. The use of a one-electron picture is validated by earlier calculations conducted at the correlated level, showing that the new subgap electron transitions appearing in the absorption spectra of singly and doubly charged oligothiophenes are characterized by a single dominant configuration [86].

The AM1-optimized geometries of the doubly oxidized oligomers indicate that the charged species is localized at the center of the molecule and is characterized by a reversal of the single and double character of the C–C bonds, while the C–S bonds are almost unaffected; the formation of bipolarons thus induces the appearance of a strong quinoidic character within the rings [86]. In contrast, the geometries of the radical cations exhibit weaker structural deformations; the AM1 results provide C–C bond lengths intermediate between those obtained for the neutral and doubly oxidized systems and thus show that the formation of polarons leads to the appearance of a semiquinoidic character along the chain [86]. As the size of the oligomer is increased, the amplitude of the geometric deformations is found to diminish when going from the center to the end of the molecule. We describe in Fig. 1.8 the typical evolution of the C–C bond lengths upon doping; this plot indicates that the bipolaron deformations can be felt over up to nine thiophene units, as also suggested by the theoretical calculations of Ehrendorfer and Karpfen [87], while a weaker spatial extension of five rings is expected for the polarons. However, these estimates have to be considered upper limits owing to the fact that the influence of counterions has been neglected.

The geometric relaxations taking place upon doping are accompanied by a strong modification of the one-electron structure of the oligomers; indeed, two molecu-

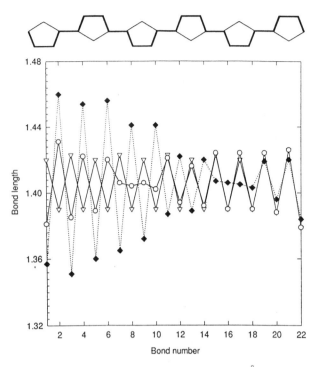

Fig. 1.8 AM1 optimized C–C bond lengths (in Å) of the 11-ring oligothiophene in the neutral state (▽), singly oxidized state (○), and doubly oxidized state (◆).

lar orbitals are shifted inside the gap almost symmetrically to give rise to new subgap absorption peaks in the optical absorption spectra [4,5]. In the case of radical cations, the VEH calculations show the appearance of *two* new subgap features that, in ascending order, originate from an electron transition between the HOMO level and the lower polaronic level (H → POL1) and between the two polaron levels (POL1 → POL2); their transition energies and intensities are reported in Table 1.3. The VEH-calculated transitions are found to be in very good agreement with experimental data measured following chemical doping [88–91], photoinduced absorption [92,93], or voltage-modulation spectroscopy [94]. Moreover, both the theoretical and experimental polaron transitions are found to evolve linearly with the inverse of the number of rings.

The absence of any electron transition between the HOMO level and the upper polaronic level is in fact directly related to the selection rules imposed by the symmetry of the oligomers. Taking account of C_{2v} symmetry, for instance, we find this transition to be strongly limited because the two levels that are involved belong to the same a_2 irreducible representation; this therefore leads to an excitation that is polarized in a direction transverse to the chain axis (the symmetry constraints are even more drastic when dealing with C_{2h} symmetry, for such a transition then becomes forbidden). On the other hand, electron transitions from the a_2 to b_1 levels (or vice versa) give rise to excitations that are polarized along the chain axis and thus present significant intensities, such as those observed in the spectra of charged oligothiophenes. It should be noted that a third symmetry-allowed transition is given by our calculations [86]; this feature, which is weak, is dominantly described by electron transitions between the lower polaronic level and the LUMO + 1 level (POL1 → L + 1) and between the HOMO-1 level and the upper polaronic level (H-1 → POL2); this third absorption peak is calculated to lie within the same energy range as the first excitation of the neutral system.

The formation of positive bipolarons leads to the appearance of a single subgap absorption peak that originates from an electron transition between the HOMO level and the lower bipolaronic level and red-shifts with increasing chain length, as shown in Table 1.3. The symmetry considerations invoked above result in a vanishing intensity for the transition between the HOMO level and the upper bipolaronic level. There is once again excellent agreement between the theoretical values and the spectroscopic data [88–91], both of which demonstrate the existence of a linear relationship between the bipolaron transition energies and the inverse number of rings. We present in Fig. 1.9 the VEH simulated spectra of the seven-ring oligomer in the various oxidation states. Note that the relative intensities of the two polaronic transitions, as obtained within the framework of the VEH one-electron picture, are significantly lowered when similar calculations are conducted at the correlated level [86].

These theoretical results do rationalize a wide range of experimental observations. However, an additional type of charged species has been recently isolated; these

Table 1.3 VEH-Calculated Polaron Transition Energies Between the HOMO Level and the Lower Polaronic Level (H → POL1) and Between the Two Polaron Levels (POL1 → POL2) and Bipolaron Excitation Energies Between the HOMO Level and the Lower Bipolaronic Level (H → BIP1) in Thiophene Oligomers Containing Three, Five, Seven, and Nine Rings[a]

Number of rings	POL1 → POL2	H → POL1	H → BIP1
3	2.72 (30.1)	1.58 (37.0)	2.01 (38.4)
5	1.64 (50.8)	1.08 (58.7)	1.30 (63.0)
7	1.39 (65.9)	0.82 (74.6)	0.99 (83.5)
9	1.26 (77.2)	0.64 (86.7)	0.81 (101.4)

[a] The relative intensities of the transitions (in arbitrary units) are reported in parentheses.

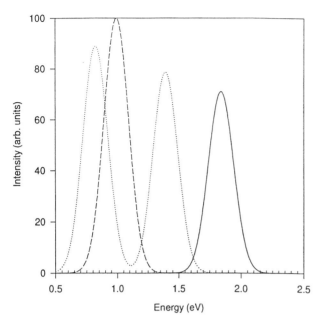

Fig. 1.9 VEH calculated absorption spectra of heptathiophene in the neutral state (——), singly oxidized state (· · ·), and doubly oxidized state (– – –). The spectra are simulated by convolution with Gaussians whose full width at half-maximum is set to 0.2 eV.

species are referred to as π-dimers and correspond to complexes formed upon interaction of two polaron-carrying oligomers [95,96]. Such defects, which are spinless, are invoked when two subgap absorption features are observed in the optical absorption spectra of lightly doped oligomers where no paramagnetic signal is detected, thus excluding the formation of polarons; the differentiation with respect to bipolarons is straightforward as the latter would lead to a single subgap feature. These considerations have led to a new interpretation of the absorption spectrum of a doped dodecamer where spinless species are generated and where the two strong subgap absorption peaks are present (these were initially assigned to bipolaronic transitions) [97]. We note that a model based on the interaction of two bipolarons on the same chain would also be consistent with such experimental data. Indeed, the formation of interacting bipolarons results in the appearance of four molecular levels inside the gap due to the splitting of the defect levels and gives rise to the appearance of two new intense subgap peaks in the spectra, as we have shown on the basis of correlated calculations [86]. However, in this dodecamer case, such an assumption is not consistent with the rates of doping reached in the experiment, namely two charges per molecule. The possible explanation is then based on the formation of a fourfold oxidized entity corresponding to a double π dimer; this hypothesis appears to be confirmed by recent cyclic voltabsorptometry measurements [98].

The existence of a single subgap absorption feature in the optical absorption spectra of doubly charged oligomers is in contrast to the typical properties of conjugated polymers supporting charged bipolarons where two intense subgap peaks are observed [3]. One reason that could be invoked in an attempt to rationalize this discrepancy is the fact that the symmetry constraints governing the nature of the optical transitions in the oligomers do not have to hold true at the scale of a long disordered polymer chain. However, the breakdown of the symmetrical relaxation of the defects does not give any significant changes in the aspect of the spectra, as revealed by calculations where we require the bipolarons to each be localized near one end of the chain [86]. As a consequence, the spectra of polymers supporting isolated bipolarons should be characterized by a dominant subgap feature at low doping level. As we mentioned above, an important aspect, however, is that the lineshape of the spectra is strongly affected as soon as interaction between the charged defects takes place, leading to the appearance of two intense subgap features such as those observed in the spectra of conjugated polymers at high doping level [99].

The signature of isolated defects should in principle be detected in photoinduced absorption experiments, since the concentration of the photogenerated charged species is expected to be very weak. So far, the two long-lived subgap absorption features observed in such polymer photoinduced absorption spectra have been assigned as bipolaronic transitions [100]. This interpretation is, however, in contrast with the results of the calculations we presented above. In order to unravel this discrepancy, we suggest either that polarons constitute more appropriate candidates to rationalize the aspect of the photoinduced absorption spectra or that long-lived bipolarons are trapped in crystalline regions where they may interact; note also that an additional origin for the photoinduced features could be the formation of π dimers in the most ordered regions of the samples.

The analysis we have presented is actually representative of any conjugated system that possesses a high degree of symmetry in its geometric structure. As a matter of fact, the same trends prevail when we look at the spectra of doped oligopyrroles [101], oligophenylenes [102,103], or oligophenylenevinylenes [104]. A single subgap peak is also induced upon double ionization of short polyenes where charge storage occurs through generation of soliton pairs [105,106]. We note that the formation of polarons in short polyenes is accompanied by the appearance of a single dominant feature. In this specific case, the one-electron picture does not hold because further symmetry considerations have to be addressed; this is due to the fact that the two possible polaron transitions possess almost the same energy [107].

IV. NONLINEAR OPTICAL RESPONSE OF DONOR–ACCEPTOR POLYENES

Organic compounds with delocalized π-electron systems are leading candidates for nonlinear optical (NLO) materials, and interest in these materials has grown tremendously in the past decade [108–118]. Reliable structure–property relationships—where property here refers to first-order (linear) polarizability α, second-order polarizability β, and third-order polarizability γ—are required for the rational design of optimized materials for photonic devices such as electro-optic modulators and all-optical switches. Here also, quantum-chemical calculations can contribute a great deal to the establishment of such relationships. In this section, we illustrate their usefulness in the description of the NLO response of donor–acceptor substituted polymethines, which are representative of an important class of organic NLO chromophores. We also show how much the nonlinear optical response depends on the interconnection between the geometric and electronic structures, as was the case of the properties discussed in the previous sections [119].

The theoretical studies we review here have been stimulated by the seminal work [120–123] of, and performed in collaboration with, Marder and coworkers [124–126]. Their main thrust has been to offer a unified picture of linear and nonlinear polarization in π-conjugated organic chromophores [124], which paves the way for the design of a wide range of novel materials. In the following, we describe how, in order to provide such guidelines, it is essential to cast in terms of simple, often well-known, models the results of more sophisticated calculations carried out at the correlated level [125,126].

The ground-state structure of donor–acceptor polymethines can be viewed as a combination of resonance forms, differing in the extent of charge separation (see Fig. 1.10). In substituted polyenes with weak donors and acceptors, the neutral resonance form dominates the ground state and the molecule has a structure with a distinct alternation in the bond lengths between neighboring carbon atoms, i.e., a high degree of bond length alternation (BLA). The contribution of the fully charge-separated resonance form to the ground state increases, and BLA decreases, when donor and acceptor substituents become stronger. The relative contributions of these resonance forms to the ground state is also controlled by the polarity of the solvent in which the chromophore is dissolved; a more polar solvent increases the ground-state polarization, which makes the charge-separated form more dominant. Experiments have clearly demonstrated that the medium influences to a large extent the molecular geometry [127–130] and, as a direct consequence, the NLO properties [131–133]. This effect of the solvent medium on the NLO properties is consistent with the linear optical solvatochromic response well documented for many organic dyes [134]. As a result, calculations on isolated molecules (i.e., in the "gas phase") are helpful but are far from being able to accurately reproduce experimental data such as the

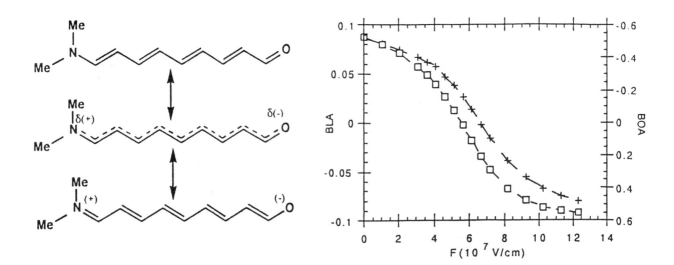

Fig. 1.10 *Left*: Canonical resonance structures for the 1-dimethylamino-1,3,5,7-octatetraen-8-al molecule, DAO. (*Top*) Neutral polyene limit; (*middle*) cyanine limit; (*bottom*) zwitterionic charge-separated limit. *Right*: Corresponding evolution of bond length alternation BLA (in Å) (x) and bond order alternation BOA (□), as a function of the applied external electric field F.

results of solvent-dependent hyperpolarizability measurements [131,132]. Consequently, an important new step in the calculation of NLO properties is to take explicitly into account the influence of the medium surrounding the NLO chromophore. Most treatments of the influence of dielectric medium on the NLO response have been based on the reaction field model of Onsager [135–139]. We return to the reaction field approach at the end of this section but first take a simpler step beyond "gas-phase" calculations by examining the influence of an external static electric field on chromophores such as linear polymethine dyes. This external static electric field has been found to be efficient in controlling the ground-state polarization of the chromophores and therefore useful in simulating the qualitative effect of the solvent medium on nuclear geometry, electronic structure, and optical properties [140]. The main advantage of this approach relative to the reaction field approach is to allow us to drive the structure all the way from the neutral polyene form to the fully charge-separated form and thus to obtain a coherent picture of the full evolution.

A. Evolution of the NLO Response upon Application of an External Field

The molecule we take as example here is 1-dimethylamino-1,3,5,7-octatetraen-8-al molecule (DAO; see Fig. 1.10). (Note that we use this unofficial nomenclature to emphasize the donor–polyene–acceptor character of the molecule.) Once again, the calculations combined several approaches based on the semiempirical intermediate neglect of differential overlap (INDO) Hamiltonian and proceeded in three steps:

1. Optimization of the molecular structure under the influence of an external homogeneous static electric field applied along the long axis of the molecule and directed in such a way as to favor charge transfer from the donor to the acceptor
2. Calculation of the transition energies, state dipole moments, and transition dipole moments, using the configuration interaction (CI) technique with single and double excitations to include electron correlation effects
3. Evaluation via the sum-over-states (SOS) expression of the polarizability α and hyperpolarizabilities β and γ

In Fig. 1.10, we follow the change in geometry for DAO when a static external electric field F is applied along the long axis of the molecule. A field strength of about 2×10^7 V/cm (in the direction that favors charge transfer from the donor end to the acceptor end) is required for the geometry to start evolving in such a way that the single bonds become significantly shorter and the double bonds longer, compared to the situation where no field is applied. Increasing the strength of F induces a continuous decrease in BLA, which eventually becomes equal to zero; such a zero-BLA situation is referred to as the *cyanine limit*. In the case of the donor–acceptor pair in the DAO molecule, the cyanine limit is reached for an electric field of approximately 6.5 $\times 10^7$ V/cm. An alternation in the single/double bond pattern resumes at higher field strengths. However, this alternation is reversed with respect to the situation in the neutral polyene limit, and the ground-state structure becomes increasingly dominated by the zwitterionic resonance form. Complete BLA reversal occurs at a field of about 10^8 V/cm. Note that the difference in π-bond orders between adjacent bonds defines a bond-order alternation (BOA), which is directly related to BLA in conjugated compounds, as can be deduced from Fig. 1.10.

In π-conjugated organic systems, as we have stressed in numerous instances, the molecular geometry and electronic structure are strongly related. Since the geometry, as probed by the BLA or BOA values, strongly evolves with the field, so does the electronic structure. When the field increases from zero (polyene limit; BOA ≈ 0.5) to 5×10^7 V/cm (cyanine limit; BOA ≈ 0), the first transition energy decreases from about 4.0 eV to 2.4 eV. That corresponds to a situation where, through the interaction with the electric field, the first excited state S_1 is preferentially stabilized with respect to the ground state S_0 (bathochromic shift). This is easily explained by the fact that the S_1 state dipole moment is larger than that in the ground state; the dipole moment thus increases during the electronic transition. This red shift in the $S_0 \rightarrow S_1$ transition takes place up to the point where the ground state becomes described by equal contributions from the polyene form and the fully charge-separated form, i.e., in the cyanine limit. Beyond this point, under the influence of an electric field $F > 5 \times 10^7$ V/cm (BOA > 0), a blue shift is observed; the ground state is then more stabilized than the first excited state through the interaction with the electric field, the dipole moment decreasing during the electronic transition. This electrochromic behavior is consistent with the closely related solvatochromic behavior observed for numerous merocyanine molecules [134].

Since the DAO molecule has a linear shape, the molecular polarizabilities are totally dominated by the longitudinal tensor components (here along the x axis). In Fig. 1.11, the evolution of the α_{xx}, β_{xxx}, and γ_{xxxx} components is plotted versus BOA (and thus as a function of F). The linear polarizability α_{xx} follows a simple evolution: it peaks at the cyanine limit (BOA≈0), with the minimum values associated with the neutral polyene and charge-separated forms, in agreement with experimental observation [141]. It is of interest to compare the fully converged SOS values, α_{xx}[SOS], to the results obtained with a simplified formula in which only a single excited state with a high one-photon absorptivity (usually S_1) is

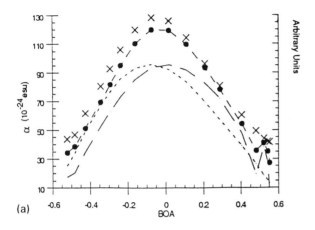

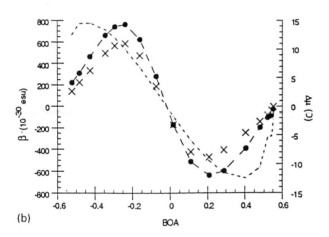

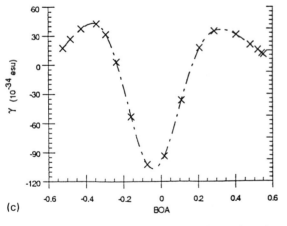

Fig. 1.11 Molecule of DAO. Evolutions plotted versus BOA. (a) α_{xx}[SOS] (x) and α_{xx}[model] (-●-) in 10^{-24} esu; the inverse of E_{ge} (---) and the transition dipole moment M_{ge} (---) in arbitrary units. (b) β_{xxx}[SOS] (x) and β_{xxx}[model] (-●-) in 10^{-30} esu; $\Delta\mu$ (---) in debye. (c) γ_{xxxx}[SOS], in 10^{-36} esu.

considered, i.e., a two-state (S_0 and S_1) model. In this case, the SOS expression for α_{xx} simplifies to

$$\alpha_{xx}[\text{model}] = 2\frac{\langle g|\mu_x|e\rangle\langle e|\mu_x|g\rangle}{E_{ge}} = 2\frac{M_{ge}^2}{E_{ge}}$$

where $|g\rangle = S_0$ and $|e\rangle = S_1$; the linear polarizability in this model, α_{xx}[model], is thus simply proportional to the square of the transition dipole moment between the ground state and the first excited state, M_{ge}, and inversely proportional to the first transition energy, E_{ge}. The electric field dependence of M_{ge} and $1/E_{ge}$ are reported in the top of Fig. 1.11, where they are plotted versus BOA. The shape and magnitude of the curve for linear polarizability versus BOA are well reproduced by the simple model. The "two-state model" thus constitutes a good approximation to describe the F dependence of the linear polarizability in such donor–acceptor polyenes. Through this model, the evolution of α can be easily understood. The first transition energy evolves with the applied electric field in such a way that, as a function of BOA, $1/E_{ge}$ peaks at the cyanine limit (Fig. 1.11a). Furthermore, the transition dipole moment M_{ge} between the ground state and S_1 evolves in a similar way as a function of BOA (note that M_{ge} is maximized at the cyanine limit since in that situation the S_0 and S_1 wave functions are most similar). The product $1/E_{ge}$ times M_{ge}^2, which constitutes the α_{xx}[model], then evolves, as a function of BOA, in such a way as to be maximized at the cyanine limit, where BOA = 0.

The electric field dependence of the first hyperpolarizability β is more complex than that of α: β is seen to first increase with increasing magnitude of the field, peak in a positive sense, decrease, pass through zero, become negative, peak in a negative sense, and finally increase again to become small for the charge-separated form of the DAO molecule (Fig. 1.11b). We can also compare the SOS results, β_{xxx}[SOS], with those of the traditional two-state model proposed by Oudar and Chemla [142,143], the β_{xxx}[model], in which only the ground state $|g\rangle$ and the first excited state $|e\rangle$ are considered. In this case, the SOS expression for β_{xxx} simplifies to

$$\beta_{xxx}[\text{model}] = 6\frac{\langle g|\mu_x|e\rangle(\langle e|\mu_x|e\rangle - \langle g|\mu_x|g\rangle)\langle e|\mu_x|g\rangle}{E_{ge}^2}$$
$$= 6\frac{M_{ge}^2 \Delta\mu}{E_{ge}^2}$$

where $\Delta\mu$ is then the difference between the dipole moments in the first excited state and the ground state. As has been discussed by Marder and coworkers [120,121], the shape for β versus BOA results from a compromise between $\Delta\mu$ on the one hand and $1/E_{ge}^2$ and M_{ge}^2 on the other hand. This model has provided design guidelines for optimization of β that have resulted in molecules with significantly improved properties [123].

The evolution of β_{xxx}[model] as a function of BOA is similar to that of β_{xxx}[SOS] (Fig. 1.11f); the model also reproduces the magnitude of β with an accuracy of $\approx 80\%$. As noted before, the $1/E_{ge}$ (and therefore $1/E_{ge}^2$) and M_{ge}^2 terms peak at the cyanine limit. The evolution of the $\Delta\mu$ term is very different and constitutes the origin of the more complex evolution of β vs. α. Starting from the polyene limit, $\Delta\mu$ first increases and reaches a maximum around BOA = -0.45 (corresponding to a degree of bond length alternation of ≈ 0.05 Å). Thereafter, it decreases and passes through zero at roughly the cyanine limit; this means that at this point S_0 and S_1 possess identical dipole moments, which both correspond to a partial charge transfer from the donor to the acceptor. $\Delta\mu$ then becomes increasingly negative and reaches a minimum around BOA = $+0.40$; after that, it increases again and tends toward a small negative value in the zwitterionic limit (Fig. 1.11b). The small $\Delta\mu$ in that limit is a consequence of large but nearly equal dipole moments in both the S_1 and S_0 states. Overall, we thus observe that the evolution of β_{xxx}[model] as a function of BOA is very similar to that of $\Delta\mu$, except that the two extrema for β are pushed closer to the cyanine limit due to the influence of the M_{ge}^2/E_{ge}^2 term peaking at that limit (the positive and negative peaks for β occur at BOA values equal to -0.25 and $+0.20$, respectively).

The experimental demonstration of the dependence of β in donor/acceptor substituted polyenes on the ground-state polarization and consequently the bond alternation has been reported [123,124]. In these experiments, solvents of increasing polarity and substituents of increasing strength are used to control the molecular structure of the chromophores; simultaneously, the $\mu\beta$ values are obtained from EFISHG measurements. The results indicate that the various regions in the β curve evolution can be mapped out by using different donor/acceptor polyenes. This allows for fine control in the design of new chromophores for second-order NLO properties.

The evolution of the second-order hyperpolarizability γ as a function of the external field F gets more complex than those of α and β (see Fig. 1.11c). The γ curve has two positive peaks associated with two intermediate forms that appear (1) between the neutral polyene and cyanine-like structures and (2) between the cyanine-like and totally charge-separated forms. At the cyanine limit, γ has a large negative peak. A negative value of γ physically corresponds to a situation where the molecular polarizability decreases with increasing laser intensity. This negative nonlinear refractive index leads to the self-defocusing of an intense beam of light. We do not detail here the origin of the γ evolution, which we have discussed at length elsewhere [125,126]; we stress, however, that we have also been able to follow a three-term model for γ that allows us to rationalize its dependence on the field. Combined with the experimental data of Marder and coworkers, the theoretical results provide structural guidelines for the optimization of γ in either a positive or negative sense.

A most important result is obtained by observing in Fig. 1.11 that the linear polarizability α, the first hyperpolarizability β, and the second hyperpolarizability γ are seen to be derivatives, with respect to the structural parameter BOA (influenced by the external field), of their next lower order polarization (for α) or polarizability (for β and γ). These derivative relations therefore provide a unified description of the linear and nonlinear polarization and of the dependence of the polarizability and hyperpolarizabilities on the structure in linear polymethine dyes [124]. A consequence to be stressed is that it is not possible, for a given compound, to find conditions that would simultaneously optimize the β and γ responses: When β reaches a peak, γ tends to a minimum.

B. Self-Consistent Reaction Field Approach

Here we discuss a joint theoretical and experimental study of the influence of solvent polarity on the second-order molecular polarizability β of p-nitroaniline and the push-pull polyene 1,1-dicyano-6-dibutylamine-hexatriene [144]. The calculations are carried out at the Hartree–Fock ab initio level on the basis of an *expanded* self-consistent reaction field approach and are compared to hyper-Rayleigh scattering measurements performed in solvents with a wide range of dielectric constants.

Given the sensitivity of the nonlinear optical response of π-conjugated molecules on solvent polarity, various theoretical approaches have been used to simulate the effect of solvent on the first hyperpolarizability β: (1) by putting point charges around the molecule [122]; (2) by considering a valence–bond charge transfer solvation model [145]; (3) as we did in the previous section, by taking account of a homogeneous external electric field [124–126]; or (4) by exploiting Onsager's reaction field theory in a simple form, *i.e.*, considering the solute to occupy a spherical cavity within the solvent and/or restricting the solute–solvent interaction to the dipolar term [133,135–139]. Although such a reaction field approach could be a priori considered to be well adapted to the problem, it must be borne in mind that most π-conjugated molecules possessing large second- or third-order NLO responses present a quasi-one-dimensional character and a complex pattern of their π-charge distribution; accordingly, the use of a spherical cavity and the dipolar approximation can prove to be inadequate. We demonstrated the validity of a more refined approach to model the effects of solvent polarity on NLO properties, in which we used an expanded self-consistent reaction field (SCRF) theory; in this case, the self-consistent solute–solvent interactions are described by multipolar terms up to 2^6 poles and the solute is taken to occupy an ellipsoidal cavity [144,146–148].

We have examined p-nitroaniline (PNA, I, see below), whose structure and NLO properties are only moderately sensitive to solvent polarity, and the push-pull polyene 1,1-dicyano-6-dibutylamine-hexatriene (DCH, II), which is much more strongly affected by solvent effects.

The use of solvents of different polarity can modulate somewhat the structure of the molecule from a neutral (form a) to a charge-separated zwitterionic structure (form b) for which the degree of bond length alternation δr is completely reversed.

I NH_2—⟨⟩—NO_2 ↔ NH_2=⟨⟩=NO_2

II NR_2—/=\—/=\—/=\—C(CN)(CN)   ↔ NR_2=/—\=/—\=/—\=C(CN)(CN)

(a)   (b)

The theoretical β values for these molecules in solvents of different polarity have been compared with experimental data from hyper-Rayleigh scattering (HRS) measurements performed at 1064 nm in dichloromethane ($\epsilon = 9.08$), 1,4-dioxane ($\epsilon = 2.23$), and mixtures of these two solvents in 20/80, 40/60, 60/40, 80/20 (v/v) proportions corresponding to ϵ values of 3.33, 4.47, 5.73, and 7.25, respectively. These solvents have been selected for their good miscibility, the high solubility of the NLO chromophores in the mixtures, and their almost equal indices of refraction (1.4224 for 1,4-dioxane and 1.4242 for dichloromethane at the sodium D line); the Lorentz local field correction factors, which depend only on the index of refraction, can therefore be assumed to be the same for the various mixtures. In the case of DCH, we have also performed HRS measurements in acetone ($\epsilon = 20.7$) and acetonitrile ($\epsilon = 37.5$). A major advantage of the HRS technique is the possibility of obtaining directly the value of β, while the more traditional electric field induced second-harmonic generation (EFISHG) measurements provide only the $\mu\beta$ value. Due to the difficulties in extracting μ in different solvents, the HRS method can thus provide more accurate β determinations.

For each value of the dielectric constant, the geometric optimization of the molecules and the second-order polarizability β calculations have been performed at the restricted Hartree–Fock ab initio level using a 3–21 G split valence basis set (which was found earlier to be satisfactory to obtain reliable trends in β values). Note that because of the size of the DCH molecule, we have considered the amino donor group in our calculations rather than the dibutylamino group. The components of β are calculated analytically via electric field derivatives of the total energy within a coupled perturbed Hartree–Fock approach.

We have compared the theoretical static β_v values to the experimental HRS data for PNA and DCH for solvents with ϵ in the range 2.23–37.5. For DCH, we observe some β enhancement for the HRS measurements due to the resonance effect as the experimental two-photon resonance occurs at 960 nm (twice l_{max}) and is not far from the wavelength used in the HRS experiment (1064 nm). As a result, the corresponding HRS β data have been extrapolated to the zero frequency values using the dispersion factor of the two-state model, as was done recently in a similar context. Although the calculated values are somewhat smaller than the HRS data (which is expected owing to the neglect of electron correlation contributions in the calculations), there is excellent agreement with the experimental trends. In particular, for PNA we find an increase in β within the range of solvent polarity. It is worth stressing that the β increase in passing from the gas phase ($\epsilon = 1$) to dioxane, characterized by a low dielectric constant ($\epsilon = 2.2$), is as strong as that in passing from dioxane to acetonitrile, which is a much more polar solvent ($\epsilon = 37.5$). In the case of DCH, *β exhibits a peak as a function of solvent polarity* (see Fig. 1.12), as was recently deduced from $\mu\beta$ determinations by EFISHG. The calculations not only reproduce this peak behavior but also provide the same peak position ($\epsilon = 7.25$). Furthermore, the bond length alternation (δr) at which this peak occurs, 0.05 Å, is in excellent agreement with predictions of the molecular structure providing an optimal β response as well as with previous semiempirical calculations in which the molecule was polarized using a static homogeneous electric field. We stress that the weak solvent sensitivity of the β value in PNA is due to the aromaticity of the benzene ring, which inhibits electron polarization in the ground state. In contrast, the ground-state structure and β of the DCH molecule are much more easily modulated by the polarity of the solvent.

In order to analyze the importance of taking account in the calculations of both an ellipsoidal shape of the cavity and a multipolar expansion of the interaction energy, we present, in Fig. 1.12, the different curves of β as a function of ϵ in the case of DCH, as calculated by considering (1) an ellipsoidal cavity and a multipolar expansion ($l = 6$), (2) an ellipsoidal cavity but a dipolar approximation ($l = 1$), and (3) a multipolar expansion ($l = 6$) but a spherical cavity shape. We observe that the latter two types of calculations lead to β values that keep growing within the solvent dielectric constant range and do not provide the peak behavior of β. Actually, *both kinds of approximations underestimate the interaction energy* between the solvent and the molecule; as a result, even in very polar solvents the medium-induced electronic and structural molecular changes re-

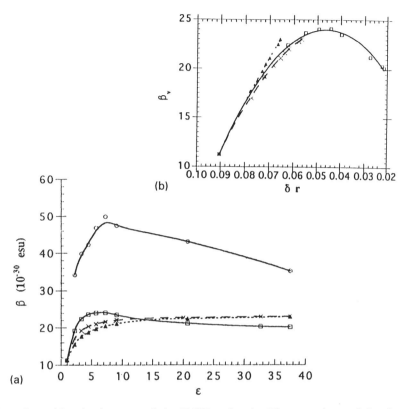

Fig. 1.12 (a) Evolution of β with ϵ in the case of the DCH molecule. The experimental β values extrapolated to zero frequency are represented as ($\bigcirc$); the 3-21G SCRF β_V values are plotted for three cases: ($\square$) $l = 6$ and ellipsoidal cavity shape; ($\times$) $l = 1$, ellipsoidal shape; ($\blacktriangle$) $l = 6$, spherical shape. (b) The same three series of theoretical β values as in (a) are given as a function of the degree of bond length alternation δr (in Å), optimized in each case for the same range of ϵ values.

main too weak for the β peak to be reached (see insert of Fig. 1.12): Using a spherical cavity results in overly large distances between the medium and the molecule in the directions perpendicular to the molecular long axis; restricting the interaction to the dipolar term makes it impossible to account for the complex π-charge distribution (presenting charge alternation) along the polyene segment.

In conclusion, the very good agreement with experiment validates the expanded SCRF theory as a powerful theoretical tool to study the solvent dependence of β; we emphasize that both a multipolar expansion of the interaction energy and the use of an ellipsoidal cavity shape in the SCRF theory are necessary [144]. It is especially relevant that this model is able to predict the nature of the solvent that maximizes the value of β; it is thus of particular interest to apply it for tuning of the NLO response in solution. Finally, we stress that our results confirm recent statements made about the effect of the medium on hyperpolarizability–structure relationships.

V. SUMMARY

We have presented an overview of some of our recent work in the field of conjugated oligomers and polymers. The examples we have chosen have allowed us to scan a wide range of topics where we feel quantum chemistry can bring significant contributions. We therefore hope that this review can stimulate further developments and applications of quantum-chemical methods in the field of advanced materials.

As should be apparent from our discussion, different problems require different sets of computational techniques to be put in play. Furthermore, if we have privileged in the examples we have given the use of Hartree–Fock-based techniques, there is no doubt that density functional theory methods are enjoying an increasing role, especially when dealing with metallic systems or when looking at dynamical properties. Hence, a major skill required of the quantum chemist is to be able to make the best judgment as to the choice of an appropriate methodology. It must be emphasized that performing the largest possible number of ab initio calculations is not necessarily the way to proceed: Simpler techniques, even though they might be less accurate on an absolute basis, can in many instances provide excellent physical insight; in the long run, such insight is often extremely valuable in the design of novel materials.

As for the future, a new challenge at the quantum-chemical level is to investigate interchain interactions, which are of major importance in real materials, for instance to optimize the functioning of light-emitting diodes or the efficiency of nonlinear optical response. This is a demanding task as it requires that the interactions between large conjugated segments that are bonded only via van der Waals interactions be taken properly into account; correlation effects are thus essential. We believe, however, that it is necessary extension of current work.

ACKNOWLEDGMENTS

We gratefully acknowledge stimulating collaborations and discussions with C. Adant, C. Dehu, R. H. Friend, A. J. Heeger, A. B. Holmes, R. Lazzaroni, S. R. Marder, J. W. Perry, B. M. Pierce, C. Quattrocchi, A. Persoons, W. R. Salaneck, D. A. dos Santos, H. Schenk, Z. Shuai, R. Silbey, E. Staring, and G. Zerbi. The conjugated polymer work reviewed in this chapter has been partly supported by the Belgian Prime Minister's Office of Science Policy (SSTC) "Pôle d'Attraction Interuniversitaire en Chimie Supramoléculaire et Catalyse," the Belgian National Fund for Scientific Research FNRS/FRFC, the European Commission [SCIENCE program (Project 0661 POLYSURF), Human Capital and Mobility program, and ESPRIT program (Project 8013 LEDFOS and Network of Excellence on Organic Materials for Electronics NEOME)], Hoechst AG (in the framework of the European Commission BRITE-EURAM program PolyLED), and an IBM Academic Joint Study.

REFERENCES

1. T. A. Skotheim (ed.), *Handbook of Conducting Polymers*, Marcel Dekker, New York, 1986.
2. J. L. Brédas, in *Handbook of Conducting Polymers* (T. A. Skotheim, ed.), Marcel Dekker, New York, 1986, p. 859.
3. R. R. Chance, D. S. Boudreaux, J. L. Brédas, and R. Silbey. in *Handbook of Conducting Polymers* (T. A. Skotheim, ed.), Marcel Dekker, New York, 1986, p. 825.
4. A. J. Heeger, S. Kivelson, J. R. Schrieffer, and W. P. Su, *Rev. Mod. Phys.* 60:781 (1988).
5. J. L. Brédas and G. B. Street, *Acc. Chem. Res. 18*: 319 (1985).
6. W. P. Su, J. R. Schrieffer, and A. J. Heeger, *Phys. Rev. Lett. 42*: 1698 (1979); *Phys. Rev. B* 22:2209 (1980).
7. W. R. Salaneck, S. Stafström, and J. L. Brédas, *Conjugated Polymer Surfaces and Interfaces: Electronic and Chemical Structure of Interfaces for Polymer Light Emitting Devices*, Cambridge Univ. Press, Cambridge, U.K., 1997.
8. J. M. André, J. Delhalle, and J. L. Brédas, *Quantum Chemistry Aided Design of Organic Polymers. An Introduction to the Quantum Chemistry of Polymers and Its Applications*, World Scientific, Singapore, 1991.
9. J. H. Burroughes, D. D. C. Bradley, A. R. Brown, R. N. Marks, K. Mackay, R. H. Friend, P. L. Burn, and A. B. Holmes, *Nature* 347:539 (1990).
10. R. W. Gymer, *Endeavour* 20:115 (1996).
11. G. Gustafsson, Y. Cao, G. M. Treacy, F. Klavetter, N. Colaneri, and A. J. Heeger, *Nature 357*:477 (1992).
12. D. G. Lidzey, M. S. Weaver, T. A. Fisher, M. A. Pate, D. D. C. Bradley, and M. S. Skolnick, *Synth. Met., 76*:129 (1996).
13. I. D. W. Samuel, B. Crystall, G. Rumbles, P. L. Burn, A. B. Holmes, and R. H. Friend, *Chem. Phys. Lett.* 213:472 (1993).
14. T. Granier, E. L. Thomas, D. R. Gagnon, F. E. Karasz, and R. W. Lenz, *J. Polym. Sci. B* 24:2793 (1986).
15. D. D. C. Bradley, R. H. Friend, H. Linderberger, and S. Roth, *Polymer* 27:1709 (1986).
16. D. Braun, E. G. J. Staring, R. C. J. E. Demandt, G. L. J. Rikken, Y. A. R. R. Kessener, and A. H. J. Venhuizen, *Synth. Met.* 66:75 (1994).
17. P. L. Burn, A. Kraft, D. R. Baigent, D. D. C. Bradley, A. R. Brown, R. H. Friend, R. W. Gymer, A. B. Holmes, and R. W. Jackson, *J. Am. Chem. Soc. 115*:10117 (1993).
18. A. R. Brown, D. D. C. Bradley, J. H. Burroughes, R. H. Friend, N. C. Greenham, P. L. Burn, A. B. Holmes, and A. Kraft, *Appl. Phys. Lett.* 61:2793 (1992).
19. P. L. Burn, A. B. Holmes, A. Kraft, D. D. C. Bradley, A. R. Brown, R. H. Friend, and R. W. Gymer, *Nature 356*:47 (1992).
20. D. A. dos Santos, C. Quattrocchi, R. H. Friend, and J. L. Brédas, *J. Chem. Phys.* 100:3301 (1994).
21. L. S. Swanson, J. Shinar, A. R. Brown, D. D. C. Bradley, R. H. Friend, P. L. Burn, A. Kraft, and A. B. Holmes, *Phys. Rev. B* 46:15072 (1992).
22. D. Braun and A. J. Heeger, *Thin Solid Films* 216:96 (1992).
23. S. Doi, M. Kuwabara, T. Noguchi, and T. Ohnishi, *Synth. Met.* 55:4174 (1993).
24. J. L. Brédas and A. J. Heeger, *Chem. Phys. Lett.* 217:507 (1994).
25. N. C. Greenham, S. C. Moratti, D. D. C. Bradley, R. H. Friend, and A. B. Holmes, *Nature* 365:628 (1993).

26. D. R. Baigent, N. C. Greenham, J. Grüner, R. N. Marks, R. H. Friend, S. C. Moratti, and A. B. Holmes, *Synth. Met.* 67:3 (1994).
27. X. C. Li, A. B. Holmes, A. Kraft, S. C. Moratti, G. C. W. Spencer, F. Cacciali, J. Grüner, and R. H. Friend, *J. Chem. Soc. Chem. Comm.* 21:2211 (1995).
28. R. E. Gill, G. G. Malliaras, J. Wildeman, and G. Hadziioannou, *Adv. Mater.* 6:132 (1994).
29. M. Berggren, G. Gustafsson, O. Inganäs, M. R. Andersson, O. Wennerström, and T. Hjertberg, *Adv. Mater.* 6:488 (1994).
30. M. Berggren, O. Inganäs, G. Gustafsson, J. Rasmusson, M. R. Andersson, T. Hjertberg, and O. Wennerström, *Nature* 372:444 (1994).
31. P. Dyreklev, M. Berggren, O. Inganäs, M. R. Andersson, O. Wennerström, and T. Hjertberg, *Adv. Mater.* 7:43 (1995).
32. F. Geiger, M. Stoldt, H. Schweizer, P. Bauërle, and E. Umbach, *Adv. Mater.* 5:922 (1993).
33. G. Horowitz, P. Delannoy, H. Bouchriha, F. Deloffre, J. L. Fave, F. Garnier, R. Hajlaoui, M. Heyman, F. Kouki, P. Valat, V. Wintgens, and A. Yassar, *Adv. Mater.* 6:752 (1994).
34. B. Xu and S. Holdcroft, *J. Am. Chem. Soc.* 115:8447 (1993).
35. M. J. S. Dewar, E. G. Zoebisch, E. F. Healy, and J. J. P. Stewart, *J. Am. Chem. Soc.* 107:3702 (1985).
36. M. J. S. Dewar and W. Thiel, *J. Am. Chem. Soc.* 99:4899 (1977).
37. C. J. Finder, M. G. Newton, and N. L. Allinger, *Acta Crystallogr. B* 30:411 (1974).
38. S. Hotta and K. Waragai, *J. Mater. Chem.* 1:835 (1991).
39. P. Tavan and K. Schulten, *J. Chem. Phys.* 85:6602 (1986).
40. J. A. Pople, D. L. Beveridge, and P. A. Dobosh, *J. Chem. Phys.* 47:2026 (1967).
41. J. Ridley and M. C. Zerner, *Theoret. Chim. Acta* 32:111 (1973).
42. B. J. Orr and J. F. Ward, *Mol. Phys.* 20:513 (1971).
43. S. Heun, R. F. Mahrt, A. Greiner, U. Lemmer, H. Bässler, D. A. Halliday, D. D. C. Bradley, P. L. Burn, and A. B. Holmes, *J. Phys.: Condens. Matter* 5:247 (1993).
44. R. J. Buenker and S. Peyerimhoff, *Theoret. Chim. Acta* 35:33 (1974).
45. D. Beljonne, J. Cornil, J. L. Brédas, R. H. Friend, and R. A. J. Janssen, *J. Am. Chem. Soc.*, 118:6453 (1996).
46. D. Beljonne, Z. Shuai, R. H. Friend, and J. L. Brédas, *J. Chem. Phys.* 102:2042 (1995).
47. J. L. Brédas, R. R. Chance, and R. Silbey, *Phys. Rev. B* 26:5843 (1982).
48. L. Salem, *Molecular Orbital Theory of Conjugated Systems*, Benjamin, New York, 1966.
49. J. Cornil, D. Beljonne, R. H. Friend, and J. L. Brédas, *Chem. Phys. Lett.* 223:82 (1994).
50. M. J. Rice and Y. N. Gartstein, *Phys. Rev. Lett.* 73:2504 (1994).
51. J. Cornil, D. Beljonne, Z. Shuai, T. Hagler, I. Campbell, D. D. C. Bradley, J. L. Brédas, C. W. Spangler, and K. Müllen, *Chem. Phys. Lett.* 247:425 (1995).
52. J. Guay, P. Kasai, A. Diaz, R. Wu, J. M. Tour, and L. H. Dao, *Chem. Mater.* 4:1097 (1992).
53. J. Grimme, M. Kreyenschmidt, F. Uckert, K. Müllen, and U. Scherf, *Adv. Mater.* 7:292 (1995).
54. J. Cornil, D. Beljonne, J. L. Brédas, F. Rohlfing, D. D. C. Bradley, A. Eberhardt, and K. Müllen, unpublished data.
55. L. S. Swanson, J. Shinar, and K. Yoshino, *Phys. Rev. Lett.* 65:1140 (1990).
56. J. C. Scaiano, R. W. Redmond, B. Mehta, and J. T. Arnason, *Photochem. Photobiol.* 52:655 (1990).
57. R. A. J. Janssen, D. Moses, and N. S. Sariciftci, *J. Chem. Phys.* 101:9519 (1994).
58. H. Chosrovian, S. Rentsch, D. Grebner, D. U. Dahm, and E. Birckner, *Synth. Met.* 60:23 (1993).
59. R. S. Becker, J. S. de Melo, A. L. Maçanita, and F. Elisei, *Pure Appl. Chem.* 67:9 (1995).
60. R. Rossi, M. Ciofalo, A. Carpita, and G. Ponterini, *J. Photochem. Photobiol. A: Chem.* 70:59 (1993).
61. Z. G. Soos, S. Ramasesha, D. S. Galvao, and S. Etemad, *Phys. Rev. B* 47:1742 (1993).
62. J. Cornil, D. A. dos Santos, D. Beljonne, and J. L. Brédas, *J. Phys. Chem.* 99:5604 (1995).
63. S. C. Moratti, D. D. C. Bradley, R. Cervini, R. H. Friend, N. C. Greenham, and A. B. Holmes, *SPIE-Int. Soc. Opt. Eng.* 2144:108 (1994).
64. B. Sjögren and S. Stafström, *J. Chem. Phys.* 88:3840 (1988).
65. C. J. Yang and S. A. Jenekhe, *Chem. Mater.* 3:878 (1991).
66. K. Pakbaz, C. H. Lee, A. J. Heeger, T. W. Hagler, and D. McBranch, *Synth. Met.* 64:295 (1994).
67. C. H. Lee, G. Yu, and A. J. Heeger, *Phys. Rev. B* 47:15543 (1993).
68. T. W. Hagler, K. Pakbaz, and A. J. Heeger, *Phys. Rev. B* 49:10968 (1994).
69. J. M. Leng, S. Jeglinski, X. Wei, R. E. Benner, Z. V. Vardeny, F. Guo, and S. Mazumdar, *Phys. Rev. Lett.* 72:156 (1994).
70. M. Chandross, S. Mazumdar, S. Jeglinski, X. Wei, Z. V. Vardeny, E. W. Kwock, and T. M. Miller, *Phys. Rev. B* 50:14702 (1994).
71. R. H. Friend, D. D. C. Bradley, and P. D. Townsend, *J. Phys. D: Appl. Phys.* 20:1367 (1987).
72. P. Gomes da Costa and E. M. Conwell, *Phys. Rev. B* 48:1993 (1993).
73. R. Kersting, U. Lemmer, M. Deussen, H. J. Bakker, R. F. Mahrt, H. Kurz, V. I. Arkhipov, H. Bässler, and E. O. Göbel, *Phys. Rev. Lett.* 73:1440 (1994).
74. J. L. Brédas, J. Cornil, and A. J. Heeger, *Adv. Mater.* 8:447 (1996).
75. C. M. Heller, I. H. Campbell, B. K. Laurich, D. L. Smith, D. D. C. Bradley, P. L. Burn, J. P. Ferraris, and K. Müllen, *Phys. Rev. B*, in press.
76. I. H. Campbell, T. W. Hagler, D. L. Smith, and J. P. Ferraris, *Phys. Rev. Lett.* 76:1900 (1996).
77. Y. Yang, Q. Pei, and A. J. Heeger, *Synth. Met.*, in press.

78. Q. Pei, G. Yu, C. Zhang, Y. Yang, and A. J. Heeger, *Science* 269:1086 (1995).
79. M. Fahlman, M. Lögdlund, S. Stafström, W. R. Salaneck, R. H. Friend, P. L. Burn, A. B. Holmes, K. Kaeriyama, Y. Sonoda, O. Lhost, F. Meyers, and J. L. Brédas, *Macromolecules* 28:1959 (1995).
80. J. Cornil, D. Beljonne, and J. L. Brédas, *J. Chem. Phys.* 103:842 (1995).
81. H. Y. Choi and M. J. Rice, *Phys. Rev. B* 44:10521 (1991).
82. Z. Shuai, S. K. Pati, W. P. Su, J. L. Brédas, and S. Ramasesha, *Phys. Rev. B*, in press.
83. R. Kersting, U. Lemmer, R. F. Mahrt, K. Leo, H. Kurz, H. Bässler, and E. O. Göbel, *Phys. Rev. Lett.* 70:3820 (1993).
84. G. Nicolas and P. Durand, *J. Chem. Phys.* 72:453 (1980).
85. J. L. Brédas, R. R. Chance, R. Silbey, G. Nicolas, and P. Durand, *J. Chem. Phys.* 75:255 (1981).
86. J. Cornil, D. Beljonne, and J. L. Brédas, *J. Chem. Phys.* 103:842 (1995).
87. C. Ehrendorfer and A. Karpfen, *J. Phys. Chem.* 98:7492 (1994).
88. J. Guay, P. Kasai, A. Diaz, R. Wu, J. M. Tour, and L. H. Dao, *Chem. Mater.* 4:107 (1992).
89. G. Horowitz, A. Yassar, and H. J. von Bardeleben, *Synth. Met.* 62:245 (1994).
90. D. Fichou, G. Horowitz, B. Xu, and F. Garnier, *Synth. Met.* 39:243 (1990).
91. S. Hotta and K. Waragai, *J. Phys. Chem.* 97:7427 (1993).
92. G. Lanzani, L. Rossi, A. Piaggi, A. J. Pal, and C. Taliani, *Chem. Phys. Lett.* 226:547 (1994).
93. J. Poplawski, E. Ehrenfreund, J. Cornil, J. L. Brédas, R. Pugh, M. Ibrahim, and A. J. Frank, *Mol. Cryst. Liq. Cryst.* 256:407 (1994).
94. M. G. Harrison, R. H. Friend, F. Garnier, and A. Yassar, *Synth. Met.* 67:215 (1994).
95. P. Bäuerle, U. Segelbacher, A. Maier, and M. Mehring, *J. Am. Chem. Soc.* 115:10217 (1993).
96. M. G. Hill, J. F. Penneau, B. Zinger, K. R. Mann, and L. L. Miller, *Chem. Mater.* 4:1106 (1992).
97. A. Yassar, D. Delabouglise, M. Hmyene, B. Nessakh, G. Horowitz, and F. Garnier, *Adv. Mater.* 4:490 (1992).
98. B. Nessakh, G. Horowitz, F. Garnier, F. Deloffre, P. Srivastava, and A. Yassar, *J. Electroanal. Chem.* 399:97 (1995).
99. T. C. Chung, J. H. Kaufman, A. J. Heeger, and F. Wudl, *Phys. Rev. B* 30:702 (1984).
100. Z. Vardeny, E. Ehrenfreund, O. Brafman, M. Nowak, H. Schaffer, A. J. Heeger, and F. Wudl, *Phys. Rev. Lett.* 56:671 (1986).
101. G. Zotti, S. Martina, G. Wegner, and A. D. Schlüter, *Adv. Mater.* 4:798 (1992).
102. R. K. Khanna, Y. M. Jiang, B. Srinivas, C. B. Smithhart, and D. L. Wertz, *Chem. Mater.* 5:1792 (1993).
103. H. Gregorius, W. Heitz, and K. Müllen, *Adv. Mater.* 5:279 (1993).
104. R. Schenk, H. Gregorius, and K. Müllen, *Adv. Mater.* 3:492 (1991).
105. E. Ehrenfreund, D. Moses, A. J. Heeger, J. Cornil, and J. L. Brédas, *Chem. Phys. Lett.* 196:84 (1992).
106. M. Lögdlund, P. Dannetun, S. Stafström, W. R. Salaneck, M. G. Ramsey, C. W. Spangler, C. Frederiksson, and J. L. Brédas, *Phys. Rev. Lett.* 70:970 (1993).
107. T. Bally, K. Roth, W. Tang, R. R. Schrock, K. Knoll, and L. Y. Park, *J. Am. Chem. Soc.* 114:2440 (1992).
108. D. J. Williams (ed.), *Nonlinear Optical Properties of Organics and Polymeric Materials*, ACS Symp. Ser. Vol. 233, American Chemical Society, Washington, DC, 1983.
109. D. S. Chemla and J. Zyss (eds.), *Nonlinear Optical Properties of Organic Molecules and Crystals*, Academic, New York, 1987.
110. A. J. Heeger, J. Orenstein, and D. R. Ulrich (eds.), *Nonlinear Optical Properties of Polymers*, Mater. Res. Soc. Symp. Proc., Vol. 109 1988.
111. J. Messier, F. Kajzar, P. N. Prasad, and D. R. Ulrich (eds.), *Nonlinear Optical Effects in Organic Polymers*, NATO-ARW Ser. E162, 1989.
112. J. L. Brédas and R. R. Chance (eds.), *Conjugated Polymeric Materials: Opportunities in Electronics, Optoelectronics, and Molecular Electronics*, NATO-ARW Ser. E182, 1990.
113. Messier, F. Kajzar, and P. N. Prasad (eds.), *Organic Molecules for Nonlinear Optics and Photonics*, NATO-ARW Ser. E194, 1991.
114. P. N. Prasad and D. J. Williams, *Introduction to Nonlinear Optical Effects in Molecules and Polymers*, Wiley-Interscience, New York, 1991.
115. S. R. Marder, J. E. Sohn, and G. D. Stucky (eds.), *Materials for Nonlinear Optics: Chemical Perspectives*, ACS Symp. Ser., Vol. 455, American Chemical Society, Washington, DC, 1991.
116. S. R. Marder and J. W. Perry, *Science* 263:1706 (1994).
117. G. Zerbi (ed.), *Organic Materials for Photonics: Science and Technology,* Elsevier Science, Amsterdam, 1993.
118. D. M. Burland (guest ed.), *Optical Nonlinearities in Chemistry*, special issue of *Chem. Rev.*, Vol. 94, January 1994.
119. J. L. Brédas, *Science* 263:487 (1994).
120. S. R. Marder, D. N. Beratan, and L. T. Cheng, *Science* 252:103 (1991).
121. S. R. Marder, C. B. Gorman, L. T. Cheng, and B. G. Tiemann, *Proc. SPIE-Int. Soc. Opt. Eng.* 1775:19 (1992).
122. C. B. Gorman and S. R. Marder, *Proc. Natl. Acad. Sci. USA* 90:11297 (1993).
123. S. R. Marder, L. T. Cheng, B. G. Tiemann, A. C. Friedly, M. Blanchard-Desce, J. W. Perry, and J. Skindhoj, *Science* 263:511 (1994).
124. S. R. Marder, C. B. Gorman, F. Meyers, J. W. Perry, G. Bourhill, J. L. Brédas, and B. M. Pierce, *Science* 265:632 (1994).

125. F. Meyers, S. R. Marder, B. M. Pierce, and J. L. Brédas, *Chem. Phys. Lett.* 228:171 (1994).
126. F. Meyers, S. R. Marder, B. M. Pierce, and J. L. Brédas, *J. Am. Chem. Soc.* 116:10703 (1994).
127. S. R. Marder, J. W. Perry, B. G. Tiemann, C. B. Gorman, S. Gilmour, S. L. Biddle, and G. Bourhill, *J. Am. Chem. Soc.* 115:2524 (1993).
128. H. G. Benson and J. N. Murell, *J. Chem. Soc., Faraday Trans. 2* 68:137 (1972).
129. R. Radeglia and S. Dähne, *J. Mol. Struct.* 5:399 (1970).
130. R. Radeglia, G. Engelhardt, E. Lippmaa, T. Pehk, K. D. Nolte, and S. Dähne, *Org. Magn. Res.* 4:571 (1972).
131. G. Bourhill, J. L. Brédas, L. T. Cheng, S. R. Marder, F. Meyers, J. W. Perry, and B. G. Tiemann, *J. Am. Chem. Soc.* 116:2619 (1994).
132. S. R. Marder, J. W. Perry, G. Bourhill, C. B. Gorman, B. G. Tiemann, and K. Mansour, *Science* 261:186 (1993).
133. K. Clays, E. Hendrickx, M. Triest, T. Verbiest, A. Persoons, C. Dehu, and J. L. Brédas, *Science* 262:1419 (1993).
134. C. Reichardt, *Solvents and Solvent Effects in Organic Chemistry*, VCH, Weinheim, 1988.
135. A. Willetts and J. E. Rice, *J. Chem. Phys.* 99:426 (1993).
136. G. Rauhut, T. Clark, and T. Steinke, *J. Am. Chem. Soc.* 115:9174 (1993).
137. K. V. Mikkelsen, Y. Luo, H. Agren, and P. Jorgensen, *J. Chem. Phys.* 100:8240 (1994).
138. J. Yu and M. C. Zerner, *J. Chem. Phys.* 100:7487 (1994).
139. S. DiBella, T. J. Marks, and M. A. Ratner, *J. Am. Chem. Soc.* 116:4440 (1994).
140. K. D. Nolte and S. Dähne, *Adv. Mol. Relax. Int. Proc.* 10:299 (1977).
141. S. Dähne and K. D. Nolte, *J. Chem. Soc. Chem. Commun.* 1972:1056.
142. J. L. Oudar and D. S. Chemla, *J. Chem. Phys.* 66:2664 (1977).
143. J. L. Oudar, *J. Chem. Phys.* 67:446 (1977).
144. C. Dehu, F. Meyers, E. Hendrickx, K. Clays, A. Persoons, S. R. Marder, and J. L. Brédas, *J. Am. Chem. Soc.* 117:10127 (1995).
145. D. Lu, G. Chen, J. W. Perry, and W. A. Goddard, *J. Am. Chem. Soc.* 116:10679 (1994).
146. J. L. Rivail and D. Rinaldi, *J. Chem. Phys.* 18:223 (1976).
147. D. Rinaldi, M. F. Ruiz-Lopez, and J. L. Rivail, *J. Chem. Phys.* 78:834 (1983).
148. D. Rinaldi, J. L. Rivail, and N. Rguini, *J. Comput. Chem.* 13:675 (1992).

2
Metal–Insulator Transition in Doped Conducting Polymers

Reghu Menon, C. O. Yoon, D. Moses, and A. J. Heeger
Institute for Polymers and Organic Solids, University of California at Santa Barbara, Santa Barbara, California

I. INTRODUCTION

The initial impetus for the plethora of work on conducting polymers was generated by the discovery in 1977 [1–3] of the increase, by nearly 10 orders of magnitude, in the electrical conductivity (σ) of polyacetylene when it was doped with iodine or other acceptors. The subsequent demonstration of the important role of nonlinear excitations, solitons, polarons, and bipolarons upon chemical doping or photoexcitation in the semiconducting regime provided a conceptual framework for understanding the electronic structure of these novel polymer semiconductors at low doping levels [4–13]. Although there has been impressive progress toward the goal of improving conductivity and achieving truly metallic polymers [14–17], parallel progress toward understanding the transport in the "metallic" state has been limited by the quality of the disordered polymer materials.

The existence of a Pauli spin susceptibility [18], a quasilinear temperature dependence of the thermopower [19], and a linear term in the specific heat [20] provided early evidence of a continuous density of states with a well-defined Fermi energy. However, the real "fingerprints" of metallic behavior were not observed—for example, finite dc conductivity as $T \to 0$, a positive temperature coefficient of the logarithmic derivative of the conductivity, $W = [\Delta \ln \sigma / \Delta \ln T]$, and metallic (Drude-like) reflectivity in the infrared. In recent years, improved homogeneity and a reduction in the degree of disorder resulting from improved synthesis and processing of conducting polymers have provided a new opportunity for investigating the nature of the "metallic" state through transport and optical measurements [21–32]. The improvement in the quality of the new generation of materials has enabled the observation of these typical metallic features in specific systems [33–35] and has resulted in a deeper understanding of the role of disorder as the limiting factor in transport and as the origin of the metal–insulator (M–I) transition in doped conducting polymers.

In first-generation conducting polymers, the electrical conductivities were limited; the maximum value observed in iodine-doped Shirakawa polyacetylene, $(CH)_x$, was on the order of 10^3 S/cm [1–3], whereas in doped polypyrrole (PPy) [36,37] and polythiophene (PT) [38–40], the best values were below 200 S/cm. Moreover, $\sigma(T)$ decreased rapidly upon lowering the temperature; i.e., the logarithmic derivative, $W = \Delta \ln \sigma / \Delta \ln T$, had a negative temperature coefficient indicative of transport on the insulating side of the metal–insulator transition. With the exception of doped $(CH)_x$ [41], a finite dc conductivity as $T \to 0$ was not observed. Although the thermopower was quasilinear with temperature, typical of that expected for metallic systems [19], the temperature dependence of the magnetic susceptibility showed a relatively large Curie contribution, observable even at room temperature and the dominant contribution at low temperatures [18,27–30]. Thus, in early measurements on conducting polymers, the presence of strong disorder masked the metallic behavior; the features observed were typical of those expected for highly disordered and inhomogeneous media.

The high electrical conductivities of doped polyacetylene reported by Naarmann et al. [14–16] in 1987 signaled the onset of a new generation of conducting polymers. Conductivities on the order of 10^4 S/cm, comparable to those of traditional metals like lead, were reported. By continuing to improve the material, Tsukomoto increased the conductivity by another order of magnitude, to 10^5 S/cm, in 1990 [17]. Correspondingly, the temperature dependence of conductivity, $\sigma(T)$, in doped $(CH)_x$ consistently became weaker as the conductivity increased.

Although values of σ on the order of 10^4 S/cm were observed for doped oriented PPV in 1990, no rigorous transport measurements were carried out [22–24]. In 1991, the low temperature electrochemical polymerization of PF_6-doped PPY, subsequently stretch oriented, yielded $\sigma \sim 10^3$ S/cm [25,26]. In this material, for the first time in doped conducting polymers, a significant positive temperature coefficient of resistivity (TCR) was observed at temperatures below 20 K.

In 1991, the development of the counterion-induced processibility of polyaniline (PANI) enhanced the conductivity of PANI to 300–400 S/cm [27–30]. These samples showed substantially weaker temperature dependence of $\sigma(T)$ than previous generation PANI [34,35]. Moreover, for the first time, a doped conducting polymer showed a significant positive TCR in the temperature range 160–300 K [34,35].

In 1992, the synthesis of regioregular polyalkylthiophenes (PAT) resulted in materials with substantially enhanced conductivity with values on the order of 10^3 S/cm [31,32].

This brief summary of the significant developments that have occurred within the last few years in the preparation and processing of doped conducting polymers indicates substantial progress; further progress is necessary, however, to reduce the microscopic and macroscopic disorder and thereby bring out the intrinsic metallic features. The systematic improvement in material quality needs to be characterized by rigorous transport measurements in order to quantify the various parameters involved in the disorder-induced localization that leads to the M–I transition. The conclusions inferred from such experimental studies will provide a deeper understanding of the microscopic parameters involved in charge transport in these metallic polymers and consequently a deeper understanding of the requirements for further improvement in the quality of the materials.

The objective of this review is to summarize the results of recent transport measurements near the M–I transition in the doped conducting polymers, $(CH)_x$, PANI, PPy, and PAT, including dc conductivity, thermoelectric power, and magnetoresistance, all as a function of temperature, pressure, and magnetic field. The results are compared with theoretical models of the M–I transition [42–45] in order to understand the microscopic transport mechanisms, especially as a function of disorder and as a function of the strength of the interchain transfer interaction.

A comparison of the conductivities of doped conducting polymers with those of conventional conductors is shown in Fig. 2.1. Although conductivities on the order of 10^5 S/cm have been observed for stretch-oriented $(CH)_x$ (along the chain direction), the absence of a positive TCR, typical of a metal, indicates that the bottleneck responsible for the "nonmetallic" temperature dependence of $\sigma(T)$ is the transport perpendicular to the chain axis. Typical conductivity values perpendicular to the chain axis for highly oriented $(CH)_x$ are on the order of 10^2 S/cm, i.e., comparable to or less than Mott's minimum metallic conductivity. Although orientation of the polymer chains by tensile drawing can enhance the spatial extent of the localized states to a few hundred angstroms, interchain electron transfer is required for delocalization into three dimensions [46].

II. EXPERIMENTAL

A brief description of the relevant experimental techniques that are used in the study of electrical transport in doped conducting polymers (in particular, in the transport laboratory at the Institute for Polymers and Organic Solids at UCSB) is presented in this section. The preparation of $(CH)_x$, PPy-PF_6, PANI-CSA, and regioregular PAT samples is described in detail in Refs. 21, 25 and 26, 27–30, and 31 and 32, respectively. Four-terminal dc resistivity, magnetoresistance, and thermopower measurements were carried out using a computer-controlled automated measuring system. Electrical contacts were made with conducting carbon paint; in cases where the paint contact adversely affects the measurements, pressure contacts were used. The linearity was checked by measuring voltage versus current, and the resistivity was obtained from the slope of the straight line [34,35].

High pressure conductivity measurements were carried out in a self-clamped beryllium-copper pressure cell [47]. After pressurizing, the cell was clamped at room temperature and then cooled to 1.2 K in a cryostat containing a superconducting magnet (0–10 T). Fluorinert was used as the hydrostatic pressure-transmitting medium. Temperature was measured with a calibrated platinum resistor (300 to 40 K), a calibrated carbon-glass resistor (40 to 1.2 K), and a calibrated cernox resistor (300 to 1.2 K). To avoid sample heating at low temperatures, the current source was adjusted at each temperature so that the power dissipated into the sample was less than 1 μW. Moreover, to avoid any sample heating, the samples were always immersed directly in liquid helium during measurements below 4.2 K.

Thermopower measurements used the differential technique [48,49] two isolated copper blocks were alternately heated with the sample mounted between the copper blocks with pressure contacts. The heating current was accurately controlled by computer. The temperature difference between the two copper blocks was measured by a chromel-constantan thermocouple and did not exceed 0.5 K for each thermal cycle. The voltage difference across the sample was averaged for one complete cycle. Any temperature difference between sample and thermocouple was less than 10% of the temperature gradient across the sample; the thermometry was carefully calibrated for the entire temperature range (5 K < T < 300 K). The absolute thermopower of the sample was obtained from the absolute scale for lead [48,49].

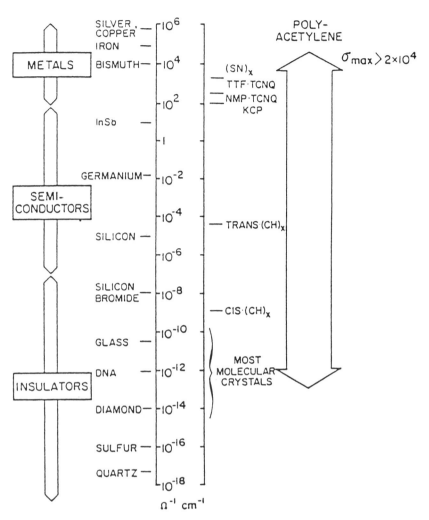

Fig. 2.1 Comparison of the conductivities of doped conducting polymers with those of conventional conductors.

III. RESULTS AND DISCUSSION

A. Disorder-Induced Metal–Insulator Transition

In 1958 Anderson [50] showed that localization of electronic wave functions can occur if the random component of disorder potential is large enough compared to the bandwidth, as sketched in Fig. 2.2. The mean free path (λ) in such a system is given by

$$1/\lambda = 0.7\,(1/a)\,(V_0/B)^2 \tag{1}$$

where a is the interatomic distance, V_0 is the random potential, and B is the bandwidth. The associated broadening of the density of states (vs, energy) due to the random potential is shown in Fig. 2.3. According to Anderson, states will become localized throughout the band for a critical value of V_0/B that is estimated to be of order unity in three dimensions. The wave functions of the localized states are of the form $e^{-r/\xi}\,\mathrm{Re}[\psi_0]$, where $\psi_0 = \sum c_n \exp(i\phi_n)\psi_n$; ($c_n$ are the real coefficients, ϕ_n are random phases), and ξ is the localization length, which tends to infinity as V_0/B tends to the critical value [42,43].

In 1966, Mott [42,43] pointed out that the states in the band tails are more susceptible to localization. Consequently, there exists a critical energy separating the localized states from the nonlocalized states, called the mobility edge (E_c). The mobility edge can be expected to play an important role in cases of significant disorder, even when V_0/B is below the critical value.

In 1970, Anderson [51] proposed that a degenerate electron gas in a random disorder potential tends to localize if the magnitude of the disorder potential is large compared with the bandwidth. In such a case all the states become localized, and the system is a "Fermi glass": an insulator with a continuous density of local-

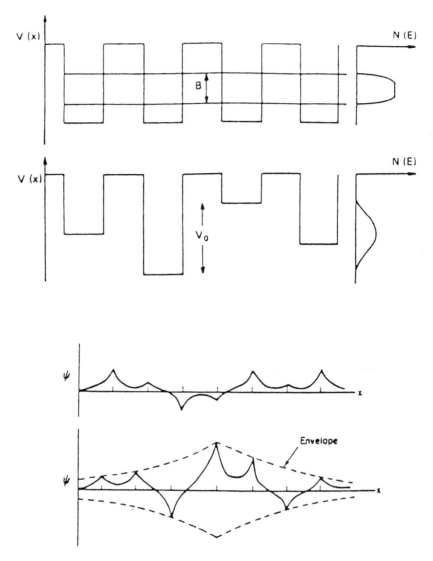

Fig. 2.2 Anderson localization: random potentials versus bandwidth.

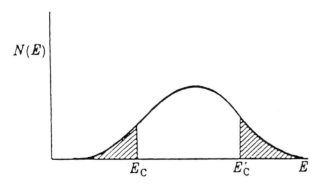

Fig. 2.3 Density of states versus random potential.

ized states (no energy gap) occupied according to Fermi statistics. When the extent of disorder is sufficiently large to induce the Fermi energy (E_F) to cross E_c (in other words, to place the Fermi energy in the region of localized states), there is a transition from the metallic state (finite value of the conductivity as $T \to 0$) to a nonmetallic state (the conductivity goes to zero as $T \to 0$). Mott called this disorder-induced M–I transition the "Anderson transition" [42,43]. The localization length increases as the Fermi energy approaches E_c, $\xi \sim a[E_0 / (E_c - E)]^\nu$, where $\nu \approx 1$ and E_0 is a constant.

Based upon the Ioffe–Regel criterion that the lower limit of mean free path is the interatomic spacing [52], Mott proposed [42,43] in 1973 that the M–I transition that occurs as $E_F \to E_c$ (from the metallic side) is discontinuous and introduced the concept of the "minimum metallic conductivity," $\sigma_{min} = 0.03\ e^2/3\hbar a$ in three dimensions. In 1979, the scaling theory of localization of Abrahams and coworkers [44,53] demonstrated, however, that the M–I transition is continuous in three dimensions. The conductivity of a metal goes smoothly to zero as $E_F \to E_c$, so that in general σ_{min} does not exist. This important progress stimulated extensive theoretical and experimental work, which has been thoroughly reviewed [42–45]. Möbius [54] pointed out, however, that the scaling theory does not completely disprove the existence of σ_{min}, since the low temperature conductivity data near the M–I transition can be explained alternatively by a combination of σ_{min} on the metallic side and the Coulomb interaction contribution on the insulating side. The concept of σ_{min} survives as a qualitative measure of the importance of localization phenomena; systems with $\sigma \sim \sigma_{min}$ should be considered as near the M–I transition; systems with $\sigma < \sigma_{min}$ should be viewed as on the insulating side of the M–I transition.

An important parameter for characterizing the disorder is the product of the Fermi wavevector (k_F) and the mean free path (λ), which is the order parameter for the disorder-induced M–I transition; for $k_F\lambda \sim 1$, $\sigma \sim \sigma_{min}$. In the Fermi glass regime, $k_F\lambda \ll 1$. Recent progress has resulted in conducting polymers on the metallic side of the M–I transition with $k_F\lambda \geq 1$ and $\sigma \geq \sigma_{min}$. In this situation, the lifetime broadening of the electronic states is less than the Fermi energy ($\epsilon_F\tau > \hbar$), so a band model can be used as a starting point [42,43,46].

In the strict one-dimensional limit, all wave functions are localized in the presence of disorder. Interchain electron transfer suppresses this extreme tendency toward localization. For a given level of disorder, the strength of the interchain coupling needed to suppress the localization depends on the coherence length along the quasi-one-dimensional chains; an electron must be able to hop to an adjacent chain prior to the resonant backscattering, which inevitably leads to localization in one dimension. Thus, although conducting polymers are correctly referred to as quasi-one-dimensional electronic systems, the interchain coupling can be sufficiently large to enable the formation of three-dimensional metals [46,55]. When such a material is oriented (e.g., by tensile drawing), the properties are those of an anisotropic three-dimensional metal. Without macroscopic orientation, the macroscopic properties of conducting polymers are isotropic, even though on a microscopic level the electronic structure (*intra*chain vs. *inter*chain) is highly anisotropic.

The transport properties of disordered systems are sensitive to the presence of both extended and localized states [42–44]. The extent of disorder determines the relative importance of the roles played by localization and by electron–electron (e–e) interactions; the extent of disorder determines the screening length and scattering processes involved in the charge transport. Among various length scales, the correlation length on the metallic side, the localization length on the insulating side, the e–e interaction length, the thermal diffusion length, and the inelastic scattering length determine the dominant mechanism(s) involved in the transport.

Recent improvements in the quality of doped conducting polymers have presented the opportunity to investigate the disorder-induced M–I transition in these materials. Disorder is an inherent feature of polymers, which often exhibit complex morphology; such systems are often partially crystalline and partially amorphous in nature [56–64]. Since disorder leads to qualitatively different charge transport mechanisms in the homogeneous and inhomogeneous limits, it is important to try to quantify these two limits. The critical parameter is the localization length, L_c; if L_c is greater than the structural coherence length (which characterizes the length scale of the crystalline regions and thus the length scale for inhomogeneity), the disorder can viewed as "homogeneous"; *the system sees an average*. On the other hand, when there are large-scale inhomogeneities, as in a granular metal, the disorder must be viewed as "inhomogeneous."

In the previous generation of conducting polymers, inhomogeneities often dominated the transport properties, and "metallic islands" models were constructed to handle such larger scale granularity [65–70].

The extent of disorder in conducting polymers can be controlled to some degree by the details of sample preparation and processing. Moreover, order can be enhanced by tensile drawing to achieve chain extension, chain orientation, and interchain order. Even in such cases, however, the disorder introduced during the doping process can be substantial. In this review, the transport properties of various doped conducting polymers having a wide range of disorder, on both sides of the M–I transition, are covered.

In the classical definition, a metal should have a positive TCR and $k_F\lambda > 1$, where

$$k_F\lambda = [\hbar(3\pi^2)^{2/3}] / (e^2\rho n^{1/3}), \tag{2}$$

ρ is the resistivity, and n is the number of charge carriers [42,43]. However, the more precise experimental definition of a metal requires that there be a finite conductivity as $T \to 0$ and that the logarithmic derivative of the temperature dependence of conductivity, $W = \Delta \ln \sigma / \Delta \ln T$, show a positive temperature coefficient [71,72]. In conducting polymers of the previous generation, these metallic features were not prominently observed (with the exception of doped polyacetylene [41,73,74]). However, in the new generation conducting polymers there are examples that exhibit a positive temperature coefficient for W and even exhibit a positive temperature coefficient for the resistivity (TCR) at temperatures above 150 K [33–35]. Even on the metallic side of the M–I transition, however, $\sigma(T)$ is dominated by the interplay between the important contributions from localization and e–e interactions at low temperatures.

On the insulating side of the M–I transition, $\sigma \to 0$ as $T \to 0$, and W shows a negative temperature coefficient [42,43,71,72]. The exponential temperature dependence, $\ln \sigma(T) \propto T^{-x}$ with $x < 1$, on the insulating side of the M–I transition is typical of hopping transport. The exponent x is determined by the extent of disorder, by the dimensionality of the system, and by the morphology, granularity, and microstructure (homogeneous or inhomogeneous). Most of the transport measurements in previous generation conducting polymers were obtained from materials in the insulating regime [65–69,75].

As noted above, in Fermi glass insulators, the Fermi level lies in an energy interval in which all states are localized [42,43,51]. The M–I transition occurs when the disorder is sufficiently weak that the mobility edges move away from the center of the band toward the band tails such that E_F lies in a region of extended states.

The new generation of conducting polymers has made possible detailed studies of the critical regime of the M–I transition, previously unexplored in doped conducting polymers. When the extent of disorder is near the critical disorder for the Anderson transition, the temperature dependence of conductivity follows a power law over a substantial range of temperatures, $\sigma_{\text{crit}} \propto T^{-\beta}$, and W is temperature-independent [76,77]; $W = \beta$. This power law behavior is universal near the critical regime of the M–I transition and does not depend on the details of the system. Thus, the power law temperature dependence plays a key role in defining the critical regime.

"Tuning" through the critical regime by varying the extent of disorder, by varying the interchain interaction through application of high external pressure, and by shifting the mobility edge through application of high magnetic fields has provided insight into the general features of the transport near the M–I transition [78]. Because of the ability to tune through the critical regime, doped conducting polymers have recently become particularly interesting systems for investigation of the transport properties near the M–I transition.

B. Models for the M–I Transition and Transport in Doped Conducting Polymers

The electrical conductivity parallel to the chain axis of a quasi-one-dimensional metallic polymer is given by [46,79]

$$\sigma_\parallel = \frac{e^2 na}{\pi \hbar} v_F \tau = \left(\frac{e^2 na^2}{\pi \hbar}\right)\left(\frac{\lambda}{a}\right) \qquad (3)$$

where n is the conduction electron density per unit volume, a is the carbon–carbon distance along the chain direction, $v_F = 2t_0 a/\hbar$ is the Fermi velocity, $t_0 \approx 1$–3 eV is the π-electron hopping matrix element, and τ is the backscattering lifetime.

The strong interconnection between the electronic system and the lattice in one-dimensional metals leads to the Peierls transition from metal to insulator and to the formation of a gap at E_F [80]; a corresponding structural distortion also occurs (which is the source of the gap). For light doping levels in such Peierls insulators, the electron–phonon coupling leads to the formation of localized quasiparticles, solitons, polarons, and bipolarons. The stability of these excitations depends on the extent of disorder, dopant–charge carrier interactions, and interchain interactions.

Kivelson and Heeger [81] proposed a first-order transition from soliton lattice to polaronic metal at a critical doping level ($\approx 8\%$) for doped polyacetylene, $(CH)_x$. Recent calculations by various groups indicate, however, that all doping levels the "metallic state" is a soliton lattice [82–86]. This dilemma can be resolved by interchain interactions that lead to true metallic behavior in three dimensions [46]. Theoretical work by Harigaya and coworkers [82,83] has shown that the energy gap can vanish at a certain doping level (e.g., near 8% for doped $(CH)_x$) even when the bond alternation pattern persist (perhaps enhanced by order among the dopant ions). The disappearance of the energy gap can explain the observed M–I transition as a function of dopant concentration as characterized by the onset of the metallic Pauli susceptibility. Calculations by Jeckelmann and Baeriswyl [87] on the one-dimensional Peierls–Hubbard model indicate that quantum fluctuations and interchain coupling tend to reduce the Peierls gap.

According to Stafstrom [85], in the lightly doped regime the dopant potentials stabilize the soliton states and increase the gap between the occupied soliton band and the conduction band. In the heavily doped regime, the dopant potential effect favors a low-band-gap system. Moreover, lattice fluctuations and disorder are not expected to completely destroy the soliton lattice.

In the metallic state of a conjugated polymer, the charged dopant ions adjacent to the chain induce Friedel-type oscillations that can enhance the local infrared-active vibrational (IRAV) modes [87]. Thus, strong IRAV modes cannot be interpreted as evidence of a Peierls gap. Salkola and Kivelson [88] proposed that the arrangement of counterions has a strong influence on the nature of the ground state. They proposed that a semimetallic phase intervenes between the nonmetallic (Peierls) and the metallic (Fermi liquid) phases. Park and coworkers [89–92] proposed a soliton–antisoliton condensation model for doped polyacetylene.

Many models have been proposed, but none are broadly accepted. Thus, although the nature of the M–I transition as a function of dopant concentration in conducting polymers has been heavily studied, the understanding remains incomplete and controversial.

We therefore adopt the point of view that the metallic state exists (quite likely because of the interchain coupling) and that there is a continuous density of states as a function of energy with a well-defined Fermi energy.

In other words, the fundamental electronic structure of heavily doped conducting polymers is that of a metal. Since, however, polymeric systems are inherently disordered, states at the band tails are always localized and there exists a mobility edge that separates the region of the extended states in the interior of the band from the region of localized states. Whether the doped system is a metal or an insulator, then, is determined by the relative position of the Fermi energy with respect to the mobility edge.

Kivelson and Heeger [46] have shown that due to the relatively large π-electron bandwidth (e.g., $4t_0 \approx 10$ eV in polyacetylene) and the relatively small number of thermally excited $2k_F$ phonons in quasi-one-dimensional conducting polymers, the intrinsic conductivity is quite large (only the $2k_F$ phonons can cause the electron backscattering that limits the conductivity); see Eq, (3). Since even relatively weak interchain coupling ($t_\perp \sim 0.1$ eV) is sufficient for three-dimensional delocalization of charge carriers, one can hope to have this quasi-one-dimensional advantage even in the limit of an anisotropic three-dimensional metal. Moreover, since the phonon frequencies are higher than in conventional metals, the intrinsic conductivity is predicted to be very high at room temperature and to increase exponentially as the temperature is lowered. For any $t_\perp$ the coherent interchain quantum diffusion of carriers becomes three-dimensional as long as the mean separation between chain breaks is sufficiently great that the chain break concentration (x) is much less than $t_\perp/t_0$, i.e., $x \ll t_\perp/t_0$. Conversely, if $x \ll (t_\perp/t_0)^2$, incoherent interchain hopping due to one-dimensional localization will limit the transport [46].

Prigodin and Efetov [93,94] proposed that the M–I transition and the critical behavior in an irregular structure of coupled metallic chains depends on the disorder, the interchain coupling, the localization length, and the number of interchain crossings. In their theory, the following parameters govern $\sigma(T)$: (1) the mean relaxation time (τ), (2) the charge carrier velocity in the direction of the chain (v_F) or the density of states $N(0) = 1/\pi v_F a^2$, and (3) the value of the interchain hopping integral. In quasi-one-dimensional systems, localization takes place if $t_\perp \tau < 0.3$, and a transition to the metallic state is observed for $t_\perp \tau > 0.3$. For quasi-one-dimensional systems with weak disorder, a crossover from one-dimensional hopping transport at high temperatures to three-dimensional transport at low temperatures was predicted.

Stafstrom [95–97] proposed that an increase in interchain interactions with increasing dopant concentration can induce three-dimensional delocalization of the electronic states; he used the multichannel Buttiker–Landauer formalism [98] to estimate the conductance and localization in a system of coupled conjugated polymer chains and found that both are strongly dependent on the interchain hopping strength.

Among the previous generation conducting polymers, metallic behavior with finite conductivity at millikelvin (mK) temperatures was observed only in doped polyacetylene. Gould et al. [41] and Thummes et al. [73,74] carried out temperature and magnetic field dependence studies of the conductivity in doped $(CH)_x$ down to millikelvin temperatures. Contributions from both weak localization and e–e interactions were observed in the disordered metallic regime. In doped PPV samples, Madsen et al. [99] also observed the e–e interaction contribution to conductivity ($\sigma \propto T^{1/2}$) in metallic samples.

The typical dependence observed for $\sigma(T)$ in previous generation conducting polymers was the exponential behavior characteristic of hopping transport [75],

$$\rho = \rho_0 \exp[T_0/T]^{1/n}$$

Various values of exponent, e.g., $\ln \rho \sim T^{-1/4}$, $T^{-1/3}$, $T^{-1/2}$, T^{-1}, have been reported, and different models have been used to interpret the data, ranging from variable range hopping in different dimensionalities d (with $n = 2$ for 1d, $n = 3$ for 2d, and $n = 4$ for 3d) to a distribution of conjugation lengths.

Sheng's fluctuation-induced tunneling (FIT) model has also been widely used [100,101]. Although this model was originally developed for granular metals (metallic particles in an insulating matrix, such as carbon black particles in insulating polymers), the FIT model has been applied even in the case of highly conducting $(CH)_x$. Voit and Buttner [102] showed that the fitting parameters in the Sheng FIT model are not consistent with the physical properties of conducting polymers. Kaiser and Graham [103] extended the FIT model for heterogeneous systems by introducing geometric factors to the insulating barriers. A different extension of the FIT model was proposed by Paasch [104]. In general, however, the multiple-parameter fitting procedure in FIT models has not been able to provide a satisfactory physical understanding of the transport in doped conducting polymers.

Conwell and Mizes argued [105] that the conduction mechanism is not due to FIT; on the contrary, in their model the conductivity is limited primarily by conjugation defects and chain breaks within the metallic regions. The increase in conductivity with temperature is due to thermal activation (the charge carriers absorb phonons and are thereby activated over the barriers).

Baughman and Shacklette [106,107] proposed that the exponential temperature dependence of the conductivity results from the effect of finite conjugation lengths. They treat the nearest-neighbor interchain hopping in terms of a random resistor network model with a wide distribution of activation energies. A similar correlation between the conjugation length, conductivity, and the exponential temperature dependence of $\sigma(T)$ was put forward by Roth [75].

Recently, Movaghar and Roth [108] proposed that the transport properties of doped conducting polymers could be explained by a percolation-type model for inhomogeneously disordered systems, a point of view that is very different from Anderson localization in a homogeneously disordered system. Andrade et al. [109] reported computer simulations in a random resistor network model; initial results indicated some correlations between microstructure and transport properties. These models assume that the conductivity decreases exponentially upon lowering the temperature, an assumption that is not true for high quality materials of the new generation of conducting polymers. For example, Ishiguro et al. [110] observed a logarithmic temperature dependence for the resistivity and attributed this weak temperature dependence to low energy excitations associated with relaxation of the molecular conformation at low temperatures.

The "metallic islands" model (a composite model consisting of high conductivity crystalline regions surrounded by insulating amorphous regions) has been used to interpret the transport properties of polyaniline (PANI) protonated by common acids (e.g., HCl, H_2SO_4) [65–69,111,112]. However, analysis of the data according to this model yields a mean free path on the order of 1–100 μm and $k_F\lambda \sim 10^5$, both of which are far too large to be reasonable for a disordered system near the M–I transition. In the "metallic islands" model, both the FIT and the charge energy limited tunneling (CELT) mechanisms [65–69,113,114] were used in attempts to understand the conductivity and thermopower data.

Phillips and Wu [115] and Dunlap et al. [116] proposed a random dimer model that has a set of delocalized conducting states (even in one dimension!) that ultimately allow a particle to move through the lattice almost ballistically. They suggested that this interesting absence of localization in one dimension, despite the disorder, might be applicable to doped conducting polymers.

Recently Zuppiroli and coworkers [117–119] proposed a polaron/bipolaron model involving correlated hopping and mutiphonon processes as an explanation for the exponential temperature dependence of conductivity.

To summarize, studies of the temperature dependence of the electrical conductivity of previous generation conducting polymers [65–69,75] have resulted in a somewhat confusing situation. In particular, the strongly activated temperature dependence that was typically observed is more characteristic of an insulator than a metal; the metallic nature anticipated for these doped materials has not emerged.

As described in detail in the following sections, the best way to analyze the temperature dependence of conductivity (or resistivity, ρ) is to plot the logarithmic derivative, $W = -(\Delta \ln \rho/\Delta \ln T)$ against T [71,72]. This plot facilitates the identification of the various regimes of transport. In early publications in the field, the data analysis was not focused on the precise identification of metallic, critical, and insulating regimes. As a result, a comprehensive model for charge transport in conducting polymers developed slowly. Only recently has the improvement in sample quality sufficiently reduced the dominant role of disorder-induced localization (although it certainly remains important even in the best materials) that genuine metallic properties can begin to be observed in transport property measurements. Studies carried out on materials with a wide range of disorder have shown that the transport properties in the metallic regime, in the critical regime, and in the insulating regime of the disorder-induced M–I transition are distinct and recognizable and that the results obtained in these regimes can be understood in terms of the well-developed theory of the M–I transition.

C. Quantification of Disorder and Identification of the Metallic, Critical, and Insulating Regimes

1. Introduction

Measurements of the resistivity, $\rho(T)$, over a wide range of temperatures (e.g., 1.4–300 K) is the simplest method to begin to identify the various regimes. Since the room temperature conductivities of heavily doped conducting polymers are often comparable to Mott's minimum metallic conductivity [42,43], disorder is the dominant factor in determining the temperature dependence and the classification of the various transport regimes. As emphasized above, since conducting polymers are quasi-one-dimensional, the interchain transfer integral is a parameter of particular importance near the M–I transition; the role of interchain hopping can be explored through application of high pressure (sufficiently high to increase the interchain transfer interaction).

The resistivity ratio,

$$\rho_r = \rho(1.4\ \text{K})/\rho(300\ \text{K})$$

is a useful empirical parameter for quantifying the extent of disorder and for sorting out the various regimes. In general, as the disorder increases, the materials become more insulating, and the conductivity decreases more rapidly upon lowering the temperature; i.e., ρ_r increases.

To explicitly describe the characteristic behavior of $\rho(T)$, we define the reduced activation energy W as the logarithmic derivative of $\rho(T)$ [71,72],

$$W = -T\left(\frac{d \ln \rho(T)}{dT}\right) = -\frac{d \ln \rho}{d \ln T} \quad (4)$$

For best results, sets of data with 50–100 data points should be extracted from the raw data, with the interval of $\ln T = 0.5$–1.0, and successive differences taken to calculate the logarithmic derivatives.

The temperature dependence of W in various regimes is as follows:

1. In the insulating regime, W has a negative temperature coefficient.

2. In the critical regime, W is temperature-independent for a wide range of temperatures.
3. In the metallic regime, W has a positive temperature coefficient.

Although the interaction–localization contributions in the metallic regime and the hopping contribution in the insulating regime were well studied in earlier work on the M–I transition [42–45], the critical regime has only recently been studied in detail.

In the insulating regime, when the resistivity follows the activated temperature dependence characteristic of variable range hopping (VRH), $\ln \rho \propto (T_0/T)^x$, the reduced activation energy becomes

$$\log_{10} W(T) = A - x \log_{10} T \quad (5)$$

where $A = x \log_{10} T_0 + \log_{10} x$. Using Eq. (5), one can determine x from the slope in the plot of $\log_{10} W$ vs. $\log_{10} T$.

For a three-dimensional system in the critical regime of the M–I transition, the correlation length is large and has a power-law dependence on $\delta = |(E_F - E_c)/E_F| < 1$ with critical exponent ν, $L_c = a\delta^{-1/\nu}$, where a is a microscopic length, E_F is the Fermi energy, and E_c is the mobility edge [42–45]. In this critical region [76], the resistivity is not activated but rather follows a power law dependence on the temperature as shown by Larkin and Khmelnitskii [77]:

$$\rho(T) \approx \frac{e^2 p_F}{\hbar^2} \left(\frac{k_B T}{E_F}\right)^{-1/\eta} = T^{-\beta} \quad (6)$$

where p_F is the Fermi momentum and e is the electron charge. The predicted range of validity includes $1 < \eta < 3$, which is consistent with the observed values $0.3 < \beta = 1/\eta < 1$ (see Section III.C on the critical regime for details).

This power law is universal and requires only that the disordered system be in the critical regime where $\delta \ll 1$. A value of $\eta > 3$ indicates that the system is just on the metallic side of the M–I transition. Although $\eta > 3$ is above the theoretical limit for the power law dependence, values for η as large as 4.5 have been reported for n-doped germanium near the critical regime [71,72]. At millikelvin temperatures, however, the system becomes either a metal or an insulator depending on the extent of disorder; extension of the power law dependence to $T = 0$ requires that the system be precisely at the critical point. Conducting polymers are particularly interesting for investigations of the critical behavior near the M–I transition because a wide range of parameters that play a significant role can be controlled (e.g., carrier concentration, interchain interaction, and extent of disorder).

Electron–electron interactions also play an important role. According to McMillan's scaling theory [76], the energy scale (correlation gap) as the system crosses over from the critical regime to the conducting or insulating regime is $\Delta_c = (\hbar D_a/a^2)(a/L_c)^\eta$, where D_a is the diffusion constant. The correlation gap is related to the characteristic crossover temperature (T_{corr}) from the power law dependence of the resistivity at high temperatures to the exponential dependence of the resistivity at low temperature (insulating regime) or to the $T^{1/2}$ dependence of the conductivity with finite $\sigma(T = 0)$ (metallic regime).

For doped conducting polymers, the approximate values of ρ_r in the critical regime of the M–I transition are the following [120]:

1. For nonoriented I-$(CH)_x$ with $\sigma(300$ K$) \approx 800$–1500 S/cm, $\rho_r \approx 50$–100.
2. For oriented I-$(CH)_x$ with $\sigma(300$ K$) \approx 5000$–8000 S/cm, $\rho_r \approx 8$–16.
3. For oriented K-$(CH)_x$ with $\sigma(300$ K$) \approx 4000$ S/cm, $\rho_r \approx 15$–30.
4. For nonoriented PPy-PF_6 with $\sigma(300$ K$) \approx 200$–400 S/cm, $\rho_r \approx 4$–8.
5. For nonoriented PANI-CSA with $\sigma(300$ K$) \approx 150$–300 S/cm, $\rho_r \approx 3$–5.
6. For nonoriented I-P3HT with $\sigma(300$ K$) \approx 1000$–1500 S/cm, $\rho_r < 45$.

Plots of W vs. T for various doped conducting polymers near the critical regime, at ambient pressure and at various magnetic fields, are shown in Fig. 2.4. The temperature-independent W (characteristic of the critical regime) is clearly observed over a significant range of temperature for these different materials and conditions.

The pressure-induced transition from the critical to the metallic regime results from enhancement of the interchain transfer interaction; high pressure is expected to decrease the interchain, van der Waals, spacing while leaving the intrachain, covalent, atomic spacing essentially unchanged. Under pressure, the increased overlap of the π-electron wave functions centered on different chains increases the conductivity both parallel and perpendicular to the chain axis. The pressure-induced crossover from the critical to the metallic regime demonstrates that charge transport perpendicular to the chain axis is an important factor in the M–I transition in quasi-one-dimensional conducting polymers, in agreement with the theoretical predictions [46].

Since an ideal one-dimensional system would not be sensitive to the magnetic field, the magnetic field induced transition from the critical regime to the insulating regime provides unambiguous evidence of the importance of interchain transport. Khmelnitskii and Larkin [121] have presented a scaling argument that indicates that the mobility edge can be shifted with respect to the Fermi energy by a magnetic field. There is, however, no detailed microscopic theory for describing the effect of a magnetic field on the M–I transition, particularly in a highly anisotropic (quasi-one-dimensional) system. The overlap of the wave functions in a quasi-one-dimen-

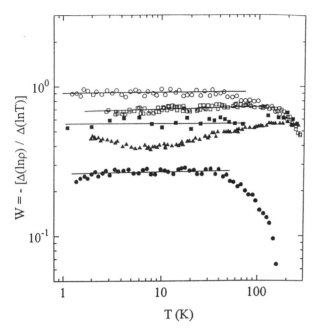

Fig. 2.4 Log-log plot of W vs. T for unoriented I-$(CH)_x$ (○); oriented I-$(CH)_x$ (□); oriented K-$(CH)_x$ (■); PPy-PF6 (▲); PANI-CSA (●). The straight lines are drawn to guide the eye.

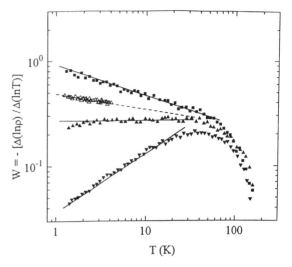

Fig. 2.5 Log-log plots of W vs. T for PANI-CSA in the metallic regime (▼); the critical regime at ambient pressure and $H = 0$ (▲) and at ambient pressure and $H = 8$ T (△); and the insulating regime (■). The lines are drawn to guide the eye.

sional metal is rather sensitive to magnetic field, since the field tends to shrink the wave functions [122,123]. The field-induced crossover from the critical regime to variable range hopping (VRH) among localized states is consistent with a field-induced shift of the mobility edge with respect to the Fermi level. The crossover occurs when the localization length and magnetic length [$(\hbar c/eH)^{1/2}$] become comparable.

2. PANI-CSA

For PANI-CSA samples in the critical regime, ρ_r falls in the range 2.5–3.5 [34,35,120]. The sample-to-sample variation of the critical exponent (η) in the power law temperature dependence correlates with ρ_r; samples with smaller ρ_r give smaller values of η, and those with larger ρ_r give larger values of η. The temperature range over which the dependence is a power law is wider for samples with smaller values of η. A log-log plot of W vs. T for PANI-CSA (nonoriented, cast from solution) is shown in Fig. 2.5. For the data in Fig. 2.5, $\rho_r = 3$ and $\eta = 3.8$. The power law regime extends from 1.4 to 40 K as shown by the temperature independence of $W(T)$, at ambient pressure, in Fig. 2.5.

Two examples are shown in Fig. 2.6, $\rho(T) \propto T^{-0.26}$, where $\eta = 3.8$, and $\rho(T) \propto T^{-0.36}$, where $\eta = 2.8$. The linearity on the log-log plot from 1.4 to 40 K demonstrates that

$$\rho(T,0) = \rho_{0m}(T/T_{0m})^{-\beta}$$

where ρ_{0m} and T_{0m} are constants.

Assuming that the resistivity in zero magnetic field follows Eq. (6) with $\eta \approx (0.36)^{-1} = 2.77$, one can compare the prefactor with that obtained from the measurements [34,35]. Using $p_F = \hbar k_F$ and $k_F \approx \pi/2c$, the prefactor is given by $\rho_{0m} \approx e^2 p_F/\hbar^2 \approx 2 \times 10^{-4}$ $\Omega \cdot$cm. From the data in Fig. 2.6 [with $E_F \approx 1$ eV in Eq. (2)], we obtain $\rho_{0m} \approx 6 \times 10^{-4}$ $\Omega \cdot$cm, in approximate agreement with the Khmelnitskii–Larkin theoretical value.

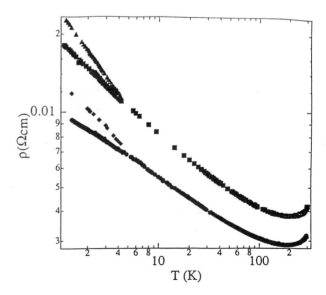

Fig. 2.6 Log-log plot of the $\rho(T)$ vs. T for two PANI-CSA samples in the critical regime: $H = 0$ T (●) and $H = 10$ T (♦) for sample with $\rho(T) \propto T^{-0.26}$; $H = 0$ T (■) and $H = 8$ T (▲) for sample with $\rho(T) \propto T^{-0.36}$.

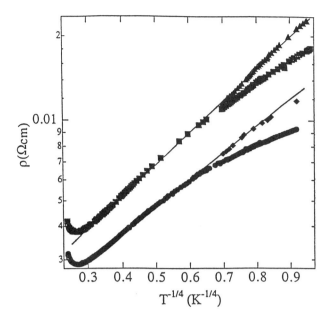

Fig. 2.7 Field-induced crossover from critical regime to VRH. $\rho(T)$ vs. $T^{-1/4}$ for PANI-CSA samples in the critical regime: $H = 0$ T (●) and $H = 10$ T (◆) for sample with $\rho(T) \propto T^{-0.26}$, and $H = 0$ T (■) and $H = 8$ T (▲) for sample with $\rho(T) \propto T^{-0.36}$

Application of a high magnetic field increases the resistivity; for $H = 8$ T, the temperature dependence is stronger than that of a power law, as shown in Fig. 2.6 [34,35]. The $\ln \rho(T, H)$ vs. $T^{-1/4}$ plots for two samples with $\eta = 3.8$ and 2.8 are compared in Fig. 2.7. In both cases, high magnetic fields induce the crossover from power law behavior to VRH. To our knowledge, this crossover as observed in PANI-CSA was not previously observed in any system. The data yield a characteristic VRH hopping temperature; T_0 ($H = 10$ T) = 27 K for the sample with $\eta = 3.8$, and $T_0(H = 8$ T) = 56 K for the sample with $\eta = 2.8$. The $T^{-1/4}$ dependence of $\ln \rho(T)$ with such low values of T_0 indicates that the magnetic field truly fine-tunes the system from the critical regime into the VRH hopping regime with very small hopping energies.

Measurements of $\rho(T)$ for a sample close to the critical point were extended down to 20 mK in order to identify the transition from the critical regime to either the insulating or metallic regime (depending upon the extent of disorder in the sample). As shown in Fig. 2.8, the power law dependence persists down to 0.4 K, below which a resistivity increase toward insulating behavior is observed.

The pressure dependence of $\sigma(T)$ of PANI-CSA is more complicated than that of doped $(CH)_x$ and PPy-PF_6 [120]. In PANI-CSA, the magnitude of the conductivity *decreases* above 2 kbar, with an associated stronger temperature dependence at high pressures, both of which might be associated with pressure-induced mor-

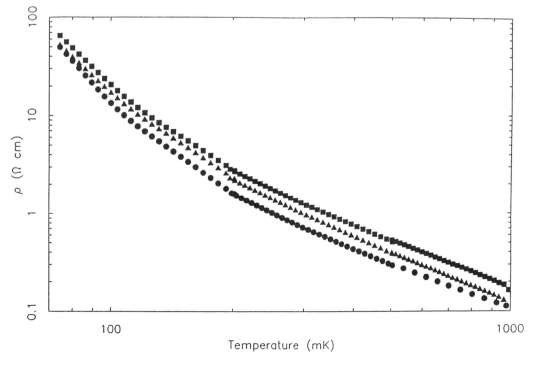

Fig. 2.8 Plot of $\rho(T)$ vs. T for the PANI-CSA sample in the critical regime at $H = 0$ (●), 2 T (▲), and 8 T (■) at millikelvin temperatures.

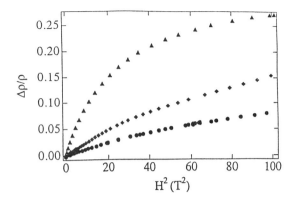

Fig. 2.9 Plot of $\Delta\rho/\rho$ vs. H^2 for PANI-CSA in the critical regime $[\rho(T) \propto T^{-0.2}]$: 4.2 K (●), 2.5 K (♦), and 1.4 K (▲).

phological changes. Thus, tuning the critical regime near the M–I transition by pressure for PANI-CSA has not proven to be as interesting as in other conducting polymers.

The magnetoresistance (MR) for PANI-CSA in the critical regime is shown in Fig. 2.9 [34,35]. The MR is positive and larger than in the metallic regime but smaller than in the insulating regime. At 4.2 K, the MR is linear in H^2 up 8 T; for lower temperatures, the initial linear dependence is followed by saturation at higher fields. The qualitative features of the MR are identical in the critical and insulating regimes, although the magnitude of the MR increases dramatically as the system moves into the insulating regime [34,35].

3. K-(CH)$_x$ and I-(CH)$_x$

Polyacetylene has charge conjugation symmetry and therefore can be n-type doped (reduced by electron donors such as potassium, K) or p-type doped (oxidized by electron acceptors such as iodine, I) into the metallic regime. The room temperature conductivities (σ_{RT}) of oriented I-(CH)$_x$ and K-(CH)$_x$ samples (in the direction parallel to the chain axis) shown in Figs. 2.10–2.12 are $\sigma_{RT} \approx 12{,}000$ and 4000 S/cm, respectively [120,124]. In both K-(CH)$_x$ and I-(CH)$_x$, the conductivity parallel to the chain axis is much higher than that typically found for systems near the M–I transition. However, since the anisotropy of the conductivity (parallel versus perpendicular to the draw axis, $\sigma_\parallel/\sigma_\perp$) is approximately 100, the conductivity perpendicular to the draw axis is comparable to Mott's minimum metallic conductivity and comparable to that of other known systems near the M–I transition [45].

The room temperature conductivities of the K-(CH)$_x$ sample at ambient pressure and 10 kbar are $\sigma_{RT} \approx 4000$ and 8000 S/cm, respectively [124]. The resistivity ratios for K-(CH)$_x$ are $\rho_r \approx 25$ at ambient pressure and $\rho_r \approx 9$ at 10 kbar. The critical regime can be identified from a log-log plot of W vs. T; Fig. 2.10 includes data obtained at ambient pressure and at 10 kbar (both at $H = 0$) and data obtained at $H = 8$ T with the sample at ambient pressure. The pressure dependence of the conductivity is shown in the inset of Fig. 2.10.

At ambient pressure (and $H = 0$), W is essentially constant from 300 to 1.2 K, implying that these oriented K-(CH)$_x$ samples are in the critical regime. At 10 kbar,

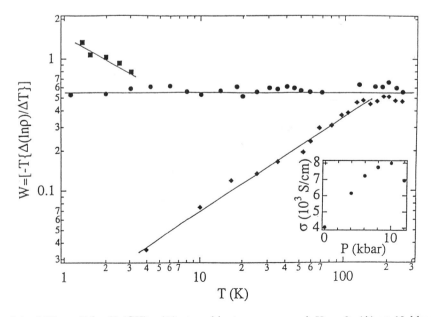

Fig. 2.10 Log-log plot of W vs. T for K-(CH)$_x$: (●) at ambient pressure and $H = 0$; (♦) at 10 kbar and $H = 0$; (■) at ambient pressure and $H = 8$ T. The inset shows the pressure dependence of conductivity. The lines are drawn to guide the eye.

the positive temperature coefficient of $W(T)$ indicates that the system has crossed over into the metallic regime, qualitatively consistent with expectations based on increased interchain transfer at high pressures.

The negative temperature coefficient of $W(T)$ at $H = 8$ tesla (see Fig. 2.10) implies a magnetic field induced crossover from the critical regime to the insulating regime (the large positive MR in K-(CH)$_x$ is consistent with the presence of strong disorder [124]). The magnetic field induced transition from power law to VRH hopping temperature dependence ($\ln \rho \propto T^{-1/4}$) is shown in Fig. 2.11; the results are similar to those obtained for PANI-CSA in the critical regime [34,35].

Log-log plots of W vs. T for I-(CH)$_x$ (current parallel to the chain axis) are shown in Fig. 2.12. The pressure dependence of the conductivity is shown in the inset of Fig. 2.12; for this sample, the room temperature conductivities parallel to the chain axis are $\sigma_{RT} \approx 11{,}000$ S/cm at ambient pressure and $\sigma_{RT} \approx 9300$ S/cm at 8 kbar [120,125,126]. Note that $\sigma(P)$ goes through a maximum at approximately 4 kbar. Although the conductivity at 8 kbar is lower than that at ambient pressure, the temperature dependence of conductivity is weaker (more nearly "metallic") at the highest pressures, an interesting result that is not understood. The values of ρ_r for I-(CH)$_x$ are $\rho_r \approx 3$ at ambient pressure and $\rho_r \approx 2$ at 8 kbar. The conductivity in the direction perpendicular to the chain axis is lower by two orders of magnitude. At 8 kbar, the

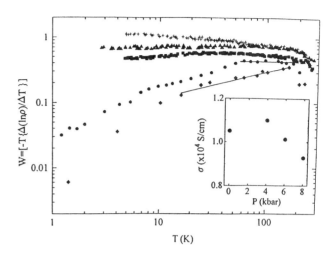

Fig. 2.12 Log-log plot of W vs. T for I-(CH)$_x$: at ambient pressure and $H = 0$, $\rho_r \approx 3$ and $\sigma_{RT} \approx 11{,}000$ S/cm (●); at 8 kbar and $H = 0$, $\rho_r \approx 2$ and $\sigma_{RT} \approx 9500$ S/cm (♦); at ambient pressure and $H = 0$, $\rho_r \approx 9$ and $\sigma_{RT} \approx 7000$ S/cm (■); at ambient pressure and $H = 0$, $\rho_r \approx 17$ and $\sigma_{RT} \approx 3500$ S/cm (▲); and at ambient pressure and $H = 0$, $\rho_r \approx 32$ and $\sigma_{RT} \approx 2450$ S/cm (+). The inset shows the pressure dependence of conductivity for sample (●). The lines are drawn to guide the eye.

enhanced interchain transport reduces the anisotropy by a factor of about 1.6 (e.g., from $\sigma_\parallel/\sigma_\perp \approx 105$ at 1 bar to $\sigma_\parallel/\sigma_\perp \approx 66$ at 8 kbar).

For the sample with $\sigma_{RT} \approx 11{,}000$ S/cm and $\rho_r \approx 3$, $W(T)$ is temperature-independent at ambient pressure over the limited temperature range from 180 to 60 K. As ρ_r gradually increases, this critical behavior crosses over to a negative temperature coefficient for $W(T)$ (insulating regime), as shown in Fig. 2.12. For $T < 60$ K, $W(T)$ shows a weak positive temperature coefficient for the samples having $\rho_r \approx 3$, indicating that the system is close to the metallic "boundary." However, at 8 kbar for samples having $\rho_r \approx 3$, $W(T)$ exhibits a strong positive temperature coefficient in the temperature range (from 1.3 to 180 K). Thus, by improving the interchain transport, it is possible to cross over into the metallic regime.

The magnetic field dependence of the conductivity in I-(CH)$_x$ is more complex than in K-(CH)$_x$ due to the interplay of weak localization, electron–electron interactions, and anisotropic diffusion coefficient contributions to the magnetoconductance as described below [125–128].

4. PPy-PF$_6$

Log-log plots of W vs. T for PPy-PF$_6$ are shown in Fig. 2.13. The room temperature conductivity of PPy-PF$_6$ samples in the critical regime is typically $\sigma_{RT} \approx 200$–300 S/cm [121] with $\rho_r \sim 3$–6. For the data in Fig. 2.13, the

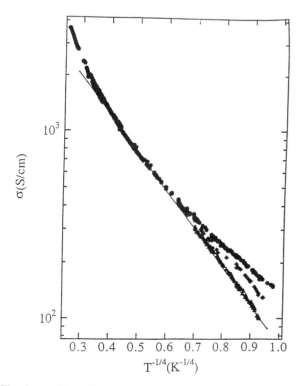

Fig. 2.11 Plot of $\ln \sigma(T)$ vs. $T^{-1/4}$ for K-(CH)$_x$ at $H = 0$ (●), 4 T (♦), and 8 T (▲) (at ambient pressure).

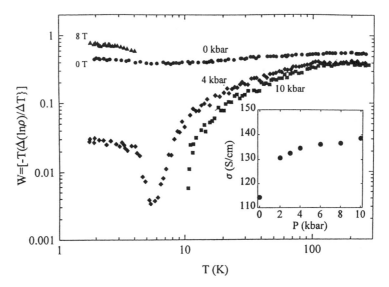

Fig. 2.13 Log-log plot of W vs. T for PPy-PF$_6$: at ambient pressure and $H = 0$ (●); at 4 kbar and $H = 0$ (♦); at 10 kbar and $H = 0$ (■); and at ambient pressure and 8 T (▲). The inset shows the pressure dependence of conductivity.

values of ρ_r are approximately 11, 2.6, and 2 at ambient pressure, 4 kbar, and 10 kbar, respectively.

The pressure dependence of the conductivity at room temperature is shown in the inset of Fig. 2.13. Although the positive temperature coefficient of $W(T)$ at 4 kbar indicates a crossover from the critical regime to the metallic regime, the temperature-independent $W(T)$ below 4 K shows that it is only marginally metallic. Thus, again, the interplay of the critical and metallic regimes can be fine-tuned by pressure. At 10 kbar, however, the temperature coefficient of $W(T)$ remains positive to 10 K, whereas below 10 K the slope changes from positive to negative, typical of that of a metal. The negative temperature coefficient of $W(T)$ at 8 T (at ambient pressure), similar to that observed in the case of K-(CH)$_x$, indicates the crossover from the critical to the insulating regime. This complex scenario of the M–I transition on both sides of the critical regime, depending upon the disorder, the pressure, and the magnetic field, indicates once again that heavily doped conducting polymers are in general in (or close to) the critical regime of the M–I transition.

Although the room temperature conductivity and ρ_r for PPy-PF$_6$ and PANI-CSA are nearly identical, the conductivity is not the same in the different temperature regimes [34,35,121,129]. The conductivity of PANI-CSA is more strongly temperature-dependent at low temperatures than that of PPy-PF$_6$. At high temperatures, the converse is true. The localization length for both PANI-CSA and PPy-PF$_6$ near the critical regime is 100–200 Å, i.e., comparable to the magnetic length at $H = 4$ T (128 Å).

5. Summary

The disorder-induced critical regime of the M–I transition has been observed in K-(CH)$_x$, I-(CH)$_x$, PPy-PF$_6$, and PANI-CSA. The critical regime can be precisely identified from log-log plots of W vs. T; $W(T)$ is temperature-independent in the critical regime, $W = \beta$, and $\sigma(T) = cT^\beta$. For all four heavily doped conducting polymer systems, the positive temperature coefficient of $W(T)$ observed at high pressures indicates the crossover to the metallic regime. For K-(CH)$_x$, PPy-PF$_6$, and PANI-CSA, at 8 T, the negative temperature coefficient of $W(T)$ indicates the crossover from the critical region into the insulating regime. Thus, in heavily doped conducting polymers, the transport can be fine-tuned from the critical regime into the metallic or insulating regimes by pressure and magnetic field, respectively.

D. Transport in the Metallic Regime Near the M–I Transition

1. Introduction

The approximate values of ρ_r for the metallic regime in doped conducting polymers are the following [120]:

1. For nonoriented I-(CH)$_x$ with $\sigma(300\text{ K}) > 1000$ S/cm, $\rho_r < 20$.
2. For oriented I-(CH)$_x$ with $\sigma(300\text{ K}) > 10{,}000$ S/cm, $\rho_r < 3$.
3. For oriented K-(CH)$_x$ with $\sigma(300\text{ K}) > 4000$ S/cm, $\rho_r < 10$.

4. For nonoriented PPy-PF$_6$ with $\sigma(300$ K$) > 200$ S/cm, $\rho_r < 3$.
5. For nonoriented PANI-CSA with $\sigma(300$ K$) > 200$ S/cm, $\rho_r < 2$.

In general, the temperature dependence of the conductivity in the metallic regime is relatively weak compared to that in either the critical or insulating regimes. Positive (metallic) TCR has been observed only for PANI-CSA [34,35] and FeCl$_3$-doped oriented (CH)$_x$ [33,130] above 160 and 220 K, respectively. In other conducting polymers, the magnitude of the negative TCR depends strongly on the extent of disorder.

Log-log plots of W vs. T precisely identify the metallic regime. If the temperature coefficient of W is positive, then the conductivity in the disordered metallic regime at low temperatures is expressed by [42–44]

$$\sigma(T) = \sigma(0) + mT^{1/2} + BT^{p/2} \qquad (7)$$

where the term $T^{1/2}$ results from thermally induced electron diffusion through states near the Fermi energy (reduced by electron–electron scattering) and the third term on the right is the correction to the zero-temperature "metallic" conductivity, $\sigma(0)$, due to disorder. The value of p is determined by the temperature dependence of the scattering rate [$\tau^1 \propto T^p$] of the dominant dephasing mechanism. For electron–phonon scattering, $p = 3$; for inelastic electron–electron scattering, $p = 2$ in the clean limit or $p = 3/2$ in the dirty limits. The calculation by Belitz and Wysokinski [132] gives $p = 1$ very near the M–I transition. In the disordered metallic regime, the conductivity depends on three length scales [42–44]: the correlation length L_c describing the M–I transition, the interaction length $L_T = (\hbar D/k_B T)^{1/2}$, and the inelastic diffusion length $L_{in} = (D\tau_{in})^{1/2}$ (where D is the diffusion coefficient and τ_{in} is the inelastic scattering time). In practice, however, it is difficult to distinguish these contributions using only the temperature dependence of the conductivity; finer details of the various contributions can be determined from magnetoconductance (MC) measurements.

Since the magnetoconductance is sensitive to the extent of disorder, it serves as an especially useful probe (compared to other transport measurements) for identifying the microscopic transport mechanisms and scattering processes. The MC in the metallic regime is determined by contributions from weak localization (positive MC for weak spin–orbit coupling, negative for strong spin–orbit coupling), and e–e interaction contributions (negative MC) [45,133,134]. In addition, there is a negative MC contribution from strongly localized states due to wave function shrinkage in a magnetic field [122], the latter being critically dependent on the extent of disorder in the system. Because doped conducting polymers are made up of light atoms with small spin–orbit coupling, one does not expect the contribution arising from spin–orbit coupling to be important. A small additional contribution to the negative MC could arise from residual hopping transport in highly disordered mesoscopic regions (this contribution is smaller than in the critical and insulating regimes). Three-dimensional calculations of the band structures of heavily doped conducting polymers are necessary for a deeper understanding (for example, scattering processes in multivalley band structures are known to yield an additional negative contribution to the MC).

In disordered metals, electron–electron interactions play an important role in the low temperature transport; $\sigma_I(T)$ can be expressed as [42–44,134]

$$\sigma_I(T) = \sigma(0) + mT^{1/2} \qquad (8a)$$

where

$$m = \alpha (4/3 - 3\gamma F_\sigma/2) \qquad (8b)$$

$$\alpha = (e^2/\hbar)(1.3/4\pi^2)(k_B/2\hbar D)^{1/2} \qquad (8c)$$

$$F_\sigma = 32[(1 + F/2)^{3/2} - (1 + 3F/4)]/3F \qquad (8d)$$

The finite temperature correction term due to electron–electron interactions in Eq. (8a) consists of exchange and Hartree contributions [44,45]. The sign of this correction depends on the relative size of the exchange and Hartree terms, which depend on the screening length. In doped semiconductors the sign of the finite temperature correction is related to various parameters such as the degeneracy of the conduction band minima in k space (valleys), intervalley scattering, and mass anisotropy. The Hartree factor (F) is the screened interaction averaged over the Fermi surface; α is a parameter that depends on the diffusion coefficient (D), and γF_σ is the interaction parameter. The value of γ depends on the band structure [135,136]. The coefficient was found to change sign as a function of disorder [44], a change that can be interpreted as being due to a sign change in the term $4/3 - 3\gamma F_\sigma/2$. Usually, the sign of m is negative when $\gamma F_\sigma > 8/9$.

Equations (8a)–(8d) are valid in zero magnetic field; at fields sufficiently high that $g\mu_B H \gg k_B T$, both the zero temperature conductivity and the coefficient of the $T^{1/2}$ term are altered:

$$\sigma_I(H,T) = \sigma(H,0) + m'T^{1/2} \qquad (9a)$$

where

$$m' = \alpha[4/3 - \gamma(F_\sigma/2)] \qquad (9b)$$

Using Eqs. (8b) and (9b),

$$\gamma F_\sigma = \left(\frac{3}{8}\right) \frac{m' - m}{3m' - m} \qquad (9c)$$

assuming that α, γ, and F_σ are not dependent on the magnetic field [86]. Thus, the parameters α and γF_σ can be estimated from the values of m and $m' = m(H)$ obtained at $H = 0$ and at $H = 8$ T by using Eqs. (8) and (9), respectively.

The MC at high fields arises mainly from the interaction contribution (weak localization contribution is less important in strong fields) [45]. In the free electron model, using the Thomas–Fermi approximation,

$$F_\sigma = x^{-1} \ln(1 + x) \qquad (10)$$

where $x = (2k_F \Lambda_s)^2$, and Λ_s is the Thomas–Fermi screening length. Equations (8d) and (10) yield $0 < F_\sigma < 0.93$ and $0 < F_\sigma < 1$ (note that $F_\sigma \approx 1$ for short-range interactions and $F_\sigma \ll 1$ for long-range interactions). Decreasing γF_σ leads to a change in the sign of m, corresponding to the divergence of screening length near the M–I transition, consistent with McMillan's prediction [76]. Kaveh and Mott [135] argued, however, that the inelastic electron–electron scattering should dominate near the M–I transition.

The magnetic field dependence of the contribution to the MC from electron–electron interactions can be written as $\Delta\Sigma(H,T) = \sigma(H,T) - \sigma(0,T)$ [42–44,133]

$$\Delta\Sigma_I(H,T) = -0.041\alpha(g\mu_B/k_B)^2 \gamma F_\sigma T^{-3/2} H^2$$
$$(g\mu_B H \ll k_B T) \qquad (11a)$$

and

$$\Delta\Sigma_I(H,T) = \alpha\gamma F_\sigma T^{1/2} - 0.77\alpha(g\mu_B/k_B)^{1/2}\gamma F_\sigma H^{1/2}$$
$$(g\mu_B H \gg k_B T) \qquad (11b)$$

Thus at low and high fields, $\Delta\Sigma_I(H,T)$ is proportional to H^2 and $H^{1/2}$, respectively.

For the low magnetic field regime, we follow Rosenbaum et al. [137,138] and assume that the contributions to $\Delta\Sigma(H,T)$ that arise from electron–electron interactions and weak localization are additive. Thus, the total low-field magnetoconductance is given by the following.

For weak spin–orbit coupling (positive contribution to MC),

$$\Delta\Sigma(H,T) = -0.041\alpha(g\mu_B/k_B)^2 \gamma F_\sigma T^{-3/2} H^2 \qquad (12)$$
$$+ (1/12\pi^2)(e/c\hbar)^2 G_0 (l_{in})^3 H^2$$

For strong spin–orbit coupling (negative contribution to MC),

$$\Delta\Sigma(H,T) = -0.041\alpha(g\mu_B/k_B)^2 \gamma F_\sigma T^{-3/2} H^2 \qquad (13)$$
$$- (1/48\pi^2)(e/c\hbar)^2 G_0 (l_{in})^3 H^2$$

where $G_0 = e^2/\hbar$ and l_{in} is the inelastic scattering length. The first term on the right-hand side of Eqs. (12) and (13) is the contribution from e–e interactions (negative MC), and the second term on the right-hand side is the contribution from weak localization. The first term can be estimated by using Eqs. (8) and (9). Then, using the slope of $\Delta\Sigma(H,T)$ vs. H^2 in the low-field region, the second term can be estimated. In this way, the value of the inelastic scattering length can be calculated at each temperature.

2. Iodine-Doped Oriented Polyacetylene

Iodine-doped oriented polyacetylene has been studied extensively by several groups [17,125–128,130,131]. The maximum room temperature conductivity parallel to the chain axis for the best quality oriented I-$(CH)_x$ is on the order of 10^5 S/cm, and the anisotropy is greater than 100.

Recently Mizoguchi et al. [139,140] showed that the main difference between Shirakawa and Naarmann polyacetylenes is the higher density and higher degree of chain orientation in the latter; the basic features of the spin dynamics are identical in the two kinds of samples. The structural and physical properties of highly conducting polyacetylene have been thoroughly reviewed by Tsukamoto [17]. The crystalline coherence lengths parallel and perpendicular to the chain axis are 120 and 50 Å, respectively. The number of charge carriers in heavily doped samples is on the order of 10^{22} cm^{-3}, and the mean free path is approximately 500 Å. The density of states at the Fermi level is approximately 0.3 state per electronvolt per coulomb.

Although resistivities as low as 10^{-5} $\Omega \cdot$cm have been reported for iodine-doped oriented polyacetylene parallel to the draw direction (chain axis), a positive temperature coefficient of the resistivity, typical of a metal, has not been observed.[1] This indicates that defects or transport perpendicular to the chain axis limit the conductivity. We conclude, therefore, that the 500 Å mean free path is not limited by phonon scattering. This implies that significantly higher room temperature conductivities will be achieved as the quality of the material is improved. The absence of any positive temperature coefficient implies that values at least an order of magnitude higher are to be expected; i.e., the intrinsic conductivity at room temperature is greater than 10^6 S/cm.

Comprehensive studies of the dc conductivity and MC in directions both parallel and perpendicular to the chain axis indicate that interchain transport is significant [125,126]. Transport measurements on oriented iodine-doped $(CH)_x$ have shown that $\rho(T)$, ρ_r, and the MC depend on the extent of disorder [120,125–128,130]. For iodine-doped oriented $(CH)_x$ samples having a room temperature resistivity of $\sim 10^{-5}$ $\Omega \cdot$cm, $\rho_r \approx 3$. Although disorder can be reduced by tensile drawing and thereby by orienting the $(CH)_x$ chains, the subsequent doping process introduces disorder [58]. The degree of anisotropy in oriented $(CH)_x$ plays a significant role in the transport properties, especially in the MC. The transport measurements were carried out on oriented samples with a maximum draw ratio of $l/l_0 \approx 5$–10 (where l is final length and l_0 is initial length) [17]. Thus, the absence of prominent anisotropic features in MC measurements and the rather low values of anisotropy in conductivity (25–50) are due to misaligned chains. Recent work on samples stretch-oriented to draw ratios of 15 showed strong anisotropy in the MC [125,126].

At room temperature, the anisotropy of the conductivity is $\sigma_\parallel/\sigma_\perp > 100$. The temperature dependences of the conductivity are nearly identical for directions parallel and perpendicular to the chain axis. The anisotropic features in the MC have been interpreted in terms of anisotropy in the diffusion coefficient [141], the electron–electron interactions, and weak localization [125,126].

As noted in Section III.D.1, the combination of disorder and anisotropy can lead to a wide range of behavior in the transport properties of disordered anisotropic metals, where both electron–electron interactions and disorder-induced localization near the M–I transition are important. The field-induced crossover from positive to negative MC results from the subtle interplay of weak localization and electron–electron interaction contributions to the MC. From the MC data it is possible to estimate the inelastic scattering length as a function of temperature; inelastic electron–electron scattering in disordered metals dominates the transport at low temperatures in I-(CH)$_x$ [125,126].

a. Pressure Dependence of Conductivity

The conductivity increases for pressures up to 4 kbar, then gradually decreases, as shown in Fig. 2.14 [125,126]. The data are consistent for both collinear four-probe and Montgomery methods. The increase in conductivity up to 4 kbar is reversible with pressure, whereas the decrease at higher pressures is not reversible. The pressure dependence of conductivity in potassium-doped oriented (CH)$_x$ [124] is nearly identical to that obtained from iodine-doped material [125,126]. The inset in Fig. 2.14 shows the pressure dependence of the conductivity perpendicular to the chain axis. The initial increase in $\sigma(P)$ is attributed to enhanced interchain transport, whereas the decrease above 4 kbar is not understood. At 8 kbar, the anisotropy has decreased by a factor of 1.6, from 110 to 67, at room temperature. The pressure dependences parallel and perpendicular to the chain axis are identical, implying that the macroscopic conductivity in both directions is limited by interchain charge transport.

b. Temperature Dependence of Conductivity

Ishiguro and coworkers have reported extensive measurements of $\sigma(T)$ for iodine-doped [127,128] and FeCl$_3$-doped [130] (CH)$_x$ down to millikelvin temperatures. Although the room temperature conductivity of FeCl$_3$-(CH)$_x$ is nearly an order of magnitude lower than that

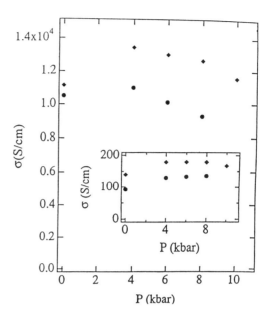

Fig. 2.14 Conductivity parallel to the chain axis versus pressure for I-(CH)$_x$ (inset shows conductivity perpendicular): (●) samples cut parallel (inset perpendicular) to the chain axis for collinear four-probe technique; and (♦) rectangular samples for Montgomery technique.

of I-(CH)$_x$, ρ_r values for FeCl$_3$-(CH)$_x$ and I-(CH)$_x$ are similar, 1.3 and 2.8, respectively, as shown in Figs. 2.15a and 2.15b. Moreover, the positive TCR for FeCl$_3$-(CH)$_x$ above 200 K implies that probably the doping is more homogeneous and that the dopant-induced interchain interaction is higher in this system than in I-(CH)$_x$. In the metallic regime, the low temperature conductivity can be explained by the localization–interaction model appropriate to disordered metallic systems. Details on the temperature dependence of the anisotropy for various samples are summarized in Table 2.1. Since doped (CH)$_x$ samples are highly susceptible to oxidative degradation, ρ_r increases with time due to defect formation, reduction in conjugation length, dedoping, etc.

As noted above, the fluctuation-induced tunneling model [100] was widely used to explain the weak temperature dependence of $\sigma(T)$ of doped polyacetylene [142]. Recent results have shown, however, that this model is not satisfactory for highly conducting polyacetylene [127,128,143]. A log T dependence of $\rho(T)$ was found for samples having $\rho_r \approx 14$ [110]. For highly conducting samples with $\rho_r < 10$, a considerable deviation from the log T dependence occurs at low temperatures. Plots of ρ versus log T and log ρ versus log T for I-(CH)$_x$ samples having $\rho_r \approx 3$ are shown in Figs. 2.16a and 2.16b. In both cases, the log T fit deviates below 20 K. Similar results were obtained for FeCl$_3$-(CH)$_x$ [130]. Thus, the log T dependence of ρ is observed only in the case of

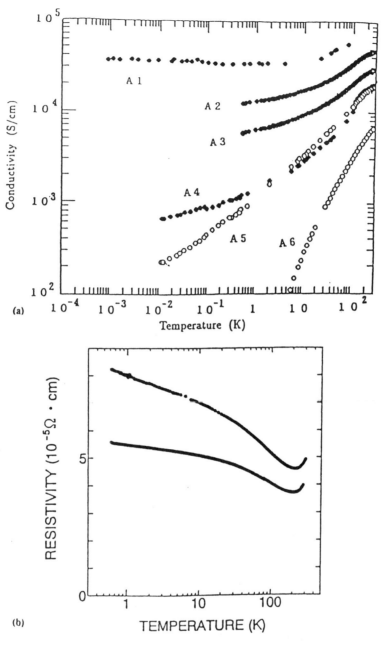

Fig. 2.15 Log-log plot of resistivity $\rho(T)$ or conductivity $\sigma(T)$ vs. T of doped $(CH)_x$ samples. (a) I-$(CH)_x$ samples (data from Ref. 110). (b) $FeCl_3$-$(CH)_x$ samples (data from Ref. 130).

Table 2.1 Conductivity and Anisotropy of Conductivity of I-$(CH)_x$ at 250 K and 1.2 K, at Ambient Pressure and 8 kbar, for Current Parallel and Perpendicular to the Chain Axis

		Temperature	
	Pressure (kbar)	250 K	1.2 K
Conductivity σ (S/cm)			
Parallel to chain axis ($\sigma_\parallel$)	0	11050	3670
Perpendicular to chain axis ($\sigma_\perp$)	0	105	34.2
Parallel to chain axis ($\sigma_\parallel$)	8	8460	3880
Perpendicular to chain axis ($\sigma_\perp$)	8	127	45
Anisotropy of conductivity ($\sigma_\parallel/\sigma_\perp$)	0	105.2	107.3
	8	66.6	86.3

M–I Transition in Doped Conducting Polymers

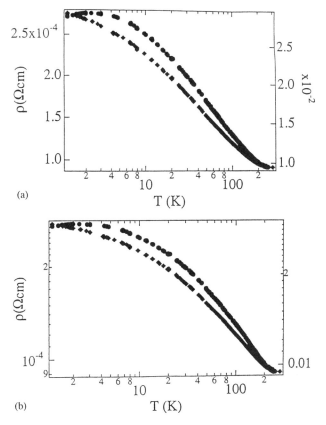

Fig. 2.16 (a) $\rho(T)$ vs. log T for I-$(CH)_x$ samples with $\rho_r < 3$ (●) parallel to the chain axis (left-hand axis) and (♦) perpendicular to the chain axis (right-hand axis). (b) Log-log plot of resistivity for I-$(CH)_x$ samples with $\rho_r < 3$ (●) parallel to the chain axis and (♦) perpendicular to the chain axis.

$(CH)_x$ samples with intermediate disorder and not for the best materials having $\rho_r < 4$. For these low values of ρ_r, the materials are in the critical regime where $\rho(T) \propto T^{-\beta}$ [120]. As emphasized above, the various regimes are most easily identified from plots of $W(T)$ vs. T.

The $W(T)$ vs. T plots for transport parallel and perpendicular to the chain axis for samples with $\rho_r \approx 3$ are shown in Fig. 2.17 [120,125,126]. Although the anisotropy is nearly 100, $W(T)$ is identical in the two directions. Moreover, for both directions, W is nearly temperature-independent from 180 to 60 K; below 60 K at ambient pressure, the positive temperature coefficient of $W(T)$ indicates that the transport is just on the metallic side of the critical regime. Note that the conductivity parallel to the chain axis is much higher than Mott's minimum metallic value; however, the conductivity perpendicular to the chain axis is on the order of 10^2 S/cm, i.e., close to the Mott value.

At high pressures, both ρ_r and the anisotropy decrease because of the enhanced interchain transport. At 8 kbar, W vs. T exhibits metallic behavior, as shown in Fig. 2.17. When ρ_r gradually increases above 3, the regime where $W(T)$ is temperature-independent extends to lower temperatures, whereas at rather high values of ρ_r (e.g., $\rho_r > 20$), the slope of $W(T)$ vs. T becomes negative as shown in Fig. 2.12. Similar behavior has been observed in highly doped inorganic semiconductors [71,72] as well as in PANI-CSA and PPY-PF_6 [120].

The $T^{1/2}$ and $T^{3/4}$ fits (below 60 K), parallel and perpendicular to the chain axis, are shown in Figs. 2.18a and 2.18b [125,126]. Although the anisotropy is nearly 100, these fits are identical for the two directions, indicating that an anisotropic three-dimensional model is appropriate for highly conducting $(CH)_x$. The linearity of the $T^{3/4}$ fits is better than that of the $T^{1/2}$ fits, implying

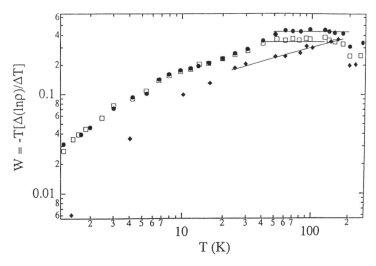

Fig. 2.17 Log-log plot of $W(T)$ vs. T for I-$(CH)_x$ (●) parallel to the chain axis at ambient pressure; (□) perpendicular to the chain axis at ambient pressure; and (♦) parallel to the chain axis at 8 kbar. The lines are drawn to guide the eye.

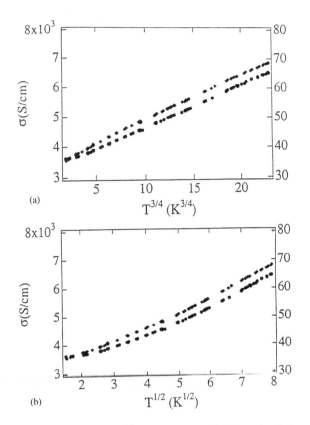

Fig. 2.18 (a) $\sigma(T)$ vs. $T^{3/4}$ for I-$(CH)_x$ and (b) conductivity vs. $T^{1/2}$ for I-$(CH)_x$ (●) parallel to the chain axis; (◆) perpendicular to the chain axis.

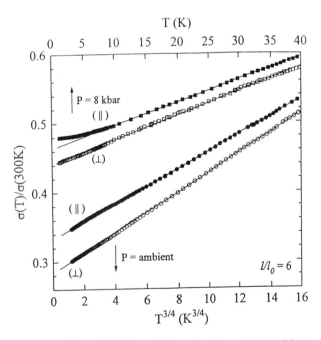

Fig. 2.19 Plots of $\sigma(T)$ vs. $T^{3/4}$ and $\sigma(T)$ vs. T at ambient pressure and at 8 kbar for I-$(CH)_x$ parallel direction to chain axis: $T < 40$ K and $\rho_r < 3$.

that the contribution from localization is dominant at higher temperatures. For temperatures where $\sigma \propto T^{3/4}$, inelastic electron–phonon scattering ($p = 3/2$) is the dominant scattering mechanism. However, when $\sigma \propto T^{1/2}$ ($T < 3$ K), e–e interactions are more important. A pressure-induced crossover from $T^{3/4}$ to $T^{1/2}$ is observed at low temperatures, as shown in Fig. 2.19. Thus, although the localization and interaction effects dominate at high and low temperatures, respectively, the two can be fine-tuned by varying the interchain interactions with pressure, orientation, etc. This has been confirmed from MC measurements as described below.

The temperature dependences of the conductivity at 8 kbar, parallel and perpendicular to the chain axis, are shown in Fig. 2.20 [125,126]. Although the room temperature conductivity decreases above 4 kbar, the temperature dependence at 8 kbar is substantially reduced. The values of ρ_r at ambient pressure and at 8 kbar, parallel to the chain axis, are 3 and 2.2 respectively, and those perpendicular to the axis are 3 and 2.8, respectively, demonstrating substantial enhancement of the interchain transport at high pressure. Nevertheless, a positive temperature coefficient of the resistivity has not been observed. Thus, even at high pressure, the combination of weak interchain transport and disorder limit the three-dimensional mean free path in this metallic quasi-one-dimensional conducting polymer.

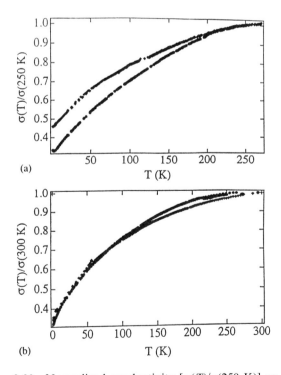

Fig. 2.20 Normalized conductivity $[\sigma(T)/\sigma(250\text{ K})]$ vs. T for I-$(CH)_x$ (a) parallel to the chain axis; (b) perpendicular to the chain axis; (●) at ambient pressure; (◆) at 8 kbar.

c. Magnetoconductance

Ishiguro and coworkers reported a detailed study of the magnetoconductance in I-$(CH)_x$ [127,128] and $FeCl_3$-$(CH)_x$ [130]; the MC features are rather different in the two cases, as shown in Figs. 2.21a and 2.21b. The MC measurements in oriented I-$(CH)_x$ ($l/l_0 \approx$ 5–10, $\sigma_\parallel/\sigma_\perp \approx$ 25–50) display a wide range of behavior, including both positive and negative MC [127,128,144]. When ρ_r decreases, the sign of the MC shifts from negative to positive. For samples with intermediate disorder ($\rho_r \approx$ 3–6), the sign of the MC was positive when the field was perpendicular to the chain axis, and negative when the field was parallel to the chain axis, at temperatures above 2 K. In both cases (H parallel or perpendicular to the chain axis), the magnitude of the positive MC decreased gradually as the temperature decreased from 10 to 1 K. This indicates that the weak localization contribution (positive MC) dominates at higher temperatures and that the contribution from e–e interactions (negative MC) becomes increasingly important at lower temperatures.

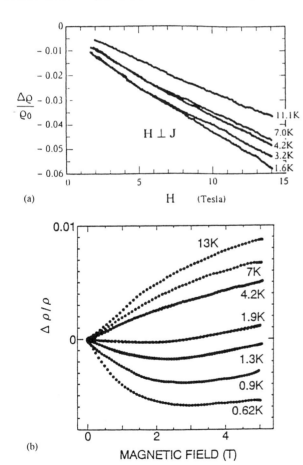

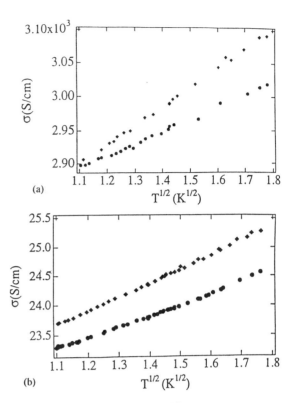

Fig. 2.22 Conductivity versus $T^{1/2}$ for I-$(CH)_x$ (a) parallel to the chain axis and (b) perpendicular to the chain axis (●) at $H = 0$ and (♦) at $H = 8$ T.

Fig. 2.21 Magnetoconductance [$\Delta\sigma = \sigma(H) - \sigma(0)$] vs. H (a) for I-$(CH)_x$ ($\rho_r < 3$) in the transverse field (data from Refs. 125 and 126) and (b) for $FeCl_3$-$(CH)_x$ ($\rho_r < 1.5$) in the longitudinal field (data from Ref. 130).

When the disorder is weaker ($\rho_r \leq 3$), the weak localization contribution dominates to lower temperatures ($T \leq$ K). Thus, the anisotropy in both conductivity and MC is related to the extent of misaligned chains (anisotropy on the molecular scale) and to the anisotropy in the diffusion coefficient.

The $T^{1/2}$ dependence of $\sigma(T)$ is shown in Fig. 2.22 for temperatures below 3 K [125,126]. Although the temperature range of the $T^{1/2}$ fit is rather narrow, this contribution is evident at very low temperatures from the enhanced negative contribution to the MC. The existence of a $T^{1/2}$ term indicates that at very low temperatures the contribution from e–e interactions in disordered metals plays a dominant role.

The values of m, $m(H)$, and γF_σ [Eqs. (8) and (9)] are summarized in Table 2.2. The values of $\sigma(0)$, extrapolated to $T \to 0$, are $\sigma_\parallel(0) \approx 3600$ S/cm, $\sigma_\perp(0) \approx 30$ S/cm at ambient pressure and $\sigma_\parallel(0) \approx 3800$ S/cm, $\sigma_\perp(0) \approx 40$ S/cm at 8 kbar. Although $\sigma_\parallel(0)$ exceeds values typical of systems near the M–I transition, values of $\sigma(0)$ as large as 4135 S/cm have been reported for Y_xSi_{1-x} [145]. Since the conductivity parallel to the chain axis (on the order of 10^3 S/cm) exceeds that typical of systems near the M–I transition, the values of m and m' are unusually large; correspondingly, the values of m, m', and σ perpendicular to the chain axis are typical of systems near

Table 2.2 Values of the Parameters m, $m(8\ T)$, α, and γF_σ, and Temperature Dependence of the Inelastic Scattering Length (Current Parallel and Perpendicular to the Chain Axis and Field Perpendicular to the Current and Chain Axis) for I-$(CH)_x$

Field (H) vs. current (I) vs. chain axis (ν)	m^a	$m(8\ T)^a$	α	γF_σ	Inelastic scattering length (Å)		
					4.2 K	2 K	1.2 K
$H \perp \nu;\ I \parallel \nu;\ H \perp I$	195	316	283	0.43	644	902	1163
$H \perp \nu;\ I \perp \nu;\ H \perp I$	1.67	2.3	1.93	0.31	134	170	209

^{a} In S/(cm·K$^{1/2}$).

the M–I transition. Thus the localization–interaction model is appropriate for the analysis of MC data in the direction perpendicular to the chain axis.

The MC for current parallel to the chain axis is shown in Figs. 2.23a–2.23c [125,126]. When the field is perpendicular (the current is parallel) to the chain direction, the MC is positive; however, the magnitude of positive

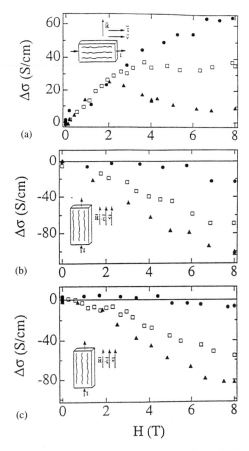

Fig. 2.23 Magnetoconductance versus H for I-$(CH)_x$ at 4.2 K (●), 2 K (□), and 1.2 K (▲). The current is parallel to the chain axis. (a) The field is perpendicular (perpendicular) to the chain axis (current); (b) the field is parallel (parallel) to the chain axis (current) at ambient pressure; (c) the field is parallel (parallel) to the chain axis (current) at 8 kbar.

contribution decreases at low temperatures, as shown in Fig. 2.23a, due to the interplay of weak localization and electron–electron interaction contributions. When the field and current are parallel to the chain direction, the sign of MC is always negative, and its magnitude increases at low temperatures, as shown in Fig. 2.23b. This anisotropy is attributed to the anisotropic diffusion coefficient. This is proven by MC measurements under high pressure. At 4.2 K, the conductivity anisotropy is 98 at ambient pressure and 77 at 8 kbar. The enhancement of the interchain interaction at high pressure reduces the anisotropy (both conductivity and diffusion coefficient). The MC at 8 kbar with field and current parallel to the chain axis is shown in Fig. 2.23c. For $H < 4$ T, the sign of the MC has reversed from negative to positive at 4.2 K and 8 kbar; at higher fields, the sign remains negative due to the dominant contribution from electron–electron interactions, as shown in Fig. 2.23c. Finally, the magnitude of negative MC is reduced under pressure at temperatures below 4.2 K, showing that the sign and magnitude of the MC depend on the anisotropic diffusion coefficient in oriented $(CH)_x$ [141].

The MC for current perpendicular to the chain axis is shown in Fig. 2.24. At 4.2 K, the MC is positive when the field and current are perpendicular to the chain direction as shown in Figs. 2.24a and 2.24b. This is due to the dominant contribution from weak localization. Since the contribution from e–e interactions increases at lower temperatures and higher magnetic fields, the sign of the MC reverses from positive to negative at 1.2 K and $H = 4$ T. For the case of current perpendicular and field parallel to the chain axis, the MC is shown in Fig. 2.24c. The sign is negative, similar to when current and field are parallel to the chain direction, as shown in Fig. 2.24b. These data show that the anisotropic MC is not caused by the direction of current with respect to the chain axis; the anisotropic MC is due to the anisotropic diffusion coefficient, which in turn is a function of the angle between the field direction and the chain axis.

Similar anisotropy in the MC has been observed by Ishiguro and coworkers [127,128] (between 4.2 and 1.6 K) for samples having $\rho_r \approx 3$–5, as shown in Figs. 2.25a and 2.25b. Although the sign did not change for samples having $\rho_r \approx 3$, the positive MC is nearly three times as large when the field is perpendicular to the chain axis

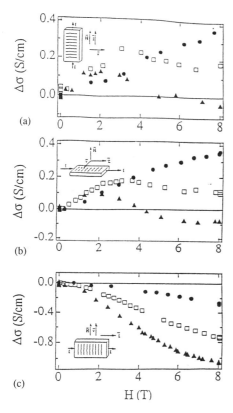

Fig. 2.24 Magnetoconductance versus H for I-(CH)$_x$ at 4.2 K (●), 2 K (□), and 1.2 K (▲). The current is perpendicular to the chain axis (n). (a) The field is perpendicular to the chain axis (parallel to the current); (b) the field is perpendicular to the chain axis and current; (c) the field is parallel to the chain axis. (perpendicular to the current).

as when it is parallel to the chain axis, indicating that the interaction contribution to the MC is larger in the latter case due to the anisotropic diffusion coefficient. The field-induced crossover from positive to negative MC shown in Figs. 2.24a and 2.24b results from the interplay of weak localization and e–e interaction contributions.

The inelastic scattering length at different temperatures can be estimated from $l_{in} = (D\tau_{in})^{1/2}$ for directions parallel and perpendicular to the chain axis. At 1.2 K, the inelastic scattering length in the directions parallel and perpendicular to the chain axis are 1163 and 210 Å, respectively. The temperature dependence of the inelastic scattering length is shown in Fig. 2.26. The $T^{-3/4}$ dependence of τ_{in}, in directions parallel and perpendicular to the chain axis, is typical of inelastic electron–electron scattering in disordered metals [42,43,125,126]. This is in agreement with the $T^{3/4}$ dependence of conductivity shown earlier in Fig. 2.18a. Thus, both the conductivity and the MC are consistent with the localization–interaction model of transport in anisotropic disordered metals.

d. Summary

The room temperature conductivity increases up to 4 kbar and then gradually decreases at higher pressures. At 8 kbar, the temperature dependence of $\sigma_\parallel$ and $\sigma_\perp$ have decreased by factors of 1.4 and 1.6, respectively, indicating that interchain transport is enhanced by pressure. $W(T)$ is temperature-independent from 180 to 60 K in directions both parallel and perpendicular to the chain axis, indicating that at ambient pressure I-(CH)$_x$ is on the metallic side of the critical regime. At 8 kbar, the system exhibits more metallic behavior due to enhanced interchain transport. The $\sigma(T) \propto T^{3/4}$ dependence (4–50 K) indicates that inelastic e–e scattering in disordered metals is the dominant scattering process.

The sign and magnitude of the MC are determined by the extent of disorder, the temperature, the degree of chain orientation, and the angle between the magnetic field and the chain axis. In samples with fewer misaligned chains (e.g., $\sigma_\parallel/\sigma_\perp > 100$) and less disorder ($\rho < 3$), the sign of MC is positive when the field is perpendicular to the chain axis and negative when the field is parallel to the chain axis. Whether the field is parallel or perpendicular to the chain axis, the e–e interaction contribution (negative MC) dominates over the weak localization contribution (positive MC) at high magnetic fields and low temperatures. The anisotropic MC arises mainly from the anisotropic diffusion coefficient, as confirmed by the crossover from negative MC to positive MC and the decrease in the magnitude of the negative MC at 8 kbar. The inelastic scattering length for transport parallel and perpendicular to the chain axis is 1160 and 200 Å, respectively. The $T^{-3/4}$ dependence of inelastic scattering length is consistent with the $T^{3/4}$ dependence of conductivity at low temperatures, indicating that the inelastic electron–electron scattering in disordered metals is the dominant scattering mechanism in I-(CH)$_x$.

3. Potassium-Doped Oriented Polyacetylene

As noted in Section III.C.3, K-(CH)$_x$ is metallic at high pressures. The positive temperature coefficient of W at higher pressures, as shown in Fig. 2.10, indicates the crossover from the critical regime to metallic behavior [120,124]. Although the large negative MC is reduced by a factor of 2 at 10 kbar with respect to that at ambient pressure, the extent of disorder is sufficiently important that the dominant contribution to the negative MC is from hopping transport.

Recently Bernier and coworkers [146,147; D. Bormann and P. Bernier, private communication] observed rather weak temperature dependence ($\rho_r < 2$) with a positive TCR below 7 K at various doping levels in K-(CH)$_x$. The correlation between the staging-induced structural transitions and the electronic properties in K-(CH)$_x$ are in the preliminary stages of investigation [Bor-

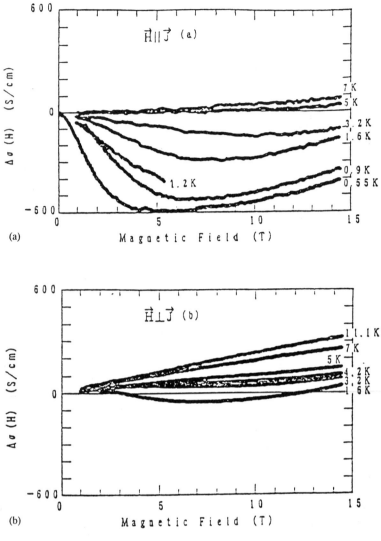

Fig. 2.25 Magnetoconductance versus H of I-$(CH)_x$ ($\rho_r \approx 3.7$) (a) for longitudinal field; (b) for transverse field. (From Ref. 127.)

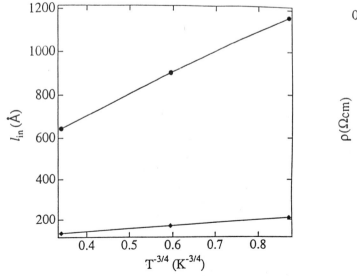

Fig. 2.26 The inelastic scattering length versus $T^{-3/4}$ for I-$(CH)_x$ (●) parallel and (♦) perpendicular to the chain axis.

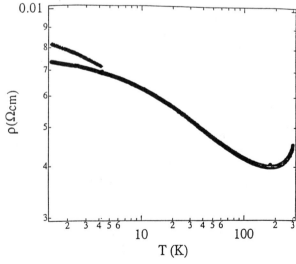

Fig. 2.27 Log-log plot of $\rho(T)$ vs. T for PANI-CSA in the metallic regime: $H = 0$ T (●) and $H = 8$ T (♦).

mann and Bernier, private communication; 146–148]. Thus more work is necessary to fully characterize the M–I transition and the metallic regime in alkali metal doped $(CH)_x$.

4. PANI-CSA

a. Temperature Dependence of the Conductivity

PANI-CSA is metallic when prepared by casting from solution in m-cresol [34,35]. The metallic regime is characterized by $\rho_r \approx 1.5$–2 and by a positive temperature coefficient of W (below 40 K), as shown in Fig. 2.27a. Although $\rho(T)$ increases at low temperature, the temperature dependence is extremely weak; the system has crossed over from the power law dependence characteristic of the critical regime to "metallic" behavior. Application of an 8 T external magnetic field increases the low temperature resistivity as shown in Fig. 2.27a. In an external magnetic field of 8 T, the temperature dependence approaches the power law regime, implying that the external field moves the system toward the critical regime ($\delta \to 0$). Even for $H = 8$ T, however, log-log plots show residual curvature, indicating that the metallic regime of PANI-CSA is robust. Moreover, the recent observation of large negative dielectric constant in the microwave regime by Joo et al. [149,150], as shown in

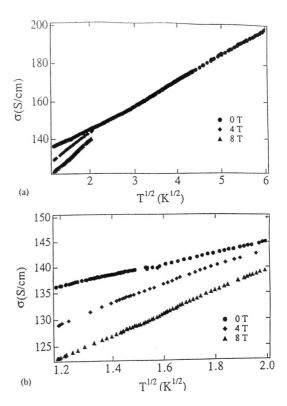

Fig. 2.29 (a) Plot of σ vs. $T^{1/2}$ for PANI-CSA in the metallic regime at $H = 0$, 4 T, and 8 T. (b) Data below 4.2 K are shown on an expanded scale.

Fig. 2.28, has confirmed the metallic nature of PANI-CSA.

The temperature dependence of the conductivity with $\rho_r = 1.6$ is plotted as $\sigma(T)$ vs. $T^{1/2}$ in Figs. 2.29a and 2.29b [34,35]. The $T^{1/2}$ dependence is in agreement with Eq. (8a) and consistent with metallic behavior near the M–I transition [the $T^{1/2}$ dependence implies that $p = 1$ in Eq. (7)]. The data yield $\sigma(T) = \sigma(0) + mT^{1/2}$ with $m(0) = 10.8$ S/(cm·K)$^{1/2}$. Figures 2.29a and 2.29b show $\sigma(T)$ vs. $T^{1/2}$ for $H = 0$, 4 T, and 8 T; the magnetic field decreases the low temperature conductivity (positive magnetoresistance). At low temperatures, shown in greater detail in Fig. 2.29b, the slope of σ vs. $T^{1/2}$ is field-dependent, as predicted by Eq. (9).

The temperature dependence of the resistivity of metallic PANI-CSA is shown for temperatures down to 75 mK in Fig. 2.30 [151]. The relatively large conductivity ($\sigma \approx 30$ S/cm) at ultralow temperatures confirms the metallic nature of PANI-CSA. The resistivity at $H = 0, 2$ T, and 8 T shows a minimum (vs. T) that could be due to the interplay of weak localization and electron–electron interaction contributions at millikelvin temperatures. The field dependence of the resistivity at these temperatures demonstrates the robust nature of the three-dimensional transport and the absence of one-dimensional localization.

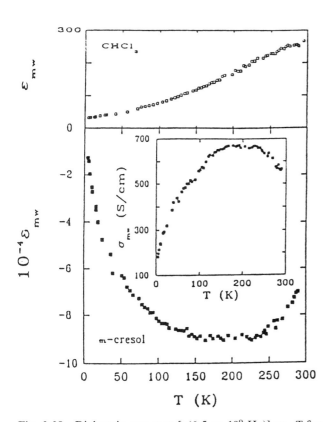

Fig. 2.28 Dielectric constant [$\epsilon(6.5 \times 10^9$ Hz)] vs. T for PANI-CSA (m-cresol). (From Ref. 149.)

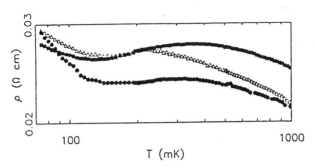

Fig. 2.30 Data of $\rho(T)$ vs. T for PANI-CSA in the metallic regime at 0 T (●), 2 T (△), and 8 T (■). (From Ref. 151.)

b. Magnetoconductance

In the metallic regime, the negative MC of PANI-CSA is relatively weak compared to that in the critical or insulating regimes [34,35]. The dependence of the MC

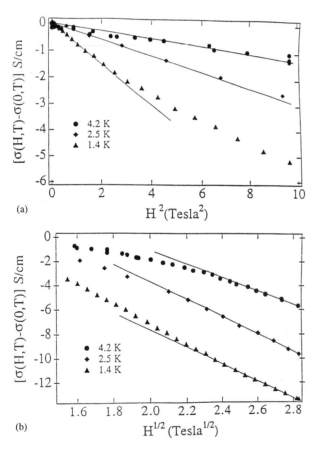

Fig. 2.31 Magnetoconductance (MC) for PANI-CSA in the metallic regime at 4.2 K, 2.5 K, and 1.4 K. (a) MC vs. H^2; the solid lines indicate H^2 dependence. (b) MC vs. $H^{1/2}$; the solid lines indicate $H^{1/2}$ dependence.

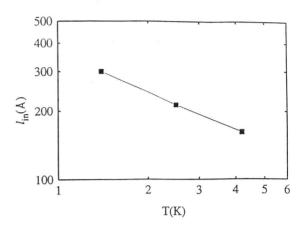

Fig. 2.32 Inelastic scattering length (l^{in}) vs. T for PANI-CSA in the metallic regime.

on H^2 and $H^{1/2}$ at low and high fields, respectively, is shown in Figs. 2.31a and 2.31b. The magnetic field decreases the zero-temperature conductivity (negative MC) and increases the slope of σ vs. $T^{1/2}$; both effects result directly from electron–electron interactions and arise predominantly from the Zeeman splitting of the spin-up and spin-down bands [44]. The values for α and γF_σ determined by using Eq. (8), (9), and (11) are $\alpha \approx 18.3$ and $\gamma F_\sigma \approx 0.5$. This relatively small value for γF_σ is comparable to that found in studies of doped semiconductors at doping levels near the M–I transition [134].

The inelastic scattering length estimated from $\tau_{in} = (l_{in})^2/D$ is shown in Fig. 2.32; $l_{in} \propto T^{-1/2}$, $\tau_{in} \propto T^{-1}$. Using the known value of $\alpha = 18.3$, D can be calculated from Eq. (8c); $D \approx 1.25 \times 10^{-2}$ cm^2/s. We conclude, therefore, that at 4.2 K, $\tau_{in} \sim 10^{-10}$ s, which is the same as that reported for potassium-doped polyacetylene [73,74] and various other amorphous metals [152]. The T^{-1} temperature dependence of τ_{in} is in agreement with the theoretical prediction of Belitz and Wysokinski [132] for systems very close to the M–I transition. The same T dependence has been observed in metallic Si:B near the M–I transition.

The MC measurements down to 20 mK at 16 T [151] are shown in Fig. 2.33. The data indicate the H^2 and $H^{1/2}$ dependence of negative MC at low and high fields, respectively. The absence of saturation of the negative MC at 20 mK and 16 T suggests the interesting possibility of an open Fermi surface for PANI-CSA.

c. Intrinsic Conductivity and Metallic Nature of PANI-CSA

The inelastic scattering length can be estimated from $\tau_{in} = (l_{in})^2/D$; $l_{in} \sim 150$–300 Å [34,35]. This enables an estimate of the intrinsic conductivity parallel to the chain

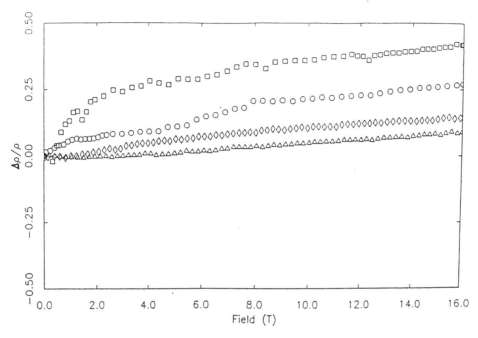

Fig. 2.33 Magnetoresistance for PANI-CSA in the metallic regime at 20 mK (□), 70 mK (○), 370 mK (◇), and 500 mK (△). (From Ref. 151.)

axis in an aligned sample:

$$\sigma_{int} = Ne^2\tau/m = (Ne^2/\hbar\ k_F)l_{in} \quad (14)$$

where N is the density of carriers and k_F is the Fermi wave number in the chain direction. Using $N \approx 2.5 \times 10^{21}$ cm^{-3} and $k_F \approx \pi/2c$, where $c \approx 7$ Å [153,154], one obtains $\sigma_{int} \approx 2.5 \times 10^4$ S/cm. In an unoriented sample, the measured value would be reduced by the anisotropy. An alternative estimate can be obtained from the observed increase in resistivity in the region of the positive temperature coefficient; $\Delta\rho \sim 3 \times 10^{-4}$ Ω·cm. This would imply $\sigma_{int} \sim 3 \times 10^3$ S/cm for a nonoriented sample; since the anisotropy would be expected to reduce the value in a nonoriented sample by about a factor of 10–100, the estimated value for $\sigma_\parallel(300$ K$)$ would be in excess of 3×10^4 S/cm. We conclude that for chain-oriented and chain-aligned PANI, the intrinsic conductivity along the chain axis at room temperature should be significantly greater than 10^4 S/cm.

The metallic nature of PANI-CSA has been confirmed from the temperature-independent Pauli susceptibility [27–30,155], the observation of modified Korringa relation from ^{13}C-NMR [156], large negative dielectric constant at 8–12 GHz [149,150], and the observation of plasma edge in infrared reflectivity measurements [157–158]. The microscopic spin dynamics in previous generation PANI (PANI-HCl) indicated that the intra/interchain spin diffusion rates are highly anisotropic and suggested that doped PANI should be considered a highly one-dimensional system [161–166]. This is in contrast to PANI-CSA, in which the electronic states are delocalized in three dimensions.

d. Summary

For metallic PANI-CSA, the conductivity at 20 mK is approximately 30 S/cm, the positive TCR extends from 300 to 150 K, and $\rho_r < 2$. In this metallic regime, the $T^{1/2}$ dependence at low temperatures points to the importance of e–e interactions. The magnitude of the inelastic scattering time, $\tau_{in} \approx 10^{-10}$ s, is typical of that of amorphous metals. The intrinsic conductivity of PANI-CSA along the chain axis is estimated to be significantly greater than 10^4 S/cm at room temperature. The magnetic and optical properties also indicate that PANI-CSA is a disordered metal.

5. PPy-PF$_6$

a. Temperature Dependence of the Conductivity

The room temperature conductivity for films grown by electropolymerization at -40°C is $\sigma_{RT} = 150$–400 S/cm [25,26,167]. Films prepared at room temperature are less conducting [$\sigma_{RT} = 10$–100 S/cm], and the temperature dependence of the resistivity is rather strong, typical of that of an insulator. Thus, one can vary the disorder by changing the polymerization temperature. The temperature dependence of normalized resistivity for various PPy-PF$_6$ samples is shown in Fig. 2.34. For samples with higher conductivity, $\sigma_{RT} > 150$ S/cm

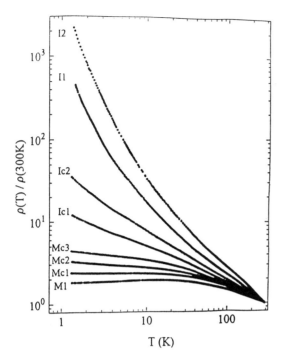

Fig. 2.34 Log-log plot of normalized $\rho(T)$ vs. T of PPy-PF_6 samples. (M = metallic regime, Mc = metallic side of critical regime, Ic = insulating side of critical regime, and I = insulating regime.) The values of $\sigma(300\ K)$ and resistivity ratios (ρ_r) are listed in Tables 2.3 and 2.4.

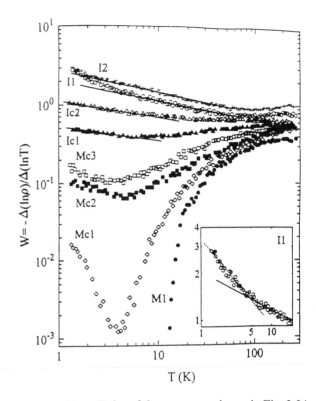

Fig. 2.35 W vs. T plot of the same samples as in Fig. 2.34. The inset shows data for sample I1 below 20 K.

(samples denoted by M), the resistivity decreases at low temperatures; there is a resistivity maximum around $T = 7$–20 K.

Again, for PPy-PF_6 the disorder can be characterized by ρ_r [167]:

1. For samples on the metallic side of the critical regime ($2 < \rho_r < 10$), $\sigma(0)$ is finite but TCR remains negative at all temperatures (denoted as Mc samples).
2. For samples in the metallic regime ($\rho_r < 2$), the TCR is positive at low temperatures with a conductivity minimum at $T = T_m$ (denoted as M samples).

Although the existence of finite $\sigma(0)$ defines the boundary of the metallic regime in PPy-PF_6 ($6 < \rho_r < 10$), the sign of the TCR changes at $\rho_r \approx 2$. The power law dependence, $\rho(T) = T^{-\beta}$, is observed at high temperatures ($T > 100$ K). The various parameters obtained from the data are listed in Tables 2.3 and 2.4 [167].

The $\rho(T)$ plots for various PPy-PF_6 samples demonstrate once again that such plots alone (Fig. 2.34) are not sufficient to identify the M–I transition. On the other

Table 2.3 Experimental Values and Parameters for PPy-PF_6 Samples in the Insulating Regime

Sample	$\sigma(300\ K)$ (S/cm)	ρ_r[a]	$\Delta\rho/\rho$[b]	x[c]	T_{Mott} (K)[d]	L_c (Å)
Ic1	114	11.6	0.40	0.19 ± 0.03 (T < 4 K)	20	269
Ic2	103	35.8	0.51	0.24 ± 0.02 (T < 5 K)	290	177
I1	52	527	—	0.24 ± 0.02 (T > 5 K)	3,700	—
I2	34.4	2590	1.78	0.29 ± 0.03 (T > 2 K)	17,500	86

[a] $\rho_r = \rho(1.4\ K)/\rho(300\ K)$.
[b] Data at $H = 8$ T and at $T = 1.4$ K.
[c] Results from data using Eq. (4).
[d] Values are obtained assuming $x = 0.25$.

M–I Transition in Doped Conducting Polymers

Table 2.4 Experimental Values and Parameters for PPy-PF$_6$ Samples in the Metallic Regime

Sample	Pressure	σ(300 K) (S/cm)	ρ_r[a]	$\Delta\rho/\rho$[b]	m [S/(cm·K$^{1/2}$)]	m' [S/(cm·K$^{1/2}$)]	$\sigma(0)$[c] (S/cm)	L_c (Å)[d]	T_m (K)
M1	Ambient	338	1.75	0.12	−7.55	+7.80	201	12.1	12
M2	Ambient	298	1.97	0.13	−3.19	+8.34	155	15.7	7.5
M2	9 kbar	330	1.33	0.05	−8.83	+0.86	261	9.3	24
Mc1	Ambient	271	2.40	0.16	+1.75	+11.6	108	22.5	—
Mc2	Ambient	313	3.22	0.21	+12.9	+25.9	82	29.4	—
Mc2	4 kbar	358	1.81	0.12	−3.98	+10.2	191	12.7	12
Mc2	10 kbar	377	1.54	0.10	−9.13	+6.22	247	9.8	19
Mc3	Ambient	192	4.45	0.23	+8.00	+12.9	34	70.9	—
Ic1	4 kbar	133	2.64	0.18	+2.05	+6.83	46	52.6	—
Ic2	10 kbar	137	2.08	0.15	−0.20	+5.11	64	37.8	—

[a] $\rho_r = \rho(1.4\ \text{K})/\rho(300\ \text{K})$.
[b] Data at $H = 8$ T and at $T = 1.4$ K.
[c] Extrapolated values from $T^{1/2}$ dependence of the conductivity.
[d] Calculated from the relation $\sigma(0) = 0.1 e^2/hL_c$.

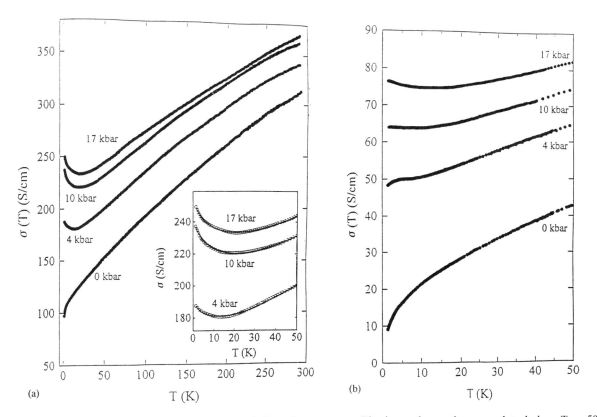

Fig. 2.36 (a) $\sigma(T)$ for PPy-PF$_6$ sample Mc2 ($\rho_r = 3.2$) under pressure. The inset shows the same data below $T = 50$ K. Solid lines in the inset are fitted curves using $\sigma(T) = \sigma(0) + mT^{1/2} + BT^{p/2}$, where $p = 2.50 \pm 0.04$ and $B = 0.4 \pm 0.01$. Note that $\sigma(0)$ and m depend on pressure. (b) The temperature dependence of the conductivity ($T < 50$ K) of the sample Ic1 ($\rho_r = 12$) under pressure.

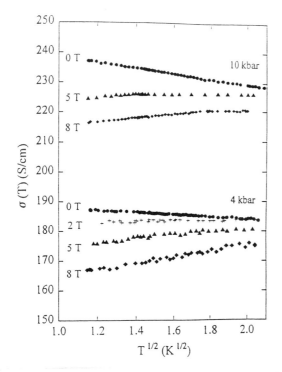

Fig. 2.37 $\sigma(T)$ at low temperature (1.3 K $< T <$ 4 K) for PPy-PF$_6$ sample Mc2 under $P =$ 4 kbar ($\rho_r =$ 1.81) and 10 kbar ($\rho_r =$ 1.54) in various magnetic fields ($H =$ 0, 2, 5, and 8 T).

hand, the W vs. T plot in Fig. 2.35 (the same data as in Fig. 2.34) brings out the subtle variations in the temperature dependence of the resistivity with clarity and precisely defines the M–I transition. For samples on the metallic side of the M–I transition, the temperature coefficient of W is positive at high temperatures, whereas for samples on the insulating side, the temperature coefficient of W is negative.

Since the coefficient m in Eq. (8a) can have either sign depending on the competition between the Hartree and exchange contributions, the positive TCR for samples in the metallic regime (sample M; $\rho_r < 2$) and negative TCR close to the transition (sample Mc; $2 < \rho_r <$ 5) are thought to be associated with a breakdown of the Thomas–Fermi screening near the M–I transition [137,138]. The tuning of the temperature dependence of conductivity by pressure on either side of the M–I transition is shown in Figs. 2.36a and 2.36b.

In the metallic regime, the $\sigma(T)$ follows Eq. (7) as shown in the inset of Fig. 2.36a. Since the conductivity of metallic PPy-PF$_6$ at 1.3 K is typically 100–300 S/cm, correlation effects are expected to play a major role in the low temperature transport. The $\sigma(T) \sim T^{1/2}$ dependence at low temperatures, shown in Fig. 2.37, indicates a significant contribution from e–e interactions. The temperature and field dependence of the conductivity of metallic PPy-PF$_6$ at millikelvin temperatures (75 mK to 1 K) are shown in Fig. 2.38 [151]. As in PANI-CSA, the high conductivity (150–200 S/cm) at 20 mK and 14 T field indicates robust three-dimensional transport and the absence of one-dimensional localization.

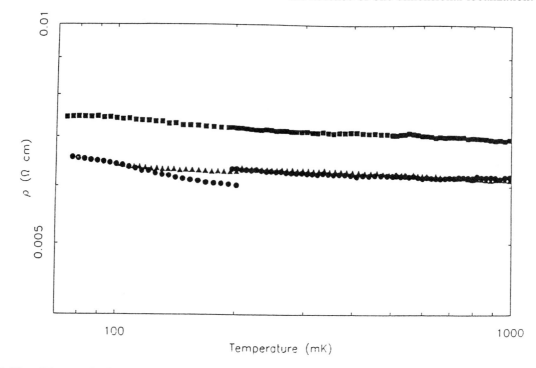

Fig. 2.38 $\rho(T)$ vs. T for PPy-PF$_6$ sample M1 at $H =$ 0 (●), 2 T (▲), and 8 T (■) at millikelvin temperatures.

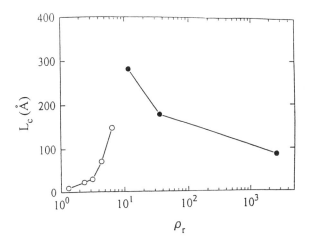

Fig. 2.39 The correlation length (L_c) obtained both from the insulating (●) and metallic (○) regimes plotted as a function of ρ_r for PPy-PF$_6$ samples. L_c is expected to diverge at the M–I transition near $\rho_r = 10$.

According to the scaling theory of localization, the correlation length L_c is expected to diverge at the M–I transition. The correlation length as a function of ρ_r is shown for PPy-PF$_6$ in Fig. 2.39. The L_c values from the metallic regime were obtained from $\sigma(0)$, using the relation $\sigma(0) = (0.1 e^2/\hbar L_c)$; those from the insulating regime were obtained from analysis of the magnetoresistance (see following section). The maximum in L_c at $\rho_r \sim 10$ indicates the transition, consistent with the transport data, which show the power law (indicative of the critical regime) over the widest temperature range at $\rho_r \approx 10$.

b. Magnetoconductance

The low and high field MC for various PPy-PF$_6$ samples near the M–I transition, at $T = 1.4$ K, are shown in Figs. 2.40a and 2.40b. The H^2 and $H^{1/2}$ dependence at low and high fields, respectively, are in agreement with the localization–interaction model as expressed in Eqs. (11)–(13). The MC in Fig. 2.40b is normalized to $\alpha\gamma F_\sigma$. The dashed line in Fig. 2.40b is the field dependence expected from Eq. (11) at 1.4 K. As ρ_r increases, however, the slope deviates from the theoretical value. This can be interpreted as due to the localization contribution, but the origin of negative MC is not yet fully understood. According to the theory of weak localization, the quantum interference between time-reversed backscattering paths is destructive when the spin–orbit scattering is strong, leading to the negative MC [44,45]. This contribution has been experimentally observed in disordered metal films and p-type doped semiconductors. However, as noted above, one does not expect strong spin–orbit effects in conducting polymers that are made up of atoms of relatively low atomic number.

The temperature dependence of inelastic scattering length (l_{in}) is shown in Fig. 2.41; the data indicate that $\tau_{in} \propto l_{in}^2 \propto T^{-p}$ with $p = 1.02 \pm 0.05$, consistent with inelastic scattering due to the Coulomb interaction close to the M–I transition [132,134]. For samples in the metallic regime ($\rho_r < 3$), we estimate that $l_{in} \sim 200$–300 Å and that it is nearly temperature-independent. In this regime, D (diffusion coefficient) ≈ 0.02–0.04 cm^2/s, and the interaction length $L_T = (\hbar D/k_B T)^{1/2} \approx 30$–40 Å ($T = 1.4$ K), i.e., much smaller than the inelastic scattering length. As the disorder increases, the system moves toward the critical regime, and the Coulomb interaction is less well screened, thereby decreasing the inelastic electron–electron scattering length. Hence the contribution due to the localization increases with ρ_r.

The MC at millikelvin temperatures is shown for metallic PPy-PF$_6$ in Fig. 2.42 [151]. The MC is positive below 300 mK. The crossover from negative to positive MC as the temperature is lowered below 1 K indicates

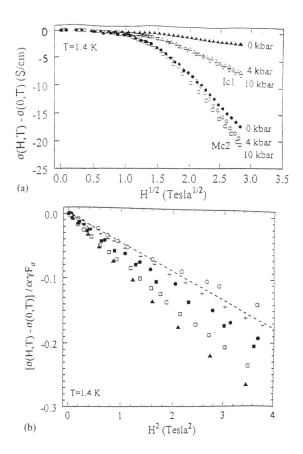

Fig. 2.40 (a) High field magnetoconductance of PPy-PF$_6$ samples plotted as a function of $H^{1/2}$ at $T = 1.4$ K and at $P = 4$ kbar and 10 kbar. (b) The low field magnetoconductance normalized by $\alpha\gamma F_\sigma$ plotted as a function of H^2 for sample M2 at $P = 9$ kbar (○), $\rho_r = 1.33$; Mc2 at $P = 10$ kbar (+), $\rho_r = 1.54$; M2 (●), $\rho_r = 1.97$; Mc2 (■), $\rho_r = 3.2$; Mc3 (■), $\rho_r = 4.5$ and Ic1 (▲), $\rho_r = 12$. The dashed line is the theoretical estimate [44,45].

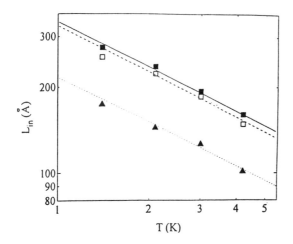

Fig. 2.41 Log-log plots of the inelastic scattering length (l_{in}) vs. T for PPy-PF$_6$ samples Mc2 (■), $\rho_r = 3.2$; Mc3 (□), $\rho_r = 4.5$; and Ic1 (▲), $\rho_r = 12$.

the competition between the weak localization and e–e interaction contributions to the MC. The enhanced negative contribution below 300 mK at fields above 8 T indicates the dominance of the e–e contribution at higher fields, as expected in the localization–interaction model. However, more work is necessary to quantify the scattering parameters below 1 K.

c. Summary

As for the other conducting polymers, ρ_r can be used to quantify the disorder; as the disorder decreases in PPy-PF$_6$, ρ_r decreases systematically, with the M–I transition at $\rho_r \approx 10$. The resistivity at high temperatures follows the power law temperature dependence with the power law exponent β decreasing from 1 to 0.3 as ρ_r decreases. As the system approaches the transition from the metallic regime ($\rho_r \sim 1-6$):

1. $\sigma(0)$ decreases continuously, and the correlation length increases.
2. The screening length increases (γF_σ decreases), and the effect of the electron–electron interaction increases.
3. The sign of TCR changes from positive to negative (at $\rho_r = 2$).
4. The temperature (T_m) of the conductivity minimum decreases with ρ_r (for $\rho_r < 2$).
5. The inelastic scattering at $T > T_m$ is due to the electron–phonon interaction ($p = 2.5$).
6. The inelastic diffusion length decreases, and the contribution due to localization increases near the M–I transition.

In metallic PPy-PF$_6$, the crossover from negative to positive MC below 300 mK, and the finite conductivity ($\sim 100-200$ S/cm) at 20 mK results from the interplay of the weak localization and e–e interaction contributions.

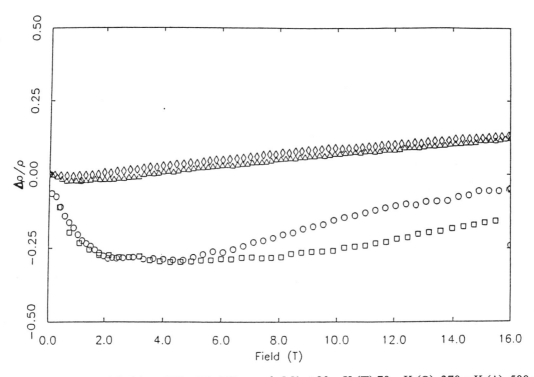

Fig. 2.42 Magnetoresistance ($\Delta\rho/\rho$) vs. H for PPy-PF$_6$ sample M1 at 20 mK (□), 70 mK (○), 370 mK (△), 500 mK (◊). (From Ref. 151.)

6. Far-Infrared Response of PANI-CSA and PPy-PF$_6$

The microwave and far-infrared dielectric response in metallic conducting polymers is typical of the usual metals [149]; the dielectric function is negative for frequencies below the plasma frequency. However, disorder and localization play an important role. Studies of the frequency response and the charge dynamics in metallic conducting polymers are in the preliminary stages of investigation. Lee et al. [157–160] observed that the infrared optical response of metallic PANI-CSA and PPy-PF$_6$ are consistent with the theory of disordered metals; the frequency-dependent optical conductivity and dielectric functions are in excellent agreement with the "localization-modified Drude model." Moreover, the value of $k_F\lambda$ estimated from the localization-modified Drude model is consistent with that obtained from transport measurements. Disorder tends to decrease $\sigma(\omega)$ and drive $\epsilon(\omega) > 0$ as $\omega \to 0$; thus, on the insulating side of the M–I transition, $\epsilon(\omega) > 0$ even in the far-infrared.

Kohlman et al. [168] and Epstein et al. [65–69] proposed a "metallic islands" model with separate contributions from crystalline and amorphous phases. They observed $\epsilon(\omega) < 0$ in the microwave regime and analyzed their data in terms of having two different plasma frequencies for the localized and delocalized electrons, respectively. Their data analysis [65–69,168] yielded $\lambda \sim$ 1–100 μm and $k_F\lambda \sim 10^5$, neither of which is consistent with the dc transport properties. Thus, the "metallic islands" model is unable to provide a quantitative description of the data with acceptable parameters.

7. Polyparaphenylenevinylene (PPV)

Although PPV is well known as the most widely used conducting polymer for light-emitting diodes [169], transport measurements on heavily doped PPV are rather few. The room temperature conductivity along the chain axis of H$_2$SO$_4$-doped PPV (PPV-HSO$_4$) is nearly 10^4 S/cm [24]. Preliminary transport measurements [170] indicate that highly doped PPV-HSO$_4$ samples are metallic ($k_F\lambda > 1$) with $\sigma(T)$ nearly temperature-independent ($\rho_r < 1.3$). Madsen et al. [99] reported $\sigma(T)$ for oriented PPV-AsF$_5$; anisotropies of 165 and 250 were found at 300 and 4 K, respectively. Park et al. [171] also observed the low temperature increase in anisotropy in doped PPV samples, unlike the temperature-independent anisotropy of doped oriented (CH)$_x$. For PPV-AsF$_5$ samples [99] having somewhat lower conductivity, $\sigma_{RT} \approx$ 2300 S/cm, the temperature dependence is rather weak; $\rho_r \approx 1.28$, and the temperature dependences parallel and perpendicular to the chain axis are nearly identical, similar to those observed in doped oriented (CH)$_x$. Although a positive TCR [99] has been observed below 6 K for PPV-AsF$_5$ samples with a relatively weak temperature dependence ($\rho_r < 1.8$), this has not been rigorously verified as in PPy-PF$_6$. The $T^{1/2}$ dependence of the conductivity due to contributions from e–e interactions at low temperatures has been reported for heavily doped and highly conducting samples [99]. Thus, it seems that high quality PPV-AsF$_5$ samples are in the metallic regime. Work is in progress to fully characterize the M–I transition in doped PPV [170].

E. Transport in the Insulating Regime Near the M–I Transition

1. Introduction

As in the critical and metallic regimes, the extent of disorder can be characterized in terms of the resistivity ratio [120]. The approximate ρ_r values for the insulating regime are the following [172]:

1. For unoriented I-(CH)$_x$ with σ(300 K) < 500 S/cm, $\rho_r > 50$.
2. For oriented I-(CH)$_x$ with σ(300 K) < 3000 S/cm, $\rho_r > 15$.
3. For oriented K-(CH)$_x$ with σ(300 K) < 4000 S/cm, $\rho_r > 25$.
4. For nonoriented PPy-PF$_6$ with σ(300 K) < 100 S/cm, $\rho_r > 10$.
5. For nonoriented PANI-CSA with σ(300 K) $\approx$ 150–250 S/cm, $\rho_r > 4$.
6. For nonoriented I-P3HT with σ(300 K) < 500 S/cm, $\rho_r > 45$.

Evidence of Mott VRH conduction [42,43] in three dimensions has been reported extensively in the literature on doped conducting polymers [75]. Deviations from the $\ln \rho \propto (T_0/T)^{1/4}$ law have also been reported; $\ln \rho \propto (T_0/T)^{1/3}$ was observed in doped polyhexylthiophene [173] and in polyacetylene doped to intermediate levels [174], where the effect of the nearest-neighbor hopping due to the finite conjugation length was suggested as the origin [106,107]. Recently, $\ln \rho \propto (T_0/T)^{1/2}$ behavior was reported for protonated polyaniline [65–69] and doped polypyrrole [117–119]. However, the strong temperature dependence ($\rho_r > 10^5$) and the nonlinear (with T) temperature dependence of the thermoelectric power indicate that such samples are deep in the insulating regime. In such samples, the extensive disorder and the formation of inhomogeneous "metallic islands" dominate the transport. The importance of homogeneity versus inhomogeneity has been demonstrated by the systematic increase in the VRH exponent from 1/4 to 1 with dilution toward the percolation threshold in PANI-CSA blends. The observation of $\ln \rho \propto (T_0'/T)^{1/2}$ below the percolation threshold [175] demonstrates clearly the crossover from "homogeneous" to "granular" behavior and demonstrates that the transport can be strongly influenced by microstructure and morphology (see following section).

2. Theory

In the insulating regime, transport occurs through variable-range hopping among localized states as described by Mott [42,43] for noninteracting carriers and by Castner [78], Efros and Shklovskii [122] when the Coulomb interaction between the electron and the hole left behind is dominant. For Mott VRH conduction in three dimensions,

$$\ln \rho \propto (T_0/T)^{1/4} \quad (15a)$$

$$T_0 = 18/k_B L_c^3 N(E_F) \quad (15b)$$

where k_B is the Boltzmann constant, L_c is the localization length, and $N(E_F)$ is the density of states at Fermi level. In the Efros–Shklovskii (ES) limit,

$$\ln \rho \propto (T_0'/T)^{1/2} \quad (16a)$$

$$T_0' = \beta_1 e^2/\epsilon k_B L_c \quad (16b)$$

where e is the electron charge, ϵ is the dielectric constant, and $\beta_1 = 2.8$ (a numerical constant).

Measurements of $\sigma(T)$ for PPy-PF$_6$, PANI-CSA, and the PATs [172] show that samples in the insulating regime, but near the critical regime, follow Mott's VRH conduction down to 1 K. Samples farther into the insulating regime, with $\rho_r = 10^2$–10^3, exhibit a crossover from Mott to ES hopping VRH conduction at $T_{cross} = 2$–10 K. The VRH parameters in Eqs. (15) and (16) have been determined from the temperature and magnetic field dependence of the resistivity. The results are consistent with strong localization theory and the expected effect of the Coulomb interaction near the disorder-induced M–I transition [78,122]. The experiments have also shown that Eq. (16) holds in a strong magnetic field ($H = 8$ T) where $T_0'(H)/T_0'(0) \propto H^p$ with $p = 1.0$–1.2.

The crossover from Eq. (15) to Eq. (16) occurs when the mean hopping energies (Δ_{hop}) in the Mott and ES limits are comparable [78,176–178]. The mean hopping energy from each theory is given by

$$\Delta_{hop} = (1/4)(k_B T)(T_0/T)^{1/4} \quad \text{(Mott, } x = 1/4\text{)} \quad (17a)$$

$$\Delta_{hop}' = (1/2)(k_B T)(T_0'/T)^{1/2} \quad \text{(ES, } x = 1/2\text{)} \quad (17b)$$

The Efros–Shklovskii VRH theory [78,122] predicts a power law energy dependence in the density of states near the Fermi level that occurs within the Coulomb gap, Δ_C:

$$\Delta_C = e^3 N(E_F)^{1/2}/\epsilon^{3/2} \quad (18)$$

where $N(E_F)$ is the unperturbed density of states at the Fermi level (i.e., in the absence of the Coulomb gap) and ϵ is the dielectric constant. Castner [78] pointed out that near the M–I transition the dielectric constant can be expressed as

$$\epsilon = \epsilon_\infty + 4\pi e^2 N(E_F) L_c^2 \quad (19)$$

where ϵ_∞ is the core dielectric constant and the second term results from the polarizability of the localized states. Note that L_c diverges as $\delta \to 0$. Thus, if the system is not too far from the M–I transition, then the second term is dominant, and $\epsilon \approx 4\pi e^2 N(E_F) L_c^2$. Assuming that the above approximation is valid, Castner [78] and Rosenbaum [176] have noted the following relations:

$$T_0/T_0' = 18(4\pi)/\beta_1 = 81 \quad (20)$$

$$\Delta_C \approx k_B T_0/18(4\pi)^{3/2} \approx k_B T_0'/\beta_1(4\pi)^{1/2} \quad (21)$$

$$T_{cross} = 16(T_0')^2/T_0 \quad \text{(if } \Delta_{hop} \approx \Delta_{hop}') \quad (22)$$

The localization length can be estimated from the expression for the weak magnetic field dependence of the VRH resistivity [122,179]:

$$\ln[\rho(H)/\rho(0)] = t(L_c/L_H)^4 (T_0/T)^{3x} \quad (23)$$

where $t = 0.0015$ for $x = 1/4$, $t = 0.0035$ for $x = 1/2$, and $L_H = (c\hbar/eH)^{1/2}$ is the magnetic length [122,179].

The following method was used for testing Castner's equations:

1. First, the values of T_0 and T_0' are obtained as accurately as possible by using Eqs. (4) and (5) and the W vs. T plots.
2. The value of L_c is determined from the slope of plots of $\ln[\rho(H)/\rho(0)]$ vs. H^2 by using Eq. (23) and the known value of T_0 or T_0'.
3. The values of $N(E_F)$ in the Mott regime can be determined by using Eq. (15b), and the values of ϵ in the ES regime can be obtained by using Eq. (16b) and the known values of T_0' (assuming that $\beta_1 = 2.8$).
4. The ratio of the experimental values of T_0 and T_0' can be used to check the validity of Eqs. (20)–(22).
5. By using the values of T_0 and T_0', it is possible to compare the theoretical value of T_{cross} with that obtained from the W vs. T plot, as shown in Tables 2.5 and 2.6.
6. Finally, the deviation of $4\pi e^2 N(E_F) L_c^2/\epsilon$ from unity is checked to identify the samples that satisfy Castner's approximation.

All the above parameters, as obtained from experiment and from theory, are listed in Tables 2.5 and 2.6 (including T_{cross} and ϵ).

The experimental ratio $T_0/T_0' = 85$–115 is close to the value predicted by Castner [78], and $\Delta_C = 0.3$–0.6 meV. The experimental values of T_{cross} are in good agreement with those estimated from the theory as shown in Tables 2.5 and 2.6. In the crossover regime ($\rho_r = 2 \times 10^2$–2×10^3), $N(E_F) = (3$–4×10^{19} states/(eV·cm^3). We have tested the validity of the approximation $\epsilon \approx 4\pi e^2 N(E_F) L_c^2$; we find $4\pi e^2 N(E_F) L_c^2/\epsilon = 0.70$–0.95, indicating that ϵ_∞ makes a relatively small contribution to ϵ. However, this approximation is not valid for the highly disordered sample (PPy-PF$_6$, $\rho_r = 2180$), which shows

Table 2.5 Experimental Values and Variable Range Hopping Parameters for PPy-PF$_6$ Samples

Symbol Fig. 2.43	σ(300 K) (S/cm)	ρ_r	T_0 (K)	T_0' (K)	T_0/T_0'	$T_{\text{cross}}^{(\text{exp})}$ (K)	$T_{\text{cross}}^{(\text{th})}$ (K)	L_c (Å)	$N(E_F)$ [×10^{19}states (eV·cm^3)]	$4\pi^2 N(E_F) L_c^2/\epsilon$
△	16.1	>10^4	1.30 × 10^5							
○	28.9	6190	3.06 × 10^4					88		
■	34.8	2180	1.05 × 10^4	5.63 × 10^1	187	4.8	4.8	108	1.6	0.43
♦	40.4	822	4.34 × 10^3	4.53 × 10^1	95	6.5	7.5	110	3.6	0.84
●	57.0	734	3.91 × 10^3	4.27 × 10^1	91	6.1	7.5	111	3.9	0.88
▲	83.4	576	3.34 × 10^3	2.91 × 10^1	115	5.4	4.1	123	3.3	0.7
+	92.8	107	1.01 × 10^3					118	22.4	

a relatively low crossover temperature in spite of its high T_0 and T_0'.

For samples farther into the insulating regime ($\rho_r > 10^3$), two different kinds of materials are known, homogeneous and inhomogeneous [172]:

By "homogeneous," we mean materials in which the localization length is greater than the disorder length scale, e.g., greater than the structural coherence length (ξ) in a polymer that has both crystalline and amorphous regions ($L_c \geq \xi$). Mott VRH conduction is recovered at low temperatures for more disordered, but relatively homogeneous, samples. In such materials, the thermoelectric power is linear with temperature and the magnitude of thermoelectric power increases slightly with ρ_r.

By "inhomogeneous" we mean inhomogeneous doping, phase segregation of doped and undoped regions, partial dedoping, and large-scale morphological disorder, etc. ($L_c \leq \xi$). Such materials are like granular metals where $\ln \rho \propto (T_0'/T)^{1/2}$ is well established [175,180,181]. Although the factors leading to the fit for granular metals are not completely understood, recent theoretical work by Cuevas et al. [181] has shown that the low temperature transport properties can be dominated by the long-range Coulomb interaction rather than by charging effects (as previously believed). In this inhomogeneous limit, hopping contributions to the temperature dependence of thermoelectric power are substantial, as shown below.

Ovchinnikov and Pronin [182,183] proposed a quasi-one-dimensional percolation model for explaining the conductivity of doped conducting polymers in the insulating regime. In this model an impurity (e.g., acceptor) captures an electron from one of the adjacent chains and forms a charged impurity center. Such a carrier can detrap by an activated process and diffuse along the chain. This polaron can recombine with another impurity center near the chain and then escape to an arbitrary chain adjacent to the second impurity center. Thus, conduction by percolation is possible in such a system if an infinite cluster of chains can be connected by impurity centers.

3. Polypyrrole

The resistivity ratio of electrochemically polymerized PPy-PF$_6$ varies over the relatively wide range from $\rho_r = 1.7$ to $\rho_r > 10^5$ depending on details of the polymerization conditions, as shown in Fig. 2.35 [167]. The M–I transition occurs at $\rho_r \sim 10$; the system becomes metallic for $\rho_r < 10$ (with finite zero temperature conductivity) and becomes an insulator for $\rho_r > 10$ [with $\rho(T)$ following

Table 2.6 Experimental Values and the Variable Range Hopping Parameters for Iodine-Doped PAT Samples

Sample	Symbol (Fig. 2.46)	σ(300 K) (S/cm)	ρ_r	T_0 (K)	T_0' (K)	T_0/T_0'	$T_{\text{cross}}^{(\text{exp})}$ (K)	$T_{\text{cross}}^{(\text{th})}$ (K)	L_c (Å)	$N(E_F)$ [×10^{19}states (eV·cm^3)]	$4\pi^2 N(E_F) L_c^2/\epsilon$
PBT		450	219	3.61 × 10^3	3.48 × 10^1	104	6.5	5.4	140	2.1	0.78
PHT-1	▲	1,170	45.1	7.60 × 10^2					129	12.7	
PHT-2	△	194	908	4.92 × 10^3	5.18 × 10^1	95	8.0	8.7	121	2.4	0.86
POT-1	●	502	244	3.29 × 10^3	2.88 × 10^1	114	5.5	4.0	149	1.9	0.71
POT-2	○	168	1,640	5.67 × 10^3	6.69 × 10^1	85	9.8	12.6	114	2.5	0.95
PDT	◇	710	47.3	7.97 × 10^2					132	11.3	
PDDT		489	250	3.38 × 10^3	3.82 × 10^1	112	6.8	6.9	147	1.9	0.71
PTDT	+	167	>10^6		4.33 × 10^2						

Eq. (15) or (16) at low temperature]. For films grown at room temperature, the electrical conductivity at 300 K is typically $\sigma(300\text{ K}) = 10–100$ S/cm with $10^2 < \rho_r < 10^5$.

The temperature dependence of the resistivity is shown in Figs. 2.43a and 2.43b as plots of $\ln \rho$ vs. $T^{-1/4}$ and $T^{-1/2}$, respectively. The sample with $\rho_r = 107$ (close to the critical regime) shows a linear dependence of $\ln \rho$ on $T^{-1/4}$ below $T = 15$ K, characteristic of Mott's VRH conduction. Samples with $\rho_r \sim 2 \times 10^2$–$2 \times 10^3$ exhibit linear $\ln \rho$ vs. $T^{-1/4}$ for 10 K $< T <$ 40 K but show a clear deviation from linearity for $T <$ 10 K. Below 10 K, the $\ln \rho (T)$ data follow a $T^{-1/2}$ dependence as shown in Fig. 2.43b. Thus, the data indicate a crossover from Mott to ES VRH conduction. This crossover is confirmed by the W vs. T plots shown in Fig. 2.44, from which the crossover temperature (T_{cross}) is accurately determined. The VRH parameters, T_0 in Eq. (15) and T_0' in Eq. (16), are determined from the slopes in Fig. 2.44. Samples with $\rho_r > 5 \times 10^3$ again show Mott's $T^{-1/4}$ behavior below 100 K down to the lowest measured temperature. The various parameters obtained from analysis of the data in terms of Eqs. (15) and (16) are listed in Table 2.5. The ratio $T_0/T_0' = 85$–115, as determined from the data, is close to that predicted using Castner's approximation.

The plot of $\ln [\rho(H)/\rho(0)]$ vs. H^2 at $T = 1.4$ K is shown in Fig. 2.45. The magnetoresistance is positive as expected for VRH conduction; the data are linear in H^2 up to $H = 1$–2 T, showing a slight deviation from quadratic dependence at higher fields. Using Eq. (23), we obtain $L_c = 110$–150 Å from the slopes of the straight

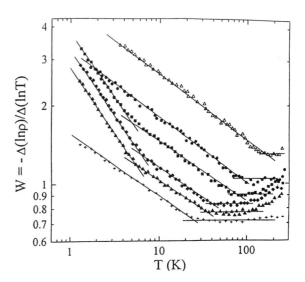

Fig. 2.44 Log-log plot of W vs. T for PPy-PF$_6$ samples in insulating regime. Solid lines represent different temperature regimes in which $x = 1/2$ (higher slopes) or $x = 1/4$ (lower slopes); the power law (zero slope) dependence can also be seen.

lines in Fig. 2.45 (see Table 2.5). The parameters $x = 1/2$ and T_0' are used in Eq. (23) except for the PPy-PF$_6$ sample following Mott's law at this temperature. The localization length increases slightly as ρ_r decreases, as expected. The relatively small difference in the values of L_c for the PPy-PF$_6$ sample with $\rho_r = 107$ and $\rho_r = 2018$ could be due to the uncertainty in numerical constant t

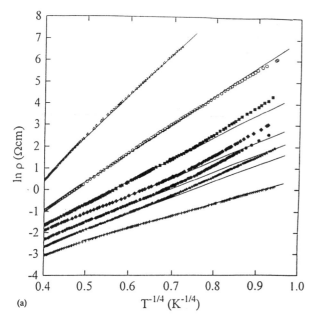

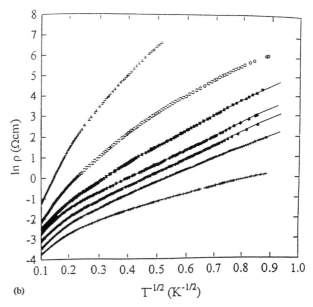

Fig. 2.43 (a) Plot of $\ln \rho$ vs. $T^{-1/4}$ for PPy-PF$_6$ samples in the insulating regime. Solid lines represent linear fit regions; (b) $\ln \rho$ vs. $T^{-1/2}$ for the same data. Solid lines represent linear fit regions. Symbols are as described in Table 2.5.

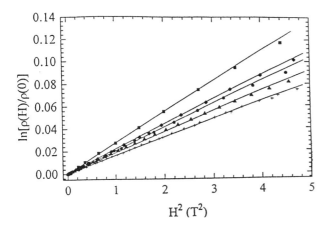

Fig. 2.45 The magnetic field dependence of the resistivity, $\ln[\rho(H)/\rho(0)]$ vs. H^2 for PPy-PF$_6$ samples in the insulating regime. The localization length was calculated from the slope (solid line) using Eq. (23).

in Eq. (23) calculated from two different theoretical models [122,179] applied for $\ln \rho \propto (T_0/T)^{1/4}$ and $\ln \rho \propto (T_0/T)^{1/2}$ dependences in the same temperature interval (1.4 K < T < 4.2 K).

Samples farther into the insulating regime ($\rho_r > 5 \times 10^3$) follow Mott VRH temperature dependence of the resistivity without showing crossover behavior. This implies that the Mott to ES crossover is restricted to the two limits, very close to the M–I transition (where the Coulomb gap may appear at millikelvin temperatures) and well on the insulating side [due to the decrease of $N(E_F)$ and L_c].

4. Polyalkylthiophenes

Advances in the synthesis of regioregular PATs have significantly improved the electronic properties by increasing the conjugation length [31,32]; the regular head-to-tail arrangement of the alkyl side chains in regioregular PATs has improved the crystalline coherence length over that in the regiorandom PATs. Moreover, side-chain-induced crystallization in regioregular PATs favors the self-assembly of well-ordered films when cast from solution.

In situ conductivity measurements were carried out during doping on several PAT samples [172]. Although the maximum conductivity obtained in regioregular PATs increases upon increasing the length of the side chain from butyl to decyl, the disorder that limits the conductivity is apparently determined by the solution casting process [31,32]. The maximum conductivities obtained for various regioregular PATs are summarized in Table 2.6. The highest room temperature conductivity, $\sigma_{RT} \sim 1200$ S/cm, was from P3HT.

Typical examples of the temperature dependence of the resistivity of iodine-doped regioregular PATs are shown in Figs. 2.46a and 2.46b [172]. Although the room temperature conductivity of some of the PATs is considerably higher than those of PPy-PF$_6$ and PANI-CSA, the temperature dependence of the resistivity is rather strong compared to the best samples of PPy-PF$_6$ and PANI-CSA. For example, PHT-1 with room temperature conductivity $\sigma_{RT} \sim 1200$ S/cm has $\rho_r \approx 50$. In general, as is the case for all other conducting polymers as well, the higher the value of σ_{RT}, the lower the value of

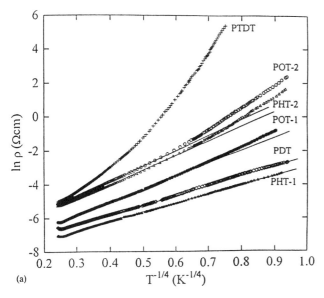
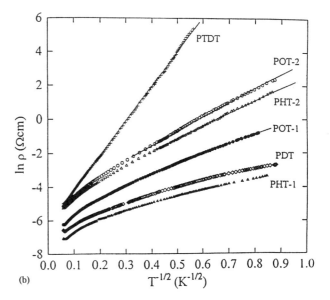

Fig. 2.46 (a) $\ln \rho$ vs. $T^{-1/4}$ for iodine-doped PAT samples. Solid lines represent linear fit regions. (b) $\ln \rho$ vs. $T^{-1/2}$ for the same data. Solid lines represent linear fit regions. Symbols are described in Table 2.6.

ρ_r, as shown in Table 2.6. The temperature dependence of the resistivity for PBT and PDDT is not shown in Fig. 2.46 since it is identical to that of POT-1.

The temperature dependence of conductivity of the PATs can be classified into three categories [172]:

1. For $\sigma_{RT} = 700$–1200 S/cm and $\rho_r < 100$, the power law dependence was observed in the temperature range 50–250 K with $\ln \rho \propto (T_0/T)^{1/4}$ behavior below 50 K.
2. For $\sigma_{RT} = 200$–600 S/cm and $2 \times 10^2 < \rho_r < 2 \times 10^3$, the crossover from $T^{1/4}$ to $T^{1/2}$ occurs (Mott to ES) at temperatures below 10 K.
3. For $\sigma_{RT} < 200$ S/cm and $\rho_r = 10^4$–10^6, $\ln \rho \propto (T_0/T)^{1/2}$ behavior was observed.

The existence of these three distinct regimes in iodine-doped PATs was confirmed from the W vs. T plots shown in Fig. 2.47. The data in regimes (1) and (2) are very similar to those obtained from PPy-PF$_6$. The power law dependence observed for PHT-1 and PDT samples above 50 K indicates that the samples are near the critical regime. The crossover from Mott to ES VRH for PBT, POT-1, and PDDT indicates that correlation effects in the insulating regime near the M–I transition are important. The VRH parameters summarized in Table 6 are consistent with results from the previous analysis of the PPy-PF$_6$ data.

The field dependence of $\rho(T,H)$ for magnetic fields up to $H = 8$ T for PPy-PF$_6$ and iodine-doped POT samples exhibiting the highest crossover temperature from Mott to ES VRH conduction is shown in Fig. 2.48 [172]. Although strong magnetic fields significantly alter the lo-

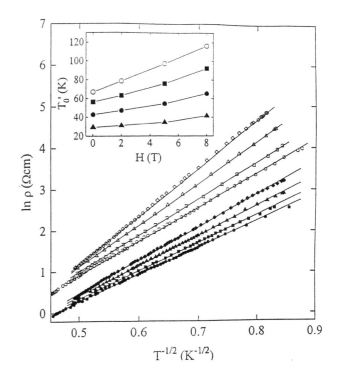

Fig. 2.48 The temperature dependence of the resistivity of PPy-PF$_6$ ($\rho_r = 734$, solid symbols) and iodine-doped POT ($\rho_r = 1640$, open symbols) in various magnetic fields plotted as $\ln \rho$ vs. $T^{-1/2}$: (●,○) $H = 0$, (■,□) $H = 2$ T, (▲,△) $H = 5$ T, and (♦,◊) $H = 8$ T. The T_0' values obtained from the slopes (solid lines) for various samples are plotted as a function of the magnetic field in the inset.

calized electronic wave functions, decreasing the overlap and increasing the hopping length, the $\ln \rho \propto (T_0'/T)^{1/2}$ law remains valid but with increased T_0'. Various VRH exponents ($x = 3/5$, $1/2$, and $1/3$) have been suggested as appropriate in a strong magnetic field in the presence of the Coulomb gap [184–187]; however, the data in Fig. 2.48 clearly indicate $x = 1/2$. The magnetic field dependences of $T_0'(H)$ obtained from the slopes in Fig. 2.48 are plotted in the inset of Fig. 2.48; the data indicate $T_0'(H)/T_0' \propto H^p$ with $p = 1.0$–1.2, consistent with Shklovskii's theory [185,186].

The $\ln \rho \propto (T_0/T)^{1/2}$ dependence for samples in regime (3), which have relatively large resistivity ratios ($\rho_r > 10^4$), is typical of granular metals. Similar behavior has been reported for polyaniline doped with conventional protonic acids [65–69] and for doped polypyrrole [117–119]. In such samples, the strong disorder and the formation of inhomogeneous "metallic islands" dominate over the $\ln \rho \propto (T_0/T)^{1/4}$ behavior expected for homogeneously disordered materials, as previously described. The $\ln \rho \propto (T_0/T)^{1/2}$ dependence observed over a wide temperature range has a different origin from the ES VRH conduction in the homogeneous limit. Phase segregation of the doped and undoped regions (i.e., the

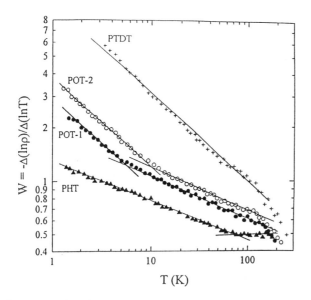

Fig. 2.47 Log-log plot of W vs. T for iodine-doped PAT samples. Solid lines represent different temperature regimes in which $x = 1/2$ (higher slopes) or $x = 1/4$ (lower slopes).

inhomogeneous doping) result in the formation of granular metals. Similar behavior was observed in microscopically disconnected networks of polyaniline in blends at concentrations below the percolation threshold (see following section for details) [175].

5. Polyaniline

Progress in the solution processing of high quality conducting PANI films showed that the electrical transport properties of PANI can be greatly improved by reduction of disorder [34,35]. In this section, we address the differences in transport properties between PANI-CSA processed from solution in *m*-cresol [27–30] and PANI doped by conventional protonic acids [65–69] such as HCl and H_2SO_4.

The temperature dependence of the resistivity of PANI-CSA is sensitive to the sample preparation conditions. This is clearly shown in the W vs. T plots, e.g., as in Fig. 2.49. The resistivity ratio for PANI-CSA is typically less than 50; in the metallic regime ($\rho_r < 3$), $\rho(T)$ approaches a finite value as $T \to 0$ [151]; in the critical regime ($\rho_r \sim 3$), $\rho(T)$ follows power law dependence; and in the insulating regime ($\rho_r > 4$), $\rho(T)$ follows Eq. (15) (Mott VRH conduction) with VRH exponent $x = 0.25 \pm 0.3$ and $T_0' = 10^1$–10^3. The Mott to ES VRH crossover noted for PPy-PF_6 and PATs has not yet been clearly observed in PANI-CSA. The systematic variation from the critical regime to the VRH regime as the value of ρ_r increases from 2.94 to 4.4 is shown beautifully in the W vs. T plot of Fig. 2.49. This is a classical demonstration of the role of disorder-induced localization in doped conducting polymers.

Plots of $\ln \rho$ vs. $T^{-1/x}$ for various PANI samples, doped with CSA and doped with conventional protonic acids (HCl and H_2SO_4), are shown in Figs. 2.50a and 2.50b. When PANI-CSA samples are treated ("washed") with acetone after casting, the behavior of $\rho(T)$ changes to $\ln \rho \propto T^{1/2}$ due to the partial deprotonation. PANI-HCl samples ("exchanged" samples) were prepared by the exchange of counterions in HCl solution after complete dedoping of PANI-CSA. For a stretch-"oriented" PANI-HCl sample with a draw ratio of 3, the resistivity data (parallel to the draw direction) are shown in Fig. 2.50b [188]. The resistivity ratio for these samples is typically $\rho_r > 10^3$, and the typical values of T_0' increase from $T_0' = 160$ K for exchanged PANI-HCl to $T_0' = 5400$ K for PANI-H_2SO_4 (see Table 2.7).

6. Summary

The transport properties of PPy-PF_6, iodine-doped PAT, and doped PANI in the insulating regime have common features; the hopping transport can be categorized as follows:

1. Very close to the M–I transition ($\rho_r < 10^2$), the low temperature resistivity follows Mott's VRH above 1 K.
2. Samples with intermediate ρ_r ($10^2 < \rho_r < 10^3$) show crossover from Mott to Efros–Shklovskii hopping (from $\ln \rho \propto T^{-1/4}$ to $\ln \rho \propto T^{-1/2}$) below 10 K, with a Coulomb gap of $\Delta_C = 0.3$–0.6 meV. The data yield $T_0/T_0' = 85$–115, $4\pi e^2 N(E_F) L_c^2/\epsilon = 0.70$–$0.95$, and values of T_{cross} that are nearly identical to those estimated from $T_{\text{cross}} = 16$

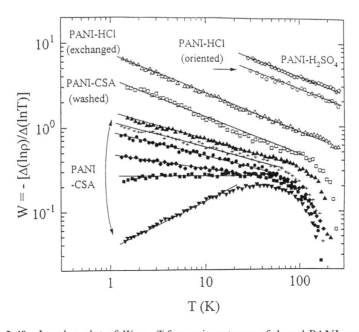

Fig. 2.49 Log-log plot of W vs. T for various types of doped PANI samples.

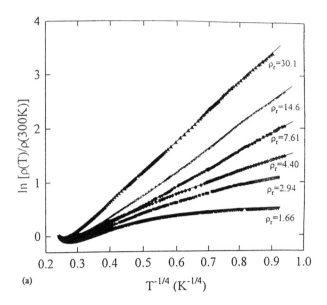
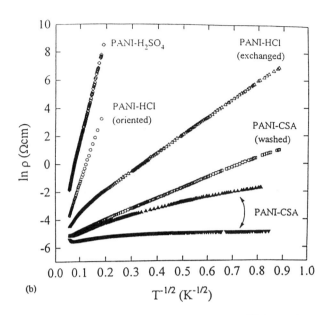

Fig. 2.50 (a) $\ln \rho$ vs. $T^{-1/4}$ for PANI-CSA samples. Solid lines represent linear fit regions. (b) $\ln \rho$ vs. $T^{-1/2}$ for various doped PANI samples.

$(T'_0)^2/T_0$; all are close to the values predicted by Castner [78]. The resistivity at low temperature in a strong magnetic field follows the $\ln \rho \propto (T'_0/T)^{1/2}$ law, where $T'_0(H)/T'_0(0) \propto H^p$ with $p = 1.0–1.2$, in agreement with the theory of Shklovskii.

3. For samples farther into the insulating regime ($\rho_r > 10^3$), $\rho(T)$ shows two distinct types of behavior. In homogeneous material ($L_c \geq$ structural coherence length), Mott VRH conduction is recovered. In inhomogeneous samples ($L_c \leq$ structural coherence length), where "metallic islands" are formed after partial dedoping or by strong morphological disorder, $\ln \rho \propto (T'_0/T)^{1/2}$, characteristic of a granular system. In this granular metal limit, there is a substantial nonlinear hopping contribution to the intrinsic diffusion thermoelectric power of doped conducting polymers.

F. Thermopower Near the M–I Transition in Doped Conducting Polymers

1. Introduction

Although the thermopower, $S(T)$, of doped conducting polymers has been studied for many years, the evolution of $S(T)$ as a function of the extent of disorder is not yet fully understood. Usually, thermopower is not as sensitive to disorder as electrical conductivity, since the latter is strongly dependent on the scattering and hopping processes involved in charge transport in the disorder-induced localized regime. Kaiser [19] analyzed the

Table 2.7 Experimental Values and Variable Range Hopping Parameters for PANI Samples

Sample	Symbol (Fig. 2.49)	$\sigma(300\ K)$ (S/cm)	ρ_r	T_0 (K)	T'_0 (K)
PANI-CSA	▼	225	1.66		
	■	322	2.94		
	◆	112	4.40	18.8	
	●	226	7.61	129	
	+	268	14.6	390	
	▲	171	30.1	1337	
PANI-CSA (washed)	□	165	486		70.6
PANI-HCl (exchanged)	△	88.4	8.8×10^4		164
PANI-HCl (oriented)	○	40.8			3060
PANI-H$_2$SO$_4$	◇	6.31			5400

thermopower in terms of the heterogeneous model in which the thermal current carried by phonons is less impeded by thin insulating barriers than the electric current carried by electrons or holes. Wang et al. [69] considered the interplay of dimensionality of coupled metallic chains. Li et al. [70] suggested that the U-shaped $S(T)$ could result from the temperature-dependent tunneling between granular metallic islands. More recently, however, Yoon et al. [172] observed that the gradual change of $S(T)$ from the positive linear temperature dependence to the negative U-shaped behavior was correlated with microstructure and indicative of negative hopping contributions in addition to the metallic diffusion thermoelectric power.

2. Theory

The diffusion thermoelectric power for a metallic system can be expressed as [189]

$$S_d(T) = \frac{\pi^2}{3}\left(\frac{k_B}{e}\right) k_B T \left[\frac{d \ln \sigma(E)}{dE}\right]_{E_F} \quad (24a)$$

or alternatively,

$$S_d(T) = +\frac{\pi^2}{3}\left(\frac{k_B}{|e|}\right)\left(k_B T \frac{z}{E_F}\right) \quad (24b)$$

where the energy dependence of $\sigma(E)$ arises from a combination of the band structure and details of the scattering mechanism, and z is a constant (of order unity), again determined from the band structure and the energy dependence of the mean scattering time. The linear temperature dependence of $S(T)$ corresponds to the characteristic diffusion thermopower of a metal. Although phonon drag often contributes to the thermopower of metals, this contribution is expected to be suppressed by disorder [190,191].

The VRH hopping contribution to the thermoelectric power depends on the details of the hopping mechanism [42,43],

$$S_{hop}(T) = \frac{1}{2}\left(\frac{k_B}{e}\right)\left(\frac{\Delta_{hop}^2}{k_B T}\right)\left[\frac{d \ln N(E)}{dE}\right]_{E_F} \quad (25)$$

where Δ_{hop} is the mean hopping energy. From Eqs. (17a) and (17b), we have $S_{hop} \propto T^{1/2}$ for $x = 1/4$ and $S_{hop} =$ constant for $x = 1/2$ [172]. Thus, in a Fermi glass with a finite density of states at E_F, $S(T)$ should have contributions from both $S_d(T)$. One finds that the hopping contribution to the total thermoelectric power of PANI-CSA samples fits well to the empirical formula

$$S(T) - AT = B T^{1/2} + C \quad (26)$$

where A is the linear slope of $S(T)$ and B and C are fitting parameters (see Section III.F.4 for details). The magnitude of hopping thermoelectric power increases with ρ_r. The origin of the positive or negative sign for hopping contributions is not understood (generally the sign depends on asymmetry corrections to the density of states with respect to the Fermi level [190].

Assuming energy-independent scattering for $S_d(T)$, the magnitude of hopping contribution can be estimated by using Eqs. (24) and (25),

$$S_{hop}/S_d \approx (3/2\pi^2)(\Delta_{hop}/k_B T)^2 = (3/2\pi^2)W^2 \quad (27)$$

where $W = \Delta_{hop}/k_B T = x(T_0/T)^x$ is the reduced activation energy. For $W < 1$, $S_{hop}/S_d \ll 1$, and the hopping contribution to the thermoelectric power is insignificant. For $x = 1/4$, Eq. (27) becomes $S_{hop}/S_d \approx (\lambda T_0/T)^{1/2}$, where $\lambda \sim 10^{-3}$; and for $x = 1/2$, $S_{hop}/S_d \approx (\lambda' T_0'/T)$, where $\lambda' \sim 10^{-1}$ [172]. The condition for the hopping thermoelectric power to be comparable to the diffusion thermoelectric power is, for example, $T_0 \geq 10^5$ K or $T_0' \geq 10^3$ K at 100 K. In the homogeneous limit ($\rho_r < 10^2$, $x = 1/4$), T_0 less than 10^3 K implies that the hopping contribution to $S(T)$ is negligible. In the inhomogeneous limit ($\rho_r > 10^3$, $x = 1/2$), the hopping thermoelectric power contributions from both the large values of $T_0 > 10^4$ and $T_0' > 10^2$ become important, and the temperature dependence of the resistivity is $\ln \rho \propto T^{-1/2}$.

3. Polyacetylene

The thermopower of doped $(CH)_x$ has been extensively studied by Kaiser [19] and by Park et al. [89–92]. The quasilinear temperature dependence is consistent with metallic contribution, while the "knee" around 50 K has been attributed to a contribution from the electron–phonon interaction. Park et al. [89–92] and Yoon [188] carried out extensive thermopower measurements on $(CH)_x$ samples doped with iodine and various transition metal halides. Park et al. [192] have also shown that the thermopower is positive for both I-$(CH)_x$ (p-type) and K-$(CH)_x$ (n-type), indicating that the sign of thermopower is not determined by the sign of the charge carrier, a surprising and unexpected result. As noted above, the sign and magnitude of thermopower in doped $(CH)_x$ are determined by the details of band structure and the dominant contributions from diffusion and hopping transport [172].

4. PANI-CSA

The thermopower data for various types of doped PANI are shown in Figs. 2.51a and 2.51b [172,193]. The room temperature value is approximately 10 μV/K with small variations (± 2 μV/K) depending on the details of the process for casting the film. The magnitude and positive sign of $S(T)$ are similar to those obtained from a number of partially doped p-type conducting polymers [89–92]. The positive sign of the thermopower is consistent with the calculated band structure of the metallic emeraldine salt, a three-quarter-filled π band with one hole per

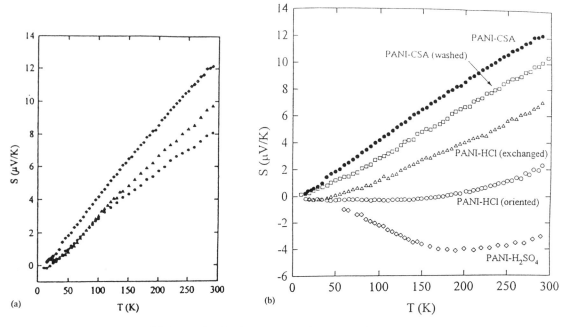

Fig. 2.51 (a) $S(T)$ vs. T of PANI-CSA samples in the (●) metallic, (▲) critical, and (■) insulating regimes. (b) $S(T)$ for various protonated PANI samples including the ones in deep insulating regime.

(—B—NH—B—NH—) repeat unit [193]. Although ρ_r for PANI-CSA samples varies by three orders of magnitude near the M–I transition, the quasilinear thermopower is relatively insensitive to ρ_r. The linear temperature dependence of $S(T)$ corresponds to the diffusion thermoelectric power, Eq. (24). The relatively large magnitude, $S(300\text{ K}) = 8\text{–}12\ \mu\text{V/K}$, indicates the diffusion of charge carriers in electronic states with relatively narrow bandwidth. Using $z = 1$, $E_F \approx 1$ eV [194,195] and $T = 300$ K, Eq. (24) yields $S(300\text{ K}) \approx 7.5\ \mu\text{V/K}$, close to the measured value. The density of states estimated from the magnitude of $S(T)$ [193] is 1.1–1.6 states per electronvolt per two rings (assuming energy-independent scattering), consistent with the value of 1 state/eV per two rings obtained from magnetic susceptibility measurements [27–30,155].

Table 2.8 Experimental Values of Thermoelectric Power and Fitting Parameters for PANI Samples

Sample	Symbol (Fig. 2.51b)	$S(300\text{ K})$ (μV/K)	B (μV/K$^{3/2}$)	C (μV/K)
PANI-CSA	●	+12.8		
PANI-CSA (washed)	□	+10.5	−0.13	−0.09
PANI-HCl (exchanged)	△	+7.0	−0.35	+0.37
PANI-HCl (oriented)	○	+2.3	−0.79	+3.01
PANI-H$_2$SO$_4$	◇	−3.0	−1.35	+6.89

The U-shaped temperature dependence $S(T)$ appears only when the hopping contribution (usually in the inhomogeneous limit) becomes significant, as shown in Fig. 2.51b. The least squares fitting parameters for B and C [Eq. (26)] are listed in Table 2.8. Neither the phonon drag effect [190], which is usually suppressed by disorder, nor the positive contribution due to the electron–phonon

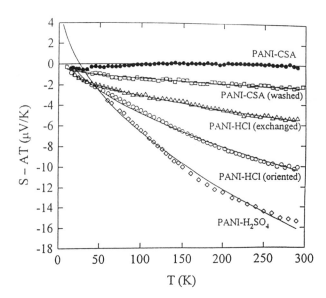

Fig. 2.52 Hopping contributions to the diffusion thermoelectric power of various doped PANI samples. Solid lines are fits to Eq. (24).

interaction [19] was observed. The data showing the hopping contribution to the thermopower for various doped PANI samples are shown in Fig. 2.52.

5. PPy-PF$_6$

The temperature dependences of $S(T)$ for various PPy-PF$_6$ samples near the M–I transition are shown in Fig. 2.53 [167]. The room temperature value is positive, with $S(300 \text{ K}) = +(9–12) \, \mu\text{V/K}$; the magnitude decreases as ρ_r decreases. The relatively large magnitude of $S(T)$ again implies that the partially filled π band is relatively narrow, in this case less than 1 eV. The density of states at the Fermi level, estimated from Eq. (24), is $N(E_F) \sim 1.0–1.6$ states/eV per four pyrrole units, assuming the ideal doping level of one dopant per four rings. As for PANI-CSA, $S(T)$ for PPy-PF$_6$ is rather insensitive to ρ_r near the disorder-induced M–I transition.

6. Summary

The thermopower of doped conducting polymers near the disorder-induced M–I transition is not as sensitive to disorder as the conductivity. Because of the continuous density of states in a Fermi glass, $S(T)$ is nearly identical in the metallic and insulating regimes. Deep in the insulating regime, however, contributions from the "metallic" diffusion thermopower (positive with $S \propto T$) and the insulating hopping thermopower become comparable; the latter can be positive or negative depending on the asymmetric corrections to the density of states with $S_{\text{hop}} \propto T^{1/2}$ for $x = 1/4$ and $S_{\text{hop}} = $ constant for $x = 1/2$ depending on whether the material is homogeneous or granular.

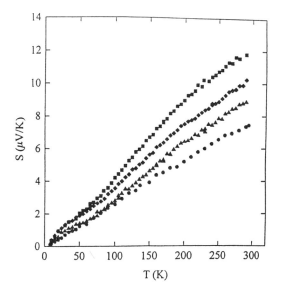

Fig. 2.53 Plots of $S(T)$ vs. T of PPy-PF$_6$ samples in various regimes: I2 (■), Ic1 (♦), Mc1 (▲), and M1 (●).

IV. TRANSPORT IN POLYANILINE NETWORKS

A. Introduction

The metal–insulator transition in doped conducting polymers can be investigated from a different perspective by blending a metallic conducting polymer into an insulating polymer as host matrix. The counterion-induced solution processing of PANI-CSA [27–30] has made possible the fabrication of conducting polymer blends in the form of films and fibers and thereby provided the opportunity to study transport in such blends [153,154,196,197].

Although classical percolating systems (where $f_c \approx 16\%$ by volume for globular conducting objects dispersed in an insulating medium in three dimensions) have been studied in detail for many years [198], the transport properties of conducting polymer blends, consisting of a network of fibrillar conducting objects, are in the early stage of investigation [199–206]. From previous experimental and theoretical studies of polymer composites filled with metal [207,208] or carbon fibers [209], it is known that the percolation threshold decreases when the aspect ratio (A) of the conducting object increases, where A is the ratio of the length to the diameter [210].

The formation of self-assembled networks in conducting polymer blends containing polyaniline provides a new class of percolating systems [153,154,175, 196,197]. Because the multiply connected, phase-separated morphology is the lowest energy configuration, the critical volume fraction of conducting material required to reach the percolation threshold can be quite low. Near the critical concentration for percolation, the conducting networks in the blends are fractal [196,197]. Transport on such fractal networks is of fundamental interest but poorly understood.

The PANI-CSA networks in PANI-CSA/PMMA blends exhibit an extremely low percolation threshold and a continuous increase of conductivity, $\sigma(f)$, while retaining the mechanical properties of the matrix polymer [153,154,175,196,197]. Homogeneous films of any size and shape can be easily fabricated either by co-dissolving the conducting PANI-CSA and a suitable matrix polymer in a common solvent and casting onto a substrate or by melt processing the blend. The networks in such blends have been directly imaged by transmission electron microscopy (TEM) [153,154,175,196,197]. The intrinsic metallic nature of PANI-CSA (as inferred from the positive temperature coefficient of resistivity) is retained upon dilution to low volume fractions (as low as $f \approx 0.003$), a feature that is not observed in other conducting polymer blends. Moreover, the positive and linear temperature dependence of thermopower remains unchanged upon dilution to 0.6% volume fraction of PANI-CSA [193,196,197]. Thus, the transport data imply the formation of a self-assembled interpenetrating fibrillar network made up of high quality PANI-CSA; the network forms spontaneously during the course of liquid–liquid phase separation.

Since the homogeneity and processibility of conducting polymer blends are superior to those of filled polymer composites, these all-polymer materials are of technological interest. The potential of such conducting polymer blends for use in a variety of applications has stimulated a more detailed study of these materials with a goal of achieving a deeper understanding of the effect of processing on the morphology and thus on the transport properties of the materials. For example, PANI-CSA networks have been demonstrated to be useful as carrier injection electrodes in conjugated polymer light-emitting diodes and as the grid in polymer grid triodes, a new architecture for "plastic" transistors [211–213].

The systematic change in the VRH exponent (x) as a function of the volume fraction of PANI-CSA suggests superlocalization of the electronic wave functions on the fractal network for concentrations near the percolation threshold [175,196,197]; x increases from 0.25 (at $f \approx 0.8$) to $x \sim 1$ upon decreasing the volume fraction of PANI-CSA to the percolation threshold. Because of the relatively high conductivity of the phase-separated PANI-CSA in PMMA, it was possible to extend the measurements well below f_c. Below the percolation threshold, $x \approx 0.5$, typical of that observed in granular metals and consistent with the disconnected granular morphology seen in the TEM micrographs [153,154,175].

B. Sample Preparation

PANI-CSA solutions are prepared by dissolving the emeraldine base form of PANI and CSA at 0.5 molar ratio of CSA to phenyl-N repeat unit in m-cresol [27–30]. This solution is then mixed in the appropriate ratio with a solution of PMMA in m-cresol. Films of thickness 20–60 μm were obtained by casting the blend solution onto a glass plate. After drying at 50°C in air for 24 h, the poly-

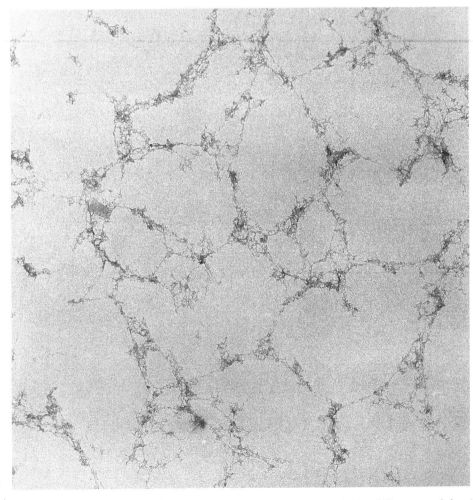

Fig. 2.54 Transmission electron micrographs of extracted PANI-CSA/PMMA polyblend films containing (a) $f = 0.005$ and (b) $f = 0.0025$ of PANI-CSA.

blend film was peeled off the glass substrate to form a free-standing film for transport measurements.

The features of the network are dependent on the molecular weight of PMMA [27–30,153,154]. The use of lower molecular weight PMMA enables greater mobility of the macromolecules during the process of liquid–liquid phase separation (carried out slowly at 50°C), thereby enhancing the diffusion of PANI-CSA in PMMA. As a result, the lower molecular weight blends more closely approach the morphology associated with the minimum energy of the PANI-CSA self-assembled network.

C. Results and Discussion

The electrical properties of PANI-CSA/PMMA blends are novel compared to those observed in more traditional systems [153,154,175,196,197]. Theoretical models of superlocalization [214–217], multifractal localization [218,219], multiple percolation [206], and high field magnetotransport [220,221] near the percolation threshold have provided additional stimulus to carefully study the transport properties of the PANI-CSA network, especially at low volume fractions near and below the percolation threshold. Improvements in material quality have enabled the temperature dependence of conductivity measurements and magnetoresistance measurements in samples containing volume fractions of PANI-CSA as low as 0.02% ($f = 0.0002$) [175].

1. Electron Microscopy and Conductivity Near the Percolation Threshold

As shown in Figs. 2.54a and 2.54b, the PANI-CSA network can be seen clearly in TEM micrographs of blends

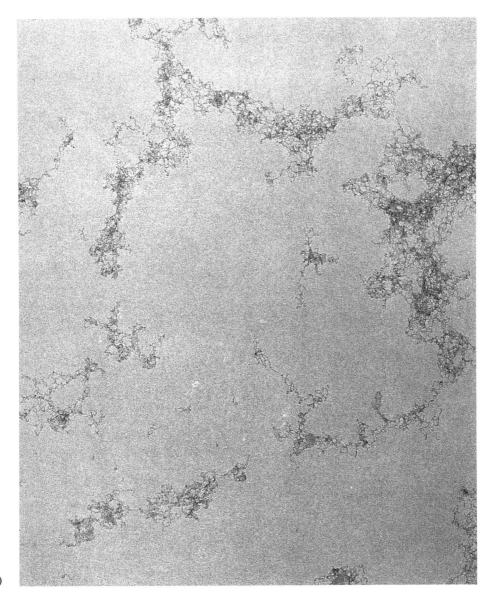

(b)

made from 0.5% and 0.25% PANI-CSA in PMMA (see Refs. 153 and 154 for details on the preparation of samples for imaging by TEM). Figures 2.54a and 2.54b resemble the typical scenario imagined for a percolating medium [198] with "links" (PANI-CSA fibrils), "nodes" (crossing points of the links), and "blobs" (dense, multiply connected regions). The distance between the nodes and the typical diameter of the blobs is assumed to be on the order of the percolation correlation length (ξ_p). The TEM photograph of the sample containing 0.5% PANI-CSA indicates that $\xi_p \sim$ 400–800 Å [153,154,175], numerous links, with diameters of about 100–500 Å, are clearly visible. The 0.25% sample is just below the percolation threshold; there are rather few links between the nodes and blobs. Thus, at these very dilute concentrations, the network becomes unstable and tends to break up; the critical concentration is approximately 0.3% PANI-CSA in PMMA.

The TEM micrographs suggest the existence of a minimum diameter for the connecting links on the order of a few hundred angstroms. Although the origin of this minimum dimension is not understood in detail, it appears that when the surface-to-volume ratio of the PANI-CSA segregated regions becomes too large, the connected network structure cannot be maintained.

The conductivity versus volume fraction of PANI-CSA is shown in Table 2.9 and in Fig. 2.55a. The σ vs. f data are dependent upon the mass distribution among the links, nodes, and blobs in the sample, which in turn are determined by various parameters involved in the

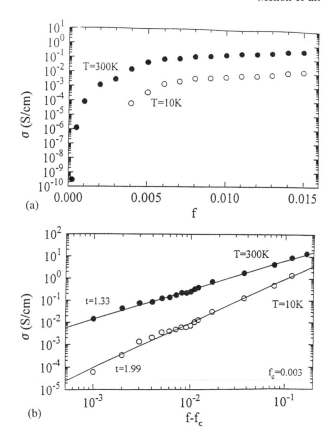

Fig. 2.55 (a) Conductivity ($\log_{10}$ scale) versus volume fraction (f) of PANI-CSA at 300 and 10 K. (b) $\log_{10}$-$\log_{10}$ plot of σ vs. $(f-f_c)$ at 300 K (●) and 10 K (○), where $f_c = 0.003$. The solid line through the points corresponds to a slope (t) of 1.99 at 10 K and 1.33 at 300 K, respectively.

Table 2.9 The Room Temperature Conductivity and Resistivity Ratio of PANI-CSA/PMMA Blends at Various Volume Fractions (f) of PANI-CSA

f	σ(300 K) (S/cm)	ρ(4.2 K) ρ(300 K)
1	200–400	1.3–10
0.8	140	11
0.67	110	13
0.5	66	18
0.33	21	19
0.12	9	30
0.08	4	60
0.04	1.8	210
0.02	0.7	710
0.015	0.4	1830
0.012	0.22	2200
0.010	0.17	2600
0.008	0.12	—
0.006	0.074	3780
0.004	0.014	5250
0.003	0.003	—
0.002	0.0012	—
0.001	10^{-4}	—
0.0005	10^{-5}	—

sample preparation—the molecular weight of PANI and PMMA, the viscosity of the polyblend solution, the solvent, the drying temperature, etc. The data in Fig. 2.55a were obtained from samples prepared by optimizing some of the above parameters with the intent of allowing the system to approach equilibrium during the liquid–liquid phase separation.

The conductivity of the sample containing 0.05% PANI-CSA is approximately 10^{-5} S/cm, several orders of magnitude higher than the typical values of conductivity obtained in filled polymer composites containing similar volume fractions of carbon black [222–224; H. B. Brom, personal communication, 1994.] or graphite particles [225, 226]. Although percolation thresholds as low as 0.1 vol % [222,223; H. B. Brom, personal communication] and 0.4 wt % [224] have been reported for carbon black/polymer composites, the conductivity near the percolation threshold for those samples is less than 10^{-7} S/cm.

Even though the TEM photographs [153,154,175] show that the macroscopic connectivity in PANI-CSA/PMMA blends is rather low for samples containing less than 0.5% PANI-CSA, a small number of residual nanoscopic connections in the phase-segregated structure persist down to volume fractions of PANI-CSA as low as 0.01%. As shown in Table 2.9, room temperature conductivities on the order of 10^{-4}–10^{-5} S/cm are observed in samples containing such extremely low volume fractions of PANI-CSA. The conductivity increases approximately four orders of magnitude (from 10^{-10} to 10^{-6} S/cm) when the PANI-CSA content is increased from $0.01 \pm 0.005\%$ to $0.025 \pm 0.005\%$, as shown in Fig. 2.55a. This suggests that nanoscopic connectivity probably does occur at these extremely low volume fractions of PANI-CSA, although such links are not clearly observable within the resolution limits of the TEM micrographs.

In order to identify the percolation threshold more precisely, the data were fit to the scaling law of percolation theory [198],

$$\sigma(f) \approx \sigma_T |f - f_c|^t \tag{28}$$

where $\sigma_T \approx (r_h)^{\zeta_R} \sum(r_h)$, which is interpreted as the conductance for each basic unit; t is the critical exponent ($t = 1$ in two dimensions and $t = 2$ in three dimensions); ζ_R is the resistivity scaling exponent ($\zeta_R = 0.975$ in two dimensions and 1.3 in three dimensions); r_h is the hopping length; $(r_h)^{\zeta_R} \sim (T_0/T)^{\zeta_R x/\zeta}$, which contains the information about the fractal network (where x and ζ are the VRH conductivity and the superlocalization exponent, described in detail below).

The fit to Eq. (28) is shown in Fig. 2.55b; at 10 K, we find $f_c = 0.3 \pm 0.05\%$ and $t = 1.99 \pm 0.04$, in agreement with the predicted universal value of $t = 2$ for percolation in three dimensions [198]. At room temperature, however, $t = 1.33 \pm 0.02$ (and $f_c = 0.3$). The smaller value of the exponent at room temperature arises from thermally induced hopping transport between disconnected (or weakly connected) parts of the network. Similar values for both the percolation threshold (0.4 wt %) and the critical exponent ($t \approx 1.3$, at room temperature) have been obtained from carbon black/polyethylene/polystyrene blends and attributed to the two-dimensionality of the system [224]. An important difference between conducting polymer blends and filled polymer systems containing metallic particles or carbon black is that the conductivity of the conducting objects in the latter are nearly temperature-independent or exhibit a metallic temperature dependence, whereas for conducting polymers the moderate temperature dependence associated with variable range hopping within the conducting polymer plays an important role.

At 10 K and at room temperature, the values of the electrical conductivity of PANI-CSA/PMMA samples with $f \approx f_c$ are on the order of 10^{-5} S/cm and 10^{-3} S/cm, respectively, values that are quite high compared to those obtained with other PANI blends. For example, polyaniline blends made by dispersing intractable polyaniline in host polymers [204,205] show percolation only at much higher levels, $f_c \approx 8.4\%$. In such dispersions, the room temperature conductivity at f_c is five orders of magnitude lower, on the order of 10^{-8} S/cm.

2. Temperature Dependence of the Network Resistivity

The temperature dependence of the resistivity for PANI-CSA/PMMA blends is shown in Fig. 2.56 for $0.002 \leq f \leq 1$ [175]. As a metallic system near the boundary of the metal–insulator transition, $\rho(T)$ in PANI-CSA is characterized by a positive temperature coefficient [34,35]. Although the positive temperature coefficient is restricted to higher temperatures upon dilution of PANI-CSA in PMMA, it is remarkable that this distinctly metallic feature is observed even in samples containing volume fractions of PANI-CSA as low as 0.3%, indicating that even at such dilution the PANI-CSA within the phase-separated network is comparable in quality to that of pure PANI-CSA.

The temperature dependence of the resistivity is relatively weak above the percolation threshold, as shown in Fig. 2.56. The data above 4.2 K show a qualitative change for samples with f below and above the percolation threshold (the resistivity measurements could not

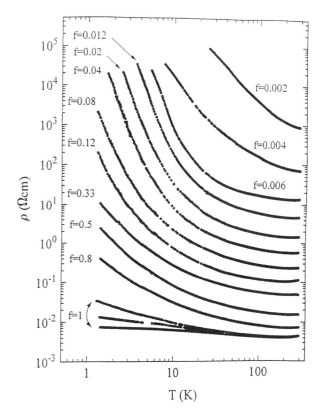

Fig. 2.56 $\log_{10}$-$\log_{10}$ plot of $\rho(T)$ vs. T of PANI-CSA/PMMA blends at various concentrations (f) of PANI-CSA ($0.002 < f < 1$).

be extended below 4.2 K for $f < 0.4\%$ due to the large contact resistance (above 20 MΩ) at very low temperatures).

Again, the subtle variations in the temperature dependence can be most clearly observed from W vs. T plots, as shown in Fig. 2.57. The temperature dependence of the resistivity of PANI-CSA/PMMA blends can be classified into three categories:

1. $0.01 \leq f \leq 1$: the VRH exponent x increases systematically from 0.25 to 1.
2. $0.006 \leq f \leq 0.01$: the VRH exponent $x \approx 1$.
3. $0.002 \leq f \leq 0.006$: $x \approx 1/2$.

a. Temperature Dependence of the Conductivity Above the Percolation Threshold

Pure PANI-CSA ($f = 1$) is at the boundary of the disorder-induced M–I transition; in the VRH regime near the transition, $x \approx 1/4$. Upon dilution of PANI-CSA with PMMA, x increases systematically from 0.25 to 1 until the PANI network breaks up at concentrations below the percolation threshold.

The $\ln \rho \propto T^{-x}$ dependence for samples containing volume fractions of PANI-CSA below 12% is shown in Figs. 2.58a–2.58c. The results are consistent with the conclusions drawn from the W vs. T plots. The systematic variation of T_0 and x as functions of the volume

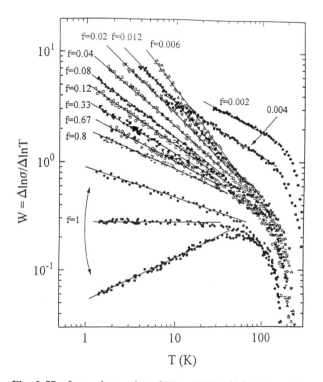

Fig. 2.57 $\log_{10}$-$\log_{10}$ plot of $W = (\Delta \ln \sigma / \Delta \ln T)$ vs T for $0.002 < f < 1$. The values of T_0 and x in Eq. (15) were determined from the straight lines using Eq. (4,5).

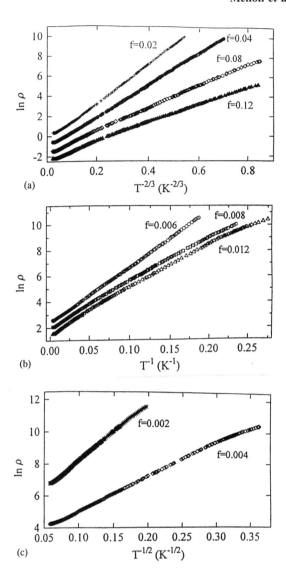

Fig. 2.58 Natural log of resistivity versus $T^{-\gamma}$ (a) $\gamma \approx 2/3$ for $f = 0.12, 0.08, 0.04$, and 0.02; (b) $\gamma \approx 1$ for $f = 0.012, 0.008$, and 0.006; (c) $\gamma = 1/2$ for $f = 0.004$ and 0.002.

fraction of PANI-CSA is shown in Fig. 2.59. Although T_0 is practically constant and x increases systematically over the wide range of volume fractions above the percolation threshold, both T_0 and x change rather abruptly when the network becomes disconnected.

The systematic increase of x from 0.25 to 1 upon dilution of PANI-CSA ($0.012 \leq f \leq 1$) is not expected in the standard VRH model. This increase in x is observed at volume fractions of PANI-CSA well above the percolation threshold ($f_c = 0.3 \pm 0.05\%$) where the system behaves as an effective medium [175,196,197]. Superlocalization is expected to play a significant role only very near the percolation threshold where the connected structure is fractal. Moreover, the theoretical models of

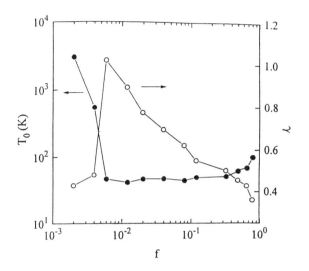

Fig. 2.59 T_0 (●) and exponent x (○) plotted as functions of the volume fraction (f) of PANI-CSA in PMMA.

superlocalization [214–217] do not predict a systematic variation of the exponent for hopping transport.

Levy and Souillard [214] have shown that in a fractal structure the wave functions for states near the Fermi level are superlocalized and decay as $\psi(r) \propto \exp[-(r/L_c)^\zeta]$, where L_c is the localization length and ζ is the superlocalization exponent, which is greater than unity (in Anderson localization $\zeta \approx 1$). Deutscher et al. [215] predicted that the temperature dependence of the electrical conductivity that results from VRH between superlocalized states would be of the form $\sigma(T) \propto \exp[-(T_0/T)^\gamma]$, where $\gamma = \zeta/(\zeta + D) \approx 3/7$. Harris and Aharony [227] predicted that γ is approximately 0.38 and 0.35 for two and three dimensions, respectively, for hopping transport in the superlocalized regime. The theory of VRH among superlocalized states was generalized by van der Putten et al. [223] to include the Coulomb interaction; van der Putten et al. obtained $\gamma \approx 0.66$ and $\zeta \approx 1.94$ consistent with the experimental results in carbon black/polymer composites. However, Aharony et al. [228,229] argued that $\zeta = 1.36$ for three dimensions and suggested that the generalized VRH equation should be used in the superlocalized regime:

$$\sigma(T) = \sigma_0 (T_0/T)^s \exp[-(T_0/T)^\gamma] \tag{29}$$

where s is the unknown exponent of the prefactor. Due to the large uncertainty [228,229] in the value of s, it is not possible to determine the values of T_0 and γ unambiguously from Eq. (29).

The fractal structure near and above the percolation threshold in PANI-CSA/PMMA blends has been demonstrated [182,183]. Aharony et al. [228,229] pointed out that the a priori requisite for applying the fractal geometry near the percolation threshold is that the length scales satisfy the condition $a \ll L_c \ll r_h \ll \xi_p$. The localization length (L_c) at 4.2 K for a 0.4% PANI-CSA sample, from the magnetic field dependence of resistivity as described below, is nearly 25 Å. The hopping length for the same 0.4% PANI-CSA sample can be estimated from the expression [175,196,197]

$$r_h(T) = (1/4) L_c (T_0/T)^{1/2} \tag{30}$$

Substituting the appropriate values for L_c and T_0 at 4.2 K gives $r_h \approx 85$ Å. The TEM micrograph for 0.5% PANI-CSA in PMMA indicates that the lower estimate for ξ_p is approximately 400 Å. The value of the smallest unit (a) is the length of the unit cell along the PANI chain direction, $a \approx 10$ Å. Hence, the criterion for applying the fractal geometry for a 0.4% PANI-CSA sample, which is near the percolation threshold, is [228,229]

$$a \ll L_c \ll r_h \ll \xi_p \rightarrow 10 \text{ Å} < 25 \text{ Å} < 85 \text{ Å} < 400 \text{ Å} \tag{31}$$

Thus, the length scales appear to satisfy the conditions presented by Aharony et al. [228,229] required for superlocalization of the electronic wave functions in a material with fractal network near the percolation threshold.

The $T^{-0.66}$ dependence of $\ln \sigma$ is valid for samples containing volume fractions $0.02 \leq f \leq 0.1$ [196,197]. Detailed studies of a wider range of samples indicate, however, that the $T^{-0.66}$ dependence of conductivity is not unique for superlocalization near the percolation threshold. The experimental results show that the exponent of the temperature dependence of conductivity in the superlocalized regime in a fractal network is sensitive to D, which in turn is determined by the morphology and the volume fraction of PANI-CSA in the system.

An alternative interpretation for the observed systematic increase in x is based on the concept of fractal character of the localized wave functions near the mobility edge [230–234]. In the standard VRH theory, the total number of states involved in the hopping conduction is assumed to be x^d times the density of states per unit volume in a region of linear dimension x, and the temperature dependence of conductivity is expressed by Eqs. (15) and (16). Aoki [230,231] and Schreiber [233] have shown that near the mobility edge the localized wave functions are fractal and that because of the fractal nature of wave functions, the number of states involved in the hopping conduction behaves like x^D where D is the fractal dimensionality (rather than x^d). The conductivity resulting from variable range hopping among such spatially fractal localized wave functions is expressed by

$$\sigma(T) \propto \exp[-(T_0/T)^{1/(D+1)}] \tag{32}$$

Since $D < d$, x would be larger than the usual values (1/4 for 3d, etc). Moreover, calculations by Schreiber and Grussbach [234] have shown that D decreases significantly with increasing disorder, yielding a further increase in the exponent of 1/T for increasing localization. The extent of disorder increases upon dilution of PANI-

CSA by PMMA since the localization length decreases for concentrations that approach the percolation threshold. However, the persistence of the positive temperature coefficient of resistivity near room temperature, even for samples near the percolation threshold, indicates that increased disorder does not drastically affect the metallic properties of PANI-CSA upon dilution.

Although the fractal wave function model offers a qualitative description for the systematic increase in γ for samples containing volume fractions of PANI-CSA above 1%, the $\ln \rho \propto 1/T$ dependence for samples near the percolation threshold (volume fractions of PANI-CSA between 0.6% and 1%) must result from other factors, since D is known to be greater than zero.

Activated conductivity ($\ln \rho \propto 1/T$ dependence) is usually observed when the dominant contribution to charge transport takes place by nearest-neighbor hopping [122,123]. Since the fibrillar links are the most resistive units in the network near the percolation threshold, charge transport through the links dominates when the fibrillar diameter becomes comparable to the hopping length. As shown above, the hopping length is nearly 100 Å, comparable to the diameter of the fibrillar links observed in the TEM photographs. The intrafibrillar transport in the highly resistive links dominates over the less resistive intrablob transport. As a result, the activated $1/T$ dependence for samples containing volume fractions of PANI-CSA between 1 and 0.5% is probably due to the dominant contribution from the nearest-neighbor hopping in the fibrillar links.

b. Temperature Dependence of the Conductivity Below the Percolation Threshold

The exponent $x(f)$ goes through a maximum at the percolation threshold; for samples containing volume fractions of PANI-CSA below 0.5%, the exponent decreases rapidly from 1 to 0.45 ± 0.05 as shown in Fig. 2.59. The $\ln \sigma \propto T^{-1/2}$ fits for samples containing 0.4% and 0.2% of PANI-CSA are shown in Fig. 2.58c.

The dramatic change in the transport properties near the percolation threshold, where the connectivity of the PANI-CSA network breaks up, is consistent with the TEM results. When the volume fraction of PANI-CSA decreases below 0.5%, the fibrillar diameter of the links between multiply connected regions decreases until the connected network cannot be sustained. Precisely at the point where the morphology changes, the charge transport undergoes a transition to that typical of granular metallic systems. The $\ln \sigma \propto T^{-1/2}$ dependence for samples containing volume fractions of PANI-CSA below the percolation threshold is typical of granular metals [175].

3. Magnetoresistance of the Network Near the Percolation Threshold

Theoretical work has shown that experimental studies of high field magnetotransport in a percolating medium can provide insight into the interrelationship between microstructure and charge transport [220,221]. The magnetoresistance data, at 4.2 K, for samples containing volume fractions of PANI-CSA from 1.5 to 0.4% are shown in Fig. 2.60. The H^2 dependence of the positive MR at low fields, typical of that observed in VRH transport, is due to the shrinkage in the overlap of the wave functions of the localized states in the presence of the magnetic field [122]. The temperature dependence of the MR for samples containing 12% and 4% PANI-CSA, as typical examples of the large difference in the MR behavior at higher and lower volume fractions of PANI-CSA, is shown in Figs. 2.61a and 2.61b. Finally, the variations of MR at 4.2 and 1.4 K as a function of the volume fraction of PANI-CSA from 100% to 0.4% are shown in Fig. 2.62.

The inset of Fig. 2.60 shows the field dependence of the resistivity for a 0.4% PANI-CSA sample at 4.2 K. The localization length (L_c) for the 0.4% PANI-CSA sample can be calculated from the slope of the straight line in the inset of Fig. 2.61 (using $T_0 \sim 800$ K obtained from the temperature dependence of the resistivity). Thus, near the percolation threshold, $L_c \approx 25$ Å at 4.2

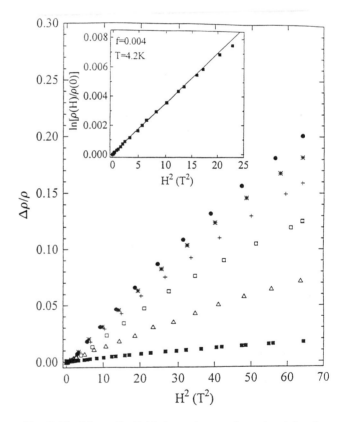

Fig. 2.60 Magnetic field dependence of conductivity for $0.015 < f < 0.004$ at $T = 4.2$ K. Magnetoresistance [$\Delta \rho/\rho(0) = \{\rho(H) - \rho(0)\}/\rho(0)$] vs. H^2 for $f = 0.015$ (●), 0.012 (*), 0.01 (+), 0.008 (□), 0.006 (△), and 0.004 (■). The inset shows $\ln \{\rho(H)/\rho(0)\}$ vs. H^2 for $f = 0.004$ at $T = 4.2$ K. The solid line is the linear fit according to Eq. (23).

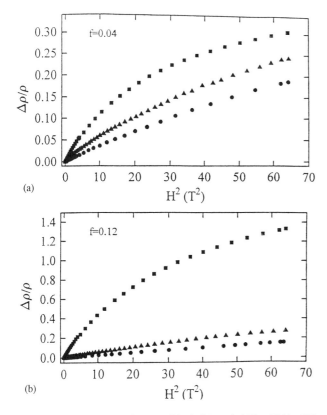

Fig. 2.61 Magnetoresistance $[\Delta\rho/\rho(0) = \{\rho(H)\text{-}\rho(0)\}/\rho(0)]$ vs. H^2 at 4.2 K (●), 2.5 K (▲), and 1.4 K (■) for (a) $f = 0.04$ and (b) $f = 0.12$.

K. Since this is below the characteristic sizes seen in the TEM micrographs, the system is expected to be in the inhomogeneous limit; and it is!

The magnitude of positive MR shows a temperature-dependent maximum upon decreasing the volume fraction of PANI-CSA. Above 4.2 K the MR is rather low for both 100% PANI-CSA and blends. At 4.2 K, the MR is maximum at 1.5% PANI-CSA; at 1.4 K, the MR is maximum at 8% PANI-CSA.

A discrete model for magnetotransport in percolating systems has been proposed [220,221]. This model, which assumes that the conducting component has a closed Fermi surface and that the MR saturates at high fields, predicts a large MR in the vicinity of the percolation threshold. This is contrary to the predictions of effective medium theory in which there is no MR near the percolation threshold [235,236]. For insulating PANI-CSA (100%), the MR tends to saturate at 8 T, but it is not known whether the Fermi surface of PANI-CSA is open or closed. In order to address this question, we have carried out MR measurements in many PANI-CSA/PMMA samples (0.4–1.5 vol %) near the percolation threshold in which the volume fraction of PANI-CSA varied by 0.1% from sample to sample. At 4.2 K, the MR increases systematically upon dilution from 100% to 1.5% PANI-CSA, whereas below 1.5% the MR de-

creases rapidly as f approaches the percolation threshold. However, the MR is much higher at 1.4 K than at 4.2 K and the maximum in MR shifts to nearly 8% PANI-CSA.

The increase in MR upon dilution is consistent with the VRH model since the wave function overlap of the localized states decreases due to superlocalization of the wave functions on the fractal network upon decreasing the volume fraction of PANI-CSA. As noted earlier, when the volume fraction of PANI-CSA decreases toward f_c, the diameter of links decreases and the interblob distance increases; consequently the hopping length and the diameter of links become rather similar. At temperatures below 4.2 K, the hopping length continues to increase, and the maximum in MR shifts to higher volume fractions of PANI-CSA (larger diameter of the fibrillar links), both of which are consistent with the observations at 1.4 K. This is also consistent with the $\ln \rho \sim 1/T$ dependence for samples containing volume fractions of PANI-CSA of 1–0.5%. The rapid decrease in MR on approaching the percolation threshold is in agreement with effective medium theory [235,236]; the data do not support the discrete model [220,221].

4. Thermopower in PANI-CSA Networks

Although the temperature dependence of electrical conductivity and that of magnetoresistance in PANI-CSA/PPMA blends are sensitive to dilution, the temperature dependence of the thermopower remains linear at high temperatures as shown in Fig. 2.63 [193,196,197]. The deviation from linearity below 100 K is obvious in Fig. 2.63, and the magnitude of this deviation increases as f decreases. For the most dilute samples studied (e.g., f

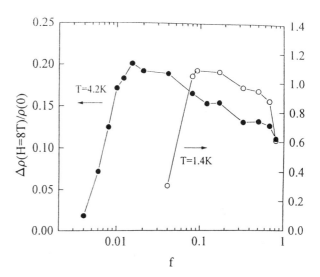

Fig. 2.62 Magnetoresistance $[\Delta\rho(H = 8\text{ T})/\rho(0) = \{\rho(H = 8\text{ T})-\rho(0)\}/\rho(0)]$ is plotted as a function of the volume fraction (f) of PANI-CSA in PMMA at 4.2 K and 1.4 K for $1 < f < 0.004$.

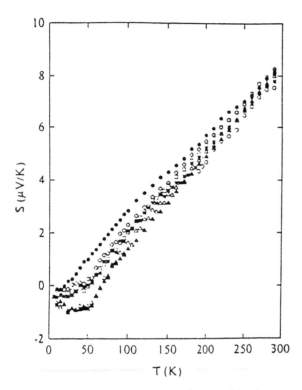

Fig. 2.63 $S(T)$ vs. T of PANI-CSA/PMMA blends at various concentrations of PANI-CSA in PMMA: y = 100% (●), 66.6% (◊), 33.3% (□), 9.09% (*), 4.76% (+), 2.43% (△), 1.24% (○).

= 0.024 PANI-CSA), the low temperature behavior is reminiscent of the characteristic U-shaped dependence well known for PANI-HCl [46,172]. Note, however, that the magnitude of the U-shaped deviation from linearity in Fig. 2.63 is much smaller than that reported for PANI-HCl samples (as shown in Fig. 2.52). The percolation threshold represents the contribution below which the conducting network breaks up into disconnected regions. In this sense, the observation of the weak U-shaped contribution to $S(T)$ at low volume fractions of PANI-CSA is consistent with the existence of large-scale inhomogeneity ("metallic islands") in PANI-HCl, and the U shape results from activated transport through the insulating regions.

The metallic temperature dependence of $S(T)$ observed at surprisingly low concentrations of PANI-CSA indicates that the microscopic conduction mechanism is not changed as PANI-CSA is diluted in PMMA. Although $\sigma(T)$ is strongly dependent on the mean free path and the number of connected pathways, $S(T)$ is rather insensitive to the change in the number of pathways once the connected paths are formed above the percolation threshold.

5. Summary

Transmission electron microscopic studies and conductivity measurements of PANI-CSA/PMMA blends indicate that the volume fraction of PANI-CSA at the percolation threshold is approximately 0.3 wt %. The formation of a self-assembled interpenetrating network of PANI-CSA results in a low percolation threshold with rather high conductivity at threshold in comparison with other percolating systems; the conductivity near percolation threshold is 0.003 S/cm at room temperature. The positive temperature coefficient of the resistivity typical of PANI-CSA remains even at volume fractions near the percolation threshold. The value of x in the $\ln \sigma \propto T^{-x}$ dependence increases systematically, from 0.25 to 1, upon dilution. This suggests that the exponent depends on the complex morphology of the network, perhaps due to superlocalization on the fractal network near the percolation threshold. For 0.4% PANI-CSA in PMMA, the length scales [($a \ll L_c \ll r_h \ll \xi_p$) → (10 Å < 25 Å < 85 Å < 400 Å)] satisfy the criteria for applying the concept of fractal geometry. For samples below 1% PANI-CSA, $x \approx 1$ until the fibrillar network breaks up at the percolation threshold. In this regime where the hopping length and the diameter of fibrillar links become similar, the $\ln \approx \propto 1/T$ dependence is typical of nearest-neighbor hopping. In the disconnected regime below the percolation threshold, the $\ln \sigma \propto T^{-1/2}$ dependence is typical of granular metals. The positive MR increases upon decreasing the volume fraction of PANI-CSA, and when the fibrillar diameter and the hopping length become comparable the MR decreases rapidly. The small MR is in agreement with the effective medium theory but contrary to the large MR predicted near the percolation threshold by the discrete model. Although the thermopower remains linear at temperatures above 100 K upon dilution, the U-shaped feature that is observed at lower temperatures for concentrations near the percolation threshold indicates the importance of the contribution from hopping thermopower.

V. SUMMARY AND CONCLUSIONS

The developments in synthesis, doping, and processing of doped conducting polymers that have occurred in the last five years have dramatically improved the structural and electronic properties of these materials [237]. These advancements have reduced the extent of disorder and inhomogeneities in dopant distribution, and consequently the conjugation length and charge delocalization have been substantially increased. Thus the intrinsic metallic nature of doped conducting polymers, previously camouflaged by disorder, is now beginning to be revealed in transport property measurements.

Detailed transport measurements and data analyses of the present generation of doped conducting polymers have enabled precise identification and detailed investigation of the metallic, critical, and insulating regimes. Highest quality doped samples of $(CH)_x$, PPy-PF$_6$, PANI-CSA, and PPV show typical features of metallic systems ($k_F \lambda > 1$).

The following are highlights of the current status of the M–I transition in doped conducting polymers.

1. The disorder-induced critical regime has been observed in K-$(CH)_x$, I-$(CH)_x$, PPy-PF_6, PANI-CSA, and PPV-HSO_4. The critical regime can be precisely identified from log-log plots of W vs. T. In the critical regime, the $\sigma(T)$ follows a power law, $\sigma(T) = cT^\beta$, and $W(T)$ is temperature-independent, $W = \beta$. The power law exponent β decreases from 1 to 0.3 as ρ_r decreases, and the system moves from the insulating to the metallic regime. At high pressures, the positive temperature coefficient of $W(T)$ observed over a wide range of temperatures indicates the crossover to the metallic regime. For K-$(CH)_x$, PPy-PF_6, and PANI-CSA, at 8 T, the negative temperature coefficient of $W(T)$ indicates a magnetic field induced crossover from the critical regime into the insulating regime. Thus the transport can be fine-tuned from the critical regime into the metallic or insulating regimes by pressure and magnetic field, respectively.

2. Metallic properties of doped conjugated polymers are observed in the temperature dependence of conductivity, magnetoresistance, thermopower, magnetic susceptibility, and infrared reflectivity. However, the materials of this class are not yet "really metallic" with long mean free paths; they remain just on the metallic side of disorder-induced M–I transition. This implies that significantly higher electrical conductivities will be obtained with continued improvement of the materials.

The resistivity ratio (ρ_r) for high quality samples of I-$(CH)_x$, K-$(CH)_x$, $FeCl_3$-$(CH)_x$, PPy-PF_6, PANI-CSA, PPV-AsF_5, and PPV-HSO_4 is less than 2. For some samples [e.g., PANI-CSA, $FeCl_3$-$(CH)_x$], the positive TCR of real metals has been observed at temperatures above 160 K; and in the case of PPy-PF_6 and PPV-AsF_5, a positive TCR appears below 10 K.

The $T^{-3/4}$ dependence of inelastic scattering length is consistent with the $T^{3/4}$ dependence of the conductivity at low temperatures (60 K to 3 K) for I-$(CH)_x$, indicating that the inelastic electron–electron scattering in disordered metals is the dominant scattering mechanism. The anisotropic MC in I-$(CH)_x$ is mainly due to the anisotropic diffusion coefficient. The positive MC [e.g., I-$(CH)_x$] is typical of that observed in disordered metals where the dominant contribution is from weak localization. Thus, in the metallic regime at low temperatures, both conductivity and MC are typical of systems near the M–I transition; the localization–interaction model is appropriate. The inelastic scattering lengths, at 1.2 K for I-$(CH)_x$, parallel and perpendicular to the chain axis are 1160 Å and 210 Å, respectively.

The $\sigma \propto T^{1/2}$ dependence observed for I-$(CH)_x$, PPy-PF_6, PANI-CSA, and PPV-AsF_5 indicates that the e–e interaction contribution is important at low temperatures. The linear temperature dependence of thermopower is typical of that of metals. The T^{-1} dependence of inelastic scattering time (τ_{in}) for PPy-PF_6 and PANI-CSA is in agreement with the prediction for systems very near the M–I transition. An inelastic scattering time (τ_{in}) on the order of 10^{-10} s is typical of amorphous metals. The finite conductivity for metallic PANI-CSA at 75 mK is nearly 30 S/cm. The intrinsic conductivity of PANI-CSA along the chain axis is nearly 10^4 S/cm.

For PPy-PF_6, ρ_r decreases systematically upon a reduction in the disorder; the M–I transition occurs at $\rho_r \approx 10$. The correlation length from the metallic side and the localization from the insulating side increase in accordance to the predictions of scaling theory. The conductivity for unoriented PPy-PF_6 at 75 mK is nearly 100–150 S/cm. The crossover from negative to positive MC below 300 mK arises from the interplay of weak localization and e–e interaction contributions in metallic PPy-PF_6.

3. In the insulating regime, the transport properties of PPy-PF_6, iodine-doped PAT, and doped PANI samples can be divided into three categories:

(a) Close to the critical regime on the insulating side ($\rho_r < 10^2$), the low temperature resistivity follows Mott's VRH conduction ($T^{-1/4}$) to very low temperatures ($T > 1$ K).

(b) In the intermediate regime ($10^2 < \rho_r < 10^3$), crossover is observed from Mott to Efros–Shklovskii VRH conduction (from $T^{-1/4}$ to $T^{-1/2}$) below 10 K. The size of the Coulomb gap (Δ_C) is about 0.3–0.6 meV. The data yield $T_0/T_0' = 85$–115, $4\pi e^2 N(E_F) L_c^2/\epsilon = 0.70$–$0.95$, and values of T_{cross} that are nearly identical to those estimated from $T_{cross} = 16(T_0')^2/T_0$; all are close to the approximate values estimated by Castner [78]. The resistivity at low temperature in a strong magnetic field follows the $\ln \rho \propto (T_0'/T)^{1/2}$ law, where $T_0'(H)/T_0'(0) \propto H^p$ with $p = 1.0$–1.2, in agreement with the theory of Shklovskii.

(c) For samples farther into the insulating regime ($\rho_r > 10^3$), $\rho(T)$ shows two distinct types of behavior. In the homogeneous limit ($L_c \geq$ structural coherence length), Mott VRH conduction is recovered at low temperature. In the inhomogeneous limit ($L_c \leq$ structural coherence length), the formation of "metallic islands" due to inhomogeneous distribution of dopants or strong morphological disorder lead to the $\ln \rho \propto (T_0'/T)^{1/2}$ behavior characteristic of granular systems.

4. The thermopower in doped conducting polymers near the disorder-induced M–I transition is not as sensitive as the conductivity; $S(T)$ is nearly identical in both the metallic and insulating regimes near the M–I transition. Deep in the insulating regime, however, $S(T)$ exhibits contributions from both the diffusion thermopower (positive and $S \propto T$) and the hopping thermopower (positive or negative and $S \propto T^{1/x}$, where $x = 4$ and 2 in the homogeneous and inhomogeneous limits, respectively).

5. Transmission electron microscopic studies and

conductivity measurements of PANI-CSA/PMMA blends indicate that the volume fraction of PANI-CSA at the percolation threshold is approximately 0.3%. The conductivity near the percolation threshold is 0.003 S/cm at room temperature. The value of x in the $\ln \sigma \propto T^{-x}$ dependence of the conductivity increases systematically, from 0.25 to 1, upon dilution, indicating that x is sensitive to the morphology of the interpenetrating network, perhaps due to superlocalization of electronic wave functions on the network near the percolation threshold where the network becomes fractal. In the disconnected regime below the percolation threshold, the $\ln \sigma \propto T^{-1/2}$ dependence is typical of granular metals. The thermopower remains linear at high temperatures upon dilution. The positive MR increases upon decreasing the volume fraction of PANI-CSA. When the fibrillar diameter and the hopping length become comparable, the MR decreases rapidly.

In conclusion, doped conducting polymers are intrinsically metallic. Disorder-induced localization changes the electronic states in the partially filled band of these systems from extended and delocalized (metallic) to localized (insulating). The recent advances in the materials science of doped conducting polymers have resulted in highly conducting polymers that are processible and environmentally stable [237]. The electronic and optical properties of these systems, on either side of the M–I transition, can be varied for use in a wide range of potential applications. Based on our current understanding, further improvement of these materials is expected to yield conducting polymers with electrical conductivities comparable to or higher than those of even the best metals.

ACKNOWLEDGMENTS

We thank Dr. Yong Cao and Professor Paul Smith for their many contributions to the materials science that enabled the transport measurements at UCSB. We thank Dr. N. S. Sariciftci and Dr. Kwanghee Lee for important discussions on the properties of conducting polymers near the metal–insulator transition. Support for the transport studies at UCSB and for the preparation of this review was obtained from the MRL Program of the National Science Foundation (NSF-DMR-9123048) and from the Office of Naval Research (Dr. Kenneth Wynne), N00014-91-J-1235.

REFERENCES

1. H. Shirakawa, E. J. Louis, A. G. MacDiarmid, C. K. Chiang, and A. J. Heeger, *J. Chem. Soc. Chem. Commun. 1977*:578.
2. C. K. Chiang, C. R. Fincher, Y. W. Park, A. J. Heeger, H. Shirakawa, E. J. Louis, S. C. Gua, and A. G. MacDiarmid, *Phys. Rev. Lett. 39*:1098 (1977).
3. C. K. Chiang, M. A. Druy, S. C. Gua, A. J. Heeger, H. Shirakawa, E. J. Louis, A. G. MacDiarmid, and Y. W. Park, *J. Am. Chem. Soc. 100*:1013 (1978).
4. A. J. Heeger in *Handbook of Conducting Polymers*, Vol. 2 (T. A. Skotheim, ed.), Marcel Dekker, New York, 1986, p. 729.
5. W. P. Su, *Handbook of Conducting Polymers*, p. 757.
6. R. R. Chance, D. S. Boudreaux, J. L. Brédas, and R. Silbey, *Handbook of Conducting Polymers*, p. 825.
7. D. K. Campbell, A. R. Bishop, and M. J. Rice, *Handbook of Conducting Polymers*, p. 937.
8. S. A. Brazovskii and N. N. Kirova, *Sov. Sci. Rev. A Phys. 5*: 99 (1984) and references therein.
9. S. Roth and H. Bleier, *Adv. Phys. 36*:385 (1987) and references therein.
10. A. J. Heeger, S. Kivelson, J. R. Schrieffer, and W. P. Su, *Rev. Mod. Phys. 60*:781 (1988) and references therein.
11. Y. Lu (ed.), *Solitons and Polarons in Conducting Polymers*, World Scientific, Singapore, 1988.
12. Kiess (ed.) *Conjugated Conducting Polymers* (Springer Ser. Solid State Sci., Vol. 102), Springer New York, 1992.
13. W. R. Salaneck, I. Lundstrom, and B. Ranby (eds.), *Conjugated Polymers and Related Compounds*, Oxford Univ. Press, London, 1993.
14. H. Naarmann and N. Theophilou, *Synth. Met. 22*:1 (1987).
15. N. Basescu, Z.-N. X. Liu, D. Moses, A. J. Heeger, H. Naarmann, and Theophilou, *Nature 327*:403 (1987).
16. Th. Schimmel, D. Glaser, M. Schwoerer, and H. Naarmann in *Conjugated Polymers* (J. L. Bredas and R. Silbey, Eds.), Kluwer Academic, Dordrecht, 1991, p. 49.
17. J. Tsukamoto, *Adv. Phys. 41*:509 (1992) and references therein.
18. Bernier in *Handbook of Conducting Polymers*, Vol. 2 (T. A. Skotheim, ed.), Marcel Dekker, New York, 1986, p. 1099 and references therein.
19. A. B. Kaiser, *Phys. Rev. B 40*:2806 (1989) and references therein.
20. D. Moses, A. Denenstein, A. Pron, A. J. Heeger, and A. G. MacDiarmid, *Solid State Commun. 36*:219 (1980).
21. J. Tsukamoto, A. Takahashi, and K. Kawasaki, *Jpn. J. Appl. Phys. 29*:125 (1990).
22. I. Murase, T. Ohnishi, T. Noguchi, and M. Hirooka, *Synth. Met. 17*:639 (1984).
23. F. E. Karaz, J. D. Capistran, D. R. Gagnon, and R. W. Lenz, *Mol. Cryst. Liq. Cryst. 118*:327 (1985).
24. T. Ohnishi, T. Noguchi, T. Nakano, M. Hirooka, and I. Murase, *Synth. Met. 41–43*:309 (1991).
25. T. Hagiwara, M. Hirasaka, K. Sato, and M. Yamaura, *Synth. Met. 36*:241 (1990).
26. K. Sato, M. Yamaura, T. Hagiwara, K. Murata, and M. Tokumoto, *Synth. Met. 40*:35 (1991).

27. Y. Cao, P. Smith, and A. J. Heeger, *Synth. Met. 48*: 91 (1992).
28. Y. Cao and A. J. Heeger, *Synth. Met. 52*: 193 (1992).
29. Y. Cao, J. J. Qiu, and P. Smith, *Synth. Met. 69*:187, 191 (1995).
30. Y. Cao, P. Smith, and A. J. Heeger, in *Conjugated Polymeric Materials: Opportunities in Electronics, Optoelectronics and Molecular Electronics*, (NATO Adv. Study Inst, Ser. E: Appl. Sci. Vol. 82) (J. L. Bredas and R. R. Chance, eds.), Kluwer Academic, Dordrecht, 1990.
31. R. D. McCullough and R. D. Lowe, *J. Org. Chem. 70*:904 (1993).
32. T.-A. Chen and R. D. Rieke, *J. Am. Chem. Soc. 114*: 10087 (1992).
33. Y. W. Park, C. Park, Y. S. Lee, C. O. Yoon, H. Shirakawa, Y. Suezaki, and K. Akagi, *Solid State Commun. 65*:147 (1988).
34. M. Reghu, Y. Cao, D. Moses, and A. J. Heeger, *Phys. Rev. B 47*:1758 (1993).
35. M. Reghu, C. O. Yoon, D. Moses, A. J. Heeger, and Y. Cao, *Phys. Rev. B 48*:17685 (1993).
36. G. B. Street, in *Handbook of Conducting Polymers*, Vol. 1 (T. A. Skotheim, ed.), Marcel Dekker, New York, 1986, p. 265.
37. P. Pfluger, G. Weiser, J. Campbell Scott, and G. B. Street, in *Handbook of Conducting Polymers*, Vol. 2 (T. A. Skotheim, ed.), Marcel Dekker, 1986, p. 1369.
38. G. Tourillon, in *Handbook of Conducting Polymers*, Vol. 1 (T. A. Skotheim ed.), Marcel Dekker, New York, 1986, p. 293.
39. A. O. Patil, A. J. Heeger, and F. Wudl, *Chem. Rev. 88*:183 (1988).
40. J. Roncali, *Chem. Rev. 92*:711 (1992).
41. C. M. Gould, D. M. Bates, H. M. Bozler, A. J. Heeger, M. A. Dury, and A. G. MacDiarmid, *Phys. Rev. B 23*:6820 (1980).
42. N. F. Mott and E. A. David, *Electronic Process in Noncrystalline Materials*, Oxford Univ. Press, Oxford, 1979.
43. N. F. Mott, *Metal–Insulator Transition*, 2nd ed., Taylor & Francis, London, 1990.
44. P. A. Lee and T. V. Ramakrishnan, *Rev. Mod. Phys. 57*: 287 (1985) and references therein.
45. A. L. Efros and M. Pollak (eds.), *Electron–Electron Interactions in Disordered Systems*, North-Holland, Amsterdam, 1985.
46. S. Kivelson and A. J. Heeger, *Synth. Met. 22*:371 (1989).
47. H. Fujiwara, H. Kadomatsu, and K. Tohma, *Rev. Sci. Instrum. 51*:1345 (1980).
48. Y. W. Park, *Synth. Met. 45*:173 (1991).
49. R. B. Roberts, *Phil. Mag. 36*:91 (1977).
50. P. W. Anderson, *Phys. Rev. 109*:1492 (1958).
51. P. W. Anderson, *Comments Solid State Phys. 2*:193 (1970).
52. A. F. Ioffe and A. R. Regel, *Prog. Semicond. 4*:237 (1960).
53. E. Abrahams, P. W. Anderson, D. C. Licciardello, and T. V. Ramakrishnan, *Phys. Rev. Lett. 42*:695 (1979).
54. A. Möbius, *J. Phys. C: Solid State Phys. 18*:4639 (1985); *Phys. Rev. B 40*:4194 (1989).
55. A. J. Heeger and P. Smith, in *Conjugated Polymers* (J. L. Bredas and R. Silbey, eds.), Kluwer Academic, Dordrecht, 1991, p. 141.
56. R. H. Baughman, N. S. Murthy, and H. Eckhardt, *Phys. Rev B 46*:10515 (1992).
57. N. S. Murthy, R. H. Baughman, and L. W. Shacklette, *Phys. Rev. B 41*:3708 (1990).
58. Y. Wada, in *Physics and Chemistry of Materials with Low-Dimensional Structures* (H. Aoki, M. Tsukada, M. Schluter, and F. Levy, eds.), Kluwer Academic, Dordrecht, 1992.
59. D. Chen, M. J. Winokur, Y. Cao, A. J. Heeger, and F. E. Karasz, *Phys. Rev. B 45*:2035 (1992).
60. M. J. Winokur, J. Maron, Y. Cao, and A. J. Heeger, *Phys. Rev. B 45*:9656 (1992).
61. T. J. Prosa, M. J. Winokur, J. Moulton, P. Smith, and A. J. Heeger, *Phys. Rev. B 51*:159 (1995).
62. P. Papanek, J. E. Fisher, J. L. Sauvajol, A. J. Dianoux, G. Mao, M. J. Winokur, and F. E. Karasz, *Phys. Rev. B 50*:15668 (1994).
63. J. P. Pouget, Z. Oblakowski, Y. Nogami, P. A. Albouy, M. Laridjani, E. J. Oh, Y. Min, A. G. MacDiarmid, J. Tsukamoto, T. Ishiguro, and A. J. Epstein, *Synth. Met. 65*:131 (1994).
64. Nogami, J. P. Pouget, and T. Ishiguro, *Synth. Met. 62*:257 (1994).
65. A. J. Epstein, J. M. Ginder, F. Zuo, H. S. Woo, D. B. Tanner, A. F. Richter, M. Angeloupolos, W. S. Huang, and A. G. MacDiarmid, *Synth. Met. 21*:63 (1987).
66. A. J. Epstein, J. Joo, R. S. Kohlman, G. Du, A. G. MacDiarmid, E. J. Oh, Y. Min, J. Tsukamoto, H. Kaneko and J. P. Pouget, *Synth. Met. 65*:149 (1994).
67. A. G. MacDiarmid and A. J. Epstein, *Synth. Met. 65*:103 (1994).
68. A. J. Epstein, J. Joo, C. Y. Wu, A. Benatar, C. F. Faisst, Jr., J. Zegarski, and A. G. MacDiarmid, in *Intrinsically Conducting Polymers: An Emerging Technology* (M. Aldissi, ed.), Kluwer, Dordrecht, 1993, p. 165 and references therein.
69. Z. Wang, A. Ray, A. G. MacDiarmid, and A. J. Epstein, *Phys. Rev. B 43*:4373 (1991).
70. Q. Li, L. Cruz, and P. Phillips, *Phys. Rev. B 47*:1840 (1993).
71. A. G. Zabrodskii and K. N. Zeninova, *Zh. Eksp. Teor. Fiz. 86*:727 (1984) [*Sov. Phys. JETP 59*:425 (1984)].
72. A. G. Zabrodskii, *Fiz. Tekh. Poluprovodn. 11*:595 (1977) [*Sov. Phys. Semicond. 11*:345 (1977)].
73. G. Thummes, U. Zimmer, F. Korner, and J. Kotzler, *Jpn. J. Appl. Phys. Suppl. 26(3)*:713 (1987).
74. G. Thummes, E. Korner, and J. Kotzler, *Solid State Commun. 67*:215 (1988).
75. S. Roth, in *Hopping Transport in Solids* (M. Pollak and B. Shklovskii, eds.), North-Holland, Amsterdam, 1991, p. 377 and references therein.

76. W. L. McMillan, *Phys. Rev. B 24*:2739 (1981).
77. A. I. Larkin and D. E. Khmelnitskii, *Zh. Eksp. Teor. Fiz. 83*: 1140 (1982) [*Sov. Phys. JETP 56*:647 (1982)].
78. T. G. Castner, in *Hopping Transport in Solids* (M. Pollak and B. I. Shkolvskii, eds.), North-Holland, Amsterdam, 1990, p. 1 and references therein.
79. L. Pietronero, *Synth. Met. 8*:225 (1983).
80. R. E. Peierls, *Quantum Theory of Solids*, Oxford Univ. Press, London, 1955.
81. S. Kivelson and A. J. Heeger, *Phys. Rev. Lett. 55*: 308 (1985).
82. K. Harigaya, A. Terai, and Y. Wada, *Phys. Rev. B 43*:4141 (1991).
83. K. Harigaya, and A. Terai, *Phys. Rev. B 44*:7835 (1991).
84. E. M. Conwell and H. A. Mizes, *Synth. Met. 65*:203 (1994).
85. S. Stafstrom, *Phys. Rev. B 47*:12437 (1993).
86. P. Phillips and L. Cruz, *Synth. Met. 65*:225 (1994).
87. E. Jeckelmann and D. Baeriswyl, *Synth. Met. 65*:211 (1994) and references therein.
88. M. I. Salkola and S. A. Kivelson, *Phys. Rev. B 50*: 13962 (1994).
89. Y. W. Park, C. O. Yoon, C. H. Lee, and H. Shirakawa, *Makromol. Chem. Macromol. Chem. Macromol. Symp. 33*:341 (1990).
90. Y. W. Park, C. O. Yoon, B. C. Na, H. Shirakawa, and K. Akagi, *Synth. Met. 41–43*:27 (1991).
91. Y. W. Park, C. O. Yoon, C. H. Lee, H. Shirakawa, Y. Suezaki, and K. Akagi, *Synth. Met. 28*:D27 (1989).
92. Y. W. Park, A. J. Heeger, M. A. Dury, and A. G. MacDiarmid, *J. Chem. Phys. 73*:946 (1980).
93. V. N. Prigodin and K. B. Efetov, *Phys. Rev. Lett. 70*:2931 (1993).
94. V. N. Prigodin and K. B. Efetov, *Synth. Met. 65*: 195 (1994).
95. S. Stafstrom, *Synth. Met. 65*:185 (1994).
96. S. Stafstrom, *Synth. Met. 69*:667 (1995).
97. S. Stafstrom, *Phys. Rev. B 51*:4137 (1995).
98. M. Buttiker, Y. Imry, R. Landauer, and S. Pinhas, *Phys. Rev. B 31*:6207 (1985).
99. J. M. Madsen, B. R. Johnson, X. L. Hua, R. B. Hallock, M. A. Masse, and F. E. Karasz, *Phys. Rev. B 40*:11751 (1989).
100. P. Sheng, *Phys. Rev. B 21*:2180.
101. P. Sheng and J. Klasfter, *Phys. Rev. B 27*:2583 (1983).
102. J. Voit and H. Buttner, *Solid State Commun. 67*: 1233 (1988).
103. A. B. Kaiser and S. C. Graham, *Synth. Met. 36*:367 (1990).
104. G. Paasch, *Synth. Met. 51*:7 (1992).
105. E. M. Conwell and H. A. Mizes, *Synth. Met. 38*:319 (1990).
106. R. H. Baughman and L. W. Shacklette, *Phys. Rev. B 39*:5872 (1989).
107. R. H. Baughman and L. W. Shacklette, *J. Chem. Phys. 90*:7492 (1989).
108. B. Movaghar and S. Roth, *Synth. Met. 63*:163 (1994).
109. J. S. Andrade, Jr., Y. Shibusa, Y. Arai, and A. F. Siqueira, *Synth. Met. 68*:167 (1995).
110. T. Ishiguro, H. Kaneko, Y. Nogami, H. Ishimoto, H. Nishiyama, M. Yamaura, T. Hagiwara, and K. Sato, *Phys. Rev. Lett. 62*:660 (1992).
111. A. Raghunathan, T. S. Natarajan, G. Rangarajan, S. K. Dhawan and D. C. Trivedi, *Phys. Rev. B 47*:13189 (1993).
112. A. K. Meikap, A. Das, S. Chatterjee, H. Digar, and S. N. Bhattacharya, *Phys. Rev. B 47*:1340 (1993).
113. P. K. Kahol, V. Pendse, N. J. Pinto, M. Traore, W. T. K. Stevenson, B. J. McCormick, and J. N. Gundersen, *Phys. Rev. B 50*:2809 (1994).
114. B. Abele, P. Sheng, M. D. Coutts, and Y. Arie, *Adv. Phys. 24*:407 (1975).
115. P. Phillips and H.-L. Wu, *Science 252*:1805 (1991).
116. D. H. Dunlap, H.-L. Wu, and P. Phillips, *Phys. Rev. Lett. 65*:88 (1990).
117. L. Zuppiroli, M. N. Bussac, S. Paschen, O. Chauvet, and L. Forro, *Phys. Rev. B 50*:5196 (1994).
118. M. N. Bussac and L. Zuppiroli, *Phys. Rev. B 49*: 5876 (1994); *47*:5493 (1993).
119. O. Chauvet, S. Paschen, L. Forro, L. Zuppiroli, P. Bujard, K. Kai, and W. Wernet, *Synth. Met. 63*:115 (1994).
120. M. Reghu, K. Vakiparta, C. O. Yoon, Y. Cao, D. Moses, and A. J. Heeger, *Synth. Met. 65*:167 (1994).
121. D. E. Khmelnitskii and A. I. Larkin, *Solid State Commun. 39*:1069 (1981).
122. B. I. Shklovskii and A. L. Efros, *Electronic Properties of Doped Semiconductors*, Springer, Heidelberg, 1984.
123. H. Bottger and V. V. Bryksin, *Hopping Conduction in Solids*, Deerfield Beach, FL, 1985.
124. K. Vakiparta, M. Reghu, M. R. Anderson, Y. Cao, D. Moses, and A. J. Heeger, *Phys. Rev. B 47*:9977 (1993).
125. M. Reghu, K. Vakiparta, Y. Cao, and D. Moses, *Phys. Rev. B 49*:16162 (1994).
126. C. O. Yoon, M. Reghu, A. J. Heeger, E. B. Park, Y. W. Park, K. Akagi, and H. Shirakawa, *Synth. Met. 69*:79 (1995).
127. Y. Nogami, H. Kaneko, H. Ito, T. Ishiguro, T. Sasaki, N. Toyota, A. Takahashi, and J. Tsukamoto, *Phys. Rev. B 43*:11829 (1991).
128. Y. Nogami, H. Kaneko, T. Ishiguro, A. Takahashi, J. Tsukamoto, and N. Hosoito, *Solid State Commun. 76*:583 (1990).
129. M. Reghu, C. O. Yoon, D. Moses, and A. J. Heeger, *Synth. Met. 64*:53 (1994).
130. H. Kaneko and T. Ishiguro, *Synth. Met. 65*:141 (1994).
131. H. Shirakawa, Y. X. Zhang, T. Okuda, K. Sakamaki, and K. Akagi, *Synth. Met. 65*:93 (1994).
132. D. Belitz and K. I. Wysokinski, *Phys. Rev. B 36*: 9333 (1987).
133. A. Kawabata, *Solid State Commun. 34*:431 (1980).
134. P. Dai, Y. Zhang, and M. P. Sarachik, *Phys. Rev. B 45*: 3984 (1992); *46*:6724 (1992).
135. M. Kaveh and N. F. Mott, *Phil. Mag. B 55*:1 (1987).

136. R. N. Bhatt and P. A. Lee, *Solid State Commun.* 48: 755 (1983).
137. T. F. Rosenbaum, R. M. F. Milligan, G. A. Thomas, P. A. Lee, T. V. Ramakrishnan, and R. N. Bhatt, *Phys. Rev. Lett.* 47:1758 (1981).
138. T. F. Rosenbaum, R. F. Milligan, M. A. Paalanen, G. A. Thomas, R. N. Bhatt, and W. Lin, *Phys. Rev. B* 27:7509 (1983).
139. K. Mizoguchi, H. Sakurai, F. Shimizu, S. Masubuchi, and K. Kume, *Synth. Met.* 68:239 (1995).
140. K. Mizoguchi, S. Masubuchi, K. Kume, K. Akagi, and H. Shirakawa, *Phys. Rev. B* 51:8864 (1995).
141. V. N. Prigodin and S. Roth. *Synth. Met.* 53:237 (1993).
142. A. B. Kaiser, *Synth. Met.* 45:183 (1991).
143. Y. Nogami, M. Yamashita, H. Kaneko, T. Ishiguro, A. Takahashi, and J. Tsukamoto, *J. Phys. Soc. Jpn.* 62:664 (1993).
144. H. H. S. Javadi, A. Chakraborty, C. Li, N. Theophilou, D. B. Swanson, A. G. MacDiarmid, and A. J. Epstein, *Phys. Rev. B* 43:2183 (1991).
145. P. Hernandez and M. Sanquer, *Phys. Rev. Lett.* 68:1402 (1992).
146. N. Foxonet, P. Bernier, and J. Voit, *J. Chim. Phys. Phys.-Chim. Biol.* 89:977 (1992).
147. N. Coustel, P. Bernier, and J. E. Fisher, *Phys. Rev. B* 43:3147 (1991).
148. M. R. Andersson, K. Vakiparta, M. Reghu, Y. Cao, and D. Moses, *Phys. Rev. B* 47:9238 (1993).
149. J. Joo, Z. Oblakowski, G. Du, J. P. Pouget, E. J. Oh, J. M. Wiesinger, Y. Min, A. G. MacDiarmid, and A. J. Epstein, *Phys. Rev. B* 49:2977 (1994).
150. J. Joo and A. J. Epstein, *Rev. Sci. Instrum.* 65:2653 (1994).
151. J. C. Clark, G. G. Ihas, A. J. Rafanello, M. W. Meisel, M. Reghu, C. O. Yoon, Y. Cao, and A. J. Heeger, *Synth. Met.* 69:215 (1995).
152. B. Beri, A. Fert, G. Creuzet, and A. Schul, *J. Phys. F* 16:2099 (1986).
153. C. S. Yang, Y. Cao, P. Smith, and A. J. Heeger, *Synth. Met.* 53:293 (1993).
154. C. S. Yang, Y. Cao, and P. Smith (to be published).
155. N. S. Sariciftci, A. J. Heeger, and Y. Cao, *Phys. Rev. B* 49:5988 (1994).
156. A. C. Kolbert, S. Caldarelli, K. F. Thier, N. S. Sariciftci, Y. Cao, and A. J. Heeger, *Phys. Rev. B* 51:1541 (1995).
157. K. Lee, A. J. Heeger, and Y. Cao, *Phys. Rev. B* 48:14884 (1993).
158. K. Lee, A. J. Heeger, and Y. Cao, *Synth. Met.* 72:25 (1995).
159. K. Lee, M. Reghu, E. L. Yuh, N. S. Sariciftci, and A. J. Heeger, *Synth. Met.* 68:287 (1995).
160. K. Lee, M. Reghu, C. O. Yoon, and A. J. Heeger, *Phys. Rev. B* 52:4779 (1995).
161. K. Mizoguchi and K. Kume, *Solid State Commun.* 89:971 (1994).
162. K. Mizoguchi, M. Nechtschein, J. P. Travers, and C. Menardo, *Synth. Met.* 29:E417 (1989).
163. J. P. Travers, P. Le Guyadec, P. N. Adams, P. J. Laughlin, and A. P. Monkman, *Synth. Met.* 65:159 (1994).
164. P. N. Adams, P. J. Laughlin, A. P. Monkman, and N. Bernhoeft, *Solid State Commun.* 91:875 (1994).
165. D. S. Galvao, D. A. dos Santos, B. Laks, C. P. de Melo, and M. J. Caldas, *Phys. Rev. Lett.* 63:786 (1989); 65:527 (1990).
166. P. A. Schulz, D. S. Galvao, and M. J. Caldas, *Phys. Rev. B* 44:6073 (1991).
167. C. O. Yoon, M. Reghu, D. Moses, and A. J. Heeger, *Phys. Rev. B* 49:10851 (1994).
168. R. S. Kohlman, J. Joo, Y. Z. Wang, J. P. Pouget, H. Kaneko, T. Ishiguro, and A. J. Epstein, *Phys. Rev. Lett.* 74:773 (1995).
169. D. Braun, A. Brown, E. Staring, E. W. Meijer, *Synth. Met.* 65:85 (1994).
170. M. Ahlskog, M. Reghu, et al. *Phys. Rev. B53*: 15529 (1996).
171. Y. W. Park, E. B. Park, K. H. Kim, C. K. Park, and J.-I. Jin, *Synth. Met.* 41–43:315 (1991).
172. C. O. Yoon, M. Reghu, D. Moses, A. J. Heeger, Y. Cao, T.-A. Chen, X. Wu, and R. D. Rieke, *Synth. Met.* 75:229 (1995).
173. K. Vakiparta, M. Reghu, M. R. Andersson, J. Moulton, and T. Taka, *Solid State Commun.* 87:619 (1993).
174. K. Ehinger and S. Roth, *Phil. Mag.* 53:301 (1986).
175. M. Reghu, C. O. Yoon, C. Y. Yang, D. Moses, P. Smith, A. J. Heeger, and Y. Cao, *Phys. Rev. B* 50:13931 (1994).
176. R. Rosenbaum, *Phys. Rev. B* 44:3599 (1991).
177. Y. Zhang, P. Dai, M. Levy, and M. P. Sarachik, *Phys. Rev. Lett.* 64:2687 (1990).
178. A. Aharony, Y. Zhang, and M. P. Sarachik, *Phys. Rev. Lett.* 68:3900 (1992).
179. A. N. Ionov and I. S. Shlimak, in *Hopping Transport in Solids* (M. Pollak and B. I. Shkolvskii, eds.) North-Holland, Amsterdam, 1990, p. 397 and references therein.
180. M. Pollak and C. J. Adkins, *Phil. Mag. B* 65:855 (1992).
181. E. Cuevas, M. Ortuno, and J. Ruiz, *Phys. Rev. Lett.* 71:1871 (1993).
182. A. A. Ovchinnikov and K. A. Pronin, *Synth. Met.* 41–43:3373 (1991).
183. A. A. Ovchinnikov and K. A. Pronin, *Sov. Phys. Solid State* 28:1666 (1986) [*Fiz. Tverd. Tela* 28:2964 (1986)].
184. H. Tokumoto, R. Mansfield, and M. J. Lea, *Phils. Mag. B* 46:93 (1982).
185. B. I. Shklovskii, *Zh. Eksp. Teor. Fiz. Pis. Red.* 36:43 (1982) [*Sov. Phys. JETP Lett.* 36:51 (1982)].
186. B. I. Shklovskii, *Fiz. Tekh. Polouprov.* 17:2055 (1983) [*Sov. Phys. Semicond.* 17:1311 (1983)].
187. I. Shlimak, M. Kaveh, M. Yosefin, M. Lea, and P. Fozooni, *Phys. Rev. Lett.* 68:3076 (1992).
188. C. O. Yoon, Ph.D. Thesis, Seoul Natl. University, 1992.
189. D. K. C. MacDonald, *Thermoelectricity: An Introduction to the Principles*, Wiley, New York, 1962.

190. I. P. Zvyagin, in *Hopping Transport in Solids* (M. Pollak and B. I. Shklovskii, eds.), Elsevier/North-Holland, Amsterdam, 1990, p. 143.
191. M. J. Burns and P. M. Chaikin, *J. Phys. C* 18:L743 (1985).
192. E. B. Park, Y. S. Yoo, J. Y. Park, Y. W. Park, et al., *Synth. Met.* 69:61 (1995).
193. C. O. Yoon, M. Reghu, D. Moses, A. J. Heeger, and Y. Cao, *Phys. Rev. B* 48:14080 (1993).
194. S. Stafstrom, J. L. Bredas, A. J. Epstein, H. S. Woo, D. B. Tanner, W. S. Huang, and A. G. MacDiarmid, *Phys. Rev. Lett.* 59:1464 (1987).
195. D. S. Boudreaux, R. R. Chance, J. F. Wolf, L. W. Shacklette, J. L. Bredas, B. Themans, J. M. Andre, and R. Silbey, *J. Chem. Phys.* 85:4584 (1986).
196. M. Reghu, C. O. Yoon, C. Y. Yang, D. Moses, A. J. Heeger, and Y. Cao, *Macromolecules* 26:7245 (1993).
197. C. O. Yoon, M. Reghu, D. Moses, A. J. Heeger, and Y. Cao, *Synth. Met.* 63:47 (1994).
198. A. Aharony and D. Stauffer, *Introduction to Percolation Theory*, 2nd ed., Taylor & Francis, London, 1993 and references therein.
199. A. Fizazi, J. Moulton, K. Pakbaz, S. D. D. Rughooputh, P. Smith, and A. J. Heeger, *Phys. Rev. Lett.* 64:2180 (1990).
200. Y. Y. Suzuki, A. J. Heeger and P. Pincus, *Macromolecules* 23:4730 (1990).
201. A. Andreatta, A. J. Heeger, and P. Smith, *Polymer Commun.* 31:275 (1990).
202. Y. Wang and M. F. Rubner, *Macromolecules* 25:3284 (1992).
203. M. Makhlouki, M. Morsli, A. Bonnet, A. Conan, A. Pron, and S. Lefrant, *J. App. Polym. Sci.* 44:443 (1992).
204. C. K. Subramaniam, A. B. Kaiser, P. W. Gilberd, and B. Wessling, *J. Polym. Sci. B* 31:1425 (1993).
205. R. Pelster, G. Nimtz, and B. Wessling, *Phys. Rev. B* 49:12718 (1994).
206. K. Levon, A. Margolina, and A. Z. Patashinsky, *Macromolecules* 26:4061 (1993).
207. B. Bridge and H. Tee, *Int. J. Electron.* 6:785 (1990) and references therein.
208. S. De Bondt, L. Froyen, and A. Deruyttere, *J. Mater. Sci.* 27:1983 (1992).
209. F. Carmona, *Physica A* 157:461 (1989) and references therein.
210. H. Munson-McGee, *Phys. Rev. B* 43:3331 (1991).
211. Y. Yang, E. Westerweele, C. Zhang, P. Smith, and A. J. Heeger, *J. Appl. Phys.* 77:694 (1995).
212. Y. Yang and A. J. Heeger, *Nature* 372:344 (1994).
213. A. J. Heeger, *Trends Polym. Sci.* 3:39 (1995).
214. Y. E. Levy and B. Souillard, *Europhys. Lett.* 4:233 (1987).
215. G. Deutscher, Y. E. Levy, and B. Souillard, *Europhys, Lett.* 4:577 (1987).
216. A. Aharony and A. B. Harris, *Physica A* 163:38 (1990).
217. A. Aharony and A. B. Harris, *Physica A* 205:335 (1994).
218. A. Aharony and A. B. Harris, *Physica A* 191:365 (1992).
219. H. Grussbach and M. Schreiber, *Physica A* 191:394 (1992).
220. A. K. Sarychev, D. J. Bergman, and Y. M. Strelniker, *Phys. Rev. B* 48:3145 (1993).
221. D. J. Bergman and A. K. Sarychev, *Physica A* 200:231 (1993).
222. M. A. Michels, J. C. M. Brokken-Zijp, W. M. Groenewoud, and A. Knoester, *Physica A* 157:529 (1989).
223. D. van der Putten, J. T. Moonen, H. B. Brom, J. C. M. Brokken-Zijp, and M. A. J. Michels, *Phys. Rev. Lett.* 69: 494 (1992); 70:4161 (1993).
224. F. Gubbels, R. Jerome, P. Teyssie, E. Vanlathem, R. Deltour, A. Calderone, V. Parente, and J. L. Bredas, *Macromolecules* 27:1972 (1994).
225. A. Quivy, R. Deltour, A. G. M. Jansen, and P. Wyder, *Phys. Rev. B* 39:1026 (1989).
226. M. Mehbod, P. Wyder, R. Deltour, C. Pierre, and G. Geuskens, *Phys. Rev. B* 36:7627 (1987).
227. A. B. Harris and A. Aharony, *Europhys. Lett.* 4:1355 (1987).
228. A. Aharony, O. Entin-Wohlman, and A. B. Harris, *Physica A* 200:171 (1993).
229. A. Aharony, O. Entin-Wohlman, and A. B. Harris, *Phys. Rev. Lett.* 70:4160 (1993).
230. H. Aoki, *J. Phys. C: Solid State Phys.* 16:L205 (1983).
231. H. Aoki, *Phys. Rev. B* 33:7310 (1986).
232. C. M. Soukoulis and E. N. Economou, *Phys. Rev. Lett.* 52:565 (1984).
233. M. Schreiber, *Phys. Rev. B* 31:6146 (1985).
234. M. Schreiber and H. Grussbach, *Phil. Mag. B* 65:707 (1992).
235. B Ya. Balagurov, *Fiz. Tverd. Tela* 28:3012 (1986) [*Sov. Phys. Solid State* 28:1694 (1986)].
236. D. Stroud and F. P. Pan, *Phys. Rev. B* 13:1434 (1976).
237. Y. W. Park and H. Lee, (eds.), *Synth. Met.*, Vol. 69, 1995 [Proc. Int. Conf. Sci. Technol. Synthetic Metals (ICSM '94), Seoul, Korea].

3
Insulator–Metal Transition and Inhomogeneous Metallic State in Conducting Polymers

R. S. Kohlman* and Arthur J. Epstein
The Ohio State University, Columbus, Ohio

I. INTRODUCTION

For the past 50 years, conventional insulating polymer systems have been increasingly used as substitutes for structural materials such as wood, ceramics, and metals because of their high strength, light weight, ease of chemical modification and customization, and processibility at low temperatures (see, e.g., Ref. 1). In 1977, the high electrical conductivity of an organic polymer, doped polyacetylene, was reported [2], spurring interest in "conducting polymers" [2–6] (for recent research activity, see, e.g., the conference proceedings listed in Ref. 5). The frequency-dependent electrical transport in doped polyacetylene was investigated intensively to understand the transport mechanisms [6,7]. The common electronic feature of pristine (undoped) conducting polymers is the π-conjugated system that is formed by the overlap of carbon p_z orbitals and alternating carbon–carbon bond lengths [4,7–9]. (In some systems, notably polyaniline, nitrogen p_z orbitals and C_6 rings are also part of the conjugation path.) Figure 3.1 shows the chemical repeat units of pristine forms of several families of conducting polymers, i.e., *trans*- and *cis*-polyacetylene [$(CH)_x$]; the leucoemeraldine base (LEB), emeraldine base (EB), and pernigraniline base (PNB) forms of polyaniline (PAN); and polypyrrole (PPy).

Due to the susceptibility of one-dimensional chains to Peierls distortions [10], charge doped into the polymer is stored in novel states such as solitons, polarons, and bipolarons, which include a charge and a lattice distortion that surrounds it [7]. Despite the strong electron–phonon coupling in conducting polymers, the conductivities of the pristine polymers can be transformed from insulating to metallic through the process of doping, with the conductivity increasing as the doping level increases. Both *n*-type (electron-donating) and *p*-type (electron-accepting) dopants have been used to induce an insulator–metal transition in electronic polymers [2–5,7–9].

An extraordinarily large range of conductivity is obtained upon doping [11–29]. Figure 3.2 shows that the conductivity of these undoped polymers can be increased by 10 orders or more. For instance, the conductivity of polyacetylene processed by various routes [11–14,26,27,30–33] when doped with iodine can be increased from $\sim 10^{-10}$ [*cis*-$(CH)_x$]–10^{-5} [*trans*-$(CH)_x$] $(\Omega \cdot cm)^{-1}$ [= S/cm] to greater than 4×10^4 S/cm [3,13,34], comparable to the conductivity of good metals (e.g., $\sigma_{dc} \sim 4.8 \times 10^4$ S/cm for lead and 5.9×10^5 S/cm for copper at room temperature [35]). Recent advances in the processing of other conducting polymer systems have led to improvements in their σ_{dc} values to the range of $\sim 10^3$ S/cm [3,5,23,25,34,36,37]. With such high conductivity, conducting polymers become useful for applications as wires, in electromagnetic shielding, and for electrostatic dissipation [3], with the advantages that conducting polymers are lightweight and facilitate recycling, compared with metallized polymers [1].

Accompanied by the enhancement in σ_{dc}, many traditional signatures of an intrinsic metallic nature have become apparent, including negative dielectric constants [3,19,34,38], a Drude metallic response [19,24,34,39,40], temperature-independent Pauli susceptibility [34,41–48], and a linear dependence of thermoelectric power on temperature [49–52]. However, the conductivities of even the new highly conducting polymers, though comparable to those of traditional metals at room

---
*Current affiliation: Los Alamos National Laboratory, Los Alamos, New Mexico.

trans-polyacetylene

cis-polyacetylene

polyaniline: leucoemeraldine (y=1), emeraldine (y=0.5) and pernigraniline (y=0)

polypyrrole

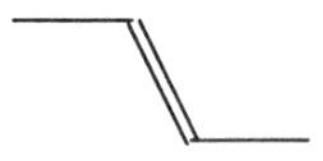

Fig. 3.1 Schematic chemical structures of polyacetylene, polyaniline, and polypyrrole.

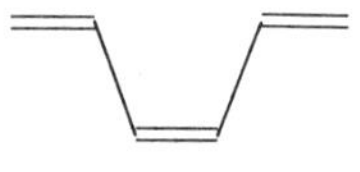

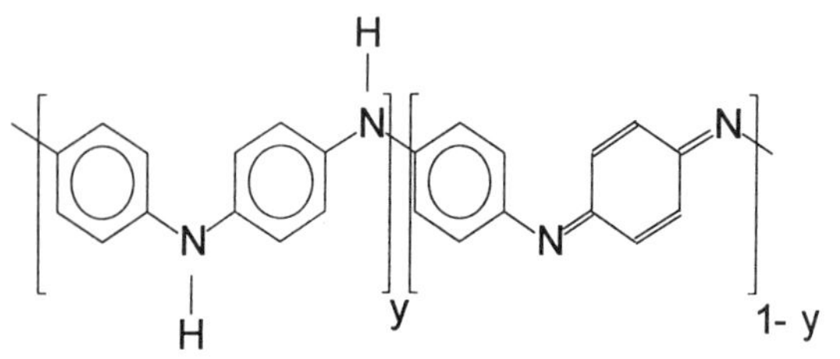

temperature, generally decrease as the temperature is lowered [3,5,34,36,37]. Some of the most highly conducting samples remain highly conducting even in the millikelvin range [36,37,40,53], demonstrating that a truly metallic state at low temperature has been attained.

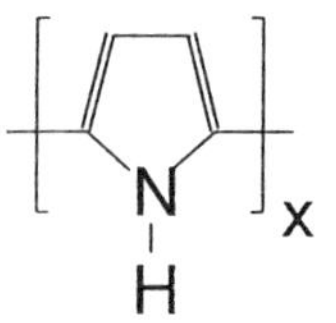

These metallic features have been reported in conducting versions of both polyaniline and polypyrrole. The ability to process PAN doped with camphorsulfonic acid (CSA) from solution [18,54] has resulted in free-standing films with high conductivity ($\sigma_{dc} \sim$ 100–400 S/cm) [39,55,56] that span the insulator–metal transition (IMT) even at low temperature [40,53]. Some samples of electrochemically prepared polypyrrole doped with hexafluorophosphate (PF_6) are metallic to low temperature [23,36,37]. However, when PPy is synthesized using different dopants or at high temperatures, the materials are more disordered and show insulating behavior [23–25,57,58]. Similar results are reported in the literature for conducting polyacetylene [36,37].

The transport properties of conducting polymers are highly dependent upon the structural disorder arising from sample quality, doping procedure, and aging [3,34]. In doped polyacetylene [36,37], polypyrrole [36,37], and polyaniline [40], studies of the millikelvin resistivity ($\rho = 1/\sigma$) showed that the conductivity is thermally activated in strongly disordered films, the resistivity appears like that of a disordered metal ($\rho \propto \log T$) for samples of intermediate conductivity, and ρ has a very weak temperature dependence for the weakly disordered state. Careful studies of the microwave dielectric contant in polyaniline as the crystallinity of the films improves demonstrate that the carriers become more delocalized as the structural order is improved [19]. Variation of the solvent from which PAN-CSA is cast yields a change in the local order in the polymer [59–63] and a concomitant change in the crystallinity of the films, with the transport properties showing stronger localization as the crys-

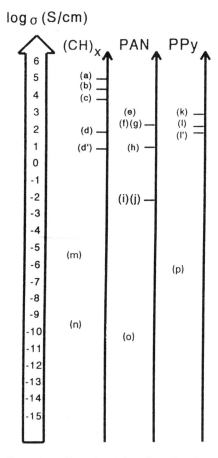

Fig. 3.2 Overview of conductivity of conducting polymers at room temperature. (a) Stretched $[CH(I_3)]_x$ (from Refs 11 and 12). (b) Stretched $[CH(I_3)]_x$ (from Ref. 13). (c) $[CH(I_3)]_x$ (from Ref. 14). (d) $[CH(I_3)]_x$ (from Ref. 15). (d') $[CH(I_3)]_x$ (from Ref. 16). (e) Stretched PAN-HCl (from Ref. 17). (f) PAN-CSA from *m*-cresol (from Ref. 18). (g) PAN-CSA from *m*-cresol (from Ref. 19). (h) PAN derivative: poly(*o*-toluidine) POT-CSA fiber from *m*-cresol (from Ref. 20). (i) POT-HCl (from Ref. 21). (j) Sulfonated PAN (from Ref. 22). (k) Stretched $PPy(PF_6)$ (from Ref. 23). (l) $PPy(PF_6)$ and (l') PPy(TsO) (from Refs. 24 and 25). (m) Undoped *trans*-$(CH)_x$ (from Ref. 26). (n) Undoped *cis*-$(CH)_x$ (from Ref. 27). (o) Undoped PAN (EB) (from Ref. 28). (p) Undoped PPy (from Ref. 29). The conductivity reported for the undoped polymers should be considered an upper limit because of the possibility of impurities.

tallinity decreases [60–63]. Even the most metallic samples are not single crystals. Doped polyacetylene has a complex fibrillar morphology with ~80–90% crystallinity, while doped PAN and doped PPy have ~50% crystallinity at best [59].

The effect of the disorder and the one-dimensionality of the polymer on the nature of the metallic state and the insulator–metal transition are still strongly debated. Disorder [64,65] and one-dimensionality [65] lead to localization of the electron wave functions. Even if the polymer chains are well-ordered, macroscopic transport is impossible unless the carrier can hop or diffuse to another chain to avoid the chain breaks and defects [66,67]. Previous theoretical calculations [38,68–70] as well as experimental work [19,21,38] have stressed the importance of interchain interaction and three-dimensionality of the electron states in highly conducting polymers to avoid one-dimensional localization. There is evidence that the metallic states are three-dimensional, though the transport properties are highly anisotropic [66,67,71–73]. The nature of the disorder is still an important question. Disorder can result in different properties depending on whether it is homogeneous or inhomogeneous.

The wave functions of the charge carriers may become localized to a few atomic sites if the disorder is strong enough. The behavior of a uniformly disordered material near an insulator–metal transition has been discussed in the context of an Anderson transition [64,65,74,75]. For a three-dimensional system, when the disorder potential becomes comparable to the electronic bandwidth, a mobility edge exists that separates localized from extended states [64,65]. If the Fermi level (E_F) lies in the region of extended states, σ_{dc} is finite [65] and the logarithmic derivative of the conductivity [$W \equiv d \ln \sigma_{dc}/d \ln T$] [76] has a positive slope at low temperature. When E_F lies in the range of localized states, the carriers have a hopping behavior (where $\sigma_{dc} \to 0$) [65] and the W function has a negative slope at low temperature [76]. Also, the dielectric function is positive owing to the polarization of the localized state. The strong disorder necessary to cause localization of three-dimensional electronic wave functions necessarily limits the mean free time (τ) to small values. As the IMT is crossed, the electronic localization length (L_{loc}) diverges and the system monotonically becomes more metallic, displaying higher σ_{dc} values with weaker temperature dependences. However, τ varies slowly with energy in crossing the IMT (V. N. Prigodin, private communication). Therefore, τ remains short close to the IMT. In fact, the Ioffe–Regel condition requires that $k_F \lambda \sim 1$ for a material near an IMT [77], where k_F is the Fermi wavevector and λ is the mean free path. This implies that the scattering time $\tau \sim 10^{-15}$ s. For such short scattering times, the frequency-dependent conductivity, $\sigma(\omega)$, is monotonically suppressed at low frequency. Also, localization corrections to the metallic Drude response result in a positive dielectric function $\epsilon(\omega)$ for short τ at low frequency, rather than the negative low frequency $\epsilon(\omega)$ of normal metals [35,78].

In contrast, if the disorder is inhomogeneous with large variations on the length scale of or larger than the characteristic electron localization lengths, then the behavior expected for a composite system may be expected. The inhomogeneous disorder model [24,34,38,39] treats the metallic state of conducting polymers as a composite system comprising metallic ordered regions (with delocalized charge carriers) coupled by disordered quasi-one-dimensional regions through which hopping transport along and between chains occurs. Localization occurs in the disordered regions owing to the one-dimensional electronic nature of the polymer chains in this region. When the polymer chains in the disordered region are tightly coiled, the in-chain localization length is short and coupling between metallic regions is poor, so that free electrons are confined within the metallic regions [3,19,34,39,60–63,79–81]. The temperature-dependent transport is then dominated by hopping and phonon-induced delocalization in the disordered regions [38] or even tunneling between metallic islands [82], depending upon the morphology. When the polymer chains in the disordered regions are sufficiently straight (i.e., larger radii of curvature or longer persistence lengths), the in-chain localization length is larger than the typical separation between metallic islands and carriers are able to diffuse macroscopically among the metallic regions [19,39,79–81]. In this case, a fraction of the carriers will percolate through the ordered paths. Just as in the Anderson transition, there is a crossover in slope for the W plot as percolation occurs, though the IMT is no longer necessarily a monotonic function of the room temperature σ_{dc} [39,79]. In this model, the magnitude of σ_{dc} depends on the number of well-coupled metallic regions across the sample. On the metallic side of the IMT, a fraction of the carriers will demonstrate free carrier response even at low temperature. Due to phonon-induced delocalization in the disordered regions, a fraction of the carriers may appear percolated at room temperature even for samples on the insulating side of the IMT [39,79].

To determine the nature of the metallic state in conducting polymers, it is necessary to use a wide variety of probes. Direct current transport measurements provide insight into the insulating or metallic nature of electrons at the Fermi level. Measurement of high frequency transport provides an important probe away from the Fermi energy to help discriminate between the homogeneously and inhomogeneously disordered metallic states [6,39]. For instance, $\epsilon(\omega)$ and the microwave dielectric constant (ϵ_{MW}) can be used to determine the presence of Drude (free carrier) dispersion in the electrical response of the sample as well as the plasma frequency of the free carriers [3,19,24,34,39]. This free carrier behavior can be monitored as a function of processing and temperature. Also, $\epsilon(\omega)$ and the optical frequency conductivity $\sigma(\omega)$ provide probes of the scattering times and mean free paths for the samples as the IMT is crossed.

In this chapter, coordinated studies of the temperature-dependent dc, optical (2 meV to 6 eV), and microwave (6.5 GHz) transport on two different polymer systems, polyaniline (PAN) and polypyrrole (PPy), are emphasized. From a comparison of this set of experiments with the reports in the literature for polyacetylene, it is determined that there is a universal type of behavior of the metallic state in conducting PAN, PPy, and (CH)x. The metallic state is characterized by a small fraction of carriers that are delocalized down to low T with long mean free times ($\tau > \sim 10^{-13}$ s) and a large majority of carriers that are more strongly localized ($\tau_1 \sim 10^{-15}$ s). These delocalized electrons result in a weak temperature dependence for σ_{dc} down to millikelvin temperatures. This metallic state is strongly dependent on structural order. More disordered polymers show strong localization and hopping (insulating) transport. We conclude that although the insulator–metal transition is due to structural disorder, it is not due to a conventional homogeneous Anderson transition. We propose that the insulator–metal transition is instead better described by the inhomogeneous disorder model.

II. MODELS FOR LOCALIZATION AND METALLIC CONDUCTIVITY

It is established experimentally that a finite density of states at the Fermi level [$N(E_F)$] can be induced by doping [3,16,34,41–45]. Even though there is a high density of conduction electrons at the Fermi level for the highly doped state, the carriers may be spatially localized so they cannot participate in transport except through hopping. The prime source of localization is structural disorder in the polymers [3,34,59]. X-ray diffraction studies of these systems show that they are generally of modest crystallinity, with regions of the material that are more ordered while other regions are more disordered [57,59,83,84]. Also, the fibrillar nature of many of the conducting polymers may lead to localization by reducing the effective dimensionality of the electrons delocalized in a bundle of polymer chains [70]. In the following sections, the effect of inhomogeneous and homogeneous structural disorder on the resulting transport properties of the disordered metallic state is discussed.

A. Inhomogeneous Disorder-Induced Insulator–Metal Transition

Conducting polymers generally display a rich variety in their morphology [59], being partially crystalline and partially disordered as shown schematically in Fig. 3.3. If the localization length in the more disordered regions

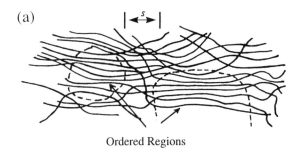

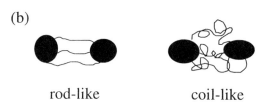

Fig. 3.3 (a) Schematic picture of the inhomogeneously disordered state of metallic and insulating conducting polymers. (b) Schematic picture of rodlike and coil-like morphology of disordered regions.

of the electrons (L_{loc}) is comparable to or smaller than the crystalline coherence lengths ($\sim$10 Å) in the polymer, then the disorder present in the conducting polymer is viewed as inhomogeneous [3,24,34,39,79]. The localization effects in the inhomogeneously disordered (partially crystalline) conducting polymers are proposed to originate from quasi-one-dimensional localization in the disordered regions [19,24,34,38] that surround the ordered regions. It is well known that for a one-dimensional metallic chain the localization of charge carriers arises for even weak disorder because of quantum interference of static backscattering [65], with L_{loc} increasing as the disorder decreases. For quasi-one-dimensional conducting polymers, the transfer integral perpendicular to the chain axis is crucial. As the polymer chains are finite, if the carrier cannot diffuse to another chain before being reflected back onto itself and thus experiencing quantum interference, then the polymer will not be a conductor. This leads to the condition $w_\perp > 1/2\tau$, where $w_\perp$ is the interchain transfer rate and τ is the mean free time, to avoid quasi-one-dimensional localization [66,67].

Prigodin and Efetov [70] studied the insulator–metal transition of these quasi-one-dimensional conducting polymers using a random metallic network (RMN) model to represent weakly connected, fibrous bundles of metallic chains. In this zero temperature model, the phase transition from insulating to metallic behavior is a function of the cross section of electronic overlap between fibers (α) and $\rho = pL_{loc}$, the product of the localization radius (L_{loc}) and the concentration of cross-links between fibers (p). The metallic state can be induced by strengthening the interchain (or interfibril) interaction (increasing α), increasing the density of cross-links between fibers (increasing p), or increasing the localization length (increasing L_{loc}). This model developed for contacts between fibers composed of parallel polymer chains can be generalized to the three-dimensional delocalization transition that occurs in inhomogeneously disordered (partially crystalline) nonfibrillar polymers [3,34]: the metallic state is induced as the strength of connection between ordered or crystalline regions (α) is increased, as the density of interconnections between ordered or crystalline regions (p) increases, and as the localization length within the disordered regions (L_{loc}) increases. L_{loc} depends on the morphology of the disordered region. Figure 3.3b schematically shows examples of rod-like and coil-like chain morphologies [54]. L_{loc} will be larger for the rod-like and smaller for the coil-like morphology. Within this model, conduction electrons are three-dimensionally delocalized in the "crystalline" ordered regions (though the effects of paracrystalline disorder may limit delocalization within these regions [85]). In order to move between ordered regions, the conduction electrons must diffuse along electronically isolated quasi-one-dimensional chains through the disordered regions where the electrons readily become localized.

Because the IMT depends on the coupling between metallic regions, the inhomogeneous disorder model resembles a percolation transition. The main difference between percolating conducting polymers and more traditional percolating system such as silver particles in potassium chloride is that the conducting polymer "crystals" do not have sharp boundaries. As a single polymer chain can be a part of both metallic "crystalline" regions and quasi-one-dimensional disordered regions, the percolating object is a fuzzy ellipsoid [3,39]. When L_{loc} is greater than the average distance between metallic regions, carriers may diffuse among the metallic regions and an IMT occurs. If L_{loc} is temperature-dependent due to phonon-assisted processes [38], it is possible to show the behavior of a percolated metallic system at high temperatures (L_{loc} greater than the average distance between metallic islands) and the behavior of an unpercolated insulating system at low temperatures (L_{loc} less than the average distance between metallic islands).

The dependence of the conductivity on transport through quasi-one-dimensional chains leads to some profound effects. For an ordered quasi-one-dimensional metal with its highly anisotropic electronic Fermi surface, σ in the chain is given direction by

$$\sigma_\parallel = \frac{e^2 n a}{\pi \hbar} v_F \tau = \frac{e^2 n a}{\pi \hbar} \left(\frac{\lambda}{a} \right) \quad (1)$$

where n is the conduction electron density per unit volume, a is the interatomic distance in the chain direction, $v_F = 2 t_0 a / \hbar$, t_0 is the electron hopping element along the chain, and τ is the mean free time. Kivelson and Heeger [86] showed that only $2k_F$ phonons can relax the momentum of the electrons for a quasi-one-dimensional chain from which the counterions are spatially removed and therefore effectively screened. Since the $2k_F$ phonons have a relatively small thermal population (owing to their high energies, ~ 0.1 eV), the conductivity is predicted [86] to be quite high at room temperature ($\sim 2 \times 10^6$ S/cm for polyacetylene) and increase exponentially upon cooling. A small interchain hopping integral $t_\perp \sim 0.1$ eV is all that is necessary to obtain three-dimensional delocalization of carriers, and the resulting metallic state may take advantage of this large quasi-one-dimensional (1-D) conductivity. The lack of effective scatterers in an ordered quasi-one-dimensional metal may lead to an anomalously long scattering time [19] compared with conventional 3-D metals.

In a real system, disorder scattering must be taken into account as the dc conductivity is limited by the least conducting (most disordered) part of the conduction path. In disordered quasi-1-D conducting polymers, static disorder scattering [with scattering time $\tau_{\text{imp}}(2k_F)$] as well as $2k_F$ phonons [with scattering time $\tau_{\text{ph}}(2k_F,T)$] can relax momentum. The effect of phonon-induced delocalization [scattering time $\tau_{\text{ph}}(0,t)$], which increases the conductivity with increasing temperature, must be taken into account. Within this model [38], the T-dependent dc conductivity is given by

$$\sigma_{\text{dc}}(T) = \frac{\Omega_p^2}{4\pi} \tau_{\text{tr}}(T) \frac{f(T) - 1}{f(T) + 1} \quad (2)$$

where $\Omega_p^2 (= 4\pi n e^2/m^*)$ is the carrier plasma frequency, τ_{tr} is the effective transport scattering time determined via Matthieson's rule

$$\frac{1}{\tau_{\text{tr}}(T)} = \frac{1}{\tau_{\text{imp}}(2k_F)} + \frac{1}{\tau_{\text{ph}}(2k_F, T)} \quad (3)$$

and

$$f^2(T) = 1 + 4 \frac{\tau_{\text{imp}}(2k_F)}{\tau_{\text{ph}} \in 0, T)} \quad (4)$$

The behavior with temperature of the scattering times and the resultant $\sigma_{\text{dc}}(T)$ [87] are shown in Fig. 3.4. In well-ordered materials, it is possible for $\tau_{\text{imp}}(2k_F)$ to be greater than $\tau_{\text{ph}}(2k_F, T)$ near room temperature. Therefore, τ_{tr} is dominated by inelastic phonon scattering near room temperature. Since $\tau_{\text{ph}}(2k_F,T)$ increases with decreasing temperature, σ_{dc} increases with decreasing T, similar to the behavior of metallic systems. When $\tau_{\text{imp}}(2k_F) < \tau_{\text{ph}}(2k_F, T)$, localization effects begin to dominate (phonon-induced delocalization is important), resulting in a suppression in σ_{dc} at lower temperatures and a consequent maximum in the diffusion rate and the corresponding $\sigma_{\text{dc}}(T)$ (Fig. 3.4). For weaker disorder [longer $\tau_{\text{imp}}(2k_F)$], this maximum in σ_{dc} is shifted to

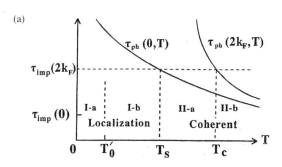

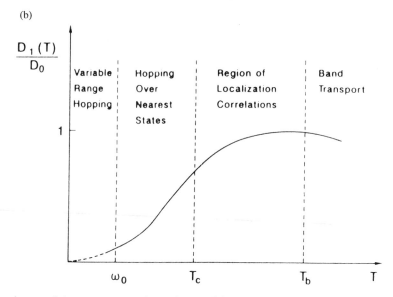

Fig. 3.4 (a) Schematic picture of the temperature dependence of the various scattering times in the inhomogeneous disorder model and the respective localization domains. (b) The corresponding diffusion constant expected for the inhomogeneous quasi-one-dimensional system. (From Ref. 87.)

lower temperature. At even lower temperature, if $\tau_{imp}(2k_F) > \tau_{ph}(0,T)$, there is a crossover to a more strongly localized hopping conductivity. This model accounts for localized behavior at low temperature despite conductivities at room temperature in excess of the Mott minimum conductivity. Only the most delocalized electrons on the percolating network contribute to the high dc conductivity. (The more localized electrons lead to a conductivity less than ~ 10 S/cm [79].) The fraction of delocalized carriers can be determined from high frequency measurements.

For a truly metallic system, the dc conductivity will remain finite as $T \to 0$ [65] and the logarithmic derivative of the conductivity ($W = d \ln \sigma_{dc}(T)/d \ln T$) will show a positive slope at low T [76]. In contrast, for an unpercolated system, the dc conductivity will decrease rapidly at low T and the W plot will have a negative slope, characteristic of hopping systems. In the inhomogeneous disorder model, the W plot will show a crossover in the low T slope for the W plot from negative to positive, though the IMT is not necessarily a monotonic function of room temperature σ_{dc} owing to its dependence on the density and strength of connections between metallic regions and the localization length of electronic states in the disordered regions.

Effective medium theories characterize the frequency-dependent transport in systems with large-scale inhomogeneities such as metal particles dispersed in an insulating matrix [88,89]. An IMT in the effective medium model represents a percolation problem where a finite σ_{dc} as $T \to 0$ is not achieved until metallic grains in contact span the sample. To understand the frequency dependence of the macroscopic material, an effective medium is built up from a composite of volume fraction f of metallic grains and volume fraction $1 - f$ of insulator grains. The effective dielectric function $\epsilon_{EMA}(\omega)$ and conductivity function $\sigma_{EMA}(\omega)$ are solved self-consistently.

The characteristic composite behavior of $\sigma_{EMA}(\omega)$ for a medium consisting of spherical particles with volume fractions f of Drude conductor and $1 - f$ of insulator is shown in Fig. 3.5. For a volume fraction f less than the percolation value ($f = 1/3$ for spheres), $\sigma_{EMA}(\omega)$ is dominated by an impurity band of localized plasmon-like excitations. As the system approaches the percolation threshold, the localized peak in $\sigma_{EMA}(\omega)$ shifts to lower frequency. Above the percolation threshold, a Drude peak corresponding to the carriers that have percolated through the composite structure occurs at low frequency. Only a fraction ($\sim (3f - 1)/2$ [89]) of the full conduction electron plasma frequency appears in the Drude peak, depending on the proximity to the percolation threshold. The same percolating free electron behavior is observable in the dielectric response $\epsilon_{EMA}(\omega)$ for the system.

This characteristic behavior for traditional composites is expected to be modified for conducting polymer composites. As mentioned earlier, the geometry of the percolating objects resembles fuzzy ellipsoids with large aspect ratios rather than spheres. Also, phonon-assisted transport in the disordered regions of the conducting polymer (which are the insulating regions) may give the appearance of percolation in $\sigma(\omega)$ and $\epsilon(\omega)$ at room temperature even though a particular system may not truly be metallic at low T [39,79]. For both $\sigma(\omega)$ and $\epsilon(\omega)$, the distinct behavior of localized and delocalized charge carriers will be evident above percolation. Specifically, for percolated samples, $\sigma(\omega)$ will increase and $\epsilon(\omega)$ will become increasingly negative with decreasing frequency below the unscreened free carrier plasma frequency Ω_p ($= \sqrt{4\pi\delta ne^2/m^*}$, where n is the full charge carrier density, δ is the fraction of the charge carrier density that is delocalized, e is the electronic charge, and m^* is the effective mass of the delocalized carriers). The response of localized carriers (the localized plasmon-like excitations) will occur for frequencies less than the "full" unscreened plasma frequency Ω_{p1} ($= \sqrt{4\pi ne^2/m_1^*}$, where m_1^* is the averaged effective mass of the carriers) [39,79]. The anomalously long scattering times possible for quasi-1-D systems may lead to huge negative dielectric functions at low frequency.

B. Anderson Disorder-Induced Insulator–Metal Transition

It is useful to describe the form of localization that occurs in a homogeneously disordered material in contrast to the model described in the previous section. The three-dimensional models described below assume that the materials are isotropic; i.e., the materials should be electrically the same in all directions. In a perfect crystal with periodic potentials, the wave functions form Bloch waves that are delocalized over the whole solid [35]. In systems with disorder, impurities and defects introduce substantial scattering of the electron wave function, which may lead to localization. Anderson demonstrated [64] that electronic wave functions can be localized if the random component of the disorder potential is large enough compared with the electronic bandwidth (Fig. 3.6). In this case, the localized wave functions have the form

$$\Psi(\mathbf{r}) \propto \exp(-\mathbf{r}/L_{loc}) \, \mathrm{Re} \, (\Psi_0) \qquad (5)$$

where L_{loc} is the localization length of the state. Later, Mott showed that band tail states are more easily localized than states in the center of the band [65]. Therefore, a critical energy called the mobility edge (E_c) exists that separates localized states in the band tails from extended states in the center of the band (Fig. 3.6). The resulting electronic behavior of a material depends on where the Fermi energy (E_F) lies relative to E_c. If E_F lies in the range of extended states, then $\sigma_{dc}(T)$ is finite as $T \to 0$ [65] and the W plot has a positive slope at low temperature [76]. If the disorder potential is strong enough to cause E_F to be in the range of localized states, then the material will be nonmetallic, with $\sigma_{dc} \to 0$ as $T \to 0$ [65] and a negative slope for the low TW plot [76], even though there is a finite density of states at E_F.

When E_F approaches E_c on the insulating side of the I–M transition, the localization length L_{loc} diverges as the electronic wave function becomes delocalized through the material. However, because of the strong disorder, the mean free path (λ) is still very short. In 1960, Ioffe and Regel [77] proposed that the lower limit

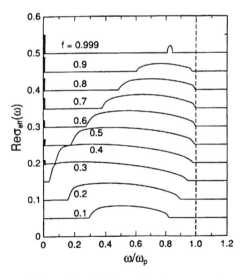

Fig. 3.5 Schematic $\sigma(\omega)$ for an insulator-metal composite made up of volume fractions f of a Drude metal and $1 - f$ of an insulator, as calculated in the effective medium theory. The heavy line at $\omega = 0$ represents the Drude peak. The integrated stength of the delta function is proportional to the height of the delta function. The scattering time is chosen to be very long so that the width of the Drude peak is too narrow to be resolved in the plot, emphasizing the behavior of the localized modes. (From Ref. 88.)

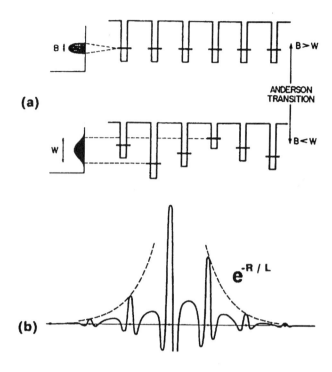

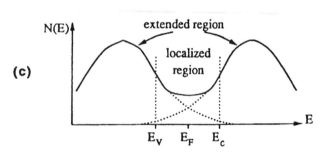

Fig. 3.6 (a) The Anderson transition and (b) the form of the localized wave function in an Anderson metal–insulator transition. (c) The Fermi glass state where the Fermi level lies in the region of localized states. (From Refs. 64,65,80.)

for the metallic mean free path is the interatomic spacing, which occurs at and on the insulating side of the IMT. This condition led Mott to propose a minimum metallic conductivity ($\sigma_{min} \sim 0.03e^2/3\hbar a \sim 10^2$ S/cm in three dimensions, where a is the interatomic spacing). The Ioffe–Regel condition implies that $k_F\lambda \sim 1$ (k_F is the Fermi wavevector) when $\sigma_{dc} \sim \sigma_{min}$. If applied to conducting polymers, this leads to a very short mean free path and scattering time. For typical repeat distances along the polymer chain in doped PAN [59] (other doped conducting polymers have similar values), the Ioffe–Regel condition requires that

$$\lambda \sim 1/k_F \sim 1/[\pi/2(10 \text{ Å})] \sim 10 \text{ Å} \quad (6)$$

and

$$\tau = \lambda/v_F \sim (10 \text{ Å})/(10^6 \text{ m/s}) \sim 10^{-15} \text{ s} \quad (7)$$

Far into the metallic regime where transport is due to diffusion of carriers in extended states, $k_F\lambda \gg 1$.

When the Fermi level lies in the localized region, the conductivity at zero temperature is zero even for a system with a finite density of states. The transport at higher temperature involves phonon-activated hopping between localized levels. The Mott variable range hopping (VRH) model is applicable to systems with strong disorder such that ΔV (disorder energy) $\gg B$ (bandwidth) [65]. The general form of the temperature-dependent conductivity of Mott's model is described as

$$\sigma = \sigma_0 \exp\left[-\left(\frac{T_0}{T}\right)^{1/(d+1)}\right] \quad (8)$$

where d is the dimensionality and, for three-dimensional systems, $T_0 = c/k_B N(E_F)L^3$ (c is the proportionality constant, k_B the Boltzmann constant, and L the localization length). In Mott's model, electron correlations are neglected as for the classical Fermi liquid. Efros and Shklovskii [90,91] pointed out that the interactions between localized electrons and holes play an important role in the hopping transport, especially at low temperature, changing the expected temperature dependence of the conductivity to

$$\sigma = \sigma_0 \exp\left[-\left(\frac{T_0'}{T}\right)^{1/2}\right] \quad (9)$$

where $T_0' = e^2/\epsilon L$ (e is the electron charge and ϵ is the dielectric constant). For materials very close to the I–M transition, instead of an exponential temperature dependence, a conductivity with a power law temperature dependence is predicted [92].

Typical behavior for the evolution of the temperature-dependent resistivity (ρ) as E_F crosses E_c is shown in Fig. 3.7 for p-doped germanium as a function of the compensation (carrier density) [65]. For the material with the largest compensation (smallest carrier density), ρ has the largest magnitude and strongest temperature dependence. As the compensation is decreased (carrier density is increased), ρ decreases monotonically in magnitude and the temperature dependence decreases monotonically. For the most metallic sample, ρ shows a very weak T dependence at low temperature. This monotonic evolution for the magnitude and T dependence of σ (= $1/\rho$) is characteristic of an Anderson IMT [65].

At high frequencies, $\sigma(\omega)$ for ordered materials is given by the Drude formula,

$$\sigma_{\text{Drude}}(\omega) = \frac{\Omega_p^2 \tau}{4\pi(1 + \omega^2\tau^2)} \quad (10)$$

where Ω_p (= $\sqrt{4\pi ne^2/m^*}$) is the plasma frequency of the free electrons, n is the full carrier density, and m^* is the effective mass. This behavior follows from the Kubo formula when the mean free time (τ) is such that $\omega\tau \gg 1$ and breaks down when $\omega\tau \sim 1$ [65,93]. For three-

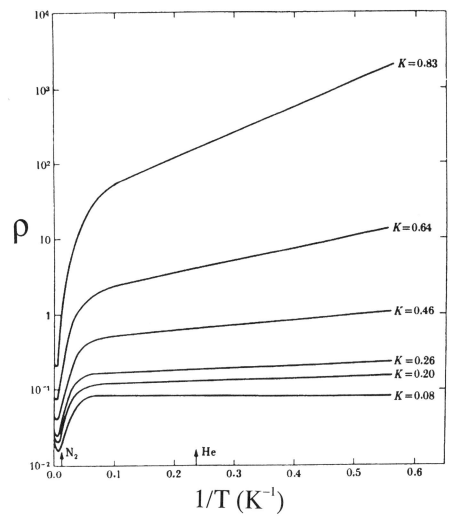

Fig. 3.7 Temperature-dependent resistivity in an Anderson transition. Experimental temperature-dependent resistivity for p-type germanium with different levels of compensation (K) leading to a variation in the electron density. (From Ref. 65.) Notice that the magnitude and temperature dependence change monotonically with increasing compensation (decreasing carrier density).

dimensional materials near an Anderson IMT, $\sigma(\omega)$ is suppressed at low frequencies relative to $\sigma_{\text{Drude}}(\omega)$ due to quantum interference of the electronic wave functions [94,95]. Localization corrections to σ_{Drude} calculated within scaling theories of the Anderson transition [65,74,75,95–98] yield the localization-modified Drude model;

$$\sigma_{\text{LMDM}}(\omega) = \sigma_{\text{Drude}}(\omega)\left[1 - \frac{C}{(k_F\lambda)^2}\left(1 - \frac{\lambda}{L(\omega)}\right)\right] \quad (11)$$

where C is an undetermined universal constant ($\sim$1), k_F is the Fermi wavevector, λ is the mean free path, and $L(\omega)$ is the distance a charge would diffuse during an oscillation of the electromagnetic wave. It is noted that

in this model at zero frequency, the high dc conductivity of a sample near the IMT is due to localized charge carriers. Since in three dimensions,

$$L(\omega) = \sqrt{D/\omega} \quad (12)$$

where $D = \lambda^2/3\tau$ is the diffusion constant, the localization-modified Drude model can be written as

$$\sigma_{\text{LMDM}}(\omega) = \sigma_{\text{Drude}}(\omega)\left[1 - \frac{C}{(k_F v_F \tau)^2} + \frac{C(3\omega)^{1/2}}{(k_F v_F)^2 \tau^{3/2}}\right] \quad (13)$$

The real part of the dielectric function $\epsilon_{\text{LMDM}}(\omega)$ corresponding to the localization-modified Drude model can be calculated using the Kramers–Kronig relations, giving

$$\epsilon_{\text{LMDM}}(\omega) = \epsilon_\infty + \frac{\Omega_p^2 \tau^2}{1+\omega^2\tau^2}\left[\frac{C}{(k_F v_F)^2 \tau^2}\left(\sqrt{\frac{3}{\omega\tau}} - (\sqrt{6}-1)\right) - 1\right] \quad (14)$$

where ϵ_∞ is the dielectric screening due to higher energy interband transitions. It is noted that these expressions are not complete as they ignore cutoffs at low frequency when $L(\omega) \sim L_T$, where L_T is the Thouless length (V. N. Prigodin, private communication).

The behavior of both $\sigma_{\text{LMDM}}(\omega)$ and $\epsilon_{\text{LMDM}}(\omega)$ for materials close to the IMT is determined in the Anderson model by the Ioffe–Regel condition, which requires $\tau \sim 10^{-15}$ s. $\sigma_{\text{LMDM}}(\omega)$ and $\epsilon_{\text{LMDM}}(\omega)$ are shown as the mean free time τ ($\sim 10^{-15}$ s) is varied in Fig. 3.8. $\sigma_{\text{LMDM}}(\omega)$ has Drude dispersion at high frequency, a maximum, and then monotonic suppression at low frequencies. As τ increases (reflecting a more ordered material), the Drude dispersion occurs over a wider frequency range and the maximum shifts to lower frequency. An important consequence of the localization-modified Drude model is that $\epsilon_{\text{LMDM}}(\omega)$ becomes positive at very low frequencies, reflecting the short τ due to strong disorder scattering. This behavior should be contrasted with the negative value of $\epsilon(\omega)$ for the Drude model at low frequency ($\omega \ll 1/\tau$),

$$\epsilon_{\text{Drude}} \simeq \epsilon_\infty - \Omega_p^2 \tau^2 \quad (15)$$

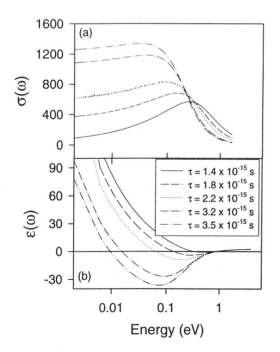

Fig. 3.8 Behavior of the localization-modified Drude model with increasing mean free time (mean free path). (a) $\sigma(\omega)$; (b) $\epsilon(\omega)$. The parameters used were $\Omega_p^2 = 2$ eV^2 and $C/(k_F v_F)^2 = 1.9 \times 10^{-30}$ s^2.

III. EXPERIMENTAL TECHNIQUES

A. Chemical Preparation

The final structural order and hence the electronic properties of conducting polymers are sensitive to preparation techniques. The high quality free-standing films of polyaniline emeraldine base (EB) doped with d,l-camphorsulfonic acid (CSA), for which data are presented here, were prepared using EB synthesized to provide different molecular weights [99]. Low molecular weight polyaniline, $\bar{M}_w < \sim 100{,}000$ was prepared by a well-known route [60–63,99]. Polyaniline of high molecular weight, $\bar{M}_w > \sim 300{,}000$, was synthesized [99] by adding lithium chloride (LiCl) to the reaction vessel and lowering the temperature of the reaction. All molecular weights were determined by gel permeation chromatography using a 0.5% w/v LiCl/N-methyl pyrrolidone (NMP) solvent and a polystyrene standard.

To obtain fully doped polyaniline salt, the molar ratio of EB to the dopant (CSA) should be 1:2, assuming that all of the dopant ions successfully protonate an imine nitrogen. For example, 1.0 g (0.00276 mol) of EB would be mixed with 1.287 g (0.00552 mol) of HCSA. The EB and HCSA was mixed in two different ways. For some films, the EB and HCSA were mixed as powders using a mortar and pestle. This doped EB-CSA powder could then be dissolved in appropriate solvent mixtures. For the data reviewed here, these powders were dissolved in m-cresol or chloroform or mixtures of these solvents. The resulting properties were shown to vary dramatically depending on which solvent was used through the effects of secondary doping [60–63,99]. The alternative way to obtain solutions of doped PAN-CSA is to separately add the EB powder into a volume of solvent and the HCSA into a volume of solvent and then mix the two solutions.

In addition, the concentration of PAN-CSA in the solvent was varied from 1 to 3 wt %, as the amount of solvent that must evaporate from the resulting film has an effect on the structural order of the film. When the solution of PAN-CSA was prepared, it was stirred overnight with a magnetic stirrer and then sonicated for ~ 2 h to completely dissolve all visible particles. The solution was then filtered and cast on a microscope slide upon which it dried under a hood. The resulting films could be peeled from the microscope slide to obtain free-standing films of approximate thickness 40–100 μm. All polyaniline samples for which data are reported here were prepared in the laboratories of Alan G. MacDiarmid at the University of Pennsylvania. The details of the effects of different processing conditions on the free-standing films are given in Ref. 99. For this study, samples of these films were chosen that illustrate the properties of PAN-CSA as the transport properties approach and cross the insulator–metal transition. The samples for this study are listed in Table 3.1 along with their specific processing conditions.

Table 3.1 Preparation Conditions for the PAN Samples of this Study[a]

Sample	EB $\overline{M}_w$	Mixed as	Solvent	Wt %	σ_{dc}(300 K) (S/cm)
A	400,000	Powder	m-Cresol	2	240
B	300,000	Powder	m-Cresol	2	110
C	50,000	Solution	m-Cresol	6.8	400
D	400,000	Solution	m-Cresol	1	120
E	50,000	Powder	30% m-cresol	1	70
F	50,000	Powder	Chloroform	1	20
G	50,000	Powder	Chloroform	1	0.2

[a] Notice that in the column describing the solvent used, when the solvent listed is a percentage of m-cresol, the balance of the solvent is chloroform. Also, samples F and G appear to have been exposed to m-cresol.

The polypyrrole films that were doped with hexafluorophosphate (PF_6) and p-toluenesulfonate were obtained by anodic oxidation of pyrrole at $-30°C$ under potentiostatic conditions. The electrolytic cell contained 0.06 mol/L of pyrrole and 0.06 mol/L of the appropriate salt in propylene carbonate containing 1 vol % of water. A glassy carbon plate was used as the anode, and a platinum foil was used as the anode. The current was adjusted to ~0.125 mA/cm². The reaction took place for ~24 h under a stream of nitrogen gas. For PPy(PF_6) and PPy(TsO), elemental analyses determined that there is approximately one unit of dopant per three pyrrole rings [23,25,36,37].

The polypyrrole film doped with sulfonated polyhydroxyether (S-PHE) [100] was synthesized on rotating stainless steel electrodes in a solution of 0.05 mol/L S-PHE, 5 vol % pyrrole, and 1 vol % water in propylene carbonate (PC) under galvanostatic conditions using a current density of 2 mA/cm². The film was peeled, washed with PC and ethanol, and then dried under vacuum at 50°C for 12 h. The molar ratio of sulfonation is defined as MR = $n/(n + m)$, where n is the number of sulfonated polyhydroxyether segments and m is the number of unsulfonated polyhydroxyether segments as shown in Fig. 3.9. For samples with low molar ratios, a large fraction of the volume of the polymer is occupied by the PHE, a saturated polymer that does not transport electric charge effectively. Thus, a large volume of insulator is added to the film, drastically reducing the conduction electron density and the interaction between chains. Data presented here are from a study of a PPy(S-PHE) film with MR = 0.125 [58,100].

In general, data presented here are for samples that are under vacuum or "pumped" conditions that substantially reduce or eliminate the effects of weakly bound moisture or solvents.

B. Transport Measurements

The dc conductivity (σ_{dc}) measurements were performed using a four-probe technique as described previously [66,67]. Four thin gold wires (0.05 mm thick and of 99.9% purity) were attached to the sample surface using a graphite paste (Acheson Electrodag 502) to improve the electrical contact. By placing the sample probe in a Janis DT dewar with helium exchange gas, $\sigma_{dc}(T)$ could be measured from 4.2 to 300 K. Measurements at millikelvin temperatures were carried out by placing the sample probe in thermal contact with the mixing chamber of a ^{3}He-^{4}He dilution refrigerator [53]. The millikelvin conductivity measurements were carried out using a low frequency (~19 Hz) alternating current and a lock-in amplifier. A magnetic field of 0–5 T was applied perpendicular to the plane of the samples.

Reflectance spectra from 2 meV to 6 eV were recorded as reported previously [24]. The high energy (0.5–6 eV) reflectance was recorded using a Perkin-Elmer λ-19 UV/Vis/NIR (ultraviolet/visible/near-infrared) spectrometer equipped with a Perkin-Elmer RSA-PE-90 reflectance accessory based on the Labsphere DRTA-9a integrating sphere. The low energy (2 meV to 1.2 eV) reflectance measurements were made on a BOMEM DA-3 Fourier transform infrared (FTIR) spectrometer equipped with a homemade cryostat that could be lifted so that a reference mirror and sample could be placed alternately in the IR beam. A Michelsen interferometer (1–12 meV) was used to record temperature-dependent far-IR reflectance spectra [40]. The measured reflectance spectra from the different spectrometers typically agreed to within 1% in the regions of overlap for samples with highly specular surfaces.

The samples were free-standing films with thickness ~40–100 μm. This thickness is greater than the electro-

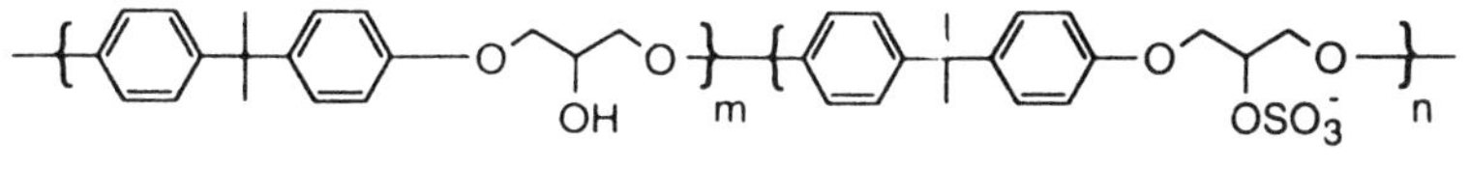

Fig. 3.9 Chemical structure for sulfonated polyhydroxyethers. (From Ref. 100.)

magnetic penetration depth [$\delta = c/\sqrt{2\pi\mu\omega\sigma(\omega)}$, where c is the speed of light and $\sigma(\omega)$ is the conductivity at frequency ω] when $\sigma_{dc} > 10$ S/cm. Samples with $\sigma_{dc} < 10$ S/cm were first checked to determine whether they transmitted far-infrared radiation. Therefore, the reflectance can be analyzed using the Fresnell reflection coefficients for semi-infinite media [78].

The optical conductivity $\sigma(\omega)$ and the real part of the dielectric function $\epsilon(\omega)$ were calculated from the reflectance spectra using the Kramers–Kronig analysis and the Fresnell relations for semi-infinite media [78]. To calculate the optical functions using the Kramers–Kronig technique, reasonable extrapolations of the experimental reflectance data at low and high frequencies is necessary. In this study, the reflectance data were extrapolated at low energy using the Hagens–Rubens relation [78]. At high energy (6–12 eV), the reflectance spectra were extrapolated with a reflectance $R \propto \omega^{-2}$ (interband) followed by an $R \propto \omega^{-4}$ (free electron) at higher energy (>12 eV).

The microwave conductivity and dielectric constant were measured using the cavity perturbation technique [66,67,80,81]. The resonant cavity used was a cylindrical TM_{010} cavity with a resonant frequency at 6.5 GHz. The whole cavity is inserted into a dewar filled with He gas to provide a temperature range of 4.2–300 K.

To understand the nature of the spins in the polymers, electron paramagnetic resonance (EPR) experiments were carried out using a Bruker ESP 300 spectrometer equipped with a TE_{102} rectangular cavity that resonates at 9.5 GHz (X band). An ESR-900 continuous flow helium cryostat from Oxford Instruments provided temperature control from 4 to 300 K.

IV. POLYANILINE

Polyaniline doped with camphorsulfonic acid has been shown to cross the insulator–metal transition and show metallic behavior at low temperature [53,55,56,101]. The proximity of the samples to the I–M transition can be controlled to some degree by the processing conditions [55,56,60–63,99]. By using the concept of secondary doping [60–63], the electrical behavior and structural order of PAN-CSA can be varied by varying the solvent from which the film is cast [60–63]. Also, the weight percent of the polymer in the solvent can be varied to change the resulting electrical properties of PAN-CSA. In the remainder of this section, the experimental results for the low and high frequency transport near the I–M transition are discussed. For simplicity, the real parts of the optical conductivity and dielectric function are referred to as $\sigma(\omega)$ and $\epsilon(\omega)$, respectively.

A. X-Ray Diffraction

Polyaniline forms a rich set of structures dependent upon the processing sequence and dopant [42,59,80, 81,83,84,102–105]. Generally, polyaniline obtained from solution in the doped (conducting salt) form exhibits a local crystalline order of the emeraldine salt I (ES-I) type. In contrast, polyaniline obtained by doping powder or films cast as the base form from solution are of the ES-II type [42,80,81,83,104,106]. Both preparation methods lead to between a few percent and about 50% crystallinity depending on the details of the processing route. In addition, there are significant differences in the type of local order that exists in the disordered regions between the crystalline ordered regions, varying from liquid benzenelike, to coil-like, to expanded coil-like, to more rodlike [60–63,106,107]. For many undoped and doped polyanilines, short-range local order in the disordered regions resembles that in the ordered regions [59,106].

Comparing the X-ray patterns for PAN-CSA cast from m-cresol with those of PAN-CSA cast from chloroform, there is a drastic difference in the structure. In PAN-CSA cast from m-cresol, the repeat distance along the chain direction is ~9.2 Å [59]. In the PAN-CSA cast from chloroform, the repeat distance along the chain increases to ~14.2 Å, possibly due to the periodicity of a snaking chain that has undergone a "rodlike to coil-like" transition [59]. There is evidence that the benzene rings of the polyaniline tend to align very well with the plane of the free-standing film, though there is less order in the stacking of layers [105]. In the most highly ordered, most conductive samples of PAN-CSA, a diffraction peak occurs at low scattering angles, suggesting a d spacing of ~21 Å [59,108], which may correspond to the distance between 2-D stacks of PAN chains separated by CSA ions [59].

For each of these systems the coherence length within the doped crystallographic regions generally is no more than 50–75 Å along the chain direction, with smaller values in the perpendicular direction. It has been proposed that these coherent crystalline regions form metallic islands and the disordered weak links between more ordered regions are areas where conduction electrons are subject to localization, as expected for charges moving through isolated one-dimensional chains. That is, for each very highly conducting polymer system studied there are regions of one-dimensional electronic character through which conduction electrons must pass [34].

It is pointed out that subtle disorder in the ordered "crystalline" regions can have drastic effects. For example, the percent crystallinity and the size of the crystalline regions of the HCl salt of poly(o-toluidine) (i.e., polymethylaniline) are similar to those of HCl-doped polyaniline, with the important difference of "paracrystallinity" introduced by two possible positions for the CH_3-substituted C_6 rings. That is, a rotation of the ring by 180° about its nitrogen–nitrogen axis changes the structure. The disorder introduced disrupts the interchain interaction, leading to relatively strong localization [21,85,104]. The disorder in HCl-doped poly(o-toluidine) is reduced (and the delocalization increased) through a processing-induced increase in structural order using fiber formation [20].

B. Magnetic Susceptibility

Magnetic susceptibility studies identify the charge storage mechanism at low doping levels as well as the density of states at the Fermi level and the density of localized "Curie" spins at higher dopant levels. The magnitude of χ_{Pauli} depends on the structural order and morphology of the polymers as these affect the uniformity of the doping.

For PAN, $N(E_F)$ is finite and has been shown to increase with the level of protonic acid doping and the volume fraction of crystalline material for both the ES-I and ES-II structures [41,42]. The $N(E_F)$ values differ for ES-I HCl and ES-II HCl, being 0.26 state/eV (C+N), and 0.083 state/[eV·(C+N)], respectively. For highly conducting PAN-CSA (m-cresol) [46], $N(E_F) \sim 0.07$ state/[eV·(C+N)]. Recently, a differently prepared stretched PAN doped with HCl was reported to have a much higher $N(E_F)$, ~ 1.4 state/[eV·(C+N)] [17]. Some solutions of PAN-CSA have been reported to have a Pauli-like susceptibility [109]. It has been proposed that there is primarily spinless bipolaron formation in the disordered regions, especially for ES-II materials [42]. It is also noted that oxygen and moisture can have a significant effect on both the Pauli and Curie susceptibility observed [110,111].

C. DC Conductivity

To establish that a material has undergone an insulator–metal transition, it is necessary to show that σ_{dc} is finite as $T \to 0$ [65] and that the logarithmic derivative of the temperature-dependent conductivity ($W = d \ln [\sigma(T)]/d \ln T$, a generalized activation energy) has a positive temperature coefficient at low temperature [76]. For a conductor very close to the insulator–metal transition, called a critical sample, the resistivity follows a power law behavior with T [92]. The plot of log W vs. log T for a critical sample approaches $T = 0$ K at a constant value, providing a dividing line between the plot of log W vs. log T for insulating hopping behavior, which increases

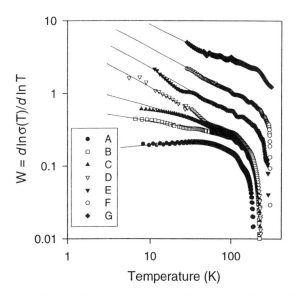

Fig. 3.10 The reduced activation energy $W = d \ln [\sigma(T)]/d \ln T$ for selected PAN-CSA samples.

with decreasing T [i.e., the slope of log W vs. log T is equal to $-\gamma$ if $\sigma \propto \exp (T_0/T)^\gamma$], and the plot of log W vs. log T for metallic samples, which decreases with decreasing T.

The W plots for PAN-CSA samples A–G are shown in Fig. 3.10. The lettering for each sample reflects the relative behavior of the W plot for the various samples. The W plot for sample A has a positive slope at low temperature, indicating that it is metallic at low temperature. Measurements of σ_{dc} for sample A at millikelvin temperature also show a large finite conductivity ($\sigma_{dc} > \sim 70$ S/cm) down to ~ 20 mK (discussed in greater detail in Section IV.D), confirming the presence of delocalized states at the Fermi level as $T \to 0$. Therefore, sample A has crossed the IMT and is metallic. However, the remaining samples show a negative slope at low temperature, varying from -0.2 to -0.7 (Table 3.2), is characteristic of the hopping transport shown by

Table 3.2 Comparison of the DC Transport Properties and Far-IR Reflectance at Room Temperature for PAN-CSA

Sample	σ_{dc}(300 K) (S/cm)	ρ_R[a]	γ[b]	T_M[c] (K)	Reflectance at 6 meV
A	230	1.7	0.08	188	0.8484
B	110	2.9	-0.2	218	0.7991
C	400	4.5	-0.3	224	0.9262
D	120	11	-0.6	240	0.8431
E	70	130	-0.5	288	0.7537
F	20	1.3×10^6	-0.7	>300	0.6915
G	0.7	—	-0.6	>300	0.3133

[a] Resistivity ratio $\rho_R = \rho(4.2 \text{ K})/\rho(300 \text{ K})$.
[b] The parameter γ is the hopping exponent determined from the low temperature W plot.
[c] T_M is the temperature at which the maximum occurs in σ_{dc}.

samples that are insulating at low temperature ($\sigma_{dc}(T) = \sigma_0 \exp[-(T_0/T)^\gamma]$). As W, the reduced activation energy, increases in magnitude, the samples become more insulating and the exponent γ grows. For samples B and C, $\gamma \sim 0.25$, indicating 3-D variable range hopping at low temperature. For samples D–G, $\gamma \sim 0.5$–0.7, suggesting that the localization is now more quasi-1-D. Therefore, as the materials become more insulating, the charge transport becomes more characteristic of hopping on isolated chains with reduced dimensionality (in the disordered regions).

Figure 3.11 shows $\sigma_{dc}(T)$ for each of the samples. Table 3.2 gives $\sigma_{dc}(300\ K)$ and the resistivity ratio $\sigma_{dc}(300\ K)/\sigma_{dc}(5\ K) \equiv \rho_R$ for each sample. A wide range is observed for ρ_R. For samples A–E, $\sigma_{dc}(T)$ has a positive temperature coefficient of resistivity near room temperature [19,34,55,56], a maximum, and then localization effects at lower temperatures. These experimental $\sigma_{dc}(T)$ values are consistent with the predictions of the inhomogeneous disorder model. The temperature at which the maximum in σ_{dc} occurs (T_M) is given for each sample in Table 3.2. The shift of the maximum in PAN-CSA to lower temperature as the materials become more metallic (as gauged by the W plot) is consistent with PAN-CSA sample A having the least disordered conduction paths. Assuming that the materials are homogeneous, sample A is the most ordered material.

Though PAN-CSA sample A shows intrinsic metallic behavior, it does not have the highest $\sigma_{dc}(300\ K)$ value compared to sample C. In fact, samples C and D each show stronger temperature dependences for σ_{dc} than materials with lower $\sigma_{dc}(300\ K)$ (Fig. 3.11). This nonmonotonic behavior of the temperature dependence with increasing $\sigma_{dc}(300\ K)$ contrasts with the monotonic behavior shown in Fig. 3.7 for p-doped germanium, which undergoes an Anderson IMT. This nonmonotonic evolution of $\sigma_{dc}(T)$ is a strong argument against a homogeneous 3-D Anderson IMT. Since PAN-CSA sample A is metallic whereas sample C is insulating at low T, $\sigma_{dc}(RT) > \sigma_{min}$, the minimum metallic conductivity, for sample C. However, sample C becomes insulating at low T.

D. Millikelvin Conductivity

The millikelvin σ_{dc} with no applied magnetic field is shown in Fig. 3.12 along with the high temperature conductivity for samples A, B, and D. Sample A has a large finite $\sigma_{dc} > \sim 70$ S/cm down to 18 mK. This indicates that there are extended states at E_F as $T \to 0$ for PAN-CSA sample A, i.e., PAN-CSA sample A is a metal at low temperature. The remaining samples show more complex behavior at low temperature, though they demonstrate hopping behavior at ~ 10 K. Sample B shows a monotonic decrease in σ_{dc} down to 18 mK. Sample D shows a turnover in σ_{dc} below ~ 200 mK. For sample D σ_{dc} then increases and saturates at lowest temperatures. A similar increase in conductivity at low temperature has been reported for samples of polypyrrole [23–25,112,113], though the turnover was seen at $T \sim 5$–20 K. In those materials, the increase at low temperature was ascribed to a tunneling mechanism [23–25,112]. This complex behavior implies that σ_{dc} does not go to zero as $T \to 0$ for samples B and D, even though σ_{dc} for these samples shows variable range hopping behavior at higher temperatures. Therefore, σ_{dc} does not follow

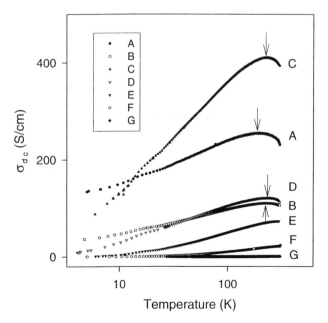

Fig. 3.11 Temperature-dependent dc conductivity for selected PAN-CSA samples.

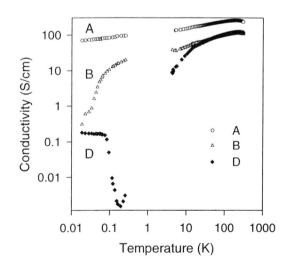

Fig. 3.12 Millikelvin σ_{dc} for selected PAN-CSA samples. The high temperature (4–300 K) σ_{dc} data for each sample are included for comparison.

I–M Transition and Inhomogeneous Metallic State

variable range hopping behavior as $T \to 0$ as predicted for materials with homogeneous disorder and Anderson localization. It is conjectured that the dip in the conductivity may be due to reduced tunneling due to long-wavelength phonon scattering, which affects the polyaniline rings, reducing the overlap integrals and decreasing the wave function overlap. This saturation of σ_{dc} is not expected at low temperature for strongly disordered materials such as PAN-CSA samples F or G and PPy(S-PHE) due to larger separation between metallic regions and shorter localization lengths for chains in the disordered regions.

The behavior of the millikelvin σ_{dc} when a magnetic field of 5 T is applied is shown in Fig. 3.13. For sample A, there is a weak *positive* magnetoconductance ($\Delta\sigma/\sigma \sim 10^{-1}$). This same behavior was reported for doped polyacetylene [36,37,51] and polypyrrole [53], which were highly conducting at millikelvin temperatures. For these samples, a positive magnetoconductance was attributed to magnetic field destruction of weak localization effects [36,37,51] (quantum interference of electronic wave functions leading to standing wave patterns or localization [74,75]). In such a model, the increase in σ_{dc} with increasing T may be seen as a result of activating parallel conduction paths through phonon scattering induced destruction of weak localization in those parallel pathways. Positive magnetoconductance is also predicted for enhanced delocalization on percolating structures due to increased magnetic energy in the metallic regions [114]. Sample B also shows a positive magnetoconductance [$\Delta\sigma(H)/\sigma \sim 10^{1}$] that is larger than that of sample A at low temperatures (~30 mK) and reduced though positive [$\Delta\sigma(H)/\sigma \sim 10^{-1}$] at higher temperatures (~200 mK). The larger $\Delta\sigma(H)/\sigma$ for sample B compared with A may result from paths in series (bottlenecks) with the main conduction path in which weak localization is reduced, resulting in high σ_{dc}. The magnetic field shifts the minimum of the conductivity to higher temperature in sample D, which also shows positive magnetoconductance at the lowest temperatures.

E. Reflectance

For free electron metals such as copper [78], silver [78], and aluminum [115], the reflectance approaches unity at low frequencies and remains high up to frequencies near the conduction electron plasma frequency (Ω_{p1}) for that system. Figure 3.14 shows the room temperature reflectance for samples A–G. At low energy, the reflectance approaches unity for PAN-CSA samples A–D. The value of the reflectance in the far-IR at 6 meV (50 cm^{-1}) scales with σ_{dc} for that sample (Table 2), providing an independent verification of the σ_{dc}(RT) for each sample. At higher energy, the reflectance for samples A–E decreases monotonically with increasing energy to a minimum near ~ 2 eV, the conduction electron plasma frequency. Two distinct peaks are observed at higher energy near 2.8 and 6 eV, attributed to interband transitions within a polaron lattice [116,117].

For samples F and G, there is a strong peak at 1.5 eV, more pronounced in the lowest σ_{dc} sample, G. When the peak at 1.5 eV is strong, the reflectance in the far-infrared is diminished. This indicates that there is a second "localization–delocalization" transition from isolated polarons localized to one or two C_6H_4N repeat

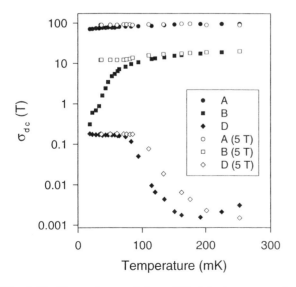

Fig. 3.13 Dependence of the millikelvin transport of selected PAN-CSA samples on applied 5 T magnetic field.

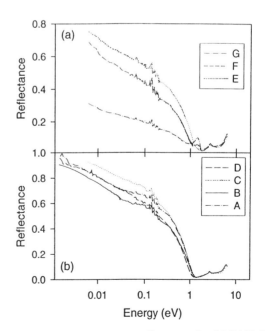

Fig. 3.14 Room temperature reflectance for (a) PAN-CSA samples E–G and (b) PAN-CSA samples A–D.

units to partially delocalized charges that are proposed [71] to extend over distances of ~10–100 Å.

The temperature-dependent reflectance data of PAN-CSA samples A and D in the far-IR are shown in Fig. 3.15. With decreasing T, the reflectance at ~2 meV is decreased relative to the room temperature reflectance while the higher energy (>~4 meV) reflectance initially increases above the room temperature value. At the lowest T, the far-IR reflectance is suppressed over the whole energy range. The difference between the metallic (sample A) and insulating (sample D) materials is the magnitude by which the reflectance is suppressed. For sample A, the far-IR reflectance is suppressed by ~3% from its value at room temperature whereas the suppression is ~10% for sample D. This suppression of the reflectance in the far-IR is consistent with the decrease in σ_{dc} at low T for each of these materials and results from the localization (trapping) at low temperature of carriers that are delocalized at room temperature.

F. Optical Conductivity

Kramers–Kronig analysis of the reflectance data provides the optical conductivity function $\sigma(\omega)$. We can obtain information about the conduction electrons in PAN-CSA by comparing this function with the Drude expression for free electrons and the localization models. The room temperature $\sigma(\omega)$ is shown in Fig. 3.16 for samples A–G. In sample G there is a strong peak at ~1.5 eV and a weaker peak at ~0.6 eV. For

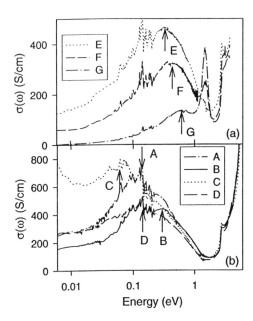

Fig. 3.16 Room temperature optical conductivity $\sigma(\omega)$ for (a) PAN-CSA samples E–G and (b) PAN-CSA samples A–D. The shift of the maximum in $\sigma(\omega)$ indicates the growth of the electron mean free path. The arrows indicate the peak (maximum) in the intraband $\sigma(\omega)$.

sample F, $\sigma(\omega)$ shows that the peak at 1.5 eV is reduced and the peak in the IR has grown and the maximum has shifted to lower energy. For the samples with higher σ_{dc}, the peak at 1.5 eV is absent, indicating that it is not an intrinsic transition in the highly conducting state. Comparison of $\sigma(\omega)$ for samples E–G shows that there is an isosbestic point at ~1.3 eV, characteristic of composite behavior [39]. This directly supports the occurrence of composite behavior in bulk polyaniline.

For samples A–E, $\sigma(\omega)$ shows Drude dispersion with decreasing energy beneath ~1.4 eV until reaching a maximum in the mid-infrared. The oscillations of $\sigma(\omega)$ in the mid-infrared correspond to the vibrational modes of doped polyaniline. $\sigma(\omega)$ is monotonically suppressed beneath the σ_{Drude} at lower energy except for sample C. For sample C, $\sigma(\omega)$ begins to increase again with decreasing energy beneath 0.02 eV. Though the increase of $\sigma(\omega)$ in the far-IR is not as rapid, the behavior is qualitatively similar to what was reported for iodine- [72,73,118] and perchlorate- [119] doped $(CH)_x$ where free electrons are reported. This type of frequency behavior for $\sigma(\omega)$ is qualitatively similar to that of metallic particles in an insulating matrix after percolation of the metallic particles [88]. It is indicative of a small plasma frequency of conduction electrons that are delocalized through the material. Also, Fig. 3.16 demonstrates that the far-IR $\sigma(\omega)$ scales with σ_{dc}, confirming the measured σ_{dc} values.

As σ_{dc} of the samples increases, the frequency at

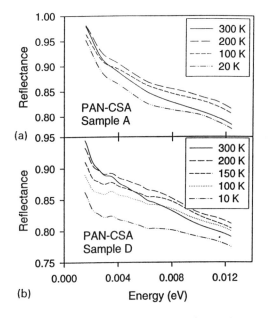

Fig. 3.15 Temperature dependence of the far-infrared (10–100 cm^{-1}) reflectance for (a) PAN-CSA sample A and (b) PAN-CSA sample D.

which the maximum in $\sigma(\omega)$ occurs (ω_{max}) shifts monotonically to lower energy (Table 3.3). This shift of the conductivity peak to low energies is the same behavior the localization-modified Drude model displays as the mean free time (and therefore the mean free path) grows and the behavior the inhomogeneous disorder model predicts as the samples near the percolation threshold. Since ω_{max} is lowest for PAN-CSA sample C, this indicates that the scattering time (mean free path) for sample C is the longest and therefore the disorder is weakest. Assuming a homogeneous material, sample C is inferred to be the most ordered sample. Recalling that the maximum in $\sigma_{dc}(T)$ implied that sample A is the most ordered, a contradiction is obtained from assuming that the materials are homogeneous. Within the inhomogeneous disorder model, there is no contradiction. This is because $\sigma_{dc}(T)$ reflects the behavior of percolated carriers in well-ordered conduction paths while the high frequency $\sigma(\omega)$ is dominated by the contribution of localized carriers, which contribute less than 10 S/cm to σ_{dc}, in disordered regions of the sample.

Drude conductivity fits to the energy range ~0.4–1.4 eV provide further information about the carriers. The plasma frequency (Ω_{p1}) and localized carrier scattering time (τ_1) for each of the samples (Table 3.3) fall into the range $\Omega_{p1} \sim 1.7$ eV and $\tau_1 \sim 10^{-15}$ s. The small scattering time ($\tau_1 \sim 10^{-15}$ s) is attributed to the disorder in the polymer. This scattering time is the magnitude predicted from the Ioffe–Regel condition for materials close to a 3-D Anderson IMT. However, the Anderson IMT model is not appropriate for these materials. Analysis of the dielectric function $\epsilon(\omega)$ (Section IV.G) demonstrates that $\sigma_{dc}(T)$ in conducting polymer systems is due to a small fraction of carriers with $\tau > \sim 10^{-13}$ s that percolate through the material. The increase in $\sigma(\omega)$ for sample C in the far-infrared is evidence of the percolated free electrons.

The "average" effective mass (m^*) of the conduction electrons can also be calculated from knowledge of

Table 3.3 Comparison of the DC and High Frequency Conductivity Parameters for Selected PAN-CSA Samples

Sample	$\sigma_{dc}(300\ K)$ (S/cm)	Ω_p[a] (eV)	τ[a] (10^{-15} s)	ω_{max}[b] (eV)
A	240	1.7	1.1	0.09
B	110	1.7	0.9	0.3
C	400	1.6	1.3	0.06
D	120	1.5	1.0	0.14
E	70	2.0	0.7	0.33
F	20	1.8	0.6	0.37
G	0.7	1.3	0.4	0.7

[a] The values for Ω_{p1} and τ_1 are obtained by fitting the Drude model to the optical conductivity near the plasma edge.
[b] ω_{max} represents the frequency where $\sigma(\omega)$ shows a maximum (ignoring phonon features).

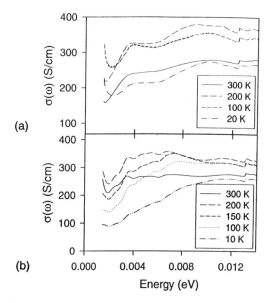

Fig. 3.17 Temperature dependence of the far-infrared optical conductivity in (a) PAN-CSA sample A and (b) PAN-CSA sample D.

Ω_{p1} ($= \sqrt{4\pi n e^2/m^*}$). From X-ray measurements [59], the volume of a 2-PAN ring repeat with a CSA counterion is $V \sim 760$ Å^3 assuming a unit cell with dimensions 10.4 Å (the repeat length along the chain), 3.5 Å, and 21 Å (which represents the repeat distance between alternating stacks of PAN and CSA). Since there is one doped charge added per two-ring repeat of PAN-CSA, $n = 1/760$ Å$^3 \sim 1.3 \times 10^{21}$ cm^{-3}. Therefore, $m^* \sim 0.6 m_e$ (m_e is the electron mass).

Temperature-dependent far-IR $\sigma(\omega)$ values for samples A and D are shown in Fig. 3.17. For both samples, the trend is clear. As the temperature is lowered, $\sigma(\omega)$ first increases and then at lower temperatures is suppressed. The resulting suppression for sample D is much stronger than for sample A, in agreement with the trend in σ_{dc}. The growth and then decrease of $\sigma(\omega)$ is consistent with the localization of charges at low temperature that are delocalized at room temperature. Some of the free electrons contributing to the Drüde plasma frequency are being frozen out at low (zero) frequency. Therefore, due to sum rules on $\sigma(\omega)$ [78], the transitions of the "frozen out" electrons must occur at higher frequency. From this point of view, sample D becomes insulating (hopping) at low temperature because the density of free electrons goes to zero at low temperatures. Sample A remains metallic at low temperatures because the free electron density is not sufficiently reduced.

G. Optical Dielectric Function

The dielectric function determined by Kramers–Kronig analysis provides information about whether the carriers are free or localized. The importance of localization is clear from $\sigma(\omega)$. As σ_{dc} increased, the mean free path clearly increased. With increasing electron delocalization, a larger polarization of the electron gas is possible. The far-IR $\epsilon(\omega)$ for samples E–G (Fig. 3.18a) is positive in the far-IR, characteristic of localized carriers. The growth of the far-IR $\epsilon(\omega)$ directly demonstrates the growing polarization of localized states with increasing σ_{dc}. For sample G, $\epsilon(\omega)$ is small and positive in the far-IR with strong dispersion near the "localized polaron" peak at 1.5 eV. The growth of the average scattering time and mean free path is also made evident by the development of a zero crossing at ~1 eV. This zero crossing is termed ω_{p1} (the screened plasma frequency Ω_{p1}). When the mean free path grows, $\epsilon(\omega)$ shows Drude dispersion to lower frequencies, so $\epsilon(\omega)$ crosses zero at the screened plasma frequency of all the conduction electrons, even though localization corrections force $\epsilon(\omega)$ positive at lower frequencies. This behavior is reminiscent of the $\epsilon(\omega)$ for the localization-modified Drude model of Section II.B. The zero crossing of $\epsilon(\omega)$ at ω_{p1} is therefore due to localized electrons.

However, the low frequency electrical response is not dominated by localized carriers for samples A–D near the insulator–metal transition. Figure 3.18b shows $\epsilon(\omega)$ for samples A–D. In addition to the dielectric reponse of localized electrons at ~1 eV, a third zero crossing of $\epsilon(\omega)$ termed ω_p is evident in the far-IR. For energies lower than ω_p, $\epsilon(\omega)$ remains negative as expected for free carriers. If the carriers are free electrons, then their frequency response is described by the Drude model.

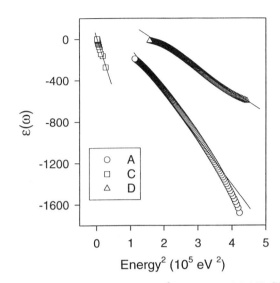

Fig. 3.19 $\epsilon(\omega)$ plotted versus $1/\omega^2$ for selected PAN-CSA samples at room temperature. The linearity confirms that the low energy carriers are free (Drude) carriers.

At sufficiently high frequencies ($\tau \gg 1/\omega$), the Drude dielectric function is given by

$$\epsilon_{\text{Drude}}(\omega) = \epsilon_b - \Omega_p^2/\omega^2 \quad (16)$$

where ϵ_b is the background dielectric response due to localized electrons and interband transitions and Ω_p is the plasma frequency for free carriers. Figure 3.19 shows $\epsilon(\omega)$ plotted against $1/\omega^2$ for samples A, C, and D. [The plot for sample B is not shown because, though $\epsilon(\omega)$ is clearly turning toward negative values, the zero crossing is not observed in the experimental frequency range. For this sample, we do not have low enough frequency data where $\epsilon(\omega)$ is dominated only by free electrons.] The plots provide straight lines where the slope is Ω_p^2. The value of Ω_p for each of these samples is given in Table 3.4. The frequency response of $\epsilon(\omega)$ for PAN-

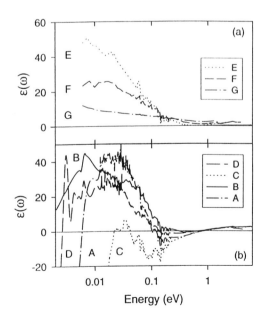

Fig. 3.18 The real part of the dielectric response $\epsilon(\omega)$ at room temperature for (a) PAN-CSA samples E–G and (b) PAN-CSA samples A–D.

Table 3.4 Free Electron Plasma Frequency Ω_p and Scattering Time τ Versus Temperature for Selected PAN-CSA Samples

Sample	T (K)	σ_{dc}(300 K) (S/cm)	Ω_p (eV)	$n_{\text{free}}/n_{\text{cond}}$[a] ($10^{-3}$)	τ[b] (10^{-13} s)
A	300	230	0.07	1.2	2.3
	200	250	0.08	1.6	2.0
	100	240	0.06	0.9	2.9
	20	170	0.05	0.6	3.4
C	300	400	0.11	3.0	1.6
D	300	120	0.04	0.4	3.7
	200	120	0.03	0.2	6.6

[a] Estimated fraction of the conduction electrons that are free.
[b] Estimated scattering time.

CSA samples A–D (i.e., the response of both localized carriers and delocalized carriers with long τ) is typical of percolating systems [88,89].

From Table 3.4, it is clear that ω_p (the plasma frequency of free electrons) scales with σ_{dc} for each sample. This supports the notion that the high dc conductivity and low frequency transport are controlled by the delocalized "free" carriers. Also, the scaling of Ω_p with σ_{dc} argues against the Anderson transition model for conducting polymers. Because sample C is insulating at low T within the Anderson model, the Fermi level for sample C lies in the region of localized (hopping) states. However, sample C has a larger free electron plasma frequency (Ω_p) at room temperature than sample A, which is metallic at low T and therefore has its Fermi level in the region of delocalized (free electron) states. In addition, the scaling of Ω_p with σ_{dc} argues against the negative $\epsilon(\omega)$ being due to the formation of an intrinsic many-body gap as proposed for polypyrrole materials [120], especially since the negative far-IR $\epsilon(\omega)$ is observed at room temperature in PAN-CSA samples.

The values obtained for Ω_p are very small compared to the full conduction electron density ($\Omega_{p1} \sim 2$ eV), suggesting that only a small fraction (n_{free}/n_{cond}) of the conduction electrons are delocalized macroscopically. This small fraction of delocalized carriers is consistent with the small number of percolation paths that occur close to the percolation threshold in composite systems. The fraction of the carriers that are delocalized can be estimated by comparing the plasma frequency of free electrons (Ω_p) with the full conduction electron plasma frequency (Ω_{p1}),

$$\frac{n_{free}}{n_{cond}} = \frac{m^*_{free}}{m^*_{cond}} \left(\frac{\Omega_p}{\Omega_{p1}}\right)^2 \quad (17)$$

This ratio is estimated in Table 3.4 for samples A, C, and D, assuming that the effective masses (m^*_{free} and m^*_{cond}) are approximately the same. In each case,

$$n_{free}/n_{cond} \sim 10^{-3} \quad (18)$$

Even assuming a tenfold increase in m^*_{free} as the free carriers may reside in a very narrow band (due to passage through the disordered regions), the fraction of delocalized carriers is only on the order of 1%. That the current is carried by a small number of percolated paths is not surprising, as high conductivities (>10 S/cm) have been reported for PAN-CSA diluted to only 10% in insulating PMMA [121].

Because the plots of $\epsilon(\omega)$ vs. $1/\omega^2$ in Fig. 3.19 do not show a tendency to saturate down to 10 cm^{-1}, the mean free time for the free electrons (τ) can be estimated as

$$\tau \gg 1/(10 \text{ cm}^{-1}) \sim 5 \times 10^{-13} \text{ s} \quad (19)$$

Such a huge mean free time in a disordered polymer is surprising, especially when compared with the typical room temperature scattering times for copper, $\tau \sim$ 10^{-13}–10^{-14} s [35]. This small fraction is delocalized sufficiently that $k_F\lambda \gg 1$. Similar values for τ have been reported previously in doped polyacetylene [73]. The large scattering time likely results from the ramifications of an open Fermi surface as expected for an anisotropic metal [19,24,34,86], and that the usual scattering centers, being off the polymer chains, are likely screened, reducing their effectiveness. If the whole density of conduction electrons were able to diffuse with $\tau \sim 10^{-13}$ s, the dc conductivity would be $\sim 10^5$ S/cm. Using this estimate, it is suggested that substantial improvements in the electrical conductivity can still be obtained.

It is also possible to estimate τ within the Drude model where

$$\sigma_{dc} = \Omega_p^2 \tau / 4\pi \quad (20)$$

Since the free electron plasma frequency Ω_p scales with σ_{dc}, the scattering time is estimated as $\tau \sim 10^{-13}$ s for samples A, C, and D (Table 3.4), assuming that only the free electrons participate in the low frequency transport. This estimate of the free electron scattering time is in good agreement with the previous estimate. This large scattering time ($\tau \gtrsim 10^{-13}$ s) compares with $\tau \sim 10^{-14}$ s at room temperature in copper [122]. The presence of carriers with $\tau \gg 10^{-15}$ s indicates that the low frequency carriers are not subject to the Ioffe–Regel condition, which requires that $\tau \sim 10^{-15}$ s in conducting polymers. This constitutes further evidence against the applicability of the Anderson IMT model.

The temperature dependence of the far-IR $\epsilon(\omega)$ explicitly demonstrates the essential difference between metallic and insulating samples. The temperature-dependent $\epsilon(\omega)$ for sample A (Fig. 3.20a) shows the crossover to negative values and the Drude dispersion in the far-IR. This free carrier dispersion is present in $\epsilon(\omega)$ from 300 K down to ~ 20 K. The presence of free carriers down to low T is consistent with the presence of high σ_{dc} down to millikelvin T. In contrast, the temperature-dependent $\epsilon(\omega)$ for sample D is shown in Fig. 3.20b. As the temperature is lowered to ~ 10 K and σ_{dc} drops, $\epsilon(\omega)$ crosses from negative to positive in the far-infrared. Therefore, the free carriers become localized at low temperature in PAN-CSA sample D. It is proposed that the localization at low T is due to the ineffectiveness of phonon-assisted delocalization in the disordered regions at low T [38]. The localization of the free carriers in PAN-CSA sample D at low T is consistent with the strong decrease of σ_{dc} at low T. Therefore, the difference between sample A, which remains metallic, and sample D, which becomes insulating at low T, is the presence of percolated free carriers down to low temperatures. The values for the low temperature Ω_p and τ [shown only for those temperatures where Drude dispersion is evident in $\epsilon(\omega)$] determined from Drude fits to σ and ϵ [40] are shown in Table 3.4. The presence of free electrons at room temperature in insulating sample D indicates that percolation behavior can occur at high T even

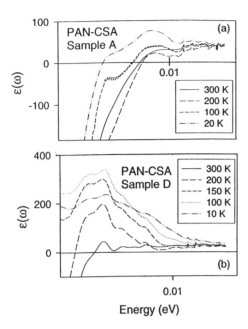

Fig. 3.20 Comparison of the temperature dependence of the dielectric response $\epsilon(\omega)$ for (a) metallic PAN-CSA sample A and (b) insulating PAN-CSA sample D. The value of the abscissa at the left hand axis is 0.002 eV.

though at low T the composite is not percolated. This is proposed to reflect the importance of phonon-induced delocalization on the electronic localization length (L_{loc}) in the disordered regions, which electronically couple "crystalline" metallic regions in inhomogeneous conducting polymers [24,34,39,79].

Figure 3.21a shows a plot of $\epsilon(\omega)$ vs. $1/\omega^2$ at various temperatures for sample A. The slope varies with temperature, indicating that Ω_p is temperature-dependent. This is further evidence that the insulator-metal transition is due to a loss of free carrier density $n(T)$. Figure 3.21b shows directly that the free carrier plasma frequency Ω_p approximately scales with σ_{dc} at each temperature for metallic sample A. This behavior contrasts with the case in conventional metals where the temperature dependence of σ_{dc} is determined solely by the temperature dependence of the scattering time. Assuming that $\sigma_{dc}(T) = \Omega_p^2 \tau / 4\pi$, the scattering time actually increases at low T (Table 3.4), as in conventional metals. Therefore, the temperature dependence of σ_{dc} in inhomogeneous metallic conducting polymers must be determined primarily by $n(T)$. Within the Anderson model, lowering the temperature would result in depopulation of more extended states above the mobility edge and therefore a decrease in the scattering time. The contradiction of the Anderson model prediction and the trend in the data argue further against the Anderson IMT. The increase in τ at low T may instead reflect the more robust percolation paths that remain at low T as the less robust ones "freeze out."

Figure 3.22 shows the dielectric function parallel and perpendicular to the stretch direction for HCl-doped, stretched PAN obtained from polarized reflectance measurements [71]. There is a strong anisotropy evident in the electrical response. Parallel to the stretch direction, the high energy (~1 eV) zero crossing of $\epsilon(\omega)$ is observed as well as strong dispersion in the IR. Perpendicular to the stretch direction, $\epsilon(\omega)$ is positive over the whole energy range, showing very weak dispersion. The anisotropy present along the chain direction must be accounted for within the model of the insulator-metal transition. The homogeneous Anderson model that has been employed in the literature [96,116,120] assumes that the material is isotropic. The fact that the anisotropic nature of the polymer transport is not included in the framework of the homogeneous Anderson IMT model also rules out this model for the conducting polymer IMT.

In summary, $\epsilon(\omega)$ is characteristic of strongly localized electrons ($\tau \sim 10^{-15}$ s) for samples E-G far from the I-M transition. For samples A-D near the I-M transition, a small fraction of the carriers become macroscopically delocalized ($\tau \gtrsim \sim 10^{-13}$ s) while the majority of carriers are still strongly localized ($\tau \sim 10^{-15}$ s). The difference between metallic sample A and insulating sample D is that the free carriers freeze out in sample D at low T. This behavior is characteristic of percolating systems. Unlike the usual percolating systems, the per-

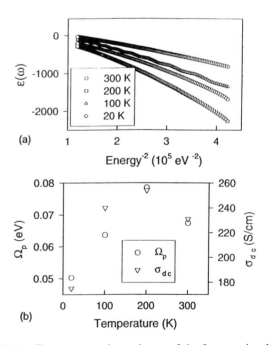

Fig. 3.21 Temperature dependence of the free carrier dielectric response for metallic PAN-CSA sample A. (a) Far-infrared $\epsilon(\omega)$ plotted against $1/\omega^2$ as a function of temperature. (b) Comparison of the temperature dependence of the free electron plasma frequency Ω_p and σ_{dc}.

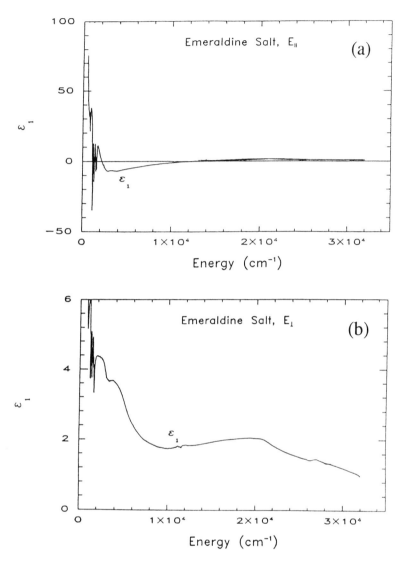

Fig. 3.22 Real part of the dielectric constant versus energy for light polarized (a) parallel and (b) perpendicular to the stretch direction for PAN doped with HCl. (From Ref. 71.)

colation for doped polymers can be temperature-dependent. The dielectric response of oriented materials clearly points out the importance of dimensionality in the electronic behavior.

H. Discussion of Conductivity and Dielectric Functions

The dielectric function and optical conductivity provide insight into the nature of the disorder in the metallic state. In this section, $\epsilon(\omega)$ and $\sigma(\omega)$ for the PAN-CSA samples are compared with the localization-modified Drude model for homogeneously disordered systems and a model for inhomogeneous disorder.

For the homogeneously disordered system, the mean free time is limited to a very short time owing to substantial disorder. For materials near the insulator–metal transition, $k_F\lambda \sim 1$ [24,95,96,116]. Due to the limitation of λ, $\sigma(\omega)$ is suppressed and $\epsilon(\omega)$ is driven positive at low energy. The frequency dependence of σ and ϵ in this model is described by the localization-modified Drude model [65,95,96,116], Section II.B, with a short scattering time.

A typical fit of the localization-modified Drude model is shown in Fig. 3.23 for sample E. To ensure that causality is satisfied, the parameters were chosen (Table 3.5) to describe both $\sigma(\omega)$ and $\epsilon(\omega)$. The experimental $\sigma(\omega)$ and $\epsilon(\omega)$ are well represented for PAN-CSA sample E. It is important, however, to determine whether the parameters obtained are reasonable. The values obtained for the localized carrier scattering time τ_1 are compara-

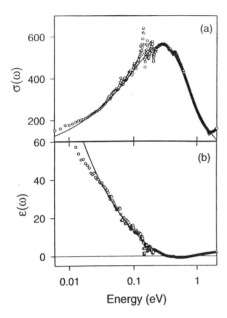

Fig. 3.23 Typical comparison of the localization-modified Drude model fits to the experimental data. The data shown are (a) $\sigma(\omega)$ and (b) $\epsilon(\omega)$ for PAN-CSA sample E.

ble to the values obtained from the Drude fits to $\sigma(\omega)$ (Table 3.3), confirming the short scattering time ($\tau_1 \sim 10^{-15}$ s). However, the "full" carrier plasma frequency Ω_{p1} is small compared to what was found for the Drude fits (Table 3.3), suggesting a sizable difference in oscillator strength between the two models. Integration of $\sigma(\omega)$ for the localization-modified Drude model shows that

$$8\int_0^\infty \sigma_{\text{LMDM}}(\omega)d\omega = \Omega_{p1}^2 \left[1 + C/(k_F v_F \tau_1)^2\right] \quad (21)$$

so that the oscillator strength is greater in the localization-modified Drude model than just Ω_{p1}^2 as in the unmodified Drude model [35,78]. For sample F, the full plasma frequency calculated using Eq. (21) and the pa-

Table 3.5 Fit Parameters for PAN-CSA Using the Localization-Modified Drude Model

Sample	Ω_{p1} (eV)	τ_1 (10^{-15} s)	$C/(k_F v_F)^2$ (10^{-30} s^2)	ϵ_{inf}
A	1.2	2.0	3.1	2.8
B	1.3	1.4	1.6	2.6
C	1.3	2.4	1.6	3.6
D	1.2	1.6	1.4	2.8
E	1.4	1.2	1.4	2.5
F[a]	1.2	1.0	1.1	2.2
G[a]	0.9	0.7	0.6	1.8

[a] An additional Lorentzian oscillator was included for samples F and G to model the 1.5 eV localized polaron peak.

rameters from Table 3.5 is ~2.0 eV, reconciling the oscillator strength differences between the Drude model and localization-modified Drude model estimates of the oscillator strength.

The parameter $C/(k_F v_F)^2$ can be used to estimate the Fermi velocity v_F, the mean free path ($= v_F\tau$), and the one-dimensional density of states [$N(E_F) = 2/\pi\hbar v_F$] [96]. To make the estimates, C is assumed to be of order unity [74,75,95,96,116]. Assuming that the sample is fully doped, k_F has the value $\pi/2c$ with $c \sim 10.2$ Å [59]. Using $\tau \sim 1.2 \times 10^{-15}$ s and $C/(k_F v_F)^2 \sim 1.4 \times 10^{-30}$ s^2, $v_F \sim 5 \times 10^7$ cm/s and $\lambda = v_F\tau \sim 7$ Å. Using these parameters, $k_F\lambda \sim 1$. This value of $k_F\lambda$ is in accord with the Ioffe–Regel condition for conductors close to an insulator–metal transition because the mean free path is of the same size as a nitrogen and benzene ring repeat in PAN (~5.2 Å). The prediction for the one-dimensional density of states at the Fermi level, $N(E_F) \sim 2$ state/eV per formula unit (two nitrogens and two rings), is also in reasonably good agreement with the measured value of $N(E_F) \sim 1$ state/eV per formula unit [46]. The consistency of these estimates obtained from the localization-modified Drude model indicates that the high frequency behavior of samples similar to E (dominated by localized excitations) is reasonably represented within a homogeneous Anderson transition picture.

The fits for $\sigma(\omega)$ and $\epsilon(\omega)$ of PAN-CSA samples E–G using the parameters in Table 3.5 are shown in Figs. 3.24 and 3.25. For samples F and G, $\sigma(\omega)$ and $\epsilon(\omega)$ are complicated by the presence of localized polarons as well. Since the localized polaron feature occurs in a nar-

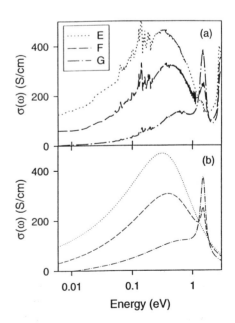

Fig. 3.24 (a) Experimental $\sigma(\omega)$ compared with (b) localization-modified Drude model fits to $\sigma(\omega)$ for PAN-CSA samples E–G.

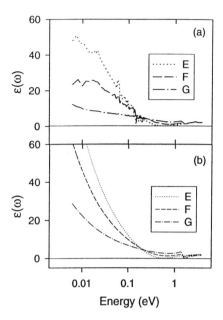

Fig. 3.25 (a) Experimental $\epsilon(\omega)$ compared with (b) localization-modified Drude model fits to $\epsilon(\omega)$ for PAN-CSA samples E–G.

row frequency window (~1.5 eV), it can be modeled by including a Lorentzian function. For sample F, the Lorentzian parameters were $\Omega_p = 0.8$ eV, $\gamma = 4.4 \times 10^{14}$ Hz, and $\omega_0 = 1.5$ eV, while $\Omega_p = 1.3$ eV, $\gamma = 5.4 \times 10^{14}$ Hz, and $\omega_0 = 1.5$ eV were used for PAN-CSA sample G.

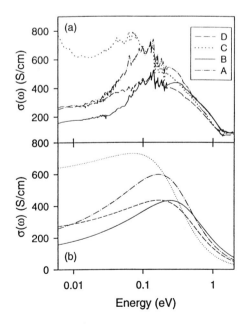

Fig. 3.26 (a) Experimental $\sigma(\omega)$ compared with (b) localization-modified Drude model fits to $\sigma(\omega)$ for PAN-CSA samples A–D.

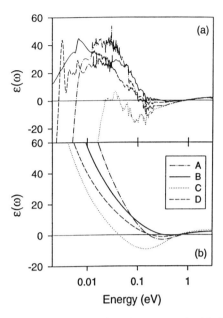

Fig. 3.27 (a) Experimental $\epsilon(\omega)$ compared with (b) localization-modified Drude model fits to $\epsilon(\omega)$ for PAN-CSA samples A–D.

One set of experimental trends is well represented by the localization-modified Drude model. τ_1 increases with increasing σ_{dc}, suggesting that the disorder decreases, consistent with the shift of ω_{max} in $\sigma(\omega)$. Samples A–D deviate in an important way from the predictions of the localization-modified Drude model in the far-IR (Figs. 3.26 and 3.27). Figure 3.26 shows that $\sigma(\omega)$ for PAN-CSA samples A–D can be reasonably modeled except for the upturn in $\sigma(\omega)$ for sample C at ~0.02 eV. On the other hand, Fig. 3.27 demonstrates that the localization-modified Drude model cannot account for the free carrier behavior in $\epsilon(\omega)$ at low frequency. This is a consequence of the assumption within a homogeneous material that all of the carriers have short mean free times ($k_F v_F \tau_1 \sim 1$) limited by strong disorder. It is proposed that this model explains well the behavior of only the relatively strongly localized electrons ($k_F \lambda \sim 1$) in conducting polymers with reasonable parameters.

The early studied PAN-CSA [96] and PPy doped with perchlorate [123] both had $\sigma_{dc} \sim 100$ S/cm, indicating that the carriers were reasonably localized. Therefore, the agreement of the optical properties with the localization-modified Drude model (with $\tau \sim 10^{-15}$ s) is expected. However, this model is unable to account for the free electron behavior observed in higher σ_{dc} samples because Drude dispersion requires that $k_F \lambda \gg 1$.

The necessity of treating the highly conducting polymers within a composite picture with inhomogeneous disorder is clear from an estimate of the mean free paths for the percolated electrons. Using the same Fermi velocity estimated from the localization-modified Drude model ($v_F \sim 5 \times 10^7$ cm/s), the mean free path for free electrons in PAN-CSA is estimated as $\lambda_{\text{free}} \sim 10^3$ Å, with $k_F \lambda_{\text{free}} \sim 10^2$ using τ determined from the infrared measurements. (As noted earlier, band narrowing due to disorder will increase m^* and decrease estimates of λ_{free}.) This estimate of λ_{free} is consistent with previous estimates of the mean free time in doped polyacetylene [73,119] and the inelastic length obtained from magnetoresistance experiments [101,124]. Since $\lambda_{\text{free}} \gg \xi$, the crystalline coherence length (~ 50 Å), these free electrons are capable of diffusing large distances among ordered regions between scattering events.

In contrast to the macroscopically homogeneous Anderson transition, effective medium models allow for macroscopic inhomogeneity and composite behavior. Therefore, there can be percolated metallic regions of the sample where the mean free time is large. For composite systems, the disorder is macroscopic and there may be two very different phases. Most calculations within the effective medium theory are for composites of an insulator and a metal [88,89]. For the conducting PAN and PPy samples, the different parts of the composite are proposed to be the more ordered (metallic) regions and the disordered regions where localization becomes important. The percolating behavior of conducting polymers is expected to be modified from traditional insulator–metal composites as the ordered and disordered regions do not have sharp boundaries; a single polymer chain may be a part of both ordered and disordered regions. Also, phonon-assisted transport occurs in the disordered regions. Nevertheless, there are many qualitative similarities with the insulator–metal composite. Figure 3.5 shows the behavior of $\sigma(\omega)$ for a system of metal particles percolating in an insulating matrix. At low volume fractions before percolation, the Drude peak in $\sigma(\omega)$ is suppressed and shows up as a localized plasmon excitation at higher energy [88,89]. The maximum in this localized plasmon band in $\sigma(\omega)$ shifts to lower frequency as the materials near percolation. Above percolation, the Drude peak is observed at low frequency, though the conductivity is only a fraction of the conductivity of the bulk metal (alternatively, the Drude peak contains only a fraction of the density of carriers). The fraction depends on the proximity of the system to the percolation threshold. Though it was not calculated in Ref. 88, the appearance of a Drude peak in $\epsilon(\omega)$ is required by causality when there is a Drude peak in $\sigma(\omega)$.

For insulating PAN-CSA samples E–G, $\epsilon(\omega)$ shows only localized excitations. The peak in $\sigma(\omega)$ shifts to lower frequency with increasing σ_{dc}, consistent with the behavior of a localized plasmon band in a composite system [88,89]. For metallic PAN-CSA samples A–D, $\epsilon(\omega)$ has a large plasma frequency, ω_{p1}, for electrons with localized behavior and a small plasma frequency for free electrons. The fraction of the free carriers that percolate was estimated as $\sim 10^{-2}-10^{-3}$. This value may seem small for percolation, but the volume fraction required for percolation depends upon the aspect ratio of the percolating item [88,89]. Conducting polymers have already been shown to percolate in insulating polymers at volume-filling fractions beneath 1% [121,125]. The experimental $\sigma(\omega)$ for PAN-CSA sample C demonstrates an upturn at low energy (~ 0.02 eV) similar to the percolating Drude peak in Fig. 3.5. Therefore, both the insulating and metallic samples are consistent with the expectations for a composite system.

The presence of two electron gases, one of which is strongly localized ($\tau \sim 10^{-15}$ s) and the other of which has percolated, showing Drude dispersion and a long mean free time ($\tau \sim 10^{-13}$ s), can be taken into account qualitatively by introducing a distribution function for scattering times into the localization-modified Drude model. This distribution function is strongly peaked with $k_F \lambda \sim 1$ but with a small tail with larger $k_F \lambda$. Such a distribution of scattering times assumes that there are some paths in the material that are more ordered so that τ is longer, taking into account the inhomogeneous nature of a polymer solid as determined from X-ray experiments [59,83]. For such a case, we assume for convenience that $\sigma_{\text{inhomo}}(\omega)$ and $\epsilon_{\text{inhomo}}(\omega)$ are given by

$$\sigma_{\text{inhomo}}(\omega) = \int_0^\infty P(\tau) \sigma_{\text{LMDM}}(\omega, \tau) \, d\tau \tag{22}$$

and

$$\epsilon_{\text{inhomo}}(\omega) = \int_0^\infty P(\tau) \epsilon_{\text{LMDM}}(\omega, \tau) \, d\tau \tag{23}$$

where $P(\tau)$ is the distribution function for the scattering times.

For analytical simplicity, we consider the representative distribution function

$$P(\tau) = \frac{2\Delta}{\pi} \left[\frac{\tau^2}{(\tau^2 - \tau_0^2)^2 + \tau^2 \Delta^2} \right] \tag{24}$$

where Δ describes the width of the spread in scattering times and τ_0 is the average scattering time. The resulting functions were fit to $\epsilon(\omega)$ and $\sigma(\omega)$ for sample C. The plots, shown in Fig. 3.28, indicate that by including a distribution that allows for carriers with long scattering times, both the localized ($\tau \sim 10^{-15}$ s) and Drude carriers ($\tau \sim 10^{-13}$ s) can be roughly fit. The parameters used were $\Omega_p = 1.0$ eV, $\tau_0 = 2.7 \times 10^{-15}$ s, $C/(k_F v_F)^2 = 11 \times 10^{-30}$ s^2, and $\Delta = 1.3 \times 10^{-15}$ s. The behavior of these model functions as the width of the distribution function is varied is shown in Fig. 3.29, which shows the crossover in behavior from insulating (with only localized carriers) to metallic (with some delocalized electrons). An increase in the width of spread of scattering times (Δ) is equivalent to having more carriers with long mean free times. In other words, an increase in Δ places

I–M Transition and Inhomogeneous Metallic State

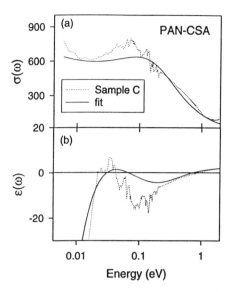

Fig. 3.28 Typical fit of the localization-modified Drude model with a distribution of scattering times to the experimental data for PAN-CSA sample C. The plot shows (a) $\sigma(\omega)$ and (b) $\epsilon(\omega)$.

the sample further above percolation so that there are more ordered paths where free electron diffusion is possible.

Qualitatively, $\sigma(\omega)$ and $\epsilon(\omega)$ for PAN samples E–G show the behavior expected for the good conductor–poor conductor composites as well as for the 3-D localization-modified Drude model with only short scattering times (corresponding to the Ioffe–Regel criterion). $\sigma(\omega)$ and $\epsilon(\omega)$ of highly conducting PAN-CSA samples A–D show Drude behavior for a small fraction of the conduction electrons that essentially percolate through the film while the remaining conduction electrons are more localized, showing the behavior expected only within the composite picture. Therefore, $\sigma(\omega)$ and $\epsilon(\omega)$ indicate that the metallic state in conducting PAN-CSA and PPy(PF$_6$) is inhomogeneous. This same experimental behavior has been reported for iodine [72,73,118] and perchlorate [119] doped polyacetylene and is shown for doped PPy in Section V.

I. Microwave Frequency Dielectric Constant

The microwave frequency dielectric constant (ϵ_{MW}) is a key probe of the delocalization of charge carriers. For delocalized Drude electrons at frequencies less than their plasma frequency, the real part of the dielectric function $\epsilon(\omega)$ is negative due to the inertia of the free electron in an alternating current field [35,78,93,122]. For a localized carrier, the charges can stay in phase with the field and $\epsilon(\omega)$ is positive at low frequencies. Thus, the sign of the microwave dielectric constant serves as a sensitive probe of the presence of free elec-

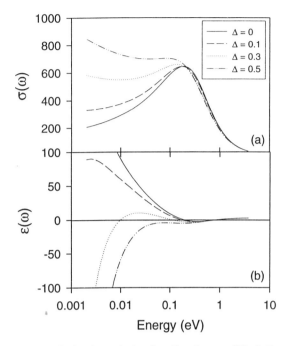

Fig. 3.29 Behavior of the localization-modified Drude model with a spread in scattering times as the width of the distribution changes. Results are shown for (a) $\sigma(\omega)$ and (b) $\epsilon(\omega)$. The parameters other than Δ are for PAN-CSA sample E, Table 3.5. The units for Δ are 10^{-15} s.

trons and provides independent verification of the infrared results.

Figure 3.30 shows ϵ_{MW} for selected samples. For PAN-CSA samples E and F, ϵ_{MW} is positive from 300 K down to ~4 K, indicating the importance of localiza-

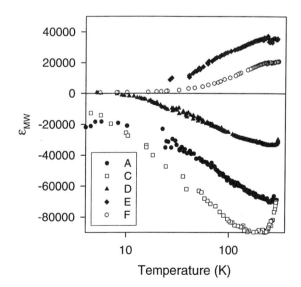

Fig. 3.30 Temperature-dependent microwave (6.5 GHz) dielectric constant for selected PAN-CSA samples. ϵ_{MW} for sample C was reported in Ref. 19.

tion in these materials at all temperatures. This also agrees well with the infrared $\epsilon(\omega)$ for samples E and F, which were positive in the far-infrared, characteristic of localized carriers. For samples A, C, and D, ϵ_{MW} is negative at room temperature, providing an independent confirmation of the free electron behavior observed in the infrared. For sample A, ϵ_{MW} remains negative down to ~4 K, demonstrating independently of the infrared data that the metallic state in conducting polymers is accompanied by free (Drude) carriers. Also, the trends observed in $\sigma_{dc}(T)$ for samples A and C are observed in $\epsilon_{MW}(T)$. Near room temperature, ϵ_{MW} for sample C is metallic (negative), with an absolute value greater than that for sample A; however, at low T, sample A shows a weaker dependence on temperature. The strong temperature dependence of ϵ_{MW} at low T for sample C indicates that ϵ_{MW} will cross over to positive values at lower T, reflecting low T insulating behavior. For PAN-CSA sample D, the crossover of ϵ_{MW} from negative to positive values is directly observed at low T (~10 K). This provides a confirmation of the crossover observed for sample D from metallic (negative values) to insulating (positive values) in the T-dependent far-IR $\epsilon(\omega)$ at ~150 K and reflects the progressively increasing localization of the charge carriers as the temperature is lowered. This changeover from delocalized (free) carriers to localized carriers at low temperature reflects the importance of phonon-induced delocalization in the disordered regions of the inhomogeneous metallic state [38,80,81]. For sample D, the crossover in ϵ_{MW} occurs at ~10 K. For other samples, this crossover has been observed at high temperatures (~20 K) [38], indicating the importance of the specific material-processing conditions and the resulting composite network.

Using the Drude model at low frequency ($\omega\tau \ll 1$),

$$\sigma \sim \omega_p^2 \tau / 4\pi \quad (25)$$

and

$$\epsilon \sim -\omega_p^2 \tau^2 \quad (26)$$

the plasma frequency (Ω_p) and scattering time (τ) can be estimated from ϵ_{MW} and σ_{MW} (not shown) (Table 3.6). The predicted Ω_p for free electrons in samples A, C, and D is in the far-infrared, in good agreement with the observed zero crossings of $\epsilon(\omega)$ in the far-infrared,

Table 3.6 Room Temperature Free Electron Plasma Frequency ω_p and Scattering Time τ for Selected PAN-CSA Samples Calculated from ϵ_{MW} and σ_{MW} Using the Drude Model

Sample	ω_p (eV)	τ (10^{-11} s)
A (300 K)	0.007	2.5
C (300 K)	0.016	1.1
D (300 K)	0.004	2.9

though the values are smaller than those obtained from the Drude fits in the far-infrared. The relative size of the plasma frequencies for the different samples [ω_p(sample C) > ω_p(sample A) > ω_p(sample D)] is in good agreement with the infrared measurements. The scattering time predicted in each case is ~10^{-11} s. The values are two orders of magnitude larger than the τ predicted in Section IV.G.

The quantitative difference between the the Ω_p and τ estimated from infrared and microwave transport measurements may be a reflection of the inhomogeneity of the percolating network [79]. From the discussion of the optical conductivity and dielectric functions, a distribution of scattering times for the conduction electrons is likely involved in the transport. Therefore, at the lowest frequencies, the carriers that are the most delocalized (with the longest scattering times, $\tau \sim 10^{-11}$ s) may dominate the transport. Ω_p appears smaller because a smaller fraction of the charge carriers have such a long mean free time ($\tau \sim 10^{-11}$ s).

The trends observed in the infrared are confirmed by the transport measurements at 6.5 GHz. The high σ_{dc} observed as $T \rightarrow 0$ in metallic PAN-CSA is accompanied by the presence of free electrons down to low temperature. In the samples where the free electron density "freezes out" at low temperature, σ_{dc} has a much stronger insulating temperature dependence.

V. POLYPYRROLE

The dc and high frequency transport in polypyrrole (PPy) prepared with different dopants has also been systematically investigated. Introducing different dopants varies the disorder in the PPy system. For instance, PPy doped with hexafluorophosphate (PF$_6$) can be highly conducting down to millikelvin temperatures [36,37,53]. When the slightly larger dopant p-toluenesulfonate (TsO) is used, the conductivity is reported to be more strongly temperature-dependent [23,24], showing three-dimensional variable range hopping behavior. When doped not with an anion but with a polyanion (sulfated polyhydroxyether), the conductivity decreases drastically from ~10 S/cm at room temperature down to ~10^{-10} S/cm at ~1.5 K [100], showing a stronger temperature dependence ($\sigma_{dc} \propto \exp[-(T_0/T)^{1/2}]$).

The key difference between PPy doped with PF$_6$ and PPy doped with TsO is the disorder introduced into the polymer during polymerization. When, for instance, PF$_6$ was chemically exchanged with TsO as the dopant for PPy in a film polymerized with PF$_6$, the magnitude and temperature dependence of the conductivity of the resulting film were nearly identical to those of the parent film, PPy(PF$_6$) [113]. The following set of experiments used the structural order of the doped PPy sample as a background to understand the difference between the metallic states in these systems.

A. X-Ray Diffraction

The X-ray diffractometer tracing for each of the films of this study is shown in Fig. 3.31. X-ray studies of PPy(PF$_6$) indicate that these films are ~50% crystalline, with a structural coherence length of ~20 Å [59,126]. The reduction in structural disorder induced during synthesis compared with that of PPy materials studied earlier has been suggested to account for the improved metallic behavior of these new films [25,57,113]. The structure assigned to the crystalline portion of PPy(PF$_6$) samples consists of two-dimensional stacks of PPy chains in the bc plane separated by an intervening layer of PF$_6$ counterions [59,126]. The PPy(TsO) for this study was ~25% crystalline, approximately half that of PPy(PF$_6$), with a crystalline domain size that is decreased to ~15 Å [24] compared to ~20 Å. The bulky TsO dopant not only increases the separation between bc layers of PPy chains by 50% but also likely enhances the conformational disorder of the PPy chain since intrachain X-ray reflections are not recorded for PPy(TsO). In contrast, the PPy(S-PHE) films show only a single broad peak reflecting disordered chains [58]. The fact that the maximum average d spacing for the disordered chains in PPy(S-PHE) is different from those of PPy(PF$_6$) and PPy(TsO) may reflect the average stacking of the S-PHE polyanions, which account for the majority of the volume in this material. The differences in crystallinity allows us to probe the effects of structural disorder on the metallic state.

B. Magnetic Susceptibility

The spin susceptibility was measured from room temperature down to 4 K for PPy doped with PF$_6$, TsO, and S-PHE. The temperature dependence provides insight into the nature of the spin states. A simple model for the total susceptibility is

$$\chi = \chi^{Pauli} + \chi^{Curie} \quad (27)$$

where χ^{Pauli} [$= 2\mu_B^2 N(E_F)$] is temperature-independent and χ^{Curie} is proportional to T^{-1} [35]. Figure 3.32 shows χT vs. T for each of the samples. The solid lines are fits to $\chi T = \chi_P T + C$, where χ_P is the temperature-independent Pauli susceptibility and C is the Curie constant.

The positive slopes in Fig. 3.32 indicate a finite $N(E_F)$ for each of the materials. $N(E_F) \simeq 0.80$ state/eV per ring for PPy(PF$_6$); $N(E_F) \simeq 0.20$ state/(eV·ring) for PPy(TsO); and $N(E_F) \simeq 0.17$ state/(eV·ring) for PPy(S-PHE). This value is comparable to the $N(E_F)$ measure for other highly conducting polymers [3]. For doped polyacetylene [34], $N(E_F) \sim 0.18$ state/(eV·C), and in PAN-CSA, $N(E_F) \sim 0.1$ state/[eV·(C + N)] [46]. The positive intercept corresponds to a finite number of uncorrelated Curie spins. PPy(TsO) [$C = 1.7 \times 10^{-3}$ emu·K/(mol·ring) ~ 1 spin/200 polypyrrole rings] and PPy(S-PHE) [$C = 6.3 \times 10^{-4}$ emu·K/(mol·ring) ~ 1 spin/500 polypyrrole rings] show a larger Curie component than PPy(PF$_6$) [$C = 3.6 \times 10^{-4}$ emu·K/(mol·ring) ~ 1 spin/10^3 polypyrrole rings]. The fact that PPy(TsO) has a larger number of Curie spins than PPy(S-PHE) per chain may reflect that bipolarons are preferentially formed in the more disordered chains of PPy(S-PHE).

The larger $N(E_F)$ and smaller density of uncorrelated spins for PPy(PF$_6$) correlates with the larger fraction of the sample having three-dimensional order [126], i.e., crystallinity stabilizes a metallic density of states. Simi-

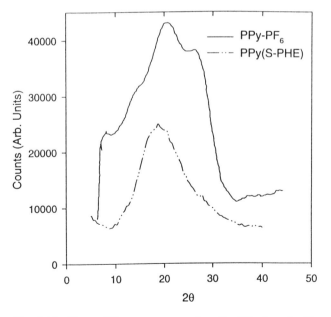

Fig. 3.31 X-ray diffractometer tracings for PPy doped with PF$_6$ (from Ref. 126) and S-PHE (from Ref. 58), showing the change in crystallinity of the PPy as the dopant is varied.

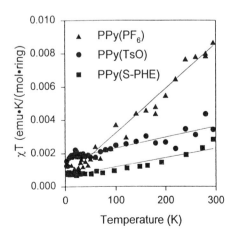

Fig. 3.32 χT vs. T for PPy(PF$_6$), PPy(TsO), and PPy(S-PHE).

lar to doped polyaniline, where the metallic Pauli susceptibility is associated with the three-dimensionally ordered regions [41,42], a large finite $N(E_F)$ is present in PPy when there are large three-dimensionally ordered (crystalline) regions. In more disordered materials, localized polarons or bipolarons predominate.

C. DC Conductivity

The reduced activation energy [76] ($W \equiv d \ln \sigma / d \ln T$) versus T is shown in Fig. 3.33 for PPy(PF$_6$), PPy(TsO), and PPy(S-PHE). For PPy(PF$_6$), the positive slope of W with increasing T implies that it is in the metallic regime. Millikelvin dc conductivity measurements on the same PPy(PF$_6$) materials show metallic behavior [36,37] and a negative magnetoresistance [53], confirming that PPy(PF$_6$) has crossed the IMT and is metallic at low T. In contrast, PPy(TsO) has a negative slope for its W plot, placing it in the localized regime. The slope of the line is ~0.25, indicating three-dimensional variable range hopping. For PPy(S-PHE), the W plot also has a negative slope (~0.5) at low temperature, characteristic of quasi-one-dimensional variable range hopping. Therefore, with increased disorder and chain isolation, doped PPy becomes more insulating, with charge transport that becomes more characteristic of reduced dimensionality hopping, just as for the PAN-CSA samples.

The same reliance on structural order is seen for σ_{dc} in Fig. 3.34. For PPy(PF$_6$) σ_{dc} very slowly decreases as temperature decreases from room temperature ($\sigma_{dc} \sim$ 300 S/cm) to ~20 K, and $d\sigma/dT < 0$ for $T < 20$ K. The increase in σ_{dc} with decreasing T at low temperature was reported previously [23,25,36,37,112,113] and attributed to a tunneling mechanism [112]. The ratio of $\sigma(RT)/\sigma(20 K)$ is <1.8, relatively small in comparison to that of other highly conducting polymers.

For more disordered PPy(TsO), $\sigma_{dc}(T)$ shows a stronger temperature dependence than PPy(PF$_6$), consistent with a 3-D variable range hopping (VRH) model;

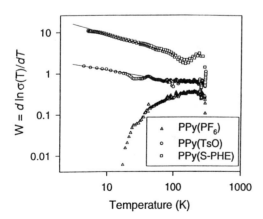

Fig. 3.33 W plot for PPy doped with PF$_6$, TsO, and S-PHE.

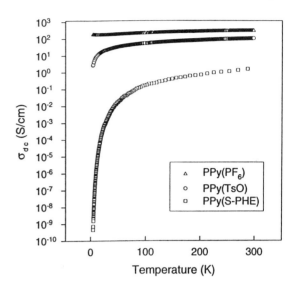

Fig. 3.34 σ_{dc} for PPy doped with PF$_6$, TsO, and S-PHE.

$\sigma_{dc}(T) = \sigma_0 \exp[-(T_0/T)^{1/4}]$, where $T_0 = 16/k_B N(E_F) L^3$ (k_B is the Boltzmann constant). Using the slope of $\ln \sigma$ vs. $T^{-1/4}$, $T_0 \simeq 4100$ K, and $N(E_F) = 0.2$ state/(eV·ring) determined from the magnetic studies, the localization length is estimated as $L \sim 30$ Å. This estimate of localization length is comparable with the X-ray crystalline coherence length ξ.

For PPy(S-PHE), which has the least structural order of the doped PPy samples studied here, $\sigma_{dc}(T)$ shows the highest degree of localization, with a large resistivity ratio [$\rho(4 K)/\rho(300 K) \sim 10^{10}$]. The temperature dependence is that of quasi-one-dimensional variable range hopping.

D. Reflectance

Similar to the results for highly conducting polyaniline, the reflectance of PPy doped with PF$_6$ and TsO, shown in Fig. 3.35, shows Drude-like behavior [78], increasing monotonically with decreasing energy from the conduction electron plasma edge at ~2.1 eV for PPy(PF$_6$) and ~1.9 eV for PPy(TsO) and approaching unity at low energy [24]. The oscillations in the mid-infrared (~0.1 eV) are due to the phonons of doped PPy. The reflectance of the S-PHE-doped PPy samples shows a more modest reflectance in the far-infrared, only approaching 40% near 0.006 eV [58]. Unlike polyaniline, there is not a separate peak at higher energy (~1.5 eV) that develops with decreasing conductivity.

The reflectance for PPy(PF$_6$) measured at 10 K (2 meV to 1.2 eV) is compared with its room temperature reflectance in Fig. 3.36. There is very little difference at high frequencies (>0.1 eV). In the far-IR, the 10 K reflectance is higher than the room temperature reflectance down to ~0.01 eV and then drops beneath the

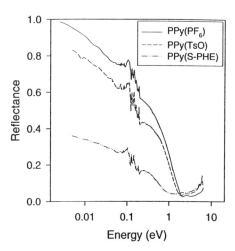

Fig. 3.35 Room temperature reflection spectra for PPy doped with PF$_6$, TsO, and S-PHE.

room temperature reflectance. The decrease of the reflectance measured in the far-IR is consistent with the drop in σ_{dc} as T decreases. The region where the 10 K reflectance rises above the room temperature reflectance is due to free carriers that are localized at low temperature, similar to what has been reported in polyacetylene [118] and earlier in this chapter for polyaniline. This decrease in the reflectance in the far-IR is inconsistent with the opening of an intrinsic gap in the electronic spectrum of PPy(PF$_6$) as reported earlier [120].

E. Absorption Coefficient

The absorption coefficients (α) for PPy doped with PF$_6$, TsO, and S-PHE obtained via Kramers–Kronig analysis of the reflectance data are shown in Fig. 3.37. For each of the samples, α is similar in character at low energy.

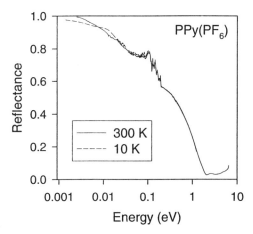

Fig. 3.36 Comparison of the reflectance of PPy(PF$_6$) at room temperature and 10 K. Note the weak T dependence at high energy and the strong T dependence in the far-infrared.

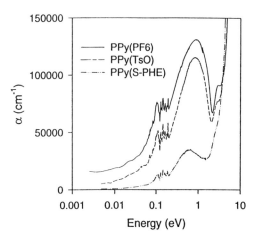

Fig. 3.37 Room temperature absorption coefficient for PPy doped with PF$_6$, TsO, and S-PHE.

There are peaks at ~1.0 eV and 2.8 eV. The absorption at higher energy depends upon the counterion that was used. Aside from the presence of a peak at 1.5 eV (attributed to isolated polarons [54]), the absorption coefficients for PAN-CSA [60–63] and for PPy doped with PF$_6$, TsO, and S-PHE are very similar, implying that highly conducting PPy forms a polaron lattice like doped PAN instead of a bipolaron lattice [127]. In contrast, a peak at ~1.5 eV was observed in lightly doped poorly conducting PPy [123].

The absorption coefficient for PPy(PF$_6$) at 10 K is compared with the room temperature α in Fig. 3.38. There is very little change in the absorption coefficient at high energy as the temperature is lowered. In the far-IR, a peak at ~0.01 eV forms due to absorption by carriers that are delocalized at room temperature but localized at 10 K. This behavior is very similar to what has been reported for iodine-doped polyacetylene with decreasing temperature [118]. The lack of change of the absorption at high energy indicates that the localized electrons do not contribute significantly to the high σ_{dc}; the temperature dependence of their conductivity contribution is different from that of the highly conducting carriers at zero frequency.

F. Optical Conductivity

Similar to the results for polyaniline, $\sigma(\omega)$ for doped PPy (Fig. 3.39) does not show Drude-like conductivity at low energies. Instead, $\sigma(\omega)$ shows Drude dispersion at high energy (~0.4–2 eV), a maximum at ω_{max}, and then suppression at low energy. The lack of structural order in PPy(S-PHE) results in a much greater suppression in $\sigma(\omega)$. The ratio $\sigma(6 \text{ meV})/\sigma(\omega_{max})$ gives an indication of the localization. For PPy(PF$_6$), this ratio is ~0.9; for PPy(TsO), ~0.3; and for the PPy doped with S-PHE, ~0.05. The fact that $\omega_{max} \sim 0.3$ eV for PPy(PF$_6$) is smaller than $\omega_{max} \sim 0.4$ eV for PPy(TsO) is consistent

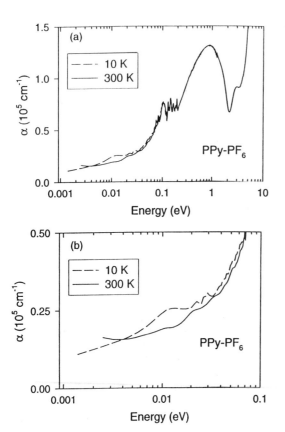

Fig. 3.38 Comparison of the absorption coefficient of PPy(PF$_6$) at room temperature and 10 K (a) for the full optical range and (b) in the far-infrared.

with the increased disorder in PPy(TsO), resulting in a smaller mean free path. The values obtained from a Drude fit to the high energy region of each sample are given in Table 3.7.

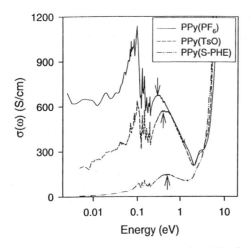

Fig. 3.39 Room temperature optical conductivity for PPy doped with PF$_6$, TsO, and S-PHE. $\sigma(\omega)$ for disordered PPy(S-PHE) is suppressed strongly in the far-infrared.

Table 3.7 Parameters from Drude Fit to High Energy (0.4–2 eV) $\sigma(\omega)$ for PPy Samples

Sample	Ω_{p1} (eV)	τ_1 (10^{-15} s)
PPy(PF$_6$)	2.5	0.53
PPy(TsO)	2.5	0.51
PPy (S-PHE)	1.5	0.43

The plasma frequency for conduction electrons is higher in PPy doped with PF$_6$, TsO, and S-PHE than in PAN-CSA. For PPy(PF$_6$), a three-pyrrole-ring repeat plus dopant has dimensions $a = 13.4$ Å, $b = 10.95$ Å, $c = 3.4$ Å [126]. For PPy(TsO), the unit cell dimensions are the same except that $a = 18.0$ Å [24] as the toluenesulfonate dopant is larger. In heavily doped PPy(PF$_6$) and PPy(TsO), there is one dopant unit added per three rings [23–25,36,37]; therefore the doping density for PPy(PF$_6$) is $n \sim 2 \times 10^{21}$ cm^{-3} and for PPy(TsO), $n \sim 1.5 \times 10^{21}$ cm^{-3}. For PPy(S-PHE), the carrier density can be approximated from the density (1.3 g/cm^3) [100] as $n \sim 2.9 \times 10^{20}$ cm^{-3}. From these conduction electron densities and the Ω_{p1} ($= \sqrt{4\pi n e^2/m^*}$) determined from Drude fits, $m^* \sim 0.5 m_e$ for PPy(PF$_6$), $m^* \sim 0.5 m_e$ for PPy(TsO), and $m^* \sim 0.2 m_e$ for PPy(S-PHE). Therefore, the effective masses (m^*) for doped PPy are lower than for PAN-CSA, implying that the overlap integrals are larger.

The scattering times determined from Drude fits (from 1.4 to ~2.4 eV) for the conduction electrons are small, $\tau_1 \sim 10^{-15}$ s. This small value is comparable to though smaller than what was found for τ_1 in PAN-CSA. The smaller values of τ for PPy compared with PAN may reflect greater disorder present in the disordered regions of PPy [59]. With increasing structural order within the PPy samples, τ increases, implying a larger mean free path for carriers in PPy(PF$_6$), in agreement with the behavior of $\omega_{\max}$.

G. Optical Dielectric Function

The percolation behavior seen in $\epsilon(\omega)$ for doped polyaniline is also observed for doped polypyrrole. For metallic PPy(PF$_6$), $\epsilon(\omega)$ shows three zero crossings (Fig. 3.40a). The dispersion of the localized electrons at high energy (>0.06 eV) and the Drude dispersion in the far-infrared are both evident. For PPy(TsO), the scattering time is reduced sufficiently by disorder that ϵ does not cross zero within the entire experimental frequency range. For PPy(S-PHE), the dispersion is very weak throughout the optical frequency range. The plot of $\epsilon(\omega)$ vs. $1/\omega^2$ for PPy(PF$_6$) is shown in Fig. 3.40b. The slope of the curve gives $\Omega_p \sim 0.17$ eV. Compared to the plasma frequency of the full conduction electron gas (~2.5 eV), only a small fraction of the conduction electrons are "free." Estimating in the same manner as for PAN-CSA [Eq. (17)],

$$n_{\text{free}} \sim (0.17/2.5)^2 \sim 5 \times 10^{-3} \qquad (28)$$

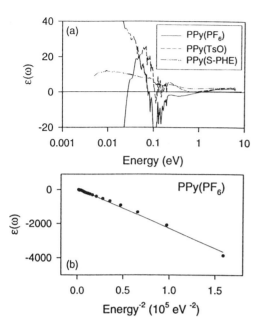

Fig. 3.40 (a) Comparison of the real part of the dielectric function $\epsilon(\omega)$ for PPy doped with PF_6, TsO, and S-PHE. (b) Far-infrared $\epsilon(\omega)$ for PPy(PF_6).

As discussed for PAN-CSA, the estimate of the fraction of delocalized electrons is dependent upon the ratio of m^* in the disordered regions to m^* in the ordered regions. Due to the linearity of $\epsilon(\omega)$ vs. $1/\omega^2$ down to 20 cm^{-1}, $\tau > \sim 1/20$ cm$^{-1} \sim 3 \times 10^{-13}$ s. Similar to the metallic state in PAN-CSA, a large majority of conduction electrons are localized ($\tau_1 \sim 10^{-15}$ s) and a small fraction of the conduction electrons are delocalized ($\tau \geq \sim 10^{-13}$ s) in metallic PPy(PF_6). With this long τ, a mean free path comparable to that of delocalized carriers in PAN-CSA is estimated. For more disordered PPy(TsO) and PPy(S-PHE), the electrical response is determined by only localized electrons. Similar to the behavior of $\epsilon(\omega)$ in PAN-CSA, $\epsilon(\omega)$ is larger in the far-infrared for more ordered, higher conductivity PPy(TsO) than for PPy(S-PHE).

For PPy(PF_6), $\epsilon(\omega)$ at room temperature is compared with $\epsilon(\omega)$ at ~ 10 K in Fig. 3.41. Two important points are observed. At high energy, there is very little difference in the dispersion of the localized carriers. $\epsilon(\omega)$ remains negative to slightly lower frequency, indicating that the scattering time changes only slightly as the temperature is lowered. This lack of a strong dependence of the scattering time on thermal processes indicates that the (static) disorder is the predominant scattering mechanism for the charge carriers. Similar results were reported for PAN doped with CSA [96]. The slight increase in the scattering time for localized electrons while σ_{dc} decreases implies that there is not a direct correlation of the localized carriers and the dc transport.

The temperature dependence of $\epsilon(\omega)$ in the far-infrared is shown in Fig. 3.41b. In contrast with the weak temperature dependence at high energy, the Drude dispersion is more strongly affected. Down to 10 K, the dielectric function remains negative in the far-IR, indicating that free carriers are present down to low T in PPy(PF_6), consistent with the high millikelvin σ_{dc} measured for this material [36,37] and also with the behavior of metallic PAN-CSA. At 10 K, the plasma frequency for free electrons (Ω_p) is 0.1 eV, decreased from 0.17 eV at room temperature. In comparison to σ_{dc}, which decreases by ~ 1.8, ω_p decreases by ~ 1.7. The scaling of Ω_p with σ_{dc} indicates that free carriers dominate the high dc conductivity for doped PPy as well as for doped PAN. The free carriers that are frozen out at low temperature give rise to the additional absorption at ~ 10 meV seen in the absorption coefficient (Fig. 3.37). As for PAN-CSA, the presence of a negative far-infrared $\epsilon(\omega)$ at room temperature excludes the possibility of an intrinsic many-body gap [120].

H. Microwave Transport

Microwave measurements for PPy samples provide independent confirmation of the results reported at infrared frequencies. Figure 3.42 compares $\sigma_{dc}(T)$ with $\sigma_{MW}(T)$ for PPy(PF_6) [24], PPy(TsO) [24], and PPy(S-PHE) [58]. The absolute values and the temperature dependence of σ in the dc and microwave frequency ranges for PPy(PF_6) are nearly identical, in agreement with the Drude theory. The stronger temperature dependence of

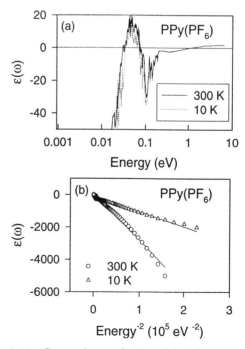

Fig. 3.41 Comparison of the dielectric response of PPy(PF_6) at room temperature and 10 K. (a) $\epsilon(\omega)$ in the range of localized carriers. (b) $\epsilon(\omega)$ vs. $1/\omega^2$ in the far-infrared.

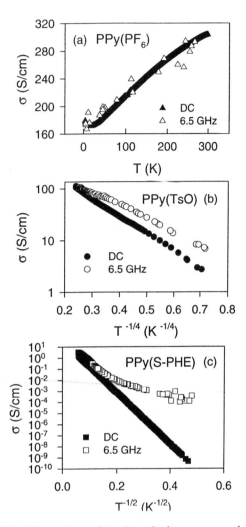

Fig. 3.42 Comparison of the dc and microwave conductivity for (a) PPy(PF$_6$), (b) PPy(TsO), and (c) PPy(S-PHE).

σ_{dc} for PPy(TsO) and PPy(S-PHE) in comparison with $\sigma_{MW}(T)$ is expected because the sample is in the localized regime [128].

Figure 3.43 shows $\epsilon_{MW}(T)$ for PPy(PF$_6$) [24]. ϵ_{MW} at 265 K is huge and negative, $\sim -10^5$, corresponding to the Drude dielectric response at microwave frequencies. ϵ_{MW} remains huge and negative down to 4.2 K ($\sim -4 \times 10^4$). The absolute value of ϵ_{MW} has a very weak T dependence and a maximum at $\sim$20 K. This behavior qualitatively agrees with $\sigma_{dc}(T)$ and $\sigma_{MW}(T)$. Using the Drude model in the low frequency limit ($\omega\tau \ll 1$), $\sigma_{MW} \simeq (\omega_p^2/4\pi)\tau$ and $\epsilon_{MW} \simeq -\omega_p^2\tau^2$. Hence, ω_p and τ ($\simeq 4\pi\epsilon_{MW}/\sigma_{MW}$) are $\sim$0.007 eV and $\sim 3 \times 10^{-11}$ s, respectively. The prediction of a small plasma frequency of free electrons with a zero crossing in the far-infrared agrees well with the infrared measurements, providing an independent confirmation of the Drude carriers. Again, as for PAN-CSA, the plasma frequency for free carriers is small and the scattering time large compared with the infrared estimates. Therefore, a distribution of scattering times is important for highly conducting PPy samples as well.

Though its room temperature σ_{MW} is one-third that of PPy(PF$_6$), ϵ_{MW} for more disordered PPy(TsO) is positive in the entire temperature range [24]. The $T \to 0$ localization length can be estimated as $\sim$25 Å using the metallic box model [66,67] and the ESR $N(E_F)$ in agreement with the crystalline coherence length ($\xi \sim 15$ Å). ϵ_{MW} linearly increases as temperature increases even though the localization length is small, which implies that the charge can easily delocalize through the disordered regions, and the phase segregation between the metallic and disordered regions is weak in the PPy(TsO) sample.

For PPy(S-PHE) also, ϵ_{MW} is positive over the entire temperature range and smaller in absolute value than ϵ_{MW} for PPy(TsO), consistent with stronger localization of charges in the PPy(S-PHE) sample [58]. For PPy(S-PHE), ϵ_{MW} does not change much from its low temperature saturation value until the temperature increases above $\sim$150 K, at which point it increases rapidly at

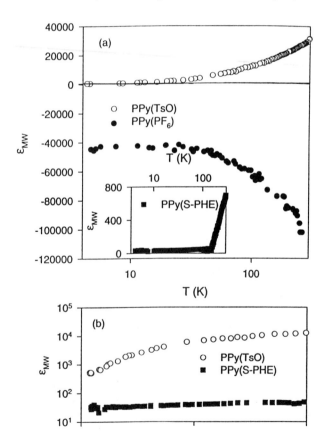

Fig. 3.43 (a) Microwave (6.5 GHz) dielectric constant for metallic PPy(PF$_6$) and insulating PPy(TsO) and PPy(S-PHE). (b) Low T dielectric functions for PPy(TsO) and PPy(S-PHE) showing their saturation.

higher temperatures. This indicates that there may be an energy barrier to hopping below 150 K. Using the metallic box model for this sample and the ESR $N(E_F)$, the $T \to 0$ localization length is estimated as ~7 Å, comparable to two pyrrole rings and smaller than that of PPy(TsO), consistent again with greater localization.

VI. POLYACETYLENE

It is noted that similar behavior for highly conducting doped polyacetylene samples has been reported, including a finite $N(E_F)$ [3,13,34,43,45], σ_{dc} finite as $T \to 0$ [36,37], large and negative ϵ_{MW} [3,34,130], and a rapid increase in the negative value of ϵ in the far-infrared [72,73,119]. Further, samples for which $\sigma_{dc} \to 0$ as $T \to 0$ generally have ϵ_{MW} positive with the magnitude related to sample conductivity [131] and a positive $\epsilon(\omega)$ in the infrared [73,119]. Transport in the poorly conducting regime for doped polyacetylene has been thoroughly reviewed [6]. Models very similar to those applied to doped polyaniline and polypyrrole can be applied to doped polyacetylene [3,34].

VII. DISCUSSIONS AND CONCLUSIONS

The experimental data for doped polyaniline, polypyrrole, and polyacetylene show a wide range of electrical behavior, varying from localized charge carrier hopping behavior for all charge carriers to Drude free carrier diffusion for a small fraction of the carriers while the remaining carriers are localized. Regardless of the family of conducting polymers, the metallic state possesses certain universal properties, including a finite density of states at the Fermi level, a high σ_{dc} as $T \to 0$, and a dielectric function that becomes negative in the infrared and remains negative to the microwave frequency range, as expected for Drude carriers. Conducting polymers exist that range from insulating to metallic and span the insulator-metal transition at low temperature.

Through analyses of the high frequency dielectric and conductivity functions, new insights have been obtained concerning the metallic state in conducting polymers. For highly conducting doped PAN and doped PPy, with decreasing energy, there are two negative-going zero crossings for the optical frequency dielectric function for metallic conducting polymers. The dielectric frequency response for highly conducting doped polyacetylene is very similar, with a high energy zero crossing at ~3 eV and a very rapid increase in the absolute value of the dielectric function in the far-infrared. These two electrical responses correspond to the plasma response of delocalized (Drude) carriers at low frequencies (far-IR, ~0.05 eV) and localized carriers at high frequencies (~1-3 eV). The delocalized carriers have an unusually long scattering time, $\tau \sim 10^{-13}$ s, which is attributed to the intrinsic anisotropy of the Fermi surface for quasi-one-dimensional conducting polymers and to the fact that most off-chain impurities are effectively screened. However, the fraction of the charge carrier density that is delocalized is estimated to be less than 1%. Analysis of the temperature dependence of the dc conductivity and dielectric function indicate that σ_{dc} is high at low T when $\epsilon(\omega)$ has free carrier behavior at low T. For materials for which the free carrier behavior in $\epsilon(\omega)$ "freezes out" at low T, σ_{dc} decreases rapidly at low T. This suggests that the high σ_{dc} in metallic conducting polymers is controlled by only the small fraction of delocalized carriers. This indicates that a large potential increase in σ_{dc} (up to ~10^5 S/cm) can be obtained if the entire charge carrier density has $\tau \sim 10^{-13}$ s.

The zero crossing (ω_{p1}) at higher energy (1-3 eV) is attributed to the majority of carriers that are localized with short mean free times ($\tau \sim 10^{-15}$ s). For conducting polymers whose dielectric response is dominated by only localized carriers, $\epsilon(\omega)$ is positive in the far-infrared and σ_{dc} has a strong temperature dependence, becoming insulating at low T.

The insulator-metal transition was shown explicitly to be controlled by disorder in PPy through structural studies and in PAN by varying the local chain conformation by casting PAN-CSA from different solvents. Though the IMT is controlled by disorder, it is not a homogeneous 3-D Anderson transition. This is asserted because

1. PAN-CSA samples with conductivity (σ_{dc}) higher than the minimum metallic conductivity (σ_{min}) become insulating at low temperature.
2. The evolution of $\sigma_{dc}(T)$ through the IMT is not monotonic.
3. Millikelvin σ_{dc} for selected insulating samples is not consistent with hopping transport.
4. The density of free electrons present in a sample scales with $\sigma_{dc}(T)$, so a sample that demonstrates metallic behavior at low temperature may have a smaller density of free electrons at room temperature than a sample that demonstrates insulating behavior at low temperature due to the nonmonotonic evolution of $\sigma_{dc}(T)$ through the IMT.
5. $\epsilon(\omega)$ and $\sigma(\omega)$ for metallic samples are consistent with macroscopically inhomogeneous models but not Anderson localization models.
6. A long scattering time ($\tau \gtrsim \sim 10^{-13}$ s) inconsistent with the Ioffe-Regel condition for homogeneous 3-D Anderson localization of carriers close to the IMT is determined.
7. Doped oriented films of polyaniline and polyacetylene demonstrate a known strong anisotropy along the chain direction, implying a strong influence of dimensionality on the entire class of polymers that is not accounted for in the Anderson IMT model.

Therefore, it is suggested that the metallic state and IMT are not controlled by homogeneous disorder.

Instead, the metallic state and IMT are proposed to be controlled by inhomogeneous disorder, consistent with the percolation (composite) behavior evident in $\epsilon(\omega)$ and $\sigma(\omega)$ at high frequencies, X-ray diffraction results, and the growth of a Pauli component of the magnetic susceptibility with increasing crystallinity. The composite nature results from the distinct transport among ordered metallic regions and the surrounding disordered regions where localization is strong. The IMT occurs when L_{loc} in the disordered regions exceeds the distance between ordered metallic regions so that carriers effectively percolate among metallic islands through the ordered paths. A small number of percolated paths could account for the small fraction of delocalized carriers. For metallic samples, a long mean free path up to $\sim 10^3$ Å (or perhaps longer if the microwave τ is used to estimate λ) is estimated (assuming that m^* is the same in the ordered and disordered regions), much larger than the crystalline domain sizes ($\sim 10^1$ Å). Such a long mean free path is consistent with percolation among many ordered regions.

For systems where the carriers have not percolated among the ordered regions, the transport is controlled by hopping and phonon-induced delocalization in the intermittent disordered regions. It is suggested that for materials with more rodlike chain conformation in the disordered regions, free carriers are present at room temperature that may "freeze out" at low T, causing a temperature-dependent IMT. For systems with more coil-like local order in the disordered regions, free carriers are not observed and transport is due to hopping among localized states, which contributes less than ~ 10 S/cm to σ_{dc}.

ACKNOWLEDGMENTS

This work was supported in part by the Office of Naval Research, the National Institute for Science and Technology under contract NIST ATP 1993-01-0149, and the National Science Foundation under contract NSF DMR-9508723. We are deeply indebted to our collaborators over many years, especially Alan G. MacDiarmid, Jean Paul Pouget, Vladimir Prigodin, Gary Ihas, Takaheko Ishiguro, Yongong Min, David Tanner, Libero Zuppiroli, and coworkers at The Ohio State University including Jinsoo Joo, Richard McCall, Yunzhang Wang, and Zhao-hui Wang. Without the cooperation of these scientists, this work could not have been done.

REFERENCES

1. J. E. Mark (ed.), *Handbook of the Physical Properties of Polymers*, AIP Press, Woodbury, CT, 1996.
2. C. K. Chiang, C. R. Fincher, Jr., Y. W. Park, A. J. Heeger, H. Shirakawa, E. J. Louis, S. C. Gau, and A. G. MacDiarmid, *Phys. Rev. Lett.* 39:1098 (1977).
3. R. S. Kohlman, J. Joo, and A. J. Epstein, Conducting polymers: electrical conductivity, in *Handbook of the Physical Properties of Polymers*, (J. E. Mark, ed.), AIP Press, Woodbury, CT, 1996, Chap. 34.
4. T. A. Skotheim (ed.), *Handbook of Conducting Polymers*, Marcel Dekker, New York, 1986.
5. Proceedings of International Conferences on Science and Technology of Synthetic Metals: ICSM '96, Snowbird, Utah, July 28–Aug. 2, 1996, *Synth. Met.*, in press; ICSM '94, Seoul, Korea, July 21–29, 1994, *Synth. Met. 69–71* (1995); ICSM '92, Goteborg, Sweden, Aug. 12–18, 1992, *Synth. Met. 55–57* (1993); ICSM '90, Tubingen, FRG, Sept. 2–9, 1990, *Synth. Met. 41–43* (1991); ICSM '88, Santa Fe, NM, June 26–July 2, 1988, *Synth. Met. 27–29* (1988).
6. A. J. Epstein, AC conductivity of polyacetylene: distinguishing mechanisms of charge transport, in *Handbook of Conducting Polymers* (T. A. Skotheim, ed.), Marcel Dekker, New York, 1986, p. 1041.
7. A. J. Heeger, S. A. Kivelson, J. R. Schrieffer, and W. P. Su, *Rev. Mod. Phys.* 60:781 (1988).
8. D. Baeriswyl, D. K. Campbell, and S. Mazumdar, in *Conjugated Conducting Polymers* (H. G. Keiss, ed.), Springer-Verlag, Berlin, 1992, p. 7.
9. E. M. Conwell, *IEEE Trans. Electr. Insul.* EI-22:591 (1987).
10. R. E. Peierls, *Quantum Theory of Solids*, Clarendon, Oxford, 1955, p. 108.
11. J. Tsukamoto, *Adv. Phys.* 41:509 (1992).
12. J. Tsukamoto, A. Takahashi, and K. Kawasaki, *Jpn. J. Appl. Phys.* 29:125 (1990).
13. H. Naarmann and N. Theophilou, *Synth. Met.* 22:1 (1987).
14. H. Shirakawa, Y.-X. Zhang, T. Okuda, K. Sakamaki, and K. Akagi, *Synth. Met.* 65:93 (1994).
15. J.-C. Chiang and A. G. MacDiarmid, *Synth. Met.* 13:193 (1986).
16. A. J. Epstein, H. Rommelmann, R. Bigelow, H. W. Gibson, D. M. Hoffman, and D. B. Tanner, *Phys. Rev. Lett.* 50:1866 (1983).
17. P. N. Adams, P. Laughlin, A. P. Monkman, and N. Bernhoeft, *Solid State Commun.* 91:895 (1994). The value of conductivity reported is for samples kindly provided by Monkman and coworkers, and measured at The Ohio State University.
18. Y. Cao, P. Smith, and A. J. Heeger, *Synth. Met.* 48:91 (1992).
19. J. Joo, Z. Oblakowski, G. Du, J. P. Pouget, E. J. Oh, J. M. Weisinger, Y. Min, A. G. MacDiarmid, and A. J. Epstein, *Phys. Rev. B* 69:2977 (1994).
20. Y. Z. Wang, J. Joo, C.-H. Hsu, J. P. Pouget, and A. J. Epstein, *Phys. Rev. B* 50:16,811 (1994).
21. Z. H. Wang, H. H. S. Javadi, A. Ray, A. G. MacDiarmid, and A. J. Epstein, *Phys. Rev. B* 42:5411 (1990).
22. J. Yue, Z. H. Wang, K. R. Cromack, A. J. Epstein, and A. G. MacDiarmid, *J. Am. Chem. Soc.* 113:2655 (1991).

23. M. Yamaura, T. Hagiwara, and K. Iwata, *Synth. Met.* 26:209 (1988).
24. R. S. Kohlman, J. Joo, Y. Z. Wang, J. P. Pouget, H. Kaneko, T. Ishiguro, and A. J. Epstein, *Phys. Rev. Lett* 74:773 (1995).
25. K. Sato, M. Yamaura, T. Hagiwara, K. Murata, and M. Tokumoto, *Synth. Met.* 40:35 (1991).
26. A. J. Epstein, H. Rommelmann, M. Abkowitz, and H. W. Gibson, *Phys. Rev. Lett.* 47:1549 (1981).
27. A. J. Epstein, H. Rommelmann, and H. W. Gibson, *Phys. Rev. B* 31:2502 (1985).
28. F. Zuo, M. Angelopoulos, A. G. MacDiarmid, and A. J. Epstein, *Phys. Rev. B* 39:3570 (1989).
29. J. C. Scott, P. Pfluger, M. T. Krounbi, and G. B. Street, *Phys. Rev. B* 28:2140 (1983).
30. T. Ito, H. Shirakawa, and S. Ikeda, *J. Polym. Sci. Polym. Chem. Ed.* 12:11 (1974).
31. K. Ito, Y. Tanabe, K. Akagi, and H. Shirakawa, *Phys. Rev. B* 45:1246 (1992).
32. J. H. Edwards and W. J. Feast, *Polym. Commun.* 21:595 (1980).
33. J. C. W. Chien, *Polyacetylene: Chemistry, Physics, and Material Science*, Academic, New York, 1984, p. 24.
34. A. J. Epstein, J. Joo, R. S. Kohlman, G. Du, A. G. MacDiarmid, E. J. Oh, Y. Min, J. Tsukamoto, H. Kaneko, and J. P. Pouget, *Synth. Met.* 65:149 (1994).
35. C. Kittel, *Introduction to Solid State Physics*, Wiley, New York, 1986, p. 157.
36. T. Ishiguro, H. Kaneko, Y. Nogami, H. Nishiyama, J. Tsukamoto, A. Takahashi, M. Yamaura, and J. Sato, *Phys. Rev. Lett.* 69:660 (1992).
37. H. Kaneko, T. Ishiguro, J. Tsukamoto, and A. Takahashi, *Solid State Commun.* 90:83 (1994).
38. J. Joo, V. N. Prigodin, Y. G. Min, A. G. MacDiarmid, and A. J. Epstein, *Phys. Rev. B* 50:12,226 (1994).
39. R. S. Kohlman, J. Joo, Y. G. Min, A. G. MacDiarmid, and A. J. Epstein, *Phys. Rev. Lett.* 77:2766 (1996).
40. R. S. Kohlman, A. Zibold, D. B. Tanner, G. G. Ihas, Y. G. Min, A. G. MacDiarmid, and A. J. Epstein, *Phys. Rev. Lett.* 78 (1997).
41. J. M. Ginder, A. F. Richter, A. G. MacDiarmid, and A. J. Epstein, *Solid State Commun.* 63:97 (1987).
42. M. E. Jozefowicz, R. Laversanne, H. H. S. Javadi, A. J. Epstein, J. P. Pouget, X. Tang, and A. G. MacDiarmid, *Phys. Rev. B* 39:12958 (1989).
43. A. J. Epstein, H. Rommelmann, M. A. Druy, A. J. Heeger, and A. G. MacDiarmid, *Solid State Commun.* 38:683 (1981).
44. P. K. Kahol, H. Guan, and B. J. McCormick, *Phys. Rev. B* 44:10393 (1991).
45. S. Ikehata, J. Kaufer, T. Woerner, A. Pron, M. A. Druy, A. Sivak, A. J. Heeger, and A. G. MacDiarmid, *Phys. Rev. Lett.* 45:1123 (1980).
46. N. S. Saricifti, A. J. Heeger, and Y. Cao, *Phys. Rev. B* 49:5988 (1994).
47. K. Mizoguchi, M. Nechtschein, J.-P. Travers, and C. Menardo, *Phys. Rev. Lett.* 63:66 (1989).
48. M. Nechtschein, F. Genoud, C. Menardo, K. Mizoguchi, J.-P. Travers, and B. Villeret, *Synth. Met.* 29: E211 (1989).
49. Y. W. Park, *Synth. Met.* 45:173 (1991).
50. Y. W. Park, A. J. Heeger, M. A. Druy, and A. G. MacDiarmid, *J. Chem Phys.* 73:946 (1980).
51. H. H. S. Javadi, A. Chakraborty, C. Li, N. Theophilou, D. B. Swanson, A. G. MacDiarmid, and A. J. Epstein, *Phys. Rev. B* 43:2183 (1991).
52. C. K. Subramaniam, A. B. Kaiser, P. W. Gilberd, C. J. Liu, and B. Wessling, *Solid State Commun.* 93: 235 (1996).
53. J. C. Clark, G. G. Ihas, A. J. Rafanello, M. W. Meisel, R. Menon, C. O. Yoon, Y. Cao, and A. J. Heeger, *Synth. Met.* 69:215 (1995).
54. A. G. MacDiarmid and A. J. Epstein, *Synth. Met.* 65:103 (1994).
55. R. Menon, C. O. Yoon, D. Moses, A. J. Heeger, and Y. Cao, *Phys. Rev. B* 48:17685 (1993).
56. R. Menon, Y. Cao, D. Moses, and A. J. Heeger, *Phys. Rev. B* 47:1758 (1993).
57. J. H. Kim, J. H. Kim, H. K. Sung, H. J. Kim, C. O. Yoon, and H. Lee, *Synth. Met.* in press.
58. R. S. Kohlman, S. M. Long, K. Bates, W. P. Lee, H. Kaneko, L. Zuppiroli, and A. J. Epstein, submitted.
59. J. P. Pouget, Z. Oblakowski, Y. Nogami, P. A. Albouy, M. Laridjani, E. J. Oh, Y. Min, A. G. MacDiarmid, J. Tsukamuto, T. Ishiguro, and A. J. Epstein, *Synth. Met.* 65:131 (1994).
60. A. G. MacDiarmid and A. J. Epstein, *Synth. Met.* 65:103 (1994).
61. A. G. MacDiarmid, J. M. Weisinger, and A. J. Epstein, *Bull. Am. Phys. Soc.* 38:311 (1993).
62. A. G. MacDiarmid and A. J. Epstein, Trans. 2nd Congresso Brazileiro de Polimeros, São Paulo, Brazil, Oct. 5–8, 1993, p. 544.
63. Y. Min, A. G. MacDiarmid, and A. J. Epstein, *Polym. Prep.* 35:231 (1994).
64. P. W. Anderson, *Phys. Rev.* 109:1492 (1958).
65. N. F. Mott and E. Davis, *Electronic Processes in Non-Crystalline Materials*, Clarendon Press, Oxford, 1979, p. 6.
66. Z. H. Wang, C. Li, E. M. Scherr, A. G. MacDiarmid, and A. J. Epstein, *Phys. Rev. Lett.* 66:1749 (1991).
67. Z. H. Wang, E. M. Scherr, A. G. MacDiarmid, and A. J. Epstein, *Phys. Rev. B* 45:4190 (1992).
68. M. I. Salkola and S. A. Kivelson, *Phys. Rev. B* 50: 13,962 (1994).
69. S. A. Kivelson and M. I. Salkola, *Synth. Met.* 44: 281 (1991).
70. V. N. Prigodin and K. B. Efetov, *Phys. Rev. Lett,* 70:2932 (1993).
71. R. P. McCall, E. M. Scherr, A. G. MacDiarmid, and A. J. Epstein, *Phys. Rev. B* 50:5094 (1994).
72. G. Leising, *Phys. Rev. B* 38:10313 (1988).
73. J. Tanaka, C. Tanaka, T. Miyamae, M. Shimizu, S. Hasegawa, K. Kamiya, and K. Seki, *Synth. Met.* 65: 173 (1994).
74. P. Lee and T. V. Ramakrishnan, *Rev. Mod. Phys.* 57:287 (1985).

75. H. Fukuyama, in *Electron–Electron Interactions in Disordered Systems* (A. L. Efros and M. Pollak, eds.), Elsevier Science, New York, 1985, p. 155.
76. A. G. Zabrodskii and K. N. Zeninova, *Zh. Eksp. Teor. Fiz.* 86:727 (1984) [*Sov. Phys. JETP* 59:425 (1984)].
77. A. F. Ioffe and A. R. Regel, *Prog. Semicond.* 4:237 (1960).
78. F. Wooten, *Optical Properties of Solids*, Academic, New York, 1972, p. 173.
79. R. S. Kohlman, D. B. Tanner, G. G. Ihas, Y. G. Min, A. G. MacDiarmid, and A. J. Epstein, Synth. Met., in press.
80. J. Joo, Charge localization and delocalization phenomena in conducting polymers, Ph.D. Thesis, The Ohio State University, 1994.
81. J. P. Pouget, to be published.
82. F. Zuo, M. Angelopoulos, A. G. MacDiarmid, and A. J. Epstein, *Phys. Rev. B* 36:3475 (1987).
83. J. P. Pouget, M. E. Jozefowicz, A. J. Epstein, X. Tang, and A. G. MacDiarmid, *Macromolecules* 24:779 (1991).
84. J. P. Pouget, C.-H. Hsu, A. G. MacDiarmid, and A. J. Epstein, *Synth. Met.* 69:119 (1995).
85. Z. H. Wang, A. Ray, A. G. MacDiarmid, and A. J. Epstein, *Phys. Rev. B* 43:4373 (1991).
86. S. Kivelson and A. J. Heeger, *Synth. Met.* 22:371 (1988).
87. V. N. Prigodin and K. B. Efetov, *Synth. Met.* 65:195 (1994).
88. D. J. Bergman and D. Stroud, in *Solid State Physics* (H. Ehrenreich and D. Turnbull, eds.), Academic, New York, 1992, Vol. 46, p. 148.
89. G. L. Carr, S. Perkowitz, and D. B. Tanner, *Infrared Millimeter Waves 13*:171 (1985).
90. A. L. Efros and B. I. Shklovski, *J. Phys. C* 8:L49 (1975).
91. B. I. Shklovski and A. L. Efros, *Electronic Properties of Doped Semiconductors*, Springer-Verlag, Heidelberg, 1984).
92. A. I. Larkin and D. E. Khmelnitskii, *Zh. Eskp. Teor. Fiz,* 83:1140 (1982) [*Sov. Phys. JETP* 56:647 (1982)].
93. P. Drüde, *Ann. Phys.* 1:566 (1900); 3:369 (1900).
94. N. F. Mott, *Metal–Insulator Transitions*, Taylor and Francis, New York, 1990.
95. N. F. Mott and M. Kaveh, *Adv. Phys.* 34:329 (1985).
96. K. Lee, A. J. Heeger, and Y. Cao, *Phys. Rev. B* 48:14884 (1993).
97. N. F. Mott, in *Localization and Interaction in Disordered Metals and Doped Semiconductors* (D. M. Finlayson, ed.), Proc. 31st Scottish Univ. Summer School in Physics, 1986.
98. N. F. Mott, in *Localization 1990*, (K. A. Benedict and J. T. Chalker, eds.), Inst. Phys. Conf. Ser. No. 108, Institute of Physics, Bristol, 1990. Paper presented at the Localization 1990 Conference held at the Imperial College, London.
99. Y. G. Min, Determination of factors promoting increased conductivity in polyaniline Ph.D. Thesis, Univ. Pennsylvania, 1995.
100. O. Chauvet, S. Paschen, L. Forro, L. Zuppiroli, P. Bujard, K. Kai, and W. Wernet, *Synth. Met.* 63:115 (1994).
101. R. Menon, C. O. Yoon, D. Moses, and A. J. Heeger, Metal–insulator transition in doped conducting polymers, in *Handbook of Conducting Polymers* (T. Skotheim, R. Elsenbaumer, and J. Reynolds, eds).
102. W. Fosong, T. Jinsong, W. Lixiang, Z. Hongfang, and M. Zhishen, *Mol. Cryst. Liq. Cryst. 160*:175 (1988).
103. Y. B. Moon, Y. Cao, P. Smith, and A. J. Heeger, *Polym. Commun.* 30:196 (1989).
104. M. E. Jozefowicz, A. J. Epstein, J. P. Pouget, J. G. Masters, A. Ray, and A. G. MacDiarmid, *Macromolecules* 25:5863 (1991).
105. C. D. G. Minto and A. S. Vaughan, *Polymer*, in press.
106. M. Laridjani, J. P. Pouget, E. M. Scherr, A. G. MacDiarmid, M. E. Jozefowicz, and A. J. Epstein, *Macromolecules* 25:4106 (1992).
107. M. Laridjani, J. P. Pouget, A. G. MacDiarmid, and A. J. Epstein, to be published.
108. L. Abell, S. J. Pomfret, E. R. Holland, P. N. Adams, and A. P. Monkman, Proc. Soc. of Plastic Engineers Ann. Technical Conf. (ANTEC 1996), 1996, p. 1417.
109. Y. Cao and A. J. Heeger, *Synth. Met.* 52:193 (1992).
110. H. H. S. Javadi, R. Laversanne, A. J. Epstein, R. K. Kohli, E. M. Scherr, and A. G. MacDiarmid, *Synth. Met.* 29:E439 (1989).
111. P. K. Kahol, A. J. Dyakonov, and B. J. McCormick, *Synth. Met.* (1997).
112. R. Menon and S. V. Subramanyam, *Solid State Commun.* 72:325 (1989).
113. M. Yamaura, K. Sato, T. Hagiwara, and K. Iwata, *Synth. Met.* 48:337 (1992).
114. B. Movaghar and S. Roth, *Synth. Met.* 63: 163 (1994).
115. H. Ehrenreich and H. R. Phillip, *Phys. Rev.* 128:1622 (1962).
116. K. Lee, A. J. Heeger, and Y. Cao, *Synth. Met.* 72:25 (1995).
117. S. Stafstrom, J. L. Brédas, A. J. Epstein, H. S. Woo, D. B. Tanner, W. S. Huang, and A. G. MacDiarmid, *Phys. Rev. Lett* 59:1464 (1987).
118. H. S. Woo, D. B. Tanner, N. Theophilou, and A. G. MacDiarmid, *Synth. Met.* 41–43:159 (1991).
119. T. Miyamae, M. Shimizu, and J. Tanaka, *Bull. Chem. Soc. Jpn.* 67:40253 (1994).
120. K. Lee, R. Menon, E. L. Yuh, N. S. Sariciftci, and A. J. Heeger, *Synth. Met.* 68:287 (1995).
121. C. O. Yoon, R. Menon, D. Moses, A. J. Heeger, and Y. Cao, *Synth. Met.* 63:47 (1994).
122. G. Burns, *Solid State Physics*, Academic, New York, 1985, p. 187.
123. K. Yakushi, L. J. Lauchlan, T. C. Clarke, and G. B. Street, *J. Chem. Phys.* 79:4774 (1983).

124. R. Menon, K. Vakiparta, C. O. Yoon, Y. Cao, D. Moses, and A. J. Heeger, *Synth. Met. 65*:167 (1994).
125. G. Du, V. Prigodin, J. Avlyanov, A. G. MacDiarmid, and A. J. Epstein, to be published.
126. Y. Nagomi, J. P. Pouget, and T. Ishiguro, *Synth. Met. 62*:257 (1994).
127. J. L. Brédas, J. C. Scott, K. Yakushi, and G. B. Street, *Phys. Rev. B 30*:1023 (1984).
128. A. R. Long, *Adv. Phys. 31*:553 (1982).
129. E. P. Nakhmedov, V. N. Prigodin, and A. V. Samukhin, *Sov. Phys. Solid State 31*:368 (1989).
130. J. Joo, G. Du, V. N. Prigodin, J. Tsukamoto, and A. J. Epstein, *Phys. Rev. B 52*:8060 (1995).
131. J. Joo, G. Du, J. Tsukamoto, and A. J. Epstein, *Synth. Met.*, in press.

4
The Dynamics of Solitons in *trans*-Polyacetylene

Christoph Kuhn
University of Pittsburgh, Pittsburgh, Pennsylvania

I. INTRODUCTION

The concept of solitons and their dynamics applied to *trans*-polyacetylene has become a paradigm in understanding conducting polymers [1] since it was introduced by Su and coworkers [2,3] using a simple Hückel molecular orbital (HMO) Hamiltonian. The understanding of these nonlinear, pseudo-one-dimensional topological excitations has been improved by applying solid-state theory (continuum limit of the SSH-HMO model [4–7], including electron correlation [8–11] and by quantum-chemical methods (MNDO [12], ab initio [13–16] and semiempirical [17,18]). Many excellent reviews on the theory of solitons in *trans*-polyacetylene have appeared [19–28]; for reviews on soliton dynamics see Refs. 20–22, 24, 25, and 27–29.

In the present chapter we attempt to model solitons in *trans*-polyacetylene and their dynamics as simply as possible. Models confined to what is really essential for a given set of phenomena are useful for grasping the physics behind these phenomena and are indispensable in the search for more complex new phenomena. For that purpose we use the step-potential model. It is the conceptually simplest approach to treat π-electrons coupled to the phonons of a σ-bonded lattice of carbon ions in unsaturated hydrocarbon molecules and a powerful alternative to the SSH-HMO theory. To get acquainted with the theory we first discuss the ground-state and excited-state properties of polyenes (which are chain molecules with even numbers of sites and alternating bond lengths), of cyanine dyes (which are chain molecules with odd numbers of sites and equal bond lengths), and of annulenes (which are ring molecules with even numbers of sites). The application of the model to larger π-electron systems such as *trans*-polyacetylene, where the alternating bonds lead to a band gap, is straightforward, and the formation of a kink-soliton as a topological defect is easily seen. The lattice dynamics within adiabatic Born–Oppenheimer approximation are derived and applied to investigate the dynamics of solitons (such as frictionless motion of a neutral kink, collision between kinks, and excitation of a defect-free long-chain polyene). The investigation should be of interest in the search for possibilities of information transfer in supramolecular systems and in tracing future prospects in molecular electronics [30].

II. STEP-POTENTIAL MODEL FOR ELECTRON–PHONON COUPLING IN π-ELECTRON SYSTEMS

In single-electron models the many-body π-electron Hamiltonian is simplified to a Hamiltonian of independent electrons moving in orbitals to be described as solutions of the three-dimensional Schrödinger equation of an electron in a potential $V(x, y, z)$ associated with the lattice sites. These orbitals have nodes in the layer plane (orthogonality with the σ-electrons). $V(x, y, z)$ is constructed from atomic contributions [31–35]. With the particular postulates defining $V(x, y, z)$, the correlation of electrons in the ground state is indirectly considered: it is assumed that $V(x, y, z)$ is the sum of the Slater potentials of the two adjacent charged lattice sites in the chain (see Appendix A). The contributions of all other sites are neglected; their charges are considered to be shielded by the residual π-electrons while nearest-neighbor sites are unshielded. In this way the interdependence of the electron under consideration with all other electrons is taken into account. The motion of this electron is considered to be correlated with the motions of the other electrons. The Schrödinger equations for such potentials $V(x, y, z)$ have been solved in some typical cases [31–35]. It was shown that a considerable simplification in the shape of this potential still leads to a reasonable description of unbranched π-electron systems and branched π-electron systems [36], i.e., the C_{60} [37], can be easily considered. The wave functions in an unbranched system can be written as the product of a function $\varphi_k(s)$ of the coordinate s along the zigzag line connecting the atoms in the chain and a function of the coordinates perpendicular to the chain, which is identical for all solutions of physical interest. The functions $\varphi_k(s)$ are solutions of the one-dimensional Schrödinger equation

$$\left[-\frac{\hbar^2}{2m}\frac{d^2}{ds^2} + V(s)\right]\varphi_k(s) = E_k\varphi_k(s) \quad (1)$$

in the potential $V(s)$, where $V(s)$ is the average of $V(x, y, z)$ taken over the coordinates perpendicular to s. $V(s)$ again has been systematically simplified in order to investigate what is crucial in describing the relevant properties of π-electrons [31–35]. In the case of a chain with bonds fixed to equal bond lengths, the most radical simplification, $V(s)$ = constant along the chain, still leads to a reasonable description of the wave functions of an electron in the potential $V(x, y, z)$ if the electron wave is assumed to extend by about one bond length beyond the centers of the sites at the chain ends [length $L = a(Z + 1)$], bond length $a = 1.40$ Å, number of sites Z). In this case $\varphi_k(s)$ is simply a sine function,

$$E_k = \frac{h^2}{8mL^2}k^2, \quad \varphi_k(s) = \sqrt{\frac{2}{L}}\sin\left(k\frac{s}{L}\pi\right), \quad (2)$$
$$k = 1, 2, 3, \ldots$$

This simplification is appropriate in the case of a cyanine dye (Fig. 4.1), where $L = 2na$. The $N = 2n$ π-electrons occupy the $n = N/2$ orbitals of lowest energy. The agreement of the excitation energies between HOMO (highest occupied molecular orbital) and LUMO (lowest unoccupied molecular orbital),

$$\Delta E = E_{n+1} - E_n = \frac{h^2}{8mL^2}(2n + 1) \approx \frac{A}{n}, \quad (3)$$
$$A = \frac{h^2}{16ma^2} = 9.57 \text{ eV}$$

with experiment shows that this simple free electron

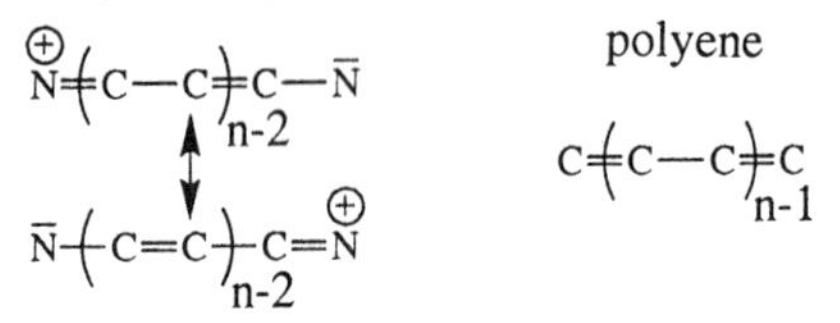

cyanine dye

$$\overset{\oplus}{N}\!\!+\!\!C\!\!-\!\!C\!\!\underset{n-2}{+}\!\!C\!\!-\!\!\bar{N}$$
$$\updownarrow$$
$$\bar{N}\!\!+\!\!C\!\!=\!\!C\!\!\underset{n-2}{+}\!\!C\!\!=\!\!\overset{\oplus}{N}$$

polyene

$$C\!\!+\!\!C\!\!-\!\!C\!\!\underset{n-1}{+}\!\!C$$

Fig. 4.1 The chemical representation (hydrogen atoms are omitted) of cyanine dyes (by resonating structures) and a polyene.

(FE) model leads to a rationalization of the color of cyanine dyes [38]. The FE model can easily be refined to the nearly free electron (NFE) model by solving the Schrödinger equation for the given potential $V(s)$, which has troughs at each atom in the chain, but this does not change the relevant wave functions essentially [39]. The troughs can easily be taken into account to check this statement. However, in the case of unequal bond lengths a refinement is indeed important in describing orbital wave functions and excitation energies [31–35,39]. $V(s)$ is lower in short bonds because the adjacent nuclei, in the average, are closer to the electron, so the Coulomb attraction is larger. We assume, for simplicity, that the potential V_i is constant along a given bond i (bond length d_i) and of constant length $a = 1.40$ Å. We call this potential, given by Appendix A, the bond potential V_i:

$$\mathbf{v}_i = \frac{V_i}{\hbar^2/2ma^2} = \beta_1\left(\frac{d_i}{a} - 1\right) - \beta_2\left(\frac{d_i}{a} - 1\right)^2 \quad (4)$$
$$(\beta_1 = 20.2; \beta_2 = 12.3)$$

($\hbar^2/2ma^2 = 1.944$ eV.). This model, the step-potential model [39], is the logically simplest extension of the free electron model. It describes this generalized situation sufficiently accurately for the present purposes, and it can easily be refined if desired for its justification. The exact wave functions for a stepwise constant potential are easily obtained by numerical evaluation.

A. Dimerization and the Electron Gap: From Small Molecules to *trans*-Polyacetylene

The step-potential model allows us to calculate the bond lengths in π-electron systems. Bond length and π-electron density in a bond are related: the π-electron cloud attracts the nuclei. Thus assuming first a σ-bonded molecular skeleton with uniform bond length a, the skeleton is elastically deformed in the field of the π-electron cloud and the cloud is deformed in the changed potential. This iterative process is repeated until self-consistency between bond length (or bond potential V_i) and π-electron density ρ_i in the middle of bond i is reached for each bond. We find

$$\mathbf{v}_i = \alpha_1(1 - \rho_i a) + \alpha_2(1 - \rho_i a)^2 \quad (5)$$
$$(\alpha_1 = 1.95; \alpha_2 = 0.52)$$

where the strength of the electron–phonon coupling, $\{\alpha_1, \alpha_2\}$, is scaled to the bond lengths of butadiene: solving the Schrödinger equation with $\mathbf{v}_i(d_i = 1.344$ Å$)$ and $\mathbf{v}_i(d_i = 1.474$ Å$)$ for the double and single bonds, respectively, in butadiene and calculating $\rho(s)$. This gives ρ_i for $d_i = 1.344$ Å and $d_i = 1.474$ Å. Note that the values of $\{\alpha_1, \alpha_2\}$ are specific to the present model; different values are derived when the atomic troughs are taken into account. This consideration explains the bond alternation in a polyene, while bond lengths in a cyanine dye are essentially equal; assuming equal bond lengths in polyenes when starting the search for self-consistency,

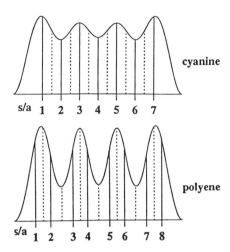

Fig. 4.2 Cyanines and polyenes in the step-potential model. Example $n = 4$. Charge density maxima at atoms 1, 3, 5, 7 (cyanines) and at bonds 1-2, 3-4, 5-6, 7-8 (polyenes).

the π-electron density is accumulated in bonds 1–2, 3–4, . . . (Fig. 4.2), causing an instability that leads to bond alternation [31–35]. In cyanine dyes, however, the density accumulations are at atoms and not in bonds (Fig. 4.2). For long-chain hydrocarbons with odd numbers of sites [40] and for long-chain cyanine dyes [41–42], upon increasing the chain length a change from equal bond lengths to alternating bonds and a soliton-like region of equal bonds are expected.

The step-potential model leads to bond equalization in benzene and higher Hückel annulenes with 10 and 14 sites and bond alternation in all other annulenes (the Hückel annulenes are rings with 6, 10, 14, 18, . . . sites). The value $\alpha_1 = 1.95$ in Eq. (5) is unambiguously given. A value $\alpha_1 > 2.6$ would yield bond length alternation in benzene. Therefore the resulting equal bond lengths in benzene and the transition from equal bonds to alternating bonds in 18-annulene are not obvious. The latter is in agreement with experimental facts (vanishing ring current in 18-annulene measured by NMR [43]) and thus an important test for the model.

We describe the ground state of *trans*-polyacetylene as a large even-numbered ring (e.g., 140 sites) with cyclic boundary condition to avoid termination effects. For the ground state we obtain the bond length alternation $\Delta d = 0.074$ Å, the potential difference for the single and double bond is $\Delta V = 2.1$ eV, which gives numerically a band gap $\Delta E = 1.34$ eV $= 0.64 \cdot \Delta V$ in agreement with an analytical solution of the first band gap of the infinite step potential given in Ref. 44. This value of the band gap is in good agreement with the experimental values $\Delta E = 1.4$–1.8 eV [26,45].

B. Comparison of the Nearly Free Electron Model with the SSH-HMO Hamiltonian

It is of interest to compare the physical implications of the SSH-HMO theory and the NFE model. Both models have the same starting point: the π-electron Hamiltonian [26],

$$H_\pi(\{r, R\}) = \sum_i \left[\frac{p_i^2}{2m} + V_\pi(r_i, \{R\})\right] + V_{\text{e-e}}^{\text{eff}}(\{r\}) + \sum_i \frac{P_i^2}{2M_{\text{CH}}} + V_n(\{R - R_0\}) \quad (6)$$

depending on the coordinate configuration $\{r\}$ of the π-electrons and on the ionic coordinate configuration $\{R\}$. p_i (P_i) is the momentum operator for the electron numbered i (CH$^+$ numbered i) respectively. The pseudopotential $V_\pi(r_i, \{R\})$ for the π-electrons in the field of the CH$^+$ ions as well as the effective electron–electron interaction $V_{\text{e-e}}^{\text{eff}}(\{r\})$ incorporate the screening and renormalization effects of core and σ-electrons. The potential $V_n(\{R - R_0\})$ in which the ions move has, in the absence of coupling to the π-electrons, a minimum for $\{R\} = \{R_0\}$. This π-electron Hamiltonian is simplified to a Hamiltonian for independent electrons moving in orbitals to be described as solutions of the Schrödinger equation of an electron in a potential $V(x, y, z)$ associated with the CH$^+$ lattice sites. These orbitals have nodes in the layer plane (orthogonality with σ-electrons). In both approaches $V(x, y, z)$ is constructed from atomic contributions. The difference between the two approaches is given by the strategy of finding approximate solutions of that Schrödinger equation.

The HMO theory and approaches beyond are based on an approximate solution of the Schrödinger equation in a given molecular potential, e.g., neglecting the overlap integrals and restricting to nearest-neighbor interaction (with the electron–hole symmetry as a consequence). This approximate solution leads to an inconsistency between the LCAO-MO wave functions and their energies [46–50]. In contrast, in the NFE model, the potential is simplified, but then the solutions of the Schrödinger equation are exact. In the following we compare specific LCAO model approximations (terminology is given in Appendix B).

1. The Electronic Energy Levels

We consider the lowest excitation energy ΔE of a long hydrocarbon chain of $2n$ π electrons and the bonds kept to length a. Within Hückel theory, $\Delta E = t_0\pi/n$; within the present approach (case of a particle in a box of length $2na$, Eq. 3) $\Delta E = A/n$. Comparison leads to $t_0 = A/\pi = 3.05$ eV. The description is analogous, but t_0 is unambiguous in the NFE model and adjusts to each particular problem in HMO: t_0 values range from $t_0 = 0.69$ eV (empirical resonance energy) to $t_0 = 4.0$ eV (UV ionization potential) [51], with a recommended value of $t_0 = 2.9$ eV [52] close to $A/\pi = 3.05$ eV.

2. The Electron–Phonon Coupling

The geometry of a hydrocarbon molecule predicted by the SSH-HMO model depends strongly on the values used for K and α_{SSH}. Benzene is predicted to have equal or alternating bond lengths depending on the choice of these parameters [53]. Treating the electron–phonon coupling within the SSH-HMO model and beyond, the minimum of the total energy is determined. This leads to Eq. (B3) for the band gap $2\Delta_{SSH}$. An inconsistency appears: the experimental value of the band gap gives the value $\lambda = 0.2$ for the parameter $\lambda = 2\alpha_{SSH}^2/\pi t_0 K_x$, but the value $\lambda = 0.08$ is calculated from the experimental values of $K_{\exp} = 47.5$ eV/Å² and of the bond lengths of graphite and benzene (giving α_{SSH}/K_x). This leads to a band gap smaller than the experimental value by a factor of $1/40$. This clear disagreement is usually assumed to be an indication of the failure of the single-particle picture [26]. However, there is no such inconsistency within the NFE model, and a comparison of the SSH model with the step-potential model suggests a different reason. We consider the change in the π-electron potential by proceeding from the reference configuration (all bonds of length a) to a configuration with alternating bond lengths ($d_i - a$ alternating between $+\Delta$ and $-\Delta$) [54,55]. In the step-potential model the potential [Eq. (4)]

$$V_i = (39.2 \text{ eV})\left(\frac{\Delta}{a}\right) - (23.9 \text{ eV})\left(\frac{\Delta}{a}\right)^2 \qquad (7)$$

alternately increases and decreases by 39.2 eV(Δ/a) [first term in Eq. (7)], and in addition the average potential drops by -23.9 eV$(\Delta/a)^2$ [second term in Eq. (7)]. This latter effect is neglected in the SSH model [second term in Eq. (B1)]. Taking it into account we have to add the electronic term -23.9 eV$(\Delta/a)^2$ to the elastic energy $(1/2)K_x(u_i - u_{i+1})^2 = (1/2)K\Delta^2 = (1/2)K_{\exp}\Delta^2 - (23.9$ eV$)(\Delta/a)^2$ and with $K_{\exp} = 47.5$ eV/Å² we obtain $K = 23.6$ eV/Å², this is in agreement with the adjusted values $K = 23$ eV/Å² to $K = 28$ eV/Å² (Appendix B), and the discrepancy is removed. Thus the first term in Eq. (B1) is not the elastic energy of the σ-bonded lattice but the sum of this elastic energy and a hithero neglected π-electron contribution due to the overproportional increase in the Coulomb attraction toward adjacent nuclei. Kakitani [56] introduces a quadratic term in the π-electron portion of the energy but gives no simple physical picture.

III. STATIC PROPERTIES OF SOLITONS IN *trans*-POLYACETYLENE

A. Kink

In an odd-numbered ring the two degenerate bond length alternation patterns called A phase and B phase unite, forming a localized defect called a neutral kink. The kink can be weak or strong, the center of the corresponding defect being shifted by one bond (see Fig. 4.3). The kink creates an intergap state (energy spectrum in Fig. 4.4). The corresponding wave function is well localized (envelope of square of wave function of weak neutral kink in Fig. 4.4). Weak and strong kinks are self-consistent solutions describing two different stable states with the center of the defect at adjacent sites. Thus the energy of a kink moving along the chain is minimal for weak and strong kinks, while intermediate states have higher energies. Considering two adjacent bonds enclosing a particular site i in the course of a kink passing over it (e.g., from left to right in Fig. 4.4), the single bond gradually turns into a double bond and vice versa for the adjacent bond according to the change of the π-electron density. The bond length alternation $\delta_i = (-1)^i(d_i - d_{i-1})$ between the two adjacent bonds proceeds from $+\delta$ to $-\delta$ with $|\delta| = 0.053a$. Hence the intermediate state of a kink has to be a self-consistent solution with the constraint of the bond length alternation δ_i fixed to a given value between the value in the weak and the strong kink enclosing the intermediate state. The transition state has a little higher energy than the two stable states

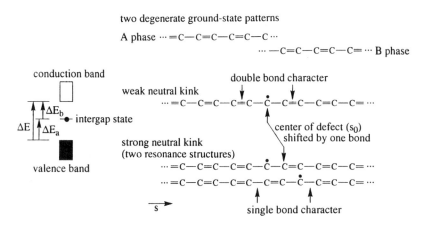

Fig. 4.3 Weak and strong neutral kinks: boundary between two degenerate alternation patterns called the A phase and B phase (schematic). Corresponding energy spectrum (schematic).

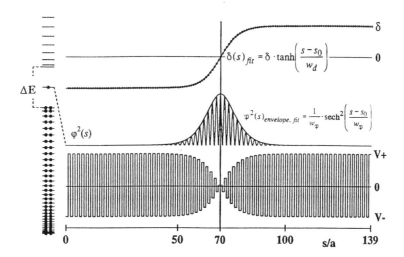

Fig. 4.4 Weak neutral kink $K^0\{w_d = w_\varphi = 9.15a\}$. Self-consistent potential along lattice coordinate s, center $s_0 = 70a$ (at atom) alternating between $V_+ = +1.1$ eV and $V_- = -1.0$ eV gives band gap $\Delta E = 1.34$ eV. Intergap state φ (square of wave function with its envelope shown): occupied by one electron (spin 1/2), electronic half-width $w_\varphi = 9.15a$, antinode at center. Bond length alternation $\delta_i = (-1)^i(d_i - d_{i-1})$ proceeds from $+\delta$ (right) to $-\delta$ (left) with $|\delta| = 0.053a$, kink of geometric half-width $w_d = 9.15a$. Formulas from continuum limit of SSH-HMO model [4], $\{w_d = w_\varphi\}$ fitted.

(this kink-pinning energy or Peierls–Nabarro energy barrier according to the present treatment is less than 0.001 eV), indicating a nearly free transition of kinks through the otherwise perfect lattice. The same value has been calculated within the SSH model [2].

The negative kink, $K^-\{w_d = 8.31a, w_\varphi = 9.02a\}$, has two electrons (spin 0); the neutral kink, $K^0\{w_d = w_\varphi = 9.15a\}$ (Fig. 4.4), has one electron (spin 1/2); and the positive kink, $K^+\{w_d = 5.62a, w_\varphi = 8.00a\}$, has no electron (spin 0) occupying the intergap state. Proceeding from the neutral to the positive kink, the single-bond character in the range of the defect is increased prominently, reducing the geometric half-width w_d of the defect more than by proceeding from the neutral kink to the negative kink. Inspection of the half-widths shows that the localized wave functions are broader than the geometric solitons consistent with ab initio calculation [16], but there the charged kinks have similar geometric widths, in contrast to the present model. Due to the electron–hole symmetry within the SSH-HMO model, both the electronic half-widths w_φ and the geometric half-widths w_d of kinks derived there have the same value of $7a$ independent of the occupation of the intergap state. Figure 4.4 confirms that the SSH-HMO model provides an excellent first-order approximation to the charge distribution of the neutral kink. However, as indicated by the results above, the Coulomb forces dominate the structural details of charged solitons. This is most remarkable: even though the step-potential model is formally a one-electron treatment, it takes care of the interdependence of the electron under consideration with all other electrons (see introduction to Section II, Section II.A, and Appendix A). In exciting the neutral kink, one has to consider two-electron terms explicitly (see below). The value $w_d = 9.15a$ in the case of the neutral kink is in good agreement with a value deduced from magnetic resonance ENDOR experiments giving $w_d = 12a$ [57], in contrast to a recent calculation based on a detailed variational study giving $w_d = 3a$ [52].

The observed additional absorption peaks of 1.35 and 0.45 eV in photoexcited *trans*-polyacetylene have been attributed to the formation of a kink–antikink pair from the excited state and a subsequent photoinduced transition [58,59]. The 0.45 eV peak was attributed to the charged kinks, and the 1.35 eV peak to the neutral kink. In the negative kink the intergap state (which lies 0.52 eV above the valence band edge when Coulomb repulsion is neglected [39]) is lifted by the Coulomb repulsion of the two electrons by $E_c \approx 0.5$ eV (assuming for simplicity that the charge is equally distributed among the $n = 15$ sites of the kink and counting only the repulsion when the two electrons are at the same site (Hubbard approximation), $U = 8$ eV [52,60–62], we obtain $E_c = U(1/n)^2 n = 0.53$ eV). Then the energy of transition from the negative kink state to the lowest state of the conduction band is 1.34 eV − (0.52 eV + 0.53 eV) = 0.29 eV. This is in reasonable agreement with the observed excitation energy of 0.45 eV. It must be taken into account that the theoretical gap 1.34 eV is smaller than the measured values (1.4–1.8 eV). In the positive kink the intergap state (which lies 0.86 eV above the valence band edge when the Coulomb attraction of the electron by the kink is neglected [39]) is lowered by the same amount, 0.53 eV, and the energy of transition from the highest state of the valence band to the positive kink state becomes 0.86 eV − 0.53 eV = 0.33 eV. In the neutral kink the transition from the midgap state to the lowest state of the conduction band and the transition from the highest state of the valence band to the midgap state are coupled: each electron is in the field of the incident light and in the field due to the polarization of the other electron. According to Appendix C [54,55,60–62], the corrected excitation energy becomes $\Delta E_2^{\text{corr}} = [(\Delta E/2)^2 + (\Delta E/2) \cdot 2J]^{1/2}$, where the coupling integral $J = 0.3$ eV is obtained by numerical integration. With the theoretical band gap $\Delta E = 1.34$ eV we obtain $\Delta E_2^{\text{corr}} = 0.9$ eV. The gap $\Delta E = 1.8$ eV gives $\Delta E_2^{\text{corr}} = 1.2$ eV. This is in reasonable agreement with the measured excitation energy of 1.35 eV.

B. Polaron

If an electron is added to the ground state of *trans*-polyacetylene, the lowest state of the conduction band is occupied by one electron. Relaxation by the electron–phonon coupling reveals two intergap states (negative polaron in Fig. 4.5). The lower state is very close to the valence band edge and is occupied by two electrons; the upper state is near the conduction band edge and is occupied by one electron. The wave functions of these two states are well localized, but the lower state is less localized than the upper state. The additional electron occupying the upper intergap state accumulates in the single bonds in the range of the defect and changes them to bonds with more double bond character. This causes a further charge density increase in these bonds at the expense of charge density in the double bonds in the defect, forming the lower intergap state. Thus a negative polaron $P^-\{w_d = 11.3a, w_\varphi^v = 15.1a, w_\varphi^c = 11.2a, 2b = 11a\}$ is formed (the indices v and c refer to the intergap states originated from the valence and conduction bands, respectively). The negative polaron P^- can be viewed as a bound kink–antikink pair with distance $2b = 11a$, one of them neutral, the other negatively charged ($K^0, \overline{K}^- \leftrightarrow K^-, \overline{K}^0$). Similarly, if an electron is removed from the ground state of *trans*-polyacetylene, then the highest state of the valence band is occupied by one electron. Relaxation reveals two intergap states: the lower state is near the valence band edge and is filled with one electron, and the upper state is very close to the conduction band edge and is empty. The wave functions of these two states are well localized; the lower state is more localized than the upper state, and a positive polaron, $P^+\{w_d = 11.1a, w_\varphi^v = 11.1a, w_\varphi^c = 17.2a, 2b = 9.5a\}$ is formed. The positive polaron P^+ can be viewed as a bound kink–antikink pair with distance $2b = 9.5a$, one of them neutral, the other positively charged ($K^0, \overline{K}^+ \leftrightarrow K^+, \overline{K}^0$). The values $w_d = 11.3a$ and $w_d = 11.1a$ for the negative and positive polaron, respectively, are in excellent agreement with the value from ^{13}C-NMR spectroscopy by comparing the

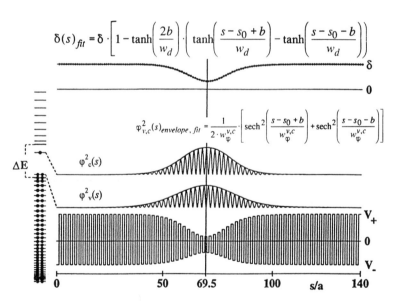

Fig. 4.5 Negative polaron $P^-\{w_d = 11.3a, w_\varphi^v = 15.1a, w_\varphi^c = 11.2a, 2b = 11a\}$. Self-consistent potential along lattice coordinate s, center $s_0 = 69.5a$ (at bond) alternating between $V_+ = +1.1$ eV and $V_- = -1.0$ eV gives band gap $\Delta E = 1.34$ eV. Two intergap states φ_v (originating from valence band) and φ_c (originating from conduction band) (square of wave functions with envelopes shown): φ_c occupied by one electron (spin 1/2), electronic half-width $w_\varphi^c = 11.2a$, antinode at center φ_v occupied by two electrons, electronic half-width $w_\varphi^v = 15.1$, node at center. P^- viewed as bound ($K^0, \overline{K}^- \leftrightarrow K^-, \overline{K}^0$) with distance $2b = 11a$. Bond length alternation $\delta_i = (-1)^i(d_i - d_{i-1})$ proceeds from $+\delta$ (right) to $+\delta$ (left) with $|\delta| = 0.053a$, polaron of geometric half-width $w_d = 11.3a$. Formulas from continuum limit of SSH-HMO model [5,6], numbers $\{w_d, w_\varphi, 2b\}$ fitted.

chemical shift of neutral and anionic polyenes of varying chain length [63].

The unbound bipolaron, where two electrons are added to or subtracted from the ground state of *trans*-polyacetylene evolves into a well-separated negative kink–negative antikink pair (K^-, $\overline{K}^-$) or positive kink–positive antikink pair (K^+, $\overline{K}^+$), respectively.

IV. LATTICE DYNAMICS

The lattice dynamics for the understanding of moving solitons in *trans*-polyacetylene are usually treated in standard adiabatic approximation using Newton's equation of motion [3,64–71]. The force required to integrate Newton's equation is obtained as the negative gradient of a potential energy surface. The equation of motion is integrated numerically using finite time steps. The result of the integration depends on the accuracy and stability of the algorithm used [72–74]. Treating the dynamics of the vibronic coupling within the step-potential model [75–77], the force acting on adjacent hydrocarbon sites is obtained from the strain within the bond, which is proportional to the deviation of the actual π-electron density in the middle of the bond from its equilibrium density. In the following we show that this approach gives results essentially identical with those obtained by the treatments mentioned above. However, these treatments yield a maximum speed of the soliton depending strongly on adjustable parameters, while in our model the result (maximum speed of the soliton three times the velocity of sound) is unambiguous and is given by the strength of the electron–phonon coupling. The dynamics in standard adiabatic approximation follow from Newton's equation.

$$M_{CH} \frac{d^2}{dt^2} x_i = F_i \qquad (8)$$

where x_i is the coordinate of site i representing a CH^+ group with mass $M_{CH} = 23888m = 2.176 \times 10^{-26}$ kg. The displacements are constrained to the molecular axis along x [3,78]; thus $d_i \sin \varphi_i = a/2$, where $d_i = (x_{i+1} - x_i)/\cos \varphi_i$ is the length of bond i (Fig. 4.6). During dynamics the molecular lattice is not relaxed. The σ-bonds in the field of the π-electron cloud are compressed according to the π-electron density in each bond. In analogy to Hooke's law, we assume that the strain arising in bond i is proportional to the deviation $\tilde{\rho}_i - \rho_i$ of the actual π-electron density ρ_i from the equilibrium density $\tilde{\rho}_i$. For

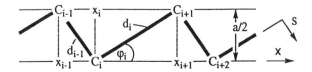

Fig. 4.6 Zigzag structure of polyacetylene along s. Coordinate x_i of the CH^+ group i moving on the track along x.

small deviations of the actual density ρ_i from its equilibrium, the equilibrium density $\tilde{\rho}_i$ is obtained by (1) applying Eq. (5) for each ρ_i to get a new configuration of bond potentials and (2) solving the Schrödinger equation with this configuration of bond potentials (thus by just one cycle of the iterative process of Section III). The force caused by the strain within bond i acts on the two adjacent sites i and $i + 1$ and is projected to the x axis by the angle φ_i. The strain within the two adjacent bonds $i - 1$ and i contributes to the force acting on site i.

$$F_i = [(\tilde{\rho}_i - \rho_i) \cos \varphi_i - (\tilde{\rho}_{i-1} - \rho_{i-1}) \cos \varphi_{i-1}]ka^2 \qquad (9)$$

Newton's equation, Eq. (8), with the force defined by Eq. (9) is approximated by a difference equation (Verlet algorithm, time step $\Delta t = 1.2 \times 10^{-15}$ s) and is simplified to Eq. (D4) describing the time evolution of the bond potentials. This is outlined in Appendix D. The value $k = 11.2$ eV/Å^2 is obtained from scaling the time unit: calculating the in-phase stretching mode of the polyene lattice (zero-momentum optical phonon frequency) by numerical integration of (D4) with the frequency $\nu_{max} = c \times 1400$ cm^{-1} (vibronic structure of the absorption band of polyenes). $u_a = \pi a \nu_{max} = 1.85 \cdot 10^4$ m/s then is the velocity of sound along s. The dynamical scaling factor $k = 11.2$ eV/Å^2 should not be confused with the force constant $K_{exp} = 47.5$ eV/Å^2 of the bare σ-bond, which would give the frequency $\omega_0 = 2\sqrt{K_{exp}/M_{CH}}$ [56,79]. The reduction of frequency [79–81], which is in our case $2\pi \nu_{max}/\omega_0 = 0.675$, is well known in solid-state physics as the Kohn anomaly [82] due to the Peierls instability [83].

A. Moving Neutral Kink

A neutral kink embedded in a ring with 139 sites initially at rest is given a velocity boost of $u_K = 10u_a$,* and the time evolution is investigated (Fig. 4.7). Immediately after the boost the velocity of the kink u_K drops to $u_K = 5u_a$ and its geometrical half-width w_d drops from initially $w_d = 9.15a$ to $w_d = 5.3a$, while the rise of a huge hump behind the soliton takes over the excess kinetic

---

* With the velocity $u_F = h/4ma = 1.3 \times 10^6$ m/s of a π-electron at the Fermi surface, the ratio $u_s/u_F = 0.014$. The localized π-electron corresponding to a kink is fast enough to fill the space given by its wave function before the defect leaves the region with a velocity on the order of u_s, and the adiabatic approximation is expected to give correct results. The validity of the adiabatic approximation must be seriously questioned for kink velocities of $\approx 10u_s$ and it surely breaks down for velocities approaching u_F. Using the translational degeneracy of a soliton by two bond lengths $2a$, its initial condition for a velocity $u_K = \eta \cdot u_a$ is assumed to be $v_i^t - v_i^{t-\Delta t} = (1/2)\eta(\pi/19.8)(v_i^t - v_{i+2}^t)$.

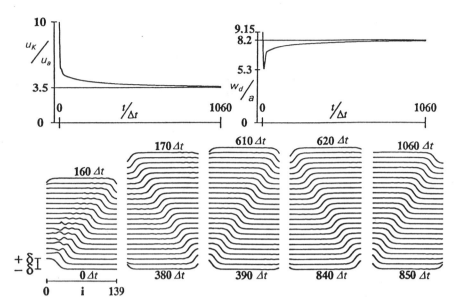

Fig. 4.7 Dynamics of a neutral kink K^0 along a ring of 139 sites with an initial velocity $u_K = 10u_a$ during 1060 time steps $\Delta t = 1.2 \times 10^{-15}$ s. Time evolution of the kink velocity u_K (top left) and the kink half-width w_d (top right). Bottom: Time evolution of bond length alternation $\delta_i = (-1)^i(d_i - d_{i-1})$ with $|\delta| = 0.053a$ (strobelight every $10\Delta t$). To avoid kink–phonon collision, phonons evolving behind the fast moving kink are relaxed by removing kinetic energy.

energy. The hump starts to oscillate, and smaller wiggles with frequency $0.92\,\nu_{\max}$ and wavelength $13a$ develop behind the kink, slowing it further and increasing its width again. The vibration is due to the fact that the soliton, as it moves over a distance given by its width, produces a maximum in the alternation and leaves a minimum behind. With progressive time evolution, the amplitude of these wiggles fades and the soliton reaches a constant width of $w_d = 8.2a$ and a constant velocity of $u_\infty = 3.5u_a$ without energy dissipation into the lattice. For this long-time study the phonons behind the soliton are suspended at some distance from the soliton to avoid soliton–phonon collision. The total energy is thus not conserved until steady state is reached. Guinea [65] and Bishop et al. [64] obtain within the SSH model a value for the maximal velocity between $u_\infty = 0.06u_a$ and $u_\infty = 4.0u_a$ by using different values of adjustable parameters. Waves of wavelength $12a$ were also described by Guinea [65] for $u_\infty = 4.3u_a$ and $u_\infty = 3.0u_a$. Note that our result $u_\infty = 3.5u_a$ depends on the electron–phonon coupling $\{\alpha_1, \alpha_2\}$ scaled to the experimental bond lengths but does not depend on the dynamical scaling factor k.

B. Excitation

Starting with the ground state of a *trans*-polyacetylene chain, one electron is excited from the highest state of the valence band to the lowest state of the conduction band (Fig. 4.8). These two states split off the band edges and meet $12\Delta t = 15 \times 10^{-15}$ s after excitation as nearly degenerate gap states (φ_v is the gap state originating from the valence band, φ_c is the gap state originating from the conduction band), and both corresponding wave functions have two localization centers close to each other (Fig. 4.9). Two hybrid states can be formed, $\varphi_{\text{left}} = (\varphi_c + \varphi_v)/\sqrt{2}$ with its localization center at the left and $\varphi_{\text{right}} = (\varphi_c - \varphi_v)/\sqrt{2}$ with its localization center at the right. This breaking of symmetry can be achieved by applying a small electric field (e.g. a potential drop to the right, where φ_v becomes localized to the right and φ_c becomes localized to the left, Fig. 4.9). Further dynamics show three "breaths" of annihilation (with overshooting, where the two gap states become separated), and creation of this "bound kink–antikink pair" is followed by its separation. $76\Delta t = 91 \times 10^{-15}$ s after excitation the two kinks move apart and a "breather" is left behind with the period of $21\Delta t = 25 \times 10^{-15}$ s. The band edges oscillate in and out of the gap, and the (reversed) bond length alternation of central bonds oscillates with the same frequency. Bishop et al. [64] obtained a breather period of 39×10^{-15} s.

Two channels are possible: a covalent channel (K^0, $\overline{K}^0$) and an ionic channel (K^+, $\overline{K}^-$ or K^-, $\overline{K}^+$). Following adiabatically the Born–Oppenheimer surface, a neutral kink K^0 and an antikink $\overline{K}^0$ move apart (Fig. 4.8) with both φ_v and φ_c singly occupied; thus there is no electron transfer [84–86]. If an electron transfer takes place [88–93] by jumping nonadiabatically to another Born–Oppenheimer surface, a positive kink K^+ (in Fig. 4.10 to the left) and a negative antikink $\overline{K}^-$ (in Fig. 4.10 to the right) move apart with doubly occupied φ_v and empty φ_c. Within the SSH-HMO model, only charged solitons are obtained after photoexcitation, which is shown by using an analogy to the hydrogen molecule [94,95]. The value of the photoproduced neutral-to-

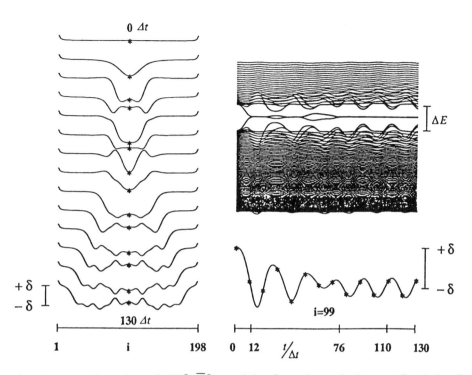

Fig. 4.8 Exciton-breather, covalent channel, $(K^0, \overline{K}^0)$, evolving from the excited state of a chain of 198 sites: ground-state geometry, but one electron excited from the highest level of the valence band to the lowest level of the conduction band. To break the symmetry, a small electric field is applied (potential drop to the right). Time evolution of δ_i (strobelight every 10 Δt). Time evolution of eigenvalues (both intergap states singly occupied) and of δ_i of central bonds.

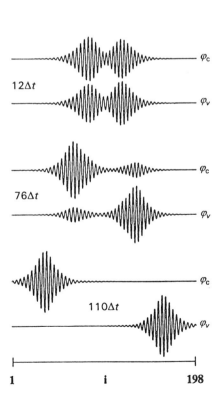

Fig. 4.9 Wave functions of lower intergap state φ_v and of upper intergap state φ_c (both singly occupied) at time indicated in multiples of Δt after excitation (Fig. 4.8).

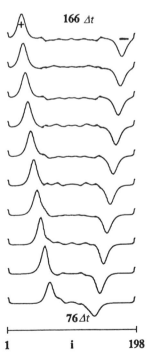

Fig. 4.10 Exciton-breather, ionic channel, $(K^+, \overline{K}^-)$, evolving $76\Delta t$ after excitation (Fig. 4.8) jumping nonadiabatically to another Born–Oppenheimer surface by electron transfer at $76\Delta t$; φ_v doubly occupied and φ_c empty (Fig. 4.9). Time evolution of $1/4(-\rho_{i-1} + 2\rho_i - \rho_{i+1}) - 1$, separation of opposite charges, K^+ moving to the left and $\overline{K}^-$ moving to the right (strobelight every $10\Delta t$).

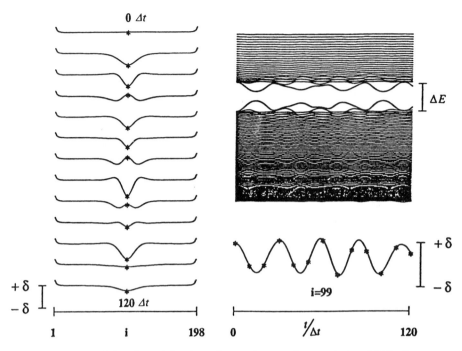

Fig. 4.11 Negative polaron P^- evolving from injection of one electron into a chain of 198 sites: ground-state geometry, but electron added to lowest state of conduction band. Time evolution of bond length alternation δ_i (strobelight every $10\Delta t$). Central bonds indicated by star. Time evolution of eigenvalues and of δ_i of central bonds.

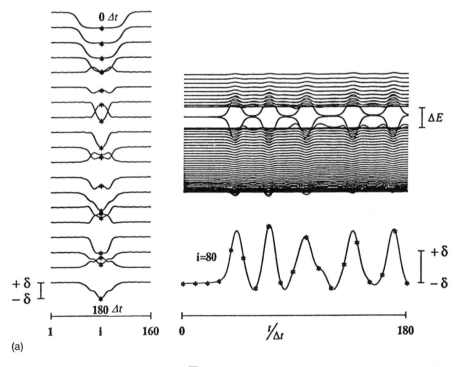

(a)

Fig. 4.12 Scattering of neutral kink K^0 and antikink $\overline{K}^0$ within a ring of 160 sites. Initial kink velocities $u_K = +3.5u_a$ and $u_{\overline{K}} = -3.5u_a$, respectively. Time evolution of bond length alternation δ_i (strobelight every $10\Delta t$). Time evolution of eigenvalues and of δ_i of central bonds and of kink velocities u_K respectively. (a) A_g occupation: lower intergap state φ_v doubly occupied, upper intergap state φ_c empty. (b) B_u occupation: both intergap states φ_v and φ_c singly occupied. (c) A_g occupation: lower intergap state φ_v empty, upper intergap state φ_c doubly occupied.

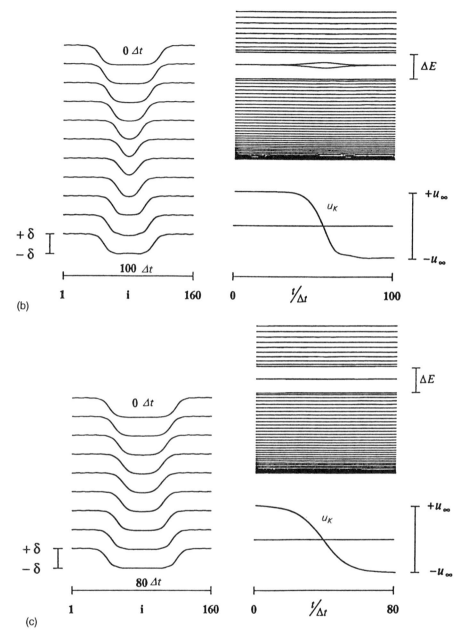

Fig. 4.12 (continued).

charged soliton branching ratio is still an open issue both experimentally [96–98] and theoretically [99,100].

C. Electron Injection

If one electron is added to the ground state of a *trans*-polyacetylene chain (Fig. 4.11), the lowest state of the conduction band is singly occupied. This state and the highest state of the valence band split off the band edges and move deep into the gap (they do not meet), while a negative polaron evolves. Later the states move back to the band edges, while the negative polaron disappears (the charge delocalizes), and this continues periodically with the period of $28\Delta t = 33.6 \times 10^{-15}$ s. Su and Schrieffer [3] report the appearance of shake-off phonons after electron injection. This we do not see in our model simulation.

If two electrons are added to the ground state of a *trans*-polyacetylene chain, the lowest state of the conduction band is doubly occupied. This state and the highest state of the valence band split off the band edges and meet as degenerate gap states. The evolving doubly negative charged defect is localized in the center of the chain. This behavior is similar to that of the covalent channel of the exciton. After three breaths a negative kink and a negative antikink (K^-, $\overline{K}^-$) move apart and a breather is left behind.

D. Kink Collision

Figure 4.12 shows the computer simulation of two neutral kinks (K^0, $\overline{K}^0$) embedded in a ring of 160 sites and approaching each other with velocities $u_K = +u_\infty = +3.5u_a$ and $u_{\overline{K}} = -u_\infty = -3.5u_a$. For the well-separated neutral kinks the spectrum shows two nearly degenerate intergap states. These states are occupied with two electrons; thus three different occupation configurations are possible:

1. The lower state is doubly occupied, and the upper state is empty (A_g symmetry). The degeneracy is removed as the two kinks collide and annihilate; the lower, doubly occupied intergap state digs into the valence band, while the upper, empty intergap state digs into the conduction band. The kinks are trapped in "bounce resonances", an oscillation between the annihilated state of a defect-free but overshot bond length alternation with an empty gap and the bound kink–antikink pair with two intergap states. The energy dissipates slowly into the lattice. These findings do not depend on the initial velocities.
2. Both states are each singly occupied (B_u symmetry). This occupation configuration has a similar course of dynamics as case 3 below.
3. The lower state is empty, and the upper state is doubly occupied (A_g symmetry). The degeneracy is removed slightly as the two kinks approach each other; the lower intergap state loses a small amount of energy, while the upper intergap state gains a small amount of energy. The kinks do not annihilate, but they stop before collision and turn backward; they repel each other, in the present case more strongly than in case 2. The degeneracy of states and the final kink velocities are restored. These findings do not depend on the initial velocities.

Guinea [65] reports within the SSH model the scattering of neutral kinks with different initial velocities: The kinks "are trapped in bounce resonances" for small initial velocities, and above a certain threshold they "bounce twice and separate to infinity with velocities significantly lower than the initial ones." Campbell et al. [101–103] report, within the ϕ^4 model, kinks "trapped in bounce resonances" for small initial velocities and above a certain threshold "n bounce collisions and separation to infinity" and "narrow windows in which kinks are trapped in bounce resonances." These phenomena do occur in our model when one changes nonadiabatically the occupation of levels of the nearly degenerate intergap states [88–93].

V. CONCLUSION

This chapter has presented a detailed study of the dynamics of nonlinear excitations of *trans*-polyacetylene chains. Neutral kinks move frictionless for velocities up to 3.5 times the velocity of sound and shake off phonons at higher velocities. Excitation leads to a localized defect, which breathes and later evolves into either a neutral or an oppositely charged kink–antikink pair moving apart. Injection of an electron leads to an oscillation between a delocalized charge distribution and the negative polaron state. Scattered neutral kinks are either trapped and form a breather or repel each other, depending on the occupation of their states.

The step-potential model is the conceptually simplest approach to treating electron–phonon coupling in π-electron systems and is the logical extension of the free electron model. Scaled to the experimental bond lengths of butadiene and benzene, the static properties of nonlinear excitations in *trans*-polyacetylene are well described. Based on standard adiabatic approximation, the extension of the step-potential model to the study of the dynamical behavior of nonlinear excitations in *trans*-polyacetylene is straightforward. The time unit is scaled to the frequency of the in-phase stretching mode of the polyene lattice. However, this does not affect the course of the dynamics shown.

Unsaturated long-chain hydrocarbon molecules with their extended π-electron system are suitable for bridging the gap between quantum chemistry and solid-state theory. A simple step-potential model of the nearly free π electron coupled to the lattice distortions is presented. The geometry of the molecule is obtained by requiring self-consistency between bond length and π-electron density in the middle of each bond. This quantum-mechanical model has no free parameter and leads to a satisfactory description of unsaturated hydrocarbon chains and rings and of defect states in *trans*-polyacetylene called solitons, such as kinks, polarons, bipolarons, and excitons.

ACKNOWLEDGMENT

I am grateful to Professor D. Beratan for encouraging me in writing this article.

APPENDIX A. POSTULATES DEFINING $V(d)$

We estimate the bond potential $V_i = V(d_i) - V(a)$ corresponding to bond i of length d_i by considering the electron as being in a Slater atomic orbital $\varphi_{\text{Slater}} \propto z \exp(-Z_{\text{Slater}} r/2a_0)$ where a_0 is the Bohr radius. The effective potential $V_{\text{eff}} = -(Z_{\text{Slater}} - 1)e^2/4\pi\epsilon_0 r$ is averaged [Eq. (A1)] over the plane perpendicular to the bond and crossing the bond in its middle (Fig. 4.A1) to give

$$V(d) = \int_{\text{plane}} 2V_{\text{eff}}\varphi_{\text{Slater}}^2 \, \rho d\rho d\varphi \Big/ \int_{\text{plane}} \varphi_{\text{Slater}}^2 \, \rho d\rho d\varphi$$

where the two adjacent C atoms are considered as point charges of $Z_{\text{Slater}} - 1$ atomic units (shielded by an electron each, $Z_{\text{Slater}} = 3.25$ for carbon). The other atoms are assumed to be shielded by the residual π-electrons [no contribution to $V(d)$].

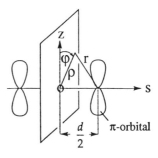

Fig. 4.A1 Averaging the effective potential V_{eff} of two adjacent C atoms.

$$\begin{aligned}
V(d) &= -\frac{2(Z_{\text{Slater}} - 1)e^2}{4\pi\epsilon_0} \left(\frac{\int_{\rho=0}^{\infty} \int_{\varphi=0}^{2\pi} \frac{\rho^2 \cos^2\varphi}{[\rho^2 + (d/2)^2]^{1/2}} \exp\left\{ -\frac{Z_{\text{Slater}}}{a_0}\left[\rho^2 + \left(\frac{d}{2}\right)^2\right]^{1/2} \right\} d\varphi \, \rho \, d\rho}{\int_{\rho=0}^{\infty} \int_{\varphi=0}^{2\pi} \rho^2 \cos^2\varphi \exp\left\{ -\frac{Z_{\text{Slater}}}{a_0}\left[\rho^2 + \left(\frac{d}{2}\right)^2\right]^{1/2} \right\} d\varphi \, \rho \, d\rho} \right) \\
&= -\frac{e^2}{4\pi\epsilon_0 2a_0} 4(Z_{\text{Slater}} - 1)a_0 \frac{\int_{r=d/2}^{\infty} \left[r^2 - \left(\frac{d}{2}\right)^2\right] \exp\left(-\frac{Z_{\text{Slater}}}{a_0} r\right) dr}{\int_{r=d/2}^{\infty} \left[r^2 - \left(\frac{d}{2}\right)^2\right] \exp\left(-\frac{Z_{\text{Slater}}}{a_0} r\right) r \, dr} \\
&= (-13.6 \text{ eV}) \frac{8(Z_{\text{Slater}} - 1)Z_{\text{Slater}}\left(2 + \frac{Z_{\text{Slater}}}{a_0}d\right)}{12 + 6\frac{Z_{\text{Slater}}}{a_0}d + \frac{Z_{\text{Slater}}^2}{a_0^2}d^2} \approx \left[39.2\left(\frac{d}{a} - 1\right) - 23.9\left(\frac{d}{a} - 1\right)^2 - 61.3\right] \text{eV}
\end{aligned} \quad (A1)$$

APPENDIX B. THE SSH–HÜCKEL–HMO HAMILTONIAN

In the SSH model the tightly bound π electron is treated by Hückel MO theory. Coupling to the lattice distortion is achieved by determining the minimum of the total energy. The Hamiltonian

$$H_{\text{SSH}} = \sum_i \frac{1}{2} K_x(u_i - u_{i+1})^2 \quad \text{(B1)}$$
$$- 2\sum_i [t_0 + \alpha_{\text{SSH}}(u_i - u_{i+1})]p_{i,j+1}$$

is the sum of

1. The elastic energy of the lattice resulting from the compression of the σ-bonds due to the (projected) displacements u from the reference configuration. The values used for the force constant $K = (4/3)K_x$ range from $K = 17.3$ eV/Å² to $K = 90$ eV/Å². Recommended values are $K = 23$–28 eV/Å² [2,25] or $K = 46$ eV/Å² [53].

2. The electronic energy of the π-electrons, where

$$p_{i,j+1} = \frac{1}{2} \sum_\sigma (c_{i,\sigma}^+ c_{i+1,\sigma} - c_{i+1,\sigma}^+ c_{i,\sigma})$$

is the bond order operator, where $c_{i,\sigma}^+$ creates and $c_{i,\sigma}$ annihilates an electron in the Wannier state ψ_i. Values for the intrinsic transfer matrix element t_0 frequently used range from $t_0 = 2.5$ eV to $t_0 = 3.0$ eV with the recommended value being $t_0 = 2.9$ eV [53]. Values for the electron–phonon coupling constant α_{SSH} range from $\alpha_{\text{SSH}} = 4.1$ eV/Å to $\alpha_{\text{SSH}} = 9.$ eV/Å. Recommended: $\alpha_{\text{SSH}} = 4.5$ eV/Å [52].

It was discussed by Peierls [83] that a one-dimensional metal is unstable against a periodic lattice distortion with the wave vector $2k_F$ (for a half-filled band $k_F = \pi/2a$) and the lattice dimerizes: the gain in electronic energy always outweighs the loss of elastic energy. Inserting the dimerization, $u_i = (-1)^i u$, $t_{d,s} = t_0 \pm 2\alpha_{\text{SSH}}u$ for single bonds (s) and double bonds (d) respectively, and the cyclic boundary condition $\psi_{2n+2} = \psi_{2n} \exp(2ika)$ into (B1), the Schrödinger equation

$$E^2 \psi_{2n} = (t_d^2 + t_s^2)\psi_{2n} + t_d t_s(\psi_{2n+2} + \psi_{2n-2})$$

gives the dispersion relation

$$E_k^2 = 4t_0^2 \cos^2(ka) + \Delta_{\text{SSH}}^2 \sin^2(ka)$$

with the gap $2\Delta_{\text{SSH}} = 8\alpha_{\text{SSH}}u$. With the elliptic integral

$$\int_0^{\pi/2} d\varphi [1 - (1 - z^2)\sin^2\varphi]^{1/2} = E(1 - z^2)$$
$$\approx 1 + \frac{1}{4}z^2 \left[2\ln\left(\frac{4}{|z|}\right) - 1 \right]$$

the total energy per site is

$$E_0(\Delta_{SSH}) = -\frac{4t_0}{\pi}\int_0^{\pi/2}\left\{1-\left[1-\left(\frac{\Delta_{SSH}}{2t_0}\right)^2\right]\sin^2(ka)\right\}^{1/2}d(ka) + 2K_x\left(\frac{\Delta_{SSH}}{4\alpha_{SSH}}\right)^2 \quad \text{(B2)}$$

$$\approx -\frac{4t_0}{\pi} - \frac{2\ln(8t_0/\Delta_{SSH})-1}{4\pi t_0}\Delta_{SSH}^2 + \frac{K_x}{8\alpha_{SSH}^2}\Delta_{SSH}^2$$

Minimizing the energy with respect to Δ_{SSH} and introducing the parameter $\lambda = 2\alpha_{SSH}^2/\pi t_0 K_x$ gives the amount of dimerization and the corresponding gap

$$2\Delta_{SSH} = 16t_0\exp\left[-\left(1+\frac{1}{2\lambda}\right)\right] \quad \text{(B3)}$$

APPENDIX C. EXCITATION INCLUDING ELECTRON CORRELATION EFFECTS

We consider the π-electron system in the perturbing electric field of the incident light and apply the time-dependent Schrödinger equation. The electron under consideration is treated as being exposed to the time-dependent field of the light and to the time-dependent field of all other electrons in the system. Omitting electron exchange (antisymmetrization requirement) in contrast to the standard approximations allows transparency, and the essential physical effects determining spectral features are easily grasped [54,55,60–62]. Electron 1 described by the time-dependent wave function $\Psi_1 = a_1(t)\psi_{A1} + b_1(t)\psi_{B1}$ is assumed to be in orbital ψ_{A1}, and it can be excited into orbital ψ_{B1} with the excitation energy ΔE_1 and the oscillator strength

$$f_1 = \frac{2m_e}{e^2}\left(\frac{\Delta E_1}{\hbar^2}\right)\left(\frac{M_1^2}{3}\right) \quad \text{(C1)}$$

where $M_1 = \int \psi_{A1}\cdot ex\cdot \psi_{B1}ds$ is the transition moment in x-direction. Electron 2 is described analogously by $\Psi_2 = a_2(t)\psi_{A2} + b_2(t)\psi_{B2}$ with excitation energy ΔE_2, transition moment $M_2 = \int \psi_{A2}\cdot ex\cdot \psi_{B2}ds$ and oscillator strength

$$f_2 = \frac{2m_e}{e^2}\left(\frac{\Delta E_2}{\hbar^2}\right)\left(\frac{M_2^2}{3}\right) \quad \text{(C2)}$$

We focus on the interaction of electron 1 with electron 2 [60–62],

$$\mathbf{H}_1\Psi_1 = i\hbar\frac{\partial}{\partial t}\Psi_1 \quad \text{(C3)}$$

$$\mathbf{H}_1 = \mathbf{H}_{10} + ex_1F_0\cos(2\pi\nu t) + \mathbf{H}_{12} \quad \text{(C4)}$$

$$\mathbf{H}_{12}(s_1) = \int g(1,2)\Psi_2\Psi_2^* \, ds_2 \quad \text{(C5)}$$

$\mathbf{H}_{10}$ relates to the time-independent Schrödinger equation for electron 1 in the field of the skeleton of the molecule. Correspondingly, $ex_1F_0\cos(2\pi\nu t)$ is the potential energy of electron 1 in the radiation field. We have to include the repulsion energy of electron 1 in the field of electron 2. This corresponds to the term $\mathbf{H}_{12}$ where $\Psi_2\Psi_2^*$ is the time-dependent probability density distribution of electron 2 and $g(1,2)$ is the repulsion energy of electrons 1 and 2 depending on the coordinates s_1 and s_2 and assuming a permittivity of $\epsilon = 2.5$ to take care of the medium and the σ electrons. Then $\mathbf{H}_{12}$ is the repulsion energy for a given coordinate s_1 of electron 1 averaged over the distribution of electron 2. A corresponding time-dependent Schrödinger equation holds for electron 2.

$$\mathbf{H}_2\Psi_2 = i\hbar\frac{\partial}{\partial t}\Psi_2 \quad \text{(C6)}$$

$$\mathbf{H}_2 = \mathbf{H}_{20} + ex_2F_0\cos(2\pi\nu t) + \mathbf{H}_{21} \quad \text{(C7)}$$

$$\mathbf{H}_{21}(s_2) = \int g(1,2)\Psi_1\Psi_1^* \, ds_1 \quad \text{(C8)}$$

We obtain the time-dependent part of the dipole moment $\mu(t) = M_1\text{Re}[b_1(t)/a_1(t)] + M_2\text{Re}[b_2(t)/a_2(t)]$ by solving the two coupled Schrödinger equations given by Eqs. (C3)–(C8) [60–62]. The problem of finding the time evolution of $\mu(t)$ from (C3)–(C8) turns out to be the same as that of asking for the time evolution of $\mu(t) = \hat{x}_1(t)q_1 + \hat{x}_2(t)q_2$, where $\hat{x}_1(t)$ and $\hat{x}_2(t)$ are the elongations of two coupled classical oscillators in the electric field of the incident light (with force constants D_1 and D_2, charges q_1 and q_2, and damping constants ρ_1 and ρ_2):

$$m_1\frac{d^2}{dt^2}\hat{x}_1 = -D_1\hat{x}_1 - \rho_1\frac{d}{dt}\hat{x}_1 + q_1F_0\cos(2\pi\nu t) - D_{12}\hat{x}_2 \quad \text{(C9)}$$

$$m_2\frac{d^2}{dt^2}\hat{x}_2 = -D_2\hat{x}_2 - \rho_2\frac{d}{dt}\hat{x}_2 + q_2F_0\cos(2\pi\nu t) - D_{21}\hat{x}_1 \quad \text{(C10)}$$

We obtain formal identity with the quantum-mechanical treatment for

$$\frac{D_{12}}{m_1} = \frac{D_{21}}{m_2} = \frac{2}{\hbar^2}\sqrt{\Delta E_1 \Delta E_2}\, J_{12} \quad \text{(C11)}$$

$$\frac{1}{2\pi}\sqrt{\frac{D_1}{m_1}} = \nu_{10} = \frac{\Delta E_1}{h}, \quad \frac{1}{2\pi}\sqrt{\frac{D_2}{m_2}} = \nu_{20} = \frac{\Delta E_2}{h} \quad \text{(C12)}$$

$$\frac{q_1^2}{m_1} = \frac{2\Delta E_1}{\hbar^2}M_1^2, \quad \frac{q_2^2}{m_2} = \frac{2\Delta E_2}{\hbar^2}M_2^2 \quad \text{(C13)}$$

where $J_{12} = \iint \psi_{A1}\psi_{B1}g(1,2)\psi_{A2}\psi_{B2}ds_1 ds_2$ is the coupling integral. The corrected transition energies

ΔE_1^{corr} and ΔE_2^{corr} and oscillator strengths f_1^{corr} and f_2^{corr} are given by the relations

$$\Delta E_1^{\text{corr}} = \sqrt{A - B\sqrt{1 + \eta^2}},$$

$$\Delta E_2^{\text{corr}} = \sqrt{A + B\sqrt{1 + \eta^2}} \tag{C14}$$

$$f_1^{\text{corr}} = \left(C\sqrt{\frac{f_1}{2}} - D\sqrt{\frac{f_2}{2}}\right)^2,$$

$$f_2^{\text{corr}} = \left(C\sqrt{\frac{f_1}{2}} + D\sqrt{\frac{f_2}{2}}\right)^2 \tag{C15}$$

$$2A = (\Delta E_2)^2 + (\Delta E_1)^2, \quad 2B = (\Delta E_2)^2 - (\Delta E_1)^2 \tag{C16}$$

$$C = \left(1 + \frac{1}{\sqrt{1 + \eta^2}}\right)^2, \quad D = \left(1 - \frac{1}{\sqrt{1 + \eta^2}}\right)^2$$

$$\eta = \frac{2}{B}\sqrt{\Delta E_1 \Delta E_2} J_{12} \tag{C17}$$

Equations (C1), (C2), (C11), (C13), and (C17) are valid for transitions from singly occupied states to empty states. They are also valid for transitions from doubly occupied states to singly occupied states, since by the Pauli principle just one electron can carry out the transition without changing its spin. For transitions from doubly occupied states to empty states the terms on the left side of Eqs. (C1), (C2), (C11), (C13), and (C17) must be multiplied by a factor of 2.

APPENDIX D. DERIVATION OF THE EQUATION OF MOTION

The dynamics of bond i are given by Eq. (D1), subtracting Eq. (8) from its nearest neighbor.

$$\frac{d^2}{dt^2}(d_i \cos \varphi_i) = \frac{1}{M_{\text{CH}}}(F_{i+1} - F_i) \tag{D1}$$

With the inverse of Eq. (4),

$$\frac{d_i}{a} - 1 = \frac{1}{\beta_1} v_i + \frac{\beta_2}{\beta_1^3} v_i^2 + o[v_i^3] \tag{D2}$$

and setting $\varphi_i(t) = \varphi$ for all i, inserting the force [Eq. (9)] with

$$\chi_i = [(\tilde{\rho}_{i+1} - \rho_{i+1}) - 2(\tilde{\rho}_i - \rho_i) + (\tilde{\rho}_{i-1} - \rho_{i-1})]a$$

Eq. (D1) can be written as

$$\frac{d^2}{dt^2} v_i = \frac{\frac{k\beta_1}{M_{\text{CH}}} \chi_i - \frac{2\beta_2}{\beta_1^2}\left(\frac{d}{dt} v_i\right)^2}{1 + (2\beta_2/\beta_1^2) v_i} \tag{D3}$$

$$\approx \frac{k\beta_1}{M_{\text{CH}}} \chi_i \left(1 - \frac{2\beta_2}{\beta_1^2} v_i\right) - \frac{2\beta_2}{\beta_1^2}\left(\frac{d}{dt} v_i\right)^2$$

To integrate Eq. (D3) we use its discrete form with $\omega^2 = k\beta_1/M_{\text{CH}}$,

$$\mathbf{v}_i^{t+\Delta t} = 2\mathbf{v}_i^t - \mathbf{v}_i^{t-\Delta t} + (\Delta t \omega)^2 \chi_i^t \left(1 - \frac{2\beta_2}{\beta_1^2} \mathbf{v}_i^t\right)$$

$$- \frac{2\beta_2}{\beta_1^2}(\mathbf{v}_i^t - \mathbf{v}_i^{t-\Delta t})^2 \tag{D4}$$

Starting with an appropriate initial condition and choosing $\Delta t\omega$ sufficiently small to reach good accuracy and stability [72–74], the evolution in time is obtained from Eq. (D4). Numerical integration with $\Delta t\omega = 0.5$ of the in-phase stretching mode of a polyene lattice gives 19.8 $\Delta t = 1/\nu_{\text{max}}$; thus with the experimental value $\nu_{\text{max}} = c \times 1400$ cm^{-1} (vibronic structure of the absorption band of polyenes), we have set the time step $\Delta t = 1.2 \times 10^{-15}$ s. The velocity of sound along s is $u_a = \pi a \nu_{\text{max}} = 0.159 a/\Delta t = 1.85 \cdot 10^4$ m/s. Conservation of total energy is proved by either demonstrating no loss of amplitude in the case above or by demonstrating constant velocity of a soliton moving with low speed.

REFERENCES

1. D. K. Campbell, in *Molecular Electronic Devices II* (F. L. Carter, ed.), Marcel Dekker, New York, 1987, p. 111.
2. W.-P. Su, J. R. Schrieffer, and A. J. Heeger, *Phys. Rev. B* 22:2099 (1980); 28:1138 (1983).
3. W.-P. Su and J. R. Schrieffer, *Proc. Natl. Acad. Sci. USA* 77:5626 (1980).
4. H. Takayama, Y. R. Lin-Liu, and K. Maki, *Phys. Rev. B* 21:2388 (1980).
5. D. K. Campbell and A. R. Bishop, *Phys. Rev. B* 24:4859 (1981).
6. D. K. Campbell and A. R. Bishop, *Nucl. Phys. B* 200:297 (1982).
7. S. A. Brazovskii and N. Kirova, *Pris'ma Zh. Eksp. Teor. Fiz* 33:6 (1981) [*JEPT Lett.* 33:4 (1981)].
8. P. Horsch, *Phys. Rev. B* 24: 7351 (1981).
9. S. N. Dixit and S. Mazumdar, *Phys. Rev. B* 29:1824 (1984).
10. D. Baeriswyl and K. Maki, *Phys. Rev. B* 31:6633 (1985).
11. S. Kivelson and W.-K. Wu, *Phys. Rev. B* 34:5423 (1986).
12. D. S. Boudreaux, R. R. Chance, J. L. Brédas, and R. Silbey, *Phys. Rev. B* 28:6927 (1983).
13. A. Karpfen and J. Petkov, *Theoret. Chim. Acta* 53:65 (1979).
14. H. O. Villar, M. Dupius, J. D. Watts, G. J. B. Hurst, and E. Clementi, *J. Chem. Phys.* 88:1003 (1988).
15. H. O. Villar, M. Dupius, and E. Clementi, *Phys. Rev. B* 37:2520 (1988).
16. J. S. Craw, J. R. Reimers, G. B. Bacskay, A. T. Wong, and N. S. Hush, *Chem. Phys.* 167:77, 101 (1992).

17. W. Förner, *Chem. Phys. 160*:173, 189 (1992).
18. J. Paldus, J. Cizek, and P. Piecuch, *Int. J. Quant. Chem. 42*:135, 165 (1992).
19. D. Baeriswyl, in *Theoretical Aspects of Band Structures and Electronic Properties of Pseudo-One-Dimensional Solids* (H. Kamimura, ed.), Reidel, Dordrecht, 1985, p. 1.
20. W.-P. Su, in *Handbook of Conducting Polymers*, Marcel Dekker, New York, 1986, (T. A. Skotheim et al., eds.), p. 757.
21. E. J. Mele, in *Handbook of Conducting Polymers*, Marcel Dekker, New York, 1986, (T. A. Skotheim et al., eds.), p. 795.
22. D. K. Campbell, A. R. Bishop, and M. J. Rice, in *Handbook of Conducting Polymers*, Marcel Dekker, New York, 1986, (T. A. Skotheim et al., eds.), p. 937.
23. S. Roth and H. Bleier, *Adv. Phys. 36*:385 (1987).
24. Y. U. Lu (ed.), *Solitons and Polarons in Conducting Polymers*, World Scientific, Singapore, 1988.
25. A. J. Heeger, S. Kivelson, J. R. Schrieffer, and W.-P. Su, *Mod. Phys. 60*:781 (1988).
26. D. Baeriswyl, D. K. Campbell, and S. Mazumdar, in *Conjugated Conducting Polymers* (H. Kiess, ed.), Springer, Heidelberg, 1992, p. 7.
27. Y. Wada (ed.), *Soliton no bussei to seigyo* (*The Properties and Control of Solitons*), Japanese/American Physical Society, Pergamon/IOP/Elsevier, New York, 1988.
28. Y. Ono, A. Terai, and H. Takayama (eds.), *Nonlinear Excitations in Quasi-One-Dimensional Materials, Soliton and Its Relevance, Prog. Theor. Phys. Suppl. 113*: (1993).
29. W. Förner, *Adv. Quantum Chem. 25*:207 (1994).
30. F. L. Carter, A. Schulz, and D. Duckworth, in *Molecular Electronic Devices II* (F. L. Carter, ed.), Marcel Dekker, New York, 1987, p. 149.
31. H. D. Försterling, W. Huber, and H. Kuhn, *Int. J. Quant. Chem. 1*:225 (1967).
32. H. Kuhn, W. Huber, G. Handschig, H. Martin, F. Schäfer, and F. Bär, *J. Chem. Phys. 32*:467 (1960).
33. H. Martin, H. Kuhn, F. Bär, W. Huber, and G. Handschig, *J. Chem. Phys. 32*:470 (1960).
34. H. Kuhn, *Prog. Chem. Org. Natural Products 17*: 404 (1959).
35. H. D. Försterling and H. Kuhn, *Int. J. Quant. Chem. 2*:413 (1968).
36. H. Kuhn, *Helv. Chim. Acta 32*:2247 (1949).
37. C. Kuhn, in *Electronic Properties of Fullerenes* (H. Kuzmany, J. Fink, M. Mehring, and S. Roth, eds.), Springer Ser. Solid State Sci. 117, Springer-Verlag, Berlin, 1993, p. 131.
38. H. Kuhn, *Helv. Chim. Acta 31*:1441 (1948).
39. C. Kuhn, *Phys. Rev B 40*:7776 (1989).
40. J. A. Pople and S. H. Walmsley, *Mol. Phys. 5*:15 (1962).
41. C. Kuhn, *Synth. Met. 41*:3681 (1991).
42. J. R. Reimers and N. S. Hush, *Chem. Phys. 176*:407 (1993).
43. L. M. Jackson, F. Sondheimer, Y. Amiel, D. A. Ben-Efraim, R. Gaoni, R. Wolovski, and A. Bothner-By, *J. Am. Chem. Soc. 84*:4307 (1962).
44. S. Flügge, *Practical Quantum Mechanics*, Vol. 1, Springer, New York, 1974.
45. J. B. Lando and M. K. Thakur, in *Molecular Electronic Devices II* (F. L. Carter, ed.), Marcel Dekker, New York, 1987, p. 237.
46. H. Kuhn, *J. Chem. Phys. 29*:958 (1958).
47. K. S. Schweizer, *J. Chem. Phys. 85*:4181 (1986).
48. Z. G. Soos and S. Ramasesha, *Phys. Rev B 29*:5410 (1984).
49. H. Suzuki, *Electronic Absorption Spectra and Geometry of Organic Molecules*, Academic, New York, 1967.
50. C. Sandorfy, *Die Elektronenspektren in der theoretischen Chemie*, Verlag Chemie, Weinheim, 1961, pp. 95–98.
51. A. Streitwieser, *Molecular Orbital Theory for Organic Chemists*, Wiley, New York, 1961, p. 192 and 241.
52. E. Jeckelmann and D. Baeriswyl, *Synth. Met. 65*: 211 (1994).
53. D. Baeriswyl, G. Harbeke, H. Kiess, and W. Meyer, in *Electronic Properties of Polymers* (J. Mort and G. Pfister, eds.), Wiley, New York, 1982, Chap. 7.
54. H. Kuhn and C. Kuhn, *Chem. Phys. Lett. 204*:206 (1993).
55. C. Kuhn and H. Kuhn, *Synth. Met. 68*:173 (1995) and references therein.
56. T. Kakitani, *Prog. Theor. Phys. 51*:656 (1974).
57. P. K. Kahol, W. G. Clark, and M. Mehring, in *Conjugated Conducting Polymers* (H. Kiess, ed.), Springer, Heidelberg, 1992, p. 217.
58. Z. Vardeny, J. Orenstein, and G. L. Baker, *Phys. Rev. Lett. 49*:1043 (1928).
59. K. Tanaka, K. Yoshizawa, and T. Yamabe, *Int. J. Quant. Chem. 40*:315 (1991).
60. W. Huber, G. Simon, and H. Kuhn, *Z. Naturforsch. 17a* 99 (1962).
61. H. D. Försterling, W. Huber, H. Kuhn, H. H. Martin, A. Schweig, F. F. Seelig, and W. Stratmann, in *Optische Anregung organischer Systeme*, 2nd Int. Farbensymp. (W. Foerst, ed.), Verlag Chemie, Weinheim, 1966, p. 55.
62. H. D. Försterling and H. Kuhn, *Int. J. Quant. Chem. 2*:413 (1968).
63. L. M. Tolbert and M. E. Ogle, *J. Am. Chem. Soc. 112*:9519 (1990).
64. A. R. Bishop, D. K. Campbell, P. S. Lomdahl, B. Horovitz, and S. R. Phillpot, *Phys. Rev. Lett. 52*: 671 (1984).
65. F. Guinea, *Phys. Rev. B 30*:1884 (1984).
66. C. L. Wang and F. Martino, *Phys. Rev. B 34*:5540 (1986).
67. A. Terai and Y. Ono, *J. Phys. Soc. Jpn. 55*:213 (1986).
68. S. R. Phillpot, D. Baeriswyl, A. R. Bishop, and P. S. Lomdahl, *Phys. Rev. B 35*:7533 (1987).

69. F. Chien, Y. Kashimori, K. Nishimoto, and O. Tanimoto, *Chem. Phys. 125*:269 (1988).
70. W. Förner, C. L. Wang, F. Martino, and J. Ladik, *Phys. Rev B 37*:4567 (1988).
71. Y. Ono and A. Terai, *J. Phys. Soc. Jpn. 59*:2893 (1990).
72. J. Baumgarte, *Comput. Methods Appl. Mech. Eng. 1*:1 (1972).
73. R. W. Hockney and J. W. Eastwood, *Computer Simulation Using Particles*, Hilger, Philadelphia, 1988.
74. H. J. C. Berendsen and W. F. van Gunsteren, in *Molecular Dynamics Simulation of Statistical-Mechanical Systems*, Proc. Enrico Fermi School (G. Ciccotti and W. G. Hoover, eds.), North-Holland, Amsterdam, 1986, p. 43.
75. C. Kuhn and W. F. van Gunsteren, *Solid State Commun. 87*:203 (1993).
76. C. Kuhn, *Synth. Met. 57*:4350 (1993).
77. C. Kuhn, in *Future Directions of Non-Linear Dynamics in Physical and Biological Systems* (P. L. Christiansen, J. C. Eilbeck, R. D. Parmentier, and A. C. Scott, eds.), Plenum, New York, 1993, p. 67.
78. C. Tric, *J. Chem. Phys. 51*:4778 (1969).
79. E. Ehrenfreund, Z. Vardeny, O. Brafman, and B. Horovitz, *Phys. Rev. B 36*:1535 (1987).
80. H. J. Schulz, *Phys. Rev. B 18*:5756 (1978).
81. M. Nakahara and K. Maki, *Phys. Rev. B 25*:7789 (1982).
82. W. Kohn, *Phys. Rev. Lett. 2*:393 (1959).
83. R. E. Peierls, *Quantum Theory of Solids*, Clarendon, Oxford, 1955, p. 108.
84. Y. Ono and A. Terai, *J. Phys. Soc. Jpn. 59*:2893 (1990).
85. M. Kuwabara and Y. Ono, *J. Phys. Soc. Jpn. 63*:1081 (1994).
86. W. Förner, *Phys. Rev. B 44*:11743 (1991).
87. W. Förner, *J. Mol. Struct. (Theochem.) 282*:235 (1993).
88. D. deVault, *Quantum Mechanical Tunneling in Biological Systems*, 2nd ed., Cambridge Univ. Press, Cambridge, UK, 1984.
89. K. Prassides, P. N. Schatz, K. Y. Wong, and P. Day, *J. Phys. Chem. 90*:5588 (1986).
90. T. Azumi and K. Maztsuzaki, *Photochem. Photobiol. 25*:315 (1977).
91. S. Larsson and M. Braga, *Chem. Phys. 176*:367 (1993).
92. E. Neria and A. Nitzan, *J. Chem. Phys. 99*:1109 (1993).
93. T. W. Hagler and A. J. Heeger, *Phys. Rev. B 49*:7317 (1994).
94. R. Ball, W. P. Su, and J. R. Schrieffer, *J. Phys. C3*:429 (1983).
95. E. R. Davidson, *J. Chem. Phys. 33*:1577 (1960); *35*:1189 (1961).
96. P. D. Townsend and R. H. Friend, *Phys. Rev. B. 40*:3112 (1989).
97. Z. V. Vardeny, *Chem. Phys. 177*:743 (1993).
98. D. Yu. Paraschuk, T. A. Kulakov, and V. M. Kobryanskii, *Phys. Rev. B 50*:907 (1994).
99. J. Takimoto and M. Sasai, *Phys. Rev. B 39*:8511 (1989).
100. O. Rubner, W. Förner, and J. Ladik, *Synth. Met. 61*:279 (1993).
101. D. K. Campbell, J. F. Schonfeld, and C. A. Windgate, *Physica 9D*:1 (1983).
102. M. Peyrard and D. K. Campbell, *Physica 9D*:33 (1983).
103. D. K. Campbell, M. Peyrard, and P. Sodano, *Physica 19D*:165 (1986).

5
Electron Spin Dynamics

Maxime Nechtschein*
Département de Recherche Fondamentale sur la Matière Condensée, CEA–Grenoble, France

I. INTRODUCTION

Electron spin dynamics occupies a special place between magnetic resonance and transport studies. As discussed in the first edition of the *Handbook of Conducting Polymers* [1], it has been used for investigating the motion of solitons in neutral polyacetylene. Since then such studies have been extended to the motion of polarons in various conductive polymers. In the case of polarons, spin carriers being also charge carriers, spin dynamics studies become a method of investigating transport properties. A unique advantage of this method is that it enables one to probe the conduction process at the polymeric chain scale. For example, the intra- and interchain conductivities can be distinguished. In other words, the anisotropy of conductivity can be estimated, even in powder or amorphous samples. Furthermore, the conductivity of isolated conducting islands can be determined.

The aim of spin dynamics studies is to specify the motion of the electron spins that are present in a given compound. This motion reflects basic properties such as magnetism or transport. We can deal with a "true motion" of the spin carriers or with just spin flips (the position of the spin remaining fixed). The latter case is concerned with magnetic insulators in which the spin orientation is modulated by exchange interactions. In such a case spin dynamics is related to magnetic properties. The case of a true motion of the spin carriers can be encountered in various categories of systems: triplet excited states in organic semiconducting salts (e.g., TCNQ salts), conduction electrons in organic conductors, solitons in neutral polyacetylene, polarons in conducting polymers. It is noteworthy that in all these examples we are dealing with pseudo-one-dimensional systems. For reasons that will appear later, spin dynamics studies are usually concerned with low-dimensional systems.

The basic quantity is $S_\lambda^\alpha(t)$, that is, the α component of the spin located on site λ at time t. Note that the value of this operator is zero if there is no spin carrier on site λ at time t. For simplicity, in the following we assume isotropic conditions, i.e., that the spin behavior is not dependent on the orientation of the applied magnetic field. Then all components are equivalent, and we shall restrict the discussion to the consideration of the z component.

In practice, we are concerned with the correlation function

$$\Gamma_\lambda(t) = \langle S_\lambda^z(t) S_0^z(0)\rangle / \langle S_0^z(0)^2\rangle \qquad (1)$$

which expresses the probability of finding S^z on site λ at time t, knowing that it was on site 0 at $t = 0$. Another convenient way to describe the spin motion is through the Fourier transform:

$$f_\lambda(\omega) = \int_{-\infty}^{+\infty} e^{i\omega t}\Gamma_\lambda(t)\, dt \qquad (2)$$

which is also called "the spectrum of the motion." In most cases the autocorrelation function ($\lambda = 0$) is sufficient to account for experiments related to spin dynamics. Then, $\Gamma_0(t)$ is nothing but the probability, $P(t)$, of the spin carrier coming back to its starting site at time t with the same spin orientation, $\Gamma_0(t) = P(t)$. The crucial point in spin dynamics studies is that the shape of the return probability, $P(t)$, as well as that of the spectrum $f(\omega)$, are significantly dependent on the motion dimensionality.

---

* Member of CNRS.

II. SPIN DYNAMICS BACKGROUND

A. Random Walk and Dimensionality

We assume a random walk, i.e., an incoherent motion, for the spin carrier. In practice this assumption is not restrictive. Coherence or incoherence of the motion is essentially a question of time scale. A motion appears coherent, i.e., ballistic, as long as it is not interrupted by any kind of collision. After a collision the memory is left and the motion appears incoherent. In spin dynamics studies the time scale to probe the motion corresponds to the Larmor periods in the applied magnetic fields, typically 10^{-8} and 10^{-11} s for the nuclear and electron spins, respectively. This is longer than the usual collision times of charge carriers in conducting materials.

Let a random walk particle start from site 0 at $t = 0$, and consider the probability $P(t)$ that this particle returns to its starting point at time t. We shall show that the shape of $P(t)$ depends on the dimensionality d of the space covered by the walker. We assume that the space is homogeneous, that is, that after a given time t the walker has the same probability of being at any point within a volume $V(t)$. We can thus write

$$P(t) \sim \frac{1}{V(t)} \tag{3}$$

The volume $V(t)$ is related to the distance $r(t)$ covered by the walker and to the space dimensionality:

$$V(t) \sim \langle r^2(t) \rangle^{d/2} \tag{4}$$

So, one has

$$P(t) \sim \langle r^2(t) \rangle^{-d/2} \tag{5}$$

For a classical diffusion process, and for a long time compared to elementary steps, the mean square distance increases linearly with time:

$$\langle r^2(t) \rangle = \tilde{D} t \tag{6}$$

where $\tilde{D}$ is the diffusion coefficient (expressed in cm^2/s). Note that it is often more convenient to use the diffusion rate (in rad/s): $D = \tilde{D}/c^2$, with c the intersite distance. Equations (5) and (6) yield

$$P(t) \sim (Dt)^{-d/2} \tag{7}$$

which shows the direct link between $P(t)$ and dimensionality.

By Fourier transform one obtains the shape of the motion spectrum for $d = 1$, 2, and 3.

$$f(\omega) \sim (D\omega)^{-1/2} \quad \text{for } d = 1 \tag{8a}$$

$$f(\omega) \sim D^{-1} \ln \frac{D}{\omega} \quad \text{for } d = 2 \tag{8b}$$

$$f(\omega) \sim \text{constant} \quad \text{for } d = 3 \tag{8c}$$

Let us note that these expressions are valid for time long compared to the elementary step time, $t \gg D^{-1}$, and for low frequency, $\omega \ll D$.

In fact, one is usually dealing with real systems, which are not purely one (or two- or three-dimensional) but present some anisotropy of their transport properties, so that three diffusion coefficients should be defined along the three axes. The diffusion equation, Eq. (9), has to be solved in the general case of anisotropic diffusion.

$$\frac{\partial}{\partial t} P(r, t) = \mathrm{D} \Delta P(r, t) \tag{9}$$

where D is a tensor.

If the diffusion coefficients along the three axes are sufficiently different from each other ($D_1 \gg D_2 \gg D_3$), the different regimes corresponding to Eqs. (8) can be observed, according to the frequency domains. Explicit forms of these typical regimes are given in Table 5.1. The shape of $f(\omega)$ is shown in Fig. 5.1.

B. One-Dimensional Motion

Equations (8) display the basic characteristics of the motion spectrum versus dimensionality. In particular, in one dimension there is a divergence of the spectrum as $\omega \to 0$. This corresponds to a slow decay of the correlation function in the time domain: $P(t) \sim t^{-1/2}$. The 1-D random walker has a significantly high probability of coming back to its starting point after a long time. This behavior can be visualized by the picture of a drunk man whose walk is restricted inside a corridor. Since he is unable to escape from the corridor there will be a significant probability that he will come back to his starting point after a long time. The divergence of the spectrum as $\omega \to 0$ also exists in 2-D systems, but it is much slower (logarithmic). This peculiar behavior is the reason, spin dynamics studies are mainly concerned with

Table 5.1 Motion Spectrum for Anisotropic Diffusion

Regime	1-D	2-D	3-D
Frequency domain	$D_1 \gg \omega \gg D_2, D_3$	$D_1, D_2 \gg \omega \gg D_3$	$D_1, D_2, D_3 \gg \omega$
$f(\omega)$	$\dfrac{1}{\sqrt{2D_1\omega}}$	$\dfrac{1}{2\pi\sqrt{D_1 D_2}} \ln \dfrac{4\pi^2 D_2}{\omega}$	$\dfrac{1}{2\pi^2 \sqrt{D_1 D_2}} \ln \dfrac{4\,e^2 D_2}{D_3}$

Electron Spin Dynamics

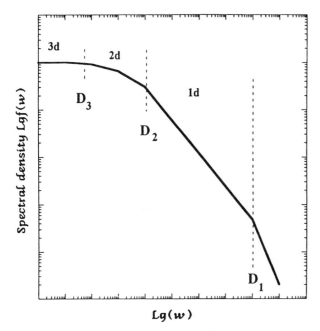

Fig. 5.1 Typical shape of the spectral density in the case of anisotropic motion.

low-dimensional, more particularly one-dimensional, systems.

However, significant differences exist between the ideal 1-D system considered by the theorist and the real material of the experimentalist. Only pseudo-one-dimensional systems can be found in nature. First, in a real system the chains are not totally isolated from each other. Second, the chains are not infinite. Third, they are not perfect; they contain defects. These different departures from ideal chains have consequences for the motion spectrum. Conversely, from the departure from the ideal 1-D motion spectrum (see Table 5.1), information can be obtained on the real material characteristics.

1. Interchain Couplings

In a real compound there are always some interchain couplings. In other words, the walker has a nonzero probability of jumping to the neighboring chain. In such a case the probability of coming back to the starting site drops rapidly. This results in a truncation of the low frequency divergence of $f(\omega)$. We feel that exploring the low frequency part of the motion spectrum can be a sensitive way to estimate the interchain couplings. A "cutoff frequency", ω_c, can be defined that is characteristic of the interchain couplings. Such behavior can be described in terms of crossover between 1-D and 3-D diffusion regimes. An intermediate 2D regime may also exist. In the case of cylindrical symmetry ($D_2 = D_3 = D_\perp$), a direct 1-D → 3-D crossover takes place, with a cutoff (or crossover) frequency on the order of $D_\perp$: $\omega_c = 1.7 \times D_\perp$ [2].

An approximate analytical expression can be given for $f(\omega)$ in a pseudo-one-dimensional system. We admit that the cutoff effect by interchain couplings can be described in the time domain by multiplying $P(t)$ by $\exp(-\omega_c t)$, which gives $f(t) \propto t^{-1/2} \exp(-\omega_c t)$. By Fourier transforming one obtains

$$f(\omega) = \frac{1}{\sqrt{4D_\parallel \omega_c}} \left[\frac{1 + \sqrt{1 + (\omega/2\omega_c)^2}}{1 + (\omega/2\omega_c)^2} \right]^{1/2} \quad (10)$$

where $D_\parallel$ is the diffusion coefficient along the chain. Note that the exponential cutoff used for the time correlation function is faster than the decay given by 3-D diffusion.

2. Finite Chain Length

Most authors have ignored the effect of chain finiteness on spin dynamics. However, it is clear that in real compounds the chains have limited length, which disturbs the spin carrier motion. The simplest picture we can give is that when a random walker reaches one end of a chain it is reflected. The disturbance occurs only after the time needed for the carrier to reach the chain end. This characteristic time varies as the square of the chain length,

$$\tau_c \sim \frac{1}{4}\left(\frac{N^2}{D_\parallel}\right)$$

where N is the number of unit cells. For large N, the disturbance due to chain ends may occur after the disturbance due to interchain couplings. The latter takes place after the interchain hopping time, $\sim D_\perp^{-1}$. According to this rough estimate, as long as $N > (2D_\parallel/D_\perp)^{1/2}$, the chain end effects can be ignored because interchain couplings are the dominant effects. In the opposite case, i.e., short chain length compared to the square root of the anisotropy, the diffusion motion will be disturbed first by the chain end [3]. To our knowledge, a complete calculation of the correlation function taking into account chain end effects has been carried out only very recently [4,5]. A simple model can be derived that has the advantage of giving an intuitive picture of the phenomenon. In this model the time domain is divided into two parts: (1) $t < \tau_c$; the walker has not reached a chain end, so $P(t) \sim (Dt)^{-1/2}$; and (2) $t > \tau_c$: the walker has reached a chain end, and $P(t)$ is close to its asymptotic value $P(t) \approx N^{-1}$. In the simple picture we admit that in the second part the fluctuations can be ignored since $P(t) \approx$ constant. Hence, only the first part has to be taken into account for the spectrum motion. The time to reach the chain end acts as a cutoff time. As $\omega \to 0$, instead of going to infinity, $f(\omega)$ goes to a finite value: $f(\omega) \to N/D$. Note that the constant value for $P(t)$ in the second part, which corresponds to a Dirac peak at $\omega = 0$ for $f(\omega)$, is not relevant for dynamic properties. More complete calculations lead to qualitatively the same conclusions [4].

3. Chain Defects

In the absence of microscopic characterization, chain defects are more difficult to model. Many kinds of defects or imperfections may exist that disturb the spin motion. So far, three parameters have been introduced to describe the spin motion: the intra- and interchain diffusion rates $D_\parallel$ and $D_\perp$ and the chain length N. Accounting for the influence of possible defects leads to the introduction of additional parameters. Many theoretical papers have been produced on the random walk problem in the presence of impurities, traps, or random hopping rates. It is out of the scope of this chapter to present a detailed discussion of this topic. Let us just envisage briefly a few typical cases.

a. Traps

Defects or impurities can have a localizing effect on the spin/charge carriers. For instance, such an effect has been invoked to explain the evolution of ESR linewidth in polyacetylene with oxygen contamination. The localizing effect is more and more pronounced as the temperature is lowered. It may appear only at low temperature if the trap energy is small compared to kT. Due to the one-dimensional character of polymeric componds, one can assert that a localizing effect will always appear if the temperature is sufficiently low. It is a matter of fact that a slowing down of the spin motion is usually observed at low temperature.

b. Disorder

Defects and imperfections produce disorder. A way to account for disorder effects is to introduce random values for the diffusion rates. It can be shown that if the diffusion rate distribution is not too peculiar the motion can be described in terms of diffusion with a diffusion rate that is the average of the distribution: $D \to \overline{D}$.

c. Chain Interruptions

If the defect interrupts the conjugation on the chain, it is equivalent to a chain length reduction. The case of partial interruption could be modeled by chain segments linked together by a diffusion coefficient $D' < D$. If there are a large number of small segments, the motion can be described as a diffusion with an average diffusion coefficient, as above.

C. Anomalous Diffusion and Fractal Network

If, instead of moving in a classical space, the random walker considered in Section II.B is moving on a fractal network, the volume visited during time t varies as $V(t) \sim \langle r^{2(t)} \rangle^{\tilde{d}/2}$, where $\tilde{d}$ is the spectral dimension [6]. Then the probability of returning to the starting site given by Eq. (5) can be generalized:

$$P(t) \sim \langle r^{2(t)} \rangle^{-\tilde{d}/2} \tag{11}$$

The regular diffusion law [Eq. (6)] is replaced by

$$\langle r^2(t) \rangle = \bar{D} t^\nu \tag{12}$$

where $\nu = \tilde{d}/d_f$, with d_f being the fractal dimension. Since $\tilde{d} < d_f$, one has $\nu < 1$. Diffusion in a fractal network is an anomalous diffusion [7]. It is less efficient than in a regular system. This is due to the fact that some paths of the ramified fractal network lead to culs-de-sac. For the spectral density of the motion, one obtains

$$f(\omega) = A\omega^{\tilde{d}/2 - 1} + B, \quad \tilde{d} < 2 \tag{13a}$$
$$f(\omega) = A\ln(B/\omega), \quad \tilde{d} = 2 \tag{13b}$$

Application of the concept of anomalous diffusion to conducting polymers has been introduced by Devreux and Lecavellier [8] to account for the nuclear relaxation in polypyrrole, which is a very disordered system (see Section V.C).

III. SPIN DYNAMICS AND MAGNETIC RESONANCE

Magnetic resonance offers various ways to access the spin motion. The basic idea is to use magnetic resonance techniques to look at spins of the sample that can be used as probes for the electron spin motion. The observed spins can be nuclear spins that are fixed probes, not participating in the motion. They can also be the electron spins themselves, which are then both the moving spins and the probes.

A. Nuclear Relaxation

Nuclear spins, usually the protons present in the material, are fixed probes, which undergo fluctuating magnetic fields produced by the nearby moving electronic spins. If the fluctuations have components at the Larmor frequencies, they will be a source of relaxation for the nuclear spins. The induced relaxation rate is then given by [9]

$$\frac{1}{T_1} = kT\chi \left[\frac{3}{5} d^2 f(\omega_N) + \left(a^2 + \frac{7}{5} d^2\right) f(\omega_e) \right] \tag{14}$$

where a and d are the averaged scalar and dipolar electronic–nuclear couplings, ω_N and ω_e are the Larmor nuclear and electronic frequencies ($\omega_{e,N} = \gamma_{e,N} H_0$, with H_0 the applied magnetic field), respectively; k is the Boltzmann constant, T the absolute temperature, and χ the reduced spin susceptibility [$\chi = \chi_{\text{molar}}/N(\hbar\gamma_e)^2$]

expressed in units of inverse energy. The spin susceptibility is commonly expressed in terms of a spin concentration c, whose susceptibility would obey the Curie law ($\chi_C \propto 1/T$).

$$k_B T \chi = (1/3) S(S+1) c \tag{15}$$

Expression (14) has been derived for a powder sample. It is noteworthy that the motion spectrum is involved at the two frequencies ω_N and ω_e. The former term implies nuclear spin flips with no change of the electronic spin orientation. The latter term corresponds to transitions of both nuclear and electronic spins, which implies quanta $\hbar(\omega_e \pm \omega_N) \approx \hbar\omega_e$. The shape of $f(\omega)$ can be extracted from data of the spin-lattice nuclear relaxation time T_1 versus frequency (or magnetic field). In fact, the motion spectrum being involved at two frequencies, as T_1 is measured versus frequency, information is obtained on $f(\omega)$ within two windows corresponding to ω_N and ω_e. The data of T_1^{-1} are usually plotted as a function of $\omega_N^{-1/2}$ to test the 1-D features of $f(\omega)$. A straight line in this plot can be a signature of 1-D diffusion of the spin carriers. The diffusion rate along the chains $D_\parallel$ is then obtained from the slope of the straight line. Departure from the straight line at low frequency may be related to interchain couplings, which result in a truncation of the low frequency divergence of $f(\omega)$. It noteworthy that a $\omega_N^{-1/2}$ dependence of T_1^{-1} could also result from a completely different process, namely nuclear relaxation due to fixed electronic spin and transmitted by nuclear spin diffusion [10]. Careful analysis of the data should be performed to discriminate between these two possible cases. A number of spin dynamics studies in low-dimensional systems based on the frequency dependence of T_1 have been carried out. In the case of conducting polymers, most of the work has been concerned with polyacetylene [11], polypyrrole [8], polyaniline [12], and polythiophenes.

B. Dynamic Nuclear Polarization

Dynamic nuclear polarization (DNP) experiments permit us to demonstrate unambiguously the existence of spin motion. More precisely, this technique is able to establish whether the motion spectrum has components at the ESR Larmor frequency: $\omega_e/2\pi \sim 10^{10}$ Hz (for an X band spectrometer). It consists of observing the NMR (at ω_N) while pumping with microwave power at ω_e. Two very different results may occur according to whether the electron nuclear coupling is static or dynamic [10]. In the static case, the electronic spin is fixed (at least its hopping frequency is less than ω_N), and forbidden transitions at $\omega_e \pm \omega_N$ can be induced, giving rise to negative or positive enhancements of the NMR signal. This is the so-called solid-state effect. On the other hand, if the electronic spin is moving such that a frequency component of motion exists at ω_e, it is possible to enhance the NMR signal by pumping at ω_e. This is the Overhauser effect.

The Overhauser effect can be simply explained as follows. By pumping at the ESR Larmor frequency, quanta at $\hbar\omega_e$ are sent into the spin system. If this results in enhancement of the NMR signal, it is evidence that $\hbar\omega_e$ quanta have been converted to $\hbar\omega_N$. In order to obey energy conservation, the energy difference $\hbar(\omega_e \pm \omega_N) \approx \hbar\omega_e$ must be supplied by some part of the system; in other words, a "motion reservoir" must exist whose spectrum contains $\sim \hbar\omega_e$ [13].

Let us stress that these statements are not model-dependent. If an Overhauser effect is actually observed upon DNP experimentation, the electronic spins do move, at least at the rate ω_e. More quantitatively, the Overhauser enhancement, as defined by

$$\rho = (\langle I_P \rangle - \langle I_0 \rangle)/\langle I_0 \rangle \tag{16}$$

(where $\langle I_P \rangle$ and $\langle I_0 \rangle$ are the nuclear magnetization under dynamic polarization and at thermal equilibrium, respectively) is proportional to the motion spectral density at ω_e [14]:

$$\rho = -(\gamma_e/\gamma_N)(\Omega_{++} - \Omega_{+-})\, 2\pi T_1\, f(\omega_e) \tag{17}$$

where Ω_{++} and Ω_{+-} denote combinations of hyperfine coupling coefficients corresponding to the spin operators $S_\lambda^+ I_\mu^+$ and $S_\lambda^+ I_\mu^-$, respectively, and T_1 is the nuclear spin-lattice relaxation time. Note that for protons $-(\gamma_e/\gamma_N) = 660$. Observation of a crossover from the Overhauser effect to the solid-state effect is evidence of a slowing down of the spin motion. Such an effect has been observed in polyacetylene, as a function of oxygen defect content [11,15] or temperature [16,17]. A quantitative derivation of the crossover from the Overhauser effect to the solid-state effect, as a function of the motion correlation time τ, has been given [18].

C. Electron Spin Relaxation

The moving electronic spins can also be used as probes for their own motion. In a similar way as for fixed nuclei, the fluctuating magnetic fields induced by the moving spins are relaxation processes themselves. An expression similar to Eq. (14) can be derived for an electronic spin-lattice relaxation time T_{1e}. For a powder average, it becomes

$$T_{1e}^{-1} = 3\gamma_e^4 \hbar^2 k_B T \chi \Sigma \left[\frac{1}{5}\hat{f}(\omega_e) + \frac{4}{5}\hat{f}(2\omega_e)\right]$$

$$= \gamma_e^4 \hbar^2 S(S+1) c \Sigma \left[\frac{1}{5}\hat{f}(\omega_e) + \frac{4}{5}\hat{f}(2\omega_e)\right] \tag{18}$$

where Σ is the average of $(r_{ij})^{-6}$ over the lattice electronic spin sites. The motion spectrum is relative to a given moving spin; consequently, $f(\omega)$ which was defined with respect to a fixed point, has been replaced by $\hat{f}(\omega)$. In the case of diffusive motion, D should be replaced by $2D$ in Eqs. (8).

In contrast to the nuclear relaxation, which involves the motion spectrum at two very different frequencies, ω_N and $\omega_e \sim 10^3 \omega_N$, the electronic spin relaxation is essentially governed by the spectrum components around ω_e. Measurements of T_{1e} at a given frequency can be performed by a pulse technique or with a continuous wave (cw) spectrometer using the saturation method. Since ESR spectrometers are usually narrowband equipment, it is quite difficult to obtain T_{1e} data over a wide range of frequency. However, outstanding spin dynamics studies have been performed by Mizoguchi and coworkers using a low frequency ESR spectrometer that covers two decades in frequency (from a few megaherz to 500 MHz). In fact, instead of T_{1e}, it is easier to measure the linewidth, which in principle should directly give the spin–spin relaxation rate,

$$\gamma_e \Delta H_{pp} = \frac{2}{\sqrt{3}} (T_{2e})^{-1}$$

for a Lorentzian lineshape. The spin–spin relaxation rate is connected to the spectrum motion by expressions similar to Eq. (18), although it also implies the spectrum at $\omega = 0$.

$$T_{2e}^{-1} =$$
$$3 \gamma_e^4 \hbar^2 k_B \chi T \Sigma \left[\frac{3}{10} \hat{f}(0) + \frac{5}{10} \hat{f}(\omega_e) + \frac{2}{10} \hat{f}(2\omega_e) \right]$$
$$= \gamma_e^4 \hbar^2 S(S+1) c \Sigma \left[\frac{3}{10} \hat{f}(0) + \frac{5}{10} \hat{f}(\omega_e) + \frac{2}{10} \hat{f}(2\omega_e) \right]$$
(19)

Equation (19) has been written under the assumption that $f(\omega)$ has a finite value as $\omega \to 0$. This would not be the case for a pure one-dimensional system [see Eq. (8a)]. In a pseudo-one-dimensional system the interchain couplings result in a truncation of $f(\omega)$ for $\omega < \omega_c \sim D_\perp$. If the cutoff frequency is smaller than the linewidth, the 1-D features of the spin dynamics reflects on the lineshape, which requires a special analysis.

For completeness the phonon modulation of the spin–orbit coupling should also be mentioned as a possible source of spin-lattice relaxation. However, the spin–orbit coupling is weak in polymers that contain only light atoms. Furthermore, in ideal 1-D systems this relaxation route is forbidden by time reversal and inversion symmetry. It was suggested by Soda et al. [19] that in pseudo-one-dimensional systems this selection rule can be overcome by interchain hopping, in which case the relaxation rate becomes proportional to the inverse interchain transfer integral $t_\perp^{-1}$.

D. ESR Line Analysis

The effect of 1-D spin dynamics on the ESR line was investigated in various 1-D paramagnetic systems in the 1970s [20–23]. Basically, if the cutoff frequency is less than the linewidth, $\omega_c < \Delta \omega$, the $\omega^{-1/2}$ divergence of $f(\omega)$ is felt on the line, which no longer has the Lorentzian shape derived from motional narrowing. The lineshape $F(\omega)$ can be analyzed either directly in the frequency domain or in the time domain by considering the total spin time correlation function, $G(t) = \langle S^+(t) S^-(0) \rangle$, with $S = \Sigma_\lambda S_\lambda$, which is the Fourier transform of $F(\omega)$:

$$F(\omega) = \frac{1}{2\pi} \int_{-\infty}^{+\infty} d\omega\, G(t)\, e^{-i\omega t}$$

It has been shown that, instead of an exponential function as in the case of 3-D system, in a pure 1-D system motional narrowing leads to $G(t) \propto \exp(-t^{3/2})$. Such a time dependence is intermediate between the static case (no motion), which gives a Gaussian $G(t) \propto \exp(-t^2)$, and the 3-D motional narrowing, $G(t) \propto \exp(-t)$. This shows that motion is less efficient in one dimension than in three dimensions in narrowing the line.

The different cases are given in Table 5.2. The "static" linewidth, i.e., the linewidth one would observe without motion, is denoted by $\Delta \omega_0$, and D is the characteristic hopping frequency. When the interchain couplings are such that the cutoff frequency is larger than the linewidth, $\omega_c > \Delta \omega$ (pseudo-1-D case), the line is

Table 5.2 Linewidth and Lineshape (in the Time Domain) in Different Cases of Motional Narrowing[a]

	"Static" case	3-D motion	1-D motion $\omega_c \ll \Delta\omega$	Pseudo-1-D $\omega_c \gg \Delta\omega$
Linewidth: $\Delta\omega$	$\Delta\omega_0$	$\dfrac{(\Delta\omega_0)^2}{D}$	$\dfrac{(\Delta\omega_0)^{4/3}}{D^{1/3}}$	$\dfrac{(\Delta\omega_0)^2}{\sqrt{(D\omega_c)^{1/2}}}$
Lineshape: expressions for $G(t)$	$\exp\left[-\dfrac{(\Delta\omega)^2 t^2}{2}\right]$	$\exp[-(\Delta\omega)t]$	$\exp\left[-\left(\dfrac{4}{3}\right)^{2/3}(\Delta\omega)\,t^{3/2}\right]$	$\exp[-(\Delta\omega)t]$

[a] $\Delta\omega_0$ is the linewidth in the absence of motion, $D\ (= D_\parallel)$ the hopping frequency, and $\omega_c \approx D_\perp$ the cutoff frequency.

Lorentzian, as in the 3-D case, but with an effective frequency for motional narrowing that is $(D\omega_c)^{1/2}$, instead of D. Since $\omega_c (\approx D_\perp) \ll D (= D_\parallel)$, it again appears that the 1-D features of the motion result in a less efficient narrowing than with a 3-D motion.

If the cutoff frequency is on the order of the linewidth, $\omega_c \gtrsim \Delta\omega$, one should observe a crossover from a Lorentzian shape at the center of the line ($|\omega - \omega_e| < \omega_c$) to the Fourier transform of $\exp(-t^{3/2})$ in the wings ($|\omega - \omega_e| > \omega_c$).

Observation of a lineshape such as $G(t) \propto \exp(-t^{3/2})$, at least in some part of the line, is a direct signature of 1-D motion. It requires that the interchain couplings be so small that the cutoff frequency is smaller than the linewidth. The latter being commonly on the order of a few 10^7 rad/s, this sets the maximum value for the interchain hopping rate. In the class of conducting polymers, to our knowledge, such behavior has only been observed in undoped *trans*-polyacetylene. It has been evidenced immediately after Fourier transformation of the usual ESR signal in the frequency/field domain [24]. More quantitatively, from a detailed analysis using the memory function formalism, the characteristic parameters $D_\parallel$ and ω_c can be extracted from the lineshape [25].

E. Electron Spin Echo

The time domain ESR line, $G(t) = \langle S^+(t)S^-(0)\rangle$, should be, in principle, directly observable using pulsed techniques, as the free induction decay signal after a $\pi/2$ pulse. The spin echo generated at time $t_E = 2\tau$ after a sequence $\pi/2 - \tau - \pi$ decays with a time constant T_2 in simple cases of fixed or slowly moving spins. For rapidly moving spins in a field gradient G there is an additional decay as $\exp[-(1/12)\gamma^2 G^2 D\, t_E^3]$ [26].

Spin echo techniques have been used successfully to demonstrate the 1-D electron spin transport in highly anisotropic organic conductors [27,29]. Such studies have been carried out on single crystals, e.g., fluoranthenyl radical cation salts. The anisotropy of spin diffusion is deduced from the time constants of the echo signal decay for the gradient field parallel and perpendicular to the chains.

As single crystals are not available in conducting polymers, the use of the spin echo technique is more complicated and interpretation of the data is quite model-dependent. However, electron spin echo studies of spin dynamics in polyacetylene were reported in the early 1980s [30]. The results on the soliton diffusion rate, which were not consistent with those obtained by the other methods, gave rise to a long controversy. They were recently carefully discussed and invalidated [31].

Note that time-resolved ESR spectroscopy has been shown to be an efficient way to distinguish between mobile and fixed spins [32].

F. Fixed Spin-Induced ESR Line Broadening

A new method has been recently proposed for spin dynamic studies in conducting polymers [33]. It basically relies on the following. Collisions between moving spins (species A) and unlike spins (species B), instead of giving rise to a motional narrowing, can result in a broadening of the line. Such an effect can be observed in two cases. First, in the slow exchange case, i.e., when the spin flip-flop frequency ω_x due to the collisions is much smaller than the difference of the Larmor frequencies of the two species: $\omega_x \ll \delta\omega_{AB} = |\omega_{eA} - \omega_{eB}|$. Second, in the fast exchange case: $\omega_x \gg \delta\omega_{AB}$, with the supplementary condition that the B species is strongly coupled to the lattice. The line broadening $\delta(\Delta\omega)$ is given by the simple expression

$$\delta(\Delta\omega) = pC_B D \quad (20)$$

where C_B is the B species concentration, D is the moving spin hopping frequency, and p is the efficiency of the collisions in flipping the spins. For fixed spin $S = 1$ (typically adsorbed molecular oxygen), p is expressed as

$$p = \frac{16}{27}\left(\frac{\alpha^2}{\alpha^2 + 1}\right)$$

with

$$\alpha = \frac{3}{2}\left(\frac{J}{\hbar}\right)t_w$$

J being the exchange integral during the collisions and t_w the collision duration. In the case of strong exchange, long collision ($\alpha \gg 1$), the flipping probability is a constant and one has $\delta(\delta\omega) \propto D$.

It is noteworthy that in the above process the linewidth increases with conductivity (since $\sigma \propto D$), which is opposite to the relationship given by the Elliott process [34,35]. The role of the relaxing spin species (B) can be played by fixed paramagnetic impurities such as adsorbed oxygen molecules. With the use of Eq. (20), determining the polaron hopping frequency is reduced to measurement of the ESR linewidth.

IV. NEUTRAL SOLITONS IN POLYACETYLENE

Polyacetylene, $(CH)_x$, is an almost ideal candidate for spin dynamics studies: it contains electronic spins [36] (about one spin per 3000 CH units in *trans*-$(CH)_x$), typical 1-D characteristics are expected, and all protons are equivalent. Spin dynamics studies in this compound started in the early 1980s [11,37]. They were rapidly proving to be successful in demonstrating the presence of highly mobile spins in *trans*-$(CH)_x$, while only localized, fixed spins were present in *cis*-$(CH)_x$. With the two isomers differing from each other by the presence of a degenerate ground state in *trans*-(CH)$_x$ but not in *cis*-(CH)$_x$, this result strongly supported the soliton picture proposed just a few years earlier [38–40]. However, if the spin carriers possess properties similar to those expected for neutral solitons in *trans*-(CH)$_x$, it is noteworthy that their origin is quite different. Solitons are described as topological defects, or "kinks," originating from nonlinear excitations of a 1-D system with a nondegenerate ground state. The spins present in pristine (CH)$_x$ do not correspond to excitations, as their concentration is not temperature-dependent. Consequently, they do not originate in excitations of the (CH)$_x$ chain. Their origin likely lies in the formation of conjugated segments with an odd number of π electrons during the cis–trans isomerization process. Therefore, we would be dealing with a kind of macromolecular π free radical with behavior similar to that of the solitons described for *trans*-(CH)$_x$. However, in the following, as is usually practiced in the conducting polymer community, we shall call this species "moving spins" or "neutral solitons," using the terms interchangeably. It is worth mentioning that "true solitons" can indeed be generated by photoexcitation or by charge injection. Essentially the same mobility has been found for the soliton-like free radicals and for the photogenerated solitons.

The presence of mobile spins, or "neutral solitons," in undoped *trans*-(CH)$_x$ was established unambiguously from the beginning by the observation of the Overhauser effect in DNP experiments. Further studies were then concerned with (1) quantitative estimates of the diffusion rate and of the anisotropy of the motion and (2) trapping of the diffusive solitons in the presence of defects or impurities or as the temperature is lowered.

A. Evidence for Spin Motion in *trans*-(CH)$_x$

The observation of a narrow ESR line ($\Delta H_{pp} \approx 1.4$ G) in *trans*-(CH)$_x$ was suggestive of motional narrowing and hence of highly mobile spins [36]. However, the narrowness of the line is not in itself a proof of the motion. It can also be due to a high degree of delocalization. If the electronic spin is delocalized over N sites, the hyperfine contribution to the linewidth goes like $N^{-1/2}$: $\Delta H_{hf} \propto N^{-1/2}$. It is a matter of fact that spin delocalization with no motion would give rise to a Gaussian lineshape, in contrast to the Lorentzian shape that was first recognized. However, after careful analysis, the line appears to deviate somewhat from Lorentzian [37]. The direct proof for the motion was supplied by the observation of the Overhauser effect in DNP experiments [11].

The data of DNP experiments carried out on *cis*- and *trans*-(CH)$_x$ are shown in Fig. 5.2. The proton NMR signal enhancement is given as a function of the electronic pumping frequency. It clearly appears that for *trans*-(CH)$_x$ a positive enhancement is obtained by pumping just at the electronic Larmor frequency ω_e (this is the Overhauser effect), whereas for the *cis* isomer negative

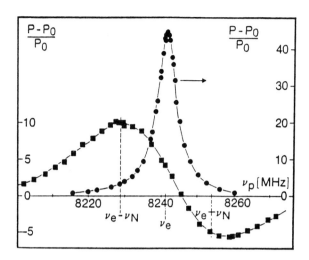

Fig. 5.2 Dynamic nuclear polarization experiment. Proton NMR signal enhancement as a function of the pumping frequency for *trans*-polyacetylene (●) and *cis*-polyacetylene (■).

and positive enhancement are obtained by pumping at $\omega_e \pm \omega_N$ (this is the solid-state effect). This result proves that spins are mobile in *trans*-(CH)$_x$, at least at the rate $\omega_{hop} \geq \omega_e \approx 10^{10}-10^{11}$ rad/s, whereas they are fixed in *cis*-(CH)$_x$, or at least the motion is so slow that $\omega_{hop} \leq \omega_N \approx 10^7-10^8$ rad/s.

B. Soliton Mobility

After having demonstrated that the spins were moving in undoped *trans*-(CH)$_x$, the aim of the researchers was to characterize the motion. Use was made of various methods described in section III. At first, as suggested by the polymeric nature of (CH)$_x$, the motion was expected to proceed essentially along the chains. Evidence for the 1-D feature of the motion was given from the $\omega^{-1/2}$ dependence of the nuclear relaxation rate: $T_1^{-1} \propto \omega^{-1/2}$. It should be noted that such a frequency dependence could also be obtained in the case of nuclear relaxation induced by fixed electronic spins and transmitted by nuclear spin diffusion.* However, this alternative explanation for $T_1^{-1} \propto \omega^{-1/2}$ was invalidated by the Overhauser effect, which proved that the electronic spins were not fixed. Let us stress that the same spins are responsible for both the Overhauser effect and the nuclear relaxation. Further data supported the 1-D nature of the motion: (1) The $\omega^{-1/2}$ dependence was also observed for the electronic spin relaxation [41] and (2) the ESR lineshape was shown to display 1-D features, namely the Fourier transform of $G(t) \propto \exp(-t^{3/2})$ [24,25,36].

---

* See Ref. 10, Chapter 9.

Electron Spin Dynamics

The first estimate of the diffusion rate was obtained from the data of proton T_1 versus frequency (see Fig. 5.3) and the use of Eq. (14) with $f(\omega) = 1/\sqrt{2D_\parallel \omega}$. This yields $D_\parallel \approx (3-6) \times 10^{13}$ rad/s with the room temperature data, which is a quite high value, close to the estimated velocity of sound in $(CH)_x$. For the soliton mobility one obtains $\tilde{D} \sim 5 \times 10^{-3}$ cm^2/s.

A large uncertainty in the estimate is related to the incomplete knowledge of the hyperfine couplings **a** and **d**. In the absence of a determination of the soliton–proton hyperfine coupling in $(CH)_x$, it was reasonable to consider the tensor coupling constant determined for unpaired electrons in $2p_z$ orbitals in other compounds [42,43] and to take the MacConnell value $|a_{CH} / \gamma_e| =$ 23.4 G [44] for **a**. Since then, the spin density distribution of unpaired electrons in polyacetylene has been determined by several groups, especially using various ENDOR methods [45–52]. Spectroscopic studies, which require the electronic spin to be immobile, have been performed on cis-$(CH)_x$ or on trans-$(CH)_x$ at low temperature. The electronic soliton wave function is the result of an intricate balance between localization and condensation energy with account taken of the electron correlation energy, which results in negative spin densities. It is now established that the spin density (1) is spread over 21 ~ 20 CH units and (2) has alternating positive and negative values. A picture of a proposed spin density distribution is given in Fig. 5.4.

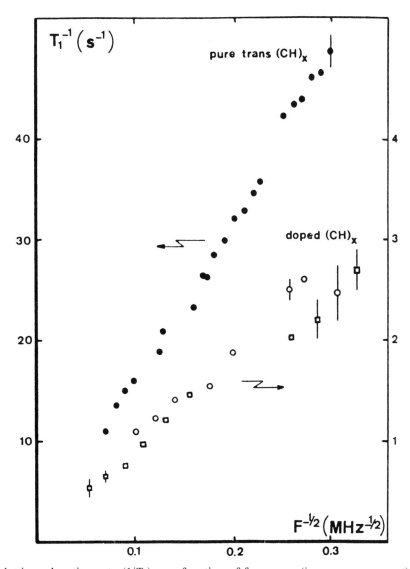

Fig. 5.3 Proton spin-lattice relaxation rate ($1/T_1$) as a function of frequency (inverse square root) for undoped (●) and heavily AsF$_5$-doped (○ and □) trans-polyacetylene. (From Ref. 11.)

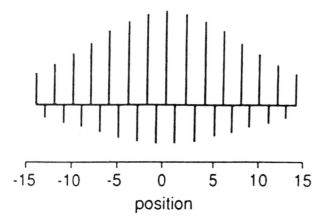

Fig. 5.4 Proposed spin density distribution in *trans*-polyacetylene according to Käss et al. (From Ref. 51.)

Complete calculation of the proton T_1, taking into account a realistic shape of the spin density distribution with alternating positive and negative values, would be much more complicated than simple use of Eq. (14), as is commonly done. In particular, cross-correlation terms should be included [54]. The result might be somewhat different from that obtained from the simplified theory.

Mizoguchi et al. have performed quantitative spin dynamics studies using the frequency dependence of the electron spin-lattice (T_{1e}) and spin-spin relaxation times (T_{2e}). Figure 5.5 shows T_{1e}^{-1} measured with a saturation technique as a function of frequency for *trans*-$(CH)_x$ and *trans*-$(CD)_x$ [41]. The $\omega^{-1/2}$ dependence is well-established. Furthermore, the higher relaxation rate in the protonated compound compared to that in the deuterated one provides evidence for a hyperfine contribution to T_{1e}^{-1}. As shown in the insert, there are both concentration-dependent and concentration-independent contributions. The latter is almost zero in the case of *trans*-$(CD)_x$. These contributions are easily assigned to the (electron–electron) dipolar and (electron–nucleus) hyperfine interactions, respectively. The dipolar contribution has been calculated from Eq. (18) by taking into account the crystallographic structure of *trans*-$(CH)_x$ [55] for the lattice summations. From the fit with the data (solid curves in Fig. 5.5), values for $D_\parallel$ have been obtained in the range $(1–5) \times 10^{13}$ rad/s, that is, in full agreement with the proton T_1 determination.

Measurement of the ESR linewidth, from which one obtains T_{2e} [$\gamma_e \Delta H_{pp} = 2/\sqrt{3}\,(T_{2e})^{-1}$ for a Lorentzian lineshape], is easier than that for T_{1e} and can be done at low temperature. Mizoguchi et al. have developed a spin dynamics methodology based on these ideas. Typical data covering the range 3–500 MHz are shown in Figs. 5.6a and 5.6b for *trans*-$(CH)_x$ and *trans*-$(CD)_x$, respectively. As the linewidth is governed only by the motional narrowing process described in Section III. C, the data should be analyzed with the use of Eq. (19). This equation is similar to Eq. (18), except that the motion spectral density is also involved at $\omega = 0$.

However, various effects may contribute to the linewidth such as (1) *g*-factor anisotropy, (2) trapping of the

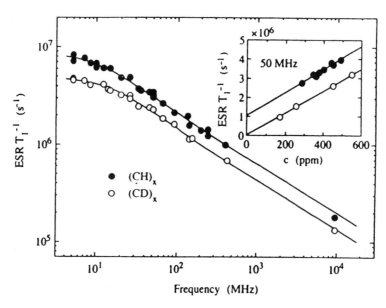

Fig. 5.5 Electron spin-lattice relaxation rate ($1/T_{1e}$) as a function of frequency for protonated (●) and deuterated (○) *trans*-polyacetylene, according to Mizoguchi et al. (From Ref. 41.) The insert gives $1/T_{1e}$ as a function of the spin concentration.

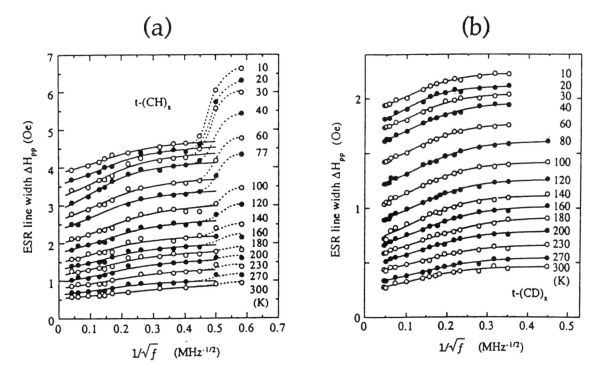

Fig. 5.6 ESR linewidth as a function of frequency (inverse square root) for (a) protonated and (b) deuterated *trans*-polyacetylene, according to Mizoguchi et al. (From Ref. 62.)

spin species, or (3) the presence of magnetic impurities. Failure to take possible extra contributions to ΔH into account can invalidate the analysis. It has been shown that oxygen contamination has a drastic effect on ΔH in undoped *trans*-$(CH)_x$ [25,56]. It has been proposed to account for this effect in terms of a two-spin model: moving solitons and trapped solitons. The nuclear relaxation is mainly governed by the moving spins, while the fixed spins may have the dominant contribution to the ESR linewidth. If one includes the roles of these two kinds of spins, the NMR and ESR data lead to quite consistent values: $D_\parallel \sim 5 \times 10^{13}$ rad/s, which yields $\tilde{D} = D \times c_\parallel^2 \sim 5 \times 10^{-3}$ cm^2/s (with the inter-CH unit distance $c_\parallel = 2.46$ Å). These values were first obtained on Shirakawa $(CH)_x$. Similar values have been found recently on Naarman–Theophilou $(CH)_x$ [57]. They compare quite well with that obtained from optical measurements, i.e., analysis of the photoinduced bleaching of the interband transition by Vardeny et al. [58]: $\tilde{D} \approx 2 \times 10^{-2}$ cm^2/s.

It is noteworthy that these spin dynamics studies, leading to fully convergent results, were performed independently in Grenoble and in Tokyo. Let us mention that a controversy concerning the value of the soliton diffusion coefficient took place in the 1980s. In contrast to the high value found by the Grenoble and Tokyo groups, other groups maintained that $D_\parallel$ was much smaller [30,59,60]: in the range 10^9–10^{11} rad/s. The latter estimate relied on various pulse ESR experiments including the spin-echo phase memory time T_M and the spin-echo-detected multiple-quantum NMR. Mizoguchi et al. [31] recently refuted the arguments given in favor of slow solitons. They have shown that, basically, the phase memory time and the multiple-quantum phase coherences are not able to detect mobile solitons, so that the data based on these techniques are essentially concerned with trapped solitons.

Not only are the solitons highly mobile, but in addition the motion is highly one-dimensional. Since $T_1^{-1} \propto \omega^{-1/2}$ throughout the experimental frequency range, the cutoff frequency is smaller than the smaller frequency used, that is, $\omega_c < 6 \times 10^7$ rad/s. Estimates of ω_c have been obtained from $T_{1\rho}$ measurements ($\omega_c \approx 4.5 \times 10^7$ rad/s), from the residual (diffusive) ESR linewidth ($\omega_c \approx 3 \times 10^7$ rad/s) [25], and from the data of ΔH versus ω [$\omega_c \approx (6-8) \times 10^7$ rad/s] [57]. The cutoff frequency ω_c is an upper limit for the transverse diffusion rate $D_\perp$, but that may be due to another process more efficient than transverse soliton diffusion. The interchain electronic dipole–dipole interactions are large enough to account for the observed ω_c. In fact, owing to the actual nature of the soliton-like species, namely $(CH)_x$ seg-

C. Soliton Trapping

After the discovery of highly mobile spins in *trans*-(CH)$_x$ at room temperature, it was soon recognized that spin dynamics were strongly affected by temperature and by oxygen contamination. Much evidence exists for a "freezing" of the spin motion at low temperatures, including (1) the increase of the ESR linewidth at low temperature, (2) the observation of a crossover from the Overhauser effect to the solid-state effect [16–18], and (3) the low temperature ENDOR spectra [45,49,51]. We also mention an effect of anomalous broadening of the ESR linewidth, observed at low temperature and low frequency ($\omega_e/2\pi < 6$ MHz) and explained in terms of a crossover from an "unlike-spin" broadening to a "like-spin" broadening [31]. This effect requires the hyperfine field to be static, that is, the electronic spins should be in a localized state.

Furthermore, it has been noted that the effect of oxygen contamination is rather similar to that of lowering the temperature; both result in an overall slowing down of the motion. For instance, both give rise to a crossover from the Overhauser effect to the solid-state effect as illustrated in Figs. 5.7a and 5.7b, and both increase the ESR linewidth (see Fig. 5.8).

However, the nuclear relaxation rate T_1^{-1} decreases at low temperature, while it is scarcely affected in the presence of oxygen. The latter result shows that the data could not all be accounted for just by introducing a diffusion coefficient dependent on both temperature and oxygen content. A comprehensive explanation of the data has been proposed in terms of a two-spin species model [56], also called the "diffusive-trap model," as follows. In a pure ideal (CH)$_x$ chain we would be dealing with diffusive solitons with a diffusion coefficient $\tilde{D}(T)$ that will be discussed later. In the presence of impurities such as adsorbed oxygen, traps are created in which the solitons can be temporarily pinned and thus localized. The total number of solitons, n, is thus $n = n_D + n_L$, where n_D and n_L are the numbers of diffusive and localized spins, respectively. The latter is temperature-dependent as $n_L = c_L(T)n$, with $c_L(T)$ being the relative concentration of localized spins. Then the contributions of diffusive and localized spins to ΔH and to T_1^{-1} can be analyzed, and these quantities can be expressed as a function of $c_L(T)$. As for ΔH, it has to be decomposed into a hyperfine part and a dipolar part: $\Delta H = \Delta H^{hf} + \Delta H^{dip}$. While the hyperfine part is simply the weighted sum of the diffusive and localized spin contributions,

$$\Delta H^{hf} = (1 - c_L) \Delta H_D^{hf} + c_L \Delta H_L^{hf} \tag{21}$$

The dipolar part contains a c_L^2 contribution that accounts for soliton–soliton dipole interactions:

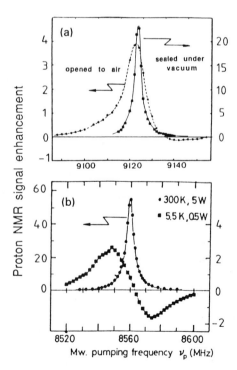

Fig. 5.7 Dynamic nuclear polarization in *trans*-polyacetylene. Proton NMR signal enhancement as a function of the pumping frequency. (a) Effect of air contamination. (From Ref. 25). (b) Effect of temperature (From Ref. 16).

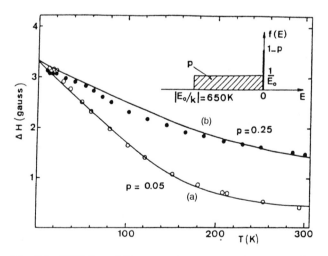

Fig. 5.8 ESR linewidth versus temperature in *trans*-polyacetylene in (a) samples sealed under vacuum and (b) air-contaminated samples. (From Ref. 56.) The theoretical curves are obtained from a trap model, with a trap energy distribution as in the insert and a trap concentration p taken as adjustable parameter.

Electron Spin Dynamics

$$\Delta H^{\text{dip}} = [1 + 2(\sqrt{2} - 1)c_L - (2\sqrt{2} - 1)c_L^2]$$
$$\times \Delta H_D^{\text{dip}} + c_L^2 \Delta H_L^{\text{dip}} \quad (22)$$

The diffusive contributions ΔH_D^{dip} and ΔH_D^{hf} are frequency-dependent and are proportional to $1/\sqrt{\omega D}$. From the data of ΔH versus temperature, the temperature dependence of the localized site concentration $c_L(T)$ can be extracted. This requires that $D(T)$ be known. In the work of Ref. 56, $D(T)$ was obtained from data of the proton relaxation rate as a function of temperature and frequency. There are also two contributions to the nuclear relaxation: from diffusive spins and from localized spins. However, due to the local hyperfine field, the protons located around the localized electronic spins are not seen by NMR. Consequently, only the diffusive contribution [$\propto (1 - c_L)$] is effective for T_1^{-1}.

The data can be accounted for as a result of trapping in trap sites whose potential is distributed from ~0 to a maximum value $V_0/kT \approx 650$ K. As for the microscopic nature of the oxygen-induced traps, it has been suggested that adsorbed oxygen can act as an electron acceptor (or oxidant), resulting in an ion pair, O_2^-/carbonium$^+$. In other words, oxygen adsorption would result in the formation of a charged soliton. As estimated by Brédas et al. [61], there exists a binding energy on the order of 0.05 eV between a neutral soliton and a charged soliton. This value compares well with the maximum

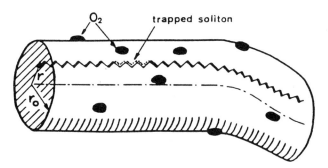

Fig. 5.9 Schematic representation of a polyacetylene fiber with O_2 impurities bonded to the surface. Soliton traps are created on $(CH)_x$ chains, with trapping energy that is a function of $r - r_0$, the chain-to-surface distance. (From Ref. 56.)

value of the trapping energy distribution, $V_0 \approx 0.06$ eV, found by spin dynamic studies. The oxygen molecules being essentially adsorbed at the surface of a $(CH)_x$ fibril, and the trapping energy of a soliton in a given chain being a decreasing function of the chain-to-surface distance (see Fig. 5.9), the overall trapping energy must be distributed from V_0 for solitons at the surface of the fibril to a very small value for solitons deep inside the fibril.

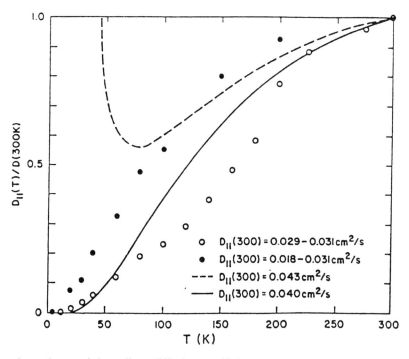

Fig. 5.10 Temperature dependence of the soliton diffusion coefficient in *trans*-polyacetylene from NMR data obtained in Grenoble (●) and ESR data obtained in Tokyo (○). Theoretical results (by Jeyadev and Conwell [67]) for phonon scattering only (dashed line) and for phonon scattering plus barriers of 0.01 eV (solid line).

The emerging picture is then as follows. The neutral solitons in *trans*-$(CH)_x$ are basically diffusive along the chains. However, they can fall into traps. In particular, oxygen contamination creates traps. The population of trapped solitons increases as the temperature is lowered. At a given temperature there is a given ratio of trapped to diffusive solitons. This can also be depicted as the ratio of the times spent by a given soliton in the trapped and diffusive states. During the time the soliton is in the trapped state it is localized and exhibits the properties of fixed spins, similar to the properties of spins in *cis*-$(CH)_x$. The trapping time τ_{tr} increases as the temperature is lowered, depending of the value of the trap potential. As the latter does not have a well-defined value but is distributed, τ_{tr} also is distributed. Various ESR data reported in the literature can be qualitatively interpreted by comparing the average trapping time $\bar{\tau}_{tr}$ to the relevant characteristic times [31].

As for those spins that remain in the diffusive state, they also undergo a slowing down of the motion. The diffusion coefficient decreases as the temperature is lowered. Experimental determinations of the temperature dependence of $D_\parallel$ have been achieved independently by the Grenoble [56] and Tokyo [62] groups from NMR and ESR data, respectively. The results, which are in qualitative agreement, can be fitted with a T^2 dependence. The theoretical problem of the soliton diffusion has been investigated by several authors. If the diffusion process is originating from a ballistic motion limited by collisions, particularly with phonons, $D_\parallel$ is expected to diverge as $T \to 0$ since there are fewer and fewer phonons [63]. The soliton random walk can also be envisaged as resulting from the shifts induced by the collisions. A T^2 law has been predicted in the low temperature limit [64,65]. An overall discussion of the mechanisms that give rise to the soliton diffusion has been given recently by Wada [66]. Also, Jeyadev and Conwell have proposed to account for the experiment in terms of collision-limited ballistic motion but including trapping effects [67]. Indeed, this approach predicts a temperature dependence for $D_\parallel$, which is consistent with the data (see Fig. 5.10), provided that the effect of barriers of 0.01 eV is included.

V. POLARONS IN CONDUCTING POLYMERS

In the case of conjugated polymers in the conducting state, the moving spin carriers are also charge carriers, and spin dynamics can be used in transport property studies. In contrast with the conventional methods for transport, spin dynamics is able to supply data at the microscopic scale. For example, the intra- and interchain conductivity can be estimated. Furthermore, the conductivity of the conducting phase can be determined in heterogeneous samples, e.g., conducting islands in an insulating matrix or a blend. Despite its unique advantage, spin dynamics has been used by only a limited number of groups. Most of the studies have been carried out by the groups in Grenoble and Tokyo. Until recently, the spin dynamics studies have been less concerned with polarons than with the neutral solitons of pristine polyacetylene. However, although the studies in pristine polyacetylene are essentially completed, there are still a number of open questions concerning the conducting state of conjugated polymers for which spin dynamics is potentially a highly valuable method.

A. Doped Polyacetylene

The first spin dynamics studies in conducting $(CH)_x$ were attempted on AsF_5-doped Shirakawa-type $(CH)_x$ [11]. Further studies of this type were presented 15 years later at ICSM'94 in Seoul [68,69]. At this time the Naarmann–Theophilou synthesis, which is known to produce much longer and more regular chains, had appeared.

In the heavily doped state ($y > 6$), polyacetylene exhibits a Pauli-type spin susceptibility corresponding to a density of states at the Fermi level of $N(E_f) \approx 0.1$ state/eV per monomer unit [70,71]. As in the work performed on various 1-D conductors [19,72], it was tempting to use spin dynamics to explore transport properties at the chain level. The works of Refs. 11 and 68 are concerned with proton spin-lattice relaxation time measurement as a function of frequency on AsF_5-doped Shirakawa-type [11] and K-doped Naarmann–Theophilou-type $(CH)_x$ [68]. In both cases the data are consistent with pseudo-one-dimensional diffusive motion of the spin carriers. From the values of $D_\parallel$, the intrachain conductivity has been deduced to be in the range $(1-5) \times 10^4$ S/cm, which is quite consistent with the value expected for doped $(CH)_x$. Furthermore, an extremely high anisotropy has been found: $\sigma_\parallel/\sigma_\perp > 10^5$, which is more difficult to understand. It cannot be excluded that other processes are responsible for the small proton spin-lattice relaxation rate. Namely, in Ref. 11 it was mentioned that the relaxation can be caused by ~5% of the remaining diffusive spins in possible undoped regions. Besides, the data of Ref. 68 could be accounted for by the presence of ~20 ppm impurities with fixed spins. The idea that the "conducting spins" are not responsible for the nuclear relaxation is supported by the data of Ref. 69, which show a correlation between the nuclear relaxation rate (at a given frequency) and the amount of Curie spins—such as residual neutral solitons confined in limited undoped regions—contained in samples heavily doped with various dopants. Indeed, the temperature dependence of T_1^{-1} does not display a metallic behavior, except for the bromine-doped sample before aging. In this case the data for $T > 100$ K can be fitted with $T_1^{-1} \propto T^{1.5}$. Such a temperature dependence is expected for a 1-D metal whose resistivity increases proportionally to temperature [69].

Since in most cases the nuclear relaxation in doped polyacetylene is likely connected to spins that are not involved in the conduction process, the following question arises: What is the effect of the "conducting spins," which certainly are present, i.e., why would these spins have a negligible contribution to the relaxation? The answer to this question is as follows. The "conducting spins" would contribute significantly to the relaxation provided that highly anisotopic 1-D behavior is included (due to the low frequency divergence of the motion spectrum in one dimension). But if the conductivity is rather three-dimensional, or if the 1-D character is not so pronounced ($\sigma_\parallel/\sigma_\perp < 10^2$), then the induced relaxation is actually much less than that observed.

B. Polyaniline

The polyaniline family (see Fig. 5.11) is known for its remarkable insulator-to-conductor transition as a function of protonation. The emeraldine base form (PAN-EB) is insulating. Upon protonation of the amine functions in an acidic medium, the emeraldine salt form (PAN-ES) is obtained and the conductivity increases by 10 orders of magnitude while the number of electrons on the polymer chains remains constant [73,74]. In addition, a temperature-independent spin susceptibility appears—the so-called Pauli susceptibility—that is essentially proportional to the protonation level. This strongly suggests that instead of being homogeneous the protonation leads to phase segregation into fully protonated domains with high spin concentration (conducting islands) embedded in an unprotonated sea [75]. It has been proposed to account for the Pauli-like spin susceptibility in terms of a "polaronic metal" [76,77] or, alternatively, in terms of a disorder-induced Fermi glass [78,79].

1. Spin Dynamics Versus Protonation Level

In preliminary work on polyaniline combining conventional transport measurements and magnetic resonance data, a frequency dependence of the proton relaxation as $T_1^{-1} \propto A + B\omega^{-1/2}$ was noted [80]. A detailed spin dynamics study was then carried out within a collaboration of the Grenoble group and Mizoguchi [12]. The study was concerned with a set of emeraldine salt samples equilibrated in aqueous HCl solutions of varying pH. It included both NMR and ESR data. A plot of proton T_1 versus frequency exhibits low-dimensional features, namely an increase of T_1^{-1} at low frequency as shown in Fig. 5.12. The data can be fitted with pseudo-one-dimensional spin diffusion. However, the cutoff frequency $\omega_c \sim D_\perp$ can be chosen in either the ω_e or the ω_N range. Furthermore, the case of anomalous diffusion [8], i.e., random work on a fractal network, can also fit the data. It is possible to discriminate between these different cases by taking into account the ESR data. The ESR linewidth versus frequency seems to be consistent only in the case of 1-D diffusion with ω_c in the ω_e range (see Fig. 5.13). The two parameters $D_\parallel$ and $D_\perp$ have been extracted from the data. They are shown in Fig. 5.14 as a function of the protonation level, determined from the ratio $y = [Cl]/[N]$. Data obtained independently from NMR and ESR are in reasonable agreement. It should be stressed that the $D_\parallel$ values are dependent on the hyperfine couplings and on the structural data for the NMR and ESR studies, respectively. Use of set of crystallographic data [81] other than the one used in Ref.

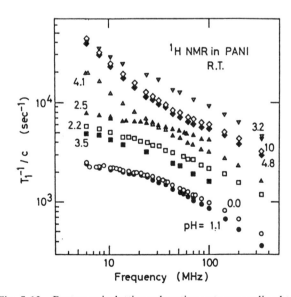

Fig. 5.12 Proton spin-lattice relaxation rate normalized to the spin concentration as a function of frequency for polyaniline equilibrated in aqueous solutions of varying pH from 0 to 10. (From Ref. 12.)

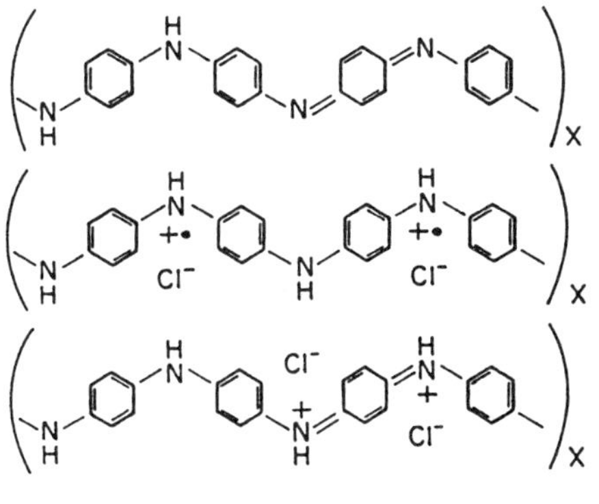

Fig. 5.11 Emeraldine forms of polyaniline. (a) Polyemeraldine base; (b) polyemeraldine salt, polaronic form; (c) polyemeraldine salt, bipolaronic form.

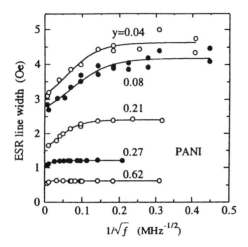

Fig. 5.13 ESR linewidth as a function of frequency (inverse square root) for polyaniline (polyemeraldine) of various protonation levels, $y = [Cl]/[N] = 0.04-0.62$. (From Ref. 5.)

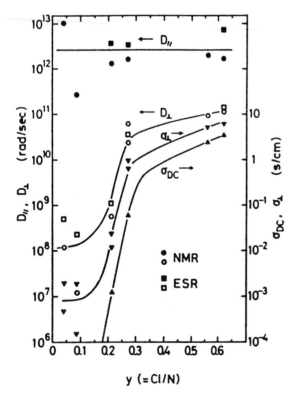

Fig. 5.14 Room temperature spin-diffusion rates, as obtained from NMR (○, ●) and from ESR (□, ■), and conductivity in polyaniline as a function of the protonation level. (From Ref. 12.) σ_{dc} is the measured conductivity, and $\sigma\perp$ has been calculated from $D\perp$, using $\sigma\perp = ne^2 D\perp / kT$, with n, the carrier concentration, proportional to y.

12 leads to higher values for $D_\parallel$. Moreover, it should be added that for the major part of polyaniline, which is amorphous, there are no structural data available. But, whatever its absolute value, it appears clearly that $D_\parallel$ is essentially independent of the protonation level. This is evidence that the intrachain hopping frequency, i.e., the polaron mobility, remains the same whatever the number of conducting chains in the sample. The polaron motion within a given chain is not influenced by the neighboring chains. In contrast, $D_\perp$, whose determination is not parameter-dependent, presents a sudden drop for $y < y_c \approx 0.2-0.3$, so that y_c appears as a kind of percolation threshold. The change of $D_\perp$ is parallel to the change of σ_{dc}, the dc conductivity, which suggests that σ_{dc} is governed by the interchain hoppings. Furthermore, the σ_{dc} value agrees with the calculated value $\sigma_\perp = ne^2 D_\perp / kT$, where n is the charge carrier concentration and e is the electron charge. (See Fig. 5.14.) To account for these data, and consistent with the conducting island picture, it has been proposed that the conducting islands are composed of single (or very few) chains [12]. The picture that emerges from spin dynamics studies is as follows. Protonation, instead of being homogeneous, would be a zip process, resulting in fully protonated chains surrounded by nonprotonated ones. Macroscopic dc conductivity across the sample is limited by the interchain transfers, so that $\sigma_{dc} \approx \sigma_\perp$. As the protonation level increases, the number of protonated (conducting) chains increases. At a certain protonation level, $y = y_c$, the conducting chains are percolating; then the interchain hoppings, and hence the conductivity, increase rapidly. Why $y_c \approx 0.2-0.3$, and not a much smaller value as expected for 1-D percolating systems, is not understood. It might be due to folding of the chains into balls, giving them the percolation features of 3-D objects. It is noteworthy that the proposal of single-chain conducting islands has been questioned by Epstein et al. [82,83]. These authors have proposed a description of the transport properties of polyaniline in terms of bundles with 3-D "metallic" features [84]. Instead of hopping, conduction would be governed by the metallic (coherent) regime, at least in the crystalline regions. This proposal is based on dc conductivity, thermopower, microwave electric constant, and ESR versus temperature data. In the following we discuss the ESR data, namely the temperature dependence of the linewidth, and show that an alternative interpretation can be proposed.

2. Spin Dynamics Versus Temperature

The relationship $\sigma_{dc} \propto D_\perp$, which was first shown as a function of the protonation level, was then tested as a function of temperature [85]. The interchain diffusion rate $D_\perp$ was extracted from the data of the frequency dependence of the ESR linewidth measured in the temperature range 20-300 K. As shown in Fig. 5.15, the temperature dependence of the resulting $D_\perp$ agrees with

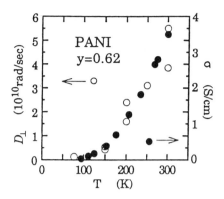

Fig. 5.15 Temperature dependence of $D_\perp$ (from ESR data) and of the measured conductivity for polyemeraldine salt, $y = 0.62$. (From Ref. 85.)

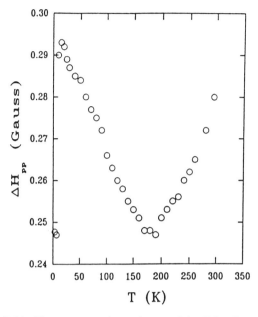

Fig. 5.16 Temperature dependence of the ESR linewidth in polyemeraldine salt. (From Ref. 84.)

that of the conductivity. This offers further support for the picture of the conductivity being limited, and therefore essentially governed, by the interchain hoppings.

An analysis of the temperature dependence of $D_\parallel$ obtained from the data of the ESR linewidth versus frequency has also been proposed [86]. It is claimed that $D_\parallel$ increases by about one order of magnitude as the temperature decreases from 300 to ~100 K. Then, below 100 K, $D_\parallel$ decreases with decreasing T. The first increase of $D_\parallel$ with decreasing T (down to ~100 K) has been interpreted as evidence of metallic behavior at the intrachain level. According to this description, the polaron mobility increases with decreasing T, as in a metal. It reaches a maximum at $T \sim 100$ K. Below that temperature, 1-D localization effects become effective, resulting in a decrease in the mobility. This interpretation is consistent with the data of $\Delta H(T)$, which display a minimum for $T = T_{min}$ in the range 100–200 K (Fig. 5.16) if one accounts for the temperature dependence of ΔH in terms of the Elliott mechanism: The linewidth originates from the electron–phonon scattering, like the resistivity, and then $\Delta H \propto \sigma^{-1}$. The decrease of ΔH as T decreases to T_{min} would reflect an increase of σ as in a metal.

However, I believe in another interpretation, which is based on the following. As mentioned in Section III.F, the relation $\Delta H \propto \sigma$, i.e., the opposite to that of the Elliott mechanism, can be encountered. Such a relationship has been observed, for instance, in amorphous silicium [87], in SiC fibers [88], and in polypyrrole as a function of aging [89]. Movaghar and Schweitzer [87] give an explanation in terms of collisions of the spin carriers, which, in the presence of spin–orbit coupling, result in a shortening of the spin state lifetime T_1 and hence in line broadening. The broadening is proportional to the collision rate, which is proportional to the spin carrier mobility. In the case of conducting polymers, the moving spins are carried by delocalized π electrons,

which have a very small spin–orbit coupling. A more efficient broadening mechanism can exist if paramagnetic impurities, strongly coupled to the lattice, are present [33]. In the case of conducting polymers, several kinds of paramagnetic impurities can be present. Oxygen contamination is certainly the most usual. It is a matter of fact that ΔH increases upon exposure to oxygen [90,91]. As the atmospheric air is pumped off, ΔH decreases to a residual value ΔH_{res} that is not perfectly reproducible from sample to sample. It is very likely that a large part of Δ_{res}, if not the major part, originates in residual, trapped, molecular oxygen or other paramagnetic impurities. Therefore, the Mizoguchi and Kume analysis of the linewidth [86], from which $D_\parallel (T)$ has been extracted, is questionable. It is noteworthy that in this analysis there was a linewidth offset (frequency-independent) $\Delta H_0 \approx 0.45$ G, unexplained in terms of a diffusion-narrowed line and suggested by the authors to be linked to oxygen contamination. In this work ΔH_0 was supposed to be temperature-independent. But, according to Eq. (20), the oxygen-induced linewidth is expected to follow the temperature dependence of the diffusion coefficient: $\gamma \Delta H(T) = p C_i D_\parallel (T)$ [33]. In terms of this process of impurity-induced line broadening, the $\Delta H(T)$ data reported by Wang et al. [84] can be explained as follows. As the temperature is lowered, two competing processes take place, giving rise to a minimum in $\Delta H(T)$. First, the usual decrease of conductivity (in contrast to the assumed "metallic conductivity") results in a decrease of the linewidth (observed for $T >$

T_{min}). Second, at low temperature, 1-D localization effects oppose motional narrowing and thus the line broadens for $T < T_{min}$.

3. Effects of Hydration, or Chain Alignment, on Spin Dynamics

Various parameters have been shown to affect the transport properties in polyaniline. For instance, upon hydration the conductivity can increase by one order of magnitude [92–96]. Besides, upon stretching, the (macroscopic) conductivity of films becomes anisotropic, and the conductivity along the stretching direction $\sigma_\parallel$ is enhanced compared to that of the unstretched material [97,98]. In both cases, spin dynamics studies have been used to probe the mechanisms at the microscopic level.

Polyaniline is hydrophilic. The PAN-ES form is able to absorb reversibly up to 0.5 H_2O molecule per aniline ring. The absorbed molecules are located on the acidic sites [99]. It turns out that the proton spin-lattice relaxation time T_1 gets longer as the hydration level increases [100]. This T_1 lengthening cannot be explained only by the increase in the number of inactive protons, since it is also observed when deuterated water is used for hydration. The frequency dependence of T_1 for the hydrated material displays 1-D spin diffusion as for the dried material, as shown in Fig. 5.17. This figure displays data for four samples. Sample A is a regular dried PAN-ES; sample B is the same as A but equilibrated in 10 torr water vapor pressure. Since proton exchange takes place between adsorbed water and the polyaniline acidic sites [101], PAN-ES sample C has been prepared in which all protons, except those of the aniline ring, have been replaced by deuterium. Sample D is the same as C but equilibrated in 10 torr water vapor pressure. The comparison of data C and D makes clear the neat effect of hydration: a proportional decrease of T_1^{-1}.

As discussed by Travers [102], this can be due to either (1) an increase in the intrachain diffusion rate or (2) a decrease in the electron–proton coupling constants. In case 1, hydration has an effect on the polaron mobility, i.e., the latter is enhanced; in case 2, hydration modifies the polaron electronic wave function. In addition, $T_{1\rho}$ shows that the low frequency contribution to the proton relaxation is not affected, which is not consistent with a change in the coupling constant. It can be concluded that the increase in the macroscopic conductivity observed on hydration is related to an increase of the on-chain polaron mobility. A possible explanation can be proposed in terms of a solvation effect of the counterions resulting in a depinning of the polarons.

Upon stretching, the conductivity along the stretching direction can increase by more than one order of magnitude. However, the effect of stretching at the microscopic level was unknown. By use of spin dynamics, Travers et al. obtained insight into the connections between the microscopic and macroscopic levels [103]. The frequency dependence of the proton relaxation rate for stretched and unstretched films of PAN-ES is shown in Fig. 5.18. It appears that the knee characteristic of the occurrence of the 3-D couplings moves toward high

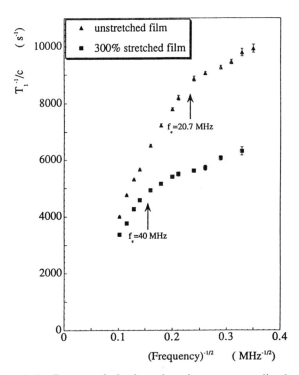

Fig. 5.18 Proton spin-lattice relaxation rate normalized to the spin concentration as a function of frequency (inverse square root) for unstretched and 300% stretched films of polyaniline. (From Ref. 103.)

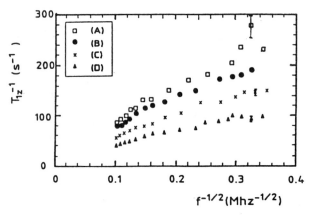

Fig. 5.17 Proton spin-lattice relaxation rate as a function of frequency for polyaniline. (A, C) dried state; (B, D) hydrated state (see text). (From Ref. 102.)

frequency after stretching. This is evidence for a net increase in the interchain hoppings (the intrachain diffusion rate also seems to increase, but the uncertainty in the $D_\parallel$ determination is much larger than for $D_\perp$). The data for the stretched and unstretched films are summarized in Table 5.3. It is noteworthy that the increase in the macroscopic anisotropy (≈ 10) is significantly higher than that of the interchain hopping rate (≈ 2). The data can be accounted for by a simple geometrical model that relies on the following picture. The material is composed of conducting grains connected together through resistive contacts. Inside a given grain the chains are oriented, which results in an anisotropic intragrain conductivity. The main effect of stretching is to orient the grains. It can be shown that as long as the microscopic anisotropy is very high ($D_\parallel \gg D_\perp$), the macroscopic anisotropy after stretching is determined only by geometrical parameters (and not by the microscopic transport characteristics). This model is able to explain, in particular, that the anisotropy does not vary with temperature, whereas large changes are observed for $\Sigma_\parallel$ and $\Sigma_\perp$.

C. Polypyrrole

Spin dynamics studies in polypyrrole-perchlorate (PPy-ClO$_4$) have been performed by Devreux and Lecavellier [8]. At first, observation of the Overhauser effect proved the existence of a direct dynamic coupling between electronic and nuclear spins [104]. The frequency dependence of the proton relaxation rate is shown in Fig. 5.19. The data can be fitted with $T_1^{-1} \propto \omega^{-1/2}$ for temperatures $T \leq 150$ K. For $T > 150$ K, the data deviate from 1-D diffusion behavior. They also cannot be fitted with the law of a pseudo-one-dimensional diffusion [Eq. (10)] with the introduction of a cutoff frequency ω_c. Instead, they can be accounted for by taking the spectral density

Table 5.3 Macro- and Microscopic Transport Parameters for Unstretched and 300% Stretched PANI-ES Films[a]

Parameter	Unstretched film	300% Stretched film
$\Sigma_\parallel$ (S/cm)	40	210
$\Sigma_\perp$ (S/cm)	40	23
$A_{macro} = \Sigma_\parallel/\Sigma_\perp$	1	9
$D_\parallel (10^{12}$ rad/s)	1.6 ± 0.9	3 ± 1.6
f_c (MHz)	20.7	40
$D_\perp (10^8$ rad/s)	1.3	2.5
$A_{micro} = D_\parallel/D_\perp$	$\approx 10^4$	$\approx 10^4$

[a] Conductivities along ($\Sigma_\parallel$), and perpendicular to ($\Sigma_\perp$) the stretching direction; spin diffusion rate along ($D_\parallel$) and perpendicular to ($D_\perp$) the chains; f_c is the cutoff frequency of the spectrum motion.
Source: Ref. 103.

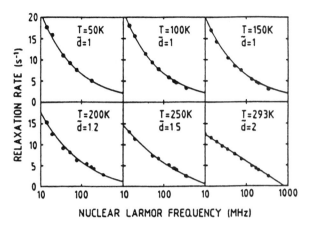

Fig. 5.19 Proton spin-lattice relaxation rate as a function of frequency for ClO$_4$-doped polypyrrole at different temperatures. (From Ref. 8.)

for diffusion on a fractal network (see Section II.C): $f(\omega) = A\omega^{\bar{d}/2 - 1} + B$, where $\bar{d}$ (<2) is the spectral dimension of the motion. It turns out that $\bar{d}$ increases with temperature from ≈ 1 at low T up to ~ 2 at room temperature.

For interpreting these results, in particular the departure from the commonly observed "pseudo-one-dimensional diffusion," it is important to recall that polypyrrole is a very disordered (fully amorphous) material [105]. The following picture has been proposed [8]. As the effective dimensionality remains locked at 1 for $T < 150$ K, it is reasonable to admit that this value reflects the confinement of the polarons on the polymer chain. At low temperature the spins undergo intrachain 1-D diffusion. At higher temperature, they are allowed to execute interchain hoppings, giving rise to an increasing effective dimensionality. In contrast to the behavior of pseudo-1-D diffusion, in which the dimensional crossover takes place at a definite frequency corresponding to a well-defined interchain hopping rate, the high degree of disorder in PPy-ClO$_4$ results in a continuous change in the dimensionality with temperature.

The above qualitative picture is supported by the data of frequency-dependent transport properties. The frequency dependence of conductivity provides another, independent way of determining the effective dimensionality $\bar{d}$. In disordered systems the conductivity as a function of frequency usually follows a power law: $\sigma \propto \omega^s$. Considering that the basic process of conduction is an anomalous diffusion, i.e., a random walk of the charge carriers on a network of effective dimensionality $\bar{d} < d$, where d is the space dimension ($d = 3$ in the present case), the exponent s can be expressed as $s = 1 - \bar{d}/d$. This expression with data of conductivity versus frequency given in the literature [106] leads to values for $\bar{d}$ that agree satisfactorily with those obtained from spin dynamics.

D. Polythiophenes and Polyalkylthiophenes

Spin dynamics studies into compounds of the polythiophene family have been started recently. The Tokyo group has investigated the temperature dependence of the spin dynamics as obtained from the ESR linewidth [107], and the Grenoble group has been concerned with the spin dynamics–structure relationship derived from room temperature data [108,109].

Mizoguchi et al. explain the ESR linewidth in terms of three contributions:

$$\Delta H = \Delta H_{\text{diff}}(T,\omega) + \Delta H_{\text{El}}(T) + \Delta H_{g\text{-ani}}(T,\omega) \tag{23}$$

where $\Delta H_{\text{diff}}(T, \omega)$ is the 1-D diffusion-narrowed dipolar linewidth, $\Delta H_{\text{El}}(T)$ is the linewidth originated in the conduction electrons via the Elliott process, and $\Delta H_{g\text{-ani}}(T, \omega)$ is the broadening due to the anisotropic g shifts (proportional to the applied magnetic field). The diffusion rates, $D_\parallel$ and $D_\perp$, are extracted from $\Delta H_{\text{diff}}(T, \omega)$, which is obtained by subtracting the two other contributions from the measured linewidth. The data for $\Delta H_{\text{diff}}(T,\omega)$ give a good fit with a pseudo-1-D behavior, provided a (large) frequency-independent term, ΔH ($\omega = 0$), is added (Fig. 5.20). Since the linewidth decreases with decreasing temperature, as in a "true metal," the frequency-independent term has been assumed to correspond to the Elliott relaxation mechanism $\Delta H(\omega = 0) = \Delta H_{\text{El}}(T)$ [110]. The latter gives a line broadening that originates from the finite lifetime of the spin states due to the electron–phonon scattering and via the spin–orbit coupling. It is expressed as [34,35]

$$\gamma \Delta H_{\text{El}}(T) = \alpha \left(\frac{\lambda}{\Delta E}\right)^2 \tau^{-1} \tag{24}$$

where λ is the spin–orbit coupling, ΔE is the interband energy, τ is the scattering time of the electron momentum, and α is a numerical factor on the order of unity, which should be zero in pure 1-D systems because of symmetry restrictions. Since the involved electron–phonon scattering is the same as for the transport process, $\Delta H_{\text{El}}(T)$ varies the same way as the resistivity: $\Delta H_{\text{El}}(T) \propto \rho(T) = \sigma^{-1}(T)$. This theoretical behavior for the Elliott mechanism is in contrast with experiment, since as T increases, $\rho(T)$ increases [$\sigma(T)$ decreases], while $\Delta H \approx \Delta H(\omega = 0)$ increases. Mizoguchi et al. have proposed, instead of taking into account the dc conductivity as usually measured, to consider the conductivity value obtained by the voltage-shorted compaction (VSC) method. The latter is supposed to avoid interfibril and interchain contributions to the resistivity and to yield the intrinsic value of the intrachain conductivity. Indeed, the VSC resistivity measured in PT(ClO$_4^-$) and in PT(AsF$_6^-$), increases as T increases and gives a good fit with $\Delta H(T)$ measured at 50 MHz [110]. Furthermore, the data are well fitted with a theoretical expression that accounts for the resistivity in a 1-D metal due to the scattering with a single phonon. Despite the remarkable fits, this approach suffers two difficulties. First, the VSC method for conductivity data is controversial. Data obtained from the conventional four-contact measurements lead to the opposite temperature dependence. Then the relation $\Delta H(T) \propto \rho(T)$ would not be true. Second, in principle the Elliott mechanism should not be efficient in one dimension for symmetry reasons. Mizoguchi et al. argue that in a five-member ring compound the symmetry restriction can be overcome because there is no inversion symmetry.

In Grenoble, the diffusion rates $D_\parallel$ and $D_\perp$ have been determined from proton T_1 measurements versus frequency in the following series of polymers of the polythiophene family: (1) polybithiophene (PBT), (2) polydimethyltetrathiophene (PDMTT), (3) polyhexylthiophene (PHT); and (4) a polystyrene-polythiophene copolymer (PS-PT). From the data shown in Fig. 5.21, the diffusion rates and the microscopic conductivities $\sigma_\parallel$ and $\sigma_\perp$ have been deduced (see Table 5.4). These values are consistent with qualitative chemical and steric considerations. The intrachain conductivities are the smallest in PDMT, which is expected to contain numerous localizing defects. The anisotropy is very large ($\sim 5 \times 10^5$) in PHT, in which the chains are very well separated from each other by the hexyl segments. The measured dc conductivity has a value intermediate between $\sigma_\parallel$ and $\sigma_\perp$. For the PS-PT copolymer, the σ_{dc}, on the order of $\sigma_\perp$, is very small (5×10^{-3} S/cm). This is consistent with the fact that in such a block copolymer the PT chains are organized in nonpercolating, rather well crystallized, grains embedded in an insulating polystyrene matrix. In

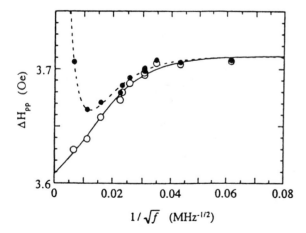

Fig. 5.20 ESR linewidth as a function of frequency (inverse square root) for ClO$_4$-doped polythiophene. (From Ref. 110.) (●) Raw data; (○) data corrected by the g-anisotropy broadening, which is proportional to the applied magnetic field.

Electron Spin Dynamics

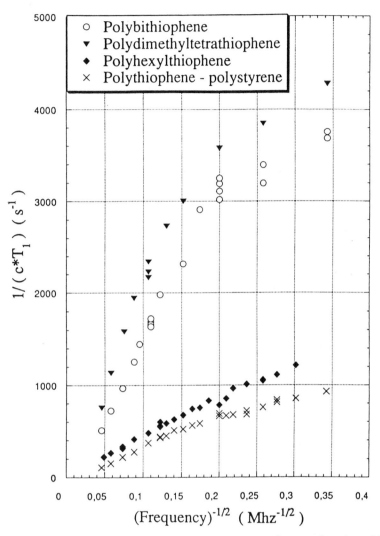

Fig. 5.21 Proton spin-lattice relaxation rate normalized to the spin concentration as a function of frequency (inverse square root) for four polythiophene family compounds. (From Ref. 109.)

Table 5.4 Spin Diffusion Rates and Microscopic Conductivities, $\sigma_\parallel$ and $\sigma_\perp$, for Four Polythiophene Family Compounds

	PBT	PDMT	PHT	PS-PT
$D_\parallel$ (10^{13} rad/s)	2.3	≈0.5	47	48
$D_\perp$ (10^8 rad/s)	2.2	3.5	0.7	1.8
$\sigma_\parallel$ (S/cm)	1.3×10^2	≈30	2×10^3	3.1×10^3
$\sigma_\perp$ (S/cm)	6×10^{-3}	1×10^{-2}	1×10^{-4}	6×10^{-3}
Anisotropy $\sigma_\parallel/\sigma_\perp$	2×10^4	3×10^3	2×10^7	5×10^5
Measured dc-conductivity (S/cm)	0.6	0.5	—	5×10^{-3}

PBT = polybithiophene; PDMTT = polydimethyltetrathiophene; PHT = polyhexylthiophene, PS-PT = polystyrene-polythiophene copolymer.
Source: P.Y. Mabboux, Thèse, Grenoble (1996).

contrast, inside the conducting grains the estimated intrachain conductivity is very high: $\sigma_\| \approx 3 \times 10^3$ S/cm. Let us stress that such intragrain transport data can be obtained only by spin dynamics methods. It is noteworthy that the diffusion rate values obtained by both groups in the case of polythiophene (PT) are in qualitative agreement.

VI. CONCLUSION

After more than 15 years of use and development, spin dynamics has proved to be an efficient method for transport property studies in conducting polymers. Numerous novel results have been obtained that would not have been obtained by conventional methods.

Among the first results was the proof that highly mobile spins are present in *trans*-polyacetylene and that the motion is highly one-dimensional. This finding provided strong support for the soliton picture. Note that since neutral solitons have no charge, they are not attainable by methods based on charge transport. This contribution of spin dynamics to the understanding of soliton behavior was discussed in the first edition of this handbook. The essential work in this field had already been done at that time. Since then, the developments have been mainly concerned with the motion of polarons in polymers in the conducting state. As illustrative examples of what can be done with spin dynamics we mention the estimate of the microscopic conductivity parameters inside nonpercolating conducting islands in partially protonated polyaniline or in block copolymers [109].

Spin dynamics as a method for transport studies is still developing. There are some fundamental points that have not yet been solved thoroughly. For instance, in contrast to the interchain transfer, which is a well-established parameter, in most cases the effect of the chain finiteness is ignored. In most of the work the chains are implicitly assumed to be infinite, which obviously does not correspond to a realistic picture of the actual material. A thorough discussion of the influence of finite chain length on spin dynamics has not yet been published.

Another important point that must be developed in the future is concerned with nuclear relaxation in the case of semidilute diffusive spins. In spin dynamics studies based on nuclear T_1 measurements, it is usually assumed that nuclear relaxation is directly induced by electronic spins. This is well verified in the case of a material homogeneously visited by mobile spins. A single T_1 can be defined. The opposite limiting case corresponds to dilute fixed electronic spins. The nuclear relaxation is essentially transferred by nuclear spin diffusion. It also results in a single T_1, defined only for long enough times. The intermediate case is the one of electronic spins diffusing only in certain regions of the material. The nuclear relaxation mechanism is much more complicated to model, but this case opens the way to various interesting compounds such as block copolymers and blends.

Spin dynamics in conducting polymers is a living domain of research, with its internal debates and open questions. One of the latter is concerned with the relationship between the polaron ESR linewidth and the conductivity. Two contradictary interpretations can be found: (1) The linewidth, or part of it, varies with the resistivity ($\Delta H \propto \rho = \sigma^{-1}$), and (2) the linewidth, or part of it, is proportional to the conductivity ($\Delta H \propto \sigma$). According to the first of these, an increase in ΔH, typically as a function of temperature, corresponds to a slowing of the motion. Such an interpretation is consistent with the Elliott mechanism for conduction electrons (line broadening and resistivity both originate from electron–phonon scattering) or with the motional narrowing process (less motion leads to less narrowing and therefore to a broader line). Conversely, in the second interpretation, the linewidth originates in the collisions of the moving spins (polarons) with fixed impurity spins or other moving spins (via spin–orbit coupling). Therefore, ΔH varies with the collision rate, which is proportional to the spin carrier mobility. It is clear that, according to which point of view is adopted, contradictory conclusions can be presented for the same set of data. Some polyaniline data were given interpretation 1 by the Tokyo group, while the Grenoble group was in favor of the second scheme. In principle, one could discriminate between the two possibilities by direct conductivity measurements; but, as spin dynamics reflects the conductivity at a nanoscopic scale, comparison with conventional measurements might be delicate.

ACKNOWLEDGMENTS

I thank all colleagues and visiting scientists who contributed to the work of the Grenoble group on spin dynamics in conducting polymers, in particular F. Devreux, F. Genoud, K. Holczer, and J.-P. Travers. I also thank K. Mizoguchi for his collaboration and stimulating discussions and P. Beadle for checking the manuscript.

REFERENCES

1. T. S. Clarke and J. C. Scott, *Handbook of Conducting Polymers*, Vol. 2, Marcel Dekker, New York, 1986, Chap. 31.
2. F. Devreux, Thèse, Grenoble.
3. W. G. Clark, K. Glover, M. D. Lan, and L. J. Azevedo, *J. Phys. (Paris)* 44(C3):1493 (1983).
4. P. Y. Mabboux, B. Beau, J. P. Travers, and Y. Nicolau, Proceedings of the ICSM '96, to appear in *Synth. Met.* (1997).

5. K. Mizoguchi, *Jpn. J. Appl. Phys.* **34**:1 (1995).
6. S. Alexander and R. Orbach, *J. Phys. (Paris) Lett.* **43**:L625 (1982).
7. Y. Gefen, A. Aharony, and S. Alexander, *Phys. Rev. Lett.* **50**: 77 (1983).
8. F. Devreux and H. Lecavellier, *Phys. Rev. Lett.* **59**: 2585 (1987).
9. F. Devreux, *Phys. Rev. B 13*:46 (1976).
10. A. Abragam, *The Principles of Nuclear Magnetism*, Oxford Univ. Press, Oxford, 1961.
11. M. Nechtschein, F. Devreux, R. L. Greene, T. C. Clarke, and G. B. Street, *Phys. Rev. Lett.* **44**:356 (1980).
12. K. Mizoguchi, M. Nechtschein, J-P. Travers, and C. Ménardo, *Phys. Rev. Lett.* **63**:66 (1989).
13. J. P. Boucher and M. Nechtschein, *J. Phys. (Paris) 31*:783 (1970).
14. J. P. Boucher, F. Ferrieu, and M. Nechtschein, *Phys. Rev. B, 9*:3871 (1974).
15. R. A. Wind, M. J. Duijvestijn, and J. Vriend, *Solid State Commun.* **56**:713 (1985).
16. K. Holczer, F. Devreux, M. Nechtschein, and J. P. Travers, *Solid State Commun.* **39**:881 (1981).
17. W. G. Clark, K. Glover, G. Mozurkewich, C. T. Murayama, J. Sanny, S. Etemad, and M. Maxfield, *J. Phys. Colloq.* **44**:C3–239 (1983).
18. W. G. Clark, K. Glover, G. Mozurkewich, S. Etemad, and M. Maxfield, *Mol. Cryst. Liq. Cryst. 117*:447 (1985).
19. G. Soda, D. Jérome, M. Weger, J. Alizon, J. Gallice, H. Robert, J. M. Fabre, and L. Giral, *J. Phys. (Paris) 38*:931 (1977).
20. Z. G. Soos, T. Z. Huang, J. S. Valentine, and R. C. Hughes, *Phys. Rev. B8*:993 (1973).
21. R. E. Dietz, F. R. Meritt, R. Dingle, D. Hone, B. G. Silbernagel, and P. M. Richards, *Phys. Rev. Lett.* **26**:1186 (1971).
22. M. J. Hennessy, C. D. McElwee, and P. M. Richards, *Phys. Rev. B7*:930 (1973).
23. J. P. Boucher, M. Ahmed-Bakheit, M. Nechtschein, M. Villa, G. Bonera, and F. Borsa, *Phys. Rev. B 13*: 4098 (1976).
24. J. Tang, C. P. Lin, M. K. Bowman, J. R. Norris, J. Isoya, and H. Shirakawa, *Phys. Rev. B* **28**:2845 (1983).
25. K. Holczer, J. P. Boucher, F. Devreux, and M. Nechtschein, *Phys. Rev. B* **23**:1051 (1981).
26. H. C. Torrey, *Phys. Rev.* **104**:563 (1956).
27. G. G. Maresh, A. Grupp, M. Mehring, J. U. V. Schultz, and H. C. Wolf, *J. Phys.* **46**:461 (1985).
28. G. G. Maresh, A. Grupp, M. Mehring, and J. U. V. Schultz, *Synth. Met.* **16**:161 (1986).
29. R. Ruf, N. Kaplan, and E. Dormann, *Phys. Rev. Lett.* **74**:2122 (1995).
30. N. S. Shiren, Y. Tomkiewicz, T. G. Tazyaka, A. R. Taranko, H. Thomann, L. Dalton, and T. C. Clarke, *Solid State Commun.* **44**:1157 (1982).
31. K. Mizoguchi, S. Masubichi, and K. Kume, *Phys. Rev. B* **51**: 8864 (1995).
32. M. Mehring, H. Seidel, W. Müller, and G. Wegner, *Solid State Commun.* **45**:1075 (1983).
33. E. Houzé and M. Nechtschein, *Phys. Rev. B53*: 14309 (1996).
34. R. J. Elliott, *Phys. Rev.* **96**:266 (1954).
35. Y. Yafet, in *Solid State Physics* Vol. 14, (H. Ehrenreich, F. Seitz, and D. Turnbull, eds.), Academic, New York, 1965.
36. I. B. Goldberg, H. R. Crowe, P. R. Newman, A. J. Heeger, and A. G. MacDiarmid, *J. Chem. Phys.* **70**: 1132 (1979).
37. B. R. Weinberger, E. Ehrenfreund, A. Pron, A. J. Heeger, and A. G. MacDiarmid, *J. Chem. Phys.* **72**: 4749 (1980).
38. W. P. Su, J. R. Schrieffer, and A. J. Heeger, *Phys. Rev. Lett.* **42**:1698 (1979).
39. W. P. Su, J. R. Schrieffer, and A. J. Heeger, *Phys. Rev. B* **22**:2099 (1980).
40. J. M. Rice, *Phys. Lett.* **71A**:152 (1979).
41. K. Mizoguchi, K. Kume, and H. Shirakawa, *Solid State Commun.* **50**:213 (1984).
42. H. M. MacConnell, C. Heller, T. Cole, and W. R. Fessenden, *J. Am. Chem. Soc.* **82**:766 (1960).
43. F. Devreux, C. Jeandey, M. Nechtschein, J. M. Fabre, and L. Giral, *J. Phys. (Paris)* **40**:65 (1979).
44. H. M. MacConnell, *J. Chem. Phys.* **24**:532 (1956).
45. H. Thomann, L. R. Dalton, Y. Tomkiewicz, N. S. Shiren, and T. C. Clarke, *Phys. Rev. Lett.* **50**:533 (1983).
46. H. Thomann, H. Kim, A. Morrobel-Sosa, L. R. Dalton, M. T. Johnes, B. H. Robinson, T. C. Clarke, and Y. Tomkiewicz, *Synth. Met.* **9**:255 (1984).
47. J. F. Cline, H. Thomann, H. Kim, A. Morrobel-Sosa, L. R. Dalton, and B. Hoffman, *Phys. Rev. B 31*:1605 (1985).
48. H. Thomann and L. R. Dalton, in *Handbook of Conducting Polymers*, Vol. 2 T. A. Skotheim, ed. Marcel Dekker, New York, 1986, p. 1157.
49. S. Kuroda and H. Shirakawa, *Solid State Commun.* **43**:591 (1982).
50. S. Kuroda and H. Shirakawa, *Phys. Rev. B 35*:9380 (1987).
51. H. Kaas, P. Hofer, A. Grupp, P. K. Kahol, G. Wiesenhofer, G. Wegner, and M. Mehring, *Europhys. Lett.* **9**:947 (1987).
52. P. K. Kahol and M. Mehring, *J. Phys. C: Solid State Phys.* **19**:1054 (1986).
53. S. Kuroda, *Int. J. Mod. Phys.* **B9**:221 (1995).
54. F. Devreux, J. P. Boucher, and M. Nechtschein, *J. Phys. (Paris)* **35**:271 (1974).
55. C. R. Fincher, C. E. Chen, A. J. Heeger, A. G. MacDiarmid, and J. B. Hastings, *Phys. Rev. Lett.* **48**:100 (1982).
56. M. Nechtschein, F. Devreux, F. Genoud, M. Guglielmi, and K. Holczer, *Phys. Rev. B* **27**:61 (1983).
57. K. Mizoguchi, H. Sakurai, F. Shimizu, S. Masubichi, and K. Kume, *Synth. Met.* **68**:239 (1995).
58. Z. Vardeny, J. Strait, D. Moses, T. C. Chung, and A. J. Heeger, *Phys. Rev. Lett.* **49**:1657 (1982).

59. B. H. Robinson, J. M. Schurr, A. L. Kwiram, H. Thomann, H. Kim, A. Morrobel-Sosa, P. Bryson, and L. R. Dalton, *J. Phys. Chem. 89:* 4994 (1985).
60. H. Thomann, H. Jin, and G. L. Baker, *Phys. Rev. Lett. 59:* 509 (1987).
61. J. Brédas, R. Chance, and R. Silbey, *Mol. Cryst. Liq. Cryst. 77:*319 (1981).
62. K. Mizoguchi, K. Kume, and H. Shirakawa, *Synth. Met. 17:*439 (1987).
63. K. Maki, *Phys. Rev. B 26:*2181 (1982).
64. Y. Wada and J. R. Schrieffer, *Phys. Rev. B 18:*38976 (1978).
65. M. Ogata, A. Terai, and Y. Wada, *J. Phys. Soc. Jpn. 55:*(7) (1986).
66. Y. Wada, *Progr. Theor. Phys. Suppl. 113:*1 (1993).
67. S. Jeyadev and E. M. Conwell, *Phys. Rev. B 36:*3284 (1987).
68. M. Nechtschein and W. Park, *Synth. Met. 69:*77 (1995).
69. F. Shimizu, K. Mizoguchi, S. Masubuchi, and K. Kume, *Synth. Met. 69:*43 (1995).
70. S. Ikehata, J. Kaufer, T. Woerner, A. Pron, M. A. Druy, A. Sivak, A. J. Heeger, and A. G. MacDiarmid, *Phys. Rev. Lett. 45:*1123 (1980).
71. T. C. Chung, F. Moraes, J. D. Flood, and A. J. Heeger, *Phys. Rev. B 29:*2341 (1984).
72. F. Devreux, M. Nechtschein, and G. Gruner, *Phys. Rev. Lett. 44:*53 (1980).
73. R. de Surville, M. Jozefowicz, L. T. Yu, J. Perichon, and R. Buvet, *Electrochim. Acta 13:*1451 (1968).
74. W. S. Huang, B. D. Humphrey, and A. G. MacDiarmid, *J. Chem. Soc. Faraday Trans. 82:*2385 (1986).
75. J. M. Ginder, A. F. Richter, A. G. MacDiarmid, and A. J. Epstein, *Solid State Commun. 63:*97 (1987).
76. A. G. MacDiarmid, J. C. Chiang, A. F. Richter, and A. J. Epstein, *Synth. Met. 18:*285 (1987).
77. F. Zuo, M. Angelopoulos, A. G. MacDiarmid, and A. J. Epstein, *Phys. Rev. B 36:*3475 (1987).
78. F. Wudl, R. O. Angus, F. L. Lu, P. M. Allemand, D. J. Vachon, M. Nowak, Z. X. Liu, and A. J. Heeger, *J. Am. Chem. Soc. 109:*3677 (1987).
79. M. Nechtschein, F. Genoud, C. Ménardo, K. Mizoguchi, J. P. Travers, and B. Villeret, *Synth. Met. 29:*E211 (1989).
80. J. P. Travers, J. Chroboczek, F. Devreux, and F. Genoud, *Mol. Cryst. Liq. Cryst. 121:*195 (1985).
81. J. P. Pouget, M. E. Jozefowicz, A. J. Epstein, X. Tang, and A. G. MacDiarmid, *Macromolecules, 24:*779 (1991).
82. A. J. Epstein, A. G. MacDiarmid, and J. P. Pouget, *Phys. Rev. Lett. 65:*664 (1990).
83. Z. H. Wang, C. Li, E. M. Scherr, A. G. MacDiarmid, and A. J. Epstein, *Phys. Rev. Lett. 66:*1745 (1991).
84. Z. H. Wang, E. M. Scherr, A. G. MacDiarmid, and A. J. Epstein, *Phys. Rev. B 45:*4190 (1992).
85. K. Mizoguchi, M. Nechtschein, and J. P. Travers, *Synth. Met. 41:*113 (1991).
86. K. Mizoguchi and K. Kume, *Solid State Commun. 89:*971 (1994).
87. B. Movaghar and L. Schweitzer, *Phil. Mag. B 37:* 683 (1978).
88. O. Chauvet, T. Stoto, and L. Zuppiroli, *Phys. Rev. B 46:* 8139 (1992).
89. F. Genoud, M. Nechtschein, M. F. Planche, and J. C. Thiéblemont, *Synth. Met. 69:*(1995).
90. M. Nechtschein and F. Genoud, *Solid State Commun. 91:*474 (1994).
91. K. Aasmundtveit, F. Genoud, E. Houzé, and M. Nechtschein, *Synth. Met. 69:*193 (1995).
92. M. Nechtschein, C. Santier, J. P. Travers, J. Chroboczek, A. Alix, and M. Ripert, *Synth. Met. 18:*311 (1987).
93. J. P. Travers and M. Nechtschein, *Synth. Met. 21:* 135 (1987).
94. M. Angelopoulos, A. Ray, and A. G. MacDiarmid, *Synth. Met. 21:*21 (1987).
95. O. N. Timofeeva, B. Z. Lubentsov, Y. Z. Sudakova, D. N. Chernishov, and M. I. Khidekel, *Synth. Met. 40:*111 (1991).
96. T. Taka, *Synth. Met. 57:*5014 (1993).
97. A. P. Monkman, P. N. Adams, P. J. Laughlin, and E. R. Holland, *Synth. Met. 69:*183 (1995).
98. H. Y. Hwang, S. W. Lee, I. W. Kim, and H. Lee, *Synth. Met. 69:*225 (1995).
99. A. Alix, V. Lemoine, M. Nechtschein, J. P. Travers, and C. Ménardo, *Synth. Met. 29:*E47 (1989).
100. J. P. Travers, M. Nechtschein, and A. Savalle, *Synth. Met 41:*613 (1991).
101. M. Nechtschein and C. Santier, *J. Phys. (Paris) 47:* 935 (1986).
102. J. P. Travers, *Synth. Met. 55:*731 (1993).
103. J. P. Travers, P. Le Guyadec, P. N. Adams, P. J. Laughlin, and A. P. Monkman, *Synth. Met 65:*159 (1994).
104. H. Lecavellier, F. Devreux, M. Nechtschein, and G. Bidan, *Mol. Cryst. Liq. Cryst. 118:*183 (1985).
105. G. B. Street, in Handbook of Conducting Polymers, Vol. 1 (T. A. Skotheim et al., eds.), Marcel Dekker (1986), Chap. 8, p. 265.
106. M. el Kadiri, J. P. Parneix, and A. Chapoton, *Ann. Phys. (Paris) 11:*89 (1896).
107. K. Mizoguchi, M. Honda, S. Masubishi, S. Kazawa, and K. Kume, *Jpn. J. Appl. Phys. 33:*L1239 (1994).
108. P. Leguennec, J. P. Travers, and Y. Nicolau, *Synth. Met. 55:* 672 (1993).
109. P. Y. Mabboux, J. P. Travers, Y. Nicolau, E. Samuelsen, P. H. Carlsen, and B. Francois, *Synth. Met. 69:*361 (1995).
110. K. Mizoguchi, M. Honda, N. Kachi, F. Shimizu, H. Sakamoto, K. Kume, S. Masubuchi, and K. Kazama, *Solid State Commun. 96:*333 (1995).

6
π-Electron Models of Conjugated Polymers: Vibrational and Nonlinear Optical Spectra

Zoltán G. Soos and Debasis Mukhopadhyay
Princeton University, Princeton, New Jersey

Anna Painelli and Alberto Girlando
Parma University, Parma, Italy

I. INTRODUCTION

Hückel theory is a simple, elegant, and general approach to π electrons in conjugated molecules [1]. Its seminal ideas and approximations have motivated a wide variety of chemical and solid-state studies, including current applications to conjugated polymers [2] discussed in this chapter. Diversity has naturally led to major differences in themes and goals within different fields. Electronic and vibrational spectra of conjugated polymers, for example, retain the original microscopic and phenomenological formulation of π-electron theory. More accurate all-electron descriptions are now available for small molecules, while more approximate microscopic parameters are encountered in quantum cell models of solids. The competition between higher accuracy and wider scope will be quite apparent in our discussion of π-electronic excitations and their vibrational signatures in polymers.

Hückel or tight-binding methods generate delocalized orbitals for noninteracting electrons with transfer integrals $\beta(R)$ in chemistry and $t(R)$ in physics, where R is a C—C bond length. The procedure is equally applicable to bands formed from atomic or molecular orbitals of any complexity, as long as $t(R)$ is taken from experiment. This yields $\beta(R_o) = -2.40$ eV at the benzene bond length $R_o = 1.397$ Å, or π-electron bandwidths $4t \sim 10$ eV in conjugated polymers. A *constant, uniform* $\beta(R_o)$ is convenient for discrete excitations of small conjugated hydrocarbons. Transferable parameters were emphasized [3–5] in extending π-electron theory to include Coulomb interactions $V(R)$ in the Pariser–Parr–Pople (PPP) model [1,6,7].

Constant transfer integrals yield a metallic ground state for the infinite polyene, however, and the question of uniform vs. alternating bond lengths was actively pursued until Longuet-Higgins and Salem (LHS) [8] showed the uniform chain to be unstable to dimerization. This manifestation of the Peierls instability [9], stated a few years earlier, has since been confirmed in polyacetylene (PA). The Su–Schrieffer–Heeger (SSH) model [10,11] has alternating transfer integrals $t(1 \pm \delta)$ for partial double and single bonds $R = R_o \mp u$. *Static* modulation of $t(R)$ generates semiconductors in Hückel theory for polymers with an even number of sp^2-hybridized carbons per repeat unit along the chain; *dynamic* modulation of $t(R)$ couples electronic and vibrational degrees of freedom.

The potential $V(R)$ introduces electron–electron (e–e) interactions, and Taylor expansion of $t(R)$ or $V(R)$ about equilibrium generates electron–phonon (e–ph) coupling. Conjugated polymers abundantly illustrate [12,13] e–ph and e–e contributions whose joint analysis is difficult mathematically. But a joint analysis will undoubtedly emerge, and this review is a step in that direction. We seek sufficiently powerful e–ph descriptions for detailed fits of vibrational spectra and sufficiently accurate correlated states to understand excitations, including a host of recent nonlinear optical (NLO) spectra. Both e–ph and e–e interactions appear naturally in models, and both lead to characteristic susceptibilities. A related issue for vibrational spectra is the precise identification of π-electronic contributions. We will emphasize the advantages of models for microscopic descriptions of conjugated polymers. To develop these themes, we recall some basic results for infinite chains with alternating single and double bonds. Although apparently restricted to PA, alternating chains are widely applicable to other conjugated polymers.

A. Idealized One-Dimensional Chain

The Hückel chain with alternating transfer integrals $t(1 \pm \delta)$ is

$$H_0(\delta) = -|t| \sum_{n\sigma} [1 - (-1)^n \delta](a_{n\sigma}^+ a_{n+1\sigma} + a_{n+1\sigma}^+ a_{n\sigma})$$
$$\equiv \sum_n 2t_n p_n \quad (1)$$

The fermion operators $a_{n\sigma}^+$ and $a_{n\sigma}$ create and annihilate an electron with spin σ at carbon n, with partial double and single bonds at sites $2n, 2n-1$, and $2n, 2n+1$ in Fig. 6.1. Coulson's mobile π-electron bond order [14] is the operator p_n defined by Eq. (1). The familiar eigenvalues of $H_0(\delta)$ are

$$\epsilon(k,\delta) = \pm 2|t|(\cos^2 k + \delta^2 \sin^2 k)^{1/2} \quad (2)$$

and form the bands in the first Brillouin zone sketched in Fig. 6.1a. The valence band is filled in the PA ground state. Vertical excitations $2|\epsilon(k,\delta)|$ give an electron and hole, with opposite wavevector k. The corresponding vibrational model in Fig. 6.1 is a chain of particles of mass M and alternating spring constants K_1, $K_2 =$ $K(1 \pm \Delta)$. The normal frequencies squared, $\omega(q)^2$, are [15]

$$M\omega(q)^2 = 2K[1 \pm (\cos^2 q + \Delta^2 \sin^2 q)^{1/2}] \quad (3)$$

A gap between the optical and acoustical branches appears at the zone edge, $q = \pi/2$, for $\Delta > 0$ in Fig. 6.1b.

The Peierls instability of PA modeled by SSH is based on $H_0(\delta)$, a harmonic lattice with $\Delta = 0$ for the σ electrons, and linear e–ph coupling constant

$$\alpha(R_0) = \left(\frac{\partial t(R)}{\partial R}\right)_{R_0} = t'(R_0) \quad (4)$$

The $\pm \delta$ ground states for $\delta > 0$ are degenerate, with the gain of electronic energy more than offsetting the strain. The elementary excitations [10,11] are solitons, or self-localized states, at lower energy than $2|\epsilon(k,\delta)|$ when the excited-state relaxation is included. These electronic excitations and their counterparts, polarons and bipolarons, in chains with nondegenerate ground states have motivated solid-state studies [16] of conjugated polymers. Mele [17] reviewed the vibrational aspects of the SSH model. The $q \sim 0$ optical phonons are strongly depressed by linear e–ph coupling and, as indicated by the dashed line in Fig. 6.1b, the PA dispersion increases rather than decreases near the zone center. The one-dimensional Hückel chain in Fig. 6.1 and linear e–ph coupling $\alpha(R_0)$ illustrate the fundamentals of both electronic and vibrational contributions.

Several major generalizations are needed to discuss vibrational and NLO spectra. The CH units represented by M in Fig. 6.1 have vibrational degrees of freedom, and the PA backbone is planar rather than one-dimensional. The amplitude mode (AM) formalism developed by Horovitz and coworkers [18,19] extends the SSH model to several coupled $q = 0$ modes, as discussed in Section II.C, with the partitioning taken from experiment. The second term in the Taylor expansion of $t(R)$ appears in force fields, as already recognized by Coulson and Longuet-Higgins [20]. While the form of $t''(R)$ need not be specified in advance, wave function overlap is usually taken to be exponential, and this fixes the curvature without additional parameters as

$$t''(R) = \alpha(R)^2/t(R) \quad (5)$$

Even at the π-electron level, conjugated polymers have additional vibrational modes, quadratic e–ph coupling, and Coulomb interactions $V(R)$ that generate other e–ph coupling constants. More seriously, $V(R)$ spoils the convenient single-particle solutions in Eq. (2) or Fig. 6.1a and alters the spectrum drastically.

B. Vibrational and Electronic Models

Two recent reviews cover the vibrational spectroscopy of linear polyenes [21] and conjugated polymers [22],

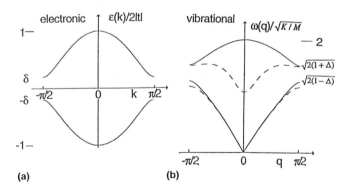

Fig. 6.1 Idealized linear chain with alternating transfer integrals $t(1 \pm \delta)$ and spring constants $K(1 \pm \Delta)$ for double and single bonds, with $\delta = \Delta = 0.2$. (a) Valence and conduction bands, Eq. (2), of the Hückel chain $H_0(\delta)$ with unit cell $2a$, $a = 1$. (b) Optical and acoustical phonon frequencies, Eq. (3), of the chain; the anomalous dispersion (dashed line) is for about 50% stronger e–ph coupling than in polyacetylene.

respectively. Many different experimental and theoretical approaches have been considered. One broad demarcation is between π-electron and all-electron descriptions. Direct quantum-chemical calculations range from ab initio to semiempirical, with many gradations in the size and quality of the basis, and recent density-functional approaches are almost as diverse. No special role is assigned to π-electrons or their fluctuations in these treatments. The ground-state potential surface, force constants, barrier heights, and systematic variations among related systems are typical goals. Vibrational models designed for conjugated polymers have also been developed. The effective conjugation coordinate (ECC) model of Zerbi and coworkers [22] adapts spectroscopic techniques to obtain force fields for conjugated polymers; a special coordinate is introduced for delocalization contributions that cannot be treated by standard methods. Kürti and Kuzmany [23], Brivio and Mulazzi [24] and coworkers have focused on finite conjugation lengths in polymers to analyze vibrational spectra as an inhomogeneous superposition of segments.

Longuet-Higgins and Salem [1,8] proposed a molecular force field for localized σ electrons and delocalized π electrons. Their scheme has become attractive for conjugated polymers [25,26], now in conjunction with semiempirical rather than simple Hückel calculations. Warshel and Karplus [27] and Hemley et al. [28] combine all-electron calculations of geometry with PPP models for π electrons (see also Ref. 21). We will focus exclusively on delocalization or π-electron contributions using linear response (LR) theory, as illustrated by the AM formalism. LR has been widely applied to vibrational spectra of charge transfer and ion-radical organic crystals [29–32]. The idea is to model *shifts* due to delocalization relative to some localized reference. The concomitant problem of localized or "other" contributions is the choice of a reference force field $\mathbf{F}^0$ discussed in Section II.B. The solid-state perspective of LR theory is quite compatible with π-electron or other models based on frontier orbitals.

Quantum-chemical calculations on conjugated hydrocarbons support the spectroscopic estimate, $\beta(R_0) = -2.40$ eV, and all-electron descriptions are appealing as soon as they become feasible. There are too many levels of theory to enumerate here, but quantitative ones are not yet applicable to conjugated polymers. Moreover, we are interested in excited states, which remain challenging even in molecules. The rationale for σ–π separability, for the Coulomb potential $V(R)$, and for the zero differential overlap (ZDO) approximation were discussed [1] in connection with the PPP model. Hubbard [33] considered the same issues for d electrons in transition metals. Quantum cell models [12,13,34] for frontier orbitals of any kind implicitly invoke ZDO to obtain two-center interactions. In many cases, the relevant transfer integrals t, Hubbard repulsion U, and intersite interactions $V(R)$ are small and hence difficult to evaluate in organic molecular crystals, transition metal oxides, or other narrow band solids.

The PPP model is formally a special case of extended Hubbard models [34], with transferable on-site and intersite interactions, although its origins are quantitative studies of conjugated molecules. An unshielded Coulomb interaction [35]

$$V(R) = e^2/(\rho^2 + R^2)^{1/2} \tag{6}$$

is introduced for C^+ or C^- sites separated by R. The radius ρ of the cell is taken from the atomic ionization potential and electron affinity [12]. We have $U = V(0) = e^2/\rho$, with $U = 11.26$ eV or $\rho = 1.28$ Å for carbon. Such precise values are unexpected in the context of Hubbard models, and variations of 10% may occur, but the emphasis is on a common molecular potential $V(R)$. The PPP model for an alternating chain is [12]

$$H_{\text{PPP}}(\delta) = H_0(\delta) + U\sum_p n_p(n_p - 1) + \sum_{p>p'} V_{pp'} q_p q_{p'} \tag{7}$$

with $V_{pp'} = V(R_{pp'})$, while $n_p = \sum_\sigma a^+_{n\sigma} a_{n\sigma}$ and $q_p = 1 - n_p$ are, respectively, the π-electron number and charge operators. Extended Hubbard models corresponds to other choices of $V_{pp'}$ in Eq. (7). We may also add site energies (Hückel α's) for inductive effects or heteroatoms. The mathematical properties of alternating chains are summarized in Section V. Although we will use Hückel, Hubbard, and PPP results, our general analysis of vibrational and electronic signatures of π electrons holds for any model.

The structure is an input for π-electron models. Once $V(R)$ and $t(R)$ have been specified, excitations and e–ph coupling can be treated comprehensively. Any structural change has both π and σ components. Conversely, e–ph analysis is not restricted to a particular $V(R)$. Although both reflect π electrons, vibrational and electronic models focus on different issues and have been developed separately. Debates about the relative importance of e–ph and e–e interactions miss their common origin, their close connection to delocalization, and the many guises of π electrons. Both e–ph and e–e interactions are in fact crucial.

C. Models, References, and Thresholds

Models have become separate research areas in statistical and solid-state physics. In addition to Hubbard models, familiar examples are two-level models, impurity models, and networks of spins in one, two, or three dimensions with isotropic, anisotropic, or frustrated interactions. Exact results for Hubbard chains with uniform t and for $s = 1/2$ Heisenberg chains with uniform exchange J are based on the Bethe ansatz, while exact results for two-dimensional Ising lattices come from transfer matrix techniques. Exact results for extended interacting systems are rare and interesting. They provide strong checks on approximations, Monte Carlo results, or oligomer calculations that can be carried out more generally on models with intersite interactions or alternating bonds. The direct connection between PPP and solid-state models is exploited in Section V and is a major reason for working with models.

Quantum cell models are introduced phenomenologically in solid-state physics to describe frontier orbitals in extended systems. They are not "derived" in the π-electron sense, although ZDO and related approximations are readily discerned. Models pose well-defined theoretical problems at the expense of direct contact with actual systems. Such contact is clearly vital for spectroscopic studies, however, of real solids or polymers. The problem is to identify contributions due to π electrons or frontier orbitals *without* detailed knowledge of the full system. For example, the oriented gas model of organic molecular crystals [36] describes small vibrational or electronic shifts and splitting relative to the gas phase. The microscopic parameters of quantum cell models can then be directly related to vibrational or electronic spectra.

Hückel and PPP theory emphasize common aspects of π electrons in molecules whose geometry is specified in advance. The backbones of the representative conjugated polymers in Fig. 6.2, for example, immediately fix the Hückel or PPP Hamiltonians. Effects due to substituents R or sulfur heteroatoms in polythiophene (PT) are approximated by site energies or neglected as perturbations of π–π^* transitions. Such a *priori* information about H_0 and $V(R)$ comes from molecules and contrasts sharply with the adjustable parameters of extended Hubbard models. The σ conjugation encountered in polysilanes (PS) also leads [37] to an alternating chain in Fig. 6.2. Now $t(1 + \delta)$ is associated with sp^3–sp^3 bonds of adjacent Si, $t(1 - \delta)$ with transfer between sp^3 orbitals of the same Si, and U is taken from the ionization potential and electron affinity of Si atoms.

The polymers in Fig. 6.2 are semiconductors in Hückel or band theory. The gap between the valence and conduction bands in Fig. 6.1a is $4t\delta$ for PA or PS. Four atoms per repeat unit in polydiacetylene (PDA) lead to two valence bands and two conduction bands. The larger alternation produced by triple bonds can be approximated [38] by an effective δ, while phenyl rings in poly(*para*-phenylenevinylene) (PPV) lead to a topological alternation [39]. The divergent density of states at band edges implies a *common* energy threshold: both triplets and singlets start at $4t\delta$ in alternating Hückel chains. They normally differ in molecules, where confinement lowers e–e repulsion for parallel spins. The energy thresholds of crystalline anthracene in Table 6.1 closely match molecular states, except for generating charge carriers, as expected for a lattice of neutral molecules in van der Waals contact. The thresholds [40] of polydiacetylene (PDA-PTS) crystals, poly(di-n-hexylsilane) (PDHS), and PPV in Table 6.1 resemble molecules rather than bands. Different thresholds for linear and two-photon absorption are particularly interesting. Mutually exclusive one and two-photon selection rules follow directly from the centrosymmetric backbones in Fig. 6.2. One-photon excitation is lower in PDHS, PPV, or PT, whereas two-photon excitation is lower in PDA or PA. The excitation thresholds in fact reflect an interplay [41] between the potential $V(R)$ in Eq. (6) and the alternation δ.

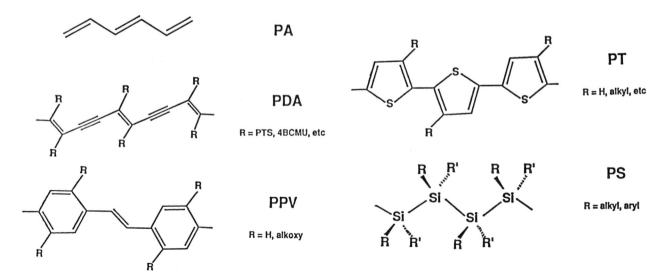

Fig. 6.2 Idealized planar backbones of conjugated polymers.

Table 6.1 Energy Thresholds (eV) of Anthracene, PDA-PTS, PDHS, and PPV

	Triplet, E_T	Singlet, $1B$	Singlet, $2A$	Charge transfer
Anthracene	1.8	3.1	3.5	4.1
PDA-PTS	~1.1	2.00	1.80	2.5
PDHS	3.4	3.5	4.2	5.2
PPV	~1.6	2.5	2.9	–

Source: Ref. 40.

D. Scope of Review

Our goal is to model quantitatively π-electronic contributions to both vibrational and electronic spectra. The general e–ph analysis introduced in Section II combines the microscopic AM formalism [18,19] with the spectroscopic ECC model [22]. The reference force field $\mathbf{F}^0$ for PA provides an experimental identification of delocalization effects. Transferable e–ph coupling constants are presented in Section III for polyenes and isotopes of *trans*- and *cis*-PA. The polymer force field in internal coordinates directly shows greater delocalization in *t*-PA, while coupling to C—C—C bends illustrates $V(R)$ participation and different coupling constants $\alpha(R_d)$ and $\alpha(R_s)$ in Eq. (3) support an exponential $t(R)$. NLO spectra of PDA crystals and films are presented in Section IV, with multiphoton resonances related to excited states of PPP models and vibronic contributions included in the Condon approximation. Linear and electroabsorption (EA) spectra of PDA crystals provide an experimental separation of vibrational and electronic contributions, and the full π–π^* spectrum is needed to model EA. We turn in Section V to correlated descriptions of electronic excitations, with particular attention to theoretical and experimental evidence for one- and two-photon thresholds of centrosymmetric backbones. The final section comments on parameters for conjugated polymers, extensions, and open questions.

We have sought throughout to combine Hückel and correlated states, vibrational and electronic spectra, and a common framework for the conjugated molecules and the polymers in Fig. 6.2. In this respect, we are guided by early π-electron discussions [1] of both molecular spectra and force fields. Vibrational analysis has been extended over the years to Raman and IR evidences of delocalization, while electronic excitations now also include two-photon and NLO spectra. We certainly recognize that the increased scope and detail of current models make it impractical to pursue vibrational and electronic applications together, except perhaps in reviews. Yet their unified microscopic description in interesting new materials such as conjugated polymers may well be the central reason for retaining π-electron models in spite of their inherently approximate nature.

II. ELECTRON–PHONON COUPLING AND REFERENCE STATE FOR CONJUGATED POLYMERS

A. Special Vibrational Features of Delocalized States

Progress in understanding the vibrational spectra of conjugated polymers has been slow in comparison with those of other polymers [42]. A traditional molecular spectroscopic approach to polymer vibrations would in fact be based on the GF formalism [43] and the solution of the secular equation

$$\mathbf{GFL} = \mathbf{L}\Lambda \qquad (8)$$

Here $\mathbf{G}$ and $\mathbf{F}$ are the mass and potential energy matrices on the internal coordinate basis $\mathbf{R}$, Λ is the diagonal matrix of the squared vibrational frequencies, and $\mathbf{L}$ is the transformation matrix to normal coordinates, $\mathbf{R} = \mathbf{LQ}$. The choice of internal coordinates is the key to the success of the GF method. Internal coordinates are directly related to chemical descriptions of molecular motions in terms of bond stretches, bends, and torsions. The internal coordinates in Fig. 6.3, for example, describe in-plane vibrations of polyenes.

Moreover, the $\mathbf{F}$ matrix of nonconjugated molecules is often quasidiagonal, with larger diagonal than off-diagonal elements and negligible interactions between coordinates of well-separated atoms. Force constants in internal coordinates are then understood as local properties, characteristic of each internal coordinate and, in this respect, transferable among similar molecular fragments. Transferability is of paramount importance in classical spectroscopy. The vibrational problem for a single molecule is in fact largely undetermined because there are many more force constants than experimental frequencies. But tractable solutions are possible with overlay techniques [44], in which experimental frequencies of related molecules are used to construct highly reliable force fields for the structures at hand with an overall fitting procedure. This approach has been applied by Schachtschneider and Snyder [45] to polyalkanes, using oligomer frequencies to determine the force field of the polymer. The classical approach to polymers treats the infinite chain as a collection of weakly interacting oscillators, each relevant to an internal coordinate, so that oligomer frequencies fall on the dispersion curve of the polymer itself. This approach necessarily fails in conjugated polymers due to the anomalous dispersion of the optical phonons in Fig. 6.1b.

Two main assumptions underlie the GF method: (1) full separation of electronic and vibrational degrees of freedom (crude adiabatic approximation) and (2) local-

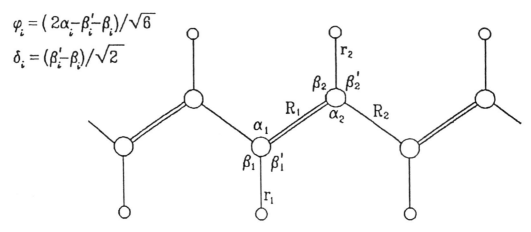

Fig. 6.3 Internal coordinates of *trans*-polyacetylene or *trans*-polyenes.

ized chemical bonds between atoms. Neither assumption holds in systems with low-lying delocalized electronic states, and the normal GF method fails. These difficulties already appear for the polyenes in Fig. 6.3. Force constants are not transferable in the $C_{2n}H_{2n+2}$ series, and interaction constants between distant bonds are needed for detailed fits [21,46,47] Striking conjugation or delocalization effects are inherent to systems whose electronic excitations vary with length, and their anomalous spectroscopic features are well known. Among the models successfully applied to vibrational spectra, the AM formalism [18,19] concentrates on the role of e–ph coupling, namely on the failure of the crude adiabatic approximation. The ECC model [22] instead handles the problem in terms of an "extended" or delocalized internal coordinate and variations of its force constant. The two approaches are therefore complementary, facing two different aspects of the same problem, but neither offers a comprehensive analysis of the overall problem.

B. Reference Force Field

The approach we recently proposed [48] overcomes these limitations by explicitly accounting for e–ph coupling and delocalization. We separate the problem by properly defining a reference state *without* e–ph coupling or delocalization. For this state we resort to general and reliable methods of traditional spectroscopy [43–45]. Then e–ph coupling and electronic delocalization are introduced in a second step, using powerful linear response techniques from solid-state physics [18,19,29,49]. We outline the method and its application to conjugated polymers in this section and specialize the treatment to the prototypical conjugated polymer, *t*-PA, in the next.

In the crude adiabatic approximation with fixed nuclei, the electronic Hamiltonian H_e does not depend on the nuclear coordinates q. The q dependence can be introduced via a Herzberg–Teller expansion of H_e, to obtain the e–ph coupling Hamiltonian [30,31]

$$H_{\text{e-ph}} = \sum_m \frac{\partial H_e}{\partial q_m} q_m + \frac{1}{2} \sum_{m,n} \frac{\partial^2 H_e}{\partial q_m \partial q_n} q_m q_n \qquad (9)$$

$H_{\text{e-ph}}$ acts as a perturbation on the q-independent eigenstates, $|F\rangle$, of H_e and yields q-dependent states. The ground-state energy $E_0(\mathbf{q})$ is the adiabatic potential for nuclear motions. In the harmonic approximation, the force constants for nuclear motions are second partial derivatives of $E_0(\mathbf{q})$. They are obtained perturbatively as

$$\Phi_{mn}^{(\text{HT})} = \Phi_{mn}^s + \langle G | \frac{\partial^2 H_e}{\partial q_m \partial q_n} | G \rangle \qquad (10)$$

$$- 2 \sum_F \langle G | \frac{\partial H_e}{\partial q_m} | F \rangle \langle F | \frac{\partial H_e}{\partial q_n} | G \rangle / E_F$$

where $|G\rangle$ and $|F\rangle$ are the ground and excited states of H_e, with excitation energy E_F relative to $E_G = 0$, and Φ_{mn}^s represents the skeleton force constants due to core electrons. In systems without delocalized electrons, the matrix Φ^s expressed in internal coordinates thus coincides with the total force field $\mathbf{F}$ in Eq. (8).

The spectroscopic signatures originate from *linear* e–ph coupling, the third term in Eq. (10), which is particularly important in the presence of electronic fluctuations and depends strongly on low-energy excitations. It is responsible for the anomalous red-shifted optical branch in Fig. 6.1b. On the other hand, contributions from quadratic coupling only involve the ground state and are conveniently included in a reference force field Φ^0 that describes the potential due to core electrons and the ground-state distribution of valence electrons but excludes electronic fluctuations.

Equations (9) and (10) were originally derived for vibrational spectra of ion-radical organic crystals [29–32]. The best characterized are based on the strong π donors D = TTF, BEDT-TTF, or TMPD and π acceptors A = TCNQ, DCNQI, or chloranil. They include organic conductors and superconductors based on substituted TTFs, as well as paramagnetic semiconductors and CT salts with a variety of structural motifs and oxidation states [34]. The effects of quadratic e–ph coupling in these systems led us to define the reference state as a collection of noninteracting molecules bearing the mean charge they have in the crystal [31]. This *experimental* reference automatically includes the first two terms of Eq. (10). The corresponding reference frequencies, at least for molecular vibrations, are either directly accessible from experiment or can be interpolated from vibrational frequencies of isolated molecular units with different charges. With respect to the reference state, linear e–ph coupling induces a frequency softening, often accompanied by "anomalous" (in the sense of classical spectroscopy) infrared and/or Raman intensity. A careful analysis of vibrational spectra rationalizes these effects in terms of linear e–ph coupling constants, here electron–molecular vibration coupling constants, that are molecular properties and consequently transferable between different systems [50].

The corresponding reference state for conjugated polymers is a hypothetical chain with frozen electronic fluctuations or, equivalently, with the same ground state as the actual polymer but with electronic excitations at infinite energy. Of course, this reference state has no experimental counterpart. As mentioned after Eq. (10), the reference force field contains both core or σ-electron contributions and ground-state π-electron contributions. It is expected to be local and thus amenable to the traditional GF approach [43,44]. Specifically, the reference force constant matrix $\mathbf{F}^0$ defined in internal coordinates can be transferred from the force fields of short oligomers whose high excitation energies minimize π-electron fluctuations.

The π–π^* excitation of *trans*-butadiene, for example, is over three times as high as in t-PA. The well-studied force field [51,52] of *trans*-butadiene is thus a suitable starting point for the reference of t-PA. To account for different bond lengths in the molecule and polymer, we estimate diagonal force constants for C—C stretches from empirical force constant vs. bond length correlations [53]. By exploiting the full symmetry of the polymer, the infinite-dimensional GF problem factors into an infinite number of finite-dimensional problems, with $q = 0$ relevant for optical probes. The strongly coupled modes of t-PA are C—C stretches that do not appear in the b_u symmetry block, which is therefore decoupled from the π system. The corresponding experimental IR frequencies [54,55] are used to refine [48] (and to validate) $\mathbf{F}^0$ by standard methods. We report in Table 6.2 the reference force field for t-PA and, for comparison,

Table 6.2 Reference Force Field $\mathbf{F}^0$ of Polyacetylene[a]

Symbol	Coordinate	Force constant[b,c]
Stretch		
K_1	R_1 (C=C)	8.10 (8.89)
K_2	R_2 (C—C)	5.80 (5.43)
K_3	r_1, r_2 (C—H)	5.00 (5.07)
Bend		
H_1	ϕ_1, ϕ_2 (C—C—C)	0.891 (0.691)
H_2	δ_1, δ_2 (C—C—H)	0.533 (0.519)
Interaction		
F_1	$R_1 R_2$	0.267 (0.400)
F_2	$R_1 \phi_1 = R_2 \phi_2$	0.170 (0.170)
F_3	$R_1 \delta_1 = -R_2 \delta_2$	0.200 (0.228)

[a] The coordinates are shown in Fig. 6.3.
[b] The units are mdyn/Å for stretches K and F_1, mdyn·Å/rad² for bends H, and mdyn/rad for stretch-bends F_2 and F_3.
[c] Butadiene values from Ref. 51 are given in parentheses. *Source*: Ref. 56.

the force field of *trans*-butadiene. The procedure is general and can be applied to other conjugated polymers. We are currently attempting to construct $\mathbf{F}^0$ for polythiophene.

C. The AM Equation for Coupled Modes

Once the reference force field $\mathbf{F}^0$ is properly derived, diagonalization of the $\mathbf{GF}^0$ matrix [see Eq. (8)] yields the reference frequencies ω_i^0 and normal coordinates Q_i^0. We use capital Q's to distinguish normal coordinates from the generic nuclear coordinates q in Eq. (9). This is a partial solution, since we still have to account for linear e–ph coupling. The expansion in Eq. (9) is conveniently carried out on the basis of the reference normal coordinates. The simplest case arises when, at least for a given symmetry subspace, only one electronic operator, θ, is coupled to the phonons,

$$H_{\text{e-ph}}^{(\text{linear})} = \frac{2}{\sqrt{N}} \sum_i Q_i^0 g_i \theta \qquad (11)$$

Here g_i is the e–ph coupling constant for mode Q_i^0. In spite of its simplicity, $H_{\text{e-ph}}$ is rather general. It accounts for the coupling of a_g molecular vibrations to CT excitations in organic ion-radical crystals [29–32] and is implicity assumed in both AM [18,19] and ECC [22] models.

The form of Eq. (11), with one electronic operator coupled to phonons, has far-reaching consequences. The total force field, Eq. (10), can be expressed on the basis of the reference normal coordinates, Q_i^0, as

$$\Phi_{ij} = \Phi_{ij}^0 - \Delta\Phi_{ij} = (\omega_i^0)^2 \delta_{ij} - \chi g_i g_j \qquad (12)$$

where Φ^0 is the diagonal matrix consisting of squared reference frequencies, δ_{ij} is the Krönecker symbol, and χ is the relevant electronic susceptibility [30,48].

$$\chi = 2 \sum_F |\langle F|\theta|G\rangle|^2/E_F \tag{13}$$

We emphasize that whenever a single electronic operator θ couples to the phonons, the $\Delta\Phi$ matrix has the special form above, with off-diagonal elements given by the geometric mean of diagonal elements.

The eigenvalues of Φ are the squared frequencies ω_i^2 of the actual system. The eigenvalue equation may be written as [48]

$$\sum_i \chi g_i^2 \prod_{j \ne i} [(\omega_j^0)^2 - \omega_p^2] = \prod_i [(\omega_i^0)^2 - \omega_p^2] \tag{14}$$

for each coupled mode p. When reliable reference and experimental frequencies are known, we have a system of linear equations for the unknown χg_i^2 and thus for the diagonal elements of $\Delta\Phi$ scaled by χ. Due to its special form, the $\Delta\Phi$ matrix is then fully determined [48].

In the particular case of t-PA, only three of the four a_g modes are appreciably coupled to π-electron fluctuations. Equation (14) is then equivalent to the AM equation. Its graphical solutions are obtained by plotting χ vs. ω, as illustrated in Fig. 6.4 for the three coupled Raman modes, and such plots summarize both AM and ECC results. In the absence of e–ph coupling the $\chi = 0$ solutions of Eq. (14) are the reference frequencies, ω_i^0. All modes shift to lower frequencies with increasing χ, and relative $\chi = 1$ is assigned to t-PA. The forced fit at $\chi = 1$ gives the coupling constant $\sqrt{\chi}\, g_i = 912, 122,$ and 355 cm^{-1}, respectively, for $i = 2,3,4$. The Raman modes of finite polyenes satisfy Eq. (14) with $\chi < 1$ in Fig. 6.4 [56]; larger excitation energies in Eq. (13) rationalize weaker e–ph coupling.

The preceding analysis holds for the a_g block of infinite extended PA. But the AM formalism also describes successfully the infrared frequencies [18,19] of chemically doped or photoexcited samples with different χ in Eq. (14). Strong IR activity is associated with localized vibrations about charged solitons, S$^+$ or S$^-$. From a phenomenological point of view, the appearance of infrared-activated vibrations (IRAV) corresponding to the Raman modes of pristine PA suggests that the original a_g modes, or closely related ones, are coupled to π-electron fluctuations. As implicitly assumed in the AM formalism, we retain the form of Eq. (11) for e–ph coupling in doped or photoexcited samples, except possibly for a different electronic operator θ. The IR frequencies of doped and photoexcited samples also fall on the curves in Fig. 6.4. Larger χ, which is now an adjustable parameter, indicates stronger coupling.

Curves similar to Fig. 6.4 have also been obtained within the AM [18,19] and ECC [22] models. The a_g frequencies in the ECC model are interpreted in terms of variations of the diagonal force constant of the special extended coordinate, without introducing a reference. No microscopic interpretation of the variation is offered, and the model is typically phenomenological. Up to this stage, our treatment is also phenomenological and formally equivalent to AM. However, whereas we derive reference frequencies from $\mathbf{F}^0$, they are adjustable parameters in the AM formalism. The precise physical meaning of that model's fitting parameters, ω_i^0 and λ_i, is left open. New parameters are consequently needed even for isotopic substitutions that are routinely included in molecules [43] by changing nuclear masses in the G matrix.

Like the AM formalism [18,19], our treatment is amenable to further microscopic analysis when the explicit form of $H_{\text{e-ph}}$ in Eq. (11) has been specified. The coupling in PA is essentially due to modulation of $t(R)$ by C—C stretches. We retain the AM choice,

$$\theta_- = \sum_n (-1)^n p_n, \qquad g_i = \sqrt{N}\frac{\partial t_n}{\partial Q_i^0} \tag{15}$$

where p_n is the π-electron bond order between sites n and $n+1$ defined in Eq. (1), t_n is the corresponding transfer integral, and N is the chain length. Thus θ_- is the staggered bond order, and the relevant phonons are totally symmetric, or a_g, modes. As shown in Section III, the assumption that only θ_- is coupled to phonons is

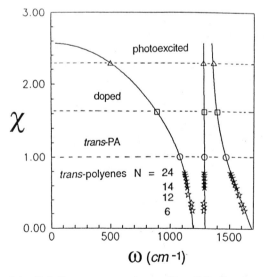

Fig. 6.4 Relative χ vs. ω curves, Eq. (14), for three a_g modes of *trans*-polyacetylene coupled to π-electron fluctuations. The $\chi = 0.0$ and 1.0 values are fixed by the reference force field and the coupling constants g_i. The IR frequencies of doped and photoexcited samples fall on the curves at larger χ, while the Raman frequencies of N-site *trans*-polyenes fit with smaller χ. (From Ref. 56.)

strictly justified for alternating chains only within Hückel theory.

The electronic susceptibility χ in Eq. (14) can now be explicitly evaluated for Hückel, PPP, or other alternating chains. Estimates of the bare e–ph coupling constants are thus possible. The g_i represent the modulation of $t(R)$ in the reference normal coordinates and can be compared with theoretical estimates [56,57]. Moreover, the dependence of g_i on isotopic substitution is completely given by the reference normal coordinates and can be treated by standard spectroscopic methods.

To conclude this section, we reiterate that our approach to e–ph coupling in conjugated polymers is strictly analogous to a widely used model for ion-radical organic solids, CT crystals, and dimers [29–32], that has recently been applied to linear Pt–halogen chains [58] and hydrogen at megabar pressures [59]. Reference frequencies for CT crystals are taken directly from experiment. For conjugated polymers, by contrast, they are estimated from a reference force field $\mathbf{F}^0$ in internal coordinates. By explicitly accounting for quadratic e–ph coupling, $\mathbf{F}^0$ is related to the (empirical) force field of short oligomers. Differences between the actual and reference frequencies of polymers then unambiguously fix the linear e–ph coupling constants. The fundamental parameters of our model, reference force constants and linear e–ph coupling constants, are determined empirically. No assumptions are needed about π-electron models, e–e correlations, or other microscopic quantities. On the other hand, the precise definition of linear e–ph coupling constants and electronic susceptibilities provides the basis for a sound microscopic interpretation. Indeed, a careful analysis of the vibrational effects provides stringent tests for π-electron models and new estimates of microscopic parameters. These implications are better illustrated through the specific example of PA, to which we now turn.

III. A CASE STUDY: POLYACETYLENE

A. Transferable Electron–Phonon Coupling Constants

Translational and unit cell symmetries greatly simplify the vibrational spectra of infinite polymers. In the case of t-PA, whose C_{2h} unit cell is shown in Fig. 6.3, we have just six in-plane, optically active $q = 0$ modes. The four a_g Raman-active modes are even under inversion at bond centers, whereas the two IR-active b_u modes are odd. Only C—C stretches couple to π electrons in Hückel theory, and they remain the strongly coupled modes of PPP models. Since C—C stretches do not appear in the b_u subspace, the IR modes are not coupled efficiently to π electrons. Formally, the symmetry of the polymer is exploited by defining a set of symmetry coordinates $\mathbf{S}$ related to the internal coordinates by the unitary transformation $\mathbf{S} = \mathbf{UR}$. For instance, the symmetry coordinates for single- and double-bond stretches in Fig. 6.3 are [60].

$$S_{d,s}^q = \left(\frac{2}{N}\right)^{1/2} \sum_{n=1}^{N/2} e^{-iq2an} R_{d,s}^n \quad (16)$$

where $N/2$ is the number of unit cells of length $2a$. Other $q = 0$ symmetry coordinates are similarly constructed from internal coordinates in Fig. 6.3. The C—C—C bend has a_g symmetry, while in- and out-of-phase combinations of C—H stretches and bends have a_g and b_u symmetry, respectively. We note that, at variance with the linear chain discussed in the Introduction, the redundant (translational) a_g mode of the actual chain is a mixture of single- and double-bond stretches and C—C—C bend [22].

In the previous section we discussed the reference force field $\mathbf{F}^0$ of t-PA (see Table 6.2) derived from the force field of butadiene. In the a_g symmetry block, the high frequency C—H stretch is decoupled from the other modes and thus from π electrons. We are left with three relevant a_g modes; their reference and experimental frequencies are reported in Table 6.3 and, as discussed in Section II, fix the $\Delta\Phi$ matrix and the $\chi(\omega)$ curves in Fig. 6.4. The $\Delta\Phi$ matrix is written on the basis of the reference normal coordinates $\mathbf{Q}^0$. It consequently depends on both the $\mathbf{G}$ and $\mathbf{F}^0$ matrices and varies with molecular or polymeric structure. The e–ph coupling constants g_i thus vary even with isotopic substitution. To define coupling constants independent of mass, we use the symmetry coordinates to solve the GF problem for the reference state. In fact, diagonalization of $\mathbf{GF}^0$ gives both ω_i^0 and the eigenvector matrix $\mathbf{L}^0$ in the $\mathbf{S}$ basis. The $\mathbf{L}^0$ matrix is used to transform $\Delta\Phi$ in Eq. (12) back to the $\mathbf{S}$ basis:

$$\Delta \mathbf{F} = [(L^0)^{-1}]^T \Delta\Phi (L^0)^{-1} \quad (17)$$

where the superscript T indicates the transpose matrix. The experimental Raman frequencies of t-PA fix only the magnitude of $\Delta\Phi$ elements, and not the signs of off-diagonal elements. The relative signs of the e–ph coupling constants are required for the transformation. The ambiguity in signs is easily resolved, however, based on physical considerations [48].

Due to the redundant (translational) mode in the a_g subspace, the 3×3 matrix $\Delta\Phi$ generates the 4×4 matrix $\Delta\mathbf{F}$ shown in Table 6.4. It has the same form as $\Delta\Phi$, with off-diagonal elements given by the geometric mean of diagonal elements. The diagonal elements are

$$\Delta F_{ii} = \chi \gamma_i^2, \quad \text{with } \gamma_i = \frac{\partial t_n}{\partial S_i} \quad (18)$$

and have a precise physical meaning: The γ_i are linear e–ph coupling constants in the $\mathbf{S}$ basis. The γ_i, and thus the ΔF matrix, are independent of nuclear mass and consequently hold for isotopes of t-PA.

Table 6.3 Reference and Experimental Frequencies (in cm^{-1}) of In-Plane Modes of *trans*-Polyacetylene

Symmetry	Mode	(CH)$_x$		(CD)$_x$		(^{13}CH)$_x$	
		Ref.	Expt[a]	Ref.	Expt[a]	Ref.	Expt[a]
a_g	ν_1	3052	—	2306	—	3038	—
	ν_2	1688	1469	1665	1364	1627	1442
	ν_3	1298	1291	1210	1199	1261	1256
	ν_4	1195	1079	861	855	1184	1054
b_u	ν_5	3022	3013	2219	2231	3013	—
	ν_6	1291	1292[b]	949	916	1287	1231?

[a] From Refs. 23 and 27 of Ref. 56.
[b] The assignment of $b_u \nu_6$ is controversial; an alternative is 1250 cm^{-1}.
Source: Ref. 56.

The reference vibrational problem for the *t*-PA derivatives (CD)$_x$ and (^{13}CH)$_x$ is defined by the same **F**^{0} matrix in Table 6.2. The **G** matrices, which depend on geometry and masses, are different and lead to new frequencies ω_i^0 and eigenvectors **L**^{0}. Once the reference vibrational problem is solved, we work backward and use again Eq. (17) to derive Δ**F** matrices for (CD)$_x$ and (^{13}CH)$_x$ from the Δ**F** of *t*-PA and the appropriate **L**^{0} matrices. Starting from *t*-PA information, *no new parameter* is needed to calculate the Raman and IRAV frequencies of isotopically substituted *t*-PA. In Fig. 6.5 we report the χ dependence of the coupled modes ω_i for (CD)$_x$ and (^{13}CH)$_x$. Since the electronic problem is independent of the nuclear masses, the relative χ values are the same as in Fig 6.4. The predicted frequencies [56] are in good agreement with experiment, the data in Fig. 6.5, with mean deviation of about 19 cm^{-1}.

We stress that five parameters (three e–ph coupling constants and two relative χ values) fixed by experimental frequencies allow us to reproduce quantitatively 24 experimental frequencies. The same parameters also give semiquantitative estimates of the relative intensities of the coupled modes in the Raman spectrum of pristine samples and in the IR spectra of doped and photoexcited samples. In first approximation, C—C stretches are responsible for both IRAV and resonance Raman intensities. Following Zerbi and coworkers [22], we assume single-bond contributions to be equal in magnitude and opposite in sign to double-bond stretches. The relative intensities are then fixed by the eigenvectors of the vibrational problem.

In fairly good agreement with experiment [61], we find the Raman intensities of the highest and lowest energy bands (ν_2 and ν_4) to be comparable in (CH)$_x$ and (^{13}CH)$_x$, while the middle band (ν_3) has negligible intensity. In (CD)$_x$, by contrast, ν_1 is ~6 times more intense than ν_4, while the ν_3/ν_4 intensity ratio is ~1/7. In agreement with experiment [62,63], the IRAV intensity of ν_3 is always very small in (CH)$_x$ or (^{13}CH)$_x$, and ν_4 acquires intensity from ν_2 in going from doped to photoexcited samples; the calculated ν_1/ν_3 intensity ratio is ~0.4 and 0.1 in doped and photoexcited samples, respectively. In (CD)$_x$, both ν_3 and ν_4 acquire intensity from ν_2, with intensity ratios $\nu_2/\nu_3 \sim 0.1$ and $\nu_2/\nu_4 \sim 0.03$ for the parameters relevant to the photoexcited spectrum.

Up to now our analysis has been phenomenological and, apart from a smaller number of adjustable parameters, equivalent to the AM model and, in some respect,

Table 6.4 Experimental Δ**F** Matrix of *trans*-Polyacetylene in the a_g Symmetry Block[a]

Internal coordinate	Symmetry coordinate	S_1	S_2	S_3	S_4
C═C	S_1	8.346	−7.721	−1.030	0.624
C—C	S_2		7.145	0.955	−0.574
C—C—C	S_3			0.125	−0.075
C—C—H	S_4				0.044

[a] Units are eV/Å^2 for stretches, eV/rad^2 for bends, and eV/(Å·rad) for bend-stretches.

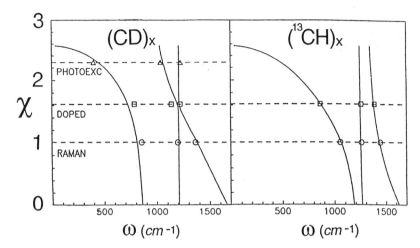

Fig. 6.5 Same as Fig. 6.4, for $(CD)_x$ and $(^{13}CH)_x$, with unchanged χ. (From Ref. 56.)

to the ECC model. The precise definition of the reference state and microscopic description offered prompts a deeper analysis. In particular, having identified π-electron contributions to vibrational spectra, we can relate them to different π-electron models in Eq. (7) and test their applicability.

B. Hückel Model and Force Field in Internal Coordinates

We start with the alternating Hückel chain in Eq. (1), which simply consists of the bond order operators $p_{d,s}$ for nearest-neighbor hopping between sites $2n$, $2n - 1$ and $2n$, $2n + 1$. Following SSH [10,11] and LHS [8], we discuss the dimerized ground state and write the linear e-ph Hamiltonian, Eq. (11), in terms of the symmetry coordinates in Eq. (16) for single- and double-bond stretches. In the $q = 0$ subspace, we have

$$H_{e-ph} = 2(2/N)^{1/2}(\alpha_d S_d \theta_d + \alpha_s S_s \theta_s)$$
$$= (2/N)^{1/2}[\alpha_d S_d(\theta_- + \theta_+) + \alpha_s S_s(\theta_- - \theta_+)] \quad (19)$$

where $\alpha_{d,s} = \alpha(R_{d,s})$ are linear e-ph coupling constants for double and single bonds and θ_- is the staggered bond-order operator defined in Eq. (15). The bond-order operator θ_+ couples to $k = 0$ electronic states. We have two coupled operators

$$\theta_\pm = (\theta_d \pm \theta_s)/\sqrt{2} \quad (20)$$

in terms of θ_d and θ_s for double and single bonds, respectively.

The electronic susceptibilities χ_+ and χ_- corresponding to the $\theta_\pm$ follow from Eq. (13), whose direct evaluation is limited to Hückel chains. The ΔF_{ii} elements in Eq. (18) are given in symmetry coordinates as [60]

$$\Delta F_{ss} = \alpha_s^2(\chi_- + 2\chi_\pm + \chi_+)/2 \quad (21a)$$
$$\Delta F_{dd} = \alpha_d^2(\chi_- - 2\chi_\pm + \chi_+)/2 \quad (21b)$$
$$\Delta F_{sd} = \alpha_s\alpha_d(\chi_- - \chi_+)/2 \quad (21c)$$

The susceptibilities are related according to $\chi_+ = \delta^2 \chi_-$ and $\chi_\pm = \delta\chi_-$ in alternating Hückel chains. Since $H_0(\delta)$ contains only terms proportional to θ_d and θ_s, their matrix elements in Eq. (13) are related [60] by eigenvector orthogonality, $\langle F|H_0(\delta)|G\rangle = 0$. Such relations are lost on adding alternating site energies, for example, to Eq. (1) that still allow exact solution. The AM formalism [18,19], which retains χ_- and neglects χ_+ and $\chi_\pm$, is restricted to small δ. The ΔF_{ii} elements of alternating Hückel chains simplify to

$$\Delta F_{ss}(\delta) = \alpha_s^2 \chi_-(\delta)[1 + \delta]^2/2 \quad (22a)$$
$$\Delta F_{dd}(\delta) = \alpha_d^2 \chi_-(\delta)[1 - \delta]^2/2 \quad (22b)$$
$$\Delta F_{sd}(\delta) = \alpha_s\alpha_d\chi_-(\delta)[1 - \delta^2]/2 \quad (22c)$$

in terms of $\chi_-(\delta)$. This matrix retains the form of Eq. (12), provided that effective coupling constants are introduced: [60] $\alpha_{d,eff} = \alpha_d(1 - \delta)$ and $\alpha_{s,eff} = \alpha_d(1 + \delta)$. The special form of ΔF is thus preserved in Hückel theory, even for two electronic operators coupled to phonons, and off-diagonal elements remain the geometric mean of diagonals. As noted above, the two susceptibilities are related and we still have just one independent electronic operator.

The electronic susceptibility $\chi_-(\delta)$ is readily found in Hückel theory [56] from the curvature of the ground-state energy per site, E_0/N, of $H_0(\delta)$,

$$\chi_-(\delta) = -\frac{1}{t^2 N}\frac{\partial^2 E_0}{\partial \delta^2} = \frac{4[(1 + \delta^2)K(p) - 2E(p)]}{\pi t(1 - \delta^2)^2} \quad (23)$$

where $p = (1 - \delta^2)^{1/2}$ and $E(p)$, $K(p)$ are complete elliptic integrals of the first and second kind. We find $\chi_-(0.18) = 0.617$ eV^{-1} for the PA alternation and bandwidth $4t = 10$ eV. The experimental $\Delta\mathbf{F}$ in Table 6.4 lead to $\alpha_s = 4.1$ eV/Å and $\alpha_d = 6.4$ eV/Å. Linear $t(R)$ with constant $\alpha(R)$ is a poor approximation, while exponential $t(R)$ implies $\alpha(R_d)/\alpha(R_s) = (1 + \delta)/(1 - \delta) = 1.44$ at $\delta = 0.18$, close to the observed ratio of 1.56. Analysis of the vibrational spectra with proper account of e–ph coupling thus points to exponential rather than linear $t(R)$, as expected on physical grounds [60,64].

The discussion so far has been restricted to the optically important $q = 0$ subspace, using symmetry coordinates for the translational symmetry of the infinite polymer. On the other hand, if we want to return to internal coordinates, we need matrices $\Delta\mathbf{F}(q)$ for all q wavevectors in the Brillouin zone. In Hückel theory, we can evaluate the q dependence of the electronic susceptibilities [60]. Figure 6.6 shows $\chi_-(q)$, $\chi_+(q)$, and $\chi_\pm(q)$ for alternation $\delta = 0.20$, close to the PA value. The behavior of $\chi_-(q)$ justifies qualitatively the anomalous frequency dispersion of C—C stretches, as sketched in Fig. 6.1 for the optic mode of a linear chain. We also see from Fig. 6.6 that θ_- and θ_+ and the corresponding susceptibilities are most useful at $q = 0$, where $\chi_-(q) \gg \chi_+(q)$, whereas they are almost the same at the zone edge.

The $\alpha(R_d)$, $\alpha(R_s)$ values found above from $q = 0$ data and $\chi_-(q)$ curves based on Hückel theory with $\delta = 0.18$ lead to the force and interaction constants listed in Table 6.5 for C—C stretches [60,64]. The C=C and C—C diagonal force constants are significantly lowered compared to $\mathbf{F}^0$, as expected, while interaction constants between C—C stretches extend over about three unit cells. At this point we have made contact with earlier work by Kakitani [65], by Mele [66], and by Piseri et al. [67], who derived force constants from real-space π-electron polarizabilities introduced by Coulson and Longuet-Higgins [20].

Table 6.5 Valence Force Fields of *cis*- and *trans*-Polyacetylene in Internal Coordinates

Constant[a]	*trans*-PA	*cis*-PA
Stretch		
K_1 (C=C)	7.50	7.88
K_2 (C—C)	4.97	5.04
K_3 (C—H)	5.00	5.00
Bend		
H_1 (CCC)	0.859	0.530
H_2 (CCH)	0.531	0.530
Interaction[b] (Stretch–stretch, at m unit cell of *trans*-PA separation)		
$m = 0$, F_1 (C=C,C—C)	0.770	0.561
$m = 1$, F_4 (C=C,C=C)	−0.289	−0.111
F_5 (C—C,C—C)	−0.116	−0.045
F_6 (C=C,C—C)	0.081	0.021
$m = 2$, F_7 (C=C,C=C)	−0.064	−0.011
F_8 (C—C,C—C)	−0.026	−0.004
F_9 (C=C,C—C)	0.021	0.002
$m = 3$, F_{10} (C=C,C=C)	−0.019	0.0
F_{11} (C—C,C—C)	−0.008	0.0
F_{12} (C=C,C—C)	0.007	0.0

[a] Units are mdyn/Å for K and F, mdyn/rad^2 for H.
[b] Bend–stretch and bend–bend interaction constants for $m = 0$ are given in Table 6.2.
Source: Ref. 64.

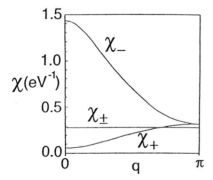

Fig. 6.6 Dispersion of the electronic susceptibilities χ in Eq. (21) as a function of wavevector q for an alternating Hückel chain with $\delta = 0.20$, hopping integral $t = 1$ and unit cell $2a = 1$. The staggered bond order χ_- dominates at $q = 0$. Source: Ref. 60.

If we combine Table 6.5 for C—C stretches with the other force constants listed in Table 6.2, the calculated frequencies are not satisfactory. The reason is that, as shown in Table 6.4, C—C—C and C—C—H bends are also coupled to π-electron fluctuation, and this weak coupling cannot be accounted for in Hückel theory. We introduced minor adjustments to bend, bend–bend, and bend–stretch force constants, as described in detail in Ref. 60, to obtain good fits for in-plane Raman and IR frequencies while preserving the Raman and IRAV fits in Figs. 6.4 and 6.5. The basic idea is to combine the dispersion of C—C stretches found in Hückel theory with dispersionless bends suggested by exact PPP results for oligomers [68]. Thus we have obtained a valence force field in internal coordinates for t-PA and its isotopes while maintaining a clear connection with the corresponding force fields of the oligomers.

We have also extended [60] the procedure described above to *cis*-polyacetylene (*cis*-PA), assuming a planar D_{2h} structure with t-PA bond lengths and angles. The *cis*-PA in-plane modes are classified as $4a_g + 4b_{3g} + 3b_{1u} + 3b_{2u}$, with gerade and ungerade modes active in Raman and IR, respectively. Two minor changes are needed for the reference force field of *cis*-PA starting from that of t-PA: an interaction constant of 0.02 mdyn/rad^2 for nearest-neighbor C—C—H bends and lowering the diagonal constant for the C—C—C bend from 0.891 to 0.550 to fit the b_{1u} ν_{16} mode in the infrared [69]. The different topology of *cis*-PA yields a different distribution of C—C stretching modes, which now appear also in the b_{3g} and b_{2u} symmetry blocks. However, we expect and indeed find only three a_g modes to be strongly coupled to π-electron fluctuations.

Since the symmetry coordinates for a_g modes are the same as in *t*-PA, we can test the transferability of e–ph coupling constants. We account for e–ph coupling in the a_g block through $\mathbf{F} = \mathbf{F}^0 - \eta \Delta \mathbf{F}$, with the same $\Delta \mathbf{F}$ as in *t*-PA, and plot the a_g frequencies as functions of η in Fig. 6.7. The best fit to the experimental Raman frequencies [70,71] of *cis*-PA occurs at $\eta = 0.56$. Within Hückel theory, η corresponds to $\chi(cis)/\chi(trans)$. The predicted intensities are again in qualitative agreement with experiment [70,71]. The calculated Raman intensities of ν_2 and ν_3 are similar in *cis*-PA and an order of magnitude larger than ν_4. In *cis*-(CD)$_x$, the calculated ν_2 intensity is ~ 10 times that of ν_3 and $\sim 10^2$ times ν_4.

The observed η leads to an artificially high alternation $\delta = 0.36$ for *cis*-PA and is an example of the limitations of Hückel theory, which does not distinguish between the two isomers [64]. In any case, the Hückel dispersion $\chi(0.36, q)$ and the *t*-PA parameters $\alpha_{d,s} = 6.4$ and 4.1 eV/Å also generate a successful valence force field for *cis*-PA. As seen in Table 6.5, interaction constants for C—C stretches are now restricted to two unit cells, in agreement with the higher localization expected in the *cis* isomer.

C. Role of Electron–Electron Interactions

Hückel theory clearly provides a useful analytical model for e–ph coupling in polyenes and PA, but some limitations are evident. The observed coupling to bending modes and the relative χ values of *cis*- and *trans*-PA require going beyond $H_0(\delta)$. The PPP model defined in Eq. (7) has a specified potential $V(R)$, but this is not necessary in general. $V(R)$ raises two major issues for our phenomenological analysis: Partial derivatives of $V(R)$ generate additional electronic operators coupled to phonons, so that the special form of the $\Delta \mathbf{F}$ matrix in Eq. (18) is no longer ensured, and the susceptibilities in Eq. (13) must be approximated in extended systems whose eigenstates $|F\rangle$ are not known. But symmetry coordinates $\mathbf{S}$ are retained, and the Herzberg–Teller expansion in Eq. (9) still holds. The force constants associated with $H(\delta) = H_0(\delta) + V(R)$ are formally [68]

$$F_{ij}^\pi = \frac{\partial^2 E_G}{\partial S_i \partial S_j} - \left\langle G \left| \frac{\partial^2 H(\delta)}{\partial S_i \partial S_j} \right| G \right\rangle \quad (24)$$

where the superscript π is a reminder that E_G and $|G\rangle$ refer to a π-electron model.

For *t*-PA, the four S_i in the a_g block are again C—C stretches and C—C—C, C—C—H bends. The C—C—C bend S_3 is seen from Fig. 6.3 to involve second neighbors, with [68]

$$R_{n,n+2} = (R_d^2 + R_s^2 - 2R_d R_s \cos \zeta)^{1/2} \quad (25)$$

for all n, where ζ is the bond angle. The linear e–ph coupling constant for S_3 is the partial derivative of $V(R_{n,n+2})$ with respect to ζ, while the electronic operator analogous to the staggered bond order in Eq. (15) is seen from Eq. (7) to be a sum over the charge operators $q_n q_{n+2}$. The choice of $V(R)$ completely fixes both linear and quadratic e–ph coupling constants due to the potential. The C—H bend S_4 remains uncoupled for any parameters in Eq. (7). The small finite coupling to C—H

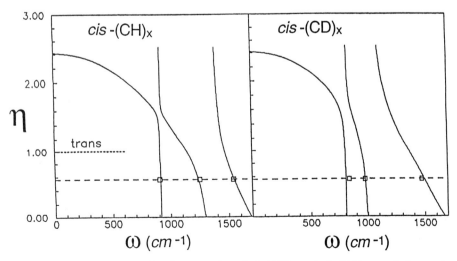

Fig. 6.7 Relative susceptibility η vs. ω curves for the a_g modes of *cis*-(CH)$_x$ and *cis*-(CD)$_x$ coupled to π-electron fluctuations. $\eta = \chi(cis)/\chi(trans) = 0.56$ gives the best fit to the indicated Raman data and indicates reduced delocalization in *cis*-PA. (From Ref. 60.)

bends in Table 6.4 is at the limits of our analysis; if accepted, it would require going beyond π-electron models.

The force constants in Eq. (24) can be found exactly for quantum cell models of oligomers, currently up to $N = 14$ sites [68]. The PPP model describes molecular excitations, and all microscopic parameters in Eq. (24) are fixed in advance. The C—C force constants $-F_{11}$, $-F_{22}$ and interaction constants F_{12} of *trans*-polyenes are extrapolated in Fig. 6.8 as $N^{-\gamma}$, with adjustable $\gamma = 0.70$ giving the best straight lines. The opposite N dependence of cyclic polyenes with $4n + 2$ sites facilitates convergence, but much longer chains are needed for quantitative polymer results. The $\alpha(R)^2$ contributions account for ~70% of the force constants for C—C stretches, with the rest due mostly to $\alpha(R)V'(R)$ terms in Eq. (24) rather than to $V'(R)^2$. Although the PPP model has reduced [68] $\alpha(R)$ compared to Hückel theory, the same e–ph coupling mechanism dominates and F_{12} is approximately the geometric mean, $(F_{11}F_{22})^{1/2}$, as found exactly for noninteracting electrons. The force constant $-F_{33}$ for the C—C—C bend is similarly constrained to 0.28 ± 0.01 eV/rad^2, while interaction constants with C—C stretches are small and weakly dependent on N. We consequently neglected the dispersion of bending modes in constructing the PA force field (Table 6.5) in internal coordinates.

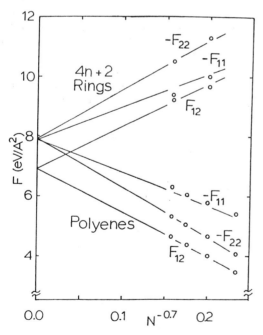

Fig. 6.8 Force and interaction constants F_{ij}, Eq. (24), in eV/Å^2 for C—C stretches of N-site linear and cyclic polyenes, based on the Pariser–Parr–Pople model with molecular potential, Eq. (6), alternation $\delta = 0.07$, and exponential $t(R)$. The indicated power law extrapolations of F_{ij} were constrained to coincide in the polymer. Source: Ref. 68.

The π-electron force constants in Eq. (24) can readily be found for butadiene, which motivated the reference force field $\mathbf{F}^0$ in Table 6.2. Since $\mathbf{F}^0$ is phenomenological and PPP parameters include some core effects, the precise *experimental* identification of π-electron contributions emphasized above does not extend to precise *theoretical* connections to PPP or other models. PPP results indicating additional delocalization beyond butadiene can simply be associated with $\Delta F(N) = F(N) - F(4)$. The staggered bond order θ_- couples to the out-of-phase combination, $S_- = (S_1 - S_2)2^{-1/2}$, of single and double bonds, and S_- is the special ECC coordinate [22] of PA. The PPP force constants for S_- are

$$-\Delta F_-(N) = \Delta F_{12}(N) - [\Delta F_{11}(N) + \Delta F_{22}(N)]/2 \quad (26)$$

The extrapolated value for the polymer, $-\Delta F_- = 10$ eV/Å^2, is the correlated counterpart of χ in Fig. 6.4. The relative χ values of finite polyenes are then calculated [68] to be $\Delta F_-(N)/10$, or 0.35, 0.44, 0.51, and 0.57 for $N = 8$, 10, 12, and 14; the fitted χ values in Fig. 6.4 are 0.30, 0.43, 0.48, and 0.57. The close agreement extends the PPP model to polyene vibrations without introducing any adjustable parameter. The change of the C—C—C force constant relative to butadiene, $-\Delta F_{33} = 0.12$ eV/rad^2, also agrees quantitatively with $\gamma_{33}\sqrt{\chi}$ obtained from the vibrational analysis.

The curvatures in Eq. (24) correspond to *static* ground-state susceptibilities for various bond-order and charge operators. There is a sum over virtual states $|F\rangle$ in Eq. (13). The NLO spectra in Section IV involve third-order *dynamic* responses with three sums. Low-energy excitations are crucial in both cases. The magnitudes of e–ph coupling constants, on the other hand, are primarily of interest for vibrational analysis. Both coupling constants and susceptibilities can be found for quantum cell models such as Eq. (7). We have sought the vibrational implications of the PPP model with standard molecular parameters. The reduced χ of *cis*-PA and the relation between π-electron models and the reference state are issues that require attention for the next level of understanding. The PPP model decisively improves the Hückel picture while retaining much of its physical simplicity. The scarcity of exact solutions for correlated models, however, suggests continued reliance on Hückel results and judicious combinations with PPP or other models.

D. Microscopic Model for Soliton Modes

Infrared spectra of doped or photoexcited t-PA samples are dominated by very strong absorptions [10,11]. The huge intensity of IRAV bands leaves no doubt about their electronic origin. As originally suggested by Mele and Rice [72], vibrations around charged solitons S^+ or S^- induce large oscillations of charge due to e–ph coupling. The AM formalism [18,19] unambiguously relates the IRAV modes in Fig. 6.4 to the same physics as Raman modes of pristine PA. But a simple justification of this counterintuitive result is still lacking, in our opinion. In terms of classical spectroscopy, a defect spread over a few sites in the chain should originate several (perhaps many) bands in both IR and Raman spectra [73], and doped or photoexcited PA spectra are then too simple. Furthermore, continuum models popular in physics [17,74–78] are too idealized for stringent experimental comparisons, while more realistic models based on discrete lattices and multimode coupling [17,79,80] are rather cumbersome. Quantum-chemical calculations of the vibrational properties of solitons have also been proposed [81–84], with different degrees of sophistication. However, the picture strongly depends on the details of the model and loses contact with the simple and elegant AM formalism.

We recall that only a_g modes have large intensity in the Raman spectrum of pristine PA. In the presence of a defect, translational symmetry is lost, and many modes could acquire Raman and/or IR activity. The success of AM formalism implies a one-to-one correspondence between the Raman modes of pristine polymers and IRAV bands. Even in the absence of translational symmetry, only a special kind of localized vibration appears to be strongly coupled to π-electron fluctuations.

To investigate the qualitative shapes of the vibrations around a soliton, we return to the idealized linear chain in Fig. 6.1. We choose* a dimerized chain with 102 sites and periodic boundary conditions to model pristine PA, which is sufficient for the infinite chain. The vibrational problem is solved on the internal coordinate basis using different C—C and C=C stretching force constants, as given in Table 6.3, and only nearest-neighbor interaction constants K_{12}. The vibrational problem maps into a Hückel chain with $\beta = K_{12}$ and alternating $\alpha = K_{11}$, K_{22}. We then consider a chain of 101 sites with soliton-like distortions u_n taken from the SSH model [10,11],

$$u_n = u_0 \tanh[(n - n_0)/\xi]\cos n\pi \quad (27)$$

Here u_n is the displacement of the nth site in a chain with uniform bond lengths $a = 1$, u_0 is the dimerization amplitude of the ground state, and 2ξ is the spatial extent of a soliton centered at $n_0 = 51$. The bond lengths are close to uniform for $|n - n_0| < \xi$, and the alternation reverses at n_0.

The vibrational problem with a fixed defect is solved by assuming a linear dependence of K_{11} and K_{22} on bond order and constant K_{12}. Depending on the assumed soliton width ξ and force constants, we find a variable number of localized modes.* In particular, a local mode al-

---

* A. Painelli, A. Girlando, and Z. G. Soos, unpublished results.

ways develops from the $q = \pi$ mode of the perfect chain (in the extended zone representation), where single- and double-bond stretches are out of phase. This mode is strongly reminiscent of the ECC coordinate, but it is a local mode, with maximum amplitude on the defect and decaying rapidly away from the soliton. This local mode is odd under reflection at n_0 and is therefore IR-active. Other IR local modes develop near $q = \pi/2$, depending on the model parameters, with pairs of adjacent single- and double-bond stretches moving out of phase. Several local Raman modes also appear around $q = \pi$ or $\pi/2$. Such a "zoo" of local modes is expected about defects, and their number and shape depend sensitively on the values of ξ, K_{12}, and $K_{11} - K_{22}$.

What is more interesting is the coupling of these modes to π electrons. We have already seen that Hückel theory accounts approximately for e–ph coupling in t-PA. By adopting a Hückel model we have found the electronic susceptibilities relevant to all normal modes in the 102- and 101-site chains, as shown in Fig. 6.9. Small deviations of the χ relevant to the soliton with respect to the perfect chain are observed around $q = \pi$ and $q = \pi/2$. However, only one mode, corresponding to the infrared-active out-of-phase combination of C—C stretches shows sizable deviations. This mode is sketched in Fig. 6.9 and is a damped ECC coordinate.

The relative IR intensities of the 101-site chain are then calculated as derivatives of the ground-state dipole moment with respect to nuclear displacements. About half of the 101 normal modes of the chain are antisymmetric and hence IR-allowed. The IR intensity of the

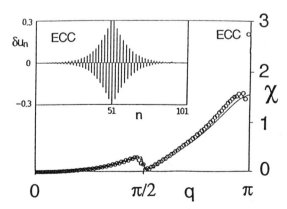

Fig. 6.9 Relative susceptibilities $\chi(q)$ vs. phonon wavevector q. Solid curve: 102-site alternating chain; open circles: 101-site chain with a charged soliton at the center. The damped ECC mode at $q = \pi$ is strongly coupled; the inset shows its normalized amplitudes for out-of-phase C—C stretches for soliton width $\xi = 7$ in Eq. (27).

damped ECC mode is three orders of magnitude larger than any other mode. Although symmetry does not select a special IR mode, e–ph coupling does. We also obtained off-resonance Raman intensities as derivatives of the ground-state polarizability and found significant intensity for a few modes around $q = \pi$. These modes have χ values very similar to those of the perfect chain, however, so no appreciable change in the Raman spectrum is expected for light doping.

These clear-cut results are for an idealized Hückel chain, but they hold for a wide range of parameters and offer a simple microscopic picture. We find that only one special combination of C—C stretches about a soliton couples effectively to π electrons and gains strong IR intensity. When the actual zigzag chain is considered, this special mode is expected to couple to C—C—C and C—H bends, thus originating the three IRAV bands in Figs. 6.4 and 6.5. For the parameters relevant to t-PA, this rough model gives semiquantitative agreement with experiment. By taking the force constants of the idealized chain to be those of t-PA in Table 6.3, and with the usual $\delta = 0.18$, we find* the ratio of the calculated χ for the damped ECC mode and the perfect chain to be 1.5, 1.8, 2.0 for $\xi = 3, 7, 15$, respectively. The adjustable χ values in Figs. 6.4 and 6.5 are 1.6 for doping and 2.3 for photoexcitation. The counterion in doped samples pins the soliton and decreases its width compared to nominally free solitons produced by photoexcitation. The total IR intensity borrowed from the soliton's absorption increases as 1, 1.8, and 3.3 for $\xi = 3, 7,$ and 15. The relative IRAV intensities of doped and photoexcited samples are roughly 1/2 [62,63].

In closing this section, we draw attention to two related points. First, the coupled vibration of the soliton is found to be the *localized* analog of the ECC. Localization is crucial, since extended states have negligible contributions from a defect localized on a small part of the chain. Second, physically sensible χ values require much longer chains than the soliton width 2ξ. Solitons are often modeled [72,82–84] on chains comparable to their width, roughly 15 sites, especially when e–e interactions make longer chains computationally very demanding. Such a procedure is justified for classical vibrational studies, where only local interactions need be considered, but is misleading for delocalized systems. Even if the soliton is quite localized (e.g., $\xi = 7$), π electrons fluctuate over much longer segments, and the calculated χ of the damped ECC mode does not saturate until $N \sim 60$ for alternation $\delta = 0.18$. From this perspective, the recent observation [85] of frequency dispersion of IRAV bands in polyenes with up to 20 double bonds ($N = 40$) need not demonstrate a highly extended defect

---

*A. Painelli, A. Girlando, and Z. G. Soos, unpublished results.

but may instead show the effects of electronic delocalization for charged defects extending over a few sites.

IV. NONLINEAR OPTICAL AND ELECTROABSORPTION SPECTRA

A. Condon Approximation for NLO Coefficients

The electronic response χ in Eq. (13) is a static susceptibility involving even-parity singlets of polymers with centrosymmetric backbones. The p_n in Eq. (1) indicate χ to be a bond-order/bond-order polarizability [20]. Since the excited states $|F\rangle$ in the sum depend on $V(R)$, different χ's are found in Hubbard or PPP models. The large NLO responses of conjugated polymers are also due to π electrons. The dipole operator in the ZDO approximation of quantum cell models is [12]

$$\mu = e\sum_n \mathbf{r}_n q_n \quad (28)$$

where $\mathbf{r}_n$ is the position of the nth site relative to an origin that is arbitrary for a neutral system and q_n is the π-electron charge operator. The linear response to an applied field **F** gives the polarizability tensor

$$\chi_{ij}^{(1)}(-\omega,\omega) \equiv \alpha_{ij}(\omega) = \sum_F \frac{\langle G|\mu_i|F\rangle\langle F|\mu_j|G\rangle}{E_F/\hbar - \omega - i\Gamma_F} + (\omega \to -\omega) \quad (29)$$

where $\omega \to -\omega$ indicates another sum with opposite sign in the denominator; i,j are Cartesian components; and Γ_F are excited-state lifetimes. The static ($\omega = 0$) polarizability for $\Gamma_F = 0$ is directly analogous to χ.

The polymers in Fig. 6.2 have centrosymmetric backbones in their ideal extended conformation. Dipole selection rules imply mutually exclusive linear absorption to odd-parity states such as $|F\rangle$ in Eq. (29) and two-photon absorption (TPA) to even-parity states. We retain the A_g and B_u designation for the C_{2h} backbone of t-PA, but g and u indices apply equally to PDA, PPV, PT, and PS. NLO coefficients [86–89] are higher order perturbations in μ, with additional sums over states (SOS), polarizations, frequencies, and lifetimes. The transition dipoles $\langle A_g|\mu|B_u\rangle$ control all NLO responses in the dipole approximation. To simplify notation, $2A$ and $1B$ refer to 2^1A_g and 1^1B_u below and correspond to excitation thresholds in Table 6.1.

The nature of the excited states $|F\rangle$ is left open in formal SOS expressions for NLO coefficients [86–89] that include all states. The same e–ph coupling governs the vibrational properties of ground and excited states, but exhaustive treatment of e–ph coupling in the latter requires more vibrational information than is presently available. Indeed, accurate excited states for conjugated polymers are difficult even at the π-electron level. Since large responses are due to dipole-allowed virtual excitations, vibronic analysis can be carried out with crude adiabatic states and the Condon approximation. In the following, we always consider products of π-electron functions $|F\rangle$ and backbone vibrations taken as harmonic oscillators. As suggested by these approximations, current NLO discussions are more preliminary than e–ph models but probe π-electronic excitations directly.

The coupled a_g modes of PA, PDA, or other polymers give Franck–Condon overlaps for different excited-state geometries. The three states in Fig. 6.10 are generic for centrosymmetric systems with an even-parity G, even and odd singlets $2A$ and $1B$, and transition dipoles μ_{GB} and μ_{AB}. The largest SOS terms [88,89] for the third-harmonic generation (THG) coefficient, $\chi^{(3)}(-3\omega; \omega,\omega,\omega)$, have all polarizations along the polymer backbone. We denote the most divergent term $\chi_{AB}(-3\omega)$ to indicate a single A,B pair and omit multiplicative constants. The Condon approximation is [90]

$$\chi_{AB}(-3\omega) = \mu_{GB}^2 \mu_{AB}^2 \times$$
$$\sum_{pqp'} \frac{F_{0p}(\mathbf{b}) F_{pq}(\mathbf{a}-\mathbf{b}) F_{qp'}(\mathbf{b}-\mathbf{a}) F_{p'0}(-\mathbf{b})}{(\omega_p - 3\omega - i\Gamma_p)(\omega_q - 2\omega - i\Gamma_q)(\omega_{p'} - \omega - i\Gamma_{p'})}$$
(30)

The sums are over vibrations $\omega_B + p\omega_v$ or $\omega_A + q\omega_{v'}$, starting from $|G,0\rangle$, to excited states whose potentials in Fig. 6.10 have displacements $\mathbf{b}$ and $\mathbf{a}$ along normal modes; F_{pq} and Γ_p, Γ_q are nuclear overlaps and excited-state lifetimes. The full THG coefficient has two sums over rB singlets and one over sA singlets, including G.

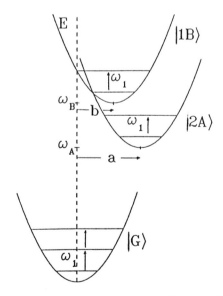

Fig. 6.10 Schematic representation of three electronic states with excitation energies ω_A, ω_B; harmonic potentials with common frequency ω_1; and displacements a,b. Dipole-allowed transitions connect the odd-parity $|1B\rangle$ and even-parity $|G\rangle$ or $|2A\rangle$ states. Two- and three-photon resonances overlap when $\omega_A/2 \sim \omega_B/3$. (From Ref. 90.)

Harmonic oscillators simplify the analysis. Normal coordinates and frequencies differ in principle from state to state and involve both state rotations and energy renormalization. In practice, unrotated ground-state coordinates are retained even when different frequencies are used in detailed fits [21,91,92]. Displaced harmonic oscillators yield analytical expressions [93,94] for F_{pq}, when ground-state frequencies ω_1 are kept in excited states. Moreover, since ω_1 is typically smaller than electronic splittings, some of the sums in Eq. (30) can be evaluated by closure. Closure can readily be checked and holds unless an energy denominator becomes small. Closure leads to Franck–Condon averages $\bar\omega_B$ and $\bar\omega_A$ in NLO coefficients [90].

Two-photon absorption (TPA) from G to A corresponds to Im $\chi^{(3)}(-\omega;\omega,-\omega,\omega)$ when $2\hbar\omega \sim E_A$. The largest SOS terms are again polarized along the chain. The Condon approximation for the most divergent term, designated as $\chi_A(-\omega)$ for a single A state, is [90]

$$\chi_A(-\omega) = \mu_{GB}^2 \mu_{AB}^2 \times$$
$$\sum_{pqp'} \frac{F_{0p}(\mathbf{b}) F_{pq}(\mathbf{a}-\mathbf{b}) F_{qp'}(\mathbf{b}-\mathbf{a}) F_{p'0}(-\mathbf{b})}{(\omega_p - \omega - i\Gamma_p)(\omega_q - 2\omega - i\Gamma_q)(\omega_{p'} - \omega - i\Gamma_{p'})}$$
(31)

When $\omega \sim \omega_A/2$ is small compared to ω_B, closure holds for the p and p' sums. This reduces Im $\chi_A(-\omega)$ to Lorentzian profiles with FWHM of Γ_q and Franck–Condon factor $F_{0q}(a)^2$ for each vibronic of A. The linear absorption is also Lorentzian, with $F_{0q}(b)^2$ for vibronics of B, as seen from the imaginary part of Eq. (29).

The Condon approximation applies term by term to any NLO coefficient. The simplest general vibronic treatment of NLO spectra thus requires frequencies and displacements of harmonic oscillators for electronic states with specified location, transition dipoles, and lifetimes. We have separate electronic and vibrational problems, and either experimental or theoretical inputs may be used in SOS expressions. To proceed, appropriate A and B excitations must be chosen. The linear absorption of conjugated polymers gives a $1B$ state or exciton in all models, while TPA clearly probes even-parity states whose locations have been debated. The choices [95] in Table 6.6 are two a_g modes and three electronic excitations, a B state and two A states, that will be applied to PDA spectra. This pattern of excitations and transition moments is based on exact PPP results [96,97] for oligomers with $\delta = 0.15$ in Eq. (7).

Nonlinear optical spectra could in principle be treated as "two-level" models, with everything adjustable. This is not practical at present resolution: even three excitations and two coupled modes generate too many parameters. As few states as possible are chosen, based on some model. The resulting fits are *local* optimizations with built-in constraints and may differ significantly for the same number of states.

Table 6.6 Excited-State Parameters for PDA-PTS Crystals and PDA-4BCMU Films

System	State $|X\rangle$	Energy $E(X)$(eV)	Width Γ (eV)	Displacement x_2	Displacement x_3	Dipole with $1B$ $\langle X|\mu|1B\rangle/\mu_{G1B}$
PTS	$1B$	2.00	0.025	0.778	0.566	
4BCMU		2.35	0.15	0.95	0.70	
PTS	$2A$	1.80	0.025	−1.00	−0.75	0.50
4BCMU		1.90	0.15	−1.05	−0.80	0.71
PTS	nA	2.70	0.10	0.8	0.6	1.22
4BCMU		3.22	0.30	0.7	0.5	1.34

Source: Ref. 95.

B. NLO Spectra of PDA Crystals and Films

Batchelder and Bloor [91] analyzed linear and resonance Raman (RR) spectra of PDA-PTS crystals. The $1B$ exciton has a 0–0 peak at 2.00 eV and resolved 0–1 sidebands at 2.20 and 2.26 eV, respectively, for double- and triple-bond stretches. The RR spectrum has additional backbone modes as well as overtones and combinations that imply corrections to harmonic potentials. The linear absorption and Raman excitation profiles (REP) of the fundamentals were fit with four a_g modes using harmonic oscillators with different ground and $1B$ frequencies but unchanged excited-state normal coordinates. The dimensionless C=C and C≡C displacements of $1B$ are accurately known to be 0.778 and 0.566 in Table 6.6.

The linear absorption of PDA-4BCMU films at 300 K is the dashed line in the upper panel of Fig. 6.11. The peaks at 2.35 and 2.55 eV are broader and 0.36 eV higher than the crystal's 0–0 line [98] at 1.99 eV, with $\Gamma \sim 0.02$ eV. The 0.2 eV spacing in Fig. 6.11 indicates strong 0–1 sidebands. Although sample-dependent, the 0–1 intensity clearly implies larger $1B$ shifts in films. The fit (solid line) has $b_2 = 0.95$ and $b_3 = 0.70$ (Table 6.6), with the Franck–Condon factors shown in the stick spectrum. The 0–0 and 0–1 intensities in Fig. 6.11 are reversed from experiment, but larger shifts broaden the NLO spectrum that is our primary objective. The blue shift of $1B$ in films suggests shorter conjugation lengths and/or altered environments [99]. Since E_{1B} is adjustable in Table 6.6, it contains conformational and environmental factors.

The lower panel in Fig. 6.11 shows the THG intensity and phase θ ($= \tan^{-1}$ Im/Re) of PDA-4BCMU films [100]. The largest SOS contribution, Eq. (30), is for A states with transition dipoles to $1B$. The prominent three-photon resonance at $\omega = E_{1B}/3\hbar$ is typical of conjugated polymers, and, as in β-carotene [101,102], resembles the linear absorption. The indicated fit [95], with scaled intensity, is the Condon approximation for two a_g modes and displacements in Table 6.6. Contributions from two-photon resonances at $E_{sA}/2$ are possible and are discussed below but are not resolved.

Lawrence et al. [103] reported the resolved TPA of PDA-PTS single crystals shown in Fig. 6.12. The 1.80 eV feature has previously been assigned [96,97] as $2A$ in PDA-PTS, based on PPP results at $\delta = 0.15$ placing $2A$ about 10% below $1B$. Far stronger TPA around 1.5

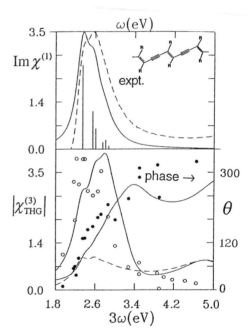

Fig. 6.11 Top: Linear absorption (dashed line) of PDA-4BCMU films at 300 K and scaled fit (solid line) to $1B$ in Table 6.6; the lines are 0–0, 0–1, and 0–2 Franck–Condon factors for C=C and C≡C vibrations. Bottom: THG intensity (open circles) and phase (closed circles) of the same films (from Ref. 100). Solid lines are based on Table 6.6; the dashed line is the THG intensity when $2A$ is deleted ($\mu_{2A1B} = 0$) (from Ref. 105).

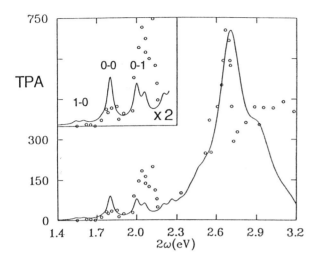

Fig. 6.12 Two-photon absorption spectrum of PDA-PTS single crystals (from Ref. 103). The fit is based on the A_g states and parameters in Table 6.6, with $2A$ magnified in the inset; the 1–0 feature has 10% excitation of the ground state (from Ref. 95).

E_{1B} was predicted, and this nA_g state or biexciton is naturally associated with the intense 2.7 eV feature at 1.35 E_{1B}. The resolved TPA in Fig. 6.12 invites more complete analysis. The indicated fit places $2A(0-0)$ at 1.80 eV and a_g modes at 0.20 and 0.26 eV for this state (Table 6.6). Since the power dependence of the 1.6 eV feature is close to I^3, we interpret it [95] as a 1–0 transition arising from anti-Stokes scattering. The $2A$ displacements $a_2 = -1.0$ and $a_3 = -0.75$ in Table 6.6 are larger than b_2 and b_3. Larger $2A$ than $1B$ displacements are directly seen in octatetraene spectra [104] and were used to model [90] the THG spectra of β-carotene. The polyene double bond resembles the in-phase combination of PDA multiple bonds shown in Fig. 6.2. Slightly better fits are found for a_2, a_3 and b_2, b_3 with opposite signs. This enhances vibronic contributions for virtual transitions μ_{2A1B} between $2A$ and $1B$, with relative displacement $\mathbf{b} - \mathbf{a}$ in Eq. (31).

The TPA fit in Fig. 6.12 has equal 0–0 and 0–1 intensities at 1.80 and 2.03 eV, respectively, contrary to experiment. Larger a_2, a_3 are not attractive because they give comparably strong 0–2 features around 2.3 eV. The 2.7 eV peak is taken as the 0–0 of nA in the fit. The dipoles μ_{2A1B} and μ_{nA1B} between $1B$ and $2A$ and between nA and $1B$ control the relative TPA intensities. The ratio of their squares is 1/6 in Table 6.6, twice the 8% in PPP calculations at $\delta = 0.15$ for oligomers up to 14 carbons. Relative intensities become important when several electronic transitions contribute, even when calculated spectra are scaled to NLO experiments that, in turn, are relative rather than absolute. The indicated fit is preliminary, and both experimental and theoretical improvements can be anticipated. Excited-state absorption is a major concern for resonant excitations, and shorter pulses can alter spectra (W. E. Torruellas and B. Lawrence, personal communication).

The magnitude and phase of nondegenerate four-wave mixing (NDFWM) spectra of PDA-4BCMU films [100] are shown in Fig. 6.13 for two frequencies, $\omega_2 = 0.651$ and $\omega_2 = 1.165$ eV. They were analyzed [100] using standard expressions for $\chi^{(3)}(-[2\omega_1 - \omega_2]; \omega_1, \omega_1, -\omega_2)$ without vibronic structure. The fits are the same response, plus vibronic contributions and a Raman resonance, for the states in Table 6.6. Since $1B$ is fixed by the linear spectrum and $2A$ and nA inputs by PDA-PTS, the only new parameters are greater widths in the film. The $1B$ blue shift from 4BCMU crystals to films places nA around 3.22 eV, or 1.37 times $E_{1B} = 2.35$ eV. The resonances in Fig. 6.13 constrain the location of nA in 4BCMU films: there is a two-photon resonance at $2\omega_1 = E_{nA}$ and a one-photon resonance at $2\omega_1 - \omega_2 = E_{1B}$. These resonances overlap in Fig. 6.13. Both the magnitudes and phases in Fig. 6.13 agree with experiment better than the few-level models discussed in Ref. 100.

The location of nA emerges from NDFWM spectra and implies resonances at $E_{nA}/2$ and $E_{2A}/2$ in the THG spectrum (Fig. 6.11). The μ_{2A1B}/μ_{nA1B} ratio now matters, as well as overlap with the resonance at $E_{1B}/3$. The dashed line in Fig. 6.11 has the same parameters as in Table 6.6, except for $\mu_{2A1B} = 0$, and shows directly the effect of removing $2A$: the THG peak is three times smaller without any change at $3\omega > 3.5$ eV. As seen from Fig. 6.10, we have overlapping two- and three-photon resonances when $E_{2A}/2 \sim E_{1B}/3$, and two terms in the

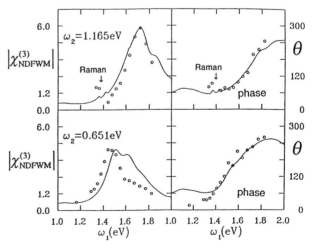

Fig. 6.13 Amplitude and phase of NDFWM spectra of PDA-4BCMU films (from Ref. 100) at $\omega_2 = 0.651$ and 1.165 eV; solid lines are based on Table 6.1. Raman resonances appear when $\omega_1 - \omega_2 = 0.20$ or 0.26 eV matches a C—C stretch. (From Ref. 95.)

denominator in Eq. (30) are simultaneously small. The resonances overlap almost perfectly in β-carotene, where the three-photon peak is enhanced fivefold [90]. Overlapping resonances make possible a large THG signal at $E_{1B}/3$ without predicting a strong (and unobserved) two-photon resonance. $2A$ contributions are otherwise modest and, for NDFWM in Fig. 6.13, are largely confined to $2\omega_1 < 1.3$ eV.

Cha et al. [100] remarked that the $\omega_2 = 1.165$ eV spectrum in Fig. 6.13 could have resonances when $\omega_1 - \omega_2$ matches a backbone vibration, but they did not pursue vibronic analysis. Raman resonances contain mainly $1B$ contributions to $\chi^{(3)}$ $(-[2\omega_1-\omega_2]; \omega_1, -\omega_2, \omega_1)$ when, as here, μ_{G1B} dominates $I(\omega)$. The response is weak [95], however, unless both ω_1 and $2\omega_1 - \omega_2$ are close to ω_B and closure is inappropriate. The calculated Raman features in Fig. 6.13 are indeed small, even for $\Gamma \sim 100$ cm^{-1} taken from RR spectra of crystals, and the data are too sparse for detailed comparison.

C. Linear and Electroabsorption Spectra of PDA Crystals

The electroabsorption spectrum in a static electric field F is conventionally recorded as $I(\omega,F) - I(\omega)$ and is formally Im $\chi^{(3)}(-\omega;\omega,0,0)$. Stark shifts are familiar in atomic, molecular, and extended systems: F mixes and shifts electronic states. We neglect mixing of vibrational states, whose transition dipoles are much smaller. The spectrum $I(\omega,F)$ contains both allowed and field-induced transitions,

$$I(\omega,F) = \sum_{r,p} I_{r,p}(\omega,F) \qquad (32)$$
$$= \sum_{r,p} \frac{\Gamma_{r,p}}{\pi} \left(\frac{\mu_r^2(F) F_{0p}^2(\mathbf{a}_r)}{[\omega_{r,p}(F) - \omega]^2 + \Gamma_{r,p}^2} \right)$$

and consists of Lorentzians. We scale all transition moments by μ_{G1B}, the dominant $F = 0$ moment. The $1B$ contribution to EA becomes [105]

$$\mathrm{EA}_{1B}(\omega,F) = \frac{2[\mu_{G1B}(F) - \mu_{G1B}]}{\mu_{1GB}} I_{1B}(\omega) \qquad (33)$$
$$+ [\omega_{1B} - \omega_{1B}(F)] \frac{\partial I_{1B}(\omega)}{\partial \omega}$$

The first term on the right is the change in transition moment, the second is the Stark shift, and they are additive to lowest (F^2) order. Similar expressions hold for other B states. Field-induced transitions at even-parity states lead to

$$\mathrm{EA}_{sA}(\omega) = [\mu_{sA}(F)/\mu_{G1B}]^2 I_{sA}(\omega) \qquad (34)$$

Here $I_{sA}(\omega)$ contains A-state vibrations, as in TPA, and the transition moment μ_{G1B}; this facilitates comparison of F^2 contributions below. The full EA spectrum is the sum over B states in Eq. (33) and A states in Eq. (34).

We emphasize that $I(\omega)$ and $I'(\omega)$ in Eq. (33) contain information about the ground and $1B$ potential surfaces sketched in Fig. 6.10. The force constants in Eq. (10) require all electrons and are formidable calculations for polyenes [21]. The F-dependent coefficients in Eq. (33), on the other hand, are due to virtual $\pi-\pi^*$ excitations. Similarly, the vibronics of two-photon excitations appear in Eq. (34), and the induced intensity depends on the $\pi-\pi^*$ spectrum. These EA expressions hold for an isolated molecule or polymer. An isotropic distribution of conjugated backbones in films also gives $I''(\omega)$ terms due to internal fields [106] or site disorder, and such profiles have been reported in PA [107] and PDA [108] films. Since EA of extended states in semiconductors [109,110] also goes as $I''(\omega)$ and disorder is poorly understood, films are more difficult to model. In crystals, Stark shifts scale as $I'(\omega)$ and induced moments as $I(\omega)$ or TPA.

Weiser [111] reported Stark shifts of 10 μeV for $F \sim 2 \times 10^4$ V/cm in PDA-R single crystals at $T \sim 10$ K, where R = PTS, DCHD, and PFBS; the singlet-exciton polarizablity is ~ 7000 Å^3. The DCHD and PTS spectra in Fig. 6.14 compare EA (top) and $I'(\omega)$ (bottom) derived from reflection and electroreflection spectra of these highly absorbing crystals with $\alpha \sim 10^6$ cm^{-1}. The PTS phase transition [112,113] leads to nonequivalent chains and doubles the spectra. The high resolution [111] of $I(\omega)$ and EA(ω) also shows small (<10 meV) variations of multiple C—C bonds as well as single-bond and low energy lattice modes. The close correspondence between EA(ω) and $I'(\omega)$ in the 0.4 eV interval above $1B(0-0)$ constitutes [105] an *experimental* demonstration that Stark shifts in these PDA crystals dominate in Eq. (33). Since the vibronics of EA(ω) and $I(\omega)$ coincide, the $1B$ potential surface governs the EA spectrum in this interval.

Stark shifts of discrete states define their static polarizabilities, Eq. (29). The shift for the singlet exciton is

$$E_{1B}(F) = \hbar\omega_{1B}(F) = E_{1B} - (1/2)\mathbf{F}\cdot(\alpha_{1B} - \alpha_G)\cdot\mathbf{F} \qquad (35)$$

for the polarizability tensors of G and $1B$. In analogy with $\chi_-(\delta)$ in Eq. (23), the static polarizability is a curvature of the ground-state energy, now with respect to F. Table 6.7 lists Stark shifts of PPP oligomers [105] used previously for NLO spectra, with *trans*-polyene geometry, $\delta = 0.15$, and $F = 1.65 \times 10^6$ V/cm applied along $\mathbf{r}_N - \mathbf{r}_1$. The proportionality to F^2 can be verified by changing F in direct solutions or by finding first-order corrections [114] to wave functions and thus generating only the F^2 part in Eq. (35). The different shifts at $N = 12$ are due to limitations of perturbation theory at large F. The 100 μeV oligomer shifts in Table 6.7 are convenient computationally. The 10 μeV shifts inferred [111] in PDA crystals are for $F \sim 2 \times 10^4$ V/cm, almost 100

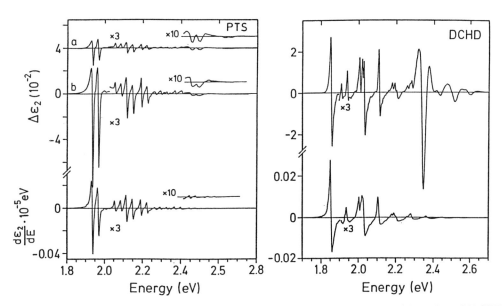

Fig. 6.14 Electroabsorption (top) and linear (bottom) spectra of PDA single crystals with PTS and DCDH side groups derived in Ref. 111 from reflectance and electroreflectance at $T \sim 10$ K. PTS signals are doubled owing to nonequivalent chains in the low temperature phase. The EA signal between 1.9 and 2.4 eV resembles the derivative of the linear absorption for a Stark shift of 10 μeV. (Adapted from Ref. 111.)

times smaller, for a 1000-fold larger polarizability along the backbone. Table 6.7 illustrates convergence problems discussed in Section V: The polymer limit does not follow from such small N.

The observed F^2 dependence of the EA intensity is important for our analysis based on large fields and oligomers. The other entries in Table 6.7 are field-induced changes in transition moments and go as F^2, as expected from first-order corrections to the wave functions. To compare Stark shifts and intensity changes in Eq. (35), we need the widths Γ of the linear absorption in Fig. 6.14, since sharp features are enhanced in the derivative and $I'(\omega)$ has maxima around Γ. In the $1B$ region, the scaling of EA(ω) depends on the ratio [105]

$$R = \frac{\mu_{1BG}\mathbf{F} \cdot (\alpha_{1B} - \alpha_G) \cdot \mathbf{F}}{(\mu_{G1B}(F) - \mu_{G1B})4\Gamma} \quad (36)$$

R is independent of F, since both terms go as F^2, and has no adjustable parameters. We have an $I'(\omega)$ spectrum for $R \gg 1$ and a bleached $I(\omega)$ profile for $R \ll 1$. The measured linewidth [111], $\Gamma < 0.010$ eV, and oligomer results in Table 6.7 lead to $R > 10$ and give an $I'(\omega)$ spectrum. As seen in Fig. 6.12 and Table 6.6, PDA-PTS crystals have both two-photon and linear absorption at 2.00 eV. Stark profiles due to overlapping vibronics depend sensitively on the precise spacing [115] and contain bleaching and induced absorptions that may approximate $I'(\omega)$.

The induced intensities to $2A$ in Table 6.7 follow first-order perturbation theory up to $N = 12$. Since $2A$ is below $1B$ in PDA, the denominators in Eq. (29) are finite for all B_u. The situation is quite different for nA because the excitation spectrum of Eq. (7) rapidly becomes congested [116] with increasing N. The mixing of closely spaced levels near nA is already beyond perturbation

Table 6.7 PPP Results for Stark Shifts and Transition Moments of N-Site Oligomers with Alternation $\delta = 0.15$, *trans*-Polyene Geometry, and Field $F = 1.65 \times 10^6$ V/cm

	$N = 8$	$N = 10$	$N = 12$	$N = 14$
$E_{1B} - E_{1B}(F)$ (μeV)	159	240	293	292
$\mathbf{F} \cdot (\alpha_{1B} - \alpha_G) \cdot \mathbf{F}/2$ (μeV)	153	232	331	—
$[\mu_{G1B} - \mu_{G1B}(F)]/\mu_{G1B}$ ($\times 10^4$)	5.69	8.55	13.3	20.6
$[\mu_{2AG}(F)/\mu_{G1B}]^2$ ($\times 10^4$)	10.8	15.9	24.1	38.1
$[\mu_{nAG}(F)/\mu_{G1B}]^2$ ($\times 10^4$)	27.3	35.0	169	—

Source: Ref. 105.

theory for $F = 1.65 \times 10^6$ V/cm, as shown by contrasting exact result and first-order corrections[105]; the latter are 10% above the $N = 8$ result in Table 6.7 for $\mu_{nA}(F)$, 30% higher for $N = 10$, and tenfold higher at $N = 12$. Although discrete, excitations for $N > 10$ are congested except near $1B$ and $2A$. We suggest the same approximation for PDA crystals, where the singlet exciton has a ~0.5 eV binding energy and $2A$ is even lower. We are pursuing the implications of Table 6.7 concerning induced absorptions in the $2A$ region [115].

We can now contrast the excitations in Table 6.6 for NLO spectra in Figs. 6.11–6.13 and the full spectrum for EA in Table 6.7. Both are based on PPP results for oligomers. The resonant states identified in Table 6.6 are augmented by vibrational inputs, and their relative positions and transition moments are adjustable. A few states are not even qualitatively adequate for Stark shifts, however. For example, Table 6.6 gives a positive Stark shift for nA, the highest state, while the full PPP result for nA has a large negative shift due to many higher energy rB states. Nor is the required compensation for mixing $1B$ and several A states borne out by the full solution. The 1:3 ratio of $[\mu_{2A}(F)/\mu_{nA}(F)]^2$ in Table 6.7 is reversed to 6:1 on considering four states with the same transition moments. Such phenomenological descriptions are no substitute for the full π–π^* spectrum.

V. CORRELATED EXCITED STATES OF QUANTUM CELL MODELS

A. Symmetries and Valence Bond Basis

The vibrational consequences of π-electron fluctuations discussed in Sections II and III drew on both molecular spectroscopy and solid-state physics. The analysis of NLO and EA spectra in Section IV combined PPP models for molecules with quantum cell models of alternating chains. We proposed at the outset to relate the conjugated polymers in Fig. 6.2 to alternating Hückel or PPP chains and have so far discussed vibrational and optical *implications* of π-electron models rather than the Hamiltonian, Eq. (7), or its mathematical properties. The analysis holds for any $H(\delta)$ with appropriate vibrational or optical susceptibilities. Equation (7) is sufficiently general to encompass Hückel, Hubbard, extended Hubbard, PPP, and other models with suitable choices of U and $V_{pp'}$. This generality is an extremely useful feature of solid-state models.

Accurate correlated *excited* states pose major theoretical challenges for extended systems, even at the π-electron level. We consider quantum cell models from several perspectives in this section, starting with symmetries and a many-electron basis. In addition to the total spin S, Eq. (7) has electron–hole (e–h) symmetry [12,117–119] for arbitrary intersite interactions $V_{pp'}$ in systems with one electron per site. The correlated singlets G and A_g have e–h index $J = 1$, while B_u singlets have $J = -1$. If electrons are added or subtracted, the spectra of positive and negative ions of any charge are identical by e–h symmetry, as first realized by McLachlan [119]. Site energies break e–h symmetry, however, and were omitted in Eq. (7) to retain this property.

To construct a many-electron basis for Eq. (7), we note that each site n can appear in only four ways: n may contain no electrons, two spin-paired electrons, or one electron with spin α or β. Each assignment $\{n_p\}$ of N_e electrons to N sites gives [120] a Slater determinant. The basis increases as 4^N, and this is slightly reduced by particle conservation or symmetry. Nevertheless, the *finite* basis of Eq. (7) for finite N is the key to exact analysis [121], while the exponential growth of the basis is the major challenge.

The total spin S is found by vector addition of p singly occupied sites. Pairing $2p$ sites into a singlet implies a linear combination of 2^p Slater determinants with fixed S_z at each site. Slater determinants and CI with single-particle states are difficult to apply to open-shell systems. We resort instead to the familiar valence bond (VB) diagrams of organic chemistry, where Heitler–London spin pairing is indicated by a line [120,121]. The polymers in Fig. 6.2, for example, have paired π electrons indicated by multiple bonds. The VB diagrams $|k\rangle$ shown are *covalent*, with one π electron ($n_p = 1$) per carbon. A C^+ or C^- site has $n_p = 0$ or 2, respectively, and any $|k\rangle$ with one or more $n_p \neq 1$ is *ionic*. The unique correspondence [121] between many-electron states and VB diagrams without crossing lines show that $|k\rangle$ gives a complete basis for Eq. (7) that conserves S. The state created by $a_{n\sigma}^+$ need not be specified beyond adding an electron with spin σ. Electron transfer between exact many-electron atoms or molecules also maps [34] into Eq. (7) when only the ground state for each oxidation state is considered. The nature of these site functions is governed by the parameters t, U, and $V_{pp'}$.

The easy visualization of VB diagrams facilitates the construction of linear combinations with inversion, e–h, and point-group symmetries [122]. Any eigenstate $|F\rangle$ of Eq. (7) can be expanded as

$$|F\rangle = \sum_k C_{Fk}|k\rangle \tag{37}$$

as originally suggested by Pauling [123,124]. The expansion is precisely defined for models where the complete basis of real molecules is not an issue. Since diagrams with identical n_p but different spin pairing are not orthogonal, the matrix representation of Eq. (7) is not symmetric. But all interaction terms are now diagonal, and off-diagonal $t(R)$ generate extremely sparse matrices [121]; each transfer gives at most a few new $|k'\rangle$ and there are only ~N bonds. There are about 2.8×10^6 and singlet 5.0×10^6 triplet VB diagrams for 14 electrons and 14 sites, and a fourfold reduction for chains with inversion and e–h symmetry. The many-electron eigenstates $|F\rangle$

in Eq. (37) are now equally accessible for any U or $V_{pp'}$. We typically find the lowest (≤ 10) states with given S, e–h, and inversion symmetry with a Silicon Graphics computer.

Exact eigenstates $|F\rangle$ for oligomers also yield transition moments, bond orders, and other matrix elements [121]. More generally, static or dynamic susceptibilities can be found exactly [114] using the finite basis and avoiding SOS expressions for NLO coefficients. The relevant state, normally $|G\rangle$, must be known, and vanishing lifetimes Γ_p are assumed. When many $|F\rangle$ are found directly, the full spectrum with adjustable lifetimes can be computed, and such PPP calculations [96,97] led to excited states and transition moments in Table 6.6. Valence bond methods are exact for oligomers with arbitrary parameters in Eq. (7). As shown by Bonner and Fisher [125] for spin 1/2 Heisenberg antiferromagnets, considerable insight can be gained from short ($N < 12$) oligomers.

B. Spin–Charge Separation and Dimer Limit

The generality of Eq. (7) is particularly advantageous for discussing exact limits. Alternating Hückel chains $H_0(\delta)$ in Eq. (1) with noninteracting electrons are the basis for linear e–ph coupling in SSH and continuum models. Self-consistent field results are similar [1]: Hückel and SCF orbitals of PPP models are often interchangeable. The large U limit of strong correlations is also understood in detail; the lowest states of Eq. (7) for one electron per site are spin excitations. Regular ($\delta = 0$) Hubbard chains yield exact results [126,127] for arbitrary U/t. Large alternation $\delta \sim 1$ can also be treated quantitatively [116] using molecular exciton theory. Limiting cases guide and constrain results for conjugated polymers with intermediate correlations and alternation.

In the atomic limit $U \gg t$, Hubbard models have a localized spin at each site. The 2^N spin states are precisely the covalent VB diagrams [128]. The Hückel spectrum in Eq. (2) is fundamentally altered as charge degrees of freedom are frozen out. The most dramatic change occurs in $\delta = 0$ chains with one site per unit cell. The band is metallic, with vanishing gap signaling the Peierls instability to dimerization. Lieb and Wu [126] demonstrated an *electronic* instability of the rigid lattice to on-site repulsion U by using the Bethe ansatz to solve the regular Hubbard chain exactly. Since $E_{1B}(U/t)$ is finite for any $U > 0$, the system is an insulator at $T = 0$ K for arbitrarily small correlations. Ovchinnikov [127] found other exact excitations of regular Hubbard chains, including triplets and homopolar singlets that are A_g states with $J = 1$, starting at $E_T = E_{2A} = 0$ for arbitrary U/t. The lowest singlet excitation S_1 of *regular* Hubbard chains is rigorously $2A$. This conclusion holds [41] for regular PPP and other chains with inversion and e–h symmetry and one fermion per site.

The atomic limit for $\delta > 0$ leads to spin chains with alternating antiferromagnetic exchange constants [129]. The static magnetic susceptibility $\chi_m(T)$ distinguishes sharply between regular and alternating chains, Takahashi [130] showed $\chi_m(T,U/t)$ of regular Hubbard chains to be finite at $T = 0$ for $U > 0$. The atomic limit is the regular Heisenberg chain with $J = 2t^2/U$ treated by Bonner and Fisher. Its $\chi_m(T)$ accurately describes both inorganic [131] and organic [129] chains. Alternating spin 1/2 chains, by contrast, have an energy gap E_T to the lowest triplet, 1^3B_u. Now $T\chi_m(T)$ decreases exponentially for $kT < E_T$ and vanishes as $T \to 0$. Triplet spin excitons and $E_T \sim 0.1$ eV are found [129,132] in ion-radical organic salts forming dimerized chains. Alternating ($\delta > 0$) chains have finite E_T in general. Since two triplets can always be combined into a $J = 1$ singlet, the $2A$ threshold is $2E_T$ for $U \gg t$.

There is another exact limit of Eq. (7), at $\delta \sim 1$, when delocalization is suppressed and the many-body problem reduces to dimers D for sites $2n, 2n - 1$. The ground state of D is an even-parity singlet. The lowest excited state is a triplet at

$$\epsilon_T(t_+, U_e) = (\sqrt{U_e^2 + 16t_+^2} - U_e)/2 \qquad (38)$$

where $t_+ = t(1 + \delta)$ and $U_e = U - V$. There is also an odd-parity singlet at $\epsilon_T + U_e$ and an even-parity singlet at $2\epsilon_T + U_e$. Any excitation, including composite states with several triplets, can readily be constructed [116] for the infinite chain. Interdimer interactions or hopping $t_- = t(1 - \delta)$ are typical perturbations in molecular exciton theory. A chain of dimers also has CT excitations leading to D^+D^- states. The complete spectrum of Eq. (7) can be developed analytically [116] in the dimer limit to follow spin–charge separation for arbitrary U, V, and t, including intermediate correlations where both band and atomic limits fail. The connection between E_{2A} and $2\epsilon_T$ is evident in this limit, as is the fundamentally different spin–charge separation of Hubbard and excitonic chains. Although dimers become increasingly approximate at alternations realized in conjugated polymers, they remain pertinent when E_{1B} and E_{2A} are comparable.

C. The 2A–1B Crossover and Fluorescence

The band and atomic limits show all alternating chains to have finite $1B$ and $2A$ thresholds, although their magnitudes are not known exactly. The $4t\delta$ band gap indicates that a criterion other than opening a gap is needed for spin–charge separation with increasing e–e correlations as the system evolves from a wideband semiconductor to a paramagnetic insulator. Covalent VB singlets have e–h index $J = 1$, while the lowest $J = -1$ singlet is ionic, with one or more C^- sites. Thus $1B$ and other $J = -1$ states in Eq. (37) have vanishing expansion coefficients for *all* covalent diagrams $|k\rangle$ for any N.

While the lowest singlet S_1 is clearly covalent in the atomic limit, we have $S_1 = 1B$ in Hückel theory for oligomers or in exciton models. Spin–charge separation involves an *excited-state* crossover [133] from $S_1 = 1B$ to $S_1 = 2A$. Photophysical properties related to S_1 change discontinuously at the crossover [41], which is a generic feature of alternating chains.

The related $E_{1B}/2E_T$ crossover is more convenient to analyze. The triplet 1^3B_u is the $S = 1$ threshold in the $J = 1$ manifold, and $2A$ evolves into two triplets for strong correlations. The relations [133]

$$E_{1B}(\delta, V(\{R\})) = 2E_T(\delta, V(\{R\})) \geq E_{2A}(\delta, V(\{R\})) \quad (39)$$

define a spin–charge crossover. The inequality indicates that repulsive interactions in infinite chains can be relieved by separating the triplets. Equation (39) summarizes band results for arbitrary δ, exact Hubbard results for $\delta = 0$ and arbitrary U/t, and $\delta \sim 1$ results for arbitrary potential $V(R)$; it is also consistent with finite Hubbard and PPP chains. For given $V(R)$, the crossover $E_{1B} = 2E_T$ fixes δ_c, while for given δ the crossover gives the critical correlations for single-particle gap $4t\delta$ and bandwidth $4t$. We have $U_c = 0$ at $\delta = 0$, where any $U > 0$ opens a gap, but finite U_c is required for finite $4t\delta$.

The crossover defined by Eq. (39) is shown in Fig. 6.15 for Hubbard and PPP oligomers [133] as a function of alternation, with $2E_T < E_{1B}$ above the lines. Both crossovers occur when $U \sim 2t_+ = 2t(1 + \delta)$ is the arithmetic mean of the band gap and bandwidth, except near regular chains. The inset in the Hubbard panel is the crossover based on the exact $E_{1B}(U/t)$ at $\delta = 0$ and $E_T = 4t\delta$ at $U = 0$. The gap is exponentially small [127]. Although Hubbard and PPP chains converge properly at small δ, they do so far too slowly to deduce the exact result. That is not the case near $\delta = 1$, however, where the exact Hubbard result $U_c = 2t_+$ holds for dimers, and the dashed line in Fig. 6.15 is the first-order correction in $t_- = t(1 - \delta)$ for infinite chains. Intersite $V_{pp'}$ persist in PPP chains even at $\delta = 1$. The first-order contribution, $\langle V(R) \rangle$, for the molecular potential in Eq. (6) and polyene geometry is the arrow in the PPP panel.

Rapid convergence at large alternation is expected on general grounds. The sufficiency of $N = 14$ oligomers for linear or TPA spectra of Hubbard and PPP chains with $\delta \geq 0.6$ follows in detail from molecular exciton theory [116]. The different shapes of PPP and Hubbard crossovers in Fig. 6.15 are due to different $1B$ states, which are in turn related to $V_{pp'}$ in Eq. (7): $1B$ is an exciton in PPP theory [134] and evolves [116] to an excited dimer at $\delta = 1$, while $1B$ is a CT state [37] D^+D^- in Hubbard chains at $\delta \sim 1$. These possibilities for $1B$ are another generic feature of Eq. (7). Molecular PPP parameters place [133] PA on the correlated side, PS and PPV on the band side.

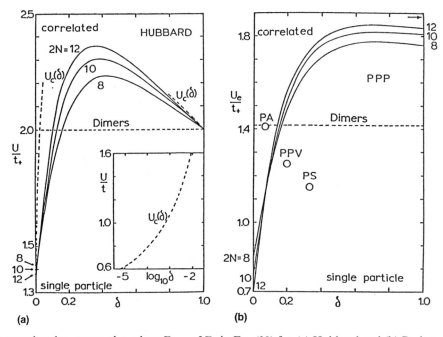

Fig. 6.15 Band-to-correlated crossover based on $E_{1B} = 2E_T$ in Eq. (39) for (a) Hubbard and (b) Pariser–Parr–Pople chains with $t_\pm = t(1 \pm \delta)$. The crossover of isolated dimers is at $U = 2t_+$ in Hubbard chains and $U_e = U - V = t_+\sqrt{2}$ in PPP chains. The oligomer curves, exact limits at $\delta = 0$ and 1, and indicated polymers are discussed in the text. (Adapted from Ref. 133.)

The crossover of infinite chains follows approximately from Fig. 6.15. The PPP curves hardly depend on N around $0.1 < \delta < 0.2$, where the N dependence reverses. In retrospect, this reversal rationalizes the success of using E_{1B} as an internal standard [96,97] for E_{2A} and other even-parity states in fitting NLO spectra. Scaling is also used experimentally for E_{1B} shifts between crystals and films whose backbone conformations are not known. The N dependence of Hubbard models reverses at smaller δ, and both the regular and dimer limits are exact.

The parameters of Eq. (7) can be varied to delineate spin–charge separation or other properties with e–e correlations. Such variations are not possible experimentally, however. The sp^2 carbons of conjugated polymers suggest similar e–e interactions in these chemically related systems. We have systems with almost *constant* $V(R)$ and variable [41] δ due to bond lengths, topology, or heteroatoms. As shown by the open circles in Fig. 6.16, the observed [135] E_{2A}/E_{1B} ratios indicate both polymers with $2A$ above and below $1B$. When $2A$ is higher, we have $S_1 = 1B$, and fluorescence is in fact typical in these families [41]. Experimental data on polyenes [136] and a PDA oligomer [137] are shown as open circles at finite $1/N$ in Fig. 6.16.

The solid points in Fig. 6.16 are exact PPP results for *trans*-polyenes, the molecular potential $V(R)$ in Eq. (6),

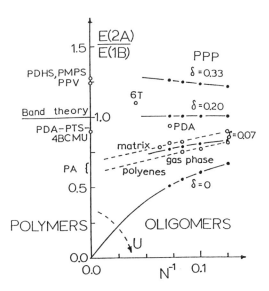

Fig. 6.16 The E_{2A}/E_{1B} ratio of the lowest two- and one-photon excitations of conjugated polymers, molecules, and N-site oligomers discussed in the text. Experimental results are shown as open circles, exact PPP calculations at alternation δ as closed circles. Electron–hole symmetry fixes the $\delta = 0$ value of regular chains, where increasing $U \gg t$ leads to $E_{2A}/E_{1B} \sim (t/U)^2$ for spin and charge excitations; the band result is $E_{2A} = E_{1B}$ for any δ. (From Ref. 135.)

and the indicated δ. We took equal bond lengths at $\delta = 0$ to facilitate extrapolation to the polymer. The weak N dependence of E_{2A}/E_{1B} around $\delta = 0.20$ is evident of π-conjugated chains. The larger alternation, strong fluorescence, and NLO spectra of polysilanes are consistent [138] with Si parameters in Eq. (6). Conjugated molecules and polymers show *intermediate* correlations and comparable one- and two-photon thresholds.

Hückel theory for PPV or other polymers with *para*-conjugated phenyls gives a simple explanation for their reversed $2A/1B$ ordering relative to polyenes [39]. We label phenyl sites in Fig. 6.2 adjacent to one bridgehead as ± 1, those adjacent to the other bridgehead as ± 2. The new orbitals

$$\varphi_{\pm 1,2} = (\varphi_{1,2} \pm \varphi_{-1,-2}) / \sqrt{2} \quad (40)$$

have $t = 0$ and $t\sqrt{2}$ at bridgeheads for the odd and even combinations, respectively. The odd combinations form a bonding molecular orbital for two of the six π electrons of each phenyl. The remaining four are in extended states of linear chains with increased $t\sqrt{2}$ at bridgeheads. A hypothetical planar poly-*para*-phenylene with equal bond lengths reduces to an alternating chain with $\delta = (\sqrt{2} - 1)/2$. An extended PPV chain has six sites per unit cell and larger band gap [39] for *equal* bonds R_0 than *t*-PA with *alternating* bonds. Topological alternation of *para*-conjugated phenyls carries over to PPP models, and exact results [139] for *trans*-stilbene, the two-ring oligomer, account for singlet, triplet, and two-photon excitations of this well-studied molecule. The $2A$–$1B$ crossover in PT is modeled [140,142] in terms of site energies $\pm \epsilon$ in Fig. 6.2 due to nonconjugated sulfur atoms, which generate a charge-density wave ground state. In contrast to topological alternation, ϵ is an adjustable parameter in PPP models of thiophene oligomers.

D. Oligomers and Polymers

Different parameter ranges of Eq. (7) are sampled by conjugated polymers, ion-radical organic crystals, oxide superconductors, heavy fermions, or exchange-coupled spin networks. Band theory and low-order CI are unequivocally preferred in extended systems with weak e–e interactions, while the atomic limit is clearly best for strong correlations. Comparable $2A$ and $1B$ thresholds in Table 6.1 and Fig. 6.16 are particularly difficult to approximate. Band or mean-field schemes must go beyond first-order CI merely to place $2A$ below $1B$, while $U \gg t$ describes covalent states. Intermediate correlations require other methods. Exact solutions are restricted to short oligomers whose extrapolations may be unreliable [141,142]. We have accordingly discussed spin–charge crossovers using symmetry and exact limits. Although rigorous, such arguments have limited usefulness for understanding particular polymers, and limitations of exact solutions to small N deserve closer examination.

Extrapolations in $1/N$ or $1/N^2$ are the traditional approach to extended systems. They are best for the ground-state energy, which often converges rapidly for cyclic boundary conditions. Even then, excitation energies vary as $1/N$, and higher energy states such as nA in Table 6.6 cannot be extrapolated directly. Bond orders illustrate both convergence and sum rules. The PPP results in Table 6.8 are average double- and single-bond orders for *trans*-polyenes with $\delta = 0.15$ [105]. They are roughly two-third of the Hückel value and clearly converge in the ground state. The polymer values are $1/N$ extrapolations of successive $N, N + 2$ pairs. The $1B$ bond orders of H_0 are smaller on exciting from a bonding to an antibonding orbital. The PPP bond orders of $1B$ in Table 6.8 are also smaller and increase with N. Their polymer limit must in fact coincide with the ground state, since G and $1B$ are connected by the one-electron operator μ.

The interaction terms of Eq. (7) are completely specified by number operators n_p and thus commute with μ in Eq. (28). Quantum cell models satisfy a sum rule for oscillator strengths based on the identity [143,144]

$$[[H_0 + V, \mu], \mu] = \sum_n e^2 t_n \mathbf{b}_n \cdot \mathbf{b}_n 2 p_n \quad (41)$$

where $\mathbf{b}_n = \mathbf{r}_{n+1} - \mathbf{r}_n$ is the bond vector. The expectation value with respect to $|G\rangle$ gives

$$\sum_{r=1} |\langle rB|\mu|G\rangle|^2 E_{rB} = e^2 b^2 \sum_n |t_n| \langle G|p_n|G\rangle \quad (42)$$

where b is the average bond length for simplicity. The oscillator strength per site, f/N, for dipole transitions originating from $|G\rangle$ is governed by the ground-state bond orders, whose convergence with N is shown in Table 6.8.

The bandwidth $4t(R_0) = 9.6$ eV leads in Eq. (42) to $f/N \sim 0.25$ and increases slowly with N in Table 6.8. The $r = 1$ term in the sum is the $1B$ contribution, $f_{1B}/N \sim 0.14$, including all vibronics. Its weak N dependence is comparable to $\langle G|p|G\rangle$, but decreases [105] slowly with N. The N dependence of $f_{1B}/N \sim 0.15$ has previously been found [145] for the same potential and the polyene alternations $\delta = 0.07$. The measured [111] f values for the 0–0 line of PDA crystals in Fig. 6.14 are around 0.6 per repeat unit, or 0.15 per π electron. The 0–0 line is several times more intense at 10 K than the 0–1 sidebands. The total oscillator strength of the exciton is $f_{1B} \sim 0.7$–0.8 per repeat unit, or 0.18–0.20 per δ electron, slightly above the PPP estimate. As there are no adjustable parameters or scale factors, oligomers with molecular $V(R)$ account directly for the polymer intensity. Local field and other corrections will be needed for more stringent comparison.

The sum rule for the singlet exciton has implications for the location of even-parity states. The expectation value of Eq. (41) with respect to $|1B\rangle$ is

$$\sum_{s=1} |\langle sA|\mu|1B\rangle|^2 (E_{sA} - E_{1B}) = e^2 b^2 \sum_n |t_n| \langle 1B|p_n|1B\rangle \quad (43)$$

with the sum now over $|sA\rangle$ states, including $|G\rangle$. Size extensivity implies that the total oscillator strengths, Eqs. (42) and (43), increase as N for sufficiently large systems and thus become equal in polymers, where the indicated bond orders per site are equal to order $1/N$. The $s = 1$ term in Eq. (43) involves G and $1B$, the dominant ground-state process, and is negative. The oscillator strengths of sA states *above* E_{1B} must exceed the ground-state intensity, Eq. (42), by the large $s = 1$ contribution.

In band theory, two-electron excitation across the gap gives an A state at $2E_{1B}$ with transition moment μ_{G1B} from $1B$, and this cancels the $s = 1$ term in Eq. (42). In correlated systems, the nA state derived from two-electron excitation [96,97] of G shifts below $2E_{1B}$ and the transition dipole μ_{nA1B} becomes larger than μ_{G1B}, as follows analytically for dimers. The sum rule compensates for decreasing $E_{nA} - E_{1B}$ with increasing μ_{nA1B}. The NLO fits in Section IV are based on Table 6.6 and place nA, slightly lower than in PPP oligomers. The indicated $\mu_{nA1B}/\mu_{G1B} \sim 1.3$ must be increased to 2 to satisfy Eq. (43) within these four states.

Table 6.8 Exact PPP Bond Orders and Oscillator Sum Rule for G and $1B^a$

N	$\langle p_d \rangle$, G	$\langle p_s \rangle$, G	f/N, G	$\langle p_d \rangle$, $1B$	$\langle p_s \rangle$, $1B$	f/N, $1B$
8	0.8928	0.2825	0.2484	0.6860	0.4615	0.2229
10	0.8879	0.2853	0.2501	0.7201	0.4425	0.2324
12	0.8845	0.2870	0.2512	0.7424	0.4272	0.2380
14	0.8821	0.2882	0.2520	0.7583	0.4145	0.2417
Polymer	0.867	0.295	0.257			

^{a} $\langle p_d \rangle$, $\langle p_s \rangle$ are the average for $N/2$ double and $N/2 - 1$ single bonds, respectively, and f is the sum rule, Eqs. (42) and (43), for G and $1B$.
Source: Ref. 105.

For Hubbard chains with $U \gg t$, the ground state becomes a spin wave, with $p_n = 0$ for all n, and dipole processes are suppressed; Eq. (42) then decreases as $(t/U)^2$, and the bond orders vanish when only virtual transfers are possible. Dipole-allowed transitions from $1B$ are still possible to A states with a C^+C^- pair, however, in the manifold of states within $\sim t$ of $E_{1B} \sim U$. Such excitations in Eq. (43) go as t^2/U in units of E_{1B}. The only even-parity excited state, mA_g, considered in the essential states model [146,147] for NLO is just above $1B$, which changes the sign of the left-hand side of Eq. (43) and implies [100] a strong two-photon resonance in the THG spectrum (Fig. 6.11) around $3\omega \sim 3$ eV. Vanishing $2A$ contributions in the essential states model also point to strong correlations, when $2A$ is a spin state based on two triplets. The intermediate nature of molecular correlations is again apparent: correlations place $2A$ and nA far below the band limit, but the intense linear absorption is far from spin waves.

VI. EXTENSIONS AND CONCLUSIONS

A. Molecular Potential for Conjugated Polymers

Conjugated polymers and molecules are chemically similar. Polymers are larger and have additional conformational degrees of freedom, and, in some cases, high electrical conductivity on doping with donors or acceptors. We have related spin–charge separation in alternating chains to the $2A$–$1B$ crossover and distinguished in Eq. (7) between models whose $1B$ is an exciton or an electron–hole pair. Since intersite interactions are needed to stabilize excitons, the different behavior of PPP and Hubbard chains is not surprising. What is remarkable, rather, is that molecular parameters developed decades ago [3–5] remain satisfactory, at least with current limitations to oligomers, for both optical and vibrational features of conjugated polymers. Molecular electronic parameters are transferable [12] to conjugated polymers, and no special size separates molecules and polymers. The joint treatment of conjugated molecules has major implications.

The strength of e–ph coupling is usually measured by the dimensionless constant [17]

$$\lambda = 2 \frac{\sum_i (g_i/\omega_i^0)^2}{\pi t} \quad (44)$$

where g_i is the linear coupling constant and ω_i^0 is the reference frequency (Section II.C). We estimate $\chi\lambda\pi t/2 = 0.389$ from the Raman frequencies of t-PA. The Hückel susceptibility $\chi(0.18) = 0.617$ eV^{-1} gives [57] $\lambda = 0.16$, slightly below the SSH value of 0.2. An even lower estimate of $\lambda = 0.12$ is found in the PPP model, with $\chi \sim 0.83$ eV^{-1}. The strength of e–ph coupling is therefore small but not negligible compared to effective $U_e/t \sim 1.5$ of dimers, with $U_e = U - V$. Another effective U_e' appropriate to PPP chains embedded in ionic lattices with $(-1)^{r-s}V_{rs}$ Coulomb interactions leads to even smaller $U_e'/t \sim 0.7$. The interplay of e–e and e–ph interactions in determining the ground-state structure of linear chains is rather subtle [148]. For conjugated polymers, however, an alternating ground state is an experimental input and t-PA has the most extensive excited-state relaxation. Polymers with nondegenerate ground states are simpler in this respect and closer to conjugated molecules. The relaxation of excited-state or molecular ion structures estimated [139] from π-electron bond orders may reduce excitation energies by 5–10%. Such refinements will become important as π–π^* spectra of conjugated polymers are more accurately measured or assigned than in current NLO studies. The potential $V(R)$ fixed by the ground-state geometry suffices for electronic states, while dynamic $t(R)$ modulation and Hückel models are the basis for more sophisticated e–ph treatments of vibrations.

Once the relevance of molecular parameters to conjugated polymers is recognized, the PPP model emerges as the best and most complete π-electron description. Kohler and coworkers [136] have documented E_{2A} to be below E_{1B} in polyenes from $N = 8$ to 16 (Fig. 6.16), and the splitting increases with N. Alternating even and odd orbitals in band theory give the opposite ordering, even at the SCF level or in first-order CI. When used in exact solutions, standard PPP parameters give the proper order and magnitudes [12]. The magnetic and optical properties of naphthalene [149], anthracene [150], and $trans$-stilbene [139] follow from the same parameters. Molecular spectra clearly support e–e correlations in general and $V(R)$ in particular. Thus Eq. (6) is a logical choice for Coulomb interactions in pristine polymers.

Hückel analysis of t-PA with $\delta = 0.18$ matches $E_{1B} \sim 1.8$ eV and suffices for most aspects of e–ph coupling in Section III, where analytical results for the infinite chain are particularly important. Smaller $\delta = 0.07$ occurs in PPP models of polyenes or PA, since $V(R)$ also contributes to E_{1B}, and yields comparable but less precise χ_- for vibrations on extrapolating to the polymer (Fig. 6.8). Different π-electron models are therefore suitable for different applications. Wider applications place more restrictions on the form and parameters of H_e. The inherently approximate nature of models, however, limits quantitative comparisons even if parameters are readjusted for each polymer, as is often done in solid-state discussions, and leaves open the range of optical, vibrational, NLO, or other spectra described by H_e. More detailed analyses and novel applications are certain in view of the generality of alternating chains and the flexibility of models.

The SSH [10,11] and continuum approaches to conjugated polymers focus on linear e–ph coupling and the resulting self-localized gap states. Although $t(R)$ and force constants K are close to molecular values, no comparisons with molecular excitations are attempted. General expectations are obtained instead for polymers with degenerate or nondegenerate ground states, or for metal–insulator transitions on doping, but different parameters are routinely assigned to every polymer. The novel electrical properties of t-PA and PTs indeed called for new ideas and approximations, especially in connection with excited-state properties. The SSH model has guided both experiment and all-electron calculations on many conjugated polymers.

Subsequent optical studies of pristine polymers amply demonstrate resemblances to molecular spectra, where shortcomings of Hückel theory have long been recognized. Yet solid-state approaches to NLO coefficients remain largely free of molecular comparisons. They emphasize instead the challenges of limited CI in longer chains of $N = 50$–100, novel methods for estimating correlated states, and Hubbard rather than Coulomb potentials. These are important issues, and accurate treatments of longer chains are clearly desirable. Molecules with 10–20 π electrons afford many opportunities for testing approximations whose validity has not been shown to improve with increasing N. Moreover, the proper choice of $V(R)$ and molecular comparisons focus attention on quantum cell models and parameter sectors appropriate for conjugated polymers.

B. Toward Force Fields and Susceptibilities of Nondegenerate Polymers

The key to our understanding of vibrational spectra of conjugated polymers is the separation of highly nontrivial effects of π-electron fluctuations, modeled in terms of linear e–ph coupling, from classical spectroscopic effects associated with the ground-state distribution of σ and π electrons. This separation parallels σ–π separation in Hückel or PPP theory or the identification of frontier orbitals in solid-state models. The basic idea is to single out the effects of π-electron fluctuations from the details of the backbone vibrational structure, which is accessible by standard methods. The interesting physics is then described in terms of simple models whose parameters can be reliably extracted from, or calibrated against, experiment and, in principle, transferred among different polymeric structures.

Our approach to vibrational spectra is general, and its success in the case of polyacetylene suggests applications to other conjugated polymers. The procedure for extracting the reference force field is standard. Major difficulties with other polymers arise from the lack of extensive vibrational data on short oligomers and/or of exhaustive vibrational assignments. On the other hand, force fields have been recently obtained [151] by adopting overlay techniques to oligomers of increasing length. Care has to be taken with these force fields: they can in fact lead to satisfactory fits, but they do not correspond to the reference because e–ph coupling is implicitly included.

The analysis of t-PA clearly shows that e–ph coupling is dominated by $t(R)$ modulation by C—C bond stretches. In this view t-PA is particularly simple: C—C stretches occur only in the a_g symmetry block, leading to only three coupled modes. The other polymers in Fig. 6.2 have more C—C bonds per unit cells and different local geometries involving sp hybridization, aromatic subunits, or heteroatoms. The possible number of coupled modes is larger, and effective coupling may not be confined to totally symmetric modes. Moreover, the larger alternation of fluorescent polymers, with $1B$ below $2A$, tends to reduce [152] the response χ by increasing the $2A$ gap in Eq. (13). Smaller softening of coupled modes is more difficult to recognize or calibrate precisely. Similar complications are apparent in the SSH model, in which all polymers with nondegenerate ground states are lumped together. The different backbones of PDA, PPV, or PT are lost, and their inclusion in quantum-chemical treatments tends to obscure connections to π electrons. The alternating chains emphasized in this review are a partial solution to modeling the polymeric excitations realistically within a common π-electron framework.

Transferability of molecular coupling constants has been demonstrated and widely applied in organic CT complexes [30,31] whose crystal structures indicate molecules or ions at, or just below, van der Waals contact. Such a goal is far more ambitious in conjugated polymers, where transferable e–ph coupling constants for strong π bonds are sought. The simplicity of the PA backbone and extensive studies of polyenes are ideally suited to demonstrate the general procedure. Transferable e–ph coupling constants must be defined in internal coordinates, and their estimates in t-PA relies on a Hückel description of the electronic system. We anticipate that e–ph coupling analysis in other polymers will rely mostly on Hückel theory, supplemented and integrated with PPP or other calculations on oligomers. As in the case of electronic excitations, the challenge is to incorporate the chemical diversity of conjugated polymers without sacrificing the simplicity of models.

The interpretation of dopant and photoinduced IR spectra of nondegenerate polymers presents other opportunities. Due to the larger number of coupled modes with respect to PA, and to the overall smaller effects of e–ph coupling, phenomenological analysis of spectra based on the AM formalism is delicate. A simple microscopic model for the IR activity of polarons is not available. We have shown that among the localized and delocalized modes of a soliton in an idealized chain, strong e–ph coupling is restricted to just one, the damped ECC mode. Infrared spectra of doped or photoexcited PA are therefore strongly dominated by three bands, corresponding to the three Raman-active a_g modes of pristine samples. Such a one-to-one correspondence is not guaranteed for polarons in nondegenerate polymers. First, the e–ph coupled modes in pristine samples occur not only in Raman but also in IR spectra. Second, the number of polaron or bipolaron modes that acquire appreciable IR intensity on doping or photoexcitation remains to be found. Fresh ideas are needed for a clear interpretation of IRAV modes in nondegenerate polymers.

As noted in Section IV, increasingly diverse and resolved NLO spectra are becoming available for conjugated polymers. These spectra confirm the importance of vibronic effects, especially for backbone modes coupled to the π system, and point to more detailed vibronic studies. Current approaches focus on the position and transition moments of resonant states in terms of crude adiabatic states and displaced harmonic oscillators. Resolved spectra are now available for PDAs, which are uniquely available as single crystals. The ground-state and $1B$ potential surfaces of conjugated molecules are amenable to detailed modeling [21]. These are also the best characterized states of conjugated polymers, where $1B$ is a singlet exciton. High resolution electroabsorption spectra in the exciton region of PDA crystals are dominated by backbone vibrations and differ from the EA profiles of polymer films. The rich excitation spectrum of quantum cell models offers many opportunities for analyzing NLO, EA, and other spectra. A unified semiquantitative description such as PPP theory for conjugated molecules remains to be fully realized.

Both vibrational and electronic consequences of delocalized π electrons were recognized and modeled in early studies of conjugated molecules [1]. Conjugated polymers and modern experimental methods have enormously expanded the scope of both vibrational and electronic spectra, as discussed above. But the fundamental susceptibilities arising from low-lying excitations still describe the responses of π systems. The simplicity of the alternating chain sketched in Fig. 6.1 is undoubtedly responsible for many of the extensions to correlated excited states, vibronic analysis of NLO coefficients, and transferable e–ph coupling constants presented in this review. We have found both dynamic modulation of $t(R)$ and Coulomb interactions $V(R)$ to be manageable in alternating chains and have used the alternation δ of transfer integrals to describe the different photophysics and properties of the polymers in Fig. 6.2. The larger unit cells of PPV or PT provide additional problems for identifying π-electron contributions. Time will tell whether their interpretation requires going beyond π-electron models.

ACKNOWLEDGMENTS

This research was made possible through NATO Collaborative Research Grant No. 900629. We thank the National Science Foundation for support of work at Princeton through NSF-DMR-9300163, and the National Research Council (CNR) and the Ministry of the University and of Scientific and Technological Research (MURST) for support of work at Parma.

REFERENCES

1. L. Salem, *The Molecular Orbital Theory of Conjugated Systems*, Benjamin, New York, 1966.
2. T. A. Skotheim (ed.), *Handbook of Conducting Polymers*, Vols. 1 and 2, Marcel Dekker, New York, 1986.
3. B. Roos and P. N. Skancke, *Acta Chem. Scand. 21*: 233 (1967).
4. K. Schulten, I. Ohmine, and M. Karplus, *J. Chem. Phys. 64*: 4422 (1976).
5. L. Labhart and G. Wagnière, *Helv. Chim. Acta 46*: 1314 (1967).
6. R. Pariser and R. G. Parr, *J. Chem. Phys. 21*: 767 (1953).
7. J. A. Pople, *Trans. Faraday Soc. 42*: 1375 (1953).
8. H. C. Longuet-Higgins and L. Salem, *Proc. Roy. Soc. Lond. A251*:172 (1959).
9. R. E. Peierls, *Quantum Theory of Solids*, Clarendon, Oxford, 1955, p. 108.
10. A. J. Heeger, S. Kivelson, J. R. Schrieffer, and W. P. Su, *Rev. Mod. Phys. 60*:81 (1988).
11. W. P. Su, J. R. Schrieffer, and A. J. Heeger, *Phys. Rev. Lett. 42*:1698 (1979): *Phys. Rev. B22*: 2099 (1980).
12. Z. G. Soos and G. W. Hayden, Excited states of conjugated polymers, in *Electroresponsive Molecular and Polymeric Systems* (T. A. Skotheim, ed.), Marcel Dekker, New York, 1988, p. 197.
13. D. Baeriswyl, D. K. Campbell, and S. Mazumdar, An overview of the theory of π-conjugated polymers, in *Conducting Polymers* (H. Kiess, ed.), Springer-Verlag, Heidelberg, 1992, p. 7.
14. C. A. Coulson, *Proc. Roy. Soc. Lond. A169*:413 (1939).
15. N. W. Ashcroft and N. D. Mermin, *Solid State Physics*, Holt-Saunders, New York, 1976, p. 433.
16. Proceedings of international conferences on the science and technology of synthetic metals: Tübingen, Germany, *Synth. Met 41–43*, 1991; Göteborg, Sweden, *Synth. Met 55–57*, 1993; Seoul, South Korea, *Synth. Met. 69–71*, 1995.
17. E. J. Mele, Phonons and the Peierls instability in polyacetylene, in *Handbook of Conducting Polymers* (T. A. Skotheim, ed.), Marcel Dekker, New York, 1986, Vol. 2, p. 795.
18. B. Horovitz, *Solid State Commun. 41*:721 (1982).
19. E. Ehrenfreund, Z. Vardeny, O. Brafman, and B. Horovitz, *Phys. Rev. B36*:1533 (1987).
20. C. A. Coulson and H. C. Longuet-Higgins, *Proc. Roy. Soc. Lond. A191*:39 (1947); *A192*:16 (1947); *A193*:447, 456 (1948).

21. G. Orlandi, F. Zerbetto, and M. Zgiersky, *Chem. Rev. 91*:867 (1991).
22. M. Gussoni, C. Castiglioni, and G. Zerbi, Vibrational spectroscopy of polyconjugated materials: polyacetylene and polyenes, in *Spectroscopy of Advanced Materials*, Adv. Spectrosc. Vol. 19 (R. J. H. Clark and R. E. Hester, eds.), Wiley, New York, 1991, p. 251.
23. J. Kürti and H. Kuzmany, *Phys. Rev. B44*:597 (1991).
24. G. P. Brivio and E. Mulazzi, *Chem. Phys. Lett. 95*: 555 (1983).
25. M. Kertész and P. R. Surján, *Solid State. Commun. 39*:611 (1981).
26. J. Kürti, P. R. Surján, and M. Kertész, *J. Amr. Chem. Soc. 113*:9865 (1991).
27. A. Warshel and M. Karplus, *J. Am. Chem. Soc. 94*: 5612 (1972).
28. R. J. Hemley, B. R. Brooks, and M. Karplus, *J. Chem. Phys. 85*:6550 (1986).
29. M. J. Rice, *Solid State Commun. 31*:93 (1979).
30. A. Painelli and A. Girlando, *J. Chem. Phys. 84*:5655 (1986).
31. C. Pecile, A. Painelli, and A. Girlando, *Mol. Cryst. Liq. Cryst. 171*:69 (1989).
32. R. Bozio and C. Pecile, Charge transfer crystals and molecular conductors, in *Spectroscopy of Advanced Materials*, Adv. Spectrosc., Vol. 19 (R. J. H. Clark and R. E. Hester, eds.), Wiley, New York, 1991, p. 1.
33. J. Hubbard, *Proc. Roy. Soc. Lond. A276*:238 (1963); *A277*: 237 (1964); *A281*: 401 (1964).
34. Z. G. Soos and D. J. Klein, Charge-transfer in solid-state complexes, in *Molecular Association*, Vol. 1 (R. Foster, ed.), Academic, New York, 1975, p. 1.
35. K. Ohno, *Theor. Chim. Acta 2*:219 (1964).
36. M. Pope and C. E. Swenberg, *Electronic Processes in Organic Crystals*, Clarendon, Oxford, 1982.
37. Z. G. Soos and G. W. Hayden, *Chem. Phys. 143*:199 (1990).
38. G. P. Agrawal, C. Cojan, and F. Flytzanis, *Phys. Rev. B17*:776 (1978).
39. Z. G. Soos, S. Etemad, D. S. Galvão, and S. Ramasesha, *Chem. Phys. Lett. 194*:341 (1992).
40. Z. G. Soos, M. H. Hennessy and D. Mukhopadhyay, Correlations in Conjugated Polymers, in *Primary Photoexcitations in Conjugated Polymers: Molecular Exciton Versus Semiconductor Band Model* (N. S. Sariciftci, ed.) World Scientific, Singapore, in press).
41. Z. G. Soos, D. S. Galvão, and S. Etemad, *Adv. Mater. 6*:280 (1994).
42. D. I. Bower and W. F. Maddams, *Vibrational Spectroscopy of Polymers*, Cambridge Univ. Press, Cambridge, 1992, p. 156.
43. E. B. Wilson, Jr., J. C. Decius, and P. Cross, *Molecular Vibrations*. McGraw-Hill, New York, 1955.
44. S. Califano, *Pure Appl. Chem. 18*: 353 (1969) and references therein.
45. J. H. Schachtschneider and R. G. Snyder, *Spectrochim. Acta 19*: 117 (1963).
46. B. Hudson, B. Kohler, and K. Schulten, Linear polyene electronic structure and potential surfaces, in *Excited States* (C. Lim, ed.), Academic, New York, 1982, Vol. 6, p. 1.
47. M. Kofranek, H. Lischka, and A. Karpfen, *J. Chem. Phys. 96*: 982 (1992).
48. A. Girlando, A. Painelli, and Z. G. Soos, *Chem. Phys. Lett. 198*: 9 (1992).
49. A. Painelli and A. Girlando, *Solid State Commun. 63*: 1087 (1987).
50. A. Painelli, A. Girlando, and C. Pecile, *Solid State Commun. 52*: 801 (1984).
51. Y. Furukawa, H. Takeuchi, I. Harada, and M. Tasumi, *Bull. Chem. Soc. Jpn. 56*: 392 (1983).
52. H. Guo and M. Karplus, *J. Chem. Phys. 94*: 3679 (1991).
53. H. J. Bernstein and S. Sunder, Resonance Raman spectra and normal coordinate analysis of some model compounds of heme proteins, in *Vibrational Spectroscopy: Modern Trends* (A. J. Barnes and W. J. Orville-Thomas, eds.), Elsevier, Amsterdam, 1977, p. 413.
54. H. Shirakawa and S. Ikeda, *Polymer J. 2*: 231 (1972).
55. H. Takeuchi, T. Arakawa, Y. Furukawa, and I. Harada, *J. Mol. Struct. 158*: 179 (1987).
56. A. Girlando, A. Painelli, and Z. G. Soos, *J. Chem. Phys. 98*: 7459 (1993).
57. A. Girlando, A. Painelli, and Z. G. Soos, *Synth. Met. 55–57*: 4549 (1993).
58. A. Girlando, A. Painelli, M. Ardoino, and C. Bellitto, *Phys. Rev. B51*: 17338 (1995).
59. Z. G. Soos, J. H. Eggert, R. J. Hemley, M. Hanfland, and H.-K. May, *Chem. Phys. 200*: 23 (1995).
60. A. Girlando, A. Painelli, G. W. Hayden, and Z. G. Soos, *Chem. Phys. 184*: 139 (1994).
61. I. Harada, Y. Furukawa, M. Tasumi, H. Shirakawa, and S. Ikeda, *J. Chem. Phys. 73*: 4746 (1980).
62. H. E. Schaffer, R. H. Friend, and A. J. Heeger, *Phys. Rev. B36*: 7537 (1987).
63. Y. H. Kim and A. J. Heeger, *Phys. Rev. B40*: 8393 (1989).
64. Z. G. Soos, A. Girlando, and A. Painelli, *Mol. Cryst. Liq. Cryst. 256*: 711 (1994).
65. T. Kakitani, *Progr. Theor. Physics 51*: 656 (1974).
66. E. J. Mele, *Mol. Cryst. Liq. Cryst. 77*: 75 (1981).
67. L. Piseri, R. Tubino, and G. Dellepiane, *Solid State Commun. 44*: 1589 (1982).
68. Z. G. Soos, G. W. Hayden, A. Girlando, and A. Painelli, *J. Chem. Phys. 100*: 7144 (1994).
69. P. Piaggio, G. Dellepiane, L. Piseri, R. Tubino, and C. Taliani, *Solid State Commun. 50*: 947 (1984).
70. L. S. Lichtman, E. A. Imhoff, A. Sarhangi, and D. B. Fitchen, *J. Chem. Phys. 81*: 168 (1984).
71. G. Lanzani, S. Luzzati, R. Tubino, and G. Dellepiane, *J. Chem. Phys. 91*: 732 (1989).
72. E. J. Mele and M. J. Rice, *Phys. Rev. Lett. 45*: 926 (1980).
73. G. Zannoni and G. Zerbi, *Chem. Phys. Lett. 87*: 50 (1982).

74. M. Nakahara and K. Maki, *Phys. Rev. B25*: 7789 (1982).
75. J. C. Hicks and G. A. Blaisdell, *Phys. Rev. B31*: 919 (1985).
76. E. J. Mele and J. C. Hicks, *Phys. Rev. B32*: 2703 (1985).
77. J. C. Hicks and E. J. Mele, *Phys. Rev. B34*: 1091 (1986).
78. A. Terai, Y. Ono, and Y. Wada, *J. Phys. Soc. Jpn. 55*: 2889 (1986).
79. J. C. Hicks and J. T. Gammel, *Phys. Rev. B37*: 6315 (1988).
80. S. Xie and L. Mei, *Phys. Rev. B47*: 14905 (1993).
81. K. A. Chao and Y. Wang, *J. Phys. C18*: L1127 (1985).
82. Y. Mori and S. Kurihara, *Synth. Met. 22*: 219 (1987); *24*: 357 (1988).
83. H. D. Viller, M. Dupuis, and E. Clementi, *J. Chem. Phys. 88*: 5252 (1988).
84. H. D. Viller, M. Dupuis, and E. Clementi, *Phys. Rev. B37*: 2520 (1988).
85. M. Rumi, A. Kiel, and G. Zerbi, *Chem. Phys. Lett. 231*: 70 (1994).
86. D. C. Hanna, M. A. Yuratich, and D. Cotter, *Nonlinear Optics of Free Atoms and Molecules*. Springer-Verlag, Berlin, 1979.
87. Y. R. Shen, *The Principles of Nonlinear Optics*, Wiley, New York, 1984.
88. B. J. Orr and J. F. Ward, *Mol. Phys. 20*: 513 (1971).
89. P. W. Langhoff, S. T. Epstein, and M. Karplus, *Rev. Mod. Phys. 44*: 602 (1972).
90. Z. G. Soos and D. Mukhopadhyay, *J. Chem. Phys. 101*: 5515 (1994).
91. D. N. Batchelder and D. Bloor, *J. Phys. C: Solid State Phys. 15*: 3005 (1982).
92. A. R. Mantini, M. P. Marzochi, and G. Smulevich, *J. Chem. Phys. 91*: 95 (1989).
93. C. Manneback, *Physica 17*: 1001 (1951).
94. P. M. Morse and H. Feshback, *Methods of Theoretical Physics*, McGraw-Hill, New York, 1953, p. 786.
95. D. Mukhopadhyay and Z. G. Soos, Vibronic analysis of NLO spectra of PDA crystals and films, in *Optical and Photonic Applications of Electroactive and Conducting Polymers* (S. C. Yang and P. Chandrasekhar, eds.), SPIE Proc. Ser. Vol. 2528, SPIE San Diego, 1995, p. 116.
96. P. C. M. McWilliams, G. W. Hayden, and Z. G. Soos, *Phys. Rev. B43*: 9777 (1991).
97. S. Etemad and Z. G. Soos, Non-linear optical spectroscopy of conjugated polymers, in *Spectroscopy of Advanced Materials* (R. J. H. Clark and R. E. Hester, eds.), Adv. Spectrosc., Vol. 19, Wiley, New York, 1991, p. 87.
98. M. J. Nowak, G. J. Blanchard, G. L. Baker, S. Etemad, and Z. G. Soos, Exciton relaxation in PDA-4BCMU: from crystals to films, in *Conjugated Polymeric Materials: Opportunities in Electronics, Optoelectronics, and Molecular Electronics* (J. L. Brédas and R. R. Chance, eds.), NATO ASI Ser. E, Vol. 182, Kluwer, Dordrecht, The Netherlands, 1989, p. 421.
99. D. J. Sandman and Y. J. Chen, *Polymer 30*: 1027 (1989).
100. M. Cha, W. E. Torruellas, G. I. Stegeman, H. X. Wang, A. Takahashi, and S. Mukamel, *Chem. Phys. Lett. 228*: 73 (1994).
101. J. B. van Beek, F. Kajzar, and A. C. Albrecht, *J. Chem. Phys. 95*: 6400 (1991).
102. J. B. van Beek, F. Kajzar, and A. C. Albrecht, *Chem. Phys. 161*: 299 (1992).
103. B. Lawrence, W. E. Torruellas, M. Cha, M. L. Sundheimer, G. I. Stegeman, J. Meth, S. Etemad, and G. L. Baker, *Phys. Rev. Lett. 73*: 597 (1994).
104. M. F. Granville, G. R. Holtom, and B. E. Kohler, *J. Chem. Phys. 72*: 4671 (1980).
105. D. Mukhopadhyay and Z. G. Soos, *J. Chem. Phys. 104*: 1600 (1996).
106. A. Horvath, G. Weiser, G. L. Baker, and S. Etemad, *Phys. Rev. B51*: 2751 (1995).
107. S. D. Phillips, R. Worland, G. Yu, T. Hagler, R. Freedman, Y. Cao, V. Yoon, J. Chiang, W. C. Walker, and A. J. Heeger, *Phys. Rev. B40*: 9751 (1989).
108. S. Jeglinski and Z. V. Vardeny, *Synth. Met. 49–50*: 509 (1992).
109. D. E. Aspnes and J. E. Rowe, *Phys. Rev. B5*: 4022 (1972).
110. D. Aspnes and N. Bottka, Electric-field effects on the dielectric function of semiconductors and insulators, in *Semiconductors and Semimetals*, Vol. 9 (R. K. Willardson and A. C. Beer, eds.), Academic, New York, 1972, p. 457.
111. G. Weiser, *Phys. Rev. B45*: 14076 (1992).
112. D. Bloor and F. H. Preston, *Phys. Stat. Sol. 39(a)*: 607 (1977).
113. V. Enkelmann, *Acta Cryst. B33*: 2842 (1977).
114. Z. G. Soos and S. Ramasesha, *J. Chem. Phys. 90*: 1067 (1989).
115. Z. G. Soos, D. Mukhopadhyay, and M. H. Hennessy, *Chem. Phys. 210*: 249 (1996).
116. D. Mukhopadhyay, G. W. Hayden, and Z. G. Soos, *Phys. Rev. B51*: 9476 (1995).
117. O. J. Heilmann and E. H. Lieb, *Trans. N. Y. Acad. Sci. 33*: 116 (1971).
118. S. R. Bondeson and Z. G. Soos, *J. Chem. Phys. 71*: 3807 (1979).
119. A. D. McLachlan, *Mol. Phys. 2*: 276 (1959).
120. S. Mazumdar and Z. G. Soos, *Synth. Met. 1*: 77 (1979).
121. Z. G. Soos and S. Ramasesha, Diagrammatic valence bond theory, in *Valence Bond Theory and Chemical Structure* (D. J. Klein and N. Trinajstic, eds.), Elsevier, Amsterdam, 1990, p. 81.
122. S. Ramasesha and Z. G. Soos, *J. Chem. Phys. 98*: 4015 (1992).
123. L. Pauling, *J. Chem. Phys. 1*: 280 (1933).
124. L. Pauling and G. W. Wheland, *J. Chem. Phys. 1*: 362 (1933).

125. J. C. Bonner and M. E. Fisher, *Phys. Rev. 135*: A640 (1964).
126. E. H. Lieb and F. Y. Wu, *Phys. Rev. Lett. 20*: 1445 (1968).
127. A. A. Ovchinnikov, *Soviet Phys. JETP 30*: 1100 (1970).
128. S. R. Bondeson and Z. G. Soos, *Phys. Rev. B22*: 1793 (1980).
129. Z. G. Soos and S. R. Bondeson, Magnetic resonance in ion-radical organic solids, in *Extended Linear Chain Compounds*, Vol 3 (J. S. Miller, ed.), Plenum, New York, 1983, p. 193 and references therein.
130. M. Takahashi, *Progr. Theor. Phys. 42*: 1098 (1970); *43*: 1619 (1970).
131. W. E. Hatfield, W. E. Estes, W. E. Marsh, M. W. Pickens, L. W. ter Haar, and R. R. Weller, The synthesis and static magnetic properties of first-row transition-metal compounds with chain structures, in *Extended Linear Chain Compounds*, Vol. 3 (J. S. Miller, ed.), Plenum, New York, 1983, p. 43 and references therein.
132. P. L. Nordio, Z. G. Soos, and H. M. McConnell, *Ann. Rev. Phys. Chem. 17*: 237 (1966).
133. Z. G. Soos, S. Ramasesha, and D. S. Galvão, *Phys. Rev. Lett. 71*: 1609 (1993).
134. D. Yaron and R. Silbey, *Phys. Rev. B45*: 11655 (1992).
135. Z. G. Soos, D. S. Galvão, S. Etemad, and R. G. Kepler, Conjugated polymer fluorescence: interplay of correlations and alternation, in *Electrical, Optical, and Magnetic Properties of Organic Solid State Materials* (A. F. Garito, A. K. Y. Chen, C. Y. C. Lee, and L. R. Dalton, eds.), MRS Symp. Proc. 328, Boston, 1994, p. 383.
136. B. E. Kohler, C. Spangler, and C. Westerfield, *J. Chem. Phys. 89*: 5422 (1988).
137. B. E. Kohler and D. E. Schilke, *J. Chem. Phys. 86*: 5214 (1987).
138. R. G. Kepler and Z. G. Soos, Electronic properties of polysilanes: excitations of σ-conjugated chains, in *Relaxation in Polymers* (T. Kobayashi, ed.), World Scientific, Singapore, 1993, p. 100.
139. Z. G. Soos, S. Ramasesha, D. S. Galvão, and S. Etemad, *Phys. Rev. B47*: 1742 (1993).
140. Z. G. Soos and D. S. Galvão, *J. Phys. Chem. 98*: 1029 (1994).
141. D. Guo and S. Mazumdar, *J. Chem. Phys. 97*: 2170 (1992).
142. P. C. M. McWilliams, Z. G. Soos, and G. W. Hayden, *J. Chem. Phys. 97*: 2172 (1992).
143. Z. G. Soos and K. S. Schweizer, Optical spectra of flexible conjugated polymers, in *Nonlinear Optical and Electroactive Polymers* (P. N. Prasad and D. R. Ulrich, eds.), Plenum, New York, 1988, p. 331.
144. S. Mazumdar and Z. G. Soos, *Phys. Rev. B23*: 2810 (1981).
145. Z. G. Soos and S. Ramasesha, *Phys. Rev. B29*: 5410 (1984).
146. D. Guo, S. Mazumdar, and S. N. Dixit, *Synth. Met. 49*: 1 (1992).
147. D. Guo, S. Mazumdar, and S. N. Dixit, *Nonlinear Opt. 6*: 337 (1994).
148. A. Painelli and A. Girlando, *Phys. Rev. B48*: 10683 (1993).
149. S. Ramasesha and Z. G. Soos, *Chem. Phys. 91*: 35 (1984).
150. S. Ramasesha, D. S. Galvão, and Z. G. Soos, *J. Phys. Chem. 97*: 2823 (1993).
151. G. Louarn, J. P. Buisson, S. Lefrant, and D. Fichou, *J. Phys. Chem. 99*: 11399 (1995).
152. Z. G. Soos, S. Ramasesha, D. S. Galvão, R. G. Kepler, and S. Etemad, *Synth. Met. 54*: 35 (1993).

7
Synthesis of Polyacetylene

Hideki Shirakawa
University of Tsukuba, Tsukuba, Ibaraki, Japan

I. INTRODUCTION

The term "polyacetylene" contains some ambiguity in that it is used not only for the name of the polymer of acetylene [—(CH=CH)$_n$—] but also as a generic term for compounds that contain more than two acetylenic bonds [—C≡C—]. Nevertheless, the former use is widely accepted in the field of polymer science. The synthesis of this polymer is the subject of this chapter.

Polyacetylene is the simplest organic polymer. Its interesting electronic properties and the ability to form highly conducting derivatives have stimulated great interest among many scientists in various fields. The two major developments in the birth of conducting organic polymers with metallic properties were the discovery by Ito et al. [1] of a synthetic route to high quality flexible films of polyacetylene and the discovery that the electrical conductivity of the films could be increased by doping them from the semiconducting to the metallic regime [2,3]. Doping has generated a great deal of interest in the fields of polymer science, condensed matter physics, and others. Since its discovery a huge number of papers have appeared dealing with the various physical, chemical, and structural properties of polyacetylene and giving a fair insight into the whole topic of conductivity in organic polymers.

A large amount of work has been devoted to the investigation of synthesis, characterization, and properties, and these results are summarized in a specialized textbook on polyacetylene by Chien [4], several textbooks [5–8], a handbook [9], and many review articles and conference proceedings [see Ref. 10] on conjugated polymers. Nevertheless, polyacetylene appears to be a quite poorly characterized material from the viewpoint of material chemistry. The most serious problem is that it is insoluble in all solvents tested so far and infusible. Thus, fundamental quantities such as molecular weight and its distribution and details of chemical structures have not been determined and their influence on the polymer's physical properties have not been evaluated. The second problem is its instability under ambient conditions. Polyacetylene is an aliphatic conjugated polymer whose double bonds are reactive to oxygen and halogens via electrophilic addition reaction. A trace amount of electrophilic reagent may cause serious damage for the long conjugation system because the addition reaction shortens the long conjugation, which is consistent with the fact that in recent years much effort devoted to improving the stability of polyacetylene was unsuccessful. Many applications have been proposed and intensive work has been done, for example, to develop a rechargeable battery system [11]; however, reactivity limits its usefulness for possible applications. In contrast to a recent decrease in the application studies of polyacetylene due to its poor stability, a large number of publications have focused on the synthesis of high quality oriented films, characterization, and various properties of polyacetylene.

The electrical conductivity of polyacetylene has been much improved since 1960 as shown in Fig. 7.1 [12]. Data reported by earlier investigators were obtained with nondoped compressed pellets prepared from the intractable black powder [13–15]; therefore, the values were between 10^{-5} and 10^{-13} S/cm depending on the processing and manipulation of the specimens. The conductivities of as-prepared free-standing films were in the same range as those of the compressed pellets, and it become clear that values changed almost linearly with *cis-trans* content. For example, the values of 76 and 15% *trans* films were 1.3×10^{-4} and 6.0×10^{-9} S/cm, respectively [16]. At the first report of doping in 1977, the values were 0.5 and 38 S/cm for films doped with bromine and iodine [2], respectively, and 560 S/cm for films doped with AsF$_5$ [3].

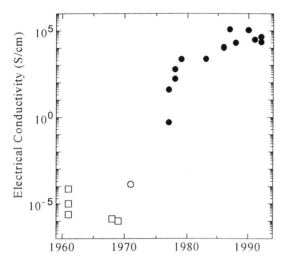

Fig. 7.1 Advances in the electrical conductivity of polyacetylene achieved since 1960. (□) Undoped compressed pellets; (○) undoped films; (●) doped films. (From Ref. 12.)

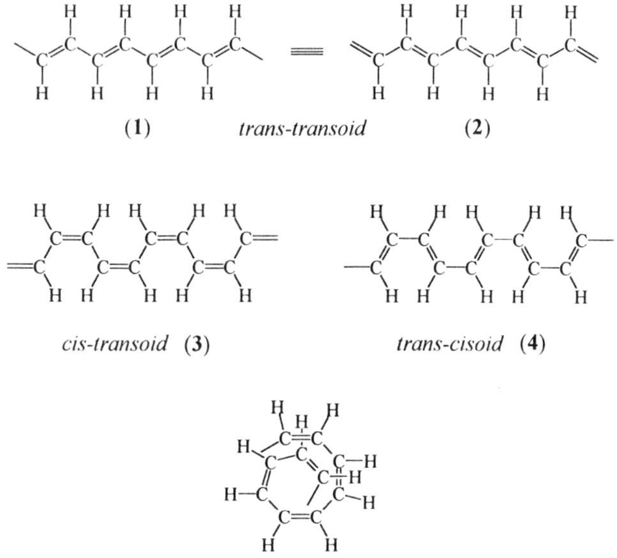

Fig. 7.2 Possible molecular structures of polyacetylene.

Continuing efforts by many investigators have been directed to the synthesis of higher quality films and oriented films. The most primitive method for orientation was a tensile stretching of as-prepared *cis*-rich films [17]. Partial orientation of the films led to a significant improvement in the electrical conductivity and to the introduction of electrical and optical anisotropy. With an elongation $l/l_0 \cong 3$, the room temperature conductivity of AsF$_5$-doped film was in excess of 2×10^3 S/cm [18]. In 1987, Naarmann and Theophilou [19] succeeded in preparing highly conducting films whose conductivity was more than 10^5 S/cm by iodine doping. Tsukamoto and coworkers [20] obtained high quality films with 10^5 S/cm after iodine doping by using a modification of the Naarmann and Theophilou method. These modifications suggest that specimens of polyacetylene so far prepared are far from ideally perfect but that many more modifications in its synthesis are still possible.

II. MOLECULAR STRUCTURES

Polyacetylene is often denoted simply by (CH)$_x$, its structural repeating unit contains only two atoms, one hydrogen and one carbon. Its chemical repeating unit is vinylene [—(CH=CH)—], and therefore two configurations, *cis* (Z) and *trans* (E), are possible for each double bond. Including two conformational isomers for single bonds, *transoid* and *cisoid*, four different isomeric structures are possible, as shown in Fig. 7.2. Among these configurations, structures **1** and **2** are the same as the *trans-transoid* form or simply the *trans* isomer. The *cis-cisoid* form **5** cannot be realized owing to large steric hindrance.

The comparison of calculated vibrations for copolymers of CH≡CH and CD≡CD with the observed infrared spectra led to the result that the structure of vinylene unit agreed with *cis*—(CD=CD)—, suggesting that *cis*-polyacetylene consists of the *cis-transoid* form (**3**)[21], consistent with the results of a mutation NMR study by Yannoni and Clarke [22]. They obtained the carbon–carbon bond lengths in polyacetylene with use of nutation NMR spectroscopy using copolymer samples with doubly ^{13}C-enriched acetylene [^{13}CH≡^{13}CH] and doubly ^{13}C-depleted acetylene [^{12}CH≡^{12}CH]. The mutation NMR spectroscopy is a method specifically designed to measure interatomic distances in amorphous solids based on the splitting due to magnetic dipole–dipole coupling in a ^{13}C NMR spectrum. In *trans*-polyacetylene, values of 1.36 and 1.44 Å were found for the double and single bond lengths, respectively. On the other hand, in *cis*-polyacetylene only one bond distance, 1.37 Å, was observed for the double bond, which strongly suggests that the polymerization mechanism of acetylene by the Ziegler–Natta catalyst leaves the original carbon pair double-bonded in the resulting polymer as shown in Fig. 7.3, in good agreement with an IR study of copolymers of acetylene and acetylene-d_2 [21]. In other words, in the polymerization of acetylene each propagation step is a cis addition of acetylene to a metal–carbon bond and forms the *cis-transoid* configuration [23]. Thus, the *trans-cisoid* form (**4**) could be ruled out by the IR and nutation NMR studies. The values of the double and single-bond lengths as deduced from the two-dimensional NMR experiment [24] were 1.38 ± 0.01 and 1.45 ± 0.01 Å, respectively, for *trans*-polyacetylene

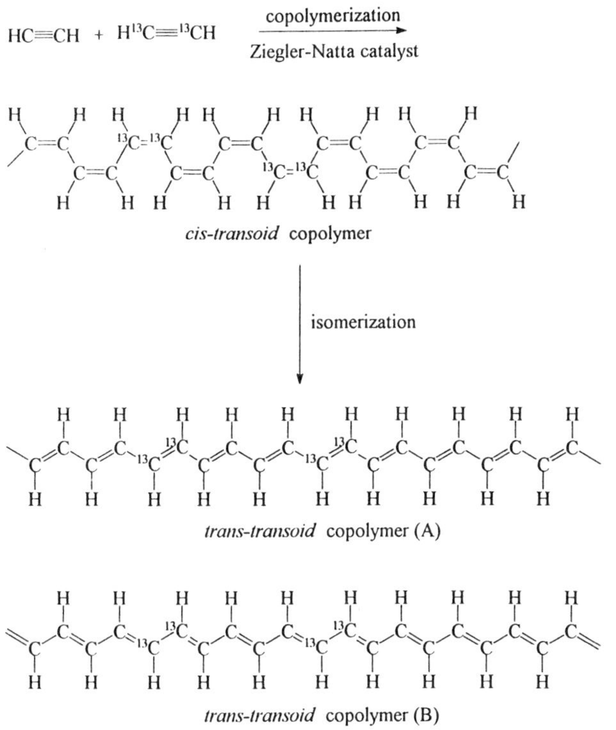

HC≡CH + H¹³C≡¹³CH $\xrightarrow[\text{Ziegler-Natta catalyst}]{\text{copolymerization}}$

cis-transoid copolymer

↓ isomerization

trans-transoid copolymer (A)

trans-transoid copolymer (B)

Fig. 7.3 Position of a given monomer unit in the conjugated sequence of polyacetylene.

cis-transoid copolymer

↓ isomerization

trans-transoid (2)   *trans-transoid* (1)

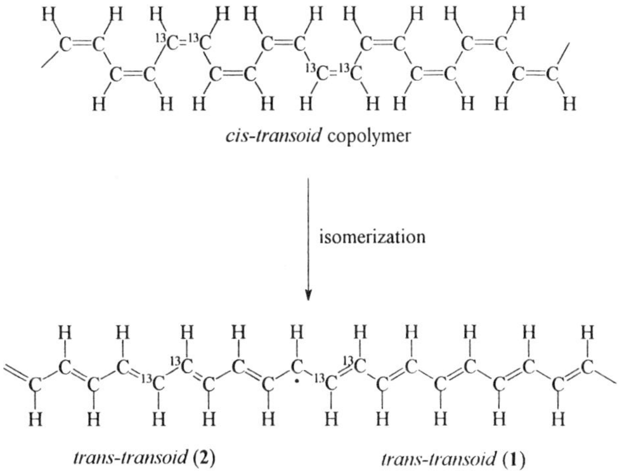

Fig. 7.4 Isomerization reaction of *cis*-polyacetylene causes the formation of a free radical and two different vinylene units.

and were in good agreement with the results of Yannoni and Clarke [22].

The reason only the double bond length was determined by nutation NMR spectroscopy in the *cis*-polyacetylene is that the energy of the *trans-cisoid* form (**4**) is slightly higher than that of the *cis-transoid* form (**3**) [25] and only *cis* ¹³C=¹³C double bonds are formed because no double-bond shift reaction is involved in the polymerization reaction. On the other hand, the *trans* isomers **1** and **2** are energetically equal and have lower energy than the *cis-transoid* form. Therefore, during the thermal *cis–trans* isomerization, a double-bond shift and some interchain reaction may produce free radicals or unpaired spins that can be stabilized between the two conjugated sequences in the linear molecule [26], as shown in Fig. 7.4. As a result, two different bonds, ¹³C=¹³C and ¹³C—¹³C, exist in *trans*-polyacetylene.

Pople and Walmsley considered that the energy required to form a pair of defects in the linear polyene molecule [H(CH=CH)$_n$H] is relatively low and should lead to a significant thermal population. They estimated the number of unpaired spins at about one per 70 carbon atoms at room temperature [26]. The observed values were much lower, 3.0×10^{19} and 2.2×10^{18} spin/g, for 97% *trans*- and 93% *cis*-polyacetylene, respectively, or one spin per 3000 and one per 10,000 carbon atoms, re-

spectively. The unpaired spin density in *cis*-polyacetylene is substantially smaller. When acetylene was polymerized at −78°C directly in an ESR sample tube, no resonance corresponding to the unpaired spin was found in the reaction mixture [27,28], suggesting strongly that the nascent *cis*-polyacetylene has no structural defect.

III. SYNTHESIS OF POLYACETYLENE

Although polyacetylene was first synthesized as early as 1958 by Natta et al. [29], it remained for some time a material of limited interest to organic chemists [30,31], polymer chemists [13–15,21,32], and theoreticians [33–35], who saw it as an infinitely long conjugated molecule in which one would expect the one-dimensional π electrons to form a half-filled band leading to metallic behavior [36,37].

A variety of routes have been proposed for the synthesis of polyacetylene. These can be classified into four categories as summarized in Table 7.1. The first is via the catalytic polymerization of acetylene. The second is noncatalytic polymerization. So far, spontaneous polymerization of acetylene has been reported under high pressure. The third type of route is the catalytic polymerization of monomers other than acetylene. The fourth is a so-called precursor route in which linear conjugated polyene chains are formed either by decomposition or by isomerization of soluble precursor polymers.

A. Polymerization of Acetylene by Ziegler–Natta Catalysts

The polymerization reactions of acetylene that produce not only a polymeric material, cuprene, but also oligo-

Table 7.1 Synthetic Routes for Preparing Polyacetylene

Method	References
1. Catalytic polymerization of acetylene monomer	
Ziegler–Natta catalysts	1, 12, 17, 19, 20, 29
Single-component catalysts	38–41
Luttinger catalysts	30, 42–46
Metathesis catalysts	47–50
Rh and Re catalysts	51, 52
Electrochemical method	53
Metal oxide surfaces	54–56
AsF_5	57, 58
2. Noncatalytic polymerization of acetylene monomer	
Polymerization of acetylene under high pressure	59–62
Radiation-induced polymerization	63
3. Catalytic polymerization of monomers other than acetylene	
Polymerization of cyclooctatetraene	64, 65
4. Indirect methods (precursor methods)	
Durham method	66–74
Isomerization of polybenzvalene	75
Dehydrochlorination of poly(vinyl chloride)	76
Dehydration of poly(vinyl alcohol)	77, 78

mers such as benzene, cyclooctatetraene, and vinyl acetylene have been known for many years [79]. Unlike the polymerization of ethylene and α-olefins, the polymerization of acetylene by organometallic catalysts is characterized by the simultaneous formation of cyclic oligomers. Among them, the cyclic trimer benzene is the most popular by-product. Generally speaking, a wide variety of Ziegler–Natta, Luttinger, and metathesis catalysts are effective in the polymerization of acetylene to form polyacetylene, but some catalysts produce predominantly benzene, which is a cyclic trimer of acetylene. For example, the main product of acetylene polymerization by tris(acetylacetonato)titanium(III) and diethylaluminum chloride [Ti(acac)$_3$—(C$_2$H$_5$)$_2$AlCl] catalyst is benzene (~80%) and gives a small amount of polymer [80]. Thus, it is important to select a catalyst system that produces polyacetylene with high selectivity. Another important factor required for the catalysts used in the synthesis of high quality polyacetylene is their solubility in organic solvents. The use of soluble catalysts is recommended, because the removal of residual catalyst enclosed in the polymer is substantially impossible when a heterogeneous catalyst is employed. Since polyacetylene is insoluble in organic solvents, the reprecipitation method, which is the most popular purification process for soluble polymers, is not applicable.

Among Ziegler–Natta catalysts derived from the transition metals of groups 4–8* in combination with organometallic compounds of groups 1–3, soluble catalysts used for polyacetylene synthesis are listed in Table 7.2.

Among these catalysts, the preferred one is the combination of Ti(O-n-C$_4$H$_9$)$_4$ and (C$_2$H$_5$)$_3$Al because it is soluble in organic solvents and highly active for acetylene polymerization to produce crystalline polyacetylene in the form of mechanically strong free-standing films and each component of the catalyst is commercially available. The synthetic process for producing the films was developed by Shirakawa et al. [1], who employed a very high concentration ([Ti] ≃ 0.3 mol/L) of the Ti(O—n—C$_4$H$_9$)$_4$ (C$_2$H$_5$)$_3$Al catalyst and allowed acetylene to polymerize on the free surface of the catalyst solution or on the wall of the reaction flask on which the catalyst solution is coated.

The minimum concentration of the catalyst for the synthesis of polyacetylene in the form of film was estimated to be ~3 mmol/L of Ti(O—n—C$_4$H$_9$)$_4$ in the case of Al/Ti = 4, a polymerization temperature at −78°C,

---

* The periodic group notation used here is in accord with actions of the IUPAC nomenclature committees.

Table 7.2 Soluble Catalysts Used for Polyacetylene Synthesis

Transition metal compound	Alkyl metal	Ref.
$Ti(OC_3H_7)_4$	$(C_2H_5)_3Al$	29
$Ti(OC_4H_9)_4$	$(C_6H_{13})_3Al$	29
$Ti(OC_3H_7)_4$	$C_5H_{11}Li$	29
$Ti(OC_4H_9)_4$	$(C_2H_5)_3Al$	13
$TiO(acac)_2$	$(C_2H_5)_3Al$	81
$VO(acac)_2$	$(C_2H_5)_3Al$	81
$Cr(acac)_3$	$(C_2H_5)_3Al$	81
$Fe(acac)_3$	$(C_2H_5)_3Al$	81
$VO(acac)_2$	$(C_2H_5)_3Al$	82
$Cr(acac)_3$	$(C_2H_5)_3Al$	82
$Co(acac)_3$	$(C_2H_5)_3Al$	82
$Ti(OC_4H_9)_4$	$(C_2H_5)_3Al$	1
$M(acac)_3$[a]	$(C_2H_5)_3Al$	1
$Ti(CH_2C_6H_5)_4$	None	48
$M(naph)_3$[b]	$(C_2H_5)_3Al$	83
$Ti(OC_4H_9)_4$	R_2Al-O-AlR_2[c]	84
$Ti(OC_4H_9)_4$	$(C_2H_5)_2Al$-S-$Al(C_2H_5)_2$	84
$Ti(OC_4H_9)_4$	$(C_2H_5)_2Al$-$N(n$-$C_4H_9)$-$Al(C_2H_5)_2$	84
$Ti(OC_4H_9)_4$	C_2H_5MgBr	85
$Ti(OC_4H_9)_4$	C_6H_5MgBr	
$Ti(OC_4H_9)_4$	n-C_4H_9Li	86[d]
$(C_5H_4)(C_5H_5)_3Ti_2$(Ti-Ti)[e]	None	87
$Cp_2Ti(PMe_3)_2$[f]	None	38, 39
$Cp_2M(PMe_3)_2$[g]	None	40
$Ti(OC_4H_9)_4$	R_3Al[h]	88
$Ti(OC_2H_5)_4$	R'_3Al[i]	88
$Ti(OR)_4$[j]	$(C_2H_5)_3Al$	88
$Ti(OR)_4$[k]	$(C_6H_{13})_3Al$	89
$Ti(OR)_4$	$(C_8H_{17})_3Al$	89
$Ti(OR)_4$	$(C_{10}H_{21})_3Al$	89
$V(acac)_3$	R_3Al[l]	90
$V(mmh)_3$[m]	R_3Al	90
$Ti(OC_2H_5)_4$	$(CH_3)_3Al$	91
$Ti(O$-n-$C_4H_9)_4$	$(C_2H_5)_3Al$	92–94[n]
$Ti(O$-n-$C_4H_9)_4$	$(C_2H_5)_3Al$	95–97[o]

[a] M = Ti, V, Fe, Cr, and Co; acac = 2,4-pentanedionato (acetylacetonato).
[b] M = Y, La, Pr, Nd, Gd, Tb, and Dy; naph = naphthenate.
[c] R = i-C_4H_9, n-C_6H_{13}.
[d] Silicone oil was used as a solvent.
[e] μ-$(\eta^1:\eta^5$-Cyclopentadienyl)-tris(η-cyclopentadienyl)dititanium (Ti-Ti).
[f] Cp = η^5-cyclopentadienyl; Me = CH_3.
[g] M = Ti, Zr; Me = CH_3.
[h] R = C_2H_5, C_6H_{13}, C_8H_{17}, $C_{10}H_{21}$.
[i] R = C_6H_{13}, C_8H_{17}.
[j] OR = OC_4H_9, OC_6H_{13}, OC_8H_{17}.
[k] OR = OC_6H_{13}, OC_8H_{17}, $OC_{10}H_{21}$.
[l] R = C_2H_5, n-C_3H_7, i-C_4H_9.
[m] mmh = 2-methyl-2,3-butanedionato.
[n] No solvent was used.
[o] Solvent evacuation method.

and acetylene pressure of ~700 torr [17]. The critical concentration depends on the catalytic activity and the preparative conditions such as the Al/Ti ratio, polymerization temperature, acetylene pressure, and aging conditions. When the concentration is extremely low, the introduction of acetylene gas causes no formation of polyacetylene on the surface of the catalyst solution, but the pale yellow solution of the catalyst gradually develops an intense color depending on the polymerization temperature, red at $-78°C$ and purple at room temperature, suggesting formation of the polymer. The reaction mixture appears to be a homogeneous solution for a while, but soon precipitation of the polymer occurs to form a flock or a powder of polyacetylene. The period depends on the catalyst concentration. The higher the concentration, the shorter the period. If a drop of the suspension is spread on a glass plate and the solvent is evaporated, the precipitate gives a very thin shiny metallic film with a golden tint. When the solvent is replaced by benzene and the benzene suspension is allowed to freeze-dry, a fluffy powder of polyacetylene can be obtained.

With increasing catalyst concentration, the polymerization occurs on the solution surface to form a thick swollen mass or a gel that grows gradually toward the bottom until the whole solution becomes a gel. When a piece of the gel is dried, it shrinks to give a black solid whose bulk density is in the range 0.1–0.5 g/cm^3. It should be noted that the solid does not swell again in any solvent once the gel is allowed to dry. When a piece of the gel is squeezed between filter papers and then pressed between mirror-polished metal plates, a film can be prepared whose surfaces are both smooth and shiny. The gel can be freeze-dried to give a porous material whose dimensions are essentially the same as those of the original gel. Thus bulk density of the polyacetylene foam is very low, for example, less than 0.1 g/cm^3, depending on the catalyst concentration. The foam can be pressed between the mirror-polished metal plates to form a high quality film.

When the catalyst concentration is sufficiently high, a film forms immediately on the surface of the catalyst solution. If its bulk density is great enough, then the diffusion of acetylene gas into the catalyst solution through the film formed becomes the rate-determining step for the polymerization. Thus, the use of a highly concentrated catalyst solution is recommended for synthesis of high quality films; however, an extremely concentrated or highly active catalyst solution results in the formation of very thin films of high density.

Aldissi and coworkers [98] demonstrated this situation with a set of three experiments. In the first, polymerization was attained by a nonsolvent catalyst consisting of 2.9 mmol of tetrabutoxytitanium and 11.8 mmol of triethylaluminum, and its concentration was the maximum ([Ti] = 1.12 mol/L, Al/Ti = 4). After aging at room temperature for 1 h, the polymerization was carried out on the quiescent surface of the catalyst solution at $-78°C$ under an initial pressure of 690 torr. Rapid consumption of acetylene occurred in the initial few minutes, and substantially no consumption of acetylene was observed after 25 min. This polymerization behavior is in contrast to that seen with catalysts at lower concentrations. In a second experiment, Aldissi et al. [98] polymerized acetylene-d_2 on the quiescent surface of a standard catalyst solution ([Ti] = 0.02 mol/L, Al/Ti = 4) at $-78°C$. After a film formed on the surface, the polymerization was interrupted by evacuating the system and then acetylene-h_2 was added into the reactor. The consumption of acetylene-h_2 suggested that a fraction of the active catalyst was included in the film formed or that acetylene gas was able to diffuse through the film to reach the catalyst layer underneath. Infrared spectroscopic study of the cross section of the film revealed that the side of the film facing the acetylene gas was a mixture of poly(acetylene-h_2) and poly(acetylene-d_2), whereas the side facing the catalyst solution was poly(acetylene-h_2), suggesting that an additional polymerization of acetylene-h_2 took place in and at the boundary of the poly(acetylene-d_2) film and in the catalyst solution. In the third experiment, the morphology of the film was observed from both sides with a scanning electron microscope (SEM), and the cross section of the film was observed by microprobe analysis. The morphology of the gas-side surface was quite different from that of the other surface. Thus, the SEM observations revealed that the fibrils are flattened to 400 Å and packed more tightly than on the side facing the catalyst solution, in good agreement with the carbon profile of the cross section of the film obtained by microprobe analysis.

B. Polymerization of Acetylene with Luttinger Catalysts

Luttinger catalysts consist of a hydridic reducing agent such as sodium borohydride plus a salt or complex of a group 8 metal such as nickel chloride. The major reaction product in experiments carried out with these catalysts was a high molecular weight polymer of acetylene, and there was no indication of the formation of benzene or cyclooctatetraene [30]. Daniels [31] reported that nickel halide–tertiary phosphine complexes themselves are effective catalysts for the polymerization of acetylene. The catalytic activity is apparently specific for nickel-phosphine complexes, cobalt and palladium complexes being completely inactive, as was a nickel-phosphine oxide complex. Bis(triphenylphosphine)nickel chloride was not a very active catalyst, whereas both its bromide and bis(tri-n-butylphosphine)nickel bromide were good catalysts, as was bis(tirphenylphosphine) nickel iodide.

It is important to note that a hydrophilic solvent such as ethanol, THF, or acetonitrile, and even water, can be used as a solvent for the Luttinger catalysts, in contrast to the requirement of rigorously dehydrated hydrocarbon solvents for the Ziegler–Natta and metathesis catalysts. Thus, the Luttinger catalysts have an advantage over the Ziegler–Natta catalysts in that they are stable in air and do not require equipment and handling with the rigorous exclusion of oxygen and moisture.

Generally the catalytic activity of the Luttinger catalysts is much lower than that of the standard Ziegler–Natta catalyst, [Ti(O—n—C_4H_9)$_4$—(C_2H_5)$_3$Al]. Therefore, special cautions are needed to synthesize high quality films. Thus, the formation of a thin film is difficult because the diffusion of acetylene into the catalyst solution precedes the polymerization reaction, resulting in the formation of a gelatinous mass throughout the solution. After washing the gel thoroughly, the gelatinous mass can be dried on a Teflon sheet under vacuum to give a high quality film. Lieser and coworkers [99] prepared a thin film by dipping a microscope slide into a catalyst solution and then rapidly transferring it into an atmosphere of acetylene. Once it is exposed to acetylene, polymerization starts in the thin liquid layer of the catalyst solution on the glass substrate within a few seconds at −30°C. Lieser et al. [99] prepared the Luttinger catalyst by adding 1 mL of a 1.0 wt% ethanol solution of Co(NO$_3$)$_2$ to a solution of 20 mg of NaBH$_4$ dissolved in 50 mL of ethanol and 50 mL of diethyl ether at −80°C.

The cis-rich polyacetylenes prepared from the standard Ziegler–Natta catalyst and the Luttinger catalyst did not differ with regard to the overall morphology and other properties, but Lieser et al. found a large difference with regard to polymer reactions. It is well known that polyacetylene reacts with chlorine and bromine to give halogenated polyacetylene, for example, [—(CHCl—CHCl)$_n$—]. They demonstrated that the cis-rich polyacetylene prepared by the Luttinger catalyst can react with Cl$_2$ at 0°C to give a completely soluble chlorinated polyacetylene whose molecular mass was generally on the order of (2–5) × 10^4 daltons, corresponding to 200–500 vinylene units per polyacetylene molecule. The polyacetylenes prepared with the Ziegler–Natta catalysts, on the other hand, could not be solubilized upon chlorination, suggesting that the materials were cross-linked during the polymerization process.

Kobryanskii [46,52] investigated acetylene polymerization with the Co(NO$_3$)$_2$—NaBH$_4$ catalyst ([Co] = 2.15 × 10^{-4} mol/L and Co(NO$_3$)$_2$/NaBH$_4$ = 1/4) in ethanol solution containing 1.5–2.8 wt % of polyvinylbutyral at −50 to +20°C. He claimed that a soluble polyacetylene-polyvinylbutyral composition was produced at polyvinylbutyral concentrations greater than 1.5 wt%. The polymerization was also carried out with a new catalyst system based on binuclear rhenium compounds with NaBH$_4$ or KBH$_4$ as a reducing reagent.

C. Polymerization of Acetylene with Metathesis Catalysts

Aldissi and coworkers demonstrated the formation of uniform films on the quiescent surface of a concentrated solution of a soluble catalyst prepared by mixing equimolar amounts of WCl$_6$ and (C$_6$H$_5$)$_4$Sn in toluene. A combination of MoCl$_5$ with (C$_6$H$_5$)$_4$Sn also gave a soluble catalyst, but the catalytic activity was too low to give a uniform film [48]. A more active tungsten catalyst was prepared by Theophilou et al. [49] who used n-BuLi as a co-catalyst instead of (C$_6$H$_5$)$_4$Sn. In a typical experiment, 5 mmol of n-BuLi was added into the reaction vessel, which contained a solution of 5 mmol of WCl$_6$ in 5 mL of toluene. After the resultant dark red-brown mixture was stirred for 22 min at room temperature, acetylene was introduced at a pressure of 670 torr. The product was obtained as a black powder. The as-prepared powder is particularly interesting, for it is doped by WCl$_6$ as indicated by the three doping-induced bands at 1400, 1290, and a broad band around 800 cm^{-1} in the IR spectrum. The relative activities for acetylene polymerization by the metathesis catalysts were in the following order:

$$Ti(O—C_4H_9)_4—(C_2H_5)_3Al \gg WCl_6—n—BuLi$$
$$> WCl_6—(C_6H_5)_4Sn > MoCl_5—(C_6H_5)_4Sn$$

D. Polymerization of Acetylene with Single-Component Catalysts

In general, Ziegler–Natta, Luttinger, and metathesis catalysts are composed of a main catalyst (transition metal compound) and a co-catalyst. However, several metal complexes alone are reported to be active for acetylene polymerization as shown in Table 7.2. Although the catalytic activity for acetylene polymerization is very small, tetrabenzyltitanium is one of them. Hsu et al. [87] synthesized polyacetylene film and bulk samples by polymerizing acetylene in the presence of μ-(η^1:η^5-cyclopentadienyl)-tris(η-cyclopentadienyl)dititanium (Ti-Ti) (6) (Fig. 7.5) [100] in hexane. Their synthetic techniques were as follows. Acetylene was polymerized by contacting the gas at 1 atm pressure with a rapidly stirred solution of the titanium complex, 30 mg in 250 mL of n-hexane. Addition of acetylene was continued for ∼24 h, until there was no further uptake. A dark reddish voluminous mass of solvent-swollen polymer was obtained. This was dried under high vacuum to yield 1–1.5 g of a metallic gray bulk polymer. At room temperature a polymer having a predominantly trans isomer was obtained. The polymerization reaction conducted at −80°C gave a cis-rich polyacetylene. Cis-rich films were prepared by treatment of unstirred solutions of the catalyst (3.4 × 10^{-3} M) in hexane at −80°C under 760 torr of acetylene pressure. After 24 h a 20 mm thick copper-colored gelatinous mass of polymer was formed. This was washed with hexane, and upon very gradual removal of solvent under vacuum at room temperature, a 0.3 mm thick polyacetylene film having a lustrous gray metallic appearance was formed.

Alt and coworkers [40] demonstrated that $Cp_2Ti(PMe_3)_2$ (**7**) reacts readily with acetylene and various substituted acetylenes $R^1C{\equiv}CR^2$ ($R^1 = R^2 =$ H, Me, Ph), yielding acetylene complexes of the type $Cp_2Ti(R^1C{\equiv}CR^2)(PMe_3)$ (**8**), titanacyclopentadienes $Cp_2Ti{-}(C_4R_2{}^2R_2{}^2)$ (**9**), and polymers as shown in Fig. 7.6. Martinez and coworkers [39] found that the titanium compound polymerizes acetylene to give predominantly *trans*-polyacetylene although the polymerization was carried out at low temperatures.

E. Polymerization of Acetylene with Rh and Re Catalysts

Chlorine-bridged Rh(I) complexes such as [Rh(COD)Cl]₂ (**10**) and [Rh(NBD)Cl]₂ (**11**), where COD is cycloocta-1,5-diene and NBD is bicyclo[2.2.1]hepta-2,5-diene, in the presence of sodium ethoxide as a cocatalyst can initiate polymerization of acetylene to give polyacetylene in the form of solid flakes [51]. Generally [Rh(NBD)Cl]₂ is so effective that it does not require the addition of sodium ethoxide to initiate the polymerization, whereas [Rh(COD)Cl]₂ needs the addition of the co-catalyst. The advantage of Rh(I) complexes over the Ziegler–Natta eatalysts is that they are stable in air and their catalytic activity is not inhibited by the presence of oxygen and/or moisture. This characteristic is completely analogous to that shown by the Luttinger catalysts. The polyacetylene freshly prepared over the Rh(I) complexes can be chlorinated as a slurry in dichloromethane by bubbling gaseous chlorine at 0°C to give soluble chlorinated polyacetylene in dichloromethane and can be recovered by precipitation in methanol. The polymerization was also carried out with a new catalyst system based on binuclear compounds of Re with NaBH₄ or KBH₄ as a reducing reagent, but further details of the Re catalyst were not reported [46,52].

F. Polymerization of Acetylene by Miscellaneous Methods

Electrochemical synthesis of polyacetylene was carried out with platinum foil as cathode and nickel foil as anode with nickel bromide in acetonitrile as an electrolyte at room temperature to precipitate in the form of powder in the cell [101]. Chen and Shy [53] observed a thin layer of black material on a surface of the platinum cathode when a voltage of 4–40 V was applied for about 50 min. During this stage, no precipitation of polyacetylene was observed in the solution. The material has the same chemical structure as those produced by the standard Ziegler–Natta catalyst as examined by elemental analysis, infrared spectroscopy, X-ray diffraction, and differential scanning calorimetry.

Although it is not a practical method for producing polyacetylene specimens for various properties, acetylene polymerization occurs on a surface of metal oxide such as rutile (TiO_2) [54], γ-alumina (Al_2O_3) [55,56], or zeolite KX [56] to give *trans*-polyacetylene.

Soga et al. [57] found a new and convenient method for synthesizing polyacetylene in the form of thin films by exposing acetylene gas to a small amount of AsF_5 in a glass reaction vessel at −75 to −198°C. Polyacetylene films formed instantaneously on the glass wall. Infrared spectra of these films showed a strong band at 740 cm⁻¹ assignable to *cis* C—H out-of-plane bending vibration and doping-induced broad bands at 1370 and 900 cm⁻¹, suggesting the formation of *cis*-polyacetylene doped by AsF_5. Aldissi and Liepins [58] carried out the polymerization in AsF_3 solvent. The product is soluble in solvent, and the infrared spectra of the methylene dichloride solution showed a strong peak at 700–740 cm⁻¹ and two small bands at 1375 and 900 cm⁻¹, suggesting a product similar to that prepared by Soga et al.

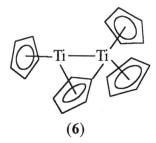

(**6**)

Fig. 7.5 A stable dititanium complex as single component catalyst for acetylene polymerization. A molecule of THF attached to the formally divalent titanium atom by coordination through oxygen is not shown.

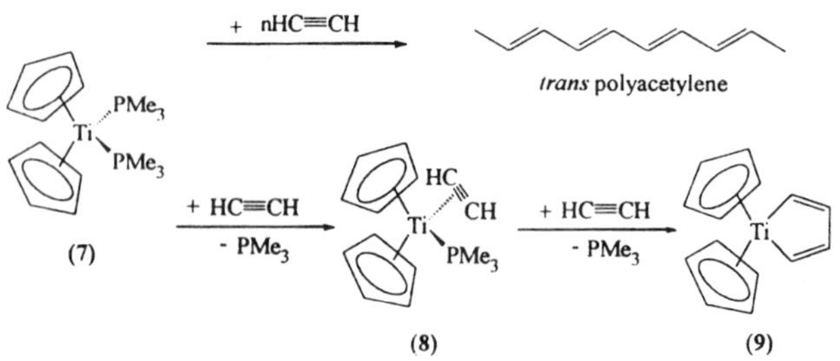

Fig. 7.6 Reaction products from $Cp_2Ti(PMe_3)_2$ and acetylene.

It has been known that the irradiation of acetylene gas with ultraviolet rays or ionizing radiation produces cuprene and benzene. Radiation-induced polymerization of acetylene was first investigated using α-rays from radon as early as the 1920s.

Tabata et al. [63] reported the polymerization of acetylene in the liquid and solid state by irradiation with γ-rays from a ^{60}Co source over the dose rate range of 3×10^4 to 1×10^5 R/h. The product obtained in the liquid-state polymerization was a relatively bulky orange-yellow powder that was soluble in aniline, isopropylamine, and DMF, and the polymer obtained by the solid-state polymerization was a rigid deep brown material and was insoluble in almost all solvents. Infrared spectra of the product were quite different from those of polyacetylene prepared with Ziegler–Natta catalysts.

Aoki and coworkers [59–62] observed spontaneous solid-state polymerization of acetylene above 3.5 GPa during the phase study of acetylene by Raman scattering. A wedge-driven diamond cell was employed for pressure generation. Acetylene gas was charged by solidification in the cell cooled in liquid nitrogen. The cell was then warmed up to room temperature for Raman scattering measurements. Through the microscopic observations, a liquid phase was obtained when the cell was warmed to room temperature. At 0.7 GPa the liquid transformed into a solid phase, which subsequently transformed into another solid phase at 0.9 GPa. Both solid phases were colorless. At about 3.5 GPa the sample began to change in color from colorless and transparent to deep red. It took a few tens of hours to complete the reaction. The Raman spectra of the deep red material at 3.5 GPa showed two strong bands at around 1000 and 1500 cm^{-1} assignable to the C—C and C=C stretching vibrations, respectively, suggesting that this material is a linear conjugated polymer with a trans configuration. However, the conclusion is in conflict with the deep red product that is characteristic of the cis-polyacetylene. The peak positions of the C—C and C=C stretching vibration modes shifted down to 1098 and 1480 cm^{-1}, respectively, with release of the pressure. The Raman bands diminished as the bleaching in the deep red phase proceeded at pressures higher than 4.5 GPa, suggesting that intermolecular cross-linking proceeded to destroy the conjugation. The molecular arrangement determined from high pressure X-ray diffraction data at the reaction pressure indicated that the monomer undergoes trans opening of the triple bond to form the trans-polyacetylene in the bc plane of the orthorhombic unit cell.

G. Polymerization of 1,3,5,7-Cyclooctatetraene

Korshak and coworkers [64] synthesized polyacetylene films by the ring-opening polymerization of 1,3,5,7-cyclooctatetraene (COT) with a metathesis catalyst, W[OCH(CH$_2$Cl)$_2$]$_n$Cl$_{6-n}$—(C$_2$H$_5$)$_2$AlCl (n = 2 or 3). The polymerization of the monomer in the form of film was carried out on a solid layer of the catalyst deposited on the glass surface by evaporating toluene from the catalyst solution under vacuum. Then the monomer was slowly condensed onto the catalyst for periods of 24 h to several days at such a rate as to cause the formation of a solid polymer film. The infrared spectrum of the polymer showed typical bands of polyacetylene: trans C—H out-of-plane vibration at 1010 cm^{-1} and cis C—H out-of-plane vibration at 740 cm^{-1}, indicating that the polymer is polyacetylene with 25% cis and 75% trans configurations with saturation defects and a partial doping with the tungsten component.

Klavetter and Grubbs [65] developed a versatile and convenient route to polyacetylene through the condensed-phase polymerization of the monomer with the well-defined metathesis tungsten-based catalysts **10** and **11**, as shown in Fig. 7.7. Shiny and silvery films with a smooth surface were prepared by dissolution of catalyst **10** in 50–150 equivalents of COT and subsequent polymerization on a glass surface at ambient temperature and pressure. Properties of these poly-COT films are nearly identical with those of polyacetylene produced with Ziegler–Natta catalysts.

H. Durham Method

So far, various routes for synthesizing polyacetylene by the polymerization of monomers such as acetylene and

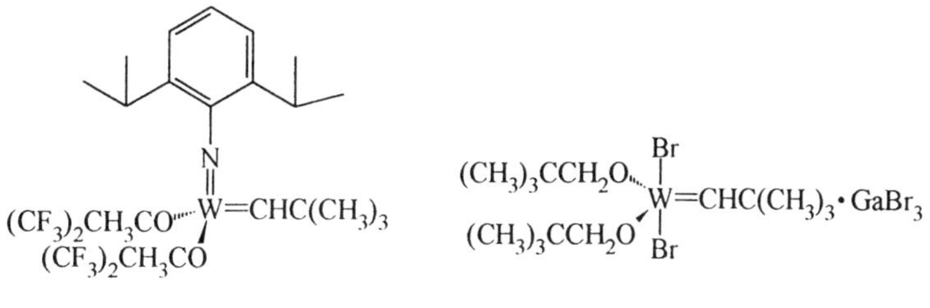

Fig. 7.7 Well-defined catalysts for the ring-opening metathesis polymerization of 1,3,5,7-cyclooctatetraene.

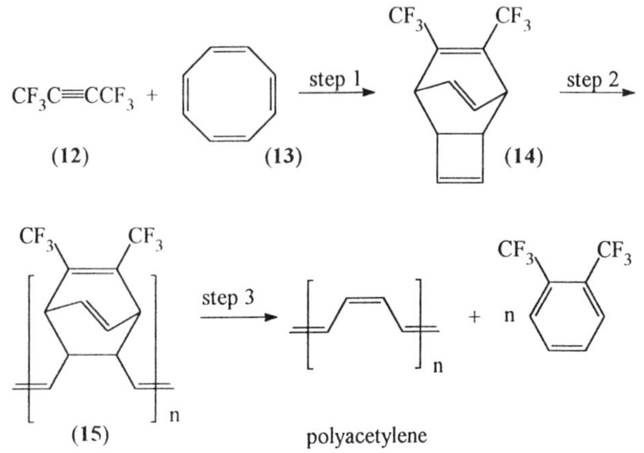

Fig. 7.8 The Durham route to polyacetylene.

COT have been described, but an alternative route for synthesis of long conjugated polyenes has been known for many years. A typical example is the base-catalyzed thermal elimination of hydrogen chloride from poly(vinyl chloride). Unfortunately, synthesis of well-characterized polyacetylene by this route was not successful. The first synthesis of high quality, well-characterized polyacetylene films was reported by Edwards and Feast [66]. The process [67–74] consists of three reaction steps as shown in Fig. 7.8.

The first step is the synthesis of the monomer, 7,8-bis(trifluoromethyl)tricyclo[4.2.2.0^{2,5}]deca-3,7,9- triene (**14**) by thermal reaction between hexafluorobut-2-yne (**12**) and COT (**13**). The second step is the ring-opening polymerization of **14** with either the metathesis catalyst, WCl_6—$(C_6H_5)_4Sn$ (W/Sn = 1/2) or the classical Ziegler catalyst, $TiCl_4$—$(C_2H_5)_3Al$ (Ti/Al = 1/2) to give the precursor polymer **15**. The former catalyst gives roughly equal amounts of *cis* and *trans* main-chain double bonds, whereas the latter gives predominantly a *trans* configuration. Since the polymer is soluble in acetone and chloroform, a thin film can be cast from an acetone solution. The polymer decomposed spontaneously on standing in the dark under an atmosphere of dry nitrogen and more rapidly in solution. The decomposition was accompanied by a sequence of color changes from yellow, through orange, red, and brown, to eventually yield a black material with a metallic luster, suggesting the evolution of long conjugation in the precursor molecule. Actually, in the third step, the thermal elimination was carried out in the range of 40–150°C for 10 min to 12 days under vacuum. A predominantly *cis* polyacetylene was prepared by slow transformation of the precursor polymer at 60°C. At higher temperature, *trans*-rich polyacetylene was obtained as a result of the thermal *cis*-to-*trans* isomerization of the *cis* polymer.

I. Isomerization of Polybenzvalene

Another example of a soluble polyacetylene precursor polymer is polybenzvalene [75], which is prepared by the ring-opening metathesis polymerization (ROMP) of the highly reactive but readily available monomer, benzvalene (**16**) as shown in Fig. 7.9. Thermal and photochemical isomerization of polybenzvalene (**17**) were unsuccessful in producing polyacetylene but successful with transition metal catalysts. Thus, solutions of $HgCl_2$, $HgBr_2$, and Ag salts in THF transformed films of polybenzvalene into shiny silvery materials resembling polyacetylene films.

IV. CONCLUSION

Synthetic methods of preparing polyacetylene are summarized in this chapter. Polyacetylene has been at center stage throughout the evolution of this field with an electrical conductivity that can be increased by doping to values comparable to those of copper and silver. However, because this polymer is insoluble in any solvent tested so far, infusible, and unstable under ambient conditions, there are serious difficulties in characterizing it at the molecular level, for example, by molecular weight or conjugation length and their distribution, or by microstructure or morphology. All of these characteristics are set at the time of synthesis. Therefore, different synthetic routes might be expected to give polyacetylenes with different conjugation lengths, defects, microstructure, morphology, and crystallinity and different species of impurities at different concentrations. The most widely used procedure for polyacetylene synthesis is the film synthesis via addition polymerization of acetylene using the soluble Ziegler–Natta catalyst system com-

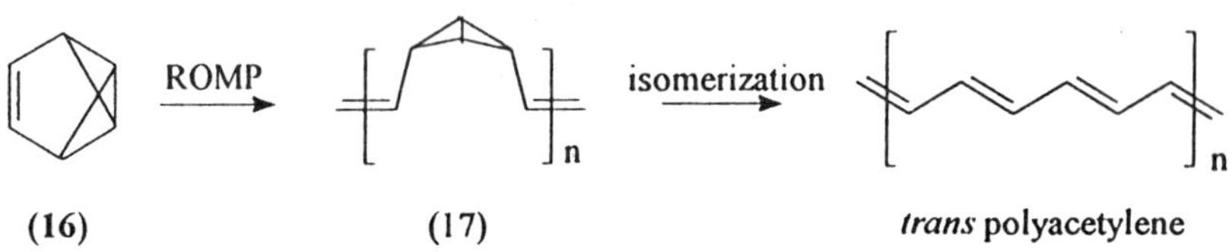

Fig. 7.9 Synthesis of polyacetylene by the isomerization of polybenzvalene.

bined with Ti(O—C$_4$H$_9$)$_4$ and (C$_2$H$_5$)$_3$Al. However, its polymerization activity is affected by concentration of the catalyst, Al/Ti ratio, solvent, aging conditions, acetylene pressure, and polymerization temperature, giving rise to different molecular weights, microstructure, morphology, defects, etc. Further developments of the field require a better understanding of catalysts and more breakthroughs in the synthesis of high quality polyacetylene.

REFERENCES

1. T. Ito, H. Shirakawa, and S. Ikeda, *J. Polym. Sci., Polym. Chem. Ed. 12*:11 (1974).
2. H. Shirakawa, E. J. Louis, A. G. MacDiarmid, C. K. Chiang, and A. J. Heeger, *J. Chem. Soc., Chem. Commun. 1977*:578.
3. C. K. Chiang, C. R. Fincher, Jr., Y. W. Park, A. J. Heeger, H. Shirakawa, E. J. Louis, S. C. Gau, and A. G. MacDiarmid, *Phys. Rev. Lett. 39*:1098 (1977).
4. J. C. W. Chien, *Polyacetylene: Chemistry, Physics, and Materials Science*, Academic, Orlando, 1984.
5. L. Yu, *Solitons and Polarons in Conducting Polymers*, World Scientific, Singapore, 1988.
6. J. L. Brédas and R. Silbey (eds.), *Conjugated Polymers*, Kluwer Academic, Dordrecht, 1991.
7. H. G. Kiess (ed.), *Conjugated Conducting Polymers*, Springer-Verlag, Berlin, 1992.
8. S. Roth, *One-Dimensional Metals: Physics and Materials Science*, VCH, Weinheim, 1995.
9. T. A. Skotheim (ed.), *Handbook of Conducting Polymers*, Vols. 1 and 2, Marcel Dekker, New York, 1986.
10. J. W. Blatchford and A. J. Epstein, *Am. J. Phys. 64*:120 (1996).
11. A. G. MacDiarmid and R. B. Kaner, in *Handbook of Conducting Polymers*, Vol. 1 (T. A. Skotheim, ed.), Marcel Dekker, New York, 1986, Chap. 20.
12. H. Shirakawa, *Synth. Met. 69*:3 (1995).
13. M. Hatano, S. Kambara, and S. Okamoto, *J. Polym. Sci. 51*:S26 (1961).
14. W. H. Watson, Jr., W. C. McMordie, Jr., and L. G. Lands, *J. Polym. Sci. 55*:137 (1961).
15. D. J. Berets and D. S. Smith, *Trans. Faraday Soc. 64*:823 (1968).
16. H. Shirakawa, T. Ito, and S. Ikeda, *Makromol. Chem. 179*:1565 (1978).
17. H. Shirakawa and S. Ikeda, *Synth. Met. 1*:175 (1980).
18. Y. W. Park, M. A. Druy, C. K. Chiang, A. G. MacDiarmid, A. J. Heeger, H. Shirakawa, and S. Ikeda, *J. Polym. Sci., Polym. Lett. Ed. 17*:195 (1979).
19. H. Naarmann and N. Theophilou, *Synth. Met. 22*:1 (1987).
20. J. Tsukamoto, A. Takahashi, and K. Kawasaki, *Jpn. J. Appl. Phys. 29*:125 (1990).
21. H. Shirakawa and S. Ikeda, *Polym. J. 2*:231 (1971).
22. C. S. Yannoni and T. C. Clarke, *Phys. Rev. Lett. 51*:192 (1983).
23. S. Ikeda, *Kogyo Kagaku Zasshi 70*:1880 (1967).
24. M. J. Duijvestijn, A. Manenscheijn, A. Smidt, and R. A. Wind, *J. Magn. Res. 64*:461 (1985).
25. T. Yamabe, K. Tanaka, H. Terama-e, K. Fukui, A. Imamura, H. Shirakawa, and S. Ikeda, *Solid State. Commun. 29*:329 (1979).
26. J. A. Pople and S. H. Walmsley, *Mol. Phys. 5*:15 (1962).
27. J. C. W. Chien, F. E. Karasz, G. E. Wnek, A. G. MacDiarmid, and A. J. Heeger, *J. Polym. Sci., Polym. Lett. Ed. 18*:45 (1980).
28. P. Bernier, M. Rolland, C. Linaya, and M. Aldissi, *Polymer 21*:7 (1980).
29. G. Natta, G. Mazzanti, and P. Corradini, *Atti Acad. Naz. Lincei, Cl. Sci. Fis. Mat. Nat. Rend. 25(8)*:3 (1958).
30. L. B. Luttinger, *J. Org. Chem. 27*:1591 (1962).
31. W. E. Daniels, *J. Org. Chem. 29*:2936 (1964).
32. H. Shirakawa, T. Ito, and S. Ikeda, *Polym. J. 4*:460 (1973).
33. Y. Ooshika, *J. Phys. Soc. Jpn. 12*:1238, 1246 (1957).
34. H. C. Longuet-Higgins and L. Salem, *Proc. Roy. Soc. A251*:172 (1959).
35. A. A. Ovchinnikov, I. I. Ukrainskii, and G. V. Kventsel, *Sov. Phys. Usp. 15*:575 (1973).
36. N. S. Bayliss, *J. Chem. Phys., 16*:287 (1948); *Quart. Rev. 6*:319 (1952).
37. H. Kuhn, *Helv. Chim. Acta 31*:1441 (1948).
38. H. G. Alt, H. E. Engelhardt, M. D. Rausch, and L. B. Kool, *J. Am. Chem. Soc. 107*:3717 (1985).
39. J. R. Martinez, M. D. Rausch, J. C. W. Chien, and H. G. Alt, *Makromol. Chem. 190*:1309 (1989).
40. H. G. Alt, H. E. Engelhardt, M. D. Rausch, and L. B. Kool, *J. Organomet. Chem. 329*:61 (1987).
41. A. Munardi, N. Theophilou, R. Anzar, J. Sledz, and H. Naarmann, *Makromol. Chem. 188*:395 (1987).
42. L. B. Luttinger, *Chem. Ind. (Lond) 1960*:1135.
43. M. Monkenbusch, B. S. Morra, and G. Wegner, *Makromol. Chem. Rapid Commun. 3*:69 (1982).
44. L. Terlemezyan and M. Mihailov, *Makromol. Chem. Rapid Commun. 3*:613 (1982).
45. V. M. Kobryanskii and E. A. Tereshko, *Synth. Met. 39*:367 (1991).
46. V. M. Kobryanskii, *Synth. Met. 46*:251 (1992).
47. M. G. Voronkov, V. B. Pukhnarevich, S. P. Sushchinksaya, V. Z. Annenkova, V. M. Annenkova, and N. J. Andreeva, *J. Polym. Sci.: Polym. Chem. Ed. 18*:53 (1980).
48. M. Aldissi, C. Linaya, J. Sledz, F. Schué, L. Giral, J. M. Fabre, and M. Rolland, *Polymer 23*:243 (1982).
49. N. Theophilou, A. Munardi, T. Aznar, J. Sledz, F. Schué, and H. Naarmann, *Eur. Polym. J. 23*:15 (1987).
50. A. J. Amass, M. S. Beevers, T. R. Farren, and J. A. Stowell, *Makromol. Chem., Rapid Commun. 8*:119 (1987).
51. F. Cataldo, *Polym. Commun. 33*:3073 (1992).
52. V. M. Kobryanskii, *Synth. Met. 49*:203 (1992).
53. S.-A. Chen and H.-J. Shy, *J. Polym. Sci.: Polym. Chem. Ed. 23*:2441 (1985).

54. V. Rives-Arnau and N. Sheppard, *J. Chem. Soc. Faraday I 76*:394 (1980).
55. N. Kurokawa, M. Tabata, and J. Sohma, *J. Polym. Sci.: Polym. Lett. Ed. 19*:355 (1985).
56. J. Heaviside, P. J. Hendra, P. Tsai, and R. P. Coony, *J. Chem. Soc. Faraday I 74*:2542 (1978).
57. K. Soga, Y. Kobayashi, S. Ikeda, and S. Kawakami, *J. Chem. Soc. Chem. Commun. 1980*:931.
58. M. Aldissi and R. Liepins, *J. Chem. Soc. Chem. Commun. 1984*:256.
59. K. Aoki, Y. Kakudate, S. Usuba, M. Yoshida, K. Tanaka, and S. Fujiwara, *Solid State Commun. 64*:1329 (1987).
60. K. Aoki, Y. Kakudate, S. Usuba, M. Yoshida, K. Tanaka, and S. Fujiwara, *J. Chem. Phys. 88*:4565 (1988).
61. K. Aoki, S. Usuba, M. Yoshida, Y. Kakudate, K. Tanaka, and S. Fujiwara, *J. Chem. Phys. 89*:529 (1988).
62. K. Aoki, Y. Kakudate, S. Usuba, M. Yoshida, K. Tanaka, and S. Fujiwara, *Synth. Met. 28*:D91 (1989).
63. Y. Tabata, B. Saito, H. Shibano, H. Sobue, and K. Oshima, *Makromol. Chem. 76*:89 (1964).
64. Y. V. Korshak, V. V. Korshak, G. Kanischka, and H. Höcker, *Makromol. Chem., Rapid Commun. 6*:685 (1985).
65. F. L. Klavetter and R. H. Grubbs, *J. Am. Chem. Soc. 110*:7807 (1988).
66. J. H. Edwards and W. J. Feast, *Polym. Commun. 21*:595 (1980).
67. D. White and D. C. Bott, *Polym. Commun. 25*:98 (1984).
68. J. H. Edwards, W. J. Feast, and D. C. Bott, *Polymer 25*:395 (1984).
69. P. D. Townsend, C. m. Pereira, D. D. C. Bradley, M. E. Horton, and R. H. Friend, *J Phys. C: Solid State Phys. 18*:L283 (1985).
70. P. J. S. Foot, P. D. Calvert, N. C. Billingham, C. S. Brown, N. S. Walker, and D. I. James, *Polymer 27*:448 (1986).
71. M. M. Sokolowski, E. A. Marseglia, and R. H. Friend., *Polymer 27*:1714 (1986).
72. C. S. Brown, M. E. Vickers, P. J. S. Foot, N. C. Billingham, and P. D. Calvert, *Polymer 27*:1719 (1986).
73. D. C. Bott, C. S. Brown, J. N. Winter, and J. Barker, *Polymer 28*:601 (1987).
74. W. J. Feast, M. J. Taylor, and J. N. Winter, *Polymer 28*:593 (1987).
75. T. M. Swager, D. A. Dougherty, and R. H. Grubbs, *J. Am. Chem. Soc. 110*:2973 (1988).
76. C. S. Marvel, J. H. Sample, and M. F. Roy, *J. Am. Chem. Soc. 61*:3241 (1939).
77. C. A. Finch, *Polyvinyl Alcohol*, Wiley, New York, 1973, p. 477.
78. K. Maruyama, M. Take, N. Fujii, and Y. Tanizaki, *Bull. Chem. Soc. Jpn. 59*:13 (1986).
79. J. A. Nieuland and R. R. Vogt, *The Chemistry of Acetylene*, Reinhold, New York, 1945, pp. 138–170.
80. H. Shirakawa and S. Ikeda, *J. Polym. Chem. Polym. Chem. Ed. 12*:929 (1974).
81. S. Kambara, M. Hatano, and T. Hosoe, *Kogyo Kagaku Zasshi 65*:720 (1962).
82. E. Angelescu and I. V. Nicolescu, *J. Polym. Sci.: Part C 22*:203 (1968).
83. Z.-Q. Shen, M.-J. Yang, M.-X. Shi, and Y.-P. Cai, *J. Polym. Chem., Polym. Lett. Ed. 20*:411 (1974).
84. M. Catellani, S. Destri, and A. Bolognesi, *Makromol. Chem. 187*:1345 (1986).
85. I. Kminek and J. Trekobval, *Makromol. Chem., Rapid Commun. 7*:53 (1986).
86. A. Munardi, N. Theophilou, R. Anzar, J. Sledz, and H. Naarmann, *Makromol. Chem. 188*:395 (1987).
87. S. L. Hsu, A. J. Signorelli, G. P. Pez, and R. H. Baughman, *J. Chem. Phys. 69*:106 (1978).
88. M. Soga, S. Hotta, and N. Sonoda, *Synth. Met. 30*:251 (1989).
89. M. Soga, in *Progress in Pacific Polymer Science 2* (Y. Imanishi, ed.), Springer-Verlag, Berlin, 1992, p. 245.
90. S. Y. Oh, K. Akagi, and H. Shirakawa, *Synth. Met. 32*:245 (1989).
91. K. Akagi, D. Hashimoto, H. Shirakawa, and J. Isoya, *Polym. Commun. 31*:411 (1990).
92. K. Akagi, K. Sakamaki, and H. Shirakawa, *Macromolecules 25*:6725 (1992).
93. K. Akagi, K. Sakamaki, and H. Shirakawa, *Synth. Met. 55*:779 (1993).
94. K. Akagi, K. Sakamaki, H. Shirakawa, and H. Kyotani, *Synth. Met. 69*:29 (1995).
95. G. Luguli, U. Pedretti, and G. Perego, *J. Polym. Sci., Polym. Lett. Ed. 23*:129 (1985).
96. G. Luguli, U. Pedretti, and G. Perego, *Mol. Cryst. Liq. Cryst. 117*:43 (1985).
97. K. Akagi, M. Suezaki, H. Shirakawa, H. Kyotani, M. Shimomura, and Y. Tanabe, *Synth. Met. 40*:D1 (1991).
98. M. Aldissi, F. Schué, L. Giral, and M. Rolland, *Polymer 23*:246 (1982).
99. G. Lieser, G. Wegner, W. Müller, and V. Enkelmann, *Makromol. Chem. Rapid Commun. 1*:621 (1980).
100. G. Pez, *J. Am. Chem. Soc. 98*:8072 (1976).
101. W. A. Kronicker, U.S. Patent, 3,474,012 (1969).

8
Synthesis of Poly(*para*-phenylene)s

A.-Dieter Schlüter
Freie Universität Berlin, Berlin, Germany

I. INTRODUCTION

Until recently, structurally defined poly(*para*-phenylene)s (PPPs) were a blank space on the map of polymer synthesis. There was not even one passable trail leading toward the synthesis of a high molecular weight polymer that could represent this class in spite of numerous attempts. Only some ill-defined oligomers and polymers, whose structures more or less resemble that of PPP, were obtained [1–13]. This lack of synthetic access is tightly associated with the unusual structure of the PPP backbone, whose relatively rigid aromatic subunits are directly linked to each other in a linear fashion (structure). Quite apart from all other aromatic main chain polymers (Vectra, Kevlar, etc.), there are no functional groups like esters or amides between the phenylene moieties, and as a consequence classical condensation procedures cannot be applied. These procedures involve bond formation between a carbon and a heteroatom, typically oxygen or nitrogen, but not between two carbon atoms. The unusual structure of PPP not only renders the synthetic access problematic, it is also responsible for the rigid-rod nature of this polymer, a property that makes it insoluble, intractable, and very difficult to process.

Over the years, both academic and industrial research have continued to be strongly interested in PPP regardless of all these difficulties with the polymer and materials derived from it. Theoretical considerations of the parent polymer and some of the derivatives [14] as well as investigations performed with the imperfect materials mentioned, created high expectations for the unknown properties that structurally homogeneous PPP might have. Ill-defined materials have been found to show high mechanical strength, excellent chemical resistance (due to the lack of functional groups), considerable thermal and oxidative stability, electrical conductivity in the oxidized state, and electroluminescence [15–23]. This last property has currently prompted special interest because of its potential impact on a future market, display technology [24]. Optical indicators and displays with excellent pixel resolution will become more and more important for various technical applications and for consumer electronics such as high definition television and new generations of computers and software. In recent years synthetic chemists have learned to use transition metal mediated C—C bond-forming processes in macromolecular chemistry. This includes such important developments as the Ziegler–Natta polymerization using metallocenes [e.g., 25] and the ring-opening metathesis polymerization (ROMP) based on metal carbenes [26]. Considering the almost unique bouquet of interesting properties of PPP, it was just a matter of time before new synthetic routes for this polymer were also opened. This chapter gives a brief overview of the most important developments of recent years and, rather than be a mere compilation of results, puts some emphasis on how and why decisions were made. This should mediate an understanding of how these developments were possible. For other recent reviews on this matter and related topics, the reader is referred to the literature [9,19,20,27,28].

II. A BRIEF GENERAL COMMENT

The attempts to synthesize PPP may be classified as either direct or indirect methods. In the first case, the monomers that are connected already contain the phenylene moiety that will become the repeating unit of the final polymer. In the second case, a precursor polymer is first synthesized from which PPP is then released, e.g., by thermal treatment. The most prominent examples are Kovacic's direct route [9] and the ICI researcher's indirect route [11] (Fig. 8.1). Unfortunately, both methods have serious drawbacks. For most of the direct syntheses, the reaction conditions are too harsh to allow a regiospecific coupling reaction to take place. Thus, linkages between wrong sites, cross-linking, and other side reactions occur. Furthermore, the molecular weights achieved are very low, which is specifically due to solubility problems. Typical degrees of polymerization (DPs) range from 5 to 15. This is the point when the growing chain simply precipitates from solution, thus inhibiting further growth.

The precursor approach is superior to all direct methods inasmuch as the molecular weights that are achieved are much higher because of the precursor's mode of synthesis. A serious limitation of this method is that the structural irregularities contained in the precursor are inevitably transplanted into the final polyarylene, not to mention that the conversion of the precursor polymer does not proceed as cleanly as desired. It is either incomplete or leads to chain fracture. Also there are only a small number of suitable precursor polymers available. From this consideration it becomes evident what chemists had to do to develop solutions for the individual problems of both approaches.

In regard to the direct approach, a C—C coupling reaction needed to be developed that, on the one hand, does not require harsh conditions and, on the other hand, tackles the solubility problem. For the indirect approach, however, a method had to be discovered that would allow polymerization of a substituted cyclohexa-1,3-diene monomer under full regio and possibly even full stereo control. Additionally, the monomer substituents had to be designed so that they were not only compatible with the polymerization conditions but could also be cleanly removed once the precursor polymer was formed. For both approaches, significant progress could be achieved. Again transition metal induced C—C bond formation reactions play a key role, which further consolidates the ever-growing importance of metal-catalyzed processes in polymer synthesis.

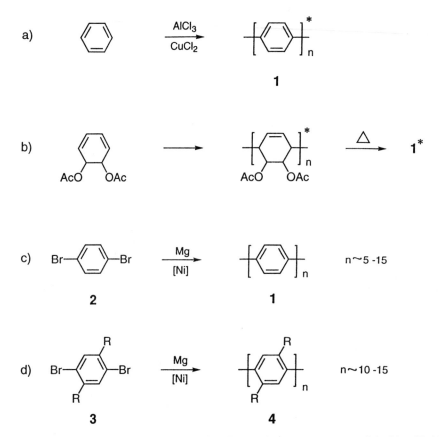

Fig. 8.1 Synthetic routes to PPP followed by (a) Kovacic, (b) ICI, (c, d) Yamamoto. Asterisks identify idealized structures. R = n-alkyl.

III. THE DIRECT APPROACH

A. Two Roots of Success: Yamamoto's and Suzuki's Work and Flexible Alkyl Chains

There is one paper in the literature for which the critical assessment of regiochemical control in the direct synthesis of PPPs does not apply. In 1978 and subsequent years, Yamamoto reported on the reaction between dihaloaromatic compounds and Mg metal in the presence of various low valency Ni catalysts [27,28]. When 1,4-dibromobenzene (2), 1 equiv of Mg, and dichloro-Ni(bipyridyl) in refluxing tetrahydrofuran is used, an exclusively para-linked, low molecular weight PPP (IR) is obtained (Fig. 8.1). The superiority of this very mild coupling process over other direct methods becomes immediately apparent if Yamamoto's material is visually compared with, e.g., Kovacic's. Whereas the former material is slightly yellow, the latter is black.

At this point it was asked whether this process could be improved to furnish polymers instead of oligomers. The obvious question that had to be answered was: How can one suppress the termination of the Ni-catalyzed polycondensation? Two very different possible reasons had to be considered. Termination may either be inherently associated with the chemistry involved or be due to precipitation of the growing polymer chain at a very early stage of growth. Evidence for both possibilities was available in the literature. First, Yamamoto and co-workers [29] proved that some termination takes place in the course of the coupling, unfortunately without quantifying the extent to which it occurs. Second, the absolute solubilities of PPP oligomers in conventional organic solvents at ambient temperatures decrease dramatically with increasing length and reach negligibly small values for a few linked benzene rings. For example, the solubility of octaphenylene in toluene at 25°C is less than 10^{-8} g/L [32]. In view of this almost unmeasurable value, the DPs of 10–15 for PPP reported by Yamamoto are surprising high and can only be explained by assuming that the precipitation of the growing chain is kinetically hindered. To differentiate between the modes of termination, monomer 3 was designed and synthesized. It is closely related to 2, but it carries flexible alkyl chains (6–12 C atoms per chain). These chains' attachments are known to drastically increase the solubility of low molecular weight compounds [33–36] as well as that of polymers [37–39]. Enhancements on the order of 10–100-fold are frequently encountered. It was therefore expected that a polymer from 3 would remain soluble up to much higher DPs than the parent Yamamoto PPP.

When monomer 3 instead of 2 was subjected to Yamamoto's reaction conditions, it apparently behaved remarkably differently, inasmuch as the reaction mixture stayed homogeneous throughout the entire process. Additionally, the material obtained upon isolation was soluble, which enabled a thorough analysis of its structure and molecular weight. High field ^{1}H- and ^{13}C-NMR spectroscopy and vapor pressure osmometry showed that this material is the corresponding, alkyl chain substituted PPP 4 with a DP of approximately 13 [40]. Variation of both the reaction conditions and the Ni catalysts always gave materials with correctly 1,4-connected structures but with molecular weights in the same range. Even though the alkyl chains obviously keep the growing polymer in solution, the molecular weights cannot be pushed beyond the magic number of DP = 15. Thus the answer to the question asked above is that the Ni-catalyzed polycondensation terminates itself. What causes termination? A number of potential reasons have to be considered, e.g., homolytic cleavage of the C—Ni bond to give a carbon-centered radical that undergoes typical radical reactions like hydrogen abstraction; heterolytic C—Ni bond cleavage, e.g., by protolysis; hydrolysis of the Mg-carrying terminus; transfer of ligands (or fragments thereof) centered at Ni to carbon; or transmetallation reactions [41–46]. Yamamoto has already observed that before work-up, Grignard carrying termini (—C_6H_4MgBr) are not present. He assumed that trace amounts of water are responsible for this. In the case of the alkyl chain substituted PPP 4, reduction of active end groups also seems to significantly contribute to termination. For example, the HPLC trace for each of the low DP fractions of an oligomeric mixture of 4 showed a "fine structure," the individual lines of which were assigned to chains with the different end-group combinations H—H, H—Br, and B–Br [40] (Fig. 8.2). Upon reaction with n-butyllithium, which is known to undergo

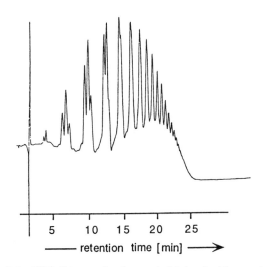

Fig. 8.2 HPLC trace of polymer 4 obtained with a methylene chloride/acetonitrile gradient at room temperature.

bromine–lithium exchange reactions when followed by the addition of water, the HPLC trace of the material obtained gave only one line per DP. It is reasonable to assume that this line reflects the H—H end-group combination.

In the initial approach to the synthesis of **4** via Ni-catalyzed Grignard coupling, chain growth termination of side reactions and difficulty in maintaining stoichiometric balance inhibited the very high reaction conversions necessary to give high DP materials. With this in mind, the search for other transition metal catalyzed coupling reactions began that would allow circumvention of these problems. Suzuki and coworkers [47] and later Miller and Dugar [48] had already described the Pd-catalyzed coupling of various bromobenzene derivatives with benzeneboronic acid. These cross-couplings, which use the less electropositive boron instead of magnesium, proceed with good to excellent yields. In some cases yields of 99% were reported. The authors also investigated the influence of substituents attached at the ortho position on this reaction, which was found to be virtually negligible. Hence it was decided to adapt this coupling reaction for the synthesis of PPP **4** and to explore its potential for the synthesis of polyarylenes and related polymers in general. Similar work using other haloaromatics and organometallic compounds based on boron and other metals was also available [49–51].

First a suitable monomer had to be prepared that would carry both the required functionalities and flexible alkyl chains. Fortunately, monomer **3**, which was used in the Yamamoto condensation, met almost all of these requirements and was abundantly available. The conversion of one of the bromo substituents into boronic acid was easily achieved. Monomer **5** was obtained in 10–20 g batches in a purity exceeding 98%. "Polycondensation grade" monomer was available by further recrystallization, whereby only small amounts of material were lost. The first experiment with the AB-type monomer **5** turned out to be a success [52]. Application of the reaction conditions reported by Suzuki's group furnished a colorless, soluble polymer whose structure was unambiguously assigned as **4** (Fig. 8.3). It became immediately apparent by comparing the aromatic regions of the ^{13}C-NMR spectra of representative samples of both materials (Fig. 8.4) that the molecular weight of **4** was higher than that of the Mg/Ni system. While the Yama-

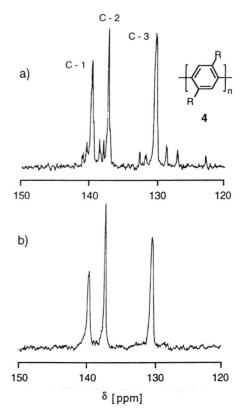

Fig. 8.4 Aromatic regions of the ^{13}C-NMR spectra of polymer **4** prepared according to (a) the Yamamoto and (b) the Suzuki procedure with assignment of the main signals.

moto polymer shows a number of end-group signals in addition to the three required main signals, the Suzuki polymer does not. The SEC traces of both materials qualitatively confirm the same trend (Fig. 8.5). Due to the lack of an appropriate standard, the SEC traces were

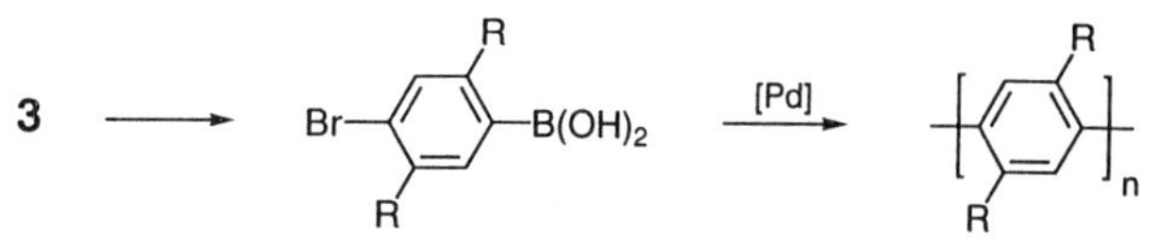

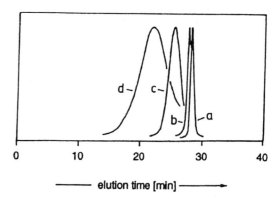

Fig. 8.5 SEC traces for samples of polymer **4** in 1,2-dichlorobenzene and related model compounds. (a) Dimer; (b) trimer; (c) DP = 12 (Yamamoto): (d) DP ≈ 28 (Suzuki).

Fig. 8.3 Synthetic route to Suzuki PPP.

not analyzed quantitatively. Vapor pressure osmometry (VPO) measurements, which provide a lower threshold of the actual molecular weight, gave approxiately DP = 30. A DP in this range is not fully satisfying from a synthetic point of view. Combined with the processibility of **4** into films, however, it is sufficient to make measurements, e.g., on optical or electrooptical effects with relevance for applications.

B. Other PPP Derivatives

1. Toward a Parent PPP

There is no doubt that PPP with a DP of 30 cannot be synthesized in a direct approach without the solubilizing groups having already been attached to the monomer. This statement also holds true for the synthesis of rigid-rod polymers in general. One should not oversee that the attachment of flexible chains does not have only positive effects. When they are attached to the parent PPP, the chemical constitution of the backbone is left unaltered, but its conformation will be changed, maybe even undesirably. In the specific case of PPP, the dihedral angle between two adjacent phenyl rings will increase the repulsive interactions between the chain on one ring and an ortho hydrogen on the other. The phenyl rings are forced even more out of coplanarity, which they would not normally attain anyway [53–56]. Although small deviations from coplanarity do not have a profound effect on electrical and optical properties, large deviations do [57]. In this context it is interesting to mention that the Suzuki PPP is absolutely colorless if trace impurities of residual catalyst are carefully removed.

With this new development in mind and the Suzuki polycondensation now at hand, it seemed feasible to try to model the parent PPP. All that had to be done was to synthesize PPPs with blocks of unsubstituted phenylene units of varying lengths that were linked to each other by a unit carrying alkyl chains. This unit, despite acting as an electronic insulator between the unsubstituted blocks, is required to mediate sufficient solubility. The syntheses were achieved according to the sequence shown in Fig. 8.6. Proper combination of one of the two boronic acids **6** or **9** with one of the dibromo compounds **3**, **7**, or **8** gave the telomer copolymers **10** with defined blocks of one to six unsubstituted phenylene units [58]. The DPs were in a typical range for this method. For example, VPO measurements of **10** ($y = 2$) gave a number-average molecular weight $M_n = 21,000$ corresponding to a DP of 37. This DP corresponds to a polymer with more than 100 phenylene rings linked to each other exclusively in para positions. The series of PPPs **10** comprises some of the longest well-characterized polyarylenes known today. The UV spectra of the first four members of the series show that the λ_{max} values undergo a significant bathochromic shift from 270 to 330 nm (in cyclohexene), as expected. A more systematic study should allow an extrapolation of these values to the parent PPP.

Clearly, this kind of strategy to model parent PPP has its limitations. As the density of the alkyl groups at the backbone is further and further decreased, the solubility of the polymers also decreases. An elegant way to circumvent this problem was recently reported from Müllen's laboratory [59]. They prepared alkylated polytetrahydropyrenes, **11**, from the corresponding dibromo monomer by applying the Yamamoto procedure (Fig. 8.7). The molecular weights reported are surprisingly high (DP = 40), which underlines the finding that this cross-coupling does not always fail to give polymers. Little steric crowding at the coupling site seems to be an important factor in this regard. The structure of polymer **11** is remarkable in that it models unsubstituted PPP very closely. The dihedral angle between adjacent phenylene rings belonging to two different repeat units resembles the one between the phenylene rings in parent PPP. It is therefore not surprising that the longest wavelength absorption of polymer **11** is bathochromically shifted compared with polymers **10** to $\lambda_{max} = 385$ nm. From a comparison of this value with the λ_{max} values of a series of related monodisperse model compounds containing up to 12 repeat units, the effective conjugation length of **11** was found to correspond to approximately 20 phenylene rings.

2. Ladder-Type PPP Derivatives

The monomers used in the cross-couplings described so far carried alkyl substituents of varying lengths, typically hexyl or dodecyl. An important characteristic of the Suzuki reaction is its compatibility with many functional groups. The use of aromatic monomers with functional groups instead of, or in addition to, alkyl chains in the cross-couplings should provide access to functionalized PPPs. Such PPPs are of interest in their own right and for further chemical modification. Scherf was one of the first to exploit this synthetic potential. He showed that, provided the nature and the position of the functional groups in the monomers are carefully designed, single-stranded PPP derivatives can be obtained and, in a subsequent step, converted into double-stranded, ladder-type PPPs by a simple chemical modification. The synthesis of polymer **15** is an important example (Fig. 8.8). Cross-coupling of monomers **6** and **12** yields PPP **13**, whose carbonyl functions provide the carbon atoms necessary to build the second strand. This second strand was tied by reducing the carbonyls to alcohols and activating the latter for electrophilic aromatic substitution with a Lewis acid. Because of the geometry implemented in the precursor **14**, each single electrophilic substitution step takes place at the nearest ortho position. This leads directly to the formation of a ladder-type structure. The regiochemical control of this polymer transformation is high, certainly the highest ever ob-

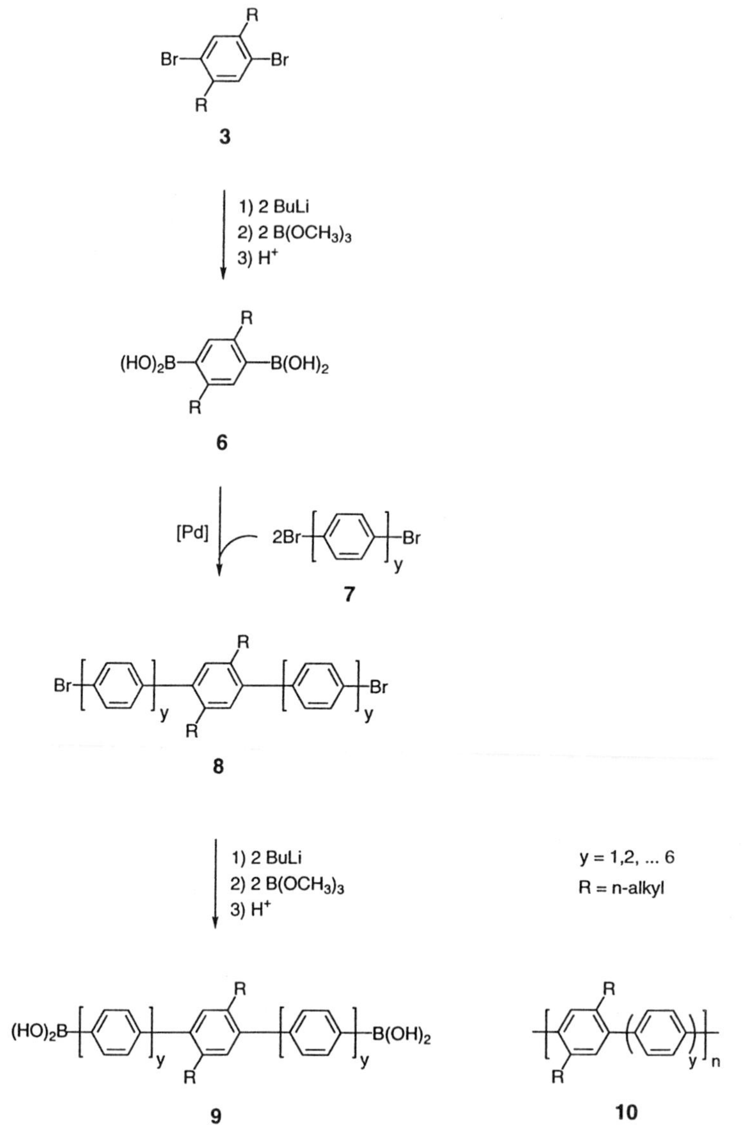

Fig. 8.6 Route to telomers **8** and **9** for the synthesis of PPP **10**, which has a reduced density of alkyl substituents.

served for a "classical" ladder polymer synthesis [60]. Besides the close proximity of the carbocation formed in situ to the ortho position of the adjacent phenylene ring, the shielding of the reaction site by the rest of the substituent contributes to this control.

The backbone of polymer **15** contains an alternating sequence of annulated five- and six-member rings. It is therefore reasonable to assume that all phenylene rings are more or less coplanar. As a consequence, the π conjugation between the aromatic rings is even better than in polymer **11**. This is reflected by a further bathochromic shift of the longest wavelength absorption from $\lambda_{max} = 385$ nm for **11** to $\lambda_{max} = 441$ nm for **15** [61]. The effective conjugation length was determined to be 11–12 phenylene rings [62]. Since the first discovery of light-emitting diodes based on polymers, an intensive search has been carried out for polyconjugated polymers that may be suited for application in such devices [24]. Polyphenylenevinylenes proved to be the best organic materials as far as emissions in the yellow and red are concerned. Difficulties arose, however, in finding polymers with blue emission. Leising and coworkers [32] first described a blue electroluminescence using conventionally prepared and thus ill-defined unsubstituted PPP as active material in the device. The use of a poorly defined material limits the impact of this interesting discovery be-

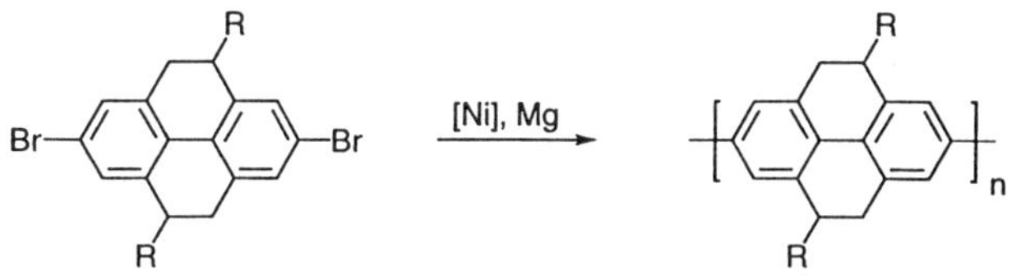

Synthesis of PPP

Fig. 8.7 Yamamoto route to PPP **11**, a model for parent PPP.

cause electro-optical effects tend to strongly depend upon purity and structural regularity. This is where ladder polymer **15** comes into play. It has a band gap of approximately 2.75 eV and should therefore be a potential candidate for a blue LED. In fact, polymer **15** shows in solution blue emission with an extraordinarily small Stokes shift [63,64]. The quantum efficiency was determined to be greater than 60%. Some problems, however, occurred with excimer formation in the solid state, resulting in a color change from blue to turquoise green during operation of the device. For a thorough description of the electro-optical properties of polymer **15** and related materials, see Chapters XY and XX in this volume.

The idea to planarize PPP through incorporation of bridges between the phenylene rings was also pursued by the groups of Swager and Tour. Their approaches, like the one by Scherf, can be categorized as classical ladder polymer syntheses whereby, in the first step, a single-stranded polymer is made that is modified in a second step to establish the second strand. Again the substitution pattern of the backbones is carefully de-

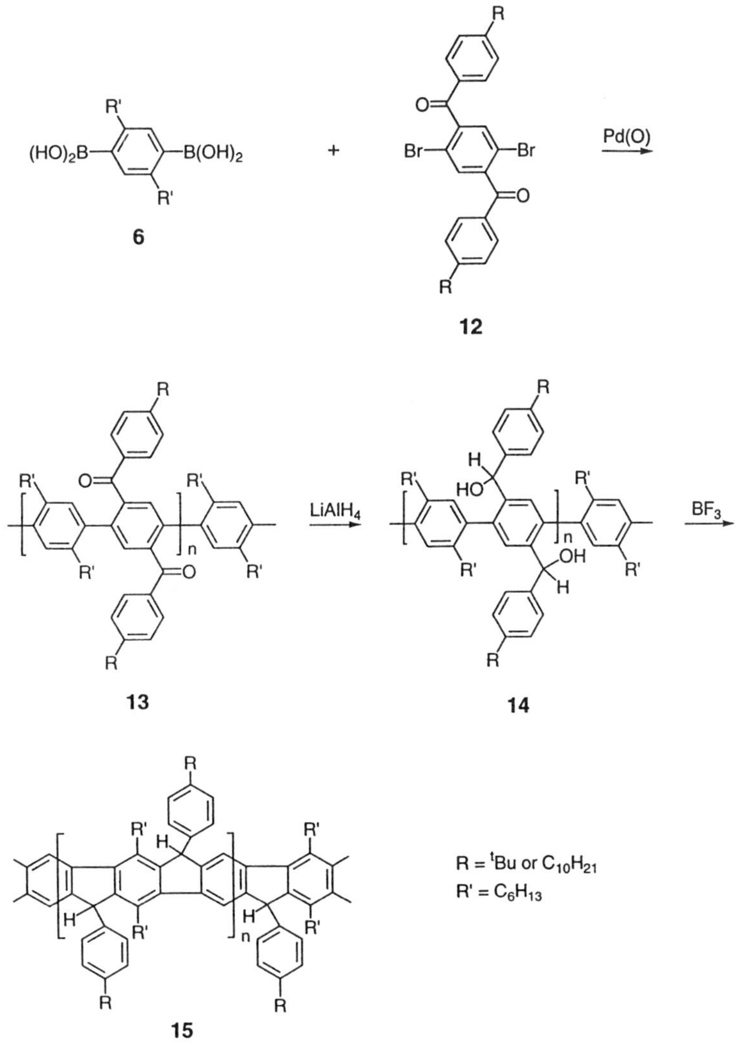

$R = {}^tBu$ or $C_{10}H_{21}$
$R' = C_6H_{13}$

Fig. 8.8 Scherf's synthesis of ladder-type PPP **15**.

signed. The repeat units of Swager's precursor polymer, **16** (Fig. 8.9), contains a phenylene ring with two alkoxys and one with phenyl acetylenes [65]. Upon addition of acid, the acetylene is protonated to give, formally speaking, a vinyl cation, which then attacks the adjacent (donor-substituted!) phenylene ring. Tour's precursor, **18**, contains keto functions at one ring and protected amines on the other [66]. Immediately upon deprotection, the amine functions attack the keto groups and cyclic imines are formed. The backbone of the resulting polymer **19** closely resembles Swager's **17**. Just two carbon atoms are replaced by nitrogen. It should be mentioned that the three ladder polymers described in this chapter can still be described as PPPs. In the case of **17**, for example, the PPP substructure becomes visible, when in a Gedankenexperiment the newly formed two-carbon bridges are considered substituents and not integral constituents of the backbone. The band gaps of the unsubstituted analog of **17** and (the nonexistent) planar PPP have been calculated by Brédas and found to be 2.86 and 2.75 eV, respectively. Since these values are very similar, the description of **17** as a PPP derivative

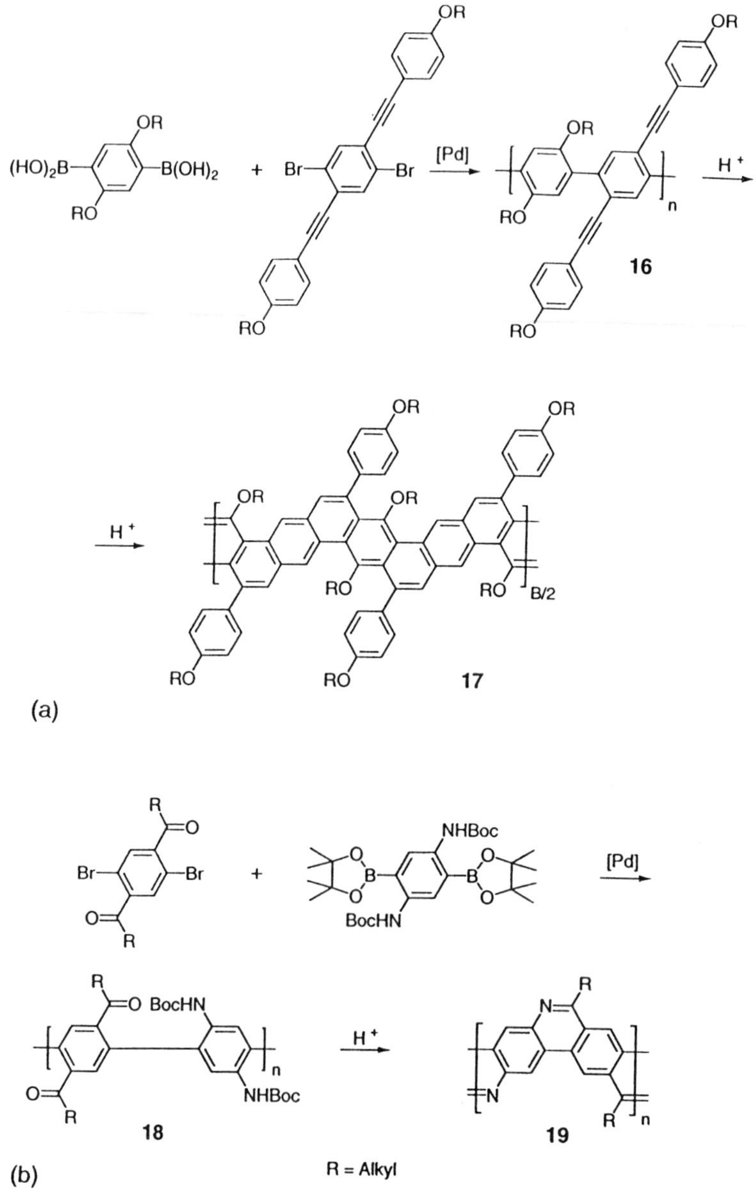

Fig. 8.9 Swager's (top) and Tour's (bottom) syntheses of the ladder-type PPPs **17** and **19**.

is not unreasonable. A polymer closely related to **17** has also been prepared by Chmil and Scherf [67].

3. PPP Polyelectrolytes

Despite the effort recently invested in the synthesis and characterization of rigid-rod polyelectrolytes, it is not yet clear in which fields of application they may be superior to the well-established representatives of this class of polymers. The present state is still best described as exploratory. Research is directed toward an understanding of their mesophase and miscibility/compatibility behavior, as well as their behavior at interphases, to name a few directions. Within this context the PPP-based polyelectrolytes **20–23** were prepared (Fig. 8.10). The PPP derivatives **20** and **22** were prepared in Novak's [68] and

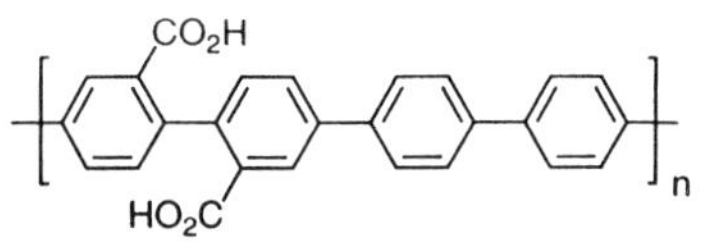

20

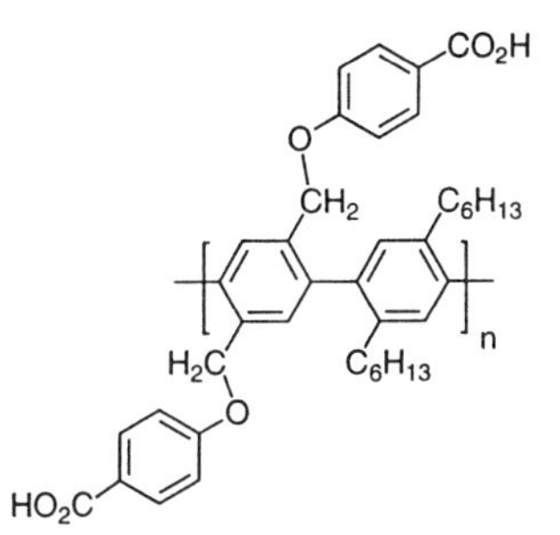

21

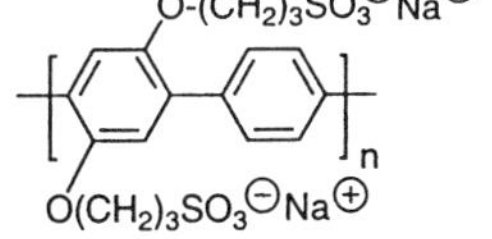

22

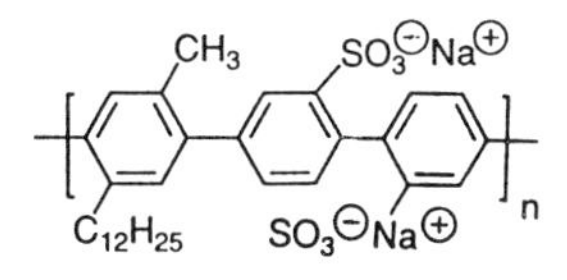

R = Alkyl **23**

Fig. 8.10 Structures of polyelectrolyte PPPs **20–23**.

Reynold's [69] laboratories under homogeneous reaction conditions. By using water–DMF mixtures and the water-soluble Casalnuovo Pd catalyst [70], the highly polar acid groups were introduced at the very beginning. This has the advantage that no further synthetic step is required and the disadvantage that the molecular weight determination becomes very difficult because of aggregation phenomena. In the preparation of **21** and **23**, Rau and Rehahn [71] and Rulkens et al. [72] followed the alternative, more work-intensive strategy in which a precursor polymer is first synthesized whose functional groups are then converted into carboxylate or sulfonate. This way the molecular weights could be reliably determined. For example, the degree of polymerization of a precursor of **23**, and thus of **23** itself, was determined to be DP = 72 by membrane osmosis. In a careful electrochemical study, Reynolds was able to show that his water-soluble PPP polyelectrolyte **22** can be both *p*- and *n*-type redox-doped, which precipitated an extensive investigation into the optoelectronic properties of this and related polymers.

C. Toward Cheaper Procedures

The PPPs described so far were synthesized using relatively expensive monomers, catalysts, and reaction conditions. Many of them are therefore destined to never gain any industrial importance. They will, however, be indispensable "toys" for all kinds of basic research studies, which will further knowledge about rigid-rod polymers in specific and add to the understanding of polymer science in general. Considering the great expectation associated with this class of polymers, it is not surprising that scientists in industry have tried to develop some ways to make certain PPPs available at a reasonable cost. The two important factors to determine were the nature of the leaving group and the catalyst. The question was: Can one replace bromine by chlorine, and Pd by Ni? The Union Carbide researchers Colon and Kelsey in 1986 made the important observation that under certain conditions chlorobenzene can be coupled to biphenylene in the presence of Ni complexes and Zn as reducing agent [74] (Fig. 8.11). In a detailed study, which also shed light on possible side and termination reactions, they found that the yield for this coupling step may exceed 98%. This was a clear indication that this C—C bond-forming reaction may be of use for step-growth polymerizations. Colon and Kwiatkowski [75] made the first polymer, the aromatic polyether sulfone **24**, with this methodology. Shortly thereafter, Maxdem Inc. launched a press release in which the commercialization of polymer **25** was announced. Even though not all the experimental details were disclosed, the researchers Marrocco and Gagné seem to have used a very similar procedure to obtain their PPP, which they named Poly-X. Presumably because of the interesting proper-

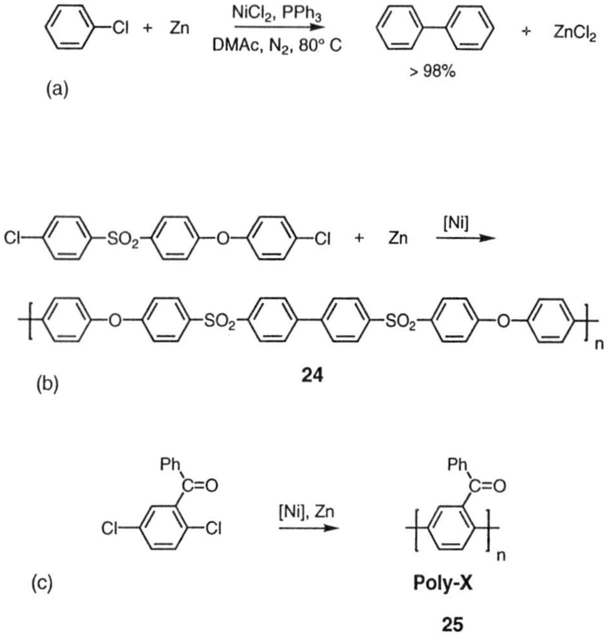

Fig. 8.11 Nickel-catalyzed couplings. (a) Colon's initial experiments: (b) the first use in polymer synthesis, (c) Maxdem's commercialized product Poly-X **25**.

ties reported by Maxdem for the thermoplastic polymer **25** (tensile modulus 0.9–2.6 msi, tensile strength 15–35 ksi, dielectric constant 3.0–3.3, moisture uptake 0.2–0.3%, moldability, etc.) two prominent university research groups started investigations into this matter. Wang and Quirk [76,77] and DeSimone's group [78] mostly concentrated on the effect that changes in the synthetic protocol, e.g., the nature of chelating ligand in the Ni-catalyzed polymerization of 2,5-dichlorobenzophenone in the presence of excess zinc, have on the microstructure of **25** and consequently on properties like solubility and glass transition temperature. Wang and Quirk found, for example, that by changing the monodentate ligand triphenylphosphine to the bidentate 2,2′-bipyridine, polymer **25** is produced with a 68° higher glass transition temperature and a λ_{max} in the UV-vis spectrum that is bathochromically shifted by 24 nm. Together with other evidence, these striking differences were attributed to more head-to-tail structures in **25** when the bidentate ligand was used.

During the early 1990s, Kaeriyama's group contributed to this field when they reported the synthesis of the ester derivative of PPP, **26**, which was used as a precursor for parent PPP [79–81] (Fig. 8.12). Hydrolysis of the ester functions in **26** afforded the free acids, which were catalytically decarboxylated using copper(II) oxide, basic copper(II) carbonate, and copper(I) oxide.

PPP was obtained as an amorphous powder. Very similar experiments were recently reported by Ueda and Yoneda [82]. BASF chemists were also active in this area during the same period but did not reach the commercialization stage. Kallitsis and Naarmann [83] made a series of PPPs with the general structure **27**. They found that the best results were obtained with 20 mol % catalyst precursor per mole of monomer. Such a high amount of "catalyst" is, of course, disadvantageous for many reasons, including financial ones. Recently a very interesting development for scientific, but maybe also industrial, reasons came from Percec's laboratory [84–86]. For a couple of years, they had tried to introduce oxygen-based leaving groups like tosylates and triflates in Ni-catalyzed cross-coupling reactions but met with only limited success with respect to the molecular weight of the PPP derivatives prepared [87,88]. They now seem to have accomplished a significant breakthrough. It was shown that mesylates, which are much cheaper than, e.g., triflates, may give good results and that high molecular weight polymers can be obtained from simple hydroquinones or bisphenols. For polymer **29**, which was prepared from the easily accessible dimesylated monomer **28**, a DP of 101 was reported. This case is admittedly the one that went best; there are many more examples, however, in the DP range of 30–45, a molecular weight regime that is already interesting for many applications. It should be mentioned that Ni/Zn-based cross-coupling has been used not only for the synthesis of PPP derivatives but also for many heteroarylenes and related aromatic polymers [89–97].

It was clear from the beginning that prices of commodity polymers are not within reach for PPP derivatives. It may especially surprise academic researchers to learn from the Maxdem Inc. press release that certain Poly-X resins are expected to sell for only about $10/lb. This shows the dramatic effect that twisting the two screws, leaving group and catalyst, can have.

IV. THE INDIRECT OR PRECURSOR APPROACH

A. Marvel's Experiment

The synthesis of Durham polyacetylene [98] and Wessling–Lenz polyphenylenevinylene [99,100] in the 1980s raised a lot of interest, and not only because of the importance of these polymers for material science. They were successful examples of the so-called indirect or precursor approach in which an insoluble target polymer is liberated from a well-characterizable precursor by some well-understood, cleanly proceeding chemistry. The polymer chemist's community, which, for good reasons, tends to be skeptical toward this kind of approach, realized that the precursor approach can be a powerful method in spite of all the problems inherently associated with it. In view of these developments, which date back

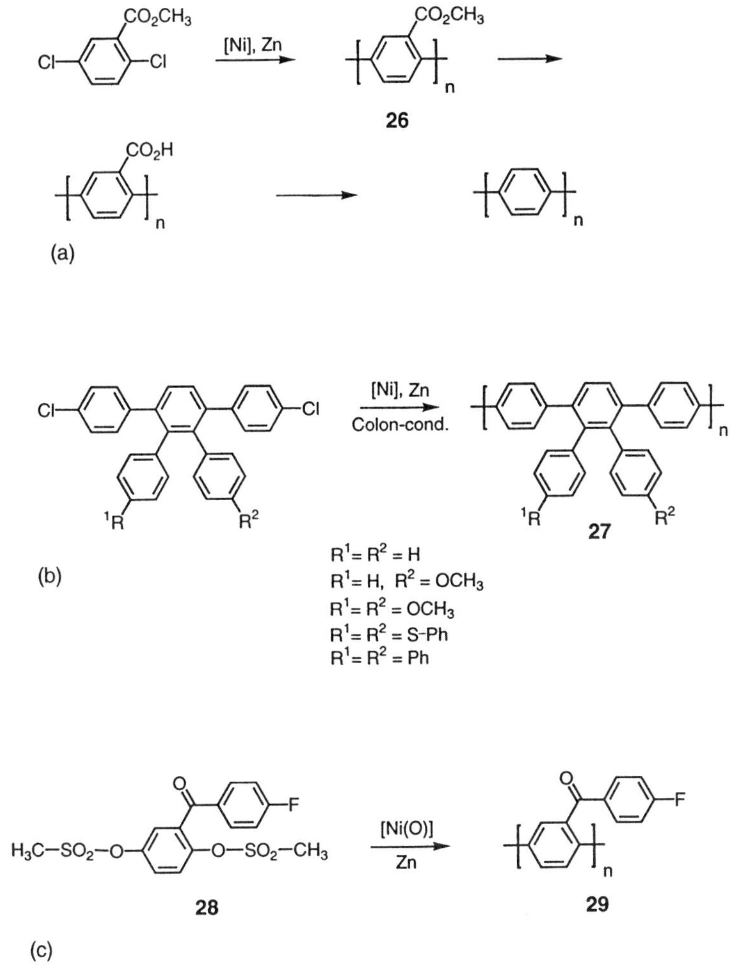

Synthesis of PPP

Fig. 8.12 Nickel-catalyzed couplings. (a) Kaeriyama's PPP synthesis; (b) Naarmann's PPP derivative **27**; (c) Percec's PPP route using mesylate as the leaving group.

only a few years, Marvel's precursor concept for the synthesis of parent PPP first published in the late 1950s was certainly a very farsighted one, even though nowadays one has to say that by and large it was not successful [101,102] (Fig. 8.13). The difficulties Marvel encountered with his strategy were (1) The low molecular weight of precursor polymer **30**; (2) poor regiochemical control in **30**; (3) the lack of stereochemical control in **30**; and (4) the poor control of the aromatization of **30** regardless of the method tried. His strategy, however, contains all the important features that are also found in modern approaches to PPP, from which we now know that they actually solved the synthetic problem. These features are the use of a cyclohexadiene monomer, its chain-growth polymerization to furnish the precursor polymer, and the precursor's subsequent aromatization to the targeted PPP. In this regard, Marvel's work is certainly an important milestone in the history of PPP synthesis.

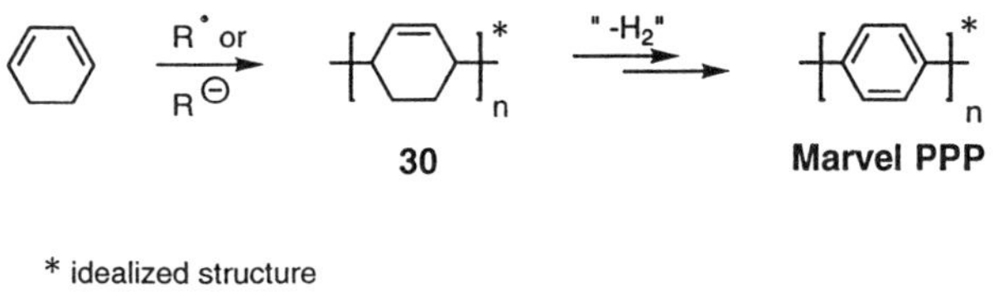

Fig. 8.13 The basic steps of Marvel's 1958 concept for the synthesis of parent PPP.

B. ICI's Biotechnological Monomer Synthesis

Marvel's failure to aromatize precursor **30** with at least some structural control led ICI researcher Ballard and coworkers to develop a new monomer based on cyclohexadiene [103,104]. They sought a monomer with functional groups that could not only be carried through the polymerization but, more important, allow for clean aromatization to take place. In organic chemistry, ester pyrolysis and related reactions are an important tool for the generation of unsaturation. One C—C double bond is formed per ester function. Such groups were therefore attractive, and Ballard decided to try the sequence shown in Fig. 8.14. It starts with monomer **31**, which carries two ester, carbonate, or related groups. This choice required the synthesis of the dihydroxycyclohexadiene **33** from which to prepare **31**. A synthesis of **33** was known at that time but involved many steps and a work-intensive procedure that could by no means be done on a technical scale. Additionally, the considerable sensitivity of **33** toward traces of acids, which inevitably leads to its dehydration to phenol, had to be taken into account. At this point the ICI chemists came up with the brilliant idea to use biotechnological techniques to circumvent the synthetic obstacles. They knew that the microorganism *Pseudomonos putida* 11767 oxidizes benzene within the bacterial cell according to the simplified picture in Fig. 8.14b. The dioxygenase E1 with the assistance of the protonated co-catalyst nicotinamide adenine dinucleotide (NADH) reacts with oxygen to form **33**. The complete oxygen molecule is used in this process. The protons are supplied by NADH. The undesired subsequent aromatization had to be suppressed. This could be done by a genetic modification of the microorganism to a new variant lacking the enzyme E2 that is responsible for the aromatization of **33**. With this new variant, compound **33** was produced on the kilogram scale and under very mild conditions. It should be mentioned that the *cis* isomer shown was the only product and that the conversion of benzene was virtually 100%. The modifications to the differently substituted monomers **31** were accomplished straightforwardly. Monomer **31** with R = methoxycarbonyl was investigated most intensely.

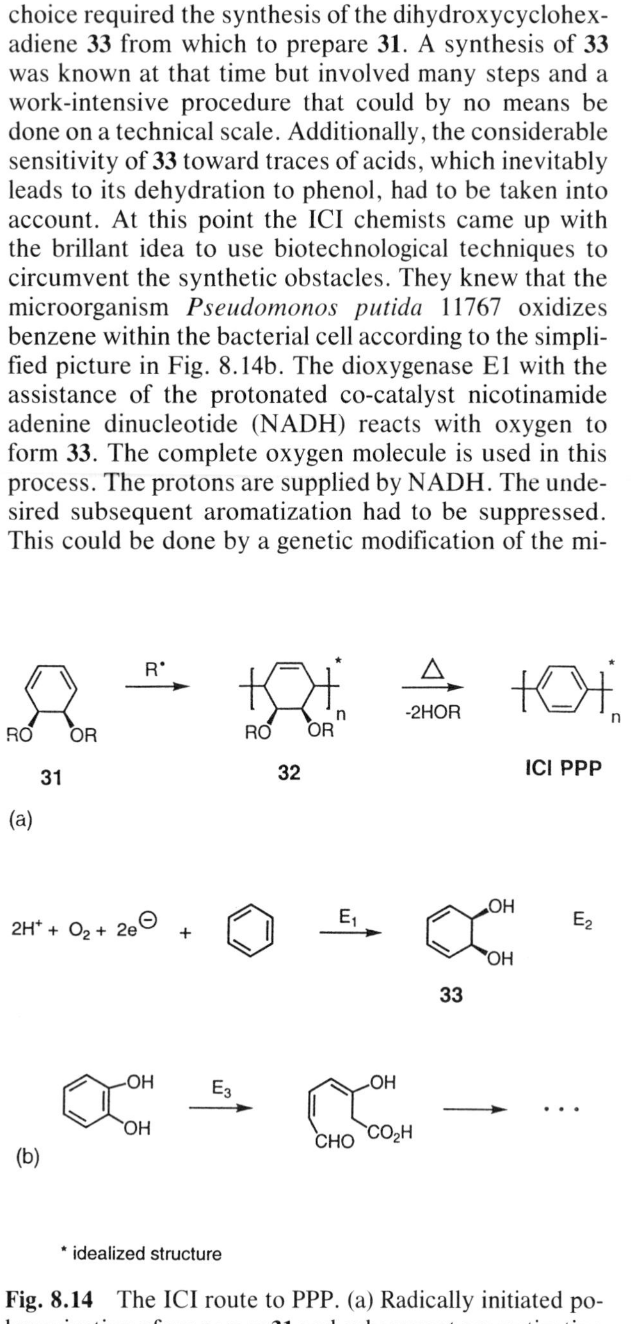

The beauty of the monomer synthesis could unfortunately be only partially transferred to the rest of the sequence. Monomer **31** was polymerized to **32** employing radical initiators [105]. The use of radical initiators, which is convenient for many, many reasons, had two consequences. First, there is the limited regio- and stereospecific control during polymerization. For example, polymer **32** contained not only 1,4 linkages but also approximately 15% 1,2 linkages. Second, it is difficult to control both molecular weight and distribution. Nonetheless, the precursor polymer obtained has a very high molecular weight that enables it to be processed into films and fibers. This was the first time that a potential precursor for PPP (better yet, polyphenylene) could be reasonably processed. The random stereochemistry of **32** and the considerable amount of 1,2 linkages caused by the choice of initiator had a disadvantageous impact on the thermally induced aromatization process. Neutron scattering experiments of ICI PPP suggested that chain fracture during the thermally induced aromatization occurs preferentially at repeat units with certain stereochemistry as well as at 1,2-linked repeat units (internal report, ICI Chemicals and Polymers, Ltd., Runcorn, UK). As a consequence, aromatization furnished a relatively ill-defined, low molecular weight phenylenic material. Even though ICI could not carry through the whole sequence to PPP fully satisfactorily, one has to arrive at the conclusion that their work has significantly pushed the whole area forward. As will be seen later, ICI PPP, in spite of its imperfect structure, is far superior to Marvel's PPP and became an important material studied by physicists in great detail.

C. Full Structure Control of the Precursor Polymer

After radical initiators had not led to the desired result, and anionic as well as cationic initiators were not likely to do the job any better, the next logical step was to consider transition metal complexes as potential initiators. This required, among other things, the selection of a catalyst that combines a regiospecific 1,4 polymerization and sufficient compatibility toward the functional groups of the monomer. Considering the strong electrophilic character of most Ziegler–Natta catalyst systems, these prerequisites were not trivial. A key observation in this whole matter was made by Teyssie's group. They

* idealized structure

Fig. 8.14 The ICI route to PPP. (a) Radically initiated polymerization of monomer **31** and subsequent aromatization of **32**. (b) Biotechnological route to **33**, a starting material for the synthesis of monomer **31**.

found that the π-allyl nickel complex **34** (Fig. 8.15) is an excellent catalyst for 1,3-butadiene [106]. It combines high activity and excellent regioselectivity (~ 1% 1,2 linkages) with an amazing stereoselectivity that can be switched from 99% *cis* to 99% *trans* depending on the ligands and reaction conditions. Additionally, it was observed a few years earlier that this catalyst also exhibits a partial tolerance to heteroatom functionalities [107].

It seemed logical to try the polymerization of monomer **31** with this catalyst. Chemistry, however, does not always do what chemists want. Grubbs and his coworkers who did these experiments found that the catalyst aromatizes a stoichiometric amount of the monomer under self-decomposition. Not discouraged by this undesired finding, the same group started an intensive search for monomers compatible with **34** and were finally successful [108]. Compound **35** turned out to be the ideal candidate (Fig. 8.15). Two factors seem likely to be responsible for this: (1) The steric effect of the TMS group prevents the oxygen atoms of the monomer from coordinating to the catalyst and interfering with the polymerization, and (2) TMS ethers are very poor leaving groups compared to esters and carbonates, a feature that inhibits aromatization of the monomer and subsequent catalyst decomposition. The bulky TMS groups may also play an advantageous role during polymerization by blocking off one face of the ring and directing the inserting monomer into the propagating chain exclusively from the unshielded side. The consideration was that this π-facial selectivity afforded by the monomer, coupled with the 1,4-regioselectivity afforded by the catalyst, should yield polymer **36** as a completely 1,4-linked, stereoregular polymer with 1,4-*SSRR* repeat units. That this actually holds true could be proved by Grubbs. The only stereochemical feature that could not be unambiguously clarified was whether the polymer is isotactic or syndiotactic. Fortunately, this feature was not as important for the aromatization process as the 1,4-regiochemistry and the geometric relationships within each repeat unit.

All that remained to be done on the way to a structurally perfect precursor polymer was to convert the TMS ether groups in **36** to the esters in **38**, which are needed for aromatization. Standard organic chemistry employing tetra-*n*-butylammonium fluoride (TBAF; "naked" fluoride) in methanol led to a clean deprotection of **36** to furnish the polyalcohol **37**, which was then converted into the "polyester" **38** with acetic anhydride and catalytic amounts of base [108]. With this result, obtained some 30 years after Marvel's first indirect "PPP" synthesis, the first well-defined and processable potential PPP precursor had been prepared—clearly an important accomplishment.

D. The Final Step: Aromatization

As the parent PPP is absolutely insoluble and intractable, the matter of converting its precursor polymer into PPP comprises two aspects that are quite different but that are both very important: chemistry and processing. On the one hand, it is not sufficient if the aromatization can be done meticulously when, at the same time, the final product can be obtained only as an amorphous powder and not in a useful form. On the other hand, it is not sufficient that the precursor polymer can be easily processed into films, fibers, and all kinds of interesting shapes when the chemical conversion into the final product cannot be accomplished satisfactorily. In both cases, the intrinsic properties of PPP cannot be used to the full extent, thus defeating the goal of the previous efforts.

As mentioned, processing of the ICI PPP precursor **32** is convenient. For example, substrates can be spin-coated using solutions of **32**. The conversion into PPP, which is often done by heating films of the precursor to approximately 340°C under vacuum, seems to lead to significant chain fracture. Nevertheless, the material obtained is still of sufficient quality to be used as the active material in electroluminescent devices [22]. In contrast to this, the Caltech PPP obtained from the stereo- and regioregular precursor **38** using acid-catalyzed bulk pyrolysis techniques has a chemically well-defined structure; chain fracture can be avoided under certain circumstances [109]. In their first publication on this matter, however, Grubbs and his coworkers were not able to obtain the PPP other than as amorphous foams. The reason for this seems to be that during the aromatization process acetic acid vapor is rapidly formed. This expands the PPP matrix into a void-filled foam. An additional disadvantage is that the Lewis and Bronsted acids used to lower the pyrolysis temperature either cannot

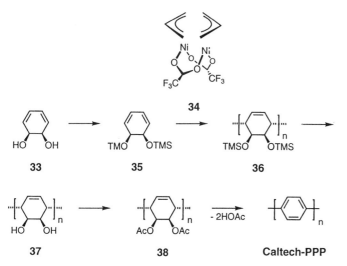

Fig. 8.15 The Caltech route to PPP using Ni complex **34** for the regio- and stereospecific synthesis of precursor polymer **38**.

be removed from the product, thus rendering it brittle, or sublime from the precursor before its conversion is completed. This is specifically important when thin films are to be prepared. It is reasonable to assume that these problems will be overcome shortly. An indication is that MacDiarmid recently succeeded in converting Grubbs PPP precursor into free-standing films of high molecular weight, structurally regular PPP by using polyphosphoric acid to catalyze the elimination [110]. Even though this looks like a breakthrough it is still too early to rest on one's laurels. The material obtained contains significant amounts of the acid. If this would just act as a plasticizer, it might be acceptable for some applications. This kind of impurity, however, strongly influences the electro-optical properties of the polymer and prevents any use of it for applications where these properties are essential.

V. SUMMARY

For some 30 years now, chemists have tried to tackle the problem of how to synthesize the simplest aromatic polymer, PPP. The main difficulty is that the repeat units are linked by carbon–carbon bonds and not by carbon–heteroatom bonds. Two approaches have been followed, the direct and the indirect, which are tightly associated with the names of the researchers Kovacic and Marvel, who did the initial experiments. Since then a whole brigade of chemists have contributed to solving the problem. Truly significant progress has been made, even though here and there improvements are still possible. The direct route has led to a whole bouquet of relatively high molecular weight and structurally well-defined PPPs. According to the material properties targeted, different substitution patterns were realized. This not only shows the variability of the synthetic strategy but will certainly also further the importance of this class of polymers. In the case of the indirect route, the achievements are also spectacular. High molecular weight parent PPP is now accessible, for example, in film form from regio- and stereochemically defined precursor polymers. Marvel's dream has come true. The PPP story is a beautiful example of what can be accomplished if chemists from various disciplines (macromolecular chemistry, organic chemistry, transition metal chemistry, etc.) and from various institutions (universities, industry, and research institutes) work toward the same goal and share their individual knowledge and expertise.

REFERENCES

1. W. Ried, and D. Freitag. *Naturwiss.* 53:305 (1966).
2. J. K. Stille and Y. Gilliams, *Macromolecules* 4:515 (1971).
3. J. K. Stille, *Makromol. Chem.* 154:49 (1971).
4. J. M. Dineen and A. A. Volpe, *Am. Chem. Soc., Polym. Div., Polym. Prepr.* 19:34 (1978).
5. W. R. Krigbaum and K. J. Krause, *J. Polym. Sci., Polym. Chem. Ed.* 16:3151 (1978).
6. J. M. Dineen, E. E. Howell, and A. A. Volpe, *J. Polym. Sci., Polym. Chem. Ed.* 23:282 (1982).
7. D. G. H. Ballard, A. Courtis, I. M. Shirley, and S. C. Taylor, *J. Chem. Soc., Chem. Commun.* 1983: 954.
8. K. Mukai, T. Teshirogi, N. Kuramoto, and T. Kitamura, *J. Polym. Sci., Polym. Chem. Ed.* 23:1259 (1985).
9. P. Kovacic and M. B. Jones, *Chem. Rev.* 87:357 (1987).
10. D. R. McKean and J. K. Stille, *Macromolecules* 20: 1787 (1987).
11. D. G. H. Ballard, A. Courtis, I. M. Shirley, and S. C. Taylor, *Macromolecules* 21:294 (1988).
12. C. B. Gorman and R. H. Grubbs, in *Conjugated Polymers* (J. L. Brédas and R. Silbey, eds.), Kluwer, Dordrecht, 1991.
13. A. F. Diaz and J. Bargon, in *Handbook of Conducting Polymers* (T. A. Skotheim, ed.), Vol. 1, Marcel Dekker, New York, 1986, p. 81.
14. J. L. Brédas, *J. Chem. Phys.* 82:3808 (1985).
15. G. K. Noren and J. K. Stille, *Macromol. Rev.* 5:385 (1971).
16. D. M. Gale, *J. Appl. Polym. Sci.* 22:1971 (1978).
17. R. H. Baughman, J. L. Brédas, R. R. Chance, R. I. Elsenbaumer, and L. W. Shacklette, *Chem. Rev.* 82: 209 (1982).
18. R. I. Elsenbaumer and L. W. Shacklette, in *Handbook of Conducting Polymers* (T. A. Skotheim, ed.), Marcel Dekker, New York, 1986, Vol. 1, Chap. 7.
19. M. B. Jones and P. Kovacic, in *Comprehensive Polymer Science* (G. C. Eastmond, A. Ledwith, S. Russo, and P. Sigwalt, eds.), Pergamon, Oxford, 1989, Vol. 5, p. 465.
20. V. Percec and D. Tomazos, in *Comprehensive Polymer Science* (S. L. Aggarwal and S. Russo, eds.), Pergamon, Oxford, 1992, Suppl. 1, p. 318.
21. T. Yamamoto, *Prog. Polym. Sci.* 17:1153 (1992).
22. G. Grem, G. Leditzky, B. Ullrich, and G. Leising, *Adv. Mater.* 4:36 (1992).
23. E. B. Stephens and J. M. Tour, *Macromolecules* 26: 2420 (1993).
24. J. H. Burroughes, D. D. C. Bradley, A. R. Brown, R. N. Marks, K. Mackay, R. H. Friend, P. L. Burn, and A. B. Holmes, *Nature* 347:539 (1990).
25. J. Okuda, *Nachr. Chem. Techn. Lab.* 41:8 (1993).
26. R. Schrock, *Acc. Chem. Res.* 23:158 (1990).
27. A.-D. Schlüter and G. Wegner, *Acta Polym.* 44:59 (1993).
28. J. M. Tour, *Adv. Mater.* 6:190 (1994).
29. T. Yamamoto, Y. Hayashi, and Y. Yamamoto, *Bull. Chem. Soc. Jpn.* 51:2091 (1978).
30. T. Yamamoto, A. Morita, Y. Miyazaki, T. Maruyama, H. Wakayama, Z. Zhou, Y. Nakamura, T.

Kanbara, S. Sasaki, and K. Kubota, *Macromolecules* 25:1214 (1992).
31. T. Yamamoto, *Prog. Polym. Sci. 17*:1153 (1992).
32. H. O. Wirth, in *Physik and Chemie der Scintillatoren* (N. Riehl and H. P. Kallmann, eds.), Thiemig Verlag, München, 1966.
33. W. Kern and H. O. Wirth, *Kunststoffe, Plastics 6*: 12 (1959).
34. W. Kern, M. Seibel, and H. O. Wirth, *Makromol. Chem. 29*:164 (1959).
35. W. Kern, H. W. Ebersbach, and I. Ziegler, *Makromol. Chem. 31*:154 (1959).
36. H. Langhals, S. Demig, and T. Potrawa, *J. Prakt. Chem. 333*:733 (1991).
37. W. Heitz, *Chem.-Ztg. 110*:385 (1986).
38. H. Ringsdorf, P. Tschirmer, O. Hermann-Schönherr, and J. H. Wendorff, *Makromol. Chem. 188*:1431 (1987).
39. M. Ballauf, *Angew. Chem. Int. Ed. Engl. 28*:253 (1989).
40. M. Rehahn, A.-D. Schlüter, G. Wegner, and W. J. Feast, *Polymer 30*:1054 (1989).
41. M. F. Semmelhack and L. S. Ryono, *J. Am. Chem. Soc. 97*:2873 (1975).
42. D. R. Fahey and J. E. Mahan, *J. Am. Chem. Soc. 98*:4499 (1976).
43. A. Zask and P. Helquist, *J. Org. Chem. 43*:1619 (1978).
44. M. Lewin, A. Aizenshtat, and J. Blum. *J. Organomet. Chem. 184*:255 (1980).
45. I. Colon, *J. Org. Chem. 47*:2622 (1982).
46. I. Colon, and D. R. Kelsey, *J. Org. Chem. 51*:2627 (1986).
47. N. Miyaura, T. Yanagi, and A. Suzuki, *Synth. Commun. 11*:513 (1981).
48. R. B. Miller and S. Dugar, *Organometallics 3*:1261 (1984).
49. V. Snieckus, *Chem. Rev. 90*:925 (1990).
50. V. N. Kalinin, *Synthesis 1992*:413.
51. A. R. Martin and Y. Yang, *Acta Chem. Scand. 47*: 221 (1993).
52. M. Rehahn, A.-D. Schlüter, G. Wegner, and W. J. Feast, *Polymer 30*:1060 (1989).
53. A. Almennigen, O. Bastiansen, L. Fernhold, B. N. Cyvin, S. J. Cyvin, and S. Samdal, *J. Mol. Struct. 128*:59 (1985).
54. G.-P. Carbonneau and Y. Delugeard, *Acta Crystallogr. B33*:1586 (1977).
55. C. P. Brock and R. P. Minton. *J. Am. Chem. Soc. 111*:4586 (1989).
56. K. N. Baker, A. V. Fratine, and W. Wade Adams, *Polymer 31*:1623 (1990).
57. J. L. Brédas, R. R. Chance, R. H. Baughman, and R. Silbey, *J. Chem. Phys. 76*:3673 (1982).
58. M. Rehahn, A.-D. Schlüter, and G. Wegner, *Makromol. Chem. 191*:1991 (1990).
59. M. Kreyenschmidt, R. Uckert, and K. Müllen, *Macromolecules 28*:4577 (1995).
60. A.-D. Schlüter, *Adv. Mater, 3*:284 (1991).
61. U. Scherf and K. Müllen, *Makromol. Chem. Rapid Commun. 12*:489 (1991).
62. J. Grimme, M. Kreyenschmidt, F. Uckert, K. Müllen, and U. Scherf, *Adv. Mater. 7*:292 (1995).
63. J. Huber, K. Müllen, J. Salbeck, H. Schenk, U. Scherf, T. Stehlin, and R. Stern, *Acta Polym. 45*:244 (1994).
64. J. Grüner, P. J. Hamer, R. H. Friend, H.-J. Huber, U. Scherf, and A. B. Holmes, *Adv. Mater. 6*:748 (1994).
65. M. B. Goldfinger, and T. M. Swager, *J. Am. Chem. Soc. 116*:7895 (1994).
66. J. J. S. Lamba and J. M. Tour, *J. Am. Chem. Soc. 116*:11723 (1994).
67. K. Chmil and U. Scherf, *Makromol. Chem. Rapid Commun. 14*:217 (1993).
68. T. I. Wallow and B. M. Novak, *J. Am. Chem. Soc. 113*:7411 (1991).
69. A. D. Child and J. R. Reynolds, *Macromolecules 27*: 1975 (1994).
70. A. L. Casalnuovo and J. C. Calabrese, *J. Am. Chem. Soc. 112*:4324 (1990).
71. I. U. Rau and M. Rehahn, *Polymer 34*:2889 (1993).
72. I. U. Rau and M. Rehahn, *Acta Polym. 45*:3 (1994).
73. R. Rulkens, M. Schulze, and G. Wegner, *Macromol. Rapid Commun. 15*:669 (1994).
74. I. Colon and D. R. Kelsey, *J. Org. Chem. 51*:2627 (1986).
75. I. Colon and G. T. Kwiatkowski, *J. Polym. Sci., Polym. Chem. Ed. 28*:367 (1990).
76. G. Y. Wang and R. P. Quirk, *Am. Chem. Soc., Div. Polym. Chem., Polym. Prepr. 35*:712 (1994).
77. Y. Wang and R. P. Quirk, *Macromolecules 28*:3495 (1995).
78. R. W. Phillips, V. V. Sheares, E. T. Samulski, and J. M. DeSimone, *Macromolecules 27*:2354 (1994).
79. V. Chaturvedi, S. Tanaka, and K. Kaeriyama, *J. Chem. Soc., Chem. Commun. 1992*:1658.
80. V. Chaturvedi, S. Tanaka, and K. Kaeriyama, *Macromolecules 26*:2607 (1993).
81. K. Kaeriyama, M. A. Mehta, V. Chaturvedi, and H. Masuda, *Polymer 36*:3027 (1995).
82. M. Ueda and M. Yoneda, *Macromol. Rapid. Commun. 16*:469 (1995).
83. J. K. Kallitsis and H. Naarmann, *Synth. Met. 44*:247 (1991).
84. V. Percec, J.-Y. Bae, M. Zhao, and D. H. Hill, *J. Org. Chem. 60*:176 (1995).
85. V. Percec, J.-Y. Bae, M. Zhao, and D. H. Hill, *J. Org. Chem.* submitted.
86. V. Percec, J.-Y. Bae, M. Zhao, and D. H. Hill, *Am. Chem. Soc., Polym. Chem. Div., Polym. Prepr. 36*(1):699 (1995).
87. V. Percec, S. Okita, and R. Weiss, *Macromolecules 25*:1816 (1992).
88. V. Percec, and S. Okita, *J. Polym. Sci., Polym. Chem. Ed. 31*:877 (1993).

89. M. D. Bezoari, P. Kovacic, S. Gronowitz, and A.-B. Hörnfeldt, *J. Polym. Sci., Polym. Lett. Ed. 19*: 347 (1981).
90. I. Khoury, P. Kovacic, and H. M. Gilow, *J. Polym. Sci., Polym. Sci., Polym. Lett. Ed. 19*:395 (1981).
91. T. Yamamoto, K. Osakada, T. Wakabayashi, and A. Yamamoto, *Makromol. Chem., Rapid Commun. 6*: 671 (1985).
92. M. Ueda and F. Ichikawa, *Macromolecules 23*:926 (1990).
93. M. Ueda, Y. Miyaji, T. Ito, Y. Oba, and T. Sone, *Macromolecules 24*:2694 (1991).
94. T. Yamamoto, A. Morita, Y. Miyazaki, T. Maruyama, H. Wakayama, Z. Zhou, Y. Nakamura, T. Kanbara, S. Sasaki, and K. Kubota, *Macromolecules 25*:1214 (1992).
95. T. Yamamoto, Z. Zhou, I. Ando, and M. Kikuchi, *Makromol. Chem., Rapid. Commun. 14*:833 (1993).
96. T. Chen, X. Wu, and R. D. Rieke, *J. Am. Chem. Soc. 117*:233 (1995).
97. T. Yamamoto, Y. Yoneda, and K. Kizu, *Macromol. Rapid Commun. 16*:549 (1995).
98. W. J. Feast and J. H. Edwards, *Polymer 21*:595 (1980).
99. R. A. Wessling, *J. Polym. Sci., Polym. Symp. 72*:55 (1985).
100. R. W. Lenz, C.-C. Han, J. Stenger-Smith, and F. E. Karasz, *J. Polym. Sci., Polym. Chem. Ed. 26*:3241 (1988).
101. C. S. Marvel and G. E. Hartzell, *J. Am. Chem. Soc. 81*:448 (1959).
102. P. E. Cassidy, C. S. Marvel, and S. Ray, *J. Polym. Sci., Part A. 3*:1553 (1965).
103. D. G. H. Ballard, A. Courtis, I. M. Shirley, and S. C. Taylor, *J. Chem. Soc., Chem. Commun. 1983*: 954.
104. D. G. H. Ballard, A. Courtis, I. M. Shirley, and S. C. Taylor, *Macromolecules 21*:294 (1988).
105. D. R. McKean and J. K. Stille, *Macromolecules 20*: 1787 (1987).
106. P. Hadjiandreon, M. Júlemont, and P. Teyssié, *Macromolecules 17*:2455 (1984).
107. F. Borge-Visse and F. Dawans, *J. Polym. Sci., Polym. Chem. Ed. 18*:2481 (1980).
108. D. L. Gin, V. P. Conticello, and R. H. Grubbs, *J. Am. Chem. Soc. 116*:10507 (1994).
109. D. L. Giu, V. P. Conticello, and R. H. Grubbs, *J. Am. Chem. Soc. 116*:10934 (1994).
110. D. L. Gin, J. K. Avlyanov, and A. G. MacDiarmid, *Synth. Met. 66*:169 (1994).

9
Regioregular, Head-to-Tail Coupled Poly(3-alkylthiophene) and Its Derivatives

Richard D. McCullough and Paul C. Ewbank
Carnegie Mellon University, Pittsburgh, Pennsylvania

I. INTRODUCTION

Over the past few years it has become clear that structure plays a critical role in determining the physical properties of conducting polymers. The most striking conclusion is that control of the regularity and order in the polymeric structure leads to remarkable enhancements in the electronic and photonic properties of these novel materials. This discovery leads to the exciting prospect that the properties of conducting polymers can be selectively engineered through synthesis and assembly. After the initial discovery of conducting organic polymers [1], chemists also prepared polymers that were soluble (and/or processible) and also exhibited reasonable electrical conductivities. The early problems concerning material intractability in polythiophenes [2–6], in particular, have been thoroughly addressed, and now efforts may be focused on the molecular engineering of structures to produce materials that possess enhanced electrical and optical properties. An effective design strategy is aimed at controlling both the microscopic and solid-state macroscopic structure, because they collectively define the resultant band structure [7] of a given material and thereby determine its electrical [8–13] and optical [14–18] properties. Molecular engineering must begin with the synthesis of homogeneous structures. This is the crucial first step toward the engineering of organic materials that will undergo controlled macroscopic assembly and possess exceptional properties.

It is clear that the ability to design and control conjugated π architectures is the key to creating new materials. Synthesis can determine the magnitude of π overlap along the backbone and eliminate structural defects. Materials assembly (and/or processing) determines interchain overlap and dimensionality. Together these concepts can be used to maximize carrier mobility and minimize electron–phonon scattering and thereby produce highly conductive organic polymers. A clear indication of how synthesis and assembly can determine the electrical and optical properties is found in Naarman polyacetylene [19]. Classical synthetic methods yield polyacetylene [1,20] with sp^3 hydridized defects at junctions of cross-links between polymer chains. Given that the degree of π overlap [8–10] along the chain directly determines the bandwidths (and band gap) and sp^3 centers decrease the conjugation, classically prepared polyacetylenes will possess larger band gaps than defect-free materials. Naarman's I_2-doped, defect-free polyacetylene exhibits electrical conductivities of up to 10^5 S/cm, which is within an order of magnitude of that of copper. The recent synthesis of regioregular head-to-tail coupled polythiophenes also clearly demonstrates that design, controlled synthesis, and directed assembly can lead to greatly enhanced properties in conjugated polymers. Work on regioregular HT polythiophenes has led to new highly conductive materials and to new insights on assembly in polythiophenes. This chapter addresses the design, synthesis, assembly, structure, and physical properties of regioregular head-to-tail coupled polythiophenes.

II. POLYTHIOPHENE AND IRREGULAR POLY(3-ALKYLTHIOPHENES)

A. Polythiophene

Early preparations of polythiophene employed both electrochemical and chemical methods [1,21–23]. Chemical preparations via oxidative coupling of thiophene or Kumada cross-coupling of Grignard reagents of dihalothiophenes have led to polythiophene (PT); electrochemical polymerization of thiophene also yields PT. The powdered PT from the chemical preparations and the films and powders generated electrochemically are insoluble and infusible and therefore are difficult to characterize. While it can be assumed that the Kumada coupling leads primarily to a 2,5-coupled polythiophene, the other synthetic methods are likely to produce the conjugation-disrupting 2,4-coupled polythiophene structure.

In polythiophenes (PTs), the thiophene rings coupled in the 2 and 5 positions allow for the conjugation of the π orbitals along the polymer chain that leads to the development of semiconductor/insulator band structure for the solid-state material. The large band gap leads to an electrical conductivity for neutral PT [1] of about 10^{-8} S/cm. When the material is oxidized (or doped), higher electrical conductivity in PT can be achieved. The consequence of the oxidation of polythiophene is a gross change in the electronic band structure manifested as new midgap states being created and a quinoidal type of resonance structure being formed (Fig. 9.1). This allows for charged carriers called bipolarons [8] to be created. The oxidation leads to an effective reduction in the band gap and an increase in the electrical conductivity of the material, with maximum conductivities ranging from 1 to 100 S/cm.

A quinodial type of resonance structure necessitates that a coplanar orientation of rings be readily accessible in order for a small band gap and high conductivity to be achieved [24]. If high energy constraints prevent the coplanar arrangement from being formed, then a low conductivity will result. In the solid state, dense packing of polymer chains can allow for overlap to occur in three dimensions. This 3-D electronic connectivity creates low resistivity pathways for electrons (or holes) to travel. Dense chain packing with overlap of the π orbitals can also lead to high electrical conductivities. In general, the processing of the polymer controls the assembly of the three-dimensional structure, determines the polymer morphology, and influences the resulting electrical conductivity [25]. Whereas unsubstituted polythiophene (PT) can form a desirable dense packed structure, that same tight packing and intermolecular chain aggregation renders unsubstituted PT completely insoluble and not melt-processable [1]. Despite the lack of processibility, the expected stability and high electrical conductivity of PT make it a highly desirable material. However, the intractability of PT makes the polymer difficult to characterize and study.

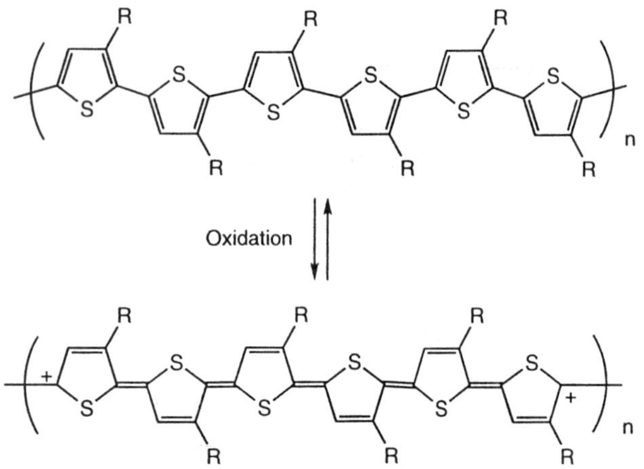

B. Irregular Poly(3-alkylthiophene)

The synthesis and development of chemically stable, soluble, poly(3-alkylthiophenes) (PATs) [2–5] was first reported by Elsenbaumer and coworkers [2,3] and further developed by a number of other groups [4,5]. Poly(3-alkylthiophene) can readily be melt- or solution-processed into films that after oxidation can exhibit reasonably high electrical conductivities [2–6]. In addition, the neutral polymers exhibit fast and large nonlinear optical responses [14–18,26–29], photo- and electroluminescence [30–37] behavior, and other band gap dependent phenomena. Conjugated polymers in general, and polythiophenes in particular, could lead to new chemical and optical sensors, light-emitting diodes and displays, electrodes, molecular devices, microwave-absorbing materials, new types of memory devices, batteries, artificial muscles, antistatic and electromagnetic shielding materials, nanoswitches, optical modulators and valves, imaging materials, smart windows, polymer electronic interconnects, nanoelectronic and optical devices, and transistors [30–47]. Progress toward practical applications of these materials critically depends on the understanding of the basic chemistry and physics that govern the optimization and stability of electronic and optical properties in organic conjugated polymers and on the development of new materials syntheses.

1. Synthesis

The first poly(3-alkylthiophenes) were prepared via Kumada cross-coupling [2,3] (Fig. 9.2), using a method similar to the one used for the chemical preparation of polythiophene [23]. In this synthesis a 2,5-diiodo-3-alkylthiophene is treated with 1 equiv of Mg in THF, generating a mixture of Grignard species. A catalytic amount of nickel(II) 1,3-bis(diphenylphosphino)propane [21] dibromide is then added, and the polymer is generated by a bromo-Grignard cross-coupling reaction. Large quantities of PATs can be prepared by this method and, though initially reported to have low molecular

Fig. 9.1 Neutral and oxidized forms of poly(3-alkylthiophene).

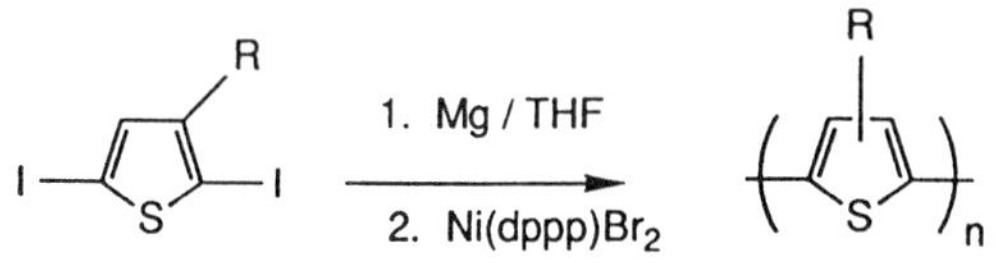

Fig. 9.2 Kumada coupling method for the synthesis of PATs.

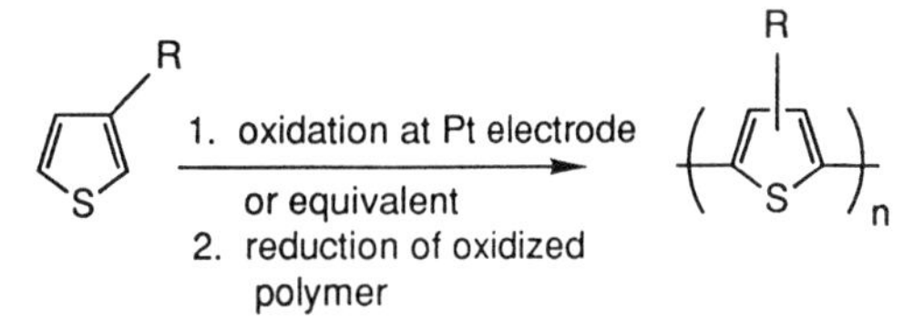

Fig. 9.4 The electrochemical method for the synthesis of PATs.

weights (e.g., M_n = 5000, polydispersity index (PDI) = 2), high molecular weights are possible [48]. No 2,4 linkages are possible using Kumada methods. Similar materials may also be prepared on a large scale by the oxidative polymerization of 3-alkylthiophene with FeCl₃ (Fig. 9.3) [49] or on a small scale by electrochemical oxidation of alkylthiophene monomers (Fig. 9.4). Materials prepared by the FeCl₃ method produce a PAT with number-average molecular weight M_n = 30,000–300,000 and polydispersities ranging from 1.3 to 5. One very good preparation of PAT reports [50] molecular weights of (62–297) × 10³ and polydispersities of 1.7–2.5.* The FeCl₃ method has been the synthetic method of choice for the synthesis of PATs. Although chemical oxidation using FeCl₃ may disrupt the conjugation of the resulting polymer by creating 2,4 linkages, these materials are more crystalline and regular than electrochemically prepared polymers because the latter contain more 2,4-coupling defects.

There are, however, several other chemical methods of synthesis that generate no 2,4 linkages. A recent method developed by Curtis and coworkers [51] couples 2,5-bis(chloromercurio)-3-alkylthiophenes using Cu powder and a catalytic amount of PdCl₂ in pyridine (Fig. 9.5). This method generates homopolymers as well as random copolymers with 3-alkyl and 3-esteric substituents. Molecular weights are reasonably high for a cross-coupling method [for poly(3-butylthiophene), M_n = 26 × 10³; PDI = 2.5], and the proportion of alkyl groups in the copolymers matched the ratio of their respective monomers in the reaction mixture. Furthermore, this method is tolerant to the presence of the carbonyl functionality. A partial list of other recent methods includes the coupling of 5,5′-dilithiobithiophenes with CuCl₂ [52], a similar coupling of a 5,5′-dilithiobithiophene with Fe(acac)₃ in refluxing THF [53], Stille coupling of 2,5′-dibromobithiophenes with 2,5′-bis(trimethylstannyl)bithiophenes using a catalytic amount of PdCl₂(AsPh₃)₂ [53], and nickel(0) dehalogenative coupling.

2. Regiochemistry in Irregular PATs

While some of the methods just mentioned reduce or eliminate 2,4 linkages, they do not address the lack of regiochemical control over head-to-tail couplings between adjacent thiophene rings. A 3-alkylthiophene is not a symmetrical molecule. Consequently, there are three relative orientations available when two thiophene rings are coupled between the 2- and 5-positions. The first of these is 2,5′ or head-to-tail coupling (referred to herein as HT), the second is 2,2′ or head-to-head coupling (HH), and the third is 5,5′ or tail-to-tail coupling (TT) (Fig. 9.6). All of the above methods afford products with three possible regiochemical couplings: HH, TT, and HT. This leads to a mixture of four chemically distinct triad regioisomers when 3-substituted (asymmetric) thiophene monomers are employed (Fig. 9.6) [54,55]. (Specific procedures with special asymmetric thiophene molecules that can lead to exceptions are discussed later.) These structurally irregular polymers are denoted here as *irregular* or *non-HT*. Irregular, substituted polythiophenes have structures in which unfavorable HH couplings cause a sterically driven twist of thiophene rings resulting in a loss of conjugation. This is illustrated in Fig. 9.7. On the other hand, regioregular, head-to-tail (HT) poly(3-substituted)thiophene can easily access a low energy planar conformation, leading to highly conjugated polymers. Increasing the torsion angles between

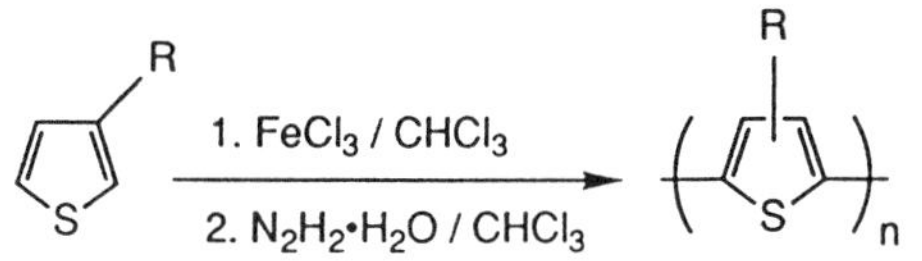

---

* The synthesis of PATs using the FeCl₃ method has been reported multiple times with variations in molecular weights being reported. A listing of all those reports is beyond the scope of this chapter.

Fig. 9.3 The "FeCl₃ method" for the synthesis of PATs.

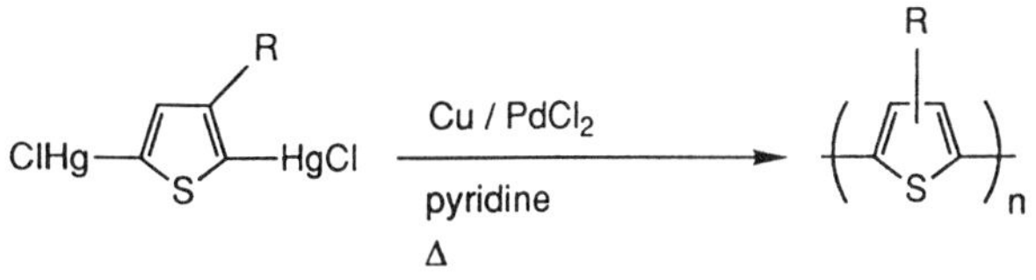

Fig. 9.5 The Curtis method for the synthesis of PATs.

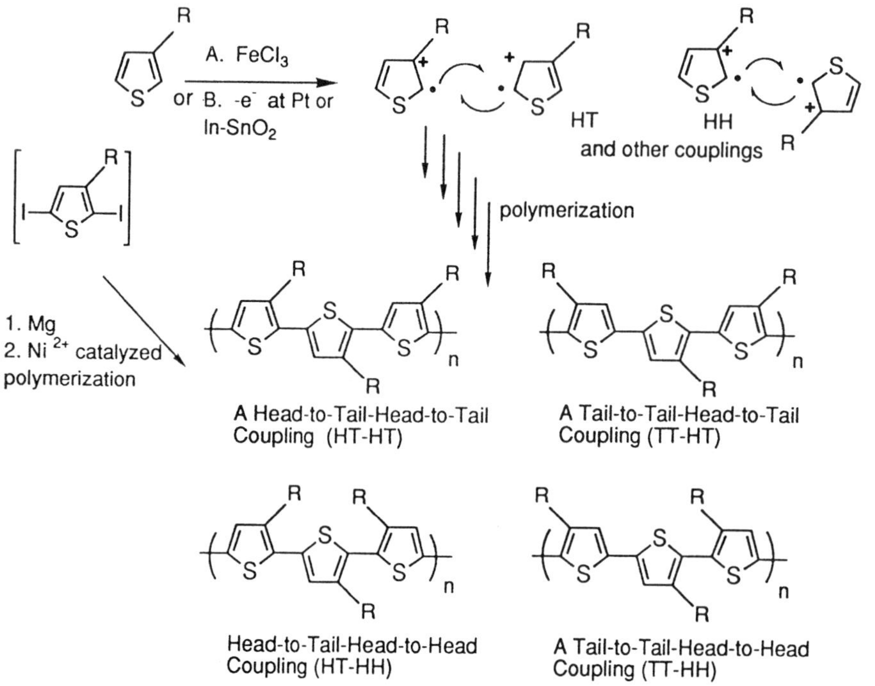

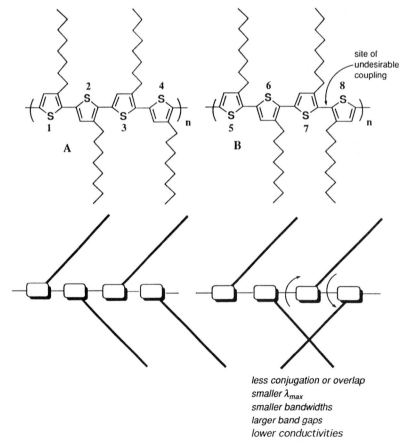

Fig. 9.6 Classical synthetic methods lead to a number of regiochemical isomers.

Fig. 9.7 Regioregular, head-to-tail polythiophene can be planar. Irregular poly(3-alkylthiophene) is not planar.

Poly(3-alkylthiophene) and Its Derivatives

thiophene rings leads to greater band gaps, with consequent destruction of high conductivity and other desirable properties. Elsenbaumer et al. [56] were the first group to study the effect of regiospecificity on the electrical conductivity in PATs. Their results clearly demonstrated that the polymerization of a 3-butylthiophene-3'-methylthiophene dimer that contains a 63:37 mixture of HT:HH couplings leads to a threefold increase in electrical conductivity over the random copolymerization of butylthiophene and methylthiophene (50:50) reaction. The detrimental consequence of HH and TT couplings in the structure and conjugation and its potential effect on the electrical conductivity in PATs and poly(3-substituted)thiophenes is examined in the next section.

III. DESIGN OF REGIOREGULAR POLY(3-ALKYLTHIOPHENE): STRUCTURE CALCULATIONS ON OLIGOMERS WITH DIFFERING REGIOREGULARITY

To investigate the conformational energetic consequences of each of the possible regioisomers that can occur in PATs, the four oligomeric triads have been modeled in the gas phase by molecular mechanics and ab initio methods [57,58]. For the HT-HT example, both methods indicate that the thiophene rings prefer a trans coplanar orientation. Structures with the rings twisted up to 20° (molecular mechanics) or up to 50° (ab initio STO-3G) from coplanarity all lie within less than 1 kcal of each other on a very flat potential energy surface and accordingly are easily accessible [57]. The advantage of HT coupling is supported by crystallographic evidence from HT-HT oligomers of 3-methylthiophene [59]. Head-to-tail trimers of 3-methylthiophene are calculated to have a torsional angle of 7–8° between conjoined rings [57]. This compares favorably with the 6–9° observed in the X-ray structure of unsubstituted α-terthienyl [60].

Introduction of a head-to-head coupling, as in the HT-HH example, dramatically alters the calculated conformation at the defective HH junction. The thiophene rings maintain a trans conformation, but they are now severely twisted approximately 40° from coplanarity [57]. Rotation from this state toward planarity encounters a rapidly increasing energy barrier, costing more than 5 kcal to achieve even a 20° twist toward planarity. The electronic consequences of reduced conjugation resulting from an HH junction include increased ionization potential, widening of the band gap, and narrowing of the bandwidth [10,11]. Significant loss of conjugation, as a result of twisting, also leads to segments with localized electronic wave functions, thereby inhibiting intrachain charge mobility [61]. Although these are gas-phase calculations on oligomers of thiophene, and packing forces to a great extent determine the solid-state structure, the calculations indicate that head-to-head couplings de-

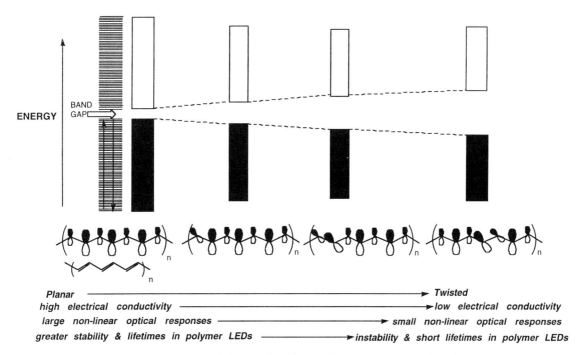

Fig. 9.8 Band gaps and electrical and optical properties vary with planarity.

stroy conjugation and can result in poor electrical conductors in polythiophenes that contain non-HT connectivity. It is important to point out that Bredas [10] reported that the π orbitals must be within 30° of coplanarity in order to achieve enough overlap to generate conducting polymer band structure. In contrast, a regioregular, head-to-tail coupled polythiophene can access very low energy structures that can possess coplanarity between 3-alkylthiophene (or 3-substituted thiophene) rings and lead to greater intrachain π overlap and small band gaps and high electrical conductivities.

The current challenge is one of understanding the solid-state chemistry of PT such that one can devise new materials with some anticipation of the resultant physical properties. Surmounting this challenge requires the construction of materials with very well defined physical structures. Only then is it apparent to what degree the substituents can affect the ordering of PT in the solid state and it becomes feasible to systematically correlate observed properties with the physical structure. However, one can speculate that as the conjugated polymer is twisted out of conjugation, electrical conductivity, nonlinear optical response, and stability will dramatically decrease (Fig. 9.8).

Structurally homogeneous PTs, denoted as regioregular PTs, can be obtained by one of two general strategies. The obvious, currently most common approach is to polymerize symmetric thiophene monomers or oligomers. The number of available publications is too large to be fairly considered here and will not be addressed [62–65]. This review is directed at the alternative strategy: the use of asymmetric coupling of asymmetric monomers to achieve regioregular HT-coupled structures of polythiophene derivatives.

IV. REGIOREGULAR, HEAD-TO-TAIL COUPLED POLY(3-ALKYLTHIOPHENES)

A. Methods of Synthesis

The first synthesis of regioregular head-to-tail coupled poly(3-alkylthiophenes) (PATs) was reported by McCullough and Lowe [66] early in 1992 (Fig. 9.9). This new synthetic method [57,66,74–81] regiospecifically generates 2-bromo-5-(bromomagnesio)-3-alkylthiophene (6, Fig. 9.10) (from monomer 3 [82–85]), which is polymerized with catalytic amounts of Ni(dppp)Cl$_2$ using Kumada [67,68] cross-coupling methods [67–73] to give PATs with 98–100% HT-HT couplings. In this approach, HT PATs were prepared in yields of 44–69% in a one-pot, multistep procedure. Molecular weights for HT poly(3-alkylthiophenes) are typically in the range of $(20-40) \times 10^3$ (PDI ≈ 1.4). [A recently prepared sample of HT poly(dodecylthiophene) had M_n = 130,000 (PDI = 2.1).]

Some key features of the synthesis are the selective metallation of 3 with LDA [86,87] to generate 5. The organolithium intermediate 5 is stable at −78°C and does not undergo metal–halogen exchange via any process, including the halogen dance mechanism [88–90]. In addition, thienyllithiums are relatively poor organolithium reagents and therefore are unlikely to undergo metal–halogen exchange reactions with 2-bromo-3-al-

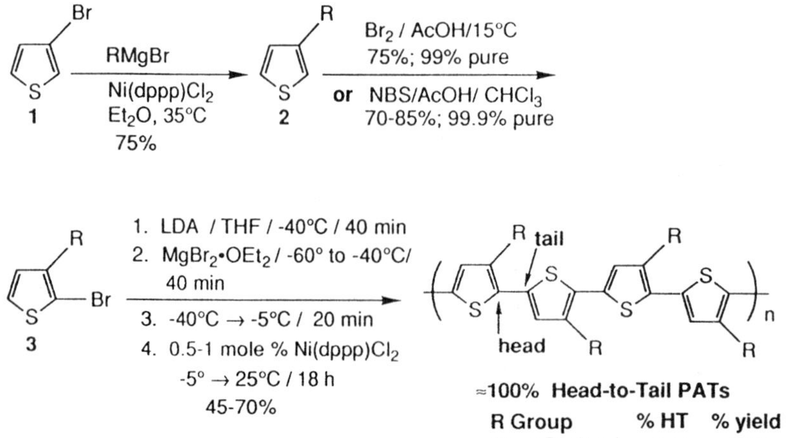

Fig. 9.9 The McCullough method for the regiospecific synthesis of poly(3-alkylthiophene)s (PATs) with 100% head-to-tail couplings. (From Refs. 57 and 66.)

Poly(3-alkylthiophene) and Its Derivatives

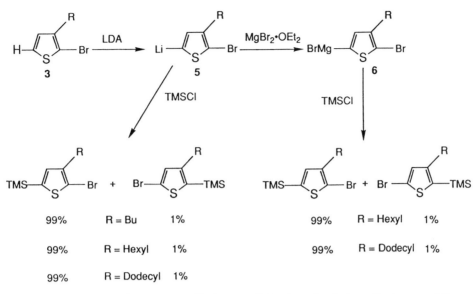

Fig. 9.10 Trapping of organometallic intermediates in order to investigate scrambling.

kylthiophenes. The intermediate **5** is then reacted with recrystallized (from Et$_2$O, in a dry box), MgBr$_2$·Et$_2$O which results in the formation of **6**, which does not rearrange at higher temperatures. Quenching studies performed on the intermediates **5** and **6** indicate 98–99% of the desired monomer and less than 1–2% of the 2,5-exchange product are observed [57] (Fig. 9.10). The subsequent cross-coupling polymerization also occurs without any scrambling. The resulting HT PAT is precipitated in MeOH, washed (fractionated) with sequential MeOH and hexane Soxhlet extractions, and then recovered by Soxhlet extraction with chloroform [50,63].

The method is very straightforward and has been used by a number of research groups to prepare regioregular HT derivatives of polythiophene as discussed later.

The second synthetic approach to HT PAT was subsequently described by Chen and coworkers [91–95]. This related cross-coupling approach differs primarily in the synthesis of the asymmetric organometallic intermediate (**8**, Fig. 9.11). In the Rieke method, a 2,5-dibromo-3-alkylthiophene is added to a solution of highly reactive "Rieke zinc" (Zn*). This metal reacts quantitatively to form a mixture of the isomers 2-bromo-3-alkyl-5-(bromozincio)thiophene (**8**) and 2-(bromozincio)-3-alkyl-5-

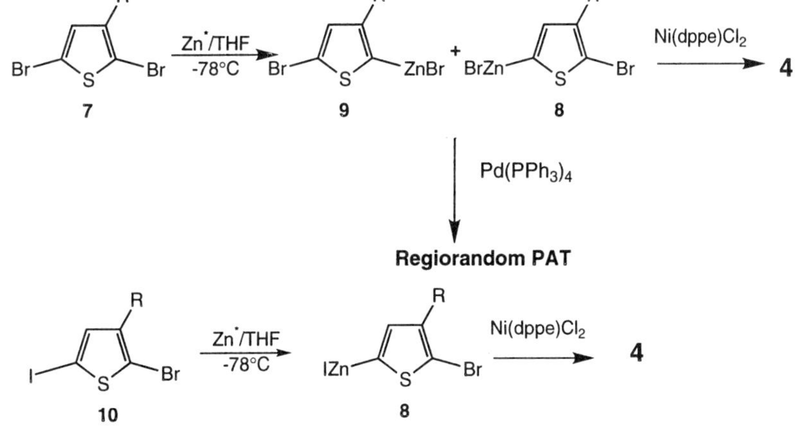

Fig. 9.11 The Rieke method for the preparation of HT PATs.

bromothiophene (9). The ratio between these two isomers is dependent on the reaction temperature and, to a much lesser extent, the steric influence of the alkyl substituent. Although there is no risk of metal–halogen exchange, cryogenic conditions must still be employed because the ratio of isomers 8 and 9 produced is affected by the temperature. The addition of a Ni cross-coupling catalyst, Ni(dppe)Cl$_2$, leads to the formation of a regioregular HT PAT, whereas addition of a Pd cross-coupling catalyst, Pd(PPh$_3$)$_4$, will result in the formation of a completely regiorandom PAT. As an alternative approach, a 2-bromo-3-alkyl-5-iodothiophene (10) will react with Rieke Zinc to form only 2-bromo-3-alkyl-5-(iodozincio)thiophene (8). This species will then react in an identical fashion to form either a regioregular HT PAT or the regiorandom equivalent, depending on what catalyst was used for the polymerization [94]. After precipitation and Soxhlet extraction, the yield for these reactions is reported to be ≈75%. Molecular weights for polymers prepared by this method are $M_n = (24–34) \times 10^3$ (with PDI = 1.4). One advantage of the Rieke method is that highly reactive Rieke zinc affords a functional group tolerant synthesis. The McCullough method and Rieke method of synthesis of regioregular HT-coupled polythiophenes produce comparable materials that are not spectroscopically distinct. Both methods appear to be generally applicable to thiophenes that are tolerant to organolithium, Grignard reagents, or zinc reagents.

Following the reports by McCullough, Rieke and co-workers, other groups found that specific oxidative conditions with a limited number of thiophene monomers can lead to an increase in the number of HT couplings in polythiophene derivatives. The Inganäs group reports [96] that the combination of a very sterically hindered 3-substituent and the slow addition of FeCl$_3$ lead to an apparent regioselective synthesis of phenyl-substituted polythiophenes with PDIs of 2.5. It was also shown by Levesque and Leclerc [97] that the preparation of poly(3-alkoxy-4-methylthiophenes) by FeCl$_3$ coupling can lead to highly regioregular materials. This may be due to an asymmetric reactivity of the oxidized monomers. Other methods generate as much as 60–70% HT-HT coupling [50,63] (electrochemical polymerizations generate ≈55% HT-HT couplings) [54,55]. Precipitation of the more crystalline HT-HT polymer from the irregular material can lead to samples with up to 75% HT-HT couplings [25,63].

B. Mechanism

The polymerizations described in Figs. 9.9 and 9.11 use a metal-catalyzed cross-coupling technique that has been investigated extensively [67–73]. The reaction is believed to proceed by (1) oxidative addition of an organic halide with a metal-phosphine catalyst, (2) transmetallation between the catalyst complex and a reactive organometallic reagent (or disproportionation) to generate a diorganometallic complex, and (3) reductive elimination of the coupled product with regeneration of the metal-phosphine catalyst. Numerous organometallic species [including organomagnesium (Grignard), organozinc, organoboron, organoaluminum, and organotin] demonstrate sufficient efficiency to be used in cross-coupling reactions with organic halides. The number of effective, regiospecific PT syntheses is expected to grow as alternatives to Mg and Zn (Figs. 9.9 and 9.11) are investigated. It should be noted that the choice of catalyst is a critical concern. It has been observed that the proportion of cross-coupling to homocoupling of the substrate, indicated by the degree of regioregularity in the product PT, can be dependent upon both the metal and the ligands used in the catalyst [91,94]. A comparison of Ni vs. Pd with monodentate (PPh$_3$) or bidentate (Ph$_2$PCH$_2$CH$_2$PPh$_2$; dppe) ligands suggested that cross-coupling selectivity was a function of the steric environment of the catalyst. The catalyst with the greatest steric congestion, Ni(dppe)Cl$_2$, produced almost exclusively cross-coupled product; the catalyst with the least congestion, Pd(PPh$_3$)$_4$, produced a random mixture of cross- and homocoupled product.

C. NMR Characterization of HT PAT

Since poly(3-alkylthiophenes) and poly(3-substituted thiophenes) are soluble in common organic solvents, ^{1}H- and ^{13}C-NMR can be used to determine their structure and regiochemistry [2,3,50,54,55,57,63,66,94]. In a regioregular PT, HT PAT for example, there is only one aromatic proton signal in the ^{1}H-NMR due to the 4-proton on the aromatic thiophene ring. Proton NMR investigations of regio*irregular*, electrochemically synthesized PATs reveal that four singlets exist in the aromatic region that can be clearly be attributed to the protons on the 4-position of the central thiophene ring in each configurational triad: HT-HT, TT-HT, HT-HH, TT-HH (Table 9.1). Sato and Morii [54,55] used the ^{1}H-^{1}H nuclear Overhauser effect and ^{1}H nuclear Overhauser effect difference spectra to assign these peaks to their respective couplings. In addition, ^{1}H-NMR was used to determine that regioirregular, electrochemically synthesized PATs contain merely 54% HT-HT couplings

Table 9.1 Four Singlets in the Aromatic Region

Configuration	Chemical shift (ppm)
HT-HT	δ 6.98
TT-HT	δ 7.00
HT-HH	δ 7.02
TT-HH	δ 7.05

Poly(3-alkylthiophene) and Its Derivatives

[54,55]. Holdcroft and coworkers [98] synthesized regioirregular poly(3-hexylthiophenes) (PHTs) with the percent of HT-HT couplings varying from 52% to 80% using Kumada coupling methods. Studies on these various samples of regioirregular PHTs clearly demonstrate the variations in the ^{1}H-NMR with regioregularity [98]. Subsequent synthesis of the four isomeric trimers by Barbarella and coworkers [58] unambiguously assigned the relative chemical shift of each triad, with each trimer being shielded by about 0.05 ppm relative to the polymer. In this analysis the HT-HT (δ = 6.98), TT-HT (δ = 7.00), HT-HH (δ = 7.03), and TT-HH (δ = 7.05) couplings are readily distinguished by a 0.02–0.03 ppm shift. These assignments are the same as the assignment by Holdcroft's group [98] and different from those proposed by Sato and Morii [54,55].

The relative ratio of HT-HT couplings to non-HT-HT couplings can also be determined by an analysis of the protons that are on the α-carbon of the 3-substituent on thiophene [56]. For example, in poly(3-dodecylthiophene), the resonance for an HH coupling is observed at δ = 2.56 ppm, and that of an HT coupling appears at δ = 2.79 ppm [57]. The relative integration of the α-methylene protons of the 3-substituent reveals the ratio of HT-HT to non-HT-HT couplings in the polymer. The same information can also be determined from the β-

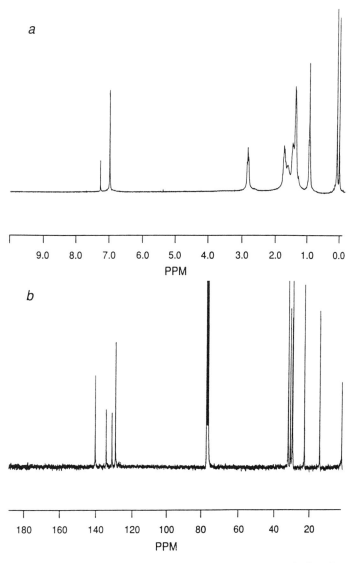

Fig. 9.12 (a) Full ^{1}H-NMR spectrum of regioregular HT PHT prepared by the McCullough method (note that extra peak near 0.0 ppm is silicon grease impurity). (b) Full ^{13}C-NMR spectrum of regioregular HT PHT prepared by the McCullough method.

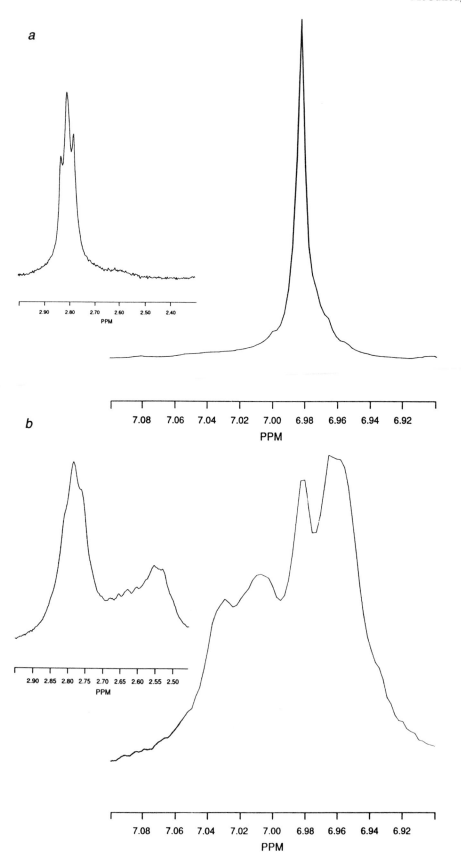

Poly(3-alkylthiophene) and Its Derivatives

methylene protons of the 3-substituent [94]. The ^{1}H-NMR resonance for the HT-coupled β-methylene proton in PHT may be observed at δ = 1.72 ppm, and that of the HH-coupled β-methylene proton appears at δ = 1.63 ppm. Clearly, the relative regioselectivity of a polymerization can be determined by ^{1}H-NMR analysis.

The comparison between regioregular HT PATs and irregular PATs is illustrated by the ^{1}H-NMR spectra of regioregular HT poly(3-hexylthiophene) (PHT) as synthesized by the McCullough method (shown in Figs. 9.12 and 9.13) [57,66]. As expected, only one of the aforementioned aromatic resonances may be observed in the ^{1}H-NMR spectra of regioregular HT PATs. The chemical shift of this peak is usually around δ = 6.98 ppm, which is typical of an HT-HT sequence (Figs. 9.12a and 9.13a). In addition, the aromatic and methylene regions of the ^{1}H-NMR spectra of regioregular HT PHT can be directly compared to those of PHTs prepared by the FeCl$_3$ method [49] (Fig. 9.13). The abundance of defective couplings in the latter case is apparent.

The degree of structural regularity is likewise apparent in the ^{13}C-NMR spectrum as shown in Figs. 9.12b and 9.14. For HT PATs, only four resonances are apparent in the aromatic region, and they are attributable to the four carbons of an HT-coupled thiophene ring (δ = 128.5, 130.5, 134.0, and 140.0 ppm). Examination of the ^{13}C-NMR spectrum of regioirregular PHT shows the resonances corresponding to the HT-HT coupled regions of the polymer as well as several peaks corresponding to non-HT regioisomers (125.2, 126.6, 127.4, 128.3, 129.6, 134.9, 135.7, 140.3, 142.9, and 143.4 ppm) [63]. With this knowledge it is possible to evaluate the ^{1}H- and ^{13}C-NMR spectra of both regioregular and irregular PATs. The NMR spectra of PATs synthesized either electrochemically or chemically have been published as part of other studies on the polymerization of dialkylbithiophenes and related work and show a large number of imperfect couplings and random regiospecificity.

In addition, the end groups of PATs have been identified by quenching of the end-group Grignards [98]. The ^{1}H-NMR resonances for the end-group thiophene residues may be readily observed in the spectrum of regiorandom poly(3-hexylthiophene) [94]. These peaks appear as two doublets (at δ = 7.15 ppm and δ = 6.93 ppm) that represent the terminal 4- and 5-protons of the chain, as well as a singlet at δ = 6.91 ppm that corre-

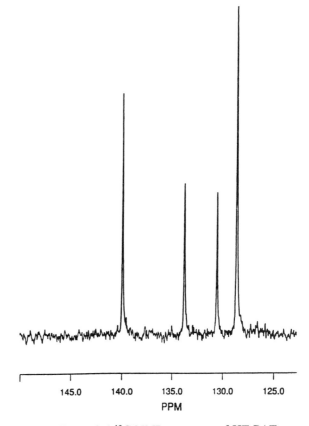

Fig. 9.14 Expanded ^{13}C-NMR spectrum of HT PAT prepared by the McCullough method.

sponds to the terminal 2-proton at the opposite end of the polymer chain (Fig. 9.15).

D. Infrared Spectroscopy Characterization of HT PAT

All linear alkyl-substituted PTs have very similar infrared spectra; thus this technique has limited diagnostic

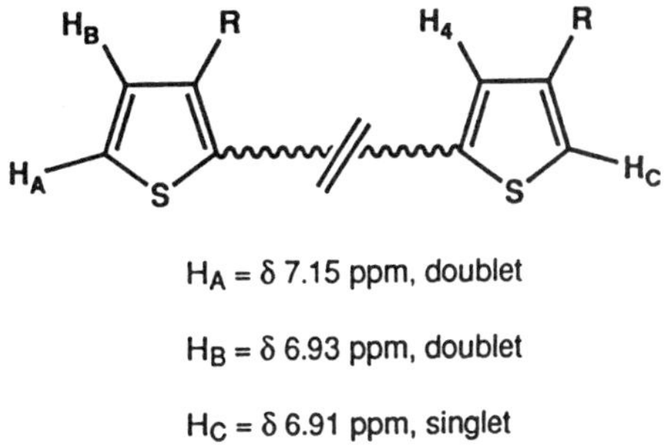

Fig. 9.13 (a) Expanded ^{1}H-NMR spectrum of regioregular HT PHT in the aromatic region prepared by McCullough method. The inset shows the expanded ^{1}H-NMR spectrum of regioregular HT PHT in the aryl methylene region. (b) Expanded ^{1}H-NMR spectrum of irregular PHT aromatic region prepared by the FeCl$_3$ method. The inset also shows the aryl methylene region for PHT prepared by the FeCl$_3$ method.

H$_A$ = δ 7.15 ppm, doublet
H$_B$ = δ 6.93 ppm, doublet
H$_C$ = δ 6.91 ppm, singlet

Fig. 9.15 End groups in PATs.

utility. However, there are two notable differences that can help distinguish an HT PAT from a regioirregular sample. The first is the C–H out-of-plane vibration that appears at 820–822 cm^{-1} for HT PAT but at 827–829 cm^{-1} for irregular PAT. The second conspicuous difference in the IR spectra is qualitatively related to the conjugation length in the polymer backbone [99]. The intensity ratio of the symmetric (~1460 cm^{-1}) to the asymmetric (~1510 cm^{-1}) ring stretches decreases with increasing conjugation length. For HT PATs this ratio is 6–9, less than half of the 15–20 value measured for regiorandom samples [94].

E. UV-Vis Characterization of HT-PAT

1. Solution UV-Vis Spectra

A prominent feature in conducting polymers is their extended conjugated π system. A qualitative measure of π-orbital overlap in solution and the solid state of both neutral and oxidized conjugated polymers can be probed by UV-vis spectroscopy [100]. The maximum absorption is the π–π* transition for the conjugated polymer backbone. In conjugated polymers the extent of conjugation directly affects the observed energy of this π–π* transition. Therefore, UV-vis spectroscopy provides a method by which the conformational state and structure of conjugated polymers may be observed [8,100].

If a qualitative comparison of the solution UV-vis spectra of PATs is made (Table 9.2), a red shift of λ_{max} is clearly evident upon comparing the regiorandom, regioirregular, and two regioregular HT PAT samples with identical substituents [57,66,74,78,79,81,91,94]. Thus the regioregular HT PATs have lower energy π–π* transitions, indicating longer conjugation length. Regioregular polythiophenes can access a more planar (or rod-like) structure on average in solution, and regiorandom PATs are likely to have a more coil-like structure [101–105]. Early indications are that regioregular HT PATs are clearly more rod-like. It appears that the regular placement of the alkyl side chains helps to drive a solution self-assembly that can induce a coil-to-rod transition [106]. The McCullough data seem to indicate that HT PATs exhibit a change in λ_{max} as a function of the alkyl side chain [57,74]. Longer conjugation lengths are observed in the PATs that possess the longer alkyl side chains, suggesting that in solution side-chain ordering improves with increasing chain length. It has been proposed [106] that solution aggregation of the alkyl side chain occurs to a greater extent in HT PDDT, giving rise to greater conjugation lengths than that of HT PBT. Although a more thorough discussion of this aggregation phenomenon is discussed later, light scattering in solution of HT PATs [106] shows that such aggregation of polymer chains occurs to a much greater extent for HT PDDT than for HT POT. Aggregation occurs to a great extent at low temperature and leads to a large thermochromic effect [100,101,107–109], marked by a deep red solution at 25°C and a dark purple solution at −40°C (Fig. 9.16). Since this aggregation can also occur to a greater extent in more concentrated solutions, direct correlation of conjugation length to λ_{max} can be difficult. Concentrated solutions of regioregular HT PATs can form supramolecular aggregates that have very broad peaks in the UV-vis spectra, obscuring the true solution absorption maxima. So, in effect, the reported λ_{max} values may be a function of concentration of the polymer. In extreme cases, the sample may actually be a colloidal suspension of aggregates [101,106]. Therefore, great care must be taken to use dilute solutions or filtered solutions of HT PATs. Due to solvatochromism in PATs, direct comparison is only feasible using the same solvent system. As an example, the λ_{max} in xylene is red-shifted relative to chloroform solutions.

2. Solid-State UV-Vis Spectra

Spectra of regioirregular PAT films contain little structure and consist of a single broad absorption for the π–π* transition. Drop-cast films of HT PAT, in contrast,

Table 9.2 Qualitative Comparison of the Solution UV-vis Spectra of PATs

	λ_{max} (nm) in solution				
3-Substituent	Random 50% HT[a]	FeCl$_3$ 70% HT[b]	Electrochem[c] HT[d]	Rieke 98–99% HT[a]	McCullough 98–99% HT[b]
Butyl	428	436	434	449	450
Hexyl	428	436	434	456	442
Octyl	428	436	440	451	446
Dodecyl	428	436	440	453	460

[a] From Ref. 94.
[b] From Ref. 57.
[c] From Ref. 40.
[d] Holdcroft and coworkers [98] have shown that 69% HT-HT poly(3-hexylthiophene) has a λ_{max} of 434 and 80% HT-HT has a λ_{max} of 440.

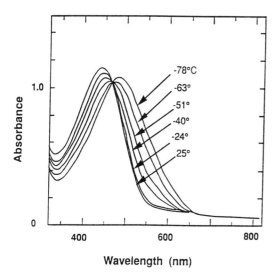

Fig. 9.16 UV-vis spectra of a THF solution of a regioregular HT poly(3-dodecylthiophene) showing solution thermochromism (3 g/L).

are red-shifted, and a great deal more resolution is apparent in the spectra (Fig. 9.17). It is interesting to note that this fine structure may range from shoulders [57] to well-defined peaks [74,94] (Fig. 9.18) with definite absorption frequencies but varying intensities. As shown in Table 9.3, films of irregular PAT polymers prepared from the FeCl$_3$ route have a λ_{max} of 480 nm. Films of the regioregular HT PATs have solid-state absorptions ranging from λ_{max} = 560 nm for "thin" films of HT poly(3-dodecylthiophene) to 500 nm for "thick" films of HT poly(3-butylthiophene) (Fig. 9.18). This difference in energy between the FeCl$_3$-coupled materials and the regioregular HT polymers is directly related to the differences in the ability of the regioregular materials to form more ordered solid-state structures with smaller band gaps [57,66,74,91,92,94]. An estimation of the Fermi level differences between the regiorandom and regioregular can be made by observing the band edge of the spectra. The band edge is a qualitative indication of the band gap in the neutral conjugated material. The band edge for regioregular PT ranges from 1.7 to 1.8 eV [57,94], a 0.3–0.4 eV improvement over the 2.1 eV reported for a regiorandom sample. Simply by eliminating coupling defects, thereby minimizing unfavorable steric interactions of the substituents, the solid-state order is markedly improved, resulting in a reduction in the energy of the band gap.

Analysis of thin films of HT PATs by UV-vis spectroscopy reveals the presence of at least three distinct peaks. For example, HT poly(3-dodecylthiophene) has a λ_{max} of 562 nm in the solid state. The lower intensity peaks appear at λ = 530 and 620 nm [57]. These non-λ_{max} peaks are quite substantial in intensity, ranging from 60 to 100% of the λ_{max} intensity (depending on film thickness) (Figs. 9.17 and 9.18). It has been observed [57] that these peaks indicate that distinct, reproducible solid-state structures with differing yet defined conjugation lengths are responsible for each of the three absorptions. The long-wavelength absorptions have been interpreted as evidence of structures possessing long-range order [57,74], and the relative intensity ratio of the three peaks is an indication of the extent of that order present in the solid-state thin films [110].

The differences in the degree of conjugation and macroscopic morphological order in HT PATs is a function of film thickness. This is evident from UV-vis studies on films of varying thicknesses (Fig. 9.18), which indicate that "thick" films ($\approx$3 μm) show distinctly different absorption spectra than "thin" films (<1 μm) (Table 9.3). A similar observation was reported by Roncali and coworkers [40,111,112], who found that regioirregular poly(3-methylthiophene) (PMT) prepared electrochemically exhibited a thickness-dependent solid-state UV-vis spectrum that correlated with the conjugation lengths and electrical conductivity. They found that 0.2 μm thin films of poly(3-methylthiophene) (PMT) had a λ_{max} of 510 nm, while the thinner films of 0.006 μm had λ_{max} values as high as 552 nm. The observation was explained by noting that as the polymer film thickens, morphological disorder can increase, leading to more disorder than in ultrathin films. These thin films of non-HT PMT had a very high degree of structural order and extended π-conjugation lengths with conductivities of up to 2000 S/cm. In much thicker films (1–3 μm) of HT PATs, the λ_{max} of POT is 559 nm and that of PDDT is 562 nm; therefore the conjugation lengths of HT PATs are comparable to those of the very highly ordered, very thin films (0.006 μm) of non-HT poly(3-methylthiophene). These results indicate that the packing arrangements and molecular structures of these PATs are ordered to the same degree as those of ultrathin films of PMT. Therefore the morphological order found in the "thin-film regime" has been greatly extended from 0.006 μm to $\approx$1–3 μm by the regular placement of the side chains in HT PATs. This order in HT PATs appears to be driven by a supramolecular ordering of the alkyl side chains through aggregation and appears to be maximized in HT PDDT.

F. Supramolecular Ordering and Self-Assembly in HT PATs in Solution and in the Solid State: Light-Scattering, Morphology, X-Ray, and Electrical Conductivity Studies

One of the most fascinating physical property differences between irregular PATs and regioregular HT PATs is that supramolecular ordering occurs in regioregular HT PATs and not in irregular PATs. Light-scattering studies and UV-vis experiments clearly show that

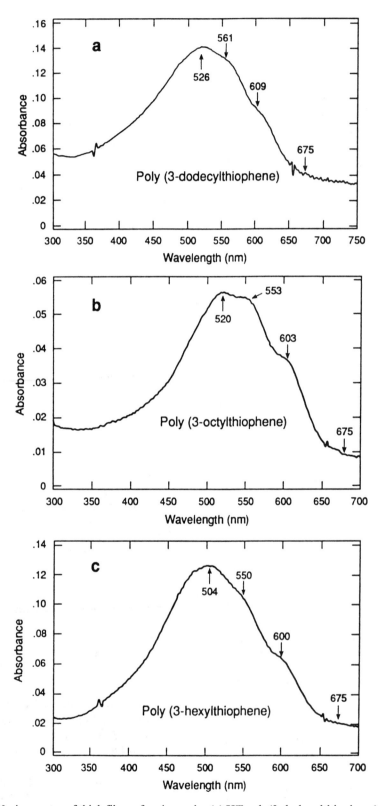

Fig. 9.17 Solid-state UV-vis spectra of thick films of regioregular (a) HT poly(3-dodecylthiophene), (b) HT poly(3-octylthiophene), and (c) HT poly(3-hexylthiophene) cast on quartz from chloroform.

Poly(3-alkylthiophene) and Its Derivatives

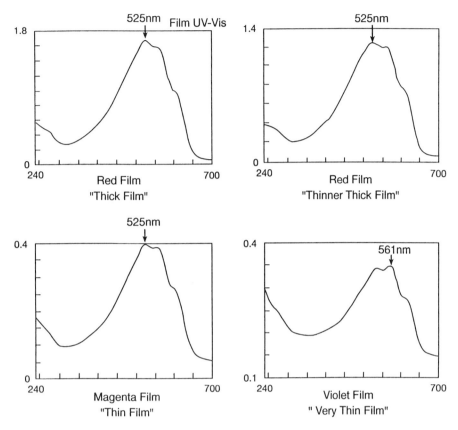

Fig. 9.18 Solid-state UV-vis spectra of films of regioregular HT poly(3-dodecylthiophene) of varying thicknesses. Note the change in the maximum absorbance as the film gets thinner.

Table 9.3 Comparison of Various Films and Absorption Wavelengths

	λ_{max} (nm) solid state			
Polymer	McCullough "thick film" >98% HT[a]	McCullough "thin film" >98% HT[b]	Rieke "thin film" >98% HT[c]	FeCl$_3$ 70% HT[d]
PBT (**4d**)	500[e]	525	522	480
	580	560[e]	556[e]	
	610	608	605	
PHT (**4c**)	504[e]	525	526	480
	550	555[e]	556[e]	
	600	610	608	
POT (**4b**)	520[e]	525	522	480
	553	559[e]	556[e]	
	603	610	608	
PDDT (**4a**)	526[e]	530	524	480
	561	562[e]	560[e]	
	609	620	610	

[a] From Ref. 74.
[b] From Ref. 80.
[c] From Ref. 94.
[d] From Ref. 57.
[e] λ_{max}.

supramolecular association of HT PATs occurs in solution [106]. Scanning electron microscopic morphology studies of HT PATs coupled with X-ray studies and electrical conductivity on solid thin films of HT PATs show macromolecular self-assembly in these conducting polymers [74].

1. Solution Studies

Self-association in solutions of HT PATs is found to be a function of concentration. In dilute solution, HT PATs act as independent chains with little to no self-association between the polymer chains [101,102]. UV-vis spectra of these dilute red solutions exhibit standard UV-vis data with $\lambda_{max} \approx 450$ nm (see Fig. 9.16). However, concentrated solutions exhibit very broad UV-vis spectra with λ_{max} values ranging from 450 to 600 nm and are deep magenta. As the concentration of HT PATs increases, supramolecular association or aggregation drives the formation of colloids and eventually macroscopic, precipitated aggregates. Precipitation of irregular PATs occurs only with poor solvents, generating particles that, by X-ray, are identical to those of disordered neutral films [101,102]. In contrast, precipitation of regioregular HT PATs occurs eventually even in good solvents due to the highly crystalline nature of HT PATs. It has been observed [106] that clear solutions of HT PATs that are allowed to sit overnight form precipitates. These solids can be (and should be) filtered out before film-casting or physical studies proceed.

The properties of solutions of irregular PATs and their derivatives have been well studied by a number of research groups [100–105]. Concepts such as thermochromism, solvatochromism, and a number of important physical properties have been examined in order to elucidate the fundamental molecular and microscopic reasons for the observed phenomena. These studies have led to the observation of reversible thermochromism and solvatochromism in solutions of PATs. The presence of an isosbestic point indicates that two conformations are possible in solution, with one possessing increased conjugation length relative to the other. The different forms are often attributed to either a conformationally disordered random coil structure or an ordered rodlike structure [100,101,107–109]. Solutions can be a mixture of the two phases (biphasic). The two forms can be distinguished by solution light scattering. In HT PATs [113,114] the chromic effects are even more pronounced than in irregular PATs, indicating that the solution order is remarkably sensitive to the regularity of the polymer.

Static and dynamic light-scattering studies have demonstrated that solutions of regioregular HT PDDT of different thermal histories and concentrations show strong intermolecular association between the polymer chains [106]. This association (or aggregation) drives main-chain conformation order, leading to a highly conjugated polymer system. Supramolecular association is present under a variety of conditions, including temperatures as high as 65°C and concentrations as low as 0.5 g/L. In fact, only at this low concentration was there indication of significant numbers of independent chains, and then only at the highest temperature studied.

The supramolecular structure may assume a variety of shapes dependent on concentration. One solution structure is a nematic liquid crystalline phase consisting of lamella of extended PT chains interspersed with lamella of alkyl chains (Fig. 9.19). The supramolecular aggregate structure is thought to have a disklike shape and possess long-range order. Supramolecular ordering increases as the concentration of the polymer increases. This behavior is consistent with the concepts embodied in the Flory phase diagram for semiflexible polymer chains. The theory predicts that with increasing concentration of semiflexible chains, these polymer solutions go from a disordered phase to a biphasic solution to an ordered phase. It is also found that at low temperature the dodecyl side chains slowly self-order, enhancing coplanarity of the thiophene rings along the polymer chain [106].

Another shape of the solution structure that can exist in HT PDDT consists of needle-like aggregates of extended chains of PT surrounded by a sheath of ordered alkyl chains. The needle-like aggregate structure does not possess a great deal of two-dimensional order, in contrast to the lamellar form (Fig. 9.19).

It is very interesting to note that the supramolecular ordering is greatest in HT PDDT. The aggregation effects are much smaller in PHT and intermediate in HT POT. This mirrors the trends in the UV-vis spectra, electrochemistry, and electrical conductivity of HT PATs [74].

2. Solid-State Order

a. Morphology Studies

The solution studies indicate that the ordered supramolecular disk structures with globally ordered extended conformations of polythiophene are desirable precursors to self-assembled PAT films, whereas the needle-like structures do not pack well and cannot give a globally ordered structure. This discovery implicitly has a profound effect on film casting and the resultant solid-state order (discussed below). The light-scattering data also suggest that solution spectra will vary with temperature, concentration, solvent composition, and the thermal history of the sample.

The observed properties of PAT films arise from the prevalent structure of the material in the solid state. In contrast to traditionally synthesized amorphous PATs, regioregular HT PATs are self-orienting polycrystalline polymers [74,94]. The observation of two distinct aggregated forms in solution implies that two (or more) phases may be generated upon film casting. Efforts to generate

Poly(3-alkylthiophene) and Its Derivatives

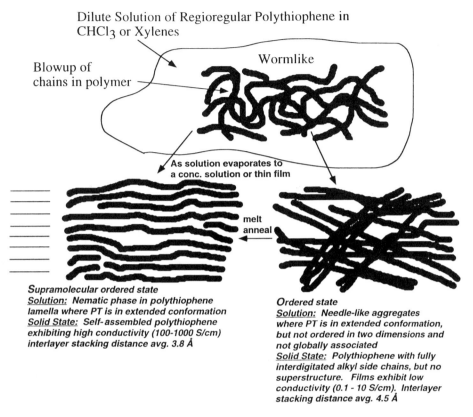

Fig. 9.19 Understanding self-assembly in regioregular HT poly(3-alkylthiophenes): metastable aggregation in solution to solid-state self-ordering.

uniform films by drop casting support this premise. Films with a variety of morphologies, ranging from brittle and cracked to sturdy and smooth, are commonly produced from the same sample of polymer under similar conditions. Additionally, trace impurities such as silicon grease can dramatically affect the film-forming properties of a given sample [80].

Scanning electron microscopic (SEM) photographs of thin films of both regioregular HT PATs and irregular PATs are presented in Figs. 9.20–9.28. Figures 9.20–9.23 show three samples of irregular PDDT synthesized via the $FeCl_3$ method that were drop-cast from xylene and chloroform. These amorphous, irregular PATs form good films with dense, smooth morphologies when cast from a number of organic solvents. The SEM pictures are examples from a large number of morphology studies on irregular PATs done in the McCullough lab. The conductivities of these iodine-doped irregular PATs are about 1 S/cm [25,40,57,74]. The SEM photos shown in Figs. 9.24–9.28 are of regioregular HT PATs drop-cast from xylene, toluene, or chloroform. These films show a wide variety of morphologies ranging from brittle, cracked films to fluffy films, to dense, uniform films. The film in Fig. 9.24 is typical of a regioregular PDDT. The film shown was cast from xylene. With some regularity, drop-cast films are brittle and contain microscopic and/or macroscopic cracks. The film in Fig. 9.25 is a regioregular HT PDDT film cast from toluene. This morphology displays horizontal microscopic cracks and fissures that cause physical discontinuity between regions of the film. While the film is not as cracked as that in Fig. 9.24, the morphology is intermediate between that of a smooth dense film and that of a macroscopically cracked film. The conductivity in this film (Fig. 9.25) after iodine doping is about 10 S/cm. The film in Fig. 9.26 shows yet another morphology. This fluffy morphology, in an HT PHT cast from xylene, again apparently lacks the polymer chain density to lead to high electrical conductivity. In contrast, the film in Fig. 9.27 is an HT PDDT film drop-cast from toluene. This film has a dense, uniform morphology and displays no macroscopic or microscopic cracks. Upon oxidation with iodine, this film exhibits an electrical conductivity of 1000 S/cm. The density of the film is critical for high electrical conductivity [112]. Figure 9.28 is of an HT PDDT cast from chloroform, which forms a film that is

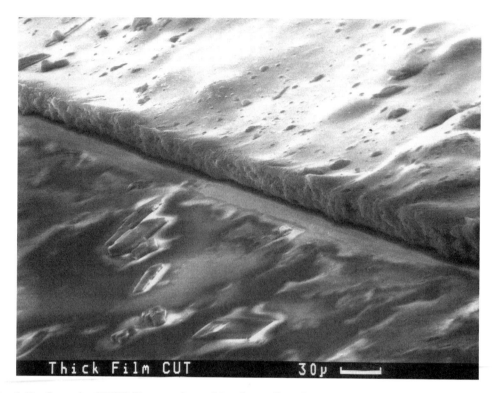

Fig. 9.20 Irregular PDDT film cast from chloroform. Sample synthesized by the $FeCl_3$ method.

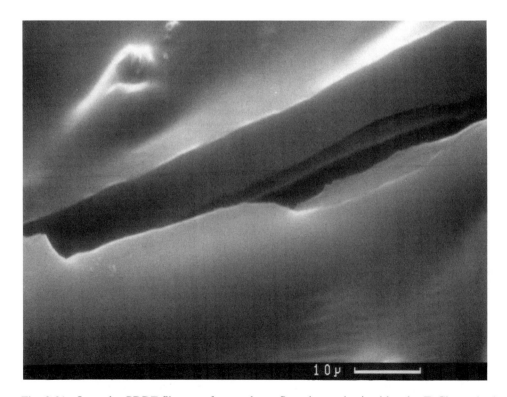

Fig. 9.21 Irregular PDDT film cast from xylene. Sample synthesized by the $FeCl_3$ method.

Poly(3-alkylthiophene) and Its Derivatives

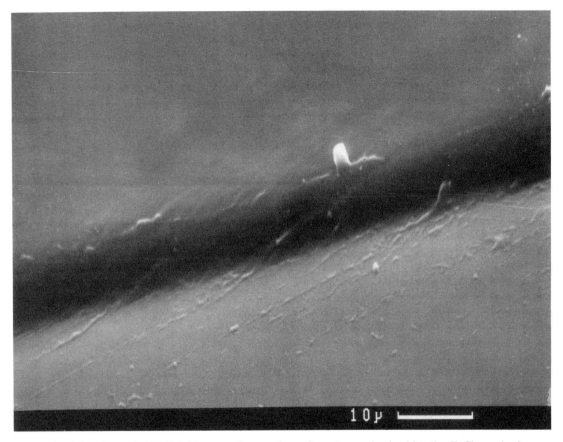

Fig. 9.22 Irregular PDDT film cast from xylene. Sample synthesized by the FeCl₃ method.

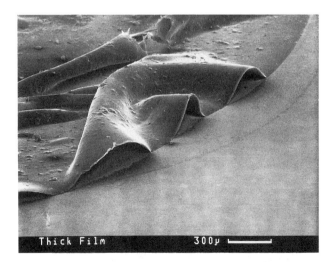

Fig. 9.23 Irregular PDDT film cast from chloroform. Sample synthesized by the FeCl₃ method.

quite dense in appearance and yet has microscopic holes. The conductivity of this film was determined to be 340 S/cm after doping with iodine. Clearly, microscopic defects can dramatically alter the macroscopic properties (i.e., electrical conductivity) in a dramatic fashion. The processing of the polymers must be explored in order to fully realize the potential of materials derived from HT PATs.

b. X-Ray Studies

Simple air evaporation of filtered 2–5% *p*-xylene solutions of PATs onto 70 μm glass coverslips leads to thin films that can be studied by X-ray scattering. These studies on thin films of HT PATs show very highly oriented structures [74,94]. Both X-ray film data, at a 5° angle of incidence to the coverslip surface, and scans of intensity vs. 2θ (Fig. 9.29) show strong intensity in three small-angle reflections that correspond to a well-ordered lamellar structure with an interlayer spacing, in the case of PHT, of 16.0 ± 0.2 Å (Fig. 9.29). A single narrow (0.1 Å half-width) wide-angle reflection is observed at a

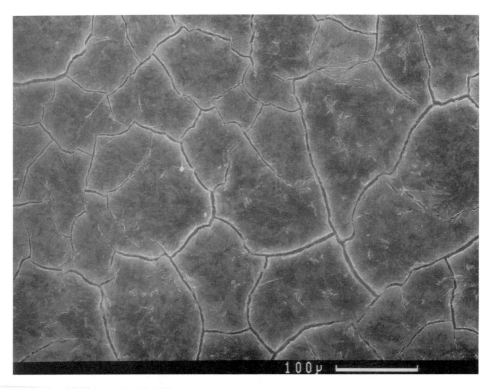

Fig. 9.24 Regioregular HT PDDT film cast from xylene. Film shows brittle cracked morphology.

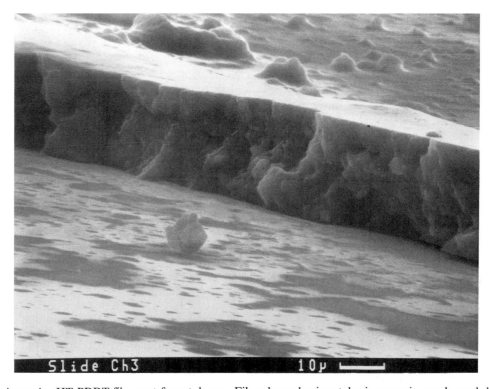

Fig. 9.25 Regioregular HT PDDT film cast from toluene. Film shows horizontal microscopic cracks and discontinuities.

Poly(3-alkylthiophene) and Its Derivatives

**Poly(3-hexylthiophene)
xylene**

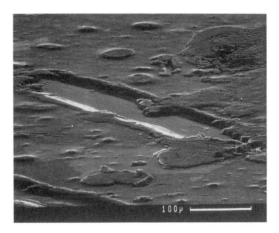

Poly (3-dodecylthiophene)
7μm σ = 1000 S cm^{-1}
PhCH$_3$

Fig. 9.27 Regioregular HT PDDT film cast from toluene. Film shows dense, uniform morphology.

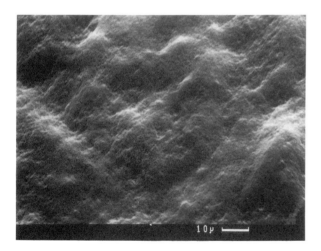

Fig. 9.26 Regioregular HT PHT film cast from xylene. Film shows fluffy morphology.

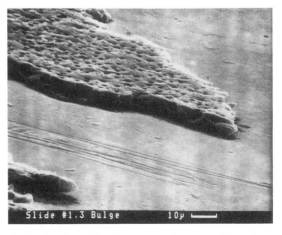

Fig. 9.28 Regioregular HT PDDT film cast from chloroform. Film shows morphology with microscopic holes.

90° angle of incidence to the coverslip (Fig. 9.29, inset), corresponding to 3.81 ± 0.02 Å stacking distance of the thiophene rings between two polymer chains [74] (Fig. 9.30). The minimum energy structure for an HT (3-hexylthiophene) tetramer can be calculated using molecular mechanics [74] and is shown in Fig. 9.30. The molecules were then arranged as shown to fit the X-ray data. These structures mirror those presented in the work of Winokur and coworkers [115,116] on stretch-oriented PATs [115] and those presented in the work of Mardalem et al. [117,118] and Chen and Ni [119]. However, an important point is that in all other PAT samples examined previ-

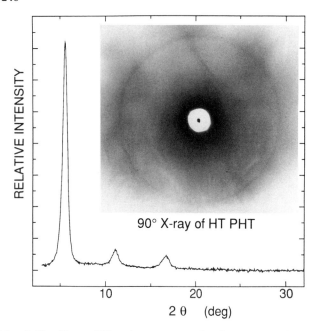

Fig. 9.29 X-ray diffraction pattern of a drop-cast regioregular HT poly(3-hexylthiophene) on a 70 μm coverslip, from scans of intensity vs. 2θ (5° incidence). Inset: X-ray film data taken at a 90° angle of incidence to the coverslip.

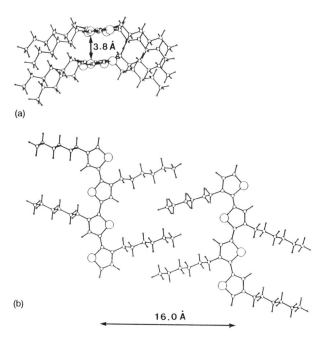

Fig. 9.30 Calculated structure for an HT 3-hexylthiophene tetramer using molecular mechanics. These globally minimized tetramers have been "docked" in an idealized manner to fit the X-ray structural parameters from PHT thin films. (a) Intermolecular π stacking between thiophene rings inferred from 90° X-ray pattern of HT-PHT film. (b) Lamellar stacking inferred from X-ray scans of intensity vs. 2θ data.

ously, the wide-angle peak is quite broad [115–119]. Our observation of a very narrow width in the wide-angle region indicates that along the polymer chain, conformational order gives rise to a single stacking distance and not a distribution of stacking distances. The narrow widths of all these dominant X-ray features indicate self-oriented, highly ordered crystalline domains in PHT [74]. It appears, however, that the samples are disordered from domain to domain, which gives rise to the wide-angle *ring* found in the 90° picture (Fig. 9.29, inset). In striking contrast, PHT prepared from FeCl$_3$ [119] shows much weaker intensity reflections in the small-angle region, with a d spacing of 17.3 ± 0.5 Å and a very broad amorphous halo centered at 3.8 Å. Similar X-ray results have been reported by Chen et al. [94] for HT PATs cast from chloroform (Table 9.4). These results show an unprecedented structural order for HT PATs, indicating that the order is induced by the regiochemical purity afforded by the synthetic method used to prepare these samples.

Examples of morphological extremes were studied by X-ray, revealing the existence of two distinct packing

Table 9.4 X-Ray Results on Thin Films of PATs

PAT	D Spacing (Å) Refs. 74 and 120	D Spacing (Å) Ref. 94	Stacking distance, all groups (Å)
PBT	—	12.63	3.8
PHT	16.0	16.36	3.8
POT (type I)	20.8	20.10	3.8
POT (type II)	14.5	—	4.47
PDT	—	23.88	3.8
PDDT (type I)	27.1	27.19	3.8[74], 3.95[94]
PDDT (type II)	19.8	—	4.47
P3TDT	—	30.48	—

structures [120]. Films of HT POT or HT PDDT were formed by the evaporation in air of 1–3% (by weight) polymer solutions in *m*-xylene solutions onto thin mica sheets both at room temperature and at 60°C. The smooth, flexible films exemplify the most common packing pattern, designated type I by Winokur and coworkers [120]. The self-assembled PATs in this structure adopt a well-ordered lamellar structure with partially interdigitated side chains. The lamellar interchain spacing is dependent upon the substituent length. For HT PDDT this distance is 27.1Å, with a stacking distance between rings of approximately 3.8 Å. For HT POT the lamellar spacing is 20.8 Å, with a stacking distance of 3.8 Å. In contrast, the brittle, powdery films are dominated by a metastable type II packing pattern in which the side chains are more interdigitated than type I. These films are cracked and very brittle as shown in Fig. 9.24. The lamellar interchain spacing is reduced by ~30% (HT PDDT, d = 19.8 Å), and the intrastack distance increases from 3.8 to 4.47 Å. (Interchain hole hopping is quite diminished with a stacking distance of 4.47 Å.) Heating of type II samples causes reversion to a type I structural form at temperatures well below that of the melt. Winokur and coworkers [120] have shown X-ray diffraction profiles of regioregular HT POT at various temperatures, demonstrating the solid-state phase change. The existence of disparate packing structures clearly indicates that the method of processing is crucial in determining the electronic and photonic properties of the material (Fig. 9.19).

c. Electrical Conductivity

The measured conductivity of HT PAT films cast from the same sample can differ markedly as a result of varying morphology from film to film. The majority of the conductivity measurements reported for PATs use I_2 as the dopant and measure conductivities via the four-point probe method. For HT PDDT the maximum values measured are around 1000 S/cm, with many more samples ranging as low as 1 S/cm because of defects incurred in film casting [56,74,78–81,91,92]. Due to the aforementioned competition between the self-assembled structure with flexible morphology (high conductivity state) and the brittle morphology with fully interdigitated side chains, the electrical conductivity of these HT PATs is highly variable (Fig. 9.19).

Upon exposure to I_2 vapor, thin films (7–12 μm thick) of HT PATs exhibit maximum electrical conductivities of 1000 S/cm for PDDT, 200 S/cm for POT, and 150 S/cm for PHT (as measured in air) [74]. Thin films of PATs prepared from the $FeCl_3$ method show maximum conductivities between 20 and 0.1 S/cm. McCullough et al. have routinely measured conductivities of 100–200 S/cm in these samples in HT PDDT. In contrast, PDDT from $FeCl_3$ gave conductivities of 0.1–1 S/cm (58–70% HT). Rieke and coworkers have reported that the electrical conductivity for their HT PATs is 1000 S/cm [91]. Rieke also reports that the average conductivity for HT PBT is 1350 S/cm, with a maximum conductivity of 2000 S/cm [92]. While the absolute values of the electrical conductivity are important, it is most critical to examine the relative values of thin films of similar thicknesses, thermal history, and morphology [74]. It is also critical to precisely measure the film thicknesses. Previous work on the conductivity of irregular PATs has reported values around 100 S/cm [25,40]; however, in some cases the film thicknesses were guessed or measured with calipers or by equally error-prone methods. It is suggested that only scanning electron microscopy or profilometry should be used to get accurate measurements of the film thicknesses. Since the resistivity (1/conductivity) is proportional to the film thickness, it is very important to have exact film thickness values. The same conductivity in all PATs prepared from $FeCl_3$ has been reported [74], whereas there is a clear trend in the electrical conductivity in HT PATs, decreasing in the series as PDDT>POT>PHT (>PBT) (Fig. 9.31).

Although HT PATs are very much more crystalline than classical PATs, the materials have good film-forming properties. However, it has been found that the formation of high quality films of these HT PATs is sensitive to grease impurities [80]. Any grease present from reaction vessels can make the formation of good films very difficult. Silicon grease can be detected by proton NMR and is found near 0 ppm (Fig. 9.12). Dissolving and reprecipitation of HT PTs in hexanes removes any grease.

G. Electrochemistry

The electrochemical properties of self-oriented HT PATs exhibit variations in oxidation potential as a function of side-chain length (Fig. 9.31). Cyclic voltammetry on thin film samples of HT PDDT exhibit two reversible oxidation potentials of $\Delta E_{1/2}(1)$ = 0.65 V and $\Delta E_{1/2}(2)$ = 0.99 V (vs. SCE); HT POT exhibits peaks at $E_{pa}(1)$ = 0.8 V and $E_{pa}(2)$ = 1.07 V; and HT PHT exhibits a broad oxidation wave that peaks at E_{pa} = 1.08 V [66,74]. These results are in contrast to PATs prepared by other methods [40,74]. It was found that HT PDDT displays the lowest oxidation potential in the alkyl series, followed by HT POT and HT PHT. It is expected that as the adjacent thiophene rings move toward coplanarity the increase in molecular orbital overlap will help to stabilize both polaronic (radical cationic) and bipolaronic (dicationic) states, thus lowering the oxidation potentials. The differences in the solid-state electrochemistry between irregular PATs and regioregular HT PATs is seen in Fig. 9.32. The HT PDDT exhibits two pseudo-reversible oxidations, while irregular PDDT prepared by the $FeCl_3$ method shows only one oxidation at a much higher potential. The difference between the oxidation potentials is an indication of the stability expected for the conjugated polymer in the conducting state.

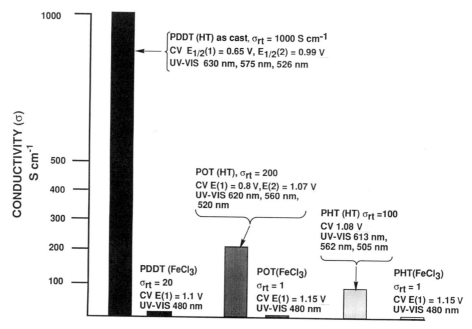

Fig. 9.31 Relative electrical conductivity in HT PATs relative to irregular PATs.

H. Photoluminescence and Electroluminescence

Solutions of HT PATs in $CHCl_3$ have shown fluorescence with a maximum emission wavelength of 570 nm [94]. This wavelength corresponds to the onset of the $\pi-\pi^*$ transition in the electronic absorption spectra. Solutions of the regiorandom poly(3-alkylthiophenes) that have been prepared by Rieke's method fluoresce at $\lambda = 550$ nm [94].

The photoluminescence of polythiophene thin films has been explored as a function of the head-to-tail ratio [121]. It was determined that intensity and emission frequency of polythiophenes are dramatically affected by the regularity of the polymer both in solution and in the solid state. A direct comparison of a regioregular PDDT film with a sample prepared by a traditional method emphasizes the differences. The emission from the regioregular sample ($\lambda_{max} = 717$ nm) is red-shifted 67 nm in comparison to the irregular sample ($\lambda_{max} = 650$ nm), indicating a smaller band gap [121]. The spectrum of the regioregular sample is also much narrower, suggesting a more rigid ordering of the polymer backbone.

The electroluminescence spectra of HT PATs were studied by the fabrication of light-emitting diodes (LEDs) [121–123]. Once again, the spectrum of the regioregular sample was found to be red-shifted 30–35 nm relative to the traditionally prepared sample because of the reduced band gap [121]. Additional benefits include a tenfold increase in light-generating efficiency, a 44% decrease in the turn-on voltage, and a longer lifetime for the LEDs [121].

V. HEAD-TO-TAIL COUPLED, RANDOM COPOLYMERS OF ALKYL THIOPHENES: SUBSTITUENT EFFECT ON ORDER

It has been established that the alkyl substituents on the backbone have a marked effect on the solution and solid-state order of HT PATs. As is apparent from the above discussions and data, the dodecyl ($C_{12}H_{25}$) side chain has the most effective side chain length for the aggregation and self-assembly in HT PATs. To investigate this theory further, HT random copolymers of PAT were prepared [76]. Head-to-tail coupled PAT random copolymers were synthesized by the route shown in Fig. 9.33. The Grignard compounds **11** and **12** were generated using the standard procedure [57] and polymerized to give polymers **13–17**, by simply mixing aliquots of **11** and **12** in direct proportion to the amount of incorporation desired. These copolymers are very soluble in typical organic solvents and possess excellent film-forming abilities. Polymers **13–15** and **17** all form violet films, whereas polymer **16** forms red films.

Using the same approach, a substituted HT polythiophene copolymer containing alkyl and ω-hydroxyalkyl side chains has been prepared [122]. The ω-hydroxyalkyl side chain was first protected with a tetrahydropyranyl (THP) ether, polymerized, and deprotected to give the random copolymer **19** (Fig. 9.34). The copolymer **19** contains a free alcohol at the end of the side chain and can be functionalized by a number of reagents to tailor the properties of the conjugated polymer.

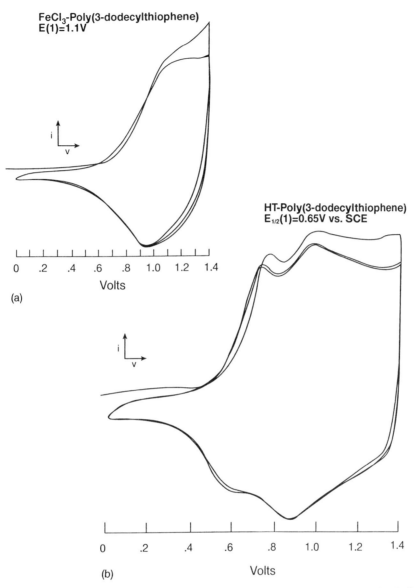

Fig. 9.32 (a) Electrochemistry of irregular poly(3-dodecylthiophene) thin films synthesized by the FeCl₃ method. (b) Electrochemistry of regioregular HT poly(3-dodecylthiophene) thin films synthesized by the McCullough method.

In Table 9.5, the physical properties of these random, HT poly(3-alkylthiophene) copolymers are given. It can seen that copolymer **17** exhibits a solution λ_{max} of 510 nm, which corresponds to one of the longest conjugation lengths known for a polythiophene in solution. It can also be seen that there is little difference in the absorption maximum between solution and the solid state (25 nm). Apparently **17** has an extended planar conformation even in solution, owing to the lack of steric hindrance caused by alkyl side chains and solution supramolecular ordering.

The solution UV-vis data for the copolymers **13–15** indicate that increasing the amount of dodecyl side chains increases the *solution* disorder, leading to a nonplanar structure [76]. However, in the solid state, the polymers with a higher percentage of dodecyl side chains self-assemble to form planar structures with long conjugation lengths. The conjugation in the solid state was greatest in polymer **13** (2:1, $C_{12}H_{25}$:CH_3; λ_{max} = 565 nm) and less in polymer **14** (1:1, $C_{12}H_{25}$:CH_3; λ_{max} = 550 nm) and polymer **15** (1:2, $C_{12}H_{25}$:CH_3; λ_{max} = 545 nm). In addition, cyclic voltammetry of thin films

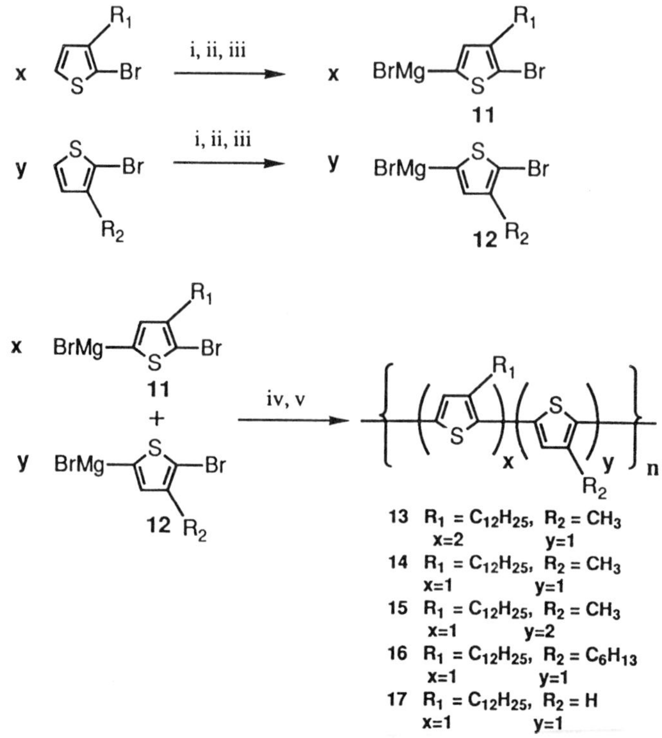
i, ii, iii
11
12
iv, v
13 R1 = C12H25, R2 = CH3 x=2 y=1
14 R1 = C12H25, R2 = CH3 x=1 y=1
15 R1 = C12H25, R2 = CH3 x=1 y=2
16 R1 = C12H25, R2 = C6H13 x=1 y=1
17 R1 = C12H25, R2 = H x=1 y=1

i, LDA, THF, -60°C, 1.5 h; ii, $MgBr_2 \cdot OEt_2$, -60°C, 0.33 h; iii, -60° → 25°C, 1.5 h; iv, 0.5 mol % Ni(dppp)Cl_2, 25°C, 14 h; v, 0.5 mol % Ni(dppp)Cl_2, 12 h

Fig. 9.33 Synthesis of random copolymers of PATs.

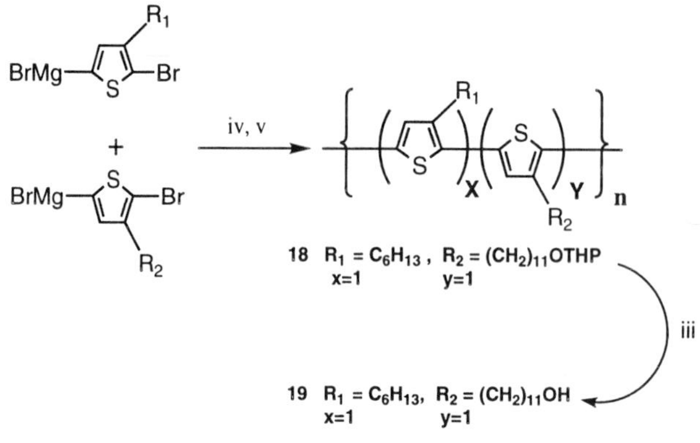
iv, v
18 R1 = C6H13 , R2 = (CH2)11OTHP x=1 y=1
iii
19 R1 = C6H13, R2 = (CH2)11OH x=1 y=1

i, 0.5 mol % Ni(dppp)Cl_2, 25°C, 14 h; ii, 0.5 mol % Ni(dppp)Cl_2, 12 h; iii, MeOH, dil. aq. HCl, 24 reflux

Fig. 9.34 Synthesis of other random HT copolymers of PATs.

Table 9.5 Properties of Random HT Copolymers of 3-Alkylthiophenes

Polymer	Solution (xylene) λ_{max}	Solid state λ_{max}	Cyclic voltammetry[a]	Electrical conductivity[b]
13	458	565[c]		
		526	0.45 V	90 avg., 190 max: (27 samples)
		610	0.85 V	
14	470	550[c]		
	600[d]	520	0.5 V	70 avg., 70 max. (11 samples)
		600	0.85 V	
15	466	545[c]		
	600[d]	510	0.65 V	—
		600	0.85 V	
16	450	556[c]		
		525	0.66 V	—
		604	0.99 V	
17	510	535[c]	0.71 V	
	575[d]	580	0.97 V	
19[e]	450	520[c]		
		540	—	—
		600		

[a] Electrochemistry on thin films in 0.2 M TBAPF$_6$ vs. SCE.
[b] Measured by four-probe method on 0.5–4 μm I$_2$-doped films cast from xylene. Films exposed to I$_2$ for 4 h. Exact film thickness measured by stylus profilometer.
[c] λ_{max}.
[d] Well-resolved peak.
[e] From Ref. 122.

of **13–15** indicates that there are longer conjugation lengths in the solid state for **13** and that the oxidation potential decreases as the proportion of dodecyl side chains decreases. The first oxidation potential decreases from **15** (1:2, C$_{12}$H$_{25}$:CH$_3$; 0.65 V) to **14** (1:1, C$_{12}$H$_{25}$:CH$_3$; 0.5 V) to **13** (2:1, C$_{12}$H$_{25}$:CH$_3$; 0.45 V), indicating that in cases where there are more dodecyl groups there is a more extended coplanar conformation [76]. This results in more stable charge carriers. This data supports the notion that the dodecyl side chains induce a planar structure along the backbone of the conducting polymer chain, presumably by aggregation [106].

Preliminary conductivity results indicate that I$_2$-doped thin films (0.5–4 μm) of polymers **13** and **15** exhibit electrical conductivities in the range of 50–200 S/cm [76]. In addition, the physical properties of these polymers are unchanged over number average molecular weights (M_n) ranging from 9000 to 28,000 (PDI = 1.6). Therefore, these properties appear not to be a function of the molecular weight of the polymer.

VI. HETEROATOMIC SUBSTITUENTS ON HT PATs

Structurally homogeneous PT makes possible the rational study of the effect of substituents on the physical properties of the parent polymer. To date the linear alkyl-substituted polymers have been most widely studied because of their simple synthesis and chemical stability. Increasingly, heteroatom substituted PTs are being explored in order to exploit particular properties of the conducting polymer backbone [40]. The many possibilities include altering the solid-state packing structure in a definable manner, inductively donating electron density to or withdrawing it from the backbone, and generating sensors that exploit the band gap sensitive torsion of thiophene rings to elicit a response. Current progress is outlined below.

A. Etheric Substituents on HT PTs

Among the many regioirregular PTs prepared to date, some ether-substituted examples are among the highest conducting materials reported [6]. It is clear that such substituents, regioregularly distributed on a PT backbone, might impart enhanced conductivity or elicit a chemoselective response to external stimuli through a binding phenomenon. Three related polymers, **26–29**, were prepared for study (Fig. 9.35) [75,77,80,81]. The substituent on **28** is too short to confer solubility on the growing polymer; therefore only low molecular weight materials are produced. Similarly, **29** yielded marginally better results (M_w = 6000; PDI = 2). Polymer **26**, determined to be >99% HT by NMR, was markedly different. The product was regularly of high molecular weight (M_w = 71,000; PDI = 2; ≈160 thiophene rings/chain). When doped with I_2, this polymer possesses very high electrical conductivity, 500–1000 S/cm on average and a maximum conductivity of 5500 S/cm for one sample of exceptional film quality [75]. These results indicate that **HT-26** exhibits higher electrical conductivities than HT PATs. This is in line with early reports on the high conductivity found in irregular **26** [6] and contrary to reports on other studies on irregular **26** [40]. It is possible that solid-state ion–dipole binding led to a highly ordered structure for irregular **26** polymerized in the presence of Bu_4N^+ [75] and hence led to a highly conductive sample. Molecular models show that the Bu_4N^+ fits well in a cavity formed by polyether arms. In addition, high conductivity in **26** may also be due to an increase in the ionic conductivity, facilitated by the etheric side chains, thereby increasing the charge carrier mobility. As was the case with the PATs, film morphology appears to be the limiting factor in reproducing high conductivity.

Polymer **26** does exhibit ion-binding properties, and an ionochromic response occurs upon exposure to Li^+ in acetonitrile and leads to a blue shift of up to 11 nm [75]. A dramatic chemoselective response to Pb^{2+} in $CHCl_3$ was discovered for **26** as indicated by a 200 nm blue shift in dilute solution (77,81,124), upon addition of $PbCl_2$. In concentrated solutions the effect is different. Concentrated solutions of **HT-26** are deep magenta without added Pb salts (λ_{max} = 575–600 nm; band edge at 700 nm). Again a striking transformation occurs upon the introduction of $Pb(BPh_4)_2$ to the solution of **HT-26**. The conjugation length immediately decreases, as indicated by a blue shift of 50–100 nm upon introduction of Pb^{2+} ion accompanied by a 50 nm blue shift in the band edge (λ_{max} = 480–550 nm; band edge at 650 nm). A film cast from **HT-26** and $Pb(BPh_4)_2$ is yellow (λ_{max} = 440 nm; band edge at 550 nm). In contrast, films cast from the salt-free solution are deep crimson (λ_{max} = 520 nm; band edge at 720 nm). There is a large (170 nm) difference in the band edge between films cast in the presence and absence of Pb^{2+}. Since films of pure $Pb(BPh_4)_2$ are colorless, the Pb^{2+} ions induced a large amount of disorder in the polymer. Comparison, by X-ray analysis, of a film of **HT-26** with a similar film that was cast in the presence of $Pb(OAc)_2$ indicates that the Pb^{2+} ions cause a significant amount of disorder in the film. X-ray diffraction shows that the very strong wide-angle reflection, which represents interchain stacking of thiophene rings, has a half-width of 0.23 Å for a film that contains Pb^{2+}. In contrast, the corresponding half-width for the uncontaminated film is 0.11 Å. In addition, iodine-doped films of **HT-26** that had been cast from a solution (in $CHCl_3$) saturated with $Pb(OAc)_2$ showed about a 10,000-fold decrease in electrical conductivity compared to similar samples that contained no Pb salts (σ ≈0.001–0.01 vs. σ ≈100–1000 for samples without Pb) [124].

Regioregular, head-to-tail poly(3[β-(p-methoxyphenoxy)hexyl]thiophene [125] has been recently prepared using the McCullough method. NMR analysis shows

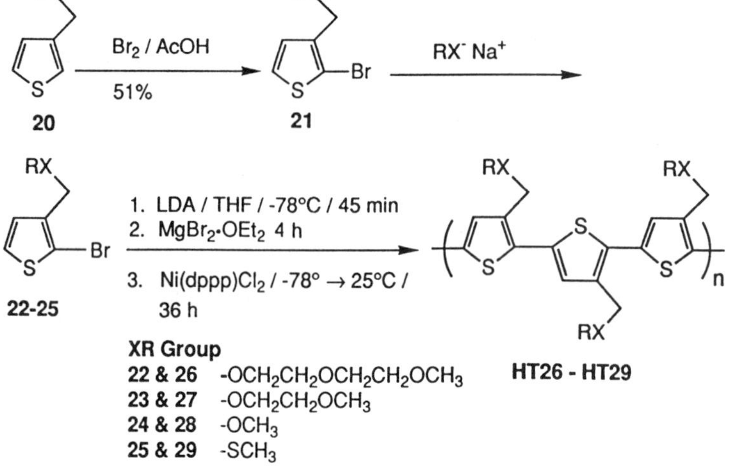

Fig. 9.35 The synthesis of head-to-tail coupled, heteroatom-substituted polythiophenes.

that the polymer is 99% head-to-tail coupled. The regioregular polymer is much more soluble than the cross-linked irregular polymer, whose preparation was also reported. The polymer is thermochromic like **HT-26**. Four-probe conductivity measurements on pressed pellet samples show that the regioregular polymer shows a 1000-fold increase in the electrical conductivity versus the irregular sample.

B. Chiral Substituents

Most polymers are known to adopt helical conformations in solution or in the solid state or both. It is possible to induce optical activity in the main chain of such a polymer by substituting with enantiomerically pure side chains. In such a system, significant chirality is induced in the backbone only when the polymer forms a well-ordered aggregate, a state that is common in self-assembled, regioregular poly(3-alkylthiophene)s [74,126]. The role of chirality and optical activity as a function of regioregularity in polythiophenes can be examined by comparing the work of Roncali with that of Meijer and Boumann. Roncali [127] prepared chiral polythiophenes by polymerizing $(S)(+)$- and $(R)(-)$-2-phenylbutyl ether of 3-propylthiophene to yield chiral polythiophenes with reported specific rotations of $[\alpha]^{22} = 3000°$ [127]. Using the McCullough method, Meijer and Boumann [113,114] synthesized an optically active regioregular polythiophene, **33** (Fig. 9.36), that exhibits a specific rotation of $[\alpha]^{22} = 140,000$ for $\lambda = 513$ nm, and at the sodium D line $[\alpha]^{22} = -9000$ for $\lambda = 589$ nm. This, of course, points to the variation in the optical rotation as a function of wavelength and regioregularity and assembly. Polymer **33** also undergoes stereomutation in the solid state. Solvatochromic studies of polymer **33** ($M_n = 16,900$; PDI = 1.4) show that varying the solvent composition dramatically affects the shape and λ_{max} of the $\pi-\pi^*$ transition by altering the distribution of disordered and ordered aggregate structures. The solid-state thermochromism in **33** is typical for a polythiophene, except that at the melting point of the polymer a complete loss of optical activity is observed. Even more interesting is the observation that when the polymer is cooled very fast from the disordered melt (by pouring the sample into a water bath at 0°C), the absorption spectrum is unchanged, but a mirror image CD spectrum (relative to the original sample before melting) is found. Therefore the regioregular chiral polythiophene **33** undergoes stereomutation. This process is thought to be driven by an aggregation effect. The effect is reversible, affording the opportunity to tune the chirality of the spectrum simply by controlling the cooling rate. Irregular chiral polythiophenes do not show this effect.

C. Thioalkyl Substituents

Substitution of a sulfur atom directly on the thiophene ring is expected to lower the oxidation potential of the conjugated polymer. Examples with varying alkyl chain length (Fig. 9.37) (polymers **HT-36**, **HT-37**, **HT-38**, and **HT-39**) have been synthesized with greater than 90% HT-HT linkages [95]. The solubility is notably poor, though, suggesting that there is a stronger affinity between polymer chains and that the lack of solubility leads to low molecular weights ($M_n = 4417$) as determined by gel-permeation chromatography. The solubil-

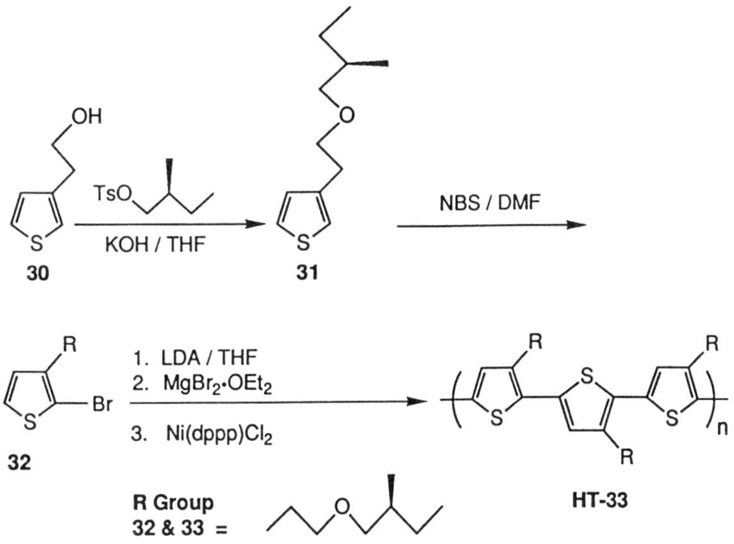

Fig. 9.36 The synthesis of polythiophenes with chiral side chains.

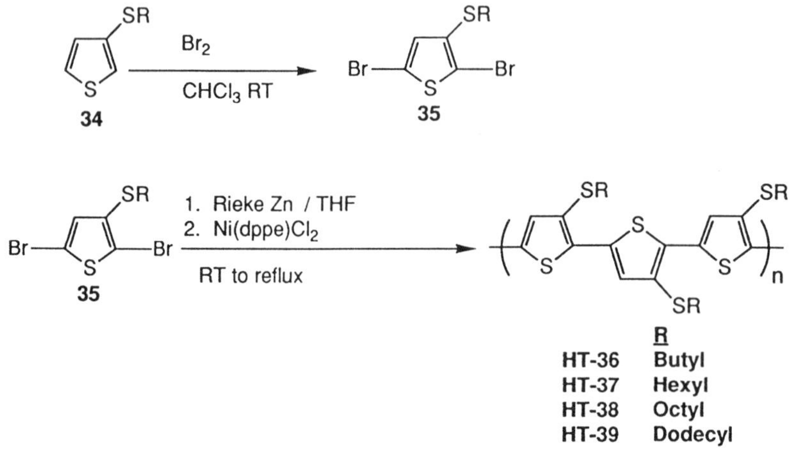

Fig. 9.37 The synthesis of polythiophenes with thioether side chains.

ity of these polymers is in contrast to that of the regioirregular polymer from 3-ethylmercaptothiophene prepared by Reynolds [128] and coworkers. This polymer is soluble in common solvents and has $M_n = 2200$ ($M_w = 13,000$) and a broad polydispersity. The solution UV-vis spectra in chloroform for **HT-37**, for example, exhibited three peaks at 263, 324, and 513 mm, with a shoulder at 605 nm. The destabilization of the HOMO reduces the HOMO-LUMO gap and leads to a red shift in these polymers in solution. However, the solid-state UV spectra and conductivity do not vary markedly from those of the alkyl-substituted model [94]. The conductivity of I_2-doped thin films of these polymers was reported to range from 450 to 750 S/cm. The conductivity in irregular poly(3-ethylmercaptothiophene) powder samples is about 10^{-3} S/cm [128].

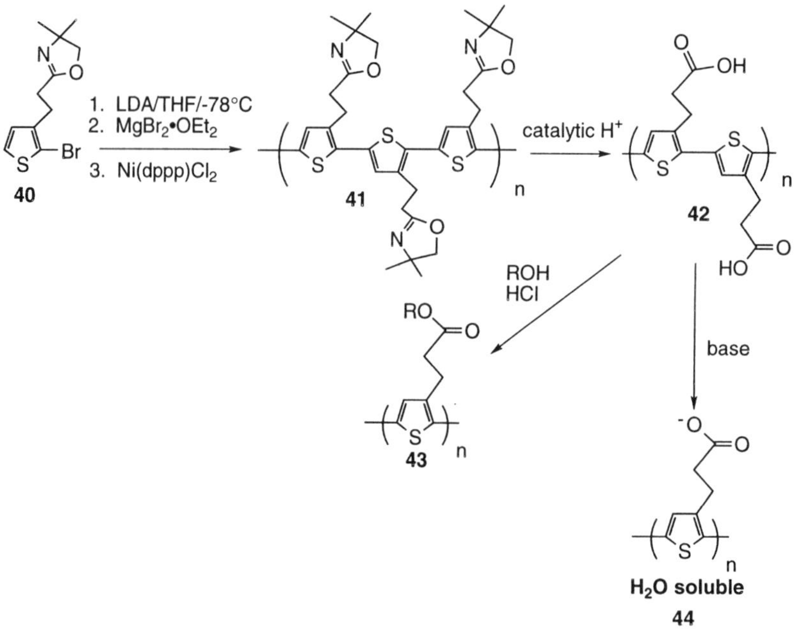

Fig. 9.38 Synthesis of water-soluble, highly ionochromic HT polythiophenes.

D. Carboxylic Acid Derivatives

It has already been demonstrated that the torsion between thiophene rings is extraordinarily sensitive to the steric interactions of the side chains. It was hypothesized that a polymer with easily varied substituent bulk might evidence a tunable band gap (Fig. 9.8) and thus afford control over all band gap dependent properties. Carboxylic acid salts were chosen as the substituent because it is trivial to dramatically change the steric demands of the function simply by changing the size of the counter cation. In addition, carboxylate-substituted polythiophene should be water-soluble.

However, by employing the sturdy oxazoline protecting group, regioregular HT polymers **41–44** were synthesized [80,129] (Fig. 9.38). Deprotection of **41** in aqueous HCl yields the desired product **42** as a dark purple precipitate. Neither of these products could be directly characterized spectroscopically because of their sparing solubility and high degree of aggregation. However, the derivative **43** was generated and fully characterized. The polymer chains are short (M_n = 4200, PDI = 2, ≈20 thiophene rings) but contain virtually no defective couplings according to NMR analysis. Irregular polythiophenes [130] containing ester side chains have been prepared with limited success. The FeCl$_3$ method leads to partial deesterification of the isolated polymer and with some ester functionalized thiophenes does not lead to polymerization. The polymer **42** is insoluble in all solvents in its native form; however, upon deprotonation, the new polymer **44** is found to be completely water-soluble. Most interesting, the predicted ionochromic response is evident. It is possible to tune the color from purple to yellow and the λ_{max} of the polymer over a 130 nm range simply by varying the counterion (Fig. 9.39) [80,129–131]. The observed chromism is not merely counterion size dependent but is also related to the hydrophobicity of the counterion. It is currently thought that a protein-like hydrophobic assembly occurs with small counterions, whereas larger counterions break up aggregation of the polythiophene chains [129]. Irregular polythiophenes carrying carboxylic acids [132] have been prepared electrochemically and do not show any dramatic ionochromism or ionic self-assembly as do the regioregular polymer **44**.

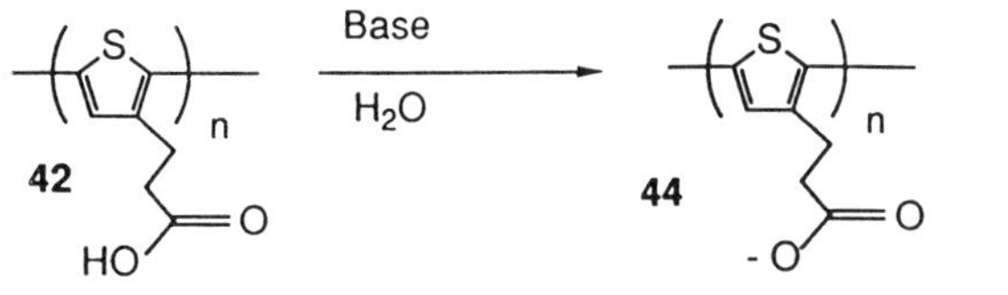

VII. SUMMARY

Comparison of the physical properties of regioregular HT polythiophenes and regiorandom polythiophenes clearly demonstrates the narrower band gap of the former, with concurrent improvement in all band gap dependent properties including higher electrical conductivities. The development of general syntheses of HT PATs affords self-assembled, self-oriented conducting polymers that were previously inaccessible. The sensitivity of these structures to torsional deconjugation makes possible the tuning of the band gap across a broad range of potentials simply by altering the nature or distribution of the substituents. Both of these discoveries have elucidated new structure–property relationships through systematic studies of structure–physical property correlations in these PATs. The optimizing of the electrical conductivity and stability in conducting polymers as is shown in HT PATs is also important in helping to make future commercial applications possible.

Base	Color	λ(max, nm, in water)
NH$_2$NH$_2$	purple	550
NH$_4$OH	light purple	543
NH$_2$CH$_2$CH$_2$OH	dark red	522
(CH$_3$)$_4$NOH	red	508
H$_2$NCH$_2$CH$_2$NH$_2$	red-orange	497
[(CH$_3$)$_2$CH]$_2$NH	orange	460
H$_2$NCH$_2$CH$_2$NHCH$_2$CH$_2$NH$_2$	orange-yellow	442
(C$_4$H$_9$)$_4$NOH	yellow	433
KOH (18-crown-6)	light yellow	414

Fig. 9.39 Ionochromism and assembly in water-soluble, HT polythiophenes.

REFERENCES

1. T. A. Skotheim (ed.), *Handbook of Conducting Polymers*, Marcel Dekker, New York, 1986.
2. R. L. Elsenbaumer, K.-Y. Jen, and R. Oboodi, *Synth. Met. 15*:169 (1986).
3. G. G. Miller and R. L. Elsenbaumer, *J. Chem. Soc., Chem. Commun. 1986*:1346.
4. M. Sato, S. Tanaka, and K. Kaeriyama, *J. Chem. Soc., Chem. Commun. 1986*:873.
5. S. Hotta, S. D. D. V. Rughooputh, A. J. Heeger, and F. Wudl, *Macromolecules 20*:212 (1987).
6. M. R. Bryce, A. Chissel, P. Kathirgamanthan, D. Parker, and N. R. M. Smith, *J. Chem. Soc., Chem. Commun. 1987*:466.
7. R. Hoffmann, *Angew. Chem. Int. Ed. Engl. 26*:846 (1987).
8. J. L. Bredas and G. B. Street, *Acc. Chem. Res. 18*:309 (1985).

9. D. O. Cowan, and F. M. Wiygul, *Chem. Eng. News* 64(29):28 (1986).
10. J. L. Bredas, *J. Chem. Phys.* 82:3809 (1985).
11. J. L. Bredas, G. B. Street, B. Themans, and J. M. Andre, *J. Chem. Phys.* 83:1323 (1985).
12. J. P. Pranata, R. H. Grubbs, and D. A. Dougherty, *J. Am. Chem. Soc.* 110:3430 (1988).
13. Y.-S. Lee, M. Kertesz, and R. L. Elsenbaumer, *Chem. Mater.* 2:526 (1990).
14. A. J. Heeger, J. Orenstein, and D. R. Ulrich (eds.), *Nonlinear Optical Properties of Polymers, Mat. Res. Soc. Symp. Proc. 109*, Materials Research Society, Pittsburgh, PA, 1988.
15. D. J. Williams, *Angew. Chem., Int. Ed. Engl.* 23:690 (1984).
16. C. P. de Melo and R. Silbey, R. *Chem. Phys. Lett.* 140:537 (1988).
17. C. P. de Melo and R. Silbey, *J. Chem. Phys.* 88:2567 (1988).
18. S. Ramasesha and Z. G. Soos, *Chem. Phys. Lett.* 153:171 (1988).
19. N. Basescu, Z.-X. Liu, D. Moses, A. J. Heeger, H. Naarmann, and N. Theophilou, *Nature (Lond.)* 327:403 (1987).
20. J. C. W. Chien, *Polyacetylene: Chemistry, Physics, and Material Science*, Academic Press, New York, 1984.
21. T. Yamamoto, K. Sanechika, and A. Yamamoto, *J. Polym. Sci., Polym. Lett. Ed.* 180:13 (1980).
22. G. Tourillon and F. Garnier, *J. Electroanal. Chem.* 135:173 (1982).
23. M. Kobayashi, J. Chen, T.-C. Chung, F. Moraes, A. J. Heeger, and F. Wudl, *Synth. Met.* 9:77 (1984).
24. R. H. Baughman, J. L. Bredas, R. R. Chance, R. L. Elsenbaumer, and L. W. Shacklette, *Chem. Rev.* 82:209 (1982).
25. J. Moulton and P. Smith, *Polymer* 33:2340 (1992).
26. J.-L. Bredas and R. Silbey (eds.), *Conjugated Polymers: Novel Science and Technology of Conducting and Nonlinear Optically Active Materials*, Kluwer, Amsterdam, 1991.
27. C. W. Spangler, P.-K. Liu, T. J. Hall, D. W. Polis, L. S. Sapochak, and L. R. Dalton, *Polymer* 33:3937 (1992).
28. P. N. Prasad and D. J. Williams, *Introduction to Nonlinear Optical Effects in Molecules and Polymers*, Wiley, New York, 1991.
29. K. G. Chittibabu, L. Li, M. Kamath, J. Kumar, and S. K. Tripathy, *Chem. Mater.* 6:475 (1994).
30. J. H. Burroughes, D. D. C. Bradley, A. R. Brown, R. N. Marks, K. Mackay, R. H. Friend, P. L. Burns, and A. B. Holmes, *Nature (Lond.)* 347:539 (1990).
31. D. Braun and A. J. Heeger, *Appl. Phys. Lett.* 58:1982 (1991).
32. P. L. Burns, A. B. Holmes, A. Kraft, D. D. C. Bradley, A. R. Brown, and R. H. Friend, *J. Chem. Soc., Chem. Commun. 1992*:32.
33. R. Kaneto and K. Yoshino *Synth. Met.* 28:C287 (1989).
34. J. Ruhe, N. F. Colareri, D. D. C. Bradley, R. H. Friend, and G. Wegner, *J. Phys.: Condens. Matter.* 2:5465 (1990).
35. L. S. Swanson, J. Shinar, L. R. Lichty, and K. Yoshino *Mat. Res. Soc. Symp. Proc.* 173:385 (1990).
36. G. G. Malliaras, J. K. Herrema, J. Wildeman, R. H. Wieringa, R. E. Gill, S. S. Lampoura, and G. Hadziioannou, *Adv. Mater.* 5:721 (1993).
37. B. Xu and S. Holdcroft, *Macromolecules* 26:4457 (1993).
38. C. H. McCoy and M. S. Wrighton, *Chem. Mater.* 5:914 (1993).
39. F. Garnier, A. Yassar, R. Hajilaoui, G. Horowitz, F. Deloffre, B. Servet, S. Reis, and P. Alnot, *J. Am. Chem. Soc.* 115:8716 (1993).
40. J. Roncali, *Chem. Rev.* 92:711 (1992).
41. P. Yam, *Sci. Am.* June 1995, p. 82.
42. F. L. Carter (ed.), *Molecular Electronic Devices*, Marcel Dekker, New York, 1982.
43. J. M. Tour, R. Wu, and J. S. Schumm, *J. Am. Chem. Soc.* 112:5662 (1990).
44. C. R. Martin, L. S. Van Dyke, Z. Cai, and W. J. Liang, *J. Am. Chem. Soc.* 112:8967 (1990).
45. W. Liang and C. R. Martin, *J. Am. Chem. Soc.* 112:9666 (1990).
46. J. M. Tour, R. Wu, and J. S. Schumm, *J. Am. Chem. Soc.* 113:7064 (1991).
47. C.-G. Wu and T. Bein, *Science (Washington, D.C.)* 264:1757 (1994).
48. S. A. Chen and C. C. Tsai, *Macromolecules* 26:2234 (1993).
49. R. Sugimoto, S. Takeda, H. B. Gu, and K. Yoshino, *Chem. Express* 1:635 (1986).
50. M. Pomerantz, J. J. Tseng, H. Zhu, S. J. Sproull, J. R. Reynolds, R. Uitz, and H. J. Arnott, *Synth. Met.* 41–43:825 (1991).
51. M. D. McClain, D. A. Whittington, D. J. Mitchell, and M. D. Curtis, *J. Am. Chem. Soc.* 117:3887 (1995).
52. A. Berlin, G. A. Pagani, and F. Sannicolò, *J. Chem. Soc., Chem. Commun. 1986*:1663.
53. M. J. Marsella and T. M. Swager, *J. Am. Chem. Soc.* 115:12214 (1993).
54. M. Sato and H. Morii, *Polym. Commun.* 32:42 (1991).
55. M. Sato and H. Morii, *Macromolecules* 24:1196 (1991).
56. R. L. Elsenbaumer, K.-Y. Jen, G. G. Miller, H. Eckhardt, L. W. Shacklette, and R. Jow, in *Electronic Properties of Conjugated Polymers* (H. Kuzmany, M. Mehring, and S. Roth, eds.), Springer Ser. Solid State Sci., Vol. 76, Springer, New York, 1987, p. 400.
57. R. D. McCullough, R. D. Lowe, M. Jayaraman, and D. L. Anderson, *J. Org. Chem.* 58:904 (1993).
58. G. Barbarella, A. Bongini, and M. Zambianchi, *Macromolecules* 27:3039 (1994).
59. G. Barbarella, M. Zambianchi, A. Bongini, and L. Antolini, *Adv. Mater.* 6:561 (1994).

60. G. Barbarella, A. Bongini, and M. Zambianchi, *Tetrahedron 48*:6701 (1992).
61. R. H. Baughman and R. R. Chance, *J. Appl. Phys. 47*:4295 (1976).
62. B. Krische, J. Hellberg, and C. Lilja, *J. Chem. Soc., Chem. Commun. 1987*:1476.
63. R. M. Souto Maior, K. Hinkelmann, and F. Wudl, *Macromolecules 23*:1268 (1990).
64. M. Zagorska and B. Krishe, *Polymer 31*:1379 (1990).
65. M. Zagorska, I. Kulszewicz-Bajer, A. Pron, L. Firlcj, P. Berier, and M. Galtier, *Synth. Met. 45*:385 (1991).
66. R. D. McCullough and R. D. Lowe, *J. Chem. Soc., Chem. Commun. 1992*:70.
67. K. Tamao, K. Sumitani, and M. Kumada, *J. Am. Chem. Soc. 94*:9268 (1972).
68. M. Kumada, *Pure Appl. Chem. 52*:669 (1980).
69. E. Negishi, *Pure Appl. Chem. 53*:2333 (1981).
70. N. Bumagin and I. P. Beletskaya, *Russ. Chem. Rev. 59*:1174 (1990).
71. J. K. Stille, *Angew. Chem., Int. Ed. Engl. 25*:508 (1986).
72. V. N. Kalinin, *Synthesis 1992*:413.
73. E. Negishi, T. Takahashi, S. Baba, D. Van Horn, and N. Okukado, *J. Am. Chem. Soc. 109*:2393 (1987).
74. R. D. McCullough, S. Tristram-Nagle, S. P. Williams, R. D. Lowe, and M. Jayaraman, *J. Am. Chem. Soc. 115*:4910 (1993).
75. R. D. McCullough and S. P. Williams, *J. Am. Chem. Soc. 115*:11608 (1993).
76. R. D. McCullough and M. Jayaraman, *J. Chem. Soc., Chem. Commun. 1995*:135.
77. R. D. McCullough, J. A. Belot, and S. P. Williams, *Mol. Eng. Adv. Mater. 456*:349 (1995).
78. R. D. McCullough and R. D. Lowe, *Polym. Prepr. 33*:195 (1992).
79. R. D. McCullough, R. D. Lowe, M. Jayaraman, P. C. Ewbank, D. L. Anderson, and S. Tristram-Nagle, *Synth. Met. 55*:1198 (1993).
80. R. D. McCullough, S. P. Williams, S. Tristram-Nagle, M. Jayaraman, P. C. Ewbank, and L. Miller, *Synth. Met. 69*:279 (1995).
81. R. D. McCullough, S. P. Williams, M. Jayaraman, J. Reddinger, L. Miller, and S. Tristram-Nagle, in *Electrical, Optical, and Magnetic Properties of Organic Solid State Materials*, Vol. 328 (L. Dalton and C. Lee, eds.), Mater. Res. Soc., Pittsburgh, PA, 1994, p. 215.
82. K. Tamao, S. Kodama, I. Naajima, M. Kumada, A. Minato, and K. Suzuki, *Tetrahedron 38*:3347 (1982).
83. C. Van Pham, H. B. Mark, Jr., and H. Zimmer, *Synth. Commun. 16*:689 (1986).
84. D. D. Cunningham, L. Laguren-Davidson, H. B. Mark, Jr., C. Van Pham, and H. Zimmer, *J. Chem. Soc., Chem. Commun. 1987*:1021.
85. G. Consiglio, S. Gronowitz, A.-B. Hornfeldt, B. Maltesson, R. Noto, and D. Spinelli, *Chem. Scripta 11*:175 (1977).
86. C. Soucy-Breau, A. MacEachern, L. C. Leitch, T. Arnason, and P. Morand, *J. Heterocycl. Chem. 28*:411 (1991).
87. A. MacEachern, C. Soucy, L. C. Leitch, T. Arnason, and P. Morand, *Tetrahedron 44*:2403 (1988).
88. C. van Phan, R. S. Macomber, H. B. Mark, Jr., and H. Zimmer, *J. Org. Chem. 49*:5250 (1984).
89. G. M. Davies and P. S. Daview, *Tetrahedron Lett. 1972*:8507.
90. S. Kano, Y. Yuasa, T. Yokomatsu, and S. Shibuya, *Heterocycles 20*:2035 (1983).
91. T.-A. Chen and R. D. Rieke, *J. Am. Chem. Soc. 114*:10087 (1992).
92. T.-A. Chen and R. D. Rieke, *Synth. Met. 60*:175 (1993).
93. T.-A. Chen, R. A. O'Brien, and R. D. Rieke, *Macromolecules 26*:3462 (1993).
94. T.-A. Chen, X. Wu, and R. D. Rieke, *J. Am. Chem. Soc. 117*:233 (1995).
95. X. Wu, T.-A. Chen, and R. D. Rieke, *Macromolecules 28*:2101 (1995).
96. M. R. Andersson, D. Selse, M. Berggren, H. Jarvinen, T. Hjertberg, O. Inganas, O. Wennerstrom, and J. E. Osterholm, *Macromolecules 27*:6503 (1994).
97. I. Levesque and M. Leclerc, *J. Chem. Soc., Chem. Commun. 1995*:2293.
98. H. Mao, B. Xu, and S. Holdcroft, *Macromolecules 26*:1163 (1993).
99. Y. Furukawa, M. Akimoto, and I. Harad, *Synth. Met. 18*:151 (1987).
100. A. O. Patil, A. J. Heeger, and F. Wudl, *Chem. Rev. 88*:183 (1988).
101. S. D. D. V. Rughooputh, S. Hotta, A. J. Heeger, and F. Wudl, *J. Polym. Sci. B: Polym. Phys. 25*:1071 (1987).
102. O. Inganas, *Trends Polym. Sci. 2*:189 (1994).
103. J.-P. Aime, in *Conjugate Polymers: Novel Science and Technology of Conducting and Nonlinear Optically Active Materials* (J.-L. Bredas and R. Silbey, eds.), Kluwer, Amsterdam, 1991, p. 229.
104. O. Inganas, W. R. Salanek, W. R. Osterholm, and J. Laakso, *Synth. Met. 22*:395 (1988).
105. O. Inganas, G. Gustafsson, W. R. Salaneck, J. E. Osterholm, and J. Laasko, *Synth. Met. 28*:C377 (1989).
106. S. Yue, G. C. Berry, and R. D. McCullough *Macromolecules 29*:933 (1996).
107. C. Roux, J.-Y. Bergeron, and M. Leclerc, *Makromol. Chem. 194*:869 (1993).
108. C. Roux and M. Leclerc, *Chem. Mater. 6*:620 (1994).
109. K. Tashiro, K. Ono, Y. Minagawa, K. Kobayashi, T. Kawai, and K. Yoshino, *Synth. Met. 41*:571 (1991).
110. A. G. MacDiarmid and A. J. Epstein, *Synth. Met. 65*:103 (1994).
111. J. Roncali, A. Yassar, and F. Garnier, *J. Chem. Soc., Chem. Commun. 1988*:581.
112. A. Yassar, J. Roncali, and F. Garnier, *Macromolecules 22*:804 (1989).
113. M. M. Bouman and E. W. Meijer, *Adv. Mater. 7*:385 (1995).

114. M. M. Bouman, E. E. Havinga, R. A. Janssen, and E. W. Meijer, *Mol. Cryst. Liq. Cryst. 256*:439 (1994).
115. T. J. Prosa, M. J. Winokur, J. Moulton, P. Smith, and A. J. Heeger, *Macromolecules 25*:4364 (1992).
116. M. J. Winokur, P. Wamsley, J. Moulton, P. Smith, and A. J. Heeger, *Macromolecules 24*:3812 (1991).
117. J. Mardalen, E. J. Samuelsen, O. R. Gautun, and P. H. Carlsen, *Solid State Commun. 80*:687 (1991).
118. J. Mardalen, E. J. Samuelsen, O. R. Gautun, and P. H. Carlsen, *Synth. Met. 48*:363 (1992).
119. S.-A. Chen and J.-M. Ni, *Macromolecules 25*:6081 (1992).
120. T. J. Prosa, M. J. Winokur, and R. D. McCullough, *Macromolecules 29*:3654 (1996).
121. F. Chen, P. G. Mehta, L. Takiff, and R. D. McCullough, *J. Mater. Chem. 6*:1763 (1996).
122. K. A. Murray, S. C. Moratti, D. R. Baigent, N. C. Greenham, K. Pichler, A. B. Holmes, and R. H. Friend, *Synth. Met. 69*:395 (1995).
123. G. Yu, H. Nishino, A. J. Heeger, T.-A. Chen, and R. D. Rieke, *Synth. Met. 72*:249 (1995).
124. R. D. McCullough, and S. P. Williams, *Chem. Mater. 7*:2001 (1995).
125. A. Iraqi, J. A. Crayston, and J. C. Walton, *J. Mater. Chem. 5*:1831 (1995).
126. S. Yue, G. C. Berry, and R. D. McCullough, *Macromolecules 29*:933 (1996).
127. M. Lemaire, D. Delabouglise, R. Garreau, A. Guy, and J. Roncali, *J. Chem. Soc., Chem. Commun. 1988*:658.
128. J. P. Ruiz, K. Nayak, D. S. Marynick, and J. R. Reynolds, *Macromolecules 22*:1231 (1989).
129. R. D. McCullough, P. C. Ewbank, and R. S. Loewe, *J. Am. Chem. Soc. 119*:633 (1997).
130. F. Andreani, P. C. Bizzari, C. D. Casa, and E. Salatelli, *Polym. Bull. 27*:117 (1991).
131. P. C. Ewbank, J. L. Reddinger, and R. D. McCullough, *Polym. Prepr. 36*:198 (1995).
132. P. Bauerle, K.-U. Gaudl, F. Wurthner, N. S. Sariciftci, H. Neugebauer, M. Mehring, C. Zhong, and K. Doblhofer, *Adv. Mater. 2*:490 (1990).

10
Recent Advances in Heteroaromatic Copolymers

John P. Ferraris and Douglas J. Guerrero*
The University of Texas at Dallas, Richardson, Texas

I. INTRODUCTION

Polymers based on heteroaromatics comprise a huge class of materials that have received considerable attention due to their interesting electrical, electrochemical, and optical properties, and a number of reviews have appeared [1–4]. These polymers have seen application in semiconductor devices such as molecular-based and field-effect transistors [5–7], energy storage devices such as batteries [8–11] and ultracapacitors [12–14], and active materials for electrochromic and light-emitting displays [15], to name a few. The properties of these materials can be varied over a wide range of conductivity, processibility, and stability depending on the type (and number) of substituents, rings, and ring fusions. As is true for other areas of polymer chemistry, the use of *copolymers* can simultaneously broaden and allow better control over these and other properties. Whether formed by copolymerization of mixtures of the individual monomers or through the use of polycyclic monomers that have the comonomer moieties already built in, heteroaromatic copolymers have aroused considerable interest from the scientific community.

A particularly important determinant of the redox and optical properties of polyheteroaromatics is the nature of the heteroatom itself. Homopolymers based on pyrrole and thiophene derivatives are the most highly studied polyheteroaromatic systems, so it should come as no surprise that copolymers based on them constitute the main emphasis of work in this area. Nevertheless, many other combinations of heteroaromatics with other (hetero)aromatic species have been reported. This review examines many of the recent advances in the field but makes no claim to be an exhaustive review of the literature. In particular, the formal copolymers of heteroaromatics with acetylene (e.g., heteroarylene-vinylene and its derivatives) are not treated, as they are the topic of other chapters in this volume.

II. COPOLYMERS OF PYRROLE

The relative ease of substitution on the N atom of pyrrole makes it an exceptionally versatile monomer with respect to its derivative chemistry. It can also be substituted at the 3- and/or 4-positions [16–19] to afford a large number of monomers amenable to oxidative polymerization. Stereoelectronic effects of substitution on the N must be appreciated when conducting copolymerizations between pyrrole and its derivatives. To a first approximation, *N*-alkyl substitution is expected to impart little electronic perturbation to the monomer's oxidation potential, because these substituents exhibit weak electronic affects in the first place and reside at a node in the monomer's highest occupied molecular orbital (HOMO). When the substituents become large (e.g., *t*-butyl, cyclohexyl) or conjugated with the pyrrole (e.g., aryls), they can influence the redox potential of the monomer as well as impede the coupling of radical cations. These factors result in considerable discrepancies between final copolymer composition and the comonomer feedstock ratios. This complication notwithstanding, a large number of copolymers between pyrrole and its N-substituted derivatives have been prepared and characterized. Several of these are discussed in the following sections.

A. Poly(pyrrole-*co*-*N*-arylpyrrole)s

Several groups have copolymerized pyrrole (**I**) with *N*-phenyl- and substituted *N*-phenylpyrroles. Unlike the

---

* *Current affiliation:* Brewer Science, Inc., Rolla, Missouri.

homopolymerization of N-(p-nitrophenyl)pyrrole (**IIa**, $R_1 = NO_2$, $R_2 = H$), which fails to afford good quality polymer, its copolymerization with **I** readily yields a copolymer with a 4:1 **I**:**II** ratio [20]. Reynolds' group copolymerized pyrrole with N-phenyl- (**IIb**, R_1, $R_2 = H$) and N-(3-bromophenyl)pyrrole (**IIc**, $R_1 = H$, $R_2 = Br$) [21,22]. The conductivity of the copoly(**I**/**IIb**) increased as the pyrrole content increased. However, elemental analysis proved unreliable for correlating the different feedstock ratios and final copolymer compositions. Copolymerization with N-(3-bromophenyl)pyrrole (**IIc**) gave more consistent elemental composition data. The redox properties of **Ib** and **IIb** were shown to be similar. The polymer composition in these pyrrole/arylpyrrole copolymers is very much controlled by the oxidation potentials of the monomers employed and less by monomer feed ratios. A maximum incorporation of 10% of the substituted monomer was achieved even when the feed ratio of **I** to **IIc** was 5:95. At this level of incorporation a conductivity drop of six orders of magnitude compared to polypyrrole was observed.

Nitroaryl groups have also been attached to the nitrogen either by chemical modification of polypyrrole by nitrobenzoyl chlorides and nitrohalobenzenes or electrocopolymerization of pyrrole in the presence of nitrostyrenes and nitrobenzaldehydes [23].

B. Poly(pyrrole-co-N-methylpyrrole)s

Early studies on the electrochemical copolymerization of mixtures of pyrrole and N-methylpyrrole (**III**) were carried out by the groups of Kanazawa et al. [24] and Inganas et al. [25]. Unlike the poly(pyrrole-co-N-arylpyrrole)s, these copolymers have higher conductivities and better mechanical properties. The final copolymer compositions paralleled the comonomer feed ratios due to the similarities in the oxidation potentials of the monomers [21,27]. Higher incorporation of N-methylpyrrole into the feed ratio decreases the conductivity of the copolymer, however [21–23,27–29]. This has been attributed to steric interactions between the H's on the N-methyl groups and the β-hydrogens on the adjacent rings, which force a twisted chain conformation of the polymer backbone [22,26]. Nevertheless, a higher incorporation of the substituted monomer into the copolymer is allowed before a comparable decrease in conductivity is obtained [22] when compared to copolymers from N-arylpyrroles (Table 10.1). Copolymers of **I** and **III** have been prepared by the simultaneous oxidation and copolymerization of mixtures of these monomers using bromine [26]. They have been characterized by X-ray photoelectron spectroscopy (XPS), and their electrical conductivities were in the $(1.2–2.3) \times 10^{-5}$ S/cm range, depending on the number of charged nitrogens on the main chain and incorporation of bromine counterions [26].

Copolymers of **I** and **III** have been studied by in situ measurements using microarray electrodes [27]. This technique allowed the copolymer films to be characterized in the absence of irreversible air oxidation and mechanical damage. Copolymers with different compositions were obtained by controlling monomer feed ratios, and the threshold potentials for conductivity varied with composition. These array devices showed reproducible responses to various chemical oxidants and reductants and hence could enjoy sensor applications. Copolymers from pyrrole/N-methylpyrrole are significantly sensitive to preparation conditions, such as dopant counterions and temperature [28]. X-ray scattering studies of films prepared using toluene sulfonate as the dopant showed structural anisotropy, whereas films prepared using perchlorate anion were isotropic [28]. Structural anisotropy and conductivity decreased with increased N-methylpyrrole in the monomer feed ratio. Recently, this phenomenon has also been observed in an organized assembly of two-dimensional conducting pyrrole/N-methylpyrrole copolymers prepared using the Langmuir–Blodgett technique [29]. These experiments revealed a high anisotropy for the in-plane/transverse conductivity ratio (10^{10}) with in-plane conductivities in the range of 0.68–0.01 S/cm.

C. Poly(pyrrole-co-pyrrole-N-alkylsulfonate)s

The poly(pyrrole-co-pyrrole-N-alkylsulfonate)s belong to a class of materials that have been termed "self-doping." They possess an anionic dopant ion covalently attached to the polymer backbone, forcing predominant *cation* movement during the doping/dedoping process. The most highly studied copolymer in this family has been prepared from pyrrole and 3-(pyrrol-1-yl)propane-

Table 10.1 Compositional Dependence of Conductivity for pyrrole-(N-substituted) pyrrole Copolymers

Copolymer	Composition	Conductivity (S/cm)	Ref.
I-co-**III**	2:8	10^{-2}	22
I-co-**IIc**	9:1	10^{-2}	22

sulfonate (**IVa**, n = 3) [30–33]. The ratio of **I** to **IVa** in this random copolymer was calculated to be 3:1 [30] or 2:1 [27,28] in different preparations. In the oxidized form, about two-thirds of the bound sulfonates are used to compensate the polymer backbone, and the remaining are associated with electrolyte or residual monomer cations. This polymer is sensitive to cation size, and charge transport is faster than for polypyrrole. A conductivity of 10^{-2} S/cm has been reported for this copolymer. A similar system with a butyl sulfonate chain (**IVb**) has also been prepared [34].

D. Poly(pyrrole-co-(3-substituted pyrrole))s

Copolymers of **I** and pyrrole derivatives containing oligo(ethyleneoxy) chains on the 3-position have recently been prepared and studied as battery electrodes [35]. The conductivity of these copolymers (**V**) was higher than that of the corresponding homopolymer, and the copolymers also exhibited larger current density and stability in cyclic voltammetric and discharge tests.

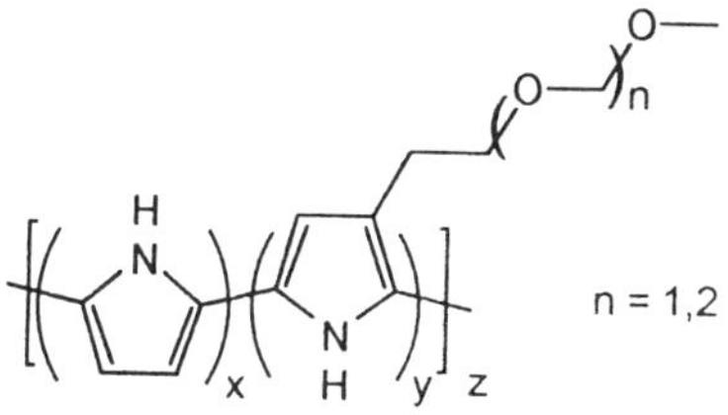

V

E. Poly(pyrrole-co-N-(redox couple)pyrrole)s

Several pyrrole copolymers containing redox active groups have been prepared in order to control electrical properties. One set is derived from the electrochemical copolymerization of pyrrole and N-(2-ferrocenylethyl) pyrrole (**VIa**) [36] or N-(4-ferrocenylbutyl)pyrrole (**VIb**)

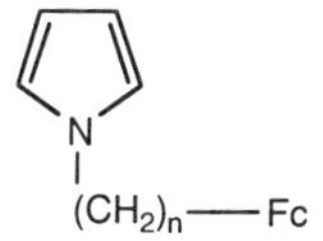

Fc = ferrocenyl

VIa (n = 3)
VIb (n = 4)

[37]. In the latter system, ferrocenylpyrrole incorporation was limited by the different redox potentials of the monomers and the slower kinetics of nucleation of the substituted monomer. In some cases the copolymer had five times greater unsubstituted pyrrole content than the feed ratio. In general these copolymers exhibited lower conductivity than pyrrole itself and inferior mechanical strength. Other copolymers have been prepared that in-

corporate electron donors (10-propylphenothiazine, **VII**), electron transfer acceptors (1,1'-dipropyl-4,4'-bipyridinium, **VIII**), or a metal-to-ligand charge transfer chromophore Ru(4-(2-pyrrol-1-ylethyl)-4'-methyl-2,2'-bipyridine) (**IX**) [38,39] attached to the nitrogen. These systems were of interest as possible avenues to spatially separated photooxidative and photoreductive sites that might mimic *pn* junctions. Similar copolymers containing anthraquinone, phenothiazine, or anthracene moieties linked to the nitrogen as well as to the 3-position have been reported [39]. The properties of these copolymers show strong dependence on the nature of the electroactive chromophore and less on polymer composition, electrolyte, or monomer concentration [39]. Recently a novel copolymer system for photoelectrical conversion was reported [40]. This material was formed by the copolymerization of pyrrole with an N-triphenylaminophenyl-substituted pyrrole (**X**). Spectral response was limited to the UV, and conversion efficiencies for the transmitted light using different film compositions was quite low, varying from 0.01% to 0.09%. Performance improvements were thought possible through optimization of the copolymerization process. In another report [41], pyrrole was reacted with quinone in the presence of polyphosphoric acid to give a material with a conductivity of 10^{-5} S/cm in the undoped state.

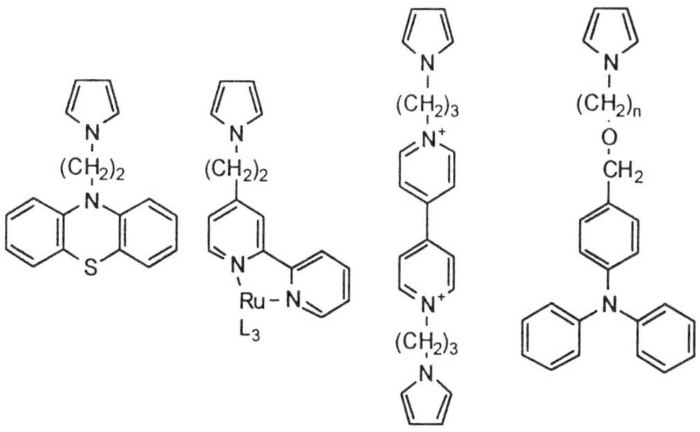

VII   **VIII**   **IX**   **X**

III. COPOLYMERS OF THIOPHENE

A. Copolymers from Thiophene and Substituted Thiophenes

1. Copoly(3-alkylthiophene)s

Thiophene (**XI**) copolymers have been prepared by Grignard coupling of 2,5-diiodo-3-methylthiophene and either 3-n-butyl-2,5-diiodothiophene or 2,5-diiodo-3-n-octylthiophene. These materials were soluble in common organic solvents, had molecular weights of about 5000, and displayed conductivities of 4–6 S/cm after doping with I_2 or $NOSbF_6$ [42]. Chemical oxidative copolymerization of monomers with longer alkyl chains [*n*-butyl (**XII**, n = 4)/*n*-octyl (**XII**, n = 8) and *n*-butyl/*n*-dodecyl (**XII**, n = 12)] has been carried out with FeCl$_3$ [43]. The copolymer compositions reflected the feedstock ratios, and the molecular weights ranged between 1.2×10^5 and 1.6×10^5. X-ray diffraction patterns on long-chain alkylthiophenes suggested an ordered layered structure similar to that of the corresponding homopolymers but with reduced coplanarity of the main chain [43]. One of the undesirable results of chemical polymerization of mixtures of 3-alkyl-substituted thiophenes is the undefined structure of the resulting copolymer. By extending the method they have developed for producing head-to-tail poly(3-alkylthiophene)s, McCullough's group synthesized highly regioregular random copolymers as well as structurally homogeneous alternating copolymers of 3-dodecyl/3-methyl- and 3-dodecyl/3-hexylthiophenes (**XII**, n = 6) [44,45]. This procedure gives 100% head-to-tail (2-5) coupling, and the copolymers exhibit higher conductivities (50–200 S/cm) [45] than their regiorandom counterparts. The monomer feed ratios were used to control the effective degree of conjugation in these copolymers.

Electrochemical polymerization of a mixture of 3-*n*-octylthiophene and 3-methylthiophene produces a processible copolymer with a 1:2.5 C$_8$:C$_1$ composition ratio as shown by NMR [46]. When chemically doped with FeCl$_3$, this polymer had a conductivity in the 5–10 S/cm range and displayed high stability in both the neutral and doped states.

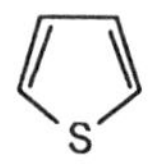

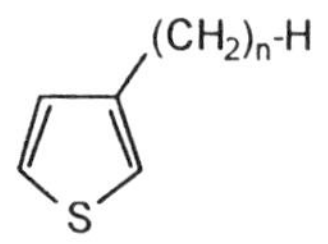

X-ray diffraction was used to examine the structures of two poly(3-*n*-hexylthiophene-*co*-3-*n*-octylthiophene) compositions. These partially crystalline materials displayed structures intermediate to those of their corresponding homopolymers, and the interstack distance varied linearly with the 3-*n*-octylthiophene mole fraction [47]. The use of specifically designed regioregular copolymers of 3-alkylthiophenes or block copolymers of oligothiophenes with organosilanylenes (**XIII** and **XIV**) to control the extent of conjugation and hence the wavelength of luminescence has recently been reported [48]. The latter afforded the desired control whereas the solution luminescence of the 3-alkylthiophene copolymers was dominated by the quinoid character of the excited state, which outweighed the steric constraints of the various substitution patterns.

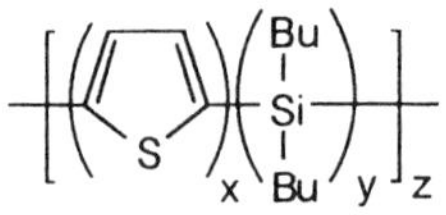

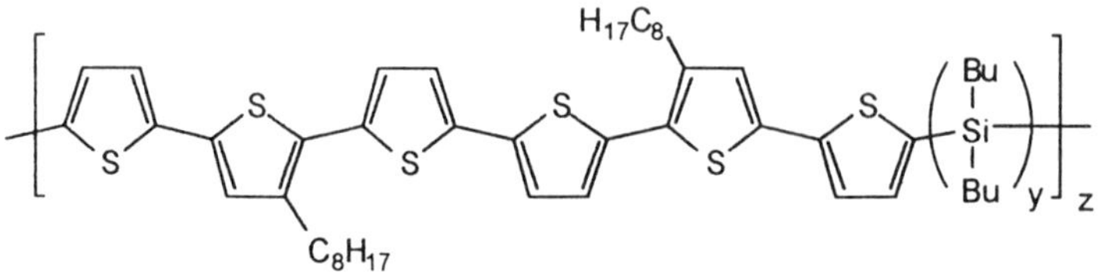

2. Other Thiophene Copolymers

Novel materials were prepared by electrochemical copolymerization of thiophene with thiophene-3-acetic acid (**XV**) [49]. The incorporation of the —COOH group allows for the loading of metal catalysts into these systems. These copolymers were soluble and had high degrees of polymerization (>2000). Copolymers between 3-methylthiophene and ferrocene esters of 2-(3-thienyl-ethanol) with varying alkyl chain lengths have been prepared (**XVI**) [50].

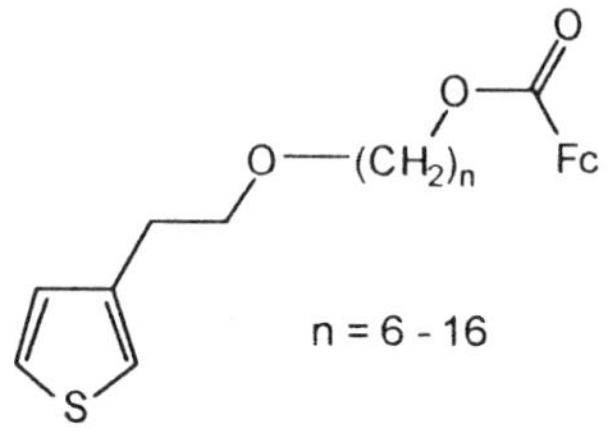

Copolymers resulting from the chemical polymerization of 3-hexylthiophene with 3-thiopheneethylacetate (**XVII**) [51], bithiophene with a crown ether linked bithiophene (**XVIII**) [52], and thiophene with 3,4-crown ether containing thiophene (**XIX**) [53] have also been reported. The last two copolymers can be n-doped by complexation of the crown ether units with alkali metal ions.

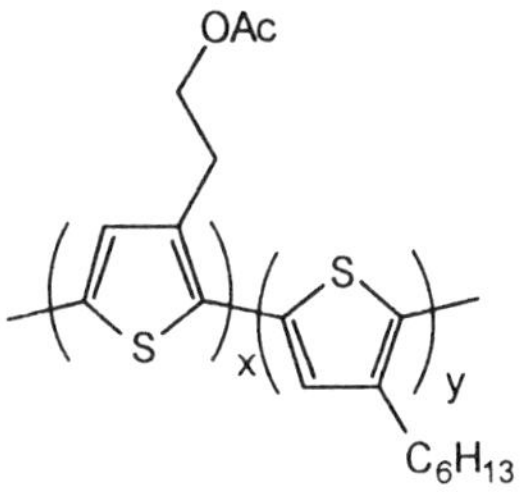

XVII

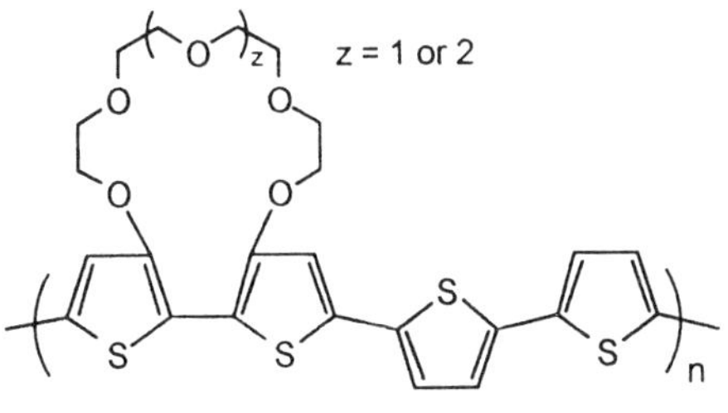

XVIII

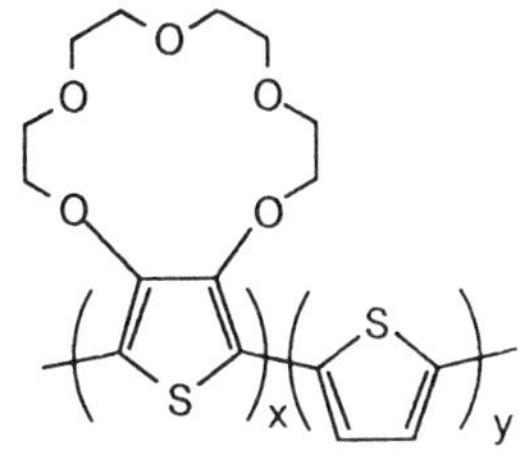

XIX

A series of polyaryleneethynylene copolymers (—Ar—C≡C—Ar'—C≡C—)$_n$, where Ar = 3-hexylthiophene, 3,4-dinitrothiophene, or selenophene and Ar' = 3-hexylthiophene, 2,5-pyridyl, or 1,4-phenylene units were prepared by Pd(0)-catalyzed coupling of the 2,5-dihalothiophene and the appropriate diacetylene [54]. These materials had MW $>10^4$, were soluble, and exhibited large third-order nonlinear optical susceptibilities and strong fluorescence. Several were n-dopable.

B. Copolymerization from Substituted Bi- or Terthiophenes

Monomers that incorporate two or three thiophenes with various substituents have been used to generate copolymers with well-defined comonomer sequences.

1. Copoly(3-alkylthiophene)s and Copoly(3-alkoxythiophene)s

Electrochemical polymerization of bi- and terthiophenes is a facile method for the synthesis of copolythiophenes due to their lower oxidation potentials relative to single-ring monomers. Some chemical polymerizations of these monomers have also been carried out. These monomers yield well-defined alternating structures and provide a means to control the effective conjugation length of the resulting polymer. Since substituents are present in every other or every third ring, steric hindrance is decreased, resulting in increased planarity of the main chain. Tables 10.2 and 10.3 show some of alkyl-substituted bi- and terthiophenes that have been polymerized.

Monomer **XXXII** can be electrochemically polymerized from acetonitrile solutions *containing no added*

Table 10.2 Substituted Bithiophene Monomers

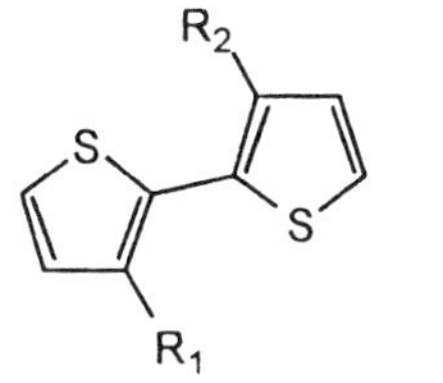	Polymerization method	Ref.
XX $R_1 = C_2H_5$, $R_2 = H$	Electrochemical	55
XXI $R_1 = C_2H_5$, $R_2 = CH_3$	Electrochemical	55
XXII $R_1, R_2 = C_2H_5$	Electrochemical	55

Table 10.3 Substituted Terthiophene Monomers

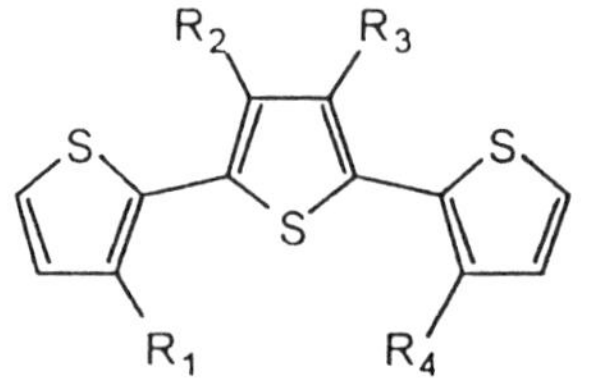

		Polymerization method[a]	Ref.
XXIII	$R_1, R_3, R_4 = H; R_2 = CH_3$	E	56
XXIV	$R_1, R_4 = H; R_2, R_3 = CH_3$	E	57
XXV	$R_1, R_4 = CH_3; R_2 = C_2H_5; R_3 = H$	E	55
XXVI	$R_1, R_3, R_4 = H; R_2 = C_4H_9$	E	57
XXVII	$R_1, R_4 = H; R_2, R_3 = C_4H_9$	C	58
XXVIII	$R_1, R_3, R_4 = H; R_2 = C_7H_{15}$	E, C	59
XXIX	$R_1, R_3, R_4 = H; R_2 = C_8H_{17}$	E	60
XXX	$R_1, R_3, R_4 = H; R_2 = C_{18}H_{37}$	E	59
XXXI	$R_1, R_4 = H; R_2, R_3 = -(CH_2)_3-$	E	57
XXXII	$R_1, R_3, R_4 = H; R_2 = -(CH_2)_3SO_3^- K^+$	E	57
XXXIII	$R_1, R_3, R_4 = H; R_2 = CH_2CH_2O\ CH_2CH_2OCH_3$	E	56

[a] E = electrochemical; C = chemical

electrolyte and affords a "self-doped" copolymer. This polymer can be cycled between the self-doped and de-doped states in electrolyte solutions containing *small* cations (H$^+$, Li$^+$, Na$^+$) but not in those containing only large cations like Bu$_4$N$^+$. Based on observations of the spectral behavior of polymers derived from **XXVI**, **XXIX**, and **XXXIII**, Roncali et al. [56] concluded that substitution on the central ring does not significantly affect the oxidation potential of the terthiophene systems but does affect the electropolymerization process and the structure of the resulting polymers. In the cases where the central ring contains an alkyl or oxyalkyl chain, the resulting polymer showed a decrease in oxidation potential and more extensively conjugated polymers, with λ_{max} increasing from 380 nm for polyterthienyl to 450 nm for the films of copolymers derived from **XXIX**. These materials were grown at current densities of 2 mA/cm^2. Previously Ferraris and Newton [59] reported that monomers **XXVIII** and **XXX** electrochemically polymerized at even lower current densities (0.5 mA/cm^2) afforded films with even higher effective conjugation lengths ($\lambda_{max} \approx 540$ nm). Compound **XXVII** was chemically polymerized by Wang et al. to yield a soluble copolymer that displayed a film with $\lambda_{max} \approx 522$ nm, slightly less than that derived from **XXVIII** and **XXX** but still indicative of a high effective degree of conjugation. The authors ascribe this to reduced steric interactions in the copolymer. Consistent with its effective conjugation length, this copolymer exhibited high electrical conductivity and p-type metallic behavior (thermopower) when doped.

A series of regiochemically defined dialkoxy-substituted thiophene oligomers (**XXXIV–XXXIX**) were polymerized electrochemically [61]. The degree of polymerization decreased as the monomer length increased. The alkoxy groups lowered the oxidation potentials of the monomers and copolymers and led to smaller energy gaps (E_{gap}) in the polymers. The lack of correlation between the alkoxy group substitution position and the effective degree of conjugation was taken to mean that the oxygen atom has a smaller steric demand than a methylene at the same site would in the corresponding polyalkylthiophene and thus could aid in the planarization of the polymer chain.

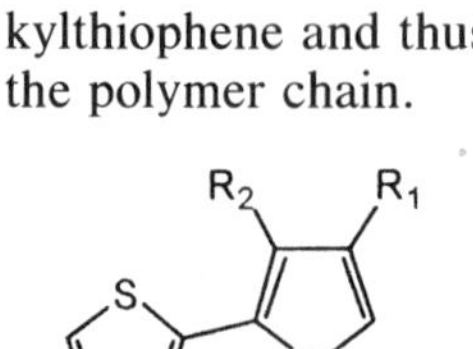

XXXIV $R_1 = OC_5H_{11}, R_2 = H$
XXXV $R_2 = OC_5H_{11}, R_1 = H$

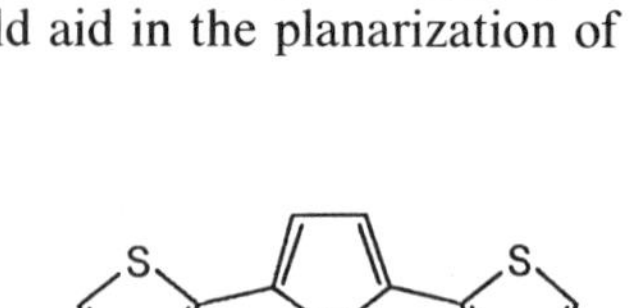

XXXVI $R_1 = OC_5H_{11}, R_2 = H$
XXXVII $R_2 = OC_5H_{11}, R_1 = H$

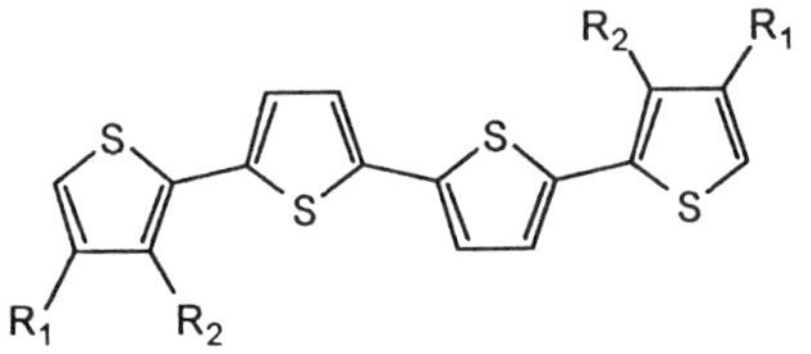

XXXVIII $R_1 = OC_5H_{11}, R_2 = H$
XXXIX $R_2 = OC_5H_{11}, R_1 = H$

2. Copoly(3-arylthiophene)s

Bithiophene and α-terthiophene monomers containing β-aryl substituents have also been polymerized. A material derived from **XL** has been used in blends with other poly(3-substituted thiophene)s to afford electroluminescent devices whose emission color depended on the applied voltage [62]. Monomers in which the central ring of α-terthiophene has been substituted with phenyl-[57,63], *p*-cyanophenyl-, *p*-methoxyphenyl-, *p*-pyridyl-, 2-thienyl-, or 3-methyl-2-thienyl [63] have been electrochemically oxidized to yield mixtures of oligomers in all cases [63].

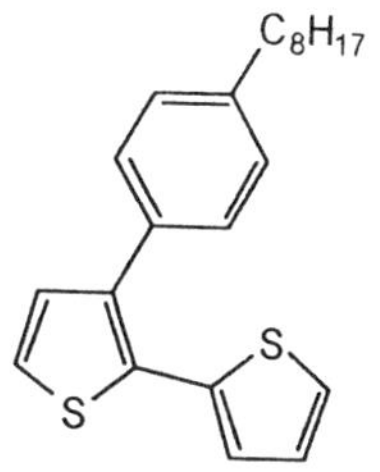

XL

3. Copolythiophenes from α-Terthiophenes Containing a Central Fused Ring System

Polymers derived from α-terthiophenes bearing a variety of fused central ring systems have been prepared. Three groups simultaneously reported on the polymerization of the monomer **XLI** (R = H) [64–66] prepared by different routes. This monomer affords the 1:1 alternating copolymer of bithiophene (BT) and isothianaphthene (ITN) with an E_{gap} of 1.7 eV. This value is intermediate to those of the corresponding homopolymers of BT (≈2.1 eV) and ITN (≈1 eV) [67]. Derivatives in which the heteroatom in the central ring was O [64–66] or N-CH$_3$ [64,66] were also prepared, as was the dimethoxy compound (**XLI**, R = OCH$_3$) [66].

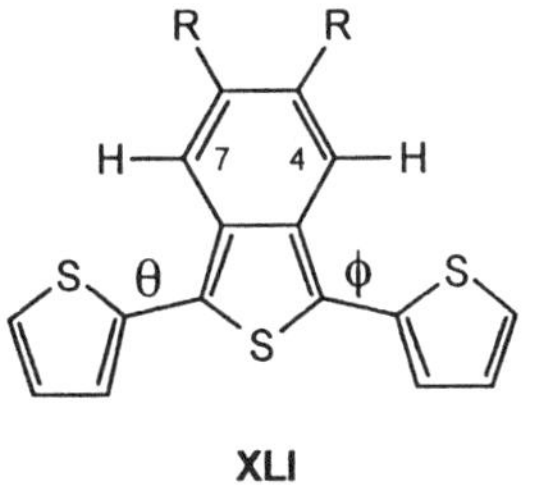

XLI

The copolymer derived from **XLI** can still experience some steric interaction between the H's on positions 4 and 7 of the fused benzo ring and the adjacent thiophenes, which can cause some rotation out of coplanarity ($\theta, \phi \approx 35$–$40°$) [66], thus raising the E_{gap}. Replacing one or both of these C—H groups with an N atom was predicted to lower this steric interaction in related systems [68]. Polymers based on compounds **XLII** [69] and **XLIII** [70] have been reported. X-ray structures of these monomers reveal θ, ϕ dihedral angles of 3° and 38° for **XLII** [69] and ≈10° for **XLIII** [70]. The E_{gap} values for their derived polymers are 1.4 eV and ≈1.0 eV, respectively. Two other members of this family are derived from the monomers incorporating the nonclassical fused thienothiadiazole **XLIV** [71] whose completely planar structure yields a polymer with $E_{gap} \approx 0.9$ eV and the extended system **XLV** [72], which has an estimated polymer $E_{gap} \approx 0.3$ eV!

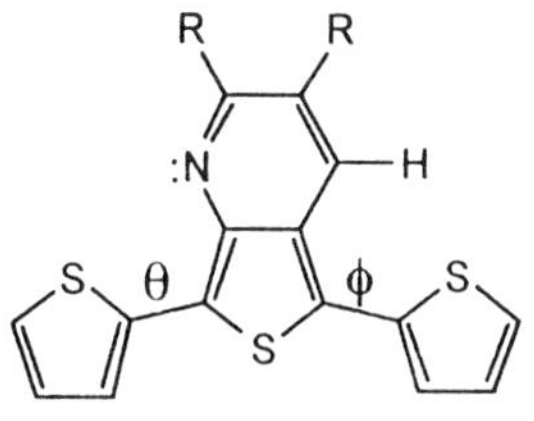

XLII 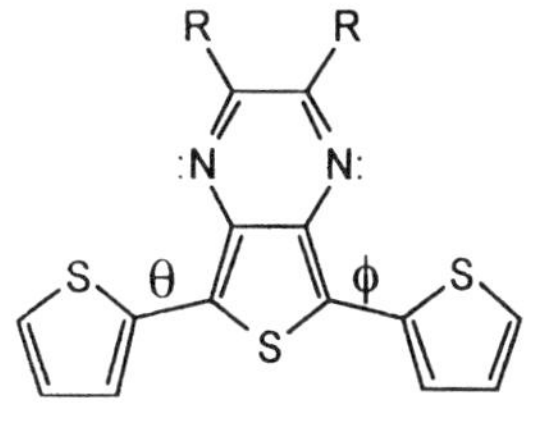   **XLIII**

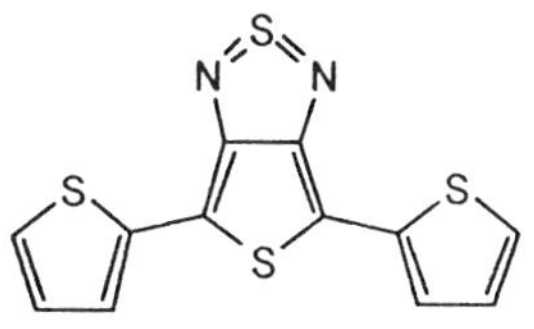

XLIV 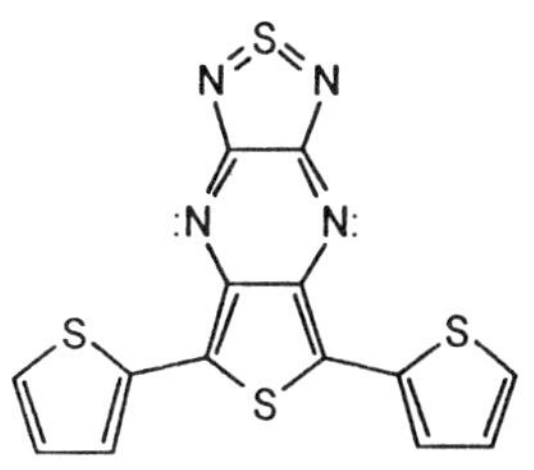   **XLV**

C. Other Copolymers

Highly colored (purple) but insoluble copolymers of 3,4-dialkoxythiophenes with thiophenes (**XLVII**) have been obtained by cyclization of the intermediate polymer **XLVI** [73] [Eq. (1)].

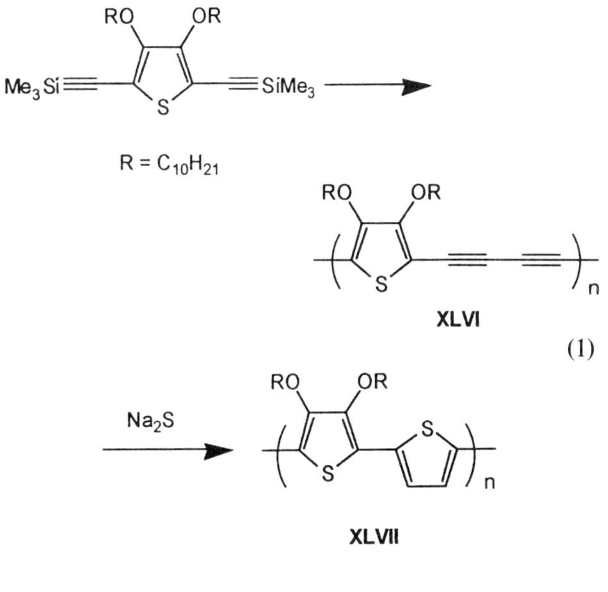

XLVI

XLVII

(1)

IV. COPOLYTHIENYLPYRROLES

Copolymerization of thiophenes and pyrroles offers the promise of combining the higher stability associated with thiophene-based polymers with the higher conductivity characteristic of polypyrroles. Three different methods have been used to prepare such copolymers. These are (1) direct copolymerization of pyrrole and thiophene monomer mixtures, (2) copolymerization of pyrrole with bi- or α-terthiophene, and (3) polymerization of α-linked pyrrole-thiophene monomers. The last case includes both two- and three-ring systems.

A. Copolymers from Pyrrole and Thiophene

The least potentially useful way to synthesize copolythienylpyrrole is the oxidative copolymerization of pyrrole and thiophene monomers. The main problem associated with this process is the great difference between the electrochemical oxidation potentials of pyrrole (0.6 V vs. SCE) and thiophene (1.6 V vs. SCE). A strategy used to carry out this copolymerization has been to "oxidize pyrrole under diffusion limiting conditions at potentials where thiophene oxidation takes place" [74]. Under these conditions pyrrole is kept at concentrations ~100 times lower than thiophene. By changing the applied potential over the range 1.37–2.07 V, copolymers rich in pyrrole or thiophene could be prepared. Cyclic voltammograms of copolymers prepared by this method with V_{appl} = 1.87 V were similar to those obtained by polymerization of 2,2'-thienylpyrrole [75]. The conductivities were reproducible and varied between 0.15 S/cm (thiophene-rich) and 44 S/cm (pyrrole-rich) [74] for the different compositions.

In another study, pyrrole was chemically polymerized with a 3-substituted thiophene [3-(methoxyethoxyethoxymethyl)thiophene], **XLVIII** [76]. This copolymer had a pyrrole/thiophene ratio of 7:1 and a conductivity of 1 S/cm. Pyrrole acted as a initiator of the polymerization and produced a material that would be less expensive while still maintaining the solubility properties of the substituted thiophene.

CH₂OCH₂CH₂OCH₂CH₂OCH₃

XLVIII

B. Copolymers from Pyrrole and Bi- or α-Terthiophene

There are several examples of the copolymerization of pyrrole with bithiophene [77–80] and pyrrole with α-terthiophene [78,81]. The lowered oxidation potentials of the thiophene dimers and trimers facilitates copolymerization with pyrrole. Nevertheless, early research showed that bithiophene content in polymers of pyrrole/bithiophene was not linearly related to the bithiophene feed ratio [77]. This was attributed to the difference in oxidation rates of the two monomers prior to oxidative coupling. Reactivity ratios for both monomers were determined at different potentials, with the conclusion that for a given feedstock ratio the copolymer composition could be varied systematically by changing the applied potential. Cyclic voltammetric data revealed three distinct oxidizable units on pyrrole/bithiophene copolymers: one due to blocks of polypyrrole, one to blocks of bithiophene, and one to random and alternate polypyrrole and polybithiophene groupings [76]. Conductivities were in the 1–20 S/cm range, with the lower conductivities associated with higher bithiophene content. Chemical polymerization of pyrrole bithiophene mixtures with copper perchlorate gave polymers whose composition could be varied by adjusting the feedstock ratios. Reactivity ratios for pyrrole and bithiophene were 3.2 and 0.2, respectively [80].

Copolymerization of pyrrole with α-terthiophene is readily accomplished due to their almost identical polymerization potentials [81]. Fox has studied random and graded copolymers of pyrrole/bithiophene and pyrrole/terthiophene as potential candidates for single charge rectification layers [78]. Graded copolymers were formed by lowering the concentration of one monomer and adjusting the applied potential. In contrast to bilayers of polypyrrole/polythiophene [78], significant rectifying effects were not observed, and no charge was trapped within random or graded films.

C. Copolymers of Pyrrole and Thiophene Derived from α-Linked Ring Systems

The first syntheses of copolythienylpyrroles using an α-linked ring system were carried out with 2,2'-thienylpyrrole (**XLIX**) as the starting monomer [75,82,83]. These films had conductivities of 3.3 S/cm when electrochemically doped with bisulfate ions [75,83].

Electrochemical polymerization of 2,5-di(2-thienyl)pyrrole (**L**) and doping with p-toluene sulfonate produced films with electrical conductivities ranging between 10^{-8} and 10^{-1} S/cm, depending on the concentration of electrolyte and voltage used [84,85]. Chemically polymerized polymers have electrical conductivities of 2.7×10^{-3} and $(0.36–9.0) \times 10^{-2}$ S/cm when doped with NOSbF₆ and NOPF₆, respectively [86]. Substitution of an octyl group on the β-position of the central ring has been carried out, and this monomer (**LI**) has been polymerized chemically and electrochemically [59] to afford soluble and insoluble products, respectively.

XLIX

L R = H
LI R = C₈H₁₇

Ferraris et al. [86–88] studied the steric effects on the optical and electrochemical properties of nitrogen-substituted (methyl, ethyl, isopropyl, heptyl, and octadecyl groups) copolythienylpyrroles. They observed that as the steric demands of the various N-substituents increased, so did the interring deviations from coplanarity. This trend was reflected in the monomer's oxidation potential and the energies of the long-wavelength absorption maxima for the monomers and polymers (Table 10.4). The conductivity of the chemically prepared N-methyl-substituted polymer was 2.4×10^{-3} and 2.7×10^{-2} S/cm when doped in methylene chloride with NOSbF$_6$ and NOPF$_6$, respectively [86]. In another study, the conductivity of the same polymer was reported to be $<10^{-5}$ S/cm when NOPF$_6$ in methanol was used as the dopant [89]. Electron paramagnetic resonance (EPR) studies of the copolymer prepared this last way demonstrated the presence of *localized* radicals rather than mobile charge carriers, possibly due to β-branching and off-chain coupling during polymerization [89], which would naturally lead to low conductivities. Electrochemically generated films from **LI** or **LII** displayed conductivities of >100 S/cm and 1 S/cm, respectively [88].

D. Other Approaches

In a synthesis similar to the one used for copolythiophenes described above, an N-phenyl copolythienylpyrrole was obtained as a brown solid [73] [Eq. (2)].

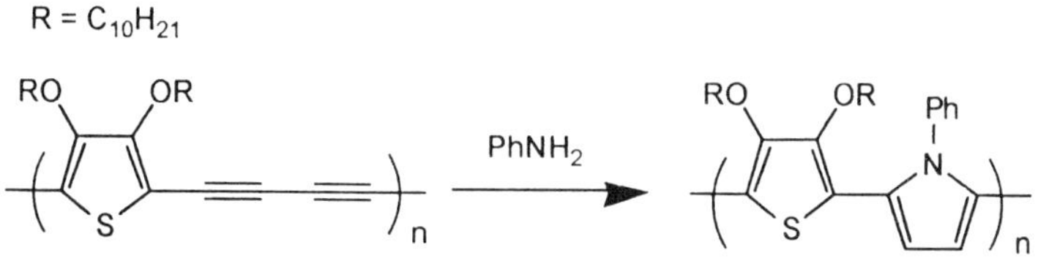

$$\text{XLVI} \xrightarrow{\text{PhNH}_2} \text{LVII} \quad (2)$$

R = C$_{10}$H$_{21}$

V. OTHER BINARY SYSTEMS

In the next two sections α-linked ring systems are designated by the heteroatom in each ring whenever appropriate. For example, the system derived from thiophene-pyrrole-thiophene is designated as SNS. Substitution on the nitrogen atoms is indicated appropriately [e.g., SN(Me)S].

A. Polyfurylpyrroles and Poly(pyrrole-benzo[b]thiophene)s

Unlike the polythienylpyrroles, there are only a few reports of copolymers containing the pyrrole and furan moieties. Synthesis from the corresponding monomers suffers both from the high oxidation potential of furan ($E_{ox}^{furan} = 1.7$ V vs. Ag/Ag$^+$) relative to substituted thiophenes or bithiophenes (≤ 1.5 V vs. Ag/Ag$^+$) [90] and from its susceptibility toward ring opening. A 1:1 copolymer has been prepared from the α-linked two-ring system 2,2′-furylpyrrole (ON) [82]. Most commonly, copolyfurylpyrroles have been synthesized by electrochemical or chemical polymerization of the α-linked three-ring systems. Cava and coworkers [89,91] synthesized ON(Me)O, which they chemically polymerized with NOPF$_6$ to afford a material with a conductivity of $<10^{-5}$ S/cm. They postulated that the low conductivity was a result of overoxidation, which created more bipo-

Table 10.4 N-substituted 2,5-di(2-thienyl)pyrrole Monomers/Polymers

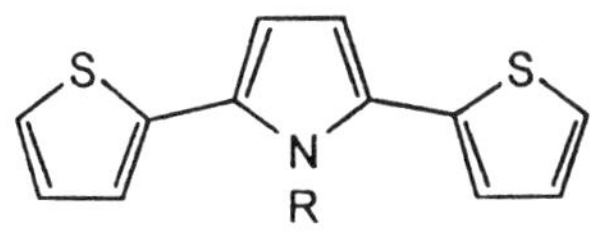

| | | Monomer | | Polymer |
	R	$E_{p,a}$(V) vs. SCE	$h\nu_{min}$ (eV)	$h\nu_{min}$ (eV)
LI	H	0.620	3.65	2.62
LII	CH$_3$	0.685	3.89	2.98
LIII	C$_2$H$_5$	0.706	3.97	3.02
LIV	C$_7$H$_{15}$	0.712	3.92	3.18
LV	C$_{18}$H$_{37}$	0.738	3.94	—
LVI	—CH(CH$_3$)$_2$	0.807	4.13	3.42

larons than polarons, thus reducing the number of charge carriers [91].

A highly conductive polymer (**LVIII**) has been prepared from the electrochemical polymerization of benzo[*b*]thiophene with pyrrole [92]. The tensile strength of the material doubled as the mole fraction of pyrrole in the copolymer increased from 20% to 70%. Doped polymer films had conductivities in the 0.4–1.2 S/cm range depending on the copolymer's composition [92].

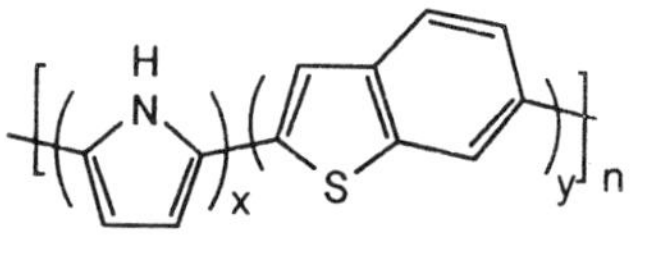

LVIII

B. Polyco(thiophene-furan)s, Polyco(thiophene-pyridine)s, Polyco(thiophene-aniline)s, and Polyco(thiophene-thiazole)s

A report on the 1:1 copolymer derived from 2,2′-thienylfuran (SO) has appeared [55]. Upon doping, this material displays an electrochromic change from brown to green. The oxidation potential for this copolymer is 240 mV lower than that of polythiophene. Most copolymers of furan and thiophenes have been synthesized from α-linked three-ring systems. Chemical polymerization of SOS gave a material with conductivities of 2.5×10^{-3} and 5.6×10^{-3} S/cm when doped with NOSbF$_6$ or NOPF$_6$, respectively [86]. In another study [89] this polymer had a reported conductivity of 0.2 S/cm. Cava et al. [89,91] examined an extensive series of α-linked heteroaromatics that included the monomers OSO, SOO, and SSO. Chemical polymerizations of these afforded copolymers with conductivities of $\leq 10^{-5}$ S/cm when doped with NOPF$_6$.

In an effort to combine the electron-donating properties of thiophene with the electron-accepting nature of pyridine, Kaeriyama and coworkers [93] electropolymerized 2,5-, 2,6-, and 3,5-[di(2-thienyl)]pyridine derivatives. Others have carried out chemical polymerization of 2-(2-thienyl)pyridine and 2,5-di(2-pyridyl)thiophene by dehalogenative polycondensation of the corresponding dihaloaromatic compounds using a zero-valent nickel complex [94,95] to form copolymers **LVIV** and **LV**, respectively. In their neutral states these polymers had low conductivities, but when doped with iodine or sodium their conductivities increased to the 4.5×10^{-4}–1×10^{-2} S/cm and 9.1×10^{-5}–1.3×10^{-4} S/cm ranges, respectively [95]. The long-wavelength absorption for **LVIV** was red-shifted from that of polythiophene, and this was taken as evidence of intramolecular charge transfer. The absorbance maximum for **LV** was intermediate to those of the corresponding homopolymers. Polymer **LVIV** exhibited facile n-doping [94].

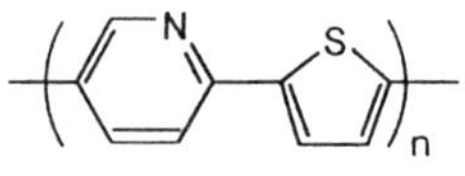

LVIV 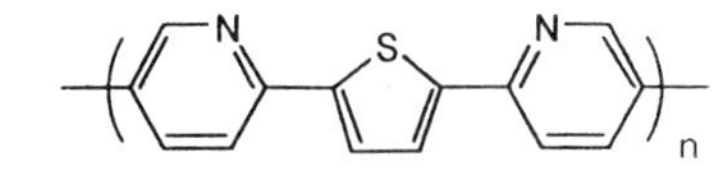   **LV**

Systems in which acetylenic spacers are placed between the pyridine and the thiophene rings have been reported (**LVI**). In these cases, the thiophene ring contained 3-hexyl or 3,4-dinitro groups [54]. These polymers can be electrochemically n-doped more easily than they can be p-doped. These materials are also of interest because of their large third-order nonlinear optical properties.

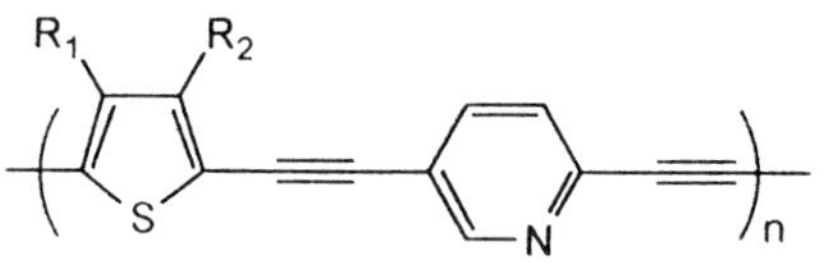

$R_1 = H; R_2 = C_6H_{13}$
$R_1 = R_2 = NO_2$

LVI

Copolymers of bithiophene and aniline have been chemically prepared using Cu(ClO$_4$)·6H$_2$O in acetonitrile [80]. The conductivities of these materials were in the 1×10^{-3}–6×10^{-1} S/cm range, depending on composition, but in all cases were lower than those of the respective homopolymers at comparable doping levels.

Copolymers of thiophene or 3-methoxythiophene with methylthiazole (**LVII**) have been electrochemically synthesized from α-linked four-ring monomers [96]. The maximum conductivity observed for the methoxy-substituted polymer was 0.01 S/cm. The authors concluded that placing an N atom in the conduction pathway had little effect on the conductivity and redox potential of the copolymers, which were comparable to those observed for poly(α-terthiophene). Incorporation of alkoxy groups had the expected effect of lowering the oxidation potentials of the monomer and its corresponding polymer.

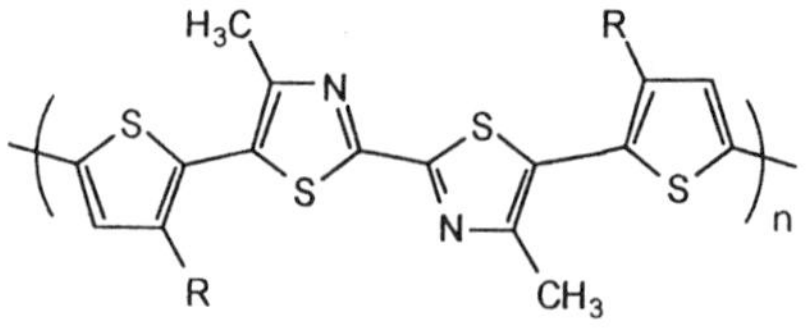

R = H or OCH$_3$

LVII

C. Other Binary Systems

Numerous other electroactive copolymers containing more complex heterocyclic rings have been prepared, and several are shown in Table 10.5. Compound **LXIII** represents an attempt to extend the work on thiophene-

Recent Advances in Heteroaromatic Copolymers

Table 10.5 Copolymers Containing Complex Heterocycles

	Structure	Ref.
LVIII	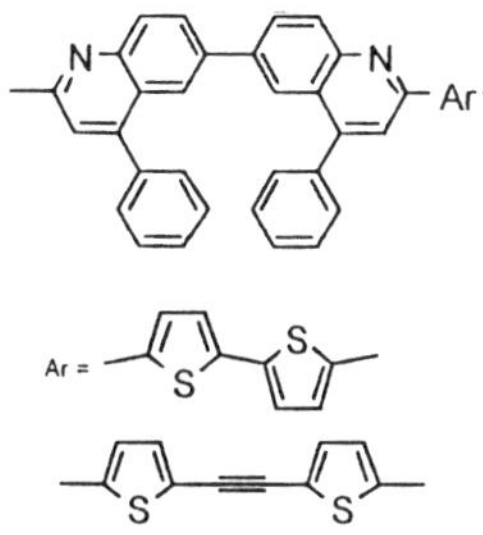 (structure with two quinoline-type rings bearing Ph groups, linked to Ar) Ar = bithiophene-methyl or bis(thienyl)acetylene	97
LVIX	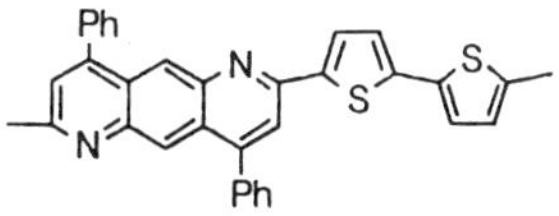 (Ph-substituted phenanthroline linked to bithiophene-methyl)	98
LX	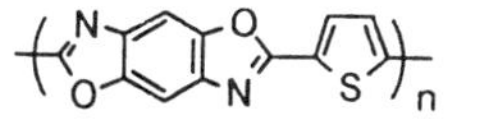 (benzobisoxazole-thiophene copolymer)	99
LXI	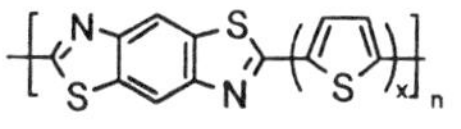 (benzobisthiazole-thiophene copolymer, x = 1, 2 or 3)	100
LXII	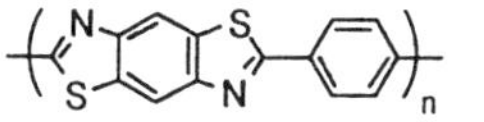 (benzobisthiazole-phenylene copolymer)	101
LXIII	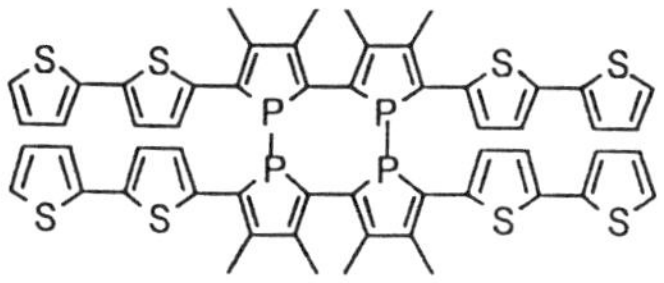 (bis-phospholyl linked bithiophenes)	102
LXIV	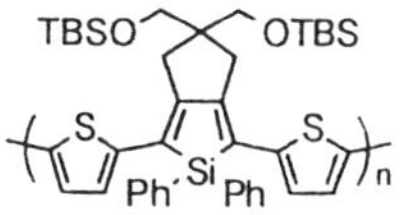 (silole-thiophene copolymer with TBSO groups, Ph₂Si, n = 2, 3 or 11)	103

based polymers to the phosphorus-containing heterocycle. This compound could be quantitatively cleaved to the bisphosphoryl anion, which is isoelectronic with sexithiophene [102]. Air-stable co-oligomers and copolymers of thiophene and diphenylsiloles (**LXIV**) have been reported [103]. The polymer had a conductivity of $<10^{-11}$ S/cm when neutral, which increased by 8 orders of magnitude upon I_2 doping. The dimethyl analog has recently been made. The UV/vis absorption red-shifted as the silole content increased, but the electrical conductivity followed the reverse order [104].

VI. COPOLYHETEROLENE-ARYLENES

A. Poly(pyrrole-*co*-arylene)s

Poly(pyrrole-*co*-phenylene)s have been synthesized electrochemically from 2-phenylpyrrole, **LXV** (σ^{doped} = 1–10 S/cm) [82], and from 1,4-bis(2-pyrrol-yl)phenylene, **LXVI** (σ^{doped} = 15 S/cm) [105]. Chemical syntheses of

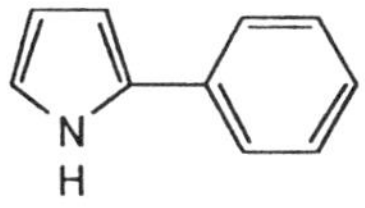

LXIV 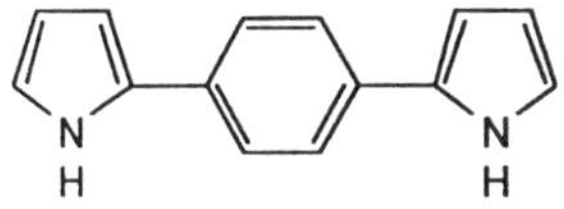   **LXV**

similar copolymers have been carried by cyclization of a polyphenylene diketone in the presence of ammonia (I_2-doped, 0.1 S/cm) [106] [Eq. (3)] and by palladium(0)-catalyzed coupling reactions of boronic acid derivatives of 1,4-bisdodecylbenzene and 2,5-dibromopyrrole [107] (**LXVI**). The copolymer where $x = 1$ was soluble and

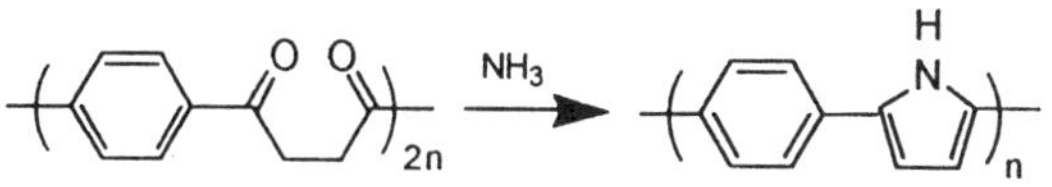

$$\left(\!\!\begin{array}{c}\text{Ph}-\text{CO}-\text{CH}_2\text{CH}_2-\text{CO}\end{array}\!\!\right)_{2n} \xrightarrow{NH_3} \left(\!\!\begin{array}{c}\text{Ph}-\text{pyrrole}\end{array}\!\!\right)_n \quad (3)$$

had a weight-average MW of 25,000, PDI = 2.3. The corresponding $x = 3$ copolymer was insoluble.

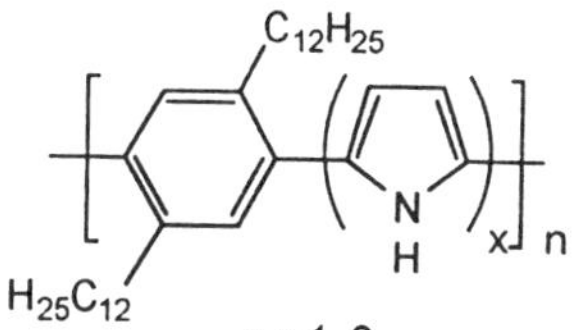

LXVI (1,4-didodecyl phenylene linked to pyrrole, n = 1, 3)

Very recently Reynolds et al. extended their work on 1,4-bis(2-furanylbenzenes) and 1,4-bis(2-thienylbenzenes) (see below) to a series of 1,4-bis(2-pyrrol-2-yl-arylenes), **LXVII**. These monomers can be polymerized chemically and electrochemically to afford copolymers that exhibit highly reversible redox chemistry and electrochromism with lifetimes of several thousand cycles [108].

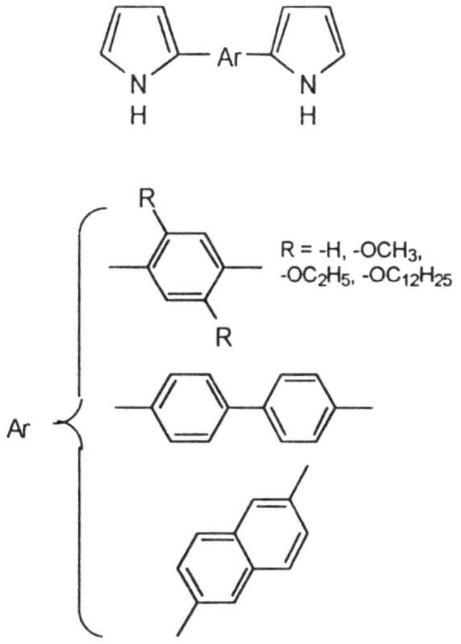

LXVII
Ar = substituted phenylene (R = -H, -OCH₃, -OC₂H₅, -OC₁₂H₂₅), biphenylene, or naphthylene

B. Poly(thiophene-co-arylene)s

Taliani et al. first studied the electropolymerization of 1,4-bis(2-thienyl)phenylene (**LXVIII**) [109,110] and structural changes of the electrochemically doped (ClO_4^-, 1.4×10^{-4} S/cm) polymer [110]. The electrochromic properties of this copolymer were studied by Haynes et al. [111], who observed good color contrast for it and lifetimes of $>10^3$ cycles. Electrochemical polymerization of the 1,3-bis(2-thienyl)phenylene (**LXIX**) and 1,4-bis(2-thienyl)biphenylene (**LXX**) isomers has also been reported [93,112]. These polymers can be electrochemically doped to a greenish color that persists in the presence of air and the absence of an applied potential. Chemical syntheses of random copoly[1,4-bis(2-thienyl)phenylene] derived from the Ni(II) coupling of continuous molar ratios of 2,5-dibromothiophene and 1,4-dibromobenzene have been carried out [113,114]. The band gaps were intermediate to those of the corresponding homopolymers, but the bandwidths were wider. Electrical conductivities of the neutral and doped materials varied as a function of composition, with the highest gain in σ upon doping occurring for the 1:1 composition [114].

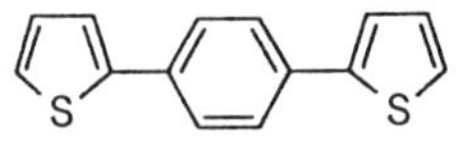

representative members of this family of copolymers are given in Table 10.6.

The Reynolds group prepared an extensive series of 2,5-disubstituted poly[1,4-bis(2-thienyl)phenylene]s (**LXI**). In most cases the phenyl ring was symmetrically substituted using groups like OCH_3, OC_7H_{15}, $OC_{12}H_{25}$, $OC_{16}H_{33}$, $OC_{20}H_{41}$ [117–120], cyclohexylmethyloxy [121], CH_3, and C_6H_{13}. Asymmetric substitution was also reported with OCH_3/OC_7H_{15} and $CH_3/OC_{12}H_{25}$ as the substituents [118]. In situ EPR showed the presence of polaron and bipolaron charge carriers on the symmetrically substituted OC_7H_{15}-doped polymer [117]. The polymers containing alkoxyphenylene groups have lower band gaps and higher conductivities due to the lower barrier of rotation between each monomeric unit relative to the alkyl-substituted polymer [119]. Chemical doping of alkoxy-substituted polymers with $NOPF_6$ gave materials with conductivities of 1–4 S/cm, whereas the alkyl-substituted derivatives were poor conductors (10^{-8}–10^{-6} S/cm) [118–120]. Chemical polymerization ($FeCl_3$) of the symmetrically substituted cyclohexylmethyloxy monomer gave a totally insoluble material, and electrochemical polymerization of the same monomer resulted in very slow film formation [121].

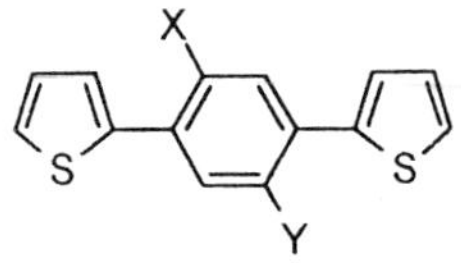

LXVIII   **LXIX**

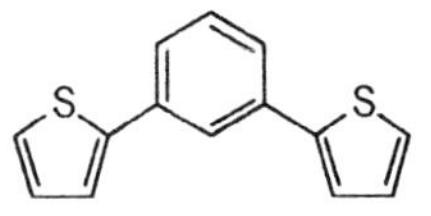

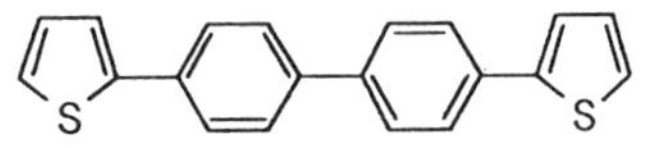

LX

Pelter and coworkers prepared a series of substituted thiophene-phenylenes by chemical polymerization of two aryl moieties [115,116] [Eq. (4)]. Conductivities for

LXI

Yamamoto et al. [122] developed a synthetic strategy to obtain poly(thienyl-co-phenylene), **LXII**, with well-defined linkages between the monomer units. This

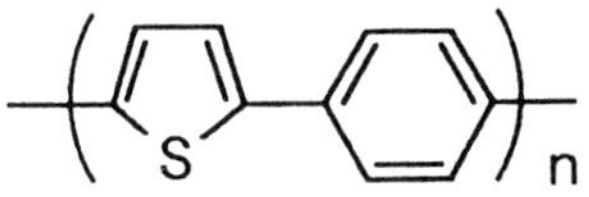

LXII

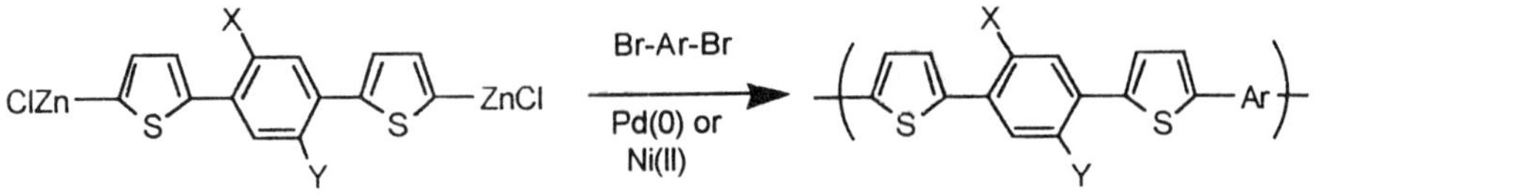

$$\text{ClZn-} \underset{S}{\bigcirc}\text{-} \underset{Y}{\overset{X}{\bigcirc}}\text{-} \underset{S}{\bigcirc}\text{-ZnCl} \xrightarrow[\text{Ni(II)}]{\text{Br-Ar-Br, Pd(0) or}} \left(\underset{S}{\bigcirc}\text{-} \underset{Y}{\overset{X}{\bigcirc}}\text{-} \underset{S}{\bigcirc}\text{-Ar} \right) \qquad (4)$$

Table 10.6 Conductivities for poly(thiophene-co-arylene)s

Polymer structure	Undoped σ (S/cm)	Maximum doping σ (S/cm)
X = Y = H; Ar = 1,4-phenyl	4.0×10^{-11}	12×10^{-4} (FeCl$_3$)
X = Y = methyl; Ar = 1,4-(2,5-dimethyl)phenyl	1.3×10^{-10}	1.9×10^{-4} (FeCl$_3$)
X = NO$_2$; Y = H; Ar = 1,4-(2-nitro)phenyl	3.3×10^{-9}	1.4×10^{-4} (FeCl$_3$)
X = Y = methoxy; Ar = 1,4-(2,5-dimethoxy)phenyl	1.0×10^{-7}	1.6×10^{-3} (I$_2$)
X = Y = butyl; Ar = 1,4-(2,5-dibutyl)phenyl	3.6×10^{-11}	1.3×10^{-3} (FeCl$_3$)
X = Y = H; Ar = 4,4'-biphenyl	2.0×10^{-11}	1.9×10^{-4} (FeCl$_3$)

polymer can be obtained in high yield by a nickel(0)-catalyzed reaction of 2,5-dihalothiophene and 1,4-dihalobenzene. The chemically and electrochemically doped polymer was a semiconductor ($\sim 10^{-5}$ S/cm). Similarly, 2,5-dibromothiophene has been copolymerized with diiododimethoxybenzene [123]. Removal of the methyl groups was reported to yield a self-doped poly(p-hydroquinone-alt-thiophene), **LXIII**, as a black insoluble solid.

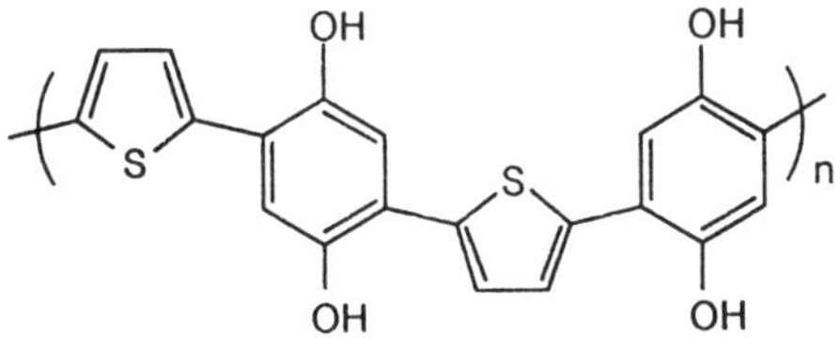

LXIII

Recently, chemical copolymerization of 2,5-(bistributyltin)thiophenes with 2,5-(bisalkoxy)dihalobenzenes (alkoxy: C_4–C_9, C_{12}, and C_{16}) in the presence of palladium(0) afforded materials with liquid crystalline properties [124,125]. Also, a new family of trithienylbenzenes (**LXIV**) have been copolymerized with 3-n-octylthiophene with the objective of developing xerogels of polyalkylthiophenes with well-defined conjugated cross-linking points. Homopolymers of these trifunctional monomers are highly cross-linked materials that cannot be doped and remain insulating. Chemically prepared [126] (FeCl$_3$) copolymers of the $f = 3$ systems with 3-n-octylthiophene gave a material that swells in polar solvents and achieves a conductivity of 0.1–0.2 S/cm provided the cross-link concentration is kept low.

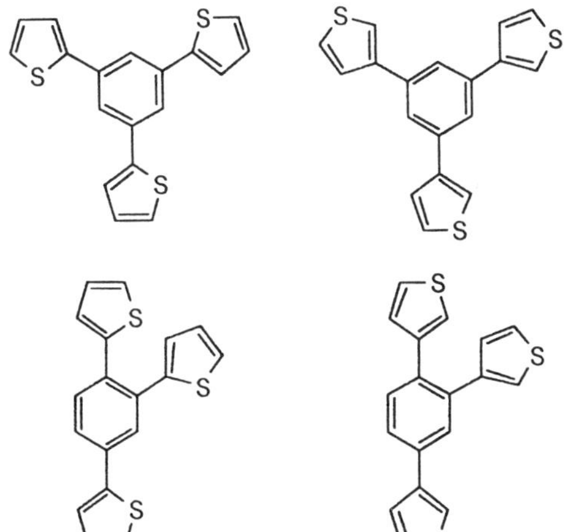

LXIV

Two other copolythienylarylenes of interest are those containing alternating bisthienyl and corenene units, **LXV** (ClO$_4^-$, 8.5×10^{-13} S/cm) [127], and one having an acetylenic group between 3-hexylthiophene and phenylene rings [54].

LXV

In a series of papers [128–133], the Reynolds group recently reported an impressive set of monomers that straddle various aryl groups with the electron-rich 3,4-ethylenedioxythiophene system, **LXVI**. These monomers offer almost unprecedented control over the E_{gap} and redox properties of their resultant polymers. Furthermore, since the majority of these polymers are soluble, the ambiguities regarding their structure are lifted. These (monomers and) polymers oxidize at relatively low potentials, making them extremely stable in their doped state. Electrochromic devices made from them can be cycled $>10^5$ times without noticeable degradation.

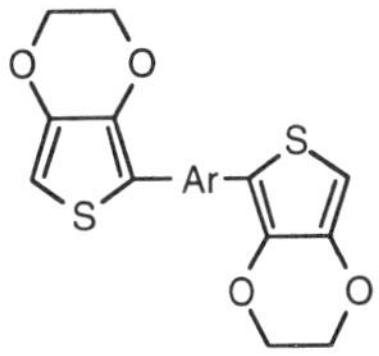

LXVI

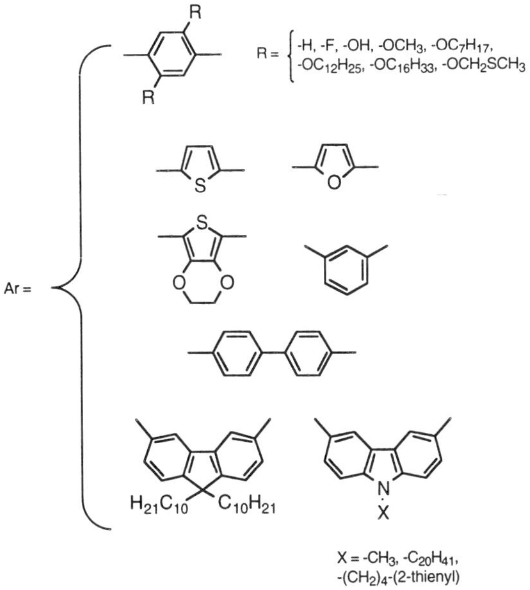

R = { -H, -F, -OH, -OCH$_3$, -OC$_7$H$_{17}$, -OC$_{12}$H$_{25}$, -OC$_{16}$H$_{33}$, -OCH$_2$SCH$_3$ }

X = -CH$_3$, -C$_{20}$H$_{41}$, -(CH$_2$)$_4$-(2-thienyl)

C. Poly(furan-co-arylene)s

Pelter and coworkers [115] chemically prepared structural isomers of poly(furan-co-phenylene)s by reacting bis-zinc chlorides of 1,4-bis(2-furanyl)phenylenes with 1,3- and 1,4-dibromobenzene. In another study [116],

Table 10.7 Conductivities of Poly(furan-co-phenylene)s

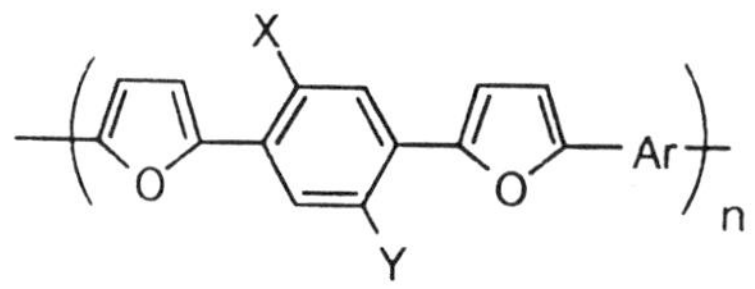

LXVII

Structure	Undoped (S/cm)	Maximum doping (S/cm)
X = Y = H; Ar = 1,4-phenyl	1.0×10^{-10}	3.24×10^{-4} (FeCl$_3$)
X = Y = methyl; Ar = 1,4-(2,5-dimethyl)phenyl	6.8×10^{-10}	4.5×10^{-2} (FeCl$_3$)
X = Y = methoxy; Ar = 1,4-(2,5-dimethoxy)phenyl	5.0×10^{-6}	1.6×10^{-3} (I$_2$)
X = Y = H; Ar = 4,4'-biphenyl	6.2×10^{-11}	5.9×10^{-5} (FeCl$_3$)

the same group prepared a number of substituted poly-(furan-co-phenylene)s with the various conductivities listed in Table 10.7.

Reynolds' group found that chemical polymerization of 1,4-bis(2-furanyl)phenylenes, symmetrically substituted in the 2- and 5-positions of the phenylene ring with methyl, methoxy, and heptyloxy and asymmetrically substituted with methoxy and heptyloxy groups, caused ring opening of the furan rings. On the other hand, electrochemical polymerization [90] of these monomers in the presence of ClO$_4^-$ gave copolymers with conductivities of 0.1–1.0 S/cm. A lowering of the optical band gap and oxidation potential of these polymers is observed with increasing electron-donating ability of the substituents [90]. Potential-dependent EPR studies on the symmetrical heptyloxy derivative showed evidence of the existence of polarons on the electrochemically doped polymer [121].

VII. COPOLYMERS OF PYRROLE AND THIOPHENE WITH NONHETEROCYCLIC OR NONCONJUGATED COMPOUNDS

Conducting copolymers of pyrrole with nonconjugated moieties are relatively few. Electrooxidative copolymerization of pyrrole and N-phenyl maleimide [134] in sulfuric acid gave a material with an electrical conductivity of 10^{-4} S/cm. X-ray analysis suggested that the copolymer was alternating and isotactic. In another study [135], various N-substituted poly(pyrrole-ethylenes), **LXVIII**, were exposed to iodine, and the conductivities of the resulting materials ranged between 1.8×10^{-4} and 2.0×10^{-3} S/cm.

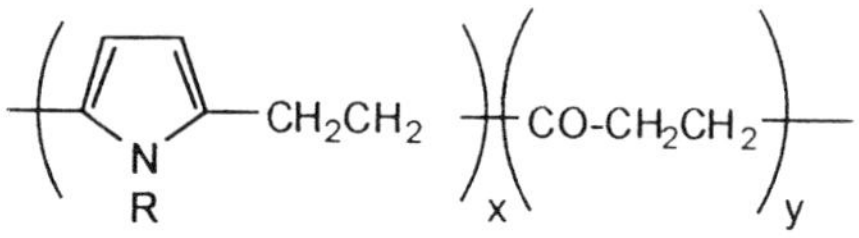

R = -C$_4$H$_9$, -C$_6$H$_5$, p-CH$_3$O-C$_6$H$_4$, p-Cl-C$_6$H$_4$

LXVIII

Early reports of a thiophene copolymer with a nonheterocyclic compound included the copolymerization of a 3,4-dicyanothiophene ring with several sulfur nitride monomers [136]. These polymers could be reversibly p-doped with bromine, giving conductivities of 0.01–0.002 S/cm. In two different reports, nickel(II)-catalyzed copolymerizations of 3-methylthiophene and methyl methacrylate have produced processible nonconjugated conductive polymers (I$_2$-doped, 0.2 S/cm or 6.5 S/cm) [137,138]. 3-Methylthiophene has also been copolymerized with methoxyocta(oxyethylene)methacrylate, **LXIX**. This copolymer exhibits both electronic and ionic conductivities [139].

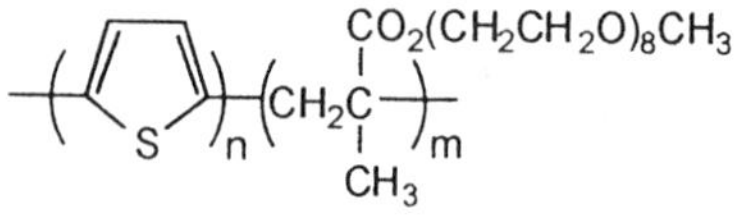

LXIX

A polyester incorporating quaterthiophene units in the main chain has been reported [140]. This material was soluble in its neutral state and had a MW = 13000. Upon I$_2$ doping it achieved conductivities as high as 0.3 S/cm.

Di(2-thienyl)silane has been copolymerized with 1,5- and 1,8-dichloroanthraquinone [141] in the presence of palladium to produce polymers, **LXX**, that upon doping with sodium naphthalenide have conductivities of 3×10^{-5} S/cm. Several of the derivatives also exhibit electrochemical n-doping depending on the substituent on the silicon atom.

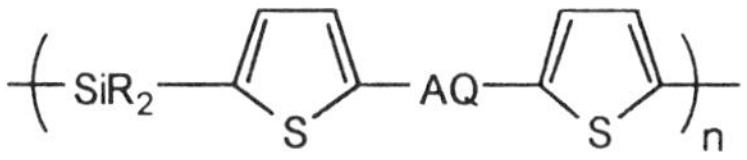

AQ = 1,5 and 1,8 anthraquinone
R = -CH$_3$, -C$_6$H$_{13}$ and -C$_6$H$_5$

LXX

Other poly(silanylene-*co*-thienylene) systems (**LXXXI**) containing one or more silanylene units and with doped conductivities in the 0.1–10 S/cm range have been reported [142,143]. Similar multiblock copolymers whose luminescence can be tuned by changing the thiophene block size have recently appeared [144].

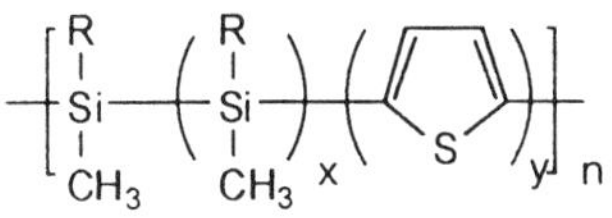

R = -CH$_3$, -C$_6$H$_5$, -*p*-CH$_3$-C$_6$H$_4$
y = 1,2 3; x = 1,2

LXXXI

Hanack et al. [145–148] have developed a series of polyarenemethylidenes that are copolymers containing mixtures of pyrrole, thiophene, thieno[*c*]thiophene, and isothianaphthene rings linked by a methine carbon (**LXXXII**). These polymers are characterized by having low E_{gap} values (~1.0 eV).

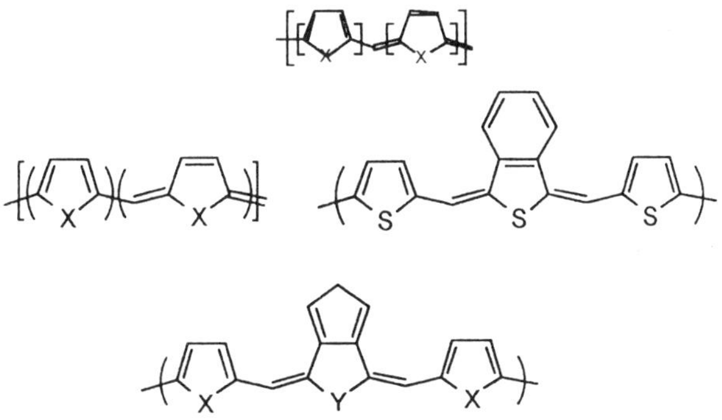

X = SO$_2$, SO, S, NR, -CH=CH- or
X = Y = Z = S

LXXXII

Jenekhe has also developed a extensive series of poly(heteroarylene-methylene)s, **LXXXIII** [149,150]. These systems also have small band gaps (<1.5 eV) and iodine-doped conductivities of 10^{-3}–10^{-1} S/cm (undoped $<10^{-7}$ S/cm) [150].

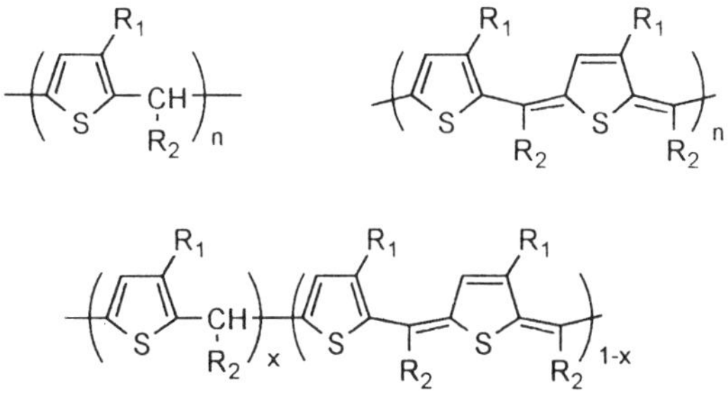

R$_1$ = -CH$_3$, -C$_4$H$_9$, -C$_6$H$_{13}$
R$_2$ = aryl, H

LXXXIII

VIII. POLYMERIZATION OF TRIHETEROCYCLIC SYSTEMS

There are only a few examples in which three different heterocycles have been incorporated into a polymer. One of those is obtained by the chemical polymerization [89,91] of an α-linked three-ring system containing thiophene, *N*-methylpyrrole, and furan rings, SN(Me)O, **LXXXIV**. The NOPF$_6$-doped polymer had a conductivity of $<10^{-5}$ S/cm.

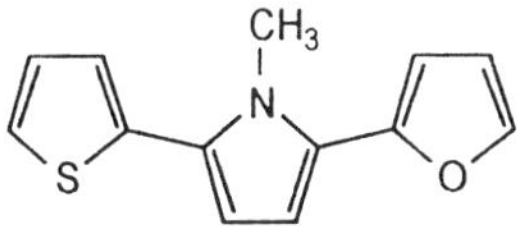

LXXXIV

IX. CONCLUSIONS

The copolymerization of pyrrole, thiophene, and furan-based heterocycles with each other and with other aromatic systems has afforded an impressive array of materials whose composition and properties can be controlled by judicious choice of comonomers and polymerization conditions. We remain confident that this is will continue to be a fertile field for exploration, development, and materials design.

REFERENCES

1. T. E. Skotheim et al. (eds.), *Handbook of Conducting Polymers*, Marcel Dekker, New York, 1986.
2. J. Heinze, *Topics Current Chem.* 152:1 (1990).
3. G. K. Chendler, and D. Pletcher, *Electrochemistry* 10:117 (1985).
4. A. Techagumpuch, H. S. Nalwa, and S. Miyata, in *Electroreponsive Molecular and Polymeric Systems* T. E. Skotheim, (ed.), Marcel Dekker, New York, 1991, p. 257.

5. A. Tsumura, H. Koezuka, S. Tsunoda, and T. Ando, *Chem. Lett. 1986*:863.
6. H. S. White, G. P. Kittlesen, and M. S. Wrighton, *J. Am. Chem. Soc. 106*:5375 (1984).
7. E. W. Paul, A. J. Ricco, and M. S. Wrighton, *J. Phys. Chem. 89*:1441 (1985).
8. K. Kaneto, M. Maxfield, D. P. Nairns, A. G. MacDiarmid, and A. J. Heeger, *J. Chem. Soc. Faraday. Trans I 78*:3147 (1982).
9. C. Arbizzani, A. M. Marinageli, M. Mastragostino, T. Hamaide, and A. Guyot, *Synth. Met. 41–43*:1147 (1991).
10. T. Osaka, K. Naoi, H. Sakai, and S. Ogano, *J. Electrochem. Soc. 134*:285 (1987).
11. F. Trinidad, J. Alonso-Lopez, and M. Nebot, *J. Appl. Electrochem. 17*:217 (1987).
12. B. E. Conway, *J. Electrochem. Soc. 138*:1539 (1991).
13. A. Rudge, I. Raistrick, S. Gottesfeld, and J. P. Ferraris, *Electrochim. Acta 39*:273–287 (1994), and references therein.
14. A. Rudge, J. Davey, I. Raistrick, S. Gottesfeld, and J. P. Ferraris, *J. Power Sources 47*:89 (1994) and references therein.
15. B. Scrosati (ed.), *"Applications of Electroactive Polymers"*, Chapman and Hall, New York, 1993.
16. H. Masuda, S. Tanaka, and K. Kaeriyama, *J. Chem. Soc., Chem. Commun. 1990*:725.
17. D. Delabouglise and F. Garnier, *New J. Chem. 15*:233 (1991).
18. T. Iyoda, M. Aiba, T. Saika, K. Honda, and T. Shimidzu, *J. Chem. Soc., Faraday Trans. 2 87*:1765 (1991).
19. D. M. Collard, and M. A. Fox, *J. Am. Chem. Soc. 113*:9414 (1991).
20. M. V. Rosenthal, T. A. Skotheim, C. Linkous, and M. I. Florit, *Polym. Prepr. 25*:258–259 (1984).
21. J. R. Reynolds, P. A. Poropatic, and R. L. Toyooka, *Synth. Met. 18*:95–100 (1987).
22. J. R. Reynolds, P. A. Poropatic, and R. L. Toyooka, *Macromolecules 20*:958–961 (1987).
23. S. J. Vigmond, K. M. R. Kallury, and M. Thompson, *Anal. Chem. 64*:2763–2769 (1992).
24. K. K. Kanazawa, A. F. Diaz, M. F. Diaz, M. T. Krounbi, and G. B. Street, *Synth. Met. 4*:119 (1981).
25. O. Inganas, B. Liedberg, W. R. Wu, and H. Wynberg, *Synth. Met. 11*:239 (1985).
26. H. S. O. Chan, E. T. Kang, K. G. Neoh, K. L. Tan, B. T. G. Tan, and Y. K. Lim, *Synth. Met. 30*:189–197 (1989).
27. M. Nishizawa, T. Sawaguchi, T. Matsue, and I. Uchida, *Synth. Met. 45*:241–248 (1991).
28. M. S. Kiani and G. R. Mitchell, *Synth. Met. 46*:293–306 (1992).
29. A. Paul, D. Sarkar, and T. N. Misra, *Solid State Commun. 89*:363–367 (1994).
30. N. S. Sundaresan, S. Basak, M. Pomerantz, and J. R. Reynolds, *J. Chem. Soc., Chem. Commun. 1987*:621–622.
31. J. R. Reynolds, N. S. Sundaresan, M. Pomerantz, S. Basak, and C. K. Baker, *J. Electroanal. Chem. 250*:355–371 (1988).
32. J. R. Reynolds, J. P. Ruiz, F. Wang, C. A. Jolly, K. Nayak, and D. S. Marynick, *Synth. Met. 28*:C621–C628 (1989).
33. J. P. Travers, P. Audebert, and G. Bidan, *Mol. Cryst. Liq. Cryst. 118*:149 (1985).
34. G. Bidan, B. Ehui, and M. Lapkowski, *J. Phys. D: Appl. Phys. 21*:1043–1054 (1988).
35. D. Moon, A. B. Padias, H. K. Hall, Jr., T. Huntoon, and P. D. Calvert, *Macromolecules 28*:6205–6210 (1995).
36. A. Haimerl, and A. Merz, *Angew. Chem. Int. Engl. 25*(2):180–181 (1986).
37. A. Merz, A. Haimerl, and A. J. Owen, *Synth. Met. 25*:89–102 (1988).
38. A. J. Downard, N. A. Surridge, T. J. Meyer, S. Cosnier, A. Deronzier, and J. G. Moutet, *J. Electroanal. Chem. 246*:321–335 (1988).
39. C. P. Andrieux, P. Audebert, and C. Salou, *J. Electroanal. Chem. 318*:235–246 (1991).
40. N. Noma, K. Namba, and Y. Shirota, *Synth. Met. 64*:227–232 (1994).
41. J. Kowalik, H. T. Nguyen, and L. M. Tolbert, *Synth. Met. 41–43*:435–438 (1991).
42. R. L. Elsenbaumer, K. Y. Jen, and R. Oboodi, *Synth. Met. 15*:169–174 (1986).
43. S.-A. Chen, and J.-M. Ni, *Macromolecules 26*:3230–3231 (1993).
44. M. Jayaraman, and R. D. McCullough, *Polym. Prepr. 35*(1):299 (1994).
45. R. D. McCullough and M. Jayaraman, *J. Chem. Soc., Chem. Commun. 1995*:135–136.
46. Q. Pei and O. Inganas, *Synth. Met. 46*:353–357 (1992).
47. H. J. Fell, E. J. Samuelsen, E. Bakken, and P. H. Carlsen, *Synth. Met. 72*:93–96 (1995).
48. P. F. Van Hutten, R. E. Gill, J. K. Herrema, and G. Hadziioannou, *J. Phys. Chem. 99*:3218–3224 (1995).
49. W. J. Albery, F. Li, and A. R. Mount, *J. Electroanal. Chem. 310*:239–253 (1991).
50. R. Back and R. B. Lennox, *Langmuir 8*:959–964 (1992).
51. J. Lowe and S. Holdcroft, *Polym. Prepr. 35*(1):297–298 (1994).
52. M. J. Marsella and T. M. Swager, *Polym. Prepr. 35*(1):271–272 (1994).
53. Y. Miyazaki and T. Yamamoto, *Chem. Lett. 1994*:41–44.
54. T. Yamamoto, W. Yamada, M. Takagi, K. Kizu, T. Marayama, N. Ooba, S. Tomaru, T. Kurihara, T. Kaino, and K. Kubota, *Macromolecules 27*:6620–6626 (1994).
55. D. D. Cunningham, A. Galal, C. V. Pham, E. T. Lewis, A. Burkhardt, L. Laguren-Davidson, A. Nkansah, O. Y. Ataman, H. Zimmer, and H. B. Mark, Jr., *J. Electrochem. Soc.: Electrochem. Sci. Technol. 135*(11):2750–2751 (1988).

56. J. Roncali, A. Gorgues, and M. Jubault, *Chem. Mater.* 5:1456–1464 (1993).
57. E. E. Havinga and L. W. van Horssen, *Makromol. Chem., Makromol. Symp.*, 24:67–76 (1989).
58. C. Wang, M. E. Benz, E. LeGoff, J. L. Schindler, C. R. Kannewurf, and M. G. Kanatzidis, *Polym Prepr.* 34(2):422–423 (1993).
59. J. P. Ferraris and M. D. Newton, *Polymer* 33:391–397 (1992).
60. M. Granstrom and O. Inganas, *Synth. Met.* 55–57:460–465 (1993).
61. G. Zotti, M. C. Gallazi, G. Zerbil, and S. V. Meille, *Synth. Met.* 73:217–225 (1995).
62. M. Berggren, O. Inganas, G. Gustafsson, J. Rasmusson, M. R. Andersson, T. Hjertberg, and O. Wennerstrom, *Nature* 372:444–445 (1994).
63. C. Visy, J. Lukkari, and J. Kankare, *Macromolecules* 27:3322–3329 (1994).
64. P. Bauerle, G. Gotz, P. Emerle, and H. Port, *Adv. Mater.* 4(9):564–568 (1992).
65. D. Lorcy, and M. P. Cava, *Adv. Mater.* 4(9):562–564 (1992).
66. S. Musmanni and J. P. Ferraris, *J. Chem. Soc., Chem. Commun.* 1993:172–174.
67. F. Wudl, M. Kobayashi, and A. J. Heeger, *J. Org. Chem.* 49:3382 (1984).
68. D. Marynick and K. Nayak, *Macromolecules* 23:2237 (1990).
69. J. P. Ferraris, A. Bravo, W. Kim, and D. C. Hmcir, *J. Chem. Soc., Chem. Commun.* 1994:991–992.
70. C. Kitamura, S. Tanaka, and Y. Yamashita, *J. Chem. Soc., Chem. Commun.* 1994:1585–1586.
71. S. Tanaka and Y. Yamashita, *Synth. Met.* 55–57:1251–1254 (1993).
72. S. Tanaka and Y. Yamashita, *Synth. Met.* 69:599–600 (1995).
73. B. R. McKellar and W. A. Feld, *Polym. Prepr.* 34(2):380–381 (1993).
74. S. Kuwabata, S. Ito, and H. Yoneyama, *J. Electrochem. Soc.* 135(7):1691–1695 (1988).
75. S. Naitoh, K. Sanui, and N. Ogata, *J. Chem. Soc., Chem. Commun.* 1986:1348–1350.
76. P. Kathirgamanathan, *J. Electroanal. Chem.* 247:351–353 (1988).
77. B. L. Funt, E. M. Peters, and J. D. Van Dyke, *J. Polym. Sci. A Polym. Chem.* 24:1529–1537 (1986).
78. W. Torres and M. A. Fox, *Chem. Mat.* 4:146–152 (1992).
79. E. M. Peters and J. D. Van Dyke, *J. Polym. Sci. A Polym. Chem.* 24:1891–1898 (1992).
80. Q. Y. Liang, K. G. Neoh, E. T. Kang, K. L. Tan, and H. K. Wong, *Eur. Polym. J.* 28:755–763 (1992).
81. O. Inganas, B. Liedberg, and W. Chang-Ru, *Synth. Met.* 11:239–249 (1985).
82. M. Aldissi, *Mol. Cryst. Liq. Cryst.* 160:121–131 (1988).
83. S. Naitoh, *Synth. Met.* 18:237–240 (1987).
84. G. G. McLeod, M. G. B. Mahboubian-Jones, R. A. Pethrick, S. D. Watson, N. D. Truong, J. C. Galin, and J. Francois, *Polymer* 27:455–458 (1986).
85. A. Berlin, W. Wernet, and G. Wegner, *Makromol. Chem.* 188:2963 (1987).
86. J. P. Ferraris and G. D. Skiles, *Polymer* 28:179–182 (1987).
87. J. P. Ferraris, R. G. Andrus, and D. Hrncir, *J. Chem. Soc., Chem. Commun.* 1989:1318–1320.
88. J. P. Ferraris and T. R. Hanlon, *Polymer* 30:1319–1327 (1989).
89. M. V. Joshi, C. Hemler, M. P. Cava, J. L. Cain, M. G. Bakker, A. J. McKinley, and R. M. Metzger, *J. Chem. Soc. Perkin Trans. 2* 1993:1081–1086.
90. J. R. Reynolds, A. D. Child, J. P. Ruiz, S. Y. Hong, and D. S. Marynick, *Macromolecules* 26:2095–2103 (1993).
91. M. V. Joshi, M. P. Cava, M. G. Bakker, A. J. McKinley, J. L. Cain, and R. M. Metzger, *Synth. Met.* 55–57:948–953 (1993).
92. M. Seki, H. An, K. Sato, and R. Yosomiya, *Synth. Met.* 26:33–39 (1988).
93. T. Mitsuhara, S. Tanaka, and K. Kaeriyama, *Makromol. Chem.* 189:1755–1763 (1988).
94. Z. H. Zhou, T. Maruyama, T. Kanbara, T. Ikeda, K. Ichimura, T. Yamamoto, and K. Tokuda, *J. Chem. Soc., Chem. Commun.* 1991:1210–1212.
95. T. Yamamoto, Z. H. Zhou, T. Maruyama, and T. Kanbara, *Synth. Met.* 55–57:1209–1213 (1993).
96. I. H. Jenkins and P. G. Pickup, *Macromolecules* 26:4450–4456 (1993).
97. A. K. Agrawal and S. A. Jenekhe, *Polym Prepr.* 33(2):349–350 (1993).
98. A. K. Agrawal and S. A. Jenekhe, *Macromolecules* 24(25):6806–6808 (1991).
99. J. H. Promislow, J. Preston, and E. T. Samulski, *Macromolecules* 26:1793–1795 (1993).
100. M. Dotrong, R. Mehta, G. A. Balchin, R. C. Tomlinson, M. Sinsky, C. Y.-C. Lee, and R. C. Evers, *J. Polym. Sci. A: Polym. Chem.* 31:723–729 (1993).
101. P. A. DePra, J. G. Gaudiello, and T. J. Marks, *Macromolecules* 21:2297–2299 (1988).
102. M.-O. Bevierre, F. Mercier, L. Ricard, and F. Mathey, *Angew. Chem. Int. Ed. Engl.* 29(6):655–657 (1990).
103. K. Tamao, S. Yamaguchi, M. Shiozaki, Y. Nakagawa, and Y. Ito, *J. Am. Chem. Soc.* 114:5867–5869 (1992).
104. K. Tamao, S. Yamaguchi, Y. Ito, Y. Matsuzaki, T. Yamabe, M. Fukushima, and S. Mori, *Macromolecules* 28:8668–8675 (1995).
105. J. R. Reynolds, A. R. Katritzky, J. Soloducho, S. Belyakov, G. A. Sotzing, and M. Pyo, *Macromolecules* 27:7225–7227 (1994).
106. K. L. Pouwer, T. R. Vries, E. E. Havinga, E. W. Meijer, and H. Wynberg, *J. Chem. Soc., Chem. Commun.* 1988:1432–1433.
107. S. Martina and A.-D. Schluter, *Macromolecules* 25:3607–3608 (1992).
108. G. A. Sotzing, J. R. Reynolds, A. R. Katritzky, J.

Soloducho, S. Belyakov, and R. Musgrave, *Macromolecules* 29:1679–1684 (1996).
109. R. Danieli, P. Ostoja, M. Tiecco, R. Zamboni, and C. Taliani, *J. Chem. Soc., Chem. Commun.* 1986: 1473–1474.
110. C. Taliani, R. Danieli, R. Zamboni, G. Ruani, and P. Ostoja, in *Electronic Properties of Conjugated Polymers: Proceedings of an Intermatinal Winter School*, Kirchberg am Wechsel, Austria, Springer-Verlag Ser. Solid-State Sci. 76 (H. Kuzmany M. Mehring, and S. Roth, eds.) Springer-Verlag, Berlin, 1987, p. 322–325.
111. D. M. Haynes, A. R. Hepburn, D. M. Goldie, J. M. Marshall, and A. Pelter, *Synth. Met.* 55–57:839–844 (1993).
112. T. Mitsuhara, K. Kaeriyama, and S. Tanaka, *J. Chem. Soc., Chem. Commun.* 1987:764–765.
113. W. Czerwinski, N. Nucker, and J. Fink, *Synth. Met.* 25:71–77 (1988).
114. W. Czerwinski, *Synth. Met.* 35:229–237 (1990).
115. A. Pelter, M. Rowlands, and I. H. Jenkins, *Tetrahedron Lett.* 28:5213–5216 (1987).
116. A. Pelter, J. M. Maud, I. Jenkins, C. Sadeka, and G. Coles, *Tettrahedron Lett.* 30(26):3461–3464 (1989).
117. A. D. Child and J. R. Reynolds, *J. Chem. Soc., Chem. Commun* 1991:1779–1781.
118. J. P. Ruiz, A. D. Child, K. Nayak, D. S. Marynick, and J. R. Reynolds, *Synth. Met.* 41–43:783–788 (1991).
119. J. R. Reynolds, J. P. Ruiz, A. D. Child, K. Nayak, and D. S. Marynick, *Macromolecules* 24:678–687 (1991).
120. J. P. Ruiz, J. R. Dharia, J. R. Reynolds, and L. J. Buckley, *Macromolecules* 25:849–860 (1992).
121. B. Sankaran, A. D. Child, F. Larmat, and J. R. Reynolds, *Polym. Prepr.* 35(1):263–264 (1994).
122. T. Yamamoto, A. Morita, Y. Miyazaki, T. Maruyama, H. Wakayama, Z. Zhou, Y. Nakamura, T. Kanbara, S. Sasaki, and K. Kubota, *Macromolecules* 25:1214–1223 (1992).
123. J. Kowalik and L. M. Tolbert, *Polym. Mat. Sci. Eng. (ACS Div. Polym. Mat.: Sci. Engl.)* 64:214–215 (1991).
124. A. Bao, R. Cai, and L. Yu, *Polym. Prepr.* 34(2): 749–750 (1993).
125. L. Yu, Z. Bao, and R. Cai, *Angew. Chem. Int. Ed. Engl.* 32(9):1345–1347 (1993).
126. E. Rebourt, B. Pepin-Donat, and E. Dinh, *Polymer* 36(2):399–412 (1995).
127. M. Thelakkat and H. Naarmann, *Synth. Met.* 68: 153–155 (1995).
128. J. R. Reynolds, G. A. Sotzing, B. Sankaran, S. A. Sapp, D. J. Irvin, J. A. Irvin, and J. L. Keddinger, *Polym. Prepr.* 37(1):135 (1996).
129. D. J. Irvin and J. R. Reynolds, *Polym. Prepr.* 37(1): 532 (1996).
130. D. J. Irvin and J. R. Reynolds, *Polym. Prepr.* 37(1): 682 (1996).
131. G. A. Sotzing, J. L. Reddinger, and J. R. Reynolds, *Polym. Prepr.* 37(1):795 (1996).
132. S. A. Sapp, G. A. Sotzing, J. L. Reddinger, and J. R. Reynolds, *Polym. Prepr.* 37(1):797 (1996).
133. F. Larmat, J. R. Reynolds, B. Reinhart, and L. L. Brott, *Polym. Prepr.* 37(1):799 (1996).
134. D. P. Amalnerkar, S. Radhakrishnan, S. R. Sainkar, S. Badrinarayanan, and S. G. Joshi, *Solid State Commun.* 84:911–915 (1992).
135. Z. Jiang and A. Sen, *Macromolecules* 25:880–882 (1992).
136. J. W. Chien, and M. Zhou, *J. Polym. Sci. A., Polym. Chem. Ed.* 24:2947–2957 (1986).
137. W.-S. Huang, and J. M. Park, *J. Chem. Soc., Chem. Commun.* 1987:856–858.
138. H. S. Nalwa, *Synth. Met.* 35:387–391 (1990).
139. J. Li and M. Khan, *Polym. Prepr.* 34(2):662–663 (1993).
140. Y. Hong and L. L. Miller, *Chem. Mater.* 7: 1999–2000 (1995).
141. S. Ho Yi, S. Ohashi, H. Sato, and H. Nomori, *Bull. Chem. Soc. Jpn.* 66:1244–1247 (1993).
142. J. Ohshita, D. Kanaya, M. Ishikawa, T. Koike, and T. Yamanaka, *Macromolecules* 24:2106–2107 (1991).
143. P. Chicart, R. J. P. Corriu, J. J. L. Moreau, F. Garnier, and A. Yassar, *Chem. Mater.* 3:8–10 (1991).
144. J. K. Herrema, P. F. Can Hutten, R. E. Gill, J. Wildeman, R. H. Wieringa, and G. Hadziioannou, *Macromolecules* 28:8102–8116 (1995).
145. M. Hanack, G. Hieber, G. Dewald, and H. Ritter, *Synth. Met.* 41–43:507–511 (1991).
146. H. Ritter, K. M. Mangold, U. Rohrig, U. Schmid, and M. Hanack, *Synth. Met.* 55–57:1193–1197 (1993).
147. M. Hanack, U. Schmid, U. Rohrig, J.-M. Toussaint, C. Adant, and J.-L. Bredas, *Chem. Ber.* 126: 1487–1491 (1993).
148. M. Hanack, G. Hieber, K. M. Mangold, H. Ritter, U. Rohrig, and U. Schmid, *Synth. Met.* 55–57:827–832 (1993).
149. W. C. Chen and S. A. Jenekhe, *Macromolecules* 28: 454–464 (1995).
150. W. C. Chen and S. A. Jenekhe, *Macromolecules* 28: 465–480 (1995).

11
Low Band Gap Conducting Polymers

Martin Pomerantz
The University of Texas at Arlington, Arlington, Texas

I. INTRODUCTION

From the very early days of the study of conducting polymers scientists envisioned that there might be a class of these polymers that would have either a zero energy band gap (a single, continuous band consisting of the valence and conduction bands) or a very low band gap (Fig. 11.1). It was envisioned that these materials might be inherent electronic conductors, that is, would conduct without the need for doping, similar to metals that have overlapping bands. If these materials were indeed like metals, they might also show the high electrical conductivity of metals. Also, conducting polymers were known to be highly colored in the nonconducting state owing to optical transitions of electrons from the valence band to the conduction band. Upon doping, for example oxidatively, new low energy transitions were observed due to the production of bipolarons, which gave rise to two new states within the gap. While transitions involving these energy states increased in intensity, the original absorptions decreased in intensity (Fig. 11.1). It was envisioned that a low band gap polymer might be colored when undoped, with the optical absorption in the visible region of the spectrum shifted to the red compared to a polymer with a higher band gap. Since, as indicated above, oxidative doping would produce new absorptions of even lower energy, these new absorptions might be pushed into the near-infrared region of the spectrum. Thus, with the new optical absorptions being in the near-infrared and the original absorptions decreasing in intensity, the conducting form of the polymer might well be very lightly colored and essentially transparent.

With these objectives in mind, new systems were developed that were designed to have low band gaps. Since the initial synthesis of one of the first truly low band gap conducting polymers by Wudl and Heeger, namely, polyisothianaphthene (**1**) [poly(benzo[c]thiophene)] [1], much theoretical and experimental work has been car-

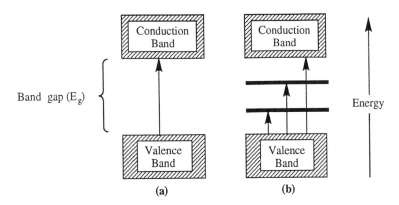

Fig. 11.1 Optical transitions in (a) an undoped and (b) an oxidatively doped (bipolaron) conducting polymer.

ried out to design and prepare new systems and to understand and predict the properties, including the band gaps, of these polymers. Prior to 1984 several systems were prepared that presumably had low band gaps, and some quantum-mechanical calculations were also performed on these polymeric molecules [2,3]. It is the purpose of this chapter to review and discuss the research on low band gap systems, and we begin with the polyisothianaphthene (1) story, just prior to the publication of the first edition of this handbook.

First, we must define what we mean by a *low band gap* polymer. The band gaps of many of the well-studied conducting polymers are >2 eV. Thus, for example, that of poly(*p*-phenylene) is 2.7 eV [4,5], that of poly(*p*-phenylene vinylene) is 2.4 eV [6,7], that of polythiophene is 2.0–2.1 eV [8,9], and that of polypyrrole is 3.2 eV [10]. Polyacetylene has a lower band gap, about 1.5 [11] to 1.7 eV [12]. I have arbitrarily decided to use a band gap of about 1.5 eV as the cutoff and define a low band gap conjugated polymer as one having a band gap of less than 1.5 eV. I have also decided not to include polyacetylene, this being a polymer whose band gap is borderline by the definition above, since it has been reviewed many times and is not considered to be a low band gap polymer.

A major problem with these low band gap materials results from the relatively high energy of the electrons in the valence band. While this, in combination with the relatively low energy of the conduction band, is what gives rise to a low band gap material, it also renders the polymer easily oxidized. These low band gap readily oxidized polymers can be doped by, and react with, ambient oxygen. Thus, low band gap polymers are usually handled under anaerobic conditions to prevent atmospheric oxidation from occurring.

II. POLYISOTHIANAPHTHENE

The first of the low band gap polymers to be prepared was polyisothianaphthene (1) [poly(benzo[*c*]thiophene)]

[1,13]. Scheme 1 shows the synthesis of **1** from monomer **3** reported by the Wudl and Heeger group in 1984 [1]. Benzo[*c*]thiophene (**3**) was prepared by the method of Cava from sulfoxide **2** [14]. Polymer **1** was prepared either electrochemically, using LiBr in CH$_3$CN, or chemically by acid-catalyzed polymerization of **3**, which produced polydihydroisothianaphthene (**4**), which could then be oxidized by any of a variety of oxidizing agents including chloranil and sulfuric acid. **1** is a blue-black, insoluble material in the nonconducting state that becomes much lighter in color (decreased optical density in the visible part of the spectrum), and a thin film is transparent yellow-green [15]. The band gap, based on the band edge at ~1100 nm [1], is about 1.0–1.2 eV [16]. That this is a low band gap polymer was confirmed by electrochemical measurements [15]. The absorption maximum, corresponding to the interband transition, was at about 886 nm (1.4 eV) [15]. Electrochemical experiments showed that a thin film of polymer **1** could be reversibly oxidized at 0.6 V (vs. SCE) and the color changed from blue-black to light yellow-green. The oxidation was reversible only if brought to about 0.8 V and brought back. At higher voltages, say 1.3 V, the scans were no longer reversible and the polymer apparently overoxidized [15]. Four-point probe conductivity measurements on film and pressed pellets electrochemically doped with chloride ion gave values of 2–7 S/cm, and doping of a film, grown on four narrow gold electrodes, with I$_2$ or perchloric acid gave conductivities of about 50 S/cm. Compensation (dedoping) of these samples with ammonia or hydrazine reduced the conductivity to about 10^{-3} S/cm, still indicating a rather good semiconductor. Interestingly, the as-grown polymer, even after ammonia compensation, exhibited an ESR signal at $g = 2.005$ with one unpaired spin per about 200 isothianaphthene units, which could not be reduced by further ammonia treatment [15].

Following this initial pioneering work, many papers appeared on studies of **1**, the search for new low band gap polymers, and theoretical investigations of these systems, including the prediction of new low band gap

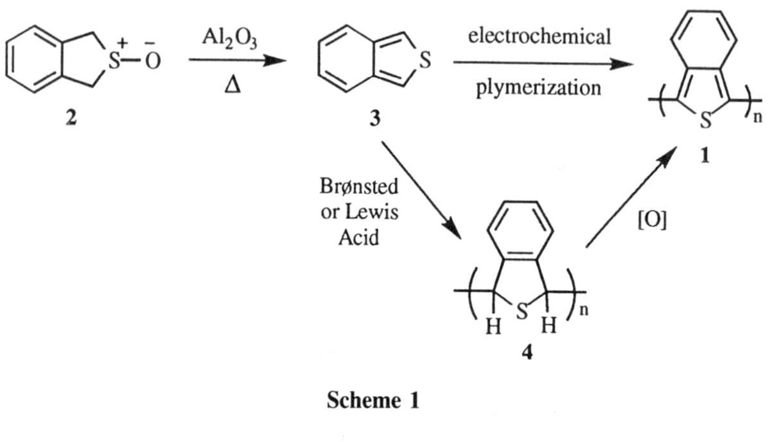

Scheme 1

polymers. Many of the subsequent investigations involving polymer **1** have concentrated on improved methods of preparation and on very careful physical and chemical studies. In 1986 Jen and Elsenbaumer [17] reported a new facile synthesis of polyisothianaphthene (**1**)

<chemical scheme: compound 5 (benzo[c]thiophene with S) → compound 1 (polyisothianaphthene) via O$_2$ and/or FeCl$_3$>

whereby the easily prepared monomer **5**, from which **2** and **3** were made, could be directly oxidized to **1** using oxygen or, even better, FeCl$_3$ in the presence of oxygen (air). This avoids the need to work with the reactive monomer **3**, but the polymer is prepared as an intractable powder. The pressed pellet conductivity (FeCl$_3$-doped) was 0.5 S/cm, essentially the same as that of a sample prepared by the Wudl–Heeger procedure. It is interesting that this polymerization can be carried out by oxygen alone and that the product is the doped polymer with a conductivity of 10^{-1}–10^{-2} S/cm. This means that O$_2$ is a strong enough oxidizing agent to p-dope the polymer [17].

Several years later, Rose and Liberto reported [18] on the formation of polyisothianaphthene (**1**) by oxidation of the dihydro polymer **4** (polydihydroisothianaphthene) prepared from isothianaphthene (**3**) with methanesulfonic acid. Films of the processible dihydro polymer **4** could be aromatized to **1** using sulfuryl chloride (SO$_2$Cl$_2$) and showed properties similar to those of the material prepared by electrochemical deposition from **3**. Pressed pellet conductivity of **1** prepared from a solution of **4** and SO$_2$Cl$_2$ was reported to be 50 S/cm [18].

It has been reported that upon photolysis in the presence of tetrabutylammonium bromide and CCl$_4$ as an electron acceptor, isothianaphthene (**3**) produces the blue-black polyisothianaphthene (**1**) as either a film or powder [19]. The conductivity was somewhat lower than had been reported previously, but films of **1** could be readily converted electrochemically (Sb-doped SnO$_2$ on glass electrode) into the dihydro polymer **4**. No polymer **1** could be obtained unless an electron acceptor or oxygen was present; however, **4** could be isolated under these conditions. Based on the gel permeation chromatographic (GPC) molecular weight determination of the **4** that was produced, the number-average molecular weight of **1** was 1.2×10^4, giving a degree of polymerization of about 300. The polymerization mechanism was suggested to involve electron transfer from a photoexcited molecule of isothianaphthene (**3**) to CCl$_4$ followed by coupling of the resulting radical cations [19].

1,3-Dihydrobenzo[c]thiophene (**5**) was reported to be electrochemically converted directly to polyisothianaphthene (**1**) [20]. What was most interesting in this report was that the polymer displayed small solubility in organic solvents. This was attributed to the lack of cross-linking in the polymer prepared in this way. It was said that polymers prepared by other (electrochemical) routes were cross-linked and hence insoluble [20]. It was found that high concentrations of the monomer were required for efficient polymerization and a counter electrode of graphite rather than ITO glass. Further, it was reported that **1** could not be prepared directly by electrochemical polymerization of the S-oxide **2**. Electropolymerization in the presence of air gave the same results and was said to be preferred for the larger runs [20]. The absorption maximum of this polymer was ~633 nm and, from the solution UV-vis-near-IR spectrum provided in the paper, the band-edge band gap would appear to be close to 1.4 eV [20]. The report by Kobayashi et al. [15] gave λ_{max} for the film at about 886 nm, which suggests that the polymer reported by Chandrasekhar et al. [20] has a relatively low conjugation length and that this soluble material may well be oligomeric. GPC molecular weight determination was not performed.

A somewhat more recent paper [21] reports on another synthesis of **1**, this time directly from 1,3-dihydrobenzo[c]thiophene-S-oxide (**2**), by oxidation with N-chlorosuccinimide or by treatment with sulfuric acid. This removes one step from the original synthesis. In addition, direct oxidation of **5** with N-chlorosuccinimide also gave polyisothianaphthene (**1**) [21] in a reaction similar to that reported by Jen and Elsenbaumer [17]. A direct one-step synthesis of **1** has been reported [22] that involves the reaction of phthalic anhydride (or phthalide) with phosphorus pentasulfide at temperatures above 120°C. Doping with NO$^+$ SbF$_6^-$ gave maximum conductivities of 10 S/cm.

<chemical scheme: phthalic anhydride → polyisothianaphthene (1) ← phthalide, via P$_4$S$_{10}$>

In a 1991 paper, Eichinger and Kritzinger [23] demonstrated that doped **1** prepared by oxidative polymerization of the dihydro monomer **5** using FeCl$_3$, air, or TCNQ contained oxidized sulfur in the form of sulfone groups. These groups, however, apparently did not affect the conductivity of **1** compared to samples prepared by other methods reported in the literature.

Shortly after the first preparation of polyisothianaphthene (**1**) and the contention that the polymer might display a higher stability and lower band gap than polythiophene because of the contribution of the quinonoid, diradical, and "nonclassical" structures, **1b–1d**, relative to the aromatic form **1a** [1,15,24], quantum-mechanical calculations began to appear [25–27]. These and subsequent calculations are discussed in the next section, but the results from a number of these calculations indicated

1a ↔ **1b** ↔ **1c** ↔ **1d**

that **1b**, the quinonoid form of **1**, was more stable than **1a**, the aromatic form of **1**, and the actual polymer might exist in the quinonoid form. This, of course, is the result of the benzene ring in **1b** being aromatic whereas in **1a** it is quinonoidal.

There have been a number of studies designed to probe whether the structure of **1** was quinonoid or aromatic. The initial studies employed resonance Raman spectroscopy coupled with vibrational frequency analysis [16,28,29], and in these studies it was concluded that the data were more consistent with the quinonoid structure (**1b**) than with the aromatic structure (**1a**). Further, the behavior of the lines at about 1500 cm^{-1} as the excitation wavelength was changed was interpreted as being due to short conjugation lengths within the polymer backbone [28]. Using ab initio scaled quantum-mechanical force field calculations (scaled to fit polythiophene), the spectrum of the quinonoid form of the polymer could be predicted, thus further supporting the idea that **1** exists in the quinonoid form [30,31].

Gelan's group [32] suggested that the Raman results [28,29] based strictly on the Raman spectra of **1** cannot adequately distinguish between the aromatic and quinonoid forms, and they have used solid-state CP/MAS ^{13}C NMR spectroscopy to characterize polyisothianaphthene as either the quinonoid (**1b**) or aromatic (**1a**) form. This was accomplished by preparing model compounds, monomers and oligomers, and comparing the solid-state ^{13}C NMR chemical shifts with those of the polymer **1**. It is necessary to use selective pulse sequences to help resolve overlapping bands in **1**. Thus, compounds such as **6**, **7**, **8** [33], and **9** were prepared and studied both in the solid state and in solution. The conclusion was that polymer **1** has the quinonoid structure [32,34–37]. The solid-state chemical shifts of polyisothianaphthene (**1**) were determined to be δ = 125 (C-4), 126 (C-3), 128 (C-5), and 139 (C-3a) and matched more closely those of compounds such as **6** and **7** [32].

Gelan and coworkers [32] also examined the Raman spectra of the model compounds, both quinonoid and aromatic, which include **6**, **7**, and **9**, and compared these to the spectrum of polyisothianaphthene (**1**). Here, the vibrational bands of **1** were shown to match more closely with those of the quinonoid structures **6** and **7** than those of the aromatic structures such as **9** that were studied. In another recent paper [38], the Raman spectra of the model compounds and that of the pristine polyisothianaphthene (**1**) are again compared. Using the effective conjugation coordinate theory to interpret the spectra, it is again concluded that the structure is quinonoid.

In 1988 Lazzaroni et al. [39] examined the X-ray photoelectron spectra of polyisothianaphthene (**1**) and compared the results to the theoretical calculations. The conclusions were that **1** is similar to other conducting polymers with a disorder in the material that is the result of the anion inclusion process. Also, the low values of the shifts between neutral and oxidized states is given as evidence of a very low band gap in the material [39].

Polyisothianaphthene (**1**) has been shown to be both p- and n-dopable electrochemically by Hillman and coworkers [40], who also reported that the best electrochemical results were obtained after the films were conditioned by a single potential cycle in tetraethylammonium tetrafluoroborate in acetonitrile. They reported that the neutral polymer had a broad UV-vis absorption at 740 nm that reversibly decreased upon p-doping at potentials up to 1 V (vs. SCE). At the same time a new absorption was growing in at a wavelength longer than 820 nm, the longest wavelength examined. On the other hand, n-doping at voltages down to −1.7 V (vs. SCE) also reduced the 740 nm absorption, but the absorption at 820 nm decreased very substantially and there was no evidence for a new long-wavelength absorption growing in. The doping and dedoping could be carried for more than 10 cycles provided they were done in the absence of oxygen. The n-doped species showed less visible absorption than the p-doped material [40]. In a paper published the following year, Onoda et al. [41] reported that p-doping of **1** resulted in a decrease of the optical absorption at around 800 nm and a new band appeared that grew with extent of doping at around 1500 nm. By combining electrochemical studies with ESR spectroscopy they concluded that the valence band in polyisothianaphthene (**1**) was 0.5 eV higher in energy than that of polythiophene and the conduction band was 0.6 eV lower than that of polythiophene. This led them to conclude that it might be possible to n-dope polymer 1, which, of course, had been published the previous year [40]. Further, in 1994 Onoda et al. [42] indicated in the introductory part of a publication that "it is very interesting to clarify whether n-type doping is possible in PITN or not," again with no reference to the previous work [40]. In this publication they did point out that the reduced polymer was colorless and also studied the photoconductivity.

Because of the intractability of polyisothianaphthene (1), it would be quite difficult to envision practical applications for this material that would require processing. The idea that it is nearly transparent in the conducting state is rather appealing and led Defieuw et al. [43] to devise a method of coating an antistatic layer of 1 from an aqueous dispersion. The monomer, 1,3-dihydrobenzo[c]thiophene (5), was polymerized in concentrated sulfuric acid and dispersed in water in the presence of λ-carrageenan acting as a surfactant. This could then be coated on poly(ethylene terephthalate) to produce a transparent antistatic film whose surface resistivity was as low as 8.5 Ω/square. Optical density measurements of the films indicated that they were extremely transparent, with an optical density of about 0.02. Polyisothianaphthene (1) has also been used in a heterojunction device with C_{60} consisting of ITO glass, 1, C_{60}, and Al, and it was shown to act as a rectifying diode both in the dark and upon irradiation with 390 nm light. The currents were enhanced somewhat under irradiation. It was argued that the LUMO of C_{60} and the HOMO of 1 are very close in energy, resulting in charge transfer and bending of the energy states and thus hindering current flow with a negative bias but allowing current flow with a positive bias applied [44].

Polyisothianaphthene (1) has recently been prepared by an interesting photochemical polymerization of 1,3-dihydroisothianaphthene (5) [45]. It was necessary to carry out the photolysis in the presence of CCl_4 or O_2 as electron acceptors. It was postulated that the reaction, prior to irradiation, formed the radical cation of 5 which lost a proton to give an intermediate radical. This radical could cause polymerization to occur when irradiated with UV light, and when the light was turned off the polymerization continued. Thus it was suggested that the polymerization proceeded through two mechanisms, one photochemical and the other cationic [45].

III. QUANTUM-MECHANICAL CALCULATIONS

In this section the quantum-mechanical calculations pertaining to low band gap polymers are reviewed. Those relating to polyisothianaphthene are discussed first, followed by calculations on other systems. In some instances the syntheses preceded the calculations, whereas in others the calculations were the impetus for the search for new polymers. The calculations must be carried out in two parts. Since the polymers are too large to do high-level calculations, small molecules are calculated first and the geometries of these are optimized. These may include monomers, dimers, trimers, and other oligomers and are generally used to obtain the structure of the repeat unit of the polymer. These calculations can be at either the semiempirical or ab initio level. Once a structure for the repeat unit is determined, the band structure is calculated using methods such as the valence effective Hamiltonian (VEH), extended Hückel, or modified extended Hückel calculations. Because of the approximate nature of these calculations, the values calculated, (e.g., for the band gap), particularly for the earliest ones, were scaled by fitting to a known polymer. In addition, in the band structure calculations discussed below, the bandwidths of the highest occupied bands were generally calculated; however, they have not been included in these discussions. The reader is directed to the original papers for these calculations and numerical values.

After the report of the first preparation of polyisothianaphthene (1), Brédas published a paper [46], based on previous work on polyacetylene, where it was known that there was a degenerate ground state and hence bond alternation, examining poly(p-phenylene) (10), polypyrrole (11), and polythiophene (12). These three polymers have nondegenerate aromatic and quinonoid states, and

Brédas discusses how the band gap changes as the contribution of the quinonoid form increases relative to the aromatic form. In these cases the structures used were from ab initio Hartree–Fock optimizations of the aromatic and quinonoid structures of tetramers, i.e., quaterphenyl, quaterpyrrole, and quaterthienyl. Band structures were obtained by using the VEH technique. The conclusion was that as the contribution of the quinonoid structure increases, the band gap decreases linearly and that by designing systems that have increasing quinonoid contribution new low band gap polymers can be obtained.

Subsequently, Brédas et al. [25] calculated the band structures of polyisothianaphthene (**1**) and the 5,6-dimethyl, 5,6-dimethoxy, and 5,6-dicyano derivatives, **13–15**, using the MNDO-optimized geometry of isothianaphthene (**3**) corresponding to the aromatic structures and the VEH band structure methodology.

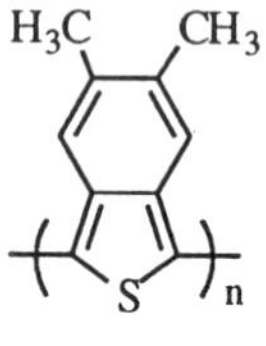

The calculated bandgap of **1** (0.54 eV) is 1.17 eV lower than that calculated for polythiophene (**12**, 1.71 eV), which is in very good agreement with the experimental data. The calculated value of 0.54 eV for **1** is about 0.5 eV too low, as is that calculated for **12**, and this is ascribed to the errors in the MNDO-optimized geometries. However, it should be pointed out again that the geometries employed in these calculations were aromatic geometries. It is also suggested that since the width of the highest occupied band in **1** is large, 2.53 eV, which is 0.17 eV larger than the width calculated for the corresponding band in polythiophene (**12**) using similar calculations, the maximum conductivities of **1** and **12** should be comparable [25]. The three derivatives **13**, **14**, and **15** were calculated to have band gaps and bandwidths very close to those of **1**, indicating there is only a minimal substituent effect. Brédas [47] also did similar calculations on polyisonaphthothiophene (**16**) and polythieno[3,4-c]thiophene (**17**), which provided band gaps of 0.01 and 1.02 eV, respectively. The band gap calculated by Lee et al. [48] for **17**, using MNDO structure and extended Hückel band calculations, was 2.00 eV, with the quinonoid form being the more stable. A calculation was reported in 1988 by Bakhshi and Ladik [49] on **17** using ab initio Hartree–Fock crystal orbital calculations, with the geometry calculated for the monomer and a single bond between the rings. This calculation gave a band gap of 1.97 eV, which was then scaled using a factor that made polythiophene fit the experimental value, to provide a band gap of ≈0.5 eV [49]

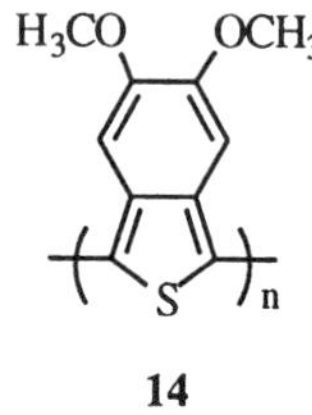

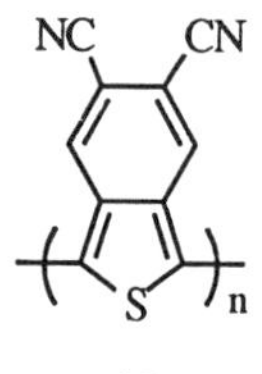

Bakhshi and Ladik [27] used ab initio Hartree–Fock crystal orbital theory and a geometry derived from that of isothianaphthene (**3**) with an inter-ring distance of 1.49 Å, to calculate the band gaps of **1** (4.52 eV) and **12** (8.14 eV), both of which are much too high, along with other electronic properties. The observation that the band gap of **1** is half that of **12** is mirrored in these calculations.

Lee and Kertesz [50] used MNDO for geometry optimization and Hückel theory for band structure calculations and applied them separately to the aromatic and quinonoid forms of polythiophene (**12**) and derivatives with fused benzene **1**, naphthalene **16**, and anthracene **18** rings. The band gaps for the quinonoid and aromatic forms are shown in Table 11.1 along with those of polythiophene (**12**). It is immediately obvious from the calculations that only for polythiophene (**12**) is the band gap for the quinonoid structure lower than that for the aromatic structure, and as a benzo, naphtho, or anthro ring is fused onto the thiophene the aromatic form of the polymer gives the lower band gap, approaching zero for **18**. This is due to the fused rings stabilizing the LUMO and destabilizing the HOMO in the aromatic structures. On the other hand, the LUMOs of the quinonoid structures are destabilized while the HOMOs are stabilized. As far as the stability of both forms is concerned, only in polythiophene (**12**) is the aromatic form more stable, whereas in the fused systems **1**, **16**, and **18**, the quinonoid form is more stable [50]. The calculated value of 1.16 eV for the band gap of **1** agrees quite well with the experimental value of 1.0–1.2 eV [1,16], as does the calculated value of 1.50 eV for **16** compared to the experimental value of ≈1.5 eV [51]. Further, although there should be some twisting around the bonds joining the two rings in the aromatic forms of **1**, **16**, and **18** due to S⋯H steric

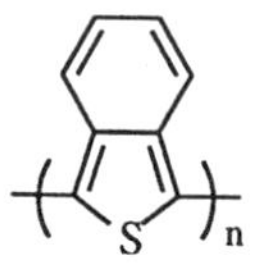

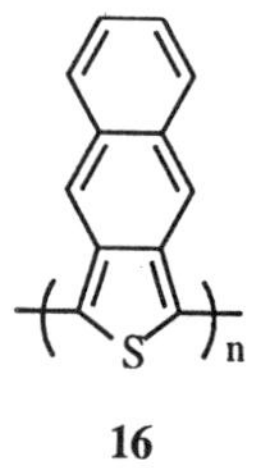

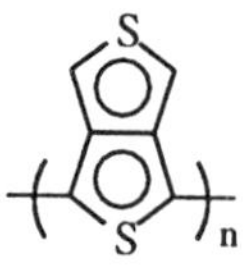

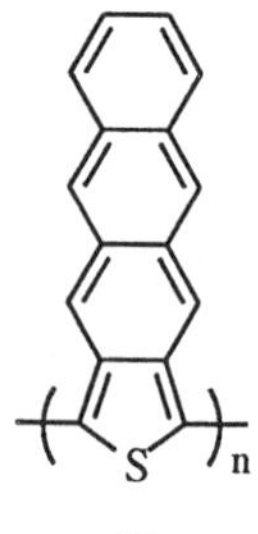

Table 11.1 Calculated Quinonoid and Aromatic Band Gaps (eV) of **12**, **1**, **16**, and **18**

	Polymer			
	12	1	16	18
Quinonoid	0.47	1.16	1.50	1.66
Aromatic	1.83	0.73	0.28	0.08

interactions, the double bond character in the quinonoid forms is enough to overcome the S····H steric interactions and the rings are coplanar in the quinonoid form. The calculated energy difference between the planar quinonoid and aromatic forms of polyisothianaphthene (**1**) is 10.55 kcal/(mol·ring), favoring the quinonoid form, while the S····H repulsion energy is 3.41 kcal/(mol·ring), not enough to cause twisting [50]. The conclusion here was that in order to form the low band gap polymers where the band gap is lower than that in **1**, the aromatic form of the fused ring polythiophenes should be stabilized. It is also predicted that in the polypyrrole series, because the calculated aromatic stabilization per ring is greater in polypyrrole [8.62 kcal/(mol·ring)] than in polythiophene [3.52 kcal/(mol·ring)], the crossover in the stability favoring the quinonoid form should occur with more rings than for polythiophene [50].

In a brief note in 1990, Kürti and Surján [52] pointed out that the more stable form of polyisothianaphthene (**1**) is quinonoidal and one must be very careful in extrapolating polymer structure from oligomer structures. Oligomers with fewer than eight repeat units have aromatic rings (Hückel calculations), whereas longer oligomers show quinonoid rings in the center of the chain and aromatic rings localized at the chain ends [52]. In 1993, Surján and Németh [53] showed that in both poly-*p*-phenylene and polythiophene, by adjusting the quinonoid vs. aromatic character of the polymer systems, band gaps could be lowered. At the appropriate inter-ring distances, using fused rings that are themselves quinonoidal, band gaps as low as about 0.4 eV were predicted [53].

In a paper in 1990, Nayak and Marynick [54] used PRDDO and ab initio calculations for geometry optimizations and extended Hückel calculations for the band structures to examine polyisothianaphthene (**1**) and related polymers. The PRDDO geometry optimization was shown to be quite good, and, since it is known that nonbonded interactions are poorly reproduced by MNDO calculations, the PRDDO method would seemingly give better results involving planar and twisted aromatic and planar quinonoid forms. The calculations (PRDDO and STO-3G ab initio) show that the rotational barriers around the bond between two isothianaphthene aromatic units are quite high (15.6 and 9.4 kcal/mol, respectively), thus leading to the conclusion that the aromatic form of **1** is nonplanar. Energy differences between aromatic and quinonoid monomers, dimers, trimers, up to hexamers were calculated and extrapolated to the infinite polymer chain. This gave an energy difference of 2.4 kcal/(mol·ring), with the quinonoid form being more stable than the aromatic form [54], in agreement with the calculations of Lee and Kertesz [50]. Band structure calculations were done on all three forms of **1**. The calculated band gap for the hypothetical planar aromatic form was 0.68 eV, which is quite low. The nonplanar aromatic form was calculated as a much higher 1.64 eV, and the planar quinonoid form, which is presumably the actual form of **1**, was calculated to be 0.80 eV [54]. This is to be compared to the experimental value of 1.0–1.2 eV [1,16].

Nayak and Marynick [54] also performed similar calculations of polynaphthothiophene (**16**), polythieno[3,4-*b*]pyrazine (**19**; R = H), and polythieno[3,4-*b*]quinoxaline (**20**), with similar results [54]. Polymers **19** (R = H) and **20** are related to **1** and **16** in that the C—H groups have been replaced by nitrogen atoms. Replacement of C—H by N was suggested as a way to eliminate the S····H nonbonded interactions present in **1** and **16** (D. S. Marynick and M. Pomerantz, private discussions, 1989). In all cases the quinonoid form was calculated to be the more stable, and in the case of **16** the aromatic form was twisted whereas in **19** (R = H) and **20** the aromatic forms were planar as a result of the elimination of the S····H nonbonded interactions. In the case of **19**

(R = H) the calculated difference in energy between the more stable quinonoid and less stable aromatic forms was calculated to be 4.2 kcal/mol per repeat unit. The calculated band gaps for polythieno[3,4-*b*]pyrazine (**19**; R = H) were 0.70 eV for the quinonoid and 0.12 eV for the aromatic form [54] compared to 0.95 eV observed for the 2,3-dihexyl derivative (**19**; R = C_6H_{13}) [55,56]. Polythieno[3,4-*b*]quinoxaline (**20**) gave calculated quinonoid and aromatic band gaps of 0.53 and 0.27 eV, respectively [54]. In a more recent paper, Hong and Marynick used a modified extended Hückel band calculation, whereby the UV-vis absorption maximum is the version of the band gap that is calculated. They obtained a value of 2.28 eV for the quinonoid form of polyisothianaphthene (**1**) [57]. They point out that the reported band gap of **1** based on optical absorptions is 1.4 eV (which corresponds to λ_{max} = 886 nm) [15], while the electrochemical band gap (of material that had λ_{max} = 633 nm) is 2.0 eV [20], and this is said to be in reasonable agreement with their calculations [57]. In addition they used PRDDO calculations to obtain geometrical parameters and modified extended Hückel band structure calculations on polythieno[3,4-*c*]thiophene (**17**) and obtained a result that indicated that the quinonoid form is 30.5 kcal/mol per repeat unit more stable than the aromatic form and that the band gap, as λ_{max}, is quite high, 3.69 eV for the quinonoid form and 2.54 eV for the high energy aromatic form [57]. This is to be contrasted with the band gap of 1.02 eV calculated by Brédas [47] mentioned above. Brédas also pointed out that the quinonoid form of this polymer is the more stable form.

Recently other calculations have been reported on various derivatives of polyisothianaphthene (**1**) with nitrogen atoms replacing CH groups at a number of positions, such as **19** (AM1 structure/VEH band calculations) [58]. Using structure **21**, several systems have been examined and the band gaps (E_g) for the aromatic and quinonoid forms are as follows: For **21a**, W = N,

21

X = Y = Z = CH, E_g = 0.60 eV and 1.31 eV for the aromatic and quinonoid forms, respectively; for **21b**, X = N, W = Y = Z = CH, E_g = 0.71 eV and 1.31 eV, respectively; for **21c**, W = X = N, Y = Z = CH, E_g = 0.27 eV and 1.51 eV, respectively; for **21d**, X = Y = N, W = Z = CH, E_g = 0.84 eV and 0.52 eV, respectively; for **21e**, W = Y = N, X = Z = CH, E_g = 1.05 eV and 1.46 eV, respectively; and for **21f** (which is the same as **19**, R = H), W = Z = N, X = Y = CH, E_g = 0.18 eV and 1.29 eV, respectively. The prediction was that in the case of **21b** and **21d** the quinonoid form is the more stable; in the case of **21c**, **21e**, and **21f** (**19**, R = H), the aromatic form was predicted to be more stable; and in **21a** the two forms were of equal energy [58]. The band gap values for polythieno[3,4-b]pyrazine (**21f**; **19**, R = H) should be compared with those for the aromatic form of 0.12 eV and the quinonoid form of 0.70 eV calculated by Nayak and Marynick [54], and the value of 0.95 eV observed for the 2,3-dihexyl derivative (**19**, R = C_6H_{13}) observed by Pomerantz and coworkers [55,56]. The most serious discrepancy between the calculations of Nayak and Marynick and those of Quattrocchi et al. is that in the former calculations, which use PRDDO and STO-3G ab initio structural calculations, the more stable structure was the quinonoid form (by 4.2 kcal/mol per repeat unit), and in the latter calculations, which use AM1 structural calculations, the more stable structure was the aromatic form (by 2.9 kcal/mol per

repeat unit). This seems strange because most calculations on these types of polymeric benzo-annulated thiophenes give the quinonoid structure as the stable form. In addition, recent ab initio calculations (using Gaussian 92) and a 3-21G basis set) [59,60] suggest that the quinonoid form is the stable one and that it is 13 kcal/mol per repeat unit more stable than the aromatic form. Further, Raman spectroscopic studies of several disubstituted derivatives of **19**, including the previously prepared **19** with R = C_6H_{13}, are completely consistent with the quinonoid ground-state geometry [59,60].

In 1986 a reported synthesis by Jenekhe [61,62] of the low band gap polymer **22** prompted Kertesz and Lee to

22

perform calculations on this interesting system [63,64]. While calculations on heterocyclic polymers suggested that low band gaps might be achieved by designing systems in which the aromatic and quinonoid forms were close in energy, this system combines both forms in a single polymer. The calculated band gaps, using MNDO geometries and Hückel band calculations, for the exactly alternating polymer **22**, R = H, x = y = 1, was 1.21 eV. For **22**, R = H, x = y = 2, E_g was calculated to be 1.13 eV, and for x = y = 3, E_g was calculated to be 1.05 eV [63,64]. The reported corresponding experimental values for R = C_6H_5 were 1.1, 0.83, and 0.75 eV, respectively, for x = y = 1; x = 1, y = 2; and x = y = 3 [61,62], which are in quite good agreement with the calculations.

Since 1987 calculations have been done on a variety of systems, some of which had previously been prepared and others that were predicted to have low band gaps. Some of these predicted systems were subsequently prepared, whereas others remain unknown. Thus, the pyrrole polymer derivative of **22** (x = y = 1), namely **23**, was calculated (MNDO geometry/Hückel band calculations) to have a band gap of 0.99 eV [64] and 1.10 eV using MNDO geometry and VEH band structure calculations [65]. A related vinylogous system (**24**) was calcu-

23   **24**

lated to have a band gap of 1.18 eV [65]. Another related structure where the conjugated backbone is somewhat different (**25**) and the protonated version **26** were calculated (MNDO geometry/VEH band calculations) to have band gaps of 1.47 eV and 0.78 eV, respectively [66]. A derivative of **24** was reported recently, but the band gap was not obtained [67].

25   **26**

In addition to calculations of polythieno[3,4-c]thiophene (**17**), Hong and Marynick [68] also calculated the

Low Band Gap Conducting Polymers

conformations (PRDDO calculations) and band structure (modified extended Hückel calculations) of polythieno[3,4-b]thiophene (**27**, R = H) and poly(3-thiabicyclo[3.2.0]cyclohepta-1,4,6-trien-2,4-diyl) (**28**) [68]. From the structure of **28** it can be predicted that the aromatic structure **28a** would be much more stable than the quinonoid structure **28q**, since the latter involves an antiaromatic cyclobutadiene ring. On the other hand, it was suggested that polymer **27** might have more nearly equal energy for the aromatic **27a** and quinonoid **27q**

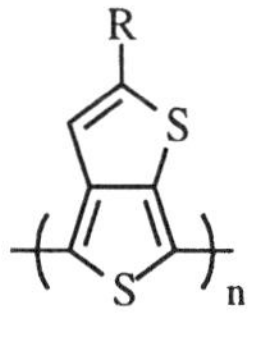

27a 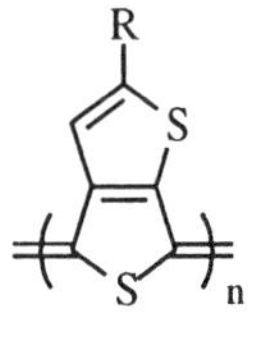   **27q** 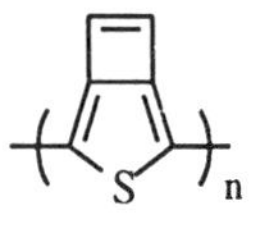   **28a** 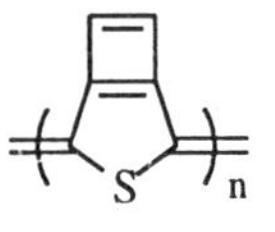   **28q**

forms, because in each form there is one aromatic thiophene ring and one quinonoid thiophene ring (Pomerantz and Marynick, private communication, 1991). In the case of polymer **28** the aromatic structure **28a** is 33.5 kcal/mol per repeat unit more stable than the quinonoid form **28q**, and the band gap for the aromatic form **28a** is a rather high 4.34 eV (λ_{max}) and for the high energy form **28q**, 0.83 eV. For **27**, as predicted, the energy difference between **27a** and **27q** is 0.3 kcal/mol per repeat unit, or they are equal in energy to within the errors of the calculations. The λ_{max} band gap was calculated to be 1.54 eV for the aromatic form **27a** and 1.63 eV for the quinonoid form **27q** [68]. Very recently, poly(4-decylthieno[3,4-b]thiophene) (**27**, R = H) was prepared, and the band gap determined from the absorption maximum at 738 nm was reported to be 1.68 eV, in excellent agreement with the calculations. The band gap obtained from the band edge was 1.2 eV (Ref. 69 and Pomerantz and Gu, unpublished observations, 1995). Lee et al. [48] also performed calculations (MNDO structure and extended Hückel band calculations) on **28a** and **28q** that gave band gaps of 2.56 and 0.97 eV, respectively.

The pyrrole analogs of the polybenzo-, polynaphtho-, and polyanthrothiophenes, **1**, **16**, and **18**, namely, **29**, **30**, and **31**, were calculated (MNDO geometry/Hückel band calculations) [70], and the more stable form in these cases is the aromatic rather than the quinonoid form. The band gaps were calculated for the planar and nonplanar forms of these polymers and, while adding benzo fusions lowers the band gap, the band gaps for the planar and nonplanar forms are nearly equal. The reason for this is that as the rings twist out of planarity the steric interactions decrease and the inter-ring C—C bond distance decreases, which act on the band gap in opposite directions. The calculated band gaps for the planar and nonplanar forms of **29** are 2.02 and 2.1 eV, respectively, those of the planar and nonplanar forms of **30** are 1.13 and 1.2 eV, respectively, and those of **31** are 0.62 and 0.8 eV, respectively.

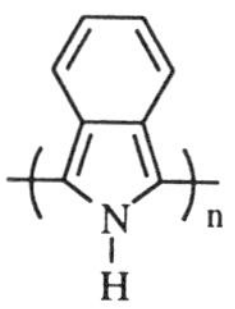

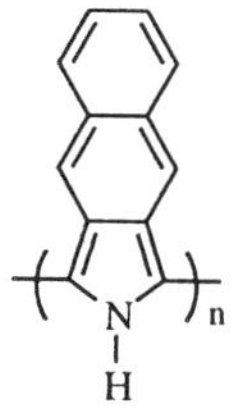

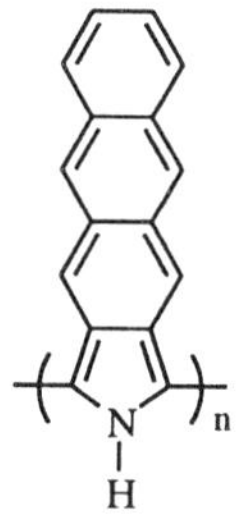

Copolymers of isothianaphthene and thiophene have been the subject of calculations, one of the ideas being that the isothianaphthene would prefer the quinonoid structure whereas the thiophene would prefer the aromatic structure and this could give rise to a lower band gap situation. The calculated bandgaps for **32**, **33**, and **34** [using the Longuet-Higgins–Salem (LHS) Hamiltonian for geometry optimization and extended Hückel band calculation] were, respectively, 1.020, 0.447 [71] (0.54 eV) [72], and 0.747 eV [71]. In a related series, **35**–**37**, with an additional thiophene in the repeat unit, the calculated band gaps were 1.293, 0.880, and 0.600 eV, respectively [71]. Calculations on block copolymers of isothianaphthene and thiophene have also been reported, but the band gaps given were not scaled and were in the range of 4.5–6.5 eV [73].

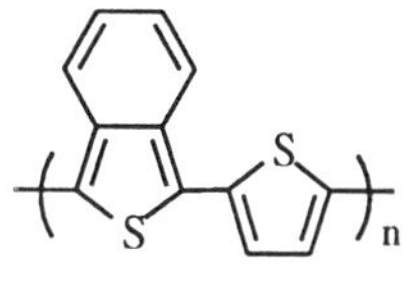

29   **30**   **31**

32 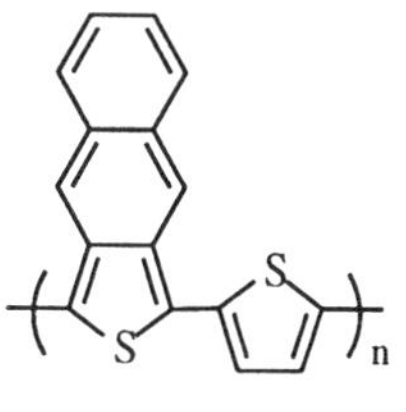   **33** 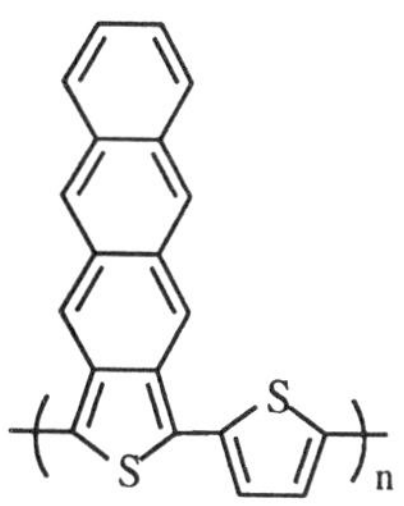   **34**

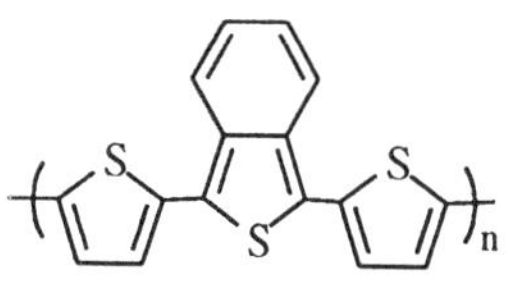

35 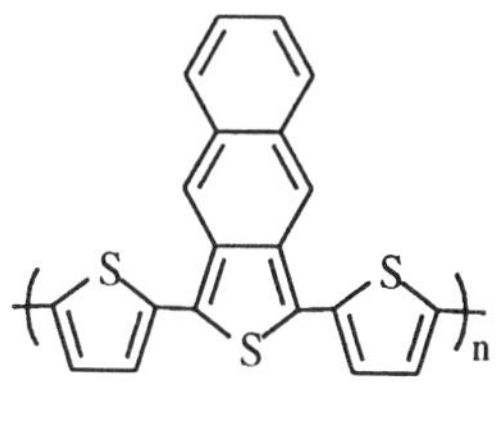   **36**

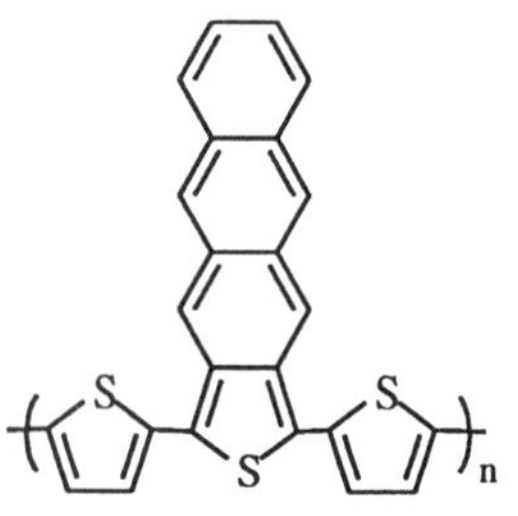

37

The observed band gap of **35**, reported by Lorey and Cava [74], was 1.58 eV, whereas Musmanni and Ferraris [75] reported an observed band gap of 1.7 eV and Bäuerle et al. [76] reported 1.77 eV for the same polymer. These values are higher than the calculated band gaps and are outside of what we consider low band gaps for the purposes of this review. Cava and Metzger and coworkers have also reported the synthesis and band gap of **36**. Interestingly, the band gap here is very low, 0.65 eV [77,78], which now is close to the calculated value. Why the band gap of **35** is so much higher than the calculated value while the band gap of **36** is about the same or even lower than the calculated value remains unclear.

Polymer **22** ($x = y = 1$; R = H), along with the bis(isothianaphthene)-methine system and its benzologues **38**–**40** have been calculated by Kürti, Kertesz and coworkers to have band gaps of 1.026 [71], 1.04 [72], or 1.21 eV [63,64] for **22** ($x = y = 1$; R = H); 0.634 [71] or 0.69 eV [72] for **38**; 0.418 eV for **39** [71]; and 0.254 eV for **40** [71].

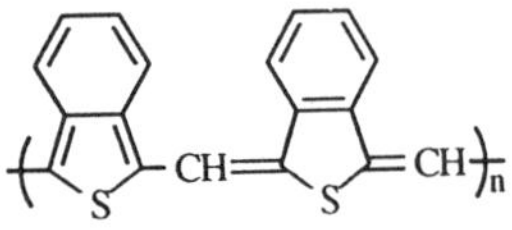

38

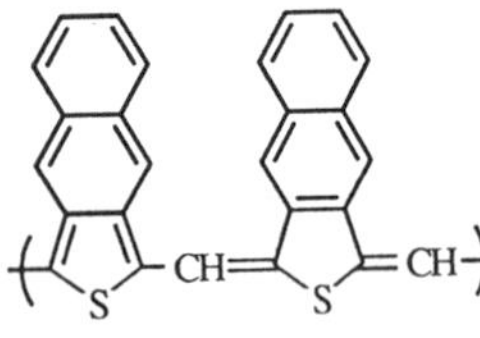

39

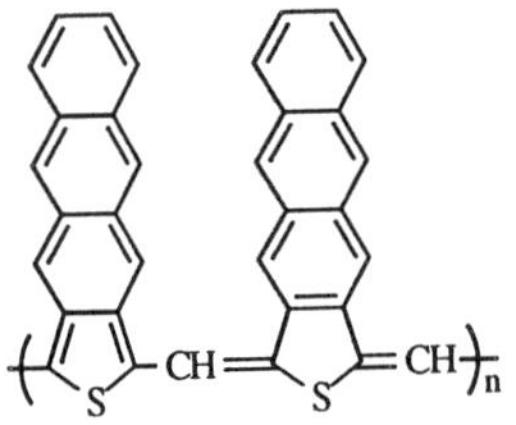

40

In 1988 Pranata et al. [79] examined polymers related to polythiophene, polypyrrole, and their benzannelated derivatives but had fulvene rings replacing the five-membered heterocycles. The calculations involved the MNDO-optimized geometries and Hückel MO and VEH band structure calculations and were done on both the fulvenoid and quinonoid structures of polymers **41**–**44**. **41** and **41a** show the fulvenoid and quinonoid structures for polyfulvene (**41**), respectively. Table 11.2 shows the calculated band gaps of these polymers obtained using the VEH methodology. Interestingly, in all cases, the fulvenoid forms have the lower band gaps, which is a

Table 11.2 Calculated (VEH) Fulvenoid and Quinonoid Band Gaps (eV) of **41**–**44**

	Polymer			
	41	42	43	44
Fulvenoid	0.87	0.98	0.65	0.05
Quinonoid	1.80	1.74	2.69	2.26

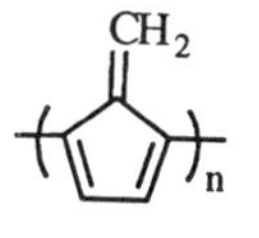

41 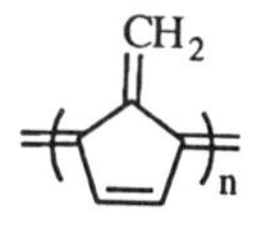   **41a** 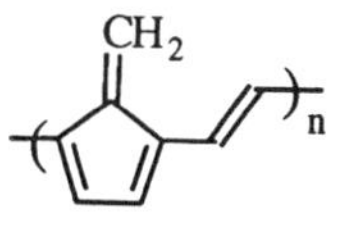   **42**

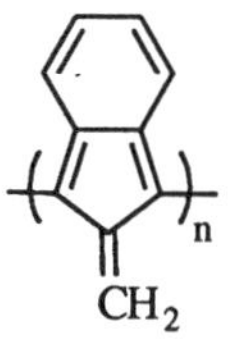

43 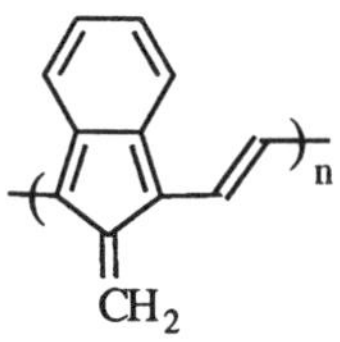   **44**

result of the symmetry of the highest occupied crystal orbitals and lowest unoccupied crystal orbitals and can be rationalized using conventional orbital mixing arguments similar to what is done with small molecular systems [79]. Steric interactions in **41** and **43** will no doubt give rise to nonplanar polymers, but with the vinylene spacers the planar species appear to be much more likely. In addition, since fulvene is not an aromatic molecule like thiophene or pyrrole, it is said that there is no overwhelming tendency for the polymers to exist in the fulvenoid forms and so in **41** the two structures may be comparable in π-electron energy [79].

Dougherty's group also calculated polycyclobutadiene-1,3-diyl (**45**) (MNDO geometry/Hückel and VEH band structure) and found a band gap of 0.68 eV for structure **45a** and a band gap of 0.00 eV for both structures **45b** and **45c**. Structure **45a** was calculated to be a little more stable than **45b** and **45c**, but a clear choice

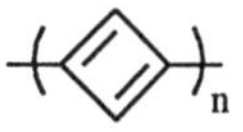

45 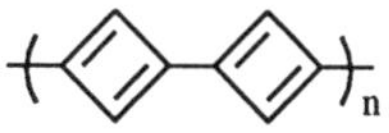   **45a**

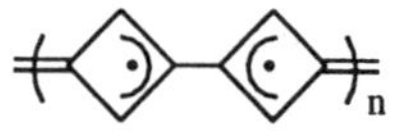

45b 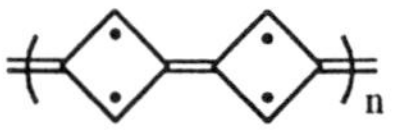   **45c**

for the preferred structure could not be made. The reason for considering structures that are radical in nature is that ab initio calculations show that the dimer **46** has the structure shown here, and MNDO-SCF calculations show the trimer **47** and the tetramer **48** to have the structures shown [80].

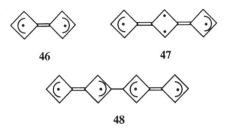

Ladder polymers have been of interest for many years, and an early paper on poly(*peri*-naphthalene) (**49**) using the VEH method combined with crystal packing calculations provided a band gap of 0.44 eV. With interchain overlap the calculated band gap decreases to 0.29 eV [81,82]. The synthesis of polymer **49** by pyrolysis of 3,4,9,10-perylenetetracarboxylic dianhydride at temperatures between 530 and 900°C gave very highly conducting undoped films (σ up to 15 S/cm) that were probably not the simple polymer **49** [82].

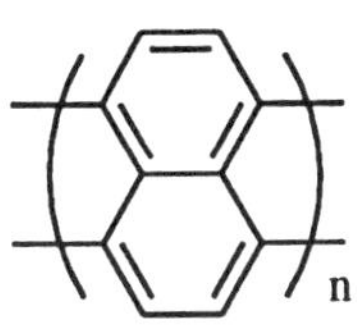

49

Pomerantz and coworkers reported in 1989 [83] that very simple perturbational molecular orbital (PMO) methods could be applied to a number of hydrocarbon aromatic oligomers and conducting polymers. These calculations provided band gaps and UV-vis absorption maxima that were in excellent agreement with both observed data, where available, and data calculated by more sophisticated procedures. Thus for polymer **49** the PMO method gave a value of 0.54 eV [83], while the VEH calculations mentioned above gave a band gap of 0.44 eV [81] and the PPP-SCF band gap calculations gave 0.73 eV [84].

Polyacene (**50**) calculations have been carried out by a number of different methods, which, depending on the geometry, have given different values for the band gap. Extended Hückel methodology gave values of 0.45 and 0.002 eV, respectively, for **50a** and **50b** [2]. In this case

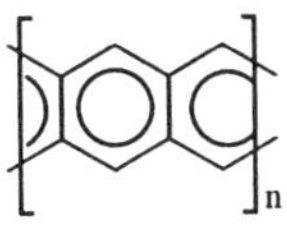

50 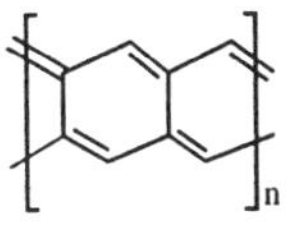   **50a** 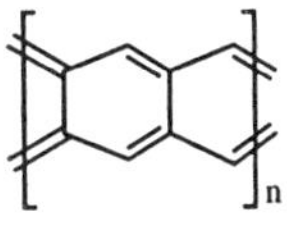   **50b**

the symmetric structure **50** was calculated to be much less stable than **50a** or **50b** [2]. The VEH methodology gave 0.0 eV for both **50** and **50b** and 0.3 eV for **50a** [85]. The PMO methodology gave a band gap of 0.09 eV [83]. Other calculations on polyacene (**50**) and related heteroatom-containing systems such as **51** with X, Y = N, O, S have provided band gaps in the 0.3–0.5 eV range: 0.32 eV for **51** with X = S, Y = N; 0.41 eV for **51** with X = NH, Y = N; and 0.46 eV for **51** with X = O, Y = N [86,87].

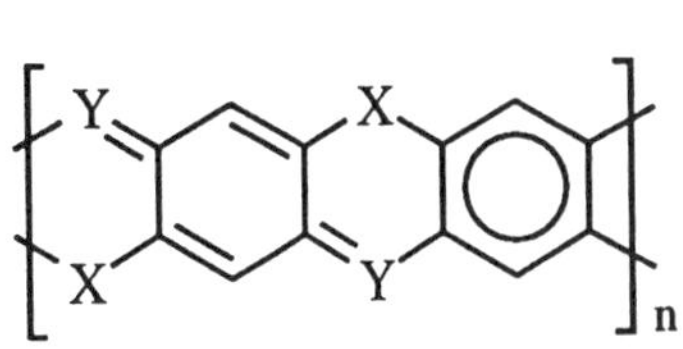

51

An early paper in which calculations on several variants of **51**, including **51a** where X = NH and Y = CH, **51b** where X = S and Y = CH, and **51c** where X = CH$_2$ and Y = CH, provided band gaps of 0.60, 1.02, and 0.97 eV, respectively, using MNDO geometry and VEH band gap calculations [88]. The non-ladder polymer related to these systems, namely **51d**, where Y = CH and X is not a group or atom, also shows a calculated low band gap of 1.17 eV [88].

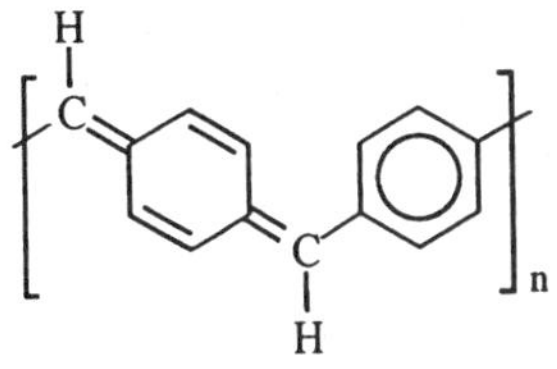

51d

An interesting ladder polymer that has recently been prepared is shown as structure **52(p,m)** (R = *p*-C$_{10}$H$_{21}$-C$_6$H$_4$ and R' = C$_6$H$_{13}$) [89,90], where indene-type ring systems are fused in what has been termed a (*p,m*) configuration [86]. The *p* and *m* (*para* and *meta*) describe the relationship of the nonbenzenoid carbons around the six-membered rings. The calculated band gap for this

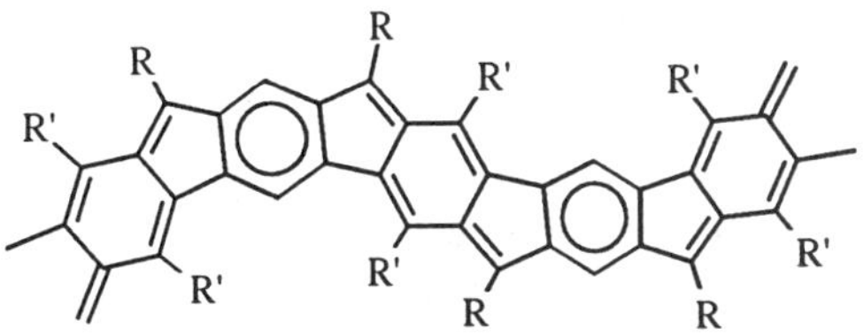

52(p,m)

species with both quinonoid and aromatic six-membered rings, was 1.19 eV and 0.98 eV using PM3 and Longuet-Higgins–Salem (LHS) methodology, respectively, for geometry optimization and Hückel calculations for band gaps [86]. This is to be compared with the band gap of the substituted molecule **52(p,m)** (R = *p*-C$_{10}$H$_{21}$-C$_6$H$_4$ and R' = C$_6$H$_{13}$), estimated from the band edge of the peak at 605 nm to be about 1.5 eV [89]. Other related isomeric systems with alternate *ortho*, *para*, and *meta* relationships were also calculated. The band gaps are predicted to generally be less than 1 eV, and some are very much lower, even 0.00 eV. Thus, for example, **52m** is calculated to have a band gap of 0.12 eV [86]. Interestingly, an isomer of **52m**, namely **52m'**, was calculated,

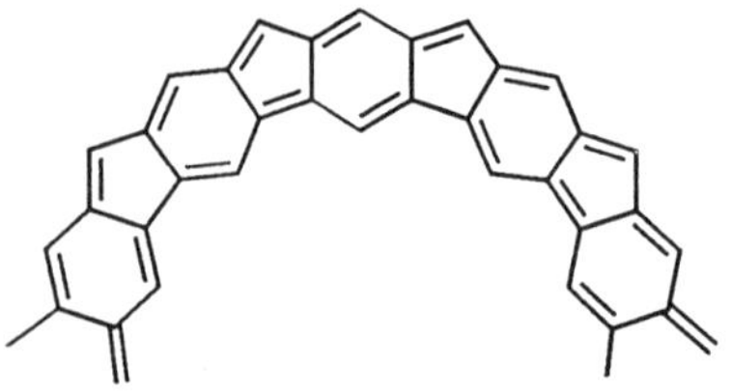

52m

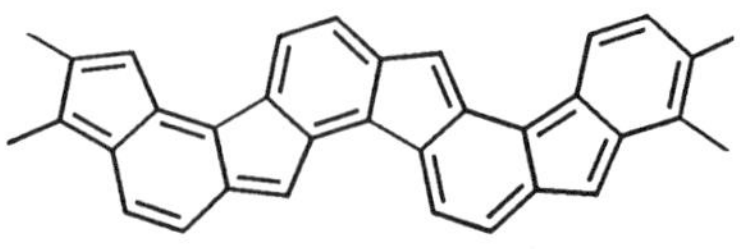

52m'

using the LHS calculated geometry, to be an intrinsic metal with a band gap of 0.00 eV [86,91]. Related nonladder polymers where the six-membered rings are bonded through a single carbon atom (again with the possibility of *o*, *m*, and *p* isomers) were also calculated, using planar geometries, and here the band gaps were generally higher, up to 1.67 eV [86]. Interestingly, a spin-unrestricted Hartree–Fock calculation using the crystal orbital version on **52m** suggests that the lowest energy state may be the one that shows a high-spin ferromagnetic ground state. It should be noted that a form of polymer **52m** can be written with free electrons on the nonaromatic carbon atoms (**53**).

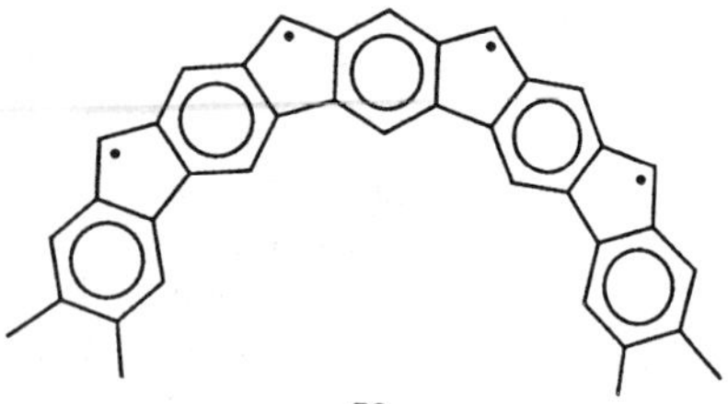

53

Another class of low band gap polymers are those whose repeat units contain two thiophene or pyrrole rings fused to a central five-membered ring that contains a substituted methylene group. The object here is to combine electron-withdrawing groups with electron-donating groups to lower the band gap. The first of these molecules to be synthesized and studied were **54** and **55** [92–94]. Using AM1 geometry optimization and VEH band gap calculations, the band gap for **55** was calculated to be 0.58 eV, which agreed quite well with the 0.8 eV observed experimentally [95]. The corresponding pyrrole derivative **56** was calculated to have an even lower band gap of 0.38 eV.

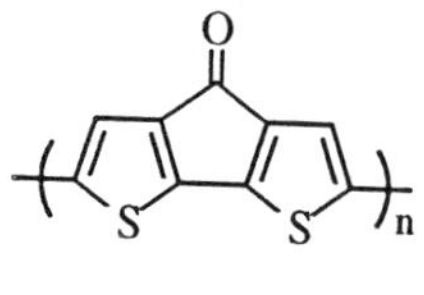

54

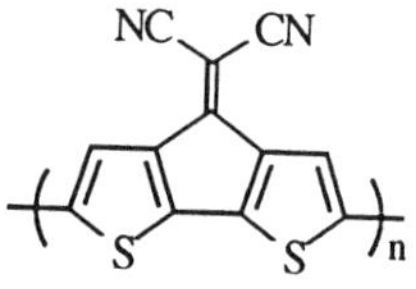

55

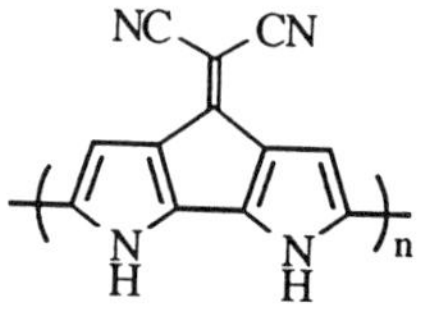

56

In 1988 polydithieno[3,4-*b*:3′,4′-*d*]thiophene (**57**) was prepared [96] and was reported to have a band gap of 1.1 eV [97]. The monomer **58** was electrochemically polymerized, and the structure of the polymer was not determined at that time. However, in 1994 it was reported [98] that the photochemical polymerization of **58** re-

57

58

sulted in the formation of some soluble oligomers whose NMR spectrum suggested that the linkage, at least in this photochemical polymerization, was at the 2- and 5-positions in one thiophene ring as shown in structure **57**. Calculations using AM1 geometry optimization and VEH band gap calculations were carried out on three different isomeric polymers, **59–61**, and provided band

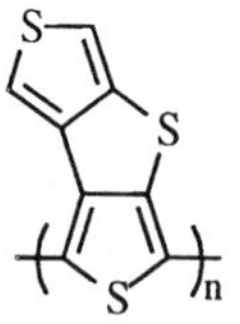

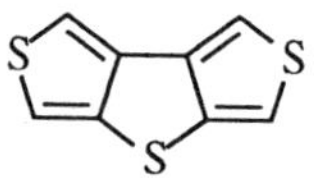

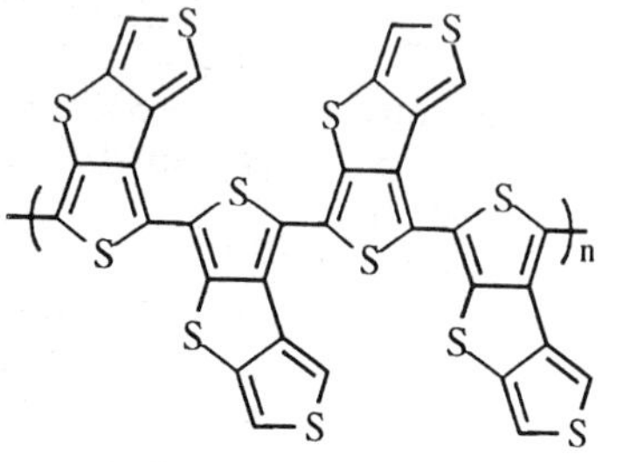

59

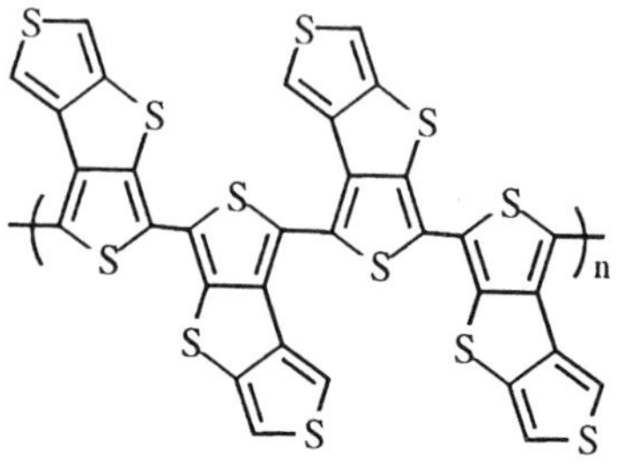

60

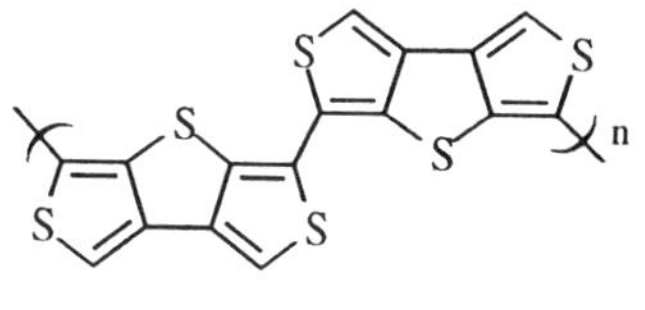

61

gaps of 1.27, 1.77, and 2.39 eV, respectively [99]. Polymers **59** and **60** differ in the relative orientation of the fused rings. The conclusion was that since the absorption of the actual polymer is broad with a maximum at 2.1 eV and a band edge at 1.1 eV, it may well be a mixture of many different sequences and the onset at 1.1 eV was due to sequences of the type shown in structure **59** [99].

A polymer related to structure **55** but with the sulfur atoms replaced by CH_2 groups, **62**, has also been recently calculated, again using AM1 geometry optimization and VEH band gap calculations. In this instance a very low band gap of 0.16 eV was calculated [100]. It was shown that this is mainly due to the large stabilization that occurs between the LUMO of the conjugated polymer skeleton and the LUMO of the electron-withdrawing cyano groups [100].

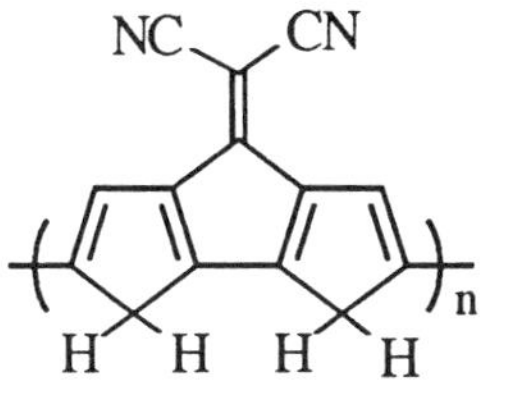

62

Three simple derivatives of polythiophene, namely the S-oxide **63** [the sulfoxide of polythiophene; poly(thiophene-1-oxide)], the S,S-dioxide **64** [the sulfone of polythiophene; poly(thiophene-1,1-dioxide)], and the S-methyl sulfonium polycation **65** [poly(1-methylthiophenium cation)], were calculated with CNDO/2 and the SCF–crystal orbital methodology. The results, after scaling the band gaps to fit for polythiophene, showed that the band gaps of **63**, **64**, and **65** were 1.12, 0.48, and

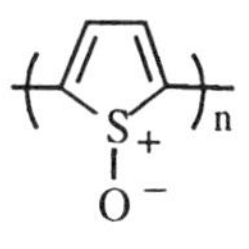

63 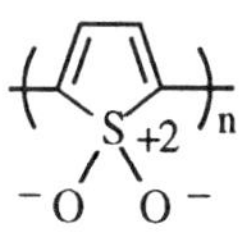   **64** 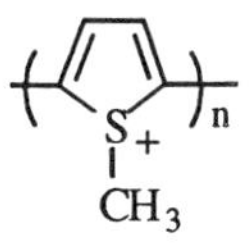   **65**

2.23 eV, respectively [101]. The band gaps of **63** and **64** have more recently been recalculated using the *ab initio* Hartree–Fock crystal orbital method using a double-zeta basis set, and the scaled band gaps calculated by this methodology are higher and reversed, namely 1.14 and 1.53 eV for **63** and **64**, respectively [102]. The reason for the discrepancy is not clear, and it is also known that the Hartree–Fock methodology overestimates band gaps.

Direct attachment of alkoxy groups onto the thiophene rings in polythiophenes has been shown to give rise to conducting polymers with considerably lower band gaps. Thus, for example, while the band gap of poly(thienylene vinylene) (**66**, R = H) is 1.74 or 1.64 eV [103,104], the band gap of the oxygen-substituted derivative poly(3-methoxythienylene vinylene) (**66**, R = OCH_3), obtained electrochemically, has been reported to be 1.32 eV and 1.67 eV from the band edge, while the ethoxy derivative, **32**, R = OC_2H_5, gave a band-edge band gap of 1.48 eV and an electrochemical band gap of 1.31 eV [104,105]. This compares with calculated values of 1.24 eV (MNDO geometry/Hückel band calculations) [70] and 1.69 eV (MNDO geometry/VEH band calculations) [104] for the methoxy derivative **32**, R = OCH_3.

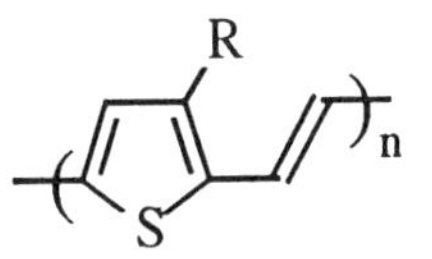

66

IV. DERIVATIVES OF POLYISOTHIANAPHTHENE

Since the preparation and study of polyisothianaphthene (**1**), the syntheses and study of a number of substituted derivatives have appeared. One of the earliest claims of the synthesis of such derivatives was in 1988, when polymers **67a**, **67b**, **68**, and **14** were prepared and the properties of **67a** were reported [106]. They were pre-

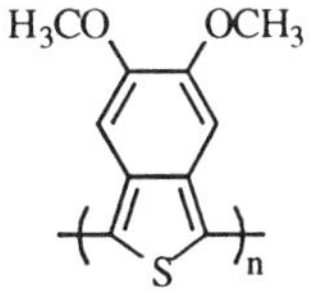

14 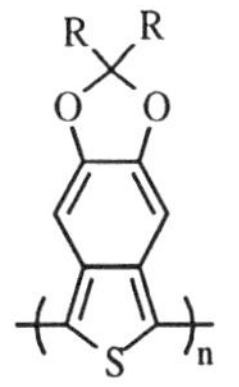   **67a**: R = H   **68**
       b: R = CH_3

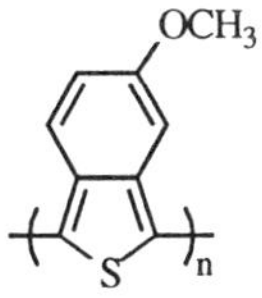

pared in a manner similar to the original preparation of the parent polyisothianaphthene, and the synthesis of **67a** is shown in Scheme 2 [106,107]. It is interesting to note that the electrochemical polymerization could be carried out using a solution of the monomer **70** in acetonitrile containing tetraphenylphosphonium chloride ($Ph_4P^+ Cl^-$), simply using a series of 1.5 V batteries as the power supply [107]. The $FeCl_3$-induced polymerization of the dihydromonomer, 1,3-dihydro-5,6-dioxymethyleneisothianaphthene (**69**), was reported in 1991 [107]. Since it was claimed that polyisothianaphthene was stable in air in the neutral state but that in the highly doped state it was unstable in air and rapidly dedoped, a more stable derivative of **1** was needed, and this was **67a** [107]. The band gap of **67a** was originally said to be

to 3×10^{-2} and 6×10^{-2} S/cm, respectively [107]. When the cyclic voltammetry of the monomer **70** was examined using acetonitrile solvent and tetraphenylphosphonium chloride ($Ph_4P^+ Cl^-$) as supporting electrolyte, it was observed that the anodic oxidation potential was the same as that of the parent monomer **3**. That is, there was no substituent effect. However, using a nonnucleophilic supporting electrolyte, $Bu_4N^+ ClO_4^-$, a lowering of the oxidation potential of 0.2 V was observed. The polymer **67a** showed an oxidation peak at 0.68 V (Ag^+/Ag) and a reduction peak at 0.26 V ($CH_3CN/Bu_4N^+ ClO_4^-$), and, contrary to expectations, there was no redox potential lowering effect of the dioxymethylene group. Optoelectrochemical studies showed that the band-edge band gap was about 1 eV, and the heavily doped polymer showed a new near-IR absorption maximum at 1290 nm. Polymer **67a** showed about the same TGA thermal stability as the parent polyisothianaphthene (**1**) [107].

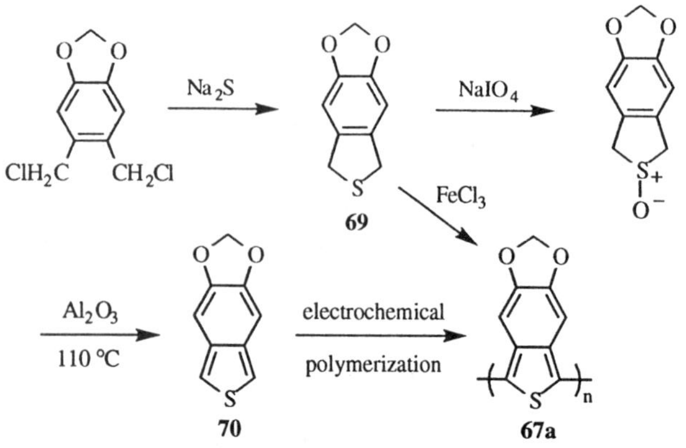

Simple alkyl derivatives of **1** have been reported and studied [56,108–111]. A 1990 patent reports the synthesis of 5-dodecylbenzo[c]thiophene by the sulfoxide route (see Schemes 1 and 2) and its electrochemical polymerization (acetonitrile/$Bu_4N^+ Cl^-$) to poly(5-dodecylbenzo[c]thiophene) (**71**). The polymer was reported to be soluble in a variety of organic solvents, and an electrochemically doped film had a four-point probe conductivity of 10^{-2} S/cm. Several other alkyl derivatives were said to have been made, and the following conductivities (σ) were reported: **72**, $\sigma = 10^{-3}$ S/cm; **73**, $\sigma = 10^{-3}$ S/cm; **74**, $\sigma = 10^{-3}$ S/cm; and **75**, $\sigma = 10^{-3}$ S/cm

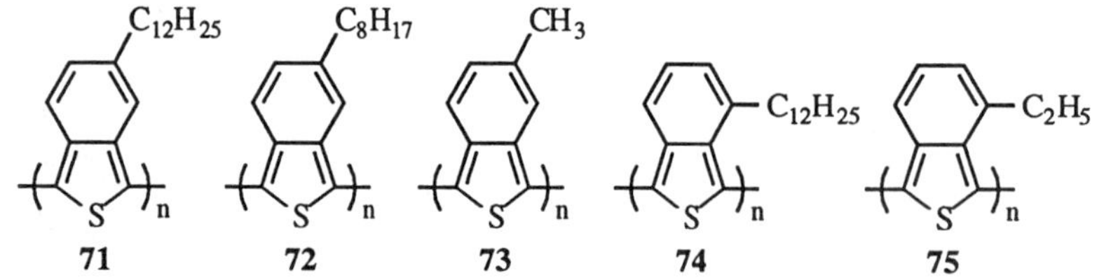

≈ 0.6 eV [106], based on the band edge, but this was subsequently modified to ≈ 1 eV [107]. The undoped material was blue-black or deep blue, and the doped polymer was gray and transparent [106]. A film of **67a** could be recycled electrochemically between oxidized and reduced forms, and the oxidized form could be compensated chemically with hydrazine, but even then it remained slightly doped. Two-point probe conductivity of pressed pellets of the electrochemically prepared polymer **67a** was 5×10^{-4} S/cm, while for the polymer prepared by $FeCl_3$ polymerization it was 4×10^{-3} S/cm. Upon doping with I_2, the conductivities rose modestly

[108]. It is quite surprising that the conductivities of these derivatives are within one order of magnitude independent of whether the alkyl groups, both small and large, are on carbon atom 4 or 5. Poly(5-decylisothianaphthene) (**76**) was synthesized by Pomerantz et al. [56] using the $FeCl_3$ oxidation of the dihydro monomer as shown in Scheme 3. It was a purple-black polymer in the undoped state and formed a dark blue-black solution in $CHCl_3$ that showed $\lambda_{max} = 512$ nm with a shoulder at 670 nm. The polymer could be doped in solution with nitrosonium fluoborate and produced a light yellow solution. A film of this material showed a band-edge band gap of 1.0–1.3 eV; scattering from the film prevented an accurate determination [56].

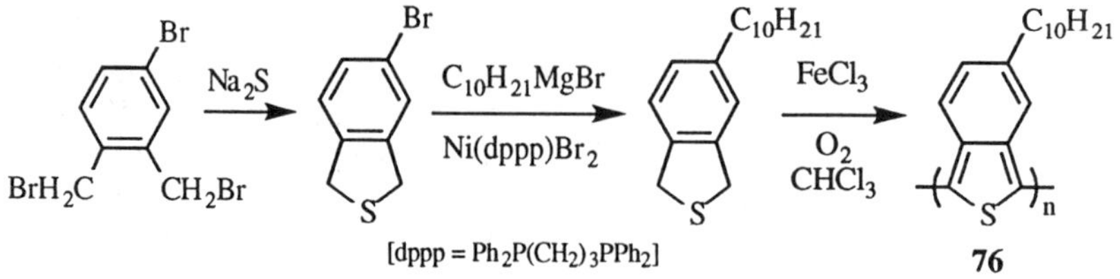

Low Band Gap Conducting Polymers

The 5-methyl derivative, poly(5-methylbenzo[c]thiophene) (**73**) was reported recently by Higgins and coworkers [109–111] and was made by the electrochemical polymerization of the monomer produced by the sulfoxide route, similar to that shown in Schemes 1 and 2. The reported band gap, obtained from the band edge, was 1.13 eV. Two anodic peaks were reported at 0.50 and 0.77 V (SCE, Et_4N^+ BF_4^-/CH_3CN), and, in addition, it was reported that the onset of *n*-doping was at −1.1 V [109–111]. This should be compared to the values for the parent **1**, where the two anodic peaks were at 0.45 and 0.85 V, respectively, and the onset of *n*-doping was at −1.1 V, which shows that there is little effect due to the methyl substituent [109–111]. However, by comparing these with fluorine-substituted polybenzo[c]thiophenes (**77**, **78**) and with poly(5,6-dichlorobenzo[c]thiophene) (**79**) it was shown that there was indeed very little substituent effect on the *p*-doping but there was a significant effect on the *n*-doping. Thus the anodic peaks for **77–79** were 0.60 and 0.80 V, 0.70 and 1.00 V, and

77 **78** **79**

0.72 and 0.98 V, respectively. The onset of *n*-doping of these three polyisothianaphthene derivatives was at −0.6, ≤−0.8, and −0.2 V, respectively, with the fluoro derivative **78** being rather more difficult than the others to *n*-dope [111]. Thus, while the total range in the *p*-doping values for **1**, **73**, and **77–79** is 0.25 V, the *n*-doping values cover a range of 0.9 V. In addition, these derivatives were reported to be transparent when *n*-doped [111]. However, the properties of **78**, such as its color, which was red-purple, compared to those of the others (**1**, **73**, **77**, and **79**), which were blue, and the very negative *n*-doping onset potential indicated that the backbone of this polymer was twisted [111]. Interestingly, the band-edge band gap for **77** was 0.95 eV where λ_{max} = 720 nm [111].

Other monosubstituted polyisothianaphthenes have been reported. In 1989 a report appeared stating that poly(5-benzoylbenzo[c]thiophene) (**80**) had been prepared by electrochemical polymerization of the monomer, 5-benzoylbenzo[c]thiophene, which had been prepared by the sulfoxide route similar to that shown in Scheme 1 for the parent polymer **1** and in Scheme 2 [28]. Unfortunately, no other data were given. A variety of other substituted derivatives were prepared by the sulfoxide route (Schemes 1 and 2) as described in a Japanese patent, and four-point probe conductivities (electrochemical doping/Bu_4N^+ Cl^-/CH_3CN) of 10^{-2} S/cm were reported for poly(5-octyloxybenzo[c]thiophene) (**81**) and poly(5-octylaminobenzo[c]thiophene) (**82**)

80 **81** **82**

[108]. Other derivatives of polyisothianaphthene containing electron-withdrawing groups have been reported in another Japanese patent [112]. The synthesis was again via the sulfoxide route, and the electrochemical synthesis of poly(5-cyanobenzo[c]thiophene) (**83**) was given. Using Bu_4N^+ Cl^- as supporting electrolyte in CH_3CN, a film of the doped polymer showed a conductivity of 10^{-2} S/cm and light transmission of 30%, but with Bu_4N^+ ClO_4^- the light transmission improved to 65%. This is better than the parent polyisothianaphthene (**1**), which showed a doped transmission of only 40% with Bu_4N^+ ClO_4^- as the electrolyte [112]. The doped (Ph_4P^+ Cl^-) conductivities and light transmittances of other substituted polyisothianaphthenes are shown in Table 11.3 [112].

83 **84** **85**

86 **87**

Another simple substituted polyisothianaphthene that was reported in 1993 was the tetrafluoro derivative, poly(4,5,6,7-tetrafluorobenzo[c]thiophene) (**88**). It was prepared by electrochemical polymerization of the monomer using Bu_4N^+ Cl^- as supporting electrolyte in CH_3CN (Scheme 4) [113] and by DDQ oxidation of the

Table 11.3 Properties of Polymers **84–87**

Polymer	Conductivity (S/cm)	% Transmittance
84	10^{-2}	60
85	10^{-2}	55
86	10^{-3}	60
87	10^{-3}	50

Source: Ref. 112.

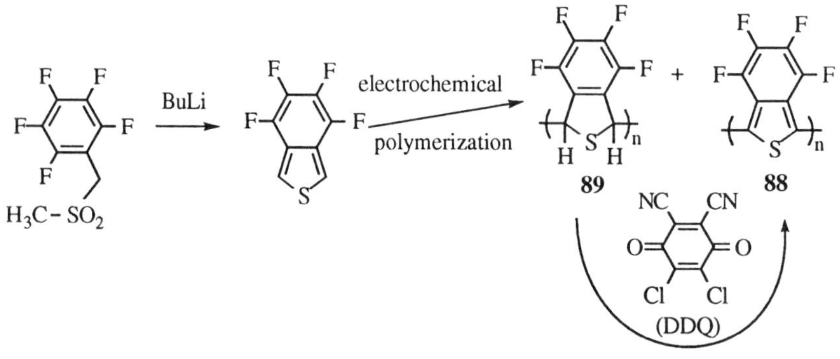

Scheme 4

precursor dihydro polymer, poly(1,3-dihydro-4,5,6,7-tetrafluorobenzo[c]thiophe ne) (**89**), which had been prepared from the monomer using Bu_4N^+ ClO_4^- as supporting electrolyte [114]. In addition, **89** was prepared by polymerization of the monomer, 4,5,6,7-tetrafluorobenzo[c]thiophene, using SbF_5-HF [115]. Polymer **88** was a red-black solid, soluble in chlorinated solvents, that showed λ_{max} = 485 in CH_2Cl_2 solution and a band gap of 2.7 eV. Thus the four fluorine substituents greatly affect the band gap, quite possibly because of steric hindrance, and this derivative of **1** is not a low band gap polymer [113]. Although polymer **88** was first prepared in 1990, because of its solubility it was misidentified [115].

Naphtho[2,3-c]thiophene was prepared by the sulfoxide route, and this monomer (which contained some 1,3-dihydronaphtho[2,3-c]thiophene) was then polymerized electrochemically (Ph_4P^+ Cl^-/CH_3CN) to poly(naphtho[2,3-c]thiophene) (**16**) as shown in Scheme 5. The oxidation potential (AgCl-Bu_4N^+ ClO_4^-/CH_3CN) was 0.74 V in the anodic process and 0.64 in the cathodic process and was reversible over many cycles. The neutral polymer film was blue and became transparent gray upon doping, but, unlike polyisothianaphthene (**1**), it did not become completely transparent. The polymer showed λ_{max} = 575 nm, and the band-edge band gap was 1.5 eV [51]. As discussed above, this is in good agreement with the calculated band gap for the quinoid form but is much higher than that calculated for the aromatic form. It has, however, been suggested that this seemingly high observed band gap may be due to structural irregularities due to bonding to the *meso* positions in the central ring [77,78].

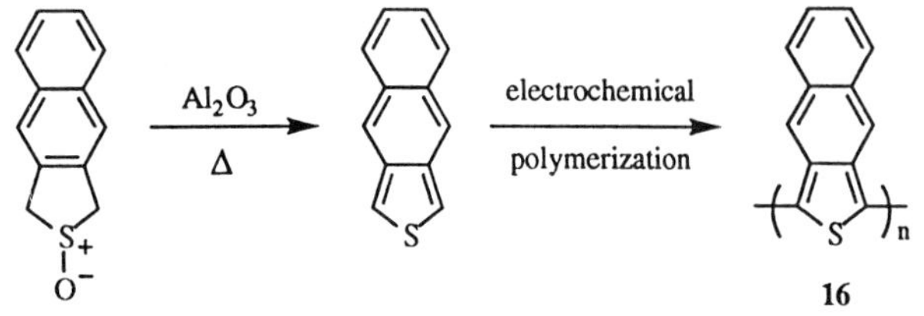

Polyphenanthro[9,10-c]thiophene (**90**) was reported to have been prepared by electrochemical polymerization (Bu_4N^+ PF_6^-/CH_3NO_2) of phenanthro[9,10-c]thiophene (**91**) (see Scheme 6). The polymer showed electrochromic behavior, going from red, formed on oxidation of **91**, to yellow in the range 0.8–0 V [Ag/AgCl/KCl(sat)], to black at potentials between 0 and -0.8 V [116]. Interestingly, the red polymer produced in acetonitrile solution had a conductivity of 105 S/cm, whereas that formed in nitromethane solvent had a conductivity of 4.4×10^{-3} S/cm. Reduction of monomer **91** at -0.8 V produced an air-sensitive brown, n-doped polymer whose conductivity was 4.5×10^{-1} S/cm (pressed pellet). Chemical polymerization of **91** with NO^+ BF_4^- in CH_3CN gave a greenish-yellow polymer with a conductivity of 1.7×10^{-1} S/cm (pressed pellet) [116]. Since the band gap was not determined, it is not known whether this is a low band gap polymer; however, it would appear that the phenanthrene rings would cause twisting in the backbone, and this would agree with the observed color changes.

Another polymer related to polyisothianaphthene (**1**) in which the two CH groups adjacent to the thiophene rings have been replaced by sterically less demanding nitrogen atoms is polythieno[3,4-b]pyrazine (**19**, R = H), and the first derivative of this structure to be prepared, the dihexyl derivative **19**, R = C_6H_{13}, was prepared by Pomerantz and coworkers [55,56]. It was expected that the replacement of the CH groups by N would cause less torsion in the backbone and would give rise to a lower band gap polymer, and, as discussed above, this was verified by quantum-mechanical calculations [54]. The synthesis of poly(2,3-dihexylthieno[3,4-b]pyrazine) (**19**, R = C_6H_{13}) is shown in Scheme 7

Scheme 5

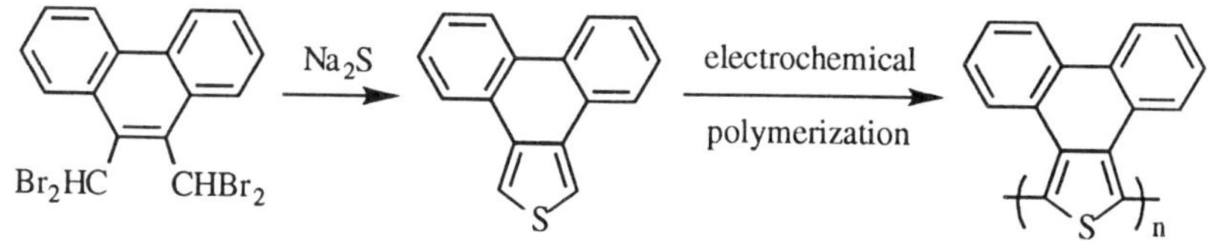

[55,56]. A cast film of the polymer showed a band-edge band gap of 0.95 eV, confirming the quantum-mechanical predictions that the band gap of **19** would be 0.1 eV lower than that of the parent polyisothianaphthene. Interestingly, a film of the polymer showed an absorption maximum in the near-infrared at 915 nm, while λ_{max} was 875 nm in CHCl$_3$ solution [55,56]. The undoped polymer was dark blue-black, became light yellow and transparent upon oxidative doping with nitrosonium fluoborate, and showed a four-point probe conductivity of 3.6×10^{-2} S/cm. TGA analysis under N$_2$ showed the onset of decomposition at about 230°C and ~50% weight loss at 600°C. Upon doping, the band at 875 nm decreased in intensity and moved to shorter wavelength as the amount of dopant was increased, and a new absorption appeared at around 1600 nm. Unfortunately, the polymer, either doped or undoped, was not stable in the air. It could easily be irreversibly overdoped, and if too much oxidative dopant was added, a new absorption with $\lambda_{max} > 2100$ nm began to appear and the conductivity decreased.

It is interesting to compare the band-edge band gaps of poly(2,3-dihexylthieno[3,4-b]pyrazine) (**19**, R = C$_6$H$_{13}$) and poly(5-fluorobenzo[c]thiophenes) (**77**), both of which are reported to have band gaps of 0.95 eV. The absorption maximum, λ_{max}, is 915 nm for a film of **19**, R = C$_6$H$_{13}$ [55,56], while that for **77** is 720 nm. The difference is clearly due to a long tail in the UV-vis-NIR spectrum of **77** [111], while the absorption in **19**, R = C$_6$H$_{13}$, is much sharper, suggesting that the band gap is lower for poly(2,3-dihexylthieno[3,4-b]pyrazine) (**19**, R = C$_6$H$_{13}$).

The parent polythieno[3,4-b]pyrazine (**19**, R = H) has recently been prepared by reaction of 2,3-pyrazinedicarboxylic anhydride (**92**) with P$_4$S$_{10}$ (Scheme 8) as a black powder whose dedoped conductivity was 10^{-6}–10^{-7} S/cm. Doping with NO$^+$ BF$_4^-$, I$_2$, or FeCl$_3$ gave conductivities of $(5–7) \times 10^{-5}$ S/cm [22]. A similar reaction using 2,3-pyridinedicarboxylic anhydride (**93**) produced polythieno[3,4-b]pyridine (**94**), whose dedoped conductivity was 10^{-9} S/cm and whose conductivities after doping with NO$^+$ BF$_4^-$, I$_2$, and FeCl$_3$ were 2×10^{-5}, 3×10^{-5}, and 6×10^{-7} S/cm, respectively [22]. No band gap data were presented for either polymer.

Other derivatives of polythieno[3,4-b]pyrazine that have been prepared [117] and studied are the 5,6-dimethyl derivative (**19**, R = CH$_3$), the 5,6-diethyl derivative (**19**, R = C$_2$H$_5$), the 5,6-diundecyl derivative (**19**, R = C$_{11}$H$_{23}$), the 5,6-ditridecyl derivative (**19**, R = C$_{13}$H$_{27}$), and the di-2-thienyl derivative (**19**, R = 2-C$_4$H$_3$S) [59,60]. Raman spectroscopic studies, combined with scaled quantum-mechanical oligomer force field (SQMOFF) calculations of the Raman spectra of the quinonoid form of the polymer, verified that the ground-state form of the polythieno[3,4-b]pyrazines is quino-

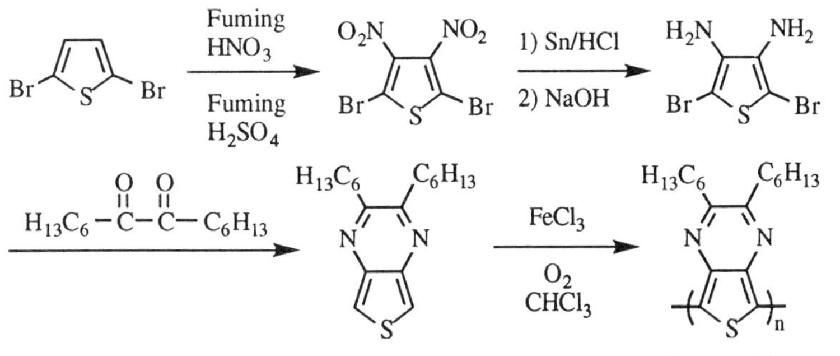

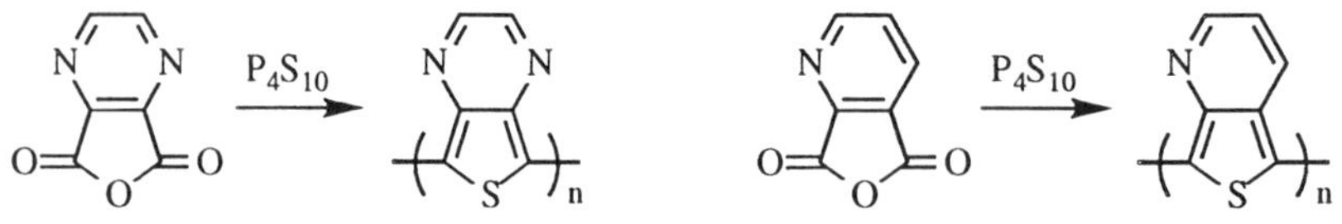

Scheme 8

noidal. Further, the band gaps were all around 0.9 eV [60,61], as had been reported earlier for the dihexyl derivative **19**, R = C_6H_{13} [55,56]. The Raman spectra of these polymers all showed ring C=C stretching frequencies at around 1520 and 1560 cm^{-1}. Upon doping with $FeCl_3$ the relative intensities of these peaks changed with respect to those of the undoped polymers, and several other lines became more intense [59]. Analysis of the Raman spectrum of the doped polymer suggested that the rings had become aromatic rather than quinonoidal as in the undoped polymer [59].

Very recently poly(4-decylthieno[3,4-b]thiophene) (**27**, R = $C_{10}H_{21}$) was prepared by Gu and Pomerantz using $FeCl_3$ chemical polymerization by the route shown in Scheme 9. The band gap, determined from the absorption maximum of the film at 738 nm. was 1.68 eV, which, as indicated above, is in excellent agreement with the calculations. The band gap obtained from the band edge of this absorption was 1.2 eV (Ref. 69 and unpublished observations, 1995). Polymer **27**, R = $C_{10}H_{21}$, ws blue-green and had a GPC molecular weight (vs. polystyrene) of M_n = 30,000 (M_w/M_n = 1.4). Upon doping with $FeCl_3$ the peak at 738 nm decreased gradually while a new peak at λ_{max} = 1052 nm was formed, and the color went from blue-green to yellow-green for the doped polymer. The doping was reversible, with NH_3 gas used to dedope the polymer films. Electrical conductivities (four-point probe) of the polymer films cast from $CHCl_3$ solution were 2.2 × 10^{-5} S/cm before doping, 8.8 × 10^{-3} S/cm after doping with I_2, and 7.2 × 10^{-2} S/cm after doping with $FeCl_3$ (Ref. 69 and Pomerantz and Gu, unpublished observations, 1995).

Polydithieno[3,4-b:3′,4′-d]thiophene (**57**) was prepared in 1988 [96] by electrochemical polymerization ($LiClO_4$/CH_3CN) of the monomer **58** and was reported to show an electrochemically doped (1.04 V vs. SCE) conductivity of 1.0 S/cm. The neutrral polymer was opaque and had λ_{max} = 590 nm, whereas the doped polymer was said to be colorless and highly transparent [96]. The band-edge band gap was reported to be 1.1 eV [97].

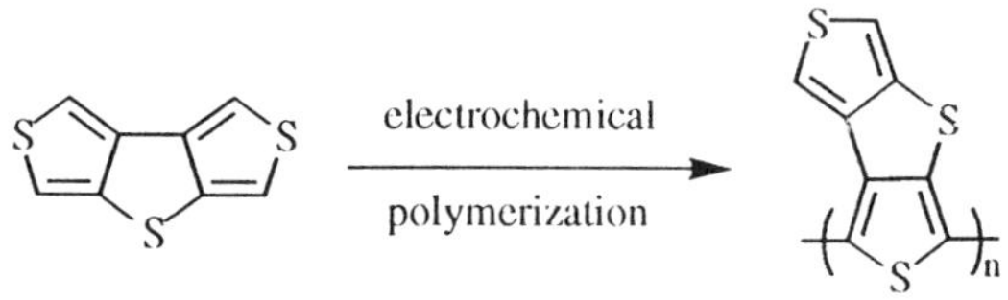

It should be noted that there are four α-positions through which polymerization can occur, and the structure of the polymer was not determined at that time. In 1992 a report appeared [118] that repeated that the doped polymer was transparent and the reduced form was opaque. The polymer was stable for a few hundred electrochemi-

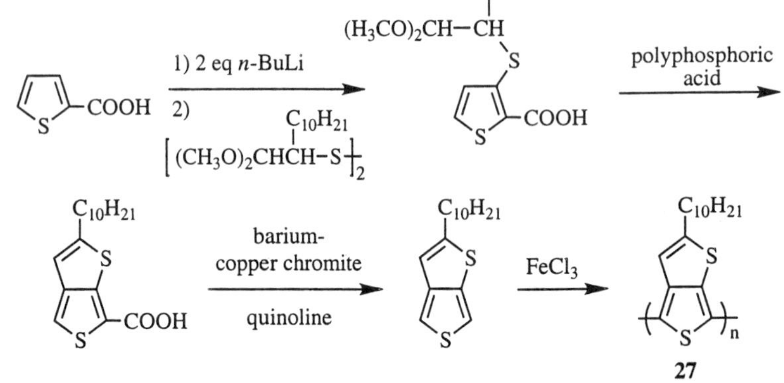

Scheme 9

cal cycles between −0.15 and 1.0 V (Ag/AgCl), but scanning to 1.35 V resulted in an irreversible oxidation. In addition it was shown that **57** also slowly degraded in the atmosphere [118]. Polymer **57** can also be prepared photochemically (254 nm/CH$_3$CN/CCl$_4$/Bu$_4$N$^+$Br$^-$) from the monomer **58**. Soluble oligomers were formed along with the insoluble polymer, and from the NMR spectrum it was tentatively concluded that the oligomers, and therefore the polymer, were bonded through the 2- and 5-positions in one thiophene ring. That is, the polymer had structure **57** [98]. Pressed pellet conductivity of the as-formed polymer (two-point probe) was 5 × 10^{-2} S/cm, and the band gap determined on films, which also can be formed by this photochemical polymerization, was confirmed to be 1.1 eV. X-ray diffraction studies indicated that there was some stacking, with an interplanar spacing of 3.5 Å [98]. In addition, it has been shown that it is possible to electrochemically *n*-dope the polymer in addition to *p*-doping for use in supercapacitors. By using tetraalkylammonium salts as supporting electrolyte in the polymerization on a carbon paper electrode, it was shown that comparable *n*- and *p*-doping levels could be achieved [119].

V. COPOLYMERS OF POLYISOTHIANAPHTHENE AND DERIVATIVES

A number of copolymers containing two thiophene rings and one benzo[*c*]thiophene or a derivative of benzo[*c*]thiophene in the repeat unit have been reported. The rationale was to combine a unit that prefers the quinonoid structure in a polymer with thiophenes that prefer the aromatic form in order to form a low band gap polymer. Lorey and Cava [74] prepared copolymer **35**, poly(benzo[*c*]thiophene-*alt*-bithiophene) by the route shown in Scheme 10, and this same polymer was prepared independently by Musmanni and Ferraris [75] and by Bäuerle and coworkers [76] by similar routes. Lorey and Cava and Bäuerle et al. prepared the 1,2-bis(2-thenoyl)benzene (**95**) from thiophthalic anhydride and phthalic anhydride, respectively, while Musmanni and Ferraris prepared **95** from *o*-phthalaldehyde (Scheme 10). Ring closure was with either Lawesson's reagent [74] or P$_4$S$_{10}$ [75]. The polymerization was carried out electrochemically. The purple (or dark violet) polymer **35** showed λ_{max} = 584 nm and a band-edge band gap of 1.58 eV [74], or λ_{max} = 534 nm and a band gap of 1.7 eV [75], or λ_{max} = 500 nm and a band gap of 1.77 eV [76], probably reflecting differences in conjugation length or molecular weight. Thus this polymer is not one with a particularly low band gap, and the observed band gap lies about two-thirds of the way between those of polyisothianaphthene (**1**) and polythiophene (**12**) [74–76].

Poly(naphtho[2,3-*c*]thiophene-*alt*-bithiophene) (**36**) was also prepared by Cava and coworkers [77,78] electrochemically (Bu$_4$N$^+$PF$_6^-$/CH$_2$Cl$_2$) as shown in Scheme 11. The polymer was gray in the neutral form and became blue when oxidized. Neutral **36** showed a broad absorption maximum at 476 nm with a shoulder in the near-infrared at 1240 nm. The band edge of this shoulder was at 1900 nm, and this provided a band gap of 0.65 eV. The electrical conductivity of an electrochemically doped film of **36** was 2 × 10^{-2} S/cm and that of a FeCl$_3$-doped film was 1 × 10^{-3} S/cm. Cyclic voltammetry of a film of polymer **36** on a platinum disk electrode showed two oxidation waves at 0.57 and 0.87 V (SCE) [77,78].

A rather interesting "nonclassical" copolymer, poly(4,6-di(2-thienyl) thieno[3,4-*c*][1,2,5]thiadiazole), **96**, was prepared electrochemically (R$_4$N$^+$ClO$_4^-$/CH$_3$CN) as shown in Scheme 12 [120]. As in the case of **19**, it was anticipated that there would be reduced torsional strain in these systems because the nitrogen atoms do not have a hydrogen atom bound to them. In addition, an X-ray structure determination of the monomer **97** showed short intermolecular S···N distances, 0.22 Å shorter than the sum of the van der Waals radii,

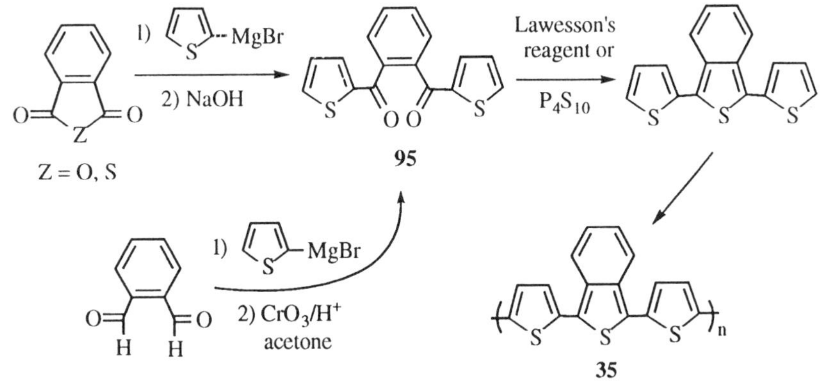

Scheme 10

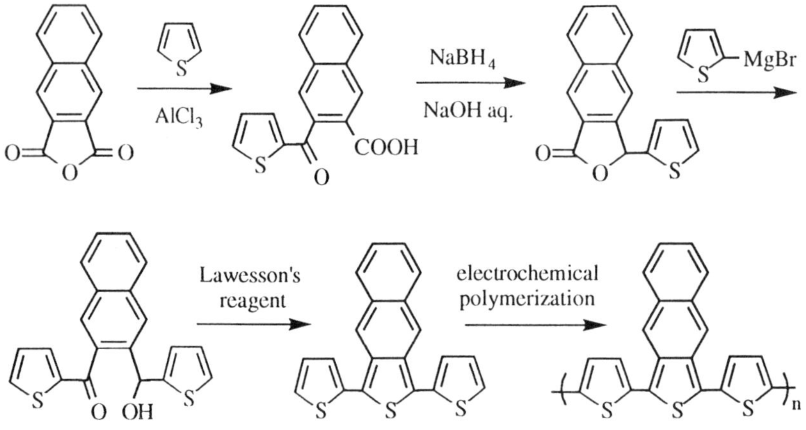

Scheme 11

suggesting that if the polymer **96** had similarly short intermolecular S····N distances it might show enhanced conductivity [120]. Electrochemical polymerization of **97** produced **96**, which could readily be both *p*- and *n*-doped. The oxidation and reduction peaks were at +0.7 and −1.1 V (SCE), respectively. From the onset of the *p*- and *n*-doping potentials the band gap was 0.9 eV, which agreed with the band-edge band gap obtained from the electronic spectrum. The absorption maximum was at 934 nm, and the band edge was at 1350 nm or 0.9 eV.

By using a sequence of reactions similar to those used in the preparation of **35**, Ferraris et al. [121] prepared the thienopyridine copolymer poly(thieno[3,4-*b*]pyridine-*alt*-bithiophene) (**98**) by electrochemical polymerization (LiBF$_4$/CH$_3$CN) of monomer **99**. This polymer showed λ_{max} = 560 nm and a band gap of 1.4 eV, about 0.3 eV lower than that of the benzo analog **35**. An X-ray structural determination of the monomer **99** showed a dihedral angle of 39° around the C—C bond joining a thiophene ring with the thienopyridine ring on the CH side and a dihedral angle of 3 ± 3° around the C—C bond joining a thiophene ring with the thienopyridine ring on the N side. This shows that in the monomer there is less steric interaction on the nitrogen side, and it is suggested that the lower band gap in **98** compared to **35** could be due to the smaller steric interactions in the polymer [121].

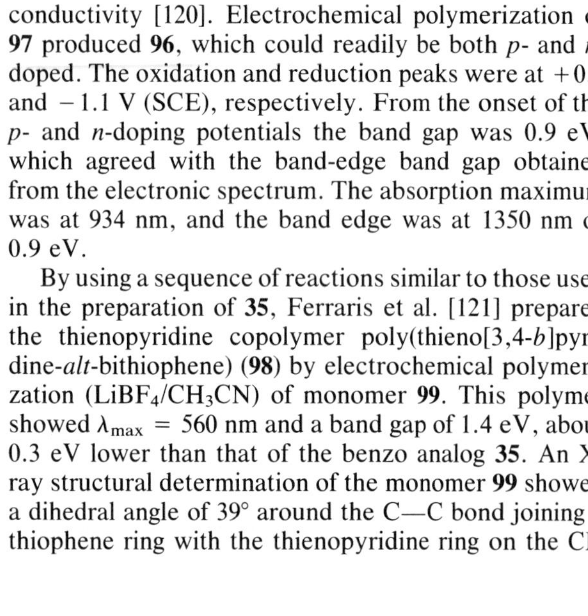

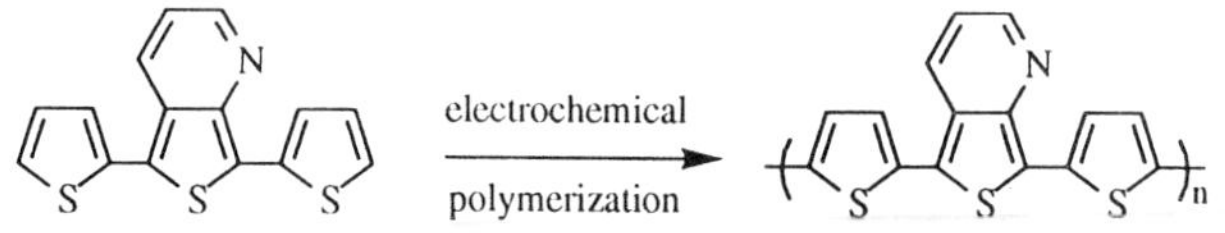

The pyrazine analog, namely poly(thieno[3,4-*b*]pyrazine-*alt*-bithiophene) (**100**, R = R′ = H), which now has both CH groups replaced by nitrogen atoms, and several derivatives containing alkyl groups on the pyrazine and/or thiophene rings (to improve processability), were prepared by Yamashita and coworkers [122]. The monomers with R = CH$_3$, R′ = H; R = C$_6$H$_{13}$, R′ = H; and R = C$_{13}$H$_{27}$, R′ = H were electrochemically polymerized (Bu$_4$N$^+$ClO$_4^-$/CH$_3$CN), while that with R = R′ = C$_{16}$H$_{33}$ was polymerized chemically with Cu(ClO$_4$)$_2$ (Scheme 13). The GPC molecular weight of this latter chemically prepared polymer was M_n = 5800 (M_w/M_n = 1.5). All polymers in the neutral state were dark blue-black and could be either *p*- or *n*-doped. The parent polymer **100**, R = R′ = H, had a band-edge band gap of 1.0 eV, which, as expected due to increased planarity, is 0.7 eV lower than that of the benzo analog **35** and 0.4 eV lower than that of the pyridino analog **98**. In CHCl$_3$ solution, **100**, R = R′ = C$_{16}$H$_{33}$, exhibited a band gap of 1.5 eV [122].

There is a 1995 report of the electrochemical preparation (Bu$_4$N$^+$ClO$_4^-$/CH$_3$CN or C$_6$H$_5$CN) of polymers **101a** and **101b** and of **102a** and **102b** from their monomers, and it is simply stated that these are low band gap

Scheme 12

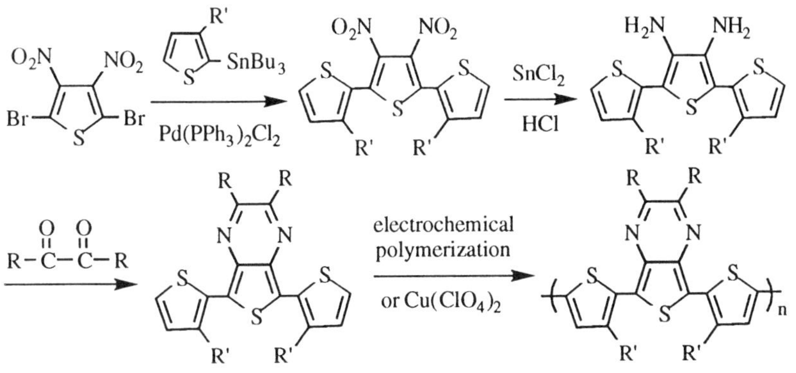

Scheme 13

polymers [123]. In this series there is one remarkable polymer (103) that shows an electrochemical band gap, determined from the onset of the p- and n-type doping, of about 0.3 eV. The dedoped state of this polymer

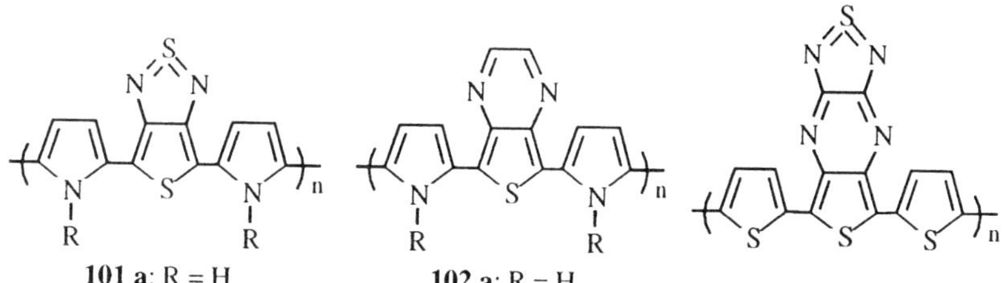

101 a: R = H
   b: R = CH$_3$

102 a: R = H
   b: R = CH$_3$

103

shows the onset of absorption below 0.5 eV; the exact onset is masked by the absorption of the ITO electrode substrate [123].

A related polymer (104), poly(benzo[1,2-c:4,5-c'] bis (1,2,5 - thiadiazole) - 4,8 - diyl - alt - bithiophene), which contains two 1,2,5-thiadiazole rings fused onto a benzene ring between two thiophene rings, has been reported to also have a band gap below 0.5 eV [124]. The synthesis using electrochemical polymerization (Bu$_4$N$^+$ BF$_4^-$/CH$_3$CN) is presented in Scheme 14. Polymer 104 can be both p- and n-doped electrochemically, and the difference in onset of these processes is small, indicative

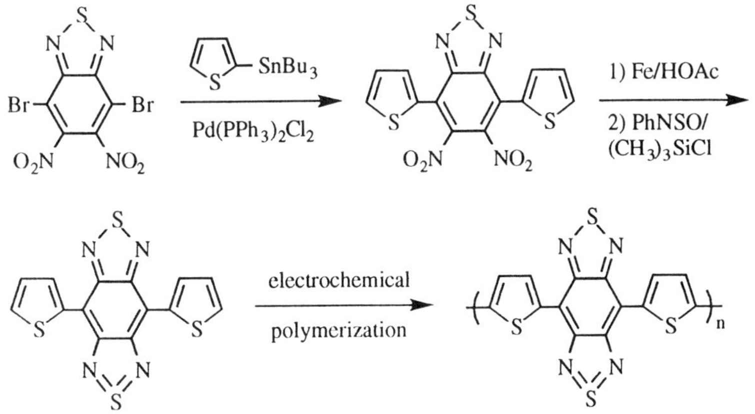

Scheme 14

of a low band gap polymer. From the absorption spectrum the band gap is below 0.5 eV, but because of the ITO absorption it cannot be determined accurately. Pressed pellet conductivities of dedoped and I_2-doped material were 5.0×10^{-5} and 5.6×10^{-3} S/cm, respectively [124].

These latter two polymers, **103** and **104**, are among the lowest band gap polymers reported to date.

A related polymer with only one thiadiazole ring, **105**, was prepared in a similar way [124] and was also found to be a low band gap polymer, with a band gap of 1.1 eV.

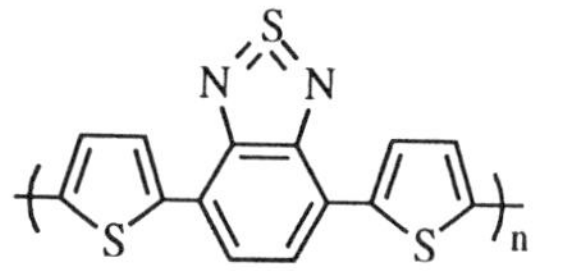

105

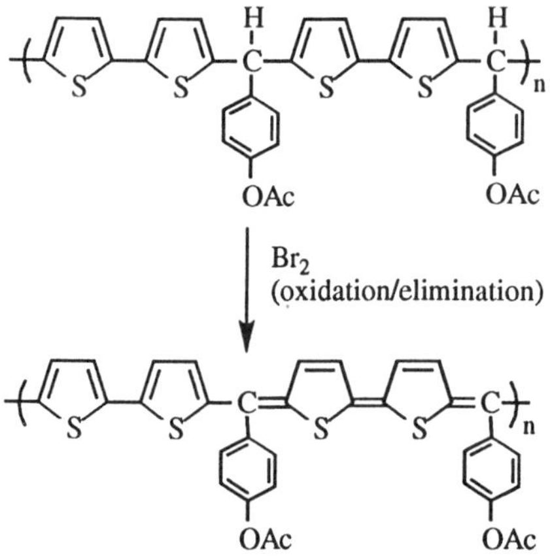

106

Scheme 15

VI. OTHER LOW BAND GAP CONDUCTING POLYMERS

One of the earliest low band gap polymers to be reported combined aromatic with quinonoid units in the backbone of the polymer [61,62]. It was prepared by oxidative elimination of a precursor polymer containing all aromatic rings, thus converting some into quinonoid rings. Scheme 15 shows the reaction to form polymer **106**, which is one of several related polymers differing in the number of thiophene and quinonoid units and substituents in the repeat unit. The syntheses involved preparation of the precursor polymer containing sp^3 carbon atoms and its reaction with bromine in an oxidation/elimination reaction [62]. The precursor polymer was blue with λ_{max} = 692 nm, and the new polymer, **106**, whose formation was followed by UV-vis-NIR and IR spectroscopy, was reported to have a band-edge band gap of 0.83 eV and was gray. Other similar thiophene-containing polymers had band gaps as low as 0.75 eV [61]. The oxidation/elimination could also be carried out with I_2, and other low band gap polymers containing thiophene, pyrrole, and carbazole rings were also prepared. In 1988 a paper appeared challenging the results and conclusions of this work and suggesting that the polymer obtained was doped and still contained bromine in the backbone [125]. Specifically, after the bromination reaction the polymer was reduced with hydrazine to remove the dopant bromine and then showed a band gap of 1.53 eV rather than 0.83 eV. Further, the NMR spectra revealed that methine hydrogens remained in the polymer. In addition, by comparing the polymer with model compounds, it was concluded that there were some dehydrogenated units, but the reaction was much more complicated than originally suggested [125].

Subsequent to these reports a number of publications have dealt with other syntheses of these types of polymers. A paper in 1991 reports that diol **107**, when treated with acid, provides the cation by loss of water, and this then forms polymeric material suggested to be **108** in the

107

108

doped state [126]. This polymer, which is still charged, showed near-infrared absorption at 1050 and 1200 nm. Treatment with base provided the dedoped material, but from the optical spectra it was said not to be a low band gap polymer. Several possible reasons for this were given, including that the material has only short chains, that it has a nonplanar backbone due to more sterically crowded conformations than is shown in structure **108**, or that the dedoping with base causes the addition of OH groups to the polymer chain, giving sp^3-hybridized carbon atoms along the backbone [126].

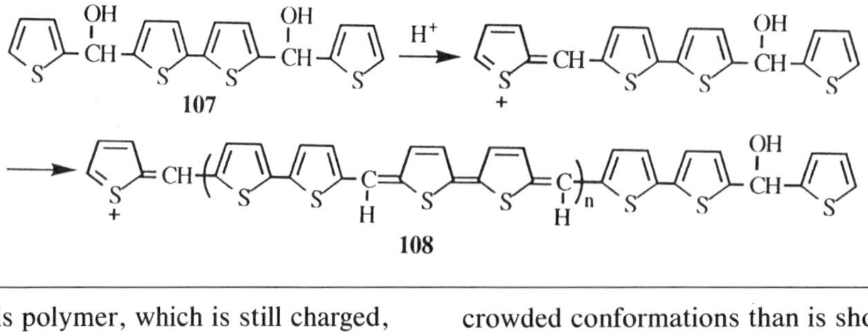

When substituents on the methine carbon were CH_3 or H, λ_{max} was about 1200 nm for the polymeric material formed in acid, while with aromatic rings as substituents on the methine carbon the absorption maximum was moved further into the near-infrared, to around 1500 nm [127]. The conductivities of these materials was in the range of 10^{-1}–10^{-10} S/cm depending on the pretreatment and the doping levels. Studies of the temperature dependence of the conductivity above 100 K gave an activation energy of 0.25–0.45 eV for **108**, but below 100 K there were deviations from the Arrhenius equation and variable range hopping dominated the charge transport [127]. In addition it was shown that the polymer could be electrochemically reduced and oxidized many times, and the reduced polymer showed much less absorption above 1000 nm. Furthermore, these polymers showed promising third-order nonlinear optical properties [127].

Monomers such as **109a** and **109b** with two aromatic thiophene rings, which were prepared by Knoevenagel-type condensation reactions (Scheme 16), were shown to form polymers **110a** and **110b** by electrochemical polymerization ($Bu_4N^+PF_6^-/CH_3CN$) [128]. Similarly, the pyrrole monomer **109c** provided polymer **110c**. This latter polymer showed a very broad near-IR absorption with a maximum around 1200–1300 nm, whereas in the reduced form this absorption disappeared and the longest wavelength feature was a shoulder at around 800 nm. From the band edge of this shoulder the band gap appeared to be around 1.3–1.4 eV [128].

In two 1995 papers Chen and Jenekhe [130,131] describe the synthesis of poly(heteroarylene-methylenes), which are precursor polymers, and then the syntheses of a series of low band gap poly(heteroarylene methines) from these. Band gaps as low as 1.14 eV were observed for these polymers [4,129]. The first paper gives the synthesis of 20 new methylene-bridged polythiophenes of

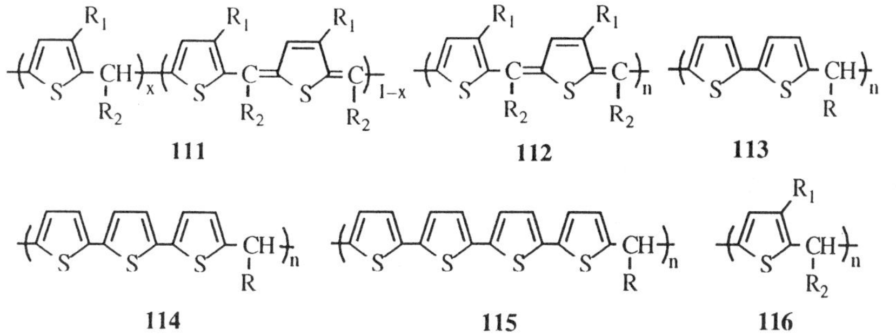

111 112 113

114 115 116

the type **111–115** [130]. Polymers **113–115** were obtained by reaction of bithienyl, terthienyl, or quaterthienyl with an aldehyde, RCH=O, in the presence of acid. However, attempts to prepare polymer **116** by reaction of an alkyl thiophene with an aldehyde in the presence of acid did not produce **116** but instead gave partially dehydrogenated material **111** and completely dehydrogenated polymer **112**. The molecular weights of the polymers varied from $M_n = 870$ to $M_n = 8600$. Interestingly, the completely conjugated polymers **112** with R_1 = H and R_2 = Ph or p-NO_2-C_6H_4 provided λ_{max} = 471 and 485 nm, respectively, whereas polymers **111** with $R_1 = CH_3$ and R_2 = Ph or p-NO_2-C_6H_4, which are less conjugated, gave maxima at 451 and 450 nm, respectively.

In the second paper [131], polymers prepared as above but that still contained saturated carbon atoms were reported to be oxidized by 2,3-dichloro-5,6-dicyano-1,4-benzoquinone (DDQ) to produce polymers **112**, **117**, and **118**, while the deep green polymer **119** was made by the treatment of diol **120** with formic acid. Of

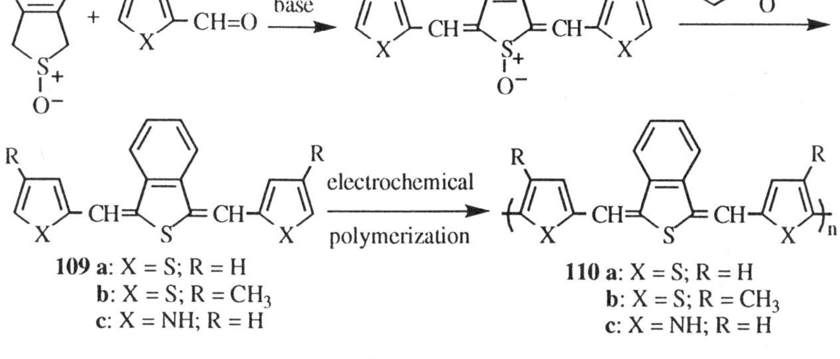

109 a: X = S; R = H
b: X = S; R = CH_3
c: X = NH; R = H

110 a: X = S; R = H
b: X = S; R = CH_3
c: X = NH; R = H

Scheme 16

the variety of polymers prepared this way, the two that showed the longest wavelength λ_{max} were **119** and **117**, where R = p-NO$_2$-C$_6$H$_4$, which provided film λ_{max} = 652 and 675 nm, respectively. The four low band gap materials ($E_g <$ 1.5 eV, determined from the band edges) and their respective band gaps were **117**, R = p-C$_7$H$_{15}$-O-C$_6$H$_4$, E_g = 1.14 eV; **117**, R = Ph, E_g = 1.27 eV; **119**, E_g = 1.31 eV; and **117**, R = p-NO$_2$-C$_6$H$_4$, E_g = 1.45 eV. The polymers were characterized by NMR, IR, and ESR spectroscopy in addition to UV-vis spectroscopy and by viscosity and thermal measurements. It was determined that the optical and electrochemical properties of these polymers depended on the size of the quinonoid unit and on the nature of the substituents. The electrochemistry of the polymers was also examined, and the electrochemical band gaps for the four low band gap polymers above (**112, 117–119**) were 1.28, 1.32, 1.28, and 1.40 eV, respectively, in close agreement with the optical values [131]. The electrical conductivity (four-point probe) of the undoped low band gap polymers was determined to be less than 10^{-7} S/cm. Clearly, these polymers are rather poor intrinsic conductors. When doped with I$_2$ vapor the conductivities of **117**, R = Ph; **117**, R = p-NO$_2$-C$_6$H$_4$, and **117**, R = p-C$_7$H$_{15}$-O-C$_6$H$_4$ were 6 × 10^{-3}, 2.2 × 10^{-2}, and 1.6 × 10^{-1} S/cm, respectively. Upon doping, **117** with R = p-C$_7$H$_{15}$-O-C$_6$H$_4$ showed a new near-IR optical absorption at 1596 nm [131].

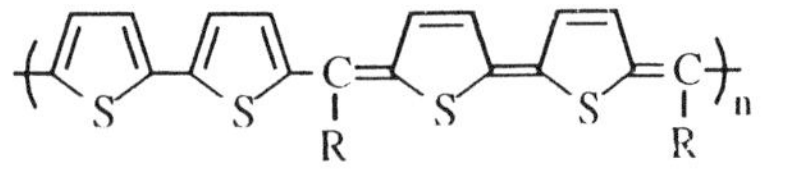

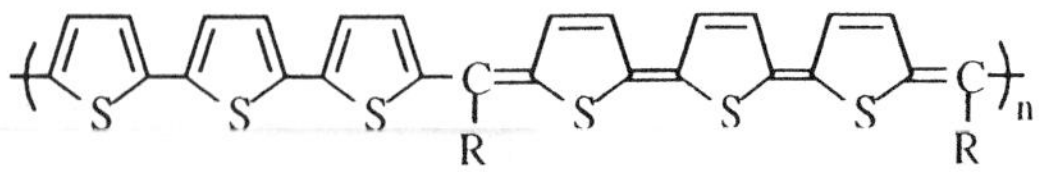

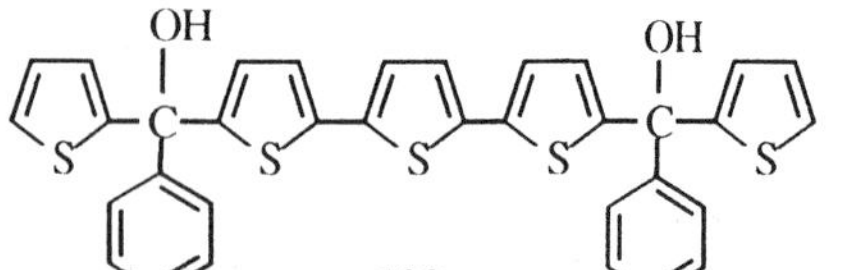

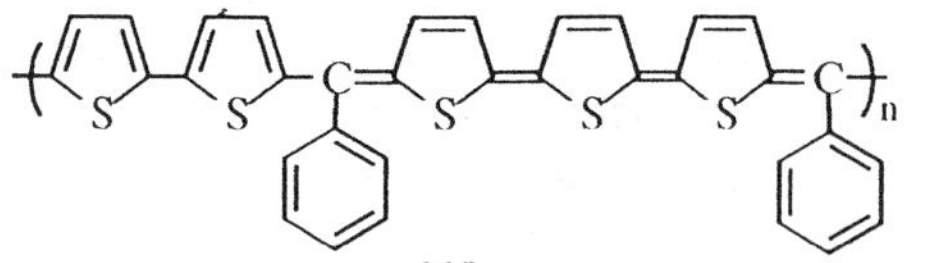

Another class of low band gap polymers first prepared and studied by Lambert and Ferraris [92–94] contain electron-deficient carbon atoms between the thiophene rings as shown in the resonance structures for **54** and **55**. These were prepared electrochemically (Ph-No$_2$/Bu$_4$N$^+$ BF$_4^-$) from the respective monomers. Polymer **54** shows λ_{max} = 740 and 425 nm, and the band-edge band gap is ≤1.2 eV. From the cyclic voltammetry of **54**, E_{pa} is +0.75 V (SCE), which is very close to that for poly(α,α'-bithiophene), indicating that the carbonyl group has little effect on the HOMO (top of the valence band), which should sit at a node. It was also suggested that the 1.2 eV (1020 nm) gap is due to the aromatic HOMO-to-quinonoid LUMO transition [92]. Similar studies on polymer **55** produced electrochemically (Ph-NO$_2$/Bu$_4$N$^+$BF$_4^-$) from its monomer showed the band-edge band gap at ≈0.8 eV [93]. This is reasonable because there should be a greater partial positive charge on the central carbon atom of the repeat unit and hence it should have a smaller HOMO–LUMO gap. Once again the cyclic voltammetry (which showed both p- and n-doping) provided E_{pa} = +0.76 V (SCE) or about the same as **54** and poly(α,α'-bithiophene), again demonstrating that the central carbon sits on a node in the HOMO.

A series of related polymers **120** and **121** were reported by Yamashita and coworkers [132], and the absorption maxima of the **120** series was 614–689 nm for films on ITO glass. The band-edge band gap of **120**, R = S(CH$_2$)$_5$CH$_3$, which was soluble and showed λ_{max} = 626 nm in CHCl$_3$ solution and 653 nm for the film, was 1.4 eV (film), demonstrating that many of the polymers in that series are indeed low band gap. The number-average molecular weight of **120**, R = S(CH$_2$)$_5$CH$_3$, was 2700 (GPC, polystyrene standards), giving a degree of polymerization of only about 11. Electrical conductivities of up to 52 S/cm for **120**, R = COOCH$_3$ (BF$_4^-$-doped), and 33 S/cm for **120**, R = R = benzo (PF$_6^-$-doped). The absorption maxima for the members of the series **121** indicate that they are not low band gap [132].

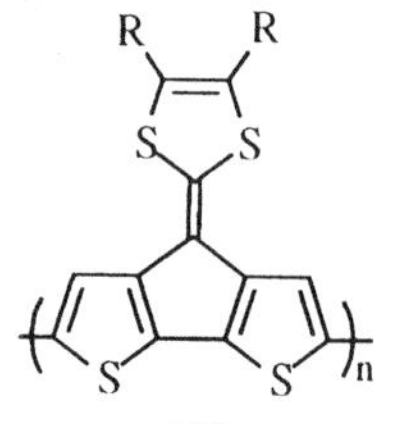

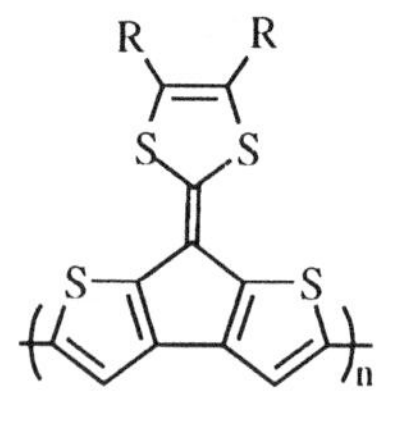

Another related polymer, poly(cyclopenta[2,1-b:4,3-b']dithiophen-4-[cyano(nonafluorobutylsulfonyl)methylidene]) (**122**) with two different highly electron-withdrawing groups, which can be both n- and p-doped, was recently reported [133]. It was prepared from the monomer by electrochemical polymerization (Me$_4$N$^+$CF$_3$SO$_3^-$/CH$_3$CN). The rather low band gap determined by the separation of the n- and p-doping regions is about 0.68 eV. This agrees with the band-edge band gap of 0.67 eV obtained from the absorption maximum at approximately 980 nm, which was said to have an onset at 1850 nm. The neutral polymer was red, and both the n- and p-doped materials were purple [133].

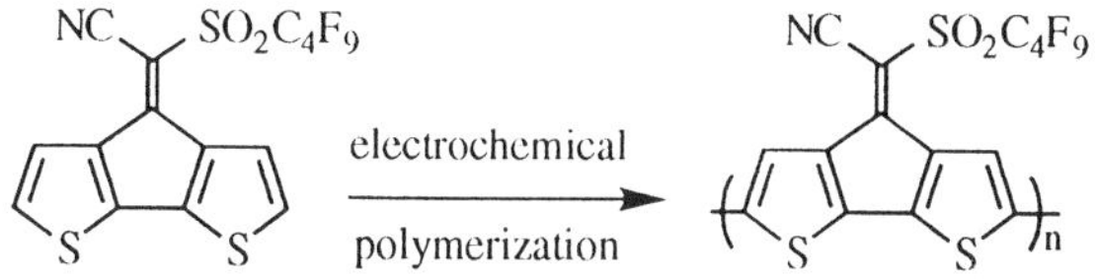

In a series of interesting papers, Roncali and coworkers [134–136] have argued that the lowering of the band gap in polymers such as **54**, **55**, and other related derivatives is due mainly to rigidifying of the polymer backbone. This was initially based on electrochemically (Bu$_4$N$^+$ PF$_6^-$/CH$_3$CN) preparing poly(4,4-ethylenedioxy-4H-cyclopenta[2,1-b:3,4-b']dithiophene) (**123**) and showing that this material, which does not have an sp^2-hybridized carbon between the rings, still has a low

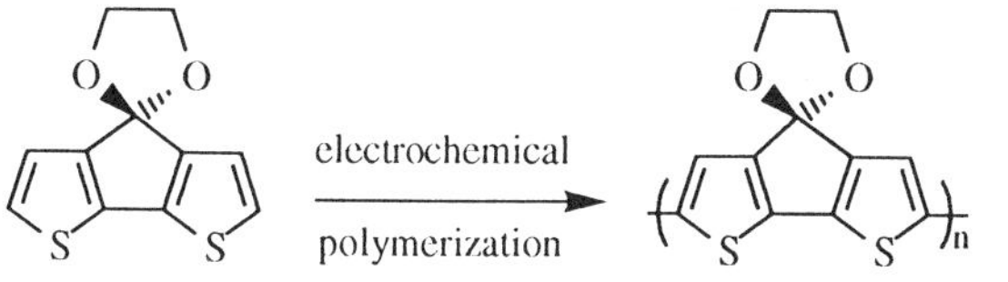

band gap. The band gap obtained from the cyclic voltammetry for n- and p-doping is 0.9 eV, and from the onset of optical absorption it is 1.2 eV. Interestingly, the monomer shows a low oxidation wave at 1.14 V (vs. SCE), compared to 1.36 V for the corresponding ketone. This suggests that the ketal group donates electrons into the bithiophene ring system. In addition, the dithioketal analog of **123** gave an electrochemical band gap of 1.40 eV [134]. As further evidence that making the polymer backbone more rigid results in band gap lowering, a comparison was made between polymers **124** and **125** [135,137,139]. Whereas polymer **124** has a band gap of 1.8 eV [139], the band gap of polymer **125** has been lowered to 1.40 eV, or a lowering of about 0.4 eV [135,137,138].

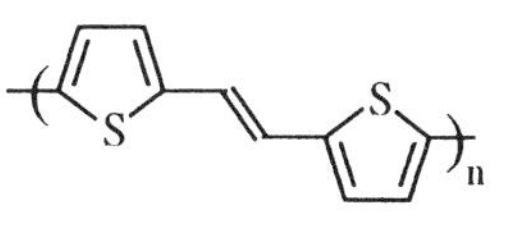

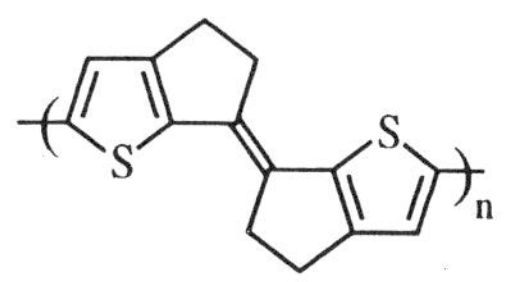

The rigid terthienyl polymer **126** was prepared electrochemically (Bu$_4$N$^+$PF$_6^-$/CH$_3$CN) from the corresponding monomer and shown to have an extremely low band gap [140–142]. Indeed, while the λ_{max} of terthienyl (**127**) is 380 nm, that of the rigidified terthienyl monomer **128** is 750 nm [135,138,143]. The incredibly low band gap obtained from the n- and p-doping in the cyclic voltammogram for polymer **126** was 0.8–1.0 eV, which agreed with that obtained from the optical spectrum, 1.0–1.1 eV [143]. The pressed pellet conductivity of the doped polymer, which was prepared by first casting a film on ITO glass and then doping electrochemically, was 0.1 S/cm, which is a high value for a polyterthienyl [143].

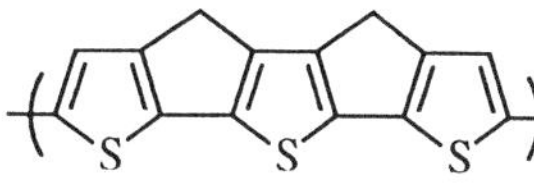

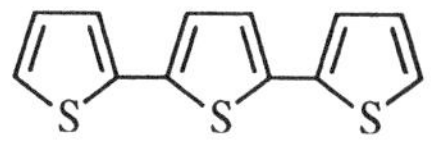

As discussed above, another class of low band gap conducting polymers that have been studied are ladder or ribbon polymers such as poly(peri-naphthalene) (**49**), polyacene (**50**), and related systems such as **51**. In 1985 a report appeared on the synthesis of poly(peri-naphthalene) (**49**) that was formed by pyrolysis of 3,4,9,10-perylenetetracarboxylic dianhydride at temperatures between 530 and 900°C as shown in Scheme 17 [82,144]. Very highly conducting undoped films were reported ($\sigma \approx 250$ S/cm) [144], but elemental analysis revealed that the formula was about C$_{10}$H$_{0.8}$, strongly suggesting partially graphitic material. Subsequent work provided polymer with $\sigma = 10^{-2}$ S/cm when the pyrolysis was carried out at 530°C and $\sigma = 15$ S/cm when the pyrolysis was carried out at 800–900°C. These were intrinsic conductivities, since they were not changed upon attempted compensation with NH$_3$. The 530°C pyrolysate showed an elemental composition with 4.1% oxygen still remaining, but the 900°C pyrolysate had an acceptable elemental composition for poly(peri-naphthalene) (**49**). X-ray diffraction studies suggested, however, that the pyrolysis did not go strictly as shown in Scheme 17 but was of a more statistical nature. The experimental band gap was not given [82].

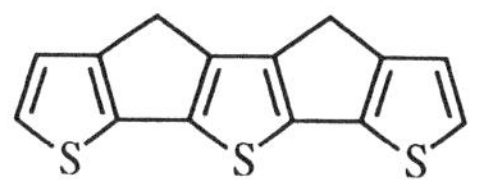

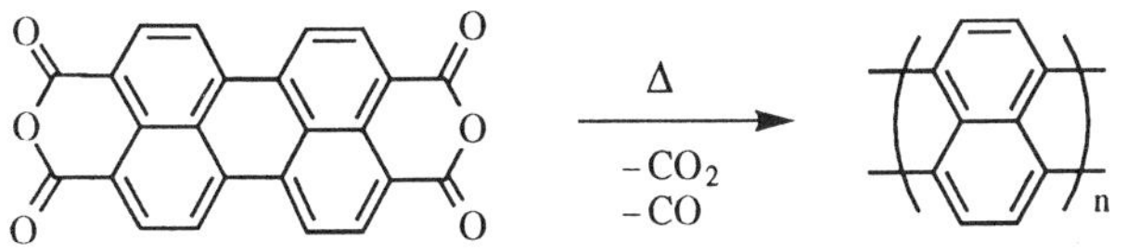

Scheme 17

Although polyacene (**50**) has not been prepared [90], related heterocyclic ladder polymers have been reported. Thus, for example, the ladder polymers poly(1,6-dihydropyrazino [2,3-*g*] quinoxaline-2,3,8-triyl - 7(2*H*)-ylidene-7,8-dimethylidene) (**51**, X = NH, Y = N), where R = H, $C_{11}H_{23}$, or $OC_{20}H_{41}$; poly(2*H*,11*H*-bis[1,4]oxazino[3,2-*b*:3′,2′-*m*]triphenodioxazine-3,12-diyl-2,11-diylidene-11,12-bis(methylidene)) (**51**, X = O, Y = N); and the corresponding polymer where sulfur has replaced oxygen, **51**, X = S, Y = N, have been prepared [145]. The band gaps were not reported, but the ESR spectra were studied using the undoped polymers, which are paramagnetic, and the relationship of these spectra to other properties such as conductivity was discussed [145].

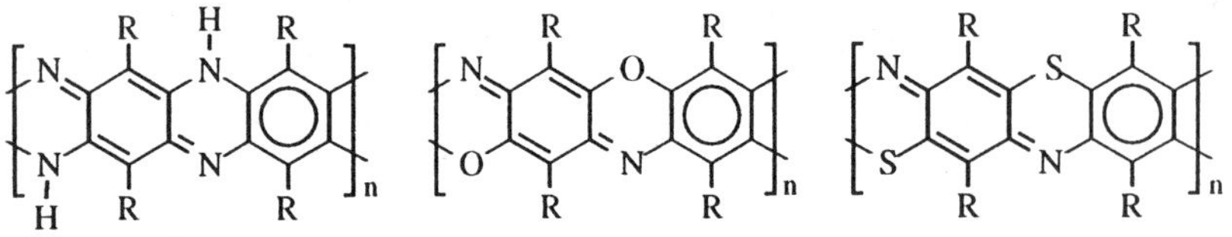

Doping of polyphenothiazine and polyphenoxazine (**51**, X = O, Y = N and **51**, X = S, Y = N, respectively) with I_2, AsF_5, H_2SO_4, and $ClSO_3H$ allowed the conductivity to go from 5.0×10^{-6} S/cm for **51**, X = S, Y = N and 1.7×10^{-8} S cm^{-1} for **51**, X = O, Y = N to conductivities up to 2.1×10^{-2} S/cm for **51**, X = S, Y = N when doped with $ClSO_3H$ [146]. Both doped and pristine polymers exhibited ESR spectra, with the number of spins increasing only moderately in going from the pristine to the doped state. Once again the band gap was not reported [146].

The ladder polymer **52(p,m)** (R = *p*-$C_{10}H_{21}$-C_6H_4; R′ = C_6H_{13}) has recently been prepared by oxidation of the reduced precursor polymer as shown in Scheme 18 [89,90]. The band-edge band gap of this substituted polymer **52(p,m)** (R = *p*-$C_{10}H_{21}$-C_6H_4, R′ = C_6H_{13}), estimated from the peak at 605 nm, was about 1.5 eV [89]. The precursor polymer was orange, and after oxidation with DDQ the polymer solution was deep blue. ^{1}H NMR spectroscopy indicated about 90% conversion, that is, the aliphatic proton at δ = 5.03 ppm decreased to about 10% of its original size. The polymer did not show an ESR spectrum and decomposed after several weeks of exposure to the air [89]. The precursor polymer had a GPC molecular weight, M_n, of 11,000 (M_w/M_n = 1.9; polystyrene standards) [147]. For the repeat unit shown in Scheme 18 and in structure **52(p,m)** (R = *p*-$C_{10}H_{21}$-C_6H_4, R′ = C_6H_{13}), this molecular weight means that there are only about seven of these units per chain.

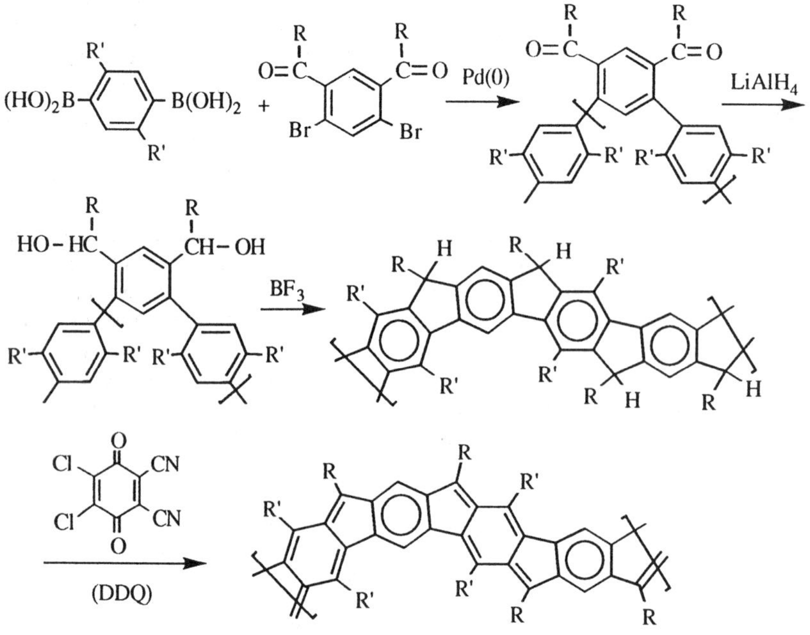

52(p,m): R = *p*-C_6H_4-$C_{10}H_{21}$, R′ = C_6H_{13}

Scheme 18

Direct attachment of alkoxy groups to the thiophene rings in polythiophenes has been shown to give rise to conducting polymers with considerably lower band gaps. Thus, for example, poly(ethylenedioxythiophene) (**129**) has a band gap of 1.5–1.6 eV, a lowering of about 0.5 eV compared to polythiophene [148,149]. This material also switches from opaque black or blue-black to transparent sky blue [149] or bronze in reflection [148]. Poly(4,4′-dimethoxybithiophene) (**130**) was reported to have a band gap of 1.6 eV [150].

When poly(thienylene vinylene) (**66**, R = H, band gap = 1.74 or 1.64 eV) [103,104] is substituted with alkoxy groups, it too shows considerably reduced band gaps. Poly(3-methoxythienylene vinylene) (**66**, R = OCH_3) was reported to have an electrochemical band gap of 1.32 eV [105] and an optical band gap of 1.37 eV [104]; poly(3-ethoxythienylene vinylene) (**66**, R = OC_2H_5) was reported to have electrochemical and optical band gaps of 1.31 and 1.48 eV, respectively [104]; and poly(3,4-dibutoxythienylene vinylene) (**131**) was reported to have a band gap of 1.62 eV [151].

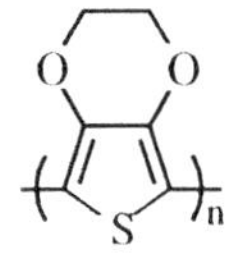

129, 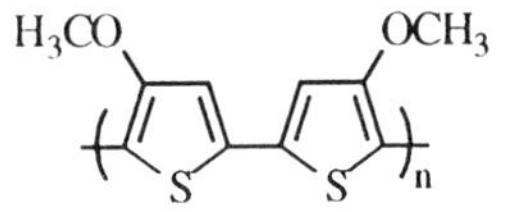 130, 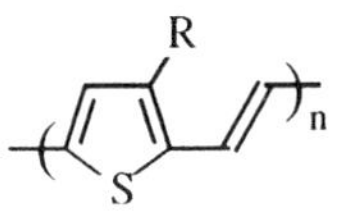 66, 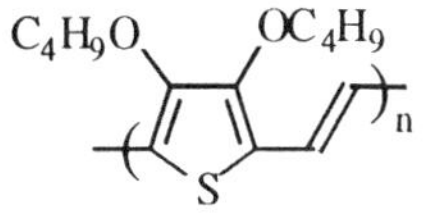 131

Recently these concepts were combined, and poly(*trans*-bis(ethylenedioxythienylene vinylene) (**132**) was reported by Sotzing et al. [152], Sotzing and Reynolds [153], and Fu and Elsenbaumer [154,155]. The polymer was prepared by electrochemical polymerization of the monomer ($Bu_4N^+ClO_4^-/CH_3CN$) or by chemical polymerization using $FeCl_3$, and the synthesis is shown in Scheme 19 [152–155]. It was reported to have a band gap of 1.4 eV [152,153] or 1.48 eV [154] from the onset of the band at 590 nm. From the onset of electrochemical *p*- and *n*-doping, the band gap was determined to be 1.17 eV [155]. It was reported that the deep purple color of the dedoped polymer goes to a transmissive blue upon oxidative doping. Further it was observed that switching from the oxidized to the reduced state took about 500 ms but the reverse was somewhat slower, 1.5 s, and after 400 cycles between oxidized and reduced forms the polymer retained about 50% of its electroactivity and absorption contrast [153]. The four-point probe pressed pellet conductivity of **132** was 20 S/cm [155].

Poly(3,4-dialkoxythienylene vinylene) (**133**) was prepared by Cheng and Elsenbaumer by an interesting chemical route through a precursor polymer containing sulfoxide groups that were thermally eliminated. The synthesis of **133** is shown in Scheme 20 [156,157]. The band-edge band gap (λ_{max} = 680 nm) of cast films of the butoxy derivative **131** was 1.22 eV and was thus considerably lower than that for other preparations of this polymer via the Grignard route or by $TiCl_4$/Zn McMurry coupling of the corresponding dialdehyde. The GPC molecular weight (M_n = 87,000; M_w/M_n = 2.2) was somewhat higher than for the other preparations, but the absorption maximum, 702 nm, was about 100 nm red-shifted compared to the others, and the $FeCl_3$-doped conductivity (σ = 15 S/cm) was more than one order of magnitude higher [156,157]. The band gap observed is consistent with the band gap of poly(3-alkoxythienylene vinylenes) (**66**, R = *O*-alkyl) reported previously [104,105]. Other polymers prepared were **133**, R = Me; **133**, R = R = $-CH_2-CH_2-$; and **133**, R = R = $-CH_2-CH(C_6H_{13})-$, and their properties were comparable to those of **131** [156,157].

In 1992 and 1993 two papers appeared [158,159] reporting the preparation of a new class of polymers with band gaps down to 0.5 eV [159]. They were polysquaraines and polycroconaines, and the low band gap arises

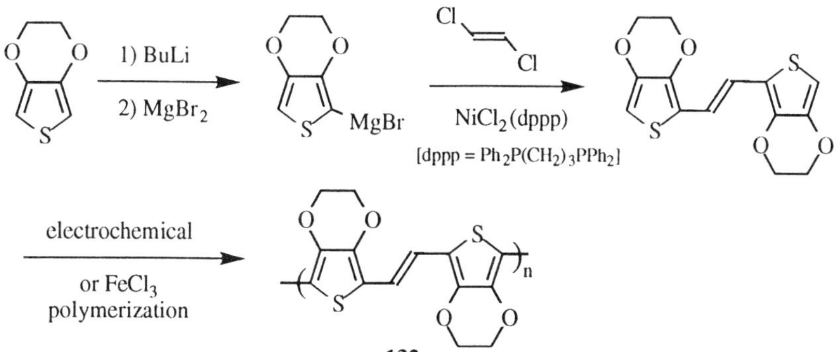

Scheme 19: Synthesis of 132 via BuLi, MgBr₂, then NiCl₂(dppp) [dppp = Ph₂P(CH₂)₃PPh₂] coupling with dichloroethylene, followed by electrochemical or FeCl₃ polymerization.

Scheme 19

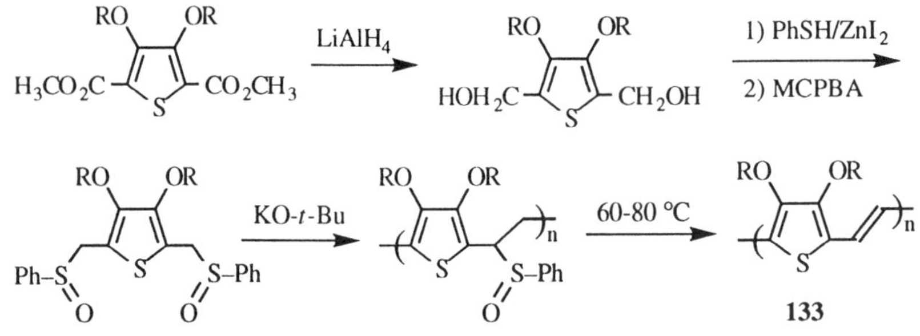

Scheme 20

from the regular alternation of strong donor and acceptor groups within the conjugated polymer backbone. Thus the strong acceptors squaric acid (**134**) and croconic acid (**135**) were incorporated into the polymers along with donor moieties **136** and **137** containing alkyl groups for solubility. The polymers **138–141** were produced upon reaction of the acceptors and donors as shown in Scheme 21 for **138** and **139** [158,159]. The unsubstituted polymers were insoluble, and the solubility could be increased with alkyl groups. GPC (polystyrene standards) of **138**, R = C_7H_{15} and **138**, R = $C_{12}H_{25}$ gave molecular weights corresponding to 15 and 25 repeat units. The band-edge band gaps of **138**, R = $C_{12}H_{25}$; **139**, R_1 = CH_3 R_2 = $C_{12}H_{25}$; **140**, R_1 = CH_3, R_2 = $C_{12}H_{25}$; and **141**, R = $C_{12}H_{25}$, were 1.15, 0.5, 0.8, and 1.2 eV, respectively, based on the vis-NIR absorption maxima of 919, 1378, 992, and 919 nm, respectively. The conductivities of the pristine polymer films were 10^{-7}, 10^{-5}, 10^{-7}, and 10^{-9} S/cm, respectively. Doping with I_2 resulted in increased conductivities, up to 1 S/cm, and doping with DDQ also gave values of approximately 1 S/cm. The polymers decomposed upon standing in air in the daylight, apparently producing lower molecular weight oligomers [158]. Related water-soluble polymers **142–144** have also been reported [160]. They are black solids that dissolve in water and methanol to produce green-blue solutions. Polymer **143** was reported to have a band-edge band gap of 0.7 eV, which agrees with the band-edge band gap obtained from the photoconductivity spectrum, since these polymers show photoconductivity. In addition, **142** and **144** have band gaps of 0.7 and 1.2 eV, respectively [160]. Interestingly, these water-soluble polymers showed better air stability than the other polymers such as **138**, R = C_7H_{15}, which dope in

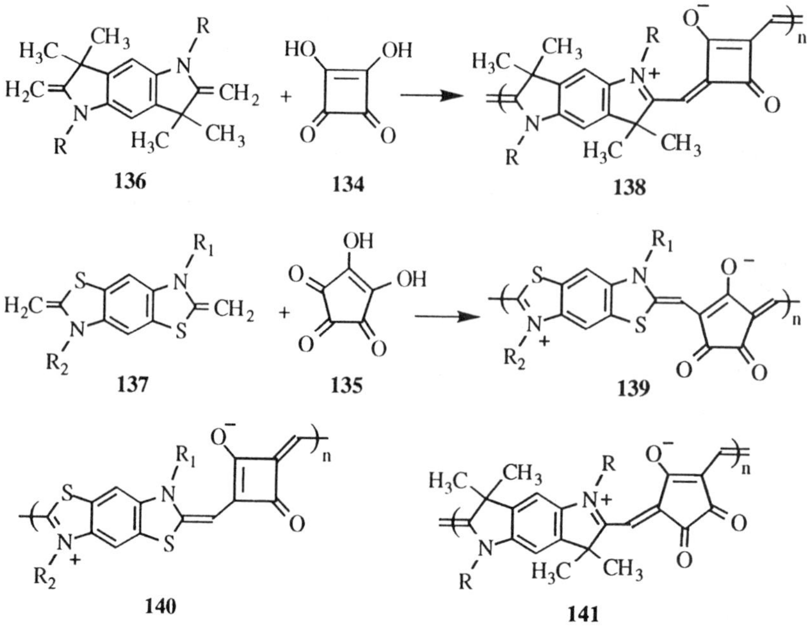

Scheme 21

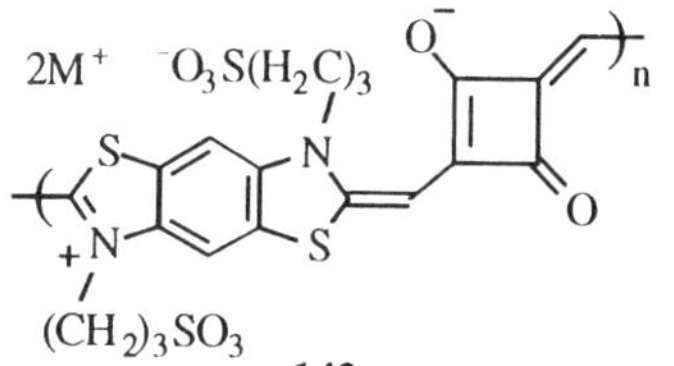

the presence of air or oxygen. Doping with I_2 or DDQ gave conductivities of up to 1 S/cm for **142–144** [160].

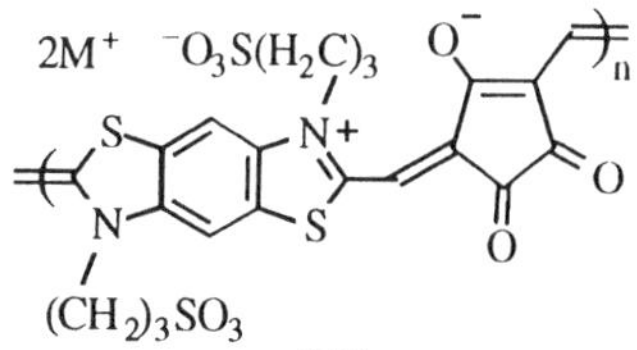

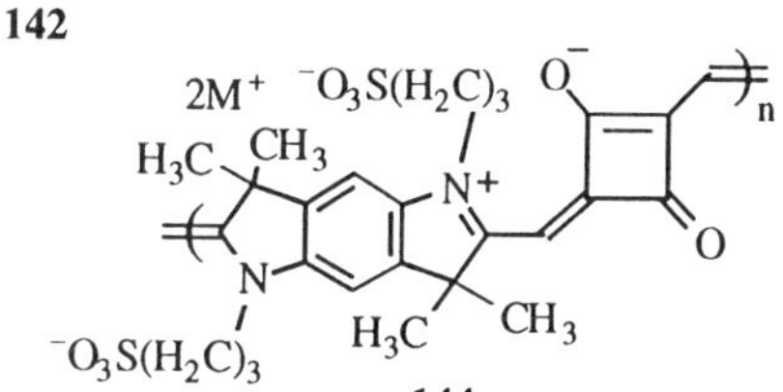

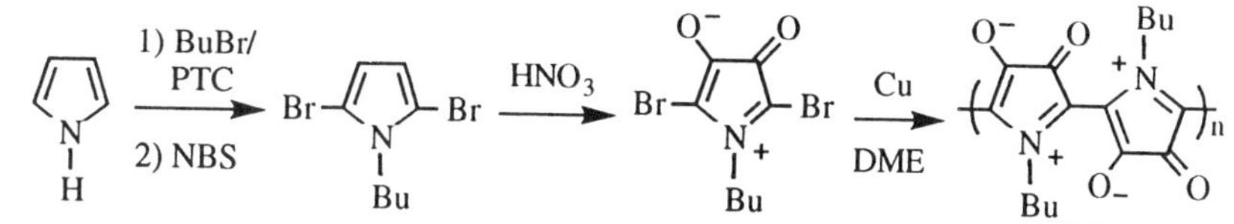

An interesting low band gap zwitterionic polypyrrole-derived polymer **145** has been reported by Brockmann and Tour [161,162]. The preparation is shown in Scheme 22. **145** shows an intrinsic pressed pellet (four-point probe) conductivity of 1.4×10^{-5} S/cm, and this increased only slightly upon iodine doping, to 4.2×10^{-4} S/cm. It showed interesting solvatochromic behavior, having λ_{max} = 520 nm in CCl_4, 512 nm in THF, 482 nm in acetone, and 498 nm in H_2SO_4. When NaOH was added to a THF solution of **145**, the red solution became pale orange and then brown with λ_{max} = 881 nm. Lewis basic solvents also gave striking red shifts to 901 nm in hexamethylphosphoric triamide (HMPA) and 746 nm in N-methylpyrrolidone (NMP). These spectral changes were reversible with the addition of HCl. It was suggested that the structure of the polymer that resulted from Lewis base reaction and showed the long-wavelength absorption indicative of a low band gap polymer is that shown in structure **146** in Scheme 23. It was also shown that polymer **145** could be partially hydrogenated to a new polymer that also showed strong solvatochromism. It was yellow in THF solution and became blue upon the addtion of NaOH, and this could be reversed upon the addition of HCl [161,162]. Polymer **145** with a dodecyl group instead of a butyl group on the nitrogen was prepared, and it showed solvatochromism comparable to that of the butyl polymer. Interestingly, a cast film of this dodecyl polymer was maroon, and the color did

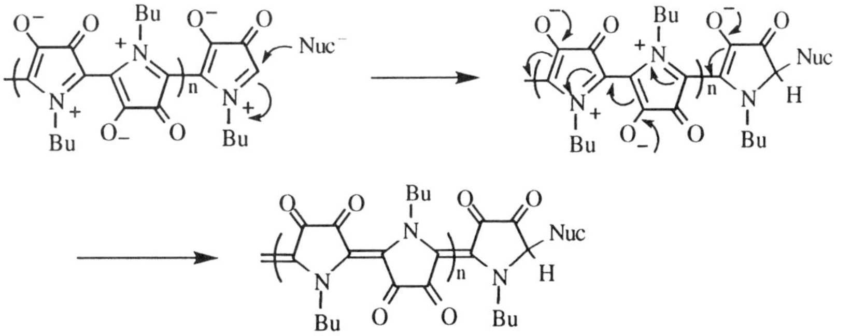

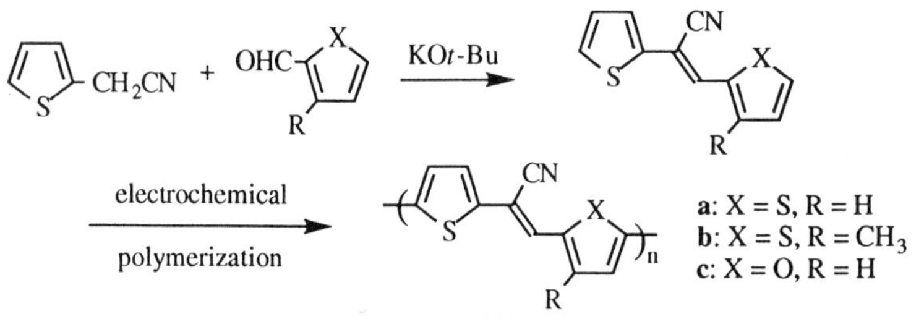

Scheme 24

not change on exposure to acid or base solutions [162]. A polyether chain was also introduced in place of the butyl group in **145**, and the effect of metal chelation was studied [162].

Another class of low band gap polymers (**147**) recently reported by Roncali and coworkers contains a cyano group on the double bond of an electrochemically ($Bu_4N^+ \ PF_6^-/CH_3CN$) polymerized E-1,2-(di-2-thienyl)ethylene or E-1,2-(di-2-furyl)ethylene. The syntheses are shown in Scheme 24 [163]. The band-edge band gaps for **147a**, **147b**, and **147c** were 0.5–0.6 eV, 1.50 eV, and 1.40 eV, respectively. It is remarkable that the cyano group on the double bond in **147a** lowers the band gap 1.2–1.3 eV from that of the parent polydithienylethylene (**124**) [139]. Polymers **147a** and **147c** showed pressed pellet conductivities (four-point probe) of 7×10^{-2} and 4.8 S/cm, respectively [163].

VII. EPILOGUE AND CONCLUSIONS

It is clear that during the past decade or so, considerable progress has been made in an attempt to obtain very low band gap conducting polymers. However, the dream is to find a polymer that will have a zero band gap and will thus be an inherent metal without the need for doping. Currently the lowest band gaps reported are about 0.5 eV, which means that there is still a way to go. Scientists have also tried to make these polymers as transparent and colorless as possible in the conducting state. A number of these polymers have been prepared that as thin films are very lightly colored and transparent. Finally, a recurring problem with many of these low band gap conducting polymers is their poor stability, either doped or undoped, in ambient air. Clearly we are not yet at the stage where these materials are stable enough for long-term applications.

ACKNOWLEDGMENTS

I wish to thank my students and postdoctoral associates whose research at the University of Texas at Arlington was referenced in this manuscript for their hard work. I also thank the Defense Advanced Research Projects Agency through a grant monitored by the office of Naval Research and the Robert A. Welch Foundation for financial support of our work.

REFERENCES

1. F. Wudl, M. Kobayashi, and A. J. Heeger, *J. Org. Chem.* 49:3382 (1984).
2. M.-H. Whangbo, R. Hoffmann, and R. B. Woodward, *Proc. Rou. Soc. Lond. A 366*:23 (1979).
3. J. L. Brédas and R. H. Baughman, *J. Polym. Sci., Polym. Lett. Ed.* 21:475 (1983).
4. G. Grem, G. Leditzky, B. Ullrich, and G. Leising, *Synth. Met.* 51:383 (1992).
5. G. Grem, G. Leditzky, B. Ullrich, and G. Leising, *Adv. Mater.* 4:36 (1992).
6. K. Yoshino, T. Takiguchi, S. Hayashi, D. H. Park, and R.-i. Sugimoto, *Jpn. J. Appl. Phys.* 25:881 (1986).
7. J. Obrzut and F. E. Karasz, *J. Chem. Phys.* 87:2349 (1987).
8. M.-A. Sato, S. Tanaka, and K. Kaeriyama, *Synth. Met.* 14:279 (1986).
9. T. C. Chung, J. H. Kaufman, A. J. Heeger, and F. Wudl, *Phys. Rev. B 30*:702 (1984).
10. J.-L. Brédas, in *Handbook of Conducting Polymers* (T. A. Skotheim, ed.), Marcel Dekker, New York, 1986, Vol. 2, p. 859.
11. A. J. Heeger, in *Handbook of Conducting Polymers* (T. A. Skotheim, ed.), Marcel Dekker, New York, 1986, Vol. 2, p. 729.
12. J. H. Burroughes and R. H. Friend, in *Conjugated Polymers: The Novel Science and Technology of Highly Conducting and Nonlinear Optically Active Materials* (J. L. Brédas and R. Silbey, eds.), Kluwer Academic, Dordrecht, The Netherlands, 1991, p. 555.
13. F. Wudl, M. Kobayashi, and A. Heeger, Eur. Patent Appl. 0 164 974 (1985).

14. M. P. Cava, N. M. Pollack, O. A. Mamer, and M. J. Mitchell, *J. Am. Chem. Soc. 36*:3932 (1971).
15. M. Kobayashi, N. Colaneri, M. Boysel, F. Wudl, and A. J. Heeger, *J. Chem. Phys. 82*:5717 (1985).
16. J. Poplawski, E. Ehrenfreund, H. Schaffer, F. Wudl, and A. J. Heeger, *Synth. Met. 28*:C539 (1989).
17. K.-Y. Jen and R. L. Elsenbaumer, *Synth. Met. 16*: 379 (1986).
18. T. L. Rose and M. C. Liberto, *Synth. Met. 31*:395 (1989).
19. T. Iyoda, M. Kitano, and T. Shimidzu, *J. Chem. Soc., Chem. Commun. 1991*:1618.
20. P. Chandrasekhar, A. M. Masulaitis, and R. W. Gumbs, *Synth. Met. 36*:303 (1990).
21. I. Hoogmartens, D. Vanderzande, H. Martens, and J. Gelan, *Synth. Met. 47*:367 (1992).
22. R. van Asselt, I. Hoogmartens, D. Vanderzande, J. Gelan, P. E. Froehling, M. Aussems, O. Aagaard, and R. Schellenkens, *Synth. Met. 74*:65 (1995).
23. K. Eichinger and F. Kritzinger, *Spectrochim. Acta 47A*:661 (1991).
24. F. Wudl, M. Kobayashi, N. Colaneri, M. Boysel, and A. J. Heeger, *Mol. Cryst. Liq. Cryst. 118*:199 (1985).
25. J. L. Brédas, A. J. Heeger, and F. Wudl, *J. Chem. Phys. 85*:4673 (1986).
26. Y. Lee and M. Kertesz, *J. Phys. Chem. 91*:2690 (1987).
27. A. K. Bakhshi and J. Ladik, *Solid State Commun. 61*:71 (1987).
28. W. Wallnöfer, E. Faulques, H. Kuzmany, and K. Eichinger, *Synth. Met. 28*:533 (1989).
29. E. Faulques, W. Wallnöfer, and H. Kuzmany, *J. Chem. Phys. 90*:7585 (1989).
30. L. Cuff, M. Kertesz, J. Geisselbrecht, J. Kürti, and H. Kuzmany, *Synth. Met. 55–57*:564 (1993).
31. L. Cuff and M. Kertesz, *Polym. Mater.: Sci. Eng. 72*:281 (1995).
32. I. Hoogmartens, P. Adriaensens, R. Carleer, D. Vanderzande, H. Martens, and J. Gelan, *Synth. Met. 51*:219 (1992).
33. Y. Okuda, M. V. Lakshmikantham, and M. P. Cava, *J. Org. Chem. 56*:6024 (1991).
34. I. Hoogmartens, D. Vanderzande, H. Martens, and J. Gelan, *Synth. Met. 41–43*:513 (1991).
35. R. Kiebooms, I. Hoogmartens, P. Adriaensens, D. Vanderzande, and J. Gelan, *Synth. Met. 52*:395 (1992).
36. I. Hoogmartens, P. Adriaensens, D. Vanderzande, J. Gelan, C. Quattrocchi, R. Lazzaroni, and J. L. Brédas, *Macromolecules 25*:7347 (1992).
37. R. Kiebooms, I. Hoogmartens, P. Adriaensens, D. Vanderzande, and J. Gelan, *Macromolecules 28*: 4961 (1995).
38. G. Zerbi, M. C. Magnoni, I. Hoogmartens, R. Kiebooms, R. Carleer, D. Vanderzande, and J. Gelan, *Adv. Mater. 7*:1027 (1995).
39. R. Lazzaroni, J. Riga, J. Verbist, J. L. Bredás, and F. Wudl, *J. Chem. Phys. 88*:4257 (1988).
40. S. M. Dale, A. Glidle, and A. R. Hillman, *J. Mater. Chem. 2*:99 (1992).
41. M. Onoda, S. Morita, H. Nakayama, and K. Yoshino, *Jpn. J. Appl. Phys. 32*:3534 (1993).
42. M. Onoda, H. Nakayama, K. Tada, S. Morita, T. Kawai, and K. Yoshino, *Mol. Cryst. Liq. Cryst. 256*: 657 (1994).
43. G. Defieuw, R. Samijn, I. Hoogmartens, D. Vanderzande, and J. Gelan, *Synth. Met. 55–57*:3702 (1993).
44. K. Tada, S. Morita, T. Kawai, M. Onoda, K. Yoshino, and A. A. Zakhidov, *Synth. Met. 70*:1347 (1995).
45. M. Kitano, T. Iyoda, and T. Shimidzu, *Polym. J. 27*: 875 (1995).
46. J. L. Brédas, *J. Chem. Phys. 82*:3808 (1985).
47. J. L. Brédas, *Synth. Met. 17*:115 (1987).
48. Y.-S. Lee, M. Kertesz, and R. L. Elsenbaumer, *Chem. Mater. 2*:526 (1990).
49. A. K. Bakhshi and J. Ladik, *Solid State Commun. 65*:1203 (1988).
50. Y. S. Lee and M. Kertesz, *Int. J. Quantum Chem.: Quantum Chem. Symp. 21*:163 (1987).
51. Y. Ikenoue, *Synth. Met. 35*:263 (1990).
52. J. Kürti and P. R. Surján, *J. Chem. Phys. 92*:3247 (1990).
53. P. R. Surján and K. Németh, *Synth. Met. 55–57*:4260 (1993).
54. K. Nayak and D. S. Marynick, *Macromolecules 23*: 2237 (1990).
55. M. Pomerantz, B. Chaloner-Gill, L. O. Harding, J. J. Tseng, and W. J. Pomerantz, *J. Chem. Soc., Chem. Commun. 1992*:1672.
56. M. Pomerantz, B. Chaloner-Gill, L. O. Harding, J. J. Tseng, and W. J. Pomerantz, *Synth. Met. 55*:960 (1993).
57. S. Y. Hong and D. S. Marynick, *J. Chem. Phys. 96*: 5497 (1992).
58. S. Y. Hong and D. S. Marynick, *Macromolecules 23*:4652 (1992).
59. C. Quattrocchi, R. Lazzaroni, R. Kiebooms, D. Vanderzande, J. Gelan, and J. L. Brédas, *Synth. Met. 69*:691 (1995).
60. J. Kastner, H. Kuzmany, D. Vegh, M. Landl, L. Cuff, and M. Kertesz, *Macromolecules 28*:2922 (1995).
61. J. Kastner, H. Kuzmany, D. Vegh, M. Landl, L. Cuff, and M. Kertesz, *Synth. Met. 69*:593 (1995).
62. S. A. Jenekhe, *Nature 322*:345 (1986).
63. S. A. Jenekhe, *Macromolecules 19*:2663 (1986).
64. M. Kertesz and Y. S. Lee, *J. Phys. Chem. 91*:2690 (1987).
65. Y.-S. Lee and M. Kertesz, *J. Chem. Phys. 88*:2609 (1988).
66. J. M. Toussaint, B. Thémans, J. M. André, and J. L. Brédas, *Synth. Met. 28*:C205 (1989).
67. J. M. Toussaint and J. L. Brédas, *Synth. Met. 41–43*: 3555 (1991).
68. L. Hevesi, G. Proess, and A. Lazarescu-Grigore, *Synth. Met. 59*:201 (1993).

69. X. Gu, Ph.D. Dissertation, The University of Texas at Arlington, 1995. See also: M. Pomerantz and X. Gu, *Synth. Met. 84*:243 (1997).
70. M. Kertesz and Y.-S. Lee, *Synth. Met. 28*:C545 (1989).
71. J. Kúrti, P. R. Surján, M. Kertesz, and G. Frapper, *Synth. Met. 55–57*:4338 (1993).
72. J. Kúrti, P. R. Surján, and M. Kertesz, *J. Am. Chem. Soc. 113*:9865 (1991).
73. A. K. Bakhshi, C.-M. Liegener, and J. Ladik, *Synth. Met. 30*:79 (1989).
74. D. Lorcy and M. P. Cava, *Adv. Mater. 4*:562 (1992).
75. S. Musmanni and J. P. Ferraris, *J. Chem. Soc., Chem. Commun. 1993*:172.
76. P. Bäuerle, G. Götz, U. Segelbacher, D. Huttenlacher, and M. Mehring, *Synth. Met. 55–57*:4768 (1993).
77. M. V. Lakshmikantham, D. Lorcy, C. Scordilis-Kelley, X.-L. Wu, J. P. Parakka, R. M. Metzger, and M. P. Cava, *Adv. Mater. 5*:723 (1993).
78. R. M. Metzger, P. Wang, X.-L. Wu, G. V. Tormos, D. Lorcy, I. Shcherbakova, M. V. Lakshmikantham, and M. P. Cava, *Synth. Met. 70*:1435 (1995).
79. J. Pranata, R. H. Grubbs, and D. A. Dougherty, *J. Am. Chem. Soc. 110*:3430 (1988).
80. J. Pranata, V. S. Marudarajan, and D. A. Dougherty, *J. Am. Chem. Soc. 111*:2026 (1989).
81. J. L. Brédas and R. H. Baughman, *J. Chem. Phys. 83*:1316 (1985).
82. Z. Iqbal, D. M. Ivory, J. Marti, J. L. Brédas, and R. H. Baughman, *Mol. Cryst. Liq. Cryst. 118*:103 (1985).
83. M. Pomerantz, R. Cardona, and P. Rooney, *Macromolecules 22*:304 (1989).
84. P. M. Lahti, J. Obrzut, and F. E. Karasz, *Macromolecules 20*:2023 (1987).
85. J. L. Brédas, R. R. Chance, R. H. Baughman, and R. Silbey, *J. Chem. Phys. 76*:3673 (1982).
86. M. Kertesz, *Macromolecules 28*:1475 (1995).
87. S. Y. Hong, M. Kertesz, Y. S. Lee, and O.-K. Kim, *Chem. Mater. 4*:378 (1992).
88. D. S. Boudreaux, R. R. Chance, R. L. Elsenbaumer, J. E. Frommer, J. L. Brédas, and R. Silby, *Phys. Rev. B 31*:652 (1985).
89. U. Scherf and K. Müllen, *Polymer 33*:2443 (1992).
90. U. Scherf and K. Müllen, *Synthesis 1992*:23.
91. M. Kertesz and T. R. Hughbanks, *Synth. Met. 69*:699 (1995).
92. T. L. Lambert and J. P. Ferraris, *J. Chem. Soc., Chem. Commun. 1991*:752.
93. J. P. Ferraris and T. L. Lambert, *J. Chem. Soc., Chem. Commun. 1991*:1268.
94. J. P. Ferraris and T. L. Lambert, *Polym. Mater.: Sci. Eng. 64*:332 (1991).
95. J. M. Toussaint and J. L. Brédas, *Synth. Met. 61*:103 (1993).
96. A. Bolognesi, M. Catellani, S. Destri, R. Zamboni, and C. Taliani, *J. Chem. Soc., Chem. Commun. 1988*:246.
97. A. Bolognesi, M. Catellani, S. Destri, C. Taliani, and R. Zamboni, *Synth. Met. 37*:134 (1990).
98. M. Catellani, T. Caronna, and S. V. Meille, *J. Chem. Soc., Chem. Commun. 1994*:1911.
99. C. Quattrocchi, R. Lazzaroni, J. L. Brédas, R. Zamboni, and C. Taliani, *Synth. Met. 55–57*:4399 (1993).
100. J. M. Toussaint and J. L. Brédas, *Synth. Met. 69*:637 (1995).
101. K. Tanaka, S. Wang, and T. Yamabe, *Synth. Met. 30*:57 (1989).
102. A. K. Bakhshi, *Solid State Commun. 94*:943 (1995).
103. K.-Y. Jen, M. Maxfield, L. W. Shacklette, and R. L. Elsenbaumer, *J. Chem. Soc., Chem. Commun. 1987*:309.
104. H. Eckhardt, L. W. Shacklette, K. Y. Jen, and R. L. Elsenbaumer, *J. Chem. Phys. 91*:1303 (1989).
105. K.-Y. Jen, H. Eckhardt, T. R. Jow, L. W. Shacklette, and R. L. Elsenbaumer, *J. Chem. Soc., Chem. Commun. 1988*:215.
106. F. Wudl, A. J. Heeger, Y. Yoshiaki, and M. Kobayashi, Eur. Patent Appl. 0 273 643 (1988).
107. Y. Ikenoue, F. Wudl, and A. J. Heeger, *Synth. Met. 40*:1 (1991).
108. F. Eiji, Jpn. Kokai Tokkyo Koho JP 02 252 726 (1990).
109. G. King and S. J. Higgins, *J. Chem. Soc., Chem. Commun. 1994*:825.
110. G. King, S. J. Higgins, S. E. Garner, and A. R. Hillman, *Synth. Met. 67*:241 (1994).
111. G. King and S. J. Higgins, *J. Mater. Chem. 5*:447 (1995).
112. F. Eiji, Jpn. Kokai Tokkyo Koho JP 02 252 727 (1990).
113. M. J. Swann, G. Brooke, D. Bloor, and J. Maher, *Synth. Met. 55–57*:281 (1993).
114. G. M. Brooke, C. J. Drury, D. Bloor, and M. J. Swann, *J. Mater. Chem. 5*:1317 (1995).
115. G. M. Brooke and S. D. Mawson, *J. Chem. Soc., Perkin Trans. 1 1990*:1919.
116. P. Kathirgamanathan and M. K. Shepherd, *J. Electroanal. Chem. 354*:305 (1993).
117. D. Vegh, M. Landl, J. Kastner, and H. Kuzmany, *Synth. Commun.* In press. See: J. Kastner, H. Kuzmany, D. Vegh, M. Landl, L. Cuff, and M. Kertesz, *Macromolecules 28*:2922 (1995).
118. M. Siekierski and J. Plocharski, *Synth. Met. 51*:81 (1992).
119. C. Arbizzani, M. Catellani, M. Mastragostino, and M. Mingazzini, *Electrochim. Acta 40*:1871 (1995).
120. S. Tanaka and Y. Yamashita, *Synth. Met. 55–57*:1251 (1993).
121. J. P. Ferraris, A. Bravo, W. Kim, and D. C. Hrncir, *J. Chem. Soc., Chem. Commun. 1994*:991.
122. C. Kitamura, S. Tanaka, and Y. Yamashita, *J. Chem. Soc., Chem. Commun. 1994*:1585.
123. S. Tanaka and Y. Yamashita, *Synth. Met. 69*:599 (1995).
124. M. Karikomi, C. Kitamura, S. Tanaka, and Y. Yamashita, *J. Am. Chem. Soc. 117*:6791 (1995).

125. A. O. Patil and F. Wudl, *Macromolecules 21*:542 (1988).
126. H. Bräunling, R. Becker, and G. Blöchl, *Synth. Met. 41–43*:1539 (1991).
127. H. Bräunling, R. Becker, and G. Blöchl, *Synth. Met. 55–57*:833 (1993).
128. M. Hanack, G. Hieber, K.-M. Mangold, H. Ritter, U. Röhrig, and U. Schmid, *Synth. Met. 55–57*:827 (1993).
129. D. R. McKean and J. K. Stille, *Macromolecules 20*:1787 (1987).
130. W.-C. Chen and S. A. Jenekhe, *Macromolecules 28*:454 (1995).
131. W.-C. Chen and S. A. Jenekhe, *Macromolecules 28*:465 (1995).
132. M. Kozaki, S. Tanaka, and Y. Yamashita, *J. Org. Chem. 59*:442 (1994).
133. J. P. Ferraris, C. Henderson, D. Torres, and D. Meeker, *Synth. Met. 72*:147 (1995).
134. J. Roncali, H. Brisset, C. Thobie-Gautier, M. Jubault, and A. Gorgues, *J. Chim. Phys. (Paris) 92*:771 (1995).
135. J. Roncali, C. Thobie-Gautier, and H. Brisset, *Phosphorus, Sulfur, Silicon 95–96*:513 (1994).
136. H. Brisset, C. Thobie-Gautier, A. Gorgues, M. Jubault, and J. Roncali, *J. Chem. Soc., Chem. Commun. 1994*:1305.
137. J. Roncali, C. Thobie-Gautier, E. H. Elandaloussi, and P. Frère, *J. Chem. Soc., Chem. Commun. 1994*:2249.
138. J. Roncali, C. Thobie-Gautier, H. Brisset, M. Jubault, and A. Gorgues, *J. Chim. Phys. (Paris) 92*:767 (1995).
139. M. Martinez, J. R. Reynolds, S. Basak, D. Black, D. S. Marynick, and M. Pomerantz, *J. Polym. Sci., Phys. Ed. 26*:911 (1988).
140. J. M. Williams, J. R. Ferraro, R. J. Thorn, K. D. Carlson, U. Geiser, H. H. Wang, A. M. Kini, and M.-H. Whangbo, *Organic Superconductors (Including Fullerenes): Synthesis, Structure, Properties, and Theory*, Prentice-Hall, Englewood Cliffs, NJ, 1992.
141. J. M. Williams, J. R. Ferraro, R. J. Thorn, K. D. Carlson, U. Geiser, H. H. Wang, A. M. Kini, and M.-H. Whangbo, *Organic Superconductors (Including Fullerenes): Synthesis, Structure, Properties, and Theory*, Prentice-Hall, Englewood Cliffs, NJ, 1992, Chapter 3.
142. I. Murase, T. Ohnishi, T. Noguchi, and M. Hirooka, *Polym. Commun. 25*:327 (1984).
143. J. Roncali and C. Thobie-Gautier, *Adv. Mater. 6*:846 (1994).
144. M. L. Kaplan, P. H. Schmidt, C.-H. Chen, and J. M. W. Walsh, *Appl. Phys. Lett. 36*:867 (1980).
145. L. R. Dalton, J. Thomson, and H. S. Nalwa, *Polymer 28*:543 (1987).
146. M. D. Pace and O.-K. Kim, *Synth. Met. 25*:333 (1988).
147. U. Scherf, A. Bohnen, and K. Müllen, *Makromol. Chem. 193*:1127 (1992).
148. M. Dietrich, J. Heinze, G. Heywang, and F. Jonas, *J. Electroanal. Chem. 369*:87 (1994).
149. Q. Pei, G. Zuccarello, M. Ahlskog, and O. Inganäs, *Polymer 35*:1347 (1994).
150. M. Dietrich and J. Heinze, *Synth. Met. 41–43*:503 (1991).
151. P. C. Van Dort, J. E. Pickett, and M. L. Blohm, *Synth. Met. 41–43*:2305 (1991).
152. G. A. Sotzing, J. R. Reynolds, A. R. Katritzky, J. Soloducho, and R. Musgrave, *Polym. Mater.: Sci. Eng. 72*:317 (1995).
153. G. A. Sotzing and J. R. Reynolds, *J. Chem. Soc., Chem. Commun. 1995*:703.
154. Y. Fu and R. L. Elsenbaumer, *Polym. Mater.: Sci. Eng. 72*:315 (1995).
155. Y. Fu, Ph.D. Dissertation, The University of Texas at Arlington, 1995.
156. H. Cheng and R. L. Elsenbaumer, *J. Chem. Soc., Chem. Commun. 1995*:1451.
157. H. Cheng, Ph.D. Dissertation, The University of Texas at Arlington, 1995.
158. E. E. Havinga, W. ten Hoeve, and H. Wynberg, *Polym. Bull. 29*:119 (1992).
159. E. E. Havinga, W. ten Hoeve, and H. Wynberg, *Synth. Met. 55–57*:299 (1993).
160. E. E. Havinga, A. Pomp, W. ten Hoeve, and H. Wynberg, *Synth. Met. 69*:581 (1995).
161. T. W. Brockmann and J. M. Tour, *J. Am. Chem. Soc. 116*:7435 (1994).
162. T. W. Brockmann and J. M. Tour, *J. Am. Chem. Soc. 117*:4437 (1995).
163. H. A. Ho, H. Brisset, P. Frère, and J. Roncali, *J. Chem. Soc., Chem. Commun. 1995*:2309.

12
Advances in the Molecular Design of Functional Conjugated Polymers

Jean Roncali
Ingénierie Moléculaire et Matériaux Organiques CNRS, Université d'Angers, Angers, France

I. INTRODUCTION

For more than 20 years, electrical conduction in organic solids has been one of the most fascinating topics for organic chemists and solid-state physicists. This considerable interest was triggered by the discovery of high electrical conductivity in molecular crystals of tetrathiafulvalene-tetracyanoquinodimethane (TTF-TCNQ) in 1973 [1] and in the oxidized form of a π-conjugated polymer, namely polyacetylene, a few years later [2].

Whereas in charge transfer complexes or cation radical salts, conduction involves a translation of mixed valence states in a direction parallel to the stacks of the π-donor molecules, in π-conjugated polymers charge transport occurs preferentially along the π-conjugated chain. Consequently, whereas the packing arrangement of donor and acceptor molecules is a crucial factor in molecular conductors, the extent of conjugation is the relevant parameter determining the electronic and electrical properties of linearly π-conjugated systems.

Polyacetylene remains the archetype of this second class of materials. However, despite a high conductivity, its instability in atmospheric conditions constitutes a major obstacle to practical applications. A milestone in the young history of conjugated polymers is the synthesis of more stable heterocycle-based polymers such as polypyrrole [3] and polythiophene (PT) [4,5] in the early 1980s, together with the discovery that these polymers can be easily obtained in their oxidized conducting form by means of a one-pot single-step electrochemical synthesis.

In this context, polythiophene has rapidly acquired a prominent position owing to its high conductivity, the environmental stability of its neutral state and, to a lesser extent, its doped conductive state, and its structural versatility, allowing the derivatization of the π-conjugated backbone at the price of moderate losses of conjugation and conductivity. As underlined in an analysis of the recent trends in the field of conjugated polymers, both the relative and absolute numbers of papers devoted to PT derivatives have risen steadily over the past decade, making PTs the most widely investigated conjugated polymers [6].

In the pioneering works of the early 1980s PTs were obtained as poorly conductive intractable materials [7–11], but they are now available in a wide diversity of forms such as films, solutions, and gels, allowing the development of bulk utilizations such as antistatic coatings and EMI shieldings. During this period, a wide range of advanced applications including electrode materials and sensors, electronic devices, nonlinear optics, electroluminescent diodes, and models for molecular logic have been proposed and investigated [11].

Although physical characterizations have played an important role in the progress of our knowledge of the structure and electronic properties of PTs, the major driving force of their development lies in the considerable research efforts that organic chemists have invested in the control of the structure and properties of thiophene-based polymeric or molecular conjugated systems.

Since the first review covering the early developments of the field up to 1985 [12]. PTs have been considered in surveys devoted to some specific aspects of conducting polymers such as electrochemistry [13–15] or optical properties [16], and their synthesis, functionalization, and applications were reviewed a few years ago [11]. Owing to the tremendous increase in the number of papers devoted to PTs, these various topics as well as the numerous characterizations of the electrical, magnetic, and optical properties of PTs cannot be discussed in the frame of a short review. Accordingly, the aim of this chapter is to present an overview of the recent advances in the chemistry of PTs.

Whereas the modification of the structure of the monomer is the essential tool of the organic chemist, this manipulation can have quite different objectives. A first angle of attack involves the use of modified precursors to achieve better control of the polymerization reaction and hence of the structural definition of the polymer chain. A second and probably more widely acknowledged research area is the synthesis of tailored precursors to create new materials in which the inherent electronic properties of the π-conjugated system are combined with specific properties afforded by covalently grafted prosthetic groups. These topics have been intensively investigated during the past decade, and recent years have witnessed the emergence of new research areas such as the control of the macromolecular and macroscopic structures of the polymer by means of substituent interactions, the revival of thiophene oligomers, both as model compounds of PT and as molecular materials, and the growing development of small band gap PTs. In this chapter, these various aspects of PTs are discussed through representative examples that emphasize the contribution of synthetic chemistry to the development of the field.

II. SYNTHESIS OF POLYTHIOPHENES

A. Polymerization of Unsubstituted Thiophenes

The ability to control the structure and properties of conjugated polymers at a molecular level is a prerequisite that conditions the validity of all subsequent physical and structural studies as well as the development of technological applications. Consequently, in addition to a considerable number of structural characterizations by a whole arsenal of physical techniques, tremendous research efforts have been continuously focused on the development of new methods of preparation allowing better control of the form, structural definition, and properties of PTs.

Polythiophenes can be synthesized by electrochemical polymerization or by chemical methods. Although, because of several distinct advantages, electropolymerization has been for a long period the preferred method at the laboratory scale, the recent development of solution-processable PTs has triggered a strong renewal of interest in chemical polymerization, which remains the most suitable method as far as industrial production is envisioned.

The chemical synthesis of PTs can be achieved by two main routes: oxidative polymerization using ferric chloride, which is probably the simplest method [17], and polycondensation of organometallic derivatives using nickel or palladium catalysts. While the coupling of Grignard compound is the most widely employed method [18,19], organometallic compounds involving other metals such as zerovalent nickel [20] or zinc [21] have also been employed.

Electrochemical polymerization presents several advantages such as rapidity, absence of catalyst, control of film thickness by the deposition charge, and direct obtainment of the polymer in the oxidized conducting form. Electropolymerization proceeds by a complex succession of electrochemical and chemical reactions based on the coupling of the cation radicals resulting from the electrooxidation of the monomer on the anode surface [13–15]. The polymers are obtained as powdery deposits or free-standing films, depending on experimental conditions, with an oxidation (or "doping") level of 0.20–0.25 electron/monomer and conductivities in the range of 10–100 S/cm [11].

The optimization of the electrosynthesis conditions (concentration of reagents, solvent, electrical conditions) has been the subject of several studies [22–24]. These works have led to significant progress in the control of the structural definition and electrical properties of PTs, and polymers with conductivities as high as 2000 S/cm have been obtained [25]. This considerable increase of conductivity has been correlated with a large extension of the effective mean conjugation length (MCL) revealed by a 200 mV decrease of oxidation potential and a 30 nm bathochromic shift of the absorption maximum of the polymer [24]. These studies and others [26–28] have contributed to put clearly in evidence the tight correlation existing between the electrical conductivity and the MCL. They have also underlined the interest of a parallel use of electrochemical and spectroscopic techniques as methods of investigation.

B. Polymerization of Tailored Precursors

In the absence of steric hindrance to planarity, the MCL is determined by the degree of polymerization and the regularity of the α–α' linkages in the polymer chain. While the MCL depends to a large extent on the synthesis conditions [11], a more straightforward control of the stereoregularity of the polymer has been searched for through tailoring of the precursor structure.

1. Terthienyls

One of the first attempts in this direction involved the use of conjugated thiophene oligomers as substrates for the electropolymerization. This approach was based on two main ideas. On the one hand, since the oxidation potential of conjugated oligomers decreases with chain length [29], electropolymerization of conjugated oligomers can be performed at lower potential than that of the monomer, thus avoiding possible undesired side reactions of the cation radical with the solvent or electrolyte or the degradation of the growing polymer at highly positive potentials [30,31]. On the other hand, since the starting molecule already contains exclusive α–α′ linkages, one may expect the resulting polymer to contain fewer α–β′ conjugation defects than that prepared from the monomer.

The electropolymerization of thiophene oligomers, in particular that of terthienyl (TT) (**1**), was investigated by several groups [32–34]. Despite a wide diversity of experimental conditions, consistent results were obtained, showing that the resulting polymers differ strongly from that prepared from the monomer. Thus, instead of free-standing films, poly(TT) is generally obtained as powdery deposits with conductivities inferior by several orders of magnitude to that of PT [32–34]. Furthermore, poly(TT) exhibits an oxidation potential 300 mV more positive than that of PT and an absorption maximum hypsochromically shifted by about 120 nm, changes that are indicative of a considerable decrease in the MCL [33]. Further studies showed that these effects are related to the composition of poly(TT), which contains essentially unreacted TT, and its dimer, i.e., sexithienyl [33,35–37].

This correlated decrease in the conductivity and MCL is related to the effects of the size of the oligomer on the reactivity and solubility of the resulting cation radical. The increase in chain length and hence in electron delocalization stabilizes the corresponding radical, which becomes less and less reactive [13–15]. This phenomenon is clearly evidenced by the comparison of the electrochemical behavior of the monomer and oligomers when electrooxidized by means of repetitive potential scans (Fig. 12.1). The monomer oxidizes irreversibly at ~2.0 V, and the new redox system emerging around 1.0 V indicates the formation of an extended conjugated system. With the trimer, oxidation occurs at 1.05 V and becomes partially reversible while a new redox system corresponding to the coupling product develops at 0.80 V. Finally, oxidation of the hexamer leads to two symmetrical oxidation waves at 0.88 and 1.10 V characteristic of the stable cation radical and the dication. This stability results in a homothetic increase in the wave intensity indicative of the electrodeposition of a dication salt and hence of the absence of coupling. However, partial electrochemical coupling can be observed in more polar solvents [38].

As stressed in a paper by Wei et al. [39], if a monomeric radical has a coupling probability of 1, this probability decreases to 2/3 for a trimer and 2/n for an oligomer of degree n since only the mesomeric forms of the radical with the unpaired electron on a terminal thiophene ring can contribute to chain growth. This implies that in the case of TT, about one-third of the cation radicals can be trapped in the film as cation radical salts. In addition to these problems of reactivity, the solubility of the precursor and of the various cationic intermediates in the electrolytic medium also plays a determining role in the maximum length of the coupling product. While solubility decreases with chain length, oxidation of the oligomer to its cation radical produces a sudden drop in solubility that favors precipitation, this phenomenon being even more dramatic in the case of the product of dimerization. Finally, the difficulty oligomeric radicals have in orienting correctly for coupling represents another important limiting factor as confirmed by the negative values of the activation entropies determined for the electrochemical dimerization of oligothiophenes [40].

More recent works have shown that the enhanced solubility resulting from the grafting of alkyl (**2**) or oxyalkyl substituents (**3**) on the central thiophene ring of TT allows a significant increase in the chain length of the coupling product. This is shown by the negative shift of the oxidation potential of the first redox system in the cyclic voltammogram (CV) of poly**2**) (0.68 V) compared to poly(**1**) (1.05 V) (Fig. 12.2). However, the presence of the starting trimer reflected by the occurrence of a second redox system in the CV of the deposited material confirms that in each case electropolymerization remains incomplete [36]. On the other hand, the increase in solubility seems to be more effective in the case of chemical polymerization, and oxidative polymerization of 3′,4′-dibutylterthienyl (**4**) and 3,3″- or 4,4″-dipentyloxyterthienyl (**5,6**) has been shown to lead to extensively conjugated polymers [41,42].

2. Bithienyls

The polymerization of symmetrical 3,3′- and 4,4′-dialkylbithiophenes (**7,8**) has been reported in several works [43,44]. While the resulting polymers, which consist in a periodic succession of head-to-head and tail-to-tail

1 **2** **3**

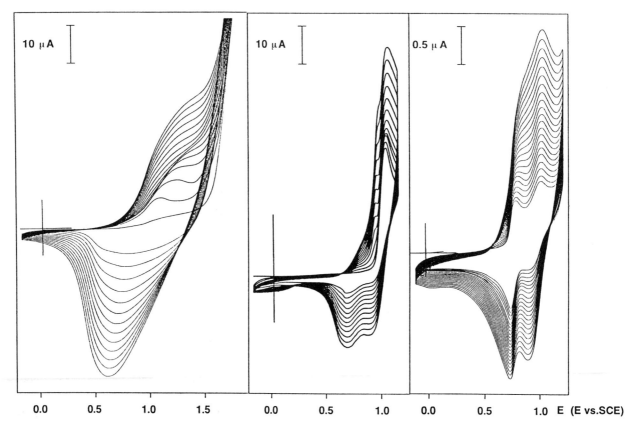

Fig. 12.1 Electrodeposition of thiophene derivatives by recurrent potential scans (v = 100 mV/s). From left to right: 0.1 M thiophene in 0.1 M Bu$_4$NPF$_6$-MeCN; 5 × 10^{-4} M 3'-methylterthienyl in 0.1 M Bu$_4$NPF$_6$-MeCN; 2 × 10^{-3} M 3',3'''''-di(3,4-dioxaheptyl)sexithienyl in 0.1 M Bu$_4$NPF$_6$-CH$_2$Cl$_2$. (From Refs. 36 and 38.)

linkages, can be formally considered as regioregular, they are considerably less conjugated and less conductive than the polymers obtained from unsubstituted bithiophene or from 3-substituted monomers, although in this latter case the polymers contain 10–15% of irregular head-to-head linkages [44,45]. Furthermore, although 3,3'- and 4,4'-disubstituted bithiophenes should lead ultimately to the same polymer, the differences observed between the electrogenerated polymers indicate that steric hindrance in 3,3'-disubstituted precursors affects the polymerization process [43].

A recent extension of this approach involves the use

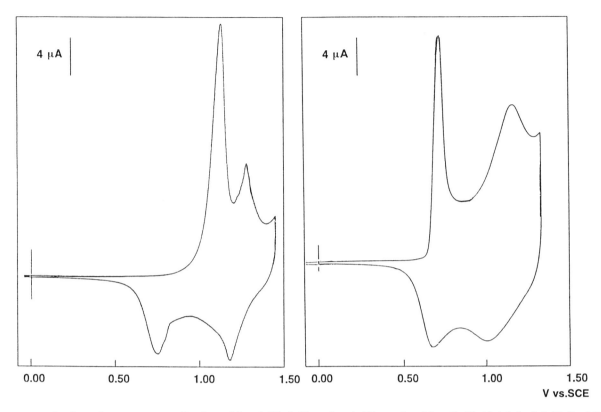

Fig. 12.2 Cyclic voltammograms of polyterthienyl (**1**) (left) and poly(3'-octylterthienyl) (**2**) (right) in 0.1 M Bu$_4$NClO$_4$-MeCN, scan rate 100 mV/s. (From Ref. 36.)

of substituted cyclopentadithiophenes (**9**) as substrate. This strategy, which combines the advantages of the previous ones without the problems related to steric hindrance, allows the obtainment of highly conducting regioregular polymers [46].

3. 3,4-Disubstituted Thiophenes

Another possible strategy to control the stereoregularity of PTs consists in the disubstitution of the 3- and 4-positions in order to prevent any possibility of α–β' coupling. However, this approach is severely limited by the steric interactions between substituents grafted on adjacent thiophene rings that distort the π-conjugated system and lead to a considerable loss of effective conjugation. Consequently, the polymers obtained from 3,4-dialkylthiophenes (**10**) have larger optical band gaps and much lower conductivities than monosubstituted polymers [47].

Various proposals have been made to solve this problem. A first strategy involves the cyclization between the 3- and 4-positions to reduce the steric hindrance. Thus, cyclopenta[c]thiophene (**11**) has been shown to produce a polymer considerably more conjugated than those obtained from 3,4-dialkylthiophenes (Fig. 12.3) [48].

Another approach consists in the grafting of alkoxy groups that cause less steric hindrance than alkyl chains. This hypothesis has been confirmed for both unsymmetrically disubstituted monomers such as poly(3-methyl, 4-methoxythiophene) (**12**) and symmetrical 3,4-dialkoxythiophenes (**13**) [49,50]. More recently the electrosynthesis and properties of poly(3,4-ethylenedioxythiophene), poly(**14**), which can be viewed as a combination of the above two strategies, has been reported [51]. Owing to its low oxidation potential the doped conducting form of this polymer shows remarkable stability, and its small optical band gap is consistent with a considera-

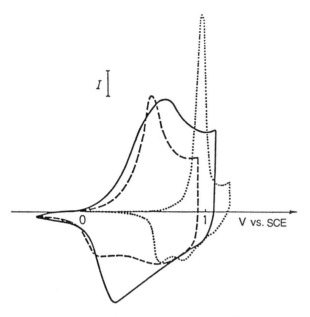

Fig. 12.3 Cyclic voltammograms of poly(3-methylthiophene) (dashed line), $l = 20$ μA; poly (**10**) (dotted line), $l = 40$ μA; and poly(**11**) (solid line), $l = 20$ μA. Electrolytic medium: 0.1 M LiClO$_4$-MeCN; scan rate 50 mV/s. (From Ref. 48.)

ble reduction of the steric hindrance. This approach was recently extended to the case of dialkylmercaptothiophenes, and it has been shown that the cyclization achieved in 3,4-ethylenedithiathiophene (**15**) allowed this monomer to be electropolymerized [52], unlike its uncyclized analog 3,4-bis(ethylmercapto)thiophene [53].

4. Activated Precursors

An alternative strategy to improve the stereoregularity of electrogenerated PTs consists in the activation of the α-positions of the thiophene ring to increase the selectivity of α–α coupling during polymerization. A first attempt in this direction involved the oxidative polymerization of 2,5-dilithiothiophene in the presence of CuCl$_2$ [54]. More recently, the use of silicon-directed reactions was proposed as a means to improve the selectivity of α–α coupling. Whereas the polymer synthesized from 2,5-bis(trimethylsilyl)thiophene (**16a**) is rather similar to that prepared with unactivated thiophene, electropolymerization of 2,5-bis(trimethylsilyl)bithiophene (**16b**) allows a significant improvement of the electrochemical and electrical properties compared to conventional polybithiophene [55].

Thus, instead of the powdery deposits generally obtained with bithiophene, electropolymerization of (**16b**) produces flexible free-standing films with a conductivity ~30 times larger than that of polybithiophene. Furthermore, the doping level was found to be 50% higher than that of PT and the voltammetric response identical to that of PT [55]. This method appears particularly interesting for substrates difficult to polymerize such as 3-chlorothiophene (**17**) [55] or terthienyl (**16c**) [56]. In a recent extension of this approach, the trimethylsily/thiophene ratio was reduced to its minimal value by the design of tetrathienylsilane as precursor (**18**). This substrate, which can be electropolymerized at very low concentrations (2.5×10^{-3} M), leads to highly conducting films showing a highly defined cyclic voltammetric response and the lowest oxidation potential reported so far [57]. Other recent works have shown that the replacement of trimethylsilyl by trialkoxysilyl groups (**19**) leads to xerogels that can be subsequently electropolymerized [58].

III. SUBSTITUTED POLYTHIOPHENES

Besides the fundamental questions posed by the mechanisms responsible for their electronic and electrochemical properties, PTs are intensively investigated with respect to their multiple technological applications extending from bulk utilizations to electronic and optoelectronic devices and selective sensors. The wide diversity of these potential uses requires the definition of methods of preparation capable of producing large modifications of the form and properties of the polymer in order to meet the specific requirement of each type of envisioned application.

While the elaboration of composite materials in which PTs are associated with organic polymers [59], metals [60], or inorganic host matrices [61] provides a first answer to this problem, the modification of the structure of the precursor remains the most appropriate strategy for controlling the structure and properties of the polymer at the molecular level. In this regard, one of the main motivations for the widespread current interest in PT lies in its structural versatility, which, under defined conditions, allows the grafting of bulky functional groups with only a moderate loss of conjugation and conductivity.

A. Structural Basis of Functionalization

Despite an apparent simplicity, the functionalization of polyaromatic conjugated polymers puts stringent requirements on the possible structural modifications of the precursor. As a matter of fact, the structure of the modified precursor must remain compatible with both the polymerization reaction and the preservation of the relevant electronic, electrochemical, and electrical properties of the π-conjugated backbone in the resulting polymer. Such conditions imply in turn a detailed knowledge of the consequences of the electronic and steric effects of substitution both on the polymerization reaction and on the structure and properties of the polymer.

Early works on commercially available substituted thiophenes made evident the electronic effects of substitution on the polymerizability of the thiophene ring. Thus, 3-substitution by the strongly electron-withdrawing cyano, carbonyl, carboxylic, or nitro groups leads to a 0.50–0.70 V increase in the oxidation potential of thiophene [62]. This leads to highly reactive cation radicals capable of reacting with the solvent and/or electrolyte and hence to a lack of polymerization. Halogen-substituted thiophenes are also difficult to polymerize and produce polymers of high oxidation potential that are therefore unstable in air and poorly conducting [62,63]. In the other extreme case, the introduction of electron-releasing methoxy or alkylthio groups decreases the oxidation potential by almost a whole volt. Consequently, the resulting cation radicals are strongly stabilized, which favors the formation of short-chain oligomers [64,65] or fully inhibits electropolymerization [48,53]. The results obtained in these rather extreme cases clearly show that in order to synthesize extensively conjugated and highly conducting substituted PTs it is suitable to keep the reactivity of the substituted monomer within precise limits.

While steric factors do not directly affect the reactivity of the monomer, they can exert considerable influence on the structure and properties of the polymer. Steric interactions between substituents on adjacent monomers induce a distortion of the π-conjugated system, which results in a shortening of effective conjugation and hence in a drop in conductivity.

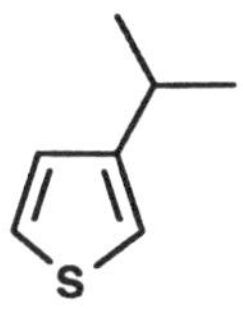

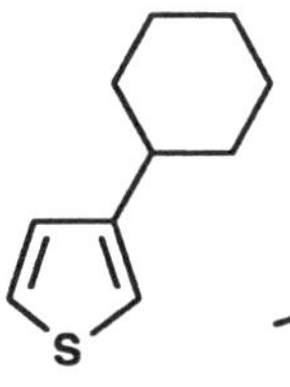

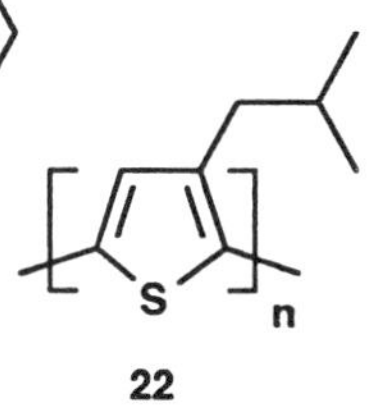

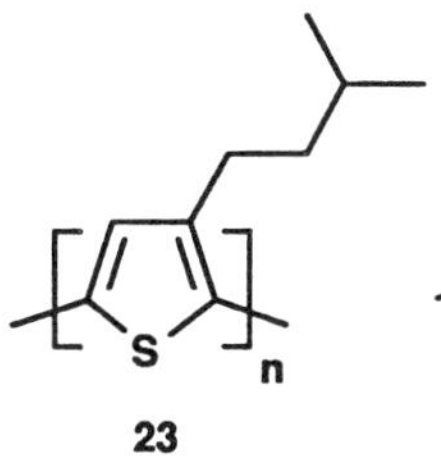

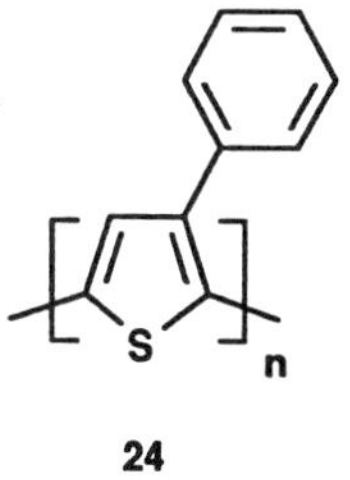

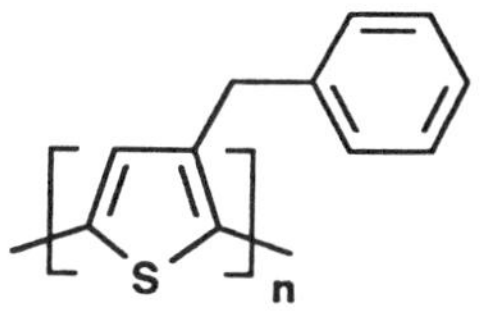

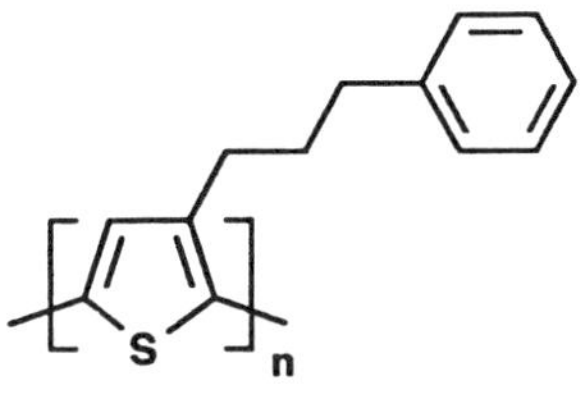

These problems have been systematically investigated by analyzing the effects of the distance between a bulky group, such as a branched alkyl chain or a phenyl group, and the thiophene ring on the structure and properties of the resulting polymer [66,67]. Steric hindrance was clearly evidenced by the failure to electropolymerize 3-isopropylthiophene (**20**) [66] and 3-cyclohexylthiophene (**21**) [68] and by the considerably shorter effective conjugation of poly(3-isobutylthiophene) (**22**) compared to its linear analog [66]. In contrast, poly(3-isoamylthiophene) (**23**) exhibits an oxidation potential and an absorption maximum similar to those of poly(3-amylthiophene). These results have been interpreted as due to a progressive decrease in the torsion angle between adjacent monomer units as the distance between the secondary carbon and the thiophene ring is increased [66]. Similar conclusions were reached by analyzing the structural conditions of introduction of phenyl rings in (**25** and **26**), and in this case an alkyl spacer involving at least three carbons was required to obtain a polymer of properties comparable to those of a poly(3-alkylthiophene) [67]. Poly(3-phenylthiophene) (**24**) was not considered in these studies because the conjugation of the phenyl and thiophene rings does not allow a clear discrimination between steric and electronic effects. This conjugation of the two cycles was recently confirmed by the good correlation observed between the oxidation potentials of a series of para-substituted 3-phenylthiophenes and σ_p Hammett constants [69].

These few examples thus confirm that the design of functional PTs requires a detailed knowledge of the structural parameters of the envisioned substituent in order for the resulting functionalized polymer to retain its relevant electrical and electrochemical properties.

B. Processible Polythiophenes

1. Poly(3-alkylthiophenes)

Since their first synthesis in 1986 [70–72], poly(3-alkylthiophenes) (PATs) have been the focus of considerable interest motivated in large part by the significant improvement of processability imparted by long alkyl chains. Thus undoped PATs have been shown to be soluble in common organic solvents such as tetrahydrofuran, dichloromethane, chloroform, benzene, toluene, and xylene [70–74]. The discovery of the solubility and fusibility [75] of PATs represents a milestone in the development of conducting polymers, and the resulting solution and melt processibility have considerably enlarged the potential technological applications of PTs.

PATs can be prepared by electropolymerization [66,70], by nickel-catalyzed Grignard coupling of 2,5-dihalogenated thiophenes [72], and by oxidative polymerization using ferric chloride (Scheme 1) [17,74].

Although it is evident that long alkyl chains increase the solubility of the PT backbone, unequivocal conclusions regarding the effects of the length of the alkyl substituent on solubility are difficult to draw as this parameter also affects the degree of polymerization (dp) [70]. Whereas in initial reports PATs were implicitly considered as fully soluble, further studies on electrogenerated polymers have shown that the solubility as well as the conjugation length and conductivity depend strongly on the synthesis conditions [34,66,76] and that this dependence increases with the length of the alkyl chain [66]. Thus in the case of poly(3-nonylthiophene), the increase of the synthesis current density from 0.5 to 10 mA/cm^2 leads to a bathochromic shift of the absorption maximum of the undoped polymer from 440 to 542 nm while a vibronic fine structure becomes discernible (Fig. 12.4). These considerable changes are correlated with a strong increase in conductivity up to 100 S/cm and with a parallel loss of solubility, which decreases down to a few percent [66]. While a wide range of dp values (20–2500) have been reported [70,72–74,76], the exact determination of this parameter is not straightforward. In fact, comparative studies performed with authentic 3-alkythiophene oligomers have shown that the polystyrene standards generally used for the determination of the molecular weight of PATs lead to an overestimation of the molecular weight by at least an order of magnitude [34].

In addition to solution processibility, the grafting of long alkyl chains leads to several important modifications of the electrochemical and optical properties of the PT backbone. Thus PATs with alkyl chains containing seven to nine carbons exhibit a significant enhancement of the mean conjugation length and electrochemical reversibility compared to polymers with shorter or longer alkyl substituents [66,77]. On the other hand, PATs have been shown to undergo thermochromism and solvatoch-

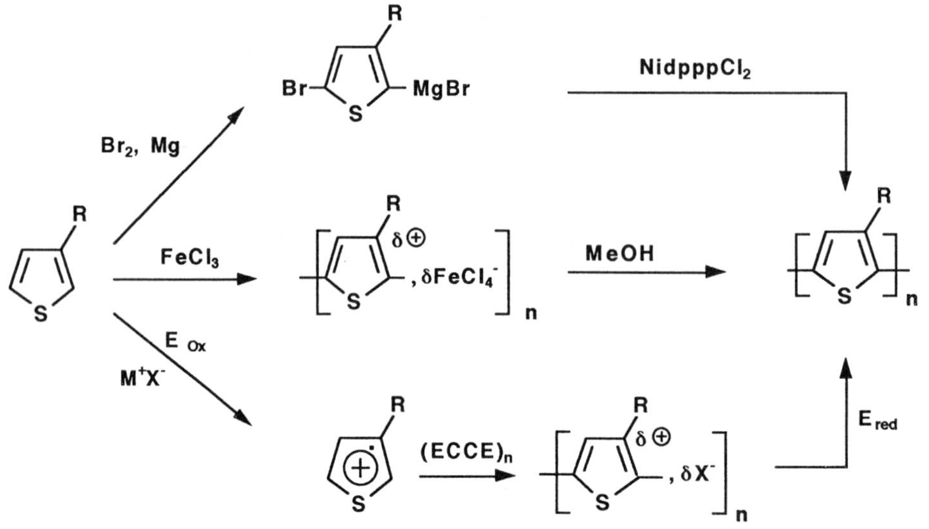

Scheme 1

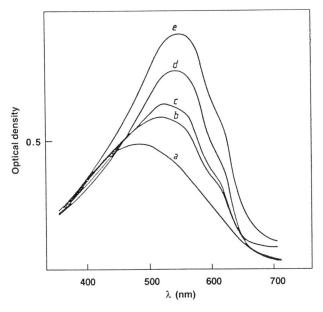

Fig. 12.4 Effect of the electrosynthesis current density (J) on the electronic absorption spectrum of undoped poly(3-nonylthiophene) films on ITO. (a) $J = 0.5$, (b) $J = 1$, (c) $J = 2$, (d) $J = 5$, (e) $J = 10$ mA/cm^2. (From Ref. 66.)

romism [78] (see also the chapter by M. Leclerc in this volume).

2. Self-Doped Polythiophenes

The concept of self-doped conducting polymers was introduced in 1987 with the synthesis of poly(3-thiophene-β-ethanesulfonate) (**27a**) and poly(3-thiophene-β-butanesulfonate) (**27c**) [79]. In these polymers the charge-compensating anion is covalently bound to the polymer backbone. Consequently, instead of anion incorporation, charge compensation upon electrochemical doping involves the expulsion of cationic species.

Furthermore, the presence of the sulfonate groups makes these polymers soluble in water in both the doped and undoped states [79,80]. Due to the failure to electropolymerize these monomers in their acidic or sodium salt forms, the polymers have been prepared by polymerization of the corresponding methyl esters and subsequent conversion into the polyelectrolytic form [79]. This problem was not observed with 3′-propanesulfonate-terthienyl (**28**), which, unlike the parent monomers, can be electropolymerized without supporting electrolyte, thus providing unequivocal evidence of self-doping [34]. More recently, poly(3-thiophene-β-propanesulfonate) (**27b**) was directly obtained by a chemical polymerization [81]. Attempts to prepare self-doped PTs from 3-(ω-carboxyalkyl)thiophenes (**30**) have also been reported [82]. These monomers are easily electropolymerized, leading to polymers with redox and spectroelectrochemical properties resembling those of PATs. While an auto-doping mechanism has been inferred from experiments performed in aqueous media, this hypothesis has not been definitively confirmed yet [82]. On the other hand, an intramolecular chemical reaction leading to the intermediate formation of a lactone has been proposed to account for the unusual electrochemical behavior of poly(3-thiopheneacetic acid) (**29**) in aqueous medium [83].

C. Polythiophenes with Specific Electrochemical Properties

1. Polyether Chains

The potentialities of PTs as materials for the elaboration of modified electrodes were recognized at an early stage and became a major focus of attention. Thus, the derivatization of PT with functional groups able to specifically interact with the physical or chemical environment allows the reversible alteration of the redox properties of the conjugated backbone. Advantage can thus be taken of these phenomena for the construction of selective electrochemical sensors. One of the first steps in this direction involved the substitution of the conjugated PT backbone by polyether chains. As a matter of fact, such substituents can lead to interesting properties such as solubility, enhanced hydrophilic character, and ionic sensitivity. Furthermore, the formation of an ether function constitutes an interesting method for covalent attachment of a functional group [84]. The electronic effects related to the position of the first oxygen atom in the side chain exert a strong influence on the formation and properties of the resulting polymer. As already discussed, the direct grafting of the oxygen atom on the thiophene ring leads to a considerable decrease in the oxidation potential (~700 mV), which has deleterious consequences for the efficiency of the electropolymerization [64]. However, this difficulty can be circumvented by increasing the length of the alkyl residue and by the use of chemical polymerization [49,50,85]. As could be expected, poly(3-alkoxythiophenes) have lower oxidation potentials than PATs and exhibit lower conductivities (~1 S/cm) [49,85,86]. On the other hand, these polymers exhibit good solubility and thermochromic properties [85,87].

The effects of the position of the first oxygen atom have been analyzed in more detail on two basic structures, namely, poly(3-butylthiophene) and poly(3-nonylthiophene) (PNT). In both cases the introduction of the oxygen after one methylene group raises the oxidation potential of the monomer by ~150 mV and decreases the conjugation length, doping level (y), and conductivity of the corresponding polymer (Table 12.1). In contrast, the introduction of an alkyl spacer involving at least two methylene groups brings the oxidation potential and absorption maximum back to values close to those of PNT and leads to more conjugated and more conductive polymers [88,89].

As the number of oxyethylene units in the side chain increases, electropolymerization becomes increasingly difficult, and complete inhibition was observed in the case of a monomer containing seven ether groups [88]. In the course of these investigations, poly(3-(3,6-dioxaheptylthiophene)) (PDHT) appeared as an interesting compromise between the conservation of extensive conjugation and high conductivity and the emergence of original properties associated with the oligooxyethylene side chains.

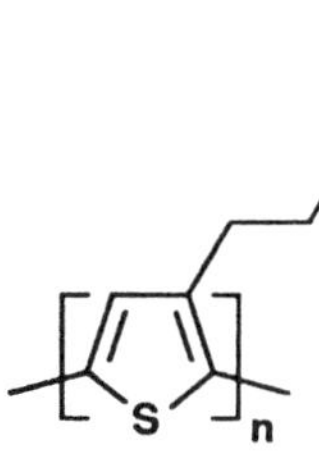

Thus, whereas the oxygen is electronically decoupled from the π-conjugated backbone, the optical spectrum of PDHT shows an absorption maximum at 556 nm and a well-resolved vibronic fine structure. Such optical features, characteristic of extensively conjugated PTs, have been observed only at low temperature on regiorandom PATs [78] or in the room temperature spectra of their regioregular head-to-tail versions [90,91]. These strong similarities suggest that electrogenerated PDHT has a rather rigid and regioregular structure [92]. The presence of the polyether side chain also produces several interesting modifications in the electrochemical properties of the polymer. Thus PDHT is highly hydrophilic and remains fully electroactive in aqueous media [93]. Furthermore, it is the first example of a PT derivative capable of being n-doped with alkali cations [94]. PDHT has been shown to exhibit solvatochromic properties in the solid state, and, as shown in Fig. 12.5, immersion of a PDHT film in a strong hydrogen bonding solvent pro-

Table 12.1 Effects of the Position of the Ether Function on the Properties of Poly(3-oxynonylthiophenes)[a]

Structure	$E_{ox\,mon}$ (V vs. SCE)	$E_{ox\,pol}$ (V vs. SCE)	y (%)	λ_{max} (nm)	σ (S/cm)
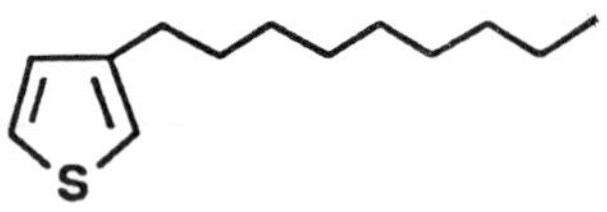 thiophene-nonyl	1.80	1.02	16	542	100
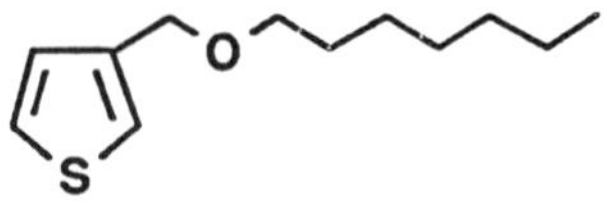 thiophene-CH$_2$-O-pentyl	1.94	1.10	6	460	0.1
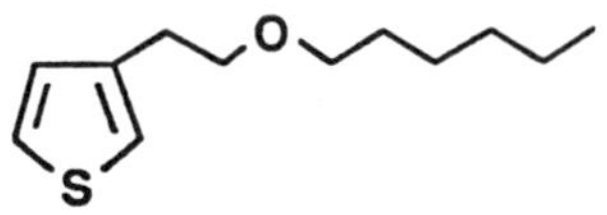 thiophene-CH$_2$-O-butyl	1.80	0.88	17	520	50
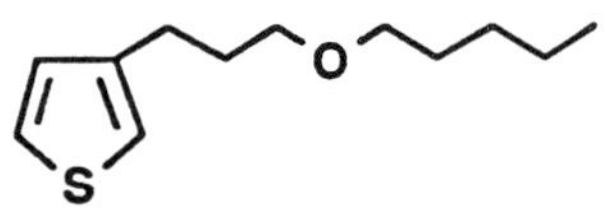 thiophene-(CH$_2$)$_2$-O-butyl	1.80	0.88	14	500	10

[a] mon = monomer; pol = polymer; y = doping level.

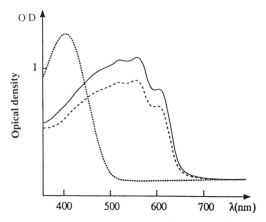

Fig. 12.5 Effect of hydrogen bonding on the in situ electronic absorption spectrum of undoped PDHT films on ITO. Solid line: Spectrum recorded at 0.0 V vs. SCE after redox cycling in 0.1 M LiClO$_4$-MeCN. Dotted line: Same film immersed in hexafluoroisopropanol. Dashed line: Same film at 0.0 V vs. SCE after redox cycling in 0.1 M LiClO$_4$-MeCN. (From Ref. 95.)

vokes a reversible dramatic loss of conjugation [95]. Specific electrochemical and spectroelectrochemical responses have been observed in the presence of Li$^+$, demonstrating that the interactions between the polyether chains and Li$^+$ can modify the geometry of the π-conjugated system [96] (Fig. 12.6). These original properties have been related to feedback control of the conformation and rigidity of the polymer backbone by the interactions between the polyether side chains and the chemical environment [96].

2. Crown Ethers

To improve the selectivity of the new class of cation-sensitive electroactive polymers, several groups have undertaken the synthesis of PTs derivatized with macrocyclic polyethers. However, the length of the spacer groups imposed by the size of such substituents could be expected to limit the possibilities of directly controlling the complexing properties of the crown ether from the redox state of the π-conjugated backbone. A first answer to this problem has consisted in the electropolymerization of a precursor involving two thiophene rings linked by a polyether chain (Scheme 2). Whereas the low potential of electropolymerization observed in the presence of Li$^+$ (~300 mV lower than PDHT) appears consistent with template-assisted ring closure [97], the cation-complexing properties of the resulting polymer have not been analyzed. A recent extension of this approach involves the chemical ring closure to afford a crown ether derivatized bithiophene precursor (Scheme 3).

Although chemical polymerization of this type of substrate leads to rather short polymer chains, the optical spectra of polymer solutions containing various alkali cations show that the absorption maximum undergoes a bathochromic shift whose magnitude depends on the nature of the cation [98].

A thiophene monomer with a crown ether moiety linking the 3- and 4-positions (**31**) has been synthesized. The corresponding polymer exhibits a short conjugation length (λ_{max} = 332 nm), probably because of the large steric hindrance between the crown ether subunits. When reacted with metallic sodium, the polymer afforded an n-doped polymer with excellent stability against oxygen [99].

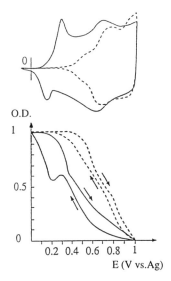

Fig. 12.6 Effects of cation on the simultaneous cyclic voltammetric (a) and cyclic voltabsorptometric (b) (λ = λ_{max} = 556 nm) responses of PDHT on ITO. Solid lines: In 1 M LiClO$_4$-MeCN. Dashed lines: In 1 M Bu$_4$NClO$_4$-MeCN. Scan rate 10 mV/s. (From Ref. 96.)

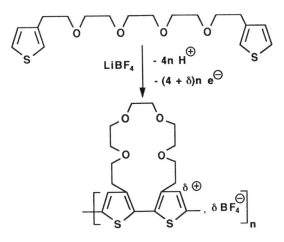

Scheme 2

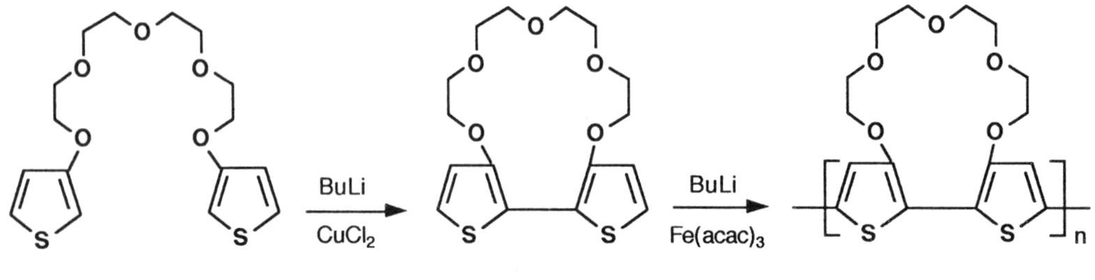

Scheme 3

With the development of 3-(ω-bromoalkyl)thiophenes [100], several new types of crown ether functionalized PTs have been synthesized. While electropolymerization of monomeric derivatives (32) failed, probably for steric reasons, bithiophene (33) and terthiophene (34) were easily electropolymerized to the corresponding crown ether derivatized PTs [101].

The analysis of the electrochemical response of these polymers in the presence of increasing concentrations of various alkali cations reveals a positive shift of the oxidation peak and a decrease in the amount of charge reversibly exchanged. Thus a linear dependence of the current versus cation concentration has been obtained, and the differences observed between the slopes corresponding to the different cations have confirmed the selectivity of these systems (Fig. 12.7).

3. Chiral Substituents

Conducting polymers containing optically active groups are potentially interesting for application as enantioselective electrochemical sensors or modified electrodes for electrocatalysis.

Several examples of chiral PTs (35–39) have been synthesized by electropolymerization of enantiomeric monomers [102,103]. In spite of the size of the attached substituents, these polymers retain high conductivities lying typically in the range of 1–10 S/cm. The high spe-

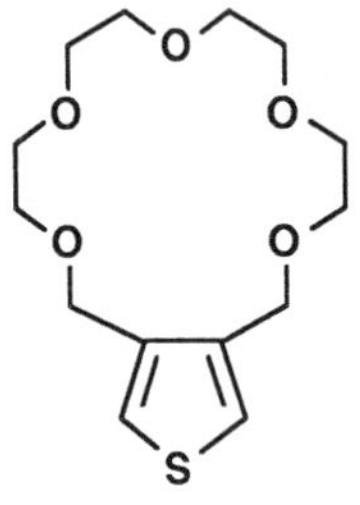

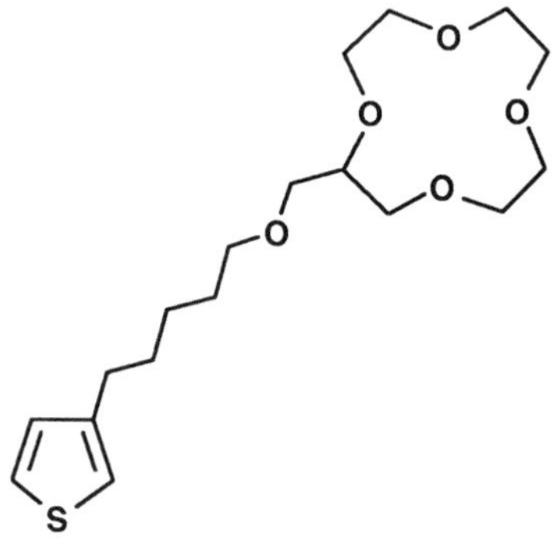

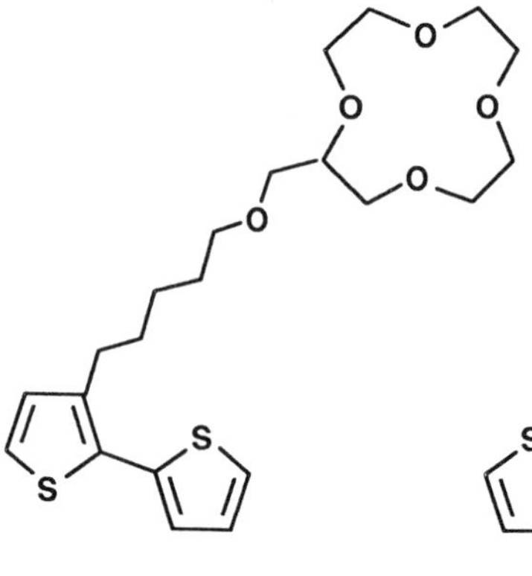

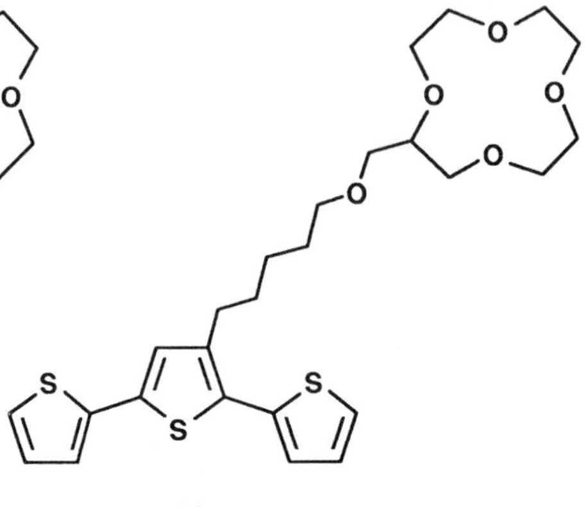

Molecular Design of Functional Conjugated Polymers

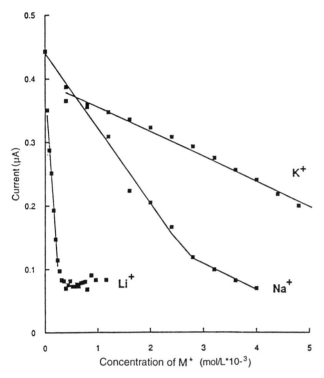

Fig. 12.7 Current versus concentration of various alkali cations. Currents were taken from each set of cyclic voltammograms of poly(**33**) at a fixed potential of 532 mV (Li^+), 590 mV (Na^+), and 570 mV (K^+). (From Ref. 101.)

cific rotations measured on the two antipodes of **37** (α_{22}^D = +3000° and −3000°) have been interpreted as due to a macromolecular asymmetry arising from a helical polymer structure [102].

Such high optical rotations have recently been confirmed on solutions of chemically synthesized **36** [104]. The electrochemical behavior of the two antipodic forms of **37** has been analyzed using d- and l-camphorsulfonates as supporting electrolyte, and the differences observed in the resulting electrochemical responses provided first evidence of enantioselective molecular recognition on chiral conducting polymers [102].

4. Redox Active Groups

In regard to the considerable number of functional polypyrroles containing redox active centers [105], examples of PTs derivatized with redox groups appear rather limited. While the higher conductivity of substituted PTs can present some advantages, in particular for electrochemical applications requiring thick films, the introduction of redox groups on the PT backbone poses several specific problems related to the more drastic structural conditions of functionalization of PT and to the higher oxidation potential of thiophene, which, in some cases, may lead to the degradation of the attached group during electropolymerization.

Conversely, the presence of a redox group of low oxidation potential on the monomer may sometimes limit or inhibit the electropolymerization. In the first steps in this direction, redox groups such as Fe(bipyridine)$_3^{2+}$ (**40**) [106], benzoquinone (**41**) [107], anthraquinone (**42**) [108], or viologen (**43**) [109] were grafted to the thiophene ring via spacers equivalent to two to five carbons. While the electrochemical signature of the attached group was observed in each case, these polymers were not characterized as bulk materials, and no conductivity data were reported.

A systematic analysis of the effects of the length of the alkyl spacer was carried out in the case of viologen-derivatized PTs [110]. This work showed that the lowest

because of the irreversible oxidation of the ferricinium groups at the potential needed for the polymerization of thiophene. This obstacle was circumvented by the use of a bithiophene precursor (**44**) of lower polymerization potential [111].

IV. CONTROL OF THE MACROMOLECULAR STRUCTURE BY SUBSTITUENT INTERACTIONS

Since the early works on functional PTs, modification of the monomer structure has been focused on obtaining either new bulk properties such as processability or polymers with specific electrochemical properties. Parallel to the continuation of these works a new research area recently emerged in which substituents having a strong tendency to self-assemble or to stack are grafted on the monomer. The purpose of this novel approach is to achieve better control of the long-range order and packing arrangement of the polymer chains in order to reach enhanced optical or charge transport properties. Until now research in this direction was essentially concerned with three classes of substituents: fluoroalkyl chains, mesogenic groups, and tetrathiafulvalenes.

A first demonstration of the feasibility of this strategy was provided by a series of PTs (**45–48**) in which substituted phenyl groups are linked to the 3-position of the thiophene ring by means of an oxabutyl spacer. The electronic and steric decoupling of the phenyl rings from the π-conjugated backbone is confirmed by the invariance of the absorption maximum at 520 nm, a value characteristic of extensively conjugated PTs [66,89]. However, as shown in Fig. 12.9, the resolution of the vibronic fine

value of the oxidation potential of the PT backbone was reached with an alkyl spacer involving 8–10 carbons (Fig. 12.8). In the case of ferrocene, electropolymerization of the derivatized monomer was inhibited, probably

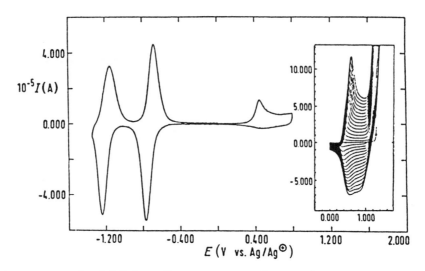

Fig. 12.8 Cyclic voltammogram of poly(**43**) ($n = 8$) ($\tau = 9.4 \times 10^{-8}$ mol/cm^2) in 0.1 M Bu$_4$NPF$_6$-MeCN, scan rate 50 mV/s. The inset shows the electropolymerization by repetitive potential scans. (From Ref. 110.)

Molecular Design of Functional Conjugated Polymers

[Structures 45, 46, 47, 48 shown: polythiophenes with -CH₂CH₂-O-CH₂-Ar side chains, where Ar = phenyl (45), 4-fluorophenyl (46), 4-trifluoromethylphenyl (47), pentafluorophenyl (48)]

structure, which is a measure of the rigidity of the conjugated system [112], exhibits a marked dependence on the substitution of the phenyl ring. This result shows that while substitution of the phenyl group has no direct effect on the π-conjugated system, the interactions between the substituents exert an indirect control on the structure and rigidity of the π-conjugated backbone. On the other hand, a correlation has been found between the resolution of the vibronic structure and the n-doping level [113]. This result together with other observations [114] confirms that the rigidity of the π-conjugated backbone is a crucial factor for efficient n-doping. An interesting implication of these results is that they strongly suggest that the spatial extension of negatively charged polarons and bipolarons could be larger than that of their positively charged counterparts.

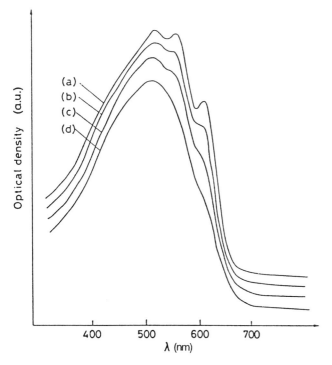

Fig. 12.9 Electronic absorption spectra of undoped polymers films on ITO. (a) Poly(**46**); (b) poly(**45**); (c) poly(**47**); (d) poly(**48**). (From Ref. 113.)

A. Fluorinated Polythiophenes

Fluorinated polymers are known for their high chemical and thermal stability, low coefficient of friction, and hydrophobicity. In addition to the fact that the association of these properties with the electrical properties of conjugated polymers can produce new materials with original properties, the propensity of fluorinated groups to self-assemble (fluorophilic effects) can constitute an interesting tool for controlling the long-range order in PTs.

While 3-halothiophenes have been known for a long time [62,63], 3-fluorothiophene (**49**) was synthesized only recently by means of a five-step synthesis [115]. Although 3-fluorothiophene has the highest oxidation potential among the 3-halothiophenes, the oxidation potential of the polymer (1.00 V vs. SCE) is the lowest within the series, probably because of smaller steric interactions.

Fluorinated 3-methylthiophene derivatives in which there is a gradual increase in the number of fluorine atoms in the side chain (**50–52**) have been reported recently [116]. While the oxidation potential increases in the order methyl > fluoromethyl > difluoromethyl by ~ 0.50 V, trifluoromethylthiophene seems to have a lower oxidation potential than mono- and difluorinated compounds. The redox potentials of the corresponding polymers are among the highest of the known PT derivatives (1.30–1.40 V vs. Ag), and these polymers are therefore unstable and spontaneously dedope when exposed to air [116].

Different strategies have been adopted to keep the oxidation potential of polyfluorinated 3-alkylthiophenes within acceptable limits. In the initial report on the subject, the effects of the length of an alkyl spacer inserted between the thiophene ring and the perfluoroalkyl moiety were analyzed, and the best results were ob-

[Structures 49, 50, 51, 52 shown: 3-substituted thiophenes with F, CH₂F, CHF₂, CF₃ respectively]

tained for an alkyl spacer involving three —(CH$_2$)— groups (**54, 55**) [117]. In these conditions, introduction of up to 50% fluorine in the polymer structure leads to polymers showing elastomeric properties, together with enhanced electroactivity and electrochemical stability [117].

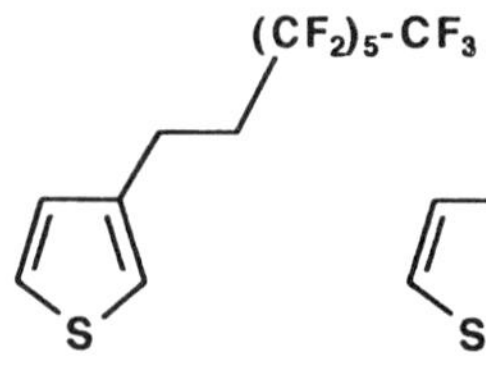

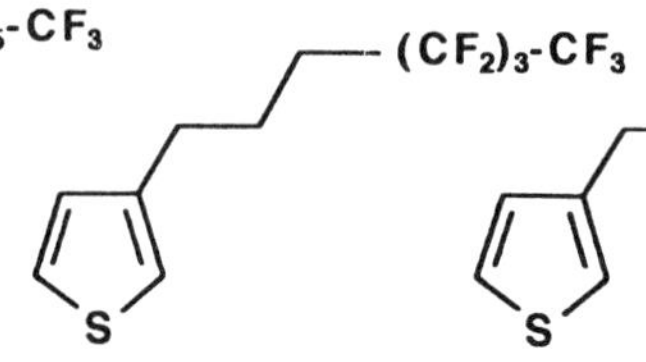

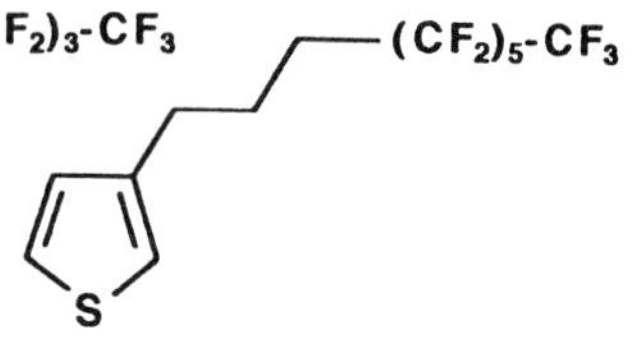

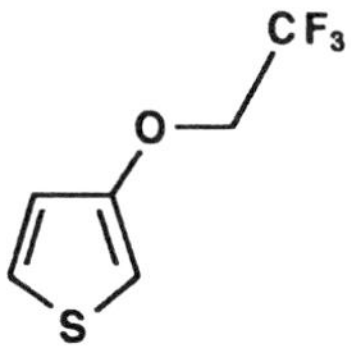

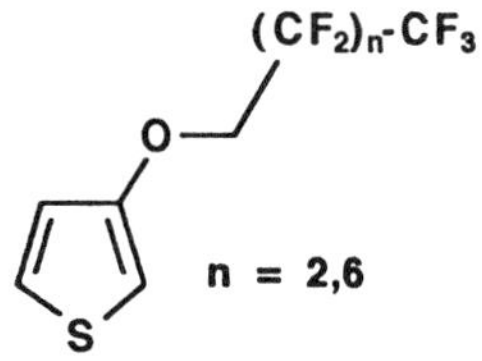

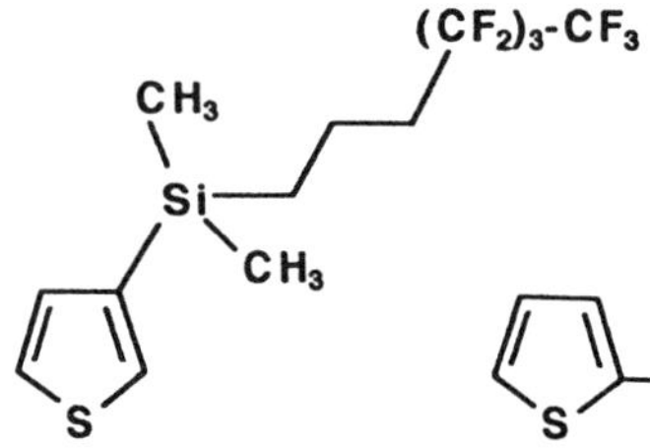

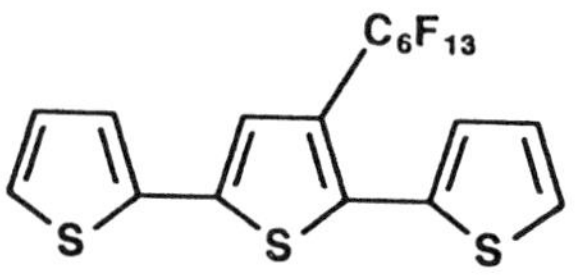

More recently, polyfluorinated PTs with rather high conductivities ($\sigma = 1$–10 S/cm) were obtained from monomers in which the oxidation potential was lowered by insertion of an alkoxy spacer between the thiophene ring and the perfluoroalkyl chain (**56, 57**) [118]. On the other hand, the direct attachment of a perfluorohexyl chain to the thiophene ring was achieved using a trimeric precursor (**59**). This substitution raises the oxidation potential of both the precursor and the polymer by 0.30–0.40 V [119].

Unlike that of the 3-alkylthiophenes, the synthesis of polyfluorinated thiophene monomers is not straightforward and requires multiple steps [120,121]. It has been shown that a polyfluoroalkyl chain could be directly attached at the 3-position of the thiophene ring by means of the single coupling of a terminal dimethylchlorosilane group to 3-lithiothiophene (**58**). Despite the steric hindrance expected from the direct grafting of the dimethylsilyl group at the 3-position, this monomer could be readily electropolymerized, suggesting that fluorophilic interactions are strong enough to counterbalance the steric hindrance [122].

B. Mesogenic Groups

The covalent attachment of mesogenic substituents on the framework of PT is another potentially powerful method for controlling the long-range order and orientation of the polymer chains. Until now this interesting approach has been considered in only a few papers. A seminal work reported the synthesis of thiophenes **60–62** [123]. Monomers **61** and **62** were claimed to exhibit mesogenic properties, but no detailed characterization was reported.

Monomer **60** was reported to undergo electropolymerization, but both the low conductivity and the yellow color of the films indicate a limited effective conjugation, probably for steric reasons. Other thiophenes containing the *p*-cyanobiphenyl mesogenic group (**63, 64**) have been synthesized, but the mesomorphic properties of the monomers and polymers were not investigated [124]. Thiophene **65** was prepared using ω-bromooctylthiophene as starting material. This compound undergoes thermotropic transitions, and textures typical of nematic and smectic phases have been observed [125]. In contrast to the previous ones, this monomer can be readily electropolymerized, leading to free-standing films with electrochemical and electrical properties comparable to those of PATs. However, the mesomorphic properties of the polymer have not yet been analyzed.

C. Tetrathiafulvalene

Organic conductors based on the cation radical salts of tetrathiafulvalene (TTF) derivatives consist of segre-

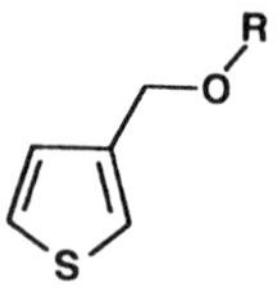

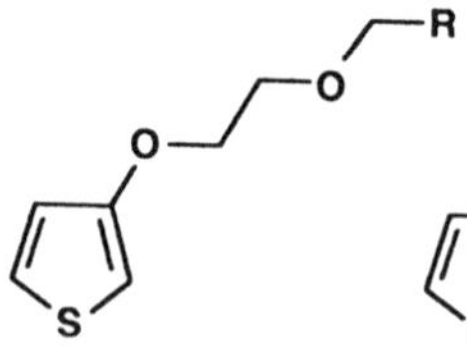

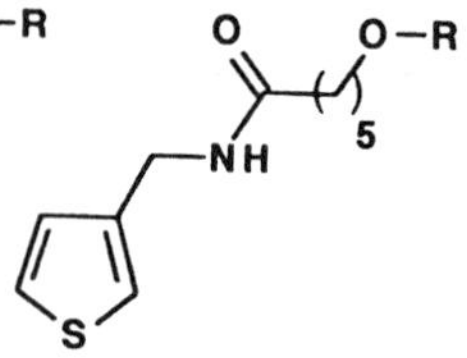

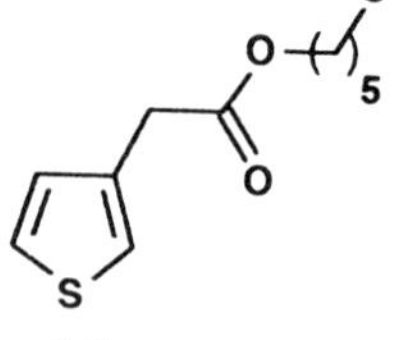

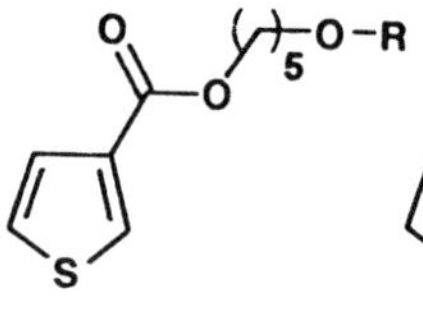

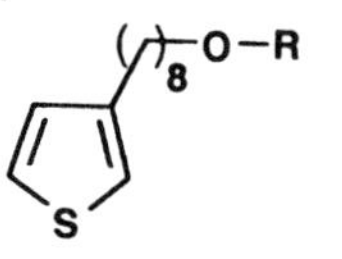

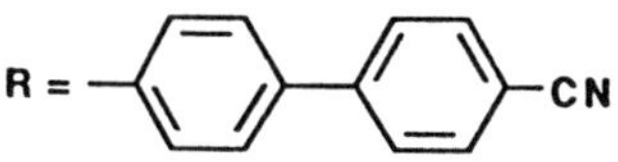

66 **67** **68**

gated stacks of π-donors and counter anions. The short interplanar distances allow significant interactions between π-molecular orbitals of neighbors, which results in a highly anisotropic conductivity along the stacks [126]. On this basis, the covalent grafting of TTF moieties on the backbone of a conjugated polymer represents a potentially very interesting strategy for controlling the orientation and long-range order of the polymer chains. Furthermore, the association of two different mechanisms of charge transport in the same material can contribute to increasing the dimensionality of the conduction and extending the potential window of high conductivity.

As a first step in this direction, a short communication [123] reported the synthesis of monomers **66** and **67**.

However, only in one case (**66**) did electrooxidation give rise to polymerization, and furthermore the orange color of the resulting doped polymer indicates a rather short conjugation length. This limited conjugation probably arises from steric interactions between TTF moieties linked to adjacent thiophene rings by too short spacer groups. Such problems were not encountered in the case of **68** with an oxadecyl spacer. This monomer readily electropolymerizes in nitrobenzene, leading to conducting dark blue films [127]. As expected, the cyclic voltammogram of the resulting film exhibits the typical two successive reversible oxidations of TTF followed by the redox system of PT (Fig. 12.10). Interestingly, the cathodic branch of the first redox system involves two components at 0.41 and 0.32 V. While the 0.41 V wave corre-

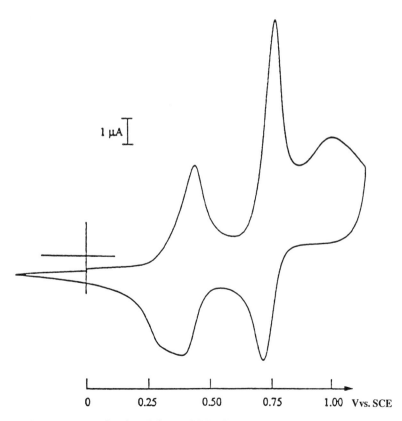

Fig. 12.10 Cyclic voltammogram of poly(**68**) in 0.1 M Bu$_4$NPF$_6$-MeCN, scan rate 10 mV/s. (From Ref. 127.)

sponds to the reduction of the TTF cation radical and of its dimer dication, the less positive component reflects the reduction of the mixed-valence TTF dimer $(TTF)_2^+$, indicating that self-assembly of the oxidized TTF moieties takes place in the material [127].

D. Three-Dimensional Polythiophenes

Whereas the above strategies tend to confer a two-dimensional character on the conjugated PT backbone, the possibility of building up three-dimensional conjugated systems has been envisioned very recently. In a first attempt in this direction, tetrahedral PT precursors in which four polymerizable monomers are linked to a central silicon atom have been synthesized. While the insolubility of **69** did not allow any polymerization (A. Guy and J. Roncali, unpublished), electrooxidation of **70** leads to the deposition of a tetrakis cation radical salt that can be subsequently converted into a material containing sexithienyl units by further electrooxidation [128]. Although reported electrochemical, spectroscopic, and EDX data are in accord with the expected 3D structure, further physical characterizations are needed to definitively confirm this conclusion.

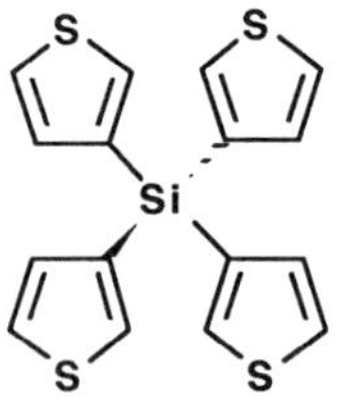

69

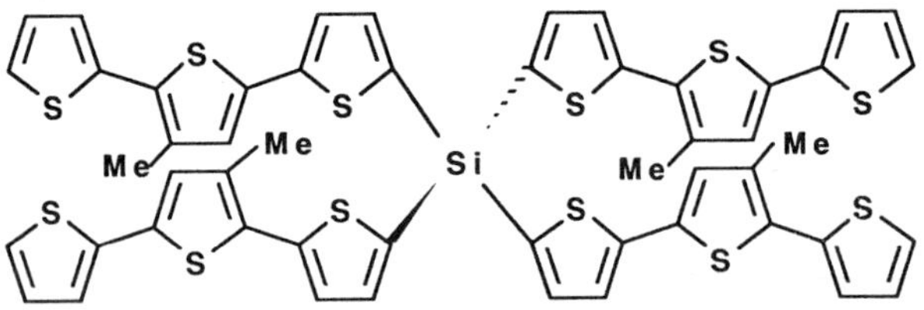

70

To summarize, the control of the organization and dimensionality of PT by substituent interactions is one of the most recent trends in the field. Although some promising preliminary results have been obtained, a critical evaluation of the real potentialities of this approach is obviously premature at this very early stage of development.

V. THE OLIGOMERIC APPROACH

Parallel to the continuing development of research on PT, recent years have witnessed a strong renewal of interest in thiophene oligomers (nTs). Owing to their processibility by vacuum sublimation techniques, nTs have been used as active components in electronic devices such as diodes [129], transistors [130], and electroluminescent diodes [131]. At a more fundamental level, because of their simple and unequivocal structure, nTs constitute interesting simplified models for the analysis of the electropolymerization mechanism of thiophene [33,35,38,40] for the study of the electronic and electrochemical properties of PT [26,132–135] (see also chapter by F. Garnier, this volume). The conjunction of these technological and fundamental potentialities has given rise to a considerable number of physical studies. A detailed discussion of these work is outside the scope of this chapter, and this section essentially focuses on some of the most representative results related to the chemistry of oligothiophenes.

A. Synthesis of Oligothiophenes

Oligothiophenes have been known for several decades [136]. Whereas initial syntheses used Ullman-type reaction of iodothiophene with copper bronze [136,137], the coupling of α-lithiated short-chain oligomers such as bi- or terthienyl in the presence of $CuCl_2$ [138] or the coupling of Grignard compounds in the presence of bis(diphenylphosphino)-1,3-propane-nickel(II) chloride ($NidpppCl_2$) [139] are presently the most widely employed syntheses. nTs can also be obtained by electrochemical duplication of shorter chain nTs [33,39], and many of the modern methods of aryl–aryl bond formation including the coupling of organozinc, organotin, or boronic acids with halogenothiophenes in the presence of nickel or palladium catalysts are also effective [140,141] (Scheme 4). One of the major problems encountered in the synthesis of unsubstituted nTs (**71**) is their very low solubility, which limits the maximum attainable length to sexithiophene, although characterization of septithiophene has been reported [140]. To obtain soluble nTs of longer chain length, several groups have developed the synthesis of nTs β-substituted at some of the thiophene rings by alkyl or oxyalkyl chains (**72**) [38,141–146].

However, the regiorandom placement of the side chains prohibited the isolation of homogeneous material. This problem has been addressed through different approaches such as the synthesis of fully symmetrical nTs [141], the introduction of 3,4-disubstituted thiophene rings in the nT chain (**73**) [145], and, more recently, the

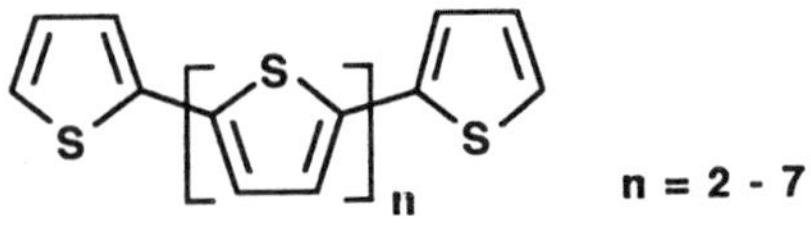

n = 2 - 7

71

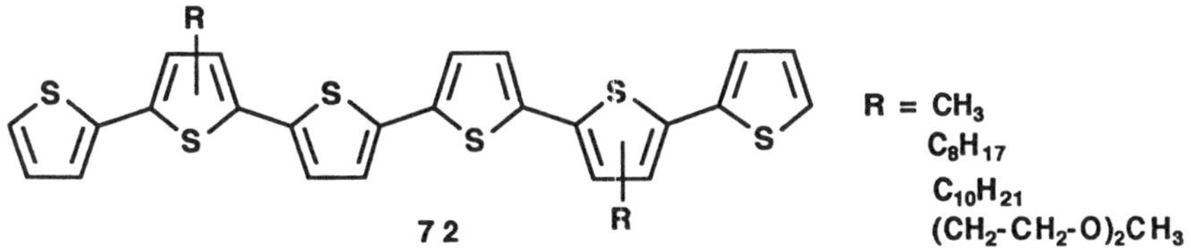

Molecular Design of Functional Conjugated Polymers

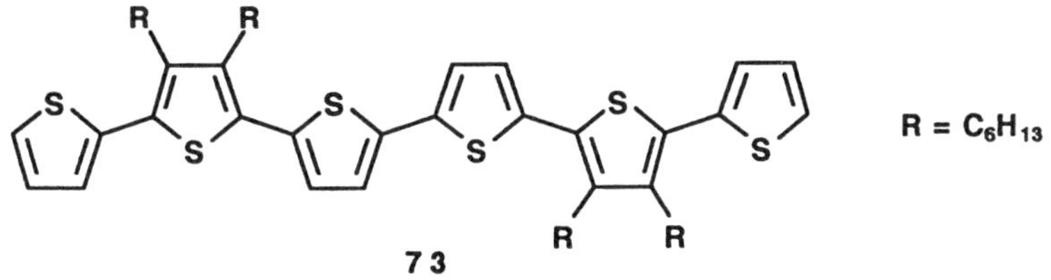

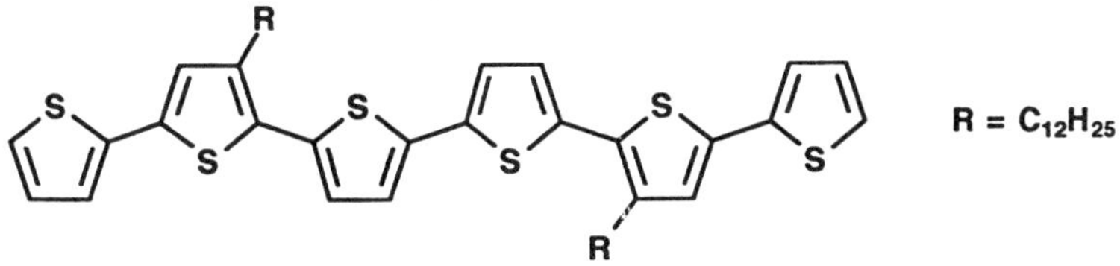

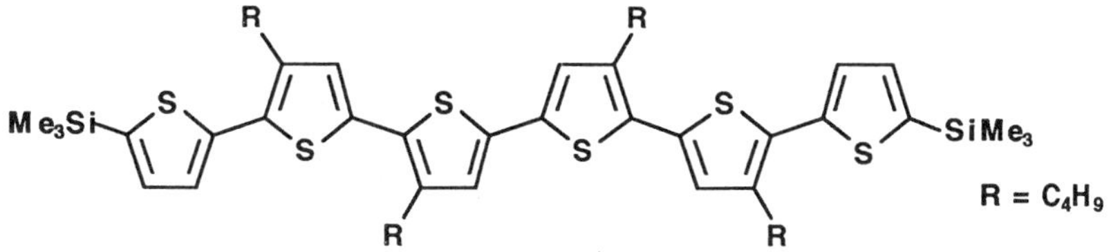

72 R = CH₃, C₈H₁₇, C₁₀H₂₁, (CH₂-CH₂-O)₂CH₃

73 R = C₆H₁₃

74 R = C₁₂H₂₅

75 R = C₄H₉

synthesis of the first isomerically pure dialkylsexithiophene (**74**) [146].

While these materials have permitted the analysis of some properties of nTs such as charge carrier generation [147] or thermochromism [145], their possible coupling when electrooxidized in polar solvents [37,143] can render their use as models of isolated PT chains rather delicate. This problem has been solved by the synthesis of nTs with one or both terminal α-positions blocked by a protecting group (**75–77**) [133,141,147].

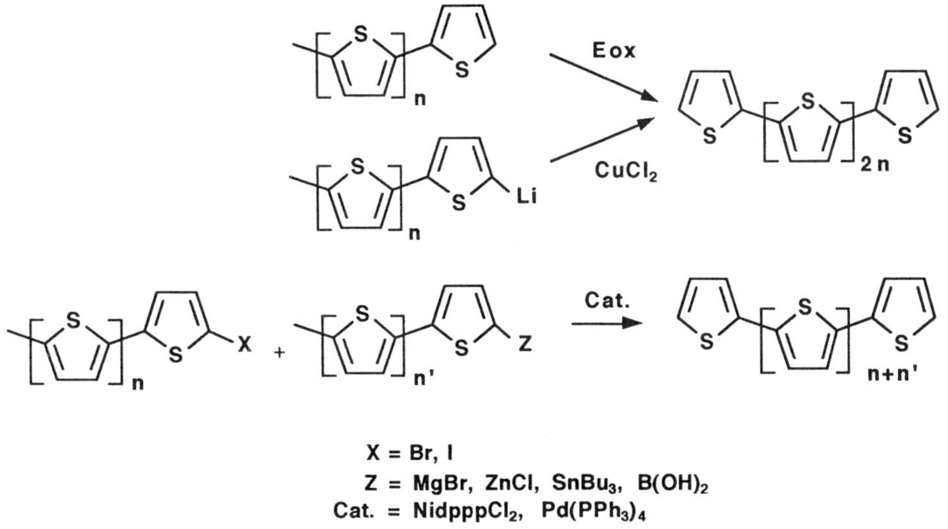

X = Br, I
Z = MgBr, ZnCl, SnBu₃, B(OH)₂
Cat. = NidpppCl₂, Pd(PPh₃)₄

Scheme 4

76

77

Table 12.2 Conductivity, Oxidation Potential, and Absorption Maximum of Polythiophene Extrapolated from Data Obtained on Oligomers

Compd	n	σ (S/cm^a)	$E_{ox}1^a$	$h\nu_{max}$ (eV)	Ref.
71	2–5	1600			26
76	2–7		0.07	2.30	133
75	3–8		0.05	2.30	134
71	2–5			2.70	132

^{a} In volts versus ferrocene.

B. Properties of Oligothiophenes

The wide current interest in nTs is largely motivated by their use as model compounds for the study of the electronic properties of the parent polymer. As a matter of fact, the uncertainty in the results of theoretical calculations and the large variations in the experimental data obtained on polymers prepared using different methods make very difficult a precise determination of parameters such as ionization potential, bandwidth, and energy gap that control the electronic properties of the polymer. In this context nTs appear as interesting model compounds for which experimental quantities such as oxidation and reduction potentials, optical transitions, or conductivity can be correlated to defined conjugation lengths and/or compared to theoretical results.

As could be expected, the conductivity of doped nTs increases with chain length, and values have been reported ranging from 2×10^{-3} S/cm for the dimer to 2 S/cm^{-1} for the pentamer [26]. However, more recent studies revealed that these values might not reflect the intrinsic conductivities of nTs and that short-chain nTs are subject to post polymerization upon doping [148]. Data obtained on longer chain lengths have shown that a conductivity of 20 S/cm, i.e., approaching that of doped PATs, could be obtained with an alkyl-substituted undecithiophene [148].

The evolution of the electrochemical, optical, and electrical properties of nTs as a function of the number of thiophene rings in the chain (n) has been the subject of several studies. As expected, increasing n leads to a steady decrease of the first oxidation potential ($E_{ox}1$). From n = 2, a second oxidation wave appears whose peak potential ($E_{ox}2$) also decreases with n. The increase in n produces a decrease in the difference $E_{ox}2 - E_{ox}1$, which reflects a reduction of the Coulombic repulsion between positive charges in the dicationic state. With the **76** series, $E_{ox}2 - E_{ox}1$ was found to cancel when $1/n$ tends to zero [133], while it was found to be constant (~0.18 V) beyond n = 6 for the **75** series [134]. Finally in the case of the **77** series, a two-electron single oxidation wave was observed for n = 8 [149], which is in satisfactory agreement with theoretical calculations [150]. These contradictory behaviors show that the possibility of directly oxidizing the nT chain to its dicationic state can be strongly influenced by substitution. On the other hand, the extrapolation of these results to an infinite chain leads to different views concerning the oxidation process of the polymer, which is considered to involve either a single multielectronic step [133] or two discrete steps [149]. There is better agreement concerning the chain-length dependence of the optical properties since both **75** and **76** series lead to an extrapolated value of 2.30 eV for the absorption maximum [133,134]. Using the same methodology, optimal values of the conductivity, oxidation potential, absorption maximum, and energy gap of an ideal defect-free PT chain have been proposed by extrapolation of experimental data obtained on nTs of increasing length (see Table 12.2).

Detailed electrochemical studies on nTs have also provided new insights into the redox properties of PTs and the charged species responsible for the electrical conduction in PTs. Thus, in the case of **74**, first evidence of reversible reduction to the mono- and dianion states was obtained (Fig. 12.11). On the other hand, analyses of the redox behavior of nTs by ESR, UV-vis absorption, and cyclic voltammetry provided evidence for the dimerization of oligomeric cation radicals (polarons) into spin-dimerized polarons, which have been proposed as

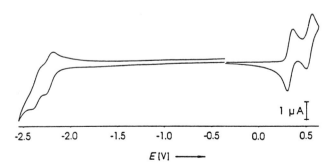

Fig. 12.11 Cyclic voltammogram of **74**. Oxidation in CH$_2$Cl$_2$, reduction in THF, scan rate 100 mV/s; potentials are referenced to ferrocene. (From Ref. 151.)

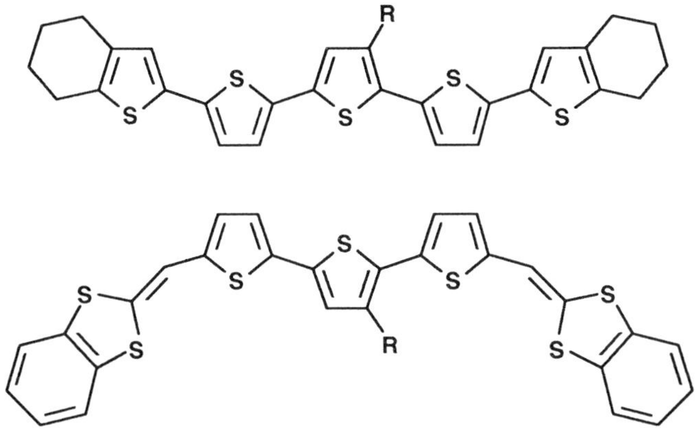

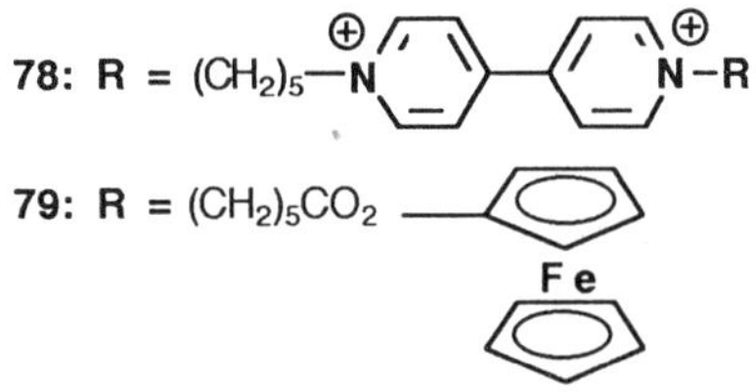

80 a: R = CH_3
b: R = C_8H_{17}
c: R = $(CH_2\text{-}CH_2\text{-}O)_2\text{-}CH_3$

an alternative to bipolarons in the mechanisms of charge transport [135,149,151,152].

C. Functional Oligothiophenes

A very recent trend concerns the use of nTs as the basic unit for building more sophisticated molecular architectures with properties specifically tailored to the development of molecular electronic devices. Thus, end-capped nTs functionalized by ferrocene (**78**) or viologen (**79**) have been described. These compounds exhibit the specific electrochemical responses of both the attached functional group and the nT chain [153]. nTs end-substituted by the 1,3-dithiole moiety (**80**) have been synthesized [154]. These substituted nTs, which can also be viewed as extensively conjugated analogs of tetrathiafulvalene, exhibit a considerable reduction of oxidation potential compared to those of their constitutive building blocks. Furthermore, the substitution of the nT spacer allows the alteration of the redox behavior, and a cation-dependent electrochemical response has been observed with a compound bearing a polyether side chain (**80c**) [155].

Spiro-fused thiophene trimers and heptamers (**81**) have been synthesized. The heptamer segments oxidize

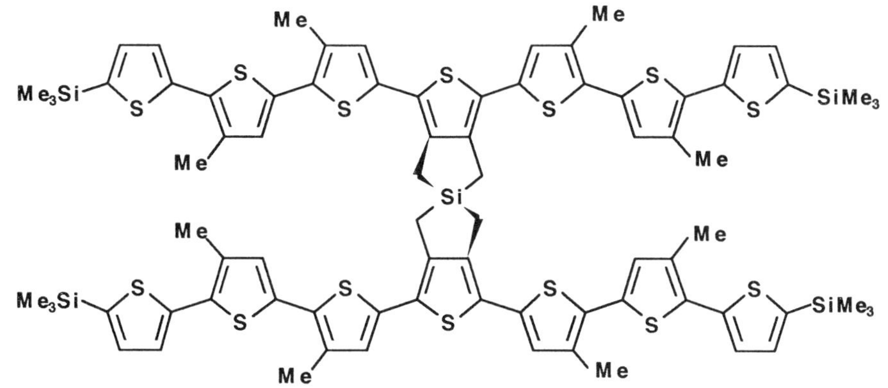

81

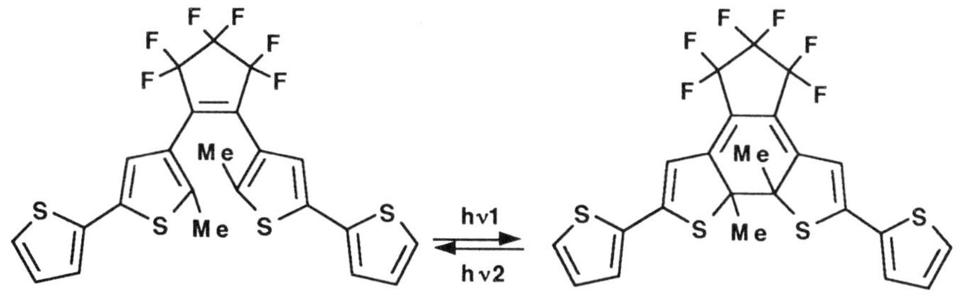

82

Scheme 5

stepwise to produce the monocation radical, the bis(cation radical), the cation radical/dication, and the bis(dication) [156]. While there is no significant interaction between the orthogonally positioned cations, a weak magnetic coupling has been suggested. Recently the synthesis of an nT-based molecular photoswitch (82) was reported. Photocyclization of the open structure by irradiation with 350 nm light leads to the formation of an extended conjugated system with an ~1.0 V lower oxidation potential and a 300 nm red-shifted absorption maximum [157]. Subsequent irradiation with 550 nm light converts the closed form into the initial open structure (Scheme 5).

VI. SMALL BAND GAP POLYTHIOPHENES

The design of narrow band gap (E_g) conjugated polymers is probably one of the most exciting challenges in the field of organic conductors. Since E_g governs the intrinsic electronic and optical properties of the π-conjugated system, it constitutes the key factor in many fundamental and technological problems. In addition to potential applications in electronics and nonlinear optics, the synthesis of narrow band gap conjugated systems can pave the way toward the development of intrinsically conducting organic materials and also contribute to a better understanding of the structural factors that condition the eventual obtainment of a superconducting polymer [158].

A. Small Band Gap Polymers

Theoretical considerations have shown that E_g should decrease as a function of the increase in the quinoid character of the π-conjugated system at the expense of its aromaticity [159]. On this basis, various strategies aimed at increasing the quinoid character of PT have been developed. The most representative example of this approach is polyisothianaphthene [poly(83)] in which annellation of a benzene ring to the thiophene nucleus leads to a decrease in E_g from 2.20 to ~1.15 eV [160]. A small band gap polymer was obtained by chemical polymerization of 2,3-dihexylthieno[3,4-b]pyrazine [poly(84)] [161]. This polymer, which exhibits an E_g value comparable to that of poly(83), has the advantage of being solution-processable.

To reduce the steric interactions between the annelled rings, trimeric precursors in which the fused ring system occupies the median position have been synthesized (85–87). Due to the stabilization of the corresponding radicals, these precursors are rather difficult to electropolymerize [162]. Nevertheless, E_g values ranging between 1.70 eV for poly(85) [163] and 0.90 eV for poly(87) [164] have been obtained (Table 12.3).

Table 12.3 Small Band Gap Polythiophenes Based on Fused-Ring Precursors

Precursor	Polymer band gap E_g (eV)	Ref.
83	 1.10	160
84	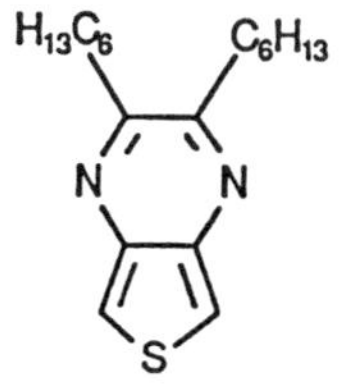 0.95	161
85	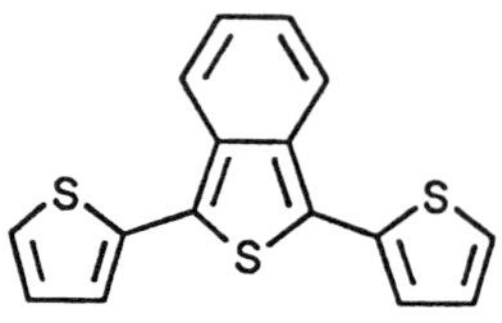 1.70	163
86	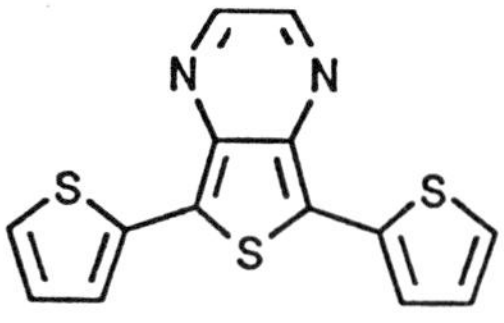 1.00	162
87	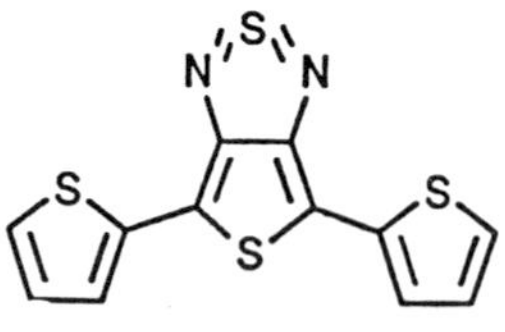 0.90	164

Another approach involves the introduction of electron-withdrawing substituents such as a keto (88) or a dicyanomethylene (89) group at an sp^2 carbon bridging the 4,4'-positions of a bithienyl precursor [165,166]. While these substituents do not greatly affect the position of the HOMO compared to bithiophene, they induce a significant lowering of the LUMO level and thus a reduction of the HOMO–LUMO gap (Table 12.4). More recently, a considerable reduction of the band gap of polybithiophene was achieved by the introduction of a dioxolane group at an sp^3 carbon bridge (91) [167]. Thus, both the optical spectra and the cyclic voltammogram of poly(91) (Fig. 12.12) indicate a decrease in E_g from

Molecular Design of Functional Conjugated Polymers

Table 12.4 Small Band Gap Polythiophenes Derived from Bridged Bithienyl Precursors

Precursor		Polymer band gap (eV)		Ref.
		$E_{g_{opt}}$	$E_{g_{el}}$	
88	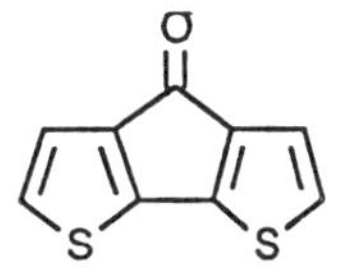 (cyclopenta-dithiophene ketone)	1.20		165
		1.20	1.10	168
89	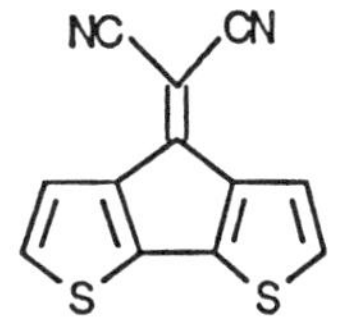 (dicyanomethylene derivative)	1.00	0.80	166
90	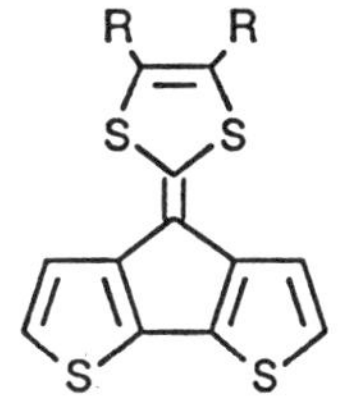 (dithiole derivative)	1.50		167
91	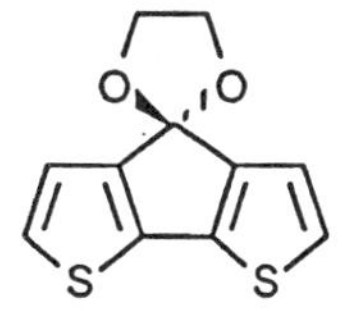 (dioxolane derivative)	1.10	0.90	168

$E_{g_{opt}}$: optical band gap.
$E_{g_{el}}$: electrochemical band gap.

2.20 to 0.90–1.00 eV. However, since the electron-withdrawing effects of the dioxolane group are significantly smaller than those of the keto or dicyanomethylene groups in **88** and **89**, it has been concluded that the increase in the overall planarity and rigidity of the π-conjugated backbone should also strongly contribute to the reduction of E_g.

As a matter of fact, several experimental and theoretical studies have shown that nTs and their cation radicals

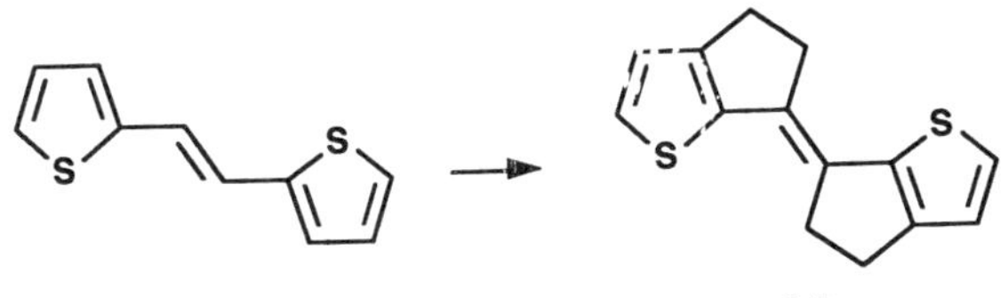

Scheme 6

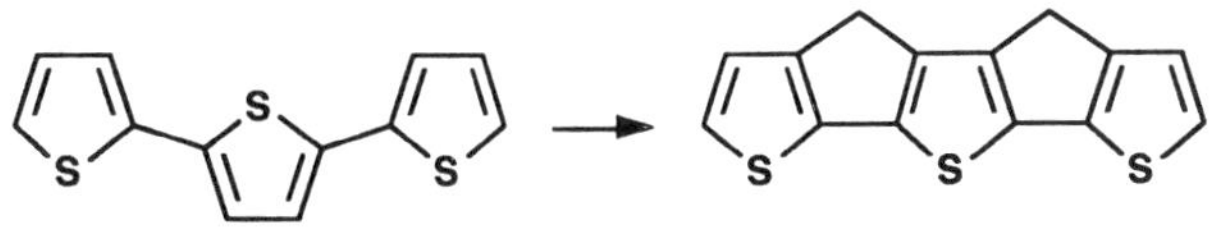

Scheme 7

are subject to interannular rotations around single bonds [168–173]. Therefore, it is evident that a statistical deviation from planarity must exert a major influence on the magnitude of the band gap. This conclusion has led to the definition of a novel strategy for the reduction of the band gap based on the rigidification of the π-conjugated system. In addition to a decrease or complete suppression of rotational disorder, the bridging of the monomer units can be expected to reduce bond-length alternation, which is in large part responsible for the existence of a finite band gap in low-dimensional π-conjugated polymers. As a first example, the application of this strategy concerned the synthesis of a rigid version (**93**) of dithienylethylene (**92**) (Scheme 6). As shown in Fig. 12.13a, rigidification leads to a strong enhancement of the resolution of the vibronic fine structure together with a bathochromic extension of the absorption to ~460 nm. As could be expected, this results in a marked reduction of the band gap, which decreases from 1.80 eV for poly(**92**) [174] to 1.40 eV for poly(**93**) (Fig. 12.13b) [175], thus demonstrating that significant band gap reduction can be achieved without resorting to mesomeric or inductive electronic effects.

A further step in this direction was accomplished by the synthesis of a first example of fully rigidified terthienyl (**94**) (Scheme 7). Although the dramatic decrease in oxidation potential and hence the strong stabilization of the resulting cation radical make this compound difficult to polymerize, the resulting polymer exhibits a band gap of ~1.00 eV (Fig. 12.14), which ranks among the smallest values reported so far [176]. On the other hand, the comparison of the cyclic voltammogram of **94** and terthienyl (TT) reveals several important differences. Both the first and second oxidation peaks shift negatively from 1.07 and 1.65 V for TT to 0.83 and 0.60 V for **94**. Furthermore, whereas TT is reduced at −2.0 V in DMF [132], reduction of **94** occurs at −0.75 V in MeCN and reversible reduction to the dianion state can be observed at −1.00 and −1.15 V in THF (Fig. 12.15). These shifts result in a considerable decrease in the potential difference between the first oxidation and reduction waves that corresponds to a reduction of the HOMO–LUMO gap (ΔE) from 3.10 to 1.35 eV. The fact that effects of such magnitude are observed on a simple terthienyl molecule confirms that the rigidification represents a powerful new approach of the problem of the synthesis of small band gap π-conjugated systems.

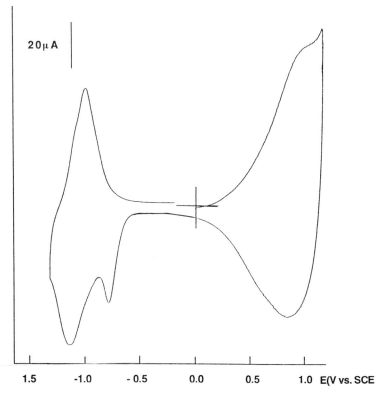

Fig. 12.12 Cyclic voltammogram of poly(**91**) on platinum. Electrolytic medium 0.1 M Bu$_4$NPF$_6$-MeCN; scan rate 50 mV/s. (From Ref. 168.)

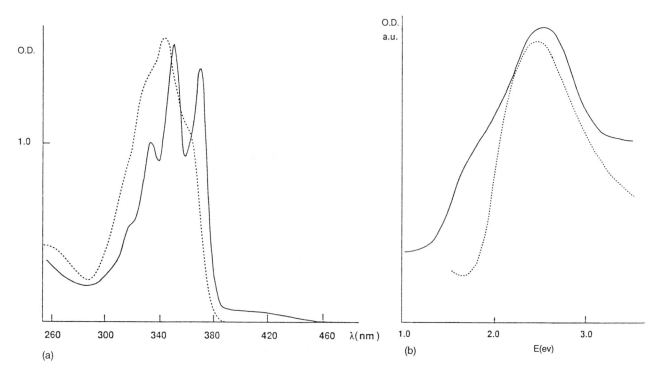

Fig. 12.13 (a) Electronic absorption spectra of **92** (dotted line) and **93** (solid line) in CH$_2$Cl$_2$. (b) Electronic absorption spectra of neutral polymer films on ITO. Dotted line, poly(**92**); solid line; poly(**93**). (From Ref. 175.)

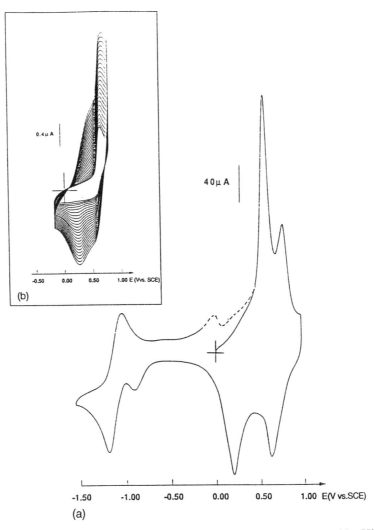

Fig. 12.14 (a) Cyclic voltammogram of poly(**94**) on Pt in 0.1 M Bu$_4$NPF$_6$-MeCN, scan rate 20 mV/s. (b) The electropolymerization by repetitive potential scans. (From Ref. 176.)

B. Small Band Gap Molecular Systems

The results just discussed underscore the interest of working on molecular systems of defined structure, especially when a detailed understanding of the effects of structural variations is the goal. Furthermore, due to the considerable deleterious consequences of rotational disorder on the electronic properties of PTs, similar or even better performance in terms of band gap might be expected with adequately tailored short-chain oligomers rather than with the polymer itself. In fact, several reports have described short-chain oligomers in which a considerable reduction of ΔE was achieved by introduction of electron-withdrawing substituents. Thus, increasing n from 1 to 4 in compounds **95** leads to a decrease from 1.81 to 1.15 V in the potential difference between the first oxidation and reduction potentials while λ_{max} shifts bathochromically from 2.95 to 1.49 eV [177]. More recently, ΔE values of 1.20–1.30 eV were reported for compounds **96** combining thiophene and cyclopentadienone rings [178].

X-ray studies on tetrathiafulvalene analogs built around nT-conjugated spacers have shown that the end parts of these molecules are stabilized by strong interactions between the sulfur atoms of the thiophene and 1,3-dithiole rings [179]. Consequently, suppression of the remaining possible rotation between the thiophene rings achieved in **97** and **98** leads to a fully rigidified π-conjugated structure. As shown in Fig. 12.16, these compounds, which can be reversibly charged to the tricationic and trianionic states, exhibit ΔE values as low as 0.90 V [180]. Furthermore, the optical spectrum recorded on a vacuum-evaporated thin film (Fig. 12.17) confirms that in the solid state the band gap of this material is close to the smallest values reported so far on conjugated polymers.

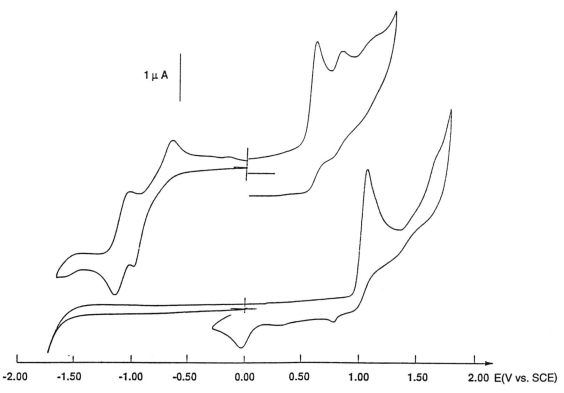

Fig. 12.15 Single-scan cyclic voltammogram of **94** (top) and terthienyl (bottom) in 0.1 M Bu$_4$NPF$_6$, scan rate 50 mV/s. Oxidation in MeCN, reduction in THF.

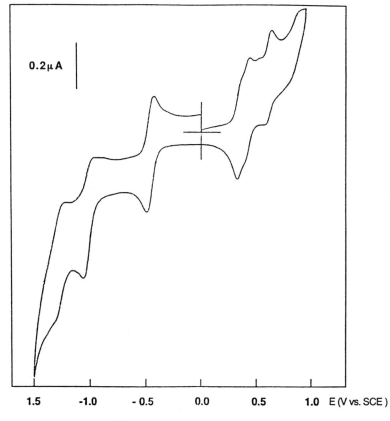

Fig. 12.16 Cyclic voltammogram of 10^{-3} M **98** in 0.1 M Bu$_4$NPF$_6$-CH$_2$Cl$_2$; scan rate 100 mV/s. (From Ref. 180.)

Molecular Design of Functional Conjugated Polymers

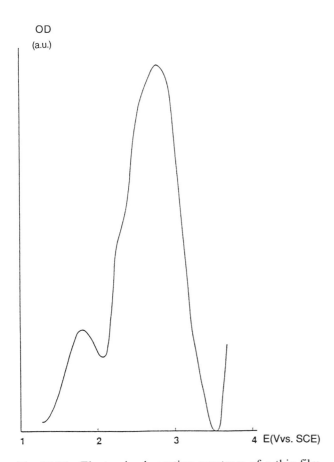

Fig. 12.17 Electronic absorption spectrum of a thin film of **98** vacuum-evaporated on glass.

VII. CONCLUSION AND PERSPECTIVES

Conjugated polymers based on thiophene derivatives have a 15 year history. This period has been marked by the considerable progress accomplished in the control of the structural definition of PTs and in the understanding of the mechanisms responsible for their electronic and electrochemical properties. Although synthetic chemistry has contributed to this progress, its major contribution has concerned the development of new materials with unusual properties. Thus, highly conducting and processible polymers are now available, allowing the development of large-scale bulk applications such as antistatics or EMI shieldings. Whereas hopes based on early envisioned applications such as rechargeable batteries or electrochromic devices have not yet been fulfilled, the derivatization of polythiophene has opened new avenues to advanced applications that were not even imagined 15 years ago, such as supercapacitors, selective sensors, and electronic or optoelectronic devices. However, in order for these applications to go beyond the laboratory scale, several conditions such as reproducibility, processability, and acceptable cost must be simultaneously fulfilled, which implies the need for further progress in the chemistry of PTs. Although these problems are still subject to intensive research effort, recent years have witnessed the emergence of new research areas. Thus thiophene oligomers are attracting renewed interest both as model compounds for basic research and for their intrinsic properties, which in many instances approach or even surpass those of the polymers. On the other hand, small band gap π-conjugated systems have become the focus of growing current interest. However, the development of this field presents some similarities with the situation of conjugated polymers 15 years ago. Thus, despite interesting perspectives opened by several recent works, some results unpredicted by theoretical considerations have underlined the rather rudimentary level of understanding of the structure–properties relationships in this field.

Linearly π-conjugated systems with small or even vanishing band gaps have an enormous potential of development. In addition to the obvious technological implications, the long-term achievement of intrinsic conductivity and perhaps of a superconducting polymer remains a quite exciting perspective. The synthesis of a "true organic metal" is of course, an ambitious objective that might well become the challenge of the next decade. Such a goal will imply both a better understanding of the structural factors that control the magnitude of the band gap and the ability to build molecular architectures capable of fulfilling these requirements, and in this regard synthetic chemists still have plenty of work to do.

ACKNOWLEDGMENTS

I thank Professor M. Lemaire (Laboratoire de Catalyse et Synthèse Organique, Lyon) and Professor A. Guy and Dr. R. Garreau (Laboratoire de Chimie Organique, CNAM, Paris) for our longstanding collaboration. I am grateful to my colleagues from Angers (Ingénierie Moléculaire et Matériaux Organiques), Professors A. Gorgues and M. Jubault, for our many stimulating discussions and Drs. P. Frère, C. Thoble, and H. Brisset for their contribution to some of the work discussed in this review.

REFERENCES

1. J. P. Ferraris, D. O. Cowan, V. J. Walatka, and J. H. Perlstein, *J. Am. Chem. Soc.* 95:948 (1973).
2. C. K. Chiang, Y. W. Park, A. J. Heeger, H Shirakawa, E. J. Louis, and A. G. MacDiarmid, *Phys. Rev. Lett.* 39:1098 (1977).
3. A. F. Diaz, K. K. Kanazawa, and G. P. Gardini, *J. Chem. Soc. Chem. Commun.* 1979:635.
4. A. F. Diaz, *Chem. Scr.* 17:142 (1981).
5. G. Tourillon and F. Garnier, *J. Electroanal. Chem.* 135:173 (1982).
6. P. Bäuerle, *Adv. Mater.* 5:879 (1993).
7. P. Kovacic, K. N. McFarland, *J. Polym. Sci. Polym. Chem. Ed.* 17:1963 (1979).
8. T. Yamamoto, K. Sanechika, and A. Yamamoto, *J. Polym. Sci. Polym. Lett.* 18:9 (1980).
9. J. W. Lin and L. P. Dudek, *J. Polym. Sci.* 18:2869 (1980).
10. G. Kossmehl and G. Chatzitheodorou, *Makromol. Chem. Rapid Commun.* 2:551 (1981).
11. J. Roncali, *Chem. Rev.* 92:711 (1992).
12. G. Tourillon, in *Handbook of Conducting Polymers* (T. Skotheim, ed.), Marcel Dekker, New York, 1986, p. 294.
13. R. J. Waltman and J. Bargon, *Tetrahedron* 64:76 (1986).
14. G. K. Chandler and D. Pletcher, *Spec. Period Rep. Electrochem.* 10:117 (1985).
15. J. Heinze, *Topics in Current Chemistry*, Springer-Verlag, Berlin, 1990, vol. 152, p. 1.
16. A. O. Patil, A. J. Heeger, and F. Wudl, *Chem. Rev.* 88:163 (1988).
17. R. Sugimoto, S. Takeda, H. B. Gu, and K. Yoshino, *Chem. Express* 1:635 (1986).
18. T. Yamamoto, K. Sanechika, and A. Yamamoto, *Bull. Chem. Soc. Jpn.* 56:1497 (1983).
19. M. Kobayashi, J. Chen, T. C. Moraes, A. J. Heeger, and F. Wudl, *Synth. Met.* 9:77 (1984).
20. T. Yamamoto, A. Morita, Y. Miyazaki, T. Maruyama, H. Wakayama, Z.-H. Zhou, Y. Nakamura, T. Kanbara, S. Sasaki, and K. Kubota, *Macromolecules* 25:1214 (1992).
21. T.-A. Chen, X. Wu, and R. D. Rieke, *J. Am. Chem. Soc.* 117:244 (1995).
22. J. Roncali and F. Garnier, *New J. Chem.* 4-5:237 (1986).
23. M. Sato, S. Tanaka, and K. Kaeriyama, *Synth. Met.* 14:279 (1986).
24. A. Yassar, J. Roncali, and F. Garnier, *Macromolecules* 22:804 (1989).
25. J. Roncali, A. Yassar, and F. Garnier, *J. Chem. Soc. Chem. Commun.* 1988:581.
26. Y. Cao, D. Guo, M. Pang, and R. Qian, *Synth. Met.* 18:189 (1987).
27. Y. Furukawa, M. Akimoto, and I. Harada, *Synth. Met.* 18:151 (1987).
28. D. Delabouglise, R. Garreau, M. Lemaire, and J. Roncali, *New J. Chem.* 12:155 (1988).
29. A. F. Diaz, J. Crowley, J. Bargon, G. P. Gardini, and J. B. Torrance, *J. Electroanal. Chem.* 121:355 (1981).
30. P. Marque, J. Roncall, and F. Garnier, *J. Electroanal. Chem.* 218:107 (1987).
31. B. Krische and M. Zagorska, *Synth. Met.* 28:C263 (1989).
32. Y. Yumoto and S. Yoshimura, *Synth. Met.* 13:185 (1985).
33. J. Roncall, R. Garreau, F. Garnier, and M. Lemaire, *Synth. Met.* 15:323 (1986).
34. E. E. Havinga and L. W. Van Horssen, *Makromol. Chem. Makromol. Symp.* 24:67 (1989).
35. G. Zotti and G. Schiavon, *Synth. Met.* 39:183 (1990).
36. J. Roncali, A. Gorgues, and M. Jubault, *Chem. Mater.* 5:1456 (1993).
37. C. Visy, J. Lukkari, and J. Kankare, *Macromolecules* 27:3322 (1994).

38. J. Roncali, M. Giffard, M. Jubault, and A. Gorgues, *J. Electroanal. Chem. 361*:185 (1993).
39. Y. Wei, C.-C. Chan, J. Tian, G.-W. Jang, and K. F. Hsueh, *Chem. Mater. 3*:888 (1991).
40. G. Zotti, G. Schiavon, A. Berlin, and G. Pagani, *Chem. Mater. 5*:430 (1993).
41. M. C. Galazzi, L. Castellani, R. A. Marin, and G. Zerbi, *J. Polym. Sci. A 31*:3339 (1993).
42. C. Wang, M. E. Benz, E. LeGoff, J. L. Schindler, J. Albritton-Thomas, C. R. Kannewurf, and M. Kanatzidis, *Chem. Mater. 6*:401 (1994).
43. B. Krische, J. Hellberg, and C. Lilja, *J. Chem. Soc. Chem. Commun. 1987*:1476.
44. R. M. Souto-Maior, K. Hinkelmann, H. Eckert, and F. Wudl, *Macromolecules 23*:1268 (1990).
45. M. Zagorska and B. Krische, *Polymer 31*:1379 (1990).
46. G. Zotti, G. Schiavon, A. Berlin, G. Fontana, and G. Pagani, *Macromolecules 27*:1938 (1994).
47. G. Tourillon and F. Garnier, *J. Electroanal. Chem. 161*:51 (1984).
48. J. Roncali, F. Garnier, R. Garreau, and M. Lemaire, *J. Chem. Soc. Chem. Commun. 1987*:1500.
49. M. Feldhues, G. Kämpf, H. Litterer, T. Mecklenburg, and P. Wegener, *Synth. Met. 28*:C487 (1989).
50. M. Leclerc and G. Daoust, *J. Chem. Soc. Chem. Commun. 1990*:273.
51. M Dietrich, J. Heinze, G. Heywang, and F. Jonas, *J. Electroanal. Chem. 369*:87 (1994).
52. C. Wang, J. L. Schindler, C. R. Kannewurf, and M. G. Kanatzidis, *Chem. Mater. 7*:58 (1995).
53. J. P. Ruiz, K. Nayak, D. S. Marynick, and J. R. Reynolds, *Macromolecules 22*:1231 (1989).
54. A. Berlin, G. Pagani, and F. Sannicolo, *J. Chem. Soc. Chem. Commun. 1986*:1663.
55. M. Lemaire, W. Büchner, R. Garreau, H. A. Hoa, A. Guy, and J. Roncali, *J. Electroanal. Chem. 281*:293 (1990).
56. J. Roncali, A. Guy, M. Lemaire, R. Garreau, and H. A. Hoa, *J. Electroanal. Chem. 312*:277 (1991).
57. J. L. Sauvajol, C. Chorro, J. P. Lère-Porte, R. J. P. Corriu, J. J. E. Moreau, P. Thépot, and M. Wong Chi Man, *Synth. Met. 62*:233 (1994).
58. R. J. P. Cornu, J. J. E. Moreau, P. Thépot, and M. Wong Chi Man, *Chem. Mater. 6*:640 (1994).
59. J. Roncali and F. Garnier, *J. Phys. Chem. 92*:833 (1988) and references therein.
60. A. Yassar, J. Roncali, and F. Garnier, *J. Electroanal. Chem. 255*:53 (1988).
61. M. G. Kanatzidis, H. O. Marcy, W. J. McCarthy, C. R. Kannewurf, and T. J. Marks, *Solid State Ionics 32/33*:594 (1989).
62. R. J. Waltman, J. Bargon, and A. F. Diaz, *J. Phys. Chem. 87*:1459 (1983).
63. H. S. Li, J. Roncali, and F. Garnier, *J. Electroanal. Chem. 263*:155 (1989).
64. A.-C. Chang, R. L. Blankespoor, and L. L. Miller, *J. Electroanal. Chem. 236*:239 (1987).
65. S. Tanaka, M. Sato, and K. Kaeriyama, *Synth. Met. 25*:277 (1988).
66. J. Roncali, R. Garreau, A. Yassar, P. Marque, F. Garnier, and M. Lemaire, *J. Phys. Chem. 91*:6706 (1987).
67. M. Lemaire, R. Garreau, D. Delabouglise, J. Roncali, H. K. Youssoufi, and F. Garnier, *New J. Chem. 14*:359 (1990).
68. W. A. Goedel, N. S. Somanathan, V. Enkelmann, and G. Wegner, *Makromol. Chem. 193*:1195 (1992).
69. D. J. Guerrero, X. Ren, and J. P. Ferraris, *Chem. Mater. 6*:1437 (1994).
70. M. Sato, S. Tanaka, and K. Kaeriyama, *J. Chem. Soc. Chem. Commun. 1986*:873.
71. M. Lemaire, J. Roncali, R. Garreau, F. Garnier, and E. Hannecart, Fr. Patent 86 04 744 (Apr. 4, 1986).
72. K. Y. Jen, G. G. Miller, and R. L. Elsenbaumer, *J. Chem. Soc. Chem. Commun. 1986*:1346.
73. S. Hotta, S. D. D. V. Rughooputh, A. J. Heeger, and F. Wudl, *Macromolecules 20*:212 (1987).
74. S. Hotta, M. Soga, and N. Sonoda, *Synth. Met. 26*:267 (1988).
75. K. Yoshino, S. Nakajima, M. Onoda, and R. Sugimoto, *Synth. Met. 28*:C439 (1989).
76. M. Leclerc, F. Martinez-Diaz, and G. Wegner, *Makromol. Chem. 190*:3105 (1989).
77. P. Marque and J. Roncali, *J. Phys. Chem. 94*:8614 (1990).
78. S. D. D. V. Rughooputh, S. Hotta, A. J. Heeger, and F. Wudl, *J. Polym. Sci. B 25*:1071 (1987).
79. A. O. Patil, Y. Ikenoue, F. Wudl, and A. J. Heeger, *J. Am. Chem. Soc. 109*:1858 (1987).
80. Y. Ikenoue, N. Uotani, A. O. Patil, F. Wudl, and A. J. Heeger, *Synth. Met. 30*:305 (1989).
81. Y. Ikenoue, Y. Saida, M. Kira, H. Tomozawa, H. Yashima, and M. Kobayashi, *J. Chem. Soc. Chem. Commun. 1990*:1694.
82. P. Bäuerle, K. W. Gaudi, F. Würthner, S. Sariciftci, H. Neugebauer, M. Mehring, C. Zhong, and K. Doblhofer, *Adv. Mater. 2*:490 (1990).
83. P. N. Bartlett and D. H. Dawson, *J. Mater. Chem. 4*:1805 (1994).
84. E. Schultz, K. Fahmi, and M. Lemaire, *Acros Chim. Acta 1*:1 (1994).
85. G. Daoust and M. Leclerc, *Macromolecules 24*:455 (1991).
86. S. A. Chen and C.-C. Tsai, *Macromolecules 26*:2234 (1993).
87. C. Roux, K. Faid, and M. Leclerc, *Makromol. Chem. Rapid Commun. 14*:461 (1993).
88. M. Lemaire, R. Garreau, J. Roncali, D. Delabouglise, H. Korri, and F. Garnier, *New J. Chem. 13*:863 (1989).
89. B. Zinger, Y. Greenvald, and I. Rubinstein, *Synth. Met. 41–43*:583 (1991).
90. T.-A. Chen, X. Wu, and R. D. Rieke, *J. Am. Chem. Soc. 117*:223 (1995).
91. R. D. McCullough, R. D. Lowe, M. Jayaraman, and D. L. Anderson, *J. Org. Chem. 58*:904 (1993).
92. H. S. Li, F. Garnier, and J. Roncali, *Macromolecules 25*:6425 (1992).

93. J. Roncali, H. S. Li, R. Garreau, F. Garnier, and M. Lemaire, *Synth. Met. 36*:267 (1990).
94. J. Roncali, R. Garreau, D. Delabouglise, F. Garnier, and M. Lemaire, *J. Chem. Soc. Chem. Commun. 1989*:679.
95. H. S. Li, F. Garnier, and J. Roncali, *Solid State Commun. 77*:811 (1991).
96. J. Roncali, H. S. Li, and F. Garnier, *J. Phys. Chem. 95*:8983 (1991).
97. J. Roncali, R. Garreau, and M. Lemaire, *J. Electroanal. Chem. 278*:373 (1990).
98. M. J. Marsella and T. M. Swager, *J. Am. Chem. Soc. 115*:12214 (1993).
99. Y. Miyazaki and T. Yamamoto, *Chem. Lett. 1994*:41.
100. P. Bäuerle, F. Würthner, and S. Heid, *Angew. Chem. Int. Ed. Engl. 29*:419 (1990).
101. P. Bäuerle and S. Scheib, *Adv. Mater. 5*:848 (1993).
102. M. Lemaire, D. Delabouglise, R. Garreau, A. Guy, and J. Roncali, *J. Chem. Soc. Chem. Commun. 1988*:658.
103. D. Kotkar, V. Joshi, and K. Gosh, *J. Chem. Soc. Chem. Commun. 1988*:917.
104. M. M. Bouman and E. W. Meijer, *Polym. Prepr. 35*:309 (1994).
105. A. Deronzier and J. C. Moutet, *Acc. Chem. Res. 22*:849 (1989).
106. R. Mirrazaei, D. Parker, and H. S. Munro, *Synth. Met. 30*:265 (1989).
107. J. Grimshaw and S. D. Perera, *J. Electroanal. Chem. 278*:287 (1990).
108. J. A. Crayston, A. Iraqi, P. Mallon, J. C. Walton, and D. P. Tunstall, *Synth. Met. 55–57*:867 (1993).
109. D. Ofer, R. M. Crooks, and M. S. Wrighton, *J. Am. Chem. Soc. 112*:7869 (1990).
110. P. Bäuerle and K.-U. Gaudl, *Adv. Mater. 2*:185 (1990).
111. P. Bäuerle and K.-U. Gaudl, *Synth. Met. 41–43*:3037 (1991).
112. H. H. Jaffé and M. Orchin, *Theory and Applications of Ultraviolet Spectroscopy*, Wiley, New York, 1962.
113. J. Roncali, H. Korri, R. Garreau, F. Garnier, and M. Lemaire, *J. Chem. Soc. Chem. Commun. 1990*:414.
114. R. M. Crooks, O. M. R. Chyan, and M. S. Wrighton, *Chem. Mater. 1*:2 (1989).
115. A. ElKassmi, F. Fache, and M. Lemaire, *J. Electroanal. Chem. 373*:241 (1994).
116. S. K. Ritter, R. E. Noftle, and A. E. Ward, *Chem. Mater. 5*:752 (1993).
117. W. Büchner, R. Garreau, M. Lemaire, and J. Roncali, *J. Electroanal. Chem. 277*:355 (1990).
118. A. ElKassmi, W. Büchner, F. Fache, and M. Lemaire, *J. Electroanal. Chem. 326*:357 (1992).
119. L. Robitaille and M. Leclerc, *Chem. Mater. 27*:1847 (1994).
120. W. Büchner, R. Garreau, J. Roncali, and M. Lemaire, *J. Fluorine Chem. 59*:301 (1992).
121. A. ElKassmi, G. Héraud, W. Büchner, F. Fache, and M. Lemaire, *J. Mol. Catal. 72*:299 (1992).
122. C. Thobie-Gautier, A. Guy, A. Gorgues, M. Jubault, and J. Roncali, *Adv. Mater. 5*:637 (1993).
123. M. R. Bryce, A. D. Chissel, J. Gopal, P. Kathirgamanathan, and D. Parker, *Synth. Met. 39*:397 (1991).
124. F. Sundholm, G. Sundholm, and M. Törrönen, *Synth. Met. 53*:109 (1992).
125. C. Thobie-Gautier, Y. Bouligand, A. Gorgues, M. Jubault, and J. Roncali, *Adv. Mater. 6*:138 (1994).
126. G. Saito and S. Kagoshima (eds.), *The Physics and Chemistry of Organic Superconductors*. Springer-Verlag, London, 1990.
127. C. Thobie-Gautier, A. Gorgues, M. Jubault, and J. Roncali, *Macromolecules 26*:4094 (1993).
128. J. Roncali, C. Thobie-Gautier, H. Brisset, J. F. Favard, and A. Guy, *J. Electroanal. Chem. 381*:257 (1995).
129. D. Fichou, Y. Nishikitani, G. Horowitz, J. Roncali, and F. Garnier, *Synth. Met. 28*:C729 (1989).
130. F. Garnier, G. Horowitz, X. Peng, and D. Fichou, *Adv. Mater. 2*:592 (1990).
131. F. Geiger, M. Stoldt, H. Schweizer, P. Bäuerle, and E. Umbach, *Adv. Mater. 5*:922 (1993).
132. D. Jones, M. Guerra, L. Favaretto, A. Modelli, M. Fabrizio, and G. Distefano, *J. Phys. Chem. 94*:5761 (1990).
133. P. Bäuerle, *Adv. Mater. 4*:102 (1992).
134. J. Guay, P. Kasai, A. Diaz, R. Wu, J. Tour, and L. H. Dao, *Chem. Mater. 4*:1097 (1992).
135. M. G. Hill, J. F. Penneau, B. Zinger, K. R. Mann, and L. L. Miller, *Chem. Mater. 4*:1106 (1992).
136. W. Steinkopft, R. Leitsmann, and K. H. Hoffmann, *Liebigs Ann. 546*:180 (1941).
137. J. W. Sease and L. Zechmeister, *J. Am. Chem. Soc. 69*:270 (1947).
138. J. Kagan and S. K. Arora, *Heterocycles 20*:1937 (1983).
139. D. D. Cunningham, L. Laguren-Davidson, H. B. Mark, Jr., C. V. Pham, and H. Zimmer, *J. Chem. Soc. Chem. Commun. 1987*:1021.
140. J. Nakayama, T. Konishi, S. Murabayashi, and M. Hoshino, *Heterocycles 26*:1793 (1987).
141. J. M. Tour and R. Wu, *Macromolecules 25*:1901 (1992).
142. W. ten Hoeve, H. Wynberg, E. E. Havinga, and E. W. Meijer, *J. Am. Chem. Soc. 113*:5887 (1991).
143. D. Delabouglise, M. Hmyene, G. Horowitz, A. Yassar, and F. Garnier, *Adv. Mater. 4*:107 (1992).
144. A. Yassar, D. Delabouglise, M. Hmyene, B. Nessak, G. Horowitz, and F. Garnier, *Adv. Mater. 4*:490 (1992).
145. K. Faid and M. Leclerc, *J. Chem. Soc. Chem. Commun. 1993*:962.
146. P. Bäuerle, F. Pfau, H. Schupp, F. Würthner, K.-U. Gaudi, M. Balparda Caro, and P. Fischer, *J. Chem. Soc. Perkin Trans. 2 1993*:489.
147. F. Garnier, F. Deloffre, G. Horowitz, and R. Hajlaoui, *Synth. Met. 55–57*:4747 (1993).
148. E. E. Havinga, I. Rotte, E. W. Meijer, W. Ten Hoeve, and H. Wynberg, *Synth. Met. 41–43*:473 (1991).

149. G. Zotti, G. Schiavon, A. Berlin, and G. Pagani, *Chem. Mater. 5*:620 (1993).
150. C. Ehrendorfer and A. Karpfen, *J. Phys. Chem. 98*:7492 (1994).
151. P. Bäuerle, U. Segelbacher, K.-U. Gaudl, D. Huttenlocher, and M. Mehring, *Angew. Chem. Int. Ed. Engl. 32*:76 (1993).
152. P. Hapiot, P. Audebert, K. Monnier, J. M. Pernaut, and P. Garcia, *Chem. Mater. 6*:1549 (1994).
153. P. Bäuerle, G. Götz, U. Segelbacher, D. Huttenlocher, and M. Mehring, *Synth Met. 55–57*:4768 (1993).
154. J. Roncali, M. Giffard, P. Frère, M. Jubault, and A. Gorgues, *J. Chem. Soc. Chem. Commun. 1993*:689.
155. J. Roncali, M. Giffard, M. Jubault, and A. Gorgues, *Synth. Met. 60*:163 (1993).
156. J. Guay, A. Diaz, R. Wu, and J. M. Tour, *J. Am. Chem. Soc. 115*:1869 (1993).
157. T. Saika, M. Irie, and T. Shimidzu, *J. Chem. Soc. Chem. Commun. 1994*:2123.
158. W. A. Little, *Phys. Rev. 134*:A1416 (1964).
159. J. L. Brédas, *J. Chem. Phys. 82*:3808 (1985).
160. F. Wudl, M. Kobayashi, and A. J. Heeger, *J. Org. Chem. 49*:3382 (1984).
161. M. Pomerantz, B. Chaloner-Gill, L. O. Harding, J. J. Tseng, and W. J. Pomerantz, *Synth. Met. 55–57*:960 (1993).
162. C. Kitamura, S. Tanaka, and Y. Yamashita, *J. Chem. Soc. Chem. Commun. 1994*:1585.
163. D. Lorcy and M. P. Cava, *Adv. Mater. 4*:562 (1992).
164. S. Tanaka and Y. Yamashita, *Synth. Met. 55–57*:1251 (1993).
165. T. L. Lambert and J. P. Ferraris, *J. Chem. Soc. Chem. Commun. 1991*:752.
166. J. P. Ferraris and T. L. Lambert, *J. Chem. Soc. Chem. Commun. 1991*:1268.
167. M. Kosaki, S. Tanaka, and Y. Yamashita, *J. Org. Chem. 59*:442 (1994).
168. H. Brisset, C. Thobie-Gautier, A. Gorgues, M. Jubault, and J. Roncali, *J. Chem. Soc. Chem. Commun. 1994*:1305.
169. A. Alberti, L. Favaretto, and G. Seconi, *J. Chem. Soc. Perkin Trans. 2 1990*:931.
170. G. Barbarella, M. Zambianchi, A. Bongini, and L. Antolini, *Adv. Mater. 5*:834 (1993).
171. G. Distefano, M. Da Colle, D. Jones, M. Zambianchi, L. Favaretto, and A. Modelli, *J. Phys. Chem. 97*:3504 (1993).
172. J. E. Chadwick and B. E. Kohler, *J. Phys. Chem. 98*:3631 (1994).
173. M. Belletête, M. Leclerc, and G. Durocher, *J. Phys. Chem. 98*:9450 (1994).
174. M. Martinez, J. R. Reynolds, S. Basak, D. A. Black, D. S. Marynick, and M. Pomerantz, *J. Polym. Sci. B 26*:911 (1986).
175. J. Roncali, C. Thobie-Gautier, E. Elandaloussi, and P. Frére, *J. Chem. Soc. Chem. Commun. 1994*:2249.
176. J. Roncali and C. Thobie-Gautier, *Adv. Mater. 6*:846 (1994).
177. K. Takahashi and T. Suzuki, *J. Am. Chem. Soc. 111*:5483 (1989).
178. K. Tamao, S. Yamaguchi, and Y. Ito, *J. Chem. Soc. Chem. Commun. 1994*:229.
179. J. Roncali, L. Rasmussen, C. Thobie-Gautier, P. Frère, M. Sallé, H. Brisset, A. Gorgues, M. Jubault, J. Becher, J. Garin, and J. Orduna, *Adv. Mater. 6*:841 (1994).
180. H. Brisset, C. Thobie-Gautier, A. Gorgues, M. Jubault, and J. Roncali, *J. Chem. Soc. Chem. Commun. 1994*:1765.

13
The Chemistry and Uses of Polyphenylenevinylenes

Stephen C. Moratti
University of Cambridge, Cambridge, England

I. INTRODUCTION

There has been a steadily growing interest in poly(*p*-phenylenevinylene) (PPV) (Fig. 13.1) and its close derivatives. Almost 200 papers are published each year on the chemistry and physics of this versatile class of polymers. There are many reasons for this attention. PPV is easily synthesized in good purity and high molecular weight. It is relatively stable and quite insoluble, yet it can be manipulated as a water-soluble precursor polymer to form films and fibers. The optical band gap of ~ 2.6 eV and bright yellow fluorescence make PPV a strong contender in applications such as light-emitting diodes and photovoltaic devices. It can be readily doped to form electrically conductive materials (although without quite the stability of polymers such as polyaniline), and its physical and electronic properties can be varied over a wide range by the inclusion of functional side groups. Even 30 years after its first synthesis, there is still much to learn about this interesting polymer.

The nomenclature used here is semisystematic, with the numbering system as in Fig. 13.1. The systematic IUPAC name is poly(1,4-phenylene-1,2-ethenylene), but this is rarely used in practice.

II. DIRECT ROUTES TO PPV

The insolubility of PPV has been a recurring problem in the synthesis of high molecular weight polymer. Step-growth coupling reactions (such as Wittig condensation) will lead to only low molecular weight species containing at most 5–10 repeat units before the product becomes insoluble and polymerization halts. This has led to two different strategies to overcome this problem. Through the incorporation of various side groups (usually alkyl, alkoxy, or phenyl), the conjugated polymer becomes soluble in organic solvents such as chloroform and toluene. This has the added effect that the optical and electronic properties of the polymer can also be significantly varied. Of course, this technique is of no use in preparing the parent, unsubstituted polymer, and invariably some kind of precursor route is used. In this method, a soluble polymer is converted by various means into the fully conjugated form, which allows thin films, fibers, and even large objects to be fabricated from PPV.

Historically, the first syntheses of PPV-type materials were achieved via step-growth polymerization. This type of coupling has many disadvantages, such as low degrees of polymerization and the presence of large amounts of unreacted end groups. Incomplete elimination or formation of the double bond is common, as well as the presence of both cis and trans isomers in the chain. In many cases, heating to 200–300°C improves the regularity and structure by isomerization of the cis double bonds and increases the amount of unsaturation [1]. An advantage of many step-polymerization reactions is that *ortho-*, *meta-*, and *para*-xylylene linkages can be incorporated in the main chain as well as a great variety of functionality on the ring. Copolymers of defined stereoregularity can also be easily made.

McDonald and Campbell [2] recognized in 1960 that PPV could be synthesized by repetitive Wittig-type couplings between an aromatic bisphosphonium salt and bisaldehyde (Scheme 1). Reaction in ethanol produced an insoluble yellow solid. Gourley et al. [3] investigated this reaction and suggested that the material was in fact an oligomer of PPV, with an average degree of polymerization of 3–9. The presence of some cis double bonds was noted, which were isomerized to the trans form by heating with iodine. Since the initial work by McDonald and Campbell, other PPV derivatives, both soluble and insoluble, have been synthesized by this route [3–7].

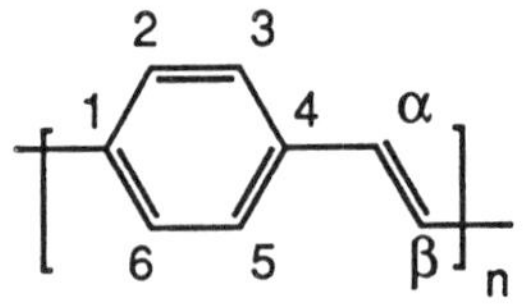

Fig. 13.1 Structure and numbering system for PPV.

Stilbenes can be produced in high yield by the coupling of benzylic halides in liquid ammonia with sodium amide. This reaction was extended to the preparation of PPV by Smith [8] and later by Hoeg et al. [9]. By using dichloroxylene under the same conditions, they produced a yellow solid polymer containing a significant amount of residual chlorine. Similar results were obtained by Höerhold and Opfermann [10] using sodium hydride in DMF. The presence of chlorine in the final product was attributed to the presence of incompletely eliminated unsaturated linkages that could be removed by heating to 300°C [3]. A related route was used by Moritani et al. [11], whereupon treatment of tetrabromo-p-xylene with two equivalents of methyllithium gave PPV. There remains doubt about the precise mechanism of some of these reactions, which could conceivably go via a quinoidal, carbenoid, or carbanionic pathway, although the former seems more likely [7,12]. PPV and derivatives can also be produced by dechlorination condensation of tetrachloroxylenes with chromium(II) acetate [13–15]. Similarly, perchloroxylene can be polymerized with iron pentacarbonyl to give a perchloro-PPV derivative [16]. These last two examples almost certainly involve a step-polymerization mechanism. Possible polymerization pathways for the coupling of benzylic halides are shown in Scheme 2.

Rehahn and Schlüter [17] were the first to synthesize a soluble PPV derivative by the McMurray deoxygenative coupling of an aromatic dialdehyde. By using an improved version of the McMurray conditions, fairly high molecular weight polymer was obtained that contained a mixture of both cis and trans linkages. Later, Cataldo [18] used terephthaldehyde under similar conditions to prepare PPV and its meta analog. However, the materials were not well characterized, and there was no attempt to estimate the degree of polymerization. Polythienylenevinylene (PTV) derivatives have also been synthesized by this method [19].

It is possible to introduce the double bond directly into the polymer chain. Griener and coworkers polymerized ethylene with a variety of aromatic dibromides via a Heck reaction to produce both PPV and soluble derivatives (Scheme 3) [20]. Reasonable molecular weights (3000–10,000) were obtained when solubilizing groups were present. The disadvantage of this route is that one of the reactants is gaseous but must be added in precise amounts. There are also many possible side reactions that have to be carefully controlled [21]. However, the reaction conditions are mild and are tolerant of many functional groups (e.g., nitro) that can be deleterious in other PPV syntheses. The Heck reaction can be extended to use functionalized styrenes [22,23], which obviates the need for a gaseous reactant but can make synthesis of the monomers much more demanding.

Metathesis catalysts are used to break and recombine double bonds. In this way, unsaturated compounds can be isomerized and rearranged to give new products. Kumar and Eichenger [24] oligomerized divinylbenzene with the Shrock catalyst to give low molecular weight PPV. This was extended by Thorn-Csányi and coworkers [25,26] to give a soluble PPV derivative of moderate molecular weight. Due to the sensitivity of the catalysts, the low degrees of polymerization obtained, and the difficulty in synthesizing the appropriate monomers, there seems little advantage to this route at present. PPV is also obtainable by metathesis of paracyclophane dienes (Scheme 4) [27]. A mixture of cis and trans double bonds is found in the product. The synthesis of substituted cyclophanes is a significant barrier to further exploitation of this idea for other PPV derivatives.

Double bonds can be synthesized by a variety of base-catalyzed condensations between an active methylene compound and a carbonyl group. In a few cases, yields are high enough to be used in polymerization reactions. Meier et al. [28,29] used the Siegrist reaction to form soluble PPV derivatives. This involved base-catalyzed self-coupling of the Schiff bases of various p-methyl-benzaldehydes. The Knoevenagel condensation between a benzylic nitrile and an aromatic aldehyde is often very high yielding and has been used by a number of research groups to give insoluble PPV derivatives (Scheme 5) [30–34]. Soluble alkoxy- and alkyl-substituted side chain derivatives have also been synthesized [35–38]. Careful optimization of the reaction conditions was needed to reduce unwanted side reactions such as hydrolysis of the nitrile group or Michael addition to the double bond.

Polyparaphenylenevinylene films have been grown electrochemically by the reductive coupling of $\alpha,\alpha,\alpha',\alpha'$-tetrabromoxylene [39]. Deposition could be accelerated by the addition of chromium or molybdenum carbonyl complexes. The material produced appeared to be contaminated with both residual chlorine and elec-

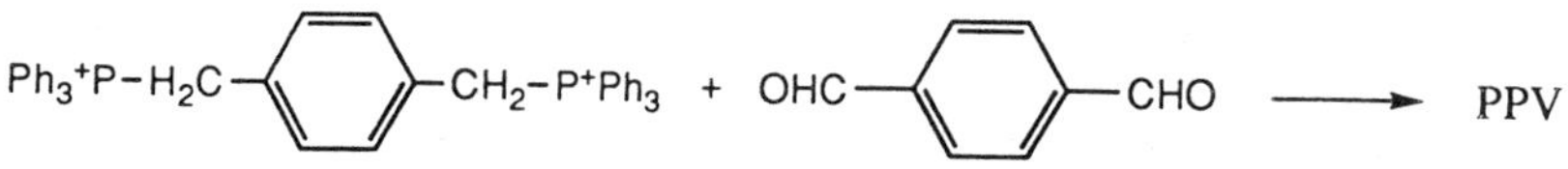

Scheme 1

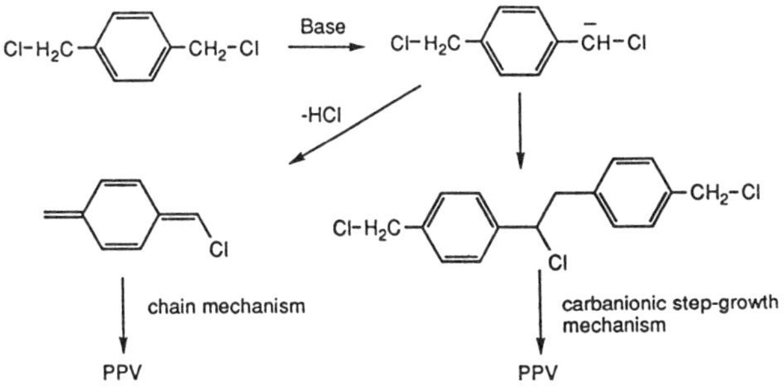

Chemistry and Uses of PPV

Scheme 2

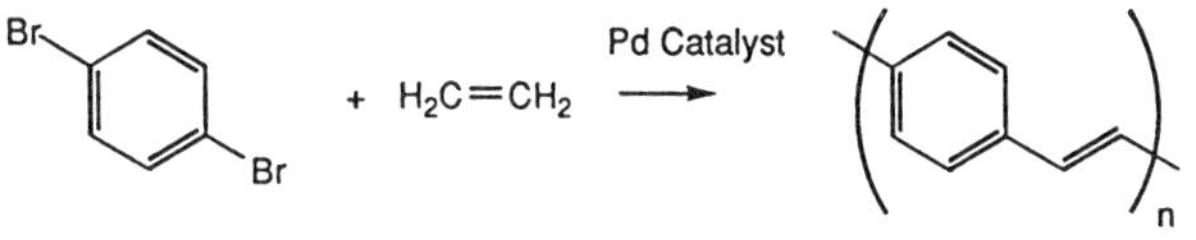

Scheme 3

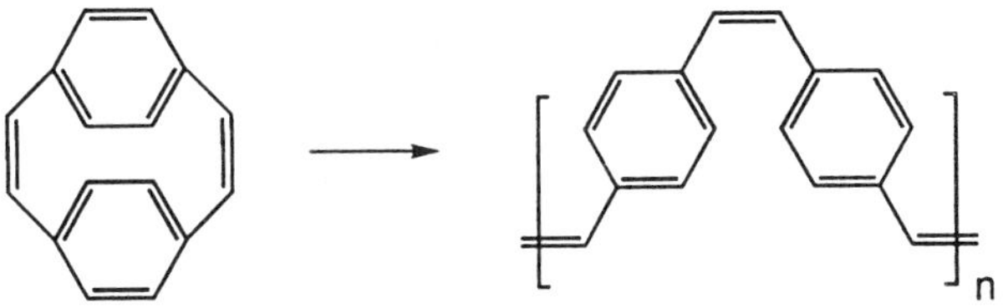

Scheme 4

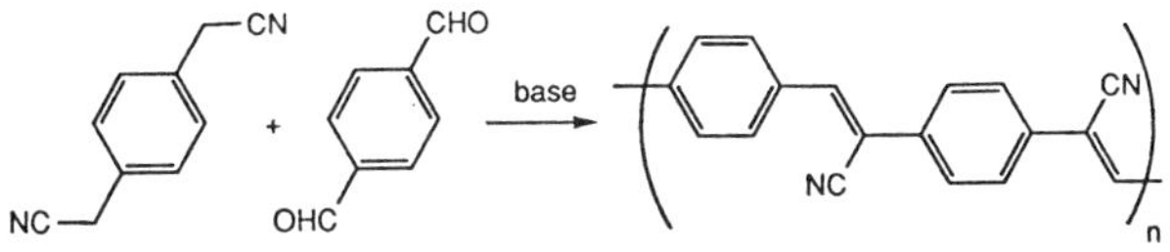

Scheme 5

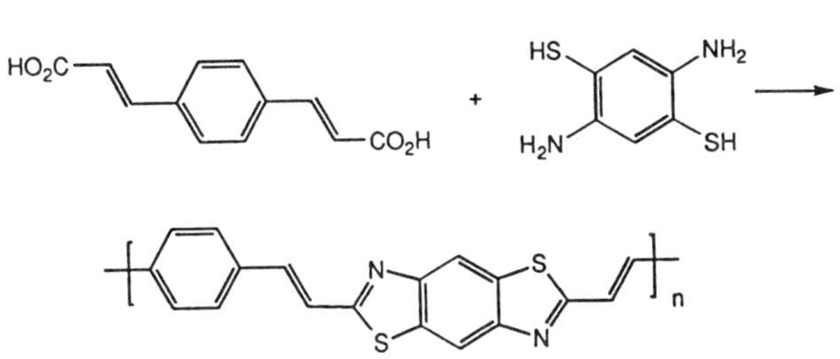

Scheme 6

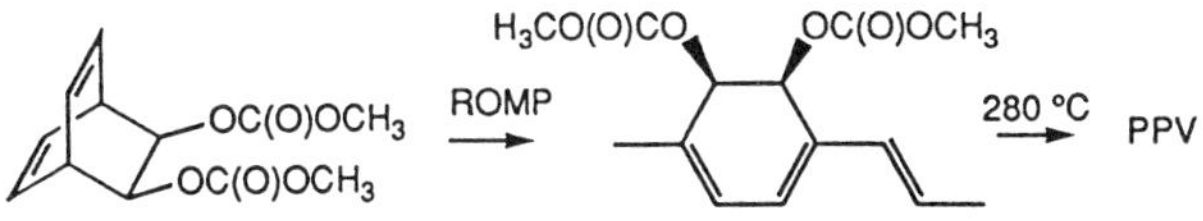

Scheme 7

trolyte. If the process is improved to remove these contaminants, it might represent an attractive route for low temperature fabrication of PPV films for optoelectronic applications.

A different approach has been used by Osaheni and Jenekhe [40] to synthesize a heterocyclic PPV derivative. Condensation of diacrylic acid with 2,5-diamino-1,4-benzenedithiol produced a benzobisthiazolevinylene-PPV copolymer (Scheme 6). The red fluorescent polymer was soluble in a nitromethane–$GaCl_3$ solvent mixture and could be cast into good quality films.

III. PRECURSOR ROUTES

A. General

Almost all of the above general routes to PPV have the disadvantage of giving insoluble, unworkable products. The need for an alternative route that could be used for forming films and fibers has driven a number of groups to design alternative syntheses of PPV. These involve intermediate, processable polymers that can be later converted into the fully conjugated material. It is perhaps remarkable that one of the simplest precursor routes to PPV was in fact first discovered by accident by Wessling and Zimmerman 30 years ago.

A novel precursor route to PPV was reported by Conticello et al. [41]. A bicyclooctadiene compound was coupled by ring-opening metathesis polymerization (ROMP) to give a precursor polymer of high molecular weight but soluble in organic solvents (Scheme 7). This polymer could be deposited as thin films and converted thermally to PPV at 280°C. Even lower conversion temperatures (200°C) could be employed in the presence of an amine catalyst.

Thin films of PPV were also produced by gas-phase-pyrolysis of a dichloroparacyclophane [42]. The active intermediate was probably a p-quinodimethane species that condensed and polymerized to form a sparingly soluble chloro precursor polymer. This intermediate polymer could be converted to PPV by heating to 300°C under nitrogen.

A modification of the Thorn-Csányi ROMP route to PPV used a silyl-substituted paracyclophane derivative that was reacted under ROMP conditions to give a soluble precursor polymer [43]. Transformation into PPV could be achieved by treating the precursor polymer with acid or by hydrolysis of the silyl group followed by thermal treatment (Scheme 8). The advantage of this

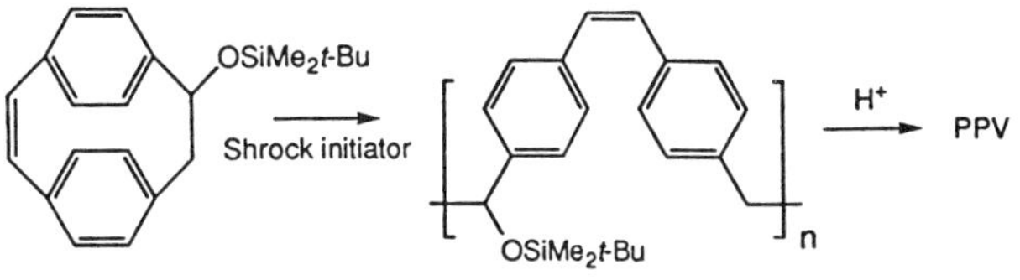

Scheme 8

method is that the polymerization is "living," and thus polymers and block copolymers of well-defined molecular weight can be prepared.

B. Sulfonium Route

1. Synthesis

The best known and most widely used precursor route was discovered fortuitously by Wessling and Zimmerman (Dow Chemical Co.) [44–46], who were trying to produce conventional polymers via alkylation of *p*-xylylene sulfonium salts. Unexpectedly, treatment of the sulfonium salt with base produced a water-soluble, viscous, yellow-green solution. This aqueous polyelectrolyte (**2**) could be efficiently purified by dialysis. Addition of more base caused the formation of a yellow solid identified as PPV. It was recognized that the intermediate polyelectrolyte could be cast and dried to form fibers and films and thermally converted into PPV (Scheme 9). It was initially hoped that this material would be useful in high temperature structural applications, but it eventually proved to have insufficient oxidative stability. Both open-chain and cyclic sulfides can be used, the latter being preferred because of fewer side reactions upon thermal elimination [47]. The Wessling–Zimmerman route is able to produce high quality PPV material cheaply and quickly. The main drawbacks are the toxicity and the very bad odor of the sulfur compounds used in the synthesis, and the instability of the intermediate polyelectrolytes.

The polymerization mechanism has been studied in some detail, and it seems clear that the first steps involve deprotonation and 1,6-elimination of a sulfide group (see probable reaction route in Scheme 10) to form a reactive *p*-xylylene derivative. Both steps are reversible, and polymerization is inhibited by excess sulfide [48]. If the reaction is carried out in water, then yields of polymer can be improved by using either an immiscible organic solvent (e.g., pentane or toluene) to extract the sulfide as it is being formed or a water-insoluble sulfide as the leaving group [49]. It is also possible to use an emulsified dispersion as the polymerization medium [50]. Exchange of the methylene protons occurs even at $-50°C$, whereas formation of the *p*-xylylene derivative occurs only at temperatures over $-40°C$ [51]. Generally, the best results are achieved at temperatures around 0°C. If the temperature is too high ($>70°C$), the base tends to attack the polymer instead, resulting in the formation of conjugated, insoluble product [46]. Excess base has been found in some circumstances to lead to nonrandom

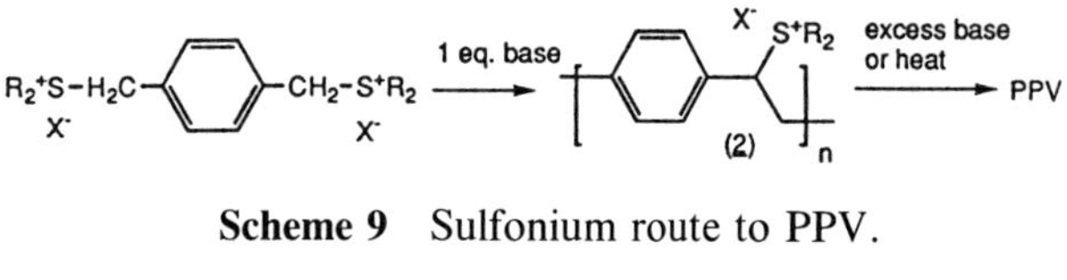

Scheme 9 Sulfonium route to PPV.

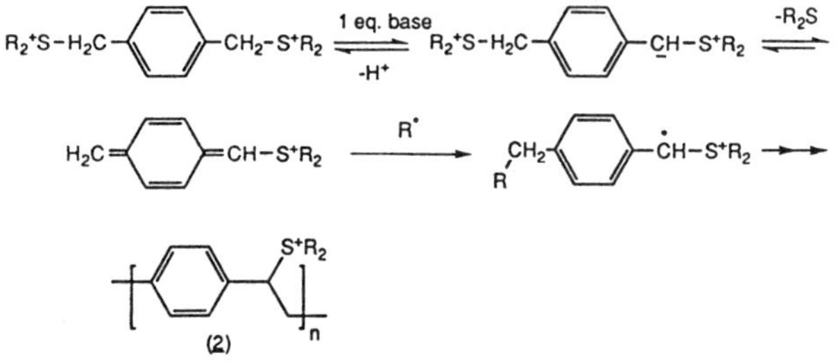

Scheme 10

elimination in dialkoxy precursor polymers [52]. Instead, isolated stilbene units were preferentially formed over longer conjugated sequences, but the reason for this behavior could not be determined.

The formation of the quinoidal *p*-xylylene intermediate can be monitored by the appearance of a peak in the UV spectrum around 310 nm. This has been used to optimize reaction conditions for polymerizations involving unreactive sulfonium salts [48]. There has been some controversy over the precise nature of the polymer coupling reaction. The initial assumption was that the polymerization was a radical-promoted process [46]. The presence of radicals was very hard to prove, and the pendulum swung for a while toward an anionic mechanism [51]. However, careful work by Lahti and coworkers [53] showed that radical trapping reagents did indeed suppress the polymerization. As an example, the addition of TEMPO to the reaction mixture not only dramatically lowered the yields and molecular weights but also caused the disappearance of the spin label. The mechanism of radical initiation is unknown; it may involve spontaneous coupling of two quinoidal *p*-xylylene intermediates to form a biradical.

A wide variety of aromatic polycyclic and heterocyclic sulfonium salts can be used in the polymerization. Formation of the quinoid intermediate is energetically unfavorable due to loss of aromaticity. In practice, if more than two aromatic ring systems have to be forced out of aromaticity when forming the *p*-xylylene intermediate, then the polymerization is unlikely to succeed (Fig. 13.2) [49]. The presence of extra double bonds outside the aromatic ring but within the main chain does not seem to prevent polymerization [54]. Sulfonium salts based on a 1,4-naphthalene system are able to polymerize, as only one ring loses aromaticity during formation of the intermediate quinoid [55]. However, because both aromatic rings would need to lose some conjugation, the 2,6 isomer has been reported not to polymerize [49] (although an earlier communication from the same group [56] reported low molecular weight polymer). On the other hand, the polymerization is inhibited if the *p*-xylylene is too stable, as is probably the case for 9,10-substituted anthracene derivatives. *Meta*-substituted phenylene systems also do not polymerize due to the impossibility of forming the necessary quinoid interme-

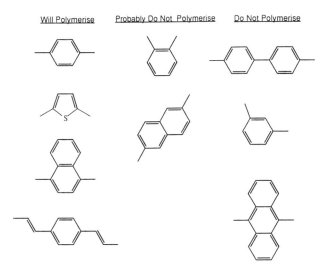

Fig. 13.2 Polymerization tendencies of various sulfonium salts $R_2S^+CH_2-Ar-CH_2S^+R_2$.

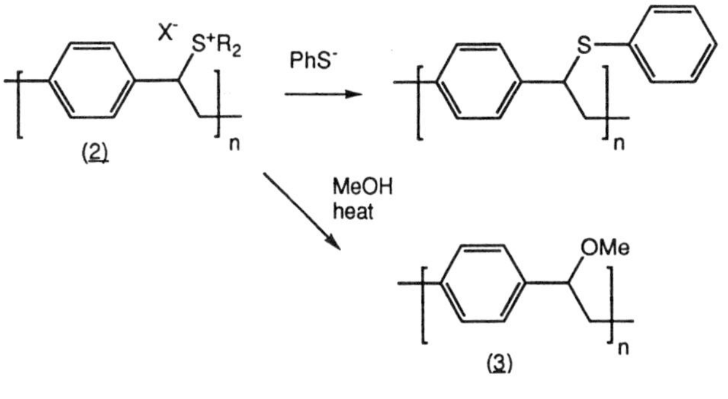

Scheme 11

diate. The situation regarding *ortho*-phenylene derivatives is not entirely clear. There is no doubt that *o*-xylylenes could be formed; however, these are very prone to other reactions such as dimerization or ring closure to form fused cyclobutanes. Successful polymerization of *ortho*-substituted compounds (halide rather than sulfonium leaving group) via phase transfer catalysis has been claimed [57], but the products were not well characterized.

A large variety of substituents can be tolerated on the aromatic ring, including aromatic [58–60], alkoxy [61–70], alkyl [44,45,61,71], silyl [72,73], halogen [74–76], sulfur [77], and amino groups [78]. Electron-poor aromatic systems (e.g., nitro- or cyano-substituted) polymerize with extreme difficulty [48,79], although polymerization proceeds more smoothly if the electron-withdrawing group is attached to the α-methylene carbon rather than the ring (S. C. Moratti, unpublished results). Mixtures of different sulfonium salts can be used to generate copolymers [74,75,80–85], which is especially useful if a particular sulfonium salt does not homopolymerize [79].

The sulfonium polyelectrolyte (**2**) is usually soluble in methanol or water, and so these are the preferred solvents for polymerization. However, the nature of the counterion (X^-) has a large influence on its solubility. Beerden et al. [86] measured the solubility of **2** with different counterions. Chloride ions give good aqueous solubility, whereas bromide sulfonium precursor polymers are often only sparingly soluble in water [55,86,87]. Gel permeation chromatography (GPC) of the aqueous polyelectrolyte usually gives irreproducible results [88], so it is more common to substitute the sulfonium groups to give an organic-soluble polymer or to change the counterion. The tetrafluoroborate precursor polymer is relatively insoluble in water [86,89] but is soluble in both acetonitrile and dimethylformamide, a fact that allows direct measurement of the molecular weight of these polymers by GPC in these solvents [88]. Substitution of the sulfonium group itself can be done by reaction with phenylthiolate ions [44,88] or more simply by extended heating in methanol (Scheme 11) [90,91]. Both treatments give polymers soluble in chloroform or tetrahydrofuran that can then be analyzed by conventional methods such as GPC [88,92].

The intermediate sulfonium precursor polymer (**2**) is not very stable, and the actual structure is probably closer to that shown in Fig. 13.3. The vinylic double bonds arise from base-catalyzed elimination during the polymerization and often give the resulting polymer a light blue-yellow tinge. The alkoxy substituent could arise from direct attack of the base (hydroxide or methoxide) during polymerization, or via S_N1 hydrolysis during storage.

With electron-rich aromatic systems (such as dialkoxy-substituted phenylene or thiophene) this hydrolysis reaction is complete within a matter of hours at room temperature. The presence of these hydroxy or methoxy groups complicates the subsequent pyrolysis step and can lead to residual unsaturation in the main chain. It is also possible to get some substitution of the sulfonium salt with the counterion, although this reaction may be more prevalent in the solid state [47] or at elevated temperatures in solution [93]. The chloride and bromide precursor polymers are not stable in the solid state at room temperature, quickly turning yellow and becoming insoluble due to partial elimination. However, they can be

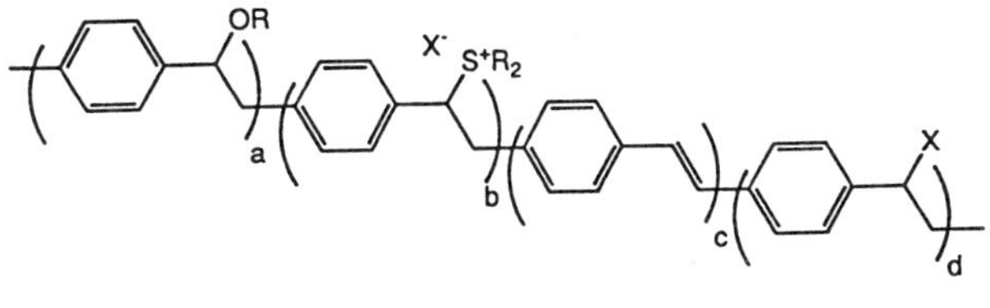

Fig. 13.3 Representative portion of sulfonium precursor polymer **2**.

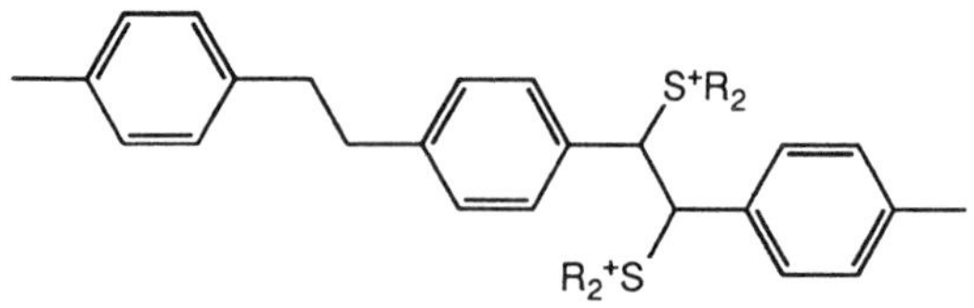

Fig. 13.4 Possible defects in the structure of sulfonium polyelectrolyte **2**. (From Ref. 30.)

stored safely for some months in solution below 0°C without appreciable change.

There are other complicating factors in the synthesis and pyrolysis of sulfonium polymers that should be mentioned. The first is the possibility of head-to-head coupling of the *p*-xylylene intermediates during polymerization, which would lead to irregularities in the polymer chain and reduction in conjugation after conversion (Fig. 13.4). This has been postulated as one reason for residual sulfur levels in some converted PPV-type polymers [30]. Small levels of carbonyl groups are often noted by infrared spectroscopy after thermal conversion, which could result from oxidation during conversion, or from side reactions involving hydroxy-substituted portions of the chain. These residual carbonyl groups may have a large effect on the optoelectronic properties of the polymer [94]. This is a general, recurring problem when results obtained from different groups are being compared. Many measurements, such as conductivity and fluorescence, depend critically on the quality of the precursor polymer and the conversion conditions. Even within the same group, there can sometimes be significant batch-to-batch variation.

2. Conversion to PPV

The sulfonium precursor polymer can be converted to fully conjugated PPV either by treatment with excess base or by heating under vacuum or nitrogen. The temperature needed for full conversion depends on the nature of the counterion. These range from ~120°C for bromide, to 200°C for chloride and fluoride, to over 300°C for acetate [86,87,95]. Partial elimination starts for all four precursor polymers at the same temperature [95], but complete conversion is complicated by the presence of side reactions. Elimination of the sulfide group and the counterion is not always synchronous over the entire temperature range of the conversion (as monitored by analysis of the gaseous products) [86,96]. This suggests that there may be competing reactions occurring (Scheme 12). The counterion could attack the proton $\beta-$ to the sulfonium group in an E2 or E1 type process to give elimination of both HX and SR_2 [path (a), Scheme 12]. Alternatively, substitution of the sulfonium group could occur first, with later loss of the counterion [path (b)]. There seems some evidence from elemental analyses that both processes operate when X = Cl [47], but more work needs to be done for other counterions for any firm conclusions to be drawn.

A side reaction that occurs with simple dialkylsulfonium salts is attack by the counterion on the alkyl group, with formation of RX and a neutral sulfide group on the main chain [path (c), Scheme 12]. This can result in incomplete conversion and moderate amounts of residual sulfur left in the polymer. Cyclic sulfonium groups do not seem to suffer from this side reaction so much (or if they do it must be reversible) and have become standard in the Wessling–Zimmerman synthesis of PPV.

Whereas in practice most sulfonium precursor polymers are converted to PPV thermally, there has been some interest in other methods. Several groups have reported that exposing films of the sulfonium polymer (**2**) to light (250–350 nm) causes partial conversion to PPV. This allows any unexposed parts of the film to be washed away (the exposed polymer being insoluble). Even longer wavelength light (514 nm) can be used if a sensitizing dye is also present [97]. By exposing the precursor polymer to extended irradiation, the opposite effect is found, wherein the exposed portions become so degraded that they do not convert into PPV upon heating. Thus, photolithography and patterning of PPV becomes an attractive and viable option [98–100].

Other methods that have been used for the conversion of sulfonium polymers to PPV include the use of strong acids [101] and microwave heating [102]. It is also possible to grow films of PPV from the sulfonium precursor polymer electrochemically. The formation of PPV is thought to arise from a local increase in the pH around the electrode due to hydrolysis reactions, leading to elimination of the precursor polymer [46].

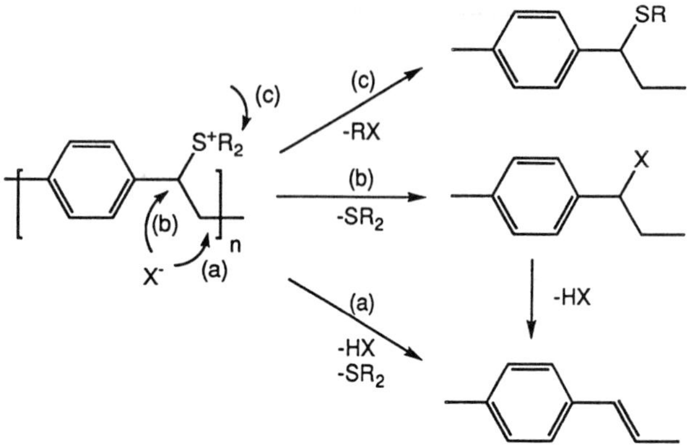

C. Gilch Route

While not necessarily a precursor route, a closely related polymerization method to the Wessling–Zimmerman

Scheme 12 Competing reactions in the thermal elimination of sulfonium groups.

synthesis of PPV was discovered independently by Gilch and Wheelwright in 1966 [12]. Instead of water-soluble sulfonium salts, they used dichloro-*p*-xylene that was polymerized with potassium *tert*-butoxide in organic solvents. Other bases have been used such as pyridine [103]. Although this method has no advantages over the sulfonium route for the synthesis of PPV itself, it allows easier entry into a large range of substituted organic-soluble PPV derivatives [15]. It can also be modified to allow the synthesis of a precursor polymer, which can be thermally eliminated at a later stage (see below). The mechanism of the reaction is believed to closely parallel that of the Wessling–Zimmerman route, via a reactive *p*-xylylene intermediate. Unlike the sulfonium route, the reaction is usually allowed to proceed to completion by the addition of excess base, solubility being ensured by the use of substituents on the aromatic ring or on the vinylic bond (Fig. 13.5). Both chlorine and bromine leaving groups have been used. Molecular weights as measured by GPC are as high if not higher than those measured for sulfonium-based precursors. However, the fully conjugated backbone of the former polymers might cause some overestimation of the molecular weight by GPC analysis.

In certain cases the fully conjugated polymer may not be soluble. It is then possible to use a precursor variant of the Gilch synthesis, where only one equivalent of base is used. This produces an intermediate soluble halide-substituted polymer that can be dehydrohalogenated thermally (Scheme 13) [104–107] or partially eliminated in an alcoholic solvent [108]. However, the intermediate halide-containing polymers may not be very stable, though this can depend on storage conditions [104].

Substituents used to give solubility include phenyl, alkoxy, and alkyl groups. Branched long-chain alkoxy and alkyl groups at least six carbons in length are especially good at improving solubility. Asymmetric substitution of the ring (where R1 ≠ R2, Scheme 11) also appears to help. Based on these principles, one of the most widely used soluble PPV derivatives synthesized by the Gilch method is poly(1-methoxy-4-(2-ethylhexyloxy)-*p*-phenylenevinylene) (MEH-PPV) [109], which possesses both asymmetry and branching. In contrast, poly(1,4-dihexoxy-*p*-phenylenevinylene), which contains a more regular side chain, is virtually insoluble when prepared under the same conditions (S.C. Moratti, unpublished results).

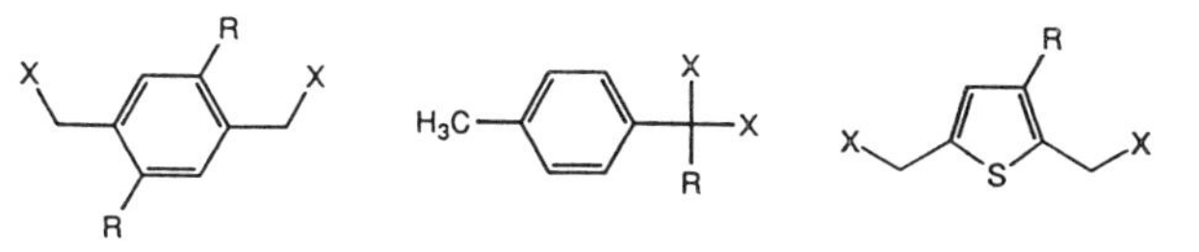

Fig. 13.5 Typical monomers used in the Gilch route to PPV derivatives. R is a solubilizing group, and X is Cl or Br.

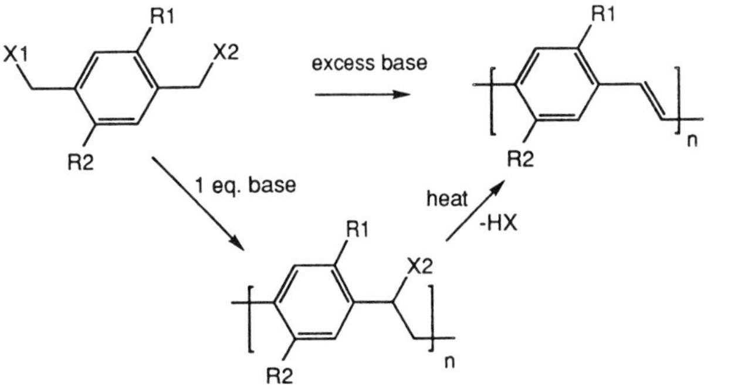

Scheme 13

A potentially very useful variant of the Gilch route to PPV uses a sulfone or sulfoxide group (i.e., X2, Scheme 13) in place of one of the halide atoms [110]. The resulting precursor polymers are soluble in organic solvents such as chloroform and can be thermally converted at low temperatures. For instance, the phenylsulfenyl-substituted precursor (X2 = SOPh) apparently undergoes conversion at temperatures of ~100°C, although higher temperatures may be needed to volatilize the eliminated sulfur compounds from the film.

Xanthates have been used as the leaving group in a Gilch-type synthesis of PPV [1]. The resulting polymer had a much higher level of cis double bonds (although these could be isomerized on prolonged heating) and as a result was amorphous rather than semi-crystalline after conversion. This may be very useful in optical devices to lessen light scattering and was shown to improve performance in light-emitting diodes.

D. Alkoxy Route

It was mentioned above that sulfonium precursor polymers can be readily converted to methoxy-substituted polymers, which can be used in molecular weight determinations. The methoxy-substituted polymer is soluble in organic solvents such as chloroform and is much more stable than the sulfonium precursor [70,90,91]. The alkoxy group can be removed with acid catalysis and/or heating to produce fully converted PPV derivatives (see Scheme 14).

There is usually little direct advantage in this route as the conversion step often requires an acid catalyst for completion. Heat alone has sometimes been used to effect elimination, and for thick samples this may be

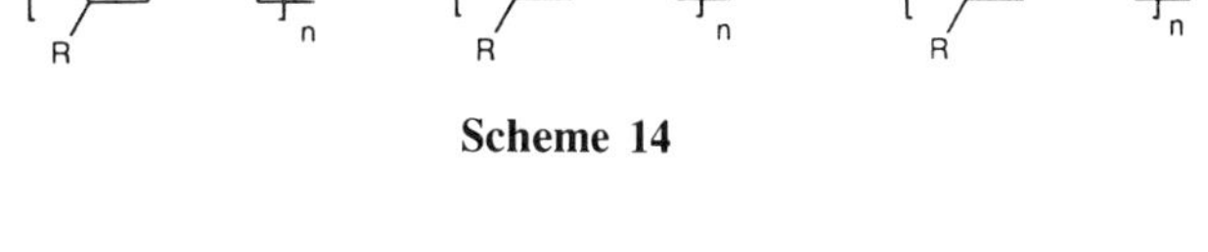

Scheme 14

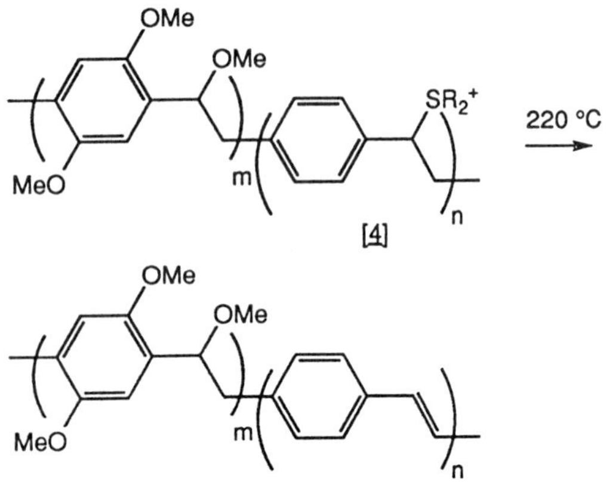

Scheme 15

methoxy group. Under normal thermal conversion conditions (thin film, 220°C, vacuum) the methoxy groups do not eliminate, and the resulting polymer is only semiconjugated (Scheme 15). Electroluminescence efficiency was increased by a factor of 3, believed to be due in part to the lower migration of excited states to quenching centers. Chemical patterning of thin films of precursor polymer 4 was accomplished by depositing aluminum on top of selected areas prior to conversion. In the masked areas, escape of HCl (formed from the elimination of the chloride sulfonium salt) was hindered and was able to catalyze the elimination of remaining methoxy groups to form a fully conjugated polymer. As a result, these areas were significantly red-shifted in absorption and emission compared to the unmasked areas [113].

IV. CHEMICAL MODIFICATION OF PPV AND DERIVATIVES

There has been much less work done on postmodification of fully conjugated PPV and derivatives. This is natural, considering the problems of achieving selectivity and high yields in chemical transformations on polymers. It has been observed that dialkoxy-PPV derivatives become insoluble if heated over 200°C [114,115], probably due to cross-linking, and this may be regarded as a crude (though potentially useful) form of modification.

A very early transformation of PPV involved reduction of the double bonds with sodium in liquid ammonia [9]. The product obtained was colorless but insoluble, suggesting that cross-linking had occurred. Poly(2-(N,N-dimethylamino)-p-phenylenevinylene) could be prepared via both the Wessling–Zimmerman and Gilch routes as an orange solid that was insoluble in organic solvents but soluble in strong acids. The polymer was treated with fuming sulfuric acid to sulfonate the aromatic ring, and a black rubbery solid was obtained with a conductivity of 10^{-2} S/cm.

Hsieh [116] attempted to prepare polyphenyleneacetylene (PPA) via bromination and dehydrobromination of PPV (Scheme 16). Although the bromination step went largely as expected, elimination of the halogen atoms proved troublesome. Base treatment gave hydrolysis and bond cleavage rather than elimination. Instead,

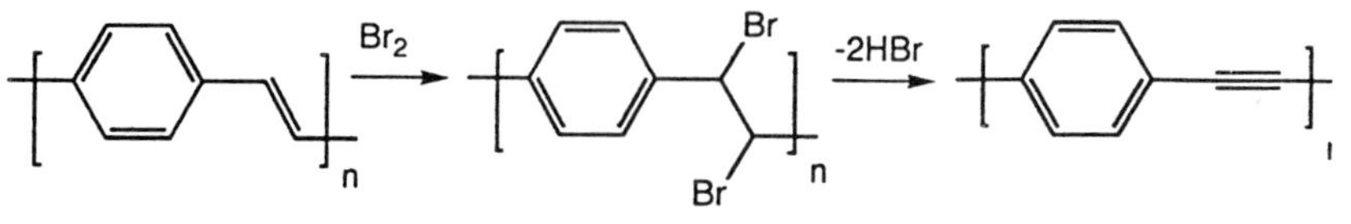

adequate as traces of acid (possibly from residual traces of sulfonium groups) may get trapped in the polymer and help catalyze the reaction. Acid catalysts may also be added to the polymer before processing and heating. Alternatively, the polymer may be heated under an acidic atmosphere (such as hydrogen chloride) [61,111].

In some cases there is little choice when using the Wessling–Zimmerman route, as the intermediate sulfonium precursor polymer can be unstable and rapidly substitute in alcohol solvents to give the alkoxy-substituted polymer. This is especially true for polymers that contain electron-rich aromatic rings such as dialkoxyphenylenes and -thiophenes, when substitution occurs within hours at room temperature. This high reactivity of the sulfonium group results from extra stabilization of the intermediate carbocation in the solvolysis process. Interestingly, it has been reported that weak bases such as pyridine are able to stabilize solutions of alkoxy-substituted sulfonium precursor polymers. Apparently, the weak base substitutes for the sulfonium groups within the polymer, which results in a more stable polyelectrolyte [112].

It is possible to chemically tune the optical properties of PPV-type polymers by the copolymerization of different sulfonium salts. This was used to advantage to increase the electro- and photoluminescence efficiency of PPV by the incorporation of up to 20% dimethoxyphenylene units within the polymer (Fig. 13.5) [113]. The sulfonium groups adjacent to a methoxyphenylene ring rapidly undergo solvolysis and become substituted with a

Scheme 16

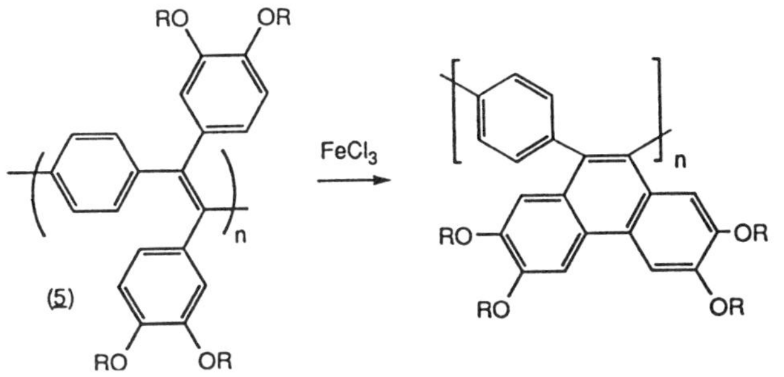

Scheme 17

the intermediate polymers were heated to 250°C for 8 h, and substantial triple bond formation did indeed occur. However, there was still a large amount of residual bromine (up to 20%) left in the products. A kinetic study has been reported for this reaction [117].

It proved possible to selectively oxidatively cyclize the two dialkoxyphenyl substituents of polymer 5 with ferric chloride to produce phenanthrene structures (Scheme 17) [118]. Similar cyclizations were performed on di(monoalkoxyphenyl)-PPVs.

Perhaps a rather extreme form of chemical modification is graphitization at high temperatures. PPV has an advantage in that films and fibers can be readily formed via a precursor route and subsequently pyrolyzed at high temperatures to produce highly conducting materials. Conductivities of 100 S/cm were reported for PPV pyrolyzed at 1000°C, rising to 1400 S/cm at 3000°C [119,120].

V. LIQUID CRYSTALLINE AND ORDERED PPV DERIVATIVES

Liquid crystallinity often occurs in polymers containing both rigid and flexible regions, for example, in PPV derivatives that have (1) a rigid, fully conjugated backbone with flexible side groups, (2) flexible linkages in the main chain, usually caused through incomplete elimination, or (3) bulky lateral substituents that hinder planarity and conjugation. Liquid crystalline (LC) behavior is potentially important as it allows some ordering of the polymer while processing, which can have significant impact on strength, conductivity, and optical properties. Simple dialkoxy-PPV derivatives do not appear to show thermotropic LC behavior, although a lyotropic phase was found for chloroform solutions of poly(1,4-dinonyl-p-phenylene vinylene) [121]. X-ray diffraction data suggested a stacked layer arrangement in which the alkoxy groups are spread out in a plane along the layer at right angles to the main chain.

While no simple alkoxy-substituted PPV derivatives have been found to be thermotropic, many LC copolymers have been discovered. Yu and Bao [114] synthesized a series of PPV copolymers exhibiting nematic phases (Fig. 13.6), often over large temperature ranges. However, cross-linking was observed at temperatures over 200°C. This was suppressed by the use of very long (16-carbon) side chains, which presumably separated the vinylene groups on neighboring chains, preventing interchain reaction.

A series of substituted PPVs and copolymers were synthesized by palladium-catalyzed coupling of aryl bromides and ethylene [115]. Bulky lateral substituents (such as phenyl) were found to improve solubility and to lower the softening points of the polymers. All the polymers with low softening points (~130–180°C) were found to be at least partially anisotropic under a polarizing microscope. However, polymers became insoluble when they were taken 50°C above their softening points, suggesting that cross-linking was occurring. In contrast, decomposition (as measured by thermal gravimetric analysis) usually occurred at temperatures over 350°C.

Partially eliminated precursor polymers may also exhibit liquid crystalline phases. It was possible to control the elimination of butanol from a butoxy-substituted precursor polymer by using weak acids with appropriate amounts of alcohol scavengers to suppress the reverse reaction (Scheme 18). Shear alignment was found above 140°C, although at temperatures over 160°C further loss of butanol occurred and liquid crystallinity was lost [122].

Langmuir–Blodgett (LB) films have been formed from PPV precursor polymers. Kim et al. [123] deposited the methoxy precursor to PPV (3, Scheme 11) by this method and converted it to PPV by thermal treatment at 300°C. Regular STM features were observed, indicating that a high degree of order was obtained both before and after conversion. The photoluminescence (PL) spectrum was blue-shifted from normal, which was

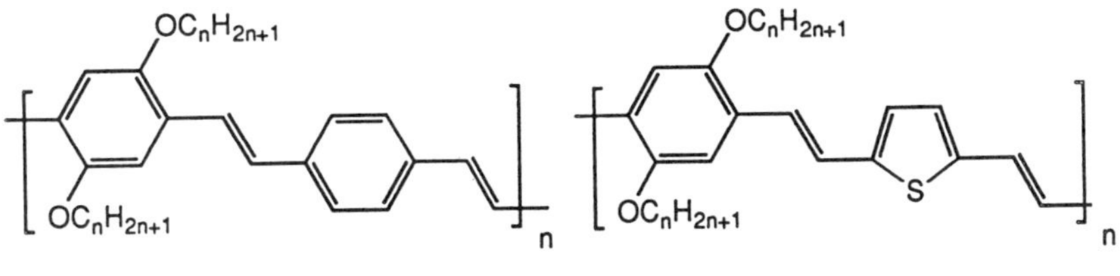

Fig. 13.6 Typical liquid crystalline PPV derivatives. (From Ref. 114.)

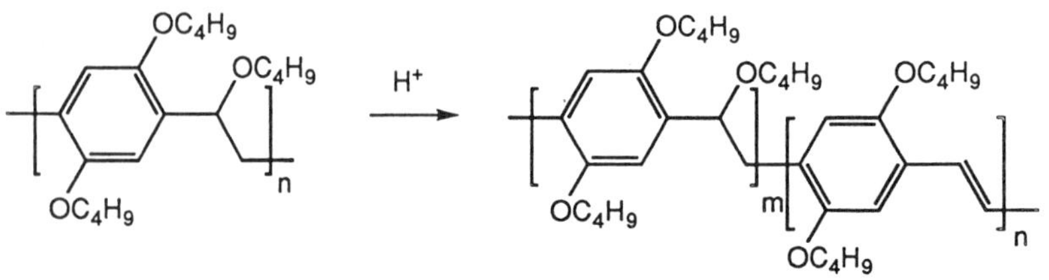

Scheme 18 Synthesis of a liquid crystalline PPV derivative by partial elimination. (From Ref. 122.)

taken to indicate that the thermal conversion was not complete.

A number of groups have formed LB films from sulfonium precursor polymers [124–127]. It was found that the structure of the counterion was important in forming good monolayers, with long-chain perfluoroalkylcarboxylic acids or alkylsulfonic acids giving the best results [124,125]. Large anisotropy was observed in the electrical conductivity of SO_3-doped PPV LB films, with measurements of 10^{-5} S/cm normal to the film as opposed to 0.5 S/cm in the plane [125].

Highly ordered PPV was obtained by the acid-catalyzed conversion of films of partially eliminated methoxy precursor polymer (**6**, Fig. 13.7) [111]. In comparison to conventionally prepared PPV, the optical absorption was strongly red-shifted (Fig. 13.8). Electron diffraction studies also indicated much greater ordering of the polymer chains [128]. It is not clear whether the increased order occurred during deposition of the precursor film by spin-coating or during thermal conversion in a liquid crystalline phase. Other studies have also shown red shifts in the absorption spectra of PPV derivatives upon stretch-induced alignment [129]. It is also possible to orient the methoxy precursor polymer (**6**) on rubbed polytetrafluoroethylene substrates to achieve high degrees of alignment in the final converted polymer [130].

VI. CONDUCTIVITY STUDIES

Even though PPV is a fully conjugated polymer, intrinsic electrical conductivity is very low, on the order of 10^{-13} S/cm [131–133]. In order to increase the conductivity, the polymer has to be doped with such compounds as iodine, ferric chloride, alkali metals, or acids. The environmental stability of these doped materials is usually low; however, alkoxy-substituted PPV polymers can be stably doped with iodine even in the presence of air. It can be difficult to reliably compare conductivity and charge transport measurements from different research groups, as these seem to be sensitive to the molecular weight, structural regularity, and the method of preparation of the polymer [68,69,134–141].

In general, unaligned, unsubstituted PPV exhibits only moderate conductivity upon doping, ranging from approximately $\ll 10^{-3}$ S/cm (I_2-doped) to 100 S/cm (H_2SO_4-doped) (Table 13.1). Alignment is usually achieved by the stretching of a precursor film before or during thermal conversion. Draw ratios of up to 10 can be obtained in this way, with increases in conductivity of the same order or more. Much lower conductivities are found upon n-doping with, e.g., sodium naphthalide. Alkoxy-substituted PPVs are more easily oxidized than the parent polymer (Table 13.2), which results in much higher conductivities, especially with dopants of low oxidizing power such as iodine. Other groups able to stabilize positive charges (such as silyl) also increase the conductivity in PPV materials after doping with iodine. For stronger oxidants such as ferric chloride, there does not seem to be any consistent difference in conductivity between alkoxy-substituted and unsubstituted PPV. However, increasing the length of the side chain lowers the

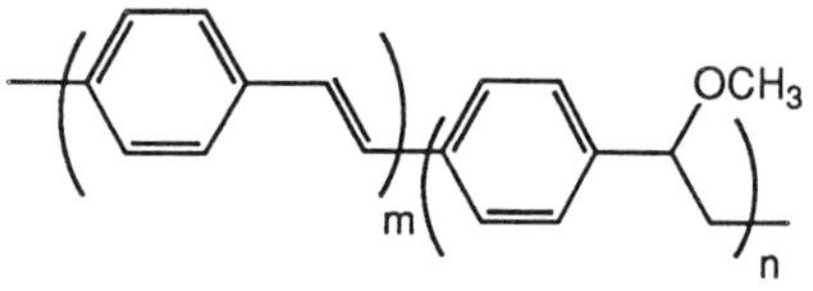

Fig. 13.7 Precursor polymer giving increased order upon conversion to PPV. (From Ref. 112.)

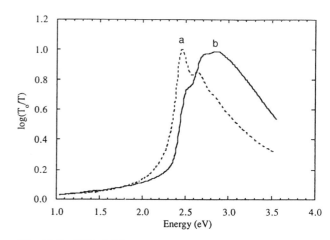

Fig. 13.8 UV-visible spectra of standard PPV (solid line) and "ordered" PPV (dashed line). (From Ref. 112.)

Table 13.1 Electrical Conductivity of Several Doped Complexes of PPV

Dopant (concentration)	Conductivity (Ω^{-1} cm^{-1})	Draw ratio L/L_o	Ref.
FeCl$_3$ (0.071 equiv)	35	1	136
FeCl$_3$ (0.39 equiv)	230	4	136
Na naphthalide (?)	2×10^{-4}	1	137
H$_2$SO$_4$ (?)	100	1	137
AsF$_5$ (?)	10	1	137
H$_2$SO$_4$ (?)	2700	6	135
I$_2$ (?)	$<10^{-3}$	1	151

conductivity [68,83], probably by hindering interchain hopping of charge carriers. Alkyl and phenyl groups also do not seem to improve conductivity much [58,71,142], possibly for the same reason. There have also been many other conductivity measurements reported on PPV [88,101,125,135,143–147], alkoxy derivatives [1,63–66, 77,84,85,122,148–155], alkyl derivatives [82,83,151, 152,156], and halogen-substituted [75,76,152,157] and heterocyclic copolymers [6,84,158–160].

It is possible to produce doped polymers during the thermal conversion process. It was found that sulfonium precursor polymers with BF$_4^-$ and AsF$_6^-$ counterions produced a black conductive material (0.1–0.3 S/cm) after heating [89]. A black converted polymer was also found with IO$_3^-$ and BrO$_3^-$ counterions, although the conductivity was not measured [86]. Self-doping was also achieved by the inclusion of a sulfonate group on a side group, though the resulting conductivity was not high (10^{-6}–10^{-2} S/cm) [161].

VII. POTENTIAL APPLICATIONS

The stability, processability, and electrical and optical properties of PPV have led to its being investigated for use in a wide variety of applications. Much of the initial work centered on the wide range of conductivity that can be achieved, covering some 20 orders of magnitude. Although PPV can indeed form highly conductive materials upon doping, the long-term stability of these materials may be a problem. It is probable that it will be the semiconductive and photoconductive properties of this polymer that will prove more useful in the future.

Organic electroluminescence (EL) has been one of the most exciting recent developments in polymer science. Materials based on organic molecules have been used in experimental EL devices for many years, but it was not until 1989 that the first polymer-based light-emitting diode (LED) was discovered, using PPV as the emissive layer [162]. Polymers are thought to have certain advantages over molecular materials in LEDs, such as ease of processing, greater thermal and mechanical stability, and reduced tendency for crystallization. This must be balanced by the often greater synthetic challenges in producing processable, highly functional polymers and by the difficulty in purifying polymers to the necessary degree. Since 1989, a large number of PPV derivatives have been synthesized and used for LED applications, with alteration of such properties as color of emission and electron affinity [35–38,108,163–184]. Although solid-state lasing has yet to be demonstrated in an organic LED, MEH-PPV has been shown to be a useful laser dye owing to its high fluorescence efficiency in solution [185].

Conjugated polymers can be used in the construction of field effect transistors (FETs) and diodes, although charge carrier mobilities tend to be too slow [$<10^{-5}$ cm^2/(V·s)] for most serious applications. Undoped PPV films often have too low a charge carrier concentration for use in metal insulator semiconductor FETs; however, both poly(2,5-dimethoxy-p-phenylenevinylene) [186] and poly(1,4-naphthalenevinylene) [187] have been used in similar devices. The charge carrier concentration can be very sensitive to the method of synthesis of the polymer, and in the case of poly(2,5-dimethoxy-p-phenylenevinylene) some extra doping of the polymer probably occurs during the acid-catalyzed conversion [186]. Carrier injection into the polymer during the operation of the FET causes new subband gap levels to appear in the absorption spectrum, and this effect was used in the fabrication of an electro-optical modulator [106,186]. It is also possible to increase the charge carrier concentration for FET operation by doping using ion implantation [102].

There has also been much research into the photoconductive properties of PPV and derivatives [188–190]. This work has direct application for use in photodiodes [191], photovoltaic cells [109,192,193], optocouplers [113], and electrophotography [194]. It was discovered that the photoconductivity of PPV derivatives can be enhanced by several orders of magnitude through the use of dopants such as C$_{60}$ compounds. [189,190,195].

The nonlinear properties of PPV derivatives may

Table 13.2 Selected Conductivities for Several Doped 2,5-Substituted PPVs

Ring substituents	Dopant	Conductivity (Ω^{-1} cm^{-1})	Draw ratio L/L_o	Ref.
H, I	FeCl$_3$	0.96	5	138
H, SiMe$_3$	I$_2$	2×10^{-2}	1	73
C$_7$H$_{15}$, C$_7$H$_{15}$	SO$_3$	2.7×10^{-3}	1	139
H, OCH$_3$	FeCl$_3$	0.38	6	140
OCH$_3$, OCH$_3$	FeCl$_3$	37	1	69
OCH$_3$, OCH$_3$	I$_2$	20	1	141
OCH$_3$, OCH$_3$	I$_2$	1200	8	141
H, OC$_8$H$_{17}$	I$_2$	2.6×10^{-3}	8	68
OCH$_3$, OC$_4$H$_9$	FeCl$_3$	2160	3.6	150

prove useful in such applications as optical switches and electromodulators. An asymmetrically substituted PPV precursor polymer was poled in an electric field during thermal conversion to produce poly(2-cyano-5-methoxy-1,4-phenylenevinylene) [196]. A resulting electro-optic coefficient, γ_{113}, of 1.2 pm/V was obtained that was stable even during prolonged annealing at 200°C. Values of $\chi^{(3)}$ for PPV of 2×10^{-11} esu (1.06 μm) [197], 7.5×10^{-11} esu (1.06 μm) [198], and 4×10^{-10} esu (0.6 μm) [199] have been reported. As PPV is highly crystalline [128,200], films tend to scatter light, which is a major problem for optical devices. One solution is to use PPV/sol gel composites, where the optical quality was found to be much improved [157,199,201,202]. Slightly lower $\chi^{(3)}$ values were found for the PPV-silica/sol gel materials than for the conjugated polymer alone [202], but values of 1×10^{-9} esu (0.6 μm) were still obtained for a poly(2-butoxy-5-methoxy-p-phenylenevinylene)/silica composite film [199]. Third-order nonlinear optical susceptibilities have been measured for other substituted polymers as well [64,68,75,203]. PPV derivatives have also been investigated for use in waveguides. Strips as narrow as 1–4 μm in a PPV copolymer could be defined by a chemical masking procedure for use as waveguide channels [204].

Polyparaphenylenevinylidene and derivatives have some promise as charge storage materials for rechargeable batteries. One study showed that an energy density of 532 W h/kg might be possible with a Li/PPV cell [205], much higher than could be achieved with polyaniline or polypyrrole. The doping level was almost one charge per repeat unit; however, the charging process was rather slow. This may be due to the compact, dense morphology of PPV, preventing diffusion of the electrolyte into the film. Diethoxy-PPV has similarly been investigated [206], and although it gave a lower estimated energy density (240 W h/kg) than PPV, little degradation was found after up to 1000 charging cycles. The general electrochemistry of PPV and derivatives has been well reviewed [207,208]. A new approach to light generation from conjugated polymers was reported recently using an electrochemical system. Chemiluminesence was observed in an electrochemical cell containing MEH-PPV, and this may represent a useful alternative technology for organic electroluminescent devices [209].

The structural and mechanical properties of PPV and derivatives have also been of interest. A tensile strength of 500 MPa was measured for uniaxially oriented PPV [143], which approaches that of many high performance polymers. Substitution on the ring with long alkoxy chains can improve processibility but at the expense of mechanical properties [155]. Doping of PPV was found to produce embrittlement and sharply reduce the mechanical properties [143]. However, fibers of poly(2,5-dimethoxy-p-phenylenevinylene) were found to retain their high tensile strength (0.7 GPa) upon doping with iodine and exhibited conductivities as high as 200 S/cm [141].

REFERENCES

1. S. Son, A. Dodabalapur, A. J. Lovinger, and M. E. Galvin, Luminescence enhancement by the introduction of disorder into poly(p-phenylene vinylene), *Science* 269:376 (1995).
2. R. N. McDonald and T. W. Campbell, *J. Am. Chem. Soc.* 82:4669 (1960).
3. K. D. Gourley, C. P. Lillya, J. R. Reynolds, and J. C. W. Chien, Electrically conducting polymers: AsF$_5$-doped poly(phenylenevinylene) and its analogues, *Macromolecules* 17:1025 (1984).
4. V. G. Kossmehl, M. Härtel, and G. Manecke, *Makromol. Chem.* 131:37 (1970).
5. V. G. Manecke and D. Zerpner, *Makromol. Chem.* 129:183 (1969).
6. W. S. Huang, K. Y. Jen, M. Angelopoulos, A. G. McDiarmid, and M. P. Cava, Electrical and electrochemical properties of some polyarylene vinylene derivatives and their mixed copolymers, *Mol. Cryst. Liq. Cryst.* 189:237 (1990).
7. H. H. Hörhold and J. Opfermann, Poly-p-xyliden: Synthesen und Beziehungen zwischen Struktur und elektrophysikalischen Eigenschaften, *Makromol. Chem.* 131:105 (1970).
8. G. H. Smith, U.S. Patent 3,110,687 (1963).
9. D. F. Hoeg, D. I. Lusk, and E. P. Goldberg, Poly-p-xylylidene, *J. Polym. Sci. Part B: Polym. Lett.* 2:697 (1964).
10. H. H. Hörhold and J. Opfermann, *Makromol. Chem.* 131:105 (1970).
11. I. Moritani, T. Nagai, and Y. Shirota, *Kogyo Kagaku Zasshi* 68:296 (1965).
12. H. G. Gilch and W. L. Wheelwright, Polymerization of α-halogenated p-xylenes with base, *J. Polym. Sci. Part A: Polym. Chem.* 4:1337 (1966).
13. H.-H. Hörhold, H. Räthe, M. Helbig, and J. Opfermann, Synthese, Photoleitfähigkeit und -verhalten von Poly(9-methylcarbazol-3,6-diyl-1,2-diphenylvinylen, *Makromol. Chem.* 188:2083 (1987).
14. H. H. Hörhold and D. Raabe, *Acta Polym.* 30:86 (1979).
15. H. H. Hörhold, J. Gottschaldt, and J. Opfermann, *J. Prakt. Chem.* 139:611 (1977).
16. F. Fors, L. Julia, J. Riera, J. M. Tura, and J. Saulo, Syntheses and properties of polymers from perchloro-p-xylene, perchlorobi-p-tolyl, and perchloro-m-xylene with Fe(CO)$_5$ as dechlorinating agent, *J. Polym. Sci. Part A: Polym. Chem.* 30:2489 (1992).
17. M. Rehahn and A.-D. Schlüter, Soluble poly(p-phenylenevinylene)s from 2,5-dihexylterephthaldehyde using the improved McMurray reagent, *Macromol. Chem. Rapid Commun.* 11:375 (1990).

18. F. Cataldo, A new method of synthesizing poly(p-phenylene vinylene), *Polym. Commun.* 32:354 (1991).
19. S. Iwatsuki, M. Kubo, and Y. Itoh, Preparation of poly(3,4-dibutoxy-2,5-thienyl-vinylene) via titanium dicarbonyl-coupling reaction of dibutoxythiophene-2,5-dicarbaldehyde, *Chem. Lett. 1993*:1085.
20. A. Greiner and W. Heitz, New synthetic approach to poly(1,4-phenylenevinylene) and its derivatives by palladium catalyzed arylation of ethylene, *Macromol. Chem., Rapid Commun.* 9:581 (1988).
21. M. Brenda, A. Greiner, and W. Heitz, Model reactions for the synthesis of poly(1,4-phenylenevinylene). The palladium catalyzed arylation of ethylene with halogenated arenes or benzoyl chlorides, *Makromol. Chem.* 191:1083 (1990).
22. M. Pan, Z. Bao, and L. Yu, Regiospecific, functionalised poly(phenylenevinylene) using the Heck coupling reaction, *Macromolecules* 28:5151 (1995).
23. W. Heitz et al., Synthesis of monomers and polymers by the Heck reaction, *Makromol. Chem.* 189:119 (1988).
24. A. Kumar and B. E. Eichinger, Synthesis of poly(1,4-phenylenevinylene) by metathesis of p-divinylbenzene, *Macromol. Chem., Rapid Commun.* 13:311 (1992).
25. E. Thorn-Csányi and P. Kraxner, Synthesis of soluble, all trans poly(2,5-diheptyl-p-phenylenevinylene) via metathesis polymerisation, *Macromol. Chem., Rapid Commun.* 16:147 (1995).
26. E. Thorn-Csányi and K.-P. Pflug, Synthesis of copolymers with a controllable sequence length of p-phenylenevinylene units, *Makromol. Chem., Rapid Commun.* 14:619 (1993).
27. E. Thorn-Csányi and H.-D. Höhnk, Ring-opening metathesis polymerisation of [2.2]paracyclophane-1,9-diene (PCPDE) to poly(p-phenylene vinylene) (PPV) and copolymers with cyclopentene, *J. Mol. Catal.* 76:101 (1992).
28. H. Meier, H. Kretzshmann, and M. Lang, Synthesis of 2-alkoxy substituted oligo and poly(1,4-phenyleneethylene)s and 2-arylbenzo[b]furanes by applying the Siegrist reaction, *J. Prakt. Chem.* 336:121 (1994).
29. H. Kretzschmann and H. Meier, A new synthesis of soluble poly(1,4-phenylenevinylene)s and poly(2,5-pyridinylenevinylene)s, *Tetrahedron Lett.* 32:5059 (1991).
30. P. M. Lahti, A. Sarker, R. O. Garay, R. W. Lenz, and F. E. Karasz, Polymerisation of 1,4-bis(tetrahydrothiopheniomethyl) - 2 - cyano - 5 - methoxybenzenedibromide: synthesis of electronically "push-pull" substituted poly(p-phenylene vinylene)s, *Polymer* 35:1312 (1994).
31. D. Debord and J. Golé, Propriétés thermiques et électriques de quelques systemes macromoléculaires aromatiques, *Bull. Chem. Soc.* 4:1401 (1971).
32. V. W. Funke and E. C. Schütze, Polykondensationreaktionen mit Xylylendicyaniden, *Makromol. Chem.* 74:71 (1963).
33. R. W. Lenz and C. E. Handlovits, Thermally stable hydrocarbon polymers: polyterephthalylidenes, *J. Org. Chem.* 25:813 (1960).
34. H. H. Hörhold, Neue Polykondensationsreaktioen zur Synthese von Polymeren mit Halbleitereigenschaften, *Z. Chem.* 12:41 (1972).
35. N. C. Greenham, S. C. Moratti, D. D. C. Bradley, R. H. Friend, and A. B. Holmes, Efficient light-emitting diodes based on polymers with high electron affinity, *Nature* 365:628 (1993).
36. S. C. Moratti, et al., High electron affinity polymers for LEDs, *Synth. Met.* 71:2117 (1995).
37. D. R. Baigent, P. J. Hamer, R. H. Friend, S. C. Moratti, and A. B. Holmes, Polymer electroluminescence in the near infra-red, *Synth. Met.* 71:2175 (1995).
38. E. G. J. Staring, et al., Electroluminesence and photoluminesence efficiency of poly(p-phenylenevinylene) derivatives, *Synth. Met.* 71:2179 (1995).
39. H. Nishihara, M. Tateishi, K. Aramaki, T. Ohsawa, and O. Kimura, Electrochemical synthesis of poly(p-phenylene vinylene) films, *Chem. Lett. 1987*:539.
40. J. A. Osaheni and S. A. Jenekhe, A new light-emitting conjugated rigid rod polymer: poly(benzobisthiazole-1,4-phenylene vinylene), *Macromolecules* 26:4726 (1993).
41. V. P. Conticello, D. L. Gin, and R. H. Grubbs, Ring-opening metathesis polymerisation of substituted bicyclo[2.2.2]octadienes: a new precursor route to poly(1,4-phenylenevinylene), *J. Am. Chem. Soc.* 114:9708 (1992).
42. S. Iwatsuki, M. Kubo, and T. Kumeuchi, New method for preparation of poly(phenylene-vinylene) film, *Chem. Lett. 1991*:1071.
43. Y.-J. Miao and G. C. Bazan, Paracyclophane route to poly(p-phenylenevinylene), *J. Am. Chem. Soc.* 116:9379 (1994).
44. R. A. Wessling and R. G. Zimmerman, U.S. Patent 3,404,132 (1968).
45. R. A. Wessling and R. G. Zimmerman, U.S. Patent 3,532,643 (1970).
46. R. A. Wessling, The polymerization of xylylidene bisalkylsulfonium salts, *J. Polym. Sci., Polym. Symp.* 72:55 (1985).
47. R. W. Lenz, C.-C. Han, J. Stenger-Smith, and F. E. Karasz, Preparation of poly(phenylene vinylene) from cycloalkylene sulfonium salt monomers and polymers, *J. Polym. Sci. Part A: Polym. Chem.* 26:3241 (1988).
48. F. R. Denton III, A. Sarker, P. M. Lahti, R. O. Garay, and F. E. Karasz, Paraxylylenes and analogues by base-induced elimination from 1,4-bis(dialkylsulfoniomethyl)arene salts in poly(1,4-arylene vinylene) synthesis by the Wessling soluble precursor method, *J. Polym. Sci. Part A: Polym. Chem.* 30:2233 (1992).
49. R. Garay and R. W. Lenz, Anionic polymerisation of p-xylenesulfonium salts, *Makromol. Chem. Suppl.* 15:1 (1989).

50. M. J. Cherry, S. C. Moratti, A. B. Holmes, P. L. Taylor, J. Grüner, and R. H. Friend, The dispersion polymerisation of poly(*p*-phenylene vinylene), *Synth. Met. 69*:493 (1995).

51. P. M. Lahti, D. A. Modarelli, F. R. Denton III, R. W. Lenz, and F. E. Karasz, Polymerization of α,α'-bis(dialkylsulfonio)-*p*-xylene intermediates: evidence for a nonradical mechanism, *J. Am. Chem. Soc. 110*:7258 (1988).

52. A. Delmotte, M. Biesemans, B. Van Mele, M. Gielen, M. M. Bouman, and E. W. Meijer, Selective elimination in dialkoxy-PPV precursors yielding polymers with isolated tetraalkoxy-stilbene units, *Synth. Met. 68*:269 (1995).

53. F. R. Denton III, P. M. Lahti, and F. E. Karasz, The effect of radical trapping upon formation of poly(α-tetrahydrothiophenio paraxylene) polyelectrolytes by the Wessling soluble precursor method, *J. Polym. Sci. Part A: Polym. Chem. 30*:2223 (1992).

54. Y. Sonada and Y. Suzuki, Preparation of *p*-phenylene-3,3-bis(1-allyltetrahydrothiophenium) dibromide and its reactions in basic solution, *J. Chem. Soc., Perkin Trans. 1*:317 (1994).

55. J. D. Stenger-Smith, T. Sauer, G. Wegner, and R. W. Lenz, Preparation, spectroscopic and cyclic voltammetric studies of poly(1,4-naphthalene vinylene) prepared from a cycloalkylene sulfonium precursor polymer, *Polymer 31*:1632 (1990).

56. J. D. Capistran, D. R. Gagnon, S. Antoun, R. W. Lenz, and F. E. Karasz, Synthesis and electrical conductivity of high molecular weight poly(arylene vinylene), *Polym. Prepr. 1984*:282.

57. L. M. Leung and G. L. Chik, Phase-transfer catalysed synthesis of disubstituted poly(phenylene vinylene), *Polymer 34*:5174 (1993).

58. R. O. Garay, F. E. Karasz, and R. W. Lenz, Synthesis and properties of phenyl-substituted arylene vinylene polymers and copolymers, *J. Macromol. Sci.-Pure Appl. Chem. A 32*:905 (1995).

59. A. Sarker, P. M. Lahti, and F. E. Karasz, Synthesis of new poly(arylene vinylene) analogues: poly(4,7-benzofuran vinylene), poly(4,7-benzothiophene vinylene), *J. Polym. Sci. Part A: Polym. Chem. 32*:65 (1994).

60. M. Pomerantz, J. Wang, S. Seong, K. P. Starkey, L. Nguyen, and D. S. Marynick, Poly(benzo[1,2-*b*:4,5-*b'*]dithiophene-4,8-diylvinylene). Synthesis, properties, and electronic structure of a new dithiophene-fused *p*-phenylenevinylene conducting polymer, *Macromolecules 27*:7478 (1994).

61. P. L. Burn, et al., Precursor route chemistry and electronic properties of poly(*p*-phenylenevinylene), poly[(2,5-dimethyl-*p*-phenylene)vinylene] and poly-[(2,5-dimethoxy-*p*-phenylene)vinylene], *J. Chem. Soc. Perkin Trans. 1*:3225 (1992).

62. R. O. Garay, H. Naarmann, and K. Mullen, Synthesis and characterization of poly(1,4-anthrylenevinylene), *Macromolecules 27*:1922 (1994).

63. H.-K. Shim, D. H. Hwang, J. I. Lee, and K.-S. Lee, Synthesis, electrical and optical properties of asymmetrically monoalkoxy-substituted PPV derivatives, *Synth. Met. 55–57*:908 (1993).

64. K.-J. Moon, K.-S. Lee, and H.-K. Shim, Synthesis, electrical and nonlinear properties of PPV derivatives containing alkoxynitrostilbene group, *Synth. Met. 71*:1719 (1995).

65. I. Murase, T. Ohnishi, T. Noguchi, and M. Hirooka, Alkoxy-substituent effect of poly(*p*-phenylene vinylene) conductivity, *Polym. Commun. 26*:362 (1985).

66. W. B. Liang, R. W. Lenz, and F. E. Karasz, Poly(2-methoxyphenylene vinylene): synthesis, electrical conductivity, and control of electronic properties, *J. Polym. Sci. Part A: Polym. Chem. 28*:2867 (1990).

67. S. H. Askari, S. D. Rughooputh, and F. Wudl, Soluble substituted-PPV conducting polymers: spectroscopic studies, *Synth. Met. 29*:E129 (1989).

68. M.-S. Jang and H.-K. Shim, Synthesis, electrical conductivity and optical properties of multifunctional poly[2-(2-ethyl-hexyloxy)-1,4-phenylenevinylene], *Polym. Bull. 35*:49 (1995).

69. K.-Y. Jen, L. W. Shacklette, and R. Elsenbaumer, Synthesis and conductivity studies of poly(2,5-dimethoxy-1,4-phenylene vinylene), *Synth. Met. 22*:179 (1987).

70. S. Tokito, T. Momii, H. Murata, T. Tsutsui, and S. Saito, Polyarylene films prepared from precursor polymers soluble in organic solvents, *Polymer 31*:1137 (1990).

71. Y. Sonada and K. Kaeriyama, Preparation of poly(2,5-diheptyl-1,4-phenylenevinylene) by sulfonium salt pyrolysis, *Bull. Chem. Soc. Jpn. 65*:853 (1992).

72. S. Höger, J. J. McNamara, S. Schricter, and F. Wudl, Novel silicon-substituted, soluble poly(phenylenevinylene)s: enlargement of the semiconductor band-gap, *Chem. Mater. 6*:171 (1994).

73. D.-H. Hwang, H.-K. Shim, J.-I. Lee, and K.-S. Lee, Synthesis and properties of multifunctional poly(2-trimethylsilyl-1,4-phenylenevinylene): a novel, silicon-substituted, soluble PPV derivative, *J. Chem. Soc. Chem. Commun. 1994*:2461.

74. R. K. McCoy and F. E. Karasz, Synthesis and characterization of ring-halogenated poly(1,4-phenylenevinylenes), *Chem. Mater. 3*:941 (1991).

75. I.-N. Kang, G.-J. Lee, D.-H. Kim, and H.-S. Kim, Electrical and third-order nonlinear optical properties of poly(2-fluoro-1,4-phenylenevinylene) and its copolymers, *Polym. Bull. 33*:89 (1994).

76. J.-I. Jin, J.-C. Kim, and H.-K. Shim, Synthesis and electrical properties of poly(2-bromo-5-methoxy-1,4-phenylenevinylene) and copolymers, *Macromolecules 25*:5519 (1992).

77. J.-I. Jin, C.-K. Park, and H.-K. Shim, Synthesis and electroconductivities of poly(2-methoxy-5-methylthio-1,4-phenylene vinylene) and copolymers, *J. Polym. Sci. Part A: Polym. Chem. 29*:93 (1991).

78. J. D. Stenger-Smith, A. P. Chafin, and W. P. Norris,

Synthesis and initial characterisation of poly[2-(*N*,*N*-dimethylamino)-1,4-phenylenevinylene], *J. Org. Chem.* 59:6107 (1994).

79. J.-I. Jin, S.-H. Yu, and H.-K. Shim, Synthesis and electrical conductivity of poly(1,4-phenylenevinylene-*co*-2-nitro-1,4-phenylenevinylene), *J. Polym. Sci. Part B: Polym. Phys.* 31:87 (1993).

80. J.-I. Jin, Y.-H. Lee, and H.-K. Shim, Synthesis and characterisation of poly(2-methoxy-5-nitro-1,4-phenylenevinylene) and poly(1,4-phenylenevinyl-co-2-methoxy-5-nitro-1,4-phenylenevinylene)s, *Macromolecules* 26:1805 (1993).

81. R. M. Gregorius, P. M. Lahti, and F. E. Karasz, Preparation and characterisation of poly(arylenevinylene) copolymers and their blends, *Macromolecules* 25:6664 (1992).

82. J.-I. Jin, Y.-H. Lee, C. K. Park, and B.-K. Nam, Synthesis and characterisation of poly[(2-(2-phenylethenyl)-1,4-phenylene)vinylene], poly[(2-methoxy-5-(2-phenylethenyl)-1,4-phenylene)vinylene] and their copolymers containing 1,4-phenylenevinylene units, *Macromolecules* 27:5239 (1994).

83. J.-I. Jin, C. K. Park, and H.-K. Shim, Synthesis and electroconductivity of poly(1,4-phenylenevinylene-*co*-2-*n*-butoxy-5-methoxy-1,4- phenylene vinylene)s and poly(1,4-phenylenevinylene-*co*-2-*n*-dodecyloxy-5-methoxy- 1,4-phenylene vinylene)s, *Polymer* 35:480 (1994).

84. H.-K. Shim, R. W. Lenz, and J.-I. Jin, Synthesis and electrical conductivity of poly(1,4-phenylenevinylene-*co*-2,5-thienylvinylene), *Makromol. Chem.* 190:389 (1989).

85. C.-C. Han, R. W. Lenz, and F. E. Karasz, Highly conducting, iodine doped copoly(phenylene vinylene)s, *Polym. Commun.* 28:261 (1987).

86. A. Beerden, D. Vanderzande, and J. Gelan, The effect of anions on the solution behaviour of poly(xylene tetrahydrothiophenium chloride) and on the elimination to poly(*p*-phenylene vinylene), *Synth. Met.* 52:387 (1992).

87. R. O. Garay, U. Baier, C. Bubeck, and K. Müllen, Low-temperature synthesis of poly(*p*-phenylene vinylene) by the sulfonium route, *Adv. Mater.* 5:561 (1993).

88. J. M. Machado, F. R. Denton III, J. B. Schlenoff, F. E. Karasz, and P. M. Lahti, Analytical methods for molecular weight determination of poly(*p*-xylylidene dialkyl sulfonium halide): degree of polymerisation of poly(*p*-phenylene vinylene) precursors, *J. Polym. Sci. Part B: Polym. Phys.* 27:199 (1989).

89. A. O. Patil, S. D. D. V. Rughooputh, and F. Wudl, Poly(*p*-phenylene vinylene): incipient doping in conducting polymers, *Synth. Met.* 29:E115 (1989).

90. P. L. Burn, D. D. C. Bradley, A. R. Brown, R. H. Friend, and A. B. Holmes, *Synth. Met.* 41–43:261 (1991).

91. T. Momii, S. Tokito, T. Tsutsui, and S. Saito, *Chem. Lett.* 1201 (1988).

92. D. A. Halliday, P. L. Burn, R. H. Friend, D. D. C. Bradley, and A. B. Holmes, Determination of the average molecular weight of poly(*p*-phenylene vinylene), *Synth. Met.* 55–57:902 (1993).

93. V. Massardier, J. C. Beziat, and A. Guyot, Kinetic study of synthesis of the poly(phenylene vinylene sulfonium) precursor through ^{1}H NMR and UV analyses, *Eur. Polym. J.* 31:291 (1995).

94. F. Papadimitrakopoulos, K. Konstadinidis, T. M. Miller, R. Opila, E. A. Chandross, and M. E. Galvin, The role of carbonyl groups in the photoluminescence of poly(*p*-phenylenevinylene), *Chem. Mater.* 6:1563 (1994).

95. J. B. Schlenoff and L.-J. Wang, Elimination of ion-exchanged precursors to poly(phenylenevinylene), *Macromolecules* 24:6653 (1991).

96. G. Montaudo, D. Vitalini, and R. W. Lenz, Mechanism of thermal generation of poly(*p*-phenylene vinylene) from poly(*p*-xylene-α-dimethylsulfonium halides), *Polymer* 28:837 (1987).

97. A. Torres-Filho and R. W. Lenz, Electrical, thermal, and photoconductive properties of poly(phenylene vinylene) precursors: I. Laser-induced elimination reactions in precursor solutions, *J. Polym. Sci. Part B: Polym. Phys.* 31:959 (1993).

98. W. Schmid, R. Dankesreiter, J. Gmeiner, T. Vogtmann, and M. Schwoerer, Photolithography with poly-(*p*-phenylene vinylene) (PPV) prepared by the precursor route, *Acta Polym.* 44:208 (1993).

99. S. S. Taguchi and T. Tanaka, U.S. Patent 4,816,383 (1989).

100. J. Bullot, B. Dulieu, and S. Lefrant, Photochemical conversion of polyphenylenevinylene, *Synth. Met.* 61:211 (1993).

101. V. Massardier, A. Guyot, and V. H. Tan, Direct conversion of sulfonium precursors into poly(*p*-phenylene vinylene) by acids, *Polymer* 35:1561 (1994).

102. A. Torres-Filho and R. W. Lenz, Electrical, thermal, and photoconductive properties of poly(phenylene vinylene) precursors: I. Microwave-induced elimination reactions on sulfonium precursor-films, *J. Appl. Polym. Sci.* 52:377 (1994).

103. B. R. Hsieh, A facile synthesis of poly(phenylene 1,2-diphenylvinylene), *Polym. Bull.* 26:391 (1991).

104. G. J. Sarnecki, P. L. Burn, A. Kraft, R. H. Friend, and A. B. Holmes, The synthesis and characterisation of some poly(2,5-dialkoxy-1,4-phenylene vinylene)s, *Synth. Met.* 55–57:914 (1993).

105. G. J. Sarnecki, R. H. Friend, A. B. Holmes, and S. C. Moratti, The synthesis of a new regioregular conjugated polymer, poly(2-bromo-5-dodecyloxy-1,4-phenylenevinylene), *Synth. Met.* 69:545 (1995).

106. W. J. Swatos and B. Gordon III, *Polym. Prepr.* 30:505 (1990).

107. B. R. Hsieh and W. A. Field, A dehydrochlorination route to poly(2,3-diphenyl-1,4-phenylene vinylene), *Polym. Prepr.* 34:410 (1993).

108. D. Braun, E. G. J. Staring, R. C. J. E. Demandt, G. L. J. Rikken, Y. A. R. R. Kessener, and A. H. J. Venhuizen, Photo- and electroluminesence effi-

ciency in poly(dialkoxy-*p*-phenylenevinylene), *Synth. Met.* 66:75 (1994).
109. F. Wudl, P.-M. Allemand, G. Srdanov, Z. Ni, and D. McBranch, in *Materials for Nonlinear Optics: Chemical Perspectives* (S. R. Marder, J. E. Sohn, and G. D. Stucky, eds.), American Chemical Society, Washington, DC, 1991, p. 683.
110. F. Louwet, D. Vanderzande, and J. Gelan, A general synthetic route to high molecular weight poly(*p*-xylylene)-derivatives: a new route to poly(*p*-phenylene vinylene), *Synth. Met.* 69:509 (1995).
111. D. A. Halliday, et al., Large changes in optical response through chemical preordering of poly(*p*-phenylenevinylene), *Adv. Mater.* 5:40 (1993).
112. C.-C. Han, K. Y. A. Jen, and R. L. Elsenbaumer, U.S. Patent 4,900,782 (1990).
113. G. Yu, C. Zhang, K. Pakbaz, and A. J. Heeger, Photonic devices made with semiconducting conjugated polymers: new developments, *Synth. Met.* 71:2241 (1995).
114. L. Yu and Z. Bao, Conjugated polymers exhibiting liquid crystallinity, *Adv. Mater.* 6:156 (1994).
115. H. Martelock, A. Griener, and W. Heitz, Structural modifications of poly(1,4-phenylenevinylene) to soluble, fusible, liquid crystalline products, *Makromol. Chem.* 192:967 (1991).
116. B. R. Hsieh, Synthesis of poly(phenylene acetylene) from poly(phenylene vinylene), *Polym. Bull.* 25:177 (1991).
117. V. L. Patel, S. P. Church, and N. Khan, An in situ infrared study of the thermal conversion of polyphenylene-1,2-dibromoethylene to polyphenyleneacetylene, *Polym. Bull.* 29:527 (1992).
118. H. H. Hörhold, A. Bleyer, E. Birckner, S. Heinze, and F. Leonhardt, A novel approach to light emitting polyarylene: cyclisation of poly(arylene vinylenes), *Synth. Met* 69:525 (1995).
119. H. Ueno and K. Yoshino, Electrical conductivity and thermoelectric power of highly graphitizable poly(*p*-phenylene vinylene) films, *Phys. Rev. B 34*:7158 (1986).
120. J.-I. Jin, J.-H. Kim, G.-H. Lee, Y.-H. Lee, and Y. W. Park, Carbonization of poly(1,4-phenylene ethynylene), *Synth. Met.* 55–57:3742 (1993).
121. M. Hamaguchi and K. Yoshino, Lyotropic behaviour of poly(2,5-dinonly-*p*-phenylene), *Jpn. J. Appl. Phys.* 33:L1689 (1994).
122. C.-C. Han and R. L. Elsenbaumer, Conveniently processable form of electrically conductive poly(dibutoxyphenylene vinylene), *Mol. Cryst. Liq. Cryst.* 189:183 (1990).
123. J. H. Kim et al. Preparation and characterization of poly(*p*-phenylene vinylene) Langmuir–Blodgett films formed via precursor method, *Synth. Met* 71:2023 (1995).
124. Y. Nishikata, M.-A. Kakimoto, and Y. Imai, Preparation of a conducting ultrathin film of poly(*p*-phenylene vinylene) using a Langmuir–Blodgett technique, *J. Chem. Soc. Chem. Commun. 1988*:1040.
125. Y. Nishikata, M.-A. Kakimoto, and Y. Imai, Preparation and properties of poly(*p*-phenylene vinylene) Langmuir–Blodgett film, *Thin Solid Films* 179:191 (1989).
126. A. Wu et al., Further investigation of the preparation process of poly(*p*-phenylene vinylene) Langmuir Blodgett films, *Thin Solid Films* 244:750 (1994).
127. M. Era, H. Shinozaki, S. Tokito, T. Tsutsui, and S. Saito, *Chem. Lett. 1988*:1097.
128. J. H. F. Martens, et al., Structural order in poly(*p*-phenylene vinylene), *Synth. Met.* 55–57:434 (1993).
129. T. Tsutsui, H. Murata, T. Momii, K. Yoshiura, S. Tokito, and S. Saito, Highly orientated polyarylene vinylene thin films, *Synth. Met.* 41–43:327 (1991).
130. K. Pichler, R. H. Friend, P. L. Burn, and A. B. Holmes, Chain alignment in poly(*p*-phenylene vinylene) on orientated substrates, *Synth. Met.* 55–57:454 (1993).
131. J. Gmeiner, S. Karg, M. Meier, W. Riess, P. Strohrieg, and M. Schwoerer, Synthesis, electrical conductivity and electroluminescence of poly(*p*-phenylene vinylene) prepared by the precursor route, *Acta Polym.* 44:201 (1993).
132. S. Tokito, T. Tsutsui, S. Saito, and R. Tanaka, Optical and electrical properties of pristine poly(*p*-phenylenevinylene) film, *Polym. Commun.* 27:333 (1986).
133. G. Kossmehl, Semiconductive conjugated polymers, *Ber. Bunsenges. Phys. Chem.* 83:417 (1979).
134. B. R. Hsieh, H. Antoniadis, M. A. A. Abkowitz, and M. Stolka, Charge transport properties of poly(phenylene vinylene) and its sulfonium precursor polymer at different degrees of conversion, *Polym. Prepr.* 33:414 (1992).
135. T. Ohnishi, T. Noguchi, T. Nakano, M. Hirooka, and I. Murase, Preparation and properties of highly conducting poly(arylene vinylene), *Synth. Met.* 41–43:309 (1991).
136. R. Mertens, P. Nagels, R. Callaerts, J. Briers, and H. J. Geise, Electrical conductivity of poly(paraphenylene vinylene) films doped with $FeCl_3$, *Synth. Met.* 55–57:3538 (1993).
137. D. R. Gagnon, J. D. Capistran, F. E. Karasz, R. W. Lenz, and S. Antoun, Synthesis, doping, and electrical conductivity of high molecular weight poly(*p*-phenylene vinylene), *Polymer* 28:567 (1987).
138. I.-N. Kang, D.-H. Hwang, and H.-K. Shim, Synthesis and electrical properties of halogen substituted PPV derivatives, *Synth. Met.* 69:547 (1995).
139. Y. Sonada, Y. Nakao, and K. Kaeriyama, Preparation and properties of poly(1,4-phenylene vinylene) derivatives, *Synth. Met.* 55–57:918 (1993).
140. W. B. Liang, M. A. Masse, and F. E. Karasz, Highly conductive crystalline (2-methoxy-*p*-phenylene vinylene), *Polymer* 33:3101 (1992).
141. S. Tokito, P. Smith, and A. J. Heeger, Highly conductive and stiff fibres of poly(2,5-dimethoxy-*p*-phenylenevinylene) prepared from soluble precursor polymer, *Polymer* 32:464 (1991).
142. Z. Yang, et al., Modification of poly(*para*-phenylene

vinylene) by introduction of aromatic groups on the olefinic carbons, *Synth. Met. 47*:111 (1992).
143. J. M. Machado, M. A. Masse, and F. E. Karasz, Anisotropic mechanical properties of uniaxial orientated electrically conducting poly(*p*-phenylene vinylene), *Polymer 30*:1992 (1989).
144. J. Briers et al., Molecular-orientation and conductivity in highly oriented poly(*p*-phenylene vinylene), *Polymer 35*:4569 (1994).
145. M. Esteghamatian and G. Xu, Effect of lithium doping on the conductivity of poly(*p*-phenylene vinylene), *Synth. Met. 63*:195 (1994).
146. Z. Yang and H. J. Geise, Preparation and electrical conductivity of blends consisting of modified Wittig poly(*para*-phenylene vinylene), iodine and polystyrene, polymethyl methacrylate or polycarbonate, *Synth. Met. 47*:105 (1992).
147. M. A. Masse, J. B. Schlenoff, F. E. Karasz, and E. L. Thomas, Crystalline phases of electrically conductive poly(*p*-phenylene vinylene), *J. Polym. Sci. Part B: Polym. Phys. 27*:2045 (1989).
148. I. Murase, T. Ohnishi, T. Noguchi, M. Hirooka, and S. Murakimi, Highly conducting poly(*p*-phenylene vinylene) prepared from sulfonium salt, *Mol. Cryst. Liq. Cryst. 118*:333 (1985).
149. J.-I. Lee, H.-K. Shim, G. J. Lee, and D. Kim, Synthesis of poly(2-methoxy-5-methyl-1,4-phenylenevinylene) and its poly(1,4-phenylenevinylene) copolymers: electrical and third-order nonlinear optical properties, *Macromolecules 28*:4675 (1995).
150. J.-I. Jin, C. K. Park, H.-K. Shim, and Y.-W. Park, Highly conducting poly(2-*n*-butoxy-5-methoxy-1,4-phenylene vinylene), *J. Chem. Soc. Chem. Commun. 1989*:1205.
151. I. Murase, T. Ohnishi, T. Noguchi, and M. Hirooka, Highly conducting poly(phenylene vinylene) derivatives via soluble precursor process, *Synth. Met. 17*:639 (1987).
152. S. Antoun, D. R. Gagnon, F. E. Karasz, and R. W. Lenz, Synthesis and electrical conductivity of AsF$_6$-doped poly(arylene vinylene), *Polym. Bull. 15*:181 (1986).
153. Y. W. Park, E. B. Park, K. H. Kim, C. K. Park, and J.-I. Jin, Electrical properties of a ring substituted PPV, *Synth. Met. 41–43*:315 (1991).
154. R. O. Garay, B. Mayer, F. E. Karasz, and R. E. Lenz, Synthesis and characterisation of poly[2,5-bis-(triethoxy)-1,4-phenylene vinylene], *J. Polym. Sci. Part A: Polym. Chem. 33*:525 (1995).
155. F. Motamedi, K. J. Ihn, Z. Ni, G. Srdanov, F. Wudl, and P. Smith, Fibers of poly(methoxy-2-ethyl-hexyloxy)phenylenevinylene prepared from the soluble, fully converted polymer, *Polymer 33*:1102 (1992).
156. R. W. Lenz, C.-C. Han, and M. Lux, Highly conducting, iodine doped arylene vinylene copolymers with dialkoxyphenylene units, *Polymer 30*:1041 (1989).
157. C. J. Wung, K.-S. Lee, P. N. Prasad, J.-C. Kim, J.-I. Jin, and H.-K. Shim, Study of third-order optical non-linearity and electrical conductivity of sol-gel processed silica: poly(2-bromo-5-methoxy-*p*-phenylene vinylene) composite, *Polymer 33*:4145 (1992).
158. J.-I. Jin, H.-K. Shim, and R. W. Lenz, Electrical conductivity of poly(1,4-phenylenevinylene-*co*-2,5-thienylenevinylene)s and polyblends of poly(1,4-phenylenevinylene) and poly(1,4-thienylenevinylene), *Synth. Met. 29*:E53 (1989).
159. R. M. Gregorius and F. E. Karasz, Conduction in poly(arylene vinylene) copolymers and blends, *Synth. Met. 53*:11 (1992).
160. H.-K. Shim, S.-K. Kim, J.-I. Jin, K.-H. Kim, and Y.-W. Park, Electrical properties of poly(1,4-phenylene vinylene-*co*-2,5-dimethoxy-1,4-phenylene vinylene)s and poly(1,4-phenylene vinylene-*co*-2,5-thienylene vinylene)s, *Bull. Korean Chem. Soc. 11*:11 (1990).
161. S. Shi and F. Wudl, Synthesis and characterisation of a water soluble poly(*p*-phenylenevinylene) derivative, *Macromolecules 23*:2119 (1990).
162. J. H. Burroughes, et al., Light-emitting diodes based on conjugated polymers, *Nature 347*:539 (1990).
163. C. Zhang, D. Braun, and A. J. Heeger, Light-emitting diodes from partially conjugated poly(*p*-phenylene vinylene), *J. Appl. Phys. 73*:5177 (1993).
164. C. Zhang, H. von Seggern, K. Pakbaz, B. Kraabel, H.-W. Schmidt, and A. J. Heeger, Blue electroluminescent diodes utilizing blends of poly(*p*-phenylphenylene vinylene) in poly(9-vinylcarbazole), *Synth. Met. 62*:35 (1994).
165. C.-C. Wu, et al., Heterostructure electroluminescent diodes prepared from poly(*p*-phenylene vinylene) and aluminum-tris(8-quinolate), *Polym. Prepr. 35*:101 (1994).
166. A. Wu, M. Jikei, M. Kakimoto, M. Imai, S. Ukishima, and Y. Takahashi, Fabrication of polymeric light emitting diodes based on poly(*p*-phenylene vinylene) LB films, *Chem. Lett. 1994*:2319.
167. S. Doi, M. Kuwabara, T. Noguchi, and T. Ohnishi, Organic electroluminescent devices having poly(dialkoxy-*p*-phenylene vinylenes) as a light emitting material, *Synth. Met. 55–57*:4174 (1993).
168. A. V. Vannikov and A. C. Saidov, Electroluminescence in ether-substituted poly(phenylenevinylenes), *Mendeleev Commun. 1993*:54.
169. M. Uchida, Y. Ohmori, T. Noguchi, T. Ohnishi, and K. Yoshino, Color variable light-emitting diode utilising conducting polymer containing fluorescent dye, *Jpn. J. Appl. Phys. 32*:L921 (1993).
170. I. D. Parker and H. H. Kim, Fabrication of polymer light-emitting diodes using doped silicon electrodes, *Appl. Phys. Lett. 64*:1774 (1994).
171. M. Onada, M. Uchida, Y. Ohmori, and K. Yoshino, Organic electroluminescence devices using poly(arylene vinylene) conducting polymers, *Jpn. J. Appl. Phys. 32*:3895 (1993).
172. M. Onada and K. Yoshino, Heterostructure electroluminescent devices prepared from self-assembled monolayers of poly(*p*-phenylene vinylene) and sulfonated polyaniline, *Jpn. J. Appl. Phys. 34*:L260 (1995).

173. H. Herold, J. Gmiener, and M. Schoewer, Preparation of light emitting diodes on flexible substrates: elimination reaction of poly(p-phenylene vinylene) at moderate temperatures, *Acta Polym.* 45:392 (1994).
174. H. Vestweber, R. Sander, A. Greiner, W. Heitz, R. F. Mahrt, and H. Bassler, Electroluminescence from polymer blends and molecularly doped polymers, *Synth. Met.* 64:141 (1994).
175. H. H. Kim, et al., Silicon compatible organic light-emitting diode, *J. Lightwave Technol.* 12:2107 (1994).
176. H. H. Kim, R. G. Swartz, Y. Ota, T. K. Woodward, M. D. Feuer, and W. L. Wilson, Prospects for silicon monolithic opto-electronics with polymer light-emitting diodes, *J. Lightwave Technol.* 12:2114 (1994).
177. P. L. Burn, A. B. Holmes, A. Kraft, D. D. C. Bradley, A. R. Brown, and R. H. Friend, Synthesis of a segmented conjugated polymer giving a blue-shifted electroluminescence and improved efficiency, *J. Chem. Soc. Chem. Commun.* 1992:32.
178. P. L. Burn, et al., Chemical tuning of the electronic properties of poly(p-phenylenevinylene)-based copolymers, *J. Am. Chem. Soc.* 115:10117 (1993).
179. B. R. Hsieh, H. Antoniadis, D. C. Bland, and W. A. Feld, Chlorine precursor route to poly(p-phenylene vinylene)-based light emitting diodes, *Adv. Mater.* 7:36 (1995).
180. T. Zyung, J.-J. Kim, and W. Y. Hwang, Electroluminescence from poly(p-phenylenevinylene) with monoalkoxy substituent on the aromatic ring. *Synth. Met.* 71:2167 (1995).
181. D. R. Baigent, R. N. Marks, N. C. Greenham, R. H. Friend, S. C. Moratti, and A. B. Holmes, Surface emitting polymer light emitting diodes, *Synth. Met.* 71:2177 (1995).
182. M. Onada, Y. Ohmori, T. Kawai, and K. Yoshino, Visible-light electroluminescent diodes using poly(arylene vinylene), *Synth. Met.* 71:2181 (1995).
183. D. Braun and A. J. Heeger, Visible light emission from semiconducting polymer diodes, *Appl. Phys. Lett.* 58:1982 (1991).
184. G. Gustafsson, Y. Cao, G. M. Treacy, F. Klavetter, N. Colaneri, and A. J. Heeger, *Nature* 357:477 (1992).
185. D. Moses, High quantum efficiency luminescence from a conducting polymer in solution: a novel laser dye, *Synth. Met.* 55–57:22 (1993).
186. I. D. Parker, R. W. Gymer, M. G. Harrison, R. H. Friend, and H. Ahmed, Fabrication of a novel electro-optical intensity modulator from the conjugated polymer, poly(2,5-dimethoxy-p-phenylene vinylene), *Appl. Phys. Lett.* 62:1519 (1993).
187. Y. Ohmori, K. Muro, M. Onada, and K. Yoshino, Fabrication and characterization of Schottky gated field-effect transistors utilizing poly(1,4-naphthalene vinylene) and poly(p-phenylene vinylene), *Jpn. J. Appl. Phys.* 31:L 646 (1992).
188. J. Obrzut, M. J. Obzrut, and F. E. Karasz, Photoconductivity of poly(p-phenylene vinylene), *Synth. Met.* 29:E103 (1989).
189. C. H. Lee, et al., Large enhancement of the transient and steady-state photoconductivity of conducting polymer/C_{60} composite films, *Synth. Met.* 70:1353 (1995).
190. K. Yoshino, et al., Marked enhancement of photoconductivity and quenching of luminescence in poly(2,5-dialkoxy-p-phenylene vinylene) upon C_{60} doping, *Jpn. J. Appl. Phys. part 2A* 32:L357 (1993).
191. G. Yu, K. Pakbaz, and A. J. Heeger, Semiconducting polymer diodes: large size, low cost photodetectors with excellent visible-ultraviolet sensitivity, *Appl. Phys. Lett.* 64:3422 (1994).
192. H. Antoniadis, B. R. Hsieh, M. A. Abkowitz, S. A. Jenekhe, and M. Stolka, Photovoltaic and photoconductive properties of aluminium/poly(p-phenylene vinylene) interfaces, *Synth. Met.* 62:265 (1994).
193. S. Karg, W. Riess, M. Meier, and M. Schwoerer, Characterization of light emitting diodes and solar cells based on poly-phenylene-vinylene, *Synth. Met.* 55–57:4186 (1993).
194. G. Drefahl, H.-H. Hörhold, and J. Opfermann, GDR Patent 75233 (1970).
195. N. S. Sariciftci, L. Smilowitz, A. J. Heeger, and F. Wudl, *Science* 258:1474 (1992).
196. J. J. Kim, S.-W. Kang, D.-H. Hwang, and H.-K. Shim, Electro-optic properties of poly(2-cyano-5-methoxy-1,4-phenylenevinylene) and para-phenylenevinylene copolymers, *Synth. Met.* 55–57:4024 (1993).
197. D. McBranch, M. Sinclair, A. J. Heeger, A. O. Patil, S. Askari, and F. Wudl, Linear and nonlinear optical studies of poly(p-phenylene vinylene) derivatives and polydiacetylene-4BCMU, *Synth. Met.* 29:E85 (1989).
198. D. D. C. Bradley and Y. Mori, Third harmonic generation in precursor route poly(p-phenylene vinylene), *Jpn. J. Appl. Phys.* 28:174 (1989).
199. K.-S. Lee, et al., Sol-gel processed conjugated polymers for optical communications, *Mol. Cryst. Liq. Cryst.* 224:33 (1993).
200. Y. B. Moon, S. D. D. V. Rughooputh, A. J. Heeger, A. O. Patil, and F. Wudl, X-ray scattering study of the conversion of poly(p-phenylene vinylene) precursor to the conjugated polymer, *Synth. Met.* 29:E79 (1989).
201. C. J. Wung, M. K. P. Wijekoon, and P. N. Prasad, Characterisation of sol-gel processed poly(p-phenylene vinylene) silica and V_2O_5 composites using waveguide Raman, Raman and FTIR spectroscopy, *Polymer* 34:1174 (1993).
202. F. W. Embs, E. L. Thomas, C. J. Wung, and P. N. Prasad, Structure and morphology of sol-gel prepared polymer-ceramic composite thin films, *Polymer* 34:4607 (1993).
203. D.-H. Hwang, H.-K. Shim, J.-I. Lee, and K.-S. Lee, Optical third-harmonic generation of poly(2-trimethylsilyl-1,4-phenylenevinylene) and poly(2,5-bis(trimethylsilyl)-1,4-phenylenevinylene), *Synth. Met.* 71:1721 (1995).
204. R. W. Gymer, R. H. Friend, H. Ahmed, P. L. Burn,

A. M. Kraft, and A. B. Holmes, The fabrication and assessment of optical waveguides in poly(p-phenylenevinylene) poly(2,5-dimethoxy-p-phenylenevinylene) copolymer, *Synth. Met. 55–57*:3683 (1993).
205. S. Tanaka and J. R. Reynolds, Electrochemical doping of poly(p-phenylene vinylene) thin films, *J. Macromol. Sci.-Pure Appl. Chem. A32*:1049 (1995).
206. T. Kawai, T. Iwasa, T. Kuwabara, M. Onada, and K. Yoshino, Secondary battery characteristics of poly(p-phenylene vinylene) derivatives, *Jpn. J. Appl. Phys. 29*:1833 (1990).
207. H.-H. Hörhold and M. Helbig, Poly(phenylenevinylenes)—synthesis and redox chemistry of electroactive polymers, *Makromol. Chem., Macromol. Symp. 12*:229 (1987).
208. M. Helbig and H.-H. Hörhold, Investigation of poly(arylene vinylene)s, 40: electrochemical studies on poly(p-phenylenevinylenes), *Makromol. Chem. 194*:1607 (1993).
209. M. M. Richter, F.-R. F. Fan, F. Klavetter, A. J. Heeger, and A. J. Bard, Electrochemistry and electrogenerated chemiluminescence of films of the conjugated polymer 4-methoxy-(2-ethylhexoxyl)-2,5-polyphenylenevinylene, *Chem. Phys. Lett. 226*:115 (1994).

14
Conjugated Ladder-Type Structures

Ullrich Scherf
Max-Planck-Institut für Polymerforschung, Mainz, Germany

I. INTRODUCTION

When the talk is of conjugated polymers, single-stranded structures like polyacetylene, poly(p-phenylene), or polythiophene are usually meant. The extent of π-conjugation present in these single-stranded systems and their characteristic optoelectronic properties, such as absorption or emission behavior, are strongly dependent on the geometric construction of the polymer backbone, in which the mutual twisting of the unsaturated subunits represents the major effect. Thus, the specific conformation present is often quite decisively influenced by the substitution pattern on the conjugated polymer backbone (configuration, tacticity), the physical state of the sample (solution, film, liquid crystalline phase), the temperature, and also by the supramolecular degree of order.

One means by which the undesired dependence of the electronic properties on the steric construction of the molecule can be largely overcome involves incorporation of a rigid, planar polymer backbone in the form of a double-stranded molecule (conjugated ladder-type structures). The characteristic electronic properties of the resulting flattened and sterically restrained π-electron systems are then largely independent of external factors such as temperature and physical state. The properties are then determined only by the particular π-topology of the system, the connection pattern of the π-orbitals of the carbon atoms and heteroatoms. The potential range possible for the HOMO–LUMO energy difference (band gap energy) of such conjugated ladder-type polymers thereby ranges from semiconductor structures [e.g., ladder-type polymers of the poly(p-phenylene) type], in which the second strand of the ladder structure serves solely to geometrically restrain the system), to so-called low band gap systems with continuous double-stranded conjugation [e.g., poly(n)acene or p-phenylenemethine ladder polymers (Scheme 1)].

Since the 1930s there have been a variety of synthetic efforts aimed at the experimental realization of such ladder-type structures. Nevertheless, the first of these attempts exclusively involved structures in which the conjugated ladder structure was only incompletely formed, either on purely statistical grounds [e.g., the case of the polymer-analogous cyclization of poly(methyl vinyl ketone); see Section II] or by the lack of continuous π-conjugation, e.g., in BBL (polybenzimidazophenanthroline; see Section II), caused by the presence of m-phenylene subunits as barriers to conjugation. A series of further examples involved not fully determined products, the structures of which were never unambiguously characterized (frequently because of the insolubility of the products) or that displayed numerous structure defects [e.g., cyclization products of polyacrylonitrile or polycondensates of tetrafunctional monomers (aromatic tetraamines + dihalo- or dihydroxybenzoquinones; see Section II].

Since the end of the 1980s, the synthesis of conjugated ladder-type structures has received new, extremely fertile impulses. Characteristic of this, on the one hand, are the attempts to prepare soluble structures in which solubilizing side groups are introduced into the macromolecule, an approach that had previously shown its value in the synthesis of other single-stranded polymers with rigid backbones. On the other hand, new types of synthetic strategies have been introduced, in which both molecular strands have been constructed in a synchronous (i.e., simultaneous) process. Concerted reactions of this sort allow the formation of structurally almost perfect ladder-type structures. These processes involve repetitive polyaddition reactions (of the 4 + 2 or 2 + 2 + 2 type), mostly employing Diels–Alder reac-

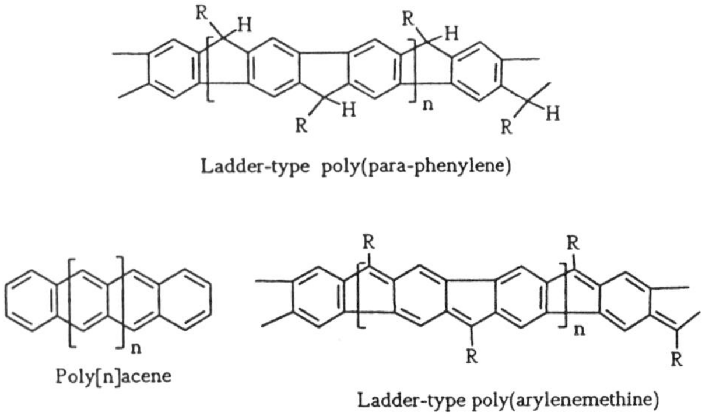

Scheme 1 Low and high band gap conjugated ladder polymers.

tions as the basis for construction of the ladder structure (see Section III). Thus, as far as the formation of double-stranded skeletons is concerned, remarkable successes were quickly obtained, such as the synthesis of soluble, structurally defined products. Nevertheless, as the resulting polymer intermediates are exclusively nonconjugated ladder structures, subsequent polymer-analogous reaction steps (removal of cyclic O-bridges, dehydrogenation, retro-Diels–Alder reactions for the cleavage of ethylene or ethano bridges) are necessary for the generation of conjugated structures. These steps in particular represented enormous difficulty, so that only partial successes could be reported. Promising results have only recently been published (Schlüter et al., in 1994; see Section III), although even here the ladder polymers were still insoluble. However, comprehensive characterization is only possible with the formation of soluble products.

Surprisingly, the synthesis of the first soluble, conjugated ladder polymer in 1991 (Scherf and Müllen; see Section II.A) was not achieved by the polyaddition route, which had been for a long time strongly preferred. These authors used the so-called classical synthetic variant, i.e., the cyclization of suitably substituted single-stranded precursors. They showed that it was quite possible, with the appropriate choice of starting materials and reaction conditions, to prepare structurally defined and soluble conjugated ladder-type polymers. To guarantee sufficient solubility, the macromolecules contained solubilizing alkyl or alkoxy side chains. The first polymer of this type was a methylene-bridged poly(p-phenylene), the planarized structure of which produces attractive optical and electronic properties (particularly regarding the absorption and emission behavior; the application as the active layer in light-emitting diodes based on polymers is highly topical; see Section II.B). The synthetic principle used successfully here was then employed for the synthesis of a second, soluble, conjugated ladder polymer that was formed exclusively from annelated, six-member aromatic rings, the first angular polyacene (Chmil and Scherf, in 1993; see Section II.C).

Using this as a basis, other groups also took up this synthetic method to develop multistep syntheses to conjugated ladder-type polymers (Tour et al., in 1993; Swager et al., in 1994; see Section II.C). This type of synthetic method currently represents the most powerful of the available routes to conjugated, double-stranded macromolecules.

Hitherto, only purely synthetic aspects of the preparation of ladder-type polymers have been discussed. Naturally, from the viewpoint of *conducting* polymers, physical properties and applications of the materials are most important. Nevertheless, a discussion of conjugated ladder-type polymers on the basis of physical parameters made little sense until today, as in most cases no clear structure–property relationships were available. Such a discussion is only possible, if at all, on the basis of unambiguously characterized materials, low in or completely free of defects. Precisely for this reason the synthesis of *defined* conjugated ladder-type structures will be the focus of this article. Thus, the chapter is organized purely according to synthetic aspects ("classical" multistep variant, Section II; synchronous routes, Section III). Where references to typical physical properties of conjugated polymers (e.g., charge transport properties, photo- and electroluminescence) are available, these are considered and appraised for the structure in question. Of particular interest are topical applications for conducting (ladder) polymers, e.g., as active (emitter) material in light-emitting diodes.

II. "CLASSICAL" MULTISTEP ROUTES

Efforts to prepare ladder-type polymers in a multistep process were first undertaken in the late 1930s. Marvel and coworkers [1,2] described the first attempt at the polymer-analogous cyclization of poly(methyl vinyl ketone) (Scheme 2, aldol condensation of the acetyl side groups). However, a degree of conversion (cyclization) only up to 86% is possible on purely statistical grounds.

Other polymer-analogous precursors employed include polyacrylonitrile (oxidative or nonoxidative cyclization [3–5]), poly(1,2-butadiene) or poly(3,4-isoprene) [6,7], and polyalkinylacetylenes [8–11] prepared by the polymerization of butadiynes. The unsaturated side

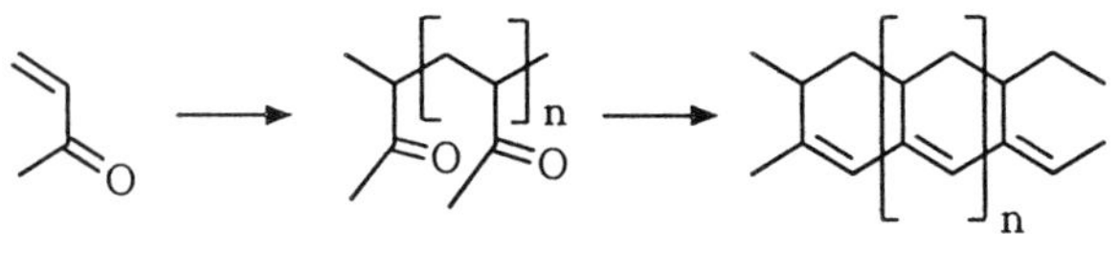

Scheme 2 Polymer-analogous cyclization of poly(methyl vinyl ketone) according to Marvel et al. [1,2].

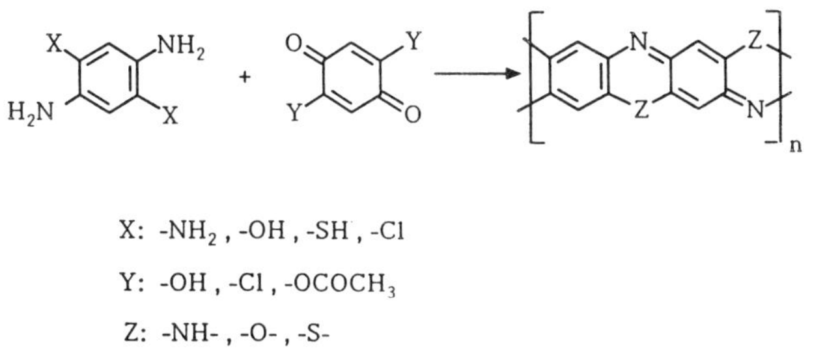

X: -NH$_2$, -OH, -SH, -Cl
Y: -OH, -Cl, -OCOCH$_3$
Z: -NH-, -O-, -S-

Scheme 3 Multifunctional condensation of tetrafunctional monomers.

chains remaining from the 1,2-polymerization of the monomers (acrylonitrile, butadiene, isoprene, butadiyne) employed here are converted in a subsequent step to form the desired ladder structure ("zipping up" cyclization). These syntheses are, however, characterized by the formation of products with poorly defined structure and with a considerable number of defects (incomplete cyclization, cross-linking, radical sites).

In the 1950s and 1960s, experiments involving the polycondensation of multifunctional monomers (e.g., tetraamino aromatics with 2,5-dihalo-1,4-benzoquinones) were carried out first by Stille and coworkers [12,13] and Marvel and coworkers [14,15] (Scheme 3). The resulting products were poorly soluble, mostly insoluble, and did not exist in neutral form (they were oxidatively doped) and contained numerous structural defects (incomplete cyclization, branching, cross-linking).

The synthesis of BBL (poly benzimidazobenzophenanthroline) from 1,2,4,5-tetraaminobenzene and 1,4,5,8-naphthalene tetracarboxylic acid dianhydride (Arnold and vanDeussen [16,17]) represented a definite step forward (Scheme 4). BBL is to some extent soluble, e.g., in strongly acidic media such as sulfuric acid or methanesulfonic acid and can be worked into transparent films and layers. Jenekhe et al. [18] published a method to solubilize and process insoluble rigid polymers like BBL by complexation with Lewis acids in aprotic solvents (nitroalkanes or nitrobenzene). Nevertheless, BBL is not a continuously conjugated ladder polymer, as meta- and para-phenylene subunits (corresponding to cis and trans connections of the condensed functionalities) are built into the polymer in purely statistical fashion.

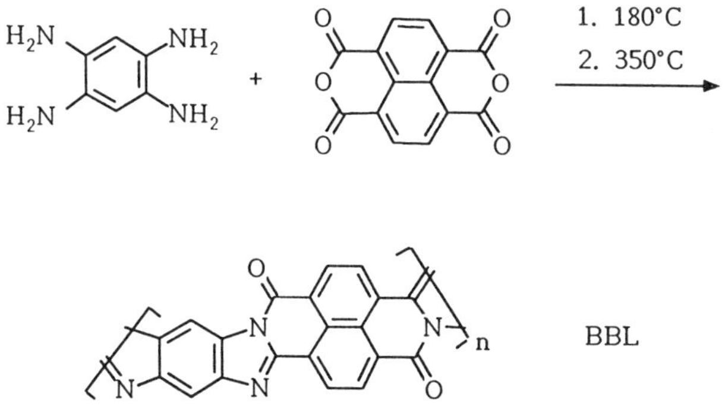

A. Methylene-Bridged Ladder Polymers of the PPP Type

In order to be able to conduct a polymer-analogous cyclization leading to a ladder polymer with a completely defined structure, the monomers employed for this purpose must contain functional groups that survive the initial coupling reaction without damage and with which the subsequent ring closure can be carried out in the desired manner (chemo- and regioselectively, quantitatively). This challenge was completely solved for the first time in 1991 with the synthesis of methylene-bridged poly(p-phenylene) (Scheme 5) [19]. A "classical" multistep synthesis was developed for the preparation of the target structure, the first step of which involves an AA/BB-type polycondensation (aryl-aryl coupling after Suzuki) of an aromatic diboronic acid with a substituted 2.5-dibromo-1.4-dibenzoylbenzene to give a single-stranded precursor polymer of the PPP type. On the one hand, the alkyl substituents of the 1.4-phenylene diboronic acid (R_1) together with the peripheral alkyl or alkoxy groups R_2 of the dibromo monomer ($R' = 1,4-C_6H_4-R_2$) guarantee sufficient solubility of the primary coupling product (Table 14.1). A reduction in the fraction of solubilizing substituents leads to insoluble or insufficiently soluble primary polycondensation products. On the other hand, the two alkyl substituents of the boronic acid component block two potential sites of attack for an electrophilic substitution on the aromatic ring, so that the subsequent, successful ring closure is forced, with no alternative, to take place at the two remaining positions (Table 14.1).

The highest number-average molecular weight attained (20,000–30,000) corresponds to a maximum degree of polymerization of 50–80 six-member aromatic rings. The polydispersity (M_w/M_n) is 1.7–2.1. Direct determination of M_n using vapor-phase osmometry confirms the molecular weights routinely measured using gel chromatography (polystyrene standards). The polymer-analogous ring closure to give the double-stranded target structure is then conducted after the formation of the single-stranded intermediate as a two-step sequence in which the keto group is first reduced to a secondary alcohol and the latter is then subjected to an intramolecular cyclization analogous to a Friedel–Crafts reaction. Reduction of the keto group was achieved smoothly with lithium aluminum hydride (LAH) in toluene–THF.

Scheme 4 Synthesis of BBL according to Arnold and van Deussen [16,17].

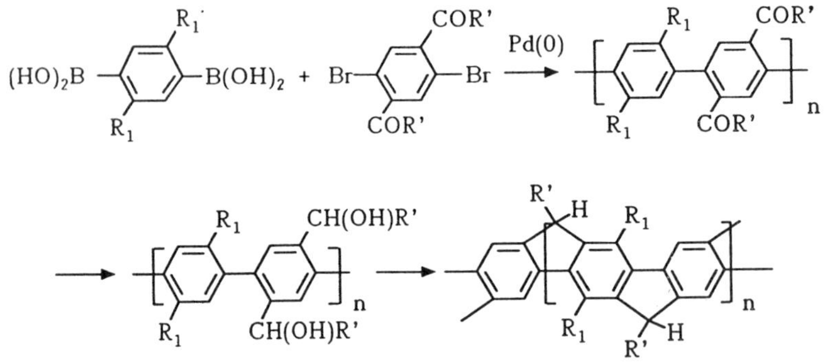

Scheme 5 Synthesis of methylene-bridged ladder polymers of the poly(*para*-phenylene) type (LPPP).

Surprisingly, if a solution of the polyalcohol was treated with boron trifluoride etherate in, e.g., dichloromethane or chloroform, the ring closure to give the ladder polymer took place within seconds and was quantitative. The resulting new type of ladder polymer, a planar methylene-bridged poly(*p*-phenylene), is deep yellow, and solutions thereof display an extremely intense blue photoluminescence.

Nuclear magnetic resonance analysis of the products [19] gave absolutely no indication of defects, which could arise, for example, from incomplete cyclization or branching. Signals from remaining keto or (secondary) alcohol groups could not be demonstrated within the limits of detection of the method.

The reaction sequence described here represents the first synthesis of a structurally defined, soluble band polymer using a multistep process. In addition, it is actually the first known conjugated ladder polymer of defined molecular structure.

The success of the polymer-analogous ring closure depends essentially on the substitution pattern of the methylene carbon atom, which, during the course of the conversion, forms the methylene bridge of the ladder polymer (Scheme 6, Table 14.2). If, for example, there are no further substituents on the hydroxymethylene carbon atom (—CH_2OH—), only colorless and insoluble (cross-linked) products are formed during the cyclization. In the presence of β hydrogen atoms in the side chain (*n*-alkyl substituents other than methyl), 1,2-elimination again takes place in competition with the polymer-analogous ring closure [21].

An aryl substituent or the substituent combination aryl–*n*-alkyl (preferably methyl) on the methylene carbon atom represents the extraordinarily favorable substitution pattern already presented for the quantitative intramolecular ring closure (Table 14.2). The cyclization can be effected completely even with weak Lewis acids or protic acids (boron trifluoride or trifluoroacetic acid).

However, if the substituent density on the methylene bridge is increased further, the steric hindrance increases to the extent that complete cyclization is no longer certain. The degree of conversion for ring closure, which cannot go to completion, is about 80–90% (NMR).

The effects to date on the ring closure illustrate that the careful choice of substitution pattern on the mono-

Table 14.1 Coupling Reaction of 2,5-Dialkyl-1,4-phenylenediboronic Acids with 2,5-Dibromo-1,4-dibenzoylbenzenes to Precursors for Ladder-Type PPPs

Diboronic acid, R_1	Dibromodiketone, R_2	Solubility of the coupling products	M_n[a]
H	*t*-Butyl	Insoluble	—
H	*n*-Decyl	Almost insoluble	—
n-Hexyl	*t*-Butyl	Chloroform	6000
n-Hexyl	*n*-Decyl *n*-Hexyl	Chloroform, dichloromethane, toluene	11,000–30,000
n-Hexyl	O-*n*-Decyl O-*n*-Hexyl	Chloroform, dichloromethane, toluene	10,000–25,000
O—$(CH_2)_{10}$—O—	*n*-Decyl	Chloroform, dichloromethane, toluene	23,000

[a] Determined by gel permeation chromatography.

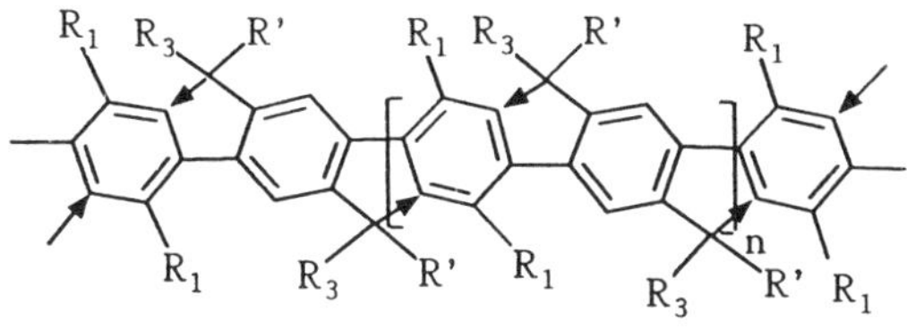

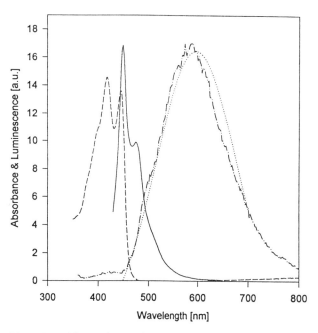

Scheme 6 Polymer-analogous cyclization of open-chain LPPP precursors.

mers employed is of considerable importance for the synthesis of defect-free ladder-type structures. Only that combination of substituents that gives both solubility and quantitative ring closure brings the desired result, the synthesis of planarized polyphenylene. The PPP-ladder polymers presented are characterized by high chemical and thermal resistance. In this respect, the characteristic of the thermal decomposition in the lower temperature range is dominated by the alkyl or alkoxy side groups whose thermal decomposition starts first and determines the upper limit of the thermal stability of the material. The optical and electronic properties of the new ladder polymers have proven themselves to be of utmost importance.

B. Optical and Electronic Properties of PPP Ladder Polymers

As already briefly mentioned, easily the most obvious change regarding optical properties appears in the transition from the open-chain precursor polymers to the planarized PPP ladder polymers (LPPP). The primary single-stranded coupling product (a benzoyl-substituted polyphenylene) from a Suzuki-type polycondensation between an alkyl- or alkoxy-substituted 2,5-dibromo-1,4-dibenzoylbenzene derivative and 2,5-dihexyl-1,4-phenylenediboronic acid is colorless with an absorption maximum λ_{max} of 264 nm [$n-\pi^*$ transition; $\epsilon = 39,000$ L/(mol · cm)] [19]. This absorption behavior supports the strongly twisted structure of the main PPP chain. The mutual distortion of the aromatic subunits reduces the conjugative interaction to a minimum, so that the typical electronic properties of a conjugated polymer are almost completely lost.

Fig. 14.1 Absorption and photo- and electroluminescence spectra of LPPP. (- - -) Solid-state absorption; (—) photoluminescence in solution (toluene); (····) solid-state photoluminescence (film); (–·–) electroluminescence of an ITO/LPPP/aluminum device.

In the cyclization to the ladder structure LPPP, the color of the reaction solution changes from colorless to an intense yellow, connected with the appearance of a considerably intense blue photoluminescence. The absorption spectrum of the resulting ladder polymer substantiates the drastic changes in the electronic structure. A very sharp, structured absorption band appears on the long-wave absorption maximum of about 450 nm ($\pi-\pi^*$ transition) [19] (see Fig. 14.1).

The experimentally determined λ_{max} values are in very good agreement with theoretical predictions for the optical excitation energies of planarized PPP. Thus, Froyer, et al. [22] predicted a value of 2.8 eV (λ_{max} 442 nm) and Toussaint and Bredas [23] a value of 2.75 eV [for comparison, poly(p-phenylene) PPP with a torsional angle in the phenylene units of 23° ~3.10 eV]. Depending

Table 14.2 The Influence of the Substitution at the Methylene Bridge on the Course of the Polymer-Analogous Cyclization Reaction

R_3	R' (-1,4—C_6H_4-R_2)	Cyclization reaction	Catalyst
H	H	Network formation	$AlCl_3$
H	n-Alkyl	Incomplete ring closure; 1,2-elimination as side reaction	$AlCl_3$
H	Aryl	Complete cyclization	BF_3, CF_3COOH
n-Alkyl	Aryl	Complete cyclization[a]	BF_3
Aryl	Aryl	Incomplete ring closure; steric hindrance to cyclization	BF_3, $AlCl_3$

[a] From Ref. 20.

on the substituents on the methylene bridge (—1,4—C_6H_4—R_2), only minor variations appear in the value of λ_{max} (438–450 nm), as there is no direct conjugative interaction with the PPP main chain. However, it is striking and quite plausible that the molar extinction coefficients are reduced with increasing internal "dilution" of the chromophore (with increasing molar fraction of solubilizing side groups). The introduction of alkoxy substituents directly on the main PPP chain resulted in a significant, albeit only moderate, bathochromic shift of λ_{max} to about 460 nm [24].

In addition to the typical absorption properties mentioned, the emission properties of PPP ladder polymers are also worthy of mention. The open-chain precursor polymers are characterized by a low-intensity yellow (polyketone intermediate) or blue (polyalcohol intermediate) photoluminescence. The transition to the planar ladder polymer is accompanied by the appearance of a very intense blue photoluminescence. Measurements of quantum yield of the photoluminescence in solution (dichloromethane) gave values for the quantum efficiency of the photoluminescence up to 85% [25,26]. The emission spectrum of LPPP in solution is the mirror image of the absorption or photoluminescence (PL) excitation spectrum, with the 0–0 emission as the most intense band in the spectrum (see Fig. 14.1). The unusually low Stokes shift between absorption and emission of $\sim 180 \pm 20$ cm^{-1} [27] is particularly striking and to date unique for a conjugated polymer. The distance between the longest wavelength absorption and the shortest wavelength emission maximum amounts to only 3–8 nm (with a low dependence on the substituents on the methylene bridge). This behavior is reflected in the incorporation of the PPP chromophore into the extremely rigid skeleton of the ladder structure. As a result of the restraint imposed by the molecular structure, only marginal changes in geometry are possible in the transition from the ground state to the excited state. Finally, the quite low overall spectral width of the photoluminescent band (<50 nm), the foundation for the emission of visual "pure blue" light, is conspicuous. Without exception, the PL characteristics just discussed refer to investigations of the light emission in *strongly diluted solution*. Thus, cooperative effects of the individual chromophores are almost completely excluded. However, the transition to the *solid state* (film) in the case of the PPP ladder polymer hides a surprise: In addition to the blue photoluminescence of the individual chromophore [λ_{max}(emission) 445–470 nm], a second PL component appears. The latter is bathochromically shifted into the yellow region of the spectrum [λ_{max}(emission) 530–610 nm] and, in addition, is broad and without structure. The λ_{max} value of the second, yellow PL component is markedly dependent on the intensity behavior of both PL components with respect to each other; with increasing fraction of yellow emission the corresponding λ_{max} value is shifted bathochromically. In turn, the intensity behavior is markedly influenced by the method used to prepare the LPPP films (pouring or spin coating, thermal posttreatment conditions) [28]. Tempering the film at a temperature of 150–200°C leads to almost complete disappearance of the blue PL component; the λ_{max} value of the yellow PL band then amounts to 600–610 nm [29], (see Fig. 14.1). The described emission characteristics of LPPP in the solid state suggest an interpretation of the yellow photoluminescence as an emission of molecular aggregates [27]. The formation of aggregates is supported by detailed studies on the photoluminescence of polymer mixtures and time-resolved photoluminescence [27]. The experiments conducted above on the photoluminescence of PPP ladder polymers suggest that these materials might be suitable as active layers in light-emitting diodes (LEDs). As the spectral distribution of photo- and electroluminescence in the overwhelming majority of cases is almost congruent, LPPP (with its very high photoluminescent quantum yield in the solid state) represents a very promising candidate for the construction of yellow- or blue-emitting LEDs based on conjugated polymers and oligomers [28–31].

For LPPP, a yellow electroluminescence was first observed in 1992 by Leising et al. ($\lambda_{max} \sim 600$ nm) [29]. In this case it was a tempered polymer film in which only the broad, unstructured yellow aggregate emission could be discerned in the photoluminescence spectrum (see Fig. 14.1). However, if the conditions used for preparation of the polymer film were altered (no tempering, preferential use of halogenated hydrocarbons as solvent), it was also possible to observe both components of the emission [λ_{max} 445–470 nm (blue); λ_{max} 530–610 nm (yellow)] in parallel in the electroluminescence [28]. The emission then appears to the eye as white. Such a spectral characteristic for light emission is also attractive for various applications (resemblance to daylight!). During operation of the unit, the fraction of blue emission decreased in favor of the yellow component. Heating of the film as a result of operating the diode (>98% of the energy is converted to heat) leads to an increase in mobility of the individual polymer segments to the extent that a more effective overlap of the π-systems was obtained (formation of aggregates). Thus, the wavelength of aggregate emission itself underwent a slight bathochromic shift, reaching a final value of about 600 nm.

Grüner et al. [25] used LPPP in a two-layer arrangement as the active layer with poly(p-phenylenevinylene) (PPV) as the hole transport layer. As a result of the higher hole mobility (by several orders of magnitude compared with that of electrons), recombination of the charge carriers takes place in the region of the (electron-injecting) metal electrodes, so that LPPP in fact functions as the active layer in the component. Quantum efficiencies for the electroluminescence of up to $\sim 1\%$ have been measured for LPPP as the active layer in LEDs. In this regard, the samples have been shown to be remarkably stable; the LEDs operated during the first

24 h with increasing efficiency and stable current/voltage values, with a radiant density of ~150 C/m². In addition, the devices survived a current density of 300 mA/cm² [25]. The emission maximum of the electroluminescence displayed a λ_{max} value of 610–625 nm. These results are remarkable. As no electron transport films have yet been used for optimization of the electron injection, further improvement in these figures should be possible. The (best internal) quantum yields of about 1% are nevertheless very encouraging, as similar efficiencies have also been measured in the application of PPV, hitherto the most preferred active layer in LEDs of similar construction.

As a result of their electronic structure, PPP band polymers are, however, very promising candidates for the realization of a blue electroluminescence. To attain this, it will be necessary to effectively suppress the undesired formation of aggregates. A first possibility in this direction is the internal dilution of the LPPP macromolecule in the solid state by its embedment in a polymer matrix [25]. A basic requirement for this is the complete miscibility of the components. If polyvinylcarbazole (PVK) is employed as the polymer matrix, the intermolecular aggregation is largely suppressed even with a mixing ratio of 10% LPPP as active component to 90% polymer matrix. In addition to the simple dilution effect, donor–acceptor interactions between PVK and LPPP molecules may also play a role here. A mixture of 10 wt % LPPP (R_2 = n-decyl, R_1 = n-hexyl, R_3 = H) and 90 wt % PVK is characterized by a pure blue light emission of the corresponding electroluminescence (EL) device. The efficiency of the light emission in these experiments amounted to about 0.16% and was quite stable, chronologically and spectrally. These results therefore stimulated further experiments into the application of LPPP blends as active material in polymer LEDs.

Apart from the work just described on the internal dilution of the active species in a suitable matrix, a synthetic strategy for structural (chemical) modification of PPP ladder polymers has also been developed and its utility for suppression of the intermolecular aggregation examined. The basic idea behind this concept to break up the intermolecular aggregations so heavily favored in LPPP was the synthesis of polymers that are no longer completely planar, so-called stepladder structures (Scheme 7). Incorporation of appropriately bulky intermediate groups ("spacers") between the PL- and EL-active planar ladder segments effects a distortion of the main polymer chain, thereby either at least markedly inhibiting the formation of molecular aggregates or at best suppressing it altogether [30,31].

One method of choice for the incorporation of the intermediate members (spacers) that give rise to distortion is the formation of copolymers in which a part of the potential bridging monomers [2,5-dibromo-1,4-bis(4-alkylbenzoyl)benzene] is replaced by another monomer, one that is incapable of forming methylene bridges in the course of the polymer-analogous cyclization. In addition, the comonomer should possess substituents that are sufficiently bulky to lead to an effective twisting of the aromatic subunits.

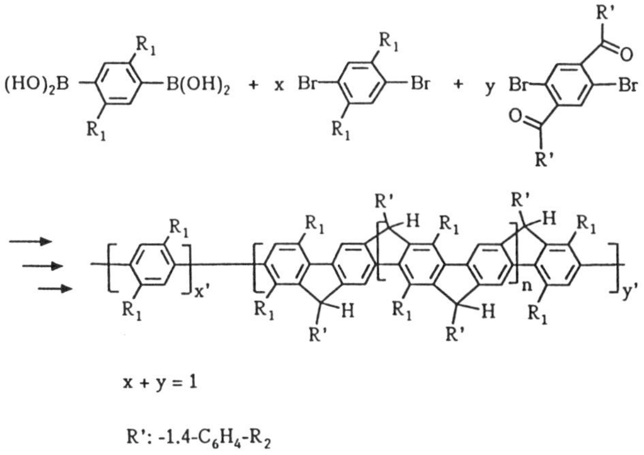

The route taken for this purpose is the execution of statistical copolycondensation using 2,5-dibromo-1,4-dihexylbenzene as comonomer. It is known from investigations by Rehahn et al. [32,33] and Schlüter and Wegner [34] that the 2,5-dihexyl-1,4-phenylene subunits in poly(2,5-dihexyl-1,4-phenylene) effect a strong mutual distortion of the aromatic phenylene subunits, so that conjugative interaction in the main chain is reduced to a minimum (λ_{max} absorption 250 nm). Thus, the 2,5-dihexyl-1,4-phenylene unit, also used for the synthesis of the copolymers, should be well suited for attaining the synthetic target described above.

The resulting statistical copolymers (Scheme 7) then consist of planar ladder segments (planar terphenyl, quinquephenyl, heptaphenyl subunits, etc.) that are connected by 2,5-dihexyl-1,4-phenylene intermediate units. The number-average molecular weight of the copolymers lies in the range 11,000–18,000.

^{1}H NMR spectra of the copolymers show that the mixing ratios of the monomer components [2,5-dibromo-1,4-bis(4-alkylbenzoyl)benzene (y)/2,5-dibromo-1,4-dihexylbenzene (x)] in the copolymer are highly reproducible. Analysis of the UV-vis is absorption spectra support the statistical (or almost statistical) course of the copolycondensation; block formation or the occurrence of an alternating copolycondensation can be excluded. The appearance of several sharp absorption maxima reflects the presence in the copolymer of planar oligophenylene segments of variable lengths. In comparison with the UV-vis spectra of the corresponding oligomeric model compounds [methylene-bridged planar ter- (n = 0), quinque- (n = 1), and heptaphenyls (n = 2)], the individual signal groups of the copolymer absorption spectra could be unambiguously assigned to individual

Scheme 7 Synthesis of stepladder copolymers of the LPPP type.

segment lengths (λ_{max} ~340 nm, methylene-bridged terphenyl ~ 370/390 nm, quinquephenyl; ~415 nm, heptaphenyl segments) [30,31] (see Fig. 14.2). The absorptions of the extended ladder segments form an unstructured band edge (~450 nm), the position of which almost corresponds to that of the homopolymer (LPPP). From the viewpoint of the absorption characteristics described here, the planar PPP segments act almost as an independent chromophore. In spite of the formally uninterrupted conjugation in the main chain, the spacer units incorporated operate as effective barriers to conjugation.

To verify this concept aimed at suppressing the formation of aggregates, it was necessary to carry out a detailed analysis of the photoluminescence behavior of the copolymers in the solid state. The PL emission appears at a λ_{max} value of about 450 nm, even with a high fraction (x) of spacer units. The emission characteristics of the copolymers thereby correspond to that of the homopolymer LPPP and are almost independent of the copolymer composition (x/y) (see Fig. 14.2). Thus, it can be assumed that a rapid intra- and intermolecular energy relaxation is possible after successful electronic excitation in the solid state, so that the emission almost always arises from states with minimal energy gaps close to the band edge, i.e., from the most extended π-segments, although their percentage fraction is very low and statistically controlled.

As anticipated, the formation of aggregates is markedly suppressed by the incorporation of spacer units. If insignificant emissions are discernible above 550 nm at moderate levels (20–50%) of spacer units, these emissions disappear completely with a higher fraction of 2,5-dihexyl-1,4-phenylene units (>50%). The color of the PL emission thereby is blue to light blue-green. The PL quantum yields of the copolymers are remarkably high (in solution, ~85%; in the film, ~25%) [26].

It was not until 1993 that a blue light-emitting diode based on these PPP "stepladder" structures was constructed by Leising and coworkers [30]. The efficiencies obtained are considerable, Grüner et al. determined quantum yields for the electroluminescence of 1.0 to 2.5% [31]. The half-life of the blue emission amounted in the first experiments to about 10 h (under constant current). The spectral changes in the emission (increasingly reduced yellow fraction) could be neglected. The results represent peak values for LEDs emitting blue light. The chronological stability and the spectral invariance of the light emission are currently being investigated in more detail.

For the preparation of ladder polymers and copolymers of the PPP type, as described in detail, the "classical" synthetic methods were adopted as the polymer-analogous cyclization of a suitably substituted single-stranded precursor polymer. In spite of frequently expressed doubts regarding the power of this method [35], it was possible to attain the impressive proof that it really does lead to defect-free band structures after careful selection of reaction centers and reaction conditions. The effect of this was a renaissance for the classical synthetic routes for ladder polymers [36–39] (see Section II.C).

The multistep routes always appear to be particularly advantageous when conjugated ladder structures are the synthetic aim, as the subsequent examples demonstrate.

C. Synthesis of an Angular Polyacene as the Second Example of the Successful "Classical" Synthesis of a Ladder Polymer

Ladder structures that are composed exclusively of aromatic carbon centers have been the focus of theoretical calculations and diverse synthetic efforts for a long time. Polyacenes are a major focus in this interest [40,41]. Polyacenes are band molecules composed exclusively of annelated six-member carbon rings. Linear polyacenes, poly(n)acenes, are, as polymers with low band gap energy, poorly accessible as a result of their high reactivity (oxidation, dimerization) [42–44]. Hitherto only oligo-(n)acenes up to hexa- and heptacene [45,46] have been known and characterized. A synthetic entry to the corresponding polymer does not yet exist, although there are

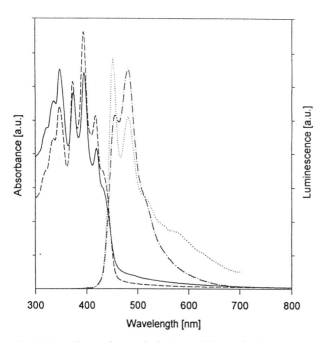

Fig. 14.2 Absorption and photo- and electroluminescence spectra of two LPPP-type stepladder copolymers A and B, (- - -) Solid-state absorption of a copolymer A with x/y = 60/40; (—) solid-state absorption of a copolymer B with x/y = 50/50; (-·-) solid-state photoluminescence (film) of copolymer A; (···) electroluminescence spectrum of an ITO/copolymer A/aluminum device.

very promising efforts based on the use of repetitive Diels–Alder polyadditions (see Section III) [35,47]. The last step in particular of a reaction series leading to poly(n)acenes, which takes place with formation of the unsaturated structure, requires very mild and selective reaction conditions. Angular polyacenes are characterized, in comparison to poly(n)acene, by an enlarged energy gap (band gap) resulting from the diverse topology of the π-electron systems [48]. The extension of the longest linear (n)-acene segment in the angular band polymer is decisive for specific values of the band gap energy.

The target structure of the following synthetic sequence is an angular polyacene that is formally distinguished from the LPPP ladder structures described in detail by the substitution of the methylene for a vinylene bridge. As an angular polyacene it contains anthrylene as the longest linear (n)-acene subunit. The synthetic sequence to these new ladder structures (Scheme 8) again represents a classical ladder polymer synthesis: a suitably substituted, open-chain precursor polymer is cyclized to a band structure in polymer-analogous fashion [36].

The first step here, the formation of the polymeric open-chain precursor structure, is an AA-type coupling of a 2,5-dibromo-1,4-dibenzoylbenzene derivative. In contrast to the syntheses previously described, the monomer is now used as the sole component in the polycondensation. For this purpose, a new type of condensation method after Yamamoto [49–53] was employed. Until now, however, Yamamoto and coworkers have described only coupling of dibromoaromatics and heteroaromatics (including 1,4-dibromobenzene [53], 2,5-dibromothiophene [53], 2,5-dibromopyridine [51], various dibromoquinolines [50], and 2,7-dibromophenanthrene [52]). The reagent employed for the dehalogenation, the nickel(0)/1,5-cyclooctadiene complex [Ni(COD)$_2$], was used in stoichiometric amounts with coreagents 2,2'-bipyridine and 1,5-cyclooctadiene, with dimethylacetamide or dimethylformamide as solvent.

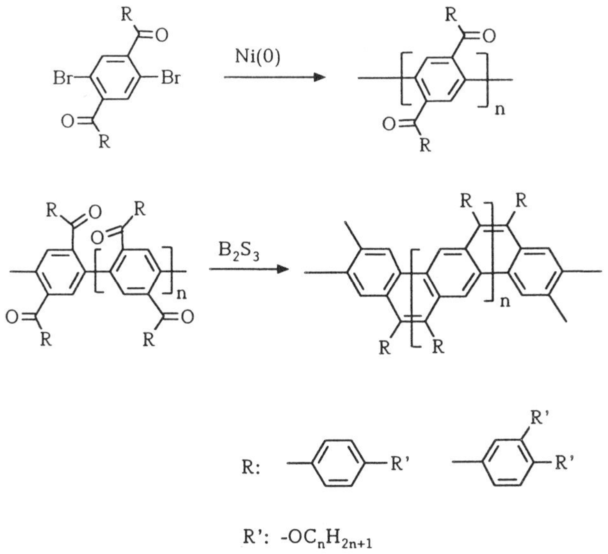

For the synthesis of the angular polyacene, the question now arose as to whether this synthetic method allows, on the one hand, an effective coupling of dibromoaromatics that contain keto functions without changes also taking place at this functional group, and on the other hand, whether a sufficiently high degree of polycondensation could be attained in spite of the sterically demanding circumstances (two *ortho* substituents in the monomer). Since the experiments that have been conducted, both questions can be answered positively [36]. Nevertheless, it is necessary to introduce solubilizing substituents in the peripheral positions to the benzoyl substituents. The primary coupling product, a poly(2,5-dibenzoyl-1,4-phenylene) derivative—a poly(*p*-phenylene) with two benzoyl substituents in each structural unit—is, as expected, very poorly soluble. The precipitation of the initial condensation product during the coupling reaction is the factor limiting the chain length in these experiments, which use 4-decyloxyphenyl substituents for solubilization (M_n ~4000; M_w ~5500). The degree of polymerization attained is only about 7–8. The quite small molecular weight distribution is a result of the coupling taking place by precipitation condensation. The transition to more highly substituted monomers [2,5-dibromo-1,4-bis(3,4-dihexyloxybenzoyl)benzene], with four solubilizing alkoxy groups per monomer unit, allows the synthesis of dramatically longer polymer chains (M_n ~12,000; M_w ~22,000). Moreover, the coupling products no longer precipitate during the reaction, so that a polydispersity (molecular weight distribution) typical of that for a polycondensation (~1.8) is obtained (Table 14.3). The number-average molecular weight (determined by gel permeation chromatography, calibrated with polystyrene standards) corresponds to the connection of about 22 phenylene units. The polycondensates are soluble in common organic solvents (particularly halogenated hydrocarbons and aromatics).

^{1}H and ^{13}C NMR spectroscopy confirm the regular 1,4 connection of the building blocks and the absence of structural defects, within the limits of detection of the method.

The open-chain precursor polymers described can now be cyclized in a polymer-analogous ring closure. A carbonyl olefination reaction, first described in 1992 by Steliou et al. [54], was used with boron sulfide generated in situ. This synthetic variant was similarly used in 1992 by Wang and Zhang [55] for the polymer-analogous generation of phenanthryl subunits in the main chain of poly(ether ketone)s with 2,2'-dibenzoylbiphenyl subunits, and the high synthetic potential of this method was demonstrated (quantitative conversion, no side re-

Scheme 8 Synthesis of an angularly annelated polyacene ribbon.

Table 14.3 Coupling Reaction of 2,5-Dibromo-1,4-dibenzoylbenzenes to Single-Stranded Precursors for an Angularly Annelated Polyacene

Monomer[a]	Solvent	Reaction time (h)	Yield (%)	Molecular weight		M_n/M_w
				M_n	M_w	
A	DMF	70	87	4.000	5.500	1.37
A	DMAc	70	76	4.100	5.500	1.34
B	DMF	70	79	12,000	22,000	1.83

[a] **A** = 2,5-Dibromo-1,4-bis(4-decyloxybenzoyl)benzene; **B** = 2,5-Dibromo-1,4-bis(3,4-dihexyloxybenzoyl)benzene.

actions). In the course of the cyclization, the corresponding thioketones are initially formed. In the next step, the C=S group undergoes dimerization (2 + 2 cycloaddition) with the formation of cyclic disulfide bridges. These intermediates stabilize themselves in turn by elimination of sulfur (S_2) to give the conjugated aromatic ladder polymer. The reaction product is actually that with the anticipated structure, as confirmed by ^{1}H and ^{13}C NMR experiments [36].

The polymer-analogous cyclization is accompanied by a remarkable change in the absorption properties. The colorless intermediate is converted to the deep yellow planar ladder polymer; connected with this is a distinct bathochromic shift of the longest wavelength absorption maximum. The polymer possesses an absorption band with well-defined vibrational fine structure and a sharp absorption edge of the 0–0 transition (see Table 14.4 and Fig. 14.3).

The angular polyacene now possesses absorption properties quite similar to those of the methylene-bridged LPPP (band gap energy LPPP, ~ 2.78 eV; angular polyacene, 2.88 eV). Thus, the absorption is, compared with that of LPPP, shifted hypsochromically by about 10–15 nm. These results are in very good agreement with calculations of Toussaint and Bredas, who found a band gap energy for planar PPP (comparable with LPPP) of 2.75 eV and for the angular polyacene prepared here [which they called poly(2,3,8,9)-benzanthracene)] a value of 2.86 eV [56].

The photoluminescence behavior of the new angular polyacenes is characterized by the appearance of a structured emission band (shortest wavelength emission maximum, 478 or 485 nm, respectively). The emission is mirror image symmetrical to that of the absorption, although the Stokes shift is enlarged compared with that of LPPP. At present, the cause of this behavior cannot be satisfactorily explained, especially when it is taken into account that a planar and rigid geometrically restrained π-system also exists here.

In the solid state, in addition to the described blue emission, there is a broad emission shoulder in the yellow region of the spectrum, which can be assigned to the formation of aggregates. The new type of angular polyacenes, such as LPPP, could be used as active material in light-emitting diodes, although the quantum efficiency of the electroluminescence is still quite low (~0.001%; single-layer construction: ITO/angular polyacene as emitter/Al): the yellow-green light emission is just perceptible to the eye in daylight.

Building on the results just described in the synthesis of angular polyacenes, efforts were undertaken by two American groups to prepare ladder polymers of the angular polyacene type using multistep routes (Tour et al. in 1993 and Swager et al. in 1994). In this respect, the synthetic methods used are closely related to the general synthetic scheme developed by Scherf and Müllen (aryl–aryl coupling according to Suzuki or Yamamoto; polymer-analogous ring closure).

Tour and Lambda [37,38] coupled (aryl–aryl coupling after Suzuki) the previously mentioned 2,5-dibromo-1,4-dibenzoylbenzene with an N,N'-(tert-butoxycarbonyl)-protected 2,5-diamino-1,4-phenylene diboronic acid (Scheme 9). In this fashion, they obtained open-chain intermediates of the poly(p-phenylene) type with a number-average molecular weight of 10,000–28,000; the higher molecular weight products show a quite high polydispersity of ~3–4. Treatment of this intermediate with trifluoroacetic acid then effected the intramolecular cyclization to the ladder structure, which represents an N-heteroanalog of the previously described angular polyacenes. Cleavage of the protective groups and formation of the ketimine should take place in parallel with the ring closure.

The resulting products are very poorly soluble; only the use of solvent mixtures such as dichloromethane–trifluoroacetic acid results in dissolution of the ladder structures. Films of the products have been prepared in

Table 14.4 Absorption and Emission Properties of the Angularly Annelated Polyacene Structures Derived from Monomers **A** and **B** as Fully Aromatic Ladder Polymers

Monomer	Absorption		Emission
	λ_{max} (nm)	ϵ [L/(mol·cm)]	λ_{max} (nm)
A	431	(4200)	478
	406, 386	(6700, 6500)	510, 545 (shoulder)
B	437	(4500)	485
	414, 390	(4600, 6600)	516, 553

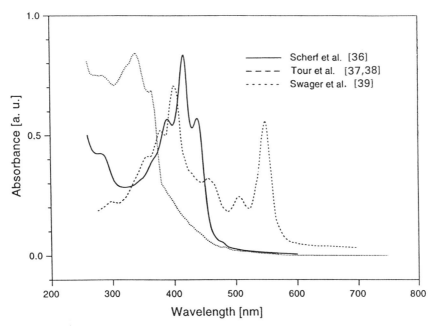

Fig. 14.3 Absorption spectra of the polymer-analogous cyclization products to angularly annelated polyacenes according to (——) Chmil and Scherf [36], (– – –) Tour and Lamba [37,38], and (- - -) Goldfinger and Swager [39].

which films of the open-chain intermediates have been treated in the solid state with hydrogen chloride/ethyl acetate followed by neutralization with trimethylamine/sodium hydroxide. In films treated in this way, carbonyl signals could no longer be detected in the Fourier transform infrared (FTIR) spectrum.

The absorption spectra of the synthesized ladder polymers (see Fig. 14.3) display two groups of signals in the visible part of the spectrum. One of these, with an absorption maximum λ_{max} of 396–402 nm (with vibrational fine structure 356 and ~375 nm) is very similar to the absorptions of the π-system of the angular polyacene structure of Scherf and Chmil described previously, although shifted somewhat (~30 nm) hypsochromically

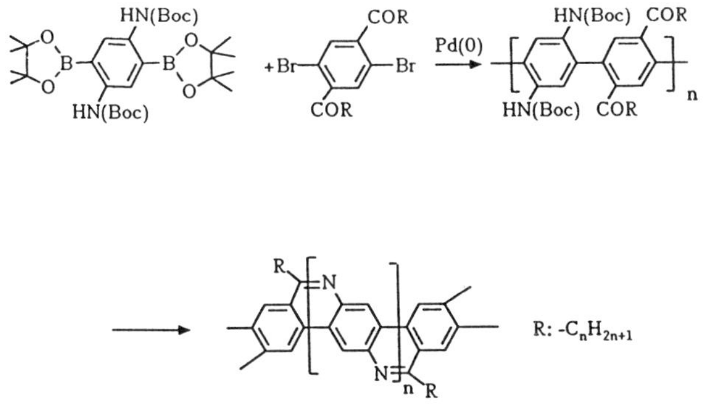

but with identical vibrational fine structure. These absorptions should therefore be attributed to the conjugated framework of the N-heteroanalog of the angular polyacene, whereby the hypsochromic shift is a result of incorporating the N-hetero unit. Surprisingly, a second group of absorptions could be detected in the products of Tour and Lambda, which cause a blue to blue-green coloration of the material. Depending on the peripheral solubilizing substituents, intense absorptions appear at 506–549 nm, which are difficult to assign to the conjugated base structure, the more so when considering the previously cited calculations of Toussaint and Bredas [56]. On the contrary, it can be assumed that these absorptions can be assigned to defects, whereby the formation of benzoquinonediimine subunits or the corresponding radical cation species is quite probable. The reason for this could be the ready oxidation of the p-amino or p-ketimine functional groups.

Goldfinger and Swager [39] described in 1994 the synthesis of an angular polyacene structure that is identical with the polymer prepared by Scherf and Chmil, in regard to the conjugated, double-stranded framework (Scheme 10). The differences are solely in the substitution pattern of the solubilizing side groups. The authors coupled, again with the assistance of the aryl–aryl coupling method of Suzuki, a 2,5-dialkyl-(C_{12}) or dialkoxy-(C_{10})-1,4-phenylene diboronic acid with a 2,5-dibromo- or 2,5-diiodo-1,4-diphenylethynylbenzene. The resulting open-chain intermediates of the polyphenylene type were obtained with a number-average molecular weight of 4000–6000, whereby a high polydispersity of about 10 is characteristic of the resulting products. For the

Scheme 9 Synthesis of a nitrogen-containing angularly annelated polyacene structure according to Tour and Lamba [37,38].

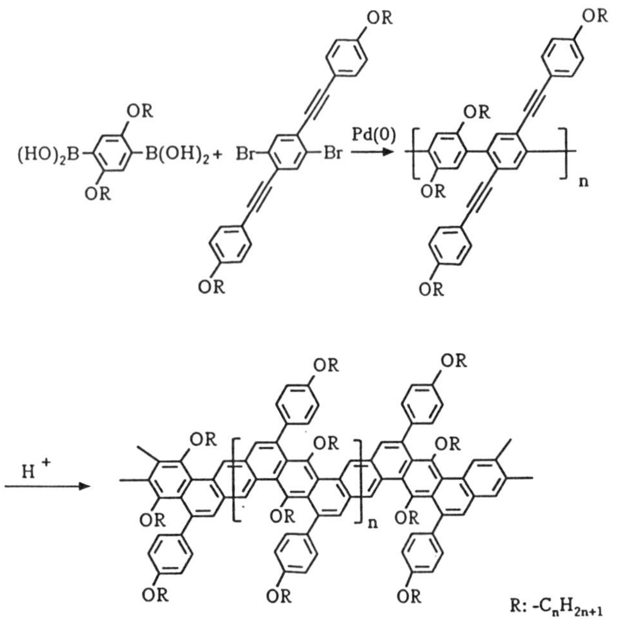

Scheme 10 Synthesis of a polyacene ribbon according to Goldfinger and Swager [39].

polymer-analogous cyclization of the axial ethynyl groups they used trifluoroacetic acid in dichloromethane, from which they obtained yellow-orange to red-brown materials. In this connection they postulated the formation of an angular polyacene structure. The IR spectra of the products show the complete disappearance of the triple bonds due to the ethynyl groups.

The absorption behavior of the resulting cyclization products (see Fig. 14.3), however, was unusual, with λ_{max}-340 nm (shoulder at ~364 nm), which was scarcely different from that of the open-chain precursor polymer (apart from a weak salient in the range above 400 nm). This absorption behavior, compared with the results of Scherf and Chmil and the calculations of Toussaint and Bredas, is difficult to reconcile with the postulated conjugated ladder structure, even considering the somewhat different substitution pattern in the peripheral substituents. It is rather the result of an incomplete polymer-analogous cyclization.

The photoluminescence spectra published by Swager and Goldfinger do not contradict this assumption. The green emission with a maximum at about 500 nm should be assigned to structural defects.

D. Nonconjugated Ladder Polymers as Precursors for Conjugated Polycations with a Ladder Framework

In addition to the new types of conjugated ladder polymers described hitherto [methylene-bridged poly(p-phenylene), angular polyacenes], multistep syntheses of other, primarily nonconjugated, ladder structures have also be achieved [57]. Here, a nucleophilic aromatic substitution of difunctional monomers with thiol and halo functions was used for the formation of the primary coupling products. The high molecular weight intermediates [dibenzoyl-substituted poly(phenylene sulfide)s] are characterized by an unusually high molecular weight (M_n up to 100,000). The intermediates are then subjected to a cyclization sequence analogous to that used for the preparation of methylene-bridged poly(p-phenylene)s (LPPP). The ladder polymers thus obtained (Scheme 11) are attractive as very high molecular weight compounds, in particular for investigations into the formation and dynamics of rigid ladder polymers in solution and in the solid state (viscometry, light scattering, and neutron scattering) and for mechanical investigations.

As a result of the extensive loss of conjugation, these ladder polymers are almost colorless products. Nevertheless, these structures are of considerable importance as precursors for conjugated polycations with ladder structure. In order to prepare polycations that are isoelectronic with the poly(n)acenes, preliminary experiments were first undertaken with model compounds. The two-step reaction sequence contained an oxidation of the methylene bridge with DDQ/water to a hydroxymethyl group, followed by the formation of cationic species in strongly acidic media.

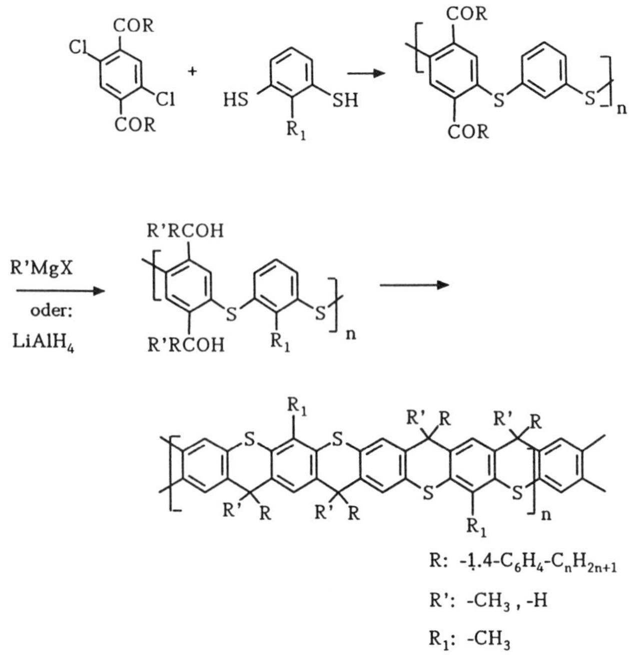

Scheme 11 Synthesis of nonconjugated sulfur-containing ladder polymers as precursors for conjugated ladder-type polycations.

Conjugated Ladder-Type Structures

III. SYNCHRONOUS ROUTES

At the beginning the 1960s, as briefly mentioned in the Introduction, typical synchronous routes to ladder structures were also intensively examined. In this regard they involved predominantly repetitive Diels–Alder reactions of suitable bisdienes and bisdienophiles [58,59]. Although the products were at first poorly soluble, the use of monomers with solubilizing substituents meant that it was also possible to prepare readily soluble, higher molecular weight oligomers and polymers and to characterize them in detail [35,60–62].

The major obstacle on the path to conjugated ladder polymers is, however, the fact that, because of the specific structure of the monomers, at first nonconjugated, polycyclic adducts with exocyclic bridges (e.g., —O— or —CH=CH—), keto functions, and saturated units were formed. It now became apparent that the polymer-analogous conversion (aromatization) of these intermediates was very difficult, particularly from the viewpoint of the required high degree of completion of the reaction as well as the regio- and chemoselectivity. The results known to date have doubtless subdued the initial euphoria at having found an entry to a new generation of conjugated, soluble ladder polymers [35].

A. Poly(n)acene Precursors

In spite of partial successes and very promising attempts, the execution of a complete aromatization sequence leading to a soluble, conjugated ladder polymer has not yet been attained [47,60].

The reactions that lead to the formation of the primary Diels–Alder polyadducts have already been extensively discussed in numerous review articles [35,47,60]. Thus, in this chapter 9 describe strategies and experiments on the polymer-analogous conversion of the initially formed structures in aromatized, i.e., conjugated, ladder polymers. In this respect, efforts have been concentrated predominantly on the synthesis of linear (and angular) polyacenes, i.e., of ladder polymers in which the double-stranded main chain is composed largely of annelated six-member aromatic rings. Particularly challenging in this respect is the synthesis of the linear poly(n)acenes, as these are characterized (as a result of their electronic structure) by a high reactivity toward electrophiles and by a high tendency to undergo intermolecular dimerization, i.e., these are chemically very unstable compounds [63]. Hitherto, only short-chain oligomers up to heptacene have been described as fully characterized neutral compounds [63]. The Diels–Alder polyaddition route now offers a simple method to prepare poly(n)acene precursor structures that are structurally accurate and of high molecular weight. Potential oligo- and poly(n)acenes of this sort have been described, e.g.,

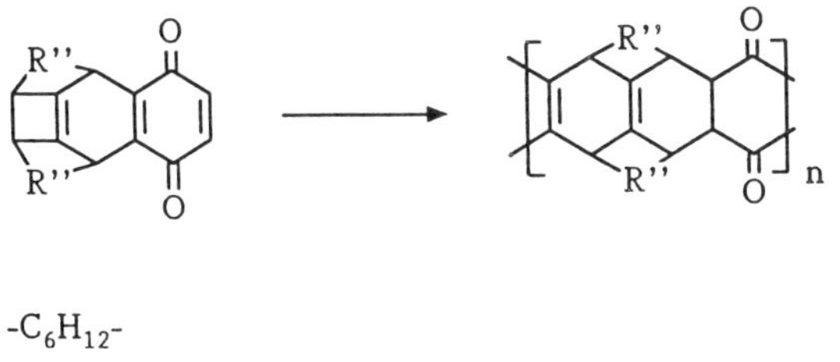

R'': -C$_6$H$_{12}$-

Scheme 12 Diels–Alder polyaddition of a AB-type diene-dienophile-monomer according to Schlüter [35].

by Schlüter and coworkers and by Müllen and coworkers.

Schlüter and Löffler [35,62] described the synthesis of a Diels–Alder polyadduct from an AB-type monomer, consisting of a 1,4-benzoquinone subunit as dienophile and a cyclobutene substructure as pseudodiene. The resulting polymers have a number-average molecular weight of up to 20,000 and are completely soluble thanks to solubilizing hexano bridges on the periphery of the ladder polymer (Scheme 12).

The precursor polymers still contain keto functions and saturated carbon centers that must be converted in the polymer-analogous transition to the conjugated ladder structures, but this has not yet been achieved for these structures.

Müllen and coworkers [47] and Wegener and Müllen [61] investigated precursor polymers (Scheme 13) formed by repetitive Diels–Alder additions of AA/BB type from tetramethylenebicyclo(2.2.2)octene as bisdiene with 1,4,5,8-diepoxyanthracene as bisdienophile. The bisdiene effects formation of barrelene substructures, which bring about an additional increase in solubility because of their angular construction. Under high pressure conditions and with long-chain peripheral alkyl substituents to solubilize the products, it is possible to condense up to 120 six-member rings (number average).

The aromatization sequence leading to conjugated structures now contains several steps. The oxy bridges are removed first by dehydration; in the next step the

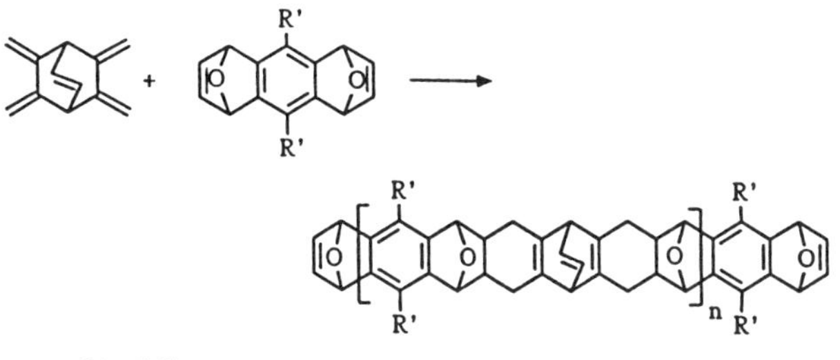

R': -C$_n$H$_{2n+1}$

Scheme 13 AA/BB-type polyaddition of a bisdiene and a bisdienophile according to Wegener and Müllen [61].

saturated carbon atoms are converted by dehydrogenation. The last step is then the removal of the vinylene bridges by means of a retro-Diels–Alder reaction (either by direct elimination of acetylene or by elimination of ethylene after hydrogenation of the vinylene bridges to ethano units). It should be possible to conduct the last step very elegantly in the solid state, i.e., in an immobilized phase, which should dramatically increase the chances of success in the aromatization, given the high instability of the poly(n)acenes. Even if the first step in the reaction sequence could be successfully solved with gaseous hydrogen chloride as dehydrating reagent [47], the second step, the dehydrogenation, would still present problems. Nevertheless, the synthetic efforts to form barrelene substructures are very promising.

To circumvent the difficulties in execution of the first two steps in the aromatization sequence, Müllen and coworkers [64,65] developed a synthesis using the application of potential bisarynes as diene component (Scheme 14).

The arynes are generated in situ, i.e., in the reaction mixture (from the corresponding dibromo compounds with n-butyllithium). The resulting precursor polymers are characterized by the advantage that they contain no oxo bridges. The steps to polymer-analogous aromatization are therefore reduced to a dehydrogenation step and the retro-Diels–Alder reaction for the removal of the vinylene bridges. For short-chain oligomers (pentacene, nonacene), the synthetic potential of this method has already been demonstrated. For the corresponding polymeric precursor structures, the polymer-analogous dehydrogenation was complete and led to structurally defined products. The complete removal of the vinylene bridges, on the other hand, is not yet fully solved, as a result of the surprisingly high temperature required for the retro-Diels–Alder reaction.

For complete success in the synthesis of poly(n)-acenes, careful optimization of the structure of the starting materials and of the reaction conditions is necessary.

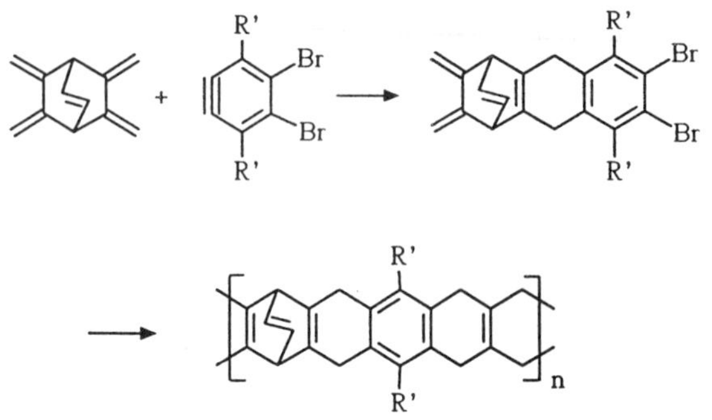

In this respect the last step, the formation of the target molecule poly(n)acenes, plays a decisive role.

B. Conjugated Ladder Polymers Containing Pyracyclene Subunits

The target structures of the hitherto described experiments were all linear polyacenes, composed exclusively of annelated six-member aromatic rings. In 1993/94, Schlüter and coworkers began to construct ladder polymers with pyracyclene subunits containing five-membered rings via the synchronous Diels–Alder route. To simplify the polymer-analogous aromatization they used previously prepared extended aromatic monomer blocks as the diene component (benzodifluoranthrenes) and treated these with a bisfuran derivative, generated in situ, as the bisdiene or pseudobisdiene components [66]. In a second synthetic effort they used a pyracyclene derivative as AB monomer for the preparation of the primary, not yet aromatized Diels–Alder adduct [67–69]. For the solubilization of these polyadducts, Schlüter and coworkers used cyclic alkano loops (dodecano bridges), which should also suppress aggregation of the unsaturated target structures. A very similar strategy was employed by Huber and Scherf [70] for the solubilization of open-chain poly(p-phenylene).

Using AB monomers (Scheme 15) [67–69], precursor polymers with a high number-average molecular weight M_n (up to 34,000) were obtained, corresponding to about 54 repeat units. The yellow-orange precursor polymer is soluble in chloroform, and the structure could be verified with the aid of ^{1}H and ^{13}C NMR. Advantageous for the further conversion into the fully aromatic ladder polymer is that the precursor polymers (Diels–Alder polyadducts) do not contain any exocyclic bridges, as carbon monoxide is eliminated in the course of the polyaddition of the olefin with the cyclopentadienone substructure.

The target of the polymer-analogous aromatization is then the conversion (dehydrogenation) of the saturated carbon centers still present. For this purpose, Schlüter et al. used DDQ as oxidant, in somewhat more than equimolar amounts. The course of the reaction can be followed by an increase in color from yellow-orange to

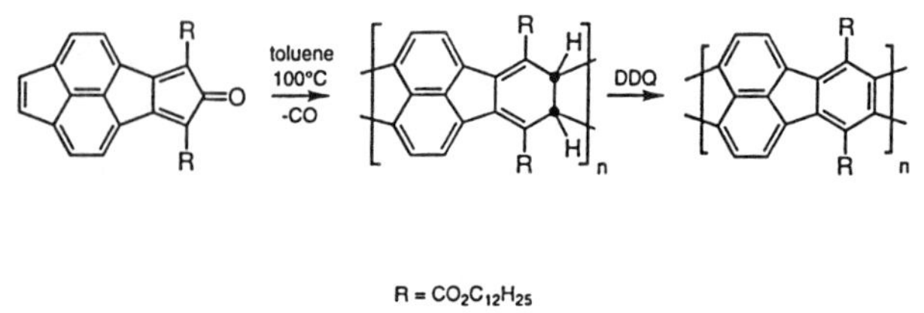

R': -C$_6$H$_{13}$

Scheme 14 AA/BB-type Diels–Alder polyaddition of a bisdiene and a pseudobisdienophile (pseudobisaryne) according to Horn et al. [64,65].

R = CO$_2$C$_{12}$H$_{25}$

Scheme 15 Diels–Alder polyaddition of an AB-type diene dienophile monomer of the pyracyclene-type according to Schlüter and coworkers [67–69].

deep red. Unfortunately, however, the products precipitate out of the reaction mixture. Only a small fraction of the oligomeric reaction products is soluble in the common organic solvents, giving a violet solution. Thus, determination of the exact degree of dehydrogenation is not possible. Solid-state ^{13}C NMR spectra suggest an extensive degree of aromatization. The UV-vis spectrum of the soluble fraction of the ladder polymer displays the longest wavelength absorption spectrum at ~510 nm, together with a broad absorption tail with several shoulders to over 600 nm, which can be traced back to the oligomeric nature of the soluble fraction.

The primary polyaddition products from AA/BB-type reactions between the bifunctional pyracyclene dienophiles and the bisfuran diene [66] led to soluble precursor polymers (Scheme 16) with a number-average molecular weight of 5000–10,000, corresponding to a degree of polycondensation of 4–9 repeat units. ^{1}H and ^{13}C NMR spectroscopy confirm the presence of the precursor polymers as the correct structure. These precursor polymers still contain exocyclic oxo bridges. Thus, a polymer-analogous dehydration leading to formation of the fully unsaturated ladder structure is necessary here for the primary polyadducts. This polymer-analogous elimination of water should lead to removal of the exocyclic oxo bridges and to unfolding (planarization) of the molecule (Scheme 16). For the dehydration Löffer et al. used toluene sulfonic acid (in excess). With this reaction an increase in color was also observed (from yellow to red); unfortunately, this time only insoluble products were formed. The "solubilizing" substituents are unable to keep the resulting polymers in solution to the desired extent. It was only possible to obtain soluble products with incomplete dehydration. In the ^{13}C NMR spectrum of the insoluble products of the (attempted) complete dehydration, no saturated structural units (carbon atoms with oxo bridges) could be detected, although one must take into account the low sensitivity of solid-state NMR spectroscopy. The resulting polymer is deep violet with an absorption maximum at about 600 nm, which corresponds to a band gap of ~2.1 eV. This value is in good agreement with theoretical calculations [68]. The work of Schlüter and coworkers presented here were accompanied by extensive studies on the corresponding low molecular weight model compounds [66–69].

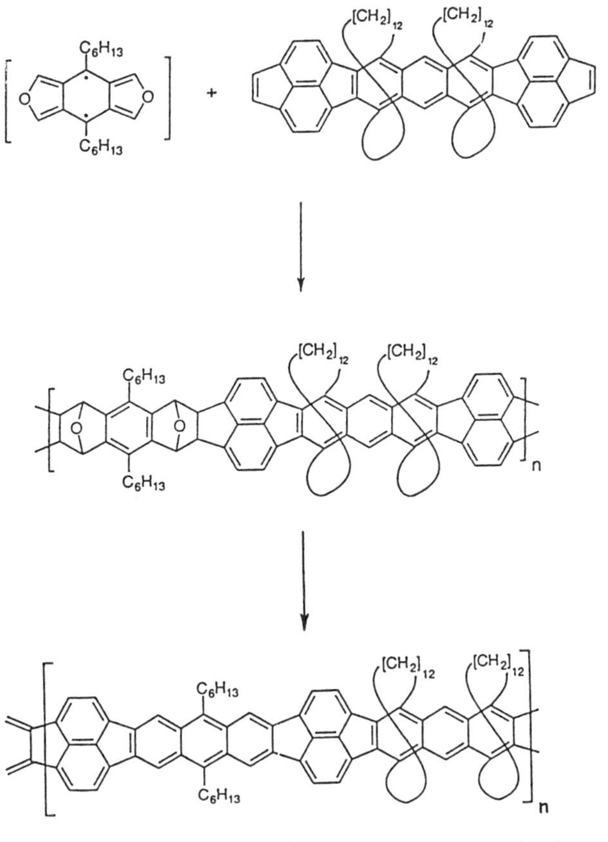

Should it be possible in the future, by means of structural variations in the solubilizing substituents, to attain fully soluble and thus well-characterized and workable materials, attractive application perspectives may emerge, e.g., as potential emitter substances in light-emitting diodes based on polymers. Nevertheless, nothing is yet known of the photoluminescence behavior of these substances. The observed high chemical stability of the materials ought to be of advantage.

The wealth of results from the last few years presented in this chapter impressively demonstrate the advance and the enormous dynamism of polymer research in the area of conjugated ladder polymers. Generally, a definite trend toward structurally well defined and easily worked materials can be ascertained. A particularly high level of interest appears from the side of potential applications (e.g., the use as emitter substances in polymer LEDs), from a direction in which structural accuracy or a minimum number of structure defects and the purity of the substances are absolute requirements for attaining optimal values. Apart from this, the availability of several powerful synthetic strategies for conjugated ladder polymers are extremely promising for the developments achieved hitherto and those anticipated in the future.

Scheme 16 AA/BB-type polyaddition of a pseudobisdiene (bisfuran) and a bisdienophile building block of the bispyracyclene type according to Löffler and Schlüter [66].

REFERENCES

1. C. S. Marvel and C. L. Levesque, *J. Am. Chem. Soc.* 60:280 (1938).
2. C. S. Marvel, J. O. Cormer, and E. H. Riddle, *J. Am. Chem. Soc.* 64:92 (1942).
3. N. Grassie and I. C. McNeil, *J. Chem. Soc. 1956:* 3929.
4. N. Grassie and R. McGuchan, *Eur. Polym. J.* 7:1357, 1503 (1971).
5. J. E. Bailey and A. J. Clarke, *Nature 234*:529 (1971).

6. R. J. Angelo, M. L. Wallach, and R. M. Ikeda, *Polym. Prepr.* 8:221 (1967).
7. R. J. Angelo, M. L. Wallach, and R. M. Ikeda, *Polym. Prepr.* 4:32 (1963).
8. A. W. Snow, *Nature* 292:40 (1981).
9. F. Bohlmann and E. Inhoffen, *Chem. Ber.* 89:1276 (1956).
10. P. Teysie and A. C. Korn-Girard, *J. Polym. Sci. A* 2:2849 (1964).
11. N. Kobayashi, M. Mikitoshi, H. Ohno, E. Tsuchida, H. Matsuda, H. Nakanishi, and M. Kato, *New Polym. Mat.* 1:3 (1987).
12. J. K. Stille and M. E. Freeburger, *J. Polym. Sci. A1* 6:161 (1968).
13. J. K. Stille and E. L. Mainen, *Macromolecules* 1:36 (1968).
14. M. Okada and C. S. Marvel, *J. Polym. Sci. A1* 6:1259 (1968).
15. R. Wolf, M. Okada, and C. S. Marvel, *J. Polym. Sci. A1* 6:1503 (1968).
16. F. E. Arnold and R. L. vanDeussen, *Macromolecules* 2:497 (1969).
17. F. E. Arnold and R. L. vanDeussen, *Polym. Lett.* 6:815 (1968).
18. S. A. Jenekhe, P. O. Johnson, and A. K. Agrawal, *Macromolecules* 22:3216 (1989).
19. U. Scherf and K. Müllen, *Makromol. Chem. Rapid Commun.* 12:489 (1991).
20. U. Scherf, A. Bohnen, and K. Müllen, *Makromol. Chem.* 193:1127 (1992).
21. U. Scherf and K. Müllen, *Macromolecules* 25:3546 (1992).
22. G. Froyer, J. Y. Goblot, J. L. Guilbert, F. Maurice, and Y. Pelours, *J. Phys. Colloq.* 44:C3-745 (1983).
23. J. M. Toussaint and J. L. Bredas, *Synth. Met.* 46:325 (1992).
24. R. Fiesel, J. Huber, U. Scherf, *Augen. Chem.* 108:2233 (1996), *Int. Ed. Engl.* 35:2111 (1996).
25. J. Grüner, H. F. Wittmann, P. J. Hamer, R. H. Friend, J. Huber, U. Scherf, K. Müllen, S. C. Moratti, and A. B. Holmes, *Synth. Met.* 67:181 (1994).
26. J. Stampfi, W. Graupner, G. Leising, and U. Scherf, *J. Lumin.*, 63:117 (1995).
27. R. F. Mahrt, U. Siegner, U. Lemmer, M. Hopmeier, U. Scherf, S. Heun, E. O. Göbel, K. Müllen, and H. Bässler, *Chem. Phys. Lett.* 240:373 (1995).
28. J. Huber, K. Müllen, J. Salbeck, H. Schenk, U. Scherf, T. Stehlin, and R. Stern, *Acta Polym.* 45:244 (1994).
29. G. Grem and G. Leising, *Synth. Met.* 55–57:4105 (1993).
30. G. Grem, C. Paar, J. Stampfl, G. Leising, J. Huber, and U. Scherf, *Chem. Mat.* 7:2 (1995).
31. J. F. Grüner, P. Hamer, R. H. Friend, J. Huber, U. Scherf, and A. B. Holmes, *Adv. Mater.* 6:748 (1994).
32. M. Rehahn, A.-D. Schlüter, G. Wegner, and W. J. Feast, *Polymer* 30:1054, 1060 (1989).
33. M. Rehahn, A.-D. Schlüter, and G. Wegner, *Makromol. Chem.* 191:1991 (1990).
34. A.-D. Schlüter and G. Wegner, *Acta Polym.* 44:59 (1993).
35. A.-D. Schlüter, *Adv. Mat.* 3:282 (1991).
36. K. Chmil and U. Scherf, *Makromol. Chem. Rapid Commun.* 14:217 (1993).
37. J. M. Tour and J. S. S. Lamba, *J. Am. Chem. Soc.* 115:4935 (1993).
38. J. S. S. Lamba and J. M. Tour, *J. Am. Chem. Soc.* 116:11723 (1994).
39. M. B. Goldfinger and T. M. Swager, *J. Am. Chem. Soc.* 116:7895 (1994).
40. T. Davidson, *Polymers in Electronics*, Am. Chem. Soc., Washington, DC. 1984, p. 227.
41. H. E. Zimmermann and D. R. Amick, *J. Am. Chem. Soc.* 95:3977 (1973).
42. E. Clar, *Polycyclic Hydrocarbons*, Academic, New York, 1964.
43. E. Clar, *The Aromatic Sextet*, Wiley, London, 1972.
44. D. Biermann and W. Schmidt, *J. Am. Chem. Soc.* 102:3163, 3173 (1980).
45. V. R. Sastri, R. Schulman, and D. C. Robert, *Macromolecules* 15:939 (1982).
46. T. Fang, Ph.D. Thesis, University of California, Los Angeles, 1986.
47. U. Scherf and K. Müllen, *Synthesis* 1992:23.
48. V. R. Sastri, R. Schulmann, and D. C. Robert, *Macromolecules* 15:939 (1982).
49. T. Yamamoto, A. Morita, Y. Miyazaki, T. Marayama, H. Wakayama, Z. Zhou, Y. Nakumura, T. Kanbara, S. Sasaki, and K. Kubota, *Macromolecules* 25:1214 (1992).
50. T. Kanbara, N. Saito, and T. Yamamoto, *Macromolecules* 24:5883 (1991).
51. T. Maruyama, K. Kubota, and T. Yamamoto, *Macromolecules* 26:4055 (1993).
52. N. Saito, T. Kanbara, T. Saito, and T. Yamamoto, *Polym. Bull.* 30:285 (1993).
53. T. Yamamoto, *Prog. Polym. Sci.* 17:1153 (1992).
54. K. Steliou, P. Salama, and X. Yu, *J. Am. Chem. Soc.* 114:1456 (1992).
55. Z. Y. Wang and C. Zhang, *Macromolecules* 25:585 (1992).
56. J. M. Toussaint and J. L. Bredas, *Synth. Met.* 46:325 (1992).
57. T. Freund, K. Müllen, and U. Scherf, *Macromolecules* 28:547 (1995).
58. W. J. Bailey, J. Economy, and M. E. Hermes, *J. Org. Chem.* 27:3295 (1962).
59. W. J. Bailey, E. J. Fetter, and J. Economy, *J. Org. Chem.* 27:3497 (1962).
60. U. Scherf and K. Müllen, *Adv. Polym. Sci.* 123.
61. S. Wegener and K. Müllen, *Chem. Ber.* 124:2101 (1991).
62. M. Löffler and A.-D. Schlüter, *GIT Fachz. Lab.* 1992:1101.
63. T. Fang, Ph.D. Thesis, University of California, Los Angeles, 1986.
64. T. Horn, U. Scherf, S. Wegener, and K. Müllen, *Polym. Prep.* 33:190 (1992).

65. T. Horn, S. Wegener, and K. Müllen, *Macromol. Chem. Phys. 196*:2463 (1995).
66. M. Löffler, A.-D. Schlüter, K. Gessner, W. Saenger, J.-M. Toussaint, and J.-L. Bredas, *Angew. Chem. 106*:2281 (1994); *Int. Ed. Engl. 33*:2209 (1994).
67. M. Löffler and A.-D. Schlüter, *Macromol. Symp. 77*:359 (1994).
68. A.-D. Schlüter, M. Löffler, and V. Enkelmann, *Nature 368*:831 (1994).
69. B. Schlicke, H. Schirmer, and A.-D. Schlüter, *Adv. Mater. 7*:544 (1995).
70. J. Huber and U. Scherf, *Macromol. Rapid Commun. 15*:897 (1994).

15

Synthesis and Properties of Conducting Bridged Macrocyclic Metal Complexes

Michael Hanack, Michael Hees, Patrick Stihler, Götz Winter, and L. R. Subramanian
Universität Tübingen, Tübingen, Germany

I. INTRODUCTION

Although only a decade has passed since the publication of the first edition of this book [1], the chemistry and practical uses of organic conducting polymers based on macrocyclic metal complexes have advanced in long strides. These polymers have attracted wide attention due to their multitude of applications in materials science. In this context, metal complexes of phthalocyanines and structurally related compounds have excelled in many ways. Presently, phthalocyanines are of particular interest in nonlinear optics [2–5], as liquid crystals [6,7], as Langmuir–Blodgett (LB) films [8–12], in optical data storage [13,14], as electrochromic substances [15–17], as low-dimensional metals [1,18–21], in rectifying devices [22], in electrocatalysis [23], as gas sensors [24,25], as photosensitizers [26], as photoconductors [27], as carrier generation materials in the near-infrared [28], and in studies related to spectral hole burning [29]. The more soluble substituted derivatives of phthalocyanines function as active components in various processes driven by visible light: in photoredox reactions and photo-oxidations in solution [30–32], photodynamic cancer therapy [30,33,34], photoelectrochemical cells [35–37], photovoltaic cells [38–40], and in electrophotographic applications [41–43]. In the solid state, phthalocyanines (hereafter abbreviated Pc's) can behave as molecular organic semiconductors [20]. The conductive properties of metal Pc's are attributed to the favorable electronic properties and morphology of the macrocyclic systems. The centrosymmetric, planar Pc molecule, a tetrabenzotetraazaporphyrin of formula $C_{32}H_{18}N_8$, and numerous metal Pc's (Fig. 15.1) occur as a beautifully crystalline series of macrocyclic organic pigments of extraordinary stability [44].

II. SYNTHESIS OF PHTHALOCYANINES AND RELATED MACROCYCLIC COMPOUNDS

A. Unsubstituted Phthalocyanines

More than 70 elements have been used as the central metal atom in the phthalocyanine moiety that controls the oxidation potential of the phthalocyaninato compounds, leading to very different electrical properties. Figure 15.2 shows some additional phthalocyaninato (PcM) derivatives containing extended π systems like 2,3-naphthalocyaninato- (2,3-NcM), 1,2-naphthalocyaninato- (1,2-NcM), phenanthrenocyaninato-, and anthracenocyanatometal (AncM) compounds [45,46]. 1,2-NcM is shown only in its symmetrical (C_{4h}) form; three other constitutional isomers are known.

Most metallophthalocyanines can be prepared from phthalodinitrile derivatives and the corresponding metals or metal salts in high boiling solvents such as 1-chloronaphthalene or quinoline (Fig. 15.3) [18,19,44–47]. They are also obtained by subsequent metal insertion into the phthalocyanine [Fig. 15.3, Eq. (5)]. The synthesis of naphthalocyaninatophenanthrenocyaninato and anthraceno systems follows similar routes from the appropriate starting materials [44–49]. A mild method to obtain phthalocyaninatometal complexes in high yields is to heat phthalodinitrile with metal salts in an alcohol (e.g., 1-pentanol) in the presence of 1,8-diazabicyclo[5.4.0]undec-1-ene [Fig. 15.3, Eq. (6)] [50–52]. Phthalocyanines can be prepared also by ring-insertion reaction from subphthalocyanines leading to unsymmetrically substituted phthalocyanines (see Section II.B). Some Pc's are more difficult to synthesize; PcRu is given as an example.

Metallophthalocyanines are insoluble in common organic solvents and can be easily purified by extracting

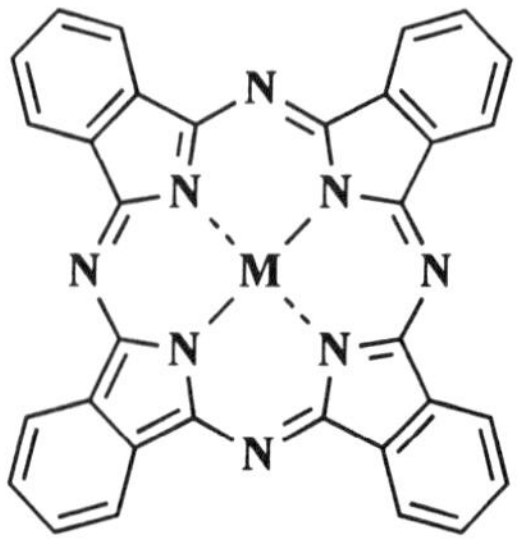

Fig. 15.1 Structure of metallophthalocyanines (M = metal).

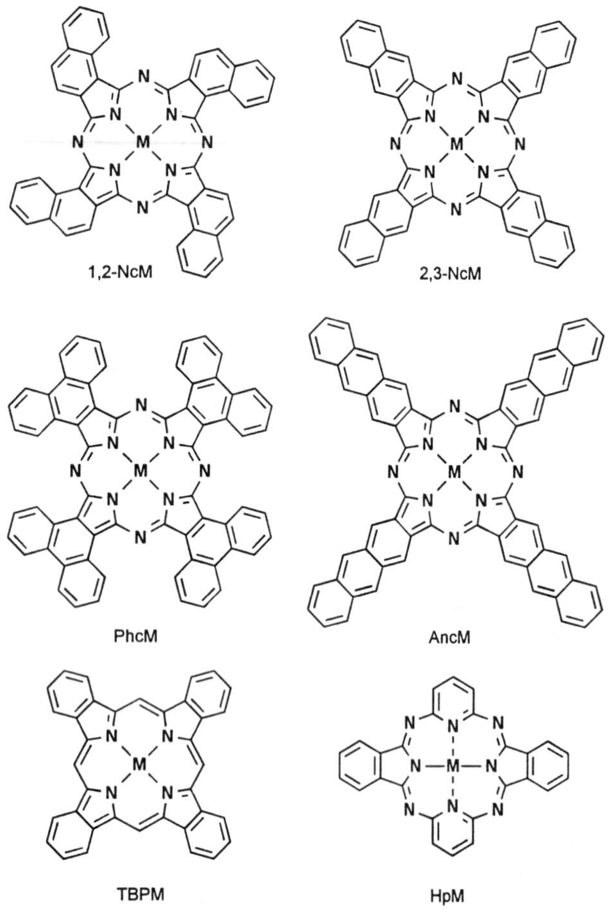

Fig. 15.2 1,2-NcM = 1,2-metallonaphthalocyanine; 2,3-NcM = 2,3-metallonaphthalocyanine; PhcM = metallophenanthrenocyanine; AncM = metalloanthracenocyanine; TBPM = metallotetrabenzoporphyrin; HpM = metallohemiporphyrazine.

the impurities with a suitable solvent. PcRu is synthesized via the axially coordinated bisisoquinoline complex PcRu(iqnl)$_2$ (iqnl = isoquinoline). The isoquinoline complex is formed from phthalonitrile and RuCl$_3$·6H$_2$O in quinoline (from the 1–2% isoquinoline present as impurity in commercial quinoline) and purified by column chromatography [53; R. Polley and M. Hanack, unpublished]. Subsequent thermal decomposition in vacuo at 250°C affords pure PcRu. Based on magnetic measurements and UV-vis spectroscopy, the structure of PcRu is reported to exist as a dimer containing a Ru–Ru double bond [54].

The stability of metallophthalocyanines depends on the central metal atom. Many of the PcMs (H$_2$, Ni, Cu, Zn) are very stable in the presence of oxygen; however, other examples are known that are not. While synthesizing, e.g., PcFe, oxygen cannot be excluded totally, leading to the formation of (PcFe)$_2$O as a by-product [55].

Phthalocyanines can also be prepared by ring insertion reactions from subphthalocyanines [56]. This reaction is suitable for obtaining unsymmetrically substituted phthalocyanines and is discussed later.

Besides the aforementioned subphthalocyanines, superphthalocyanines (SPc's) containing five isoindolenine units have also been synthesized [57]. Heating a mixture of phthalonitrile with anhydrous uranyl chloride in DMF at 170°C affords SPc(UO$_2$) [58].

B. Peripherally Substituted Phthalocyanines

1. Tetra- and Octasubstituted Phthalocyanines

The solubility of phthalocyanines increases when substituents are introduced in the periphery of the macrocycle. These substituents are responsible for a larger distance between the inclined stacked phthalocyanines and enable their solvation. Introducing bulky groups like *tert*-butyl [59] was an early approach to obtain phthalocyanines that are soluble in organic solvents. Recently substituted phthalocyanines have shown to be quite soluble in common organic solvents and therefore are being intensively investigated. Some examples of tetrasubstituted and octasubstituted metallophthalocyanines are given in Figs. 15.4 and 15.5, respectively.

Due to the location of the substituents in either the 2, 3, 9, 10, 16, 17, 23, 24- or 1, 4, 8, 11, 15, 18, 22, 25-positions, the tetra- and octasubstituted metallophthalocyanines shown in Figs. 15.4 and 15.5 are called 2,3- and 1,4-substituted phthalocyanines, respectively, in the following discussion.

Tetrasubstituted 2,3-phthalocyanines can be synthesized from 4-substituted phthalonitriles, and tetrasubstituted 1,4-phthalocyanines are obtained from 3-substituted phthalonitriles as shown in Fig. 15.6. Monosubstituted phthalonitriles always give a mixture of four constitutional isomers of tetrasubstituted phthalocyanines. These isomers are depicted in Fig. 15.7 for the 1,4-substituted systems; similar isomers are obtained for the 2,3-substituted phthalocyanines.

Several trials have been made to separate these four isomers [60,61]. Initial attempts were concentrated on the separation of the four isomers of 2,3-substituted

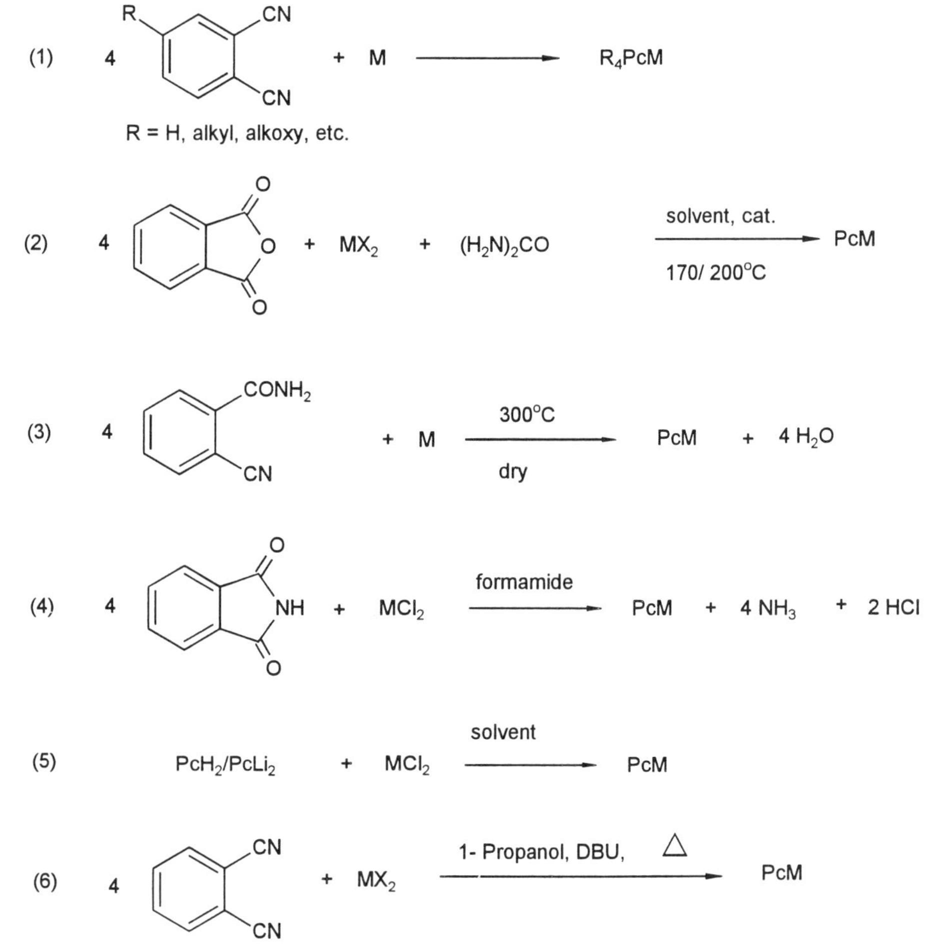

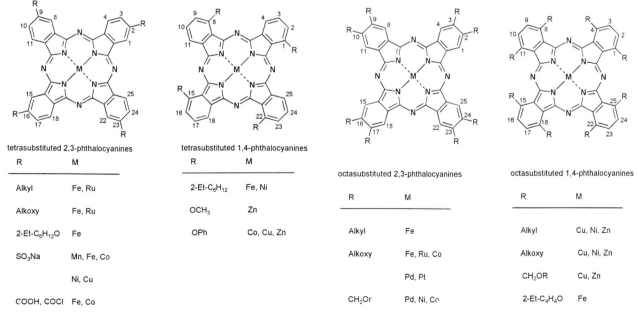

Fig. 15.3 Synthesis of metallophthalocyanines.

tetrasubstituted 2,3-phthalocyanines

R	M
Alkyl	Fe, Ru
Alkoxy	Fe, Ru
2-Et-$C_6H_{12}O$	Fe
SO_3Na	Mn, Fe, Co
	Ni, Cu
COOH, COCl	Fe, Co

tetrasubstituted 1,4-phthalocyanines

R	M
2-Et-C_6H_{12}	Fe, Ni
OCH_3	Zn
OPh	Co, Cu, Zn

Fig. 15.4 Tetrasubstituted phthalocyanines.

octasubstituted 2,3-phthalocyanines

R	M
Alkyl	Fe
Alkoxy	Fe, Ru, Co
	Pd, Pt
CH_2Or	Pd, Ni, Co

octasubstituted 1,4-phthalocyanines

R	M
Alkyl	Cu, Ni, Zn
Alkoxy	Cu, Ni, Zn
CH_2OR	Cu, Zn
2-Et-C_4H_4O	Fe

Fig. 15.5 Octasubstituted phthalocyanines.

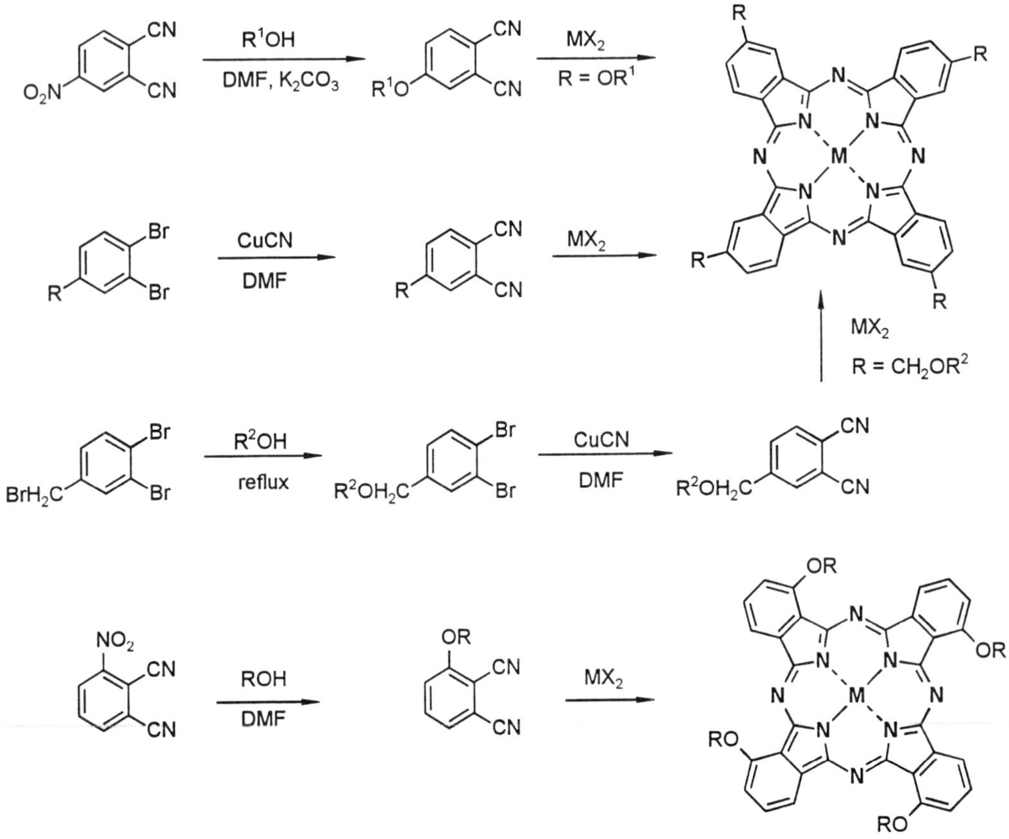

Fig. 15.6 Syntheses of tetrasubstituted phthalocyanines (only C_{4h} isomers are shown).

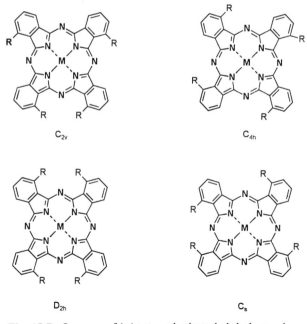

Fig. 15.7 Isomers of 1,4-tetrasubstituted phthalocyanines.

tetra-*tert*-butylphthalocyaninatonickel by MPLC (medium pressure liquid chromatography) and HPLC methods [60]. Although it was not possible to isolate the four isomers, the C_{2v} and the C_s isomers could be enriched to the extent of 84% and 93%, respectively, by HPLC separation. These separations are also possible with phthalocyaninatonickel complexes carrying other substituents in the 2,3-position (R = e.g., OC_6H_{13} or OC_6H_{17}, linear or branched) using specially developed chromatographic columns (M. Sommerauer and M. Hanack, unpublished). Recently, the separation (and full characterization by ^{1}H NMR spectroscopy) of all four isomers of the 1,4-substituted (2-ethylhexyloxy)phthalocyaninatonickel (Fig. 15.7, R = OC_8H_{17}) by MPLC and HPLC methods was reported [61; M. Sommerauer and M. Hanack, unpublished]. 1,4-Tetrasubstituted phthalocyanines are easier to separate than 2,3-tetrasubstituted phthalocyanines, since they are quite distinguished by different steric interactions in the various constitutional isomers, which can be especially seen in the D_{2h} isomer (Fig. 15.5). The alkoxy substituents (R = ethylhexyloxy) are not in the plane of the phthalocyanine ring; one is located above and the other under the plane of the macrocycle [62]. Hence there can be different interactions between the various constitutional isomers and the stationary phase of the chromatographic column.

Concerning the distribution of the four isomers, optically active substituents show interesting effects. While the isomer with the highest symmetry (C_{4h}) in phthalocyanines with racemic substituents is formed only in small amounts, it is the main product with an optically active substituent. During the synthesis, a directing effect of the enantiomerically pure substituents is observed leading to the phthalocyanine with the highest symmetry (G. Schmid and M. Hanack, unpublished).

2, 3, 9, 10, 16, 17, 23, 24-Octaalkyl-substituted [63] and 1, 4, 8, 11, 15, 18, 22, 25-octaalkyl-substituted phthalocyanines are synthesized similarly to the tetrasubstituted analog starting from 4,5- or 3,6-disubstituted phthalonitriles.

Usually tetrasubstituted phthalocyanines (Fig. 15.4) show higher solubilities in organic solvents than the octasubstituted ones (Fig. 15.5). There are mainly two reasons for this behavior: The isomer mixture of four tetrasubstituted phthalocyanines leads to a lower order in the solid state compared to symmetrically octasubstituted phthalocyanines, and tetrasubstituted phthalocyanines possess a higher dipole moment caused by the unsymmetrical arrangement of the substituents in the periphery of the macrocycle.

Besides substituted phthalocyanines, substituted naphthalocyanines [64–67] and tetra-*tert*-butyl-substituted anthracenocyanines [45,46] were also synthesized. Most of the soluble substituted phthalocyanines show insulating behavior; however, they are important building blocks for the synthesis of soluble bridged transition metal phthalocyanines. The electrical properties of these polymers have been investigated (see below).

A particular attribute of octasubstituted phthalocyanines is their liquid crystalline behavior. Phthalocyanines substituted in peripheral positions by long-chain substituents form, depending on the structure (alkyl or alkoxy), liquid discotic (from the disklike shape of the molecules) crystalline phases [68–71] or columnar mesophases [72]. Liquid crystalline phases are found for side chains longer than C_4 or C_6 (depending on the metals and substituents used) [73]. In these mesophases the phthalocyanine moieties stack in columns with arrangements of different symmetry (Fig. 15.8), depending on the nature of the side chains [71,74]. Only chains connected with a heteroatom X (X = O, S) to the aromatic macrocycle show an arrangement with the molecular planes perpendicular to the column axis.

The transition from the solid to liquid crystalline phase (melting) corresponds to the melting of the flexible chains; the aromatic cores retain positional and orientational order. The transition from the mesophase to the isotropic liquid (clearing) corresponds to the destruction of the columns. There are two possible means of influencing the transition temperatures: extension or branching of the aliphatic chains depresses the transition tem-

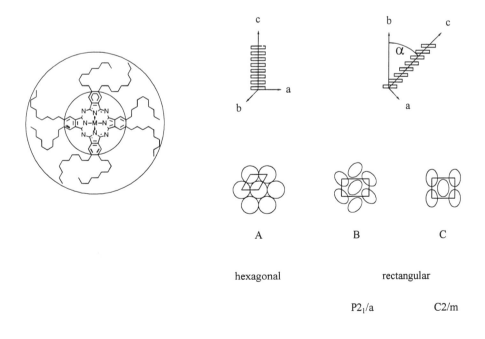

Fig. 15.8 Arrangements of phthalocyaninato mesophases.

peratures. Branching in the middle of the chains depresses the melting point [75,76]. Branching of the chains near the aromatic core, however, depresses the clearing point [75,77].

When R_4PcM compounds with R = OC_5H_{11}, OC_8H_{17}, $OC_{10}H_{21}$ and M = Ni, Pd, Pt are compared, the melting points of these compounds show that the metallophthalocyanines containing pentyloxy substituents do not exhibit liquid crystalline behavior [78,79]. Those with octyloxy chains show a transition temperature between 100 and 84°C (M = Ni, 100°C; M = Pd, 92°C; M = Pt, 84°C). The melting points of the decyloxy series are between 86 and 64°C (M = Ni, 86°C; M = Pd, 74°C; M = Pt, 64°C). As can be seen from these data, the longer the chain the lower the transition temperature.

The electrical properties of liquid crystalline phthalocyanines have been determined, for example, from complex impedance spectroscopic measurements [76]. Thus it was possible to observe that liquid crystalline phthalocyanines show a slight increase in conductivity when going from the solid phase to the mesophase. The conductivities of these compounds are low ($\sim 5 \times 10^{-10}$ S/cm) [80]. Higher conductivities are obtained if crown ether phthalocyanines (CEPcM) are aggregated by adding metal picrate salts (M = K^+, Rb^+, Cs^+; $\sigma_{RT} \sim 10^{-6}$–10^{-7} S/cm) [81]. In these aggregates the cations are sandwiched between two stacked crown ether rings. The picrate anion is located in the neighborhood of an alkali metal ion, forming an ion pair [81]. So far the most often used crown ethers (CE) have been 15-crown-5, 18-crown-6, and 21-crown-7. These CEPcMs (M = H_2, Zn, Cu, Co, Ni, Fe) show high solubility in chloroform and dichloromethane and lower solubility in acetone, DMF, DMSO, toluene, and benzene [82–84]. Elemental analysis of the metal picrate complexes of CEPcCu (18-crown-6) show a host/guest ratio of 1:4. A potassium ion just fits into the 18-crown-6 ring. Ultraviolet-visible experiments, using anhydrous solvents, suggested the aggregation of the phthalocyanine units upon addition of potassium picrate [82–84]. Seebeck measurements revealed CEPcCu to be a p-type electronic conductor as observed earlier for PcCu [85,86], whereas the K^+, Rb^+, and Cs^+ picrate complexes of CEPcCu exhibit n-type electronic conductivity. For PcCu the conduction pathway of charge carriers is formed by π–π overlap of the macrocyclic rings [87]. With the addition of metal picrate salts to CEPcCu, the conduction pathway changes from ligands centered by holes to metals centered by electrons, as a result of interaction of d orbitals of the Cu^{2+} centers [88–90].

As has been recognized, the closepacked aromatic macrocycles are capable of transporting charge and energy along the stacking axis due to interactions between single macrocycles (one-dimensional conduction) [81,91]. With the increased tractability of liquid crystalline materials, these one-dimensional systems could serve in future molecular electronic devices. Further information about phthalocyanine liquid crystals is available in Ref. 92.

2. Unsymmetrically Substituted Phthalocyanines

In contrast to symmetrically substituted (tetra- or octasubstituted) phthalocyanines, reports on phthalocyanines with lower symmetry have rarely appeared, mainly because of the problems associated with their preparation [93,94]. Phthalocyanines with lower symmetry show interesting properties in NLO and are furthermore important materials for LB films [73] or ladder polymers [94]. They can also be useful in understanding the nature of phthalocyanines; for example, fine-tuning of the absorption bands of phthalocyanines can be achieved by stepwise adjustment of the size of the π-conjugated macrocyclic system [95].

The preparation of unsymmetrical phthalocyanines can follow several routes:

Statistical condensation routes [93,96–98]
Polymer support route [99–101]
Subphthalocyanine route [102–104]

The statistical condensation route has already been covered in Fig. 15.6. The formation of tetrasubstituted phthalocyaninatometal compounds according to Fig. 15.6 always leads to a mixture of constitutional isomers (cf. Fig. 15.7) that subsequently must be separated by chromatographic methods.

The strategy based on the statistical condensation reaction of two different phthalonitriles (A and B) leads to six kinds of phthalocyanines (Fig. 15.9) [93].

Phthalocyanines containing four similar units (AAAA or BBBB)
Phthalocyanines containing three similar units and one different unit (AAAB or BBBA)
Phthalocyanines containing two different units (two isomers: AABB or ABAB)

The number of products can be reduced by using a phthalonitrile bearing bulky groups, e.g., phenyl groups (no BBBB compound is obtained), in the 3,6-positions [93,105]. The statistical condensation is carried out by reaction of equimolar amounts of 3,4,5,6-tetraphenylphthalonitrile and 4,5-dipentoxyphthalonitrile with nickel acetate in n-pentanol in the presence of catalytic amounts of DBU at 140°C for 24 h. The reaction products are shown in Fig. 15.9. All reaction products can be isolated by column chromatography [93].

Because of their different symmetry these phthalocyanines show different UV-vis spectra [18,19]. The lower the symmetry of the phthalocyanine, the larger the split in the Q band in the UV-vis spectrum.

The polymer support route is based on the fixation of a phthalonitrile (A) on a polymer, while a different

Conducting Bridged Macrocyclic Metal Complexes

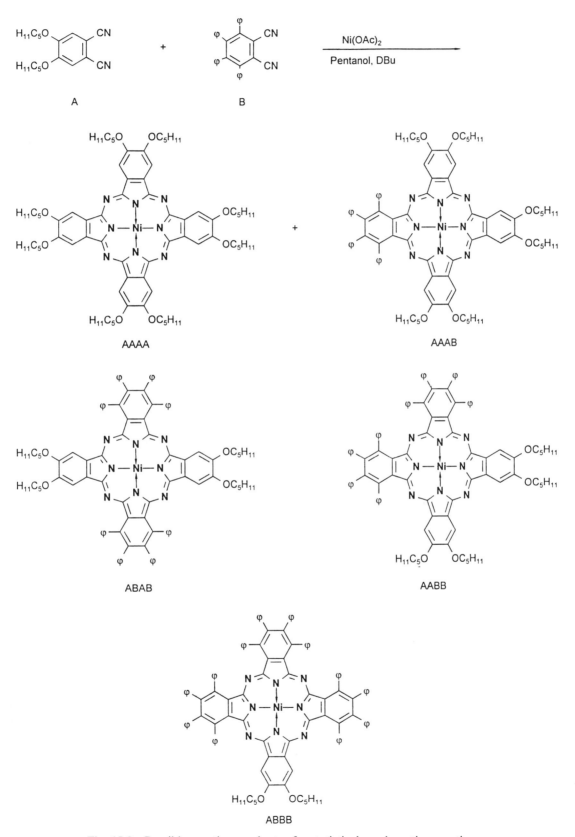

Fig. 15.9 Possible reaction products of a statistical condensation reaction.

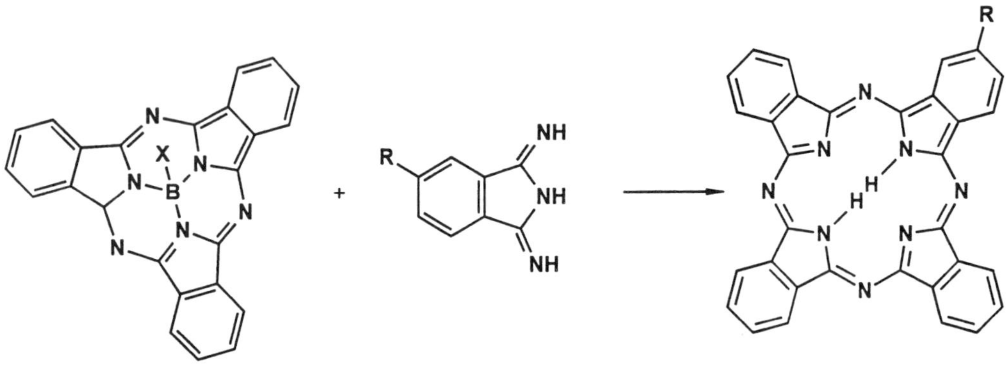

Fig. 15.10 Synthesis of an unsymmetrical substituted phthalocyanine via a subphthalocyanine.

phthalonitrile (B) is added forming a 1:3 product, which can be isolated by removing it from the polymer.

Figure 15.10 shows another route that in principle can lead to the formation of an unsymmetrically substituted phthalocyanine, in this case a monosubstituted one. In this route a subphthalocyanine is reacted with a monosubstituted isoindolenine with formation of the monosubstituted phthalocyanine [102,106]. This route has been applied for the preparation of a variety of unsymmetrically substituted phthalocyanines [102]; however, the yields of the monosubstituted phthalocyanines are low and other side products are formed as well. Subphthalocyanines, shown in Fig. 15.10, are a class of compounds that have found much interest recently, mostly due to the possibility of preparing unsymmetrically substituted phthalocyanines [107,108].

Unsymmetrically substituted water-soluble phthalocyanines and naphthalocyanines as well as water-soluble tetrasubstituted phthalocyanines [R_4PcM, R = COOH, SO_3H; M = Zn, Al(C)] have been applied in photodynamic cancer therapy [109].

III. ELECTRICAL PROPERTIES

For achieving good semiconducting or even conducting properties, a special spatial arrangement of the macrocycles, namely either planar or stacked, is a necessary condition. While phthalocyanines very seldom crystallize in a stacked cofacial arrangement that does not favor the formation of a conducting band by $\pi-\pi$ overlap of the macrocycles, methods have been constructed to obtain oligomeric bridged Pc's with good semiconductivity. For example, oxygen-bridged Pc metal complexes have to be doped to obtain semiconducting behavior (see Section III.A), while coordination polymers prepared from phthalocyaninato transition metal complexes and a bridging ligand, e.g., s-tetrazine, show intrinsic semiconducting properties (see Section III.C).

In the following, results obtained on oligomeric bridged Pc metal complexes since the publication of the first edition of this handbook are described under appropriate subheadings.

A. Oxo-Bridged Macrocyclic Main Group Metal Complexes

One of the best investigated types of cofacially linked stacked polyphthalocyaninatometalloxanes [PcMO]$_n$ (M = Si, Ge, Sn) is shown in Fig. 15.11. The synthesis of [PcMO]$_n$ (M = Si, Ge, Sn) was carried out starting from PcMCl$_2$ prepared by standard procedures [110–116]. Hydrolysis of PcMCl$_2$ led to PcM(OH)$_2$. The polycondensation was achieved by dehydration, either by heating PcM(OH)$_2$ in vacuo at 325–440°C or by heating it in refluxing 1-chloronaphthalene or quinoline (Fig. 15.11). A topotactic polymerization mechanism was proposed by Marks [116]. Random reaction of all end groups with each other was also discussed [117].

Dehydration in vacuo provides the highest degree of polymerization. Both [PcSiO]$_n$ and [PcGeO]$_n$ have high thermal and chemical stabilities. In fact, [PcSiO]$_n$ can be recovered without any change after being dissolved in a strong concentrated acid such as H_2SO_4 or CF_3COOH. Estimation of the average molecular weights for [PcMO]$_n$ (M = Si, Ge, Sn) by IR end group analysis, tritium labeling, and laser light scattering experiments yields a degree of polymerization of 70–140 subunits for [PcSiO]$_n$ and lower values for [PcGeO]$_n$ and [PcSnO]$_n$. Longer polymerization reaction times and higher temperatures seem to increase the molecular weight [118].

Based on a model structure for [PcSiO]$_n$, a one-dimensional stacking of the PcM^{2+} subunits linked by O^{2-} bridges is derived from powder diffraction data by computer simulation techniques. The M–O–M distances vary from 333 pm for M = Si to 353 pm for M = Ge and 382 pm for M = Sn. The best fit of the data indicates

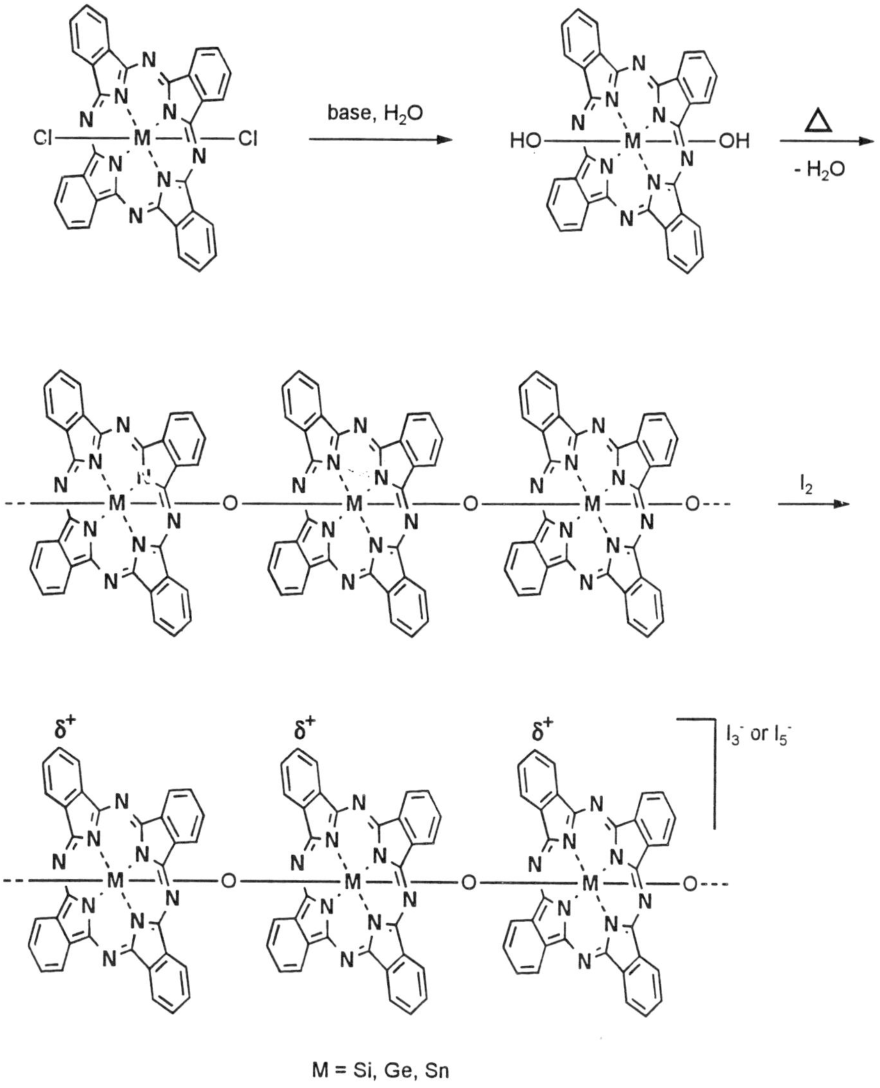

Fig. 15.11 Oxo-bridged phthalocyaninato metal complexes.

a staggering angle of the macrocycles of 39° for [PcSiO]$_n$ and 0° for [PcGeO]$_n$ and [PcSnO]$_n$. These data are confirmed by ^{13}C CP-MAS-NMR spectroscopy [119,120], electron diffraction [121], and electron transmission spectroscopy [122].

To achieve conductivity, charge carriers must be generated, either by oxidation (*p*-doping) or reduction (*n*-doping). Very little experimental material is available for *n*-doped phthalocyanines [123]. The most frequently used doping method is oxidation with iodine, which can be carried out heterogeneously with iodine vapor, by treatment with iodine solutions, or by grinding both components together. Partial oxidation is also achieved with other electron acceptors, e.g., chlorine, bromine, quinones, and nitrosyl compounds [124], or by electrochemical methods [117,124]. Oxidation leads to well-defined, air-stable conducting polymers of relatively high thermal stability up to 120°C.

The most thoroughly investigated iodine-doped compounds lead to stoichiometries [(PcMO)I$_y$]$_n$ (M = Si, Ge) with a y_{max} of about 1.1, which represents a degree of maximum oxidation of 1/3. For peripherally substituted polymers, e.g., [((*t*Bu)$_4$PcSiO)I$_y$]$_n$, a y_{max} of up to 2.0 is possible (Table 15.1). As in the case of iodine doping of PcNi, iodine is thereby reduced to I$_3^-$ or I$_5^-$, which was proved by resonance Raman spectroscopy and ^{129}I Mössbauer studies. Chains of these counterions are disordered in channels parallel to the crystallographic *c* axis. Increasing the interring distance leads to the formation of I$_5^-$ counterions [125]. In the case of [PcSnO]$_n$, doping leads to the destruction of the polymeric structure [126]. It is also possible to use sulfur as

Table 15.1 Room Temperature Electrical Conductivity Data of Doped and Undoped Polycrystalline μ-Oxophthalocyaninatometal(IV) Compounds [PcMO]$_n$ and [R$_4$PcMO]$_n$

Compound	y	σ_{RT} (S/cm)
[PcSiOI$_y$]$_n$	0	5.5×10^{-6}
	1.1	6.7×10^{-1}
[PcGeOI$_y$]$_n$	0	2.2×10^{-10}
	1.1	1.1×10^{-1}
[PcSnOI$_y$]$_n$	0	1.2×10^{-9}
	1.1	2.2×10^{-6}
[(t-Bu)$_4$PcSiOI$_y$]$_n$	0	8.0×10^{-9}
	2.0	2.0×10^{-3}
[(t-Bu)$_4$PcGeOI$_y$]$_n$	0	6.0×10^{-11}
	1.9	1.0×10^{-3}

Source: Ref. 19.

bridging ligand, for instance in [PcGeS]$_n$ prepared by heating PcGe(OH)$_2$ with H$_2$S in an autoclave at 130°C, but the Ge–S bond is cleaved on doping [127]. However, the undoped polymer shows photoconductivity (see below).

Electrochemical oxidation using counterions as BF$_4^-$ or ClO$_4^-$ has not led to higher conductivities so far [117]. Analysis of powder diffraction data of [(PcMO)I$_y$]$_n$ (M = Si, Ge) that were compared with those of the model compound PcNiI shows similarities to this compound. In all cases a staggered arrangement of the macrocycles with a staggering angle of 39–40° and the above-described parallel chains of disordered I$_3^-$ counterions is evident. Doping also decreases the interring distances in the range of 3–5 pm [110–113]. Single-crystal data are not available for lack of crystals of suitable size.

The room temperature conductivities of polycrystalline samples of [(PcMO)I$_y$]$_n$ (M = Si, Ge) for various stoichiometries are given in Table 15.1. The nature of the dopant (iodine, bromine, quinones, etc.) has no significant effect on the conductivities [125,128–130]. In addition to the main charge transport mechanism via the phthalocyanine system, other mechanisms, e.g., percolation theory and fluctuation-induced carrier tunneling through potential barriers separating metal-like regions, have also been discussed [131].

The presence of bulky substituents in the periphery of the phthalocyanine ring system increases their solubility in common organic solvents. The synthesis of the peripherally alkylated μ-oxo polymers [R$_4$PcMO]$_n$ (R = t-Bu, Si(CH$_3$)$_3$; M = Si, Ge, Sn) was described by Metz et al. in 1983 [132]. The formation of alkoxymethylene- and alkoxy-substituted polymeric phthalocyaninatometalloxanes [R$_8$PcSiO]$_n$ (R = CH$_2$OC$_{12}$H$_{25}$, OC$_{12}$H$_{25}$) was also reported [133–135].

The synthesis of these complexes [R$_x$PcSiO]$_n$ (x = 4, 8) from the corresponding monomeric phthalocyaninatodihydroxysilanes R$_x$PcSi(OH)$_2$ was carried out with trifluoroacetic anhydride followed by thermal polymerization at 200°C [134]. The dihydroxides R$_x$PcSi(OH)$_2$ can be obtained by following routes similar to the unsubstituted derivatives. The undoped materials show electrical conductivities that are similar to or somewhat lower than those of the peripherally unsubstituted [PcMO]$_n$ (M = Si, Ge, Sn). The doped polymer [(R$_4$PcSiO)I$_y$]$_n$ [R = t-Bu, Si(CH$_3$)$_3$] is thermally stable up to 140°C. Above this temperature a loss of the doping agent occurs, giving pure [R$_4$PcSiO]$_n$ as the residue (380°C). Independently of the doping procedure, all silicon and germanium samples [R$_4$PcMO]$_n$ (M = Si, Ge) exhibit the characteristic features reported for the conducting [(PcSiO)I$_y$]$_n$ materials. In the case of μ-oxoocta(alkoxy)phthalocyaninatosilicon, the degree of polymerization was determined to be 140 [134]. These types of oligomeric or polymeric phthalocyaninatosiloxanes exhibit liquid crystalline properties and also form LB films [136].

Tetrabenzoporphyrinatogermanium dihydroxide [TBPGe(OH)$_2$] (cf. Fig. 15.1) could be polymerized to [TBPGe(O)]$_n$ by heating at 350°C and 0.001 torr [137]. X-ray powder diffractometer measurements showed that [TBPGe(O)]$_n$ is microcrystalline and isocrystalline with [PcGe(O)]$_n$ and exhibits electrical conductivity of 1×10^{-6} S/cm, which is four orders of magnitude higher than [PcGe(O)]$_n$. As expected, the conductivity increased strongly on doping with iodine [138] with a maximum value of $\sigma = 5 \times 10^{-2}$ S/cm for [TBPGe(O)]I$_{0.75}$]$_n$.

B. Fluorine- and Alkynyl-Bridged Macrocyclic Metal Complexes

Other examples of stacked phthalocyanines are the fluorine-bridged phthalocyaninatometal complexes [PcMF]$_n$ (M = Al, Ga, Cr) [139–143]. For their preparation, e.g., PcAlCl or PcGaCl was converted into PcAl(OH) or PcGa(OH), respectively. These PcM(OH) (M = Al, Ga) complexes react with concentrated hydrofluoric acid to form the polymeric [PcMF]$_n$, which can be purified by sublimation in vacuo [140–142]. As shown by single-crystal X-ray analysis, [PcGaF]$_n$ crystallizes in stacks of nearly eclipsed macrocycles connected by linear Ga–F–Ga bridges, with an interring distance of 387 pm [140,141,144]. All known polymeric fluorophthalocyaninatometal compounds [PcMF]$_n$ (M = Al, Ga, Cr) can be doped with iodine to yield the partially oxidized [(PcMF)I$_y$]$_n$, with a y between 0.012 and 3.3, depending on the central metal. These doped systems contain I$_3^-$ or I$_5^-$ as counterions [140,141].

The conductivities of the [(PcMF)I$_y$]$_n$ compounds (M = Al, Ga, Cr) increase with increasing iodine content. The highest conductivity is observed for [(PcAlF)I$_{3.3}$]$_n$ with a room temperature conductivity of 5 S/cm and an activation energy of 0.017 eV. This complex is prepared from sublimed [PcAlF]$_n$. Due to the increasing interring distances when gallium is used instead of aluminum, the π-orbital overlap between the cofacially arranged macrocycles decreases and therefore the conductivity decreases [140,141].

In general the fluoro-bridged [(PcMF)I$_y$]$_n$ complexes (M = Al, Ga, Cr) show lower thermal stability, with regard to the loss of iodine, compared to [(PcSiO)I$_y$]$_n$. [PcMF]$_n$ (M = Al, Ga) can also be oxidized with nitrosyl salts, for instance with NO$^+$ BF$_4^-$ to give [(PcMF)(BF4)$_y$]$_n$. The conductivities are ~0.3 S/cm [143].

Besides the phthalocyaninatometal complexes [PcMO]$_n$, [PcGeS]$_n$, and [PcMF]$_n$, complexes with covalently linked alkynyl ligands have also been prepared, e.g., μ-ethynylphthalocyaninatosilicon [PcSi(C≡C)]$_n$, which is obtained by treating PcSiCl$_2$ with bisbromomagnesiumacetylene as Grignard reagent. [PcSi(C≡C)]$_n$ is an insulator with a powder conductivity of less than 10^{-12} S/cm. When doped under the conditions described above, the polymer decomposes [145–149].

A more detailed description of the electrical properties of the polymetalloxanes [PcMO]$_n$ (M = Si, Ge, Sn), the bridged fluorophthalocyaninato metal complexes [PcMF]$_n$ (M = Al, Ga, Cr), and related compounds is given in the literature [18].

C. Bridged Macrocyclic Transition Metal Complexes

1. Bridged Transition Metal Complexes with Bidentate Ligands

A stacked arrangement of phthalocyaninato and naphthalocyaninato transition metal compounds that leads to coordination polymers where the macrocycle, the central metal atom, and the bridging ligand can be varied systematically was developed a few years ago by our group [19]. The stacking is achieved by bisaxially connecting the central transition metal atoms of the macrocycles with bidentate bridging ligands (L) (Fig. 15.12). We have synthesized and investigated such bridged macrocyclic metal compounds [MacM(L)]$_n$ in detail with respect to their physical properties [1,18,19].

The bridging ligands (L) are linear π-electron-containing organic molecules, e.g., pyrazine (pyz), p-diisocyanobenzene (dib) and substituted p-diisocyanobenzenes, tetrazine (tz), substituted tetrazines, and others (see below). If the oxidation state of the central metal atom is +3 (e.g., Co^{3+}, Fe^{3+}), charged bridging ligands such as cyanide (CN$^-$), thiocyanate (SCN$^-$), and others can also be used. As macrocycles, phthalocyanines, substituted phthalocyanines, 1,2- and 2,3-naphthalocyanines, phenanthrenocyanines, tetrabenzoporphyrines, and tetranaphthoporphyrins have been used [1,18,19]. Diisocyanobenzene as bridging ligand in [MacM(L)]$_n$ complexes leads to a larger interring distance of about 1190 pm compared to the pyrazine bridged polymer with a distance of about 680 pm.

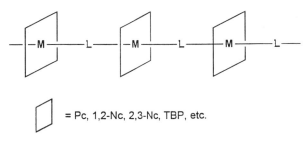

= Pc, 1,2-Nc, 2,3-Nc, TBP, etc.

M = transition metal (+II: Fe, Ru, Os; +III: Co, Rh)

L = pyz, tz, bpy, dib, me$_2$dib, CN$^-$, SCN$^-$

Fig. 15.12 Construction of bridged macrocyclic metal complexes.

Many of the bridged macrocyclic metal complexes [MacM(L)]$_n$ ("shishkebab" polymers) can be prepared in high yields and high purity by treating the metallomacrocyclic MacM with the pure ligand or with the ligand in an appropriate solvent such as acetone or chlorobenzene. For example, [PcFe(pyz)]$_n$ was prepared as a dark violet solid from PcFe and pyrazine in benzene or directly by reacting PcFe in a melt of pyrazine [150,151]. [PcFe(dib)]$_n$ was formed quantitatively by the reaction of PcFe with 1,4-diisocyanbenzene in acetone under reflux. Powder diffraction data indicate a high crystallinity for many of the coordination polymers obtained in this manner [150,151].

The bridged structure shown in Fig. 15.12 has been proven with many compounds using a variety of physical methods [1,18,19] including thermogravimetry, detailed IR investigations [152], Mössbauer spectroscopy [30], ^{1}H and ^{13}C NMR spectroscopy [153] for soluble systems (see below), and scanning tunnel microscopy (STM) [154]. The crystal structure of a very similar compound [DMGCo(pyz)]$_n$ (DMG = dimethylglyoximato) shows that the pyrazine molecules within the chain are all arranged within a plane perpendicular to the plane of the planar DMGCo units [155]. The mechanism of the formation for this type of coordination polymers has also been investigated using spectroscopic methods [156].

The powder conductivities of most of the bridged phthalocyaninato transition metal complexes [PcM(L)]$_n$ for M = Fe, Ru, Os, Co, Rh, for example, and L = pyz, dib, etc. are low, in the range of 10^{-6}–10^{-7} S/cm. However, they can be doped either chemically or electrochemically. Table 15.2 shows a list of compounds that have been doped chemically (with iodine) or electrochemically (with BF$_4^-$, PF$_6^-$, HSO$_4^-$, ClO$_4^-$), leading to good semiconducting properties ($\sigma_{RT} = 10^{-5}$–10^{-1} S/cm) with thermal stabilities up to 120–130°C. The chemically doped compounds [PcM(L)I$_y$]$_n$ are obtained by heterogeneous doping either in benzene or in CHCl$_3$.

Table 15.2 Room Temperature Powder Conductivities of Undoped and Doped [MacML]$_n$

Compound	σ_{RT} (S/cm)
Chemically doped compounds	
[PcFe(pyz)]$_n$	1×10^{-6}
[PcFe(pyz)I$_{2.5}$]$_n$	2×10^{-1}
[PcFe(dib)]$_n$	2×10^{-5}
[PcFe(dib)I$_{1.4}$]$_n$	7×10^{-3}
[PcFe(dib)I$_{3.0}$]$_n$	3×10^{-2}
[PcFe(Me$_4$dib)]$_n$	1×10^{-7}
[PcFe(Me$_4$dib)I$_{1.5}$]$_n$	1×10^{-3}
[PcFe(Me$_4$dib)I$_{3.0}$]$_n$	2×10^{-2}
[PcRu(pyz)]$_n$	1×10^{-7}
[PcRu(pyz)I$_{2.0}$]$_n$	2×10^{-2}
[PcRu(dib)]$_n$	2×10^{-6}
[PcRu(dib)I$_{1.0}$]$_n$	1×10^{-3}
[PcRu(dib)I$_{2.0}$]$_n$	7×10^{-3}
Electrochemically doped compounds	
[PcFe(pyz)]$_n$	1×10^{-6}
[PcFe(pyz)(BF$_4$)$_{0.45}$]$_n$	5×10^{-2}
[PcFe(pyz)(PF$_6$)$_{0.5}$]$_n$	4×10^{-2}
[PcFe(pyz)(HSO$_4$)$_{0.4}$]$_n$	1×10^{-5}
[PcFe(pyz)(ClO$_4$)$_{0.3}$]$_n$	3×10^{-3}

Source: Ref. 19.

[PcFe(pyz)]$_n$ can also be doped electrochemically [19]. With X = BF$_4^-$, PF$_6^-$, HSO$_4^-$, ClO$_4^-$, stable compounds [PcM(L)X$_y$]$_n$ are obtained. The composition of the doped polymers was established by elemental analysis and TG/DTA [18]. The results of ^{57}Fe Mössbauer spectroscopy prove that doping does not destroy the bridged structure of the polymers. Isomer shifts and quadrupole splitting of chemically and electrochemically doped compounds show values nearly identical to those found for [PcFe(pyz)]$_n$ [30]. Therefore the oxidation does not take place at orbitals centered at the metal atoms. As can be seen from Table 15.2, the conductivities can be increased to 10^{-3}–10^{-1} S/cm via doping depending on the iodine content. Despite the fact that the interring distance in the dib oligomers is larger, their conductivities are in the same range ($\sim 10^{-5}$ S/cm) as those of the pyz oligomers. Peripheral substitution of the phthalocyanine ring by electron-donating substituents has little effect on the conductivity. Only substitution of the phthalocyanine with chlorine leads to a decreased conductivity of 3×10^{-11} S/cm for [Cl$_{16}$PcFe(dib)]$_n$ [157].

In general, the complexes [MacM(L)]$_n$ are practically insoluble in organic solvents. However, soluble oligomers [R$_4$PcM(L)]$_n$ and [R$_8$PcM(L)]$_n$ have been prepared using substituted metallomacrocycles R$_4$PcM in which

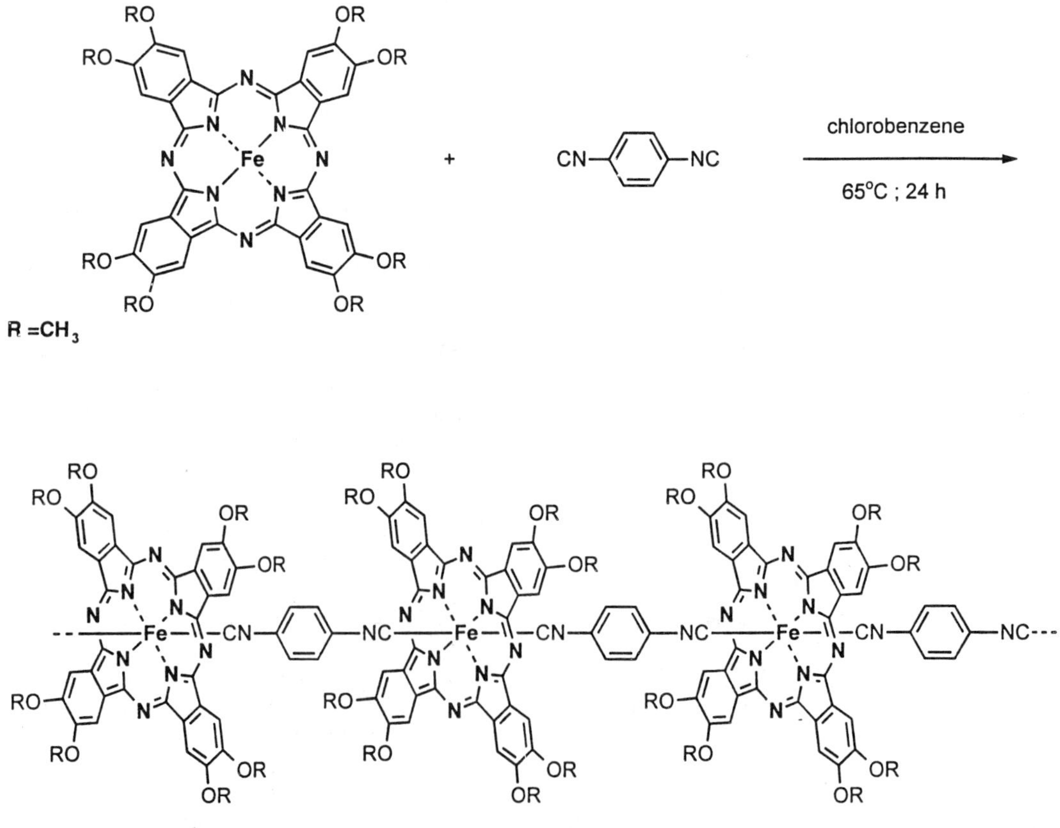

Fig. 15.13 Synthesis of substituted phthalocyanine polymers.

Conducting Bridged Macrocyclic Metal Complexes

Table 15.3 Room Temperature Powder Conductivities of Undoped and Doped Soluble Substituted Bridged Complexes [R$_4$MacML]$_n$ and [R$_8$MacML]$_n$

Compound	σ_{RT} (S/cm)
[t-Bu$_4$PcRu(dib)]$_n$	2×10^{-7}
[t-Bu$_4$PcRu(dib)I$_{1.2}$]$_n$	1×10^{-4}
[t-Bu$_4$PcRu(Me$_4$dib)]$_n$	1×10^{-11}
[t-Bu$_4$PcRu(Me$_4$dib)I$_{1.3}$]$_n$	1×10^{-4}
[(EHO)$_4$PcFe(pyz)]$_n$	$< 10^{-12}$
[(EHO)$_4$PcFe(pyz)I$_{1.3}$]$_n$	6×10^{-5}
[(n-C$_5$H$_{11}$)$_8$PcFe(Me$_4$dib)]$_n$	$< 10^{-12}$
[(n-C$_5$H$_{11}$)$_8$PcFe(Me$_4$dib)I$_{1.8}$]$_n$	9×10^{-7}
[(EHO)$_4$PcFe(Me$_4$dib)]$_n$[a]	$< 10^{-12}$
[(EHO)$_4$PcFe(Me$_4$dib)I$_{2.5}$]$_n$[a]	2×10^{-10}

[a] Substituents in the 1,4-position.
Source: Ref. 19.

R = t-Bu, Et, OR' (R' = C$_5$H$_{11}$–C$_{12}$H$_{25}$) and R$_8$PcM in which R = C$_5$H$_{11}$–C$_{12}$H$_{25}$, OC$_5$H$_{11}$–OC$_{12}$H$_{25}$ and M = Fe or Ru [53,73,158–163]. These types of oligomers are soluble in most common organic solvents, thereby allowing the determination of the chain lengths by the usual methods, e.g., ^{1}H NMR spectroscopy [164,165]. Depending on the preparation method, oligomers that contain 20–50 MacM subunits can be obtained. Figure 15.13 shows one of the routes to synthesize soluble bridged coordination polymers [163].

Doping of the soluble oligomers with iodine also leads to semiconductive systems as can be seen from Table 15.3. Similar to the peripherally unsubstituted systems, the composition of the doped polymers was determined by elemental analysis [166]. Spectroscopic investigations show that the polymers were not destroyed by the doping process. The conductivities of the doped polymers are in the range of 10^{-7}–10^{-4} S/cm, except for that of the 1,4-tetrasubstituted [(EHO)$_4$PcFe(Me$_4$dib)I$_{2.5}$]$_n$. From the few data obtained so far, it can be seen (cf. Table 15.3) that the conductivities of the peripherally substituted doped chain compounds are somewhat lower than those of the unsubstituted systems. Furthermore, the size and location of the substituents at the macrocycle seem to play a role that affects the conductivities.

The doping process has been studied very carefully using not only different macrocycles and central metal atoms but also different bridging ligands with comparatively low oxidation potentials. The question was studied whether or not the doping process is possible not only at the macrocycle leading to a radical cation, but also at the bridging ligand. If, e.g., in the oligomer [PcFe(L)]$_n$ a ligand with a comparatively low oxidation potential such as 1,4-diisocyanoanthracene (dia) is used, doping with iodine indeed leads to an oxidation of the bridging ligand, thereby increasing the powder conductivity from 3×10^{-7} S/cm in [PcFe(dia)]$_n$ to 8×10^{-3} S/cm in [PcFe(dia)I$_{2.6}$]$_n$ [167].

Another doping process of a bridging ligand is shown in Fig. 15.14: Phthalocyaninatoruthenium (PcRu) reacts with p-phenylenediamine (ppd) to form the corresponding oligomer [PcRu(ppd)]$_n$. Doping of this oligomer is

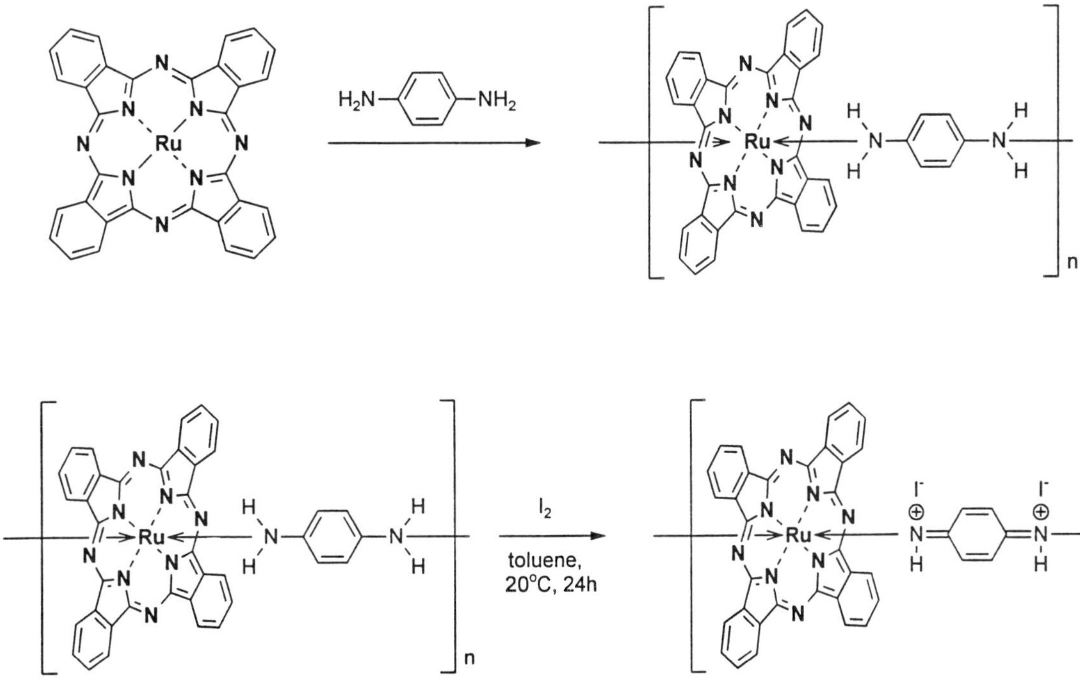

Fig. 15.14 Preparation of [PcRu(ppd)]$_n$.

possible under the conditions given in Fig. 15.14. As in the case of polyaniline [168], quinoid structures are formed in the bridging ligand and an increase in conductivity of five orders of magnitude in comparison with undoped [PcRu(ppd)]$_n$ is observed [169] (cf. Section III.C).

Variation of the macrocycle also changes the properties of the corresponding bridged transition metal polymers. Theoretical calculations of the electronic properties of annulated phthalocyanines show that linear annulation of benzene rings in phthalocyanines produces a continuous destabilization of the HOMO level and a narrowing of the HOMO–LUMO energy gap [170]. One-dimensional stacks of linear annulated phthalocyanines are calculated to have lower oxidation potentials and lower VB–CB (VB = valence band, CB = conduction band) energy gaps than angular annulated systems [171]. These theoretical results are confirmed by our studies of the redox potentials of PcFe, 1,2-NcFe, 2,3-NcFe, TBPFe, and 2,3-TNFe (Table 15.4) as well as the powder conductivities of their corresponding 1,4-diisocyanobenzene bridged polymers (Table 15.5).

Phthalocyaninatoiron (PcFe) possesses almost the same oxidation potentials ($E_{1/2}^1$ and $E_{1/2}^2$) as the unsymmetrical 1,2-naphthalocyaninatoiron, 1,2-NcFe. 2,3-Naphthalocyaninatoiron, however, exhibits lower oxidation potentials, which can also be observed with TBPFe and 2,3-TNPFe (see Table 15.4). Due to the low oxidation potentials of the 2,3-Nc and 2,3-TNP macrocycles, the corresponding dib-bridged polymers are already doped by air oxygen. This results in higher powder conductivities of these compounds compared to those of the "undoped" phthalocyanine, 1,2-naphthalocyanine, or tetrabenzporphyrin systems. The oxidation potentials given in Table 15.4 can be directly related to the powder conductivities of the corresponding bridged systems listed in Table 15.5.

2. "Intrinsic" Semiconductive Polymers

PcFe, PcRu, PcOs, and 2,3-NcFe react easily with tz (tetrazine) and me$_2$tz (2,5-dimethyltetrazine) to form the corresponding monomers MacM(L)$_2$ (L = tz, me$_2$tz)

Table 15.5 Room Temperature Powder Conductivities of Chain Compounds [MacM(dib)]$_n$

Compound	σ_{RT} (S/cm)
[PcFe(dib)]$_n$	2×10^{-5}
[1,2-NcFe(dib)]$_n$	6×10^{-10}
[2,3-NcFe(dib)]$_n$	2×10^{-3}
[TBPFe(dib)]$_n$	2×10^{-6}
[2,3-TNPFe(dib)$_-$]$_n$	5×10^{-3}

Source: Ref. 19.

and under somewhat different conditions to form the bridged systems [MacM(L)]$_n$ (L = tz, me$_2$tz) [172]. The tetrazine bridged macrocycles, in contrast to other bridged compounds [MacM(L)]$_n$ (M = Fe, Ru, or Os and L = pyz or dib) show good semiconducting properties without external oxidative doping (σ_{RT} = 0.05–0.3 S/cm).

The powder conductivities of a selection of monomeric and bridged macrocycles (mostly phthalocyanines) in the nondoped state are listed in Table 15.6. All the bridged complexes [MacM(L)]$_n$ (M = Fe, Ru, Os; L = pyz, tz, dabco) consist of cofacially arranged macrocycles that are separated by approximately the same distance (about 600 pm) (see Fig. 15.12).

Systematic investigations of the influence of the bridging ligands on the semiconducting properties in [MacM(L)]$_n$ reveal that changing L from diazabicyclo[2.2.2]octane (dabco) to pyrazine (pyz) to s-tetrazine (tz) leads to a steady increase of the semiconducting properties without external oxidative doping. Powder conductivities on the order of 0.1 S/cm can be reached by using s-tetrazine as bridging ligand [172]. Substituted tetrazines (Me$_2$tz, Cl$_2$tz), p-diaminotetrazine, triazine, and others were also used for the formation of complexes of this type [173].

While the monomeric complexes PcM(L)$_2$ (L = pyz, dabco, tz; M = Fe, Ru, Os) show insulating behavior (Table 15.6), it can be seen that the ligand L has a significant effect on the conductivity of the bridged complexes

Table 15.4 Redox Potentials (V vs. SCE) of Some Iron Macrocyclics in Pyridine/Bu$_4$NClO$_4$

	$E_{1/2}^1$	$E_{1/2}^2$	$E_{1/2}^3$	$E_{1/2}^4$	$E_{1/2}^5$
PcFe	1.10	0.69	−1.085	−1.39	−1.93
1,2-NcFe	1.01	0.68	−0.95	−1.21	−1.80
2,3-NcFe	0.81	0.43	−1.09	−1.32	−1.86
TBPFe	0.82	0.34	−0.90	−1.59	−1.87
2,3-TNPFe	0.61	0.15	−0.90	−1.49	−1.87
Assignment	Mac^{-2}/Mac^{-1}	Fe^{II}/Fe^{III}	Fe^I/Fe^{II}	Mac^{-3}/Mac^{-2}	Mac^{-4}/Mac^{-3}

Source: Ref. 19.

Table 15.6 Room Temperature Powder Conductivities of Monomeric and Bridged Macrocyclic Transition Metal Complexes[a]

Compound	σ_{RT} (S/cm)	Compound	σ_{RT} (S/cm)
PcFe(dabco)$_2$	1×10^{-10}	[2,3-NcFe(tz)]$_n$	3×10^{-1}
PcFe(pyz)$_2$	3×10^{-12}	Me$_8$PcFe(pyz)$_2$	3×10^{-9}
[PcFe(pyz)]$_n$	1×10^{-6}	[Me$_8$PcFe(pyz)]$_n$	9×10^{-6}
PcFe(tz)$_2$	$< 10^{-9}$	[Me$_8$PcFe(tz)]$_n$	1×10^{-2}
[PcFe(tz)]$_n$	2×10^{-2}	[(CN)$_4$PcFe(pyz)]$_n$	5×10^{-9}
[PcFe(Me$_2$tz)]$_n$	4×10^{-3}	[(CN)$_4$PcFe(tz)]$_n$	1×10^{-6}
[PcRu(pyz)]$_n$	1×10^{-7}	[t-Bu$_4$PcFe(pyz)]$_n$	5×10^{-11}
PcRu(tz)$_2$	$< 10^{-11}$	[t-Bu$_4$PcFe(tz)]$_n$	9×10^{-9}
[PcRu(tz)]$_n$	1×10^{-2}	[Et$_4$PcFe(pyz)]$_n$	8×10^{-9}
[PcRu(tri)]$_n$	2×10^{-4}	[Et$_4$PcFe(tz)]$_n$	2×10^{-4}
[PcRu(NH$_2$)$_2$tz)]$_n$	4×10^{-3}	[Et$_4$PcFe(tri)]$_n$	5×10^{-9}
[PcRu(Cl$_2$tz)]$_n$	3×10^{-3}	[(EHO)$_4$PcFe(pyz)]$_n$	$< 10^{-12}$
[PcRu(Me$_2$tz)]$_n$	4×10^{-3}	[(EHO)$_4$PcFe(tz)]$_n$	3×10^{-6}
[PcRu(p-(NH$_2$)$_2$C$_6$H$_4$)]$_n$	5×10^{-9}	[Et$_4$PcRu(pyz)]$_n$	5×10^{-10}
[PcRu(CN)$_2$C$_6$F$_4$)]$_n$	1×10^{-3}	[Et$_4$PcRu(tri)]$_n$	2×10^{-8}
[PcOs(pyz)]$_n$	1×10^{-6}	[Et$_4$PcRu(tz)]$_n$	5×10^{-9}
PcOs(tz)$_2$	4×10^{-8}	[t-Bu$_4$PcRu(pyz)]$_n$	7×10^{-8}
[PcOs(tz)]$_n$	1×10^{-2}	[t-Bu$_4$PcRu(tz)]$_n$	1×10^{-6}
[2,3-NcFe(pyz)]$_n$	5×10^{-5}		

[a] Pressed pellets, 10^8 Pa.
Source: Ref. 19.

[MacM(L)]$_n$. As dabco is a ligand containing no π orbitals to interact with the metallomacrocycle, the complex [PcFe(dabco)]$_n$ is an insulator. An increase in conductivity is observed for the pyrazine-bridged compounds [MacM(pyz)]$_n$, which exhibit conductivities in the low semiconducting region. However, by changing the bridging ligand from pyrazine to s-tetrazine, the conductivity is increased by three to five orders of magnitude without external oxidative doping [174].

One of the factors responsible for the electrical conductivities in bridged macrocyclic transition metal complexes [MacM(L)]$_n$ is the band gap, which has been shown by theoretical calculations to be determined by the energy difference between the LUMO of the bridging ligand and the HOMO of the transition metallomacrocycle [87,175]. Therefore, to achieve semiconducting properties, the metallomacrocycle should contain a high-lying HOMO, whereas the bridging ligand should have a low-lying LUMO. The ligands used for the preparation of the bridged macrocyclic transition metal complexes listed in Table 15.6 have been selected under consideration of possessing a low-lying LUMO. In addition to s-tetrazine [172,174] and its derivatives, some other interesting attempts were also made to achieve intrinsic conductivity in such systems. The 3,6-diamino-1,2,4,5-tetrazine bridged polymer [PcRu(NH$_2$)$_2$tz] shows a powder conductivity that is about six orders of magnitude higher than that of [PcRu(p-(NH$_2$)$_2$C$_6$H$_4$)]$_n$ [173], due to the fact that p-phenylenediamine contains no heteroatoms in the aromatic ring (Table 15.6) [169]. Also, soluble (peripherally substituted) tetrazine-bridged phthalocyaninato transition metal complexes have been synthesized using a variety of tetraalkyl- and alkoxy-substituted phthalocyaninatoiron and -ruthenium complexes. These tetrazine-bridged systems again show higher conductivities than the corresponding pyrazine analogs, but to a lesser extent than the peripherally unsubstituted ones (Table 15.6).

Triazine has also been used for the preparation of bridged phthalocyaninatoruthenium complexes [173]. The corresponding bridged coordination polymers also show good semiconducting properties. In addition to nitrogen-containing heterocycles, other ligands can be used for the preparation of intrinsically conductive phthalocyaninato transition metal polymers: Tetrafluoroterephthalic acid dinitrile (CN)$_2$C$_6$F$_4$ reacts with PcRu to give the bridged compound [PcRu(CN)$_2$C$_6$F$_4$]$_n$, which exhibits a powder conductivity of $\sigma_{RT} = 10^{-3}$ S/cm without external oxidative doping. Fumarodinitrile (NC–CH=CH–CN), dicyanoacetylene, and even cyanogen have been used to prepare the corresponding bridged phthalocyaninatoruthenium complexes. All compounds show intrinsic conductivities of about $\sigma_{RT} = 10^{-3}$ S/cm [173,176; J. Pohmer and M. Hanack, unpublished].

Electron-withdrawing substituents in the peripheral positions of the phthalocyanine macrocycle show the expected effect; for example, [(CN)$_4$PcFe(tz)]$_n$ exhibits a conductivity that is at least three orders of magnitude less than the conductivities of other tetrazine-bridged compounds investigated so far [177] (Table 15.6).

The low band gaps of all tetrazine-bridged coordination polymers with group VIII transition metals [MacM(L)]$_n$ [Mac = Pc, and M = Fe, Ru, Os or Mac = 2,3-Nc and M = Fe; L = e.g., tz, Me$_2$tz, (NH$_2$)$_2$tz] can be demonstrated by physical properties that are not shown by the corresponding systems [MacM(L)]$_n$ with L = pyz or dib. All polymers with group VIII transition metals containing tetrazine or related compounds as bridging ligands [MacM(L)]$_n$ [L = tz, me$_2$tz, tri, (CN)$_2$C$_6$F$_4$, NC–CH=CH–CN, p-(NH$_2$)$_2$tz] show broad bands in the UV-vis-near-IR spectra between 1250 and 2500 nm with different maxima, e.g., for [PcFe(tz)]$_n$ at 1650 nm (0.75 eV) and for [PcRu(tz)]$_n$ at about 1300 nm (0.95 eV). The corresponding pyrazine-bridged systems [PcM(pyz)]$_n$ (M = Fe, Ru, Os) exhibit "normal" UV-vis spectra with Soret and Q bands between 245 and 700 nm. [PcRu((p-NH$_2$)$_2$tz)]$_n$ shows a broad band in the UV-vis-near-IR region between 1000 and 2000 nm with two maxima at 1525 (0.81 eV) and 1340 nm (0.93 eV). The corresponding p-phenylenediamine-bridged system, [PcRu((p-NH$_2$)$_2$C$_6$H$_4$)]$_n$, shows a "normal" UV-vis spectrum.

Related systems like the 2,3-naphthalocyanines have attracted much attention because of their potential use as semiconducting materials, in nonlinear optics, as laser dyes, and in photodynamic therapy [19]. Owing to intermolecular interactions between the macrocycles, peripherally unsubstituted metallo-2,3-naphthalocyanines are practically insoluble in common organic solvents such as chloroform or toluene. Solubility, however, is necessary for many applications. Although it has been shown that soluble compounds are formed by inserting side chains in the periphery of phthalocyaninato metal complexes, very little is known about the synthesis and properties of peripherally substituted 2,3-naphthalocyanines. The first synthesis of pure tert-butyl-substituted phthalocyaninato- and 2,3-naphthalocyaninatoruthenium [(t-Bu)$_4$-MacRu] by thermal decomposition of (t-Bu)$_4$-MacRuL$_2$ (L = 3-chloropyridine, ammonia) has been recently described [178]. (t-Bu)$_4$PcRu is highly soluble in common organic solvents, whereas the influence of the tert-butyl group in (t-Bu)$_4$-2,3-NcRu is insufficient to obtain noticeable solubility.

The monomeric complexes (t-Bu)$_4$PcRu(3-Clpy)$_2$ and (t-Bu)$_4$-2,3-NcRu(3-Clpy)$_2$ were prepared by reaction of stoichiometric amounts of 4-tert-butylphthalonitrile and 6-tert-butyl-2,3-dicyanonaphthalene [178], respectively, with RuCl$_3$·3H$_2$O in 2-ethoxyethanol in the presence of an excess of 3-chloropyridine and small amounts of DBU. The bisammonia complex (t-Bu)$_4$PcRu(NH$_3$)$_2$ was obtained by refluxing a mixture of stoichiometric amounts of 5-tert-butyl-1,3-dihydro-1,3-diiminoisoindole and RuCl$_3$·3H$_2$O in 2-ethoxyethanol that was saturated with NH$_3$.

For (t-Bu)$_4$PcRu and (t-Bu)$_4$-2,3-NcRu an electrical conductivity of 3.4×10^{-7} and 1.2×10^{-5} S/cm, respectively, was measured [176,178]. The conductivities are significantly higher than those observed for monomeric and generally nonconducting metallophthalocyanines and metallonaphthalocyanines. The high conductivities found for (t-Bu)$_4$PcRu, (t-Bu)$_4$-2,3-NcRu, and also for PcRu ($\sigma_{RT} = 2.0 \times 10^{-5}$ S/cm [179]) and 2,3-NcRu ($\sigma_{RT} = 2.0 \times 10^{-1}$ S/cm [176]) may be caused by the dimeric structure in the solid state [54].

The magnetic and spectroscopic properties found for (t-Bu)$_4$PcRu are very similar to those observed for PcRu. These results led us to the conclusion that (t-Bu)$_4$PcRu, like PcRu, forms a dimeric structure in the solid state. The magnetic and spectroscopic behavior of 2,3-NcRu [176] also points to a dimeric structure. Hence we assume that (t-Bu)$_4$-2,3-NcRu also forms a dimer.

3. Bridged Macrocyclic Complexes with Trivalent Transition Metal Ions

As described above, the central transition metals within the macrocycles may also have the oxidation number +3. An octahedral configuration of the metal in the macrocycles is also possible, for instance, for Fe^{3+}, Co^{3+}, and Rh^{3+}. For the formation of the corresponding polymers, CN$^-$, SCN$^-$, and N$_3^-$ are suitable bridging ligands [180–183].

Figure 15.15 schematically shows a cyano-bridged intrinsic semiconductive phthalocyanine complex that can be synthesized by the displacement of the axial anion X$^-$ by CN$^-$ in a coordinatively unsaturated compound PcMX. This synthesis has been used for the preparation of [PcMn(CN)]$_n$ [184], [PcFe(CN)]$_n$ [185], and [2,3-NcFe(CN)]$_n$ [186]. The starting materials were synthesized by oxidative chlorination of the appropriate phthalocyanines with thionyl chloride or oxygen and concentrated aqueous hydrochloric acid. The chlorides PcMCl (M = Fe, Mn) were converted into the bridged complexes [PcM(CN)]$_n$ in aqueous or ethanolic alkali metal cyanide solutions.

The appropriate precursor of the type PcMCl for this reaction path is not known in the case of cobalt phthalocyanine. The earlier reported PcCoCl [187] is indeed an impure PcCoCl$_2$. In PcCoCl$_2$ the oxidation state of the macrocycle is assumed to be −1 and that of the metal +3. But although it does not have the right stoichiometry, pure PcCoCl$_2$ can also be converted into the polymer by treatment with an aqueous alkali metal cyanide solution.

Another route also leading to cyano-bridged polymers [PcM(CN)]$_n$ is the elimination of alkali metal cyanide from dicyanophthalocyaninato transition metal(III) complexes M'[PcM(CN)$_2$] (M' = Na, K; M = Co, Rh, Fe, Mn, Cr). This material can be easily obtained either by reaction of the chloro compounds PcMCl$_2$ (M = Co, Cr) with an excess of alkali metal cyanide in ethanol or

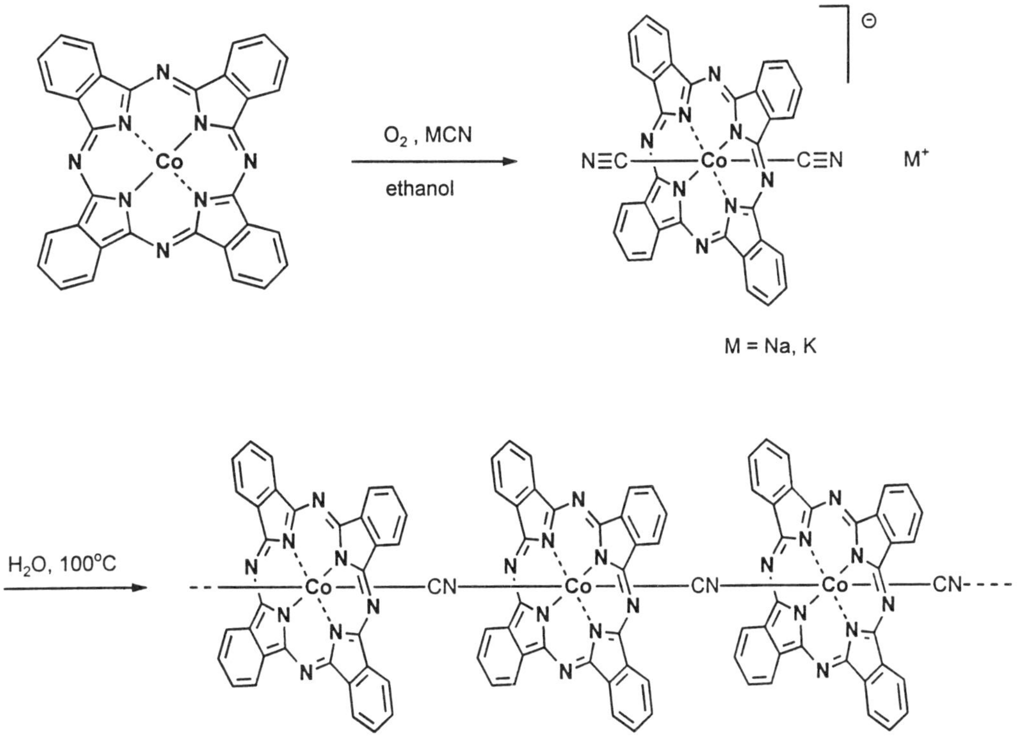

Fig. 15.15 Synthesis of [PcCoCN]$_n$.

by oxidation of PcM (M = Co, Fe, Mn) in the presence of cyanide with atmospheric oxygen. From the formed mononuclear alkali metal dicyanophthalocyaninato transition metal(III), alkali metal cyanide is split off by treatment with boiling water to give the polymer [PcM(CN)]$_n$ in nearly quantitative yield.

The alkali metal dicyano(phthalocyaninato)cobalt-(III) complexes M′[PcM(CN)$_2$] were characterized by infrared spectroscopy and in the case of peripherally substituted phthalocyanines also by UV-vis and ^{1}H NMR spectroscopy. The ^{1}H NMR spectra of the complexes recorded in acetone-d_6 show the expected number and intensities of signals that confirm the proposed structures. The fact that ^{1}H NMR spectra are obtained supports the proposed oxidation state +3 and the octahedral ligand field of the Co atom. Otherwise a paramagnetic substance would result.

The UV-vis spectra of the complexes M′[PcM(CN)$_2$] recorded in DMF show a bathochromic shift of the Q band in comparison to the corresponding PcCo due to the increase of the oxidation state of the cobalt ion from +2 to +3 (see Table 15.7).

Other metals (e.g., Fe, Rh) and macrocycles (e.g., TBP, 1,2-Nc, 2,3-Nc) were used analogously [188,189]. In the case of peripherally substituted cobalt phthalocyanines, soluble [R$_x$PcCo(CN)]$_n$ (R = t-Bu for x = 4 and R = C$_7$H$_{15}$ for x = 8) polymers have also been prepared in this way [189]. The crystal structure of K[PcCo(CN)$_2$]$_2$ that was obtained by oxidative electrocrystallization of K[PcCo(CN)$_2$] is known [190,191], and the bisaxial coordination of the CN groups was proved.

Soluble polymers can also be prepared in one step starting with peripherally substituted cobalt phthalocyanine by reaction with cyanogen (CN)$_2$ in CHCl$_3$ at room temperature [183]. Cyanogen is used as the oxidant, and

Table 15.7 UV-Visible Data of Cyanophthalocyanine Compounds

Compound	λ_{max} (nm)
(t-Bu)$_4$PcCo	664
Na[(t-Bu)$_4$PcCo(CN)$_2$]	676
[(t-Bu)$_4$PcCo(CN)]$_n$	675
PcCo	658
Na[PcCo(CN)$_2$]	670
[PcCo(CN)]$_n$	668
(t-Bu)$_4$–2,3-NcCo	748
K[(t-Bu)$_4$–2,3-NcCo(CN)$_2$]	765
[(t-Bu)$_4$–2,3-NcCo(CN)]$_n$	767
2,3-NcCo	—[a]
K[2,3-NcCo(CN)$_2$]	761
[2,3-NcCo(CN)]$_n$	762

[a] Insoluble in DMF.
Source: Ref. 19.

the cyanide (CN$^-$) formed coordinates the Co^{3+} ion. This reaction takes place only if the precursor cobalt phthalocyanine is soluble in CHCl$_3$. Peripherally unsubstituted PcCo reacts only if a coordinating solvent like pyridine is used. But in this case no polymer but the complex PcCo(CN)(py) is formed.

The polymers [PcM(CN)]$_n$ were characterized by IR and UV-vis spectroscopy, by elemental analyses, and, in the case of peripherally substituted soluble compounds, by ^{1}H NMR spectroscopy. The ν_{CN} valence frequency of the unsubstituted [PcM(CN)]$_n$ is shifted 15–30 cm^{-1} to higher energy. This increase is evidence for the presence of a cyano bridge in the polymer. Treating the polymer with a competing ligand such as pyridine or piperidine destroys the polymeric structure of [PcCo(CN)]$_n$. Monomeric complexes with the composition PcCo(L)(CN) are formed by this reaction [180].

The peripherally substituted polymers could also be characterized by their solubility in common organic solvents. In solution the polymer structure of the compounds is destroyed and they are converted into monomers. This is the reason no ^{1}H NMR spectrum was obtained in CDCl$_3$ because of the paramagnetism of the samples. ^{1}H NMR spectra of soluble polymers could be recorded in pyridine-d_5. But it must be considered that in this solvent not the polymers but monomeric complexes of the type MacCo(CN)(py-d_5) are measured.

Although the main emphasis of spectroscopic and chemical characterization of [PcM(CN)]$_n$ has been directed toward the cobalt derivative, the results can be generalized. The compounds with other transition metals are found by X-ray diffraction studies to be isostructural to [PcCo(CN)]$_n$. The electrical conductivities without doping of the complexes are given in Table 15.8. When the polymeric structure was destroyed by treatment with a competing ligand to form PcCo(L)(CN), the conductivity was diminished by 6–10 orders of magnitude.

On comparing the conductivities of [PcCo(CN)]$_n$ and [(t-Bu)$_4$PcCo(CN)]$_n$, a decrease of 10^5 is observed. Generally, the conductivities of axial bridged substituted phthalocyanines are lower than those of the unsubstituted systems; steric hindrance of the bulky or long-chain groups in the peripheral positions of the macrocycles prevents the intercalation between the polymeric chains that is necessary for the charge transport by the hopping mechanism. In (t-Bu)$_4$-2,3-Nc the influence of the bulky *tert*-butyl groups seems to be lower. The *tert*-butyl substituent is relatively small compared with the large heteroaromatic π system of the naphthalocyanine ring, hence leading to good semiconducting properties of [(t-Bu)$_4$-2,3-NcCo(CN)]$_n$ without losing the advantage of high solubility in common organic solvents (Table 15.8).

IV. PHOTOCONDUCTIVITY

Photoconductivity is a typical phenomenon for many organic semiconductors. When a photoconductor is illuminated by light that it is able to absorb, charge carriers are produced and consequently the conductivity of the material increases [193]. Several processes are responsible for the magnitude of a photocurrent within an organic solid:

1. For the amount of charge carriers generated, the number of excited charge carriers per absorbed photon is decisive. The absorption of photons can cause electronic transitions from occupied states of the valence band either to the conduction band or to free defect levels (e.g., impurities, adsorbed gases) within the gap or alternatively from defect levels to the conduction band. Indirect excitation processes (e.g., formation of charge carriers from excitons) can also contribute to the photocurrent. An applied high external voltage may cause injection of electrons or holes from the electrodes into the sample and thus a current flow.

Table 15.8 Room Temperature Electrical Conductivity Data for Mono- and Polynuclear Cyanometal Complexes

Compound[a]	Conductivity (S/cm)
M′[PcCo(CN)$_2$]	—
M′[PcRh(CN)$_2$]	3 × 10^{-9}
K[PcCr(CN)$_2$]	6 × 10^{-9}
K[PcMn(CN)$_2$]	5 × 10^{-6}
K$_2$[PcFe(CN)$_2$]	—
H[TBPCo(CN)$_2$]	—
[PcCo(CN)]$_n$	2 × 10^{-2}
[1,4,8,11,15,18,22,25-(C$_7$H$_{15}$)$_8$PcCo(CN)]$_n$	< 10^{-12}
[(t-Bu)$_4$PcCo(CN)]$_n$	9 × 10^{-8}
[2,3,9,10,16,17.23,24-(C$_7$H$_{15}$)$_8$PcCo(CN)]$_n$	3 × 10^{-9}
[(t-Bu)$_4$-2,3-NcCo(CN)]$_n$	8 × 10^{-2}
[PcFe(CN)]$_n$	6 × 10^{-3}
[2,3-NcFe(CN)]$_n$	1 × 10^{-3}
[PcCr(CN)]$_n$	3 × 10^{-6}
[PcMn(CN)]$_n$	1 × 10^{-5}
[HPcFe(CN)]$_n$	8 × 10^{-4}
[TBPCo(CN)]$_n$	4 × 10^{-2}
[2,3-TNPCo(CN)]$_n$	2 × 10^{-3}
[2,3-NcCo(CN)]$_n$	1 × 10^{-1}

[a] M′ = Na or K.
[b] Obtained with four-probe or two-probe technique.
Source: Ref. 19.

2. The transport of these charge carriers can be described by a band model or a hopping mechanism.
3. The photoelectric behavior of a solid is further determined by deactivation processes and thus the lifetime of the charge carriers. Deactivation can take place via direct recombination of electrons and holes. This process occurs especially when the charge carrier concentration is high and when it is accompanied by the release of a large number of photons. Electrons or holes can also be trapped by defects first. Then they recombine with the holes of the valence band or the electrons of the conduction band.

Photoelectrical properties of phthalocyanine powders and their films on a variety of substrates have been the subject of numerous investigations [193,194]. Photoconductivity is affected by the nature of the phthalocyanine substrate as well as by its crystal modification, doping agents employed, and by the way the substrate is prepared.

The conductivity changes of phthalocyaninatometal complexes most probably originate from the electronic influences of substituents as observed for phthalocyaninatozinc and -copper complexes.

The following sequence of photoelectrical sensitivity for the substituted macrocyclic was found: CN ≫ H > OCH$_3$ > CH$_3$. Despite the differences in photoelectrical sensitivity, electron-withdrawing and electron-donating substituents on the macrocyclic rings hardly change the value of the photocurrent measured against the wavenumber.

Unsubstituted PcCu shows a high sensitivity in the near-IR region compared to substituted phthalocyaninatocopper complexes. This is explained by a disarrangement of the skipped stacked arrangement characteristic for PcCu caused by substituents.

The oxo-bridged phthalocyaninatometal complexes

Table 15.9 Photosensitizable Mononuclear Phthalocyaninatometal Complexes of the Type PcM

Compound	Ref.
(CH$_3$O)$_8$PcZn	192
(CN)$_8$PcZn	193
(CH$_3$O)$_8$PcCu	192
PcCu	193
(CN)$_8$PcCu	193
PcTiO	195
(C$_7$H$_{15}$)$_8$PcTiO	195
(C$_5$H$_{11}$)$_8$PcTiO	195
(C$_4$H$_9$)$_8$PcTiO	195
(C$_3$H$_7$)$_8$PcTiO	195

Table 15.10 Photosensitizable Phthalocyaninatometal Complexes Polymerized via Bridging Ligands, [MacML]$_n$

Compound	Ref.
[PcGeO]$_n$	193
[(t-Bu)$_4$PcGeO]$_n$	193
[(Me$_3$Si)$_4$PcGeO]$_n$	193
[PcGeS]$_n$	193
[PcCrCN]$_n$	193
[PcMnCN]$_n$	193
[PcCoCN]$_n$	197
[PcFe(tz)]$_n$	193
[Me$_8$PcFe(dib)]$_n$	193
[Me$_8$PcFe(Cl$_4$dib)]$_n$	193
[Cl$_{16}$PcFe(pyz)]$_n$	193
[PcFe(bpy)]$_n$	193
[PcRu(dib)]$_n$	193
[PcRu(pyz)]$_n$	193
[HpFeO]$_n$	193
[taaFe(pyz)]$_n$	198, 199
[TPPFe(bpy)]$_n$	198, 199

maintain a stacked arrangement even if the compound is substituted by bulky groups. Thus the infrared absorption of the photoconduction is not diminished. Tables 15.9 and 15.10 show some photosensitizable oxygen-bridged phthlocyaninatometal complexes. In addition to the electronic influences of substituents on the macrocycle on the photoconductivity parameters, electronic influences of substituents on the bridging ligands have to be taken into account when bridged phthalocyaninato complexes are investigated. Bulky groups reduce the size of the photocurrent. Changing the bridging ligand from oxygen to sulfur leads to a reduction in photoconductivity. Unsubstituted μ-pyrazine phthalocyaninato-iron(II) [PcFe(pyz)]$_n$ can be compared with the corresponding perchloro-substituted complex. Photoconductivity is found to be two orders of magnitude larger for [PcFe(pyz)]$_n$ than for [Cl$_{16}$PcFe(pyz)]$_n$. Chloride substitution on the bridging ligand, diisocyanobenzene, also decreases the photoconductivity as is evident when comparing the corresponding values of [Me$_8$PcFe(dib)]$_n$ and [Me$_8$PcFe(Cl$_4$dib)]$_n$. The bridging ligand can also influence the photoresponse via its size. For this purpose, dib- and bpy-linked iron and ruthenium complexes were investigated, and a decrease in photoconductivity from tz to bpy was noticed. The oligomeric complexes of Fe and Ru with pyz and dib demonstrate the influence of the central metal atom. The iron oligomers were found to be the better photoconductors. The photoelectrical sensitivity of cyanobridged phthalocyaninato complexes decreases when cobalt is replaced with chromium and manganese: [PcCoCN]$_n$, [PcCrCN]$_n$, [PcMnCN]$_n$. The influence of the macrocyclic ligand on the photoconductivity was tested by comparing [taaFepyz]$_n$ and [TPPFebpy]$_n$ with the respective phthalocyaninato complexes [PcFepyz]$_n$ and [PcFebpy]$_n$. The taa complexes seem to be the most efficient and the TPP complexes the least efficient photoconductors. For several iron complexes, e.g., [PcFetz]$_n$, electrons were found to be the major charge carriers in the illuminated materials [196,197].

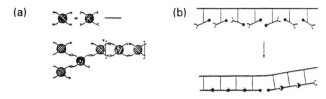

Fig. 15.16 Schematic representation of the synthesis of ladder polymers. See text for details.

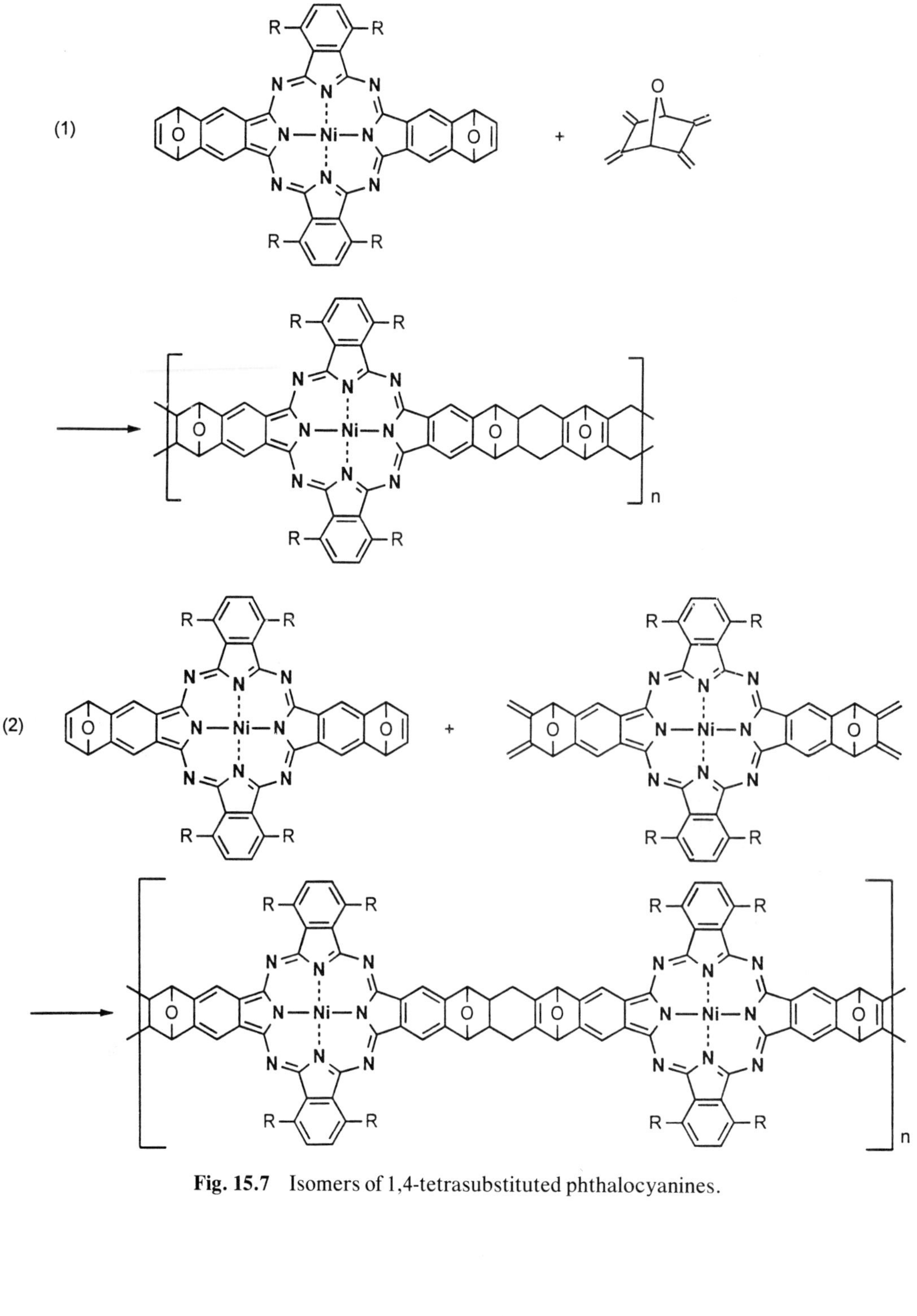

Fig. 15.7 Isomers of 1,4-tetrasubstituted phthalocyanines.

V. LADDER POLYMERS BASED ON PHTHALOCYANINES

Conjugated polymers with ladder structures have become more significant in the last few years with respect to preparative and theoretical aspects [198,199]. They combine the advantage of high stabilities with interesting electrical and optical properties. The most suitable method to synthesize these polymers and their precursor molecules (oligomers) is the repetitive Diels–Alder reaction [200]. Ladder polymers, which include phthalocyanine units, are difficult to synthesize because only soluble phthalocyanines with D_{2h} symmetry are suitable as precursors. Such substituted Pc's are difficult to obtain because they have to be separated from the mixture of isomers formed during the synthesis. A convenient macrocyclic metal complex for repetitive Diels–Alder reactions is the hemiporphyrazine (HpM) system (cf. Fig. 15.2). It is easier to synthesize hemiporphyrazine compounds, which always exhibit D_{2h} symmetry.

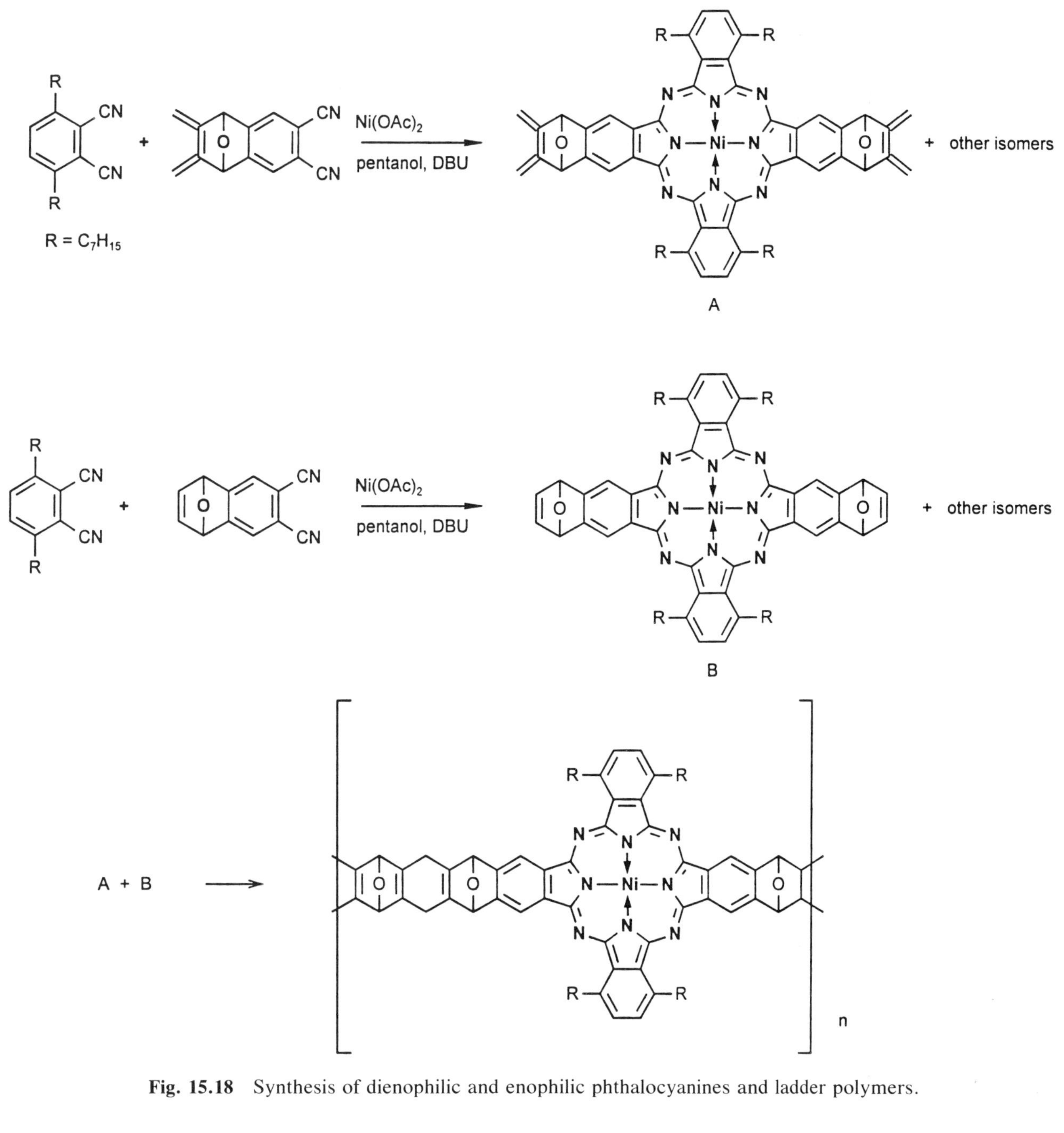

Fig. 15.18 Synthesis of dienophilic and enophilic phthalocyanines and ladder polymers.

However, most of the known syntheses of ladder polymers fall into two categories.

1. Two tetrafunctional monomers are reacted with each other, whereby—assuming the reaction proceeds in the desired way—the double-stranded structure grows from the very beginning [201–210]. (Fig. 15.16a).
2. Single-stranded polymers are synthesized that carry the required functionalities at defined, regular distances along the chain. These functionalities are then used to generate the second strand in a series of polymer-analogous reaction steps [211–215] (Fig. 15.16b).

Both strategies have serious disadvantages. In the first case it is difficult to see which factors, under homogeneous reaction conditions, should force the monomers to react exclusively in the desired way. One wrong linkage, as indicated in Fig. 15.16a, inevitably leads to inter-ladder cross-linking and is one of the reasons for the insolubility of most of the known ladder polymers. In addition to this topological issue, problems arise from the condensation reactions that are most often used and specifically in this case are difficult to drive to completion. This is because, due to the rigidity of the ladder polymers, the reaction mixture becomes very viscous and polymerization stops at relatively low conversions.

The second strategy looks very elegant on paper but could be successful only if the polymer-analogous reactions would not proceed randomly but rather start at one terminus of the single-stranded prepolymer and proceed from there along the chain one after the other like a zipper. Such a reaction is statistically unfavorable and therefore very unlikely to take place.

The repetitive Diels–Alder reaction [216,217] can be achieved by linking soluble nonidentically substituted

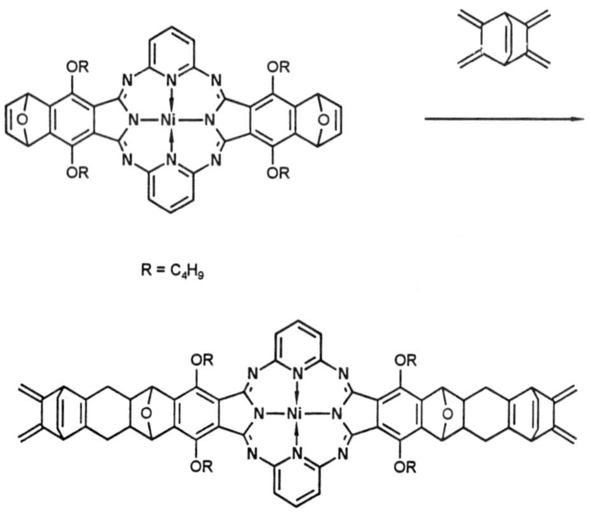

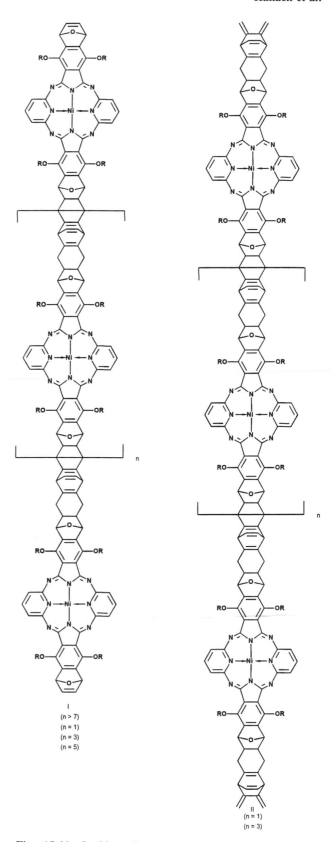

Fig. 15.19 Synthesis of precursors for obtaining ladder polymers based on phthalocyanines.

Fig. 15.20 Ladder oligomers based on hemiporphyrazines.

phthalocyanines over suitably functionalized bridging units, for example, by reacting a dienophilic phthalocyanine with a bifunctional diene in a Diels–Alder reaction [Fig. 15.17, Eq. (1)]. Ladder polymers can also be obtained by Diels–Alder reaction of dienophilic and enophilic phthalocyanines [Fig. 15.17, Eq. (2)]. The resulting oligomers must then be converted into a system containing conjugated π electrons [218; M. Hanack and P. Stihler, unpublished].

One way to obtain such dienophilic and enophilic phthalocyanines is by statistical synthesis starting from two correspondingly substituted phthalonitriles (Fig. 15.18). The D_{2h} isomer can be separated by column chromatography from the other five isomers [218; M. Hanack and P. Stihler, unpublished].

Similar to the above-described phthalocyanines, the hemiporphyrazinatometal complexes can be functionalized with dienophilic peripheral substituents, thereby enabling an oligomerization by repeated Diels–Alder reactions [219,220]. To increase the solubility of the resulting ladder polymers, additional peripheral substituents can be introduced into the monomeric subunits. The hemiporphyrazinatometal ladder polymers obtained by Diels–Alder couplings must also be converted into a π-electron-containing system to achieve electrical conductivity.

By repetitive Diels–Alder reactions starting from suitable precursors (Fig. 15.19), oligomers I and II (Fig. 15.20) were prepared and characterized [221]. For achieving full conjugation, the oxygen-bridged and saturated rings have to be aromatized.

REFERENCES

1. M. Hanack, A. Datz, R. Fay, K. Fischer, H. Keppler, J. Koch, J. Metz, M. Metzger, O. Schneider, and H.-J. Schulze, in *Handbook of Conductive Polymers*, Vol. 1 (T. A. Skotheim, ed.), Marcel Dekker, New York, 1986, pp. 133–204.
2. P. N. Prasad and D. J. Williams, *Introduction to Nonlinear Optical Effects in Molecules and Polymers*, Wiley, New York, 1991.
3. J. Zyss (ed.), *Molecular Nonlinear Optics*, Academic, New York, 1993.
4. J. L. Brédas, C. Adant, P. Tackx, A. Persoons, and B. M. Pierce, *Chem. Rev.* 94:243 (1994).
5. H. S. Nalwa, *Adv. Mater.* 5:5 (1993).
6. P. G. Schouten, J. M. Warman, M. P. de Haas, C. F. van Nostrum, G. H. Gelinck, R. J. M. Nolte, M. J. Copyn, J. W. Zwikker, M. K. Engel, M. Hanack, Y. H. Chang, and W. T. Ford, *J. Am. Chem. Soc.* 116:6880 (1994).
7. J. Simon and C. Sirlin, *Pure Appl. Chem.* 61:1625 (1989).
8. R. H. Tredgold, *Order in Thin Organic Films*, Cambridge Univ. Press, Cambridge, UK, 1994.
9. A. Ulman, *Ultrathin Organic Films*, Academic, New York, 1991.
10. M. J. Cook, A. J. Dunn, M. F. Daniel, R. C. O. Hart, R. M. Richardson, and S. J. Roser, *Thin Solid Films* 159:395 (1988).
11. M. A. Mohammad, P. Ottenbreit, W. Prass, G. Schnurpfeil, and D. Wöhrle, *Thin Solid Films* 213:285 (1992).
12. W. Yan, Y. Zhou, X. Wang, W. Chen, and X. Xi, *J. Chem. Soc., Chem. Commun.* 1992:872.
13. J. E. Kuder, *J. Imag. Sci.* 32:51 (1988).
14. A. Manivannan and L. A. Nagahara, *Thin Solid Films* 226:6 (1993).
15. G. Corker, B. Grant, and C. Clecak, *J. Electrochem. Soc.* 126:1339 (1979).
16. G. C. S. Collins and D. J. Schiffrin, *J. Electrochem. Soc.* 132:1835 (1985).
17. M.-T. Rion and C. Clarissse, *J. Electroanal. Chem.* 149:181 (1988).
18. H. Schultz, H. Lehmann, M. Rein, and M. Hanack, *Struct. Bonding (Berlin)* 74:41 (1991).
19. M. Hanack and M. Lang, *Adv. Mater.* 6:819 (1994); *Chemtracts—Org. Chem.* 8:131 (1995).
20. J. Simon and J. J. André, *Molecular Semiconductors*, Springer, Berlin, 1985, Chapter 3.
21. T. J. Marks, *Angew. Chem. Int. Ed. Engl.* 29:857 (1990).
22. K. Abe, H. Saito, T. Kimura, Y. Ohkatsu, and T. Kusano, *Makromol. Chem.* 190:2693 (1989).
23. B. Staiger, C. Shi, and F. C. Ansan, *Inorg. Chem.* 32:2107 (1993).
24. T. A. Temofonte and K. F. Schock, *J. Appl. Phys.* 65:1350 (1989).
25. H.-Y. Wang and J. B. Lando, *Langmuir* 10:790 (1994).
26. M. Kato, Y. Nishioka, K. Kaifu, K. Kawamura, and S. Ohno, *Appl. Phys. Lett.* 46:196 (1985).
27. P. Haisch, G. Winter, M. Hanack, L. Lüer, H. J. Egelhaaf, and D. Oelkrug, *Adv. Mat.* 9:316 (1997).
28. K.-Y. Law, *Chem. Rev.* 93:449 (1993).
29. M. Ehrl, F. W. Deeg, C. Brächle, O. Franke, A. Sobbi, G. Schulz-Eklott, and D. Wöhrle, *J. Phys.* 98:47 (1994).
30. M. Hanack, U, Keppeler, A. Lange, A. Hirsch, and R. Dieing, in *Phthalocyanines, Properties and Applications* (A. B. P. Lever, ed.), VCH, Weinheim, 1993, p. 43.
31. J. R. Darwent, P. Douglas, A. Harriman, G. Porter, and M. C. Richoux, *Coord. Chem. Rev.* 44:83 (1982).
32. D. Wöhrle, J. Gitzel, G. Krawczyk, E. Tsuchida, H. Ohno, and T. Nishisaka, *J. Macromol. Sci. Chem.* A 25:1227 (1988).
33. B. A. Anderson and T. J. Dougherty, *Photochem. Photobiol.* 55:145 (1992).
34. D. Wöhrle, M. Shopova, S. Müller, A. D. Milev, V. N. Mantareva, and K. K. Krastev, *Photochem. Photobiol.* 21B:155 (1993).
35. T. J. Klofta, J. Danziger, P. Lee, J. Pankow, K. W. Nebesny, and N. R. Armstrong, *J. Phys. Chem.* 91:5646 (1987).
36. D. Schlettwein, M. Kaneko, A. Yamada, D. Wöhrle, and N. I. Jaeger, *J. Phys. Chem.* 95:1748 (1991).

37. D. Schlettwein, D. Wöhrle, F. Karmann, and U. Melville, *Chem. Mater.* 6:3 (1994).
38. R. O. Loutfy and J. H. Sharp, *J. Chem. Phys.* 71:1211 (1979).
39. C. W. Tang, *Appl. Phys. Lett.* 48:183 (1986).
40. D. Wöhrle and D. Meissner, *Adv. Mater.* 3:129 (1991).
41. S. Takano, T. Enokida, and A. Kabata, *Chem. Lett. 1984*:2037.
42. R. O. Loutfy, A. M. Hor, C. K. Hsiao, G. Baranyl, and P. Kazmaier, *Pure Appl. Chem.* 60:1047 (1988).
43. R. O. Loutfy, C. K. Hsiao, A. M. Hor, and G. J. Di Paola-Baranyl, *J. Imaging Sci.* 29:148 (1985).
44. F. H. Moser and A. L. Thomas, *The Phthalocyanines*, CRC Press, Boca Raton, FL, 1983, Vols. I and II.
45. M. Hanack and G. Renz, *Chem. Ber.* 123:1105 (1990).
46. M. Hanack, R. Dieing, and U. Röhrig, *Chem. Lett. 1993*:399.
47. D. Wöhrle and G. Mayer, *Kontakte 1985*:38.
48. S. Deger and M. Hanack, *Isr. J. Chem.* 27:347 (1986).
49. M. Hanack, G. Renz, J. Strähle, and S. Schmid, *J. Org. Chem.* 56:350 (1991).
50. H. Tomoda, S. Saito, and S. Shiraishi, *Chem. Lett. 1983*:313.
51. H. Tomoda, S. Saito, S. Ogawa, and S. Shiraishi, *Chem. Lett. 1980*:1277.
52. D. Wöhrle, G. Schnurpfeil, and G. Knothe, *Dyes Pigm.* 18:91 (1992).
53. M. Hanack, J. Osio-Barcina, E. Witke, and J. Pohmer, *Synthesis 1992*:211.
54. A. Capobianchi, A. M. Paoletti, G. Pennesi, G. Rossi, R. Caminiti, and C. Ercolani, *Inorg. Chem.* 33:4635 (1994).
55. R. Dieing, G. Schmid, E. Witke, C. Feucht, M. Dressen, J. Pohmer, and M. Hanack, *Chem. Ber.* 128:589 (1995).
56. N. Kobayashi, in *The Phthalocyanines, Properties and Applications* (C. C. Leznoff and A. B. P. Lever, eds.), VCH, Weinheim, 1993, Vol. 2, Chapter 3.
57. J. E. Bloor, J. Schlabitz, C. C. Walden, and A. Demerdache, *Can. J. Chem.* 42:2201 (1964).
58. T. J. Marks and D. R. Stojakovic, *J. Am. Chem. Soc.* 100:1645 (1978).
59. S. A. Mikhalenko, S. U. Barkanova, O. L. Lebedev, and E. A. Luk'yanets, *J. Gen. Chem. USSR* 41:2770 (1971).
60. M. Hanack, D. Meng, A. Beck, M. Sommerauer, and L. R. Subramanian, *J. Chem. Soc., Chem. Commun. 1993*:58.
61. M. Hanack, G. Schmid, and M. Sommerauer, *Angew. Chem.* 105:1540 (1993); *Angew. Chem. Int. Ed. Engl.* 32:1422 (1993).
62. M. J. Cook, J. McMurdo, and A. K. Powell, *J. Chem. Soc., Chem. Commun. 1993*:903.
63. K. Otha, L. Kaquemin, C. Sirlin, L. Bosio, and J. Simon, *New J. Chem.* 12:751 (1988).
64. M. J. Cook, A. J. Dunn, S. D. Howe, A. J. Thomson, and K. J. Harrison, *J. Chem. Soc., Perkin Trans. 1 1988*:2453.
65. E. I. Kovshev, V. A. Puchnova, and E. A. Luk'yanets, *Zh. Org. Khim* 1:369 (1971).
66. M. Karayose, S. Tai, K. Kamijima, H. Hagiwara, and N. J. Hayshi, *J. Chem. Soc., Perkin Trans. 2 1992*:403.
67. R. Polley and M. Hanack, *J. Org. Chem.* 60:8278 (1996).
68. A.-M. Giroud-Godquin and P. M. Maitlis, *Angew. Chem.* 103:370 (1991); *Angew. Chem. Int. Ed. Engl.* 30:375 (1991).
69. C. Piechocki, J. Simon, A. Skoulios, D. Guillon, and P. Weber, *J. Am. Chem. Soc.* 104:5245 (1982).
70. P. Espinet, M. A. Esternelas, L. A. Oro, J. L. Serrano, and E. Sola, *Coord. Chem. Rev.* 117:215 (1992).
71. M. K. Engel, P. Rassoul, L. Bosio, H. Lehmann, M. Hanack, and J. Simon, *Liquid Cryst.* 15:709 (1993).
72. S. Chandrasekhar and G. S. Rangnath, *Rec. Prog. Phys.* 53:570 (1990).
73. M. J. Cook, M. F. Daniel, J. Harrison, N. B. McKnowen, and A. J. Thomson, *J. Chem. Soc., Chem. Commun. 1987*:1086.
74. P. Weber, D. Guillon, and A. Skoulios, *Liq. Cryst.* 9:369 (1991).
75. D. M. Collard and C. P. Lillya, *J. Am. Chem. Soc.* 113:8577 (1991).
76. P. G. Schouten, J. F. van der Pol, J. W. Zwikker, W. Drenth, and S. J. Picken, *Mol. Cryst. Liq. Cryst.* 195:291 (1991).
77. A. N. Cammidge, M. J. Cook, K. J. Harrison, and N. B. McKeown, *J. Chem. Soc., Perkin Trans. 1 1991*:3053.
78. P. Haisch, Ph.D. Dissertation, University of Tübingen, Germany, 1994.
79. P. Haisch and M. Hanack, *Synthesis 1995*:1251.
80. J. H. van der Linden, J. Schoonman, R. J. M. Nolte, and W. Drenth, *Rec. Trav. Chim. Pays-Bas* 103:260 (1984).
81. O. E. Sielcken, H. C. A. van Lindert, W. Drenth, J. Schoonman, J. Schram, and R. J. M. Nolte, *Ber. Bunsenges. Phys. Chem.* 93:702 (1989).
82. V. Ahsen, A. Gürek, E. Muslouglu, and Ö. Bekaroglu, *Chem. Ber.* 122:1073 (1993).
83. V. Ahsen, E. Yilmazer, M. Ertas, and Ö. Bekaroglu, *J. Chem. Soc., Dalton Trans. 1988*:401.
84. O. E. Sielcken, M. M. van Tielborg, M. T. M. Roks, R. Hendriks, W. Drenth, and R. J. M. Nolte, *J. Am. Chem. Soc.* 109:4261 (1987).
85. P. E. Fiedling and S. Guttman, *J. Chem. Phys.* 2:411 (1957).
86. P. Gomez-Romero, Y.-S. Lee, and M. Kertesz, *Inorg. Chem.* 127:3672 (1988).
87. E. Canadell and S. Alvarez, *Inorg. Chem.* 23:573 (1984).
88. N. Kobayashi and A. B. P. Lever, *J. Am. Chem. Soc.* 109:7433 (1987).

89. L. S. Grigoryan, W. Hilezer, and M. Krupski, *Ferroelectrics 80*:11 (1988).
90. M. Abkowitz and A. R. Monahan, *J. Chem. Phys. 88*:2281 (1973).
91. D. Markovitsi, I. Lécuyer, and J. Simon, *J. Phys. Chem. 95*:3620 (1991).
92. J. Simon and P. Bassoul, in *The Phthalocyanines—Properties and Applications* (C. C. Leznoff and A. B. P. Lever, eds.), VCH, Weinheim, 1993, Vol. 2, pp. 223–299.
93. T. Linssen and M. Hanack, *Chem. Ber. 127*:2051 (1994).
94. C. Feucht, T. Linssen, and M. Hanack, *Chem. Ber. 127*:113 (1994).
95. M. J. Stillway and T. Nyokong, in *The Phthalocyanines—Properties and Applications* (C. C. Leznoff and A. B. P. Lever, eds.), VCH, Weinheim, 1989, Vol. 1, pp. 133–289.
96. Y. Ikeda, H. Konami, M. Hatano, and K. Mochizuki, *Chem. Lett. 1992*:763.
97. H. Konami and M. Hatano, *Chem. Lett. 1988*:1359.
98. C. Piechocki and J. Simon, *J. Chem. Soc., Chem. Commun. 1985*:259.
99. C. C. Leznoff, P. I. Svirskaya, B. Khouw, R. L. Cerny, P. Seymour, and A. B. P. Lever, *J. Org. Chem. 56*:82 (1991).
100. Wöhrle and G. Krawczyk, *Polym. Bull. 15*:193 (1986).
101. C. C. Leznoff and T. W. Hall, *Tetrahedron Lett. 23*:3023 (1982).
102. N. Kobayashi, R. Kundo, S. Nakajima, and T. Osa, *J. Am. Chem. Soc. 112*:9640 (1990).
103. E. Mosluoglu, A. Gürek, V. Ahsen, A. Gül, and Ö. Bekaroglu, *Chem. Ber. 125*:2337 (1992).
104. K. Kasuga, T. Idehara, M. Hande, and K. Isa, *Inorg. Chim. Acta 196*:127 (1992).
105. N. J. Kobayashi, T. Ashida, and T. Osa, *Chem. Lett. 1992*:2031.
106. F. Plenzig, *Diploma Thesis*, University of Tübingen, Germany, 1994.
107. M. Hanack and M. Geyer, *J. Chem. Soc., Chem. Commun. 1994*:2253.
108. M. Hanack and J. Rauschnabel, *Tetrahedron Lett. 36*:1629 (1995).
109. J. Spikes, *Photochem. Photobiol. 43*:691 (1986).
110. R. D. Joyner and M. E. Kenney, *J. Am. Chem. Soc. 82*:5790 (1960).
111. R. D. Joyner and M. E. Kenney, *Inorg. Chem. 1*:236 (1962).
112. R. D. Joyner and M. E. Kenney, *Inorg. Chem. 1*:717 (1962).
113. C. W. Dirk, T. Inabe, K. F. Schoch, Jr., and T. J. Marks, *J. Am. Chem. Soc. 105*:1539 (1983).
114. D. W. DeWulf, J. K. Leland, B. L. Wheeler, A. J. Bard, D. A. Batzel, D. R. Dininny, and M. E. Kenney, *Inorg. Chem. 26*:266 (1987).
115. X. Zhou, T. J. Marks, and S. H. Carr, *Polym. Mater. Sci. Eng. 51*:651 (1984).
116. T. J. Marks, *Science 227*:881 (1985).
117. E. Orthmann, V. Enkelmann, and G. Wegner, *Makromol. Chem., Rapid Commun. 4*:687 (1983).
118. T. J. Marks, K. F. Schoch, and B. R. Kundalkar, *Synth. Met. 1*:337 (1979/80).
119. P. J. Toscano and T. J. Marks, *J. Am. Chem. Soc. 108*:437 (1986).
120. B. Wehrle, H.-H. Limbach, T. Zipplies, and M. Hanack, *Angew. Chem. Adv. Mater. 101*:1783 (1989).
121. X. Zhou, T. J. Marks, and S. H. Carr, *J. Polym. Sci. 23*:305 (1985).
122. X. Zhou, T. J. Marks, and S. H. Carr, *Mol. Cryst. Liq. Cryst. 118*:337 (1985).
123. J. A. Ibers, L. J. Pace, J. Martinsen, and B. M. Hoffman, *Struct. Bond. 50*:47 (1982).
124. J. G. Gaudiello, M. Almeida, T. J. Marks, W. J. McCarthy, J. C. Butler, and C. R. Kannewurf, *J. Phys. Chem. 90*:4917 (1986).
125. C. W. Dirk, T. Inabe, K. F. Schoch, Jr., and T. J. Marks, *J. Am. Chem. Soc. 105*:1539 (1983).
126. O. Schneider, J. Metz, and M. Hanack, *Mol. Cryst. Liq. Cryst. 81*:273 (1982).
127. M. Hanack and K. Fischer, *Chem. Ber. 116*:1860 (1983).
128. T. Inabe, M. K. Moguel, T. J. Marks, R. Burton, J. W. Lyding, and C. R. Kannewurf, *Mol. Cryst. Liq. Cryst. 118*:349 (1985).
129. W. Dirk, E. A. Mintz, K. F. Schoch, Jr., and T. J. Marks, *J. Macromol. Sci. Chem. A 16*:275 (1981).
130. T. Inabe, M. K. Moguel, T. J. Marks, R. Burton, J. W. Lyding, and C. R. Kannewurf, *Mol. Cryst. Liq. Cryst. 118*:349 (1985).
131. P. Sheng, *Phys. Rev. R21*:2180 (1980).
132. J. Metz, G. Pawlowski, and M. Hanack, *Z. Naturforsch. 38b*:378 (1983).
133. C. Sirlin, L. Bosio, and J. Simon, *Mol. Cryst. Liq. Cryst. 155*:231 (1988).
134. W. Caseri, T. Sauer, and G. Wegner, *Macromol. Chem., Rapid Commun. 9*:651 (1988).
135. E. Orthmann and G. Wegner, *Angew. Chem., Int. Ed. Engl. 25*:1105 (1986).
136. T. Sauer, T. Arndt, D. N. Batchelder, A. A. Kalachev, and G. Wegner, *Thin Solid Films 187*:357 (1990).
137. M. Ilanack and T. Zipplies, *J. Am. Chem. Soc. 107*:6127 (1985).
138. U. Keppeler, O. Schneider, W. Stöffler, and M. Hanack, *Tetrahedron Lett. 25*:3679 (1984).
139. J. P. Linsky, T. R. Paul, R. S. Nohr, and M. E. Kenney, *Inorg. Chem. 19*:3131 (1980).
140. P. M. Kuznesof, R. S. Nohr, K. J. Wynne, and M. E. Kenney, *J. Macromol. Sci. Chem. A16*:299 (1981).
141. R. S. Nohr and K. J. Wynne, *J. Chem. Soc., Chem. Commun. 1981*:1210.
142. K. J. Wynne and R. S. Nohr, *Mol. Cryst. Liq. Cryst. 81*:243 (1983).
143. P. Brant, R. S. Nohr, K. J. Wynne, and D. C. Weber, *Mol. Cryst. Liq. Cryst. 81*:255 (1982).
144. K. J. Wynne, *Inorg. Chem. 24*:1339 (1985).
145. M. Hanack, K. Mitulla, G. Pawlowski, and L. R.

Subramanian, *Angew. Chem., Int. Ed. Engl. 18*:322 (1979).
146. K. Mitulla and M. Hanack, *Z. Naturforsch. 35b*:1111 (1980).
147. M. Hanack, K. Mitulla, G. Pawlowski, and L. R. Subramanian, *J. Organomet. Chem. 204*:315 (1981).
148. M. Hanack, W. Kobel, J. Metz, M. Mezger, G. Pawlowski, O. Schneider, H.-J. Schulze, and L. R. Subramanian, *Mater. Sci.* 185 (1981).
149. M. Hanack, K. Mitulla, and O. Schneider, *Chim. Scr.* 17 (1981).
150. O. Schneider and M. Hanack, *Chem. Ber. 116*:2088 (1983).
151. O. Schneider and M. Hanack, *Angew. Chem. 95*:804 (1983); *Angew. Chem. Int. Ed. Engl. 22*:784 (1983).
152. J. Metz, O. Schneider, and M. Hanack, *Spectrochim. Acta 38A*:1265 (1982).
153. U. Keppeler, W. Kobel, H.-U. Siehl, and M. Hanack, *Chem. Ber. 118*:2095 (1985).
154. R. Aldinger, M. Hanack, K.-H. Herrmann, A. Hirsch, and K. Kasper, *Synth. Met. 60*:265 (1993).
155. F. Kubel and J. Strähle, *Z. Naturforsch. B 36*:441 (1981).
156. M. Hanack and U. Keppeler, *Chem. Ber. 119*:3363 (1986).
157. O. Schneider, Ph. D. Thesis, University of Tübingen, Germany, 1983.
158. M. Hanack, A. Hirsch, and H. Lehmann, *Angew. Chem. 102*:1499 (1990); *Angew. Chem. Int. Ed. Engl. 29*:1467 (1990).
159. M. Hanack, A. Beck, and H. Lehmann, *Synthesis 1987*:703.
160. M. Hanack, J. Osio-Barcina, E. Witke, and J. Pohmer, *Synthesis 1992*:211.
161. M. Hanack, A. Gül, A. Hirsch, B. K. Mandal, L. R. Subramanian, and E. Witke, *Mol. Cryst. Liq. Cryst. 187*:365 (1990).
162. G. Schmid, E. Witke, U. Schlick, S. Knecht, and M. Hanack, *J. Mater. Chem. 5*:855 (1995).
163. E. Witke, Dissertation, Universität Tübingen, 1993.
164. M. Hanack and P. Vermehren, *Synth. Met. 32*:257 (1989).
165. P. Vermehren, Ph. D. Thesis, University of Tübingen, 1989.
166. B. N. Diel, T. Inabe, N. K. Jaggi, J. W. Lyding, O. Schneider, M. Hanack, C. R. Kannewurf, T. J. Marks, and L. H. Schwarz, *J. Am. Chem. Soc. 106*:3207 (1984).
167. M. Hanack and H. Ryu, *Synth. Met. 46*:113 (1992).
168. A. G. MacDiarmid and A. J. Epstein, in *Science and Application of Conducting Polymers* (W. R. Salaneck, D. T. Clark, and E. J. Samuelsen, eds.), Adam Hilger, Bristol, 1991, p. 117.
169. M. Hanack and Y.-G. Kang, *Synth. Met. 48*:79 (1992).
170. E. Orti, J. L. Brédas, M. C. Piqueras, and R. Crespo, *Chem. Mater. 2*:110 (1990).
171. E. Orti, J. L. Brédas, M. C. Piqueras, and R. Crespo, *Synth. Met. 41*:2647 (1991).

172. U. Keppeler, S. Deger, A. Lange, and M. Ilanack, *Angew. Chem. 99*:349 (1987); *Angew. Chem. Int. Ed. Engl. 26*:344 (1987).
173. J. Pohmer, M. Hanack, and J. Osio Barcina, *J. Mat. Chem. 6*:957 (1996).
174. S. Hayashida and M. Hanack, *Synth. Met. 52*:241 (1992).
175. W. Koch, Ph. D. Thesis, University of Tübingen, 1984.
176. M. Hanack, R. Polley, S. Knecht, and U. Schlick, *Inorg. Chem. 34*:3621 (1995).
177. R. Grosshans, Ph. D. Thesis, University of Tübingen, 1990.
178. M. Hanack, S. Knecht, and R. Polley, *Chem. Ber. 128*:929 (1995).
179. W. Kobel and M. Hanack, *Inorg. Chem. 25*:103 (1986).
180. J. Metz and M. Hanack, *J. Am. Chem. Soc. 105*:828 (1983).
181. C. Hedtmann-Rein, M. Hanack, K. Peters, E.-M. Peters, and H. G. v. Schnering, *Inorg. Chem. 19*:787 (1987).
182. M. Hanack, C. Hedtmann-Rein, A. Datz, U. Keppeler, and X. Münz, *Synth. Met. 19*:787 (1987).
183. M. Hanack and R. Polley, *Inorg. Chem. 33*:3201 (1994).
184. A. Datz, J. Metz, O. Schneider, and M. Hanack, *Synth. Met. 19*:787 (1987).
185. O. Schneider and M. Hanack, *Z. Naturforsch. B39*:265 (1984).
186. U. Ziener, N. Fahmy, and M. Hanack, *Chem. Ber. 126*:2559 (1993).
187. J. F. Myers, G. W. R. Canham, and A. B. P. Lever, *Inorg. Chem. 14*:46 (1975).
188. M. Hanack and X. Münz, *Synth. Met. 10*:357 (1985).
189. M. Hanack and R. Fay, *Rec. Trav. Chim. Pays-Bas 105*:427 (1986).
190. T. Inabe and Y. Maruyama, *Chem. Lett. 1989*:55.
191. K. Morimoto and T. Inabe, *J. Mat. Chem. 5*:1749 (1995).
192. C. Hamann, J. Hein, and H. Burghardt, *Organische Leiter, Halbleiter und Photoleiter*, Vieweg, Braunschweig, 1981.
193. H. Meier, W. Albrecht, and E. Zimmerhackl, *Polym. Bull. 13*:43 (1985).
194. H. Meier, *Research Report No. I/61 635*, Stiftung Volkswagenwerk, Wolfsburg, Germany.
195. G. Winter, Diplomarbeit, Tübingen, 1995.
196. H. Meier, W. Albrecht, E. Zimmerhackl, M. Hanack, and J. Metz, *Synth. Met. 11*:333 (1985).
197. H. Meier, W. Albrecht, M. Hanack, and J. Koch, *Polym. Bull. 16*:75 (1986).
198. U. Scherf and K. Müllen, *Synthesis 1992*:23.
199. M. B. Goldfinger and T. B. Scrager, *J. Am. Chem. Soc. 116*:7895 (1994).
200. B. Schlicke, H. Schirmer, and A. D. Schlüter, *Adv. Mater. 7*:544 (1995).
201. F. E. Arnold and R. L. van Deusen, *Macromolecules 2*:497 (1969).

202. M. Kurihara, *Macromolecules* 3:722 (1970).
203. J. Szita, L. H. Brannigan, and C. S. Marvel, *J. Polym. Sci., Polym. Chem. Ed.* 9:691 (1971).
204. W. J. Bailey and J. H. Feinberg, *Polym. Prepr., ACS, Div. Polym. Chem. Ed.* 13:287 (1972).
205. K. Imai, M. Kurihara, L. Mathias, J. Wittmann, W. B. Alston, and J. K. Stille, *Macromolecules* 6:158 (1973).
206. C. G. Berry, *J. Polym. Sci., Polym. Symp.* 65:143 (1978).
207. V. R. Sastri, R. Schulman, and D. C. Roberts, *Macromolecules* 15:939 (1982).
208. B. H. Lee and C. S. Marvel, *J. Polym. Sci., Polym. Chem. Ed.* 21:83 (1983).
209. Z. Iqbal, D. M. Ivory, J. Marti, J. L. Bredas, and R. H. Baughman, *Mol. Cryst. Liq. Cryst.* 118:103 (1985).
210. X.-T. Bi and M. H. Litt, *Polymer* 28:2346 (1987).
211. U. Kador and P. Mehnert, *Makromol. Chem.* 144:37 (1971).
212. V. N. Salaurov, Y. G. Kryazhev, T. L. Yushmanova, T. I. Vakul'skaya, and M. G. Voronkov, *Makromol. Chem.* 175:757 (1974).
213. K. Messmer and P. Mehnert, *Kolloid Z. Z. Polym.* 252:97 (1974).
214. H. Kämmerer and G. Hegemann, *Makromol. Chem.* 185:499 (1984).
215. J. Z. Ruan and M. H. Litt, *Synth. Met.* 15:237 (1986).
216. A. D. Schlüter, *Nachr. Chem. Tech. Lab.* 1990:38.
217. A. D. Schlüter, *Adv. Mater.* 3:282 (1991).
218. T. G. Linssen, Ph. D. Thesis, University of Tübingen, 1994.
219. M. Rack, B. Hauschel, and M. Hanack, *Chem. Ber.*, 129:237 (1996).
220. B. Hauschel, D. Ruff, and M. Hanack, *J. Chem. Soc., Chem. Commun.*, 1995:2449.
221. M. Rack and M. Hanack, *Angew. Chem.* 106:1712 (1994); *Angew. Chem. Int. Ed. Engl.* 33:1646 (1994).

16
Template Polymerization of Conductive Polymer Nanostructures

Charles R. Martin
Colorado State University, Fort Collins, Colorado

I. INTRODUCTION

Nanochemistry is an emerging subdiscipline of the chemical and materials sciences that deals with the development of methods for synthesizing nanoscopic particles of a desired material and with scientific investigations of the nanomaterial obtained [1–4]. Nanomaterials have numerous possible commercial and technological applications, including use in electronic, optical, and mechanical devices [3–7], drug delivery [8], and bioencapsulation [9]. In addition, this field poses an important fundamental philosophical question: How do the properties (electronic, optical, magnetic, etc.) of a nanoscopic particle of a material differ from the analogous properties for a macroscopic sample of the same material?

My research group has been exploring a method for preparing nanomaterials called template synthesis. (For recent reviews see Refs. 1 and 2.) This method entails synthesizing the desired material within the pores of a nanoporous membrane or other solid. The membranes employed have cylindrical pores of uniform diameter (Fig. 16.1). In essence, we view each of these pores as a beaker in which a piece of the desired material is synthesized. Because of the cylindrical shape of these pores a nanocylinder of the desired material is obtained in each pore (Fig. 16.2). Depending on the material and the chemistry of the pore wall, this nanocylinder may be solid (a fibril) or hollow (a tubule).

The template method has a number of interesting and useful features. First, it is a very general approach; we have used this method to prepare tubules and fibrils composed of electronically conductive polymers [9–16], metals [14,17–23], semiconductors [24], carbons [25], and other materials. Furthermore, nanostructures with extraordinarily small diameters can be prepared. For example, Wu and Bein [26] recently used this method to prepare conductive polymer (polyaniline) nanofibrils with diameters of 3 nm (30 Å). It would be difficult to make nanowires with diameters this small using lithographic methods. In addition, because the pores in the membranes used have monodisperse diameters, analogous monodisperse nanostructures are obtained. Finally, the tubular or fibrillar nanostructures synthesized within the pores can be freed from the template membrane and collected (Fig. 16.2A). Alternatively, an ensemble of nanostructures that protrude from a surface like the bristles of a brush can be obtained (Fig. 16.2B).

We began our template synthesis work in 1985 by electrochemically synthesizing the electronically conductive polymer polypyrrole within the pores of a nanoporous polycarbonate filtration membrane [10]. Since then, we [9,11–16] and others [26–35] have explored, in some detail, the electrochemical, electronic, and optical properties of template-synthesized conductive polymers. I review this work in this chapter. Topics discussed include the membranes used to do template synthesis, the electronic properties of template-synthesized conductive polymer fibrils and tubules, and the morphology of the template-synthesized conductive polymers.

II. MEMBRANES USED

A. "Track-Etch" Membranes

A number of companies (e.g., Nuclepore and Poretics) sell micro- and nanoporous polymeric filtration membranes that have been prepared via the "track-etch" method [36]. As shown in Fig. 16.1A and 16.1B, these membranes contain cylindrical pores of uniform diameter. The pores are randomly distributed across the mem-

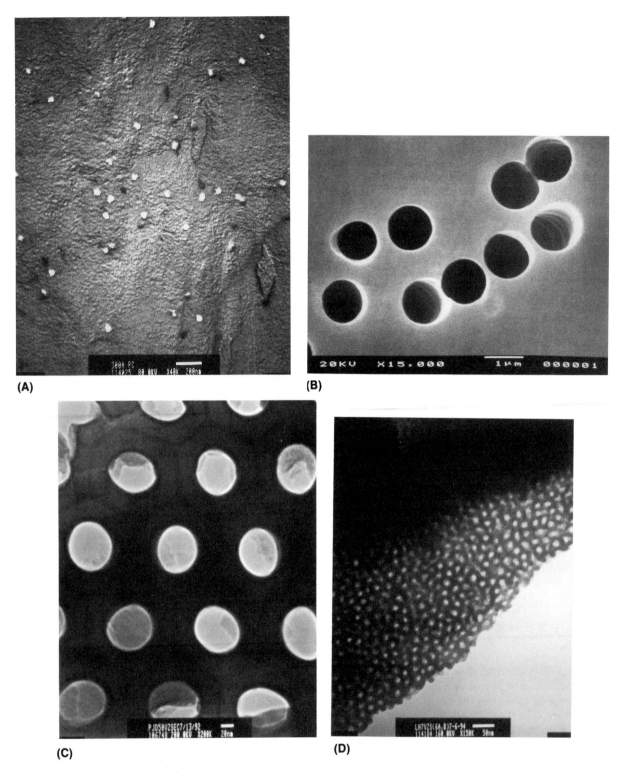

Fig. 16.1 Electron micrographs of polycarbonate (A, B) and alumina (C, D) template membranes. For each type of membrane, an image of a larger pore membrane is presented (a, c) so that the characteristics of the pores can be clearly seen. An image of a membrane with extremely small pores is also presented in (B) and (D). (A) Scanning electron micrograph of the surface of a polycarbonate membrane with 1 μm diameter pores. (B) Transmission electron micrograph (TEM) of a graphite replica of the surface of a polycarbonate membrane with ∼30 nm diameter pores. The pores appear "ragged." This is an artifact of the graphite replica. (C, D) TEMs of microtomed section of alumina membranes with ∼70 nm (C) and ∼10 nm (D) diameter pores.

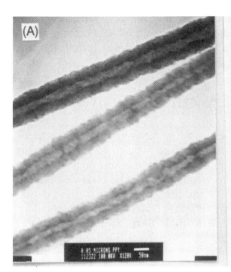 

Fig. 16.2 (A) Transmission electron micrograph of three polypyrrole nanotubules. (B) Scanning electron micrograph of an array of capped polypyrrole microtubules.

brane surface. Membranes with a wide range of pore diameters (down to 10 nm) and pore densities approaching 10^9 pores/cm^2 are available commercially. The material most commonly used to prepare membranes of this type is polycarbonate; however, a broad range of materials are amenable to the track-etch process [36].

B. Porous Alumina Membranes

Porous alumina membranes are prepared electrochemically from aluminum metal [37]. As indicated in Fig. 16.1C, the pores in these membranes are arranged in a regular hexagonal lattice. Pore densities as high as 10^{11} pores/cm^2 can be achieved [38]. Although such membranes are sold commercially, only a limited number of pore diameters are available. We have prepared membranes of this type with a broad range of pore diameters [19,21,22]. We have made membranes with pores as small as 5 nm, and we believe that even smaller pores can be prepared.

C. Other Nanoporous Materials

Tonucci et al. [39] describe a nanochannel array glass membrane. Membranes of this type containing pores with diameters as small as 33 nm and densities as high as 3×10^{10} pores/cm^2 were prepared. Beck et al. [40] prepared a new large pore diameter zeolite. Wu and Bein [26,35] used the pores in these materials as templates to synthesize polyaniline and graphitic nanofibrils. Douglas et al. [41] have shown that the nanoscopic pores in a protein derived from a bacterium can be used to transfer an image of these pores to an underlying substrate.

Finally, Ozin [3] discusses a wide variety of other nanoporous solids that could be used as template materials.

III. TEMPLATE METHODS

Most of our work has focused on polypyrrole, poly(3-methylthiophene), and polyaniline. These polymers can be synthesized via oxidative polymerization of the corresponding monomer. This can be accomplished either electrochemically [10,12,42] or by using a chemical oxidizing agent [43–45]. We have adapted both of these approaches so that they can be used to do template synthesis of conductive polymers within the pores of our nanoporous template membranes. The easiest way to do electrochemical template synthesis is to coat a metal film onto one surface of the membrane and then use this film to electrochemically synthesize the desired polymer within the pores of the membrane [12]. Chemical template synthesis can be accomplished by simply immersing the membrane into a solution of the desired monomer and its oxidizing agent [9,16,44].

In developing these template synthetic methods, we made an interesting discovery. When these polymers are synthesized (either chemically or electrochemically) within the pores of the track-etched polycarbonate membranes, the polymer preferentially nucleates and grows on the pore walls [11,14,46]. As a result, polymeric tubules are obtained at short polymerization times (Fig. 16.2A). These tubular structures have been quite useful in our fundamental investigations of electronic conductivity in the template-synthesized materials (see below). In addition, tubular structures of this type have a number of proposed technological applications [47]. For example, we have shown that capped versions of our tubules can be used for enzyme immobilization [9].

The reason the polymer preferentially nucleates and grows on the pore walls is straightforward [14]. Although the monomers are soluble, the polycationic forms of these polymers are insoluble. Hence, there is a solvophobic component to the interaction between the polymer and the pore wall. There is also an electrostatic component because the polymers are cationic and there are anionic sites on the pore wall [14].

Finally, by controlling the polymerization time, conductive polymer tubules with thin walls (short polymerization times) or thick walls (long polymerization times) can be obtained. This point is illustrated by the transmission electron micrographs shown in Fig. 16.3 [14]. For polypyrrole, the tubules ultimately close up to form solid fibrils. By controlling the polymerization time we can make hollow polypyrrole tubules or solid fibrils. In contrast, the polyaniline tubules will not close up, even at long polymerization times [44].

IV. ENHANCED CONDUCTIVITY

The easiest way to measure the conductivities of our template-synthesized fibrils is to leave them in the pores of the template membrane and measure the resistance across the membrane [11,15]. Provided the number and diameter of the fibrils are known, the measured transmembrane resistance can be used to calculate the conductivity of a single fibril. Conductivity data obtained in this way for polypyrrole fibrils are shown in Fig. 16.4 [15]. While the large diameter fibrils have conductivities

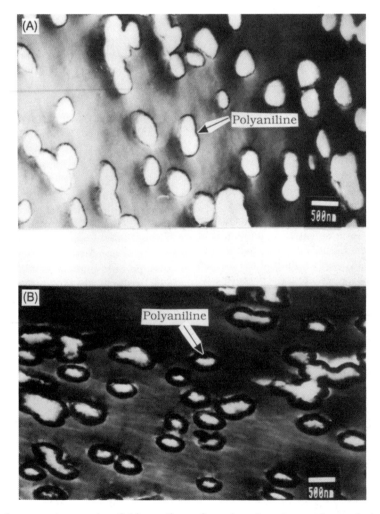

Fig. 16.3 Transmission electron micrographs of thin sections of a polycarbonate membrane during the template synthesis of polyaniline. Polymerization time was 30 min (A) and 6 h (B). Note thickness of polyaniline layer on pore walls in (A) vs. (B).

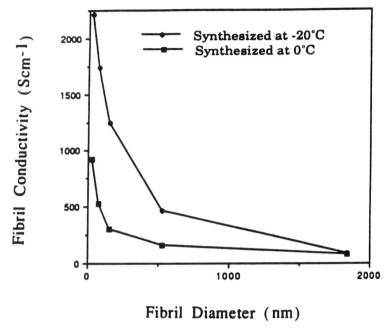

Fig. 16.4 Conductivity of template-synthesized polypyrrole tubules vs. diameter of the tubules. Data for two different synthesis temperatures are shown. (From Ref. 15.)

comparable to those of bulk samples of polypyrrole (e.g., electrochemically synthesized films), the conductivity of the smallest diameter fibrils is over an order of magnitude higher. Similar enhancements in conductivity have been observed for template-synthesized polyaniline [16,44] and poly(3-methylthiophene) [11]. Analogous results have been obtained by Granström and Inganäs [32].

This transmembrane conductivity method is a two-point measurement. Contact resistance is always a worry with such measurements; indeed, we must apply substantial pressure across the membrane (7×10^3 psi) to obtain reproducible resistance data [15]. We recently developed a method for forming thin films from our template-synthesized nanostructures [16,44]. This has allowed us to make four-point conductivity measurements on these nanomaterials (without applying pressure during the measurement). The thin films are prepared by dissolving the template membrane, collecting the conductive polymer nanofibrils or nanotubules by filtration to form a film across the surface of the filter, and then compacting this film in an IR pellet press [16,44].

Figure 16.5 shows scanning electron micrographs of cross sections of thin films prepared in this way from polyaniline and polypyrrole nanotubules. Note that because of the high pressure used in the compaction step (6×10^4 psi) the tubular structure can no longer be seen in the cross section of the polyaniline film (Fig. 16.5A). In contrast, the tubular structure is still clearly evident in the film prepared from the polypyrrole nanotubules (Fig. 16.5B). That the polypyrrole nanotubules can survive the high pressures used during the film compaction step is quite remarkable and indicates that these tubules are either very strong or very resilient.

Table 16.1 shows four-point conductivity data for films prepared from polyaniline tubes of various diameters [16,44]. In complete agreement with the two-point data (Fig. 16.4), conductivity increases as the diameter of the tubules used to prepare the film decreases. The conductivity for the film prepared from the narrowest tubes is over five times that of a film prepared, under the same conditions, from bulk polyaniline [16,44]. Conductivities of films prepared from polypyrrole tubules show the same trend (increasing conductivity with decreasing tubule diameter) but are lower than those obtained by the two-point method because, as Fig. 16.5B clearly shows, these films are not space-filling.

Table 16.1 Conductivity as a Function of Tubule Diameter for Films Made from Template-Synthesized Polyaniline Tubules

Tubule diameter (nm)	Conductivity (S/cm)
100	50 ± 4
200	14 ± 2
400	9 ± 2

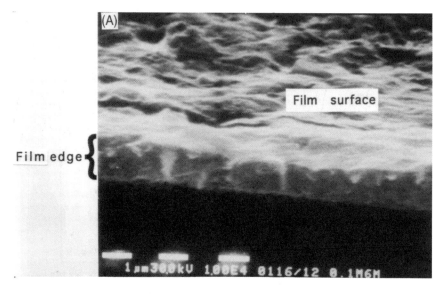

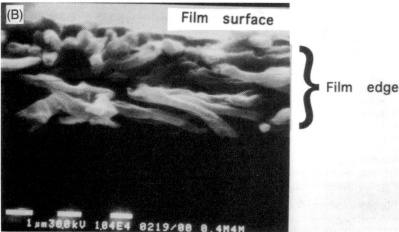

Fig. 16.5 Scanning electron micrographs of cross sections of films made from template-synthesized conductive polymer tubules. (A) Polyaniline tubules; (B) polypyrrole tubules.

V. MOLECULAR AND SUPERMOLECULAR STRUCTURE

We and others have shown that the polymer chains in the template-synthesized materials are preferentially aligned (enhanced supermolecular order) and that these chains contain fewer conjugation-interrupting defects (improved molecular structure) [13,15,16,32,44]. These improvements in molecular and supermolecular structure are responsible for the enhancements in conductivity discussed above.

A. Supermolecular Structure: Chain Alignment

We have used X-ray diffraction and polarized infrared absorption spectroscopy (PIRAS) to prove that the template-synthesized materials show enhanced supermolecular order [13,15,16,44]. PIRAS entails measuring the absorbance by the sample of two orthogonal polarizations of IR radiation [48,49]. In our case, one polarization is perpendicular to the axes of our fibrils ($I_\perp$) and the other has a component that is parallel to the fibril axes ($I_\parallel$). In general, if $I_\perp$ and $I_\parallel$ are absorbed to the same extent, the polymer chains show no preferred orientation. Unequal absorbance indicates that the polymer chains are preferentially aligned. PIRAS data can be quantified by calculating a parameter called the dichroic ratio, R. An R value of unity means no preferential chain alignment. For our studies, an R value less than 1 means some degree of alignment, and the lower the R, the greater the extent of alignment [13,15,16,44].

Table 16.2 shows typical R data for polypyrrole fibrils synthesized at two temperatures [15]. While the largest

Template Polymerization

diameter fibrils show no preferential chain alignment, the narrowest fibrils show strong dichroism, indicating that the polymer chains are aligned. The PIRAS and conductivity data tell a consistent story about the template-synthesized fibrils. The narrowest fibrils have higher conductivities (Fig. 16.4) because the polymer chains are strongly aligned (Table 16.2). Furthermore, fibrils synthesized at low temperatures are more conductive (Fig. 16.4) because the extent of chain alignment is higher for the low temperature material (Table 16.2).

Recall from Fig. 16.3 that by controlling the polymerization time we can prepare tubules with very thin walls (Fig. 16.3A) or tubules with thick walls (Fig. 16.3B). By obtaining PIRAS data as a function of polymerization time, we can explore the extent of polymer chain alignment in the layer of conductive polymer that is deposited directly onto the polycarbonate (short polymerization times) and in subsequently deposited layers (long polymerization times). Figure 16.6 shows the results of such an experiment. We find that the layer of polypyrrole that is deposited directly on the pore wall is ordered (low dichroic ratio) but that the extent of order decreases in subsequently deposited layers (dichroic ratio approaches unity). Analogous results were obtained with polyaniline tubules [44].

These data show that a template-synthesized conductive polymer fibril or tubule has a layer of ordered polymer chains at its outer surface and that the extent of this chain order decreases toward the center of the nanostructure. This "anatomy" is shown schematically in Fig. 16.7. The narrowest template-synthesized tubules and fibrils have the highest conductivities because they contain proportionately more of the ordered (and less of the disordered) material (Fig. 16.7).

The template-synthesized conductive polymer nanostructures have this interesting anatomy because the polycarbonate template membranes are stretch-oriented during processing [44]. As a result, the polycarbonate chains on the pore walls are aligned. When conductive polymer is grown on these aligned chains, the conductive polymer chains also become aligned. This concept of using the polymer chains in a prealigned substrate to align conductive polymer chains synthesized on the surface of the substrate has been demonstrated for other systems [51]. The disordered central core results be-

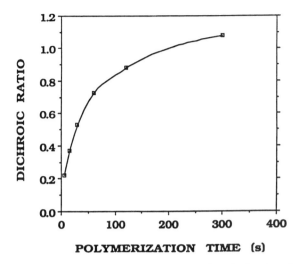

Fig. 16.6 Dichroic ratio for polypyrrole tubules (synthesized in a polycarbonate membrane with 400 nm diameter pores) as a function of polymerization time. Since polymerization time controls tubule wall thickness (see Fig. 16.3), the x axis is also a wall thickness axis. The band at 1560 cm^{-1} (see Fig. 16.8A) was used to obtain the data.

cause the order-inducing influence of the pore wall is ultimately lost in subsequently deposited layers. An analogous effect occurs when polypyrrole is electrochemically deposited on electrode surfaces [52]; i.e., the first layer of polymer chains lies parallel to the electrode surface, but this preferential orientation is lost in subsequently deposited layers.

Finally, it is worth noting that in the two-point conductivity method (Fig. 16.4) we measure conductivity in

Table 16.2 Dichroic Ratios (R) for Polypyrrole Fibrils Synthesized at Two Temperatures

Diameter (nm)	R	
	0°C	25°C
600	—	1.01 ± 0.04
400	0.95 ± 0.04	0.97
50	0.40	0.61
30	0.11	0.17

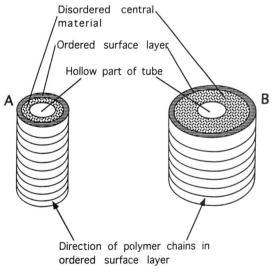

Fig. 16.7 Anatomy of the polypyrrole tubules. (A) Small and (B) large outside diameters. (From Ref. 50.)

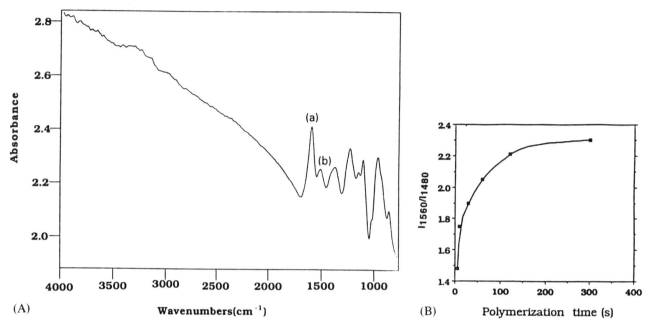

Fig. 16.8 (A) Infrared spectrum of polypyrrole showing the 1560 cm^{-1} (a) and 1480 cm^{-1} (b) bands. (B) Plot of I_{1560}/I_{1480} vs. polymerization time. Since polymerization time controls tubule wall thickness (see Fig. 16.3) the x axis is also a wall thickness axis. Membrane as in Fig. 16.6.

a direction perpendicular to the chain alignment direction. Nevertheless, conductivity is enhanced relative to that of bulk polypyrrole. MacDiarmid and Epstein [53] observed an analogous enhancement in conductivity in a direction perpendicular to the chain axis in stretch-oriented polyaniline. If we could obtain a thin film sample in which the axes of all the tubes were aligned, we could measure the conductivity in a direction parallel to the chain axis. In analogy to the MacDiarmid–Epstein data, this should yield even higher conductivities and high conduction anisotropies. We are currently attempting to prepare such uniaxial films.

B. Molecular Structure: Extended Conjugation

Defects that interrupt conjugation include sp^3 carbons, carbonyls, and twists and bends in the polymer chain [15,43,54–56]. We have used a variety of methods, including X-ray photoelectron spectroscopy (XPS) and UV-vis/near-IR and Fourier transform infrared (FTIR) spectroscopies to identify and quantify defect sites within conductive polymers [15,43,54–56]. Perhaps the most useful of these methods is a new IR-based approach, developed in these laboratories, for probing the conjugation length in polypyrrole [43]. This method is based on theoretical work by Tian and Zerbi [57,58]. We have used this method to show that conjugation lengths in our template-synthesized polypyrroles are longer than in conventional polypyrrole.

Tian and Zerbi [57,58] conducted a theoretical analysis of the vibrational spectra of polypyrrole. This theory successfully predicts the number and position of the main IR bands and also predicts how the intensities and positions of these bands change with the extent of delocalization along the polymer chains (i.e., with the conjugation length). The bands at 1560 and 1480 cm^{-1} (Fig. 16.8A) are especially affected by changes in the conjugation length. These changes can be most easily visualized by taking the ratio of the integrated absorption intensity of the 1560 cm^{-1} band to the integrated absorption intensity of the 1480 cm^{-1} band; we call this ratio I_{1560}/I_{1480} [43].

According to Tian and Zerbi's analysis [57,58], I_{1560}/I_{1480} is inversely proportional to the extent of delocalization. To test this prediction we chemically synthesized a family of polypyrroles that ranged from being relatively defect-free to having high concentrations of defect sites along the polymer chains; and we used various chemical and instrumental methods of analysis to identify and quantify the various defect sites [43]. As might be expected, polymers with high concentrations of defect sites showed low conductivities, and polymers with lower defect concentrations showed the highest conductivities [43]. More important, we found [43] that Zerbi's prediction concerning the effect of extent of delocalization on I_{1560}/I_{1480} was correct—polymers with high concentrations of defect sites (short conjugation lengths) showed high values of I_{1560}/I_{1480} whereas polymers with low concentrations of defect sites (long conjugation lengths) showed low values of I_{1560}/I_{1480}.

Figure 16.8B shows the application of this method to template-synthesized polypyrrole tubules [1]. Like the PIRAS data in Fig. 16.6, FTIR data were obtained at various times during the polymerization of polypyrrole tubules within a template membrane that contained 400 nm diameter pores. Note that I_{1560}/I_{1480} increases with polymerization time (i.e., with wall thickness). This clearly shows that the layer of polypyrrole that is deposited directly on the pore wall has extended conjugation relative to subsequently deposited layers. Hence, these data mirror the PIRAS data shown in Fig. 16.6.

Figure 16.9 shows another application of this method to the template-synthesized materials. We have previously shown that materials synthesized at low temperatures are more conductive (Fig. 16.4). The relative intensities of the ring-stretching bands in Fig. 16.9 show that the low temperature material has longer conjugation lengths. Hence, the higher conductivity in the low temperature material is attributable to both greater chain alignment (PIRAS, Table 16.2) and extended conjugation (FTIR, Fig. 16.9). Finally, the effect of synthesis temperature on conductivity can be explained as follows. We and others have shown that polypyrrole synthesized at low temperature has a longer conjugation length than polypyrrole synthesized at higher temperature [56]. As a result, the low temperature synthesized material is more conductive. Low temperature synthesized polypyrroles have extended conjugation because defect-forming reactions have higher activation energies than the desired α–α linking reaction [56].

These analyses of conductivity and extent of supermolecular and molecular order have led us to an important generalization; Alignment of polymer chains (improvement in supermolecular order) typically produces an increase in conjugation length (improvement in molecular level order). This observation makes sense because in order to align, the polymer chains must be linear (i.e., straight), and linear chains will have fewer kinks and bends that interrupt conjugation. Finally, by improving molecular and supermolecular order in this way, we obtain materials with enhanced electronic conductivities.

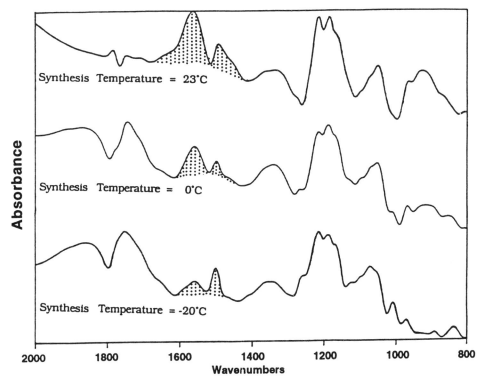

Fig. 16.9 FTIR data for 30 nm diameter polypyrrole fibrils synthesized at three temperatures. Bands at 1560 cm^{-1} and 1480 cm^{-1} are highlighted.

VI. CONDUCTION MECHANISM IN THE TEMPLATE-SYNTHESIZED MATERIALS

How might the unique "anatomy" (Fig. 16.7) of the template-synthesized nanostructures affect the mechanism of electronic conduction in thin films prepared from these materials? We have been exploring this question in collaboration with Professor H. D. Hochheimer in the Physics Department here at Colorado State University and Dr. P.-H. Hor of the Texas Center for Superconductivity [50,59–63]. These investigations (which are very much ongoing) entail measurements of both the temperature and pressure dependence of conductivity in thin films prepared from our template-synthesized nanostructures. Preliminary results of the effect of temperature on conductivity are briefly reviewed here.

According to the Mott variable-range hopping (MVRH) model for conduction in solids, the temperature (T) dependence of electronic conductivity (s) is given by

$$s = K_0 T^{-1/2} \exp[-(T_0/T)^{1/n}] \quad (1)$$

where K_0 and T_0 are constants and $n = 4, 3,$ or 2 for three-dimensional (3D), 2D, and 1D conduction, respectively [64,65]. According to Eq. (1), resistance (R) data for conductive polymer films can be analyzed via plots of $\ln(RT^{-1/2})$ vs. $T^{-1/n}$. The dimensionality of conduction can be obtained from the value of n that yields the best linear plot. The Mott temperature parameter, T_0, can be obtained from the slope. T_0 is directly proportional to the density of states at the Fermi level and inversely proportional to the localization length. By conducting such analyses at various applied pressures, information about the effect of pressure on extent of delocalization can be obtained.

Figure 16.10A shows a plot of $\ln(RT^{-1/2})$ vs. $T^{-1/4}$ for a film prepared from 400 nm diameter polypyrrole tubules [50]. The points are the experimental data, and the solid curve is the least-squares best fit line. The $\ln(RT^{-1/2})$ data for this film show a $T^{-1/4}$ dependence over a temperature range from about 256 to 20 K. These data indicate that 3D MVRH is appropriate for conduction in this film over this temperature range. To prove this point, we ratioed the measured resistance (R_m) to the best-fit resistance (R_f) at each temperature. These data were then plotted as $\ln(R_m/R_f)$ vs. T, and, as indicated in Fig. 16.10B, this analysis was done using R_f values calculated using the 3D, 2D, and 1D MVRH models. This analysis clearly shows that only the 3D model fits the experimental data for this sample. Identical results (i.e., 3D MVRH) were obtained for films prepared from 400 nm diameter polyaniline tubules [50]. The 3D model has been shown to be applicable to polypyrrole and polyaniline samples prepared by conventional synthetic methods [66–68].

Figure 16.11A shows analogous data for a film pre-

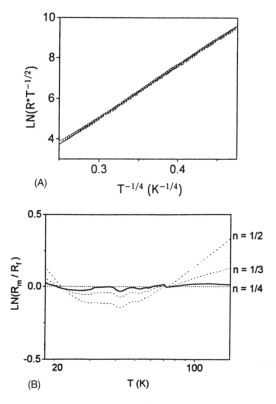

Fig. 16.10 (A) Resistance data for films prepared from 400 nm diameter polypyrrole tubules plotted according to 3D MVRH. (B) Residual analysis showing that only the 3D model fits the data.

pared from 50 nm diameter polypyrrole tubules. In this case, the data show a $T^{-1/3}$ dependence, indicating 2D MVRH (see Fig. 16.11B for proof). Identical results (i.e., 2D MVRH) were obtained for films prepared from small diameter polyaniline tubules [50]. These data show that the dimensionality of conduction can be predictably and reproducibly changed in template-synthesized conductive polymer samples.

Knotek and coworkers [69,70] investigated the dimensionality of conduction in amorphous semiconductor films. When the film was thick (>400 nm), 3D conduction was observed, whereas in thinner films conduction was two-dimensional. They suggest that when the film thickness becomes small relative to the hopping distance (thin films), conduction is constrained to two dimensions. In contrast, when the film thickness is large relative to the hopping distance (thick films), conduction occurs in three dimensions.

We have used analogous arguments to explain the change in dimensionality of conduction in our materials [50]. As indicated in Fig. 16.7, our nanostructures consist of a highly ordered (and highly conductive) surface skin surrounding a disordered (and lower conductivity) core. Because the small diameter tubules have a larger

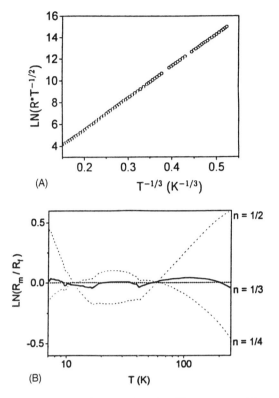

Fig. 16.11 (A) Resistance data for films prepared from 50 nm diameter polypyrrole tubules plotted according to 2D MVRH. (B) Residual analysis showing that only the 2D model fits the data. (From Ref. 50.)

proportion of the ordered material (Fig. 16.7A), conduction in the surface layer predominates. Because this surface layer is thin (~ 5 nm), conduction is constrained to two dimensions (Fig. 16.11). Epstein et al. [71] observed an analogous 2D conduction process in films composed of polyacetylene fibrils. They suggest that 2D conduction results because only a thin layer at the surface of the fibril is doped.

In contrast, the large diameter tubules have a larger proportion of disordered material (Fig. 16.7B). As a result, conduction in the disordered phase predominates despite its lower intrinsic conductivity. Because the disordered layer is thick, conduction can occur in three dimensions (Fig. 16.10). While this proposed explanation is consistent with the experimental data, it is clear that we have just scratched the surface in our investigations and analyses of conduction in these unique polymeric materials. These preliminary data set the stage, and point the direction, for future research.

VII. CONCLUSIONS

We have learned that template synthesis provides a route for controlling the extent of molecular and supermolecular order in electronically conductive polymers. This allows us to predictably vary not only the magnitude of the conductivity but also the conduction mechanism within the material. For these reasons, our template-synthesized nanostructures are proving to be useful materials for exploring the fundamentals of the conduction process in conductive polymers. Such investigations are under way in our laboratories and in a number of other labs around the world.

In addition, while I have not been able to discuss it here, these template-synthesized nanostructures have a number of possible technological and commercial applications. For example, we have recently shown that capped versions of our polypyrrole microtubules can be used for enzymatic bioencapsulation to make a new type of microbioreactor [9]. This tubule-based microbioreactor might find applications in biosensors [30,72,73]. In addition, the groups headed by Professors Inganäs and Granström of the Department of Physics, University of Linköping have recently shown that the template approach can be used to prepare nanoscopic polymeric light-emitting diodes [33].

Finally, as indicated in the Introduction, the template approach is a universal method for preparing nanomaterials. We and others have shown that this method can be used to make nanotubules and fibrils of polymers, metals, semiconductors, carbons, and other materials. This creates the interesting possibility of preparing nanowires and nanotubules that are composed (in a spatially controlled fashion) of more than one material. The simplest example is a nano-Schottky barrier composed of a nanowire segment of a semiconducting material in contact with a nanowire segment of an appropriate metal. We have recently described template-synthesized devices of this type [24].

Cylindrical junctions (i.e., concentric tubules of different materials) should also be possible. For example, consider a nanocapacitor that is 50 nm in diameter and consists of an outer tube of conductive polymer surrounding an inner tube of an insulating plastic (e.g., polystyrene) surrounding a solid nanowire of a second conductive polymer. One of our alumina template membranes (Fig. 16.1C) could contain 10^{10} of these nanocapacitors per square centimeter of membrane area. We have recently prepared such ensembles of nanocapacitor-type devices (74). Analogous ideas will lead to nanobatteries, nano fuel cells and nanoelectronic and electro-optical devices. This next generation of work will be one focus of our research efforts well into the 21st century.

ACKNOWLEDGMENTS

First, I would like to acknowledge the invaluable help of my professional colleagues Prof. H.D. Hochheimer of the Department of Physics, Colorado State University;

Drs. P.-H. Hor and J. Bechtold of the Texas Center for Superconductivity, University of Houston; and Dr. S.W. Tozer of the National High Magnetic Field Laboratory, Florida State University. Second, this work would not have been possible without the efforts of a number of hardworking and highly motivated graduate students and postdoctoral students. They include Vinod P. Menon, Zhihua Cai, Junting Lei, Wenbin Liang, Ranjani V. Parthasarathy, Gabor L. Hornyak, Leon S. Van Dyke, and Reginald M. Penner. Finally, financial support from the Office of Naval Research and the National Science Foundation is also gratefully acknowledged.

REFERENCES

1. C. R. Martin, *Acc. Chem. Res.* 28:61–68 (1995).
2. C. R. Martin, *Science* 266:1961–1966 (1994).
3. G. A. Ozin, *Adv. Mater.* 4:612–649 (1992).
4. *Science* Nov. 29, 1991. This issue contains a series of articles and features under the title "Engineering a Small World: From Atomic Manipulation to Microfabrication."
5. R. T. Bate, *Sci. Am.* 259:96–100 (1988).
6. M. H. Devoret, D. Esteve, and C. Urbina, *Nature* 360:547–553 (1992).
7. J. L. Jewell, J. P. Harbison, and A. Scherer, *Sci. Am.* 265:86–94 (1991).
8. R. Gref, Y. Minamitake, M. T. Peracchia, V. Trubetskoy, V. Torchilin, and R. Langer, *Science* 263:1600–1603 (1994).
9. R. Parthasarathy and C. R. Martin, *Nature* 369:298–301 (1994).
10. R. M. Penner and C. R. Martin, *J. Electrochem. Soc.* 133:2206–2207 (1986).
11. Z. Cai and C. R. Martin, *J. Am. Chem. Soc.* 111:4138–4139 (1989).
12. L. S. Van Dyke and C. R. Martin, *Langmuir* 6:1123–1132 (1990).
13. W. Liang and C. R. Martin, *J. Am. Chem. Soc.* 112:9666–9668 (1990).
14. C. R. Martin, *Adv. Mater.* 3:457–459 (1991).
15. Z. Cai, J. Lei, W. Liang, V. Menon, and C. R. Martin, *Chem. Mater.* 3:960–966 (1991).
16. C. R. Martin, R. Parthasarathy, and V. Menon, *Synth. Met.* 55–57:1165–1170 (1993).
17. R. M. Penner and C. R. Martin, *Anal. Chem.* 59:2625–2630 (1987).
18. C. J. Brumlik and C. R. Martin, *J. Am. Chem. Soc.* 113:3174–3175 (1991).
19. C. A. J. Foss, G. L. Hornyak, J. A. Stockert, and C. R. Martin, *J. Phys. Chem.* 1992:7497–7499 (1992).
20. C. J. Brumlik, C. R. Martin, and K. Tokuda, *Anal. Chem.* 64:1201–1203 (1992).
21. C. A. J. Foss, G. L. Hornyak, J. A. Stockert, and C. R. Martin, *Adv. Mater.* 5:135–136 (1993).
22. C. A. J. Foss, G. L. Hornyak, J. A. Stockert, and C. R. Martin, *J. Phys. Chem.* 98:2963–2971 (1994).
23. C. J. Brumlik, V. P. Menon, and C. R. Martin, *J. Mater. Res.* 9:1174–1183 (1994).
24. J. D. Klein, R. D. I. Herrick, D. Palmer, M. J. Sailor, C. J. Brumlik, and C. R. Martin, *Chem. Mater.* 5:902–904 (1993).
25. R. V. Parthasarathy and C. R. Martin, *Adv. Mater.*, 7:896–897 (1995).
26. C.-G. Wu and T. Bein, *Science* 264:1757–1759 (1994).
27. S. N. Atchison, R. R. 'Burford, T. A. Darragh, and T. Tontong, *Polym. Int.* 26:261–266 (1991).
28. R. P. Burford and T. Tongtam, *J. Mater. Sci.* 26:3264–3270 (1991).
29. W. Cahalane and M. M. Labes, *Chem. Mater.* 1:519 (1989).
30. R. Czajka, C. G. J. Koopal, M. C. Feiters, J. W. Gerritsen, R. J. M. Nolte, and H. Van Kempen, *Biochem. Bioenerg.* 29:47–57 (1992).
31. C. G. J. Koopal and R. J. M. Nolte, *J. Chem. Soc. Chem. Commun.* 1991:1691.
32. M. Granstrom and O. Inganas, *Synth. Met.* 55–57:460–465 (1993).
33. M. Granstrom, M. Berggren, and O. Inganas, *Science* 267:1479–1481 (1995).
34. T. Kyotani, L.-F. Tsai, and A. Tomita, *Chem. Mater.* 7:1427–1432 (1995).
35. C.-G. Wu and T. Bein, *Science* 266:1013–1015 (1994).
36. R. L. Fleischer, P. B. Price, and R. M. Walker, *Nuclear Tracks in Solids*, Univ. California Press, Berkeley, 1975.
37. A. Despic and V. P. Parkhutik, in *Modern Aspects of Electrochemistry* (J. O. Bockris, R. E. White, and B. E. Conway, eds.), Plenum, New York, 1989, Vol. 20, Chap. 6.
38. D. AlMawiawi, N. Coombs, and M. Moskovits, *J. Appl. Phys.* 70:4421–4425 (1991).
39. R. J. Tonucci, B. L. Justus, A. J. Campillo, and C. E. Ford, *Science* 258:783–785 (1992).
40. J. S. Beck, J. C. Vartuli, W. J. Roth, M. E. Leonowicz, C. T. Kresge, K. D. Schmitt, C. T.-W. Chu, D. H. Olson, E. W. Sheppard, S. B. McCullen, J. B. Higgins, and J. L. Schlenker, *J. Am. Chem. Soc.* 114:10834–10843 (1992).
41. K. Douglas, G. Devaud, and N. A. Clark, *Science* 257:642–644 (1992).
42. R. M. Penner, L. S. Van Dyke, and C. R. Martin, *J. Phys. Chem.* 92:5274–5282 (1988).
43. J. Lei, Z. Cai, and C. R. Martin, *Synth. Met.* 46:53–69 (1992).
44. R. V. Parthasarathy and C. R. Martin, *Chem. Mater.* 6:1627–1632 (1994).
45. J. Lei, V. P. Menon, and C. R. Martin, *Polym. Adv. Tech.* 4:124–132 (1992).
46. C. R. Martin, L. S. Van Dyke, Z. Cai, and W. Liang, *J. Am. Chem. Soc.* 112:8976–8977 (1990).
47. R. Pool, *Science* 247:1410 (1990).

48. L. Monnerie, in *Developments in Oriented Polymers* (I. M. Ward, ed.), Elsevier, London, 1987, Vol. 2, p. 199.
49. R. Zbinden, *Infrared Spectroscopy of High Polymers*, Academic, New York, 1964, p. 186.
50. J. P. Spatz, B. Lorenz, K. Weishaupt, H. D. Hochheimer, V. P. Menon, R. V. Parthasarathy, C. R. Martin, J. Bechtold, and P.-H. Hor, *Phys. Rev. Lett. 50*: 14,888–14,892 (1994).
51. J. C. Wittmann and P. Smith, *Nature 352*:414–417 (1991).
52. Z. Cai and C. R. Martin, *J. Electroanal. Chem. 300*: 35–50 (1991).
53. A. G. MacDiarmid and A. J. Epstein, *Synth. Met. 65*: 103 (1994).
54. J. Lei, W. Liang and C. R. Martin, *Synth. Met. 48*: 301–312 (1992).
55. J. Lei and C. R. Martin, *Synth. Met. 48*:331–336 (1992).
56. W. Liang, J. Lei, and C. R. Martin, *Synth. Met. 52*: 227–239 (1992).
57. B. Tian and G. Zerbi, *J. Chem. Phys. 92*:3886 (1990).
58. B. Tian and G. Zerbi, *J. Chem. Phys. 92*:3892 (1990).
59. H. D. Hochheimer, B. Lorenz, S. W. Tozer, V. Menon, R. Parthasarathy, C. R. Martin, J. Bechtold, and P.-H. Hor, *Bull. Am. Phys. Soc. 40*:345 (1995).
60. J. P. Spatz, B. Lorenz, H. D. Hochheimer, V. Menon, R. Parthasarathy, C. R. Martin, J. Bechtold, and P.-H. Hor, *Bull. Am. Phys. Soc. 39*:622 (1994).
61. J. P. Spatz, B. Lorenz, H. D. Hochheimer, V. Menon, R. Parthasarathy, C. R. Martin, J. Bechtold, and P.-H. Hor, *Bull. Am. Phys. Soc. 39*:160 (1994).
62. B. Lorenz, J. P. Spatz, H. D. Hochheimer, V. Menon, R. Parthasarathy, C. R. Martin, J. Bechtold, and P.-H. Hor, *Phil. Mag. B 71*:929–940 (1995).
63. B. Lorenz, J. P. Spatz, H. D. Hochheimer, V. Menon, R. Parthasarathy, and C. R. Martin, in *High Pressure in Materials Science and Geoscience* (J. Kamarad, Z. Arnold, and A. Kapicka, eds.), Proc. 32nd Euro. High Pressure Research Group Conf. Brno, Czech Republic, 1994, p. 73.
64. J. C. W. Chien, *Polyacetylene*, Academic, New York, 1984, pp. 485–488.
65. N. F. Mott and E. A. Davis, *Electronic Processes in Noncrystalline Materials*, 2nd ed., Clarendon Press, Oxford, 1979, pp. 32–37.
66. A. K. Meikap, A. Das, S. Chatterjee, M. Digar, and S. N. Bhattacharyya, *Phys. Rev. B47*:1340 (1993).
67. R. Menon, Y. Cao, D. Moses, and A. J. Heeger, *Phys. Rev. B47*:1758 (1993).
68. K. Sato, M. Yamaura, T. Hagiwara, K. Murata, and M. Tokumoto, *Synth. Met. 40*:35 (1991).
69. M. L. Knotek, M. Pollak, T. M. Donovan, and H. Kurtzman, *Phys. Rev. Lett. 30*:853 (1973).
70. M. L. Knotek, *Solid State Commun. 17*:1431 (1975).
71. A. J. Epstein, H. W. Gibson, P. M. Chailin, W. G. Clark, and G. Gruner, *Phys. Rev. Lett. 45*:1730 (1980).
72. C. G. J. Koopal, R. J. M. Nolte, and B. De Ruiter, *J. Chem. Soc. Chem. Commun. 1991*:1691.
73. S. Kuwabata and C. R. Martin, *Anal. Chem. 66*:2757 (1994).
74. V. M. Cepak, J. C. Hulleen, G. Che, K. B. Jirage, B. B. Lakshmi, E. R. Fisher, and C. R. Martin, *Chem. Matr.*, in press.

17
Colloidal Dispersions of Conducting Polymers

Steven P. Armes
University of Sussex, Brighton, East Sussex, England

I. INTRODUCTION

Organic conducting polymers (OCPs) are usually prepared in the form of intractable films, gels, or powders that are insoluble in common organic solvents. For example, polyacetylene is typically synthesized by the polymerization of gaseous acetylene monomer using a soluble Ziegler–Natta catalyst system such as $Ti(OBu^t)_4$-$4AlEt_3$ in nonpolar solvents such as toluene or n-hexane [1,2]. Depending on the reaction conditions, the polyacetylene is obtained as an insoluble film, precipitate, or gel. On the other hand, polypyrrole and polyaniline are usually synthesized using chemical oxidants [e.g., $FeCl_3$ or $(NH_4)_2S_2O_8$] in either acidic aqueous media [3,4] or certain nonaqueous solvents such as ethers [5], esters [5], alcohols [6], or acetonitrile [7]. In each case the conducting polymer is invariably obtained as an insoluble bulk powder, with compressed pellet conductivities lying in the range 1–200 S/cm. Scanning electron microscopy studies have confirmed that these powders have a fused pseudospherical or globular morphology with submicrometer features [8,9]. Typical BET surface areas for these powders are 10–20 m^2/g, which correspond to grain sizes of 200–500 nm [8–10]. These observations suggest that, provided agglomeration of the precipitating conducting polymer nuclei can be prevented, it should be possible to obtain discrete, microscopic conducting polymer particles.

Over the last decade or so there has been enormous worldwide interest in improving the processing characteristics of OCPs, with many hundreds of research papers published in this area.* This chapter summarizes the recent progress made in the synthesis of processible conducting polymers in the form of colloidal dispersions.

II. COLLOID STABILITY

In general, a colloidal system is one that comprises a dispersed phase (solid, liquid, or gas) within a continuous phase (solid, liquid, or gas) with one or more dimensions of the dispersed phase lying in the size range 1–1000 nm [11,12]. All the conducting polymer colloid systems discussed in this article are solid/liquid colloids, i.e., insoluble, microscopic conducting polymer-based particles dispersed in a liquid phase (usually water). These colloids are generally prepared via dispersion polymerization [13]. Initially the monomer and oxidant species are dissolved in a homogeneous reaction solution; as the polymerization proceeds, the system becomes heterogeneous owing to precipitation of insoluble, microscopic conducting polymer particles. Such particles are usually stabilized with respect to aggregation by either a steric or a charge stabilization mechanism. This approach is necessary because conducting polymers, like conventional metals, have high Hamaker constants [14], i.e., strong interchain attractive forces exist in these materials. Steric stabilization involves the use of a suitable polymeric stabilizer that adsorbs onto the conducting polymer particles to form a thick solvated layer. This outer layer of adsorbed polymer provides a steric barrier to particle aggregation [15], thus resulting in a stable dispersion of colloidal conducting polymer particles. Unlike metals [16] and semiconductors [17], conducting polymers synthesized in aqueous media do not appear to form charge-stabilized colloidal particles spontaneously, even at high dilution. This is probably due to electrolyte-induced flocculation owing to the relatively high concentration of chemical oxidants

---
* See the Proceedings for the biennial International Conferences on Synthetic Metals (ICSM) for 1988, 1990, 1992, and 1994, all published in *Synth. Meth.*

such as $FeCl_3$ or $(NH_4)_2S_2O_8$ used in such syntheses. However, it is now possible to prepare charge-stabilized conducting polymer-based nanocomposite particles using ultrafine silica sols (see Section VI).

III. STERICALLY STABILIZED POLYACETYLENE COLLOIDS

The Bristol group have published two papers describing the synthesis of polyacetylene colloids via dispersion polymerization in n-hexane–THF mixed solvent media using two different tailormade block copolymer stabilizers: poly(t-butylstyrene-b-ethylene oxide) and poly(styrene-b-isoprene) [18,19]. Both block copolymers were synthesized via anionic polymerization techniques using sequential monomer addition. In this approach one part of the block is designed to be solvatophobic [e.g., the poly(ethylene oxide) and polystyrene components, respectively] and is adsorbed onto the surface of the polyacetylene particles; the other part of the block is solvated and provides the steric barrier that prevents particle aggregation. The polyacetylene colloids produced from these syntheses had spherical morphologies but were rather polydisperse, with particle diameters in the range 40–200 nm. Unfortunately, the well-documented air sensitivity of polyacetylene is exacerbated by its synthesis as a high surface area colloid; routine cleanup operations such as centrifugation and redispersion are both tedious and problematic for such dispersions [20]. A further problem is that the steric stabilizer appears to compete with the polyacetylene core for the iodine dopant during the postpolymerization doping step [19,20]. This results in relatively low doping levels and poor conductivities ($<10^{-4}$ S/cm). In view of these difficulties, and given the general shift in research activity from polyacetylene toward more airstable conducting polymers such as polypyrrole and polyaniline, it is rather doubtful whether any further studies on polyacetylene colloids can be justified.

IV. STERICALLY STABILIZED POLYPYRROLE COLLOIDS

Polypyrrole colloids can be easily synthesised via dispersion polymerization using a wide range of commercial water-soluble nonionic polymers such as methylcellulose [21], poly(vinyl alcohol) [22,23], poly(N-vinyl pyrrolidone) [14,20–22], poly(vinyl methyl ether) [24,25], and poly(ethylene oxide) [26–28]. Cationic [29,30] and anionic [31] polyelectrolytes have also been successfully used as polymeric stabilizers. The water-soluble polymer is physically adsorbed onto the surface of the growing polypyrrole particles, probably via a hydrogen bonding mechanism in many cases (see Fig. 17.1). The physicochemical properties of this steric stabilizer layer can have a profound influence on the colloidal stability of the dispersions. For example, both methylcellulose- and poly(methyl vinyl ether)-stabilized polypyrrole particles can be reversibly flocculated in hot aqueous solution due to the inverse-temperature solubility behavior exhibited by these adsorbed polymeric stabilizers [21,24]. Similarly, poly(vinyl pyridine)-stabilized polypyrrole particles can be reversibly flocculated or redispersed simply by altering the pH of the aqueous solution [29,30].

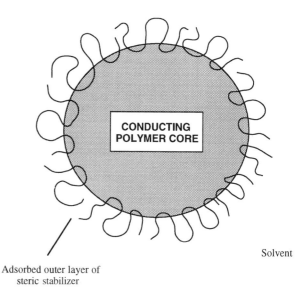

Fig. 17.1 Schematic representation of an isolated, sterically stabilized conducting polymer particle.

The morphology of polypyrrole particles prepared in aqueous media is usually spherical and generally fairly monodisperse as evidenced by transmission electron microscopy (see Fig. 17.2). The particle size is strongly dependent on the nature of the water-soluble polymer stabilizer, for example, Armes et al. [28] reported that the average particle diameter can be varied over the range 66–300 nm. Thus, the use of poly(vinyl alcohol) stabilizers usually leads to relatively small polypyrrole particles of ~ 66–100 nm diameter depending on the stabilizer molecular weight, whereas the use of poly(ethylene oxide) yields rather larger particles (~ 300 nm). The Bristol group reported that polypyrrole particles as small as 30 nm and as large as 445 nm can be prepared given the correct stabilizer-oxidant combination [26]. Dynamic light scattering studies by Rawi et al. [32] and Stejskal et al. [33] indicate that the polypyrrole particles have a high degree of dispersion and that the solvated, adsorbed layer of steric stabilizer is relatively thin (<25 nm). There have been only two reports on the use of tailormade block copolymers as stabilizers for polypyrrole particles. Beadle et al. [34] reported that a poly(N,N'-dimethylaminoethyl methacrylate-b-n-butyl methacrylate) block copolymer was a rather inefficient stabilizer that yielded weakly flocculated polypyrrole particles. Similarly, Munk and coworkers [35] used poly-

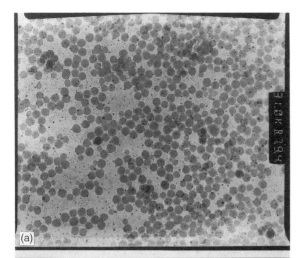
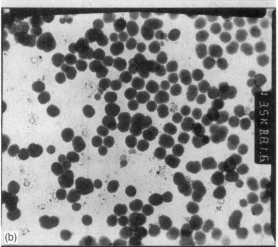
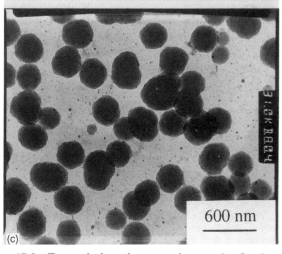

Fig. 17.2 Transmission electron micrograph of polypyrrole particles of (a) 100 nm diameter, (b) 200 nm diameter, and (c) 300 nm diameter synthesized in aqueous media using a poly(vinyl alcohol) stabilizer, a poly(2-vinyl pyridine-*co*-*n*-butyl methacrylate) stabilizer, and a poly(ethylene oxide) stabilizer, respectively. Note the uniform spherical morphology. (Reproduced with permission from Ref. 28.)

(styrene-*b*-ethylene oxide) block copolymer stabilizers and suggested that in this case pyrrole polymerization occurs within the swollen block copolymer micelles.

The stabilizer content of the polypyrrole particles can vary from 3% to more than 50% by mass depending on the particle size and stabilizer type. Stabilizer contents have been determined using direct methods such as elemental microanalyses of the dried colloid [24,25,31]; for polymeric stabilizers that contain no nitrogen, the reduced nitrogen content of the colloid relative to polypyrrole "bulk powder" allows the polypyrrole, and hence the stabilizer, content to be calculated. This method can also be used if the nitrogen content of the stabilizer is nonzero, provided that there is a sufficient difference between the nitrogen content of the stabilizer and the polypyrrole components [34]. Alternatively, indirect methods based on UV-vis, Raman, or FTIR spectroscopic assays of the postreaction supernatant solution for nonadsorbed polymeric stabilizer have been developed [24,29,30,36].

Electrical conductivities are usually reported for compressed pellets made from the dried colloidal particles using the four-point probe technique [37]. The observed conductivity is invariably at least an order of magnitude lower than the conductivity of the corresponding conducting polymer bulk powder. This is not really surprising considering the presence of electrically insulating polymeric stabilizer and the increased number of resistive interparticle contacts at the submicrometer level. Nevertheless, conductivities of up to 10 S/cm can be achieved with the addition of co-dopants [38]. Naively, it might be expected that those dispersions that contained the least electrically insulating stabilizer would be more conductive, but this does not necessarily appear to be the case. Thus Armes et al. [28–30,34] reported that a polypyrrole colloid that contained only 3% poly(ethylene oxide) stabilizer had a significantly *lower* conductivity (by two to three orders of magnitude) than various polypyrrole colloids containing 10–50% stabilizer. It is quite plausible that the spatial distribution of the stabilizer layer on the polypyrrole particles may affect the solid-state conductivity values. Thus lower conductivities might be expected for a continuous stabilizer layer, whereas a "patchy" layer would allow direct electrical contacts between adjacent polypyrrole particles in the solid state. It is noteworthy that Epron et al. [39] carried out microwave measurements on aqueous dispersions of methylcellulose-stabilized polypyrrole particles and obtained electrical conductivities of up to 3 S/cm, which is comparable to the conductivity of polypyrrole bulk powder synthesized in the absence of a steric stabilizer. This "zero-current" technique is probably a more reliable indication of the intrinsic conductivity of the polypyrrole component than conventional four-point probe measurements on pressed pellets, since in this latter approach the overall macroscopic conductivity is almost certainly limited by inefficient charge transport at the microscopic level due to interparticle resistances.

Cawdery et al. [26] described the preparation of polypyrrole colloids in nonaqueous media by an indirect route. Polypyrrole particles were first synthesized in aqueous media using a poly(ethylene oxide) stabilizer and then transferred into 1,4-dioxan via methanol by repeated centrifugation/redispersion cycles. It was claimed, on the basis of elemental microanalyses, that the poly(ethylene oxide) stabilizer was "stripped off" the polypyrrole particles during this solvent exchange. Further indirect experimental evidence for "stripping" was the relatively poor colloid stability of the final dispersion. In a follow-up paper, Markham et al. [14] determined the Hamaker constant for polypyrrole based on the critical coagulation concentrations of electrolyte required to flocculate the "bare" polypyrrole particles. The value obtained was much nearer the Hamaker constants found for metals than those of conventional polymers, which is consistent with the high conductivity of the polypyrrole component. The first "single-step" synthesis of polypyrrole colloids in nonaqueous media was reported by Armes and Aldissi [40] using a poly(vinyl acetate) stabilizer with anhydrous $FeCl_3$ in various alkyl ester solvents. Rather broad, trimodal particle size distributions were obtained, and solid-state conductivities were relatively low (10^{-6}–10^{-1} S/cm). More recently, the same steric stabilizer was used to prepare polypyrrole particles in both 2-methoxyethanol [36] and acetonitrile–methanol [41] mixtures. Finally, an Indian group led by Mandal published several papers on the synthesis of polypyrrole [24,25] and also polyaniline [42] colloids in 50:50 ethanol–water solvent mixtures using a poly(methyl vinyl ether) steric stabilizer.

The vast majority of sterically stabilized polypyrrole colloid syntheses have been carried out using $FeCl_3$ as an oxidant for pyrrole polymerization. This oxidant is known to polymerize pyrrole smoothly to high yield under a wide range of conditions at room temperature [3,43]. Yamamoto's group [44] reported the use of the milder $H_2O_2/Fe^{3+}/HBr$ oxidant system, but there have been very few reports on the use of the $(NH_4)_2S_2O_8$ oxidant [26,31]. According to Bjorklund [45], this latter oxidant polymerizes pyrrole very rapidly compared to the $FeCl_3$ oxidant. Presumably this causes uncontrollably fast nucleation, which subsequently leads to macroscopic precipitation of the conducting polymer.

V. STERICALLY STABILIZED POLYANILINE COLLOIDS

The various commercial water-soluble polymers used for the synthesis of polypyrrole colloids are rather ineffective for the preparation of sterically stabilized polyaniline colloids and usually fail to prevent macroscopic precipitation [46]. One exception to this rule is poly(vinyl alcohol), but even here a relatively high stabilizer concentration is required for the quantitative formation of stable polyaniline dispersions [42,47–50]. On the other hand, tailormade *reactive* copolymers that are capable of chemically grafting (rather than merely physically adsorbing) onto the surface of the growing polyaniline particles are much more efficient steric stabilizers. For example, Armes and coworkers have favored the use of reactive stabilizers that contain pendant aniline groups as graft sites [46,51–56]. These stabilizers are easily synthesized by the free-radical statistical copolymerization of 4-aminostyrene with 2-vinylpyridine, 4-vinylpyridine, N-vinylpyrrolidone, N-vinylimidazole, etc. Alternatively, poly(vinyl alcohol)-based stabilizers can be synthesized via chemical modification [53]. All of these copolymer stabilizers are rather ill-defined; they

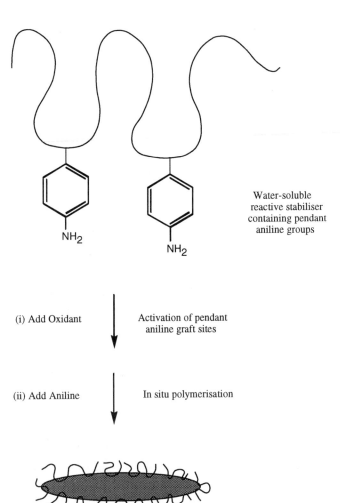

Fig. 17.3 Schematic representation of the formation of sterically stabilized "rice grain" polyaniline particles using a water-soluble reactive steric stabilizer.

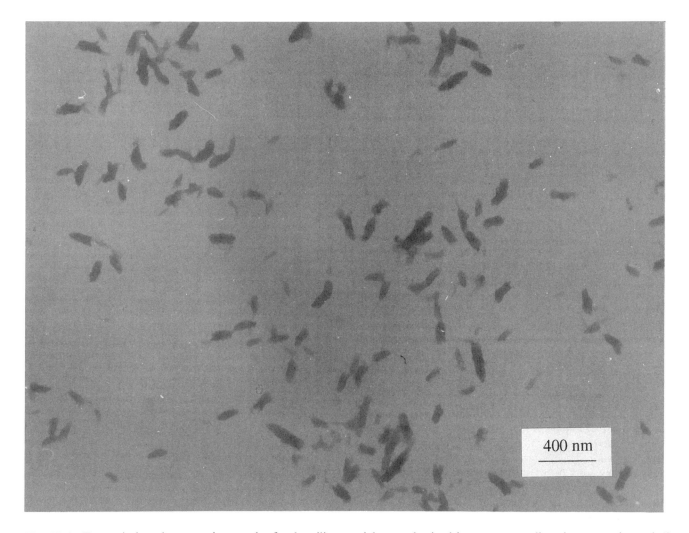

Fig. 17.4 Transmission electron micrograph of polyaniline particles synthesized in aqueous media using a reactive poly(2-vinyl pyridine-co-4-aminostyrene) steric stabilizer. Note the polydisperse "rice grain" morphology. (Reproduced with permission from Ref. 46.)

have rather broad molecular weight distributions ($M_w/M_n > 2.0$) and almost certainly suffer from significant compositional heterogeneity. The average graft site content of these stabilizers is typically determined by ^{1}H NMR spectroscopy and can vary from less than 1 mol % up to 40 mol % [51–56]. There is no doubt that these graft sites are essential for the formation of stable polyaniline dispersions, since in their absence a macroscopic precipitate of the conducting polymer is always obtained. Visible absorption spectroscopy studies on various reactive stabilizers dissolved in aqueous oxidant solutions in the absence of aniline monomer have confirmed that in each case the pendant aniline groups on such stabilizers are readily oxidized [52,56]. Thus it was suggested that such activated stabilizers inevitably become grafted onto the growing polyaniline nuclei during the dispersion polymerization of aniline (see Fig. 17.3). Other reactive stabilizers with terminal methacrylate groups or pendant glycidyl groups have been reported by the Bristol group, but in this case no experimental evidence was offered to support the postulated chemical grafting mechanism [57].

The particle size and morphology of the resulting polyaniline particles is very dependent on the synthesis conditions; "rice grain" morphologies are most common [51–54,56,57] (see Fig. 17.4), but needles [57] and polydisperse spheres [55,57] have also been reported. Precise control over the morphology of conducting polymer colloids can be useful. Cooper and Vincent [58] showed that dispersing needle-shaped polyaniline particles within an acrylic latex film produces a significantly lower conductivity percolation threshold than that obtained using spherical polypyrrole particles. The critical synthesis parameters that determine the polyaniline morphology have not yet been identified, but it is well known that subtle changes in reaction conditions can have dramatic effects. For example, Vincent and Waterson [57] reported that the use of poly(ethylene oxide)-based homopolymer, graft copolymer, and reactive graft copolymer stabilizers can alter the particle morphology from needles to rice grains to polydisperse spheres. Similarly, DeArmitt and Armes [55] showed that changing the chemical oxidant can also profoundly affect the particle morphology. Using a reactive statistical copolymer based on poly(N-vinylpyrrolidone), these workers obtained a rice grain morphology using the KIO$_3$ oxidant but spherical particles with the (NH$_4$)$_2$S$_2$O$_8$ oxidant. Surprisingly, this is one of the few stabilizers to date that is compatible with both types of oxidants. The (NH$_4$)$_2$S$_2$O$_8$ oxidant tends to form insoluble complexes with most cationic stabilizers, e.g., those based on poly(2- or 4-vinylpyridine)s, poly(N-vinylimidazole), etc., and is ineffective at preventing precipitation with the poly(vinyl alcohol)-based stabilizers. On the other hand, the KIO$_3$ oxidant cannot be used successfully with poly(ethylene oxide)-based stabilizers (see below). These observations may be at least partly due to differences in aniline polymerization kinetics between these two oxidants: our in situ ^{1}H NMR spectroscopic studies have confirmed that the (NH$_4$)$_2$S$_2$O$_8$ oxidant polymerizes aniline significantly faster than KIO$_3$ [59]. In general, the conductivities of sterically stabilized polyaniline colloids lie in the range 10^{-3}–10^1 S/cm, perhaps more typically 10^{-1}–10^0 S/cm. The stabilizer content of the polyaniline particles can vary from 11 to 50 wt % depending on the stabilizer type.

The Sussex group have used near-monodisperse, low molecular weight ($M_w/M_n < 1.15$; $M_n < 8800$) poly(ethylene oxide)s containing a single tertiary aromatic amine graft site on each polymer chain as reactive stabilizers for polyaniline colloids [56,59]. However, the performance of such "model" stabilizers proved to be somewhat disappointing; relatively high stabilizer concentrations were required for stable polyaniline dispersions, presumably due to the low level of graft sites, and the particle morphology remained largely unaffected under various synthesis conditions, with only polydisperse "rice grain" particles being obtained. Thus there appear to be no obvious advantages in using well-defined stabilizers of narrow molecular weight.

DeArmitt and Armes have attempted to develop a nonaqueous route to polyaniline particles (personal communication; see also Ref. 60). The most promising results were obtained using a poly(vinyl acetate)-based reactive stabilizer in acetonitrile using the Cu(ClO$_4$)$_2$·6H$_2$O oxidant first reported by Inoue and coworkers [7]. However, the Cu(I) by-products resulting from the use of this oxidant crystallized out during the dispersion polymerization and tended to destabilize the "rice grain"-shaped polyaniline particles. In addition, only substoichiometric oxidant/monomer molar ratios could be used, which in turn led to rather low pressed pellet conductivities (~10^{-4} S/cm).

The Sussex group are currently exploring the use of reactive polymeric stabilizers containing thiophene or pyrrolic graft sites for the synthesis of polypyrrole particles in aqueous media [61]. Like the syntheses of sterically stabilized polyaniline colloids described above, this approach appears to be rather general and, in principle, can be used to prepare polypyrrole particles using new types of water-soluble polymers as steric stabilizers. In principle this approach should allow polypyrrole particles to be synthesized using functional stabilizers, which might offer some advantage(s) in certain biomedical applications such as immunodiagnostic assays (see Section IX).

VI. CONDUCTING POLYMER-SILICA NANOCOMPOSITES

In 1992 the Sussex group reported a serendipitous but remarkable discovery. Polymerization of aniline in the presence of ultrafine silica particles of 20 nm diameter

Colloidal Dispersions of Conducting Polymers

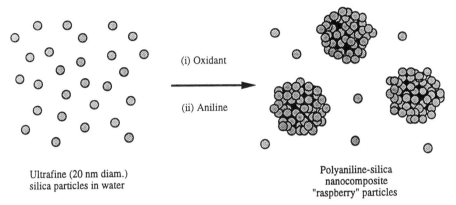

Fig. 17.5 Schematic representation of the formation of polyaniline-silica nanocomposite particles obtained by the in situ polymerization of aniline in the presence of ultrafine silica particles.

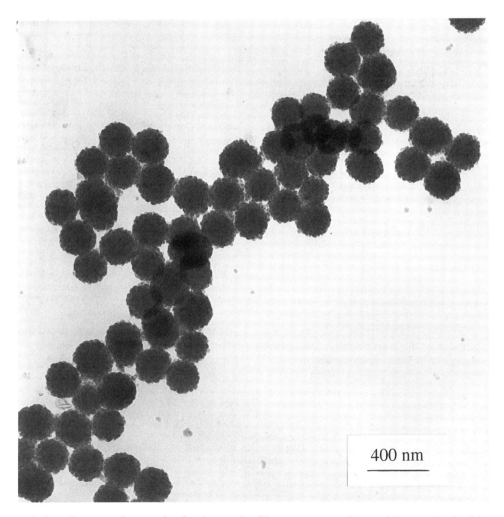

Fig. 17.6 Transmission electron micrograph of polypyrrole-silica nanocomposite particles synthesized in aqueous media using a commercial 20 nm silica sol. Note the unusual "raspberry" morphology. (Reproduced with permission from Ref. 64.)

results in the formation of stable polyaniline colloids in the *absence* of any added polymers or surfactants [62]. Subsequently, it was confirmed that these "polyaniline" colloids actually consist of submicrometer aggregates of the ultrafine silica particles "glued" together by the conducting polymer component (see Fig. 17.5) [63]. The same synthetic approach has been used to synthesize the analogous polypyrrole-silica particles [64,65]. These two systems have been studied extensively over the last few years. The conducting polymer loading within the particles can be easily varied from 20 to 80 wt% simply by varying the synthesis conditions (e.g., silica size, initial silica concentration, type of chemical oxidant). The particles have an unusual "raspberry" morphology (see Fig. 17.6), and their particle diameters can be readily controlled over the 100–500 nm size range (Refs. 62–65 and Lascelles and Armes, unpublished results). The excess ultrafine silica particles are easily separated from the larger "raspberry" particles by repeated centrifugation-redispersion cycles. Nonspherical, elongated, or "stringy" silica particles can also be used to prepare stable conducting polymer-silica dispersions [66].

Terrill et al. [67] reported a small-angle X-ray scattering study on polyaniline-silica particles. Their results indicate an average silica–silica separation distance of approximately 4 nm within the "raspberries" [67]. Thus this study confirms that these particles are true nanocomposites (since both the silica and the conducting polymer components have nanoscale dimensions) and suggests that the polyaniline chains are dispersed within the silica clusters at, or near, the molecular level. Complementary small-angle neutron scattering studies to examine directly the nanomorphology of the conducting polymer component are in progress.* Recently Maeda and coworkers [68] used X-ray photoelectron spectroscopic studies to show that the surfaces of both the polypyrrole-silica and the polyaniline-silica "raspberry" particles are distinctly silica-rich. These observations are consistent with a charge stabilization mechanism and account for the excellent long-term colloid stability exhibited by these systems. BET surface areas of these conducting polymer-silica "raspberries" are up to six times higher than those of the conventional conducting polymer bulk powders [10]. Thus these nanocomposites undoubtedly represent a high surface area form of polypyrrole and polyaniline. Furthermore, the measured surface areas are always significantly larger than those estimated on the basis of the TEM particle size, even allowing for the considerable surface roughness of the raspberry particles. Thus the BET data indicate that the raspberries have significant microporosity [10]. Recently Maeda and Armes [69] showed that certain ultrafine tin(IV) oxide sols can be used instead of silica sols to prepare the analogous polypyrrole-tin(IV) oxide raspberry particles. This observation indicates that "raspberry" formation is not dependent on some unique aspect(s) of the surface chemistry of the silica particles used in the original syntheses. On the other hand, many other ultrafine aqueous oxide sols such as magnetite, titania, zirconia, and alumina do *not* lead to the formation of stable raspberry colloids [69]. Finally, the Sussex group recently showed [70,71] that silica-coated magnetite particles can be used to prepare polypyrrole-silica-magnetite dispersions, i.e., conductive magnetic particles.

The polypyrrole-silica particles are rather more resistant to pH- or electrolyte-induced flocculation than the polypyrrole-tin(IV) oxide particles [68]. This is probably due to the well-documented hydrophilic nature of the solvated silanol surface [72,73]. More surprisingly, polypyrrole-silica raspberry particles exhibit excellent colloid stability in the pH range 1–9, whereas polyaniline-silica particles become flocculated at pH >2. It is difficult to account for such marked differences in behavior in view of the XPS data reported by Maeda et al. [68], which confirm that the surfaces of both types of dispersions are silica-rich. Since the polypyrrole-silica dispersions remain colloidally stable at physiological pH, this particular system might prove to be suitable as highly colored "marker" particles for visual agglutination immunodiagnostic assays [74,75]. Such applications often require hydrophilic particles with carboxylic acid or amine surface groups for the covalent binding of proteins, antigens, etc. Accordingly, the Sussex group has devoted considerable effort to the synthesis of surface-functionalized polypyrrole-silica particles. Carboxylated particles can be readily synthesized by copolymerizing pyrrole with 1-(2-carboxyethyl)pyrrole in the presence of silica particles [75]. On the other hand, copolymerization appears to be unsuitable for the synthesis of aminated polypyrrole-silica particles, which are best prepared by other methods [76,77]. The color intensity (absorbance) of one of the carboxylated polypyrrole-silica dispersions was compared with that of a commercial carboxylated, extrinsically dyed polystyrene latex (Polymer Laboratories) using visible absorption spectrophotometry [75]. The conducting polymer-based particles were 2–4 times as absorbing at all wavelengths across the visible spectrum (400–800 nm). Thus the carboxylated polypyrrole-silica dispersions appear to have real potential as high performance "marker" particles in diagnostic assays.

VII. CONDUCTING POLYMER-COATED PARTICLES

In 1986 Jasne and Chiklis [78] reported that the electrochemical synthesis of polypyrrole in aqueous media in the presence of carboxylated latex particles led to the

---

* R. D. Wesley, T. Cosgrove, G. P. McCarthy, M. D. Butterworth, and S. P. Armes, unpublished results.

formation of polypyrrole-latex composite films at the electrode surface. Subsequently Dufort et al. [79] claimed that the use of polychloroprene and poly(styrene-butadiene) latexes with surface carboxylic acid groups in similar electrochemical syntheses led to the formation of 50–500 μm composite films with conductivities in the range 10^{-3}–10^{-1} S/cm and improved mechanical properties.

Several research groups have described the chemical synthesis of conducting polymer-coated particles. For example, Garnier and coworkers [80] polymerized pyrrole using chemical oxidants such as $FeCl_3$ in the presence of commercial sulfonated polystyrene latexes to obtain polypyrrole-polystyrene composites. Similarly, Yamamoto's group [81] coated carboxylated styrene-butadiene-methacrylate latex particles with polypyrrole, polyaniline, and poly(3-methylthiophene). The oxidant used by the Japanese group was H_2O_2/HBr with a catalytic amount of Fe^{3+}. This redox system was selected because it was considered less likely to cause electrolyte-induced flocculation of the anionic latex particles than highly charged cations such as Fe^{3+}. Remarkably, these conducting polymer-coated latexes were apparently colloidally stable with respect to particle aggregation, though no attempts were made to assess the degree of dispersion of the particles. The Sussex group has spent considerable time attempting to repeat both the French and Japanese groups' studies using identical latex particles obtained from the same commercial sources, but their syntheses have invariably resulted in agglomerated, rather than stable, dispersions [82]. However, these syntheses are apparently not trivial to reproduce successfully (T. Yamamoto, personal communication, 1994).

Partch and coworkers at Clarkson [83] used inorganic oxide particles such as hematite and ceria as particulate (rather than soluble) oxidants for pyrrole polymerization. This approach ensures that the locus of polymerization occurs exclusively at the surface of the oxide particles and results in the formation of polypyrrole-inorganic oxide particles (see Fig. 17.7). Only relatively low conductivities (10^{-3}–10^{-5} S/cm) were obtained even at conducting polymer loadings as high as 40 wt %. This is almost certainly due to the rather high reaction temperature (~100°C) used for the conducting polymer syntheses. As expected, these coated oxide particles are colloidally unstable due to the "stickiness" (high Hamaker constant) of the polypyrrole overlayer and are prone to precipitation unless a polymeric stabilizer such as poly(vinyl alcohol) is added to the reaction solution (R. Partch, personal communication, 1994).

Wiersma et al. at DSM Research have shown that sterically stabilized latex particles can be coated with polypyrrole or polyaniline in aqueous media to form conducting polymer-coated latexes with good colloid stability [84,85]. These workers emphasize that the use of an adsorbed nonionic polymeric stabilizer such as poly(ethylene oxide) or hydroxymethylcellulose in these syntheses is critical for producing stable colloidal dispersions at high Fe^{3+} concentrations; all control experiments carried out in the absence of such nonionic stabilizers resulted in macroscopic precipitation [85]. Examination of the coated particles by transmission electron microscopy provided direct evidence for a "core–shell" morphology; this observation was supported by both aqueous electrophoresis and dielectric measurements [84]. Presumably the conducting polymer is formed as a thin layer at the surface of the latex particles without significantly interfering with the steric stabilization mechanism conferred by the solvated outer layer of nonionic polymer (see Fig. 17.8). The Dutch group have focused on coating low T_g latexes of 50–500 nm diameter based on polyurethane, poly(vinyl acetate), and alkyd resins. Unlike the sterically stabilized polypyrrole or polyaniline colloids discussed in Sections IV and V, the conducting polymer-coated latex particles exhibit remarkably good

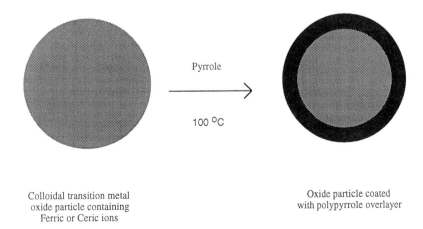

Colloidal transition metal oxide particle containing Ferric or Ceric ions

Oxide particle coated with polypyrrole overlayer

Fig. 17.7 Schematic representation of the synthesis of conducting polymer-coated inorganic oxide particles according to Ref. 83.

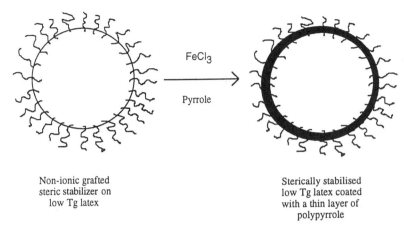

Fig. 17.8 Schematic representation of the DSM protocol for coating sterically stabilized latexes with a thin overlayer of polypyrrole according to Refs. 84 and 85.

film-formation properties at room temperature, despite encapsulation of the low T_g latex component by an outer layer of high T_g conducting polymer. Conductivities lie in the range 10^{-5}–10^1 S/cm. The Sussex group have successfully coated similar latexes with polypyrrole using the DSM protocol [86] and are currently engaged in the synthesis, characterization, and evaluation of these fascinating core–shell particles. At the present time very little is known regarding their surface composition, colloid stability, solid-state conductivity, or the mechanism of film formation.

VIII. OTHER CONDUCTING POLYMER COLLOIDS

Eisazadeh et al. [87,88] have shown that polymer-stabilized polypyrrole and polyaniline colloids can be synthesized electrochemically using a flow-through electrochemical cell. Only rather low yields of colloid were obtained; the authors suggest that this approach may be a useful alternative to chemical syntheses because, in principle, any dopant anion can be incorporated. It is likely that further studies in this area may be useful in studying the mechanism of conducting polymer particle formation, e.g., the role played by soluble oligomers. The Sussex group showed [89,90] that small-molecule surfactants such as sodium dodecylbenzenesulfonate (SDBS) can be used instead of high molecular weight polymeric stabilizers to synthesize surfactant-stabilized polypyrrole and polyaniline colloids; Gan and coworkers [91,92] subsequently reported similar results. Jonas and coworkers at Bayer [93,94] reported that 3,4-dioxyethylenethiophene (EDOT) can be chemically polymerized in aqueous solution in the presence of low molecular weight poly(styrenesulfonic acid) stabilizers to produce poly(EDOT) colloids that can be used for low color antistatic coatings. Finally, the polyacetylene block copolymers reported by Aldissi [95] and other workers [96,97] in the 1980s have since been shown to exist as highly aggregated micellar structures with a poorly solvated polyacetylene core rather than true solutions [98]. Similar results were obtained by a French group working with poly(p-phenylene)-based and polythiophene-based block copolymers [99,100]. Perhaps all these materials are best classified as "colloidal" rather than "soluble" forms of conducting polymers [101,102]. However, since all soluble synthetic polymers are lyophilic colloids, the real issue is not whether a given "soluble" conducting polymer system is actually "colloidal" (in the most general sense they all are!), but whether the conducting polymer component is actually solvated to any appreciable extent.

IX. APPLICATIONS AND PROSPECT

Although the colloidal approach has significantly improved the processability of OCPs, their electrical conductivities are only comparable to much cheaper materials such as carbon black. Conductivity alone, then, is not likely to justify the future commercial exploitation of OCP colloids. Instead, their future technological success (or failure) will depend critically on whether they offer additional "value-added" properties. Such properties may well include well-defined colloidal dimensions or morphologies, intense coloration, biocompatibility, high surface area, efficient radiation absorption, potential surface functionalization, good film-forming properties, etc. Specific applications that have recently been suggested include electrochromic displays [27], microwave welding [103,104], conductive paints [58], anticorrosion and antistatic coatings [84,85,105,106], electrochromatography [107], and immunodiagnostic assays [108–110]. The Sussex group is evaluating polypyrrole-based particles for use as novel visual "marker" particles in this latter application. Tarcha's group at Abbott Laboratories has already shown that sterically stabilized polypyrrole colloids have considerable potential in this field, particularly if they are derivatized with surface functional groups such as carboxylic acids or amines prior to use [108,109,111]. It is rather ironic that while this particular application requires the particles to exhibit most of the value-added properties cited above, it does not actually require them to be electrically conductive!

ACKNOWLEDGMENTS

I thank all the current and former students in my research group at Sussex and my many collaborators, both in academia and in industry. I am grateful to the EPSRC, Royal Society, Nuffield Foundation, Society for Chemical Industry, and the British Council for their generous financial support, and to the following industrial sponsors: Courtaulds Research, Zeneca Resins, ICI, Cabot Plastics, the New Oji Paper Co., Synthetic Chemicals Ltd., and the Defence Research Agency. Finally I thank the referees of this manuscript for their insightful comments.

REFERENCES

1. H. Shirakawa and S. Ikeda, *Polym. J.* 2:231 (1971).
2. C. K. Chiang, C. R. Fincher, Y. W. Park, A. J. Heeger, H. Shirakawa, E. J. Louis, S. C. Gau, and A. G. MacDiarmid, *Phys. Rev. Lett.* 39:1098 (1977).
3. S. P. Armes, *Synth. Met.* 20:367 (1987).
4. J. C. Chiang and A. G. MacDiarmid, *Synth. Met. 13*: 193 (1986).
5. R. E. Myers, *J. Elect. Mater.* 2:61 (1986).
6. J. A. Walker, L. F. Warren, and E. F. Witucki, *J. Polym. Sci., Polym. Chem.* 26:1285 (1988).
7. M. M. Castillo-Ortega, M. B. Inoue, and M. Inoue, *Synth. Met.* 28:65 (1989).
8. T. H. Chao and J. March, *J. Polym. Sci., Polym. Chem.* 26:743 (1988).
9. S. Maeda, D.Phil. Thesis, Sussex University, 1994.
10. S. Maeda and S. P. Armes, *Synth. Met.* 73(2):151 (1995).
11. D. J. Shaw, *Introduction to Colloid and Surface Chemistry*. 4th ed., Butterworth-Heinemann, Oxford, UK, 1992.
12. R. J. Hunter, *Foundations of Colloid Science*, Vol. 1, Clarendon Press, Oxford, UK, 1987.
13. K. E. J. Barrett (ed.), *Dispersion Polymerisation in Organic Media*, Wiley, New York, 1975.
14. G. Markham, T. M. Obey, and B. Vincent, *Colloids Surf.* 51:239 (1990).
15. D. H. Napper, *Polymeric Stabilization of Colloidal Dispersions*, Academic, London, 1983.
16. J. Turkevich, P. C. Stevenson, and J. Hiller, *Discuss. Faraday Soc.* 11:55 (1951).
17. Y. Nosaka, N. Ohta, T. Fukuyama, and N. Fujii, *J. Colloid Interface Sci.* 155:23 (1993).
18. J. Edwards, R. Fisher, and B. Vincent, *Makromol. Chem., Rapid Commun.* 4:393 (1983).
19. S. P. Armes and B. Vincent, *Synth. Met.* 25:171 (1988).
20. S. P. Armes, Ph.D. Thesis, University of Bristol, 1987.
21. R. B. Bjorklund and B. Liedberg, *J. Chem. Soc., Chem. Commun.* 1986:1293.
22. S. P. Armes and B. Vincent, *J. Chem. Soc., Chem. Commun.* 1987:288.
23. S. P. Armes, J. F. Miller, and B. Vincent, *J. Colloid Interface Sci.* 118:410 (1987).
24. M. L. Digar, S. N. Bhattacharyya, and B. M. Mandal, *J. Chem. Soc., Chem. Commun.* 1992:18.
25. M. L. Digar, S. N. Bhattacharyya, and B. M. Mandal, *Polymer 35*:377 (1994).
26. N. Cawdery, T. M. Obey, and B. Vincent, *J. Chem. Soc., Chem. Commun.* 1988:1189.
27. R. Odegard, T. A. Skotheim, and H. S. Lee, *J. Electrochem. Soc.* 138:2930 (1991).
28. S. P. Armes, M. Aldissi, G. C. Idzorek, P. W. Keaton, L. J. Rowton, G. L. Stradling, M. T. Collopy, and D. B. McColl, *J. Colloid Interface Sci. 141*:119 (1991).
29. S. P. Armes, M. Aldissi, and S. F. Agnew, *Synth. Met.* 28:837 (1989).
30. S. P. Armes and M. Aldissi, *Polymer 31*:569 (1990).
31. P. M. Beadle, S. P. Armes, S. Greaves, and J. F. Watts, *Langmuir,* 12:1784 (1996).
32. Z. Rawi, J. Mykytiuk, and S. P. Armes, *Colloids Surf.* 68:215 (1992).
33. J. Stejskal, P. Kratochvil, N. Gospodinova, L. Terlemezyan, and P. Mokreva, *Polym. Commun. 33*: 4857 (1992).
34. P. M. Beadle, L. Rowan, J. Mykytiuk, N. C. Billingham, and S. P. Armes, *Polymer 34*:1561 (1993).
35. E. Arca, T. Cao, S. E. Webber, and P. Munk, *ACS Polym. Prepr.* 35(1):334 (1994).
36. M. Beaman and S. P. Armes. *Colloid Polym. Sci.* 271:70 (1993).
37. F. M. Smits, *Bell. Syst. Tech. J.* 37:711 (1958).
38. E. Destryker and E. Hannecart, U.S. Patent 5,066,706.
39. F. Epron, F. Henry, and O. Sagnes, *Makromol. Chem., Macromol. Symp. 35/36*:527 (1990).
40. S. P. Armes and M. Aldissi, *Synth. Met.* 37:137 (1990).
41. M. French, N. C. Billingham, and S. P. Armes, *Synth. Met.* 55–57:3556 (1993).
42. P. Banerjee, M. L. Digar, S. N. Bhattacharyya, and B. M. Mandal, *Eur. Polym. J.* 30:499 (1994).
43. J. Lei, Z. Cai, and C. R. Martin, *Synth. Met.* 46:53 (1992).
44. C.-F. Liu, D.-K. Moon, T. Maruyama, and T. Yamamoto, *Polym. J.* 25:775 (1993).
45. R. B. Bjorklund, *J. Chem. Soc., Faraday Trans. 83*: 1507 (1987).
46. S. P. Armes and M. Aldissi, *J. Chem. Soc., Chem. Commun.* 1989:88.

47. J.-M. Liu and S. C. Yang, *J. Chem. Soc., Chem. Commun. 1991*:1529.
48. N. Gospodinova, P. Mokreva, and L. Terlemezyan, *J. Chem. Soc., Chem. Commun. 1992*:923.
49. N. Gospodinova, L. Terlemezyan, P. Mokreva, J. Stejskal, and P. Kratochvil, *Eur. Polym. J. 29*:1305 (1993).
50. J. Stejskal, P. Kratochvil, N. Gospodinova, L. Terlemezyan, and P. Mokreva, *Polym. Int. 32*:401 (1993).
51. S. P. Armes and M. Aldissi, *Mat. Res. Soc. Symp. Proc. 173*:311 (1989).
52. S. P. Armes, M. Aldissi, S. F. Agnew, and S. Gottesfeld, *Langmuir 6*:1745 (1990).
53. S. P. Armes, M. Aldissi, S. F. Agnew, and S. Gottesfeld, *Mol. Cryst. Liq. Cryst. 6*:1745 (1990).
54. R. F. C. Bay, S. P. Armes, C. J. Pickett, and K. S. Ryder, *Polymer 32*:2456 (1991).
55. C. DeArmitt and S. P. Armes, *J. Colloid Interface Sci. 150*:134 (1992).
56. P. Tadros, S. P. Armes, and S. Y. Luk, *J. Mater. Chem. 2*:125 (1992).
57. B. Vincent and J. Waterson, *J. Chem. Soc., Chem. Commun. 1990*:683.
58. E. C. Cooper and B. Vincent, *J. Phys. D 22*:1580 (1989).
59. M. Gill, S. Chapman, F. L. Baines, S. P. Armes, J. G. Stamper, C. M. Dadswell, N. C. Billingham, and G. A. Lawless, manuscript in preparation.
60. C. DeArmitt, D.Phil. Thesis, in preparation, Sussex University, 1996.
61. M. Simmons, P. A. Chaloner, and S. P. Armes, *Langmuir, 11*:4222 (1995).
62. M. Gill, J. Mykytiuk, S. P. Armes, J. L. Edwards, T. Yeates, P. J. Moreland, and C. Mollett, *J. Chem. Soc., Chem. Commun. 1992*:108.
63. M. Gill, S. P. Armes, D. Fairhurst, S. N. Emmett, T. Pigott, and G. C. Idzorek, *Langmuir 8*:2178 (1992).
64. S. Maeda and S. P. Armes, *J. Colloid Interface Sci. 159*:257 (1993).
65. S. Maeda and S. P. Armes, *J. Mater. Chem. 4*:935 (1994).
66. R. Flitton, J. Johal, S. Maeda, and S. P. Armes, *J. Colloid Interface Sci. 173*:135 (1995).
67. N. J. Terrill, T. Crowley, M. Gill, and S. P. Armes, *Langmuir 9*:2093 (1993).
68. S. Maeda, M. Gill, S. P. Armes, and I. W. Fletcher, *Langmuir 11*:1899 (1995).
69. S. Maeda and S. P. Armes, *Chem. Mater. 7*:171 (1995).
70. M. D. Butterworth, S. P. Armes, and A. W. Simpson, *J. Chem. Soc., Chem. Commun. 1994*:2129.
71. M. D. Butterworth, S. P. Armes, and A. W. Simpson, *J. Colloid Interface Sci, 183*:91 (1996).
72. R. K. Iler, *The Chemistry of Silica*, Wiley-Interscience, Chichester, UK, 1979.
73. H. E. Bergna (ed.), *The Colloid Chemistry of Silica*, Adv. Chem. Ser. No. 234, Am. Chem. Soc., Washington, DC, 1994.
74. S. Maeda and S. P. Armes, *ACS Polym. Prepr. 35*: 217 (1994).
75. S. Maeda, R. Corradi, and S. P. Armes, *Macromolecules 28*:2905 (1995).
76. M. D. Butterworth, R. Corradi, J. Johal, S. F. Lascelles, S. Maeda, and S. P. Armes, *J. Colloid Interface Sci. 174*:510 (1995).
77. B. Newby, J. Johal, G. McCarthy, and S. P. Armes, manuscript in preparation.
78. S. J. Jasne and C. K. Chiklis, *Synth. Met. 15*:175 (1986).
79. M. Dufort, C. Levassort, and L. Olmedo, *Synth. Met. 41–43*:3063 (1991).
80. A. Yassar, J. Roncali, and F. Garnier, *Polym. Commun. 28*:103 (1987).
81. C.-F. Liu, T. Maruyama, and T. Yamamoto, *Polym. J. 25*:363 (1993).
82. P. M. Beadle, Ph.D. Thesis, Sussex University, 1995.
83. R. Partch, S. G. Gangolli, E. Matijevic, W. Cai, and S. Arajs, *J. Colloid Interface Sci. 144*:27 (1991).
84. A. E. Wiersma, L. M. A. vd Steeg, and T. J. M. Jongeling, poster presentation at the ICSM '94 conference in Seoul, South Korea, 1994.
85. A. E. Wiersma and L. M. A. vd Steeg, Eur. Pat. 589,529.
86. S. F. Lascelles and S. P. Armes, *Adv. Mater. 7*:864 (1995).
87. H. Eisazadeh, G. Spinks, and G. G. Wallace, *Polymer 35*:3801 (1994).
88. H. Eisazadeh, Ph.D. Thesis, University of Wollongong, Australia, 1994.
89. C. DeArmitt and S. P. Armes, *Langmuir 9*:652 (1993).
90. S. Y. Luk, M. Keane, W. Lineton, C. DeArmitt, and S. P. Armes, *J. Chem. Soc., Faraday Trans. 91*:905 (1995).
91. L. M. Gan, C. H. Chew, H. S. O. Chan, and L. Ma, *Polymer Bull. 31*:347 (1993).
92. H. S. O. Chan, L. M. Gan, C. H. Chew, L. Ma, and S. H. Seow, *J. Mater. Chem. 3*:1109 (1993).
93. F. Jonas and W. Krafft, Eur. Patent 90124841.9 (1990).
94. F. Jonas and G. Heywang, *Electrochim Acta 39*:1345 (1994).
95. M. Aldissi, *J. Chem. Soc., Chem. Commun. 1984*: 1347.
96. A. Bolognesi, M. Catelloni, and S. Destri, *Mol. Cryst. Liq. Cryst. 117*:29 (1985).
97. S. P. Armes, B. Vincent, and J. W. White, *J. Chem. Soc., Chem. Commun. 1986*:1525.
98. L. Dai and J. W. White, *J. Polym. Sci., Polym. Phys. 31*:3 (1993).
99. B. Francois and T. Olinga, *Synth. Met. 57*:3489 (1993).
100. G. Widawski, M. Rawiso, and B. Francois, *J. Chim. Phys., Physico-Chem. Biol. 89*:1331 (1992).
101. B. Wessling, *Synth. Met. 41*:907 (1991).
102. M. Aldissi, *Adv. Mater. 5*:60 (1993).

103. A. J. Epstein, J. Joo, C.-Y. Wu, A. Benatar, C. F. Faisst, J. Zegarski, and A. G. MacDiarmid, in *Intrinsically Conducting Polymers: An Emerging Technology* (M. Aldissi, ed.), Kluwer, Amsterdam, The Netherlands, 1993, pp. 165–178.
104. P. Kathirgamanathan, *Polymer 34*:3105 (1993).
105. B. Wessling, *Adv. Mater.* 6:226 (1994).
106. V. G. Kulkarni, J. C. Campbell, and W. R. Mathew, *Synth. Met.* 57:3780 (1993).
107. H. Ge, P. R. Teasdale, and G. G. Wallace, *J. Chromatog.* 544:305 (1991).
108. P. J. Tarcha, D. Misun, M. Wong, and J. J. Donovan, *ACS PMSE Prepr.* 64:352 (1991).
109. P. J. Tarcha, D. Misun, M. Wong, and J. J. Donovan, in *Polymer Latexes: Preparation, Characterisation and Applications* (E. S. Daniels, E. D. Sudol, and M. S. El-Aassar, eds.), ACS Symp. Ser. No. 492, ACS, Washington, DC, 1992, Vol. 22, p. 347.
110. B. Miksa and S. Slomkowski, *Colloid Polym. Sci.* 273(1):47 (1995).
111. M. R. Pope, S. P. Armes, and P. J. Tarcha, *Bioconjugate Chem.* 7:436 (1996).

18

Solution Processing of Conductive Polymers: Fibers and Gels from Emeraldine Base Polyaniline

Richard V. Gregory
Clemson University, Clemson, South Carolina

I. SOLUTION PROCESSING OF CONDUCTIVE POLYMERS

A. Introduction

Since the early work of Shirakawa and subsequent work of MacDiarmid and Heeger, inherently conductive polymers (ICPs) have demonstrated potential for use in many different applications where the characteristics of an organic polymer and the electronic and electrical properties of semiconductors and/or metals are desired [1]. Although there has been substantial research involving ICPs over the last several years, practical uses of these polymers in real-world applications are still limited by two major obstacles: (1) lack of processibility and (2) instability to electrical and environmental decay. The lack of processibility results, in most cases, from the structural form of the doped polymer systems, which has inherent in its nature a high degree of conjugation along the polymer chain. This conjugation imparts stiffness to the chain, which in turn leads to intractability from a processing point of view [2]. The lack of environmental stability results from, among other considerations, the formed conducting polymer existing in an energy state that is susceptible to attack by oxygen and other agents such as acids, which results in either chain scission or in saturation of some of the double bond systems, thus destroying the conjugation path along the polymer chain. A thorough discussion of stability, both electronic and environmental, is left to other areas of this handbook. This chapter is primarily devoted to the formation of fibers from polyaniline (PANI) and to a discussion of the gelation behavior of PANI base in organic solvents. A brief review of efforts to improve the processibility of various conductive polymers is included as well as a discussion of some of the problems encountered along the way.

Since the discovery of conductive polyacetylene films in 1977, there has been substantial interest in the scientific and engineering communities in understanding these materials and finding applications for their unique properties [3]. Early on, applications such as lightweight conductors, battery electrodes, solar cells, semiconductors, and molecule-sized electronic devices and nanostructures were envisioned [4]. The implementation of ICPs in these applications has lagged behind the early predictions of widespread near-term usage due primarily to the inability to process these materials using conventional polymer processing technologies. Although substantial progress has been made in the formation of films of these materials for small-scale devices by spin cast techniques and other in situ processes, efforts to process ICPs into long continuous filament or film form were not realized as industrially feasible processes until quite recently. Continuous processing of ICPs into film and fiber form is preferable to small-scale batch processes for large-scale production [5].

B. Processibility Improvements

1. Solubility

In their neat form most ICPs are not melt-processible and must be formed from various solution casting or spinning techniques. Unfortunately, most ICPs are also insoluble in their doped or conductive form unless special conditions are met. The difficulty in solubilizing ICPs originates from their inherent delocalized π-electronic structure, which is the same characteristic that gives rise to their unique optical and electronic properties. The delocalized π-electronic structure leads to (1) large electronic polarizability and (2) a rigid polymer backbone. Both of these properties are causes for the low solubility of ICPs in most common polymer solvents. High electronic polarizability leads to large interchain π–π attraction (dispersive force), which favors aggregation rather than solvation. Also, the rigidity of the polymer backbone formed by the delocalization of the electron density contributes to an unfavorable entropy of solution. Additionally, ICPs are composed of ionic dopants complexed with the polymer chain, which also contains hydrophobic organic segments. It is difficult to find solvents to simultaneously solvate both hydrophilic and hydrophobic segments in one chain. Add to this difficulty the problems of polarizability and the unfavorable chain conformation in solution, and the problem of solvation becomes quite complex. To overcome these difficulties various methods for the preparation of conductive polymers with good tractability have been explored [6]. The methodologies involved can basically be divided into four groups:

1. Preparation of colloidal dispersions and coated latexes
2. Preparation of substituted derivatives
3. Use of processible precursor polymers for later conversion to ICPs
4. Others

Some of these approaches can yield conductive polymers with varying degrees of tractability but usually at the expense of conductivities. Colloidal dispersions are discussed elsewhere; therefore only the last three approaches are briefly discussed here [7–9].

2. Substituted Derivatives

A well-established technique for rendering an intractable polymer soluble in certain solvents is to add solubilizing substituents to the monomer unit of the polymer structure. This technique has also been used in solubilizing conductive polymers [10]. The addition of substituents onto the polymer chain may have steric consequences that reduce the ability of the polymer to form delocalized π-electron density over the chain, thus reducing the level of achievable conductivity. Another popular method for solubilization is the inclusion of solubilizing monomeric structures along the polymer chain, rendering a modified chain that includes some percentage of different monomers used to enhance solubility. This method is not particularly useful for ICPs where reasonable levels of conductivity are needed, because the inclusion of these structures at a level necessary for solubility may significantly reduce the conductivity of the polymer system due to these charge-carrying interruptions along the chain. The morphology and the crystalline domains of the polymer may also be significantly affected by incorporation of these groups [11].

An example of the monomeric substituted systems is phenyl-substituted polyphenylene, which is soluble in common solvents such as chloroform. Polypyrrole and polythiophene derivatized with solubilizing groups have also been extensively investigated, and large amounts of soluble derivatized species have been synthesized [2,3,10]. Polypyrrole derivatives containing alkyl or alkoxy substituents at either N-, 3, or 4 positions are found to be soluble in several common organic solvents such as chloroform, tetrahydrofuran (THF), and o-dichlorobenzene [12]. However, these substituents, which improve solubilization, also induce a steric interference that unfavorably alters the planarity of the polymer structure, thus reducing the degree of orbital overlap and the levels of obtainable conductivity. Planar structures are necessary for achieving high levels of electrical conductivity in polypyrrole and polythiophene [13,14]. It was reported that the inclusion of solubilizing groups at the β position reduces the conductivity of the formed polymer depending on the size of the pendent group. Methyl substituents at the β position reduce the conductivity, and larger groups may completely destroy it [15,16]. Substitution at the nitrogen also reduces the conductivity of the formed polymer. Substitution at one of the α positions results in α–β-linked monomeric units in the polymer backbone, which also destroys the planar structure. In addition, many of the solutions of these derivatized polypyrroles are unstable over time and precipitation of the polymer occurs. This solution instability makes processing quite difficult [10].

Derivatized polythiophenes are very different from the pyrroles and are unique in that the introduction of substituents on the polymer chain has only a minor effect on the planarity of the formed polymer. Polythiophene derivatives such as the 3-alkyl-substituted system demonstrate reasonable solubility while maintaining good levels of conductivity [17]. Work by Elsenbaumer and others [18–21] showed that the poly(3-aklylthiophenes) can be dissolved in common solvents such as chloroform and methylene chloride, and films prepared from these solutions have conductivities ranging from 1 to 1000 S/cm depending on the substituent size. Poly(3-dodecylthiophene) can even be melt spun into fibers with good mechanical properties and when doped with iodine can achieve conductivities up to 55 S/cm. Despite the fact that these alkyl-substituted thiophenes have good mechanical and electrical properties in the film and fiber form, general applications are still limited because the tedious synthesis of the modified monomers makes large-scale production difficult. Finite solubility, particularly of the doped form, also reduces the amount that can be solubilized for certain technologies, further limiting processibility. In general the literature has been somewhat sparse over the last two decades concerning the effects of polymer–solvent interactions, although some reports have detailed specific interactions for certain systems [22–27]. Detailed studies concerning the

Solution Processing of Conductive Polymers

thermodynamics of solubility for various solvents and solvent mixtures with regard to solvation parameters, polymer–polymer interactions, solvent–polymer interaction forces, etc. are beginning to be reported with more frequency by various groups due to the increased interest in processing of these materials.

3. Precursor Polymers

Another method explored recently for making conductive polymers into useful end products of various geometries is the precursor polymer approach. In this type of processing a precursor polymer is prepared and formed into the desired end product geometry and then converted to the final dopable ICP, usually by the application of heat. Poly(p-phenylenevinylenes) (PPVs) are representative of this type of approach. Figure 18.1 shows the basic scheme for this type of synthesis [10]. The common characteristic of most of the precursor polymers used for this type of processing is the polyelectrolyte structure, which is highly water-soluble. Usually films and fibers can be prepared from the precursor polymer by spin casting or wet spinning, respectively, and the films and fibers so produced are thermally converted to the final undoped ICP. For example, the PPV precursor polymer is converted to the undoped PPV by heating at temperatures above 200°C with the elimination of sulfur compounds and acid gases. The converted PPV is then doped with solutions of acids and oxidizing agents to conductivities of 100 S/cm. Murase et al. [28] prepared a PPV film from a drawn (oriented) precursor PPV polymer with an elastic modulus of 9.2 GPa, a tensile strength of 274 MPa, and an elongation of 7%. The conductivity of this film is on the order of several thousand Seimens per centimeter.

This type of processing involves several steps including the preparation of a precursor polymer for subsequent processing. Although films and fibers can be produced by this process, the synthesis of the precursor polymers can be difficult and the production of the precursor on a large scale may not be economically feasible. A major advantage of these materials is that the precursor polymer can be oriented by existing draw technologies prior to doping, thus providing the formed materials with relatively good mechanical strength and excellent electrical conductivity.

4. Other Methods

Copolymerization is a technique frequently used to combine diverse properties of differing polymers into one polymeric system. This process is widely used in the fiber industry to impart dyeability to nonreactive fiber-forming polymers such as acrylics [29]. The polymer is copolymerized with monomeric groups that form anionic moieties at the proper pH and can be readily dyed with cationic dyes. Using this approach one might anticipate that the solubility properties of one type of polymer might be incorporated with the electrical properties of an ICP by copolymerization to give a final soluble conductive polymer. Generally there are four different methods used in the copolymerization process: (1) in situ polymerization of both monomers to form a random copolymer, (2) grafting of the solubilizing polymer onto the backbone of the intractable polymer, (3) grafting of the intractable polymer onto the backbone of the soluble

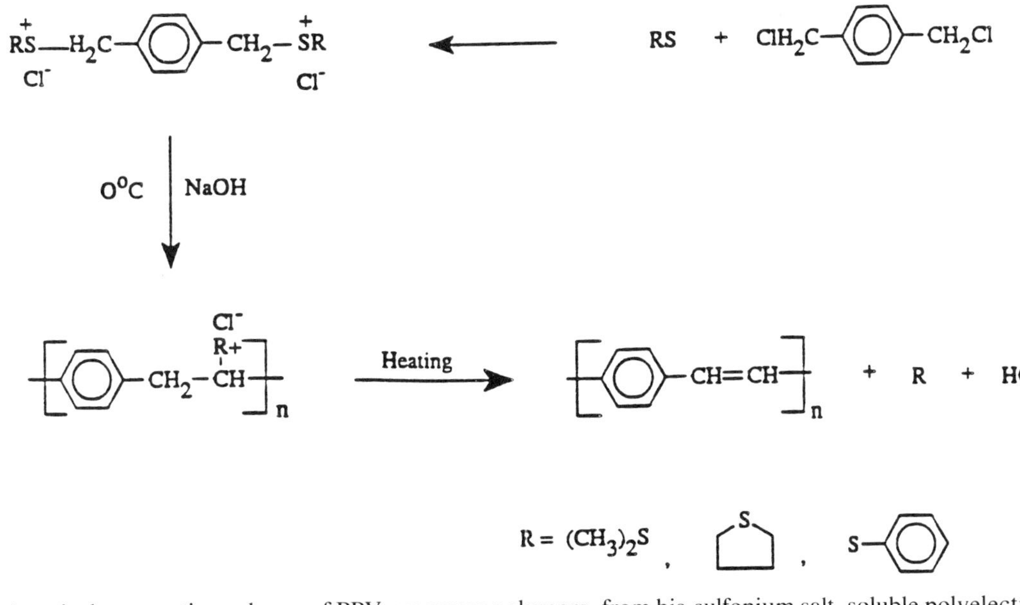

Fig. 18.1 A typical preparation scheme of PPV precursor polymers, from bis-sulfonium salt, soluble polyelectrolyte precursor, to heat-converted PPV. (From Ref. 10.)

polymer, and (4) combination of both soluble and intractable polymer to form a block copolymer by various synthetic techniques. These techniques generally result in substantially lower electrical properties as previously discussed. One notable exception to this is the counterion-induced processible conductive polyaniline, where the dopant acts as a pseudograft on the polymer backbone, giving the PANI chain a degree of solubility in the doped form. This form of processing is discussed in Section I.C.

Another widely used approach is the in situ polymerization of an intractable polymer such as polypyrrole onto a polymer matrix with some degree of processibility. Bjorklund [30] reported the formation of polypyrrole on methylcellulose and studied the kinetics of the in situ polymerization. Likewise, Gregory et al. [31] reported that conductive fabrics can be prepared by the in situ polymerization of either pyrrole or aniline onto textile substrates. The fabrics obtained by this process maintain the mechanical properties of the substrate and have reasonable surface conductivities. In situ polymerization of acetylene within swollen matrices such as polyethylene, polybutadiene, block copolymers of styrene and diene, and ethylene-propylene-diene terpolymers have also been investigated [32,33]. For example, when a stretched polyacetylene-polybutadiene composite prepared by this approach was iodine-doped, it had a conductivity of around 575 S/cm and excellent environmental stability due to the encapsulation of the ICP [34]. Likewise, composites of polypyrrole and polythiophene prepared by in situ polymerization in matrices such as poly(vinyl chloride), poly(vinyl alcohol), poly(vinylidine chloride-co-trifluoroethylene), and brominated poly(vinyl carbazole) have also been reported. The conductivity of these composites can reach up to 60 S/cm when they are doped with appropriate species [10].

Work reported by Liu and Gregory [35] on alkyl-substituted polythiophenes synthesized with the addition of a urethane group on the β-substituted alkyl chain shows excellent blending characteristics with hydrogen bonding polymers such as urethanes or polyamides such as nylon. These blends have significantly reduced percolation thresholds and are solution and melt spinnable.

C. Counterion-Induced Processibility of Conducting PANI

Cao et al. [36] reported that difficulties in the processing of PANI in the conductive form from reasonably high molecular weights could be overcome by the use of a functionalized protonic acid. This functionalized system dopes PANI and simultaneously renders the conductive PANI complex soluble in common organic acids. In this study the authors defined a "functionalized protonic acid" as $H^+(M^-$—$R)$ in which the counterion anionic species, $(M^-$—$R)$, contains an R functional group chosen to be compatible with nonpolar or weakly polar organic solvents. An example of this is dodecylbenzenesulfonic acid (DBSA)

R—⟨phenyl⟩—SO$_3$H

R = C$_{12}$H$_{25}$

These workers surmise that the long alkyl chains on the phenyl ring lead to solubility in common organic solvents such as toluene, xylenes, and chloroform and the anionic part of the molecule (SO$_3$) dopes the PANI, forming a complex that is conductive and soluble. Table 18.1 reports their findings on the solubility and conductivity of protonated emeraldine salt using a variety of functionalized counterions. As can be seen in this table, films prepared with DBSA and camphor sulfonic acid (CSA) exhibit initial conductivities of better than 100 S/cm. These authors conclude that one can "design" the conducting PANI complex to be soluble in specific solvents and cite as an example the use of CSA as the functionalized protonic acid. After doping and complexation with CSA, the conducting PANI complex is soluble in m-cresol. The CSA-PANI complex, due to the excellent solvation effects of m-cresol, can be used to make films of high optical quality. Use of the soluble PANI complex in making appropriate blends with other nonconductive polymers soluble in the same solvent is also cited as a use of these PANI complexes as well as preparation of the neat conducting polymer. Since the polymer is processed in the conductive form, no postprocessing chemical treatment is necessary.

Work by Menon et al. [37] on PANI complexed with CSA reported on the temperature and magnetic field dependence of formed films. The authors noted that the intrinsic metallic nature of this PANI complex can be observed from the positive temperature coefficient of resistivity in the temperature range of 180–300 K. Some sign of hopping transport was also observed below 180 K, which they interpreted as indicative of marginal metallic behavior with disorder-induced localizations at low temperature. Resistivity ratios as low as 1:3 (300 to 1.2 K) were found. A typical electronic localization length in these PANI-CSA complexes was on the order of 100–150 Å based on evaluation of data obtained from the dependence of resistivity on temperature and magnetic fields. Menon et al. report that the PANI-CSA complex is on the metal/insulator boundary and a magnetic field can shift the mobility edge and cause the complex to cross over from a disordered metal to an insulator. The intrinsic conductivity (σ_{int}) of PANI-CSA complexes

Solution Processing of Conductive Polymers

Table 18.1 Solubility and Conductivity of Protonated Emeraldine Salt with (SO_3^-—R) Counterion

R	σ (S/cm)		Solubility[b]				
	Pellet	Film[a]	Xylene	CH_3Cl	m-Cresol	Formic acid	DMSO
C_6H_{13}	10		○	○			
C_8H_{17}	19		○	○			
$C_8F_{17}COOH$	2.7		—[c]				
C_8F_{17}	3.7		—[c]				
(L, D)-Camphor	1.8	100–400		⊕	⊕	⊕	○
4-Dodecylbenzene	26.4	100–250	⊕	⊕	⊕	○	
o-Anisidine-5-	7.7×10^{-3}				○	○	○
p-Chlorobenzene	7.3				○	○	○
4-Nitrotoluene-2-	5.7×10^{-2}				○	○	○
Dinonylnaphthalene	1.8×10^{-5}		○	⊕	⊕		
Cresol red	2.2×10^{-4} [d]				○		
Pyrogallol red	1.2×10^{-1} [d]				○		
Pyrrocatechol violet	1.9×10^{-1} [d]				○		

[a] Films were cast from concentrated solution.
[b] ○ = soluble at room temperature; ⊕ = very soluble at room temperature.
[c] Soluble in perfluoroalkanes, e.g., perfluorodecaline.
[d] Pressed at 165°C.
Source: Ref. 36.

was calculated to be $\approx 2.5 \times 10^4$ S/cm for an oriented sample and $\approx 3 \times 10^3$ S/cm for a nonoriented film.

Cao et al. [38] prepared blends from PANI produced by counterion processes. Partial orientation of the PANI complex was accomplished upon drawing of the blend film [38]. An example cited by the authors is a PANI-DBSA polyethylene composite film that upon drawing at 105°C to a draw ratio of 40 increases in electrical conductivity by two orders of magnitude at surprisingly low levels of PANI complex (less than 10%) in the blend. X-ray studies by these authors demonstrate that the PANI complex indicates a relatively high degree of orientation of the PANI chains, no doubt resulting in the increased conductivity. This is consistent with the results of MacDiarmid and coworkers [39] on polyaniline oriented films. The mechanical properties of the polyblend material are also quite good, with no substantial loss of mechanical properties at modest levels of the PANI complex that give conductivities of less than 10 S/cm.

Further work by Osterholm's group in conjunction with Cao and coworkers determined that a PANI complex of a proprietary nature called POLARENE, developed as a commercial product by Neste Oy in Finland, could be processed with various other thermoplastic polymers such as polystyrene, PVC, and others including thermoplastic elastomers [40]. These blends could be processed by existing techniques such as injection and blow molding. Conductivities in the semiconductor range can be achieved at low loading levels in the 1–2% range. Due to the low loading levels of Polarene, the mechanical properties of the blend are excellent, approaching those of the neat host polymer and in some cases exceeding them.

D. Fiber Overview

1. Fiber Spinning

According to Andrzej Ziabicki, noted author of *Fundamentals of Fiber Spinning*, the term "spinnability" in man-made fiber technology has no precise meaning [41]. The term is usually used to mean "fiber-forming," i.e., suitable for the manufacture of fibers. Essentially, a polymeric material is deemed to be spinnable if it is capable of assuming large irreversible deformations when subjected to a uniaxial stress. Spinnability is a necessary but not the sole condition for the formation of fibers. Examples of spinnable fluids that do not form fibers are honey and highly viscous mineral oils. In order to successfully spin a fiber the spinning solution (spin dope) must not only undergo the deformations required with the application of uniaxial stress but must also be of significant molecular weight and be capable of forming some type of chain–chain interactions such as hydrogen bonding (nylon), dispersion interactions (polyethylene), mechanical entanglement of the polymer chains, or some combination of these or other mechanisms.

Generally fiber spinning technologies can be subdivided into two basic areas, melt spinning and solution spinning. There are many subsets of these such as air gap spinning, but for the most part all spinning areas can be described as either melt or solution. In melt spinning the polymer material must soften and interchain interactions are minimized on heating, thus allowing the necessary deformation without deterioration of the polymer chains. These types of materials are referred to as thermoplastic. In solution spinning, the polymer is dissolved into a solvent and spun into a coagulating solvent bath in which the original polymer is not soluble. In both cases the fiber-forming material is spun through a device called a spinnerette, which is a highly engineered metal or ceramic plate containing many small diameter holes. The cross-sectional shape of these holes determines the fibers' cross-sectional shape—circular, trilobal, etc. In melt spinning the fluid is typically delivered to the spinnerette by a metering device called a gear pump. The gear pump is fed material by a long metallic rotating screw housed in an extruder barrel whose purpose is to transport the polymer through several heating, melting, and mixing zones prior to interfacing with the gear pump and subsequently the spinnerette. In solution spinning the dissolved polymer in reasonably high concentration (normally 15% w/w) is pumped to the spinnerette and into a coagulation bath.

In both of these cases, knowledge of the rheology of the spinning fluid during the spinning process is an absolute necessity for forming fiber with reproducibly uniform mechanical, and, in conductive polymer fibers, electrical properties. For the spinning process, constitutive equations are developed that describe the mass balance and momentum of the fluid during the spinning process. In solution spinning the thermodynamics of the coagulation bath and the kinetics of diffusion of solvent into and out of the forming fiber must be considered and understood. This problem is not trivial, as the concentrations and mass balance are in a state of flux during the fiber formation process.

Most inherently conductive polymers that are stable under ambient conditions and whose monomers are available in large commercial quantities, for example pyrrole, aniline, and thiophene, are not thermoplastic and cannot be easily spun into fibers. Polyaniline, however, can be solution spun into fiber in its base form with relative ease owing to its solubility in solvents such as N-methylpyrrolidione (NMP). Methods for the formation of polyaniline fibers from the base form of the polymer are discussed at length in Section III.

2. PANI Fibers

Angelopoulos et al. [42] reported that N-methylpyrrolidione (NMP) could dissolve PANI in the base form. Subsequent work demonstrated that PANI solutions formed in NMP at solids concentrations above 6% formed gels after a short period of time [43–45]. The formation of these gels at concentrations necessary for the formation of fibers and films by solution spinning technologies ($\geq$10%) prohibited the use of this solvent without modification to a binary solvent system.

Hsu et al. [46] reported that fibers could be spun from NMP solutions at fiber spinning concentrations at which they would normally gel if certain amines were added to the polymer solution. These workers report that a PANI/1,4-diaminocyclohexane solution provides a stable spin dope and that fibers could be extruded by continuous dry-jet spinning with an initial tenacity of 0.8 gram per denier (gpd). The formed fiber could then be drawn at 215°C with a final tenacity of 3.9 gpd, an elongation of 9.3%, and modulus of 83 gpd. Assuming that the density of the fiber is 1.2 g/cm^3, the increase in strength on drawing is in the range 80–400 MPa.

The spin dope for these fibers is prepared by the addition of gaseous ammonia or pyrrolidine to the NMP solution of PANI base in concentrations great enough for fiber spinning. The gelation of the PANI is reversed, and a viscous solution is formed for use as the spin dope at concentrations of 20% solids. Hsu et al. [46] speculate that the gelation process is related to the interaction of microcrystalline regions that the new binary solvent disrupts. They observed that when the solution is stirred at low speeds, on the order of 15–20 rpm, it remains smooth and stable, but on increasing the stirring to 50–70 rpm the solution becomes lumpy. It is suggested that this phenomenon is the result of shear-induced crystallization.

Doping of the heat-drawn fibers produced from the spin dopes in 1 M aqueous HCl reduced the tensile strength by approximately 64%. This reduction was attributed to the microscopic incorporation of chain defects resulting from the exposure to HCl, such as chain scission due to acid hydrolysis. The highest conductivity reported for the HCl-doped fiber was $\approx$157 S/cm. When similarly prepared fibers were doped with sulfuric acid the reported conductivity was $\approx$320 S/cm. The authors did not speculate on the difference in observed conductivity between HCl and H_2SO_4 or report if reduction in fiber strength was similar to that of the HCl-doped fibers.

Recent work by Wang et al. [47] reported on the formation of fibers spun from CSA-doped PANI and poly(o-toluidine) (POT) in m-cresol. This study details the transport properties (i.e., temperature dependent dc conductivity, thermopower, and microwave frequency dielectric constant) of the fibers so produced. The authors report that the results for the PANI-CSA fibers are much like those reported for PANI-CSA films although the fibers contain some areas of poorly conducting linkages. The fibers formed from the derivatized PANI, poly(o-toluidine), show much higher electrical conductivity and weaker temperature dependence than similarly prepared HCl-doped powders and films of this polymer. The higher conductivity is assumed to be due to the CSA-doped poly(o-toluidine) polymer being much closer to the localization/delocalization boundary. The tenacity reported for the PANI/CSA and POT/CSA fibers was 0.2 gpd, with the elongation to break and the modulus being 8.4%, 7.3 gpd and 3.0%, 9.7 gpd, respectively. A negative microwave frequency dielectric constant was noted for the PANI/CSA fibers, indicating the formation of three-dimensional metallic states as discussed by Joo et al. [48]. These authors did not speculate on the morphology of the as-spun fibers or the possibility of voids or other anomalies contributing to the very low mechanical properties.

E. Solution Processing of PANI

A goal of some research groups is to develop a method to process PANI in its conductive form [36,37,40,46,49]. Processing in this form removes subsequent doping steps, thus speeding production, and may be less expensive. Work along these lines resulted in the counterion processible PANIs and other conductive fiber and film processing methods discussed previously [40,46]. Other groups, including ours, look to maximize the mechanical properties of the PANI base by elucidating the structure of the polymer in solvents that are stable and "good" in a thermodynamic sense [50]. Both of these approaches have validity; however, processing PANI in the base form, where the polymer is not a rigid rod, is attractive because existing spin technology can be employed if proper solvent systems can be identified. In the proper solvent, under the right conditions, the polymer exists more as a solvated coil rather than being "bunched" in on itself [51]. That is, the difference between the strength of the polymer–polymer interactions more closely matches that of the polymer–solvent interactions. If the polymer is relaxed in solution, the probability of chain entanglement for the purpose of developing fiber and film strength on extrusion is enhanced. This entanglement should remain after doping to produce a conductive fiber or film with reasonable mechanical properties [41,51]. In addition, if the PANI chains are entangled throughout the matrix, electron transport between polymer chains should be enhanced, giving better segment-to-segment electrical reproducibility. By investigating different solvent systems and studying the rheology and polymer–solvent interactions of these systems, spin dopes of PANI, and for that matter other ICPs, can be developed in which the solution is stable owing to better polymer–solvent interactions and a decrease in polymer–polymer attraction. The stability of the solution is necessary in order to produce fiber or films with segment-to-segment reproducible mechanical and electrical properties on extrusion [41]. Interaction of the polymer and the solvent must be quantified for differing systems if process parameters are to be defined for the production of fibers and films with reproducible properties.

A significant study by Shacklette and Han [24] pointed out that the solubility of undoped polyaniline could be understood from the point of view of standard solubility parameters that measure the likelihood of a solvent and a polymer engaging in dispersive, polar, or hydrogen bonding. These forces are usually referred to as secondary bonding forces when employed to hold polymer chains together to provide strength, as in the case of nylons, or as the interactive forces between solvent and polymer for the purpose of solvation. These workers determined the total solubility factor (δ) for undoped polyaniline to be 22.2 MPa$^{1/2}$, which represents the contribution from the polar interaction δ_p of 8.1 MPa$^{1/2}$, the dispersive interaction δ_d of 17.4 MPa$^{1/2}$, and the hydrogen bonding interaction δ_h of 10.7 MPa$^{1/2}$. In addition, these workers report the solubility parameter for doped PANI to be 23.6 MPa$^{1/2}$. In the doped form the contribution from the hydrogen bonding term is significantly higher than in the PANI base and is 13.7 MPa$^{1/2}$. This work characterized the solution parameters of polyanilines, both doped and undoped, and polyacetylenes and compared them with several conventional synthetic polymers. These studies were carried out using empirical methods and group additive calculations so that the modifying influence of the dopant ion need not be considered. Without this modifying influence doped PANI compositions were found to be more polar and hydrogen-bonding than the undoped PANI base. Table 18.2 summarizes the findings of Shacklette and Han for PANI and polyacetylene and also lists the values for interaction parameters for some conventional polymers for comparison. The results obtained by these authors is consistent with our work on blends of conductive PANI, within hydrogen-bonding matrices such as polyurethane and nylons, that show peak shifts in the infrared indicating hydrogen bond formation between the PANI salt and its host polymer [52,53].

These solution studies of polymer–solvent interactions demonstrate that the strong propensity for hydrogen bond formation in doped PANI systems may be responsible for PANI self-associating in solution and phase separating from its host or solvent, forming microdispersions as suggested by Wessling and Polk [54,55]. In the undoped form, association due to hydrogen bonding may well be the basis for the formation of gels in solvents such as NMP. Although the δ_h is not as large as in the undoped base form it still has a reasonably high value. Work in our laboratory has determined on the basis of the rheology of these systems, that PANI gels formed in NMP and other solvents are thixotropic. Thixotropic gels do not involve primary valence crosslinks, as will be discussed in Section II.B.1.

F. Stability and Rheological Characterization of PANI/NMP and PANI/DMPU

Although there have been several recent studies involving solvent systems for ICPS, and particularly PANI, most of these studies to date, with the exception of those mentioned above and a few others, have not focused efforts on the interactions between the solvent and the

Table 18.2 Summary of Polymer Properties

Polymer	ϵ	γ (mN/m)	δ (MPa$^{1/2}$)	δ_d (MPa$^{1/2}$)	δ_n (MPa$^{1/2}$)	δ_h (MPa$^{1/2}$)
Polyanilines						
Emeraldine base (EB)	5	49.6	22.2	17.4	8.1	10.7
Emeraldine salt (ES)		69.6	23.6	17	8.9	13.7
Leucoemeraldine (LE)			23–25	21.1	5.6	7.3
Polyacetylenes						
cis-Polyacetylene	3.5	40.1	17.5	17.5	0	0
t-Polyacetylene-iodine		44.2	—	—	—	—
Conventional polymers						
Polyethylene (LLDPE)	2.28	31	17	17	0	0
Polystyrene (PS)	2.55	33	18.5	19.2	0.9	2.1
Poly(methyl methacrylate)	3.4	39	19–23	18.6	10.5	7.5
Nylon 6	3.5	42	22–26	16.1	7.2	12.9
Poly(ethylene terephthalate)	3.3	43	21.6	19.5	3.5	8.6
Poly(vinyl chloride) PVC	3	39	21–22.5	18.2	7.5	8.3

Source: Ref. 24.

polymer. For instance, work reported by Cao et al. [56] considered the case for fully protonated PANI (H$^+$/PhN = 0.5) and found that it can be dissolved in organic solvents only when the PANI-(acid)$_{0.5}$ complex is solvated by at least an additional 0.5 mole of solvating agent per PANI unit (PhN). For solvents with appropriately strong hydrogen bond interactions this solvating agent can be the solvent itself. It was found that with the appropriate choice of solvent and cosolvent the resultant electrical conductivity of films cast from solution could be much higher or lower than that of the initial PANI powder. This work and several others developed the functional relationship between final film conductivities and solubilities, temperatures, etc.; however, the authors do not address the "goodness" of the solvent systems in a thermodynamic sense nor do they evaluate the systems by rheological characterizations.

A thorough understanding of the solvent–polymer interactions as well as the polymer–polymer interactions is necessary if we are to be able to predict the solution processing behavior of ICPs. An example of such studies is the recent work in our laboratories addressing the behavior of PANI base in N-methyl-2-pyrrolidione (NMP) and N,N'-dimethylpropylene urea (DMPU), two solvents used for processing films and fibers, at or below spinnable concentrations. The solvent–polymer interactions are studied using Huggins and Kreamer relationships employing dilute solution viscometry to determine the intrinsic viscosity ([η]) of the solved PANI and shear studies to determine its stability [57].

Figure 18.2 shows the values obtained for the intrinsic viscosity [η] of PANI in DMPU and NMP, prepared by the method of Cao, using the Huggins and Kreamer equations. (Experimentally determined η_{red}/η_{inh} at 0 concentration = [η].) The value of [η] in DMPU is higher than in NMP, suggesting a more expanded coil due to polymer–solvent compatibility as suggested by Mangaraj et al. [58] in their study of differing polymer–solvent systems. Table 18.3 reports the values obtained for the constants K_H and K_K and the intrinsic viscosities. In addition, the results of shear studies shown in Fig. 18.3 clearly show the development of a second Newtonian region in 17.5% (w/w) PANI/DMPU after shear thinning that is absent in 10% PANI/NMP, indicating that the PANI solution is relatively stable at higher concentrations in DMPU. Figure 18.4 details the rheological stability of PANI in NMP and DMPU at a shear rate of 0.75 s^{-1} over time, at concentrations of 10% PANI/NMP and 17.5% PANI/DMPU, and shows that the instability of the PANI/NMP solution may well be due to the initial association of PANI chains bunching in on themselves and with other PANI chains, which eventually leads to the formation of gels at a concentration of 10% or more. The nature of these gels is discussed in the next section.

This study as well as those of Andretta et al., Cao et al., and several other authors help to define the environment of the PANI coil in solution and the solvent effect on the final mechanical and electrical properties of the formed fiber or film [56,59]. Studies of the characteristics of the PANI polymer solutions, including fundamental solvation phenomena and rheological characterization, will lead to the development of constitutive equations and to subsequent processes for forming large quantities of reproducible materials by continuous processing.

G. Summary

Although many different routes have been explored to produce processible conductive polymers with acceptable mechanical properties, the tedious methods of syn-

Solution Processing of Conductive Polymers

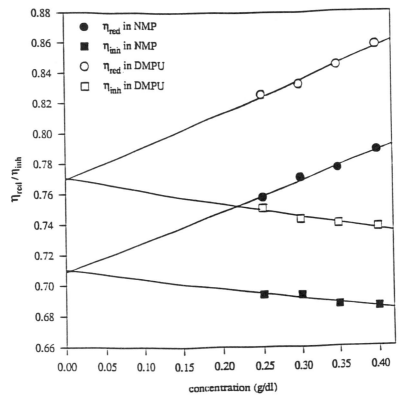

Fig. 18.2 Evaluation of intrinsic viscosity of PANI in NMP and DMPU. (From Ref. 57.)

thesis, availability of monomer in sufficient quantity for large-scale processes, lack of appropriate continuous process ranges, and a variety of other drawbacks has led to renewed efforts to process those ICPs developed from monomers already produced in large quantities. These efforts have focused primarily on forming polypyrroles, polythiophenes, and polyanilines in their neat or nonderivatized forms.

There have been some notable efforts in the past several years to process these particular ICPs into fibers and other commercial end products where large volumes are necessary. Work by Gregory et al. [31] demonstrated that conductive textiles could be developed by the in situ polymerization of either pyrrole or aniline monomer on the surface of the material, resulting in conductive fabrics with little or no change in the general mechanical properties of the fabric substrate. Further work by CHILD and Kuhn [60] showed that high levels of surface conductance could be obtained and substantial increases in environmental stability resulted if the proper dopants were chosen and the process was correctly carried out. These materials are in effect a thin film of the conductive polymer on the host polymer. Recent work by Nowak

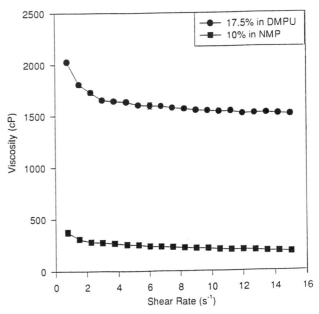

Fig. 18.3 Effect of solvent on the rheology of polyaniline solution in NMP and DMPU. (From Ref. 29.)

Table 18.3 Intrinsic Viscosities, Huggins Constants (K_H), and Kraemers Constants (K_K) of PANI in NMP and DMPU

Solvent	$[\eta]$ (dL/g)	K_H	K_K
NMP	0.71	0.384	−0.133
DMPU	0.77	0.371	−0.138

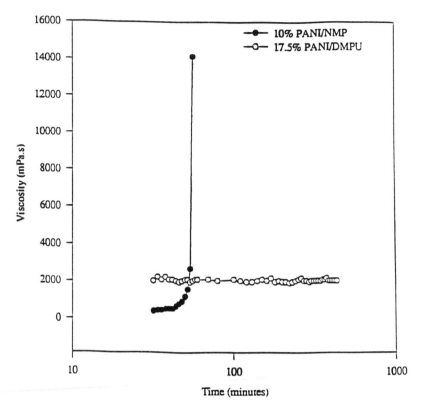

Fig. 18.4 Rheological stability of PANI solutions in NMP and DMPU at a constant shear rate of 0.75 s^{-1}. (From Ref. 57.)

[61] indicates that there is a direct interaction between certain substrates and the ICP film that results in strong adhesions. These interactions may eventually form the basis for new solvent systems. The interactions are strongest on nylons and other polymers with active areas or end groups that interact as secondary bonding sites. Quantification of these interactions may be helpful in determining necessary solvent parameters. Other host polymers such as PET, which do not rely on the same type of secondary bonding force interactions, have a reduced level of ICP–host polymer adhesion. The interaction is still reasonably strong due most likely to van der Waals interactions of one type or another.

Thiophenes that have not been altered to enhance their processibility form intractable powders that are not soluble in either the undoped or doped form. The high oxidation potential of the unsubstituted monomer also prohibits processing in most solvents. Derivatization of this monomer with appropriate pendent groups seems to be the only way at present to solubilize these polymers for processing on a large scale [10]. Of the ICPs whose monomers are produced in commercial quantities and that are easily polymerized, only polyaniline is soluble to some degree in organic solvents in its base form. In this form it can be processed into films and fibers and later doped to some level of conductivity.

Although there is considerable ongoing effort to produce easily processed ICPs from monomers constructed to produce melt- or solution-processible polymers, commercial production on a large scale has not yet been realized. Processing of polyaniline in its base form where the polymer is not considered a rigid rod and the molecular chain has a reasonable flexibility has recently been shown to be a viable technique for the production of PANI fibers [62]. These fibers can then be doped to reasonable levels of conductivity and maintain acceptable mechanical properties. Also, gels formed from PANI base can be formed into structures that are flexible and that upon doping have mechanical and electrical properties that may prove useful for various applications.

II. SOLUTION PROCESSING AND GEL FORMATION OF EMERALDINE BASE POLYANILINE

A. Introduction

PANI was originally considered to be an intractable nonprocessable intrinsically conductive polymer. The nonprocessibility of PANI results from the delocalization of the electron density along the polymer chain resulting in a stiff backbone that, from a processing point of view, is undesirable [63]. Although in some polymer systems a rigid backbone prohibiting processing may be overcome at elevated temperatures, this is not the case for PANI, as it decomposes prior to softening and therefore is not considered thermoplastic. The T_g (glass transition temperature) for PANI was recently reported to be $\approx 256°C$, and a T_m (melt temperature) is not observed for this polymer [64]. Other investigators have reported differing values for the T_g for PANI, with the different reported values thought to result from differences in solvents and film casting techniques [65,66]. These differences are also a function of the differing formation processes resulting in varying amounts of residual solvents, differing oxidation states, etc. In general the higher the oxidation state of the PANI the higher the T_g. Early on, PANI researchers noted that in its nonconductive emeraldine base form, where delocalization of electron density is no longer a problem, this polymer could be processed into films and then doped to a conductive state while maintaining reasonable mechanical properties [42,67]. These films were usually cast from a 1–2 wt % solution because at higher concentrations the solutions normally gelled [44,67,68]. Casting from a weakly concentrated solution presents problems of solvent removal and is not considered an industrially feasible method on a large scale [5]. With present solution spinning fiber technology, a 1–2 wt % solution is not considered spinnable. Additionally, fibers and films produced from low concentration solutions experience substantial mechanical stress during the evaporation or removal of the solvent, which may adversely alter the mechanical properties of the formed fiber or film [5,41]. When solvent is removed too quickly, void spaces may result that not only affect the mechanical properties but may also reduce the density of the formed structures and adversely affect the electrical properties by decreasing the interchain hopping ability of the conductive regimes within the PANI matrix [62]. The formation of large void spaces is obvious under microscopic investigations; however, the formation of microvoids is most likely the most damaging to the electrical properties of the formed PANI.

Angelopoulos et al. [42] reported that the emeraldine base form of PANI could be dissolved in polar solvents such as NMP and DMSO. Fibers spun from or films cast from these solvents displayed reasonable mechanical properties; however, at concentrations greater than 6 wt % the solutions tended to gel after a relatively short time [44,45]. This solution instability makes it difficult to form fibers because the amount of time available for spinning is reduced and solvent–polymer interactions in the spin dope change the rheology of the system. However, useful structures may be formed from PANI gels once the gels are obtained. Strong, flexible, thick films made from PANI gels can be prepared and doped to significant levels of conductivities.

B. Gelation of Emeraldine Base Polyaniline Solutions

Spinning of emeraldine base fiber from solution in NMP was reported by MacDiarmid and coworkers in 1989 [69]. This work, however, did not attempt to characterize the PANI solution from which the fiber was spun. Further studies on this system by these and other workers later demonstrated that this polymer solution gelled after some period of time [44,70,71]. Extensive work on the gel systems indicated that the gelation may result from interactions or cross-linking between the microcrystalline areas of the PANI polymer [52]. The question still remained as to whether the gelation is physical or chemical in nature, that is, whether or not the cross-links result from hydrogen bonds or from other secondary bonding phenomena or primary valence cross-links.

Gelation caused by primary valence cross-linkages in a polyfunctional system is irreversible [72]. This is due to the immobilization of the polymer structure resulting from primary bond formation between the polymer chains. This type of behavior is best demonstrated by polyethylenes, which can be chemically cross-linked by radiation or chemical agents such as peroxides [72, pp. 705–706]. In some systems where primary valence cross-links are responsible for the networking and gel formation, the actual cross-link percentage may be quite small, on the order of 1–2%. Cross-links at low percentages are difficult to determine by spectroscopic methods if the polymers in question are of significant molecular weight [73]. Another type of gelation results from physical cross-links resulting from secondary bonding forces such as hydrogen bond formation, van der Waals forces, or other types of interactions that do not involve primary valence bond formation [72, pp. 121–123]. This type of gelation normally results from conditions in which the polymer–polymer interactions are stronger than the polymer–solvent interaction. To some extent the polymer gel formed by these types of cross-links can be broken by simple stirring where shear forces are strong enough to break the physical cross-links. This type of gelation phenomenon is called thixotropy [76, pp. 390–391]. A thixotropic gel once redissolved by stirring will re-form upon sitting. Gels formed by physical cross-links such as hydrogen bonds may be thermally reversible, but this is not always the case. Nylons, for instance, show some degree of hydrogen bonding even in the molten state [74].

1. Thixotropic PANI Gels

One of the characteristics of a thixotropic gel is that it can be broken up by the addition of a second solvent of appropriate character. A study by Hsu et al. [46] indicated that such behavior was noted in a solution of PANI base and NMP. This study reported that a concentrated (17% w/w) solution of PANI in NMP gelled after a short period of time but could be redissolved when exposed to gaseous ammonia and formed a viscous solution.

Thixotropic gels also demonstrate a characteristic shear thinning rheology. Shear viscosities, as a function of shear rate, for PANI solutions of 7.5% and 10% aged for a period of 4 h, are shown in Fig. 18.5. The 7.5%

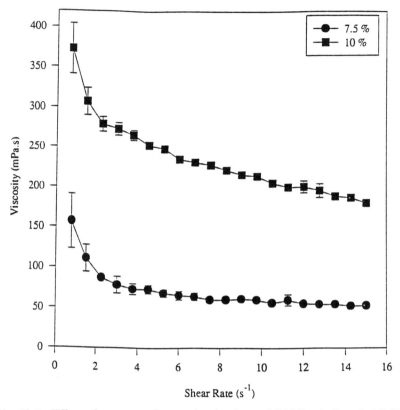

Fig. 18.5 Effect of concentration on the rheology of PANI solutions in NMP.

solution shows shear thinning behavior followed by a second Newtonian region. The first Newtonian region occurs at very low shear rates and is not accessible using the instrumentation available in our laboratory. Measurement of this region is of little consequence to the present study. A 10% solution of PANI base in NMP shows shear thinning behavior at a low shear rate (<3 s^{-1}), as did the 7.5% solution, but in this case there is a continuous decrease in the viscosity at higher shear rates. This continuing viscosity decrease may well be due to the breaking of loosely associated regions in the early stages of thixotropic gel formation. When the 7.5% solution was aged for a longer period of time, rheological behavior similar to that of the 10% solution was observed. Results from dilute solution viscometry studies of PANI/NMP solutions support the assumption that NMP may not be a "good" solvent for PANI base in a thermodynamic sense [57,75]. These studies demonstrate that the polymer–polymer interactions are stronger in NMP than in some other solvents that may be better candidates for solution processing of PANI base. These studies are detailed in Section III.

The results of the work of Hsu and that carried out in our laboratory indicate that the gel formation in NMP results from physical interactions because the material can be redissolved in appropriate solvents and demonstrates continuous shear thinning behavior. However, we cannot at this time categorically state that chemical primary cross-links do not form under appropriate thermodynamic conditions.

C. Flexible, Mechanically Strong, Electrically Conductive Structure Formed from PANI Base Gels

The gelation time for an 8% PANI base/NMP solution prepared by the method of MacDiarmid and aged at various temperatures is shown in Table 18.4 [44]. It clearly shows that the onset of gelation increases with tempera-

Table 18.4 Gelation Times of an 8 wt % PANI Base/NMP Solution at Various Formation Temperatures

Temperature (°C)	−22	4	22	37	72
Gelation time (h)	150 <	8.7 <,< 9.2	3.2 <,< 3.5*	1.0 <,< 1.1*	0.4 <,< 0.5

Source: Ref. 44.

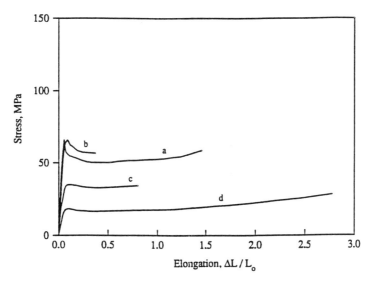

Fig. 18.6 Stress–strain curves of PANI films. (a) Undoped base film; (b) HCl-doped; (c) TSA-doped; (d) MeSO$_3$H-water/acetic acid–doped.

ture. This temperature dependence is typical of thixotropic gels and allows processing of a PANI/NMP solution after preparation and storage for some time at reduced temperatures. The solution can then be exposed to elevated temperatures and the gel allowed to form. Once formed, the gel can either be doped in the gel form and cast into a film or first cast into a film and then doped [44,71]. Film formed from the material doped prior to film formation exhibits a discontinuous structure, whereas the film doped after formation is smooth and featureless. The discontinuous structure is composed of small spheroids approximately 100 nm in diameter and very uniform in distribution. A variety of doping organic acids were used to dope the gels prior to film formation. All those that were tried gave conductivities in the range of 0.1–0.4 S/cm. The films were brittle and could not be properly mounted for mechanical testing without breaking. The films cast from the undoped gels prior to doping, however, maintained their smooth features after doping with the same organic acids and were easily prepared for mechanical testing. The films were smooth to the touch and quite flexible with surprising strength. Fig. 18.6 shows the stress–strain curves for the films prepared prior to doping and then doped with HCl, TSA (toluene sulfonic acid), or MeSO$_3$H and also for the undoped base film. Since PANI base films prepared from NMP still contain 15–20% of the solvent it is not unrealistic for the undoped base film to demonstrate reasonable stretching with a yield stress of 62 MPa and an elongation of 150% while the HCl-doped gel film is more rigid, having a yield stress of 66 MPa and an elongation of only 40%. The undoped films are, in effect, being plasticized by the NMP, a phenomenon observed for many polymer systems. It is surprising and quite interesting, however, that the film doped with MeSO$_3$H is quite flexible, with a low yield stress of 18 MPa and an elongation of 280%. TGA results indicate that for this film the NMP solvent content is almost negligible. This is consistent with NMP not acting as a plasticizer in the doped film, as there is no significant weight loss below 250°C (NMP bp ≈ 81°C). Moisture may play an important role in the mechanical behavior of this film, because all doped PANI films are hygroscopic. Table 18.5 reports the yield stress and elongation of films doped in

Table 18.5 Yield Stresses and Elongations for Dried and Undried (Moisture-Containing) PANI Films Doped with Various Dopants

Film dopant	Yield stress (MPa)		Elongation (%)	
	Dry	Undried	Dried	Undried
TSA/acetic acid	102 ± 4%	34 ± 4%	30 ± 10%	80 ± 10%
1 M HCl	140 ± 4%	66 ± 4%	40 ± 10%	40 ± 10%
MeSO$_3$H/acetic acid	62 ± 4%	18 ± 4%	80 ± 10%	280 ± 10%
None	78 ± 4%	66 ± 4%	150 ± 10%	150 ± 10%

Table 18.6 Moisture Regains of PANI Films Doped with Various Dopants Equilibrated at 25°C and 65% RH for 24 h

Doping solution	Base film	EtSO$_3$H	TSA	MeSO$_3$H	1 M HCl
Regain (%)	2.1 ± 2%	14.4 ± 2%	5.5 ± 2%	20 ± 2%	13.7 ± 2%

different acids and of the undoped film both in the moist state and after moisture has been removed. This table clearly shows the effect of moisture in acting as a plasticizer for the formed films. The films' mechanical properties are substantially reduced when moisture is removed; however, these properties can be easily restored by simply placing the films on the bench top and allowing them to regain atmospheric moisture. Table 18.6 gives the results of a moisture regain study at room temperature for 24 h and a relative humidity of 62–68%. It is easily seen that undoped PANI base films regain only ≈ 2% moisture in air and have a corresponding negligible loss in yield stress or elongation as seen in Fig. 18.7. The MeSO$_3$H-doped films, however, show substantial differences in yield strength and elongation between the moist and dry films. This result is not surprising, as it is well known that nylon is effectively plasticized by water [76].

D. Dopant Effects on Gel Film Conductivities

Table 18.7 reports the measured conductivities of PANI gel films doped in various doping solutions. It clearly shows that the films doped from an MeSO$_3$H solution and an EtSO$_3$H/acetic acid solution have the highest conductivities and thermal stabilities of the films listed. Acetic acid seems to be the best solvent for the doping of films using MeSO$_3$H or EtSO$_3$. Acetic acid superior to water for these dopants owing to the inability of aqueous solutions of these dopants to effectively wet the PANI films. The contact angle for undoped PANI gel films is 85°. This contact angle is substantially reduced in acetic acid as reported by Liu et al. [77]. As can be seen in Table 18.7, water, although having poor wettability on PANI, is still adequate for some dopants. Films doped from purely aqueous solvents demonstrate a reduced thermal stability, however, compared to the acetic acid solutions. The superiority of the acetic acid solvent is most likely due to its acidity (pK_a 4.75), which can compensate for the doping level, and also its good solvency for PANI, which can help the dopant molecules penetrate the PANI films. NMP solvent might also be expected to be a good carrier of dopant molecules into the PANI films, but conductivity data do not support this. This may well be attributable to the higher pK_a values of the dopants in this solvent. An example is MeSO$_3$H, which has a pK_a value of −1.97 in water but a value of 1.76 in DMSO [78].

The size and shape of the dopant molecules have a significant influence on ion mobility within the films. We examined three different dopant molecules with respect to their size differences and geometries. The dopants selected were MeSO$_3$H, HCl, and TSA. The sizes of the examined molecules decrease in the order TSA$^-$ > MeSO$_3^-$ > Cl$^-$, and the doping times decrease in the

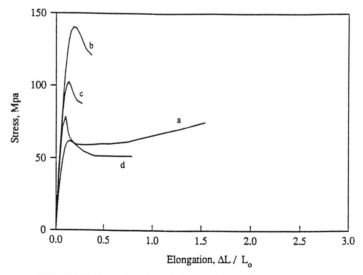

Fig. 18.7 Stress–strain curves of PANI films as in Fig. 18.6 but dried prior to testing. (a) Undoped base film; (b) HCl-doped; (c) TSA-doped; (d) MeSO$_3$H-water/acetic acid–doped. (From Ref. 29.)

Table 18.7 Conductivities of Doped Films of PANI Under Various Drying Conditions

Dopant solution	Conductivity (S/cm)		
	Air-dried	Vacuum-dried	145°C, 4 h
MeSO$_3$H/acetic acid	54 ± 3%	60 ± 3%	44 ± 3%
EtSO$_3$H/acetic acid	51 ± 3%	56 ± 3%	40 ± 3%
TSA/acetic acid	46 ± 3%	32 ± 3%	10 ± 3%
SSUA/acetic acid	—	0.2 ± 3%	—
SSA/acetic acid	—	18 ± 3%	—
NDSA/acetic acid	—	7 ± 3%	—
1 M HCl	50 ± 3%	5 ± 3%	0.4 ± 3%
MeSO$_3$H/H$_2$O	56 ± 3%	51 ± 3%	24 ± 3%
TSA/H$_2$O	0.44 ± 3%	0.4 ± 3%	—
MeSO$_3$H/NMP	—	<0.01	<0.01
TSA/NMP	—	<0.01	<0.01
80% acetic acid	—	0.3 ± 3%	—

same order, which is consistent with smaller dopant ions having greater mobility in the PANI gel film matrix.

E. Summary: Gel Films

Films prepared from PANI gels prior to doping demonstrate smooth homogeneous surfaces at high magnifications. Upon subsequent doping with appropriate dopants in solvents such as acetic acid, the gel films demonstrate good to excellent mechanical properties. In addition, significant levels of conductivity, approaching 70 S/cm in the case of films doped with MeSO$_3$H in acetic acid solution, can be obtained. The conductivity of these films can be increased to 200 S/cm or better if the films are drawn by mechanical force. Films of any thickness can be prepared or the gels can be coated onto substrates and subsequently doped. Fabrics and other structures coated with gel films and then doped show a very uniform level of surface conductance and, as in the case of the Kuhn materials, have little fiber–fiber bonding for thin films formed on textile substrates [31].

III. FIBER FORMATION FROM EMERALDINE BASE POLYANILINE

A. Introduction

Although films formed from PANI gels have good mechanical properties and subsequent conductivity, fibers cannot be formed from previously set gels [5,51]. Since PANI is not thermoplastic, it cannot be processed in the pure state by melt-spinning technologies. However, if suitable solvents can be found that allow the formation of rheologically stable PANI base solutions of ≈15% w/w, then it is possible to spin PANI base into fiber form and subsequently dope the fibers. Although MacDiarmid and coworkers successfully spun fibers from NMP solution, long-term solution stability studies were not carried out other than to point out the aforementioned gelation behavior of solutions whose concentrations were greater than 6% w/w [2,42,69,70], Wei et al. [79] and others [80] did show, however, that solutions at higher concentrations demonstrate bimodal distributions of molecular weights in NMP. The addition of LiCl to the solution reduced the bimodal distribution, with the removal of the higher molecular weight fraction indicating that fraction may well be an agglomeration of polymer due to interactive secondary forces between the polymer chains. The addition of a salt such as LiCl is a well-known technique used in spinning technologies for the reduction of "clustering" of polymers in solution [41]. Work at IBM by Angelopoulos et al. [64] has quantified the interaction of the Li$^+$ ion in reducing the effect of hydrogen bonding interactions between the PANI chains, thus reducing the clusters of associated polymer. This work is consistent with our previous conclusions that PANI forms a thixotropic gel in reasonably concentrated solutions. The clustering of PANI polymer in solution most likely continues to increase until the gel "sets."

Work in our laboratory on the rheology of NMP solutions of PANI and PANI/LiCl showed that the solution characteristics of NMP and NMP/LiCl do not provide a stable window for spinning of PANI fiber because the rheological profile of the solution is changing. NMP/LiCl is better than NMP alone but is still not satisfactory for fiber production with reproducibility from fiber segment to fiber segment. Figure 18.8 shows a plot of the normalized viscosities of 3%, 6%, and 8% w/w solutions of PANI/NMP at 25°C versus time and clearly indicates that at concentrations of 8% the viscosity quickly increases and the solution is not stable. Figure 18.9 is a similar study of PANI/NMP/LiCl solutions (0.5 wt %

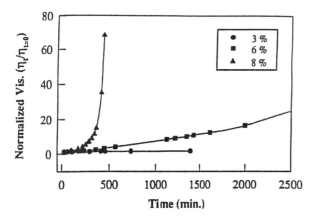

Fig. 18.8 Normalized viscosities vs. time of PANI/NMP solutions at 3, 6, and 8 wt % at 25°C. $\eta_{t=0}$ is the viscosity at time 0.

LiCl) at 6%, 8%, and 10% w/w PANI. This plot shows that although the solution viscosity is more stable it still increases at an undesirable rate at a concentration of 10%. For fiber production purposes a concentration of 10% is considered the lowest spinning concentration, and fibers formed at this low concentration tend to have undesirable defects and reduced mechanical properties. Figure 18.10 further demonstrates the effect of time on a 6% NMP/PANI solution. This plot of normalized viscosities versus shear rate clearly shows that at a solution aging time of 181 min the solution behaves as a Newtonian fluid and follows the general Newtonian relation, $\tau_{yx} = -\mu(dU_x/dy)$, for the shearing of a fluid element by a velocity gradient. However, at longer aging times there is a significant deviation from Newtonian behavior, and finally, in the solution aged for 2887 min prior to study, the solution shows a power law type of behavior that obeys the general power law expression for non-Newtonian fluids, $\mu(dU_x/dy) = k\,|dU_x/dy|^{n-1}$, where k is a

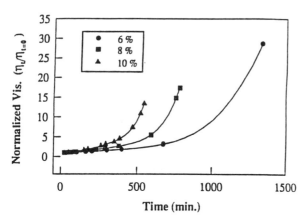

Fig. 18.9 Normalized viscosities vs. time of PANI solutions in 0.5 wt % LiCl/NMP at concentrations of 6, 8, and 10 wt %, at 25°C.

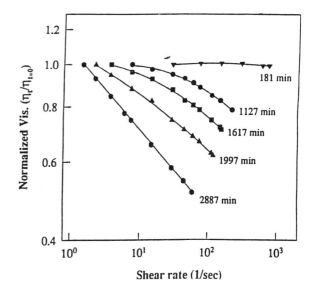

Fig. 18.10 Normalized viscosities vs. shear rate of a 6 wt % PANI/NMP solution at different aging times.

proportionality constant relating the shear stress to the nth power of the shear rate. This behavior indicates that the NMP solutions will behave initially as Newtonian fluids but deviate to a non-Newtonian fluid with the passage of time. Since the solution is in a state of flux, spinning a fiber with consistency over an extended period of time would prove difficult.

Observing the viscosity changes over a period of time is a useful method to study polymer solution stability and determine which solvents may provide a window of opportunity from which to spin fiber. To reduce polymer–polymer interactions, several solvents were studied for their ability to effectively solvate PANI at concentrations acceptable for wet spinning of emeraldine fibers. These solvents were selected from typical polymer solvents used for extraction and solvation of differing polymers and those with characteristics similar to NMP. Examples of solvents included N-methylpyrrolidione, N,N'-dimethylurea, dimethyl formamide, trifluoroethanol, tetramethylurea, formic acid, and other well-known polymer solvent systems. Of the solvents studied, N,N'-dimethylpropylene urea (DMPU) showed promise for reducing the likelihood of gel formation for an extended period of time and providing solvent stability for fiber spinning under applied shear. Figure 18.11 is a plot of the normalized viscosity of PANI/DMPU solutions at 10%, 15%, and 20% w/w vs. time at 25°C. This plot demonstrates a much higher solution stability for PANI/DMPU solutions, and, in fact, a 10% solution is stable to viscosity changes for over 2500 min. Even at a higher concentration of 20%, more than adequate for fiber processing, the viscosity has increased by only a factor of 4, whereas that of an 8% PANI/NMP solution as shown in Fig. 18.12 has increased by a factor of 35.

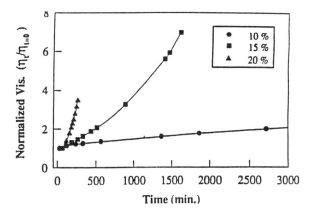

Fig. 18.11 Normalized viscosities of PANI/DMPU solution at concentrations of 10, 15, and 20 wt % at 25°C.

The importance of a good spinning solvent cannot be overstated. If the polymer–polymer interactions are strong enough for the polymer to fold in on itself, resulting in areas of high polymer density ("bunched" coils) separated by areas mostly composed of solvent, the likelihood of polymer chain entanglement necessary for strength in the formed fiber is reduced. The results of viscometry and shear studies as reported above indicate that in DMPU the PANI emeraldine chains are most likely more relaxed than in NMP. This results in the polymer chains not being "bunched" or coiled in on themselves, and the likelihood of chain entanglement is much greater upon extrusion. This chain entanglement not only gives superior mechanical properties but should also facilitate interchain electron hopping, thus giving a better overall macro conductivity. Although the very nature of the extrusion process results in some chain orientation along the extrusion channel axis due to shear forces, extension will be substantially less for polymers that are bunched in on themselves, or self-associated,

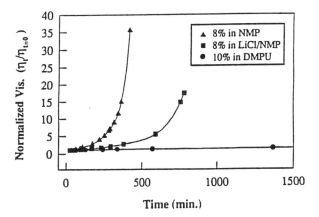

Fig. 18.12 Normalized viscosities vs. time of PANI solutions in NMP (8 wt %), LiCl/NMP (8 wt %), and DMPU (10 wt %).

than for those polymers in good solvents that are more solvated and extended. Also, polymers in poorer solvents will tend to rebunch during the die swell phenomenon occurring at the spinnerette exit, thereby minimizing chain entanglement in the formed fiber [5,27,41,62].

It should be noted that the mechanical strength of fibers formed during extrusion processes may not necessarily be due only to entanglement of molecular chains. Polymers that are of sufficiently high molecular weight and are highly crystalline may interact via secondary bonding, such as dispersive forces, to form high strength materials. An example of this is the solution-spun Spectra polyethylene fiber. The molecular weight of the polymer in this fiber is on the order of several million, and the crystallinity is around 85%. Although there is still considerable debate as to the actual molecular weight of PANI produced by various methods, it is reasonable to assume that the molecular weight and the degree of crystallinity are significantly less than that of Spectra fiber, and therefore the mechanical properties of the formed fiber to date result primarily from chain entanglement. PANI base films and fiber have been shown to be amorphous and to have a molecular weight (M_W) below 100,000. X-ray patterns (wide-angle) for the PANI base fibers prior to and after doping further support the noncrystalline amorphous structure of the PANI base as spun fiber. On the basis of these and other previously reported observations, we believe that the initial values for the tenacities, etc. result from chain entanglement, although upon doping dispersive interactions between microcrystalline regions may also play a role in the observed mechanical properties. When the fibers are drawn at a later stage there is an increase in crystallinity as reported by MacDiarmid and other workers. Dispersive interactions as well as hydrogen bonding interactions pointed out by Shacklette may well be greater in the case of the doped drawn fiber [24].

B. Spinning of Emeraldine Base Fiber

Solutions at concentrations high enough to allow for the production of PANI base fibers have been prepared using DMPU as the solvent. This "spin dope" was found to be stable for extended periods of time, and, in fact, if kept refrigerated at 10°C, the solution remains free of particulate matter, indicating that gels have not formed after months of storage. The PANI base is made according to the methods described by MacDiarmid, although special care is taken in the production of the PANI base by dedoping to ensure that all acid has been removed. Residual acid has been shown to catalyze gel formation [71]. The solution is filtered to remove any solid particles that could interfere with the spinning process due to clogging of the spin pack filters or spinnerette openings. After prolonged storage the spin dope may be refiltered to ensure that agglomeration has not occurred. In almost all cases we have seen little evidence of particulate formation upon prolonged storage. At Clemson University, we have constructed solution fiber spinning equipment for use in the production of PANI emeraldine base fibers. This equipment uses a pressurized canister to feed the PANI spin dope to a gear pump that provides the necessary pressure and flow to the spinnerette. A design of this type allows for the spinning of small amounts of fiber or prolonged runs producing thousands of meters of reproducible PANI base fibers. Interchangeable spinerettes allow for the formation of multifilament or monofilament fiber. The denier (linear density; mass/unit length) can also be controlled by the take-up speed in combination with the spinnerette hole diameter. A gear pump is used to deliver the spin dope to the spinnerette. The use of a gear pump provides a constant pressure that is easily monitored and can be controlled simply by adjusting the speed of the pump.

The polymer solution used for the production of the PANI filaments discussed in this chapter was a 15% w/w DMPU/PANI spin dope having a viscosity of 21,000 cP at a shear rate of 1.5 s^{-1} at a temperature of 25°C. A Brookfield cone-and-plate rheometer was used for the viscosity determination. The polymer solution is placed in the spin dope tank and pressurized with N_2, providing a steady flow of the spin dope to the gear pump. The solution is then pumped through a 0.1 mm circular spinnerette ($L/D = 2.0$) into a coagulation bath consisting of NMP and water. The fiber is continuously taken up on a 5 in. diameter take-up roll system equipped with a digital controller for precise take-up speed control and is washed with copious amounts of water. Depending on the speed of the take-up roll compared to the rate at which fiber is produced at the spinerette, some initial orientation can be achieved, producing a partially oriented yarn (POY) prior to actual drawing. The concentration of the coagulation bath is an important factor in the final mechanical and electrical properties of the formed fiber. Most of the fibers formed in our laboratories for initial studies are formed in a coagulation bath approximately 2 m long with an NMP/H_2O concentration of 25–40%. The effects of changing coagulation bath concentrations and other process conditions on the initially formed PANI fibers are outlined in the next subsection. In an effort to completely characterize the electrical and mechanical properties of PANI fibers formed from emeraldine base, the initial properties and the effect of process conditions on these properties must be evaluated prior to annealing and drawing the fibers. Maximizing the initial mechanical and structural properties prior to draw should result in fibers with excellent mechanical and electrical properties.

C. Effect of Process Conditions on the Mechanical and Electrical Properties of PANI Base As-Spun Fibers

Because of the many process steps involved in the formation of wet-spun PANI fibers, the effect of these various stages must be studied to determine an optimal set of conditions to maximize the desired mechanical and electrical properties. The development of constitutive equations for the process depends on an understanding of the effect of varying the spin conditions. Since the PANI fibers are initially undrawn, the molecular orientation and therefore the resultant properties are expected to be rather low. However, as previously stated, it is generally assumed that with the initial stages of the fiber spinning process optimized, the later stages of processing (i.e., drawing, annealing, etc.) will be more efficient in approaching the intrinsic limits of the material.

PANI fibers formed according to the method previously described were found to have both the initial modulus and the tenacity strongly dependent on both take-up speed and the coagulation bath concentration. The tenacity of the formed fiber increases by nearly 50% as the take-up speed is increased from 2 to 10 m/min. Figure 18.13 shows this relationship quite clearly. These tenacities are somewhat lower than one might expect on the basis of the molecular structure of PANI. Figure 18.14 shows increasing initial modulus as a function of take-up speed. The increase in both the initial modulus and the tenacities is not surprising, as the increased take-up speed certainly imparts some degree of orientation to the formed fiber. Although tenacities of 0.3 gpd may be

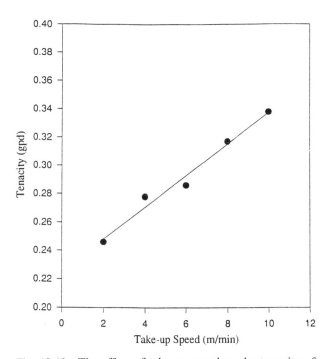

Fig. 18.13 The effect of take-up speed on the tenacity of as-spun PANI fibers. Coagulation bath concentration: 35 wt % NMP/water.

Solution Processing of Conductive Polymers

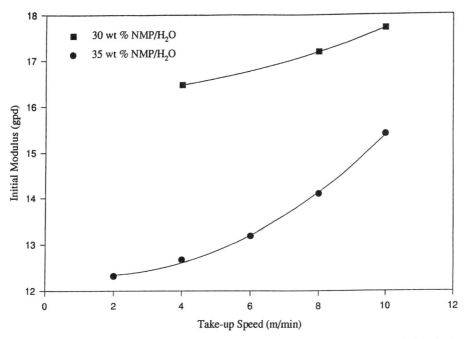

Fig. 18.14 The effect of take-up speed on the initial modulus of as-spun PANI fibers. Coagulation bath concentration: 35 wt % NMP/water.

acceptable for some specialized applications, they are not sufficient for most large-scale industrial processes. Subsequent drawing of the formed fiber will increase tenacities to an acceptable level. Tenacities as high as 3.2 gpd for the doped fibers are observed after drawing and annealing the fiber.

Figure 18.15 shows the relationship between the conductivity and the take-up speed of the formed fiber upon doping with methane sulfonic acid. As can be seen, the conductivity increases with increasing take-up speed, demonstrating the effect of fiber orientation on conductivity. This observation is consistent with those of MacDiarmid and other workers, who have noted that the drawing of PANI films and fibers increases conductivity [81]. It is unclear as of the time of this writing whether doping and drawing in unison or in separate steps provides the highest and most reproducible conductivities. Again it is important to state that the initially formed fiber must be as reproducible as possible on a segment-by-segment basis if subsequent drawing and/or doping are to be uniform.

From fiber diameter measurements and measured linear densities (denier), the fiber density can be calculated. Figure 18.16 shows the relationship between the calculated fiber density of the as-spun fibers and the conductivity of the doped fiber using $MeSO_3H$ as the dopant. Fibers obtained at differing take-up speeds had different calculated densities, and the conductivity increases as a function of fiber density. The fiber densities range between 0.25 and 0.6 g/cm³. These densities are much lower than expected on the basis of the density of PANI itself, which is approximately 1.3. One infers from Fig. 18.16 that if fibers were formed at higher densities the conductivity of the as-spun fiber upon doping would be substantially increased. The nature of the low

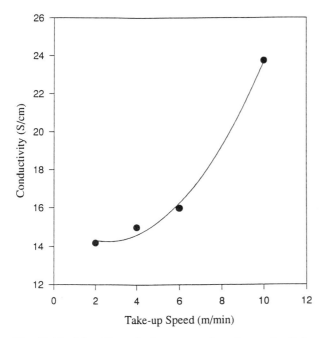

Fig. 18.15 The effect of take-up speed on the conductivity (1 M $MeSO_3H$-doped) of as-spun PANI fibers. Coagulation bath concentration: 35 wt % NMP/water.

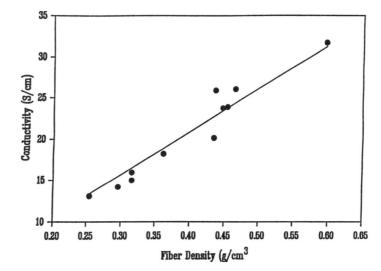

Fig. 18.16 The effect of fiber density on the conductivity (1 M MeSO$_3$H-doped) of as-spun PANI fibers. (From Ref. 84.)

densities was determined to be the result of large and small void spaces within the formed fiber. Figure 18.17 is a typical SEM photo of the undoped fiber cross section and is characterized by several interesting features. The most prominent of these are the large void spaces in the fiber, which explains the low densities. At higher takeup speed these void spaces are reduced in size due to the orientation (i.e., stretching) of the fibers, and this results in higher fiber density. Several other important features are the circular fiber cross section and the grainy structure of the "solid" material. The circular cross section is a confirmation that the spinnerette is properly machined and that the coagulation process is not too fast. More important, however, it indicates that the polymer solution (i.e., spin dope) is homogeneous,

giving further evidence that DMPU is a reasonable spin solvent. If the solution were not homogeneous, shrinkage would not be uniform during fiber formation and noncircular (crescent or dogbone-shaped) cross sections would result. The grainy surface indicates that very small voids may be present as well as the larger ones.

Since the conductivity–density relationship is approximately linear as shown in Fig. 18.16, the voids may be the major factor affecting both the mechanical and electrical properties of the as-spun fiber. In order to more thoroughly investigate this possibility, the electron microscopic images were digitally captured and were analyzed using a binary threshold to distinguish void from solid. The void spaces were numbered, and cross-sectional areas were calculated. This allowed the formation of histograms showing the distribution of void sizes resulting from various processing conditions. Figure 18.18 shows a series of histograms for fibers spun into a coagulation bath of 35% NMP in water. It clearly shows that the fiber spun at 4 m/min has larger voids than the one spun at 10 m/min. The largest void space in the 4 m/min fiber is approximately 430 μm^2, and the largest void size in the fiber spun at 10 m/min is 150 μm^2. Although the sizes may vary from cross section to cross section, the fibers spun at higher take-up speeds consistently have smaller void spaces than those at lower speeds.

The total cross-sectional area of the voids within each fiber is shown in Figs. 18.19 and 18.20 related to both the fiber tenacity and the conductivity. In both of these figures the measured property is low and shows little dependence on the total void area above 3000 μm^2. However, as the total void area decreases, the dependence of the measured properties becomes very pronounced. This suggests that small decreases in total void area produce large changes in tenacity and conductivity upon doping. According to the Griffith failure criteria,

Fig. 18.17 Typical scanning electron micrograph of an as-spun PANI emeraldine base fiber cross section (900×).

Solution Processing of Conductive Polymers

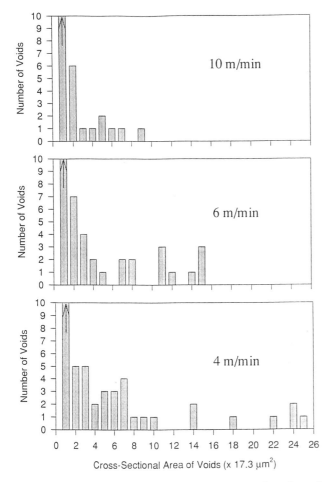

Fig. 18.18 Histogram of void dimensions as a function of take-up speed in as-spun PANI fibers. Coagulation bath concentration 35% NMP/water.

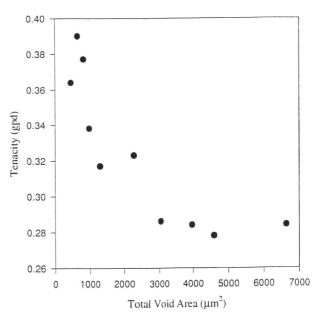

Fig. 18.19 The effect of total void cross-sectional area on the tenacity of as-spun PANI fibers. (From Ref. 84.)

the material strength is proportional to $c^{1/2}$, where c is the defect size [82]. Therefore as the defect size decreases, the strength increases, and this behavior is clearly seen in these plots. As the data in Fig. 18.20 suggest, the conductivity of MeSO$_3$H-doped as-spun fibers, prior to any external drawing other than take-up speed, is dependent on the total void area. The highest conductivity achieved for these fibers was approximately 32 S/cm, corresponding to a take-up speed of 10 m/min and the lowest total void area. When this fiber is drawn prior to doping, the conductivity increases to better than 350 S/cm on average with good reproducibility on a segment-by-segment basis, representing a tenfold increase in the conductivity of the as-spun fiber. Most likely the increase in conductivity of the as-spun doped fiber results from minimizing the very small or micro void spaces, allowing better chain-to-chain electrical proximity.

It is clear from the foregoing that in order to optimize the properties of the as-spun fibers, the void sizes must be reduced. The effects of two processing parameters, coagulation bath concentration and take-up speed, have recently been studied in our laboratory for their effects on fiber microstructure. Figure 18.21 shows the total void space area, determined by the previously discussed method, versus coagulation bath concentration for two

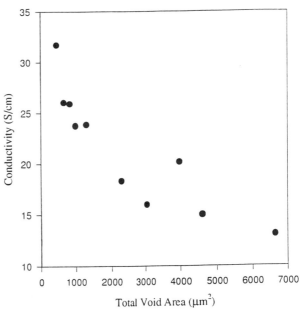

Fig. 18.20 The effect of total void cross-sectional area on the conductivity of MeSO$_3$H-doped PANI fibers.

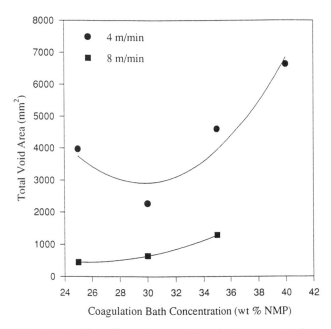

Fig. 18.21 The effect of coagulation bath concentration and take-up speed on the total void cross-sectional area of PANI base fibers.

different take-up speeds. At low speeds, the total area of the void spaces goes through a minimum at a coagulation bath concentration of 30%. This behavior is not unexpected, since it is common to observe a minimum in the diffusion coefficient and a maximum in fiber density versus concentration curves for various other well-characterized wet spin systems. Figure 18.21 also shows that the total void space area for a take-up speed of 8 m/min is considerably less than the area for a 4 m/min take-up speed. It also indicates that at speeds of 8 m/min, the dependence on coagulation bath concentration is significantly decreased relative to the concentration dependence at speeds of 4 m/min. This seems to indicate that in order to reduce the void size significantly, thus increasing the mechanical properties of the PANI base fiber, higher take-up speeds at lower coagulation bath concentrations of NMP in water are appropriate.

Figure 18.22 is a scanning electron micrograph of the drawn undoped fiber. The grainy fracture surface of the undrawn fiber in Fig. 18.17 has been replaced by a smooth cross section, indicating that the microvoids have been substantially reduced. Some microvoids are still present, however, in this fiber, indicating that the intrinsic limit of conductivity for the doped fiber and the mechanical properties have probably not been reached. These studies indicate that the elimination of both the large- and small-scale voids is necessary to form a fiber that maximizes both the mechanical and electrical properties and approaches an intrinsic limit. Reduction of the void spaces should produce a fiber with a fiber density close to that of the polymer itself. In fact, it may well be possible to correlate the measured fiber density with the areas of the void spaces within the fiber.

D. Fiber Properties

Initial X-ray data obtained for the as-spun fibers indicate that they are amorphous at slow take-up speeds, and some slight orientation is observed at higher speeds. This is consistent with the data obtained by MacDiarmid and Epstein on their films and fibers after orientation. Slight orientation occurs during higher take-up speeds. There is considerable scatter in the small-angle X-ray films, further confirming the presence of microvoid spaces on the order of 60 nm across within the fiber that affect both the electrical and mechanical properties. The grainy fracture surface observed in the SEM scans also is indicative of microvoids. Work being completed at this time in our laboratories suggests that the fiber should be doped immediately after or during the draw stage and not doped after a prolonged time period to ensure a more homogeneous doping. Table 18.8 compares some of the properties of PANI/DMPU fibers after drawing with those of other synthetic fibers. These values are reproducible on a segment-by-segment basis. Much higher conductivities of 500 S/cm have been obtained on single filaments, but it is difficult at this time to produce thousands of meters of this fiber with these conductivities on a segment-by-segment basis. After resolution of the void space problem, by determination of the proper mass balance and thermodynamic coagulation bath parameters, production of uniform fibers at these higher conductivities with level segment-to-seg-

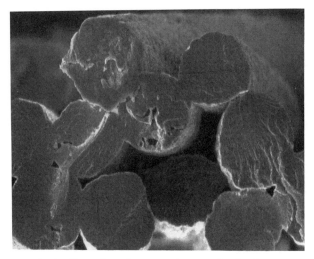

Fig. 18.22 Scanning electron micrograph of a drawn undoped PANI fiber (3000×).

Table 18.8 Mechanical Properties of Some Commercial Fibers and PANI/DMPU

Fiber	Tenacity (gpd)	Elongation (%)	Modulus (gpd)
PANI/DMPU (EB) (undoped drawn)	1.5–3.0	12–18	8–12
Nylon 6	4.0–7.2	17–45	18–23
PET	2.8–5.6	24–42	10–30
Glass (e-type)[a]	15.3	4.5	320
Rayon[b]	1.8–2.3	20–25	

[a] DuPont dacron, single-filament regular tenacity.
[b] North American rayon.
Source: Physical Properties of Textile Fibers, Ciba-Geigy Inc.

ment reproducibility should not prove difficult. Table 18.9 lists the processing parameters for formation of the PANI base as-spun fibers. The spin dope solution of 15% w/w of PANI base in DMPU with a viscosity of 21,000 cP is stable, and thousands of meters of fiber can be easily spun with negligible changes in the morphology of the as-spun fiber. This is a direct result of the stability of the PANI/DMPU spin dope.

E. Recent Advances

To more thoroughly investigate the effect of chain interactions in the spinning solution prior to fiber extrusion on the properties of the formed fiber, the spinning solution must be characterized. One method of doing this is to perform mechanical measurements commonly referred to as oscillatory shear measurements.

To spin a fiber either by melt or solution spinning technologies, the viscoelastic properties of the fluid must be known in order to obtain reproducible mechanical properties of the formed fiber. In addition, the electrical properties are dependent on the final polymer structure in the condensed phase and will also be dependent on the viscoelastic characteristics of the fiber and the fiber-spinning solution. The fiber-forming solution (spin dope) must be stable and cannot agglomerate or cross-link during spinning. Likewise, the polymer-spinning solution should not form cross-links, either chemical of physical, during storage, and no significant degree of interaction should occur between polymer chains during the spinning prior to extrusion at the spinnerette interface. Only certain specialized fibers employ cross-linking or gelation (i.e., gel spinning) after extrusion from the spinnerette. As mentioned in previous sections, PANI base solutions do tend to cross-link and form thixotropic gels at higher solids concentrations, and these interactions must be minimized if reproducible fiber is to be formed on an industrial scale. To study the changes in PANI base solutions with respect to time and shear, viscoelastic studies were carried out on PANI base solutions.

Several workers recently characterized the nature of the interactions as being due to secondary bonding (i.e., hydrogen bonding) between the oxidized portion of the PANI chain (imine nitrogens) and amine protons on neighboring chains [83–85]. In order to more thoroughly understand the nature and consequences of the chain–chain interactions and their effect on fiber and film formation, we decided to study the viscoelastic properties of the leucoemeraldine form of polyaniline. The leucoemeraldine form is the reduced form of PANI and contains significantly fewer imine nitrogens. Since imine nitrogens show a much greater propensity for the formation of hydrogen bonds, the degree of secondary bonding between chains should be substantially reduced. These studies were carried out by viscoelastic oscillatory shear measurements.

1. Polymer Solution Characterization: Viscoelasticity Theory

Application of small deformations displaces the equilibrium configuration of polymer chains, and the resulting recoil and elastic effects can easily be observed due to the slow relaxation nature of polymer liquids. Given an arbitrary strain history, corresponding stress history can be calculated using Boltzmann's constitutive equation of linear viscoelasticity [86],

$$\sigma(t) = \int_{-\infty}^{t} G(t - t') \frac{d\gamma(t')}{dt} dt'$$

where $s(t)$ is the shear stress at a given time t, $d\gamma(t')$ is the strain at time t', and $G(t - t')$ is called the relaxation modulus. When a sinusoidal strain $\gamma(t) = \gamma_0 \sin \omega t$ is imposed on a viscoelastic material, the Boltzmann equation becomes

$$\sigma(t) = \gamma[G'(\omega) \sin \omega t + G''(\omega) \cos \omega t]$$

where G' (storage modulus) and G'' (loss modulus) are the in-phase and out-of-phase components of the dynamic modulus, respectively. For a perfectly elastic material, $G'' \sim 0$, and for a perfectly viscous material, $G' \sim 0$. By measuring frequency dependence and time dependence of dynamic moduli, the elastic/viscous nature and evolution of chain interactions can be identified.

Table 18.9 As-Spun Fiber Processing Parameters

Spinning solution	15 wt % PANI/DMPU (either EB or LEB); $\eta = 21,000$ cP
Spinnerette	$D = 0.004$ in.; $L/D = 2.0$
Coagulation bath	25–40% NMP/H_2O
Take-up speed	2–10 m/min (LEB solutions up to 100 m/min)

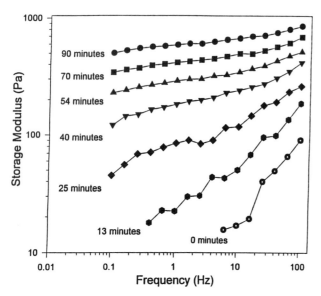

Fig. 18.23 Storage modulus (G') of a 14 wt % PANI/DMPU solution at various time intervals (min).

2. Viscoelastic Characterization of Concentrated PANI Emeraldine (EB) and Leucoemeraldine (LEB) Base Solutions

The viscoelastic characteristics of polyaniline (emeraldine base)/DMPU [PANI(EB)/DMPU] solution (14% w/w) and polyaniline (leucoemeraldine base)/DMPU [PANI(LEB)/DMPU] (14% w/w) were studied using a controlled stress rheometer in the frequency range of 1–100 Hz. The polyaniline solution was subjected to an oscillatory shear as a function of frequency at various time intervals. Figure 18.23 shows G' (storage modulus) as a function of frequency at various intervals of time. The modulus increases with frequency; however, the frequency dependence decreases with increasing time, suggesting a transformation from a viscoelastic liquid to a viscoelastic solid. The storage modulus increased significantly with time, indicating a development of elastic interaction between polyaniline base chains. Figure 18.24 shows the storage and loss moduli of a 14 wt % polyaniline (EB)/DMPU solution (PANI-C1) as a function of time at a frequency of 100 Hz.

It can be seen that at times less than 20 min the loss modulus G'' is greater than the storage modulus G', indicating that the viscous contribution is greater than the elastic contribution. There is, however, a pronounced increase in the storage modulus with time, suggesting that some type of network structure forms as a consequence of chain interactions resulting in the increased elasticity. Figure 18.25 shows the dynamic moduli of the PANI(LEB)/DMPU solution (PANI-C2). In contrast to the C1 solution, the loss modulus remains greater than the storage modulus throughout the experiment, indicating that formation of an elastic or gel structure is inhibited when PANI is in the reduced form. The increased chain flexibility due to the absence of cross-links is reflected in the order-of-magnitude difference in moduli between the EB and LEB solutions.

The loss tangent is the ratio of energy lost to energy stored in a cyclic deformation and is a measure of the fluidity of the solution (tan $\delta = G''/G'$). The loss tangent for both the C1 and C2 solutions is presented in Fig. 18.26. The loss tangent is very low for a cross-linked

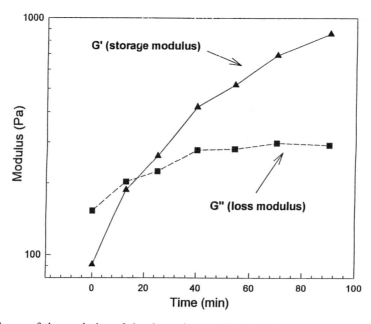

Fig. 18.24 Time dependence of the evolution of the dynamic moduli of a 14 wt % PANI(EB)/DMPU solution at 100 Hz.

Solution Processing of Conductive Polymers

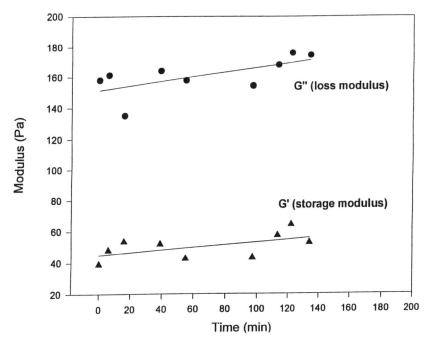

Fig. 18.25 Time dependence of the evolution of dynamic moduli of a 14 wt % PANI(LEB)DMPU solution at 100 Hz.

material, whereas it is very high for a dilute solution. The very low values of loss tangent for the C1 solution compared to the C2 solution and its decrease over time indicates the presence of elastic cross-links in the C1 solution, while few if any exist in the C2 solution. Scatter in the C2 data is due to the limited sensitivity of the rheometer in low elasticity solutions.

The dramatic changes in viscoelasticity when EB solution is converted to LEB solution motivated us to study the chain conformation. We compared the molecular weight of EB and LEB solutions by gel permeation chromatography because molecular weight determination by GPC is based on chain size and shape. Gel permeation chromatographic (GPC) data were obtained for

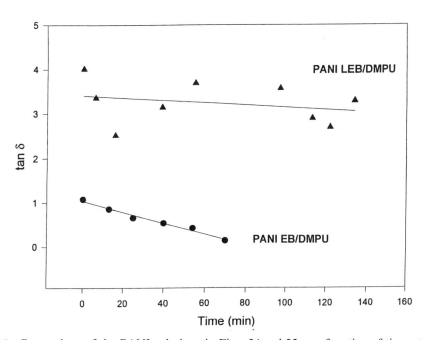

Fig. 18.26 Comparison of the PANI solutions in Figs. 24 and 25 as a function of time at 100 Hz.

Table 18.10 Mechanical and Electrical Properties of PANI C2 Fibers from Leucoemeraldine Base/DMPU Spin Dope

Property	As-spun (2×)	Drawn doped	As-spun doped	Drawn (2×)
Tenacity (gpd)	1.1	3.6	0.8	1.9
Modulus (gpd)	57	89	15	41
Elongation (%)	51	15	75	23
Conductivity (S/cm)	—	—	20	150

both C1 and C2 materials. The molecular weight of the C1 material was estimated to be $M_w = 85,000$ with a polydispersity index of 2.4, while the molecular weight and polydispersity index of the C2 material were 25,000 and 1.6, respectively.

Preliminary studies indicate that the difference in the observed molecular weights is due primarily to a difference in hydrodynamic volume between the oxidized and reduced forms of the polymer and may not be due solely to degradation as reported in the literature [87]. The minimized chain interaction and changes in chain conformation are reflected in the processing flexibility. Both C1 and C2 solutions in DMPU were solution-spun into fibers. Maximum attainable take-up speeds and draw ratios were greater for the C2 solutions than for the C1 solutions. As previously discussed, the C1 fibers exhibit a pronounced void structure. Fibers spun from leucoemeraldine base in DMPU had substantially fewer large area void spaces than did the C1 fibers. Variation of the oxidation state of the spin dope will allow the fiber spinning process to be tailored to obtain a wide range of fiber properties (electrical, mechanical, morphological, etc.).

3. Characterization of PANI Fibers from EB and LEB Solutions

Thermal characteristics of polyaniline fibers were determined in order to understand the effect of changes in oxidation state on morphology. In the first DSC scan of C2 fiber, the temperature was ramped to 400°C. A broad glass transition was observed at $T_g \sim 190°C$, and a broad melting endotherm was observed at $T \sim 355°C$, indicative of imperfect crystals with low sample crystallinity. The sample was quenched and remeasured under the same conditions. During the second scan, the glass transition was observed, but the melting peak was absent. In order to understand this melting transition, a second sample was heated to 350°C at 5°C/min to anneal, followed by quenching. A DSC scan of the annealed sample revealed $T_g \sim 190°C$ and a sharper endotherm at $T \sim 365°C$, indicating more perfect crystals. A second scan revealed the same glass transition temperature but no melting endotherm. The disappearance of the endotherm on the second heating may be due to the cross-linking of melted chains, preventing their recrystallization during second heating, or the recrystallization kinetics may be sufficiently slow that the melting temperature is reached before crystals can be formed. If the behavior is due to cross-linking, one would expect an exothermic reaction accompanied by an increase in the glass transition temperature. The data indicate an exothermic trend for $T > 300°C$, but the second scan shows no increase in the glass transition temperature.

Mechanical and electrical properties of the fibers were measured and are presented in Table 18.10. These properties should be compared with those of the PANI C1 fibers (EB/DMPU) and other fibers detailed in Table 18.8. Table 18.11 is a direct comparison of PANI C1 and PANI C2 (EB and LEB) fibers. The mechanical properties of the C2 fibers are significantly higher than those of the C1 fibers, as can be seen by comparing the values in Tables 18.8, 18.10, and 18.11. Even after doping with methane sulfonic acid (MSA) or HCl, these new fibers demonstrate mechanical properties strong enough for processing using existing technologies for yarn spinning.

Electrical conductivities for the C2 fibers after oxidation and doping gave reproducible values of 150 S/cm, which is lower than the 350 S/cm observed for C1 fibers. The presence of voids in C1 fibers most likely facilitates easier diffusion of dopants throughout the fiber, resulting in higher conductivities, while lesser voids and nonuniform oxidation give lower conductivity values for C2 fibers. Draw ratios in excess of 2 should dramatically increase the conductivity, and we are investigating methods to uniformly oxidize the C2 fibers to attain maximum conductivity. However, these conductivity values

Table 18.11 Tenacity and Conductivity of Doped PANI C1 (EB) and PANI C2 (LEB) Fibers

Fiber	Tenacity (gpd)	Conductivity (S/cm)	Doping solution
C1 As-spun	10–32	<0.2	1 M MeSO$_3$H, aq. acetic acid
C1 4× drawn	<1.0	350	1 M MeSO$_3$H, aq. acetic acid
C2 As-spun	15	0.8	1 M MeSO$_3$H, aq. acetic acid
C2 2× drawn	1.9	140	1 M HCl

are high enough for many applications where high mechanical properties are desired. Processing in reduced form also allows preparation of fibers from blends of polyaniline with other insulating and conducting polymers [88; Chacko and Gregory, unpublished results].

IV. CONCLUDING REMARKS

Fibers and films can be processed from emeraldine base polyaniline and then doped to appreciable levels of conductivity. Likewise, fibers and films can be formed from doped PANI as in the case of counterion-processible materials. Regardless of the methods used to produce the final conductive PANI there is still much work to be done if fibers and films are to approach the calculated intrinsic conductivities of the doped PANI polymer [37,89]. Dreams of a "plastic wire" to replace heavier metallic motor windings or to provide transmission lines are yet to be realized. In fact, they may well not be realized by PANI-type systems but may depend on new ICPs whose monomers are yet to be synthesized. As detailed in Section III, the final electrical and mechanical properties of PANI and other polymeric conductors will be based on the final microstructure and morphologies developed during processing [62,90]. In the solution spinning of fibers from PANI, the reduction in void spaces and corresponding increase in density should yield a fiber in which the interchain electron transport capability is enhanced, and there should be an increase in the mechanical strength as well. Reduction of the void spaces is a function of the mass transport of the solvent from the coagulating polymer. The coagulation bath must be such that the fiber skin remains swollen in order to allow the spin dope solvent to be removed, and the fiber formation kinetics must be predictable and controllable for reproducible fibers [5,41]. The rate of solvent diffusion out of the coagulating fiber is the predominant factor in forming uniform fibers from wet spinning techniques.

Methods for processing ICPs into the desired form for various applications will have to be developed whether it be by melt or solution processing. Likewise, new technologies for the production of these materials will need to be developed such as the electrostatic spinning of PANI fiber developed by Reneker and coworkers at the University of Akron [91]. In the electrospinning process, a high electric field is applied to a hanging droplet of polymer solution in a capillary tube. The applied electric field overcomes the surface tension, and a charged jet of the solution is ejected. Fibers are then formed by removal of the solvent and coagulation in a coagulation bath as in wet spinning. This process produces very highly oriented fibers with large surface areas and small diameters. Reneker's group has successfully spun conductive PANI by such a process. The use of chaotic mixing and supercritical fluids for doping and fiber formation will also need to be explored as possible alternatives to the present methods to maximize electrical and mechanical properties.

Recently, Mattes and coworkers at Los Alamos National Laboratories reported that high quality PANI fiber could be spun from emeraldine base using gel formation inhibitors [92]. These workers state that gel inhibitors prevent secondary bonding from occurring between the imine nitrogens on one chain and the amine nitrogens on a neighboring chain. An example they cite as being representative of a gel inhibitor is pyrrolidine. The role of these inhibitors is to associate with the unbound electrons on the nitrogens and prevent them from hydrogen bonding with neighboring chains. Mattes et al. [92] report that the mole ratio of the gel inhibitor used to the PANI and the use of high molecular weight PANI solutions are necessary to the formation of quality fibers. The density of the as-spun fibers is comparable to that of the fibers spun in our group from EB/DMPU (C1) solutions, and the as-formed fibers show no evidence of crystallinity, which is also similar to the C1 fibers spun in the Clemson laboratories. The fibers spun by the Los Alamos group are also much thicker and may have many useful applications, especially in the area of hollow fiber production for gas separations, etc. The use of these gel inhibitors allows stable spin dope solutions to be formed that contain a high concentration of solids, which, as previously discussed, is necessary for solution spinning of PANI fibers.

In order for successful industrial scale production of inherently conductive organic polymers to become a reality, the effects of processing on the final desired properties must be quantified. Since processing of polymeric materials has a direct effect on such properties as orientation, crystallinity, mechanical strength, and polymer morphology in the condensed state, processing can be used to tune or maximize certain desired properties. For example, certain organic polymers can be drawn to many times their original length, thus increasing their mechanical strength. This process may also facilitate interchain charge transport mechanisms by orientation and elongation of the polymer chain. Likewise, the optical and semiconductor properties of polymers used for nonlinear optical and electronic device applications will be dependent on processing of the polymer. Since many of the properties are process-dependent, advances in processing techniques unique to conductive polymers will have to be developed.

Simply put, a conductive organic polymer, or for that matter any organic polymer, is not a metal, although many of the theories used to explain the electrical nature of ICPs are based on our understanding of conduction mechanisms in metals and semiconductors. Band theory, polarons, bipolarons, thermopower, etc., are but a few examples of the models borrowed from solid-state chemistry and physics to explain the observed electronic behavior of these materials. Lattice distortions that occur in polymers as a result of polarons or other charge-carrying defects along the chain are not necessary to the conduction mechanism of metals. Polymers, unlike metals, are composed of individual organic moieties that are joined together with some identifiable chemical repeat unit in most cases. There are some exceptions to this, such as plasma-induced polymerization, where the starting monomeric material may not be identifiable [93]. In almost every case synthetic polymers are composed of a distribution of molecular weights with polydispersities (M_w/M_n) greater than 2. The degree of orientation and crystallinity in synthetic polymers will vary according to the methods used to process these materials and are a function of many differing variables during the initial polymerization and the subsequent processing steps. Since the basic properties of most polymeric systems are defined by these steps, efforts to produce conductive polymers with reproducible mechanical and electrical properties must consider the effects of processing on the morphology of the desired end product. Because of the defects inherent in any polymer chain, many of the desired mechanical and electrical properties cannot be fully realized. Defects in high molecular weight materials can be minimized by studying and quantifying the processes used to form the desired geometries of the solid-state polymer but cannot be completely eliminated. Controlling the molecular weight distribution in the formed polymer, along with accurately controlling the final polymer microstructure, are and will be necessary steps in the production of reproducible materials for various applications.

Advances in polymer processing over the last 50 years have brought us many useful techniques that can be used to enhance the properties of ICPs. Techniques of blending, spinning, injection molding, etc. can be employed to produce ICP solid-state structures for a variety of applications ranging from conductive fibers to nonlinear optical devices. Optimization of the process parameters and the development of constitutive equations for various processes are necessary to predicting and enhancing conductive polymer properties. In addition, the use of computer modeling for large-scale processes will also become more and more necessary to the production of materials with segment-to-segment reproducibility.

ACKNOWLEDGMENTS

I thank Drs. Koutai Tzou, Mingjun Liu, Claudia Nowak, Bin Huang, Steve Hardaker, and Antony Chacko and also Mssrs. James Moreland and Rajiv Jain for their contributions to the fiber formation and gelation work. I also gratefully acknowledge the financial support of the Department of Commerce under a grant administered by the National Textile Center supporting the fiber formation work.

Finally, on behalf of all those involved in the laboratories of the School of Textiles, Fiber and Polymer Science at Clemson University I dedicate this chapter to the memory of colleague and fellow graduate student Hanfeng Huang, who left us so suddenly and whose dedication to his work and friends is sorely missed.

REFERENCES

1. H. Shirakawa, E. J. Lewis, A. G. MacDiarmid, C. K. Chiang, and A. J. Heeger, *J. Chem. Soc. Chem. Commun. 1977*:578–580.
2. G. L. Baker, Progress toward processable, environmentally stable conducting polymers, in *Electronic and Photonic Applications of Polymers* (M. J. Bowden and R. J. Turner, eds.), *Adv. Chem. Ser.* Vol. 218, ACS, Washington, DC, 1989, pp. 271–296.
3. T. A. Skotheim (ed.), *Handbook of Conducting Polymers*, Vols. 1 and 2, Deckker, New York, 1986.
4. M. G. Kanatzidis, Conductive polymers, *Chem. Eng. News*, Dec. 3, 1990, pp. 36–54.
5. Z. Tadmore, *Principles of Polymer Processing*, Wiley-Interscience, New York, 1979.
6. A. Andreatta, S. Tokito, J. Moulton, P. Smith, and A. J. Heeger, *Science and Applications of Conducting Polymers*, IOP Publishing, Bristol, UK, 1990.
7. S. P. Armes, M. D. Aldissi, and S. F. Agnew, *Synth. Met. 28*:C837 (1989).
8. S. P. Armes, Conducting polymer colloids: a review, in *Colloidal Polymer Particles* (J. W. Goodwin and R. Buscall, eds.), Academic, New York, 1995, pp. 207–231.
9. S. Maeda, R. Corradi, and S. P. Armes, *Macromolecules 28*(8):2905–2911 (1995).
10. J. R. Reynolds and M. Pomereantz, Processable electronically conducting polymers, in *Electroresponsive Molecular and Polymeric Systems* (T. Skotheim, ed.), Marcel Dekker, New York, 1989, pp. 187–256.
11. G. Odian, *Principles of Polymerization*, Wiley-Interscience, New York, 1991, pp. 453–523.
12. J. Ruhe, T. Ezquerra, and G. Wegner, *Synth. Met. 28*:C177 (1989).
13. A. G. MacDiarmid, J.-C. Chaing, A. E. Richter, N. L. D. Somasiri, and A. J. Epstein, in *Conductive Polymers* (L. Alcacer, ed.), Reidel, Dordrecht, Holland, 1975.
14. S. Wang, T. Kawai, K. Yoshino, K. Tanaka, and T. Yamabe, *Jpn. J. Appl. Phys. 29*(10):2010–2013 (1990).
15. J. Ruhe, T. Ezguerra, and G. Wegner, *Polym. Bull. 18*:277–281 (1987).
16. Y. Wei, D. Yang, and J. Tian, U.S. Patent 5120807 (1992).
17. K.-Y. Jen, T. R. Jow, and R. E. Elsenbaumer, *J. Chem. Soc. Chem. Commun. 1987*:1113.
18. M. R. Bryce, A. Chissel, P. Kathirgamanathan, D. Parker, and N. R. M. Smith, *J. Chem. Soc. Chem. Commun. 1987*:466.
19. K.-Y Jen, M. Maxfield, L. W. Shacklette, and R. L.

Elsenbaumer, *J. Chem. Soc. Chem. Commun.* 1987: 309.
20. R. L. Elsenbaumer, K.-Y Jen, G. G. Miller, H. Echkardt, L. W. Shacklette, and R. Jow, in *Electronic Properties of Conjugated Polymers* (J. Kazmany, M. Mehring, and S. Roth, eds.), Springer-Verlag, Berlin, 1987, p. 400.
21. K.-Y Jen, R. L. Elsenbaumer, and L. W. Shacklette, PCT Int. Patent WO 88 00,954 (1988); *Chem. Abstr.* *109*:191098q (1988).
22. S. Li, Y. Cao, and Z. Xue, *Synth. Met.* 20:141–149 (1987).
23. Y. Cao and P. Smith, *Synth. Met.* 69:191–192 (1995).
24. L. W. Shacklette and C. C. Han, *Mat. Res. Soc. Symp.* 328 (1994).
25. Y. Cao, P. Smith, and A. J. Heeger, *Synth. Met.* 48: 91 (1992).
26. L. W. Shacklette and R. H. Baughman, *Mol. Cryst. Liq. Cryst.* 189:193 (1990).
27. K. T. Tzou and R. V. Gregory, *Synth. Met.* 69: 109–112 (1995).
28. T. Murase, T. Ohnishi, and Noguchi, *Chem. Abstr.* *104*:169124v (1990).
29. R. Aspland, *Application of Basic Dye Cations to Anionic Fiber: Dyeing Acrylic & Other Fibers with Basic Dyes, A Series on Dyeing*, Cpt. 12, Vol. 25, No. 6, 1993, pp. 21–26.
30. R. B. Bjorkland, *J. Chem. Soc. Faraday Trans.* 83: 1507 (1987).
31. R. V. Gregory, W. C. Kimbrell, and H. H. Kuhn, *Synth. Met. 28* (1989).
32. M. E. Galvin and G. E. Wnek, *Polym. Commun. 23*: 795 (1982).
33. M. E. Galvin and G. E. Wnek, *J. Polym. Sci., Polym. Chem. Ed.* 1983:21.
34. G. E. Wnek, in *Handbook of Conducting Polymers*, Vol 1 (T. A. Skotheim, ed.), Marcel Dekker, New York, 1986, pp. 205–212.
35. M. Liu and R. V. Gregory, *Synth. Met.* 72:45–49 (1995).
36. Y. Cao, P. Smith, and A. J. Heeger, *Synth. Met.* 48: 91–97 (1992).
37. R. Menon, Y. Cao, D. Moses, and A. J. Heeger, *Phys. Rev. B 47*: (1993).
38. Y. Cao, P. Smith, and A. J. Heeger, *Synth. Met.* 48: 91 (1992).
39. J. E. Fisher, X. Tang, E. M. Scherr, V. B. Cajipe, and A. G. MacDiarmid, *Synth. Met.* 661–664 (1991).
40. O. T. Ikkala, J. Laakso, K. Vakiparta, E. Virtanen, H. Ruohonen, H. Jarvinen, T. Taka, P. Passiniemi, J.-E. Osterholm, Y. Cao, A. Andreatta, P. Smith, and A. J. Heeger, *Synth. Met.* 69:97–100 (1995).
41. A. Ziabicki, *Fundamentals of Fibre Formation, Wet and Dry-Spinning from Solutions*, 1976, pp. 249–350.
42. M. Angelopoulos, C. E. Asturier, S. P. Ermer, E. Ray, M. E. Scherr, A. G. MacDiarmid, A. M. Akhtar, Z. Kiss, and A. J. Epstein, Polyaniline: solutions, films, and oxidation state, *Mol. Cryst. Liq. Cryst.* 160: 151–163 (1988).
43. O. Oka, S. Morita, and K. Yoshino, *Jpn. J. Appl. Phys.* 29:L679–L682 (1990).
44. K. T. Tzou and R. V. Gregory, *Synth. Met.* 55–57: 983 (1993).
45. E. J. Oh, Y. Min, S. K. Manohar, A. G. MacDiarmid, and A. J. Epstein, *Abstr. Am. Phys. Soc. Mtg.* 1992: M30-8.
46. C.-H. Hsu, J. D. Cohen, and R. F. Tietz, *Synth. Met.* 59:37–41 (1993).
47. Y. Z. Wang, J. Joo, C.-H. Hsu, and A. J. Epstein, *Synth. Met.* 68:208–211 (1995).
48. J. Joo, Z. Oblakowski, J. P. Pouget, E. J. Oh, J. M. Wiesinger, Y. Min, A. G. MacDiarmid, and A. J. Epstein, *Phys. Rev. B* 49:2977 (1994).
49. J. D. Cohen and R. F. Tietz, Eur. Patent Appl. 0446943 A2 (1991).
50. R. Jain, Masters Thesis, Clemson University, August 1995.
51. F. Grulke, *Polymer Process Engineering*, Prentice-Hall, Englewood Cliffs, NJ, 1994, pp. 363–434.
52. M. J. Liu, C. K. Nowak, and R. V. Gregory, *Polym. Mater. Sci. Eng.* 72:306–307 (1995).
53. M. J. Liu, C. K. Nowak, and R. V. Gregory, *Polym. Mater. Sci. Eng.* 72:308–309 (1995).
54. B. Wessling and H. Polk, *Mol. Cryst. Liq. Cryst.* 160: 205–220.
55. B. Wessling, *Polym. Mater. Sci. Eng.* (1991).
56. Y. Cao, J. Qui, and P. Smith, *Synth. Met.* 69:187–190 (1995).
57. R. Jain and R. V. Gregory, *Synth. Met.* in press.
58. D. Mangaraj, S. K. Bhatnagar, and S. B. Rath, *Makromol. Chem.* h67:5 (1963).
59. A. Andreatta, S. Tokito, P. Smith, and A. J. Heeger, *Mol. Cryst. Liq. Cryst.* 189:169–182 (1990).
60. A. D. Child and H. H. Kuhn, *Synth. Met.* 71: 2139–2142 (1995).
61. C. Nowak, Ph.D. Dissertation, Clemson University, 1995.
62. R. V. Gregory, *Proc. Soc. Plast. Eng.* 1683–1686 (1995).
63. R. Fried, Commodity thermoplastics and fibers, *Polym. Sci. Technol.* (1995).
64. M. Angelopoulos, Y. H. Liao, B. Furman, D. Lewis, and T. Graham, *Proc. Soc. Plast. Eng.* 41:1678–1671 (1995).
65. Y. Wei, G. W. Jang, K. F. Hsueh, E. M. Scherr, A. G. MacDiarmid, and A. J. Epstein, *Polymer 33*:314–322 (1992).
66. K.-Y. Jen, M. Drzewinski, H. H. Chin, and G. Boara, in *Electrical, Optical and Magnetic Properties of Organic Solid State Materials* (L. Chiang, A. Garito, and D. Sandman, eds.), MRS Symp. Proc. Vol. 247, 1991, pp. 687–692.
67. A. G. MacDiarmid and A. J. Epstein, *Faraday Discuss. Chem. Soc.* 88:17 (1989).
68. M. Angelopulos, A. Ray, A. G. MacDiarmid, and A. J. Epstein, *Synth. Met.* 21:(1987).
69. X. Tang, M. E. Scherr, A. G. MacDiarmid, and A. J. Epstein, *Bull. Am. Phys. Soc.* 34:583 (1989).

70. A. G. MacDiarmid and A. J. Epstein, in *Science and Application of Conducting Polymer* (W. R. Salaneck, D. T. Clark, and E. J. Samuelson, eds.), IOP Publishing, Bristol, 1990, p. 117.
71. K. T. Tzou, *Ph.D. Dissertation*, Clemson University, May 1994.
72. G. Odian, *Principles of Polymerization*, Wiley-Interscience, New York, 1991, pp. 17–19.
73. J. E. Mark, *Physical Properties of Polymers*, 2nd ed., ACS, Washington, DC, 1995, pp. 3–59.
74. L. R. Schroeder and S. L. Cooper, *J. Appl. Phys. 47*(10): (1976).
75. P. J. Flory, *Principles of Polymer Chemistry*, Cornell Univ. Press, Ithaca, NY, 1953.
76. G. J. Kettle, *Polymer 18*:742 (1977).
77. M. J. Liu, K. T. Tzou, and R. V. Gregory, *Synth. Met. 63*:67–71 (1994).
78. R. L. Benoit and C. Buisson, *Electrochim. Acta 20*: 105 (1973).
79. Y. Wei, K. F. Hsueh, and G. W. Jang, *Macromolecules 27*(2):518 (1993).
80. F. J. Oh, Y. Min, J. M. Wiesinger, S. K. Manohar, E. M. Scherr, P. J. Prest, A. G. MacDiarmid, and A. J. Epstein, *Synth. Met. 55–57*:977 (1993).
81. E. M. Scherr, A. G. MacDiarmid, S. K. Manohar, J. G. Masters, Y. Sun, X. Tang, V. B. Cajipe, J. E. Fischer, K. R. Cromack, M. E. Jozefowicz, J. M. Ginder, R. P. McCall, and A. J. Epstein, *Synth. Met. 41–43*:735–738 (1991).
82. A. Kelly and N. H. MacMillan, *Strong Solids*, 3rd ed., Clarendon Press, Oxford, UK, 1985.
83. M. Angelopulous, Y. H. Laio, B. Furman, and G. Teresits, *Macromolecules 29*:1996.
84. S. S. Hardaker, B. Huang, A. P. Chacko, and R. V. Gregory, *Proc. Mater. Res. Soc. 1996*:413.
85. S. S. Hardaker, B. Huang, A. P. Chacko, and R. V. Gregory, *SPE ANTEC Proc.* 1996.
86. J. D. Ferry, *Viscoelastic Properties of Polymers*, Wiley, New York, 1970.
87. Y. Wei, K. F. Hseuh, and G. W. Jang, *Macromolecules 27*:(1994).
88. A. P. Chacko, S. S. Hardaker, R. J. Samuels, and R. V. Gregory, *Synth. Met.* in press (1996).
89. A. J. Epstein and A. G. MacDiarmid, *Synth. Met. 69*: 179–182 (1995).
90. A. G. MacDiarmid and A. J. Epstein, in *Electrical, Optical and Magnetic Properties of Organic Solid State Materials* (L. Chiang, A. Garito, and D. Sandman, eds.), MRS Symp. Proc. Vol. 247, 1991, pp. 565–576.
91. G. Srinavasan and D. H. Reneker; *Polym. Int.* 36, 1995.
92. B. R. Mattes, H. L. Wang, D. Yanga, Y. Zhu, W. Luenenthal, and M. Hundley, *Synth. Met.* in press (1996).
93. H. Yasuda, *Plasma Polymerization*, Academic, New York, 1985.

19
Dispersion as the Key to Processing Conductive Polymers

Bernhard Wessling
Zipperling Kessler & Company, Ormecon Chemie, Ahrensburg, Germany

I. INTRODUCTION

It has long been known that it is possible to modify the electrical properties of polymers by means of conductive admixtures of many kinds or, rather, to make them conductive. Despite this, electrically conductive polymers are still not of any great economic importance as materials. They do, however, play an indispensable role in many specialized technical applications. They are mostly manufactured by incorporating electrically conductive carbon black or carbon fibers in a variety of polymers, especially thermoplastic polymers. Other conductive additives, such as steel fibers, aluminum flakes, metal-coated carbon fibers, metal-coated hollow glass spheres, and low melting metallic alloys have not so far played any decisive part in the development of electrically conductive polymers.

Today, however, there are signs of new ways and means of achieving electrical conductivity in plastics. New kinds of polymers, known as intrinsically conductive polymers (ICPs), have been developed whose electrical conductivity is no longer due to additives but to their chemical and crystalline structure, or morphology.

The crucial prerequisite for a full understanding of the properties of these substances and for using ICPs as materials is a research approach that integrates chemistry, physics, and processing.

This chapter describes the development of the hitherto successful scientific approach of ICP materials research and the results to date.

A. History of ICP Materials Research

It was in 1910 that Green and Woodhead [1] first reported the discovery of an aniline polymer that displayed a marked increase in electrical conductivity when treated with acetic acid. This phenomenon was soon forgotten, however. In the succeeding decades it was the dream of chemists and physicists to create organic substances, preferably polymers, with metallic conductivity. This conductivity, the researchers thought, ought to be made possible by π orbitals. Unfortunately, the "ideal model," polyacetylene, which was synthesized in the late 1950s and early 1960s by Natta et al. [2] and by Luttinger [3], did not exhibit any conductive properties. Moreover, the black powder also appeared to be nonformable and therefore completely useless. Not until 1978, when Shirakawa and Ikeda [4] succeeded in polymerizing polyacetylene (PAc) in the form of a film, were intrinsically conductive polymers with a relatively high conductivity available as a result of an oxidation process on the molecular chain. "Intrinsically conductive polymers" is taken to mean polymers that are conductive merely as a result of their chemical and physical properties. This breakthrough discovery gave this field of research a great boost and paved the way for detailed study. Now there were free charge carriers available that were mobile within the conductivity band of the polymers.

The years that followed saw the synthesis of a bewildering number of conjugated polymers and copolymers from a wide variety of aromatic, heteroaromatic, and acetylene monomers with and without substituents, that were converted into their respective conductive forms with the aid of equally numerous oxidizing agents or Brønsted acids. These highly conductive polymers include the polyacetylenes (PAc), polypyrroles (PPy), polythiophenes (PTh), polyphenylenes (PPP), and polyanilines (PAni). The conductivities achieved range from $\approx 10^{-5}$ S/cm (intrinsic semiconductors) to $\approx 10^4$ S/cm.

For many years the instability of many conductive polymers remained a problem. Polyacetylene is especially sensitive to oxygen. Even when stored in an inert atmosphere and at low temperatures, polyacetylene loses its original properties. After a while, for example, it can no longer be oriented. Even when stored in an inert atmosphere, complexed polyacetylenes almost completely lose their excellent electrical properties after a short time. These phenomena are attributed to oxidative degradation and to cross-linking processes. A similar lack of stability—albeit not to such a dramatic extent and partly for other reasons—is displayed by PPy, PTh, and many representatives of the PAni family. If they are to be used as materials, therefore, it is essential to solve the mystery of their chemical and physical instability.

Right from the start, special importance was attached to investigating the structure of the polymers. Clarification of their morphology, first with PAc, then with PAni and PPy, was important for us, both for questions of planned further processing and for the mechanism of their conductivity. This aspect was therefore given special priority within our developing materials research in the field of ICPs.

Conductive polymers occur as polycrystalline powders, filmlike agglomerates, or lumps of primary particles that, it was thought at the time, were not capable of further deformation. Since they are neither soluble nor fusible, it was a major advance when Shirakawa and Ikeda [4] succeeded, by means of interfacial polymerization, in producing self-supporting though very thin films whose properties were comparable to those of thin polymer films. This formed the basis for the idea formed by many groups of not bothering with attempts to process them but to polymerize ICP directly in the desired form as a film or in a blend.

We, by contrast, favored the polymerization of powders, with a view to following it up with a flexible range of processing options.

The great impacts that processing, including manufacture, has on the properties of all materials had long been known in the case of polymers and of carbon-black-filled conductive compounds. For conductive polymers too we at Zipperling* developed in the following years a systematic research into the interrelationships between polymerization, processability, and the resulting properties. Among other things this included studies of the interactions between chemical structure, morphology, processability, structure formation during processing, and electrical, mechanical, and rheological properties.

In view of the fact that conductive polymers are insoluble and infusible, it is not possible to manufacture formed parts from them by means of molding processes (molding by extrusion, injection molding, etc. [5]). To get around these processing problems, we attempted in our research approach to achieve processability through "dispersion." It was this that paved the way for making ICPs available in all forms. At the same time this unexpectedly proved to be the integrative basis for a comprehensive ICP materials research that solved the problems mentioned above and permitted a comprehensive understanding of ICPs.

B. Strategy and Function of ICP Materials Research

In spite of the obvious possibilities that intrinsically conductive polymers have to offer, it seems that what began some time ago as a phase of euphoria lasting several years among scientists and the media has now reached a critical stage. A number of scientific journals have altered their attitude with respect to conductive polymers from the initial euphoria to sarcastic criticism. Research into materials and processing is still regarded by most scientists as a necessary evil and one they do not generally want to be bothered with themselves.

In the countries engaged in research in this field, the consequences of this have already become apparent. In Japan, MITI did not renew a 10 year research fund, thereby forcing private companies to bear the entire financial responsibility for ICP research in industry and at the universities. In the United States too, continuity of finance for research would seem to be a subject of much uncertainty. The present situation in Germany is no different. A number of leading companies that had spent more than 10 years researching in the field of conductive polymers have mostly cut back their research activities or even discontinued them. Even the development of batteries based on PAni, PPy, or lithium-ICP composites has been stopped entirely in Europe and the United States. Attempts at commercialization in Japan proved a failure.

But what is the cause of this crisis affecting research into conductive polymers? Why is there this crisis that some scientists do not believe exists? There may be several reasons for it.

On the one hand, fundamental research concentrated too early on the mere use of conductive polymers in batteries or accumulators on the basis of PAc, PAni, or PPy, instead of first trying to reach a more profound understanding of the properties of the material. On the other hand, the quest for "solitons" dominated basic research in physics.

Too many scientists are still trying to understand conductivity as a microscopic property, reduced to properties they assume to exist in a single molecular chain. But how can the macroscopic properties of a material be reduced to molecular properties alone? Does one Fe atom constitute a metal? No, it does not.

To this day, the debate is characterized by a lack of cooperation and communication between the various areas and viewpoints of ICP research. The

---

* Zipperling Kessler & Co., Ahrensburg, Germany.

result of this is that only a very few scientists from other research fields (metals or semiconductors, solids, polymer science, etc.) have taken an interest in this particular field of research.

Processing of conductive polymers after polymerization with the aim of obtaining a reproducible morphology and lattice structure and hence reproducible material properties continues to be regarded by most of those active in the field of ICP research as of minor importance or is even totally overlooked.

It seems to us that the reasons people have not yet succeeded in implementing significant applications for conductive polymers include the following.

1. For many years research focused on rechargeable batteries on the basis of PAc, PAni, or PPy. Apart from the exception mentioned above, however, the hoped-for success has not materialized, for various reasons—technical reasons since the performance attainable mostly falls well short of the battery systems available on the market, and commercial reasons because even the battery with the new technologically attractive Li-PPP composite electrode is not (yet) capable of being produced and sold at a profit [6].
2. Owing to this one-sided concentration, electrochemical synthesis of self-supporting and nonprocessible films was adequate, with the result that very little work was put into the (chemical) polymerization of powders, because they did not appear to be usable in batteries.
3. As a result there was (outside our laboratory) no moldable ICP material available for broadly based applications-oriented research, which meant that ideas for applications (for which films could not be used) either did not emerge, were not investigated, or were confined to all too obvious areas such as "substitution of carbon black in antistatic applications."

Contrary to prevailing opinion, our strategic approach is based on the conviction that the macroscopic properties of a material cannot be explained in terms of molecular properties alone. It takes the correct chemical structure, the appropriate morphology, and finally the interaction between the polymer chains and the primary morphological units right up to the macroscopic morphology level to make up the totality of the properties of a material including its electrical properties.

Thus a comprehensive science of the processing of conductive polymers requires the integration of studies at the molecular level with studies at the level of the primary particles and more highly aggregated structures. What is needed, therefore, is information about chemical, crystalline, and morphological structure; about macroscopic transport properties and mechanical properties, their possible applications, and a fundamental theoretical understanding of them.

Until the community of ICP researchers as a whole succeeds in pursuing an integrated materials science approach, applications using ICPs cannot be expected to materialize on a broad front.

C. Processing as a Pillar of ICP Materials Science

What alternative ways and means of processing conductive polymers exist or are being sought? What are the advantages and disadvantages of the following processes?

1. Polymerization directly into the finished product, in the form of films or blends, on the surface of molded components or in their morphology. Here there are limitations on practical flexibility and purity, constraints on the shape or design of the finished products and their reproducibility and the chemical process. This approach is not possible for converters (for lack of chemical infrastructure) and is unsuitable even for basic research because of poor reproducibility.
2. Modification of the chain with side chains or by synthesis of block copolymers with a view to achieving fusibility and/or solubility. One argument against this approach is the fact that to date doped forms of even these modified forms are neither soluble nor fusible. "Postdoping," i.e., a chemical process involving solution or fusing, is required after conversion. Poor flexibility, reproducibility, poor stability, and poor homogeneity have been observed.
3. Dispersing the ICPs in convertible polymers (dispersion hypothesis). This approach makes available from a single raw material a wide range of application and conversion options for various products. A disadvantage here is that the polymerization process is complicated and the dispersion difficult. On the other hand, the process does result in flexible convertibility with regard to diverse applications and the reproducibility and purity of the products.

D. Success of the Dispersion Hypothesis

Since 1982 we have been advocating drawing up a materials science strategy for the development of ICPs that includes not only chemistry, physics, and theory but also processing science and processing itself as a fourth element. Each of these elements should be an "equal rights partner" in the foundations of a materials science. In the many years of research during which we have pursued this strategy, the "dispersion hypothesis" we developed proved to be an excellent tool for investigat-

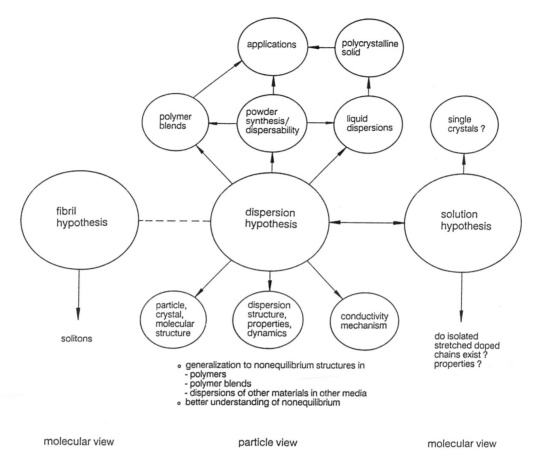

Fig. 19.1 Structure of the main question: What can the dispersion hypothesis contribute to a new integrated materials science of conductive polymers? (From Ref. 8.)

ing and describing conductive polymers (Fig. 19.1). The dispersion hypothesis is able to integrate formerly separated chemical, physical, theoretical, and processing approaches. In the process we were able to find both experimental and theoretical support for the new theory.

In 1983–1984 we succeeded for the first time in dispersing PAc. The synthesis and dispersion of PPy and PAni followed. In 1986 we were able to draw up complete concentration curves for conductivity between 0% and maximum filling in a thermoplastic polymer matrix. 1988 brought a breakthrough in the polymerization technology of dispersible ICPs. In 1990 we produced pure thin layers of PAni from organic dispersions. Transparent antistatic coatings with PAni in paint carriers followed. First applications of PAni coatings in the electronics industry were achieved. The anticorrosive effect of a coating containing PAni, recognized in 1988, was finally understood in 1992–1993 and at last became really practicable thanks to a new process. As this chapter is being written, the market launch of a first anticorrosive system on this basis is taking place by Ormecon Chemie, a subsidiary of Zipperling Kessler & Co., Ahrensburg, Germany.

Since the beginning of 1982 we have been carrying out research into the polymerization of processible powders and into structures, processing, applications, and materials. Our approach consists in identifying the chemical structure (purity, freedom from defects), crystalline lattice structure, and morphology and optimizing them by means of suitable measures (process steps) in polymerization and dispersion. This also led to the development of a new theory of thermodynamics in 1991 [7], which we have since exploited commercially and corroborated scientifically. The entire materials science of multiphase polymer systems and of ICP systems can benefit from this newly developed "nonequilibrium theory." In the end we began to understand the conductivity properties by analogy with the behavior of mesoscopic metals.

Finally it must be said here that finding the fundamental interrelationships of structures, morphology, interactions, and microscopic and macroscopic properties is an ambitious task of basic research and indispensable for

Dispersion as the Key to Processing

a broadly based introduction of ICPs in technical applications. Conversely, the excessively one-sided strategic orientation of research to date and the time lost as a result means that there is now an urgent need to implement significant technical applications if ICP research is not to be relegated to the role of an exotic outsider but receive the adequate financial backing and "moral support" that it deserves.

Practical use of our dispersions could be a new start in this direction.

II. BACKGROUND TO FINDINGS AND THEORIES IN THE PROCESSING OF ICPs

This section explains a number of terms and phenomena necessary for an understanding of intrinsically conductive polymers (ICPs) and their processing.

A. Percolation Theory

The term "percolation" [8] has been used to describe in particular the discontinuous conductivity behavior of carbon-black-filled compounds. It is a widely known fact that the conductivity of a compound filled with carbon black does not increase in a linear fashion with the concentration of conductive carbon black added. What in fact happens is that there is a sudden increase in conductivity that does not take place until a certain characteristic concentration, the "critical volume concentration," is reached, whereas continued addition of carbon black after this point once again brings only a gradual increase.

This behavior has been called percolation (Figs. 19.2 and 19.3). The "percolation behavior" of carbon blacks shows a dependence on the specific surface area. Carbon blacks with a large surface area display a much lower percolation point (which can be expressed in percentage by weight or volume, wt % or vol %) [9].

In an attempt to explain this phenomenon of the sudden leap in conductivity at a critical concentration, it was originally assumed that the conductivity of a compound was due to the fact that above the percolation point the conductive particles statistically distributed in the compound (secondary agglomerates) were sufficiently close together or in contact, resulting in continuous current paths. Since a statistical distribution of spherical particles would not result in adequate contact (and hence conductivity) until their concentrations were between about 27 and 52 vol %, depending on the geometrical model used, it was assumed that the carbon blacks had a "structure" that deviated from the spherical form. The critical volume concentrations of less than 20 vol % (corresponding to around 35–40 wt %) observed even in carbon blacks without a very marked structure would demand a complicated geometrical arrangement that would seem unlikely to form under the conditions prevailing in the compound.

Carbon blacks with a more marked structure display critical concentrations of around 10%.

Carbon black manufacturers in particular therefore worked on the basis of a model in which the primary particles measuring ≈ 30 nm arranged themselves preferentially in a linear fashion to form a secondary agglomer-

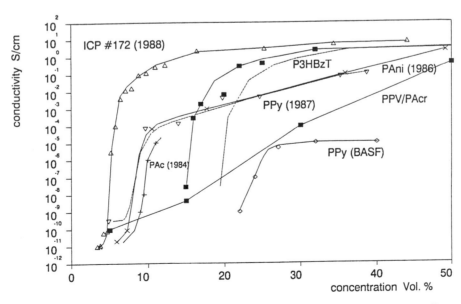

Fig. 19.2 Critical volume concentration of various ICP qualities showing the relation between, dispersibility and critical volume concentration: very bad dispersibility with PPy (BASF), leading to 25 vol% ϕ_c; no explicit dispersion behavior for PPV in polyacrylate; no critical volume concentration behavior; increasing dispersibility from polyacetylene (1984) to PAni No. 172 (1988). (From Ref. 8.)

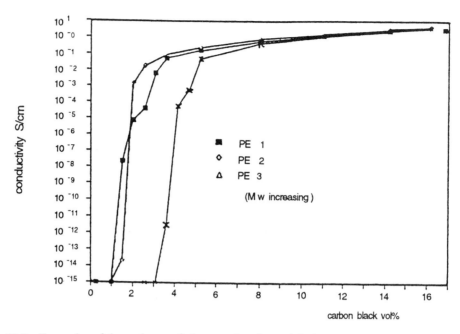

Fig. 19.3 Examples of dependence of Φ_c on molecular weight in polyethylene. (From Ref. 8.)

ate with a structure that remained at least partly intact even when incorporated into the plastic. This promised adequate contact in the supposed statistical distribution. The idea thus became established that the structure of the conductive particles was degraded during compounding [10] and that overprocessing [11] of the conductive substance during the incorporation and further processing of the compound (e.g., injection molding) should be avoided. It is important to note that the percolation theory is a mathematical approach based entirely on statistical (geometrical) considerations. It can be used only for filled compounds (or other heterogeneous systems) in cases where the particles are statistically distributed and interfacial forces do not play any significant role. It is possible to deduce from the theory a number of predictions for testing its validity, such as the following:

1. For a given type of carbon black the percolation point should be independent of the properties (polarity, viscosity) of the polymer matrix.
2. Any dependence on viscosity should mean that "overprocessing" with highly viscous polymers results in a higher percolation point than with low viscosity polymers;
3. A tempering process without shearing after compounding and/or processing ought to result in correction of any deviation from the statistical distribution, which would mean a deterioration in the conductivity of the compound.

These predictions, and hence the theory, were not confirmed but were, rather, negated by our own studies, which indicated that the phenomenon was one of a primarily interfacial nature [12,13].

The essential aspect here is that a compound with carbon black is controlled by the dynamics of the interfacial energies, namely adhesion (polymer–carbon black) and cohesion (carbon black–carbon black). First the carbon black is completely dispersed, i.e., the entire agglomerate structure is totally destroyed, and in the process adhesion interfaces are created. Above the critical concentration, however, it becomes energetically advantageous (from a thermodynamic point of view) to form more and more carbon black/carbon black cohesion interfaces; a network with a predominantly linear structure is built up. The formation of these structures is thermodynamically induced and kinetically controlled.

The scanning electron micrographs of carbon blacks with different specific surface areas shown in Fig. 19.4 [13] were interpreted as follows. Whereas the carbon black used for coloring purposes exhibits a smooth particle surface (Fig. 19.4a) and its primary particles are densely packed and highly agglomerated (Fig. 19.4b), in the case of conductive blacks the more suitable they are for imparting conductive properties, the more strongly structured their particle surface is (Figs. 19.4c, 19.4e); their primary particles can be more clearly differentiated (Fig. 19.4f) and evidently have preferably only point contact with each other. In our view this is in line with the better dispersibility of conductive blacks. It should be noted that the spherical primary particles identifiable under the scanning electron microscope as the smallest morphological units are not the primary units of graphitic structure; the latter are in fact formed from the morphological primary particles measuring ≈50 nm.

Thus after laboratory tests conducted with special precision we found that complete dispersion of the conductive black in various plastics yielded widely differing percolation points: The higher the surface tension of the matrix polymer, the higher the critical concentration (Fig. 19.5). In other words, the wetting of the carbon black by the matrix must play a dominant role. The better the matrix wetted (dispersed) the carbon black, the more carbon black was necessary to bring about the leap in conductivity. Conductivity evidently occurs at a concentration at which the matrix can no longer disperse the entire carbon black completely—in other words, its "wetting capacity" is exhausted. This observation, in conjunction with the results of other experiments, later gave rise to the "flocculation model," to the interpretation of the structures as "dissipative structures," and to the nonequilibrium thermodynamics of multiphase polymer systems.

In this view, carbon blacks with a larger specific surface area exhibit lower percolation points only because they possess a larger accessible wettable surface, i.e., they are more readily dispersed. They therefore reach a maximum of adhesion interface energy earlier, after which the system, its wetting capacity being exhausted, forms other structures (which are advantageous for conductivity). It will readily be seen why a reduction in particle size (increase in surface area) results in a reduction in the critical volume concentration, something that does not apply to even the most complicated geometrical model. Thus one can obtain a mixture with stable conductivity or one with reproducible conductivity only if the carbon black is fully dispersed (wetted) and if this concentration level is above the critical volume concentration. The conductivity bridges or carbon black networks then form inevitably and automatically for thermodynamic reasons.

At this point it is important to note that if ICPs are incorporated (dispersed) in polymer systems there is likewise a sudden leap in conductivity at a characteristic concentration ϕ_c just as in systems filled with carbon black. We found that the more readily dispersible we managed to make the ICP, especially PAni, the lower the critical point ϕ_c was. Thus here too we are concerned not with a percolation phenomenon but with a thermodynamic phenomenon involving interfacial energy. The term "percolation" should be avoided in connection with the phenomenon discussed here.

B. The Fibril Hypothesis

Even today, despite the evidence of our findings, a large number of materials scientists working on conductive polymers still regard the structure of ICPs as fibrillar. This fibril hypothesis has its origins in the early studies by Shirakawa and coworkers [15], who in 1974 were the first to succeed in producing self-supporting polyacetylene films. Initial morphological studies of this material clearly revealed fibrils with a diameter of some 20 nm in a fiberlike structure.

Subsequent electron microscopic studies (SEM and TEM) [14, Chap. 4] came to the same conclusion. To avoid changing polymer morphology as a result of handling the samples, thin polymer films were applied directly to EM grids in a layer thickness of about 100 nm. The SEM/TEM images showed a large number of thin strings, the fibrils, with an average diameter of 20 nm. In certain areas these aggregated to form bundles of parallel aligned fibers.

Investigation of thinner *cis*-polyacetylene films revealed large numbers of even smaller fibrils, microfibrils, with a diameter of 2–3 nm. On the basis of the crystal structure of the unit cell of *cis*-polyacetylene, the number of polyacetylene chains contained in and therefore making up the microfibrils was deduced. This ranged from about 13 chains (for 2 nm microfibrils) to 60 (for 3 nm microfibrils). Since then, these microfibrils have been regarded as the basic morphological unit of the fibril hypothesis. On this basis the fibrils discovered earlier are made up of straight microfibrils.

According to Chien [14], the origin of the microfibrils lies in the formation of the polymer systems. Since polyacetylene can be formed by various processes, the morphology of the polymer should be dependent on the nature of the crystallization process.

When microfibrils aggregate to form fibrils of larger diameter, a further morphological development is assumed to occur. For example, the appearance of a globular morphology, e.g., in acetylene, has been explained in terms of poor heat distribution during polymerization in combination with a smooth film surface. The spheres that formed in this case were accordingly assumed to be an accumulation of fibrils.

For highly conductive PAc (hc PAc) in particular (for a review see Ref. 16), the fibril hypothesis was used (even though globular morphology was also observed for hc PAc synthesized by a "nonsolvent" method (Ref. 16, p. 98). For other ICPs as well, also in the light of the "granular metal" concept, the fibril hypothesis and the understanding that conductivity is based on single-chain properties and their arrangement in fibrils (see Ref. 2, Fig. 2) plays a dominant role.

C. Rheological Phenomena

One of the most striking properties when processing carbon-black-filled conductive compounds (and also ICP blends) is the massive increase in melt viscosity compared with the unfilled matrix polymers and with other filled compounds. It is known that filled thermoplastics

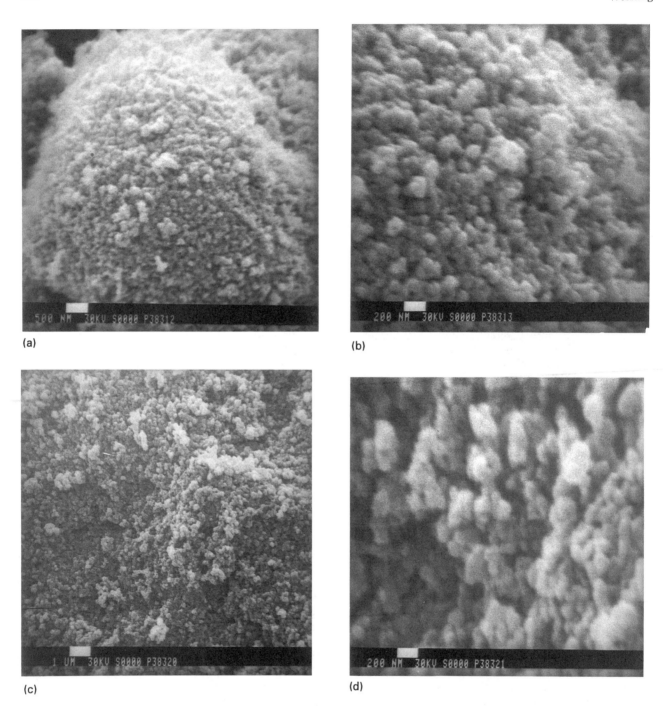

Fig. 19.4 Scanning electron microscopic pictures showing carbon black powders with different specific surface areas. (a, b) Pigment carbon black (Monarch 880); (c, d) carbon black for conductivity applications with medium quality (Philblack P); (e, f) carbon black for improved conductivity applications (Philblack XE-2). (From Ref. 13.)

Dispersion as the Key to Processing

(e) (f)

Fig. 19.4

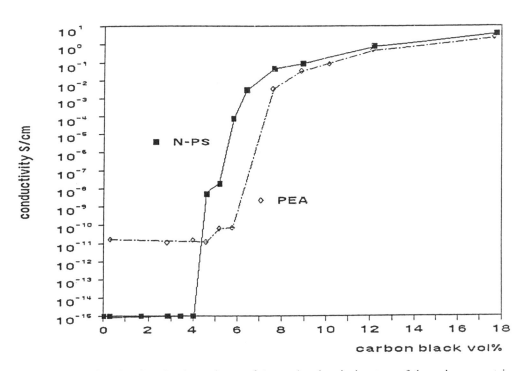

Fig. 19.5 Examples showing the dependence of ϕ_c on the chemical nature of the polymer matrix.

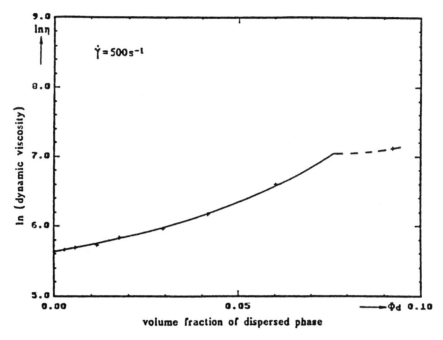

Fig. 19.6 Nonlinear dependence of the viscosity of a polymeric compound on the carbon black content (crystalline polystyrene with carbon black printex XE-2). (From Ref. 7.)

display increased viscosity (Fig. 19.6). Two equations are commonly used to describe this phenomenon, the Einstein equation,

$$\eta = \eta_0(1 + \mu\phi) \quad \text{(Ref. 17)} \quad (1)$$

and the Mooney equation,

$$\ln\left(\frac{\eta}{\eta_0}\right) = \frac{\nu\phi}{1 - \phi/\phi_0} \quad \text{(Ref. 18)} \quad (2)$$

Neither equation contains any terms that tell us anything about the interactions between fillers (dispersed phase) and matrix. They have not proved their value in forecasting the viscosity of filled systems and polymer blends. For calculations on polymer blends used as phonograph record molding compounds, we analyzed on the basis of systematic studies the influence of

1. Copolymers compatible with PVAc-VC copolymers
2. Homopolymers partially compatible with PVAc-VC copolymers
3. Fillers (chalk, carbon black)
4. Plasticizers

on the viscosity of a PVC-VAc copolymer (the matrix polymer). From this we derived an equation of general validity that for the first time takes into account the interactions between matrix polymer and dispersed phase:

$$\ln \eta = \ln \eta_0 + \chi_N \ln \eta_N + \chi_A a + \chi_B b \quad \text{(Wessling)} \quad (3)$$

Here χ_A, χ_B, χ_N are the percentages by volume of the components of the mixture; η_N the viscosity of the matrix polymers N and the polymers compatible with it; a, b were formerly described by us as "thermodynamic interaction factors" and are parameters that are constant over a wide concentration range and appear in the equation as apparent (logarithmic) viscosities. The value of this equation is first that it makes it possible to accurately predict melt viscosities of multiphase polymer compounds and blends at any shear rate. It is sufficient for this purpose to measure (1) the viscosity of the matrix polymer in question and (2) the effect of components planned to be blended with the matrix polymer in any other comparable polymer system (i.e., it is not necessary to analyze the effect of a given component in all the different polymer matrices, and it is sufficient to measure only two or three concentrations over a realistic range).

With these basic measurements one can derive the viscosity η_0 of the matrix polymer, the viscosity contribution η_N of a fully compatible polymer component N, and the viscosity-increasing or -decreasing effect a or b of an incompatible dispersed component A or B. With these starting values, the viscosity prediction can be performed with sufficient accuracy.

Dispersion as the Key to Processing

This equation has proved very successful in practical work to date. It has been supported by various works and publications [19], and we now understand it thermodynamically (see Section V).

The question arises as to the nature of the interactions and/or structures that influence the viscosity of the matrix polymer in this way and why very different substances exert such a comparable influence. From the structures underlying conductivity it may be assumed that it is largely, if not completely, the size of the specific surface area of the disperse phase rather than the nature (strength) of the interaction between dispersed phase and matrix that causes the change in viscosity.

Decisive factors here seem to be the formation of the firmly adsorbed phase (i.e., the reduction in freely flowing volume), the phase separation ("seams"), the flocculation structures (chain formation), and the interactions between the adsorbed phase and the matrix (see Fig. 19.7). Of these, the formation of the seams and chains appears to be particularly important. These are extremely thin and very extensive structures that have the effect of increasing viscosity. At the same time it must be remembered that the dispersed phase does not interact directly with the matrix but only, in a very intensive fashion, through the shell of matrix material firmly adsorbed on it.

D. Mechanical Properties

Other widely known phenomena are the influence of the incorporation of carbon black or ICP and the influence of processing on the mechanical properties of products. The outstanding features here are the massive increase in the modulus of elasticity and the decrease in percentage elongation at break, breaking strength, and notched-bar impact strength.

To influence these phenomena and compensate for them insofar as possible, the approach most commonly adopted was the empirical one of incorporating the appropriate conductive additives and then compensating for the "lost" impact strength and breaking strength by means of impact modifiers. In most cases, however, this made the compounds too soft, too low in heat deflection temperature, though still too inelastic or too brittle. Optimization is a balancing act. The goal is to obtain a compound with properties that, apart from the gain in conductivity, are still comparable to those of the original

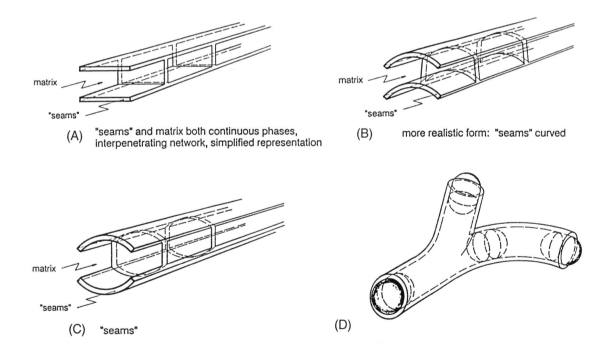

Fig. 19.7 Schematic models of possible continuous seam structures in a (continuous) polymer matrix. (A, B) Seams and matrix, both continous phases, interpenetrating network, simplified representation; (C, D) seams.

matrix material. For this purpose an understanding of the thermodynamic fundamentals of structure formation in such systems is a great advantage.

It is not yet possible to give an exact description of the reasons for the massive adverse effects of the incorporated conductive additives on the mechanical properties of the polymer. The following aspects play an important part.

1. Conductive black or PAni actually takes up about 3–6 times its own volume in the compound volume, owing to the firmly adsorbed shell. In a compound with 9 vol % carbon black, for example, only about 60% of the matrix is still freely mobile.
2. Polymers seem to adsorb differently and/or the changes in the properties of the chain as a result of absorption appear to vary in intensity, with the result that the interactions between the adsorbed shell and the free matrix are of varying degrees of "poorness."
3. The structures (within a seam, above all the su-

Table 19.1 Overview on Zipperling Conductive Carbon Black Compounds and Concentrates[a]

Polymer base	Designation		Processing	Resistance (Ω)	Impact strength[b]
LD-PE	PE	101 CON-EP	Injection/blow molding	10^3–10^4	100
HD-PE	PE	102 CON	Blow molding/pipe extrusion	10^3–10^4	100
LD-PE/EVA	PE	103 CON-EP	Injection molding	10^2–10^4	100
LD-PE/EVA	PE	104 CON-EP	Injection molding	10^3–10^4	100
HD-PE	PE	105 CON-EP	Extrusion	10	30
EVA	EVA	106 CON	Inject molding/extrusion	10^4–10^6	100
HD-PE	PE	109 CON	Blow molding	10^3–10^4	100
LD-PE/EVA	PE	110 CON	Film extrusion	10^3–10^4	100
PP	PP	203 CON	Extrusion	10^3–10^4	100
PP	PP	204 CON	Injection molding	10^3–10^4	100
PP	PP	205 CON	Injection molding	10^6–10^9	100
PP	PP	207 CON-EP GF	Injection molding	10^3–10^4	95
PP	PP	208 CON-EP	Injection molding	10^3–10^4	25
PP	PP	209 CON-EP	Injection molding	10^4–10^7	100
HI-PS	PS	304 CON	Injection molding	10^3–10^4	80
HI-PS	PS	306 CON	Injection molding	10^3–10^4	50
HI-PS	PS	307 CON-EP	Extrusion	10^3–10^4	80
HI-PS	PS	308 CON	Injection molding	10^6–10^{10}	20
HI-PS	PS	309 CON-EP blue	Injection molding	10^3–10^4	60
			Injection molding	10^3–10^5	20
Rigid PVC	H-PVC	401 CON	Extrusion	10^3–10^4	100
Rigid PVC	H-PVC	402 CON	Sheet extrusion	10^3–10^4	95
Rigid PVC	H-PVC	403 CON	Profile extrusion	10^4–10^6	95
Flexible PVC	W-PVC	404 CON	Extrusion	10^3–10^4	100
Flexible PVC	W-PVC	405 CON-FR	Profile extrusion	10^3–10^4	100
TPU	TPU	502 CON	Inject. molding/extrusion	10^3–10^6	100
TPU	TPU	503 CON	Inject. molding/extrusion	10^3–10^6	100
ABS concentrate	ABS	601 CON-EP	Inject. molding/extrusion	10^3–10^9	30–75
PA concentrate	PA	801 CON	Inject. molding/extrusion	10^3–10^9	50–90
PA concentrate	PA	802 CON-GF	Inject. molding/extrusion	10^3–10^9	40–80
PA	PA	803 CON-EP	Injection molding	10^3–10^5	95
PC	PC	901 CON-EP	Injection molding	10^3–10^4	65
PETG	PETG	902 CON	Extrusion	10^3–10^4	65
PVDF	PVDF	904 CON-EP	Inject molding/extrusion	10^3–10^5	65
PSU	PSU	905 CON-EP	Injection molding	10^3–10^5	30
PBTC	PBTC	906 CON-EP	Extrusion	10^3–10^4	95
PBT	PBT	907 CON-EP	Injection molding	10^3–10^4	55
PSU	PSU	908 CON-CF-EP	Injection molding	10^3–10^5	35

[a] Note the high values of the impact strength.
[b] In percent of the base polymer.

perstructure of the seams) probably have a very great influence on the deterioration in the mechanical properties. The "seams" form long continuous paths that may be present as open (Fig. 19.7, A and B), semiopen (Fig. 19.7, C) or closed (Fig. 19.7, D) structures.

The forms in Fig. 19.7 appear to merge into one another depending on the polymer matrix and the processing conditions. This explains why samples from the same compound display differences in modulus of elasticity (usually higher after tempering) and in percentage elongation at break (lower after tempering) depending on their shear and temperature history. While forms A and B as an interpenetrating network are probably more favorable for mechanical properties, further curvature would lead to the formation of tubes and ultimately of closed cellular or foamlike structures. The compartments seem to have diameters ranging from several tens of micrometers up to 100 μm (with a wall thickness of ≈ 0.1 μm). If the structures we have observed are representative, this could pave the way to explaining the mechanical behavior of the compounds: Parts of the matrix would no longer have any contact with one another!

This raises the question of how such structures arise and how they could be optimized in the interests of processing and mechanical properties.

With our present understanding we have succeeded in developing a product family of conductive compounds with carbon black, each with maximum impact strength, covering the whole range (see Table 19.1 and Figs. 19.8 and 19.9).

E. Conductivity: Molecular or Lattice Structure Phenomenon?

The mechanism of conductivity in conductive polymers has long been the subject of theoretical controversy. Whereas Wegener drew attention at an early stage to the fact that the morphological structure and the interactions between the chains had a decisive influence on conduc-

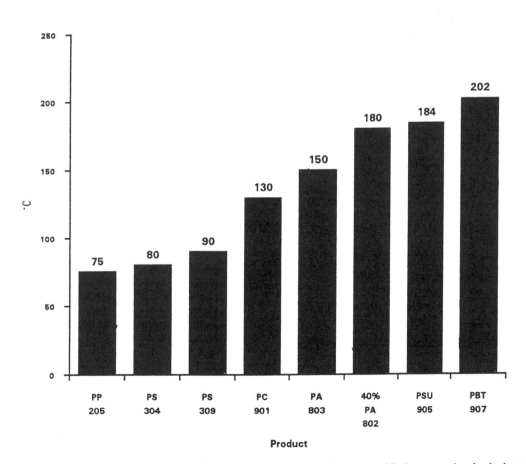

Fig. 19.8 The Vicat softening point shown here is the temperature at which a specified penetrating body has penetrated 1 mm deep vertically into the test piece under a specified force. The Vicat softening point was determined in accordance with DIN 53 460 (Method B 50: force 49.05 N, temperature increase 50 K/h).

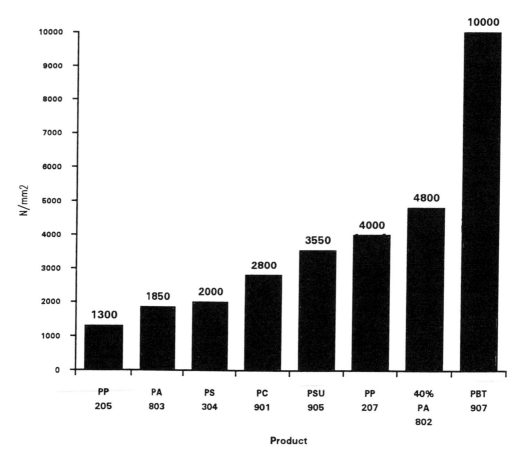

Fig. 19.9 The modulus of elasticity in tension shown here is the quotient of the tension (σ) and the elongation (ε) in the case of bar elongation with unhindered transverse shrinkage. It was determined in accordance with DIN 53457.

tivity [20–22] the greater part of worldwide research work devoted to explaining the conductivity mechanism is still geared to isolated consideration of a single chain. For this reason models of solitons and polarons have been very popular. The model proposed by Epstein and MacDiarmid [23] in which polyaniline is regarded as a "granular metal" was a very important step toward an understanding of the conductivity mechanism in ICPs.

The different kind of conductivity compared with metals has hitherto been largely associated only with defects at the molecular level (interruption of the conjugation length). However, this approach alone undoubtedly does not do justice to the problem. Conductivity is rather a supramolecular phenomenon in which the properties of the atomic (in metals) or the molecular (in organic metals) building blocks and their morphology, i.e., their interactions, give rise to the macroscopically measurable effect. This a common view of things in the materials sciences. In 1989, Wegener and coworkers analyzed the contribution of conductivity perpendicular to the direction of the chain, taking PPy derivatives [22] as an example. With increasing chain distance, the conductivity decreased, showing at least a very important interchain contribution to the conductivity. The influence of morphology and its alteration have hitherto been underestimated. Although there have been theoretical approaches in which the marked effect of orientation of the neutral PAc on subsequent oxidation becomes apparent by comparison with oxidation of nonoriented PAc, they have attributed the high conductivity solely to the lack of molecular defects and the improved chain alignment. They did not look in any detail at the morphology and its probable alteration during orientation. In the majority of cases the properties of the individual chain were regarded as the decisive factors for conductivity.

Since the character of the conductivity even in highly conductive PAc is "only" semimetallic, however, it appeared increasingly important to investigate more systematically the influence of chain arrangement, morphology, and purity, and also of processing. Why should conductive polymers, which are only halfway to being

Dispersion as the Key to Processing

a materials category in their own right, behave in a fundamentally different way from metals or thermoplastic polymers, in which purity, degree of crystallization, morphological and lattice structure, and processing history have, both qualitatively and quantitatively, a decisive impact on electrical and mechanical properties? A number of examples would appear to be classic in this respect:

1. When amorphous metals crystallize, the temperature dependence of their conductivity alters.
2. In high purity form, copper is more than 15 times as conductive as in the usual quality.
3. Polyethylene (HDPE) occurs as a material with a modest modulus of elasticity or as a high modulus fiber, depending on its processing history but regardless of its chemical structure.
4. Depending on the crystalline morphology in which some charge transfer salts crystallize, e.g., TTF-TCNQ, they may be insulating or highly conductive. Today CT salts are also known with metallic temperature dependence of conductivity.
5. Mesoscopic metals display a conductivity that is several powers of 10 smaller with decreasing particle size [Section VI and Ref. 24] (quantum-mechanically limited conductivity), and below 1μm they also display a nonmetallic temperature dependence. Conversely, it is not yet known what order of magnitude of clusters of metal atoms represents the threshold for a metallic character. This is not yet the case for up to 500 atoms, i.e., a cluster size of several nanometers.

Evidently the structures that are more favorable for conductivity tend to form preferentially during tempering. In 1986 we first drew attention to an interesting effect: In carefully formulated compounds and after optimal manufacture, conductive compounds display an increase in conductivity after tempering (see Figs. 19.2 and 19.23). Frequently this effect cannot be observed in compounds with the wrong composition and processing history, since disperse structures that are normally generated before "percolation" do not form in such compounds until the tempering stage.

We have already seen that conductive polymers are insoluble and infusible. These two properties can be linked with the property of conductivity (cf. [100]).

Conductive polymers as conjugated polyradical cationic salts are rigid and highly polar molecules; they bear negative charges arranged in a nonrotationally symmetrical fashion that compensate for the positive charges on the chain. The charge distribution is therefore asymmetrical (Fig. 19.10A). During polymerization, this will result in the attachment of this chain to the more positive side of another chain. This will continue until the chain conglomerate has a sufficiently large surface area whose net external charge can be stabilized by ad-

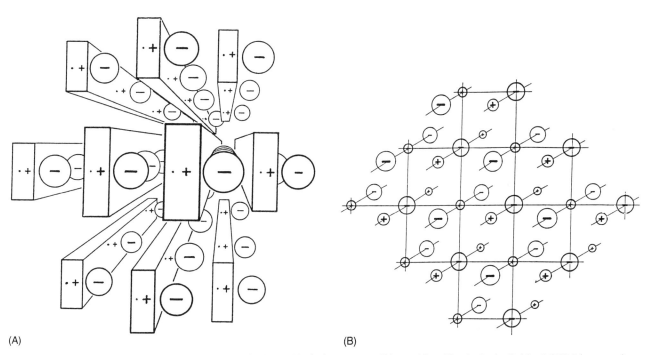

Fig. 19.10 In a spherical primary particle of ICPs (A), it is not possible to identify single individual ICP$^+$/counterion$^-$ units; in the comparable case of a Na$^+$ Cl$^-$ crystal (B), one cannot identify specific individual Na$^+$Cl$^-$ units.

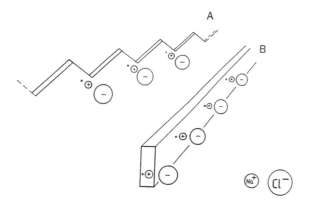

Fig. 19.11 Isolated doped chains of polyacetylene or other ICPs (A) cannot exist in this form like a single individual in Na$^+$Cl$^-$ unit (B), which also does not exist in this form.

sorbates (e.g., water). This configuration also allows the major interchain charge transfer contribution, if not the crucial contribution to conductivity. Organic solvents are incapable of disrupting such chain arrangements. At any rate the interactions between charged chains and solvents appear to be weaker than between the chains themselves. Any disturbance of this configuration in an attempt to achieve solubility by means of copolymerization or side chains results in loss of conductivity. Even a disklike arrangement that imitated the solubility of 3-alkylthiophenes would not be a genuine solution but a colloidal structure comparable with soaps. A special case could exist in cases where the solvent is a strong acid and both "dopes" (complexes, oxidizes) the polymer and dissolves it. Such solutions are assumed to be the case for PPS in AsF$_3$.

Thus, conductivity and insolubility belong together; if you want solubility, you lose conductivity. The situation with regard to fusibility is similar [100].

What this all means, therefore, is that ICPs cannot be understood in terms of the individual molecule. They can only be regarded as an arrangement of a relatively large number of chains, with several positive charges on them and several negative counterions beside them (Fig. 19.10A). Neither is it possible to allocate an arbitrarily chosen negative charge to a given chain. Consequently it is not possible either to regard a specific combination of positive charges and negative countercharges as the smallest metallic unit. This distinction, in fact, belongs to the primary particles. Ideal ICP particles may have an anisotropic charge distribution according to Fig. 19.10A (e.g., with an excess of negative counterions on the "right" side); this leads to a high tendency of irregular agglomeration or (more regular) arrangement into "fibrils" or "transparent thin layers." The conductivity and related properties are again directed by the particle interactions and their arrangements (cf. Section VI).

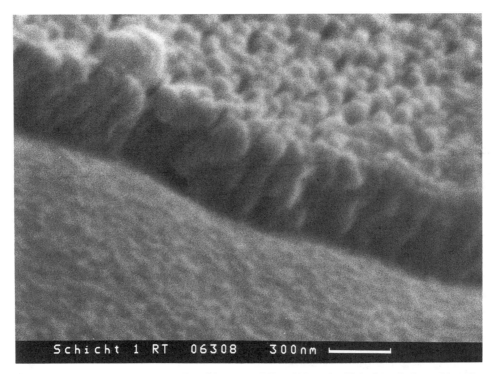

Fig. 19.12 Cross section through a transparent PAni layer on PET (300 nm). Globules of about 50–100 nm diameter are visible as constructive units (SEM).

Dispersion as the Key to Processing

The analogy of the NaCl system (Fig. 19.11B) may help to understand the problem. Individual NaCl molecules, which consist of a positively charged sodium ion and a negatively charged chlorine ion, do not occur in a solid or gaseous form; in an NaCl crystal it is neither possible nor correct to describe an individual NaCl cell as the smallest unit of the crystal (Fig. 19.10B).

Unlike NaCl, which is readily dissolved in water though without creating a single NaCl molecule, ICPs cannot be dissolved by introducing a single charged chain with its countercharges into a solvent. Thus ICPs are not soluble in the media that are of interest for processing but form discrete phases consisting of one or more primary particles.

The conductivity of the ICP, and hence of the ICP blend, and the associated phenomena are determined primarily by the conductivity properties of the primary particles, their arrangement, and their mutual interactions.

The very high charge density is the reason for a very high surface tension (we have measured more than 69 mN/m) [25], which results, in conjunction with the anisotropic arrangement of the counterions, in a strong tendency to absorb impurities. The high surface tension and the strong propensity to aggregate are the reasons for the extremely great problems involved in dispersing ICPs in other media.

F. Structures and Structural Studies of ICPs

For a long time there was no consensus of opinion about the morphology of ICPs. On the basis of scanning electron micrographs, some research groups favored a fibrillar structure for PAc (polyacetylene) produced by the Shirakawa method [15]. This would, it was thought, be an explanation for an anisotropy of electrical conductivity that was observed following orientation of the material [26].

The morphological studies of PAc films performed under the scanning electron microscope initially confirmed this opinion and then led to a fibril theory that held that polyacetylene congregates to form long fibers, i.e., fibrils, which in turn form crystalline domains in the direction of the fiber. According to this theory an electric current should flow along the fiber axis in these fibrils after doping. By contrast, further studies that we carried out on a sample of PAc that was shown by TEM methods to be "fibrillar" clearly indicated a spherical morphology (see Figs. 19.12–19.21). This suggests that the PAc exists in a spherical form and aggregates to form chains. The aggregation of these spherical particles is only physical; they are not chemically linked. Only at a suitable scale do these chains appear to be fibrils [28]. This explains the original opinion that proceeded from a fibrillar structure.

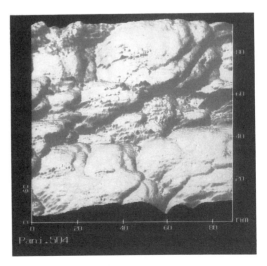

Fig. 19.13 Scanning transmission microscopic images of 600 × 600 nm showing terraces at low magnification. (From Ref. 27.)

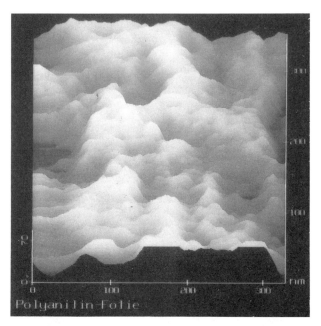

Fig. 19.14 STM image (100 × 100 nm) of a film deposited on a gold substrate consisting of spherical particles of about 5–15 nm (see Fig. 19.13).

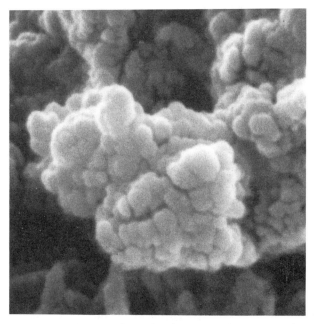

Fig. 19.16 Aggregation of primary particles to tertiary structures resembling a bunch of grapes (From Ref. 28.)

III. UNDERSTANDING THE PROCESSING PROPERTIES OF INTRINSICALLY CONDUCTIVE POLYMERS

Since their discovery in 1977, intrinsically conductive polymers (ICPs) have been regarded as unprocessible.

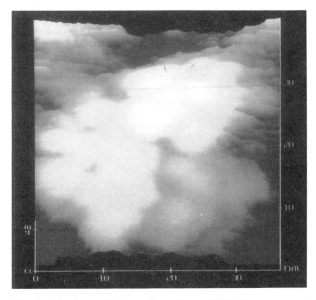

Fig. 19.15 Area of about 40 × 40 nm of STM images showing spherical primary particles of 5–10 nm (see Fig. 19.13).

This was a great disadvantage for applications-oriented research and possible applications. We have developed methods that make it possible to process ICP raw material after polymerization, both in pure form and as dispersions in other matrices.

After several years of working with numerous different ICPs (PAc, PPy, PTh, PPP, etc.), we have concentrated since 1988 entirely on PAni. We have optimized the chemistry (composition) of PAni, its synthesis, and its processibility in a parallel interactive approach. In this work the focus was on purity, dispersibility, and stability (during processing, in use, resistance to temperature, light, etc.).

A. Processing Pure ICPs: Thermoductility

The earlier experiments that we made with the aim of turning pure raw ICPs into a technically interesting form at relatively high temperatures and under high pressure resulted in products whose morphology displayed marked changes compared with the original material [26]. Spherical primary particles became rodlike. The products were very rigid, and their surface had a metallic sheen. It transpired, however, that these were not homogeneous and in some cases were not completely "melted."

Figures 19.22a and 19.22b show some scanning electron micrographs. The original globular morphology has disappeared and is replaced by a new compact, highly oriented, and obviously rodlike morphology. X-ray

Dispersion as the Key to Processing

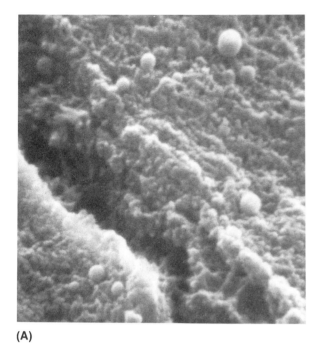

(A)

(B)

Fig. 19.17 Magnification of a tear due to sample preparation of the polyacetylate film. Primary (better: secondary) particles are clearly differentiated. The "grapes" consist of about 15 secondary particles. (From Ref. 28.)

⟶

Fig. 19.19 SEM picture of the same film as shown in Fig. 19.18, showing the aggregation of much smaller primary (secondary) particles than those shown in Fig. 19.16 (powder PAc). (From Ref. 28.)

Fig. 19.18 TEM picture of the same "fibrillar" PAc film as in Figs. 19.17, 19.19, and 19.20. (From Ref. 28.)

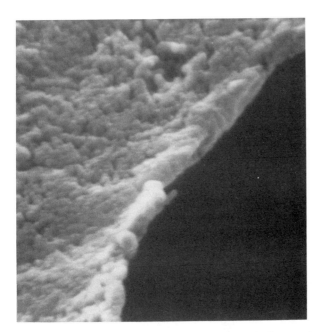

Fig. 19.20 Tear edge of the fibrillar sample of Fig. 19.18. In the lower part of the picture as well as in the upper left area of the film, clearly differentiated spherical particles are to be seen. (From Ref. 28.)

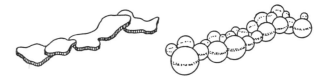

Fig. 19.21 Comparison of Wegner's earlier "lamella model" with my own spherical primary particle model for explaining the fibrillar structure.

structural analyses show an increasing crystallinity as a result of the deformation.

Regardless of their processability under high pressure and high temperature, ICPs are not thermoplastic materials. The necessary processing conditions are much more far-reaching, and the process is therefore considerably more complicated than that used for thermoplastic polymers. Deformation involves flow processes, which is why we advocate describing ICPs in general as "thermoductile."

So far studies of homopolymeric ICPs have not observed any glass transition temperature or even a melting point. To date ICPs have been generally classified as solids or at least regarded as such. We have always proceeded from the assumption that the process we observed was "genuine" flow, i.e., that the results or products that we obtained were not only a result of a sintering process. Instead of a flow of initially disorganized polymer chains consisting of different primary particles (i.e., they slide relative to one another), the particles crystallize in the process. One interpretation of this process could be to regard ICPs as undercooled liquids that have solidified in amorphous form, i.e., as "glasses." During the pressure-shear process a pressure-induced crystallization takes place during the flow.

It remains to be investigated whether this concept is correct, and if so whether the resulting flow is that of a solid (similar to ductility in metals or creep in polymers) or that of other (extremely viscous) fluids [12]. Until this

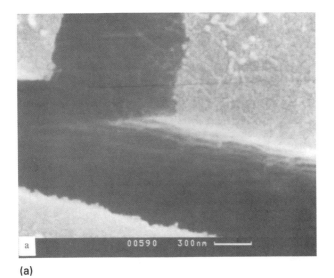

(a)

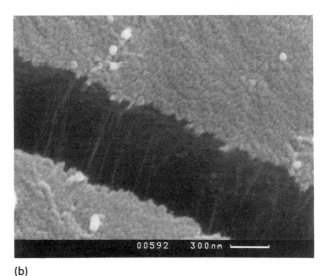

(b)

Fig. 19.22 The original globular primary morphology has disappeared and is replaced by a new compact, highly oriented, and obviously rodlike morphology.

point is clarified, we assume that the term "thermoductility" can be used to describe the flow properties of ICPs.

B. ICP Dispersions in Liquids of Low Molecular Weight

Using special dispersion techniques it is possible to make dispersions of PAni in several mainly polar solvents and, more recently, in water. The most efficient solvents for dispersing PAni are NMP and DMSO. Such solvents are used by many other groups for "dissolving" neutral PAni (emeraldine base, EB). We have shown by membrane filtration and by photon correlation spectroscopy (see Section IV.K) that the EB in NMP or DMSO, which are transparent clear blue liquids, are in fact also dispersions and not solutions.

The scientific community should be careful when using terms that are clearly defined in other scientific areas. The term "solution" means a complete solvent shell around an atom, a molecule, or a single polymeric chain *on the molecular scale*. Such systems contain only a single phase and are homogeneous solutions. The term "dispersion"—used by colloidal scientists—means that *particles* (in colloidal chemistry and physics, particles are understood to be atomic or molecular arrangements bigger than about 1 nm) are surrounded by and embedded in the medium (here the solvent or dispersant). Such systems are not homogeneous but consist of at least two phases—the pure solvent phase containing no dissolved molecules, and the pure dispersed phase. The phases contact each other at the interfaces (which might be of complex structure).

This is even more true, and may be less doubted, of dispersions of PAni in NMP or DMSO. Such dispersions require much more sophisticated dispersion techniques. Most of the trials in other laboratories were unsuccessful; these then preferred to make PAni "solutions" in, e.g., sulfuric acid [29]. They look like solutions (clear green), but they are dispersions. They are in most cases unstable, that is,

1. They decrease in viscosity over time (especially when stored at elevated temperature).
2. They exhibit precipitation of flocculated PAni.
3. They lose color strength (or absorption strength) over time.

We have succeeded in preparing clear green dispersions in NMP or DMSO and other solvents that are "stable" (more precise, metastable), i.e., they do not change their properties under ambient conditions (especially no flocculation, no discoloration, no viscosity change). But it is very important not to exceed certain critical concentrations (depending on particle size and dispersion medium) around 0.5 vol %, because above this concentration sudden gelation or flocculation occurs. The preparation procedures and their reproducibility allow the supply of such dispersions on a commercial scale.

In contact with other substrates (plastic films, glass sheets, etc.), PAni is deposited on the surface and builds transparent coatings of pure PAni (50–300 nm thickness). This deposition is due to the metastability of these dispersions.

Recently we also succeeded in preparing water-based dispersions, which are now being thoroughly investigated.

C. Polymer Blends with Intrinsically Conductive Polymers

The manufacture of polymer blends from insulating conventional polymers (regardless of whether these are thermoplastic or nonthermoplastic polymers) and intrinsically conductive (homo-) polymers (especially PAni) by physical methods begins after the polymerization of the individual components. This has not only practical but also fundamental theoretical aspects.

It is much more complicated to disperse ICP powder in conventional thermoplastic polymers than is the case with, for example, pigments or conductive carbon black. ICPs form very large, extremely dispersion-resistant aggregates, and we therefore started by dispersing them at low concentrations.

On the basis of the percolation model, the objective of polymerizing an ICP powder suitable for blends had to be to create especially highly structured fine aggregates (this was also attempted by other working groups, e.g., Bayer). By contrast, we set out to polymerize spherical dispersible ICP powders that, it was widely assumed, would "never" percolate in a polymer blend (and, because we would "destroy" the fibrils, conductivity would not be achievable).

Surprisingly enough, we nevertheless observed as long ago as 1984 a conductivity leap after dispersion, first with PAc powder, then with other ICP powders, and with the progress of our polymerization techniques and the development of new blending methods we were able to achieve better and better dispersion [13]. Thanks to further development of this process, we obtained, as the years went on, ever lower critical concentrations, down to as low as 1.5 vol %, and ever lower saturation concentrations, of ≈35 down to 25 vol % of ICP (especially PAni) in the polymer blends (Fig. 19.23).

It is not possible on the basis of percolation theory to explain the percolation behavior of the spherical ICP primary particles that exist in totally disperse form below the percolation point. The percolation theory is based on the assumption that particles that do not enter into any interaction with the matrix are statistically distributed within it. This enabled the theoreticians to arrive at a mathematical description of the percolation behavior from a purely statistical and mechanistic point

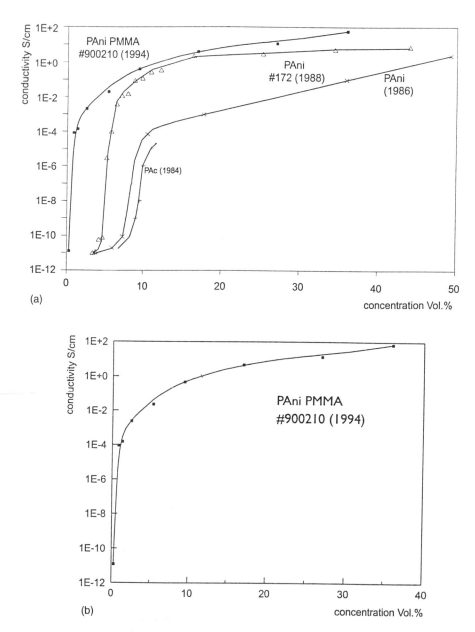

Fig. 19.23 Data obtained in 1994 on PAni dispensibility, with a critical volume concentration of about 1.5 vol %, compared with 1986 and 1988 data on PAni and 1984 results with PAc.

of view. In the case of polymer compounds containing conductive carbon black, this theory demands that highly structured aggregates of irregular and nonspherically symmetrical form are statistically distributed in the matrix.

This would mean that percolation of completely dispersed ICP globules is theoretically not permissible or would not take place until well above 30–50%. The fact that it nevertheless took place, and at very low concentrations, encouraged us to polymerize powders of conductive polymers (PAni) that were increasingly easy to disperse. Finally we tested this by means of internally standardized dispersion tests capable of quantitative evaluation.

By using the new theory we can now obtain processible and flexible polymer blends with conductivities of up to 20 S/cm for research and development work on a scale of greater than 100 kg. Even blends with conductivities of ≈100 S/cm have become feasible, although not yet on a large scale.

IV. EXPERIMENTAL BACKGROUND TO THE PROCESSING OF ICPs

In the preceding sections we have taken a closer look at the phenomenon of the sudden existence of electrical conductivity in heterogeneous polymer systems when the volume concentration of the conductive phase (ϕ_c) in a matrix polymer surrounding it reaches a characteristic level. In the past this has often been explained in terms of the percolation theory. What is not generally known, however, is that this theory is unable to interpret the experimental results obtained, for example, with carbon-black-filled polymer systems. A critical assessment of the percolation theory with regard to the interpretation of the conductivity leap has been given in Sections I–III. Our studies have not been able to support the use of this theory for this problem either. We have therefore developed a new theory that is based on the interpretation of dynamic interactions between the dispersed phase and the continuous phase.

In this section we put forward a number of quantitative confirmations that have been obtained with the aid of various experiments conducted on ICPs and ICP-polymer blends and in comparison with carbon-black-filled compounds. These experiments included quantitative investigations of the critical volume concentration, measurements of density, the dependence of conductivity on temperature history, scanning electron microscopic studies, pyrolysis, and Brunauer–Emmett–Teller measurements, in each case in a variety of polymer matrices. These experiments have provided support for the theory that above the critical volume the disperse phase forms new dissipative structures that are responsible for the occurrence of conductivity in blends and compounds.

A. Background to Experiments

The experimental basis for and the application of the percolation theory to conductivity in polymer systems derive mainly from studies of metal powders in polymers. From these a number of (inadmissible) conclusions have been deduced, among them,

1. The distribution of the carbon black particles in the matrix is thought to be uniform and statistical.
2. These particles are asymmetrically structured aggregates, because otherwise the percolation concentration would be much higher.
3. The original particle structure is preserved during compounding, and the better the structure is preserved, the better are the results of percolation (smaller ϕ_c).
4. The interactions between the carbon black particles and the matrix are negligible.

Table 19.2 Critical Volume Concentration ϕ_c and Maximum Possible Content of Carbon Black ϕ_{max} for Selected Carbon-Black Filled Polymer Compounds[a]

Polymer	ϕ_c (exp.)	ϕ_{max}
PE MI 70	0.013	0.29
MI 20	0.018	0.28
MI 1	0.04	0.28
PS N 2000	0.04	
N 5000	0.063	
PMMA	0.037	0.26
CPE	0.049	0.3
PVC K 50	0.04	—[b]
K 57	0.045	—
PEA	0.07	0.28
PU-PE, ϕ_c 1	0.11	0.26
ϕ_c 2	0.145	

[a] See also Figs. 19.24–19.38.
[b] Not determined—too unstable.

On the basis of new methods we developed, which permitted better dispersion, we performed a series of concentration experiments with carbon black or an ICP as a conductive phase in polymers of different chemical compositions and different molecular weights. We performed all these experiments several times to ensure and demonstrate reproducibility. Selected results are described in the following pages.

B. Dependence of the Critical Volume Concentration ϕ_c on the Properties of the Matrix of the Polymer

The design and conditions of the experiments, with the exception of PVC, have to a large extent been described in other publications [30]. In the case of PVC a mixture of conventionally stabilized PVC powder with the planned quantities of carbon black was made in a high speed mixer. The dry blend was extruded in a counterrotating twin-screw extruder ($L/D = 24:1$ with 45 mm screw diameter). The electrical conductivity of all materials obtained was measured on pressed panels by the four-point method (down to values of $\approx 10^{-3}$ S/cm). Samples with even lower conductivity were measured by using ring electrodes in accordance with DIN 53482.* The measurements on the pressed panels were made before and after artificial aging or tempering [31], but only the readings obtained after aging were used [30]. Table 19.2 shows results published in 1988 and 1991. Information is also given about the maximum incorporatable

---
* Determination of volume resistivity.

quantity of conductive carbon black (ϕ_{max}) for which the mixtures remain free-flowing. At a volume concentration as little as 1 % above ϕ_{max}, a concentration is reached at which the compound is no longer capable of thermoplastic processing. Table 19.2 shows values found, in some cases with different molecular weights, in a variety of polymer matrices such as polyethylene (PE), polystyrene (PS), and poly(vinyl chloride) (PVC). It is interesting to note that some carbon black/polymer composites in which the matrix polymer is a two-phase system exhibited two conductivity leaps, a phenomenon never previously observed. Figure 19.24 shows the dependence of conductivity on the concentration of the conductive filler, taking a polyurethane–polyethylene glycol blend (PUR-PEG) as an example. Here two separate phase transition points are clearly recognizable (ϕ_{c1} and ϕ_{c2}).

The most important conclusion is that the properties of the polymer matrix, i.e., the interactions between matrix and disperse phase, have, contrary to the predictions of the percolation theory, a crucial and decisive influence on the critical volume concentration ϕ_c and the maximum filler loading content ϕ_{max}.

C. The Nonlinear Increase in Density

When measuring the densities of the compounds in the above-mentioned test series we noticed for the first time that the increase in density resulting from the increasing admixture of carbon black stagnates in the region of the critical volume concentration [30,32]. Untempered samples displayed a stagnation of density at the same concentration of ϕ_c, but after tempering or artificial heat aging, the point of density stagnation shifts parallel to ϕ_c in the direction of a lower concentration. It thus looks as if this effect is associated with the effect of the conductivity leap. This conclusion is also supported by the

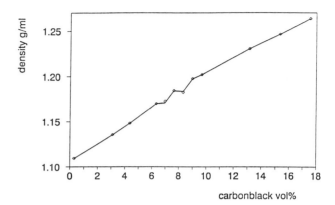

Fig. 19.25 Two density stagnation points in the blend of Fig. 19.24.

occurrence of a two-stage density stagnation (Fig. 19.25) for the system with the two ϕ_c points shown in Fig. 19.24.

Figure 19.26 shows the density dependence curves for various systems. These are polyethylene (PE), polystyrene (PS), PVC, and poly(methyl methacrylate) (PMMA) as the polymer matrix with various concentrations of carbon black. The critical volume concentration ϕ_c of these systems is shown in Table 19.2.

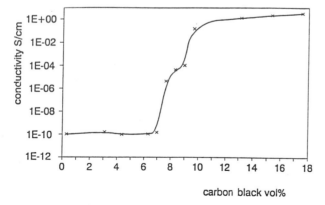

Fig. 19.24 Two critical volume concentration in a carbon-black-filled PUR-PEG blend.

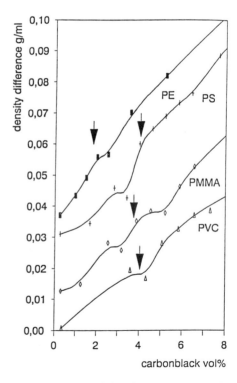

Fig. 19.26 Dependence of density on concentration in the systems of Table 19.2. (Arrows mark Φ_c.)

Dispersion as the Key to Processing

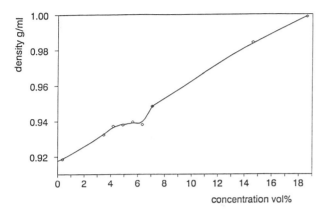

Fig. 19.27 Dependence of density on concentration in a pigment 13/PE system.

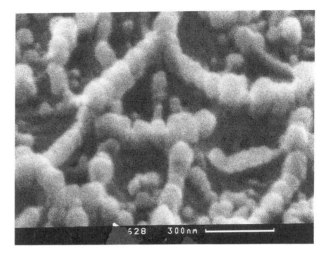

Fig. 19.29 Scanning electron micrograph of break surface in a carbon black compound at concentrations at the flocculation point (6%).

To determine whether this phenomenon is associated solely with the increase in conductivity, we investigated a pigmented polyethylene system. This system does not become conductive at a critical volume concentration. The aim was to determine whether in these circumstances as well there would be a phase transition comparable with that found in the systems filled with carbon black. And indeed Irgalith yellow BAWP dispersed in a polyethylene matrix also displays a stagnation of density on the same scale as is found with carbon black in a variety of polymer matrices (Fig. 19.27).

We therefore conclude that the conductivity at and above ϕ_c comes about because of a sudden change in the structure (arrangement) of the disperse phase in the compound. The structural change is not confined to systems with a conductive disperse phase but is probably a general phenomenon.

D. Morphological Investigations with Scanning and Transmission Electron Microscopy

Various studies of the different compounds were made by means of direct scanning electron microscopic (SEM) investigations below and in the region of the critical volume concentration [33]. The SEM images in Figs. 19.28–19.30 are taken from a broadly based investigation of carbon black structures in a compound. They were

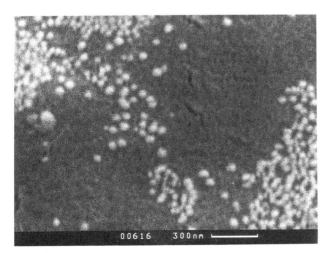

Fig. 19.28 Scanning electron micrograph of break surface in a carbon black PS compound at concentrations below the flocculation point (0.5%).

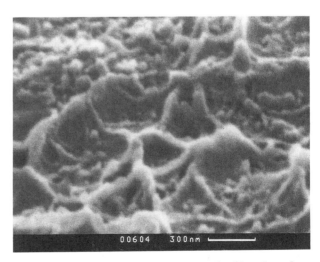

Fig. 19.30 Scanning electron micrograph of break surface in a carbon black compound at concentrations at the flocculation point (tempered).

obtained from fracture surfaces of mixtures containing various concentrations of carbon black.

These three pictures are characteristic of the entire study of the structure of the various compounds. They show clearly that below the region of ϕ_c carbon black is present in the form of isolated spherical particles. At the critical volume concentration one can see networks of long drawn-out and branching chains (flocculates) consisting of spherical particles, plus some dispersed particles that continue to exist in isolation. An analysis of the carbon black concentration found at the fracture surface revealed that in all the samples examined this was considerably larger than was to be expected on the basis of the quantity of carbon black incorporated.

E. Interim Status of Model Development

When developing the qualitative part of the nonequilibrium theory we first worked out fundamental features of a qualitative model of dispersion and structure formation. Several elements of this model were derived directly from the experiments.

1. The crucial role of the interfacial forces and the nature of the polymers for the minimum concentration required for conductivity (ϕ_c) is based on results of the conductivity studies performed on the various concentration series.
2. The phase transition of the dispersion state of carbon black at the critical concentration is based on the SEM studies.
3. That this phase transition from the dispersed to the flocculated structure of the chains is accompanied by the formation of cavities or voids between the carbon black particles was deduced from the results of the investigations of density change behavior. This aspect of the model also derives from the "gas absorption" experiments performed by Tanioka et al. [33]; they found a sudden increase in the CO_2 sorption of carbon-black-filled compounds at the critical volume concentration. They did not, however, offer any explanation for this phenomenon. The new model, which includes explanations for the creation of the cavities during the flocculation process, can be used to explain the experimental results.

The experiments described below were conducted to confirm further elements of the new model that until then had only been formulated as hypotheses. This proved to be necessary in the initial stages of working with the new theory in order to explain interactions between the matrix and the dispersed phase and the formation of the networklike flocculates. The subject of these studies was the layers adsorbed on the filler particles, their character, and to some extent the layer structure of the "seams" (Fig. 19.7), because the idea of a formation of layer arrangements is suggested only by the results of the SEM studies and the observations of phase separation [30,32,34]. The SEM observations do not point to the existence of the layer structures—the seams—as the fractures might also have propagated themselves at the "wrong," i.e., suboptimally mixed, parts of the samples.

A further aim was to determine by means of new experiments whether there was any justification for the idea of an oscillation of the carbon black particles between the dispersed and the flocculated stage (reversible phase transition) within the adsorbed shell.

F. Pyrolysis

Pyrolysis is a thermal degradation in the absence of oxygen. In this case it is carried out to determine the carbon black content of a compound and is supplemented as a matter of routine by pyrolysis in the presence of oxygen to determine the ash content in accordance with DIN 53585.

Although work was carried out under very carefully controlled conditions, it was noticed that some of the compounds analyzed in this way did not display the carbon black content that was to be expected on the basis of their original composition. In fact, there was found to be an additional pyrolysis residue. The smaller the incorporated concentration of carbon black or ICP, the greater was the amount of this residue. These observations led to the assumption that the adsorbed shell underwent incomplete pyrolytic reactions that were different from those entered into by the free polymer matrix.

Figure 19.31 shows results of pyrolysis of compounds with different concentrations of carbon black, while Fig. 19.32 shows the corresponding experiments with ICP

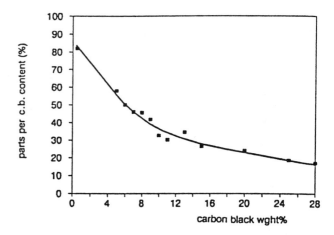

Fig. 19.31 Pyrolytic residue from carbon-black-filled PS with various concentrations. c.b. = carbon black.

Dispersion as the Key to Processing

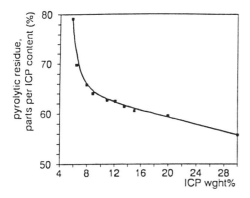

Fig. 19.32 Pyrolytic residue from a PAni/polyester blend with various concentrations.

polymer blends. These findings supported the idea of an adsorbed shell consisting of matrix molecules. After adsorption these undergo a very marked change in properties (ultimately in their capacity to be fully decomposed by pyrolysis). In these cases the adsorbed shell is not joined by chemical bonds to the surfaces of the carbon black or ICP (here PAni) particles, as can be shown by extraction; before pyrolysis an extraction with suitable solvents results in the total removal of the matrix polymer. The curves in Figs. 19.31 and 19.32 show that with increasing concentration of carbon black and ICP particles in the compound in question a small portion of the adsorbed shell remains on the surface of the particles during pyrolysis. This represents further strong support for the assumption that a change in the state of dispersion is connected with the increase in the concentration of the incorporated particles, and the more dispersed particles are built into the flocculate structure, the less matrix is present in adsorbed form on the disperse phase.

A consideration of the dispersion/flocculation model suggests that during flocculation parts of the polymer matrix can no longer be adsorbed on the surface of the conductivity phase but act as a connecting sleeve that surrounds the particles, like the skin of a snake that has swallowed a number of golf balls (as shown later in Figs. 19.39 and 19.40).

G. Evaluation of SEM Observations: The Layer Structure

Figures 19.33 and 19.34 show scanning electron microscopic images of the structure of mixtures with different contents of carbon black obtained by pyrolysis under nitrogen. They were carefully broken to obtain observable surfaces. None of these samples showed an unstructured and irregular arrangement of carbon black aggregates of the kind typical of pure and unprocessed carbon black [31].

Fig. 19.33 Scanning electron micrograph of carbon black (c.b.) structure after pyrolysis of a 6% carbon-black-filled PS.

Fig. 19.34 Scanning electron micrograph of carbon black structure after pyrolysis of a 15% carbon-black-filled PE showing wave-like "seams."

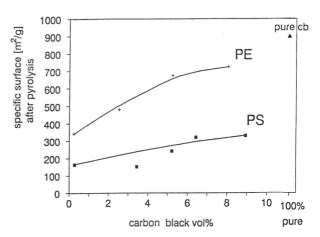

Fig. 19.35 Specific surface area (BET) of carbon black after pyrolysis of c.b.-filled PE and PS compounds.

In none of these samples was it possible to find structures that were comparable with the structures previously published by other authors [35]. In their investigations, carbon black was liberated from the matrix by solution and subsequent filtration. In contrast, we find different structures that support the assumption of a concentration of the carbon black in layers. The structures found after pyrolysis by studies with the scanning electron microscope may be flat (Fig. 19.33), wavelike, tubular, or foamlike (Fig. 19.34). These observations can be interpreted with the aid of the model of an interpenetrating network of layers with different structures, as has already been proposed in Ref. 36.

H. Brunauer–Emmett–Teller Adsorption Isotherm Studies

We carried out Brunauer–Emmett–Teller (BET) measurements to determine whether with the adsorption of a layer of the matrix polymer different quantities of "inner" surface became accessible to the nitrogen. Figure 19.35 shows the figures obtained in this way for the BET surface of the carbon black after pyrolysis: The more carbon black or PAni present in the compound to be pyrolyzed, the greater was the specific surface area measured after pyrolysis. This too suggests a different distribution or dispersion state or different adsorption of matrix polymer layers as a function of the quantity of carbon black incorporated.

The more completely the carbon black or PAni particles are surrounded by the adsorbed shell, the less the nitrogen contained in the mixture after pyrolysis can penetrate to the inner surface of the carbon black particles during the BET measurements. On the other hand, the more flocculated structures and voids present, the greater the proportion of the inner surface that is accessible to the nitrogen after pyrolysis.

I. Temperature Dependence of Electrical Conductivity

A study of the temperature dependence of the electrical conductivity of carbon-black-filled compounds after storage at various elevated temperatures [37] led to the finding that the polymer matrix used initially displays an increase in specific resistance (similar to the PTC effect for crystalline matrices) or more or less stable behavior depending on the degree of crystallization.

After passing through a certain critical temperature the specific resistance begins to show a marked reduction. This can be attributed to increased formation of flocculated network structures at increased temperature. Figure 19.36 shows this situation in terms of the

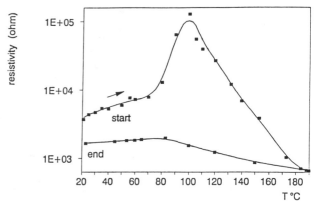

Fig. 19.36 Temperature–conductivity behavior of a PAni-polyester blend during tempering and recooling.

Dispersion as the Key to Processing

curve for electrical resistance as a function of temperature. The effect, after one tempering cycle, is completely reversible. In other words, an increased supply of energy to the system does not promote improvement in dispersion or statistical distribution or the occurrence of isolated dispersed particles, but phase separation (layer formation) and flocculation. This agrees with the fact that dispersions in thermoplastic polymers are thermodynamically unstable or nonequilibrium systems and that the Gibbs free energy for the manufacturing process is positive.

J. Summary of Results and Their Implications for the Models

The experiments just described provide strong foundations for the new nonequilibrium theory for understanding the sudden leap in electrical conductivity in heterogeneous polymer systems. As a result it is necessary to refrain from using the term "percolation" for this phenomenon at a certain critical volume concentration of the conductive phase. The new theory, by contrast, is able to provide a logical interpretation of all experimental results obtained by various groups researching into heterogeneous polymers. It also explains the phenomenon of "overdispersion" [30,38] and the increase in con-

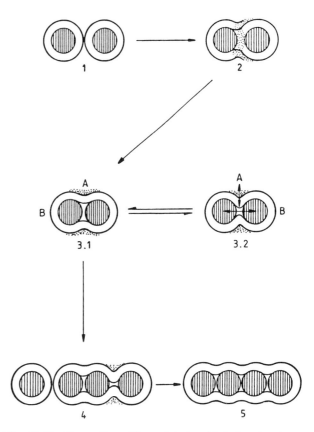

Fig. 19.38 Formation of flocculated and tubular structures above critical concentration from physically dispersed carbon black particles.

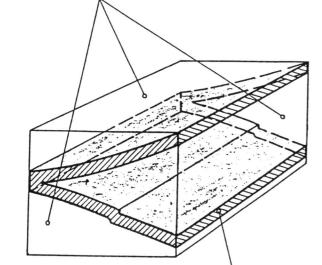

Fig. 19.37 Dispersed carbon black forming layers ("seams").

ductivity in heterogeneous systems in which an intrinsically conductive polymer (ICP) has been incorporated as a conductive phase in the polymer matrix [32].

The most important conclusion that can be drawn from the new theory is that the phenomenon of the rise in conductivity must be regarded as a phase transition phenomenon. The conductive phase suddenly changes, at least partially, from a completely dispersed state to a flocculated state. The critical concentration at which this phase transition takes place is crucially dependent on the properties of the interface between the conductive particles and the molecules of the polymer matrix (the interfacial energy) and on temperature. The underlying mechanism and the resulting structures are shown in Figs. 19.37–19.40. The existence of adsorbed shells and the changes in the properties of the matrix molecules adsorbed in the conductive phase have been conclusively demonstrated. The new theory has made it possible to determine the thickness of these adsorbed layers (monomolecular, $\approx$ 15–20 nm) and also the interfacial energy between layer and conductive phase [38].

In summary it may be said that these experiments do not yield any confirmation of the percolation theory.

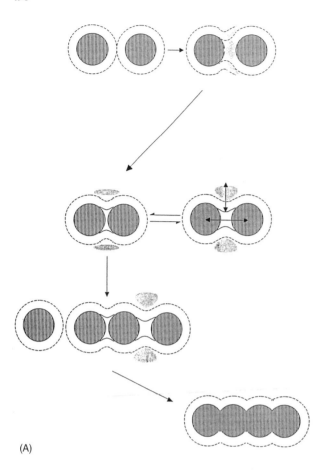

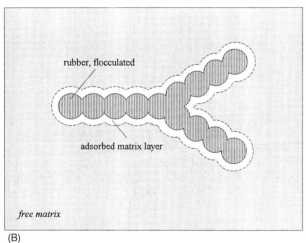

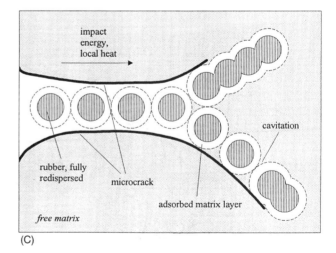

Fig. 19.39 Fracture mechanism and energy dissipation. The flocculated rubber network becomes redispersed by the impact energy; the crack propagates at the interface between free matrix and adsorbed polymer matrix layer; cavitation occurs at the point where redispersion begins. (A) Formation of flocculated and tubular structures from physically dispersed rubber particles. In contrast to carbon black particles, the rubber particles may fuse in tubular form. (B, C) Fracture mechanism and energy dissipation. The flocculated rubber network becomes redispersed by the impact energy; the crack propagates at the interface between the free matrix and the adsorbed polymer matrix layer; cavitation occurs at the point where redispersion begins.

Rather they tend to confirm the validity of our new theory. The general statements that can be derived from the experiments can be briefly summarized as follows.

1. The nonlinear behavior of ICPs and that of carbon black at a characteristic critical concentration are basically comparable despite the differences in the structures of their aggregates.
2. The critical volume concentration (for a defined conductive phase) is crucially dependent on the type of polymer matrix.
3. The critical volume concentration ϕ_c also depends on the molecular weight of the polymer matrix.
4. ϕ_c also depends on the specific surface area of the conductive phase.

Dispersion as the Key to Processing

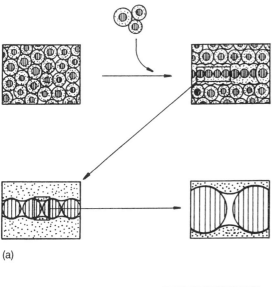

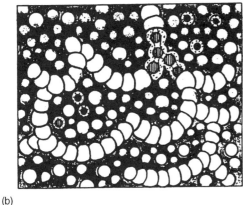

Fig. 19.40 Flocculated particles, forming chains and networks.

5. The gas absorption capacity of the carbon-black filled polymer compounds changes at the critical volume concentration.
6. The density does not show a linear increase with the concentration of the conductive phase but stagnates at the critical volume concentration.
7. Dispersed particles are fully dispersed below ϕ_c and form networklike flocculate structures at and above ϕ_c.

It is not possible to explain these experimental results in terms of the percolation theory.

K. Structures in Colloidal Dispersions

At almost the same time as we found the structures and dynamics in our polymeric colloidal systems, workers in other laboratories discovered a comparably rich variety of structures and structure dynamics in microemulsions, which are also colloidal systems but free of polymers. They consist of water, oil, and a surface-active agent, for example. As Strey and coworkers have shown [39,40], they form bicontinuous lamellar structures (i.e., structures consisting of two continuous phases), which are very sensitively history-dependent and exhibit phase transitions (i.e., sudden structural changes) at certain critical temperatures and/or concentrations.

The comparison of colloidal systems with polymer matrices and water-based microemulsions leads directly to the question of whether other dispersions are also able to form dissipative structures. Interesting systems for the study of such phenomena could possibly be dispersions of intrinsically conductive polymers (e.g., polyaniline) in organic solvents. Clear PAni–solvent systems are actually still thought to be solutions [41], although we have put forward several arguments for the hypothesis that conductive polymers are not soluble and hence clear PAni–solvent systems, which "look" like solutions, are in fact dispersions. Also, other groups have observed instabilities such as sudden gelation or a decrease in viscosity with time [41], which is characteristic of colloidal systems that aggregate anisotropically, but they still interpret these phenomena using the "solution hypothesis."

In fact, most of these so-called soluble conductive polymers are not conductive, but the authors refer only to the neutral, insulating base form of a doped conducting and insoluble form (e.g., polyalkylthiophenes). Only a few doped and really conductive polymers were reported to be soluble. On the other hand, dispersions in various solvents and polymers that "look" like solutions could be prepared using almost any kind of ICP powder polymerized in a proper way, in either doped or neutral form [36].

How can we decide whether the system is a solution or a dispersion? Colloid scientists would still propose the light-scattering effect as a simple way to differentiate between dispersions and solutions. According to Maxwell's theory, light is scattered at inhomogeneities of the dielectric constant and magnetic permeability of the medium it traverses. An electromagnetic wave impinging on matter generates secondary electromagnetic waves in all directions, which are observed as scattered light. The intensity of scattered light increases with the size of the scattering center, so particles with a larger radius have greater scattering power than particles with a smaller radius or even completely solvated atoms or molecules. Photon correlation spectroscopy uses this scattering phenomenon to give quantitative information about the sizes of particles in a dispersion.

Photon correlation spectroscopy experiments have been carried out in completely clear and filtered (membrane filters) dispersions (0.1% concentration) of polyaniline in isopropanol-based organic solvents (Fig. 19.41) [42].

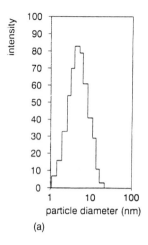 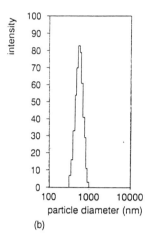 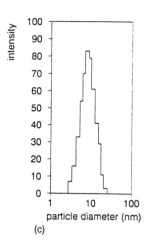 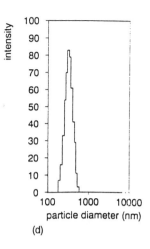

Fig. 19.41 (a, b) Results of photon correlation spectroscopy of a neutral PAni dispersion. (c, d) Result of photon correlation spectroscopy of a dispersion of doped PAni. (See Ref. 35 and Fig. 19.42.)

The first correlation signal to be detected occurs at around 5 nm and 8 nm for neutral and doped PAni, respectively. Surprisingly, a second signal can be observed in both systems at a much larger particle size. We consider these correlation signals to arise from elongated structures analogous to a structure found in microemulsions (see Fig. 19.42). It would be interesting to compare other photon correlation spectroscopic results from microemulsions and conductive polymer "solutions" with our results.

Additionally, the following experiments have been performed with PAni protonated with HCl, which visually would be called solutions.

1. Determination of Particle Size

A dispersion* of compensated (neutral) PAni in DMSO (0.5% PAni) can be filtered through the 20 nm sieve (pressure filter) without discoloration. Using membranes of different sizes, the membrane with 5 nm pores does not allow the dispersion to pass colored, whereas the 10 nm pore membrane does. This indicates that this is also a dispersion, with a particle size between 5 and 10 nm. PAni-HCl is dissolved, when filtered through 5 nm pores, and stays completely colored through a 20 nm filter. Using 10 nm, partial decoloration occurs, which shows that "doped" PAni occurs in somewhat bigger particles around 10 nm. These could be primary particles, which are bigger than the neutral particles because of the counterions. For many dispersions of various other ICPs prepared by our method, we found a particle size between 20 and 100 nm.

---

* All these dispersions are clear and do not show phase separation after even long-term centrifugation.

2. Viscosity

The viscosity of dispersions with various PAni contents was determined using a Couette viscosimeter. The results are shown in Fig. 19.6 (see Section II.C). The viscosity–concentration dependence obeys Relationship (3),

$$\ln \eta_{res} = \ln \eta_{matrix} + x\alpha$$

where η_{res} is the resulting viscosity, η_{matrix} is the viscosity of the matrix, x is the volume content of the dispersed phase, and α is the logarithmic viscosity increment.

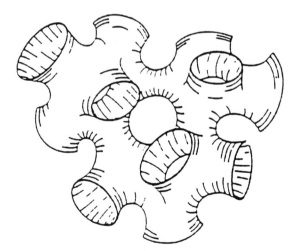

Fig. 19.42 Schematic picture of a structural element of the bicontinuous monolayer in a microemulsion.

3. Thin Layers: SEM Structure

By contracting substrates (films, glass, fibers, etc.) with such dispersions, thin layers of ICPs such as PAni-HCl, can be deposited whose properties are comparable, if not identical, with layers chemically grown on substrates. The use of dispersions is much simpler, so we developed a technically applicable procedure and evaluated the properties of the thin layers.

The specific conductivity of these dispersions ranges around about 5 S/cm. In cases in which the layer has been deposited very carefully, one can reach about 200 S/cm and more. The most important observation is that the layer is formed before the dispersion has been taken away and dried off. The SEM evaluation reveals a very smooth and even surface. At imperfect sites one can estimate the layer thickness to be 50–80 or 300–400 nm, depending on contact time. Closer examination of torn and bent samples shows the layers to be composed of globular particles that seem to overlay exactly one on top of the other.

4. Surface Tension

Poly(ethylene terephthate) PET and PC films have been coated with a thin layer ($\approx$ 250 nm) of PAni-HCl. These have been measured by Springer and coworkers using the contact angle technique with water and CH_2I_2. The data have been evaluated with the harmonic mean method.

Solvent used	Surface tension (mN/m)	Contact x	Surface angle Θ(deg)	Polarity PAni-HCl	tension PAni-HCl
H_2O	72.8	0.696	21.7 ± 3.3		
				69.6 ± 1.9	0.58 ± 0.03
CH_2I_2	50.8	0.132	33.77 ± 5		

Using the equation $\gamma = 0.2575 \delta^2$ [43], a solubility parameter of 17.5 $(cal/cm^3)^{1/2}$ is obtained. A discussion of solubility parameters of PAni can be found in Ref. 44, although it is based on the assumption that real solutions are under investigation; Shacklette did not consider the fact that these are dispersions (cf. also [100]).

The value of 17.5 is an extremely high value compared with conventional polymers, which range between 8 (PE) and 14 (PA). It is to be expected that even higher values for PAni may be found in the course of further studies, because we do not yet know the degree of the surface purity. We expect that low surface energy materials may have been adsorbed onto the surface, thereby lowering the real value to the level we found. We expect that we might have measured the surface tension of adsorbed water, which in consequence makes even higher surface tension and solubility parameter values logical and understandable [100].

These results and many other complex characteristics of polymers, such as viscosity or impact strength, or phenomena observed in practical applications, e.g., deposition buildup, "chalking," and streaks, can be understood using the same principle—the principle of a sudden self-organization of dissipative structures.

L. Conclusions and Outlook

The new theory opens up hitherto unexplored approaches to researching and interpreting the dispersion process, the prevailing energetic and kinetic processes, and the resulting morphology. It must be stressed that the theory put forward here for electrical conductivity in multiphase polymer systems is not a special case for describing special structures in systems where the dispersed or flocculated phase is a conductive phase (see also Figs. 19.43 and Fig. 19.44); it is rather the case that such dispersion and flocculation structures are typical of all (thermoplastic) polymer systems. A first pillar for this generalization is the nonlinear density curve in the polyethylene–Irgalith yellow system investigated (Section IV.C). Further confirmation can be derived from the viscosity data: The viscosity of the polymer blend dispersions and of dispersions of any kind follows the

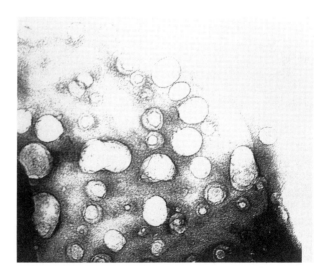

Fig. 19.43 Transmission electron micrographic image of a 47% PAni-PVC blend; the sample was prepared by microtoming, which may account for variations in the background but also for the visibility of PAni particles of different layers ("seams"). Nevertheless, at least one nice flocculate chain (in the middle of the picture, leading in bent form to the upper left corner) is to be seen. Note the distance between the particles.

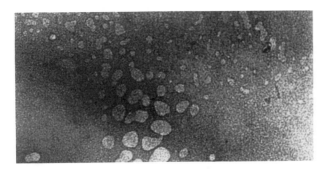

Fig. 19.44 Transmission electron micrographic image of a 40% PAni-PET blend. In comparison to the blend in Fig. 19.43 it is obvious that more than one particle in diameter is forming the chain, maybe due to the high concentration.

same general concentration dependence [36] as can be applied to conductive multiphase polymer systems. It can be shown that the relationship is valid over the entire concentration range of conductive carbon-black-filled polymer blends, other polymer blends, filled polymers, and even liquid dispersions.

It is surprising to find, however, that the universal applicability of the new theory goes even further. The modification of polymers to achieve better impact strength by means of dispersed rubber particles or similar means has hitherto been explained by assumptions that essentially interpret crack propagation and energy absorption in terms of primarily geometrical and topological factors, e.g., Wu's "critical particle distance model" [45]. Years ago Borggreve and coworkers [46] reported observations regarding the transition from brittle to tough in the impact modification of nylon. They presented their results as being compatible with the "critical particle distance model," but they stated that the physical explanation of this model (overlapping of force fields) cannot claim any validity. In other words, Borggreve et al. were unable to explain their observations.

To explain their findings we made a critical analysis of the data supplied by Borggreve et al. [46], which indicate a sharp brittle-to-tough transition at certain temperatures for certain admixture quantities of EPDM rubber. In fact, something that was not originally recognized, the curves strongly suggest the existence of phase transformations. In light of the model discussed here it seems logical to interpret the reported dependences as indications of phase transitions from a dispersed to a flocculated rubber phase. Such a phase transition would lead to structures that were comparable with the arrangements described for conductive phases [30,34].

Neither did Borggreve and coworkers elaborate the transition from brittle to tough as a function of the EPDM concentration incorporated in the nylon. Instead they were inclined to suspect a linear dependence, although this dependence is very similar to the dependence of conductivity in a heterogeneous polymer system. The temperatures they noted, which are reproduced in Ref. 46 and at which, depending on the concentration of rubber, a certain notched-bar energy is necessary to provoke the breakage of rod-shaped samples, all point to a phase transition behavior. On this basis, Borggreve's system must undergo its dispersion/flocculation transition at $\approx$ 12 vol % EPDM [47].

In the meantime we succeeded in simulating the flocculation process by computer [101].

V. THERMODYNAMICS

Only a very few of the countless different polymer systems we meet in daily life are straight single-phase systems. Most real polymers consist of two phases. For most of them, many surprising and nonlinear phenomena are well known and have been the subject of continuous work for decades. Impact modification, viscosity, and conductivity, in particular, display a nonlinear dependence on a given parameter. The property–parameter relationship can be described by an S-shaped curve. Theories that are in principle based on considerations of equilibrium thermodynamics have been developed and are currently being used to explain these phenomena. These theories include the Flory–Huggins theory, the percolation theory, the nearest-neighbor model, and also the constitutive and related equations for the description of rheological phenomena.

In contrast to this, on the basis of experimental results, we developed a new theory [7] that is applicable to all heterogeneous polymer systems [47,48]. Our main principle is to define the nonequilibrium character of colloidal dispersions in polymeric matrices and to interpret the experimental findings (phase separation, dispersion–flocculation phase transition [42]) as "dissipative structures." This term was introduced by Prigogine [49,50] for self-organizing structures in nonlinear systems far from equilibrium. In the case of dispersions in polymers, properties like conductivity and impact strength are measured as a result of the "frozen dissipative structure," whereas rheological phenomena (viscosity, etc.) are to be considered dynamic nonequilibrium phenomena. Dissipative structures can generally survive only under continuous entropy export. This is the case under dispersion conditions, in which a huge amount of high value energy is pumped through the system. It continues to be the case during melt flow in the molten stage but not after the multiphase system has been quenched; then the high viscosity energy barrier prevents the generated dissipative structures from falling apart.

Since colloidal dispersions in polymeric matrices are isothermal processes, differences of temperature are neglected in the following; this means it is always assumed that $\Delta T = 0$.

In this section the following abbreviations are used.

A	area
a	rheological interaction parameter
d	distance
G	Gibbs free enthalpy of mixing; $G = U + pV - TS$
H	dispersion or mixing enthalpy
J_i	flux, flow
k	Boltzmann constant
n	number of particles
P	probability
Q	heat
S	entropy
T	temperature
t	time
W	work, energy
X_i	force
γ	surface tension
η	viscosity
μ	chemical or dispersion state potential
τ	viscous strain
χ	interaction parameter
Φ	volume fraction

A. Thermodynamic Considerations Concerning the Present Theory

The new "nonequilibrium thermodynamic theory of heterogeneous polymer systems" [7] is aimed at providing a basis for an integrated description of the dynamics of dispersion and blending processes, structure formation, phase transition, and critical phenomena.

According to Prigogine [49] and Nicolis and Prigogine [50], dissipative structures are to be expected if in open systems the distance from thermodynamic equilibrium exceeds some critical value, that is, if $d_i > d_{i,\text{crit}}$. In that region, the relations between flows (fluxes) and forces are nonlinear and the standard Prigogine principle, which is valid only in the linear regime, is to be replaced by the Glansdorff–Prigogine evolution criterion

$$d_x P = \sum J_k dX_k \leq 0 \quad (4)$$

It is well known that none of the multiphase systems is spontaneously formed. The formation processes are all endergonic.

Therefore the entropy change of the (irreversible) dispersion process can be calculated according to irreversible thermodynamics of nonlinear processes. In principle, diffusion and dispersion can be treated in a similar way; they apparently lead to formally similar structures. Both processes are irreversible and nonlinear.

According to irreversible thermodynamics in the neighborhood of equilibrium, the internal entropy change (production) would be the product of all fluxes and forces:

$$P = \frac{d_i S}{dt} = \sum J_k X_k \quad (5)$$

The entropy change (production) is then

$$\frac{d_i S}{dt} = \frac{dn_k}{dt} \left(\frac{(\gamma_{12e} - \gamma_{12})}{d} \right) \left(\frac{1}{T} \right) \quad (6)$$

This is the difference between an "excess interfacial energy," $\gamma_{12e} A_{12}$, caused by the work needed for dispersing the particles, W_{disp}, and the surface energy of component 2 before dispersion, $\gamma_2 A_{12}$ (the term $\gamma_1 A_1$ can be neglected), divided by the average dispersion path length.

This can be estimated empirically with a good theoretical basis by stating that $\gamma_{12e} A_{12}$ is the energy that is necessary to provide 1 m² of matrix polymer and force it to wet 1 m² of carbon black or 1 m² of an ICP, during which the interface A_{12} is formed. This is the minimum amount of work W_η necessary to force the volume V_{syst} of the developing polymer dispersion with the viscosity η at the shear rate s_r to flow:

$$\gamma_{12e} A_{12} > W_\eta = \eta V_{\text{syst}} s_r \quad (7)$$

whereby (see Ref. 7, footnote 4d):

$$\ln \eta = \Phi_p \ln \eta_p + (1 - \Phi_p) a = \Phi_m \ln \eta_p + (\Phi_d + \Phi_c) \ln \eta_{\text{app}} \quad (8)$$

$$\ln \nu_{\text{app}} = \ln \eta_{\text{ads}} + \frac{\gamma_{12e}}{\gamma_{12}} \quad (9)$$

where γ_{12} is the equilibrium surface tension [Eq. (7) in Ref. 7].

In reality, W_η is obviously only a minimum (ideal) value for $\gamma_{12e} A_{12}$. A better experimental basis is the real dispersion work, W_{disp}, which can be experimentally determined.

According to our experience, the total enthalpy needed for preparing such systems is about 2 MJ/kg, a value large enough to suggest that these systems are being driven far from equilibrium.

A rough estimation can show that one of the conditions for order under nonequilibrium, a significant amount of negative entropy change, is fulfilled [7]. The next steps in analyzing the nonequilibrium properties are to prove the nonlinearity of the process and to determine whether the distance to equilibrium is supercritical. Therefore we consider similarities to and differences from the irreversible diffusion process, which shows a positive entropy production.

To arrive at a solution for the dispersion law, entropy development during the diffusion process must be considered. This entropy develops over time as an exponential function approaching a saturation value. This behavior is characteristic of an irreversible process in the direction of thermodynamic equilibrium, and we are still in the thermodynamic branch.

The solution of a "dispersion law" is a function like

$$n = \left(\frac{r_u}{r}\right)^3 n_0 \qquad (10)$$

where r_u, n_o are the radius and number of the undispersed particles, respectively.

The "dispersion law" might therefore have a special solution like

$$n = \left(\frac{r_u}{r_d}\right)^3 n_0 \exp\left[-\frac{e}{r_d t s_r}\right] \qquad (11)$$

Two new factors have been introduced. r_d describes the particle radius after dispersion, and c is a constant describing dispersibility. It is used with the dimension $[c] = [m]$.

The expression c/r_d has the dimension 1, where

$$[c] = [m] = [(m^2 \cdot N)/(N \cdot m)]$$

$$= [\text{surface tension/pressure}]$$

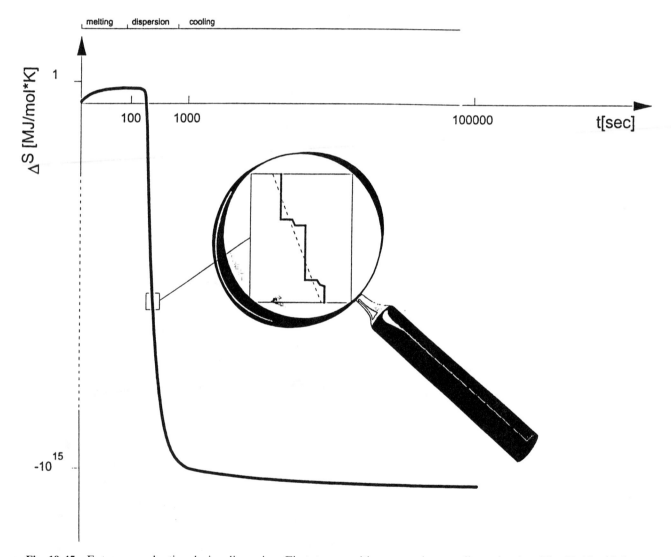

Fig. 19.45 Entropy production during dispersion. First stage, melting; second stage, dispersion (see Fig. 19.46); third stage, cooling. During the melting phase, positive entropy production takes place; during dispersion, the entropy export is occuring ($\cong$ negative entropy production). The magnified area models the rhythmic compression and decompression in an extruder. (From Ref. 7.)

Dispersion as the Key to Processing

This correctly reflects the fact that dispersion is the result of a stress (N/m²), a "dispersion pressure" induced through the polymer by W_{disp} being applied against a surface tension (N/m) of the material to be dispersed, which works as a counterpressure induced by interfacial forces between the particles and the matrix and is directed against dispersion. The "dispersion stress" is transferred to the agglomerates to be dispersed via the shear force applied to the polymer matrix ($W_\eta = \eta V s_r$).

There is some evidence from the arguments given above that $dS/dt < 0$ and also that $d_x P < 0$ for the dispersion process itself. The entropy function is more complicated (visualized in the detailed diagrams of (Figs. 19.45 and 19.46), because in every dispersion the extruder shear (leading to dispersion) is not continuous but intermittent, followed by flow relaxation with phase separation, a further small negative entropy change step ($dS/dt < 0$).

Table 19.3 Dependence of the Degree of Dispersion on the Shear Rate, s_r [a]

s_r	n [b]	Dispersion degree (%) after 100 s residence time
1	0	—
10	0	—
100	10^{-43}	—[c]
200	5×10^{-22}	—
300	9×10^{-15}	—
500	6×10^{-9}	
600	1.5×10^{-7}	6 ppb
700	1.6×10^{-6}	1 ppm
1,000	10^{-4}	—[d]
1,300		0,1
1,800	0.01	1
3,100	0.1	10
9,100		90
10,000	1	100

[a] See Fig. 19.48.
[b] n = number of dispersed particles during dispersion, normalized to 1.
[c] This shear rate would require 2.78 h for complete dispersion.
[d] This shear rate would require 1000 s = 16 min time for complete dispersion.

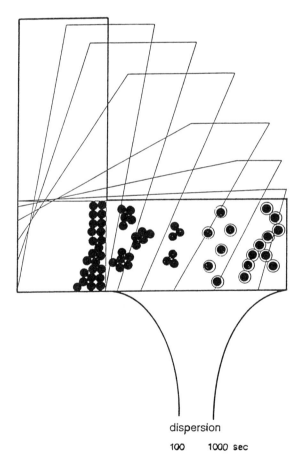

Fig. 19.46 Schematic representation of the dispersion process. The polymer matrix is sheared and transported. The particles are first distributed and destroyed, followed by dispersion (adsorption of a monomolecular layer of matrix polymer onto the particle surface), and later separation. (From Ref. 7.)

So we see that dispersion provides enough negative entropy flow (entropy export) to force the system far away from equilibrium and allow it to build up "dissipative structures." The distance from equilibrium is very large, i.e., we are beyond the thermodynamic branch.

B. Critical Shear Rate at Bifurcation Point

This problem is related to the question of whether there is a minimum work input required before dispersion begins to take place. Or, in nonequilibrium terms: What is the critical parameter at which "bifurcation" (Fig. 19.47) occurs, and what is the value needed to make the system leave the thermodynamic branch?

The dispersion law and its possible solution, Eq. (11), describing the dynamics of dispersion, could enable us to find the instability. With the values

$$c = 0.1 \text{ m}, \quad r_d = 10^{-7} \text{ m}, \quad t = 100 \text{ s}$$

we analyzed the evolution of the particle numbers with changing s_r (shear rate); see Table 19.3. It can be seen that an appreciable initial degree of dispersion after a residence time of 100 s will be found only if $s_r > 1300$ s^{-1} (dispersion degree $> 0.1\%$). Therefore it can be concluded that the shear rate is the critical parameter, and its critical value above which dispersion takes place or dissipative dispersion structures are created is around 1000 s^{-1}.

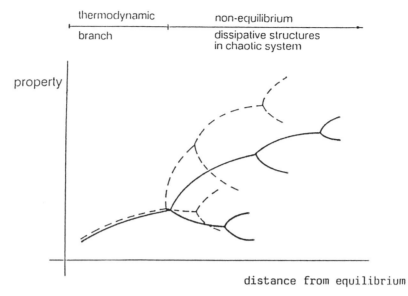

Fig. 19.47 Thermodynamic branch and series of bifurcations above a certain critical point. (After Refs. 7 and 49.)

It seems helpful to reformulate the exponent in Eq. (11), starting with the considerations about the dimension of c given there:

$$\frac{c}{r_d t\, s_r} = \begin{cases} \dfrac{\gamma_{12}}{\eta\, s_r\, r_d} & \text{with} \quad \dfrac{c}{t} = \dfrac{\gamma_{12}}{\eta} \\ \dfrac{\gamma_{12}}{r_d\, \tau} & \text{with} \quad \tau = \eta s_r \end{cases} \quad (12)$$

Empirically we know that there is no dispersion to be detected under pure pump extrusion conditions (pure melting and conveying screw design). It is known that a minimum shear stress has to be applied to obtain a significant degree of dispersion, such as is needed for pigments (see Fig. 19.48). Figure 19.48 shows the development of the color intensity (or color strength) of any kind of pigment in a polymer, which can be measured according to DIN 53234.* With increasing dispersion degree, the color strength [represented by an increasing color strength value†] also increases. This can be achieved either by increasing dispersion time (at a given supercritical shear rate) or by increasing shear rate (for a given residence time). Also, certain types of carbon black are used as pigments.

Rwei et al. [52] compared such a carbon black dispersed in three low viscosity media (water, squalene, polydimethylsiloxane) and showed that dispersion takes place only above a critical shear rate, and the lower the viscosity, the higher the necessary critical shear rate. Moreover, Ref. 52 is the only report available with a quantitative description of this qualitatively known dependence. But even Rwei et al. [52] do not supply an answer to the question: What is this critical shear rate in physical terms?

Introducing the experimentally observed critical boundary for first occurrence of dispersion as τ_{crit}, it follows that

$$\frac{c}{r_d t(s_r - S_{r,\text{crit}})} = \frac{\gamma_{12}}{r_d t(\tau - \tau_{\text{crit}})} \quad \text{for } \tau > \tau_{\text{crit}} \quad (13)$$

This means that there is no mathematical information about n [in Eq. (11)] for $\tau \leq \tau_{\text{crit}}$, and Eq. (11) with the exponent as shown in Eq. (13) is not applicable. This is in accordance with our own experiments in low shear extrusion and with the results published in Ref. 52, which indicate that if $\tau \leq \tau_{\text{crit}}$, no dispersion occurs.

Another approach to describing the observed behavior is to introduce the above-mentioned definition of γ_{12e}. The value of $\gamma_{12e}A_{12}$ is at least as large as the work W_η necessary to overcome the viscous strain of the polymer before it is able to wet the dispersed phase [see Eq. (7)]. Introducing this into Eq. (12), it follows that, assuming that $\gamma_{12e}A_{12} = W_\eta$ and taking $r_d A_{12} = V_{\text{disp}}$ (V_{disp} = volume of the dispersed phase),

$$\frac{c}{r_d t s_r} = \frac{\gamma_{12} V_{\text{syst}}}{\gamma_{12e} r_d A_{12}} = \frac{\gamma_{12} V_{\text{syst}}}{\gamma_{12e} V_{\text{disp}}} \quad (14)$$

where V_{syst} is the total volume of the matrix polymer–dispersed phase system.

---

* DIN 53234, Determination of the relative tinting strength, 1972.

† Color strength is measured in percent against a given standard material.

Dispersion as the Key to Processing

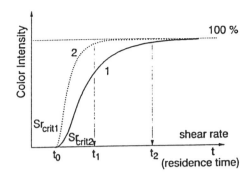

Fig. 19.48 Color intensity development. The color strength of a pigment is dependent on time (at a given shear rate in different polymers 1 or 2) or on shear rate with $t =$ constant, polymer 2 being the more viscous polymer. (From Ref. 51.)

At the present stage, in which we are now just entering a nonequilibrium thermodynamic description of multiphase polymer systems, an ab initio theoretical derivation of Eqs. (10) and (11) is still lacking. The foregoing thoughts and reformulations may at least lead to some important conclusions:

1. The critical shear rate above which Eq. (11) results in a first physically appreciable degree of dispersion (>0.1%) is in the neighborhood of what is known to cause "melt fracture" ($s_r \geq 1000$ s^{-1}; $\tau \geq 10^5$ N·m). This leads to the hypothesis that dispersion can occur and be observed only under conditions of melt fracture, a widely known rheological instability [53].* ("Melt fracture" can be observed at the die of an extruder or melt rheometer as a sudden change in the surface aspect of the emerging melt beyond a critical point in the vicinity of $\tau \sim 10^5$ N·m or $s_r \sim 1000$ s^{-1}. Under given extrusion conditions it suddenly appears at a certain critical point during continuously increasing output. It can be viewed as analogous to turbulent flow in low viscosity media. However, "melt fracture" structures can be frozen by simply cooling the melt strand or the produced film, which exhibits irregular wave and/or fish-scale patterns. These patterns will again suddenly change to new patterns after a second critical point in response to a further increase in output.)
2. The nonlinear behavior of the melt is then well reflected in the exponent [Eq. (13)], which leads to a definition of $n = f(t)$ only for $\tau > \tau_{crit}$ (above melt fracture).
3. Independently of this approach, Eq. (14) tell us about two other nonlinear phenomena:
   a. The nonlinear dependence of V_{syst} on dispersed phase concentration (cf. the density nonlinearity) [54].
   b. The relation of γ_{12}/γ_{12e} (the "structure factor"), which behaves nonlinearly according to Fig. 19.49.
      γ_{12e} is not defined for a dispersed phase concentration of zero; for low viscosity systems, γ_{12e} will be identical to γ_{12} for a certain concentration regimen. For "easy-to-disperse" fillers, γ_{12e} will differ from γ_{12} only above a certain concentration. In general, two-phase systems cannot reach the γ_{12e} level from the γ_{12} equilibrium level without experiencing a (nonlinear) jump; there is no continuity between γ_{12} and γ_{12e}.

These results, combining the widely known instability (and dissipative structure!) phenomenon of melt fracture with the new nonequilibrium description of multiphase polymer systems, will hopefully stimulate more experimental and theoretical work devoted to these (frozen) dissipative structures. It still remains an open question which property of the melt may be responsible for its suddenly occurring capability to disperse fillers (pigments, carbon black, etc.) or other incompatible polymers above melt fracture conditions. We can only speculate that, in particular, the creation of microvoids (inner surfaces) and a sudden increase in gas solubilization capability at and above "melt fracture" allow the polymer melt to wet the surface of the material that is to be dispersed. This means that a polymer melt might have completely different (supercritical) properties above melt fracture than we usually observe.

This new theoretical view of dispersion in polymer systems and the important critical parameter can probably be used to describe any other colloidal systems [48] in an analogous way.

We can investigate microemulsion systems in the same way. The first attempts of Strey and coworkers [55] to explain microemulsion structures were still based on equilibrium thermodynamic considerations. But structure generation is equivalent to an entropy decrease, which makes $(-T \Delta S)$ a large positive term. Microemulsion scientists are thus forced to believe that the enthalpy of mixing is so large and negative that the total free energy of microemulsion formation is negative. However, it is legitimate to doubt whether this is a general phenomenon or even whether it occurs at all. So, for some reasons, we can propose to consider microemulsions and their structure to also be the result of a supercritical energy input and entropy export, leading to self-organized dissipative structures.

---

*Melt fracture itself is a very interesting instability and nonequilibrium phenomenon, exhibiting interacting "dissipative structures" and worth detailed thermodynamic analysis. For a qualitative description see, e.g., Ref. 53.

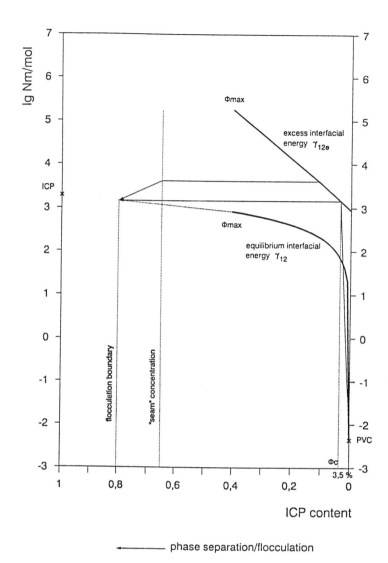

Fig. 19.49 Graphic analysis of the different energy levels responsible for dissipative structures ("excess interfacial energy") and flocculation ("equilibrium interfacial energy"), calculated for a conductive polymer ("ICP")-PVC blend. ICP concentrations in separated phases: $X_{ICP} = 0.65$. ICP concentration in these phases after flocculation: $X_{ICP} = 0.8$. A full phase separation ($X_{ICP} = 1$; $X_{PVC} = 1$) of ICP and PVC is not possible owing to the high viscosity work barrier; therefore, the region between $X_{ICP} = 1\ 0.8$ and the excess interfacial energy region above γ_{12e} for $\Phi_{max} = 0.4$ are not accessible. The value of γ_{12} for $X_{ICP} = 0.8$ is practically a trap, because a further energy gain, e.g., by a phase separation along the γ_{12} curve, is theoretically and practically not possible, because one of the phases would have to cross the inaccessible region. (From Ref. 7.)

VI. THE METALLIC PROPERTIES OF POLYANILINE AND ITS DISPERSIONS

A. Original Ideas About the Conductive Nature of ICPs

Research into the conductivity phenomena of the new materials category of "intrinsically conductive polymers" proceeded from the paradigm that the transport phenomena were linked to the properties of the polymer chain. It began with the hypothesis that polyacetylene was an ideal candidate for the search for solitons. It was thought that oxidation of the chain (which was incorrectly called "doping") made possible a kind of "folding over" of the conjugated double bonds, thereby setting in motion the flow of electrons.

This idea was then linked with the supposedly discovered morphology of PAc, which appeared to be fibrillar. According to this view the PAc chains exhibited a more or less tidy parallel arrangement in the fibrils and were axially oriented. The transport phenomena were tied to the chain and therefore took place along the length of the fibrils. We shall therefore call the entire paradigm the "fibril hypothesis." A thorough description can be found, for example, in Chien's monograph [14].

This paradigm was also applied to all other conductive polymers, and fibrils were found everywhere. These ideas appeared to receive strong confirmation with the discovery of "high conductivity PAc" ("Naarmann-Polyacetylen," which displayed a particularly high conductivity that was attributed to stretching and orientation processes and to minimal defects in the PAc chains. For an overview, see Ref. 56).

B. The "Granular Metal" Hypothesis

With the discovery of partly metallic behavior (Pauli susceptibility [57], thermopower [58,59]) and the extension of crystalline domains in polyaniline (PAni) and other ICPs [60], Epstein and coworkers [61] developed the hypothesis of a "granular metal." Although this concept still involved the idea that the chains were arranged along the fibrils (and that defects in the chains were the cause of poor metallic conductivity), the sites of metallic transport properties were now associated with the crystalline domains, which had a diameter of several nanometers. Conversion processes such as "dissolving" or "orientation," however, were still understood as acting on the individual chains [62]. This model has certain similarities with the ideas developed by Kaiser [63]. He pointed out the similarities between ICPs and amorphous metals.

The current versions of a combination of the granular metal hypothesis and the fibril paradigm proceed from the assumption that the chains of ICPs are arranged along the fibrils and are present in proportions of 30–50%, or a maximum of 80%, in crystalline phases that are interlinked by amorphously bundled chains. The fibrils are thought to be linked to one another by quasi-one-dimensional contacts [64]. Most groups of workers within the worldwide community engaged in research into conductivity polymers continue to attribute the conductivity phenomena primarily to the properties of individual chains [62,65].

C. The "Sphere" Hypothesis

As long ago as the early 1980s we developed the idea that conductive polymers possessed, as the morphological primary unit to which one could attribute the essential properties of an ICP, not fibrils (or chains) but small spheres [66]. Investigations using the scanning electron microscope (cf. Section IV.D) had shown that PAc fibrils that were recognizable as such under the transmission electron microscope proved in the same sample under the scanning electron microscope (with careful sputtering) to consist of spherical particles with a diameter of 50–200 nm. Later [34] we suspected that these in turn consisted of even smaller subunits, which proved to be correct.

Dispersions using PAni in binderless solvents today display a particle size of 10 nm. Thermoplastic polymers or paints contain (secondary) agglomerates with a size of about 100 nm.

Another hypothesis voiced at an early stage [34], according to which the primary particles are the "active units" not only of dispersion and hence of conversion but also of the electrical properties, has also proved viable. Unlike the "granular metal" hypothesis [the crystallites (coherence length 3–5, max. 7.5 nm [67]) of which, incidentally, have dimensions similar to those of the primary particles we found], we conclude from our findings that there are no material links between the metallic units (no chains creating chemical or physical links between the individual particles). In fact, the particles have physical contact only with one another (cohesion) and display interactions of an electronic nature that are the basis for the electrical transport phenomena.

D. Mesoscopic Metals

In 1944 Landau published a theory that it should be possible to observe quantum-mechanical effects in metals if the dimensions of the metal were in the region of or less than the wavelength of the electrons [68]. For many years attempts were made to confirm or refute this theory by creating thin layers of metals. From today's viewpoint these experiments were unsuitable because only one of three dimensions of the metal was brought below the critical limit, with the result that the electrons were still able to spread unhindered.

The credit for finally proving this old theory experimentally goes to Nimtz et al. [69], who succeeded in creating mesoscopic metals (i.e., metals with particle sizes from less than 1 μm to $\approx$20 nm) and investigating their conductivity as a suspension or colloidal dispersion in a suitable test design with the aid of microwaves. They found that metals consisting of submicrometer particles behave differently from the macroscopic form.

The conductivity of mesoscopic metals can be measured only by noncontact means. For this reason the particles were embedded in an insulating matrix. The manufacture of the (indium) particles was generally achieved by condensation from the gas phase in a rotating oil film [69]. This method yielded metal particles of about 20 nm that were (colloidally) dispersed in the oil matrix. By means of thermal coalescence, particles with a diameter of up to several hundred nanometers were obtained. Thus the effective dielectric function (DF) of the heterogeneous oil–indium system was measured. At constant volumetric filling ratio it was possible to measure the dielectric response as a function of particle size during the growth process.

A similar procedure was adopted with other metals (e.g., silver).

The experiments were performed at frequencies between 10 and 10^{10} Hz using a test design that made it possible to measure the microwave absorption of the system to be measured in one beam by comparison with a parallel beam. The design chosen permitted noncontact measurement of both the electric and magnetic contributions to the total electromagnetic response. An effective medium analysis permits extraction of the metallic components from the total DF of the heterogeneous system.

At low frequencies what one obtains from the DF ($\epsilon = \epsilon_1 + i\epsilon_2$) is practically the direct current conductivity of the free charge carriers, since the equation

$$\sigma = \epsilon_2 \epsilon_0 \omega$$

applies, in which ϵ_0 is free space permeability and ω is the measuring frequency.

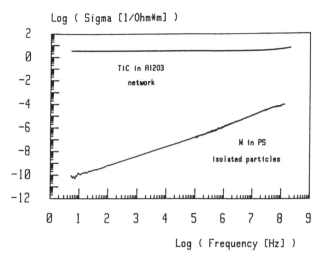

Fig. 19.51 Conductivity σ vs. frequency ω. Isolated particles have σ proportional to ω, whereas networks of particles have a frequency-independent conductivity like a bulk metal. (W particles in PS: $f = 0.11$, $d = 400$ nm; TiC particles in Al_2O_3: $f = 0.17$, $d = 800$ nm, where f is the volume filling factor). (From Ref. 70.)

The result—initially surprising, but understandable in the light of the Landau theory—is that the DF, and hence also the conductivity σ, depends on the particle size. Figure 19.50 shows this for indium. Comparable situations were demonstrated for a number of other metals such as Cu, Ag, Pt, Ni, and TiC.

It is interesting to note that the temperature dependence of the conductivity of mesoscopic metals is different from that of the same metals in bulk form. Whereas it is well known that temperature dependence is characterized as metallic by the fact that conductivity increases with falling temperature, metals in mesoscopic form exhibit nonmetallic behavior in that their conductivity declines, albeit only slightly, as the temperature drops [71].

No significant dependence of ϵ_2 on frequency was found, so $\sigma \sim \omega$. The results of various experiments are shown in Fig. 19.51 [72].

This function is completely different, however, if the particles aggregate or even form a continuous (percolating) network with the result that the system displays a limited degree of direct current conductivity. The frequency dependence of conductivity then disappears completely and displays a metallic function with σ = constant and $\epsilon_2 \sim \omega^1$.

E. ICPs: A Family of "Natural" Mesoscopic Metals

Finally, in 1989, we suggested a first joint study with the Cologne group with the aim of finding out whether our polyaniline might perhaps display behavior similar

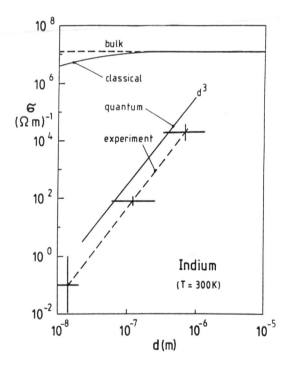

Fig. 19.50 Size dependence of the quasi-dc conductivity $\sigma(x)$ of isolated indium particles measured at 10 GHz. The accuracy of the experimental data is determined by rather broad particle size distribution of the ensembles available. Quantum size calculations have predicted a particle conductivity proportional to the third power of the diameter d. Calculated values for the classical surface scattering are also shown for comparison. (From Ref. 70.)

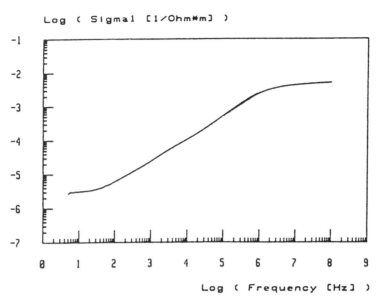

Fig. 19.52 Conductivity of σ vs. frequency ω of PAni layers at room temperature. Layer thickness 100 nm. (From Ref. 70.)

to that of an artificially produced mesoscopic metal. Our hypothesis was that owing to its spherical primary particles of ≈ 10 nm diameter, PAni ought to display properties similar to those of a metal with particles of 10 nm diameter. This was indeed the case [70].

First we investigated PAni layers on polycarbonate films (100 μm). The PAni coating was applied to the PC from binderless organic dispersions (see Section III.B) and was 100 or 250 nm thick. The conductivity of the layer was ≈ 1 S/cm. For the measurements 78 coated films (for the 100 nm PAni layers) or 72 (for the 250 nm layers) were arranged in nonconducting fashion one above the other. The conductivity stagnated at frequencies in the megahertz range and up to 10 GHz (Fig. 19.52).

In the region of the saturation conductivity of $\sigma(\omega)$ we observed a pronounced relaxation process in both samples, as can be seen from Fig. 19.53.

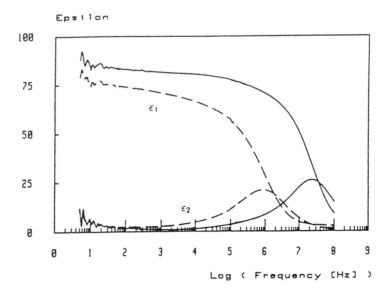

Fig. 19.53 Complex dielectric function $\varepsilon = \varepsilon_1 + i\varepsilon_2$ of two samples of PAni layers. The solid line was measured with layers 250 nm thick, the broken line was measured with layers 100 nm thick. (From Ref. 70.)

It is clear that the peak frequency of ϵ_2 depends on the properties of the two samples. No change in the relaxation frequency was observed when we increased the measuring temperature to 340 K.

Agglomerated metal particles also displayed a relaxation process [73]. The relaxation frequency increased with the degree of agglomeration. The authors interpret this as a charge and discharge process of the gap between the agglomerated particles (Maxwell–Wagner relaxation). This interpretation may also apply to the processes observed in the PAni layers on PC, as we had identified the morphology of the pure PAni layers as being that of individual spheres built up one on top of the other to form pillars [74].

Other features that mesoscopic metals and conductive polymers (here PAni) have in common are the frequency dependence of conductivity,

$$\sigma \approx \omega^\alpha$$

where $\alpha = 0, \ldots, 1$, and the positive nonmetallic temperature dependence of conductivity.

Following this, in a joint project lasting several years, we investigated over a broad concentration range the conductivity of polyaniline dispersed in a thermoplastic matrix. The results confirmed, in quantitative terms as well, the metallic properties and the dimensions (9.6 nm) of the PAni primary particles, which contain an 8 nm metallic core [75].

The investigations were performed with a series of PAni dispersions in PETG (an amorphous copolyester) in concentrations ranging from 3.1 to 37.7 vol % (plus 0% and 100%) PAni, where 0 and 100% represent the PAni-free PETG matrix and the pure PAni, respectively. Here again two network analyzers were used with measuring ranges of 5 Hz to 200 MHz and 200 MHz to 2 GHz.

The critical volume concentration for the blends used here was in the region of 8 vol %. In agreement with direct current measurements by the four-point method, this was determined in the test design used here by analyzing the complex dielectric function (DF) $\sigma = \epsilon_2 \epsilon_0 \omega$ at 5 Hz (see Fig. 19.54).

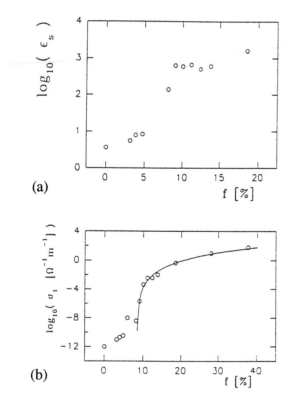
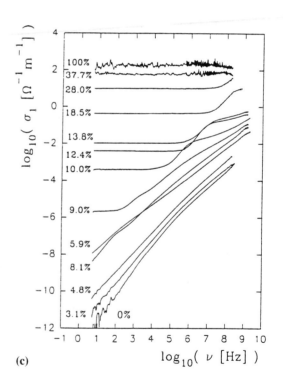

Fig. 19.54 (a) Extrapolated static dielectric constant and (b) conductivity at 5 Hz versus filling factor of PAni-PETG blends in semilog plot ($T = 24°C$). Above $f \cong 8\%$ the measured s values equal the dc conductivity. Best fit $\sigma = a(f-f_c)\mu$ with $a = 8230 \, \Omega^{-1} \, m^{-1}$, $f_c = 0.084$, and $\mu = 4.28$. Obviously, since $\mu > 2$, there is no statistical distribution of polyaniline above f_c. (c) Conductivity vs. frequency of PAni-PETG blends with different volume filling factors in double-log plot ($T = 24°C$) Above $f \cong 8\%$ the constant low frequency values equal dc conductivity. For $f \gtrsim 37\%$ measured conductivities are determined by contact resistance of the samples. The bulk value for pure PAni (compressed Ormezon™) is $\sigma \cong 2000 \, \Omega^{-1} \, m^{-1}$. (From Ref. 75.)

(An interesting side result can be observed. In Fig. 19.54b, the best approximation to the measured values can be achieved with the relationship

$$\sigma = a(c - c_{\text{crit}})^\mu \qquad (15)$$

where $\mu = 4.28$, in other words $\mu > 2$, which according to the percolation theory means that there is no statistical distribution. This is precisely what we found in our investigations described in the preceding sections and have theoretically underpinned with the new nonequilibrium thermodynamic description. Once again the percolation theory proves to be an inadequate description of the causes of the sudden increase in conductivity above the critical volume concentration.)

Figure 19.54c shows conductivity as a function of frequency.

Below the critical volume concentration a relaxation was again observed, but this must be ascribed to the β relaxation of polyester. The analysis of DF(ϵ_∞) for the concentration range below the critical volume concentration showed that here the relaxation frequency of the interfacial polarization would be outside the range of measurement, i.e., that there is no energy dissipation due to free charge carriers.

Using the effective medium formula of Looyenga [76], it was possible to calculate the dielectric constant

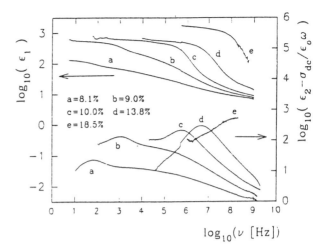

Fig. 19.56 Dielectric function of blends ($f \geq f_c$, 24°C) in a double-log plot (ε_1: upper curves, left y axis; ε_2: lower curves, right y axis). The dc contribution $\sigma_{\text{dc}}/\varepsilon_0 \omega$ has been subtracted from the imaginary part to show the relaxation process more clearly. (From Ref. 75.)

for the dispersed PAni as

$$\epsilon_{\text{PAni}}(\omega\tau \ll 1) \approx 400 \qquad (\text{cf. Fig. 19.55}).$$

According to information in the literature, undoped (neutral, insulating) PAni (emeraldine base) had a dielectric constant of 4–10; i.e., the quasistatic permitivity of the dispersed PAni was markedly increased by interfacial polarization. According to calculations by Cini and Ascarelli (see Ref. 75), this depends on the dimensions of the conductive phases. The assessment of the number of free charge carriers as 4×10^{27} m^{-3} yielded a figure of ≥ 6.3 nm for the extent of the metallic phase [a figure that agrees in principle with the figures for the coherence length in highly conductive ICPs (see Section VI.B)].

The analysis of the measurements above the critical volume concentration now made it possible to say something about whether the transport mechanism was dominated by these mesoscopic metallic regions (of which we now definitely know that they are not interlinked by bridges of material or amorphous loops of chains but at most display point-type physical contact above the critical volume concentration).

Above the critical volume concentration it is possible to observe a process (see Fig. 19.56) that displays a reaction time $\tau = 1/2\pi\nu_{\text{max}}$ that is related to the effective conductivity of the blend. For $c > 18.5\%$, the relaxation frequency has become shifted to magnitudes outside the range of measurement, so that only frequency-independent conductivity can be observed.

With declining temperature and with declining fill factor, both the direct current conductivity and the relaxation frequency grow smaller. For more than 7 orders of magnitude the known conduction current relaxation

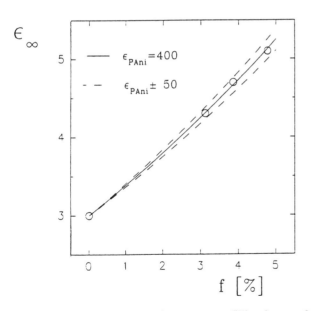

Fig. 19.55 High frequency value ε_∞ versus filling factor of blends with $f < f_c$ at 24°C (extrapolated from Cole–Cole diagrams, i.e., from plots of the imaginary part of the DF vs. the real part). The solid line is calculated from the Looyenga formula with $\varepsilon_{\text{PAni}} = 400$ (the uncertainty is about 20). To show the accuracy of the evaluation, the dashed lines are calculated with $\varepsilon_{\text{PAni}} = 350$ and $\varepsilon_{\text{PAni}} = 450$, respectively. (From Ref. 75.)

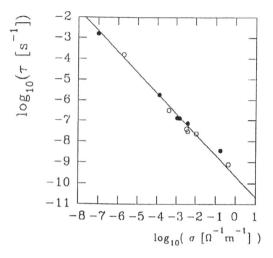

Fig. 19.57 Log-log plot of relaxation time vs. dc conductivity of the process shown in Fig. 19.56. $\tau(f,T) \propto 1/\sigma_{dc}(f,T)$ holds. (○) $T = 24°C$; (●) $T = 172°C$. (From Ref. 75.)

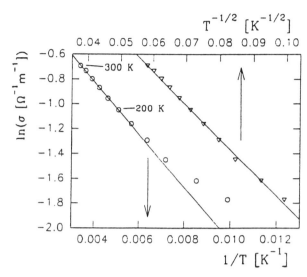

Fig. 19.58 Temperature dependence of the dc conductivity of a blend with $f = 18.5\%$. (○) Arrhenius plot (lower x axis). (▽) $\ln(\sigma_{dc})$ vs. $1/\sqrt{T}$ (upper x axis). (From Ref. 75.)

relation [77] applies (see Fig. 19.57):

$$\tau(c,T) = f[1/\sigma(c,T)] \quad (16)$$

This behavior is characteristic of partial interfacial polarization induced by conductive paths that are not arranged parallel to the electric field. The same phenomenon has been observed in heterogeneous combinations of mesoscopic metallic particles in insulating matrices and (see above) in thin layers of pure PAni on PC.

For further analysis we examined the temperature dependence of conductivity and of the high frequency and low frequency values of ϵ_∞ and ϵ_s. At temperatures between 200 and 300 K we observed Arrhenius behavior (Fig. 19.58) (with an activation energy on the order of $k_B T$), whereas at lower temperatures the conductivity follows the relation

$$\sigma = \sigma_0 \exp[(-T_0/T)^{1/2}] \quad (17)$$

which was also observed in the investigations of the thermopower of comparable PAni-PVC blends down to 10 K (cf. Ref. 78). The corresponding parameters are given in Table 19.4.

It follows from considerations outlined above, which resulted from SEM studies, pyrolysis, and density measurements, that conductivity takes place along paths that consist of pure PAni (flocculated PAni particles). For these the size of the localization centers can be determined independently of the question of whether the transport mechanism corresponds to a three-dimensional hopping model or intermolecular hopping between "metallic" chains. From considerations related to the activation energy $W = f(1/d_T)$, which is determined by the charge energy (cf. Ref. 75), we arrive at

$$d_T \sim 8 \text{ nm}$$

Table 19.4 Parameters of Blends Above f_c (100–300 K) and to $\sigma = \sigma_1 \exp(-W/k_B T)$ (200–300 K) [a]

f (%)	σ_0 ($\Omega^{-1}m^{-1}$)	T_0 (K)	σ_1 ($\Omega^{-1}m^{-1}$)	W (meV)	ϵ_∞	d_T (nm)
9.02	8.9×10^{-3}	4848	1.1×10^{-5}	49.6		
10.03	3.6×10^{-3}	1154	1.1×10^{-3}	20.8	6.5	10.65
11.15	2.2×10^{-2}	931	8.8×10^{-3}	21.2	10.0	6.8
12.40	2.1×10^{-2}	697	9.0×10^{-3}	17.5	10.5	7.8
13.80	5.1×10^{-2}	674	2.3×10^{-2}	18.2	11.5	6.9
18.49	2.2	687	1.0	18.0		
27.99	37	509	18	14.7		

[a] For $f > 14\%$, the high frequency value of the DF, ϵ_∞, could not be determined.

Dispersion as the Key to Processing

From this it follows in turn that transport is determined by 3D "hopping" between mesoscopic metallic phases. The figure quoted here also agrees with the figure of 6.3 nm mentioned earlier, which was obtained from measurements below the critical volume concentration.

These mesoscopic metallic phases are quite evident in the morphological primary particles of ≈10 nm diameter found by other methods.

In this view the barriers between the (partly) crystalline metallic phases in the PAni primary particles also consist of PAni and not of components of the insulating matrix. This is probably amorphous PAni, whereas the metallic phase consists of (partly) crystalline PAni. Assuming a crystallinity of 30%, this results in a barrier thickness of 1.6 nm, giving a particle size of 9.6 nm, which approximates closely to the 10 nm estimated by other methods in earlier works by Wessling.

Figure 19.59 shows the transport scenario. This is intended to indicate that the "hopping" is not to be seen as the hopping of an individual electron but as a phenomenon of wave mechanics. The primary particles evidently possess a genuine metallic core in which the electrons may be freely mobile (electron gas). The dimensions of the metallic phase do not, however, allow any desired wavelength for the electrons, but only certain wavelengths that are a whole multiple of the diameter of the correlation length (see Fig. 19.60).

This is the reason the conductivity found in mesoscopic metals is not the conductivity familiar from bulk metals but a much lower conductivity. Nimtz has referred to this in the past as SIMIT (size-induced metal-to-insulator transition). We do not know yet whether this interpretation can be applied entirely to PAni and other ICPs. To date the measurements have permitted statements only about transport from particle to particle.

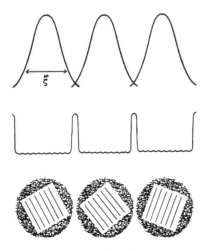

Fig. 19.59 Schematic view of (a) electronic wave functions (ξ = localization length); (b) potential; and (c) morphology of crystalline metallic regions and amorphous barriers.

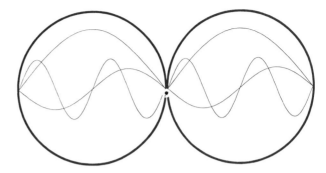

Fig. 19.60 Mesoscopic metallic particles (e.g., PAni with 19 nm diameter) allow only certain electron wavelengths to occur. This is the reason for their limited conductivity.

We do not yet know the conductivity and transport mechanisms within the particles themselves.

The macroscopically observed transport is based on the quantum-mechanical interaction between the electron waves in neighboring "quantum metal particles," which is probably facilitated by the Maxwell–Wagner polarization and thermally stimulated. Instead of "hopping," the expression "tunneling" may approximate more closely the quantum-mechanical fundamentals of this process.

F. Thermopower of Polyaniline Dispersions

It was recognized at an early stage [63] that ICPs display a temperature dependence of thermopower that identifies them as metals (according to Kaiser [63], with similarities to amorphous metals). It was therefore interesting for us to know what the thermopower of blends of nonconductive matrix polymers with PAni was like. According to our studies (cf. Sections I–IV), these are formed from a dispersion (complete separation of the spherical primary particles) of PAni or other ICPs in the matrix polymer by flocculation (formation of chainlike branched aggregates) and thus display no material links between the (metallic) primary or secondary particles. Although the individual particles in the flocculates are in close proximity, they are nevertheless separated from one another by several angstroms (i.e., the particles can be reversibly dispersed and flocculated).

As early as 1987 [79], therefore, we measured the thermopower of a PAni blend (at saturation conductivity concentration, at that time around 35%) with PVC and recognized its temperature dependence as linear, though in view of our limited experimental facilities in this field we were only able to measure a ΔT of ≈100 K and were above all not able to measure sufficiently far in the direction of absolute zero.

It was only thanks to many years of cooperation with Kaiser that we succeeded in making temperature measurements over the entire temperature range down to 2.7 K and in a broad range of PAni concentrations in a variety of blends [78,80]. The matrix polymer used in the first study was PVC; in the second, polyester and PMMA.

The PVC blend was prepared and investigated in two concentrations, 20% and 47%. Its composition corresponded in principle to a commercially available (though not commercially successful) blend, although an average concentration of 35% was planned for the industrial applications. For comparison purposes pure PAni powder was measured (as a molding made at room temperature and 10 kbar). The thermopower and conductivity and their temperature dependence were measured using standard techniques [78,80]. Conductivity as a function of temperature is shown in Fig. 19.61.

As expected, conductivity at room temperature (or σ_∞ extrapolated to $T \to \infty$) was higher for pure PAni (σ

Table 19.5 Parameters T_0 and σ_0 in Eq. (18) Deduced from Conductivity Data on Polyaniline Blends and Unblended Polyaniline

% PAni	σ_0 (S/cm)	T_0 (K)	Ref.
20	2.0	194	
47	15.5	52	
100	132	938	
100		6000	70
100		4300	73
100		2930	74

Source: Ref. 78.

= 20 S/cm) than for the PVC blends, especially for the sample containing 20% PAni (1 S/cm; 47%, 10 S/cm). Over a broad temperature range the temperature dependence of σ follows the formula

$$\sigma = \sigma_0 \exp(-T_0/T)^\gamma \quad (18)$$

where a better fit was obtained with $\gamma = 1/2$ than with $\gamma = 1/4$, although with serious deviations (see below). The figures for σ_0 and T_0 are given in Table 19.5.

Surprisingly, the conductivity of the PVC-PAni blends at low temperatures was one order of magnitude greater than for pure PAni. At 50 K the 47% sample still possesses half its room temperature conductivity and the 20% sample nearly 40%, but the pure PAni has less than 10% of its initial conductivity.

As regards the temperature dependence gradient, the 47% and 20% samples are much more similar (although their original conductivity differs by a factor of 10) than the 47% sample and the pure PAni (whose original conductivity differs by only a factor of 2). Moreover, the PAni used by us had a much lower temperature dependence of σ than that used in comparable studies of pure PAni (for literature references see Refs. 78 and 80).

It is also interesting to note the situation shown in Fig. 19.61b. The pure PAni displays a marked conductivity maximum, whereas the 47% sample shows only a weak maximum and the 20% sample shows none at all. A later study [81] showed that with even better dispersion in a PMMA blend, a higher and broader conductivity maximum is to be observed, and hence a more metallic behavior for the blend than for the pure PAni. Above room temperature the first two samples—both the pure PAni and the 47% PAni blend—behave like metals in a fashion also observed in numerous highly conductive polyacetylenes. Thus in this respect there is no difference between these materials (PAni, PAni blend, and PAc).

The above-mentioned formula for describing temperature dependence therefore cannot be used in this temperature range.

The thermopower of the relevant samples is shown in Fig. 19.62. From the start it is particularly striking to note the similarity with traditional metallic behavior,

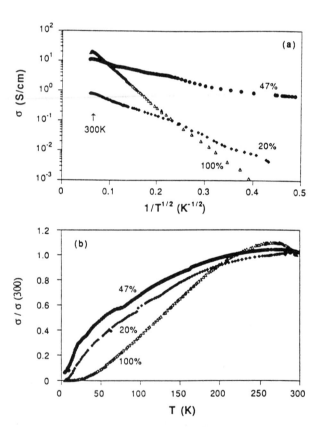

Fig. 19.61 Conductivity σ of PAni-PVC blends between the temperatures of 4.2 and 300 K compared to that for unblended PAni (the figures next to the curves give the percentage of PAni by weight). (a) Log σ plotted versus $T^{-1/2}$ to show the agreement with Eq. (18) over a wide temperature range. (b) Conductivity normalized by the conductivity $\sigma(300)$ at 300 K, plotted using linear scales. (From Ref. 78.)

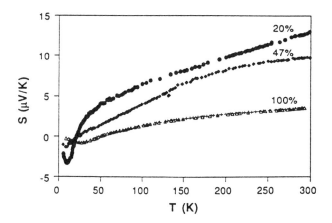

Fig. 19.62 Thermopower S of PAni-PVC blends (with curves labelled by percentage of PAni). For unblended PAni, the triangles and crosses show data taken on two different runs. (From Ref. 78.)

namely the small size of the thermopower figures and the increase with rising temperature. Above 50 K all values are positive (which might suggest, but does not prove, that the conductivity is due to holes rather than electrons). Below 50 K the values for all three samples are negative (conductivity contribution by electrons?). Earlier measurements by other groups on different forms of PAni yielded mainly negative values of S, while positive values tended to be found for samples with lower conductivity; a sharp transition to positive values was observed above 300 K. For literature references see Refs. 78 and 80.

Although the PAni-PVC blends display a greater thermopower than the pure PAni, the figures are still well below what can be expected for semiconductors. Above 100 K the temperature dependence of S is described relatively well by a linear approach (which when extrapolated to $T = 0$ results in a positive value of S_0). The blends exhibit a more marked negative thermopower peak, which is considerably sharper for the 20% sample than for the 47% sample, at 10 K, whereas the pure PAni displays a weaker minimum, namely at 25 K.

The investigations show that the fundamental transport mechanism is the same for pure PAni and for (dispersed/flocculated) blends. The interpretation of the transport mechanism set out in the previous section is supported by these studies.

The differences in low temperature conductivity between pure and dispersed/flocculated PAni could be explained either by "purification effects" during the dispersion process (e.g., a reduction in the barriers through removal of superficially absorbed impurities) or by an improved correlation between the primary and/or secondary particles.

The practically pure metal behavior of thermopower can easily be understood in terms of the fact that a thermal gradient plays a more important part here than an electrical potential gradient. The barriers between the metallic phases therefore play only a secondary role in thermopower. The negative peaks at low temperatures are still waiting for a convincing explanation but do at least suggest that the conductivity may not necessarily be due (solely) to holes but (at least partially) to electrons as well. They may also indicate an increase in the importance of charge energy, quantum effects (SIMIT), or localization at low temperatures.

These interpretations were supported by further studies of PAni-PMMA and PAni-PET blends [80]. The polyester blends (with 40% and 15% PAni) were appreciably less conductive (presumably owing to hydrolysis effects during production, which would result in partial compensation of PAni), while the PMMA blends (40% and 33% PAni) were much more conductive than the PVC blends previously studied. The PMMA blends were produced using an improved dispersion technology and displayed at only 33% a conductivity similar to the earlier 47% blends with PVC, whereas the 40% sample, with a figure of 20 S/cm, was more conductive at room temperature than the pure original material. At 50 K this sample still displayed 60% of its original conductivity and also, similar to the pure PAni, an (albeit broader) maximum of conductivity at (slightly below) room temperature and metallic behavior thereafter. This most conductive PAni blend to date exhibits behavior that practically agrees with that of highly conductive polyacetylenes (apart from the absolute conductivity figure).

The thermopower of the 40% blend (cf. Fig. 19.63) is also more similar to that of highly conductive polyacety-

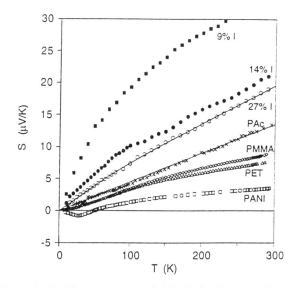

Fig. 19.63 Thermopower of PAni, PMMA, and PET blends with 40% PANI, and of polyacetylene doped with $MoCl_5$ (PAc) and iodine (9, 14, and 27 at % I). The lines are fits for metallic diffusion thermopower. (From Ref. 80.)

lene, in that it is linear over an even larger temperature range and, unlike the PVC blends described above and pure PAni, does not display a negative peak at low temperatures.

An important difference from iodine-doped PAc lies in the fact that in the latter the thermopower rises sharply as the iodine concentration falls, whereas even at low concentrations (low conductivity) the PAni blends still exhibit metallically low values of S. This indicates that the metallic nature of PAni remains uninfluenced in low dispersion concentrations (there is no change in chemical composition, crystallinity, or particle size, but the possibility of interactions with other PAni particles is reduced), whereas different iodine concentrations in PAc represent completely different chemical substances, which will be reflected in the distribution, size, and crystallinity of the metallic phases and also in the nature of the insulating barriers.

The fundamental similarity of (dispersed/flocculated) PAni in blends to pure PAni and even to highly conductive PAc is, however, clearly demonstrated by the above studies.

We recently succeeded in analyzing the two different contributions to the conductivity, i.e., the intrinsic metallic part and the tunneling part [81]. This analysis allows for a perfect fit of all curves $\sigma = f[T]$ that we measured in various PAni blends. It can be seen that the metallic character becomes more obvious and a greater influence on the $\sigma[T]$ dependence, the more perfect the dispersion.

G. The New Picture of Conductive Polymers

The conclusions that have to be drawn from the morphological studies, the dispersion experiments and associated analyses, and the investigations into the metallic character of conductivity are clear. Even if ultimate proof of a common nature of all ICPs has yet to be obtained and the majority of arguments are based on studies with PAni, the various comparisons with PAc, hcPAc, PPy, etc. described in the preceding sections suggest that the following is highly probable:

1. Intrinsically conductive polymers (ICPs) are, as far as we have found, made up of spherical primary particles ≈ 10 nm in diameter that contain a (partially) crystalline metallic core (≈ 8 nm).
2. The primarily spherical structure has been demonstrated for all ICPs that we have investigated; even highly conductive PAc has (before orientation) a spherical structure (Ref. 56, p. 99). Although the morphology after orientation is described as a fibrillar structure, in light of our earlier investigations one cannot rule out the possibility that these fibrils merely consist of differently arranged globules; my group, however, works on the basis of PAc chains oriented along the fibril axis and advocate the "fibril hypothesis."
3. The high charge density (one positive charge on two monomer units and the corresponding counterions) on the polymer chain results in an extremely high solubility parameter and thus in total insolubility even at short chain lengths [100]. (It should be noted that only the "doped" form of the ICP is meant here, since the neutral form is not an ICP, not a conductive polymer, but "only" a conjugated polymer and possesses a completely different composition, structure, and physical behavior).
4. An idealized structure (Fig. 19.64) shows the ICPs must have an anisotropic (bipolar) structure, which explains a strong tendency to form aggregates, especially linearly arranged fibrillar aggregates to screen the charges.
5. ICPs behave like traditional metals existing in mesoscopic particle sizes. The Maxwell–Wagner relaxation found in both cases could be facilitated by the anisotropic (bipolar) structure.
6. The conductivity phenomena are primarily of a metallic nature. This means that ideas based on individual electrons on individual chains cannot adequately describe the phenomena. Instead, the metallic phase in the primary particle is filled with an "electron gas," a superimposition of waves of numerous charge carriers (some 10^{27}/m^3, with $d_{particle}$ = 9.6 nm $\Rightarrow V_{particle} = d^3\pi/6$ = 4.63×10^{-25} m$^3 \Rightarrow 2.16 \times 10^{-24}$ particles/m$^3 \Rightarrow \approx 5 \times 10^3$ per particle). Their mobility is limited, however, by the fact that the dimensions of the metallic phase are of magnitude similar to that of the wavelength of the electrons. This results in a quantum-mechanically induced reduction in room temperature conductivity and in a transport mechanism that is characterized by tunneling (interaction of the electron gas waves between the primary particles) that is thermally stimulated. In blends with PAni, in pure PAni (insofar as it meets our dispersibility specifications), and in highly conductive PAc, purely metallic behavior is observed around room temperature. We are dealing with "organic metals."
7. If the electronic correlation between the primary metallic phases could be extended without barriers, there would be a dramatic increase in conductivity. This appears to be the case in oriented highly conductive PAc.
8. The primary particles in turn are frequently aggregated to spherical secondary particles. These are the particles present in high polymeric matrices, whereas in binderless aqueous and organic dispersions it is the 10 nm primary particles that are present.

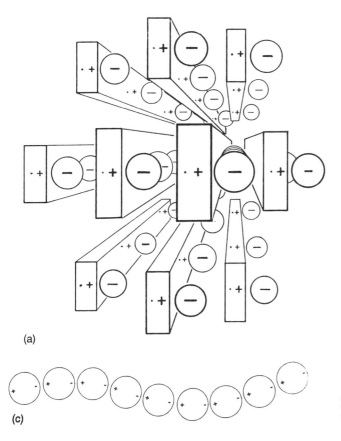

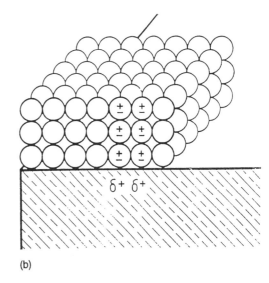

Fig. 19.64 Due to their anisotropic structure, ICP particles tend to agglomerate, forming fibril-like arrangements.

9. Dispersion of ICPs in solvents or polymer matrices does not alter the mechanisms underlying conductivity. Changes may occur, however (we have observed above all marked improvements up to a factor of 20), in the interactions between the particles, presumably as a result of better arrangement of the particles in relation to one another. This could in principle approximate to the process of orientation in highly conductive PAc.
10. The primary spherical structure of ICPs with mesoscopic metallic phases and their dispersibility (albeit with great difficulty owing to high surface energy) is the basis for processing through dispersion. Applications of ICPs are largely possible and to be expected in cases where the unusual combinations of properties that ICPs have to offer are called for.
11. The nonequilibrium thermodynamics description of multiphase dispersions (with PAni or other substances as the disperse phase) and the resulting explanation of the formation of dissipative structures is, on the basis of present knowledge, an adequate theoretical foundation for describing the phenomena observed.

This concise summary makes it clear that with the integrated materials science approach bringing together all disciplines we have been able to develop a logical and comprehensive picture of conductive polymers that meets present-day requirements. It has successfully accompanied the chemical work, helped in the strategy and interpretation of the physical investigations, and directed and interpreted the chemical, physical, and engineering aspects of dispersion, processing, and applications research. The successful forecasts of the early versions of this materials science approach and its logical coherence, which can be felt right through to the applications already being implemented, are a good basis for further work devoted to solving the questions still outstanding and aimed at finding exciting and socially useful new applications.

VII. PRACTICAL APPLICATIONS OF INTRINSICALLY CONDUCTIVE POLYMERS

On the basis of the innovative technical possibilities resulting from research into the processing properties of ICPs and the development of better and better processing technologies, new areas of application have been and are being opened up for the materials category of ICPs and their blends. This makes possible solutions to technical problems that are currently impracticable or unsatisfactory with conventional materials.

A large number of applications of conductive polymers are already in commercial use, although still in low volume, undergoing precommercial trials, or at the specific development stage. These include the following:

1. Dispersion paints containing polyaniline. These are used to create transparent electrically conductive or antistatic coatings on films or injection-molded products (see Section VII.A).
2. Anticorrosion paints containing polyaniline. As the first actively passivating anticorrosion system, a newly developed PAni primer could perform the important environmental protection function of conserving energy-intensive and raw-material-intensive assets (Section VII.B).
3. Transparent and semitransparent functional layers containing polyaniline, made from dispersions of pure polyaniline in organic or aqueous solvents. By this means it is possible to create highly transparent and ultrathin (200 nm) layers of polyaniline and apply them to a variety of substrates (Section VII.C).
4. EMI shielding. The high specific conductivity at the saturation concentration makes for a high degree of shielding from electromagnetic waves (Section VII.D).
5. Coatings used for through-plating of printed circuit boards.

Further applications using PAni dispersions, paints, and blends are conceivable or have already been tested experimentally, including

6. Membranes for gas separation.
7. "Intelligent" windows.
8. Electrochromism. When an electric potential is applied (and the corresponding current flows), polyaniline changes color between green, blue, and yellow/colorless.
9. Sensors.

A. Transparent Antistatic Coatings

By means of transparent PAni dispersion paints it is possible to coat injection-molded articles (by spray coating) and films, products, or semifinished products (by spray coating or immersion, continuous or discontinuous), resulting in transparent green antistatic products [82]

Depending on the PAni concentration it is possible to attain surface resistances of 10^9–10^2 Ω; coated films can be deep-drawn or thermoformed. The layer thickness is usually 1–2 μm or up to 10–20 μm. In this way it is possible to obtain (depending on the product to be coated) almost any desired slightly green tint or (green) transparency for products with antistatic or conductive properties (Fig. 19.65) (see also Chapter 38).

Waterborne dispersion paints leading to clear transparent antistatic coatings have been developed and are presently undergoing practical evaluation [84].

B. Corrosion Control

The use of conductive polymers in the particularly important field of protecting metal surfaces from corrosion has developed into another extremely important and very interesting future application for polyaniline.

1. Development of Corrosion Protection Using ICPs

Over 10 years ago DeBerry [85] recognized the possibility of anodic protection of metals, in 1985 he found that polyaniline applied electrochemically to steel prepassivated in a strongly acid solution gave a kind of anodic protection to the passivated metal. This substantially reduced the rate of corrosion in a solution containing sulfate and acid. Before the electrochemical deposition of the PAni there was already a passive oxide layer on the surface of the steel. The doped PAni layer was in electrochemical contact with the surface of the steel and evidently stabilized the oxide layer against dissolution or conversion to rust.

Subsequent studies by Troch-Nagels et al. [86] on electrochemical deposition of PAni and polypyrrole on mild steel showed that PAni, unlike polypyrrole, did not offer protection against corrosion. This finding was made more precise by Sekine et al. [87], who found in their investigations that PAni electrochemically deposited on the surface of steel offered only very limited protection against corrosion.

2. Corrosion Protection with Doped Polyaniline

We ourselves found in 1987 [88] that paint formulations containing PAni offered a certain improved protection against corrosion. Although further work with, among others, Elsenbaumer resulted in improvements [89], it did not bring a breakthrough.

Thompson and coworkers [83,90] found a reduction in corrosion in samples of steel that had been coated with neutral polyaniline (EB) and exposed to a salt solution (3.5% NaCl) and an acid environment (0.1 N HCl) after "doping" and applying a top coat. They found, however, that they were able to achieve good corrosion protection with only certain dopants; moreover, adhesion problems prevented them from obtaining reproducible and acceptable results.

All previous studies had suffered not only from a lack of understanding of a possible protection mechanism but also from a lack of practical feasibility. We therefore

Dispersion as the Key to Processing

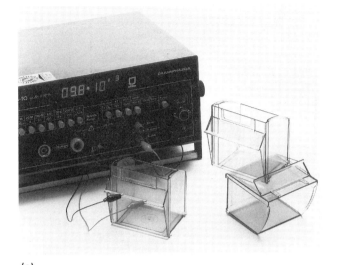
(a)

(b)

(c)

Fig. 19.65 (a) Using ultimately dispersed PAni in acylate coating systems, any plastic product can be coated with a transparent antistatic layer, here with spray coating, leading to a surface resistivity of about 10^4 Ω. (b) A polyester film continously coated on a film coating machine, then cut and dedrawn. (c) High performance loudspeaker using a PAni-acrylate coating on a polyester film for about a 1.5 m^2 sound generating area. (From Ref. 83.).

wanted both to clarify the mechanism (chemism) and to develop paints (primers) for practical use that would offer optimal protection against corrosion.

By means of repeated immersion of steel specimens in dispersions of doped polyaniline (Versicon), we succeeded [91] in coating the bright metal surfaces of various types of steel, and also copper and aluminum, with a layer of PAni and then bringing about targeted passivation. What makes this different from previous studies is the fact that in this case the polyaniline coating was applied to a bright metal surface that had not been electrochemically treated or treated with acid (passivated) and that this was done using a prepared dispersion of doped PAni and not by electrochemical means. We repeated the immersion coating process 5–20 times to obtain a polyaniline layer of suitable thickness. More detailed investigations revealed that the cause of the passivation lay not only in the polyaniline coating itself but also in the resulting formation of an oxide layer between the steel and the PAni coating. This became evident after removal of the entire polyaniline layer from the surface. This left behind a matte gray surface, and the passivation properties continued to exist without the polyaniline coating. The PAni layer alone brings about a "surface ennoblement" due to its potential.

These apparently astonishing results provided the motivation to take a closer look at the mechanism behind the corrosion protection behavior of doped and undoped polyaniline.

3. Studies of the Corrosion Process

To investigate the corrosion process we took various specimens of steel coated with polyaniline and exposed them for a considerable period to two different corrosive environments: a salt solution (a neutral 3.5% NaCl solu-

Table 19.6 Summary of Visual Observations During the Corrosion Test

	Initial stage (2 h to 3 days)	Intermediate stage (1–5 wk)	Final stage (8 wk)
In 0.1 N HCl			
D, e/s	Surface became black at 1–2 days	Considerable rust and filiform corrosion	Underfilm rusting
D, e/dP/s	Shiny surface	Still shiny	Bare steel covered by deep gray-black thin film with some shiny steel surfaces
D, e/nP/s	Dim gray	Small brown spots at several areas	Mild rust
In 3.5 % NaCl			
D, e/s	Rust at 3 h	Serious corrosion over epoxy film	Voluminous rust
D, e/dP/s	Light brown/black film	Very rusty	Mild attack to epoxy layer
D, e/nP/s	Shiny with white-gray film	Thin light brown film over bare steel	Mild attack on bare steel; some attack at epoxy surface

Source: Ref. 83.

tion saturated with carbon dioxide) and dilute hydrochloric acid (0.1 N HCl) [92].

Table 19.6 shows some of the results obtained at the end of the measuring period by means of visual examination of the three types of specimens. It emerged from the studies that of the specimens exposed to the HCl environment, those coated with polyaniline were clearly the best protected against corrosion and rust. In the case of the NaCl environment the corrosion protection afforded by polyaniline appeared to be less effective than in the specimens exposed to the HCl environment. An untreated control specimen corroded within only a few hours.

Together with the results in Table 19.6, electrochemical measurements of the corrosion potential and the corrosion current led to the realization that the cause of the corrosion protection is indeed the presence of the polyaniline on the metal.

The measurements and the results shown in Figs. 19.66 and Figs. 19.67 make it clear that there is a fundamental reduction in the corrosion rate in the specimens with doped polyaniline in HCl and for the specimens with neutral polyaniline in an NaCl environment.

All the studies of the corrosion process suggest that the corrosion protection is the result of the formation of a passivating oxide layer on the surface of the coated

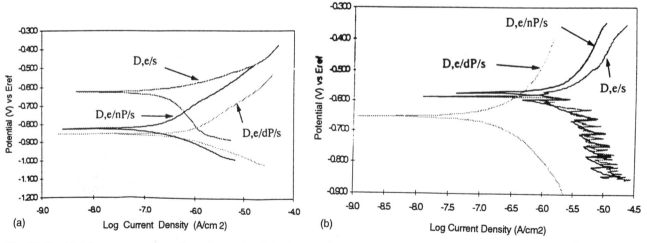

Fig. 19.66 Tafel measurements of specimens in of steel HCl (a) and NaCl (b). The diagrams show the corrosion rate curves for three different specimens (see text). (From Ref. 83.) The Tafel measurement method makes it possible to obtain information about the corrosion rate curve. Large cathodic and anodic polarizations result in curves for cathodic and anodic polarization in the individual corrosion process [93]. Extrapolation of these curves at their intersection yields both the corrosion potential and the corrosion curve.

Dispersion as the Key to Processing

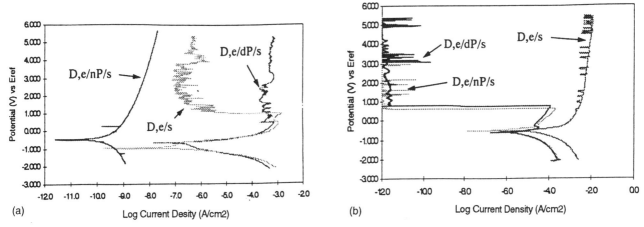

Fig. 19.67 Potentiodynamic polarization measurements of three specimens of PAni-coated steel in HCl (a) and in NaCl (b) (see text). (From Ref. 92.) Potentiodynamic polarization measurements reveal the tendency of an alloy to undergo active or passive behavior. In the specimens with polyaniline, (b) clearly shows the formation of a passive oxide layer. In a control specimen (uncoated) there is no sign of a passivation tendency.

material. This formation of the oxide layer reduces the corrosion rate by more than two orders of magnitude. It was found that the oxide layer is made up of $\gamma\text{-}Fe_2O_3$ on a very thin layer of Fe_3O_4, between the latter and the free steel [92,94]. The thickness of this layer is about 1 μm. This area of protection typically has an extent of up to 2 mm in the case of exposed areas (scratches).

All in all, it looks as if the metal oxide layer in combination with the polyaniline layer is responsible for the corrosion protection. Here the polyaniline acts as a cata-

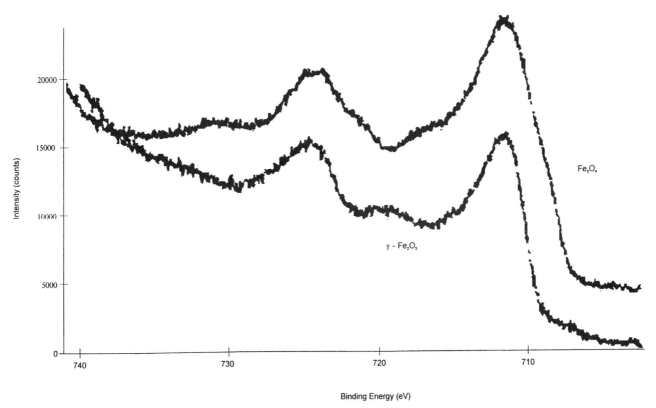

Fig. 19.68 X-ray photoelectron spectroscopic (XPS) analysis of passive iron oxide layers form in the presence of doped polyaniline. (From Ref. 83.)

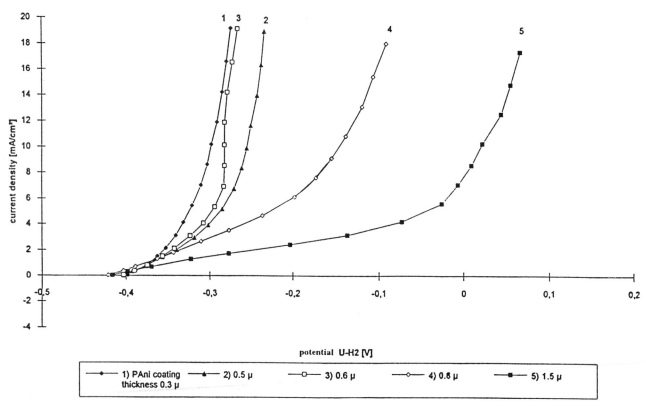

Fig. 19.69 Change in the course of corrosion as revealed by measurements of iron specimens coated with various thickness of polyaniline. (From Ref. 90.)

lyst in the formation and constant renewal of the protective layer during corrosive attack.

4. Summary of Initial Findings

Suitable coating of metals with pure polyaniline from dispersions and the subsequent interaction with the metal surface results in a significant shift in the corrosion potential of the coated metals in the direction of noble metals ("ennoblement"). This entire process can also be regarded as a "passivation of the metal by polyaniline" (see Figs. 19.68–19.72).

After the formation of the protective metal oxide layer, however, the metal specimen is durably passivated only as long as this metal oxide layer is present. The continued presence of the polyaniline coating on the metal surface permits ongoing repair of the metal oxide interface with the aid of the regenerative polyaniline layer.

5. Reaction Scheme for the Passivation of Metals by Polyaniline

In a further phase of research, we analyzed the chemism of the passivation mechanism [95]. We suspected, for example, that the passivation reaction that leads to the formation of the oxide layer might be due to a catalytic redox reaction of PAni. The reaction sequence responsible for the passivation process and the formation of the oxide layer was successfully evaluated.

Spectroscopic experiments (ultraviolet and infrared) and visible observations were carried out on two different kinds of coated polypropylene samples* in two different solutions. They showed that after a coated film of polypropylene in a quartz cuvette was placed in the solutions, the color of the polyaniline changed very quickly in the presence of iron and water.

The change in the color of polyaniline after contact with iron in the presence of water from the green to the yellow form (from green to yellowish-green and then to a pale yellow) can be understood as the reduction of polyaniline from the emeraldine salt to the leucoemeraldine salt. This requires an equivalent oxidation of iron to Fe^{2+}. The back-reaction in a neutral (IM NaCl) aqueous

---

* UV and visible spectrum of polypropylene films, 480 m, coated with PAni from dispersion—some containing 5% suspended iron powder, 10 μm diameter "pro analysi" made by reduction, others containing no iron powder—in a quartz cuvette, using two different solutions.

Dispersion as the Key to Processing

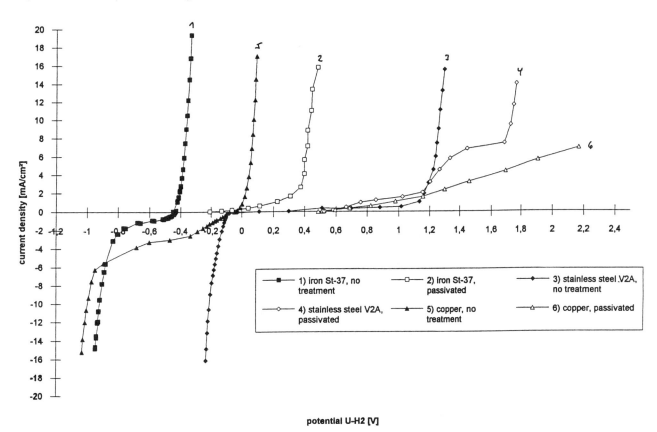

Fig. 19.70 Development of curves of corrosion density versus corrosion potential for various metals, untreated (original) and passivated. A marked shift in the corrosion potential can be seen, largely in connection with a reduction in the corrosion progress. The corrosion shift here is at least 100–200 mV. The most reproducible figures are up to 800 mV for iron and around 300 mV for stainless steel. In the case of copper the shift is even more pronounced and, as can be seen from the shallow cuve following passivation, the ennobling and passivation are almost ideal. (From Ref. 83.)

Fig. 19.71 An untreated iron plate (right) exhibits rust after a short time in salt water or after one anodic corrosion current measurement. The passivated plate (left; passivation was performed only on the lower half of the plate and the polyaniline layer was removed after passivation) exhibits no rust even after a current density measurement. (From Ref. 90.)

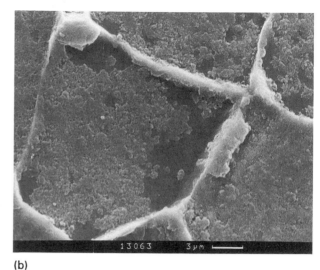

Fig. 19.72 (a) Scanning electron microscopic image showing that the first phase of passivation seems to be an etching step. Grain boundaries and crystallite orientation of the steel become visible. (b) SEM image showing that the second phase seems to be the deposition of an oxide, initially extremely thin. (c) SEM image showing that the passivation is achieved with an oxide layer about 1 μm thick coating the entire metal surface. (From Ref. 90.)

solution leads back to the normal polyaniline with its green color, although with a certain bluish component (emeraldine base or nigraniline) in the neutral environment. This back-reaction is performed with the help of oxygen. In an aqueous acid solution (pTs: 0.1 M p = toluene sulfonic acid–water solution) the back-reaction leads to the original nice PAni green. Without any additional iron powder, no color change can be seen.

In a degassed water solution, we also observed the reoxidation of the leucoemeraldine into the emeraldine salt, but, very interestingly, we could see gas formed at the surface of the film, which we suppose was hydrogen.

Further observations using paints containing polyaniline have shown that the occurrence of hydroxyl ions is very probable. Therefore the reaction scheme shown in Figs. 19.73 (reaction) and 19.74 (back-reaction) describes the entire passivation reaction sequence of polyaniline on the iron surface, showing that polyaniline acts as a redox catalyst and as a noble metal with respect to iron (or copper, stainless steel, aluminum, zinc, or others).

6. Practical Approach

In practice, coatings of pure PAni are unsuitable, largely because of problems with the adhesion of top coats (not so much because of adhesion to the metal surface) but also because of the coating technology, which is unusual and uneconomical in practical corrosion protection.

We have therefore developed various paint systems that are applied as primer (with a thickness of $\approx$ 20 μm) and satisfy the corrosion protection requirements in the

Dispersion as the Key to Processing

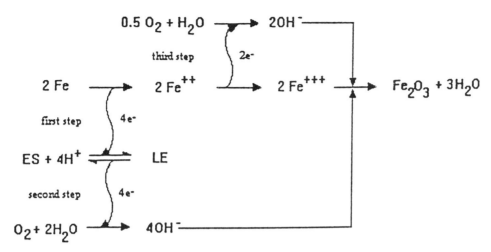

Fig. 19.73 Reaction scheme. The major steps of the passivation are af follows. Iron (FE) is oxidized by the metallic form of polyaniline, the "emeraldine salt" (ES), which in return is reduced to the "leucoemeraldine" (LE). Salt, at which time Fe^{2+} is further oxidized by oxygen to Fe^{3+}, oxygen also reoxidizes the LE form back to the ES form, which is thereby enabled to continue to passivation, hence is acting as a catalyst.

various areas of application. One- and two-component systems, i.e., physically and chemically hardening, are available. To this end we have also developed and tested suitable top coat combinations. The primers perform the function of ennoblement and passivation, the top coats that of surface sealing. Adhesion on the metal and to top coats is in most cases very good; undermining and blistering are no longer observed, and corrosion protection is excellent. This has also been confirmed by independent studies [96].

A more detailed discussion of the practical aspects can be found in Ref. 97.

C. Conductive Layers of Pure PAni

Because of the possibility of coating substrates of various kinds with pure polyaniline by means of a simple process, it became possible to produce thin conductive layers (100–400 nm). The conductive layer consists of an intrinsically conductive polymer (specific conductivity ≈1–10 S/cm). Owing to the small thickness of the coating this results in a surface resistance of 10^3–10^8 Ω. This coating displays very good transparency with a slight green tint.

For coating purposes this polymer is dispersed in small concentrations in an organic or aqueous solvent system, yielding a thin dispersion that enables the materials to be coated by spraying or immersion. The coating process is completed by subsequent drying, during which the solvent evaporates completely. This process is outstanding for its simple technology and the high yield of the dispersion.

With only 1 L of the dispersion it is possible to coat about 100 m^2 of the surface of various semifinished or finished products such as films, fibers, panels, sections, textiles, and paper, made of a wide variety of materials. There may be differences in bonding strength and abrasion resistance. Abrasion resistance on PC, PET copolymers, and ABS is good, whereas on PET homopolymers, PVC, PE, or aluminum it has to be described as moderate.

Applications of conductive coatings of pure PAni are initially to be seen in basic research and possibly later in gas separation, in sensors, and in electrochromic systems.

D. Electromagnetic Interference Shielding

Both the metallic properties of electrically conductive plastics and an uncomplicated processing technology in the form of paints, thin films, or even the possibility of direct processing to electrically conductive plastic housings suggest the possibility of using these plastics as materials for shielding against electromagnetic interference (EMI).

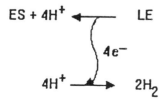

Fig. 19.74 Back-reaction under oxygen-free conditions.

1. Various Methods of EMI Shielding

Two fundamentally different methods of obtaining electrically conductive plastic housings with a sufficiently high conductivity to provide EMI shielding have been developed to date. One method consists in subsequent metallization, e.g., by flame spraying of zinc. The high cost and length of time involved makes this method unattractive.

An alternative to this would be to make a thermoplastic polymer conductive by using metal flakes or fibers measuring 10–50 μm. Here it proved difficult to ensure uniform distribution of the metal particles during processing. The resulting deficit of conductive admixture at the corners, edges, and surfaces of the polymers leads to defective shielding, which made it unattractive for commercial application.

A vast improvement in shielding performance would be achieved, however, by thermoplastic blends containing a uniformly distributed conductive phase. For most technical applications; however, the conductivity would have to be increased considerably and the mechanical properties optimized. The attempts to produce such a conductive plastic for EMI shielding by incorporating carbon black in thermoplastic polymer systems have been abandoned owing to inadequate conductivity. Such compositions do not possess sufficient conductivity or appropriate mechanical properties for industrial applications.

To circumvent these disadvantages and nevertheless obtain useful shielding by plastics, highly conductive polymer blends using polyaniline have been developed for EMI shielding purposes. By incorporating ICPs in a matrix polymer, e.g., PVC, PMMA, or polyester, conductivity figures of around 20 S/cm, and in some cases up to 100 S/cm, can be achieved. Such conductivity figures, which are higher than has so far been achieved by incorporating carbon black in polymers, promise a very high standard of EMI shielding. The shielding effect is up to 25 dB higher than with carbon black compounds and lies, depending on the frequency of the electromagnetic interference, in the region of 40–75 dB for both near and far field. A considerable improvement in mechanical values is needed, however, and, as shown below, preferably conductivity levels that are higher by one to two orders of magnitude.

2. Effectiveness of EMI Shielding

We systematically investigated the extent to which the shielding effect depends on conductivity and coating thickness [98]. To obtain information about the shielding effectiveness of a conductive material, the test material is placed between a suitable source of radiation, a source of smooth electromagnetic waves, and a suitable detector. The relationship between the received and the transmitted radiation is then calculated.

In general the shielding effectiveness (SE) of a material is defined as the ratio of the transmitted (P_o) to the arriving (P_i) radiation energy.

$$\text{SE} = 10 \log \left(\frac{P_i}{P_0}\right) = 20 \log \left(\frac{E_i}{E_0}\right) \quad [\text{dB}] \quad (19)$$

Useful shielding effectiveness values for commercial applications in electronic housings are in excess of 40 dB at frequencies of 1 GHz. For military applications and for near-field shielding requirements, even better performance, in the region of 80–100 dB, is required.

When discussing shielding effectiveness it is necessary to consider two radiation zones: the near-field and the far-field (or smooth wave) zones. The distinction between them lies in the distance from the source of radiation. If the distance from source to shielding is less than one-sixth of the free path wavelength of the radiation to be shielded, the radiation is dominated by the smaller multipole components of the source field and is described as "within the near-field zone." Above this zone it is in the far-field zone. A more detailed consideration of the problem can be found in Ref. 98.

As long as the conductive component is uniformly and well dispersed in the polymer matrix, the shielding effectiveness proves in theory and in practice (Fig. 19.75) to be a function of conductivity and thickness. Shacklette et al. [98] developed a correlation function for this.

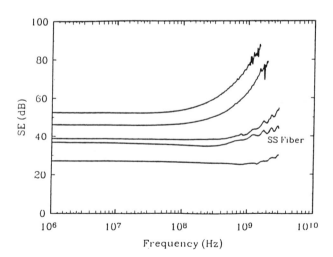

Fig. 19.75 Measurements of shielding effectiveness under various conditions on four different samples of compounds of conductive polymers (all samples contain dispersed ICP as the conductive phase; from top to bottom, Nos. 4, 10, 7, and 9 from Table 19.7). They are compared with a 10% stainless steel fiber compound. (From Ref. 95.)

3. Effectiveness of Near-Field and Far-Field Shielding

The conductivity of conductive plastic coatings is in the region of 0.1–10 S/cm. The full expression for far-field shielding effectiveness therefore possesses two special limits in the region of megahertz frequencies. These depend on whether the frequency of the radiation to be shielded is higher or lower than the frequency at which the coating thickness d is equal to the classic surface depth δ. The classic surface depth is the depth to which the radiation of frequency ω penetrates into the material and thereby undergoes a reduction in intensity to the $(1/e)$th part of its original strength. This surface depth is described by

$$\delta = \left(\frac{2}{\mu_0 \omega \sigma}\right)^{h} \quad (20)$$

where $\mu_0 = 4\pi \times 10^{-7}$ H/m is the permeability of the charge-free space.

A sample is described as "electrically thin" if $d \ll \delta$ and as "electrically thick" if $d \gg \delta$. The crossover frequency ω_c at which $d = \delta$ is determined from Eq. (20) as

$$\omega_c = \frac{2}{\delta \mu_0 d^2} \quad (21)$$

For frequencies that are very much lower than $\omega_c = 2/\delta \mu_c d^2$, the shielding effectiveness in the case of an electrically thin shielding becomes independent of the frequency, and Eq. (19) is reduced to

$$SE = 10 \log\left(\frac{1 + Z_0 \sigma d}{2}\right) = 20 \log \left(1 + \frac{Z_0}{2R_s}\right) \quad [dB] \quad (22)$$

Here Z_0 is the impedance of the free space (377 Ω) and R_z is the surface resistivity ($= 1/\sigma d$). For frequencies higher than ω_c ($d > \delta$), the effectiveness of far-field shielding can be approximated by

$$SE = 10 \log\left(\frac{\sigma}{16 \omega \epsilon_0}\right) + 20 \frac{d}{\delta} \log(e) \quad [dB] \quad (23)$$

where $\epsilon_0 = 10^7/4\pi c^2$ is the dielectric constant of the free space (c is the speed of light).

The first term in Eq. (23) is the contribution to shielding, taking account of the simple reflection wave that impinges on the front and rear of the surface of the envelope. The second term represents the attenuation of this wave by absorption during its passage through the envelope. At high frequencies ($\omega \gg \omega_c$), the second term dominates and the shielding capacity increases with frequency.

For near-field shielding, just as for far-field shielding, it is possible to approximate a general expression by means of the limits of the electrically thin and electrically thick samples. At the limit of an electrically thick shielding ($d/\delta \gg 1$ or $\omega \gg \omega_c$), near-field shielding effectiveness can be approximated as

$$SE = 10 \log\left(\frac{c^2 \sigma}{16 \epsilon_0 \omega^3 r^2}\right) + 20 \frac{d}{\delta} \log(e) \quad [dB] \quad (24)$$

The first term on the right-hand side of Eq (24) is interpreted as the shielding due to reflection, and the second as the shielding due to absorption. It is expected that the shielding due to reflection will decrease by about 30 dB per decade of frequency increase. The entire near-field shielding will naturally never be smaller than the far-field shielding. As the wavelength decreases, the near-field shielding zone approaches the far-field shielding.

Here too we find a difference in frequency dependence for electrically thick and electrically thin samples. The approximation $d/\delta \ll 1$ (or $\omega \ll \omega_c$) leads to

$$SE = 20 \log\left(\frac{c}{2\omega r} Z_0 \sigma d\right)$$
$$= 20 \log\left[\frac{c}{2\omega r}\left(\frac{Z_0}{R_s}\right)\right] \quad [dB] \quad (25)$$

Unlike Eq. (24), this equation shows a decrease of 20 dB per decade of frequency increase and 20 dB per decade of surface resistivity, $R_s = 1/\sigma d$. It may therefore be concluded that the most effective near-field screening can be expected in the region of 20–30 dB for every decade of frequency increase and 10–20 dB for every decade of increase in specific resistance.

Figures 19.76 and 19.77 show theoretically expected values of shielding effectiveness [after Eqs. (24) and (25)] compared with measured values. Figure 19.76 shows that the highest and therefore best value for shielding effectiveness was obtained for far-field shielding. At low frequencies the measured points and the predicted curve agree exactly. At high frequencies, by contrast, the measured shielding capacity (data points) tends to increase with frequency earlier than theoretically predicted (curves). Since this deviation is very small, it may be assumed that the shielding effectiveness of this sample increases at higher frequencies. One can in fact expect a frequency-dependent shielding when a conductive material is dispersed in a matrix.

Also shown is a comparison of the results of the impacts of a near field and a far field on the same sample. It will be seen that, as already outlined in the theoretical consideration, the shielding effectiveness is higher for the near-field radiation. The shallow curve that approximates to the results of the near-field measurements is a

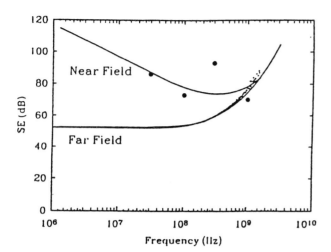

Fig. 19.76 Theoretically expected values (curves) of far-field and near-field shielding compared with actually measured values (points) of sample 4. (After Ref. 98.)

Table 19.7 Near-Field Shielding Effectiveness (SE$_n$) of Various PVC/ICP Blends 3.2 mm thick. Measured at the Four ASTM Standard Frequencies (30, 100, 300 MHz, and 1 GHz) Using the Two-Chamber Box Method and Compared with Far-Field Shielding Effectiveness (SE$_f$) at 1 GHz

Sample No.	σ (S/cm)	SE (dB)				SE$_f$ (dB) at 1 GHz
		30 MHz	100 MHz	300 MHz	1 GHz	
1	0.53	71	60	60	26	—
2	0.58	70	59	62	27	—
3	2.0	75	63	72	36	—
4	7.45	86	73	93	70	78.5
5	6.63	83	77	93	72	71
6	1.42	74	62	72	36	48.4
7	0.98	73	60	66	39	41.3
8	0.59	65	60	62	25	—
9	0.18	56	52	61	17	25.6
10	3.74	85	77	88	65	63

plot of a semiempirical relationship that was calibrated to fit the data obtained by the ASTM method. To this end the effective distance of the shielding from the source (r) was taken as the parameter to be adapted. The data from several samples could be fit relatively well by assuming a value of $r = 2.9$. Near-field data for various ICP blends obtained from a two-chamber box are summarized in Table 19.7 [95].

Figure 19.77 illustrates that the measured conductivity ranges from 0.3 to 8 S/cm. The data generally fit the combined semiempirical expressions. The tendency to display a sizable anomaly at 300 MHz is evident. There is likewise a tendency to show a lower value than expected at 1 GHz. At low frequencies (30 and 100 MHz) the value of the measured near-field shielding lies closer to the expected value. The mean gradient of the lines drawn through the two points on the basis of the data from Table 19.7 is -18.2 dB per decade of frequency (with a standard deviation of 6.6 dB). This value is remarkably close to the figure of -20 dB per frequency decade obtained from Eq. (25). Similarly, the data at 30 MHz yield the hoped-for dependence of surface resistivity of -19.4 dB per decade (with a standard deviation of 1.1) [95].

Near-field data of various ICP blends are summarized in Table 19.7. The near-field shielding values display the expected behavior. The anomaly of a relatively high figure at the frequency of 300 MHz is an exception. This, however, is not confined to our data alone. This effect is well known from the literature on measurements with the two-chamber box with various materials. It may be assumed that this is an effect that is due to the specific choice of dimensions of the "standard" two-chamber box used.

It was expected that at high frequencies the value of the near-field shielding would approach the figure for far-field shielding. In almost all samples for which we investigated far-field shielding, the shielding at 1 GHz was slightly higher than was measured for the near field at the same frequency. It is probable that the reason for this deviation lies in a weak electric contact between the sample and the corners of the test arrangement within the two-chamber box used [95].

4. Consequences

It emerged from the theoretical and practical studies [98] that it is possible to achieve very high values for shielding performance by using intrinsically conductive poly-

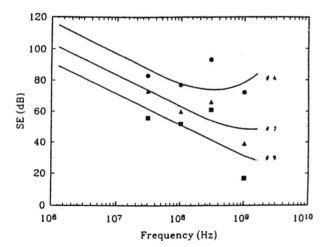

Fig. 19.77 Near-field shielding effectiveness data for samples 4, 7, and 9, as theoretically predicted for the ranges $\omega > \omega_c$ and $\omega < \omega_c$ and as measured in the two-chamber box.

mers incorporated in thermoplastic polymers. This was made possible by improving the electrical contacts and reducing the zones of poor conductivity in the ICP blends developed, resulting in outstanding electrical conductivity.

However, in view of the blends available today with conductivities of 10–100 S/cm, the technically desirable shielding figures still require coating thickness of 2–3 mm. It is therefore our task to develop more conductive blends with conductivity levels that are better by one to two orders of magnitude that also display the mechanical properties of technical polymers.

E. Final Remarks: Applications of PAni via Integrated Materials Science

From all the possible applications of conductive polymers that are already being implemented or are foreseeable and are constantly being researched and improved, one can get a good idea of the potential that awaits this category of materials. They show that research into this materials category is making an important contribution to the development of new and modern materials and that the possible applications are far from being exhausted. If these (theoretical and practical) findings are accepted by materials researchers, and they *will* become accepted, ICPs have a great future ahead of them.

In view of the unusual combinations of properties displayed by ICPs, and above all the commercially available PAni, it is to be expected that further applications will emerge that are not even under discussion at present.

REFERENCES

1. A. G. Green, and A. E. Woodhead, *J. Chem. Soc. 1910*:2388.
2. G. Natta, G. Mazzanti, and P. Corradine, *Alti Accad. Naz. Lincei Rend. Sci. Fis. Mat. Nat. 25*:2 (1958).
3. L. B. Luttinger, *Chem. Ind. (Lond.) 1960*:1135; *J. Org. Chem. 27* (1962).
4. H. Shirakawa, and S. Ikeda, *Polym. J. 2*:231 (1971).
5. Saechtling, *Kunststoff Taschenbuch*, 22nd ed., Carl Hanser, Berlin 1983, p. 58.
6. L. W. Shacklette, T. R. Jow, and M. Maxfield, *Synth. Met. 28*(1,2):C655 (1988).
7. B. Wessling, *Synth. Met. 45*:119–149 (1991).
8. D. Stauffer, *Introduction to Percolation Theory*, Taylor, and Francis, London, 1985.
9. R. G. Gilg, in *Elektrisch leitende Kunststoffe* (H. J. Mair and S. Roth, eds.), Carl Hanser, Berlin, 1986, pp. 55–76.
10. N. Probst, Conductive compounds on the move, *Eur. Rubber J.*, Sonderdruck, (1984).
11. E. C. Ketjenblack, *Technisches Merkblatt*, Akzo Chemie, Chicago, 1984.
12. B. Wessling, and H. Volk, *Synth. Met. 15*:183–193 (1986).
13. B. Wessling, Electrically conductive polymers, *Kunststoffe/Ger. Plast.* 10 (1986).
14. J. C. W. Chien, *Polyacetylene—Chemistry, Physics, and Material Science*, Academic, New York, 1984.
15. T. Ito, H. Shirakawa, and S. Ikeda, *J. Polym. Sci. Polym. Chem. Ed. 12*:11 (1974).
16. H. Shirakawa, Y.-X. Zhang, T. Okuda, K. Sakami, and K. Akagi, *Synth. Met. 65*:93–101 (1994).
17. B. Vollmert, *Grundriss der Makromolekularen Chemie, Vol. II*, Karlsruhe, 1982, p. 55.
18. M. Mooney, *J. Collatl. Sci. 6*:162 (1951).
19. S. Bhaguwan, O. Tripathy, S. De, S. Sharma, and K. Ramamurthy, *Polym. Eng. Sci. 28*(10):648 (1988).
20. C. Kröhnke, V. Enkelmann, and G. Wegner, *Angew. Chem. 92*:941 (1980).
21. G. Wegner, *Angew. Chem. 93*:352 (1981).
22. J. Rühe, T. A. Ezquerra, M. Mohammadi, V. Enkelmann, F. Kremer, and G. Wegner, *Synth. Met. 28*(1,2):C127 (1989).
23. A. J. Epstein, and A. G. MacDiarmid, *Mol. Cryst. Liq. Cryst. 160*:165 (1988).
24. G. Nimtz, P. Marquardt, and B. Mühlschlegel, *Solid State Commun. 65*:539–542 (1988).
25. B. Wessling, *Synth. Met. 41–43*:1200–1216 (1991).
26. Y. Park, M. Druy, C. Chiang, A. G. McDiarmid, and A. Heeger, *J. Polym. Lett. Ed. 17*:195 (1979).
27. B. Wessling, Hiesgen, and Meissner, *Acta Polym. 44*:132–134 (1993).
28. B. Wessling, *Makromol. Chem. 185*:1265–1275 (1984).
29. A. Andreatta, Y. Cao, J. C. Chiang, A. J. Heeger, and P. Smith, *Synth. Met. 26*:383 (1988).
30. B. Wessling, Br. Patent Appl. GB-OS 2 214 511, (1989).
31. B. Wessling, *Kunststoffe/Ger. Plast. 76*:930 (1986).
32. B. Wessling, *Synth. Met. 27*:A83 (1988).
33. A. Tanioka, A. Oobayashi, Y. Kageyama, K. Miyasaka, and K. Ishikawa, *J. Polym. Sci., Polym. Phys. Ed. 20*:2197 (1982).
34. B. Wessling, H. Volk, W. R. Mathew, and V. G. Kulkarni, *Mol. Cryst. Liq. Crist. 160*:205 (1988).
35. E. K. Sichel (ed.), *Carbon Black Polymer Composites*, Marcel Dekker, New York, 1982.
36. B. Wessling, in *Elektrisch leitende Kunststoffe* (H. J. Mair and S. Roth, eds.) Carl Hanser, Berlin, 1989, p. 483.
37. B. Wessling, and H. Merkle, unpublished results, graduate paper H. Merkle, FH (Technical College), Hamburg, 1988.
38. W. Y. Hsu, W. G. Holtje, and J. R. Barkley, *J. Mater. Sci. Lett. 7*:459 (1988).
39. R. Strey, G. Porte, and P. Bassereau, *Langmuir 6*:1635–1639 (1990).
40. M. Kahlweit, R. Strey, and G. Busse, *J. Phys. Chem. 94*:3881 (1990).
41. A. Andreatta, Y. Cao, J. C. Chiang, A. J. Heeger, and P. Smith, *Synth. Met. 26*:383 (1988).
42. B. Wessling, *Synth. Met. 41–43*:1057–1062 (1991).
43. S. Wu, *Polymer Interface, and Adhesion*, Marcel Dekker, New York, 1982, p. 96.

44. L. Shacklette, *Synth. Met.* 65:123–130 (1994).
45. S. Wu, *J. Polym. Sci., Polym. Phys. Ed.* 21:699 (1983).
46. R. J. M. Borggreve, R. J. Gaymans, J. Schuijer, and J. F. Ingen Housz, *Polymer* 28:1489 (1987).
47. B. Wessling, *Macromol. Symp.* 78:71–82 (1991).
48. B. Wessling, *Adv. Mater.* 5(4):300–305 (1993).
49. I. Prigogine, *Angew. Chem.* 90:704 (1978).
50. G. Nicolis, and I. Prigogine, *Self-Organization in Non-Equilibrium Systems*, Wiley, New York, 1977.
51. B. Wessling, *Z. Physik. Chem.* 191:119–135 (1995).
52. S. Rwei, S. Horwatt, I. Manas-Zloczower, and D. Feke, *Int. Polym. Proc.* VI:98–102 (1991).
53. *Encyclopedia of Polymer Science, and Engineering* Vol. 13, Wiley, New York, 1988, p. 441.
54. B. Wessling, *Polym. Eng. Sci.* 31(16):1200–1206 (1991).
55. M. Kahlweit, R. Strey, and G. Busse, *J. Phys. Chem.* 94:3881 (1990).
56. H. Shirakawa, Y. Zhang, T. Okuda, K. Sakamaki, and K. Akagi, *Synth. Met.* 65(2–3):93–101 (1994).
57. A. Epstein, J. Jos, R. Kohlman, G. Du, A. G. MacDiarmid, E. Oh, J. Min, J. Tsuhanoto, H. Kaneko, and J. Payet, *Synth. Met* 65:149–157 (1994) and references therein.
58. Y. Park, C. Yoon, B. Na, H. Shirakawa, and K. Akagi, *Synth. Met.* 41:27 (1991).
59. Y. Park, *Bull. Am. Phys. Soc.* 36:638 (1991).
60. H. Kaneko, and T. Ishiguro, *Synth. Met.* 65:141–148 (1994).
61. Z. Wang, C. Li, E. Scher, A. MacDiarmid, and A. Epstein, *Phys. Rev. Lett.* 66(13):1745–1748 (1991).
62. Proc. Workshop on the Metallic Phase of Conducting Polymers, *Synthetic Metals.* 65(2–3), 1994.
63. A. Kaiser, in *Electronic Properties of Conjugated Polymers*, Vol. 2, H. Kuzmany et al., eds.), Springer, Berlin, 1987, pp. 2–11.
64. A. G. MacDiarmid, and A. J. Epstein, *Synth. Met.* 65(2–3):150 (1994).
65. A. G. MacDiarmid, and A. J. Epstein, *Synth. Met.* 65(2–3):103 (1994).
66. B. Wessling, *Makromol. Chem.* 185:1265–1275 (1984).
67. A. J. Epstein et al., *Synth. Met.* 65(2–3):151 (1994).
68. P. Marquardt, G. Nimtz, and P. Mühlschlegel, *Solid State Commun.* 65:539–542 (1988).
69. G. Nimtz, P. Marquardt, and H. Gleiter, *J. Cryst. Growth* 86:66 (1986).
70. G. Nimtz, A. Enders, P. Marquardt, R. Pelster, and B. Wessling, *Synth. Met.* 45:197–201 (1991).
71. P. Marquardt, and G. Nimtz, *Phys. Rev. B* 43:12,245 (1991).
72. R. Pelster, G. Galeczki, G. Nimtz, and P. Pissis, *J. Non-Cryst, Solids* 131:238 (1991).
73. R. Pelster, P. Marquardt, G. Nimtz, A. Enders, H. Eifert, K. Friederich and F. Petzoldt, *Phys. Rev. B* 45:8929–8933 (1992).
74. B. Wessling, *Synth. Met.* 45:119 (1991).
75. R. Pelster, G. Nimtz, and B. Wessling, *Phys. Rev. B* 49(18):12,718–12,723 (1994).
76. H. Looyenga, *Physika* 31:401 (1965).
77. K. Yamamoto and H. Namikawa, *Jpn. J. Appl. Phys.* 28:2523 (1989); 31:3619 (1992).
78. C. K. Subramaniam, A. B. Kaiser, P. W. Gilberd, and B. Wessling, *J. Polym. Sci.* B31:1425–1430 (1993).
79. B. Wessling, in *Electronic Properties of Conjugated Polymers*, Vol. 3 (H. Kuzmany et al. eds.), Springer, Berlin, 1989, p. 453.
80. A. B. Kaiser, C. K. Subramaniam, P. W. Gilberd, and B. Wessling, *Synth. Met.* 69:197–200 (1995).
81. C. K. Subramaniam, A. B. Kaiser, P. W. Gilberd, C.-J. Liu, and B. Wessling, Conductivity and thermopower of blends of polyaniline with insulating polymers (PETG and PMMA), *Sol. State Commun.*, 97(3):235 (1996).
82. B. Wessling, Elektrisch leitfähige polymere, *Kunststoffe 1990*:3.
83. D. A. Wroblewski, B. C. Benicewicz, K. G. Thompson, and C. J. Bryan, *Polym. Reprints* 35:265 (1994).
84. Zipperling Kessler & Co, unpublished results, 1995.
85. D. W. DeBerry, *J. Electrochem. Soc.* 132:1022 (1985).
86. G. Troch-Nagels, R. Winand, A. Weymeersch, and L. Renard, *J. Appl. Electrochem. Soc.* 139:756 (1992).
87. I. Sekine, K. Kohara, T. Sugiyama, and M. Yuasa, *J. Electrochem. Soc.* 139:3090 (1992).
88. Zipperling Kessler & Co, Patent Appl. WO 88/00 798 (1987).
89. Allied Signal/Zipperling Kessler Patent Appl. WO 93/00543.
90. K. G. Thompson, C. J. Bryan, B. C. Benicewicz, and D. A. Wroblewski, Los Alamos Natl. Lab. Rep. LA-UR-92-360.
91. B. Wessling, *Adv. Mater.* 6:226 (1994).
92. W.-K. Lu, R. L. Elsenbaumer, and B. Wessling, *Synth. Met.* 71:2163–2166 (1995).
93. D. A. Jones, *Principles and Prevention of Corrosion*, Macmillan, New York, 1992.
94. K. Asami and K. Hashimoto, *Corros. Sci.* 17:559 (1977).
95. B. Wessling, *Mater. & Corros,* 47:439 (1996).
96. Dechema, Tests on "Passivated" Steel Specimens, Test report, Dechema, Frankfurt, Germany.
97. Ormecon Chemie, Data sheets on Corrpassiv and Correpair.
98. L. W. Shacklette, N. F. Colaneri, V. G. Kulkarni, and B. Wessling, EMI shielding of intrinsically conductive polymers, Proc. 49th Annual Tech. Conf. SPE.
99. D. E. Stutz, T. Atterbury, and W. F. Scahrenburg, *Methods and apparatus for electrically* contacting a material specimen to the conductors of a coaxial cable, U.S. Patent 4,281,284 (July 28, 1981).
100. A detailed discussion of the insolubility of conductive polymers and their thermodynamic background: B. Wessling (to be published); Internet http://www.ormecon.de/Research/soludisp
101. B. Wessling, *J. Phys. II France* 6:395–404 (1996).

20
Electrochemistry of Conducting Polymers

Karl Doblhofer
Fritz-Haber-Institut der Max-Planck-Gesellschaft, Berlin, Germany

Krishnan Rajeshwar
The University of Texas at Arlington, Arlington, Texas

I. INTRODUCTION

A. General Considerations

Conducting polymers contain electronic states [1–6] that can be reversibly occupied and emptied with electrochemical techniques. In fact, the electrochemistry of conducting polymers has enormous potential for wide-ranging practical applications, e.g., in batteries, displays, etc. These reasons have rendered the electrochemistry of conducting polymers a most actively investigated research field [7–15].

Electrochemistry provides methods for preparing conducting polymers in the convenient form of thin films on electrode (metal, possibly also carbon or semiconductor) surfaces. These preparative aspects are not discussed in this chapter. Instead, the resulting system consisting of the coated electrode immersed in an electrolyte, i.e., the arrangement metal (Me)/polymer(poly)/electrolyte (S), constitutes the basis for our discussions. Thus, the aim may be to oxidize or reduce the polymer film itself (which may be done in an inert electrolyte by electron exchange with the metal electrode), or the polymer film may act as a mediator for the oxidation or reduction of a depolarizer dissolved in the contacting electrolyte. The two approaches are illustrated in Fig. 20.1.

B. The Morphology of Polymeric Phases

In the discussion that follows, the general assumption is made that the polymers are homogeneous phases. This assumption is frequently an oversimplification of the true situation. Crystalline phases, fibrils, etc. [7,16–18] may be present that require appropriate modification of the models and methods presented.

C. Redox vs. Conjugated Polymers

The accepted mechanism of electron transfer across the redox sites in the polymer is redox electron hopping [19–26], which is discussed in Section IV.A. The electronic donor and acceptor states that are involved may have identical energy levels (disregarding the thermal fluctuations). Typically, these are polymers with bound redox couples, such as $Os(bipy)_3^{2+/3+}$. Such polymers are termed redox polymers. In other systems, the energies of the states are spread over a range of levels. This

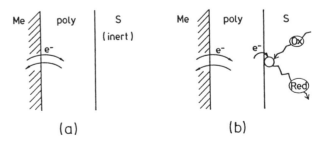

Fig. 20.1 (a) Electrochemical oxidation/reduction (e^-) of a conducting polymer layer (poly) on a metal electrode (Me); S is an inert electrolyte. (b) Electron transfer from a metal electrode to a depolarizer ("Ox", "Red") in the electrolyte, mediated by the conducting polymer.

is normally the case with "conjugated polymers" such as polypyrrole (PPy/PPy$^+$), where the electronic energy levels depend on the conjugation length and the site density [1–5]. Examples of a redox polymer and conjugated polymer are presented in Fig. 20.2.

D. Charge-Compensating Ion Fluxes

During electrochemical oxidation/reduction of the redox sites, the principle of electroneutrality coupling requires that a flux of ions into or out of the polymer film provide for internal charge compensation [7,8,27–31]. This process is shown schematically in Fig. 20.3 for the reduction of the Ru(bipy)$_2^{3+}$ polymer of Fig. 20.2. Note that this polymer constitutes an example of an anion exchanger. Thus, charge compensation for the injected electronic charge is likely to proceed by ejection of the counterions present at a large concentration in the polymer matrix. However, this is not necessarily always and/or completely the case. In principle, charge compensation could be provided as well by an influx of cations from the electrolyte. In particular, in nonequilibrium situations, e.g., following a potential step, the ions with the larger mobility will initially provide charge compensation, while ion-partitioning and ion-exchange equilibria will be established later. Note that the conjugated polymer, Fig. 20.2b, is dominantly a cation exchanger. Charge compensation following oxidation/reduction of the polymer matrix will thus be largely associated with cation ejection or influx. The concentration and mobility of mo-

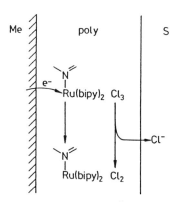

Fig. 20.3 Schematic representation of the ejection of an anion (Cl$^-$) associated with the electrochemical reduction of the redox polymer of Fig. 20.2a.

bile ions in the conducting polymer are thus matters of great importance and are discussed further below.

Next, we briefly review the electrochemical methods that have been applied to the study of conducting polymer films on electrode surfaces.

II. SURVEY OF ELECTROCHEMICAL METHODS

The arsenal of electrochemical techniques that can be applied to the study of conducting polymer films on electrode surfaces is indeed broad and versatile, as the family tree in Fig. 20.4 indicates. The scope of these techniques is further enhanced when a spectroscopic probe is added as in the (thin-layer) spectroelectrochemical approach discussed later. We briefly examine each of these techniques in turn. With the amount of material involved here, it is not feasible to treat each topic in an exhaustive manner. Instead, only the underlying principles, typical experimental arrangements, and the type of information obtained in each instance are reviewed. The reader is referred to the literature cited for more extensive discussions.

A. Electrochemical Instrumentation and Cells

As with any experiment, careful attention must be paid to cell design, electrode placement details, etc. In general, electrochemical experiments have become user-friendly thanks to the advent of computers and the availability of commercial instruments and accessories. Nonetheless, some general guidelines can be given as to the choice of a particular experimental arrangement. In the vast majority of cases, we are interested only in the processes at the conducting polymer film/support electrode. This is called the working electrode (WE) in what follows. For techniques wherein only a negligible

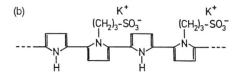

Fig. 20.2 (a) A redox polymer consisting of a poly(vinyl pyridine) matrix; a fraction of the pyridyl groups are covalently bound to a redox ion, Ru(bipy)$_2^{2+/3+}$. The remaining pyridyl groups are quaternized. The Cl$^-$ ions symbolize the charge-compensating counterions. (b) The conjugated polymer poly(N-sulfopropylpyrrole-co-pyrrole).

Electrochemistry of Conducting Polymers

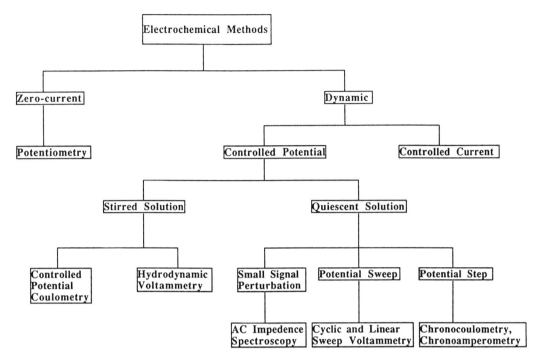

Fig. 20.4 A family tree for electroanalytical measurement techniques.

current is drawn across the cell (e.g., potentiometry), two electrodes suffice to complete the electrochemical cell. The second electrode is chosen to be an ideal nonpolarizable electrode of known potential called a reference electrode (RE). A variety of such electrodes are commercially available. For use with aqueous electrolytes, a saturated calomel electrode (SCE) or an Ag/AgCl/satd KCl electrode can be used. In nonaqueous solvents, a silver wire in contact with 0.1 M $AgNO_3$ dissolved in the particular solvent (e.g., acetonitrile) can be used and denoted as the Ag/Ag^+ reference. Figure 20.5 illustrates the relationship of these reference electrode potentials to the standard hydrogen electrode (SHE), which is arbitrarily assigned a value of 0.000 V on the redox potential (standard reduction potential) scale.

For the dynamic techniques, especially when rather large currents flow in response to the potential perturbation, a three-electrode cell arrangement is needed. In this arrangement, the current is passed between the working electrode and an auxiliary (or counter) electrode (CE). This auxiliary electrode should have a large area relative to the working electrode. For example, a Pt mesh or gauze can be used. In the three-electrode arrangement, the RE will not polarize because little current flows into it (see below).

The properties of interest in an electrochemical experiment are summarized in Table 20.1. Thus, in a potentiometry experiment, we are monitoring the cell potential (strictly E_{WE}) as a function of the WE composition and solution characteristics. In a dynamic

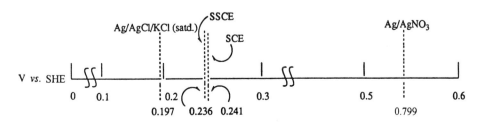

Fig. 20.5 Relationship between various reference electrode potentials expressed on a common scale.

Table 20.1 Variables Controlled, Measured Response, and Information Content in Electroanalytical Techniques

Technique	Controlled variable, symbol	Measured response, symbol	Other parameters	Information obtained
Potentiometry	—	Potential, E	Solution composition; film composition	Redox potential, ion-exchange capacity
Voltammetry	Potential, E	Current, i	Solution composition; film composition	Redox potential, charge transfer kinetics
Coulometry	Potential, E	Charge, Q	Solution composition; film composition	Electron stoichiometry
AC impedance spectroscopy	Potential, E	Current i, or impedance Z	Solution composition; film composition, frequency	Charge transfer and ion transport kinetics

electrochemistry (e.g., voltammetry) experiment, the WE potential is controlled (i.e., swept) and the current is measured. The resultant i–E trace is called a voltammogram. In a coulometry exercise, the potential is again controlled and the charge Q is measured. Controlled current experiments are not widely employed except in electrosynthetic procedures, as for example when a conducting polymer film is grown on a support electrode from a monomer solution.

In all the experiments considered in Fig. 20.4, it is important to (1) minimize ohmic drops in the electrolyte and (2) define the mass transport conditions at the WE/solution interface. A supporting electrolyte (usually at 0.1 M or higher) is added to the solution to meet both these objectives. The supporting electrolyte is chosen such that its components are electroinactive over the potential range encompassed in the electrochemical experiment. Solutions of alkali metal cations and polyatomic anions (e.g., KNO_3, $KClO_4$) meet these criteria. In nonaqueous solvents, salts comprising bulky cations (e.g., tetrabutylammonium perchlorate) are normally employed. An exception to the use of high ionic strength media for electrochemical experiments is when ultramicroelectrodes are employed. These are considered later in the discussion.

What sort of instrumentation would be needed for electrochemical experiments? A potentiometry experiment requires little more than a pH meter. A potentiostat or galvanostat can be used for the controlling potential or current in an experiment. In a coulometric procedure, a device to integrate the current (i.e., a coulometer) would also be needed. A hydrodynamic voltammetry [e.g., a rotating disk electrode (RDE)] experiment would require an electrode rotor (to spin the electrode at a precisely known rotation speed), and the rotating ring-disk or RRDE refinement (see below) would necessitate the use of a bipotentiostat so that the disk and ring potentials can be independently controlled. An ac impedance measurement involves the use of a sine-wave oscillator and a lock-in amplifier to separate the components of the impedance (see below). Alternatively, techniques based on imposition of noise and correlation methods can be employed as exemplified by the EG&G (Princeton Applied Research) and Solartron instruments, respectively.

Table 20.2 presents a compilation of manufacturers of electrochemical instruments. It is emphasized that this listing is only representative and is not meant to imply endorsement by the authors, editors, or publishers of this volume.

The cell designs vary widely and depend on the objectives of the particular experiment. To avoid interferences from the products of electrolysis at the CE/solution interface, twin-compartment cells are commonly used as exemplified by the design in Fig. 20.6. The so-called uncompensated resistance, R_u [32,33] is minimized by use of a fine capillary tip called a Luggin–Haber probe wherein the reference electrode is positioned very close to the WE surface (*not* shown in Fig.

Table 20.2 Manufacturers of Electroanalytical Equipment in the United States and Europe

Bioanalytical Systems, 2701 Kent Avenue, W. Lafayette, IN 47906

Cypress Systems, 2500 W. 31st St., Suite D, Lawrence, KS 66047

EG&G, Princeton Applied Research, P.O. Box 2565, Princeton, NJ 08543

Pine Instrument Co., 101 Industrial Dr., Grove City, PA 16127

Radiometer America Inc., 810 Sharon Dr., Westlake, OH 44145

Solartron Instruments, 2 Westchester Plaza, Elmsford, NY 10523

Tacussel, 27 rue d'Alsace, F-69627 Villeurbanne, France

Electrochemistry of Conducting Polymers

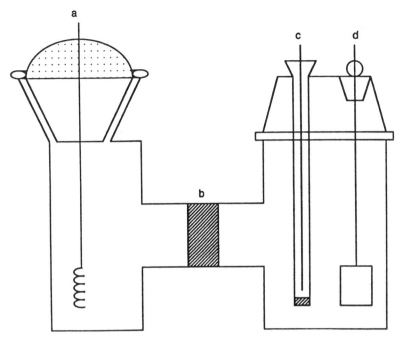

Fig. 20.6 Schematic of a twin-compartment electrochemical cell: a, counterelectrode; b, frit; c, reference electrode; and d, working electrode. Arrangements for cell sparging and other details are not shown in this simplified diagram.

20.6). Provisions are also made in the cell for solution sparging (usually with ultrapure N_2 gas). This is because the presence of dissolved air (or O_2) causes interference with voltammetric measurements. Thin-layer cells are designed to maximize the electrolysis efficiency and are especially suited for spectroelectrochemistry experiments. These are considered later.

B. Working Electrodes

As mentioned earlier, the working electrode structure in the present context comprises the conducting polymer films on top of a support electrode (the latter often referred to as a substrate). We consider the materials and configurations for the support electrode in this section.

In cases where electrolysis of the solution bulk is not desired (as in voltammetry or chronoamperometry), the working electrode area is kept reasonably small (nominal dimensions of square millimeters). Such an electrode is termed a microelectrode. On the other hand, in electrolysis procedures (as in thin-layer cells or coulometry), the ratio of electrode area to solution volume (A/V) must be maximized. Large-area working electrodes [grids or porous electrodes such as reticulated vitreous carbon (RVC)] are used in such cases.

The electrochemical techniques have a nominal time resolution only in the millisecond domain. This is a rather severe handicap relative to spectroscopic probes that now routinely access the femtosecond (10^{-15} s) regime. The two main obstacles for decreasing the time window of electrochemical techniques are both related to the uncompensated resistance, R_u. The time constant of the electrochemical cell, $R_u C_d$ (where C_d is the double-layer capacitance at the WE/solution interface), for nominal values of R_u and C_d (1 Ω and 20 μF, respectively) is in the microsecond domain. The cell therefore effectively acts as a low-pass filter with respect to the applied voltage [34]. Distortion of the applied voltage results in a distorted voltammogram.

Both of these deleterious effects can be reduced by decreasing the electrode dimension. Since C_d and the total current i are proportional to the WE surface area, i.e., to d^2 (d = electrode diameter), both $R_u C_d$ and iR_u decrease as the electrode is made smaller. For example, a decrease of the electrode diameter from 1 mm to 10 μm causes a reduction in the response time and the iR drop by a factor of 100 [35]. Thus, ultramicroelectrodes in the micrometer dimension have become exceedingly popular in the electrochemical community, and, as we shall see later, their use has not unsurprisingly permeated to the study of conducting polymer electrochemistry also.

The advantages in the use of ultramicroelectrodes are severalfold. The extremely small currents (nano- to picoamperes) mean that largely undistorted current–voltage curves can be recorded in low ionic strength media typical of conditions encountered in spectroscopy. Second, very high potential scan rates (on the order of 10^6 V/s) are possible [36], which means that (heterogeneous) rate constants of up to several centimeters per second can be accessed. Third, redox potentials of chemically unstable couples can be measured even for follow-up reactions with rate constants of 10^7 s^{-1} (first-order) or 10^9 M^{-1} s^{-1} (second-order) [37]. These facts translate to an equivalent time constant in the nanosecond regime as opposed to the millisecond domain for millimeter-sized electrode counterparts.

Developments in the use of ultramicroelectrodes and instrumentation advances in high speed voltammetry have been the subject of several review articles and book chapters [38–44]. Rather than monitoring the current with concomitant problems with double-layer charging at short time scales, a rather ingenious approach is to combine the use of ultramicroelectrodes with spectroelectrochemistry [45–47]. Very fast reactions (with bimolecular rate constants up to $\sim 10^7$ M^{-1} s^{-1}) can be thus accessed for the electrogenerated species in the diffusion layer region.

Even finer electrodes in the nanometer size regime have been claimed [48,49]. Whether such *nanodes* can be reproducibly made and used remains an open question, at least at the time of this writing. To our knowledge, they have not yet been used for the electrochemical study of conducting polymers.

What materials are to be used for the working electrodes? The common practice is to use electrochemically inert materials such as the noble metals (Pt and Au) or various forms of carbon. However, reactive metal supports (e.g., Cu) have indeed been used in the study of conducting polymer films (e.g., Refs. 21 and 22). Even purportedly stable electrode materials such as Au have undergone electrodissolution in the media used for polymer growth, with interesting results as discussed in a subsequent section.

Microelectrodes and ultramicroelectrodes are commercially available from several vendors. Table 20.3 contains a representative listing of such sources.

C. Potential-Step Techniques

The imposition of a sudden change in E_{WE} can reveal much about the electrode process. The simplest scenario is one wherein the solution initially contains only the reduced species, R, at its bulk concentration, C_R^*. At the instant $t = 0$, the electrode potential is suddenly changed from its rest value to E, at which the anodic

Table 20.3 Suppliers of Electrodes and Electrode Materials

Bioanalytical Systems, 2701 Kent Avenue, West Lafayette, IN 47906
Cypress Systems, 2500 W. 31st St., Suite D, Lawrence, KS 66047
EG&G, Princeton Applied Research, P.O. Box 2565, Princeton, NJ 08543
Electrosynthesis, 72 Ward Rd., Princeton, NJ 14086
Electrode Corporation, 100 Seventh Ave., Suite 300, Chardon, OH 44024
Pine Instrument Co., 101 Industrial Dr., Grove City, PA 16127

process R → O + ne^- occurs (Fig. 20.7a), where O is the oxidized species. Assume that the reaction is reversible and the Nernst equation is applicable:

$$\frac{C_R^* \sqrt{D_R}}{C_O^* \sqrt{D_O}} = \exp\left[\frac{nF}{RT}(E^h - E)\right] \quad (1)$$

It can be shown that if the potential is large enough ($\sim 130/n$ mV positive of E^h), so that the current is almost completely controlled by diffusion, the i–t transient is described by the Cottrell equation:

$$i = nFA\, C_R^* \left(\frac{D_R}{\pi t}\right)^{1/2} \quad (2)$$

The current thus decays with time (Fig. 20.7b). An experiment in which i is monitored as a function of t in response to a potential step is termed chronoamperometry. When electron transfer kinetics are important (i.e., a non-Nernstian regime), we obtain the Smutek equation:

$$\frac{i}{i_d} = \pi^{1/2} \lambda \exp(\lambda^2)\, \text{erfc}(\lambda) \quad (3)$$

where

$$\lambda = \frac{kt^{1/2}}{D^{1/2}} \quad (4)$$

i_d is the (diffusion) current in the absence of kinetic complications, and k is the potential-dependent rate constant for electron transfer.

Thus, in the Nernstian regime, a plot of i vs. $t^{-1/2}$ will be linear, and useful information about the parameters n and D_R can be obtained from its slope for the electrode process of interest. (Double potential step experiments similarly afford information about the reverse process, reduction.) Likewise a plot of $it^{1/2}$ vs. $t^{1/2}$ (Fig. 20.7c) yields kinetics information for a non-Nernstian process. The horizontal region at large values of $it^{1/2}$ corresponds to the Cottrell regime, whereas the short-time data are

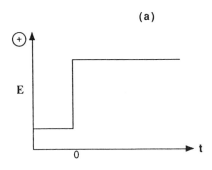

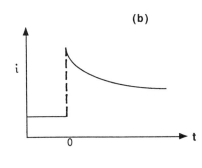

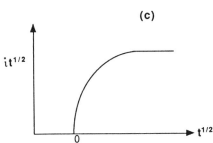

Fig. 20.7 A chronoamperometric experiment and associated data analysis. (a) The potential step; (b, c) current response and analysis of data in terms of a plot of $it^{1/2}$ vs. $t^{1/2}$.

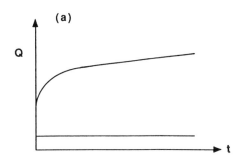

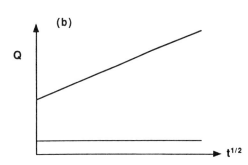

Fig. 20.8 (a) Chronocoulometric data and (b) data analyses. The top and bottom curves in each figure correspond to the test solution and the background electrolyte, respectively.

be calculated for a range of values of $t^{1/2}$. A working curve of Eq. (3) is then prepared, and from such plots of i/i_d vs. λ, k can be estimated. The importance of accessing short-time scales in electrochemical experiments to obtain information on fast kinetic processes can perhaps now be appreciated.

In a chronocoulometry experiment (see Fig. 20.4), the total charge is monitored as a function of time. A representative Q vs. t transient and the associated analysis of such data are contained in Fig. 20.8. Integration of the Cottrell equation shows that Q will vary with $t^{1/2}$:

$$Q = \frac{2nFA\, D_R^{1/2}\, C_R^*\, t^{1/2}}{\pi^{1/2}} \tag{5}$$

The advantage of chronocoulometry over chronoamperometry is that long-time data are less contaminated by noise. (Note that the signal in this case grows with time rather than decaying as in the chronoamperometric case.) The charge at short times is still distorted by the double-layer charging process, but its influence on the total charge rapidly becomes negligible.

In cases where we have a polymer film on the electrode surface (as in a conducting polymer study), the situation becomes rather different from a solution case with respect to charge and mass transport. The critical parameter here is $D_{ct}\tau/d^2$ (where d is the polymer film thickness and T is the time after the potential step). When $D_{ct}\tau/d^2 \ll 1$, semi-infinite diffusion prevails, and the i–t transient should conform to the Cottrell equation [42]:

$$i = \frac{nFA\, D_{ct}^{1/2}\, C}{\pi^{1/2}\, t^{1/2}} \tag{2a}$$

The concentration C now pertains to the electroactive sites in the polymer film, and the parameter D_{ct} is a film property and not a solution characteristic like its counterpart D_R in Eq. (2). Thus, plots of i vs. $t^{-1/2}$ (Cottrell plots) may be used to compute D_{ct} if C can be reasonably estimated (e.g., from voltammetry data, see below) and film resistance effects are negligible. Analogously, chronocoulometric plots of Q vs. $t^{1/2}$ can be used for this purpose.

At long times, the concentration profile within the polymer film impinges on the film/solution boundary, i.e., $D_{ct}\tau/d^2 \gg 1$. The chronoamperometric current will be less than that given by the Cottrell equation analog [Eq. (2a)]. We now have a finite (i.e., bounded) diffusion regime.

A major difficulty with the applicability of these (simplified) equations to the conducting polymer cases is the neglect of migration as a transport mode within the polymer film. This is a rather serious omission, as we shall see later.

D. Bulk (Exhaustive) Electrolysis

In the preceding section, we mostly considered cases wherein only a thin segment of the electroactive region (whether the solution or the film phase) was electrochemically altered. This situation must be contrasted with those in which exhaustive electrolysis is involved. An example is constant-potential coulometry (Fig. 20.4) wherein the entire solution contained within the cell is electrolyzed. As mentioned earlier, this is ensured by the use of a large A/V ratio and efficient solution agitation. The underlying coulometric equation derives from Faraday's law of electrolysis and can be expressed as

$$Q = nFN \qquad (6)$$

The measurement of the electrolysis charge Q thus affords a route to n (if N is known) or to N (if n is separately determined, e.g., from a potential-step experiment, see above).

Another route to exhaustive electrolysis again maximizes the A/V ratio by shrinking the electrolyte volume. A variety of such *thin-layer cell* configurations have been described [50–55]. The solution volume typically is few microliters in these configurations as opposed to a few milliliters in the usual scenarios. As long as the cell thickness is smaller than the diffusion layer thickness for a given experimental time $[(2 D\tau)^{1/2}]$, mass transfer within the cell can be neglected.

A polymer film (nominally of micrometer thickness) on the electrode surface can be considered to be a "solid-state" electrochemical analog of the thin-layer cell. Thus, electrochemical processes within the polymer film can be modeled similarly to their thin-layer solution counterparts (with some important distinctions as discussed in Section I earlier and in what follows). Indeed, many of the theoretical developments in the electrochemistry of polymer-modified electrode surfaces were inspired by the earlier developments (in the 1960s and early 1970s) in thin-layer electrochemistry.

E. Potential-Sweep Techniques

In potential-sweep techniques, the current flowing at the WE/solution interface is monitored as a function of the potential applied to it. We consider three such voltammetric techniques: linear sweep voltammetry (LSV), cyclic voltammetry (CV), and hydrodynamic voltammetry (Fig. 20.4). The voltammogram obtained in each case may be regarded as the electrochemical equivalent of a spectrum obtained in a spectrophotometric technique. Indeed, the term "electrochemical spectroscopy" has been applied to CV [56], and it is worth noting that the independent variable in both cases is related to energy—wavelength in the case of spectroscopy and potential in the case of CV. The potential is swept linearly at v V/s so that the potential at any time is $E(t) = E_i \pm vt$.

1. Cyclic and Linear Sweep Voltammetry

We can proceed as before from the Nernstian regime to kinetically controlled cases and progress from solution to thin-layer cells to polymer-borne electrodes in turn. However, Eq. (1) now becomes (for an initial reduction case)

$$\frac{C_O(0, t)}{C_R(0, t)} = f(t) = \exp\left[\frac{nF}{RT}(E_i - vt - E^h)\right] \qquad (1a)$$

The ratio is now a function of time (as well as potential)—a significant complication because the Laplace transformation can no longer be applied to obtain closed-form solutions to the master (flux) equation. Recourse has been sought instead to numerical solutions [57–61] or convolutive (semi-integral) techniques [62–66]. Cyclic voltammograms have also been simulated for a wide variety of mechanistic variations using finite-difference computational algorithms [67]. For the present discussion, we merely present the salient features of the information content for each type of voltammetric technique. The distinction between LSV and CV pertains to the perturbation waveform. In LSV, only a "forward" potential scan (in either the oxidation or reduction direction depending on the characteristics of the test redox system) is used. CV, on the other hand, uses a triangular ramp, and thus both directions of the redox process O + $ne^- \rightleftharpoons$ R are probed. Both LSV and CV employ quiescent solutions; hydrodynamic voltammetry, on the other hand, involves electrode rotation (i.e., solution agitation) (Fig. 20.4). In solutions of high ionic strength, the dominant mass transport mode is diffusion in LSV and CV, whereas it is diffusion-convection in the rotating electrode case.

Assume again a Nernstian (electrochemically reversible) situation, and assume that an initial reduction is carried out starting with oxidized species only and no R initially present. At a rest potential, no current flows and the concentration profile is homogeneous across the WE/solution interface (Fig. 20.9, curve a). As the potential is swept past the standard reduction potential, $E^{\circ\prime}$ (or E^h), electrolysis at the interface converts O to R at a fast rate (i.e., with no kinetic hindrance) and the current continues to rise. In fact, at this characteristic potential, half of the original O species has been converted to R (Fig. 20.9,b). At a potential several millivolts past $E^{\circ\prime}$ (or E^h), the surface concentration of O becomes virtually zero. Thus, the current peaks at this point (point c), and further excursion of the potential causes a current decay [as $t^{-1/2}$, cf. Eq. (2a)] as the diffusion layer begins to grow (Fig. 20.9, d and e). The peak current, i_p, is given by the Randles–Sevcik equation,

$$i_p = (2.69 \times 10^5) A n^{3/2} C_O^* D_O^{1/2} v^{1/2} \quad (7)$$

Thus, i_p increases as $v^{1/2}$ for Nernstian redox couples.

The reverse sweep can be analogously modeled with the aid of concentration–distance profiles as in Fig. 20.9. In fact, the forward and reverse S-shaped i–E profiles should exactly overlap for reversible redox couples.

In the presence of kinetic complications, the waves become drawn out. Figure 20.10 shows three simulated cyclic voltammograms for the reversible (Nernstian), quasi-reversible, and irreversible charge transfer cases. Obviously, kinetic information is contained in the CV wave shapes, and quantitative estimates of $\vec{k}$, $\overleftarrow{k}$, and k_0 are possible if account is taken of the effects due to R_u [61].

What happens in the thin-layer cell case? In the Nernstian ideal case, the peak current i_p occurs at $E = E^{\circ\prime}$ and is given by

$$i_p = \frac{n^2 F^2 v V C_O^*}{4RT} \quad (8)$$

A representative thin-layer cyclic voltammogram for the Nernstian ideal case is illustrated in Fig. 20.11. Note that i_p is now proportional to v (instead of $v^{1/2}$), but the total charge under the i–E curve is independent of v. In fact, this provides a direct route to assay of the electroactive species (Fig. 20.11). Again, extension to irreversible charge transfer cases has been made, and reference is made to the original literature [53–55] for a discussion of these aspects.

The thin-layer cell case bears strong analogy to immobilized redox polymer films as in the potential-step cases discussed in a preceding section. Thus, the charge encompassed under the i–E curve affords a route to Γ (mol/cm^2), the total quantity of attached electroactive sites (Fig. 20.11). It is important that such an analysis be performed at small enough v (a few millivolts per second) so that kinetics effects and complications due to R_u do not interfere. The most sensitive criterion for Nernstian

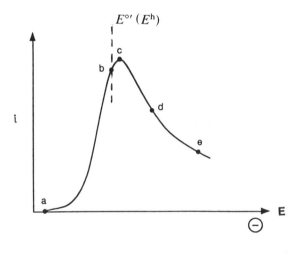

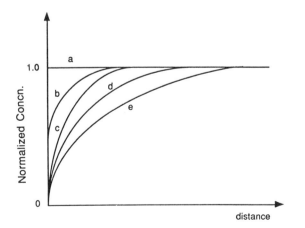

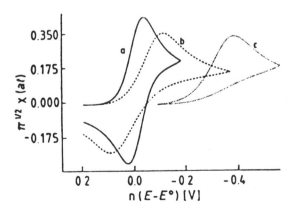

Fig. 20.9 (Top) A linear sweep voltammogram and (bottom) concentration profiles for the electroactive species at the electrode/electrolyte interface. Refer to text for significance of the points denoted a–e.

Fig. 20.10 Simulated cyclic voltammograms for an electrochemically reversible (a), quasi-reversible (b), and irreversible (c) redox couple. The ordinate is a normalized current function. (Reproduced with permission from Ref. 56.)

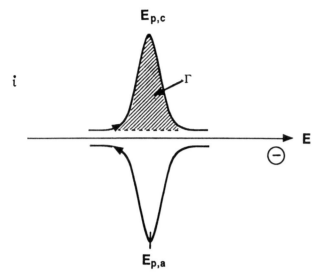

Fig. 20.11 Thin-layer cyclic voltammogram.

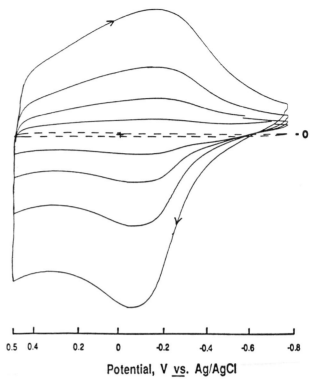

Fig. 20.12 Representative cyclic voltammograms as a function of scan rate for an electronically conductive polymer film, polypyrrole. The scan rate increases from bottom to top, and the dashed lines refer to the voltammogram for the bare (support) electrode (e.g., glassy carbon).

response is $\Delta E_p = 0$ and scan rate invariance of ΔE_p. Departure of ΔE_p from these trends and a concomitant curvature in i_p vs. v plots signify the onset of rate limitations due to charge transfer and other complications. Again, the critical parameter is $D_{ct}\tau/d^2$, with τ now controlled by the scan rate. When $D_{ct}\tau/d^2 \gg 1$ (slow scan rate), the film is completely equilibrated and the above trends hold. Conversely, when $D_{ct}\tau/d^2 < 1$ (fast scan rate or slow D_{ct} value), the CV peak exhibits a "diffusional tail" and we approach (semi-infinite) diffusion conditions akin to the bulk solution case considered earlier. Now $i_p \propto v^{1/2}$ [52].

Conducting polymer films yield cyclic voltammograms often resembling those expected for immobilized redox moieties (see below). However, the waves often are superimposed over a "capacitive" envelope (cf. Fig. 20.12) arising from the microporosity of these films. In fact, a cyclic potential sweep applied to a capacitive electric circuit yields a current envelope (Fig. 20.13a) in which the steady-state current is proportional to vC. Conversely, a ramp applied to a resistive circuit yields a sloping (ohmic) i–E "voltammogram" (Fig. 20.13b). Such voltammograms are typical of very thick polymer films or those that are highly resistive and poorly electroactive.

2. Hydrodynamic Voltammetry

When the electrode (RDE) is spun at a radial velocity ω, a Prandtl boundary layer is established with thickness $\delta_H = 3.6 \, (\nu/\omega)^{1/2}$. This roughly represents the thickness of the layer of liquid dragged by the rotating disk (Fig. 20.14). For example, for water, $\nu \approx 0.01$ cm²/s, and at a ω value of 10^4 s^{-1}, δ_H is 3.6×10^{-3} cm. Figure 20.14 also illustrates the relative magnitude of δ_H and the

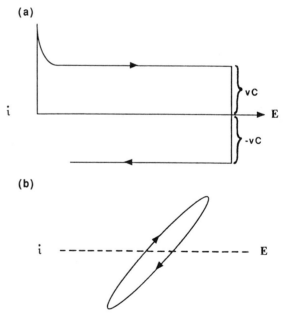

Fig. 20.13 Current flow in response to a potential ramp (scan rate v V/s) for (a) a capacitive circuit and (b) a resistive circuit.

Electrochemistry of Conducting Polymers

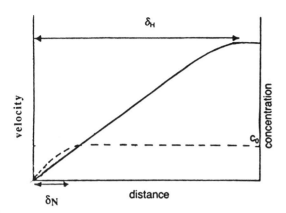

Fig. 20.14 Schematic diagram of the relative dimensions of the hydrodynamic layer (δ_H) and the diffusion layer (δ_N) at an electrode/electrolyte interface. The electrode plane is located at zero distance in this diagram, and the solution bulk is to the right. (Adapted with permission from J. Wang, *Analytical Electrochemistry*, VCH, New York, 1994.)

Nernst diffusion layer thickness δ_N and shows why we have to consider a mixed convection–diffusion mass transport mode in this case.

Again, considering a Nernst regime and at the limiting current condition (where $C_O = 0$), the Levich equation is obtained:

$$i_L = 0.620\ nFA\ D_O^{2/3} \omega^{1/2} \nu^{-1/6} C_O^* \tag{9}$$

A family of hydrodynamic voltammograms are presented in Fig. 20.15a. Thus a Levich plot of i_L vs. $\omega^{1/2}$ will be linear with zero intercept. It is important in these experiments to scan the potential rather slowly ($v = 1\text{–}5$ mV/s) so that a steady-state situation is established (at a fixed ω).

In the presence of kinetic limitations at the RDE/solution interface, curvature is seen in the Levich plot (Fig. 20.15b). An inverse Levich plot (Koutecky–Levich plot) of $1/i$ vs. $1/\omega^{1/2}$ may be constructed from which the kinetic current i_k can be extracted from the intercept (Fig. 20.16):

$$\frac{1}{i} = \frac{1}{i_k} + \frac{1}{i_L} \tag{10}$$

Obviously, when $i_k \to \infty$ (Nernstian regime), Eq. (10) collapses to the Levich case [Eq. (9)].

The range of ω is established at the lower bound by the dimension of δ_H relative to the disk radius r. When δ_H approaches the latter, the approximations inherent in the derivation of Eq. (9) break down. For $\nu = 0.01$ cm²/s and $r = 0.1$ cm, ω should be larger than $10\ \text{s}^{-1}$. Similarly, the upper bound is dictated by the onset of turbulent flow. The latter limit occurs at a dimensionless Reynolds number ($\omega r^2/\nu$) of 2×10^5, which translates

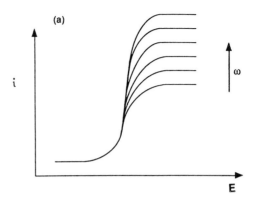

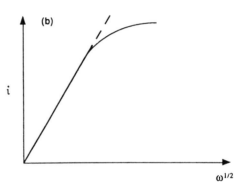

Fig. 20.15 (a) Representative hydrodynamic voltammograms and (b) the corresponding Levich plot. The electrode rotation speed increases from bottom to top in (a).

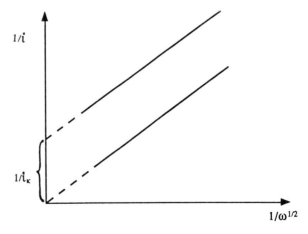

Fig. 20.16 Inverse Levich or Koutecky–Levich plot of $1/i$ or $1/\omega^{1/2}$. The intercepts afford a value for the kinetic current i_k.

to a ω value of 2×10^5 s^{-1}. Usually, the practical limit occurs at much lower values of ω because of imperfections in the RDE surface and cell geometry limitations. Thus, the range of ω (in rad/s) is given by $10 < \omega < 1000$ or in terms of rpm from 100 to 10,000.

Further in-depth discussions of RDE and the related RRDE techniques can be found in Refs. 68–70.

Hydrodynamic (RDE) voltammetry is particularly useful for the study of redox catalysis at polymer film modified electrode/solution interfaces. The catalytic scheme is illustrated in Fig. 20.17 and hinges on the charge transport being *mediated* by the polymer-confined redox sites. The transport of the substrate to the polymer film/solution interface is given by the Levich flux [Eq. (9)]. Note that in the following discussion of mediated charge transfer the depolarizer (cf. Fig. 20.1), i.e., the reacting redox species from the electrolyte, will be termed "substrate." Assume that the bimolecular reaction between the polymer sites and the substrate is rate-limiting. Then the rate law for this step is given by

$$\frac{i_{cr}}{nFA} = \frac{d\Gamma}{dt} = k_{cr} \Gamma C_s^* \quad (11)$$

where Γ is the polymer site coverage, C_s^* is the substrate concentration in the solution bulk, and k_{cr} is the bimolecular rate constant. The Koutecky–Levich plot in this instance becomes

$$\frac{1}{i} = \frac{1}{nFAk_{cr}\Gamma C_s^*} + \frac{1}{0.62nFAD_s^{2/3}\nu^{-1/6}\omega^{1/2}C_s^*} \quad (10a)$$

Thus, plots similar to those in Fig. 20.16 may be used to extract the kinetic parameter $k_{cr}\Gamma$ (cm/s).

In more complicated situations, the rates of permeation of the substrate into the polymer film or of the charge transport within the polymer film itself may limit

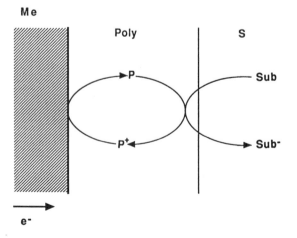

Fig. 20.17 Electron transfer mediation between the (underlying) support electrode and the (solution) substrate by polymer-confined redox counters. A catalytic reduction is assumed as a specific example.

the overall current. Theory for these situations is available and has been formulated both for the "pure" rate regimes of catalytic reaction (R), electron transport through the polymer (E), and substrate partitioning in the polymer film (S) and for "mixed" (e.g., SR, ER) cases [52,71–73]. Figure 20.18 is a schematic representation of the concentration profiles for mediator B and substrate S for various cases [74]. Table 20.4 lists the characteristic currents shown in Fig. 20.18.

A similar theoretical approach has been developed [75] except that reaction layer thicknesses are considered instead of characteristic currents. The location of these reaction layers reflects the relative rates of charge and substrate transport within the film, and the thickness of the layers conveys an idea of the fraction of the film usefully employed in the redox catalysis. Note that the two extremes here are the scenarios that the charge transport through the polymer is very fast (i.e., not rate-limiting) such that the reaction zone is essentially confined to the Nernst diffusion layer (Fig. 20.19a) and the opposite extreme that the reaction zone is located at the support electrode/polymer film boundary (Fig. 20.19b).

We shall have further opportunities to explore the applicability of these ideas to the specific case of conducting polymer films in Section III.B.

F. Electrochemical Quartz Crystal Microgravimetry

As Table 20.1 indicates, the parameters usually measured in electrochemical experiments are electrical in nature. Many electrochemical processes (e.g., dissolution, corrosion, electrodeposition), however, are accompanied by a change in mass at the working electrode/solution interface. In such cases, it is obviously beneficial to combine an electrochemical experiment with mass measurement. The classical approach to this problem was to monitor the mass changes ex situ. Such electrogravimetric experiments indeed have a long history. The use of a quartz resonator permits in situ and real-time probes of mass changes and various other processes and considerably enhances the information content, especially in studies of thin films on electrode surfaces [76–78]. A metal film deposited on the quartz surface simultaneously permits electrochemistry to be carried out on the resonator itself. Figure 20.20 is a schematic diagram of such an electrochemical quartz crystal microgravimetric (EQCM) apparatus.

The relationship between the frequency change of the quartz resonator Δf and the mass change Δm is embodied in the Sauerbrey equation [79]

$$\Delta f = -[2f_0^2 (A \sqrt{\mu_2 \rho_2})^{-1} \Delta m] \quad (12)$$

Note the negative sign; i.e., an increase of mass causes a decrease of the frequency from its original resonance condition.

Mass loading of the quartz resonator has been routinely employed for film thickness measurements by the

Electrochemistry of Conducting Polymers

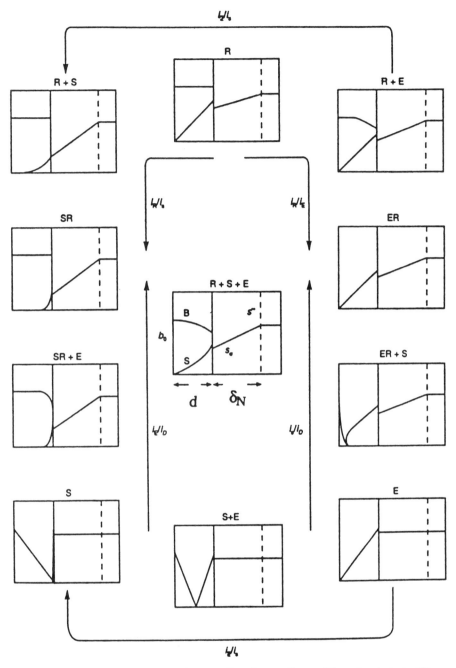

Fig. 20.18 Schematic representation of concentration profiles for mediator (B) and substrate (S) according to the Andrieux–Saveant model. The characteristic currents are also shown for the various case designations, R, S, E, etc. (Reproduced with permission from Ref. 74.)

high vacuum community. The applicability of the Sauerbrey equation when one face of the resonator is exposed to the liquid has been thoroughly explored in the literature; reviews of these studies are given in Refs. 76–78 and 80. These investigations have shown that assumptions of "ideal rigid layer" behavior may not always hold, especially in the case of polymer films on electrode surfaces. Nonidealities include viscoelastic behavior due to changes in polymer morphology and swelling upon electrochemical cycling, high mass loadings, surface stress and interfacial slippage, electrode surface roughness and liquid trapping within interfacial pores,

Table 20.4 Characteristic Currents in the Andrieux–Saveant Model for Redox Catalysis at Polymer-Modified Electrodes

1. Substrate diffusion in Nernst diffusion layer:

$$i_L = \frac{nFAD_s C_s^*}{\delta_N}$$

2. Substrate diffusion in polymer matrix:

$$i_s = \frac{nFAD_p \kappa C_s^*}{d}$$

3. Charge percolation through polymer matrix:

$$i_E = \frac{nFAD_{ct} \Gamma}{d^2}$$

4. Catalytic (cross) reaction between mediator and substrate:

$$i_{cr} = nFAk_{cr} \Gamma C_s^*$$

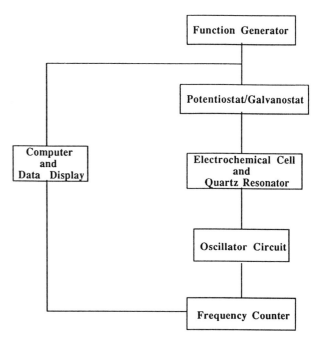

Fig. 20.20 Schematic diagram of an electrochemical quartz crystal microgravimetry (EQCM) apparatus.

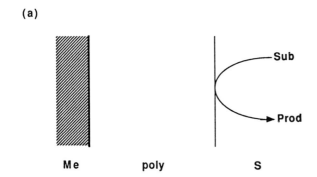

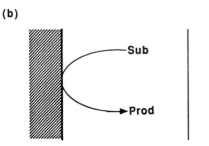

Fig. 20.19 Two limiting scenarios representing (a) facile catalysis and (b) slow charge transport (and/or facile substrate permeation) through the polymer coating.

and bubble formation as a result of water electrolysis. Film plasticization and solvent swelling effects can be conveniently probed by simultaneous monitoring of the conductance spectrum of the quartz resonator circuit [78,81]. Departures from rigid layer behavior are frequently accompanied by broadening of the resonance peak under these conditions.

Figure 20.21 demonstrates how cyclic voltammetry can be effectively combined with EQCM for the study of a complex electrochemical process such as the reduction of Te(IV) [82]. Voltammetry alone would have provided only a limited glimpse into the subtleties of the reduction process. Combination of EQCM with coulometry [i.e., combination of Eqs. (6) and (12)] affords a sensitive probe of the electron stoichiometry and Faradaic (Coulombic) efficiency for polymer deposition processes. For example, Fig. 20.22 and Table 20.5 contain assays of the platinum content thus obtained for polypyrrole-Pt composites [83] considered later in this chapter.

Electrochemical quartz crystal microgravimetry can be applied to conducting polymers for the in situ and real-time study of ion-exchange processes, and this comprises a powerful realm of application of this versatile technique. We further explore this aspect when specific conducting polymer systems are discussed later.

Improvements continue to be made in the use of QCM for electrochemical applications including advances re-

Electrochemistry of Conducting Polymers

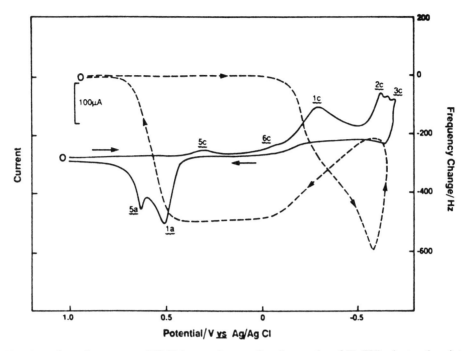

Fig. 20.21 Combined cyclic voltammetry–EQCM experiments for the study of Te(IV) electrochemistry in an aqueous medium. The dashed lines are the EQCM data. The numbers designate various electrochemical processes, the details of which are contained in Ref. 82. (Reproduced with permission from Ref. 82.)

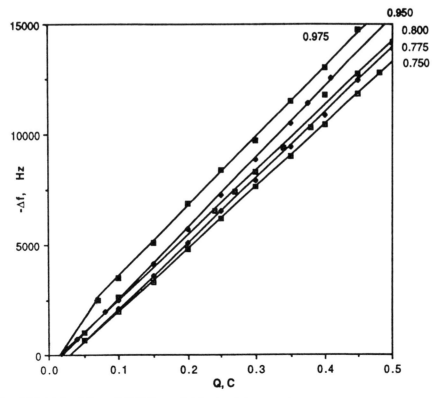

Fig. 20.22 Plots of $-\Delta f$ (obtained from EQCM) vs. Q (obtained from coulometry) for the compositional assay of polypyrrole-platinum composites. The analytical results are contained in Table 20.5. (Reproduced with permission from Ref. 83.)

Table 20.5 Platinum Content of Polypyrrole-Pt Composite Films as Measured by Electrochemical Quartz Crystal Microgravimetry and Coulometry

Growth potential[a] (mV vs. Ag/AgCl)	Slope of $-\Delta f$ vs. Q plot[b] (Hz/C)	Platinum content (wt%)
750	28,928	30.0
775	29,643	19.7
800	30,000	17.5
950	32,250	16.0
975	31,071	9.5

[a] Growth medium: 0.1 M KCl containing 0.05 M pyrrole and 3 mM platinum.
[b] See Fig. 20.22.
Source: Adapted with permission from Ref. 83.

lated to instrumentation [84,85–89], data processing (T. Hepel, Elchema Inc., Potsdam, NY, private communication, 1994), equivalent circuit modeling [90,91], and understanding of interfacial processes at the quartz/liquid interface and their consequences [92–94].

G. AC Impedance Spectroscopy

We have already examined how a working electrode/solution interface responds to various types of perturbations including potential steps and potential sweeps. These perturbations are usually of large amplitude (Fig. 20.4), and they generally drive the working electrode to a condition far from equilibrium. Another approach is to perturb the cell with an alternating (usually sinusoidal) signal of small amplitude (nominally a few millivolts peak to peak) and observe the manner in which the system follows the perturbation at steady state. A major advantage with this ac impedance spectroscopic technique is that the response may be theoretically treated via linearized current–potential (i–E) characteristics. This leads to important modeling simplifications in matters related to diffusion and charge/ion transport kinetics. Detailed knowledge about i–E curves over wide ranges of overpotential is not needed, and, an important point, high precision measurements are possible because of the steady-state nature of the response and the intrinsic multiplexing (data-averaging) capability. Indeed, multiplex data processing methods such as the fast Fourier transform (FFT) have been profitably employed in the acquisition and analysis of ac impedance spectroscopic data (e.g., Ref. 95).

The parameter (electrical) impedance Z is the ac analog of the resistance, R for dc circuits and expresses the relationship between a sinusoidal signal and the corresponding response:

$$e = E \sin \omega t \quad (13a)$$
$$i = I \sin (\omega t + \phi) \quad (13b)$$
$$Z = e/i \quad (13c)$$

The phase angle ϕ is negative for capacitive circuits and is 90° for a "pure" capacitor. The impedance, then, is obviously a vector quantity and, as usual, we can employ both rectangular and polar coordinates to denote a vector. In the former format, the vector Z is given by $R - jX_c$, where $j = \sqrt{-1}$ and X_c is termed the capacitive reactance (equal to $1/\omega C$, $\omega = 2\pi f$, where f is the ac frequency in hertz). Simply put, X_c is a frequency-sensitive "variable" resistor that switches from ∞ when $\omega \rightarrow 0$ (dc) to 0 when $\omega \rightarrow \infty$ (high-frequency). The magnitude of **Z** ($|Z|$) is $(R^2 + X_c^2)^{1/2}$, and the phase angle is given by

$$\tan \phi = \frac{X_c}{R} = \frac{1}{\omega RC} \quad (14)$$

In polar coordinates, **Z** can be written in Euler form as

$$\mathbf{Z} = |Z|e^{j\phi} \quad (15)$$

Separation of the impedance components into "real" and "imaginary" components is a bookkeeping measure and simply embodies the fact that there is a phase lag between the applied signal and the measured response. For dc circuits, this phase lag obviously is absent. Thus, we can model the systems response in terms of complex plane impedance plots and expect the response from a purely resistive circuit to be distributed along the abscissa. On the other hand, a "pure" capacitor will manifest a response along the ordinate (in the negative direction). Intermediate values of ϕ are expected for other RC circuits. These ideas are illustrated in Fig. 20.23.

Representative complex plane plots are illustrated in Figs. 20.24a–20.24c for various types of RC circuits including one with multiple relaxation times. Figure 20.24d contains an alternative data representation format that is favored by the electrical engineering community and is termed a Bode plot. The complex plane and Bode formats are entirely equivalent, and it is simply a matter of taste whether one type of plot or the other is used. The complex plane plot has also been variously named a Nyquist, Argand, or Cole–Cole plot, and these should generally be regarded as synonymous terms. A physicochemical system that can be represented by an electric circuit is said to be modeled by an *equivalent circuit* with the message that each of the circuit components can be associated with a portion or process in the test system.

A prototypical equivalent circuit for electrochemical systems is the Randles circuit illustrated in Fig. 20.25. The circuit is contained in Fig. 20.25a, and the corresponding complex plane diagram is shown in Fig. 20.25b. This particular circuit shares some of the fea-

Electrochemistry of Conducting Polymers

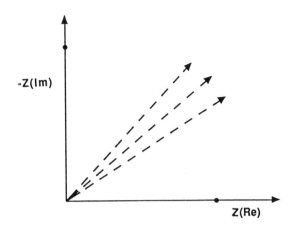

Fig. 20.23 The impedance shown in terms of a complex plane plot with purely resistive and purely capacitive behavior extremes represented by the dots on the axes. Real systems containing both R and C elements manifest behavior in the "complex plane" intermediate between these extremes.

tures associated with its counterpart in Fig. 20.24b. The major difference is the inclusion of a diffusional impedance termed the Warburg impedance, Z_W. Unlike most components of an equivalent circuit, which are frequency-invariant, the Warburg impedance is frequency-dependent:

$$Z_W = \frac{\sigma}{\omega^{1/2}} - j\frac{1}{\sigma\omega^{1/2}} \quad (16)$$

where

$$\sigma = \frac{RT}{n^2 F^2 A \, (2D)^{1/2}} \left(\frac{1}{C_O^*} + \frac{1}{C_R^*}\right) \quad (16a)$$

It is worth noting that diffusional (random walk or stochastic) phenomena manifest themselves in terms of a square root relationship, such as $t^{-1/2}$ (potential-step), $v^{1/2}$ (potential-sweep), or $\omega^{1/2}$ (small-amplitude ac perturbation).

Figure 20.26 illustrates that simple models such as those described above often indeed describe, rather

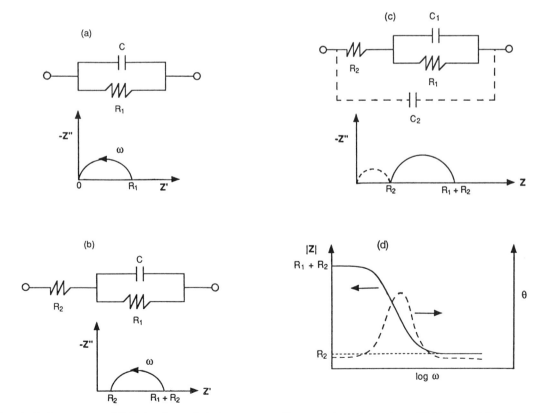

Fig. 20.24 (a–c) Various types of *RC* circuits and their corresponding complex plane plots. (d) Bode plot for circuit in (b).

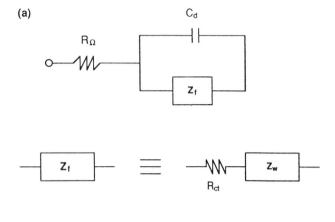

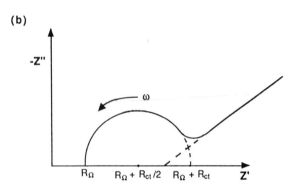

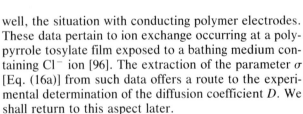

Fig. 20.25 (a) Randles circuit and (b) the corresponding complex plane plot. R_Ω, C_d, R_{ct}, and Z_w are the (uncompensated) series resistance, double-layer capacitance, charge transfer resistance, and Warburg impedance, respectively.

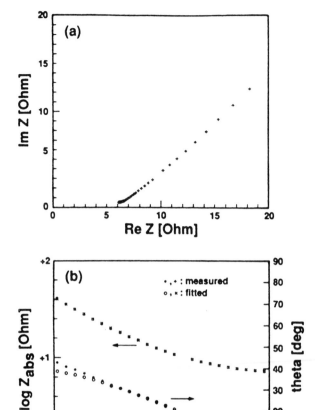

Fig. 20.26 Alternating current impedance data for a polypyrrole tosylate film in 1 M KCl in (a) complex plane format and (b) Bode representation. (Reproduced with permission from Ref. 96.)

well, the situation with conducting polymer electrodes. These data pertain to ion exchange occurring at a polypyrrole tosylate film exposed to a bathing medium containing Cl^- ion [96]. The extraction of the parameter σ [Eq. (16a)] from such data offers a route to the experimental determination of the diffusion coefficient D. We shall return to this aspect later.

Turning to experimental details, clearly some sort of a device to separate the system response into real and imaginary components is needed (cf. Section II.A). One such phase-sensitive detector is the lock-in amplifier. The classical approach based on the use of various types of ac bridge circuits has largely been supplanted by more "dynamic" and automation-friendly approaches. As mentioned earlier in Section II.A and in a preceding paragraph, methodologies based on noise, correlation, and time domain analyses have all been profitably employed in ac impedance spectroscopy, and many of these approaches have been incorporated in commercial instruments. Various types of data acquisition and analysis software have also been developed as exemplified by the IONICS [97], EQIVCT [98], and Z-PLOT (Scribner Associates, Inc, Charlottesville, VA) routines; some of these software packages are commercially available and are adaptable to a range of ac impedance spectroscopy instruments.

Complexities associated with the simple equivalent circuit models described above include phenomena (some interrelated) associated with depressed arcs and constant phase angles [99], porous electrodes and electrodes with rough surfaces [100–102], pseudoinduction behavior associated with electrosorbed intermediates [103], and nonidealities associated with (polymer) coatings on electrode surfaces. We now briefly explore this last aspect while also noting that general reviews of ac impedance spectroscopy are available [104–109].

The boundary conditions for the Warburg impedance, Z_W, previously discussed were such that semi-infinite diffusion prevails. However, as we have already seen in connection with voltammetry and other techniques for film-modified electrodes, diffusion in these cases is bounded and is restricted to a thin layer of thickness d. This problem has been independently addressed by three different groups [110–113] and leads to essentially the same end result, namely that the phase angle begins to increase at very low frequencies due to the onset of finite length effects. Figure 20.27a illustrates the complex plane impedance plot obtained in this instance.

The low frequency impedance in this case is given by the expression [112]

$$Z = \frac{RT}{j\omega n^2 F^2 C^* d} \qquad (17)$$

with the redox capacity of the film being given by

$$C_L = \frac{n^2 F^2 C^* d}{RT} \qquad (17a)$$

The limiting (low frequency) resistance R_L is obtained from the intercept of the impedance plot with the real axis as illustrated in Fig. 20.27a. In real-life systems, however, the semicircle is not centered on the real axis (i.e., depressed arc phenomenon) nor is the capacitive branch parallel to the imaginary axis as exemplified in Fig. 20.27b. The microscopic origins of these effects are not very well understood at present but are believed to be related to "surface inhomogeneities" and "roughness" effects (see above). Other analyses of the impedance of polymer-modified electrodes are available, including effects related to migration within the film and (lateral) redox-site interactions [114,115].

H. Spectroelectrochemistry

Historically, electrochemists' reliance on the measurement of current, potential, or charge (as illustrated by the preceding section) has resulted in an evolution of the discipline that would appear to be somewhat less molecularly oriented than other areas of chemistry such as spectroscopy. Indeed, this recognition has led to the assimilation of spectroscopic probes in an electrochemical measurement context resulting in the advent of spectroelectrochemical techniques. Virtually any spectroscopic technique can be thus incorporated, as Table 20.6 illustrates. The combination of thin-layer cell methodology (Section II.D) and a spectroscopic probe results in a wide range of spectroelectrochemical methods for

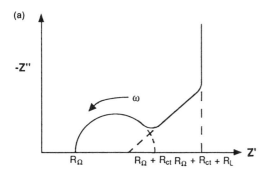

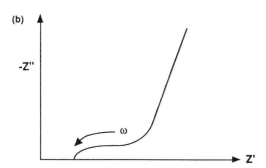

Fig. 20.27 (a) Ideal and (b) observed ac impedance behavior for film electrodes, illustrating the manifestations of bounded diffusion, depressed arc effects, and the consequence of surface roughness.

Table 20.6 Spectroscopic Measurements That Have Been Performed In Situ in an Electrochemical Cell

Technique	Ref.
UV-visible transmission spectrophotometry	116–122
Ellipsometry	123
Specular reflectance spectroscopy	124
Attenuated internal reflectance spectroscopy	125
Fourier transform IR spectroscopy	126, 127
Fourier transform Raman spectroscopy	128
Surface and resonance enhanced Raman spectroscopy	129, 130
Electron spin resonance spectroscopy	131
Mass spectrometry	132
Surface plasmon spectroscopy	133
Luminescence spectroscopy	134
Mössbauer spectroscopy	135
X-ray diffraction	136
X-ray absorption fine structure	137

studying the electrogenerated species. The earliest example of this approach consisted of an optically transparent thin-layer electrode (OTTLE) and a transmission UV-vis spectrophotometric measurement [116–120]. The electrochemical cells in these experiments used either a metal (e.g., Au) minigrid [121] or transparent conducting glass (e.g., Sn-doped indium oxide) [122].

In the present context, we are interested in spectroscopic changes undergone by the conducting polymer film (or the film/solution interface, see below) in response to a potential perturbation. However, there are some inherent experimental difficulties in the use of spectroscopic tools for studying polymer films. For example, conducting polymer films absorb strongly in the IR wavelength range, resulting in rather poor signal-to-noise ratios. Potential modulation (i.e., monitoring of $\Delta R/R$, where R is the reflectance) or the use of polarized light are effective solutions in this regard. Luminescence

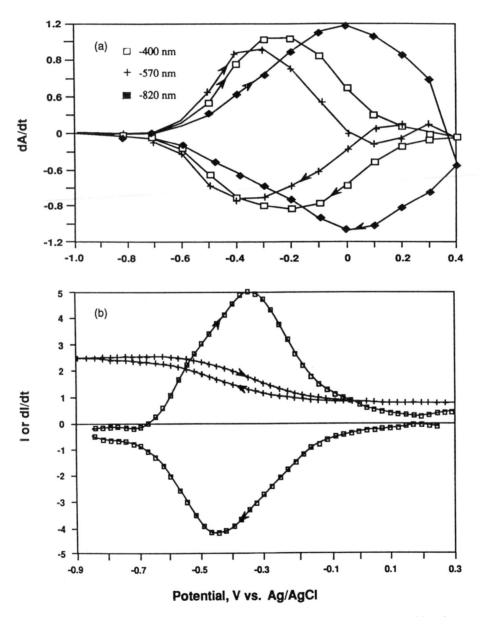

Fig. 20.28 (a) Derivative UV-vis absorbance and (b) Raman scattering signal (expressed in either integral or differential form) versus potential as a polypyrrole film is scanned through its redox transition. The Raman signal was monitored at 1576 cm^{-1}, and the electrolyte was N$_2$-saturated 0.1 M KCl. (Reproduced with permission from Ref. 144.)

measurements have high sensitivity but are fraught with difficulties from quenching phenomena and inner-filter effects, especially in the case of conducting polymer films [138]. Such measurements, however, have been effective for the study of redox polymer films (Table 20.6).

Many of the difficulties with measurements on the polymer phase can be obviated by moving the measurement zone to the solution phase. Thus, electrochemistry can be used to trigger the uptake or release of a species of interest into or from the polymer film from or into the solution phase. These (ionic) species can be either chosen for spectroscopic properties that make them amenable to tagging or derivatized in situ with a dye or a luminophore [139–143]. Use of a thin-layer cell in these instances enhances the analytical sensitivity of the method by shrinking the solution volume.

The spectroscopic measurement can be made in response to either a potential (or current) step or a potential sweep. Examples of the latter approach are contained in Fig. 20.28a and 20.28b for UV-vis and Raman scattering measurements on a polypyrrole film electrode in aqueous media [144]. The dA/dt and dI/dt (A = optical absorbance, I = Raman scattered light intensity) derivative signals versus potential have the shapes of cyclic voltammograms with the important difference that the spectroscopic signals are species-selective.

Another effective approach is to use an ac modulation for the potential and measure an optical transmittance or reflectance response signal expressed as $\Delta T/\Delta E$ and $\Delta R/\Delta E$, respectively [145–147]. Figure 20.29 shows an equivalent circuit for a thin-film modified electrode (Fig. 20.29a) and the corresponding complex plane impedance plots of the electrical and optical ac responses (Fig. 20.29b) [147]. The optical response is seen to differ from the electrical response at high frequencies. On the other hand, the shapes of electrical and optical responses are similar at low frequencies in the limiting film capacity region (see above). Thus, the electrical and optical probes respond differently to various electrochemical processes such as Faradaic charge transfer, absorption, and double-layer charging. AC perturbation has also been employed for probe beam deflection [148] as discussed in the next section.

I. Miscellaneous Techniques

In this section, we briefly review other electrochemical (and related) methods that have proven useful for the study of conducting polymers.

1. Dual-Electrode Voltammetry

The technique of dual-electrode voltammetry [149] uses a "sandwich" configuration wherein vapor deposition of a thin porous metal (usually Au) film is used over a

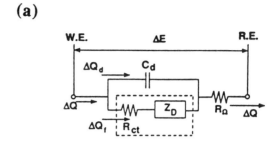

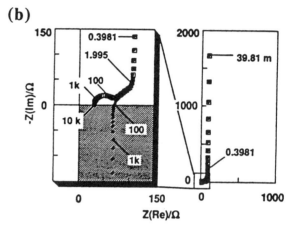

Fig. 20.29 (a) Equivalent circuit and (b) the electro-optical complex plane plot for a polypyrrole/polystyrenesulfonate composite film. Z_D is a charge transport impedance within the polymer film (modeled as a transmission line circuit). (□) Electrical data; (♦) optical data. ΔE is the applied ac potential, and the total charge ΔQ is divided into double-layer charging (ΔQ_d) and Faradaic (ΔQ_f) components. (Reproduced with permission from Ref. 144.)

polymer-coated Pt disk electrode. The porous electrode is contacted by the solvent. The gold film also covers an adjacent Pt disk that affords potential control at the polymer/solution interface. With the electrode immersed in an electrolyte solution, a four-electrode potentiostat is used to maintain a small constant potential difference (nominally 10 mV) across the polymer film while its average potential is slowly scanned. The steady-state voltammogram thus obtained can be converted to a conductivity vs. potential plot using Ohm's law.

2. Probe Beam Deflection

Refractive index changes in a quiescent electrolyte produced by an electrochemical reaction at a working electrode/solution interface can be detected either interferometrically [123] or by probe beam deflection (PBD) [148]. The "mirage" deflection is due to both a temperature gradient and a concentration gradient in the electrolyte [150]. It is the latter effect that has been exploited for the study of conducting polymers [151–156]. For example, the expulsion of ions from the polymer film into the solution bulk (cf. Section I.D) will give rise to a concentration increase of the latter in the solution phase and a corresponding refractive index gradient. Representative "cyclic deflectogram" data are graphed in Fig. 20.30 and compared to the corresponding cyclic voltammograms for polypyrrole tosylate in sodium tosylate electrolyte [156]. Note that PBD can be monitored in response to a potential step, a potential sweep (as in Fig. 20.30b), or even an ac perturbation as discussed in an earlier section.

3. Scanning Electrochemical Microscopy

Recent innovations in microscopy include the use of scanning probes that sense changes either in the tunneling current [scanning tunneling microscopy (STM)] or in the interfacial force [atomic force microscopy (AFM)]. The electrochemical counterpart of STM uses the perturbation in Faradaic current flow at a tip electrode by the presence of the test surface for imaging purposes. Figure 20.31 presents a schematic diagram of the operational modes of this scanning electrochemical microscopy (SECM) setup [157]. The "feedback" process is an important feature of the SECM method. The direction of the current feedback (i.e., positive or negative) indicates the electrical nature of the test surface (i.e., whether electronically conducting or insulating), while the magnitude of the signal gives an indication of the distance of the tip from the test surface or, alternatively, an indication of the rate of species turnover at the test surface. More recently, this technique has been used as an in situ probe of thickness and morphological changes at a film-covered electrode/solution interface [158].

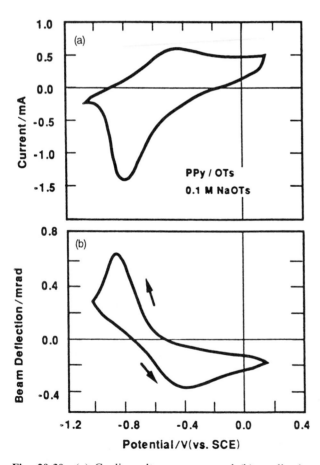

Fig. 20.30 (a) Cyclic voltammogram and (b) cyclic deflectogram polypyrrole tosylate in sodium tosylate electrolyte. (Reproduced with permission from Ref. 156.)

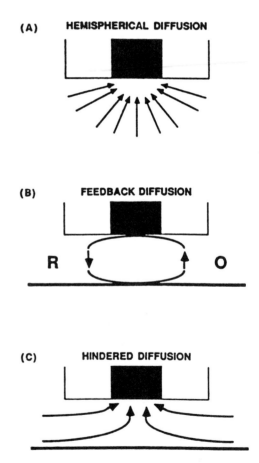

Fig. 20.31 Basic principles of scanning electrochemical microscopy (SECM). (A) Hemispherical diffusion far from electrode; (B) feedback diffusion near a conductive electrode; (C) hindered diffusion near an insulating electrode surface. The probe's electroactive area is shaded in the diagrams. (Reproduced with permission from Ref. 157.)

4. Volta Potential Difference Measurements

The state of the polymer surface and the polymer/electrolyte interface can be probed by measurement of the work function after initial polarization of the polymer-coated electrode in a liquid electrolyte. The latter is then withdrawn from the electrolyte ("emersed") under potential control, and then the work function measurement is performed [159,160]. The "emersion" procedure is schematized in Fig. 20.32. A widely used strategy for the work function measurement is based on the vibrating capacitor plate or the Kelvin probe method [161]. These methods rely on the fact that between two conducting and connected surfaces there exists a contact potential difference (CPD) because of the difference in work functions [162]. Changes in capacitance are induced by periodic vibration of one of the capacitor plates around its equilibrium position. The resulting ac current i is detected with a phase-sensitive detector and is given by

$$i = \Delta\psi \frac{dC_m}{dt} \quad (18)$$

The Volta-potential difference $\Delta\psi$ between the test electrode and the "reference" electrode is given by the sum of the CPD ($\Delta\Phi/e$) and the voltage V_c of the variable voltage source. At measurement null,

$$V_c = -\Delta\Phi/e \quad (18a)$$

The operating principle is schematized in Fig. 20.33, and Fig. 20.34 illustrates how the working electrode emersion cell is coupled with the Kelvin vibrator [159].

Analysis of work function data on polymer film covered test electrodes affords valuable insights into the electrical surface potential and thus the surface structure in the emersed state. Additionally, the equilibrium potential across the polymer/solution interface (the Donnan potential difference) is experimentally accessible (see Chapter 3 of Ref. [8]).

5. Radioisotopic Labeling

Labeling a target ion with a radioactive tag element and counting either the modified surface or the solution in the cell in response to an electrochemical perturbation is a rather classical procedure for monitoring ion fluxes at the film/solution interface [163,164]. Thus, ^{35}S-labeled

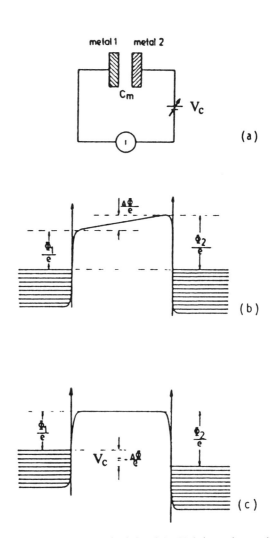

Fig. 20.33 Operating principle of the Kelvin probe method for measuring the work function.

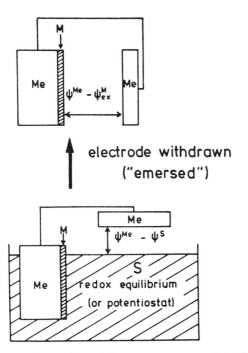

Fig. 20.32 Illustration of the measurements of the Volta potential differences by emersion of the electrode from the bathing solution medium. M symbolizes the polymer layer. For details, see Chapter 3 of Ref. [8].

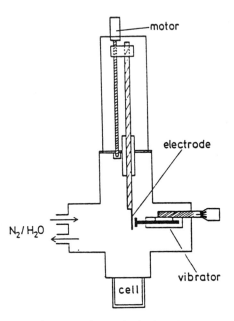

Fig. 20.34 The experimental device for measurement Volta potential differences. The position of the Kelvin vibrator is adjusted with a micrometer drive.

sulfuric acid and ^{36}Cl-labeled HCl and $HClO_4$ can be used for monitoring HSO_4^-, Cl^-, and ClO_4^- ion fluxes at polypyrrole/solution interfaces (see below).

III. THE ELECTROCHEMICAL EQUILIBRIUM STATE OF CONDUCTING POLYMERS

A. Basic Concepts

1. The Equilibrium State Neglecting Charge Compensation

Consider the polymer-coated electrode in contact with a redox electrolyte (Fig. 20.1b). When the system is undisturbed, electrons will exchange across both the Me/poly and poly/S interfaces until the state of equilibrium is attained. This state is characterized by the condition that the partial molar free energy of electrons, i.e., the electrochemical potential of electrons $\bar{\mu}_e$ in the three phases, is the same [165–167]:

$$\bar{\mu}_e^{Me} = \bar{\mu}_e^{poly} = \bar{\mu}_e^{S} \qquad (19)$$

The electrochemical potential of electrons in the metal corresponds to the Fermi level in the metal. It is defined by

$$\bar{\mu}_e^{Me} = \mu_e^{Me} - F\phi^{Me} \qquad (20)$$

where μ_e^{Me} is the chemical potential of the electrons in the considered metal, F is Faraday's constant, and ϕ is the electric potential of the phase characterized by the superscript.

For the polymer phase, the redox equilibrium between the electroactive sites (ox, red) of the polymer defines $\bar{\mu}_e^{poly}$:

$$\bar{\mu}_e^{poly} = \bar{\mu}_{red}^{poly} - \bar{\mu}_{ox}^{poly} \qquad (21a)$$

$$= (\mu_{red}^{0(poly)} - \mu_{ox}^{0(poly)}) + RT \ln \frac{a_{red}^{poly}}{a_{ox}^{poly}} - F\phi^{poly} \qquad (21b)$$

where μ^0 is the standard chemical potential and a the activity of the indicated species. The electrochemical potential of electrons in the redox electrolyte is defined analogously:

$$\bar{\mu}_e^{S} = \bar{\mu}_{red}^{S} - \bar{\mu}_{ox}^{S} \qquad (22a)$$

$$= (\mu_{red}^{0(S)} - \mu_{ox}^{0(S)}) + RT \ln \frac{a_{red}^{S}}{a_{ox}^{S}} - F\phi^{S} \qquad (22b)$$

2. The Ion-Partitioning Equilibrium

Every mobile species present, including the solvent, will be partitioned between the polymer and the solution phase [7,8,168–170]. At interfacial equilibrium, Eq. (23) will be valid for each species i:

$$\bar{\mu}_i^{S} = \bar{\mu}_i^{poly} \qquad (23)$$

Accordingly, each ionic species adjusts its partitioning equilibrium according to

$$a_i^{poly} = a_i^{S} \exp\left(-\frac{\Delta\mu_i^0}{RT}\right) \exp\left[-\frac{z_iF}{RT}(\phi^{poly} - \phi^S)\right] \qquad (24a)$$

where

$$\Delta\mu_i^0 = \mu_i^{0(poly)} - \mu_i^{0(S)} \qquad (25)$$

The ion-partitioning equilibrium, Eq. (24a), is usually formulated as

$$a_i^{poly} = a_i^{S} k_{part} \exp\left[-\frac{z_iF}{RT}(\phi^{poly} - \phi^S)\right] \qquad (24b)$$

where k_{part} is the partitioning equilibrium constant defined by

$$k_{part} = \exp\left(-\frac{\Delta\mu_i^0}{RT}\right) \qquad (26)$$

3. Electroneutrality

In the polymer and in the electrolyte phases the condition of electroneutrality may be formulated as

$$\sum_i z_i c_i = 0 \qquad (27)$$

where z_i is the ionic charge of species i and c_i is its concentration.

On the basis of these three concepts the electrochemical equilibrium behavior of the electrodes coated with conducting polymer films is discussed in the following.

B. Redox Polymers in Inert Electrolytes

1. General Considerations

In the following, the equilibrium state of an electroactive polymer in an electroinactive electrolyte is considered (Fig. 20.1a). Using an inert electrolyte reduces the complexity of the system; it is of interest also because the oxidation/reduction of the polymer in inert electrolytes is a subject of great technical potential (batteries, etc.). For characterizing these systems, one has to consider both the ion-partitioning (ion-exchange) equilibrium across the polymer/solution interface [Eqs. (23)–(27)] and the electronic equilibrium between the electrode and the polymer phase [Eqs. (19)–(22)].

Polymers of the type shown in Fig. 20.2a are considered typical redox polymers. They contain bound ionic sites in the reduced and oxidized states. Thus, they fall into the category of fixed-charge or ion-exchange polymers [170–176]. Relevant aspects of the behavior of such polymers are discussed in the following.

Some redox polymers contain electroactive nonionic sites (e.g., ferrocene units). In this case, membrane concepts apply that are discussed along with the neutral conjugated polymers (see Section III.C.2.

2. Ion Partitioning Between Electrolyte and Polymer Containing Bound Ionic Sites

Consider that the electrolyte contains the salt K_mX_n that dissociates into the cations K^{z_K} and the anions X^{z_X}. The electroneutrality condition in the polymer phase is

$$z_K c_K^{poly} + z_X c_X^{poly} + z_f c_f = 0 \qquad (28)$$

where the subscript f characterizes the nonexchangeable (bound) ionic sites. Stoichiometric considerations indicate that the density of bound ionic charge will typically have values on the order of ≈ 1 M. According to Eqs. (27) and (28), the polymer must contain at least the same concentration of counterions. Therefore, these polymers can usually be considered to constitute hydrophilic polyelectrolytes [170–176], as opposed, for instance, to the hydrophobic neutral conjugated polymers that are discussed in Section III.C.

For analyzing this situation, consider first Eq. (24a), written in an explicit form in terms of free energies, whereby the activity coefficients (γ_i of species i in the phase defined by superscript) are introduced:

$$z_i F(\phi^{poly} - \phi^S) + RT \ln \frac{c_i^{poly}}{c_i^S} + RT \ln \frac{\gamma_i^{poly}}{\gamma_i^S} + \Delta \mu_i^0 = 0 \qquad (29)$$

Practically, for such polyelectrolytes, separation of the nonelectrostatic free energy of transfer into a "medium effect," characterized $\Delta \mu_i^0$, and an activity coefficient term, $RT \ln(\gamma_i^{poly}/\gamma_i^S)$, becomes ambiguous. For describing partitioning equilibria of ions between liquid electrolytes and polyelectrolyte gels it is therefore customary and convenient [7,170,177] to set $\Delta \mu_i^0 = 0$. Any deviations from ideality of i will be reflected in the activity coefficients γ_i^S and γ_i^{poly}. The corresponding reference states are the (possibly hypothetical) infinitely diluted polymer and electrolyte, i.e., $\mu_i^{0(poly)} = \mu_i^{0(S)}$. The partitioning equilibrium of i between the electrolyte and such ionic polymers can consequently be formulated as

$$a_i^{poly} = a_i^S \exp\left[-\frac{z_i F}{RT}(\phi^{poly} - \phi^S)\right] \qquad (30)$$

or, in terms of concentrations of i,

$$c_i^{poly} = c_i^S \frac{\gamma_i^{poly}}{\gamma_i^S} \exp\left[-\frac{z_i F}{RT}(\phi^{poly} - \phi^S)\right] \qquad (31)$$

Thus, the free energy of transfer of i from electrolytes to polyelectrolytes is defined by

$$\Delta \mu_i^T = RT \ln \frac{\gamma_i^{poly}}{\gamma_i^S} \qquad (32)$$

where $\Delta \mu_i^T$ might be termed the "local relaxation" free energy of transfer.

When aqueous solutions of some uniunivalent salts (e.g., alkali chlorides or fluorides) are considered, one may assume that the ions do not specifically interact with the polymer. In such cases it is acceptable to disregard the differences in the activity coefficients in the two phases, i.e., the assumption may be made that $\Delta \mu_i^T \approx 0$. One obtains the equation for the interfacial ionic equilibrium essentially in a form as derived originally by Donnan [178]:

$$RT \ln \frac{c_i^M}{c_i^S} + z_i F \Delta \phi_D = 0 \qquad (33)$$

whereby the equilibrium potential difference across the interface is now termed the Donnan potential, $\Delta \phi_D = (\phi^{poly} - \phi^S)$.

The dependence of the Donnan potential on the electrolyte concentration can be calculated using Eq. (33) in combination with the electroneutrality condition within the membrane phase. For a $(1,-1)$-valent electrolyte of concentration c^S, it is given by the equation

$$(\phi^{poly} - \phi^S) = \Delta \phi_D = \frac{z_f RT}{|z_f|F} \ln\left[\frac{|z_f|c_f}{2c^S} + \left\{1 + \left(\frac{z_f c_f}{2c^S}\right)^2\right\}^{1/2}\right] \qquad (34)$$

Equation (34) shows that the sign of $\Delta \phi_D$ is determined by the sign of the bound ionic sites and that the absolute value of $\Delta\phi_D$ is a function of the ratio $z^f c^f / c^S$. Equation (33) and consequently Eq. (34) must fail whenever specific interactions or large differences in the solvation energy lead to strong differences of the activity coefficients in the two phases [170,174]. Despite these limitations, the above equations constitute the most practical concept for describing the equilibria across the interface between liquid electrolyte and fixed-charge polymer.

This concept is used now to illustrate the interfacial situation that is expected to prevail between the Ru(bipy)$_2^{3+}$ polymer of Fig. 20.2 and a noninteracting aqueous (1,1)-electrolyte. The interfacial situation is shown schematically in Fig. 20.35 [189]. Note the excess of anions ($\ominus$) that accumulate in the solution near the interface in response to the electric potential distribution shown schematically in the lower part of this illustration.

The values of $\Delta\phi_D$ calculated with Eq. (34) and the corresponding concentrations of univalent counterions, c_-^{poly}, and co-ions, c_+^{poly}, are shown in Fig. 20.36 for a typical total positive bound-charge density of 1 M, i.e., for the polymer with $z_f = +3$, the value of c_f is 0.33 M.

The corresponding features of polymers with negative bound charges can be derived analogously from Eqs. (33) and (34). Considerations as represented in Figs. 20.35 and 20.36 are of central relevance for an understanding of the electrochemistry of redox polyelectrolytes. At electrolyte concentrations above the fixed-charge density, the polymer becomes "solution-like." It contains both co-ions and counterions at substantial concentrations. The effect of $\Delta\phi_D$ becomes insignificant. On the other hand, as the electrolyte concentration decreases relative to $|z_f c_f|$, the effect of the Donnan potential becomes significant. The concentration of counterions is then essentially constant and corresponds to

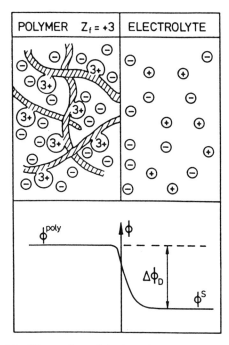

Fig. 20.35 Illustration of the interface between a polymer containing bound sites of valence $z_f = +3$ and a $(1,-1)$-electrolyte [($\oplus$) Mobile cations; ($\ominus$) mobile anions constituting the counterions]. ϕ is the electric potential, $\Delta\phi_D$ is the resulting (positive) Donnan potential.

the fixed-charge concentration, while the co-ion concentration becomes very small and varies by one decade per 59 mV change in $\Delta\phi_D$ (for $z = \pm 1$). This is the condition known as "Donnan exclusion" of co-ions.

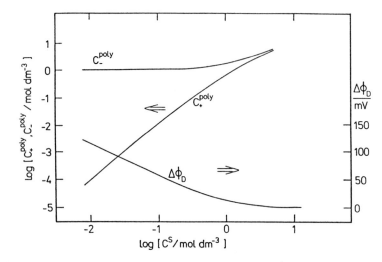

Fig. 20.36 The equilibrium concentrations of mobile univalent anions (c_-^{poly}) and cations (c_+^{poly}) partitioned from the electrolyte of concentration c^S into the polymer phase that contains a total bound-charge concentration of $z_f c_f = 1$ M; $\Delta\phi_D$ is the Donnan potential.

3. The Electrochemical Equilibrium State of Redox Polymers

Since the electrolyte is inert, redox equilibrium according to Eq. (19) can be established only between the metal electrode and the polymer redox sites. Equation (19) can thus be written in the following appropriate form:

$$\phi^{Me} - \phi^{poly} = \frac{\mu_e^{Me} - \mu_{red}^{0(poly)} + \mu_{ox}^{0(poly)}}{F} + \frac{RT}{F} \ln\left(\frac{a_{ox}^{poly}}{a_{red}^{poly}}\right) \quad (35)$$

which is a form of the Nernst equation for this electrochemical equilibrium. Note that normally the electrode potential E of a coated electrode is measured versus a reference electrode in the contacting electrolyte:

$$E = (\phi^{Me} - \phi^{poly}) + (\phi^{poly} - \phi^S) - \Delta\phi^{ref} \quad (36)$$

where $\Delta\phi^{ref}$ is a constant defined by the electrode metal and by the type of reference electrode used. One arrives at the important conclusion that at an equilibrium state characterized by a certain ratio of the activities of redox sites in the polymer ($a_{ox}^{poly}/a_{red}^{poly}$ = constant) the measurable electrode potential E varies in parallel with $\phi^{poly} - \phi^S$. In the case of the considered fixed-charge polymers, $\phi^{poly} - \phi^S = \Delta\phi_D$, i.e., the equilibrium potential E varies with the electrolyte concentration [179–183].

One can discuss this in a quantitative way by inserting appropriate expressions defining $\phi^{Me} - \phi^{poly}$ and $\Delta\phi_D$. Consider, for instance, the situation obtained with a noninteracting (1,1)-electrolyte of a concentration that is smaller than the bound-charge density. Then Eq. (34) simplifies to

$$\phi^{poly} - \phi^S = \Delta\phi_D \approx \frac{z_f RT}{|z_f| F} \ln\left(\frac{|z_f| c_f}{c^S}\right) \quad (37)$$

so that one may formulate for this system

$$E = \left(\frac{\mu_e^{Me} - \mu_{red}^{0(poly)} + \mu_{ox}^{0(poly)}}{F}\right) + \frac{RT}{F} \ln\left(\frac{a_{ox}^{poly}}{a_{red}^{poly}}\right) + \frac{z_f RT}{|z_f| F} \ln\left(\frac{|z_f| c_f}{c^S}\right) - \phi^{ref} \quad (38)$$

The experimental results summarized in Fig. 20.37 illustrate the behavior expected from Eq. (38). Note that in this experiment the electrode has been coated with an inorganic redox polymer of the Prussian blue type [182,183].

Of course, one might formulate an equation for the electrochemical equilibrium such that the half-cell includes the incorporation or expulsion of charge-compensating ions. For example,

$$Ox^{+(poly)} + X^{-(poly)} + e^- \rightleftharpoons Red^{poly} + X^{-(S)} \quad (39)$$

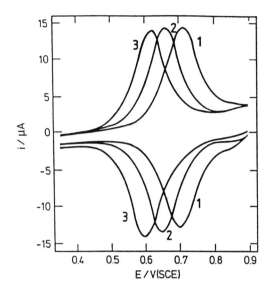

Fig. 20.37 Cyclic voltammograms of glassy carbon electrodes coated with a copper hexacyanoferrate film. Electrolytes: Aqueous KCl of concentration 1 M (1); 0.1 M (2); 0.01 M (3). (Reproduced with permission from Ref. 182.)

which leads to the appropriate Nernst equation that describes the shift of the voltammetric wave on the potential axis as a function of the electrolyte concentration in a way similar to Eq. (38):

$$E = E^{0(poly)} + \frac{RT}{F} \ln\left(\frac{a_{ox}^{poly} a_X^{poly}}{a_{red}^{poly} a_X^S}\right) \quad (40)$$

C. Conjugated Polymers in Inert Electrolytes

1. General Considerations

There are two major differences between conjugated and redox polymers. These differences are of a practical rather than a fundamental nature. First, the native conjugated polymers can be brought into a state in which they do not contain charged sites (neutral polypyrrole, polythiophene, etc.). In this state the polymer is rather hydrophobic; oxidation/reduction to an ionized state involves a significant structural change. Second, the polymeric redox sites (polarons, etc.) are less clearly defined than redox polymer sites, in particular with respect to their electronic energy levels.

2. The Membrane State of Conjugated Polymers Not Containing Bound Ions

Conjugated polymers in the electroneutral (hydrophobic) state are now considered. The contacting electrolyte contains ions in a polar solvent, e.g., water. In this case, the ions will be stabilized in the electrolyte rather than in the polymer phase, i.e., the standard free energy of transfer of the ionic species, $\Delta\mu_i^0$, will be positive [Eq. (25)].

The requirement of electroneutrality has the consequence that across the polymer/electrolyte interface a "distribution potential" drop builds up that is defined by [30,184]

$$\phi^{poly} - \phi^S = \frac{\Delta\mu_X^0 - \Delta\mu_K^0}{F(z_K - z_X)} \quad (41)$$

The sign of this potential drop is such that it causes the more hydrophobic ion to be rejected from the hydrophobic polymer phase while the more hydrophilic one is pulled into the polymer. Thus, the difference in hydrophobicity of the ions is compensated for, and equal anionic and cationic charge densities in the polymer result. Of course, in the case of a hydrophobic polymer and a polar solvent, the concentration of (hydrophilic) ions in the polymer may be extremely small [7].

The distribution of the electric potential across the interface between the coated electrode and an inert electrolyte is illustrated in Fig. 20.38 for three applied electrode potentials, $E_1 > E_2 > E_3$.

The applied electrode potential generates an electrical double layer at the metal/polymer interface, with excess ionic charge accumulating in the polymer phase near the interface (we assume that the polymer is neither oxidized nor reduced in the potential range E_1–E_3). According to the concepts governing the distribution of ions in electrical double layers, the diffuse "ion cloud," and thus the potential drop, will penetrate more deeply into the polymer phase when the bulk ion concentration in the polymer is smaller. This is illustrated in Fig. 20.38 by the electrical double layers 1 and 2.

The equilibrium concentrations of ions in polymers can be estimated on the basis of an electrostatic model, where the organic phase is assumed to be a homogeneous dielectric characterized by its dielectric constant ϵ^{poly} (see, e.g., Section 4.2.3 of Ref. 7). From such considerations it follows that the equilibrium ion concentration in the polymer may easily be so low that the electrical double layer would extend beyond the thickness of the polymer film. Eventually, the film may be considered free of ions, acting as a dielectric layer between the metal and the electrolyte. This situation is illustrated in Fig. 20.39.

The situation of Fig. 20.39 is characterized by the fact that the measurable interfacial capacitance C is defined by the parallel-plate capacitor model:

$$C = A \frac{\epsilon^{poly}\epsilon_0}{d} \quad (42)$$

where A is the electrode area, ϵ_0 is the permittivity of free space, and d is the film thickness.

3. Conjugated Polymer in an Ionic State

When the conjugated polymers are oxidized or reduced, ionic sites (polarons, etc.) are formed by extracting electrons from (or injecting electrons into) the π system. These sites are mobile, particularly along the π system, and render the polymer electronically conductive. However, they are confined to the polymer phase, i.e., they constitute bound charges that are nonexchangeable. Thus, they require counterions for charge compensa-

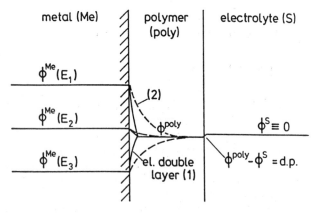

Fig. 20.38 Illustration of the electric potential distribution (ϕ) across the interface between a metal electrode (Me) coated with a neutral polymer (poly) film and an electrolyte solution containing ions that are partitioned into the polymer. E_1, E_2, E_3 are three applied electrode potentials at which the polymer is neither oxidized nor reduced. The electrical double layers (1) and (2) correspond to two different bulk concentrations of ions in the polymer. d.p. is the distribution potential.

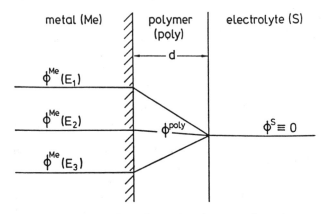

Fig. 20.39 Illustration of the electric state of a polymer-coated electrode, as in Fig. 20.38, but where the concentration of ions in the polymer is so small that the film acts as a dielectric layer.

tion. In this regard, conjugated polymers in the oxidized/reduced state constitute ion exchangers similar to ionic redox polymers.

Clearly, the transition of the conjugated polymer from the hydrophobic to the hydrophilic state, which is associated with oxidation/reduction, is a dramatic case. It is difficult to treat in a quantitative way (see Section IV). This is one reason it has been considered advantageous to introduce electroinactive bound cationic or anionic sites [186–196], such as the sulfonate groups of the polymer shown in Fig. 20.2b. If such groups are present, the effective density of bound charge (defining the ion-partitioning equilibrium) is the algebraic sum of the polaronic and the electroinactive fixed charge density. If the concentration of bound inactive ions is much larger than the maximum density of polaronic charge, the film will have at all oxidation states practically the same ion-partitioning properties. On the other hand, if the density of inert bound ions is relatively small and opposite in sign to the polaronic charge, an electronically conducting (oxidized/reduced) state of the polymer can be obtained in which the polaronic charge is exactly internally compensated. By oxidizing this polymer across this point of internal charge compensation, the polymer is transformed from an anion exchanger to a cation exchanger or vice versa. Such switchable membranes are presently a subject of great interest [172,173,181,197,198]. In Fig. 20.40, the switchable membrane consisting of a polypyrrole (PPy) film with incorporated (practically immobile) dodecylsulfate ions ($ROSO_3^-$) on a gold substrate is represented [173]. The transition from the cation exchanger (at negative potential) to an anion exchanger (at positive potentials) is shown in the upper diagram, representing the potential distribution, in which the change of the sign of the Donnan potential is shown schematically.

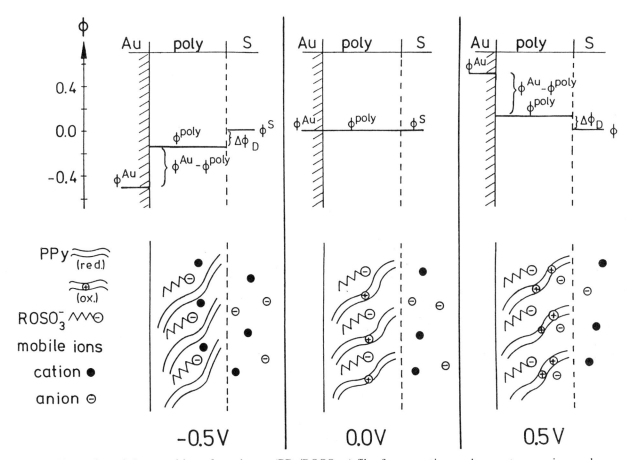

Fig. 20.40 Illustration of the transition of a polymer (PPy/$ROSO_3^-$) film from a cation exchanger to an anion exchanger phase, associated with the progressing electrochemical oxidation of the polymer (from $E = -0.5$ V to $+0.5$ V). The transition is characterized by a change in sign of the Donnan potential, $\Delta\phi_D$. $ROSO_3^-$ are the dodecyl sulfate ions that constitute practically fixed negative charges. Au is the gold electrode, ϕ the electric potential of the phase defined by the superscript.

The fact that the Donnan potential indeed varies in the way shown in Fig. 20.39 is strongly supported by the results of Volta potential measurements [172, 173,195]. The composition of the film, corresponding to the scheme of Fig. 20.40, has been determined analytically, for example with EDAX (energy-dispersive analysis of X-ray emission), Fig. 20.41 [173].

The thermodynamic aspects of electrochemical oxidation/reduction of the ionic conjugated polymer can be described in principle by the same concept that has been used with redox polymers, Eqs. (35)–(40). The question is, How should the polymer redox sites be associated with the actual polymer structure? The simplest way is to combine a number of monomer units (usually four or five) and consider them as the redox site. Consider, for example, the redox reaction for polypyrrole $(Py)_m$:

$$[(Py)_5^+]_n + nX^{-(poly)} \rightleftharpoons [(Py)_5]_n + nX^{-(S)} \quad (43)$$

where $m = 5n$. This representation is convenient in discussing internal charge compensation, for example, during oxidation of poly[(pyrrole)$_9$-co-(N-sulfopropylpyrrole)] [172]. The three characteristic states obtained when this polymer is oxidized can be represented in the way shown in Fig. 20.42 (cf. Figs. 20.2 and 20.40).

Thus the corresponding Nernst equation, analogous to Eq. (40), can be formulated:

$$E = E^{0(poly)} + \frac{RT}{F} \ln\left(\frac{a_{(py5)}^{poly} + a_X^{poly}}{a_{(py5)}^{poly} a_X^S}\right) \quad (44)$$

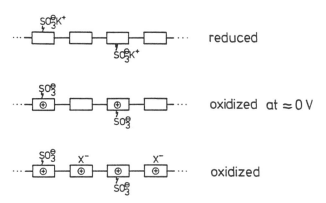

Fig. 20.42 The three characteristic states of the polymer poly[(pyrrole)$_9$-co-(N-sulfopropylpyrrole)] obtained by progressively oxidizing the coating from the completely reduced state in an experiment analogous to the one represented in Fig. 20.40. The rectangles symbolize groups of five pyrrole units.

As indicated before, the electronic energy levels in the conjugated polymers (as estimated, for instance, from pseudoreversible voltammetric oxidation/reduction currents) are generally distributed over a considerable range. Thus, in order to match voltammetric waves obtained on the basis of Eq. (44) with (quasi-reversible) experimental ones, activity coefficients are necessary

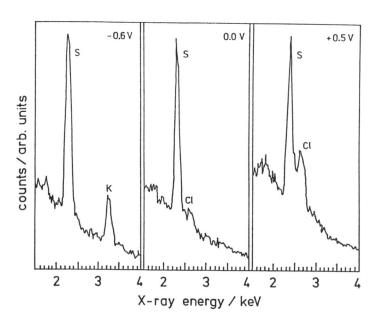

Fig. 20.41 Energy-dispersive analysis of X-ray emission (EDAX) spectra of the PPy/ROSO$_3^-$-coated gold electrode of Fig. 20.40. For the analyses the electrode was withdrawn from the electrolyte (KCl) at the indicated electrode potentials. The elements corresponding to the X-ray emission peaks from the polymer are indicated by their chemical symbols.

that are difficult to rationalize in a quantitative way. Thus, other formalisms have also been employed to describe the equilibrium state of the system, e.g., "equilibrium charging curves," see Refs. 184 and 199.

Of course, in the same way as with the redox polymers [Eqs. (36)–(40)], the equilibrium potential E (measured versus a reference electrode in the electrolyte) varies with the potential drop $\phi^{poly} - \phi^S$. Consider that the oxidation or reduction of a neutral polymer is associated with the transition of $\phi^{poly} - \phi^S$ from an electric distribution potential to a Donnan potential. It is clear that the quantitative description of such reactions is not an easy task. More is said on this subject in Section IV.

D. Coated Electrodes in Equilibrium with Redox Electrolytes

The situation of coated electrodes in equilibrium with redox electrolytes is important for electrocatalysis, where the oxidation/reduction of a solution species is mediated by polymeric redox sites, as illustrated in Fig. 20.1b. The equilibrium condition for this system can be readily understood on the basis of the concepts presented above, Eqs. (19)–(22). Unfortunately, the kinetics of the mediated electron transfer processes are more complicated, as shown below.

IV. THE DYNAMICS OF ELECTROCHEMICAL REACTIONS

A. Redox Polymer Oxidation/Reduction in Inert Electrolytes

As shown above, the typical redox polymer contains an appreciable density of mobile ionic species. Thus, the ionic conductivity of the polymer phase is rather high. An electrical double layer can form at the metal/polymer interface, i.e., the electric potential drops sharply at the polymer side of the interface (see Fig. 20.38). Consequently, the polymer-bound redox sites that contact the metal electrode surface may be oxidized/reduced by a conventional electrochemical mechanism.

Starting from the electrode surface, the reduction/oxidation of the bulk polymer phase proceeds by the mechanism of redox electron hopping. Since the redox sites are solvated, electron exchange between neighboring redox sites, as known from redox systems in liquid solutions, proceeds. An analysis of electron transport based on this type of electron exchange in electrolytes was first given in 1968 [200]. Consider a phase in which the concentration gradients of Ox and Red in the x direction constitute the driving force for electron transfer. At two positions, x_1 and $x_1 + \lambda$, where λ is the distance across which electron exchange proceeds, the concentrations of the considered species, c_i, differ by $\lambda(dc_i/dx)$. Electron exchange between the redox sites at the two positions produces "forward" and "back" electronic current. The resulting net electronic current density, i_e, is

$$i_e = Fk_{ex}\lambda^2 \left(c_{ox} \frac{dc_{red}}{dx} - c_{red} \frac{dc_{ox}}{dx} \right) \quad (45)$$

where k_{ex} is the homogeneous second-order rate constant for electron exchange between Ox and Red. Under the condition that the sites are fixed and the total concentration of sites is constant, the concentration gradients of Ox and Red are approximately equal. The current density is then

$$i_e = Fk_{ex}\lambda^2(c_{ox} + c_{red}) \frac{dc_i}{dx} \quad (46)$$

Equation (46) resembles Fick's first law, and by comparing the coefficients one can define an "electronic diffusion coefficient," D_e:

$$D_e = k_{ex}\lambda^2(c_{ox} + c_{red}) \quad (47)$$

This concept was later developed [201] and applied successfully to electronic charge transport via redox sites ("redox hopping") in polymers [19–26]. The model is now generally accepted for this process; it is represented schematically in Fig. 20.43. Experimental values of the electronic diffusion coefficients can be obtained, e.g., via electrochemical impedance measurements [199] or probably best in steady-state experiments [20]. Typically, values of $D_e \approx 10^{-10}$ cm^2/s or smaller are found. The diffusion coefficients of small counterions are typically larger than that by several orders of magnitude. Thus, the counterion motion associated with the hopping process will not be limiting; rather, the observed Faradaic current will reflect the electron hopping rate.

However, it should be noted that values of D_e calculated on the basis of Eq. (47) can be at variance with experimental values by orders of magnitude [21]. This is understandable, because in a real system electron transfer is not likely to proceed across only one value of λ but over a distribution of distances. At each distance a different rate constant k_{ex} will be effective. Usually, neither the exact distribution of redox sites in the polymer nor the dependence of k_{ex} on λ in the polymer phase is known.

Furthermore, a variation of redox-site density (associated with the oxidation/reduction of the film or with preparing films with different total site concentrations) leads normally to changes of the density of osmotically active ions. For example, during the oxidation of the Ru(bipy)$_2^{2+/3+}$ sites in the redox polymer of Fig. 20.3, the fixed-charge density, and thus the concentration of charge-compensating counterions, rise by 50%! This leads to differences in the degree of swelling. Furthermore, it is possible that the redox centers in the reduced state interact differently with the polymer matrix than in the oxidized state; for example, Fe(CN)$_6^{3-}$ interacts

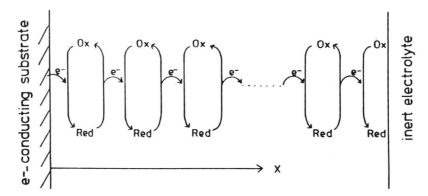

Fig. 20.43 Illustration of the electron hopping process, which leads to reduction of a redox polymer film containing the bound redox sites Ox and Red. The process is associated with a flux of charge-compensating ions into or out of the polymer phase (not shown). See also Fig. 20.19.

more strongly with a polyvinylpyridinium matrix than $Fe(CN)_6^{4-}$ [202]. Such structure changes tend to affect the ionic mobilities and the electron transfer rate [203,204]. Special experimental arrangements have been designed to minimize this problem [203,204].

Another difficulty is the fact that it is not always clear if the observed currents are indeed defined by redox hopping or if the redox sites move themselves. Diffusion-type current–time and current–voltage relations are obtained for both mechanisms. The situation is particularly serious with systems like $Os(bipy)_3^{2+/3+}$ or $Ru(bipy)_3^{2+/3+}$ in Nafion. The hydrophobic interactions between the bipyridyl ligands and the hydrophobic polymer matrix lead to small physical diffusion coefficients that may just be of the same order of magnitude as the D_e values [21,205,206].

Even in cases where the redox sites are immobilized by chemical bonding to the matrix, they will still have a certain local mobility. One can therefore describe the electron transport by assuming that electron exchange between the redox sites does not take place over the distance λ, as assumed above, but that the donor and acceptor sites come into intimate contact for electron transfer [26,207]. This mechanism was proposed to be operative, for example, with iron redox sites embedded in a plasma polymer matrix [26].

B. Oxidation/Reduction of Conjugated Polymers in Inert Electrolytes

1. General Consideration

Both the transition between a hydrophobic and a hydrophilic state (see above) and the high electronic conductivity of the conjugated conducting polymers lead to considerable practical differences in the dynamics of redox switching between these systems and redox polymers.

2. Transition Between the Hydrophobic and Hydrophilic States

When an electrochemical potential step is applied to the electrode coated with the conjugated polymer film in the neutral state, which is in contact with a polar solvent, the film may act in a way closely similar to that of the insulating barrier film shown in Fig. 20.39. The observed current flow, characterizing the rate of film oxidation, shows typically a nucleation-type behavior. The current transients of Fig. 20.44a are the result of stepping the electrode potential from a value of $E = -0.8$ V vs. SCE, at which the polymer (polypyrrole) is in the neutral, hydrophobic state, to $E = +0.4$ V vs. SCE, where the polymer is strongly oxidized and contains a high concentration of ions.

Quantitative description of such transients is difficult [208–212]. It has been proposed that the oxidation does indeed start from a weak spot in the film, from where the oxidation front grows radially, parallel to the metal surface [212].

3. The Polymer Matrix Contains Mobile Ionic Charges

When the oxidation/reduction of the film starts from a state in which it contains mobile ions [186–196], the reaction proceeds without the discussed inhibition. Clearly, the transient of Fig. 20.44b represents a significantly more desirable behavior as far as application in displays, etc. is concerned. To introduce electroinactive fixed sites, one can use polymers of the type shown in Fig. 20.2b or a mixture of the desired conjugated polymer with an electroinactive fixed-charge polymer such as the polystyrenesulfonate used in the experiment whose results are shown in Fig. 20.44b.

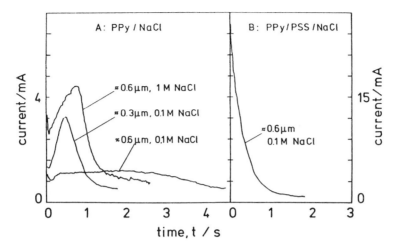

Fig. 20.44 Anodic current transients following potential steps from −0.8 V vs. SCE to +0.4 V, applied to electrodes that were coated with films of (A) polypyrrole or (B) polypyrrole/polystyrenesulfonate of the indicated thickness.

4. The Polymer as a Mixed Conductor

The high conductivity of the conjugated polymers is caused by the combined effect of both the high mobility of charge carriers along the polymer chain and a high rate of electron exchange (large k_{ex}) between the conjugated chains [213]. The electronic conductivity (σ_e) is, in fact, normally higher than the ionic conductivity in the polymer phase. This means that the rate of oxidation/reduction of the polymer is usually determined by the rate of counterion movement required for electroneutrality coupling between the electronic and ionic charges.

When a polymer of $\sigma_e \gg \sigma_{ion}$, such as polypyrrole in the oxidized state, is subjected to changes of the applied electrode potential, during the transient state electric fields develop in the polymer matrix and then disappear as electronic and ionic charge carriers migrate to new equilibrium positions [19,27,31,184]. The analyses may be based on the concepts derived for redox polymers under the condition that the hopping mobility of the electrons exceeds the counterion mobility. It has been shown [31] that in this case the system behavior is again "diffusional" in character. The coated electrode behaves like a porous metal electrode with pores of limited depth. Numerous experimental reports on this behavior of conducting polymers have appeared in the literature; the first was probably that of Bull et al. in 1982 [214].

5. Voltammetric Behavior of Conjugated Polymers

Consider for the following that electron transport across the polymer is fast. Furthermore, the polymer contains a reasonably large concentration of mobile ions. One might then anticipate that under conditions of cyclic voltammetry one should obtain reversible and easily interpreted current–voltage–time curves. However, this is generally not the case. Figure 20.45 shows the results obtained with linear polyphenylene deposited as a thin film on a platinum electrode [215]. The coated electrode was subjected to slow voltammetric potential scans. The film is solvated well enough that ionic charge transport required for electroneutrality coupling does not determine the rate of the redox process.

This voltammogram demonstrates two remarkable features. First, the anodic and cathodic voltammetric peaks are separated by 380 mV, which means that the

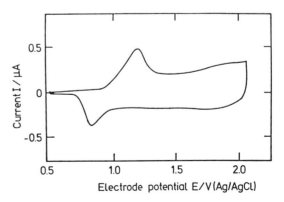

Fig. 20.45 Cyclic voltammogram of a polyphenylene-coated electrode in CH_2Cl_2/0.1 M TBAPF$_6$. (Reproduced with permission from Ref. 215.)

electrochemical redox reaction is "irreversible" in the electrochemical sense. Second, large anodic and cathodic currents of nearly constant value are observed over a potential range on the order of 1 V. Both these features have aroused considerable discussion [165,216–223].

The most probable and plausible explanations at present are as follows. Note that the large peak separation is essentially not a function of the potential sweep rate. This fact points to a chemical reaction following charge transfer [218,219]. Upon oxidation the molecules stabilize themselves from the twisted to a partially planar structure with better conjugation. In the case of aromatic polymers this is the transition from a more benzenoid structure to a quinoid structure. This interpretation of the observed irreversibility is supported by systematic work with the corresponding oligomers [215–218]. It has been shown that the voltammograms obtained with various oligomers dissolved in liquid solutions react in an electrochemically reversible way. However, when the same oligomers are deposited as films on an electrode, i.e., when the free rotation is hindered by the solid matrix, electrochemical irreversibility reappears. For the case of polythiophene the same group has presented a further explanation for the irreversibility (J. Heinze and K. Meerholz, private communication). They assumed that during oxidation crystalline domains are formed by π interactions between polymer chain segments, which stabilize the system. This view is supported by experiments with thiophene oligomers in solution, which show that the cationic species form dimeric π complexes [224].

These oligomer results also yield a plausible explanation for the constant-current region in the cyclic voltammogram of Fig. 20.45. Note that in the usual preparation procedures for conducting polymers little attention is given to the degree of polymerization. Normally one obtains a mixture of oligomers and polymers of different chain lengths and possibly different cross-linking. Considering the results of Fig. 20.45, it appears likely that numerous close-neighbor redox potentials are effective in an actual polymer. They are distributed over a considerable electrode potential range. The typical voltammogram is the superposition of all these redox waves. Based on this interpretation it is immediately clear that slight variations in the conditions of film preparation and of the electrolyte have considerable consequences for the resulting voltammograms.

It should be noted that truly irreversible electrochemistry in the chemical sense is a well-known and annoying feature with conjugated polymers. It may lead to cross-linking by further oxidative coupling or to nucleophilic attack of aromatic rings, typically introducing OH or halogen groups [225]. Such reactions are often termed "overoxidation" of the polymer [225,226]. They normally reduce the redox capacity of the polymer films in a most undesirable way.

C. Electrochemical Reactions Mediated by Conducting Polymer Films

1. General Considerations

One of the most promising aspects of polymer-modified electrodes was the prospect of finding coatings that improve the electrocatalytic properties of the electrode. It was hoped that redox systems that are known as homogeneous catalysts could be useful as electrocatalysts when attached to the electrode, for instance, in the form of a redox polymer [227,228]. The conjugated conducting polymers were expected to be catalytically active [229] for reactions such as the cathodic oxygen reduction [230].

At an early stage in the development of redox-polymer-modified electrodes, models for describing the mechanism of mediated electron transfer reactions were proposed (Section II). They have since been further developed and now constitute an accepted basis for analyzing electrocatalytic reactions on redox-polymer-coated electrodes. Considerable fundamental work has also been done in analyzing electron transfer mediation by conjugated polymer coatings [165,166,231–237]. From this perspective, an interesting system appears to be the combination of conjugated polymers with embedded catalytic sites [69,238–244].

2. Mediation of Electrochemical Reactions by Redox Polymers

In the following, the polymeric redox sites are termed P/P^+ to distinguish them from the solution redox species, O/R, which may, of course, also enter into the polymer phase. The mediated (catalytic) reaction is the chemical electron transfer reaction between the redox system from the electrolyte and the polymeric redox sites:

$$O + P \rightleftharpoons R + P^+ \tag{48}$$

The reacting polymer sites, P/P^+, are regenerated by interfacial electron transfer at the electrode and via the mechanism of redox electron hopping.

The driving force for the mediated reaction is a shift of the electrode potential E from the equilibrium to a nonequilibrium value, e.g., by the potential step from E_2 to E_3 shown in Fig. 20.46. For simplifying the situation, it is assumed that the mediated reaction [Eq. (30)] takes place at the polymer/electrolyte interface.

To understand the mechanism of this process, assume that the Donnan potential is not affected by the potential change from E_2 to E_3 (e.g., large concentration of electrolyte). Then the conditions represented in Eq. (25) and Fig. 20.38 prevail, i.e., the electric potential of the polymer phase remains unchanged. The new value of $\bar{\mu}_e$ adjusts by a change of the chemical potential of electrons, μ_e. That means that the ratio between the concentrations

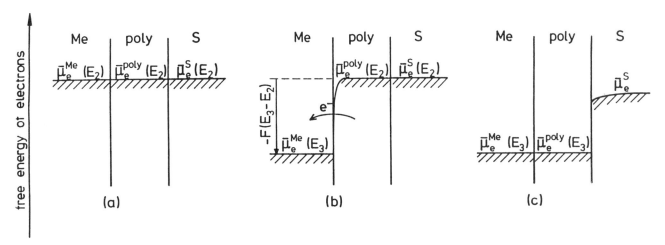

Fig. 20.46 Nonequilibrium states of a redox polymer coated electrode (Me/poly) that mediates the oxidation of a solution (S) species (e.g., R → O). The chemical charge transfer reaction takes place at the poly/S interface. $\bar{\mu}_e$ is the electrochemical potential of electrons in the considered phase. (a) The system at equilibrium at electrode potential E_2 [cf. Eq. (19)]. (b) The (nonequilibrium) situation shortly after changing E from E_2 to E_3. (c) Possible further developments, depending on the system parameters and time.

of the polymeric redox sites readjusts; in the case of Fig. 20.46, this occurs by redox electron hopping from the reduced polymer sites (P) to the electrode, until the new equilibrium concentrations of P^+ and P are achieved. The states of the Me/poly interface corresponding qualitatively to the states of Figs. 20.46a–c are represented schematically in Fig. 20.47.

As soon as the redox hopping front across the polymer reaches the polymer/electrolyte interface, the mediated redox reaction can start. Depending on the ratio between the rate of redox hopping transport across the polymer and the rate of the mediated electron transfer reaction, different gradients of the chemical potentials of electrons in the polymer and the electrolyte phase will develop. In the state characterized in Fig. 20.47b the polymer reaction front has hardly started to move. The situation of Fig. 20.47c corresponds to a relatively slow mediated reaction. Note that so far the assumption has

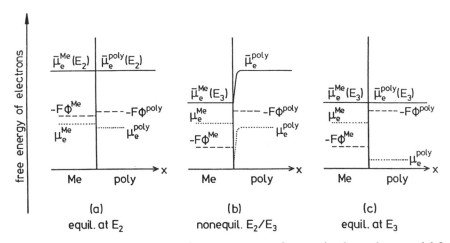

Fig. 20.47 Variation of the electronic energy terms in consequence of a step in electrode potential from E_2 to the more positive E_3. (a) Equilibrium at E_2; (b) nonequilibrium; (c) equilibrium at E_3. The polymer has a high mobile ion density, and the Donnan potential is assumed not to be affected by the film oxidation.

been made that the mediated reaction proceeds only at the poly/S interface. If the redox species enters (diffuses) into the polymer, so that Eq. (48) proceeds also inside the bulk polymer phase, the quantitative description of the reaction and transport rates becomes rather complex (see Fig. 20.18). The interested reader is referred to the literature [227].

3. Charge Transfer Mediated by Conjugated Polymer Coatings

In the conducting state, conjugated polymers normally contain a significant ion concentration. This means that the adjustment of an ionic conjugated polymer phase to a new equilibrium state corresponding to a new value of E proceeds in the same way as discussed with the redox polymers, i.e., by readjustment of the activity ratio of P^+/P, rather than by a new value of ϕ^{poly} (Figs. 20.38 and 20.40). It is therefore appropriate to consider the mechanism of electrocatalysis basically in a way similar to that of redox polymers. The high electronic conductivity of the conjugated polymers suggests that the situation involving a relatively slow catalyzed reaction might be well represented by Fig. 20.46c.

Note, however, that not only the concentration but also the energies of the polymer electronic states are functions of the degree of oxidation/reduction. For this reason, but in particular also in the case when the mediated reaction involves the conjugated polymer in the neutral state, a clearer description is obtained when the polymer is considered a semiconductor and the corresponding concepts of charge injection and transport are used, see, e.g., Ref. 245.

Basic understanding of such systems can still be derived from the redox polymer model. Consider the fact that the density of states varies according to a Nernst-type equation. Thus, on a linear concentration scale, a coating of the polypyrrole type may be practically completely in the reduced state, independent of the actually applied electrode potential. The rate of reduction of a depolarizer according to Eq. (48) may be defined by the current i_x [165,166],

$$i_x = -FAk_x \int_0^d c_P(x)c_O(x)dx \qquad (49)$$

Clearly, if the concentration of polymer sites is not a function of E, the rate of mediated reduction must also be constant. This behavior is easily demonstrated experimentally [245]. If one reduces $Fe(CN)_6^{3-}$ cathodically on an electrode coated with a poly(N-methylpyrrole) film, one obtains the rotation-dependent, but largely potential-independent, reduction currents represented in Fig. 20.48.

It is clear from such experiments that the behavior of the polymer differs significantly from that of a metal and that the term "synthetic metal" is to be used with caution.

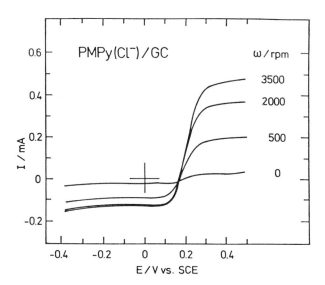

Fig. 20.48 Pseudostationary current–voltage curves obtained with a poly(N-methylpyrrole) (PMPy)-coated glassy carbon electrode in a deaerated aqueous electrolyte containing 1 mM $K_3/K_4Fe(CN)_6$ and 0.1 M KCl. Electrodes rotated at the indicated rate.

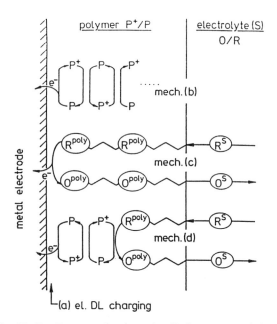

Fig. 20.49 Four mechanisms (a–d) that may each lead to current flow when the polymer-coated electrode is polarized in the redox electrolyte O/R. The polymer contains the electroactive sites P^+/P; "el. DL." is the electrical double layer.

4. The Overall Picture

Consider again a potential step applied to the coated electrode that induces mediated electron transfer as discussed above. It should be noted that the current flow following the potential step has contributions from other processes that are considerable. The reactions involved are summarized in Fig. 20.49, along with the mediated electron transfer reaction (mechanism d).

We consider the polymer to have a significant ion concentration. Thus an electrical double layer will form at the Me/poly interface, leading to capacitive charging current (mechanism a). Second, in dynamic measurements the current associated with the oxidation/reduction of the polymer redox sites P^+/P (mechanism b) is often very large. Third, it should be noted that the reaction O/R may proceed as a regular electrochemical charge transfer reaction at the metal/polymer interface. This was found to be the main mechanism in several systems, e.g., H_2/H^+ on polypyrrole-coated metal electrodes [235,246].

V. SURVEY OF EXPERIMENTAL DATA

In this section, we attempt to collate the corpus of data available to date on the electrochemical behavior of conducting polymers. This information is presented in capsule format in the form of tables. In keeping with the nature of this volume (i.e., handbook), every attempt has been made to be reasonably comprehensive, but it is entirely possible that some important references have been inadvertently left out. We apologize in advance for these omissions. Data are presented here within the framework of the discussion in the preceding sections of this chapter and are grouped according to a particular technique and/or topic.

A. Chronoamperometry, Chronocoulometry, and Chronopotentiometry

Measurements of the current or charge transient response to a potential step have been made in many in-

Table 20.7 Summary of Potential-Step and Current-Step Experiments on Conducting Polymer-Coated Electrodes

Conducting polymer	Comments	Reference(s)
Polypyrrole	Chronocoulometric charging and discharging of films studied in basic $AlCl_3$/1-methyl-(3-ethyl)-imidazolium chloride molten salts and acetonitrile.	247
	Simultaneous ESR, current, and charge measurements made versus time for films in the above molten salt electrolyte.	247
	The effect of uncompensated solution resistance on chronoamperometric transients discussed.	248
	Potential–time transients analyzed using small-amplitude current pulse experiments.	223, 249
	Transient measurements used to probe either diffusion coefficients of solution species or gas permeabilities.	250, 251
	Chronoamperometric response used to verify propagation model for the conductive domain during redox switching of polypyrrole.	252, 253
Polypyrrole and polythiophene	Analysis of Q–E profiles using both step and scan techniques to estimate various thermodynamic parameters for the redox switching reaction.	254
Polypyrrole and polyaniline	Ultramicroelectrodes employed for chronoamperometric experiments.	212, 255
Polypyrrole	Resistance and dimensional changes to potential steps probed.	256, 257
Polythiophene	Chronocoulometry used to analyze ion transport at poly(3-methylthiophene) films.	258
	Coupling and redox response of end-capped thiophene oligomers studied by potential-step methods.	259, 260
Polyaniline	Chronocoulometry used for partitioning the nitrogen involvement in the two redox processes.	261, 262
	Potential-step measurement used for measuring the diffusion coefficients of BF_4^- ion during undoping of the polymer.	263, 264
	Ultramicroelectrodes employed for low temperature measurements.	265
	Galvanostatic charge injection employed in chronopotentiometric measurements.	266

stances on polypyrrole and other conducting polymer film covered electrodes. Table 20.7 contains examples of such studies. Several types of electrolyte media including aqueous solutions, organic solvents, and molten salts have been employed. Both conventional and ultramicro working electrodes have been used. Current pulses have also been used for excitation in these measurements. Finally, other parameters such as resistance and dimensional changes have also been measured in response to a potential step.

B. Voltammetry

Table 20.8 contains a compilation of literature entries on the voltammetry of conducting polymer films. The scope of these studies is similar to that of the transient experiments discussed in Section V.A in terms of the types of electrodes and media employed. Both cyclic and hydrodynamic voltammetry have been used as shown in Table 20.8. Other aspects under discussion include the mathematic modeling of cyclic voltammograms [277,278], the occurrence and origin of prewaves in the cyclic voltammograms [319], the use of very fast scan rates [220], structural relaxation effects and their manifestation in voltammetry [304,317,320], the inactivation of polymer electroactivity when driven to extreme potentials, and the so-called polythiophene paradox [225,226,306,321]. Unusual media and cryogenic temperatures have also been employed for the voltammetric observation of doping phenomena [322–325]. Dual-electrode voltammetry (Section II.I) has been performed on derivatized polypyrrole [290] in an attempt to deconvolute the electronic and ionic contributions to the overall conductivity of the sample as a function of electrode potential. Finally, voltammetry has been carried out in the "solid state", i.e., in the absence of electrolyte solutions [215,323].

Other aspects of the voltammetry of conducting polymer films are reviewed in Refs. 261 and 326–330.

C. Electrochemical Quartz Crystal Microgravimetry

As already discussed in Section II.F, the EQCM technique provides information on the mass changes (and in some instances, film morphology changes) that accompany an electrochemical process. Thus, the technique can be profitably applied for the study of polymer film growth at a support electrode surface as a result of monomer oxidation (or reduction). However, this topic is beyond the scope of this chapter. Germane to the discussion are the redox reactions responsible for the insulator-to-conductor transitions in conducting polymer films, and much has been accomplished on this topic using the EQCM technique. Table 20.9 provides a summary of the EQCM studies on conducting polymer films. Starting with the first reported study on polypyrrole around 1984 [331], many aspects of ion transport, solvent uptake, and movement of neutral (salt) species during the redox "switching" reaction have been addressed. The studies have included both conducting polymers containing simple dopant ions (such as ClO_4^- or Cl^-) and copolymer films in which the conducting polymer film was grafted in the presence of a polymeric anion (e.g., polystyrenesulfonate).

D. AC Impedance Spectroscopy

Table 20.10 provides a compilation of ac impedance studies on conducting polymer films. The aspects under investigation include modeling of the ac impedance response of these materials [348–350,354,368,372], the separation of ionic and electronic contributions to the total conductivity [290,368], overdoping [352], the relative contribution of Faradaic and capacitive components to the total measured charge [221,351], the computation of diffusion coefficients associated with the oxidation of these polymers and the transport of dopant ions

Table 20.8 Summary of Voltammetry Studies on Conducting Polymer Film Electrodes

Conducting polymer/voltammetry technique	Ref.
Polypyrrole	
Cyclic (or linear sweep) voltammetry	142, 199, 214, 247, 250, 267–288
Hydrodynamic voltammetry	279, 289
Dual-electrode voltammetry	290
Polythiophene	
Cyclic voltammetry	222, 226, 258, 260, 291–305
Pulse voltammetry	306
Polyaniline	
Cyclic voltammetry	307–318
Polarography	310

Table 20.9 Studies of Conducting Polymer Films by Electrochemical Quartz Crystal Microgravimetry

Conducting polymer	Ref.
Polypyrrole	331–337
Polythiophene	338–341
Polyaniline	342–344
Copolymers	
Polypyrrole/polystyrene sulfonate	189, 345, 346
Polypyrrole/polyvinyl sulfonate	345
Polypyrrole/sulfated poly (β-hydroxy ether)	347
Polyaniline/Nafion	192

[96,353,365], in situ conductance changes during polymer film growth [359], the mediated oxidation of solution species [364] and the study of composite films [233,369–371]. As mentioned earlier in Section II.A (Fig. 20.29), ac impedance spectroscopy has also been profitably combined with measurements of optical absorbance changes of conducting polymer films [145–147].

E. Spectroelectrochemistry and Miscellaneous Techniques

Table 20.11 demonstrates that a vast array of in situ spectroscopic and other tools have been used to complement the electrochemical study of polypyrrole. Table 20.12 contains a similar compilation for the other two conducting polymers that have been extensively studied, polythiophene and polyaniline. Finally, Table 20.13 contains a summary of studies that have been oriented toward a study of the ion (and neutral species) transport that accompanies redox switching of conducting polymer films. Again, a vast array of techniques have been brought to bear on this important problem. A major distinction between the studies represented in Tables 20.11 and 20.12 vis-à-vis those in Table 20.13 is that in many of the latter cases the films were examined ex situ after various stages of ion exchange with the bathing medium. On the other hand, all the studies in Tables 20.11 and 20.12 were performed in situ, i.e., under potential control of the film either in contact with the electrolyte or in the emersed state.

F. Functionalized Conducting Polymers and Redox Catalysis

Table 20.14 lists instances of studies wherein the conducting polymer had been chemically modified and the electrochemical behavior of the resulting derivative studied. This compilation is by no means comprehen-

Table 20.10 Studies of Conducting Polymer Films by AC Impedance Spectroscopy

Conducting polymer	Ref.
Polypyrrole	96, 219, 221, 348–355
Polythiophene	219, 258, 356–359
Polyaniline	219, 360–365
Polyacetylene	362
Copolymers and modified polymers	
Polypyrrole/sodium dodecylsulfate	366
Polypyrrole/poly(vinyl chloride)	367
Poly-[1-methyl-3-(pyrol-1-ylmethyl)	368
Poly-(3-methyl pyrrole-4-carboxylic acid)	290
Polypyrrole/polystyrene sulfonate	369–371
Polyaniline/Nafion	233

Table 20.11 Spectroelectrochemical and Other Measurements on Polypyrrole Films[a]

Technique	Ref.
UV-visible absorption/reflection	144, 277, 373–382
Electron spin resonance	247, 369, 375, 383, 384
Raman spectroscopy	144, 385–389
Fourier transform infrared spectroscopy	390, 391
Probe beam deflection	156, 392, 393
Scanning tunneling microscopy	394, 395
Scanning electrochemical microscopy	396, 397
Work function measurements	172, 173
In situ electrical conductivity	256, 289, 290, 398–401

[a] All measurements were performed in situ under potential control of the polypyrrole electrode.

Table 20.12 Spectroelectrochemical and Other Measurements on Polythiophene and Polyaniline Films[a]

Polymer and technique	Ref.
Polythiophene	
UV-visible absorption	402–410
ESR	409–412
IR absorption	412, 413
In situ conductivity	359, 400, 414, 415
Polyaniline	
UV-visible absorption	378, 416–422
ESR	423, 424
In situ conductivity	424–426
Raman spectroscopy	427, 428
Fluorescence spectroscopy	429
Scanning electrochemical microscopy	430

[a] All these measurements were performed in situ under potential control of the polymer film electrode.

Table 20.13 Studies of Ion Exchange and Neutral Species Transport During Reduction/Oxidation of Conducting Polymer Films

Conducting polymer/technique	Ref.
Polypyrrole	
Voltammetry	142, 172, 173, 176, 184, 186, 188–190, 193–195, 198, 431–442
Chronoamperometry	443
Electrochemical quartz crystal microgravimetry	331–337, 444–450
AC impedance spectroscopy	95, 353
Spectroelectrochemistry	138–143, 451
Potentiometry	452
Radiotracer labeling	453–455
Miscellaneous methods	156, 392, 393, 396, 397, 456–468
Polythiophene	293, 339, 341, 469–475
Polyaniline	263, 264, 343, 344, 365, 421, 430, 476–487
Polyacetylene	453, 488

Table 20.14 Examples of Functionalized Polypyrrole Films

Type of functionality	Ref.
Dye	490–492
Ferrocene	493–495
Quinones	496–501
Oxometallates	437, 502, 503
Biological molecules	504–506
Chelates	507, 508
Macrocycles	388, 509–518
Electron donor/acceptor (luminophore/quencher)	519–534

sive, and previous reviews [186,190,229,489] can be consulted for further details. Table 20.15 contains a summary of the use of such chemically modified electrodes for redox catalysis applications. That is, these conducting polymer films have been used as *mediators* for the sustained electrochemical conversion of solution substrates (cf. Section IV.C). This aspect has also been reviewed in Refs. 8, 77, 166, 229, and 245.

Aside from the examples contained in Table 20.15, conducting polymer matrices have also proven to be a fertile area for the study of immobilized model redox species such as $Fe(CN)_6^{3-/4-}$ [554–560]. Further, the immobilization of species such as glucose oxidase or glucose dehydrogenase [504,561–566], cytochrome c [567], and flavins [568] have laid the ground for the use of chemically modified conducting polymer films as sensors (see below). The current response of such electrodes has also been theoretically analyzed from a sensor perspective [569,570].

G. Composites Based on Conducting Polymers

In the preceding section, conducting polymer films containing various types of functionalities were considered. Table 20.16 contains a further compilation of conducting polymer films that were copolymerized with a variety of other polymers. Some of these candidates have been discussed earlier in other contexts, particularly from the perspective of ion transport behavior. A major consideration in the design of these copolymers is the resultant improvement in mechanical properties relative to those of the parent conducting polymer film. The compilation in Table 20.16 is only meant to be illustrative of the vast range of polymers that can be grafted with polypyrrole. Nor are all the studies contained in the partial listing electrochemically oriented. Some of these polymer blends are at an advanced stage of commercial development as exemplified by Versicon and Incoblend. These products are polyaniline-poly(vinyl chloride) blends and were jointly developed by Zipperling Kessler & Co. in Germany and Allied Signal Inc. and Americhem, Inc. in the United States.

Another important category of conducting polymer-based composites includes films containing metal, metal oxide, clay, carbon black, semiconductor, and other types of particles (Table 20.17). These composites can be synthesized electrochemically by polymerizing the parent monomer (pyrrole, thiophene, or aniline) in a dis-

Table 20.15 Examples of Redox Catalysis with Chemically Modified Conducting Polymer Films

Parent conducting polymer	Substrate(s)	Ref.
Polypyrrole	Metallocenes	236, 289
	Metal bipyridyls	289, 401
	Dioxygen	83, 230, 241, 243, 535, 536
	Hydrogen	537–539
	Protons	83, 243, 246, 247, 535, 536, 539
	Ascorbic acid	194
	Methanol	541, 542
	Cyclohexanone	540
	Benzyl alcohol	543
	Quinones	544, 545
Polythiophene	Hexachloroacetone	546
	Protons	547
Polyaniline	Dioxygen	233, 239, 542, 548–550
	Formic acid	551
	Iron ferrocyanide	551
	Hydrazine	233, 552
	Nitric acid	553

Table 20.16 Copolymers of Conducting Polymers with Other Polymeric Materials

Conducting polymer	Copolymer partner	Ref.
Polypyrrole	Nafion	187, 191, 571–574
	Polystyrenesulfonate	188, 189, 345, 346, 369–371
	Polyvinylsulfonate	345
	Sulfated poly(β-hydroxyether)	347
	Poly(vinyl chloride)	367, 575, 576
	Polyurethane	577, 578
	Poly(vinyl alcohol)	579
	Poly(methyl methacrylate)	580
	Nylon 6	581–583
	Stepantane A	584
	Poly(p-phenyleneterephthalamido)propanesulfonate	585
	Polyacetylene	586
	Polyazulene	587
	Polyaniline	588
	Poly(N-vinylcarbazole)	589
	Melanin	590, 591
	Poly(vinyl sulfate)	592, 593
	Poly(N-phenylmaleimide)	594
	Polyethylene	595
Polythiophene	Poly(vinyl chloride)	297
	Polytetrahydrofuran	596
	Poly(methyl methacrylate)	597
	Polypyridine	598
	Polysilole	599
	Polyisothianaphthene	600
	Nafion	191, 601
	Poly(p-phenylene)	602
Polyaniline	Nafion	191, 192, 233, 601, 603
	Polydiphenylamine	604

Table 20.17 Conducting Polymer-Based Film Composites

Conducting polymer	Film additive (examples)	Targeted application(s)	Ref.
Polypyrrole	Noble metals (Pt, Pd)	Electrocatalysis, sensors	69, 238, 243, 536, 539, 605 541, 542, 606–609
	Metal oxides (TiO$_2$, WO$_3$, RuO$_2$, MnO$_2$, PbO$_2$, Fe$_2$O$_3$)	Electrochromism, photoelectrochemical devices, electrocatalysis, magnetic devices	610–620
	Clays	Ion exchange, electrocatalysis	187, 621
	Carbon black	Environmental pollution abatement, electrocatalysis	622–626
	Metal sulfides (CdS)	Sensors	627
Polythiophene	Metal clusters (Pt, Ag, Cu)	Electrocatalysis	547, 628–630
Polyaniline	Metal clusters (Pt, Ag)	Electrocatalysis, sensors	551, 631–634
	Metal oxides (TiO$_2$)	Electro- and photochromism	635
	Silica and clay	Electrocatalysis	636–638
Poly(2-hydroxy-3-aminophenazine)	Metal clusters (Au)	Electrocatalysis	639, 640

persion containing the particles to be entrapped within the polymer matrix, or they can be generated in powder form by chemical polymerization of the monomer in a suspension containing the (colloidal) particles and an oxidant such as $FeCl_3$ or $(NH_4)_2S_2O_8$. Of course, the former (film) configuration falls directly within the purview of the present discussion. The resultant materials have enhanced properties (relative to the parent polymer) with respect to their catalytic properties, photoactivity, ion-exchange capacity, conductivity, magnetic susceptibility, and so on.

H. Devices Based on the Electrochemistry of Conducting Polymer Films

The electrochemical behavior of conducting polymer films, as elaborated in Sections III and IV of this chapter, lends itself to the construction of a number of useful devices for practical applications. Some of these form the topics for other chapters in this handbook. Table 20.18 contains a further listing of devices that have been described in the literature.

Another important category of devices results from the superposition of two separate conducting polymer films (one atop the other) on a suitable support surface. Such bilayers, first described for redox polymer counterparts [149,700–706], have useful trapping properties. These arise from two characteristics: First, the two component layers are chosen to have disparate redox potentials. For example, if the inner layer has a higher redox potential than the outer one, then the charge resident on the outer layer is trapped because the inner layer becomes insulating over a potential regime wherein this accumulated charge can be otherwise bled through to the underlying support. A second interesting possibility occurs with conducting polymer-based bilayers [707]. If the outer layer is impervious to counterions that are needed for charge compensation in the inner layer, then the charge can be effectively trapped in the bilayer [707,708]. Table 20.19 contains examples of bilayer devices that have been constructed with conducting polymer films.

I. Miscellaneous Conducting Polymers

This chapter features mainly the electrochemistry of films of polypyrrole, polythiophene, and polyaniline on various support electrode surfaces. In a historical sense, polyacetylene predates these conducting polymer candidates. Electrochemically oriented studies on this material have been discussed in previous reviews [326–328],

Table 20.18 Devices Based on the Electrochemistry of Conducting Polymer Films[a]

Device category	Ref.
Battery	641–654
Sensor	504, 564, 655–664
Transistor	665–674
Diode	675–677
Solar cell	678–688
Display and imaging	689–694
Ion and gas transport	695–699

[a] Other chapters in this volume feature devices not considered herein, e.g., supercapacitors.

Table 20.19 Bilayer Devices Constructed from Conducting Polymers and Redox Polymers and Combinations Thereof

Inner layer	Outer layer	Ref.
A. *Conducting polymer/redox polymer bilayers*		
Poly [Ru(vbpy)$_3$] (ClO$_4$)$_2$	Polypyrrole	709, 710
Polythiophene	Polyxylylviologen	711–715
B. *Conducting polymer/conducting polymer bilayers*		
Polypyrrole	Polybithiophene	711
Polybithiophene	Polypyrrole	711
Polypyrrole	Poly(3-octylthiophene)	716
Poly(*N*-methylpyrrole)	Polypyrrole	708
Polypyrrole perchlorate	Polypyrrole or polystyrenesulfonate	370, 371
Polybithiophene	Polypyrrole	717
Poly(3-bromothiophene)	Polypyrrole	717
Polyacetylene	Poly(*N*-methyl pyrrole)	718
Polyaniline	Polypyrrole	719, 720
Polypyrrole	Polyaniline	719, 720
Polypyrrole	Polybithiophene or Poly(3-methylthiophene)	721
Polypyrrole	Polythiophene	722

Table 20.20 Monomers (Other Than Pyrrole, Thiophene, Aniline, Acetylene, and Their Derivatives) That Have Been Considered for the Preparation of Conducting Polymer Films

Monomer	Structure	Ref.
o-Phenylenediamine	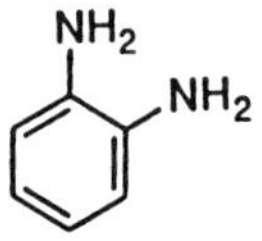	703, 724–726
Benzene		215, 727–734
p-Phenylenevinylene	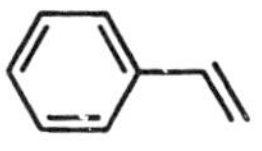	735, 736
Thianaphthene		723, 737, 738
o-Aminophenol	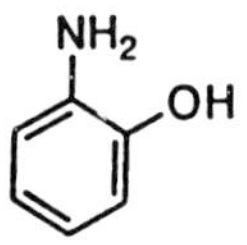	739–741
Indole (and derivatives)	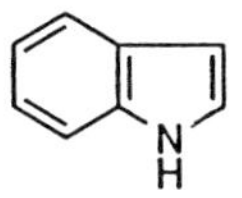	723, 742, 743
Furan		742, 743
Benzofuran		723
Carbazole	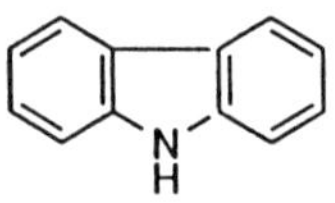	744, 745
Azulene	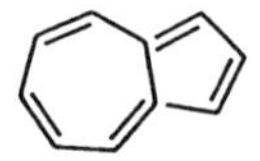	291, 745, 746
Toluene	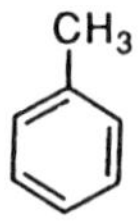	747
Phenazine (and derivatives)	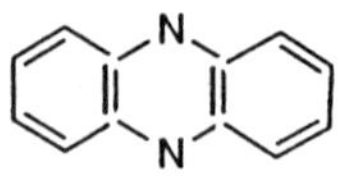	639, 640
Pyridine	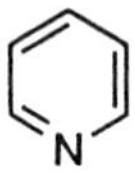	748

and little new has been done since then. It is interesting that other heterocyclic monomers had been considered for the generation of conducting polymer films fairly early in the history of these materials [e.g., 291,723]. However, further studies on these alternative polymer precursors have been sporadic. More recently, monomers such as toluene, phenol, and phenylenediamine have been investigated in terms of their proclivity toward conductive film formation. These and the earlier monomer candidates are listed in Table 20.20 along with leading literature entries to the corresponding studies. Only time will tell as to whether any of these (or other materials to emerge) will supplant the three conducting polymers featured in this review in terms of technological applicability.

VI. CONCLUDING REMARKS

This review shows that a vast literature already exists on both fundamental and practical aspects of the electrochemistry of conducting polymers. This trend is expected to extend in the future to encompass other new polymer candidates in addition to polypyrrole, polythiophene, and polyaniline. Concomitantly, electrochemical studies should provide the impetus for discovering new application areas (e.g., environmental pollution abatement and information and communication technologies) for these interesting materials.

REFERENCES

1. C. C. Ku and R. Liepins (eds)., *Electrical Properties of Polymers*, Hanser, Munich, 1987.
2. J. L. Brédas and R. Silbey (eds.), *Conjugated Polymers*, Kluwer, Dordrecht, 1991.
3. H. J. Mair and S. Roth (eds.), *Elektrisch leitende Kunststoffe*, Hanser, München, 1989.
4. H. Kuzmany, M. Mehring, and S. Roth (eds.), *Electronic Properties of Conjugated Polymers*, Vol. III, *Basic Models and Applications*, Springer, Berlin, 1989.
5. H. G. Kiess (ed.), *Conjugated Conducting Polymers*, Springer, Berlin, 1992.
6. J. L. Brédas, R. R. Chance, and R. Silbey, *Phys. Rev. B 26*:5843 (1982).
7. K. Doblhofer, in *Electrochemistry of Novel Materials* (J. Lipkowski and P. N. Ross, eds.), VCH, New York, 1994, pp. 141–205.
8. M. E. G. Lyons (ed.), *Electroactive Polymer Electrochemistry*, Plenum, New York, 1994.
9. R. W. Murray, in *Molecular Design of Electrode Surfaces* (R. W. Murray, ed.), Wiley, New York, 1992, pp. 1–48.
10. M. Kaneko and D. Wöhrle, *Adv. Polym. Sci. 84*: 143–228 (1988).
11. H. D. Abruña, in *Electroresponsive Molecular and Polymeric Systems* (T. A. Skotheim, ed.), Marcel Dekker, New York, 1988, pp. 98–160.
12. I. Rubinstein, in *Applied Polymer Analysis and Characterization* (J. Mitchell, Jr., ed.), Hanser, Munich, 1992, Vol. II, pp. 233–258.
13. G. P. Evans, in *Advances in Electrochemical Science and Engineering* (H. Gerischer and C. W. Tobias, eds.), VCH, Weinheim, 1990, Vol. 1, pp. 1–74.
14. J. Heinze, *Top. Curr. Chem. 152*:2–47 (1990).
15. T. A. Skotheim (ed.), *Handbook of Conducting Polymers*, Marcel Dekker, New York, 1986, Vols. 1 and 2.
16. M. Laridjani, J. P. Pouget, E. M. Scherr, A. G. MacDiarmid, M. E. Jozefowicz, and A. J. Epstein, *Macromolecules 25*:4106 (1992).
17. C. R. Martin and L. S. Van Dyke, in *Electrical Properties of Polymers* (C. C. Ku and R. Liepins, eds.), Hanser, Munich, 1987, pp. 403–424.
18. A. E. Woodward, *Understanding Polymer Morphology*, Hanser, New York, 1995.
19. M. Majda, in *Electrical Properties of Polymers* (C. C. Ku and R. Liepins, eds.), Hanser, Munich, 1987, pp. 159–206.
20. E. F. Dalton, N. A. Surridge, J. C. Jernigan, K. O. Wilbourn, J. S. Facci, and R. W. Murray, *Chem. Phys. 141*:143 (1990).
21. H. S. White, J. Leddy, and A. J. Bard, *J. Am. Chem. Soc. 104*:4811 (1982).
22. R. J. Mortimer and F. C. Anson, *J. Electroanal. Chem. 138*:325 (1982).
23. S. M. Oh and L. R. Faulkner, *J. Electroanal. Chem. 269*:77 (1989).
24. R. J. Forster, A. J. Kelly, J. G. Vos, and M. E. G. Lyons, *J. Electroanal. Chem. 270*:365 (1989).
25. T. Ohsaka, H. Yamamoto, and N. Oyama, *J. Phys. Chem. 91*:3775 (1987).
26. K. Doblhofer, W. Dürr, and M. Jauch, *Electrochim. Acta 27*:677 (1982).
27. R. P. Buck, *J. Phys. Chem. 92*:4196 (1988).
28. R. P. Buck, *J. Electroanal. Chem. 271*:1 (1989).
29. R. P. Buck, *Mater. Res. Soc. Symp. Proc. 135*:83 (1989).
30. R. P. Buck and P. Vanysek, *J. Electroanal. Chem. 292*:73 (1990).
31. C. P. Andrieux and J. M. Saveant, *J. Phys. Chem. 92*:6761 (1988).
32. D. Britz, *J. Electroanal. Chem. 88*:309 (1978).
33. D. K. Roe, in *Laboratory Techniques in Electroanalytical Chemistry* (P. T. Kissinger and W. R. Heineman, eds.), Marcel Dekker, New York, 1984, Ch. 7, pp. 193–234.
34. J. O. Howell and R. M. Wightman, *Anal. Chem. 56*: 524 (1984).
35. C. P. Andrieux, P. Hapiot, and J.-M. Saveant, *Chem. Rev. 90*:723 (1990).
36. C. P. Andrieux, P. Hapiot, and J.-M. Saveant, *Electroanalysis 2*:183 (1990).
37. J. Heinze, *Angew. Chem. Int. Ed. Engl. 30*:170 (1991).

38. R. M. Wightman, *Anal. Chem. 53*:1125A (1981).
39. M. Fleischmann and S. Pons, in *Ultramicroelectrodes* (M. Fleischmann, S. Pons, and P. P. Schmidt, eds.), Datatech, Morgandon, NC, 1987, pp. 1–63.
40. R. M. Wightman and D. O. Wipf, *Electroanal. Chem. 15*:267–353 (1989).
41. R. M. Wightman and D. O. Wipf, *Acc. Chem. Res. 23*:64 (1990).
42. B. R. Scharifker, in *Modern Aspects of Electrochemistry* (J. O'M. Bockris, B. E. Conway, and R. E. White, eds.), Plenum, New York, 1992, pp. 467–519.
43. J. Heinze, *Angew. Chem. Int. Ed. Engl. 32*:1268 (1993).
44. J. F. Cassidy and M. B. Foley, *Chem. Br.*, Sept. 1993, p. 764.
45. R. S. Robinson and R. L. McCreery, *Anal. Chem. 53*:997 (1981).
46. R. S. Robinson, C. W. McCurdy, and R. L. McCreery, *Anal. Chem. 54*:2356 (1982).
47. R. S. Robinson and R. L. McCreery, *J. Electroanal. Chem. 182*:61 (1985).
48. R. M. Penner, M. J. Heben, T. L. Longmire, and N. S. Lewis, *Science 250*:1118 (1990).
49. R. M. Penner and N. S. Lewis, *Chem. Ind.* Nov. 1991, p. 788.
50. P. Hülser and F. Beck, *J. Electrochem. Soc. 137*:2067 (1990).
51. N. R. de Tacconi, Y. Son, and K. Rajeshwar, *J. Phys. Chem. 97*:1042 (1993).
52. R. W. Murray, *Electroanal. Chem. 13*:191–369 (1984).
53. A. T. Hubbard, *J. Electroanal. Chem. 22*:165 (1969).
54. A. T. Hubbard and F. C. Anson, *Electroanal. Chem. 4*:129 (1970).
55. A. T. Hubbard, *CRC Crit. Rev. Anal. Chem. 2*:201 (1973).
56. J. Heinze, *Angew. Chem. Int. Ed. Engl. 23*:831 (1984).
57. J. E. B. Randles, *Trans. Faraday Soc. 44*:327 (1948).
58. A. Sevcik, *Collect. Czech. Chem. Commun. 13*:349 (1948).
59. H. Matsuda and Y. Ayabe, *Z. Elektrochem. 59*:494 (1955).
60. W. H. Reinmuth, *J. Am. Chem. Soc. 79*:6358 (1957).
61. R. S. Nicholson and I. Shain, *Anal. Chem. 36*:706 (1964).
62. J. C. Imbeaux and J. M. Saveant, *J. Electroanal. Chem. 44*:1969 (1973).
63. K. B. Oldham and J. Spanier, *J. Electroanal. Chem. 26*:331 (1970).
64. K. B. Oldham, *Anal. Chem. 44*:196 (1972).
65. K. B. Oldham, *Anal. Chem. 45*:39 (1973).
66. J. C. Myland and K. B. Oldham, *J. Electroanal. Chem. 153*:43 (1983).
67. M. Rudolph, D. P. Reddy, and S. W. Feldberg, *Anal. Chem. 66*:589A (1994), and references therein.
68. V. G. Levich, *Physicochemical Hydrodynamics*, Prentice-Hall, Englewood Cliffs, NJ, 1962.
69. W. J. Albery and M. L. Hitchman, *Ring-Disc Electrodes*, Clarendon Press, Oxford, 1971.
70. A. C. Riddeford, *Adv. Electrochem. Electrochem. Eng. 4*:47 (1966).
71. C. P. Andrieux, J.-M. Dumas-Bouchiat, and J.-M. Saveant, *J. Electroanal. Chem. 131*:1 (1982).
72. C. P. Andrieux, J.-M. Dumas-Bouchiat, and J.-M. Saveant, *J. Electroanal. Chem. 169*:9 (1984).
73. C. P. Andrieux and J.-M. Saveant, *J. Electroanal. Chem. 171*:65 (1984).
74. M. E. G. Lyons, *Analyst 119*:805 (1994).
75. W. J. Albery and A. R. Hillman, *J. Electroanal. Chem. 170*:27 (1985).
76. R. Schumacher, *Angew. Chem. Int. Ed. Engl. 29*:329 (1990).
77. D. A. Buttry, *Electroanal. Chem.* 1991, pp. 1–85.
78. D. A. Buttry and M. D. Ward, *Chem. Rev. 92*:1355 (1992).
79. G. Sauerbrey, *Z. Phys. 155*:206 (1959).
80. M. Thompson, A. L. Kipling, W. C. Duncan-Hewitt, L. V. Rajakovic, and B. A. Cavic-Vlasak, *Analyst 116*:881 (1991).
81. A. Glidle, A. R. Hillman, and S. Bruckenstein, *J. Electroanal. Chem. 318*:411 (1991).
82. E. Mori, C. K. Baker, J. R. Reynolds, and K. Rajeshwar, *J. Electroanal. Chem. 252*:441 (1988).
83. C. S. C. Bose and K. Rajeshwar, *J. Electroanal. Chem. 333*:235 (1992).
84. S. Bruckenstein, M. Michalski, A. Fensore, Z. Li, and A. R. Hillman, *Anal. Chem. 66*:1847 (1994).
85. G. C. Kamplin, F. Schleifer, and W. J. Pietro, *Rev. Sci. Instrum. 64*:1530 (1993).
86. W. Koh, W. Kutner, M. T. Jones, and K. M. Kadish, *Electroanalysis 5*:209 (1993).
87. D. W. Paul, S. R. Clark, and T. L. Beeler, *Sensors Actuators B17*:247 (1994).
88. J. H. Teuscher and R. L. Garrell, *Anal. Chem. 67*:3372 (1995).
89. D. M. Soares, C. Fruböse, K. Doblhofer, and W. Kautek, *Ber. Bunsenges. Phys. Chem.*, submitted.
90. C. Fruböse, K. Doblhofer, and D. M. Soares, *Ber. Bunsenges. Phys. Chem. 97*:475 (1993).
91. S. J. Martin, V. E. Granstaff, and G. C. Frye, *Anal. Chem. 63*:2272 (1991).
92. A. C. Hillier and M. D. Ward, *Anal. Chem. 64*:2539 (1992).
93. W.-W. Lee, H. S. White, and M. D. Ward, *Anal. Chem. 65*:3232 (1993).
94. S. Bruckenstein, A. Fensore, Z. Li, and A. R. Hillman, *J. Electroanal. Chem. 370*:189 (1994).
95. G. S. Popkirov and R. N. Schindler, *Rev. Sci. Instrum. 63*:5366 (1992).
96. E. W. Tsai, T. Pajkossy, K. Rajeshwar, and J. R. Reynolds, *J. Phys. Chem. 92*:3560 (1988).
97. S. Bhatnagar, S. Gupta, and K. Shahi, *Solid State Ionics 31*:107 (1985).
98. B. A. Boukamp, *Solid State Ionics 20*:31 (1986).
99. W. H. Mulder and J. H. Sluyters, *Electrochim. Acta 33*:303 (1988).
100. R. de Levie, *Adv. Electrochem. Electrochem. Eng. 6*:329 (1967).
101. R. de Levie, *J. Electroanal. Chem. 261*:1 (1989).

102. R. de Levie, *J. Electroanal. Chem. 281*:1 (1990).
103. L. Bai and B. E. Conway, *J. Electrochem. Soc. 138*: 2897 (1991).
104. M. Sluyters-Rehbach and J. H. Sluyters, in *Comprehensive Treatise of Electrochemistry* (E. Yeager et al., eds.), Plenum, New York, 1984, Vol. 9, pp. 177–292.
105. D. E. Smith, *Electroanal. Chem. 1*:1 (1966).
106. R. D. Armstrong, M. F. Bell, and A. A. Metcalf, *Electrochemistry 6*:98 (1978).
107. J. R. MacDonald, *J. Electroanal. Chem. 223*:25 (1987).
108. J. R. MacDonald, *Impedance Spectroscopy*, Wiley, New York, 1987.
109. F. B. Growcock, *Chemtech*, Sept. 1989, p. 564.
110. C. Ho, I. D. Raistrick and R. A. Huggins, *J. Electrochem. Soc. 127*:343 (1980).
111. O. Contamin, E. Levart, G. Magner, R. Parsons, and M. Savy, *J. Electroanal. Chem. 179*:41 (1984).
112. R. D. Armstrong, *J. Electroanal. Chem. 198*:177 (1986).
113. R. D. Armstrong, B. Lindholm, and M. Sharp, *J. Electroanal. Chem. 202*:69 (1986).
114. C. Gabrielli, F. Huet, M. Keddam, and O. Haas, *Electrochim. Acta 33*:1371 (1988).
115. M. F. Mathias and O. Haas, *J. Phys. Chem. 96*:3174 (1992).
116. T. Kuwana and N. Winograd, *Electroanal. Chem. 7*: 1 (1974).
117. T. Kuwana and W. R. Heineman, *Acc. Chem. Res. 9*:241 (1976).
118. W. R. Heineman, *Anal. Chem. 50*:390A (1978).
119. J. Robinson, *Electrochemistry 9*:101 (1984).
120. W. R. Heineman, F. M. Hawkridge, and H. N. Blount, *Electroanal. Chem. 13*:1–113 (1984).
121. R. W. Murray, W. R. Heineman, and G. W. O'Dom, *Anal. Chem. 39*:1666 (1967).
122. N. R. Armstrong, A. W. C. Lin, M. Fujihira, and T. Kuwana, *Anal. Chem. 48*:741 (1976).
123. R. H. Muller, *Adv. Electrochem. Electrochem. Eng. 9*:167 (1973).
124. J. D. E. McIntyre, *Adv. Electrochem. Electrochem. Eng. 9*:61 (1973).
125. W. N. Hansen, *Adv. Electrochem. Electrochem. Eng. 9*:1 (1973).
126. A. Bewick and S. Pons, in *Advances in Infrared and Raman Spectroscopy* (R. J. H. Clark and R. E. Hester, eds.), Wiley-Heyden, London, 1985, Vol. 12, pp. 1–67.
127. R. J. Gale (ed.), *Spectroelectrochemistry—Theory and Practice*, Plenum, New York, 1988.
128. P. J. Hendra (ed.), *Applications of Fourier Transform Raman Spectroscopy II*, *Spectrochim. Acta 47A* (1991) (special volume).
129. M. Fleischmann, P. J. Hendra, and A. J. McQuillan, *Chem. Phys. Lett. 26*:163 (1974).
130. R. P. Van Duyne, in *Chemical and Biological Applications of Lasers* (C. B. Moore, ed.), Academic, New York, 1979, Vol. 4, p. 101.
131. I. D. Goldberg and T. M. McKinney, in *Laboratory Techniques in Electroanalytical Chemistry* (P. T. Kissinger and W. R. Heineman, eds.), Marcel Dekker, New York, 1984, pp. 675–728.
132. B. Bittins-Cattaneo, E. Cattaneo, P. Königshoven, and W. Vielstich, *Electroanal. Chem. 17*:181–220 (1991).
133. G. Flätgen, K. Krischer, B. Pettinger, K. Doblhofer, H. Jenkes, and G. Ertl, *Science 269*:668 (1995).
134. M. Majda and L. R. Faulkner, *J. Electroanal. Chem. 169*:77 (1984).
135. W. E. O'Grady, *J. Electrochem. Soc. 127*:555 (1980).
136. M. Fleischmann, P. Graves, I. R. Hill, A. Oliver, and J. Robinson, *J. Electroanal. Chem. 150*:33 (1983).
137. H. Abruna (ed.), *Electrochemical Interfaces*, VCH, New York, 1991.
138. E. W. Tsai, L. Phan, and K. Rajeshwar, *J. Chem. Soc., Chem. Commun. 1988*:771.
139. G.-W. Jang, E. W. Tsai, and K. Rajeshwar, *J. Electrochem. Soc. 134*:2377 (1987).
140. G.-W. Jang, E. W. Tsai, and K. Rajeshwar, *J. Electroanal. Chem. 263*:383 (1989).
141. Y.-H. Ho, S. Basak, E. W. Tsai, and K. Rajeshwar, *J. Chem. Soc. Chem. Commun. 1989*:1078.
142. V. Krishna, Y.-H. Ho, S. Basak, and K. Rajeshwar, *J. Am. Chem. Soc. 113*:3325 (1991).
143. S. Basak, Y.-H. Ho, E. W. Tsai, and K. Rajeshwar, *J. Chem. Soc., Chem. Commun. 1989*:462.
144. Y. Son and K. Rajeshwar, *J. Chem. Soc. Faraday Trans. 88*:605 (1992).
145. T. Amemiya, K. Hashimoto, and A. Fujishima, *J. Phys. Chem. 97*:4187 (1993).
146. T. Amemiya, K. Hashimoto, and A. Fujishima, *J. Phys. Chem. 97*:4192 (1993).
147. T. Amemiya, K. Hashimoto, and A. Fujishima, *J. Phys. Chem. 97*:9736 (1993).
148. J. M. Rosolen, M. Fracastoro-Decker, and F. Decker, *J. Electroanal. Chem. 346*:119 (1993).
149. P. G. Pickup, W. Kutner, C. R. Leidner, and R. W. Murray, *J. Am. Chem. Soc. 106*:1991 (1984).
150. M. Fracastoro-Decker and F. Decker, *J. Electroanal. Chem. 266*:215 (1989).
151. C. Barbero, M. C. Miras, O. Haas, and R. Kötz, *J. Electrochem. Soc. 138*:669 (1991).
152. C. Barbero, M. C. Miras, R. Kötz, and O. Haas, *J. Electroanal. Chem. 310*:437 (1992).
153. M. C. Miras, C. Barbero, R. Kötz, O. Haas, and V. M. Schmidt, *J. Electroanal. Chem. 338*:279 (1992).
154. M. C. Pham, J. Moslih, C. Barbero, and O. Haas, *J. Electroanal. Chem. 316*:143 (1991).
155. P. Novak, R. Kötz, and O. Haas, *J. Electrochem. Soc. 140*:37 (1993).
156. V. M. Schmidt, C. Barbero, and R. Kötz, *J. Electroanal. Chem. 352*:301 (1993).
157. A. J. Bard, G. Dennault, C. Lee, D. Mandler, and D. O. Wipf, *Acc. Chem. Res. 23*:357 (1990).
158. C. Wei and A. J. Bard, *J. Electrochem. Soc. 142*: 2523 (1995).
159. M. Cappadonia, K. Doblhofer, and M. Jauch, *Ber. Bunsenges. Phys. Chem. 92*:903 (1988).

160. M. Cappadonia and K. Doblhofer, *Electrochim. Acta 34*:1815 (1989).
161. W. A. Zisman, *Rev. Sci. Instru. 3*:367 (1932).
162. A. Many, Y. Goldstein, and N. B. Grover, *Semiconductor Surfaces*, North-Holland, Amsterdam, 1965.
163. G. Horanyi, *Electrochim. Acta 25*:43 (1980).
164. G. Inzelt and G. Horanyi, *J. Electroanal. Chem. 200*:405 (1986).
165. K. Doblhofer, *J. Electroanal. Chem. 331*:1015 (1992).
166. K. Maksymiuk and K. Doblhofer, *Electrochim. Acta 39*:217 (1994).
167. H. Gerischer, D. M. Kolb, and J. K. Sass, *Adv. Physics 27*:437 (1978).
168. D. W. Van Krevelen, *Properties of Polymers, Correlations with Chemical Structure*, Elsevier, Amsterdam, 1972, Chap. 18.
169. J. Comyn (ed.), *Polymer Permeability*, Elsevier, New York, 1988.
170. F. Helfferich, *Ionenaustauscher*, Band I, Verlag Chemie, Weinheim, 1959; English translation: *Ion Exchange*, McGraw-Hill, New York, 1962.
171. P. Burgmayer and R. W. Murray, in *Handbook of Conducting Polymers* (T. A. Skotheim, ed.), Marcel Dekker, New York, 1986, Vol. 1, pp. 507–523.
172. C. Zhong, W. Storck, and K. Doblhofer, *Ber. Bunsenges. Phys. Chem. 94*:1149 (1990).
173. C. Zhong and K. Doblhofer, *Electrochim. Acta 35*:1971 (1990).
174. H. Braun, W. Storck, and K. Doblhofer, *J. Electrochem. Soc. 130*:807 (1983).
175. H. Braun, F. Decker, K. Doblhofer, and H. Sotobayashi, *Ber. Bunsenges. Phys. Chem. 88*:345 (1984).
176. C. Zhong, K. Doblhofer, and G. Weinberg, *Faraday Discuss. Chem. Soc. 88*:307 (1989).
177. N. Lakshminarayanaiah, *Membrane Electrodes*, Academic, New York, 1976, Chapter 3.
178. F. G. Donnan, *Z. Elektrochem. 17*:572 (1911).
179. P. Ugo and F. C. Anson, *Anal. Chem. 61*:1802 (1989).
180. R. Naegeli, J. Redepenning, and F. C. Anson, *J. Phys. Chem. 90*:6227 (1986).
181. A. Fitch, *J. Electroanal. Chem. 284*:237 (1990).
182. D. Engel and E. W. Grabner, *Ber. Bunsenges. Phys. Chem. 89*:982 (1985).
183. P. J. Kulesza and K. Doblhofer, *J. Electroanal. Chem. 274*:95 (1989).
184. K. Doblhofer and M. Vorotyntsev, in *Electroactive Polymer Electrochemistry* (M. E. G. Lyons, ed., Plenum, New York, 1994 pp. 375–442.
185. S. Ohki and H. Oshima, in *Electrical Double Layers in Biology* (M. Blank, ed.), Plenum, New York, 1986, pp. 1–16.
186. A. Deronzier and J.-C. Moutet, *Acc. Chem. Res. 22*:249 (1989).
187. F.-R. F. Fan and A. J. Bard, *J. Electrochem. Soc. 133*:301 (1986).
188. L. L. Miller and Q. X. Zhou, *Macromolecules 20*:1594 (1987).
189. C. K. Baker, Y.-J. Qiu, and J. R. Reynolds, *J. Phys. Chem. 95*:4446 (1991).
190. G. Bidan, B. Ehui, and M. Lapkowski, *J. Phys. D: Appl. Phys. 21*:1043 (1988).
191. T. Hirai, S. Kuwabata, and H. Yoneyama, *J. Electrochem. Soc. 135*:1132 (1988).
192. D. Orata and D. Buttry, *J. Electroanal. Chem. 257*:71 (1988).
193. G. E. Asturias, G.-W. Jang, A. G. MacDiarmid, K. Doblhofer, and C. Zhong, *Ber. Bunsenges. Phys. Chem. 95*:1181 (1991).
194. H. Mao and P. Pickup, *J. Electroanal. Chem. 265*:127 (1989).
195. C. Zhong, K. Doblhofer, and G. Weinberg, *Faraday Discuss. Chem. Soc. 88*:307 (1989).
196. B. Lindholm and M. Sharp, *J. Electroanal. Chem. 198*:37 (1986).
197. Y. J. Qiu and J. R. Reynolds, *J. Electrochem. Soc. 137*:900 (1990).
198. V. M. Schmidt and J. Heitbaum, *Synth. Met. 41*:425 (1991).
199. M. A. Vorotyntsev, L. I. Daikhin, and M. D. Levi, *J. Electroanal. Chem. 332*:213 (1992).
200. H. Dahms, *J. Phys. Chem. 72*:362 (1968).
201. I. Ruff and V. J. Friedrich, *J. Phys. Chem. 75*:3297 (1971).
202. H. Braun, F. Decker, K. Doblhofer, and H. Sotobayashi, *Ber. Bunsenges. Phys. Chem. 88*:345 (1984).
203. A. R. Hillman, D. C. Loveday, and S. Bruckenstein, *J. Electroanal. Chem. 274*:157 (1989).
204. G. Inzelt, *Electrochim. Acta 34*:83 (1989).
205. M. Sharp, B. Lindholm, and E.-L. Lind, *J. Electroanal. Chem. 274*:35 (1989).
206. I. Rubinstein, J. Rishpon, and S. Gottesfeld, *J. Electrochem. Soc. 133*:729 (1986).
207. S. M. Oh and L. R. Faulkner, *J. Am. Chem. Soc. 111*:5613 (1989).
208. K. Aoki, J. Cao, and Y. Hoshino, *Electrochim. Acta 39*:2291 (1994).
209. D. A. Kaplin and S. Qutubuddin, *Electrochim. Acta 40*:1149 (1995).
210. S. Bruckenstein, C. P. Wilde, and A. R. Hillman, *J. Phys. Chem. 97*:6853 (1993).
211. C. Lee, J. Kwak, and A. J. Bard, *J. Electrochem. Soc. 136*:3720 (1989).
212. L. M. Abrantes, J. C. Mesquita, M. Kalaji, and L. M. Peter, *J. Electroanal. Chem. 307*:275 (1991).
213. G. Wegner and J. Rühe, *Faraday Discuss. Chem. Soc. 88*:333 (1989).
214. R. A. Bull, F.-R. F. Fan, and A. J. Bard, *J. Electrochem. Soc. 129*:1009 (1982).
215. K. Meerholz and J. Heinze, *Synth. Met. 43*:2871 (1991).
216. J. Heinze, K. Meerholz, and R. Bilger, *Electronic Properties of Conjugated Polymers* (H. Kuzmany, M. Mehring, and S. Roth, eds.), Springer, Berlin, 1989, Vol. 3, p. 146.
217. J. Heinze, R. Bilger, and K. Meerholz, *Ber. Bunsenges. Phys. Chem. 92*:1266 (1988).
218. J. Heinze, M. Störzbach, and J. Mortensen, *Ber. Bunsenges. Phys. Chem. 91*:960 (1987).
219. W. J. Albery, Z. Chen, B. Horrocks, A. R. Mount,

P. J. Wilson, D. Bloor, A. T. Monkman, and C. M. Elliott, *Faraday Discuss. Chem. Soc. 88*:247 (1989).
220. C. P. Andrieux, P. Audebert, P. Hapiot, M. Nechtschein, and C. Odin, *J. Electroanal. Chem. 305*:153 (1991).
221. J. Tanguy, N. Mermilliod, and M. Hoclet, *J. Electrochem. Soc. 134*:795 (1987).
222. S. Servagent and E. Vieil, *Synth. Met. 31*:127 (1989).
223. Z. Cai and C. R. Martin, *J. Electroanal. Chem. 300*: 35 (1991).
224. M. G. Hill, K. R. Han, L. L. Miller, and J. F. Penneau, *J. Am. Chem. Soc. 114*:967 (1992).
225. F. Beck, P. Braun, and M. Oberst, *Ber. Bunsenges. Phys. Chem. 91*:967 (1987).
226. E. W. Tsai, S. Basak, J. P. Ruiz, J. R. Reynolds, and K. Rajeshwar, *J. Electrochem. Soc. 136*:3683 (1989).
227. C. P. Andrieux and J.-M. Saveant, *Molecular Design of Electrode Surfaces* (R. W. Murray, ed.), Wiley, New York, 1992, pp. 207–270.
228. C. R. Leidner, *Molecular Design of Electrode Surfaces* (R. W. Murray, ed.), Wiley, New York, 1992, pp. 313–332.
229. D. Curran, J. Grimshaw, and S. D. Perera, *Chem. Soc. Rev. 20*:391 (1991).
230. R. C. M. Jacobs, L. J. J. Janssen, and E. Barendrecht, *Electrochim. Acta 30*:1085 (1985).
231. M. D. Levi, M. A. Vorotyntsev, A. M. Skundin, and V. E. Kazarinov, *J. Electroanal. Chem. 271*:193 (1989); *319*:243 (1991).
232. J. Desilvestro and O. Haas, *Electrochim. Acta 36*: 361 (1991).
233. M. Fabrizio, G. Mengoli, M. M. Musiani, and F. Paolucci, *J. Electroanal. Chem. 300*:23 (1991).
234. C. Deslouis, M. M. Musiani, M. El Rhazi, and B. Tribollet, *Synth. Met. 60*:269 (1993).
235. M. M. Lohrengel, J. W. Schultze, and A. Thyssen, *Elektrisch leitende kunststoffe* (H. J. Mair and S. Roth, eds.), Hanser, Munich, 1989, pp. 377–401.
236. H. Mao and P. G. Pickup, *Chem. Mater. 4*:642 (1992).
237. P. N. Bartlett, D. H. Dawson, and J. Farrington, *J. Chem. Soc. Faraday Trans. 88*:2685 (1992).
238. G. K. Chandler and D. Pletcher, *J. Appl. Electrochem. 16*:62 (1986).
239. G. Bidan, E. M. Genies, and M. Lapkowski, *J. Chem. Soc., Chem. Commun. 1988*:533.
240. F. Bedioui, M. Voisin, J. Devynck, and C. Bied-Charreton, *J. Electroanal. Chem. 297*:257 (1991).
241. K. Masashi, A. Sakawaki, and T. Sato, *Bull. Chem. Soc. Jpn. 67*:2323 (1994).
242. J. E. Sheats, C. E. Carraher, Jr., and C. U. Pittman, Jr. (eds.), *Metal Containing Polymeric Systems*, Plenum, New York, 1985.
243. C. S. C. Bose, S. Basak, and K. Rajeshwar, *J. Electrochem. Soc. 139*:L75 (1992).
244. C. C. Chen, C. S. C. Bose, and K. Rajeshwar, *J. Electroanal. Chem. 350*:161 (1993).
245. K. Doblhofer and C. Zhong, *Synth. Met. 43*:2865 (1991).
246. K. Maksymiuk and K. Doblhofer, *J. Chem. Soc. Faraday Trans. 90*:745 (1994).
247. J. F. Oudard, R. D. Allendoerfer, and R. A. Osteryoung, *J. Electroanal. Chem. 241*:231 (1988).
248. C. D. Paulse and P. G. Pickup, *J. Phys. Chem. 92*: 7002 (1988).
249. R. M. Penner, L. S. Van Dyke, and C. R. Martin, *J. Phys. Chem. 92*:5274 (1988).
250. F. T. A. Vork, B. C. A. M. Schuermans, and E. Barendrecht, *Electrochim. Acta 35*:567 (1990).
251. V. M. Schmidt, D. Tegtmeyer, and J. Heitbaum, *Adv. Mater. 4*:428 (1992).
252. K. Aoki and Y. Tezuka, *J. Electroanal. Chem. 267*: 55 (1989).
253. Y. Tezuka and K. Aoki, *J. Electroanal. Chem. 273*: 161 (1989).
254. G. Zotti, G. Schiavon, and N. Comisso, *Electrochim. Acta 35*:1815 (1990).
255. M. Kalaji, L. M. Peter, L. M. Abrantes, and J. C. Mesquita, *J. Electroanal. Chem. 274*:289 (1989).
256. R. John, A. Talaie, G. G. Wallace, and S. Fletcher, *J. Electroanal. Chem. 319*:365 (1991).
257. Q. Pei and O. Inganäs, *J. Phys. Chem. 96*:10507 (1992).
258. F.-J. Pern and A. J. Frank, *J. Electrochem. Soc. 137*: 2769 (1990).
259. G. Zotti, G. Schiavon, A. Berlin, and G. Pagani, *Adv. Mater. 5*:551 (1993).
260. G. Zotti, G. Schiavon, A. Berlin, and G. Pagani, *Chem. Mater. 5*:430 (1993).
261. E. M. Genies, A. Boyle, M. Lapkowski, and C. Tsintavis, *Synth. Met. 36*:139 (1990).
262. E. M. Genies and C. Tsintavis, *J. Electroanal. Chem. 195*:109 (1985).
263. K. Kanamura, S. Yonezawa, S. Yoshioka, and Z. Takehara, *J. Phys. Chem. 95*:7939 (1991).
264. K. Kanamura, Y. Kawai, S. Yonezawa, and Z. Takehara, *J. Phys. Chem. 98*:2174 (1994).
265. M. Vuki, M. Kalaji, L. Nyholm, and L. M. Peter, *J. Electroanal. Chem. 332*:315 (1992).
266. M. Kalaji, L. Nyholm, and L. M. Peter, *J. Electroanal. Chem. 325*:269 (1992).
267. A. F. Diaz, J. I. Castillo, J. A. Logan, and W.-Y. Lee, *J. Electroanal. Chem. 129*:115 (1981).
268. A. F. Diaz, *Chem. Scripta 17*:145 (1981).
269. M. Salmon, M. E. Carbajal, J. C. Juarez, A. Diaz, and M. C. Rock, *J. Electrochem. Soc. 131*:1802 (1984).
270. G. Bidan and M. Guglielmi, *Synth. Met. 15*:49 (1986).
271. S. W. Feldberg, *J. Am. Chem. Soc. 106*:4671 (1984).
272. S. Kuwabata, H. Yoneyama, and H. Tamura, *Bull. Chem. Soc. Jpn. 57*:2247 (1984).
273. M. Saloma, M. Aguilar, and M. Salmon, *J. Electrochem. Soc. 132*:2379 (1985).
274. N. Oyama, T. Ohsaka, K. Chiba, H. Miyamoto, T. Mukai, S. Tanaka, and T. Kumagai, *Synth. Met. 20*: 245 (1987).
275. W. Wernet and G. Wegner, *Makromol. Chem. 188*: 1465 (1987).

276. Y. Tezuka, K. Aoki, and K. Shinozaki, *Synth. Met. 30*:369 (1989).
277. T. Yeu, T. V. Nguyen, and R. E. White, *J. Electrochem. Soc. 135*:1971 (1988).
278. T. Yeu, K.-M. Yin, J. Carbajal, and R. E. White, *J. Electrochem. Soc. 138*:2869 (1991).
279. A. Witkowski, M. S. Freund, and A. Brajter-Toth, *Anal. Chem. 63*:622 (1991).
280. A. Witkowski and A. Brajter-Toth, *Anal. Chem. 64*:635 (1992).
281. F. Beck, U. Barsch, and R. Michaelis, *J. Electroanal. Chem. 351*:169 (1993).
282. A. F. Diaz, J. Castillo, K. K. Kanazawa, J. A. Logan, M. Salmon, and O. Fajardo, *J. Electroanal. Chem. 133*:233 (1982).
283. D.-S. Park, Y.-B. Shim, and S.-M. Park, *J. Electrochem. Soc. 140*:609 (1993).
284. M. C. Anglada, J. Claret, and J. M. Ribo, *Synth. Met. 59*:181 (1993).
285. M. Sak-Bosnar, M. V. Budimir, S. Kovac, D. Kukuli, and L. Duic, *J. Polym. Sci. Part A: Polym. Chem. 30*:1609 (1992).
286. J.-M. Ko, H. W. Rhee, S.-M. Park, and C. Y. Kim, *J. Electrochem. Soc. 137*:905 (1990).
287. D. J. Walton, C. E. Hall, and A. Chyla, *Analyst 117*:1305 (1992).
288. E.-L. Kupila, J. Lukkari, and J. Kankarc, *Synth. Met. 74*:207 (1995).
289. H. Mao and P. G. Pickup, *J. Am. Chem. Soc. 112*:1776 (1990).
290. X. Ren and P. G. Pickup, *J. Electrochem. Soc. 139*:2097 (1992).
291. R. J. Waltman, A. F. Diaz, and J. Bargon, *J. Electrochem. Soc. 131*:1452 (1984).
292. M. Sato, S. Tanaka, and K. Kaeriyama, *J. Chem. Soc., Chem. Commun. 1986*:873.
293. P. Marque, J. Roncali, and F. Garnier, *J. Electroanal. Chem. 218*:107 (1987).
294. J. Roncali, F. Garnier, R. Garreau, and M. Lemaire, *J. Chem. Soc., Chem. Commun. 1987*:1500.
295. J. Roncali and F. Garnier, *Nouv. J. Chim. 10*:237 (1986).
296. J. Roncali, A. Yassur, and F. Garnier, *J. Chem. Soc., Chem. Commun. 1988*:581.
297. J. Roncali and F. Garnier, *J. Phys. Chem. 92*:833 (1988).
298. L. Laguren-Davidson, C. V. Pham, H. Zimmer, H. B. Mark, Jr., and D. J. Ondrus, *J. Electrochem. Soc. 135*:1406 (1988).
299. M. R. Bryce, A. D. Chissel, N. R. M. Smith, D. Parker, and P. Kathirgamanathan, *Synth. Met. 26*:153 (1988).
300. J. Roncali, H. K. Youssoufi, R. Garreau, F. Garnier, and M. Lemaire, *J. Chem. Soc., Chem. Commun. 1990*:414.
301. J. Roncali, P. Marque, R. Garreau, F. Garnier, and M. Lemaire, *Macromolecules 23*:1347 (1990).
302. H. Harada, T. Fuchigami, and T. Nonaka, *J. Electroanal. Chem. 303*:139 (1991).
303. T. F. Otero and J. Rodriguez, *J. Electroanal. Chem. 310*:219 (1991).
304. C. Odin and M. Nechtschein, *Synth. Met. 44*:177 (1991).
305. Z.-G. Xu and G. Horowitz, *J. Electroanal. Chem. 335*:123 (1992).
306. M. Gratzel, D.-F. Hsu, A. M. Riley, and J. Janata, *J. Phys. Chem. 94*:5973 (1990).
307. A. F. Diaz and T. C. Clarke, *J. Electroanal. Chem. 111*:115 (1980).
308. P. J. Nigrey, A. G. MacDiarmid, and A. J. Heeger, *Mol. Cryst. Liq. Cryst. 83*:309 (1982).
309. N. Comisso, S. Daolio, G. Mengoli, R. Salmaso, S. Zecchin, and G. Zotti, *J. Electroanal. Chem. 255*:97 (1988).
310. K. M. Choi, K. H. Kim, and J. S. Choi, *J. Phys. Chem. 93*:4659 (1989).
311. R. Greef, M. Kalaji, and L. M. Peter, *Faraday Discuss. Chem. Soc. 88*:277 (1989).
312. G. Inzelt and G. Horanyi, *Electrochim. Acta 35*:27 (1990).
313. G. Inzelt, *J. Electroanal. Chem. 279*:169 (1990).
314. Y. Sun, A. G. MacDiarmid, and A. J. Epstein, *J. Chem. Soc., Chem. Commun. 1990*:529.
315. J.-Y. Bergeron, J.-W. Chevalier, and L. H. Dao, *J. Chem. Soc., Chem. Commun. 1990*:180.
316. J. Tang and R. A. Osteryoung, *Synth. Met. 44*:307 (1991).
317. C. Odin, M. Nechtschein, and P. Hapiot, *Synth. Met. 47*:329 (1992).
318. L. Nyholm and L. M. Peter, *J. Chem. Soc. Faraday Trans. 90*:149 (1994).
319. S. Gottesfeld, A. Redondo, I. Rubinstein, and S. W. Feldberg, *J. Electroanal. Chem. 265*:15 (1989).
320. C. Odin and M. Nechtschein, *Phys. Rev. Lett. 67*:1114 (1991).
321. B. Kritsche and M. Zagorska, *Synth. Met. 28*:C263 (1989).
322. H. B. Tatistcheff, I. Fritsch-Faules, and M. S. Wrighton, *J. Phys. Chem. 97*:2732 (1993).
323. S. Chao and M. S. Wrighton, *J. Am. Chem. Soc. 109*:6627 (1987).
324. D. Ofer and M. S. Wrighton, *J. Am. Chem. Soc. 110*:4467 (1988).
325. D. Ofer, L. Y. Park, R. R. Schrock, and M. S. Wrighton, *Chem. Mater. 3*:573 (1991).
326. G. K. Chandler and D. Pletcher, *Spec. Period. Rep. Roy. Soc. 10*:117 (1985).
327. A. F. Diaz, J. F. Rubinson, and H. B. Mark, Jr., *Adv. Polym. Sci. 84*:113 (1988).
328. J. O'M. Bockris and D. Miller, in *Conducting Polymers: Special Applications* (L. Alcacer, ed.), Reidel, Dordrecht, 1987, pp. 1–36.
329. J. Heinze, *Synth. Met. 41–43*:2805 (1991).
330. A. A. Syed and M. K. Dinesan, *Talanta 38*:815 (1991).
331. J. N. Kaufman, K. K. Kanazawa, and G. B. Street, *Phys. Rev. Lett. 53*:2461 (1984).
332. Y.-J. Qiu and J. R. Reynolds, *Polym. Eng. Sci. 31*:417 (1991).

333. C. S. C. Bose, S. Basak, and K. Rajeshwar, *J. Phys. Chem. 96*:9899 (1992).
334. J. R. Reynolds, N. S. Sundaresan, M. Pomerantz, S. Basak, and C. K. Baker, *J. Electroanal. Chem. 250*:355 (1988).
335. S. Basak, C. S. C. Bose, and K. Rajeshwar, *Anal. Chem. 64*:1813 (1992).
336. A. J. Kelly, K. Naoi, T. Ohsaka, and N. Oyama, *Bunseki Kagaku 40*:835 (1991).
337. V. M. Schmidt and J. Heitbaum, *Electrochim. Acta 38*:349 (1993).
338. A. R. Hillman, D. C. Loveday, M. J. Swann, R. M. Eales, A. Hammett, S. J. Higgins, S. Bruckenstein, and C. P. Wilde, *Faraday Discuss. Chem. Soc. 88*:151 (1989).
339. A. R. Hillman, M. J. Swann, and S. Bruckenstein, *J. Electroanal. Chem. 291*:147 (1990).
340. A. R. Hillman, M. J. Swann, and S. Bruckenstein, *J. Phys. Chem. 95*:3271 (1991).
341. A. R. Hillman, D. C. Loveday, M. J. Swann, S. Bruckenstein, and C. P. Wilde, *J. Chem. Soc., Faraday Trans. 87*:2047 (1991).
342. D. Orata and D. A. Buttry, *J. Am. Chem. Soc. 109*:3574 (1987).
343. R. M. Torresi, S. I. Cordoba de Torresi, C. Gabrielli, M. Keddam, and H. Takenouti, *Synth. Met. 61*:291 (1993).
344. M. C. Miras, C. Barbero, R. Kötz, and O. Haas, *J. Electroanal. Chem. 369*:193 (1994).
345. M. Lien, W. H. Smyrl, and M. Morita, *J. Electroanal. Chem. 309*:333 (1991).
346. K. Naoi, M. M. Lien, and W. H. Smyrl, *J. Electroanal. Chem. 272*:273 (1989).
347. K. Takeshita, W. Wernet, and N. Oyama, *J. Electrochem. Soc. 141*:2004 (1994).
348. W. J. Albery, C. M. Elliott, and A. R. Mount, *J. Electroanal. Chem. 288*:15 (1990).
349. W. J. Albery and A. R. Mount, *J. Electroanal. Chem. 305*:3 (1991).
350. W. J. Albery and A. R. Mount, *J. Chem. Soc. Faraday Trans. 90*:1115 (1994).
351. J. Tanguy, N. Mermilliod, and M. Hoclet, *Synth. Met. 18*:7 (1987).
352. J. Tanguy and N. Mermilliod, *Synth. Met. 21*:129 (1987).
353. R. M. Penner and C. R. Martin, *J. Phys. Chem. 93*:984 (1989).
354. J. Tanguy, *Synth. Met. 41–43*:2991 (1991).
355. L. L. Madsen, K. Carneiro, B. N. Zaba, A. E. Underhill, and M. J. van der Sluijs, *Synth. Met. 41–43*:2931 (1991).
356. T. F. Otero and E. DeLarreta, *J. Electroanal. Chem. 244*:311 (1988).
357. J. Bobacka, A. Ivaska, and M. Grzeszczuk, *Synth. Met. 44*:21 (1991).
358. J. Tanguy, J. L. Baudoin, F. Chao, and M. Costa, *Electrochim. Acta 37*:1417 (1992).
359. J. Kankare and E.-L. Kupila, *J. Electroanal. Chem. 322*:167 (1992).
360. S. H. Glarum and J. H. Marshall, *J. Electrochem. Soc. 134*:142 (1987).
361. I. Rubinstein, E. Sabatini, and J. Rishpon, *J. Electrochem. Soc. 134*:3078 (1987).
362. D. B. Swanson, A. G. MacDiarmid, and A. J. Epstein, *Synth. Met. 41–43*:2987 (1991).
363. P. Ferloni, M. Mastragostino, and L. Meneghello, *Electrochim. Acta 41*:27 (1996).
364. C. Deslouis, M. M. Musiani, M. El Rhazi, and B. Tribollet, *Synth. Met. 60*:269 (1993).
365. C. Deslouis, M. M. Musiani, and B. Tribollet, *J. Phys. Chem. 98*:2936 (1994).
366. S. Panero, P. Prosperi, and B. Scrosati, *Electrochim. Acta 37*:419 (1992).
367. A. M. Waller, A. N. S. Hampton, and R. G. Compton, *J. Chem. Soc. Faraday Trans. I 85*:773 (1989).
368. P. G. Pickup, *J. Chem. Soc. Faraday Trans. 86*:3631 (1990).
369. C. M. Elliott, A. B. Kopelove, W. J. Albery, and Z. Chen, *J. Phys. Chem. 95*:1743 (1991).
370. X. Ren and P. G. Pickup, *J. Phys. Chem. 97*:3941 (1993).
371. X. Ren and P. G. Pickup, *J. Phys. Chem. 97*:5356 (1993).
372. S. Fletcher, *J. Electroanal. Chem. 337*:127 (1992).
373. K. Yakushi, L. J. Lauchlan, T. C. Clarke, and G. B. Street, *J. Chem. Phys. 79*:4774 (1983).
374. E. M. Genies, G. Bidan, and A. F. Diaz, *J. Electroanal. Chem. 149*:101 (1983).
375. J. H. Kaufman, N. Colaneri, J. C. Scott, and G. B. Street, *Phys. Rev. Lett. 53*:1005 (1984).
376. E. M. Genies and J. M. Pernault, *Synth. Met. 10*:117 (1984/1985).
377. E. M. Genies and J. M. Pernault, *J. Electroanal. Chem. 191*:111 (1985).
378. G. Zotti and G. Schiavon, *Synth. Met. 30*:151 (1989).
379. K. Aoki, Y. Tezuka, K. Shinozaki, and H. Sato, *Denki Kagaku 57*:397 (1989).
380. Cs. Visy, J. Lukkari, T. Pajunen, and J. Kankare, *Synth. Met. 33*:289 (1989).
381. Cs. Visy, J. Lukkari, T. Pajunen, and J. Kankare, *Synth. Met. 39*:61 (1990).
382. T. Amemiya, K. Hashimoto, A. Fujishima, and K. Itoh, *J. Electrochem. Soc. 138*:2845 (1991).
383. F. Genoud, M. Guglielmi, M. Nechtschein, E. Genies, and M. Salmon, *Phys. Rev. Lett. 55*:118 (1985).
384. C. J. Zhong, Z. Q. Tian, and Z. W. Tian, *J. Phys. Chem. 94*:2171 (1990).
385. T. Inoue, I. Hosoya, and T. Yamase, *Chem. Lett. 1987*:563.
386. H. R. Virdee and R. E. Hester, *Croat. Chem. Acta 61*:357 (1988).
387. J. Bukowska and K. Jackowska, *Synth. Met. 35*:143 (1990).
388. C. S. Choi and H. Tachikawa, *J. Am. Chem. Soc. 112*:1757 (1990).
389. N. R. de Tacconi, Y. Son, and K. Rajeshwar, *J. Phys. Chem. 97*:1042 (1993).
390. C. Zhong and K. Doblhofer, *Synth. Met. 38*:117 (1990).

391. P. Novak, B. Rasch, and W. Vielstich, *J. Electrochem. Soc. 138*:3300 (1991).
392. C. Lopez, M. F. M. Viegas, G. Bidan, and E. Vieil, *Synth. Met. 63*:73 (1994).
393. P. Novak, R. Kötz, and O. Haas, *J. Electrochem. Soc. 140*:37 (1993).
394. R. Yang, W. H. Smyrl, D. F. Evans, and W. A. Hendrickson, *J. Phys. Chem. 96*:1428 (1992).
395. S. E. Creager, *J. Phys. Chem. 96*:2371 (1992).
396. M. Arca, M. V. Mirkin, and A. J. Bard, *J. Phys. Chem. 99*:5040 (1995).
397. M. Arca, B. R. Horrocks, and A. J. Bard, to be published.
398. B. J. Feldman, P. Burgmayer, and R. W. Murray, *J. Am. Chem. Soc. 107*:872 (1985).
399. L. Olmedo, I. Chonteloube, A. Germain, M. Petit, and E. M. Genies, *Synth. Met. 30*:159 (1989).
400. G. Schiavon, S. Sitran, and G. Zotti, *Synth. Met. 32*:209 (1989).
401. J. Ochmanska and P. G. Pickup, *J. Electroanal. Chem. 297*:197 (1991).
402. K. Kaneto, Y. Kohno, and K. Yoshino, *Solid State Commun. 51*:267 (1984).
403. T. C. Chung, J. H. Kaufman, A. J. Heeger, and F. Wudl, *Phys. Rev. B 30*:702 (1984).
404. M. J. Nowak, S. D. D. V. Rughooputh, S. Holta, and A. J. Heeger, *Macromolecules 20*:965 (1987).
405. S. N. Hoier, D. S. Ginley, and S.-M. Park, *J. Electrochem. Soc. 135*:91 (1988).
406. P. Marque and J. Roncali, *J. Phys. Chem. 94*:8614 (1990).
407. M. Lapkowski, M. Zogorska, I. Kulszewicz-Bajer, K. Koziel, and A. Pron, *J. Electroanal. Chem. 310*:57 (1991).
408. J. Roncali, L. H. Shi, and F. Garnier, *J. Phys. Chem. 95*:8983 (1991).
409. Cs. Visy, J. Lukkari, and J. Kankare, *J. Electroanal. Chem. 319*:85 (1991).
410. Z. W. Sun and A. J. Frank, *J. Chem. Phys. 94*:4600 (1991).
411. A. D. Child and J. R. Reynolds, *J. Chem. Soc., Chem. Commun. 1991*:1779.
412. D. Fichou, G. Horowitz, and F. Garnier, *Synth. Met. 39*:125 (1990).
413. J. H. Wang, *Surf. Interfac. Anal. 15*:635 (1990).
414. J. Ochmanska and P. G. Pickup, *J. Electroanal. Chem. 297*:211 (1991).
415. G. Zotti, A. Berlin, G. Pagani, G. Schiavon, and S. Zecchin, *Adv. Mater. 7*: 48 (1995).
416. R. J. Cushman, P. M. McManus, and S. C. Yang, *J. Electroanal. Chem. 291*:335 (1986).
417. E. M. Genies and M. Lapkowski, *J. Electroanal. Chem. 220*:67 (1987).
418. D. E. Stilwell and S.-M. Park, *J. Electrochem. Soc. 136*:427 (1989).
419. L. H. Dao, J. Guay, and M. LeClerc, *Synth. Met. 29*:E283 (1989).
420. K. Shimazu, K. Murakoski, and H. Kita, *J. Electroanal. Chem. 277*:347 (1990).
421. M. Lapkowski and E. M. Genies, *J. Electroanal. Chem. 284*:127 (1990).
422. K. Kanamura, Y. Kawai, S. Yonezawa, and Z. Takehara, *J. Phys. Chem. 98*:13011 (1994).
423. S. H. Glarum and J. H. Marshall, *J. Phys. Chem. 90*:6076 (1986).
424. S. H. Glarum and J. H. Marshall, *J. Phys. Chem. 92*:4210 (1988).
425. P. M. McManus, R. J. Cushman, and S. C. Yang, *J. Phys. Chem. 91*:744 (1987).
426. R. Holze and J. Lippe, *Synth. Met. 38*:99 (1990).
427. M. Wicker, G. Popkirov, and R. N. Schindler, *Synth. Met. 41–43*:3005 (1991).
428. A. Hugot-LeGoff and M. Bernard, *Synth. Met. 60*:115 (1993).
429. Y. Son, H. H. Patterson, and C. M. Carlin, *Chem. Phys. Lett. 162*:461 (1989).
430. M. H. T. Frank and G. Denuault, *J. Electroanal. Chem. 354*:331 (1993).
431. M. W. Espencheid and C. R. Martin, *J. Electroanal. Chem. 188*:73 (1985).
432. T. Osaka, K. Naoi, H. Sakai, and S. Ogano, *J. Electrochem. Soc. 134*:285 (1987).
433. T. Shimidzu, A. Ohtani, T. Iyoda, and K. Honda, *J. Electroanal. Chem. 224*:123 (1987).
434. P. G. Pickup, *J. Electroanal. Chem. 225*:273 (1987).
435. Y. Li and R. Qian, *Synth. Met. 28*:C127 (1989).
436. H. Mao, J. Ochmanska, C. D. Paulse, and P. G. Pickup, *Faraday Discuss. Chem. Soc. 88*:165 (1989).
437. B. Keita, D. Bonaziz, L. Nadjo, and A. Deronzier, *J. Electroanal. Chem. 279*:187 (1989).
438. S. Basak, K. Rajeshwar, and M. Kaneko, *Anal. Chem. 62*:1407 (1990).
439. H. Mao and P. G. Pickup, *J. Phys. Chem. 96*:5604 (1992).
440. F. Beck and M. Dahlhaus, *J. Electroanal. Chem. 357*:289 (1993).
441. T. Naguoka, M. Fujimoto, H. Nakao, K. Kakuno, J. Yano, and K. Ogura, *J. Electroanal. Chem. 364*:179 (1994).
442. H. K. Youssoufi, F. Garnier, A. Yassar, S. Baiteche, and P. Srivastava, *Adv. Mater. 6*:755 (1994).
443. B. Herberg and J. P. Pohl, *Ber. Bunsenges. Phys. Chem. 92*:1275 (1988).
444. G. Schiavon, G. Zotti, N. Comisso, A. Berlin, and G. Pagani, *J. Phys. Chem. 98*:4861 (1994).
445. R. Bilger and J. Heinze, *Synth. Met. 41–43*:2893 (1991).
446. J. Heinze and R. Bilger, *Ber. Bunsenges. Phys. Chem. 97*:502 (1993).
447. J. R. Reynolds, M. Pyo, and Y.-J. Qiu, *Synth. Met. 55–57*:1388 (1993).
448. R. C. D. Peres, M. A. DePaoli, and R. M. Torresi, *Synth. Met. 48*:259 (1992).
449. M. Pyo and J. R. Reynolds, *J. Chem. Soc., Chem. Commun. 1993*:258.
450. M. Pyo, G. Maeder, R. T. Kennedy, and J. R. Reynolds, *J. Electroanal. Chem. 368*:329 (1993).
451. R. Qian, Q. Pei, and Y. Li, *Synth. Met. 61*:275 (1993).

452. E. M. Genies and A. A. Syed, *Synth. Met. 10*:21 (1984/85).
453. J. B. Schlenoff and J. C. W. Chien, *J. Am. Chem. Soc. 109*:6269 (1987).
454. G. Inzelt and G. Horanyi, *J. Electroanal. Chem. 230*:257 (1987).
455. G. Inzelt and G. Horanyi, *J. Electrochem. Soc. 136*:1747 (1989).
456. T. Shimidzu, A. Ohtani, T. Iyoda, and K. Honda, *J. Chem. Soc., Chem. Commun. 1986*:1414.
457. Q.-X. Zhou, C. J. Kolaskie, and L. L. Miller, *J. Electroanal. Chem. 223*:283 (1987).
458. L. S. Curtin, G. C. Komplin, and W. J. Pietro, *J. Phys. Chem. 92*:12 (1988).
459. L. S. Curtin, M. McEllistrem, and W. J. Pietro, *J. Phys. Chem. 93*:1637 (1989).
460. S. Gosnier, A. Deronzier, J.-C. Moutet, and J. F. Roland, *J. Electroanal. Chem. 271*:69 (1989).
461. H. Mao and P. G. Pickup, *J. Phys. Chem. 93*:6480 (1989).
462. G. L. Duffitt and P. G. Pickup, *J. Phys. Chem. 95*:9634 (1991).
463. G. L. Duffitt and P. G. Pickup, *J. Chem. Soc. Faraday Trans. 88*:1417 (1992).
464. H. Zhao, W. E. Price, and G. G. Wallace, *Polymer 34*:16 (1993).
465. A. C. R. Hogervorst, *Synth. Met. 62*:27 (1994).
466. L. L. Miller and Q. X. Zhou, *Macromolecules 20*:1594 (1987).
467. L. L. Miller, B. Zinger, and Q.-X. Zhou, *J. Am. Chem. Soc. 109*:2267 (1987).
468. Q.-X. Zhou, L. L. Miller, and J. R. Valentine, *J. Electroanal. Chem. 261*:147 (1989).
469. G. Tourillon and F. Garnier, *J. Phys. Chem. 87*:2289 (1983).
470. R. L. Blankespoor and L. L. Miller, *J. Chem. Soc., Chem. Commun. 1985*:90.
471. A.-C. Chang, R. L. Blankespoor, and L. L. Miller, *J. Electroanal. Chem. 236*:239 (1987).
472. H. Reiss and D. Kim, *J. Phys. Chem. 90*:1973 (1986).
473. D. Kim, H. Reiss, and H. M. Rabeony, *J. Phys. Chem. 92*:2673 (1988).
474. Y. Ikenoue, J. Chiang, A. O. Patil, F. Wudl, and A. J. Heeger, *J. Am. Chem. Soc. 110*:2983 (1988).
475. Y. Ikenoue, N. Uotani, A. O. Patil, F. Wudl, and A. J. Heeger, *Synth. Met. 30*:305 (1989).
476. P. Snauwaert, R. Lazzaroni, J. Riga, and J. J. Verbist, *Synth. Met. 16*:245 (1986).
477. B. Pfeiffer, A. Thyssen, and J. W. Schultze, *J. Electroanal. Chem. 260*:393 (1989).
478. H. Reiss, *Synth. Met. 30*:257 (1989).
479. A. G. MacDiarmid, J. C. Chiang, A. F. Richter, and A. J. Epstein, *Synth. Met. 18*:285 (1987).
480. H. Reiss, *J. Phys. Chem. 92*:3657 (1988).
481. J. C. Chiang and A. G. MacDiarmid, *Synth. Met. 13*:193 (1986).
482. H. S. O. Chan, P. K. H. Ho, E. Khor, M. M. Tan, K. L. Tan, B. T. G. Tan, and Y. K. Lim, *Synth. Met. 31*:95 (1989).
483. H. Shinohara, J. Kojima, and M. Aizawa, *J. Electroanal. Chem. 266*:297 (1989).
484. K. Hyodo and M. Oomae, *Electrochim. Acta 35*:827 (1990).
485. K. Shimazu, K. Murakoshi, and H. Kita, *J. Electroanal. Chem. 277*:347 (1990).
486. J.-Y. Sung and H.-J. Huang, *Anal. Chim. Acta 246*:275 (1991).
487. J. Yue and A. J. Epstein, *J. Chem. Soc., Chem. Commun. 1992*:1540.
488. J. B. Schlenoff and J. C. W. Chien, *J. Chem. Soc., Chem. Commun. 1987*:1429.
489. F. Bedioui, J. Devynck, and C. Bied-Charreton, *Acc. Chem. Res. 28*:30 (1995).
490. O. Ikeda and H. Yoneyama, *J. Electroanal. Chem. 265*:323 (1989).
491. Z. Gao, J. Bobacka, A. Lewenstam, and A. Ivaska, *Synth. Met. 62*:117 (1994).
492. D. Matthews, A. Altus, and A. Hope, *Aust. J. Chem. 47*:1163 (1994).
493. M. Velazquez-Rosenthal, T. Skotheim, and J. Warren, *J. Chem. Soc., Chem. Commun. 1985*:342.
494. J. G. Eaves, H. S. Munro, and D. Parker, *Synth. Met. 16*:123 (1986).
495. T. Inagaki, M. Hunter, X. Q. Yang, T. A. Skotheim, and Y. Okamoto, *J. Chem. Soc., Chem. Commun. 1988*:126.
496. P. Audebert, G. Bidan, and M. Lapkowski, *J. Chem. Soc., Chem. Commun. 1986*:887.
497. B. Zinger, *Synth. Met. 30*:209 (1989).
498. J. Grimshaw and S. D. Perara, *J. Electroanal. Chem. 281*:125 (1990).
499. C. P. Andrieux, P. Audebert, and C. Salon, *J. Electroanal. Chem. 318*:235 (1991).
500. H. Yoneyama, Y. Li, and S. Kuwabata, *J. Electrochem. Soc. 139*:28 (1992).
501. A. B. Kon, J. S. Foos, and T. L. Rose, *Chem. Mater. 4*:416 (1992).
502. G. Bidan, E. M. Genies, and M. Lapkowski, *J. Electroanal. Chem. 251*:297 (1988).
503. T. Shimidzu, A. Ohtano, M. Aiba, and K. Honda, *J. Chem. Soc., Faraday Trans. I 84*:3941 (1988).
504. N. C. Foulds and C. R. Lowe, *Anal. Chem. 60*:2473 (1989).
505. J. Grimshaw and S. D. Perera, *J. Electroanal. Chem. 278*:279 (1990).
506. A. Boyle, E. Genies, and M. Fouletier, *J. Electroanal. Chem. 279*:179 (1990).
507. G. Bidan, B. Divisia-Blohorn, J.-M. Kern, and J.-P. Sauvage, *J. Chem. Soc., Chem. Commun. 1988*:723.
508. G. Bidan, B. Divisia-Blohorn, M. Lapkowski, J.-M. Kern, and J.-M. Sauvage, *J. Am. Chem. Soc. 114*:5986 (1992).
509. K. Okabayaksi, O. Ikeda, and H. Tamura, *J. Chem. Soc., Chem. Commun. 1983*:684.
510. T. Skotheim, M. Velazquez-Rosenthal, and C. A. Linkous, *J. Chem. Soc., Chem. Commun. 1985*:612.
511. F. Bedioui, C. Bongars, J. Devynck, C. Bied-Char-

reton, and C. Hinnen, *J. Electroanal Chem. 207*:87 (1986).
512. M. Velazquez-Rosental, T. Skotheim, and C. A. Linkous, *Synth. Met. 15*:219 (1986).
513. J.-P. Collin and J.-P. Sauvage, *J. Chem. Soc., Chem. Commun. 1987*:1075.
514. A. Deronzier and J.-M. Latour, *J. Electroanal. Chem. 224*:295 (1987).
515. P. Moisy, F. Bedioui, J. Devynick, L. Salmon, and C. Bied-Charreton, *New J. Chem. 13*:511 (1989).
516. P. N. Bartlett, L.-Y. Chung, and P. Moore, *Electrochim. Acta 35*:1051 (1990).
517. P. N. Bartlett, L.-Y. Chung, and P. Moore, *Electrochim. Acta 35*:1273 (1990).
518. C. Armengaud, P. Moisy, F. Bedioui, J. Devynick, and C. Bied-Charreton, *J. Electroanal. Chem. 227*:197 (1990).
519. G. Bidan, A. Deronzier, and J.-C. Moutet, *J. Chem. Soc., Chem. Commun. 1984*:1185.
520. G. Bidan, A. Deronzier, and J.-C. Moutet, *Nouv. J. Chim. 8*:50.1 (1984).
521. S. Cosnier, A. Deronzier, and J.-C. Moutet, *J. Phys. Chem. 89*:4895 (1985).
522. J. G. Eaves, H. S. Munro, and D. Parker, *J. Chem. Soc., Chem. Commun. 1985*:684.
523. F. Daire, F. Bedioui, J. Devynick, and C. Bied-Charreton, *J. Electroanal. Chem. 205*:309 (1986).
524. L. Corke, A. Deronzier, and J.-C. Moutet, *J. Electroanal. Chem. 198*:187 (1986).
525. A. Deronzier, M. Essakalli, and J.-C. Moutet, *J. Chem. Soc., Chem. Commun. 1987*:73.
526. A. J. Downard, N. A. Surridge, T. J. Meyer, S. Cosnier, A. Deronzier, and J.-C. Moutet, *J. Electroanal. Chem. 246*:321 (1988).
527. A. Deronzier, M. Essahalli, and J.-C. Moutet, *J. Electroanal. Chem. 244*:163 (1988).
528. S. Cosnier, A. Deronzier, and J.-C. Moutet, *Inorg. Chem. 27*:2390 (1988).
529. S. Cosnier, A. Deronzier, and J.-F. Roland, *J. Electroanal. Chem. 284*:133 (1990).
530. A. Deronzier and M. Essakalli, *J. Chem. Soc., Chem. Commun. 1990*:242.
531. J. P. Collin, A. Jouaiti, and J. P. Sauvage, *J. Electroanal. Chem. 286*:75 (1990).
532. A. J. Downard, N. A. Surridge, S. Gould, T. J. Meyer, A. Deronzier, and J.-C. Moutet, *J. Phys. Chem. 94*:6754 (1990).
533. J. Ochmanska and P. G. Pickup, *J. Electroanal. Chem. 271*:83 (1989).
534. J. Ochmanska and P. G. Pickup, *Can. J. Chem. 69*:603 (1991).
535. K. Rajeshwar and C. S. C. Bose, U.S. Patent 5,334,292 (Aug. 2, 1994).
536. P. G. Pickup and R. A. Osteryoung, *J. Electroanal. Chem. 195*:271 (1985).
537. R. C. M. Jakobs, L. J. J. Janssen, and E. Barendrecht, *Electrochim. Acta 30*:1433 (1985).
538. F. T. A. Vork, L. J. J. Janssen, and E. Barendrecht, *Electrochim. Acta 31*:1569 (1986).
539. F. T. A. Vork and E. Barendrecht, *Electrochim. Acta 35*:135 (1990).
540. I. M. F. De Oliveira, J.-C. Moutet, and N. Vlachopoulos, *J. Electroanal. Chem. 291*:243 (1990).
541. D. J. Strike, N. F. DeRooij, M. Koudelka-Hep, M. Ulmann, and J. Augustynski, *J. Appl. Electrochem. 22*:922 (1992).
542. M. Ulmann, R. Kostecki, J. Augustynski, D. J. Strike, and M. Koudelka-Hep, *Chimia 46*:138 (1992).
543. S. Cosnier, A. Deronzier, and A. Llobet, *J. Electroanal. Chem. 280*:213 (1990).
544. R. C. M. Jakobs, L. J. J. Janssen, and E. Barendrecht, *Electrochim. Acta 30*:1313 (1985).
545. N. S. Sundaresan and K. S. V. Santhanam, *Ind. J. Technol. 24*:11 (1986).
546. L. Coche and J.-C. Moutet, *J. Electroanal. Chem. 24*:313 (1988).
547. G. Tourillon and F. Garnier, *J. Phys. Chem. 88*:5281 (1984).
548. G. Mengoli, M. M. Musiani, G. Zotti, and S. Valcher, *J. Electroanal. Chem. 202*:217 (1986).
549. Z. Li and S. Dong, *Electrochim. Acta 37*:1003 (1992).
550. L. Doubova, G. Mongoli, M. M. Musiani, and S. Valcher, *Electrochim. Acta 34*:337 (1989).
551. M. Gholamian and A. Q. Contractor, *J. Electroanal. Chem. 289*:69 (1990).
552. L. Doubova, M. Fabrizio, G. Mengoli, and S. Valcher, *Electrochim. Acta 35*:1425 (1990).
553. G. Mengoli and M. M. Musiani, *J. Electroanal. Chem. 269*:99 (1989).
554. M. Zagorska, H. Wycislik, and J. Przyluski, *Synth. Met. 20*:259 (1987).
555. J. Przyluski, M. Zagorska, A. Pron, Z. Kucharski, and J. Suwalski, *J. Phys. Chem. Solids 48*:635 (1987).
556. S. Dong and G. Lian, *J. Electroanal. Chem. 291*:23 (1990).
557. G. Lian and S. Dong, *J. Electroanal. Chem. 260*:127 (1989).
558. W. Breen, J. F. Cassidy, and M. E. G. Lyons, *J. Electroanal. Chem. 297*:445 (1991).
559. C. C. Chen, C. Wei, and K. Rajeshwar, *Anal. Chem. 65*:2437 (1993).
560. F. Rourke and J. A. Crayston, *J. Chem. Soc., Faraday Trans. 89*:295 (1993).
561. N. C. Foulds and C. R. Lowe, *J. Chem. Soc., Faraday Trans. 82*:1259 (1986).
562. P. N. Bartlett and R. G. Whitaker, *J. Electroanal. Chem. 224*:37 (1987).
563. M. Umana and J. Waller, *Anal. Chem. 58*:2979 (1986).
564. D. Belanger, J. Nadreau, and G. Fortier, *J. Electroanal. Chem. 274*:143 (1989).
565. Y. Kajiya, H. Matsumoto, and H. Yoneyama, *J. Electroanal. Chem. 319*:185 (1991).
566. Y. Kajiya, H. Sugai, C. Iwakura, and H. Yoneyama, *Anal. Chem. 63*:49 (1991).
567. M. Caselli, M. Della Monica, and M. Portacci, *J. Electroanal. Chem. 319*:361 (1991).

568. F. Battaglini, C. Bonazzola, and E. J. Calvo, *J. Electroanal. Chem. 309*:347 (1991).
569. M. E. G. Lyons, D. E. McCormack, and P. N. Bartlett, *J. Electroanal. Chem. 261*:51 (1989).
570. M. E. G. Lyons, P. N. Bartlett, C. H. Lyons, W. Breen, and J. F. Cassidy, *J. Electroanal. Chem. 304*: 1 (1991).
571. G. Nagasubramanian, S. DiStefano, and J. Moacanin, *J. Phys. Chem. 90*:4447 (1986).
572. P. Aldebert, P. Audebert, M. Armand, G. Bidan, and M. Pineri, *J. Chem. Soc., Chem. Commun. 1986*: 1636.
573. R. M. Penner and C. R. Martin, *J. Electrochem. Soc. 133*:310 (1986).
574. H. Yoneyama, T. Hirai, S. Kuwabata, and O. Ikeda, *Chem. Lett. 1986*:1243.
575. O. Niwa and T. Tamamura, *J. Chem. Soc., Chem. Commun. 1984*:817.
576. M. A. DePaoli, R. J. Waltman, A. F. Diaz, and J. Bargon, *J. Chem. Soc., Chem. Commun. 1984*: 1015.
577. J. Bao, M. L. Daroux, M. Litt, and E. B. Yeager, *J. Polym. Sci. 8*:149 (1990).
578. B. Zinger, D. Behar, and D. Kijel, *Chem. Mater. 5*: 778 (1993).
579. S. E. Lindsey and G. B. Street, *Synth. Met. 10*:67 (1984).
580. O. Niwa, M. Hikita, and T. Tamamura, *Appl. Phys. Lett. 46*:444 (1985).
581. D. Kelhar and N. V. Bhat, *Polymer 34*:986 (1993).
582. S. W. Byun and S. S. Im, *Synth. Met. 57*:3501 (1993).
583. S. S. Im and S. W. Byun, *J. Appl. Polym. Sci. 51*: 1221 (1994).
584. L. F. Warren and D. A. Anderson, *J. Electrochem. Soc. 134*:101 (1987).
585. M. B. Gieselman and J. R. Reynolds, *Macromolecules 23*:3118 (1990).
586. B. Krische and G. Ahlgren, *Mol. Cryst. Liq. Cryst. 121*:325 (1985).
587. K. Naoi, K. Oeyama, T. Osaka, and W. H. Smyrl, *J. Electrochem. Soc. 137*:494 (1990).
588. E. Dalas, *J. Mater. Sci. 27*:453 (1992).
589. U. Geissler, M. L. Hallensleben, and L. Toponi, *Synth. Met. 40*:239 (1991).
590. M. Hepel, *Ceram. Trans. (Adv. Compos. Mater.) 19*: 389 (1991).
591. M. Hepel, Z. Fijalek, and L. Dentrone, *Polym. Prepr. 33*:106 (1992).
592. A. Ohtani, M. Abe, H. Higuchi, and T. Shimidzu, *J. Chem. Soc., Chem. Commun. 1988*, 1545.
593. T. Shimidzu, A. Ohtani, and K. Honda, *J. Electroanal. Chem. 251*:323 (1988).
594. D. P. Amalnerkar, S. Radhakrishnan, S. R. Sainkar, S. Badrinarayanan, and S. G. Joshi, *Solid State Commun. 84*:911 (1992).
595. K. Yoshino, X. H. Yin, S. Morita, Y. Nakanishi, S. Nakagawa, H. Yamamoto, T. Watanaki, and I. Isa, *Jpn. J. Appl. Phys. 32*:979 (1993).
596. M. A. Druy, *J. Electrochem. Soc. 133*:353 (1986).
597. J. Roncali and F. Garnier, *J. Chem. Soc., Chem. Commun. 1986*:783.
598. Z. Zhou, T. Maruyama, T. Kanbara, T. Ikeda, K. Ichimura, T. Yamamoto, and K. Tokuda, *J. Chem. Soc., Chem. Commun. 1991*:1210.
599. K. Tamao, S. Yamaguchi, M. Shiozaki, Y. Nakagawa, and Y. Ito, *J. Am. Chem. Soc. 114*:5867 (1992).
600. D. Lorcy and M. P. Cava, *Adv. Mater. 4*:562 (1992).
601. G. Bidan and B. Ehui, *J. Chem. Soc., Chem. Commun. 1989*:1568.
602. J. R. Reynolds, J. P. Ruiz, A. D. Child, K. Nayak, and D. S. Marynick, *Macromolecules 24*:678 (1991).
603. J.-Y. Sung and J.-J. Huang, *Anal. Chim. Acta 246*: 275 (1991).
604. N. Comisso, S. Daolio, G. Mengoli, R. Salmaso, S. Zecchin, and G. Zotti, *J. Electroanal. Chem. 255*:95 (1988).
605. Z. Qi and P. G. Pickup, to be published (courtesy preprint).
606. F.-T. A. Vork, L. J. J. Janssen, and E. Barendrecht, *Electrochim, Acta 32*:1187 (1987).
607. S. Holdcroft and B. L. Funt, *J. Electroanal. Chem. 240*:89 (1988).
608. S. Edge, A. E. Underhill, P. Kathirgamanathan, P.-O'Connor, and A. J. Dent, *J. Mater. Chem. 1*:103 (1991).
609. A. Leone, W. Marino, and B. R. Scharifker, *J. Electrochem. Soc. 139*:438 (1992).
610. R. Noufi, *J. Electrochem. Soc. 130*:2126 (1983).
611. K. Kauni, N. Mihara, S. Kuwabata, and H. Yoneyama, *J. Electrochem. Soc. 137*:1793 (1990).
612. H. Yoneyama and Y. Shoji, *J. Electrochem. Soc. 137*:3826 (1990).
613. F. Beck, M. Dahlhaus, and N. Zahedi, *Electrochim. Acta 37*:1265 (1992).
614. S. Maeda and S. P. Armes, *J. Mater. Chem. 4*:935 (1994).
615. S. Maeda and S. P. Armes, *Chem. Mater. 7*:171 (1995).
616. H. Yoneyama, A. Kishimoto, and S. Kuwabata, *J. Chem. Soc., Chem. Commun. 1991*:986.
617. D. Velayutham and M. Noel, *J. Appl. Electrochem. 23*:922 (1993).
618. G. Bidan, O. Jarjayes, J. M. Fruchart, and E. Hannecart, *Adv. Mater. 6*:152 (1994).
619. O. Jarjayes, P. H. Fries, and G. Bidan, *J. Magn. Magn. Mater. 137*:205 (1996).
620. O. Jarjayes and P. Auric, *J. Magn. Magn. Mater. 138*:115 (1996).
621. W. E. Rudzinski, C. Figueroa, C. Hoppe, T. Y. Kuromoto, and D. Root, *J. Electroanal. Chem. 243*:367 (1988).
622. A. Van der Patten, W. Visscher, and E. Barendrecht, *J. Electroanal. Chem. 195*:63 (1985).
623. W. A. Wampler, C. Wei, and K. Rajeshwar, *J. Electrochem. Soc. 141*:213 (1994).
624. W. A. Wampler, C. Wei, and K. Rajeshwar, *Chem. Mater. 7*:585 (1995).
625. W. A. Wampler, K. Rajeshwar, R. G. Pethe, R. C.

Hyer, and S. C. Sharma, *J. Mater. Res. 10*:1811 (1995).
626. W. A. Wampler, S. Basak, and K. Rajeshwar, *Carbon, 34*:747 (1996).
627. M. Hepel, E. Seymour, D. Yogev, and J. H. Fendler, *Chem. Mater. 4*:209 (1992).
628. F. Garnier and G. Tourillon, *J. Phys. Chem. 90*:5561 (1986).
629. G. Tourillon, F. Dartyge, H. Dexpert, A. Fontaine, A. Jacha, P. Lagarde, and D. E. Sayers, *J. Electroanal. Chem. 178*:357 (1984).
630. M. Inoue, M. Sotelo, L. Maoki, M. B. Inoue, K. W. Nebesny, and Q. Fernando, *Synth. Met. 32*:91 (1989).
631. M. Gholamian, J. Sundaram, and A. Q. Contractor, *Langmuir 3*:741 (1987).
632. Z. Q. Tian, Y. Z. Lian, J. Q. Wang, S. J. Wang, and W. H. Li, *J. Electroanal. Chem. 308*:357 (1991).
633. R. Kostecki, M. Ulmann, J. Augustynski, D. J. Strike, and M. Koudelka-Hep, *J. Phys. Chem. 97*:8113 (1993).
634. K. M. Kost, D. E. Bartak, B. Kazee, and T. Kuwana, *Anal. Chem. 60*:2379 (1988).
635. S. Kuwabata, N. Takahashi, S. Hirao, and H. Yoneyama, *Chem. Mater. 5*:437 (1993).
636. H. Inoue and H. Yoneyama, *J. Electroanal. Chem. 233*:291 (1987).
637. M. Gill, J. Mykytiuk, S. P. Armes, J. E. Edwards, T. Yeates, P. J. Moreland, and C. Mollett, *J. Chem. Soc. Chem. Commun. 1992*:108.
638. N. J. Terrill, T. Crowley, M. Gill, and S. P. Armes, *Langmuir 9*:2093 (1993).
639. G. Kokkinidis, A. Papoutsis, and I. Poulios, *J. Electroanal. Chem. 379*:379 (1994).
640. A. Kelaidopoulou, A. Papoutsis, G. Kokkinidis, and E. K. Polychroniadis, *J. Electroanal. Chem.* (in press).
641. G. C. Farrington, B. Scrosati, D. Frydrych, and J. DeNuzzio, *J. Electrochem. Soc. 131*:7 (1984).
642. N. Mermilliod, J. Tanguy, and F. Petiot, *J. Electrochem. Soc. 133*:1073 (1986).
643. T. Osaka, K. Naoi, S. Oguno, and S. Nakamura, *J. Electrochem. Soc. 134*:2096 (1987).
644. P. Novak, O. Inganas, and R. Bjorklund, *J. Electrochem. Soc. 134*:1341 (1987).
645. T. R. Jow and L. W. Shacklette, *J. Electrochem. Soc. 136*:1 (1989).
646. K. Shinozaki, A. Kabumoto, H. Sato, K. Watanabe, H. Umemura, and S. Tanemura, *Synth. Met. 38*:135 (1990).
647. P. Novak and W. Vielstich, *J. Electrochem. Soc. 137*:1681 (1990).
648. S. Panero, P. Prosperi, and B. Scrosati, *Electrochim. Acta 32*:1465 (1987).
649. B. Scrosati, P. Prosperi, S. Panero, M. Mastragostino, and A. Corradini, *J. Power Sources 19*:27 (1987).
650. S. Panero, P. Prosperi, and B. Scrosati, *Electrochim. Acta 32*:1461 (1987).

651. G. Casalbore-Miceli, G. Ciro, G. Beggiato, P. G. DiMarco, and A. Geri, *Synth. Met. 41–43*:1119 (1991).
652. T. Sotomura, H. Uemachi, K. Takeyama, K. Naoi, and N. Oyama, *Electrochim. Acta 37*:1851 (1992).
653. T. Matsunaga, H. Daifuku, T. Nakajima, and T. Kawagoe, *Polym. Adv. Technol. 1*:33 (1990).
654. H. Tsutsumi, S. Fukuzawa, M. Ishikawa, M. Morita, and Y. Matsuda, *J. Electrochem. Soc. 142*:L168 (1995).
655. J. J. Miasik, A. Hooper, and B. G. Tofield, *J. Chem. Soc., Faraday Trans. I 82*:1117 (1986).
656. J. G. Rabe, G. Bischoff, and W. F. Schmidt, *Jpn. J. Appl. Phys. 28*:518 (1989).
657. M. Hirata and L. Sun, *Sensors Actuators A: Phys. 40*:159 (1994).
658. P. Foot, T. Ritchie, and F. Mohammad, *J. Chem. Soc., Chem. Commun. 1988*:1536.
659. S. Dong, Z. Sun, and Z. Lu, *J. Chem. Soc., Chem. Commun. 1988*:993.
660. C. G. Z. Koopal, B. de Ruiter, and R. J. M. Nolte, *J. Chem. Soc., Chem. Commun. 1991*:1691.
661. Y.-B. Shim, D. E. Stilwell, and S.-M. Park, *Electroanalysis 3*:31 (1991).
662. J. M. Slater and E. J. Walt, *Analyst 116*:1125 (1991).
663. K. Rajeshwar, J. G. Ibanez, and G. M. Swain, *J. Appl. Electrochem. 24*:1077 (1994).
664. K. Rajeshwar and J. G. Ibanez, *Environmental Electrochemistry*, Academic, New York, in press.
665. G. P. Kittlesen, H. S. White, and M. S. Wrighton, *J. Am. Chem. Soc. 106*:7389 (1984).
666. G. P. Kittlesen, H. S. White, and M. S. Wrighton, *J. Am. Chem. Soc. 107*:7373 (1985).
667. E. W. Paul, A. J. Ricco, and M. S. Wrighton, *J. Phys. Chem. 89*:1991 (1985).
668. J. W. Thackeray, H. S. White, and M. S. Wrighton, *J. Phys. Chem. 89*:5133 (1985).
669. E. P. Lofton, J. W. Thackeray, and M. S. Wrighton, *J. Phys. Chem. 90*:6080 (1986).
670. D. Belanger and M. S. Wrighton, *Anal. Chem. 59*:426 (1987).
671. M. J. Natan, T. E. Mallouk, and M. S. Wrighton, *J. Phys. Chem. 91*:648 (1987).
672. S. Chao and M. S. Wrighton, *J. Am. Chem. Soc. 109*:2197 (1987).
673. C.-F. Shu and M. S. Wrighton, *J. Phys. Chem. 92*:5221 (1988).
674. X. Peng, G. Horowitz, D. Fichon, and F. Garnier, *Appl. Phys. Lett. 57*:2013 (1990).
675. M. J. Sailor, F. L. Klavetter, R. H. Grubbs, and N. S. Lewis, *Nature 346*:155 (1990).
676. N. Leventis, M. O. Schloh, M. J. Natan, J. J. Hickman, and M. S. Wrighton, *Chem. Mater. 2*:568 (1990).
677. J. Lei, W. Liang, C. J. Brumlik, and C. R. Martin, *Synth. Met. 47*:351 (1992).
678. R. Noufi, A. J. Frank, and A. J. Nozik, *J. Am. Chem. Soc. 103*:1849 (1981).
679. F. R. F. Fan, B. L. Wheeler, A. J. Bard, and R. N. Noufi, *J. Electrochem. Soc. 128*:2042 (1981).

680. T. Skotheim, L.-B. Petersson, O. Inganas, and I. Lundstrom, *J. Electrochem. Soc.* 129:1737 (1982).
681. R. Noufi, D. Tench, and L. F. Warren, *J. Electrochem. Soc.* 127:2310 (1980).
682. K. Honda and A. J. Frank, *J. Phys. Chem.* 88:5577 (1984).
683. G. Horowitz and F. Garnier, *J. Electrochem. Soc.* 132:634 (1985).
684. D. Gningue, G. Horowitz, and F. Garnier, *J. Electrochem. Soc.* 135:1695 (1988).
685. G. Horowitz and F. Garnier, *Solar Energy Mater.* 13:47 (1986).
686. A. J. Frank, S. Glenis, and A. J. Nelson, *J. Phys. Chem.* 93:3818 (1989).
687. M. J. Sailor, E. J. Ginsburg, C. B. Gorman, A. Kumar, R. H. Grubbs, and N. S. Lewis, *Science* 249:1146 (1990).
688. D. Gningue, G. Horowitz, J. Roncali, and F. Carrier, *J. Electroanal. Chem.* 269:337 (1989).
689. T. Kobayashi, H. Yoneyama, and H. Tamura, *J. Electroanal. Chem.* 161:419 (1984).
690. F. Garnier, G. Tourillon, M. Gazard, and J. C. Dubois, *J. Electroanal. Chem.* 148:299 (1983).
691. M. Akhtar, H. A. Weakliem, R. M. Paiste, and K. Gaughan, *Synth. Met.* 26:203 (1988).
692. J. H. Burroughes, D. D. C. Bradley, A. R. Brown, R. N. Marks, K. Mackay, R. H. Friend, P. L. Burns, and A. B. Holmes, *Nature* 347:539 (1990).
693. M. S. A. Abdan, G. A. Diaz-Guijada, M. I. Arroyo, and S. Holdcroft, *Chem. Mater.* 3:1003 (1991).
694. H. Yoneyama, *Adv. Mater.* 5:394 (1993).
695. P. Burgmayer and R. W. Murray, *J. Am. Chem. Soc.* 104:6139 (1982).
696. P. Burgmayer and R. W. Murray, *J. Phys. Chem.* 88:2515 (1984).
697. M. Okano, A. Fujishima, and K. Honda, *J. Electroanal. Chem.* 185:393 (1985).
698. J. Yue and A. J. Epstein, *J. Chem. Soc., Chem. Commun.* 1992:1540.
699. S. Kuwabata and C. R. Martin, *J. Membrane Sci.* 91:1 (1994).
700. P. Denisevich, K. W. Willman, and R. W. Murray, *J. Am. Chem. Soc.* 103:9727 (1981).
701. H. D. Abruna, P. Denisevich, M. Umana, T. J. Meyer, and R. W. Murray, *J. Am. Chem. Soc.* 103:1 (1981).
702. P. G. Pickup, C. R. Ledner, P. Denisevich, and R. W. Murray, *J. Electroanal. Chem.* 164:39 (1984).
703. C. R. Leidner, P. Denisevich, K. W. Willman, and R. W. Murray, *J. Electroanal. Chem.* 164:63 (1984).
704. H. D. Abruña, *Coord. Chem. Rev.* 86:135 (1988).
705. C. R. Leidner and R. W. Murray, *J. Am. Chem. Soc.* 107:551 (1985).
706. K. W. Willman and R. W. Murray, *J. Electroanal. Chem.* 133:211 (1982).
707. K. Maksymiuk, *J. Electroanal. Chem.* 373:97 (1994).
708. K. Maksymiuk, *Electroanalysis*, in press.
709. K. Murao and K. Suzuki, *J. Chem. Soc., Chem. Commun.* 1984:238.
710. K. Murao and K. Suzuki, *J. Electrochem. Soc.* 135:1415 (1988).
711. A. R. Hillman and E. F. Mallen, *J. Electroanal. Chem.* 281:109 (1990).
712. A. R. Hillman and E. F. Mallen, *J. Chem. Soc., Faraday Trans.* 87:2209 (1991).
713. A. R. Hillman and E. F. Mallen, *J. Electroanal. Chem.* 309:159 (1991).
714. A. R. Hillman and E. F. Mallen, *Electrochim. Acta* 37:1887 (1992).
715. A. R. Hillman and A. Glidle, *J. Electroanal. Chem.* 379:365 (1994).
716. S. Demoustrier-Champagne, J. R. Reynolds, and M. Pomerantz, *Chem. Mater.* 7:277 (1995).
717. W. Torres and M. A. Fox, *Chem. Mater.* 2:306 (1990).
718. H. Koezuka, H. Hyodo, and A. G. MacDiarmid, *J. Appl. Phys.* 58:1279 (1985).
719. Z. Guo, J. Bobacka, and A. Ivaska, *Synth. Met.* 55:1477 (1993).
720. J. Bobacka, Z. Guo, and A. Ivaska, *J. Electroanal. Chem.* 364:127 (1994).
721. S. Miyauchi, Y. Goto, I. Tsubata, and Y. Sorimachi, *Synth. Met.* 41–43:1051 (1991).
722. M. Aizawa, H. Shinohara, T. Yamada, K. Akagi, and H. Shirakawa, *Synth. Met.* 18:711 (1987).
723. R. J. Waltman, A. F. Diaz, and J. Bargon, *J. Phys. Chem.* 88:4343 (1984).
724. K. Chiba, T. Ohsaka, and N. Ogura, *J. Electroanal. Chem.* 217:239 (1987).
725. J. Yano, A. Shimoyama, and K. Ogura, *J. Chem. Soc., Faraday Trans.* 88:2523 (1992).
726. D. Sazou, I. Poulios, and G. Kokkinidis, *Synth. Met.* 32:113 (1989).
727. I. Rubinstein, *J. Electrochem. Soc.* 130:1506 (1983).
728. I. Rubinstein, *J. Polym. Sci. Polym. Chem.* 21:3035 (1983).
729. M. Satoh, K. Kaneto, and K. Yoshino, *J. Chem. Soc., Chem. Commun.* 1985:1629.
730. J. H. Ye, Y. Z. Chen, and Z. W. Tian, *J. Electroanal. Chem.* 229:215 (1987).
731. J.-F. Fauvarque, M.-A. Petit, A. Digua, and G. Frayer, *Makromol. Chem.* 188:1833 (1987).
732. P. Soubiran, S. Aeiyach, J. J. Aaron, M. Delamar, and P. C. Lacaze, *J. Electroanal. Chem.* 251:89 (1988).
733. K. Meerholz and J. Heinze, *Angew. Chem. Int. Ed. Engl.* 29:692 (1990).
734. P. Soubiran, S. Aeigach, and P. C. Lacaze, *J. Electroanal. Chem.* 303:125 (1991).
735. J.-Il Jin, C. K. Park, H.-K. Shim, and Y.-W. Park, *J. Chem. Soc., Chem. Commun.* 1989:1205.
736. P. L. Burn, A. B. Holmes, A. Kraft, D. D. C. Bradley, A. R. Brown, R. H. Friend, *J. Chem. Soc., Chem. Commun.* 1992:32.
737. F. Wudl, M. Kobayashi, and A. J. Heeger, *J. Org. Chem.* 49:3382 (1984).
738. M. Kobayashi, N. Colaneri, M. Baysel, F. Wudl, and A. J. Heeger, *J. Chem. Phys.* 82:5717 (1985).

739. T. Ohsaka, S. Kunimura, and N. Oyama, *Electrochim. Acta 33*:639 (1988).
740. S. Kunimura, T. Ohsaka, and N. Oyama, *Macromolecules 21*:894 (1988).
741. C. Barbero, J. J. Silker, and L. Sereno, *J. Electroanal. Chem. 263*:333 (1989); *291*:81 (1990).
742. G. Tourillon and F. Garnier, *J. Electroanal. Chem. 135*:173 (1982).
743. F. Garnier, G. Tourillon, M. Gazard, and J. C. DuBois, *J. Electroanal. Chem. 148*:299 (1983).
744. J. F. Ambrose and R. F. Nelson, *J. Electrochem. Soc. 115*:1159 (1968).
745. J. Bargon, S. Mohmand, and R. J. Waltman, *IBM J. Res. Dev. 27*:330 (1983).
746. J. Bargon, S. Mohmand, and R. J. Waltman, *Mol. Cryst. Liq. Cryst. 93*:279 (1983).
747. K. Ashley, D. B. Parry, J. M. Harris, and S. Pons, *J. Chem. Soc., Chem. Commun. 1988*:1253.
748. G. Schiavon, G. Zotti, and G. Bontempelli, *J. Electroanal. Chem. 194*:327 (1985).

21
Ion Implantation Doping of Electroactive Polymers and Device Fabrication

André Moliton
Université de Limoges, Limoges, France

I. INTRODUCTION AND GENERALITIES

From the point of view of electrical properties, polymers are essentially used as insulators. They usually correspond to macromolecular materials with saturated chains, which can be linear (e.g., polyethylene) or can have aromatic dangling groups (polystyrene) or nonaromatic dangling groups [poly(methyl methacrylate); PMMA]. Until now, the study [1] of ion beam effects has essentially been carried out on these types of polymers as well as on polymers with saturated chains on which aromatic groups are located (polyimide); these groups give a very high stability to the polymer, which becomes a conductor (conductivity $\sigma \geq 10^2$ S/cm) after irradiation with sufficiently energetic ions (energy $E \geq 300$ keV). The study of the origin of this metallic type of conductivity gives rise to much research [2,3] and is not explained here. We devote ourselves to the study of ions' effect on unsaturated polymers that present potentially electroactive properties. These polymers are called electroactive polymers or, more specifically, conductive polymers [4] because they can be either intrinsic insulators or intrinsic semiconductors, or semiconductors or conductors after doping. The best known doping processes are chemical doping in the liquid and gaseous phases (AsF_5, for example) and electrochemical doping with counterion migration. Until recently the existence of doping by ion implantation has been contested (see p. 373 of Ref. 1). But with recent results, I attempt to show in this chapter that ion implantation can dope polymers. The stability of conductivity after doping is one of the advantages of this process. One disadvantage is the concomitant presence of degradation during implantation with too high parameters, producing metallic type conductivity that masks doping contribution to semiconductivity.

Concerning implantation parameters, three values are essential: the ion beam current density (generally contained between 0.1 and 1 $\mu A/cm^2$), the fluence D (or the dose, the number of ions deposited on a unit surface), and the energy of the implanted ions.

One major advantage of ion implantation is the control over the doping level, which, with the help of the fluence, we determine exactly [5]. The usual range is from 10^{15} to 10^{16} ions/cm², which implies for polyparaphenylene $[(C_6H_4)_n$, density $\rho = 1.2$ g/cm³] a doping level of 1–10% with regard to number of monomers or 0.1–1% with regard to total number of atoms.

For the implantation energy, we can usually [6] distinguish three ranges:

Low energy range: 10 keV $< E <$ 90 keV, which confines implanted ions near the surface.
Intermediate energy range: 50 keV $< E <$ 300 keV, which is, by continuity with the study of implantation in silicon, the most explored range even in the case of polymers.
High energy range: $E \approx 2$ MeV, which allows ion implantation throughout the thickness of the polymer film.

In this chapter, which is dedicated to the doping of electroactive polymers by the ion implantation process, I essentially analyze results obtained by low energy implantation, and we study the macroscopic and microscopic characteristics of these materials.

First of all, by briefly summarizing all the important results obtained by implantation with the energy E greater than 100 or 200 keV, I show that damage occurs mainly in this relatively high energy range. In fact, this study was carried out on poly(paraphenylene sulfide) (PPS), as a basic polymer, at the MIT laboratory.

During the first implantation study [7] made with energy equal to 100 keV, it was shown that

1. Arsenic (As) ions and krypton (Kr) ions, having nearly the same mass, produce similar effects on conductivity (Fig. 21.1), because of structural rearrangemenets brought about by ion-implantation–induced disorder. A saturation of the effects occurs for a dose $D \approx 10^{16}$ ions/cm^2.
2. Halogen ions (e.g., Br) seem to produce a specific chemical doping that is the main process for high doses, even if this effect does not appear clearly on the curve (Fig. 21.1).

More precisely, a study was carried out with energies between 200 keV and 2 MeV, on various polymers such as PPS, PAN (polyacrylonitrile), PPO (polydimethylphenylene oxide), and PMMA. In a first article [8], Dresselhaus et al. showed that for doses between 10^{14} and 5 × 10^{15} ions/cm^2, similar behavior was observed for a large number of polymers at the same dose level, the dominant effect being defect production (breaking and cross-linking). The production of broken bonds is combined with the release of volatile species and with the production of free radicals and double bonds. It was found, even in this case, that there is only a slight dependence with implanted ion species, showing that chemical changes to the polymer chains are rather insignificant in comparison to defect production. Moreover, the conductivity occuring after implantation with the highest doses is similar to that observed in amorphous carbon;

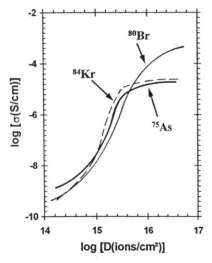

Fig. 21.1 Log-log plot of dc conductivity vs. fluence characteristics of PPS films implanted with arsenic, krypton, and bromine ions.

the authors concluded from this that the main implantation effect is to selectively remove the non-carbon atoms from the initial polymer.

In a second article [9] it was suggested that the conduction process takes place by hopping according to a one-dimensional model. The thermoelectric coefficient is very low ($S < 3 \mu V/K$) and can only be specific to a metallic-type behavior even if its sign is positive (which means a p-type semiconductor) after Br implantation through PPS or negative (n-type semiconductor) in the case of PAN. Afterwards, Wasserman [10,11] explained these transport mechanisms as being due to the creation of a percolation phase linked to a juxtaposition of clusters (produced by damage) where hopping conduction occurs.

In another study [12] on PPS, high conductivity was reached by using ion implantation with ion energy in the megaelectronvolt range. This conductivity was attributed by the authors to the high value of electronic stopping power. With some polydiacetylenes, Elman et al. [13] found a clear increase in conductivity, although it is quite difficult to chemically dope these materials. Dann et al. [14] showed that after implantation of iodine ions through phthalocyanines (energy between 100 and 200 keV), these ions produce the decomposition of the macrocycle. As established by Robinet et al. [15], these materials are fragile under implantation and become easily oxidized, even with different parameters. In our laboratory, we found for PPP [16] high values of conductivity (up to $\sigma \approx 100$ S/cm) following implantation in this energy range ($E \approx 200$ keV).

As described above, conductivity of electroactive polymers after ion implantation in this energy range is rather high and is essentially a metallic type of conductivity more linked to defect production than to the generation of real doping due to the chemical nature of the implanted ion. Conductivity levels are independent of the electroactive character of materials; conductivities up to around 100 S/cm have been reached, as in the case of polyimides [1,17].

Now we will study the effect of implantation into electroactive polymer targets at lower energies, which are necessary to induce semiconductive properties in these polymers. First I present the band scheme my group has suggested [18] to describe physical properties and notably transport phenomena; I particularly show how it is possible to balance (with a sort of compensation law) the conduction levels associated with damage and doping.

II. BAND SCHEME

The specific doping and conduction processes in electroactive polymers, which are widely described [4,19], depend mainly on three characteristics of these materials.

1. Local interactions between charges introduced by dopant atoms, and lattice deformation due to this ion bombardment. It follows that quasiparticles appear (solitons [20], polarons [21], bipolarons) that are characteristic of the type of dopant. Dopant atoms are not incorporated into the lattice at substitutional sites as covalent semiconductors (Si, Ge), but at interstitial sites. The specific electronegativity of these atoms (usually alkali or halogen atoms) is used to transfer charge to the polymer chain. Due to a Peierls transition that is energetically favorable, this polymer is intrinsically insulating as polyacetylene [$(CH)_x$, with a trans configuration and alternative single and double bonds) or polyparaphenylene (PPP with a benzenoid structure characterized by single bonds between benzene rings). After the charge transfer and alongside the chain, quasiparticles appear: solitons [$(CH)_x$] and polarons or bipolarons (PPP, polyparaphenylene sulfide, PPS, polypyrrole, polythiophene).

2. The dimensional structure of these materials. If we consider that there is no reticulation, these materials can appear one-dimensional in relation to both charge transfer and vibrational energy. This apparent anisotropy should be considered only local (at a short distance) because the study of conduction phenomena shows an isotropy of transport properties, which is quite well explained with the theory of interchain hopping mechanisms (Kivelson model [22] for solitons; polaron lattice introduced by Bredas and coworkers [23,24]). I will show that the heterogeneous polymers model may also be used.

3. Long-range disorder inside polymers, even if local order of the crystallites forming fibrils can be observed [25]. These polymers can be macroscopically considered as amorphous materials [26,27] characterized by localized states in band tails and with random distribution of potential wells where electronic sites are linked (Anderson model).

In fact [28], except for the case of unidimensional materials, disorder and electron–lattice interactions (small polarons localized within a length scale of about the interatomic spacing) do not give a strict additive effect but act together with a synergy that tends to produce one localization [29–31]. From the point of view of the energy, this localization can occur in the energy gap (polarons) or in the mobility gap (amorphous semiconductors). The polypyrrole band model established by Pfluger et al. [32] is used for these localized states, and associated to them is a density-of-states function (Fig. 21.2):

1. The introduced band tails will be all the larger, as disorder is high (zones c_5 and c_6).

2. The contribution of electron–lattice interactions will appear all the more through a polaronic or bipolaronic band as these interactions (and therefore the doping level) play a major role (zones c_3 and c_4).

With regard to transport mechanisms, the carriers can be thermally activated into either the localized states (carrier hopping between these states assisted by phonons) or extended states (zones c_1 and c_2). The necessary energy increases with the previously described order of these phenomena, and corresponding processes occur with increasing temperatures. In the case of intrinsic materials (PPP, for example [33]), this energy is equal to half the energy of the band gap from the valence band (π) to the conduction band (π^*).

Furthermore, conduction mechanisms related to the states near the Fermi level (conduction in degenerate states) can exist; indeed, structural defects can appear that are either intrinsic or induced by the implantation process [6,11]: vacancies, dangling bonds, or ends of chains. These defects induce localized states (zone c_7) near the Fermi level, as established by Mott in the case of amorphous semiconductors. Geometrical fluctuations (bond angles, for example) produce a band broadening of these levels, which split up into two bands [34]. The lower band represents the density of states due to neutral structural defects that are usually singly occupied by the electron coming from the dangling bond. The higher band, an acceptor-like band [35], corresponds to a defect that can receive one additional electron (producing a doubly occupied level, as allowed by the Pauli principle). Coulomb repulsion between the two electrons shifts this band from the first one by an amount equal to the correlation energy (Hubbard bands, as shown in Fig. 21.2, zone c_7).

In fact, as indicated elsewhere [36], two types of defect bands can appear with conjugated polymers. Let us consider the energy level formation during the bonding between two carbon atoms in the hybridization state $2sp^2\, 2p_z$. On the one hand we see (Fig. 21.3a) the formation of σ bonding and σ^* nonbonding orbitals, and on the other hand π bonding and π^* nonbonding orbitals appear.

If M denotes the monomer that contains such carbon atoms, π and π^* orbitals give rise (Fig. 21.3b) to the HOMO (valence band, VB) and LUMO (conduction band, CB) bands during the formation of polymer M_n with alternate single and double bonds; the density of states is then as represented in Fig. 21.3c. During the formation of a defect linked with the removal of one electron (as, for example, in the case of a bond breaking produced by ion implantation and giving a dangling bond), localized states depend on the type of orbitals on which the electron was primarily located. According to whether they come from a π orbital (p_z electrons) or a σ orbital (sp^2 electrons), the localized states are respec-

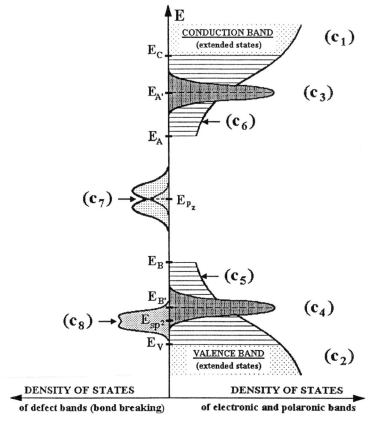

Fig. 21.2 Band diagram showing the defects band linked to C—H bond breaking. (c_1). Extended states of the conduction band; (c_2) extended states of the valence band; (c_3, c_4) localized states of the polaronic bands; (c_5, c_6) localized states of the band tails near the conduction and valence bands; (c_7) localized states in midgap; (c_8) band linked to C—H bond breaking.

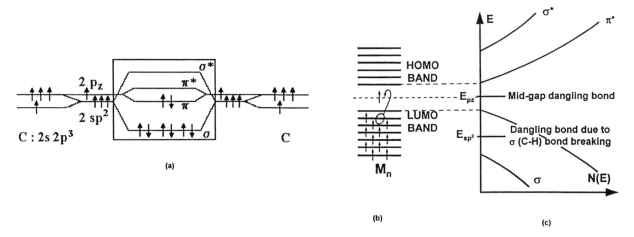

Fig. 21.3 Evolution of the energy levels due to interaction of (a) two carbon atoms, (b) n monomers giving rise to HOMO and LUMO bands, (c) resulting states density diagram. (From Ref. 18.)

Ion Implantation Doping

tively located

1. Just in the middle gap between the HOMO and LUMO bands; until now, only these states were considered [6,11].
2. Below the p_z level with a distribution focused on the sp^2 state.

By taking into account the relations [36] $E_p - E_s = 8.8$ eV and $E_{sp^2} = (E_s + 2E_p)/3$, we obtain

$$E_p - E_{sp^2} = (E_p - E_s)/3 \approx 2.9 \text{ eV}$$

Finally, in the PPP case where the top of the valence band is located 1.5 eV below the E_p level, we note that the E_{sp^2} localized state (associated with C—H bond breaking) is located 1.4 eV below the valence band; charge transfers from the top of the π band are allowed toward these localized states. Then this charge transfer pulls down the Fermi level to the valence band edge and induces a p-type conduction. In the band scheme of Fig. 21.4 are indicated the evolution of energy levels during doping [polaronic bands produced by chemical ions giving n-type doping (alkali ions) or p-type doping (halogen ions)] and during damage. (These dangling bonds produced by damages can induce charge transfer with hole generation; these carriers can be considered positive po-

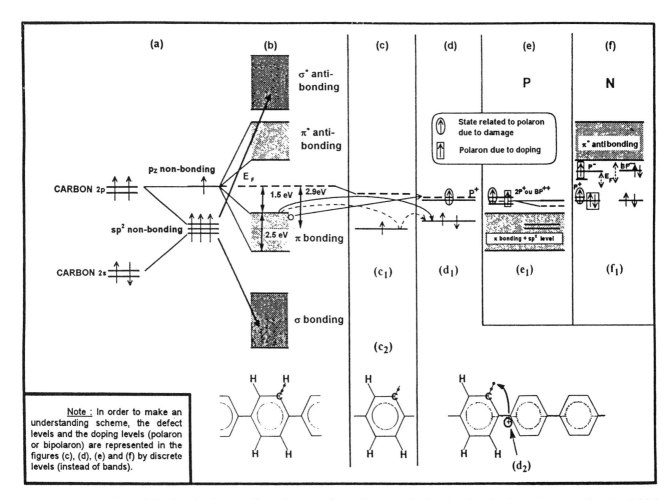

Fig. 21.4 Evolution of the band scheme of an electroactive polymer submitted to ion implantation by using the initial energy levels of carbon atoms. (c_1) Dangling bond; (c_2) neutral localized state ($q = 0$, $s = 1/2$); (d_1) effect of π-electron transfer to localized sp^2 level, with a possible appearance of Coulomb repulsion; charged localized states ($q = -e$, $s = 0$) are then produced and Fermi level goes down to the π-bonding band by producing one hole; (d_2) charge $+e$ delocalized in π system onto some phenyl groups (polaron > 0) (the electron localized on the sp^2 level is charged like the counterion during chemical doping); (e_1) appearance of polaronic band during doping process: p-type (halogen ions) in the figure. (f_1) By producing an additional n-type doping, some of the charges ($-e$) may fall on the sp^2 level (or create a C—alkali bond). The effect produced by n-type doping (P^- and BP^{2-}) competes with p doping linked to damage of the material. The conductivity and the thermopower laws can then be written $\sigma = \sigma_n + \sigma_p$ and $S = (\sigma_n S_n + \sigma_p S_p)/(\sigma_n + \sigma_p)$.

larons with low mobility because of their strong coulombic lattice interactions.) In fact, the mobility is reduced in both cases because of the disorder induced by the lattice rearrangement necessary to the introduction of the dopant into the bulk of the material.

Finally, the state density must contain (Fig. 21.2) both the defect band located at midgap (zone c_7) and a defect band located at a lower energy level (around the E_{sp^2} level, zone c_8), which is of great importance because of the main C—H bond breaking in polymers.

By comparison with amorphous semiconductors (a:Si and a:SiH) and in the case of large defect bands, we can think a priori [35] that conduction will occur by hopping near the Fermi level (pulls down to the degenerate band produced by many defects). This mechanism tends to play a major role in the low temperature range, where it generally involves a variable range hopping process (VRH) whereas it becomes thermally activated at higher temperatures (HNN: hopping to nearest neighbor). A great limitation of the number of these defects (by hydrogenation in the case of a:SiH) prevents a high contribution from this process, and conduction can take place either in band tails or in polaronic and bipolaronic bands (in the case of electroactive polymers), these bands being associated with doping. For polymers, we will show that this limitation seems to be achieved by reducing implantation parameters and the energy in particular.

III. OPTICAL CHARACTERIZATIONS (ULTRAVIOLET-VISIBLE, INFRARED, AND RAMAN)

Optical characterizations must allow us to determine the states in the energy gap and the changes undergone by chemical bonds inside the polymer. The essential difficulty in analyzing the spectra is due to the fact that the thickness of the implanted layer is quite often much less than that of the unimplanted region (virgin layer) of the material. This is all the more true since the energy and therefore the projected range (R_p) are low. To minimize this disadvantage, the thickness of the films studied was fixed at about 2000 Å, the thinnest films being reserved for implantation with lowest energies ($E < 100$ keV).

A. UV-Visible Range

From UV-visible analysis carried out on PPP made by an electrochemical process [38], Fig. 21.5 illustrates the evolution of the spectra (between 300 and 2000 nm) as a function of implantation energy. The other implantation parameters are ion implanted, cesium (Cs^+); dose, $D = 2 \times 10^{15}$ ions/cm^2; and current density $j = 0.5$ μA/cm^2. For the lowest implantation energies ($E = 40$ keV, Fig. 21.5a) we note that the absorption edge (π–π^* transi-

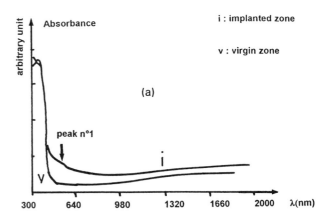

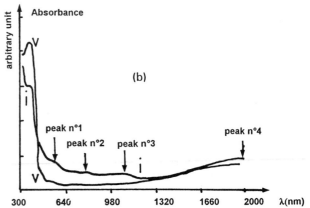

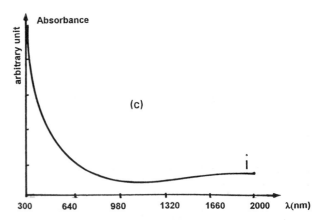

Fig. 21.5 Optical absorption spectra for cesium-implanted PPP films with various ion energies [(a) $E = 40$ keV; (b) $E = 100$ keV; (c) $E = 250$ keV]. The other implantation parameters are $D = 2 \times 10^{15}$ ions/cm^2, $j = 0.5$ μA/cm^2.

tion) is well defined and corresponds to an optical gap of 3 eV. As implantation energy increases (to 100 keV, Fig. 21.5b, and 250 keV, Fig. 21.5c), this edge becomes less clear and rather characteristic of a metallic state for higher energies (250 keV), with a very important de-

Ion Implantation Doping

crease of $\pi-\pi^*$ interband transitions, which were initially symbolized by a peak at 3.4 eV. In the case of implantation at 100 keV (an intermediate energy value), we can see in Fig. 21.6 the two essential modifications of the films [39]: the appearance of doping levels inside area B (corresponding probably to polaronic or bipolaronic levels in Fig. 21.7a) and the birth of states due to defects in the middle gap: zone C with peaks BC_1 and BC_2 (corresponding to transitions illustrated in Fig. 21.7b).

Finally [39], doping levels seem to appear after implantation with the lowest energies ($E \leq 100$ keV), while for the highest energies levels are diluted in the overlap of degenerate bands due to defects (graphitization). Such behavior is also observed with PPV [40].

In the case of polythiophene implanted with F^+ (fluorine) ions at 25 keV, Isotalo et al. [41] observed a slight decrease in the optical gap, but they did not find peaks due to doping because of the insufficient thickness of the implanted layer compared to the total thickness of the films.

A more sensitive method developed by the group of R. H. Friend [42] at the Cavendish Laboratory (UK) is based on modification of optical transmission during the

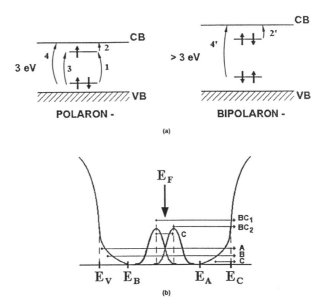

Fig. 21.7 (a) Band diagram scheme for n-type doped PPP according to the polaronic–bipolaronic model. (b) Transitions in the Mott–Davis model of an amorphous semiconductor.

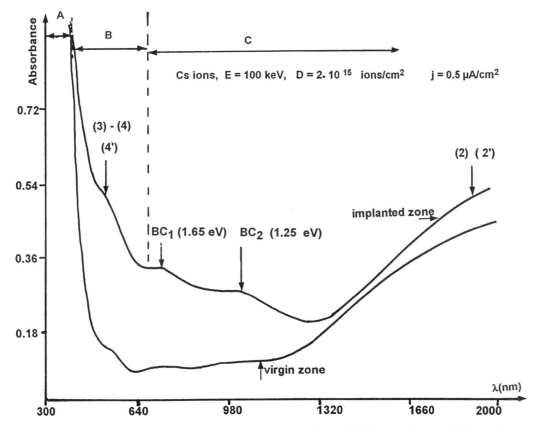

Fig. 21.6 Attribution of the absorption bands (implanted PPP film, Fig. 21.5b) in accordance with the polaron model (Fig. 21.7a) or the Mot–Davis model (Fig. 21.7b).

diffusion of the majority carriers (giving a modification of the polaronic band filling) produced by a bias voltage on a Schottky or MIS structure (with p doping, the depletion layer gives a transmission enhancement of the optical transition between the valence band and polaronic levels). With the help of this technique, it has been possible to show [43] indisputably the doping effect by charge transfer after implantations with alkali or halogen ions of low energy. In Fig. 21.8a, obtained with a PPV sample implanted with I^+ ions ($E = 30$ keV, $D = 10^{13}$ ions/cm^2), we note a strong induced absorption feature at around 0.7 eV and a weaker feature at 1.8 eV; these are consistent with transitions 2 and 4, respectively, of Fig. 21.8b with an optical gap around 2.5 eV (PPV case).

In contrast to the previous conventional optical spectra, the well-resolved structure in these spectra provides clear evidence of the doping effect despite damage effects we have in fact limited with the help of low implantation parameters.

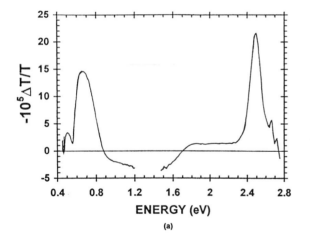

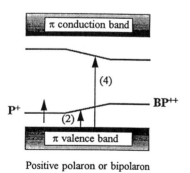

Fig. 21.8 (a) Difference in optical transmission at various gate voltages as a function of photon energy for an MIS structure with iodine (10^{13} cm^{-3}) implanted ($E = 30$ keV) PPV film as semiconductor. (b) Transition type involved in the spectrum depicted in (a).

B. Infrared and Raman Spectroscopies

By infrared spectroscopy of PPP films (made processible by using a copolymer [44]) spun on KBr substrates, we have noticed [45] that with the usual parameters ($j \approx 0.1$ μA/cm^2) the implanted area does not seem to be damaged by sodium or cesium ions at low ion energies (Figs. 21.9a and 21.9b) and absorption bands disappear only for $E \geq 250$ keV (Fig. 21.9c). Since the ratio of the thickness of the implanted layer to the thickness of the unimplanted layer is low, like the implanted ion concentration (in comparison with concentrations reached by chemical doping), it was not possible for us to induce the appearance of new bands due to doping. Nevertheless, in the case of low energy implantation with higher dose ($D = 4 \times 10^{15}$ ions/cm^2, Figs. 21.9d and 21.10a), we discovered a band at about 750 cm^{-1} whose shape and position depend on the mass of the implanted ion (band located at 745 cm^{-1} for sodium ion implantation, Fig. 21.10b; larger band at 735 cm^{-1} for cesium implantation, Fig. 21.10a). According to bibliographical data [46] this band seems to be due to the monosubstitution of a benzene molecule by a single element such as a nitrogen or oxygen atom. The slight shift of this band may mean that monosubstitution by sodium or cesium atom has occurred. Moreover, even in the case of higher energies, the characteristic band of the C=C bond does not appear (at about 1600 cm^{-1}) and characteristic bands of PPP are not shifted, which means that up to 150 keV chain length is not modified.

All the spectroscopic observations described in the case of PPP are similar to those of polyparaphenylenevinylene [40]. A band appears at 757 cm^{-1} (Figs. 21.10c and 21.10d) after cesium ion implantation with low energy ($E = 30$ keV) and the same dose level as with PPP ($D = 4 \times 10^{15}$ ions/cm^2). This behavior is typical of electroactive polymers doped by ion implantation.

Furthermore, some studies have been carried out by a Raman diffusion process on PPP and (CH)$_x$ samples. Note that with this spectroscopy nearly the entire implanted layer's thickness can be analyzed.

By using resonant Raman spectroscopy on a series of implanted (CH)$_x$ samples (ion implantation parameters: Ions implanted, sodium; energies, 50, 100, and 150 keV; doses, 10^{15}, 10^{16}, and 10^{17} ions/cm^2), Wada et al. [47] established (Fig. 21.11A) that structural defects are made only with high implantation parameters: For $E > 100$ keV and $D > 10^{16}$ ions/cm^2, a wide asymmetrical peak centered at around 1550 cm^{-1} is found, corresponding to a carbon layer being made near the surface. A deep Raman analysis of an implanted (CH)$_x$ sample (Na$^+$ ions, $E = 150$ keV, $D = 10^{17}$ ions/cm^2) shows (Fig. 21.11B) that the area between 0 and R_p (mean projected range) is extremely disrupted with the presence of this peak centered at about 1550 cm^{-1}. Beyond R_p, the initial Raman spectrum of unimplanted (CH)$_x$ is found again, which may indicate that damage is due to electronic

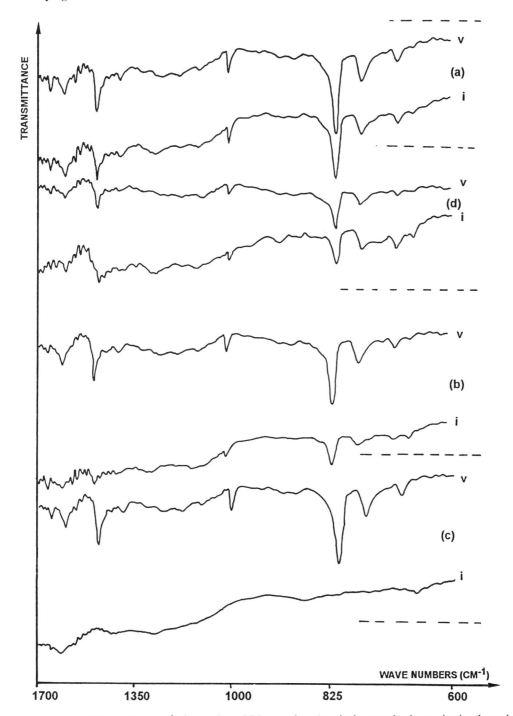

Fig. 21.9 Infrared spectra (1700 to 600 cm^{-1}) for various PPP samples. (v, virgin sample; i, cesuim-implanted sample). (a) $E = 30$ keV, $D = 10^{15}$ ions/cm^2; (b) $E = 150$ keV, $D = 10^{15}$ ions/cm^2; (c) $E = 250$ keV, $D = 10^{15}$ ions/cm^2; (d) $E = 30$ keV, $D = 4 \times 10^{15}$ ions/cm^2. (From Ref. 45.)

stopping power, which is the main process at higher energies. Damage is not visible in the area beyond R_p, corresponding to the area where the ion has lost almost all its energy and then where sodium concentration becomes lower. In the case of PPP made by the electrochemical process [38], similar results were found by Le Meil [48]; PPP spectra showed three main bands at 1600, 1280, and 1220 cm^{-1}, with the first band being insensitive to low energy implantation but disappearing when both high energy ($E \approx 250$ keV) and high dose ($D \approx 10^{16}$

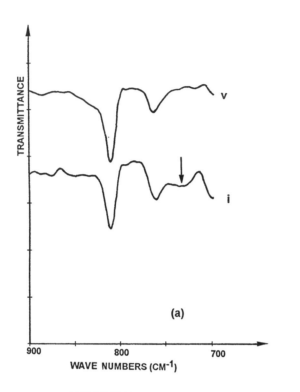
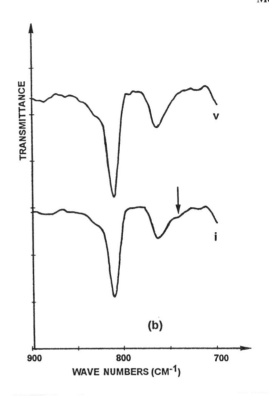
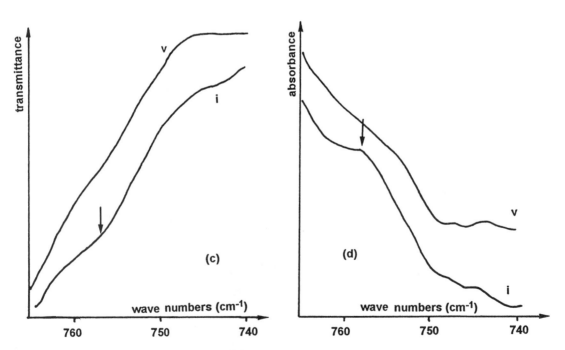

Fig. 21.10 Transmission spectra around 750 cm^{-1} for virgin (v) and implanted (i) PPP films with cesium (a) or sodium (b) ($E = 30$ keV and $D = 4 \times 10^{15}$ ions/cm^2). With the same parameters and Cs implantation, transmission and absorption spectra are reported for PPV films in (c) and (d), respectively. (From Ref. 45.)

Ion Implantation Doping

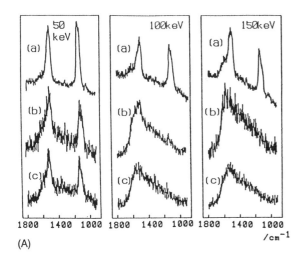

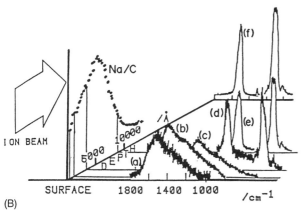

Fig. 21.11 (A) Resonant Raman spectra for sodium-implanted HD-$(CH)_x$ as a function of the energy and the fluence. a, $D = 10^{15}$ ions/cm^2; b, $D = 10^{16}$ ions/cm^2; c $D = 10^{17}$ ions/cm^2. (From Ref. 47.) (B) Depth profile of resonant Raman scattering for sodium-implanted HD-$(CH)_x$ with $D = 10^{17}$ ions/cm^2 and $E = 150$ keV at the surface (a), at depths of (b) 200 Å, (c) 500 Å, (d) 3000 Å, (e) 8200 Å (obtained by Ar$^+$ beam sputtering); (f) pristine $(CH)_x$. (From Ref. 47.)

ions/cm^2) were selected. The intensity ratio of the two other bands changed as both the energy and the dose were increased; this shows chain length shortening and then a deterioration of the material as implantation parameters increased. For the same dose levels ($D = 10^{16}$ ions/cm^2, $E = 30$ keV), cesium implantation seems to produce a much weaker Raman spectrum change than in the case of sodium ions. This behavior may be due to the smaller penetration depth of Cs$^+$ ions, allowing the large unimplanted area to produce an unaltered Raman contribution. In the case of PPP [44] made from PPP/polystyrene copolymer, modification thresholds of Raman spectra are reached with lower parameters, which may show that that polymer is more fragile under implantation [49] than PPP synthesized by the electrochemical process.

IV. ANALYSIS OF THE IMPLANTED LAYER

Studies of the surface (up to a 10 nm depth with XPS) and depth (RBS and SIMS) of the implanted layer were carried out with the help of electron spectroscopy for chemical analysis (ESCA) techniques. (XPS = X-ray photoelectron spectroscopy; RBS = Rutherford backscattering; SIMS = secondary ion mass spectroscopy.)

A. ESCA Analysis

When the first low energy implantations in $(CH)_x$ were made by Allen et al. [50] (Br$^+$ ions, $E = 25$ keV, $D = 3.4 \times 10^{16}$ ions/cm^2), ESCA analysis was used. Comparison of Br $3d$ spectra of $(CH)_x$ implanted with Br$^+$ ions and spectra of $(CH)_x$ doped chemically with Br$_2$ shows that if the envelopes of the peaks are very similar, the peak of the implanted sample is located at one bond energy higher (1 eV) than that of the chemically doped sample. The implanted bromine would appear to be linked to the carbon atom by a covalent bond but not in the Br$_2$ or Br$^-$ form. In the case of fluorine implantation ($E = 25$ keV, $D = 1 \times 10^{17}$ or 3.1×10^{17} ions/cm^2), the sharpness of the F $1s$ peak made Weber et al. [51] conclude that the implanted atom (e.g., Cl, Br, or I atom) can be in only one chemical form inside the layer below the polymer film surface. Recent studies carried out by Koshida and coworkers [52,53] on $(CH)_x$ gave major results: After the low energy implantation of sodium ions (between 8 and 10 keV), the $1s$ energy peak [at 284.4 eV for $(CH)_x$] of carbon remains unchanged (Fig. 21.12a); at the very most the full-width half-maximum can be slightly increased by 0.2 eV and occurs inside a layer analyzed in depth by etching. The material does not seem to be damaged by low energy implantation with these small ions. The slight shift after implantation of the Na $1s$ peak (Fig. 21.12b) leads us to think that a particular chemical form of sodium is produced that is localized near CH entities.

In the case of electrochemically produced PPP implanted with iodine ions (E increasing from 30 to 150 keV and doses from 10^{15} to 5×10^{16} ions/cm^2), we have shown [49] that the spectrum of the $1s$ carbon ray is broadened after implantation (Fig. 21.13) and its intensity decreases as both energies and dose increase. Moreover, it is important to note that there is a small decrease in the ratio of the number of carbon atoms in the implanted polymer to the number of carbon atoms in the

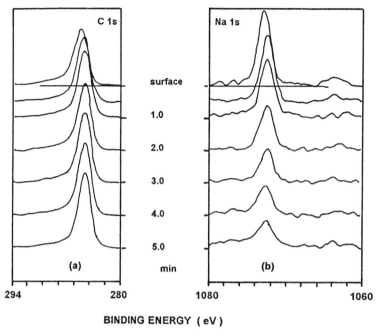

Fig. 21.12 ESCA spectra profile of (a) C $1s$ and (b) Na $1s$ for 8 keV Na$^+$-implanted (CH)$_x$, as a function of Ar$^+$ etching time. (From Ref. 53.)

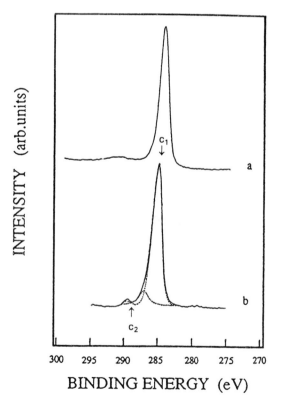

Fig. 21.13 XPS spectra C $1s$ region photoemission line of (a) pristine PPP and (b) I$^+$-implanted PPP. ($D = 10^{15}$ ions/cm^2; $E = 50$ keV, $j = 0.2$ μA/cm^2.) (From Ref. 49.)

virgin polymer depending on the energy (Fig. 21.14a) and the dose (Fig. 21.14b).

The presence of oxygen has also been discovered (insets in Figs. 21.14a and 21.14b) (characteristic peak of the C=O bond situated at 532.8 eV and that of the C—O bond peak at 533.8 eV). The ratio of carbonyl-type atoms (c_2 peak located at 289 eV) to aromatic-type atoms (c_1 peak located at 285 eV; see also Fig. 21.13a) decreases as implantation energies increase but rises with the dose. The layer analyzed close to the surface is more oxidized when ions are deposited inside a very thin layer that is buried only slightly into the polymer target. This oxidation becomes significant for higher doses. This characteristic will help us to understand some behaviors of conductivity according to the implantation parameters.

B. SIMS and RBS Analyses

In-depth analysis of implanted samples has been carried out by RBS (Rutherford backscattering) and SIMS (secondary ion mass spectroscopy). The RBS experiment on (CH)$_x$ was used by Davenas et al. [54] for the determination of oxygen concentration. Samples implanted with iodine ions ($E = 100$ keV; $D = 5 \times 10^{16}$ ions/cm^2) would seem to be less sensitive to oxidation from the air than virgin samples. Two oxidation steps would arise; the first step would concern the surface (during the first 4 h of exposure to the air), the other the volume (and would come from redistribution of oxygen atoms, which

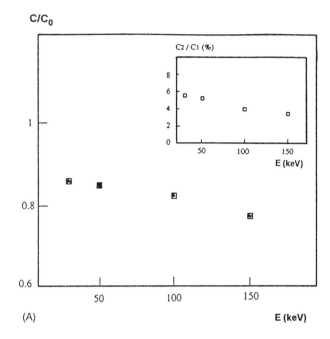

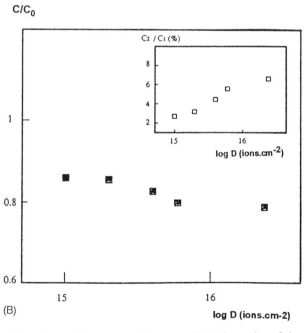

Fig. 21.14 (a) Variation of the normalized intensity of the C $1s$ line versus implantation energy ($D = 10^{15}$ ions/cm^2 and $j = 0.2\,\mu$A/cm^2). C_0 represents C $1s$ intensity in pristine PPP. Inset: Variation of the carbonyl aromatic ratio versus implantation energy. (b) Variation of the normalized intensity of the C $1s$ line versus fluence ($E = 50$ keV, $j = 0.2\,\mu$A/cm^2). Inset: Variation of the carbonyl aromatic ratio versus fluence. (From Ref. 49.)

would later diffuse inside the volume). The oxygen presence inside the volume was also confirmed by Lin et al. [55] after RBS analysis [in implanted and virgin (CH)$_x$] but without further details. Profiles of implanted ions are also presented by these authors; they have deep distribution tails, and their width is nearly 10 times the theoretical width (Fig. 21.15). Such behavior, already observed by Isotalo et al. [41] in the case of polythiophene, has been attributed to the porous nature of the material.

In the case of PPP made by following the electrochemical route (laid on an ITO substrate and called PPP/ITO) as well as by using a copolymer (PPP/PS), Ratier et al. carried out a study by ionic microanalysis [56] and discovered the presence of oxygen and nitrogen atoms within both implanted and virgin layers.

From nitrogen 15 and oxygen 18 implantations into initially virgin films, we determined the nitrogen and oxygen mean concentrations:

In the first case, we found nitrogen concentrations in the region of 0.9% and 0.2% in relation to total number of atoms for PPP/ITO and PPP/PS, respectively (which corresponds to [N]/[C] ratios of 1.5% and 0.3%).

In the second case, the oxygen concentrations were in the region of 5% and 3% in relation to total number of atoms for PPP/ITO and PPP/PS, respectively ([O]/[C] equal to 8% and 5%, respectively).

After the implantation of chemically active ions (such as Cs, Na, and I), the profiles of implanted ions present distribution tails that are more pronounced when the implantation energy is low (see Figs. 21.16 and 21.17 for cesium). Comparison with theoretical profiles shows deviations that are wider when the mass of the implanted ion is high and its energy low. In the case of cesium atoms and for $E = 30$ keV, the full-width half-maximum of the theoretical profile (Fig. 21.18) is 5 or 10 times

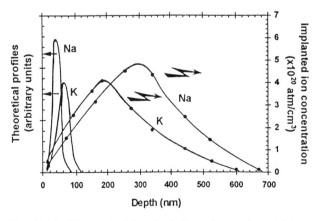

Fig. 21.15 Theoretical (on the left) and experimental (on the right) depth profiles for 24 keV K- and Na-implanted (CH)$_x$ samples.

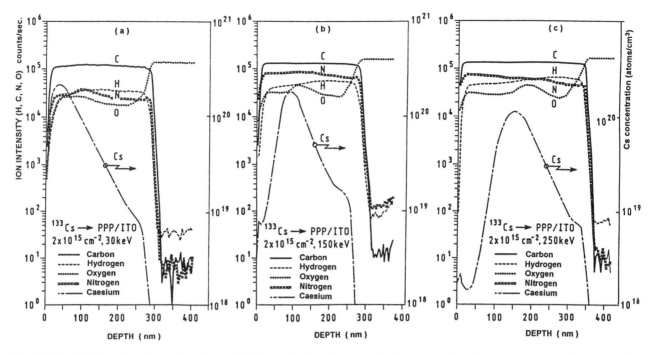

Fig. 21.16 SIMS profiles for electrochemical PPP samples implanted with Cs^+ ions and various parameters. Cesium profiles are obtained with oxygen bombardment; units are atoms/cm³; $^1H^-$, $^{12}C^-$, $^{26}CN^-$, and $^{16}O^-$ are obtained with cesium primary beam and are in relative units. (From Ref. 56.)

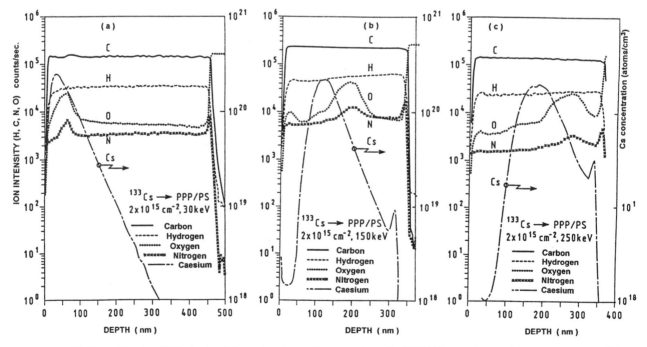

Fig. 21.17 SIMS profiles for PPP obtained from the thermal conversion of a PPP/PS copolymer. Measurement conditions are the same as in Fig. 21.16. (From Ref. 56.)

Ion Implantation Doping

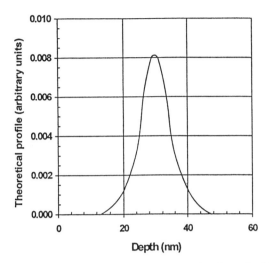

Fig. 21.18 Theoretical profiles for PPP implanted with 30 keV Cs ions. (From Ref. 56.)

lower than that of the experimental profile (Figs. 21.16 and 21.17). This result is in accord with experiments carried out by Isotalo et al. [41] and Lin et al. [55] on polythiophene and $(CH)_x$. It has been established [57,58] that the surface of PPP films presents a spongy morphology, which means that the surface density is much lower than the density of the bulk. The closer to the surface the ions are localized, the wider the gaps between the experimental and theoretical profiles.

By taking as a reference the intensity of both the nitrogen and oxygen signals at the very bottom of the unimplanted layer of the PPP/ITO sample, one can note that for all implanted ions (Fig. 21.16 in the case of cesium ions), nitrogen and oxygen concentrations increase within the implanted area (between 0 and R_p) with the appearance of both nitrogen and oxygen peaks, more markedly for the latter at $R_p + \Delta R_p$ (R_p, projected range; ΔR_p, standard deviation). The interface between the implanted zone and the unimplanted zone (which is more crystalline) must favor the accumulation of defects, particularly the oxygen localization.

Note that this effect is much more intense in the case of PPP/PS (Figs. 21.17a–21.17c), showing that this material must be much better organized with a crystallinity rate much higher than that of PPP/ITO. The analysis of Figs. 21.16 and 21.17 indicates that, particularly in the region (R_p, $R_p + \Delta R_p$), the ratio of integrated values (representative of concentrations) of the cesium and oxygen profiles decreases as the energy used for implantation increases. This major observation is useful for the understanding of the decrease in the cesium doping effect when implantation energy is increased (by increasing the relative oxygen concentration, which can produce two effects: a chemical compensation effect and the growth of a defect band).

Furthermore [58], ion diffusion both across the surface and in depth was studied from SIMS profiles measured in the three dimensions. This experiment was carried out on samples directly implanted with Na^+, Cs^+, or I^+ ions at $E = 30$ keV and also on samples that had

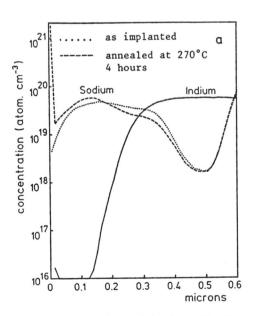
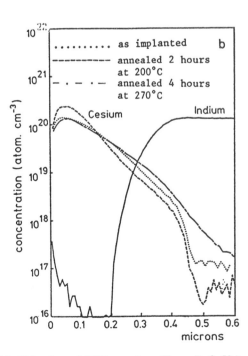

Fig. 21.19 Experimental depth profiles for various 30 keV implanted PPP samples. (From Ref. 58.)

been held at 543 K for 4 h. Whereas the annealing effect on the sodium implantation profile (chosen as an example in Fig. 21.19a) is extremely low in depth, we note toward the surface an exodiffusion amplified by the small size of the selected ion. In the case of Cs^+ (Fig. 21.19b) and I^+ ions, this exodiffusion is much more limited [58] and does not seem to be related to a migration toward defects made in the surface by ion implantation. This high mobility (of Na^+ ions in particular) at the surface is confirmed by lateral profile studies (Fig. 21.20) showing that after annealing, implanted ions (Na^+, Cs^+, I^+) diffuse toward the virgin area. One can think that there are two diffusion mechanisms related to the polyparaphenylene structure: high density areas that are locally very crystalline would allow only a slight diffusion, whereas inside low density areas, between crystallites, high diffusion would take place. Now one can understand why the stability of doping made by ion implanta-

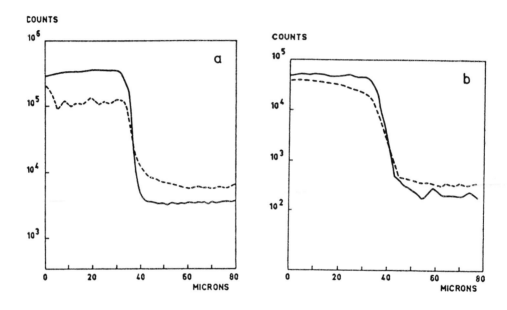

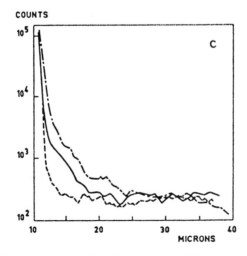

Fig. 21.20 Experimental lateral profiles for various 30 keV implanted PPP samples. (a) Sodium; (b) iodine. (———) As implanted; (---) annealed at 270°C for 4 h. (c) Cesium. (---) Nonannealed; (———) annealed at 200°C for 2 h; (-·-·-) annealed at 270°C for 4 h. (From Ref. 58.)

tion (which reaches all the sites) is much higher than the stability of chemical doping, which would be active only in low density zones reached by diffusion in the gaseous phase. This hypothesis was confirmed by the observations of Koshida and Hirayama [53] (SEM micrography), who show that morphological changes of fibrils are not uniform during chemical doping of polyacetylene, contrary to the case of doping by ion implantation.

V. STUDY OF IMPLANTATION PARAMETER EFFECTS ON ELECTRICAL PROPERTIES OF ELECTROACTIVE POLYMERS

A. Introduction and Review of Possible Transport Phenomena

1. Introduction and Results for Chemical Doping

The effect of doping on electrical properties such as (1) direct current conductivity σ_{dc} (dc conductivity), (2) thermoelectric power characterized by the Seebeck coefficient S, and (3) alternating current conductivity σ_{ac} (ac conductivity) has until recently essentially been studied in the case of electroactive polymers doped by chemical or electrochemical processes.

For the case of polyacetylene Fig. 21.21a illustrates both σ_{dc} and S measurements as a function of the chemical doping rate (Ref. 59, p. 734). We note that for a doping rate higher than about $y = 1\%$, the material becomes degenerate and shows a quasi-metallic character with S small (only a few microvolts per kelvin) and σ_{dc} high ($\sigma_{dc} > 10^2$ S/cm). In fact, an abrupt change happens for an iodine doping rate of $y \approx 1\%$, as shown in Fig. 21.21b (Ref. 59, p. 1071). As mentioned by Chiang et al. [60], the degeneracy is quickly reached, and a large decrease in the thermal activity of the conduction process is observed when this doping increases.

2. Moderate or High Doping

In fact, the interpretation of both direct conductivity and thermoelectric power behaviors in these chemically doped electroactive polymers gives rise to many discussions [61,62]. To account for these behaviors, several representations using models of heterogeneous polymers (having two or three domains) have been proposed. In the case of two domains (Fig. 21.22a) initially exposed, one of the areas (index 1) being characteristic of the fibrils, the other (index 2) representing the gaps separating (by an electric barrier) the fibrils, the resulting conductivity is equal to [63]:

$$\sigma^{-1} = f_1\sigma_1^{-1} + f_2\sigma_2^{-1}$$

where the form factor $f_i = L_iA/pLA_i$, p being the number of conducting paths characterized here by the same L_i/A_i ratio. L and A represent the sample length and sample cross section, respectively.

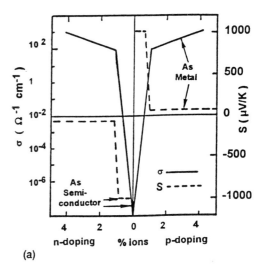

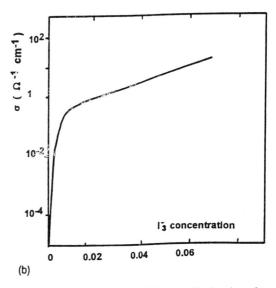

Fig. 21.21 Chemical doping. (a) General behavior of transport parameters according to the dopant concentrations; (b) experimental evolution of the dc conductivity according to I_3^- concentration.

The thermoelectric power can be written as [64]

$$S = \frac{W_1}{W} S_1 + \frac{W_2}{W} S_2$$

when the heat follows the same path toward the fibrils as the electric current. (W_i is the thermal resistance of the ith domain, W is the total thermal resistance, and S_i represents the thermoelectric power of the ith domain.)

For moderate doping, where $\sigma_1 \gg \sigma_2$, the variation of σ depends essentially on σ_2 (since $\sigma \approx f_2^{-1}\sigma_2$), which means that the conduction is due to transport phenom-

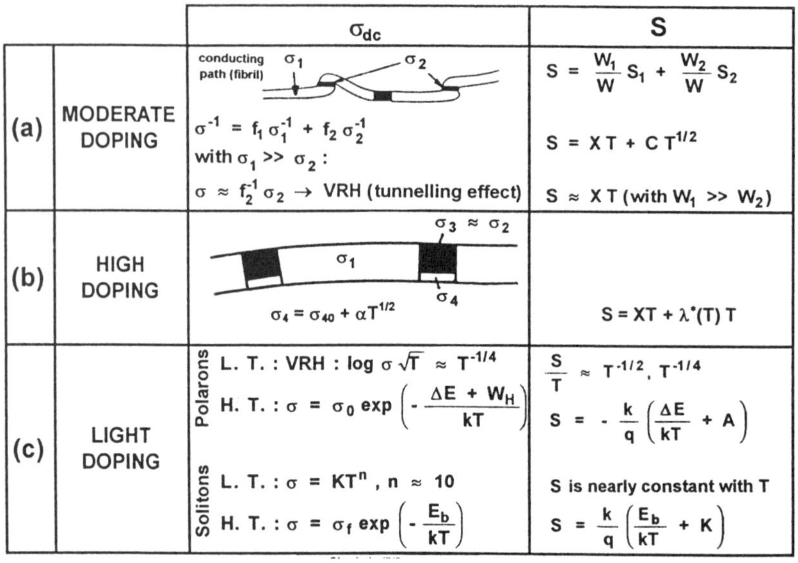

		σ_{dc}	S
(a)	MODERATE DOPING	conducting path (fibril) σ_1 — σ_2 $\sigma^{-1} = f_1\sigma_1^{-1} + f_2\sigma_2^{-1}$ with $\sigma_1 \gg \sigma_2$: $\sigma \approx f_2^{-1}\sigma_2 \rightarrow$ VRH (tunnelling effect)	$S = \dfrac{W_1}{W}S_1 + \dfrac{W_2}{W}S_2$ $S = XT + CT^{1/2}$ $S \approx XT$ (with $W_1 \gg W_2$)
(b)	HIGH DOPING	$\sigma_3 \approx \sigma_2$ σ_1 σ_4 $\sigma_4 = \sigma_{40} + \alpha T^{1/2}$	$S = XT + \lambda^*(T)T$
(c)	LIGHT DOPING	Polarons: L.T.: VRH: $\log \sigma\sqrt{T} \approx T^{-1/4}$ H.T.: $\sigma = \sigma_0 \exp\left(-\dfrac{\Delta E + W_H}{kT}\right)$ Solitons: L.T.: $\sigma = KT^n$, $n \approx 10$ H.T.: $\sigma = \sigma_f \exp\left(-\dfrac{E_b}{kT}\right)$	$\dfrac{S}{T} \approx T^{-1/2}, T^{-1/4}$ $S = -\dfrac{k}{q}\left(\dfrac{\Delta E}{kT} + A\right)$ S is nearly constant with T $S = \dfrac{k}{q}\left(\dfrac{E_b}{kT} + K\right)$

Fig. 21.22 Main formulas summarizing transport mechanisms in chemically doped polymers. (From Ref. 95.)

ena assisted by a tunneling effect. In particular at low temperatures, the hopping mechanism (VRH) proposed by Mott is satisfactory. Moreover, we notice that σ is high because of the f_2^{-1} term, which is proportional to L/L_2, where L_2 represents the narrow interfibril barrier. In agreement with the previous law of thermoelectric power, Kaiser [63] proposed the expression

$$S = XT + CT^{1/2}$$

where the first term is characteristic of metallic conduction alongside the fibrils, and the second takes into account the "interbarrier hoppings," which give a T^m law with $m \approx 0.5$ in the case of variable range hopping as established by Mott and Davis [26]. If the barriers are narrow and therefore conduct heat much better than the electricity, we can use the simplifying hypothesis that $W_1 \gg W_2$, and the metallic behavior (linear in T) of S (small) becomes preponderant.

For heavy doping, and to take into account both the finite value of σ when temperature tends to 0 K and the increase of σ with T at low temperatures, Kaiser and Graham [65] proposed to modify the representation of the previous interfibril domain (index 2). They replace it (Fig. 21.22b) by two parallel domains; one, whose conductivity is equal to σ_3, continues to represent the interfibril hoppings; the other, an amorphous metal type, introduces into the conductivity formula a new component σ_4 such that $\sigma_4(T) = \sigma_{40} + \alpha T^{1/2}$, where σ_{40} and α are constant. For high doping levels, the metallic evolution of thermoelectric power with temperature displays a distortion at low temperatures (at around 50 K); Kaiser and Graham [63–66] take this into account by introducing a nonlinear term $\lambda^*(T)$, which implies a diffusion effect due to electron–phonon interactions (phonon drag).

3. Light Doping

In the case of light doping (summarized in Fig. 21.22c), use of heterogeneous polymer models has not yet been planned, although one must still distinguish the interchain and intrachain mechanisms. In fact, the electrical conductivity is related to hopping mechanisms of quasiparticles (solitons, polarons, bipolarons) produced by doping, and their effective mass is 6 times larger than that of the free electron. Kivelson [22,67] considers, in the case of polyacetylene, the solitons moving along the chain at high temperature: the conductivity is simply activated (from the bond energy E_b of the charged soliton with the counterion):

$$\sigma = \sigma_f \exp\left(-\frac{E_b}{kT}\right)$$

As in the case of classical doped semiconductors, the thermoelectric power depends greatly on temperature and has a high value [61,67]:

$$S = \frac{k}{q}\left(\frac{E_b}{kT} + \text{const}\right)$$

Another mechanism, much less activated, seems to play a major role at low temperature. It is due to phonon-assisted interchain hopping between electronic states associated with solitons. The most likely hoppings start from a charged soliton and end on a neutral soliton located near a charged impurity (hopping energy minimized). A simplification of this model, although not very satisfactory at low temperature, leads to conductivity in the form $\sigma = KT^n$ (with $n \approx 10$) and S = constant whatever the temperature. But we can make some criticism of this solitonic interchain hopping phenomenon because it needs a total rearrangement of both the chain where the soliton was primarily and the chain that receives the soliton. Bredas and coworkers [23] proposed a mechanism of transport by bipolarons that requires only a small local reorganization of the polymer (on the scale of the localization length of the bipolarons); this model is also as valid for degenerate polymers [$(CH)_x$] as for nondegenerate polymers (PPP in particular).

The conductivity due to conduction by polarons (small polarons) in a disordered system depends greatly on the temperature [68]. At low temperature the polaron moving requires, in fact, two concomitant tunneling effects [28,69]: one related to the charge displacement and the other to the transfer of atomic sites that are associated with the deformation. Because of the mass of atoms, which gives a very low overlap factor, the latter transport mechanism is negligible.

At sufficiently high temperature, the lattice vibrations can create a configuration such that the charges have exactly the same energy level at the initial atomic site (associated with the deformation) as at the final atomic site. The polaron displacement (brought about by a thermal energy W_H nearly equal [29,69] to half the binding energy of the polaron, E_b) occurs after charge transfer by tunneling between two sites that are momentarily equivalent.

If ΔE represents the activation energy required for thermal generation of carriers, we have [34,70]

$$\sigma = \sigma_0 \exp\left(-\frac{\Delta E + W_H}{kT}\right)$$

and

$$S = -\frac{k}{q}\left(\frac{\Delta E}{kT} + A\right)$$

where A is equal to zero [30] when there is no vibrational energy transfer during the hopping (as is the case in the simplest model envisaged) or varies between 1 and 10 according to the shape of the density-of-states function used in the hopping mechanism theory.

An alternative equation was proposed by Emin [29] for cases when the carriers are located in a narrow band (smaller than kT) of degenerate states. If c is the carrier concentration by site, we obtain, in agreement with Kosarev [71] and Chaikin and Beni [72],

$$S = -\frac{k}{q}\ln\left[\frac{2-c}{c}\right]$$

This Heikes formula [73] assumes weak coulombic interactions [72]; it is used by Nagels [74] in polaronic situations where thermoelectric power is nearly temperature-independent; this case involves a weak vibrational energy transfer during hopping and temperature independence for the small polaron density. If N is the number of available sites during hopping and n is the number of carriers, S becomes [72]

$$S = -\frac{k}{q}\ln\left[\frac{N}{n}\right]$$

In the low temperature range (lower than room temperature) where the thermoelectric power is small and nearly temperature-independent, the variable range hopping (VRH) mechanism [26,68] is still quite satisfactory in polaronic situations (p. 324 of Ref. 34). In fact, S relations related to the VRH process vary among authors [70,75]. For example, the variation law is consistent with the form (p. 280 of Ref. 68) $S \approx T^{1/2}$ or is given by [27,76]

$$\frac{S}{T} = -\frac{\pi^2 k^2 \gamma}{12q}\left(\frac{T_0}{T}\right)^{1/4}$$

In this expression, γ is defined from the linear variation of the density of states near the Fermi level E_F:

$$N(E) = N(E_F)[1 + \gamma(E - E_F)]$$

We also have

$$T_0 = 7.64 \frac{\alpha^3}{kN(E_F)}$$

(with $1/\alpha$ equal to the localization length)

and the conductivity is given by

$$\sigma = \sigma_2 \exp\left[-\left(\frac{T_0}{T}\right)^{1/4}\right]$$

Such formulas agree with VRH mechanisms in conjugated polymers [77].

Relative to these transport mechanisms (summarized in Fig. 21.22 and with the curves of Fig. 21.23a for the conductivity and those of Fig. 21.23b for S) and as explained by Epstein [78], we must keep in mind that a small change of local order (even on the millielectronvolt energy scale) modifies the density-of-states function $N(E)$, and, in particular at around E_F, the transport phenomena are quickly affected. This situation will have a tendency to occur during ion implantation, one of whose effects is to produce dangling bonds; one associates with them a defect band [18] with a density of states (localized near E_F when states are degenerate) that may depend on the implantation parameters. The hopping mechanisms between these states can be described by the previously discussed Mott's law (VRH) relative to direct conductivity, and they can be observed by the study of ac conductivity.

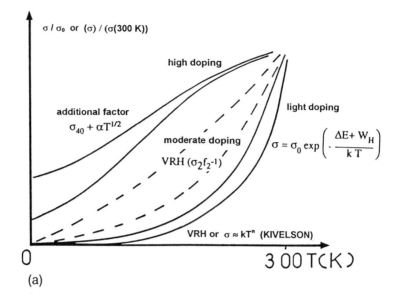

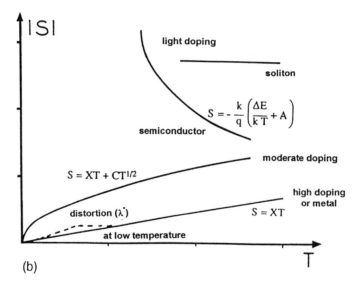

Fig. 21.23 Curves summarizing transport phenomena versus temperature. (a) Normalized conductivity; (b) thermopower. (From Ref. 95.)

4. AC Conductivity

In all cases, ac conductivity for hopping mechanisms follows a law dependent on frequency: $\sigma_{ac} \approx \omega^s$ (with $s \leq 1$) (Austin and Mott [79]); this law agrees also with polaronic tunneling considered sometimes (see p. 346 of Ref. 69). Moreover [26,27], as a function of temperature, this law is either thermally activated (conduction process in band tails) or linear (conduction in levels near E_F).

Total conductivity σ_T is given by $\sigma_T = \sigma_{ac} + \sigma_{dc}$. It will depend strongly on the contribution of σ_{ac} at high frequencies, especially when the contribution of σ_{dc} (independent of the frequency) is small (low dopings, for example). Epstein et al. [80] carried out a study of polyacetylene doping by iodine in the gaseous phase. The curve shapes are plotted in Fig. 21.24. We note that if the σ_{ac} value does not change upon doping, σ_{dc} increases with it. The values of ac conductivity measurements at various sample temperatures have been analyzed suc-

Ion Implantation Doping

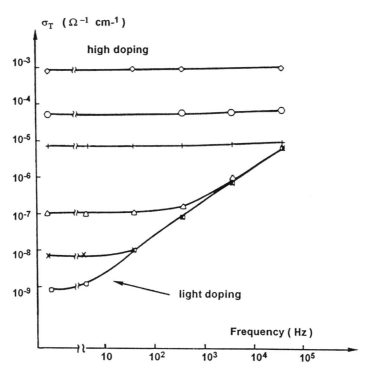

Fig. 21.24 Total conductivity versus frequency for various doping levels in cis-$(CH)_x$ doped with iodine vapor.

cessively on the basis of the Kivelson model [81] (transport by solitons) and the Emin model [31] (transport by mobile polarons).

From these results obtained by chemical doping, we will try to study the effect of ion implantation on doping compared to chemical doping.

Notes on the variation of ac conductivity at low and high frequency

At low frequencies, the carriers have plenty of time to cross the sample. If there is a conducting path (percolation path, for example) inside the sample, the conduction occurs along this path according to the law controlling transport along the path. If there is no conducting path, the only possible movement of the carriers is that of hopping from one site to another; many sites may be involved as the spacing between these sites may be large (because of the time allocated to the carrier at these low frequencies). The corresponding tunneling effect introduces the term $\exp(-2\alpha R)$, which gives a very small contribution because R is high. Moreover, the hopping will take place only with a very low frequency (excitations at low frequency), and the resulting warming of the material, and hence the material's conductivity, will be very weak.

At high frequencies, if the carriers are located on a percolation path where their mobility is relatively low, they will not have enough time to cross the sample (hence a drop in this term). On the other hand, the hopping mechanism between neighboring sites may take place at very high frequency, which is directly related to the excitation. This term, of the form $A\omega^s = \sigma$ (where $s \leq 1$ according to the Austin–Mott law), becomes dominant at high frequencies.

5. Conclusion About Transport Phenomena

Finally, in lightly doped electroactive polymers and for a sufficiently high temperature, one can consider that conduction takes place along the conducting paths by sufficiently mobile polarons to obtain a frequency-independent law (in the low frequency range) that is thermally activated:

$$\sigma_T = \sigma_0 \exp\left(-\frac{\Delta E + W_H}{kT}\right)$$

At sufficiently low temperatures, the contribution of this term is small and gives place to nonthermally activated phenomena; the variable range hopping law leads to the Austin–Mott law in the case of ac conductivity. At high frequencies, this law will generally play a major role. We generally write [78]:

Thermally activated conduction associated with mobile carriers (frequency-independent)
↓
$\sigma_{total}(\omega, T) = \sigma_{dc}(T) + \sigma_{ac}(\omega, T)$
↑
Nonthermally activated conduction in the case of hopping mechanism with tunneling effect

The curves in Figs. 21.23a and 21.23b describe σ_{dc} and S at various temperatures. Figure 21.25 illustrates the band scheme and the related theoretical laws. The frequency laws at room temperature are displayed in Fig. 21.26.

B. Effect of the Implantation Parameters on DC Conductivity and on Thermoelectric Power S

The effect of the implantation parameters on dc conductivity [82] and on thermoelectric power [83] are physical characteristics that give us both the value of conductiv-

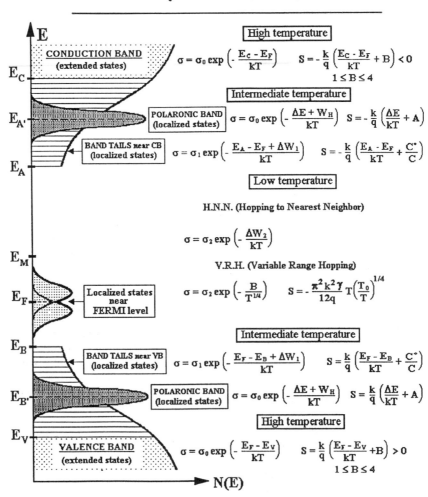

Fig. 21.25 DC conductivity σ and thermopower S main formulas for successive levels of the band diagram. (From Ref. 18.)

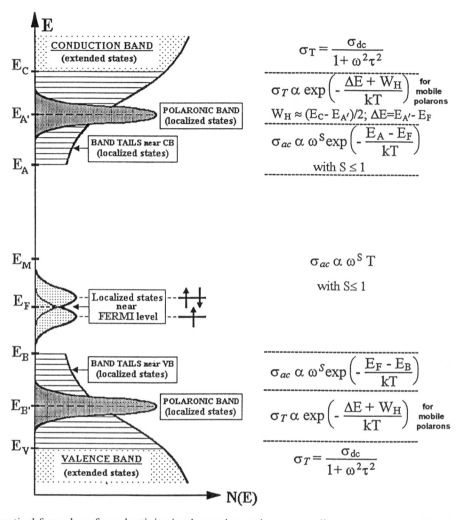

Fig. 21.26 Theoretical formulas of conductivity in alternative regimes according to temperature T and angular frequency ω for successive levels of the band diagram. (From Ref. 18.)

ity reached and the final conduction type (n or p, which give respectively $S < 0$ or $S > 0$); these results are essential to reaching conclusions about the nature of the doping. The curves in Fig. 21.27, which is a plot of the logarithm of conductivity against implantation energy are nearly linear for all chemically inert ions (noble gas ions). Only the mass seems to have an effect because the conductivity increases slightly as the ion mass rises. When the material is sufficiently conducting, it becomes possible to measure the Seebeck coefficient (implantation of Xe^+). Its sign is positive (p-type carriers), and its value is always very low: $S < 3$ $\mu V/K$ (metallic type conduction in degenerate states).

During the implantation of chemically active ions (such as alkali and halogen ions) we observe (Fig. 21.28a) for the conductivity at highest energies (100 keV $< E <$ 300 keV) the same behavior as with noble gas ions; at the same time thermoelectric power tends to remain positive with very low values (Fig. 21.28b). This means that for this energy range the chemical effect of the ion does not play a major role. In this energy range and at room temperature, one must notice that transport by variable range hopping, as introduced by Mott and Davis [26], is particularly applicable. According to the VRH conductivity law,

$$\sigma \approx \sigma_2 T^{-1/2} \exp\left[-\frac{T_0}{T}\right]^{1/4}$$

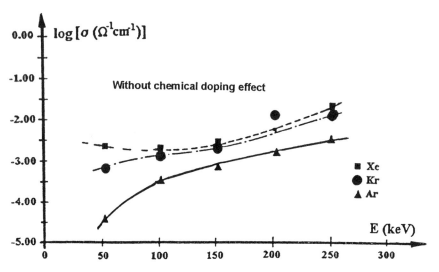

Fig. 21.27 DC conductivity variation with the ion energy of PPP samples implanted ($D = 10^{16}$ ions/cm^2) with inert gases.

T_0 decreases when σ_{dc} rises; at the same time, the absolute value of S,

$$\frac{S}{T} = -\frac{\pi^2 k^2 \gamma}{12q}\left(\frac{T_0}{T}\right)^{1/4}$$

must also decrease. We have effectively observed these behaviors in our measurements, at least up to 200 keV with alkali ions.

On the other hand, if one considers the low energy range ($E < 100$ keV), one notices with alkali and halogen ions a specific increase in conductivity that becomes higher as energy decreases. At the same time, the sign of S is positive with halogen ions (p-type doping) and negative with alkali ions (n-type doping). Doping occurs now, and we note (Fig. 21.28b) that the modulus of S is higher (semiconducting character dominates) when both the implantation energy is low and the ion mass is light. Moreover, one can note that when σ increases, the modulus of S equally rises. This indicates that the VRH mechanism is not applicable, and chemical doping starts to play a major role. This role is obvious in Figs. 21.29a and 21.29b, which present the plots of σ_{dc} and S as a function of the dose at a low energy ($E = 50$ keV) if the dose is between 10^{15} and 10^{16} ions/cm^2 ($S > 0$ for halogen ions and $S < 0$ for alkali ions; see Fig. 21.29b). For the highest doses, as we have shown by ESCA (Section IV.A), degradation occurs, producing degeneracy of the material, which presents a low absolute value of thermoelectric power, whereas the decrease in conductivity is explained by the formation of a superficial oxidized layer due to this degradation. This decrease occurs when the dose is lower and the ion is heavy.

We have studied the influence of current density j on the doping level, and we have noticed that at maximal values of j (j_{max}), degradations appear that mask the chemical effects due to active implanted ions and depend on the nature of the material. In the case of polymeric films [84], we have found that $j_{max} \leq 0.5$ μA/cm^2, whereas for compacted PPP pellets [82], $j_{max} = 1.5$ μA/cm^2.

C. Effect of the Implantation Parameters on AC Conductivity

The effect of implantation parameters on ac conductivity has been studied [85] for implanted films of PPP with chemically active ions (Cs$^+$ ions, $D = 2 \times 10^{15}$ ions/cm^2, $j = 0.2$ μA/cm^2, and $E = 250$, 100, or 30 keV) and is plotted in Fig. 21.30. The superficial conductivity σ_T ($\sigma_T = \sigma_{dc} + \sigma_{ac}$) measured between 20 Hz and 1 MHz increases when the implantation energy decreases toward the lowest energies ($E \approx 30$ keV). This behavior is in perfect harmony with results given in the previous subsection: the standard law, $\sigma_{ac} = A\omega^s$, appears at higher frequencies as the doping level, which produces a high σ_{dc}, rises (at low energy). In the case of virgin material, only a σ_{ac} component arises. According to the Emin model [31], this residual contribution is related to the dielectric lattice and produces at the lowest frequencies a low and relatively constant component [86].

Moreover, the curves in Fig. 21.30b representing the variation of log(σ) with log(frequency) for various doses

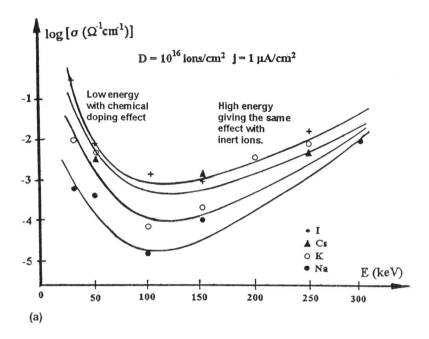

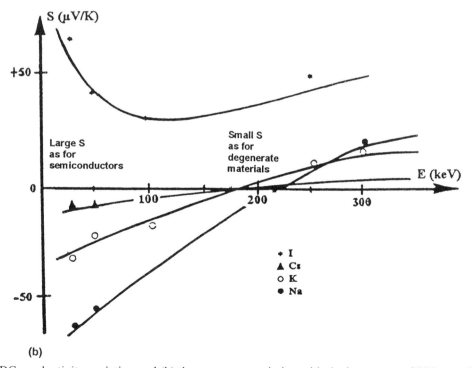

Fig. 21.28 (a) DC conductivity variation and (b) thermopower variation with the ion energy of PPP samples implanted ($D = 10^{16}$ ions/cm^2) with alkali ions and iodine.

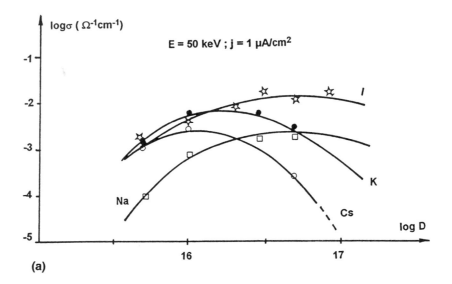

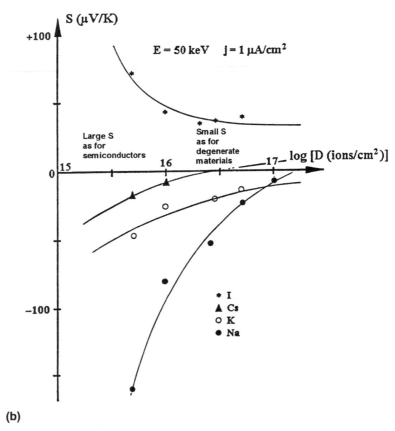

Fig. 21.29 (a) DC conductivity variation and (b) thermopower variation with the fluence of PPP samples implanted ($E = 50$ keV) with alkali ions and iodine.

Ion Implantation Doping

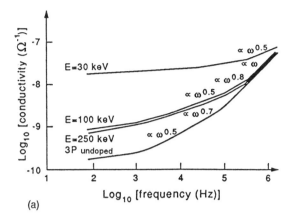

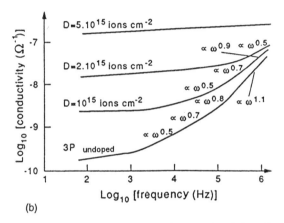

Fig. 21.30 (a) Superficial alternative conductivity of PPP films at various energies (Cs ions; $D = 2 \times 10^{15}$ ions/cm^2; $j = 0.2$ μA/cm^2). (b) Superficial alternative conductivity of PPP films at various fluences (Cs ions; $E = 30$ keV; $j = 0.2$ μA/cm^2). (From Ref. 85, with the kind permission of Elsevier Science Ltd.)

show that σ_T increases with the dose. These plots can only be fitted well by the ω^S law at lower frequencies when the doping level is low. This behavior indicates that the conducting clusters (local graphitization) cannot appear in the insulating matter [78] at this low energy ($E = 30$ keV); the doped regions are multiply interconnected and thus are not isolated (i.e., not like conducting clusters). I show in Section VI.D that the conductivity corresponding to this σ_T component is thermally activated.

D. Effect of Implantation Temperature

The effect of implantation temperature in nonelectroactive polymers has been studied by various authors [1]. Davenas [87] established that the metallic conduction in polyimide seems to be due to a double effect: the effect of degradation produced by ion implantation temperature lower than 400°C and the effect of pyrolysis obtained by thermal effects at higher temperature. The measured conductivity does not increase uniformly with implantation temperature and presents a slope at around 200°C that is probably due to local annealing of the structure. From our measurements [curves log $\sigma = f(E)$, $S = f(E)$, log $\sigma = f(D)$, and $S = f(D)$], we made a comparison [88] between implantation at low temperature and implantation at room temperature. In the case of σ (Fig. 21.31) and S plotted against energy, we did not notice any appreciable improvement of semiconducting properties in the low energy range. At low temperature implantation (LT), conductivity decreases (in relation to room temperature implantation, RT) by less than one order of magnitude (except for Na implantation, where the decrease reaches nearly two orders of magnitude) while the increase in the modulus of S is significant only

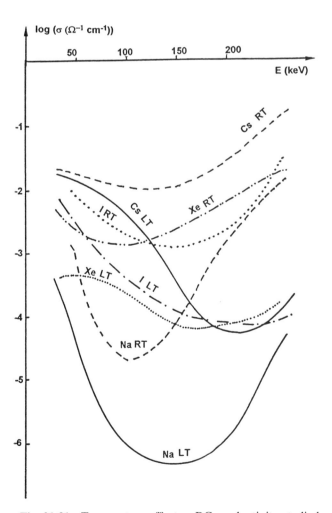

Fig. 21.31 Temperature effect on DC conductivity studied versus energy for PPP samples implanted with various ions.

in the case of sodium implantation. The conductivities produced by high energy implantations, on the other hand, are much lower (by two or three orders of magnitude) at low temperature than at room temperature; however, the modulus of S remains low. It appears, finally, that if temperature does not play an essential role (except perhaps with Na ions) on the doping level reached, it is not the same behavior in the case of conduction induced by the degradation of this material; the resulting metallic conductivity is obtained by combining the effects due to both high temperatures and high implantation energies.

Furthermore [88], during low energy implantation, we noticed that the effect of implantation temperature on the evolution of both conductivity and thermoelectric power, according to the dose, is not important with heavy ions (such as Cs and I) but is stronger for sodium ions (at low temperature, σ decreases by two orders of magnitude whereas $|S|$ increases appreciably by one order of magnitude, except at the highest doses, $D > 10^{16}$ ions/cm^2, where $|S|$ remains low and much lower than 10 μV/K).

VI. CHARACTERIZATION OF TRANSPORT PHENOMENA

We carried out specific studies to characterize the nature and the origin of the transport mechanisms discussed in the preceding section.

A. Stability of the Resulting Conductivity and the Effect of Exposing the Samples to Open Air After Ion Implantation

We know that doping by chemical or electrochemical processes is unstable and that the conductivity decreases quickly when (CH)$_x$ [59] or PPP [59,89] samples are taken to the open air. This is why we studied with PPP pellets the kinetics of conductivity decrease according to implantation parameters such as the size of implanted ions, the dose, and the current density j (Figs. 21.32a–21.32c, respectively).

In general, when the sample is kept under vacuum ($\approx 10^{-6}$ torr) for the first 20 minutes after the end of ion implantation, we observe a decrease in conductivity that can be attributed to a thermal effect. This effect is due to the fact that implantation warms the sample (generally [10,87] to more than 100°C), and the return to room temperature causes conductivity to decrease with temperature. One can note that this conductivity drop is faster when the conduction mechanism is thermally activated, which means that the material is a semiconductor and consequently was implanted using sufficiently low parameters. The importance of the conductivity drop is increased in the case of sodium implantation because the implanted layer is too well buried; the thermal exchange with the outside is limited, and the sample temperature is higher at the end of the implantation [91].

Ninety minutes after the end of this process, we open the "air admit" valve of the irradiation chamber where the samples are kept under vacuum. The conductivity falls again, and its decrease is more important when the implantation of the surface has been carried out with high parameters (ion size, j and D values). The sudden presence of air surely brings about an oxidation of the surface, which is stronger when the current density j (Fig. 21.32c) or the dose (Fig. 21.32b) are high. This is in perfect harmony with the analysis of the layer (ESCA study according to dose level) discussed in Section IV.A.

We equally notice (Fig. 21.32a) that the conductivity increases or decreases according to the nature of the ions (halogen or alkali ions, respectively). All these results were obtained at low energy ($E = 30$ keV), which produces a real doping. One observes during the entrance of air a disappearance of doping by compensation in the case of implantation with alkali ions (annihilation of n-type doping produced with alkali ions, by p-type doping created by oxygen diffusion). The more damaged by implantation the layer is (large size of the alkali atoms), the more important this phenomenon becomes (the decrease is much greater with cesium than with sodium). In the case of initial implantation with iodine ions, first one observes a slight increase in the conductivity, which can be due to an enhancement of p-type doping by oxygen diffusion; the weak decrease observed afterward can be attributed to a light exodiffusion of iodine atoms implanted into low density areas also reached by chemical doping: the instability of the chemical doping is found again. One can think that this weak phenomenon of implanted ion diffusion occurs in the case of implantation with cesium ions even if it is very limited; at the beginning [91] we did not take it into account. In the case of sodium ions, this phenomenon is almost imperceptible because these ions are implanted into a layer that is more deeply buried, located further from the surface, as previously stated. In this particular case, the role of oxygen is also minimized.

Finally, after the successive progress of the two phenomena described above (return to thermal equilibrium and return to fresh air), we noticed by our experiments on PPP samples that 2 h after taking the samples to the air the measured conductivity subsequently remained particularly stable for at least 1 year.

B. Reversibility of Doping*

We reproduced in the case of implantation doping the experiment that exhibits the doping effect in the chemi-

---

*This section is largely based on Ref. 91.

Ion Implantation Doping

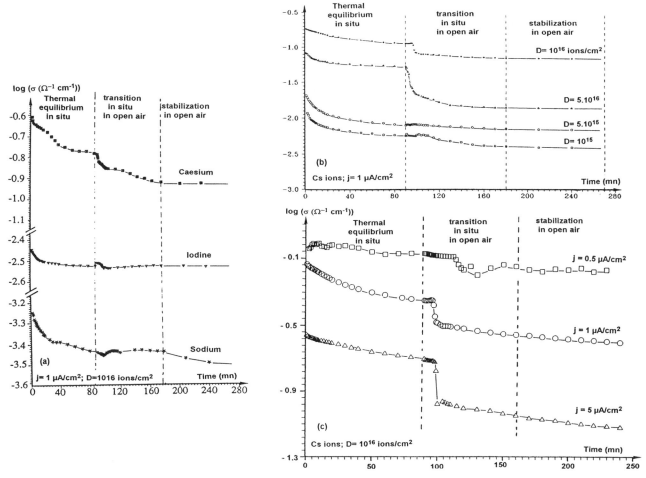

Fig. 21.32 Conductivity kinetics during return to equilibrium state after low energy ($E = 30$ keV) implantation. (a) For various ion sizes; (b) for various fluences; (c) for various current densities. (From Ref. 90.)

cal case [92]. The compensation law is found again, particularly with the thermoelectric parameter.

We studied, according to ion dose, the evolution of both the conductivity (measured inside the irradiation chamber and in open air) and the thermoelectric power (measured only outside the chamber in open air [93] for convenience) of samples previously implanted with halogen and alkali ions, respectively.

In the case of initial implantation with iodine ions giving a p-type doping, one can observe (curves 2 and 3 in Fig. 21.33a; curve 2 in Fig. 21.33b) that compensation phenomena due to cesium ions quickly appear (decrease of σ on curve 3, Fig. 21.33a) and the thermoelectric power changes sign for a cesium ion dose ($D \leq 10^{15}$ ions/cm^2) lower than the value generated initially by implantation of iodine ($D = 5 \times 10^{15}$ ions/cm^2). It appears that n = type doping due to alkali ions is much more efficient than p-type doping made by implanting halogen ions (yet this doping is favored by the contribution due to the presence of oxygen).

This result is confirmed by the study of curves 1 and 4 in Fig. 21.33a and curve 1 in Fig. 21.33b. These curves describe the evolution of σ and S under the effect of iodine ions, after an initial implantation with cesium ions (n-type doping). The slow decrease of conductivity is followed by a sign change of the thermoelectric power only at the highest doses ($D > 10^{16}$ I$^+$ ions/cm^2, much higher than the initial dose, which was 5×10^{15} Cs$^+$ ions/cm^2). The absolute value of S that is finally reached is particularly small and is a characteristic of metallic conductivity due to the considerable damage produced by these high doses.

Such behaviors show that the only way to change the conduction type is to start with a p-type doping (I$^+$ ions) and subsequently introduce Cs$^+$ ions leading to n-type doping. The two penetration volumes are superimposed

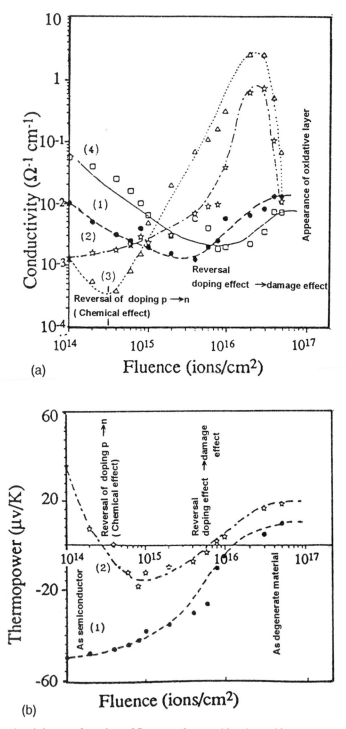

Fig. 21.33 (a) Variation in conductivity as a function of fluence of second implanted ion measured after an initial implantation ($E = 30$ keV, $D = 5 \times 10^{15}$ ions/cm^2, $j = 0.5$ μA/cm^2). Curve 1: Measured in open air, initial implantation Cs$^+$ ($\sigma = 1.6 \times 10^{-2}$ Ω^{-1} cm^{-1}), second implantation I$^+$. Curve 2: Measured in open air, initial implantation I$^+$ ($\sigma = 1.3 \times 10^{-3}$ Ω^{-1} cm^{-1}), second implantation Cs$^+$. Curve 3: Measured in situ, initial implantation I$^+$ ($\sigma = 1.35 \times 10^{-3}$ Ω^{-1} cm^{-1}), second implantation Cs$^+$. Curve 4: Measured in situ, initial implantation Cs$^+$ ($\sigma = 5.7 \times 10^{-2}$ Ω^{-1} cm^{-1}), second implantation I$^+$. (b) Variation in thermopower as a function of fluence of second implanted ion measured after an initial implantation ($E = 30$ keV, $D = 5 \times 10^{15}$ ions/cm^2, $j = 0.5$ μA/cm^2); Curves 1 and 2: Same experimental conditions as in (a).

under the experimental conditions used, and this phenomenon may not be explained in terms of differential diffusion of the two ions as argued by Lin et al. [55] with polyacetylene. It seems that the only explanation for doping reversibility ($p \to n$) and doping irreversibility ($n \to p$) is a higher "doping efficiency" of Cs^+ ions under our experimental conditions, since a D_{Cs} dose (with $D_{Cs} = 3 \times 10^{14}$ Cs^+/cm^2) is able to change the sign of the thermoelectric power, which is initially positive with a D_I dose ($D_I = 5 \times 10^{15}$ I^+/cm^2). To explain this difference, we propose the following idea based on charge transfer between implanted atoms and the polymer chain:

$Cs \to Cs^+$ (counterion) + 1 e^- (negative electron transferred locally to the polymer chain in polaronic form)

$I_2 + I \to I_3^-$ (counterion) + 1 h^+ (positive hole transferred locally to the polymer chain in polaronic form, as established during chemical doping [94])

This might account for the higher "doping efficiency" in the case of cesium because only one atom is needed whereas with iodine three atoms should meet together to give I_3^-, the probability of this being far below 100% as the atoms are located at random. $D_I/D_{Cs} \approx 17$ is much higher than 3 and would indicate that one hole is created for 17 iodine atoms.

C. Study of the Variations of DC Conductivity and Thermoelectric Power with the Temperature*

A study of the variations in dc conductivity and thermoelectric power with temperature should specify the conduction mechanisms produced by ion implantation. It is sometimes rather difficult for us to make some electrostatic measurements because of the very small conductivities reached at the lowest temperatures ($\sigma \approx 10^{-5}$ S/cm). Consequently the measure of thermoelectric power is then disturbed by electrostatic influences that cannot be easily negated, as it is almost impossible to then produce an equipotential between the two probes [93] when they are at the same temperature ($\Delta T = 0$).

First, and with PPP samples, we studied the shape of $\log \sigma = f(T)$ curves as a function of various sodium implantation parameters:

(a) Low energy range ($E \approx 30$ keV) and dose equal to 5×10^{15} ions/cm^2
(b) Low energy range ($E \approx 30$ keV) and higher dose, 10^{16} ions/cm^2
(c) High energy ($E \approx 250$ keV) and dose level of 10^{16} ions/cm^2

In case (a), the shape of the curve $\log \sigma = f(T)$ (Fig. 21.34a) is similar to that of the one obtained after light

---

*This section follows from Refs. 95 and 96.

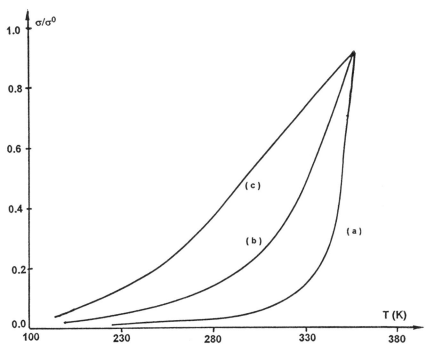

Fig. 21.34 Temperature dependence of the normalized conductivity for Na$^+$-implanted PPP samples with various parameters. (a) $E = 30$ keV, $D = 5 \times 10^{15}$ ions/cm; (b) $E = 30$ keV, $D = 10^{16}$ ions/cm^2; (c) $E = 250$ keV, $D = 10^{16}$ ions/cm^2. (From Ref. 95.)

chemical doping (Fig. 21.23a). With a dose equal to 10^{16} ions/cm^2 (Fig. 21.34b) but still at low energy ($E = 30$ keV), the doping level increases slightly while approaching near to the moderate chemical doping values. In case (c) with the energy equal to 250 keV, the curve concavity changes (Fig. 21.34c); this variation corresponds to that proposed by Kaiser (Fig. 21.23a) for heavy dopings, for which it has been supposed that there would be some conducting "bridges" between the fibrils. In our case, we think that the production of conducting grains (very carbonaceous) is due to bombardment by highly energetic ions. The conductivity variation proposed (proportional to $T^{1/2}$ for amorphous metal) masks the one related to doping (its contribution becoming negligible).

To study the transport mechanisms more carefully for cases (b) and (c), the curves are graphed for $\log \sigma = f(1/T)$ and $\log \sigma \sqrt{T} = f(T^{-1/4})$ in Figs. 21.35 and 21.36, respectively, and for $S = f(T)$ and $S/T = f(T^{-1/4})$ in Figs. 21.37 and 21.38, respectively.

First we note that after sodium implantation at low energy, the sign of S is negative (being characteristic of n-type doping by alkali ions, Fig. 21.37a) but becomes

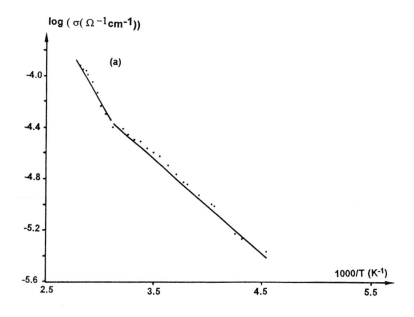

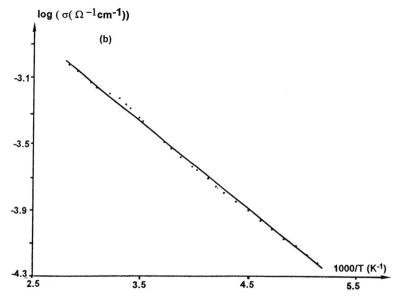

Fig. 21.35 $\log \sigma = f(1/T)$ for Na$^+$-implanted PPP sample. (a) $E = 30$ keV; (b) $E = 250$ keV. $j = 1$ μA/cm^2; $D = 10^{16}$ ions/cm^2. (From Ref. 93.)

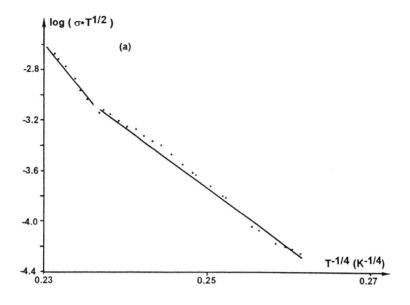

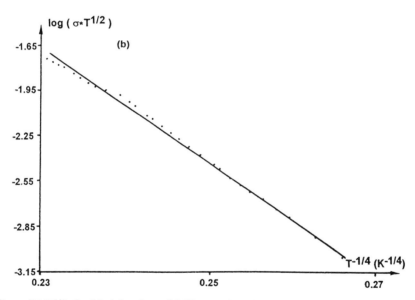

Fig. 21.36 $\log(\sigma T^{1/2}) = f(1/T^{1/4})$ for Na$^+$-implanted PPP sample. (a) $E = 30$ keV; (b) $E = 250$ keV. $j = 1$ μA/cm^2; $D = 10^{16}$ ions/cm^2. (From Ref. 93.)

positive with a much lower modulus (Fig. 21.37b) after high energy implantation (250 keV). In the latter case, especially at the lowest temperatures, the conductivity is quite well described by the variable range hopping (VRH) mechanism (Fig. 21.36b), which is compatible with the evolution of the Seebeck coefficient (Fig. 21.38b). In Fig. 21.36b, the gap between the curve and the linear law (straight line) at higher temperatures is explained by the occurrence of the conduction process in metallic amorphous interdomains (described by the $T^{1/2}$ law).

The transition of S observed at $T \approx 312$ K for a low energy implantation (Fig. 21.37a) is accompanied by a change in the conduction mechanism at low temperature (which is probably the VRH law, Fig. 21.36a) to a thermally activated conduction mechanism in localized states related to polaronic states (or band tails), the activation energy being higher for this mechanism (Fig. 21.35a). Koshida and Suzuki [97] showed that for (CH)$_x$ this energy decreases when the sodium dose increases, as observed during chemical doping. We have noticed the same behavior for PPP [98] and PPV [99] polymers.

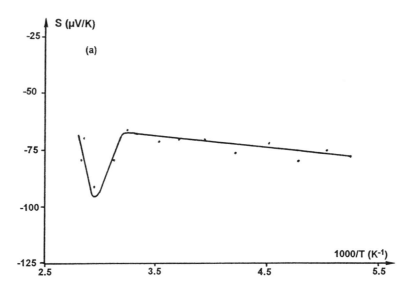

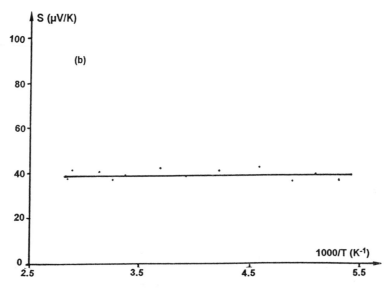

Fig. 21.37 $S = f(1/T)$ for Na$^+$-implanted PPP sample. (a) $E = 30$ keV; (b) $E = 250$ keV. $j = 1$ μA/cm^2; $D = 10^{16}$ ions/cm^2. (From Ref. 93.)

It must also be noted [96] that a negative component of the thermoelectric power appears around 400 K whose shape depends on the dominance of all transport phenomena. This component can be assigned to a migration of negative oxygen ions. Indeed, on the one hand we have exhibited an oxygen layer in the implanted zone, and on the other hand this component cannot be attributed to a chemical doping effect because it also appears in nonelectroactive polymers (polyimide).

D. Evolution of AC Conductivity with Different Temperatures

The curves $\log \sigma = f(\omega)$ for various temperatures are plotted for the virgin material in Fig. 21.39a and for the sample implanted at high energy in Fig. 21.40a and at low energy in Figs. 21.41a and 21.42a. We observed from these curves that the σ_{dc} component, which is independent of the frequency and related to the doping level, is

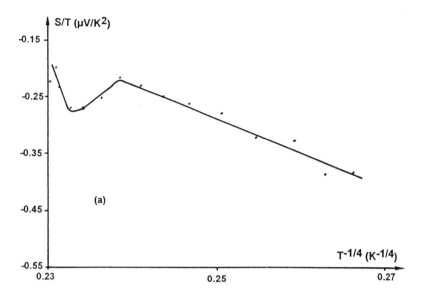

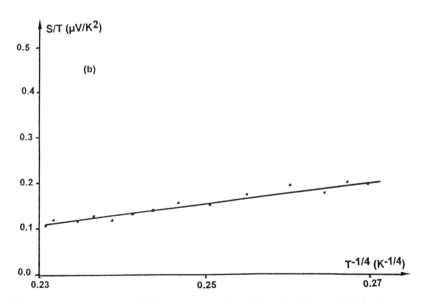

Fig. 21.38 $S/T = f(1/T^{1/4})$ for Na$^+$-implanted PPP sample. (a) $E = 30$ keV; (b) $E = 250$ keV. $j = 1$ μA/cm^2; $D = 10^{16}$ ions/cm^2. (From Ref. 93.)

higher when the implantation is carried out at both low energy and high dose and when the sample temperature is high as measurements are made.

The transport phenomenon in the virgin material is nonthermally assisted (Fig. 21.39b) for all the frequencies. The conduction may be linked only to the dielectric lattice, as established by Emin and Ngai [31]. In the case of the high energy implantation, $\sigma(\omega)$ studied at low temperature presents a similar component with the same dependence on frequency of the form $\sigma(\omega) \approx \omega^S$. The contribution due to doping is extremely small at low temperature but becomes higher at higher temperatures.

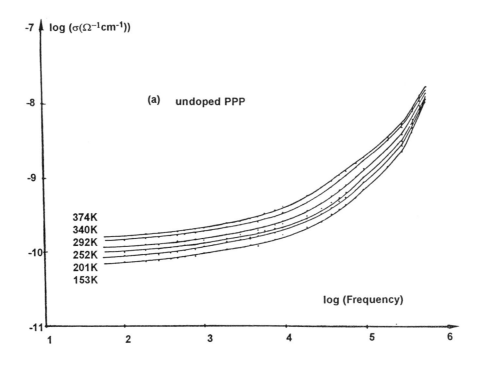

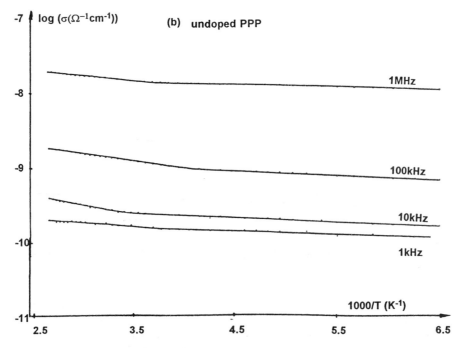

Fig. 21.39 (a) Curves of log $\sigma_T = f(\log \nu)$ for undoped PPP at various temperatures. (b) Curves of log $\sigma_T = f(1/T)$ for undoped PPP at various frequencies.

Ion Implantation Doping

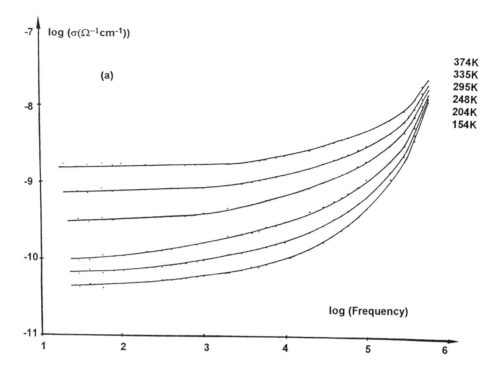

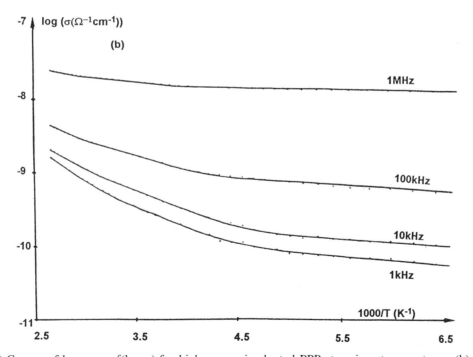

Fig. 21.40 (a) Curves of log $\sigma_T = f(\log \nu)$ for high energy implanted PPP at various temperatures. (b) Curves of log $\sigma_T = f(1/T)$ for high energy implanted PPP at various frequencies.

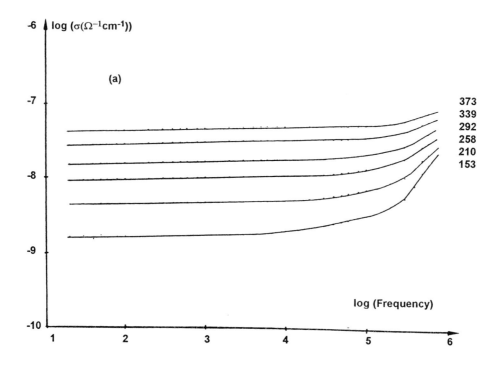

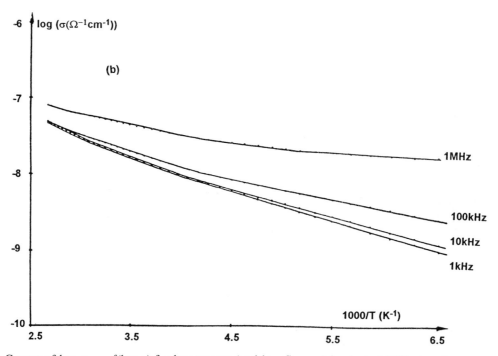

Fig. 21.41 (a) Curves of $\log \sigma_T = f(\log \nu)$ for low energy (and low fluence) implanted PPP at various temperatures. (b) Curves of $\log \sigma_T = f(1/T)$ for low energy (and low fluence) implanted PPP at various frequencies.

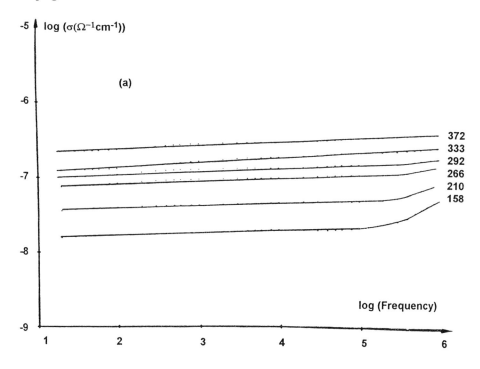

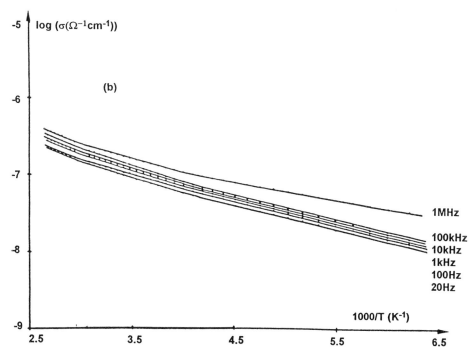

Fig. 21.42 (a) Curves of log $\sigma_T = f(\log \nu)$ for low energy (and moderate fluence) implanted PPP at various temperatures. (b) Curves of log $\sigma_T = f(1/T)$ for low energy (and moderate fluence) implanted PPP at various frequencies.

However, this effect is quickly masked by the ω^S law, which becomes dominant at a frequency that falls with temperature. In Fig. 21.40b, we notice that at $\nu = 1$ MHz only the nonthermally activated hopping mechanism appears, whereas the thermally activated process occurs at both the highest temperatures and lowest frequencies ($\nu \leq 10$ kHz).

As previously mentioned, we have also noticed that after low energy implantation (Figs. 21.41a and 21.42a), the σ_{dc} component appears in a frequency range that is wider when the doping is heavy. The conduction is thermally activated at almost all the studied frequencies ($\nu \leq 1$ MHz) for the most heavily doped sample (Fig. 21.42b). For the most lightly doped sample (Fig. 21.41b), the conductivity is no longer thermally activated at the highest frequencies ($\nu \geq 10^5$ Hz), and now the hopping mechanism [$\sigma(\omega) \approx \omega^S$] becomes preponderant.

Finally, the similarities between the ac conductive behavior of the chemically doped materials [31,78] and that of the materials doped by using the ion implantation process with a low energy lead us to think that conduction mechanisms are the same in both cases. At sufficiently high temperature, the conduction is thermally activated in localized states linked to polaronic states (or band tails). This component increases with doping and becomes independent of the frequency in the low frequency range ($\nu \leq 10^5$ Hz) because the associated mobility is sufficient. As pointed out by Epstein et al. [80], the doped regions (even with light doping) are multiply interconnected and do not form conducting clusters. The nonthermally activated hopping mechanism is independent of the doping process and appears at the highest frequencies (where the two activated and nonactivated conductivity components tend to be superimposed); for implantation at high energy, this mechanism plays a major role and must be linked to defect states.

E. Discussion: Shape of State Density Versus Implantation Parameters and Origin of Transport Phenomena

According to the general Kubo–Greenwood formula [26], compensation effects between various transport mechanisms appear in the thermoelectric power expression [18]

$$S = \frac{\sigma_{nl}S_{nl} + \sigma_{pl}S_{pl} + \sigma_H S_H}{\sigma_{nl} + \sigma_{pl} + \sigma_H}$$

The indices nl and pl are respectively related to n and p polaronic dopings, whereas the index H describes the VRH process between states induced by the defects and generally located near E_F. Indeed, these states involve both localized states induced by dangling bonds and polaronic states (with a low mobility) produced by charge transfer between the π band and E_{sp^2} localized states (Fig. 21.4).

The previously discussed density of states linked with damage is shown in Fig. 21.2, whereas the amplitude (described in Figs. 21.43a, d, and g) varies with the implantation parameter intensity. The energy levels are filled up, in agreement with the captions of Figs. 21.43b, c, e, f, h, and i. Generally, σ_H is smaller when implantation parameters are weaker as described in Fig. 21.43, where we see, from top to bottom, that S tends toward S_{pl} (halogen implantation) or S_{nl} (alkali implantation) and in this way becomes more and more characteristic of the doping effect alone. This limit is most closely approached during low energy implantation with small size ions (sodium ions) where S is large enough at a sufficiently high temperature that S is thermally activated:

$$S = -\frac{k}{q}\left(\frac{\Delta E}{kT} + A\right)$$

With large ions (e.g., cesium), a nonnegligible contribution of the defect band can appear. This contribution becomes dominant with high implantation parameters and then S tends toward

$$S_H = -\gamma \frac{\pi^2 k^2}{12q} T \left(\frac{T_0}{T}\right)^{1/4}$$

As

$$\gamma = \frac{1}{N(E_F)} \frac{dN(E)}{dE}$$

S becomes smaller in modulus when $N(E_F)$ is large, that is to say when the defect density located near E_F is large (metallic behavior with S very small, approximately a few microvolts per kelvin). Moreover, the sign of S varies as $dN(E)/dE$; we emphasize that quantitative theoretical study of S is difficult because the uncertainty on $dN(E)/dE$ whose only sign can be estimated ($S > 0$ for p doping, $S < 0$ for n doping).

Furthermore [36], it seems satisfactory to consider that the superposition of states density resulting from both disorder (induced notably by defects) and electron–lattice interactions (polaronic states) leads to the scheme in Fig. 21.44 obtained with a light doping (giving slight peaks at polaronic bands). The π or π^* band tails generally enclose the Fermi level located either in the degenerate band linked with the defects or close to the band edges as expected with the polaronic model in lightly doped polymers [100]. An evaluation of the density of state with the help of the VRH approximation in a moderate temperature range leads to values on the order of 2×10^{19}–3×10^{20} states/(eV·cm^3). We can consider these as physically reasonable values to be expected in the tails of the π or π^* band; we can then consider that for a moderate temperature range (where polarons are not thermally activated; see Section V.A

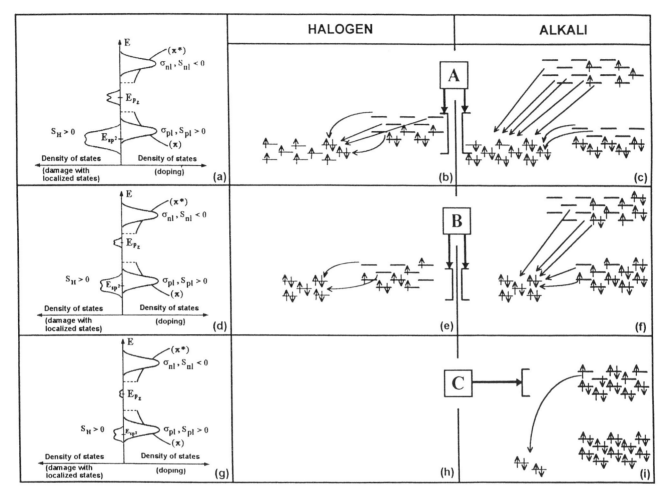

Fig. 21.43 (a) States density diagram of electroactive polymers after ion implantation with high parameters (heavy dose, $D > 2 \times 10^{16}$ ions/cm^2, and high energy, $E > 100$ keV). (b) Filling of energy levels for halogen ion implantation with high implantation parameters. A represents the zone of major energy levels where $dN/dE < 0$ and E_F must be located inside. (c) Filling of energy levels for alkali ion implantation with high implantation parameters. (d) States density diagram of electroactive polymers after ion implantation of big ions (iodine and cesium) with small parameters. (e) Filling of energy levels for iodine ion implantation with small implantation parameters. B represents a significant zone of level density. (f) Filling of energy levels for cesium ion implantation with small implantation parameters. B represents a significant zone of level density. (g) States density diagram of electroactive polymers after ion implantation with small parameters of small ions. (h) No obvious effect with small halogen ions (F and Cl). (i) Filling of energy levels for sodium ion implantation with small implantation parameters. C represents a significant zone of levels density where E_F is located. (From Ref. 18.)

and p. 339 of Ref. 69), on the whole, S verifies the VRH process.

In conclusion, the high temperature range and a sufficiently high doping give rise to thermally activated processes linked with the usual polaronic states (S_{nl} or S_{pl} components). For the other cases, when the temperature range is sufficiently low that polarons are not thermally activated and with light doping, S is consistent with the VRH law; the S modulus becomes small and its sign is negative ($dN(E)/dE > 0$) during n doping; its sign is positive ($dN(E)/dE < 0$) during p doping and in the case of slow polarons arising from the charge transfer between the top of the π band and the E_{sp^2} localized levels (defect levels). In fact, this p-type conduction must be considered as general in electroactive polymers and appears in all cases where defects are present (dangling bonds linked with the breaking of C—H bonds), without relation with their origin. In unimplanted materials, this

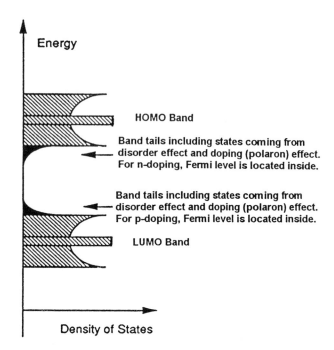

Fig. 21.44 General scheme for the density of states of ion-implanted electroactive polymers, with band edges (near LUMO and HOMO bands) resulting from both disorder (and defects) and doping (polaronic levels).

mechanism can arise from thermal treatment inducing these types of defects.

VII. APPLICATION EXAMPLE: FABRICATION OF POLYMER DEVICES

A. Introduction

From the beginning of the development of conducting polymers, it has been planned to use them in electronics. Among all the new devices, we can mention

The organic–inorganic junction (PTCDA–SiP) [101]
The heterojunction [polyacetylene/poly(N-methylpyrrole)] [102]
The Schottky junctions (polypyrrole–indium) [103], (polyparaphenylene/indium) [104], [poly(3-alkylthiophene/indium] [105], (polyacetylene/In, Al, Pb, Sb) [106]
Junctions using the phthalocyanines and their derivatives [107]
The field effect transistor using a polymer film (polythiophene [108] film, for example) or some oligomers [109]

It is important to note that these devices include only one semiconduction type related to intrinsic p-type doping (generally due to catalytic residues of synthesis). The applications are inevitably limited, and the fabrication of pn junctions is quite difficult because of the instability of n-type doping due to following the chemical route (cf. pp. 599–600 of Ref. 106). We are going to show that the use of ion implantation at low energy, even if it does not solve all the problems, allows us to make all-polymer devices of various types (Schottky junctions, pn junctions, field effect transistors).

However, we must point out that the doping effect produced by ion implantation was initially questioned [1] after only one attempt to make pn junctions by sodium ion implantation into a p-type polyacetylene substrate with a high dose (3×10^{16}–3×10^{17} ions/cm^2). In fact, we have already observed (Fig. 21.29a) that doses leading to doping can be higher when the ion size is small (which is precisely the case with sodium). For cesium, such doses are effectively forbidden.

We believe that the arguments put forth in the preceding sections must remove this doubt. While knowing that new improved devices have been made, we have carried out research work more on the electrical properties of implanted electroactive polymers than on the development of these devices (their performances are equally related to the improvement of these polymeric materials and to their processibility).

B. Devices (PN Junctions) Made by Ion Implantation

As mentioned above, the first pn junctions were made by Koshida and Wachi [110] and Wada et al. [111] after implanting sodium ions at large fluence and low energy into (CH)$_x$ substrates that were p-type intrinsically or lightly chemically doped with iodine.

Presented in Fig. 21.45 are the characteristics obtained by Koshida and Wachi after implantation of Na$^+$ ions (with $E = 12$ keV and $D = 3 \times 10^{16}$ ions/cm^2). The rectification ratio is around 100 at 5 V, but the junction deteriorates quite quickly (this ratio is equal to 15 after 37 days); moreover, this coefficient seems to be higher when the implantation energy is low (at 50 keV, it is only around 10). The junctions made by Wada et al. at higher energy ($E \geq 50$ keV) and with a higher dose ($D \approx 10^{17}$ ions/cm^2) are stable for only 1 week, with a rectification ratio of 12.

With polythiophene electrogenerated and reduced electrochemically in a solution of tetrabutylammonium perchlorate in acetonitrile, we confirmed [112] that the initial p-type doping of the film ($\sigma \approx 10^{-6}$ S/cm and $S \approx 600$ μV/K) could be compensated by implantation of alkali ions (with Cs$^+$ ions, $E = 30$ keV, and $D = 10^{16}$ ions/cm^2 we get $\sigma \approx 10^{-2}$ S/cm and $S \approx -5$ μV/K). The junctions made finally by Cs$^+$ implantation through a mask with a diameter of 0.5 mm produce a pn junction type of characteristic, with a perfect blocking current

Ion Implantation Doping

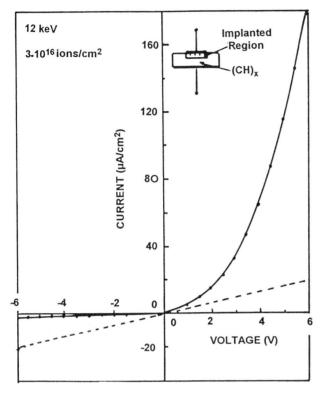

Fig. 21.45 Characteristic of a *pn* junction produced by one implantation (Na giving *n* doping) in an intrinsically *p*-type substrate, (CH)$_x$. The dashed line is the ohmic characteristic obtained on the unimplanted zone. (From Ref. 110.)

between 0 and -5 V (Fig. 21.46). Like Koshida and Suzuki [97], we checked the linearity of the curve $1/C^2 = f(V)$ characteristic of a capacity (C) evolution of the depletion layer according to the bias voltage; it is further proof of the doping effect, which leads to the realization of a *pn* junction effect (Fig. 21.47).

The evolution of such structures (heterojunctions made by implantation into intrinsically insulating substrate) over time was studied by Koshida and coworkers [52,97]. They consider that two successive effects take place inside the (CH)$_x$ sample. The first effect is a self-annealing of defects produced by implantation. First of all, the shape of the *I–V* characteristics is improved (during the first 15 days after sodium ion implantation) [52]. The second effect is due to a migration process of implanted ions and leads to the deterioration of the junction. The latter effect is particularly perceptible during implantation of lithium (Li$^+$) ions, which produces extremely unstable junctions. In the case of (CH)$_x$ implanted with bigger ions such as potassium ions (because of their size, they ought to diffuse less and produce a more efficient doping in agreement with their own electronegativity), Koshida et al. [52] notice a particularly small direct current and a very fast deterioration of the junction. One can attribute this behavior to an important production of defects due to the large size of the implanted ions. Since (CH)$_x$ is very sensitive to oxygen, we think that the *p*-type doping produced by this atom compensates quickly the initial *n*-type doping made by implantation of alkali ions. Lin et al. [55] also produced

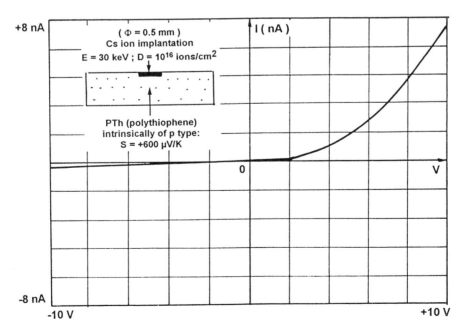

Fig. 21.46 Characteristic of a *pn* junction produced by one implantation (Cs giving *n* doping) in an intrinsically *p*-type substrate, polythiophene.

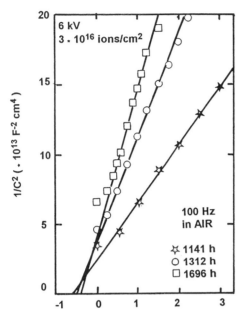

Fig. 21.47 Curves for $1/C^2 = f(V)$ characteristic of the capacity evolution of the depletion layer according to the bias voltage. (The different curves are obtained at various times after ion implantation.) (From Ref. 97.)

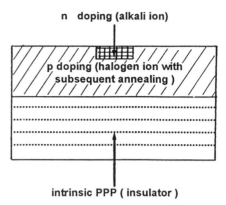

Fig. 21.48 Schematic diagram of a substrate cross section after implantation with halogen ion and subsequent annealing (p-doping) followed by alkali ion implantation through a mask (n doping). (From Ref. 113.)

a junction effect in $(CH)_x$ by using implantation of potassium ions at 27 keV with $D = 10^{17}$ ions/cm^2.

As for us, we have been able to verify that n-type doping made by implantation of the most electronegative alkali ions into materials less sensitive to oxygen (such as PPP or polythiophene) can produce a pn junction effect, the implantation having been carried out with much lower doses ($D \approx 10^{16}$ ions/cm^2) than that used for sodium ions ($D \approx 10^{17}$ ions/cm^2). Moreover, by using the results of the study of doping reversibility with implantation into PPP, we have made heterojunctions by successive implantations of halogen ions (iodine) and then alkali ions (cesium) into the intrinsically insulating substrate. In accordance with results presented in Section VI.B, we were able to verify, in particular, that the junction effect is achieved only when the implantation order is strictly obeyed (alkali ions after halogen ions). The reverse order leads to $I-V$ characteristics that are symmetrical and linear.

This shows that the junction effect is not produced by a contact phenomenon (between probes and material). Furthermore, this checking is done systematically by measuring $I-V$ characteristics on a p-type substrate, this substrate being intrinsically p type [like polythiophene and $(CH)_x$; see the dotted curve in Fig. 21.45], or after preliminary iodine ion implantation in the case of double implantation into PPP.

According to the configuration of the heterojunction (Fig. 21.48) made by double implantation into compacted PPP pellets, we present in Figs. 21.49a–21.49c the characteristics obtained [113] finally with the parameters of Table 21.1. Only the I–Cs combination leads to a satisfactory result, although the currents under forward bias are relatively small. The values of these currents were improved [114] (around 1000-fold) by using PPP films [38] while decreasing the doses used (Fig. 21.50). The development of polymers that can be now more ordered [115] must lead to an improvement of these devices.

C. Field Effect Transistor Obtained by Ion Implantation

Whereas metal–insulator–semiconductor field effect transistors (MISFETs) of conventional type with an inorganic semiconductor layer can work in various manners with a depletion layer or with an accumulation layer, in the case of an organic semiconductor layer only the last structure (accumulation) is satisfactory. The channel becomes conducting with an appropriate source-gate bias voltage (negative with a p-type semiconductor) to accumulate majority carriers [116]. On this principle, various MISFETs have been built with polythiophene [108], polyacetylene [117], polyparaphenylenevinylene [118], poly(3-octylthiophene) [119], and various oligomers [120]. Generally, carriers are p-type and often come from catalytic residues (cf. Section 4.2 of Ref. 117). In material with very low residual dopant concentration and low conductivity ($\sigma < 10^{-10}$ S/cm), the field effect is absent in MISFET devices: sexiphenyl [120], pristine PPV [116] with a polycrystalline, highly dense and compact morphology. It is with this last material that we were able to exhibit a field effect after ion implantation. The MIS structure presented in Fig. 21.51 is realized with a gate in degenerate silicon and thermally

Ion Implantation Doping

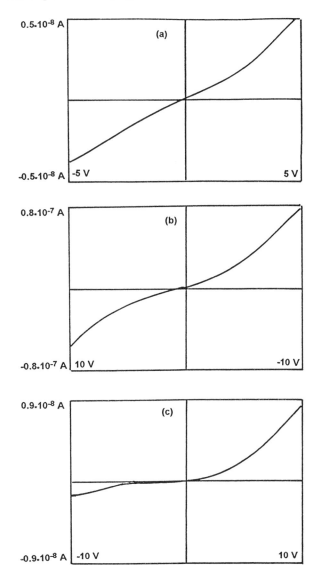

Fig. 21.49 $I = f(V)$ characteristics of halogen–alkali metal ion implanted PPP pellet. (a) Br–Li combination; (b) Br–K combination; (c) I–Cs combination. (From Ref. 113.)

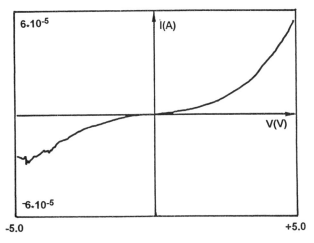

Fig. 21.50 $I = f(V)$ characteristic of halogen–alkali metal ion implanted PPP film. I–Cs combination. Iodine implantation parameters: $E = 40$ keV; $D = 2 \times 10^{15}$ ions/cm^2; $j = 0.1$ μA/cm^2. Cesium implantation parameters: $E = 30$ keV; $D = 6 \times 10^{15}$ ions/cm^2; $j = 0.2$ μA/cm^2.

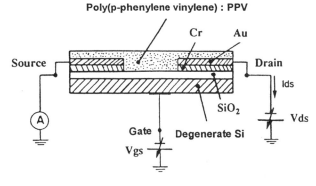

Fig. 21.51 Cross section of a MISFET structure with an implanted electroactive polymer as semiconductor.

Table 21.1 Ion Implantation Parameters

Alkali				Halogen			
Ion	E (keV)	D (ions/cm^2)	j (μA/cm^2)	Ion	E (keV)	D (ions/cm^2)	j (μA/cm^2)
Li	50	10^{16}	1	Br	50	5×10^{15}	0.25
K	50	2×10^{16}	0.5	Br	50	5×10^{15}	0.25
Cs	50	3×10^{16}	0.5	I	50	2×10^{15}	0.25

grown SiO_2 that acts as an insulator (thickness 150 nm). To obtain source and drain contacts, an interdigitated electrode in Au is deposited on Ni-Cr, which acts as an adhesion layer; the channel length is 5 µm, and the channel width is 90 cm. The PPV film was spun onto the previous substrate and then the polymer was implanted with 30 keV iodine ions; the projected range (R_p) is about 31 nm, and we obtain an ohmic contact between Au interdigitated electrodes and p-type doped polymer. With an initial dose of 4×10^{10} ions/cm^2 (doping ion concentration on the order of 8×10^{15} ions/cm^3) the source/drain current versus source/drain voltage for various gate voltages is shown in Fig. 21.52. The measurements were taken in the dark and inside the irradiation chamber to prevent photoconductivity and oxygen contamination. A very similar response is obtained with a dose increased to 10^{12} ions/cm^2. In comparison, the same MIS structure with the PPV layer doped chemically gives nearly the same behavior; in both cases, the field effect mobility is on the order of 10^{-7} cm^2/(V·s) in saturation regimes. Furthermore, it is clear [116] that the device obtained by ion implantation exhibits a more characteristic MISFET behavior with a linear regime at low source/drain voltages. This result is encouraging for further studies based on improved organic semiconductors doped by ion implantation.

VII. CONCLUSION

We have shown that, contrary to early assessments, doping effects can be obtained in conjugated polymers by the implantation technique. We have attempted to take into account all the effects and to demonstrate that both doping effects and damage appear with various implantation parameters. With high implantation parameters, damage plays a major role whereas at low energy and sufficiently low fluence, organic semiconductors can be achieved with all the advantages of ion implantation, the major one being control of the in-depth doping. Further developments of competitive devices based on implanted polymers will require higher mobilities and also improved polymers (oriented polymers, oligomers [121]). For optical applications [42] (LEDs), n doping locally realized by ion implantation will probably be able to reinforce the electron injection rate. Moreover, doping techniques must be improved by means of

1. Sweat implantation with the aim of limiting defect generation. This can be obtained with a simultaneous [122] low energy beam (for the deposition of chemically active ions) and vapor deposition (with oligomers). With respect to the field effect conditions, the ion concentration must still remain low, about 10^{16} ions/cm^3.
2. Posttreatment to saturate the dangling bonds induced by the ion implantation process with hydrogen. This technique, already developed with silicon samples [123] has been used with some significant results in the sexiphenyl case [124].

As expressed, extensive investigations are proposed to give the implantation technique the same place in organic materials (conjugated polymers) as it usually has in inorganic materials.

ACKNOWLEDGMENTS

The work reviewed in this chapter was largely carried out in Limoges (Lepofi Laboratory) with J. L. Duroux,

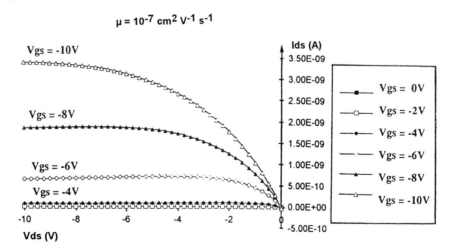

Fig. 21.52 MISFET characteristic (source/drain current vs. source/drain voltage for various gate voltages) of the device obtained after PPV implantation with iodine ions ($E = 30$ keV and $D = 4 \times 10^{10}$ ions/cm^2).

B. Ratier, B. Lucas, and C. Moreau. I thank particularly M. Moreau for his help during the writing of this chapter. We have also worked closely with R. H. Friend and his group at Cambridge, and also with G. Froyer (Nantes University) and B. Francois (Strasbourg University). This work was partly supported by the European Commission (project Napoleo under the Brite-Euram program).

REFERENCES

1. T. Venkatesan, L. Calcagno, B. S. Elman, and G. Foti, Ion beam effects in organic molecular solids and polymers, in *Ion Beam Modification of Insulators* (P. Mazzoldi and G. W. Arnold, eds.), Elsevier, Amsterdam, 1987, Chap. 8.
2. J. Davenas and G. Boiteux, Ion beam modified polyimide, *Adv. Mat.* 2(11):521 (1990).
3. L. Merhari, C. Belorgeot, and J. P. Moliton, Ion irradiation induced effects in polyamidoimide, *J. Vac. Sci. Technol.* B9(5):2511 (1990).
4. H. Kiess, *Conjugated Conducting Polymers*, Springer Ser. Solid State Sci. Vol. 102, Springer Verlag, Berlin, 1992.
5. P. N. Favennec, *L'implantation Ionique*, Masson, Paris, 1993.
6. M. S. Dresselhaus, Ion implantation in polymers and semimetals, Proceedings of the 5th Philip Morris Science Symposium, 1985.
7. H. Mazurek, D. R. Day, E. W. Maby, J. S. Abel, S. D. Senturia, M. S. Dresselhaus, and G. Dresselhaus, Electrical properties of ion implanted poly(*p*-paraphenylene sulfide), *J. Polym. Sci.* 21:537 (1983).
8. M. S. Dresselhaus, B. Wasserman, and G. Wuck, Ion implantation of polymers, *Mater. Res. Soc. Symp. Proc.* 27:413 (1984).
9. B. Wasserman, G. Braunstein, M. S. Dresselhaus, and G. E. Wuieck, Implantation induced conductivity of polymers, *Mater. Res. Soc. Symp. Proc.* 27:423 (1984).
10. B. Wasserman, Fractal nature of electrical conductivity in ion implanted polymers, *Phys. Rev. B* 34:1926 (1986).
11. B. Wasserman, Transport properties of ion implanted polymers, Thesis, Massachussetts Institute of Technology, 1985.
12. J. Bartho, B. U. Hall, and K. F. Schoch, Highly conductive poly(phenylene sulfide) prepared by high-energy ion irradiation, *J. Appl. Phys.* 59(4):1111 (1986).
13. B. S. Elman, D. J. Sandman, S. K. Tripathy, and M. K. Thakur, Ion implanted polydiacetylenes, U.S. Patent 4,647,403 (1987).
14. A. J. Dann, M. R. Fahy, M. R. Willis, and C. Jeynes, Ion implantation of polymeric phthalocyanines, *Synth. Met.* 18:581 (1987).
15. S. Robinet, M. Gauneau, M. Salvi, C. Clarisse, M. Delamar, and M. Leclerc, Spectroscopic and chemical study of the damage induced by ion implantation in scandium diphthalocyanine films, *Thin Solid Films* 200:385 (1991).
16. J. L. Duroux, A. Moliton, G. Froyer, and F. Maurice, Resultats expérimentaux sur la conductivité du PPP dopé par des ions accélérés, *Ann. Phys. Fr., Coll. 1*, 11:117 (1986).
17. M. L. Kaplan, S. R. Forest, P. H. Schmidt, and T. Venkatesan, Optical and electrical properties of ion beam irradiated films of organic molecular solids and polymers, *J. Appl. Phys.* 55(3):732 (1984).
18. A. Moliton, C. Moreau, B. Lucas, R. H. Friend, and G. Froyer, Effets de compétition entre les phénomènes de dopage et d'endommagement dans les polymères électroactifs implantés, *J. Phys. III Fr.* 4:1689 (1994).
19. K. Harigaya, Doping in conducting polymers: electronic states and metal–insulator transition, in *Molecular Electronics and Molecular Electronic Devices*, Vol. 2 (K. Sienicki, ed.), CRC Press, Boca Raton, FL, 1993, Chap. 2.
20. A. J. Heeger, S. Kivelson, J. R. Schriefer, and W. P. Su, Solitons in conducting polymers, *Rev. Mod. Phys.* 60(3):781 (1988).
21. J. L. Bredas, R. R. Chance, and R. Silbey, Comparative theoretical study of the doping of conjugated polymers: polarons in polyacetylene and polyparaphenylene, *Phys. Rev. B* 26(10):5843 (1982).
22. S. Kivelson, Electron hopping in a soliton band: conduction in lightly doped $(CH)_x$, *Phys. Rev. B* 25(6):3798 (1982).
23. R. R. Chance, J. L. Bredas, and R. Silbey, Bipolaron transportion in doped conjugated polymers, *Phys. Rev. B* 29(8):4491 (1984).
24. S. Stasfröm and J. L. Bredas, Band structure calculations for the polaron lattice in the highly doped regime of polyacetylene, polythiophene, and polyaniline, *Mol. Cryst. Liq. Cryst.* 160:405 (1988).
25. A. Boudet and P. Pradere, Morphology and structure of PPP as revealed by electron microscopy, *Synth. Met.* 9:491 (1984).
26. N. F. Mott and E. A. Davis, *Electronic Processes in Non Crystalline Materials*, Clarendon Press, Oxford, 1979.
27. A. Moliton and B. Ratier, Propriétés électroniques et schémas de bandes dans les semiconducteurs amorphes, *Ann. Phys. Fr.* 16:261 (1991).
28. D. Emin, Basic issues of electronic transport in insulating polymers, in *Handbook of Conducting Polymers*, Vol. 2 (T. A. Skotheim, ed.), Marcel Dekker, New York, 1986, Chap. 26.
29. D. Emin, The formation and motion of small polarons, in *Linear and Non Linear Electron Transport in Solids* (J. T. Devreese and V. E. Van Doren, eds.), Plenum, New York, 1976, p. 409.
30. D. Emin, Aspects of the theory of small polarons in disordered materials, in *Electronic and Structural Properties of Amorphous Semiconductors* (P. G. Le Comber and J. Mort, eds.), Academic, London, 1973, Chap. 7, p. 318.

31. D. Emin and K. L. Ngai, Hopping conduction in lightly-doped polyacetylene, *J. Phys. Fr., Coll. 3* 44(6):C3–C471 (1983).
32. P. Pfluger, G. Weiser, J. Campbell Scott, and B. Street, Electronic structure and transport in the organic "amorphous semiconductor" polypyrrole, in *Handbook of Conducting Polymers*, Vol. 2 (T. A. Skotheim, ed.), Marcel Dekker, New York, 1986, Chap. 38.
33. F. Maurice, G. Froyer, M. Minier, and M. Gaunneau, Correlation between d.c. bulk conductivity and impurities in undoped poly(*p*-phenylene), *J. Phys. Lett. 42*:L425 (1981).
34. S. R. Elliott, *Physics of Amorphous Materials*, 2nd ed., Longman/Wiley, New York, 1990.
35. R. A. Street, *Hydrogenated Amorphous Silicon*, Cambridge Univ. Press, Cambridge, UK, 1991.
36. A. Moliton, B. Lucas, C. Moreau, R. H. Friend, and B. François, Ion implantation in conjugated polymers: mechanisms for generation of charge carriers, *Phil. Mag. B 69*(6):1155 (1994).
37. J. Davenas, Ion beam modification of organic materials, *Solid State Phenom., 30–31*:317 (1993).
38. G. Froyer, Y. Pelous, F. Maurice, M. A. Petit, A. Digna, and J. F. Fauverque, Optical studies of PPP thin films prepared by electroreduction, *Synth. Met. 21*:241 (1987).
39. B. Ratier, Y. Pelous, G. Froyer, and A. Moliton, Ion implantation in PPP thin films: optical investigations, *Synth. Met. 41–43*:1483 (1991).
40. C. Moreau, R. H. Friend, G. J. Sarnecki, B. Lucas, A. Moliton, B. Ratier, and C. Belorgeot, Spectroscopic studies of ion implanted PPV films, *Synth. Met. 55–57*:224 (1993).
41. H. Isotalo, H. Stubb, and J. Saarilahti, Ion implantation of polythiophene, in *Electronic Properties of Conjugated Polymers* (H. Kuzmany, M. Mehring, and S. Roth, eds.), Springer-Verlag, Berlin, 1987, p. 285.
42. R. H. Friend, Semiconductor devices fabricated with conjugated polymers: novel optoelectronic properties, in *Conjugated Polymers and Related Materials* (W. R. Salaneck, I. Lundström, and B. Ranby, eds.), Oxford Univ. Press, Oxford, UK, 1993, Chap. 21.
43. R. H. Friend, Caractéristiques optiques et électro-optiques, Final Tech. Rep., Brite/Euram Program, Contract No. Breu-0148+C-NAPOLEO, European Communities, 1993.
44. X. F. Zhong, Soluble polystyrene (PS)-polyparaphenylene (PPP) block copolymers, *Synth. Met. 29*:E35–E40 (1989).
45. C. Belorgeot, B. Lucas, B. Ratier, A. Moliton, and B. François, Etude infrarouge et diélectrique de films de PPP vierges ou implantés, *J. Chim. Phys. 89*:1291 (1992).
46. M. Margoshes and V. A. Fassel, The infrared spectra of aromatic compounds, *Spectrochim. Acta 7*:14 (1955).
47. T. Wada, A. Takeno, M. Iwaki, and H. Sasabe, Ion beam modification of conducting polymers, *Synth. Met. 18*:585 (1987).
48. J. M. Le Meil, Etude des défauts créés dans le PPP implanté, Internal Report, CNET, Lannion, October 1989.
49. T. P. Nguyen, S. Lefrant, G. Froyer, Y. Pelous, B. Ratier, and A. Moliton, XPS and Raman characterization of ion implanted PPP thin films, *Synth. Met. 41–43*:291 (1991).
50. W. N. Allen, P. Brant, C. A. Carosella, J. J. De Corpo, C. T. Ewing, F. E. Saafeld, and D. C. Weber, Ion implantation studies of $(SN)_x$ and $(CH)_x$, *Synth. Met. 1*:151 (1979/80).
51. D. C. Weber, P. Brant, C. Carosella, and L. G. Banks, A new method for the chemical modification of polymers, *J. Chem. Soc. Chem. Commun. 130*:522 (1981).
52. N. Koshida, Y. Suzuki, and T. Aoyama, Low energy ion implantation studies of polyacetylene films, *Nucl. Inst. Meth. B37/39*:708 (1989).
53. N. Koshida and N. Hirayama, Ion implantation induced change in fibrillar morphology, *Nucl. Inst. Meth. B59/60*:1292 (1991).
54. J. Davenas, X. L. Xu, M. Maitrot, M. Gamoudi, G. Guillaud, J. J. André, B. François, and C. Mathis, Stability of polyacetylene doped by ion implantation, *J. Phys. Fr. 44*:C3-183 (1983).
55. S. Lin, K. Sheng, J. Bao, T. Rong, Z. Zhou, L. Zhang, and D. Zhu, Ion beam modification of polyacetylene films, *Nucl. Inst. Meth. B39*:778 (1989).
56. B. Ratier, M. Gauneau, A. Moliton, G. Froyer, R. Chaplain, C. Le Hüe, and J. P. Moliton, Etude par microanalyse de films de PPP et des phénomènes induits par implantation ionique, *J. Phys. III, Fr. 2*:1757 (1992).
57. A. Aboulkassim, Y. Pelous, C. Chevrot, and G. Froyer, Voltammetric studies on poly(paraphenylene) films obtained by electroreduction, *Polym. Bull 19*:595 (1988).
58. G. Froyer, Y. Pelous, M. Gauneau, R. Chaplain, A. Moliton, and B. Ratier, Ion implanted PPP: modifications of lateral and in depth concentration profiles upon annealing, in *Conjugated Polymeric Materials: Opportunities in Electronics, Optoelectronics and Molecular Electronics* (J. L. Bredas and R. R. Chance, eds.), Kluwer, Dordrecht, 1990, p. 263.
59. T. A. Skotheim, *Handbook of Conducting Polymers*, Marcel Dekker, New York, 1986.
60. C. K. Chiang, Y. W. Park, A. J. Heeger, H. Shirakawa, E. J. Louis, and A. G. MacDiarmid, Conducting polymers: halogen doped PA, *J. Chem. Phys. 69*(11):5098 (1978).
61. Y. W. Park, Structure and morphology: relation to thermopower properties of conductive polymers, *Synth. Met. 45*(2):173 (1991).
62. A. B. Kaiser, Metallic behaviour in highly conducting polymers, *Synth. Met. 45*(2):183 (1991).
63. A. B. Kaiser, Thermoelectric power and conductivity of heterogeneous conducting polymers, *Phys. Rev. B 40*(5):2806 (1989).

64. A. B. Kaiser, Electronic transport in conducting polymers, in *Electronic Properties of Conducting Polymers* (H. Kuzmany, M. Mehring, and S. Roth, eds.), Springer, Heildelberg, 1989, p. 2.
65. A. B. Kaiser and S. C. Graham, Temperature dependence of conductivity in "metallic" polyacetylene, *Synth. Met. 36*:367 (1990).
66. A. B. Kaiser, "Metallic" polymer: interpretation of the electronic transport properties, *Synth. Met. 41–43*:3329 (1991).
67. S. Kivelson, Electron hopping conduction in the soliton model of polyacetylene, *Phys. Rev. Lett. 46*(20): 1344 (1981).
68. H. Böttger and V. V. Bryksin, *Hopping Conduction in Solids*, VCH, Berlin, 1985, p. 280.
69. A. Moliton and B. Lucas, Propriétés électroniques dans les solides désordonnés, *Ann. Phys. Fr. 19*:299 (1994).
70. P. Nagels, Electronic transfer in amorphous semiconductors, in *Amorphous Semiconductors* (M. H. Brodsky, ed.), Springer Verlag, Berlin, 1985, Chap. 5.
71. V. V. Kosarev, Thermopower of lightly doped semiconductors in the hopping conduction region, *Sov. Phys. Semicond. 8*(7):897 (1975).
72. P. M. Chaikin and G. Beni, Thermopower in the correlated hopping regime, *Phys. Rev. B 13*(2):647 (1976).
73. R. R. Heikes, Narrow band semiconductors, ionic crystals, and crystals, in *Thermoelectricity* (R. R. Heikes and R. W. Ure, eds.), Interscience, New York, 1961, Chap. 4.
74. P. Nagels, Polaronic conduction in oxide glasses containing V_2O_5, in *Hopping and Related Phenomena* (H. Fritzsche and M. Pollak, eds.), World Scientific, Singapore, 1990, p. 385.
75. Z. H. Wang, A. Ray, A. G. MacDiarmid, and A. J. Epstein, Electron localization and charge transport in poly(*o*-toluidine): a model polyaniline derivative, *Phys. Rev. B 43*(5):4373 (1991).
76. P. Nagels, M. Rotti, and R. Gevers, Thermoelectric power due to variable-range hopping, *J. Non-Crystal. Solids 59–60*:65 (1983).
77. P. Nagels, H. Krikor, and M. Rotti, Electrical properties of a poly(*meta*-phenylene) network, *Synth. Met. 29*:E-29 (1989).
78. A. J. Epstein, A.C. conductivity of polyacetylene, in *Handbook of Conducting Polymers* (T. A. Skotheim, ed.), Marcel Dekker, New York, 1986, Chap. 29.
79. I. G. Austin and N. F. Mott, Polarons in crystalline or non-crystalline materials, *Adv. Phys. 18*:41 (1969).
80. A. J. Epstein, H. Rommelmann, and H. W. Gibson, Frequency dependent conductivity of lightly doped *cis*-polyacetylene, *Synth. Met. 9*:103 (1984).
81. A. J. Epstein, H. Rommelmann, M. Abkowitz, and H. W. Gibson, Anomalous frequency-dependent conductivity of polyacetylene, *Phys. Rev. Lett. 47*(21):1549 (1981).
82. J. L. Duroux, A. Moliton, and G. Froyer, Influence of implantation parameters on the poly(paraphenylene) electrical conductivity, *Nucl. Inst. Meth. B 34*: 450 (1988).
83. A. Moliton, B. Ratier, B. Guille, and G. Froyer, Thermoelectric power versus temperature in implanted PPP, *Mol. Cryst. Liq. Cryst. 186*:223 (1990).
84. A. Moliton, J. L. Duroux, B. Ratier, G. Froyer, and Y. Pelous, Etude comparative de la conductivité dans des films et des plaquettes de PPP et de PPS dopés par implantation ionique, *Makromol. Chem. Macromol. Symp. 20/21*:589 (1988).
85. B. Lucas, A. Moliton, and B. François, Etude de la conductivité alternative superficielle de films implantés de PPP obtenus à partir d'un copolymère séquencé, *Eur. Polym. J. 27*(9):911 (1991).
86. A. K. Jonsher, Frequency dependence of conductivity in hopping systems, *J. Non Cryst. Solids 8–10*: 295 (1972).
87. J. Davenas, Influence of the temperature on the ion beam induced conductivity of polyimide, *Appl. Surf. Sci. 43*:218 (1989).
88. B. Ratier, A. Moliton, B. Guille, and G. Froyer, Effet de la tempéraure d'implantation sur les propriétés électriques du PPP, *J. Chim. Phys. 89*:1313 (1992).
89. J. Y. Goblot, Contribution à l'étude du dopage du PPP par AsF_5, Thesis, University of Rennes, 1986.
90. C. Le Hüe, C. Moreau, A. Moliton, B. Guille, and G. Froyer, Conductivity kinetics and conductivity levels of ion implanted PPP pellets, *Synth Met. 55–57*:4986 (1993).
91. C. Le Hüe, A. Moliton, B. Lucas, and G. Froyer, Compensation phenomena by ion implantation doping of an electroactive polymer, poly(paraphenylene), *Adv. Mater. Opt. Electron. 1*:173 (1992).
92. C. K. Chiang, S. C. Gau, C. R. Fincher, Y. W. Park, A. G. MacDiarmid, and A. J. Heeger, Polyacetylene, $(CH)_x$: *n*-type and *p*-type doping and compensation, *Appl. Phys. Lett. 33*(1):18 (1978).
93. A. Moliton, B. Ratier, C. Moreau, and G. Froyer, Appareillage automatisé de mesure simultanée du pouvoir thermoélectrique et de la conductivité électrique. Application à l'étude de couches polymères semi-conducteurs, *J. Phys. Fr. III 1*:809 (1991).
94. S. Roth, Hopping conduction in electrically conducting polymers, in *Hopping Transport in Solids* (M. Pollak and B. I. Shklovskii, eds.), North-Holland, Amsterdam, 1991, Chap. 11.
95. A. Moliton, C. Moreau, J. P. Moliton, and G. Froyer, Transport phenomena in implanted electroactive polymers, *Nucl. Inst. Meth. B 80/81*:1028 (1993).
96. C. Moreau, B. Ratier, A. Moliton, and B. François, High temperature behaviour of thermoelectric power of implanted polymer films, *Radiation Effects and Defects in Solids, 137*:129 (1995).
97. N. Koshida, and Y. Suzuki, Electrical properties of ion-implanted polyacetylene films, *J. Appl. Phys. 61*(12):5487 (1987).
98. C. Moreau, Caractérisations thermoélectriques et optiques de polymères dopés par implantation ionique, Thesis, University of Limoges, 1984.
99. B. Lucas, B. Ratier, A. Moliton, C. Moreau, and

R. H. Friend, Transport properties of ion implanted PPV, *Synth. Met. 55–57*:4912 (1993).
100. J. H. Burroughes, D. D. C. Bradley, A. R. Brown, R. N. Marks, K. Mackay, R. H. Friend, P. L. Burns, and A. B. Holmes, Light-emitting diodes based on conjugated polymers, *Nature 347*:539 (1990).
101. S. R. Forrest, M. L. Kaplan, P. H. Schmidt, W. L. Feldmann, and E. Yanowski, Organic-on-inorganic semiconductor contact barrier devices, *Appl. Phys. Lett. 41*:90 (1982).
102. H. Koezuka, K. Hyodo, and A. G. MacDiarmid, Organic heterojunctions utilizing two conducting polymers: poly(acetylene)/poly(N-methylpyrrole) junctions, *J. Appl. Phys. 58*:1279 (1985).
103. S. Miyauchi, A. Fueki, Y. Kushihi, H. Abiko, and Y. Sorimachi, Shottky barrier formation between polypyrrole and indium, *Synth. Met. 18*:689 (1987).
104. L. M. Goldenberk, V. I. Krinichnyi, and I. B. Nazarova, The Schottky device based on doped poly(paraphenylene), *Synth. Met. 44*:199 (1991).
105. H. Tomozawa, D. Braun, S. Philips, A. J. Heeger, and H. Kroemer, Metal-polymer Schottky barrier on cast films of soluble poly(3-alkylthiophenes), *Synth. Met. 22*:63 (1987).
106. J. Kanicki, Polymeric semiconductor contacts and photovoltaic applications, in *Handbook of Conducting Polymers* (T. A. Skotheim, ed.), Marcel Dekker, New York, 1986, Chap. 17.
107. J. J. Andre, J. Simon, E. Even, B. Boudjena, G. Guillaud, and M. Maitrot, Molecular semiconductors and junction formation: phthalocyanine derivatives, *Synth. Met. 18*:683 (1987).
108. H. Koezuka, A. Tsumura, and T. Ando, Field-effect transistor with polythiophene thin films, *Synth. Met. 18*:699 (1987).
109. G. Horowitz, X. Peng, D. Fichou, and F. Garnier, The oligothiophene-based field effect transistor: how it works and how to improve it, *J. Appl. Phys. 67*(1):528 (1990).
110. N. Koshida and Y. Wachi, Application of ion implantation for doping of polyacetylene films, *Appl. Phys. Lett. 45*(4):436 (1984).
111. T. Wada, A. Takeno, M. Iwaki, H. Sasabe, and Y. Kobayashi, Fabrication of a stable *p-n* junction in a polyacetylene film by ion implantation, *J. Chem. Soc. 1985*:1194.
112. B. Lucas, B. Ratier, A. Moliton, J. P. Moliton, T. F. Otero, C. Santamaria, E. Angulo, and J. Rodriguez, Thick layers of PPy and PTh: electrosynthesis and modification of electrical parameters by ion implantation. *p-n* junctions, *Synth. Met. 55–57*:1459 (1993).
113. A. Moliton, J. L. Duroux, B. Ratier, and G. Froyer, pn^+ junction in an implanted electroactive polymer: poly(paraphenylene), *Elect. Lett. 24*(7):383 (1988).
114. A. Moliton, B. Ratier, and G. Froyer, Etude comparative de jonctions polymères (PPP) obtenues sur des films ou des pastilles par implantation avec le couple alcalin-halogène, *J. Chim. Phys. 86*(1):249 (1989).
115. M. Fahlman, J. Rasmusson, K. Karriyama, D. T. Clark, G. Beaumson, and W. R. Salaneck, Epitaxy of poly(2,5-diheptyl-*p*-phenylene) on ordered polytetrafluoroethylene, *Synth. Met. 66*:123 (1994).
116. K. Pichler, C. P. Jarret, R. H. Friend, B. Ratier, and A. Moliton, Field-effect transistor based on poly(*p*-phenylene vinylene) doped by ion-implantation, *J. Appl. Phys. 77*(7):3523 (1995).
117. J. H. Burroughes and R. H. Friend, The semiconductor device physics of polyacetylene, in *Conjugated Polymers* J. L. Bredas, and R. Silbey, eds.), Kluwer, Dordrecht, 1991, p. 555.
118. Y. Ohmori, K. Muro, M. Onoda, and K. Yoshino, Fabrication and characteristics of Schottky gated field effect transistors utilizing poly(1,4-naphthalene vinylene) and poly(*p*-phenylene vinylene), *Jpn. J. Appl. Phys. 31*:L646 (1992).
119. K. F. Voss, D. Braun, and A. J. Heeger, Temperature dependence of polymer field effect transistors, *Synth. Met. 41–43*:1185 (1991).
120. G. Horowitz, D. Fichou, X. Peng, and F. Garnier, Thin-film transistors based on alpha-conjugated oligomers, *Synth. Met. 41–43*:1127 (1991).
121. F. Garnier, Conjugated oligomers as model compounds for the charge transport process: towards molecular electronics, in *Conjugated Polymers and Related Materials* (W. R. Salaneck, I. Lundström, and B. Ranby, eds.), Oxford Univ. Press, Oxford, 1993, Chap. 20.
122. J. M. E. Harper, J. J. Cuomo, and H. R. Kaufman, Technology and applications of broad-beam ion sources used in sputtering, *J. Vac. Sci. Technol. 21*(3):737 (1982).
123. J. I. Pankove and N. M. Johnson, *Hydrogen in Semiconductors*, Academic, New York, 1991.
124. B. Ratier, L. Athouël, G. Froyer, A. Moliton, and X. L. Xu, Effect de l'hydrogène sur les propriétés électriques et optiques de films de polyparaphénylène et de parasexiphényl implantés, *J. Chim. Phys., 92*:891 (1995).

22
Optical Probes of Photoexcitations in Conducting Polymers

Zeev Valy Vardeny and X. Wei*
University of Utah, Salt Lake City, Utah

I. INTRODUCTION

The π-conjugated polymers have been intensively studied during the last 20 years. They form a new class of electronic materials with potential applications such as light-emitting diodes (LEDs) [1], thin-film transistors (TFTs) [2], and optical switches and modulators [3]. As polymers, these materials have a highly anisotropic quasi-one-dimensional electronic structure that is fundamentally different from the structures of conventional inorganic semiconductors. This has two consequences: First, their chainlike structure leads to strong coupling of the electronic states to conformational excitations peculiar to the one-dimensional (1D) system [4], and second, the relatively weak interchain binding allows diffusion of dopant molecules into the structure (between chains), whereas the strong intrachain carbon–carbon bond maintains the integrity of the polymer [4]. In their neutral form, these polymers are semiconductors with an energy gap of $\simeq 2$ eV. However, they are easily doped with various p and n dopants, increasing their conductivity by many orders of magnitude; conductivities in the range of $\simeq 10^3$–10^4 S/cm are not unusual [5].

The simplest example of the class of conducting polymers is polyacetylene, $(CH)_x$, which is depicted in Fig. 22.1. It consists of weakly coupled chains of CH units forming a pseudo-1D lattice. The stable isomer is *trans*-$(CH)_x$, in which the chain has a zigzag geometry; the *cis*-$(CH)_x$ isomer, in which the chain has a backbone geometry, is unstable at room temperature or under high illumination. Simple conducting polymers like polyacetylene are planar, with three of the four carbon valence electrons forming sp^2 hybrid orbitals (σ bonds), while the fourth valence electron is in a π orbital perpendicular to the plane of the chain. The σ bonds form the skeleton of the chain and are responsible for the strong elastic force constant of the chain. The π orbitals form the highest occupied molecular orbitals (HOMOs) and the lowest unoccupied molecular orbitals (LUMOs), which together span an energy range of $\simeq 10$ eV [4]. *trans*-$(CH)_x$ is a semiconductor with a gap $E_g \simeq 1.5$ eV, which has two equivalent lowest energy states having two distinct conjugated structures [4]. Other polymers shown in Fig. 22.1, such as *cis*-$(CH)_x$, polythiophene (PT), and polyparaphenylenevinylene (PPV), are nondegenerate in the ground state, and this can be formally described by adding an extrinsic gap component to the Peierls gap [4].

The properties and dynamics of optical excitations in conducting polymers are of fundamental interest because they play an important role in the potential applications. However, in spite of intense studies of the linear and nonlinear optical properties, the basic model for the proper description of the electronic excitations in conducting polymers is still controversial. One-dimensional semiconductor models [6], in which electron–electron (e–e) interaction has been ignored, have been successfully applied to interpret a variety of optical experiments in π-conjugated polymers [4]. In these models the strong electron–phonon (e–p) interaction leads to rapid self-localization of the charged excitations, the so-called polaronic effect. Then optical excitations across E_g, which is now the Peierls gap, are entirely different from the electron–hole pairs of conventional semiconductors. Instead, the proper description of the quasiparticles in *trans*-$(CH)_x$ is that 1D domain walls or solitons (S) separate the two degenerate ground-state structures [6]. As a result of their translational invariance, solitons in *trans*-$(CH)_x$ are thought to play the role of energy- and charge-carrying excitations. Since the soliton is a topological defect, it can be either created in soliton–antisoliton

---

* Current affiliation: Los Alamos National Laboratories, Los Alamos, New Mexico.

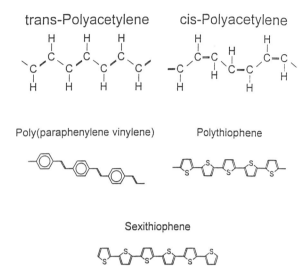

Fig. 22.1 Backbone structure of some conducting polymers and oligomers.

(S–$\bar{S}$) pairs or be created in polyacetylene chains with odd numbers of CH monomers upon isomerization from cis-$(CH)_x$. The same Hamiltonian that has predicted soliton excitations in $trans$-$(CH)_x$ predicts polarons as a distinct solution when a single electron is added to the $trans$ chain [4]. For the nondegenerate ground-state (NDGS) polymers, adding an extrinsic gap component to the electron–phonon Hamiltonian results in polarons and bipolarons as the proper descriptions of their primary charge excitations. Singlet and triplet excitons have been also shown to play a crucial role in the photophysics of conducting polymers [7–12]. However, their existence in theoretical studies can be justified only when electron–electron interaction and correlation effects are added to the Hamiltonian [13,14].

This interaction is extremely important even in the simplest example of $trans$-$(CH)_x$, where the $2A_g$ excited state (determined experimentally [15,16] by two-photon absorption) is located below the first optically allowed excited state, $1B_u$. The electron–electron interaction cannot be ignored in any conjugated polymer, and it has significant effects on various optical properties, such as photoluminescence (PL), electroabsorption (EA), and third-order optical susceptibilities. In cases where the bond alternation is relatively small, the ordering of the odd and even symmetry lowest excited states is $E(2A_g) < E(1B_u)$ [17]. When the "effective" bond alternation is relatively large, the ordering of these states is reversed, resulting in strong PL. In PPV, for instance, the benzene ring in the backbone structure gives rise to a large "effective" bond alternation for the extended π electrons [17] and therefore to high PL efficiency and improved LED devices. Nevertheless, the Coulomb interaction among the π electrons, even when it is not dominant, leads to behavior qualitatively different from the prediction of single-particle Hückel or SSH models.

The soliton excitation in t-$(CH)_x$ is an amphoteric defect that can accommodate zero, one, or two electrons [4,6,18]. The neutral soliton (S^0, spin 1/2) has one electron; positively and negatively charged solitons ($S^\pm$, spin 0) have zero or two electrons, respectively. Within the framework of the Su–Schrieffer–Heeger (SSH) model [6] Hamiltonian, which contains electron–phonon (e–p) interactions but does not contain electron–electron or 3D interactions, it has been shown that a photoexcited electronhole (e–h) pair is unstable toward the formation of a soliton–antisoliton (S–S) pair [19]. Subsequently, it was demonstrated [20] that as a consequence of the Pauli principle and charge conjugation symmetry in t-$(CH)_x$, the photogenerated soliton and antisoliton are oppositely charged. The study of photoexcited $trans$-$(CH)_x$, however, has revealed several unexpected phenomena, which were not predicted by the SSH model of the soliton (for a review, see Ref. 21). Most important, an overall neutral state as well as charged excitation have been observed; this neutral state has been recently correlated with S^0 transitions [22]. This finding together with the absence [in undoped $trans$-$(CH)_x$] of optical transitions at midgap, where transitions of neutral and charged solitons should have appeared according to the SSH picture, have shown that electron–electron interaction in $trans$-$(CH)_x$ cannot be ignored. In this case the electronic gap in $trans$-$(CH)_x$ is partially due to electron correlations rather than entirely due to electron–phonon interaction as in the SSH model. Under these circumstances, the nature of the photoexcitations in $trans$-$(CH)_x$ may be very different from that predicted by the SSH Hamiltonian. These findings have stimulated photophysical research in *all* conducting polymers, mostly the NDGS polymers. The previous picture of photogeneration of bound soliton–antisoliton pairs in NDGS polymers has been modified. On the contrary, photoexcitation of singlet and triplet excitons and/or polaron pairs [23,24] has been recently demonstrated in many such conducting polymers.

The origin of the branching process that determines the relative photoproduction of neutral vs. charged photoexcitations in the class of conducting polymers is not very well understood. One possible explanation of the branching process is afforded by the Onsager theory, which has successfully explained charge photoproduction in disordered materials and in molecular crystals [25]. The difficulty with this approach is that conducting polymers are quasi-1D semiconductors for which the Onsager theory based on the e–h Coulomb attraction may not be applicable. In addition, the application of this theory for 1D semiconductors results in negligible quantum efficiency for charge photoproduction under weak electric fields, contrary to experimental results. To solve this problem, it was suggested that the 1D–3D interplay is important in the photophysics of conducting polymers [26]. In the proposed model, *intrachain* excitation results in a neutral state (which is most probably an exciton), whereas *interchain* excitation may produce separate charges on neighboring chains (most probably a polaron pair) if the interchain geminate recombination is overcome. A demonstration [27] of the important role of interchain excitation was the observation that long-lived charged excitations in oriented films are more efficiently photogenerated with light polarized perpendicular to the polymer chain direction than with light polarized parallel to it. Another important experiment [28] in the picosecond time domain demonstrated that most of the intrachain charged solitons in $(CH)_x$ quickly recombine. However, charged solitons generated by a delayed interchain process are long-lived, and therefore interchain photogenerated charged solitons become increasingly more important at longer times. On the other hand, the recent demonstration that charged polarons and bipolarons are photogenerated in isolated PT chains [29] in both liquid and solid forms may challenge the common view of the unique importance of 3D interaction for charge photoproduction in conducting polymers. Do disorder and 3D effects play a crucial role in charge photogeneration? Are most conducting polymers excitonic or bandlike in nature? These are important questions associated with charge generation that are still unresolved.

In recent years it has been recognized that many characteristics of the excited states of conjugated polymers are in essence very similar to those of the corresponding finite oligomers. *trans*-β-Carotene, for example, is a conjugated molecule that consists of a backbone of 11 double bonds, similar to *trans*-$(CH)_x$, which has a fixed length of about 22 carbons. Upon doping it was observed that the charge is stored in a spinless stable configuration accompanied by structural relaxation [30]. Another example is the thiophene oligomers. Here, several oligomers ranging from bisthiophene through sexithiophene [31,32] to octa- and decathiophene were studied [33]. At low doping levels, the radical ions of all the oligomers show the characteristics of polarons (i.e., charged defect, with spin 1/2 accompanied by bond relaxation). Upon further doping, the di-ions show the characteristics of bipolarons (i.e., doubly charged spinless defects with stronger bond relaxation). In a recent work [34], the photoexcited defects in sexithiophene (Fig. 22.1) were identified as polarons because of the similarity of their absorption spectrum to that of the radical cations. It thus appears natural to use oligomers of various lengths to characterize the excited states of the longer polymers.

In this chapter we review the studies of photoexcitations in *trans*-$(CH)_x$, the representative of the degenerate ground state polymers, and NDGS polymers such as PT derivatives as well as those in the important PT oligomer α-sexithiophene; the backbone structure of these materials is schematically depicted in Fig. 22.1.

We studied photoexcitations in such polymers in a broad time interval from femtoseconds to milliseconds and spectral range from 0.1 to 2.4 eV. However, in this chapter we review only our continuous wave (cw) studies, where the photoexcitations are generated in quasi-steady-state conditions. The main experimental technique described herein is photomodulation (PM), which gives information complementary to that obtained by photoluminescence (PL), which is limited to radiative processes, or photoconductivity (PC), which is sensitive to high mobility photocarriers. The PM method, in contrast, is sensitive to nonequilibrium excitations in *all* states.

Among the various powerful techniques that have been used to investigate the photoexcitation properties in conducting polymers, perhaps the most controversial is the technique of light-induced electron spin resonance (LESR), which in principle has the potential to measure the spin state of photoexcitations. The sensitivity of ESR machines is usually too low for LESR; the overall minimum number of spins that can be measured is only 10^{11}. A back-of-the-envelope calculation using realistic dimensions of the type of μ-wave cavities used in LESR and the high absorbance of conducting polymers (light penetration depth of ≈1000 Å) limits the minimum steady-state density of spin-carrying photoexcitations that can be detected in LESR to 10^{17} cm^{-3}. This relatively high photoexcitation density can be photogenerated if excitation lifetime is longer than about 10^{-2} s and the generation quantum efficiency is higher than about 10%. These conditions are not easy to satisfy for most types of photoexcitations in conducting polymers. If attainable, they further require high laser excitation intensity, and this leads to sample heating, which is the second problem associated with LESR: Conducting polymers usually contain spin 1/2 defects (dangling bonds, ends of chains, etc.) of substantial density, on the order of 10^{17} cm^{-3} or larger. The modulation of the laser excitation intensity in LESR measurements causes temperature modulation, which in turn modulates the Curie susceptibility, leading to temperature artifacts in the magnetic resonance spectra. Double modulation, where the μ-wave intensity and the laser excitation intensity are simultaneously modulated, does not help to reduce the heating problem, since the LESR and the thermal lifetimes are similar. A sophisticated scheme [35] was tried in *trans*-$(CH)_x$ to avoid laser heating effects in LESR, but the results were not conclusive due to lack of spectral information associated with the LESR signal.

In this chapter we discuss the novel technique of photoabsorption-detected magnetic resonance (ADMR) [22], in which spin states and their correlated spectra in PM can be simultaneously measured; this technique is far superior to LESR. The main idea in ADMR is the detection of changes in steady-state photoexcitation density induced by μ-wave absorption in resonance with

Zeeman split sublevel electronic states. Since the changes are induced in the density of photoexcitations rather than in μ-wave absorption, the effect can be measured with detectors in the visible to near-IR spectral ranges, with improved sensitivity of up to five orders of magnitude over that of μ-wave detectors. Moreover, spectral information can also be gained in ADMR measurements, as the optical probe can be easily tuned.

II. OPTICAL TRANSITIONS OF PHOTOEXCITATIONS IN CONDUCTING POLYMERS

Perhaps the best way to detect and characterize photoexcitations in the class of π-conjugated polymers is to study their optical absorption. As a consequence of their localization they give rise to gap states in the electron and phonon level spectra, respectively. The scheme of our experiments is the following. We photoexcite with above-gap light and then probe the optical absorption of the sample in a broad spectral range from IR to visible. Essentially we obtain difference spectra, i.e., the difference in the optical absorption ($\Delta\alpha$) of the polymer when it contains a nonequilibrium carrier concentration and that in the equilibrium ground state. Therefore the optical transitions of the various photoexcitations are of fundamental importance. In this section we discuss and summarize the states in the gap and the associated electronic transitions of various photoexcitations in conducting polymers; the IR-active vibrations (IRAVs) related to the charged excitations are summarized elsewhere in this book.

Rather than discussing the various electronic states in conducting polymers in terms of bands (valence and conducting bands, for example), which might be the proper description of the infinite chains, or discrete levels with proper symmetries, which should be used for oligomers or other finite chains, we prefer the use of HOMO (highest occupied molecular orbital), LUMO (lowest unoccupied molecular orbital), and SOMO (singly occupied molecular orbital). In the semiconductor description of the infinite chain, HOMO is the top of the valence band, LUMO is the bottom of the conduction band, and SOMO is a singly occupied state in the forbidden gap. In this case E_g = LUMO − HOMO. On the other hand, in the excitonic description of the correlated infinite chain, HOMO is the $1A_g$ state and LUMO is the $1B_u$ exciton [we deal only with luminescent NDGS polymer in this review, in which $E(1B_u) < E(2A_g)$]. In this case $E_g = E(1B_u)$. In finite chains HOMO and LUMO are discrete (isolated) levels with definite symmetries. Among them the subscript g stands for even (gerade) parity and u stands for odd (ungerade) parity. These symbols are extremely important for possible optical transitions, since one-photon absorption can take place between states of *opposite* representations, such as g → u or u → g. This is true in the singlet manifold as well as in the triplet manifold.

We discuss separately excitations in degenerate and nondegenerate ground-state polymers.

A. Optical Transitions of Solitons in *trans*-(CH)$_x$

The semiconductor model for the PM spectrum associated with the soliton (S) transitions is shown in Fig. 22.2 [18]. The amphoteric S defect has ground state S^0, negatively charged state S^-, and positively charged state S^+; both S^+ and S^- are spinless. Charge conjugation symmetry is also assumed. The charged states are unrelaxed (S^-, S^+) during a time shorter than the relaxation time of the lattice around the defects; at longer times they are relaxed (S_r^-, S_r^+). The energy of the unrelaxed S^- state differs from the energy of the S^0 by the (bare) electron correlation energy $U = E^- - E^0$; E^0 is the energy of S^0, and E^- and E^+ are the energies of the unrelaxed states S^- and S^+, respectively (Fig. 22.2). The relaxed states differ from the unrelaxed states by the relaxation energy $\Delta E_r^- = E^- - E_r^-$, $\Delta E_r^+ = E_r^+ - E^+$. The relaxed state energy E_r and the ground-state energy E^0 differ by the effective correlation energy $U_{\mathrm{eff}} = E_r^- - E^0 = U - \Delta E_r^-$. The optical transitions of the soliton defects are therefore $\delta S^{\mp}$ from S_r^- (at E_r^-) to the LUMO level and from the HOMO level into S^+, and δS^0 from S^0 at E^0 into the LUMO and from the HOMO into S^- at E^-. If E_g of *trans*-(CH)$_x$ is known ($E_g \approx 1.5$ eV), we can determine all the energy levels in Fig. 22.2 from the optical transitions δS^0 and $\delta S^{\mp}$, respectively. In particular, U_{eff} (Fig. 22.2) can be directly determined from the relation

$$U_{\mathrm{eff}} = U - \Delta E_r = \delta S^0 - \delta S^{\mp} \tag{1}$$

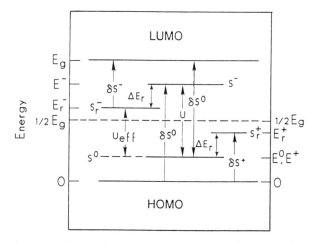

Fig. 22.2 Energy levels and associated optical transitions of charged ($S^{\mp}$) and neutral (S^0) soliton excitations in *t*-(CH)$_x$. The parameters U, U_{eff}, and ΔE_r are defined in the text. (From Ref. 18 with permission.)

Optical Probes of Photoexcitations

Also, ΔE_r can be readily calculated using the relation

$$\delta S^0 + \delta S^{\mp} = E_g + \Delta E_r \qquad (2)$$

Two additional equations can be written for the soliton transitions δS (Fig. 22.2).

$$\delta S^+ + \delta S^- = E_g + \Delta E_r - U_{\text{eff}} \qquad (3)$$

$$2\delta S^0 = E_g + U \qquad (4)$$

It is seen from Eq. (4) that we can determine U from only a single transition (δS^0) if E_g is known and charge conjugation symmetry exists. This transition should be associated with the SOMO level in the gap in the more general case in which S^0 is replaced by a polaron level. We use this relation later to determine U in NDGS polymers from the optical transitions associated with the polaron levels in the gap.

B. Optical Transitions of Charged Excitations in NDGS Polymers

The proper description of charged excitations in NDGS polymers has been the polaron ($P^{\mp}$), which carries spin 1/2, and the spinless bipolaron ($BP^{2\mp}$). However, a third possible excitation has recently gained interest, namely the π dimer, or the interchain BP [32].

1. Polaron

The states in the gap and the associated optical transitions for P^+ are shown in Fig. 22.3a. The two polaron energy states in the gap are the SOMO and LUMO, respectively, separated by $2\omega_0(P)$. Then three optical transitions, P_1–P_3, are possible. In oligomers, the parity of the HOMO, SOMO, LUMO, and LUMO + 1 levels alternate; i.e., they are g, u, g, and u, respectively. Therefore the transition vanishes in the dipole approximation, and the polaron excitation is then characterized by the appearance of two correlated optical transitions below E_g. Even for long chains in the Hückel approximation, transition P_3 is extremely weak, and therefore the existence of two optical transitions upon doping or photogeneration indicate that polarons were created. Unfortunately, polaron transitions have not been calculated for an infinite *correlated* chain. This and a possible disorder-induced relaxation of the optical selection rules cause a slight ambiguity as to the number of optical transitions associated with polarons in "real" polymer films.

2. Bipolaron

The states in the gap and the possible optical transitions for a BP^{2+} are given in Fig. 22.3b. There are now two unoccupied energy states separated by $2\omega_0(BP)$: the LUMO and LUMO + 1, which are deeper in the gap than the corresponding states for P^+. Two optical transitions are then possible: BP_1 and BP_2. In short oligomers,

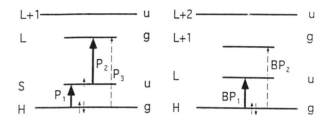

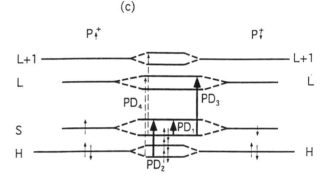

Fig. 22.3 Energy levels and associated optical transitions of (a) positive polaron, P^+; (b) bipolaron, BP^{2+}; and (c) π dimer, PD^{2+}. The full and dashed arrows represent allowed and forbidden optical transitions, respectively. H, S, and L are HOMO, SUMO, and LUMO levels, respectively, and u and g are odd (ungerade) and even (gerade) parity representations, respectively. $2\omega_0(P)$ and $2\omega_0(BP)$ are assigned; note that $\omega_0(P) > \omega_0(BP)$.

again the parity of HOMO, LUMO, and LUMO + 1 alternate (g, u, and g, respectively), and therefore the BP_2 transition vanishes. In this case the BP is characterized by a single transition below E_g. We note that even in the Hückel approximation for an infinite chain, the BP_2 transition is very weak. Electron correlation and disorder-induced relaxation of the optical selection rule, however, may cause the BP_2 transition to gain intensity, and therefore BPs with one strong transition at low energy and a second, weaker transition at higher energy should not be unexpected in "real" films.

The polaron and bipolaron transitions shown in Fig. 22.3 can also be used to calculate the important parameters U, U_{eff}, and ΔE_r defined earlier in Fig. 22.2, using Eqs. (1)–(4). This is possible because the polaron's SOMO level is singly occupied and is therefore equivalent to S^0 in Fig. 22.2, whereas BP levels are either unoccupied (BP^{2+}) or doubly occupied (BP^{2-}) and therefore are equivalent to the states S^+ and S^-, respectively in Fig. 22.2. However, since there are two levels in the gap for $P^{\mp}$ and $BP^{2\mp}$ excitations, we have to take into account their respective separation $2\omega_0(P)$ and $2\omega_0(BP)$ (Figs. 22.3a and 22.3f). The SOMO level of P^- is pushed up and that of P^+ is pushed down by $\omega_0(P)$, and the HOMO level of the BP^{2-} is pushed up and the LUMO level of BP^{2+} down by $\omega_0(BP)$. Then $P_3 - \omega_0(P)$ is equivalent to transition δS^0 in Fig. 22.2, and similarly $BP_1 + \omega_0(BP)$ is equivalent to transition δS^- in Fig. 22.2. Now within the model presented in Section II.A we find

$$U_{\text{eff}} = P_1 - BP_1 + \Delta\omega_0 \quad (5)$$

$$P_3 + BP_1 = E_g + \Delta E_r + \Delta\omega_0 \quad (6)$$

where ΔE_r is the relaxation energy of the BP with respect to the polaron and $\Delta\omega_0 = \omega_0(P) - \omega_0(BP)$. Equations (5) and (6) are equivalent to Eqs. (1) and (2), respectively, but for NDGS polymers. Another useful relation follows from Eq. (4):

$$2P_1 + P_2 = E_g + U \quad (7)$$

In Eqs. (5)–(7) we used $P_3 = P_1 + P_2$.

3. π Dimer

When two polarons come together on the same chain, then theory (Hückel approximation) predicts that they are unstable toward the formation of a bipolaron:

$$P^{\mp} + P^{\mp} \rightarrow BP^{2\mp} \quad (8)$$

Whether this model is true in "real" polymers is still an open question. Nevertheless, BPs must appear in conducting polymers upon heavily doping, regardless of whether or not Eq. (8) is exothermic. When heavily doped, however, another type of charge excitation can be formed: an *extrinsic* bipolaron (as opposed to an intrinsic BP), or π dimer ($PD^{2\mp}$), in which two polarons at *different* chains are coupled. The PD excitations are spinless (Fig. 22.3c), and therefore their formation has been recently proposed to explain the dramatic decrease in unpaired spins observed in FeCl$_3$-doped α-sexithiophene at low temperatures [32]. Then the possibility that PD charge excitations can be photogenerated cannot be discarded.

The energy levels of a PD^{2+}, which are formed from the energy levels of two coupled P^+ polarons, are shown in Fig. 22.3c. As is clearly seen, each P^+ level is split into two levels due to the coupling, where the SOMO splitting is related to the charge transfer overlap integral t_{CT}. There are now four possible transitions, PD_1–PD_4, but transition PD_4 is strictly forbidden in oligomers (similar to transition P_3 of polarons in oligomers). Again we expect this transition to be weak in real polymers. We therefore conclude that $PD^{2\mp}$ is characterized by three strong transitions: the charge transfer transition PD_1 at the lowest energy and two other transitions, PD_2 and PD_3, that are *blue-shifted* with respect to the two equivalent polaron transitions P_1 and P_2.

C. Optical Transitions of Neutral Excitations in NDGS Polymers

Upon excitation, a bound e–h pair or an exciton (X) is immediately generated. By definition the exciton is a neutral, spinless excitation of the polymer. Following photogeneration the exciton may undergo several processes: It may recombine radiatively by emitting light in the form of fluorescence (FL), which is the light source of LED devices. It may also recombine nonradiatively through recombination centers by emitting phonons. Excitons may be also trapped (X_t), either by a self-trapping process undergoing energy relaxation (this can be envisioned as a local ring rotation, etc.) or by a trapping process at defect centers. Excitons may also undergo an intersystem crossing into the triplet manifold, creating a long-lived triplet (T) state. Finally, an exciton may disassociate into a polaron pair (PP) either on the same chain (but on two different segments) or on two different chains. Since we are interested in long-lived excitations in this chapter, we deal here only with trapped singlet excitons (X_t), triplets (T), and polaron pairs (PP); their energy levels and possible optical transitions are shown in Fig. 22.4. For X_t and T we adopt here the correlated picture in which the notations for different many-body exciton levels, which follow the group theory representations, are A_g and B_u, respectively.

1. Singlet Excitons

Two important singlet exciton levels ($1B_u$ and mA_g) and a double excitation level (BX) are shown in Fig. 22.4a [36]; their electron configurations are also shown for clarity. We consider mA_g to be an excited state above the $1B_u$ level, whereas the BX level is a biexciton state, i.e., a bound state of two $1B_u$ excitons. The mA_g level is known to have strong dipole moment coupling to the $1B_u$, as deduced from the various optical nonlinear spectra of conducting polymers analyzed in terms of the "four essential states" model [37]. We therefore expect two strong optical transitions to form following the $1B_u$ photogeneration: X_1 and X_2, as shown in Fig. 22.4a. Due to exciton self-trapping, however, we do not know whether X_1 would maintain its strength, since the $1B_u$ relaxed state may no longer overlap well with the mA_g state. X_2, on the other hand, will be always strong regardless of the $1B_u$ relaxation, since there is always room for a second exciton photogeneration following the formation of the first one. We also note that the BX binding energy composed of X_t, being mostly Coulombic in nature, would not differ much from that composed of free X.

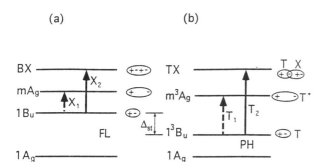

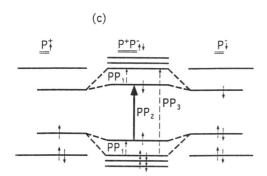

Fig. 22.4 Energy levels, optical transitions, and emission bands associated with (a) singlet and (b) triplet excitons and (c) a polaron pair. The symbols $1A_g$, $1B_u$, mA_g, and BX are the ground state, lowest allowed, and forbidden excitons, and biexciton, respectively, and T is for triplet. Full and dashed arrows are for allowed and forbidden transitions, respectively, and wiggly arrows are for emission bands. FL is fluorescence and PH is phosphorescence.

From two-photon absorption and electroabsorption spectra in PT and PPV polymers we know that mA_g is about 0.7 eV above $1B_u$ [36,38,39]. We therefore expect the X_1 transition to be in the mid-IR spectral range, at about 0.7 eV. The BX level, on the other hand, has not as yet been directly identified in conducting polymers. We may, however, estimate the biexciton binding energy $E_b(BX)$, from simple considerations, based on the Coulomb repulsive interaction and the binding energy $E_b(X)$ of a single exciton [37]:

$$E_b(BX) \simeq E_b(X) - 2e^2/2R \qquad (9)$$

where R is the extension of the exciton wave function. Equation (9) is based on the BX electronic configuration shown in Fig. 22.4a. The BX actually contains three possible e–h pairs, one e–e pair, and one h–h pair. The Coulomb binding energy of the BX can be therefore estimated to be $3E_b(X)$ due to the e–h pairs, minus the Coulomb repulsion associated with the two like pairs. Then the BX binding energy in excess of two separated excitons leads to Eq. (9). We estimate the repulsive energy in Eq. (9) to be on the order of 0.4 eV, or about $(1/2)E_b(X)$. We therefore expect transition X_2 to be lower than the fundamental transition $(1A_g \rightarrow 1B_u)$ by about $(1/2)E_b(X)$; this would place X_2 in the near-IR spectral range for the polymers studied in this chapter.

2. Triplet Excitons

The most important electronic states in the triplet manifold are shown in Fig. 22.4b. The lowest triplet level is 1^3B_u, which is lower than $1B_u$ by the singlet–triplet energy splitting Δ_{ST}. In principle, 1^3B_u can directly recombine to the ground state by emitting photons (leading to phosphorescence, PH) or phonons. But this transition is dipole-forbidden and therefore extremely weak, leading to the well-known long triplet lifetime. The other two levels shown in Fig. 22.4 are the m^3A_g level, which is equivalent to mA_g in the singlet manifold, and the TX level, which is a complex composed of a triplet exciton and a singlet exciton bound together; their electronic configuration is also shown in Fig. 22.4b for clarity.

As in the case of singlet excitons, we expect for triplets two strong transitions T_1 and T_2 (Fig. 22.4b). T_1 is into the m^3A_g level, and from Figs. 22.4a and 22.4b it is clear that is possible to estimate Δ_{ST} from the relation

$$\Delta_{ST} = T_1 - X_1 \qquad (10)$$

since the m^3A_g and mA_g levels should not be far from each other. Δ_{ST} has never been directly measured before in conducting polymers, due to lack of phosphorescence, and therefore Eq. (10) is quite important. Unfortunately, we do not know whether T_1 is indeed strong, since, as for singlet excitons, the relaxed triplet may not well overlap with the m^3A_g state, leading to a decrease in T_1 intensity. In contrast, it is quite certain that transition T_2 into the TX level is strong, because it is always possible to photogenerate a second (singlet) exciton close to a previously formed triplet exciton. Then, following the same arguments that have led before to estimate X_2 transition energy for singlet excitons [Eq. (9)], we expect T_2 to be quite close to X_2 but on the lower energy side. This is true because the TX state also contains three e–h pairs and two pairs of like charges, which are slightly tighter than the pairs in the BX state discussed previously.

3. Polaron Pairs

A polaron pair (PP) [24] is a bound pair of two oppositely charged polarons, P^+ and P^-, formed on two adjacent chains. In this respect it is similar to the π dimer discussed previously except that the binding energy is mainly Coulombic in the case of PP, in contrast to lattice relaxation for PD. The stronger overlap leads to larger splitting of the P^+ and P^- levels compared to those in PD, as shown in Fig. 22.4c. Following the same arguments as those given before for polaron transitions, we expect three strong transitions, PP_1–PP_3. For a loosely bound PP these transitions are not far from transitions P_1–P_3 of polarons. However, for a tightly bound PP we expect a *single* transition, PP_2, to dominate the spectrum, as PP_1 is considered to be intraband with traditional low intensity and PP_3 is close to the fundamental transition and therefore difficult to observe. In this case we have mainly two states in the gap, and the excitation is also known as a neutral BP (BP^0) or a polaronic exciton. We note, however, that the PP_2 transition is close in spirit to transition X_2 discussed above for excitons, as a second electron is also promoted to the excited level in the case of PP. Then from the experimental point of view, it is not easy to identify and separate in PM spectra the transitions of a trapped exciton (X_t) from those of a tightly bound PP (or BP^0). They may differ, however, in their ADMR spectra, as discussed below in Sections III.B.2 and III.B.4.

III. EXPERIMENTAL TECHNIQUES

A. Photomodulation

1. Experiment

Photomodulation [40] uses two beams, one (pump) for producing the excited states and the other (probe) for measuring the correlated changes ΔT in the transmission T. The experimental setup for measuring ΔT in the spectral range 0.1–2.5 eV is shown in Fig. 22.5. The pump beam is an argon ion laser (sometimes used to pump a dye laser for excitation spectra measurements) modulated at a frequency f of about 150 Hz. In a typical experiment the absorbed flux is about 2×10^{18} photons/(cm^2·s) at $\hbar\omega = 2.41$ eV. The transmission T and its photoinduced changes ΔT are measured using a tungsten lamp or a Perkin-Elmer Opperman source as probe beams. An infrared monochromator and several semiconductor detectors are used to span the entire energy range with optimal sensitivity. The sample temperature can be varied from 10 to 300 K. At low temperatures, the photoluminescence (PL) signal is inevitably collected with the ΔT signal; we measure the PL separately by blocking the probe beam, and in the correction of the data we assume simple superposition of ΔT and PL. The data are plotted as $-\Delta T/T = d\,\Delta\alpha$, where $\Delta\alpha$ is the change in the absorption coefficient α and d is the sample thickness (assuming $d > 1/\alpha$).

2. General Relations

Photoinduced changes in transmission ΔT are of two kinds, photoinduced absorption (PA) and photoinduced bleaching (PB). Photoinduced absorption is associated with transitions from initial states when their occupation by electrons is enhanced by the illumination or, alternatively, with transitions whose final states are emptied by the illumination. Photoinduced bleaching is associated with initial states occupied in the dark that the illumination empties or with final states filled by the illumination. We will see that because of this variety of

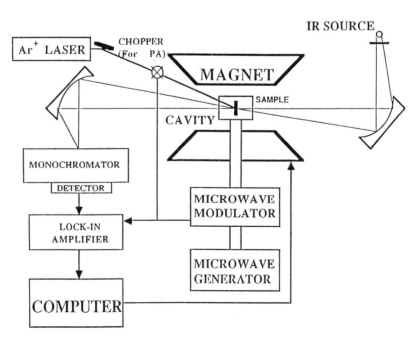

Fig. 22.5 Schematic of the PM and ADMR experimental setup. (From Ref. 78, with permission.)

photoinduced changes, the PM spectra often contain more spectroscopic information than the fundamental optical spectra (although for the interpretation of the PM spectra, knowledge of the fundamental absorption spectra seems necessary).

Since the illumination does not change the total density of electrons, it follows from the sum rule for α that [18]

$$\int \Delta\alpha(\omega)d\omega = 0 \qquad (11)$$

This equation shows that if the PM spectrum contains spectral intervals in which $\Delta\alpha > 0$ (PA), it must also contain intervals in which $\Delta\alpha < 0$ (PB). Both PA and PB are the fundamental features of any PM spectrum.

An important example that demonstrates the potential of PM spectroscopy is shown in Fig. 22.6 for a *trans*-$(CH)_x$ film ($d \simeq 1000$ Å) kept at 210 K [18]. Due to the 1D character of $(CH)_x$, the photoinduced soliton bands are sharply defined. Two well-defined bands, one PA band and one PB band, are seen, and the sum rule for $\Delta\alpha$ [Eq. (11)] is approximately obeyed. This also shows that the two bands share a common origin. The additional modulation around the PB peak is probably caused by vibronic side bands, and we identify the shoulder at 1.4 eV as the zero-phonon transition. Associated with the PA band ($\delta S^{\mp}$), which peaks at 0.5 eV at this temperature, is a narrow photoinduced IR-active vibration at 0.17 eV [41,42], which shows that the PA band is due to photoinduced charged defects ($S^{\mp}$). A sharp feature in the PB band that could be associated with the bleaching of an IR-active vibration does not exist, and therefore the PB band is due to the bleaching of the transition of a neutral state. Charge conjugation symmetry is evident from the PM spectrum of Fig. 22.6, since we do not observe separate PA bands arising from S^+ and S^-. We will come back later to this PM spectrum for further discussions.

A useful relation for the analysis of PM spectra is

$$-\frac{\Delta T}{T} = \Delta\alpha d = N_{SS}\sigma d \qquad (12)$$

where $\Delta\alpha$ is the change in the absorption coefficient α, d is the sample thickness, N_{SS} is the steady-state photoexcitation density, and σ is their optical cross section. For charged excitations in conducting polymers, σ was determined to be on the order of 10^{-16} cm^2 [41,42]. From this and Eq. (12) we can, for example, easily estimate N_{SS} in the PM spectrum of *trans*-$(CH)_x$ (Fig. 22.6). We get $N_{SS} \simeq 10^{17}$ cm^{-3}, much smaller than the density of free radicals $N_r \simeq 4 \times 10^{19}$ cm^{-3} in typical films of *trans*-$(CH)_x$. This fact is of paramount importance in the analysis of PM spectra in polyacetylene.

3. Recombination Kinetics

The steady-state photoexcitation density, N_{SS}, is determined by a rate equation that contains generation (G) and recombination (R) rates. In our studies we have frequently dealt with two types of recombination processes. One is linear or "monomolecular," and the other is nonlinear or "bimolecular." The rate equation for the monomolecular recombination process is

$$\frac{dN}{dt} = \eta G - RN \qquad (13)$$

where N is the photoexcitation density and η is the quantum efficiency (QE) for their generation ($\eta \leq 1$). The steady-state solution (N_{SS}) of Eq. (13) can be readily obtained by equating the right-hand side to zero. We then get $N_{SS} = \eta G/R$. The transient solution of Eq. (13), when G is turned off, is $N(t) = N(0)\exp(-t/\tau)$, where $N(0)$ is the photoexcitation density at $t = 0$ and the lifetime $\tau = 1/R$. From this we can express N_{SS} through τ, $N_{SS} = \eta G\tau$, showing that N_{SS} increases linearly with both the excitation intensity $I_L(\approx G)$ and the photoexcitation lifetime τ.

In many cases, in using the PM technique we do not measure N_{SS} because the laser modulation frequency f is too high and steady-state conditions are not reached. Then the PA Fourier component measured at the lock-in amplifier is proportional to $N(\omega)$, where $N(\omega)$ is the solution of Eq. (13) for $G(t) = G_0[1 + \cos(\omega t)]$, and $\omega = 2\pi f$. The in-phase component [$N_i(\omega)$] and out-of-phase component [$N_o(\omega)$] of $N(\omega)$ are

$$N_i(\omega) = \frac{N_{SS}}{1 + \omega^2\tau^2}, \qquad N_o(\omega) = \frac{N_{SS}\omega\tau}{1 + \omega^2\tau^2} \qquad (14)$$

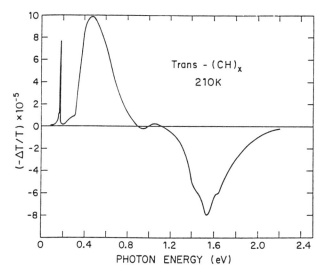

Fig. 22.6 Photomodulation spectrum of *trans*-$(CH)_x$ at 210 K showing the photoinduced absorption (PA) and photoinduced bleaching (PB) bands, respectively, associated with $S^{\mp}$ excitations. (From Ref. 18, with permission.)

It is seen from Eq. (14) that at very low f ($\omega \to 0$), $N_i(\omega)$ approaches N_{SS}, whereas $N_o(\omega)$ vanishes. The absolute density $N(\omega)$ can be readily calculated from Eq. (14), giving $N(\omega) = N_{SS}(1 + \omega^2\tau^2)^{-1/2}$; $N(\omega)$ decreases with ω at high ω as ω^{-1}. If, on the other hand, a distribution $g(\tau)$ of carrier lifetimes is assumed, then $N(\omega)$ would decrease with ω as a power law $N(\omega) \approx \omega^{-p}$, where $p \leq 1$ [34].

The solutions for the bimolecular case are much more complicated and are only briefly mentioned here. The bimolecular rate equation is

$$\frac{dN}{dt} = \eta G - bN^2 \quad (15)$$

where b is the bimolecular recombination rate. The steady-state solution of Eq. (15) is $N_{SS} = (\eta G/R)^{1/2}$, which shows that PA depends sublinearly on I_L for the bimolecular case. The transient solution of Eq. (15), when G is turned off, is $N(t) = N(0)(1 + t/\tau)^{-1}$, where the lifetime $\tau = [bN(0)]^{-1}$. This demonstrates that τ depends inversely on the laser intensity [through $N(0)$] and therefore cannot be precisely defined. If steady-state conditions are not attained, then PA is proportional to $N(\omega)$. To get $N(\omega)$ we have to solve Eq. (15) with $G(t) = G_0[1 + \cos(\omega t)]$; the solutions, however, are quite complicated and are given in Ref. 29.

4. Related Experiments

One additional technique, which is directly related to PM, is the PA excitation (PAE) spectrum [36]. In this technique the wavelength of the excitation source is continuously changed while the intensity of a particular PA band in the PM spectrum is monitored. This is readily done using a powerful xenon lamp coupled to a monochromator equipped with several long- and short-pass filter combinations. The PAE spectra are usually used for studying the excited electronic states of the polymer samples rather than the photoexcitations themselves [36]. They can yield information, for example, on the location of the lowest allowed exciton ($1B_u$), the continuum threshold, and other important electronic states above it. This complementary information is useful in relation to PL and its excitation spectrum.

Another related technique is electroabsorption (EA) spectroscopy [36]. In this method the polymer absorption spectrum $\alpha(\lambda)$ is modulated by applying an external electric field F. The changes in α, $\Delta\alpha(\lambda)$, are then measured by phase-sensitive techniques. The EA spectrum can sometimes be observed in the PM spectrum itself, for probe photon energies $\hbar\omega > E_g$ [43]. This happens because in PM the long-lived photogenerated carriers create strong local electric fields, which in turn give rise to photoinduced electroabsorption. Prior knowledge of the polymer EA spectrum may therefore assist in resolving the EA features in the PM spectrum.

One can use the PM technique with pulse excitations and study the time evolution of the spectra or the decay of the total oscillator strength of the bands. In this way one can obtain information about electronic relaxation processes. However, in this chapter we restrict ourselves to steady-state PM spectroscopy. We also note that instead of using illumination, the state occupation can be changed by changing the bias of a junction, as used in deep-level transient spectroscopy junction capacitance techniques. This method, dubbed charge-induced absorption, or CIA [44], has already been used in conducting polymers and is discussed in another chapter of this book.

B. Absorption-Detected Magnetic Resonance

During the past three decades, there have been many successful applications of optical detection of magnetic resonance (ODMR). The main idea in most of these applications is that the population redistribution among magnetic substates, in passage through magnetic resonance in the ground or excited state of a paramagnetic center, produces a change in some aspect of either emitted or absorbed light associated with the center. The change in light intensity, in turn, is used as a detector of magnetic resonance, replacing direct observation of microwave power absorption by the paramagnetic species. The scaling up of the ESR detection from the microwave to the optical region makes the ODMR methods extremely sensitive, an improvement of up to 10^5 times over conventional ESR.

Among the optical techniques involved in ODMR, detection of photoluminescence changes (PL-ODMR) has been the most widely used. However, there are disadvantages in PL-ODMR: (1) It cannot be applied to systems in which the PL is weak or absent, and (2) it provides only limited spectral information. Using PA to detect the magnetic resonance would remedy the above problems, and this technique has been dubbed absorption-detected magnetic resonance (ADMR) [22]. Although the idea was proposed long ago, the large amount of noise introduced by the fluctuations in the probe light intensity has hampered the use of the ADMR technique. The earliest measurements were described in 1972 by Clarke and Hayes [45], who observed microwave-induced changes in the intensity of triplet-triplet absorption in quinoline. Hirabayashi and Morigaki [46] applied the ADMR technique to films of hydrogenated amorphous silicon in 1983; they observed that the PA intensity decreases at resonance of trapped hole centers and dangling bond centers. In 1986, Robins et al. [7] observed the triplet excited state in single crystals of poly(diacetylene toluene sulfonate), using photoinduced reflection to detect the magnetic resonance. Improvement in the probe light stability should enable the ADMR technique to be used more often in the future.

1. Experimental

The experimental setup for the ADMR technique is illustrated in Fig. 22.5. It consists of a regular PM apparatus, described before, and a magnetic resonance part. The sample is mounted in a microwave cavity equipped with windows for optical transmission, between the poles of a superconducting dc magnet. The cavity is in a liquid helium cooled cryostat. As in photomodulation, the sample is illuminated by both pump and probe beams. The pump beam produces photoexcitations in the sample film. After being modulated, the moderately strong microwaves (μ waves) at 3 GHz are introduced into the cavity through a waveguide. Then the changes (δT) in transmission (T) of the probe beam associated with the microwave-induced changes in the photoexcitation recombination kinetics in the dc magnetic field are detected with the lock-in amplifier, using a phase-sensitive technique. The changes δT are proportional [via Eq. (12)] to δN, the μ-wave-induced change in the photoexcitation density N, produced by the pump beam. δN, in turn, is induced by transitions in the μ-wave range that change spin-dependent recombination rates. Two types of ADMR spectra are usually obtained: the H-ADMR spectrum in which δT is measured at a fixed probe wavelength λ while sweeping H, and the λ-ADMR spectrum, in which δT is measured at a constant H, in resonance, while λ(probe) is changed. With suitable signal averaging, the system $\delta T/T$ sensitivity is 3×10^{-8} in the visible to near-IR range and 3×10^{-7} in the mid-IR spectral range.

The ADMR spectrometer can also be used without the probe beam to detect changes in PL (δI) induced by μ-wave transitions, or PL-ODMR, a technique complementary to the ADMR technique. The sensitivity $\delta I/I$ of PL-ODMR using our ODMR apparatus, where I is the PL intensity, is 10^{-5} for PL in the visible range.

ADMR is an excited-state dynamic technique that is not based on spin states in thermal equilibrium. In this respect it is not similar to ESR, whose signal follows the Boltzmann statistics. In ADMR we must actively provide spin polarization, or different populations of spin sublevels of the photoexcitations. This can be achieved during either the generation process or recombination or both. We discuss spin 1/2, spin 1, and spin 1/2 correlated pairs separately, all under the assumption that the spin-lattice relaxation time is much longer than the photoexcitation lifetime. This is usually the case at low temperatures.

2. Spin 1/2 ADMR

Two spin 1/2 particles produce pairs with spins either parallel (P) or antiparallel (AP) to each other. In the case of a "geminate" pair, the photoexcited negative and positive particles are correlated following photon absorption; hence their spins are in an AP configuration.

The reason for that is that the ground state is a spin singlet and the photon absorption process conserves spin. Spin polarization is achieved, therefore, by generation: μ-Wave absorption flips the spin of one carrier, reducing the recombination rate of the pair to the ground state. Consequently, $\delta N > 0$ and the H-ADMR may contain two δN bands if the respective g values for the negative and positive carriers are sufficiently different from each other. In the case of an uncorrelated carrier, the so-called distant pair model, there are pairs with P or AP spin configuration, which are initially generated with equal probability. With time, the pairs with P configuration prevail, since their rate of recombination to the ground state is smaller. Spin polarization is therefore achieved by recombination, and any μ-wave-induced spin-flip increases recombination and consequently $\delta N < 0$. Again, two δN bands can be detected in resonance at magnetic fields H close to $H_0 \simeq 1070$ G, which is the resonant H related to species with $g = 2$ and spin 1/2, using μ waves at ≈ 3 GHz ($H_0 = h\nu/g\beta$).

These general ideas can be quantitatively formulated as follows. Consider P and AP pairs with respective generation rates G_P and G_{AP} and recombination rates R_P and R_A. The respective steady-state pair densities N_P and N_{AP}, when the μ-waves are turned off, are given by Eq. (13): $N_P = G_P/R_P$ and $N_{AP} = G_{AP}/R_{AP}$. Under μ-wave absorption saturation, the new P and AP pair densities $\tilde{N}$ are equal:

$$\tilde{N}_P = \tilde{N}_{AP} = \frac{G_1 + G_2}{R_1 + R_2} = \frac{\Sigma G}{\Sigma R}$$

Then the change δN in the total pair density N is

$$\delta N = \tilde{N} - N = (\tilde{N}_P + \tilde{N}_{AP}) - (N_P + N_{AP}) \tag{16}$$

The figure of merit in ADMR is $\delta N/N$, which can be readily calculated from the above relations:

$$\frac{\delta N}{N} = \left(\frac{\Delta N}{N}\right)\left(\frac{\Delta R}{\Sigma R}\right) \tag{17}$$

where $\Delta N = N_{AP} - N_P$ and $\Delta R = R_{AP} - R_P$. We note that whereas ΔR is always positive (since $R_P < R_{AP}$), ΔN can be either positive or negative, causing δN to be of variable sign. Again we deal with two cases.

1. *Geminate Recombination.* In this case $G_P = 0$, since the spin pairs are all generated in the AP configuration. Then $N_P = 0$, $\Delta N = N = N_{AP}$, and from Eq. (17) we get $\delta N/N = \Delta R/\Sigma R > 0$. We conclude therefore that $\delta N > 0$ for geminate pairs. We note that whereas regular ESR (in thermal equilibrium) may give a very small signal at intermediate temperatures, ADMR may be quite large and can easily exceed 5%, which is the largest $\delta N/N$ that we have ever observed in conducting polymers (see Section IV.B).

2. *Distant Pair Recombination.* In this case $G_P = G_{AP}$ and $\Delta N/N = \Delta R/\Sigma R$. Then from Eq. (17) we get $\delta N/N = -(\Delta R/\Sigma R)^2$. This shows that $\delta N < 0$ and, in general, is smaller than in the geminate pair case.

3. Spin 1 ADMR

a. General Relations

The spin Hamiltonian, $\mathcal{H}$, for spin 1 (triplet) states is more complex than that of doublets owing to the spin–spin interaction [47–49]:

$$\mathcal{H} = \mathcal{H}_0 + g\beta \overline{H} \cdot \overline{S} + D\left(S_z^2 - \frac{1}{3}S^2\right) + E(S_x^2 - S_y^2) \quad (18)$$

where $\mathcal{H}_0$ denotes the atomic Hamiltonian, $\overline{S} = \overline{S}_1 + \overline{S}_2$, where S_1 and S_2 are the spins of the two spin = 1/2 particles, $g\beta \overline{H} \cdot \overline{S}$ is the magnetic Zeeman splitting term, and D and E are the zero field splitting (ZFS) parameters. The triplet spin states are hence split into three sublevels (denoted τ_x, τ_y, and τ_z) even in the absence of $\overline{H}$. The three respective energies E_x, E_y, and E_z can be readily found by solving Eq. (18) for $\overline{H} = 0$. They are related to the ZFS parameters as follows:

$$E_x = \frac{1}{3}D + E; \quad E_y = \frac{1}{3}D - E; \quad E_z = \frac{-2}{3}D \quad (19)$$

The ZFS parameters D and E can be estimated from the triplet wave function Ψ_τ, using the single-particle operators $x_{1,2} = x_1 - x_2$, $y_{1,2} = y_1 - y_2$, $z_{1,2} = z_1 - z_2$, and $R_{1,2}^2 = (x_{1,2})^2 + (y_{1,2})^2 + (z_{1,2})^2$, as follows:

$$D = \frac{3g^2\beta}{4}\left\langle \Psi_\tau \left| \frac{r_{12}^2 - 3z_{12}^2}{r_{12}^5} \right| \Psi_\tau \right\rangle \quad (20)$$

$$E = \frac{3g^2\beta}{4}\left\langle \Psi_\tau \left| \frac{y_{12}^2 - x_{12}^2}{r_{12}^5} \right| \Psi_\tau \right\rangle$$

A useful approximation for the triplet wave function extent, $R_{1,2}$, can then be derived [47]:

$$(R_{1,2})^3 = 2.78 \times 10^4/D \quad (21)$$

where $R_{1,2}$ is measured in angstroms and D is in gauss.

The three resonant magnetic fields $H_{0,i}(i = 1-3)$ for the triplet satisfy the relation

$$\frac{2D\left(E^2 - \frac{D^2}{9}\right) \pm \frac{1}{\sqrt{3}}\left[h^2\nu^2 - \frac{D^2}{3} - E^2 - (g\beta H_0)^2\right] \times [4(g\beta H_0)^2 - h^2\nu^2 + 4(D^2/3 + E^2)]^{1/2}}{(g\beta H_0)^2}$$

$$= D(1 - 3\cos^2\theta) + 3E\sin^2\theta \cos 2\phi \quad (22)$$

where θ and ϕ are the Euler angles of the principal axis for the triplet's D tensor with respect to $\overline{H}$ and $h\nu/g\beta = H_0$. Normally, there are three solutions of H_0 in Eq. (22), corresponding to two $\Delta m_S = 1$ (full field) transitions and one $\Delta m_S = 2$ (half-field) transition, and they depend strongly on θ and ϕ, i.e., $H_0 = H(\theta, \phi)$. For a polycrystalline or amorphous sample, the D tensors are randomly oriented with respect to the applied field. Thus the magnetic resonance spectrum, referred to as a triplet powder pattern, is an average over the resonance conditions for all possible orientations:

$$S(H) = \frac{1}{4\pi}\int_0^{2\pi} d\phi \int_0^\pi \sin\theta \, d\theta \, P(\theta, \phi)\delta[H - H(\theta, \phi)] \quad (23)$$

where $S(H)$ and $P(\theta, \phi)$ denote the magnetic resonant absorption intensity at field H and the transition probability, respectively, and $\delta[H - H(\theta, \phi)]$ is a delta function that is usually replaced by a normalized Gaussian function (or a Lorentzian distribution function) in practical calculations. Under thermal equilibrium conditions (Boltzmann distribution), also called thermalized spin distribution, shoulder and divergence singularities exist in the triplet powder pattern, as given below. For full-field resonances, the powder pattern consists of

Steps at

$$H_{1,2} \cong \frac{1}{g\beta}(h\nu \pm D) \quad (24a)$$

Shoulders at

$$H_{3,4} \cong \frac{1}{g\beta}\left(h\nu \pm \frac{D + 3E}{2}\right) \quad (24b)$$

Divergences at

$$H_{5,6} \cong \frac{1}{g\beta}\left(h\nu \pm \frac{D - 3E}{2}\right) \quad (24c)$$

and the full-field resonance is distributed between H_1 and H_2 defined above.

For half-field resonances, the powder pattern of thermalized spins consists of

A divergence at

$$H_{\min} = \frac{1}{2g\beta}\left[(h\nu)^2 - \frac{4}{3}(D^2 + 3E^2)\right]^{1/2} \quad (25a)$$

A divergence at

$$H_A = \frac{1}{2g\beta}[(h\nu)^2 - (D + E)^2]^{1/2} \quad (25b)$$

A shoulder at

$$H_B = \frac{1}{2g\beta}[(h\nu)^2 - (D - E)^2]^{1/2} \quad (25c)$$

A step at

$$H_{\max} = \frac{1}{2g\beta}[(h\nu)^2 - 4E^2]^{1/2} \quad (25d)$$

and the half-field resonance is distributed between $H_{\min}$ and $H_{\max}$ defined above, where both $H < (1/2)H_0$.

b. Triplet Spin Polarization

As in the case of spin 1/2, spin polarization among the three triplet sublevels at $\bar{H} \neq 0$—(+), (0), and (−), respectively—is achieved by generation and/or recombination kinetics. The respective generation rates G_+, G_0, and G_- and recombination rates R_+, R_0, and R_- can be calculated from the generation and recombination rates of the three triplet states at $\bar{H} = 0$—G_x, G_y, and G_y, and R_x, R_y, and R_z, respectively—using the transformation coefficients $c_{\gamma,i}$ defined as follows:

$$|\tau_\gamma(\theta, \phi, H)\rangle = \sum_i c_{\gamma,i}(\theta, \phi, H)|\tau_i\rangle \quad (26)$$

where $\gamma = +, 0, -$; $i = x, y, z$; θ, ϕ are Euler angles of the magnetic field $\bar{H}$ with respect to the principal axes; and $c_{\gamma,i}(\theta, \phi, H)$ are coefficients. We propose that

$$R_\gamma(\theta, \phi, H) = \sum_i |c_{\gamma,i}(\theta, \phi, H)|^2 R_i \quad (27)$$

and

$$G_\gamma(\theta, \phi, H) = \sum_i |c_{\gamma,i}(\theta, \phi, H)|^2 G_i \quad (28)$$

Using equations similar to those determining $\delta N/N$ for spin 1/2 [Eq. (17)], we get the changes δN in the triplet population for each triplet resonance $\Delta \gamma$, between γ sublevels, as a function of θ, ϕ, and H:

$$\frac{\delta N}{N}(\theta, \phi, H) = \left(\frac{\Delta N}{N}\right)\left(\frac{\Delta R}{\Sigma R}\right) \quad (29)$$

Here $\Delta \gamma$ is a double index that stands for full-field $\Delta \gamma (+, 0)$ and $(0, -)$ and half-field $(+, -)$ resonances, respectively; ΔN is their respective steady-state population difference ($\Delta N_{+,0} = N_+ - N_0$, etc.) for μ wave off; and ΔR is the difference in the respective recombination rates ($\Delta R_{+,0} = R_+ - R_0$, etc.). From Eq. (29) we see that δN can be of either sign or even zero, depending on ΔN and ΔR. Due to the above complexity of unthermalized spin polarization, LESR, ADMR, and PL-ODMR triplet powder patterns are much more complicated than their ground-state (thermalized) ESR counterparts. Enhancing and reduction features and even derivative-like line shapes may exist in ODMR triplet powder patterns.

A good example of such complex powder patterns in PL-ODMR and ADMR is shown in Figs. 22.7a and 22.7b, respectively, for the triplet state in poly(3-octylthiophene) (P3OT), a derivative of polythiophene (see

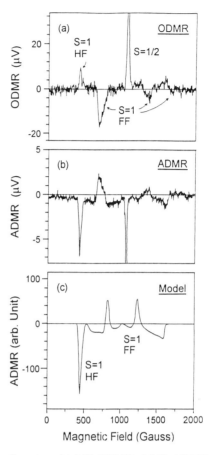

Fig. 22.7 Spectra of (a) H-ODMR, (b) H-ADMR, and (c) its model simulation, of a thin film of poly(3-octylthiophene) at 4 K. HF and FF are "half-field" and "full-field" triplet powder patterns, respectively.

later). The H-ADMR at 1.4 eV (Fig. 22.7b) shows a sharp $s = 1/2$ resonance and $s = 1$ full-field and half-field powder pattern resonances, respectively. We note, however, that whereas the half-field resonance at 410 G is negative ($\delta N < 0$), two full-field resonances at 700 and 1400 G are positive and two resonances at 600 and 1500 G are negative. The same is true in the PL-ODMR spectrum (Fig. 22.7a), which is almost a mirror image of the H-ADMR spectrum. We also show our successful computer simulation of the H-ADMR spectrum in Fig. 22.7c. This was achieved with the following fitting parameters: ZFS parameters $D = 500$ G (0.055 cm^{-1}) and $E = 75$ G (0.008 cm^{-1}), and zero-field relative recombination and generation rates: $(G_x; G_y; G_z = 0.8; 0.3; 0.5)$ and $(R_x; R_y; R_z) = (0.7; 0.3; 0.4)$. From Eq. (21) we can now estimate the triplet wave function extent $R_{1,2}$. We get $R_{1,2} \simeq 3.7$ Å, which shows that the triplet in P3OT is highly localized (molecular Frenkel exciton).

It is still instructive to find the steps and divergencies of Eq. (24) in Fig. 22.7. We find that the full-field powder pattern extends from $H_2 = 550$ G to $H_2 = 1650$ G, consistent with $H_1 - H_2 = 2D \simeq 1060$ G obtained in our model. We also find that the main divergency (positive peak in Fig. 22.7b), which occurs at about 700 G, is consistent with a singularity at $H_4 = H_0 - (D + 3E)/2$, where $H_0 = 1070$ G, $D = 500$ G, and $E = 80$ G, in agreement with the parameters extracted from our model. We also identify the other divergency at 1400 G as H_3 and note that $H_3 - H_4 = D + 3E$ [Eq. (24)] is consistent with $D = 500$ G and $E = 70$ G, also in good agreement with our model.

4. Exchange-Correlated Spin 1/2 Pairs

For the spin 1/2 case discussed in Section III.B.2, the spin 1/2 P and AP pairs were thought to be loosely bound. However, because of sample morphology and/or additional gain in relaxation energy, spin pairs can also be tightly bound, as in π dimers (PD) or polaron pairs (PP) discussed in Section II. In such a case an exchange interaction term [49] is usually added to the Zeeman Hamiltonian:

$$\mathcal{H} = \beta \overline{H} \cdot \overline{\overline{g}} \cdot (\overline{S}_1 + \overline{S}_2) + \overline{S}_1 \cdot \overline{\overline{J}} \cdot \overline{S}_2 \quad (30)$$

where $\overline{\overline{g}}$ and $\overline{\overline{J}}$ are the g tensor and exchange-coupling tensor, respectively, and $\overline{S}_1$ and $\overline{S}_2$ are spin operators for the spin 1/2 particles composing the pair. In Eq. (30) we assume that $g_1 = g_2 = g$ for the spins in both PD and PP excitations of conducting polymers. Assuming also that the principal axes of the two tensors are the same and that $\overline{H}$ is along $\hat{z}$, we get from Eq. (30)

$$\mathcal{H} = \beta g_z H_z (S_{1z} + S_{2z}) + J_x S_{1x} S_{2x} + J_y S_{1y} S_{2y} + J_z S_{1z} S_{2z} \quad (31)$$

where we assumed that the $\overline{\overline{g}}$ tensor is identical for the two spins. This is true for π dimers and is also an excellent approximation for polaron pairs in conducting polymers. We discuss separately the cases for isotropic and anisotropic J.

a. Isotropic Exchange

In the case of isotropic exchange, $J_x = J_y = J_z = J$ and Eq. (31) can be simplified to

$$\mathcal{H} = \beta g_z H_z T_z + J/2(T^2 - 3/2) \quad (32)$$

where $\overline{T} = \overline{S}_1 + \overline{S}_2$. The four solutions Ψ_s, Ψ_-, Ψ_0, and Ψ_+ of Eq. (32) can be readily found [49] and are divided into two classes: a singlet state at $E_s = -(3/4)J$ and a triplet state that is split at $E_0 = J/4$ and $E_{\mp} = J/4 \mp \beta g_z H_z$, respectively. μ-Wave transitions are strictly forbidden between the singlet (Ψ_s) and all triplet (Ψ_T) states, whereas among the triplet sublevels $\Psi_{\mp}$ and Ψ_0, μ-wave transitions are allowed only between states Ψ_0 and $\Psi_{\mp}$, respectively. Since, however, $\Delta E_{0,+} = \Delta E_{0,-}$, we expect only a single μ-wave transition to occur at H_0 such that $h\nu = \beta g_z H_0$. This is exactly the same H_0 as for an individual spin 1/2 resonance with $g = g_z$ and shows that the H-ADMR spectra of tightly bound and loosely bound pairs are indistinguishable for isotropic exchange; the PM spectrum, of course, is quite different (Section II). We note that the energy order of the singlet and triplet solutions, E_s and E_t, respectively, depends on J; for $J > 0$, $E_s < E_T$, whereas for $J < 0$, $E_s > E_T$. We expect J, however, to be on the order of kT at the temperatures used in ADMR measurements and that the sublevel spin populations will be unthermalized, i.e., determined by their respective generation and recombination rates rather than by the Boltzmann distribution at the sample temperature.

b. Anisotropic Exchange

When $J_x \neq J_y \neq J_z$ we can still simplify the exchange Hamiltonian given in Eq. (31). Let us use the notation $J_x = J'_x + J_0$, etc., where $J_0 = (1/3)(J_x + J_y + J_z)$; we can regard J_0 as the "isotropic" part of the interaction and J' the "anisotropic" part, for which the trace is zero:

$$J'_x + J'_y + J'_z = 0 \quad (33)$$

Then the Hamiltonian in Eq. (31) becomes [49]

$$\mathcal{H} = \beta g_z H_z T_z + (1/2) J_0 (T^2 - 3/2) + (1/2)(J'_x T_x^2 + J'_y T_y^2 + J'_z T_z^2) \quad (34)$$

The last part of Eq. (34) can be also written in a form equivalent to the spin–spin interaction term in Eq. (18), where the effective ZFS parameters D' and E' of the triplet state built from the three symmetric solutions of $\mathcal{H}$ in Eq. (34) are

$$D' = \frac{3}{4} J'_z, \qquad E' = \frac{1}{4}(J'_x - J'_y) \quad (35)$$

μ-Wave transitions are again forbidden between Ψ_s and Ψ_T, and the resonances associated with the triplet sublevels are exactly the same as for a regular triplet with ZFS parameters given in Eq. (35). These are in the form of two triplet powder patterns: one at half-field at H given by Eq. (25) at $H_{1/2} \simeq (1/2)H_0[1 - (1/2)(D'^2/H_0^2)]$, and the other at full field covering H between $H_0 \mp D'$. As in regular triplets, the different triplet sublevels G and R will cause the two powder patterns to be much different than for thermalized spins, with singularities of both δN signs.

5. PL-ODMR

We mentioned in Section III.B.1 that our ADMR setup can also be used for measuring PL-ODMR. Here we briefly summarize the principal relations in PL-ODMR (only an H-ODMR spectrum is possible) and its correlation with H-ADMR.

Photoluminescence is the result of radiative recombination, and its intensity, I, is given in steady-state conditions by

$$I = R'N_{SS} \tag{36}$$

where R' is the radiative rate and N_{SS} is the density of radiative photoexcitations in steady state. Magnetic resonance alters I (δI) due to spin-dependent recombination processes in the radiative and nonradiative channels, respectively. We therefore deal with these two channels separately.

a. PL-ODMR Related to Radiative Channels

In the case of radiative channels, we assume that the spin-carrying species themselves are radiative and can photoluminesce. This emission is in the form of fluorescence and delayed photoluminescence for spin 1/2 species and phosphorescence for spin triplets. Therefore, we separate our discussion into spin 1/2 and spin 1, respectively.

Spin 1/2. As in Section III.B.2 for ADMR, we can calculate the PL-ODMR figure of merit, $\delta I/I$, on the basis of μ-wave saturation conditions and spin-lattice relation time much longer than the photoexcitation lifetime. Following the same arguments as those leading to Eq. (17) for ADMR, we can readily calculate $\delta I = I_{on} - I_{off}$, where I_{on} and I_{off} are the respective PL intensities for a μ wave turned on and off. Denoting R'_P and R'_A as the radiative rates for pairs with parallel and antiparallel spin configurations, respectively (obviously $R'_P \ll R'_{AP}$), we get

$$I_{off} = R'_P N_P + R'_{AP} N_{AP} \quad \text{and} \quad I_{on} = R'_P \tilde{N}_P + R'_{AP} \tilde{N}_{AP} = \frac{\tilde{N}}{2}(R'_P + R'_{AP})$$

since at μ-wave saturation $\tilde{N}_P = \tilde{N}_{AP}$. Then from Eqs. (16) and (17) we get

$$\frac{\delta I}{I} = -\frac{R_{AP} R_P \Delta N \Delta r}{(\Sigma R)I} \tag{37}$$

where $\Delta N = N_{AP} - N_P$, $\Delta r = r_{AP} - r_P$, where r is the radiative probability given by $r = R'/R$. Combining Eqs. (17) and (37) we can then get a simple relation between PL-ODMR and ADMR, namely,

$$\frac{\delta I}{\delta N} = -R_{AP} \cdot R_P \frac{\Delta r}{\Delta R} \tag{38}$$

Since Δr and ΔR are both positive, we conclude from Eq. (38) that PL-ODMR is opposite in sign to ADMR. In Section III.B.2 we determined that $\delta N > 0$ for geminate pairs and $\delta N < 0$ for distant pairs. Equation (38) then shows that $\delta PL < 0$ for geminate pairs and $\delta PL > 0$ for distant pairs.

A useful relation can be written if we assume that $R'_P = 0$. This assumption is quite justified because this transition is dipole-forbidden anyhow. We can then simplify Eq. (37) to get $\delta I/I = -R_P/\Sigma R$ for geminate pairs and $\delta I/I = \Delta R/\Sigma R$ for distant pairs. The latter relation is particularly powerful, as $\delta N/N$ in ADMR for distant pairs is given by $-(\Delta R/\Sigma R)^2$. We therefore have for the distant pair model quite a useful relation between PL-ODMR and ADMR, namely,

$$\frac{\delta N}{N} = -\left(\frac{\delta I}{I}\right)^2 \tag{39}$$

Equation (39) can be used to check different models presented in the literature to explain spin 1/2 PL-ODMR in conducting polymers.

Spin triplet. The phosphorescence intensity I is given by the triplet sublevel radiative rates R'_γ and their steady-state population N_γ, where $I_\gamma = R'_\gamma \cdot N_\gamma$ and γ is a sublevel at (θ, ϕ, H). Using equations similar to those for ADMR with $S = 1$, we get the changes δI in phosphorescence intensity for each triplet resonance $\Delta \gamma$, as a function of θ, ϕ, and H, similar to Eq. (29):

$$\left(\frac{\delta I}{I}\right)_{\Delta\gamma}(\theta, \phi, H) = -\frac{R_\gamma R_{\gamma'} \Delta N_{\Delta\gamma} \Delta r_{\Delta\gamma}}{(R_\gamma + R_{\gamma'})(I_\gamma + I_{\gamma'})} \tag{40}$$

where $\Delta\gamma$ is a double index, which stands for the coupled triplet sublevels γ and γ' [(-1, 0) and (0, $+1$) for full-field resonances, and (-1, 1) for half-field resonance], $\Delta N_{\Delta\gamma} = N_{\gamma'} - N_\gamma$, $\Delta r_{\Delta\gamma} = r_{\gamma'} - r_\gamma$, and $r = R'/R$. From Eq. (40) we see that $\delta I_{\Delta\gamma}$ can be of either sign or zero, depending on ΔN and Δr of the two coupled levels. Also, when comparing Eq. (29) for triplet ADMR to Eq. (40) for PL-ODMR, we see that δI and δN are not related to each other by a constant factor but are given by a different combination of parameters, such as $\Delta r_{\Delta\gamma}$ in PL-ODMR and $\Delta R_{\Delta\gamma}$ in ADMR, which have different dependences on θ, ϕ, and H. Therefore the triplet powder patterns in PL-ODMR and H-ADMR should be very different from each other, for both the full-field and half-field cases. A situation such as the one in Fig. 22.7, where the PL-ODMR and H-ADMR powder patterns in P3OT are mirror images of each other, therefore cannot be explained by changes occurring in the radiative channel; this apparent symmetry must be then explained by spin-dependent changes in the nonradiative channels.

b. PL-ODMR Related to Nonradiative Channels

In the nonradiative channel model we assume that the excitons (X) also recombine through channels containing other excitations generated by the absorption of the same laser beam. These may be polarons (P), bipolarons (BP), and triplets (T). As an example, exciton recombination via BP is

$$X + BP^{2+} \to h^+ + P^+ \tag{41}$$

This recombination process is extremely efficient and happens because $X = e^- + h^+$ and the e^- quickly recombines with the BP^{2+} to give P^+. The rate equation for exciton recombination therefore contains other types of excitations with densities N_i such that

$$\frac{dN_x}{dt} = \eta G - R_x N_x - \sum_{i=1}^{n} b_i N_i N_x \tag{42}$$

where N_x is the exciton density and b_i are the bimolecular recombination rates between N_x and N_i. From Eq. (42) we get the steady-state solution for N_x, $N_x = \eta G/\Sigma R$, where $\Sigma R = R_x + \Sigma b_i N_i$. The PL intensity I is then given by $R'N_x$:

$$I = \frac{\eta G R'}{R_x + \Sigma b_i N_i} \tag{43}$$

Due to magnetic resonance we monitor changes δN_i in ADMR; then from Eq. (43) we can easily get the PL-ODMR figure of merit:

$$\frac{\delta I}{I} = -\frac{\Sigma b_i \delta N_i}{\Sigma R} \tag{44}$$

Equation (44) shows that (1) the sign of the ODMR is opposite that of the ADMR and (2) the H-ADMR spectrum for a particular spin species is directly reflected in the PL-ODMR spectrum, since in Eq. (44) δN_i is multiplied by a constant factor independent of $\overline{H}$. This is very important for triplet resonances. If triplets indirectly influence exciton photoluminescence, then we expect the PL-ODMR and the H-ADMR powder patterns to be almost identical but of opposite sign. This important finding can explain the triplet ODMR spectrum in P3OT (Fig. 22.7a). From the similarity of ODMR and ADMR (Figs. 22.7a and 22.7b, respectively) we conclude that the ODMR in P3OT is due to changes induced in spin-dependent recombination of nonradiative channels with spin 1/2 and 1, respectively. In this case the information obtained in ADMR essentially covers that in PL-ODMR also. In other words, the PL-ODMR resonances do not directly detect excitons but are sensitive to recombination centers in the sample, which can be more directly studied by ADMR.

It is also interesting to note that although excitons are spinless quasiparticles, PL-ODMR of excitons still may show strong spin 1/2 and spin 1 resonances due to "cross-recombination channels" with other quasiparticles that do possess spins. It is quite common in PL-ODMR and ADMR spectroscopies that a spinless excitation may show magnetic resonances. Therefore, we should not discard the possibility that spinless excitations in conducting polymers [$S^\mp$ in t-$(CH)_x$ and BP^{2+} in NDGS polymers] can show spin 1/2 resonances. This does not mean that these charged excitations carry spin;

it merely means that their generation and/or recombination processes may be affected by other excitations that do carry spin.

IV. PHOTOEXCITATIONS IN POLYACETYLENE

A. Trans-$(CH)_x$

1. PM Spectrum

The photomodulation (PM) spectrum of trans-$(CH)_x$ at 4 K is shown in Fig. 22.8; the trans-$(CH)_x$ sample used here was a thin film ($d \simeq 1000$ Å) polymerized on a sapphire substrate by the Shirakawa method. The PM spectrum consists of two main PA bands with $\Delta\alpha > 0$ and bleaching ($\Delta\alpha < 0$) for $\hbar\omega > 1.65$ eV. The low energy (LE) PA band peaks at 0.43 eV, and it originates from photoinduced charged defects; this has been concluded from its correlation with the photoinduced IRAVs (Fig. 22.6). The intensity of the LE band easily saturates with increased excitation power and is sample-dependent. Moreover, as was demonstrated in oriented $(CH)_x$ films, the LE band is induced preferentially with light polarized perpendicular to the chain's direction [27]. Therefore it seems that the LE band is stabilized by extrinsic defects of the polymer chain. The LE band is now considered to be due to charged soliton $\delta S^\mp$ transitions, which are pushed away from midgap (predicted by the SSH theory) due to electron correlation [18], as discussed in Section II.A.

The high energy (HE) band peaks at 1.35 eV and is due to an overall neutral state that was tentatively identified as a bound S–$\overline{S}$ pair [50]. This PA band does not depend on sample treatment nor does it saturate at high laser illumination. Therefore it is intrinsic to the $(CH)_x$ chain. The bleaching spectrum is opposite to $\alpha(\omega)$, indicating uniform bleaching of the interband transitions. However, the bleaching spectrum associated with the LE band peaks at about 1.45 eV (Fig. 22.6). Therefore it has been proposed that the long-lived $S^\mp$ are stabilized by neutral defects that are not intrinsic to the polymer chains [21]. By applying the ADMR technique, we determined that the HE photoexcitation has spin 1/2 and consequently concluded that it is due to photogenerated neutral solitons [22].

2. ADMR Spectrum

The H-ADMR spectrum at a fixed probe energy of 1.35 eV (at the HE peak) is shown in Fig. 22.8 (inset). The spectrum was taken at 4 K, with a resonant microwave frequency of 2.991 GHz and microwave power of 100 mW, modulated at 500 Hz, pump wavelength at 4880 Å, with an intensity of 500 mW/cm^2. A reduction of photoabsorption (i.e., $\delta N < 0$), with $\delta N/N \simeq -3 \times 10^{-3}$, is observed at 1067 G, which corresponds to spin 1/2 with $g \simeq 2.003$. The resonant lineshape is Lorentzian, with

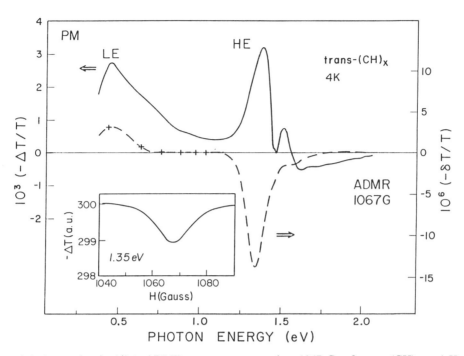

Fig. 22.8 Photomodulation and spin 1/2 λ-ADMR spectra measured at 1067 G, of *trans*-(CH)$_x$ at 4 K. The low energy and high energy PA bands are assigned. The inset shows the H-ADMR spectrum at $\hbar\omega = 1.35$ eV. (From Ref. 22, with permission.)

a full-width at half-maximum (FWHM) of 16.3 G. Except for the spin 1/2 resonance, no other signal was observed in any of our *trans*-(CH)$_x$ samples. No microwave power saturation was observed, and the microwave modulation frequency dependence of the ADMR signal was the same as the pump modulation frequency dependence of the HE band in the PM spectrum.

The H-ADMR spectrum at a fixed probe energy of 0.45 eV (at the LE peak), however, shows an increase in photoabsorption ($\delta N > 0$), with $\delta N/N \approx 10^{-3}$; it has the same spectrum as that of the high energy (HE) H-ADMR spectrum. Both the high and low energy ADMR signals have the same pump intensity and temperature dependence, which are similar to that of the high energy photoabsorption band, indicating that they are all correlated. The λ-ADMR spectrum of *trans*-(CH)$_x$ at a fixed field of 1067 G is also shown in Fig. 22.8. It consists of a relatively large reduction δN in the HE photoexcitation density ($\delta N/N \approx -3 \times 10^{-3}$), a smaller increase in the $S^{\mp}$ population at the LE band ($\delta N/N \approx 10^{-3}$), and an EA oscillation that is barely observable in the λ-ADMR spectrum, in agreement with the small value of $\delta N/N$(LE).

In principle, the spin 1/2 ADMR signal of the HE band could arise from three possible sources: (1) spin-dependent generation of spinless excitation; (2) spin-dependent recombination of spinless excitation; or (3) spin-dependent recombination of spin 1/2 excitation. The first two possibilities can be excluded as follows.

1. The HE band is generated within 3×10^{-13} s (300 fs), and its spectral shape remains unchanged even under steady-state conditions [51,52]. Even if there were spin 1/2 intermediate species, which decay to form spinless HE, the lifetime of such spin 1/2 species should have been less than 300 fs, causing an enormous "lifetime broadening" of their associated magnetic resonance line.

2. Considering that there are spin 1/2 native S^0 defects in polyacetylene, which could act as recombination centers for the spinless HE excitation, then the second possibility seems quite plausible. However, Levey et al. [35] conclusively showed photogeneration of spin 1/2 species in *trans*-(CH)$_x$, and one has to ask where these spin 1/2 particles are in the PM spectrum. Furthermore, we note that the spin 1/2 ADMR signal has the same spectral shape, temperature, and intensity dependencies as the high energy PA band. In addition, $\delta N/N$ can be as high as 5% in *cis*-rich samples (see later sections of this chapter), much larger than the change in thermalized spin polarization of neutral soliton defects under magnetic resonance conditions.

We therefore conclude that the reduction in the HE photoexcitation density is caused by enhanced recombination associated with the spin 1/2 high energy excitations themselves, with unthermalized spin polarization. This can be understood in the context of the distant pair model discussed in Section III.B.2.

3. S^0 Photogeneration

That there is magnetic resonance of only spin 1/2 excitations eliminates most of the theoretical models for the HE band in $trans$-(CH)$_x$. These include singlet excitons [53,54] and breather modes [55,56] for which $S = 0$, as well as triplet excitons [57] for which $S = 1$. Of all the models for the HE photoabsorption band in the literature, only S^0 has spin 1/2. Thus, the only viable explanation for the HE band in the photomodulation spectrum is photogenerated S^0. However, S^0 cannot be directly photogenerated because the photoexcited singlet $1B_u$ state cannot directly decay into an S^0–$\bar{S}^0$ pair. There exist two other possibilities to explain S^0 photoproduction in $trans$-(CH)$_x$ [58,59]. The first is through the triplet manifold following intersystem crossing (ISC). It has been shown that the triplet state in $trans$-(CH)$_x$ is unstable toward the formation of an S^0–$\bar{S}^0$ pair and therefore the process $T \rightarrow \bar{S}^0 + S^0$ is expected to occur following intersystem crossing. The problem with this model is that the HE PA band is photogenerated within 300 fs and there is no reason to believe that singlet-to-triplet intersystem crossing is so fast in $trans$-(CH)$_x$. The second scenario for S^0 photogeneration is via the $2A_g$ state. It has been experimentally and theoretically shown that $E(2A_g) < E(1B_u)$ in $trans$-(CH)$_x$, and this may explain its weak PL emission [60,61]. This indicates that a fast internal conversion from $1B_u$ to $2A_g$ occurs following photon absorption. The interactions that allow the $1B_u$ state to decay into the $2A_g$ state are not very well understood, but they must involve charge conjugation symmetry breaking as well as parity violation [58]. It has been shown that the $2A_g$ state in $trans$-(CH)$_x$ is unstable toward the formation of two S^0–$\bar{S}^0$ pairs and therefore the process $2A_g \rightarrow 2S^0\bar{S}^0$ is expected to occur within a picosecond.

In our ADMR measurements we found $\delta N < 0$, which can be explained within the distant pair model discussed in Section III.B.2. As explained in that section, the distant pair model can be used only for spin pairs generated with equal probability of parallel (P) or antiparallel (AP) spin configuration. This is certainly not the case for the photogenerated S^0. As concluded above, they are thought to be produced in pairs ($\bar{S}$–S) from the singlet manifold and hence presumably with preference of an AP spin configuration. To solve this apparent conflict, either a spin-flip mechanism must occur or a spin–spin interaction exists between the solitons within the S–$\bar{S}$ pair. Such a spin-flip process may happen via the native soliton defects, S_r^0 (spin 1/2 radicals), which exist in $trans$-(CH)$_x$ with a concentration N_r of about 4×10^{19} cm^{-3}. We note that an S–$\bar{S}$ pair is quite mobile, and both intrachain and interchain soliton pair diffusion have been shown to exist. Then exchange-coupling interaction may occur between the individual solitons composing the S–$\bar{S}$ pair and S_r^0, and this may randomize the spin configuration of the S–$\bar{S}$ pair. The problem with this model is that this spin randomization process may compete with the spin polarization process necessary for distant pair ADMR, causing the ADMR signal to diminish. In addition, if an exchange interaction between S–$\bar{S}$ and S_r^0 does occur, then there is no reason to disregard such an interaction between the photogenerated S^0 composing the S–$\bar{S}$ pair; after all, they are closer to each other than to S_r^0. This may actually be the solution to the conflict. If the soliton–antisoliton pair exchange-coupling constant $J < 0$, then, as discussed in Section III.B.4, its symmetric or triplet spin configuration is more stable than the singlet configuration. For isotropic J it is expected, however, that only one resonance occurs at H_0, satisfying the relation $\hbar\nu = g\beta H_0$. We therefore conclude that the high energy photoabsorption band is due to exchange-coupled S^0–$\bar{S}^0$ pairs. The photogeneration of such an excitation in $trans$-(CH)$_x$ was discussed by Orenstein [21], and our results confirm the proposed model. As shown before, this type of excitation is generated in $trans$-(CH)$_x$ with quantum efficiency QE close to 1 for excitation photon energy $\hbar\omega \geq 1.4$ eV. From our discussion above we thus conclude that in $trans$-(CH)$_x$, $E(2A_g) \leq 1.4$ eV. Actually, $E(2A_g)$ has been determined in $trans$-(CH)$_x$, by two-photon absorption (TPA) and EA spectroscopies, to be at ≈ 1.2 eV. We also note that our explanation for the high energy spin 1/2 ADMR signal does not exclude similar phenomena in other NDGS polymers.

4. Soliton Energetics

We have established the existence of soliton transitions in the PM spectrum of $trans$-(CH)$_x$: At 4 K the $\delta S^\pm$ band (LE) peaks at 0.45 eV and the δS^0 band (HE) peaks at 1.35 eV. As mentioned in Section II.A, we can now calculate all the important energy parameters associated with the soliton excitation in $trans$-(CH)$_x$. From Eq. (1) and the values of δS^0 and $\delta S^\pm$ we get $U_{\text{eff}} = \delta S^0 - \delta S^\pm \simeq 0.9$ eV. ΔE_r is calculated via Eq. (2), and we also need the value of E_g ($\simeq 1.5$ eV). We get $\Delta E_r \simeq 0.3$ eV; such a large relaxation energy should not be unusual for soliton excitations. Then from $U = U_{\text{eff}} + \Delta E_r$ we get the bare electron correlation $U = 1.2$ eV. This parameter can be directly related to the on-site Coulomb interaction constant U_H (Hubbard U) via the extent of the soliton wave function l [62],

$$U_{\text{eff}} \simeq \frac{a}{l} U_H \quad (45)$$

where a is the average C—C distance. Using Eq. (45)

and a reasonable value for l, $l \simeq 5a$, we get $U_H \simeq 6$ eV. This shows that electron correlation effects cannot be neglected in any theoretical model for trans-$(CH)_x$. Actually, such a large U_H favors the exciton model over the band model for the proper description of the electronic states in t-$(CH)_x$.

5. Soliton Recombination Processes

We found a positive ADMR signal at 0.45 eV that is associated with the photogenerated $S^{\mp}$. As mentioned in Section III.B.5, however, a situation where a spinless particle shows an ADMR signal is not uncommon in ODMR spectroscopies. This comes about via spin-dependent recombination processes of companion excitations having a "cross-recombination channel" with the spinless particle. The experimental correlation found between $\delta N(HE)$ (<0) and $\delta N(LE)$ (>0) may show that an apparent conversion process from S^0–$\bar{S}^0$ into S^+–$\bar{S}^-$ pairs occurs in trans-$(CH)_x$ even though S^0–$\bar{S}^0$ is the more energetically favorable pair. An important possibility is a fusion process of two S^0–$\bar{S}^0$ pairs into an excited S^+–$\bar{S}^-$ pair: $2S^0$–$\bar{S}^0 \rightarrow S^+$–$\bar{S}^-$, similar to the fusion process of two triplet excitations into an excited singlet exciton, commonly found in molecular crystals [63]. Then μ-wave absorption spin flips one S^0–$\bar{S}^0$ pair, and this may increase the conversion efficiency into an S^+–$\bar{S}^-$ pair.

A different explanation for the positive $\delta N(LE)$ signal, which does not involve the energetically unfavorable conversion of neutral to charged solitons, may be that photogenerated S^0–$\bar{S}^0$ pairs act as recombination centers for the long-lived S^-–$\bar{S}^+$ pairs, promoting their conversion into S^0–$\bar{S}^0$ pairs or enhancing their recombination to the ground state. This may happen, for example, via electron exchange between a pair of $(S^0$–$\bar{S}^0)$ pairs and a pair of $(S^-$–$\bar{S}^+)$ pairs:

$$(S^-\bar{S}^+) + (S^0\bar{S}^0) \rightarrow (S^0\bar{S}^+) + (S^-\bar{S}^0) \quad (46)$$

where we postulate an electron hopping from S^- in the charged soliton pair into S^0 of the neutral pair. The reaction products $(S^0\bar{S}^+)$ and $(S^-\bar{S}^0)$ are considered in trans-$(CH)_x$ physics to be just positive and negative polarons, respectively. Unlike the soliton excitations, these are not topological defects; hence they are quite mobile, and their formation enhances both the $S^{\mp}$ recombination rate to the ground state and the reaction rate to form S^0–$\bar{S}^0$ pairs via a second electron hopping process:

$$(\bar{S}^+S^0) + (\bar{S}^0S^-) \rightarrow 2(\bar{S}^0S^0) \quad (47)$$

Then fewer S^0–$\bar{S}^0$ pairs, caused by microwave absorption, may consequently reduce the charged-to-neutral soliton conversion, resulting in a correlated $\delta N(LE) > 0$ signal.

B. Trans in Cis-Rich $(CH)_x$

To extend our studies of photoexcitations in polyacetylene, we applied the PM and ADMR techniques to cis-rich $(CH)_x$ ($\approx 80\%$ cis) films, where the trans segments contain a large fraction of short chains. We also studied polyacetylene samples with hydrogen, $(CH)_x$, and deuterium, $(CD)_x$. The partially isomerized $(CH)_x$ and $(CD)_x$ samples were also in the form of thin films ($d \simeq 1000$ Å) polymerized onto sapphire substrates by the Shirakawa technique. The films contain about 80% cis and 20% trans, as determined by Raman scattering measurements [64,65].

The PM spectra of cis-rich polyacetylenes are similar to that of pure trans-$(CH)_x$. As in trans-$(CH)_x$, the PM spectrum in cis-rich samples also contain two photoabsorption bands: a weak LE band at 0.5 eV and an intense HE band, now at 1.49 eV. The relatively weak LE band shows either that long-lived $S^{\pm}$ solitons in short chains are not generated with high quantum efficiency or that their lifetime is long compared to the laser modulation frequency used in our PM measurements. The HE band, on the other hand, does not differ much from that in trans-$(CH)_x$, except for the blue shift from 1.35 to 1.49 eV, which is probably due to the relatively short trans chains in the film having higher E_g. No long-lived PA bands associated with the cis segments in the film have been detected.

The H-ADMR spectrum of the cis-rich $(CD)_x$ at 1.49 eV (HE band) at 4 K is shown in Fig. 22.9a. Two δN features were observed [66]: an intense $\delta N < 0$ band with $\delta N/N \approx 5 \times 10^{-2}$, which peaks at 1071 G, and a much weaker derivative-like half-field (HF) band with maximum $\delta N/N$ of 10^{-5} and with zero crossing at 538 G. As discussed in the previous section, the intense δN band at 1071 G shows that spin 1/2 excitations are correlated with the HE PA band and confirm, therefore, that S^0–$\bar{S}^0$ pairs are also generated in short trans chains. The reason $\delta N/N$ is much larger is cis-rich films than in pure trans films may be related either to a longer spin-lattice relaxation time in cis-rich samples or to a faster recombination rate of the S^0–$\bar{S}^0$ pairs with AP spins due to their confinement in the short trans chains in the sample. Both of these processes increase spin polarization, and this in turn enhances the ADMR signal (Section III.B).

Careful studies of the ADMR band at 1071 G reveal (Fig. 22.9b) that it contains two δN components [66] with different dynamics, as measured by their different response to the μ-wave modulation frequency: a narrow (N) component with FWHM of 7.6 G for $(CD)_x$ and 16.3 G for $(CH)_x$, and a slower and broader (B) component with FWHM of 31.6 G, which remains the same for $(CD)_x$ and $(CH)_x$. To check whether the B component was caused by unintentional doping, the samples were treated with ammonia (NH$_3$); the B component remained unchanged, excluding this possibility. In contrast, as seen in Fig. 22.9b, the H-ADMR spectrum at 0.45 eV shows only the N component; there is neither a B component nor a half-field feature at 0.45 eV. We also found that the two δN components do not have the same λ-ADMR spectra. The λ-ADMR associated with

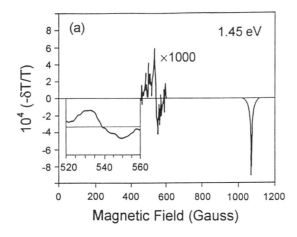

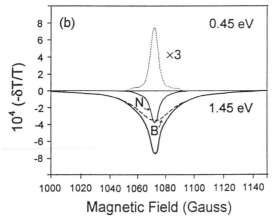

Fig. 22.9 H-ADMR spectra of *trans* segments in *cis*-rich $(CD)_x$ (a) from 0 to 1200 G at $\hbar\omega = 1.45$ eV; (b) for 1000–1145 G at $\hbar\omega = 0.45$ and 1.45 eV. The decomposition of the $S = 1/2$ resonance at 1067 G into narrow (N) and broad (B) components is shown in (b) for $\hbar\omega = 1.45$ eV. (From Ref. 66, with permission.)

not. Since PL-ODMR of excitons is due to an indirect process associated with other, long-lived photoexcitations (Section III.B.5), the lack of PL-ODMR for the *cis* segments shows that such long-lived excitations are absent in *cis* chains. This confirms that the cw PM spectrum of *cis*-rich polyacetylene is due to only the *trans* segments in the film. As in ADMR, we found that the PL-ODMR contains two δI bands: a derivative-like high frequency band with zero crossing at 537 G and a $\delta I > 0$ band at 1071 G that has the same lineshape and frequency dependence as the B component in H-ADMR.

To explain the two ADMR components and their correlation with the PL-ODMR, we consider that the distribution of *trans* chains in *cis*-rich polyacetylene films contains two main groups: one group with very short chains and the other with intermediate-length chains. This model is in agreement with the conclusions drawn from resonant Raman scattering (RRS) studies of *cis*-rich $(CH)_x$ films [67]. Then the B component is due to the short chains in the film, whereas the N component is associated with the longer chains. This is justified because the λ-ADMR of the B component peaks at a higher photon energy (1.55 eV) than the N component, indicating that chains with higher gaps are associated with the B component. The photogenerated $S^0-\bar{S}^0$ pairs in the longer chains are similar to those of pure *trans* chains, i.e., their reduction under magnetic resonance gives negative ADMR for the HE PA band (δS^0), which is correlated with a positive ADMR at the LE PA band ($\delta S^\mp$), and no PL-ODMR is observed. The photogenerated $S^0-\bar{S}^0$ pairs in the short chains, however, are quite isolated and hence lack a cross-recombination channel with photogenerated $\bar{S}^+-S^-$ pairs. This explains the reason why a B component with $\delta N > 0$ has not been observed at the LE PA band.

The $\bar{S}^0-S^0$ confinement in the short *trans* chains gives rise to a stronger and more anisotropic exchange coupling J between the solitons. As discussed in Section III.B.4, an anisotropic J for a pair of spins 1/2 results in two ADMR powder patterns: a "full field" and a "half field," the same as for regular triplets. This is indeed observed for the B component (Fig. 22.9). In this case we estimate the ZFS parameter $D \simeq 15$ G (0.0016 cm^{-1}) from the full-field powder pattern [Eq. (24)]. As mentioned in Section III.B.4, $D = (1/2)(J_z - J_0)$ and therefore J_0 cannot be directly evaluated from D. However, if D were due to a bipolar spin interaction rather than to exchange coupling, then from Eq. (21) we could have estimated the triplet wave function extent $R_{1,2}$. From $D = 15$ G we determined $R_{1,2} \simeq 12$ Å, i.e., 10 monomers, which is consistent with the model of solitons in short *trans* chains. Also, from the derivative-shaped high frequency powder pattern of the B component, we can deduce the $S^0-\bar{S}^0$ triplet sublevel generation and recombination rates. They are $G_y \simeq G_z$, $G_x \ll G_z$ and $R_x \simeq R_z$, $R_y \ll R_z$. The implication of such rates for the photogenerated $\bar{S}^0-S^0$ pair is not fully understood at this time.

the N component peaks at 1.49 eV and contains a secondary oscillation at 1.61 eV, whereas the λ-ADMR associated with the B component peaks at 1.55 eV and does not have any additional oscillation. It is obvious, therefore, that the two $\delta N < 0$ components do not share a common origin. We also found that the λ-ADMR spectrum at 535 G (HF) is similar to the λ-ADMR spectrum of the B component, indicating that the much weaker derivative-like HF band (Fig. 22.9a) is correlated with the B component. This correlation was also observed to exist in subsequent PL-ODMR measurements.

Previous PL studies [60,61,64,65] of *cis*-rich $(CH)_x$ and $(CD)_x$ films revealed two PL bands: a visible PL band at about 1.97 eV, due to the *cis* segments in the film and a near-IR PL band at about 1.3 eV, which was attributed to the *trans* segments. We found that whereas the IR PL band has PL-ODMR, the visible PL band does

We assume that PL emission is possible in *trans* segments with short chains only [60,61,64,65]. This explains the paradoxical dependence of the IR PL band with the percentage of the *trans* segments in (CH)$_x$ samples: It decreases with increasing *trans* isomerization and yet it originates from *trans* chains. As is well established from RRS studies, however, the percentage of short *trans* chains in the film *decreases* with increasing *trans* isomerization, which is, in our interpretation, in agreement with the decrease in the IR PL band [64,65]. In this case, only photogenerated $\overline{S}^0$–S^0 pairs in short *trans* chains may influence the IR PL band. It is not surprising, therefore, that the PL-ODMR shows only a B component; only the B component is due to $\overline{S}^0$–S^0 pairs in short chains. As discussed in Section III.B.5, however, we expect no additional information from the PL-ODMR spectrum; it is just a mirror image of the ADMR B component.

We have seen in this section that both $S^{\mp}$ and S^0 can be generated in relatively short *trans* chains. This indicates that short oligomers should not show completely different PM and ADMR spectra than their respective polymer films. In the next section we study an oligomer and its polymer counterpart of one important NDGS polymer, polythiophene (PT). Actually, since the PA bands of the oligomer are better defined than those of the polymer films, we discuss the PM spectroscopic results in the oligomer first, followed by the results in polythiophene.

V. PHOTOEXCITATIONS IN POLYTHIOPHENE

The backbone structure of polythiophene is shown in Fig. 22.1. It was the first NDGS polymer that was studied [68–71] using the PM technique, and therefore it is natural to discuss it next. As explained in Section II.B, NDGS polymers do not support soliton excitations. We therefore expect to observe photoabsorption related to polarons, bipolarons, and perhaps π dimers, as the dominant charge excitations, whereas trapped singlet excitons (or PP) and triplet excitons would be the dominant neutral excitations. The separation between charge and neutral excitations in the PM spectra can be relatively easily made by studying a possible correlation between the PA bands and the photoinduced IRAVs, regarding their frequency, temperature, and intensity dependencies. Photoabsorption bands of neutral excitations are not correlated, whereas those of charged excitations are correlated with the IRAVs [41,42]. We start with α-sexithiophene, a model oligomer of polythiophene.

A. Photoexcitations in α-Sexithiophene

The backbone structure of unsubstituted α-sexithiophene, α-6T, is also shown in Fig. 22.1. It consists of six thiophene rings and can thus serve as a model oligomer for PT. The sample films ($d \simeq 1000$ Å) were prepared by vapor deposition of well-purified α-6T molecules on sapphire substrates. This preparation technique [72] is known to result in high quality films containing microcrystallites on the order of 300 Å [73]. It has recently been shown that such films make very good thin film transistors with measured field effect mobilities on the order of 0.1 cm^2/(V·s) [2].

α-6T and its alkyl-substituted derivative didodecylsexithiophene, D6T, were shown to support polaron- and bipolaron-like excitations when oxidized or reduced by various dopants. Fichou et al. [31] showed that the absorption spectrum of the radical cation 6T$^{\cdot +}$ consists mainly of two strong bands (each with a pronounced shoulder on its high energy side), which collapse into a single band (again with a high energy pronounced shoulder) when the oligomer is further oxidized to its 6T^{2+} spinless form. Later, Bäuerle et al. [32] observed similar behavior for the reduced forms of D6T$^{\cdot -}$ and D6T^{2-}. It therefore appears that for the six-ring oligomer the radical ions (positive and negative polarons) are characterized mainly by two electronic transitions, whereas the divalent ions (positive and negative bipolarons) show only a single absorption band. This observation is in apparent contrast with the generally assumed spectra of polarons (P) and bipolarons (BP) in long polymers, namely, that polarons are characterized by three optical transitions and the absorption spectrum of bipolarons is composed of two bands [74]. But, as discussed in Section II.B, it is perfectly legitimate for $P^{\mp}$ and $BP^{2\mp}$ excitations in well-defined oligomers to show double and single absorption bands, respectively. In order to examine polaron/bipolaron excitations without the presence of counterions (as is the case in the doped oligomers mentioned above), we studied the PM spectra of photoexcited neutral α-6T [75].

1. Photomodulation Spectroscopy

In Fig. 22.10a we show the PM spectrum of α-6T in the energy range of 0.7–2.1 eV at 10 K and pump modulation frequency of $\simeq 200$ Hz [75]. The spectrum exhibits three main features: the two bands marked P_1 and P_2 at 0.80 and 1.54 eV, respectively, and the band BP_1 at 1.1 eV. As can be seen in Fig. 22.10b, the BP_1 band almost completely disappears in the out-of-phase spectrum, proving that the photoexcitations responsible for it have much shorter lifetimes than those responsible for P_1 and P_2. Furthermore, P_1 and P_2 have a different temperature dependence than BP_1. Each of these three peaks is accompanied by a phonon replica at 0.97, 1.27, and 1.63 eV, respectively. We note that each of the main PA peaks in the PM spectrum, namely P_1, P_2, and BP_1, corresponds to the respective band in the doping-induced absorption mentioned above [30,75]. We therefore tentatively identify P_1 and P_2 at 0.8 and 1.54 eV,

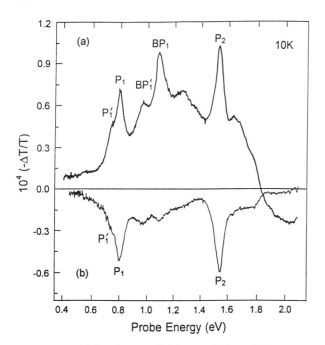

Fig. 22.10 (a) In-phase and (b) out-of-phase PM spectra of α-6T film at 10 K. Various PA bands are assigned. (From Ref. 75, with permission.)

respectively, as the PA bands associated with photogenerated polarons in α-6T and the single band BP_1 at 1.1 eV as the PA band related to photogenerated BPs in the film. This is in agreement with Fig. 22.3 for the origin of the various PA bands for P and BP charged excitations.

We note the existence of a secondary shoulder on the low energy side of P_1 (at $P_1' = 0.75$ eV) and BP_1 (at $B_1' = 0.95$ eV) (Fig. 22.10). We found that these shoulders are correlated with their respective PA peaks at higher energies; they have the same temperature, frequency, and intensity dependencies. We therefore conclude that each PA shoulder is associated with its respective higher energy PA band and consequently is correlated with their photoexcitations: P_1' is thus due to polarons, whereas BP_1' is due to bipolarons. These findings can be rationalized by assuming breaking of charge conjugation symmetry (CCS) in α-6T. This was already postulated in studies of doping-induced absorption of α-6T but for different n-type and p-type dopings [32]. In photogeneration, however, both P^+ and P^- and BP^{2+} and BP^{2-} are generated with equal densities; hence both signs of the charge excitations should be observed in photomodulation. From the doping-induced absorption studies mentioned above [32], it is known that the lower energy transitions are due to the negative-charge species. Hence we identify P_1' and BP_1' shoulders as due to P^- and BP^{2-}, respectively, whereas P_1 and BP_1 peaks are due to P^+ and BP^{2+}, respectively. The apparent asymmetry observed in the intensities of P_1' and P_1 and also for BP_1' and BP_1 (Fig. 22.10) is caused by the asymmetric lineshape of each individual PA band due to the phonon replica on its high energy side; it is not caused by an asymmetry in the optical absorption cross sections of the opposite charges. Also, from the different photon energy of the peaks and respective shoulders—BP_1' and BP_1, P_1' and P_1—we conclude that the difference in the relaxation energy of BP^{2+} with respect to BP^{2-} is larger than that of P^+ with respect to P^-. It is worth noting that we have not found a low energy companion shoulder for the P_2 PA band in the PM spectrum. This is expected for polaron transitions (Fig. 22.3), since P_2 is an intrapolaron transition, which does not change if the polaron energy shifts with respect to the HOMO and LUMO levels. The interband P_1 transition, which is HOMO to SOMO for P^+ and SOMO to LUMO for P^- (see Fig. 22.3), on the other hand, does shift with increasing polaron relaxation energy.

2. ADMR Spectroscopy

The H-ADMR spectrum of α-6T at 4 K at a probe photon energy of 1.54 eV (at the peak of the P_2 band in Fig. 22.10) is shown in Fig. 22.11a. This spectrum clearly reveals two types of spins, $S = 1$ and $S = 1/2$ photoexcitations [75]. The weak HF resonance at 420 G indicates the presence of a triplet state with $S = 1$ (we have also observed its full-field resonance). The narrower line at 1070 G, in contrast, is due to $S = 1/2$ photoexcited species.

The λ-ADMR spectra of the $S = 1/2$ resonance at 1070 G and the high frequency spin triplet resonance at 420 G are shown in Fig. 22.11b. The $S = 1/2$ spectrum consists of two relatively narrow $\delta N < 0$ bands at 0.8 and 1.52 eV and a broad $\delta N > 0$ band centered at ≈1.2 eV. The two narrow bands coincide in energy with P_1 and P_2 in the PM spectrum (Fig. 22.10). We therefore conclude that P_1 and P_2 are due to spin 1/2 polarons, which recombine via a "distant pair" process. This is in agreement with the conclusion drawn above from the comparison of the photomodulation with the doping-induced spectra.

In order to spectrally resolve the λ-ADMR spectrum associated with the positive δN component in Fig. 22.11b, we proceeded in two steps. First we spectrally separated the photoabsorption associated with polarons from that associated with bipolarons. This was done by appropriate scaling and subtracting the in-phase and out-of-phase PM spectra, respectively, shown in Figs. 22.10a and 22.10b. Then we scaled and subtracted the spin 1/2 λ-ADMR spectrum from the PA spectrum of polarons. The remainder of the λ-ADMR spectrum was then found to be quite similar to the BP spectrum in photomodulation as deduced from Figs. 22.10a and 22.10b. We therefore conclude that the broad $\delta N > 0$ feature in the spin 1/2 λ-ADMR of Fig. 22.11b is the

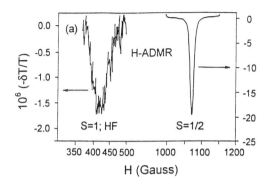

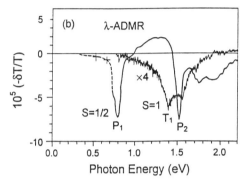

Fig. 22.11 (a) H-ADMR spectrum of α-6T at $\hbar\omega = 1.5$ eV, showing the $S = 1$ high frequency resonance at 420 G and the $S = 1/2$ resonance at 1068 G (b) λ-ADMR of α-6T at 4 K measured at 1068 G ($S = 1/2$) and 420 G ($S = 1$). Various δT bands are assigned. (From Ref. 75, with permission.)

result of a combination of $\delta N < 0$ due to the P_1 band (and its associated phonon replica) and $\delta N > 0$ due to the BP_1' and BP_1 bands (and their associated phonon replicas). In any case, the appearance of the $\delta N > 0$ feature in the λ-ADMR spectrum is due to BPs.

The reason $\delta N > 0$ for spinless BPs in α-6T is similar to the reason given in Section IV.A for $\delta N > 0$ for spinless $S^{\mp}$ in trans-$(CH)_x$. In α-6T, $BP^{2\mp}$ recombines via $P^{\mp}$; for example, we can write

$$BP^{2+} + P^- \to P^+ \tag{48}$$

and a similar reaction can be written for the recombination of BP^{2-} with P^+. Equation (48) for BP^{2+} in NDGS polymers is in the same spirit as Eq. (46) for $S^{\mp}$. The reduction ($\delta N_p < 0$) of P^+ in α-6T under magnetic resonance conditions eliminates recombination centers for bipolaron pairs, thus causing an increase in the BP steady-state density ($\delta N_{BP} > 0$). Similarly, in t-$(CH)_x$, the reduction of S^0-$\bar{S}^0$ pairs has caused an increase ($\delta N_{S^{\mp}} > 0$) in $\bar{S}^+$-S^- pair density. We conclude that the spin 1/2 ADMR results in α-6T are in complete agreement with the PM spectrum and show both $P^{\mp}$- and $BP^{2\mp}$-related δN bands.

There is another spectral feature in the spin 1/2 λ-ADMR spectrum, however, that is absent in doping-induced absorption. This is the $\delta N < 0$ band peaking at 1.85 eV. Since it does not appear in doping-induced absorption, we tentatively identify it as due to neutral excitations. α-6T does not support S^0 excitations; we therefore assign the 1.85 eV PA band as due to excitons. Triplet excitons are clearly observed in spin 1 λ-ADMR, which does not show a PA band at 1.85 eV. We thus believe that this neutral excitation is due to an overall spinless state, yet it shows spin 1/2 ADMR. We measured $\delta N < 0$ at 1.85 eV, and thus the spin 1/2 ADMR signal is not indirectly caused by cross-recombination channels with $P^{\mp}$, as is the case for $BP^{2\mp}$, which show, however, $\delta N > 0$ (Fig. 22.11b). We conclude that the 1.85 eV is caused by a neutral, overall spinless species with *direct* spin 1/2 ADMR. We are then led to believe that the spin 1/2 ADMR is caused by the internal spin structure of this exciton, as in the case of the S^0-$\bar{S}^0$ pairs discussed in Section IV.A.

Long-lived spinless excitons are trapped in defect centers, and this may cause separation into P^+-P^- (PP) pairs. As discussed in Section II.C.3, a tightly bound PP may have a single PA band in the visible spectral range, which is the equivalent of the X_2 transition (into BX) of a free exciton. This PA may show a strong spin 1/2 ADMR signal due to the exchange-coupling interaction between the two spin 1/2 polarons composing the PP pair. For an isotropic J coupling constant, we still expect a single "spin 1/2–like" resonance to occur for a PP excitation at H_0 (Section III.B.4), similar to the case of the S^0-$\bar{S}^0$ pair discussed in Section IV.A. We thus identify the 1.85 eV $\delta N < 0$ band as due to PP excitation, unique for NDGS polymers.

Finally we discuss the spin 1 excitations. As is clearly seen in Fig. 22.11b, the spin 1 λ-ADMR shows a prominent $\delta N < 0$ band with a double peak at 1.4 and 1.55 eV. This PA band is due to T_1, T_2 (or both) transitions in the triplet manifold [76], as discussed in Section II.C.2 and shown in Fig. 22.4b for the triplet excitations. We also note an unusually long energy tail accompanying this band, extending down to 0.75 eV. This shows that although the T_2 transition (the lowest triplet to the triplet biexciton, Section II.C.2) dominates the triplet PA, there are also other, weaker transitions in the PA spectrum and therefore T_2 may not be a transition to the second lowest level in the triplet manifold. In fact, the long tail may be due to T_1 transitions into excited triplet levels (Section II-C.2). From the half-field and full-field $S = 1$ H-ADMR powder patterns (Fig. 22.11a), we can calculate the triplet ZFS parameters in α-6T. They are $D = 575$ G (0.064 cm^{-1}) and $E = 46$ G (0.005 cm^{-1}), very close to the D and E extracted in Section III.B.5 for the triplets in P3OT polymer. Using Eq. (21) we can now estimate the wave function extent $R_{1,2}$ of triplets in α-6T; we find $R_{1,2} \approx 3.6$ Å, which shows quite a localized exciton.

Table 22.1 Photon Energies (in eV) of Various PA Bands in PT Films Identified by PM and ADMR Spectroscopies[a]

Film	P_1	P_2	BP_1	PP_2	T_2
α-6T	0.8	1.55	1.1	1.8	1.4
e-PT	0.45	1.25	0.85	1.8	
P3BT	0.55	1.4	1	≈1.9	1.5

[a] P_1 and P_2 are associated with $P^{\mp}$; BP_1 is due to $BP^{2\mp}$; PP_2 is for P^+–P^- excitations; and T_1 is for triplet excitons (see Figs. 22.3 and 22.4).

The various PA bands in α-6T that have been identified by our PA and ADMR techniques are summarized in Table 22.1. From the values of P_1, P_2, and BP_1 we can calculate the important charged defect parameters using Eqs. (5)–(7) and $E_g = 2.3$ eV in α-6T [77]. Assuming $\Delta \omega_0 \le 0.2$ eV, we get $U = 0.9$ eV, $\Delta E_r = 1.1$ eV, and therefore $U_{eff} = -0.2$ eV for BP excitations. The negative U_{eff} shows that $BP^{2\mp}$ excitations are more stable than $P^{\mp}$ excitations in α-6T, in agreement with Eq. (8). We note that although $U_{eff} < 0$, $U > 0$ and is close to U in $(CH)_x$ ($U = 1.2$ eV). This shows that U_H, the "Hubbard U" is similar in these polymers, indicating that e–e interaction plays an important role in both degenerate and NDGS π-conjugated polymers.

B. Photoexcitations in Polythiophene Polymers

Following the discussion of photoexcitation properties and dynamics in α-6T, the model oligomer for PT polymers, it is relatively easy now to extend the previous analysis to photoexcitations in various PT polymers. We will discuss two types of PT films: an electrochemically polymerized PT (e-PT) film, in the photophysics of which defects play a major role, and a soluble PT derivative, the chemically prepared poly(3-butylthiophene) (P3BT). These films were polymerized on sapphire and KBr substrates, respectively. We also studied photoexcitations in P3OT films [78], and the results were not different from those of P3BT films.

1. Photoexcitations in e-PT

The PM spectrum of e-PT at 4 K is shown in Fig. 22.12. It contains three PA bands: P_1 at 0.45 eV, P_2 at 1.25 eV, and PP_2 at 1.8 eV [79]. These PA bands do not share a common origin. We have measured their dependencies on the laser excitation intensity, modulation frequency,

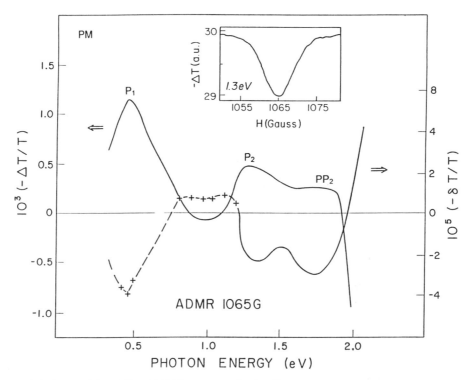

Fig. 22.12 Photomodulation and $S = 1/2$ λ-ADMR spectra of e-PT film at 4 K. Various PA bands are assigned. The inset shows the $S = 1/2$ H-ADMR spectrum at $\hbar\omega = 1.3$ eV. (From Ref. 70, with permission.)

temperature, and decay time after laser pulsed excitation; we found that the P_1 and P_2 bands are correlated with each other, whereas the PP_2 band is not. In particular, the PP_2 band decays much faster than the two other PA bands and therefore is associated with excitations with much shorter lifetime [79].

The H-ADMR spectra measured at the peaks of these three PA bands all show similar properties. As shown in the inset of Fig. 12, the ADMR is negative, with a Lorentzian line shape that peaks at 1067 G ($g \simeq 2.003$) with FWHM of about 9 G. This indicates that all three PA bands are associated with spin 1/2 excitations; spin 1 ADMR has not been detected in e-PT films.

The λ-ADMR spectrum at 1065 G (Fig. 22.12) contains the same three bands as in the PM spectrum but with different intensities [79]. The relative microwave-induced changes in the photoexcitation density $\delta N/N$ is -3×10^{-2} for both the P_1 and P_2 bands, whereas $\delta N/N$ is -7×10^{-2} for the PP_2 band. As in α-6T (Fig. 22.11), we also found a positive spin 1/2 ADMR signal ($\delta N > 0$) around 1 eV. We applied an analysis similar to that above for the $\delta N > 0$ signal in α-6T (Section V.A) and found that the $\delta N > 0$ band actually shows a peak at about 0.85 eV. From the similarity of the λ-ADMR spectra of e-PT (Fig. 22.12) and α-6T (Fig. 22.11a), we assign the P_1 and P_2 bands in e-PT as due to polarons ($P^{\mp}$), the $\delta N > 0$ band (now BP_1) to bipolarons ($BP^{2\mp}$), and the PP_2 band to polaron pairs (P^+-P^-). This is somewhat in variance with the original assignment of these PA bands [70,79] but in agreement with the new results in α-6T, which definitely show long-lived P, BP, and PP excitations, with photoinduced transitions similar to those of e-PT.

The various bands in e-PT and α-6T are compared in Table 22.1. As we can see from this table, although P_1 and P_2 bands are red-shifted by about 0.3 eV in e-PT compared to these bands in α-6T, their separation, about 0.8 eV, remains the same. This is a strong argument in favor of the identification of $P^{\mp}$ as the source of P_1 and P_2 PA bands in e-PT. We also note that BP_1 redshifts in e-PT with respect to BP_1 in α-6T by the same amount (0.3 eV) as P_1 and P_2; this is also a strong argument in favor of our assignment of the various PA bands in e-PT. The red shift of the charged excitation transitions in e-PT may be caused by the existence of chains longer than six rings in this film.

We note that we have not identified a PA band associated with the $P^{\mp}$ transition P_3 in e-PT (Fig. 22.12), and we thus conclude that P_3 remains very weak, even for PT chains longer than six rings as in α-6T. We can infer, however, P_3 from P_1 and P_2, since for $P^{\mp}$ excitations $P_3 = P_1 + P_2$ (Fig. 22.3); we get $P_3 = 1.7$ eV in e-PT. Then from Eqs. (5)–(7) and $E_g \simeq 2$ eV in e-PT [80], we can calculate the important charged defect parameters U_{eff}, ΔE_r, and U. Assuming $\Delta \omega_0 \simeq 0.1$ eV, we get $U = 0.2$ eV, $\Delta E_r = 0.5$ eV, and consequently $U_{eff} = -0.3$ eV. The negative U_{eff} shows that $BP^{2\mp}$ excitations are more stable than polarons in PT polymers, in agreement with a large body of other experiments [70,79]. To calculate U_H we use the relation $U = U_H a/3l$, where l is the BP wave function extent [62]. Using $l = 5a$ for BPs in PT, we get $U_H = 3$ eV. This U_H is intermediately large and shows again that e–e interaction should not be neglected in PT.

We also note from Table 22.1 that the PP_2 transition stays the same (1.8 eV) in e-PT and α-6T. It seems, therefore, that the PP_2 transition does not depend on chain length, in contrast to transitions P_1, P_2, and BP_1 above. In agreement with our assignment, this indicates that the PP excitation is more tightly bound than the $P^{\mp}$ or $BP^{2\mp}$ excitations in π-conjugated polymers.

2. Photoexcitations in P3BT

We also studied photoexcitations in solution-cast P3BT films [81]. These films are much more luminescent than e-PT, and although they have higher E_g (and therefore shorter chain length) we expect fewer defects than in e-PT films.

As seen in Fig. 22.13a, the PM spectrum is dominated by a PA band (T_2) at 1.45 eV; there is a second, weaker band (P_1) at 0.55 eV. From the correlation (or lack thereof) with the photoinduced IRAVs, seen in Fig. 22.13a at $\hbar\omega < 0.15$ eV, we inferred that P_1 is due to charged excitations whereas T_2 is caused by a neutral excitation [81].

Using ADMR spectroscopy we identified [81] both $S = 1/2$ and $S = 1$ excitations; their λ-ADMR is shown in Fig. 22.13b. We observed two strong $S = 1/2$ $\delta N < 0$ bands at 0.55 and 1.4 eV, a weaker shoulder at 1.8–1.9 eV, and an $S = 1/2$ $\delta N > 0$ band at about 1 eV. The $S = 1$ λ-ADMR, in contrast, contains a single $\delta N < 0$ band at 1.5 eV (Fig. 22.13b). From the discussion in Section V.B.1, we identify the strong negative $S = 1/2$ ADMR bands as due to P_1 and P_2 transitions associated with $P^{\mp}$ excitations, the $\delta N < 0$ shoulder at 1.9 eV as due to PP excitations, the $\delta N > 0$ band as BP_1 associated with $BP^{2\mp}$, and the $S = 1$ $\delta N < 0$ band as due to triplet excitations. These various PA bands are also given in Table 22.1 for comparison with the PA bands in α-6T and e-PT. It is worth noticing the potential of the ADMR spectroscopy to separate these various PA bands from such a relatively featureless PM spectrum (Fig. 22.13a). From Table 22.1 we see that the PA bands associated with the charged excitations ($P^{\mp}$ and $BP^{2\mp}$) are intermediately red-shifted (≈ 0.2 eV) compared to those in α-6T, but the PA bands due to the neutral excitations (T_1 and PP_2) stay put, showing that they are tightly bound, independent on the PT chain length. Again the P_1–P_2 separation (≈ 0.8 eV) is the same in P3BT as in e-PT and α-6T films, in agreement with their similar assignment. We can also go through an analysis similar to that of e-PT to obtain comparable defect parameters in P3BT.

The reason, however, to show and discuss P3BT (and P3OT) films is the strong PA band T_2 associated with

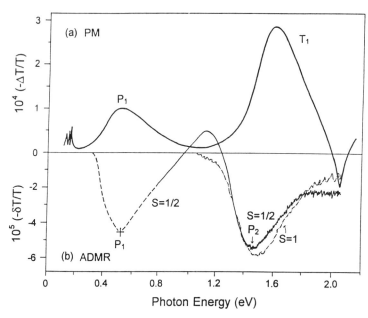

Fig. 22.13 (a) Photomodulation and (b) λ-ADMR of $S = 1/2$ (at 1068 G) and $S = 1$ (at 420 G) spectra of a P3BT film at 4 K. Various PA bands are assigned. (From Ref. 81, with permission.)

triplet excitons in the PM spectra of these films. We note that both P3BT and P3OT films have strong PL emission, and this shows that the excitons, which are the primary excitations in π-conjugated polymers, live longer in these PT derivatives than the excitons in e-PT films. In fact, the excitons in P3BT and P3OT live long enough to allow intersystem crossing into the triplet manifold. This may explain why the PM spectrum in these two polymers (and in PPV derivatives as well) is dominated by long-lived triplet excitations. The excitons in polymer films with more defects (such as e-PT) quickly disassociate before the ICS process can occur, explaining both the weak triplet PA band and strong PP band in the PM spectra of these polymer films [79].

VI. CONCLUSION

In this chapter we have reviewed the main cw techniques used to study photoexcitations in undoped conducting polymers. In particular, we have given an in-depth description, both experimental and theoretical, of the PM and ADMR techniques. We have used these techniques to study photoexcitations in two systems of conducting polymers—trans-$(CH)_x$ with its unique twofold degeneracy and polythiophene, a representative of the NDGS polymers. We have identified long-lived charged and neutral photoexcitations in both systems and verified their spin states. We have also studied the relations between the various photoexcitations and the PL emission in these two polymer systems.

The long-lived photoexcitations in degenerate and NDGS polymers have more common properties than thought before. In both polymer classes we have identified charged and neutral excitations.

1. The charged photoexcitations in trans-$(CH)_x$ are solitons, $S^{\mp}$, with a *single* PA band in the gap, whereas in polythiophene they form polarons, $P^{\mp}$, with *two* PA bands in the gap. However, we also identified in NDGS polymers photogenerated bipolarons, $BP^{2\mp}$, which have a *single* PA band below the gap. Bipolarons may play an important role in more ordered polymer films, where two polarons with like charges may more easily recombine to form a bipolaron, $BP^{2\mp}$. Then, from the experimental point of view, it is difficult to separate degenerate and nondegenerate ground-state polymers, since the long-lived charged carriers in both polymer classes are spinless excitations having a single PA band.

2. Moreover, the long-lived neutral photoexcitations in both polymer classes are also quite similar. In the singlet manifold, neutral photoexcitations form soliton pairs, S^0–$\bar{S}^0$, in trans-$(CH)_x$ and polaron pairs, P^+–P^-, in polythiophene. Both pairs show a single, strong PA band below the gap, with an associated negative spin 1/2 ADMR signal, which is caused by the exchange coupling interaction between the spin 1/2 particles composing the pair. Also, both SS and PP pairs can be generated in the picosecond time do-

main directly from the allowed exciton bands of the polymers. Then again it is difficult to separate degenerate and NDGS polymers on the basis of their long-lived singlet photoexcitations.

3. We also identified triplet excitations in both trans-$(CH)_x$ and polythiophene. In trans-$(CH)_x$, triplets are photogenerated in short chains, where there exists a strong exchange coupling interaction between the spin 1/2 neutral solitons composing the SS pair. In polythiophene, on the other hand, triplet excitons are formed via intersystem crossing from the singlet manifold. From their ZFS parameters we have concluded that the triplet excitations in NDGS polymers are more localized than the triplet formed in short trans-$(CH)_x$ chains.

We conclude that one should be very careful in claiming structure degeneracy in polymers on the basis of their long-lived photoexcitations or doping-induced charges [82]. Also, more caution has to be exercised before reaching definite conclusions about the photoexcitations in a particular conducting polymer, since their properties are not unique. For example, IRAVs are correlated with charged excitations of any kind [83], spin 1/2 signals may also occur for spinless excitations [84], and spin 1 signatures may be also observed for spin 1/2 pairs [66]. All of these fascinating properties, however, make photophysical studies of conducting polymers so much fun.

ACKNOWLEDGMENTS

We thank D. Moses for the polyacetylene samples, E. Ehrenfreund for the P3BT and P3OT samples, A. Frank and M. Ibrahim for the α-6T films, and C. Taliani for the e-PT film. We thank W. Ohlson and C. Taylor for their help with the ADMR measurements. We also wish to mention a list of collaborators over the years 1990–1996, without which this work would have never been completed. The list includes B. C. Hess, R. E. Benner, J. M. Leng, G. G. Kanner, S. Jeglinski, P. A. Lane, and M. Leiss from Utah, J. Poplawski and Y. Greenwald from Technion, S. Mazumdar from Arizona, J. Shinar and T. Barton from Ames Laboratories, and M. Ozaki and K. Yoshino from Osaka. This work was supported in part over the years 1989–1996 by the Department of Energy, grant No. FG-03-93 ER 45490, and by Office of Naval Research grant No. N00014-94-1-0853.

REFERENCES

1. J. H. Burroughes, D. D. C. Bradley, A. R. Brown, R. N. Marks, K. Mackay, R. H. Friend, P. L. Burns, and A. B. Holmes, *Nature* 335:137 (1988).
2. F. Garnier, R. Hajlaoui, A. Yassar, and P. Srivastava, *Science* 265:1684 (1994).
3. S. Etemad, G. L. Baker, and Z. G. Soos, in *Molecular Nonlinear Optics: Materials, Physics and Devices* (J. Zyss, ed.), Academic, San Diego, 1994, p. 433.
4. A. J. Heeger, S. Kivelson, J. R. Schrieffer, and W. P. Su, *Rev. Mod. Phys.* 60:781 (1988).
5. T. Skotheim (ed.), *Handbook of Conducting Polymers*, Marcel Dekker, New York, 1986.
6. W. P. Su, J. R. Schrieffer, and A. J. Heeger, *Phys. Rev. Lett.* 42:1698 (1979); *Phys. Rev. B* 22:2099 (1980).
7. L. Robins, J. Orenstein, and R. Superfine, *Phys. Rev. Lett.* 56:1850 (1986).
8. L. S. Swanson, J. Shinar, and K. Yoshino, *Phys. Rev. Lett.* 65:1140 (1990).
9. X. Wei, B. C. Hess, Z. V. Vardeny, and F. Wudl, *Phys. Rev. Lett.* 68:666 (1992).
10. R. Kersting, U. Lemmer, R. F. Mahrtt, K. Leo, H. Kurz, H. Bassler, and E. O. Gobel, *Phys. Rev. Lett.* 70:3820 (1993).
11. J. M. Leng, S. A. Jeglinski, X. Wei, R. E. Benner, Z. V. Vardeny, F. Guo, and S. Mazumdar, *Phys. Rev. Lett.* 72:156 (1994).
12. M. Yan, L. J. Rothberg, F. Papadimitrakopoulos, M. E. Galvin, and T. M. Miller, *Phys. Rev. Lett.* 72:1104 (1994).
13. Z. G. Soos and S. Ramasesha, *Phys. Rev. B* 29:5410 (1984).
14. P. Tavan and K. Schulten, *Phys. Rev. B* 36:4337 (1987).
15. F. Kajzar, S. Etemad, G. L. Baker, and J. Messier, *Synth. Met.* 17:563 (1987).
16. B. E. Kohler, C. Spangler, and C. Westerfield, *J. Chem. Phys.* 89:5422 (1988).
17. Z. G. Soos, S. Ramasesha, and D. S. Galvao, *Phys. Rev. Lett.* 71:1609 (1993).
18. Z. V. Vardeny and J. Tauc, *Phys. Rev. Lett.* 54:1844 (1985); 56:1510 (1986); *Phil. Mag. B* 52:313 (1985).
19. W. P. Su and J. R. Schrieffer, *Proc. Natl. Acad. Sci. U.S.A.* 77:5626 (1980).
20. R. Ball, W. P. Su, and J. R. Schrieffer, *J. Phys.* 44:C3-429 (1983).
21. J. Orenstein, in *Handbook of Conducting Polymers*, (T. Skotheim, ed.), Marcel Dekker, New York, 1986, p. 1297.
22. X. Wei, B. C. Hess, Z. V. Vardeny, and F. Wudl, *Phys. Rev. Lett.* 68:666 (1992).
23. M. Yan, L. J. Rothberg, F. Papadimitrakopoulos, M. E. Galvin, and T. M. Miller, *Phys. Rev. Lett.* 72:1104 (1994).
24. H. A. Mizes and E. M. Conwell, *Phys. Rev. B* 50:11243 (1994).
25. D. M. Pai and R. C. Enck, *Phys. Rev. B* 11:5163 (1975).
26. A. S. Siddiqui, *J. Phys. C* 17:683 (1984).
27. P. D. Townsend and R. H. Friend, *Synth. Met.* 17:361 (1987).
28. L. Rothberg, T. J. Jedju, S. Etemad, and G. L. Baker, *Phys. Rev. B* 36:7529 (1987).
29. C. Botta, S. Luzzatti, R. Tubino, D. D. C. Bradley, and R. H. Friend, *Phys. Rev. B* 48:14809 (1993).

30. E. Ehrenfreund, D. Moses, A. J. Heeger, J. Cornil and J. L. Bredas, *Chem. Phys. Lett.* 196:84 (1992).
31. D. Fichou, G. Horowitz, B. Xu, and F. Garnier, *Synth. Met.* 39:243 (1990).
32. P. Bäuerle, U. Segelbacher, K.-U. Gaudl, D. Huttenlocher, and M. Mehring, *Angew. Chem. (Int. Ed. Engl.)* 32:76 (1993).
33. G. Zotti, G. Schiavon, A. Berlin, and G. Pagani, *Chem. Mater.* 5:620 (1993).
34. J. Poplawski, E. Ehrenfreund, J. Cornil, J. L. Bredas, R. Pugh, M. Ibrahim, and A. J. Frank, *Mol. Cryst. Liq. Cryst.* 256:407 (1994).
35. C. G. Levey, D. V. Lang, S. Etemad, G. L. Baker, and J. Orenstein, *Synth. Met.* 17:569 (1987).
36. J. M. Leng, S. A. Jeglinski, X. Wei, R. E. Benner, Z. V. Vardeny, F. Guo, and S. Mazumdar, *Phys. Rev. Lett.* 72:156 (1994).
37. G. Guo, S. Mazumdar, S. N. Dixit, F. Kajzar, F. Jarka, Y. Kawabe, and N. Peyghambarian, *Phys. Rev. B* 48:1433 (1993).
38. C. J. Baker, O. M. Gelsen, and D. D. C. Bradley, *Chem. Phys. Lett.* 201:127 (1993).
39. N. Periasamy, R. Danieli, G. Ruani, R. Zamboni, and C. Taliani, *Phys. Rev. Lett.* 68:919 (1992).
40. P. O'Connor and J. Tauc, *Phys. Rev. B* 25:2748 (1982).
41. Z. V. Vardeny, J. Orenstein, and G. L. Baker, *Phys. Rev. Lett.* 50:2032 (1983).
42. Z. V. Vardeny, J. Orenstein, and G. L. Baker, *J. Phys.* 44:C3-325 (1983).
43. D. D. C. Bradley and O. M. Gelsen, *Phys. Rev. Lett.* 67:2589 (1991).
44. K. E. Ziemelis, A. T. Hussain, D. D. C. Bradley, R. H. Friend, J. Ruhe, and G. Wegner, *Phys. Rev. Lett.* 66:2231 (1991).
45. R. H. Clarke and J. M. Hayes, *J. Chem. Phys.* 57:679 (1972).
46. I. Hirabayashi and K. Morigaki, *Solid State Commun.* 47:469 (1983).
47. J. E. Wertz and J. R. Bolton, *Electron Spin Resonance*, Chapman and Hall, New York, 1986.
48. C. P. Poole, *Electron Spin Resonance: A Comprehensive Treatise on Experimental Techniques*, Wiley, New York, 1983.
49. A. Abragam and B. Bleaney, *Electron Paramagnetic Resonance of Transition Ions*, Clarendon Press, Oxford, 1970.
50. J. Orenstein and G. L. Baker, *Phys. Rev. Lett.* 49:1043 (1982).
51. C. V. Shank, R. Yen, R. L. Fork, J. Orenstein, and G. L. Baker, *Phys. Rev. Lett.* 49:1660 (1982); *Phys. Rev. B* 28:6095 (1983).
52. J. Orenstein, Z. V. Vardeny, G. L. Baker, G. Eagle, and S. Etemad, *Phys. Rev. B* 30:786 (1984).
53. M. J. Rice and I. A. Howard, *Phys. Rev. B* 28:6089 (1983).
54. C. L. Wang and F. Martino, *Phys. Rev. B* 34:5540 (1986).
55. A. R. Bishop, D. K. Campbell, P. S. Lomdahl, B. Horovitz, and S. R. Phillpot, *Phys. Rev. Lett.* 52:671 (1984).
56. A. R. Bishop, D. K. Campbell, P. S. Lomdahl, B. Horovitz, and S. R. Phillpot, *Synth. Met.* 9:223 (1984).
57. W. P. Su, *Phys. Rev. B* 34:2988 (1986).
58. P. Tavan and K. Schulten, *Phys. Rev. B* 36:4337 (1987).
59. G. W. Hayden and E. J. Mele, *Phys. Rev. B* 34:5484 (1986).
60. E. A. Imhoff, D. B. Fitchen, and R. E. Stahlbush, *Solid State Commun.* 44:329 (1982).
61. K. Yoshino, S. Hayashi, Y. Inuishi, K. Hattori, and Y. Watanabe, *Solid State Commun.* 46:583 (1983).
62. D. K. Campbell, D. Baeriswyl, and S. Mazumdar, *Synth. Met.* 17:197 (1987).
63. M. Pope and C. E. Swenberg, *Electronic Processes in Organic Crystals*, Clarendon Press, Oxford, 1982.
64. D. L. Weidman and D. B. Fitchen, Proc. 10th Int. Conf. on Raman Spectroscopy, Eugene, OR, 1986.
65. D. L. Weidman, Ph.D. Thesis, Cornell University, 1987.
66. X. Wei, Z. V. Vardeny, E. Ehrenfreund, and D. Moses, *J. Synth. Met.* 54:321 (1993).
67. S. Lefrant, E. Faulques, G. P. Brivio, and E. Mulazzi, *Solid State Commun.* 53:583 (1985).
68. F. Moraes, H. Schaffer, M. Kobayashi, A. J. Heeger, and F. Wudl, *Phys. Rev. B* 30:2948 (1984).
69. T. Hattori, W. Hayes, K. S. Wong, K. Kaneto, and K. Yoshino, *J. Phys. C* 17:L803 (1984).
70. Z. V. Vardeny, E. Ehrenfreund, O. Brafman, M. Nowak, H. Schaffer, A. J. Heeger, and F. Wudl, *Phys. Rev. Lett.* 56:671 (1986).
71. K. Kaneto, F. Uesugi, and K. Yoshino, *J. Phys. Soc. Jpn.* 56:3703 (1987).
72. J. Kagan and S. K. Arora, *Heterocycles* 20:1937 (1983).
73. W. Porzio, S. Destri, M. Mascherpa, S. Rossini, and S. Bruckner, *J. Synth. Met.* 55–57:408 (1993).
74. K. Fesser, A. R. Bishop, and K. K. Campbell, *Phys. Rev. B* 27:4804 (1983).
75. P. A. Lane, X. Wei, Z. V. Vardeny, J. Poplawski, E. Ehrenfreund, M. Ibrahim, and A. J. Frank, *J. Synth. Met.* 76:57 (1996).
76. R. A. J. Janssen, L. Smilowitz, N. S. Sariciftci, and D. Moses, *J. Chem. Phys.* 101:1787 (1994).
77. L. M. Blinov, S. P. Palto, G. Ruani, C. Taliani, A. A. Tevosov, S. G. Yudin, and R. Zamboni, *Chem. Phys. Lett.* 232:401 (1995).
78. X. Wei and Z. V. Vardeny, *J. Synth. Met.* 54:99 (1993).
79. G. S. Kanner, X. Wei, L. Chen, and Z. V. Vardeny, *Phys. Rev. Lett.* 69:538 (1992).
80. G. S. Kanner, S. Frolov, and Z. V. Vardeny, *Phys. Rev. Lett.* 74:1685 (1995).
81. Y. Greenwald, J. Poplawski, X. Wei, Z. V. Vardeny, E. Ehrenfreund, and S. Speiser, *Mol. Cryst. Liq. Cryst.* 242:145 (1994).
82. Q.-X. Ni, L. Swanson, J. Shinar, T. Barton, and Z. V. Vardeny, *Phys. Rev. B* 44:5939 (1991).
83. B. Horowitz, *Solid State Commun.* 41:729 (1982).
84. P. A. Lane, X. Wei, and Z. V. Vardeny, *Phys. Rev. Lett.* 77:1544 (1996).

23
Electronic and Chemical Structure of Conjugated Polymers and Interfaces as Studied by Photoelectron Spectroscopy

M. Lögdlund, Per Dannetun,* and W. R. Salaneck
Linköping University, Linköping, Sweden

I. INTRODUCTION

Photoelectron spectroscopy (PES) is a technique based on the photoelectric effect, which was first documented in 1887 by Hertz and explained in 1905 by Einstein. The use of soft X-ray sources led to the development of X-ray photoelectron spectroscopy (XPS), originally known as electron spectroscopy for chemical application (ESCA) [1], indicating the applicability of the method to chemical analysis. More or less simultaneously with the development of XPS, ultraviolet photoelectron spectroscopy (UPS) [2], i.e., the use of ultraviolet photon sources in PES, was applied to study gases and bulk and surface electronic structures of solids. The increasing use of the continuous spectral distribution of synchrotron radiation [3,4] as a photon source, however, has made the historical terminology less meaningful. Today, PES is widely used for studying the bulk and surface chemical and electronic structure of condensed matter and gases. In particular, the method is very useful for studies of the chemical and electronic structure of conjugated polymers and interfaces with polymers because (1) it provides a maximum amount of chemical and electronic information within a single technique, (2) it is essentially nondestructive to most organic systems, and (3) the method is extremely surface-sensitive.

The purpose of this chapter is to illustrate the type of information that can be obtained about organic conjugated polymers, semiconducting and conducting, through the use of photoelectron spectroscopic measurements. Thus it presents a background to the application of photoelectron spectroscopy for conducting polymers; a brief description of the interpretation of PES data with the help of quantum-chemical calculations; some illustrative examples of specific polymer systems that incorporate features that are important and representative; a discussion, including a few examples, of the use of model molecular systems for conducting polymers; and an up-to-date selection of general references as well as citations of some recent work in the area. Since this handbook deals with all aspects of conjugated polymers, no justification for the study of conjugated polymers themselves is given in this chapter. Only a very brief background on conjugated polymers is given to the extent necessary for subsequent discussions; the use of conjugated molecules as model systems for conjugated polymers is discussed in more detail. Conjugated molecules are, of course, interesting in their own right because of their position as possible active component materials in future molecular electronics applications.

Finally, some examples included in this chapter deal with doping-induced chemical and electronic structural changes in conjugated systems. Many of the PES studies on doping of conjugated polymers reported in the literature have been devoted to p-type doping, as, for example, in systems like AsF_6^- and poly(p-phenylenevinylene) [5], AsF_5- and polyacetylene [6], ClO_4^- and polyacetylene [7,8], and $NOPF_6$- and poly(3-hexylthiophene) [9]; the possible new doping-induced states are unoccupied and thus not detectable by PES. The examples included in this chapter deal only with n-type doping, which allows for a direct measurement of new doping-induced states in the previously forbidden energy gap. It is assumed here that the reader is familiar with some of the unique electronic structural issues in the physics of conjugated polymers, i.e., the concepts of solitons, polarons, and bipolarons. Also, the lowest en-

---
*Current affiliation: Swedish Research Council for Engineering Sciences, Stockholm, Sweden.

ergy electronic transition will be used operationally as "the band gap," since the focus here is on molecular solids.

II. BACKGROUND

Although conjugated polymers are finding ever-increasing use as electroactive materials, they thus far have received relatively little attention compared to inorganic semiconductors. The investigation of the interaction between conjugated polymer materials and various metals has become an important issue, for both interfacial characteristics and doping-induced effects. Thus, examples are included that demonstrate information obtainable from photoelectron spectroscopy on pristine conjugated polymers as well as the modification of the electronic and chemical structure of conjugated systems induced upon interaction with different metals such as sodium, aluminum, and calcium. In addition, some examples are given of the use of model molecules for the conjugated polymers. These can be used to help in the interpretation of spectra obtained on conjugated polymers and even to predict their interactions with doping species or metal atoms.

Studies of the interface characteristics are of great importance in connection with various device applications. Investigations of initial stages of both metal-on-polymer and polymer-on-metal interface formation are of importance in the context of how the interfaces may affect the actual device performance. In other words, when a few isolated atoms of an active metal are applied in some way to a polymer surface, they will react in a very different way than when an isolated polymer chain is applied in some way to the clean surface of an otherwise three-dimensional metal substrate with or without the presence of an oxide layer [10].

The examples presented, taken mainly but not exclusively from our own work, illustrate the kind of information about the electronic and chemical structure that can be obtained from photoelectron spectroscopic data. We also demonstrate that quantum-chemical calculations can be helpful, and are very often mandatory, in interpreting spectra, yielding an in-depth detailed description of a particular system. The combined experimental and theoretical approach provides the opportunity of yielding information at a level not possible by either approach alone.

III. CONJUGATED SYSTEMS

A. Conjugated Polymers

The synthesis of soluble conjugated polymers during the past decade has led to improvement in the quality of samples prepared by solvent techniques and thus the possibility of obtaining detailed information about, for example, electronic structure through the use of photoelectron spectroscopy. Photoelectron spectroscopy has been an important method for the investigation of the overall quality of materials during the development of conjugated polymers. Today, the high quality of substituted conjugated polymers with various side groups allows for detailed investigations of the influence of side groups on the electronic structure [11]. To demonstrate the level of information obtainable from PES (mostly UPS), some of the most commonly used conjugated polymers are presented in this chapter. The polymers chosen for discussion are polyacetylene, poly(p-phenylenevinylene)s, and polythiophene. Results from XPS studies of pristine as well as charge-transfer-doped polypyrrole and various forms of polyanilines have been reviewed by Kang et al. [12] and are not discussed here. The basic chemical structures of the conjugated polymers discussed in this chapter are displayed in Fig. 23.1 for subsequent reference.

B. Model Molecules

The approach of using conjugated molecules or oligomers of conjugated polymers as models for the conjugated polymers has been used extensively in recent years. For example, series of oligomers with increasing numbers of monomer units have been studied with the purpose of extrapolating various physical properties to the corresponding conjugated polymer [13,14]. Also, many studies have been devoted to the investigation of the interaction between metals and conjugated molecules, including oligomers of conjugated polymers [15–22]. These studies are helpful for understanding the early stages of interface formation between metals and conjugated polymers as well as the possible charge storage configurations in conjugated molecules induced by charge transfer. Because the model molecular films may be prepared essentially without impurities, the UPS spectra obtained are almost always of higher quality (higher resolution) than those for the corresponding polymers. Knowledge obtained from such studies of model molecular systems might then be used to interpret the results of similar studies on real polymer surfaces or even to predict the behavior of real polymer interfaces.

The possibility of incorporating conjugated molecules as active components in molecular electronic devices, for example, in all-organic transistors [23] and in light-emitting diodes [24–26], demonstrates the importance of studies of conjugated molecules in their own right.

From a spectroscopic point of view, one of the most important advantages in using molecular materials is the possibility of preparing samples in situ, in ultrahigh vacuum (UHV), under clean and well-defined conditions ideal for spectroscopic studies. Such films are usually amorphous or very polycrystalline, but very homogeneous and pure, because of the nature of the preparation

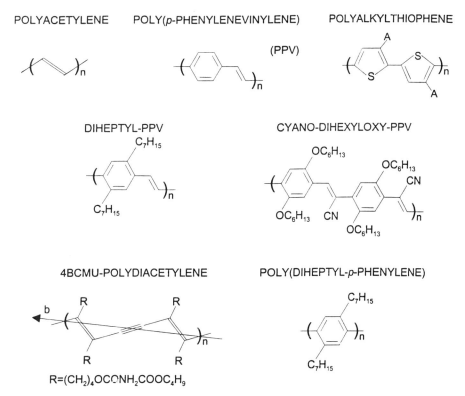

Fig. 23.1 Chemical structures of some of the conjugated polymers discussed in the chapter.

method. The conjugated molecules chosen as illustrative examples in studies of doping processes and interface formation are displayed in Fig. 23.2.

IV. PHOTOELECTRON SPECTROSCOPY

The basis of photoelectron spectroscopy and some specific features in a photoelectron spectrum that can be helpful for the understanding of the different examples exposed in this chapter are discussed in this section. Since this chapter deals with condensed conjugated systems, the main emphasis in the background discussion of PES is placed on the molecular solids aspect. For a more in-depth discussion on the technique relative to organic polymeric systems, interested readers can find more details in, for example, Refs. 27–29.

The fundamental process in molecular photoionization is represented by

$$M_0 + h\nu \rightarrow M_+^* + e^- \qquad (1)$$

where M_0 represents the isolated neutral molecule, $h\nu$ is the ionizing photon, M_+^* represents the positive molecular ion in the excited state, and e^- is the photoelectron that carries kinetic energy E_k. If E_k is sufficiently large, then the escaping electron and the molecular ion are not strongly coupled. The energy balance according to Eq. (1) is then given by

$$E_0 + h\nu = E_+^* + E_k \qquad (2)$$

where E_0 is the total energy of the neutral molecule in the ground state and E_+^* is the total energy of the positive molecular ion in an excited state. Thus, the basic equation used in interpreting photoelectron spectra can be written as

$$E_B = E_+^* - E_0 = h\nu - E_k \qquad (3)$$

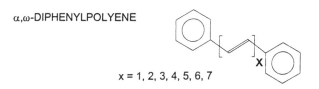

Fig. 23.2 Chemical structures of the conjugated molecules discussed in the chapter.

The photon energy $h\nu$ is known, and the photoelectron kinetic energy distribution is measured to deduce the binding energy E_B^V. The binding energy is thus not equal to the binding energy of the electrons in the neutral ground state of the molecule but corresponds to the energy difference between the initial ground state and various final excited states. The electronic structures of M_0 and M_+^* are usually modeled in terms of single-electron states, with as much physics built in as possible. Several schematic single-electron molecular configurations are shown in Fig. 23.3. The panels illustrate the neutral molecule in the ground state, M_0, where all of the electrons in the molecule occupy only the lowest allowed energy levels (V_i), while the V_i^* levels are empty; the generalized excited state of the (photoionized) molecular cation, M_+^*, where an electron has been removed form the core level, C_i, but the remaining electrons do not occupy the lowest possible single-electron states (the special case of ionization from the HOMO level is illustrated as M_+); and the optically excited neutral molecule, M_0^* (where one electron has been excited from the HOMO to the LUMO in the illustration). In addition, a case of particular interest in some of the examples included here is that of M_+^{**}. The case of M_+^{**} illustrates the phenomenon of "shake-up," where an electron is excited across the band gap simultaneously with a core-electron ionization event [1,30–32].

All of the various possible final states corresponding to a given initial state can in principle be observed in a photoelectron spectrum. In practice, however, most of the intensity usually resides in one main line. The main line corresponds to a particular simple final excited state of the molecular ion that is formed by the direct removal of a single electron from a specific initial state of the molecule. Other types of final states usually have low intensities and can be seen only as weak satellites on the main line. An idealized photoelectron spectrum, or energy distribution curve (EDC), presented as the number of electrons emitted per time unit versus the binding energy, with the corresponding molecular levels, is shown in Fig. 23.4. Within this one-electron picture there is a one-to-one correspondence between the peaks in a photoelectron spectrum and the one-electron molecular levels in the neutral molecule. Also illustrated in Fig. 23.4 is a shake-up satellite, the small feature appearing on the low kinetic energy side of the main peak labeled C_2. The escaping electron loses kinetic energy, ΔE_1, as a result of an excitation from an occupied level, V_3, to an unoccupied level, V^*, simultaneously with the electron emission from a core level, C_2. There are relax-

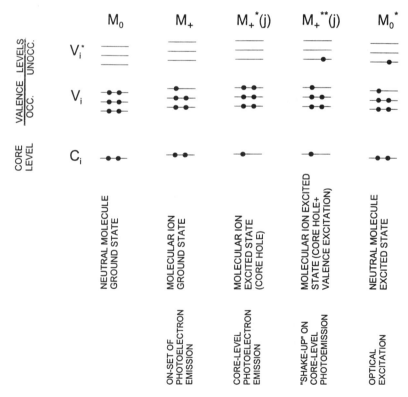

Fig. 23.3 Schematic illustration of some single-electron molecular configurations discussed in the text.

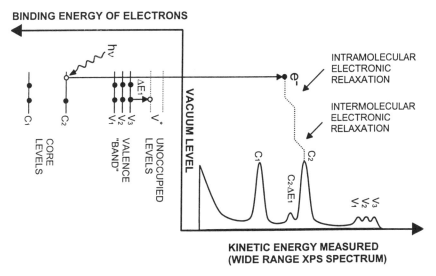

Fig. 23.4 An idealized photoelectron spectrum with the corresponding one-electron molecular levels.

ation effects that occur during the photoelectron emission event that can be rationalized, in a *classical picture,* as follows. An electron in a photoelectron emission event leaves a molecule typically within about 10^{-15} s. The nuclear geometric relaxation time is around 10^{-13} s (i.e., which corresponds to an optical phonon frequency) while the corresponding electronic relaxation time is about 10^{-16} s [33]. Thus, the main line corresponds to the binding energy of an electron in a ionized molecule where the electrons have had time to relax, i.e., the hole is fully screened, but the nuclei are frozen during the process; this is referred to as the adiabatic peak. The electronic relaxation effects are usually divided into intra- and intermolecular relaxation effects.

The *intra*molecular electronic relaxation effects occur in response to the creation of a hole state in photoelectron spectroscopy of an isolated molecule in the gas phase. In the solid state, there are also *inter*molecular relaxation effects that are due to electronic and atomic polarization of the molecules surrounding the particular molecule on which the hole state is created. The *inter*molecular relaxation energy is given to the escaping electron, and thus the kinetic energy is increased, i.e., the corresponding peak in the photoelectron spectrum appears at lower binding energy than photoelectron emission from the same level in the gas phase. This energy difference, i.e., the polarization energy, is on the order of 1–3 eV [34] throughout the valence region. The relaxation energy for core levels, however, can be much larger due to a higher degree of localization of the core holes [32,33].

A. X-Ray Photoelectron Spectroscopy

With XPS, it is possible to perform qualitative and even semiquantitative analysis of chemical composition in the near-surface region of a solid sample. Although the core electrons are not involved in the chemical bonds, core-level XPS gives information about the chemical environment [1]. Changes in the valence electron density will be reflected as small but significant shifts in the core-level binding energies. XPS can, of course, be used (and has been extensively used) for studies of the valence band region. The photoionization cross section for the valence photoelectrons in XPS is approximately one order of magnitude lower than that for the core photoelectrons, however, which leads to more time-consuming experiments. Also, the photoionization cross sections are such that it is not very convenient to study conjugated polymers, where interest is in the states near the valence band edge, which are very weak in XPS valence band spectra.

It is possible to obtain useful information about the valence π-electronic structure of a molecular ion through the so-called shake-up structures that appear as weak satellites on the low binding energy side of the main line. The shake-up structure reflects the spectrum of the one-electron–two-hole states generated in connection with photoionization and can sometimes be used as "fingerprints" of certain molecular constituents by comparison with shake-up spectra of known systems.

In solid specimens, like those dealt with here, the measurements are performed with reference to the

Fermi level of the photoelectron spectrometer since the Fermi level of the sample lines up with that of the spectrometer. Thus, in cases when the Fermi level shifts because of some chemical modifications of the sample, i.e., changes in the intercalation of graphite or other layered compound [35] or in the doping of conjugated polymers [36], it is necessary to account for the change in the Fermi energy level before interpreting the spectra. This can be done, for example, by combining the XPS core-level spectra with UPS valence band spectra, as described in more detail in the next section.

B. Ultraviolet Photoelectron Spectroscopy

Although synchrotron radiation is in extensive use today as a photon source, the most commonly used laboratory photon source for UPS is the helium resonance lamp. For valence band spectroscopy, UPS has two advantages over XPS. First, the usual in-house photon sources have high energy resolution, a full width at half-maximum (FWHM) of ~30 meV for the He lines. However, the natural linewidths in condensed molecular solids, from both homogeneous and inhomogeneous broadening effects, can approach 1 eV at room temperature [28,37,38], and thus the photon energy resolution advantage is not fully realized. The second advantage of UPS is the higher photoionization cross section for electrons in the valence region.

As mentioned above, the PES measurements of solid specimens are performed relative to the Fermi level. However, sometimes it is desirable to derive the binding energies relative to the vacuum level from a photoelectron spectrum. This is particularly important before interpreting different core-level chemical shifts induced, for example, by doping or interaction with metals. In fact, in some cases the binding energy shift can appear to be in the wrong direction before the Fermi energy shift is taken into account [9,36]. The Fermi energy shift can be deduced by measuring the position of the zero kinetic energy cutoff of the secondary electron distribution, as illustrated in Fig. 23.5. The vacuum level can be located simply by adding the photon energy to the secondary electron cutoff energy; measuring down from the vacuum level energy to the known position of the Fermi energy (E_F) determines the work function (Φ_S) of the substrate [39].

In connection with Fig. 23.5, it is worth pointing out that the reference level, i.e., the Fermi level of the spectrometer, is determined by performing a UPS measurement on a metallic specimen in equilibrium with the spectrometer, taking Au as in the example here. The Fermi level of the metal, by definition, lines up with the Fermi level of the spectrometer. Since, in equilibrium, the position of the Fermi energy is constant throughout the thickness of the thin films under discussion in this chapter, the position of the Fermi energy on the UPS spectrum is equal in analyzer voltage to the location of the Fermi level in the metallic substrate, as observed prior to the deposition of the organic overlayer. This positioning of the Fermi level, coincident with that in the substrate, has been observed in all studies reviewed in this work as well as in other molecular overlayer systems, on both metals and semiconductor substrates [29,40–43].

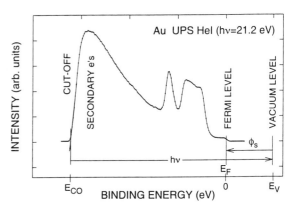

Fig. 23.5 UPS spectrum of Au illustrating the procedure to obtain vacuum level referenced spectra.

Valence band spectra provide information about the electronic and chemical structure of the system, because many of the valence electrons participate directly in chemical bonding. The experimental UPS spectra are sometimes evaluated by the "fingerprint" method, that is to say, in comparison with known standards. A much better approach is to use comparison with the results of appropriate model quantum-chemical calculations [29]. In combination with quantum-chemical calculations it is possible to specify the electronic structure in terms of atomic or molecular orbitals or in terms of band structure. The experimental valence band spectra in many of the examples included in this chapter are interpreted with the help of quantum-chemical calculations. Some basic considerations and some procedures are outlined in the next section.

C. Interpretation of Valence Bands Through Theoretical Modeling

As discussed above, the interpretation of photoelectron spectroscopic results relies either on a comparison with photoelectron spectra on reference materials or on theoretical model calculations. Along with the progress in theoretical methods and, in particular, the dramatic increase in computational power over the past decade, access to quantum-chemical calculations has become more available. Currently, the use of quantum-chemical calculations in the interpretation of experimental UPS valence band spectra is almost standard [13,14,44,45].

Interpretation of the UPS spectra of molecular systems discussed below commonly relies on the one-electron picture of the neutral molecules; in this case there exists a one-to-one correspondence between the major peaks in the photoelectron spectrum and the one-electron molecular orbitals (see Fig. 23.4). Usually the numerical values of the calculated binding energies of the peaks are set to the Hartree–Fock eigenvalues of the molecular orbitals in the neutral ground state of the molecule, e.g., employing the Koopman theorem. Important effects to consider when relying on quantum-chemical calculations are the various relaxation effects that occur during a photoelectron emission event. It is usually necessary to put in "by hand" (or in some more sophisticated theoretical fashion) corrections for the relaxation phenomena that account for differences between the molecular orbitals of the neutral molecule and the molecular ion.

Theoretical model calculations for the interpretation of UPS valence band spectra of molecular solids are usually performed on isolated molecules, i.e., not taking into account any *inter*molecular interaction. This is justified by the observation that for organic molecules there is a one-to-one correspondence between the UPS valence band spectra of the molecular solid and molecules in the gas phase, caused by the generally weak *inter*molecular van der Waals interaction between the molecular species in organic molecular solids; there is a rigid shift throughout the valence region on the order of 1–3 eV [46,47]. The shift is due to polarization of the surrounding molecules, which stabilizes the final ion state and leads to a decrease in binding energy [32].

The examples included in this chapter use some different theoretical models for the interpretation of, primarily, UPS valence band data, for pristine and doped systems as well as for the "initial stages" of interface formation for metals on conjugated molecules. Among the theoretical methods used in the examples are semiempirical Hartree–Fock methods such as the modified neglect of diatomic overlap (MNDO) [48,49] and Austin Model 1 (AM1) [50], the nonempirical valence effective Hamiltonian (VEH) pseudopotential method [51,52], ab initio Hartree–Fock techniques, and the local spin density (LSD) approximation [53,54].

A common strategy for calculating the density of valence states for large conjugated systems, for both pristine and charged (doped) systems, is to use a two-step procedure. The ground-state geometry is first optimized by applying a semiempirical method (e.g., AM1 or MNDO); the optimized geometries then serve as input for electronic structure calculations performed with the VEH method. The AM1 and MNDO methods are known to yield reliable geometries for large organic molecules. The applicability of the VEH model in interpreting photoelectron valence band spectra of conjugated systems is well established [29,55–58]; the VEH parameters have been determined by fitting to double-zeta quality *ab initio* results on model molecules [51,52].

Calculations on conjugated molecules using *ab initio* Hartree–Fock or LSD methods must, of course, be restricted to relatively small systems due to the dramatically increased computational efforts. However, in some cases one can gain reliable detailed information for a particular material system by performing calculations on an adequate model system. This has been done, for example, in studies of the interaction between different metals and conjugated molecules, on both the *ab initio* Hartree–Fock and LSD levels. The LSD approximation has been shown to give reliable estimates of geometric and electronic structures of pristine π-conjugated molecules and metal/π-conjugated molecule complexes [59–61]. The LSD calculations included in some of the examples in this chapter have been performed using the DGauss program [62] with the Vosko–Wilk–Nusair (VWN) analytic expression for the LSD exchange-correlation potential [63] and with a split valence basis set including p- and d-type polarization functions.

V. PRISTINE SYSTEMS

A. Molecules

1. Diphenylpolyenes

The capping phenyl groups on the diphenylpolyenes enhance the environmental stability with respect to the unsubstituted polyene molecules. Since the phenyl groups do not contribute significantly to the first few frontier orbitals, the diphenylpolyenes are suitable as models for short polyenes [64–66]; the separation of the frontier orbitals in terms of polyene- and phenyl-dominated molecular orbitals can be deduced from a comparison between experimental UPS spectra and theoretical quantum-chemical calculations as well as from experimental "fingerprints" in the XPS C(1s) core-level shake-up spectra, as demonstrated below for a diphenylpolyene with seven carbon–carbon double bonds in the polyene part of the molecule, i.e., α,ω-diphenyltetradecaheptaene (DP7) [66]. The chemical structures of the diphenylpolyenes are displayed in Fig. 23.2.

The UPS He I and He II valence band spectra of DP7 are compared with the density of valence states (DOVS) as convoluted from VEH calculations in Fig. 23.6 the corresponding eigenvalues are displayed at the bottom. The initial states of the photoelectrons, in terms of molecular orbitals, can be obtained from calculations. The electrons contributing to the low binding energy part, i.e., from about -5 eV to the valence band edge near -2 eV, originate from the six highest (i.e., lowest binding energy) occupied molecular orbitals. The two peaks at lowest binding energy, peaks B and C, correspond to

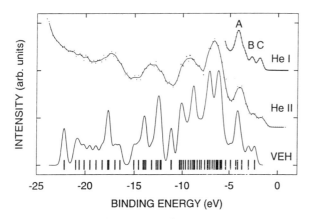

Fig. 23.6 UPS He I and He II valence band spectra of DP7 are compared with the DOVS as convoluted from VEH calculations (From Ref. 66.)

electrons originating from the highest and second highest occupied molecular orbitals, respectively, which are mainly localized to the polyene part of the molecule. Peak A corresponds to electrons in four π-orbitals, two of which are doubly degenerate and completely localized to the phenyl groups of the molecule while the other two are more evenly spread out over the molecule. In total, however, peak A is dominated by electrons corresponding to π orbitals mostly localized to the phenyl groups.

In Fig. 23.7 is shown a portion of the C(1s) XPS spectrum for trans-polyacetylene, DP7, and condensed benzene [66]. The main C(1s) peaks are not shown, only the relatively weak shake-up (s.u.) satellite features, which appear on the high binding energy side of the main C(1s) XPS peak. Fig. 23.7 indicates that the s.u. spectrum of DP7 can be considered as the superposition of the s.u. spectra of benzene and trans-polyacetylene. This separation in energy of the s.u. peaks comes directly from the fact that the frontier orbitals (both the occupied and unoccupied, by symmetry) are separated into polyene-like and benzene-like. Note that relative to trans-polyacetylene, the major (lowest energy) s.u. peak of DP7 is at higher binding energy relative to the main C(1s) and appears to be more intense, consistent with the fact that the optical band gap in DP7 is larger than that in polyacetylene.

The low binding energy part of the experimental UPS He I valence band spectra for a series of diphenylpolyenes with from one to seven C=C double bonds in the polyene part are compared with the DOVS calculated from the VEH in Fig. 23.8 [67]. The molecules are abbre-

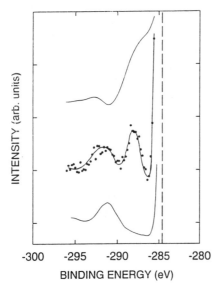

Fig. 23.7 Shake-up satellite features in the XPS C(1s) spectra for *trans*-polyacetylene (top), DP7 (middle), and benzene (bottom). (From Ref. 66.)

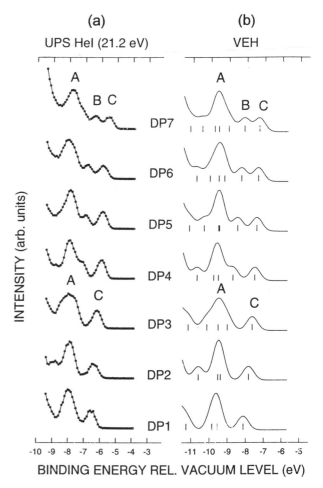

Fig. 23.8 The low binding energy part of (a) experimental UPS spectra and (b) calculated DOVS of pristine DP1–DP7. (From Ref. 67.)

viated DPx, where DP stands for diphenylpolyene and x is the number of carbon–carbon double bonds in the polyene part. The energy scales are relative to the vacuum level. The energy levels as obtained from VEH are indicated by bars in the figure. The peaks at about −8 eV (labeled A) in the UPS spectra (about −9.5 eV in the VEH simulation) have, as discussed above, the most significant contributions related to the phenyl groups. Recall that the e_{1g} degenerate HOMO orbitals of benzene split upon connection to the polyene segment. One of the degenerate e_{1g} orbitals has a node at the carbon site attached to the polyene, resulting in orbitals localized to the phenyl groups in the diphenylpolyene. Since peak A is dominated by such contributions, this peak is unaffected by the length of the attached polyene. The peaks at lower binding energies, i.e., the two peaks labeled B and C in the cases of DP7–DP4 and the peak labeled C for DP3–DP1, depend on the polyene length because the corresponding molecular orbitals are mainly localized to the polyene part of the molecules. This series of diphenylpolyenes nicely demonstrates the good agreement between theory and experiments that can be obtained for well-defined systems. In particular, note the movement of the peak labeled B; starting from DP7 the peak shifts toward higher binding energy, appears as a shoulder on peak A for DP4, and is completely merged with peak A in the case of DP3.

2. Oligothiophenes

Among the results reported from PES studies on oligomers of thiophenes can be found, for example, studies devoted to the investigation of the electronic structure of pristine [13,14] and doped [21,68,69] oligothiophenes as well as the interface formation with metals such as aluminum [17]. Fig. 23.9 shows the experimental UPS valence band spectra of a thin condensed film of pristine sexithienylene, α-6T, recorded with He I (21.2 eV) and He II (40.8 eV) radiation, compared with the density of valence states (DOVS) derived from the VEH calculation [69]; the chemical structure of α-6T is displayed in Fig. 23.2. The agreement between theory and experiments makes it possible to relate the experimental peaks to the calculated molecular energy states. The three low binding energy peaks labeled A, B, and C are of particular interest because these frontier orbitals are expected to be the orbitals mainly effected upon interaction with, for example, metal atoms. The two peaks at lowest binding energies, C and B, correspond to electrons from the highest and second highest occupied molecular orbitals, HOMO and HOMO-1, respectively. These orbitals are delocalized over the carbon backbone of the molecule, with negligible coefficients on the sulfur atoms. Peak A is related to electrons in seven different π orbitals of which six are localized mainly to the sulfur atoms and the β carbons. The lowest unoccupied molecular orbital (LUMO) has significant coefficients on the sulfur atoms.

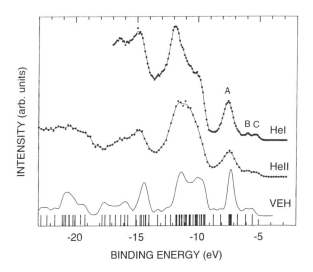

Fig. 23.9 UPS valence band spectra of pristine α-6T compared with DOVS derived from VEH calculations. (From Ref. 69.)

B. Polymers

1. Polyacetylene

trans-Polyacetylene (trans-PA) has so far been one of the most studied conjugated polymers in the family of intrinsically conducting polymers. This is because polyacetylene is the polymer that actually opened the field of "conducting polymers" and can be considered the prototype of conjugated polymers. trans-PA has a particularly simple structure (see Fig. 23.1), a planar zigzag configuration with alternating single and double bonds of 1.44 and 1.36 Å [70]. The band structure is very simple and is well suited for demonstrating the relationship between band structure, density of valence states (DOVS), and experimental valence band spectra.

In Fig. 23.10a are shown, from top to bottom, the experimental valence band spectra of trans-PA recorded with photon energies of 27 and 50 eV and the DOVS convoluted from the VEH calculation. Figure 23.10b shows the energy band diagram from which the DOVS was calculated [45]. The energy scale was determined experimentally, i.e., with respect to the Fermi level in the UPS spectra. The input geometry for the VEH calculations was taken from AM1 geometry optimizations on an oligomer of trans-PA. Differences when using experimental values instead are small and are not found to affect any of the results (the AM1-optimized bond lengths are 1.44 and 1.35 Å).

The electronic structure of trans-PA can be rationalized in the following manner [52]. In the absence of dimerization, i.e., with equal C–C and C=C bond lengths, polyacetylene would be a "regular polyene," and the

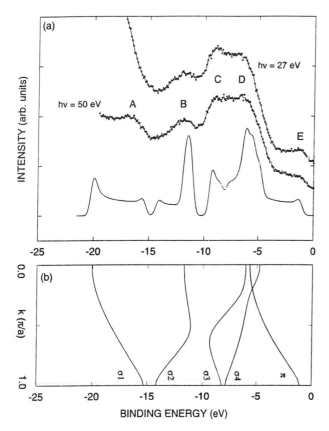

Fig. 23.10 Valence band spectra of *trans*-polyacetylene, recorded using synchrotron radiation at 27 eV and 50 eV photon energy, and the corresponding DOVS derived from VEH calculations. The VEH band structure is shown in (b). (From Ref. 45.)

unit cell would consist of a single –CH– group. At least within one-electron theory, there would exist three occupied valence bands, corresponding, in localized bond orbital terminology, to approximately a C–C σ band, a C–H σ band, and a π band, the latter being derived from the remaining p_z atomic orbital on each C atom (the z axis being perpendicular to the molecular plane). Since there would be one electron in each p_z atomic orbital on each C atom, the occupied π band would be only half filled and the electronic structure of the polymer would have a metallic character. Because of a Peierls transition (or, equivalently in molecular terminology, a Jahn–Teller distortion) and electron correlation effects, the system is dimerized with alternating single and double carbon–carbon bonds. The unit cell in trans-PA contains two carbon atoms, corresponding to CH=CH units. In the unit cell with two C atoms, the one-dimensional Brillouin zone is only one-half as wide as in the case of the (hypothetical) regular polyene; therefore each band of the regular polyene is split into two bands. In the dimerized unit cell, there are thus four occupied σ bands and one occupied π band. Due to the presence of dimerization, there are energy gaps at the Brilluoin zone edge, near –8 and –15 eV in Fig. 23.10, where each pair of bands from the regular polyene structure is folded back to form the dimerized chain band structure. The Peierls gap in the π band, which occurs at the Brilluoin zone edge, corresponds to the forbidden energy gap of approximately 1.5 eV in trans-PA. In addition, at the points where two σ bands of like symmetry intend to cross, there is an avoided band crossing.

The different peaks in the UPS spectra can be assigned to different bands from the details in the calculation: Peak A in the spectra corresponds to electrons from a band derived almost exclusively from C(2s) atomic orbitals, i.e., the C–C backbone. Peak B is derived from the flat portion of the σ_2 band near the zone center. Peaks C and D correspond to electrons from the σ_3 and σ_4 bands derived from combinations of C–C and C–H bands at points in the Brilluoin zone where there is a high density of states, i.e., at the zone center near –9 eV and at the zone edge near –8 eV. The broad peak E has its origin in the strongly dispersed π band, which also contributes to peak D. The valence band edge corresponds to the highest portion of the occupied π band, at the zone edge (at about -1 eV in the calculated band structure of Fig. 23.10).

As mentioned above, the σ bands can loosely be referred to as C–C and C–H. In polyethylene, the upper bands are essentially C–H and the lower C–C [52]. Because of the large dispersion in polyacetylene, however, this separation in energy does not occur. This is demonstrated in Fig. 23.11, where the evolution of the two σ bands is shown, going from an unfolded band structure, i.e., corresponding to a situation where the unit cell consists of a single –CH– group, to a situation where the energy gaps at the Brillouin zone edges and the avoided-band-crossing effects are included. The two σ bands cross at about –15 eV, as seen in Fig. 23.11a. A folding of the σ bands into the first Brillouin zone, Fig. 23.11b, causes another band crossing at about –18 eV. When the avoided band crossings and the energy splitting at the Brillouin zone edges, Fig. 23.11c, are taken into account; the actual σ-band structure of trans-polyacetylene can be fully rationalized.

2. Poly(p-phenylenevinylene)

The poly(p-phenylenevinylene)s, or PPVs, are among the most studied of the conjugated polymers, because they are used as active layers in light-emitting devices. The emitted color can be tuned by choosing different substituting side groups. The attachment of side groups has the added advantage of increasing the solubility of the polymers. The chemical structure of PPV is displayed in Fig. 23.1.

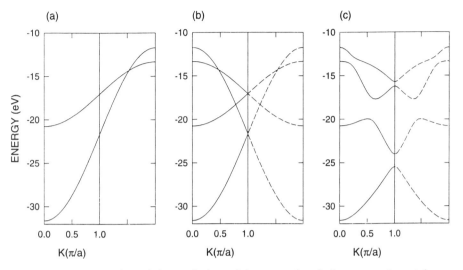

Fig. 23.11 Illustration of the evolution of the two σ-bands in *trans*-polyacetylene.

In Fig. 23.12, the experimental UPS He I and He II valence band spectra of PPV prepared by the tetrahydrothiophenium precursor route are compared with the DOVS as calculated from the VEH band structure [45]. XPS core-level spectroscopy (not shown) indicates a complete conversion of the precursor polymer. The corresponding band structure is shown in Fig. 23.12b. The energy scale is fixed relative to the experimental Fermi level. The PPV is taken to be planar in the VEH calculations used here, since neutron diffraction measurements on oriented PPV at room temperature have shown that the ring torsion angles, i.e., the twist of the phenyl rings out of the vinylene plane, are on the order of 7° ± 6° [71]. Such small torsion angles result in negligible effects on the calculated electronic band structure compared with that for the fully coplanar conformation.

In the UPS spectra of PPV, the different peaks are assigned from the details in the VEH calculations: Peaks A, B, and C originate from electrons in σ bands; peak D is built up from contributions from the four highest σ bands, the lowest π band, and a small portion from the relatively flat part of the second π band; peak E is derived from the next highest π band, which is extremely flat because it corresponds to electronic levels fully localized on the bonds between ortho carbons within the phenyl rings. In general, a flat band results in a high intensity peak in the DOVS, since there are many states per unit energy just at the flat band. There also are small contributions to peak E from the second and fourth π bands. Finally, peak F is derived from the top part of the highest π band. The larger dispersion of the top π band results in lower intensity in the UPS data.

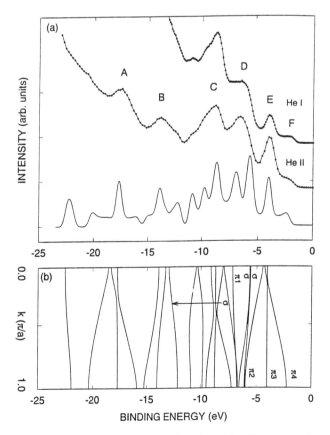

Fig. 23.12 He I and He II UPS valence band spectra of PPV compared with DOVS derived from VEH calculations. The VEH band structure is shown in (b). (From Ref. 45.)

The unit cell of PPV consists of a styrene-like unit. Upon attaching a vinylene group to a benzene molecule, the doubly degenerate outermost π orbitals split into two orbitals as a result of the lifting of degeneracy [72–74]. The π3 band is localized on the ortho carbons, i.e., it consists of localized states. The wave function, which has a nonzero intensity at the para carbon atoms (where the neighboring vinylene groups are bonded), interacts with the π state of the vinylene group, leading to the generation of the dispersed π bands π2 and π4. The top of the π4 band has almost equal contributions from the vinylene and phenylene groups. Due to the lack of symmetry within the unit cell, an avoided band crossing occurs between the π3 and π4 bands. The difference between the top of band π4 and the flat band π3 is about 2.3 eV. The flat band π3 appears only about 0.3 eV above what would have been the bottom part of the wide π4 band in the absence of an avoided band crossing. The bandwidth of π2 is about 2.0 eV, i.e., almost the same degree of delocalization as π4. Finally, the bandwidth of π1 is about 0.6 eV. This band is related to the $1a_{2u}$ orbital in benzene.

3. Polyalkylthiophene

One breakthrough in applications of conjugated polymers was the development of soluble polymers, as in the case of polythiophene [75], by substitution with, for example, alkyl side chains [76,77]. The polyalkylthiophenes exhibit properties such as solvatochromism and thermochromism [78]. A large variety of substituted polythiophenes with various band gaps exist [79].

Two He II UPS spectra of a polyalkylthiophene, namely poly(3-hexylthiophene) or P3HT, and the DOVS derived from VEH band structure are compared in Fig. 23.13 [80]. The chemical structure of P3HT is sketched in Fig. 23.1. The spectra were recorded at +190°C and −60°C, respectively, and the DOVS was derived from VEH calculations on a planar conformation of P3HT. From comparison with the details in the results from the VEH calculations, it was found that the presence of the hexyl side chains mainly influences the UPS spectra by adding intensity in the energy region from about −5 eV to −12 eV. The energy region below −5 eV is derived entirely from electrons in the π bands. Thus, the frontier orbitals, which are important in, for example, metal–polymer interactions, are unaffected by the hexyl chains. On the other hand, the XPS signal from electrons in the C(1s) core levels of carbon atoms in the thiophene units will be obscured by the signal from the carbon atoms in the side chains; this is discussed further in connection with metal–polymer interactions presented below. The fine details in the low binding energy part of the spectrum indicate that the −60°C spectrum corresponds to the planar situation, and not the +190°C spectrum, as determined from the calculations. This is in agreement with optical absorption spectroscopy, where

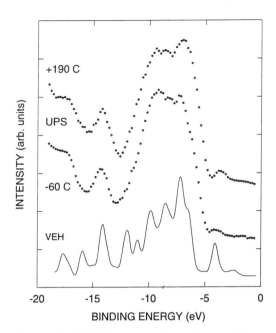

Fig. 23.13 He II UPS spectra of poly(3-hexylthiophene), recorded at -60°C and 190°C, compared with DOVS derived from VEH calculations. (From Ref. 80.)

it was found that the optical band gap increases with increasing temperature [78].

In a UPS study by Fujimoto et al.[81], the effects of air and temperature on the electronic structure of chemically and electrochemically prepared solution-cast poly(3-alkylthiophene) films were investigated. It was found that the threshold energy was almost identical for the different solution-cast films. However, electrochemically, as-prepared films had a smaller threshold value by about 0.3 eV. Changes in the work functions (of about 0.2 eV) of the samples were observed upon heating in vacuum. Subsequent recovery upon exposure to air was attributed to dedoping and doping effects caused by oxygen.

VI. INTERACTION WITH METALS

The examples included here demonstrate the information obtainable from studies of the early stages of interface formation between metals and conjugated molecules or polymers. Although the examples are specific, related or similar studies are referenced. The examples deal only with interfaces formed by evaporating metals *on top of* a polymer or molecular film. Although it is expected that the interface formed by depositing a conjugated system on top of a metal electrode behaves differently, we are not aware of any results reported from *ideal* studies of conjugated polymers on metals. The task of using real conjugated polymers on metals is a difficult one, since the polymer molecules deposited must be in the monolayer coverage region (~10 Å) in order to detect the interface region by UPS. On the other hand, in principle it is possible to reduce the thickness of the polymer film by successive removal of layers by, for example Ar$^+$ bombardment, as used in a study of PPV on chromium [82]. Unfortunately, the ion bombardment destroys the polymer molecules in the process, so the information obtained is only on the "atomic composition level." Another possible method is to use model molecules for conjugated polymers, as discussed above, allowing controlled in situ deposition of molecules onto clean metal surfaces. This technique has been used in, for example, model studies of the polyimide-on-copper interface [10].

A. Metals on Poly(*p*-phenylenevinylene)s

1. Calcium on Poly(2, 5-diheptyl-*p*-phenylenevinylene)

In connection with the progress made in the fabrication of light-emitting diodes (LEDs) from conjugated polymers, some of the most promising results have been obtained with poly(*p*-phenylenevinylene) as the light emission medium, with (hole-injecting) ITO glass as the transparent substrate and with a calcium or aluminum electrode as the (electron-injecting) metallic contact.

Fig. 23.14 presents the low binding energy parts of the UPS He I valence band spectra of poly(2,5-diheptyl-*p*-phenylenevinylene) (DHPPV) following several stages of deposition of approximately a monolayer of calcium (at each step). The chemical structure of DHPPV is displayed in Fig. 23.1. The heptyl side chains do not contribute to the intensity below −8 eV [83]. The energy scale is relative to the vacuum level and the Fermi energy for each spectrum, as indicated. The bottom-most spectrum corresponds to pristine DHPPV, while the uppermost curve corresponds to the presence of about one Ca atom per DHPPV monomer unit, as determined from the relative intensities of the XPS Ca(2*p*) and C(1*s*) core levels. The binding energy of the Ca(2*p*) level indicates that Ca is in the form of Ca^{2+} ions. Initially, Ca atoms deposited on the surface lead to the complete doping of the top layer (probed by UPS). Upon further deposition, the amount of Ca at the immediate surface (seen by UPS) does not continue to increase, i.e., saturation doping in the surface region has been reached. Additional Ca atoms then diffuse somewhat into the bulk. Finally, the C(1*s*) spectrum broadens and shifts slightly toward lower binding energy as the deposition of calcium continues, indicating charge transfer from the calcium to the carbon atoms. As a result of the deposition of Ca, the work function of the sample changes from 4.0 (± 0.1) eV to 3.0 eV, which is close to that of metallic calcium.

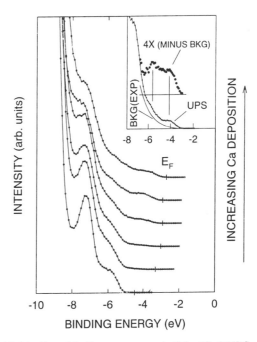

Fig. 23.14 Low binding energy part of the He I UPS spectra of DHPPV recorded during successively greater deposition of calcium. The inset shows the fully doped (one Ca atom per monomer) DHPPV with a simple estimate of the inelastic electron background to emphasize the calcium-induced structures. (From Ref. 83.)

Clear signatures of charge transfer (doping) versus covalent chemical bonding can be distinguished in UPS spectra. A simple subtraction of the inelastic electron background signal facilitates the interpretation of the new induced states in DHPPV, as shown in the inset of Fig. 23.14. It can be seen that upon doping two new electronic states above the π-band edge of the pristine polymer are induced near −4.0 and −5.8 eV relative to the vacuum level. The details, i.e., the position within the otherwise forbidden energy gap, as well as the relative splitting of the two new states are similar to those in the case of sodium on PPV [84], where two new (bipolaron) states within the gap were observed.

Also, through the XPS data, a limited diffusion of the dopant ions is indicated. The Ca atoms are not localized at the immediate surface but rather diffuse into the near-surface region. Although it would be extremely difficult to determine an exact diffusion depth profile, it is obvious from the XPS(θ) data that the Ca atoms are confined to within the depth observable by XPS. Therefore the upper and lower limits to the diffusion distance (which is undoubtedly not a sharp diffusion depth) are defined. In this way it can be estimated that the Ca atoms are distributed in some way within roughly 25 Å of the surface of the DHPPV.

In an XPS study on Ca deposited on PPV prepared by the tetrahydrothiophenium route with a surface composition of the as-converted sample consisting of 82% C, 17% O, and 1% S, it was found that the C(1s) binding energy was fixed in binding energy up to a Ca coverage of about 8 Å and shifted toward lower binding energy by 0.4 eV upon further deposition [85]; both the oxygen and the sulfur moieties reacted strongly with the deposited Ca atoms. The sudden shift of the C(1s) binding energy was suggested to be due to the exhaustion of the impurity layer and the beginning of interaction of the deposited Ca atoms with the PPV substrate. The shift of the C(1s) level was attributed to the formation of new interface states, resulting in band bending in the PPV substrate and Schottky barrier formation at the interface. Although the polymer contained sufficient oxygen, such studies are important to the understanding of the influence of "dirt" at the interfaces of electronic devices. In addition, according to Schott [86], band bending should not occur in films of undoped conjugated polymers of finite (of LED dimensions) thickness.

2. Aluminum on Poly(p-phenylenevinylene)

Results of PES studies of the interface between aluminum and poly(p-phenylenevinylene) have been reported for XPS studies [82,87,88] and UPS studies [87]. In general it is found that oxygen may be present in various degrees, either as contamination at the surface of the as-prepared PPV films or in the metal source itself. The presence of oxygen will, of course, affect the characteristics of the interface [82,88]. The XPS results obtained by Ettedgui et al.[88] were interpreted in terms of a delay of Schottky barrier formation due to formation of a buffer layer, induced by the surface impurities, which prevents the PPV substrate from interacting with the deposited aluminum layer.

In Fig. 23.15 is shown the low binding energy region of the UPS He I valence band spectra of PPV recorded during the early stages of interface formation with aluminum [87]. The pristine system is shown at the bottom, and spectra are increasingly higher in the diagram for increasing deposition of Al. The pristine PPV samples were found to be clean, with only a few atomic percent of oxygen at the surface, as determined by XPS. Although the spectrum of the pristine PPV is not as well resolved as that reported by Lögdlund et al. [45], the overall structure, e.g., the peaks at -2, -4, and -6.5 eV, which correspond to the major structure of the π system, is reproduced. The π structure is strongly affected by the Al deposition, and the band edge is essentially destroyed. This can also be seen in the XPS C(1s) core level shake-up spectra as displayed in Fig. 23.16 [87]. Two satellite peaks, at about -288 and -292 eV, can be detected on the high binding energy side of the main line (not shown). As aluminum is deposited, approximately in monolayer steps, the region above -290

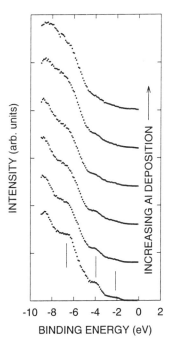

Fig. 23.15 Low binding energy part of the He I UPS spectra of PPV recorded during successively greater exposure to aluminum. (From Ref. 87.)

eV exhibits an increasing background intensity due to electrons from the C(1s) level in the atoms of PPV that are inelastically scattered in the Al overlayer. However, various fingerprints in the full C(1s) spectra indicate that the intensity of the -288 eV peak decreases, as can be seen in the successively higher curves in Fig. 23.16. Since the valence band edge of PPV is dominated by contributions from the π bands delocalized over the vinylene moieties, and the -288 eV shake-up feature corresponds approximately to electrons removed from the vinylene groups, the results have been interpreted in terms of the vinylene groups being the preferable sites for interaction with Al atoms during the initial stages of interface formation. The interpretation of the experimental results is supported by results obtained from theoretical studies of the interaction between PPV and Al atoms, where it was found that Al preferentially interacts with the vinylene moieties and strongly modifies the geometrical structure by the formation of covalent bonds and sp^3 hybridization, which leads to interruption of the π conjugation [89,90]. These results are in contrast to the case of calcium on DHPPV, where the results indicate charge transfer doping, i.e., the bondings being mainly ionic in character [83].

3. Sodium on Poly(p-phenylenevinylene)s

Some UPS results from p-type doping of polyacetylene, using potassium as a doping species, have been reported

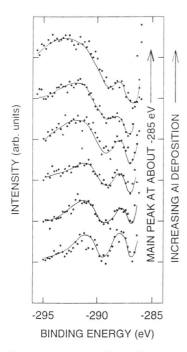

Fig. 23.16 Shake-up portion of the C(1s) spectra of PPV recorded during successively greater exposure to aluminum. (From Ref. 87.)

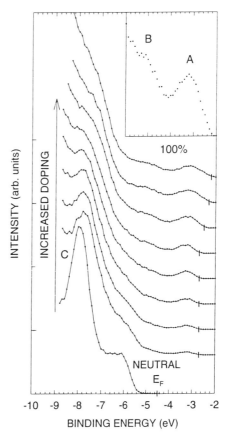

Fig. 23.17 The low binding energy part of the He I UPS spectra of PPV recorded during successively greater exposure to sodium. (From Ref. 84.)

by Tanaka et al. [8]. The first direct measure of multiple resolved gap states in a doped conjugated polymer were reported by Fahlman et al. in the case of sodium-doped poly(p-phenylenevinylene) [84]. Upon doping, the UPS spectra indicated a slight decrease in the work function at the first doping steps. At about 40% doping, defined as the Na/monomer ratio, a large change of about 1.2 eV occurred, followed by a slight decrease as the doping level approached 100%, i.e., one sodium atom per "monomer" repeat unit.

Simultaneously with the 1.2 eV change in the work function, i.e., at intermediate doping levels, two new states appear in the previously empty energy gap; the intensities of the new gap states increase uniformly with the doping as shown in Fig. 23.17. The separation between the two new peaks is about 2.0 eV at the maximum doping level, i.e., near 100%, with the lower binding energy peak positioned at about −3.2 eV. At this doping level, the charges in this nondegenerate ground-state polymer can be accommodated either in two polaron bands or in two bipolaron bands.

From model calculations performed using the VEH technique for a 100% doping level, the new states appearing in the previously forbidden energy gap are assigned to two doping-induced bipolaron bands. Also, the lack of significant density of states at the Fermi level, as would be expected for a polaron situation, indicates that the formation of bipolaron bands is most likely.

In a more recent study, the interaction between sodium and and a cyano-substituted poly(dihexyloxy-p-phenylenevinylene) (CNPPV) was investigated [91]. As expected, the evolution is very reminiscent of that for the unsubstituted PPV; two new states appear in the previously forbidden energy gap, and no density of states is detected at the Fermi level, which is consistent with a bipolaron band model. However, there are some differences between the results of Na-doped CNPPV and PPV. The experimental peak-to-peak splitting of the two bipolaron peaks is about 1.05 eV in CNPPV compared to about 2.0 eV for the sodium-doped PPV; the region where the doping-induced states appear in the UPS spectra, recorded at saturation doping, for the two systems is shown in Fig. 23.18. The difference in splitting between the gap states is caused by a stronger confinement of the bipolaron wave functions in CNPPV; AM1 calculations show that the bipolarons in CNPPV are confined on cyano-vinylene-ring-vinylene-cyano segments along the polymer backbone; the phenyl rings included in those segments can accept nearly twice as

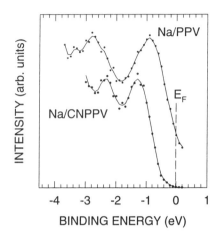

Fig. 23.18 He I UPS spectra of the band gap states of CNPPV and PPV at 100% doping level. (From Ref. 91.)

(bottom curve) indicates that only one type of sulfur site is present in P3OT.

The effect of the Al deposition can be clearly observed, as a new component is gradually growing on the low binding energy side of the S($2p$) line. Note that for most of the samples no shoulder was found on the C($1s$) level. However, a very weak shoulder could be found in some of the samples at the highest Al coverage, as seen in the C($1s$) spectrum (at the top) in Fig. 23.19a. The C($1s$) spectrum of P3OT is dominated by the aliphatic carbon atoms of the side chains, while aluminum is expected to interact preferentially with the carbon atoms of the conjugated backbone [92]. Thus, any new carbon species might be small in numbers, not only because the aliphatic carbons represent the major contribution to the C($1s$) line, but also because only a part of the aromatic carbons might be affected by the Al deposition. The main line of the Al($2p$) shifts slowly toward lower binding energy as the thickness of Al increases. This shift is consistent with the fact that the Al is growing clusterlike on the surface. The binding energy of the Al($2p$) is -74.0 eV at the first Al deposition and shifts slowly to -73.3 eV for an Al layer equivalent to a few angstroms in thickness, which is still somewhat different from the bulk metal value of -72.7 eV. A low intensity component is present on the high binding energy side of the Al($2p$) peak (corresponding to Al atoms where the electron densities are lower than in unaffected atoms). A small amount of oxygen was detected in the P3OT films, about one oxygen per 10 thiophene rings, indicating that this component might be due to oxidized aluminum.

much charge as the phenyl rings outside the sequence, which are almost unperturbed [91]. The bipolaron levels appear deeper in the gap as a result of the confinement of the charge carriers.

B. Metals on Thiophenes

1. Aluminum on Poly(3-octylthiophene) and Sexithienylene

The evolution of the C($1s$), S($2p$), and Al($2p$) lines, upon Al deposition onto poly(3-octylthiophene) films (P3OT), is shown in Fig. 23.19 [17]. The shoulder on the high binding energy side of the S($2p$) peak is due to spin–orbit coupling. Therefore, the spectrum of the pristine system

More detailed information on the Al/thiophene system can be obtained from data on studies of the interaction between aluminum and an oligothiophene, α-sexithienylene (α-6T). The two main advantages in using a model system like α-6T are (1) the lack of aliphatic side chains that can obscure the C($1s$) signal from the carbon

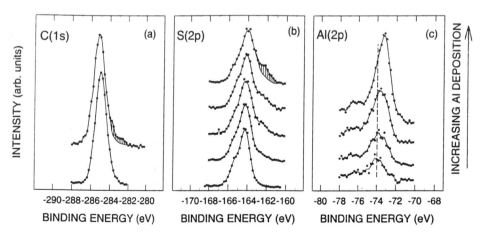

Fig. 23.19 XPS C($1s$), S($2p$), and Al($2p$) spectra of the Al/P3OT interface for increasing Al coverage. The C($1s$) and S($2p$) spectra of the pristine system are at the bottom, with increasing Al coverage upwards. (From Ref. 17.)

atoms in the backbone and (2) the possibility of in situ preparation of films, i.e., obtaining contamination-free (oxygen-free) samples. The evolution of the C(1s), S(2p), and Al(2p) core levels upon Al deposition onto α-6T is displayed in Fig. 23.20 [17]. In this system, it is more clear that a shoulder appears on the low binding energy side of the C(1s) main line upon Al deposition. The evolution of the S(2p) line is essentially identical to the evolution of the S(2p) observed for the Al/P3OT system. For the Al(2p) line, the evolution of binding energy is similar to what is observed for Al/P3OT. However, no high binding energy component is detected, indicating that this feature seen in Al/P3OT is indeed related to the formation of AlO_x species.

From ab initio quantum-chemical calculations on α-trithienylene (α-3T) interacting with aluminum atoms, it is found that the Al atoms preferentially form bonds with the α carbons of the thiophene rings [93]. The same type of structure is derived for the Al/α-6T complexes calculated using the semiempirical MNDO method. The Al–C bond leads to a redistribution of the charge density in the Al-oligothiophene complex relative to the separate partners. As expected, Al becomes strongly positive and electron density is transferred to the α carbons and the adjacent sulfur atoms, while the atomic charge of the β carbons remains unaffected. Also, the charges on the thiophene units without aluminum are not affected. The amount of negative charge transferred from the Al to the C and S atoms differs somewhat depending upon the computational level, but all of the calculations point to a charge transfer from the Al to the α carbon and the adjacent S and to only these sites. As the Al forms a bond with the α carbon, it also changes the hybridization of that carbon atom from sp^2 to sp^3. This means that the conjugation will be broken at these sites. The selectivity of the Al bonding to the α carbons can be explained in terms of the electronic structure of the lowest unoccupied states of the pristine molecule. These states are found to hybridize with the Al valence levels to form the highest occupied states of the complex. Since these states have a large contribution from the α carbon, it is expected that the bonding takes place preferentially at these sites. Reaction of the Al, which is an electron donor, directly with the sulfur atoms is less likely owing to the inherently high electron density on those sites.

Thus, the interpretation of the combined experimental and theoretical results indicates that the large chemical shift observed for the S(2p) core level, about 1.6 eV, should be due to a secondary effect from the attachment of Al atoms to the adjacent carbon atoms. This interpretation is fully consistent with ab initio Hartree–Fock ΔSCF calculations of the chemical shifts in aluminum-oligothiophene complexes [94]. For an Al_2/α-3T complex where the two Al atoms are attached to the α carbons on the central thiophene unit, the chemical shift of the S(2p) level for the central sulfur atom is found to be 1.65 eV, in close agreement with the experimental value of 1.6 eV [17]. Although several different Al/thiophene complexes were tested in the ΔSCF calculations, no stable structure where an Al atom binds directly to a S atom was found.

Upon Al deposition on the conjugated thiophene systems, only small visible changes occur in the positions of the peaks in the UPS spectra. The general intensity decreases rapidly upon Al deposition. This indicates that a metallic overlayer is formed, because the cross sections for the Al(3p) or Al(3s) are much lower than those for the C(2p) or S(3p) orbitals. This also is consistent with the Al(2p) XPS spectra discussed above. Therefore only the spectra for the pristine form of α-6T along with that corresponding to four Al atoms per α-6T molecule are shown in Fig. 23.21. Also shown in this figure are the DOVS from the calculation of α-3T and of α-3T interacting with two Al atoms. Even though cross-section

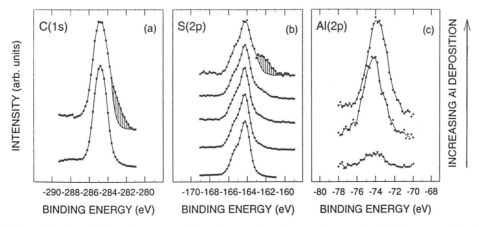

Fig. 23.20 XPS C(1s), S(2p) and Al(2p) spectra of the Al/α-6T interface for increasing Al coverage. The C(1s) and S(2p) spectra of the pristine system are at the bottom, with increasing Al coverage upwards. (From Ref. 17.)

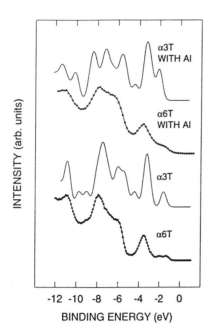

Fig. 23.21 Calculated DOVS (——) compared with the experimental UPS spectra (—·—). The theoretical curves were calculated for a trimer (α-3T) and the Al$_2$/α-3T complex; the experimental data correspond to the clean α-6T surface and to the interface with Al$_4$/α-6T stochiometry. (From Ref. 17.)

effects are not included in the calculations and the α-3T is a smaller molecule, the DOVS of the α3T in the pristine state and Al$_2$/α-3T globally agree well with the experimental UPS results: (1) the intensity of the UPS band at -3.8 eV increases upon interaction with Al relative to the most intense component of the spectrum, and (2) the large peak at -7.8 eV loses intensity in favor of the neighboring band around -6 eV, which leads to a broadening in this region.

In another study on the interaction between metals and polythiophene, the coinage metals Cu, Ag, and Au were studied by XPS during successive deposition onto poly(3-hexylthiophene) [95]. For the three metals, only the Cu(2p) line indicates a small high binding energy component upon interaction with P3HT. For all three metal–polymer systems, the C(1s) core level is unaffected, i.e., neither a shift in the binding energy position nor a new feature associated with metal–carbon formation appears. Finally, the S(2p) level is unaffected in the case of Au, while new low binding energy features are detected in the cases of Ag and Cu. The shift is on the order of 2 eV for the latter, i.e., about 0.4 eV larger than what was found in the case of aluminum deposited on P3OT. From these results, it was suggested that Au does not interact with P3HT and that Ag and Cu react exclusively with the sulfur atoms in the thiophene rings.

C. Metals on Diphenylpolyenes

1. Aluminum on α,ω-Diphenyltetradecaheptaene

The interaction between aluminum atoms and DP7 has been studied with both XPS and UPS [96]. In Fig. 23.22 are shown the He I UPS spectra obtained during the first stages of the interface formation with vapor-deposited aluminum. The strong feature A at about -8 eV (associated with molecular wave functions almost totally localized on the phenyl end groups) is essentially unaffected by the interaction with aluminum. The two peaks B and C (associated with molecular wave functions highly localized on the polyene chain) are strongly affected. Note also that the overall intensity decreases as the aluminum coverage increases, indicating the formation of an Al overlayer. The intensity decrease derives from the fact that the photoionization cross section of aluminum is lower than of that of the molecular system studied.

The main C(1s) XPS core level peak (not shown here), which occurs at about -284.6 eV, decreases in intensity upon Al deposition, again indicating the formation of an aluminum overlayer. Fig. 23.23 shows the shake-up (s.u.) portion of the C(1s) spectra corresponding to the UPS spectra of Fig. 23.22. Notice that the s.u. feature near -288 eV, which is associated with the polyene chain, is strongly affected by the presence of Al, while the higher energy feature near -292 eV, associated mostly with the phenyl end groups, is essentially unaffected. The binding energy of Al(2p) in bulk aluminum is -72.7 eV. During the deposition of aluminum, however, the Al(2p) signal first appears at about -74 eV and shifts gradually to about -73 eV for the largest coverage obtained. This shift is due to the increased electron screening of the core hole on the Al(2p) level as the Al layer, or Al clusters, grow thicker. It indicates that the initial

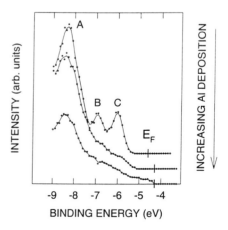

Fig. 23.22 Low binding energy part of He I UPS spectra of DP7 without and with aluminum. (From Ref. 61.)

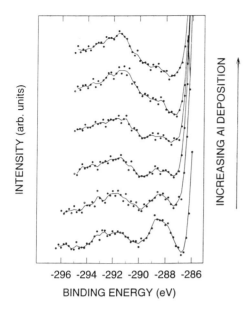

Fig. 23.23 The shake-up portion of the C(1s) spectra recorded during successive Al deposition onto DP7. (From Ref. 61.)

amount of Al deposited is in the "monolayer coverage range."

In the corresponding model calculations, performed at the LSD approximation level, two Al atoms were allowed to interact with a tetradecaheptaene molecule. The Al atoms form covalent bonds with the polyene, and the Al(3p) and Al(3s) orbitals hybridize with the molecular π orbitals [59]. The Al(3p) orbitals are found to overlap with the π orbitals of two adjacent carbon atoms in the polyene without changing the planarity of the system. The double bond character in the vicinity of the Al atom, however, is lost.

The results presented above indicate that during the initial stages of the interface formation, the Al atoms react preferentially with the polyene chain and not significantly with the phenyl end groups. The inertness of the phenyl ring toward Al deposition observed here is consistent with what is observed in other polymers. For instance, in polyethyleneterephthalate, the Al reacts initially with the ester groups, while the phenyl groups are affected only at a later stage [97].

2. Sodium and Calcium on Diphenylpolyenes

In contrast to aluminum, where a metal layer is formed almost immediately, it is possible to deposit both sodium and calcium on the DP7 system without building a metal overlayer, i.e., the metal atoms diffuse easily into the bulk. In XPS, the intensity of the Ca(3p) or Na(1s) signal increases monotonically with metal deposition, while in UPS changes occur only during the first few depositions of metal atoms. This behavior can be explained in the following way: The initial metal atoms deposited lead to the complete doping of the top layer (probed by UPS); upon further deposition, the amount of Ca or Na at the "surface" does not continue to increase, i.e., saturation doping has been reached in the surface region. Additional Ca or Na atoms then diffuse into the bulk, increasing the overall doping level as probed by XPS. When determining the doping level corresponding to a given UPS spectrum, it is necessary to use a surface-sensitive, i.e., angle-dependent, XPS mode to match the depth probed by the UPS to that probed by XPS.

The UPS He I valence band spectra for DP7 for various degrees of Na doping are shown in Fig. 23.24, the curve for the neutral DP7 being at the top and with increasing doping downwards [61,98]. The evolution of the spectra upon doping shows that two new peaks D and E appear above the valence π-band edge of the pristine molecule. These peaks correspond to the formation of soliton–antisoliton pairs confined to the polyene part of the molecule, resulting in new states in the otherwise forbidden energy gap between the HOMO and the

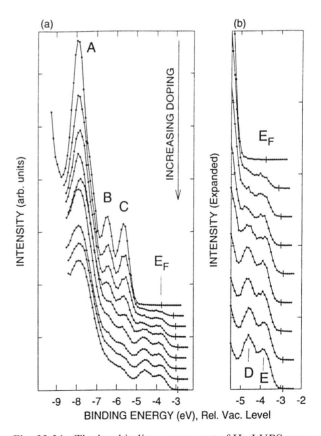

Fig. 23.24 The low binding energy part of He I UPS spectra of DP7 as a function of sodium doping. The neutral DP7 is at the top and increasing doping downwards. (From Ref. 98.)

LUMO of the pristine system. The wave functions of the two solitons are forced to overlap, because of confinement, resulting in a bonding–antibonding pair [98]. The intensities of peaks B and C decrease upon increasing doping, in good agreement with the fact that the formation of a soliton–antisoliton pair leads to the removal of states from the valence band edge (and conduction band edge) of the neutral molecule to form states within the band gap. The doping level at which the two new peaks can first be detected is about 0.4 Na per DP7, as determined by the relative intensities of the XPS C(1s) and Na(1s) core levels. The final (saturation) spectrum corresponds to two Na atoms per DP7, and the binding energy of Na(1s) is ionic, indicating that the sodium atoms donate two electrons to each DP7 molecule.

The evolution of the s.u. spectra upon sodium doping is shown in Fig. 23.25. The spectrum corresponding to the pristine system is at the bottom, and the spectra for increasing doping are successively higher. As described above, the two s.u. peaks can be assigned to the polyene part (−288 eV peak) and to the phenyl groups (−292 eV peak). Upon increasing the doping, the main effect is that the region between the main C(1s) peak and the polyene s.u. satellite becomes filled up, corresponding to new low energy electronic transitions involving the new states within the previously forbidden energy gap. The benzene contribution is unaffected by sodium deposition, since the charge transfer resulting in the generation of the soliton pairs is confined to the polyene part of the molecule.

At the initial low sodium doping levels of the DPx molecules, only doubly charged molecules could be detected for DP7 and DP6. For DP5–DP1, however, signs of singly doped molecules were found [67]. The doping levels at saturation doping are estimated to be about two sodium atoms per molecule for all of the DPx, i.e., the molecules accommodate two excess electron charges and the charge storage states can be described as interacting soliton pairs. The low binding energy part of the experimental UPS valence band spectra for the different DPx, at saturation doping, are shown in Fig. 23.26. The soliton–antisoliton wave functions are mainly confined to the polyene part of the molecules for DP7–DP3. For the smaller polyenes, the solitons are forced out somewhat onto the phenyl groups; significant LCAO coefficients can be found on the phenyl rings [99]. The interaction between the two solitons leads to the splitting of the two otherwise "degenerate" electronic states within the forbidden energy gap. It is clearly seen that the energy splitting between the two interacting soliton states, peaks D and E, increases as the solitons are forced together due to the decreasing polyene lengths.

The UPS valence band spectra of DP7, as a function of increasing Ca deposition, are shown in Fig. 23.27 [61]. In contrast to sodium/DP7, it appears that only one new

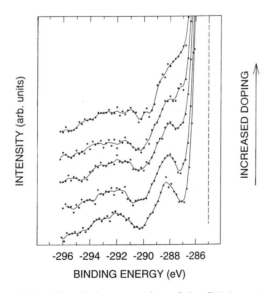

Fig. 23.25 The shake-up portion of the C(1s) spectra of DP7 recorded during successive doping with sodium. (From Ref. 61.)

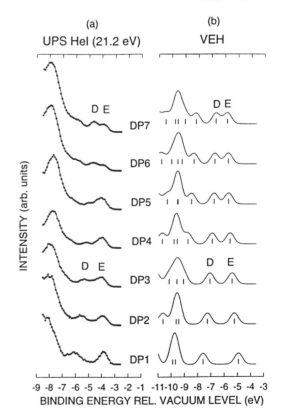

Fig. 23.26 The low binding energy part of the He I UPS spectra of the (a) fully charged and (b) calculated DOVS of DP1–DP7. (From Ref. 67.)

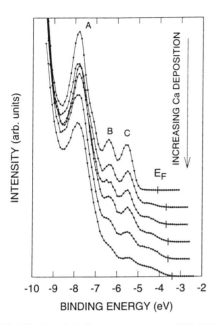

Fig. 23.27 The low binding energy part of He I UPS spectra of DP7 recorded during successively greater doping with calcium. The neutral DP7 is at the top and increasing doping downward. (From Ref. 61.)

Figure 23.28 presents a comparison of the UPS valence band spectra of DP7, for both the pristine molecule and the molecule interacting with Ca, Na, or Al atoms, and the DOVS for the corresponding tetradecaheptaene system as derived from LSD calculations. Since the tetradecaheptaene has no capping phenyl groups, only the upper part of the spectra are shown, and the contribution from the phenyl groups is omitted in the following discussion [61].

For the cases of Na and Ca on DP7, the spectra shown are for the saturated systems Ca^{2+} $DP7^{2-}$ and Na_2^{+} $DP7^{2-}$. For Al/DP7, the spectrum shown is for a few monolayers (equivalent) coverage, since the Al atoms remain essentially on the surface. All spectra are well reproduced by the DOVS derived from the results of the LSD calculations. The metal atom–induced gap states (for Ca and Na) are well resolved in both the experimental and theoretical spectra. In particular, the splitting between the soliton and antisoliton states in Na_2^{+} $DP7^{2-}$ is 0.75 eV in the experimental spectrum and 0.8 eV in the theoretical spectrum. For Ca^{2+} $DP7^{2-}$, on the other hand, because of the electrostatic interaction of

state is induced within the original forbidden energy gap (actually, two new states appear within the band gap, as discussed in more detail below). Note that upon doping, the Fermi energy decreases, consistent with filling the band gap with new occupied electronic states. For Ca/DP7, the doping-induced new states are closer to the Fermi level than in the case of Na/DP7. From the doping level, as determined by XPS, it is observed that DP7 can accommodate one Ca per DP7. The binding energy of the Ca(2p) peak corresponds to Ca^{2+}, indicating that two electrons are transferred from each Ca atom to each DP7 molecule.

From LSD calculations on the Ca/DP7 system, it was found that the charge storage states, accommodating two excess electrons per molecule, can be described as bipolaron-like states [61]. In the modeling of an isolated molecule with two negative charges, two new occupied states appear above the HOMO of the pristine molecule. When the Ca counterion is included, however, the electrostatic interaction of the large Ca^{2+} ion with the $DP7^{2-}$ leads to the stabilization of the two new states, so that the lower state (HOMO-1 in Ca/DP7) lies at the same energy as the HOMO of the pristine system. Therefore, only one new state (the HOMO of Ca/DP7) is expected to appear as a new feature in the energy gap of the pristine molecule, in agreement with the UPS spectra of Fig. 23.27.

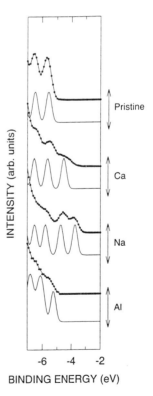

Fig. 23.28 The DOVS of tetradecaheptaene and metal/tetradecaheptaene systems derived from LSD calculations in comparison with the corresponding UPS spectra of DP7 and metal/DP7 systems as defined in the text. (From Ref. 61.)

the Ca^{2+} ion, the new states are shifted to a higher binding energy. Thus the HOMO-1 in the Ca/tetradecaheptaene system is at about the same energy as the HOMO of the pristine system, in good agreement with the experimental data.

It is of some interest to point out some of the more or less practical applications of the results discussed above. First, in connection with polymer-based light-emitting diodes using poly(p-phenylenevinylene) films with calcium electrodes, there is always a nonideal contact [83,85,100] between the calcium electrode and the polymer layer. Either an oxide is formed when the surface of the PPV contains a large concentration of oxygen-containing species [85] or a doped conducting polymer region is formed when the PPV surface is very clean [83,100]. But in no case is the interface ever without some intermediate layer, which must ultimately be included in models of charge injection in such devices. Second, studies of the vapor deposition of calcium in the presence of O_2 indicate an optimum in device yield for a pressure of about 10^{-6} mbar. This indicates that not only is UHV not necessary in the fabrication of polymer-based LEDs, but it is actually detrimental [101].

VI. FUTURE DIRECTIONS

One unique feature for organic LEDs based on conjugated polymers compared to conventional LEDs is the possibility of obtaining polarized light, as demonstrated in Ref. 102. The various techniques developed for aligning the conjugated polymers [102–104] also make it possible to obtain new information from PES by taking advantage of the inherent polarized light from synchrotron radiation. By combining the polarized light source with angle-resolved UPS (ARUPS), it is possible to measure the band dispersion along a polymer chain. This technique has been applied to different organic molecules aligned by using, for example, LB techniques, as in the work by Seki et al. [105]. The preliminary results from two ARUPS studies of oriented conjugated polymers, using synchrotron radiation, are discussed below. For a discussion of the ARUPS technique itself, the interested reader is referred to, for example, Refs. 39, 105, and 106.

The first example of ARUPS measurements demonstrates the first results from "band mapping" of a real polymer, 4BCMU-polydiacetylene (PDA) [107]. The chemical structure of PDA is displayed in Fig. 23.1. The PDA samples was prepared by electron beam radiation of 4BCMU monomer crystals as described previously [108]. In Figure 23.29 presents the low binding energy part, which corresponds to the top part of the highest occupied π band, of two UPS spectra of PDA. The spectra were recorded with a photon energy of 85 eV and correspond to two different geometries; the electric field component, E, parallel to the crystal b axis, i.e., the

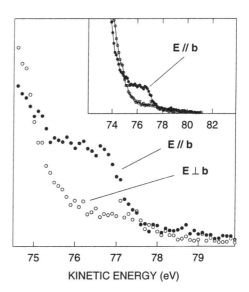

Fig. 23.29 The π-band region of the UPS spectra of 4BCMU-PDA on two different expanded scales. (From Ref. 107.)

axis along the polymer backbone (E ∥ b), and the electric field component perpendicular to the b axis (E⊥b). The π-band intensity disappears completely for E⊥b and is at a maximum for E ∥ b; the E vector of the light only excites electrons in π states with electron momentum k for k ∥ E (or k parallel to a component of E). Thus, it is possible to determine the direction of the dispersed π band since the maximum intensity of the π band is to be found when the polarization of the incoming light is parallel to this direction.

For a fixed geometry of the sample versus the incoming light, the dispersion of the k component parallel to the surface, k∥, can be determined from the relation [106]

$$k_{\parallel} = \sin(\theta) \times (E_{kin})^{1/2} \times 0.51 \text{ Å}^{-1} \qquad (4)$$

$k_{\parallel}$ is the momentum of the electrons of interest and E_{kin} is the kinetic energy of the emitted electrons measured for different emission angles (θ). In Fig. 23.30 the experimental ARUPS data points are compared with a VEH band structure. The VEH results indicate that below the dispersed π2 band there is a band of less dispersion corresponding to the transverse π orbitals that involves the triple bonds in the backbone, and a flat (nondispersed) band corresponding to the C=O groups in the 4BCMU side groups. Note that the experimental data and the calculated results do not agree for only one point close to the first Brillouin zone edge; this discrepancy is not fully understood. Also, because of the high kinetic energy of the emitted electrons used in the measurements, the Brillouin zone is covered by a relatively small spread

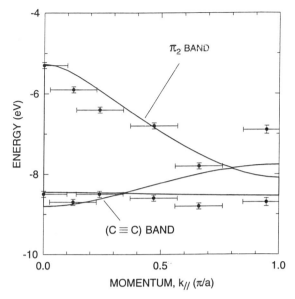

Fig. 23.30 A portion of the VEH band structure of 4BCMU-PDA, covering the energy region of the upper π band, in comparison with the ARUPS results. (From Ref. 107.)

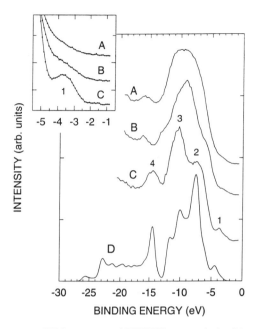

Fig. 23.31 UPS spectra of DHPPV, recorded with 40 eV photons, compared with DOVS derived from VEH calculations (D). Spectra A–C correspond to different experimental configurations as defined in the text. (From Ref. 104.)

in angles, hence the relatively few distinguishable data points in the first Brillouin zone.

It has been shown that it is possible to obtain alignment of organic polymers and molecules by depositing the materials onto aligned tetrafluoroethylene (PTFE) substrates [103]. This is nicely demonstrated in the second example, where aligned poly(2,5-diheptyl-p-phenylene) (DHPPP) films were prepared by solution casting onto aligned PTFE substrates [104]. The chemical structure of DHPPV is displayed in Fig. 23.1. Fig. 23.31 shows three valence band spectra of DHPPP compared with the DOVS derived from VEH calculation for an inter-ring torsion angle of 42° [104]. The inset shows the low binding energy part in more detail. The spectra were recorded with 40 eV photon energy for three different orientations of the sample relative to the polarization direction of the incoming radiation; the incoming light (A) parallel to, (B) at 45° to, and (C) perpendicular to the aligned chains of the PTFE substrate. From the calculations, peak 1 can be assigned to electrons removed from the two highest lying π bands: $\pi 3$ and $\pi 2$. The $\pi 3$ band is derived from molecular orbitals delocalized along the polymer backbone and has significant dispersion. The $\pi 2$ band is a flat band derived from molecular orbitals localized on the ortho carbons in the phenyl rings. These molecular orbitals correspond to one of the doubly degenerate e_{1g} molecular orbitals in benzene, with a linear combination of atomic orbital (LCAO) coefficients close to zero on the carbons connecting to the neighboring phenyl ring. As seen in the inset of Fig. 23.31, the intensity of peak 1 is a maximum for configuration C and completely disappears for configuration A. Configuration B shows something in between A and C, which would be expected if the polymer chains were almost completely aligned. These results indicate that (1) the DHPPP chains are ordered (parallel to one another), which has also been confirmed by measurements using polarized microscopy [109], and (2) the DHPPP chains are aligned perpendicular to the PTFE chains (in configuration C, the polarization of the incoming light is perpendicular to the PTFE chains but couples most strongly to the dispersed band along the polymer backbone).

VI. SUMMARY

The usefulness of photoelectron spectroscopy of polymer surfaces and interfaces has been outlined. A variety of examples have been supplied, mostly but not exclusively from our own work. The main points developed throughout this review concern

1. The usefulness of surface-sensitive techniques in the study of surfaces and interfaces of conjugated polymers.

2. The advantages of using model molecules as comparative systems for conjugated polymers.
3. The significant advantages of employing a combined experimental-theoretical approach to the study of conjugated polymer surfaces and interfaces, whereby more information may be obtained than by using either alone.

ACKNOWLEDGMENTS

Research on conjugated polymers and molecules in Linköping is supported by grants from the Swedish Natural Science Research Council (NFR), the Swedish Research Council for Engineering Sciences (TFR), the Swedish National Board for Industrial and Technical Development (NUTEK), the Neste Corporation, Finland, and Philips Research, NL (within Brite/EURAM Poly Research). We also acknowledge the important input from all of our many collaborators; there is not space enough here for them all to be mentioned by name. We do acknowledge specifically, however, the long-time theoretical and experimental collaborations with S. Stafström and coworkers in Linköping and, in particular, J. L. Brédas, R. Lazzaroni, and coworkers in Mons, Belgium. The Linköping–Mons collaboration is supported by the European Commission program SCIENCE (Project 0661 POLYSURF), the ESPRIT Network of Excellence, NEOME, and the ESPRIT Basic Research Action LEDFOS 8013. One of us (P.D.) is also partly supported by the Wenner-Gren Foundation Center in Sweden.

REFERENCES

1. K. Siegbahn, C. Nordling, G. Johansson, J. Hedman, P. F. Heden, V. Hamrin, U. Gelius, T. Bergmark, L. O. Werme, R. Manne, and Y. Baer, *ESCA Applied to Free Molecules*, North-Holland, Amsterdam, 1971.
2. D. W. Turner, C. Baker, A. D. Baker, and C. R. Brundle, *Molecular Photoelectron Spectroscopy*, Interscience, London, 1970.
3. H. Winick and S. Doniach (eds.), *Synchrotron Radiation Research*, Plenum New York, 1980.
4. R. Z. Bachrach, (ed.), *Synchrotron Radiation Research (Ad. Surface Interface Sci. Vol 1)*, Plenum, New York, 1992.
5. M. J. Obrzut and F. E. Karasz, X-ray photoelectron spectroscopy of neutral and electrochemically doped poly(p-phenylenevinylene), *Macromolecules* 22:458 (1989).
6. W. R. Salaneck, H. R. Thomas, C. B. Duke, A. Paton, E. W. Plummer, A. J. Heeger, and A. G. MacDiarmid, Photoelectron spectra of AsF$_5$-doped polyacetylenes, *J. Chem. Phys.* 71:2044 (1979).
7. K. Kamiya, H. Inokuchi, M. Oku, S. Hasegawa, C. Tanaka, J. Tanaka, and K. Seki, UPS of new type polyacetylene, *Synth. Met.* 41–43:155 (1991).
8. J. Tanaka, C. Tanaka, T. Miyamae, K. Kamiya, M. Shimizu, M. Oku, K. Seki, J. Tsukamoto, S. Hasegawa, and H. Inokuchi, Spectral characteristic of metallic state of polyacetylene, *Synth, Met.* 55-57:121 (1993).
9. R. Lazzaroni, M. Lögdlund, S. Stafström, W. R. Salaneck, and J. L. Brédas, The poly-3-hexylthiophene/NOPF6 system: a photoelectron spectroscopy study of electronic structural changes induced by the charge transfer in the solid state, *J. Chem. Phys.* 93:4433 (1990).
10. W. R. Salaneck, S. Stafström, J.-L. Brédas, S. Andersson, P. Bodö, and J. Ritsko, Pthalimide on copper: a model system to address certain site-specific interactions at the polymide-copper interface, *J. Vac. Sci. Technol. A* 6:3134 (1988).
11. M. Fahlman, M. Lögdlund, S. Stafström, W. R. Salaneck, R. H. Friend, P. L. Burn, A. B. Holmes, K. Kaeriyama, Y. Sonoda, O. Lhost, F. Meyers, and J. L. Brédas, Experimental and theoretical studies of the electronic structure of poly(p-phenylenevinylene) and its ring-substituted derivatives, *Macromolecules* 28:1959 (1995).
12. E. T. Kang, K. G. Neoh, and K. L. Tan, *X-Ray Photoelectron Spectroscopic Studies of Electroactive Polymers (Adv. Polym. Sci., 106)*, Springer-Verlag, Berlin, 1993.
13. D. Jones, M. Guerra, L. Favaretto, A. Modelli, M. Fabrizio, and G. Distefano, Determination of the electronic structure of thiophene oligomers and extrapolation to polythiophene, *J. Phys. Chem.* 94:5761 (1990).
14. H. Fujimoto, U. Nagashima, H. Inokuchi, K. Seki, Y. Cao, H. Nakahara, J. Nakayama, M. Hoshino, and K. Fukuda, Ultraviolet photoemission study of oligothiophenes: π-band evolution and geometries, *J. Chem. Phys.* 92:4077 (1990).
15. M. G. Ramsey, D. Steinmüller, and F. P. Netzer, Explicit evidence for bipolaron formation: Cs-doped biphenyl, *Phys. Rev. B* 42:5902 (1990).
16. L. M. Tolbert and J. A. Schomaker, Alkali-metal doping of polyacetylene model compounds: α,ω-diphenylpolyenes, *Synth. Met.* 41-43:169 (1991).
17. P. Dannetun, M. Boman, S. Stafström, W. R. Salaneck, R. Lazzaroni, C. Fredriksson, J. L. Brédas, R. Zamboni, and C. Taliani, The chemical and electronic structure of the interface between aluminum and polythiophene semiconductors, *J. Chem. Phys.* 99:664 (1993).
18. C. Fredriksson and J. L. Brédas, Metal/conjugated polymer interfaces: a theoretical investigation of the interaction between aluminum and trans-polyacetylene oligomers, *J. Chem. Phys.* 98:4253 (1993).
19. C. Tanaka and J. Tanaka, Molecular and electronic structure of model compounds of doped polyacetylene, *Bull. Chem. Soc. Jpn.* 66:357 (1993).
20. M. G. Ramsey, D. Steinmüller, M. Schatzmayr, M. Kiskinova, and F. P. Netzer, n-Phenyls on transition

and alkali metal surfaces: from benzene to hexaphenyl, *Chem. Phys. 177*:349 (1993).
21. D. Steinmüller, M. G. Ramsey, and F. P. Netzer, Polaron and bipolaronlike states in *n*-doped bithiophene, *Phys. Rev. B 47*:13323 (1993).
22. P. Dannetun, M. Lögdlund, M. Fahlman, C. Fauquet, D. Beljonne, J. L. Brédas, H. Bässler, and W. R. Salaneck, The evolution of charge-induced gap states in degenerate and non degenerate conjugated molecules and polymers as studied by photoelectron spectroscopy, *Synth. Met. 67*:81 (1994).
23. F. Garnier, G. Horowitz, X. Peng, and D. Fichou, An all-organic "soft" thin film transistor with very high carrier mobility, *Adv. Mater. 2*:592 (1990).
24. G. Grem, V. Martin, F. Meghdadi, C. Paar, J. Stampfl, J. Sturm, S. Tasch, and G. Leising, Stable poly(*p*-phenylene) and their application in organic light emitting devices, *Synth. Met. 71*:2193 (1995).
25. J. Stampfl, S. Tasch, G. Leising, and U. Scherf, Quantum efficiencies of electroluminescent poly(*p*-phenylenes), *Synth. Met. 71*:2125 (1995).
26. F. Geiger, M. Stoldt, H. Schweizer, P. Bäuerle, and E. Umbach, Electroluminescence from oligothiophene-based light-emitting devices, *Adv. Mater. 5*: 922 (1993).
27. D. Briggs and M. P. Seah (eds.), *Practical Surface Analysis*, Wiley, Chichester, 1983.
28. W. R. Salaneck, Photoelectron spectroscopy of the valence electronic structure of polymers, *CRC Crit. Rev. Solid State Mater. Sci. 12*:267 (1985).
29. W. R. Salaneck, S. Stafström, and J. L. Brédas, *Conjugated Polymer Surfaces and Interfaces*. Cambridge Univ. Press, Cambridge, UK, 1996.
30. H. J. Freund, E. W. Plummer, R. W. Salaneck, and R. W. Bigelow, An XPS study of intensity borrowing in core ionization of free and coordinated CO, *J. Chem. Phys. 75*:4275 (1981).
31. G. Wendin, *Structure and Bonding 45: Breakdown of the One-Electron Pictures in Photoelectron Spectra*, Springer-Verlag, Berlin, 1981.
32. H. J. Freund and R. W. Bigelow, Dynamic effects in VUV- and XUV-spectroscopy of organic molecular solids *Phys. Scripta 17*:50 (1987).
33. C. S. Fadley, Basic concepts of X-ray photoelectron spectroscopy, in *Electron Spectroscopy: Theory, Techniques and Applications* (C. R. Brundle and A. D. Baker, eds.), Academic, London, 1978, p. 1.
34. N. Sato, K. Seki, and H. Inokuchi, Polarization energies of organic solids determined by ultraviolet photoelectron spectroscopy, *J. Chem. Soc. Faraday Trans. 277*:1621 (1981).
35. G. K. Wertheim, P. M. T. M. van Attekum, and S. Basu, Electronic structure of lithium graphite, *Solid State Commun. 33*:1127 (1980).
36. W. R. Salaneck, R. Erlandsson, J. Prejza, I. Lundström, and O. Inganäs, X-ray photoelectron spectroscopy of boron fluoride doped polymer, *Synth. Met. 5*:125 (1983).
37. W. R. Salaneck, C. B. Duke, W. Eberhardt, E. W. Plummer, and H. J. Freund, Benzene, *Phys. Rev. Lett 45*:280 (1980).
38. C. B. Duke, W. R. Salaneck, T. J. Fabish, J. J. Ritsko, H. R. Thomas, and A. Paton, The electronic structure of pendant-group polymers: molecular ion states and dielectric properties of poly(2-vinyl pyridine), *Phys. Rev. B18*:5717 (1978).
39. E. W. Plummer and W. Eberhardt, Angle-resolved photoemission as a tool for the study of surfaces, in *Advances in Chemical Physics* (I. Prigogine and S. A. Rice, Eds.), Wiley, New York, 1982, p. 533.
40. T. R. Ohno, Y. Chen, S. E. Harvey, G. H. Kroll, J. H. Weaver, R. E. Haufler, and R. E. Smalley, C_{60} bonding and energy-level alignment on metal and semiconductor surfaces, *Phys. Rev. B 44*:13747 (1991).
41. G. Gensterblum, K. Hevesi, B. Y. Han, L. M. Yu, J. J. Pireaux, P. A. Thiry, D. Bernaerts, S. Amelinckx, G. Van Tendeloo, G. Bendele, T. Buslaps, R. L. Johnson, M. Foss, R. Feidenhans'l, and G. Le Lay, Growth mode and electronic structure of the epitaxial C_{60}(111)/GeS(001) interface, *Phys. Rev. B 50*:11981 (1994).
42. A. J. Maxwell, P. A. Brühwiler, A. Nilsson, and N. Mårtensson, Photoemission, autoionization and X-ray absorption spectroscopy of ultrathin-film C_{60} on Au(110), *Phys. Rev. B 49*:10717 (1994).
43. P. Rudolf and G. Gensterblum, Comment on "Adsorption of C_{60} on Ta(110): photoemission and C K-edge studies", *Phys. Rev. B 50*:12215 (1994).
44. W. R. Salaneck and J. L. Brédas, Electronic band structure of conjugated polymers, in *Organic Materials for Electronics* (J. L. Brédas, W. R. Salaneck and G. Wegner, eds.), North-Holland, Amsterdam, 1994, p. 15.
45. M. Lögdlund, W. R. Salaneck, F. Meyers, J. L. Brédas, G. A. Arbuckle, R. Friend, A. B. Holmes, and G. Froyer, The evolution of the electronic structure in a conjugated polymer series: polyacetylene, poly(*p*-phenylene), and poly(*p*-phenylenevinylene), *Macromolecules 26*:3815 (1993).
46. W. R. Salaneck, Intermolecular relaxation energies in anthracene, *Phys. Rev. Lett. 40*:60 (1978).
47. N. Sato, K. Seki, and H. Inokuchi, Polarization energies of organic solids determined by ultraviolet photoelectron spectroscopy, *J. Chem. Soc. Faraday Trans. 277*:1621 (1981).
48. M. J. S. Dewar and W. Thiel, Ground states of molecules. 38. The MNDO method. Approximations and parameters, *J. Am. Chem. Soc. 99*:4899 (1977).
49. M. J. S. Dewar and W. Thiel, Ground states of molecules. 39. MNDO results for molecules containing hydrogen, carbon, nitrogen and oxygen, *J. Am. Chem. Soc. 99*:4907 (1977).
50. M. J. S. Dewar, E. G. Zoebisch, E. F. Healy, and J. J. P. Stewart, AM1: a new general purpose quantum mechanical molecular model, *J. Am. Chem. Soc. 107*:3902 (1985).

51. J. L. Brédas, R. R. Chance, R. Silbey, G. Nicolas, and P. Durand, A nonempirical effective Hamiltonian technique for polymers: application to polyacetylene and polydiacetylene, *J. Chem. Phys.* 75:255 (1981).
52. J. M. André, J. Delhalle, and J. L. Brédas, *Quantum Chemistry Aided Design of Organic Polymers* (World Scientific Lecture and Course Notes in Chemistry, Vol. 2), World Scientific, Singapore, 1991.
53. P. Hohnenberg and W. Kohn, Inhomogeneous electron gas, *Phys. Rev.* 136:B 864 (1964).
54. W. Kohn and L. J. Sham, Self-consistent equations including exchange and correlation effects, *Phys. Rev.* 140:A 1133 (1965).
55. J. L. Brédas and W. R. Salaneck, Electronic-structure evolution upon thermal treatment of polyacrylonitrile: a theoretical investigation, *J. Chem. Phys.* 85:2219 (1986).
56. E. Orti and J. L. Brédas, Electronic structure of metal-free phtalocyanine: a valence effective Hamiltonian theoretical study, *J. Chem. Phys.* 89:1009 (1988).
57. E. Orti and J. L. Brédas, Photoelectron spectra of phthalocyanine thin films: a valence band theoretical interpretation, *J. Am. Chem. Soc.* 114:8669 (1992).
58. J. L. Brédas and W. R. Salaneck, Characterization of the interfaces between low workfunction metals and conjugated polymers in light-emitting diodes, in *Organic Electroluminescence* (D. D. C. Bradley and T. Tsutsui, eds.), Cambridge Univ. Press, Cambridge, UK, in press.
59. C. Fredriksson, R. Lazzaroni, J. L. Brédas, A. Ouhlal, and A. Selmani, Metal/conjugated polymer interfaces: a local density functional study of aluminum/polyene interactions, *J. Chem. Phys.* 100:9258 (1994).
60. C. Fredriksson and S. Stafström, Metal/conjugated polymer interfaces: sodium, magnesium, aluminum and calcium on *trans*-polyacetylene, *J. Chem. Phys.* 101:9137 (1994).
61. P. Dannetun, M. Lögdlund, R. Lazzaroni, C. Fauquet, C. Fredriksson, S. Stafström, C. W. Spangler, J. L. Brédas, and W. R. Salaneck, Reactions of low workfunction metals, Na, Al, and Ca, on α,ω-diphenyltetradecaheptaene: implications for metal/polymer interfaces, *J. Chem. Phys.* 100:6765 (1994).
62. J. W. Andzelm, DGauss: density functional-Gaussian approach: implementation and applications, in *Density Functional Methods in Chemistry* (J. K. Labanowski and J. W. Andzelm, eds.), Springer, New York, 1991, p. 155.
63. S. H. Vosko, L. Wilk, and M. Nusair, Accurate spin-dependent electron liquid correlation energies for local spin density calculations: a critical analysis, *Can. J. Phys.* 58:1200 (1980).
64. B. S. Hudson, J. N. A. Ridyard, and J. Diamond, Polyene spectroscopy. photoelectron spectroscopy of the diphenylpolyenes, *J. Am. Chem. Soc.* 98:1126 (1976).
65. K. L. Yip, N. O. Lipari, C. B. Duke, B. S. Hudson, and J. Diamond, The electronic structure of bond-alternating and nonalternant conjugated hydrocarbons: diphenylpolyenes and azulene, *J. Chem. Phys.* 64:4020 (1976).
66. M. Lögdlund, P. Dannetun, B. Sjögren, M. Boman, C. Fredriksson, S. Stafström, and W. R. Salaneck, The electronic structure of α,ω-diphenyltetradecaheptaene, a model molecule for polyacetylene, as studied by photoelectron spectroscopy, *Synth. Met.* 51:187 (1992).
67. P. Dannetun, M. Lögdlund, C. W. Spangler, J. L. Brédas, and W. R. Salaneck, Evolution of charge-induced gap states in short diphenylpolyenes as studied by photoelectron spectroscopy, *J. Phys. Chem.* 98:2853 (1994).
68. D. Oeter, C. Ziegler, and W. Göpel, Doping and stability of ultrapure α-oligothiophene thin films, *Synth. Met.* 61:147 (1993).
69. M. Lögdlund, P. Dannetun, C. Fredriksson, W. R. Salaneck, and J. L. Brédas, Theoretical and experimental studies of the interaction between sodium and oligothiophenes, *Phys. Rev. B*, 53:16327 (1996).
70. C. S. Yannoni and T. C. Clarke, Molecular Geometry of *cis*- and *trans*-polyacetylene by nutation NMR spectroscopy, *Phys. Rev. Lett.* 51:1191 (1983).
71. G. Mao, J. E. Fischer, F. E. Karasz, and M. J. Winokur, Nonplanarity and ring torsion in poly(*p*-phenylenevinylene). A neutron-diffraction study, *J. Chem. Phys.* 98:712 (1993).
72. R. A. W. Johnstone and F. A. Mellon, Photoelectron spectroscopy of sulphur-containing heteroaromatics and molecular orbital calculations, *J. Chem. Soc. Faraday Trans. II* 69:1155 (1973).
73. J. P. Maier and D. W. Turner, Steric inhibition of resonance studied by molecular photoelectron spectroscopy, *J. Chem. Soc. Faraday Trans. II* 69:196 (1973).
74. J. W. Rabalais and R. J. Colton, Electronic interaction between the phenyl group and its unsaturated substituents, *J. Electron Spectry. Rel. Phenom.* 1:83 (1972/1973).
75. T. Yamamoto, K. Sanechika, and A. Yamamoto, Preparation of thermostable and electric-conducting poly(2,5-thienylene), *J. Polym. Sci., Polym. Lett. Ed.* 18:9 (1980).
76. M. Sato, S. Tanaka, and K. Kaeriyama, Soluble conducting polythiophenes, *J. Chem. Soc., Chem. Comm.* 1986:873.
77. K. Y. Yen, G. G. Miller, and R. L. Elsenbaumer, Highly conducting, soluble, and environmentally-stable poly(3-alkylthiophenes), *J. Chem. Soc., Chem. Comm.* 1986:1346.
78. O. Inganäs, W. R. Salaneck, J.-E. Österholm, and J. Laakso, Thermochromic and solvatochromic effects in poly(3-hexylthiophene), *Synth. Met.* 22:395 (1988).

79. M. Berggren, O. Inganäs, G. Gustafsson, J. Rasmusson, M. R. Andersson, T. Hjertberg, and O. Wennerström, Light-emitting diodes with variable colours from polymer blends, *Nature* 372:444 (1994).
80. W. R. Salaneck, O. Inganäs, B. Thémans, J. O. Nilsson, B. Sjögren, J.-E. Österholm, J. L. Brédas, and S. Svensson, Thermochromism in poly(3-hexylthiophene) in the solid state: a spectroscopic study of temperature-dependent conformational defects, *J. Chem. Phys.* 89:4613 (1988).
81. H. Fujimoto, K. Iwasaki, and S. Matsuzaki, Photoemission study of poly(3-alkylthiophene)s: the effects of air and temperature on the electronic structure, *Synth. Met.* 66:99 (1994).
82. T. P. Nguyen, V. Massardier, V. H. Tran, and A. Guyot, Studies of the polymer-metal interface in metal-PPV-metal devices, *Synth. Met.* 55-57:235 (1993).
83. P. Dannetun, M. Fahlman, C. Fauquet, K. Kaerijama, Y. Sonoda, R. Lazzaroni, J. L. Brédas, and W. R. Salaneck, Interface formation between poly(2,5-diheptyl-*p*-phenylenevinylene) and calcium: implications for light emitting diodes, *Synth. Met.* 67:113 (1994).
84. M. Fahlman, D. Beljonne, M. Lögdlund, R. H. Friend, A. B. Holmes, J. L. Brédas, and W. R. Salaneck, Experimental and theoretical studies of the electronic structure of Na-doped poly(*p*-phenylenevinylene), *Chem. Phys. Lett.* 214:327 (1993).
85. Y. Gao, K. T. Park, and B. R. Hsieh, X-ray photoemission investigations of the interface formation of Ca and poly(*p*-phenylenevinylene), *J. Chem. Phys.* 97:6991 (1992).
86. M. Schott, Undoped (semiconducting) conjugated polymers, in *Organic Conductors: Fundamentals and Applications* (J. P. Farges, ed.), Marcel Dekker, New York, 1994, p. 539.
87. P. Dannetun, M. Lögdlund, W. R. Salaneck, C. Fredriksson, S. Stafström, A. B. Holmes, A. Brown, S. Graham, R. H. Friend, and O. Lhost, New results on metal-polymer interfaces, *Mol. Cryst. Liq. Cryst.* 228:43 (1993).
88. E. Ettedgui, H. Razafitrimo, K. T. Park, Y. Gao, and B. R. Hsieh, An X-ray photoemission spectroscopy study of the role of sample preparation on band bending at the interface of Al with poly(*p*-phenylenevinylene), *J. Appl. Phys.* 75:7526 (1994).
89. C. Fredriksson, R. Lazzaroni, J. L. Brédas, P. Dannetun, M. Lögdlund, and W. R. Salaneck, Theoretical studies of the aluminum/poly(*p*-phenylene vinylene) interface, *Synth. Met.* 55-57:4632 (1993).
90. M. Lögdlund and J. L. Brédas, Semiempirical studies of the interaction between metals and π-conjugated polymers: sodium on diphenylpolyenes and aluminum on poly(*p*-phenylenevinylene) and derivatives, *Int. J. Quant. Chem.: Quant. Chem. Symp.* 28:481 (1994).
91. M. Fahlman, P. Bröms, D. A. dos Santos, S. C. Moratti, N. Johansson, K. Xing, R. H. Friend, A. B. Holmes, J. L. Brédas, and W. R. Salaneck, Electronic structure of pristine and sodium-doped cyano-substituted poly(2,5-dihexyloxy-*p*-phenylene-vinylene): a combined experimental and theoretical study, *J. Chem. Phys.* 102:8167 (1995).
92. J. J. Pireaux, Private communication.
93. M. Boman, S. Stafström, and J. L. Brédas, Theoretical investigations of the aluminum/polythiophene interface, *J. Chem. Phys.* 97:9144 (1992).
94. M. Boman, H. Ågren, and S. Stafström, A Δ-SCF study of core electron binding energies of model molecules for the aluminum/polythiophene interface, *J. Phys. Chem* 99:16597 (1995).
95. A. Lachkar, A. Selmani, E. Sacher, M. Leclerc, and R. Mokliss, Metallization of polythiophenes, I. Interaction of vapor-deposited Cu, Ag, and Au with poly(3-hexylthiophene) (P3HT), *Synth. Met.* 66:209 (1994).
96. P. Dannetun, M. Lögdlund, C. Fredriksson, M. Boman, S. Stafström, W. R. Salaneck, B. E. Kohler, and C. Spangler, Chemical and electronic structure of the early stages of interface formation between aluminum and α,ω-diphenyltetradecaheptaene, in *Polymer-Solid Interfaces* (J. J. Pireaux, P. Bertrand, and J. L. Brédas, eds.), IOP Publishing, Bristol, UK, 1992, p. 201.
97. M. Bou, J. M. Martin, T. Le Mogne, and L. Vovelle, Chemistry of the interface between aluminum and polyethyleneterephtalate, *Appl. Surf. Sci.* 47:149 (1991).
98. M. Lögdlund, P. Dannetun, S. Stafström, W. R. Salaneck, M. G. Ramsey, C. W. Spangler, C. Fredriksson, and J. L. Brédas, Soliton pair charge storage in doped polyene molecules: evidence from photoelectron spectroscopy studies, *Phys. Rev. Lett.* 70:970 (1993).
99. M. Lögdlund and J. L. Brédas, Theoretical analysis of the charge storage states in diphenylpolyenes with one to seven double bonds, *J. Chem. Phys.* 100:6543 (1994).
100. W. R. Salaneck and J. L. Brédas, The metal-on-polymer interface in polymer light emitting diodes, *Adv. Mater.* 8:48 (1996).
101. P. Bröms, J. Birgersson, N. Johansson, M. Lögdlund, and W. R. Salaneck, Calcium electrodes in polymer LED's, *Synth. Met.* 74:179 (1995).
102. P. Dyreklev, M. Berggren, O. Inganäs, M. R. Andersson, O. Wennerström, and T. Hjertberg, Polarized electroluminescence from an oriented substituted polythiophene in a light emitting diode, *Adv. Mater.* 7:43 (1995).
103. J. C. Wittmann and P. Smith, Highly oriented thin films of poly(tetrafluoroethylene) as a substrate for oriented growth of materials, *Nature* 352:414 (1991).
104. M. Fahlman, J. Rasmusson, K. Kaeriyama, D. T. Clark, G. Beamson, and W. R. Salaneck, Epitaxy of poly(2,5-diheptyl-*p*-phenylene) on ordered polytetrafluoroethylene, *Synth. Met.* 66:123 (1994).
105. K. Seki, N. Ueno, U. O. Karlsson, R. Engelhardt,

and E.-E. Koch, Valence bands of oriented finite linear chain molecular solids as model compounds of polyethylene studied by angle-resolved photoemission, *Chem. Phys. 105*:247 (1986).
106. F. J. Himpsel, Angle-resolved measurements of the photoemission of electrons in the study of solids, *Adv. Phys. 32*:1 (1983).
107. W. R. Salaneck, M. Fahlman, C. Lapersonne-Meyer, J.-L. Fave, M. Schott, M. Lögdlund, and J. L. Brédas, Electronic structure of 4-BCMU polydiacetylene studied by angle-dependent photoelectron spectroscopy, *Synth. Met. 67*:309 (1994).
108. J. Berrehar, C. Lapersonne-Meyer, and M. Schott, Polydiacetylene single crystal thin films, *Appl. Phys. Lett. 48*:630 (1986).
109. H. Witteler, Substituierte poly(*p*-phenylene): synthese, struktur und phasenverhalten, *Ph.D. Thesis, University of Mainz*, (1993).

24
Conformation-Induced Chromism in Conjugated Polymers

Mario Leclerc and Karim Faïd
University of Montreal, Montreal, Quebec, Canada

I. INTRODUCTION

Conjugated polymers (e.g., polyacetylenes, polypyrroles, polythiophenes, polyanilines) are well known for their high electrical conductivity in the doped (oxidized, reduced, or protonated) state. The delocalized electronic structure of these polymers is in part responsible for the stabilization of the various charge carriers (radical cations, dications, and their negative analogs) created upon doping. In many cases, the conjugated electronic structure also involves a relatively small $\pi-\pi^*$ band gap with strong absorption in the UV-visible range. The coupling between the electronic structure and the UV-vis absorption of these polymers has led to the development of interesting electrochromic devices based on reversible doping–dedoping redox processes [1].

On the other hand, the UV-vis absorption spectrum of neutral conjugated polymers can also be altered by varying their conformational structure. Even using the simple Hückel theory, which relates the HOMO–LUMO transition to the conjugation length, it is evident that any twisting between adjacent units will reduce the effective conjugation length [2], leading to a blue shift of the UV-vis absorption. More sophisticated calculations performed by Brédas et al. [3,4] have clearly described the dependence of the electronic structure of conjugated molecules on backbone conformation. An illustration of this effect can be seen in the strong blue shift of the UV-vis absorption maximum of nonplanar polyacetylene [5,6] and polypyrrole [7] derivatives compared to their unsubstituted parent polymers.

These substituted conjugated polymers were prepared to develop processable electroactive materials because most unsubstituted conjugated polymers are not melt- or solution-processable owing to strong interchain interactions and chain stiffness. However, as mentioned above, the presence of bulky substituents in polyacetylenes and N-substituted polypyrroles has forced a twisting of the backbone that has led to processable but poorly conjugated materials with reduced electrical properties. A major breakthrough in this field occurred with the synthesis of highly conjugated, conducting, and processable poly(3-alkylthiophene)s [8–13] (Fig. 24.1a). In the solid state, an anti-coplanar (and highly conjugated) conformation of the backbone can accommodate an alkyl substituent at the 3-position. Highly conjugated and fully substituted poly(3-alkoxy-4-methylthiophene)s have also been developed [14–16] (Fig. 24.1b). The presence of a second substituent on each repeat unit in these conjugated polythiophenes was made possible by the introduction of a small oxygen atom in the vicinity of the thiophene backbone [15]. Moreover, some of these processable poly(3-alkylthiophene)s [17–30] and poly(3-alkoxy-4-methylthiophene)s [31–33] have shown striking reversible chromic effects upon heating (solid-state and solution thermochromism) or when the solvent quality was altered (solvatochromism). Although no definitive mechanism has yet been established, these interesting optical effects are believed to be related to a reversible "transition" between a coplanar (highly conjugated)

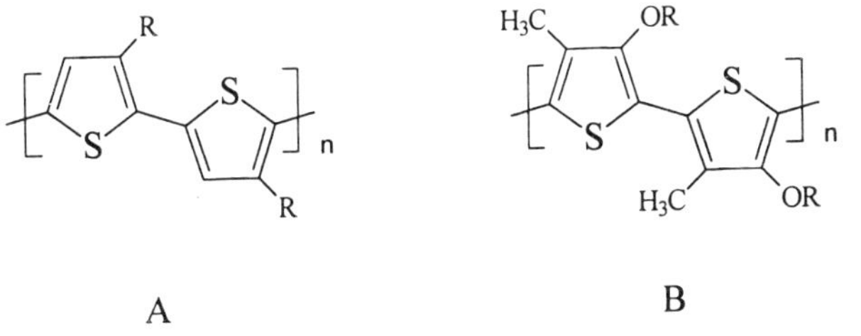

Fig. 24.1 Repeat unit of (a) poly(3-alkylthiophene)s and (b) poly(3-alkoxy-4-methylthiophene)s.

form and a nonplanar (less conjugated) conformational structure of the backbone.

In this chapter, a detailed description is given of these intriguing optical properties of certain polythiophene derivatives, and their analogy with chromic effects found in other π-conjugated (polydiacetylenes) and σ-conjugated polymers (polysilanes) is discussed. In order to explain these "transitions," various molecular mechanisms are described. Finally, potential applications of these interesting optical effects are presented.

II. CHROMIC POLYTHIOPHENES

A. Solid-State Chromism

As mentioned in the Introduction, certain neutral poly(3-alkylthiophene)s [17–30] and poly(3-alkoxy-4-methylthiophene)s [31–33] exhibit strong thermochromism in the solid state. As an example of poly(3-alkoxy-4-methylthiophene)s, Fig. 24.2 shows the temperature dependence of the absorption spectrum of a thin film of poly(3-(2-(2-methoxyethoxy)ethoxy)-4-methylthiophene). At room temperature, this polymer is highly conjugated with an absorption maximum at 548 nm, the two other peaks at 516 and 604 nm being related to a vibronic fine structure [16]. Upon heating, a new absorption band appears at 398 nm while the intensity of the 548 nm band decreases. This strong blue shift of the maximum of absorption upon heating could be related to a conformational transition of the main conjugated backbone from a highly conjugated (coplanar or nearly planar, red-violet) form at low temperatures to a less conjugated (nonplanar, yellow) conformational structure at high temperatures. These optical effects are reversible (although some hysteresis is present in this phenomenon) and cannot be attributed to any degradation of the polymer. This thermochromic transition was also studied as a function of time, and no variation in the absorption spectrum of the polymer was observed for a fixed temperature. A clear isosbestic point is also observed that indicates that these conformational defects are created in a cooperative way, involving long sequences of thiophene units. It is impossible, however, to determine whether these two "phases" coexist on different parts of the same polymer chain or on different polymer chains.

In correlation with these optical effects, differential scanning calorimetric (DSC) measurements have revealed well-defined thermal transitions (Fig. 24.3). Upon heating, in addition to a glass transition near 40°C, a clear endotherm is observed between 75 and 135°C, a temperature range where most important optical changes take place. Good agreement between the calorimetric and spectroscopic measurements is also observed during the cooling process. Temperature-dependent X-ray diffraction measurements have revealed a low degree of crystallinity for this polymer that is not strongly modified upon heating (Fig. 24.4). All these data and those obtained for poly(3-octyloxy-4-methylthiophene) [31] and poly(3-fluoroalkoxy-4-methylthiophene) [32] seem to indicate that a cooperative conformational transition is responsible for the thermochromic properties in "amorphous" poly(3-alkoxy-4-methylthiophene)s.

Similarly, semicrystalline poly(3-alkylthiophene)s exhibit a significant blue shift of their maximum of absorption upon heating (Fig. 24.5). An isosbestic point is also observed that indicates the coexistence of two distinct "chromophores" (two distinct conformational structures). However, DSC, FTIR, and X-ray analyses on these semicrystalline polymers have revealed a fairly good correlation between the melting of the polymer and the thermochromic effects [21–29]. It is even possible to modulate the temperature range of the thermochromic effects by varying the melting temperature through side-chain modification [20]. From all these results, it is evident that similar cooperative twistings of the main polythiophene chain can be observed in both the amorphous and crystalline phases. Nevertheless, it is worth

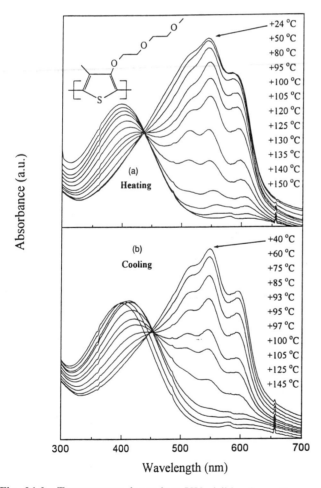

Fig. 24.2 Temperature-dependent UV-visible absorption spectra of poly(3-(2-(2-methoxyethoxy)ethoxy)-4-methylthiophene) in the solid state.

Conformation-Induced Chromism

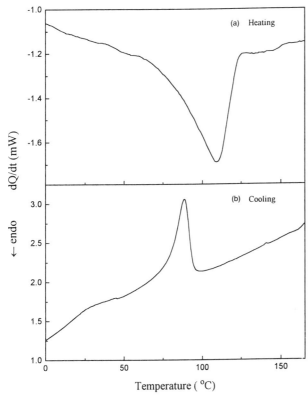

Fig. 24.3 Differential scanning calorimetric thermograms of poly(3-(2-(2-methoxyethoxy)ethoxy)-4-methylthiophene). (From Ref. 33.)

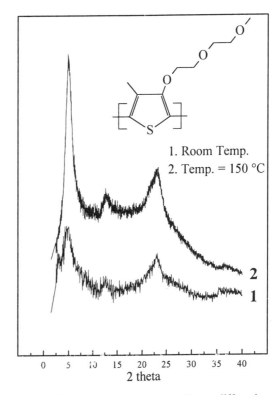

Fig. 24.4 Temperature-dependent X-ray diffraction diagrams of poly(3-(2-(2-methoxyethoxy)ethoxy)-4-methylthiophene).

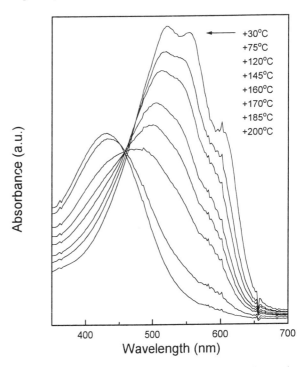

Fig. 24.5 Temperature-dependent UV-visible absorption spectra of poly(3-dodecylthiophene) in the solid state (heating scan). (From Ref. 33.)

noting that for a given polymer both the thermochromic transition temperature and the transition speed of the crystallites are expected to deviate from those of the amorphous region. These effects can explain the absence of a clear isosbestic point during the thermochromic transition of some semicrystalline polythiophene samples [28].

It has also been found that the optical absorption spectrum of various polythiophene derivatives is pressure-dependent [19,28]. For instance, the nonplanar (high temperature) form of poly(3-dodecylthiophene) is red-shifted with increasing pressure (Fig. 24.6). These results seem to indicate that the conformational change of the polymer and the increase of the band gap induced by heating can be compensated for by the applied pressure.

B. Solution Chromism

Thermochromic phenomena have also been observed with poly(3-alkoxy-4-methylthiophene)s and poly(3-alkylthiophene)s in solution. As shown in Fig. 24.7,

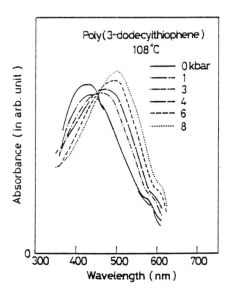

Fig. 24.6 Absorption spectra of poly(3-dodecylthiophene) at 108°C under various pressures. (From Ref. 19.)

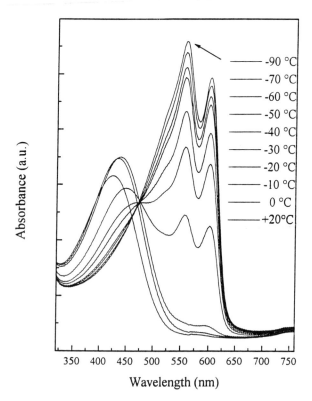

Fig. 24.7 Temperature-dependent UV-visible absorption spectra of poly(3-(2-(2-methoxyethoxy)ethoxy)-4-methylthiophene) in tetrahydrofuran (cooling process). (From Ref. 33.)

poly(3 - (2 - (2 - methoxyethoxy)ethoxy) - 4 - methylthiophene) exhibits a well-defined thermochromic (from yellow to red-violet) transition in tetrahydrofuran (THF) solution upon cooling. Similar optical effects have also been obtained with poly(3-alkylthiophene)s dissolved in 2,5-dimethyltetrahydrofuran [17] and tetrahydrofuran [33]. Recent neutron [34] and light scattering [35] measurements have indicated that poly(3-alkylthiophene)s exist as isolated flexible-coil (nonplanar) chains in dilute THF solution at room temperature. These results explain their absorption maximum near 445 nm in THF at this temperature. Upon cooling, a more conjugated (planar or nearly planar) conformation is adopted by the polymer chains. Once again, the presence of an isosbestic point in temperature-dependent optical absorption measurements indicates the coexistence of long sequences of nonplanar and planar substituted thiophene units. In dilute solutions, no dependence upon the polymer concentration was observed, which is suggestive of a single-chain phenomenon [17,33]. However, aggregation and precipitation of the polymers can be observed at very low temperatures and after a long period of time [17,36].

Strong optical changes in polythiophene solutions can also be induced by varying the solvent composition [17,18]. For instance, the addition of methanol (poor solvent) to a solution of poly(3-hexylthiophene) in chloroform (good solvent) results in a yellow to red-violet color change. This phenomenon is again related to a cooperative planar/nonplanar conformational transition of the backbone induced by a variation of the solvent quality.

C. Structural Effects

All results reported up to now involve well-defined head-to-tail (80–98%) poly(3-alkylthiophene)s or poly(3-alkoxy-4-methylthiophene)s obtained with relatively specific synthetic procedures [10,16,37–40]. The head-to-tail structure refers to couplings between the 2-position of a first thiophene unit and the 5-position of a second repeat unit. These couplings will lead to the structures shown in Fig. 24.1. However, different types of couplings can be made between thiophene units, giving materials with different regiochemical structures (triads, see Fig. 24.8) and physical properties (UV-vis absorption, crystallinity, conductivity, etc.) [16,32,37–40]. The thermochromic properties of these different regiochemical structures are also different. For instance, due to strong steric interactions, neutral poly(3,3'-dihexyl-2,2'-bithiophene) (0% head-to-tail) cannot adopt a coplanar conformation under any conditions, and therefore this polymer is not thermochromic [41]. Nonregioregular (50% head-to-tail) poly(3-alkylthiophene)s exhibit weak and monotonic shifts of their absorption maximum upon heating without the presence of any isosbestic point (Fig. 24.9 and 24.10), whereas 70% head-to-tail poly(3-pentylthio-

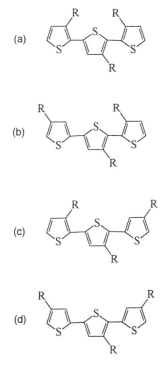

Fig. 24.8 Different regiochemical structures in poly(3-alkylthiophene)s. (a) Head-tail/Head-head; (b) tail-tail/head-head; (c) head-tail/head-tail; (d) tail-tail/head-tail.

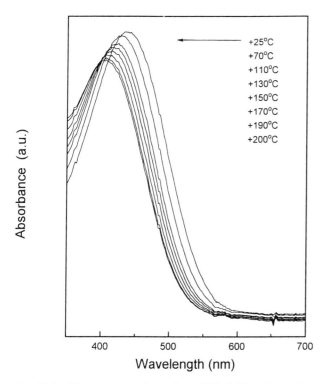

Fig. 24.9 Temperature-dependent UV-visible absorption spectra of nonregioregular (50% head-to-tail) poly(3-hexylthiophene) in the solid state (heating scan). (From Ref. 33.)

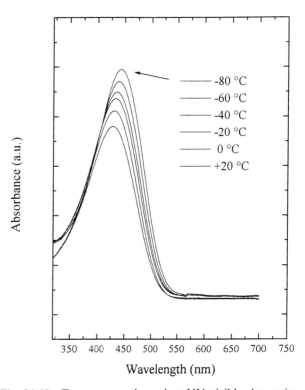

Fig. 24.10 Temperature-dependent UV-visible absorption spectra of nonregioregular (50% head-to-tail) poly(3-hexylthiophene) in tetrahydrofuran (cooling process). (From Ref. 33.)

phene) [24] and poly(3-alkylfluorothiophene) [32] do show cooperative thermochromic transitions.

Similarly, nonregioregular polybithiophenes such as poly(3-dodecyl-2,2'-bithiophene) [36] and poly(3-butoxy-3'-decyl-2,2'-bithiophene) [42] allow only the formation of localized conformational defects along the polymer backbone leading to a continuous blue shift of their absorption maximum upon heating. Although all these chromic effects are possibly related to a modification of the conformation of the conjugated backbone (chromophore), these data indicate clearly the strong dependence of these optical effects in polythiophenes on their substitution pattern, a regioregular chemical structure allowing cooperative twisting of the conjugated backbone.

III. OTHER CHROMIC POLYMERS

A. Polydiacetylenes

Chromic effects in conjugated polymers are not exclusive to polythiophenes but were indeed first observed in polydiacetylenes [43]. As reported for polythiophenes, the addition of flexible side chains had given various processible and chromic polydiacetylenes [44–48]. Among them, chromism (induced by changes in temperature, solvent quality, or pressure) in polydiacetylenes bearing urethane substituents (Fig. 24.11) have been extensively investigated [44,48–54]. However, it is worth noting that chromic effects have also been observed in polydiacetylenes with no possibility of hydrogen bonding between the side chains [45–48]. As an example, Fig. 24.12 shows the UV-visible absorption spectrum of poly(4BCMU) in toluene at various temperatures. The presence of a clear isosbestic point indicates the coexistence of two different conformational structures with different absorption features (a yellow and a red form). Similar chromic effects have been observed upon the addition of hexane to a poly(4BCMU) chloroform solution [44]. Although the determination of the exact mechanism involved in these chromic transitions is a subject of debate (see Section IV), these optical phenomena could be explained in terms of a planar–nonplanar conformational transition of the main conjugated backbone [44,48–54]. Light and neutron scattering measurements in dilute solutions have shown evidence of a wormlike coil conformation (with continuous deformation of the conjugated backbone) for the less conjugated form (high temperature phase) [55], whereas the more conjugated form is believed to be related to a rigid-rod type of conformation.

B. Polysilanes

Owing to the extensive delocalization of σ electrons along the silicon backbone, polysilanes display an intense absorption band in the near-UV region. This absorption band is also strongly coupled with the polymer conformation. Evidence of thermochromism in the solid

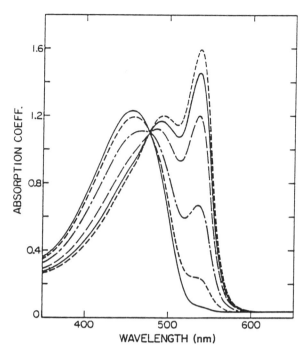

Fig. 24.12 UV-visible absorption spectra for poly(4BCMU) in toluene at 72.0, 70.5, 67.5, 63.0, 59.2, and 56.0°C. The highest temperature data are represented by the solid curve peaking at 450 nm; the lowest temperature data are shown in the dashed curve peaking at 530 nm. (From Ref. 50.)

state and in solution was first reported for poly(di-n-hexylsilane) [56,57] (Fig. 24.13). In the solid state, good correlations have been found between side-chain melting and disordering of the polymer backbone from a planar to a nonplanar conformation [58–61]. Various symmetrically substituted polysilanes have been prepared, and most of them exhibit abrupt chromic (thermochromic, solvatochromic, and piezochromic) transitions with the presence of a clear isosbestic point [62–65]. Noticeable exceptions to this trend are poly(di-n-butylsilane) and poly(di-n-pentylsilane) [66]. These two polymers show abrupt thermochromic effects in solution but not in the solid state due to a helical crystalline structure at low temperatures that is significantly different from the all-*trans* conformation usually found in poly(di-n-alkylsilane)s.

On the other hand, unsymmetrically substituted polysilanes generally display a continuous red shift of their absorption maximum with decreasing temperature until a limiting wavelength is reached [67]. This general behavior is not followed by certain alkoxy- and alkyl-substituted polysilanes [68,69], which have revealed an abrupt red shift of their absorption maximum upon cooling. Indeed, these amorphous and chromic polysilanes

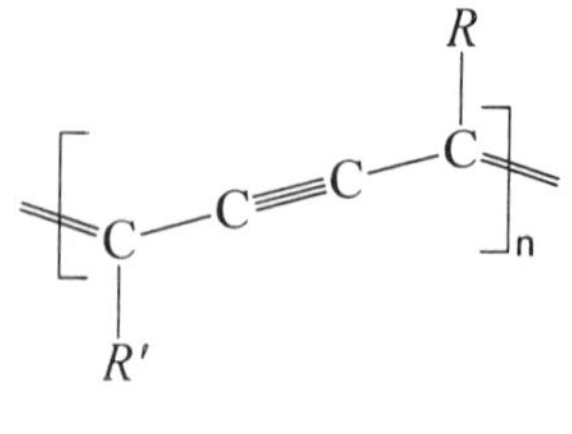

poly(3BCMU): $R=R'= -(CH_2)_3-O-\overset{O}{\underset{\|}{C}}-NH-CH_2-\overset{O}{\underset{\|}{C}}-O-(CH_2)_3-CH_3$

poly(4BCMU): $R=R'= -(CH_2)_4-O-\overset{O}{\underset{\|}{C}}-NH-CH_2-\overset{O}{\underset{\|}{C}}-O-(CH_2)_3-CH_3$

Fig. 24.11 Repeat unit of polydiacetylenes.

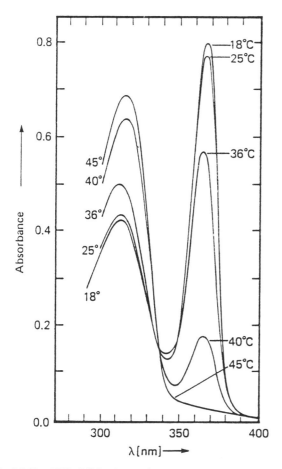

Fig. 24.13 UV-visible absorption spectra of poly(di-*n*-hexylsilane) at various temperatures between 18 and 45°C. (From Ref. 60.)

show cooperative conformational transitions very similar to those reported previously for poly(3-alkoxy-4-methylthiophene)s.

IV. MECHANISM AND DRIVING FORCE

On the basis of all these results, a striking parallelism seems to exist between the chromic phenomena observed in polythiophenes, polydiacetylenes, and polysilanes. It is even possible that the same driving force could explain most of the optical features observed in these conjugated materials. As reported in the previous sections, these chromic phenomena could be described in terms of a planar–nonplanar conformational transition of the main conjugated backbone. However, it is unclear whether this conformational transition is related to an intrachain or interchain mechanism. Results obtained in the solid state are not particularly useful in attempts to solve this problem; however, the independence of the thermochromic phenomena of dilute polythiophene solutions from the polymer concentration is suggestive of a single-chain phenomenon [17,33]. Moreover, careful spectroscopic and kinetic studies on highly dilute solutions ($<10^{-7}$ M) of polydiacetylenes [51] and polysilanes [67] have shown that aggregation is not necessarily responsible for the thermochromic transition although it occurs at later times. These results have been interpreted in terms of an intramolecular collapse of isolated polymer chains. This model is in agreement with light and neutron scattering measurements on polysilanes, which have revealed, at temperatures very near the thermochromic transition but prior to the onset of aggregation, a decrease in the radius of gyration of the polymers upon cooling [70]. This intramolecular collapse would induce a more "ordered" assembly (and a more conjugated form than the wormlike coil conformation) with distinctly different optical properties.

In the previous sections, it was reported that regioregular polythiophenes and symmetrically substituted polysilanes show a cooperative conformational transition of their main chain (involving long sequences of repeat units) as a function of temperature. In contrast, nonregioregular and partially substituted polythiophene derivatives as well as certain unsymmetrically substituted polysilanes allow only the formation of localized conformational defects along the polymer backbone leading to a continuous and monotonic red shift of their absorption maximum upon cooling. The proposed collapse could also explain this strong dependence of the thermochromic properties on the substitution pattern. To a first approximation, in highly dilute solutions the formation of intramolecular aggregates in regioregular polymers could consist in a cooperative self-assembly of the macromolecule made possible by the presence of a regioregular substitution pattern. This self-assembly of the regioregular macromolecule would drive the nonplanar–planar conformational transition of the conjugated backbone, a nonregioregular substitution pattern leading only to localized interactions with no cooperative effect. A similar mechanism can be proposed for semidilute solutions and in the solid state that involves both intramolecular and intermolecular interactions. These assemblies would be reversibly broken through side-chain disordering [33]. This assumption is in aggreement with many spectroscopic studies that have indicated an increase in disorder in both the side chains and the main chain upon heating [21–26,31,47,58,59].

Another possible explanation for these optical phenomena was developed by Schweizer [71,72], who proposed that the thermochromic phenomena can arise from the dispersion interaction of delocalized electrons along the polymer backbone with the surrounding polarizable medium (including side chains). According to this model, the conformational transition of the conjugated backbone is controlled by competition between attractive dispersion interactions (V_D) for the fully ordered chains and the rotational defect energy (ϵ). Depending on the V_D/ϵ ratio, either cooperative twisting or weak and localized rotational defects can be created along the conjugated backbone. The formation of excitons has also been mentioned recently as a possible cause of the chromic effects in conjugated polymers [73]. Although all these models can give a good description of chromic properties of conjugated polymers, further theoretical and experimental studies (in particular on well-defined oligomers) are needed to obtain more information about the mechanisms responsible for these transitions.

V. APPLICATIONS AND PERSPECTIVES

On the basis of all the examples given above, it seems that the UV-visible absorption spectrum of certain neutral conjugated polymers is dramatically altered by modifications of their conformational structure. These striking optical effects can obviously be applied for the development of various display devices, detectors, and sensors [30,74]. In particular, Charych and coworkers [75,76] developed an interesting biosensor for influenza A virus by preparing a Langmuir–Blodgett film of a polydiacetylene derivative bearing a recognition site (sialic

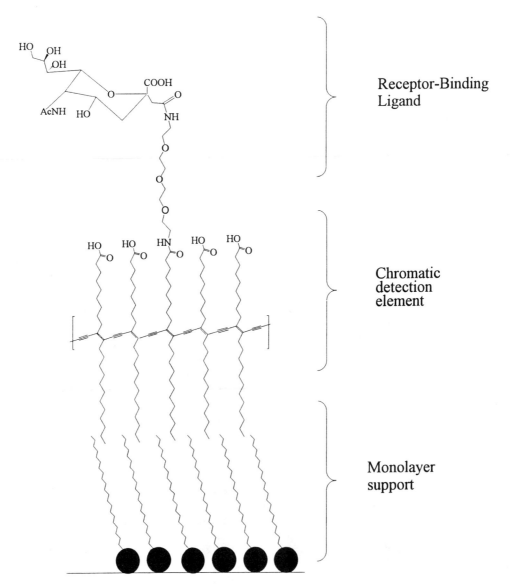

Fig. 24.14 Schematic diagram of the polymerized bilayer assembly prepared from mixtures of diacetylenes. (From Ref. 75.)

acid) for hemagglutinin (a surface group on the influenza virus) (Fig. 24.14). Binding of hemagglutinin to the polydiacetylene thin films induces a strong disorder within both the side chains and the main chain, permitting a colorimetric detection (and possibly quantification) of the virus (Fig. 24.15). Such ligand-functionalized films may also provide a simple means of screening for potential drug candidates [76]. A similar approach was recently developed with the use of functionalized polythiophenes [77].

Moreover, the conformation (and the effective conjugation length) of ether- and crown ether–containing polythiophenes in solution have been found to be sensitive to the presence of metal ions [33,78–82]. Noncovalent interactions between these ether-substituted polythiophenes and ionic species modify the side-chain conformation, which then affects the conformation of the polymer backbone and absorption in the visible range. Ionochromic effects in regioregular poly(3-oligo(oxyethylene)-4-methylthiophene) can even be used to determine optically the concentration of alkali metal ions in solution (Fig. 24.16) [81,82].

On the other hand, the control of the absorption properties of these conjugated polymers can also modify their luminescence properties and lead to the development of light-emitting diodes with tunable colors [83]. The generation of the different types of conformational defects may also create new opportunities for the fabrication of novel molecular devices based on the propagation of such conformational defects along conjugated molecules or on the resulting modification of their refractive index. In principle, it is also possible to develop novel photochromic materials by the incorporation of photoisomerizable substituents (such as azo and stilbene derivatives).

From all these examples, it is firmly believed that chromism in neutral substituted polythiophenes, polydiacetylenes, and polysilanes can be a powerful tool to develop novel smart materials. In this respect, it is worth also noting the recent development of a new class of thermochromic conjugated polymers, based on poly(alkylbithiazole)s [84]. In all these materials, a planar–nonplanar conformational transition of the backbone can be induced via a large range of external stimuli, leading to various chromic (thermochromic, solvatochromic, piezochromic, ionochromic, affinitychromic, photochromic, etc.) effects with very promising applications.

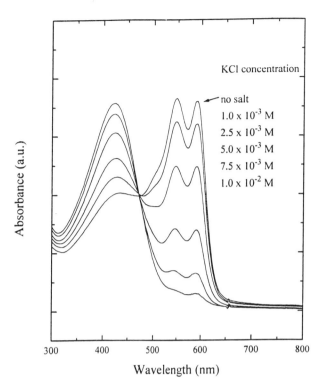

Fig. 24.16 UV-visible absorption spectra of poly(3-oligo(oxyethylene)-4-methylthiophene) in methanol with different KCl concentrations at room temperature.

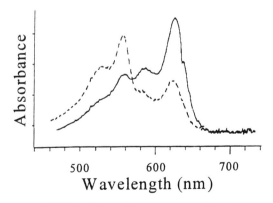

Fig. 24.15 Visible absorption spectrum of the bilayer assembly depicted in Fig. 24.16 prior to and (———) after incubation with the influenza virus. (From Ref. 75.)

REFERENCES

1. A. O. Patil, A. J. Heeger, and F. Wudl, Optical properties of conductive polymers, *Chem. Rev.* 88:183 (1988).
2. I. N. Levine, *Quantum Chemistry*, Allyn and Bacon, Boston, 1974, p. 421.
3. J. L. Brédas, G. B. Street, B. Thémans, and J. M. André, Organic polymers based on aromatic rings (polyparaphenylene, polypyrrole, polythiophene): evolution of the electronic properties as a function of the torsion angle between adjacent rings, *J. Chem. Phys.* 83:1323 (1985).
4. B. Thémans, W. R. Salaneck, and J. L. Brédas, Theoretical study of the influence of thermochromic effects

on the electronic structure of poly(3-hexylthiophene), *Synth. Met. 28*:C359 (1989).
5. M. Leclerc and R. E. Prud'homme, Conformational analysis of substituted polyacetylenes, *J. Polym. Sci., Polym. Phys. Ed. 23*:2021 (1985).
6. M. Leclerc and R. E. Prud'homme, Theoretical analysis of model compounds of substituted poly(acetylenes): conformation versus electronic properties, *Polym. Bull. 18*:159 (1987).
7. A. F. Diaz, J. Castillo, K. K. Kanazawa, J. A. Logan, M. Salmon, and O. Fajardo, Conducting poly-N-alkylpyrrole polymer films, *J. Electroanal. Chem. 133*:233 (1982).
8. R. L. Elsenbaumer, K. Y. Jen, and R. Oboodi, Processible and environmentally stable conducting polymers, *Synth. Met. 15*:169 (1986).
9. M. A. Sato, S. Tanaka, and K. Kaeriyama, Soluble conducting polymers, *J. Chem. Soc., Chem. Commun. 1986*:873.
10. R. Sugimoto, S. Takeda, H. B. Gu, and K. Yoshino, Preparation of soluble polythiophene derivatives utilizing transition metal halides as catalysts and their property, *Chem. Express 1*:635 (1986).
11. S. Hotta, S. D. D. V. Rughooputh, A. J. Heeger, and F. Wudl, Spectroscopic studies of soluble poly(3-alkylthienylenes), *Macromolecules 20*:212 (1987).
12. J. Roncali, R. Garreau, A. Yassar, P. Marque, F. Garnier, and M. Lemaire, Effects of steric factors on the electrosynthesis and properties of conducting poly(3-alkylthiophenes), *J. Phys. Chem. 91*:6706 (1987).
13. M. R. Bryce, A. Chissel, P. Kathirgamanathan, D. Parker, and N. M. R. Smith, Soluble, conducting polymers from 3-substituted thiophenes and pyrroles, *J. Chem. Soc., Chem. Commun. 1987*:466.
14. M. Feldhues, G. Kampf, H. Litterer, T. Mecklenburg, and P. Wegener, Polyalkoxythiophenes, soluble electrically conducting polymers, *Synth. Met. 28*:C487 (1989).
15. M. Leclerc and G. Daoust, Design of new conducting 3,4-disubstituted polythiophenes, *J. Chem. Soc., Chem. Commun. 1990*:273.
16. G. Daoust and M. Leclerc, Structure–property relationships in alkoxy-substituted polythiophenes, *Macromolecules 24*:455 (1991).
17. S. D. D. V. Rughooputh, S. Hotta, A. J. Heeger, and F. Wudl, Chromism of soluble polythienylenes, *J. Polym. Sci., Polym. Phys. Ed. 25*:1071 (1987).
18. O. Inganäs, W. R. Salaneck, J.-E. Osterholm, and J. Laakso, Thermochromic and solvatochromic effects in poly(3-hexylthiophene), *Synth. Met. 22*:395 (1988).
19. K. Yoshino, S. Nakajima, M. Onada, and R. Sugimoto, Electrical and optical properties of poly(3-alkylthiophenes), *Synth. Met. 28*:C349 (1989).
20. O. Inganäs, G. Gustafsson, and W. R. Salaneck, Thermochromism in thin films of poly(3-alkylthiophenes), *Synth. Met. 28*:C377 (1989).
21. M. J. Winokur, D. Spiegel, Y. Kim, S. Hotta, and A. J. Heeger, Structural and absorption studies of the thermochromic transition in poly(3-hexylthiophene), *Synth. Met. 28*:C419 (1989).
22. K. Tashiro, K. Ono, Y. Minagawa, M. Kobayashi, T. Kawai, and K. Yoshino, Structure and thermochromic solid-state phase transition of poly(3-alkylthiophene), *J. Polym. Sci., Polym. Phys. Ed. 29*:1223 (1991).
23. G. Zerbi, B. Chierichetti, and O. Inganäs, Thermochromism in poly(alkylthiophenes): molecular aspects from vibrational spectroscopy, *J. Chem. Phys. 94*:4646 (1991).
24. P. O. Ekeblad and O. Inganäs, Thermochromism in poly(3-pentylthiophene), *Polym. Commun. 32*:436 (1991).
25. S.-A. Chen and J.-M. Ni, Structure/properties of conjugated conductive polymers. 1. Neutral poly(3-alkylthiophene)s, *Macromolecules 25*:6081 (1992).
26. T. J. Prosa, M. J. Winokur, J. Moulton, P. Smith, and A. J. Heeger, X-ray structural studies of poly(3-alkylthiophenes): an example of an inverse comb, *Macromolecules 25*:4364 (1992).
27. S.-A. Chen and J.-M. Ni, Structure/properties of the conjugated conductive polymers. 3. Copolymers of 3-alkylthiophenes, *Macromolecules 26*:3230 (1993).
28. K. Iwasaki, H. Fujimoto, and S. Matsuzaki, Conformational changes of poly(3-alkylthiophene)s with temperature and pressure, *Synth. Met. 63*:101 (1994).
29. H.J. Fell, E. J. Samuelsen, J. Mardalen, and P. H. J. Carlsen, Structural and thermochromic properties of mixtures of poly(3-hexylthiophene) and poly(3-octylthiophene), *Synth. Met. 63*:157 (1994).
30. O. Inganäs, Conformational flexibility, electronic structure and chromism in conjugated polyalkylthiophenes, *Trends Polym. Sci. 2*:189 (1994).
31. C. Roux and M. Leclerc, Rod-to-coil transition in alkoxy-substituted polythiophenes, *Macromolecules 25*:2141 (1992).
32. L. Robitaille and M. Leclerc, Synthesis, characterization and Langmuir–Blodgett films of fluorinated polythiophenes, *Macromolecules 27*:1847 (1994).
33. K. Faïd, M. Fréchette, M. Ranger, L. Mazerolle, I. Lévesque, M. Leclerc, T. A. Chen, and R. D. Rieke, Chromic phenomena in regioregular and non-regioregular polythiophene derivatives, *Chem. Mater. 7*:1390 (1995).
34. J. P. Aimé, P. Garrin, G. L. Baker, S. Ramakrishan, and P. Timmins, Physical origin of the local rigidity of conjugated polymers in good solvents, *Synth. Met. 41*:859 (1991).
35. G. W. Heffner and D. S. Pearson, Molecular characterization of poly(3-hexylthiophene), *Macromolecules 24*:6295 (1991).
36. C. Roux and M. Leclerc, Thermochromic properties of polythiophene derivatives: formation of localized and delocalized conformational defects, *Chem. Mater. 6*:620 (1994).
37. M. Leclerc, F. M. Diaz, and G. Wegner, Structural analysis of poly(3-alkylthiophenes), *Makromol. Chem. 190*:3105 (1989).
38. R. M. Souto Maior, K. Hinkelmann, H. Eckert, and F. Wudl, Synthesis and characterization of two regio-

chemically defined poly(dialkylbithiophenes): a comparative study, *Macromolecules 23*:1268 (1990).
39. R. D. McCullough, R. D. Lowe, M. Jayaraman, and D. L. Anderson, Design, synthesis, and control of conducting polymer architecture: structurally homogeneous poly(3-alkylthiophenes), *J. Org. Chem. 58*: 904 (1993).
40. T. A. Chen, X. Wu, and R. D. Rieke, Regiocontrolled synthesis of poly(3-alkylthiophenes) mediated by Rieke zinc: their characterization and solid-state properties, *J. Am. Chem. Soc. 117*:233 (1995).
41. C. Roux, J. Y. Bergeron, and M. Leclerc, Thermochromic properties of polythiophenes: structural aspects, *Makromol. Chem. 194*:869 (1993).
42. C. Roux, K. Faïd, and M. Leclerc, Thermochromic properties of polythiophenes: cooperative effects, *Makromol. Chem., Rapid Commun. 14*:461 (1993).
43. G. J. Exarhos, W. M. Risen, and R. H. Baughman, Resonance Raman study of the thermochromic phase transition of a polydiacetylene, *J. Am. Chem. Soc. 98*:481 (1976).
44. G. N. Patel, R. R. Chance, and J. D. Witt, A planar–nonplanar conformational transition in conjugated polymer solutions, *J. Chem. Phys. 70*:4387 (1979).
45. C. Plachetta, N. O. Rau, A. Hauck, and R. C. Schulz, Some soluble polydiacetylenes, *Makromol. Chem., Rapid Commun. 3*:249 (1982).
46. S. D. D. V. Rughooputh, D. Phillips, D. Bloor, and D. J. Ando, Chromism of a polydiacetylene with weakly interacting sidegroups, *Polym. Commun. 25*:242 (1984).
47. N. Mino, H. Tamura, and K. Ogawa, Analysis of color transitions and changes on Langmuir–Blodgett films of a polydiacetylene derivative, *Langmuir 7*:2336 (1991).
48. G. Wenz, M. A. Muller, M. Schmidt, and G. Wegner, Structure of poly(diacetylenes) in solution, *Macromolecules 17*:837 (1984).
49. K. C. Lim, A. Kapitulnik, R. Zacher, and A. J. Heeger, Conformation of polydiacetylene macromolecules in solution: field induced birefringence and rotational diffusion constant, *J. Chem. Phys. 82*:516 (1985).
50. K. C. Lim and A. J. Heeger, Spectroscopic and light scattering studies of the conformational (rod-to-coil) transition of poly(diacetylene) in solution, *J. Chem. Phys. 82*:522 (1985).
51. M. A. Taylor, J. A. Odell, D. N. Batchelder, and A. J. Campbell, The coil to rod transition in polydiacetylenes: a kinetic study, *Polymer 31*:1116 (1990).
52. B. Chu and R. Xu, Chromatic transition of polydiacetylene in solution, *Acc. Chem. Res. 24*:384 (1991).
53. A. D. Nava, M. Thakur, and A. E. Tonelli, ^{13}C NMR structural studies of a soluble polydiacetylene poly(4BCMU), *Macromolecules 23*:3055 (1990).
54. B. F. Variano, C. J. Sandroff, and G. L. Baker, High-pressure optical absorption studies of poly(4BCMU), *Macromolecules 24*:4376 (1991).
55. M. Rawiso, J. P. Aimé, J. L. Fave, M. Schott, M. A. Muller, M. Schmidt, H. Baumgartl, and G. Wegner, Solutions of polydiacetylenes in good and poor solvents: a light and neutron scattering study, *J. Phys. Paris 49*:861 (1988).
56. P. Trefonas III, J. R. Damewood, Jr., R. West, and R. D. Miller, Organosilane high polymers: thermochromic behavior in solution, *Organometallics 4*: 1318 (1985).
57. L. Harrah and J. M. Zeigler, Rod-to-coil transition in solutions of poly(di-n-hexylsilane), *J. Polym. Sci., Polym. Lett. Ed. 23*:209 (1985).
58. J. F. Rabolt, D. Hofer, R. D. Miller, and G. N. Fickes, Studies of chain conformational kinetics in poly(di-n-alkylsilanes) by spectroscopic methods. 1. Poly(di-n-hexylsilane), poly(di-n-heptylsilane), and poly(di-n-octylsilane), *Macromolecules 19*:611 (1986).
59. A. J. Lovinger, F. C. Schilling, F. A. Bovey, and J. M. Zeigler, Characterization of poly(di-n-hexylsilane) in the solid state. 1. X-ray and electron diffraction studies, *Macromolecules 19*:2657 (1986).
60. H. Kuzmany, J. F. Rabolt, B. L. Farmer, and R. D. Miller, Studies of chain conformational kinetics in poly(di-n-alkylsilanes) by spectroscopic methods. 2. Conformation and packing of poly(di-n-hexylsilane), *J. Chem. Phys. 85*:7413 (1986).
61. F. C. Schilling, F. A. Bovey, A. J. Lovinger, and J. M. Zeigler, Characterization of poly(di-n-hexylsilane) in the solid state. 2. ^{13}C and ^{29}Si magic angle spinning NMR studies, *Macromolecules 19*:2660 (1986).
62. R. D. Miller and J. Michl, Polysilane high polymers, *Chem. Rev. 89*:1359 (1989).
63. F. C. Schilling, F. A. Bovey, A. J. Lovinger, and J. M. Zeigler, Structures, phase transitions, and morphology of polysilylenes, *Adv. Chem. 224*:341 (1990).
64. K. Song, R. D. Miller, G. M. Wallraff, and J. F. Rabolt, Studies of chain conformational kinetics in poly(di-n-alkylsilanes) by spectroscopic methods. 4. Piezochromism in symmetrical poly(di-n-alkylsilanes), *Macromolecules 24*:4084 (1991).
65. R. D. Miller and R. Sooriyakumaran, Alkoxy-substituted poly(diarylsilanes): thermochromism and solvatochromism, *Macromolecules 21*:3120 (1988).
66. F. C. Schilling, A. J. Lovinger, J. M. Zeigler, D. D. Davis, and F. A. Bovey, Solid-state structures and thermochromism of poly(di-n-butylsilylene) and poly(di-n-pentylsilylene), *Macromolecules 22*:3055 (1989).
67. R. D. Miller, G. M. Wallraff, M. Baier, P. M. Cotts, P. Shukla, T. P. Russell, F. C. De Schryver, and D. Declercq, The solution and solid-state thermochromism of unsymmetrically substituted polysilanes, *J. Inorg. Organomet. Chem. 1*:505 (1991).
68. C.-H. Yuan and R. West, Abrupt thermochromic transitions in alcohol-soluble ethoxypentyl-substituted polysilanes, *Macromolecules 26*:2645 (1993).
69. C.-H. Yuan and R. West, Side-chain effect on the nature of thermochromism of polysilanes, *Macromolecules 27*:629 (1994).
70. P. Shukla, P. M. Cotts, R. D. Miller, T. P. Russell, B. A. Smith, G. M. Wallraff, M. Baier, and P. Thiya-

garajan, Conformational transition studies of organosilane polymers by light and neutron scattering, *Macromolecules 24*:5606 (1991).
71. K. S. Schweizer, Order–disorder transitions of π-conjugated polymers in condensed phases. I. General theory, *J. Chem. Phys. 85*:1156 (1986).
72. K. S. Schweizer, Theory of order–disorder transitions and optical properties of flexible conjugated polymers, *Synth. Met. 28*:C565 (1989).
73. D. J. Sandman, Semiconducting polymers and their solid-state properties, *Trends Polym. Sci. 2*:44 (1994).
74. C. Roux, K. Faïd, and M. Leclerc, Polythiophene derivatives: smart polymers, *Polym. News 19*:6 (1994).
75. D. H. Charych, J. O. Nagy, W. Spevak, and M. D. Bednarski, Direct colorimetric detection of a receptor–ligand interaction by a polymerized bilayer assembly, *Science 261*:585 (1993).
76. W. Spevak, J. O. Nagy, and D. H. Charych, Molecular assemblies of functionalized polydiacetylenes, *Adv. Mater. 7*:85 (1995).
77. K. Faid and M. Leclerc, Functionalized regioregular polythiophenes: towards the development of biochromic sensors, *J. Chem. Soc., Chem. Commun. 1996*: 2761.
78. R. D. McCullough and S. P. Williams, Toward tuning electrical and optical properties in conjugated polymers using side chains: highly conductive head-to-tail heteroatom-functionalized polythiophenes, *J. Am. Chem. Soc. 115*:11608 (1993).
79. M. J. Marsella and T. M. Swager, Designing conducting polymer-based sensors: selective ionochromic response in crown ether containing polythiophenes, *J. Am. Chem. Soc. 115*:12214 (1993).
80. R. D. McCullough and S. P. Williams, A dramatic conformational transformation of a regioregular polythiophene via a chemoselective, metal-ion assisted deconjugation, *Chem. Mater. 7*:2001 (1995).
81. I. Lévesque and M. Leclerc, Ionochromic effects in regioregular ether-substituted polythiophenes, *J. Chem. Soc., Chem. Commun. 1995*:2293.
82. I. Lévesque and M. Leclerc, Ionochromic and thermochromic phenomena in a regioregular polythiophene derivative bearing oligo(oxyethylene) side chains, *Chem. Mater. 8*:2843 (1996).
83. M. Berggren, G. Gustafsson, O. Inganas, M. R. Andersson, O. Wennerstrom, and T. Hjertberg, Thermal control of near-infrared and visible electroluminescence in alkylphenyl substituted polythiophenes, *Appl. Phys. Lett. 65*:1489 (1994).
84. J. I. Nanos, J. W. Kampf, M. D. Curtis, L. Gonzalez, and D. C. Martin, Poly(alkylbithiazoles): a new class of variable-bandgap, conjugated polymer, *Chem. Mater. 7*:2232 (1995).

25
Structural Studies of Conducting Polymers

Michael J. Winokur
University of Wisconsin, Madison, Wisconsin

I. INTRODUCTION

The nature of a conducting polymer's intra- and intermolecular structure and the associated structural phase behavior are fundamental issues that strongly impact the physical properties manifested by this unique class of materials. Even relatively small changes in the specific chemical architecture and/or processing procedure can lead to significant variations in the resultant structural forms and in their physical properties. Ultimately a deeper understanding of the various structure–property interrelationships will, in part, form the foundation for future efforts that require even more specialized conducting polymer structures with highly specific properties.

The term "structure" may be used to describe the intrinsic unit construction on a myriad of length scales. Although the hierarchical organization of conjugated polymer structures (and polymers in general) is an exceedingly important and complex issue, it is beyond the limited scope of this chapter. For brevity and conciseness, the text that follows is restricted to reviewing the structure and structural response of prototypical conducting polymer host systems at molecular length scales (ranging from approximately 2 to 200 Å). Even within this rather narrow range there is an immense diversity with respect to the structural forms and phase behavior.

Conducting polymers have some similarities to conventional polymeric materials, but it is clearly the extensive main-chain π conjugation and its implicit "stiffness" with respect to chain bending and twisting that most influences the overall physical behavior. As a direct consequence, virtually all linearly unsubstituted conducting polymers, as shown in Fig. 25.1 are found to be intractable and infusible. These model systems also tend to form crystalline phase structures with many common features. Hence these compounds may be conveniently lumped together to form one basic "class" of conducting polymer materials that have, loosely speaking, similar structural characteristics.

In the quest for improved performance, better processibility, and novel applications [1], a wealth of newer compounds with specific chemical architectures have been synthesized. The three most common approaches that have been implemented are greater main-chain flexibility (e.g., polyaniline [2]), side-chain substitution (e.g., poly(3-alkylthiophenes) [3]), and fabrication from a soluble precursor polymer (e.g., poly-(p-xylene-α-dimethylsulfonium chloride) to yield poly(p-phenylenevinylene) [4]). All of these modifications can produce materials with a range of structural forms and, in some instances, striking new physical behavior.

In one conducting polymer, polyaniline (PANI), the added main-chain flexibility surrounding the amine/imine nitrogen linkages is ostensibly responsible for a myriad of structural effects. There is an overall reduction in the effective intrachain conjugation length and a marked decrease in the tendency to form crystalline phases [5]. This also implies, indirectly, that there are implications for the electronic transport properties that may be achieved. However, within amorphous films there is a unique ability to control the conducting polymer molecular matrix and, concomitantly, the mass transport properties so that high performance separation membranes may be fabricated [6].

Side-chain substitutions are now routinely used to enhance solvent solubility and fusibility so that various conventional polymer processing methods may be used. Some of these polymeric materials, e.g., the poly(3-alkylthiophenes), may contain only a small volume fraction of electrically active regions (i.e., the π-conjugated main chain). Still these materials can exhibit extremely high dc conductivities [7] after doping (or more precisely, intercalation) by a guest species. This intercala-

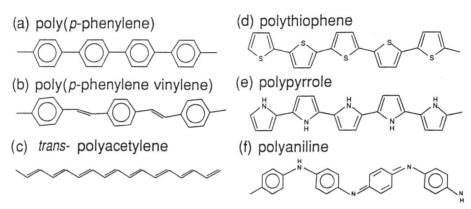

Fig. 25.1 Chemical diagrams for five "stiff" linearly unsubstituted conducting polymers (a–e) and one semiflexible host system (f).

tion process can also provoke significant dimensional changes in the molecular unit construction. This latter property enables the fabrication of novel self-structuring devices [8].

Even more exotic synthesis and processing procedures have further expanded the horizons within the structure–property interrelationships and the potential for new applications. Langmuir–Blogett monolayer film deposition techniques have been applied successfully to yield thin film multilayer heterostructures [9] for use as electronic devices. Chemical coupling of various ion-selective crown ethers [10] can create conducting polymer hosts in which the extent of main-chain π conjugation is directly influenced by the solution concentration of a specific ion when immersed. In all these examples it is often the subtle interplay between the molecular level structural ordering and the electroactive nature of the conducting polymer host that gives rise to these properties. Hence an intimate knowledge of the structure and its structural phase behavior is an integral component of conducting polymer research.

II. LINEAR UNSUBSTITUTED CONDUCTING POLYMERS

A. Rigid-Rod Polymers (Polyacetylene, Polythiophene, etc.)

Historically, rigid-rod polymers were among first conducting polymers synthesized, and they remain the best understood with respect to their structural ordering [11]. Specific examples of these polymers are shown in Fig. 25.1 and include polyacetylene (PA), poly(p-phenylenevinylene) (PPV), poly(p-phenylene) (PPP), polythiophene (PT), and polypyrrole (PPy). The majority of these compounds exhibit crystalline phases that adopt a herringbone equatorial packing (a packing motif common to many conventional linear polymers such as poly-

ethylene and polypropylene) within an orthorhombic or monoclinic unit cell as shown in Fig. 25.2. Hence there are two polymer chains per projected two-dimensional (2D) equatorial unit cell with p2gg symmetry. The average angular orientation of the polymer chains' major axis, with respect to an equatorial lattice vector (typically a), is specified by the setting angle ϕ. Table 25.1 summarizes representative reported values for the various unit cell parameters. The ratio $a/(b \sin \alpha)$ is particularly noteworthy, for it often falls near a value of $\sqrt{3}$, especially for PA and PPV. Hence the lattice formed by the equatorial projection of the polymer chain centers closely approximates that of a triangular grid.

For some conducting polymers the net crystallinity can be quite large, and values exceeding 80% have been reported [21]. Coherence lengths, i.e., the distances over which Bragg periodicity is maintained, are rather modest, with typical values ranging from 50 to 200 Å depend-

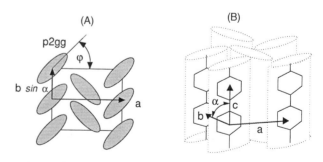

Fig. 25.2 (A) Two-dimensional equatorial projection of the most representative chain packing for linear unsubstituted conducting polymers. This structure has p2gg symmetry and two polymer chains per cell with general lattice parameters of a, $b \sin \alpha$, and ϕ (the setting angle). (B) Three-dimensional construction of the unit cell.

Table 25.1 A Partial List of Reported Crystal Structures and Lattice Constants for Various Linear Conducting Polymer Hosts

Polymer	Space group	Unit cell dimension (Å)			α (deg)	ϕ (deg)	$a/(b \sin \alpha)$	Reference
		a	b	c				
trans-PA	$P2_1/a$	7.34	4.18	2.42	90.5	57	1.76	Leising et al. [12]
trans-PA	$P2_1/n$	7.32	4.24	2.45	91.5	55	1.73	Fincher et al. [13]
trans-PA	Pnam	7.33	4.09	2.45	90.0	51	1.79	Begin et al. [14]
trans-PA		7.26	4.24	2.47	90.5	55	1.71	Sokolowski et al. [15]
PPV		8.30	6.05	6.58	123.0	58	1.63	Granier et al. [16]
PPV		8.07	6.05	6.54	123.0	49	1.59	Chen et al. [17]
PT		7.81	5.52	7.7		31	1.41	Brückner et al. [18]
PPP		7.79	5.51			65	1.41	Sasaki et al. [19]
PANI	Pbcn	7.65	5.75	10.2	90.0	0	1.3	Jozefowicz et al. [20]

ing on the specifics of the polymer sample [22], the particular synthesis route, and any additional processing procedures. In addition to these characteristics there can be pronounced fluctuations in the axial chain-to-chain registry (parallel to the chain axes) between neighboring chains so that there is appreciable paracrystallinity [23]. This type of disorder is typically inferred from systematic diffuse components along scattering profiles within the various nonequatorial ($l \neq 0$) layer lines [16]. These properties, when combined with the powder averaging by polycrystalline mats that typify as-prepared films or powders, severely reduce the overall distinguishability of individual diffraction peaks. In many cases, further processing can introduce uniaxial, biaxial, or higher order orientation of the crystallites [24]. Despite these complications, standard crystallographic analysis methods are possible [24], and significant amounts of structural detail can be discerned from pristine and doped hosts.

1. Polyacetylene

The best known member within this family of linear polymers, polyacetylene (PA) [25,26], has been the subject of numerous structural studies. One of the earliest reported structural references to polyacetylene [27] appeared in 1958. With the advent of the Shirakawa synthesis [28], high quality fibrillar PA films (s-PA) first became available. In general these films (and PA samples prepared using alternative processes*) are found to be predominantly crystalline. Even the early studies of these films identified a rich interplay in the structural ordering. When s-PA is synthesized at relatively low temperatures, the PA polymer chains are found to exist primarily in a cis conformation, as shown in Fig. 25.3, within an orthorhombic Pnma unit cell [30–33]. Of the two possible cis forms, cis-transoid or trans-cisoid, structural studies indicate a preference for the cis-transoid isomer. Thermal annealing [34] at $\approx 150°C$ or doping induces a solid-state transformation to the highly conducting trans conformational form with associated changes in the crystal structure.

---

*This includes PA films synthesized via the Durham, Akagi, Naarman, etc. methods. For a review, see Ref. 29.

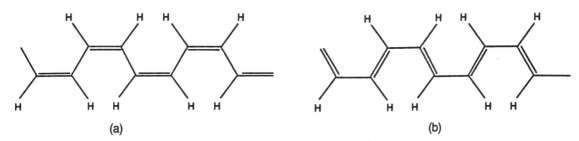

Fig. 25.3 The two possible isomers for cis-polyacetylene: (a) cis-transoid; (b) trans-cisoid.

Scattering studies of *trans*-PA have given rise to numerous debates concerning its three-dimensional (3D) ground-state crystal structure. All experimental studies are consistent with the 2D *p2gg* equatorial packing of the PA chains. With respect to the 3D unit cell, three different space groups, *P2₁/α* [12], *P2₁/n* [13, 35, 36], and *Pnam* [14], have all been proposed. The difficulty in reaching a definitive answer can be attributed to the smallness of the monoclinic distortion, $\beta \approx 92°$, incomplete thermal *cis–trans* conversion of test samples, limitations of the PA structural ordering within the films, and the subtle differences among the proposed structures.

In particular, there has been considerable controversy, both experimental and theoretical [37–40], concerning the relative phase relationship between the two chains that contribute to the PA unit cell and the degree of bond alternation between the carbon–carbon single and double bond lengths. From a theoretical standpoint PA is a very attractive model compound because the small number of CH units (four) that make up the unit cell enables comprehensive total energy calculations. The $P2_1/\alpha$ structure has the two chains in an in-phase relationship, while the $P2_1/n$ structure yields an out-of-phase arrangement as shown in Fig. 25.4. An early analysis of the nonequatorial scattering intensities by Fincher et al. [13] found strong evidence for $P2_1/n$ symmetry with double and single bond lengths of 1.44 and 1.36 Å, respectively. Although other claims have been made in the interim period, the most recent studies available [36] continue to support this model with bond lengths close to those just stated.

2. Poly(*p*-phenylene vinylene)

Poly(*p*-phenylenevinylene) is the only other member within this family of conducting polymers that has been successfully processed in film form to yield oriented samples with extremely high levels of crystallinity. Thus the structure of PPV has also been extensively studied [16, 21, 41–44]. Since PPV is typically prepared during thermal conversion of a highly processable precursor polymer, it is possible to control the degree of structural order and film morphology [45]. The unit cell is reported to have $P2_1/a$ symmetry with a large monoclinic angle of $\alpha = 123°$. This angle produces a nesting of the PPV chains along the $b \sin \alpha$ direction (in the equatorial plane) so that the phenylene rings along one chain sit centered on the vinyl linkage of the two nearest-neighbor chains as shown in Fig. 25.5. In the perpendicular direction, along the *a*-axis, there is very poor axial ordering, and this leads to substantial smearing of the scattering signal in the nonequatorial data [41].

The presence of *para*-bonded phenylene rings in the main-chain construction increases the complexity of the intrachain structure. Extended π conjugation is best achieved if the polymer chains adopt a fully planar conformation. However, the steric repulsion between, in the case of PPV, adjacent vinylene and phenylene ring hydrogen atoms favors a nonplanar construction. In PPP these effects become even more pronounced. To further quantify the degree of nonplanarity, structure factor refinements of X-ray and neutron scattering data have been used. In PPV [43] and PPP [10] the room temperature mean ring deviation from planarity, defined by θ_T in Fig. 25.5, is found to be approximately 5° and 10°, respectively.

B. Semiflexible Rods (Polyaniline)

Polyaniline (PANI), also shown in Fig. 25.1, is the best known example of semiflexible rod conducting polymer

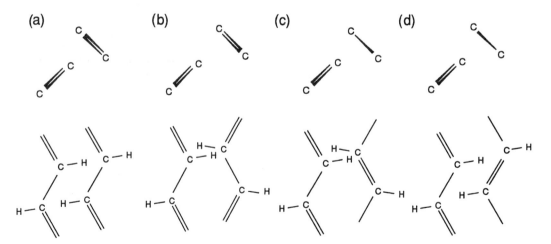

Fig. 25.4 Four "possible" equilibrium *trans*-PA structures. (a) The two chains are in phase; (b) the chains are in phase but the second chain (on the right) is rotated by π; (c) the chains are out of phase and the second chain is rotated by π; (d) the chains are out of phase. (Adapted from Ref. 37.)

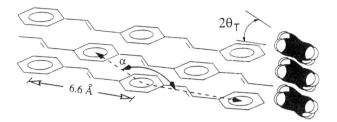

Fig. 25.5 Schematic of the PPV structure within the bc unit cell plane showing the staggered nesting of the vinylene linkage between phenylene rings of the two neighboring chains is responsible for the large monoclinic angle α. Right: Edge-on view of the PPV chains depicting the mean deviation from planarity as defined by θ_T.

systems [2]. Other related polymers include poly(p-phenylene oxide) (PPO) and poly(p-phenylene sulfide) (PPS), although their best reported conductivities are very limited. All three of these polymers have been shown to contain large equilibrium phenylene ring torsional displacements (upward of $\pm 30°$) out of the plane defined by the ring bridging atoms (amine/imine nitrogens in the case of PANI, sulfur for PPS, etc.). The most common crystalline form has an average structure best described within the orthorhombic $Pbcn$ space group [20].

The chemical and structural flexibility surrounding the amine/imine nitrogen linkages in PANI creates enormous diversity. Unlike the rigid-rod examples of the preceding section, PANI has limited solubility in a number of solvents and may be cast to yield free-standing films or precipitated into powders [46]. The chemical formula that most broadly describes this polymer is represented by

$$([(-C_6H_4-NH-C_6H_4-NH-)_{1-x}]$$
$$\times [(-C_6H_4-N=C_6H_4=N-)_x])_n$$

and it may be prepared in a variety of oxidation states ranging from leucoemeraldine to emeraldine to perniganiline with x fractions of 0.0, 0.5, and 1.0, respectively. In addition to this oxidation chemistry, the imine nitrogen can be reversibly protonated using simple acid–base chemistry.

In terms of structure, the most heavily studied form of PANI is emeraldine base (EB, $\chi = 0.5$) either as synthesized or after various acid/base treatments [5,20,47–54]. PANI-EB samples can range from fully to partially crystalline films or powders. The degree of crystallinity [52] is typically quite low; even in the best samples, it never exceeds 50%. The coherence lengths are also quite small, with values of 70 Å or less typically reported. In the approximate structure of Jozefowicz et al. [20] (see Fig. 25.6, the deviation from planarity by the phenylene rings is sequentially alternated between $+30°$ and $-30°$ as one moves along the backbone. In actuality the chemical repeat is four aniline monomer units long with three benzoidal rings and one quinoidal ring. Single-crystal studies of various oligomeric PANI analogs [53] and molecular modeling calculations [54] of the EB intrachain structure suggest a more complex ring torsion structure in which the quinoidal ring is seen to be more planar than the benzoidal rings. Recent data [51] from a direct scattering study of the amorphous polymer are also consistent with this more complex intrachain ring structure. Even within the amorphous state there is evidence of least two locally different interchain structural organizations [5] depending on the exact preparation of the PANI sample.

C. Temperature-Induced Variations

A limited number of studies have investigated the structural response of conducting polymers to changes in the ambient temperature and/or pressure [55]. While there is no evidence of a state phase change among the linear unsubstituted hosts, significant microscopic effects have been detected, and these observations provide important clues toward understanding fundamental issues of conducting polymer behavior.

The most obvious response to thermal and pressure variations is the large fractional change in the crystal

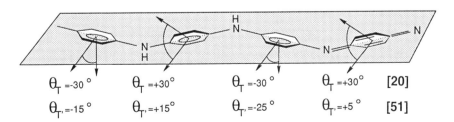

Fig. 25.6 Schematic diagram of the four-monomer repeat in PANI emeraldine base explicitly showing the deviations from planarity of the four rings, three benzoidal and one quinoidal, according to the approximate structure of Jozefowicz et al. [20] for crystalline EB and the quoted values from structural analysis of amorphous PANI EB by Maron et al. [51].

lattice repeats in the equatorial direction. Thus the "free" volume within which individual chains can travel is a strongly varying quantity. Any electronic transport properties that are dependent on the interchain and intrachain wave function overlap will, in principle, be strongly affected. In addition, rotational displacements of the host main chain about the chain axes can also have an important consequence [56]. The presence of strong torsional motions about the chain axis will alter the level of main-chain π conjugation, thus limiting intrachain transport and altering the band structure. This latter response implies that there is also a dynamical component to this effect. Long-wavelength collective torsional oscillations alter the intrachain atom-to-atom π overlap only slightly and would be expected to have a minimal influence. Pronounced local torsions resulting from phenylene ring librations and ring flips about the axis defined by the two *para*-bonded carbon atoms would be expected to have a more significant impact.

Nuclear magnetic resonance and inelastic neutron scattering have both been used to study the local dynamical response of the ring motions in PANI [57,58] and in PPV films [59,60]. The NMR experiments are able to detect the two aforementioned types of motion, ring flips and librations. At lower temperatures the ring motion is almost exclusively librational with modest torsional excursions about an equilibrium position. At elevated temperatures there are pronounced flips, approaching a full 180° of rotation.

Thermal studies of these hosts also impact a secondary side issue. For short-chain oligomeric model compounds there exists a liquid crystalline phase as seen Fig. 25.7, that is referred to as the rotator phase. In this state the underlying translational symmetry (i.e., hexagonal) of the chain packing is retained, but there is no long-range orientational ordering of the chains. In this context conducting polymers may be viewed as examples of an "infinite" model host system. Diffraction studies reveal an interesting effect that distinguishes the phenylene ring–containing compounds. Systems containing phenylene rings and hence the possibility of local ring motion (i.e., PPV and PANI) show little or no evolution toward the rotator phase, whereas systems lacking in this motion (PA and the conventional polymer polyethylene) exhibit clear signatures indicative of a transition toward this phase [61].

D. Doped Phases and Their Structural Evolution

All known conducting polymers are semiconductors in their pristine unperturbed state. Photoexcitation, charge injection, and/or doping are required to create the local electronic excitations necessary for charge transport. Of these three mechanisms, only the process of doping, or more precisely intercalation, yields a permanent transition to the conductive state. Doping can be accomplished through chemical, electrochemical, and vapor methods, and, unlike the relatively light doping concentrations typical of conventional semiconductor compounds, the dopant levels in conducting polymer hosts may approach one dopant ion per monomer unit. This process involves essentially unit transfer of charge between the polymer host backbone and the guest dopant species. To guarantee overall charge neutrality there must be interdiffusion of the guest species into the host matrix. Thus there can be massive local structural reorganizations within the polymer matrix to accommodate the uptake of dopant.

The peculiar anisotropy of the polymer host, covalent bonding along the backbone with weaker interchain interactions in the two orthogonal directions, in combination with the local structural ordering introduces considerable complexity into the doping-induced structural evolution. Even before the availability of detailed scattering data, indirect measurements found strong evidence for the existence of multiple guest–host structural phases exhibiting periodic structures in the directions both parallel and perpendicular to the polymer chain axis

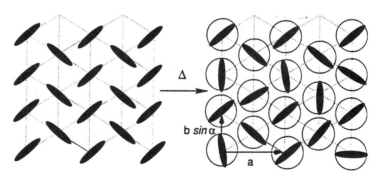

Fig. 25.7 Schematic 2D model depicting the thermal transformation (Δ implies heating) of a herringbone packed (*p2gg* symmetry) array of "rotors" to the orientationally disordered rotator phase at which point the ratio $a/(b \sin \alpha)$ becomes exactly $\sqrt{3}$.

[62]. The earliest scattering studies showed the stepwise existence of high symmetry structures in close analogy to the well-known stage-n, n = 1,2,3, ..., transformations of quasi-two-dimensional layered materials [63–65]. In conducting polymers the actual response can be significantly more complex. For layered materials, intercalation by a guest species requires only dilation of the host matrix to enable the formation of intercalant galleries. On the other hand, as seen in Fig. 25.8, intercalation of conducting polymer hosts potentially involves rotational and translation motions of the individual constituents and also in the host lattice itself.

The final quoted compositions are also highly variable. In general the relative guest ion concentrations are given in terms of their mole weight with respect to a monomer unit basis. PA has a relatively short skeletal c-axis repeat (2.45 Å), and so the nominal mole weights are considerably less than those of, for instance, PPV (with a 6.6 Å c-axis repeat).

1. Channel Structures

To first order, the equatorial structures perpendicular to the polymer chain axis can be treated independently from those parallel to the polymer chains. Two very different structural motifs, channels and layers, are characteristically seen. The most commonly reported channel structures for small dopants are characterized by a single

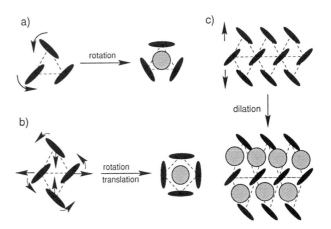

Fig. 25.8 Schematic 2D model depicting the three most common structural responses to conducting polymer intercalation. (a) Transformation to a threefold channel structure; (b) transformation to a fourfold channel structure; (c) transformation to a layer compound.

quasi-one-dimensional array of the guest species enclosed by some number of polymer main chains [12,66–73] as shown in Fig. 25.8. This phase transformation is often dominated by cooperative rotational motions of the host polymer chains about their chain axes to form the individual channel sites. Fig. 25.9 is arranged

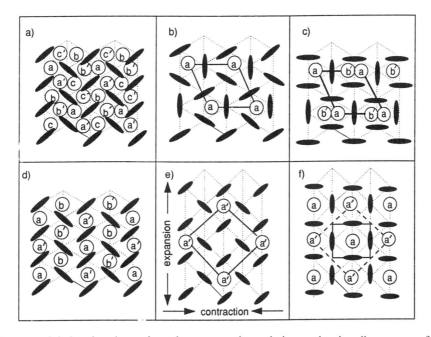

Fig. 25.9 Schematic 2D model showing the various degenerate channel sites and unit cell structures for various threefold and fourfold channel packing motifs that are appropriate for alkali metal intercalation. (a)–(c) Sequential progression of high symmetry structures from the pristine herringbone phase, HB (a) to the intermediate $\sqrt{3} \times \sqrt{3}$ (or 120° [74]) phase (b) to the distorted-120° phase (c). (d)–(f) Sequential progression of high symmetry structures from the pristine herringbone phase, HB (d) to the intermediate "stage 2" phase (e) to the final "stage 1" phase.

so that the sequential evolution of threefold and fourfold channel structures, appropriate for intercalation of alkali metal ions into PA, PPV, and PPP, can be clearly viewed. Figures 25.9a and 25.9d show the herringbone lattice of the undoped polymer in juxtaposition with the various degenerate candidate channel sites for threefold and fourfold structures, respectively. Panels (b) and (e) depict two intermediate structures that form when only a single channel sublattice is filled. The relative polymer chain/ion channel ratios are 3:1 and 4:1, respectively. Panels (c) and (f) illustrate the subsequent structural phases that form when a second sublattice site becomes filled. Thus these polymer chain/ion channel ratios are twice those of the previous two panels. Example scattering spectra for intercalation of alkali metals into PPV are shown in Fig. 25.10.

The actual local structural details are found to be extremely sensitive to the specific treatment, host polymer, and guest species. For instance, the stage 2 structure of Fig. 25.9e has been clearly observed only during electrochemical dedoping of K-PA [75]. The reported structural phases for crystalline PANI doped with halogen acids [20] are significantly different from those just introduced in the preceding paragraphs. The template structures of Fig. 25.9 also ignore much of the self-consistent structural relaxation by the polymer host. While Fig. 25.9e identifies the major translational response of the polymer lattice, a majority of these structural phases [67,75–78] have been shown to exhibit either symmetry-lowering distortions of the unit cell or pronounced displacements of the various unit cell constituents away from high symmetry positions. In one instance [73] there is experimental evidence for the presence of an even more complex ordering phenomenon. In addition, these doped structures often contain further structural ordering between guest ions in the immediately adjacent channels to give tetragonal [70,71,79], trigonal [72], or orthorhombic [78] 3D structures.

To complement these experimental studies there has been an effort to develop an understanding from a theo-

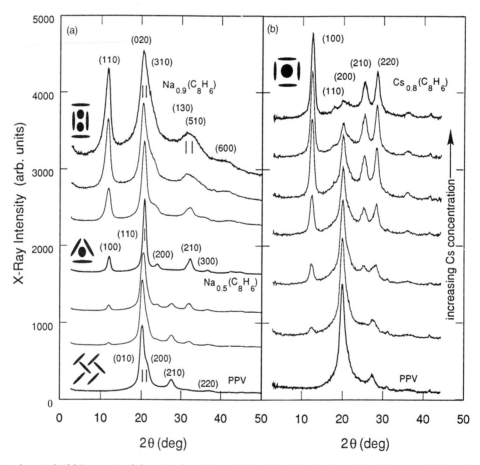

Fig. 25.10 Experimental ($hk0$) equatorial scattering data obtained in situ during (a) sodium and (b) cesium vapor doping (intercalation) of poly(p-phenylenevinylene) showing the structural progression HB → 120° → d-120° of Fig. 25.9 for sodium and the HB → "stage 1" phase transformation for cesium. ($\lambda = 1.542$ Å.)

retical perspective. The large number of atomic constituents in combination with the complexity of the structures that characterize the experimental systems precludes the use of realistic models in theoretical treatments except in certain limiting cases [80]. Hence a true fundamental understanding of the full range of structural behavior is lacking. It is possible, however, to reproduce much of the complexity in this structural phase behavior using much simplified models [74,81,82]. Choi et al. [82,83] studied a model Hamiltonian that embodies many essential aspects of the competing guest–host interactions that exist within the conducting polymer matrix. In their work the polymer host is reduced to a 2D triangular lattice of interacting planar rotors and the dopants are allowed to occupy the high symmetry sites similar to the model of Fig. 25.9a. Depending on the relative strengths of the interaction energies and that of the dopant chemical potential, a rich array of equilibrium structures are possible [74].

2. Layered Structures

The layered guest–host structures that can also develop undergo a structural evolution more reminiscent of the intercalation process in 2D layered materials (e.g., graphites or silicates). In this case, quasi-2D galleries open up between stacked sheets of the polymers chain as shown in Fig. 25.8. This type of ordering has often been reported in structural studies [84–89] using molecular dopants. Systems showing evidence of layered structures include iodine-doped PA [86] and AsF_5-doped PPV [87]. These layered structures may occur either alone or in combination with channel formations [85,89].

The details associated with the presence of layered structures are far more sketchy than for the channel phases. There are a number of compromising factors that complicate a comprehensive analysis of these experimental data. The most problematic characteristic is the significant loss of the sample coherence lengths and the resultant broadening of the scattering peak widths in the doped phase that occurs at the onset of dopant uptake. This pathological behavior is clearly captured in the in situ scattering data of Fig. 25.11 for iodine intercalation into PA. In general, few if any individual scattering features can be resolved and subsequently identified as unique lattice reflections. In addition, disproportionation reactions can yield a highly variable mixture of molecular guest ions (e.g., I_3^-, I_5^-). Detailed and rigorous structural factor refinements that test these various proposed models are rare. In view of these severe limitations, many of the structural details for large molecular dopants remain incompletely understood.

3. Structure Parallel to the Polymer Chain Axis

The ordering parallel to the polymer chain axes has also been closely studied, with both the polymer main-chain

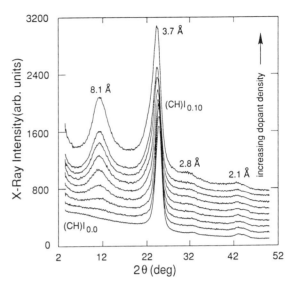

Fig. 25.11 Experimental ($hk0$) equatorial scattering data obtained in situ during iodine vapor doping of *trans*-polyacetylene. The large monotonically decreasing background clearly seen in the $(CH)I_{0.0}$ scan is an artifact due to air scatter. (λ = 1.542 Å.) (Adapted from Ref. 89)

and dopant structures receiving attention. Although the changes in the respective c-axis structure are not as dramatic as those in the equatorial plane, there are some interesting effects.

Doping of conducting polymers always involves significant charge transfer either to or from the π-conjugated orbitals that make up the skeletal backbone. This process is accompanied by a self-consistent relaxation of the electronic states and the structural ordering. n-Type doping (in which charge is donated to the polymer backbone) is associated with an overall expansion in the c-axis repeat, whereas p-type doping (in which charge is withdrawn from the backbone) initially generates a nominal reduction in the polymer lattice repeat. The magnitude of these variations can approach 1–2% of the repeat distance. Analogous behavior is seen during intercalation of graphites [90]. A number of studies [91–93] have observed these effects in conducting polymers. There are some anomalous features to this lattice relaxation. During n-type alkali metal doping of PA, it has been observed that the rate of change in the c-axis expansion abruptly increases at about 6 mol% w of the alkali metal. Detailed total energy calculations [80,94] have shown that this effect is essentially coincidental because the changes in the average bond angles and bond lengths arising from the change transfer conspire to nearly cancel out at the lighter doping levels.

The repeat spacings for the dopant ions filling the quasi-one-dimensional channels have also been measured. In samples having channels containing alkali metal ions, a modest distribution of intrachannel ion–ion spacings is obtained. These spacings range from approximately 4 to 5 Å. Most of these values are typically incommensurate with respect to the repeat distance of the surrounding polymer chains (i.e., four times the nominal 1.24 Å projected c-axis CH–CH repeat to yield 4.96 Å as shown in Fig. 25.12a). There are, however, a few reports of samples containing commensurate alkali metal ion repeat distances [67]. In all these cases the scattering spectra yield alkali metal intrachannel coherence lengths, as determined from the angular peak widths, significantly less (typically a factor of 2) than those of the surrounding polymer chains [78].

The analysis of the c-axis structure for molecular dopants is considerably more challenging. Iodine-doped PA is a prime example of a p-type doped conducting polymer system whose c-axis dopant ion structure has been probed [85,95]. In general there is an approximate 3.1 Å repeat of the iodine atoms parallel to the PA chains. However, these iodine atoms may belong to various odd-order polyiodide ions (see Fig. 25.12b) whose relative proportion is a function of both the overall doping composition and the evolution of time [85]. The presence of these mixed arrays of differing iodine species produces large repeat distances and anomalous broadening effects in the individual peak widths of scattering spectra taken along the meridional direction (i.e., parallel to the chain axis) in uniaxially oriented PA samples. Although complicated, these systemic effects have been analyzed and effectively reproduced in model calculations of representative structures.[95].

III. POLYMERS CONTAINING FLEXIBLE SIDE CHAINS

Addition of flexible side chains to the stiff π-conjugated polymer backbone of the host polymer has proven to be an extremely effective procedure for obtaining tractable and fusible materials. This approach is not limited to conducting polymers but has also attracted considerable attention in more conventional rigid-rod-like polymers [96–99] including various polyimides and polyamides and in other novel polymers such as the polysilanes [100]. Not only does this modification enable the utilization of conventional polymer processing methods, it can also, in some cases, create new materials that exhibit enhanced electronic properties in comparison to the linearly unsubstituted parent polymer. These side-chain-substituted conducting polymers also exhibit properties that do not exist in the unsubstituted hosts discussed in

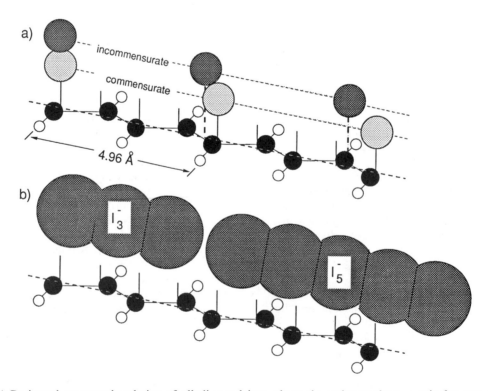

Fig. 25.12 (a) Projected structural ordering of alkali metal ions along the polyacetylene c axis for commensurate and incommensurate spacings. (b) Projected structural ordering of two different polyiodide ions along the polyacetylene c axis, highlighting the possibility of larger iodine repeat distances.

Structural Studies of Conducting Polymers

Section II. Some of these attributes include a pronounced thermo- and solvatochromism, thermotropic and lyotropic liquid crystallinity, and structural self-assembly.

Most of the side-chain-containing conducting polymers synthesized to date employ alkyl, alkoxy, or phenylalkyl side chains of varying lengths that are chemically substituted at various hydrogen atom sites along the polymer backbone. Alternatively, one can incorporate these side chains using unusual dopant molecules that have specific functionalities. A few example systems employing both of these approaches are shown schematically in Fig. 25.13.

There is, relatively speaking, considerably less definitive structural information available from the side-chain-substituted materials. In addition to the scattering from the structural components forming the main chain, there are necessarily contributions by the side-chain constituents. Depending on the specifics of the polymer sample and its processing history, a combination of crystalline, semicrystalline, and amorphous scattering signatures can be superimposed in the experimental data. As was the case for the linear unsubstituted conducting polymer hosts, considerable effort has been made and significant details concerning the detailed structure and overall phase behavior can be discerned.

A. Direct Chemical Substitution

A vast array of polymer model hosts have been synthesized and studied. Side-chain-substituted PPV, PPP, PANI, PT, and various copolymer derivatives have all appeared in the literature. From a structural standpoint, the most important defining characteristic is the presence of two chemically dissimilar building blocks: the nonpolar flexible side chains and the stiff π-conjugated main chains. Given the opportunity, these polymers will adopt structures that produce a phase separation of the two fundamental units restricted by the chemical linkage that binds them together and any additional steric packing constraints. The net result is that solvent-cast films or precipitated powders commonly form layered semicrystalline structures whereby the alkyl side chains appear as spacers between nested stacks of the main chains [101] as seen by the three differing structural motifs illustrated in Fig. 25.14.

1. Poly(3-alkylthiophene) Homopolymers

Of the many example side-chain-substituted hosts synthesized so far, one of the most heavily studied families of these compounds has been the poly(3-alkylthiophene) or P3ATs. This popularity is based in part on its versatil-

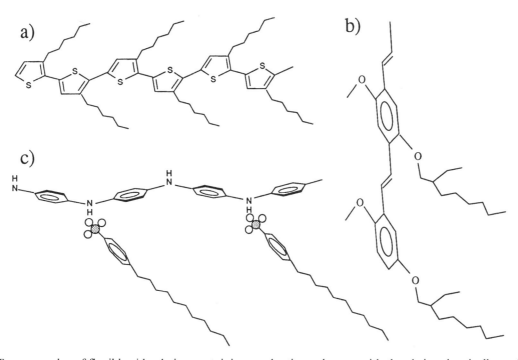

Fig. 25.13 Two examples of flexible side chains containing conducting polymers with the chains chemically anchored. (a) Monosubstituted poly(3-hexylthiophene); (b) doubly substituted poly(2-methoxy-5-(2'-ethylhexoxy)-phenylene-vinylene); and (c) an example of a linear unsubstituted host having a functionalized dopant, polyaniline-dodecylbenzenesulfonate.

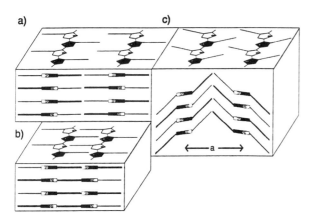

Fig. 25.14 Various possible lamellar structures appropriate for poly(3-alkylthiophenes) and other flexible side-chain-containing conducting polymer hosts. (a) Schematic model exhibiting no main-chain or side-chain tilts and no interlayer alkyl chain intermixing. (b) Schematic model exhibiting no main-chain or side-chain tilts and the maximum interlayer alkyl chain intermixing. (c) Schematic model exhibiting main-chain (i.e., a nonzero setting angle) and side-chain tilts with no interlayer alkyl chain intermixing.

ity in terms of available synthesis procedures, variety of molecular designs, and range of potential applications. Much of the structural behavior seen in the P3ATs extends to all other model systems. It is also worth noting that in most cases the molecular weights and molecular number are rather modest, and so these results pertain to P3AT samples in which the polymer chain length is relatively limited.

Diffraction studies [101–111] of the longer side-chain-containing P3ATs (butyl, pentyl, hexyl, etc.) almost universally find at least one well-defined peak at a smaller scattering angle or, equivalently, wave vector [according to the formula $k = 4\pi/\lambda \sin\theta$, where the X-ray wavelength is fixed at $\lambda = 1.542$ Å]. Scattering spectra from three representative P3AT homopolymers are displayed in Fig. 25.15. In many hosts, this low angle scattering feature has a relatively narrow 2θ width, implying scattering coherence over a significant distance [up to ~300 Å using the Scherrer relationship, $L = 0.9\lambda/(\Delta 2\theta \cos\theta)$, where L is the coherence length]. This intense feature often appears in combination with two to four less intense, higher order reflections located at integer multiples of the fundamental wavevector. These so-called $(h00)$ reflections are indicative of the lamellar structural phase alluded to previously. At somewhat larger scattering angles there is always a very broad, modest intensity peak centered near 2θ values of 22° (1.55 Å^{-1}) and, often, sharper features superimposed on this scattering at angles close to 24° (1.7 Å^{-1}). Scans out to much higher angles [112] are able to resolve additional weak, broad

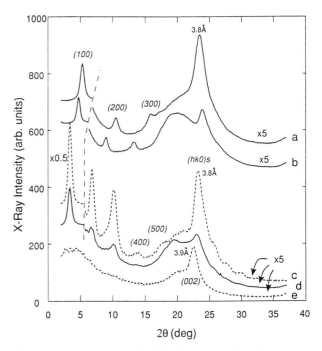

Fig. 25.15 Typical θ–2θ P3AT X-ray diffraction spectra for unoriented (powder) samples of (a) poly(3-hexylthiophene), (b) poly(3-octylthiophene) and (d) poly(3-dodecylthiophene). The two additional scans are spectra from a stretch-oriented (6:1 draw ratio) P3DDT sample (c) in the $(hk0)$ equatorial plane (perpendicular to the chain axis) and (e) along the $(00l)$ meridional direction (parallel to the polymer chain axis).

scattering features at 2θ angles near 40° (2.8 Å^{-1}), 64° (4.3 Å^{-1}), and 77° (5.1 Å^{-1}). The broad amorphous peak near 22° is primarily correlated with inter- and intrachain molecular disorder of the side chains, while the broad higher angle features are considered to be representative of the short-range intrachain structure of the polymer main chains. The presence of a sharper scattering peak near 24° implies the possibility of additional long-range order both along the polymer main chains (at half the nominal 7.8 Å chain axis repeat) and perpendicular to the main chains within individual lamellae. In some P3AT samples there are a number of well-defined steps on the high angle side of this 24° scattering feature that give evidence of even further structural ordering. Finally, in the best uniaxially oriented samples the nonequatorial scattering can be distinguished out to very high order [101] (c^* Miller indices of $l = 6$ and beyond).

The interlayer d spacing is directly correlated with the length of the side chains in accordance to the models of Fig. 25.14. However, as shown in Fig. 25.16, this functionality is somewhat less (or more) than would be expected if the side chains were oriented normal to the

Structural Studies of Conducting Polymers

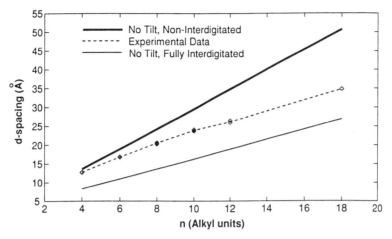

Fig. 25.16 The large interlayer d spacing as a function of alkyl chain size, $CH_3(CH_2)_{n-1}$, using the data of Table 25.2 in combination with the estimated distances from two specific structural models as shown in Figs. 25.14a and 25.14b.

main-chain axis and to the layers themselves, assuming no main-chain tilt and no interdigitation (or full interdigitation). This indicates that these side chains either are conformationally disordered, are strongly tilted away from the normal, or experience significant overlap with the side chains from the neighboring stack. In contrast, the 2θ position of the sharper features located near 24° are essentially independent of side-chain length. Studies of uniaxially oriented samples, as seen in Fig. 25.15 curves c and e, find scattering in this vicinity both parallel and perpendicular to the draw (chain c axis) direction. Hence this scattering is correlated with an extended main-chain conformation along the c axis and an intrastack chain-to-chain repeat of approximately 3.8 Å.

These qualitative features can be complemented by more rigorous analysis of the experimental data. A variety of planar orthorhombic [101,106,110] and monoclinic unit cells [104,107] have been proposed. Table 25.2 contains a partial list of reported room temperature lattice constants for some of the more common P3AT com-

Table 25.2 A Partial List of Reported Lattice Parameters for Various P3AT Homopolymers[a]

Polymer	Unit cell dimension (Å)			α (deg)	ϕ (deg)	Reference
	a	b	c			
Poly(3-butylthiophene)	12.8	7.6		90	13	[101]
	12.7	7.52	7.77	90	0	[113]
Poly(3-hexylthiophene)	16.8	7.66	7.7	90	0	[114]
	16.72	7.57	7.77	90	0	[113]
	16.8	7.6			0	[106]
	16.90	4.85	7.84	50.6	0	[104]
Poly(3-hexylthiophene)	20.65	7.63	7.7	90	6	[114]
	20.53	7.56	7.77	90	0	[113]
	20.2	7.6			0	[106]
	20.4	4.8	7.85	52	0	[104]
Poly(3-decylthiophene)	24.06	7.53	7.77	90	0	[113]
	23.7	7.6			0	[106]
Poly(3-dodecylthiphene)	25.95	7.74	8.0	90	13	[114]
	26.43	7.73	7.77	90		[113]
Poly(3-octyldecylthiophene)	34.75	8.25	7.77	90		[113]
Poly(3-octylphenylthiophene)	28.4	5.06				[115]

[a] Note that the monoclinic unit cell of Mardalen et al. [104] is only half the size of the (nonprimitive) orthorhombic structures. Also note that ϕ is the quoted setting angle (i.e., tilt) for the *trans*-planar main chain and is exclusive of the side-chain orientations.

pounds. One current structural model [114] that is found to be consistent with the full range of experimental data (infrared and UV-vis absorption, X-ray and electron diffraction, etc.) and detailed modeling calculations [116–118] (crystal structure factor refinements, molecular simulations, etc.) requires an average tilting of the side chain away from the layer normal with a minimum of side-chain interdigitation between stacks as shown in Fig. 25.13c. Other proposed P3AT models invoke variations in the chain structure including a periodic up-up-down-down doubling of the *trans* planar repeat [103], a nonplanar conformation of the main chain [119], and even a helical structure [120] [specifically for poly(3-methyl thiophene)].

2. Polydiacetylenes

Polydiacetylenes (PDAs) were among the first side-chain-substituted conducting polymers studied and are deserving of special mention. In many ways the structural properties of these PDAs are very similar to those of P3AT and other more recent systems with one remarkable exception: many PDAs can be prepared in essentially single-crystal form [121,122] through solid-state polymerization. The actual process is rather straightforward. Single crystals of the monomer are prepared by slow precipitation of the monomer in solution. Subsequent exposure to intense ultraviolet, X-ray, or γ-ray sources initiates the reaction process, which directly couples the conjugated diacetylenic linkages to form a single polymer chain. The resulting polymers are both soluble and fusible.

The thermal behavior of the materials is also quite interesting [123,124]. At temperatures somewhat below the isotropic melt temperature, T_m, the main chains and side chains can undergo a reversible conformational transition. However, if heated above T_m the single-crystal structure can no longer be recovered, and another solid-state structure is obtained. A similar result is obtained if the resultant polymers are cast from solution.

B. Functionalized Dopants

The incorporation of flexible side chains need not be restricted to synthetic methods that chemically anchor these units to the main chains at various bonding sites. The possibility also exists of using specific dopant molecules that can contribute this functionality indirectly. A well-known prototypical conducting polymer preparation employing this approach is the doping of polyaniline EB using various surfactants [125,126] including dodecylbenzenesulfonate (DBSA) (shown in Fig. 25.13) or camphor sulfonic acid. In some instances substantially improved conductivities are obtained compared to EB samples treated with only simple inorganic halogen (HX, X = F, Cl, Br, I) acids.

Studies are also possible for PANI samples having a combination of both chemically anchored side chains and functionalized dopants [127]. The structures of these and of host polymers containing dopant only are found to be qualitatively similar to those of the side-chain-substituted material discussed previously. In general, large length scale d spacings develop, indicative of a lamellar phase. Initial studies [126] of the systematic absences in the diffraction data from PANI-DBSA polymers suggests a *Pmnm* space group with lattice parameters of $a = 11.8$ Å, $b = 17.9$ Å, and $c = 7.2$ Å. Detailed structural factor calculations that would further test the validity of this model are not yet available. The quality of scattering data from other existing compounds is typically poorer than for PANI-DBSA, and this permits identification of only the most pronounced d spacings.

C. Thermotropic Behavior

A number of interesting effects are observed during the thermal cycling of the various side-chain-substituted conducting polymer hosts. The most pronounced response is the distinctive thermochromic transition that is seen upon heating [128,129]. In the P3ATs, this attribute is correlated with a structural transformation to an isotropic melt state and a loss of the π-orbital conjugation length. In this state all of the relatively sharp scattering peaks are lost and only broad, liquid-like features remain [101,130]. Cooling the samples to temperatures to well below the respective P3AT "melt" temperature usually returns the polymer to its original state, although the actual dynamics of this process appear to be rather sluggish.

There are many more subtle intermediate features that have also been observed. In addition to the endothermic behavior of the isotropic melt, calorimetric measurements [110] detect a second endothermic peak at somewhat lower temperatures. This new feature corresponds to (on heating) a transition to a liquid crystalline (LC) phase wherein the individual stacked layers become decoupled from one another and are essentially free-floating. The main-chain planarity is only somewhat affected by this transition, while the alkyl-containing side groups show a significant increase in the relative proportion of *gauche* conformers as the transition is approached from a lower temperature.

There are also detailed structural changes that can be resolved in scattering experiments [110]. Equatorial scattering peaks arising from the interlayer repeat are found to sharpen significantly in the LC phase of the longer alkyl P3ATs (octyl and larger), indicating either an increase in the crystallite size or a reduction in the strain. On cooling the ($h00$) reflections are found to broaden once more. This implies, indirectly, that strain is the predominant effect. In samples containing ($hk0$) reflections, this LC phase is marked by the loss of these

($hk0$), $k \neq 0$, peaks. Significant variations also occur in the nonequatorial scattering. These effects are indicative of changes in the side-chain and/or main-chain structure and/or conformation. In particular, the meridional scattering near the (002) reflection is seen to shift its 2θ position and reduce its intensity. While this may indicate a change in the thiophene repeat distance and in the skeletal main conformation, modest changes in the side chain structure can also strongly impact the closely spaced ($hk2$) scattering intensities near the meridional position.

D. Sensitivity to Side-Chain Structure

There are a multitude of secondary effects directly related to side-chain placement and chemical architecture that have a tremendous impact on the final structural ordering and phase behavior. For example, in P3AT homopolymer hosts it has been observed that varying degrees of regioregularity with respect to the side-chain chemical binding site can alter many physical properties. Monosubstituted polythiophenes, containing an average of one alkyl chain per thiophene ring, often exist in a *trans*-planar main-chain conformation that has a doubled monomer unit c-axis repeat of ~7.8 Å. There are three possible monomer–monomer alkylthiophene couplings, head-to-tail (HT), head-to-head (HH), and tail-to-tail (TT) as shown in Fig. 25.17. Depending on the specific synthetic approach, the regioselectivity, with respect to only the percentage of HT configurations, can span the full range from 0% to nearly 100%. At HT regioselectivities of close to 0%, there are considerable steric interactions that frustrate the formation of both a planar main-chain conformation and the lamellar construction [118]. Initial studies [131] of poly(4,4'-didecyl-2,2'-bithiophene) (PDOBT), with an all-[-(HH-TT)-]$_n$ repeat, indicate 40° torsional angles between adjacent thiophene rings. Unlike the *trans*-planar P3ATs discussed above, PBOBT is not extensively π-conjugated. Therefore it does not exhibit thermochromism. Surprisingly, cast film of this polymer can still exhibit a fairly well ordered lamellar phase. HT regioselectivities ranging near 75% enable a high degree of structural ordering in this phase and are thermochromic but yield only modest conductivities (after doping). Highly regioselective samples, 90% and above, yield similar structures but sharply enhance the measured conductivities [7]. Moreover, there is some evidence that these highly regular P3ATs can be prepared in metastable states in which there is nearly full interdigitation between alkyl side chains from immediately adjacent stacks [107].

In addition to regioregularity, the synthetic method can be adjusted to yield a variety of polyalkylthiophene "copolymers" [112]. In these host systems the side-chain substitution is modified to incorporate two or more side-chain units using various isomers or units of differing lengths. The resulting copolymers may be at random or highly regular. The structural response is also intriguing [131]. Stereoirregular alkylthiophene(A)-thiophene(T), -(A-T)-(A-T)-, or -(T-A-T)-(T-A-T)- polymers typically exhibit relatively poorly ordered lamellar structures, whereas stereoregular -(A-T-A)-(A-T-A)- polymers exhibit reasonably well ordered structures. In some of these cases it is possible to simultaneously have a *trans*-planar backbone conformation and a *poorly* ordered lamellar structure. The PPV side-chain derivative shown in Fig. 25.13b is tailored for the express purpose of suppressing interchain molecular ordering by having two very different side-chain lengths, with one of the side chains branched. There is clearly a very subtle interplay at the molecular level that gives rise to this disparity of manifested structure–property relationships.

E. Doping-Induced Structural Changes

Doping by various guest species, as noted previously, induces a full-scale co-operative structural reorganization within the host matrix. Since these side-chain-containing conducting polymer structures are initially more complex than those of the linearly unsubstituted materials, the overall structural response and location of the dopant ions is even more difficult to assess. Nevertheless, recent scattering studies have found an enormous range of detailed physical behavior.

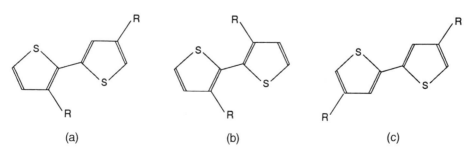

Fig. 25.17 Schematic of three types of regioisomers for the 3-alkyl-substituted thiophenes. (a) Head-to-tail; (b) head-to-head; (c) tail-to-tail.

In particular, the structural evolution of various P3ATs after p-type doping has been probed [113,132–135]. The most dramatic response is an unprecedented variation in the large interlayer d spacing. P3ATs are found to undergo upward of 20% expansions followed by ~25% reductions in their interlayer repeats [132,135]. There are also pronounced changes in the nonequatorial scattering profiles in both the peak positions and their intensities. In contrast there are only minimal changes in the intrastack polymer–polymer repeat. These structural characteristics aside, there are other surprising properties. Both DBSA-PANI [136] and regioregular iodine-doped poly(3-dodecylthiophene)s [7] can achieve remarkably high conductivities despite the limited proportion of electrically active regions. To understand the origin of these effects, determining the relative locations of the various constituents and the subsequent structural evolution is an issue of preeminent interest. In quasi-2D layered hosts, such as graphite intercalation compounds, the intercalant/dopant can be situated only between graphite layers. In these side-chain-substituted polymers the guest ions could conceivably intercalate between individual chains or equally as well lie off to the side of main-chain stacks and nested among the side chains.

A number of models [132,135,137] have been proposed that address the most significant changes in the scattering data. Although these models differ in their respective details, three key features appear to be essential for replicating the structural evolution seen in the P3ATs:

1. The molecular dopants do not overtly disrupt the stacking of the polythiophene main-chain backbones.
2. The dopants occupy sites that alter the overall orientation of the flexible side chains.
3. There are translational displacements by the polymer chains that are parallel to the chain axis.

A schematic model depicting these characteristic changes is shown in Fig. 25.18. The first feature preserves the nominal intrastack spacing at distances close to the nominal 3.8 Å, which in turn facilitates the availability of π-obital wave function overlap and hopping transport in a direction perpendicular to the main-chain axes. The second characteristic is responsible for the dramatic changes in the interlayer repeat. The third feature functions to create quasi-one-dimensional dopant ion galleries within the individual layers and enhances the uptake of the dopant ions. Detailed structural factor calculations have been used to further test and validate these defining characteristics in the iodine P3AT complexes [135]. Analogous studies are not yet available for the whole range of model compounds.

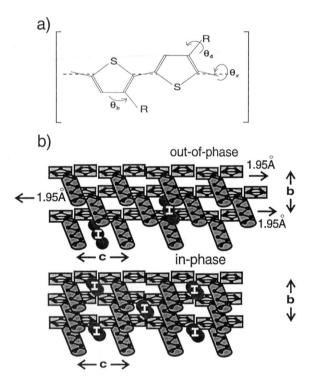

Fig. 25.18 Schematic models showing the local structural degrees of freedom necessary for iodine intercalation (doping) of poly(3-alkylthiophene) homopolymers. (a) Single-chain structural parameters: main-chain tilt (i.e., setting angle), θ_r; side-chain tilt, θ_d; and side-chain angular orientation, θ_b. (b) Intralayer c-axis translations that produce quasi-one-dimensional ion channels perpendicular to both the chain axis and the interlayer spacing (along the a axis as shown in Fig. 25.14).

IV. COMPLEX STRUCTURAL FORMS

The two previous sections focused exclusively on the most general structural properties that occur as a natural consequence of conducting polymer synthesis and the subsequent treatments of bulk samples. In fact, there are a variety of even more evolved conducting polymer structures that may be generated though highly specialized processing and/or synthesis procedures. This section briefly discusses only one of the many possible novel structural architectures that have been envisioned.

Langmuir–Blodgett (LB) monolayer film deposition techniques have been employed for many years in the fabrication of multilayer thin film structures. The basic technique involves the preparation of various monolayer films at an air/liquid interface and the subsequent stepwise transfer of these monolayers to an appropriate substrate, thereby building up a supermolecular assembly.

Typically, chemical compounds incorporating hydrophilic and hydrophobic moeties are used to stabilize the monolayers on an aqueous subphase. To enable the incorporation of conducting polymer hosts [9] into the production of these LB multilayer assemblies, a number of reaction and processing schemes have been employed. These include the direct manipulation of soluble surface-active derivatives, the transfer of various surface-active monomers followed by a secondary polymerization of the multilayer films, and, finally, the incorporation of various mixed phases that contain both conducting polymers and conventional LB agents (e.g., steric acid agents).

Alternatives to LB techniques for the preparation of multilayer thin films are also available. Control of the microscopic ordering can be achieved through structured self-assembly methods. Lamellar conjugated polymer structures have been prepared [105] using in situ electrochemical polymerization of heteroarene-containing surfactants from micellar solutions. This method avoids the highly controlled conditions necessary for monolayer film formation.

The resulting multilayer films can exhibit extremely high degrees of structural ordering and anisotropy. Diffraction studies of these films typically exhibit a series of evenly spaced low-angle 2θ reflections in analogy to the lamellar phases of the P3ATs discussed in Section III. Unlike the P3ATs, these layers are highly parallel to the substrate. The in-plane anisotropy is, at present, more difficult to control, and the full set of details concerning the structural ordering and structure–property relationship are still incomplete.

V. CONCLUSIONS

In this chapter I have briefly touched upon many, but certainly not all, of the historical developments and current issues concerning conducting polymer structure and the associated structural phase behavior. Clearly the level of knowledge is most advanced for conducting polymer hosts that have extensive crystallinity and high symmetry structures. Unfortunately, very few model host systems fulfill this criterion. Still, in the less ordered materials, the most general features of the microscopic structure are reasonably well understood.

The minute details of the local intramolecular structure, especially in terms of side-chain and main-chain orientational ordering, remain poorly defined. By their very nature conducting polymers exhibit an intimate coupling between the electronic states and the local structural degrees of freedom. In many instances direct quantitative measures of these attributes would contribute immensely to the level of understanding.

Classic diffraction techniques will continue to play an important role, but other methodologies are necessary if this local structural response is to be fully described.

For the former, i.e., diffraction, structural refinement techniques that wholly address both the crystalline and diffuse portions of the scattering profiles would be a tremendous advance. The latter component will entail heavier reliance on a variety of powerful probes of local structure including nuclear magnetic resonance, scanning tunneling microscopy, anomalous scattering methods, and radial distribution function analysis [21]. As the number of chemical/structural architectures available continues to increase, determining the fundamental nature of the structure–property relationships will become an ever more important issue.

REFERENCES

1. W. R. Salaneck, D. T. Clark, and E. J. Samuelsen (eds.), *Science and Applications of Conducting Polymers*, Adam Hilger, Bristol, 1991.
2. E. M. Genies, A. Boyle, M. Laplowsi, and C. Tsintavis, *Synth. Met. 36*:139 (1990).
3. T. Yamamoto and K. Sanechika, *Chem. Ind. 1*:301 (1982).
4. D. R. Gagnon, F. E. Karasz, E. L. Thomas, and R. W. Lenz, *Synth. Met. 20*:85 (1987).
5. J. P. Pouget, M. E. Jozefowicz, A. J. Epstein, X. Tang, and A. G. MacDiarmid, *Macromolecules 24*: 779 (1991).
6. R. B. Kaner, H. Reiss, B. R. Mattes, and M. R. Anderson, *Science 252*:1412 (1991).
7. R. D. McCullough and R. D. Lowe, *J. Chem. Soc., Chem. Commun. 1*:70 (1992).
8. E. Smela and O. Inganas, *Science 268*:1735 (1995).
9. I. Watanabe, J. H. Cheung, and M. Rubner, *J. Chem. Phys. 92*:444 (1990).
10. T. M. Swager, M. J. Marsella, Q. Zhou, and M. B. Goldfinger, *J. Macromol. Sci. A31*:1893 (1994).
11. J. P. Pouget et al., *Synth. Met. 193*:131 (1994).
12. G. Leising, O. Leitner, and H. Kahlert, *Mol. Cryst. Liq. Cryst. 117*:67 (1985).
13. C. R. Fincher, C. E. Chen, A. J. Heeger, A. G. MacDiarmid, and J. B. Hastings, *Phys. Rev. Lett. 48*:100 (1982).
14. D. Begin, F. Saldi, M. LeLaurain, and D. Billaud, *Solid State Commun. 76*:591 (1990).
15. M. M. Sokolowski, R. H. Friend, and E. A. Marseglia, *Polymer 27*:1714 (1986).
16. T. Granier, E. L. Thomas, and F. E. Karasz, *J. Polym. Sci., Polym. Phys. Ed. 26*:65 (1988).
17. D. Chen, M. J. Winokur, M. Masse, and F. E. Karass, *Phys. Rev. B 41*:6759 (1990).
18. S. Brückner, W. Porzio, and F. Wudl, *Makromol. Chem. 89*:961 (1989).
19. S. Sasaki, T. Yamamoto, T. Kanbara, A. Morita, and T. Yamamoto, *J. Polym. Sci. Polym. Phys. Ed. 30*:293 (1992).
20. M. E. Jozefowicz, R. Laversanne, H. H. S. Javadi, A. J. Epstein, J. P. Pouget, X. Tang, and A. G. MacDiarmid, *Phys. Rev. B. 39*:12958 (1989).

21. M. A. Masse, Ph.D. Thesis, University of Massachusetts, 1989.
22. D. Djurado, J. Ma, N. Theophilou, and J. E. Fischer, *Synth. Met. 30*:395 (1989).
23. R. Hosemann and S. N. Bagchi, *Direct Analysis of Diffraction by Matter*, North-Holland, Amsterdam, 1962.
24. L. E. Alexander, *X-Ray Diffraction Methods in Polymer Science*, Wiley-Interscience, New York, 1969.
25. A. J. Heeger, J. R. Schriefer, and W.-P. Su, *Rev. Mod. Phys. 40*:3439 (1988).
26. J. Tsukamoto, *Adv. Phys. 41*:509 (1992).
27. P. Corradini, *Atti Accad. Nazl. Lincei, Rend., Classe Sci. Fis., Mat. Nat. 25*:517 (1958).
28. H. Shirakawa and S. Ikeda, *J. Polym. Sci. 2*:3 (1971).
29. J. Tsukamoto, *Adv. Phys. 41*:509 (1992).
30. R. H. Baughman, S. L. Hsu, G. P. Pez, and A. J. Signorelli, *J. Chem. Phys. 68*:5405 (1978).
31. K. Shimamura, F. E. Karasz, J. A. Hirsch, and J. W. Chien, *Makrom. Chem., Rapid Commun. 2*:473 (1981).
32. J. W. Chien, F. E. Karasz, and K. Shimamura, *Macromolecules 15*:1012 (1988).
33. G. Perego, G. Lugli, and U. Pedretti, *Mol. Cryst. Liq. Cryst. 117*:59 (1985).
34. R. L. Elsenbaumer, P. Delannoy, G. G. Miller, C. E. Forbes, N. S. Murthy, H. Eckhardt, and R. H. Baughman, *Synth. Met. 11*:251 (1985).
35. Y. B. Moon, M. J. Winokur, A. J. Heeger, J. Barker, and D. C. Bott, *Macromolecules 20*:2457 (1987).
36. Q. Zhu, J. E. Fischer, R. Zuzok, and S. Roth, *Solid State Commun. 83*:179 (1992).
37. S. Stafström, *Phys. Rev. B 32*:4060 (1985).
38. P. Vogl and D. K. Campbell, *Phys. Rev. Lett. 62*:2012 (1989).
39. C. Fredriksson and S. Stafström, *Chem. Phys. Lett. 190*:407 (1992).
40. M. Boman and S. Stafström, *Phys. Rev. B 46*:4551 (1992).
41. T. Granier, E. L. Thomas, D. R. Gagnon, J. R. W. Lens, and F. E. Karasz, *J. Polym. Sci., Polym. Phys. Ed. 24*:2793 (1986).
42. D. D. C. Bradley, R. H. Friend, T. Hartmann, E. A. Marseglia, M. M. Sokolowski, and P. D. Townsend, *Synth. Met. 17*:473 (1987).
43. D. Chen, M. J. Winokur, M. A. Masse, and F. E. Karasz, *Polymer 33*:3116 (1992).
44. G. Mao, J. E. Fischer, F. E. Karasz, and M. J. Winokur, *J. Chem. Phys. 98*:712 (1993).
45. D. D. C. Bradley, *J. Phys. D: Appl. Phys. 22*:1389 (1987).
46. A. Andreatta, S. Tokito, J. Moulton, P. Smith, and A. J. Heeger, in *Science and Applications of Conducting Polymers* (W. R. Salaneck, D. T. Clark, and E. J. Samuelsen eds.), Adam Hilger, Bristol, UK, 1991, p. 105.
47. B. K. Annis, A. H. Narten, A. G. MacDiarmid, and A. F. Richter, *Synth. Met. 22*:191 (1988).
48. Y. B. Moon, Y. Cao, P. Smith, and A. J. Heeger, *Polymer 30*:196 (1989).
49. J. P. Pouget, M. E. Jozefowicz, A. J. Epstein, J. G. Masters, A. Ray, and A. G. MacDiarmid, *Macromolecules 24*:5863 (1991).
50. M. Laridjani, J. P. Pouget, E. M. Scherr, A. G. MacDiarmid, M. Jozefowicz, and A. J. Epstein, *Macromolecules 25*:4106 (1992).
51. J. Maron, M. J. Winokur, and B. R. Mattes, *Macromolecules 28*:4475 (1995).
52. J. E. Fischer, X. Tang, E. M. Scherr, V. B. Cajipe, and A. G. MacDiarmid, *Macromolecules 27*:5094 (1994).
53. L. W. Shacklette, J. F. Wolf, S. Gould, and R. H. Baughman, *J. Chem. Phys. 88*:3955 (1988).
54. J. Libert, J. L. Bredas, and A. J. Epstein, *Phys. Rev. B. 51*:5711 (1995).
55. J. Ma, J. E. Fischer, Y. Cao, and A. J. Heeger, *Solid State Commun. 83*:395 (1992).
56. D. Baranowski, H. Bütten, and J. Voit, *Phys. Rev. B 45*:10990 (1992).
57. S. Kaplan, E. M. Conwell, A. F. Richter, and A. G. MacDiarmid, *Macromolecules 22*:1669 (1989).
58. J. L. Sauvajol, D. Djurado, A. J. Dianoux, N. Theophilou, and J. E. Fischer, *Phys. Rev. B 43*:14305 (1991).
59. J. H. Simpson, D. M. Rice, and F. E. Karasz, *Polymer 32*:2340 (1991).
60. P. Papanek, J. E. Fischer, J. L. Sauvajol, G. Mao, M. J. Winokur, and F. E. Karasz, *Phys. Rev. B 50*:15668 (1994).
61. J. Ma, J. E. Fischer, E. M. Scherr, A. G. MacDiarmid, M. E. Jozefowicz, A. J. Epstein, C. Mathis, B. Francois, N. Coustel, and P. Bernier, *Phys. Rev. B 44*:11609 (1991).
62. L. W. Shacklette and J. E. Toth, *Phys. Rev. B 32*:5892 (1985).
63. J. C. W. Chien and F. E. Karasz, *J. Polym. Sci., Polym. Lett. Ed. 20*:97 (1982).
64. S. Flandrois, C. Hauw, and B. Francois, *J. Phys. (Paris) 44*:C3 (1983).
65. R. Baughman, L. W. Shacklette, N. S. Murthy, G. G. Miller, and R. L. Elsenbaumer, *Mol. Cryst. Liq. Crys. 118*:253 (1985).
66. R. H. Baughman, N. S. Murthy, and G. G. Miller, *J. Chem. Phys. 79*:515 (1983).
67. F. Saldi, M. LeLaurain, and D. Billaud, *Solid State Commun. 76*:595 (1990).
68. S. Flandrios, C. Hauw, and B. Francois, *Mol. Cryst. Liq. Cryst. 117*:91 (1985).
69. F. Saldi, J. Ghanbaja, D. Begin, M. LeLaurain, and D. Billaud, *C. R. Acad. Sci. 309*:671 (1989).
70. D. Djurado, J. E. Fischer, P. A. Heiney, and J. Ma, *Phys. Rev. B 41*:2971 (1990).
71. D. Djurado, J. E. Fischer, P. A. Heiney, J. Ma, N. Coustel, and P. Bernier, *Synth. Met. 34*:683 (1990).
72. N. S. Murthy, L. W. Shacklette, and R. H. Baughman, *Phys. Rev. B 40*:12550 (1989).
73. M. J. Winokur, Y. B. Moon, A. J. Heeger, J. Barker, D. C. Bott, and H. Shirakawa, *Phys. Rev. Lett. 35*:2329 (1987).
74. A. B. Harris, *Phys. Rev. B 50*:12441 (1994).

75. P. A. Heiney, J. E. Fischer, D. Djurado, J. Ma, D. Chen, M. J. Winokur, N. Coustel, P. Bernier, and F. E. Karasz, *Phys. Rev. B 44*:2507 (1991).
76. J. P. Aime, M. Bertault, P. Delannoy, R. L. Elsenbaumer, G. G. Miller, and M. Schott, *J. Phys. Lett. 46*:L379 (1985).
77. F. Saldi, D. Billaud, and M. LeLaurain, *Mater. Sci Forum 91*:363 (1992).
78. D. Chen, M. J. Winokur, Y. Cao, A. J. Heeger, and F. E. Karasz, *Phys. Rev. B 45*:2035 (1992).
79. D. Billaud, F. Saldi, J. Ghanbaja, D. Begin, and M. LeLaurain, *Synth. Met. 35*:113 (1990).
80. R. Baughman, N. Murthy, H. Eckhardt, and M. Kertesz, *Phys. Rev. B 46*:10515 (1992).
81. H.-Y. Choi, E. J. Mele, J. Ma, and J. E. Fischer, *Synth. Met. 27*:A75 (1988).
82. H.-Y. Choi, A. B. Harris, and E. J. Mele, *Phys. Rev. B 40*:3766 (1989).
83. H.-Y. Choi and E. J. Mele, *Phys. Rev. B 40*:3439 (1989).
84. G. Wieners, R. Weizenhofer, M. Monkenbusch, M. Stamm, G. Lieser, V. Enkelmann, and G. Wegner, *Makromol. Chem., Rapid Commun. 6*:425 (1985).
85. N. S. Murthy, G. G. Miller, and R. H. Baughman, *J. Chem. Phys. 89*:2523 (1988).
86. R. H. Baughman, N. S. Murthy, G. G. Miller, and L. W. Shacklette, *J. Chem. Phys. 79*:1065 (1983).
87. M. A. Masse, D. C. Martin, F. E. Karasz, and E. L. Thomas, *Proc. Am. Eng. Soc. Div. PMSE 57*:441 (1987).
88. T. Yamamoto, A. Morita, Y. Miyazaki, T. Maruyama, H. Wakayama, Z. Zhou, Y. Nakamura, T. Kanbara, and K. Kubota, *Marcromolecules 25*:1214 (1992).
89. M. J. Winokur, J. Maron, Y. Cao, and A. J. Heeger, *Phys. Rev. B 45*:9656 (1992).
90. C. T. Chan, W. A. Kamitakahara, and K. M. Ho, *Phys. Rev. Lett. 58*:1528 (1987).
91. N. S. Murthy, L. W. Shacklette, and R. H. Baughman, *J. Chem. Phys. 87*:2346 (1987).
92. M. J. Winokur, Y. B. Moon, A. J. Heeger, J. Barker, and D. C. Bott, *Solid State Commun. 68*:1055 (1988).
93. D. Begin, M. Lelaurain, and D. Billaud, *Mater. Sci. Forum 91*:357 (1992).
94. S. Y. Hong and M. Kertesz, *Phys. Rev. Lett. 64*:3031 (1990).
95. P.-A. Albouy, J.-P. Pouget, J. Halim, V. Enkelmann, and G. Weger, *Makromol. Chem. 193*:853 (1992).
96. A. Biswas, K. Deutscher, J. Blackwell, and G. Wegner, *Proc. Am. Chem. Soc., Polym. Prepr. 33*(1):286 (1992).
97. M. Ballauff, R. Rosenau-Eichin, and E. W. Fischer, *Mol. Cryst. Liq. Cryst. 155*:211 (1988).
98. M. Ballauff and K. Berger, *Mol. Cryst. Liq. Cryst. 157*:109 (1988).
99. G. C. Rutledge, U. W. Suter, and C. D. Papaspyrides, *Macromolecules 24*:1934 (1991).
100. F. C. Schilling, A. J. Lovinger, J. M. Zeigler, D. D. Davis, and F. A. Bovey, *Macromolecules 22*:3055 (1989).
101. M. J. Winokur, D. Spiegel, Y. H. Kim, S. Hotta, and A. J. Heeger, *Synth. Met. 28*:C419 (1989).
102. M. Leclerc, F. M. Diaz, and G. Wegner, *Makromol. Chem. 190*:3105 (1989).
103. J. Mardalen, E. J. Samuelsen, O. R. Gautun, and P. H. Carlsen, *Solid State Commun. 77*:57 (1991).
104. J. Mardalen, E. J. Samuelsen, O. R. Gautun, and P. H. Carlsen, *Solid State Commun. 80*:687 (1991).
105. D. M. Collard and M. S. Stoakes, *Chem. Mater. 6*:855 (1994).
106. G. Gustafsson, O. Inganas, H. Osterholm, and J. Laakso, *Polymer 32*:1574 (1991).
107. S. Brüchner and W. Porzio, *Makromol. Chem. 89*:961 (1989).
108. W. Porzio, A. Bolognesi, S. Destri, M. Catellani, and B. Bajo, *Synth. Met. 41*:537 (1991).
109. K. Tashiro, K. Ono, Y. Minagawa, K. Kobayashi, T. Kawai, and K. Yoshino, *Synth. Met. 41*:571 (1991).
110. K. Tashiro, K. Ono, Y. Minagawa, K. Kobayashi, T. Kawai, and K. Yoshino, *J. Polym. Sci., Polym. Phys. Ed. 29*:1223 (1991).
111. G. Gustafsson, O. Inganas, S. Stafstrom, H. Osterholm, and J. Laakso, *Synth. Met. 41*:593 (1991).
112. H. J. Fell, Ph.D. Thesis, Universitet I Trondheim, Norges Tekniske Hogskole, 1994.
113. T. Kawai, M. Nakazono, R. Sugimoto, and K. Yoshino, *J. Phys. Soc. Jpn. 6*:3400 (1992).
114. T. J. Prosa, M. J. Winokur, J. Moulton, P. Smith, and A. J. Heeger, *Macromolecules 25*:4364 (1992).
115. H. J. Fell, E. J. Samuelsen, J. Mardalen, and M. R. Andersson, *Synth. Met. 69*:283 (1995).
116. K. Tashiro, K. Kobayashi, K. Morita, T. Kawai, and K. Yoshino, *Synth. Met. 69*:397 (1995).
117. S. A. Chen and S. T. Lee, *Polymer 36*:1719 (1995).
118. T. A. Chen, X. Wu, and R. D. Rieke, *J. Am. Chem. Soc. 117*:233 (1995).
119. W. Luzny and A. Pron, *Synth. Met. 69*:337 (1995).
120. F. Garnier, G. Tourillon, J. Y. Barraud, and H. Dexpert, *J. Mater. Sci. 20*:2687 (1985).
121. R. R. Chance, G. N. Paterl, and J. D. Witt, *J. Chem. Phys. 71*:206 (1979).
122. V. Enkelmann and J. B. Lando, *Acta Crystallogr. B 34*:2352 (1978).
123. C. Plachetta, N. O. Rau, and R. C. Scultz, *Mol. Cryst. Liq. Cryst. 155*:281 (1988).
124. H. Tanaka, M. Thakur, M. A. Gomez, and A. E. Tonelli, *Macromolecules 20*:3094 (1987).
125. Y. Cao, P. Smith, and A. J. Heeger, *Synth. Met. 48*:91 (1992).
126. C. Y. Yang, P. Smith, A. J. Heeger, Y. Cao, and J. Osterholm, *Polymer 35*:1143 (1994).
127. W. Y. Zheng, K. Levon, J. Laakso, and J. Osterholm, *Macromolecules 27*:7755 (1994).
128. O. Inganas, W. R. Salaneck, H. Osterholm, and J. Laakso, *Synth. Met. 22*:395 (1988).
129. S. Hotta, S. D. D. V. Rughooputh, A. J. Heeger, and F. Wudl, *Macromolecules 20*:212 (1987).
130. M. Nakazono, T. Kawai, and K. Yoshino, *Chem. Mater. 6*:864 (1994).
131. J. Mardalen, H. J. Fell, E. J. Samuelsen, E. Bakken,

P. H. Carlsen, and M. R. Andersson, *Macromol. Chem. Phys. 196*:557 (1995).
132. M. J. Winokur, P. Wamsley, J. Moulton, P. Smith, and A. J. Heeger, *Macromolecules 24*:3812 (1991).
133. M. J. Winokur, T. J. Prosa, J. Moulton, P. Smith, and A. J. Heeger, *Proc. Am. Chem. Soc. Polym. Prepr. 33*(1):296 (1992).
134. T. Kawai, M. Nakazono, and K. Yoshino, *J. Mater. Chem. 2*:903 (1992).
135. M. J. Winokur, T. J. Prosa, J. Moulton, P. Smith, and A. J. Heeger, *Phys. Rev. B 51*:159 (1995).
136. M. Reghu, Y. Cao, D. Moses, and A. J. Heeger, *Phys. Rev. B 47*:1758 (1993).
137. K. Tashiro, Y. Minagawa, K. Kobayashi, S. Morita, T. Kawai, and K. Yoshino, *Polym. Prepr. Jpn. 41*:4595 (1992).

26
Second-Order Nonlinear Optical Materials

J. I. Chen, S. Marturunkakul, L. Li, J. Kumar, and Sukant K. Tripathy
University of Massachusetts—Lowell, Lowell, Massachusetts

I. INTRODUCTION

Second-order nonlinear optical (NLO) polymers for photonic technology have shown considerable promise, and research activity in this field has been prolific in the past decade. Photonics, an analog of electronics, is a technology in which photons instead of electrons are the carriers of information [1,2]. The importance of photonics arises from its advantages over electronics in that photonics provides larger bandwidths, faster response times, and less noise from extrinsic electromagnetic fields. Photonics exhibits high potential in many areas of present and future applications for data and image processing technologies [3–7]. The NLO phenomena are expected to provide key functions necessary for the photonic technology. The ability to change the attributes of light such as its frequency, phase, amplitude, or transmission characteristics, when the light passes through an NLO-active medium is a phenomenon that is potentially useful in photonics. Thus, advancements in the field of NLO materials will have a direct effect on the progress of photonic technology.

Currently, only inorganic crystals are commercially used in second-order NLO applications. For example, crystals of lithium niobate and potassium dihydrogen phosphate are used for electro-optic (EO) modulation and frequency conversion, respectively [2,4,6]. Over the past decade, organic materials have been identified as the materials of choice because of their design flexibility. A number of NLO organic crystals and polymers possess significantly larger second-order nonlinearities than that of traditional inorganic NLO materials [8]. Polymeric materials for second-order NLO applications hold a number of other advantages over inorganic and organic crystals. For example, the ease of processing polymers in conjunction with thin film technologies offers the unique opportunity for them to be used in integrated optic and EO applications. Moreover, the design, synthesis, and fabrication of NLO polymers are more flexible, facile, and cost-effective than those of inorganics [9]. Since NLO chromophores usually consist of electron donor and acceptor groups separated by conjugated bonds, their absorption characteristics can easily and controllably be engineered by changing the molecular structure or functional groups. Various synthetic routes can be adopted to obtain polymers with a desired molecular architecture. Inorganic and organic crystals, on the other hand, do not lend themselves easily to such manipulation.

The large dielectric constants of the inorganic materials (almost an order of magnitude larger than those of organics) lead to the large capacitance associated with the electrodes and severely limit the bandwidth for EO modulation. The lower dielectric constants of polymers make it easier to design traveling wave EO modulators due to the close match of velocity between the microwave and optical wave. Therefore, when the other characteristics of both inorganic and polymeric NLO materials are equal, the dielectric constant difference will make the polymers a better choice for use in high speed EO modulation.

Polymers for second-order NLO applications have been extensively investigated during the last ten years. However, significant progress in material development efforts toward practical applications has been made only in recent years. This chapter discusses the current state of the art in the development of second-order NLO polymers. We start with a brief description of the NLO phenomena at both the molecular and bulk levels. The design considerations and property optimization of chromophores are associated with the molecular level optical nonlinearity. The induced noncentrosymmetric dipolar orientation of chromophores leads to the bulk second-order NLO properties, which are most commonly characterized by their second harmonic generation (SHG) and EO coefficients. This discussion is followed by a brief review of second-order NLO polymeric systems, which include polymers for both EO and frequency doubling as well as photorefractive materials. Researchers have also been using various characterization techniques to understand the relaxation process of the chromophore orientation resulting in the decay of the optical nonlinearity. We focus on the SHG technique for such investigation. In the final section the evaluation of the overall performance of the materials required and optimization of the properties for potential applications are discussed.

II. NONLINEAR OPTICAL RESPONSES

The NLO processes occur when an electromagnetic field interacts with a medium. These NLO phenomena can be described at both the molecular and bulk levels. Molecular polarization in an intense electric field is field-dependent and is usually described as follows [2,10]:

$$\mu(E) = \mu_0 + \alpha E + \beta EE + \gamma EEE + \cdots \quad (1)$$

where μ is the molecular dipole moment, μ_0 is the intrinsic dipole moment of the molecule, and E is the electric field vector. The coefficients α, β, and γ represent the polarizability, hyperpolarizability, and the second hyperpolarizability of the molecule, respectively. The bulk polarization of a material can be described in a similar way:

$$P(E) = P_0 + \chi^{(1)}E + \chi^{(2)}EE + \chi^{(3)}EEE + \cdots \quad (2)$$

in which $\chi^{(1)}$, $\chi^{(2)}$, $\chi^{(3)}$ are known as the first-, second-, and third-order susceptibilities and P represents the bulk polarization of the medium. P_0 is the permanent polarization of the material. Nonlinear responses in bulk media are described by $\chi^{(2)}$, $\chi^{(3)}$, and other higher order terms. The hyperpolarizabilities and susceptibilities are tensor quantities, and expressions (1) and (2) imply tensor product yielding a vector quantity.

Based on Eq. (2), all materials, including gases, liquids, and solids, can show third-order NLO effects. In contrast to the third-order NLO materials, second-order NLO materials require a noncentrosymmetric alignment of NLO molecules. In the case of SHG, the relationship between the molecular and bulk NLO properties can usually be described as [11]

$$\chi^{(2)}(-2\omega; \omega, \omega) \propto N \beta f(\omega)f(\omega)f(2\omega)(\mu_0 E_p/kT) \quad (3)$$

where N is the number density of the NLO chromophore, $f(\omega)$ and $f(2\omega)$ are local field factors where ω is the frequency of an optical field, and E_p and T are the poling field and temperature, which are discussed in Section IV. Therefore, concentration of chromophore, $\mu\beta$ values of dye, and alignment of these dipoles are the three main factors that will have a direct impact on bulk NLO responses.

Another expression for the NLO phenomenon is the nonlinear index of refraction of the medium. When the medium is subjected to a dc field $E(0)$ and an optical field $E(\omega)$, the total field to which the medium is subjected can be described as [2,10]

$$E = E(0) + E(\omega) = E(0) + E_0 \cos(\omega t - kz) \quad (4)$$

By proper mathematical derivation, the following equation can be obtained.

$$n = n_0 + \frac{4\pi\chi^{(2)}}{n_0}E(0) + \frac{6\pi\chi^{(3)}}{n_0}E(0)^2 + \frac{3\pi}{2n_0}\chi^{(3)}E_0^2 + \cdots \quad (5)$$

where n_0 and n are the linear and nonlinear refractive indices, respectively. This equation indicates that the nonlinear effects arise from field-induced modulation of the refractive index. The second term on the right-hand side of the equation is responsible for the second-order NLO effect or the linear EO (Pockels) effect. When a second-order NLO material possessing a large $\chi^{(2)}$ value is subjected to a field $E(0)$, the refractive index of this material is changed according to Eq. (5). As a consequence, the properties of the incident optical field are changed as it passes through the medium.

III. CHROMOPHORES

An NLO chromophore acts as an active molecular component that gives rise to the second-order NLO property in a polymer. Generalized structures for NLO chromophores that exhibit molecular second-order NLO properties are shown in Fig. 26.1. These molecules posses donor–acceptor groups attached to an aromatic ring system for increased charge transfer through π-electron delocalization [12]. Such dye molecules are characterized by intramolecular charge transfer giving rise to large ground-state and excited-state dipole moments (μ) and molecular hyperpolarizability β. The value of β can be experimentally determined by electric field–induced

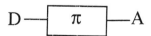

Fig. 26.1 NLO chromophores are composed of electron donor–acceptor groups separated by a π bridge. D = electron donor, e.g., $-NH_2$, $-NR_2$. A = electron acceptor, e.g., $-NO_2$, $-CN$, $-C(CN)=C(CN)_2$. π bridge: e.g., benzene, azobenzene, stilbene.

second harmonic generation [13–15], hyper-Rayleigh scattering [16], and solvatochromic [17] techniques. A few of the electron donor and acceptor groups that are commonly used are listed in the legend to Fig. 26.1. Some examples of NLO chromophores and their properties are illustrated in Table 26.1.

The molecular level NLO responses of chromophores are usually characterized by the magnitude of their $\mu\beta$ values. The linear and nonlinear optical properties of a chromophore are determined by the types of donor–acceptor groups and π bridges selected to construct the dye molecule. The typical ground-state dipole moment values of NLO chromophores are between 5 and 10 debye. The hyperpolarizability of different chromophores, however, can vary by as much as an order of magnitude. Recently, values of $\mu\beta$ in a number of chromophores have been greatly increased with the employment of heteroaromatic rings [18,19] or derivatives of barbituric acid [20]. Some of these chromophores possess $\mu\beta$ values 40 times as large as that of 4-(N,N-dimethylamino)-4'-nitrostilbene (DANS), a commonly used NLO dye [20].

The design and synthesis of NLO chromophores require simultaneous optimization of properties such as $\mu\beta$ values, absorption characteristics, thermal stability, and processibility. In general, there is a trade-off between transparency and nonlinearity for NLO dye molecules. The increase of nonlinearity of chromophores often involves a red shift of the absorption maximum. From a practical point of view, however, it is desirable that absorption at wavelengths of interest be avoided. A careful selection of the electron-donating and -withdrawing groups in conjunction with the conjugated (π-bridge) system can result in enhanced nonlinearity and a fine-tuning of the absorption. Numerous theoretical studies to understand and optimize the properties of the NLO chromophores have been reported [10,21].

Thermal stability of NLO chromophores is a very important property that cannot be overlooked. Since the chromophores are responsible for the optical nonlinearity, they must be quite stable at the elevated temperatures that are required during processing as well as under ambient conditions during the operation of the device. Thermal stability of the dyes is usually characterized by means of differential scanning calorimetry or thermogravimetric analysis or by monitoring the change in absorption characteristics of chromophores after thermal treatment. The thermal stability of a series of nitrobenzene chromophores has been studied systematically [22]. It was reported that the thermal decomposition temperatures of these types of NLO dyes can be significantly increased by replacing alkyl-substituted amino donors with aryl-substituted ones, without sacrificing their optical nonlinearities [22]. Even though it is of immense importance to understand the structure–property relationship that is associated with the thermal stability of a chromophore, only a limited number of detailed studies providing such information are available.

IV. NONCENTROSYMMETRIC DIPOLAR ORIENTATION

In centrosymmetric media, every element of the second-order susceptibility tensor $\chi^{(2)}$ is zero. To impart bulk

Table 26.1 Molecular Structures and Properties of Several NLO Chromophores

D - π - A	$\mu\beta$ (10^{-30} D cm^5/esu)
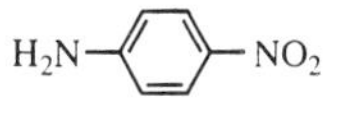 H_2N—⟨⟩—NO_2	75[a]
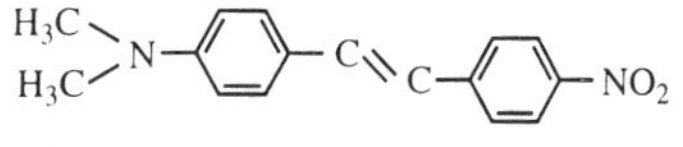 $(H_3C)_2N$—⟨⟩—C≡C—⟨⟩—NO_2	760[a]
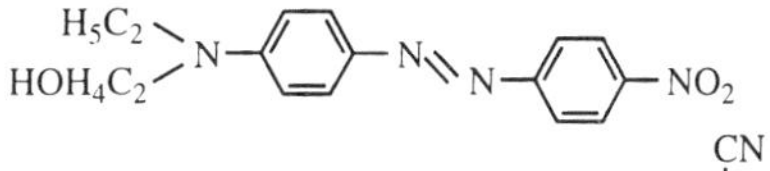 $(H_5C_2)(HOH_4C_2)N$—⟨⟩—N=N—⟨⟩—NO_2	1090[a]
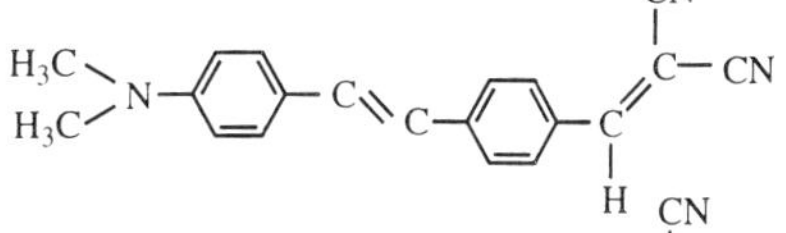 $(H_3C)_2N$—⟨⟩—C≡C—⟨⟩—CH=C(CN)—C(CN)=CN	2650[a]
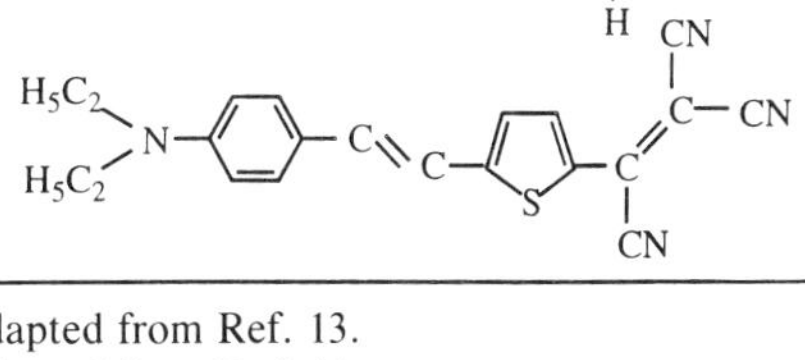 $(H_5C_2)_2N$—⟨⟩—C≡C—⟨S⟩—C=C(CN)—C(CN)=CN	6200[b]

[a]Adapted from Ref. 13.
[b]Adapted from Ref. 14.

second-order NLO properties, the NLO molecules in a material need to be aligned in a noncentrosymmetric fashion. Techniques that can be used to achieve noncentrosymmetric alignment of NLO chromophores are discussed below.

The Langmuir–Blodgett (LB) technique is a unique method that allows the preparation of very thin films with precise control of layer thickness and molecular orientation [23]. Typically, polymers are designed and synthesized to possess hydrophilic and hydrophobic groups that can assist the acentric alignment of NLO moieties at the air/water interface in an LB trough. The aligned NLO polymers are transferred onto a suitable substrate. Each layer of molecules can be organized according to the transferring process. An LB film with alternating layers where one or both layers possess a nonlinear moiety needs to be assembled in such a way that little cancellation of the nonlinearity occurs in order to exhibit second-order NLO properties [24–29]. The drawbacks of using the LB technique are that it is a time-consuming process, films tend to be fragile, and the nonlinear component can be diluted by the noncontributing aliphatic chains.

The self-assembly technique offers an alternative route to prepare monolayer or multilayer thin films of highly ordered multifunctional compounds in a more effective and time-saving manner. This technique presents some advantages over the LB technique in that each layer in the self-assembled films can be covalently bonded. Moreover, no excess deposition can take place as it is limited by the reactive sites on the layer surface [30]. A number of self-assembled films exhibiting second-order NLO effects have been prepared and studied [30–32].

Recently, a soluble unsymmetrically substituted polydiacetylene (Fig. 26.2) has been reported to exhibit an SHG signal from spin-coated or cast films without recourse to poling [33,34] (see next paragraph for the definition of poling). The polymer film showed appreciable NLO coefficients, and the NLO responses are quite stable even at elevated temperatures. It is suggested that the spontaneous orientation of the polymer chains during the film formation is responsible for the acentric alignment of the polymer. LB films prepared from such a polymer also exhibit the SHG signal.

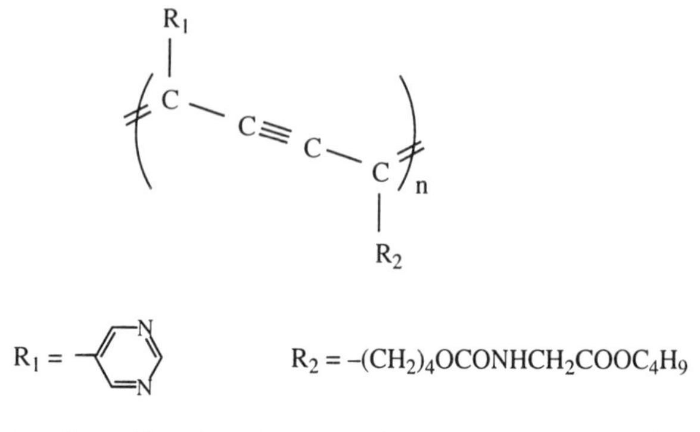

The technique that is probably the most commonly used to achieve the noncentrosymmetric alignment is called poling. Interaction between an applied electric field and the dipole moment of NLO-active dye molecules leads to the noncentrosymmetric alignment of chromophores in the polymer matrix. Both electrode [11,35] and corona poling [11,36,37] methods are widely used. Figure 26.3 shows the experimental setup for corona poling. The heating stage provides thermal control during the poling process. The poling is usually carried out in the vicinity of the glass transition temperature (T_g) where molecules have relatively large mobility. The two electrodes in the poling apparatus can be arranged in either wire-to-plane or needle-to-plane configurations [11,36,37]. A large electric potential (several kilovolts) is applied across the two electrodes to induce a corona discharge. The corona source ionizes the molecules in the air and provides an electric field to move the ions to the sample surface. The deposited ions induce image charges on the grounded electrode, forming a large static electric field across the sample. This static electric field is on the order of 10^6 V/cm, which leads to the alignment of NLO molecules with respect to the electric field. A corona current can be measured during the poling process to optimize the poling conditions. After the application of the electric field for a period of time, the material temperature is returned to well below T_g where the molecular motion is severely restricted. As a consequence, the alignment of the dipoles can be preserved. The acentric arrangement of NLO molecules leads to the optical nonlinearities.

For electrode poling, polymer samples are sandwiched between two electrodes. For example, a polymer thin film can be prepared using the spin coating technique on an indium tin oxide (ITO)–coated glass slide. The ITO layer serves as the bottom electrode. Subsequently, a metal electrode can be deposited onto the surface of the polymer film to form the sandwich structure. Poling can be carried out by applying a large dc potential at an elevated temperature. Other types of configurations for electrode poling, such as in-plane poling [35], have also been used. To obtain good poling efficiency, the applied electric field is the highest it can be without causing damage to the polymer film (usually on the order of 10^6 V/cm). In general, it is important to remove ionic impurities in the materials and avoid dust particles trapped in the polymer films as they may lower the poling efficiency and lead to the breakdown of samples during poling. The poled samples, containing a noncentrosymmetric ensemble of NLO-active molecules, are represented by the $C_{\infty v}$ symmetry [11].

$R_1 = $ (pyrimidine) $\quad R_2 = -(CH_2)_4OCONHCH_2COOC_4H_9$

Fig. 26.2 Chemical structure of an asymmetrically substituted polydiacetylene. This polymer can be readily dissolved in chloroform.

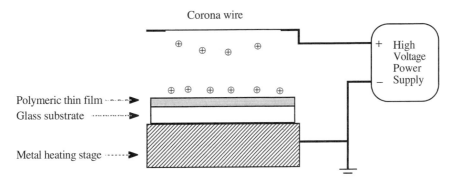

Fig. 26.3 Schematic diagram for a corona poling setup with a wire-to-plane configuration. The tungsten wire is placed above and parallel to the sample. For the needle-to-plane arrangement, the end of a tungsten wire is pointed toward the polymer film.

Optical poling of second-order NLO materials has been demonstrated in polymers containing azo dye [38]. This technique while very interesting, is not applicable to all chromophores.

V. CHARACTERIZATION

The second-order NLO properties of the poled polymers can be measured by a variety of experimental techniques. We will describe only two very commonly used methods, the SHG and EO techniques. These are the most direct techniques for the determination of the optical nonlinearities of poled polymers.

For the SHG measurement, Kleinman symmetry [39] is often assumed to hold. This assumption is approximately valid when there is no absorption of the material at the fundamental wavelength λ and at the second harmonic (SH) wavelength $\lambda/2$. For the poled polymer, the SHG coefficients, d_{31} and d_{33}, are defined as $d = \chi^{(2)}/2$. The d_{31} and d_{33} coefficients of a poled polymer depend on the noncentrosymmetric arrangement of the NLO-active molecules. Under the low poling field approximation, the relation $d_{31} = d_{33}/3$ is satisfied.

To measure the d coefficients of a poled polymer sample, the Maker fringe SHG experiment [40] is frequently carried out. The experimental arrangement is shown in Fig. 26.4. A polarized laser beam from a Q-switched Nd:YAG laser ($\lambda = 1.064 \ \mu\text{m}$) or other IR laser source is employed. The fundamental laser beam at a frequency ω passes through the poled polymer sample and an appropriate IR blocking filter. The SHG signal is selected with an interference filter at the corresponding frequency 2ω, detected by a photomultiplier tube, and measured with a boxcar integrator. The SHG intensity, $I(2\omega)$, can be expressed as

$$I(2\omega) = \frac{(512\pi/A)t_\omega^4 T_{2\omega} d^2 t_0^2 p^2 I^2(\omega) \sin^2\Psi(\theta)}{n_\omega^2 - n_{2\omega}^2} \quad (6)$$

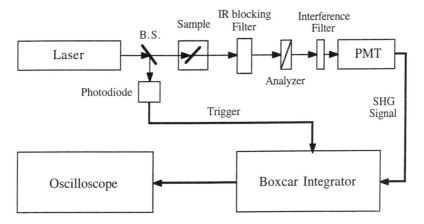

Fig. 26.4 Schematic of the SHG experimental setup.

where $I(\omega)$ is the intensity of the fundamental laser beam; A is the area of the laser beam; θ is the incident angle; t_0, t_ω, and $T_{2\omega}$ are the transmission factors; p is a projection factor; n_ω and $n_{2\omega}$ are the refractive indices of the sample at ω and 2ω, respectively; and $\Psi(\theta)$ is an angular factor related to the sample thickness, the fundamental wavelength, and the refraction angles. A Y-cut quartz crystal with $d_{11} = 0.49$ pm/V (at 1.064 μm) [41] is used as the reference. By comparing the SHG intensity from the poled polymer sample with that from the quartz crystal, the d coefficients of the poled polymer sample are determined.

A number of methods have been used to measure the EO coefficients r_{33} and r_{31} of the poled polymer samples. These EO measurements are made by detecting the change in refractive index of the poled polymer sample when a modulating electric field is applied to the sample. Mach–Zehnder [42,43], Fabry–Perot [44,45], and Michelson [46] interferometric techniques have been used to evaluate the EO coefficients. Other techniques, such as an attenuated total reflection technique [47,48] and an ellipsometric technique [49–51], have also been employed to determine the r coefficients.

Among these methods, the ellipsometric technique is the simplest and the most widely used. An experimental setup of this technique is shown in Fig. 26.5. The polymer film to be measured is sandwiched between two electrodes consisting of a transparent ITO layer and a metal coating. A laser beam at the wavelength of interest is incident on the sample, with its polarization set at 45° to the plane of incidence, so that the perpendicular (s-wave) and parallel (p-wave) components are equal in magnitude. This method measures the change in phase retardation between the s and p waves in the reflected beam while a modulating electric field is applied across the sample. The static phase shift between the s and p waves is adjusted by a Soleil–Babinet compensator positioned between the sample and the analyzer. For the measurement, the intensity of the reflected beam is set at I_c (half of the maximum intensity). The resulting modulation of the reflected beam, I_m, is detected with a lock-in amplifier. The EO coefficient r_{33} is then calculated using the equation

$$r_{33} = \frac{3\lambda I_m}{4\pi V_m I_c n^2}\left(\frac{(n^2 - \sin^2\theta)^{3/2}}{n^2 - 2\sin^2\theta}\right)\left(\frac{1}{\sin^2\omega}\right) \quad (7)$$

where λ is the laser wavelength, n is the refractive index, V_m is the modulating voltage, and θ is the angle of incidence. It is noted that $r_{33} = 3r_{31}$ is valid under the low field approximation.

VI. POLYMERIC SYSTEMS

The development of polymeric second-order NLO materials for practical applications has been pursued since the 1980s and still is a very active research field. To summarize some of the materials that have been developed during the past decades, it is convenient to classify them according to their potential applications.

A. Polymers for Electro-Optic Applications

Large second-order optical nonlinearity and stable NLO responses are the two principal properties required in polymers that have potential for EO applications. Various types of polymers have been developed to enhance the performance of these materials. The development of EO polymers is described as follows.

A conventional technique and perhaps the simplest for the preparation of a poled polymer is to incorporate an NLO chromophore into an amorphous polymer matrix. Polymer samples prepared in this manner are known as guest–host systems (see Fig. 26.6). Meredith et al. [52] reported on the first guest–host system prepared by dissolving DANS in a liquid crystalline poly-

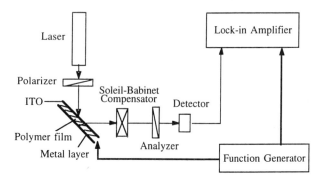

Fig. 26.5 The experimental arrangement for measurement of the EO coefficient using the ellipsometric technique.

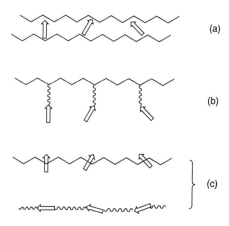

Fig. 26.6 Schematic for the (a) guest–host, (b) side-chain, and (c) main-chain polymeric systems.

mer. This system exhibited rather weak second-order NLO effects that diminished quite rapidly at room temperature. A large number of guest–host systems have been developed based on different NLO chromophores and polymer matrices. Examples for typical NLO chromophores used in these guest–host systems are p-nitroaniline, 2-methyl-4-nitroaniline, Disperse Red 1, and Disperse Orange 3. Poly(methyl methacrylate) is the most frequently used polymer host [53–55] because of its excellent optical transparency and compatibility with NLO chromophores. Styrene-acrylonitrile copolymer has been investigated to improve the solubility of chromophores [56]. Preparation of typical guest–host samples is simple, as most of the polymers and chromophores are readily available. However, the solubility of the dye molecules in a polymer host is usually limited. High loading of the dyes can cause recrystallization and phase separation, which deteriorate the optical quality of the sample. Such a restriction of chromophore density in the system also limits the magnitude of the NLO response. Moreover, T_g values of the polymeric systems are usually decreased as the chromophore concentration increases due to the plasticizing effect. This phenomenon results in the relaxation of the poled order that is evidenced from the decay of the second-order NLO response even at room temperature.

Guest–host systems based on high T_g polymers such as polyimides are promising for the enhancement of the temporal stability of the optical nonlinearity. The first polyimide guest–host system was prepared by dissolving Eriochrome Black-T in Pyralin 2611D, a DuPont polyimide. The polyimide guest–host film was poled during the imidization process at 250°C for 1 h. EO response of the poled imidized guest–host polyimide was found to be stable at 150°C for 12 h [35]. These reports have renewed interest in guest–host systems in general as they offer simplicity in sample preparation. The exploitation of high T_g polyimides was subsequently reported in a number of publications [57–60]. The temporal stability of the NLO responses was significantly improved in these polyimide-based guest–host systems. The increase of optical nonlinearity has also been demonstrated in this type of polymer. An NLO chromophore with a large $\mu\beta$ value (6200×10^{-30} cm^5 D/esu) has been dissolved in a polyimide matrix. A chromophore loading of 12 wt % gave rise to a relatively large NLO response ($r_{33} = 10.8$ pm/V at 1.52 μm) [59].

An approach to alleviating the relaxation problems encountered in the poled polymers is to covalently attach the NLO chromophores onto polymer chains. The chromophores can be either pendant groups known as side-chain polymer systems or parts of the polymer backbones known as main-chain polymer systems (see Fig. 26.6). High chromophore concentrations are possible in these systems without recrystallization or phase separation. However, poling of the NLO moieties in side-chain and main-chain polymers may be less efficient than in the guest–host samples. This is because the movement of the NLO molecules in these systems involves a larger energy barrier associated with the movement of polymer chains. For the same reason, the temporal stability of the second-order NLO effects of both the side-chain and main-chain polymers is better than that of the guest–host systems.

Side-chain polymers are usually prepared in the form of copolymers with NLO-active moieties attached on the backbone via flexible spacers such as methylene units. Examples are copolymers of methyl methacrylate and chromophore-substituted methacrylate monomers [53,61–63], poly(styrene-co-acrylic acid ester) [54], and alternating styrene-maleic-anhydride copolymer [64], and there are many others.

Main-chain polymeric systems exhibit better temporal stability than the side-chain polymers. However, it can be problematic to obtain efficient poling in these polymers. Some main-chain polymers are synthesized in such a fashion that the dipole moments of NLO chromophores are arranged head-to-tail along the polymer chains [65–68]. Others are prepared to have the dipole aligned head-to-head, and the chromophores are connected with bridges that allow the backbone to fold like an accordion [69,70]. Main-chain polymers with the dipole moments of the chromophores aligned perpendicular to the polymer backbone have also been reported [71–73]. It is easier to bring about chromophore alignment in this type of main-chain polymers than in the head-to-head or head-to-tail main-chain polymers.

NLO chromophores have recently been incorporated in polyimides as side-chain and main-chain systems [74–76]. These samples in general exhibit excellent temporal stability at elevated temperatures (e.g., long-term stability of over 1000 h at 120°C [75]). Relatively large NLO responses have also been reported (e.g., r_{33} of 13 pm/V at 1.3 μm [74]) as the materials no longer suffer from limited solubility. The excellent chemical and physical properties of polyimides in conjunction with high chromophore loading make these polyimide-based materials very promising for EO applications.

Significant progress in the enhancement of temporal stability at room temperature as well as at high temperatures arose from the incorporation of NLO chromophores in three-dimensional polymer networks. These advancements were pioneered by Hubbard et al. [77] and Eich et al. [78]. The first-generation polymer networks are based on thermally cross-linkable epoxy resins. NLO guest molecules were introduced into an epoxy-based polymer network [77]. Better stability has been observed in these cross-linked systems at room temperature than in the uncross-linked samples. A polymer network based on a bisepoxide compound and an NLO-active diamino compound was reported by Eich et al. [78]. Poling was carried out simultaneously with the curing process. Significant enhancement of the temporal stability at 85°C was achieved. Other epoxy-based

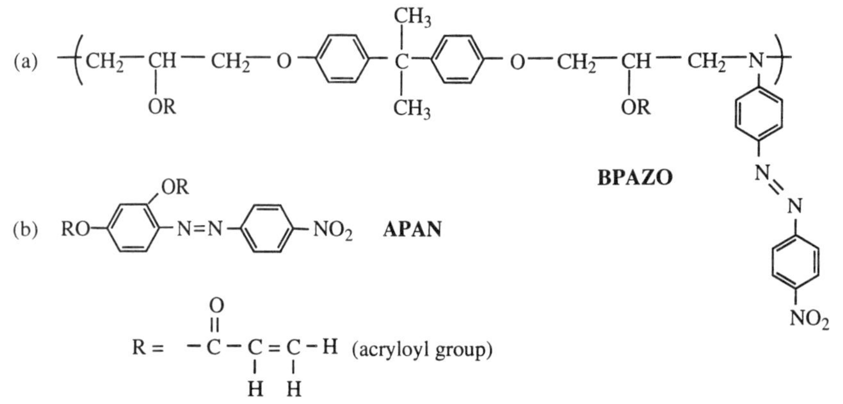

Fig. 26.7 Chemical structures of (a) a thermally cross-linkable NLO polymer, BPAZO, and (b) a thermally cross-linkable NLO chromophore, APAN.

cross-linked systems have also been developed [79–81]. Other types of polymer networks such as polyurethane [82,83], aromatic polyamides [84], polybismaleimides [85,86], and polyimide [87] have also been reported.

Figure 26.7 shows the chemical structures of an NLO chromophore (APAN) and an epoxy-based polymer (BPAZO) where NLO moieties are attached to the backbone [81]. Both the dye and the polymer are functionalized with thermally cross-linkable acryloyl groups. As the dye-doped polymer is subjected to heat as part of the simultaneous poling/curing process, the inter- and intramolecular cross-linking reactions occur simultaneously (Fig. 26.8). The T_g of the cross-linked polymer–dye network is lower than that of the undoped polymer network because of the plasticizing effect of the dissolved dye. However, the temporal stability at 100°C of the polymer–dye network is better than that of the undoped polymer network (Fig. 26.9) as a direct result of the increased cross-linking density in the cross-linked guest–host system. Therefore, the addition of the thermally cross-linkable NLO dye not only increases the NLO coefficient (16 pm/V in the undoped polymer and 28 pm/V for the doped polymer at 1.064 μm) but also enhances the temporal stability at elevated temperatures.

Cross-linking processes can also be induced by photochemical reactions. Examples of photo-cross-linkable functionalities that have often been employed are methacryloyl [88,89] and cinnamoyl groups [90,91]. A guest–host photo-cross-linked system in which both ends of the chromophore were tied onto the network was developed [90]. This system was based on a poly(vinyl cinnamate) host and an active dye guest functionalized with cinnamoyl groups. The cross-linking reaction occurred between the chromophores and polymer host via a 2 + 2 cycloaddition reaction of the cinnamoyl groups upon UV irradiation subsequent to corona poling (Fig. 26.10), resulting in the enhancement of temporal stability.

The approach to preparing cross-linked NLO polymers in which both ends of the chromophores are tied onto the network has been extensively expanded [89,92].

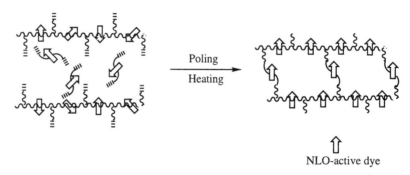

Fig. 26.8 Schematic for the formation of cross-linked network via an intermolecular cross-linking reaction between the polymer and the NLO chromophore. An electric field applied during the course of the curing process results in the noncentrosymmetric alignment of the NLO moieties.

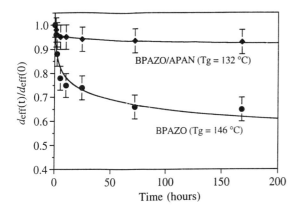

Fig. 26.9 Temporal stability of the effective second harmonic coefficients of two samples monitored at 100°C. BPAZO/APAN represents a poled/cured polymer film that was prepared from an APAN-doped BPAZO polymer solution. For comparison, a poled/cured BPAZO sample was investigated concurrently.

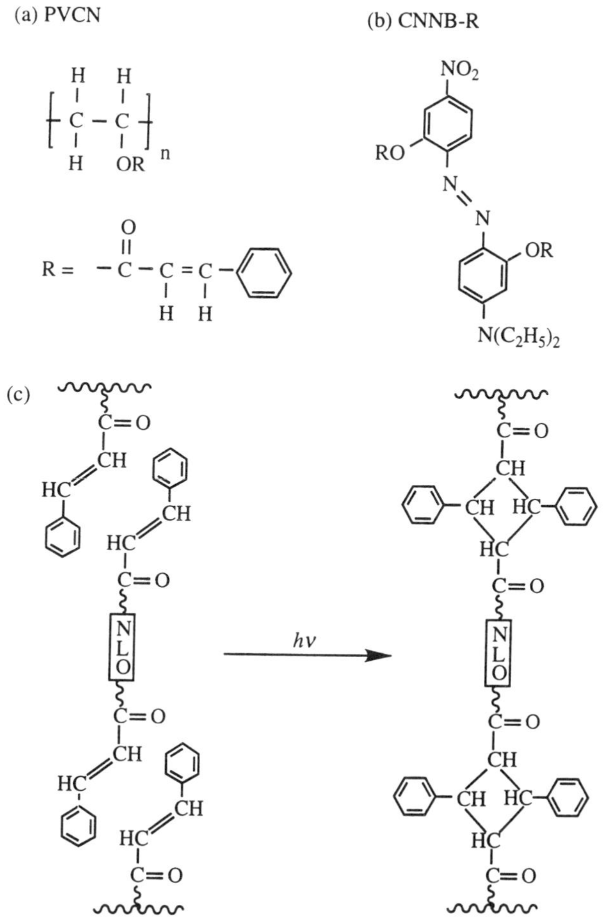

This was accomplished by inducing a cross-linking reaction between the reactive pendant groups of a side-chain NLO polymer and the reactive end groups of the chromophore. Stable NLO response at 90°C for more than 2000 h was demonstrated after a small initial decay [92].

Sol gel technology has been used to develop NLO-active organic/inorganic composite networks. The basic sol gel process involves the sequential hydrolysis and polycondensation of silicon alkoxide [93]. Various types of sol gel matrices, such as SiO_2 [94], SiO_2-TiO_2 [95], SiO_2-ZrO_2 [96,97], and Al_2O_3 [98], have been prepared to possess optical nonlinearity. Chromophores are either doped in the sol gel matrices [94] or tethered on the inorganic glassy networks [99–101]. The NLO-active sol gel glasses have been prepared mostly by traditional methods using tetraalkoxysilane as starting materials. Jeng et al. [102,103] employed a low molecular weight organosilicon oligomer as a precursor to prepare inorganic based guest–host networks. By using the oligomer, the processibility of the materials is improved, and better temporal stability has been observed compared to other sol gel systems.

The sol gel concept has been extended to the polymerization of a multifunctional organic compound and an NLO alkoxysilane dye forming a phenoxysilicon polymer network (Fig. 26.11). The poled/cured phenoxysilicon polymer [104] exhibited a d_{33} value of 77 pm/V at 1.064 μm due to its high NLO chromophore concentration (approximately 30% by weight). After a small initial decay, the NLO response was relatively stable at 105°C (T_g of the polymer is 110°C). Excellent long-term stability at temperatures close to the T_g of the system is a direct consequence of the extensively cross-linked network.

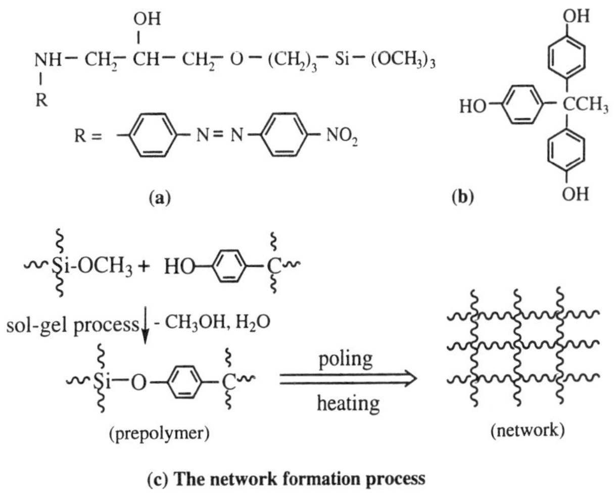

Fig. 26.10 (a,b) Chemical structures of (a) poly(vinyl cinnamate) (PVCN) and (b) photo-cross-linkable NLO chromophore, CNNB-R. (c) Schematic for the photo-cross-linking reaction between the cinnamoyl groups.

Fig. 26.11 Chemical structures of (a) an alkoxysilane functionalized NLO chromophore, ASD, and (b) a multihydroxyl organic compound, THPE. (c) The NLO network was prepared through a sol gel process.

Another class of cross-linked networks incorporating NLO chromophores based on a semi-interpenetrating polymer network (semi-IPN) [105] and fully interpenetrating polymer networks (full-IPN) [106–108] has been investigated. These systems were developed to enhance temporal stability by realizing the ability of an IPN structure to improve the viscoelastic properties of polymers. The semi-IPN system was prepared by combining a high T_g polyimide and a sol gel network [105]. The poled semi-IPN sample has a d_{33} of 28 pm/V at 1.064 μm. After over 168 h at 120°C, a retention of 73% of the d_{33} value was observed.

Full-IPN systems are based on two cross-linked networks that are formed via different reaction mechanisms that do not interfere with each other. One example of the NLO-active IPN combined the network of epoxy-based polymer, BPAZO (Fig. 26.7), and the phenoxysilicon polymer network (Fig. 26.11). The formation of the networks takes place during the poling process. This IPN, which was poled and cured at 200°C, exhibited a d_{33} of 33 pm/V at 1.064 μm and an r_{33} of 5 pm/V at 1.3 μm [106]. No measurable decay of the SH intensity of the IPN could be observed after the sample was treated at 110°C for over 1000 h. The temporal stability of this IPN system is better than that of both parent polymers. Synergistic enhancement of the temporal stability of NLO responses has also been consistently observed in other IPN systems [107,108].

Since the initiation of the investigation on poled polymers in the mid-1980s, most research has been focused on enhancing the optical nonlinearity and its temporal stability at elevated temperatures. So far, no particular system has been identified as the material of choice for use in the EO applications. With the rapid progress in research in recent years, polymeric materials exhibiting large and stable NLO properties for real applications may be anticipated in the near future. Other properties such as optical loss, optical power handling [109], processibility, and reproducibility are also of importance. Successful development of a material for NLO application will require optimization of properties in all aspects.

B. Polymers for Frequency-Doubling Applications

Poled polymers with large second-order optical nonlinearity and negligible absorption at both fundamental and SH frequencies are desired for frequency-doubling applications. NLO chromophores that exhibit large second-order susceptibilities often have longer cutoff wavelengths (visible or near-IR regions). This leads to substantial loss from absorption when the doubled frequency is in this region. This problem must be avoided as most practical NLO waveguide devices involve large optical power density. Therefore, even slight absorption can cause intolerable damage to the material [2].

Figure 26.12 shows the chemical structure for an epoxy-based polymer that was used in prototype devices for frequency-doubling applications [46,110]. The absorption maximum of the NLO moiety in this polymer is at 380 nm. The polymer exhibited a d_{33} of 13 pm/V at 1.064 μm. Cerenkov SHG of blue and green light at 450 and 532 nm has been demonstrated. However, only a 0.03%/watt conversion efficiency in the waveguide was obtained. An attempt to reduce the cutoff wavelength to 340 nm by using 4-aminobenzonitrile chromophore in a similar polymer resulted in a rather small NLO coefficient (d_{33} = 4 pm/V at 1.064 μm) [111].

A highly transparent polymer was synthesized based on an aromatic polyurea [112]. The colorless polymer film shows a cutoff wavelength at 307 nm. The poled polymer sample exhibits a d_{33} of 5.5 pm/V at 1.064 μm. This polymer is expected to be useful in frequency doubling for the emission of light in the UV region. Another approach using "blue window" chromophores for frequency doubling is also promising [113]. These chromophores are designed to have absorption peaks in the red region and very little absorption in the blue region. This type of chromophore possesses larger $\mu\beta$ values while still providing a transparent blue window at 2ω.

The major problem in the development of polymers for frequency doubling is the trade-off between the large optical nonlinearity and the absorption in the visible region. Stringent phase-matching considerations impose further limitations on enhancing the doubling efficiency. The enhancement of optical properties and the increase of SHG efficiency (to the order of a few percent) will make the poled polymer a stronger contender for use in frequency-doubling applications.

C. Photorefractive Polymers

Optical materials with large nonlinear indices of refraction and fast response times are essential components in any optical data processing scheme. Photorefractive materials have demonstrated the possibility of exhibiting large index changes at modest intensities [114]. In addition, they seem to be ideally suited for real-time holographic applications. These materials exhibit a light-in-

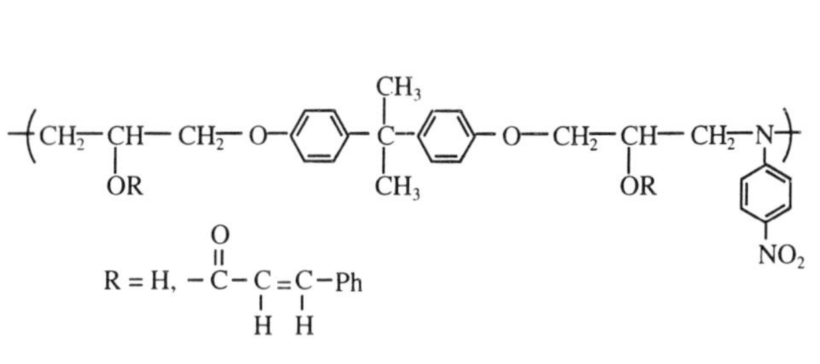

Fig. 26.12 Molecular structure of an epoxy NLO polymer bearing nitroaniline chromophores.

duced change in the index of refraction. A light pattern creates a redistribution of charges in the material because photoexcited charges drift or diffuse from bright areas to darker areas, where they are trapped. A sinusoidal light pattern can be created, for example, by the interference of two laser beams. Charge diffusion and drift (if an external field is applied) result in large electric fields (space charge fields) inside the material. This space charge field alters the refractive index of the material through the Pockels effect. Due to the possibility of nonlocal change in the refractive index (i.e., the intensity and the refractive index patterns are phase-shifted), two beam-coupling (TBC) phenomena will occur, i.e., one of the interfering beams may gain energy while the other loses the same amount of energy.

Photorefractive materials based on second-order NLO polymers have been studied only recently. In 1991, Ducharme et al. [115] first observed the photorefractive effect in an epoxy-based NLO polymer doped with a hole-transporting agent. Diffraction efficiencies on the order of 10^{-4} were achieved in this polymer when an external field of approximately 100 kV/cm was applied. This photorefractive polymer demonstrated an important proof of principle that polymeric materials with NLO properties, photoconductivity, and a charge-trapping mechanism could be photorefractive. Shortly after this observation, a number of photorefractive polymers were developed with substantial improvements in diffraction efficiency as well as TBC gain [116]. Approaches to the development of these photorefractive polymers include (1) adding a charge transport agent to epoxy-based NLO polymers [116,117], (2) doping a photoconductive polymer [e.g., poly(N-vinylcarbazole) (PVK)] with NLO chromophores and sensitizers [116,118–121], and (3) synthesizing a fully functionalized material with all functional components attached to the polymer backbone [116,122,123]. Epoxy-based materials with reasonable photorefractive properties have been studied [117,118]. A high performance epoxy-based photorefractive polymer with a large TBC gain of 56 cm^{-1} has been reported [117]. PVK-based composite materials doped with a high concentration of NLO chromophores and a small amount of sensitizers have recently been investigated. Large diffractive efficiencies and net TBC gains have been obtained in these materials [118–121]. A photorefractive polyimide system fully functionalized with an NLO-active moiety, a charge generator, and a charge-transporting agent has been synthesized [123]. The second-order NLO response of the poled polymer was exceptionally stable at high temperatures as there was no decay of the nonlinearity observed at 150°C. A TBC gain of 22.2 cm^{-1} was measured in this poled polyimide under zero external electric field.

Among all polymeric photorefractive materials developed so far, PVK-based composite materials showed the best overall performance [120,121]. Diffraction efficiency near 100% and large net TBC gain of about 200 cm^{-1} were achieved in a PVK-based composite material with a very high NLO chromophore loading (50 wt %) [120]. These large photorefractive properties were attributed to significant orientational enhancement [118,120]. However, the photorefractive response time for all reported polymeric materials is relatively large (~100 ms or longer). The slow response could be attributed to the small carrier mobilities associated with these materials leading to slow buildup of the space charge field.

Photorefractive materials based on conjugated polymers have shown potential to present a faster photorefractive response. This is because the conjugated polymers intrinsically possess much larger carrier mobilities than all reported photorefractive polymers due to the delocalized electron distribution along the conjugated bonds. Functionalized photorefractive conjugated polymers based on polyphenylenethiophenes have been reported [116,122]. A TBC gain of 5.9 cm^{-1} was measured in a fully functionalized polymer under a zero external electric field. Li and coworkers [124,125] reported on the investigation of a conjugated copolymer of thiophene derivatives (Fig. 26.13). The second-order NLO property of this polymer arises upon poling of the pendant NLO moieties and the guest photo-cross-linkable NLO molecules. The response time of the photocurrent was observed to be on the order of 0.1 s in the presence of a photosensitizer. A plasticizer was added to this polymer to enhance the poling efficiency. The TBC gain of this composite material was measured to be 24.5 cm^{-1}. The plasticizer, however, might have been responsible for the unexpectedly long photorefractive response time of 15 s in this composite material.

Progress in the research on polymers for photorefractive applications during the past 4 years has been encouraging. In addition to the material properties addressed above, however, photorefractive polymers must demonstrate other properties such as good optical quality, reproducibility, stability, processibility, and long grating

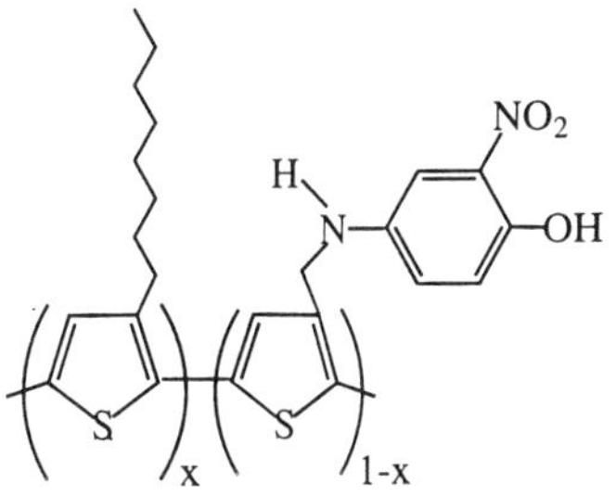

Fig. 26.13 Chemical structure of a thiophene–thiophene derivative copolymer. NLO chromophores are covalently attached to the conjugated polymer.

VII. CHROMOPHORE RELAXATION

The noncentrosymmetric alignment of the NLO chromophores in a polymer induced by the poling process is essential for materials to exhibit bulk second-order NLO properties. However, the noncentrosymmetric dipolar orientation of the NLO moieties is not thermodynamically favored. The poled order tends to relax to a random orientation in the absence of the poling field. Since the ability to control the relaxation process is the key to the development of stable second-order NLO materials, the study and understanding of relaxation behavior have become very important. A number of techniques have been used to study the degree of chromophore orientation in poled polymers and their relaxation behaviors. These techniques include SHG [58,152,127], EO measurement [35], UV-visible spectroscopy [36], dielectric measurement [71,72,128], dynamic mechanical spectroscopy [128], infrared spectroscopy [129], thermally stimulated discharge current technique [130], and waveguiding experiments [131]. Among these techniques, SHG is the most widely used because it directly monitors the decay of the poled order.

As discussed earlier, the SHG coefficient, which is proportional to the square root of the second harmonic (SH) intensity, is directly related to the alignment of the NLO dye molecules with respect to the poling field. Once the field is removed, this poled order starts to decay as the dipoles relax and lose their orientation. This leads to a decrease in the SH intensity. A plot of SH intensity as a function of time has been used to study the relaxation of oriented NLO dye dispersed in amorphous polymers. Hampsch et al. [126] reported the effect of temperatures, both above and below the T_g, on the dipole rotation and polymer relaxation by observing the decay of the SH intensity. The same group found [132] that the samples that were aged during corona poling showed improved temporal stability of the SHG signal. This was due to a decrease in the local free volume and mobility of the dipoles as a result of physical aging [132,133]. Another study [134] suggested that the relaxation of the poled order is attributed to two relaxation mechanisms. The fast decay behavior is associated with the effect of reorientation of the NLO dipole and the third-order optical nonlinearity. The slow decay is due to the rotational diffusion of the NLO dipole.

The decay of NLO response for a series of guest–host systems as a function of time at different temperatures was studied using the SHG technique [58,135]. These polymer samples were prepared based on a number of polymers and NLO chromophores. The relaxation of second-order optical nonlinearity can be reasonably well described by the Kohlrausch–Williams–Watts (KWW) equation,

$$\frac{d_{\text{eff}}(t)}{d_{\text{eff}}(0)} = e^{-\left(\frac{t}{\tau}\right)^{\beta}}, \qquad 0 < \beta < 1 \quad (8)$$

The parameter β is a measure of the width of the distribution of the relaxation times about some central value. $\beta = 1$ corresponds to the case of simple exponential decay. d_{eff} is an effective SHG coefficient. The temperature dependence of the relaxation time τ has been shown to be described by an equation of the type

$$\tau = \tau_0 \exp[B/(T_g - T)] \quad (9)$$

This equation is similar to the Williams–Landel–Ferry (WLF) expression to describe the viscoelastic behavior of polymers above T_g. To improve the correlation between τ and T_g, especially when the temperature is close to T_g, it was suggested that one can plot $\log \tau$ vs. $1/(T_g + 50 - T)$ instead of $1/(T_g - T)$. The fitting parameters B and τ_0 have been empirically determined. A master curve (see the dashed line in Fig. 26.14) can be derived from this study where the relaxation time of guest–host polymers can be predicted based on their T_g values. For example, by assuming similar values of the fitting parameters, one can estimate that for relatively stable operation over a period of many years at 100°C, the polymer must have a T_g of at least 200°C in a standard guest–host system. The relaxation of side-chain polymers has also been studied in a similar fashion [135]. It was found that the relaxation behavior of guest–host polymers and that of side-chain polymers are very similar when the temperature is below and near T_g. The decay of side-chain polymers starts to deviate from the master curve of the guest–host systems when the temperature is below and far (approximately 50°C or more) from T_g.

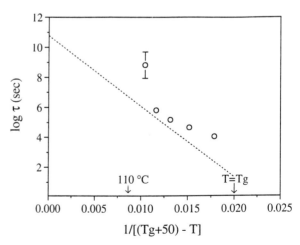

Fig. 26.14 (o) Relaxation time τ as a function of $1/[(T_g + 50) - T]$ for IPN samples at 130, 140, 150, 160, and 170°C. The dashed line is the master curve that represents the relaxation behavior of typical guest–host systems.

This master curve established a general trend between the relaxation time and T_g of polymers for various guest–host and side-chain polymers. In these polymers the relaxation process is dominated solely by the T_g of the polymer. However, according to the cross-linked system BPAZO–APAN described in Section VI [81], cross-linked networks contributed significantly to the enhancement of the temporal stability of NLO response. Therefore, in addition to the T_g of a polymeric material, the degree of cross-linking in the material may be equally important for stabilizing the poled order. The decay of SH intensity as a function of time and temperature of an NLO polymer based on an IPN has also been investigated [136]. Figure 26.14 shows such a plot for the IPN system (open circles). The dashed line refers to the master curve that was established for guest–host systems. The results show that the relaxation behavior of the IPN is significantly different from those of the guest–host and side-chain polymers. The stability of the d_{eff} coefficient for the IPN samples at low temperatures as well as temperatures close to T_g is substantially superior to that of the guest–host samples. The decay of nonlinearity of the IPN system at temperatures close to T_g is inherently a slower process than that of the guest–host and side-chain systems. This behavior suggests that a cross-linked network will further enhance the stability of the poled order during the high temperature spikes that may be encountered during device fabrication.

VIII. PROPERTY CONSIDERATIONS AND OPTIMIZATION

In addition to the linear and nonlinear optical properties as well as the temporal stability, several other material properties have to be optimized to match the requirements for different applications. Second-order NLO materials that will be used in different types of devices are required to possess specific key properties according to their applications. A promising material will have to possess properties that satisfy both the manufacturing and end-use requirements.

Polymers for applications such as EO modulation must maintain a substantial degree of poled order and associated bulk nonlinearity (e.g., EO coefficient > 30 pm/V at 830 nm has been suggested) at processing temperatures that may exceed 250°C for short periods of time as well as at operating temperatures of up to 80°C for a few years. Long-term stability at 100°C for more than 1000 h has been demonstrated in a number of polymeric systems [137]. However, the ability to withstand short-term excursions to higher temperatures while the polymer still exhibits large optical nonlinearity is still a challenge. In addition, the materials should have facile processability and low optical losses at wavelengths of interest (<1 dB/cm at 830 nm) [138,139].

Second-order NLO polymers for frequency conversion offer the possibility of obtaining a blue laser for optical data storage and retrieval. It has been suggested that materials for frequency-doubling applications should have $\chi^{(2)}$ values on the order of 60 pm/V [137]. These materials must be transparent at the wavelengths of interest, i.e., the chromophores must not absorb light at both the fundamental and SH frequencies. For practical applications, the conversion efficiency has to be improved significantly. Materials also require good processibility for incorporation into a device configuration.

To date, there has been no commercially available device made of second-order NLO polymers whereas some inorganic materials are already being used in frequency doubling of lasers and EO devices. Prototype devices made of polymeric materials for frequency doubling and EO devices are already being studied, and the properties of NLO polymers are continuously being improved. The rapid progress in materials development in recent years is a cause for optimisms and it is expected that polymeric material with strong and stable NLO properties will be made available in the near future. However, optimization of other properties such as optical quality, processibility, and reproducibility is considered to be equally important. In addition, problems associated with device design and fabrication will still need to be addressed and resolved before these polymers can find use in realistic applications.

REFERENCES

1. G. D. Stucky, M. L. F. Phillips, and T. E. Gier, *Chem. Mater. 1*:492 (1989).
2. P. N. Prasad and D. J. Williams, *Introduction to Nonlinear Optical Effects in Molecules and Polymers.* Wiley, New York, 1991.
3. M. Thakur and S. Tripathy, in *Encyclopedia of Polymer Science and Engineering*, (J. I. Kroschwitz, ed.), 2nd ed., Wiley, New York, 1986, Vol. 5, pp. 756–771.
4. S. Tripathy, E. Cavicchi, J. Kumar, and R. S. Kumar, *Chemtech 19*:620 (1989).
5. S. Tripathy, E. Cavicchi, J. Kumar, and R. S. Kumar, *Chemtech 19*:747 (1989).
6. D. F. Eaton, *Science 253*:281 (1991).
7. D. F. Eaton, *Chemtech 22*:308 (1992).
8. J. F. Nicoud and R. J. Twieg, in *Nonlinear Optical Properties of Organic Molecules and Crystals*, (D. S. Chemla and J. Zyss, eds.), Academic, Orlando, FL, 1987, Vol. 1, Chap. II-3.
9. S. Miyata (ed.), *Nonlinear Optics: Fundamentals, Materials and Devices*, Elsevier, New York, 1992.
10. J. Zyss (ed.), *Molecular Nonlinear Optics: Materials, Physics, and Devices*, Academic, San Diego, CA, 1994.
11. K. D. Singer, J. E. Sohn, and S. J. Lalama, *Appl. Phys. Lett. 49*:248 (1986).
12. S. J. Lalama and A. F. Garito, *Phys. Rev. A 20*:1179 (1979).

13. K. D. Singer, J. E. Sohn, L. A. King, and H. M. Gordon, *J. Opt. Soc. Am. B* 6:339 (1989).
14. K. Y. Wong and A. K.-Y. Jen, *J. Appl. Phys.* 75:3308 (1994).
15. C. C. Teng and A. F. Garito, *Phys. Rev. B* 28:6766 (1983).
16. K. Clay and A. Persoons, *Phys. Rev. Lett.* 66:2980 (1991).
17. J. L. Oudar and J. Zyss, *Phys. Rev. A* 26:2016 (1982).
18. A. K.-Y. Jen, V. P. Rao, K. Y. Wong, and K. J. Drost, *J. Chem. Soc., Chem. Commun.* 1993:90.
19. B. A. Reinhardt, R. Kannan, J. C. Bhatt, J. Zieba, and P. N. Prasad, *Polym. Prepr., Am. Chem. Soc. Div. Polym. Chem.* 35:166 (1994).
20. S. R. Marder, L.-T. Cheng, B. G. Tiemann, A. C. Friedli, M. Blanchard-Desce, J. W. Perry, and J. Skindhøj, *Science* 263:511 (1994).
21. D. R. Kanis, M. A. Ratner, and T. J. Marks, *Chem. Rev.* 94:195 (1994).
22. R. Twieg, V. Lee, R. D. Miller, C. Moylan, R. B. Prime, and G. Chiou, *Polym. Prepr., Am. Chem. Soc. Div. Polym. Chem.* 35:200 (1994).
23. G. Roberts (ed.), *Langmuir–Blodgett Films*, Plenum, New York, 1990.
24. I. R. Girling, N. A. Cade, P. V. Kolinsky, J. D. Earls, G. H. Gross, and I. R. Peterson, *Thin Solid Films* 132:101 (1988).
25. B. L. Anderson, J. M. Hoover, G. Lindsay, B. G. Higgins, P. Stroeve, and S. T. Kowel, *Thin Solid Films* 179:413 (1989).
26. T. Takahashi, P. Miller, Y. M. Chen, L. Samuelson, D. Galotti, B. K. Mandal, J. Kumar, and S. K. Tripathy, *J. Polym. Sci., Polym. Phys.* 31:165 (1993).
27. P. Hodge, Z. Ali-Adib, D. West, and T. King, *Macromolecules* 26:1789 (1993).
28. G. J. Ashwell, Y. Gongda, D. Lochun, and P. D. Jackson, *Polym. Prepr., Am. Chem. Soc. Div. Polym. Chem.* 35:185 (1994).
29. C. Bosshard, A. Otomo, G. I. Stegeman, M. Küpfer, M. Flörsheimer, and P. Günter, *Appl. Phys. Lett.* 64:2076 (1994).
30. D. Li, B. I. Swanson, J. M. Robinson, and M. A. Hoffbauer, *J. Am. Chem. Soc.* 115:6975 (1993).
31. D. Li, M. A. Ratner, T. B. Marks, C. H. Zhang, J. Yang and G. K. Wong, *J. Am. Chem. Soc.* 112:7389 (1990).
32. P. M. Lundquist, S. Yitzchaik, T. Zhang, D. R. Kanis, M. A. Ratner, T. J. Marks, and G. K. Wong, *Appl. Phys. Lett.* 64:2194 (1994).
33. S. K. Tripathy, W. H. Kim, B. Bihari, D. W. Cheong, and J. Kumar, *Mater. Res. Soc. Symp. Proc.* 328:433 (1994).
34. W. H. Kim, B. Bihari, R. A. Moody, N. B. Kodali, J. Kumar, and S. K. Tripathy, *Macromolecules* 28:642 (1995).
35. J. W. Wu, J. F. Valley, S. Ermer, E. S. Binkley, J. T. Kenney, G. F. Lipscomb, and R. Lytel, *Appl. Phys. Lett.* 58:225 (1991).
36. M. A. Mortazavi, A. Knoesen, S. T. Kowel, B. G. Higgins, and A. Dienes, *J. Opt. Soc. Am. B* 6:733 (1989).
37. M. Eich, H. Looser, D. Y. Yoon, R. J. Twieg, G. C. Bjorklund, J. Baumert, *J. Opt. Soc. Am. B* 6:1590 (1989).
38. F. Charra, F. Kajzar, J. M. Nunzi, P. Raimond, and E. Idiart, *Opt. Lett.* 18:941 (1993).
39. D. A. Kleinman, *Phys. Rev.* 126:1977 (1962).
40. J. Jerphagnon and S. Kurtz, *J. Appl. Phys.* 41:1667 (1970).
41. J. Jerphagnon and S. Kurtz, *Phys. Rev. B.* 1:1739 (1970).
42. K. D. Singer, M. G. Kuzyk, and J. E. Sohn, *J. Opt. Soc. Am. B* 4:968 (1987).
43. M. Sigelle and R. Hierle, *J. Appl. Phys.* 52:4199 (1981).
44. R. Meyrueix, J. Lecomte, and G. Tapolsky, *Nonlinear Opt.* 1:201 (1991).
45. H. Uchiki and T. Kobayashi, *J. Appl. Phys.* 64:2625 (1988).
46. X. Zhu, Y. Chen, L. Li, R. Jeng, B. Mandal, J. Kumar, and S. Tripathy, *Opt. Commun.* 88:77 (1992).
47. M. Dumont, Y. Levy, and D. Morichere, in *Organic Molecules for Nonlinear Optics and Photonics* (J. Messier, ed.), Kluwer, Boston, 1991, p. 194.
48. D. Morichere, V. Dentan, F. Kajzar, P. Robin, Y. Levy, and M. Dumont, *Opt. Commun.* 74:69 (1989).
49. C. C. Teng and H. T. Man, *Appl. Phys. Lett.* 56:1734 (1990).
50. J. S. Schildkraut, *Appl. Opt.* 29:2839 (1990).
51. K. Clays and J. Schildkraut, *J. Opt. Soc. Am. B* 8:2274 (1992).
52. G. R. Meredith, J. G. VanDusen, and D. J. Williams, *Macromolecules* 15:1385 (1982).
53. K. D. Singer, M. G. Kuzyk, W. R. Holland, J. E. Sohn, S. J. Lalama, R. B. Comizzoli, H. E. Katz, and M. L. Schilling, *Appl. Phys. Lett.* 53:1800 (1988).
54. L. M. Hayden, G. F. Sauter, F. R. Ore, P. L. Pasillas, J. M. Hoover, G. A. Lindsay, and R. A. Henry, *J. Appl. Phys.* 68:456 (1990).
55. R. Levenson, J. Liang, E. Toussaere, A. Carenco, and J. Zyss, *Proc. SPIE* 1560:251 (1991).
56. B. K. Mandal, Y. M. Chen, R. J. Jeng, T. Takahashi, J. C. Huang, J. Kumar, and S. Tripathy, *Eur. Polym. J.* 27:735 (1991).
57. S. Ermer, J. F. Valley, R. Lytel, G. F. Lipscomb, T. E. Van Eck, and D. G. Girton, *Appl. Phys. Lett.* 61:2272 (1992).
58. M. Stähelin, C. A. Walsh, D. M. Burland, R. D. Miller, R. J. Twieg, and W. Volksen, *J. Appl. Phys.* 73:8471 (1993).
59. A. K.-Y. Jen, K. Y. Wong, V. P. Rao, K. Drost, Y. M. Cai, B. Caldwell, and R. M. Mininni, *Mater. Res. Soc. Symp. Proc.* 328:413 (1994).
60. S. F. Hubbard, K. D. Singer, F. Li, S. Z. D. Cheng, and F. W. Harris, *Appl. Phys. Lett.* 65:265 (1994).
61. H. E. Katz, K. D. Singer, J. E. Sohn, C. W. Dirk,

L. A. King, and H. M. Gordon, *J. Am. Chem. Soc. 109*:6561 (1987).
62. S. Matsumoto, K. Kubodera, T. Kurihara, and T. Kaino, *Appl. Phys. Lett. 51*:1 (1987).
63. Y. M. Chen, M. Rahman, T. Takahashi, B. Mandal, J. Lee, J. Kumar, and S. Tripathy, *Jpn. J. Appl. Phys. 30*:672 (1991).
64. M. Ahlheim and F. Lehr, *Makromol. Chem. 195*:361 (1994).
65. G. D. Green, H. K. Hall, J. E. Mulvaney, J. Noonan, and D. J. Williams, *Macromolecules 20*:716 (1987).
66. H. E. Katz and M. L. Schilling, *J. Am. Chem. Soc. 111*:7554 (1989).
67. F. Fuso, A. Padias, and H. K. Hall, *Macromolecules 24*:1710 (1991).
68. J. D. Stenger-Smith, J. W. Fischer, R. A. Henry, J. M. Hoover, M. P. Nadler, R. A. Nissan, and G. A. Lindsay, *J. Polym. Sci. A 29*:1623 (1991).
69. G. A. Lindsay, J. D. Stenger-Smith, R. A. Henry, J. M. Hoover, R. A. Nissan, and K. J. Wynne, *Macromolecules 25*:6075 (1992).
70. M. E. Wright and S. Mullick, *Macromolecules 25*:6045 (1992).
71. D. Jungbauer, I. Teraoka, D. Y. Yoon, B. Reck, J. D. Swalen, R. J. Twieg, and C. G. Willson, *J. Appl. Phys. 69*:8011 (1991).
72. I. Teraoka, D. Jungbauer, B. Reck, D. Y. Yoon, R. Twieg, and C. G. Willson, *J. Appl. Phys. 69*:2568 (1991).
73. M. W. Becker, L. S. Sapochak, R. Ghosen, C. Xu, L. R. Dalton, Y. Shi, W. H. Steier, and A. K.-Y. Jen, *Chem. Mater. 6*:104 (1994).
74. C. R. Moylan, R. J. Twieg, V. Y. Lee, and R. D. Miller, *Proc. SPIE 2285*:17 (1994).
75. W. Sotoyama, S. Tatsuura, and T. Yoshimura, *Appl. Phys. Lett. 64*:2197 (1994).
76. D. Yu, A. Gharavi, and L. Yu, *Macromolecules 28*:784 (1995).
77. M. A. Hubbard, T. J. Marks, J. Yang, and G. K. Wong, *Chem. Mater. 1*:167 (1989).
78. M. Eich, B. Reck, D. Y. Yoon, C. G. Willson, and G. C. Bjorklund, *J. Appl. Phys. 66*:3241 (1989).
79. D. Jungbauer, B. Reck, R. Twieg, D. Y. Yoon, C. G. Willson, and J. D. Swalen, *Appl. Phys. Lett. 56*:2610 (1990).
80. J. Park, T. J. Marks, J. Yang, and G. K. Wong, *Chem. Mater. 2*:229 (1990).
81. R. J. Jeng, Y. M. Chen, J. Kumar, and S. Tripathy, *J. Macromol. Sci., Pure Appl. Chem. A29*:1115 (1992).
82. Y. Shi, W. H. Steier, M. Chen, L. Yu, and L. R. Dalton, *Appl. Phys. Lett. 60*:2577 (1992).
83. M. Chen, L. R. Dalton, L. P. Yu, Y. Q. Shi, and W. H. Steier, *Macromolecules 25*:4032 (1992).
84. L. Yu, W. Chan, and Z. Bao, *Macromolecules 25*:5609 (1992).
85. J. T. Lin, M. A. Hubbard, T. J. Marks, W. Lin, and G. K. Wong, *Chem. Mater. 4*:1148 (1992).
86. G. Tapolsky, J.-P. Lecomte, and R. Meyrueix, *Macromolecules 26*:7383 (1993).
87. G. H. Hsiue, J. K. Kuo, R. J. Jeng, J. I. Chen, X. L. Jiang, S. Marturunkakul, J. Kumar, and S. K. Tripathy, *Chem. Mater. 6*:884 (1994).
88. D. R. Robello, C. S. Willand, M. Scozzafava, A. Ullman, and D. J. Williams, in *Materials for Nonlinear Optics: Chemical Perspectives* (S. R. Marder, J. E. Sohn, and G. D. Stucky, eds.), ACS Symp. Ser. 455, American Chemical Society, Washington, DC, 1991, p. 279.
89. C. Xu, B. Wu, L. R. Dalton, Y. Shi, P. M. Ranon, and W. H. Steier, *Macromolecules 25*:6714 (1992).
90. B. K. Mandal, J. Kumar, J. Huang, and S. Tripathy, *Makromol. Chem. Rapid Commun. 12*:63 (1991).
91. R. J. Jeng, Y. M. Chen, B. K. Mandal, J. Kumar, and S. K. Tripathy, *Mater. Res. Soc. Symp. Proc. 247*:111 (1992).
92. Y. Shi, P. M. Ranon, W. H. Steier, C. Xu, B. Wu, and L. R. Dalton, *Appl. Phys. Lett. 63*:2168 (1993).
93. C. J. Brinker and G. W. Scherer, *Sol-Gel Science*, Academic, Orlando, FL, 1990.
94. M. Nakamura, H. Nasu, and K. Kamiya, *Non-Cryst. Solids 135*:1 (1991).
95. Y. Zhang, P. N. Prasad, and R. Burzynski, *Chem. Mater. 4*:851 (1992).
96. G. Pircetti, E. Toussaere, I. Ledoux, and J. Zyss, *Polym. Prepr., Am. Chem. Soc. Div. Polym. Chem. 32*:61 (1991).
97. W. E. Torruellas, D. Neher, R. Zanoni, G. I. Stegeman, F. Kajzar, and M. Leclerc, *Chem. Phys. Lett. 175*:11 (1990).
98. Y. Kobayashi, S. Muto, A. Matsuzaki, and Y. Kurokawa, *Thin Solid Films 213*:126 (1992).
99. J. Kim, J. L. Plawsky, R. LaPeruta, and G. M. Korenowski, *Chem. Mater. 4*:249 (1992).
100. C. Claude, B. Garetz, Y. Okamoto, and S. K. Tripathy, *Mater. Lett. 14*:336 (1992).
101. S. Kalluri, Y. Shi, W. H. Steier, Z. Yang, C. Xu, B. Wu, and L. R. Dalton, *Appl. Phys. Lett. 65*:2651 (1994).
102. R. J. Jeng, Y. M. Chen, A. K. Jain, S. K. Tripathy, and J. Kumar, *Opt. Commun. 89*:212 (1992).
103. R. J. Jeng, Y. M. Chen, A. Jain, J. Kumar, and S. K. Tripathy, *Chem. Mater. 4*:972 (1992).
104. R. J. Jeng, Y. M. Chen, J. I. Chen, J. Kumar, and S. K. Tripathy, *Macromolecules 26*:2530 (1993).
105. R. J. Jeng, Y. M. Chen, A. Jain, J. Kumar, and S. K. Tripathy, *Chem. Mater. 4*:1141 (1992).
106. S. Marturunkakul, J. I. Chen, L. Li, R. J. Jeng, J. Kumar, and S. K. Tripathy, *Chem. Mater. 5*:592 (1993).
107. S. Marturunkakul, J. I. Chen, L. Li, X. L. Jiang, R. J. Jeng, J. Kumar, and S. K. Tripathy, *Mater. Res. Soc. Symp. Proc. 328*:541 (1994).
108. S. Marturunkakul, J. I. Chen, L. Li, X. L. Jiang, R. J. Jeng, S. K. Sengupta, J. Kumar, and S. K. Tripathy, *Polym. Prepr., Am. Chem. Soc. Div. Polym. Chem. 35*:134 (1994).

109. M. A. Mortazavi, H. N. Yoon, and C. C. Teng, *J. Appl. Phys. 74*:4871 (1993).
110. X. Zhu, Y. M. Chen, M. Kamath, R. J. Jeng, J. Kumar, and S. K. Tripathy, *Mol. Cryst. Liq. Cryst. Sci. Technol. Sec. B 4*:175 (1993).
111. M. Kamath, C. E. Masse, R. J. Jeng, M. Cazeca, X. L. Jiang, J. Kumar, and S. K. Tripathy, *J. Macromol. Sci., Pure Appl. Chem. A31*:2011 (1994).
112. H. S. Nalwa, T. Watanabe, A. Kakuta, A. Mukoh, and S. Miyata, *Appl. Phys. Lett. 62*:3223 (1993).
113. M. Szablewski, G. H. Cross, and J. Cole, *Polym. Prepr., Am. Chem. Soc. Div. Polym. Chem. 35*:174 (1994).
114. Topics Appl. Phys. P. Günter and J.-P. Huignard (eds.), *Photorefractive Materials and Their Applications I*. Vol. 61, Springer-Verlag, Berlin, 1988.
115. S. Ducharme, J. C. Scott, R. J. Twieg, and W. E. Moerner, *Phys. Rev. Lett. 66*:1846 (1991).
116. W. E. Moerner and S. M. Silence, *Chem. Rev. 94*:127 (1994) and references therein.
117. M. Liphardt, A. Goonesekera, B. E. Jones, S. Ducharme, J. M. Takacs, and L. Zhang, *Science 263*:367 (1994).
118. W. E. Moerner, S. M. Silence, F. Hache, and G. C. Bjorklund, *J. Opt. Soc. Am. B11*:320 (1994).
119. Y. Zhang, C. A. Spencer, S. S. Ghosal, M. K. Casstevens, and R. Burzynski, *J. Appl. Phys. 76*:671 (1994).
120. K. Meerholz, B. L. Volodin, Sandalphon, B. Kippelen, and N. Peyghambarian, *Nature 371*:497 (1994).
121. M. E. Orczyk, B. Swedek, J. Zieba, and P. N. Prasad, *J. Appl. Phys. 76*:4995 (1994).
122. W.-K. Chan, Y. M. Chen, A. Peng, and L. Yu, *J. Am. Chem. Soc. 115*:11735 (1993).
123. Z. Peng, Z. Bao, and L. Yu, *J. Am. Chem. Soc. 116*:6003 (1994).
124. L. Li, K. G. Chittibabu, J. Kumar, and S. K. Tripathy, *Proc. SPIE 2042*:376 (1993).
125. K. G. Chittibabu, L. Li, Z. Chen, J. Kumar, and S. K. Tripathy, *MSE Proceedings 73*:475 (1995).
126. H. L. Hampsch, J. Yang, G. K. Wong, and J. M. Torkelson, *Macromolecules 23*:3640 (1990).
127. K. Singer, *Polym. Prepr., Am. Chem. Soc. Div. Polym. Chem. 32(2)*:98 (1991).
128. S. Marturunkakul, J. I. Chen, R. J. Jeng, S. Sengupta, J. Kumar, and S. K. Tripathy, *Chem. Mater. 5*:743 (1993).
129. J. I. Chen, S. Marturunkakul, Y. M. Chen, R. J. Jeng, J. Kumar, and S. K. Tripathy, *Eur. Polym. J. 30*:1357 (1994).
130. W. Köhler, D. R. Robello, P. T. Dao, C. S. Willand, and D. J. Williams, *J. Chem. Phys. 93*:9157 (1990).
131. R. H. Page, M. C. Jurich, B. Reck, A. Sen, R. J. Twieg, J. D. Swalen, G. C. Bjorklund, and C. G. Willson, *J. Opt. Soc. Am. B7*:1230 (1990).
132. H. L. Hampsch, J. Yang, G. K. Wong, and J. M. Torkelson, *Macromolecules 23*:3648 (1990).
133. G. A. Lindsay, R. A. Henry, J. M. Hoover, A. Knoesen, and M. A. Mortazavi, *Macromolecules 25*:4888 (1992).
134. T. Goodson and C. H. Wang, *Macromolecules 26*:1837 (1993).
135. C. A. Walsh, D. M. Burland, V. Y. Lee, R. D. Miller, B. A. Smith, R. J. Twieg, and W. Volksen, *Macromolecules 26*:3720 (1993).
136. J. I. Chen, S. Marturunkakul, L. Li, R. J. Jeng, J. Kumar, and S. K. Tripathy, *Macromolecules 26*:7379 (1993).
137. D. M. Burland, R. D. Miller, and C. A. Walsh, *Chem. Rev. 94*:31 (1994).
138. R. Lytel and G. F. Lipscomb, *Mater. Res. Soc. Symp. Proc. 247*:17 (1992).
139. D. M. Burland, *Chem. Rev. 94*:1 (1994).

27

The Influence of Charge-State Incorporation on the Nonlinear Optical Properties of Conjugated Polyenes

Charles W. Spangler*
Northern Illinois University, DeKalb, Illinois

I. INTRODUCTION

During the past decade there have been dramatic advances in the design of small molecules, oligomers, and polymers with large optical nonlinearities [1–3]. The development of design criteria has depended, to a large extent, on detailed structure–property relationships that have explored such structural features as (1) conjugation length dependence [4–6], (2) mesomeric electron-donating and -withdrawing influences [4,7,8], (3) the influence of aromatic versus heteroaromatic rings [9–11], (4) π-system dimensionality [9], (5) charge-state incorporation in the conjugation sequence [12,13], and, most recently, (6) the effect of bond length alternation (BLA) features on polarization [14–16]. Several of these topics were been reviewed in detail in an issue of *Chemical Reviews* dedicated to the topic of optical nonlinearities in chemistry [17]. This chapter focuses on the influence of charge-state incorporation, either polaron-like radical ions or bipolaron-like diions, on third-order optical nonlinearity in small model polyenes, polymer guest–host composites, and formal copolymers in which one of the repeat units is a polyene or arylene-vinylene segment of known conjugation length [18].

II. NONLINEAR OPTICAL PROPERTIES OF POLYENES

The impetus for the surge in research in nonlinear optical (NLO) materials in recent years has been supplied by the rapid developments in the new field of photonics and speculation on how best to envision the requirements for optical switching, frequency modulation, waveguiding, and eventually practical all-optical computing. A detailed introduction and discussion of the theoretical underpinnings of this new field are beyond the scope of this chapter, and the previously cited reviews of this emerging field should be consulted for more detailed discussion [1–3,17]. In brief, at the molecular level the microscopic polarization of a material resulting from an external laser-induced field can be expressed as a series expansion,

$$p(E) = \alpha E + \beta EE + \gamma EEE + \cdots \quad (1)$$

where α is the linear polarizability and β and γ the first and second hyperpolarizabilities, respectively. In a bulk material such as a crystal or polymer film, the nonlinearity that originates from the polarization response of the electrons can be represented as

$$P(E) = \chi^{(1)}E + \chi^{(2)}EE + \chi^{(3)}EEE + \cdots \quad (2)$$

where the linear optical susceptibility $\chi^{(1)}$ and the nonlinear susceptibilities $\chi^{(n)}$ are tensor quantities. At the molecular level, molecules must be noncentrosymmetric for $\beta \neq 0$. In bulk materials (crystals, polymer films, etc.) the ensemble must also be noncentrosymmetric. In recent years, rapid strides have been made in the design of new materials with large $\mu\beta$ values, where μ represents the dipole moment, and high $\chi^{(2)}$ polymers wherein the required noncentrosymmetry is induced by electric field poling have been produced with ever-increasing thermal and temporal stability. However, progress in the design of molecules possessing large γ values and/or high $\chi^{(3)}$ polymers has not kept pace with that of $\chi^{(2)}$ materials. The balance of this section concentrates on recent efforts to improve third-order nonlinear optical response.

Current affiliation: Montana State University, Bozeman, Montana.

A. Conjugation Length Dependence of γ and $\chi^{(3)}$

Early theoretical studies of the relationship between γ and organic structural parameters recognized that the extent of electron delocalization in easily polarized π-electron clouds was a key factor in designing molecules and polymers with enhanced third-order optical nonlinearity. For example, it was predicted that $\chi^{(3)}$ should be proportional to the reciprocal of the band gap raised to the sixth power [19]. Thus, any structural change resulting in a red shift in oscillator strength could result in increased $\chi^{(3)}$. One method of increasing γ is therefore simply to increase the length of the π-conjugation sequence. However, various investigators have shown that γ/N, where N is the number of repeat units, does not increase continuously and will eventually level off with increasing N. Beratan et al. [20] showed that γ increases rapidly for *trans*-polyenes as conjugation increases to 10–15 repeat units and then more slowly up to 40 repeat units. This suggests that very long conjugation sequences (e.g., a fully conjugated high molecular weight π-electron polymer) may not be required to maximize third-order NLO response.

Hurst and coworkers [21] also calculated second hyperpolarizability tensors via ab initio coupled-perturbed Hartree–Fock theory for a series of polyenes up to $C_{22}H_{24}$. They found that γ_{xxxx} was proportional to chain length, with a power dependence of 4.0, but that this dependence tapered off as N increased. Garito and coworkers [22] calculated a power law dependence of γ_{xxxx} on chain length on the order of 4.6 ± 0.2. They also suggested that large $\chi^{(3)}$ values could be obtained with conjugation sequences of intermediate length (≈ 100 Å). Prasad concurs with the conclusion that γ/N levels off with increasing N. In addition, Prasad and coworkers measured γ for a series of polythiophene and poly(*p*-phenylene) oligomers by degenerate four-wave mixing (DFWM) in solution and found a power law dependence for γ of 4. However, in the former series $\chi^{(3)}$ begins to level off at the hexamer level, whereas in the latter case $\chi^{(3)}$ levels off at the terphenyl level, $N = 3$ [23,24]. Prasad concluded that for NLO purposes the concept of *effective conjugation length*, which might be less than the overall length of the conjugation sequence, may be an important criterion for determining the effective $\chi^{(3)}$ in any system. Thus it is important to consider structural features that tend to distort the π system and disrupt the conjugation. For example, the twisting out of plane between successive phenyl rings in the polyphenylene oligomers is probably responsible for the leveling off of $\chi^{(3)}$ at the terphenyl level.

Pucetti and coworkers [6] carried out a detailed study of both symmetrically (donor–donor and acceptor–acceptor) and asymmetrically (donor–acceptor) substituted polyenes and found very large values of γ for the polyenes bearing strong electron donor groups with no saturation of γ up to ≈ 40 Å. More recently, Samuel and coworkers [25] reported a study of γ as a function of chain length in very long chain model polyene oligomers with up to 240 double bond repeat units synthesized by living polymerization techniques (**1**). They observed saturation of γ/N, but at much longer chain lengths than predicted by theory. In their study γ/N does not saturate until $N \approx 120$ [25]. At the current time, it is not clear why saturation in this system is seen at chain lengths considerably longer than predicted by theory.

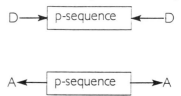

1

B. Mesomeric Substituent Effects

While the effect of mesomerically interactive electron donor (D) and acceptor (A) groups on β and $\chi^{(2)}$ has been well documented [1,2], there were few comprehensive studies on the corresponding effect on γ and $\chi^{(3)}$ until the past few years. Reinhardt, Prasad, and coworkers have shown that alkoxy group substitution on rigid rod model compounds significantly increase γ [26]. Prasad had also previously observed increases in γ when acceptor groups were incorporated on the terminal thiophene units of polythiophene oligomers [27]. What is most intriguing about such observations is that γ can be increased by either supplying additional electron density through donor groups or withdrawing electron density toward the ends of the π sequence via acceptor groups:

D⟶ | p-sequence | ⟵D

A⟵ | p-sequence | ⟶A

Until recently, relatively few D–A substituted π-electron systems had been studied systematically to determine their combined effect on γ. Garito and coworkers [22] had suggested that such substitution on polyene chains would give rise to a new type of virtual excitation process that becomes allowed when centrosymmetry is broken by the asymmetric substitution pattern. Since this new process involves a diagonal matrix element dependent on the dipole difference between ground and excited states, we can expect that γ_{xxxx} will be dramatically enhanced by the D–A pairing. As is the case with β, the greatest γ enhancement will come from the pairing of the strongest D and A groups, illustrated in Scheme

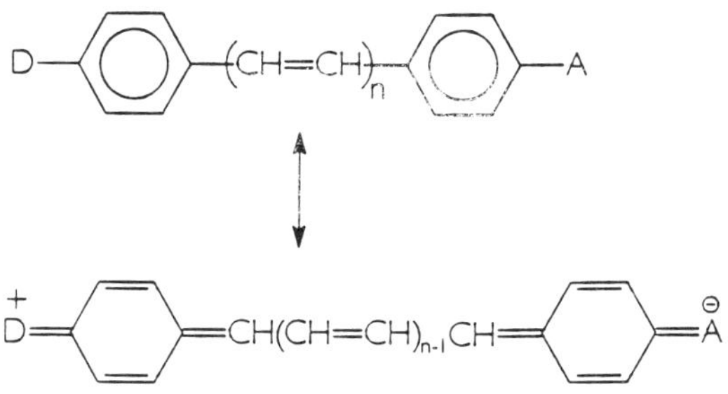

Scheme 1 D-A pairing for a typical D,A-substituted diphenyl polyene.

1 for a typical D,A-substituted diphenyl polyene. Note, however, that such delocalization requires both of the aromatic rings to assume quinoid character. More will be said about this in later sections of this chapter.

In Table 27.1, γ values for a variety of D–D, A–A, and D–A pairings in both diphenyl and dithienyl polyenes of varying conjugation lengths are presented. As can be seen from these data, γ increases with increasing conjugation length irrespective of the D–D, A–A, or D–A pairing. Also, γ increases as the strength of the D or A group increases. Finally, for comparable D and A groups and conjugation lengths, γ increases in the order D–A > A–A > D–D of interaction. For these relatively small molecules, this confirms the prediction of Garito et al. [22] that D–A-substituted polyenes should have the largest γ. However, more recent calculations by Meyers and Brédas [28], while confirming the ordering of substituent effects for short conjugation sequences ($N < 5$), predict that D–D substitution will lead to larger γ values than A–A substitution. Even more recently, Puccetti et al. [6] showed experimentally that for long polyene sequences, for example **2** and **3**, ($N = 10$), γ increases the fastest when both end groups are donors. These workers estimated that the critical length for the reversal of the relative magnitudes of D–D and A–A γ values occurs between $N = 5$ and $N = 6$ (where N is estimated to be n, the number of C=C repeat units, plus 1.5 for each phenyl ring).

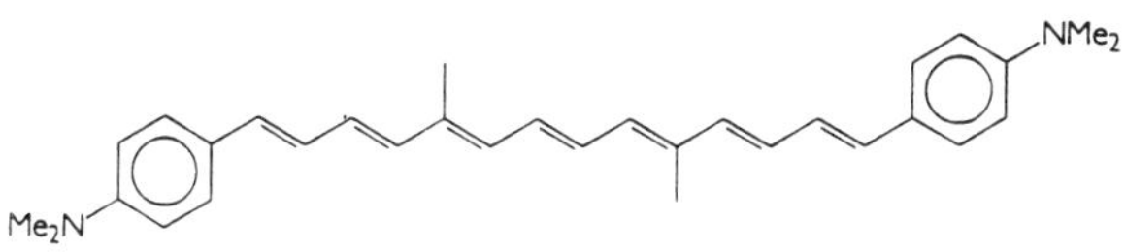

2

γ_{SHG} 12000 ± 1200 @ 1.3 μm

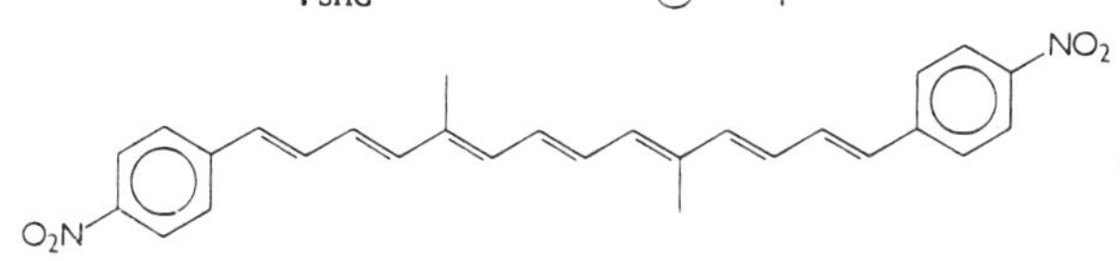

3

γ_{SHG} 3800 ± 600 @ 1.3 μm

C. Influence of Aromatic vs. Heteroaromatic End Groups

In the design of NLO chromophores with large $\mu\beta$, the influence of the choice of aromatic end group can be significant (see, e.g., **4** vs. **5**). Aromatic heterocyclic five-membered rings have six π electrons spread over

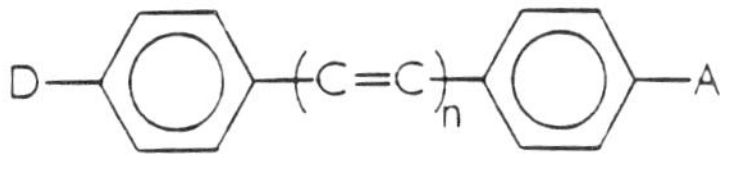

4

vs.

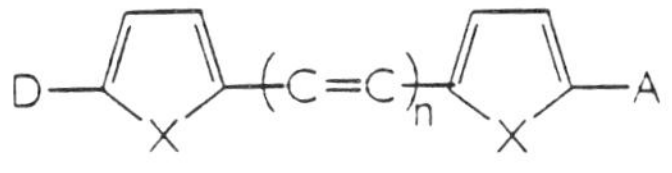

5

five atoms instead of six as in phenyl rings, thus giving an excess of electron density at each carbon. Jen and coworkers at EniChem America [29,30] showed for simple second-order NLO chromophores that substitution of a thiophene ring for phenyl has a dramatic effect on $\mu\beta$. A few selected examples (see Table 27.2) illustrate this point. Prasad and Reinhardt [9] demonstrated similar effects for third-order nonlinearity (Table 26.2). In general, it would appear that future design criteria for molecules and/or polymers with high $\chi^{(3)}$ should consider incorporating heteroaromatic end groups or repeat units to maximize the optical nonlinearity.

D. π-System Dimensionality

Most of the π-conjugation sequences discussed so far are quasi-one-dimensional systems whose third-order nonlinearity is dominated by the tensor component corresponding to the π-chain sequence, γ_{XXXX}. Prasad and coworkers [26] showed that increasing system dimensionality in model rigid-rod materials (**19, 20**) from quasi-one dimensional to quasi-two-dimensional π networks can increase γ by an order of magnitude for a given conjugation length.

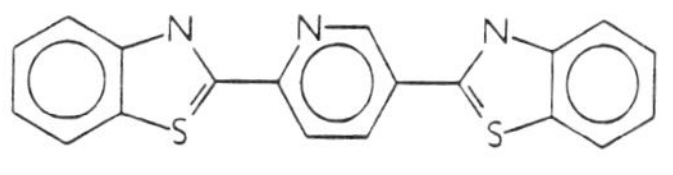

$\gamma = 1.9 \times 10^{-34}$ esu
19

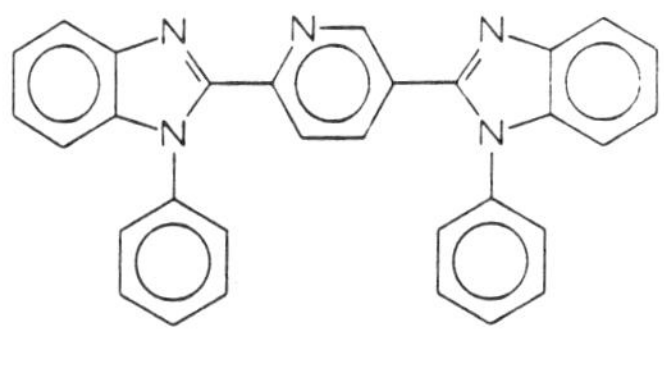

$\gamma = 1.1 \times 10^{-33}$ esu
20

Table 27.1 Second Hyperpolarizability (γ) Values for Substituted Diphenyl and Dithienyl Polyenes

D—⟨C₆H₄⟩—(CH=CH)$_n$—⟨C₆H₄⟩—A
4 a-h

Compound	n	D	A	$\gamma(10^{-36}$ esu$)^a$	Ref.
4a	1	MeO	CN	54	4
4b	2	MeO	CN	122	4
4c	3	MeO	CN	234	4
4d	1	MeO	NO$_2$	93	4
4e	2	MeO	NO$_2$	130	4
4f	3	MeO	NO$_2$	230	4
4g	1	MeS	NO$_2$	206	4
4h	1	Me$_2$N	NO$_2$	225	4

X—⟨C₆H₄⟩—(CH=CH)$_n$—⟨C₆H₄⟩—Y
6 a-l

Compound	n	X	Y	$\gamma(10^{-36}$ esu$)^b$	Ref.
6a	3	C$_8$H$_{17}$O	C$_8$H$_{17}$O	218	5
6b	4	C$_8$H$_{17}$O	C$_8$H$_{77}$O	251	5
6c	3	Me$_2$N	Me$_2$N	252	5
6d	4	Me$_2$N	Me$_2$N	307	5
6e	2	NO$_2$	NO$_2$	158	5
6f	3	NO$_2$	NO$_2$	486	5
6g	4	NO$_2$	NO$_2$	651	5
6h	5	NO$_2$	NO$_2$	1384	5
6i	1	C$_8$H$_{17}$O	NO$_2$	158	5
6j	2	C$_8$H$_{17}$O	NO$_2$	305	5
6k	3	C$_8$H$_{17}$O	NO$_2$	530	5
6l	4	C$_8$H$_{17}$O	NO$_2$	831	5

X—⟨thienyl⟩—(CH=CH)$_n$—⟨thienyl⟩—Y
7 a-e

Compound	n	X	Y	$\gamma(10^{-36}$ esu$)^b$	Ref.
7a	4	C$_{10}$H$_{21}$S	C$_{10}$H$_{21}$S	450 ± 60	7
7b	5	C$_{10}$H$_{21}$S	C$_{10}$H$_{21}$S	400 ± 150	7
7c	6	C$_{10}$H$_{21}$S	C$_{10}$H$_{21}$S	500 ± 250	7
7d	3	C$_{10}$H$_{21}$S	C$_{10}$H$_{21}$SO$_2$	250 ± 50	7
7e	4	C$_{10}$H$_{21}$S	C$_{10}$H$_{21}$SO$_2$	300 ± 50	7

[a] THG measurements at 1.9 μm in CHCl$_3$
[b] THG measurements at 1.9 μm in THF; C$_8$H$_{17}$ = 2-ethyl-1-hexyl, C$_{10}$H$_{21}$ = 1-decyl.

Table 27.2 Comparison of Effect of Aromatic vs. Heteroaromatic Structures on $\mu\beta$ and γ

Compound		$\mu\beta(10^{-48}$ esu$)^a$	Ref.
8	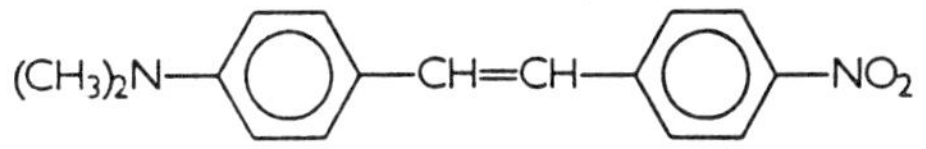 (CH₃)₂N—⟨phenyl⟩—CH=CH—⟨phenyl⟩—NO₂	580	29
9	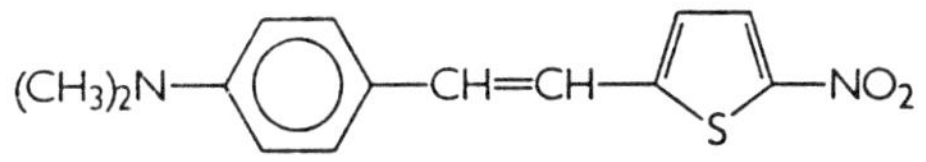 (CH₃)₂N—⟨phenyl⟩—CH=CH—⟨thiophene⟩—NO₂	600	29
10	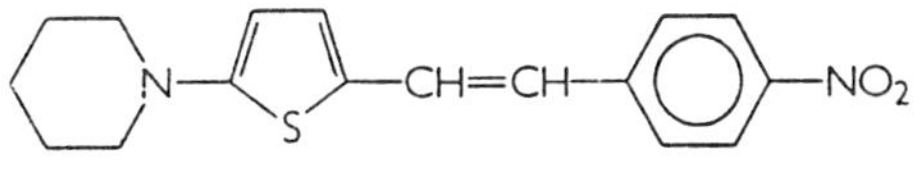 piperidinyl—⟨thiophene⟩—CH=CH—⟨phenyl⟩—NO₂	660	29
11	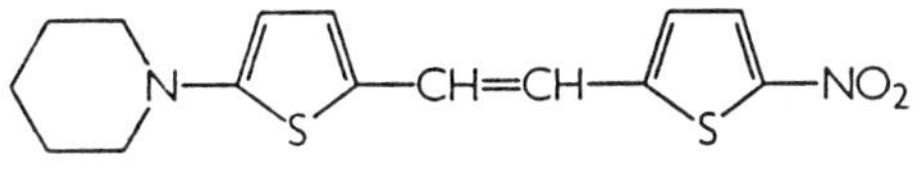 piperidinyl—⟨thiophene⟩—CH=CH—⟨thiophene⟩—NO₂	1040	29
12	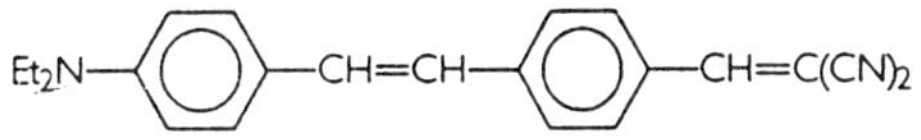 Et₂N—⟨phenyl⟩—CH=CH—⟨phenyl⟩—CH=C(CN)₂	1100	11
13	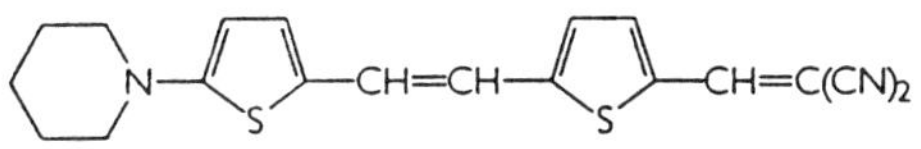 piperidinyl—⟨thiophene⟩—CH=CH—⟨thiophene⟩—CH=C(CN)₂	2600	11

Compound		$\gamma(10^{-36}$ esu$)$	Ref.
14	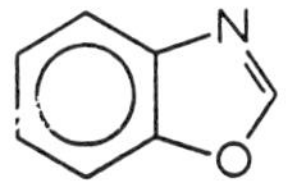 benzoxazole	16	9
15	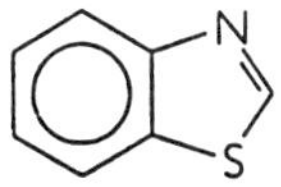 benzothiazole	26	9
16	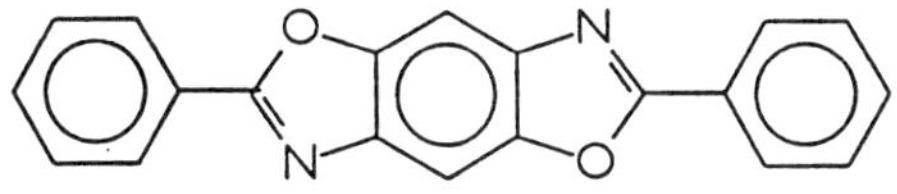 phenyl-bisoxazole-phenyl	71	26
17	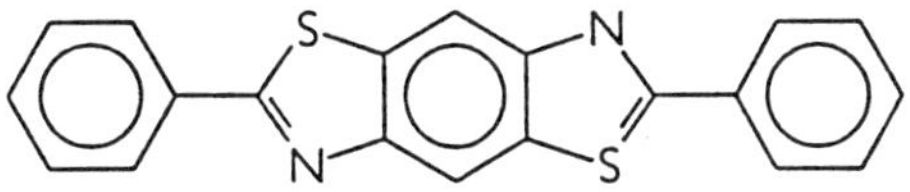 phenyl-bisthiazole-phenyl	210	26
18	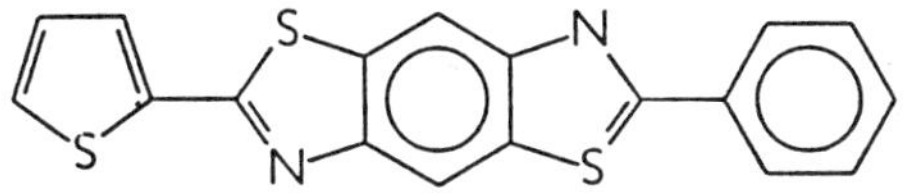 thienyl-bisthiazole-phenyl	1100	26

^{a} SHG at 1.907 μm in dioxane.

Phthalocyanines are another example of quasi-two-dimensional systems with the added advantage of variation of the central metal atoms and derivatization of the outer phenyl rings. Shirk et al. [30] measured γ for a number of derivatized phthalocyanines, obtaining γ values in the range $1 \times 10^{-32} \times 1 \times 10^{-31}$ esu and a corresponding $\chi^{(3)}$ of 2×10^{-10} esu for the Pt-containing material. At the current time, however, this approach to enhancing the optical nonlinearity of polyene systems has not been exploited.

E. Charge-State Incorporation in the Conjugation Sequence

DeMelo and Silbey calculated [31–33] the effect of charge state incorporation on γ_{xxxx} in linear conjugated polyene chains. They considered both the effect of solitons on odd-numbered chains and that of polarons (either $+ \cdot$ or $- \cdot$) and bipolarons ($+ +$ or $- -$) on even-numbered chains. Using a perturbative density matrix treatment described by a PPP Hamiltonian, they found a power law dependence of 6.3 and 6.1 for P and BP incorporation, respectively, of the γ^π, the orientational average (6.6 and 6.0 for γ_{xxxx}). The largest γ values were for the incorporation of P charge states, whose values are negative. The sign of γ for BP incorporation is positive. More recently, Birge and coworkers [34] calculated the effect of incorporation of BP charge states on γ_{xxxx} for an oligomeric series of disubstituted diphenyl polyenes. Using an INDO-PSDCI finite perturbation procedure, which includes 200 single and 400 double excitations, they also predicted BP enhancement of γ_{xxxx} but with a lower power dependency than that calculated by deMelo and Silbey [33]. Birge's calculation shows a difference in power dependency between γ_{xxxx}(neutral) and γ_{xxxx}(BP) of 0.4, which corresponds to a threefold increase in γ_{xxxx} for a 20-carbon chain. Recent experimental determination of γ_{xxxx}(neutral) and γ_{xxxx}(BP) in bisanthracenyl polyenes [12] and a triphenodithiazine model [13] have confirmed that bipolaronic enhancement of $\chi^{(3)}$ occurs, but measurement under nonresonant conditions is difficult.

F. Bond-Length Alternation Considerations in Relation to Maximizing Molecular Hyperpolarizabilities

Marder and coworkers developed new insight into the design of molecules with enhanced optical nonlinearity by consideration of the degree of bond-length alternation (BLA) that is desired in order to maximize either β or γ in conjugated molecules [14–16,35,36]. Consider the changes that occur in the structures **21** and **22**. Measurement of γ_{THG} at 1.907 μm in solvents ranging from CCl_4 to CH_3OH indicated that γ was highly solvent-dependent [37]. For highly bond-length-alternated molecules ($\Delta r \sim 0.1$ Å), such as most polyenes, γ is positive. However, as Δr decreases, γ first goes through a positive maximum and then decreases, passing through 0, before showing a large negative value at $\Delta r \sim 0$. Compound **22** exhibits the solvent dependence of γ as follows: -135×10^{-36} esu in CH_3OH, -120×10^{-36} in CH_3CN, -50×10^{-36} in CH_2Cl_2, -25×10^{-36} in dioxane, and $+40 \times 10^{-36}$ in CCl_4. Compound **21**, however, shows quite different behavior: $+73 \times 10^{-36}$ esu in CH_3OH, $+113$ in CH_3CN, $+105$ in CH_2Cl_2, and $+40$ in dioxane. This is consistent with a highly BLA polyene structure for **21** in nonpolar solvents and a cyanine-like structure for **22**, with little BLA in polar solvents. These results are totally consistent with an electric field dependence of hyperpolarizability and point to the conclusion that

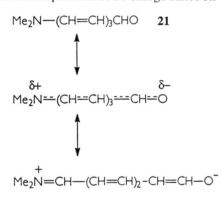

BLA is an important design criterion in attempts to maximize third-order nonlinearity.

III. GENERATION OF POLARON-LIKE RADICAL CATIONS AND BIPOLARON-LIKE DICATIONS IN POLYMER OLIGOMERS

A. Phenyl and Anthracenyl End-Capped Conjugated Polymers

Although polyacetylene may be considered to be an ideal one-dimensional polymer for theoretical study, in many ways it is not ideal to study experimentally because of its lack of processability and the difficulty of attaching electron-donating and -withdrawing substituents. Polyacetylene oligomers are also difficult to study, due to their ease of polymerization via oxidative cross-linking in air. Phenyl end groups stabilize polyacetylene toward air oxidation, and diphenyl polyenes up to $n = 8$ (23) are well defined and easily characterized crystalline solids. Polyene sequences longer than the octaene are

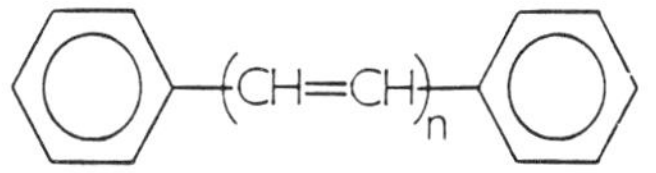

Ph—(CH=CH)$_n$—Ph $n = 1$-8 **23**

extremely insoluble in all common solvents. Polyacetylene can be oxidatively doped with iodine; however, the diphenyl oligomers have higher oxidation potentials. α,ω-Diaryl polyenes can be synthesized by either Wittig or Horner–Emmons–Wadsworth methodology from either mono- or bisphosphoranes. The bisphosphorane methodology is particularly suited to the synthesis of symmetrically substituted polyenes, while asymmetrically substituted D–A polyenes are usually produced via single condensation from monophosphoranes. These approaches are outlined in Scheme 2.

Oxidation of diphenylpolyenes can be carried out with either $FeCl_3$ or $SbCl_5$ in CH_2Cl_2 solution [39–41]. In all cases an immediate bleaching of the parent π–π^* absorption is observed with the simultaneous appearance of new red-shifted absorption bands that can be assigned to either polaron-like radical cations (**34**) or bipolaron-like dications (**35**), as illustrated in Scheme 3. In some but not all cases, the transitory polaron can be observed spectroscopically, but in the presence of an excess of dopant the bipolaron is formed exclusively. Other aromatic (e.g., anthracenyl) or heteroaromatic (e.g., thienyl) end groups yield similar results. The dications are significantly stabilized by electron-donating substituents, and this stabilization parallels the electron-donating ability of the substituent groups:

$R_2N > RS > RO > R >$ halogen $> H$

While the BP formed from diphenyloctatetraene is stable for several hours at room temperature, the p,p'-N,N-dimethylamino substituted tetraene is stable for several days in solutions in contact with air. The formation of the P $(+\cdot)$ and BP $(++)$ charge states for a variety of substituted diphenylpolyenes is reviewed in Ref. 18, where the origin of the observed transitions is discussed in detail. Dianthracenylpolyenes form extremely stable bipolaronic dications that are stable for several months in solution or as optically transparent polycarbonate composites [7,12]. This dramatic stability enhancement has been rationalized on the basis of aromatic stabilization of the quinoid intermediate and is illustrated in Scheme 4.

In general, electron-donating substituent stabilization of the BP dications results in dramatic shifts in oscillator strength toward longer wavelength. In the case of longer conjugation lengths or increased stabilization due to aromatic stabilization (anthracenyl), this shift can extend to the near-infrared (NIR). The spectral shifts observed during the formation of P $(+\cdot)$ and BP $(++)$ are shown in Table 27.3. In both the diphenyl- and dianthracenylpolyene series, the longer polyenes are quite insoluble at room temperature, even in DMF, for RO or R_2N substituents. Long-chain alkyl substituents (octyl, decyl, etc.) do not improve the solubility of the longer polyenes significantly. However, branched-chain R groups, such as 2-ethylhexyl, do improve solubility dramatically. Unfor-

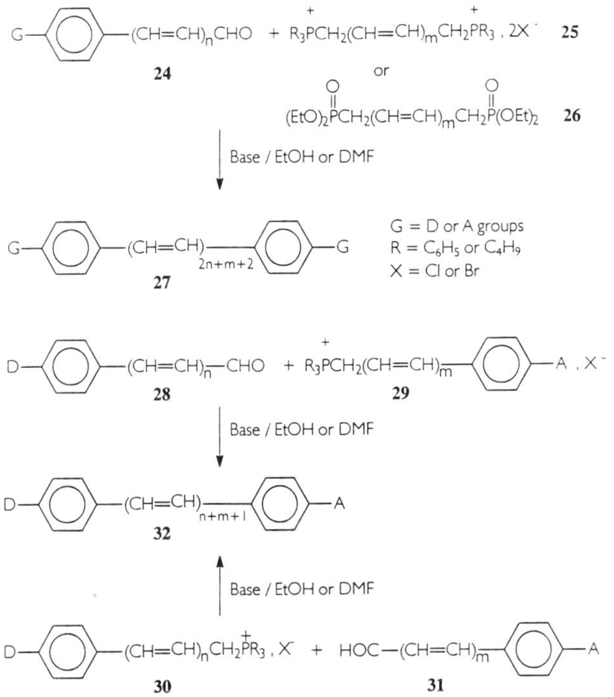

Scheme 2 Synthesis of D-D, A-A, and D-A substituted phenyl end-capped polyenes.

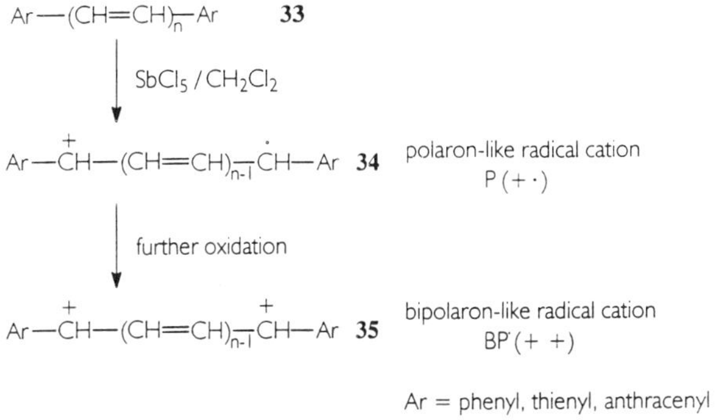

Scheme 3 Oxidative doping of diaryl polyenes.

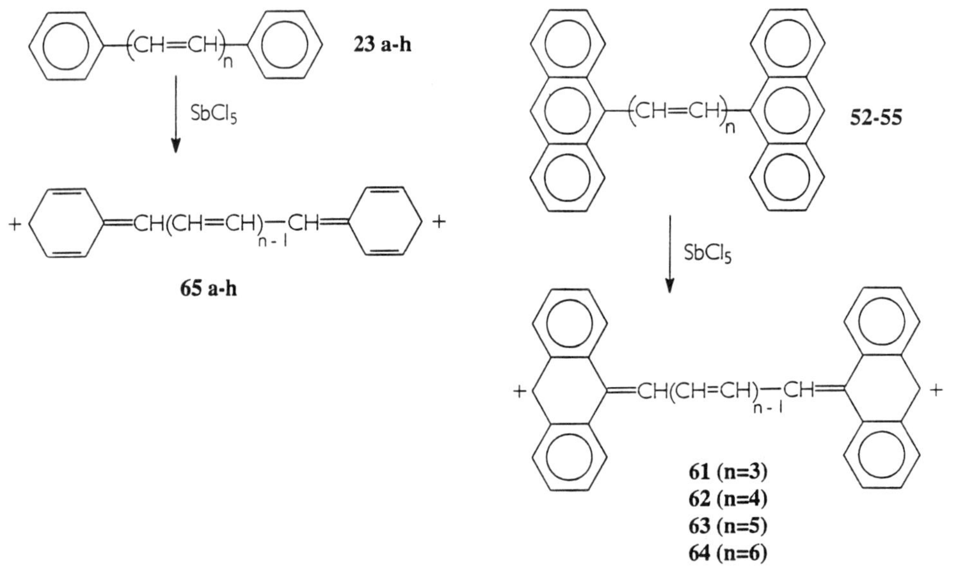

Scheme 4 Bipolaron-like dictations formed from diphenyl vs. dianthracenyl polyenes.

tunately they also increase the susceptibility of these materials to irreversible air oxidation, probably via oxidative cross-linking through the formation of free radicals at the C–H tertiary center. These materials degrade at room temperature, even in the absence of light, to yield bleached insoluble material. Probably the most significant result in these studies is the greatly increased stability of the dianthracenylpolyenes with respect to their potentially enhanced third-order nonlinearity.

B. Thienyl End-Capped Conjugated Polyenes

Dithienylpolyenes have some inherent advantages over the diphenylpolyene series. First, dithienylpolyenes are more soluble in common organic solvents, and the ease of incorporating long-chain alkyl substituents in the β or β' positions (see structure) has been shown to greatly enhance solubility in the poly(3-alkylthiophenes) [42]. Dithienylpolyenes can be synthesized by Wittig methodology similar to that previously described for the diphenylpolyene series to yield polyenes substituted in the α positions of the terminal rings [43–45].

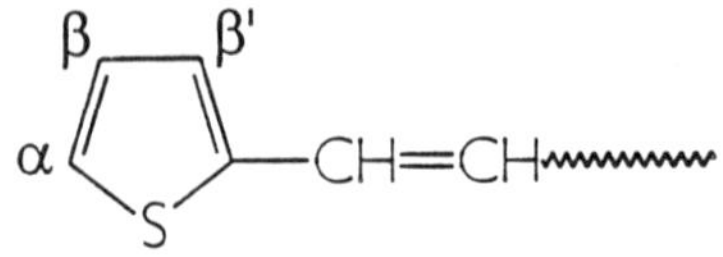

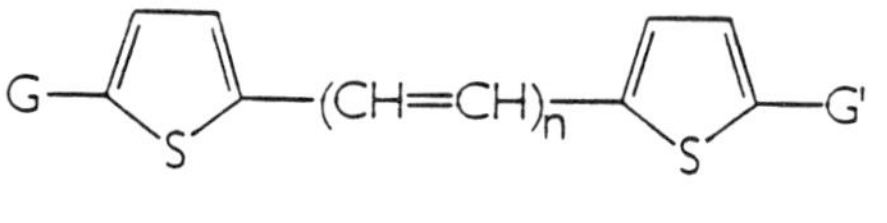

G = H, CH_3, CH_3S, CH_3O, n-$C_{10}H_{21}S$

Bipolaron formation in these series is similar to that described previously for the corresponding diphenyl series. However, when G and G' are n-$C_{10}H_{21}S$, the increased solubility and additional resonance stabilization provided by the substituent and ring sulfur atoms allows extremely stable bipolarons to be produced up to the octamer level. This is illustrated in Scheme 5. At the

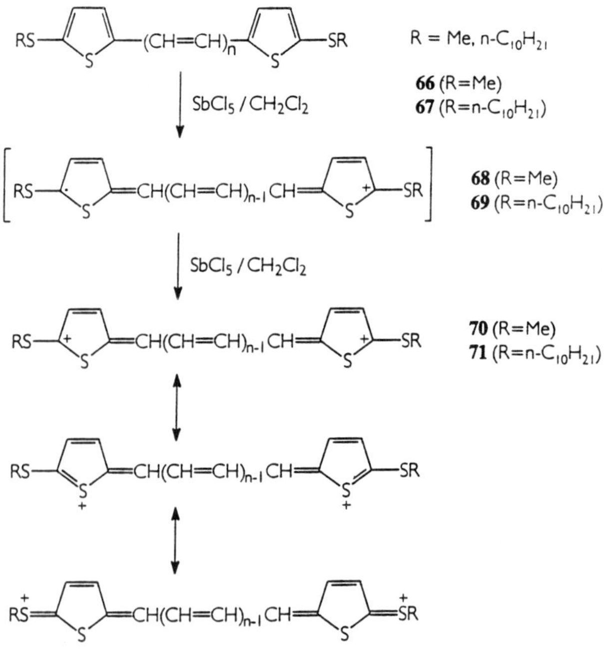

R = Me, n-$C_{10}H_{21}$

66 (R=Me)
67 (R=n-$C_{10}H_{21}$)

68 (R=Me)
69 (R=n-$C_{10}H_{21}$)

70 (R=Me)
71 (R=n-$C_{10}H_{21}$)

Scheme 5 Resonance stabilization of dithienyl polyene bipolarons stabilized by alklthio substituents.

Table 27.3 P(+ •) and BP(+ +) Formation in Diphenyl and Dianthracenyl Polyene

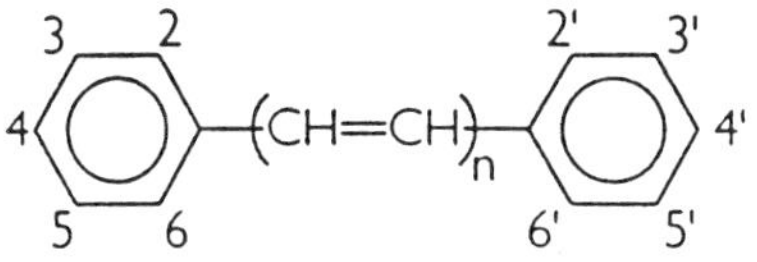

Substituent	No.	n	λ_{max} π–π* (nm)[a]	λ_{max} P(+•) (nm)	λ_{max} BP(++) (nm)
None	23e	5	374, <u>394</u>, 418	[717][b]	612, <u>564</u>
None	23f	6	393, <u>414</u>, 438	[770][b]	615, <u>685</u>
4,4'-(OMe)$_2$	36	4	370, <u>390</u>, 413	<u>740</u>, 927, 1113	593, 647, <u>700</u>
4,4'-(OMe)$_2$	37	5	388, <u>410</u>, 435	<u>797</u>, 1073, 1200	627, 692, <u>755</u>
4,4'-(OMe)$_2$	38	6	402, <u>426</u>, 452	<u>853</u>, 1175, 1300	680, 741, <u>818</u>
4,4'-(NMe$_2$)$_2$	39	4	425	—[c]	613, <u>667</u>, 723
4,4'-(NMe$_2$)$_2$	40	5	<u>445</u>, 470	—[c]	646, <u>713</u>, 773
4,4'-(NMe$_2$)$_2$	41	6	<u>458</u>, 485	—[c]	700, 748, <u>833</u>
4,4'-(F)$_2$	42	5	380, <u>398</u>, 427	[720][b]	567, <u>615</u>
4,4'-(F)$_2$	43	6	398, <u>408</u>, 438	[<u>733</u>, 1123][b]	687, <u>727</u>
4,4'-(Cl)$_2$	44	5	385, <u>405</u>, 430	[727][b]	567, <u>622</u>
4,4'-(Cl)$_2$	45	6	400, <u>422</u>, 450	<u>787</u>, 1120, 1240	630, <u>687</u>
4,4'-(Br)$_2$	46	5	381, <u>401</u>, 431	<u>740</u>, 1033, 1127	587, <u>640</u>
4,4'-(Br)$_2$	47	6	396, <u>419</u>, 446	<u>780</u>, 1113, 1273	640, <u>693</u>
4,4'-(SMe)$_2$	48	3	<u>385</u>, 403	592, <u>743</u>	733, <u>815</u>
4,4'-(SMe)$_2$	49	4	<u>400</u>, 422	808, 985, 1217	777, <u>869</u>
4,4'-(SMe)$_2$	50	5	<u>417</u>, 443	855, 1083, 1338	823, <u>920</u>
4,4'-(SMe)$_2$	51	6	<u>433</u>, 461	<u>904</u>, 1189, 1400	864, <u>966</u>

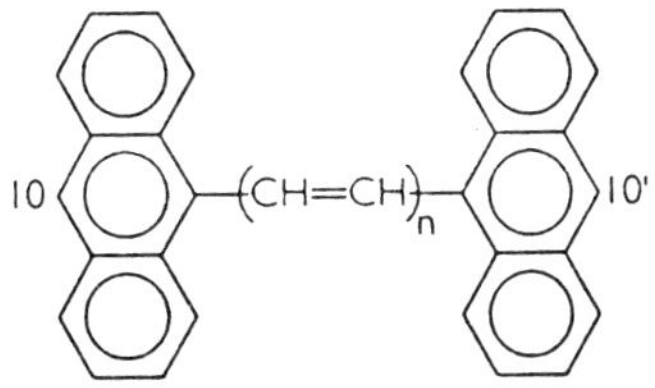

Substituent	No.	n	λ_{max} π–π* (nm)[a]	λ_{max} P(+•) (nm)	λ_{max} BP(++) (nm)
None	52	3	413	—[c]	878
None	53	4	423	—[c]	940
None	54	5	428	—[c]	978
None	55	6	438	—[c]	1000
10,10'-(n-C$_6$H$_{13}$)$_2$	56	3	429	—[c]	927
10,10'-(n-C$_6$H$_{13}$)$_2$	57	4	438	—[c]	951
10,10'-(OC$_{10}$H$_{21}$)$_2$	58	3	436	—[c]	728
10,10'-(SC$_{10}$H$_{21}$)$_2$	59	3	435	—[c]	832
10,10'-(OC$_{10}$H$_{21}$)$_2$	60	4	444	—[c]	886

Note: Underlined peaks represent peaks of maximum absorption.
[a] CH$_2$Cl$_2$ solution.
[b] Absorption spectra decay to BP(+ +) very fast; only unambiguous assignable absorption.
[c] Not observed on spectrometer scanning time scale.
Source: Portions reprinted from Ref. 71.

time this research was accomplished, the octaene was the longest model compound in which the BP delocalization could be reasonably said to extend over the length of the polyene (26 atoms).

Spangler and He [46,47] synthesized a series of 3,4-dibutylthienyl end-capped polyenes that are extremely soluble in a wide variety of solvents, including even hexane. Bipolarons for this series have been obtained up to the decamer level; however, even though polyenes containing more than 10 double bonds can be synthesized in this series, both the neutral and bipolaron forms begin to lose stability. It has been suggested that once the decaene limit has been exceeded, the stabilizing influence of the thienyl end groups declines as the molecules become more polyene (polyacetylene)-like [46,47]. The nonlinear optical properties of this series of polyenes, in both neutral and doped forms, are quite extraordinary and are discussed in greater detail in Section IV. The synthesis of this polyene series is illustrated in Scheme 6.

The decaene bipolaron delocalization in these series extends over a maximum of 30 atoms and is the longest model polyene yet studied. The absorption characteristics for both the neutral and oxidatively doped dithienylpolyenes are listed in Table 27.4. As can readily be seen in the more recent studies [47], the bipolaron absorption has been shifted into the near-infrared for the decaene oligomer. What is equally interesting, and more important for potential applications in nonlinear optics, these bipolarons are exceptionally stable. Solutions of the doped polyenes show little decline in optical absorption even after several months, thus approximating the stability achieved in the dianthracenylpolyenes but with increased processibility. Spectra for both neutral and doped hexaene oligomers (**72** and **73**) for these polyene series are illustrated in Figs. 27.1 (X = H) and 27.2 (X = BuS).

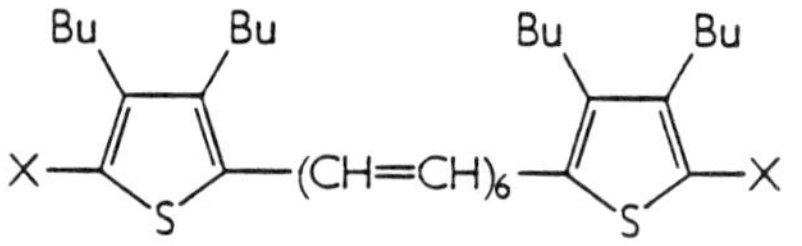

X—[thienyl(Bu,Bu)]—(CH=CH)$_6$—[thienyl(Bu,Bu)]—X   X = H, BuS

72 (X = H)
73 (X = BuS)

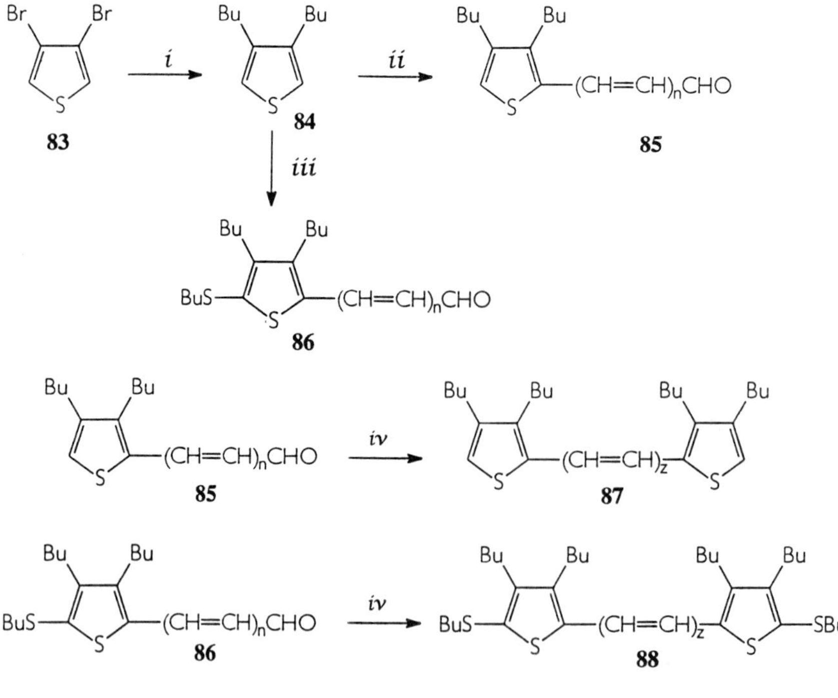

Scheme 6 Preparation of dithienyl polyenes incorporating solubilizing substituents. *i*, BuMgBr/Ni(dppp)Cl$_2$; *ii*, [dioxolane]—CH$_2^+$PBu$_3$ Br$^-$/NaOEt/DMF; 10% HCl/THF; repeat *n* times; *iii*, BuLi/TMEDA; S; BuLi; then *ii*; *iv*, Bu$_3^+$PCH$_2$(CH—OH)$_n$CH$_2^+$PBU$_3$, 2X$^-$/NaOEt/EtOH.

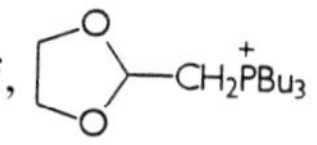

Table 27.4 P(+•) and BP(++) Formation in Dithienylpolyenes

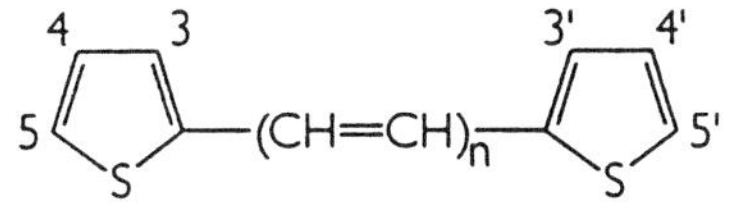

Substituent	n	λ_{max} π–π* (nm)[a]	λ_{max} P(+•) (nm)	λ_{max} BP(++) (nm)
None	5	<u>416</u>, 443	<u>705</u>, 797, 1084	653, <u>713</u>
None	6	<u>432</u>, 461	<u>760</u>, 853, 1154	713, <u>776</u>
5,5'-(Me)$_2$	5	<u>425</u>, 450	[808][b]	660, <u>710</u>
5,5'-(Me)$_2$	6	<u>441</u>, 469	<u>888</u>, 1167, 1580	720, <u>776</u>
3,3'-(Me)$_2$	5	<u>422</u>, 449	[795][b]	650, <u>699</u>
3,3'-(Me)$_2$	6	<u>440</u>, 469	<u>881</u>, 1154, 1574	713, <u>769</u>
5,5'-(OMe)$_2$	3	<u>398</u>, 420	[<u>731</u>, 1076][b]	<u>520</u>, (545)[d]
5,5'-(OMe)$_2$	4	<u>416</u>, 442	<u>806</u>, 1103, 1299	<u>577</u>, (605)[d]
5,5'-(SMe)$_2$	3	404	<u>792</u>, 1009, 1240	<u>610</u>, (643)[d]
5,5'-(SMe)$_2$	4	<u>422</u>, 444	<u>859</u>, 1114, 1348	<u>663</u>, (702)[d]
5,5'-(SMe)$_2$	5	<u>434</u>, 460	—[c]	<u>717</u>, (775)[d]
5,5'-(SMe)$_2$	6	<u>451</u>, 478	—[c]	<u>773</u>, (835)[d]
5,5'-(SC$_{10}$H$_{21}$)$_2$	7	<u>462</u>, 493	—[c]	<u>834</u>, (905)[d]
5,5'-(SC$_{10}$H$_{21}$)$_2$	8	<u>475</u>, 507	—[c]	<u>884</u>, (970)[d]
3,4,3',4'-(Bu)$_4$	3	422, <u>399</u>, 380	—[c]	593, <u>655</u>
3,4,3',4'-(Bu)$_4$	4	443, <u>418</u>, 396	—[c]	600, <u>661</u>
3,4,3',4'-(Bu)$_4$	5	462, <u>435</u>, 412	—[c]	679, <u>715</u>
3,4,3',4'-(Bu)$_4$	6	480, <u>450</u>, 426	—[c]	<u>719</u>, 809
3,4,3',4'-(Bu)$_4$	7	496, <u>464</u>, 439	—[c]	790, <u>849</u>
3,4,3',4'-(Bu)$_4$	8	510, <u>477</u>, 450	—[c]	855, <u>914</u>
3,4,3',4'-(Bu)$_4$	9	521, <u>489</u>, 461	—[c]	892, <u>971</u>
3,4,3',4'-(Bu)$_4$	10	534, <u>499</u>, 471	—[c]	950, <u>1022</u>
3,4,3',4'-(Bu)$_4$; 5,5'-(BuS)$_2$	3	424	—[c]	643
3,4,3',4'-(Bu)$_4$; 5,5'-(BuS)$_2$	4	439	—[c]	657
3,4,3',4'-(Bu)$_4$; 5,5'-(BuS)$_2$	5	451	—[c]	705
3,4,3',4'-(Bu)$_4$; 5,5'-(BuS)$_2$	6	466	—[c]	754
3,4,3',4'-(Bu)$_4$; 5,5'-(BuS)$_2$	7	476	—[c]	801
3,4,3',4'-(Bu)$_4$; 5,5'-(BuS)$_2$	8	488	—[c]	849

Note: Underlined peaks represent peaks of maximum absorption.
[a] CH$_2$Cl$_2$ solution.
[b] Absorption spectra decay to BP(++) very fast; only unambiguous assignable absorption.
[c] Not observed on spectrometer scanning time scale.
[d] Absorptions shown in parentheses represent shoulders.
Source: Portions reprinted from Ref. 71.

C. Arylene-Vinylene Oligomers

Arylene vinylene polymers, such as poly(p-phenylenevinylene) (PPV) and poly(2,5-thienylenevirylene) (PTV) have been studied extensively over the past 10 years due to the relative ease of producing high quality thin films via the soluble precursor route [48–52]. PPVs in particular have been shown to be true multifunctional materials, displaying enhanced third-order NLO properties as well as applications as light-emitting diodes [53,54]. PPV prepared by the soluble precursor route (**74** → **75**) has a broad absorption with $\lambda_{max} \approx 405$ nm and an absorption

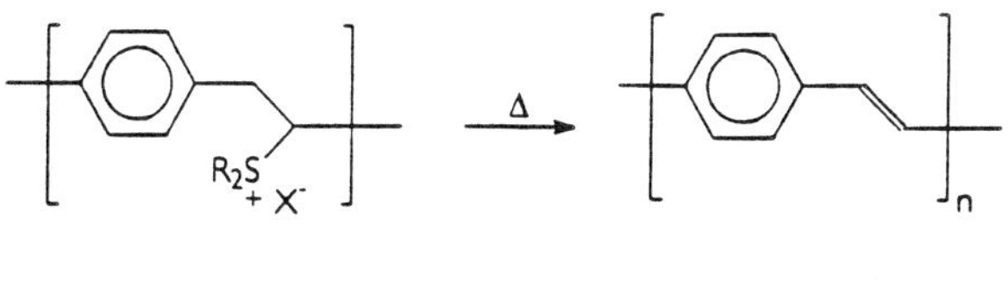

74   **75**

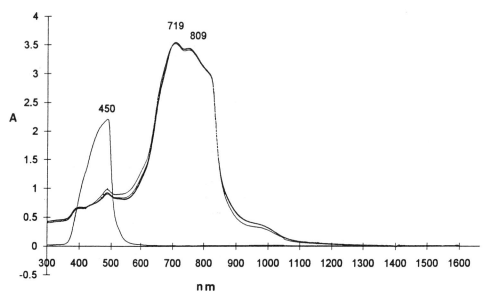

Fig. 27.1 Absorption spectra for neutral and doped dithienyl hexaene oligomer **72d** ($x = 5$).

band edge around 512 nm [55,56]. Electrochemical doping shows the appearance of two new bands at 1372 and 537 nm that Bradley et al. [55] assign to valence band–bipolaron level transitions. The band gap of PTV is lower than that of PPV, and photoinduced absorption characteristic of bipolarons is found in both, 0.6 and 1.6 eV in PPV and 0.44 and 1.0 eV in PTV [57].

We previously reported both the syntheses and doping studies on series of PPV and PTV oligomers [18,58,59]. Doping of oligomers up to the pentamer level (PTV) corresponds well to the formation of bipolarons, with the pentamer approaching the photoexcitation absorption values of the fully conjugated polymer. PPV oligomers are extremely insoluble, and BP spectra beyond the trimer cannot be obtained in solution. However, incorporation of substituent groups (e.g., RO) increases the solubility of the oligomers. At the current time, doping studies have been completed only for the

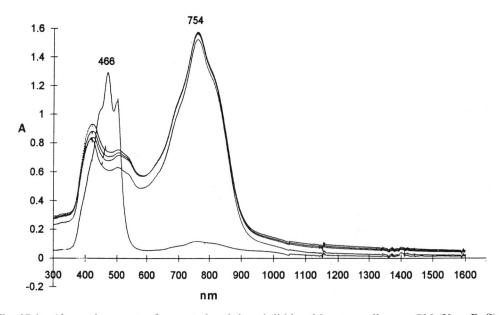

Fig. 27.2 Absorption spectra for neutral and doped dithienyl hexaene oligomer **73d** (X = BuS).

dimer series, which are listed in Table 27.5. One interesting observation is the profound effect of alkoxy substitution in the 4 and 4' positions (**76–81**). While a strong red shift is observed for the 4,4'-(OMe)₂ substitution compared to the parent, long-chain (n-C₈H₁₇O) and branched chain (2-ethylhexyl O) give much greater shifts in oscillator strength, indicating that polarization stabilization of the developing positive charge may be a factor in designing absorption and emission characteristics for PPV oligomers.

Although PTV oligomers display greater solubility than PPV oligomers of equivalent length, they are soluble in appreciable quantity only in powerful solvents such as DMF. For a variety of photonic applications, increased processibility and the capability of incorporating these oligomer lengths in copolymer formulations are highly desirable. Incorporating solubilizing substituents in the 3 and 4 positions of each thienyl unit might be expected to produce this desired result (**82**). This approach has been used most recently by Spangler and He [60] and is discussed in greater detail in Section III.D, which follows.

D. Copolymers Incorporating Polyene or Arylene Vinylene Repeat Units

Fully conjugated electroactive polymers such as polythiophene or poly(p-phenylenevinylene) absorb in the UV-vis over a wide range, primarily due to the number of different conjugation lengths in the polymer as well as varying effective conjugation lengths related to steric effects, internal rotations out of conjugation, and/or intrinsic polymer defects arising from the synthetic pathway. Nonlinear second or third harmonic generation is affected by resonance enhancement from absorption of the harmonic radiation, and for this reason nonresonant $\chi^{(2)}$ and $\chi^{(3)}$ values may be difficult to obtain except at very long irradiation wavelengths. For example, the third harmonic of the 1.9 μm irradiation commonly used in THG measurements occurs at 637 nm, a value well within the absorption envelope of several electroactive polymers. However, as discussed in Section II.A, long conjugation sequences may not be necessary for high $\chi^{(3)}$ in certain systems. Thus, an attractive alternative to the fully conjugated polymer is a copolymer design approach, in which NLO-active repeat units (NLO-phores) are alternated with saturated spacer groups:

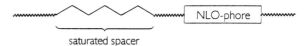

This approach has several advantages. First, the conjugation length, and thus the primary absorption characteristics of the NLO-phore, can be synthetically controlled and manipulated. Second, the spacer length and chemical identity are decoupled from the NLO-phore identity and thus can be designed to enhance such secondary polymer properties as solubility and processibility. One must keep in mind, however, that for relatively small NLO-phore conjugation lengths (<40–50 atoms), γ and $\chi^{(3)}$ increase with increasing length. Thus enhanced third-order response and the concomitant transparency–nonlinearity trade-off are still linked in the copolymer design approach.

We have previously outlined the advantages of this approach in the design of third-order NLO polymers, with particular emphasis on the subsequent incorporation of polaronic or bipolaronic charge states [61,62]. A wide range of copolymers can be synthesized by interfacial polymerization, as outlined in Scheme 7 (**89–93**). Polyamides are a particularly versatile class of copolymers that can be synthesized by this approach using either neopentyl diamine or 1,6-diaminohexane. For some monomers, however, solubility limitations require that a copolyamide approach be used in which the NLO-phore bisacyl halide is mixed with a long-chain saturated bisacyl halide [e.g., ClCO(CH₂)₁₀COCl] to yield a polymer soluble enough to be spin-coated for thin film $\chi^{(3)}$ measurement [61–63]. Typical $\chi^{(3)}$ values obtained by

Table 27.5 Stable Bipolarons Formed from Oxidative Doping of PPV Dimers

Substituent	Compound	λ_{max} π–π* (nm)[a]	λ_{max} BP (nm)
None	76	390	580, 951
4,4'-(OMe)₂	77	392	657, 1196
4,4'-(OC₈H₁₇)₂[b]	78	394	671, 1244
4,4'-(OC₈H₁₇)₂[c]	79	394	671, 1244
2,2',5,5'-(OMe)₄	80	400	594, 1399
3,3',4,4',5,5'-(OMe)₆	81	389	671, 1399

[a] CH₂Cl₂ solution.
[b] n-C₈H₁₇O.
[c] 2-ethylhexyl O.
Source: Portions reprinted from Ref. 71.

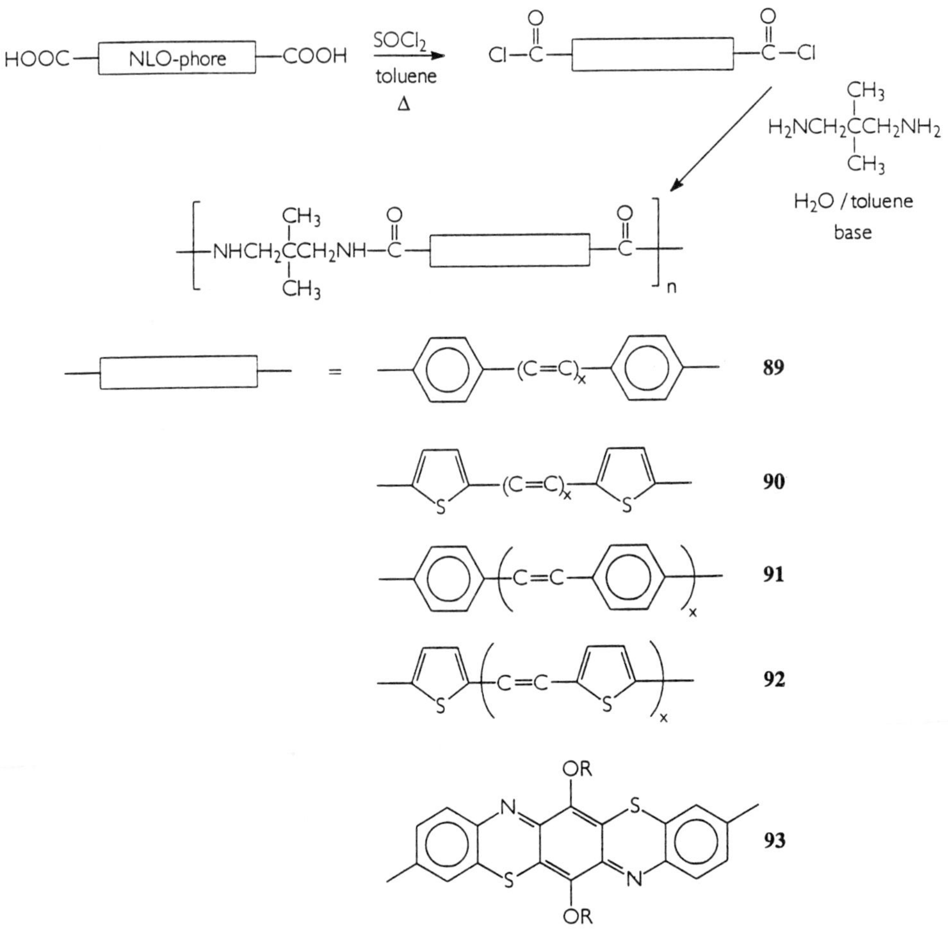

Scheme 7 Synthesis of NLO-active copolymers.

degenerate four-wave mixing studies are shown with structures **94–97**.

Mates and Ober [65] obtained similar results for a series of polyesters that incorporate distyrylbenzene segments. THG measurements obtained at 1.9 μm indicate $\chi^{(3)}$ values similar in magnitude to those shown for **94–97**. It is interesting to note that the order of magnitude of $\chi^{(3)}$ in these materials is approximately the same

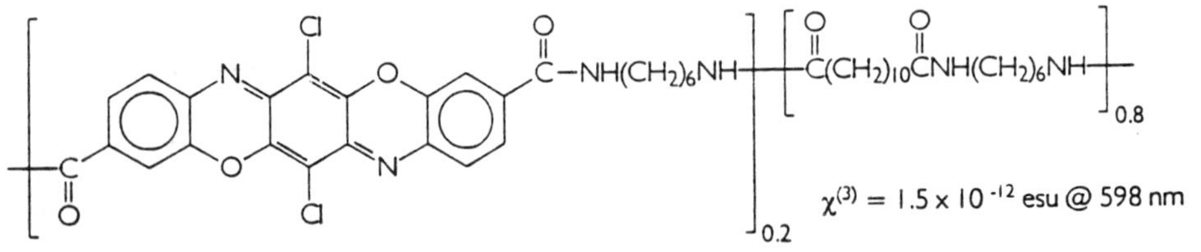

$\chi^{(3)} = 1.5 \times 10^{-12}$ esu @ 598 nm

$\chi^{(3)} = 6 \times 10^{-10}$ esu @ 532 nm (resonant)

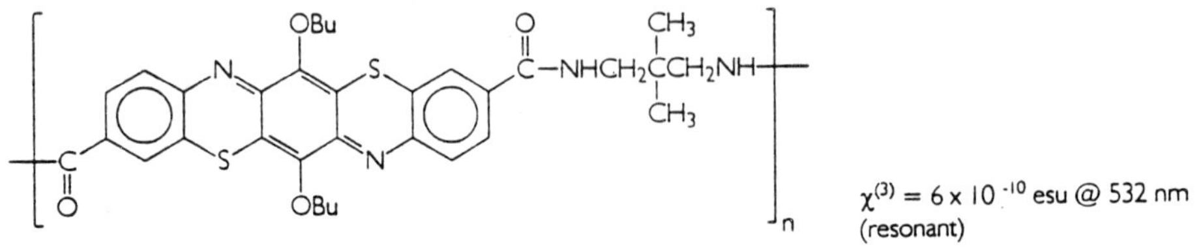

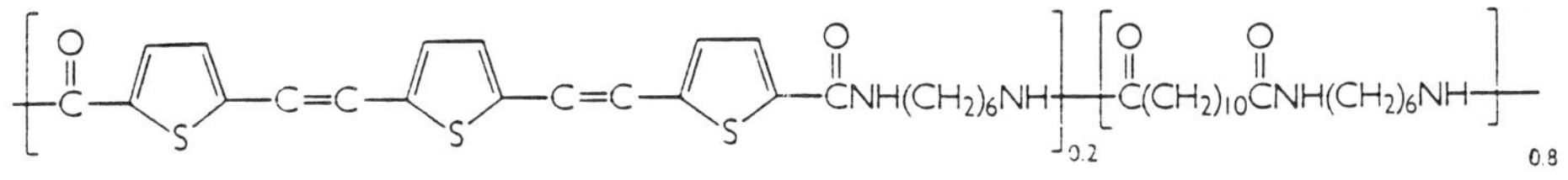

96

$\chi^{(3)} = 2 \times 10^{-12}$ esu

97

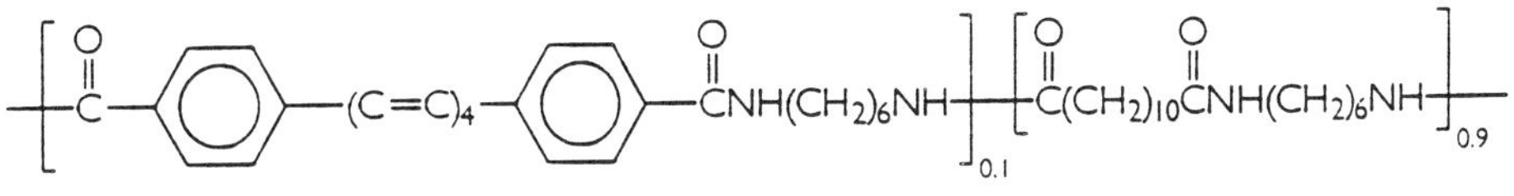

$\chi^{(3)}/\alpha = 1.4 \times 10^{-13}$ esu-cm

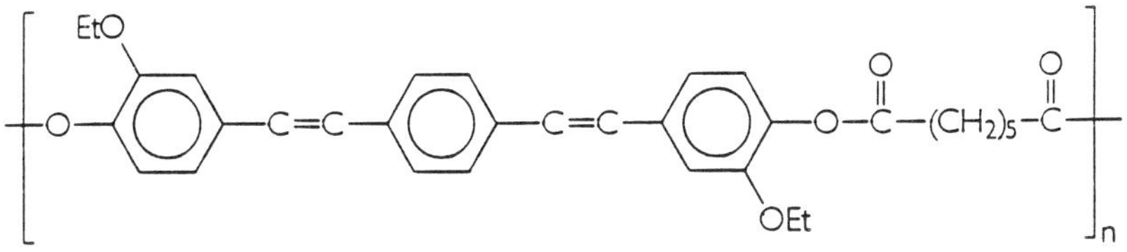

98

$\chi^{(3)}_{THG} = 8 \times 10^{-13}$ esu @ 1.9 μm

for equivalent conjugation lengths, regardless of NLO-phore structure.

For copolymers with formal NLO-phore repeat units, the intriguing question with respect to charge-state incorporation is whether each NLO-active subunit can be oxidized (or reduced) to polaronic radical ions or bipolaronic diions.

While this question is difficult to answer for homopolymers such as PPV or PTV, it can be addressed more directly for model compounds representing copolymer structures. Series of bisthienylpolyenes linked by saturated methylene spacers were studied by Spangler and Liu [45]. Absorption spectra of these model structures (**99–103**) before and after oxidative doping show clearly that both dithienyl polyene segments are oxidized to bipolarons and that their signature spectra are essentially indistinguishable from the model monomer, as shown in Table 27.6. While these model dithienylpolyenes are not soluble enough to form high molecular weight polymers *via* condensation polymerization, it seems clear that each NLO-phore segment in such structures can be oxidized to the bipolaronic dication.

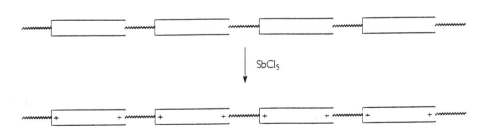

IV. EFFECT OF CHARGE-STATE INCORPORATION ON THIRD-ORDER OPTICAL NONLINEARITY

In Section II.E, theoretical predictions regarding polaronic or bipolaronic enhancement of γ and $\chi^{(3)}$ were reviewed [32–34]. Experimental verification of such enhancement, however, has been difficult to obtain for a variety of reasons. One of the earliest experimental attempts to verify bipolaronic enhancement of $\chi^{(3)}$ was that of Kaino's group [66]. These workers attempted to oxidatively dope poly(2,5-thienylenevinylene) (PTV) with I_2. They found that the "doped" sample had about

Table 27.6 Bipolaron Formation in Model Copolymers

Compound[a]	λ_{max} BP (nm)[b]
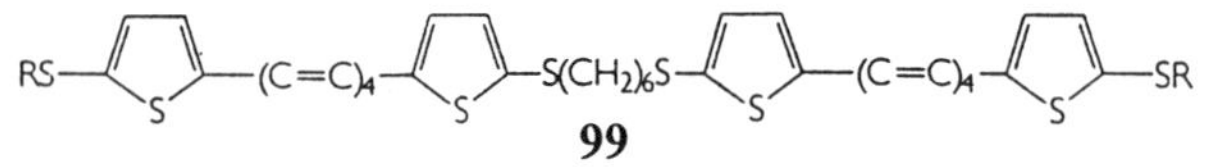 **99** RS—[thiophene]—(C=C)₄—[thiophene]—S(CH₂)₆S—[thiophene]—(C=C)₄—[thiophene]—SR	<u>420</u>, 445
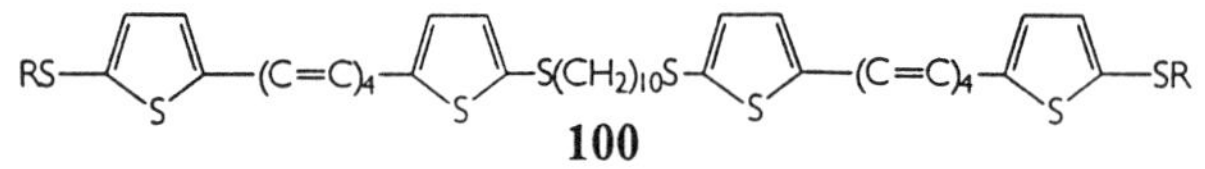 **100** RS—[thiophene]—(C=C)₄—[thiophene]—S(CH₂)₁₀S—[thiophene]—(C=C)₄—[thiophene]—SR	<u>420</u>, 444
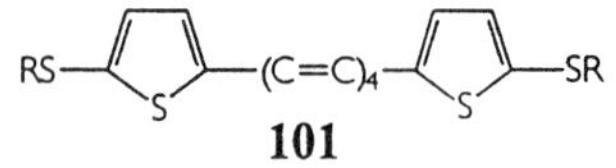 **101** RS—[thiophene]—(C=C)₄—[thiophene]—SR	<u>418</u>, 443
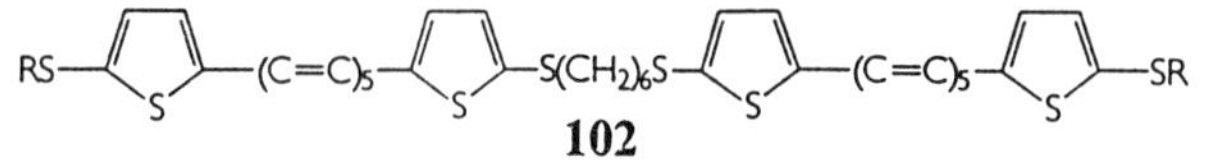 **102** RS—[thiophene]—(C=C)₅—[thiophene]—S(CH₂)₆S—[thiophene]—(C=C)₅—[thiophene]—SR	<u>435</u>, 463
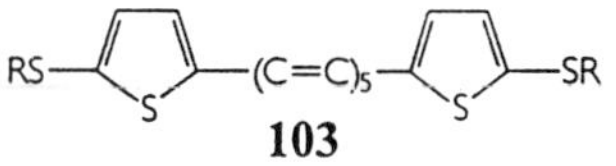 **103** RS—[thiophene]—(C=C)₅—[thiophene]—SR	<u>435</u>, 462

Note: Underlined peaks represent peaks of maximum intensity.
[a] R = n-$C_{10}H_{21}$
[b] 10^{-5} M solutions in CH_2Cl_2.

the same $\chi^{(3)}$ as the "neutral" polymer. However, they also recognized that they observed no spectral signature for bipolaron formation in the absorption spectra of their sample, and they concluded that the BP concentration might have been too low to affect $\chi^{(3)}$. Prasad and coworkers [67] carried out more detailed studies on poly(3-dodecylthiophene) in which they actually observed a decrease in $\chi^{(3)}$ at 602 nm. However, since the original $\chi^{(3)}$ at this wavelength is resonance-enhanced (~1 × 10⁻⁹ esu), the observed decrease in $\chi^{(3)}$ may be associated with the shift to longer wavelength (790 nm) of the BP absorption and a shift to lower wavelength of the residual π–π* transitions (470 nm).

Cao et al. [68] provided the first direct evidence for $\chi^{(3)}$ enhancement by charge-state incorporation by oxidizing a ladder polymer model compound by the Spangler group is methodology ($SbCl_5/CH_2Cl_2$). The magnitude of the enhancement correlated well with the predictions of Birge and coworkers [34], albeit for a different system. Evidence was also obtained in this study for enhancement by photogenerated bipolarons in the polymer samples. However, the first conclusive comparison between $\chi^{(3)}$ (neutral) and $\chi^{(3)}$ (doped) for almost equivalent resonance enhancement was obtained by Nickel et al. [12]. Those workers synthesized a series of bisanthracenylpolyenes (**52–60**) substituted in the 10,10′ positions with electron-donating substituents to stabilize the developing positive charge. The parent system (R = H, n = 3, 4, 5, 6) formed exceptionally stable bipolarons

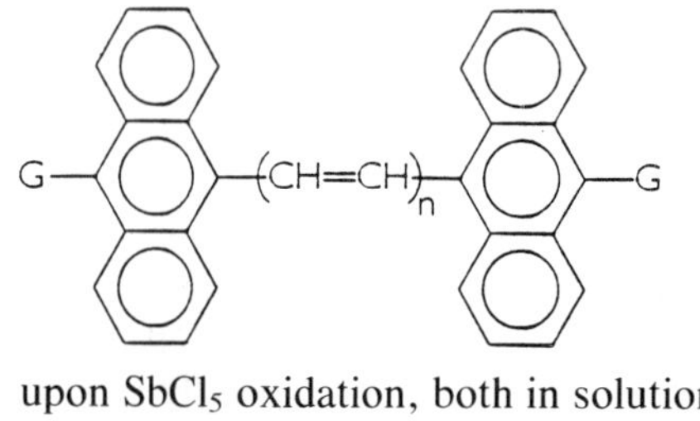

G—[anthracene]—(CH=CH)ₙ—[anthracene]—G   G = H, RS, RO, R
                                            n = 3, 4, 5, 6
52-60

upon $SbCl_5$ oxidation, both in solution (CH_2Cl_2) and as polycarbonate composites. The composites were air-stable under ambient benchtop conditions for several months and produced the largest oxidative shift in oscillator strength yet observed for such short conjugation lengths. These shifts are compared to the corresponding diphenylpolyenes in Table 27.7. The remarkable red

Table 27.7 Absorption Spectra of Bisanthracenyl Polyenes and Polyene Bipolarons in Solution and as Polycarbonate Composites Compared to Diphenylpolyenes

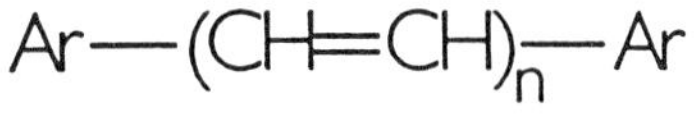

Ar—(CH=CH)$_n$—Ar

Ar[a]	n	Compound	Polyene λ_{max} (nm)		PB λ_{max} (nm)	
			CH_2Cl_2	PC composite	CH_2Cl_2	PC composite
Ph	4	**23d**	378	—	Unstable	—
Ph	5	**23e**	398	—	644	—
Ph	6	**23f**	417	—	676	—
An	4	**53**	423	435	940	924
An	5	**54**	428	443	978	950
An	6	**55**	438	447	1000	934

[a] Ph = phenyl; An = anthracenyl.

shift in oscillator strength in going from phenyl to anthracenyl end caps is attributed to the phenyl moieties' loss of aromaticity upon oxidation (quinoid formation), whereas the anthracenyl moieties retain two aromatic rings and the resulting developing positive charge is benzylic in character and thus mesomerically stabilized.

However, attempts to obtain accurate $\chi^{(3)}$ measurements for both neutral and doped species were extremely difficult due to the relative insolubility of this compound series.

We had previously shown that long-chain alkyl substituents increased the solubility of both parent and BP species in solution with bisthienyl polyenes [44]. Therefore bisanthracenyl polyenes incorporating n-$C_{10}H_{21}S$ or n-C_6H_{13} substituents in the 10 and 10' positions were synthesized [12]. These samples proved soluble enough to study $\chi^{(3)}$ changes during doping by degenerate four-wave mixing at 532 and 1064 nm in solution. These results are shown in Table 27.8. The $\chi^{(3)}$ values for the neutral polyenes are of the same order of magnitude as the solvent. Upon oxidation, significant enhancement of $\chi^{(3)}$ is observed (up to two orders of magnitude). However, this enhancement could also be caused in part by dispersion of γ near the BP absorption, resonant contri-

Table 27.8 $\chi^{(3)}$ Values for Neutral and Oxidatively Doped Bisanthracenyl Polyenes

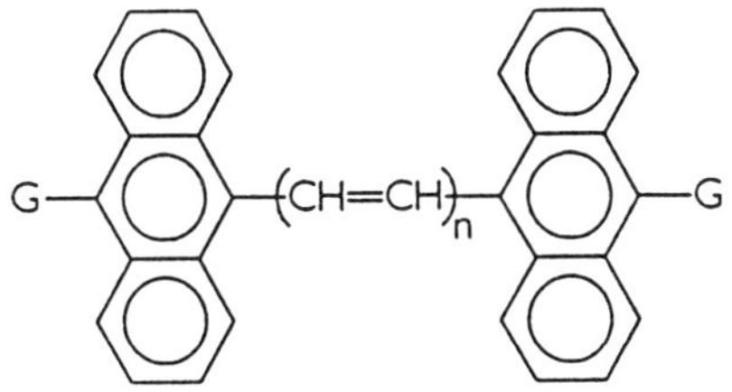

G	Compound	n	wt %	Irrad. λ (nm)	$\chi^{(3)}_{1111}(-\omega;\omega,\omega,-\omega)$ (esu)	
					Neutral	BP
$C_{10}H_{21}S$	**59**	3	0.17	1064	7×10^{-14} [a]	1.9×10^{-13}
$C_{10}H_{21}S$	**104**	4	0.21	1064	7×10^{-14}	67×10^{-13}
C_6H_{13}	**56**	3	0.14	1064	7×10^{-14}	13×10^{-13}
C_6H_{13}	**56**	3	0.14	532	2.3×10^{-14}	3.1×10^{-13}

[a] Same order of magnitude as CH_2Cl_2.

butions due to band edge absorption at 1064 nm, or some combination of these effects. When the bis-n-hexylpolyene was examined at both 532 and 1064 nm (532 nm corresponding to the band edge of the neutral polyene), $\chi^{(3)}$ enhancement by the BP was still observed, even though the degree of band edge absorption was approximately equal. Therefore we ascribe the increased $\chi^{(3)}$ to the charge-state influence.

More recently, bipolaronic enhancement of $\chi^{(3)}$ has also been observed in the dithienyl polyene series (**72,73**)

```
  Bu    Bu              Bu   Bu
   \   /                 \   /
    \ /                   \ /          72 (G = H)      a (n = 3)
G─── ─── ─(CH=CH)ₙ─── ─── ───G                        b (n = 4)
    / \                   / \                         c (n = 5)
   S                     S                 73 (G = BuS)  d (n = 6)
                                                      e (n = 7)
                                                      f (n = 8)
```

[46,47,69]. In this series, the use of Bu solubilizing substituents allowed DFWM studies up to the octaene. These values are shown in Table 27.9. Given the fixed irradiation wavelengths used in this study (532 and 1064 nm), a definitive statement regarding bipolaronic enhancement as a function of conjugation length remains elusive. Only for the $n = 5$ polyene (**72c**) can the neutral and doped results be compared with any confidence that one or the other measurement is not significantly resonance-enhanced. In this sample, bipolaronic enhancement by a factor of 2.6–2.8 is observed. This correlates well with the theoretical predictions of Birge and coworkers [34]. For $n = 7$ and 8 (**72e, 72f**), $\chi^{(3)}$ neutral shows increasing resonance enhancement. More recent measurements for a series of bisthienyl polyenes where X = BuS show enhancement of $\chi^{(3)}$ neutral over X = H for similar n, in keeping with the predictions of Meyers and Brédas [28]. Thus for $n = 5$ (**73c**), $\chi^{(3)}$ neutral = 48 × 10^{-13} esu, an enhancement of ≈18 over X = H. In addition, the power law dependence of $|\chi^{(3)}|$ for both neutral and doped species follows n^b, where $b \approx 5.5$ at 532 for the neutral series for $3 \leq n \leq 9$, and $b \approx 14$ at 1064 nm for the bipolaron series where $6 \leq n \leq 9$.

V. SUMMARY

In this chapter we have presented an overview of the various parameters that define and control the optical nonlinearity of conjugated polyenylic systems, with particular emphasis on the generation and stabilization of polaronic and bipolaronic charge states in these systems. At the current time, it is probably safe to say that the theoretical predictions of bipolaronic enhancement of third-order nonlinearity have been confirmed in a number of well-defined systems. However, the magnitude of this enhancement is less clear due to the difficulties in identifying the relative contributions of resonance enhancement to either neutral or oxidized polyene nonlinearity. This uncertainty could possibly be resolved in the future by detailed third harmonic generation studies using an irradiation wavelength whose third harmonic did not lie within the absorption envelope of either neutral or doped species. To date, these measurements have not been carried out in any series of compounds that I know of. In addition, there are no systems for which polaronic enhancement of third-order nonlinearity has been confirmed. This is due in part to the difficulty of stopping at the polaronic radical cation during chemical oxidation without having at least *some* bipolaronic dication present. What is certain at this time is that both radical cations and dications can be produced in a wide variety of extended conjugated systems with relative ease, and that these polaron and bipolaron model structures can be quite stable. Given the importance of charge-state formation in conducting polymers, light-emitting diodes, and other electronic and photonic devices, the questions still remaining with respect to their importance in nonlinear optics will in time be resolved.

ACKNOWLEDGMENTS

I am indebted to the large number of graduate and undergraduate students in my research group who have contributed to the results presented in this chapter: Dr. Tom J. Hall, Dr. Pei-kang Liu, Dr. Kathleen O. Havelka, Dr. Linda Sapochak, Dr. Eric G. Nickel, and Dr. Mingqian He. I also wish to thank members of the departments of chemistry and physics at the University of Southern California, who have collaborated with me over the past eight years, for their aid in obtaining degenerate four-wave mixing measurements on a variety of compound series: Prof. Larry R. Dalton, Prof. Robert Hellwarth, Dr. David Polis, and Dr. Joyce Laquindanum. I also acknowledge collaborations with Prof. Robert Birge (Syracuse University), Dr. Lap-tak Cheng (DuPont),

Table 27.9 $\chi^{(3)}$ Values for Neutral and Doped Dithienylpolyenes

G	n	Compd	Irrad. λ (nm)	$\chi^{(3)}$ neutral (10^{-13} esu)	$\chi^{(3)}$ doped (10^{-13} esu)
H	5	**72c**	532	2.7	7.8
H	7	**72e**	532	43.0	11
H	8	**72f**	532	258.0	14
H	5	**72c**	1064	0.54	1.4
H	7	**72e**	1064	0.85	3.2
H	8	**72f**	1064	0.66	2.7

and Dr. Robert Norwood in determining β, γ, and $\chi^{(3)}$ values for a variety of compounds discussed in this chapter. Over the years I have had many fruitful discussions on structure–property relationships with a large number of scientists too numerous to mention here; however, I would like to thank Prof. Paras Prasad (SUNY-Buffalo), Bruce Reinhardt (Wright Laboratories), Prof. J.-L. Brédas (Université de Mons-Hainaut), Dr. M. Blanchard-Desce (Collège de Paris), and Dr. Seth Marder (Beckman Institute, California Institute of Technology), for their special contributions to my knowledge of NLO processes. Finally, I thank the sponsors of this continuing project, particularly the Air Force Office of Scientific Research and the donors of the Petroleum Research Fund administered by the American Chemical Society, for their generous financial support.

REFERENCES

1. S. R. Marder, J. E. Sohn, and G. D. Stucky (eds.), *Materials for Nonlinear Optics: Chemical Perspectives*, ACS. Symp. Ser. 455, Am. Chem. Soc., Washington, DC, 1991.
2. P. N. Prasad and D. J. Williams, *Introduction to Nonlinear Optical Effects in Molecules and Polymers*, Wiley-Interscience, New York, 1991.
3. R. A. Hann and D. Bloor (eds.), *Organic Materials for Non-Linear Optics*, Vols. I and II and G. J. Ashwell and D. Bloor (eds.), Vol. III, Royal Society of Chemistry, London, 1989, 1991, 1993.
4. L.-T. Cheng, W. Tam, S. R. Marder, A. E. Stiegmann, G. Rikken, and C. W. Spangler, Experimental investigations of organic molecular nonlinear optical polarizabilities. 2. A study of conjugation dependences, *J. Phys. Chem.* 95:10643 (1991).
5. C. W. Spangler, K. O. Havelka, M. W. Becker, T. A. Kelleher, and L.-T. Cheng, Relationship between conjugation length and third-order nonlinearity in bis-donor substituted diphenyl polyenes, *Proc. SPIE* 1560:139 (1991).
6. G. Pucetti, M. Blanchard-Desce, I. Ledoux, J.-M. Lehn, and J. Zyss, Chain-length dependence of the third-order polarizability of bis-substituted polyenes. Effects of endgroups and conjugation length, *J. Phys. Chem.* 97:9385 (1993).
7. C. W. Spangler, P.-K. Liu, T. A. Kelleher, and E. G. Nickel, Substituent effects in the design of new organic NLO materials, *Proc. SPIE* 1626:406 (1992).
8. L.-T. Cheng, W. Tam, S. H. Stevenson, G. R. Meredith, G. Rikken, and S. R. Marder, Experimental investigations of organic molecular nonlinear optical polarizabilities. 1. Methods and results on benzene and stilbene derivatives, *J. Phys. Chem.* 95:10631 (1991).
9. P. N. Prasad and B. Reinhardt, Is there a role for organic materials in chemistry in nonlinear optics and photonics?, *Chem. Mater.* 2:660 (1990).
10. K. Y. Wong, A. K.-Y. Jen, V. P. Rao, K. Drost, and R. M. Mininni, Experimental and theoretical studies of heterocyclic nonlinear optical materials, *Proc. SPIE* 1775:74 (1992).
11. V. P. Rao, K. Y. Wong, A. K.-Y. Jen, and R. M. Mininni, Optimization of second-order nonlinear optical properties of push-pull conjugated chromophores using heteroaromatics, *Proc. SPIE* 2025:156 (1993).
12. E. G. Nickel, C. W. Spangler, N. Tang, R. Hellwarth, and L. Dalton, Bipolaron enhancement of $\chi^{(3)}$ in substituted bis-anthracenyl polyenes, *Nonlinear Opt.* 6:135 (1993).
13. J. Swiatkiewicz, M. E. Orczyk, P. N. Prasad, C. W. Spangler, and M. He, Resonant third-order optical nonlinearity of the neutral and the dication molecules of a triphenodithiazine model compound, *Proc. SPIE* 2025:400 (1993).
14. S. R. Marder, D. N. Beratan, and L.-T. Cheng, Approaches for optimizing the first electronic hyperpolarizability of conjugated organic molecules, *Science* 252:103 (1991)
15. S. R. Marder, J. W. Perry, B. G. Tiemann, C. B. Gorman, S. Gilmour, S. L. Biddle, and G. Bourhill, Direct observation of reduced bond length alternation in donor/acceptor polyenes, *J. Am. Chem. Soc.* 115:2524 (1993).
16. S. R. Marder, C. B. Gorman, F. Meyers, J. W. Perry, G. Bourhill, J.-L. Brédas, and B. M. Pierce, A unified description of linear and nonlinear polarization in organic polymethine dyes, *Science* 265:632 (1994).
17. *Optical Nonlinearities in Chemistry, Chemical Reviews*, Vol. 94, (J. Michl, Ed.) 1994, pp. 1–278.
18. C. W. Spangler, P.-K. Liu, and K. O. Havelka, The formation of charge states in organic molecules, oligomers and polymers for applications in molecular electronic and photonic devices, in *Molecular Electronics and Molecular Electronic Devices*, Vol. III (K. Sienicki, ed.), CRC Press, Boca Raton, FL, 1994, pp. 97–115.
19. C. Sauteret, J.-P. Herman, R. Frey, F. Pradere, J. Ducuing, R. H. Baughman, and R. R. Chance, Optical nonlinearities in one-dimensional conjugated polymer crystals, *Phys. Rev. Lett.* 36:956 (1976).
20. D. N. Beratan, J. N. Onuchic, and J. W. Perry, Nonlinear susceptibilities of finite conjugated organic polymers, *J. Phys. Chem.* 91:2696 (1987).
21. G. J. B. Hurst, M. Dupuis, and E. Clementi, Ab initio analytic polarizability, first and second hyperpolarizabilities in large conjugated organic molecules: applications to polyenes C_4H_6 to $C_{22}H_{24}$, *J. Chem. Phys.* 89:385 (1988).
22. A. F. Garito, J. R. Heflin, K. Y. Wong, and O. Zamani-Khamiri, Enhancement of nonlinear optical properties of conjugated linear chains through lowered symmetry, in *Organic Materials for Non-Linear Optics* Vol. I, (R. A. Hann and D. Bloor, eds.), Roy. Soc. Chem. London, 1989, pp. 16–28.
23. P. N. Prasad, Studies of ultrafast third-order non-linear optical processes on polymer films, in *Organic Materials for Non-Linear Optics*, Vol. I (R. A. Hann

and D. Bloor, eds.), Roy. Chem. Soc., London, 1989, pp. 264–274.
24. P. N. Prasad, E. Perrin, and M. Samoc, A coupled anharmonic oscillator model for optical nonlinearities of conjugated organic structures, *J. Chem. Phys. 91*: 2360 (1989).
25. I. D.-Samuel, I. Ledoux, C. Dhenaut, J. Zyss, H. H. Fox, R. R. Schrock, and R. J. Silbey, Saturation of cubic nonlinearity in long-chain polyene oligomers, *Science 265*:1070 (1994).
26. B. A. Reinhardt, M. R. Unroe, R. C. Evers, M. Zhao, M. Samoc, P. N. Prasad, and M. Sinsky, Third-order optical nonlinearities of model compounds containing benzobisthiazole, benzobisoxazole and benzobisimidazole units, *Chem. Mater. 3*:864 (1991).
27. M.-T. Zhao, M. Samoc, B. P. Singh, and P. N. Prasad, Study of third-order microscopic optical nonlinearities in sequentially built and systematically derivatized structures, *J. Phys. Chem. 93*:7916 (1989).
28. F. Meyers and J.-L. Brédas, Theoretical investigation of static third-order polarizabilities in push-push, pull-pull and push-pull polyenes, in *Organic Materials for Nonlinear Optics*, Vol. III (G. J. Ashwell and D. Bloor, eds.), Roy. Soc. Chem., London, 1993, pp. 1–6.
29. V. P. Rao, A. K. Jen, K. Y. Wang, K. Drost, and R. M. Mininni, Functionalized heteroaromatics for second order nonlinear optical applications, *Proc. SPIE 1775*:32 (1992).
30. J. S. Shirk, J. R. Lindle, F. J. Bartoli, Z. H. Kafafi, and A. W. Shaw, Nonlinear optical properties of substituted phthalocyanines, in *Materials for Nonlinear Optics: Chemical Perspectives* (S. R. Marder, J. E. Sohn, and G. D. Stucky, eds.), ACS Symp. Ser. 455, Am. Chem. Soc., Washington, DC, 1991, pp. 626–634.
31. C. P. de Melo and R. Silbey, Non-linear polarizabilities of conjugated chains: regular polyenes, solitons and polarons, *Chem. Phys. Lett. 140*:537 (1987).
32. C. P. de Melo and R. Silbey, Variational-perturbational treatment for the polarizabilities of conjugated chains. I. Theory and linear-polarizabilities results for polyenes, *J. Chem. Phys. 88*:2558 (1988).
33. C. P. de Melo and R. Silbey, Variational-perturbational treatment for the polarizabilities of conjugated chains. II. Hyperpolarizabilities of polyenylic chains, *J. Chem. Phys. 88*:2567 (1988).
34. J. R. Tallent, R. R. Birge, C. W. Spangler, and K. O. Havelka, Theoretical analysis of third order polarizability enhancement of disubstituted diphenyl polyenes via oxidative doping, in *Molecular Electronics—Science and Technology* (A. Aviram, ed.), Am. Inst. Phys., New York, 1992, pp. 191–203.
35. C. B. Gorman and S. R. Marder, An investigation of the interrelationships between linear and nonlinear polarizabilities and bond-length alternation in conjugated organic molecules, *Proc. Natl. Acad. Sci. USA 90*:11297 (1993).
36. J. W. Perry, G. Bourhill, S. R. Marder, D. Lu, G. Chen, and W. A. Goddard III, Hyperpolarizabilities of push-pull polyenes: experimental results and a new two-state model, *Polym. Prepr. 35(2)*:148 (1995).
37. S. R. Marder, J. W. Perry, G. Bourhill, C. B. Gorman, B. G. Tiemann, and K. Mansour, Relation between bond-length alternation and second electronic hyperpolarizability of conjugated organic molecules, *Science 261*:186 (1993).
38. C. W. Spangler, R. K. McCoy, A. A. Dembek, L. S. Sapochak, and B. D. Gates, Preparation of p,p'-disubstituted, α,ω-diphenyl polyenes, *J. Chem. Soc., Perkin Trans. 1*:779 (1991).
39. C. W. Spangler, L. S. Sapochak, and B. D. Gates, Polaron and bipolaron formation in model extended π-electron systems: potential nonlinear optics applications, in *Organic Materials for Non-Linear Optics*, Vol. I (R. A. Hann and D. Bloor, eds.), Roy. Soc. Chem., London, 1989, pp. 57–63.
40. C. W. Spangler and K. O. Havelka, Polaron and bipolaron stabilization via substituent effects and increasing conjugation length in polyacetylene oligomers, *Polym. Prepr. 31(1)*:396 (1990).
41. K. O. Havelka, Stabilization of polaronic and bipolaronic charge states in electroactive oligomers and polymers, Ph.D. Dissertation, Northern Illinois University, 1991.
42. M. A. Sato, S. Tanaka, and K. Kaeriyama, Soluble conducting polymers by electrochemical polymerization of thiophenes having long alkyl substituents, *Synth. Met. 18*:229 (1987).
43. C. W. Spangler, P.-K. Liu, A. A. Dembek, and K. O. Havelka, Preparation and oxidative doping of α,ω-dithienyl polyenes, *J. Chem. Soc., Perkin Trans. 1*: 799 (1991).
44. C. W. Spangler, P.-K. Liu, and K. O. Havelka, Comparison of bipolaron-like charge-state generation in the oxidative doping of α,ω-dithienyl and α,ω-diphenyl polyenes stabilized by methoxy and methylthio substituents, *J. Chem. Soc., Perkin Trans. 2*: 1207 (1992).
45. C. W. Spangler and P.-K. Liu, Preparation and oxidative doping studies of α,ω-dithienyl polyenes stabilized by long chain alkylthio substituents, *J. Chem. Soc., Perkin Trans. 2*:1959 (1992).
46. C. W. Spangler and M. Q. He, Synthesis of oligomeric polyacetylenes stabilized by thienyl end groups and formation of stable bipolaron-like dications, *Polym. Prepr. 35(1)*:317 (1994).
47. C. W. Spangler and M. Q. He, Preparation and oxidative doping studies of dithienyl polyenes stabilized by alkyl group substitution, *J. Chem. Soc., Perkin Trans. 1*:715 (1995).
48. J. D. Capistan, D. R. Gagnon, S. Antoun, R. W. Lenz, and F. E. Karasz, Synthesis and electrical conductivity of high molecular weight poly(arylene vinylenes), *Polym. Prepr. 25(2)*:282 (1984).
49. F. E. Karasz, J. D. Capistan, D. R. Gagnon, and R. W. Lenz, High molecular weight polyphenylene vinylene, *Mol. Cryst. Liq. Cryst. 118*:327 (1985).

50. D. R. Gagnon, J. D. Capistan, F. E. Karasz, R. W. Lenz, and S. Antoun, Synthesis, doping and electrical conductivity of high molecular weight poly(p-phenylene vinylene), *Polymer* 28:587 (1987).
51. K.-Y. Jen, M. Maxfield, L. W. Schacklette, and R. L. Eisenbaumer, Highly conducting poly(2,5-thienylene vinylene) prepared via a soluble precursor route, *J. Chem. Soc., Chem. Commun.* 309 (1987).
52. K.-Y. Jen, R. Jour, L. W. Schacklette, M. Maxfield, H. Eckhardt, and R. L. Eisenbaumer, The optical electrochemical and structure/property relationships of poly(heteroaromatic vinylenes), *Mol. Cryst. Liq. Cryst.* 160:69 (1988).
53. B. P. Singh, P. N. Prasad, and F. E. Karasz, Third order nonlinear optical properties of oriented films of poly(p-phenylene vinylene) investigated by femtosecond degenerate four wave mixing, *Polymer* 29:1940 (1988).
54. J. H. Burroughes, D. D. C. Bradley, A. R. Brown, R. N. Marks, K. Mackay, R. H. Friend, P. L. Burn, and A. B. Holmes, Light-emitting diodes based on conjugated polymers, *Nature* 347:539 (1990).
55. D. D. C. Bradley, G. D. Evans, and R. H. Friend, Characterization of poly(p-phenylene vinylene) by infrared and optical absorption, *Synth. Met.* 17:651 (1987).
56. J. Obrzut and F. E. Karasz, Ultraviolet and visible spectroscopy of poly(p-phenylene vinylene), *J. Chem. Phys.* 87:2349 (1987).
57. R. H. Friend, D. D. C. Bradley, and P. D. Townsend, Photoexcitation in conjugated polymers, *J. Phys. D, Appl. Phys.* 20:1367 (1987).
58. C. W. Spangler and T. J. Hall, Oxidative doping studies of PPV oligomers, *Synth. Met.* 44:85 (1991).
59. C. W. Spangler and P.-K. Liu, Oxidative doping studies of PTV oligomers, *Synth. Met.* 44:259 (1991).
60. C. W. Spangler and M. Q. He, unpublished results.
61. C. W. Spangler and K. O. Havelka, Design of new nonlinear optic-active polymers: use of delocalized polaronic or bipolaronic charge states, in Materials for Nonlinear Optics (S. R. Marder, J. E. Sohn, and G. D. Stucky, eds.), ACS Symp. Ser. 455, Am. Chem. Soc., Washington, DC, 1991, pp. 661–671.
62. C. W. Spangler, P.-K. Liu, J. Laquindanum, L. S. Sapochak, L. R. Dalton, and R. S. Kumar, Incorporation of ladder subunits in formal copolymers for third order NLO applications, in *Frontiers of Polymers and Advanced Materials* (P. N. Prasad, ed.), Plenum, New York, 1994, pp. 205–210.
63. C. W. Spangler, P.-K. Liu, and E. G. Nickel, New copolymers for nonlinear optics application which incorporate electroactive subunits with well-defined conjugation lengths, in *Frontiers of Polymer Research* (P. N. Prasad and J. K. Nigam, eds.), Plenum, New York, 1991, pp. 149–156.
64. C. W. Spangler, P.-K. Liu, T. J. Hall, D. W. Polis, L. S. Sapochak, and L. R. Dalton, The design of new copolymers for $\chi^{(3)}$ applications, *Polymer* 33:3937 (1992).
65. T. E. Mates and C. K. Ober, Model polymers with distyrylbenzene segments for third-order nonlinear optical properties, in *Materials for Nonlinear Optics* (S. R. Marder, J. E. Sohn, and G. D. Stucky, eds.), ACS Symp. Ser. 455, Am. Chem. Soc., Washington, DC, 1991, pp. 497–513.
66. T. Kaino, K. Kubodera, H. Kobayashi, T. Kurihara, S. Saito, T. Tsutsui, S. Tokito, and H. Murata, Optical third harmonic generation from poly(2,5-thienylene vinylene) thin flims, *Appl. Phys. Lett.* 53(21):2002 (1988).
67. P. Logsden, J. Pfleger, and P. N. Prasad, Conductive and optically non-linear polymeric Langmuir–Blodgett films of poly(3-dodecylthiophene), *Synth. Met.* 26:369 (1988).
68. X. F. Cao, J. P. Jiang, R. W. Hellwarth, L. Yu, M. Chen, and L. R. Dalton, Bipolaronic enhanced third order nonlinearity in organic ladder polymers, *Proc. SPIE 1337*:114 (1990).
69. C. W. Spangler, M. Q. He, J. Laquindanum, L. Dalton, N. Tang, J. Partanen, and R. Hellwarth, Bipolaron formation and nonlinear optical properties in bis-thienyl polyenes, *Mater. Res. Soc. Symp. Proc. 328*: 655 (1994).
70. N. Tang, J. P. Partanen, R. W. Hellwarth, J. Laquindanum, L. R. Dalton, M. Q. He, and C. W. Spangler, Studies of optical nonlinearity in bis-thienyl polyenes, *Proc. SPIE 2285*:1861 (1994).
71. K. Sienicki (ed.), *Molecular Electronics and Molecular Electronic Devices*, Vol. III, CRC Press, Boca Raton, FL (1994).

28

Molecular and Electronic Structure and Nonlinear Optics of Polyconjugated Materials from Their Vibrational Spectra

M. Del Zoppo, C. Castiglioni, P. Zuliani, and G. Zerbi
Politecnico, Milan, Italy

I. EXPERIMENTAL VIBRATIONAL SPECTROSCOPIC OBSERVABLES

The localized character of many vibrational motions in a complex molecule has allowed the development of a very large and systematic correlative body of work on group frequencies that has formed the basis for chemical and structural analysis for many years [1–3]. Most of the chemical functional groups vibrate with characteristic frequencies that show some dependence on their chemical and structural environment, thus providing essential information to synthetic chemists. Routine NMR techniques have pushed instrumental chemical diagnosis much further, allowing chemical characterization in greater detail.

Next step in molecular spectroscopy was the detailed understanding of the vibrational spectra with a meticulous assignment of the vibrational (or vibrorotational or ro-vibronic) transitions in terms of group theory, molecular and/or lattice dynamics, molecular harmonic or anharmonic force fields, etc. [4–7]. These works aimed at the detailed understanding of the molecular dynamics but were limited to relatively small and highly symmetrical molecules.

Finally one has to face the challenging problem of understanding the vibrational spectra of structurally complex systems [8–11]. The understanding of the spectrum is not limited to a spectroscopic game per se but is required by materials science for the discovery of structural aspects and/or molecular processes that modulate the macroscopic properties of particular materials with possible technological relevance.

Theoretical tools are now available to carry out a comprehensive vibrational analysis in terms of both frequency and intensity spectroscopy. These two observables can be easily and routinely obtained with the presently available instrumental technology.

In this chapter frequency correlations are discussed only marginally. We report the state of the art of frequency and intensity spectroscopy, which can help in the understanding of the electronic phenomena in polyconjugated low band gap molecules that form the basis for relevant macroscopic electrical and optical properties.

A. Frequency Spectroscopy

The physics behind molecular [5–7] and/or lattice [7,8] dynamics is well known and has been treated in great detail. The harmonic approximation is necessarily adopted in the treatment of the dynamics of very large molecules. The present study refers to pure vibrational spectra and ignores vibrorotational spectra. Some references to vibronic spectra are made when necessary.

The $3n - 6$ vibrational degrees of freedom of an isolated n-atom nonlinear molecule can be described in terms of a set of internal coordinates R_i representing vibrational displacements that, in Wilson's treatment, correspond to bond stretchings, angle bending, and conformational distortions [5,6]. The vibrational harmonic potential can be written in matrix notation as

$$2V = \tilde{\mathbf{R}}\mathbf{F}_R\mathbf{R} \tag{1}$$

where $\mathbf{F}_R$ is the potential energy matrix that collects the quadratic force constants:

$$f_{ij} = \left(\frac{\partial^2 V}{\partial R_i \partial R_j}\right)_{eq} \tag{2}$$

which are the second-order terms in the Taylor expansion of the potential energy function evaluated at the equilibrium molecular structure.

The kinetic energy of the system can be written in terms of the velocities $\dot{R}_i$ as

$$2T = \dot{\mathbf{R}}(\mathbf{G}_R)^{-1}\dot{\mathbf{R}} \qquad (3)$$

where $\mathbf{G}_R$ is the kinetic energy matrix, which can be expressed in terms of the atomic masses and molecular geometry.

The vibrational frequencies can be calculated by the solution of the eigenvalue equation

$$\mathbf{G}_R \mathbf{F}_R \mathbf{L}_R = \mathbf{L}_R \mathbf{\Lambda} \qquad (4)$$

where $\mathbf{\Lambda}$ is the diagonal matrix of the $3n - 6$ vibrational frequency parameters with $\lambda_i = 4\pi^2 c^2 \nu_i^2$ where ν_i (cm^{-1}) is the vibrational frequency (expressed in wavenumbers), $\mathbf{L}_R$ is the matrix of the eigenvectors, and each $\mathbf{L}_i$ describes the ith normal mode in terms of a set of R coordinates. Numerical algorithms are presently available for the solution of the eigenvalue equation [Eq. (4)], also for very large systems [10].

The molecules treated in this chapter are indeed large systems with complex chemical structures. Moreover, in going from the oligomers to the polymers it becomes necessary to consider the systems as one-dimensional crystals. The optical transitions (both vibrational and electronic) are determined by one-dimensional periodicity and translational symmetry. Collective motions (phonons) that extend throughout the chain need to be considered and are characterized by the wave vector $\mathbf{k}$, and their frequencies show dispersion with $\mathbf{k}$. Lattice dynamics in the harmonic approximation are well developed [12,13], and vibrational frequency spectroscopy has reached full maturity and has been widely applied in polymer science [8,9,14].

Conceptually the GF method of Wilson has been applied routinely to polymers. $\mathbf{k}$-dependent $\mathbf{F}$ and $\mathbf{G}$ matrices similar to those defined in Eqs. (1) and (3) can be defined, thus bringing to the solution of the $\mathbf{k}$-dependent dynamical matrix,

$$\mathbf{G}_R(\mathbf{k})\mathbf{F}_R(\mathbf{k})\mathbf{L}_R(\mathbf{k}) = \mathbf{\Lambda}(\mathbf{k})\mathbf{L}_R(\mathbf{k}) \qquad (5)$$

When structural defects occur (and this is usually the case in nature) in the 1-D polymer chain (with a given population and distribution), translational periodicity is lost and the treatment of the exact dynamical problem becomes extremely difficult. Numerical algorithms are available for the exact treatment of the dynamical problems even of very long chains containing configurational, conformational, and structural disorder [10]. The vibrational density of states $g(\nu)$ can be calculated and compared with the experimental $g(\nu)$ from inelastic neutron scattering experiments or with the experimental infrared and Raman spectra (after dipole or polarizability weighting).

The key issue in frequency spectroscopy is the knowledge of the quadratic force constants. In the past an enormous effort was made to obtain reliable sets of quadratic force constants conceptually by solving the inverse dynamical problem, i.e., using in Eq. (4) experimentally known frequencies and, as unknowns, the quadratic force constants [15,16].

In spite of the many approximations and difficulties (discussed in detail in specialized publications [6,17,18]), this work has been successful for simple molecules containing σ bonds or isolated double bonds. The problem becomes difficult and still unsolved systematically when a vibrational potential must be described for systems containing conjugated double bonds. The problem has not been easy even for the simple molecule of benzene [19–21] and becomes very difficult when polyconjugated molecules must be treated.

The conclusion generally accepted, even for simple organic molecules and particularly in the case of polyconjugated systems, is that mere frequency fitting by trial and error or by modification of a few arbitrarily chosen force constants is no longer acceptable and must be considered a game with numbers with no physical and/or structural meaning. Any structural consequence derived should be taken with great caution. Prototypical cases of physically unacceptable information derived from force field calculations have been presented in the literature, even for simple molecules [22,23].

Another way to approach the problem of the intramolecular potential in polyconjugated systems is to carry out quantum-mechanical calculations that provide quadratic force constants to be used in classical dynamical calculations as in Eq. (4). Also in this field arbitrary decisions must be made, thus undermining the importance and the faith in ab initio calculations. There are two types of approximations generally involved in ab initio calculations:

1. In the definition of the internal coordinates, the existence of redundant sets of coordinates is unavoidable [24–27]. Ways to reduce the number of redundancies or arbitrary decisions to avoid them have been proposed and are normally used in the literature [28].
2. The ab initio calculated force constants need to be scaled to bring the theoretical model close to the actual spectrum of any molecule. The criteria for the calculation of the scaling factors are still left to personal choice [28].

In spite of all these approximations, ab initio calculations provide extremely interesting and useful information, especially when vibrational intensities are also included in a given study. Moreover, the very recent applications of the density functional method (e.g., Ref. 29) seem to provide very reliable results for simple systems.

B. Intensity Spectroscopy

Vibrational intensities have been generally overlooked or neglected in the analysis of the vibrational spectra even if they provide very valuable information, especially in the field of polyconjugated molecules. Only recently have quantum chemists turned their attention to this experimental spectroscopic observable, and automatic computing programs include intensity as a property to be routinely calculated. However, most users of these computing programs are blindly calculating intensities together with vibrational frequencies to predict spectra. We think that, especially in the field of polyconjugated systems, intensities can provide ways to reveal interesting details of the physics of these systems.

First we treat the case of infrared intensities, which provides a direct approach to molecular electronic properties. Let the intensity of an infrared band associated with the ith mode be defined as [23]

$$A_i = \frac{1}{Cl} \int_{\text{band}} \ln\left(\frac{I^\circ}{I_i}\right) d\nu \tag{6}$$

where C is the concentration of the sample in mmol/cm^3, l the pathlength in cm, $d\nu$ is in cm^{-1}, and A_i is measured in darks (cm/mmol). Other commonly used units of measure are km/mol = 10^{-2} darks.

When the double harmonic approximation is adopted, A_i is related to molecular properties by the relation

$$A_i = \frac{N g_i \pi}{3 c^2 \nu_i} \left|\frac{\partial \mathbf{M}}{\partial Q_i}\right| \tag{7}$$

where N is Avogadro's number, c the speed of light in vacuo, g_i the degeneracy of the ith normal mode Q_i, and $\mathbf{M}$ the total molecular electric dipole moment. Since $\mathbf{M}$ depends on the distribution of electrons within the molecule, the direct relationship between vibrational intensities and $\mathbf{M}$ opens the way to the use of infrared spectra as probes of electronic properties.

Under the assumption that the instantaneous total molecular dipole moment $\mathbf{M}$ can be expressed in terms of effective electrical localized atomic charges q_p:

$$\mathbf{M} = \sum_p q_p \mathbf{r}_p \tag{8}$$

where the sum extends over all p atoms.

It has been shown [30–36] that the total dipole moment change $\partial \mathbf{M}/\partial Q_i$ appearing in Eq. (7) can be expressed as a function of equilibrium charges and charge fluxes (ECCF model):

$$\frac{\partial \mathbf{M}}{\partial Q_i} = f\left(q_p^\circ, \frac{\partial q_p}{\partial R_t}, \frac{\partial R_t}{\partial Q_i}\right) \tag{9}$$

where q_p° measures the equilibrium atomic charge of atom p, $\partial q_p/\partial R_t$ the charge flux on atom p when the tth internal coordinate is activated, and $\partial R_t/\partial Q_i = (L_R)_{ti}$ is the component of the eigenvector matrix [Eq. (4)] that represents the dynamical terms in the expression of the intensity.

The ECCF model has been shown to be capable of describing inductive and mesomeric effects in many organic systems [33–36]. For the molecular systems treated in this chapter the charge delocalization along a delocalization path in pristine molecules and the charge transfer in doped systems can be easily described with the ECCF model on the basis of experimental intensities.

The case of absolute Raman intensities (Raman cross sections) of polyconjugated molecules requires a separate discussion because it forms the basis for a whole new field of optical spectroscopy and molecular dynamics of low band gap materials. The various aspects are dealt with in detail in Section IV.

II. THE VIBRATIONAL SPECTRA OF POLYCONJUGATED MOLECULES

The vibrational spectra of polyconjugated materials have been, and are still, the subject of many experimental and theoretical works aimed at establishing structure–property correlations. These materials have been studied in the pristine (i.e., undoped), doped, and photoexcited states. The spectroscopic manifestations observed in these three states are dramatically different and have required the development of new theoretical models for their interpretation. The theoretical aspects are outlined in this chapter.

Some of the spectroscopic features are common to all the materials studied and are obviously uniquely related to the existence of a network of π electrons delocalized along the one-dimensional lattice. The discovery of such relations has been matter of experimental and theoretical work by many people.

The main spectroscopic manifestations that will be a matter of discussion later in this chapter are summarized below. The structural variable that originates and modulates the spectroscopic manifestations discussed in this chapter is the conjugation length (hereafter referred to as CL). CL is often referred to in the experimental or theoretical analysis of this chapter.

Infrared and Raman Spectra of Pristine Systems
1. The infrared spectra of undoped oligomers and polymers do not show unusual features and can be interpreted with traditional group frequency correlations or with more detailed vibrational analysis. In contrast to what is observed in the Raman spectra, the infrared spectra show the expected vibrational transitions common to organic molecules. Vibrational frequencies are generally independent

of the number of chemical units in the oligomeric molecules. Since intramolecular coupling turns out to be very limited, phonon dispersion curves for our one-dimensional systems are expected to be generally flat. For this reason only a few bands may show a small red or blue shift. The spectra of oligomers with increasing N seemingly become simpler because vibrations localized at the end of the chain (end group modes) decrease in intensity relative to the vibrations of the "bulk" of the chain. The chain length dependence of the intensity of end-group modes provides one means of identifying them. The intensity ratio I_i(end group)/I_j(bulk) is the probe for the determination of the chain length of an unknown sample [37].
2. As is usual in polymers, the infrared spectrum in polarized light of stretch-oriented samples shows dichroism [8]. For PA [38], an anomalous dichroic behavior has been observed and interpreted.
3. The Raman spectra excited with various lines from the visible to the near-IR of even structurally very complex polyconjugated oligomers and polymers are extremely simple and show very few lines. The strongest characteristic lines occur in the C=C stretching range very approximately near 1500–1400 cm^{-1} and in the C—C stretch + CH wag range very approximately near 1100–1000 cm^{-1}.
4. The strongest characteristic Raman lines generally show dispersion with the number of conjugated units N in the chain. Generally the wavenumbers decrease with increasing N (softening). In some cases the dispersion with N is very large; in other cases the dispersion is observed to be small or even very small [39].
5. The Raman cross sections are very large (relative to σ-bonded systems) and show dispersion with N. For the few cases actually measured (see, e.g., Ref. 40), the Raman cross section increases superlinearly with N.

Infrared and Raman Spectra of Doped (or Photoexcited) Polyconjugated Materials
1. The infrared spectra of doped (photoexcited) molecules are greatly different from those of the pristine molecules. New extremely strong groups of bands with complex structures (often generating a broad and ill-defined spectral pattern) are observed in the range $\approx$1450–700 cm^{-1}. The doping-induced infrared (DIIR) spectrum is generally independent of the doping species (if frequencies are considered; very little is known quantitatively about absorption intensities).
2. In a few series of oligomers the wavenumbers of the DIIR and photoinduced infrared (PIIR) spectra decrease with increasing N.
3. For some molecules (e.g., *trans*-PA), the PIIR spectrum is red-shifted with respect to the DIIR spectrum whereas for most of the other systems the two are superimposable.
4. In some cases the complexity of the DIIR spectral pattern is reduced and a cleaner spectrum is obtained.
5. The DIIR and PIIR spectra of stretch-oriented samples in polarized light show clear dichroism (generally parallel to the stretching direction).
6. The Raman spectra of doped molecules are extremely weak, broad, and often almost unobservable unless conditions of resonance enhancement with the intragap levels are reached.
7. Usually the Raman spectra of the slightly doped materials show the same features as the undoped species.

In the past a great effort was made to find a unified interpretation of the observed spectra and to rationalize trends of variations in the relevant spectral features presented by polyconjugated systems [41,42]. The important conclusion reached is that the vibrational spectra of such molecules can be explained only if one takes explicitly into account the fact that the structure (equilibrium nuclear geometry and its nuclear vibrational displacements) of the conjugated backbone is strongly coupled with the π-electron distribution. This means that any modification of the π-electron distribution induced by various mechanisms such as

1. Electronic excitation (both radiation-induced transition and static field polarization)
2. Chemical substitutions (e.g., the introduction of electron-donating or electron-accepting groups)
3. Local field effects (e.g., solvent effects, doping)

implies a relevant modification of the geometry of the conjugated portion of the molecule. Conversely, structural variations (e.g., the geometric distortions that take place during the vibrational excitations) induce a relevant charge rearrangement of the π-electron cloud.

For vibrational spectroscopists there are two main consequences:

1. It becomes impossible to account for the vibrational spectra of these systems without considering the role of π electrons.
2. The vibrational spectra offer a unique opportunity to obtain information not only on the molecular geometry (as usual) but also on the electronic structure.

Structure and Nonlinear Optics

As discussed in Section X, the most relevant application of this idea is the possibility of measuring nonlinear optical responses (which are properties based, at least in principle, on mechanisms that involve electronic charge polarization), through the measurement of infrared intensities and Raman cross sections.

A nonnegligible advantage in the treatment of the vibrational problem of polyconjugated systems derives from the fact that for a very large class of compounds it is possible to extract a unique and well-defined nuclear coordinate as the only one necessary to describe the coupling between nuclei and π electrons. It can be shown that this simplification is a consequence of the fact that it is possible to obtain a correct description of the electronic properties of these systems under the assumption that only one excited electronic state is relevant.

The hypothesis that only one coordinate in the vibrational space is important in the description of electron–phonon coupling forms the basis of the model of the "effective conjugation coordinate" Я [43], which has been the starting point in the treatment of the vibrational dynamics of conjugated systems.

In the discussion that follows, various series of molecules are considered. First polyacetylene and polyene systems are treated, because they are the prototype models on which the interpretive scheme is based. Then the concepts are generalized to the more complex case of aromatic and heteroaromatic systems.

III. THE VIBRATIONAL SPECTRA OF OLIGOENES AND POLYACETYLENE: GENERAL ASPECTS

The optical and vibrational spectra of PA in its two modifications (*cis*-PA and *trans*-PA) have been the subject of many experimental and theoretical studies for a long time, and their interpretation has laid the ground for the understanding of the physics of the whole class of polyconjugated materials. Most of the peculiar features of the IR and Raman spectra listed above are indeed observed in the case of *trans*-PA, which has become the prototypical system, seemingly the simplest and the one tackled first in all theoretical and/or experimental studies.

The scientific literature on the chemistry and physics of *trans*-PA and *cis*-PA is very rich in both books and reviews [41,42,44–46]. The discussion here is restricted to the information that has been obtained from frequency and intensity vibrational spectroscopy since the publication of Refs. 41 and 42.

A. Structural Data on *cis*-PA and *trans*-PA from the Vibrational Spectra

It is well known that PA is first obtained in the *cis* form, and by thermal isomerization in the solid state it is isomerized to the *trans* isomer. The mechanism at the molecular level of the thermal isomerization is not yet known. Some insight into such mechanisms has been obtained from the time-dependent infrared spectra during thermal isomerization at various temperatures [47–49]. Let us consider the characteristic strong group frequency absorption assigned to the out-of-plane C—H motion that is observed at room temperature at 735 cm^{-1} for the *cis* isomer and at 1015 cm^{-1} for the *trans* isomer (Fig. 28.1). If the polymers are considered as one-dimensional lattices [9,10], the frequencies at 735 and 1015 cm^{-1} can be considered the limiting $k = 0$ phonon frequencies. Time-dependent spectra (taken every 2 s) at $T = 140°C$ (Fig. 28.2a) show that the absorption of the *cis* isomer decreases in intensity with time and shifts from 735 cm^{-1} at $t = 0$ s to 741 cm^{-1} at $t \approx 200$ s while the band at 1015 cm^{-1} at $t = 0$ increases in intensity with time and shifts to 1019 cm^{-1} at $t \approx 200$ s. The

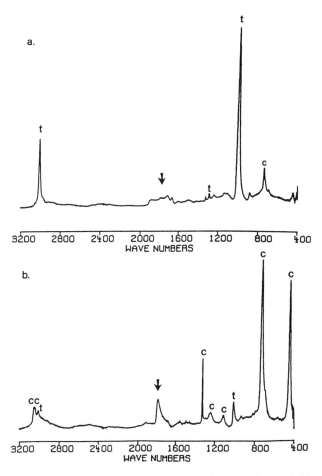

Fig. 28.1 Infrared spectra of (a) mostly *trans*-PA and (b) mostly *cis*-PA. *t* indicates the bands of *trans* configuration and *c* indicates those of *cis* configuration. (The arrows indicate two-phonon bands, see Section III.C.) (From Ref. 41.)

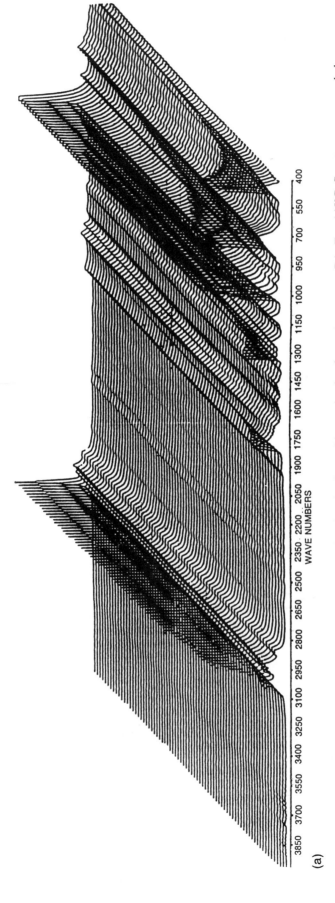

Fig. 28.2 (a) Time-dependent FTIR spectrum showing the thermal isomerization from *cis*- to *trans*-PA. $T = 145°C$. Spectra were recorded every 2 s. (b) Time dependence of CH out-of-plane wavenumber of *cis*-PA during thermal isomerization at 145°C. (c) Time dependence of CH out-of-plane wavenumber of *trans*-PA during termal isomerization at 145°C. (From Ref. 48.)

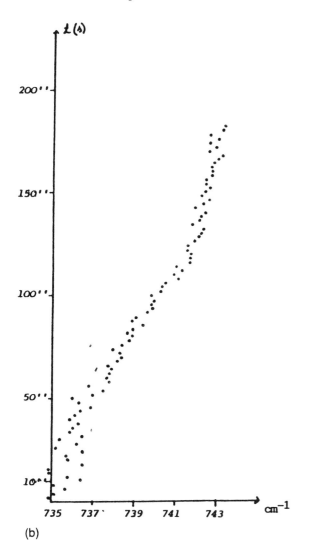

(b)

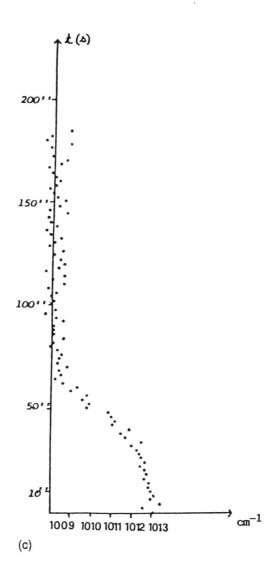

(c)

experimental observations on the out-of-plane modes are as follows.

For the *cis* isomer (band at 735 cm^{-1}) (Fig. 28.2b):

1. Almost constant wavenumber (± 2 cm^{-1}) from $t \approx 2$ s to $t \approx 50$ s
2. Wavenumber shift linear with time from $t \approx 50$ s to $t \approx 130$ s ($\Delta \nu \approx 5$ cm^{-1})
3. Almost constant wavenumber from $t \approx 130$ s to $t \approx 200$ s

Similar behavior is observed for the band assigned to the C—H in-plane wag near 1327 cm^{-1} (where the total wavenumber shift is only ≈ 2 cm^{-1}).

For the *trans* isomer (band at 1015 cm^{-1}) (Fig. 28.2c):

1. From $t \approx 0$ s to $t \approx 60$ s, the band shifts linearly with time from 1015 to 1008 cm^{-1}.
2. From $t \approx 60$ s to $t \approx 200$ s the wavenumber remains constant.

Definitely the vibrational spectra show the existence of different molecular and/or supermolecular "phases" during the thermal isomerization process. From the infrared spectra of short oligoenes one can observe a dispersion of the C—H out-of-plane motions near 1000 cm^{-1} with increasing chain length (hexatriene 941 cm^{-1}, decapentaene 980 cm^{-1}). These wavenumber changes are related to the dispersion of the C—H out-of-plane $k \neq 0$ phonons of short chains whose limiting $k \approx 0$ phonon approaches 1015 cm^{-1} (calculated A_u phonon for infinite *trans*-PA). The observed strong transition corresponds to the phonon wave with the smallest k compatible with the length of the chain. This is a classical interpretation of frequency dispersion that does not require the introduction of a conjugation length (CL)-dependent intramolecular potential (see Section IV.C).

Further proof that this interpretation is correct comes from the study of the progression of combination bands in the infrared spectra of polyenes and *trans*-PA (see Section III.C), which show that the C—H out-of-plane motion is not affected by CL.

The decrease in the wavenumbers observed from $t = 0$ s seems to indicate that the mean chain length of *trans*-PA produced in the early stages of the reaction is very long and then decreases linearly with time during the first 60 s and stabilizes. The observed plateau reached after 60 s may indicate the existence of a limiting shorter chain length dictated by the generation of a network of cross-links.

On the other hand, for the *cis* isomer the nature of the molecules probed by the C—H out-of-plane motion or C—H in-plane wag certainly differs with time. There is no information on oligomers that allows the establishment of a correlation between wavenumber and chain length (or number of oscillators) nor is there any knowledge of the shape of the phonon dispersion curve for the 735 cm^{-1} phonon. It can be proposed that during stage 1 most of the molecules maintain their original average structure, during stage 2 chains release *trans*-PA and the original *cis* chains evolve with time toward a residual material, and by the end of the process (stage 3) the chain structure is definitely different from the starting one.

This interpretation considers isolated chains and ascribes the frequency changes to intramolecular phenomena. Intermolecular interactions may also play a role. So far there are no reliable data for taking intermolecular π–π interactions into account. Only some preliminary data on the spectra of PA samples under high pressure are available [51,52].

Intensity spectroscopy has been also applied during thermal isomerization [53,54]. It must be remembered that for compounds containing CH groups the ratio $\sum A_{\text{(C-H in-plane)}} / \sum A_{\text{(C-H out-of-plane)}}$ decreases with the increase in charge fluxes generated by an angular deformation during C—H waggings [53]. The time-dependent evolution of the intensity ratio A_{1291}/A_{1015} for the nascent *trans*-PA forms a plateau until $t \approx 50$ s and then smoothly decreases, indicating that after 50 s the delocalization extends along longer delocalization paths. On the other hand, for the reactive *cis*-PA the intensity ratio $A_{\text{(C-H stretch)}}/A_{\text{(C-H out-of-plane)}}$ decreases noticeably. This can be interpreted in terms of a decrease in charge fluxes induced by C—H stretchings along the carbon–carbon backbone. This supports the idea that electron localization increases reasonably as a result of a further conformational twisting of the chain, which reduces the intramolecular electron hopping (i.e., delocalization). This can be taken as information of interest with respect to the molecular mechanism of thermal isomerization.

The evolution of the infrared spectra with time and temperature during isomerization is obviously related to the evolution of the molecular structure. While the structure of *trans*-PA is definitely and clearly established, the structure of the *cis* isomer has received relatively less attention [55] and does not seem to be definitely settled.

From X-ray diffraction [56,57] and solid-state NMR [58] experiments, the molecular chain of *trans*-PA is certainly *trans*-planar and dimerized with alternating C—C and C═C bonds. The bond lengths $d_{\text{C-C}} = 1.44$ and $d_{\text{C=C}} = 1.36$ indicate that delocalization of π electrons takes place along the chain. From the values of bond lengths (or from the values of bond orders) the delocalization does not seem to be very large. The choice between a dimerized (C_{2h}) structure and an undimerized or metallic (D_{2h}) structure can be made on the basis of the vibrational spectra. A study of the activity of Raman bands supported the choice of a dimerized structure [59]. Later this indication found unquestionable proof in the interpretation of the anomalous dichroic behavior of the infrared spectrum in polarized light of stretch-oriented samples of *trans*-PA [38]. The same experiment provided qualitative evidence that a large charge flux takes place along the skeletal backbone during C—H stretchings. A quantitative estimate of the atomic charges and charge fluxes in *trans*-PA is obtained from infrared intensity spectroscopy following ECCF theory, namely, $q_H^o = 0.134$ e, $q_C^o = -0.134$ e, $\partial q_H/\partial r_{CH} = -0.207$ e/Å, $\partial q_C/\partial r_{CH} = -0.241$ e/Å [53]. The value of q_H^o is in good agreement with that calculated from ab initio methods on some polyenes (e.g., $q_H^o = 0.127$ e) from Mulliken population analysis with a 6.31 G basis set [60].

The molecular structure of *cis*-PA is still an open question that is generally neglected in the literature. First is the question of the distinction between the *cis*-transoid or *trans*-cisoid configurations, and second is the question of coplanarity versus helicity or nonplanarity. The issue of electronic configuration has never been settled either, neither on theoretical grounds by ab initio methods nor experimentally by X-ray diffraction or solid-state NMR experiments. The latter report the value of only one carbon–carbon bond distance equal to 1.34 Å [58], very short indeed if the C═C bonds were conjugated. An additional complication arises in ab initio calculations because they are generally made for molecules *in vacuo*, while these systems are always in the solid phase where strong π–π intermolecular interactions may take place.

From intensity spectroscopy of *cis*-PA the atomic charges on the hydrogen atoms ($q_H^o = 0.160$ e) turn out to be larger than in *trans*-PA, and the charge fluxes with deformation modes are certainly much smaller than in *trans*-PA [53]. It has to be concluded that π electrons are mostly localized over a limited molecular domain. Such electron localization can take place only if the hopping integral between p_z orbitals is decreased because of conformational distortions.

Steric repulsions between the hydrogen atoms facing each other in the *cis* structure are very likely to occur (Fig. 28.3), thus forcing a conformational twisting that may generate a helical conformation. A helical structure based on conformational energy calculations was first suggested by Cernia and D'Ilario [55] but was generally undermined by the scientific community. An overall view of the vibrational data suggests that the possibility of a helical or nonplanar structure for *cis*-PA cannot be overlooked.*

B. Dynamics of *trans*-PA: Experimental Data and Phonon Dispersion Curves

A comprehensive study of the vibrational features of polyacetylene can be carried out only if a reliable intramolecular potential is obtained. For this purpose it is very useful to analyze the spectra of oligomers, i.e., polyenes with well-defined chain length.

---

* This idea was first presented by I. Bozovic. G. Z. acknowledges interesting personal discussions with him.

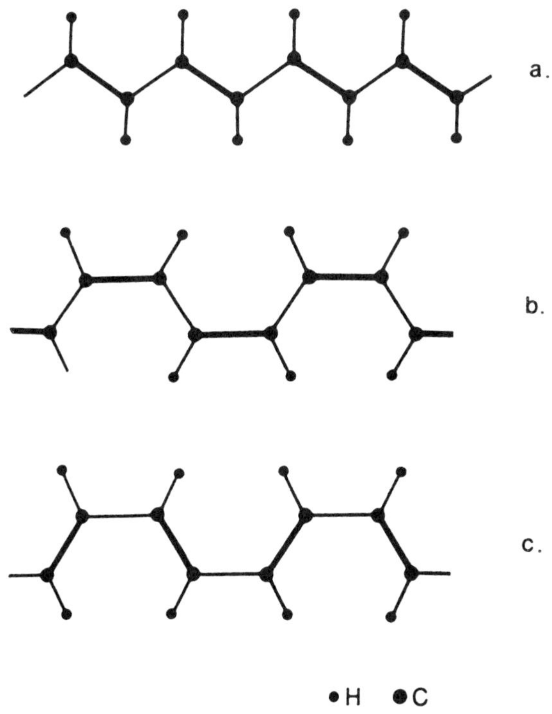

Fig. 28.3 Structure of polyacetylene. (a) *trans*-Transoid; (b) *cis*-transoid; (c) *trans*-cisoid.

Model molecules of *trans*-PA have been synthesized, and their properties have been repeatedly studied both experimentally and theoretically. Efforts have been made by Kiehl [61] and Kiehl et al. [62] to synthesize oligoenes capped at either end with *t*-butyl groups to increase their solubility and allow to study the properties of these materials in solution. The availability of natural and synthetic carotenoids with a variety of chemical modifications suggests strong correlations between the physics of polyenes or *trans*-PA and the physics of the chromophores of relevance in biophysics and bioelectronics [63–65]. Fox et al. [66], using the metathesis method, contributed to the science of polyenes by the synthesis of very long (up to ≈ 250 C=C bonds) soluble and relatively stable polyenes.

Following the classical approach of vibrational dynamics (as in the case of σ-bonded polymers) [67], the experimental data from the oligomers could in principle be used to obtain the phonon dispersion curves of the ideally infinite polymer. In the case of π-bonded systems, complications arise because the skeletal modes are strongly CL-dependent and cannot be treated in the standard way. The rationalization of the behavior of the totally symmetric Raman-active normal modes leads to the definition of a CL-dependent intramolecular potential on which the effective conjugation coordinate (ECC) theory is based (see Section IV).

An independent experimental way to derive information on the phonon dispersion curves is the recording of the incoherent inelastic neutron scattering (IINS) spectra. One-phonon density of vibrational states $g(\nu)$ from IINS spectra have been reported by Machonnachie et al. [68] and were used by Takeuchi et al. [69] for their studies of the phonon dispersion curves of *trans*-PA. Several of the singularities in the experimental $g(\nu)$ correspond to the infrared or Raman one-phonon transitions whose assignment has been widely treated in many papers. Very recently Hirata et al. [63,64] presented ab initio second-order Møller–Plesset perturbation calculations on *trans*-octatetraene, *trans*-decapentaene, and *trans*-PA. Calculations included the analysis of the force field, of dispersion curves, and of amplitude-weighted density of states.

On this basis, one could think that at present there is an approximate, but satisfactory, experimental knowledge of the extent of dispersion of the in-plane and out-of-plane phonons for *trans*-PA considered as an infinite dimerized single chain. Actually this is correct only for phonon branches corresponding to in-plane modes of B_u symmetry (at $\mathbf{k} = 0$) and to out-of-plane modes that can be successfully treated in terms of an intramolecular potential independent of CL. The actual $\mathbf{k} = 0$ frequencies and the shape of the other phonon branches are strongly CL (and thus sample) dependent. Moreover, any theoretical approach is heavily affected by the choice of the basis set. It follows that a precise determination of the vibrational force field is not easily derived experimentally. This problem is further treated in Section V.C.

C. Independence of Out-of-Plane Motion from Conjugation Length: Two-Phonon States

Further insight into the coupling of vibrational motions and delocalized π electrons can be reached from the unusual broad and weak infrared absorption observed in the frequency range 1990–1700 cm^{-1}. The infrared spectrum of *trans*-PA shows an extremely broad continuum between 1910 and 1720 cm^{-1} reminiscent of a dipole-weighted density of states $g(\nu)$ with two stronger singularities at either end of such a continuum (Fig. 28.1a). A similar feature (but with a different pattern) is observed in the IR spectrum of *cis*-PA (Fig. 28.1b). Takeuchi et al. assigned this broad absorption to the joint (or two-phonon) density of states $g^{(2)}(\nu)$ [69]. On the other hand, if a real sample of PA is considered as a polydisperse system (both in chemical length and in conjugation length), the observed band can be considered as the convolution of many binary combination transitions originating from the C—H out-of-plane fundamental modes.

The spectra were recently analyzed by Rumi and Zerbi [70], who used the infrared spectra of several oligomers as model compounds. The infrared spectra of *t*-butyl-capped dodecatetraene and *t*-butyl-capped hexadecahexaene show four lines at 1723, 1785, 1847, and 1887 cm^{-1} and six lines at 1712, 1749, 1796, 1840, 1873, 1896 cm^{-1}, respectively, which can be easily assigned to a progression of bands (Fig. 28.4) associated with the combination tones of the C—H out-of-plane fundamental motions, which are very strong in the infrared. Similar progressions with the right number of components have been observed for model molecules such as diphenylpolyenes Φ-(CH)$_n$-Φ with $n = 2, 3$, and 4.

For the interpretation of the observed band progressions, the vibrational frequencies of a molecular chain with fixed ends can be classified in terms of the phonon wavevector **k**:

$$k = m\pi/d(N + 1)$$

where N is the number of repeating units, d the repeat distance, and m an integer ($m = 1, 2, \ldots, N$). The binary sums could be evaluated if the phonon dispersion curves of *trans*-PA were known. In the analysis of the progression bands, Rumi and Zerbi [70] used the phonon dispersion curve of the C—H out-of-plane motions calculated by Takeuchi et al. [69]. The two optical out-of-plane phonon branches are responsible for the combination tones observed: the one corresponding to the **k** = 0 phonon of A_u species (IR-active, observed at 1015 cm^{-1}) and the **k** = 0 phonon of B_g species (Raman-active, observed as a very weak Raman line at 884 cm^{-1}).

The results obtained confirm that out-of-plane motions turn out to be almost independent of conjugation length. The same observation can be made for the out-of-plane motions of other polyconjugated systems.

One fact can be taken for granted: whenever a molecule contains —CH═CH— groups, the C—H out-of-

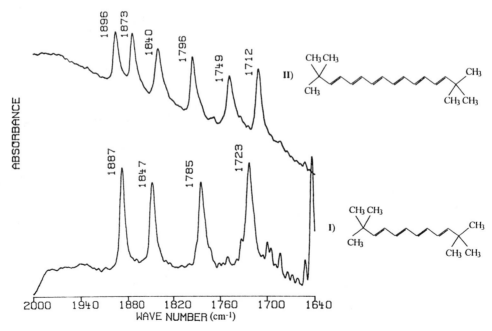

Fig. 28.4 Infrared spectra of the compounds sketched in the frequency range 2000–1640 cm^{-1}. (From Ref. 70.)

plane deformation modes generate a few very selective absorption bands due to combination tones in the 2000–1650 cm^{-1} range. Following Whiffen [71], it must be concluded that electrical anharmonicity and not mechanical anharmonicity is the dominant factor, which can be related to the nature of localized or delocalized π electrons. New data of general validity are available for further theoretical studies.

D. Vibronic Spectrum of *trans*-PA

Another source of experimental information on vibrational motions are vibronic spectra. The use of data from oligoenes allows a reliable extrapolation to *trans*-PA. One must first gain an overall description of the structure of the electronic states and of the shapes of oligoenes in the ground state and at least the first excited electronic state. The electronic spectra of simple oligoenes have been systematically studied by Kohler and coworkers [72–74], who were able to resolve the fine vibronic structure of absorption and fluorescence spectra of oligoenes (from three to seven C=C bonds) in solid inert solutions. In addition, in order to locate precisely the HOMO–LUMO A_g–B_u energy gap, Kohler et al. showed unambiguously that an excited state of A_g symmetry is lower in energy than the B_u band gap state. While the dipole-allowed one-photon transition A_g–B_u occurs as strong absorption at energies almost linearly dependent on $1/N$, the A_g–A_g symmetry-forbidden (at first order) one-photon dipole absorption is made weakly allowed by e–ph coupling. It can be directly observed as a weak one-photon absorption or as a two-photon absorption. The extrapolation to the fact that *trans*-PA may also have an A_g state with energy less than that of the B_u state has been considered straightforward and has introduced an important element into the understanding of the photophysics of transient states of *trans*-PA widely studied in the last few years [75,76].

As to the vibrational problem, as predicted by theory [77], from the vibronic structure it turns out that the totally symmetric modes are those that first occur in the vibronic spectrum of polyenes (and in general in the vibronic spectra of other polyconjugated molecules). The spacing of the vibronic components in the observed vibronic spectrum turns out to be determined mostly by the frequency of the most intense band of the Raman spectrum. This band is assigned to a longitudinal collective vibration involving stretching of C=C bonds and shrinking of C—C bonds (Я mode, see Section IV.B).

The study of the vibrational fine structure in the absorption spectrum of all-*trans*-octatetraene unquestionably shows that in going from the planar centrosymmetric ground-state equilibrium structure to the first dipole-allowed excited-state equilibrium geometry, only the single and double bond lengths are significantly changed [72]. This was already suggested by earlier semi-empirical quantum-chemical calculations; the reliability of these calculations was, however, put in question by the fact that approximate molecular orbital calculations or semiempirical calculations can fail in describing the experimentally observed excited electronic state ordering or dependency of E_g on chain length [78]. More recently, semiempirical calculations on long polyenes [79,80] confirmed both electronic states ordering and geometric changes in the excited states. All this information is of basic importance in supporting the assumption on which ECC theory is based.

IV. THE POLYENE SYSTEMS AS A STARTING POINT FOR THE DEVELOPMENT OF ECC THEORY

A. Raman Spectra

The Raman spectrum of *trans*-PA is characterized by a very simple pattern that consists of two very strong bands (1462, 1072 cm^{-1}) assigned to A_g normal modes, involving skeletal stretching and C—H wag. Both bands exhibit satellite peaks on the high energy side that undergo an appreciable frequency shift with laser excitation, merge in the principal peaks with long-wavelength visible excitation (λ_{exc} = 647 nm), and completely disappear in the FT-Raman spectrum (λ_{exc} = 1064 nm) (Fig. 28.5a).

Comparison between the Raman spectra of *trans*-PA and those of its oligomers shows that the two-band pattern is the signature of the occurrence of a *trans*-polyenic conjugated chain. In Figs. 28.5b and 28.5c, spectra of short and medium-length polyenes are reported for comparison. Comparison of the wavenumbers of the two relevant Raman bands (ν_1 and ν_3) for a series of polyenes with increasing chain length (Table 28.1; Fig. 28.6) shows a large wavenumber shift (dispersion) of both bands ($\Delta \nu_1$ = 164 cm^{-1}; $\Delta \nu_3$ = 119 cm^{-1}) from hexatriene to *trans*-PA.

This observation first suggested an interpretation of the frequency dispersion with excitation wavelength of the satellite peaks in the Raman spectrum of PA samples. The observed frequency dispersion is the experimental proof that the sample is polydisperse with respect to conjugation length. The Raman experiment with variable excitation in the visible range selectively probes chains with a particular conjugation length (CL) whose Raman-active vibrations are strongly enhanced by resonance effects.

The dispersion of ν_1 and ν_3 with CL has been the subject of extensive studies by several groups for the understanding of the CL heterogeneity of real samples of *trans*-PA [81–83]. The derivation of the distribution of CL (multimodal distributions) from the Raman line shapes as functions of the excitation wavelength has been a matter of numerous studies. Unfortunately the "multimodal distributions" proposed have always been

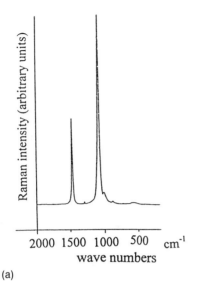

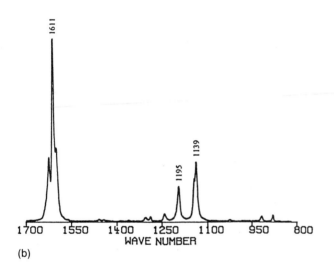

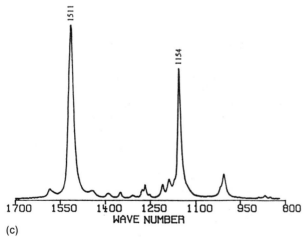

Fig. 28.5 FT-Raman spectra ($\lambda_{exc} = 1064$ nm) of (a) *trans*-PA; (b) *tert*-butyl-end-capped *trans*-octatetraene; (c) β-carotene.

Table 28.1 Experimental Wavenumbers ν_1 and ν_3 (see text) for Polyene-Like Molecules Sketched in Fig. 28.7[a]

Eff. chain length	Molecule	ν_1 (cm^{-1})	ν_3 (cm^{-1})
$N = 3$	A	1626	1191
	B	1639	1136
	C	1636	1193
	I	1593	1172
	II	1585	1183
	III	1600	1185
	IV	1586	1120
	V	1530	1200
	VI	1507	1184
	VII	1503	1181
$N = 4$	A	1613	1179
	B	1614	1142
	C	1625	1142
	I	1588	1143
	II	1581	1143
	V	1510	1160
$N = 5$	B	1587	1145
	C	1595	1144
	D	1577	1161
	I	1577	1143
	II	1573	1144
	IV	1582	1148
$N = 6$	B	1571	1144
	C	1575	1141
	III	1567	1186
	IV	1575	1150
$N = 7$	B	1556	1144
	IV	1558	1147
$N = 8$	B	1542	1140
$N = 9$	B	1531	1135
	D	1523	1156
$N = 10$	B	1521	1130
	III	1521	1156
$N = 11$	B	1514	1125
$N = 12$	B	1506	1122
$N = 13$	D	1503	1149
$N = 17$	D	1496	1144

[a] N is an "effective" chain length (see text) defined as $N = n_1 + n_2$, where n_1 is the number of double bonds and n_2 the number of phenyl rings (end groups).

handled by physicists who have never tried to critically justify their results in terms of the chemistry of the polymerization processes.

On the other hand, the large frequency dispersion observed both for the Raman bands of oligomers and for the polymer show the relevance of CL to the dynamics of polyene systems. The main conclusion derived from *only frequency analysis* of the Raman spectra is that the

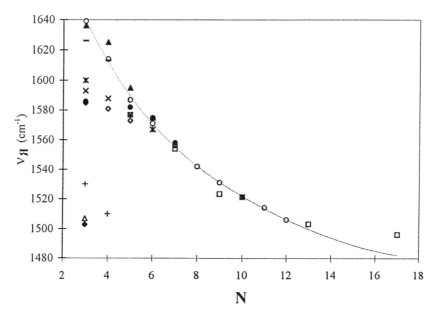

Fig. 28.6 Experimental Raman frequencies of ν_1 ("Я mode", see text) vs. N (number of double bonds in the chain) for the polyene systems sketched in Fig. 28.7. (—) *trans*-polyenes (A); (○) *tert*-butyl-end-capped *trans*-polyenes (B); (▲) diphenylpolyenes (C); (□) carotenoids (D); (×) diphenylpolyenes (I); (◇) diphenylpolyenes (II); (★) phenylpolyenes (III); (●) polyenovanillins (IV); (+) (molecule V); (△) (molecule VI); (◆) (molecule VII)

intramolecular potential of polyenes (at least for in-plane motions) is heavily affected by the degree of π-electron delocalization along the chain. Various authors have dealt with the determination of the force fields for in-plane modes of *trans*-PA and its oligomers [84–89]. A comprehensive discussion and a comparative analysis of the different force fields are reported in Ref. 41 (see also Refs. 63 and 64). The main problem in the treatment of the dynamics of polyenes is the description of the long-range carbon–carbon bond stretching interactions along the conjugated backbone due to the occurrence of delocalized and easily polarized π electrons. The determination of the values of the interaction force constants (and of the distance at which the interactions become negligible) is the main problem in an attempt to derive an empirical force field (which contains several terms) from the (very few) experimental frequency data. On the other hand, the results obtained from completely theoretical ab initio force fields are also severely limited because the effect of π-electron delocalization may be heavily underestimated in computations [86,90]. This problem can be partially overcome with the use of suitably parameterized semiempirical calculations [79,80].

Since different sets of force constants can give very similar results in terms of the predicted frequencies, a way to check the validity of a force field is to critically discuss the assignments while also taking into account the intensity data.

B. Raman Intensity

Absolute experimental Raman cross sections can be obtained for samples in solution from intensity ratios with some reference band of the solvent, whose absolute Raman intensity has been carefully measured. Measurement of FT-Raman intensities of diphenylpolyenes and carotenoids of different lengths dissolved in CCl_4 have been made with excitation at $\lambda_{exc} = 1064$ nm. Intensity values obtained for the two most relevant Raman-active lines (ν_1 and ν_3) are reported in Table 28.2. In all cases examined, unusually large intensity values were obtained in spite of the fact that the Raman experiments were made with laser excitation well below the energy gap of the systems considered ($\lambda_{exc} = 1064$ nm) and thus far from resonances.

Comparison of the intensity of the strongest line of *t*-butyl-capped dodecahexaene ($I = 1.3 \times 10^{-6}$ cm^4/g) with the Raman intensity integrated over the whole C—H stretching region ($\approx$2850–2950 cm^{-1}, the most intense lines in the spectrum) of an *n*-dodecane ($I = 1.45 \times 10^{-7}$ cm^4/g), gives an idea of the strength of the Raman-active transition in the spectra of polyenes. Comparison of ab initio calculated Raman intensities for *n*-alkanes and *trans*-polyenes gives parallel information [91,92]. Note that when the number of carbon atoms in the chains increases, the Raman intensities of saturated and conjugated chains differ remarkably in their behav-

Table 28.2 Experimental Raman Intensities and Infrared Intensities of Ħ Modes (see text) for Polyene-Like Molecules Sketched in Fig. 28.7

Eff. chain length[a]	Molecule	ν_1		ν_3	
		I_{Raman} (cm^4/g)	I_{IR} (km/mol)	I_{Raman} (cm^4/g)	I_{IR} (km/mol)
$N = 3$	C	1.7×10^{-7}	—	3.6×10^{-8}	—
	I	5.9×10^{-7}	328	1.4×10^{-7}	37
	II	8.6×10^{-7}	501	2.9×10^{-7}	250
	III	3.0×10^{-7}	833	2.1×10^{-8}	29
	IV	7.2×10^{-8}	142	1.2×10^{-8}	224
	V	4.1×10^{-7}	429	6.7×10^{-7}	309
	VI	7.7×10^{-7}	1827	5.9×10^{-7}	1754
	VII	7.5×10^{-7}	1879	7.1×10^{-7}	2000
$N = 4$	C	5.9×10^{-7}	—	7.7×10^{-8}	—
	I	1.4×10^{-6}	463	2.8×10^{-7}	64
	II	1.9×10^{-6}	614	3.7×10^{-7}	93
	V	2.7×10^{-6}	3297	2.9×10^{-6}	1917
$N = 5$	C	2.1×10^{-6}	—	3.2×10^{-7}	—
	D	9.3×10^{-7}	263	8.7×10^{-8}	69
	I	3.6×10^{-6}	491	6.6×10^{-7}	104
	II	5.5×10^{-6}	585	9.6×10^{-7}	114
	IV	9.6×10^{-7}	358	2.9×10^{-7}	217
$N = 6$	B	1.2×10^{-6}	—	4.2×10^{-7}	—
	III	4.1×10^{-6}	419	1.2×10^{-6}	579
	IV	2.4×10^{-6}	350	7.9×10^{-7}	209
$N = 7$	IV	3.7×10^{-6}	416	1.6×10^{-6}	413
$N = 9$	D	1.2×10^{-5}	—	7.0×10^{-6}	—
$N = 10$	III	1.8×10^{-5}	345	6.5×10^{-6}	125
$N = 13$	D	2.1×10^{-5}	—	9.7×10^{-6}	—
$N = 17$	D	3.6×10^{-5}	—	1.7×10^{-5}	—

[a] See footnote to Table 28.1.

ior with N. Whereas Raman intensities of saturated chains increase linearly with the number of carbon atoms, a superlinear increase is predicted for polyene chains. A superlinear increase of the Raman cross sections ν_1 and ν_3 is indeed also found from the experiments reported in Table 28.2 and in Fig. 28.8.

The relative intensities of the two strong and intense Raman lines are found to be similar when excited at 1064 or 514 nm (near resonance condition) (see Fig. 28.9).

The very important fact that Raman spectra of polyenes have unusually large intensities and always show the same two-band pattern allows us to use a simplified analytical model for the interpretation of the intensity data.

From second-order perturbation theory, an analytic expression for the Raman cross section can be derived. The general expression for the (Raman) polarizability tensor is

$$\alpha_{uv} = \sum_{nk} \frac{\langle 0j|e_u M|nk\rangle \langle nk|e_v M|0i\rangle}{E_{nk} - E_{0j} - h\nu_L} \\ + \frac{\langle 0j|e_u M|nk\rangle \langle nk|e_v M|0i\rangle}{E_{nk} - E_{0j} + h\nu_L} \quad (10)$$

where $\langle 0j|$ is the final eigenstate representing the ground electronic and jth vibrational levels, $\langle nk|$ is a vibronic level of electronic quantum number n and vibrational quantum number k, $\langle 0i|$ is the initial eigenstate representing the ground electronic and the ith vibrational levels; M is the molecular dipole moment operator, e_u and e_v are unit vectors that account for the state of polarization of the incident and scattered radiation; E_{nk}, E_{0j} are the energies of the eigenstates involved in the process; and $h\nu_L$ is the energy of the exciting laser.

More explicit expressions have been worked out by several authors (for a comprehensive review see Ref.

Structure and Nonlinear Optics

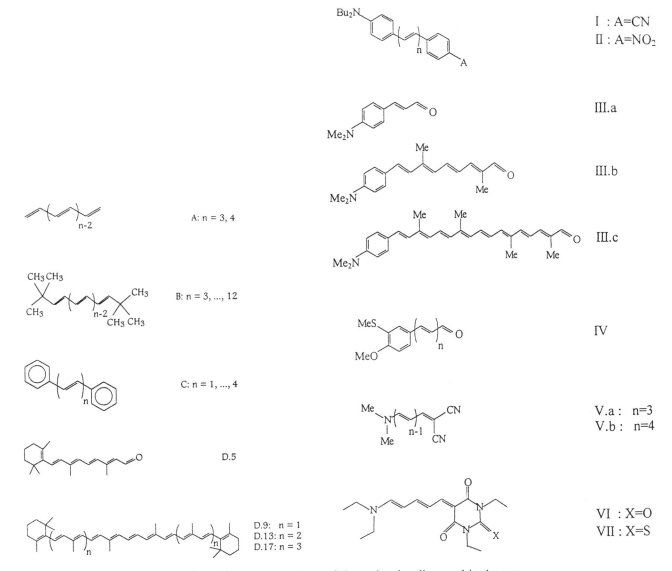

Fig. 28.7 Chemical structures of the molecules discussed in the text.

93) starting from Eq. (10). Expressions for Raman cross sections under resonance or preresonance conditions have also been derived [94,95], and resonant Raman spectra of a variety of compounds have been interpreted accordingly [96–98]. The difference between the general analytic expression for the Raman polarizability and that relative to the resonant case is determined by the fact that the sum over n reduces to the terms containing the resonant electronic state. This means that, under resonance conditions, the molecule behaves as if only two electronic states (the ground state and the excited state selected by the energy of the laser excitation) were relevant. Further elaboration of this simplified formula for Raman scattering allows us to obtain more explicit analytic expressions for resonant Raman intensities that have been used for the interpretation of the spectral pattern and to discuss excitation profiles [96–98].

Resonant Raman spectra are generally very selective, and in most cases the normal modes belonging to totally symmetric species dominate the Raman spectrum. This fact can be rationalized, according to the analytic treatment mentioned above, in terms of "normal mode displacements" relative to the transition from the ground state to the only excited state, as first suggested by Inagaki et al. [84] for the specific case of *trans*-PA.

The fact that nonresonant FT-Raman spectra of polyenes exhibit the same pattern as that obtained under resonance conditions suggests that far from resonance

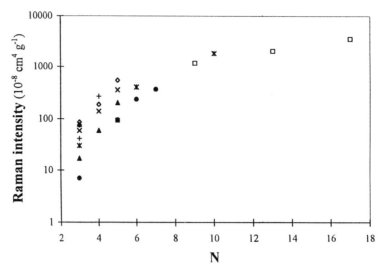

Fig. 28.8 Experimental absolute Raman cross sections vs. N (the number of double bonds in the chain) for the polyene systems sketched in Fig. 28.7 in cm^4/g. Symbols as in Fig. 28.6.

also it may be possible to treat the Raman tensor [Eq. (10)] as if only one of the electronic excited states were important [99]. Under these conditions it becomes possible to treat Raman scattering cross sections with low energy excitation under the hypothesis of a two-state model. On the other hand, the two-state model may be justified not only on the basis of a small denominator [small HOMO–LUMO transition energy in Eq. (10)] but also by the fact that the HOMO–LUMO transition of polyenes has a large oscillator strength. Following the analytic development suggested by Ting [100],

$$\frac{\partial \alpha_{uv}}{\partial Q_k} = (4\pi^2 \nu_k^2 c^2)\, (\Delta Q^{\mathrm{eg}}\, |M_{uv}^{\mathrm{ge}}|^2) \qquad (11)$$

$$\times \left[\frac{1}{(E_{\mathrm{g}} - h\nu_{\mathrm{L}})^2} + \frac{1}{(E_{\mathrm{g}} + h\nu_{\mathrm{L}})^2} \right]$$

which in the limit of a static field becomes

$$\frac{\partial \alpha_{uv}}{\partial Q_k} = (8\pi^2 \nu_k^2 c^2)\, \frac{\Delta Q^{\mathrm{eg}}\, |M_{uv}^{\mathrm{ge}}|^2}{E_{\mathrm{g}}^2} \qquad (12)$$

Note that in Eqs. (11) and (12) the variation of the polarizability tensor with respect to the normal mode is explicitly introduced by making use of the electrical harmonic approximation. Intensity values reported in Table 28.2 correspond to the mean of the tensor $\frac{\partial \alpha}{\partial Q_k}$, as obtained in an experiment with polarization of the scattered light parallel to that of the incident beam [$I_k = (1/45)(45\bar{\alpha}_k^2 + 4\beta_k^2)$], where $\bar{\alpha}_k$ and β_k^2 are the invariants of the polarizability tensor defined as

$$\bar{\alpha}_k = \frac{1}{3}\left(\frac{\partial \alpha_{xx}}{\partial Q_k} + \frac{\partial \alpha_{yy}}{\partial Q_k} + \frac{\partial \alpha_{zz}}{\partial Q_k} \right) \qquad (13)$$

$$\beta_k^2 = \frac{1}{2}\Bigg[\left(\frac{\partial \alpha_{xx}}{\partial Q_k} - \frac{\partial \alpha_{yy}}{\partial Q_k} \right)^2 + \left(\frac{\partial \alpha_{xx}}{\partial Q_k} - \frac{\partial \alpha_{zz}}{\partial Q_k} \right)^2$$
$$+ \left(\frac{\partial \alpha_{xy}}{\partial Q_k} - \frac{\partial \alpha_{zz}}{\partial Q_k} \right)^2 + 6\left(\frac{\partial \alpha_{xy}}{\partial Q_k^2} + \frac{\partial \alpha_{xz}}{\partial Q_k^2} + \frac{\partial \alpha_{yz}}{\partial Q_k^2} \right) \Bigg]$$

$$(14)$$

Equations (11) and (12) are obtained under the hypothesis of negligible vibronic coupling, and they correspond to considering only the Condon factors in the development by Tang and Albrecht [93]. The hypothesis that other terms (Herzberg–Teller terms) are negligible is fully consistent with the choice of a two-state model. The possible existence of other nonnegligible contributions to Raman intensities is irrelevant for the qualitative discussion that follows.

Expressions such as Eq. (11) or (12) are very useful in the discussion of the relative band intensities of the Raman spectra. The values of ΔQ_k^{eg} that appear in the numerator completely determine the intensity pattern of the spectrum and are the origin of the high selectivity for few normal vibrations. ΔQ_k^{eg} represents the variation of the equilibrium position of the normal mode Q_k in going from the ground to the electronic excited state. According to Eq. (12), if the exact path of nuclear equilibrium structure variation from the ground to the excited state

Structure and Nonlinear Optics

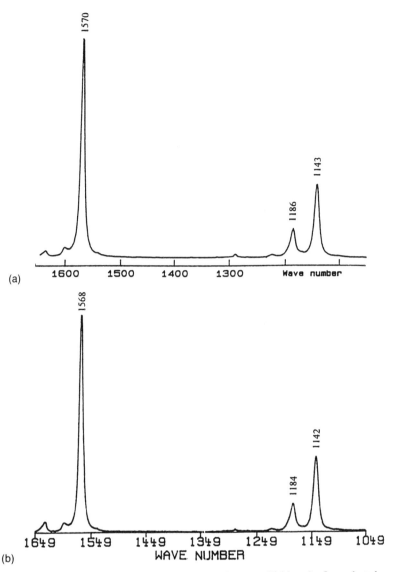

Fig. 28.9 (a) Raman ($\lambda_{\text{exc}} = 514$ nm) and (b) FT-Raman spectra ($\lambda_{\text{exc}} = 1064$ nm) of *tert*-butyl-end-capped *trans*-dodecaesaene.

is known, it is also known whether a given normal mode has a large Raman intensity. In particular, only normal modes that have an appreciable projection along the coordinate that describes such geometric variation (ΔX^{eg}) may have relevant Raman intensity in the spectrum. Moreover, if no changes take place in the molecular symmetry during the electronic transition from the ground to the excited state, only totally symmetric normal modes can have nonvanishing ΔQ_k^{eg}; accordingly, only A_g modes in all *trans*-polyenes and *trans*-PA can be observed in the Raman spectra, as the symmetry point group of the molecule (C_{2h}) is the same in the $1A_g$ and $1B_u$ electronic states. Among the modes belonging to the correct symmetry species (A_g), only a few normal modes have an appreciable projection along ΔX^{eg}; indeed, only two intense bands are observed.

Knowledge of the geometry of the relevant excited state is the tool for obtaining a correct assignment of the Raman bands. Experiments [72] and theoretical calculations [79,80,101] indicate that the structural parameter heavily modified during the first allowed $1A_g$–$1B_u$ transition in polyenes is the degree of bond alternation along the chain. More precisely, in going from the ground to the excited state, the (shorter) double bonds lengthen while the (longer) single bonds shorten. Thus the relevant structural parameter involved in the

HOMO–LUMO transition is the dimerization amplitude $u = d_{C=C} - d_{C-C}$. This parameter is directly related to the CL and is particularly useful in discussing the conjugational properties of *trans*-PA and oligoenes.

A variation of u can be described by a suitable coordinate in the vibrational space. In the case of an ideally infinite chain (*trans*-PA) one can define the vibrational coordinate

$$Я = (1/2^{1/2})(R_{C=C} - R_{C-C}) \quad (15)$$

where $R_{C=C}$ and R_{C-C} represent vibrational displacements of the C=C and C—C bond lengths from their equilibrium values. In the case of the oligomers (*trans*-polyenes) with N double bonds, the definition of Я becomes

$$Я = \frac{1}{(2N-1)^{1/2}} \sum_{j=1,N-1} (R_{2j-1} - R_{2j} + R_{2N-1}) \quad (16)$$

The variation of the equilibrium structure (variation in the u parameter) induced by the transition from the ground state to the relevant excited state can be expressed as a displacement of the nuclei along the positive Я direction (Δ Я):

$$\Delta X^{ge} \cong \Delta Я^{ge} \quad (17)$$

Equations (15) and (16) define the so-called effective conjugation coordinate of ECC theory. The important conclusion that can be derived from Eqs. (11) and (17) is that *only normal modes that contain an oscillation of the dimerization amplitude (Я vibration) can have relevant Raman cross sections*. Then the two relevant lines of the Raman spectrum of polyenes are necessarily assigned to normal modes that involve a large contribution by the Я oscillation (Я modes in the ECC theory). More precisely, the treatment of the dynamical problem of polyenes in terms of ECC theory assigns the two strong Raman lines to two different combinations (in-phase and out-of-phase) of the Я oscillation with C—H wagging vibration.

C. The Dynamics

The analysis of the intensity pattern in the Raman spectra of polyenes makes it possible to select one particular degree of freedom in the vibrational space (the Я coordinate) that plays the leading role in determining the Raman response. ECC theory requires that this particular degree of freedom be explicitly introduced in the treatment of the dynamical problem. Я is thus chosen as one of the vibrational coordinates that form the basis for the description of the vibrational modes of the molecule. The key role of the Я coordinate is due to the fact that the coupling between π-electron excitation and nuclear configuration acts mainly along the Я direction.

It is then reasonable to think that changes in the intramolecular potential related to changes of the π-electron distribution take place mainly when the Я coordinate is activated. It then becomes possible to give a simplified description of the CL-dependent intramolecular potential; a common potential for polyenes of any length and *trans*-PA was determined where all force constants but $F_Я$ were kept fixed for all molecules. $F_Я$ represents the "effective" diagonal force constant relative to the Я coordinate; it can be written as a linear combination of the more familiar valence force constants relative to carbon–carbon bond stretching coordinates, according to the expression

$$F_Я = (1/2)(K_{C=C} + K_{C-C}) - 2f^1_{C=C,C-C} \\ + \sum_{j=1,J} (f^{2j}_{C=C} + f^{2j}_{C-C} - 2f^{2j+1}_{C=C,C-C}) \quad (18)$$

where $K_{C=C}$ and K_{C-C} are the diagonal force constants relative to the valence stretching coordinates of the double and the single carbon–carbon bonds, respectively; $f^1_{C=C,C-C}$ is the first-neighbor interaction between C=C and C—C stretchings; the terms in the sum are interaction force constants between stretching of CC bonds at any distance along the chain. The relevant parameter CL finds a precise definition as the length J such that for $j > J$ the terms in the sum practically vanish. As first empirically suggested by Crawford and Califano [19] and later analytically proven by Kakitani [102], both ab initio and semiempirical calculations agree in predicting the sign of the various terms that appear in Eq. (18): interaction force constants between single and double bonds (odd indices in the sum) are positive, and interactions between bonds of the same kind (even indices in the sum) have negative values [79,80,90,102]. According to Eq. (18), an increase in the values of any interaction and/or an increase in the number of terms in the sum (increase in J) implies a decrease in the value of the collective force constant $F_Я$ (softening of $F_Я$).

The advantage of having introduced the collective force constant $F_Я$ consists in the fact that the frequency dispersion of the Raman lines of polyenes can be fully described in terms of only $F_Я$. Indeed, the relevant changes in the intramolecular potential due to the increase in CL result in a modulation of only $F_Я$.

This is shown in Fig. 28.10, where the calculated wavenumbers for three A_g modes* (ν_1, ν_2, and ν_3) of an ideally infinite polyene chain are plotted as functions of $F_Я$.

The two very intense Raman lines observed in the spectra of polyenes (ν_1 and ν_3) can be made to correspond with the curves in Fig. 28.10, which show a re-

---

* Carbon–hydrogen bond stretching of A_g symmetry is omitted because its frequency (well separated in energy from the other A_g normal modes) is not affected by changes in $F_Я$.

Structure and Nonlinear Optics

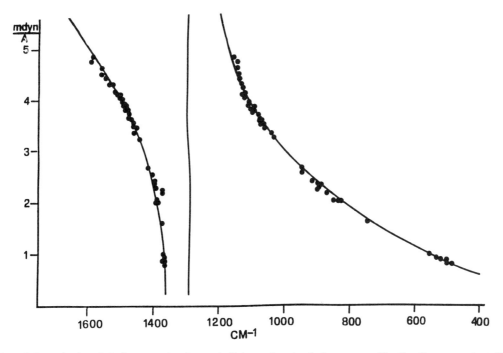

Fig. 28.10 Plot of the calculated A_g frequencies for an infinite polyenic chain versus effective force constant F_R (solid line). Data points are the experimental frequencies (both for primary and satellite peaks) of different samples of PA with different exciting laser lines; F_R values lower than 3 mdyn/Å correspond to infrared doping induced or photoinduced bands. (Figure and data from Ref. 41.)

markable dispersion with F_R: the calculated eigenvectors L_1, L_3 both show a large content of R vibration. In contrast, ν_2 practically does not shift by changing F_R: the L_2 eigenvector corresponds, indeed, to a vibration involving in-phase stretching of double and single carbon–carbon bonds in the chain, thus moving the skeletal carbon atoms orthogonal to the R vibration.

The data points in Fig. 28.10 correspond to experimental wavenumbers from the Raman spectra of different samples of PA excited with different excitation energies; both satellite and principal peak wavenumbers are reported. The experimental wavenumbers relative to infrared-doping-induced and photoinduced bands lie on the theoretical curves for F_R values of less than 3 mdyn/Å (see Section V).

In Fig. 28.11 the experimental Raman wavenumbers of oligoenes and polyene-like systems are reported on the same theoretical curves as in Fig. 28.10 obtained for

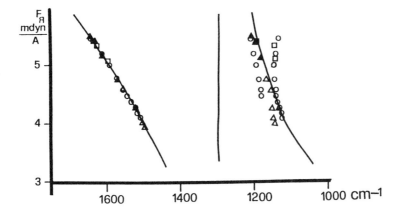

Fig. 28.11 A portion of the same theoretical curves as in Fig. 28.10 but with experimental data relative to oligomers: (▲) *trans*-Polyenes; (○) *tert*-butyl-end-capped polyenes; (□) diphenylpolyenes; (△) carotenoids. (Data from Table 28.1.)

PA. In this way it is possible to obtain for any polyene-like molecule considered a value of F_A that describes its degree of conjugation.

Generally the chain length in polyene systems is described by N, the number of carbon–carbon double bonds. However, most of the polyene chains are capped at either end by groups containing π electrons that take part in the delocalization path (e.g., α, ω-diphenylpolyenes). The values of Raman frequencies of the A modes (ν_1 and ν_3) and of F_A suggest that the effect of adding a phenyl group to the chain is similar to that of adding one carbon–carbon double bond. In Tables 28.1 and 28.2 (and Figs. 28.6 and 28.8) the chain length is defined accordingly; e.g., in the case of α, ω-diphenylpolyenes, $N = n + 2$, where n is the number of double bonds in the polyene chain.

It is worth noticing that in the case of *trans-tert*-butyl end-capped polyenes of short length (from three to six or seven double bonds), the ν_3 mode is mechanically coupled with carbon–carbon bond stretchings localized on end groups, thus giving rise, in the Raman spectra, to an extra line near 1200–1100 cm^{-1}. Frequency data relative to both peaks of the doublet are reported on the plot. Note that when the number of double bonds increases (from three to six), the lower frequency peak shifts toward the theoretical curve; correspondingly, the experimentally observed intensity of the higher energy peak decreases and disappears in the spectra of longer molecules.

The results reported in Figs. 28.10 and 28.11 show that the use of ECC theory allows the dynamics of polyene systems to be treated in a unified way, giving a correct prediction of the frequency shifts of the relevant Raman bands with conjugation length. Moreover, it gives a measure (through F_A) of the extent and strength of interaction between carbon–carbon bonds along the polyene chain. The Raman wavenumber of the A modes can thus be used as a tool to characterize quantitatively the degree of delocalization of π electrons along the chain. A further check of the validity of ECC theory can be obtained with the prediction of the evolution with F_A of the Raman intensity.

If the A coordinate is explicitly introduced in the treatment of the vibrational problem it is possible to write the parameter ΔQ_k^{eg}, which appears in Eqs. (11) and (12) as

$$\Delta Q_k^{eg} = L_{k\text{A}}^{-1} \Delta \text{A}^{eg} \qquad (19)$$

This implies that the element $L_{k\text{A}}^{-1}$ of the eigenvector relative to the kth normal mode rules the intensity distribution among the Raman-active lines. In particular it is possible to predict the intensity ratios between the ν_3 and ν_1 bands from the ratio $(L_{3\text{A}}/L_{1\text{A}})^2$. The variation of this intensity ratio with increasing polyene length has been predicted for *tert*-butyl end-capped polyenes, in nice agreement with experimental findings [99].

ECC theory does not give any direct information on the absolute intensity values of the Raman bands and on their variations with increasing chain length. However, Eqs. (11) and (12) provide a relevant theoretical insight.

Using the definition of A given in Eq. (16) it is possible to show that $\Delta Q_k^{eg} \approx L_{k\text{A}}^{-1}(2N)^{1/2}\Delta\langle u\rangle^{eg}$, where $\langle u\rangle$ is the average difference between the length of adjacent double and single bonds for a finite polyene chain. Calculations indicate that the equilibrium structures of polyenes in the ground and excited states reach a plateau for small N [79,80,90]; it is then reasonable to assume that $\Delta\langle u\rangle^{eg}$ does not change much when N increases. Because of Eq. (12) one would then expect Raman intensities to increase linearly with N. However, when N in-

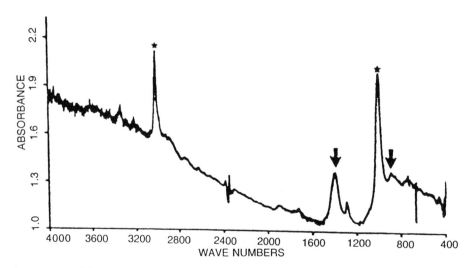

Fig. 28.12 Infrared spectrum of doped PA (at very small doping, 0.0017 I_3). The arrows indicate the doping-induced bands. (From Ref. 41.)

creases, experimental Raman intensities of ν_1 increase superlinearly with a power law of the kind N^α, with $\alpha \approx 2.7$. Such superlinear increase must then be ascribed to the modulation induced by the conjugation on other terms in Eqs. (11) and (12). First the energy gap that appears in the denominator shows a marked decrease with CL (linear with $1/N$); moreover, the terms M^2 contribute to an increase in the Raman cross section with increasing N [103].

The results reported in this section indicate that the introduction of the concept of effective conjugation coordinate is the key to the interpretation of the relevant and peculiar spectral features observed for polyenic systems. Other theoretical approaches have been independently developed with the same aim. Of fundamental importance is the amplitude mode theory (AMT) proposed by Ehrenfreund et al. [104], where the treatment of the vibrational problem of *trans*-PA is given in terms of the Green's function formalism. It has been shown that a precise correspondence can be established between ECC theory and AMT through a relationship between F_R and the adimensional electron–phonon coupling parameter $\bar{\lambda}$ [41,43]. More recently a theory that contains aspects of both AMT and ECC theory was proposed by Girlando et al. [105] based on the Hubbard Hamiltonian. The common concept of the three models is the fact that they can explain the peculiar vibrational spectra of polyene systems in terms of only the coupling between π electrons and nuclear structure. It is believed that one of the advantages of ECC is that the concept of electron–phonon interaction is translated into an "effective" force constant that can be formally handled as one of the usual empirical quadratic force constants close to molecular concepts. Moreover, ECC theory can be easily used to discuss the case of the oligomers and can be simply and immediately extended to other more complex polyconjugated systems containing aromatic and heteroaromatic units (see Sections VIII and IX).

V. DOPING-INDUCED AND PHOTOINDUCED VIBRATIONAL SPECTRA OF POLYACETYLENE

A. The Infrared Spectra of Doped Polyacetylene

Upon being doped with electron donors or acceptors, *trans*-PA becomes an electrical conductor [44,106]. The degree of conductivity is strongly dopant- and sample-dependent. Even at relatively low doping levels the electrical conductivity increases by several orders of magnitude. Recently chemistry and morphology were suitably modified to reach conductivity on the order of that of copper [107,108]. The doping levels depend on the dopant. The highest doping level for PA is $\approx 6\%$ with I_2 doping and ≈ 16–18% with alkali metal doping.

The infrared spectrum of doped PA shows the usual doping-induced infrared spectrum (DIIR) already mentioned in Section II. Three new bands appear near 1370, 1270, and 900 cm^{-1} (Fig. 28.12).

As already discussed in Refs. 41 and 42:

1. The wavenumbers of DIIR do not depend on whether the dopant is p or n (this is true for PA but may not be true for shorter oligomers).
2. DIIR shift to lower wavenumbers when ^{12}C is substituted with ^{13}C or H with D [81]. This shows that DIIR must be assigned to vibrational modes and not to electronic transitions.
3. DIIR show parallel dichroism [109].
4. The strong band at 900 cm^{-1} splits into several components.
5. The intrinsic intensity of DIIR is at least two orders of magnitude larger than that of the infrared lines of pristine *trans*-PA.
6. Even at a very high doping level ($\approx 18\%$) the frequencies of DIIR do not change, whereas their intensities change linearly with concentration [110].
7. Recent spectroscopic work on K-doped oligoenes (with 4, 6, 12, and 17 C=C bonds) have shown that DIIR show frequency dispersion with chain length [111].
8. Using special techniques of preparation from K-doped polyenes, polarons have been obtained and their IR spectra recorded with suitable experimental techniques [111].

Relevant complementary spectroscopic observations are the following:

1. The process of doping is accompanied by the appearance of gap levels in the electronic spectrum at low energies (near IR) at various λ_{max} with a simultaneous blue shift and bleaching of the HOMO–LUMO (π–π^*) strong electronic transition [61,62].
2. The electronic spectra of polarons, bipolarons, and multiple ionic species have been identified [61].
3. Pauli susceptibility during Na doping of PA is zero up to 5% concentration of dopant and then increases abruptly, showing a sort of phase transition between a diamagnetic and a paramagnetic phase, whereas conductivity increases by orders of magnitude and reaches a plateau at approximately 10–12% doping level [112,113] which is the maximum doping level with Na.

Since doping is a complex chemical and structural phenomenon and the physical measurements are often sample-dependent, any molecular modeling must allow for some flexibility and approximations. Bearing this in mind, attempts are presented below to describe the structure of the charge carrier that may be consistent with the infrared observations listed above.

The fundamental assumption accepted here as a working hypothesis is that charge carriers in PA are not clustered along the chain. It follows that the average distance D between the centers of the defects at the paramagnetic-to-diamagnetic transition is approximately 10 C=C bonds. At chain length $N < D$ (hence in the short oligomers studied) only polarons exist, whereas for lengths $N > D$ charged solitons or bipolarons may be generated (see Fig. 28.13). The size of the charge carrier generated by K doping in *trans*-PA must be larger than ≈ 20 C=C units, in agreement with quantum-chemical calculations. It is quite conceivable (in agreement with quantum-chemical calculations) that a large perturbation of the carbon–carbon bonds takes place in the center of the defect and that the perturbation dies off fairly quickly at either side along the chain.

B. The Raman Spectra of Doped Polyacetylene

The experimental observations are the following (see Ref. 41):

1. At low doping levels the Raman spectra are very similar to that of the pristine material. However, the intensities of the bands decrease with increasing doping level and become difficult to observe, just opposite to what is observed in the DIIRS.
2. The very weak Raman spectrum of doped PA can be enhanced if resonance conditions are reached by choosing an exciting line close to the transition to the gap state of the defect [115–117].
3. At large doping levels the weak Raman spectrum is very different from that of the pristine material. New lines at ≈ 1600 and 1270 cm^{-1} dominate the spectrum.
4. The Raman spectrum of doped PA shows dispersion with excitation wavelength [118].

C. The Vibrational Spectra of Segmented Polyacetylene

In 1980 Soga et al. first reported [119] that K-doped *trans*-PA films after treatment with methanol become partly hydrogenated and show a blue shift of the electronic spectrum as if the effective conjugation length has been shortened by the introduction of sp^3 carbon atoms, i.e., CH$_2$ groups. This important issue was tackled with detailed spectroscopic work (infrared and Raman) on Na-doped *trans*-PA by Furukawa et al. [120], who showed that the CH$_2$ groups in the PA chain may not be isolated but may also occur as longer sequences (ranging from two to five CH$_2$ groups). The electrical conductivity of iodine-doped segmented samples has been measured, and it is shown to decrease rapidly with increasing concentration of CH$_2$ groups.

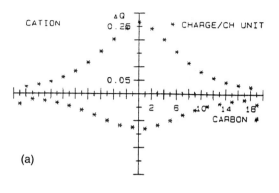

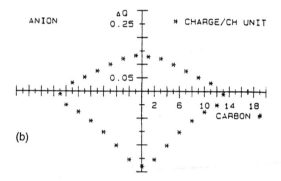

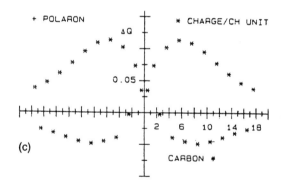

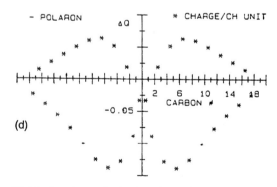

Fig. 28.13 Lattice polarization (net charge per CH unit) for (a) positive soliton, (b) negative soliton, (c) positive polaron, (d) negative polaron in C$_{40}$H$_{42}$ from MNDO calculations. (From Ref. 114.)

The issue of partly hydrogenated *trans*-PA (generally labeled segmented *trans*-PA) was the center of interest for a few years and then faded without a detailed final agreement between the communities of physically oriented and chemically oriented researchers.

From the dispersion with the excitation wavelengths of the Raman spectra (Я mode) of samples obtained with increasing doping concentration and subsequent washing with methanol, it is unquestionably clear that the average distribution of CL moves toward shorter CL. Furukawa et al. [120] even tried to derive a relationship between $\nu_Я$ and conjugation length and provided estimates of the average length of polyene segments isolated between segments of CH_2 groups. The important conclusion is that Na atoms upon doping do not attack the polyene chain at random but are clustered in segments separated from one another by distances estimated by Furukawa et al. to be in the range of 8–55 double bonds. The $\nu_Я$ versus chain length relationship may be questioned in the light of the present knowledge on PA, but the scenario is still acceptable.

Further iodine doping of segmented PA generates the usual DIIR spectra already discussed in Section V.A. The DIIR spectrum of iodine-doped segmented *trans*-PA published by Furukawa can be analyzed on the basis of ECC theory. ν_1 and ν_3 show dispersion with iodine concentration; that is, the larger the concentration of dopant, the shorter the average length of the doped segment.

The electrical conductivity of segmented PA was also measured by Shaefer-Siebert et al. [121], while the shortening of the conjugation length in the Raman spectra was again confirmed by independent Raman scattering experiments [122].

The question that came under consideration is whether and to what extent sp^3 defects introduced into a polyene chain can interrupt the delocalization of π electrons. Some authors, on the basis of measurements of magnetic susceptibility and electronic and infrared spectra, claim that the existence of sp^3 defects in the PA chain has little effect on the delocalization [123,124]. These conclusions obviously did not find a warm welcome in the chemical community. Several semiempirical [125,126] or ab initio calculations [127] were immediately generated to model the phenomena claimed in Refs. 123 and 124. It was first shown by MNDO calculations [126] or by the STO-3G basis set [127] on segmented models that the delocalization of π electrons does not cross over the defect but is strongly localized within the segments having sp^3 defects at either end. To overcome the contradiction between localization among defects and the unusual conductivity, it was proposed that the perturbations on the electronic structure induced by the doping of segmented PA are such that the perturbed orbitals of the charge transfer state cross over the C sp^3 [128]. Calculations on model molecules of segmented PA in which one electron has been extracted to simulate doping show again that sp^3 defects act as strong barriers to delocalization [129]. In a further attempt to justify delocalization between short segments it was proposed that some mechanism of hydrolysis takes place in the preparation of segmented PA with the formation of C=O defects adjacent to the CH_2 group followed by a tautomeric equilibrium that removes the separation between conjugated segments [128]. These conclusions need to be substantiated by more detailed spectroscopic observations and theoretical calculations in parallel with careful chemical experiments.

The issue of the structure of segmented PA is not marginal as it involves the general problem of the mechanism of doping and the possible subsequent "staging" or diffusion of the dopant inside the sample.

D. Photoinduced Infrared Spectra of Polyacetylene

Illumination of the sample by a light whose energy is greater than E_g and simultaneous recording of the infrared and/or electronic spectra allow probing of the structure and properties of the electronic excited state. The complex photophysics of photoexcitation and subsequent relaxation phenomena have been studied in detail by various groups, and excellent reviews are available [75,76].

In this section attention is focused on the vibrational aspects of the phenomena that can be probed at low temperatures by simple "pump and probe" techniques, even without using very fast (femtosecond or picosecond) spectroscopy. The concepts developed for *trans*-PA can also be used for understanding of the photophysics of other polyconjugated materials. The main interest is to show that, from the viewpoint of infrared spectroscopy, photoexcitation corresponds to some kind of (clean) photodoping that introduces a displacement of charges (charge carrier) within domains of the chain, thus giving rise to photoinduced infrared (PIIR) spectra. It is then expected that DIIR and PIIR spectra may show common features, as was indeed found in photoexcited PA and in many other systems [41].

The generation of charge carriers is justified by the fact that PA and other materials show photoconductivity. When the pristine material is illuminated by a short pulse of light, an electrical response is recorded with at least two response times, namely a very fast one (fs–ps) and a slower one (ms) [130,131]. It has been suggested that the photogenerated carrier moves into the material about 400 Å before being trapped [132,133]. Similar experiments have been made on polyenes as model molecules [130].

Using the techniques previously developed by Orenstein and Baker [134,135], Blanchet et al. [136] presented for the first time the infrared spectrum of photoexcited PA. The experiments were then repeated by other groups. In the classical range of infrared spectroscopy, the PIIR spectrum of *trans*-PA shows the usual three features: ν_1 strong, ν_2 (weak), and ν_3 (strong, very broad, and structured) very similar to the DIIR spectrum of *trans*-PA. The only remarkable difference between DIIR and PIIR spectra is that ν_3 of the PIIR spectrum occurs at a wavenumber much smaller than the DIIR ν_3.

The basic photophysics of *trans*-PA is conceptually the following. Upon excitation with an energy larger than E_g a transient electronic state is reached within the conduction band. Within a few femtoseconds the system relaxes in various ways in a new state located in the energy gap whose nature is variously described by different authors [75,76].

A great contribution to an understanding of the photophysics of PA has been given by the experimental observations (followed by some theoretical modeling) of an anisotropic photoconductive response in stretch-oriented PA.

These data can be obtained when, for a given orientation of the static field, the polarization of the incident beam is changed. The photocurrent obtained with polarization perpendicular to the chain direction is higher than in the parallel case. The concept of bimolecular recombination processes for the solitons has been introduced and has given rise to new experiments and theories [137].

E. ECC and the Vibrational Spectra of Doped and Photoexcited Species

The infrared spectrum of doped and photoexcited polyacetylene can be nicely interpreted with ECC theory.

Normal mode calculations on the short defect domain in the polyene chain indicate that strongly infrared active vibrations localized in the defect domain are characterized by a very large content of dimerization amplitude oscillation (Я vibration) [89]. These results suggest that the doping-induced infrared band can be interpreted as being due to the activation in the infrared of the two relevant A_g modes that show very strong Raman activity in the neutral species. The large intensities of these infrared-active modes can be ascribed to large charge fluxes along the carbon–carbon bonds, induced by the Я oscillation [41], which generate a very large molecular dipole moment variation $\partial M/\partial Я$ directed along the chain axis.

The Я oscillation can thus be seen as a mechanism able to generate a net electronic charge transfer along the doped chain within the doped domain. This mechanism is made possible by the breakdown of the electrical symmetry of carbon–carbon bonds induced by doping or photoexcitation. It must be remembered that the Я oscillation is IR-inactive by symmetry (A_g mode) in the neutral species.

Doping-induced infrared bands have lower frequencies than the two corresponding Raman bands of the pristine compound. As seen in Section IV.C, by fitting the experimental frequencies on the curves of Fig. 28.10, it is possible to obtain the proper $F_Я$ parameter. The lower values of the $F_Я$ parameter derived from infrared experiments on doped and photoexcited species indicate that the degree of delocalization within the defects is larger than along the unperturbed chains.

This result is in agreement with data from theoretical structural studies [138], which indicate the occurrence of a domain within the chain where the bond alternation is greatly reduced (almost equalized CC bonds).

An additional observation is relevant: The wavenumbers of PIIR spectra are lower than those of DIIR spectra and correspond to even lower values of $F_Я$. This fact is consistent with the idea that in the doped materials the counterions generate a kind of pinning potential [104] that confines the electrons involved in the charge carrier within a relatively small defect domain (a few CC units). In contrast, the excess charge generated by photoexcitation of the neutral chains is less confined, thus implying a larger delocalization of π electrons and a further softening of the vibrational frequency.

A relevant concept, already presented by Horowitz and coworkers [104], is that DIIR spectra are independent of the charge configuration within the conjugational defect. In other words, *DIIR spectra cannot be taken as a proof of the existence of solitons but can equally be observed when polarons or bipolarons occur in a sample.*

It has already been stated that the infrared spectra of doped PA show a complex structure, especially on the lower frequency side of ν_3. This fact reflects (1) the existence of different conjugation lengths in the chains hosting the defects or (2) the existence of a variety of chemical defects inside the material.

The influence of CL distribution on the dynamics of the charge defects is even more apparent in the case of photoexcited samples. Vardeny *et al.* [131] reported frequency dispersion of the PIIR spectra for samples of *cis*-rich PA. In this case the average CL of the *trans* fraction is not very large, and the observed photoinduced infrared bands show higher wavenumbers very close to that of the doped species. This is consistent with the confinement of the defect within the boundaries created by the *cis* segments.

An independent indication that frequency dispersion of DIIR and PIIR spectra is due to the CL distribution comes from the extensive calculations carried out by Mori and Kurihara [139,140] and Mori et al. [141]. These authors calculated by MNDO the infrared spectrum of $(C_nH_{n+2})^+$ ($n = 17$–37), which corresponds to a charged soliton embedded in a chain with increasing CL [139–141]. Even with such a semiempirical method of calculation, one of the modes, namely ν_3, shows strong frequency and intensity dispersion with chain length. On the other hand, frequency and intensity dispersion of ν_3 is confirmed by the use of ab initio calculations with large basis sets (6–31 G) on soliton-containing polyenes [from $(C_5H_7)^+$ to $(C_{11}H_{13})^+$, from $(C_5H_7)^-$ to $(C_{11}H_{13})^-$] and on bipolaron-containing chains [from $(C_4H_6)^{2+}$ to $(C_{12}H_{14})^{2+}$] [90].

While from DIIR spectra we have derived information on the charge carrier, from the Raman spectra of doped species we can derive information on the part of the molecule that is not strongly involved in the charge carrier.

At a low doping level, the Raman spectrum of PA is practically unchanged in comparison to the pristine case. Upon increasing doping concentration unperturbed segments on average become shorter and the A modes are shifted toward higher wavenumbers, as expected. The changes in the distribution of CL with doping level are clearly probed by the spectra obtained by varying the excitation lines [118]. At very high doping levels the two main A_g peaks characteristic of the neutral sample disappear and all the Raman scattering becomes very weak.

These findings can be interpreted as if a kind of selection rule holds for doped samples: namely, that the A oscillation of polarized domains in the defect regions has very high infrared activity but is silent in the Raman spectrum. This behavior is opposite to what is observed for the pristine sample where a precise selection rule, dictated by symmetry, states that A_g vibrations (and A vibration) can be observed only in the Raman.

It follows that at low doping levels the A modes of the undoped chains (or portions of chains) are observed in the Raman spectrum while the A modes of the doped domains are observed in the infrared. In a very high doping regime, all chains contribute to strong activity in the infrared whereas the off-resonance Raman spectrum is strongly weakened.

In conclusion, the observed Raman spectra of doped samples show the existence of a fraction of the material that remains undoped. This fraction may consist either of pristine chains that have not been reached by the dopant or segments of pristine chains between two charge carriers. Moreover, the whole Raman spectrum necessarily becomes weaker when doping increases because the concentration of the "pristine" material decreases. In the spectrum of Fig. 28.14, the intensity of the line near 1285 cm^{-1} (approximately associated with ν_2) seems to grow when the doping level increases. For samples of PA highly doped with Li, the intensity of the two relevant lines (ν_1 and ν_3) of the pristine material becomes comparable to that of ν_2 (generally very weak). For such samples the Raman spectrum of the doped chains is no longer selective, as discussed in Ref. 41.

An alternative interpretation of the Raman spectra of doped PA has been given by Kim et al. [143] on the basis of experiments with the exciting line in resonance with the defect state, they assign the observed Raman lines to vibrations localized on the charge defect.

VI. THE CASE OF PUSH–PULL POLYENES

An independent proof that polarization of the carbon–carbon bonds in polyenic chains makes the A mode very active in the infrared was recently obtained from a spectroscopic study on push–pull polyenes [143–148]. The experimental data obtained for these systems provide further proof of the generality of ECC theory.

Because of the relevance of push–pull molecules in today's basic science and technology, a more detailed analysis of the spectroscopy of these materials is presented in this section. The discussion that follows can also be taken as a new and worked-out guided tour for the application of ECC theory.

Push–pull polyenes were recently proposed [143–148] as promising candidates for applications in the

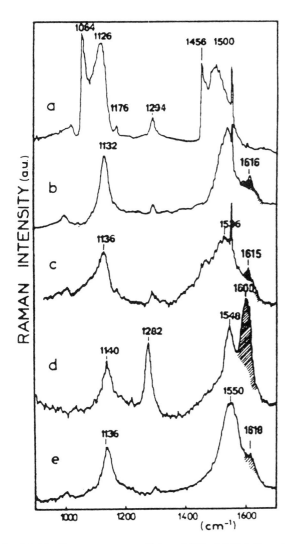

Fig. 28.14 Raman spectra of doped PA. (a) Pristine, λ_{exc} = 458 nm; (b) 2% doped with iodine, λ_{exc} = 458 nm; (c) 7% doped with ClO$_4$, λ_{exc} = 458 nm; (d) highly doped with Li, λ_{exc} = 458 nm; (e) 13% doped with FeCl$_3$, λ_{exc} = 413 nm. (From Ref. 142.)

field of nonlinear optical (NLO) devices based on large first-order NLO response. From the molecular viewpoint this corresponds to searching for molecules with large first-order molecular hyperpolarizability, β. The structure of a typical organic material with large β consists of a segment of conjugated π electrons (polyene chain or conjugated aromatic rings) connecting two chemical groups that act as electron donor (D) and electron acceptor (A) according to the scheme

A—Π—D

The conjugated π electrons of the bridge can be polarized, thus allowing a partial charge transfer from D to A. The first consequence is the appearance of a molecular dipole moment (generally large), directed from A to D, whose value is determined by the strength of the end groups and by the characteristics of the bridge (length of the bridge and its chemical nature).

The second consequence directly involves the geometry of the bridge. Let us consider the case where the bridge is a polyene chain, i.e., the case of push–pull polyenes. The electronic structure of the polyene chain can be described as a linear combination of two canonical structures, namely a neutral polyene structure (a), with alternate single and double bonds,

(a)

and a zwitterionic structure (b) where an electron is transferred from D to A:

(b)

The equilibrium average dimerization amplitude $\langle u \rangle$ of the polyene chain is modulated by the weights of the two canonical structures; moreover, the degree of stability of form (b) determines the real structure of the molecule. At least in principle, by a suitable choice of A, D, and the length of the chain, $\langle u \rangle$ can be modulated from negative to positive values.

According to the definition of the $\mathcal{R}$ coordinate, this procedure corresponds exactly to a nuclear displacement along the $\mathcal{R}$ coordinate. We thus expect that in the case of push–pull polyenes also the vibrational coordinate $\mathcal{R}$ plays a special role, since it is directly involved when the electronic structure is modified (e.g., by changing the end groups). It is then expected that the concepts developed with ECC theory will be of great help in the interpretation of the vibrational spectra of such molecules, as has indeed been found in the many experiments and ab initio calculations recently made on push–pull polyenes with a variety of end groups and different chain lengths. (Fig. 28.7). The experimental or theoretical data obtained have been discussed and compared with those relative to apolar, centrosymmetric polyenes. The more relevant observations of general validity for this class of compounds are the following:

1. Raman spectra are very strong and very selective. Only a few bands are observed with large intensities, of the order of magnitude of those relative to centrosymmetric polyenes of the same length (see Table 28.2 and Fig. 28.15a for comparison).
2. In a number of cases differences can be observed between the Raman pattern of a push–pull polyene and that of an apolar polyene of the same length. In general there are more than two relevant Raman lines, but the spectral pattern becomes simpler when polyene chains become longer. In all the cases considered, the relevant spectral features of push–pull polyenes can be fully accounted for by ECC theory. Ab initio calculations of the normal modes of some push–pull

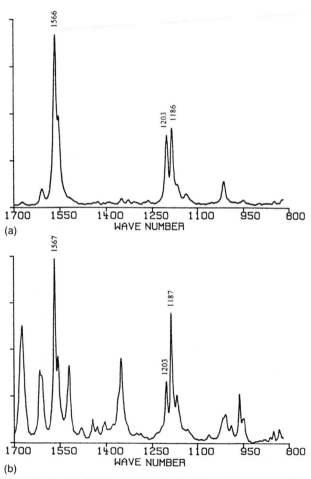

Fig. 28.15 (a) FT-Raman and (b) infrared spectra of a push–pull phenylpolyene (**III.b**, Fig. 28.7).

polyenes show that the occurrence of additional lines must be ascribed to the fact that the collective Я coordinate can couple with carbon–carbon stretching vibrations belonging to the end groups.

3. Because of the lowering of molecular symmetry, the inversion center is lost and the most relevant Raman lines (Я mode) become active in the infrared spectrum (see Fig. 28.15). The values of the absolute infrared intensities of Я modes are very large, even two orders of magnitude larger than the values usually measured in infrared intensity spectroscopy.

The origin of the intensity enhancement of infrared modes in push–pull polyenes is certainly the same as that which explains the appearance of very strong infrared bands in doped polyacetylene. In both cases the carbon–carbon bonds of the chain are strongly polarized. [Figure 28.16 reports the values of the partial effective charges calculated by ab initio methods for a model of charged soliton defect (A) and for one of the push–pull molecules considered here (B).] Due to the strong bond polarization induced by the charge transfer that takes place between the dopant and the chain in the former case and between donor and acceptor groups in the latter, adjacent CH units in the chain possess partial charges with opposite signs. Polarized carbon–carbon bonds can develop relevant charge fluxes (along the polyene axis) during the Я vibration; such fluxes are responsible for the anomalously high intensity of the infrared bands [41].

To compare spectroscopic data relative to different push–pull polyenes and centrosymmetric polyenes it is useful to select the normal mode with the largest Я content, which corresponds to the ν_1 mode in apolar polyenes, and to apply ECC theory considering both frequency and intensity. Also for push–pull polyenes it is often possible to find another normal mode, with large Я character, in the wavenumber range 1000–1200 cm^{-1}, which corresponds to ν_3 of polyenes (Tables 28.1 and 28.2). However, ν_3 in many cases is heavily affected by coupling with the vibrations of the end or side groups. For this reason, comparison between different molecules is made simpler by focusing only on ν_1.

Comparing frequency data (Table 28.1) of ν_1 for push–pull molecules of the same length (e.g., $N = 3$), it becomes apparent that changes of the end groups cause ν_1 to shift within a range of about 150 cm^{-1}. This range of variation is comparable to that covered by apolar polyenes with chain length increasing from hexatriene to PA.

Following ECC theory, ν_1 shifts because the value of $F_Я$ changes because it is modulated by the strength of the polar end groups. The stronger the donor and the acceptor, the smaller is the value of $F_Я$, which causes

A)

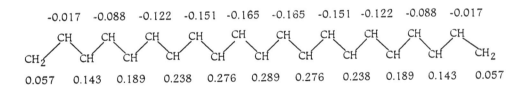

B)

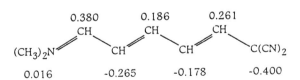

Fig. 28.16 Theoretical charge distribution of (A) C$_{21}$H$_{23}^{+}$ ion (model for charged soliton) and (B) push–pull polyene with strong polar end groups (V). The values reported are partial charges on the CH units obtained from ab initio calculation according to the definition of "corrected Mulliken charges" (see Ref. 149). Basis set used are 6–31 G in (A) and 3–21 G in (B). Units are electrons.

the ᴀ mode to shift toward lower wavenumbers. The behavior of ν_1 as a function of chain length (N) for push–pull polyenes is reported in Fig. 28.6. The reference dotted curve $\nu_\text{ᴀ}$ vs. N was drawn as the best fit of the wavenumbers obtained for end-capped *tert*-butyl polyenes. All the other apolar polyenes have wavenumbers that lie close to this curve; in contrast, short push–pull polyenes show wavenumbers that in many cases are well below the reference curve. However, when the length of the chain increases, ν_1 of push–pull polyenes approaches the values typical of apolar polyenes. In Fig. 28.6, four of the experimental data lie far from the reference curve. They belong to push–pull polyenes (compounds **V**, **VI**, and **VII**) with very strong donor and acceptor groups. For these molecules a small value of $\langle u \rangle$ is expected because the energy of the zwitterionic form [structure (b)] is very small [150].

From Eq. (18) it is possible to rationalize the fact that a reduction in the bond alternation makes the value of $F_\text{ᴀ}$ smaller. Equalization of adjacent bonds tends to lower the value of $K_{\text{C}=\text{C}}$ and to raise the value of $K_{\text{C}-\text{C}}$, giving an average value $(K_{\text{C}=\text{C}} + K_{\text{C}-\text{C}})/2$ practically independent of the degree of bond alternation. The diagonal term in Eq. (18) does not contribute to the lowering of $F_\text{ᴀ}$. In contrast, the nearest-neighbor interaction ($f^1_{\text{C}=\text{C},\text{C}-\text{C}}$) (which is positive) certainly increases by equalization of adjacent bonds, inducing softening of the ᴀ frequency [see Eq. (18)]. It thus follows that ν_1 becomes a structural probe for push–pull polyenes: Its variation ($\Delta\nu_1$), and correspondingly the variation of $F_\text{ᴀ}$ with respect to values characteristic of apolar polyenes of the same length, is a measure of the degree of bond equalization in the polyene bridge induced by the introduction of polar end groups. It will be seen in the following that the conclusions based on frequency data can be supported and improved by the parallel analysis of the intensity (Raman and infrared) of the ᴀ modes.

The fact that in the series **I**, **III**, and **IV**, the wavenumbers of the longer chains practically lie on the reference curve of the apolar polyenes shows that in the longer molecules the polyenic segment is less affected by end effects, thus keeping a strongly alternating polyene-like structure in the middle of the bridge. This suggests the existence of a critical chain length N^c above which the interaction between donor and acceptor groups is inhibited. A critical chain length can be found in every series of push–pull polyenes. For chains with $N > N^c$ the bond polarization is confined within a few bonds close to the polar end groups; the bonds in the central section of the chain are practically apolar and exhibit a structure very similar to that of a dimerized polyene chain.

The value of N^c is determined by the electron donor or acceptor strength of the end groups. From Fig. 28.6 it is apparent that the molecules of series **IV** (polyenovanillins) behave as if they were apolar polyenes (with the only exception being the shorter molecule $N = 3$).

This behavior reflects the fact that the D group of polyenovanillins is very weak. Unfortunately, data on long polyenes with very efficient end groups are not yet available. From the large wavenumber shifts of the ᴀ mode shown by the short molecules available (**V**–**VII**), it is expected that in these cases $N^c > 5$ or 6 a critical value that can be derived from ν_1 of other medium strength push–pull polyenes (Table 28.1).

As to Raman intensities, several, sometimes competitive, effects must be taken into account. The introduction of polar end groups generally causes an intensity enhancement. Also in this case the data relative to the weakest push–pull systems (**IV**) are very similar to those of the corresponding apolar polyenes (carotenoids). Trends of Raman intensities with N (Fig. 28.8) show a superlinear increase for all the series considered. Equation (11) provides a simple justification of the increase of $\partial\alpha/\partial Q_1$ in the case of push–pull polyenes. At equal chain length the experimental E_g values of push–pull polyenes are smaller than those of the apolar polyenes due to the occurrence of a low energy charge transfer state with a large contribution from the zwitterionic structure (b). The reduction of the energy gap (Fig. 28.17) also leads to an increase in the oscillator strength of the dipole transition, with a consequent increase in the Raman cross section [Eqs. (11) and (12)].

The term ΔQ_k^{eg} plays the opposite role; the polarization by the end groups induces a reduction of the bond alternation in both the ground and excited states [151,152], thus producing more similar geometrical structures in the two relevant electronic states. The fact that the intensities are even larger with respect to those of the apolar species indicates that the structure is far from a perfectly equalized CC chain even in the case of very efficient end groups. Notice, however, that the zwitterionic structure can be stabilized by solvent effects [151–153]. It is shown in Section X.C that the chain structure of compounds **VI** and **VII** can be equalized by the choice of a suitable polar solvent; in this case a marked decrease of the Raman cross section (by virtue of the reduction of ΔQ_k^{eg}) is observed. This finding is very relevant, especially if compared with the experimental data from Raman spectra of heavily doped polyacetylene, where a drastic reduction in the Raman cross section was observed (Section V).

In Fig. 28.18 infrared intensities of the ν_1 band of push–pull polyenes are reported as functions of chain length. Absolute values are very large and can be correlated with the degree of bond polarization of the chains. The more efficient the end groups, the stronger the intensities. In the case of weak and medium strength end groups, infrared intensities are practically constant with N. This means that the polarizing effect by the end groups is confined within a few bonds. On the other hand, the two compounds of series **V** show that the introduction of just one additional double bond induces an

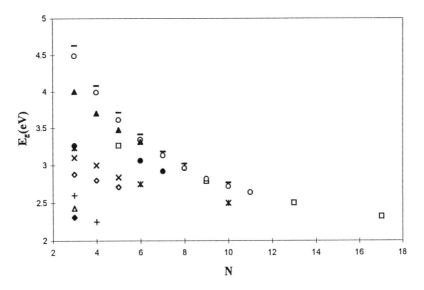

Fig. 28.17 Experimental energy gaps (λ_{max}) for the molecules of Fig. 28.7 vs. N (number of double bonds in the chain). Symbols as in Fig. 28.6.

increase in intensity of approximately one order of magnitude. In this case (strong end groups) the introduction of other units into the chain should yield even greater enhancement.

As usual, data relative to compounds **VI** and **VII** strongly deviate from the values obtained for the other molecules examined, owing to the presence of very efficient end groups.

In conclusion, a critical reading of the data reported in Table 28.1 and 28.2 and Figs. 28.6, 28.8, and 28.18 shows that the physics of the polyene systems can be interpreted in a unified way by the introduction of the concept of "effective conjugation coordinate" strongly coupled with π electrons. Moreover, with the help of ECC theory, it is possible to extract relevant information on the electronic structure and molecular geometry from

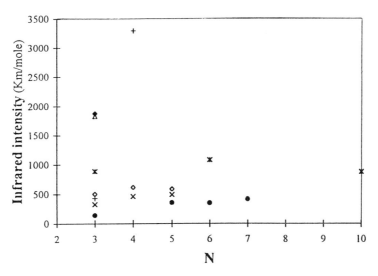

Fig. 28.18 Experimental infrared intensities of ν_1 for push–pull polyenes vs N (number of double bonds in the chain). Data refer to the molecules reported in Fig. 28.7. Symbols as in Fig. 28.6.

vibrational spectroscopic observables. In Section X these concepts are used in the field of nonlinear optics as a powerful tool for the development of structure–property correlations.

VII. ECC THEORY: GENERALIZATION TO THE CASE OF POLYAROMATICS AND POLYHETEROAROMATICS

In this section the vibrational spectra of polyaromatic and polyheteroaromatic molecules are discussed starting from their oligomers.

The prototypes of a large class of polyaromatic molecules that have been found to possess relevant electrical and/or optical properties are polyparaphenylene (PPP) and its oligomers (PP_n), polyparaphenylenevinylene (PPV) and its oligomers (PV_n), and the series of rylenes (Ry_n). In the class of polyheteroaromatics, we consider oligopyrroles (Py_n) and polypyrrole (PPy), oligothiophenes (Th_n) and polythiophene (PTh), and some derivatives.

The concepts developed in Sections IV–VI for the interpretation of the peculiar spectral features of polyenes and push–pull polyenes can be extended to the study of these more complex polyconjugated systems. The observation of a very strong and selective Raman spectrum justifies in this case also the introduction of an effective conjugation coordinate (ECC). According to the discussion in Section IV, the Я vibration is defined as an oscillation of the nuclei along the trajectory that atoms must follow in going from the ground-state equilibrium structure to that of the first excited electronic state. For all these systems the ground state (HOMO) has an essentially "aromatic" structure while the LUMO excited state has a relevant contribution from a "quinoidal" structure. The Я coordinate represents an oscillation between these two limiting structures and is defined as a collective stretching and shrinking of carbon–carbon bonds. Definitions of the Я coordinates for some of the systems considered are reported in Fig. 28.19. As discussed in Section IV, electron–phonon coupling is large along the Я mode.

Я can be easily and uniquely defined for the class of PPP, PPV, rylenes, PPy, and PTh; in these cases the spectroscopic phenomena are expected to be similar to those predicted by ECC theory and observed in PA. The case of condensed aromatic systems (i.e., anthracene, phenanthrene, tetracene, and pentacene), instead, turns out to be different because the topology of p_z orbitals does not allow a unique and absolute definition of the Я coordinate. Experimentally [154] it was found that the Raman spectra of the oligomers of the latter type cannot be accounted for in terms of ECC theory. Thus the topology of p_z orbitals becomes the determining factor for the possibility of defining the Я mode as the

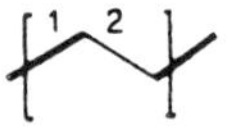

(a) Я = $1/\sqrt{2}(R_1 - R_2)$

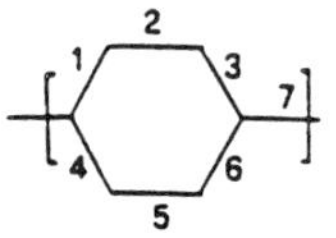

(b) Я = $1/\sqrt{7}(R_1 - R_2 + R_3 + R_4 - R_5 + R_6 - R_7)$

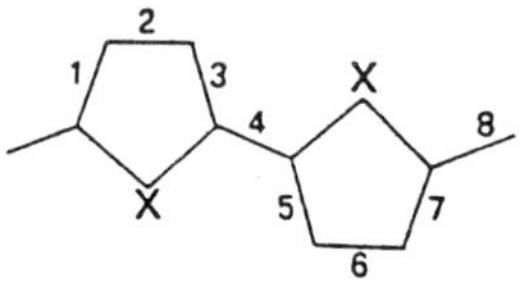

(c) Я = $1/\sqrt{8}(R_1 - R_2 + R_3 - R_4 + R_5 - R_6 + R_7 - R_8)$

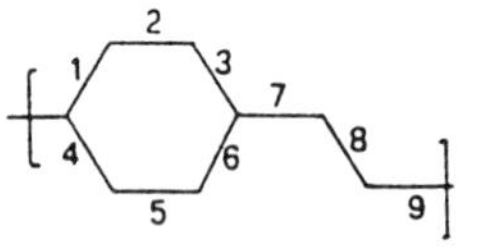

(d) Я = $1/\sqrt{9}(R_1 - R_2 + R_3 + R_4 - R_5 + R_6 - R_7 + R_8 - R_9)$

Fig. 28.19 Definition of the Я modes for a few common polyconjugated molecules. R defines the stretching of the carbon–carbon bonds. (a) Polyacetylene; (b) polyparaphenylene; (c) polypyrrole (X=NH) and polythiophene (X=S); and (d) polyparaphenylenevinylene.

preferential direction in the vibrational space along which electron–phonon coupling is maximum. This observation carries important consequences for the discussion of the nonlinear optical properties of these systems related directly to their infrared and Raman spectra [40].

In the search for the Я mode in polyaromatic and polyheteroaromatic systems, attention should be paid to "end effects" that affect both normal modes (as shown by the vibrational spectra) and electronic distribution (as shown in the electronic spectra) localized at either end of the oligomer chain. Dimers cannot be used as model molecules for longer chains; the analysis of the spectra must necessarily consider only the oligomers from the trimer on. Frequency and intensity spectroscopy of the trimer starts showing the Я mode generated in the middle unit. The Я mode of the central units dominates over the end group modes when the chain lengthens.

In spite of the large amount of work done on these systems, a comprehensive comparison of the properties is not yet possible, because complete sets of comparable data are not yet available. The following topics still need to be studied in a systematic way:

1. Structural properties of all the systems in solution and in the solid state. This knowledge is necessary to understand the relevant issue of the conformational dependence of π-electron delocalization.
2. Confinement of π electrons in the aromatic and heteroaromatic rings: competition between intraring and interring delocalization.
3. Conformational flexibility of molecules in solution and their effect on the vibrational and electronic spectra, which in turn determine the nonlinear optical responses.
4. Structure of doped materials both in solution and in the solid state, and the mechanism of conductivity as a function of the molecular environment.

For aromatic or heteroaromatic systems joined by single C—C bonds, torsional flexibility is a common structural variable that may change with the phase [155]. Consequently, optical properties become conformation-dependent. In some cases, following Boltzmann statistics, the same optical properties may then show temperature dependence.

Common structural variables (and the associated potential V) modulate the conformation of these systems [155]:

1. Steric repulsion forces between neighboring H atoms (V_{steric}) would drive the two adjacent rings to be orthogonal to each other ($\theta = 90°$).
2. Inter-ring electron hopping (delocalization) (V_{deloc}) would force planarity ($\theta = 0°$) and would increase the contribution by the quinoidal canonical structure in the resonance scheme.
3. In the crystalline phase, intermolecular packing forces (V_{lattice}) may further modify the conformation of the molecule from the gas (or liquid) phase, forcing the molecule to be planar or quasiplanar.

The intramolecular potential of the molecule in the gas phase can be written as

$$V_{\text{in vacuo}} = V_{\text{steric}} + V_{\text{deloc}}$$

and can be evaluated by ab initio calculations. The geometrical parameters in the gas phase are generally measured by electron diffraction experiments.

When the molecule is placed in an ordered lattice,

$$V_{\text{total}} = V_{\text{in vacuo}} + V_{\text{lattice}}$$

No theoretical calculations are yet available on V_{lattice} for the systems of interest in polyconjugated materials. In contrast, many structural studies with X-ray diffraction have been presented, and many structural data are available. Generally V_{lattice} does not bring the molecule to full planarity, and most of the time torsional angles $\theta \approx 5°$ have been measured [156,157].

For vibrational spectroscopy a twisting angle $\theta \approx 5°$ is irrelevant, and all molecules with such slight distortions can be considered fully planar.

For the classes of molecules that allow conformational flexibility about the inter-ring C—C bond, the problem of the effect of conformational freedom on the dynamics and vibrational or electronic optical properties is not yet fully understood. In general the frequency data do not seem to be affected much by conformational changes, whereas intensities seem to be a very sensitive probe [158]. Much more work is needed for a systematic understanding and structural use of intensity spectroscopy.

Ab initio quantum-chemical calculations have been widely applied for the understanding of the structural and spectroscopic properties of polyaromatics. Recently calculations have been extended to the interpretation of the infrared and Raman spectra of these systems in the pristine and doped states.

In some recent works it was proposed that one way to mimic the structure of polyaromatics in the "quinoidal structure" (which may be found in the doped state) is that of adding two CH_2 groups, one at either end, which necessarily force the molecule to become quinoidal in its ground state. Raman and infrared spectra have been calculated [159]. For the reasons discussed in Section V, it can be concluded that such structures are misleading and cannot reasonably mimic the structure and associated physics of the doped systems. As matter of fact, since in the quinoid models chosen no charge transfer occurs, charge displacement does not occur, carbon–carbon bonds are not polarized, and the local electronic symmetry is not drastically lowered. No infrared intensity enhancement of the Я mode can then occur; no weakening of the Raman (but eventually strong enhancement) can be expected.

VIII. POLYHETEROAROMATICS

As an introduction to the discussion of the vibrational spectra of polyheteroaromatics a first observation of general validity can be made on the monomers Fu, Py, and Th, all of which have C_{2v} symmetry. The Raman spectra of these three monomers show an unusually strong polarized Raman line at 1597 cm^{-1} (Fu), 1562 cm^{-1} (Py), and 1460 cm^{-1} (Th) assigned to the totally symmetric ring stretching of A_1 symmetry. Dynamical calculations [160] show that the heteroatoms do not move during this normal mode; thus they do not seem to give a kinetic contribution (mass effect) to this mode. The observed frequency shift must then be ascribed to the vibrational potential, which reflects electronic effects. The problem of the electronic effects of the heteroatoms was discussed by Coulson [161] and later by Streitwieser [162]; the latter introduced in his calculations an empirical parameter h that accounts for the Coulomb integral that is related to the electronegativity of O, N, and S. h takes the values 2, 1.5, and 0 for O, N, and S, respectively. The plot of the strongest Raman line versus h gives a remarkably straight line that can be extrapolated to the observed frequency of selenophene (1440 cm^{-1}), yielding a value of $h \approx -0.4$. For the purpose of the discussion that follows, this comparison allows us to focus on the electronic effects of the heteroatoms as the origin of many spectroscopic phenomena that are observed when ring-stretching modes of oligo- and polyheteroaromatic structures are considered.

As a further comparison of the electronic properties of heteroaromatics as reflected by the vibrational spectra, let us consider the C—H stretching modes as "local modes" (i.e., isolated oscillators decoupled from the neighboring CH oscillators). The wavenumbers for the stretching of the C—H groups in the α and β positions can be easily distinguished, and their numerical values can be easily read with great accuracy from the spectra. From McKean's studies of "vibrational chemical shifts" [163,164] it is known that vibrational frequencies are extremely sensitive to changes of bond lengths (and atomic charges). A bond length change of approximately 0.007 Å produces a vibrational shift of ≈ 100 cm^{-1}.

The reported bond lengths were obtained from microwave studies [165–167]. The linear relationship of ν_{CH} versus C—H bond length is remarkable (Fig. 28.20) and suggests that the study of the "vibrational chemical shifts" of the C—H oscillators may be a valuable structural probe. Another contribution to the characterization of the electronic properties of the parent chemical units in the class of heteroaromatics comes from intensity spectroscopy. The integrated intensity of the in-plane (A_{inpla}) and out-of-plane (A_{opla}) C—H modes can be easily measured. It is known that the ratio $B = A_{\text{inpla}}/A_{\text{opla}}$ decreases when the charge flux increases, which occurs within the molecule when the CH groups wag in the molecular plane [53]. Then B represents a measure of the mobility of the π-electronic charge within the ring and can be related to the "aromaticity" of the molecule. For Th, Py, and Fu the experimental values of B are 0.25, 0.56, and 0.73, respectively.

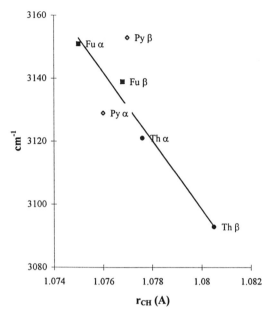

Fig. 28.20 C—H stretching frequencies (considered as local modes) vs. experimental C—H bond length in furane (Fu), thiophene (Th), and pyrrole (Py). α and β label the position of the C—H with respect to the heteroatom.

A. Polypyrrole

Polypyrrole (PPy) was the center of great interest during the early development of the science and technology of polyconjugated materials [168]. Science has been somehow hindered by the complex chemistry associated with the synthesis of the polymer. On the other hand, because of its stability the technological applications of doped PPy have spanned a wide range of possible applications and in some cases have reached the commercial level.

PPy in the doped state is chemically very stable whereas its pristine form is very unstable, can be easily oxidized or decomposed, and/or reacts with unwanted side-chain reactions. It has therefore been difficult to carry out clean spectroscopic work on chemically reliable samples [169]. Indeed the spectra of the pristine material are not reproducible and differ from one laboratory to the other. On the other hand, all the DIIR spectra of PPy are identical. This shows that the chemistry of the material is not yet under control and also that DIIR spectra present selective structural features that do not reflect the entire chemical diversity of the various samples.

The uncertainty in the chemical nature of pristine PPy makes any interpretation of the vibrational spectra either uncertain or approximate and calls for further work on clean model compounds. This has been the justification of the research into the chemistry and vibrational spectroscopy of PPy in the past few years.

Structure and Nonlinear Optics

The state of the art in the vibrational spectroscopy and structural studies of pristine and doped PPy until approximately 1990 is reviewed in Ref. 42, where also the theoretical frequency and intensity dispersion of the totally symmetric modes with $F_\mathcal{R}$ are calculated and their relevance discussed. The studies that followed concentrated on the synthesis and characterization of chemically clean samples. Martina et al. [170] reported the synthesis of well-defined oligo(2,5-pyrrole)s and the corresponding polymer with DP ≈ 20 by a Pd-catalyzed coupling reaction between N-protected activated monomers and subsequent removal of the protecting group. The structural, electrochemical, and spectroscopic characterization on these previously unknown materials are described in Ref. 170.

First the "protected" oligopyrroles of general formula (Boc)Py$_n$ with n = 3, 5, 7, 9 and "protected" polypyrrole (Boc)PPy (DP ≈ 20), where Boc indicates that all nitrogen atoms carry a *tert*-butoxycarbonyl protecting group, were studied. The synthesis and the structures from X-ray diffraction studies are reported in Ref. 171.* The materials with n = 3 and 5 are crystalline. The molecules have a *syn* configuration with a torsional angle $\varphi \approx 70°$, and as their chain length increases they form a helical structure. The absorption maxima in the electronic spectrum (in solution) are observed at 282, 295, 298, 303, and 306 nm for n = 3, 5, 7, 9, and (Boc)PPy, respectively. It is immediately clear that the small red shift of the HOMO–LUMO transition indicates that the extent of π-electron delocalization is not large.

Raman spectra were recorded [173] with excitation at 514 nm, thus certainly off resonance. The dispersion of the strongest line certainly associated to the $\mathcal{R}$ mode is observed at 1597, 1589, 1583, 1580, and 1579 cm^{-1} for n = 3, 5, 7, 9, and (Boc)PPy showing that inter-ring delocalization does occur in spite of the large conformational distortion. The observed frequency dispersion amounts to $\Delta \nu_\mathcal{R} \approx 19$ cm^{-1} (Fig. 28.21).

Other important spectral features are the lines associated with the vibrations of the end groups, which decrease as chain length increases and are not observed in (Boc)PPy. In particular, the line near 1530 cm^{-1} can be used as a marker for the determination of the chain length of the oligomers. The relevant information is that the Raman spectrum of the protected molecules arises mainly and selectively from the vibrations of the pyrrole backbone. This makes the Raman spectrum a useful structural probe of the heteroaromatic backbone, unlike the infrared spectra, which are very crowded and inextricable [173].

Next the Raman spectra of "deprotected" oligopyrroles and polypyrrole [174] need to be considered. For

---
* The X-ray diffraction data on Py$_2$ and Py$_3$ were kindly provided by Dr. G. B. Street in a private communication. See also Ref 168.

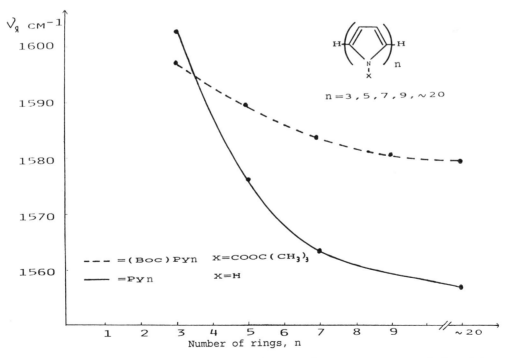

Fig. 28.21 Frequency dispersion of the $\mathcal{R}$ mode for oligopyrroles and polypyrrole. (———) Unsubstituted; (– – –) protected (Boc)Py$_n$. (From Ref. 173.)

the sake of simplicity, in this discussion let us consider the dynamics of PPy assumed as fully coplanar chains with the pyrrole units in the *anti* configuration. The one-dimensional lattice belongs to the D_{2h} point group, and its $16 \times 3 - 4$ normal modes can be classified as follows:

In-plane modes: 8 A_g(Raman-polarized Я-dependent) + 8 B_{1g}(Raman-dep) + 7 B_{2u}(IR) + 7 B_{3u}(IR)

Out-of-plane modes: 3 A_u(inactive) + 4 B_{1u}(IR) + 3 B_{2g}(Raman dep) + 4 B_{3g}(Raman-dep)

Following ECC theory we expect first to observe in the Raman spectra eight totally symmetric A_g lines. Of these, one N—H and two C—H stretching modes will occur in the 3500–3000 cm^{-1} wavenumber range (since they are not coupled with the Я coordinate, they should be vanishingly weak). We are left with five A_g modes expected in the range 1600–0 cm^{-1}, three of which should be appreciably strong in the 1600–1000 cm^{-1} range. Indeed, they are observed at 1557, 1312, and 1037 cm^{-1}. Among these the Я mode is identified as the line at 1557 cm^{-1}. It shows dispersion toward lower wavenumbers when the chain lengthens as observed in the case of the oligomers purposely synthesized by Martina et al. [170]. As expected, $\nu_Я$ values of the oligomers are observed at 1602, 1576, 1563, and 1557 cm^{-1} for Py$_3$, Py$_5$, Py$_7$, and PPy, respectively. The wavenumber range spanned by the dispersion with the number of pyrrole units ($\Delta\nu_Я = 45$ cm^{-1}) is rather large compared with other series of polyheteroaromatic molecules [39]. The increase of π-electron delocalization with chain length is thus confirmed and is certainly larger than in the case of protected (Boc)Py$_n$ and (Boc)PPy (see Fig. 28.21). The dispersion with CL shows that $\nu_Я$ reaches a plateau for $n \approx 7$ Py rings, thus showing that CL, as judged from frequency spectroscopy, turns out to be $\approx$7–9. No Raman intensity dispersion with CL has yet been measured.

The search for the vibrational spectra of truly pristine PPy is not yet completed. The infrared spectra of Py$_n$ and PPy obtained by deprotection of the corresponding Boc-protected parent molecules may not be the final ones because they were somehow exposed to air during spectroscopic manipulation.

The infrared spectra of deprotected Py$_n$ show bands associated with the vibrations of the end groups. The intensity of these bands decreases with increasing chain length, while the modes associated with the rings inside the chain increase in relative intensity and become the strongest lines in the spectra of the longer oligomers and in PPy since they represent the $k = 0$ phonons for the infinite PPy. The fact that the IR spectrum of PPy obtained by Martina et al. [170] shows a residual absorption of the end groups means that the polymer chain is not very long but is certainly longer than $\approx$9–11 units.

B. Polythiophene

Because of their potential or real applications in various fields of technology, oligothiophenes (Th$_n$) and polythiophene (PTh) have become the center of great interest in many laboratories ranging from chemical synthesis to device manufacturing [175,176]. The vibrational spectra of Th$_n$ and PTh and of their innumerable functionalized derivatives have been recorded either for routine chemical characterization or for more detailed structural studies. Reference 42 provides a rather complete review of a few years of vibrational spectroscopy of these materials.

The overall general information on PTh (details on many derivatives are also given in specialized references) is the following. PTh has been prepared by chemical and electrochemical methods [175]. The pristine material has a band gap of 2.0–2.5 eV as measured from the electronic spectrum. Upon doping, PTh reaches an electrical conductivity of $\sigma \approx 50$–100 S/cm [175]. Two interband transitions are observed in the electronic spectrum upon doping [177]. The charge carrier carries a spin at dopant concentrations below $\approx$3% [178–180]. Strong DIIR spectra appear upon doping [177,181–183], and for doping levels >3% the DIIR spectrum is generally assigned to spinless bipolarons [184]. A short-lived molecular species has been generated by photoexcitation, and its electronic and vibrational spectra have been recorded but still await a systematic interpretation [182,185]. PIIR spectra have been recorded and are practically superimposable with the DIIR spectra.

A whole class of new functionalized oligomers and polymers of thiophene have been prepared in the past few years. The discussion in this chapter mainly deals with their vibrational spectra in order to obtain the most recent and relevant information in this field to be compared with and added to previously gathered information.

1. The Structure of Oligothiophenes and Polythiophene

In Ref. 42 the complex issue of the structures of Th$_n$ and PTh in the solid state is presented in detail. Oligothiophenes in the solid state possess an *anti* configuration and are planar [186,187]. A possible slight out-of-plane distortion is irrelevant for the spectroscopy of these materials. PTh is fully planar [187]. The first thoughts of the French school that PTh could be helical [188] and that the fine structure observed in the electronic spectra could originate from privileged molecular conformers [189,190] did not find any further support.

The fine structure observed in both the solid and solution phases is ascribed to "phonon side bands," i.e., to the "vibronic structure." The electronic spectra of several alkyl-substituted PTh's have been studied [191,192], showing that the separation of the phonon side bands in both absorption and luminescence spectra is 0.182 eV (= 1460 cm^{-1}), the same as wavenumbers observed in the resonance Raman spectrum of the same material [192] (which has to be assigned to the Я mode, as discussed below). The final proof that the fine structure of the electronic spectra is vibronic comes from the careful experimental work of Birnbaum and Kohler [193], who studied the high resolution fluorescence excitation and emission spectra of 2,2′:5′,2″-terthiophene as solid solutions in n-decane at low temperatures.

The dependence of the vibrational spectra of oligo- and polythiophene (and in general of any polyaromatic oligomer or polymer) on the conformation of the chain is still an unsolved problem. Even for the simplest case of a highly symmetrical molecule such as biphenyl [194], the search for spectroscopic probes of conformational twisting about the inter-ring C—C bonds has not been very successful. Frequency spectroscopy does not seem to be a good probe for conformational distortion, because normal frequencies do not noticeably change with changes in conformation. Intensity spectroscopy is still in the early stage of development, and many more systematic studies are needed. The first results make intensity spectroscopy very promising [158].

The use of vibrational spectra as a probe for delocalization based on the analysis of the so-called Я mode of oligo- and polythiophenes enjoyed rapid success at first [42], but at present it is going through a critical reanalysis for reasons discussed below.

At present the number of oligomeric model molecules that have been analyzed for a more comprehensive study of their dynamics and spectra is very large. In the early stages of the analysis of the spectra of oligothiophenes starting from the infrared and Raman spectra published by Akimoto et al. [195] attention was focused on the dispersion of the Raman spectra as a function of CL.

D_{2h} is the symmetry point group of an infinite chain of PTh considered as a perfect one-dimensional crystal in which thiophene units are in the *anti* configuration and are coplanar. The corresponding structure of the irreducible representation of the $14 \times 3 - 4$ normal modes is the following:

In-plane modes: 7 A_g(Raman-polarized Я-dependent) + 7 B_{1g}(Raman depol) + 6 B_{2u}(IR $\perp$) + 6 B_{3u}(IR $\parallel$)

Out-of-plane modes: 3 B_{2g}(Raman depol) + 3 B_{3g}(Raman depol) + 3 A_u(inactive) + 3 B_{1u}(IR $\perp$)

The directions of the transition dipole moments as indicated above refer to polymer chains oriented along the chain axis by suitable stretching.

As seen in the case of PPy, we expect first in the Raman spectra seven totally symmetric A_g modes, two of which originate from C—H stretching vibrations that are generally vanishingly weak as explained in the case of PPy. Five lines are expected in the frequency range 1600–0 cm^{-1} and are indeed observed. However, some problems in the interpretation of the spectra remain unsolved.

Starting from the Raman spectra of Th$_n$ first published by Akimoto et al. [195], five lines were indeed observed with ν_1 unusually weak and showing unquestionable dispersion with chain length (Th$_3$, 1530; Th$_4$, 1519; and Th$_6$, 1507 cm^{-1}). This was taken as the Я mode [196], which, as usual, softens with increasing CL. When chain length increases, the softening of ν_1 turns out to be linear with the changes in E_g. In turn, E_g is almost linear with $1/N$. Two experimental observations remain unexplained: (1) the unusually weak intensity of the ν_1 line, which shows dispersion, and (2) the chain length–independent behavior of ν_2 near 1500 cm^{-1}, which is the strongest line in the Raman spectra of PTh and Th$_n$.

Since the assignment of the Я mode to the ν_1 line, which softens with increasing chain length, was initially proposed by the group in Milan [196], great care has been taken to reconsider the whole problem. This has required an analysis of the infrared and Raman spectra of more than 50 model molecules of oligothiophenes and polythiophenes in an attempt to solve the two problems mentioned above. Such correlative spectroscopic studies came to the following conclusions [197]. The characteristic lines common to most of the Raman spectra of the model molecules studied are listed below (see Fig. 28.22 for a prototypical case) along with their interpretation.

Line A. This line is generally very weak near 1600 cm^{-1} (sometimes with very weak satellites) and shows unquestionable dispersion toward lower frequencies when chain length increases. Along with the downward frequency shift, its relative intensity weakens and becomes almost unobservable with relatively long chain length (Th$_6$ or PTh). Line A has to be ascribed to a ring deformation of the end rings with some contribution by the central units. As the chain length increases, the relative weight of the deformation of the end rings becomes smaller and the A line shifts toward line B associated with ring deformation of central units. This is why line A is seen to shift and lose intensity with increasing chain length. This line is not expected for an infinite polymer for which cyclic boundary conditions hold and no end-group modes can exist.

This is in full agreement with the theoretical results presented in the work of Negri and Zgierski [198] and is substantiated by recent ab initio calculations on oligothiophenes as model molecules [197].

Of more direct interest for this work is that in the previous analysis ν_A was erroneously assigned to $\nu_Я$ and its softening was associated with the CL of the oligomer. The new assignment is critically discussed and justified in Ref. 197.

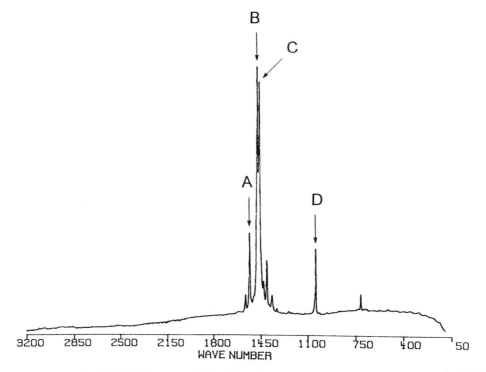

Fig. 28.22 Raman spectrum of solid 4,4‴-dipentoxy-2,2′:5′,2″:5″,2‴ tetrathiophene (λ_{exc} = 1064 nm) with the characteristic lines labeled as discussed in the text. (From Ref. 197.)

Line B. This line is the common feature of all oligo- and polythiophene molecules with or without substituents. It is always strong and dominant but, unlike other heteroaromatic systems, its wavenumber is almost independent of chain length. It shows somewhat different wavenumbers from one chemical series to another within the class of oligo- and polythiophenes, but within each series it is almost invariably strong and unshifted.

The chain length independence of line B is a theoretical puzzle. From dynamical calculations it is seen that line B does indeed correspond to the Я mode (of the units in the bulk of the chain, i.e., the inner units) as usually defined, but its wavenumber does not shift with conjugation length. Within the frame of ECC theory this means that thiophene units along the chain do not greatly interact with each other. The constancy of $\nu_Я$ indicates the existence of a strong confinement or "pinning" of the π electrons within each thiophene ring or within a very few Th rings. This does not prevent an increase of the Raman intensity, which, according to the discussion of Section IV.B, must increase because of the lowering of the energy gap and the increase in the transition moments associated with the increase in N.

Line C. This line appears at the lower frequency side of line B only for a few classes of molecules and shows intensity enhancement with increasing chain length. Line C appears in some α,ω-substituted oligomers. Since lines B and C occur at close wavenumbers and belong to the same symmetry species, mechanical coupling with intensity borrowing can occur, modulated by the CL-dependent coupling force constant.

Line D. This line occurs as a sharp line (sometimes a doublet) near 1050–1080 cm^{-1} for almost all the molecules studied. Line D has an origin similar to that of ν_3 observed for *trans*-PA and is associated with the C—H in-plane wagging coupled with the Я mode. The C—H in-plane wag borrows intensity from the Я mode.

2. Infrared Intensity Spectroscopy of Oligo- and Polythiophenes

Experimental infrared intensity spectroscopy has been applied to a class of alkoxythiophenes and to several model molecules, and new information that may seed new relevant studies has been derived [199].

The great chemical stability of polyalkoxythiophenes has initiated a systematic experimental spectroscopic study of the charge injection by the oxygen atom into the thiophene ring. The absolute infrared intensity of solutions of many regiospecific oligoalkoxythiophenes has been measured.

The inspiring concept lies directly in the expression for the infrared intensity [Eq. (7)] that refers to the dipole moment changes during specific vibrational normal modes. The dipole changes (hence the experimental intensities) have also been expressed in terms of equilibrium atomic charges and charge fluxes that take place during a given vibrational motion. Thus the electronic properties peculiar of polyconjugated molecules such as electron transfer, electron delocalization, and bond charges may be probed by infrared intensity spectroscopy. We take from the work by Villa et al. [199] the prototypical case of 3-methoxythiophene (3MOTh), whose ground-state electronic structure can be described by a linear combination of at least two canonical structures, **VIII** and **IX**.

(VIII)   (IX)

In structure **IX** the oxygen atom has injected an electron into the thiophene ring and has necessarily induced polarization of the electron cloud. The measure of the intensity of the infrared-active ring-stretching mode near 1600 cm^{-1} should reveal such strong polarization, which does not take place in a similar molecule with no alkoxy group in position 3 (3-methylthiophene, 3MTh). The measured intensities for the spectral range 1600–1300 cm^{-1} are 43.17 km/mol for 3MTh to be compared with the value of 200.27 km/mol for 3MOTh. This shows that the charge has been injected into the ring and that a strong polarization of the molecule has occurred. Moreover, this simple experiment indicates the sensitivity of infrared intensities to charge displacements within molecules and offers a new tool for the study of polyconjugated materials.

Such kinds of intensity studies have allowed us to monitor inter- and intraring delocalization and to determine the extent of the topologically dependent perturbations induced by charge injection by the oxygen atoms [199]. Moreover, indications have been found suggesting the existence of "through-space" S···O interactions that increase the conformational rigidity of the thiophene chain.

3. The "*trans* Effect" and the Case of Isothionaphthene

Polyisothianaphthene (PITN) is representative of an interesting class of polyheteroaromatic molecules that have attracted the attention of several workers [200,201]. Its molecular structure and electronic properties are still a matter of debate, and final agreement has not yet been reached. ECC theory may contribute to the knowledge in this field and provide structural probes of general validity for this class of molecules.

The structural dilemma that needs to be solved is whether the structure of PITN is "aromatic" (**X**) or "quinoidal" (**XI**).

(X)   (XI)

Quantum-chemical calculations supported either one or the other of the two structures [202–204], and other spectroscopic experiments were equally inconclusive [205]. A contribution to the solution of this problem came from the group at the University of Limburg, where several model molecules were prepared [206–208] and their infrared and Raman spectra were studied [209] (see Fig. 28.23). Model molecules such as the one shown in Fig. 28.23 do not allow the formation of quinoidal structures, whereas models such as those in Fig. 28.24 do force the quinoidal structure. A simple comparison of the Raman spectra of a few model molecules (Fig. 28.24) and the justification of the observed spectra in terms of ECC theory give a strong indication that PITN in the solid state has a quinoidal structure.

An observation of general validity that has been made for other polyconjugated systems can be made here for PITN and derivatives. Whenever three conjugated double bonds organized in the *trans* configuration exist in a molecule, a collective mode of the Я type takes place with the generation of typical Raman scattering, which is the basis of ECC theory. This *trans effect* may become a useful structural spectroscopic probe.

4. Doping-Induced and Photoexcited Infrared Spectra of Oligo- and Polythiophenes

The cleanest DIIR spectra of doped PTh have been obtained by spectroelectrochemical reflection methods using ClO_4^- as counterion [210]. The DIIR spectra of doped oligomers and polymers have been reported by Cao and Renyuan [211]. These spectra show very strong bands near 1340 and 1030 cm^{-1} and a band near 1200 cm^{-1}. Weaker satellite absorptions occur at approximately 841, 729, 679, and 644 cm^{-1}. For the doped oligomers a pattern similar to that of the polymer is observed for Th$_5$, and it becomes more difficult to correlate the spectra for shorter chains.

The vibrational assignment of the strongest bands in the DIIR spectrum of doped PTh follows the general

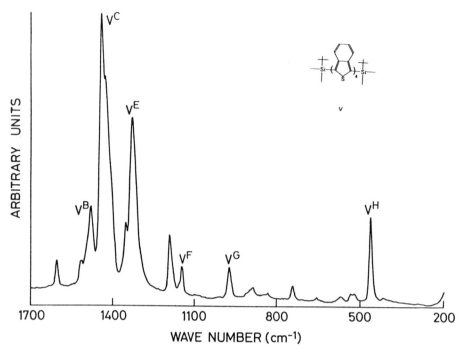

Fig. 28.23 Raman spectra (λ_{exc} = 1064 nm) of the model compounds sketched. (From Ref. 209.)

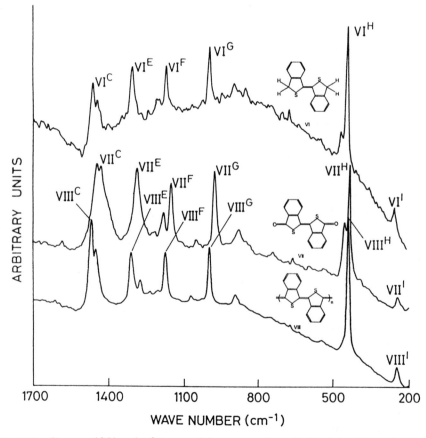

Fig. 28.24 Raman spectra (λ_{exc} = 1064 nm) of two model compounds and of polyisothionaphthene. (From Ref. 209.)

rules provided by ECC theory and previously discussed in the literature and also discussed here in Section VIII.B.1. The originally inactive Ᾱ mode for the pristine material becomes strongly infrared-active because of the lowering of the electrical symmetry caused by the polarization of the bonds that follows the process of charge transfer. The frequency of the infrared-activated Ᾱ mode shows a red shift because of the changes in the force constants of the central reference unit. Moreover, the values of the interaction force constants change while the domain of the interaction is extended to longer distances from the central reference unit.

The appearance of weaker lines throughout the spectral range from 1300 cm^{-1} downward and the complexity of the spectrum, especially on the lower frequency side of the band near 1200 cm^{-1}, can be associated, as in the case of *trans*-PA, with a polydispersivity in CL, which generates a convolution of DIIR spectra. This implies that the Ᾱ-containing modes of the doped materials show dispersion with chain length. Another possible way to explain the complexity of DIIR spectra is that because of the intrinsic chemical and structural disorder of the molecule a distribution of sites with a different environment may exist such that a distribution of "charge transfer sites" is generated, each with its own DIIR spectrum. This concept was used recently in the study of the composites of doped polythiophene derivatives and enzymes codeposited either chemically or electrochemically with the main aim of developing biosensor devices [212].

Because of such chemical and structural complexity, the verification of whether the DIIR spectra of doped polythiophenes show dispersion with CL is not easily carried out on doped samples. PIIR spectra of suitably photoexcited oligothiophenes and/or polythiophenes may provide some cleaner information. With this aim a large series of variously substituted oligo- and polythiophenes have been photoexcited and their PIIR spectra recorded and analyzed [213].

The PIIR spectra of variously alkyl-substituted oligothiophenes give the following information:

1. PIIR spectra show dispersion with conjugation length (see Fig. 28.25).
2. The complex fine structure is strongly reduced because the samples are monodisperse in CL and their chemical or structural disorder is certainly strongly reduced.
3. The PIIR spectra of polydithienyls (with suitable intraring distortion in order to modulate intraring delocalization) show frequency dispersion with the distortion angle and show the usual complexity due to polydispersivity in CL or disorder of the doped sites [213].

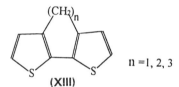

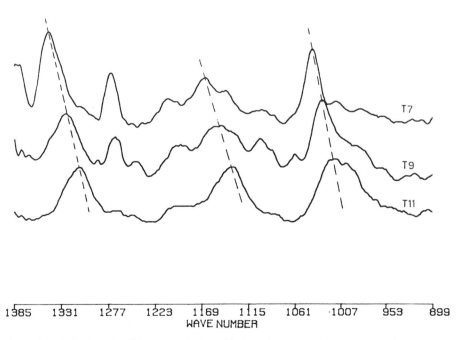

Fig. 28.25 Dispersion with chain length of the photoinduced infrared spectrum for a series of oligoalkylthiophenes illuminated by a laser light at 514 nm (M. Veronelli and G. Zerbi, to be published).

In the same experiments of photoexcitation [213], the power dependence of PIIR spectra was investigated for Th$_6$ and for polyoctylthiophene at both 20 and 150 K. In all four cases a root square dependence of the photoexcited signal from the density power on the sample was observed, suggesting a bimolecular recombination mechanism. This indicates that the charge carriers generated are bipolarons at every temperature.

As a final check of the concepts introduced by ECC theory, the PIIR spectra of stretch-oriented polyoctyl- and polyhexylthiophenes were recorded in polarized light and were analyzed and compared with the PIIR spectra of unoriented samples. As fully discussed in detail in Ref. 214 for the case of the Raman spectra of the same two polymers in their pristine state, ECC theory is fully verified, because the bands induced by photoexcitation correspond to the normal modes, which contain a large contribution by the Ᾰ mode that moves atoms along the molecular axis. The longitudinal character of the dipole moment changes are verified by the PIIR spectra of the stretch-oriented samples in polarized light. It is also apparent, as theory predicts, that the very many modes associated with the alkyl substituents or with other vibrations of the chains are not observed because they must be extremely weak, since they are not coupled to and do not borrow intensity from the Ᾰ mode.

5. Searching for Chemically Stable Conducting Polythiophenes—Use of the Infrared Spectrum

Any real technological application of conducting polymers implies that the materials show great stability in air and in a temperature range suitable for a particular application. A search for doped polythiophenes with great stability has been carried out in recent years. The suggestion arose from simple quantum-chemical calculations that the position in space of the dopant relative to the polymer molecule is an important factor in determining the chemical stability of these systems [215]. If the dopant can sit near the molecule in an energetically favorable position, the charge transfer bond becomes stronger and favors stability. The steric hindrance of the side group attached to the backbone chain then plays the dominant role in affecting the strength, which accounts for the stability of the charge transfer bond. Following this idea, Pei et al. [216,217] synthesized random copolymers of alkyl-substituted polythiophenes with the idea that the spacing between side chains could leave room for better interaction between the dopant and the polymer backbone. The observed stability was thus greatly improved [216,217].

To increase stability along these conceptual lines, Gallazzi and coworkers [218] prepared series of regiospecific polyalkylthiophenes that were doped with FeCl$_3$ or iodine. For the study of stability DIIR spectra were used for the first time in a quantitative way [218].

The existence of bipolarons (BP^{2+} or dications) was easily established from the electronic spectrum even when their concentration in the samples prepared as films could not be known. The usual DIIR spectral pattern was observed for the films of the doped polymers, and its changes with time and temperature were taken to follow the dedoping process. The measurements could be made quantitative by ratioing the observed DIIR spectra to the weak signals arising from the C—H stretching modes of the alkyl substituents. These modes

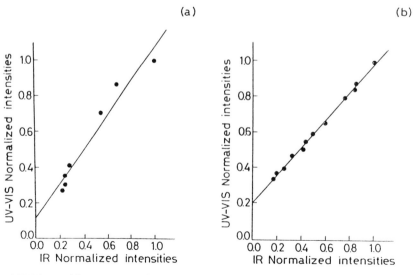

Fig. 28.26 Normalized UV intensities vs. normalized IR intensities for two poly(3,3″-dihexyl-2,2′:5′,2″-terthiophene) samples FeCl$_3$ doped and heated, respectively, (a) at 97°C and (b) at 100°C. (From Ref. 218.)

Structure and Nonlinear Optics

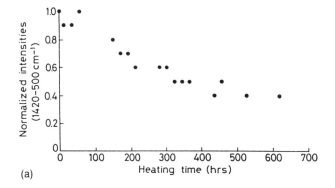

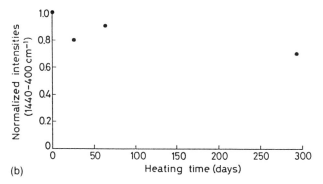

Fig. 28.27 Normalized intensities of the FeCl₃ doping-induced infrared spectrum of poly(3,3″-dihexyl-2,2′:5′,2″-terthiophene) vs. time (a) at $T = 70°C$, (b) at room temperature. (From Ref. 218.)

are observed as weak bands around 3000 cm⁻¹, but their intensities can always be easily measured.

To be sure that the intensity of DIIR spectra is directly related to the concentration of charge carriers, the normalized UV intensities of the bipolaron bands were plotted against normalized IR intensities of DIIR spectra for doped samples. An unquestionably straight line was obtained (Fig. 28.26).

The results are very pleasing. Figure 28.27 shows the time dependence of the relative intensity of DIIR spectra for two samples doped with FeCl₃, one heated to 70°C and one at room temperature. It can be seen that after 300 days the concentration of charge carriers at room temperature remained practically constant.

It is evident that the use of doping-induced infrared spectra in the study of the stability of doped samples is an easy and useful alternative to the use of the electronic spectrum.

IX. AROMATIC MOLECULES

A. Oligoparaphenylenes and Polyparaphenylene

Oligoparaphenylene (PP$_n$) and polyparaphenylene (PPP) and many derivatives have been synthesized [219–222], but, after a strong show of interest during the early stages of this research, many turned to other classes of polyconjugated materials.

For this class of compounds the problem of conformational flexibility around inter-ring carbon–carbon bonds is very important and is not yet completely settled. It was found that the torsional angle for biphenyl is $\theta \approx 45°$ [223], $\theta \approx 20°$ [224], and $\theta \approx 0°$ [225–227], in the gas, solution, and solid phases, respectively. The structure of PP$_n$ in the solid phase changes with temperature and shows the existence of strongly twisted chains in the solid at low temperature [228–230]. The spectroscopic behavior of this class of materials parallels that of Th$_n$ and PTh (Section VIII.B). Experimental E_g values change linearly with $1/N$, and ν_A is practically constant with CL [39]. Raman intensities for variously alkyl-substituted oligomers and polymers increase superlinearly either with the number of rings or with the number of p_z orbitals [40], thus showing that in spite of possible conformational twisting, inter-ring delocalization takes place.

This is also confirmed by the values of the experimental intensity ratio of the C—H deformation modes, $A_\text{in-plane}/A_\text{out-of-plane}$, for PP$_3$, PP$_4$, PP$_5$, PP$_6$, and PPP. The decrease in this ratio with increasing chain length indicates that the charge flux along the chain increases because of the increase in the delocalization of π electrons [231].

The spectra recorded for a large class of variously substituted oligomers and polymers recently prepared show the same characteristics [40] and do not yet provide a quantitative estimate of the role played by the conformation in determining the spectroscopic and optical properties.

Using vibrational chemical shifts of C—H stretching for the oligomers, three main bands are located at 3080, 3060, and 3034 cm⁻¹ (3028 cm⁻¹ for PPP) to be assigned to the CH groups of type α, β, and γ, respectively. From the correlation of ν_CH vs. C—H bond length, d_CH can

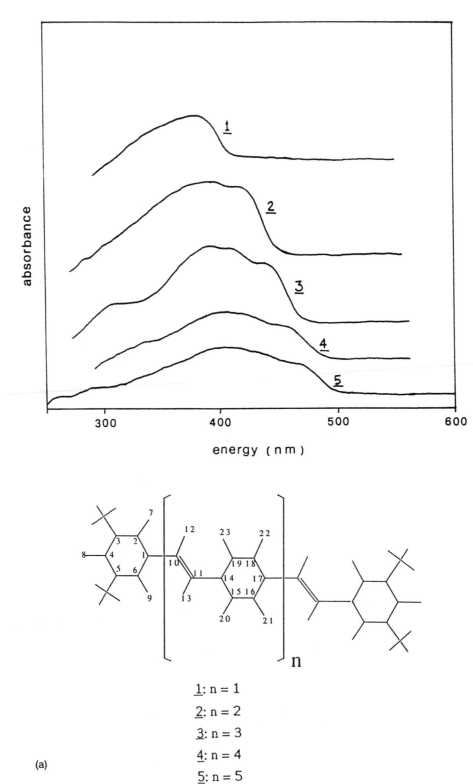

Fig. 28.28 (a) Diffuse reflectance electronic spectra of compounds 1–5 in the solid phase; (b) electronic absorption spectra of the same compounds in solution of CH_2Cl_2. (From Ref. 234.)

B. Oligo- and Polyparaphenylenevinylene

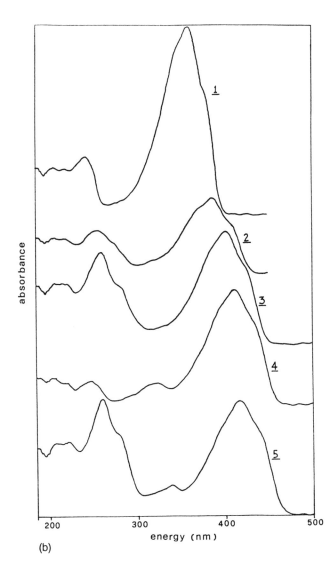

The interest in the oligomeric models of polyparaphenylenevinylene (PPV) is justified by the fact that this class of molecules (and their derivatives) are very relevant as active materials in the development of organic light-emitting diodes (see, e.g., Ref. 232). To increase solubility, t-butyl-substituted oligomers were prepared by Schenk et al. [233]. The vibrational infrared and Raman spectra of the pristine t-butyl-PV$_n$ oligomers were studied and recorded by Tian et al. [234]. The infrared spectra of the same materials doped with K were presented and discussed by Zerbi et al. [235]. Sakamoto et al. [115–117] studied the Raman spectra of other model molecules doped with Na.

trans-Stilbene is the simplest model molecule that has been the subject of many studies. For this molecule in the gas phase the two benzene rings are twisted with a torsional angle $\theta \approx 31 \pm 5°$ at 190°C [236]. In the solid phase, $\theta \approx 5°$ [237].

The first relevant information was derived from the electronic spectra of the materials in solution or in the solid phase (Fig. 28.28) [234]. In going from $n = 1$ to $n = 5$ a red shift in the electronic absorption spectra is observed, showing that the HOMO–LUMO energy gap closes with increasing CL [39]. As each model molecule passes from the solid state to the solution, its spectrum is blue-shifted, indicating an opening of E_g either because molecules in solution are conformationally more distorted than in the solid or because of the so-called solvent effect.

The analysis of the infrared spectrum of this class of materials in the pristine state shows, as expected, little effect of chain length on the IR-active vibrations. From the infrared spectra of the oligomers, end-group modes have been located and, as a consequence, $k = 0$ phonons of PPV as a one-dimensional lattice have been clearly identified. The frequency shifts from the oligomers to the polymer are extremely small, thus showing that (1) intramolecular coupling as shown by the IR-active modes is small, (2) phonon dispersion curves must be substantially flat, and (3) force constants are not dependent on chain length [234].

The Raman-active modes (including the strong Я mode) are practically CL-independent, and dispersion of the Raman intensities is not yet known. Comparison of the Raman spectra of a few oligomers and their selectively deuterated derivatives proves that the normal modes are confined either in the ring or on the C=C bonds [235].

Doping-induced infrared spectra of potassium-doped PPV and potassium-doped t-butyl-PV$_n$ have been studied in detail [235]. The infrared experiments required the development of special procedures for avoiding contact between the samples and even traces of oxygen or water. The evidence so far collected is that in solid samples, mostly polarons are generated [61].

The DIIR spectra of t-butyl-PV$_n$ models and PPV show strong bands at $\sim$1500 cm^{-1} (broad and ill-defined), $\sim$1300 cm^{-1} (possibly a doublet), and $\sim$1150 cm^{-1}, which have been identified as characteristic of the charge carrier and as modes containing the Я coordinate activated in IR because of symmetry breaking due to doping. DIIR spectra of the oligomers studied ($n = 1$–5) are slightly different in shape and frequency from those observed for PPV [235]. Furthermore, the DIIR spectrum is almost identical with the PIIR spectrum in PPV and in the oligomers studied [42].

No dispersion with CL is observed in DIIR spectra of K-doped oligomers, strongly suggesting that either the charge carrier is confined to a very limited molecular domain or, as usual, the dynamic properties of the oligomer molecules as revealed by the vibrational frequencies do not reveal long-range interactions.

Two different approaches have been adopted in the interpretation of the spectra of K- or Na-doped t-butyl-PV_n oligomers and polymers. On the basis of ECC theory we ascribe the observed spectra to collective vibrations that may involve sections of the oligomer/polymer chain [235]. On the other hand, the Japanese school seems to consider that the infrared (and Raman) spectra originate from strongly localized vibrations (almost group frequencies) [115–117].

The relevant structural information that can be derived from the combination of infrared, Raman, and ESR-ENDOR spectra is the following:

1. The charge carriers are confined over a rather short domain, which turns out to consist of a very few monomeric units according to vibrational spectroscopy but more according to ESR-ENDOR experiments [238].
2. The doping with electron donors perturbs short segments of the chain consisting approximately of one benzene unit and the two adjacent C=C

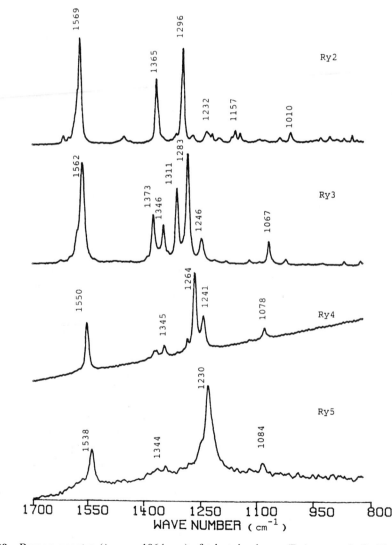

Fig. 28.29 Raman spectra (λ_{exc} = 1064 nm) of t-butyl-rylenes $(Ry)_n$, n = 2–5. (From Ref. 40.)

double bonds. The benzene rings are most perturbed.
3. The absorption to be associated with the "paraquinoid structure" in the doped systems is the infrared band near 1150 cm^{-1}. This band can be taken as characteristic of the existence of a charge carrier containing benzene rings in the quinoidal structure.

C. Rylenes

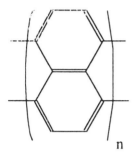

Only recently have the vibrational spectra of the first elements of the series of t-butyl-capped rylenes (Ry$_n$) been studied in detail [154] (Fig. 28.29). It is useful to discuss the data obtained from frequency and intensity spectroscopy of these systems as a prototypical case of a conformationally rigid system to be compared with the conformationally flexible PP and PV oligomers and polymers discussed in Sections IX.A and IX.B, respectively.

The infrared spectra of the t-butyl-rylenes are crowded by many bands, their interpretation is difficult, and the derivation of any structural information becomes uncertain. In contrast, the Raman spectra of oligomers starting from the "trimer" turn out to be extremely simple, showing only the usual two modes (which show dispersion with CL) associated with the Я coordinate and are very simple for pentarylene [154]. The Raman intensities of the same systems in solution increase superlinearly; the consequences for molecular nonlinearities are discussed in Section X [40].

These peculiar and very relevant optical properties can be accounted for in the light of the concepts of general validity discussed in previous sections of this chapter. The molecules are fully planar and rigid, thus favoring the largest overlap of p_z orbitals, i.e., favoring the largest degree of intramolecular delocalization. The structure allows the generation of a low energy quinoidal structure, thus causing large electron–phonon coupling, which is revealed by the frequency dispersion with CL. The topology of p_z orbitals, which allows the formation of the collective Я mode, the large polarizability of the π electrons, and the conformational rigidity of the molecules make these systems very suitable for further development in the field of materials with large nonlinear optical responses [40].

X. NONLINEAR OPTICS AND INTENSITY SPECTROSCOPY OF CONJUGATED ORGANIC MATERIALS—A NEW APPROACH

The existence of highly polarizable π electrons in polyconjugated molecules makes them interesting objects with possible large nonlinear optical responses when they are placed in strong electromagnetic fields.

For this reason the nonlinear optics of organic materials has become and still is a very active field of work of interest in both basic and applied science. Several theoretical and experimental efforts have been devoted to the evaluation of molecular hyperpolarizabilities and to the study of the mechanisms that make organic materials suitable for optoelectronic and photonic applications [239–242].

The basic relationship that rules the nonlinear optical response of a macroscopic sample under the action of an intense external electromagnetic field (**E**) is

$$\mathbf{P} = \chi^{(1)} \mathbf{E} + \chi^{(2)} \mathbf{E} \cdot \mathbf{E} + \chi^{(3)} \mathbf{E} \cdot \mathbf{E} \cdot \mathbf{E} + \cdots \quad (20)$$

where $\chi^{(1)}$, $\chi^{(2)}$, and $\chi^{(3)}$ are the bulk electric susceptibilities and **P** is the volume polarization. The corresponding microscopic equation is

$$\boldsymbol{\mu} = \alpha \, \mathbf{E} + \frac{1}{2!} \beta \, \mathbf{E} \cdot \mathbf{E} + \frac{1}{3!} \gamma \, \mathbf{E} \cdot \mathbf{E} \cdot \mathbf{E} + \cdots \quad (21)$$

In the case of molecular organic materials, where intermolecular interactions are weak, the nonlinear optical behavior is essentially determined by Eq. (21). In what follows we focus on the molecular hyperpolarizabilities of the first and second order (β and γ).

Bulk and molecular nonlinear optical properties have been measured by laser optical techniques such as second and third harmonic generation (SHG, THG), electric field–induced second harmonic generation (EFISH), and degenerate four-wave mixing (DFWM). Molecular NLO responses can also be calculated by quantum-mechanical (ab initio and semiempirical) methods, and suitable computing programs are being developed.

A completely new approach to molecular hyperpolarizabilities has recently been presented in terms of vibrational intensity spectroscopy. It has been shown both theoretically [243] and experimentally [244–246] that in the case of polyconjugated low band gap materials, molecular hyperpolarizabilities can be determined uniquely from spectroscopic observables.

It can be shown [247] that the vibrational contribution to the NLO response can be expressed in terms of spectroscopic observables according to

$$\beta^r_{nmp} = \frac{1}{4\pi^2 c^2} \sum_k \left(\frac{1}{\nu_k^2}\right)$$
$$\times \left[\left(\frac{\partial \mu_n}{\partial Q_k}\right)\left(\frac{\partial \alpha_{mp}}{\partial Q_k}\right) + \left(\frac{\partial \mu_m}{\partial Q_k}\right)\left(\frac{\partial \alpha_{np}}{\partial Q_k}\right) + \left(\frac{\partial \mu_p}{\partial Q_k}\right)\left(\frac{\partial \alpha_{nm}}{\partial Q_k}\right)\right]$$
$$(22)$$

$$\gamma^r_{nmps} = \frac{1}{4\pi^2 c^2} \sum_k \left(\frac{1}{\nu_k^2}\right) \left[\left(\frac{\partial \mu_n}{\partial Q_k}\right)\left(\frac{\partial \beta_{mps}}{\partial Q_k}\right)\right.$$
$$+ \left(\frac{\partial \mu_m}{\partial Q_k}\right)\left(\frac{\partial \beta_{nps}}{\partial Q_k}\right) + \left(\frac{\partial \mu_p}{\partial Q_k}\right)\left(\frac{\partial \beta_{nms}}{\partial Q_k}\right) + \left(\frac{\partial \mu_s}{\partial Q_k}\right)\left(\frac{\partial \beta_{nmp}}{\partial Q_k}\right)$$
$$+ \left(\frac{\partial \alpha_{nm}}{\partial Q_k}\right)\left(\frac{\partial \alpha_{ps}}{\partial Q_k}\right) + \left(\frac{\partial \alpha_{np}}{\partial Q_k}\right)\left(\frac{\partial \alpha_{ms}}{\partial Q_k}\right) + \left.\left(\frac{\partial \alpha_{ns}}{\partial Q_k}\right)\left(\frac{\partial \alpha_{mp}}{\partial Q_k}\right)\right]$$
(23)

These expressions are directly related to quantities that can easily be obtained from vibrational spectra, namely the vibrational frequency of the kth normal mode, ν_k; the variation of the molecular dipole moment, $\partial \mu_n / \partial Q_k$; and the variations of the first and second molecular polarizabilities, $\partial \alpha_{nm} / \partial Q_k$ and $\partial \beta_{nmp} / \partial Q_k$ with respect to normal coordinate Q_k. These parameters can be derived from infrared, Raman, and hyper-Raman intensities, respectively.

It must be noted that under the hypothesis of both mechanical and electrical harmonicity, Eqs. (22) and (23) are the same as those obtained by other authors with quantum-mechanical treatments [248,249].

The method that stems from the above formulas allows us to routinely measure β^r and γ^r quickly and inexpensively.

Since both the physical concepts and the experimental approach to NLO properties introduced with this method are new, it had to overcome the obvious resistance of the scientific community. The strongest conceptual difficulty to be overcome was due to the generalized belief that β^r and γ^r are small (or even very small) compared with the corresponding electronic contributions. Contrary to all expectations it is presently well assessed both theoretically and experimentally that in the case of low band gap polyconjugated compounds both β^r and γ^r are very large. The basic concept is that β^r and γ^r are very large because they correspond to a relevant projection of the electronic quantities of interest to vibrational space.

Reference is made to the discussion in Sections IV and V, where it was shown that intensity spectroscopy of polyconjugated systems provides a wealth of information on their structure, dynamics, and electronic properties. An exhaustive rationalization of the observed spectroscopic properties has been given in terms of ECC theory.

The key phenomenon is the existence, in these low band gap π systems, of strong electron–phonon coupling. The variation of the polarizability tensor $\partial \alpha / \partial Q_k$ is particularly large for those vibrational coordinates (Q_k) that oscillate in the directions along which this coupling is maximum. This explains why only a few Raman modes with extremely large intrinsic activity appear in the spectra [244–246].

The infrared spectra can also be interpreted within this framework. It has been shown that in the case of conjugated systems [34,35] infrared intensities are largely dominated by charge fluxes, i.e., by electronic charge redistributions induced by vibrational motions. Once again charge fluxes are particularly relevant for those normal modes that oscillate along the directions of maximum electron–phonon coupling.

In other words, in both infrared and Raman spectra, vibrational coordinates along these particular directions induce charge redistributions in the electronic cloud that are reflected in the vibrational intensities. Vibrational hyperpolarizabilities, which according to Eqs. (22) and (23) are expressed in terms of vibrational intensities, are obviously greatly affected.

In the attempt to evaluate the optical nonlinearities that can be obtained with the vibrational method, both theoretical and experimental approaches have been undertaken [244–246,250]. In the first case, quantum-chemical ab initio calculations were used to compute $\beta^e(0;0,0)$ and $\gamma^e(0;0,0,0)$ using standard derivative techniques and β^r and γ^r using calculated vibrational intensities. On the other hand, experimental Raman and infrared absolute intensities were used to measure β^r and γ^r. Experimental values for β^e and γ^e determined with traditional methods (EFISH, THG, etc.) can be found in the literature. In both cases (theoretical and experimental), the agreement between the NLO coefficients determined in the traditional way and their vibrational counterparts is astonishingly good.

The comparison has always been made, whenever possible, between static values. In the case of experimental β^e, the extrapolation to the static case was done in terms of a two-state model; for γ^e, where the extrapolation is not straightforward, values of the fundamental wavelength were chosen.

The results obtained in a number of meaningful test cases are presented below.

A. Third-Order Hyperpolarizabilities: γ

Centrosymmetric conjugated molecules (e.g., polyenes, thiophenes, phenylenes) are well known for their significant NLO coefficients. γ^r has been measured for several such molecules. In Fig. 28.30 the experimental and theoretical γ^r values obtained for polyene systems are compared with experimental and theoretical γ^e. The agreement in both values and trends is very good. Similar results on γ^r were obtained with other polyconjugated systems and are reported in Table 28.3 for various classes of compounds. The different values obtained for the NLO response of each class of molecules reflect the different Raman intensities measured for each compound. As discussed in Section IV, different absolute intensities correspond to different degrees of overall electronic delocalization within the molecule. A stronger charge localization implies a smaller intensity of those normal modes that couple the electronic distribution

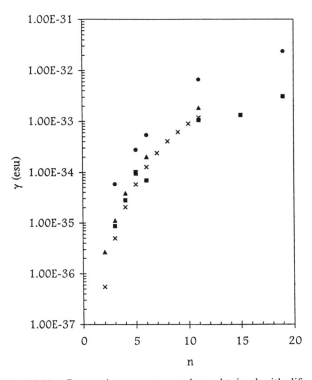

Fig. 28.30 Comparison among γ values obtained with different methods for polyene systems of increasing chain length (see Ref. 244). (▲) γ^r from calculated (ab initio 6–31G) Raman intensities; (×) γ from ab initio (6–31G) calculations; (■) γ^r from experimental Raman cross sections; (●) γ from THG measurements.

with vibrational motions (e.g., thiophenes, phenylenes) and hence yields a smaller γ^r. The existence of some localization of the electronic charge distribution within the heteroaromatic ring contrasting conjugation along the chain has been rationalized with the introduction of a "pinning" potential [39] that is a measure of the tendency of charge distribution to be confined within the single repeat unit. The case of rylenes is particularly interesting [40]; extremely high γ^r values are obtained. It must be noted that the number of π electrons involved in one repeat unit is much larger than that characterizing the other systems. Indeed, when γ^r values are normalized to the number of π electrons in the molecule, the results are those reported in Fig. 28.31. The most striking observation is that rylenes are much more similar to polyenes than to phenylenes. This might seem contradictory unless it is accepted that the certainly planar geometry of rylene compounds forces a delocalization in the plane of the molecules, in contrast to the tendency to confinement within the benzene unit [40].

B. First-Order Hyperpolarizability: β

Several classes of organic compounds have been studied for their promising first-order optical nonlinearities. At the very beginning of the NLO era of organic molecules, attention was focused on "dipolar" molecules, in which a donor group and an acceptor group are linked by a conjugated π system [144,251,252]. Various systems were studied, and the influence of the donor and acceptor groups and that of the kind of π systems on the NLO

Table 28.3 Experimental γ^r Values Obtained from Raman Intensities for Specified Series of Molecules

Number of repeating units	γ^r (esu)			
	Polyenes	Thiophenes	Rylenes	Phenylenes
1			1.88×10^{-36}	
2			3.72×10^{-35}	
3	8.69×10^{-36}	1.71×10^{-35}	1.27×10^{-34}	5.37×10^{-36}
4	2.76×10^{-35}		4.40×10^{-33}	
5	1.02×10^{-34}		7.64×10^{-33}	1.61×10^{-35}
6	6.91×10^{-35}	1.44×10^{-34}		
7				4.44×10^{-35}
9		3.71×10^{-34}		4.14×10^{-35}
10				3.23×10^{-35}
11	1.05×10^{-33}			
12		5.95×10^{-34}		
15	1.33×10^{-33}			
19	3.07×10^{-33}			

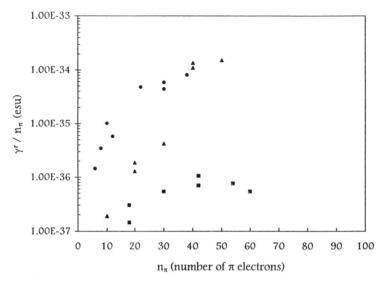

Fig. 28.31 Experimental γ^r for (●) polyene systems, (▲) rylenes, and (■) phenylenes normalized with respect to the number of π electrons (n_π). (Data from Ref. 40.)

behavior were ascertained. These studies led to the well-known and widely used "two-state model" [253]. Later, in the attempt to overcome the problems related to the dipolar nature of these compounds (centrosymmetric crystallization, transparency–NLO response trade-off, etc.) new synthetic strategies were devised. One of the most interesting outcomes was the introduction of the so-called octupolar systems proposed by Zyss [254].

Finally, only very recently, enormous interest has developed in asymmetric push–pull polyenes. These systems are particularly promising because they exploit jointly the traditional properties of dipolar systems and the efficiency, in terms of the mobility of π electrons, of the polyene bridge [144–148].

The vibrational method for estimating β has been used for all these classes of compounds. The results obtained for some prototypical molecules of each class compare favorably with other determinations that can be found in the literature [244–246].

In this work we focus on push–pull polyenes and consider the effect of changing the end groups and increasing the chain length.

In Fig. 28.32 are collected the experimental values measured for several series of push–pull molecules using the vibrational approach. Recalling Eq. (22), it becomes possible to rationalize these results in terms of infrared and Raman intensity patterns.

The ultimate goal in this field is the optimization of the molecular coefficients β. This is the result of a delicate balance between the strength of the polar end groups and the chain length. It is not easy to outline a simple relation between structural parameters and NLO behavior because the different factors that affect the properties can be competitive. This means that there must be an optimum compromise (different for each molecule) between chain length and polar end groups. The analysis of the intensity pattern can help in elucidating how changes in the structural parameters modulate β values. To optimize β values, infrared and Raman intensities must be simultaneously maximized for the same normal modes. The criteria leading to enhanced infrared and Raman intensities have been presented and discussed in Sections IV.B and VI.

As a general observation it can be stated that for a fixed chain length (e.g., $N = 3$), higher β values are

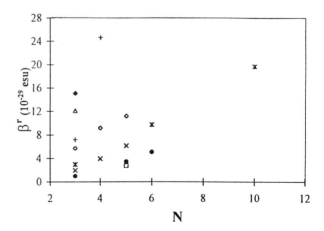

Fig. 28.32 Experimental β^r for push–pull molecules. Molecules and symbols as in Fig. 28.6. (Data from Refs. 244–246.)

obtained by increasing the strength of the polar end groups as shown by the data relative to push–pull polyenes with increasingly stronger end groups [polyenovanillins (**IV**) and diphenylpolyenes (**I** and **II**)]. This behavior can be ascribed to the increase in Raman intensity, which is associated with an electrical asymmetrization of polyene systems (see Section III). At the same time, infrared intensities are enhanced by polar substituents. An intermediate behavior is exhibited by push–pull phenylpolyenes (**III**). In this case the aldehydic C=O at one end has a lower electron-accepting capability than the cyano (CN) or nitro (NO_2) groups and should provide a smaller β value. An intermediate value (between **I** and **II**) is found (for class **III**), and this is justified by the fact that the polar group is directly grafted onto the polyene chain. In contrast, in series **I** and **II**, there is a phenyl ring between the polyene chain and the polar group. This means that the insertion of a phenyl ring has a sort of shielding effect that makes the induced polarization less effective and partially hinders charge delocalization along the chain (see Fig. 28.18). In fact, Raman intensities of compound **IIIa** ($N = 3$) are intermediate between those of compounds **Ia** and **IIa**. The infrared intensity of the aldehydic compound (**IIIa**) is much higher. This reflects the fact that the infrared spectrum is much more sensitive than the Raman to the degree of polarization of carbon–carbon bonds in the chain and this is decreased by the phenyl rings in compounds **II** and **III**. Moreover, the great intensity enhancement is due to the polarization of only a few terminal carbon–carbon bonds. This observation finds further support in the fact that the longer compounds of the aldehydic series ($N =$ 6, 10) have an almost constant infrared intensity.

An overall analysis of the plots in Fig. 28.32 indicates that substantially linear behavior is found when chain length increases as shown by the power law dependence ($\beta = kN^a$) with exponent values a very close to 1. This is not surprising if one observes that the Raman intensities become more and more "polyene-like" ("carotenoid-like" behavior). This means that there exists for each series a threshold chain length (N_c, see Section VI) beyond which the polarizing effect does not perturb the central part of the molecule. Under these conditions only Raman intensities can contribute to the increase in β values and the growth rate with N cannot be superlinear. In the polyene regime, to further increase the growth rate it is necessary to move toward more polar end groups, whereas there is no advantage to increasing chain length.

Totally different behavior is shown by compounds **VI**, **VII**, and **VIII**. In this case really strong end groups are present and relatively short chain lengths have been considered such that the polarization effect is necessarily felt by the whole chain. In the case of molecule **VI** the addition of just one double bond leads to a great increase in β. Unfortunately, only two chain lengths were available, but the trend shown makes one believe that the power law will be superlinear. From the data shown (Raman intensities) it is clear that the polyene regime has not yet been reached, indicating that it may be possible to greatly enhance the NLO response by further lengthening the molecule. Even greater is the enhancement of infrared intensities, suggesting that the induced polarization takes advantage of a more extended conjugation path. Finally, the observation of the infrared intensities of the strongest push–pull molecules makes one think that in this case the addition of other double bonds should lead to an even more striking increase of the NLO response. The practical suggestion to the chemist is that in these cases it is still meaningful to increase the chain length.

C. Solvent Effects

The intensity changes discussed can be correlated with structural modifications of the polyconjugated chains. The basic ideas underlying this observation have been outlined in Section VI. The substitution of different polar groups and/or chain lengthening induces a modification of the electronic structure of the chain by modulating the average bond length alternation, which has been shown to be the most significant structural parameter (see Section IV.B). This means that whenever bond length alternation can be modified one can expect variation of the vibrational intensities and hence of the NLO response. This is particularly important because in some cases it is possible to induce such a structural modification by working on the molecular environment without changing the chemical nature of the molecule. A typical way in which this can be done is to dissolve the molecule in various solvents of different polarity; more generally any intermolecular interaction (interactions in the solid state, inclusion in suitable stabilizing matrices or channels, etc.) can be envisaged.

The most easily accessible experimental approach to this kind of modulation is offered by the study of solvent effects. This problem has recently been successfully faced by several authors [153,255,256]. In Fig. 28.33 it is possible to observe the modulation of the NLO responses obtained by changing the polarity of the solvent. It can be shown that a similar modulation is observed when the vibrational method is used to estimate molecular hyperpolarizabilities. Obviously, not every push–pull polyene exhibits solvent-dependent behavior. The analysis of the intensity behavior will reveal which systems are more suitable for modulation of bond alternation. Again it is the acceptor–donor pair that makes the molecule more responsive to the action of the solvent. In the "polyene regime" the chain structure is less affected by the interaction with the solvent. This can be rationalized if one remembers the two limiting canonical structures whose weighted combination reproduces the molecular structure. The interaction with the solvent is reflected in stabilization of the zwitterionic structure.

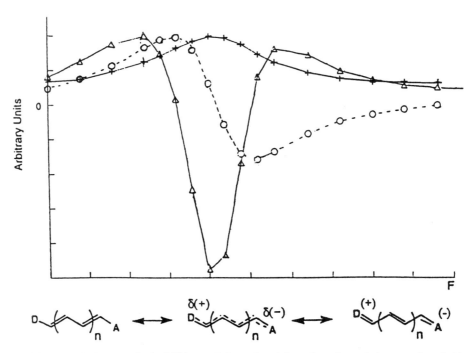

Fig. 28.33 Polarizability α ($\times$), hyperpolarizabilities β ($\bigcirc$), and γ ($\triangle$) as function of the reaction field F that modulates the dimerization parameter (u) as sketched in the figure. (From Ref. 255.)

This implies a reduction of the difference between the two wave functions and hence a reduction of bond alternation. If the starting structure (apolar solvent, unperturbed molecule) is polyene-like, the two canonical forms have very different energies and it becomes difficult to make them interact. Indeed the experimental determinations of β for the most polyene-like molecules (polyenovanillins and diphenyl push–pull molecules) do not show any solvent-dependent NLO behavior. If the structural situation is such that the two forms have more similar energies, an external perturbation (the polar solvent) can influence the mixing by changing the relative weight. A sort of fine-tuning of the structure by the solvent can thus be envisaged.

It has already been pointed out that a modulation of the structure is related to a modulation of the intensity pattern. Starting from a centrosymmetric polyene structure (alternated), it is known that modes have selectively enhanced Raman intensities and vanishing infrared activities. As soon as electrical symmetry is broken, a variation of the dimerization is induced, thus determining an enhancement of the Raman intensity and the simultaneous activation of the modes in the infrared spectrum. This means that the molecule becomes β-active and its NLO response increases as the degree of alternation decreases. Thus β reaches a maximum value after which the Raman intensity starts decreasing again as a consequence of the decreasing of ΔQ^{eg} [see Eq. (11)] until for an equalized chain structure the Raman spectrum is no longer selective and is almost vanishing. As a result β decreases and finally becomes zero for a perfectly equalized chain. This is just the symmetry requirement that must be verified by a tensor of rank 3 for a centrosymmetric molecule.

Obviously if one considers only one molecule, it is not possible, with the available solvents, to span the entire range of structural parameters from completely alternated to equalized geometry. It is then necessary to consider various molecules that have unperturbed structures in apolar solvents with different degrees of bond alternation $(\langle u \rangle)$. The experimental results obtained for molecules **V**, **VI**, and **VII** are reported in Fig. 28.34. The three molecules have the same chain length ($N = 3$), but their polar end groups are increasingly stronger. The modulation induced by the solvent is different for each molecule. A comprehensive analysis of the data shows that the three molecules taken together reproduce all the features predicted for the evolution of β as a function of the dimerization parameter [244–246]. The NLO response reaches a maximum value and then decreases until it vanishes and changes sign. This kind of behavior is exactly what is predicted on the basis of independent ab initio theoretical calculations [50, 150, 257].

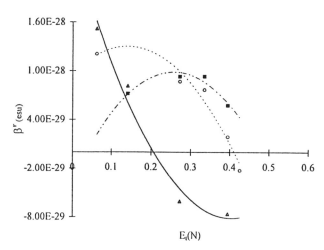

Fig. 28.34 Experimental β^r values for molecules **V** (■), **VI** (○), and **VII** (▲) in various solvents (C_6H_6, CCl_4, $CHCl_3$, CH_2Cl_2, CH_3CN, CH_3NO_2 in order of increasing polarity). Lines are meant only as a guide to the eye. (From Ref. 245.)

D. Theoretical Justification

So far the present discussion has considered a large number of experimental β values obtained with the vibrational method for various systems under different conditions. The experimental values and trends are in good agreement with other determinations based on more traditional techniques. This is quite an achievement, as it appears that the nuclear (or vibrational) contribution is the only one determining the mechanisms that lead to molecular NLO responses. As already repeatedly pointed out, in the particular class of materials we are considering (i.e., low band gap materials), the separation between electronic excitations and nuclear motions is no more rigorous. The strong electron–phonon coupling that is so important in the description of the physics of these systems makes this distinction infeasible.

It has been theoretically proven that the same quantities that appear in the two-state model expression for β^e are also present in the two-state model of β^r. This theoretical model proves that when strong electron–phonon coupling is effective, β^r represents a non-negligible projection of the electronic β^e onto the vibrational space. Moreover, this projection approaches an exact value more closely as the coupling energy becomes comparable with the energy gap [172,243]. The physical consequence is that it is possible to obtain the same final state of π polarization both by direct electronic excitation and by modification of the electronic charge distribution induced by the normal coordinate (taken in the direction of electron–phonon coupling).

XI. APPLICATION OF ECC THEORY TO BIOLOGICALLY RELEVANT MATERIALS

The theoretical and experimental knowledge gained from vibrational spectroscopy of polyene systems through ECC theory is applied to the case of biologically relevant systems whose properties are determined by polyene molecules suitably embedded in and interacting with a protein environment. These biological systems are also becoming relevant in technology.

Attention is focused on carotenoids and polyenes, which are known to be chemically very unstable as isolated entities but to acquire great stability when they are suitably surrounded by a protein cage and become the active elements in the mechanism of vision and photosynthesis. The CL dependence of the *in situ* Raman spectra of the carotenoids as naturally occurring pigments in bird feathers was studied by Veronelli et al. [65]. Later attention was focused on the bacterial membrane protein bacteriorhodopsin (bR), a small protein (~26,000 daltons) whose potential application in optical and electro-optical devices has been explored by many authors. The justification of such interest lies in the fact that bR contains all-*trans* retinal, which acts as a light-absorbing center and makes bR a naturally reversible photochromic system. All-optical switching can be achieved by proper illumination of bR with yellow or blue light.

The chromophore is covalently bonded to the protein by a protonated Schiff base linkage (PSB) and is located in the pocket resulting from the arrangement of the single polypeptide chain across the bacterial membrane. In spite of the numerous structural studies on bR, atomic resolution of the structure of the chromophore embedded in the protein has not yet been achieved. Although the sequence of amino acids comprising the retinal binding site has been identified, the protonation state of the nearby amino acids and the location of some water molecules or charged ions have not yet been established.

Vibrational spectroscopy through ECC has been shown to be an useful tool for the description of the structure of the retinal chromophore in the so-called light-adapted state of bR. For this study the Raman spectra of several model molecules have been studied (see illustrated structures).

all trans Retinal

Schiff Base (SB)

Protonated Schiff Base (PSB)

The first striking observation (Fig. 28.35) fully consistent with ECC theory is that with excitation at $\lambda_{exc} =$ 1064 nm the Raman spectra of all-*trans*-retinal and bR are very similar, thus showing that the Raman spectrum of bR is dominated by the chromophore even if the molecular weight ratio is $\sim 1/100$.

Next we refer to Section X, where it was shown that nonlinear optical properties of noncentrosymmetric polyene systems can be modulated by changing the push–pull strength of the end groups and/or by changing CL or the environment. The strength of the acceptor end group of retinal and the environment have been

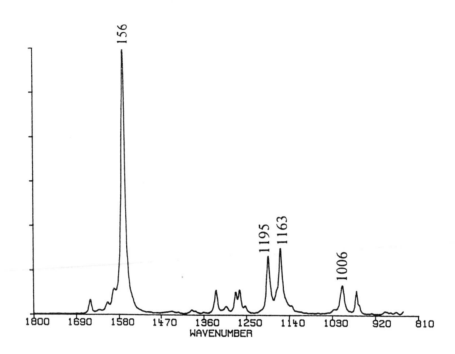

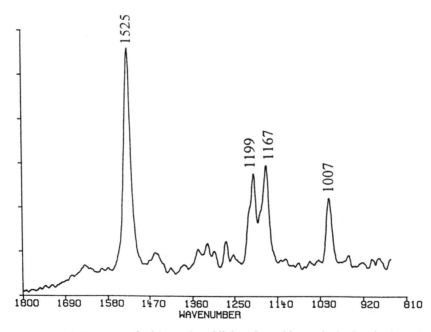

Fig. 28.35 FT Raman spectra of all-*trans*-retinal (upper) and light-adapted bacteriorhodopsin (lower) molecules in the solid state ($\lambda_{exc} = 1064$ nm). Vibrational frequencies (cm^{-1}) of the relevant bands are given.

changed by nature in such a way that the protein favors a protonated Schiff base (PSB) linkage with the chromophore, thus causing an extremely large red shift of the electronic absorption maximum. An estimate of the extent of π-electron delocalization in the all-*trans*-retinal molecule and its unprotonated (SB) and protonated (PSB) Schiff bases and in the bR-bound chromophore can be obtained from the analysis of the data reported in Fig. 28.36. This figure shows the increasing values of λ_{max} in going from SB to all-*trans*-retinal, to PSB and to bR plotted versus the experimental values of $\nu_я$. An unquestionably linear dependence is found between λ_{max} and $\nu_я$, thus making the Raman spectrum a useful probe for the local structure of protein-bound retinal.

Comparison with the spectra of other model compounds [e.g., polyenovanillins (**IV**)] indicates that the average dimerization parameter $\langle u \rangle$ of the PSB is negative, i.e., the carbon skeleton is strongly dimerized (alternated). It has been further observed that in going from the isolated PSB to the protein-linked chromophore, (1) the Raman intensity decreases remarkably, (2) $\nu_я$ decreases, and (3) λ_{max} is strongly red-shifted. Using the concepts previously elaborated in this chapter it can then be concluded that *the effect of the protein cage that surrounds the retinal-protonated Schiff base is that of generating a large degree of delocalization, almost reaching bond equalization* $\langle u \rangle \approx 0$. This is a new and relevant result from spectroscopy.

The effect of the protein matrix on the bR chromophore can be described in terms of the existence of an effective electric field at the retinal binding site originated by the charges and location of the amino acid residues that form the pocket surrounding the chromophore. The effective electric field acting along the molecular axis stabilizes an almost equalized structure of double bonds (cyanine-like) very different from the polyene-like backbone ($\langle u \rangle < 0$) found in the isolated PSB.

The degree of bond length alternation determined from spectroscopy for the bR chromophore with the use of Fig. 28.33 allows prediction of the values of the molecular nonlinear optical responses β and γ. Indeed, in the cyanine-like range ($u \approx 0$) β values are small and cross zero with negative slopes and γ can have negative values. The positive contributions to γ have been measured according to ECC applied to NLO responses.

XII. LOOKING AHEAD

The purpose of this chapter was to report on the state of the art in vibrational spectroscopy in the field of polyconjugated materials. The intent was to go beyond the approach of technique-oriented spectroscopists to show nonspecialists the large quantity of unique information that can be derived from optical spectra. Many examples have been presented as a guide to the nonspecialists on how to use spectroscopic data. Classical vibrational chemical correlations have been shown to be almost useless for deriving structure–property correlations. A large amount of new physics has been developed for the understanding of the peculiar spectra of these materials in the pristine, doped, and photoexcited states.

One of the most relevant items of information people would like to derive from the spectra is the extent of π-electron delocalization in the various classes of materials that have been produced and characterized. We believe that intensity spectroscopy will in the future be a unique tool for learning about long-range electron correlations in such highly polarizable materials.

It has also been shown that a new area of physics has been developed that uses vibrational infrared and Raman intensities for deriving nonlinear optical data for polyconjugated materials. The NLO data already collected and discussed show the relevance of the method for materials science.

The implications of these studies in the field of biophysics and biochemistry have been only touched upon but appear to be very promising.

ACKNOWLEDGMENTS

The many new theoretical concepts and new data presented and discussed in this chapter are the results of years of friendly, active collaboration and fruitful discussion with many members of the group at Milan. We are particularly grateful to Dr. M. Gussoni, who inspired the beginning and helped the development of this line of thought.

The work reported was supported by the Consiglio Nazionale delle Ricerche (Progetti Finalizzati) and the Italian government.

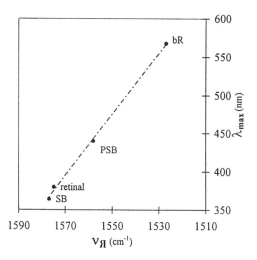

Fig. 28.36 Electronic absorption maxima versus frequencies of the Я mode (see text) for retinal analogs and bacteriorhodopsin.

REFERENCES

1. R. N. Jones and C. Sandorfy, in *Chemical Applications of Spectroscopy: Techniques of Organic Chemistry*, Vol. 9 (A. Weissenberger ed.), Interscience, New York, 1956, Ch. IV, p. 247.
2. L. Bellamy, *The Infrared Spectra of Complex Molecules*, Wiley, New York, 1958.
3. K. Nakamoto, *Infrared Spectra of Inorganic and Coordination Compounds*, Wiley, New York, 1963.
4. G. Herzberg, *Infrared and Raman Spectra of Polyatomic Molecules*, Van Nostrand, Princeton, NJ, 1959.
5. E. B. Wilson, J. C. Decius, and P. C. Cross, *Molecular Vibrations*, McGraw-Hill, New York, 1955.
6. S. Califano, *Vibrational States*, Wiley, New York, 1976.
7. M. V. Volkenstein, M. A. Eliashevich, and B. Stepanov, *Kolebaniya Molekul II*, Moscow, 1949.
8. R. Zbinden, *Infrared Spectroscopy of High Polymers*, Academic, New York, 1964.
9. G. Zerbi, in *Vibrational Spectra of High Polymers*, Vol. 2 (E. G. Brame, ed.), Appl. Spectrosc. Rev., Marcel Dekker, New York, 1963, p. 193.
10. G. Zerbi, in *Advances in Infrared and Raman Spectroscopy* (R. J. H. Clark and R. E. Hester, eds.), Wiley/Heyden, New York, 1984, p. 301.
11. G. Zerbi, *Adv. Chem. Ser.* 203:487 (1983).
12. A. A. Maradudin, E. W. Montroll, and G. H. Weiss, *Solid State Phys.*, Suppl. 3:1 (1963).
13. L. Piseri and G. Zerbi, *J. Mol. Spectrosc.* 26:254 (1968).
14. P. C. Painter, M. M. Coleman, and J. L. Koenig, *The Theory of Vibrational Spectroscopy and Its Applications to Polymeric Materials*, Wiley, New York, 1982.
15. W. T. King and B. L. Crawford, *J. Mol. Spectrosc.* 5:421 (1960).
16. E. Curtis, Thesis, University of Minnesota, Minneapolis, 1956.
17. G. Zerbi, in *Vibrational Spectroscopy—Modern Trends* (A. J. Barnes and W. J. Orville-Thomas, eds.), Elsevier, Amsterdam, The Netherlands, 1977, p. 379.
18. G. Zerbi, in *Vibrational Intensities in Infrared and Raman Spectroscopy* (W. Person and G. Zerbi, eds.), Elsevier, Amsterdam, The Netherlands, 1984, Ch. 3, p. 23.
19. B. L. Crawford and S. Califano, *Spectrochim. Acta* 16:889 (1960).
20. J. R. Scherer and J. Overend, *Spectrochim. Acta* 17:719 (1961).
21. L. Goodman, A. G. Ozkabak, and K. B. Wiberg, *J. Chem. Phys.* 91:2069 (1989).
22. J. Overend and J. R. Scherer, *Spectrochim. Acta* 16:733 (1960).
23. W. Person and G. Zerbi, *Vibrational Intensities in Infrared and Raman Spectroscopy*, Elsevier, Amsterdam, The Netherlands, 1984.
24. M. Gussoni and G. Zerbi, *Rend. Accad. Nazionale Lincei*, Ser. VIII 40:5, 6 (1966).
25. I. M. Mills, *Chem. Phys. Lett.* 3:267 (1969).
26. S. J. Cyvin, in *Molecular Structures and Vibrations* (S. J. Cyvin, ed.), Elsevier, Amsterdam, The Netherlands, 1972, Ch. 1, p. 23.
27. M. V. Volkenstein, L. A. Gribov, M. A. Eliashevich, and L. I. Stepanov, *Kolebaniya Molekul*, Moscow, 1972.
28. P. Pulay, G. Fogarasi, G. Pongor, J. E. Boggs, and A. Vargha, *J. Am. Chem. Soc.* 105:7037 (1983).
29. R. G. Parr and W. Yang, *Density Functional Theory of Atoms and Molecules*, Oxford Univ. Press, New York, 1989.
30. J. C. Decius, *J. Mol. Spectry.* 57:384 (1975).
31. A. J. Van Straten and W. Smit, *J. Mol. Spectrosc.* 62:297 (1976).
32. M. Gussoni, in *Advances in Infrared and Raman Spectroscopy* (R. J. H. Clark and R. H. Hester, eds.), Heyden, London, 1979, Vol. 6, p. 96.
33. M. Gussoni, C. Castiglioni, and G. Zerbi, *J. Mol. Struct.* 198:475 (1989).
34. M. Gussoni, C. Castiglioni, M. N. Ramos, M. Rui, and G. Zerbi, *J. Mol. Struct.* 224:445 (1990).
35. M. Gussoni, C. Castiglioni, and G. Zerbi, *J. Phys. Chem.* 88:600 (1984).
36. C. Castiglioni, M. Gussoni, and G. Zerbi, *J. Chem. Phys.* 82:3534 (1985).
37. B. Tian, G. Zerbi, R. Schenk, and K. Müllen, *J. Chem. Phys.* 95:3191 (1991).
38. C. Castiglioni, G. Zerbi, and M. Gussoni, *Solid State Commun.* 56:863 (1985).
39. V. Hernandez, C. Castiglioni, M. Del Zoppo, and G. Zerbi, *Phys. Rev. B* 50:9815 (1994).
40. C. Rumi, G. Zerbi, K. Müllen, and M. Rehahn, *J. Chem. Phys.*, 106:24 (1997).
41. M. Gussoni, C. Castiglioni, and G. Zerbi, in *Spectroscopy of Advanced Materials* (R. J. H. Clark and R. E. Hester, eds.), Wiley, New York, 1991, p. 251.
42. G. Zerbi, in *Conjugated Polymers* (J. L. Brédas and R. Silbey, eds.), Kluwer, Amsterdam, The Netherlands, 1991, p. 435.
43. C. Castiglioni, M. Gussoni, J. T. Lopez-Navarrete, and G. Zerbi, *Solid State Commun.* 65:625 (1988).
44. T. A. Skotheim (ed.), *Handbook of Conducting Polymers*, Vols. 1 and 2, Marcel Dekker, New York, 1986.
45. J. C. W. Chien, *Polyacetylene Chemistry: Physics and Material Science*, Academic, London, 1984.
46. J. L. Brédas and R. Silbey (eds.), *Conjugated Polymers*, Kluwer, Amsterdam, The Netherlands, 1991.
47. M. Miragoli, Thesis, University of Milan, 1985.
48. G. Zerbi and G. Dellepiane, *Gazz. Chim. Ital.* 117:591 (1987).
49. G. Zerbi, in *New Perspectives on Vibrational Spectroscopy in Material Science* (M. W. McKenzie, ed.), Wiley, New York, 1988, Chap. 6.
50. F. Meyers, S. R. Marder, B. M. Pierce, and J.-L. Brédas, *J. Am. Chem. Soc.* 116:10703 (1994).

51. K. Aoki, Y. Kakudate, M. Yoshida, S. Usuba, K. Tanaka, and S. Fujimara, *Synth. Met.* 28:D91 (1989).
52. A. Brillante, M. Hanfland, K. Syasson, and J. Hocker, *Physica 139*:533 (1986).
53. M. Gussoni, C. Castiglioni, M. Miragoli, G. Lugli, and G. Zerbi, *Spectrochim. Acta 41A*:37 (1985).
54. C. Castiglioni, M. Gussoni, M. Miragoli, and G. Zerbi, *Mol. Cryst. Liq. Cryst. 117*:295 (1985).
55. E. Cernia and L. D'Ilario, *J. Polym. Sci. 21*:2168 (1983).
56. H. Kalhert, O. Leitner, and G. Leising, *Synth. Met. 17*:467 (1987).
57. G. Perego, G. Lugli, and U. Pedretti, *Mol. Cryst. Liq. Cryst. 117*:59 (1985).
58. T. C. Clarke, M. T. Krounbi, V. Y. Lee, and G. B. Street, *J. Chem. Soc., Chem. Commun. 1981*:384.
59. G. Zannoni and G. Zerbi, *Chem. Phys. Lett. 87*:50 (1982).
60. R. Rabaioli, Thesis in Physics, University of Milan, 1990.
61. A. Kiehl, Ph.D. Thesis, Max-Planck-Institut für Polymerforschung, Mainz, 1993.
62. A. Kiehl, A. Eberhardt, M. Adam, V. Enkelmann, and K. Müllen, *Angew. Chem. Int. Ed. (Engl.) 31*: 1588 (1992).
63. S. Hirata, H. Yoshida, H. Torii, and M. Tasumi, *J. Chem. Phys. 103*:8955 (1995).
64. S. Hirata, H. Torii, and M. Tasumi, *J. Chem. Phys. 103*:8964 (1995).
65. M. Veronelli, G. Zerbi, and R. Stradi, *J. Raman Spectry. 26*:683 (1995).
66. H. H. Fox, M. O. Wolf, R. O'Dell, B. L. Lin, R. R. Schrock, and M. S. Wrighton, *J. Am. Chem. Soc. 116*:2827 (1994).
67. R. G. Snyder and J. H. Schachtschneider, *Spectrochim. Acta 19*:85, 117 (1963).
68. A. Machonnachie, A. J. Dianoux, H. Shirakawa, and M. Tasumi, *Synth. Met. 14*:323 (1986).
69. H. Takeuchi, T. Arakawa, Y. Furukawa, I. Harada, and H. Shirakawa, *J. Mol. Struct. 158*:179 (1987).
70. M. Rumi and G. Zerbi, *Chem. Phys. Lett. 242*:639 (1995).
71. D. H. Whiffen, *Spectrochim. Acta 7*:253 (1955).
72. B. E. Kohler, in *Electronic Properties of Polymers and Related Compounds* (H. Kuzmany, M. Mehring, and S. Roth, eds.), Solid State Sci. 63, Springer, Heidelberg, 1985, p. 100.
73. M. F. Granville, G. R. Holton, and B. E. Kohler, *J. Chem. Phys. 72*:4671 (1980).
74. M. F. Granville, B. E. Kohler, and J. B. Snow, *J. Chem. Phys. 75*:3765 (1981).
75. T. Kobajashi (ed.), *Relaxation in Polymers*, World Scientific, Singapore, 1993.
76. T. Kobajashi, M. Yoshizawa, U. Stamm, M. Taiji, and M. Hasegawa, *J. Opt. Soc. Am. B 7*:1558 (1990).
77. M. Gussoni, C. Castiglioni, M. Del Zoppo, and G. Zerbi, in *Organic Materials for Photonics, Science and Technology*, (G. Zerbi, ed.) North Holland, Amsterdam, The Netherlands, 1993, p. 27.
78. K. Schulten and M. Karplus, *Chem. Phys. Lett. 14*: 305 (1972).
79. F. Zerbetto, M. Z. Zgierski, F. Negri, and G. Orlandi, *J. Chem. Phys. 89*:3681 (1988).
80. F. Negri, G. Orlandi, F. Zerbetto, and M. Z. Zgierski, *J. Chem. Phys. 91*:6215 (1989).
81. I. Harada, Y. Furukawa, M. Tasumi, H. Shirakawa, and S. Ikeda, *J. Chem. Phys. 73*:4746 (1980).
82. H. Kuzmany, *J. Phys. (Paris) 44*:C3-255 (1983).
83. G. P. Brivio and E. Mulazzi, *Phys. Rev. B B30*:876 (1984).
84. F. Inagaki, M. Tasumi, and T. Miyazawa, *J. Raman Spectrosc. 3*:335 (1975).
85. H. Takeuchi, Y. Furukawa, J. Harada, and H. Shirakawa, *J. Chem. Phys. 84*:2882 (1986).
86. M. Kafranek, H. Lishka, and A. Karpfen, *J. Chem. Phys. 96*:982 (1992).
87. F. B. Schugerl and H. Kuzmany, *J. Chem. Phys. 74*: 953 (1981).
88. R. M. Gavin and S. A. Rice, *J. Chem. Phys. 55*:2675, (1971).
89. G. Zannoni and G. Zerbi, *J. Mol. Struct. 100*:485, 505 (1983).
90. H. O. Villar, M. Dupuis, J. D. Watts, G. J. B. Hurst, and E. Clementi, *J. Chem. Phys. 88*:1003 (1987).
91. M. Del Zoppo, C. Castiglioni, M. Rui, M. Gussoni, and G. Zerbi, *Synth. Met. 51*:135 (1992).
92. M. Del Zoppo, C. Castiglioni, and G. Zerbi, *Synth. Met. 68*:295 (1995).
93. J. Tang and A. C. Albrecht, in *Raman Spectroscopy*, Vol. 2 (H. Szymanski, ed.), Plenum, New York, 1970, p. 33.
94. R. J. H. Clark and T. J. Dines, *Angew. Chem. Int. Ed. (Engl.) 25*:131 (1986).
95. A. Warshel and P. Dauber, *J. Chem. Phys. 66*:5477 (1977).
96. W. L. Peticolas and C. Blazej, *Chem. Phys. Lett. 63*:604 (1979).
97. W. L. Peticolas, *Ber. Bunsengesl. Phys. Chem. 85*: 481 (1981).
98. F. Inagaki, M. Tasumi, and T. Miyazawa, *J. Mol. Spectrosc. 50*:286 (1974).
99. C. Castiglioni, M. Del Zoppo, and G. Zerbi, *J. Raman Spectrosc. 24*:485 (1993).
100. C. H. Ting, *Spectrochim. Acta 24A*:1177 (1968).
101. A. C. Lasaga, A. J. Aerni, and M. Karplus, *J. Phys. Chem. 87*:925 (1983).
102. T. Kakitani, *Progr. Theor. Phys. 50*:17 (1973).
103. W. Kauzmann, *Quantum Chemistry*, Academic, New York, 1957.
104. E. Ehrenfreund, Z. Vardeny, O. Brafman, and B. Horowitz, *Phys. Rev. B 36*:1535 (1987).
105. A. Girlando, A. Painelli, and Z. G. Soos, *J. Chem. Phys. 98*:7459 (1993).
106. S. Roth and H. Bleier, *Adv. Phys. 36*:385 (1987).
107. H. Naarman, *Synth. Met. 17*:223 (1987).
108. H. Naarman, in *Science and Applications of Conducting Polymers* (W. R. Salaneck, D. T. Clark, and

E. J. Samuelsen, eds.), Adam Hilger, Bristol, UK, 1991, Ch. 4, p. 81.
109. P. Piaggio, G. Dellepiane, L. Piseri, R. Tubino, and C. Taliani, *Solid State Commun.* 50:947 (1984).
110. D. B. Tanner, G. L. Doll, A. M. Rao, P. G. Eklund, G. A. Arbuckle, and A. G. McDiarmid, *Synth. Met.* 28:D141 (1989).
111. M. Rumi, A. Kiehl, and G. Zerbi, *Chem. Phys. Lett.* 231:70 (1994).
112. T. C. Chung, F. Moraes, J. D. Flood, and A. J. Heeger, *Phys. Rev. B* 29:2341 (1984).
113. J. Chen, T. C. Chung, F. Moraes, and A. J. Heeger, *Solid State Commun.* 53:757 (1985).
114. R. Chance, D. S. Bondreaux, J. L. Brédas, and R. Silbey, in *Handbook of Conducting Polymers*, Vol. 2 (T. A. Skotheim, ed.), Marcel Dekker, New York, 1986, p. 825.
115. A. Sakamoto, Y. Furukawa, and M. Tasumi, *J. Phys. Chem.* 98:4365 (1994).
116. A. Sakamoto, Y. Furukawa, and M. Tasumi, *Synth. Met.* 55–57:593 (1993).
117. A. Sakamoto, Y. Furukawa, and M. Tasumi, *J. Phys. Chem.* 96:3870 (1992).
118. H. Eckhardt, L. W. Schaklette, J. S. Szobota, and R. H. Baughman, *Mol. Cryst. Liq. Cryst.* 117:401 (1985).
119. K. Soga, S. Kawasaki, H. Shirakawa, and S. Ikeda, *Makrom. Chem. Rapid Commun.* 1:643 (1980).
120. Y. Furukawa, T. Arakawa, H. Takeuchi, I. Harada, and H. Shirakawa, *J. Chem. Phys.* 81:2907 (1984).
121. D. Schaefer-Siebert, C. Boudrowski, H. Kuzmany, and S. Roth, *Synth. Met.* 21:285 (1987).
122. H. Kuzmany and J. Kurti, *Synth. Met.* 21:95 (1987).
123. F. Zuo, A. J. Epstein, X. Q. Yang, D. B. T. Tanner, G. Arbuckle, and A. G. McDiarmid, *Synth. Met.* 17:433 (1987).
124. X. Q. Yang, D. B. T. Tanner, G. Arbuckle, A. G. McDiarmid, and A. J. Epstein, *Synth. Met.* 17:277 (1987).
125. K. Tanaka, S. Yamanaka, S. Nishio, and T. Yamabe, *Synth. Met.* 28:D267 (1989).
126. J. T. Lopez Navarrete and G. Zerbi, *Solid State Commun.* 69:289 (1989).
127. J. L. Brédas, J. M. Toussaint, G. Hennico, J. Delhalle, J. M. André, A. J. Epstein, and A. G. McDiarmid, in *Electronic Properties of Conjugated Polymers* (H. Kuzmany, M. Mehring, and S. Roth, eds.), Springer, Berlin, 1987, p. 48.
128. H. Kuzmany, M. Mehring, and S. Roth (eds.), *Electronic Properties of Conjugated Polymers*, Springer Ser. Solid State Sci., Springer, Berlin, 1987.
129. J. T. Lopez Navarrete and G. Zerbi, *Solid State Commun.* 64:1183 (1987).
130. H. Bleier, W. Göhring, and S. Roth, in *Electronic Properties of Polymers and Related Compounds* (H. Kuzmany, M. Mehring, and S. Roth, eds.), Solid State Sci. 63, Springer, Heidelberg, 1985, p. 96.
131. Z. Vardeny, E. Ehrenfreund, O. Brafman, B. Horovitz, H. Fujimoto, J. Tanaka, and M. Tareka, *Phys. Rev. Lett.* 57:2995 (1986).
132. H. Schaffer, R. H. Friend, and A. J. Heeger, *Phys. Rev. Lett. B*36:7537 (1983).
133. Z. Vardeny, H. T. Grahn, L. Cheng, and G. Leising, *Synth. Met.* 28:D167 (1989).
134. J. Orenstein and G. Baker, *Phys. Rev. Lett.* 49:1043 (1980).
135. J. Orenstein and M. Mastner, *Phys. Rev. Lett.* 46:1421 (1981).
136. G. B. Blanchet, C. R. Fincher, T. C. Chung, and A. J. Heeger, *Phys. Rev. Lett.* 50:1958 (1983).
137. R. Tubino, R. Dorsinville, A. Walser, A. Seas, and R. R. Alfano, *Synt. Met.* 28:D175 (1989).
138. D. S. Boudreaux, R. R. Chance, J. L. Brédas, and R. Silbey, *Phys. Rev.* 28:6927 (1983).
139. Y. Mori and S. Kurihara, *Solid State Commun.* 60:201 (1987).
140. Y. Mori and S. Kurihara, *Synth. Met.* 24:357 (1988).
141. Y. Mori, H. Tabei, and F. Ebisawa, *Synth. Met.* 17:447 (1987).
142. S. Lefrant, E. Falques, and A. Chentli, in *Electronic Properties of Conjugated Polymers* (H. Kuzmany, M. Mehring, and S. Roth, eds.), Springer Verlag, Berlin, 1987, p. 122.
143. J. Y. Kim, S. Ando, A. Sokamoto, Y. Furukewa, M. Tesumi, *Synth. Metals*, in press.
144. L. T. Cheng, W. Tam, S. R. Marder, A. E. Stiegman, G. Rikken, and C. W. Spangler, *J. Phys. Chem.* 95:10643 (1991).
145. M. Blanchard-Desce, J.-M. Lehn, M. Barzoukas, I. Ledoux, and J. Zyss, *Chem. Phys.* 181:281 (1994).
146. S. R. Marder, L.-T. Cheng, B. G. Tiemann, A. C. Friedli, M. Blanchard-Desce, J. W. Perry, and J. Skindhoj, *Science* 263:511 (1994).
147. M. Blanchard-Desce, V. Bloy, J.-M. Lehn, C. Runser, M. Barzoukas, A. Fort, and J. Zyss, *SPIE Proc.* 2143:20 (1994).
148. M. Blanchard-Desce, C. Runser, A. Fort, M. Barzoukas, J.-M. Lehn, V. Bloy, and V. Alain, *Chem. Phys.* 199:253 (1995).
149. M. Gussoni, M. N. Ramos, C. Castiglioni, and G. Zerbi, *Chem. Phys. Lett.* 142:515 (1987).
150. S. R. Marder, C. B. Gorman, F. Meyers, J. W. Perry, G. Bourhill, J.-L. Brédas, and B. M. Pierce, *Science* 265:632 (1994).
151. D. Lu, G. Chen, J. W. Perry, and W. A. Goddard III, *J. Am. Chem. Soc.* 116:10679 (1994).
152. G. Chen, D. Lu, and W. A. Goddard III, *J. Chem. Phys.* 101:5860 (1994).
153. S. R. Marder, J. W. Perry, G. Bourhill, C. B. Gorman, B. G. Tiemann, and K. Mansour, *Science* 261:186 (1993).
154. C. Landuzzi, Thesis in Physics, University of Milan, 1991.
155. J. T. Lopez Navarrete, B. Tian, and G. Zerbi, *Synth. Met.* 38:299 (1990).
156. G. J. Visser, G. J. Heeres, J. Wolters, and A. Vos, *Acta Cryst.* B24:467 (1968).

157. F. Van Bolhuis, H. Wynberg, E. E. Havinga, E. W. Meijer, and E. G. J. Stirling, *Synth. Met.* 30:381 (1989).
158. M. C. Rumi and G. Zerbi, *J. Chem. Phys.*, submitted.
159. L. Cuff, C. X. Cui, and M. Kertesz, *J. Am. Chem. Soc. 116*:9269 (1994).
160. D. W. Scott, *J. Mol. Spectrosc. 31*:451 (1969); *37*:77 (1971).
161. C. A. Coulson, *Valence*, 2nd ed., Oxford Univ. Press, London, 1961.
162. A. Streitwieser, *Molecular Orbital Theory for Organic Chemists*, Wiley, New York, 1961.
163. D. C. McKean, *Chem. Soc. Rev. 7*:399 (1978).
164. D. C. McKean, *J. Mol. Struct. 113*:251 (1984).
165. B. Bak, D. Christensen, L. Hansen-Nygaard, and J. Rastrupp-Andersen, *J. Mol. Spectrosc. 7*:58 (1961).
166. B. Bak, D. Christensen, L. Hansen-Nygaard and J. Rastrupp-Andersen, *J. Chem. Phys.*, 24:720 (1956).
167. B. Bak, D. Christensen, W. B. Dixon, L. Hansen-Nygaard, J. Rastrupp-Andersen and M. Schottländer, *J. Mol. Spectr.*, 9:124 (1962).
168. G. B. Street in *Handbook of Conducting Polymers* (T. J. Skotheim, ed.), Marcel Dekker, New York, 1986, vol. 1, p. 256.
169. B. Tian and G. Zerbi, *J. Chem. Phys.*, 92:3886 (1990); *ibid.* 3982.
170. S. Martina, V. Enkelman, A. D. Schlüter, G. Wegner and G. Zerbi, *Synth. Met.*, 55–57:1096 (1993).
171. S. Martina, V. Enkelmann, G. Wegner and A. D. Schlüter, *Synt. Met.*, 51:299 (1992).
172. M. Del Zoppo, C. Castiglioni, and G. Zerbi, manuscript in preparation.
173. G. Zerbi, M. Veronelli, S. Martina, A. D. Schlüter, and G. Wegner, *Adv. Mater. 6*:385 (1994).
174. G. Zerbi, M. Veronelli, S. Martina, A. D. Schlüter, and G. Wegner, *J. Chem. Phys. 100*:987 (1994).
175. G. Tourillon, in *Handbook of Conducting Polymers* (T. J. Skotheim, ed.), Marcel Dekker, New York, 1986, Vol. 2, p. 293.
176. M. C. Gallazzi and G. Zerbi, in *The Polymeric Materials Encyclopedia—Synthesis, Properties and Applications* (J. C. Solamone, ed), CRC, Boca Raton, FL, 1996.
177. A. O. Patil, A. J. Heeger, and F. Wudl, *Chem. Rev. 88*:183 (1988).
178. K. Kaneto, S. Hayashi, S. Ura, and K. Yoshino, *J. Phys. Soc. Jpn. 54*:1146 (1985).
179. K. Kaneto and K. Yoshino, *Synth. Met. 18*:133 (1987).
180. G. Harbeke, D. Baeryswyl, H. Kiess, and K. Kobel, *Phys. Scripta T1*:302 (1986).
181. W. Hayes, F. J. Pratt, K. S. Wong, K. Kaneto, and K. Yoshino, *J. Phys. C18*:L555 (1985).
182. H. E. Shaffer and A. J. Heeger, *Solid State Commun. 59*:415 (1986).
183. Y. Cao and Q. Renyuan, *Solid State Commun. 54*:211 (1986).
184. S. Straftstron and J. L. Brédas, *Phys. Rev. B 38*:4180 (1988).
185. Z. Vardeny, E. Ehrenfreund, O. Brafman, A. J. Heeger, and F. Wudl, *Synth. Met. 18*:183 (1987).
186. G. J. Visser, G. J. Heeres, J. Wolters, and A. Vos, *Acta Cryst. B24*:467 (1968).
187. S. Bruckner and W. Porzio, *Makromol. Chem. 189*:961 (1988).
188. F. Garnier, G. Tourillon, J. Y. Barraud, and H. Dexpert, *J. Mater. Sci. 20*:2687 (1985).
189. D. Fichou, F. Garnier, F. Charra, F. Kajzar, and J. Messier, in *Organic Materials for Nonlinear Optics* (R. A. Hann and D. Bloor, eds.) Royal Society of Chemistry, London, 1988, p. 176.
190. E. E. Havinga, I. Rotte, E. W. Meijer, W. ten Hoeve, and H. Wynberg, *Synth. Met.*, in press.
191. B. Hudson and B. E. Kohler, *Synth. Met. 9*:241 (1984).
192. M. Sundberg, O. Inganäs, S. Stafström, G. Gustafsson, and B. Sjögren, *Solid State Commun. 71*:435 (1989).
193. D. Birnbaum and B. E. Kohler, *J. Chem. Phys. 90*:3506 (1989).
194. G. Zerbi and S. Sandroni, *Spectrochim. Acta 24A*:483, 511 (1968).
195. M. Akimoto, Y. Furukawa, H. Takeuchi, and I. Harada, *Synth. Met. 15*:353 (1986).
196. J. T. Lopez Navarrete and G. Zerbi, *J. Chem. Phys. 94*:957 (1991).
197. E. Agosti, M. L. Rivola, V. Hernandez, and G. Zerbi, to be published.
198. F. Negri and M. Z. Zgierski, *J. Chem. Phys. 100*:2571 (1994).
199. E. Villa, E. Agosti, C. Castiglioni, M. C. Gallazzi, and G. Zerbi, *J. Chem. Phys.*, 105:9461 (1996).
200. F. Wudl, M. Kobayashi, and A. J. Heeger, *J. Org. Chem. 49*:3382 (1984).
201. R. Lazzaroni, J. Riga, J. Verbist, J. L. Brédas, and F. Wudl, *J. Chem. Phys. 82*:5717 (1985).
202. J. L. Brédas, *J. Chem. Phys. 82*:3808 (1985).
203. Y. S. Lee and M. J. Kertesz, *J. Chem. Phys. 88*:2609 (1988).
204. I. Hoogmartens, P. Adriaensens, D. Vanderzande, J. Gelan, C. Quattrocchi, R. Lazzaroni, and J. L. Brédas, *Macromolecules 25*:7347 (1992).
205. E. Faulques, W. Wallnöfer, and H. Kuzmany, *J. Chem. Phys. 90*:7585 (1989).
206. I. Hoogmartens, P. Adriaenses, R. Carleer, D. Vanderzande, H. Martens, and J. Gelan, *Synth. Met. 51*:219 (1992).
207. R. Kiebooms, I. Hoogmartens, P. Adriaensens, D. Vanderzande, and J. Gelan, *Synth. Met. 52*:395 (1992).
208. I. Hoogmartens, P. Adriaenses, D. Vanderzande, and J. Gelan, *Anal. Chim. Acta 41A*:1025 (1993).
209. G. Zerbi, M. C. Magnoni, I. Hoogmartens, R. Kiebooms, R. Carleer, D. Vanderzande, and J. Gelan, *Adv. Mater. 7*:1027 (1995).
210. H. Neugebauer, A. Neckel, and N. Brinda-Konopik,

in *Electronic Properties of Polymers and Related Compounds* (H. Kuzmany, M. Mehring, and S. Roth, eds.), Solid State Sci. 63, Springer, Heidelberg, 1985, p. 227.
211. Y. Cao and Q. Renyuan, *Solid State Commun.* 54: 211 (1986).
212. E. Agosti and G. Zerbi, *Synth. Met.* 79:107 (1996).
213. M. Veronelli, M. C. Gallazzi, and G. Zerbi, *Synth. Met.* 55:545 (1993).
214. M. Veronelli, M. C. Gallazzi, and G. Zerbi, *Acta Polym.* 45:127 (1994).
215. J. T. Lopez Navarrete and G. Zerbi, *Chem. Phys. Lett.* 175:125 (1990).
216. Q. Pei, O. Inganäs, J. E. Österholm, and J. Laakso, *Polymer* 34:247 (1992).
217. Q. Pei and O. Inganäs, *Synth. Met.* 46:353 (1992).
218. M. C. Magnoni, M. C. Gallazzi, and G. Zerbi, *Acta Polym.* 47:228 (1996).
219. A. Bohnen, K.-H. Koch, W. Lüttke, and K. Müllen, *Angew. Chem. Int. Ed. Engl.* 29:525 (1990).
220. K.-H. Koch and K. Müllen, *Chem. Ber.* 124:2091 (1991).
221. M. Rehahn, A.-D. Schlüter, and G. Wegner, *Makromol. Chem.* 191:1991 (1990).
222. M. Rehahn and P. Galda, *Synthesis*, in press.
223. O. Bastiansen, *Acta Chem. Scand.* 3:408 (1949).
224. H. Suzuki, *Bull. Chem. Soc. Jpn.* 32:1340 (1959).
225. G. Robertson, *Nature* 191:593 (1961).
226. A. Hargreaves and S. H. Rizvi, *Acta Cryst.* 15:365 (1962).
227. J. Trotter, *Acta Cryst.* 14:1135 (1961).
228. Y. Delugeard, J. Desuche, and J. L. Baudour, *Acta Cryst. Sect. B* 32:172 (1976).
229. J. L. Baudour, Y. Delugeard, and H. Cailleau, *Acta Cryst. Sect. B* 32:150 (1976).
230. J. L. Baudour, H. Cailleau, and W. B. Yelon, *Acta Cryst. Sect. B* 33:1773 (1976).
231. S. Sala, Thesis in Physics, University of Milano, 1985.
232. R. Friend, in *Conjugated Polymers and Related Materials* (W. R. Salaneck, I. Lundström, and B. Rånbi, eds.), Oxford Univ. Press, London, 1993, p. 285.
233. R. Schenk, H. Gregorius, K. Meerholz, J. Heinze, and K. Müllen, *J. Am. Chem. Soc.* 113:2643 (1991).
234. D. B. Tian, G. Zerbi, and K. Müllen, *J. Chem. Phys.* 95:3198 (1991).
235. G. Zerbi, E. Galbiati, M. C. Gallazzi, C. Castiglioni, M. Del Zoppo, and K. Müllen, *J. Chem. Phys.* 105:2509 (1996).
236. M. Traetteberg, E. B. Fransten, F. C. Mijlhoff, and A. Hoekstra, *J. Mol. Struct.* 26:57 (1975).
237. A. Hoekstra, P. Meertens, and A. Vos, *Acta Cryst.* B31:2813 (1975).
238. M. Baumgarten and K. Müllen, *Topics in Current Chemistry*, Vol. 169, Springer-Verlag, Berlin, 1994.
239. D. S. Chemla and J. Zyss (eds.), *Non-linear Optical Properties of Organic Molecules and Crystals*, Academic, London, 1987.
240. R. A. Hann and D. Bloor (eds.), *Organic Materials for Nonlinear Optics*, Royal Soc. Chem. London, 1989, Vol. 69.
241. S. R. Marder, J. E. Sohn, and G. D. Stucky (eds.) *Materials for Nonlinear Optics*, ACS Symp. Ser. 455, Am. Chem. Soc. Washington, DC, 1991.
242. J. Zyss (ed.), *Molecular Nonlinear Optics—Material Physics and Devices*, Academic, San Diego, 1994.
243. C. Castiglioni, M. Del Zoppo, and G. Zerbi, *Phys. Rev. B*, 53:13319 (1996).
244. C. Castiglioni, M. Del Zoppo, P. Zuliani, and G. Zerbi, *Synth Met.* 74:171 (1995).
245. P. Zuliani, M. Del Zoppo, C. Castiglioni, G. Zerbi, S. R. Marder, and J. W. Perry, *J. Chem. Phys.* 103: (1995).
246. M. Del Zoppo, C. Castiglioni, P. Zuliani, A. Razelli, G. Zerbi, and M. Blanchard-Desce, *J. Appl. Polym. Sc.*, in press.
247. C. Castiglioni, M. Gussoni, M. Del Zoppo, and G. Zerbi, *Solid State Commun.* 82:13 (1992).
248. D. M. Bishop, *Rev. Mod. Phys.* 62:343 (1990).
249. C. Flytzanis, *Phys. Rev.* B6:1264 (1972).
250. M. Del Zoppo, C. Castiglioni, M. Veronelli, and G. Zerbi, *Synth. Met.* 57:3919 (1993).
251. L.-T. Cheng, W. Tam, S. H. Stevenson, G. R. Meredith, G. Rikken, and S. R. Marder, *J. Phys. Chem.* 95:10643 (1991).
252. A. E. Stiegman, E. Graham, K. J. Perry, L. R. Kundkar, L.-T. Cheng, and J. W. Perry, *J. Am. Chem. Soc.* 113:7658 (1991).
253. J. L. Oudar, *J. Chem. Phys.* 67:446 (1977).
254. J. Zyss, *J. Chem. Phys.* 98:6583 (1993).
255. F. Meyers, S. R. Marder, J. W. Perry, G. Bourhill, S. Gilmour, L.-T. Cheng, B. M. Pierce, and J. L. Brédas, *Nonlin. Opt.* 9:59 (1995).
256. G. Bourhill, J.-L. Brédas, L.-T. Cheng, S. R. Marder, F. Meyers, J. W. Perry, and B. G. Tiemann, *J. Am. Chem. Soc.* 116:2619 (1994).
257. C. B. Gorman and S. R. Marder, *Proc. Natl. Acad. Sci. USA* 90:11297 (1993).

29
Electroluminescence in Conjugated Polymers

Richard H. Friend and Neil C. Greenham
University of Cambridge, Cambridge, England

I. INTRODUCTION

The initial interest in conjugated polymers dates from the report of metallic properties in chemically doped polyacetylene [1]. Polyacetylene and many of the other polymers that were studied at an early stage, such as polypyrrole [2] and polyaniline [3], have proved to be particularly suitable for these properties but for a number of reasons were less promising as semiconductors. The development of conjugated polymers with useful semiconducting properties took almost another decade; although there were early reports of diodes made with polyacetylene [4-6], the first reports of devices with potentially useful properties date from the mid-1980s, with field effect transistors (FETs) reported by Koezuka et al. [7] based on electrochemically deposited poly(3-methylthiophene). With the availability of solution-processed conjugated polymers, the scope for investigation of conjugated polymers with respectable semiconducting properties advanced rapidly; for example, FETs with improved performance were reported by several groups [7-9].

Though the metallic properties of the doped conjugated polymers are very clearly distinguishable from the properties of other organic materials, as semiconductors they must be compared with molecular semiconductors such as anthracene [10]. Molecular semiconductors have been extensively studied from the 1960s and have found very important applications as the active components for charge photogeneration and transport in many of the xerographic copiers and laser printers now made [11]. We comment on the parallels between the conjugated polymers and molecular semiconductors in the course of this chapter.

Electroluminescence—the generation of light by electrical excitation (other than blackbody radiation)—is a phenomenon that has been seen in a wide range of semiconductors and was first reported for an organic semiconductor by Pope et al. in 1963 [12]. They observed emission from single crystals of anthracene (see Fig. 29.1), a few tens of micrometers in thickness, using silver paste electrodes and required large voltages to get emission, typically 400 V. Similar studies were made by Helfrich and Schneider [13] in 1965 using liquid electrodes. The large voltages were necessary to obtain sufficient electric fields at the electrode/semiconductor interfaces, and in fact the fields necessary in many of the recent electroluminescent devices made with the polymers are not so different, even though the operating voltages are much lower because the film thicknesses are much smaller. Note that the studies on these devices had established that the process responsible for electroluminescence requires the injection of electrons from one electrode and holes from the other, the capture of one by the other (so-called recombination), and the radiative decay of the excited state (exciton) produced by this recombination process.

Development of organic thin-film electroluminescence advanced in the 1970s with the study of thin-film devices. Vincett et al. [14] made devices using films of anthracene sublimed onto oxidized aluminum electrodes, with thermally evaporated semitransparent top gold electrodes, and were able to reduce the drive voltage very considerably—down to, for example, 12 V. These devices are reported to suffer from relatively poor lifetimes and relatively poor efficiencies. Interest in molecular organic materials was revived by Tang and Van-Slyke [15], who demonstrated efficient electroluminescence in two-layer sublimed molecular film devices comprising a hole-transporting layer of an aromatic diamine and an emissive layer of 8-hydroxyquinoline aluminum (Alq_3). Indium-tin oxide (ITO) is used as hole-injecting electrode and a magnesium-silver alloy as electron-injecting electrode. Since the work of Tang and

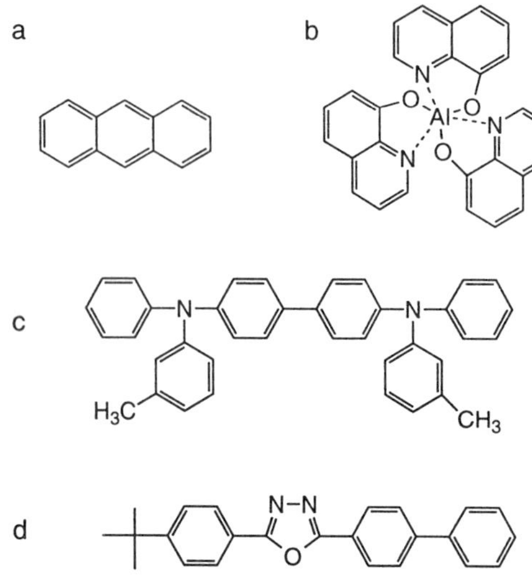

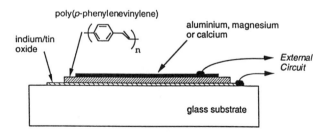

Fig. 29.1 Structures of some molecular semiconductors that have been used in thin-film electroluminescent devices. Anthracene (a) was used in early EL diodes; Alq$_3$ (b) is used as an electron transport and emissive layer, TPD (c) as a hole-transport layer, and PBD (d) as an electron-transport layer.

energy gap between π and π^* states of about 2.5 eV and produces luminescence in a band below this energy, as shown in Fig. 29.3. PPV is an intractable material with a rigid-rod microcrystalline structure, so it is infusible and insoluble in common solvents. These are excellent properties for a film of polymer once formed, but processing cannot be carried out directly with this material. PPV can, however, be obtained conveniently by in situ chemical conversion of films of a suitable precursor polymer that is itself processed from solution by spin coating. Of these, the route most commonly used is the sulfonium precursor [24–28], which is conveniently processed from solution in methanol and is converted to PPV by thermal treatment at temperatures of 200–300°C. The alternative strategy for polymer processing is to attach flexible side groups to the polymer chain so that the polymer is directly processable from solution. This is particularly convenient in that a single process step is required, but the trade-off is that the polymer film is softer and less stable thermally. The derivative of PPV that is most frequently used is poly(2-methoxy-5-(2′-ethyl)hexyloxy-p-phenylenevinylene) (MEH-PPV) [29]; its structure is shown in Fig. 29.4.

coworkers, a large number of other molecular materials have been used as the charge-transporting or emissive layer in LEDs. The structures of Alq$_3$ and TPD (a commonly used hole-transporting material) are shown in Fig. 29.1, along with that of 2-(4-biphenylyl)-5-(4-*tert*-butylphenyl)-1,3,4-oxadiazole (PBD), an oxadiazole derivative commonly used as an electron-transporting layer in organic LEDs. There has been a great deal of activity, particularly in Japan, on the development of devices of this general type, and very high levels of performance have been reported, with quantum efficiencies (photons out per charge injected) of several percent [16–22].

Electroluminescence from conjugated polymers was first reported in 1990 [23] using poly(p-phenylenevinylene) (PPV) as the single semiconductor layer between metallic electrodes, as illustrated in Fig. 29.2. PPV has

Operation of an LED is achieved when the diode is biased sufficiently to achieve injection of positive and negative charge carriers from opposite electrodes. Capture of oppositely charged carriers within the region of the polymer layer can then result in the formation of the singlet exciton, which is generated by photoexcitation across the π–π^* gap, and this can then decay radiatively to produce the same emission spectrum as that produced by photoexcitation. The absorption and emission spectra for PPV are shown in Fig. 29.3 Note that the absorption rises rapidly above the onset of the π–π^* threshold and that the emission spectrum appears on the low en-

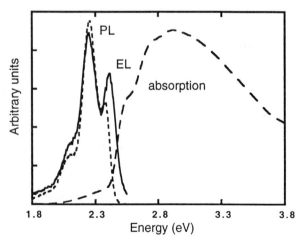

Fig. 29.2 Schematic structure of a single-layer polymer LED device formed with poly(p-phenylenevinylene) (PPV).

Fig. 29.3 Photoluminescence (PL), electroluminescence (EL), and absorption spectra for PPV as measured in a device of the type shown in Fig. 29.2.

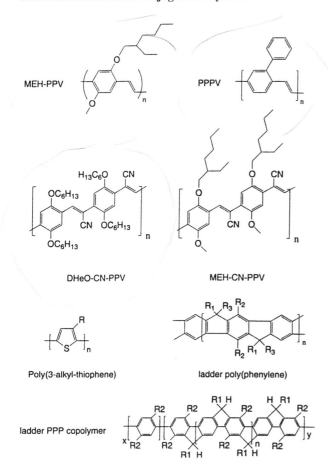

Fig. 29.4 Structures of some conjugated polymers that have been used as emissive layers in EL diodes.

ergy side of the absorption. Both absorption and emission spectra show broadening due to vibronic coupling, as is characteristic for optical transitions in molecular semiconductors where the excited state is a singlet exciton (see Section II). Note that the similarity of the emission spectra produced by photoexcitation and by charge injection establishes that the excited state responsible for light generation in the LED is the same as that produced by photoexcitation.

The levels of efficiency of the first, simple LEDs based on PPV, which were fabricated with aluminum negative electrodes, were relatively low, on the order of 1 photon generated within the device per 10^4 charges injected (an internal quantum efficiency of 0.01%) [23]. External quantum efficiencies are strongly affected by the refractive index of the emissive layer, and the relationship between the two has been discussed by Greenham et al. [30]. These values have risen rapidly over the past few years as improved understanding of the operation of these devices, aided in considerable measure by parallel developments made with sublimed molecular film devices, has allowed considerable optimization of the device characteristics. The use of negative electrodes with lower work functions was shown to improve efficiency to as high as 1% in devices made with ITO/MEH-PPV/Ca [31].

Having briefly introduced electroluminescence in conjugated polymers is this section, in the rest of the chapter we survey some of the aspects of the physics of the device operation that have been established since 1990. We do not attempt to provide a summary of all the contributions made by the many groups that have been active in the field, and we have often selected work that we are most familiar with—that originating from Cambridge.

II. BACKGROUND TO ELECTROLUMINESCENCE IN CONJUGATED POLYMERS

A. Electronic Excitations in Conjugated Polymers

1. Electron–Lattice Coupling

The combination of metallic levels of conductivity in doped polymers and optical and magnetic properties that are very far from those of traditional metals provided the direct evidence that the description of electronic excitations in conjugated polymers is very different from that of three-dimensionally bonded materials. The anisotropic bonding in the polymers allows a local rearrangement of the chain geometry to better accommodate an electronic excitation without requiring a large lattice strain in the directions perpendicular to the chain and allows, in general, formation of "polaronic" excited states, both for charged excitations and for neutral excitations (excitons).

The character of these polaronic states was first established for the case of polyacetylene, for which, in the *trans* isomer, the degeneracy with respect to the sense of bond alternation causes these states to take the form of topological, soliton-like chain excitations, which have associated with them a nonbonding π level that is situated at the middle of the π–π^* semiconductor gap [32–34]. In polymers with a nondegenerate ground state, such as PPV, the two alternative senses of bond alternation do not have equivalent energies; the charged excitations of a nondegenerate ground-state polymer are termed polarons or bipolarons and represent localized charges on the polymer chain with an accompanying local rearrangement of bond alternation. The neutral excited state shows similar structural reorganization, as shown in Fig. 29.5. These states may be considered equivalent to a confined soliton pair, and in this model the two nonbonding midgap "soliton" states form bonding and antibonding combinations, thus producing two gap states symmetrically displaced about the midgap as shown in Fig. 29.6. These levels can be occupied by 0, 1, 2, 3, or 4 electrons, giving a positive bipolaron (bp^{2+}),

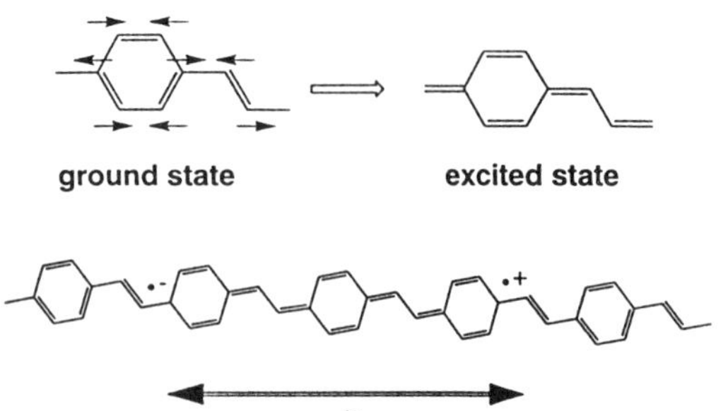

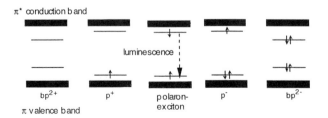

Fig. 29.5 Illustration of the geometrical relaxation of the PPV chain in response to electronic excitation, showing the movement from an aromatic structure toward a quinoidal structure. Also illustrated is a schematic representation of an intrachain exciton, showing the region of relaxed chain geometry.

positive polaron (p^+), polaron exciton, negative polaron (p^-), or negative bipolaron (bp^{2-}), respectively.

In the continuum models of Brazovskii and Kirova [35] and Fesser et al. [36] (the FBC model), the preferred sense of bond alternation is introduced through an extrinsic contribution to the gap parameter Δ, as $\Delta = \Delta_0 + \Delta_e$, where Δ_0 is the contribution to the gap due to the Peierls mechanism and Δ_e is the extrinsic contribution. It is useful to define a confinement parameter γ, as $\gamma = \Delta_e/2\lambda\Delta$, where λ is the effective electron–phonon coupling constant such that $\Delta = \Delta_0 \exp[\gamma]$.

The extent of the polaron along a polymer chain and the position in the gap of the localized energy levels are determined in these one-electron models by the strength of γ. The case of near degeneracy allows the polarons to be relatively extended with electronic levels near the center of the gap, while strong breaking of the degeneracy keeps them much more compact with levels near the band edges. Hence, the degree of nondegeneracy

Fig. 29.6 Polaron, bipolaron, and singlet exciton energy levels in a nondegenerate ground-state polymer (energy gap 2Δ). Luminescence from the singlet exciton is also shown.

can be conveniently parameterized by an experimentally accessible ratio ω_0/Δ, where $2\omega_0$ is the separation between the intragap polaron levels and 2Δ is the energy gap. The singly charged polaron is expected to have gap states that are more closely tied to the band edges than is the case for a bipolaron or exciton, with a limiting separation of $\pm\Delta/\sqrt{2}$ about the center of the gap for $\gamma = 0$. The greater degree of relaxation of the bipolaron in comparison to the polaron stabilizes the coalescence of two like-charged polarons to form a bipolaron.

The success of these models lies in their ability to account for the appearance of new optical absorption bands within the semiconductor gap on chemical doping (e.g., at 0.6 eV and 1.6 eV for the case of PPV [37]) and at the same time to account for low paramagnetic response (bipolarons being low spin). The quantum-chemical techniques that have been used to describe the ground-state electronic structure have also been used with considerable success to model the electronic structure and chain geometry of charged excited states [38,39]. In general these results have provided confirmation that the simple parameterizations used in the Hückel methods are useful.

There are, however, several areas where the Hückel models are seen to fail. Many of these problems are due to the neglect of exchange and correlation effects due to electron–electron interactions. In particular, the simple model predicts that the energy of photons emitted by luminescence should be equal to the spacing between the two bipolaron levels. Experimentally, the luminescence is found to occur at significantly higher energy than the bipolaron level spacing as determined from photoinduced absorption [37]. According to the Fesser et al. [36] model, the first excited state of a conjugated polymer is coupled radiatively to the ground state only through a one-photon transition, implying that all the photoexcited states should decay radiatively. Nonradiative decay, which occurs to some extent in all conjugated polymers, requires the breaking of charge conjugation symmetry. This can be induced by impurities or by other defects in the solid state and can also be achieved by the introduction of symmetry-breaking terms such as electron–electron interactions into the Hamiltonian. Some conjugated polymers, including *trans*-polyacetylene, exhibit no luminescence, implying solely nonradiative decay. It has been suggested that this is due to the lowest energy excited state having the wrong parity to couple radiatively to the ground state [40], contrary to the predictions of the Hückel model.

The effects of interchain interactions on the stabilization of polarons in conjugated polymers have been considered in density functional three-dimensional calculations performed on PPV [41]. These calculations indicate that the polaronic effects as modeled in the isolated chain models may be considerably less strong when interchain interactions are taken into account, and the same concern has been raised by Emin [42,43].

2. Excitons

Though the description of the excited states of the conjugated polymer chain has been couched in the framework of a noninteracting electron model with coupling to the lattice, we have already indicated that electron–electron interactions play a very important role in description of the electronic structure, both ground and excited. The effects of electron–electron interactions are to a considerable extent taken into account in the description of the ground state by the empirical parameterization of these Hückel models. However, this breaks down in the description of the excited states, most particularly for the neutral excited states, which are usually excitonic.

Excitonic effects are well known in many solid-state systems, including molecular organic crystals [10]. These effects become particularly important when the exciton binding energy is large, as predicted for PPV in the presence of strong electron–electron interactions [44]. An exciton can be considered a bound electron–hole pair and may be classified as a Frenkel exciton if the electron–hole pair is located on one molecular unit or as a Mott–Wannier exciton if it extends over many molecular units. The intermediate case, where the exciton extends over a few adjacent molecular units, is termed a charge transfer exciton.

The description of the absorption and emission from a conjugated molecule, as illustrated in Fig. 29.6, is inherently that of a Frenkel exciton. The excited state is not usefully described as a bound electron–hole pair. Since the molecular orbitals are necessarily confined to a small region (the size of the molecule), the density of the excited state wave function on the carbon–carbon bonds is high, so that there can be significant rearrangement of the molecular geometry in this excited state. The strong coupling of the electronic and vibrational excitations that is illustrated in Fig. 29.5 therefore follows.

Molecular crystals formed with molecules such as anthracene have high effective masses (since intermolecular contacts are not large) and relatively low relative permittivities. They are therefore not described by Mott–Wannier theory, the excitons being substantially confined to individual molecular units. The case of the conjugated polymers is interesting in that the effective mass for motion along the polymer chain is relatively low. The Mott–Wannier treatment has been applied to PPV, which is treated as an anisotropic semiconductor, with anisotropic effective masses ($\mu_\parallel = 0.0421\ m_e$, $\mu_\perp = 2.66\ m_e$) and anisotropic relative permittivities ($\epsilon_\parallel = 8$, $\epsilon_\perp = 3$), by Gommes da Costa and Conwell [45]. The results of this calculation give an exciton with an anisotropic ellipsoid, of extension about 2 nm along the chain and 0.4 nm transverse to the chain and binding energy of about 0.4 eV.

This value of binding energy and degree of localization give a strongly "molecular" character to the exciton, which is usually assumed to be confined to a single chain. Therefore, electron–lattice interactions are also strong, so that absorption and emission spectra show vibronic coupling. The inclusion of electron–lattice interactions is made using the Fesser–Bishop–Campbell one-electron description of polarons [36], and the neutral excited state (commonly termed the polaron-exciton) is as shown in Fig. 29.6. However, there are serious problems in getting quantitative agreement with this model [37].

A more complete description of the exciton must therefore take into account both the Coulomb and the electron–lattice interactions, and several computations have taken both into account. One consequence of the electron–electron interaction is that singlet and triplet excitons are no longer of the same energy nor of the same size. The expectation that the triplet exciton is considerably more localized than the singlet exciton (as a result of the reversed sign of the exchange interaction) is confirmed in calculations for PPV and its oligomers by Shuai et al. [44] and by Beljonne et al. [46]. Beljonne et al. use a semiempirical intermediate neglect of differential overlap (INDO) Hamiltonian with configuration interaction techniques to show that the triplet exciton in PPV is stabilized by 0.65 eV with respect to the singlet exciton and is localized over not much more than a single polymer repeat unit, whereas the singlet is considerably more extended. They are also able to calculate higher excited states, including a higher lying triplet that is measured experimentally in photoinduced absorption. The arrangement of ground and excited state energies for PPV is shown in Fig. 29.7.

We have limited discussion so far to the description of low-lying excitations. It is now established that the

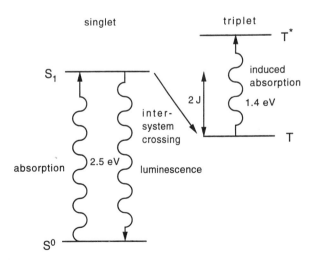

Fig. 29.7 Electronic transitions in PPV, showing both singlet and triplet states. Values for S^0–S^1 and T^1–T^* are as determined experimentally. 2J has been estimated by Beljonne et al. [46] to be 0.65 eV.

modeling of higher energy excitations is realistic only in models that include electron–electron interactions. Thus, the UV absorption spectrum of PPV can be described using configuration interaction methods [47] or Pariser–Parr–Pople methods [48]. Chandross et al. [48] consider that strong Coulomb interactions are necessary to describe these properties and propose a binding energy as high as 0.9 eV for the singlet exciton.

We have indicated that the ability of the exciton to select its extension along the polymer chain allows it to minimize its energy without recourse to extension to neighboring chains. This does seem to be appropriate for those polymers that have smaller $\pi - \pi^*$ energy gaps, but polymers with gaps in the blue or UV have poor intrachain delocalization and can show changes in optical properties as they are brought together in the solid. Interchain interactions are therefore important for these materials and lead to the formation of more extended excited states, which would be described as charge transfer excitons within the framework of molecular semiconductors. If the intermolecular contacts are optimized via a geometrical change following excitation, such excitons are described as excimers (where the exciton extends over identical molecular units) or exciplexes (where the exciton extends over two or more different molecular units).

Excitons are mobile within the solid, and their motion, either coherent or diffusive, plays a very important part in the photophysics of conjugated polymers. The theory of Frenkel excitons in molecular organic crystals is traditionally developed by the tight-binding method [10] using as a basis a set of wave functions that each have the excited state localized on a different molecule. Introduction of nearest-neighbor interactions between molecules produces an exciton band in which the exciton is a delocalized Bloch state of the crystal. If there is significant electron–phonon interaction, the exciton rapidly loses its coherence, and it is more appropriate to think of it as a localized state that moves by hopping between sites. In practice, disorder in site energies plays a very important part in causing this loss of coherence. Hopping between sites can still occur, either by tunneling (requiring direct overlap of initial and final wave functions) or by the resonant process known as Förster transfer. Förster transfer is mediated by a near-field dipole–dipole interaction and does not require the overlap of initial and final wave functions. The original theory by Förster [49] has since been applied and extended by several authors to study exciton diffusion in organic systems [10]. Note that the Förster process operates most effectively when the exciton has a strong dipole matrix element to the ground state and contributes very significantly to the diffusion of singlet excitons in polymers such as PPV. In contrast, this process is ineffective for triplet excitons, which therefore hop by the tunneling mechanism.

B. Materials Synthesis—Processing and Control of Electronic Properties

The improved scope for processing of conjugated polymers from solution is reviewed in several other chapters in this handbook. In the context of electroluminescence, it is clearly desirable to control the energy of the photon emitted in the luminescence process by control of the $\pi-\pi^*$ energy gap, and there are now a wide range of processible polymers that offer useful levels of luminescence efficiency over the whole of the visible spectrum, extending to the near infrared and to the ultraviolet. The properties of some of the polymers studied are summarized in Table 29.1, where we list the wavelength at the peak emission. PPV gives emission in the yellow-green; the emission color can be moved toward the red by the substitution of electron-donating groups such as alkoxy chains at the 2- and 5-positions on the phenyl ring [31]. Substituents can also cause changes in the energy gap through steric, rather than electronic, effects by disrupting the conjugation along the chain. The substitution of bulky cholestanoxy groups, for example, has been used to obtain green emission in soluble polymers [50]. Other polymer systems, not based on PPV, can also be used for EL; the polyalkylthiophenes, for example, conveniently give emission in the red region of the spectrum [51–53].

Polymer LEDs emitting in the blue part of the spectrum are of importance. One strategy to obtain blue-shifted emission is to introduce nonconjugated units into a PPV backbone in order to reduce the average conjugation length. Using a random copolymer of PPV and the methoxy leaving-group precursor to dimethoxy-PPV, Burn et al. [54] obtained blue-green emission. Partly because of the problem of exciton migration, however, the blue shift compared to standard PPV is not large. The introduction of nonconjugated units is also found to give an increase in the PL efficiency [55–57]. This improvement is usually attributed to the suppression of the diffusion of singlet excitons to quenching sites where they can decay nonradiatively. It is relatively easy to find molecular materials, for example oligomers of PPV, that show blue emission. These materials tend to lack the morphological stability of the conjugated polymers and therefore present some problems for device stability. Several groups used dispersions of molecular materials in inert polymer matrices to inhibit recrystallization [58,59]. This work led to the idea of attaching the molecular emitter directly to an inert polymer, either as a pendant side group [60] or incorporated into the polymer main chain [61].

Blue emission can also be produced in different conjugated polymer systems. Blue electroluminescence has been reported in poly(p-phenylene) (PPP) [62], polyalkylfluorene [63], fluorinated polyquinoline [64], and PPP-based ladder copolymers [65]. One problem that is found for the larger gap polymers is that although the

Electroluminescence in Conjugated Polymers

Table 29.1 Properties of Some Electroluminescent Diodes Made with Conjugated Polymers as the Emissive Layer

Polymer	Emission peak (nm)	Negative electrode	Transport layers	Quantum efficiency (%)	Notes	Ref.
PPV	565	Al	—	0.001–0.01		23, 70, 71
		Ca	—	0.1		72
			PBD	1.0		72
MEH-PPV	605	Ca	—	1.0		31
RO-PPV	~580	In or Mg		?	$R = C_5H_{11}$–$C_{14}H_{29}$	73
	590	Ca		0.4	$R_1 = CH_3$; $R_2 = C_{10}H_{21}$	56
	570	Ca		0.3	R = cholestanoxy	50
PPPV	560	Al		0.001		58, 74
	560	Al		0.1	Blended with PS	
	495	Ca		0.16	Blended with PVK	75
Cyano-substituted dihexyloxy-PPV	695	Al	—	0.2		76
			PPV	4		
Cyano-substituted PPV/PTV copolymer	720	Al	PPV	0.2	PL peaks at 840 nm	77
Poly(3-alkylthiophene)	690	Mg/In	—	?	$R = C_{12}H_{25}$–$C_{22}H_{45}$	51
		Ca or In	—	<0.0025	$R = C_8H_{17}$	52
		Ca	—	0.05	$R = C_6H_{13}$	53
				0.2	$R = C_{12}H_{25}$	53
Interrupted PPV copolymer	Blue-green	Ca	—	>0.1	PPV-based	55
	590		—		RO-PPV-based	56
	~530			0.07–0.21	R-PPV-based	57
Partially converted PPV	550	Ca	—	0.75		55, 78
PPP	465	Al	—	0.01		62
		Ca	PBD	1		68
PPP copolymers	420	Ca	PPV	0.5	RO-substituted	69
	400	Ca, In, or Al	—		R-substituted	68, 79
Polyalkylfluorene	470	Mg/In	—	?		63
Fluorinated polyquinoline	450	Ca	—	0.003		64
			PVK + PBD	4		64
Polydiphenylquinoxaline	490	Mg/Ag	various	?		80
Ladder polymers and derivatives	600	Ca or Al	—	?		68, 81
		Ca	PPV	0.6		65
	450	Ca	—	0.9	Copolymer	66
Blue oligomers dispersed in polymer matrix (a selection)	475	Ca	—	0.2		82
	475	Al	—	?		58
	450	Mg:Ag	Alq$_3$	Various		59
Oligomers in main chain	465	Al	—	?		61
Oligomers as side chains	450	Ca	—	0.3		60
Regioregular substituted polythiophenes	460–560	In	—	?		83
Polyphthalidylinearylene	480	Cu or Cr	—	0.01		84
Polyphenylacetylene	600	Ca	—	0.1–0.5		85

luminescence from isolated polymer chains (in solution or in solid solution) may be blue, solid films often show red-shifted emission. This is observed for the PPP-ladder polymers, for which the dominant emission band is in the yellow part of the spectrum [66], and it has been established that this is due to the formation of aggregates in the solid film that support excitons that extend over more than the single chain [67]. The tendency to form aggregates can be controlled by introducing disorder, as in the ladder-PPP copolymers shown in Fig. 29.4, which can produce good blue emission. Soluble copolymers of PPP and alkyl- or alkoxy-substituted PPP have also been used to produce blue electroluminescence [62,68,69]. A major problem in obtaining electroluminescence from materials with a high energy gap is that the barriers to electron injection, hole injection, or both are inevitably larger than in materials with lower energy gaps. This makes the problem of balancing electron and hole injection even more difficult and tends to lead to devices with high driving voltages and low efficiencies.

In addition to band gap control, it has proved to be very important to control the values of the electron affinity and ionization potentials of some of these polymers. The majority of those that had previously been well studied are known to be more easily p-type doped than n-type doped and therefore have relatively low values of ionization potential and electron affinity. This is the case for PPV and its alkoxy derivatives, and also for the polythiophenes, and these have been found to be suitable for use as hole-injecting and hole-transporting layers. Injection and transport of electrons, in contrast, have proved more difficult. There are several routes to chemical modification of conjugated polymers to improve electron injection. These include the synthesis of polymers with electron-withdrawing groups attached to the chain, such as the cyano derivative of PPV shown in Fig. 29.4 [76,86], which show an increase of both ionization potential and electron affinity by at least 0.5 eV in comparison with the dialkoxy-PPV formed without the cyano side groups. Another approach is the use of molecular units known to perform well in sublimed molecular film LEDs as electron transport layers, such as PBD as shown in Fig. 29.1. These can be attached as side groups of nonconjugated chains [87] or incorporated within the conjugated sequence of the main chain [88,89].

C. Model for Process of Electroluminescence

The basic structure of a single-layer polymer LED was briefly introduced in Section I and is shown in Fig. 29.2. The current–voltage (I–V) and luminance–voltage (L–V) characteristics of a typical ITO/PPV/Ca device are shown in Fig. 29.8. The luminance is approximately proportional to the current, indicating that the quantum efficiency is approximately constant over a wide range of currents. The voltage required to produce a given

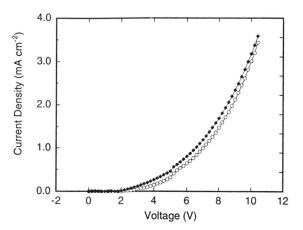

Fig. 29.8 (♦) Current density and (○) luminance (arbitrary units) versus voltage for an ITO/PPV/Ca LED.

current density has been found to be proportional to the thickness of the device, implying that the current density is simply a function of the average electric field across the device [90]. The I–V characteristics show diode-like behavior, with rectification ratios usually in excess of 10^3 between forward and reverse bias.

As mentioned above, the operation of the device requires that electrons and holes be injected from opposite electrodes. Electrons are injected into the conduction band states of the polymer, and holes into the valence band states, and for a diode formed with a polymer such as PPV, a schematic energy level diagram as shown in Fig. 29.9 is considered appropriate. Note that there are barriers at the electrodes for injection of both electrons and holes from the aluminum and indium-tin oxide electrodes, respectively. It is difficult to make accurate predictions about the barriers to electron and hole injection (ΔE_e and ΔE_h, respectively, in Fig. 29.9). Making reasonable assumptions about the polymer ionization potential and electrode work functions, it is clear that the barrier to injection of electrons from aluminum must be significantly larger than the barrier to injection of holes from ITO [90]. The majority of the current is therefore expected to be due to holes. Electroluminescence, however, requires the simultaneous injection of electrons, and the quantum efficiency will therefore depend strongly on the barrier to electron injection.

Electrons and holes capture one another within the polymer film and form either singlet or triplet excitons. Of these, the singlet excitons may decay radiatively, giving out light that is observed through one of the electrodes, which must be semitransparent. The internal quantum efficiency η_{int}, defined as the ratio of the number of photons produced within the device to the number of electrons flowing in the external circuit, is given by

$$\eta_{\text{int}} = \gamma r_{\text{st}} q \qquad (1)$$

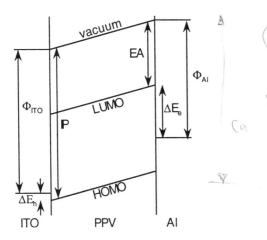

Fig. 29.9 Schematic energy level diagram for an ITO/PPV/Al LED, showing the ionization potential (IP) and electron affinity (EA) of PPV, the work functions of ITO and Al (Φ_{ITO} and ϕ_{Al}), and the barriers to injection of electrons and holes (ΔE_e and ΔE_h).

where γ is the ratio of the number of exciton formation events within the device to the number of electrons flowing in the external circuit, r_{st} is the fraction of excitons that are formed as singlets, and q is the efficiency of radiative decay of these singlet excitons. To achieve efficient luminescence, it is therefore necessary to have good balancing of electron and hole currents, efficient capture of electrons and holes within the emissive layer, and efficient radiative decay of singlet excitons. As shown in Fig. 29.3, the electroluminescence spectrum of a PPV LED is very similar to the PL spectrum for PPV, indicating that radiative decay of the same singlet exciton is responsible for the emission in both cases.

In summary, the overall performance of this type of diode depends on the factors contained in Eq. (1): (1) the balance of electron and hole currents, (2) the recombination process to form singlet or triplet excitons, and (3) the radiative decay of singlet excitons. We consider issues relating to the first factor in Section III and to factors 2 and 3 in Section IV.

III. CHARGE INJECTION AND TRANSPORT

A. Charge Injection and Transport in Single-Layer Devices

Injection of charge from most electrode materials requires that charges surmount or tunnel through a barrier at the interface. This is expected on examination of the positions of the electrode metal work functions and the positions of the π (highest occupied) and π^* (lowest unoccupied) molecular orbitals in the polymer. For the case of PPV, ITO provides a relatively good match for hole injection, though there is a barrier of around 0.2 eV [72]. However, electron injection is more difficult to achieve without the use of low work function reactive metals such as calcium (for which the barrier as determined by the difference in work function of the metal and electron affinity of the polymer is of a similar magnitude).

Modeling of the electrical characteristics of polymer LEDs is a challenging task, and at present there is no consensus as to the most appropriate model to apply. We note that the energy levels, mobilities, diffusivities, doping levels, and detailed structure of interfaces in polymer LEDs are poorly known at present. It is therefore difficult to distinguish between the effect of the bulk and the effect of the interfaces on the electrical behavior. It is also likely that many of the important parameters vary considerably between different polymers and even between the different samples of nominally the same polymer.

The improvement in efficiency in changing the electron-injecting electrode from aluminum to calcium is consistent with the reduction in barrier height for electron injection as the work function of the electrode is decreased. The I–V characteristics for PPV LEDs have been studied in detail by Marks et al. [91] for various electrode materials and temperatures. These results show a power-law dependence at low voltages, which is attributed to a space-charge-limited current with significant trapping. The effective mobilities obtained by fitting these data are between 10^{-9} and 10^{-7} cm^2/(V·s), indicating that the injection may be filamentous or that the current is due to injection of carriers with genuinely low mobility. At higher voltages, Marks et al. found that the current was predominantly an interface-limited hole current. The injection mechanism proved difficult to fit with simple models, and it was tentatively proposed that the current was limited by thermionic emission through an interfacial barrier. The importance of interfacial barriers was also identified by Vestweber et al. [58], who concentrate particularly on oxide layers at the interface between polyphenylphenylenevinylene and Al.

The characteristics of MEH-PPV LEDs have also been studied in detail by Parker [92]. He chose electrode materials to give currents that are either almost entirely due to holes or almost entirely due to electrons. At high fields, he obtained good fits to the Fowler–Nordheim theory for tunneling through a triangular barrier, with barrier heights that are in reasonable agreement with the values predicted by the electrode work functions. The model used to fit these data assumes that the bands are flat, which requires that there be negligible space charge within the device and that the doping concentration be low enough for the width of any Schottky barriers formed to be significantly greater than the width of the device. The weak temperature dependence of the current suggests that tunneling is the predominant injection mechanism, but no attempt is made to fit the data to tunneling through different kinds of barriers, and it is therefore difficult to rule out the possibility of tunneling through a Schottky barrier rather than a triangular barrier. For devices with two injecting electrodes, the current is found to be significantly greater than the sum of the currents in the equivalent single-carrier devices. This indicates that space charge effects must be significant in limiting the single-carrier currents, which contradicts the assumptions made in the model used to fit the data. Although the barrier heights extracted from the Fowler–Nordheim plots are reasonable, the observed currents are several orders of magnitude lower than those predicted by theory, indicating further problems with the simple tunneling model [93].

Riess and coworkers at the University of Bayreuth proposed a Schottky barrier model for the operation of ITO/PPV/Al LEDs [70,71,94]. They argue that the current is predominantly carried by holes and that this hole current is limited by the Schottky barrier formed at the PPV/Al interface rather than by any barrier at the ITO/PPV interface. They model the current–voltage characteristics using the equation for thermionic emission across a Schottky barrier from a semiconductor into a metal. It should be noted that the doping levels estimated for these devices are in the range 10^{16}–10^{17} cm^{-3}, considerably higher than the values estimated by Marks et al. for their devices.

Detailed understanding of the nature of the injection process will require a microscopic model for the chemistry at the polymer/metal interface, and considerable progress has been made on the nature of the polymer/cathode interface, experimentally by the Linköping group, and in terms of the quantum-chemical modeling by the Mons group [95–97]. Photoemission measurements from clean polymer surfaces with low coverages of metals evaporated in situ reveal that alkali metals such as sodium [97] and the group II metal calcium [96] reductively dope the surface layer of PPV to form a conductive charge transfer complex. In the case of calcium it appears that the metal does not diffuse far into the PPV layer (it provides a relatively stable electrode and long-lived devices [98]), so that the barrier for electron injection may in fact be between the calcium-doped surface layer of the polymer and the undoped polymer beneath it. In contrast, aluminum bonds covalently with the polymer, attacking the vinylic carbons to saturate the bonds at these positions on the chain and thereby produces barriers for charge injection [95].

We comment also on the use of chemically doped conjugated polymers as injection electrodes. Several p-doped conjugated polymers, including polypyrrole, polythiophene, and polyaniline, show good environmental stability, and such materials have been used as hole-injecting electrodes. This was first reported by Hayashi et al.[99], who used doped poly(3-methylthiophene) in combination with a sublimed film of perylene as the emissive layer. For the conjugated polymers, this was first performed with polyaniline as the hole-injecting electrode by Nakano et al.[100] and subsequently by Yang and Heeger [101], who demonstrate that doped polyaniline can show reasonable optical transparency and report lower barriers for hole injection in comparison with ITO. Pei et al.[102] recently reported that with the addition of a salt and an ion-transporting material such as poly(ethylene oxide) to the conjugated polymer, very low barriers for charge injection are found. They attribute this to the formation of p-doped regions in the polymer near the positive electrode and an n-doped region near the negative electrode, with the doping arising from electrochemical reactions that occur under drive conditions.

Carrier mobilities in polymers of this type are not well known, but there is good evidence that holes are considerably more mobile than electrons in the polyphenylenevinylenes, as is expected for polymers with relatively low ionization potentials and electron affinities [90,103]. It is therefore harder to get electrons to move into the bulk of the polymer film, and this may be responsible for poor device efficiency if excitons are generated too close to the negative electrode so that nonradiative decay becomes enhanced at the electrode [104]. Poor electron mobility, possibly resulting from charge trapping by included oxygen, is considered to play a major role in the operation of diodes of the structure as shown in Fig. 29.1 when operated in a photovoltaic or photoconductive mode [105].

We consider that the most significant problem in the context of charge injection and transport is to ensure that the currents of electrons and holes injected at opposite electrodes are balanced, so that there is not a preponderance of one charge type that passes current from one electrode to the other without encountering the carrier of opposite sign within the bulk of the polymer. Two strategies for controlling this have been reported. The first is to match the barriers for electron and hole injection by selection of the work functions of the electrode metals; as mentioned above, this has proved to be very successful when calcium has been used in place of aluminum. However, this suffers from the disadvantage that a reactive metal with a low work function must be used (calcium has proved to be the favored material) and this will constrain the application of these devices. The second method is to form a device with at least two semiconductor layers, making use of the heterojunction between them to confine electrons and holes traveling in opposite directions. This strategy is discussed in Section III.B, which follows.

B. Heterostructure Devices

Injection and transport of holes from the positive electrode into the bulk of the polymer film must be matched by injection and transport of electrons from the opposite electrode. The use of two-layer structures to control injection rates of electrons and holes by introducing barriers for charge transport at the heterojunction between the two semiconductor layers is now well established, both for devices made with sublimed molecular films [15,16] and more recently for devices containing only conjugated polymers [76]. Most of the conjugated polymers that are readily available, including the PPVs and polythiophenes, are suitable as hole-injecting and transporting materials, and until recently it has been necessary to use electron-transporting layers made either as sublimed molecular films [106] or as blends of electron-transporting molecular semiconductors in polymer hosts [72,107].

The synthesis of a family of solution-processible polycyanoterephthalylidenes, which are derivatives of PPV with nitrile groups attached to the vinylic carbons, has, however, provided the materials necessary to complement the existing hole-transporting PPVs [76,86,108]. The polymers that have been studied in greatest detail are illustrated in Fig. 29.4. The alkoxy side groups are chosen to ensure that the polymer is soluble in convenient solvents, and the dihexyloxy polymer DHeO-CN-PPV [76,86,108] can be processed from solution in chloroform. MEH-CN-PPV has asymmetric side chains (chosen to be the same as in MEH-PPV) and is also soluble in toluene. Both show optical properties similar to those of the polymers without the nitrile groups present, though in detail there are important differences, which we discuss in Section IV.B. The electron-withdrawing effect of the nitrile groups is calculated to increase the binding energies of both occupied π and unoccupied π^* states, while maintaining a similar $\pi-\pi^*$ gap [109]. The increase in binding energy is measured experimentally by cyclic voltammetry to be about 0.5 eV [108], and these polymers are found to dope readily with electron donors such as sodium [110]. As mentioned in Section IV.B, these polymers are found to be highly fluorescent and therefore useful both for electron transport and as emissive layers.

Fabrication of two-layer devices is shown in Fig. 29.10. PPV is processed via the standard precursor polymer onto ITO on a glass substrate, and the nitrile (or cyano) derivative can then be spin-coated directly on top of this intractable PPV layer. The final process step is to thermally evaporate top electrodes, for which purpose aluminum is found to be convenient. The band scheme for a two-layer device of this type is shown in Fig. 29.11. At the interface between the two polymers there are sizeable offsets in the energies of the π (highest occupied) molecular orbitals (HOMOs) for the PPV and the CN-PPV and also between the π^* (lowest unoccupied) molecular orbitals (LUMOs) of the PPV and CN-PPV layers. Under forward bias, as illustrated in Fig. 29.11, injection of holes from the ITO into the HOMOs of the PPV layer results in transport of holes to the heter-

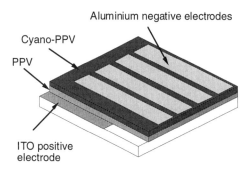

Fig. 29.10 Two-layer polymer LED.

ojunction, at which they are confined by the potential barrier, which must be surmounted if they are to progress into the CN-PPV layer. Similarly, electrons injected from the negative electrode are confined at the heterojunction by the potential step required to get transfer into the PPV LUMOs. This results in the setting up of space charge on either side of the heterojunction as indicated in Fig. 29.11. Tunneling across one or the other barrier will allow electron/hole capture and electroluminescence. We expect tunneling across the lower of the two barriers, which should be the barrier for holes. Note that it is now possible to get easy injection of electrons into the CN-PPV by using higher work function metals, such as aluminum, in place of the calcium necessary for the single-layer devices based on PPV.

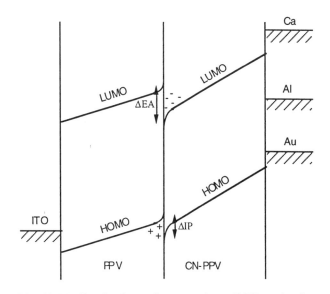

Fig. 29.11 Band scheme for a two-layer LED under forward bias. Positions of the Fermi energies for the electrode metals with respect to the π (highest occupied) molecular orbitals (HOMOs) and the π^* (lowest unoccupied) molecular orbitals (LUMOs) are illustrative.

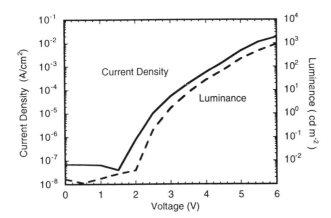

Fig. 29.12 Current and luminance versus forward bias for a heterostructure device of the type shown in Fig. 29.5, formed with PPV and MEH-CN-PPV (shown in Fig. 29.3). The polymer layers were of similar thicknesses, and the total thickness of the polymer layers was about 75 nm. (From Ref. 111.)

Experimental results for a device made with PPV and the MEH-CN-PPV polymer are shown in Fig. 29.12 [111]. Note that the turn-on voltage is close to 2 V and that the light output scales with the current through the device over many orders of magnitude of current. These devices exhibit a quantum efficiency a factor of 2 higher than that reported previously for similar devices made with DHeO-CN-PPV [76], and using the procedures reported by Greenham et al.[30] an external quantum efficiency for light emitted in the forward direction of 2.5% and an internal quantum efficiency in excess of 10% are estimated. With the peak output at 600 nm, the forward luminous intensity is now 3.5 cd/A at high current densities. Note that the luminous intensity reaches more than 1 cd/m² at 3 V bias and exceeds 100 cd/m² at 5 V bias.

The performance of these devices in terms of quantum efficiency is very high, indicating that the matching of electron and hole currents injected from opposite electrodes is probably achieved.

IV. ELECTRON–HOLE RECOMBINATION AND LUMINESCENCE

A. Electron–Hole Recombination

The process of electron/hole capture in these devices is not well studied. However, to get efficient capture in these very thin structures (typically 100 nm total thickness of polymer layers), it is necessary that one or the other charge carrier be of very low mobility so that the local charge density is sufficiently high to ensure that the other charge carrier will pass within a collision capture radius of at least one charge [90,112]. This is certainly enhanced in the heterostructure devices discussed below, where confinement at the heterojunction causes a buildup in charge density.

Electron/hole capture is expected to produce excitons with spin wave functions in the triplet and singlet configurations in the ratio 3:1. For these polymeric semiconductors, there is firm evidence that the triplet exciton is strongly bound with respect to the singlet, indicating that at least the triplet (and probably also the singlet) excitons thus produced are spatially localized to one polymer chain and to a confined segment of the chain. That there is limited crossover from triplet to singlet is evident in the very different lifetimes of the two species, with the singlet exciton decaying within typically 300 ps [113] and the triplet exciton surviving for up to 1 ms at low temperatures [114]. In this model, we expect to lose 75% of the electron–hole pairs to triplet excitons, which do not decay radiatively with high efficiency.

Experiments to detect the formation of triplet excitons have been carried out in two ways. First, measurements of electroluminescence-detected electron spin resonance have shown strong half-field resonances at the singlet emission energy [115]. Quenching of singlet states by triplets is the likely cause of this resonance. Second, the allowed optical transition between the lowest lying triplet and a higher lying triplet at 1.4 eV, which is identified in Fig. 29.7 and has been characterized in photoexcitation experiments [114,116], can be measured directly under conditions of forward bias in a diode structure. This is shown in Fig. 29.13, where the induced absorption due to photoexcitation is compared with than

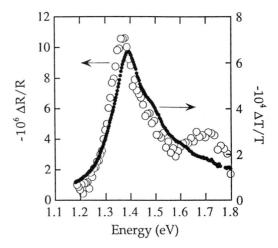

Fig. 29.13 Induced absorption response due to optical transitions within the triplet manifold of PPV, (○) arising from forward current in an ITO/PPV/Ca LED device and (●) following photoexcitation. Both were measured at low temperatures, near 20 K. (From Ref. 117.)

due to forward current in a polymer LED [117]. Brown et al.[117] consider that the strength of the observed absorption band in relation to the light emitted from radiative decay of singlet excitons is consistent with the 3:1 branching ratio expected within this simple model.

B. Radiative and Nonradiative Decay of Singlet Excitons

Radiative decay of the singlet exciton thus produced is required for light emission, and it is necessary to find conjugated polymers that show efficient luminescence in the solid state. Although measurements of photoluminescence efficiency for conjugated polymers in dilute solution have been known for some time [113,118,119], measurements for thin solid films of polymer are harder to make, and values obtained using an integrating sphere have only recently been reported [57,104,120]. Luminescence efficiency in the solid state tends to be lower than that measured for isolated molecules, as a result of exciton migration to quenching sites and also through interchain interactions, which produce lower energy excited states that are not strongly radiatively coupled to the ground state [121]. An important source of quenching sites is provided by chemical doping, and many of the first conjugated polymers to become available were sufficiently doped that luminescence yields were very low.

There has been considerable progress, however, in improving luminescence yields in conjugated polymers, both through synthesis of higher purity polymers and through the use of copolymers formed with segments of chain with different π–π^* gaps [55,122]. These are arranged so that excitons are trapped in lower gap regions of the chain and are thus unable to move easily to quenching sites. Measured values of solid-state photoluminescence efficiency for a range of polyphenylenevinylenes are now high. PPV prepared by the sulfonium precursor route in Cambridge shows an efficiency of 27% [104], and some of the soluble derivatives such as the cyano-substituted polymers shown in Fig. 29.4 and copolymers [56,120] show efficiencies of around 50%.

There has been considerable evidence that interchain interactions can significantly modify the energetics of exciton formation in conjugated polymers. For example, interchain interactions present in the ground state of some ladder polyphenylenes give strong red shifts of the luminescence through the formation of "dimer" or "aggregate" states [67,123]. For these materials, such interchain interactions seem to produce a reduction in quantum yield for luminescence, and the red shift is undesirable. Thus, polyphenylene ladder polymers of the type shown in Fig. 29.4 show blue emission in solution or in dilute solid-state solution but show only yellow emission in the solid state. Fig. 29.14 shows the EL spectra for one of these polymers in solid solution with varying amounts of polyvinylcarbazole, which acts as a hole-transporting material. Note that at 30% weight fraction of ladder polymer the emission is yellow, but at 1% the emission is substantially blue [66]. The tendency toward formation of interchain states that cause the red shift of the luminescence can be controlled by preventing an ordered structure, and the ladder copolymer of the type shown in Fig. 29.4 can show substantially blue emission even without dilution [65]. Direct evidence that the yellow emission is due to an interchain state present in the ground-state geometry (dimer or aggregate), rather than in only the excited state (excimer), is seen in the photoconductivity and photoluminescence excitation spectra, and Fig. 29.15 shows the presence of a weak optical absorption in the region of the yellow emission that gives rises to the photocurrent peak near 2.3 eV for diodes formed as ITO/LPPP/Al [67].

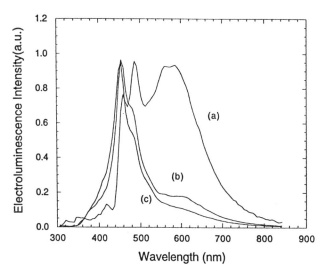

Fig. 29.14 Electroluminescence spectra for the ladder polyphenylene of the type shown in Fig. 29.4, with $R_1 = C_6H_4–C_{10}H_{21}$, $R_2 = C_6H_{13}$, and $R_3 = H$. The ladder polymer is diluted with polystyrene, so the weight fraction of the ladder polymer is (a) 30%, (b) 10%, and (c) 1%. Note that blend (a) shows yellow emission and blend (c) shows substantially blue emission. (From Ref. 66.)

The situation for PPV is controversial at present. The Cambridge measurements of photoluminescence efficiency (27%) together with decay rate (0.32 ns) are consistent with initial photogeneration of singlet intrachain excitons, which then decay both radiatively (lifetime on the order of 1 ns) and nonradiatively [104]. However, there is evidence for the formation of interchain charge-separated states in samples investigated by Rothberg and coworkers [124], though this might be associated with partially oxidized polymer, for which carbonyl groups act as electron traps, thus facilitating charge separation [125].

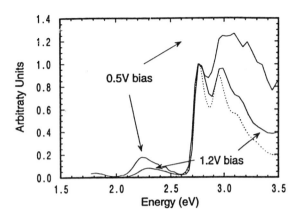

Fig. 29.15 Spectral response (solid lines) in a device in a sandwich structure of ITO/LPPP/aluminum under forward bias (ITO positive with respect to Al). For comparison, the absorption spectrum of this film is also shown (dashed line). Note that under forward bias there is an additional photocurrent peak in the yellow part of the spectrum (2.3 eV). (From Ref. 67.)

It has been found that the behavior of the cyano-PPVs shown in Fig. 29.4 is unusual and indeed provides evidence for strong interchain interactions, although a high photoluminescence efficiency is still obtained. This is illustrated by the data for absorption and emission of solid films of three cyano-PPV polymers that have different alkoxy side chains, as shown in Fig. 29.16, and the data for emission from solutions of these polymers, which are contrasted with the solid film data in Fig. 29.17. Baigent et al. [126] note that the Stokes shift for the solutions is small, comparable to that of PPV (Fig. 29.3), and similar for all three polymers. In contrast, there is a strong Stokes shift for the solid films, and this is very different for the three polymers, so that the emission maxima range from near 600 nm for MEH-CN-PPV to near 750 nm for the polymer shown in Fig. 29.16. Baigent et al. consider that this red shift is due to the formation of an excition that is delocalized over more than one chain (and thus described as an excimer), and we consider that the strong variation of the luminescence spectrum with selection of the alkyl side chain results from different packing arrangements for these polymers, which control the strength of the interchain coupling. Further evidence to support this has been provided by Samuel et al. [127], who report that the luminescence decay rate (0.8 ns) in solutions of DHeO-CN-PPV that show a photoluminescence efficiency of 52% suggest a radiative decay rate of 1.7 ns, much as for PPV, but in contrast the solid films of this polymer were measured to show a photoluminescence efficiency of 35% and a lifetime of 5.5 ns, implying a radiative decay rate of 16 ns. This factor of 10 increase in lifetime is consid-

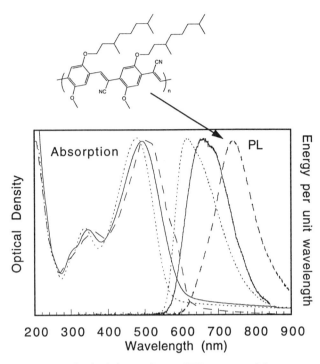

Fig. 29.16 Optical absorption and PL spectra of the cyano-PPVs shown in Fig. 29.3. (---) MEH-CN-PPV; (———) DHeO-CN-PPV; (- - -) CN-PPV. (From Ref. 126.)

ered to arise from the modified dipole matrix elements for radiative transitions of the interchain excited state.

C. Microcavities

It is well established that the spontaneous emission rate, which controls luminescence, from an excited state such

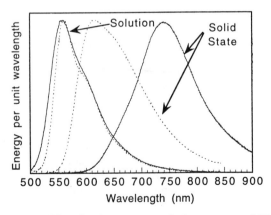

Fig. 29.17 Photoluminescence emission spectra of MEH-CN-PPV (broken line) and the CN-PPV shown in Fig. 29.16 (solid line), in dilute solution and in the solid state. (From Ref. 126.)

as the excitons of interest here is dependent on the nature of the allowed electromagnetic modes into which the photon will couple, and this can provide an important means for controlling spectral linewidth and angular dependence of the emission spectrum [128–133]. According to Fermi's golden rule, the rate of radiative spontaneous emission k_R can be written as [134]

$$k_{rad} \propto |<m>|^2 \delta(\nu)$$

where $\langle m \rangle$ describes the matrix element of the optical transition and $\delta(\nu)$ the density of optical final states as a function of frequency ν. The parameter $\langle m \rangle$ is an inherent property of the material, whereas $\delta(\nu)$ can be altered by structuring the density of optical states, e.g., by placing the luminescent thin film in an optical resonator, where only selected values of ν are allowed. For a planar configuration of such a microcavity structure, the solid angle of the modified density of optical states is large, and a large part of the spontaneous emission is affected. Radiative decay is enhanced for those emitters in the cavity for which optical transitions are compatible with the resonance condition and is reduced otherwise. For the forward emission of a sufficiently optimized cavity structure, a significant enhancement of the mode density compared to the free-space spectrum is expected. By adjusting the optical parameters of the cavity, namely the wavelength and the spectral width, the color of the cavity emission can be tuned over the whole polymer spectrum. The planar configuration of a polymer device leads also to a spatial narrowing of the emission of a cavity mode.

A number of groups have reported measurements on microcavity devices made with conjugated polymers as the emissive materials [135–138]. A typical structure is shown in Fig. 29.18 [138]; the emissive layer of PPV is formed between two electrodes, of ITO and either aluminum or silver, and the microcavity is formed by the distributed Bragg reflector (behind the ITO electrode) and the aluminum or silver electrodes, which function as a semitransparent mirror. The photoluminescence spectra (electroluminescence spectra are similar) are shown in Fig. 29.19. Note that there is a very considerable spectral narrowing, as expected for the high quality factor of this cavity, and the full-width half-maximum bandwidths are 5 and 6.8 nm, respectively, for the silver and aluminum structures. Grüner et al. [138] modeled the response of this structure and showed that the intensity of emission in the forward direction within the allowed cavity mode is strongly enhanced over that for the free-space emission of the polymer. This is particularly marked for the device with a silver electrode because the absorption losses in the silver film are lower than for the aluminum film. An important issue is whether the performance of these cavity structures is enhanced over that achieved by placing a similar Fabry–Perot filter outside the emissive layer or whether there is efficient redistribution of emission into the allowed cavity

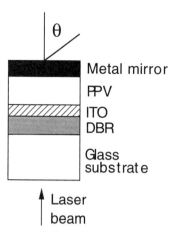

Fig. 29.18 Structure of a microcavity LED as investigated in Ref. 138. The substrate consisted of DBR on glass [Balzers: BD 100 143 03] (see Fig. 29.2), with an ITO film of thickness about 96 nm deposited by rf sputtering. The ITO acts as the bottom electrode and is treated as a transparent spacer with an index of refraction of $n = 1.8$ in the cavity device. The PPV layer was of thicknesses 160 ± 10 nm.

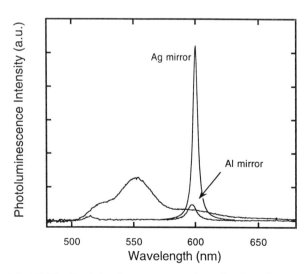

Fig. 29.19 Photoluminescence spectrum for the microcavity structure shown in Fig. 29.18, for a structure made with both aluminum and silver cathodes. The free-space emission from the PPV layer is also shown for comparison (as shown, for example, in Fig. 29.3). (From Ref. 138.)

mode through modification of the spontaneous emission rates. Grüner et al. show by measurements of the total luminescence emission that there is a real enhancement.

V. PHOTOCONDUCTIVE DIODES

Although this chapter is concerned with light-emitting diodes, it is nevertheless of interest to mention here the properties of these and related devices operating in photovoltaic or photoconductive modes. The first problem that has to be considered with the use of organic molecular semiconductors is that the excited states produced by photon absorption are usually excitons that have relatively high binding energies and do not dissociate to give electrons and holes. Exciton ionization in the bulk is therefore not a promising method to follow. However, interfaces between molecular semiconductors or with electrodes can provide the correct energetics to allow charge separation, and one of the more promising routes to follow is to use a two-layer cell similar to the two-layer LED structures of the type illustrated in Fig. 29.10. This approach has been followed by several groups, and Tang [139] obtained some of the best results, using a cell formed on glass coated with indium-tin oxide on which layers of a copper phthalocyanine (hole-collecting semiconductor) and a perylene tetracarboxylic derivative are formed by sublimation, capped with an evaporated silver electrode. Excitons that diffuse to the interface between these two semiconductors ionize, and electrons and holes are then collected. Cells of this type can show power conversion efficiencies of 1% and, if sufficiently thin, good fill factors (0.65).

Two difficulties are commonly encountered. First, carrier mobilities are low, so that power loss is likely to be problematic at high levels of illumination. Second, although the absorption coefficients for organic semiconductors are very high ($>10^5$ cm^{-1}), the absorption depth is usually greater than the diffusion range of the excitons created by the absorption process. Thus, only a fraction of the excitons generated are able to find the interface between the two semiconductors at which ionization can occur. A strategy to improve performance is to use high purity, highly crystalline molecular semiconductors, which can show greater diffusion ranges for excitons [140]. A different approach is to arrange a structure in which there is a very large surface area, so that all absorbing regions lie close to an interface at which ionization can occur. This approach was demonstrated by O'Regan and Gratzel [141], who used a sintered electrode of TiO_2 onto which they surface-adsorbed a layer of an organic dye (ruthenium bipyridinium complex). Absorption in the ruthenium complex results in electron transfer into the TiO_2, and the circuit is completed via an iodine/iodide redox couple to a back electrode to give a photoelectrochemical cell with a reported energy conversion efficiency of 7–12%.

The recent interest in the properties of electroluminescent diodes has regenerated interest in the photoconductive and photovoltaic responses of conjugated polymers, and Reiss and coworkers [70,142] were the first group to report the photovoltaic response of devices that also functioned at LEDs. Results for diodes fabricated as shown in Fig. 29.2 have been reported by several groups [70,105,142,143]. Diodes formed with PPV between ITO and Mg demonstrate an open-circuit voltage that rises to 1.2 V at high illumination intensities (approximately equal to the work function difference between the two electrodes) and a short-circuit current that gives a quantum efficiency (electrons collected per incident photon) of up to 1%. At high forward and reverse biases, however, the quantum efficiency can reach very high values; Marks et al. [105] report an efficiency of up to 500% under forward bias, and several groups have reported efficiencies above 50% under reverse bias, particularly for devices that make use of Ca and ITO electrodes [105,144,145]. These high efficiencies under bias (typically in the range up to 10 V) have generated some interest in the potential use of this type of structure as a large-area photodetector [144].

The application to the case of the polymers of the bilayer heterojunction structure that has been used successfully with molecular semiconductors [139] has recently been reported. The fullerene C_{60} provides a useful high electron affinity semiconductor, and it is found that blends of this with a range of soluble PPV and polythiophene derivatives show very efficient charge separation following photoexcitation [146–148]. Two-layer diodes of C_{60} and MEH-PPV were reported initially to show poor efficiencies [149], but photovoltaic efficiencies (short circuit) up to 9% have now been obtained for PPV/C_{60} diodes of the type shown in Fig. 29.20 [150]. Charac-

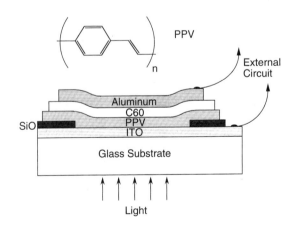

Fig. 29.20 Schematic diagram of a two-layer photovoltaic device formed with PPV and C_{60}. The thermally evaporated SiO strips define the active area of the device to be 3 mm^2. (From Ref. 150.)

teristics for a diode of this type are shown in Fig. 29.21. Halls et al. [150] were able to model the excitation spectrum of the photovoltaic short-circuit current using a value for the diffusion range of the singlet exciton in PPV of around 8 nm.

The principle demonstrated by O'Regan and Grätzel [141] of using a large effective area to ensure that all photons are absorbed close to an interface between electron-accepting and hole-accepting semiconductors has recently been applied to the polymers by Halls et al. [151] and by Yu and Heeger [152]. Similar structures using derivatized fullerene as the electron-accepting component have also been reported [153]. Blends of MEH-PPV and DHeO-CN-PPV (Fig. 29.4) are expected to form interpenetrating networks when cast from solution and correctly annealed, and films formed in this way give high photoconductive responses when formed between Al and ITO electrodes.

To investigate the phase segregation of these polymer blends, Halls et al. used a combination of transmission electron microscopy (TEM), scanning transmission electron microscopy (STEM), and parallel electron-energy-loss spectroscopy (PEELS). TEM bright field image and STEM annular dark field images of the blends clearly demonstrate the presence of an interpenetrating network structure on the scale of 10–100 nm, as expected for films of this thickness. Evidence of photoinduced charge transfer is commonly seen from quenching of photoluminescence. Halls et al. [151] reported values for the photoluminescence quantum efficiency for MEH-PPV and DHeO-CN-PPV of 10% and 32%, respectively, similar to values reported previously [104], and found quenching to a level of around 2–5% for the blends in the composition range 1:4 to 4:1. This substantial but incomplete quenching is as expected for an exciton diffusion range of 10 nm given the scale of phase segregation observed in the electron micrographs. Halls et al. propose that excitons are dissociated at the dispersed interfaces by transfer of electrons to the CN-PPV and of holes to the MEH-PPV.

Photovoltaic cells were prepared with a polymer layer sandwiched between electrodes with different work functions, as in polymer-based LEDs (Fig. 29.2). In the dark, the device exhibits a rectification ratio of 10^3 at ± 3.5 V. Under illumination, devices of this type show a strong photoresponse, with open-circuit voltages of 0.6 V and short-circuit currents that correspond to quantum efficiencies of up to 6%. Under forward or reverse biases, the quantum efficiencies rise rapidly, reaching 15% at a reverse bias of 3.5 V, 40% at 10 V, and considerably higher values under forward bias. These performance figures are very much better than those for devices made with aluminum electrodes and either MEH-PPV or CN-PPV alone. Thus, Halls et al. report values for the quantum yield of the short-circuit photocurrent in the MEH-PPV of 0.04% at the peak response wavelength (2.2 eV) and on the order of 10^{-3}%

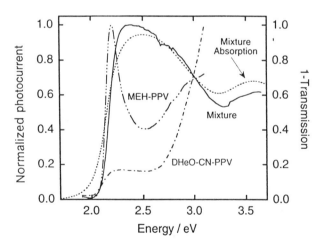

Fig. 29.22 The spectral responses of the short-circuit photocurrent through an ITO/polymer/Al device, where the polymer layers are formed as layers of pure MEH-PPV, pure DHeO-CN-PPV, and a mixture of MEH-PPV and DHeO-CN-PPV (see Fig. 29.4 for polymer structures). Currents have been corrected for the lamp–monochromator system response and scaled to give a peak current of unity for all devices. The absorption spectrum of the blend film, presented as (1 − transmission) for a film of the same thickness is also shown. Intensities of incident light were on the order of 0.1 mW/cm² at 520 nm. The polymer blend here consisted of equal masses of the two components; however, very similar results were found for blends of different ratios. (From Ref. 151.)

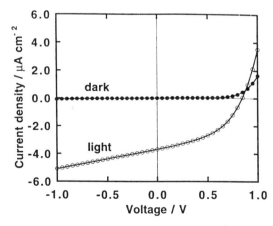

Fig. 29.21 The current–voltage characteristics of an ITO/PPV (65 nm)/C_{60} (40 nm)/Al photodiode, as shown in Fig. 29.20, in the dark and under illumination at a wavelength of 492 nm and an intensity of 0.25 mW/cm². The quantum efficiency under short-circuit conditions for diodes of this type is 9%. (From Ref. 150.)

for the DHeO-CN-PPV at 2.8 eV. (Note that better efficiencies are seen for single-polymer devices if more electronegative cathodes are employed, such as Ca and Mg [105,154].)

The spectral response of photodiodes of this type provides detailed information about the device operation. Fig. 29.22 shows the spectral dependence of the short-circuit current for devices made with the polymer blend and with the individual polymers. Halls and coworkers find that the spectral response of the blend device matches the absorption (also shown in the figure) very closely and consider that this is excellent evidence of efficient charge generation and transport to the device electrodes. In contrast, the devices made with the individual polymers show more complicated spectral responses. For example, the peak in response for the MEH-PPV at the absorption edge is attributed to electron trapping in the photogeneration region [105].

VI. SUMMARY

In this chapter we have reviewed some aspects of the electronic processes that control the properties of electroluminescent diodes that employ conjugated polymers as the active semiconductor, and we have made brief mention of the properties of diodes that function as efficient photovoltaic or photoconductive devices. The present state of the art is that very satisfactory levels of performance have been achieved, and this development has gone hand-in-hand with the development of understanding of the underlying semiconductor physics. However, there is much that remains poorly understood, and there will no doubt be important progress over the next few years. There is a high level of industrial interest and activity in these devices, not least because there is considerable demand for large-area solid-state emissive displays. The process of commercialization will be controlled by the performance of these structures according to other criteria, principally durability under drive and under storage conditions and demonstrably cheap manufacture [155]. The indications at present are very encouraging.

ACKNOWLEDGMENTS

This chapter has drawn on the work of many people in the Cambridge group, both in the department of physics and in the department of chemistry, and those in the several groups elsewhere with which we have worked. We are grateful to all those whose work and ideas have influenced the contents of this chapter, particularly those whose work we have cited. One of us (NCG) thanks Clare College, Cambridge for support.

REFERENCES

1. C. K. Chiang, C. R. Fincher, Y. W. Park, A. J. Heeger, H. Shirakawa, E. J. Louis, S. C. Gau, and A. G. MacDiarmid, Electrical conductivity in doped polyacetylene, *Phys. Rev. Lett. 39*:1098 (1977).
2. G. B. Street, Polypyrrole: from powders to plastics, in *Handbook of Conducting Polymers*, Vol. 1 (T. J. Skotheim, ed.), Marcel Dekker, New York, 1986, p. 265.
3. A. J. Epstein, J. Joo, C. Y. Wu, A. Benatar, C. F. Faisst, J. Zegarski, and A. G. MacDiarmid, Polyanilines: recent advances in processing and applications to welding of plastics, in *Intrinsically Conducting Polymers: An Emerging Technology* (M. Aldissi, ed.), NATO ASI Series E, Vol. 246, Kluwer, Dordrecht, 1993, p. 165.
4. P. M. Grant, T. Tani, W. D. Gill, M. Kroubni, and T. C. Clarke, Schottky diodes based on Shirakawa polyacetylene, *J. Appl. Phys. 52*:869 (1981).
5. E. Ebisawa, T. Kurokawa, and S. Nara, Polyacetylene metal-insulator-semiconductor devices, *J. Appl. Phys. 54*:3255 (1983).
6. J. Kanicki, Polymeric semiconductor contacts and photovoltaic applications, in *Handbook of Conducting Polymers*, Vol. 1 (T. J. Skotheim, ed.), Marcel Dekker, New York, 1986, p. 544.
7. H. Koezuka, A. Tsumara, and T. Ando, Field-effect transistor with polythiophene thin film, *Synth. Met. 18*:699 (1987).
8. A. Assadi, C. Svensson, M. Willander, and O. Inganäs, Field-effect mobility of poly(3-hexyl thiophene), *Appl. Phys. Lett. 53*:195 (1988).
9. J. H. Burroughes, C. A. Jones, and R. H. Friend, Polymer diodes and transistors: new semiconductor device physics, *Nature 335*:137 (1988).
10. M. Pope and C. E. Swenberg, *Electronic Processes in Organic Crystals*, Clarendon Press, Oxford, 1982.
11. P. M. Borsenberger and D. S. Weiss, *Organic Photoreceptors for Imaging Systems*, Marcel Dekker, New York, 1993.
12. M. Pope, H. Kallmann, and P. Magnante, Electroluminescence in organic crystals, *J. Chem. Phys. 38*:2042 (1963).
13. W. Helfrich and W. G. Schneider, Recombination radiation in anthracene crystals, *Phys. Rev. Lett. 14*:229 (1965).
14. P. S. Vincett, W. A. Barlow, R. A. Hann, and G. G. Roberts, Electrical conduction and low voltage blue electroluminescence in vacuum-deposited organic films, *Thin Solid Films 94*:476 (1982).
15. C. W. Tang and S. A. VanSlyke, Organic electroluminescent diodes, *Appl. Phys. Lett. 51*:913 (1987).
16. C. W. Tang, S. A. VanSlyke, and C. H. Chen, Electroluminescence of doped organic thin films, *J. Appl. Phys. 65*:3610 (1989).
17. C. Adachi, S. Tokito, T. Tsutsui, and S. Saito, Organic electroluminescent device with a three-layer structure, *Jpn. J. Appl. Phys. 27*:713 (1988).

18. C. Adachi, S. Tokito, T. Tsutsui, and S. Saito, Electroluminescence in organic films with three-layer structure, *Jpn. J. Appl. Phys. 27*: 269 (1988).
19. C. Adachi, T. Tsutsui, and S. Saito, Organic electroluminescent device having a hole conductor as an emitting layer, *Appl. Phys. Lett. 55*:1489 (1989).
20. C. Adachi, T. Tsutsui, and S. Saito, Blue light-emitting organic electroluminescent devices, *Appl. Phys. Lett. 56*:799 (1990).
21. C. Adachi, T. Tsutsui, and S. Saito, Confinement of charge carriers and molecular excitons within 5-nm-thick emitter layer in organic electroluminescent devices with a double heterostructure, *Appl. Phys. Lett. 57*:531 (1990).
22. C. Adachi, K. Nagai, and N. Tamoto, Molecular design of hole transport materials for obtaining high durability in organic electroluminescent diodes, *Appl. Phys. Lett. 66*:2679 (1995).
23. J. H. Burroughes, D. D. C. Bradley, A. R. Brown, R. N. Marks, K. Mackay, R. H. Friend, P. L. Burn, and A. B. Holmes, Light-emitting diodes based on conjugated polymers, *Nature 347*:539 (1990).
24. R. A. Wessling and R. G. Zimmerman, U.S. Patent 3,401,152 (1968).
25. R. A. Wessling and R. G. Zimmerman, U.S. Patent 3,706,677 (1972).
26. I. Murase, T. Ohnishi, T. Noguchi, and M. Hirooka, Highly conducting poly(p-phenylene vinylene) prepared from a sulphonium salt, *Polym. Commun. 25*: 327 (1984).
27. D. R. Gagnon, J. D. Capistran, F. E. Karasz, and R. W. Lenz, Conductivity anisotropy in oriented poly(p-phenylene vinylene), *Polym. Bull. 12*:293 (1984).
28. P. L. Burn, D. D. C. Bradley, R. H. Friend, D. A. Halliday, A. B. Holmes, R. W. Jackson, and A. M. Kraft, Precursor route chemistry and electronic properties of poly(1,4-phenylenevinylene), poly(2,5-dimethyl-1,4-phenylenevinylene), and poly(2,5-dimethoxy-1,4-phenylenevinylene), *J. Chem. Soc. Perkin Trans 1:1992*:3225.
29. F. Wudl, P. M. Allemand, G. Srdanov, Z. Ni, and D. McBranch, Polymers and an unusual molecular crystal with nonlinear optical properties, in *Materials for Nonlinear Optics: Chemical Perspectives* (S. R. Marder, J. E. Sohn, and G. D. Stucky, eds.), Vol. 455, Am. Chem. Soc., Washington, DC, 1991, p. 683.
30. N. C. Greenham, R. H. Friend, and D. D. C. Bradley, Angular dependence of the emission from a conjugated polymer light-emitting diode: implications for efficiency calculations, *Adv. Mater. 6*:491 (1994).
31. D. Braun and A. J. Heeger, Visible light emission from semiconducting polymer diodes, *Appl. Phys. Lett. 58*:1982 (1991); erratum, *59*:878 (1991).
32. W.-P. Su, J. R. Schrieffer, and A. J. Heeger, Solitons in polyacetylene, *Phys. Rev. Lett. 42*:1698 (1979).
33. W.-P. Su, J. R. Schrieffer, and A. J. Heeger, Solitons in polyacetylene, *Phys. Rev. B. 22*:2099 (1980); erratum, *28*:1138 (1983).
34. M. J. Rice, Charged π-phase kinks in lightly doped polyacetylene, *Phys. Lett. A 71*:152 (1979).
35. S. Brazovskii and N. Kirova, Excitons, polarons and bipolarons in conducting polymers, *Pis'ma Zh. Eksp. Teor. Fiz. 33*:6 (1981) [*JEPT Lett. 33*:4–8 (1981)].
36. K. Fesser, A. R. Bishop, and D. K. Campbell, Optical absorption from polarons in a model of polyacetylene *Phys. Rev. B 27*:4804 (1983).
37. R. H. Friend, D. D. C. Bradley, and P. D. Townsend, Photoexcitation in conjugated polymers, *J. Phys. D: Appl. Phys. 20*:1367 (1987).
38. J. L. Brédas, R. R. Chance, and R. Silbey, Comparative theoretical study of the doping on conjugated polymers: polarons in polyacetylene and polyparaphenylene, *Phys. Rev. B 26*:5843 (1982).
39. J. L. Brédas, B. Thémans, J. G. Fripiat, J. M. André, and R. R. Chance, Highly conducting polparaphenylene, polypyrrole and polythiophene chains: an *ab initio* study of the geometry and electronic-structure modifications upon doping, *Phys. Rev. B 29*:6761 (1984).
40. Z. G. Soos, D. S. Galvao, and S. Etemad, Fluorescence and excited-state structure of conjugated polymers, *Adv. Mater. 6*:280 (1994).
41. P. Gommes da Costa, R. G. Dandrea, and E. M. Conwell, First-principles calculation of the three-dimensional band structure of poly(phenylene vinylene), *Phys. Rev. B 47*:1800 (1993).
42. D. Emin, Self-trapping in quasi-one-dimensional solids, *Phys. Rev. B 33*:3973 (1986).
43. D. Emin, Basic issues of electronic transport in insulating polymers, in *Handbook of Conducting Polymers* (T. J. Skotheim, ed.), Vol. 2, Marcel Dekker, New York, 1986, Chap. 26, p. 915.
44. Z. Shuai, J. L. Brédas, and W. P. Su, Nature of photoexcitations in poly(paraphenylene vinylene) and its oligomers, *Chem. Phys. Lett. 228*:301 (1994).
45. P. Gommes da Costa and E. M. Conwell, Excitons and the band gap in poly(phenylene vinylene), *Phys. Rev. B 48*:1993 (1993).
46. D. Beljonne, Z. Shuai, R. H. Friend, and J. L. Brédas, Theoretical investigation of the lowest singlet and triplet states in poly(para phenylene vinylene) oligomers, *J. Chem. Phys. 102*:2042 (1995).
47. J. Cornil, D. Beljonne, R. H. Friend, and J. L. Brédas, Optical absorptions in poly(paraphenylene vinylene) and poly(2,5-dimethoxy-1,4-paraphenylene vinylene) oligomers, *Chem. Phys. Lett.*, 82 (1994).
48. M. Chandross, S. Mazumdar, S. Jeglinski, X. Wei, Z. V. Vardeny, E. W. Kwock, and T. M. Miller, Excitons in poly(para-phenylenevinylene) *Phys. Rev. B 50*:14702 (1994).
49. T. Förster, Zwischenmolekulare Energiewanderung und Fluoreszenz, *Ann. Phys. VI 2*:55 (1948).
50. C. Zhang, S. Hoger, K. Pakbaz, F. Wudl, and A. J. Heeger, Yellow electroluminescent diodes utilizing poly(2,5-bis(cholestanoxy)-1,4-phenylene vinylene), *J. Electron. Mater. 22*:413 (1993).

51. Y. Ohmori, M. Uchida, K. Muro, and K. Yoshino, Visible light electroluminescent diodes utilising poly(3-alkylthiophene), *Jpn. J. Appl. Phys. 30*:L1938 (1991).
52. D. Braun, G. Gustafson, D. McBranch, and A. J. Heeger, Electroluminescence and electrical transport in poly(3-octylthiophene) diodes, *J. Appl. Phys. 72*:564 (1992).
53. N. C. Greenham, A. R. Brown, D. D. C. Bradley, and R. H. Friend, Electroluminescence in poly(3-alkylthienylene)s, *Synth. Met. 55–57*:4134 (1993).
54. P. L. Burn, A. B. Holmes, A. Kraft, D. D. C. Bradley, A. R. Brown, and R. H. Friend, Synthesis of a segmented conjugated polymer chain giving a blue-shifted electroluminescence and improved efficiency, *J. Chem. Soc., Chem. Commun.* 32 (1992).
55. P. L. Burn, A. B. Holmes, A. Kraft, D. D. C. Bradley, A. R. Brown, R. H. Friend, and R. W. Gymer, Chemical tuning of electroluminescent copolymers to improve emission efficiencies and allow patterning, *Nature 356*:47 (1992).
56. D. Braun, E. G. J. Staring, R. C. J. E. Demandt, G. L. J. Rikken, Y. A. R. R. Kessener, and A. H. J. Venhuizen, Photo- and electroluminescence efficiency in poly(dialkoxy-*p*-phenylenevinylene), *Synth. Met. 66*:75 (1994).
57. E. G. J. Staring, R. C. E. Demandt, D. Braun, G. L. J. Rikken, Y. A. R. R. Kessener, T. H. J. Venhuizen, H. Wynberg, W. ten Hoeve, and K. J. Spoelstra, Photo- and electroluminescence efficiency in soluble poly(dialkoxy-*p*-phenylenevinylene), *Adv. Mater. 6*:934 (1994).
58. H. Vestweber, R. Sander, A. Greiner, W. Heitz, R. F. Mahrt, and H. Bässler, Electroluminescence from polymer blends and molecularly doped polymers, *Synth. Met. 64*:141 (1994).
59. J. Kido, K. Hongawa, K. Okuyama, and K. Nagai, White light-emitting organic electroluminescent devices using the poly(*N*-vinylcarbazole) emitter layer doped with three fluorescent dyes, *Appl. Phys. Lett. 64*:815 (1994).
60. J.-K. Lee, R. R. Schrock, D. R. Baigent, and R. H. Friend, A new type of blue-light emitting electroluminescent polymer, *Macromolecules 28*:1966 (1995).
61. I. Sokolik, Z. Yang, F. E. Karsz, and D. Morton, Blue-light electroluminescence from *p*-phenylene-based copolymers, *J. Appl. Phys. 74*:3584 (1993).
62. G. Grem, G. Leditzky, B. Ullrich, and G. Leising, Realisation of a blue light emitting device using poly(para phenylene), *Adv. Mater. 4*:36 (1992).
63. Y. Ohmori, M. Uchida, K. Muro, and K. Yoshino, Blue electroluminescent diode utilizing poly(alkylfluorene), *Jpn. J. Appl. Phys. 30*:L1941 (1991).
64. I. D. Parker, Q. Pei, and M. Marrocco, Efficient blue electroluminescence from a fluorinated polyquinoline, *Appl. Phys. Lett. 65*:1272 (1994).
65. J. F. Grüner, R. H. Friend, U. Scherf, J. Huber, and A. B. Holmes, A high efficiency blue light-emitting diode based on novel step-ladder poly(para-phenylene)s, *Adv. Mater. 6*:748 (1994).
66. J. Grüner, H. F. Wittmann, P. J. Hamer, R. H. Friend, J. Huber, U. Scherf, K. Müllen, S. C. Moratti, and A. B. Holmes, Electroluminescence and photoluminescence investigations of the yellow emission of devices based on ladder-type oligo(para-phenylene)s, *Synth. Met. 67*:181 (1994).
67. A. Köhler, J. Grüner, R. H. Friend, U. Scherf, and K. Müllen, Photovoltaic measurements on aggregates in ladder-type poly(phenylene), *Chem. Phys. Lett. 243*:456 (1995).
68. G. Leising, G. Grem, G. Leditzky, and U. Scherf, Electroluminescence devices with poly(paraphenylene) and derivatives as the active material, *Proc. SPIE 1910*:70 (1993).
69. W. X. Jing, A. Kraft, S. C. Moratti, J. Grüner, F. Cacialli, P. J. Hamer, A. B. Holmes, and R. H. Friend, Synthesis of a polyphenylene light-emitting copolymer, *Synth. Met. 67*:161 (1994).
70. S. Karg, W. Riess, V. Dyakonov, and M. Schwoerer, Electrical and optical characterisation of poly-phenylene-vinylene light emitting diodes, *Synth. Met. 54*:427 (1993).
71. W. Reiss, S. Karg, V. Dyakonov, M. Meier, and M. Schwoerer, Electroluminescence and photovoltaic effect in PPV Schottky diodes, *J. Luminescence 60–61*:906 (1994).
72. A. R. Brown, J. H. Burroughes, N. Greenham, R. H. Friend, D. D. C. Bradley, P. L. Burn, A. Kraft, and A. B. Holmes, Poly(*p*-phenylene vinylene) light-emitting diodes: enhanced electroluminescence efficiency through charge carrier confinement, *Appl. Phys. Lett. 61*:2793 (1992).
73. S. Doi, M. Kuwabara, T. Noguchi, and T. Ohnishi, Organic electroluminescent devices having poly(dialkoxy-*p*-phenylene vinylenes) as a light-emitting material, *Synth. Met. 55–57*:4174 (1993).
74. H. Vestweber, J. Oberski, A. Greiner, W. Heitz, R. F. Mahrt, and H. Bässler, Electroluminescence from phenylenevinylene-based polymer blends, *Adv. Mater. Opt. Electron 2*:197 (1993).
75. C. Zhang, H. von Seggern, K. Pakbaz, B. Kraabel, H. W. Schmidt, and A. J. Heeger, Blue electroluminescent diodes utilizing blends of poly(*p*-phenylene vinylene) in poly(9-carbazole), *Synth. Met. 62*, 35 (1994).
76. N. C. Greenham, S. C. Moratti, D. D. C. Bradley, R. H. Friend, and A. B. Holmes, Efficient polymer-based light-emitting diodes based on polymers with high electron affinities, *Nature 365*:628 (1993).
77. D. R. Baigent, P. J. Hamer, R. H. Friend, S. C. Moratti, and A. B. Holmes, Polymer electroluminescence in the near infra-red, *Synth. Met. 71*, 2175 (1995).
78. C. Zhang, D. Braun, and A. J. Heeger, Light-emitting diodes from partially conjugated poly(*p*-phenylene vinylene), *J. Appl. Phys. 73*:5177 (1993).
79. G. Grem and G. Leising, Electroluminescence of "wide-bandgap" chemically tunable cyclic conjugated polymers, *Synth. Met. 55–57*:4105 (1993).
80. T. Yamamoto, T. Inoue, and T. Kanbara, Polymer

light-emitting diodes with single and double-layer structures using poly(2,3-diphenylquinoxaline-5,8-diyl), *Jpn. J. Appl. Phys. 33*:L250 (1994).
81. J. Huber, K. Mullen, J. Salbeck, H. Schenk, U. Scherf, T. Stehlin, and R. Stern, Blue-light-emitting diodes based on ladder polymers of the PPP type, *Acta Polym. 45*:244 (1994).
82. A. R. Brown, Ph.D. Thesis, Cambridge, 1992.
83. R. E. Gill, G. G. Malliaras, J. Wildeman, and G. Hadziioannou, Tuning of photo and electroluminescence in alkylated polythiophenes with well-defined regioregularity, *Adv. Mater. 6*:132 (1994).
84. I. L. Valeeva, A. N. Lachinov, V. A. Antipin, and M. G. Zolotukhin, Electroluminescence in thin polymer films with a nondegenerate ground state, *JEPT 78*:83 (1994).
85. L. S. Swanson, F. Lu, J. Shinar, Y. W. Ding, and T. J. Barton, Poly(*p*-phenyleneacetylene) (PPA)-based light-emitting diodes, *Proc. SPIE 1910*:101 (1993).
86. S. C. Moratti, D. D. C. Bradley, R. H. Friend, N. C. Greenham, and A. B. Holmes, Molecularly engineered polymer LEDs, *Mater. Res. Soc. Symp. Proc. 328*:371 (1994).
87. X. C. Li, F. Cacialli, M. Giles, J. Grüner, R. H. Friend, A. B. Holmes, S. C. Moratti, and T. M. Yong, Charge transport polymers for light-emitting diodes, *Adv. Mater. 7*:898 (1995).
88. X. C. Li, A. B. Holmes, A. Kraft, S. C. Moratti, G. C. W. Spencer, F. Cacialli, J. Grüner, and R. H. Friend, Synthesis and optoelectronic properties of aromatic oxadiazole polymers, *J. Chem. Soc. Chem. Commun. 1995*:2211.
89. J. Grüner, R. H. Friend, J. Huber, and U. Scherf, A blue-luminescent ladder-type poly(para-phenylene) copolymer containing oxadiazole groups *Chem. Phys. Lett. 251*:204 (1996).
90. A. R. Brown, N. C. Greenham, J. H. Burroughes, D. D. C. Bradley, R. H. Friend, P. L. Burn, A. Kraft, and A. B. Holmes, Electroluminescence from multilayer conjugated polymer devices—spatial control of exciton formation and emission, *Chem. Phys. Lett. 200*:46 (1992).
91. R. N. Marks, D. D. C. Bradley, R. W. Jackson, P. L. Burn, and A. B. Holmes, Charge injection and transport in poly(*p*-phenylene vinylene) light-emitting-diodes *Synth. Met. 57*:4128 (1993).
92. I. D. Parker, Carrier injection and device characteristics in polymer light-emitting diodes, *J. Appl. Phys. 75*:1656 (1994).
93. N. C. Greenham and R. H. Friend, Semiconductor device physics of conjugated polymers, in *Solid State Physics* (H. Ehrenreich and F. A. Spaepen, Eds.), Academic, San Diego, Vol. 49, p. 2.
94. J. Gmeiner, S. Karg, M. Meier, W. Riess, P. Strohriegl, and M. Schwoerer, Synthesis, electrical conductivity and electroluminescence of poly(*p*-phenylene vinylene) prepared by the precursor route *Acta Polymer. 44*:201 (1993).
95. P. Dannetun, M. Lögdlund, M. Fahlman, M. Boman, S. Stafström, W. R. Salaneck, R. Lazzaroni, C. Fredriksson, J. L. Brédas, S. Graham, R. H. Friend, A. B. Holmes, R. Zamboni, and C. Taliani, The chemical and electronic structure of the interface between aluminium and conjugated polymers or molecules, *Synth. Met. 55–57*:212 (1993).
96. P. Dannetun, M. Fahlman, C. Fauquet, K. Kaerijama, Y. Sonoda, R. Lazzaroni, J. L. Brédas, and W. R. Salaneck, Interface formation between calcium and poly(2,5-diheptyl-*p*-phenylenevinylene): Implications for light emitting diodes, *Synth. Met. 67*:133 (1994).
97. M. Fahlman, D. Beljonne, M. Lögdlund, A. B. Holmes, R. H. Friend, J. L. Brédas, and W. R. Salaneck, Experimental and theoretical studies of the electronic structure of Na-doped poly(para phenylene vinylene) *Chem. Phys. Lett. 214*, 327 (1993).
98. F. Cacialli, R. H. Friend, S. C. Moratti, and A. B. Holmes, Characterization of properties of polymeric light emitting diodes over extended periods, *Synth. Met. 67*:157 (1994).
99. S. Hayashi, H. Etoh, and S. Saito, Electroluminescence of perylene films with a conducting polymer as an anode, *Jpn. J. Appl. Phys. 25*:773 (1986).
100. T. Nakano, S. Doi, T. Noguchi, T. Ohnishi, and Y. Iyechika, Organic electroluminescence device, *Euro. Patent Appl.* 91301416.3 (1991).
101. Y. Yang and A. J. Heeger, Polyaniline as a transparent electrode for polymer light-emitting diodes: lower operating voltage and higher efficiency, *Appl. Phys. Lett. 64*:1245 (1994).
102. Q. Pei, G. Yu, C. Zhang, Y. Yang, and A. J. Heeger, Polymer light-emitting electrochemical cells, *Science 269*:1086 (1995).
103. B. R. Hsieh, H. Antoniadis, M. A. Abkowitz, and M. Stolka, Charge transport properties of poly(*p*-phenylene vinylene) and its sulfonium precursor polymers at different degrees of conversion, *Polym. Prepr. 33*:414 (1992).
104. N. C. Greenham, I. D. W. Samuel, G. R. Hayes, R. T. Phillips, Y. A. R. R. Kessener, S. C. Moratti, A. B. Holmes, and R. H. Friend, Measurement of absolute photoluminescence quantum efficiencies in conjugated polymers, *Chem. Phys. Lett. 241*:89 (1995).
105. R. N. Marks, J. J. M. Halls, D. D. C. Bradley, R. H. Friend, and A. B. Holmes, The photovoltaic response in poly(*p*-phenylene vinylene) thin film devices *J. Phys. Condensed Matter 6*:1379 (1994).
106. S. Doi, T. Nakano, T. Noguchi, and T. Ohnishi, Conjugated polymer light-emitting diodes, *Polym. Prepr. Jpn. 40*:3594 (1991).
107. P. Burn, A. B. Holmes, A. Kraft, A. R. Brown, D. D. C. Bradley, and R. H. Friend, Light emitting diodes based on conjugated polymers: control of colour and efficiency, *Mater. Res. Soc. Symp. Proc. 247*:647 (1992).
108. S. C. Moratti, D. D. C. Bradley, R. Cervini, R. H. Friend, N. C. Greenham, and A. B. Holmes, Light-emitting polymer LEDs, *SPIE Proc. Ser. 2144*:108 (1994).

109. J. L. Brédas and A. J. Heeger, Influence of donor and acceptor substituents on the electronic characteristics of poly(paraphenylene vinylene) and poly(p-araphenylene), *Chem. Phys. Lett. 217*:507 (1994).
110. P. Bröms, M. Fahlman, K. Z. Xing, W. R. Salaneck, P. Dannetun, J. Cornil, D. A. dos Santos, J. L. Brédas, S. Moratti, A. B. Holmes, and R. H. Friend, Optical absorption studies of sodium doped poly(cyanoterephthalylidene), *Synth. Met. 67*:93 (1994).
111. D. R. Baigent, N. C. Greenham, J. Grüner, R. N. Marks, R. H. Friend, S. C. Moratti, and A. B. Holmes, Light-emitting diodes fabricated with conjugated polymers—recent progress, *Synth. Met. 67*:3 (1994).
112. H. Vestweber, H. Bässler, J. Grüner, and R. H. Friend, Spatial extent of the recombination zone in a PPV/polymer blend light-emitting diode, *Chem. Phys. Lett. 256*:37 (1996).
113. I. D. W. Samuel, B. Crystall, G. Rumbles, P. L. Burn, A. B. Holmes, and R. H. Friend, The efficiency and time-dependence of luminescence from poly(p-phenylene vinylene) and derivatives, *Chem. Phys. Lett. 213*:472 (1993).
114. N. F. Colaneri, D. D. C. Bradley, R. H. Friend, P. L. Burn, A. B. Holmes, and C. W. Spangler, Photoexcited states in poly(p-phenylene vinylene): comparison with *trans,trans*-distyrylbenzene, a model oligomer, *Phys. Rev. B 42*:11671 (1990).
115. L. S. Swanson, J. Shinar, A. R. Brown, D. D. C. Bradley, R. H. Friend, P. L. Burn, A. Kraft, and A. B. Holmes, Electroluminescence-detected magnetic-resonance study of polyparaphenylenevinylene (PPV)-based light-emitting diodes, *Phys. Rev. B 46*:15072 (1992).
116. X. Wei, B. C. Hess, Z. V. Vardeny, and F. Wudl, Studies of photoexcited states in polyacetylene and poly(paraphenylenevinylene) by absorption detected magnetic resonance: the case of neutral photoexcitations, *Phys. Rev. Lett. 68*:666 (1992).
117. A. R. Brown, K. Pichler, N. C. Greenham, D. D. C. Bradley, R. H. Friend, and A. B. Holmes, Optical spectroscopy of triplet excitons and charged excitations in poly(p-phenylenevinylene) light-emitting diodes, *Chem. Phys. Lett. 210*:61 (1993).
118. L. Smilowitz, A. Hays, A. J. Heeger, G. Wang, and J. E. Bowers, Time-resolved photoluminescence from poly(2-methoxy,5-(2'-ethyl-hexyloxy)-p-phenylenevinylene): solutions, gels, films, and blends, *J. Chem. Phys. 98*:6504 (1993).
119. C. L. Gettinger, A. J. Heeger, J. M. Drake, and D. J. Pine, A photoluminescence study of poly(phenylene vinylene) derivatives: the effect of intrinsic persistence length, *J. Chem. Phys. 101*:1673 (1994).
120. D. Braun, E. G. J. Staring, R. C. J. E. Demandt, G. L. J. Rikken, Y. A. R. R. Kessener, and A. H. J. Venhuizen, Photoluminescence and electroluminescence efficiency in poly(dialkoxy-p-phenylenevinylene), *Synth. Met. 66*:75 (1994).
121. J. W. P. Hsu, M. Yan, T. M. Jedju, L. J. Rothberg, and B. R. Hsieh., Assignment of the picosecond photoinduced absorption in phenylene vinylene polymers, *Phys. Rev. B 49*:712 (1994).
122. P. L. Burn, A. Kraft, D. R. Baigent, D. D. C. Bradley, A. R. Brown, R. H. Friend, R. W. Gymer, A. B. Holmes, and R. W. Jackson, Chemical tuning of the electronic properties of poly(p-phenylenevinylene)-based copolymers, *J. Am. Chem. Soc. 115*:10117 (1993).
123. U. Lemmer, S. Heun, R. F. Mahrt, U. Scherf, M. Hopmeier, U. Siegner, E. O. Göbel, K. Müllen, and H. Bässler, Aggregate fluorescence in conjugated polymers, *Chem. Phys. Lett. 240*:373 (1995).
124. M. Yan, L. J. Rothberg, F. Papadimitrakopolous, M. E. Galvin, and T. M. Miller, Spatially indirect excitons as primary photoexcitations in conjugated polymers, *Phys. Rev. Lett. 72*:1104 (1994).
125. M. Yan, L. J. Rothberg, F. Papadimitrakopoulos, M. E. Galvin, and T. M. Miller, Defect quenching of conjugated polymer luminescence, *Phys. Rev. Lett. 73*:744 (1994).
126. D. R. Baigent, A. B. Holmes, S. C. Moratti, and R. H. Friend, Electroluminescence in conjugated polymers: excited states in cyano-substituted derivatives of poly(p-phenylenevinylene), *Synth. Met. 80*:119 (1996).
127. I. D. W. Samuel, G. Rumbles, and C. J. Collison, Efficient inter-chain photoluminescence in a high electron affinity conjugated polymer, *Phys. Rev. B 52*:11573 (1995).
128. K. H. Drexhage, Interaction of light with monomolecular dye layers, *Prog. Opt. 12*:163 (1974).
129. G. Bjork, On the spontaneous lifetime change in an ideal planar microcavity—transition from a mode continuum to quantized modes, *IEEE J. Quant. Electron. 30*:2314 (1994).
130. A. Dodabalapur, L. J. Rothberg, T. M. Miller, and E. W. Kwock, Microcavity effects in organic semiconductors, *Appl. Phys. Lett. 64*:2486 (1994).
131. A. Dodabalapur, L. J. Rothberg, and T. M. Miller, Electroluminescence from organic semiconductors in patterned microcavities, *Electron. Lett. 30*:1000 (1994).
132. N. Takada, T. Tsutsui, and S. Saito, Strongly directed emission from controlled-spontaneous-emission electroluminescent diodes with europium complex as an emitter, *Jpn. J. Appl. Phys. 33*:L863 (1994).
133. T. Tsutsui, N. Takada, S. Saito, and E. Ogino, Sharply directed emission in organic electroluminescent diodes with an optical-microcavity structure, *Appl. Phys. Lett. 65*:1868 (1994).
134. A. Peres, *Quantum Theory: Concepts and Methods*, Kluwer, Dordrecht, 1993.
135. U. Lemmer, R. Hennig, W. Guss, A. Ochse, J. Pommerehne, R. Sander, A. Greiner, R. F. Mahrt, H. Bässler, J. Feldmann, and E. O. Göbel, Microcavity effects in a spin-coated polymer 2-layer system, *Appl. Phys. Lett. 66*:1301 (1995).

136. H. F. Wittmann, J. Grüner, R. H. Friend, G. W. C. Spencer, S. C. Moratti, and A. B. Holmes, Microcavity effect in a single-layer polymer light-emitting diode, *Adv. Mater. 7*:541 (1995).
137. T. A. Fisher, D. G. Lidzey, M. A. Pate, M. S. Weaver, D. M. Whittaker, M. S. Skolnick, and D. D. C. Bradley, Electroluminescence from a conjugated polymer microcavity structure, *Appl. Phys. Lett. 67*: 1355 (1995).
138. J. Grüner, F. Cacialli, and R. H. Friend, Emission enhancement in single-layered conjugated polymer microcavities, *J. Appl. Phys. 80*:207 (1996).
139. C. W. Tang, Two-layer organic photovoltaic cell, *Appl. Phys. Lett. 48*:183 (1986).
140. N. Karl, A. Bauer, J. Holzäpfel, J. Marktanner, M. Möbius, and F. Stölzle, Efficient organic photovoltaic cells, *Mol. Cryst. Liq. Cryst. 252*:243 (1994).
141. B. O'Regan and M. Grätzel, A low-cost, high-efficiency solar cell based on dye-sensitized colloidal TiO_2 films, *Nature 353*:737 (1991).
142. S. Karg, W. Riess, M. Meier, and M. Schwoerer, Electrical and optical characterization of light-emitting poly(phenylenevinylene) diodes, *Mol. Cryst. Lig. Cryst. A 234*:619 (1993).
143. H. Antoniadis, L. J. Rothberg, F. Papadimitrakopoulos, M. Yan, M. E. Galvin, and M. A. Abkowitz, Enhanced carrier photogeneration by defects in conjugated polymers and its mechanism, *Phys. Rev. B 50*:14911 (1994).
144. G. Yu, K. Pakbaz, and A. J. Heeger, Semiconducting polymer diodes: large size, low cost photodetectors with excellent visible-ultraviolet sensitivity, *Appl. Phys. Lett. 64*:1 (1994).
145. X. Wei, M. Raikh, Z. V. Vardeny, Y. Yang, and D. Moses, Photoresponse of poly(para-phenylenevinylene) light-emitting diodes, *Phys. Rev. B 49*:17480 (1994).
146. N. S. Sariciftci, L. Smilowitz, A. J. Heeger, and F. Wudl, Photoinduced electron transfer from a conducting polymer to buckminsterfullerene, *Science 258*:1474 (1992).
147. N. S. Sariciftci, L. Smolowitz, A. J. Heeger, and F. Wudl, Semiconducting polymers (as donors) and buckminsterfullerene (as acceptor): photoinduced electron transfer and heterojunction devices, *Synth. Met. 59*:333 (1993).
148. B. Kraabel, C. H. Lee, D. McBranch, D. Moses, and N. S. Sariciftci, Ultrafast photoinduced electron-transfer in conducting polymer buckminsterfullerene composites, *Chem. Phys. Lett. 213*:389 (1993).
149. N. S. Sariciftci, D. Braun, C. Zhang, V. I. Srdanov, A. J. Heeger, G. Stucky, and F. Wudl, Semiconducting polymer-buckminsterfullerene heterojunctions: diodes, photodiodes and photovoltaic cells, *Appl. Phys. Lett. 62*:585 (1993).
150. J. J. M. Halls, K. Pichler, R. H. Friend, S. C. Moratti, and A. B. Holmes, Exciton diffusion and dissociation in a poly(*p*-phenylenevinylene)/C_{60} heterojunction photovoltair cell, *Appl. Phys. Lett. 68*:3120 (1996).
151. J. J. M. Halls, C. A. Walsh, N. C. Greenham, E. A. Marseglia, R. H. Friend, S. C. Moratti, and A. B. Holmes, Efficient photodiodes from interpenetrating polymer networks, *Nature 376*:498 (1995).
152. G. Yu and A. J. Heeger, Charge separation and photovoltaic conversion in polymer composites with internal donor/acceptor heterojunctions, *J. Appl. Phys. 78*:4510 (1995).
153. G. Yu, J. Gao, J. C. Hummelen, F. Wudl, and A. J. Heeger, Polymer photovoltaic cells: enhanced efficiencies via a network of internal donor-acceptor heterojunctions, *Science 270*:1789 (1995).
154. G. Yu, C. Zhang, and A. J. Heeger, Dual-function semiconducting polymer devices: light-emitting and photodetecting diodes, *Appl. Phys. Lett. 64*:1540 (1994).
155. P. May, Polymer electronics—fact or fantasy, *Phys. World*, p. 52, March 1995.

30

Fundamentals of Electroluminescence in Paraphenylene-Type Conjugated Polymers and Oligomers

Günther Leising, Stefan Tasch, and Willhelm Graupner
Technische Universität Graz, Graz, Austria

I. INTRODUCTION

Electroluminescence is the electromagnetic radiation emitted from the solid when a thin film of a suitable solid is placed between two electrodes and an electric field is applied that drives a current through the solid. Thus electroluminescence results from the deexcitation of a solid that has been excited by the electric current passing through it. This effect is in contrast to thermal radiation, where the emitted electromagnetic radiation is related to the temperature of the sample, which is generated in every solid by the Joule heating of a passing current. In the case of electroluminescence there is in principle no limitation on the wavelength of the emitted light, so the electroluminescence emission could be in any wavelength range from the ultraviolet to the infrared.

The effect of electroluminescence has been known for a long time in inorganic semiconductors [1] as well as in organic materials [2]. It was discovered in 1907 by T. Round when he observed yellow light emission from a silicon carbide device [3] that had actually been built as a detector for radio waves. Inorganic light-emitting devices [light-emitting diodes (LEDs)] have advanced from laboratory experimental objects to industrial products with a variety of applications in a very short time. A good overview of inorganic LEDs is given in Ref. 4. In the field of conducting polymers there was research on polymeric devices from the very beginning, but it was mainly dedicated to electronic devices (diodes, field-effect transistors, etc., and some work on solar cells) [5]. In 1990 our colleagues at Cambridge University demonstrated green-yellow electroluminescence emission from the conjugated polymer polyparaphenylenevinylene (PPV) [6] during experiments on polymeric sandwich structures [7]. The emission performance of PPV electroluminescence devices has been improved [8], and the discovery of the important blue emission was reported for polyalkylfluorene [9] and for polyparaphenylene [10]. In the meantime a number of groups are working worldwide on this interesting subject. As indicated in the title of this chapter, our primary topic here is paraphenylene-based conjugated polymers and oligomers as the active material, but at some points we also cover organic molecular materials, which are used mostly as transport or buffer layers but are also used as active species as side chains in polymers, as organic thin films, and dissolved in a solid polymer matrix.

The aim of this chapter is to discuss the principles of organic electroluminescence devices based mainly on conjugated polymers and oligomers of the paraphenylene type, their function, and the related processes. In Section II we introduce the paraphenylene-type materials such as the stable oligomer parahexaphenyl, the polyparaphenylene polymers, the planar ladder-type polyparaphenylene, and copolymers of substituted polyparaphenylene and ladder-type polyparaphenylene, which are believed to represent the polymeric analogs to the quantum well structures in inorganic semiconductors. In Section III we learn about the principal possibilities for building electroluminescence devices. In Section IV we discuss how carriers, electrons, and holes are injected into the solid, and Section V is dedicated to the transport phenomena of the injected carriers within the device. How the excited states are formed from the injected carriers is discussed in Section VI, and how the emission color can be influenced or even tuned is outlined in Section VII. We pay special attention to a new technique for obtaining the basic colors red, green, and blue—color conversion using intense blue-light-emitting devices and producing other visible colors via photoluminescence of special dye and filter layers. The electroluminescence efficiency is a technically very important parameter and is discussed in detail in Section VIII along with most of the significant parameters, which can reduce (quench) the photoluminescence as well as the electroluminescence efficiency. The final section, Section IX, is dedicated to some remarks on the stability and lifetimes of the existing electroluminescence devices and is kept relatively general because it is here that the most intense industrial research is currently going on.

II. PARAPHENYLENE-TYPE POLYMERS AND OLIGOMERS

Paraphenylene-type polymers and oligomers can be described as one-dimensional semiconductors in an idealized picture. This is due to the fact that the electronic properties in the visible and ultraviolet are mainly influenced by the π electrons of the carbon atoms and the overlap of their wave functions. This overlap is called conjugation because it leads to a sequence of alternating double and single bonds. Both the strength and the spatial extent of this overlap are determined by (1) the chemical constitution of the molecules, (2) their length, and (3) defects that disturb the perfect arrangement and therefore break conjugation. Examples of all three effects are given in the following sections.

The sequence of phenyl rings leads to a rather weak conjugation; i.e., the overlap of the π-electron wave functions is weaker than, for example, in polyacetylene, which is the simplest model system for conjugated polymers. Therefore paraphenylene-type polymers and oligomers show an energetic difference between the highest occupied molecular orbital (HOMO) and the lowest unoccupied molecular orbital (LUMO) of about 3 eV. This value of 3 eV depends on the length and chemical constitution of the polymer or oligomer and may vary by a few tenths of an electronvolt, compared to only 1.4 eV in polyacetylene. The high HOMO–LUMO separation makes paraphenylene-type polymers and oligomers appear yellow or transparent in the visible spectral range; the absorbance spectrum is plotted in Fig. 30.1. Consequently their photoluminescence emission is in the blue range, as also shown in Fig. 30.1.

There are two additional facts that distinguish paraphenylene-type polymers and oligomers from polyacetylene. In polyacetylene the order of double bonds does not change the total energy of the polymer chain; therefore polyacetylene is called a degenerate ground-state system. In polyphenyls, however, the lowest total energy is reached by single bonds between neighboring phenyl rings (aromatic configuration). A double bond between two phenyl rings renders a quinoidal system with a higher total energy; therefore the aromatic configuration is preferred, and the material belongs to the

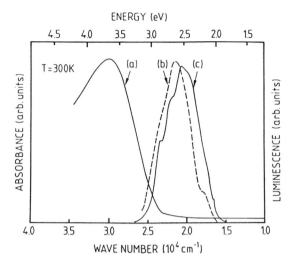

Fig. 30.1 (a) Absorption, (b) electroluminescence, and (c) photoluminescence emission of precursor PPP at 300 K.

class of nondegenerate ground-state systems. The second distinct difference between polyacetylene and polyparaphenylene lies in the order of their excited states. As demonstrated by Soos et al. [11], the energetic order of excited states determines whether or not a conjugated polymer shows strong photoluminescence. Polyparaphenylenes show a lowest singlet excitation, which is dipole-allowed and gives rise to strong photoluminescence, in contrast to polyacetylene, where the lowest excitation is two-photon-allowed but dipole-forbidden and no photoluminescence is observed.

Several polymer systems have been identified that emit blue light. They include polyparaphenylenes (PPPs) [10,12–14] derivatives of polyparaphenylenevinylenes (PPVs) [15], polyalkylfluorenes [9], polythiophenes [16], and others. Among these materials PPP and its derivatives show a high photoluminescence efficiency [17,18], which made oligophenyls interesting as laser dyes several decades ago [19]. Figure 30.1 depicts the electroluminescence emission spectrum of the first published PPP LED device [10] together with the photoluminescence emission and the optical absorption spectrum of the PPP used in this LED device. Table 30.1 summarizes the photoluminescence quantum yields obtained for different PPPs and oligophenyls. These materials are briefly discussed in detail in the following sections.

A. Synthesis of Paraphenylene-Type Polymers and Oligomers

In this section we briefly describe how the different synthetic routes influence the physical properties of the products. For a more complete review see Chapters 8

Table 30.1 Photoluminescence Quantum Yields of Oligo- and Polyparaphenyls in the Solid State and in Solution

Material[a]	PLQY (%)		Reference
	Film	Solution	
para-Terphenyl (P3)	—	98	19
para-Quaterphenyl (P4)	—	89	19
para-Hexaphenyl (P6)	30		17
DO PPP	35	85	18
EHO PPP	40	85	18
CN PPP	46	85	18
s-LPPP	10	72	17
l-LPPP	9	61	17
m-LPPP	30	≈100	20
CPLPPP	21	84	17

[a] DO, EHO, and CN stand for decyloxy, ethylhexyloxy, and cyano-methyl-heptyloxy [16]; s- and l-LPPP exhibit degrees of polymerization of 20 and 50 phenyl rings and have a hydrogen atom as substituent at the Y position in Fig. 30.2, while m-LPPP has a degree of polymerization of 50 phenyl rings and a methyl group at the Y position.

and 14 by A. D. Schlüter and U. Scherf, respectively. The term paraphenylene-type polymers or oligomers refers to strands of phenyl rings linked via *para* positions. In Fig. 30.2 we show several members of this species. The simplest representative, Ph_n (see Fig. 30.2a), is very hard to synthesize. First attempts to synthesize PPP yielded oligomers with approximately 10–15 phenylene units in each chain [21–24]. Polymers of this type obtained by the traditional routes have serious disadvantages. Oxidative coupling of benzene [25] leads to branched, infusible, and insoluble low molecular weight products. The synthesis of PPP via reductive conversion of poly(cylohexa-1,3-diene) [26] is hampered by the polymer-analogous dehydrogenation step. These materials exhibited conductivities up to 200 S/cm upon oxidation by AsF_5 [27], which stimulated attempts to improve molecular weight, purity, and processability.

Processibility can be obtained via precursor polymer routes to PPP that rely on a processible precursor, which is converted into PPP. The ability to bacterially oxidize benzene, using the microorganism *Pseudomonas putida*, to form 5,6-*cis*-dihydroxycyclohexa-1,3-diene led researchers at ICI to polymerize it as a PPP precursor, which yields PPP after pyrolysis. High molecular weights and processibility as well as a very compact morphology are observed for the PPPs [28]. Also one of the first blue LEDs based on a conjugated polymer was built using these materials [10]. However, after thermal conversion to PPP the films are completely intractable [29] and contain *ortho*-substituted phenylene segments, i.e., 1,2 defects. To suppress the formation of these defects, a synthetic route that yields highly stereoregular PPP was introduced by Grubbs and coworkers [30]. Dialkyl-substituted PPPs overcome the problem of insolubility [31–33]. Substituted PPPs of this type can be prepared via coupling reactions according to Yamamoto and Yamamoto [34]. However, such a reaction leads to significant torsion angles. Torsion about inter-ring bonds does, however, substantially reduce conjugation [35] and leads to an increase of the HOMO–LUMO distance [36–38].

One way to obtain molecular systems with well-defined configuration and conformation is to incorporate the backbone of interest (PPP) into a planar, rigid polyaromatic ribbon molecule [39]. This strategy was exploited for PPPs by Scherf et al. (see Chapter 14) via a bridging of functionalized PPP precursors [40]. Figure 30.2c shows the chemical structure of the LPPP molecule. In addition to their high structural regularity these materials also exhibit solubility, which is achieved by hexyl and decyl side groups. Joining planarized PPP segments with PPP segments that allow ring torsion to occur (Fig. 30.2d, CPLPPP) results in a sequence of alternating higher and lower π-electron overlaps. This leads to a local variation of the energy gap, and therefore such a molecule resembles a quantum well structure [41] and also shows its typical properties such as enhanced photoluminescence quantum efficiency [42].

B. Electronic Properties

The optical properties of materials are determined by the so-called dielectric function. This dielectric function was determined for PPP as a result of first principles band structure calculations. In Fig. 30.3 we depict one of the main results, namely the dependence of the imaginary part of the dielectric function (which is proportional to the optical absorption coefficient) on the orientation parallel (ϵ_c) and perpendicular (ϵ_a, ϵ_b) to the chain axis. From comparison with the experiment one can see that the optical absorption in the visible and ultraviolet range is mainly determined by the dielectric function parallel to the polymer chain. This is shown in Fig. 30.4, where the calculated absorption coefficient along the chain is compared with experimental data. The observed π–π^* transition is therefore the result of optical excitation with the dipole moment parallel to the PPP chains. Another experimental proof for this are the absorbance properties of hexaphenyl single crystals [43]. The absorbance spectra for thin films with the hexaphenyl chains oriented perpendicular and parallel to the substrate plane are shown in Fig. 30.5 together with the photoluminescence (PL) and electroluminescence (EL) emission spectra. The above-described anisotropy of the optical absorption is also valid for photoluminescence emission and can be used to obtain polarized EL spectra from Langmuir–Blodgett films made from soluble PPPs [44]

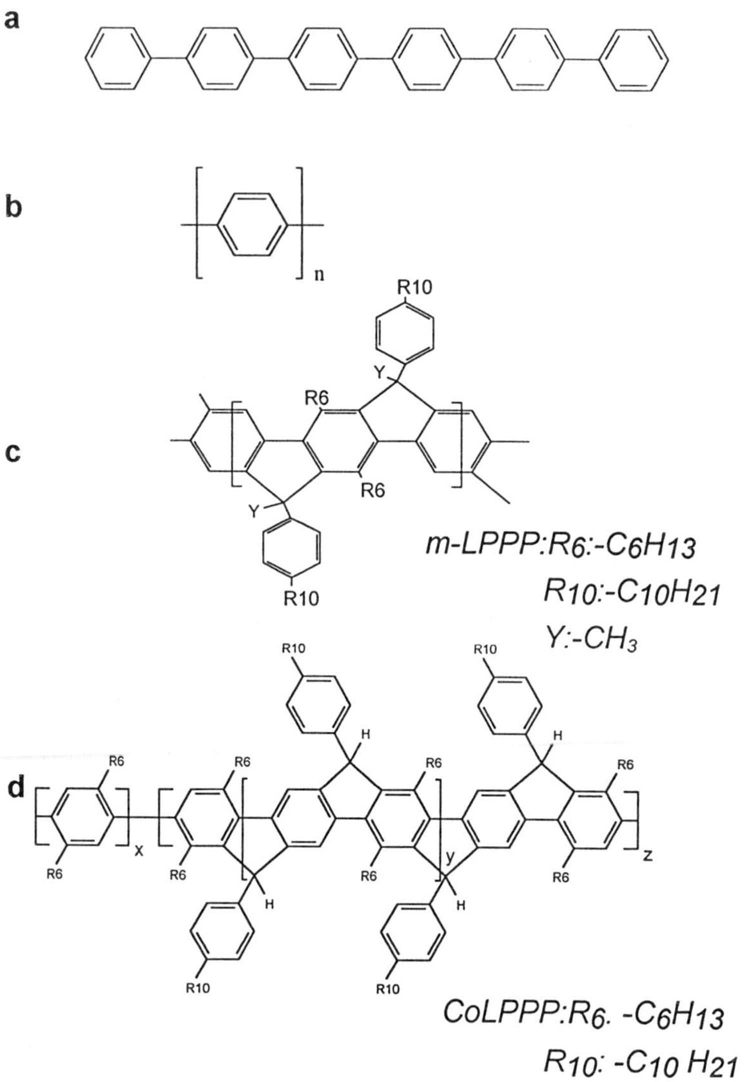

Fig. 30.2 Chemical composition of various paraphenylene systems. (a) Parahexaphenyl (PHP); (b) polyparaphenylene (PPP); (c) paraphenylene-type ladder polymer (LPPP); and (d) a copolymer of PPP and LPPP (CoLPPP). The Y substituent of the material LPPPs is either a hydrogen atom or a methyl group (m-LPPP).

or from oriented hexaphenyl films [45]. In a simple model [46,47] that describes the π electrons of conjugated polymers as particles in a box, the energetic HOMO–LUMO distance shows a linear dependence on the inverse chain length of the molecule. Since the HOMO–LUMO distance is probed by optical absorption spectroscopy, this model can be applied to the absorbance spectra of oligophenyls, and it leads to the relation [48,49]

$$E_a = \left[3.36 + \frac{3.16}{n}\right] \text{eV} \qquad (1)$$

where E_a is the energy of the peak of the absorbance and n is the number of phenyl rings along the chain. The electronic properties of conjugated polymers depend significantly on the structural regularity of the materials. Since it is known that both the type of substitution and the planarity of the molecules influence the photoluminescence quantum yield of PPPs (see Table 30.1), chemical defects like branching, cross-linking, or *meta* and *ortho* substitutions and the degree of torsion between neighboring phenyl rings are important parameters [50,51]. Since all these defects reduce the overlap of the π-electron wave functions, they also reduce the degree of conjugation. Therefore PPP molecules of a degree of polymerization as high as 100 phenyl rings can show

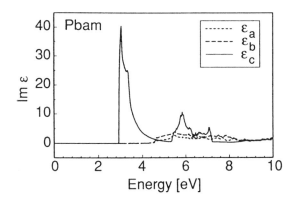

Fig. 30.3 Calculated imaginary part of the dielectric tensor in crystalline PPP in the *Pbam* phase. The ϵ_c is parallel to the chain axis; ϵ_a and ϵ_b are perpendicular to the chain axis.

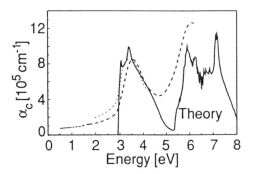

Fig. 30.4 Calculated absorption coefficient of crystalline PPP along the chain compared to the experimental data of Refs. 36 (dotted line) and 37 (dashed line).

optical properties equivalent to those of short oligomers. The inversion of Eq. (1) can be used to determine the effective conjugation length n_{eff} via the absorbance peak. Typical values for n_{eff} are 9–10 phenyl rings for PPP [52].

To illustrate the influence of the chemical composition of the PPP chains, we compare the results obtained for simple PPPs (Fig. 30.2b) to the LPPPs shown in Fig. 30.2c. To obtain the equation equivalent to Eq. (1) for the planar LPPPs, one has to use data for planarized oligomers [53], which yield

$$E_a = \left[2.56 + \frac{3.11}{n}\right] \text{eV} \quad (2)$$

The effect of planarization is to red shift the absorbance peak by about 0.8 eV (2.56 instead of 3.36 eV), but the dependence of the peak on chain length is only slightly different (3.11 eV/ring vs. 3.16 eV/ring as the slope of the regression curves). Again, typical values for the effective conjugation length for planar ladder-type PPPs are around 10 phenyl rings [53,54].

From crystallographic data on oligomers of PPP in the solid-state, PPPs with only one single bond between adjacent phenyl rings (i.e., non-ladder-type) are known

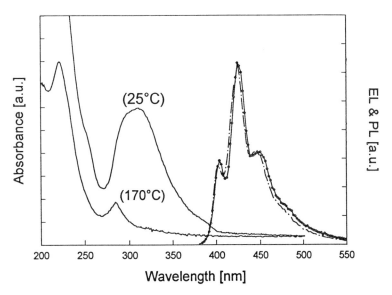

Fig. 30.5 Photoluminescence and electroluminescence emission spectra of hexaphenyl. Absorbance of vacuum-deposited hexaphenyl oriented perpendicular to the substrate (deposition temperature 170°C) and randomly oriented (deposition temperature 25°C).

to exhibit torsion angles of around 23° [35]. Since the energy gap of PPPs strongly depends on this torsional angle [36–38] the red shift of the absorbance in planar LPPPs with respect to PPP is to be expected. Examining the absorbance spectra of PPPs and LPPPs in the UV-Vis region yields a well-resolved vibrational structure in the absorption spectra of LPPP and CPLPPP as shown in Fig. 30.6, in contrast to the unstructured absorbance of PPP (see Fig. 30.1) and hexaphenyl (see Fig. 30.5). The energetic spacing of this structure corresponds to the stretching vibrations of the conjugated PPP backbone. The appearance of this structure in the LPPPs and CPLPPP originates in the suppression of the rotational degree of freedom between neighboring phenyl rings by the methine bridge. This rotational degree of freedom smears out the vibrational splitting for nonplanar PPPs [55]. Therefore the apparent shift of the absorption peak from PPP to LPPP of about 0.8 eV is partly due to a more dominant 0–0 transition in the LPPP.

However, besides planarizing the polymer backbone the nature of the synthetic route to the LPPPs induces many fewer defects in the polymer than other routes (see Scherf, this volume). As a result the LPPPs show a high degree of intrachain order. One signature of this high intrachain order is the dominance of the principal electronic transition in the optical absorption spectrum. This transition involves no phonons and is denoted as the 0–0 transition. This effect was also observed in improved PPV [56,57] as well as in highly ordered PPV in polyethylene. The high intrachain order in LPPPs is also revealed by very narrow triplet–triplet absorbance [58] and the shape of the photoinduced absorption spectrum in the infrared region [54]. However, even within the group of LPPPs the degree of intrachain order can be different depending on the degree of polymerization and the nature of the substituent attached to the conjugated backbone [54].

C. Processing

Considering processability, the goal for organic light-emitting devices is to achieve thin, large homogeneous films. Short unsubstituted oligomers are the materials of choice if single crystals or single crystalline films are to be produced. This is usually done via vapor-phase deposition, where processing temperatures in this case are significantly below those of inorganic technologies. A proper choice of the deposition parameters such as substrate material, source, and substrate temperature allows control of the crystalline arrangement of the oligomers and the optical properties of these films [43]. The obtained oligomer films show the highest mobility values in the field of organic devices [59–61] of about 0.01 cm^2/(V.s).

Substituted PPPs or PPP precursors are solution-processible and can therefore be used to obtain thin layers by drop casting, where droplets of the polymer solution are cast onto a substrate and the polymer film is formed during evaporation of the solvent. To improve the film quality and homogeneity, techniques like spin casting, dip casting, or a doctor blade process are used. A more sophisticated technique was applied to substituted PPPs with alkoxy side groups: Langmuir–Blodgett films were obtained with these polymers, which consist of up to 100 monolayers and give rise to highly anisotropic optical properties and polarized electroluminescence [44].

III. ELECTROLUMINESCENCE DEVICES

Electroluminescence (EL) devices based on conjugated polymers can be realized in several different configurations. The processes, which are the basics of EL devices, are the following:

1. Charge carriers of opposite sign are introduced into the active layer.
2. Positive and negative charge carriers move in the active layer under the applied voltage, i.e., electric field.
3. Charge carriers of different sign interact with each other to form excited species. One decay channel of these excited species into the ground states is dipole-allowed (see Section VI), so that a radiative transition occurs with emission of light.

The simplest light-emitting devices with conjugated polymers (PLEDs) are realized in such a manner that the polymer is arranged between two metal electrodes (see Figs. 30.7 and 30.8) [6,10]. The mechanism of the charge carrier injection into the polymer is discussed later in this section and in Section IV, whereas the

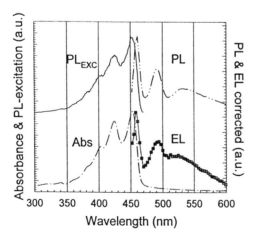

Fig. 30.6 Photoluminescence excitation (PL$_{EXC}$) and emission spectra, absorbance and electroluminescence emission spectra of m-LPPP.

Electroluminescence

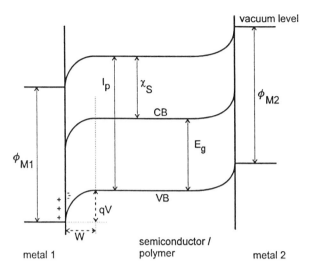

Fig. 30.7 Energy diagram of a metal/semiconductor/metal Schottky barrier (Φ, workfunction; χ_S, electron affinity; IP, ionization potential; E_g, band gap; W, depletion width).

charge transport of conjugated polymers used in PLEDs is the main topic of Section V.

Depending on the charge carrier concentration (intrinsic or extrinsic) of the conjugated polymer, this arrangement can be better described as a metal/semiconductor/metal (MSM) or metal/insulator/metal (MIM) device [5,6]. For arrangements of polymers with a high charge carrier concentration ($n > 10^{17}$ cm^{-3}), which can be caused by the chemical synthesis, the preparation technique, or doping, band bendings occur, which is characteristic of MSM structures [4,5]. The conjugated polymer that is applied in PLEDs should have a high photoluminescence (PL) quantum efficiency, as the PL quantum efficiency is directly proportional to the EL quantum efficiency (see Section VIII). A high impurity concentration in the polymer is a drawback for attaining a high PL efficiency, as impurities act as centers of nonradiative recombination for the excited states. For that reason it is very important to use pure polymers, which have a low defect and/or trap concentration. Consequently these polymers have a very low carrier concentration n (the intrinsic n for the m-LPPP is less than one carrier per cubic centimeter; see Section V), so in this case rigid band conditions—typical of MIM structures—are observed [6–8,62]. Rigid band conditions means that the bands are not bent at the interface due to space charge regions (see below) and the electric field drops continuously over the polymer layer thickness (see Fig. 30.9).

When additional charge transport layers are arranged between the polymer and the electrodes, the EL quantum efficiency (which is the number of emitted photons per injected electron) and the stability of the PLEDs can be significantly improved (see below) [7,62].

The other way to produce EL devices is based on a *pn* junction [63,64]. A *pn* junction with conjugated polymers was realized by electrochemical doping of the polymer in an electrochemical cell [65]. The working principle and the performance of such a light-emitting electrochemical cell (LEC) is described at the end of this section.

A. EL Devices from Conjugated Polymers with a High Defect Concentration

A conjugated polymer of a high charge carrier concentration ($n > 10^{17}$ cm^{-3}) arranged between two metal electrodes can be understood as an MSM structure [5]. A semiconductor can be described as a material with a Fermi energy (highest occupied level) located within the energy gap region for any temperature. If a semiconductor is brought into electrical contact with a metal, either an ohmic or a rectifying Schottky contact is formed at the interface. The nature of the contact is determined by the work function (Φ, which is the energetic difference between the Fermi level and the vacuum level) of the semiconductor relative to the metal [63,64]. As the hole conductivity in a conjugated polymer is predominant, conjugated polymers of a high defect concentration can be approximately described as p-type semiconductors. A rectifying contact between a metal and a p-type semiconductor is formed if $\Phi_{\text{semiconductor}} > \Phi_{\text{metal}}$. When the two materials are electrically connected, electrons flow from the metal into the semiconductor until the Fermi levels of the two materials are equal, so that between the metal and the semiconductor a built-in potential ($V_{\text{bi}} = \Phi_{\text{semiconductor}} - \Phi_{\text{metal}}$) is created. The electrons, which are the minority carriers in a p-type semiconductor, recombine with the holes in the semiconductor, so that a negative space-charge region is created in the semiconductor at the interface with the metal. When an

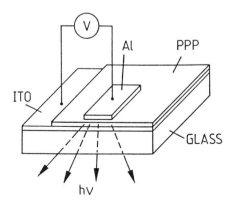

Fig. 30.8 Scheme of a polymer LED based on a indium-tin oxide (ITO)–PPP–aluminum design.

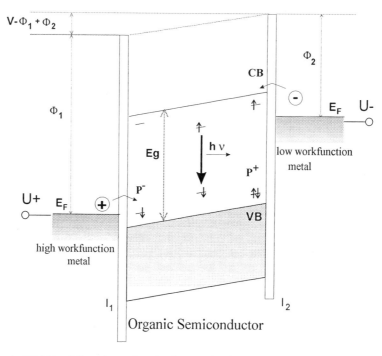

Fig. 30.9 The EL process in PLEDs. VB, valence band; CB, conducting band; V, potential; U, bias voltage; I_1, I_2, interface layers; E_g, band gap; P^+, P^-, positive and negative polarons.

external bias voltage is applied, the voltage drops over this high resistance region. This depletion width of free charge carriers is reduced when the semiconductor is made positive with respect to the metal, whereas the depletion width increases when the semiconductor is negative relative to the metal. Thus this contact is a rectifying metal–semiconductor contact (see Fig. 30.7).

When a metal and a p-type semiconductor where $\Phi_{\text{semiconductor}} < \Phi_{\text{metal}}$ are electrically connected, electrons flow from the semiconductor into the metal until the Fermi levels of the two materials are again equal. In this case no charge depletion region is formed, but the number of free positive charge carriers in the semiconductor is even increased. This contact is ohmic, and the resistance of this junction is independent of the applied voltage.

In a light-emitting MSM structure the two metal electrodes are chosen in such a way that the work functions of the electrodes are near the edge of the valence band (VB) and the conducting band (CB) of the semiconductor, respectively, so that oppositely charged carriers are injected from opposite electrodes. An ohmic contact and a rectifying contact are therefore formed in the MSM structure (see Fig. 30.7).

The charge carrier depletion width W at the rectifying contact, which forms a Schottky barrier, can be calculated using the equation [63,64]

$$W = \left[\frac{2\epsilon\epsilon_0}{qN_D}\left(\frac{\Phi_S - \Phi_M}{q} - V - \frac{kt}{q}\right)\right]^{1/2} \quad (3)$$

where ϵ is the dielectric constant, Φ_S and Φ_M are the work functions of the semiconductor and the metal, N_D is the charge density in the depletion width, T the absolute temperature, and V the potential at the junction.

For a junction of a conjugated polymer that has an energy gap of around 3 eV (around the value of the polyphenylenes) and a carrier concentration $n > 10^{17}$ cm^{-3} with a low work function metal ($\Phi_M = 2.7$ eV, the value of Ca), a depletion width of <30 nm is obtained. When Eq. (3) is used to calculate the depletion width for similar junctions with conjugated polymers with a low carrier concentration, such as pure soluble PPPs and LPPPs, values for W are obtained that are far greater than 100 nm under operating conditions. However, the thickness of the active polymer layer is usually of the same magnitude, so W is extended over the entire active polymer layer. For that reason the MSM model is not applicable for contacts of a metal with a conjugated polymer of a low intrinsic charge carrier concentration. In fact, polymers with a low charge carrier concentration sandwiched between two metal electrodes behave more like MIM structures.

B. EL Devices from Conjugated Polymers with a Low Defect Concentration

1. Electroluminescence in Single-Layer Structures

A conjugated polymer with a low defect and trap concentration ($n < 10^{17}$ cm^{-3}) that is arranged between two electrodes can be well described as an MIM structure [8,62]. In this MIM arrangement, rigid band conditions exist within the conjugated polymer (Fig. 30.9), in contrast with MSM structures, where the bands are bent at the polymer/metal interface (Fig. 30.7). To obtain optimum device performance the metal electrodes are chosen in such a way that the energetic difference between the band edges of the polymer and the work functions of the electrodes are minimal. Several difficulties are encountered in realizing this setup:

1. The electron affinity (χ_s), which is defined as the energetic difference between the lower edge of the CB and the vacuum level, of the conjugated polymers used in PLEDs is usually below 3 eV, so that metals with a very low work function are favored. Unfortunately, these metals (e.g., alkaline metals and alkaline earth metals) are known to be rather reactive and unstable under ambient conditions. Even if they can be applied in PLEDs under an inert atmosphere, chemical reactions between the polymer and the electrodes cannot be avoided [66]. Therefore, more stable metals with a higher work function (between 3.5 and 4.5 eV) are usually used. The most common metals in PLEDs used as "low" work function electrodes are Al, Ag, Mg, In, and several alloys of these materials.

2. The other electrode should have a rather high work function, as the VB edge of the most common conjugated polymers used in EL devices is in the range 5–5.9 eV, so that the use of noble metals should be most suitable. However, one of the electrodes in this sandwich configuration should be highly transparent so that the EL light generated within the polymer layer is negligibly attenuated when emitted from the active layer in the device. For that reason, indium-tin oxide (ITO) or doped Zn oxides or In oxides, which are highly transparent and also highly conducting degenerate semiconductors with a work function of up to ≈5.1 eV are usually applied as the "high" work function contact. Very thin layers of doped conjugated polymers with metallic conductivity can also be applied for the positive electrode. With doped polyaniline (emeraldine salt) as a high work function contact, levels of performance similar to those of devices using ITO were achieved but with considerably better device performance [67,68]. One advantage is the mechanical flexibility of these EL devices [67], which can be achieved by casting the conjugated polymer electrode onto a polyethylene (PET) substrate instead of a glass substrate.

The EL emission process in the MIM structures can be described as follows (see Fig. 30.9). When an electric field is applied in the forward direction (the high work function contact is positive), positive charge carriers are injected into the layer from the high work function contact and electrons are injected from the low work function contact (see Section IV). These carriers move as polarons (see Section V) under the force of the electric field and meet each other within the layer. Pairs of positive and negative charge carriers can form weakly or strongly bound states depending on the nature of the electronic structure of the active layer. These bound neutral excited states can either be singlet (spin 0) or triplet (spin ±1) states, which determines if the decay of these states into the ground state is dipole-allowed (radiative decay results in emission of light) or dipole-forbidden (nonradiative decay results in energy dissipation as heat). The dipole-allowed radiative recombination of these excited states finally causes the electroluminescence. As depicted in Fig. 30.9, the applied electric field E drops linearly through the polymer layer. This internal electric field can be calculated from the applied bias voltage U as

$$E_{\text{int}}^{\text{calc}} = \frac{U}{d_{\text{polymer}}} \quad (4)$$

where d_{polymer} is the thickness of the active polymer layer.

However, the effective internal electric field in the device deviates from the field calculated with Eq. (4) for several reasons:

1. In the PLED a small internal electric field also exists if no external bias voltage is applied. This built-in electric field is created due to the coincidence of the Fermi levels of the electrodes (see above) by the electrical connection of the metal electrodes via the polymer layer (see Fig. 30.9).

2. In impedance spectroscopic measurements it has been observed that the capacitance of PLEDs increases with increasing voltage in the forward direction [69]. This is attributed to charging of defect-induced or intrinsic states in the polymer near the metal electrodes. In the forward direction the internal electric field is therefore decreased compared to the field calculated via Eq. (4).

3. At the boundaries between the polymer layer and the electrodes, interface layers usually exist [66,70]. On the surfaces of all solids, insulating monolayers of adsorbed atoms or molecules (e.g., oxide) can be observed [71]. The time until an atomically clean surface is covered by a mono-

layer depends on the temperature and the atmospheric pressure. At ambient pressure a monolayer is formed immediately. In a vacuum compartment at a base pressure of about 10^{-6} mbar a clean surface is covered with a monolayer after only 1 s. When a PLED is being built up, the surface layers on the electrodes and the polymer become interface layers in the device arrangement. When thin insulating layers are inserted between the metal and the polymer in a PLED (metal/insulator/polymer: MIP structure) they can significantly reduce the onset electric field necessary for the electroluminescence. This has been demonstrated by intentionally producing insulating interface layers of several nanometers in thickness between the electrode and the polymer. In PLEDs, which contain such insulating layers, the charge carrier injection (see Section IV) is significantly improved compared to LEDs prepared without such layers [72–74].

Some of the molecules that are adsorbed on surfaces (especially oxygen and water) strongly interact with the polymer, so chemical reactions can occur at the interface [66]. Surface states are formed that can also cause band bending at this interface region and therefore affect the internal electric field [75]. This interface layer seems to hinder the diffusion of gases and metal atoms further into the polymer and the subsequent chemical reaction of the electrode material with the polymer and can therefore be quite important for the operation of the EL devices [76]. The diffusion of the oxygen from the electrode into the polymer that was observed for ITO devices [77] can also be prevented by introducing a thin layer of polyaniline [78]. For real systems the metal/polymer junction can therefore be understood as an Mx polymer structure, where x represents an interfacial layer of currently undefined nature.

A typical single-layer PLED is usually constructed in a very simple way [6]. The polymer is cast onto the transparent hole-injecting electrode and the electron injection contact is evaporated onto the polymer (see Fig. 30.8). In addition to this most common arrangement, PLEDs can also be produced in several other configurations [79,80].

2. Electroluminescence in Multilayer Structures

The overall performance of the PLEDs can be dramatically improved when charge transport layers are inserted at the interface between the polymer and the electrodes (see Fig. 30.10) [7,8,62].

a. Working Principle of a PLED with an Electron Transport Layer

In single-layer El devices, the region where the electroluminescence mainly occurs (emissive zone) is usually located near the cathode because of better mobility of positive polarons in conjugated polymers (see Section V). This concentration of positive charges near the cathode is a drawback for producing efficient electroluminescence, as near the cathode the probability for nonradiative recombination at defect sites is quite high. These quenching effects can be decreased when a dielectric layer is introduced between the cathode and the polymer. This layer can significantly improve the EL quantum efficiency when its LUMO is located below the CB edge of the polymer, so that the potential barrier at the interface to the low work function electrode is decreased (see Fig. 30.10a). The electron injection is then facilitated, and the balance in the charge injection can be improved. This layer, which should have good electron transport and moderate hole transport properties, is called the electron transport layer (ETL). The HOMO of the ETL is located below the VB of the polymer, so that holes accumulate at the potential barrier at this interface. As the LUMO of the ETL is located slightly below the CB of the polymer, electrons are also concentrated at this interface. Thus both electrons and holes are confined at the ETL/polymer interface, where they can effectively interact with each other to form excited states.

b. Working Principle of a PLED with a Hole Transport Layer

To improve the stability and efficiency of the PLEDs it is very important to decrease the operating bias voltages, i.e., fields, so that the power consumption and hence the Joule heating of the devices is reduced. The stability of the PLEDs is strongly influenced by malfunctions introduced by the high local temperatures [81,82] that occur during operation of the devices. These high temperatures are also a drawback for achieving high EL quantum efficiencies due to thermal quenching effects of the excited species [83]. A reduction of the operating fields is crucial not only to reduce temperature effects in the polymer but also to decrease field quenching effects of the excited species (see Section VIII).

The bias voltage, i.e., the electric field required for the onset of the current, and the electroluminescence in the forward direction decrease when the potential barrier on the polymer/hole injection contact interface is reduced (see Fig. 30.10b). This is due to the fact that the current in PLEDs is mainly dominated by positive charge carriers, as the mobility of the positive polarons in the conjugated polymers is much higher than the mobility of the negative polarons (in contrast to many organic compounds [84]) (see Section V). A reduction of the barrier height can be obtained by introducing a hole transport layer (HTL) between the anode and the polymer. To lower the barrier height, the HOMO level of the HTL, which has a high hole mobility, has to be located above the VB of the polymer and below the Fermi level of the metal (see Fig. 30.10b).

Electroluminescence

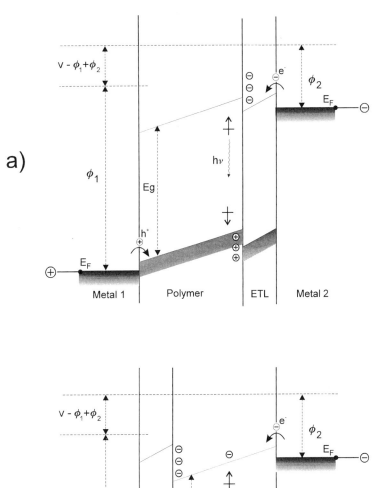

Fig. 30.10 Working principle of (a) an electron transport layer and (b) a hole transport layer.

The influence of charge transport layers in polyphenylene devices can be well described by looking at the properties and performance of single- and multilayer EL devices based on the oligomer parahexaphenyl (PHP). In air stable single-layer EL devices based on PHP with ITO and Al as electrodes, highly efficient blue LEDs (EL quantum efficiency up 1%) are obtained [85], although the band discontinuities at the interfaces in this configuration are on the order of 1 eV (when building the difference between the literature values for the work function of the electrodes [86] and band edges of the oligomer).

By applying an oligoazomethine (OAM) as HTL between the ITO and PHP, the electric field required for the onset of the current in the forward direction is decreased. For the ITO/OAM/PHP/Al structures the onset of the electric field occurs at values of around 0.3 MV/cm compared to 0.8 MV/cm obtained for single-layer PHP devices, independent of the PHP thickness (see Fig. 30.11). The electric field–induced tunneling through the potential barrier at the oligomer/electrode interface is the dominant injection mechanism for PHP devices for electric fields in excess of 0.1 MV/cm (see Section IV). The oligoazomethine layer is produced via a new polycondensation reaction of successively deposited monomers, which form highly oriented OAM films on polar surfaces with a high hole mobility (see Fig. 30.12a). The decrease of the onset electric field with the application of OAM as the HTL can be attributed to a reduction of the barrier height at the anode/HTL interface. The increased EL quantum efficiency (up to 2%) that is observed for ITO/OAM/PHP/Al devices can be attributed to better carrier injection conditions and reduced quenching effects compared to the single-layer devices, as described above.

When a diaminooctofluorobiphenyl (DOB) layer (see Fig. 30.12b) is applied between the ITO and the PHP, the operational lifetime of the EL devices is significantly improved compared to single-layer PHP devices.

In a multilayer structure applying electron and hole transport layers, the advantages that are contributed by each layer can be combined [87]. In the case of PHP devices, the best EL quantum efficiencies (>2%) and the lowest electric threshold field (≈0.25 MV/cm) are obtained in the multilayer structure as ITO/DOB/PHP/OAM/Al (see Fig. 30.11) [88].

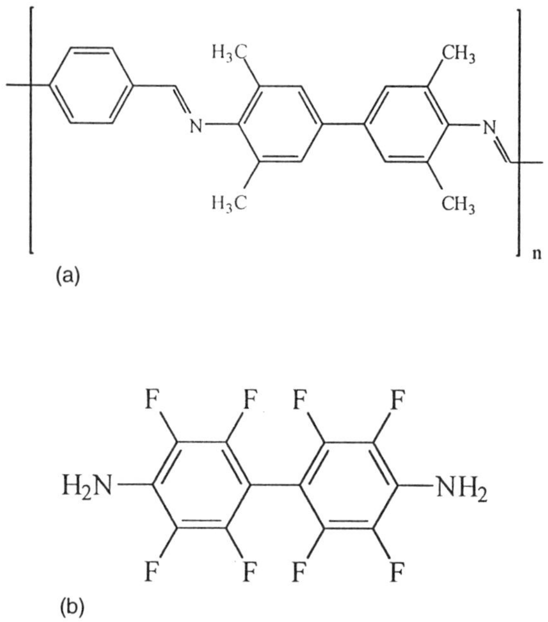

Fig. 30.12 Chemical structure of (a) oligoazomethine (OAM) and (b) diaminooctofluorobiphenyl (DOB).

C. Electroluminescence from an Electrochemical Cell

An attractive idea to realise a *pn* junction with conjugated polymers is to create this *pn* junction in situ by electrochemical doping. Bipolar light-emitting *pn* junction devices can be produced by applying the conjugated polymer in a solid-state electrochemical cell [89,90].

The working principle of the polymer light-emitting electrochemical cell is shown in Fig. 30.13. The conjugated polymer is blended with materials having good ion transport properties. This polymer blend is then arranged between two electrodes in either a planar or sandwich structure (see Fig. 30.13a). When a sufficiently high voltage is applied to the electrodes, reduction of the polymer and an injection of negative carriers occur at the cathode, and oxidation and injection of positive carriers takes place at the anode. An *n*- and a *p*-type layer are created adjacent to the electrodes on opposite sides of the cell (Fig. 30.13b). At the interface between these two layers a *pn* junction is formed. Under the applied field the oppositely charged carriers drift to the respective electrodes and meet in the region of the electrochemically induced *pn* junction (Fig. 30.13c). By the interaction of positively and negatively charged carriers, excited neutral states are formed, which subsequently can decay radiatively (see Section VI).

The EL light is emitted from the *pn* junction, which has a width of only a few micrometers. However, the

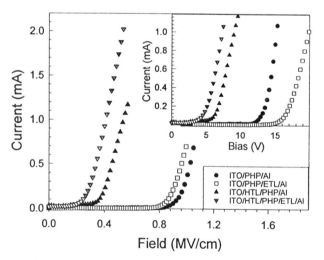

Fig. 30.11 Current vs. electric field characteristics for PHP single- and multilayer EL devices. Inset: Current vs. bias voltage.

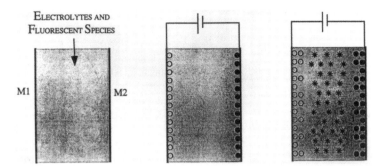

Fig. 30.13 Schematic diagram of the EL processes in an electrochemical cell. (Reproduced from Ref. 90.) **(a)** Cell before applying a voltage; **(b)** doping opposite sides as n- and p-type; **(c)** charge migration and radiative decay. M1, M2, electrodes. ○, oxidized (p-type doped) species; ●, reduced (n-type doped) species; ∗, electron–hole pair;

arrangement of the LECs offers two advantages compared to the PLEDs in an MSM or MIM arrangement:

1. The onset voltage of the LEC is quite low (2–3 eV), only slightly larger than the E_g of the active polymer layer,
2. The electrodes form ohmic contacts with the polymer. The electrode materials therefore have to possess a work function higher than the Fermi level of the conjugated polymer, which is fulfilled for most air-stable metals. Band matching between the polymer and the electrodes is not necessary.

IV. CARRIER INJECTION

The nature of the junctions between metals and conjugated polymers is described in Section III. In this section we want to focus on the charge injection mechanism in PLEDs of conjugated polymers of a low charge carrier density n (like the polyphenylenes). The metal/polymer/metal contacts with a polymer of low n can be described similarly to MIM structures (see Section III). The description of a polyphenylene type of polymer as an insulator is a good approximation for the MIM model, but with respect to the energy gap and general properties, the PPP-type materials belong to the class of wide energy gap semiconductors. For that reason we maintain the rigid band model and describe the PLEDs as metal/polymer/metal (MPM) devices. This is important insofar as in real EL devices interfacial layers exist between the polymer and the electrodes. These interfacial layers often have insulating characteristics, so the metal–polymer contact in this case has to be described as an MIP contact. Additionally, since chemical reactions can occur between the polymer and the interfacial layers, the nature of the interfacial layer can become dominant; therefore this junction can be in general terms understood as an MxP contact.

Depending on the nature of the polymer/metal junction and on the magnitude of the height of the barrier that exists at the interface, several mechanisms are responsible for the charge injection into the polymer. In the following we give an overview of the basic injection mechanisms.

a. Thermal Emission

The kinetic energy of charge carriers in a solid increases with increasing temperature, and therefore the probability that a charge carrier passes a given potential barrier also increases. The thermally induced current flow of the charge carriers from a metal contact into a polymer film can therefore occur by thermionic emission over a potential barrier or by tunneling through a potential barrier. The current flow induced by thermionic emission can be calculated with the Richardson equation

$$J = A^*T^2 \exp(q\Phi/kT) \qquad (5)$$

where Φ is the potential barrier, q the charge of the carrier, T the absolute temperature, and A^* the material-dependent effective Richardson constant. This equation, which was derived from the temperature-induced emission of hot charge carriers from a metal surface, considers that the emitted carriers are pulled away from the surface by an electric field. When a metal–polymer transition is described with this formula, the back flow of the injected charge carriers into the contact has to be taken into account [63,64]. In MPM structures a characteristic temperature dependence of the current versus voltage (I–V) curves, which is characteristic for thermionic emission, often occurs for very low applied fields and/or for MP contacts with a low barrier height. In MIP structures the potential barrier heights from the metal to insulator are usually very high and the contribution of the thermionic emission process to the current flow in the device is negligible.

2. Field-Induced Injection

Under high applied electric fields, electrons can also surmount a potential barrier at very low temperatures. This process is based on field-induced tunneling of the charge carriers through a potential barrier. The probability for the tunneling depends on the height and width of the potential barrier.

The field-induced tunneling through a potential barrier can be described using the Fowler–Nordheim equation [91]

$$I = \frac{A^*T^2}{\Phi}\left(\frac{qE}{kT\kappa}\right)^2 \exp\left[-\frac{2\kappa\Phi^{2/3}}{3qE}\right] \quad (6)$$

where E is the electric field and κ is a parameter that depends on the shape of the barrier.

When the barrier is assumed to be triangular, the parameter κ can be calculated from the potential barrier Φ as follows [63,64]:

$$\kappa = \frac{4\pi\sqrt{2m^*}}{h} \quad (7)$$

where m^* is the effective mass of the charge carriers and h is Planck's constant. When the electric field induced tunneling currents are analyzed in a log (I/E^2) vs. $1/E$ plot, a straight line should be obtained. From the slope of this straight line the height of the potential barrier can be derived using Eqs. (6) and (7).

In MIP structures large barrier heights exist at the metal/insulator interface. The charge carrier injection therefore can be well described by field-induced tunneling. The current versus electric field characteristics of MP structures can be well fitted by a straight line at high applied fields in a Fowler–Nordheim plot, when the barrier heights at the interfaces are large compared to kT. This condition is characteristic of field-induced tunneling [92].

However, the values for the current that are obtained with the actual device parameters using the Fowler–Nordheim equation are several orders of magnitude higher than the values for the measured current in real devices. This is due to the fact that the I–V characteristics of PLEDs are determined not only by the injection mechanisms but also by the charge transport mechanism in the active polymer layer (see Section V). The discrepancy between the measured and calculated values for the current in the model for field-induced tunneling can be accounted for by a backflow current of the injected charge carriers into the injection contact. This effect then reduces the net device current and seems to be especially important in low mobility conjugated polymers [93].

The bulk conductivity also influences the I–V characteristics of PLEDs based on PPV-type polymers with low barrier heights, where results from device modelings suggest that the current flow is dominated by a space charge or trap-limited current and not by the charge carrier injection [94,95].

The bulk conductivity strongly influences the I–V characteristics of PLEDs based on the LPPP polymer. In PLEDs the bulk conductivity depends not only on the carrier concentration, as outlined in Section III, but also on ordering phenomena within the active polymer layer. For conjugated polymers $\sigma_\parallel$ parallel to the chains is usually much larger than $\sigma_\perp$ (≈ 100 times [89]), so that the worst conductivity through the device is obtained when the polymer main chains are aligned parallel to the plane of the electrodes in the EL device. This orientation of conjugated polymers in the plane often occurs in thin, spin-coated films of rigid-rod polymers such as m-LPPP [97] and also in soluble polyphenylenevinylenes [98]. For the top layers of thicker films (>100 nm) this orientation diminishes more and more and the case of a statistically disordered bulk is approached. For increasing film thickness the bulk conductivity therefore increases. For that reason the I–V characteristics of the EL devices based on m-LPPP show a dependence on the film thickness [20]. The onset electric field decreases for increasing film thickness from values of around 2 down to 0.5 MV/cm (see Fig. 30.14).

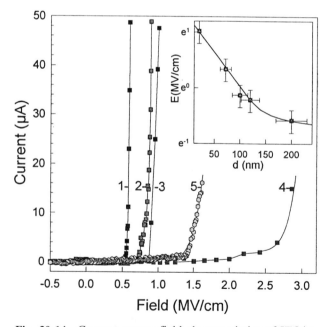

Fig. 30.14 Current versus field characteristics of ITO/m-LPPP/Al devices of polymer thickness d (1, $d = 200$ nm; 2, $d = 120$ nm; 3, $d = 100$ nm; 4, $d = 20$ nm) and a ZnO:Al/LPPP/Al device (5, $d = 170$ nm). Inset: Dependence of the onset field on the polymer thickness d; the solid line is a guide to the eye.

V. CARRIER TRANSPORT

As discussed in previous sections, PPPs and their oligomers can be treated as wide-gap one-dimensional semiconductors to describe many of their properties. The electrical conductivity is determined by the product of carrier concentration and carrier mobility. In conjugated polymers both entities depend on the material and the polarity of the charge carriers. Electrons or holes placed on a conjugated polymer lead to a relaxation of the surrounding lattice, forming so-called polarons, which can be positive or negative. Therefore the conductivity σ is a sum of the conductivity of positive polarons (P^+) and negative polarons (P^-):

$$\sigma = en_{P^-}\mu_{P^-} + en_{P^+}\mu_{P^+} \tag{8}$$

where n and μ are the concentration and mobility of the respective species P^+ and P^-, and e denotes the elementary charge. In contrast to carrier injection, which depends on the properties of the polymer/electrode interfaces, carrier transport is a bulk property of the polymer layer. In organic electroluminescent devices, both interface and bulk properties are essential for the performance. The region of radiative recombination, for example, is determined by the different transport properties of negative and positive polarons [99].

A. Carrier Concentration

Three types of contributions determine the carrier concentration in organic semiconductors used as active layers in an electronic device:

1. Intrinsic carrier concentration due to thermal excitation across the gap
2. Carriers due to ionized defects
3. Injected carriers

1. Intrinsic Semiconductors

For an intrinsic semiconductor the effective density of states $N_{C,V}$ of the valence (V) or conduction (C) band is given by [63,64]

$$N_{C,V} = 2.4 \times 10^{19} \left(\frac{m^*_{C,V}}{m}\right)^{3/2} \left(\frac{T}{300 \text{ K}}\right)^{3/2} \text{ cm}^{-3} \tag{9}$$

where $m^*_{C,V}$ is the effective mass of the charge carriers, m is the electron mass, and T is the temperature of the semiconductor in kelvins. Taking $m^*_{C,V} = m$ and $T = 300$ K we obtain a value of 2.4×10^{19} cm^{-3} for the density of states. To obtain the intrinsic carrier concentration $n_i = n_p$ due to the thermal generation of charge carriers at a given temperature, the value of the energy gap E_g has to be known as well:

$$n_i = n_p = 2.4 \times 10^{19} \left(\frac{m^*_C m^*_V}{m^2}\right)^{3/4}$$

$$\left(\frac{T}{300 \text{ K}}\right)^{3/2} \exp\left[-\frac{E_g}{2kT}\right] \text{ cm}^{-3} \tag{10}$$

Again taking the effective masses to be equal to the electron mass, using 300 K for room temperature and 2.7 eV for the energy gap of the LPPPs, we obtain an intrinsic carrier concentration of 10^{-4} cm^{-3}. This value will rise to 41 cm^{-3} upon increasing the temperature to 400 K. These values can be compared to those for common inorganic semiconductors at room temperature: 2.7 carriers/cm^3 for GaP with an energy gap of 2.24 eV and 1.8×10^6 carriers/cm^3 for GaAs with a gap of 1.43 eV.

To determine the conductivity contribution from intrinsic charge carriers at 300 K the carrier concentration of 10^{-4} cm^{-3} has to be multiplied by the elementary charge and the mobility, which is found around 10^{-4} cm^2/(V.s) in conjugated polymers [100]. This leads to a value of 1.6×10^{-27} Ω^{-1} cm^{-1}. This value is in accordance with experiments by Chiu et al. [101], who deduced the conductivity of pure PPP to be less than 10^{-17} Ω^{-1} cm^{-1}. The conductivity of pure LPPP was determined to be 5×10^{-15} Ω^{-1} cm^{-1} [102]. The experimental values together with the calculated value for the conductivity due to intrinsic charge carriers show that the contribution of these carriers to charge transport is negligible. For completeness we add here that the previous results were obtained in the conventional semiconductor model developed for inorganic semiconductors. However, the weak coupling between the structural elements in nondegenerate conjugated polymers and oligomers and the moderate dielectric constant favor the description of conjugated materials in an exciton model. This will even further reduce the calculated concentration of charge carriers.

2. Charge Carriers Due to Defects

With the exception of some high purity molecular crystals such as anthracene, which can be zone-refined [103], most organic materials exhibit transport properties determined by impurities. Chemical defects on the polymer chain may serve as sources for charge carriers. In PPP it was shown that annealing samples at temperatures up to 250°C resulted in higher conductivities [101]. This effect was attributed to the increase of crystallinity [22,104] and therefore better mobility of charge carriers. However, increasing the annealing temperature beyond this value leads to the release of residual chlorine. Since the chlorine acted as a weak electron acceptor, its release led to a decrease of conductivity, stressing the role of defects as sources of charge carriers. Instead of relying on residual impurities, charges can also be added to PPP by redox reactions; the introduction of a finite charge carrier concentration by a redox reaction started the whole field of conducting conjugated polymers [105]. When it is doped with strong oxidizing or reducing agents, PPP is converted to a metallic conductor [27,106]. Ionic processes were excluded as the dominating charge transport mechanism, and the electronic nature was established by a series of experiments comprising the thermal annealing of PPP [101]. Unfortunately, the strong doping required to achieve conductivities as high as 200 S/cm [27] totally quenches luminescence. Therefore, intentional doping to control conductivity was used only in one case [107] in light-emitting devices. However, the defects present in the "pure" materials still occur at concentrations high enough to influence charge transport. In the following paragraphs we summarize the kinds of defects that are believed to occur in conjugated polymers and their influence on charge transport.

From picosecond transient photoconductivity measurements on PPP films [102] we know that mobile charged states decay within 110 ps. In conventional routes to PPPs, defects like branched chains and large torsion angles of neighboring rings are known to occur. These defects act as shallow or deep traps for positive and negative polarons [108,109], which limit the mobility of charge carriers [110]. The synthetic route to the PPP-type ladder polymers prevents such defects and leads to a trap concentration of about one trap per 1000 monomer units [111], whereas substituted PPV, for example, exhibits trap concentrations of one trap per monomer [112]. The former value corresponds to a concentration of 10^{17} traps/cm^3, which further emphasizes the fact that intrinsic carrier concentrations of 10^{-4} cm^{-3} are negligible for electrical transport in the PPPs.

For a quantitative treatment analogous to Eq. (10) we have to take the defect concentration N_d and the energetic defect-band distance E_d into account:

$$n_d = \sqrt{2N_d} \left(\frac{m_e kT}{2\pi\eta^2}\right)^{3/4} \exp\left[-\frac{E_d}{2kT}\right] \text{cm}^{-3} \quad (11)$$

Trapping of charge carriers plays an important role in these materials, since the I–V curves are also influenced by trapped charges [113,114]. These space charges distort the fields in the charge injection region at low voltages. From time-of-flight measurements for PPVs it is known that negative polarons, in contrast to positive ones, are severely trapped [110], which results in unbalanced charge transport. The influence of unbalanced charge transport on the device operation was shown for PPV, where the limitation of the charge transport by traps leads to a bias-dependent efficiency [95]. At low bias the injected electrons, which convert into negative polarons, are trapped and do not contribute to the radiative recombination process. With increasing bias the traps are filled, giving rise to a strong increase in the number of free polarons, which participate in recombination, and hence the device efficiency increases.

Since the intrinsic electronic structure of PPV does not show any preference for holes or electrons [115], oxygen-related impurities such as molecular oxygen or carbonyls with strong electron-accepting character and reduction potential lower than that of PPV may act as the predominant P$^-$ traps and limit the range of negative polarons [110]. This is in accordance with investigations of these traps in PPV, showing that shallow traps up to a few tenths of an electronvolt seem to be intrinsic to the samples and storage in air creates deeper traps [116]. Therefore rigorous exclusion of oxygen from polymer LEDs will be necessary [117]. In addition to oxygen, other chemical agents, which are formed during heat treatment of PPV samples obtained via precursor routes, are also discussed as origins for polaronic traps [118].

Charge transport is also influenced by the degree of crystallinity of the organic films. In polycrystalline unsubstituted PPVs, the release times of those carriers that determine the transport properties are typically more than three orders of magnitude longer than comparable release times in substituted PPV. This behavior is attributed to grain boundaries, which act as efficient traps in organic photoconductors [119] and can form in the crystalline PPV while substituents inhibit crystallization. The same behavior can be expected from PPPs and might explain in part the low trap concentration in LPPPs, which also have bulky substituents preventing crystallite formation.

3. Injected Charges

For a comprehensive treatment of charge injection and the importance of this process in device operation we direct the reader to Section IV, where these issues have already been described.

B. Mobility

Mobility values in conjugated polymers are known to depend on field, temperature, and type of charge carrier [120]. The most striking fact concerning mobility in conjugated polymers is the difference in the mobilities of positive and negative polarons, which are higher by orders of magnitude for the positive polarons. In inorganic semiconductors similar differences are observed, but usually the negative polarons are more mobile than the positive ones. Typical values obtained for P$^+$ mobilities of conjugated polymers are 10^{-4} cm^2/(V·s) [110]; by careful preparation of oligomer films this value can be increased to 10^{-2} cm^2/(V·s) [100,121]. These mobility values have to be compared to 1300 and 500 cm^2/(V·s) for electrons and holes, respectively, in silicon. For LPPP the high mobility of positive polarons was demonstrated by modeling the I–V characteristics of homolayer devices of LPPP between different metal electrodes.

The most effective approach to overcoming the problem of the very different mobilities of positive and negative polarons and to enhance the EL efficiency is the use of multilayer structures, with polymer layers of different electronegativities, so that recombination is confined at the heterojunction between the two polymers [122–124]. This strategy was also applied to PPPs [18,125] with considerable success. The introduction of the transport layer also allows the use of thinner active layers, which reduces the device operating voltage without creating excessively high leakage currents in the device [18]. Blending the electroluminescent polymer with an electron or hole transport material also lowers the operating voltage [18]. The various design approaches for multicomponent devices of this kind were discussed in Section III.

Summarizing the results obtained for charge transport in conjugated polymers, the Poole–Frenkel model seems to be a good description [126,127]. This model of transport assumes the existence of two types of carrier movement, one in conductive states (positive polarons) and the other via traps (negative polarons).

VI. FORMATION OF EXCITED SPECIES

When charges are injected into conjugated polymers they initially form charge carriers with a relaxation of the surrounding lattice. Subsequently these carriers give rise to a variety of species, most of which are important for device operation but not all of them acting in favor of a stable, highly efficient light-emitting device. In this section we discuss the behavior of the charge carriers and their conversion into other species based on the lifetimes and rate constants as depicted in Fig. 30.15, interspersing short comments concerning photoexcitation in conjugated polymers. Up to now, much higher excitation densities have been obtainable via optical absorption than via charge injection. In addition, ultrafast time-resolved spectroscopy is limited to the optical domain.

When charges are injected into conjugated polymers they do not continue to exist as free electrons or holes [128]. Parallel to thermalization of the injected charges the electron–lattice coupling leads to a relaxation of the polymer nuclei in the vicinity of the introduced charge, forming polarons. These polarons are charged states within the gap region (see Fig. 30.15), which give rise to absorbance below the onset of the π–π^* transition (see Fig. 30.16). In addition, the charge distribution deviates from its ground-state conditions, giving rise to additional infrared-active vibrational modes. Therefore the presence of these polarons during the operation of light-emitting devices can be detected via their absorbance.

Driven by the electric field, positive and negative polarons (P^+ and P^-) approach each other. Upon meeting

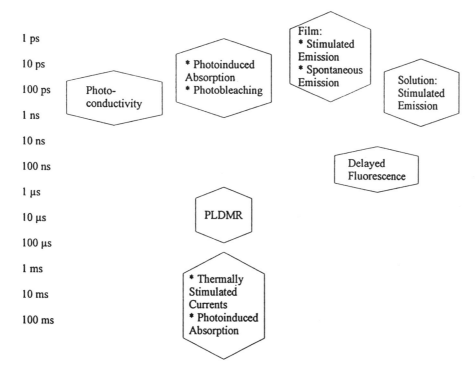

Fig. 30.15 Time scale of experiments probing photoexcited states in PPP-type polymers. PLDMR = photoluminescence detected magnetic resonance.

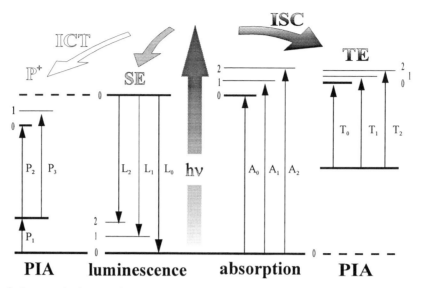

Fig. 30.16 Scheme of photoexcited states in PPP-type polymers. The symbols represent singlet exciton absorption (A_0, A_1, A_2) and luminescence emission (L_0, L_1, L_2) as well as both polaronic (P_1, P_2, P_3) and triplet exciton (T_0, T_1, T_2) absorption as detected by photoinduced absorption experiments. Except for P_1, the numerical subscript indices denote vibronic progression. (Reproduced from Ref. 137.)

they form neutral states, which are either singlet excitons or triplet excitons. In both cases the constituting charges forming the exciton are bound to their oppositely charged partner by a binding energy that is a few tenths of an electronvolt for conjugated polymers [129]. On the basis of the relative spin degeneracy of the singlet and triplet excitons, 75% of the excitons formed by P^+–P^- fusion are triplet excitons. Upon photoexcitation, triplet excitons are formed only via a weak intersystem crossing path, which leads to a triplet population of a few percent at most. Since only the singlet excitons are able to recombine radiatively, the maximum EL efficiency was predicted to be one-fourth of the PL quantum efficiency [122]. Aside from this principal reason, which reduces the maximum EL efficiency, other influences determining how many singlet excitons decay *radiatively* have to be considered. An increase in temperature leads to a decrease in the efficiency of luminescence. This is the reason for power efficiency curves that initially increase with operating voltage, show a peak, and decrease again [52,130]. Device heating and subsequent decrease in luminescence efficiency is the reason for this behavior. This clearly emphasizes the need for highly efficient devices that do not heat up when they produce the bright emission intensities required for applications. High operating fields across the active layers also have a negative effect on luminescence efficiency. As will be discussed in Section VIII, exciton dissociation is observed under high fields, which forms isolated charges, namely polarons, and prevents emission.

Photo- or electroluminescence emission from conjugated polymer films is usually unpolarized. However, this is not an intrinsic property of the films but a consequence of the preparation process. In most cases, macroscopically, more or less isotropic films are formed. Considering the experimental and theoretical evidence of the anisotropy of absorbance (for PPPs, see Refs. 36–38 and 43), the anisotropic nature of the electronic properties of conjugated polymers on a microscopic scale is well established. If one can introduce order or orientation on a macroscopic scale also, it is possible to obtain polarized luminescence emission. Indeed LEDs with polarized emission have been produced by using stretch-oriented films [131], by deposition of molecules with a preferential orientation [45], and by applying the Langmuir–Blodgett technique [44].

In addition to the emissive singlet excitons, which are created only with a 25% probability by polaronic recombination processes, triplet excitons are also present in light-emitting devices (see Fig. 30.16). The role of the triplets in this context is poorly understood, but two effects are discussed. First, the high triplet concentration may lead to faster degradation of polymer LEDs [117] compared to photoexcited samples if the oxidation proceeds through the triplet state, since no intersystem crossing is necessary as in photoexcitation experiments. Second, the triplets may also act as singlet quenching centers in LEDs [132]. Like the polarons generated by charge injection, the triplets can also be seen in absorbance [133] or optically detected magnetic resonance experiments [134] of LEDs.

On their way to recombination, a substantial portion of P^+ and P^- are trapped at defect sites, influencing device operation in many ways. First, their contribution to transport and recombination is temporarily lost. They do contribute to both processes, however, after thermal release from their trap states. In some polymers the trapping/detrapping process completely dominates their charge transport [112]. Second, they form space charge regions that affect charge transport and charge injection as discussed in Sections IV and V. Third, trapped polarons also give rise to absorption. In most polymers one of the absorption bands lies in the region of PL emission and would therefore impede the operation of a laser diode based on a polymeric LED; as discussed below, this is not the case in LPPP. The fourth and most problematic effect of trapped polarons, however, is their role as quenching centers for singlet excitons [132], which results in reducing the efficiency of electroluminescent devices.

Since none of these four effects is favorable for device operation, materials with a very low trap concentration are desirable. The LPPPs form such a class of materials as their synthesis virtually excludes defects. Evidence for this is found in their polaron absorption, which is either low or below the detection limit [54]. Upon exposing the LPPPs to the combined influence of strong UV-Vis radiation and oxygen, the defect concentration and hence the polaron absorbance can be increased, emphasizing the role of defects in extrinsically stabilizing polarons. In LPPPs their absorbance is identical to the spectrum of photogenerated polarons and polarons due to doping, and, due to the high intrachain order of the LPPP, shows a clearly resolved vibronic progression, which is usually observed only for conjugated oligomers [49,135]. The trap concentration in LPPPs is found to be on the order of one trap in 1000 monomer units [111], in contrast to other polymers which show one trap per monomer unit [112].

For completeness we note that the formation of trapped polarons by photoexcitation also depends on the quantum energy of the exciting photons. From both photoluminescence-detected magnetic resonance [136] and photoinduced absorption [137,138] experiments, we know that photons with quantum energies a few tenths of an electronvolt above the absorbance edge of the LPPP create trapped polarons less efficiently than excitation near the absorbance edge. Another way to reduce the influence of traps and space charges is to reach a balanced injection and transport of charge carriers. If polarons of one sign dominate in a certain region, both their radiative and nonradiative recombination will be delayed. Hence the probability of trapping with all its negative consequences becomes higher.

From both photoexcitation and charge injection experiments it is well known that in addition to luminescent species, absorbing species, namely polarons and triplet excitons, are formed. This absorption occurs in the same spectral region as EL and PL emission and also competes with the observed stimulated emission in pure PPVs [139] and would preclude the fabrication of a solid-state polymer laser from these materials. There are two strategies to circumvent this problem. First, the PPVs can be blended, because the absorbing species, which are sometimes denoted as polaron pairs [139], are not generated efficiently in these blends [140]. The second strategy is to find highly luminescent materials in which photoinduced or charge-induced absorbance is spectrally separated from the emission region. The LPPPs fulfill this criterion both for the occurring triplet–triplet absorption and for the absorption from polarons; dioctyloxy-p-phenylenevinylene also shows this spectral separation [141]. In agreement with this expectation and the high photoluminescent quantum yields we found strong stimulated emission from m-LPPP films [142] and solution [143] without any competition from photoinduced absorption.

With respect to the emissive singlet excitons, LPPPs show another striking phenomenon under photoexcitation: a hyperlinear dependence of the PL emission intensity on the excitation density [144]. These facts make the LPPPs a material of choice for building a polymer solid-state laser.

VII. EMISSION COLOR

The visible spectral range of the human eye spans from 380 to 750 nm. Each wavelength of light corresponds to a certain emission color. The light emission of conjugated polymers is usually spectrally distributed over more than 100 nm, so a relation to a certain pure color is difficult. One formalism used to characterize emitted light with respect to color has been established by the Commission Internationale de l'Eclairage (CIE) [145,146]. In this formalism the colors are related to CIE coordinates representing a dot in a two-dimensional plot (see Fig. 30.17).

The emission process in conjugated polymers occurs from neutral excited states, which are intrachain singlet excitons in most cases. They are generated either by direct photoexcitation or by recombination processes from other excited (band) states. The emission process is determined by the probability for the radiative transition from the excited state to the ground state. The emission color then depends on the shape and the energetic position of the emission spectrum. The same excited species are responsible for EL and PL emission; therefore, the intrinsic components of their spectra are identical (see previous sections).

The emission colors of the most common conjugated polymers used in PLEDs—e.g., PPP, polyvinylenevinylene, and polythiophenes—range from blue to red, so that the realization of any emission color with PLEDs is possible [6,10,147]. Moreover, in contrast to inorganic semiconductors, simple methods exist to change the emission color of a certain conjugated polymer by chemical means. Chemical modifications that change the emission color of the polymer are the following:

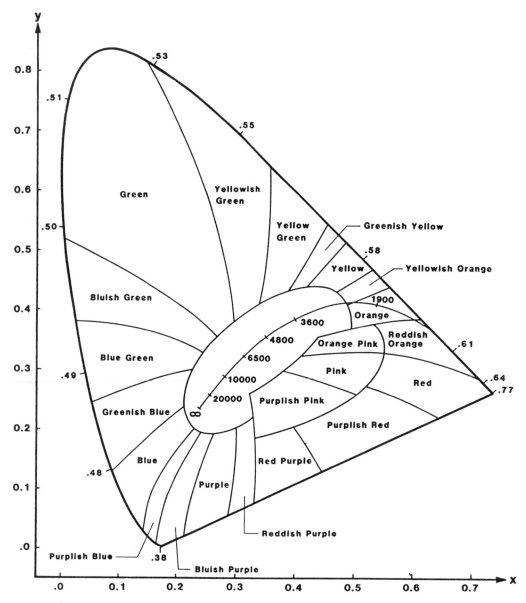

Fig. 30.17 CIE color coordinates diagram. The solid line with the numbered ticks represents the coordinates of the color of a blackbody emitter at the indicated temperatures. The wavelengths in micrometers of the pure colors are indicated at the periphery of the diagram. (Reproduced from Refs. 145 and 146.)

1. Bulky side chains influence the steric configuration of the main chain and therefore the effective conjugation length I_{eff} (which is defined as the number of undistorted sequences of single and double bonds) when they are attached to the backbone. The steric hindrance of the side chains usually causes an increase in the torsion angle between the monomer and the backbone of the polymer chains, so that I_{eff} decreases. The energy gap and the onset of emission, which strongly depend on I_{eff}, are therefore blue-shifted compared to the conjugated polymer without side chains.

2. When electronegative atoms or side groups (e.g., O, N, F, Cl, CF_3) are incorporated in the polymer either on the side or main chain, the electronic properties of the polymer are strongly altered. This electronic modification includes a change in the energetic position of the valence band (VB) and conducting band (CB) of the conjugated poly-

mer [148] and can result in a change in the emission color.

3. The emission color of a homopolymer is drastically changed when subunits other than the monomers of the homopolymer are arranged in the backbone so that a block copolymer is formed. The block copolymer consists of at least two different subunits that have different electronic properties. Due to the different positions of the molecule orbitals of the subunits, the energy along the backbone is modulated. An excitation on the polymer chain will move to subunits with the minimal energy (excitation energy transfer), so it is very probable that the emission process occurs from these subunits. The diffusion to quenching sites is therefore reduced or even suppressed. An advantage of this one-dimensional "quantum well" structure is that the photoluminescence quantum efficiency is usually increased compared to homopolymers [17,41,42,149,150]. The emission color in the case of the copolymer is blue-shifted as the emission originates from sequences with oligomeric character.

The possibility of adjusting the emission color of conjugated polymers by controlling the geometry of the main chain can be well demonstrated with polythiophenes. By attaching various substituents of different length and character on the main chain of the polythiophenes, the geometry of the backbone can be controlled, and hence the emission color of the polymer can be changed from ultraviolet to red (see Fig. 30.18) [16,151].

The influence of side chains on the geometric structure of the polymer can be excluded only when the backbone geometry of the polymer is completely planarized by connecting the monomers to each other via an additional bridge. The planarization process leads to a reduction of the energy gap and hence the onset of emission of a conjugated polymer, which is incorporated in such a ladder-type structure, due to an increase of the effective conjugation length. This effect can be well seen by looking at the ladder-type PPP (LPPP), whose absorption maximum is around 0.5 eV red-shifted compared to the PPP (see Section II). The photoluminescence emission color of the LPPP can also be influenced by extrinsic effects such as an applied external field (see Section VIII).

In addition to the possibility of instrinsically changing the emission color of a polymer by chemical means, the emission color can also be altered by doping the host polymer with miscible guest materials such as dyes, organic complexes, or other conjugated polymers [152]. Let us assume that the $\pi-\pi^*$ levels of the guest are located in between the VB and CB of the host polymer; then excitation energy is transferred from the host polymer to the guest. In this case the emission originates from the guest molecule and not from the host polymer.

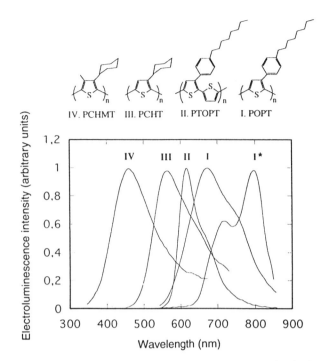

Fig. 30.18 Chemical structure and EL spectra of substituted polythiophenes. Top: Chemical structure of polymeric emitters I–IV. The devices used are of the ITO/polymer/Ca-Al type. Polymer I exists in two forms and may be thermally converted from form I to form I*. (Reproduced from Ref. 16.)

In PLEDs this effect can be used to change the EL emission color and also to increase the EL quantum efficiency when the emission stems from a guest material with a high PL quantum efficiency (see Section VI) [153].

The described modifications allow adjustment of the electronic properties of the conjugated polymers so that emission colors from the infrared [154] to the ultraviolet [155] can be realized. As many of these modifications are suitable for application in PLEDs, EL emission through the whole visible range can be achieved. However, the number of conjugated homopolymers allowing blue PLEDs to be obtained is restricted [9,10,12–14,16,18,20,52,156–159]. Blue PLEDs are of special importance in flat color displays, as discussed below.

Instead of homopolymers, certain (block or graft) copolymers can be applied to produce blue PLEDs [42,160–164]. As mentioned, emission from the copolymers stems from shorter subunits (in case of the graft copolymer, from light-emitting side chains), and therefore the emission is blue-shifted compared to that of the homopolymer.

The emission color of certain PLED arrangements can be controlled by the applied electric field, which is very interesting not only from the scientific point of view but also for many applications. The working principles of some selected device arrangements that allow the emission colors to be changed by the applied electric field are discussed below.

1. When blends of conjugated polymers with different emission colors are used as active layers in EL devices, the emission color depends on the magnitude of the electric field, as has been shown with substituted polythiophenes [16,165]. For an electric field just above the threshold field, the emission stems solely from the polymer with the lowest energy gap, whereas when the electric field is further increased, the emission intensity of the higher energy gap polymer steadily increases relative to the emission of the other component. The precondition for this effect to occur is that the polymers in the blend are phase-separated, so that excitation energy transfer to the low energy gap polymer is blocked. The field control of the emission color can be described in a model where polymers of different energy gaps are represented as parallel diodes with different voltage–luminance characteristics [16]. By using blends of three different substituted polyalkylthiophenes, PLEDs can be produced in which the emission color changes from reddish to bluish with increasing bias voltage [165].

2. The emission color of EL devices can be changed by shifting the emissive zone within the active polymer layer. The emissive zone in PLEDs is located adjacent to the cathode for electric fields slightly above the threshold field due to the higher mobility of holes in conjugated polymers (see Sections III and V). When the electric field is increased, the emissive zone is pushed away from the anode. This phenomenon can be exploited to control the emission color in a PLED, where only a small region of the active layer is doped with a dye possessing another emission color. The position of the emissive zone can then be shifted into or out of the doped region with the applied electric field [166]. In PLEDs with LPPP, field control of the emissive zone has also been used to tune the emission color (see Fig. 30.19). The EL spectrum of the LPPP is strongly influenced by self-absorption effects due to the strong overlap between the absorption and emission spectra (see Section II). The self-absorption effect changes the EL spectrum more dramatically than the PL because the EL mainly originates from an emissive zone near the cathode (Al), and due to the construction of the device the emitted light has to pass through the entire polymer film. When the emissive zone is shifted away from the cathode, the self-absorption decreases and the intensity of the peaks at 461 nm increases so the emission color can be controlled in the blue-green spectral range [20].

3. PLEDs in a special multilayer arrangement in which a charge-blocking layer is arranged between two

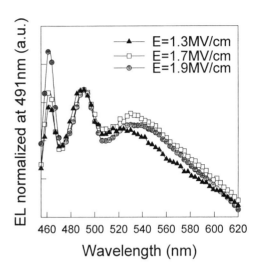

Fig. 30.19 Dependence of the electroluminescence on the applied electric field for an ITO/m-LPPP/Al device ($d_{\text{m-LPPP}} \approx 100$ nm).

active layers show symmetric I–V and EL–V characteristics. In this special configuration the EL emission stems from one of the two active layers depending on the polarity of the EL device. The EL emission can therefore be tuned between at least two emission colors [167].

The possibility of producing large-area flat devices with a high brightness (comparable to fluorescent tubes) makes the PLEDs promising candidates for application in self-emitting color flat panel displays. The multicolor dots building up a color display can be realized via several techniques:

1. *Blue, Green, and Red Organic Light-Emitting Diodes* (PLEDs). Although the emission colors that have been observed in PLEDs span the entire visible spectral range, their spectrally broad emission is a drawback for the realization of pure blue, green, and red emission colors, which are necessary to obtain the color mixture in color screens. One way to obtain a spectral narrowing of the emission spectra with minimal intensity loss is to use microcavity [168] structures or interference filters (see below).

2. *PLEDs Tunable from Blue to Red Emission Controlled by Device Parameters.* Several device parameters, e.g., temperature, current, and external (magnetic or electric) fields, are known to influence the emission color. As illustrated, the magnitude and polarity of the applied electric field (i.e., voltage) can be used to change the emission color of PLEDs. However, the realization of red-green-blue (RGB) emission colors obtained via voltage control is currently quite far

away from the CIE coordinates [165] that are required for RGB colors in color screens.

3. *Large-Area EL Devices with White Emission, Covered with Red, Green, or Blue Absorption Filters.* To produce white PLEDs, several active layers of copolymers or polymer blends have to be applied, as there is no homopolymer yet known that emits pure (white) light over the entire visible spectral range. The realization of LEDs that white light emit is possible by using multilayer structures or blends of several organic and/or polymeric materials [165,169,170]. The maximal intensity of one of the RGB emission colors obtained by covering the LEDs with absorption filters is at most one-third of the white light intensity because two-thirds of the incident light is dissipated in the filter.

4. *Blue Large-Area EL Devices Covered with Filters and Appropriate Dye Layers That Convert the Blue EL Emission into the Required Emission Color.* One of the most promising methods for the production of flat color screens is the use of blue PLEDs covered with dye layers to convert the EL emission into the RGB colors and, if necessary, with filters to purify the emission light [171–174] (see Fig. 30.20). In principle, the internal quantum efficiency of this technique for color conversion from blue into any other emission color can approach 100%.

The highly efficient EL devices based on parahexaphenyl (PHP) [52] are very suitable for the latter color conversion technique because their emission is deep blue with an emission maximum located at 425 nm (see Section II). The CIE coordinates for the PHP emission are $x = 0.15$, $y = 0.06$ (see Fig. 30.17). A fundamental advantage of the blue PHP emission light is that it can be converted into any other visible color, as the blue emission is located at the high energy side of the visible spectrum (see Figs. 30.5 and 30.21). Efficient color conversion can be attained by covering the EL device with appropriate dye layers of high fluorescence quantum efficiency. A dye absorbing in the blue spectral range is excited when it is illuminated with the PHP EL emission and subsequently emits PL light in a lower energetic range. A certain emission color is then obtained when a dielectric mirror (filter) is applied to spectrally purify the emission light. This filter possesses high transmittance for the required emission light from the dye and high reflectance for the exciting light, so only light matching the highly transparent spectral region of the filter can pass through the filter, whereas any other wavelengths of the incident light are reflected back into the dye layer. To obtain emission colors in the red-orange spectral range, the red dye layer usually has to be coupled with a green dye layer to transform the excitation light into the green spectral range, where the absorption of the red dye is much higher than in the blue range.

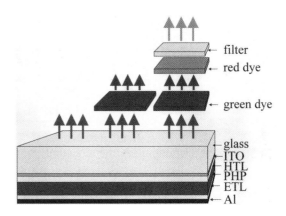

Fig. 30.20 Construction of the PLED: Glass substrate/ transparent electrode indium-tin oxide (ITO)/hole transport layer (HTL)/active layer (parahexaphenyl; PHP)/electron transport layer (ETL)/aluminum (Al). The blue PHP electroluminescence emission light is converted by covering the PHP OLED with a green dye layer (to give a green emission color) and a red dye layer together with a suitable dielectric filter (to give a red emission color).

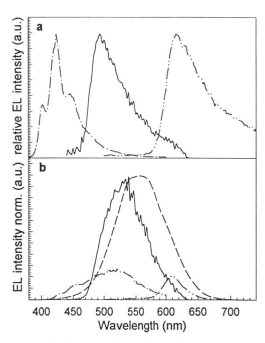

Fig. 30.21 (a) Spectra of the electroluminescence of the PHP OLED (-··-), the green (———), and the red (-···-) emission. (b) Blue (-··-), green (———), and red (-···-) emission normalized to the spectral eye sensitivity at daylight (---). Area under the spectra is proportional to the brightness of the emission colors.

Applying this color conversion technique, emission colors (e.g., green, yellow, orange, magenta, and red) throughout the whole visible range can be obtained by covering the PHP EL devices with dye layers of different concentrations and, if necessary, suitable filters.

For colored flat displays, efficient production of the two other basic colors (besides blue)—green and red—is of great interest. With this all-organic color conversion technique, RGB emission colors with CIE coordinates (green: $x = 0.27$, $y = 0.62$; red: $x = 0.65$, $y = 0.34$) similar to the basic colors that are commonly used in commercial color TVs were produced (see Fig. 30.17) [173,174]. The spectra and photographs of the blue PHP EL emission and the produced green and red emission light are presented in Figs. 30.21 and 30.22. The green and red light emitting areas were produced as follows.

Bright green emission is obtained by using a layer of the dye coumarin 102 in poly(methyl methacrylate) (PMMA) of high optical density (>3). Coumarin 102 is a very suitable dye for the color conversion owing to the strong overlap between the coumarin absorption and the PHP EL emission spectrum and also its efficient green PL light emission. The absolute external conversion efficiency from the blue PHP EL emission into the green was recorded to be around 40% [174]. In this case the green emission (luminance $\approx$ 1100 cd/m^2) appears about three times brighter to the human eye than the blue pumping light from the PHP (360 cd/m^2). This is due to the fact that human eye sensitivity is much higher in the green spectral range than in the blue (see Fig. 30.21b).

The red emission, which is more difficult to achieve, can be obtained by covering the coumarin/PMMA layer with a layer of the red dye Lumogen F300 in PMMA. This red dye strongly absorbs in the green spectral range and therefore converts the green excitation into red emission via PL emission. To obtain an emission color that is located at the pure red position ($x = 0.65$, $y = 0.34$; Fig. 30.17), the spectrum can be purified with a bandpass filter (see Fig. 30.20). The efficiency of conversion from the blue (via the green) to the red was determined to be 10% [174]. The observed intensity losses are mainly due to waveguiding within the red dye layer.

The external conversion efficiency can be significantly increased when the phase matching of the dye/matrix layers to the EL devices is improved so that waveguiding losses are reduced. When the small air gap between the color conversion films and the EL device are eliminated, e.g., by depositing the color-conversion layer directly onto the device, quantum efficiencies for color conversion from blue to green of around 90% and for blue to red of around 40% are obtained [175].

VIII. ELECTROLUMINESCENCE EFFICIENCY

The efficient formation of singlet excitons from the positive and negative polarons that are injected via the metallic contacts and the efficient radiative recombination of these singlet excitons are crucial processes for the efficient functioning of electroluminescence devices. Immediately after the formation of a singlet exciton by the combination of a positive and a negative polaron, it is impossible to distinguish such a singlet exciton created

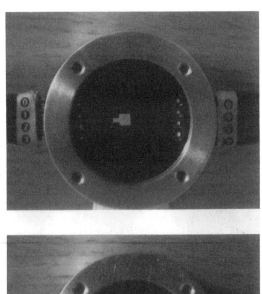

Fig. 30.22 Photographs of the RGB emission colors, all obtained from the same LED. (Light-emitting area = 9 mm^2.)

by charge carriers from a singlet exciton formed by photoexcitation. So the photoluminescence quantum yield or photoluminescence quantum efficiency is one important parameter in the electroluminescence process. A quite simple expression for the electroluminescence quantum efficiency $\eta\Phi$ was introduced by Tsutsui and Saito [176]:

$$\eta\Phi = \gamma\,\eta_r\Phi_f \qquad (12)$$

where γ is a double charge injection factor, η_r quantifies the efficiency of the formation of a singlet exciton from a positive and a negative polaron, and Φ_f is the photoluminescence quantum efficiency. Although simple, this expression gives a good qualitative picture of the EL device functions and optimization. The parameter γ is determined by the processes of carrier injection as outlined in Section IV and is maximal ($\gamma = 1$) if a balanced charge injection (number of injected positive carriers = number of injected negative carriers) into the emission layer of the device is achieved. The parameter η_r represents a number of processes such as carrier transport (number of carriers and their mobility) as described in Section V but also processes that account for the trapping and/or detrapping of carriers and the probability of the formation of a singlet exciton if two polarons of the same charge meet in a certain volume element of the emission layer.

According to Eq. (12), to obtain a high EL quantum efficiency, the active material must have a high photoluminescence quantum efficiency. In the following we discuss the photoluminescence quantum efficiency under ambient conditions. The effects that become important under the operational conditions of an EL device, e.g., high electric fields, elevated temperature, and high charge carrier concentration due to the device current, are discussed subsequently.

There are a number of different techniques for measuring the photoluminescence quantum yield of solid and solutions [177]. The easiest way to determine the quantum yield Φ_f of a substance is to compare the integral photoluminescence intensity, which is emitted from the sample under test due to excitation at an appropriate excitation wavelength to the integral photoluminescence intensity of a fluorescence standard with a known quantum yield (e.g., coumarin 102 [178], which has a quantum yield of $\Phi_f = 0.93$ in ethanol solution [179]). To minimize experimental error using this technique, the measuring geometries for the standard and for the sample should be identical. This can be easily achieved if solutions are investigated, but the situation becomes quite difficult if the quantum yield of solid samples is to be measured. To mitigate this problem one can use circular cuvetes, which can hold solutions as well as solid films. So we have a very similar optical geometry, and the quantum yield values obtained by this technique are satisfyingly accurate with measuring errors of 5% at maximum [17].

The photoluminescence quantum yield can be calculated by using the formula [177,180,181]

$$\Phi_f = \Phi_r(A_s/A_r)(\alpha_r/\alpha_s)(n_s^2/n_r^2) \qquad (13)$$

where Φ_r is the quantum yield of the reference; A_s and A_r are the areas under the photoluminescence spectra as a function of the emission wavelength; α_r and α_s denote the optical absorption of reference and sample, respectively; and n_s and n_r are the refractive indices. The photoluminescence quantum yields for the polyphenylene materials parahexaphenyl, ladder-type polyparaphenylene (with different chain lengths; the short-chain ladder polymer s-LPPP has a molecular weight that corresponds to 20 phenyl rings, and the long-chain polymer l-LPPP consists of an average of 52 phenyl rings along the backbone), and the segmented stepladder polyparaphenylene are listed in Table 30.1 (the solubility of parahexaphenyl in organic solvents is too low to measure Φ_f in solution) [179].

The accuracy of the experimental values of the photoluminescence quantum yield can be increased if the standard and the sample (either solid or film) are placed in an integrating sphere, which collects nearly the entire emitted photoluminescence intensity and is therefore less sensitive to the particular shape of the sample or the reference [182,183]. Recent comparative experiments with the circular cuvet geometry and also with an integrating sphere confirmed the quantum yield data presented in Table 30.1 within about the same experimental error [179,183].

In all samples of Table 30.1, the quantum yield in the films is significantly lower than in the corresponding solution. In intrinsically pure materials, basically two effects can be considered to be responsible for luminescence quenching in films [184]: static quenching by formation of aggregates in the unexcited ground state and collisional quenching due to interaction between singlet excitons in the excited state. We propose that collisional quenching is mainly responsible for the decrease in the quantum yield of LPPPs going from solution to the solid state. The absorption spectra of LPPP films do not differ from the absorption spectra in solution (except for a small shift caused by the dipole interaction with the solvent). This indicates that negligible aggregation, which is necessary for static quenching, takes place in the films. The temperature dependence of the luminescence in a film of l-LPPP plotted in Fig. 30.23 shows that the quantum yield increases by a factor of 2.3 (see Fig. 30.24) between 5 and 300 K as is qualitatively expected for collisional quenching [184], since higher temperatures invoke a higher collision probability. For static quenching one would expect the quantum yield to decrease at lower temperatures. Intramolecular quenching effects like internal conversion (IC) and intersystem crossing (ISC) would lead to an Arrhenius type of temperature behavior [185], in contrast to the experimental

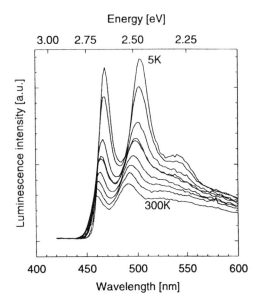

Fig. 30.23 Photoluminescence of l-LPPP [excitation at λ = 400 nm (3.1 eV)] in the temperature range between 5 and 300 K.

results in Fig. 30.24 with a linear temperature dependence of the integrated luminescence intensity.

The fast kinetics of the photoluminescence emission, picosecond and femtosecond relaxation, in ladder-type polyparaphenylenes is notable [142,143,186], since such measurements reveal stimulated emission and therefore promising aspects for LPPPs as a suitable material for optically pumped solid-state homopolymer lasers as well as for the realization of electrically pumped solid-state polymer lasers [144]. To illustrate the situation we depict the photoluminescence intensity as a function of the excitation power (excitation wavelength λ = 295 nm) for picosecond pulses (t = 4 ps) in Fig. 30.25 for an m-LPPP film at room temperature. The photoluminescence emission intensity (shown for the emission maximum at 495 nm and the long-wavelength emission above 520 nm, see Fig. 30.6, sample temperature 300 K) remains unsaturated up to very high excitation power values of 1 mJ per pulse, and the highly purified LPPP is also very stable against degradation and photooxidation if kept under inert conditions. So we conclude that LPPP is a good candidate for a homopolymer laser emitting in the blue-green at 495 nm.

All three parameters of Eq. (12) can be quite strongly influenced under extreme conditions like those present in an operating EL device. In the following we discuss the influence of the electric field on the photoluminescence quantum yield in the case of ladder-type polyparaphenylenes.

The field dependence of the PL was extensively investigated for another conjugated polymer—a soluble polyphenylphenylenevinylene—in polystyrene blends (PPV/PS) by Bässler and coworkers [187–190]. Femtosecond time-resolved PL measurements revealed that the field-induced luminescence quenching is not an instantaneous process [187]. This shows that the formation of singlet excitons is hardly affected by the field, whereas their radiative decay strongly depends on the external field. Thus the magnitude of absorption in the range of the excitation energy of the singlet excitons changes only slightly with applied electric field [191]. The lifetime of singlet excitons is drastically reduced by the electric field, i.e., by opening nonradiative decay pathways, within the first picoseconds after excitation, resulting in a strong quenching under an electric field [190].

We outline below details of the dependence of the steady-state PL of the m-LPPP homopolymer on static

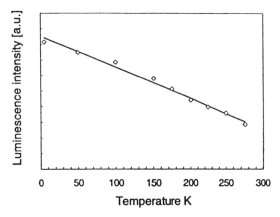

Fig. 30.24 Integral luminescence intensity of the l-LPPP film depending on temperature.

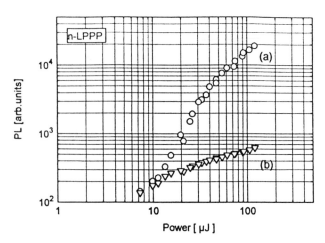

Fig. 30.25 Photoluminescence emission intensity (a) at 495 nm and (b) below 520 nm for LPPP thin films at room temperature.

electric fields. A maximum PL quenching of 76% is obtained for an applied field of 4.5 MV/cm. When the electric field exceeds a magnitude of 2 MV/cm, the dominant peak in the PL spectrum (see Fig. 30.6), located at 461 nm, is less efficiently quenched than the two other peaks at 491 and 530 nm. Therefore the color of the PL emission can be controlled by the external field in the blue-green spectral range.

The absorption and excitation spectra of the conjugated polymer m-LPPP are characterized by a steep onset at 2.69 eV due to a π–π^* transition and by well-resolved maxima (see Fig. 30.6). The PL and the EL spectra are homologous to the absorption spectrum, and their dominant maximum is very slightly Stokes-shifted to the absorption maximum. Due to the strong overlap between absorption and PL spectra, the relative intensity and location of the dominant PL and EL peak (λ = 461 nm) is strongly influenced by self-absorption effects. If the self-absorption is taken into account via a correction of the PL spectrum considering the geometry of the experiment and the optical density of the sample, this correction yields a Stokes shift of 0.036 eV, which is of the same order of magnitude as the thermal energy kT at room temperature.

To investigate the electric field effect on the PL intensity, the PL is recorded from a typical device (ITO/m-LPPP/Al) on which a bias voltage was applied. In the forward direction (ITO as anode) the onset of the current and also the EL emission occurs when the electric field exceeds a threshold field (usually between 0.5 and 1.5 MV/cm [20]). To prevent an overlap between the EL emission and the PL emission, the PL measurements are usually performed by applying reverse bias voltages or small forward bias voltages (<5 V). Moreover, one must always verify that no EL emission is detectable before the field-dependent PL measurements are performed. The m-LPPP polymer layer is not charge-depleted by applying reverse fields up to 3 MV/cm (in contrast to the case when forward electric fields are applied) [192,193]. Therefore space charge regions that would drastically affect the internal electric fields can be neglected for reverse fields. The internal electric field deviates slightly from the external field due to a built-in electric field that results from the difference of the work functions of the electrodes. For the ITO/m-LPPP/Al device configurations, the built-in voltage is determined to be around 0.7 eV, which corresponds to a built-in field of 0.7 MV/cm for an active layer with a thickness of 100 nm.

The dependence of the PL spectrum on the applied reverse external electric field E is shown in Fig. 30.26. The PL intensity (I_{PL}) and hence the PL quantum efficiency strongly decrease for an increasing applied field, whereas the location of the PL peaks remains unchanged. The relative change of the PL intensity associated with the applied electric field is spectrally dependent. The PL quenching (Q_{PL}) of the integrated PL and of the PL peak components as a function of the external electric field E, which is defined as

$$Q_{PL}(E) = \frac{I_{PL}(0) - I_{PL}(E)}{I_{PL}(0)} \quad (14)$$

where $I_{PL}(0)$ is the intensity of the zero field PL, is shown in Fig. 30.27. The decay of the PL can be divided into three sections: The PL intensity slowly decreases for increasing external electric fields up to around 1.5 MV/cm (the decay in this region can be approximated by $I \approx E^{0.2}$); then the integrated PL emission decreases rapidly down to 33% of its zero field value when the fields are increased to 3.1 MV/cm ($I \approx E^{0.54}$). Then when the applied field is further increased, the quenching of the PL starts to saturate (24% at 4.45 MV/cm; $I \approx E^{0.12}$), as was also observed for inorganic semiconductors [194].

The intensity of the peak at 461 nm is less efficiently quenched than the intensity of the other peaks in the spectrum. No correlation is observed between the excitation wavelength (380–440 nm) and the field-dependent PL.

Several mechanisms associated with the application of an electric field may contribute to the PL quenching (field-induced quenching) in the device:

1. Due to a leakage current passing through the device (see Fig. 30.28), polarons, which provide the

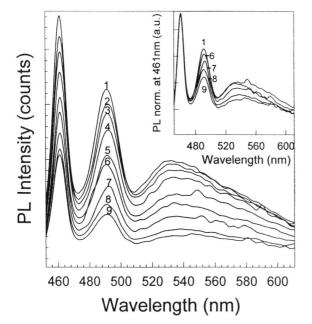

Fig. 30.26 Photoluminescence (excited at 420 nm) depending on the applied reverse external electric field of (3) 1.8 MV/cm; (4) 2.0 MV/cm, (5) 2.2 MV/cm, (6) 2.4 MV/cm, (7) 2.9 MV/cm, (8) 3.3 MV/cm, (9) 4.4 MV/cm). Inset: Spectra normalized to the intensity of the peak at 461 nm.

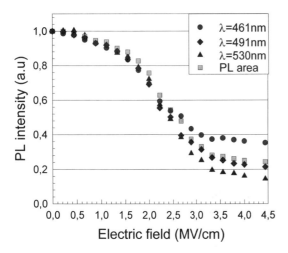

Fig. 30.27 Integrated photoluminescence: Intensity of the peak at 461, 491, and 530 nm as a function of the external electric field.

charge transport in the polymer, might act as effective quenching sites for singlet excitons [136,195,196].

2. The applied voltage and the leakage current cause electric power dissipation in the sample. The sample temperature therefore increases and enhances thermal quenching by collisional quenching mechanisms or by opening nonradiative decay paths of the excited species [17] as outlined above.

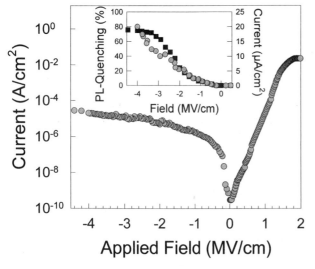

Fig. 30.28 Dark current vs. electric field characteristics of the ITO/m-LPPP/Al. Inset: (■) PL quenching and (●) current depending on the applied electric field.

3. The kinetics of the singlet excitons are changed so that the excitons might be swept to the electrodes by the applied field [197]. The probability for nonradiative recombination of the singlet excitons at quenching sites or at the electrodes is therefore enhanced.
4. The electric field could influence the exciton–exciton annihilation processes [198].
5. Quenching is also caused by field-induced annihilation of the emitting species.

Current-induced quenching was assumed to be responsible for PL quenching effects observed in organic field effect transistor (FET) devices [199,200]. However, to explain the PL quenching solely in terms of current-induced quenching mechanisms is not possible in this case, as the magnitude of the PL quenching we obtained at low leakage currents of around 10 μA/cm^2 is much higher than the PL quenching observed in FET devices, which was around 5%. In particular, the plateau at low applied fields (<1.2 MV/cm) and the saturation of the PL quenching at high fields (>3.5 MV/cm) cannot be satisfactorily described with current-induced quenching, as the current–field characteristic does not coincide with the quenching–field dependence (see Fig. 30.28). Furthermore, most recent results show that the PL quenching for forward electric fields above 1 MV/cm is around 50% lower than the PL quenching for reverse fields of the same magnitude, although the current in the forward direction is several orders of magnitude higher than in the reverse direction.

The displacement of the singlet excitons and the temperature effects, which should increase with the field or the electric power, also fail to describe the observed field dependence of the PL. The effect of temperature quenching on the photoluminescence is described below.

Therefore one can propose that the field-induced annihilation of the emitting species is the most dominant process for quenching, which also explains the spectrally nonuniform PL quenching that is observed above a magnitude of the applied electric field of 2 MV/cm (see Figs. 30.26 and 30.27). The intensity of the dominant PL peak at 461 nm and the area under the peak decrease in a less pronounced fashion than the intensity of the peak at 491 nm and the broad emission peak at approximately 530 nm. To interpret this phenomenon by field-induced annihilation, we discuss how the PL emission process in the m-LPPP polymer evolves. Time-resolved PL spectroscopy is a powerful method for studying the nature and dynamics of emission processes. Streak camera measurements on the m-LPPP polymer (for details see Ref. 144) reveal that two excited species with different lifetimes contribute to the PL emission (a behavior that is also observed in Ref. 142).

The fast component ($\tau < 25$ ps) is attributed to the radiative decay of quasi-free intrachain singlet excitons

(SE). The emission peaks located at 461, 491, and 530 nm result from this spontaneous emission into the vibronic side bands of the SE states [201]. The long-living component ($\tau \approx 800$ ps) can be discussed in terms of the radiative recombination of self-trapped excitons (STE) [202], leading to a rather broad emission peak centered at approximately 545 nm. In the time-integrated PL, the broad emission masks the vibronic peak at 530 nm, while it contributes as an intensive background to the peak at 491 nm and is negligible for the dominant peak at 461 nm (see Fig. 30.6).

When electric fields are applied, processes competitive with the radiative decay of the excitons, such as their field-induced dissociation, usually occur. The Onsager theory is often used to describe the field-dependent decay of the PL intensity (I_{PL}) due to charge separation:

$$I_{PL}(E) = \frac{k_r}{k_r - k_n}[1 - \eta_F \Omega(F)]I_{ex} \quad (15)$$

where k_r and k_n are the rate constants for radiative and nonradiative decay, respectively, Ω is the probability for the dissociation of electron–hole pairs into free charge carriers, and I_{ex} is the building rate of the excitons. In organic photoconductors a slightly different dependence of I_{PL} on the electric field is often observed, which is attributed to field-induced formation of trapped excitons, which may subsequently return to the primary excited state or be dissociated into free carriers from this state [203]. In semiconductors, the Poole–Frenkel effect, which describes the field-assisted thermal dissociation of excitons, is often successfully applied to describe the observed PL quenching.

In the case of m-LPPP, the PL quenching can be well described using the Onsager model for field-induced dissociation of SE external electric fields up to 2 MV/cm. For electric fields above 2 MV/cm, the more strongly bound STEs are also dissociated; therefore, the PL intensity resulting from the spontaneous decay of the SE decreases in a less pronounced way than the broad emission peak that stems from the radiative decay of the STE at approximately 530 nm (see Figs. 30.26 and 30.27). The nature of the plateau of the PL quenching at high electric fields is not fully understood but might be explained by deviations of the internal electric field from the external field due to the creation of space charge zones at such high applied electric fields.

A crude estimation of the magnitude of the exciton binding energy in m-LPPP can be obtained from the observed PL quenching by assuming the following scenario. The electric field changes the Coulomb potential in which the excitons are confined. The effective potential consists of a superposition of the Coulomb potential with the linear electric potential. One branch of the asymmetric effective potential a maximum comes into existence that is lowered by ΔW,

$$\Delta W = 2e\left(\frac{eE}{\epsilon\epsilon_0}\right)^{1/2} \quad (16)$$

(where ϵ is the dielectric constant) compared to the free ionization energy. When an external electric field E is applied the excitons will effectively dissociate when ΔW is equal to the binding energy of the exciton E_B. When we take a value for the electric field of 0.9 MV/cm at which the onset of the effective PL quenching occurs and use the anisotropic intrachain dielectric constant of the m-LPPP polymer ($\epsilon_{LPPP} \approx \epsilon_{PPP} = 10$ [204]), Eq. (16) yields a value of 0.8 eV for E_B. In this simple model, some effects contributing to the observed PL quenching (e.g., tunneling, temperature, or trapping of the dissociated excitons) were neglected; therefore, this value gives an upper limit for the E_B of free excitons. From electroabsorption experiments we obtain a value of E_B of ≈ 0.5 eV [191].

As the m-LPPP polymer is a low-dimensional system, some aspects, such as additionally quantized exciton energy (due to an increase in the overlap between the electron and the hole building up the exciton), have to be taken into consideration. The sample temperature also influences the shape of the PL spectrum. In the case of temperature quenching, the short-wavelength part is more effectively quenched than the long-wavelength part (just opposite to electric field quenching). This change in the shape of the PL spectrum is accompanied by a decrease in the integral PL intensity due to overall thermal quenching.

IX. STABILITY AND LIFETIME

The operational lifetime of LEDs based on organic materials is a crucial parameter in the likelihood of their being competitive with other materials (e.g., inorganic systems based on GaN, SiC, and others) for industrial applications such as flat color screens. The requirements range between operational lifetimes of 10,000 and 100,000 h depending on the particular application.

The LEDs that were fabricated in the period right after the discovery of the electroluminescence in conjugated polymers by the Cambridge group [6] survived for from only a few seconds up to a few minutes under acceptable operational conditions, i.e., the application of a substantial forward driving voltage (usually up to several tens of volts, depending on the types of polymer and electrodes used), producing EL emission clearly visible under moderate daylight conditions, which corresponds to luminance values of ≈ 1 cd/m^2 and above. Since that time, understanding of the physical processes operating in LEDs and fabrication techniques have significantly improved [205, 206], most recently leading to quite promising values of the operational lifetime—up to 6000 h for ITO/PPV/Al homolayer devices emitting 1 cd/m^2 and operated under ambient conditions [207].

The highest performance in blue organic EL devices was claimed by Hosokawa et al. [208] with a multiheterostructure device in which the active layer was a 1,4-bis(2,2-diphenylvinyl)biphenyl (DPVBi) doped with an amine. This group obtained a half-life (time in which luminance has decreased to 50% of its initial value) of more than 5000 h for devices driven initially at 100 cd/m^2. The efficiency of this type of device is 5 lm/W, corresponding to an EL quantum efficiency of about 5%. These are the highest values obtained so far.

ACKNOWLEDGMENT

We thank Joseph Shinar, Guglielmo Lanzani, Klaus Müllen, and Ullrich Scherf for stimulating and fruitful discussions.

REFERENCES

1. C. H. Gooch, *Injection Electroluminescent Devices*, Wiley, New York, 1973.
2. A. H. Kitai, *Solid State Electroluminescence—Theory, Materials and Devices*, Chapman and Hall, London, 1993.
3. T. Round, *Electr. World*, 309 (1907).
4. M. H. Pilkuhn and W. Schairer, Light emitting diodes, in *Handbook of Semiconductors*, Vol. 4 (C. Hilsum, ed.), North-Holland, Amsterdam, 1993.
5. J. Kanicki, Polymeric semiconductor contacts and photovoltaic applications, in *Handbook of Conducting Polymers* (T. Skotheim, ed.), Marcel Dekker, New York, 1986.
6. J. H. Burroughes, D. D. C. Bradley, A. R. Brown, R. N. Marks, K. Mackay, R. H. Friend, P. L. Burns, and A. B. Holmes, *Nature 347*:539 (1990).
7. R. H. Friend, Polymer LED's, this volume.
8. D. Braun and A. J. Heeger, *Appl. Phys. Lett. 58*: 1982 (1991).
9. V. Ohmori, M. Uchida, K. Muro, and K. Yoshino, *Jpn. J. Appl. Phys. 30*:L1941 (1991).
10. G. Grem, G. Leditzky, B. Ullrich, and G. Leising, *Adv. Mater. 4*:36 (1992).
11. Z. G. Soos, D. S. Galvao, and S. Etemad, *Adv. Mater. 6*:280 (1994).
12. G. Grem and G. Leising, *Synth. Met. 57*:4105 (1993).
13. J. Grüner, H. F. Wittmann, P. J. Hamer, R. H. Friend, J. Huber, U. Scherf, K. Müllen, S. C. Moratti, and A. B. Holmes, *Synth. Met. 67*:181 (1994).
14. M. Hamaguchi and K. Yoshino, *Jpn. Appl. Phys. Lett. 34*:L587 (1995).
15. C. Zhang, H. von Seggern, K. Pakbaz, B. Kraabel, H. W. Schmidt, and A. J. Heeger, *Synth. Met. 62*: 35 (1994).
16. M. Berggren, O. Inganäs, G. Gustafsson, J. Rasmusson, M. R. Andersson, T. Hjertberg, and O. Wennerström, *Nature 372*:444 (1994).
17. J. Stampfl, S. Tasch, G. Leising, and U. Scherf, *Synth. Met. 71*:2125 (1995).
18. Y. Yang, Q. Pei, and A. J. Heeger, *Appl. Phys. Lett. 79*:934 (1996).
19. H. W. Furumoto and H. L. Ceccon, *IEEE J. Quant. Electron. 6*:262 (1970).
20. Tasch, A. Niko, G. Leising, and U. Scherf, *Appl. Phys. Lett. 68*:1090 (1996).
21. P. Kovacic and C. Wu, *J. Polym. Sci. 47*:448 (1960).
22. P. Kovacic, M. B. Feldman, J. P. Kovacic, and J. B. Lando, *J. Appl. Polym. Sci. 12*:1735 (1968).
23. C. E. Brown, P. Kovacic, C. A. Wilkie, J. A. Kinsinger, R. E. Hein, S. I. Yaniger, and R. B. Cody, *J. Polym. Sci., Polym. Chem. Ed. 24*:255 (1986).
24. T. Yamamoto, Y. Hayashi, and A. Yamamoto, *Bull. Chem. Soc. Jpn. 51*:2091 (1978).
25. P. Kovacic and A. Kyriakis, *J. Am. Chem. Soc. 85*: 454 (1963).
26. C. S. Marvel and G. E. Hartzell, *J. Am. Chem. Soc. 81*:448 (1959).
27. L. W. Shacklette, R. R. Chance, D. M. Ivory, G. G. Miller, and R. H. Baughman, *Synth. Met. 1*:307 (1979).
28. G. Leising, T. Verdon, G. Louarn, and S. Lefrant, *Synth. Met. 41–43*:279 (1991).
29. J. R. Reynolds and M. Pomerantz, *Electroresponsive Molecular and Polymeric Systems*, Vol. 2 (T. A. Skotheim, ed.), Marcel Dekker, New York, 1991, p. 187.
30. D. L. Gin, V. P. Conticello, and R. H. Grubbs, *J. Am. Chem. Soc. 114*:3167 (1992).
31. M. Rehahn, A. D. Schlüter, and G. Wegner, *Makromol. Chem. 191*:1991 (1990).
32. M. Rehahn, A. D. Schlüter, G. Wegner, and W. J. Feast, *Polymer 30*:1054 (1989).
33. M. Rehahn, A. D. Schlüter, G. Wegner, and W. J. Feast, *Polymer 30*:1060 (1989).
34. T. Yamamoto and A. Yamamoto, *Chem. Lett. 1977*: 353.
35. M. Lögdlund, W. R. Salaneck, F. Meyers, J. L. Bredas, G. A. Arbuckle, R. H. Friend, A. B. Holmes, and G. Froyer, *Macromolecules 26*:3815 (1993).
36. C. Ambrosch-Draxl, J. A. Majewski, P. Vogl, and G. Leising, *Phys. Rev. B 51*:668 (1995).
37. B. Tieke, C. Bubeck, and G. Lieser, *Makromol. Chem. 3*:261 (1982).
38. G. Leising, H. Kreimaier, and R. H. Grubbs, unpublished results.
39. A. D. Schlüter, *Adv. Mater. 3*:284 (1991).
40. U. Scherf and K. Müllen, *Makromol. Chem., Rapid Commun. 12*:489 (1991).
41. D. Bloor, *Nature 356*:19 (1992).
42. G. Grem, C. Paar, J. Stampfl, G. Leising, J. Huber, and U. Scherf, *Chem. Mater. 7*:2 (1995).
43. A. Niko, F. Meghdadi, C. Ambrosch-Draxl, P. Vogl, and G. Leising, *Synth. Met. 76*:177 (1996).
44. G. Wegner, D. Neher, M. Remmers, V. Cimrova, and M. Schulze, *Mater. Res. Soc. Symp. Proc. 413*: 23 (1996).

45. M. Era, T. Tsutsui, and S. Saito, *Appl. Phys. Lett.* 67:2436 (1995).
46. H. Kuhn, *Fortschr. Chem. Org. Naturst.* 16:169 (1958).
47. H. Kuhn, *Fortschr. Chem. Org. Naturst.* 17:404 (1959).
48. G. Leising, K. Pichler, and F. Stelzer, *Springer Ser. Solid State Sci.* 91:100 (1989).
49. R. K. Khanna, Y. M. Jiang, B. Srinivas, C. B. Smithhart, and D. L. Wertz, *Chem. Mater.* 5:1792 (1993).
50. A. Heim, G. Leising, and H. Kahlert, *J. Lumin.* 31/32:573 (1984).
51. A. Heim and G. Leising, *Mol. Cryst. Liq. Cryst.* 118:309 (1985).
52. W. Graupner, G. Grem, F. Meghdadi, Ch. Paar, G. Leising, U. Scherf, K. Müllen, W. Fischer, and F. Stelzer, *Mol. Cryst. Liq. Cryst.* 256:549 (1994).
53. J. Grimme, M. Kreyenschmidt, F. Uckert, K. Müllen, and U. Scherf, *Adv. Mater.* 7:292 (1995).
54. W. Graupner, S. Eder, M. Mauri, G. Leising, and U. Scherf, *Synth. Met.* 69:419 (1995).
55. F. Momicchioli, M. C. Bruni, and I. Baraldi, *J. Phys. Chem.* 76:3983 (1972).
56. D. A. Halliday, P. L. Burn, D. D. C. Bradley, R. H. Friend, O. M. Gelsen, A. B. Holmes, A. Kraft, J. H. F. Martens, and K. Pichler, *Adv. Mater.* 5:40 (1993).
57. K. Pichler, D. A. Halliday, D. D. C. Bradley, R. H. Friend, P. L. Burn, and A. B. Holmes, *Synth. Met.* 55–57:230 (1993).
58. W. Graupner, S. Eder, K. Petritsch, G. Leising, and U. Scherf, *Synth. Met.* in print.
59. A. Dodabalapur, H. E. Katz, L. Torsi, and R. C. Haddon, *Science* 269:560 (1995).
60. F. Garnier, F. Z. Peng, G. Horowitz, and D. Fichou, *Adv. Mater.* 2:592 (1990).
61. F. Garnier, R. Hajlaoui, A. Yasser, and P. Srivastava, *Science* 265:1684 (1994).
62. A. R. Brown, D. D. Bradley, J. H. Burroughes, R. H. Friend, N. C. Greenham, P. L. Burn, A. B. Holmes, and A. Kraft, *Appl. Phys. Lett.* 61:2793 (1992).
63. M. S. Sze, *Physics of Semiconductor Devices*, Wiley-Interscience, New York, 1981.
64. B. L. Sharma (ed.), *Metal-Semiconductor Schottky Barrier Junctions and Their Applications*, Plenum, New York, 1984.
65. Q. Pei and A. J. Heeger, *Science* 269:1086 (1995).
66. P. Dannetun, M. Lögdlund, C. Fredriksson, R. Lazzaroni, C. Fauquet, S. Stafström, C. W. Spangler, J. L. Bredas, and W. R. Salaneck, *J. Chem. Phys.* 100:6765 (1994).
67. G. Gustafsson, Y. Cao, G. M. Treacy, F. Klavetter, N. Colaneri, and A. J. Heeger, *Nature* 357:477 (1992).
68. Y. Yang and A. J. Heeger, *J. Appl. Phys.* 64:1245 (1994).
69. I. H. Campbell, D. L. Smith, and J. P. Ferraris, *Appl. Phys. Lett.* 66:3030 (1995).
70. T. P. Nguyen, H. Ettaik, S. Lefrant, and G. Leising, in *Polymer-Solid Interface* (Prieaux, Bertrand, and Brédas, eds.), IOP Publication, 1992.
71. G. A. Somorjai, *Chemistry in Two Dimensions*, Cornell Univ. Press, Ithaca, NY, 1981.
72. Marks, F. Biscarini, T. Virgili, M. Muccini, R. Zamboni and C. Taliani, *Phil. Trans. Ron Soc. Lond. A* (submitted).
73. H. H. Kim, T. M. Miller, E. H. Weterwick, Y. O. Kim, E. W. Kwock, M. D. Morris, and M. Cerullo, *J. Lightwave Technol.* 12:2107 (1994).
74. Y. Kim, H. Park, and J. Kim, *Appl. Phys. Lett.* 69:1 (1996).
75. E. Ettedgui, H. Razafitrimo, Y. Gao, and B. R. Hsieh, *Appl. Phys. Lett.* 67:2705 (1995).
76. P. Bröms, J. Birgersson, N. Johnsson, M. Löglund, and W. R. Salaneck, *Synth. Met.* 74:179 (1995).
77. J. S. Scott, J. H. Kauffmann, P. J. Brock, R. DiPetro, J. Salem, and J. A. Goitia, *J. Appl. Phys.* 79:2745 (1996).
78. S. Karg, J. S. Scott, J. R. Salem, and M. Angelopoulos, *Synth. Met.* (in press).
79. E. Westerweele, P. Smith, and A. J. Heeger, *Adv. Mater.* 7:788 (1995).
80. U. Lemmer, D. Vacar, D. Moses, A. J. Heeger, T. Ohnishi, and T. Noguchi, *Appl. Phys. Lett.* 68:3007 (1996).
81. R. K. Kasim, M. Pomerantz, and R. L. Elsenbaumer, *Synth. Met.* (in press).
82. J. R. Sheats, H. Antoniadis, M. Hueschen, W. Leonard, J. Miller, R. Moon, D. Roitman, and A. Stocking, *Science* 273:884 (1996).
83. S. Tasch, O. Ekström, T. Jost, G. Leising, and U. Scherf, *Synth. Met.* (in press).
84. C. W. Tang and S. A. VanSlyke, *Appl. Phys. Lett.* 51:913 (1987).
85. G. Leising, G. Köpping-Grem, F. Meghdadi, A. Niko, S. Tasch, W. Fischer, L. Pu, M. W. Wagner, R. H. Grubbs, L. Athouel, G. Froyer, U. Scherf, and J. Huber, *Proc. SPIE* 2527:307 (1995).
86. D. R. Lide, *Handbook of Chemistry and Physics*, CRC, Boca Raton, FL, (1995).
87. C. Adachi, T. Tsutsui, and S. Saito, *Appl. Phys. Lett.* 57:531 (1990).
88. F. Meghdadi, S. Tasch, B. Winkler, W. Fischer, and G. Leising, *Synth. Met.* (in press).
89. Q. Pei and F. Klavetter, U.S. Patent 08/268763 (1994).
90. Q. Pei, G. Yu, C. Zang, Y. Yang, and A. J. Heeger, *Science* 269:1086 (1995).
91. R. H. Fowler and L. Nordheim, *Proc. Roy Soc. Lond. Ser. A* 119:173 (1928).
92. I. D. Parker, *J. Appl. Phys.* 75:1657 (1994).
93. P. S. Davids, S. M. Kogan, I. D. Parker, and D. L. Smith, *Appl. Phys. Lett.* (submitted).
94. H. Vestweber, R. Sander, A. Greiner, W. Heitz, R. F. Mahrt, and H. Bässler, *Synth. Met.* 64:141 (1994).
95. P. W. M. Blorn, M. J. M. Jong, and J. J. M. Vleggaar, *Appl. Phys. Lett.* 68:3308 (1996).

96. W. Ottinger, G. Leising, and H. Kahlert, *Springer Ser. Solid State Sci.* 63:63 (1985).
97. J. Sturm, S. Tasch, A. Niko, G. Leising, E. Toussaere, J. Zyss, T. C. Kowalczyk, and K. D. Singer, *Thin Solid Films* (in press).
98. D. McBranch, I. H. Campbell, D. L. Smith, and J. P. Ferraris, *Appl. Phys. Lett.* 66:1175 (1995).
99. A. R. Brown, N. C. Greenham, J. H. Burroughes, D. D. C. Bradley, R. H. Friend, P. L. Burn, A. Kraft, and A. B. Holmes, *Chem. Phys. Lett.* 200:46 (1992).
100. A. Dodabalapur, L. Torsi, and H. E. Katz, *Science* 268:270 (1995).
101. H. T. Chiu, T. Tsutsui, and S. Saito, *Polym. Commun.* 26:61 (1985).
102. G. Leditzky, Ph.D. Thesis, Technische Universität Graz, 1994.
103. N. Karl, in *Numerical Data and Functional Relationships in Science and Technology*, Landolt-Börnstein Springer, Berlin, 1985.
104. G. Leising, O. Leitner, F. Aldrian, and H. Kahlert, *Synth. Met.* 17:635 (1987).
105. C. K. Chiang, C. R. Fincher, Y. W. Park, A. J. Heeger, H. Shirakawa, E. J. Louis, S. C. Gau, and A. G. MacDiarmid, *Phys. Rev. Lett.* 39:1098 (1977).
106. D. M. Ivory, G. G. Miller, J. M. Sowa, L. W. Shacklette, R. R. Chance, and R. H. Baughman, *J. Chem. Phys.* 71:1506 (1979).
107. D. B. Romero, M. Schaer, L. Zuppiroli, B. Cesar, and B. Francois, *Appl. Phys. Lett.* 67:1659 (1995).
108. D. Moses and A. J. Heeger, in *Relaxation in Polymers* (T. Kobayashi, ed.), World Scientific, Singapore, 1993.
109. B. R. Hsieh, H. Antoniadis, M. A. Abkowitz, and M. Stolka, *Polym. Prepr.* 33:414 (1992), and references therein.
110. H. Antoniadis, M. A. Abkowitz, and B. R. Hsieh, *Appl. Phys. Lett.* 65:2030 (1994).
111. W. Graupner, G. Leditzky, G. Leising, and U. Scherf, *Phys. Rev. B* 54:7610 (1996).
112. R. I. Frank and J. G. Simmons, *J. Appl. Phys.* 38:832 (1967).
113. H. Meyer, D. Haarer, H. Naarmann, and H. H. Hörhold, *Phys. Rev. B* 52:2587 (1995).
114. M. Kryszewski and A. Szymanski, *Macromol. Rev.* 4:245 (1970).
115. P. Gomes da Costa and E. M. Conwell, *Phys. Rev. B* 48:1993 (1993).
116. W. Bruetting, E. Buchwald, G. Egerer, M. Meier, K. Zuleeg, and M. Schwoerer, *Synth. Met.*
117. L. J. Rothberg, M. Yan, S. Son, M. E. Galvin, E. W. Kwock, T. M. Miller, H. E. Katz, R. C. Haddon, and F. Papadimitrakopoulos, *Synth. Met.* 78:231 (1996).
118. M. Onoda, D. H. Park, and K. Yoshino, *J. Phys. C: Solid State Phys.* 1:113 (1989).
119. D. D. C. Bradley, T. Hartmann, R. H. Friend, E. A. Marseglia, H. Lindenberger, and S. Roth, *Springer Ser. Solid State Sci.* 76:308 (1987).
120. M. Van der Auweraer, F. C. De Schryver, P. M. Borsenberger, and H. Bässler, *Adv. Mater.* 6:199 (1994).
121. B. Servet, G. Horowitz, S. Reis, O. Lagorsse, P. Alnot, A. Yassar, F. Deloffre, P. Srivastava, R. Hajlaoui, P. Lang, and F. Garnier, *Chem. Mater.* 6:1809 (1994).
122. A. R. Brown, D. D. C. Bradley, J. H. Burroughes, R. H. Friend, N. C. Greenham, P. L. Burn, A. B. Holmes, and A. Kraft, *Appl. Phys. Lett.* 61:2793 (1992).
123. N. C. Greenham, S. C. Moratti, D. D. C. Bradley, R. H. Friend, and A. B. Holmes, *Nature* 365:628 (1993).
124. S. Aratani, C. Zhang, K. Pakbaz, S. Hoger, F. Wudl, and A. J. Heeger, *J. Electron. Mater.* 22:745 (1993).
125. F. Meghdadi, G. Leising, W. Fischer, and F. Stelzer, *Synth. Met.* 76:113 (1996).
126. A. Y. Kryokov, A. C. Saidov, and A. V. Vannikov, *Thin Solid Films* 209:84 (1992).
127. T. P. Nguyen, H. Ettaik, S. Lefrant, and G. Leising, *Synth. Met.* 44:45 (1991).
128. K. Fesser, A. R. Bishop, and D. K. Campbell, *Phys. Rev. B* 27:4804 (1983).
129. J. L. Bredas, J. Cornil, and A. J. Heeger, *Adv. Mater.* 8:447 (1996).
130. F. Geiger, M. Stoldt, H. Schweizer, P. Bäuerle, and E. Umbach, *Adv. Mater.* 5:922 (1993).
131. P. Dyreklev, M. Berggren, O. Inganäs, M. R. Andersson, O. Wennerström, and T. Hjertberg, *Adv. Mater.* 7:43 (1995).
132. J. Shinar, *Synth. Met.* 78:277 (1996), and references therein.
133. N. C. Greenham, R. H. Friend, A. R. Brown, D. D. C. Bradley, K. Pichler, P. L. Burn, A. Kraft, and A. B. Holmes, *SPIE Proc.* 1910:84 (1993).
134. L. S. Swanson, J. Shinar, A. R. Brown, D. D. C. Bradley, R. H. Friend, P. L. Burn, A. Kraft, and A. B. Holmes, *Phys. Rev. B* 46:15072 (1992).
135. G. Horowitz, A. Yassar, and H. J. von Bardeleben, *Synth. Met.* 62:245 (1994).
136. W. Graupner, J. Partee, J. Shinar, G. Leising, and U. Scherf, *Phys. Rev. Lett.* 77:2033 (1996).
137. K. Petritsch, W. Graupner, G. Leising, and U. Scherf, *Synth Met.* in press.
138. K. Petritsch, Diploma Thesis, Technische Universität Graz, 1996.
139. M. Yan, L. J. Rothberg, F. Papadimitrakopoulos, M. E. Galvin, and T. M. Miller, *Phys. Rev. Lett.* 72:1104 (1994).
140. M. Yan, L. J. Rothberg, E. W. Kwock, and T. M. Miller, *Phys. Rev. Lett.* 75:1992 (1995).
141. S. V. Frolov, W. Gellermann, Z. V. Vardeny, M. Ozaki, and K. Yoshino, *Synth. Met.*, in press.
142. W. Graupner, G. Leising, G. Lanzani, M. Nisoli, S. de Silvestri, and U. Scherf, *Phys. Rev. Lett.* 76:847 (1996).
143. W. Graupner, G. Leising, G. Lanzani, M. Nisoli, S. de Silvestri, and U. Scherf, *Chem. Phys. Lett.* 246:95 (1995).

144. G. Kranzelbinder, H. Byrne, S. Hallstein, S. Roth, and G. Leising, *Synth. Met.* in press.
145. W. N. Sproson, *Colour Science in Television and Display Systems*, Adam Hilger, Bristol, England, 1983.
146. J. D. Rancourt, *Optical Thin Films User Handbook*, SPIE, Washington, DC, 1996.
147. Y. Ohmori, M. Uchido, K. Muro, and K. Yoshino, *Jpn. J. Appl. Phys. 30*:L1938 (1991).
148. J. L. Bredas and A. J. Heeger, *Chem. Phys. Lett. 217*:507 (1994).
149. P. L. Burn, A. B. Holmes, A. Kraft, D. D. C. Bradley, A. R. Brown, R. H. Friend, and R. W. Gymer, *Nature 356*:47 (1992).
150. D. Braun, E. G. J. Staring, R. C. J. E. Demandt, G. L. J. Rikken, Y. A. R. Kessener, and A. H. J. Venhuizen, *Synth. Met. 66*:75 (1994).
151. M. Catellani, C. Botta, P. C. Stein, S. Luzzati, and R. Consonni, *Synth. Met. 69*:375 (1995).
152. V. M. Agranovich and M. D. Galanin, *Electronic Excitation Energy Transfer in Condensed Mater*, North-Holland, Amsterdam, 1982.
153. G. Yu, H. Nishino, A. J. Heeger, T.-A. Cehn, and R. D. Dieke, *Synth. Met. 72*:249 (1995).
154. S. C. Moratti, R. Cervini, A. B. Holmes, D. R. Baigent, R. H. Friend, N. C. Greenham, J. Grüner, and P. J. Hamer, *Synth. Met. 71*:2117 (1995).
155. M. Berggren, M. Granström, O. Inganäs, and M. Andersson, *Adv. Mater. 7*:900 (1995).
156. I. D. Parker, Q. Pei, and M. Marrocco, *Appl. Phys. Lett. 65*:1272 (1994).
157. T. Zyung, D.-H. Hwang, I.-N. Kang, H.-K. Shim, W.-Y. Hwang, and J.-J. Kim, *Chem. Mater. 7*:1499 (1995).
158. J.-K. Lee, R. R. Schrock, D. R. Baigent, and R. H. Friend, *Macromolecules 28*:1966 (1995).
159. D. D. Gebler, Y. Z. Wang, J. W. Blatchford, S. W. Jessen, L. B. Lin, T. L. Gustafson, H. L. Wang, T. M. Swager, A. G. MacDiarmid, and A. J. Epstein, *J. Appl. Phys. 78*:4264 (1995).
160. I. Sokolik, Z. Yang, F. E. Karasz, and D. C. Morton, *J. Appl. Phys. 74*:3584 (1993).
161. P. Hesenmann, H. Vestweber, J. Pommerehne, R. F. Mahrt, and A. Greiner, *Adv. Mater. 7*:388 (1995).
162. Q. Pei and Y. Yang, *Adv. Mater. 7*:559 (1995).
163. A. Hilberer, H.-J. Brouwer, B.-J. van der Scheer, J. Wildeman, and G. Hadziioannou, *Macromolecules 28*:4525 (1995).
164. C. Hochfilzer, S. Tasch, G. Leising, and B. Winkler, *Synth. Met.* in press.
165. M. Granström and O. Inganäs, *Appl. Phys. Lett. 68*:147 (1996).
166. A. Fujii, M. Yoshida, Y. Ohmori, and K. Yoshino, *Jpn. J. Appl. Phys. 34*:499 (1995).
167. M. Hamaguchi and K. Yoshino, *Appl. Phys. Lett. 69*:143 (1996).
168. T. Tsutsui, N. Takada, S. Saito, and E. Ogini, *Appl. Phys. Lett. 65*:186 (1994).
169. A. Dodabalapur, L. J. Rothberg, and T. M. Miller, *Appl. Phys. Lett. 65*:2308 (1994).
170. J. Kido, M. Kimura, and K. Nagai, *Science 267*:1322 (1995).
171. H. Tokailin, C. Hosokawa, and T. Kusomoto, U.S. Patent 5,126,214 (1992).
172. C. W. Tang, D. J. Williams, and J. C. Chang, U.S. Patent 5,294,870 (1994).
173. Matsura, H. Tokailin, M. Eida, C. Hosokawa, Y. Hironaka, and T. Kusumoto, *Proc. Asia Display '95*: 269 (1995).
174. S. Tasch, C. Brandstätter, F. Meghdadi, G. Leising, G. Froyer, and L. Athouel, *Adv. Mater. 1* (1997).
175. A. Niko, C. Brandstätter, and S. Tasch, G. Leising, unpublished.
176. T. Tsutsui and S. Saito, *NATO ASI Ser. E:Appl. Sci. 246*:123 (1993).
177. J. N. Demas and G. A. Crosby, *J. Phys. Chem. 75*:991 (1971).
178. I. B. Berlman, *Handbook of Fluorescence Spectra of Aromatic Molecules*, 2nd ed., Academic, New York, 1971.
179. A. N. Fletcher and D. E. Bliss, *Appl. Phys. 16*:289 (1978).
180. Y. S. Liu, P. Mayo and W. Ware, *J. Phys. Chem. 97*:5995 (1993).
181. W. Melhuish, *J. Phys. Chem. 65*:229 (1961).
182. A. Heim, Ph.D. Thesis, Technische Universität Graz, 1996.
183. N. C. Greenham, I. D. W. Samuel, G. R. Hayes, R. T. Philips, Y. A. R. R. Kessener, S. C. Moratti, A. B. Holmes, and R. H. Friend, *Chem. Phys. Lett. 241*:89 (1995).
184. J. R. Lakowicz, *Principles of Fluorescence Spectroscopy*, Plenum, New York, 1983.
185. C. Burgdorff and H.-G. Löhmannsröben, *J. Lumin. 59*:201 (1994).
186. T. Pauck, R. Hennig, M. Perner, U. Lemmer, U. Siegner, R. F. Mahrt, U. Scherf, K. Müllen, H. Bässler and E. O. Göbel, *Chem. Phys. Lett. 244*:171 (1995).
187. R. Kersting, U. Lemmer, H. J. Bakker, R. F. Mahrt, H. Kurz, V. I. Arkhipov, H. Bässler, and E. O. Göbel, *Phys. Rev. Lett. 73*:1440 (1994).
188. U. Lemmer, S. Karg, M. Scheidler, M. Deussen, W. Riess, B. Cleve, P. Thomas, H. Bässler, M. Schwoerer, and E. O. Göbel, *Synth. Met. 67*:169 (1994).
189. V. I. Arkhipov, H. Bässler, M. Deussen, E. O. Göbel, R. Kersting, H. Kurz, U. Lemmer, and R. F. Mahrt, *Phys. Rev. B 52*:4932 (1995).
190. Deussen, M. Scheidler, and H. Bässler, *Synth. Met. 73*:123 (1995).
191. G. Meinhardt, A. Horvath, G. Weiser, G. Leising, *Synth. Met.* (submitted).
192. I. H. Campbell, D. L. Smith, and J. P. Ferraris, *Appl. Phys. Lett. 66*:3030 (1995).
193. S. Tasch, R. Andreaus, G. Leising, and U. Scherf, unpublished.

194. S. Yi, C. Palusule, S. Gangopadhyay, U. Schmidt, and B. Schröder, *J. Non-Cryst. Solids 164–166*:591 (1993).
195. D. D. C. Bradley and R. H. Friend, *J. Phys.: Condens. Matter 1*:3671 (1989).
196. N. C. Greenham, J. Shinar, J. Partee, P. A. Lane, O. Amir, F. Lu, and R. H. Friend, *Phys. Rev. B53*: 13528 (1996).
197. H. J. Lozykowski, A. K. Alshawa, and I. Brown, *J. Appl. Phys. 76*:4836 (1994).
198. R. G. Kepler and Z. G. Soos, *Phys. Rev. B 47*:9253 (1993).
199. K. E. Ziemelis, A. T. Hussain, D. D. C. Bradley, R. H. Friend, J. Rühe, and G. Wegner, *Phys. Rev. Lett. 66*:2231 (1991).
200. P. Dyreklev, O. Inganäs, J. Paloheimo, and H. Stubb, *Appl. Phys. Lett. 71*:2816 (1992).
201. F. X. Bronold, A. Saxena, and A. R. Bishop, *Phys. Rev. B 48*:13162 (1993).
202. M. Furukawa, K. Mizuno, A. Matsui, S. D. D. V. Rughooputh, and W. C. Walker, *J. Phys. Soc. Jpn. 58*:2976 (1989).
203. J. Kalinowski, W. Stampor, and P. DiMarco, *J. Electrochem. Soc. 143*:315 (1996).
204. J. Fink, N. Nücker, B. Scheerer, W. Czerwinski, A. Litzelmann, and A. vom Felde, *Springer Ser. Solid State Sci. 76*:70 (1987).
205. F. Cacialli, R. H. Friend, S. C. Moratti, and A. B. Holmes, *Synth. Met. 67*:157 (1994).
206. W. Riess, S. Karg, V. Dyakonov, M. Meier, and M. Schwoerer, *J. Lumin. 60/61*:906 (1994).
207. K. Pichler, Cambridge Display Technology at the Discussion Meeting on Electronics with Molecular Materials: From Synthesis to Device, London, England, June 1996.
208. C. Hosokawa, M. Matsuura, M. Eida, Y. Hironakaa, and T. Kusumoto, Proceedings 49th Annu. Conf. Soc. for Imaging Science and Technology, 1996, p. 388.

31
Corrosion Inhibition of Metals by Conductive Polymers

Wei-Kang Lu, Sanjay Basak, and Ronald L. Elsenbaumer
The University of Texas at Arlington, Arlington, Texas

I. INTRODUCTION

Corrosion is the destructive result of chemical reactions between a metal or metal alloy and its environment. Corrosion impacts many aspects of our lives. Essentially anything made of metal is subject to corrosion, and when corrosion is evident, significant economic consequences may follow. This is true for structural items that comprise the infrastructure of society, such as pipelines, storage tanks, bridges, and airplanes, and for the personal items we own such as automobiles, water heaters, and metal lawn furniture. The cost of corrosion in the United States alone is estimated to be in excess of $70–100 billion per year. In addition to the economic costs, there are times when corrosion is responsible for the even greater costs of human life and safety. Several airplane accidents have occurred in which part of the fuselage tore away during flight, killing passengers. Sections of bridges have collapsed, resulting in injury and loss of life. There is a high probability that these mechanical failures occurred as the result of stress corrosion cracking due to atmospheric corrosion [1–4]. Hence, considerable expense has gone and will continue to go into the development of corrosion prevention techniques and products.

Corrosion science is the study of the chemical and metallurgical processes that occur during the corrosion process. Corrosion engineering, on the other hand, involves the design and application of methods to prevent corrosion. An understanding of both the science and engineering of corrosion is necessary to develop useful methods and materials for the prevention and control of corrosion [2–4].

Since corrosion is an electrochemical process, it is necessary to have an understanding of the electrochemical techniques used to characterize corrosion processes. These are discussed in some detail in Section III. The reader is also directed to several excellent textbooks on the subject [2–4].

The general forms of corrosion that are important to understand are uniform or general corrosion, galvanic corrosion, crevice corrosion, pitting corrosion, environmentally induced cracking, hydrogen damage, intergranular corrosion, dealloying, and erosion corrosion [3]. In this chapter we are concerned with general (uniform) corrosion, galvanic corrosion, and pitting corrosion.

General corrosion involves the uniform removal of metal from the exposed surfaces in the corrosion environment. With this type of corrosion it is generally easy to predict useful service lifetimes from measured corrosion rates.

Galvanic corrosion occurs when two dissimilar metals are electrically coupled in the same corrosive environment. Individually, these two metals usually have different corrosion potentials, but when coupled together the metal with the lower corrosion potential will preferentially corrode, while the metal with the higher corrosion potential will be protected from corrosion. A common galvanic couple is zinc on steel (galvanized steel); in a corrosion environment the zinc, with its lower corrosion potential, preferentially corrodes (becomes a sacrificial anode) while protecting the steel (which becomes the cathode). Galvanic corrosion is an important concept when dealing with conductive polymers (a form of a metal) coupled to (coated on) metals such as steel, aluminum, and copper since the two materials usually have quite different electrochemical potentials.

Pitting corrosion is a highly nonuniform type of corrosion wherein there is highly localized attack in a small area that results in the formation of pits on the surface of the corroding metal. Pitting corrosion is very unpredictable and can be quite rapid, leading to early failures. It is a form of corrosion that it is very desirable to avoid (prevent).

The most common way to prevent a metal from corroding or retard its corrosion is to provide an impervious coating over it. If a perfect barrier layer is applied to the surface of a metal exposed to a corrosive environment, then neither oxygen nor water can reach its surface and corrosion will be prevented. Unfortunately, most coatings, such as paints, are not perfect barriers, and all coating systems eventually fail, either through existing pinholes in the coating or by diffusion of oxygen and water through it. For this reason, it is highly desirable to find secondary methods of corrosion protection that can back up the primary barrier coating technology. Common secondary methods for corrosion protection involve both chemical and electrochemical techniques, such as chemical inhibitors, cathodic protection, and anodic protection. Electrically conductive polymers are capable of providing all three types of protection and as such are materials of considerable practical interest for corrosion inhibition.

II. REVIEW OF CHEMICAL CORROSION INHIBITION METHODS

A. Inhibition by Small Organic Molecules

Pyrrole and substituted pyrroles were found to be effective corrosion inhibitors for iron and aluminum alloys in both hydrochloric and sulfuric acid solutions [5]. When halide salts were added to acids containing pyrrole, a synergistic effect resulted in regard to inhibition of iron dissolution. The protective mechanism is believed to be associated with chemical adsorption of the pyrrole molecule on the metal surface involving a variation in charge transfer between the two phases. The increase in the π-electron density at the N-heteroatom is attributed to the observed increase in the absorptivity and anticorrosion ability. The corrosion rate of low carbon steel in acids containing pyrrole with and without the addition of NaI or NaCl was determined [6]. In the pyrrole–halide inhibitor system, pyrrole primarily inhibits the cathodic reaction, whereas the halide ions (Cl^- and I^-) seem to strongly inhibit the anodic reaction. Polarization studies suggest that for uninhibited acid and acid containing only pyrrole, the corrosion reaction is mostly controlled by the cathodic reaction. Addition of halides to either uninhibited acid or acid containing pyrrole changed the corrosion mechanism to anodic control. When pyrrole and inorganic halides were added to sulfuric acid, a remarkable synergistic effect resulted in the inhibition of iron dissolution. The weight loss of iron specimens was reduced by $\approx 99\%$, and the hydrogen concentration in electrolytes also decreased.

The effect of the type and the position of the substituent groups in pyrrole derivatives (N-arylpyrroles) on the corrosion inhibition of iron in strong acid pickling solution (HCl) was investigated using electrochemical methods [7]. The protection efficiencies in 5 M HCl containing 0.01 M 1-(2-fluorophenyl)-2,5-dimethylpyrrole, 5 M HCl containing 0.01 M 1-(3-fluorophenyl)-2,5-dimethylpyrrole, and 5 M HCl containing 0.01 M 1-(4-fluorophenyl)-2,5-dimethylpyrrole were 67.5%, 93.5%, and 97.6%, respectively.

Many reports have appeared on the use of aniline for corrosion protection of metals in different electrolytes. Aniline was found to be a good inhibiting additive for mild steel in various concentrations of hydrochloric acid [8]. It is seen that, in general, the efficiency of inhibition increases with the molecular weight of aniline derivatives. An adsorbed film of inhibitor builds up on the surface of the metal and is responsible for the delay of attack of the metal by acid. The electrochemical behavior of tin in H_2SO_4 solution using aniline and its ring derivatives was studied by several corrosion measurement techniques [9]. The structure–inhibition correlation indicated that Cl substitution on the aniline ring increased inhibition efficiency. It was proposed that π-electron density was related to protective ability. Desai and Desai [10] reported that N-substituted anilines can act as corrosion inhibitors for aluminum in hydrochloric acid. Results obtained with N-substituted aromatic amines showed the beneficial action of longer alkyl chains. Alkylaniline oligomers exhibit very good initial inhibition of metal corrosion in aqueous environments [11]. These kinds of amines adsorb on the metal surface through the amino group, and their hydrocarbon chains extend into the aqueous phase to form a protective monolayer film at the metal surface to interfere with either the cathodic or anodic reactions occurring at the adsorption sites.

Among the thiophene inhibitors, substitution of one hydrogen atom on the thiophene ring by any substituent, irrespective of its electrophilic or nucleophilic character, increases the inhibitory efficiency and influences both the cathodic and anodic processes as evidenced in the Tafel plots [12]. Also, the introduction of a nucleophilic (electron-donating) substituent increases the density of π electrons in the thiophene ring and improves the protective ability of the inhibitor. Hence, in the case of thiophene derivatives with nucleophilic substituents, the density of π electrons in the molecule is the decisive factor that determines the protective ability of the given compound. It was shown that the inhibitor molecules lie flat on the metal surface and the adsorption of these compounds on the metal/electrolyte interface results from the interaction of π electrons of the thiophene ring with the surface metal atoms and the influence of inhibitors on the anodic branch of the Tafel curves is a bit more pronounced. Therefore, the anodic sites are privileged sites for the reactivity of thiophene containing inhibitors. On the contrary, thiophene derivatives with electrophilic (electron-withdrawing) substituents have lower π-electron density in the ring but have higher protective ability than thiophene. This can be explained by the equilibrium of thiophene with its protonated form in strong acid solutions:

$$\underset{S}{\langle\!}}\!\!-R \;+\; H^+ \;\rightleftharpoons\; \left[\underset{\underset{H}{|}}{\underset{S}{\langle\!\!\!\!}}\!\!-R\right]^+$$

The introduction of the electron-withdrawing substituent, R, increases the positive charge of the cationic form and thus enhances its adsorption to the electron-rich metals due to increased Coulombic interaction. Introduction of strong polar substituents leads to an increase in the dipole moment of the molecule. An increase in the dipole moment of a given compound also relates to its increased protective ability. All the studied thiophene derivatives have been found to influence both the cathodic and anodic processes.

Newer experimental information now suggests that the attachment of aniline and thiophene inhibitors on the metal surface occurs through the heteroatom and not the delocalized π electrons of the aromatic rings [12,13].

B. Inhibition by Electrically Conductive Polymers

The possibility that steel and other metals could be anodically protected by conducting polymers was proposed over 10 years ago by MacDiarmid [14]. The initial findings dealt with the measurements of anodic current from a corroding iron surface by potentiostatic techniques. The experimental results indicated that the measured cell current readings were reduced when the iron was coated with polyaniline. However, the results were not precise and neglected the influence of surface coverage at the anode, and therefore no definitive conclusion could be reached about quantitative corrosion protection by conductive polymers. In 1985 DeBerry [15] found that polyaniline electrochemically deposited on ferritic stainless steels (410 and 430) provided a form of anodic protection that significantly reduced corrosion rates in sulfuric acid solutions. The electrochemical deposition of polyaniline was preceded by the formation of a passive oxide layer on the steel surface. The doped polyaniline layer in electrochemical contact with the steel stabilized the passive oxide layer against dissolution and reduction. The coatings appear to be deposited over the passive metal oxide film but can undergo electron transfer with the metal. The electron transfer exchange with the metal may be partially responsible for the ability of polyaniline to maintain the passivity of the stainless steel. This protective mechanism was then thought to be anodic protection that maintains the native passive film on the metal. More recent studies by Troch-Nagels et al. [16] on the electrochemical deposition of polyaniline and polypyrrole on mild steel showed that polyaniline provided no corrosion protection, whereas polypyrrole provided a significant amount of protection. These authors used potentiokinetic polarization curves (potential versus current density plots) to support their results. However, the major problem in this study was the adhesion of electropolymerized polymer coating to the metal surface. The mechanical and adhesive properties of these tested films do not meet practical requirements. Scanning electron micrographs showed a brittle and powdery film on the steel surface. It is hard to find sufficient experimental evidence to support the claims of the authors. Sekine et al. [17] also found that electrochemically deposited polyaniline provided very little corrosion protection to mild steel surfaces. There are also reports of other corrosion resistance studies related to vinyl- and aryl-type polymers such as polythiophene, polyacrylamide, polyphenylene oxide, and their derivatives in the Sekine et al. paper. Cathodic polarization studies showed that the current density values of coated steel were larger than those of uncoated steels. It is possible that the corrosion of coated steel was accelerated because the film was coated heterogeneously and the coated film acted as a cathodic element in the corrosion process. The most important experimental finding in this study was a decrease in double layer capacitance (C_{dl}) at the metal surface due to the decrease in the dielectric constant in the vicinity of the anode surface because of the polymer. Sathiyanarayanan et al. [18] found that soluble polyethoxyaniline inhibits the corrosion of iron in 1 N HCl solution. Corrosion inhibition by the polymer in solution was about 8 times as effective as aniline at concentrations of 75–100 ppm. The authors implied that the availability of π electrons can enable the conducting polymers to be good corrosion inhibitors. Double-layer capacitance studies indicated that there was strong adsorption of the polymer to the metal surface. These studies further showed that an increase in inhibitor concentration in electrolytes can increase the value of charge transfer resistance of metal surface and thereby increase corrosion inhibition. Ortho-substituted polyanilines with better solubility in common solvents provide improved corrosion inhibition for iron in acidic chloride solution.

Significant corrosion inhibition of polyaniline-coated mild steel exposed to saline (3.5% NaCl) and acidic (0.1 N HCl) environments was reported by Thompson et al. [19]. These polyaniline coatings seemed to promote corrosion protection even where scratches existed in the protective coating. But no reliable quantitative characterization was provided in this report, and only a qualitative explanation was made from electrochemical alternating current impedance measurements or direct current linear polarization experiments. Several other conducting polymers including poly(3-hexylthiophene), poly(3-octylthiophene), poly(3-thienylmethylacetate) and poly(3-thienylacetate) with at least 25 different dopants were used in corrosion testing on steel coupons. Various tests including both immersion and atmospheric testing such as racket outfield tests, ultraviolet radiation, and accelerated salt fog chamber tests were performed in the study. It was found that three dopants—*p*-toluenesulfonic acid, zinc nitrate, and tetracyanoethylene—gave the best results. Wessling [20] recently reported that mild steel, stainless steel, and copper were all found to passivate when clean surfaces of the metals were repeatedly dipped into dispersions of doped polyaniline (commercial trade name Versicon). The dip coating process was repeated 5–20 times to increase the thickness of the polyaniline layer. Passivation was found to occur by the formation of an oxide layer on mild steel induced by contact with the polyaniline. This was shown by removal of the polyaniline layer, which revealed a gray matted surface of γ-iron oxide with persistent passivated behavior. The protection was considered to be a form of anodic protection provided by the high oxidation state of polyaniline. Further, UV/vis spectrophotometric data showed that polyaniline in contact with iron in the presence of water containing 1 M NaCl or 0.1 M *p*-toluenesulfonic acid was quickly reduced from the emeraldine salt to leucoemeraldine. The corresponding oxidation was the formation of passive oxide on iron. The leucoemeraldine was reoxidized back to emeraldine in 24 h. The reoxidation of the polymer can be facilitated by bubbling oxygen into the medium. The authors also found that commercially available polyaniline primer Corrpassive™ coated on steel outperformed epoxy resin primer and other metal-reactive two-component primers in corrosion inhibition when subjected to (1) a crevice and pitting corrosion test (ASTM G-48 in $FeCl_3$) and (2) a salt spray test (DIN 50021) [21,22].

Poly(3-methylthiophene) (P3MT) films on both platinum and 430 stainless steel were evaluated for their effectiveness in stabilizing steel in a passive state [23]. Galvanic experiments between film-coated phosphated 430SS coupled with a 430SS electrode indicated that the P3MT films, which were stable within the periods of study, stabilized a 430SS rotating disk electrode (RDE) in 1 N sulfuric acid solution. Evidently, the film affects the working electrode potential and supplies current to the working electrode only during the short time required to establish a passive oxide layer. The P3MT film then stabilizes the passive layer by providing a transient current to heal small holes inside the passive film before they expand, a mechanism first proposed by DeBerry [15]. With an area ratio of 1:1, the coupled 430SS RDE was galvanically passivated up to 8 h in the same solution. P3MT films protected 430SS RDEs galvanically from corrosion in both air- and nitrogen-saturated 1 N sulfuric acid solutions. In a long-term corrosion test, phosphated 430SS RDE was coated with P3MT on one side, and half of the coated area was then scraped clean. This sample was left at ambient temperature in a stationary 1 N sulfuric acid solution for 30 days with no measurable corrosion. A higher degree of unsaturation in a compound (such as a conductive polymer) often shows a higher corrosion protection efficiency, which probably arises from more π bonds, which will form denser and tighter bonds to anodic metal sites. The loss of insulating and barrier properties under the destructive influence of the corrosive environment is tantamount to diminished protection. It is well known that materials that effectively limit metal dissolution in acid solution also tend to limit hydrogen adsorption, although there are some exceptions. Chemisorption will appear at the metal/polymer interface with the presence of unfilled metal electronic orbitals and low ionization potential of the polymer. According to Singh [24] chemical changes occur in the inhibitor molecules leading to an increment in the electron densities at the adsorption centers of the molecule causing an improvement in inhibitor efficiency. Thus the efficacy of an inhibitor involves dipolar, topographical, and electron donor properties [25].

Corrosion prevention of metals by monomeric and polymeric additives in solutions has been clearly demonstrated. Therefore, the practical application of conducting polymers for corrosion control seems reasonable based on previous studies. Indeed the corrosion protection of steel by conducting polymers as evidenced by fundamental electrochemical and corrosion techniques is now apparent [15,20,26]. Several similar experiments by Jasty and Epstein [27] and Wei et al. [28] using almost identical techniques confirm these initial findings. It was concluded [26] that a passivating oxide film on the iron surface consisting mainly of γ-ferric oxide was affected by the presence of a polyaniline film galvanically coupled to the metal. The equilibrium open-circuit potential of polyaniline-coated cold-rolled steel was found to be more noble (more positive) than that of an uncoated sample during immersion in 0.1 M HCl. However, when the same sample was immersed in 3.5% NaCl, the open-circuit potential decreased in the order neutral polyaniline coated steel > fully HCl-doped polyaniline > uncoated sample. Somewhat surprisingly, it was found that the base form of polyaniline provided the largest increase in polarization resistance and corrosion potential as well as the largest decrease in corrosion current in this medium [26,29].

More recently Ahmad and MacDiarmid [30] published a fundamental study on corrosion inhibition of steel with chemically deposited polyaniline. The study concludes that the corrosion protection for iron provided by this conducting polymer mostly arises from anodic protection, which requires much less current density (micro- to milliamperes per square centimeter) than a polymer can provide to the metal. Anodic protection of steel is achieved by the formation of a passive oxide layer, which can be induced by applying an anodic potential of about 0.1 V vs. SCE. The study showed that the minimum potential required for passivation of a sample of iron in a given corrosive medium is crucial for a polymer to function as a corrosion inhibitor for iron. It was found that a conducting polymer (such as polyaniline) with an open-circuit potential, V_{oc}, a little more positive than the minimum passivation potential for steel in some corrosive medium (e.g., 0.4 N H_2SO_4) can provide excellent protection for steel. If the V_{oc} of the polymer in sulfuric acid medium is less than 0.4 V vs. SCE, then the reduced form of the polymer (e.g., leucoemeraldine) formed as a consequence of the passivation of the stainless steel (430) will be reoxidized by air. In that case, the polymer can work as a redox catalyst and provide continuous corrosion protection as long as the mechanical integrity of the polymer film remains intact. On the other hand, if the V_{oc} of the polymer in the medium is higher than 0.4 V vs. SCE, it will protect stainless steel until it is consumed (completely reduced). The reduced polymer cannot be reoxidized by air and thus the polymer cannot function indefinitely. The study also showed that chemically deposited polyaniline could not provide corrosion protection for stainless steel in chloride-containing acid media for a sufficiently long time. However, pretreatment of the stainless steel surface with various chemicals (e.g., phosphoric acid and chelating agents such as alizarin sulfonate salt or chromotropic acid) prior to the chemical deposition of polyaniline on stainless steel significantly improves the corrosion protection in chloride-containing acid media.

Krstajic et al. [31] investigated the corrosion behavior of polypyrrole electrodeposited mild steel in 0.1 M sulfuric acid by electrochemical impedance spectroscopy. They concluded polypyrrole did not provide anodic protection of mild steel. However, reduced polypyrrole lowered the corrosion velocity of mild steel by about a factor of 20. The total impedance increased with exposure time after an initial period.

There are reports of the use of carbon-filled polymers for cathodic protection of metal surfaces [32]. Results of comparisons of galvanostatic polarization and impedance studies of conductive composites of ethylene-propylene-diene monomer (EPDM) with or without carbon black have been reported. It was found that the content of carbon black in the composite can significantly change the charge transfer resistance of the anode. Also, the surface pretreatment and adhesion of conducting polymers are important factors for control of corrosion.

III. CHARACTERIZATION METHODS

A. Electrochemical Corrosion Processes

Corrosion is a natural chemical process in which a metal or metal alloy is destroyed in its environment by slow oxidation and erosion. Corrosion reactions in aqueous solutions are electrochemical and involve electron transfer between the metal and the environment. For corroding metals, the anodic reaction (oxidation) is of the form

$$M \rightarrow M^{n+} + ne^- \qquad (1)$$

In the corresponding cathodic reaction (reduction), solution species like dissolved oxygen [Eqs. (2a), (2b)], hydrogen ion [Eq. (3)], or water [Eq. (4)] are reduced by accepting electrons from the metal.

$$O_2 + 2H_2O + 4e^- \rightarrow 4OH^- \qquad (2a)$$

$$O_2 + 4H^+ + 4e^- \rightarrow 2H_2O \qquad (2b)$$

$$2H^+ + 2e^- \rightarrow H_2 \qquad (3)$$

$$2H_2O + 2e^- \rightarrow H_2 + 2OH^- \qquad (4)$$

For corrosion of iron, the overall reaction can be written as

$$4Fe + 3O_2 + 2H_2O \rightarrow 2Fe_2O_3 \cdot xH_2O \text{ (rust)} \qquad (5)$$

When an excess of electrons is supplied to the metal, it is always observed that the rate of corrosion, expressed by the anodic reaction, Eq. (1), is reduced, while the rate of hydrogen evolution, Eq. (3), is increased. Thus, the application of a negative potential to the metal decreases the corrosion rate. This is the basis of cathodic protection for the mitigation of corrosion [3].

1. Polarization or Overpotential

Electrochemical reactions such as (1) and (3) occur at finite rates. If excess electrons are available for reaction (3), such as by applying an external negative potential (electrochemical cathodic polarization) or by significantly reducing the concentration of dissolved oxygen for reactions (2a) and (2b) (concentration polarization), the surface potential becomes more negative, suggesting that excess electrons with their negative charges accumulate at the metal/solution interface waiting for reaction. This means that the reaction is not fast enough to accommodate all the available electrons. This negative potential change is called cathodic polarization. Similarly, a deficiency of electrons in the metal released in reaction (1) at the interface as the result of an external applied positive potential (electrochemical anodic polarization) produces a positive potential change called anodic polarization. The relationship between cathodic polarization or overpotential, η_c, and corrosion rate given by i_c (cathodic reaction current density) is [2–4]

$$\eta_c = \beta_c \log \frac{i_c}{i_0}$$

and the relationship for anodic polarization or overpotential, η_a, and corrosion rate given by i_a (anodic reaction current density) is

$$\eta_a = \beta_a \log \frac{i_a}{i_0}$$

where β_c and β_a are the Tafel constants for the cathodically and anodically polarized reactions, respectively, and i_0 is the corrosion current density for the unpolarized reaction. As the anodic polarization becomes greater, the corrosion tendency generally becomes greater. So, anodic polarization provides a driving force for corrosion by accelerating the anodic reaction (1).

2. Passivation

Certain metals, for example iron, titanium, aluminum, chromium, and molybdenum, exhibit the ability to form thin protective oxide layers on their surface. This oxide or hydrated oxide layer is a corrosion product that tightly adheres to the metal surface and acts as a barrier to further anodic metal dissolution (corrosion). It is observed that these passivating oxide layers form on the metal surface at some critical potential, E_p, above the corrosion potential. These oxide layers can be quite protective even when very thin (20 nm). Although the metal corrodes at a relatively high rate at potentials just below E_p, the corrosion rates above E_p after the formation of the oxide layer can be many orders of magnitude less than the normal corrosion rate. By holding the metal's potential above E_p (or adjusting the oxidizing power of the corrosion environment), a metal can remain in the passive state and be considerably protected from corrosion as long as the passivating oxide layer remains intact or self-heals. The process of passivating a metal by strong anodic polarization and barrier oxide formation is referred to as anodic protection [3].

Passivation can sometimes create problems too. The breakdown of the thin and fragile passive oxide film can cause localized forms of corrosion, including pitting, crevice corrosion, and embrittlement.

B. DC Electrochemical Monitoring Techniques

1. Tafel Extrapolation Methods

Corrosion on metals occurs at a reaction rate determined by opposing electrochemical reaction equilibria established between the metal and an electrolyte solution. As described earlier, one reaction is the anodic reaction, in which the metal is oxidized, releasing electrons from its surface. The other is the cathodic reaction, in which solution species like O_2 or H^+, or even the protective coatings and oxide films that cover the metal, are reduced, attracting electrons from the metal. In a corrosion system, the Tafel equations for both cathodic and anodic reactions can be combined into the Stern–Geary [33] equation,

$$i = i_{corr}\left\{\exp\left[\frac{2.303(E - E_{corr})}{\beta_a}\right] - \exp\left[\frac{2.303(E - E_{corr})}{\beta_c}\right]\right\} \quad (6)$$

where

i is the measured cell current in amperes
i_{corr} is the corrosion current in amperes, which is a measure of the corrosion rate
E_{corr} is the corrosion potential in volts
E is the applied electrode potential in volts
β_c and β_a are the cathodic and anodic Tafel beta coefficients in volts/decade

Figure 31.1 shows a classic electrochemically measured Tafel polarization diagram [33]. The Tafel analysis is performed by extrapolating the linear portions of both cathodic and anodic curves on a log (current) versus potential plot to their point of intersection. This intersection point provides both the corrosion potential E_{corr} and the corrosion current density for the system unperturbed. This is a very simple yet powerful technique for quantitatively characterizing a corrosion process. The Tafel equation can be simplified to provide Eq. (7) by approximation using a power series expansion,

$$i_{corr} = \frac{1}{R_p}\left(\frac{\beta_c\beta_a}{2.303(\beta_c + \beta_a)}\right) \quad (7)$$

where R_p, defined as the polarization resistance, is the slope $(dE/di)_{E\to 0}$ at the origin of the polarization curve [3].

The linear regions of the current–potential curve above and below E_{corr} provide an indication of the kinetic control under which the corrosion process is occurring (i.e., anodic polarization and/or cathodic polarization). Nonlinearities in the Tafel plot can arise owing to differing concentrations of reactive species, the presence of surface oxide films, or preferential dissolution of one component in an alloy, as well as a mixed control process arising from the simultaneous occurrence of more than one cathodic and anodic reaction. The Tafel extrapolation method can be performed in a sealed electrochemical cell under nitrogen or oxygen at ambient or higher temperature.

2. Potentiodynamic Methods

The potentiodynamic technique is used to examine the passivation behavior of a metal or alloy in an electrochemical system. During the potentiodynamic scan, the metal surface may undergo several different electrochemical reactions, wherein the anodic current may vary over several orders of magnitude [34]. Generally, analysis of the anodic curve can provide potentials for active, passive, transpassive, and repassive zones; a rough estimation of corrosion current and corrosion potential; and a measure of the stability of passivity. Moreover, one can determine whether the passivation is spontaneous or needs to be polarized to induce passivation. In addition, one can determine whether the electrochemical system can induce a spontaneous transition from passive

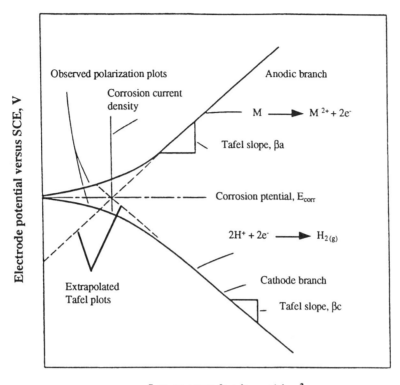

Fig. 31.1 Hypothetical Tafel extrapolation diagram for cathodic and anodic polarization. (From Ref. 33.)

to active behavior. The degree of passivation and the stability of the passive film can be assessed by measuring the passive region current and the transpassive region potential. The potentiodynamic scan must be run slowly enough to ensure steady-state behavior. More experimental details and principles can be found in ASTM Standard G3 [33]. Figure 31.2 shows hypothetical cathodic and anodic polarization plots for a material exhibiting typical passive behavior. The potentiodynamic anodic scan uses a potential scan typically starting at -1.5 V versus E_{oc} and scanning in a positive direction, usually to a potential positive enough to oxidize the electrolytes. The final potential in most studies is set at $+4$ V versus E_{oc}. The scan rate is set at 5 mV/s.

3. Cyclic Polarization

The principles of cyclic polarization are similar to that of cyclic voltammetry (CV). The cyclic polarization technique is used to qualitatively measure pitting tendencies of a metallic sample in a corrosive solution environment. ASTM Standard G61 [35] describes the experimental procedures typically used in this study. The beginning potential scan is toward the anode from a potential in the vicinity of E_{corr}. When the measured current exhibits a large increase or reaches a specified value, the scan direction is reversed to the cathodic direction. The potential at which the measured current increases dramatically on the forward sweep is considered to be the pitting potential (E_{pit}). The potential where the loop closes on the reverse scan is the protection potential (E_{prot}). New pits on the metal surface initiate only at potentials above the pitting potential. Between the pitting potential and the protection potential, new pits cannot initiate but the old ones can still grow. The hysterysis loop between the forward and backward scans is an indication of pitting formation. The larger the resulting loop, the greater the pitting tendency. If the loop does not close, E_{prot} can be determined by extrapolating the reverse scan to an infinitesimal current value. When the loops for E_{pit} and E_{prot} are close together, the pitting tendency is small. If E_{prot} is greater than E_{pit}, there may be no tendency to pit [3,35].

4. Galvanic Coupling

The galvanic corrosion technique is used to study quantitatively the corrosion reactions that occur when two dissimilar metal specimens are immersed in the same corrosive electrolyte and electrically coupled together.

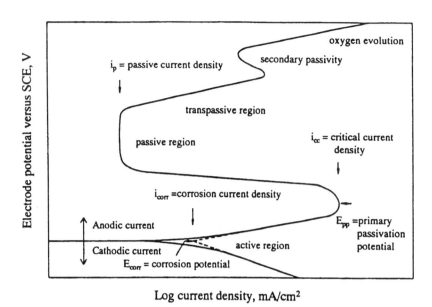

Fig. 31.2 Typical potentiodynamic anodic polarization plot. (From Ref. 33.)

One metallic sample will generally become the anode, and the other will become the cathode. The instrument used in galvanic corrosion measurements is a zero resistance ammeter (ZRA). It measures the current passing from one electrode to another and simultaneously measures the potential of the electrode. Measurement of the current or current density flowing between the counter electrode and the working electrode can give an indication of galvanic attack, and this can be monitored with time. The potentiostat can be converted into a zero resistance ammeter by connecting the counter electrode and the reference electrode together and setting them to the same potential as the working electrode. The polarization behavior of both the anode and the cathode usually vary with time along with the couple current. Thus, continuous adjustments of the potentiostat to change the measured current are necessary to maintain the potential difference at zero between anode and cathode. The potentiostat senses a difference between the reference and counter terminals of the instrument and controls the difference at a preset value by automatically varying the current between the working and counter terminals. The potentiostat will control the potential between the anode and cathode at any specified value. If the value is set at zero, the circuit will continuously read the coupled current at short circuit on the ammeter. Graphic analysis of the results of current versus time and voltage versus time can provide valuable information about predicted effectiveness of a cathode protection scheme or the extent of a galvanic coupling problem. Passive cathodic protection takes advantage of the galvanic corrosion reaction. It protects one metal by coupling it to a sacrificial anode. A familiar system is zinc on steel, described in Section I.

C. AC Monitoring Techniques: Electrochemical Impedance Spectroscopy

Alternating current electrochemical impedance spectroscopy (EIS) is a valuable tool for understanding the kinetics of an electrochemical corrosion system, especially for coating applications. Several methods are helpful in determining corrosion kinetics, corrosion mechanisms, and important physical parameters. EIS is used to characterize the interface between a metal and an electrolyte. The potentiostat applies both a dc potential and a small superimposed ac excitation to a specimen immersed in an electrolyte. In the experiments, alternating current and ac potential are measured as the excitation frequency varies over several orders of magnitude. The cell voltage and cell current are then converted into a complex impedance by a signal processing instrument such as a lock-in amplifier or a frequency response analyzer. The plot of complex impedance versus frequency can generate information that is difficult to obtain by other types of electrochemical techniques [3,36,37].

Electrochemical impedance spectroscopy is useful in the evaluation of coatings, the elucidation of transport phenomena in electrochemical systems, and the determination of corrosion mechanisms and rates. Bode and Nyquist plots are the most common data output formats, and an example of each for a simple parallel-connected resistance–capacitance circuit are shown in Figs. 31.3, and 31.4, respectively. The Bode plot format shows the

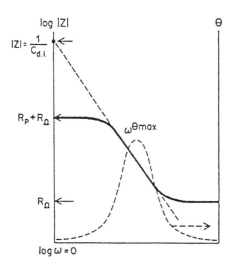

Fig. 31.3 A typical Bode plot of an electrochemical system. (From Ref. 36.)

absolute impedance, $|Z|$, and the phase shift, θ, of the impedance, each as a function of frequency f in cycles per second. The $\log |Z|$ versus $\log \omega$ (where $\omega = 2\pi f$) curve can provide values of R_p (polarization resistance) and R_Ω (solution resistance) from the horizontal plateau at low and high frequencies, respectively. At intermediate frequencies, the curve is a straight line with a slope of -1, and extrapolation of this line to the $\log |Z|$ axis yields the value of C_{dl} (double layer capacitance) from the relationship $|Z| = 1/C_{dl}$. The Bode plot format also shows the variations of phase angle θ with $\log \omega$. At the high and low frequency limits, where the behavior is resistor-like, the phase angle is nearly zero. At intermediate frequency ranges, the phase angle increases as the imaginary component of the impedance increases. The expression for Z is composed of a real and an imaginary part: $Z = Z' + jZ''$. In the Nyquist plot shown in Fig. 31.4, the real part is plotted on the X axis and the imaginary part on the Y axis. This plot has been annotated to show that low frequency data are on the right side of the plot and higher frequencies are on the left. On the Nyquist plot the overall impedance can be represented as a vector (arrow) of length $|Z|$, and the angle between this vector and the x axis is θ, where $\theta = \arg(Z)$. At very high frequency, the imaginary component, Z'', disappears, leaving only the solution resistance, R_Ω. At very low frequency, Z'' again disappears, leaving a sum of R_Ω and the Faradaic reaction resistance or polarization resistance, R_p. Both the Nyquist and Bode plots should give analogous results of R_Ω and R_p.

The Faradaic resistance or polarization resistance R_p is inversely proportional to the corrosion rate. It is evident from the Nyquist plot that the solution resistance, R_Ω, measured at high frequency can be subtracted from the sum of R_p and R_Ω at low frequency to give the value of R_p corrected for ohmic interferences from solution resistance. For processes controlled by diffusion in the electrolyte (concentration polarization) or in a surface film or coating, an additional resistive element called the Warburg impedance, W, must be included in the circuit. The Warburg impedance appears at low frequencies on the Nyquist plot as a straight line superimposed at 45° (slope = 1) to both axes, as shown in Fig. 31.5.

The measurement of polarization resistance, R_p, from impedance data has become quite popular in the study of electrochemical corrosion [38–40]. The value of R_p determined from the Nyquist plot has been shown [40] to have the usual inverse proportionality to corrosion rate determined by conventional weight loss methods. A computerized curve-fitting procedure permits extrapolation of the relatively high frequency semicircular data to R_p at the zero-frequency limit. Deviation at the low frequency side of the semicircle may occur due to the

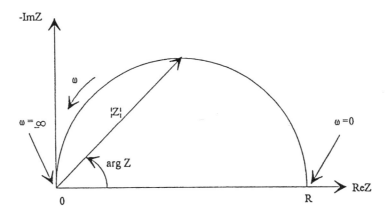

Fig. 31.4 Nyquist plot for a simple electrochemical system with impedance vector. (From Ref. 37.)

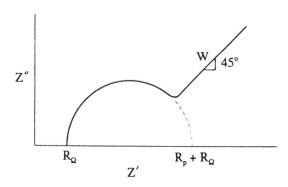

Fig. 31.5 Nyquist plot for a simple electrochemical system showing the Warburg impedance. (Ref. 3, Reprinted by permission of Prentice-Hall, Inc., Upper Saddle River, N.J.)

change in the impedance of the corroding surface during the time of the impedance measurements at low frequency. Sometimes, low frequency data are also masked by the Warburg impedance line.

Depression of the center of the semicircle below the horizontal Z' axis in the Nyquist plot often occurs for measurements at a corroding surface. This behavior is fairly typical and has been modeled [38] assuming metal dissolution under a corrosion product film with oxygen reduction within the film pores.

The electrochemical impedance technique can be extended for modeling the behavior of a metal coated with polymer. The equivalent circuit generally suitable for such systems is shown in Fig. 31.6. R_Ω, R_{cp}, R_p, C_c, and C_{dl} are solution resistance, coating pore resistance, polarization resistance, coating capacitance, and double layer capacitance at a metal surface, respectively. The parallel C_{dl}/R_p circuit, simulating the metal/solution interface, is incorporated in a second parallel circuit involving the coating parameters R_{cp} and C_c. The resistance R_{cp} has been interpreted as the pore resistance due to electrolyte penetration [41] and at damaged areas of the film where more rapid solution uptake occurs. The capacitance C_c has been interpreted as the capacitance of the capacitor consisting of the metal and the electrolyte, with the coating film as the dielectric, or simply as the capacitance of the intact film.

The Nyquist plot in this case will have two semicircles with time constants for the coated film and metal. Generally, the semicircle occurring at high frequencies will be due to the coated film. The shape of the plot will appear as two distinct semicircles if (1) the two diameters are not very different from each other and (2) maximum frequency values for the semicircles are not too close to each other. If these criteria are not both met, then the two semicircles will interact with each other, creating difficulties in separating the components of the equivalent electric circuit. In such cases, the Nyquist plot may have only one semicircle with a small bump at the high frequency region.

When both conditions are met, the corresponding Bode plot will have two distinct phase angle maxima, one at higher frequencies having coated film information, and one at lower frequencies having metal substrate information. If the conditions are not met, the two phase angle maxima may merge toward each other, showing one maximum or at best one bump associated with it in the frequency curve.

Owing to the complexity of the actual experimental system, irregularities may occur in both Bode and Nyquist plots. Therefore the plots should be inspected carefully in deriving the components of the equivalent circuit model. The Bode plot is more sensitive than the Nyquist plot for identifying the presence of plot shape irregularities. If clear separation into two time constants does not occur, then the plots may require more sophisticated and complicated data analysis to extract values of all components.

Additionally, in many cases of coated metal/solution interface, the equivalent circuit model must be modified to account for the diffusion processes within the pores in the film. These are modeled by the inclusion of a Warburg impedance, W, placed in series with R_p, as indicated by the appearance of a diffusion tail at an angle of 45° with the real Z' axis at low frequencies in the Nyquist plot.

As values of polarization resistance R_p and double layer capacitance C_{dl} are directly and inversely proportional to the corroding area, corrections may be needed to the impedance data of samples showing corrosion breakdown due to the usually localized nature of the breakdown. The increase in corroding area probably explains why the low frequency semicircle diameter in the Nyquist plot decreases with time, implying an increase in corrosion rate. Such behavior is opposite to that expected for uncoated metals.

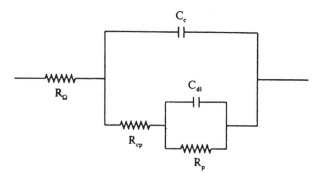

Fig. 31.6 A typical equivalent circuit providing a typical electrochemical impedance spectrum. (Ref. 3, Reprinted by permission of Prentice-Hall, Inc., Upper Saddle River, N.J.)

The impedance plot shape can vary considerably over the immersion lifetime of a coated metal sample. Initially, when the coated film shows mainly capacitive behavior, the Nyquist plot rises sharply up the imaginary Z'' axis and may bend over toward the real Z' axis in a semicircular arc at low frequencies. The corresponding Bode log Z plot is a straight line with a slope close to -1. At longer immersion times, water uptake within the coated film, with or without substrate corrosion, can modify this predominantly capacitive response. Nyquist plots with either one or two semicircles, the latter indicative of the first separation of coating and metal substrate properties, may develop. With continued immersion, the Nyquist semicircle diameter generally decreases, signifying a decrease in coating film resistance, an increase in corrosion rate, or an increase in corrosion area. At even longer times, the rate of diffusion of species through the pores of the film may become comparable to the charge transfer processes, and a low frequency diffusion tail may appear in the Nyquist plot. These can eventually dominate the plot shape as the coating film blistering and/or substrate corrosion become more severe. In some cases, the lower frequency behavior cannot be accurately resolved into individual components or can be resolved only at longer immersion times due to the initial highly capacitive behavior of the coated metal system.

The use of impedance methods for the evaluation of corrosion performance of polymer-coated metals is overall a good technique, but it is not without its limitations. There are limitations on the samples and limitations in the measuring technique itself—the frequency range, instrumentation, and data analysis capability. Finally, there are difficulties in interpreting the impedance data, particularly when the equivalent circuit model is unknown.

D. Surface Spectroscopic Techniques

1. X-Ray Photoelectron and Auger Spectroscopy

X-ray photoelectron spectroscopy (XPS) and Auger electron spectroscopy (AES) are very useful analytical techniques to determine the electronic states of conducting polymers under different chemical environments and the quantitative chemical composition of metal oxides and peroxides. But these are ex situ spectroscopic methods, and the real compositions may differ from the measured ones.

2. X-Ray Diffraction

X-ray diffraction (XRD) can be used to do a quick and easy determination of the crystalline structure of metal oxides by comparing the diffraction angle (2θ) with that of metal oxides of known composition. However, this is a qualitative and bulk method.

3. Scanning Electron Microscopy

Scanning electron microscopy (SEM) with energy-dispersive X-ray analysis (EDAX or EDS) is a powerful tool for studying the morphology and chemical composition of corroding surfaces. With proper magnification, the defect size, extended protection distance by conducting polymers on the exposed bare metal surface, and particle size of metal oxides can be precisely determined.

IV. CORROSION INHIBITION OF STEELS USING POLYANILINE AND POLYPYRROLE

Many studies have provided strong indications that polyaniline coatings on steel alter its corrosion behavior. In most of them, coated test samples were exposed to various corrosive environments and evaluated by visual inspection. Surprisingly, it was noted in these studies that samples of steel coated with polyaniline were protected from corrosion even where there were scratches in the coatings and bare steel was exposed to the corrosive environment. In an attempt to understand this mechanism of corrosion protection in regions extending from the coated areas to adjacent bare steel areas, we undertook a set of detailed experiments. These experiments were designed not only to validate the visual observations made by others, but also to provide mechanistic information on corrosion protection and to provide a quantitative measure of the corrosion inhibition possible in different corrosive environments. The concept was to study the corrosion on coated steel samples that had well-defined regions of exposed bare steel. These well-defined regions were made by drilling precision holes through the coatings, exposing areas of bare steel with precisely known surface areas. In this fashion, not only corrosion rates but also corrosion rates as a function of exposed surface area could be determined. Also, a limiting size to the areas of exposed steel that would be protected by a conductive polymer coating could be determined as a function of metal, conductive polymer, and corrosion environment.

A. Test Samples and Configurations

1. Test Materials

Two types of steel test coupons were used for these studies, C1010 carbon steel and A316 stainless steel, both of which are common structural metals. The elemental composition of these two steels can be found in common metal handbooks [42]. For the conductive polymers used in this work, an explanation and an application procedure for the metal surfaces are given here along with a shorthand abbreviation for each.

zk. An organic primer paint formulation obtained from Zipperling Kessler Co. comprising a dispersion of polyaniline particles (80–100 nm) in an organic matrix and an organic vehicle. After applying the polyaniline primer onto the metal coupon surface, the standard drying time was 20 min. The primer solution was cast onto steel surfaces to provide 4.2–6.4 μm thick coatings after drying. The coated coupons were then immersed in 1 M NaCl solution for 5–7 h. After this, the steel coupons were washed with distilled water and then dried at room temperature for 1 day. The adhesion of the conductive polymer coating on the steel coupons was good.

dp. Doped pure polyaniline coating. Dark green polyaniline powders obtained from Allied Signal (Versicon) were treated with dilute ammonium hydroxide solution to provide neutral polyaniline (emeraldine base). The neutral powder of polyaniline was filtered, washed, partially dried, and then pulverized. It was dissolved in 1-methyl-2-pyrrolidine (NMP), and the filtered solution was then cast onto steel surfaces to provide coatings of 12–25 μm thickness after drying. The resulting adhesion on the steel surfaces was moderate. The dried PAni coatings were then doped with 1 M p-toluenesulfonic acid (PTSA) or 0.1 M zinc nitrate (only for potentiodynamic studies) by immersing the coated steel samples into dopant solution for 5–10 min, followed by rinsing with distilled water and air drying for 1–2 days.

np. Pure neutral polyaniline coatings on steel prepared as above but not doped.

py. Waterborne fast cure colloidal polypyrrole/polyurethane composite coating formulation obtained from DSM. This is a core shell polypyrrole on Uraflex 401 UZ polyurethane dispersion in aqueous solution (2.5 % w/w) made by DSM Resins (solids content 40%) that was diluted with the same amount (volume) of distilled water [43]. This was then coated onto steel samples, dried, and cured at 80–90°C for 10 min to give 9–12 μm thick coatings. The adhesion and wear-resistant properties of this coating are excellent.

e. An epoxy topcoat that is composed of two parts: Ciba-Geigy liquid bisphenol-A GY2600 epoxy and XU265 cycloaliphatic/aliphatic amine as the hardener. The curing procedure involved keeping the coated steel coupons in a vacuum oven for 5–7 h at 70–85°C.

2. Coated Test Sample and Cell Configurations

Test sample configurations used in this study are shown in Fig. 31.7.

The sample configurations shown in Fig. 31.7 are represented by the following notation:

s	C1010 carbon steel corrosion test coupon (2 × 2 × 1/16 in.)
ss	A316 stainless steel corrosion test coupon (2 × 2 × 1/16 in.)
py/s	Polypyrrole/polyurethane composite film coated on carbon steel
zk/s	Polyaniline dispersion in organic vehicle coated on carbon steel
D, e/s	Epoxy topcoated on a steel coupon that has a 1.2 mm diameter hole that penetrates through the epoxy layer to the topmost surface of the steel
D,e/dp/s	Doped polyaniline coated on steel then topcoated with a layer of epoxy and drilled with a 1.2 mm diameter pinhole to the steel surface
D,e/np/s	Neutral polyaniline coated on steel then topcoated with a layer of epoxy and drilled with a 1.2 mm diameter pinhole to the steel surface
D,e/zk/s	Polyaniline primer coated on steel then topcoated with a layer of epoxy and drilled with a 1.2 mm diameter pinhole to the steel surface
D,e/py/s	Polypyrrole/polyurethane coated on steel then topcoated with a layer of epoxy and drilled with a 1.2 mm diameter pinhole to the steel surface

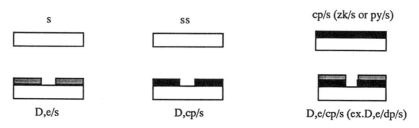

Fig. 31.7 Sample coating configurations used for corrosion experiments.

Corrosion Inhibition of Metals

The electrochemical test cell configuration used for these studies is shown in Fig. 31.8. Typically, corrosion tests were performed during an 8 week exposure period. Two corrosive environments were evaluated, 0.1 N HCl and 3.5% NaCl under normal ambient aeration.

B. Observations

1. Single-Sized Bare Steel Area Exposure Tests

Figure 31.9 shows the exposed mild steel surfaces of D,e/dp/s, D,e/np/s, and D,e/s samples, all with 1.2 mm diameter areas of exposed bare steel in both HCl and NaCl environments after 8 weeks immersion. The upper left coupon is the control sample D,e/s that was immersed in the HCl environment, and corrosion underneath the coating is visible. The lower left coupon is the control sample that was immersed in NaCl, and voluminous corrosion product with scattered yellow-brown rust can be seen over the exposed area. The accumulated rust compound started to grow from the second day of immersion. The upper middle test coupon is D,e/dp/s after 8 weeks of HCl immersion, and most of the exposed bare steel surface remained shiny for 7 weeks, after which a whitish-gray oxide film started to grow. The D,e/np/s coupon (lower middle) in NaCl shows whitish-brown corrosion products at the defect site that became visible after the second week of immersion. The upper right sample is D,e/np/s in HCl. The defect of this specimen was covered by brownish-black thin film during the 8 week immersion test. However, under an optical microscope, several rust pits can be seen at the edges and the central zone of the hole. The one at the lower right is D,e/np/s in 3.5% NaCl. Solid white crystals accumulated over the bare steel site, and the attack of epoxy layer was also observed. However, the amount of rust on this specimen was small.

Polypyrrole/polyurethane blend–coated mild steel samples with the same area of exposed steel as that used

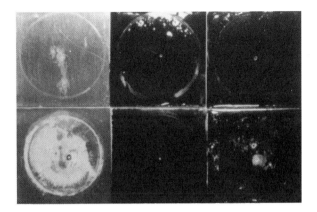

Fig. 31.9 Photograph of six steel test coupons after 8 weeks of immersion. Upper left, D,e/s in HCl; upper middle, D,e/dp/s in HCl; upper right, D,e/np/s in HCl; lower left, D,e/s in NaCl; lower middle, D,e/dp/s in NaCl; lower right, D,e/np/s in NaCl.

with the polyaniline-coated samples were studied in the same corrosive environments: 0.1 M HCl and 3.5% NaCl. Surprisingly, the color and shape of corrosive products and the oxidative film are similar to those observed with polyaniline-coated samples. However, polypyrrole/polyurethane, coated samples in NaCl showed no traces of rust at the end of the entire immersion test.

2. Variable-Sized Bare Steel Area Exposure Test

Drilled defects 1–12 mm in diameter were chosen for comparison purposes. The micrograph in Fig. 31.10

Fig. 31.10 Scanning electron micrograph showing the corrosion area just off left center of a defect site for a D,e/dp/s sample after 8 weeks exposure in 0.1 M HCl. Left side is region 1; center is region 2; and right side is region 3.

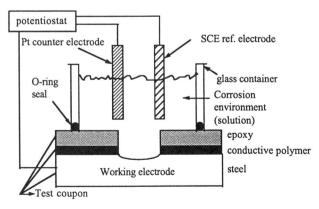

Fig. 31.8 A schematic of the corrosion cell used for corrosion experiments in various corrosion solutions.

shows the exposed area on a D,e/dp/s sample with an exposed bare steel area 6 mm in diameter after 8 weeks immersion in 0.1 M HCl. The area can be divided into three corrosion regions of different morphologies. A fully protective area is found adjacent to the edge of the coating (region 1), which shows essentially no change compared to an unexposed sample. The bare steel exposed area adjacent to this region and further from the edge of the coating toward the center of the defect region showed that there was partial protection in this area (region 2) with some obvious dendritic corrosion structures present. At the center of the bare steel defect (region 3), rust growth is evident as no protection is provided in this region. The distance from the edge of the defect to the margin of the fully protective area is important because it represents the throwing power of the conductive polymer inhibitor. In tests performed on samples with different defect sizes, a 2 mm critical protection distance from the edge of the drilled hole was consistently observed for polyaniline-coated samples in HCl environments. This indicates that conducting polymers have the ability to provide "extended protection" to uncovered (scratched) metal surfaces.

3. Aeration and Deaeration Tests

Nitrogen and oxygen gas purging tests were performed to determine the effect of oxygen concentration in electrolytes on corrosion behavior. The partial pressure of dissolved oxygen did affect the corrosion rate compared to those under ambient experimental conditions. The corrosion rate under a pure oxygen atmosphere for zk/s and dp/s samples in 0.1 M HCl increased in the initial exposure period by a small amount, then remained at a rate comparable to that of samples exposed only to a nitrogen atmosphere. However, corrosion rates increased considerably under an oxygen atmosphere for samples immersed in NaCl solutions. Apparently, the dominating corrosion mechanism in acidic solution is hydrogen evolution, whereas in NaCl it is oxygen reduction with iron dissolution. These experiments demonstrate that the nature of the corrosive environment can considerably alter corrosion rates and mechanisms.

C. Measurement of Corrosion Parameters

1. Corrosion Potential (E_{corr}) and Open-Circuit Potential (E_{oc})

The relative position of E_{corr} and E_{oc} can give a general thermodynamic corrosion tendency of an electrochemical system. Generally, if E_{corr} is more positive than E_{oc}, the electrochemical system will tend to show a slow corrosion process. If E_{corr} is more negative than E_{oc}, the electrochemical cell may become a spontaneous active corrosion system. Most of the samples investigated that had conductive polymer coatings can be divided into three major categories:

1. Cases where E_{oc} is always more positive than E_{corr} for the entire immersion period (poor or almost no protection).
2. Cases where E_{corr} is close to E_{oc} during the initial exposure period, but after a certain time E_{oc} becomes more positive than E_{corr} (moderate protection).
3. Cases where E_{corr} remains close to E_{oc} during the entire time of exposure in the corrosive environment (good protection).

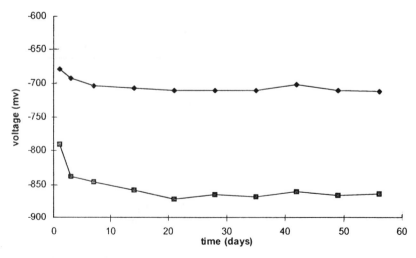

Fig. 31.11 Plots of the corrosion potential and open-circuit potential of zk/s in 3.5% NaCl during an 8 week immersion test. Category I sample in which $E_{oc} > E_{corr}$ during the entire exposure period (poor protection).

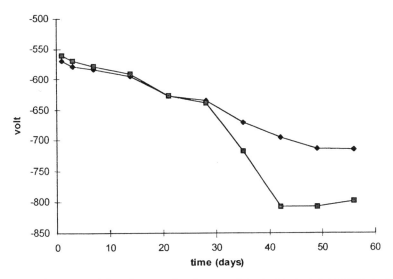

Fig. 31.12 Plots of the corrosion potential and open-circuit potential of zk/s in 0.1 M HCl during an 8 week immersion test. Category II sample in which $E_{oc} \approx E_{corr}$ initially, then $E_{oc} > E_{corr}$.

In general, mild steel samples coated with polyaniline in hydrochloric acid exhibited better corrosion protection than those exposed in sodium chloride solution under the same coating conditions as indicated simply by comparison of E_{oc} and E_{corr} as shown in Figs. 31.11–31.13. Corrosion samples tested in this study for all seven conductive polymers can be classified into one of the above three categories.

2. Tafel Plots as a Function of Exposure Time

Complete anodic and cathodic polarization curves are very useful in elucidating inhibitor properties. From the shift of Tafel plots with time, one can determine the changes in corrosion potential as well as the relative polarization tendencies of cathode and anode after applying an inhibitor or coating. The most important aspect

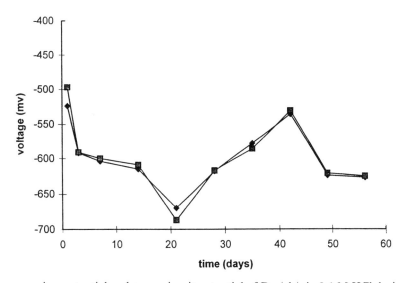

Fig. 31.13 Plots of the corrosion potential and open-circuit potential of D,e/zk/s in 0.1 M HCl during an 8 week immersion test. Category III sample in which $E_{oc} \approx E_{corr}$ during the entire exposure period.

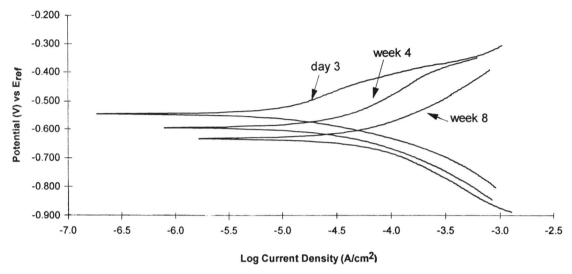

Fig. 31.14 Tafel plots for py/s at various immersion times in 0.1 M HCl.

of this technique is the measurement of corrosion rates. Figure 31.14 indicates that the polypyrrole layer in the py/s sample provided a steady cathodic polarization tendency but also showed an increasing anodic polarization with time. In addition, the corrosion current density increased by less than a factor of 10 within the 8 week testing period even though E_{corr} shifted to a more negative value.

The Tafel slopes of both cathodic and anodic curves of a py/s sample (no epoxy overcoat) in 3.5% NaCl (Fig. 31.15) showed no apparent variation even after long immersion times, indicating that corrosion stabilization is achieved in this system. E_{corr} was only slightly more positive than E_{oc}, and there was no tendency for corrosion current to increase over the entire immersion time. This indicates that polypyrrole suppresses the corrosion of mild steel in 3.5% NaCl even when the coating is porous, as was observed by optical microscopy.

The corrosion potential of a D,e/py/s sample of 0.1 M HCl shifted to more positive values after moving to more negative values in the beginning of exposure, but the corrosion current and rate remained at the same level, as shown in Fig. 31.16. The metal surface reached a repassive stage during the intermediate immersion pe-

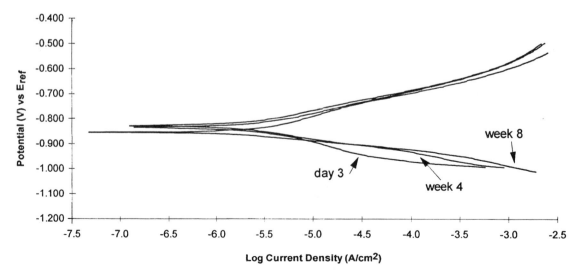

Fig. 31.15 Tafel plots for py/s at various immersion times in 3.5% NaCl.

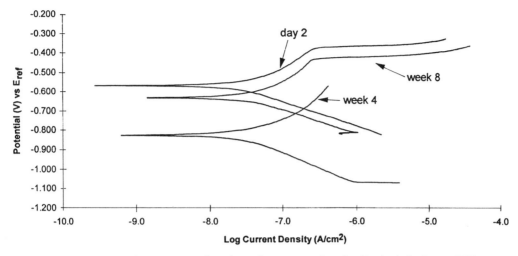

Fig. 31.16 Tafel plots as a function of exposure time for D,e/py/s in 0.1 M HCl.

riod, and the corrosion potential moved closer to the initial potential. Tafel constants of both cathode and anode stayed at the same level, indicating good corrosion prevention by doped polypyrrole for bare exposed mild steel in a low pH electrolyte.

However, it was a different story for the same sample in a saline environment (Fig. 31.17). Anodic polarization increased with time, showing strong polarization on the steel surface. This indicates that more oxidation processes are occurring on anodic sites. At the same time, the Tafel constant of the cathode increased. Even though the corrosion current went up by a factor of 10, it was low compared to that of an identical sample without polypyrrole (D,e/s). This case can be viewed as a case of metastable passivity.

3. Tafel Constant Ratios

From Tafel experiments, determination of the ratios of the cathodic Tafel constant to the anodic Tafel constant (β_c/β_a) for corrosion samples during the corrosion experiment can provide valuable information on the mechanism of corrosion inhibition and provide an indication of how corrosion mechanisms may change during the exposure time. With conductive polymers on steeel, this information is particularly revealing. Table 31.1 provides data for conducting polymer systems on mild steel in HCl. The first column (D,e/s sample) provides information on how this ratio changes with exposure time for a sample containing no conducting polymer. Initially this sample exhibits mild anodic polarization ($\beta_a > \beta_c$), but

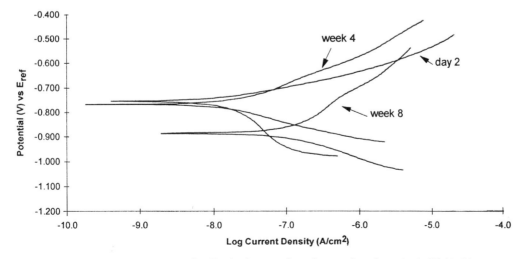

Fig. 31.17 Tafel plots for D,e/py/s at various immersion times in 3.5% NaCl.

Table 31.1 Tafel Constant Ratios (β_c/β_a) for D,e/cp/s Samples in 0.1 M HCl

Time	Sample				
	D,e/s	D,e/np/s	D,e/zk/s	D,e/dp/s	D,e/py/s
Day 1	0.78	0.82	3.17	0.97	2.37
Day 3	0.76	0.63	1.4	0.97	0.71
Day 7	0.94	1.49	1.94	1.56	0.71
Day 14	1.23	0.59	2.21	0.33	0.94
Day 21	1.43	0.59	1.36	0.28	0.93
Day 28	2.07	2.16	1.33	0.35	0.86
Day 35	2.64		1.54	1.4	0.66
Day 42	2.02		1.27	0.85	
Day 49	2.42	1.02	1.43	0.73	
Day 56	2.01	1.77	1.35	0.82	0.54
Protection	Reference	Poor	Fair	Good	Excellent

after day 14 of exposure, cathodic polarization dominates the corrosion process ($\beta_c > \beta_a$).

Samples with conductive polymer coatings showed the opposite behavior. The ranges of Tafel constant ratios for D,e/cp/s samples in HCl for good corrosion control were greater than 1 in the initial immersion stages but decreased with time to values below 1. Thus, the corrosion protection mechanism seems to have changed from cathodic control to strong anodic control. Initially, a large cathodic polarization occurs, which implies that the polarization of the conducting polymer affects the hydrogen evolution reaction (hydrogen overvoltage). The conducting polymer remains electroactive in an acidic environment, which may explain this higher polarization tendency for the cathode reactions ($\beta_c > \beta_a$). It also appears that the conductive polymer initially provides cathodic protection to the bare steel exposed surface for samples with exposed defect regions in the coatings. With continued exposure time this ratio decreased, indicating strong anodic polarization on the steel surface during the final period of immersion. At the end of immersion, anodic protection provided by the appearance of an oxide film was evident from both electrochemical measurements and observations by SEM pictures of the exposed sites, which showed that a thick oxide film had formed on exposed steel surfaces.

Table 31.2 indicates that the desired range of Tafel constant ratios (β_c/β_a) for samples exhibiting good protection in NaCl is similar to that in HCl (Table 31.1). Samples that exhibited Tafel constant ratios around 0.5 in NaCl seem to provide good protection. In HCl, protection seems to be provided by strong cathodic polarization for hydrogen evolution at early stages of immersion. The Tafel constant ratio for best corrosion protection is close to 1 at intermediate immersion time. At the end of immersion, a low Tafel constant ratio indi-

Table 31.2 Tafel Constant Ratios (β_c/β_a) for D,e/cp/s Samples in 3.5% NaCl

Time	Sample				
	D,e/s	D,e/zk/s	D,e/dp/s	D,e/np/s	D,e/py/s
Day 1	4.54	0.4	2.61	0.54	4.42
Day 3	1.8	0.24	2.45	1.64	3.83
Day 7	3.51	0.41	0.46	2.51	2.91
Day 14	3.86	0.47	2.35	2.57	0.97
Day 21	7.69	0.49	1.62		0.88
Day 28	6.88	0.76	0.48	0.87	0.78
Day 35	5.63	0.96	0.46	0.46	0.78
Day 42	5.58	0.96	0.4	0.7	
Day 49	6.99	0.98	0.62	0.64	0.46
Day 56	5.41	1.14	0.63	0.55	0.42
Protection	Reference	Poor	Fair	Good	Excellent

Table 31.3 Calculated Corrosion Rates for Coated Mild Steel Immersion Tests in 0.1 M HCl with a 1.2 mm Diameter Exposed Bare Steel Surface Present

Time	Sample				
	D,e/s	D,e/dp/s	D,e/np/s	D,e/zk/s	D,e/py/s
Day 1	0.387	0.049	0.155	0.014	0.22
Day 3	0.797	0.168	0.476	0.021	0.02
Week 1	1.775	0.487	1.367	0.034	0.02
Week 4	7.206	0.084	1.326	0.143	0.02
Week 8	8.201	0.187	9.509	0.386	0.03

cates that the corrosion prevention is dominated by anodic protection. In summary, if a conducting polymer can influence a small area of exposed steel and maintain high polarization tendency in the cathodic reaction at the initial stage, then good corrosion protection can be predicted even though the mechanism changes from cathodic to anodic protection with time.

4. Corrosion Rate and Inhibition Efficiency

The conversion of corrosion current from the Tafel plot to corrosion rate can be obtained by the method of ASTM Standard G102 [44] and is given by

$$R_{corr} = i_{corr} \times 1.288 \times 10^5 \, M_{EW}/Ad \qquad (8)$$

where

R_{corr} is the corrosion rate on exposed metal surface (unit: mpy milli-inches per year)
i_{corr} is the corrosion current (A)
M_{EW} is the equivalent weight of the metallic sample (g/equiv)
d is the density of the metallic sample (g/cm^3)
A is the exposed sample area (cm^2)

The percentage inhibitor efficiency, P or PIE, is defined as

$$P = \frac{i_b - i_c}{i_b} \times 100\%$$

where i_b is the corrosion rate in an uninhibited solution and i_c is the corrosion rate in an inhibited solution.

The corrosion rates of mild steel coated with conductive polymers and overcoated with a protective coating of epoxy and having a 1.2 mm diameter defect area of exposed bare steel in dilute HCl environments are listed in Table 31.3. For similar substrates, the corrosion rates in Table 31.4.

The PIE values obtained from corrosion rates suggest that some conductive polymers have good corrosion inhibitions (Table 31.5). The PIE values of some pyrrole- and aniline-related polymers the tested can be as high as 75–99$^+$% for samples with small areas of exposed steel in 0.1 M HCl. In 3.5% NaCl, fair to good protection can be achieved, with the best performance occurring during the early stages of exposure. Steel samples coated with doped polyaniline with 1.2 mm diameter exposed steel surfaces exhibited a PIE of nearly 98% after 8 weeks exposure, while a similar sample of doped polypyrrole exhibited a PIE of over 99%. Also, notice that the polypyrrole-containing sample exhibited this level of protection by the third day of exposure and maintained that level throughout the remaining exposure time. The sample containing neutral polyaniline showed erratic behavior in HCl, and eventually, after 8 weeks of exposure, exhibited an excellent corrosion rate compared with the control sample. Samples in neutral NaCl showed considerably different behavior. Both doped and neutral polyaniline-coated samples exhibited very

Table 31.4 Calculated Corrosion Rates for Mild Steel Immersion Tests in 3.5% NaCl with a 1.2 mm Diameter Exposed Bare Steel Surface Present

Time	Sample				
	D,e/s	D,e/dp/s	D,e/np/s	D,e/zk/s	D,e/py/s
Day 1	0.018	0.002	0.002	1.402	0.006
Day 3	0.015	0.001	0.002	1.221	0.007
Week 1	0.059	0.004	0.003	0.817	0.013
Week 4	0.125	0.011	0.016	0.722	0.009
Week 8	0.208	0.361	0.117	0.906	0.082

Table 31.5 PIE Values Compared to the Control Sample D,e/s in 0.1 M HCl and 3.5% NaCl

	Sample					
	D,e/dp/s		D,e/np/s		D,e/py/s	
Time	0.1 M HCl	3.5% NaCl	0.1 M HCl	3.5% NaCl	0.1 M HCl	3.5% NaCl
Day 1	87.4%	90.8%	59.9%	87.4%	43.1%	67.1%
Day 3	78.9%	90.3%	40.2%	89.4%	98.1%	51.1%
Week 1	72.5%	87.0%	23.1%	91.6%	98.8%	63.8%
Week 4	99.0%	91.2%	84.6%	87.4%	99.7%	93.0%
Week 8	97.7%	−73.8%	−16.2%	43.5%	99.6%	60.7%

good PIE values during early stages of exposure (up to 4 weeks); then the protection clearly broke down. In contrast, samples coated with polypyrrole only showed moderate PIE values at the early stages compared to the polyaniline samples, but they maintained this steady level of protection throughout the 8 week exposure time.

Aniline and ethoxyaniline as inhibitor additives in 1 N HCl also provide moderate protection for iron, as shown in Table 31.6 [18]. The PIE values increase with an increase in concentration of the additives. The PIE values for polyethoxyaniline suggest that the solutions containing monomers or oligomers are not as good corrosion inhibitors as polymeric materials.

D. Passivation, Pitting, and Galvanic Coupling Behavior

1. Passivation Measurements

Potentiodynamic methods can provide information about the possible potential range of passivation and passivation breakdown potentials by applying an external potential over a wide potential range to the corrosion sample in a short time interval. The purpose of using this technique is to obtain the passivation potential and passivity breakdown potential and to observe the influence of evolved oxygen and hydrogen gas on the surface passivity of the metal.

The rate of oxygen and hydrogen evolution from the steel surface is influenced by the presence of a conductive polymer coating during potentiodynamic scans. Oxygen evolution from the polyaniline and hydrogen evolution from the mild steel surface must occur in the corrosive environment to passivate the steel surface. In some cases samples were exposed in an environment in which oxygen evolution was the main electrode reaction after exceeding a determined potential value. In some other cases, at high potentials chlorine evolution took place.

The potentiodynamic scans shown in Fig. 31.18 indicate that the D,e/dp/s and D,e/np/s samples exhibited extremely good passivation in 0.1 M HCl after oxygen evolution, but the polyaniline primer (zk) and polypyrrole (py) cannot stabilize the passivation of mild steel under oxygen evolution conditions. However, the anodic current indicating metal dissolution was still lower than the control sample specimen (D,e/s). Quick oxidation measurements on doped polyaniline (dp) and neutral polyaniline (np) as the primer layer showed that these samples gave the best passivation properties, as evidenced by larger passivation potential zones and extremely low metal dissolution currents. Polypyrrole/polyurethane coating (py) gave weak passivity, showing a medium passive zone and narrow transpassivation/repassivation region toward the end of the potentiodynamic experiment. Nevertheless, every sample of steel containing a conductive polymer showed better results than the reference epoxy-only coated sample in 0.1 M HCl. These results are consistent with the direct measurements of corrosion rates on these samples. The D,e/cp/s samples have lower corrosion rates and higher tendencies to maintain the passivation state.

Table 31.6 Inhibition Efficiency of Iron in 1 N HCl Using Aniline, Ethoxyaniline (EAn), and Polyethoxyaniline Doped with HCl [PEA(Cl)]

Inhibitor	Concn. (ppm)	PIE
Aniline	100	17%
	1,000	25%
	10,000	54%
	20,000	59%
EAn	1,000	17%
	5,000	46%
	10,000	67%
	20,000	87%
PEA(Cl)	10	46%
	50	84%
	100	86%
	200	89%

Source: Ref. 18.

Corrosion Inhibition of Metals

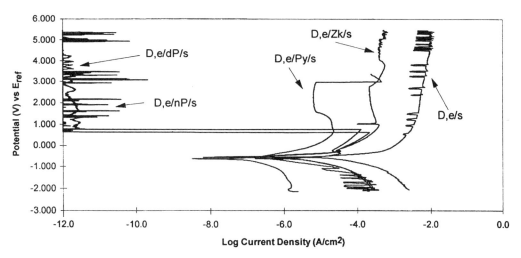

Fig. 31.18 Potentiodynamic scans for D,e/cp/s in 0.1 M HCl.

The potentiodynamic scans shown in Fig. 31.19 indicate that there is strong passivation by the polypyrrole/polyurethane composite on mild steel in 3.5% NaCl. There was fast and stable passivation on the exposed steel area. In the case of neutral polyaniline on mild steel, no passive appearance was noted but a decrease of corrosion current was observed. This result is also very consistent with the measured corrosion rates discussed earlier where some of the conducting polymers have better corrosion protection (D,e/py/s and D,e/np/s) to exposed steel areas compared to others (D,e/dp/s and D,e/zk/s) in this corrosion environment after a long exposure time.

Changing the dopant in the polyaniline coatings from toluencesulfonic acid (TSA) to zinc nitrate also influenced the anodic passivation results as evidenced by the potentiodynamic scans shown in Fig. 31.20. Changing the dopant to zinc nitrate improved the corrosion protection performance of polyaniline coatings on mild steel in NaCl compared to TSA dopant. Zinc nitrate–doped polyaniline coatings on steel exposed to HCl provided passivation, but apparently not as well as TSA in this case. Figure 31.21 shows the potentiodynamic behavior of the two different dopants for polyaniline coated on stainless steel. The results indicate that zinc nitrate can improve the corrosion inhibition performance of these coatings for stainless steel in NaCl-containing corrosion environments.

A mild steel surface covered by a polyaniline coating (without any epoxy topcoat, zk/s) showed excellent passivation and low corrosion current upon exposure to 0.1 M HCl as shown by the potentiodynamic scan in Fig. 31.22. However, similar doped polypyrrole-coated substrate showed almost no tendency for passivation com-

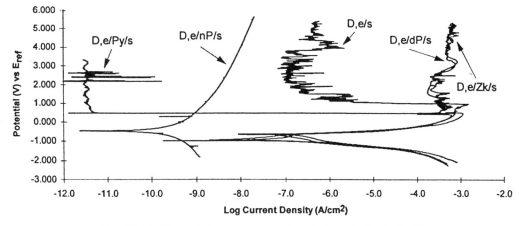

Fig. 31.19 Potentiodynamic scans for D,e/cp/s in 3.5% NaCl.

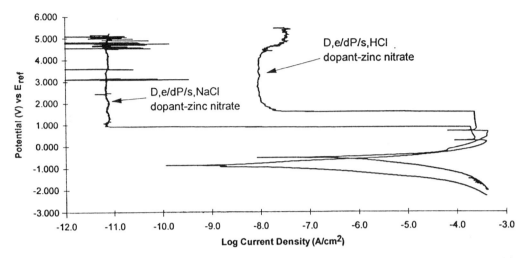

Fig. 31.20 Potentiodynamic scans for D,e/dp/s samples using zinc nitrate as dopant in polyaniline.

pared to bare steel. Both polymers showed some evidence of providing passive behavior in an NaCl environment. Interestingly, the zk/s sample had the lowest corrosion current density and best passivation behavior in both 0.1 M HCl and 3.5% NaCl environments (Figs. 31.22 and 31.23)

The potentiodynamic results illustrated clearly demonstrate the differences exhibited by samples exposed to the same corrosive environments but with different conductive polymers. This is especially true for samples containing the same conductive polymer but with different sample configurations (topcoat or no topcoat, defect areas or no defects, etc.).

2. Pitting Corrosion

The cyclic polarization technique is used to identify the protective potential (repassivation potential), pitting potential, and tendency for pitting on a metallic surface. Table 31.7 lists the pitting potential and protection potential for various samples relative to the pitting tendency of mild steel and stainless steel samples protected only by a polyaniline coating. These results show that a polyaniline coating can minimize the pitting tendency in HCl of both mild steel and stainless steel because the protection potentials were found to be more negative than the pitting potential. For most cases, even the sam-

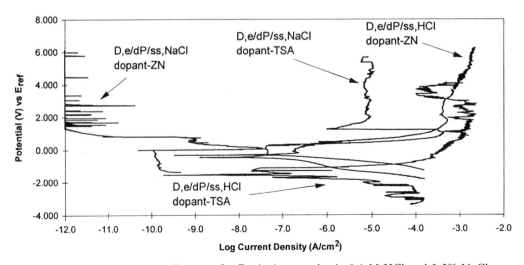

Fig. 31.21 Potentiodynamic scans for D,e/cp/ss samples in 0.1 M HCl and 3.5% NaCl.

Corrosion Inhibition of Metals

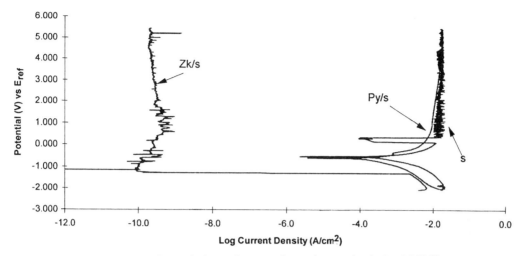

Fig. 31.22 Potentiodynamic scans for cp/s samples in 0.1 M HCl.

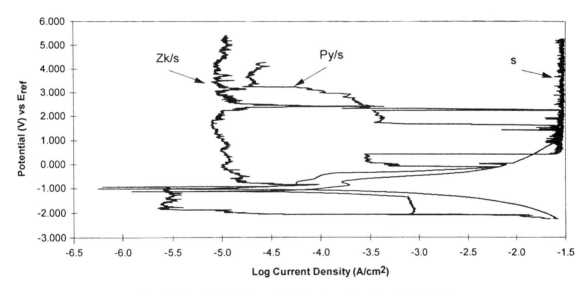

Fig. 31.23 Potentiodynamic behavior of cp/s in 3.5% NaCl.

Table 31.7 Pitting Corrosion Results of Polyaniline Primer on Steel and Stainless Steel in Different Configurations

	0.1 M HCl				3.5% NaCl			
	zk/s	zk/ss	D,zk/s	D,zk/ss	zk/s	zk/ss	D,zk/s	D,zk/ss
E_{oc}	−0.62	−0.02	−0.58	0.00	−0.58	−0.01	−0.58	−0.01
E_{corr}	−0.62	−0.11	−0.58	−0.07	−0.73	−0.08	−0.69	−0.09
E_{pit}	−0.42	+0.23	−0.20	+0.31	−0.47	+0.36	−0.67	+0.28
E_{prot}	−0.60	+0.09	−0.63	−0.11	−0.55	+0.55	−0.54	−0.14
Trend[a]	Small	Medium	Small	Large	Small	Neg[b]	Small	Large

[a] Pitting tendency.
[b] Neg = negative.

ples with multilayer coatings showed pitting corrosion when E_{pit} was more positive than E_{prot}. The initial traces of pitting can be identified by scanning electron and optical microscopy.

3. Galvanic Coupling

When the potential of a coupled system shifts to more negative potential and the current flows from the counter electrode to the working electrode, the material at the counter electrode can exhibit cathodic protection relative to the working electrode. When both D,e/cp/s and D,e/s electrodes were connected together in the same corrosive environment, the electrode bearing a conducting polymer always became the cathode and the system's coupling potential shifted to a more negative value as illustrated in Table 31.8. This table shows that polypyrrole can be used for cathodic application in both HCl and NaCl environments, but the polyaniline primer (zk) provides cathodic protection to D,e/s specimens only in HCl environments. The table also indicates that polypyrrole can be used for cathodic application on mild steel as opposed to stainless steel in both HCl and NaCl environments.

E. AC Impedance Measurements

The sum of charge transfer resistance and polarization resistance of a protective film on a metal can be estimated from the difference between the overall impedance at low frequency (e.g., 0.01 Hz) and high frequency (e.g., 100 kHz) in the Nyquist plot [36]. This provides an index of corrosion resistance for metallic materials covered by coatings. The double layer capacitance on a metal surface can be estimated from extrapolation of the absolute impedance curve at intermediate frequencies to the impedance axis in the Bode plot.

Figure 31.24 shows Bode plots for zk/s in 0.1 M HCl and demonstrates that the impedance of the zk/s sample in HCl dropped only 5% (from 200 Ω to 190 Ω) after 8 weeks immersion. After 4 weeks immersion, the impedance decreased to 70% of its original value but recovered to nearly the original impedance value at the end of the immersion period. Also, the double layer capacitance increased with time. Figure 31.25 shows that the charge transfer resistance of the zk/s sample in HCl decreased from 195 Ω to 135 Ω after a 4 week immersion. However, this specimen regained the resistance at the end of immersion even though the double layer capacitance increased. This indicates that the steel surface remained in a passivated state.

The situation changed when the same specimen was immersed in a sodium chloride solution (Fig. 31.26). The double layer capacitance increased with increasing exposure time. This indicated an increase in the oxidation reaction rate at the interface, which in turn caused the iron dissolution to increase.

By using the model-fitting utility of the commercial CMS 300 software [37], proposed equivalent circuit models that represent the corresponding electrochemical systems were determined. The equivalent circuit model and analytical electrical properties of the zk/s sample in 0.1 M HCl are illustrated in Fig. 31.27 and Table 31.9, respectively. The notations R_{sol}, R_p, W, and CPE are electrolyte resistance, polarization resistance,

Table 31.8 Galvanic Coupling Results of Two Dissimilar Electrodes

Working electrode	Counter electrode	Electrolyte	$E_{\text{couple},i}$	$E_{\text{couple},e}$	Q_{accuml}
D,e/py/s	D,e/ss	HCl	−0.48	−0.52	2.03×10^{-1}
D,e/py/s	D,e/ss	NaCl	−0.59	−0.62	8.25×10^{-3}
s	py/s	HCl	−0.53	−0.52	-2.95×10^{-2}
s	py/s	NaCl	−0.62	−0.7	1.87×10^{-1}
D,e/zk/s	D,e/py/s	HCl	−0.52	−0.56	1.12×10^{-2}
D,e/zk/s	D,e/py/s	NaCl	−0.55	−0.59	-1.62×10^{-3}
zk/s	py/s	HCl	−0.52	−0.53	8.69×10^{-1}
zk/s	py/s	NaCl	−0.67	−0.69	-1.87×10^{-1}
py/s	ss	HCl	−0.48	−0.48	1.89×10^{0}
py/s	ss	NaCl	−0.65	−0.67	2.22×10^{-1}
D,e/zk/s	D,e/s	HCl	−0.57	−0.60	8.52×10^{-3}
D,e/zk/s	D,e/s	NaCl	−0.53	−0.58	9.71×10^{-4}
D,e/py/s	D,e/s	HCl	−0.53	−0.57	-1.78×10^{-2}
D,e/py/s	D,e/s	NaCl	−0.60	−0.63	1.86×10^{-2}

$E_{\text{couple},i}$ = initial coupling potential.
$E_{\text{couple},e}$ = ending coupling potential.
Q_{accuml} = accumulated charge flow amount during 5 h coupling time.

Corrosion Inhibition of Metals

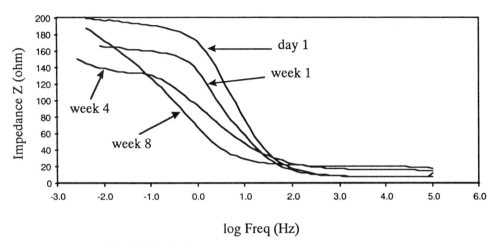

Fig. 31.24 Bode plots for zk/s samples in 0.1 M HCl.

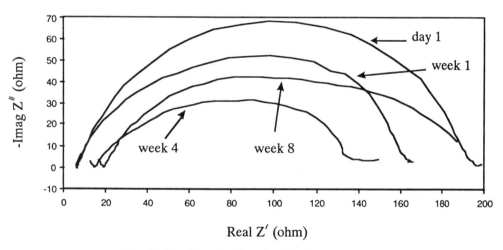

Fig. 31.25 Nyquist plots for zk/s in 0.1 M HCl.

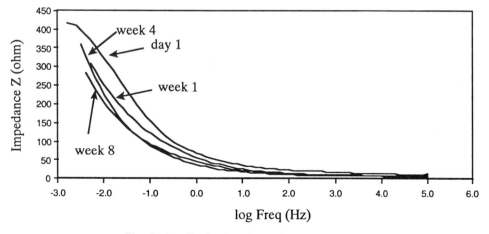

Fig. 31.26 Bode plots for zk/s in 3.5% NaCl.

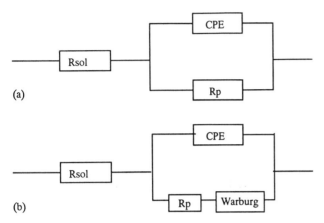

Fig. 31.27 Equivalent circuits for modeling impedance data of (a) zk/s in 0.1 M HCl and (b) zk/s in 3.5% NaCl.

Warburg impedance, and constant phase element, respectively. Warburg impedance (W) is a value used to account for mass transfer limitations due to diffusion processes adjacent to the electrode. Capacitors in EIS experiments often do not behave ideally. The double layer on an electrode surface often behaves like a CPE and not like a capacitor.

The constant phase element can be found for electrochemical behavior in both electrolytes. The unique identification came from the depressed semicircle in Nyquist plots [45]. The existence of a Warburg impedance in the 3.5% NaCl immersion test indicated diffusion control (mass transfer) in that medium. The signature of Warburg impedance is found in Nyquist plots exhibiting a 45° curve in the low frequency region (see Fig. 31.5). Clearly, the polarization resistance did not decrease much for zk/s in 0.1 M HCl even after 8 weeks of immersion. However, there was a significant decrease in 3.5% NaCl. The double layer capacitance for the sample containing a polyaniline film on mild steel was always smaller in 0.1 M HCl environments than in 3.5% NaCl.

Generally, the better the conductance of the sample surface, the smaller the electrode polarization. The addition of a conductive polymer layer on a mild steel surface has an advantageous effect. The Nyquist plot of polymer-coated samples may contain two semicircles. The first semicircle, found in the high frequency part of the spectrum, does not depend on the polarization current density. Its size depends mainly on the conductance of the working electrode. The second semicircle, found in the low frequency zone of the spectrum, depends on the polarization current density and is related to the electrochemical reactions occurring on the exposed steel surface in the small defect area. Its diameter, overlapping with the real axis, determines the resistance of the partial anodic reaction, a function of the rate of the electrode process. This semicircle shows a decaying tendency for most cases studied. Spectra with two semicircles characterize the presence of both resistance and capacitance of a coated corroding electrode. To show the correctness of the assumed circuit, an analysis has been carried out of all impedance spectra with the equivalent circuit program. The proposed equivalent circuit allows one to determine the resistance changes in the polymeric layer and oxidation layer as a result of the functioning of the anode. It also determines various kinetic parameters of electrode processes occurring on these surfaces. EIS cna be a useful source of significant information on the properties of conuctive polymers and on electrode processes occurring on the surfaces. The charge transfer resistance of polypyrrole on mild steel in 0.1 M HCl remained within a narrow range of resistance (150–250 Ω) during the entire immersion time (Fig. 31.28). The spectra indicate that a polypyrrole film can protect the steel surface consistently. The charge transfer resistance at the end of immersion returned to the value observed at the initial immersion period.

Figure 31.29 depicts an increasing charge transfer resistance for the py/s sample in NaCl owing to passivation of the steel surface. The same figure shows a gradual loss of capacitance behavior, which means that the surface coverage of the metal oxide passive film is increasing. Shown in Fig. 31.30 are Nyquist plots of D,e/py/s in HCl that form asymmetric semicircles. This indicates an energetic kinetic barrier associated with the rate-determining step where the capacitance of the passivating metal oxide film changes. After a 2 week immersion, the system develops inductor properties in the high frequency range, then exhibits two rate-determining steps in the low frequency range. As shown by the intermediate plot of Fig. 31.30b, the polarization resistance of this system dropped to two-thirds of its original value, but charge transfer resistance characteristics then emerged. At the final immersion stage, the system exhibited strong inductor behavior.

The D,e/py/s sample in 3.5% NaCl showed an inductor response in the low frequency region (Fig. 31.31). This indicates the presence of an adsorbate in the system. By the end of immersion, the polarization resistance decreased to half its initial value.

Table 31.9 Physical EC Parameters from Model Fitting for zk/s in 0.1 M HCl and 3.5% NaCl

	0.1 M HCl		3.5% NaCl	
Parameter	Day 1	Week 8	Day 1	Week 8
R_{sol} (Ω)	6.8	18.8	11.5	4.86
R_p (Ω)	189.5	165.2	672.3	241.6
$1/C_{dl}$ (Rc:F^{-1})	2280	197.7	144.6	101.2
α	0.79	0.64	0.46	0.42
W (Ω)	—	—	0.01	56.3

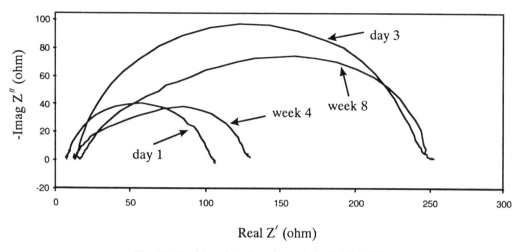

Fig. 31.28 Nyquist plots for py/s in 0.1 M HCl.

Figure 31.32 shows a two time constant system for D,e/dp/s in HCl that changes to a one time constant system at the final immersion stages. The charge transfer resistance and polarization resistance were reduced to one-tenth of their original values at the intermediate period.

Figure 31.33 indicates an unstable electrochemical system for D,e/dp/s in NaCl that tends to show a diffusion impedance response at initial immersion. And this system shows a gradual depressed semicircle, implying low surface coverage on the exposed steel surface.

Figure 31.34 shows that D,e/np/s in NaCl had high charge resistance and lower polarization resistance with strong absorption on an exposed bare mild steel surface at the beginning of immersion. Gradually, the system exhibited low charge transfer resistance and high polarization resistance.

The EIS spectra for these samples indicate that very different and often multiple processes occur during the course of the corrosion experiments (exposure time) for different polymers, corrosive environments, and sample configurations.

F. Passivating Metal Oxide Layers and Electronic Structures of Conductive Polymers

The composition of the passivating iron oxide film on steel usually has properties of an n-type semiconductor with excess negative charge carriers [46]. The electronic

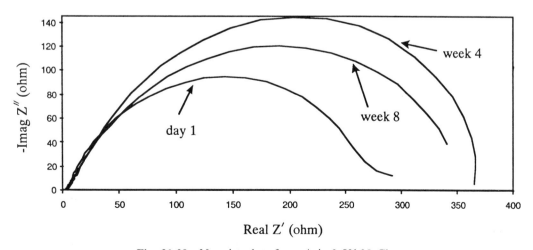

Fig. 31.29 Nyquist plots for py/s in 3.5% NaCl.

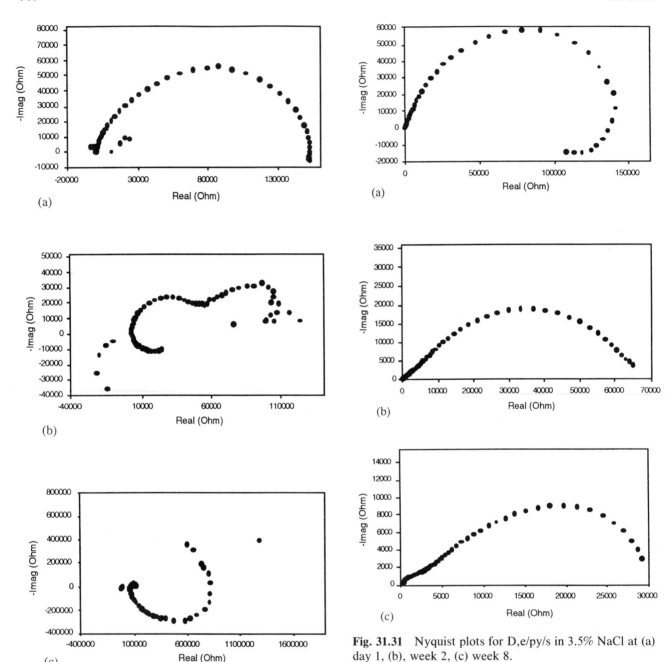

Fig. 31.30 Nyquist plots for D,e/py/s in 0.1 M HCl at (a) day 1, (b) week 2, (c) week 8.

Fig. 31.31 Nyquist plots for D,e/py/s in 3.5% NaCl at (a) day 1, (b) week 2, (c) week 8.

structure of nitrogen in polypyrrrole and polyaniline has been widely studied by XPS [47–50]. The nitrogen XPS peak locations of quinonoid, benzenoid, and positively charged ammonium nitrogen are at 398.8, 399.3, and 401.8 eV, respectively. The electronic states of nitrogen sites in the polyaniline primer coatings on mild steel changed after immersion in HCl as indicated in Table 31.10. However, there were no significant changes noted for the polyaniline coatings in NaCl. Polyaniline coatings polarized the mild steel in HCl and showed a lower —N=/—NH— (imine/amine) ratio, implying that it contains more of the reduced form of benzenoid amine in the polyaniline chains. After immersion in an acidic electrolyte (0.1 M HCl), the benzenoid amine content increased. However, the imine/amine ratio of the polyani-

Corrosion Inhibition of Metals

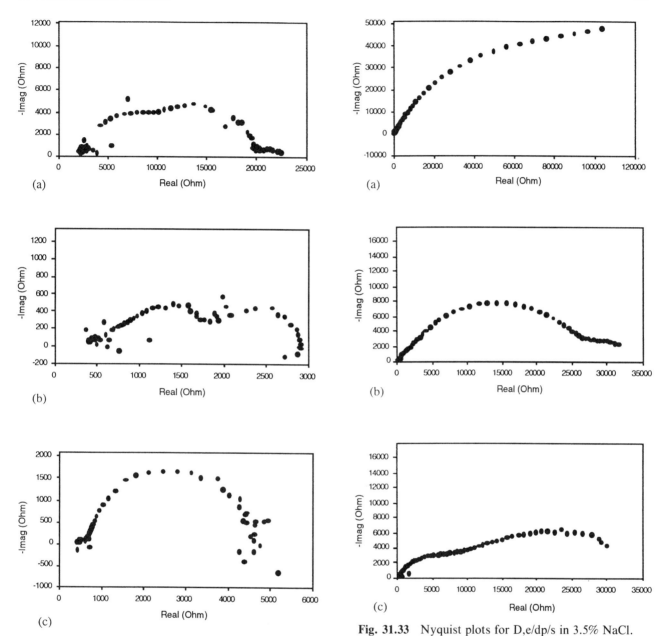

Fig. 31.32 Nyquist plots for D,e/dp/s in 0.1 M HCl.

Fig. 31.33 Nyquist plots for D,e/dp/s in 3.5% NaCl.

line primer on mild steel immersed in 3.5% NaCl did not show much change even after a long immersion. With stainless steel as the coupon material, the imine/amine ratio showed no change at all after immersion in either HCl or NaCl environments. The transitions of N(1s) binding energy of polyaniline after immersion provide information on the redox behaviors of polyaniline during the corrosion processes. However, the same analysis for polypyrrole, listed in Table 31.11, indicates that polypyrrole had stable imine/amine ratios on glass and mild steel surfaces. The polypyrrole film had an intermediate electronic structure of nitrogen binding energy around 399.7 eV on glass and 399.1 eV on mild steel after immersion in both HCl and NaCl solutions.

The iron oxides have also been studied thoroughly by XPS [51]. From the XPS peaks obtained after deconvolution of Fe($2p_{3/2}$) and Fe($2p_{1/2}$) core level spectra, the identify of the passive oxide films on the mild steel surfaces was determined. Detailed results are listed in Table 31.12 and illustrated in Figs. 31.35 and 31.36.

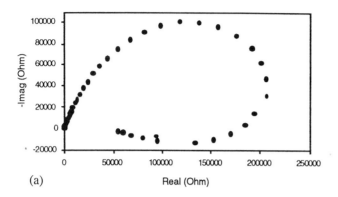

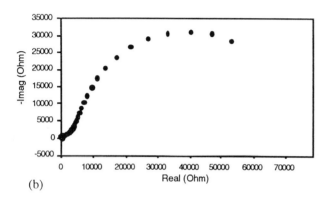

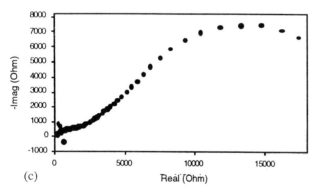

Fig. 31.34 Nyquist plots for D,e/np/s in 3.5% NaCl.

These match three different kinds of oxidation products from six structures of iron oxide passive films proposed in the literature [52].

V. CORROSION INHIBITION OF OTHER METALS WITH CONDUCTIVE POLYMERS

A. Aluminum

In addition to carbon steels and stainless steels, there are other commodity metals such as aluminum that are also used extensively in structural and industrial applications. In practice, the corrosion behavior of aluminum alloys is determined essentially by the nature of the oxide film that readily forms on the metal surface by electrochemical and chemical reactions. There are two commonly encountered forms of corrosion of Al alloys in aqueous environments: pitting corrosion and general corrosion. Pitting is of considerable concern. Many substituted anilines have been studied as inhibitors to aluminum-copper alloy corrosion in phosphoric acid [53]. It was found that the major inhibitive actions of anilines are associated with neutralizing the acid formed in the localized corrosive environment in a pit as well as strong absorption to the metal surface by chemisorption through the unshared electron pair on nitrogen. Alneami

Table 31.10 XPS Nitrogen(1s) Core Level Spectra and Stoichiometries of Polyaniline Primer Films Under Various Exposed Conditions

Sample	Peak position	Area (%)	Intensity
zk/glass	399.35	76.32	164
	401.36	23.68	42
zk/s	399.22	22.72	13
	400.27	55.45	38
	401.59	21.82	14
zk/s,HCl	399.12	87.93	30
	400.64	12.07	3
zk/s,NaCl	398.08	30.16	10
	399.61	61.9	27
	400.41	7.94	3
zk/ss	399.24	86.26	107
	400.61	13.74	17
zk/ss,HCl	399.43	86.9	58
	401.86	13.1	9
zk/ss,NaCl	399.34	90.53	47
	401.4	9.47	6

Table 31.11 XPS Nitrogen(1s) Core Level Spectra and Stoichiometries of Polypyrrole Films Under Various Exposed Conditions

Sample	Peak position (eV)	Area (%)
py/glass (G)	398.92	13.20
	400.89	86.80
py/G,HCl	399.78	100.00
py/G,NaCl	399.71	100.00
py/s	398.99	18.75
	400.90	81.25
D,e/py/s,HCl	399.14	100.00
D,e/py/s,NaCl	399.16	100.00

G = glass

Table 31.12 Chemical Compositions of Iron Oxide Passive Films Under Various Experimental Conditions

Sample	Electrolyte	Composition of passive film
D,e/py/s	HCl	γ-FeOOH mixed with γ-Fe_2O_3
	NaCl	γ-FeOOH mixed with γ-Fe_2O_3
py/s	HCl	Top layer γ-Fe_2O_3; bottom layer γ-FeOOH
	NaCl	Top layer γ-Fe_2O_3; bottom layer γ-FeOOH
D,e/zk/s	HCl	Top γ-Fe_2O_3; bottom γ-FeOOH
	NaCl	γ-Fe_2O_3
D,e/zk/ss	HCl	Fe_3O_4 mixed with Cr_2O_3
	NaCl	γ-Fe_2O_3 mixed with Cr_2O_3
zk/s	HCl	γ-Fe_2O_3
	NaCl	Top layer γ-Fe_2O_3; bottom layer Fe_3O_4

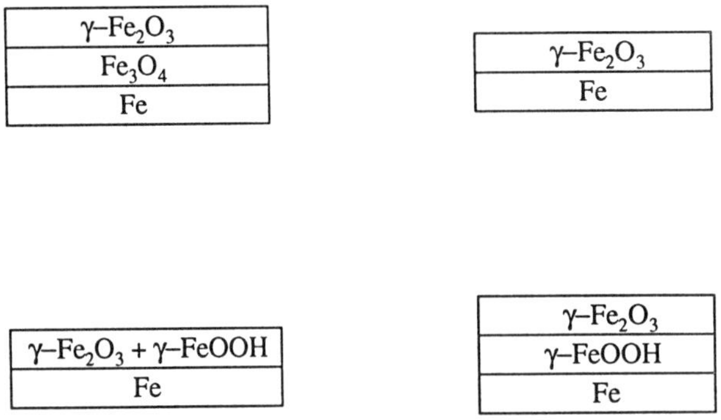

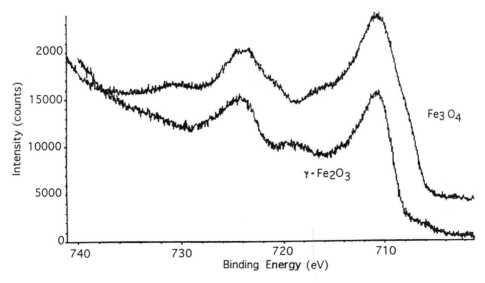

Fig. 31.35 Structures determined for passive iron oxide films formed on the surface of steel.

and Fouda [54] demonstrated that the addition of pyrrole and its derivatives can prevent the corrosion of aluminum in hydrochloric acid and sodium hydroxide. The inhibition efficiencies are determined by the electrophilic character of the substituent groups on the aniline ring and their relative adsorption thermodynamics with the metal surface. Beck et al. [55] deposited polypyrrole on aluminum anodically by direct electropolymerization. From galvanostatic measurements, complex active–passive corrosion behavior was found. The various pretreatments used on the aluminum surface provided tremendous changes in resistance determined electrochemical impedance studies. In addition, pitting corrosion sites on the aluminum surface were found to be filled with polypyrrole after electrochemical deposition, indicating some possible passivation interaction. More recently, Rabicot et al. [56] studied the use of polyaniline for anodic protection of aluminum alloy 7075-T6 in sodium chloride using cyclic polarization and electrochemical impedance techniques. The cyclic polarization results indicate a reduced tendency for pitting corrosion with polyaniline coatings. Moreover, from the observation of impedance spectra, the charge transfer resistance of doped polyaniline-coated aluminum alloy derived from the lower frequency part is greater than for the cases using neutral, nonconductive polymer and bare aluminum reference specimens.

We have conducted preliminary corrosion studies of aluminum alloys 2024, 7075, and 6061 coated with poly-

Fig. 31.36 XPS spectrum of the passive iron oxide layer formed on steel under doped polyaniline coating. (From Ref. 26.)

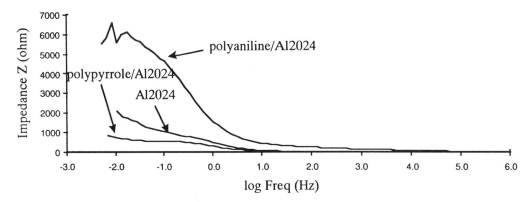

Fig. 31.37 Bode plots of aluminum alloy 2024 uncoated, coated with polyaniline, and coated with polypyrrole and exposed to 3.5% NaCl.

aniline and polypyrrole coatings, both with and without a small defect present in the coatings. The sample configurations, experimental setup, and conductive polymer materials used were the same as those described earlier for studies using steel and stainless steel (see Sections IV.A.1 and IV.A.2). Electrochemical techniques used included EIS and potentiodynamic techniques, cyclic polarization, and Tafel extrapolation.

The Bode plots shown in Fig. 31.37 obtained from EIS studies on aluminum alloy 2024 coupons with doped polyaniline and doped polypyrrole coatings (with no topcoat) indicate that the polyaniline coatings can increase the charge transfer resistance three fold whereas the polypyrrole coatings reduce the charge transfer resistance to one-half the value observed for a bare corroding aluminum sample in 3.5% NaCl.

Potentiodynamic scans for conductive polymer-coated samples of aluminum alloy 7075 in 3.5% NaCl corrosive environments are shown in Fig. 31.38. This figure shows that doped polyaniline and polypyrrole coatings on aluminum induce markedly different corrosion behavior for the metal alloy. Polyaniline coatings significantly shift the corrosion potential to more positive values but substantially reduce the corrosion rate of the metal relative to an uncoated specimen. Also significant is the fact that no passive behavior on the aluminum surface was observed even at high anodic potentials.

In contrast, polypyrrole coatings on this alloy shift the corrosion potential to more negative values and induce a strong passivation tendency at high potentials with much reduced corrosion currents relative to the reference sample. The same observations were made with coated aluminum alloy 6061.

With aluminum alloy 2024, both polyaniline and polypyrrole were found to increase the corrosion potential while providing normal Tafel behavior. Samples of this alloy coated with polypyrrole and polyaniline over-

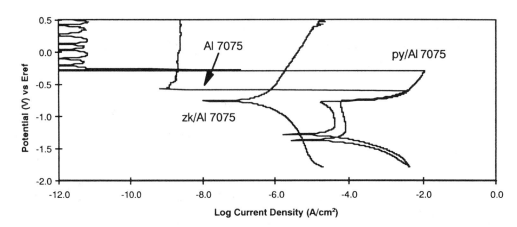

Fig. 31.38 Potentiodynamic scans of aluminum alloy 7075 coated with polyaniline or polypyrrole and exposed to 3.5% NaCl.

coated with epoxy and having small (1.2 mm diameter) bare exposed aluminum defect areas showed lower corrosion current densities than samples coated with just a conductive polymer and no epoxy topcoat. However, as can be seen in Fig. 31.39, neither showed any tendency for strong passivation.

Cyclic polarization studies on aluminum alloy 7075 coated with conductive polymer overcoated with epoxy and having small areas of exposed bare aluminum exhibited strong suppression of pitting corrosion in the exposed aluminum site. Polypyrrole depressed the pitting tendency but provided only small potential differences between the pitting potential and the passivation potential for all the aluminum alloys. Surprisingly, in the absence of a defect site in the coating, neither conductive polymer (no epoxy topcoat) showed much tendency for providing pitting corrosion protection. Quantitative results from Tafel extrapolation experiments indicated that polyaniline could stabilize an aluminum alloy surface to corrosion at the first stages of exposure but then force increased corrosion rates with time, which means that this polymer does not provide long-term passive stability. Polypyrrole provided higher corrosion rates (less protection) than polyaniline on aluminum and higher corrosion rates relative to the reference aluminum alloy in the initial periods of exposure but then stabilized the surface after an intermediate immersion time.

Visual observations and spectroscopic investigations on the various aluminum alloys coated with doped polyaniline and doped polypyrrole indicate that there is no long-range electrode polarization phenomenon like that observed on mild steel–coated samples. The corrosion protection of aluminum using conducting polymers is affected by the composition of the aluminum alloy. Corrosion inhibition is controlled solely by anodic protection. Doped polyaniline coatings provide better corrosion protection performance than polypyrrole for aluminum in 3.5% NaCl, especially in the early stages of exposure. The most significant corrosion inhibition phenomenon provided by conductive polymers on aluminum is the suppression of pitting corrosion [56].

B. Copper and Silver

Polarization resistance measurements made by Wessling [20] for copper samples coated with a dispersion of doped polyaniline in dilute acid environments showed that there was both a shift in corrosion potential and increased polarization resistance compared to samples without the doped polyaniline coatings.

The IBM group led by Brusic et al. [57,58] also studied the use of polyaniline derivatives for corrosion protection of copper as well as silver. The unsubstituted polyaniline, in neutral base form, provided good corrosion protection both at open-circuit potential and at high anodic potentials. The dissolution of metal (both Cu and Ag) was decreased by a factor of 100 when the metal surface was completely covered by the neutral polyaniline. However, polyaniline doped with dodecylbenzenesulfonic acid (the conductive form of the polymer) increased the corrosion rate of Cu and Ag in water. The doped polymer in contact with the metal is spontaneously reduced at a rate faster than the oxygen reduction rate. The faster cathodic process in turn increases the overall rate of the anodic reaction, which is the dissolution of Cu and Ag, as opposed to the formation of a passive oxide layer.

It was also found that poly-o-pheneditine (poly-o-ethoxy-aniline) acts as a better corrosion inhibitor than unsubstituted polyaniline. With the neutral base form, the oxygen reduction and hence the metal dissolution was completely inhibited. The anodic protection in-

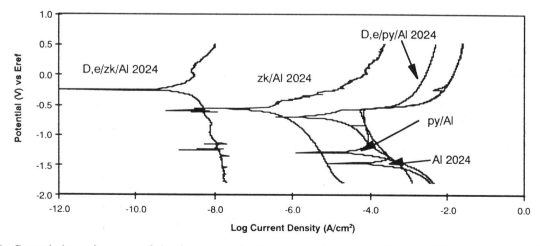

Fig. 31.39 Potentiodynamic curves of aluminum coated with polypyrrole or polyaniline with different sample configurations in 3.5% NaCl.

creases with an increase in film thickness. It is believed that the polymer network acts as a sieve for the metal ions and thus restricts the dissolution of the ions into the solution. Unlike doped polyaniline, HCl-doped poly-o-phenetidine provided excellent corrosion protection. The superior performance of poly-o-phenetidine is attributed to its better solubility and its ability to form more homogeneous and adherent coatings than polyaniline. Also, this substituted polyaniline is found to be a better corrosion inhibitor for Cu and Ag than the most commonly used inhibitor, benzotriazole (BTA), especially at high anodic potentials and under temperature/humidity/bias conditions. Further discussion is found in Chap. 32, Section IV, p. 937–939.

VI. REACTION PATHS AND CORROSION INHIBITION MECHANISMS

A. Electrochemical Corrosion Reactions

Undoubtedly, the unique corrosion-altering effects exhibited by conductive polymers in electrical contact with metals such as steel arise from the fact that both the conductive polymer and the iron are electrochemically active in the corrosive environments. The many different types of conductive polymers exhibit varied electrochemical behavior that depends on the chemical makeup of the polymer and the applied potential. One electrochemical process that is common to all conductive polymers is the reversible doping–dedoping process that is attributed to reversible oxidation and reduction, respectively, of the conjugated polymer backbone (p-type doping). However, this process occurs at very different electrochemical potentials for structurally different conjugated polymers (e.g., polypyrrole vs. polyphenylene). Polymers that oxidatively dope at higher potentials have correspondingly higher chemical potentials in their doped state and are thus stronger oxidizing agents. When in contact with metals, p-doped polymers will provide a galvanic couple whose electrochemical potential will be determined by the chemical potential of the doped polymer and the metal. Doped polymers with high chemical potentials can strongly anodically polarize metals with which they are in contact. This in turn can influence the degree to which a passivating metal may form a passivating oxide layer and "heal" (re-form) this layer when it is scratched or damaged. Also important to consider are the reversible and irreversible electrochemical processes that can occur in the conductive polymer during the corrosion process of the metal in this galvanic couple. The extent to which irreversible processes may occur will determine the useful lifetime of the conductive polymer as a corrosion-preventive element.

Because the oxidative potential of polypyrrole and hydrogen are very close, the electrochemical p-doping and dedoping processes can be simultaneously accompanied by the reduction/oxidation of hydrogen in aqueous environments. Which process dominates can depend on many factors such as pH, nature of the electrolyte, additives, and dopants. Also to be considered when the polymer is coupled with a passivating metal are two possible anodic reactions: oxidation of the polymer backbone and polarization of the metal surface to produce a metal oxide film. Another possibility is an irreversible anodic overoxidation of the polymer that can occur at potentials more positive than that of the reversible doping–dedoping process.

Oxidation of polypyrrole with the addition of OH^- or Cl^- or water to the polymer backbone leads to the formation of overoxidized products as shown below [59]:

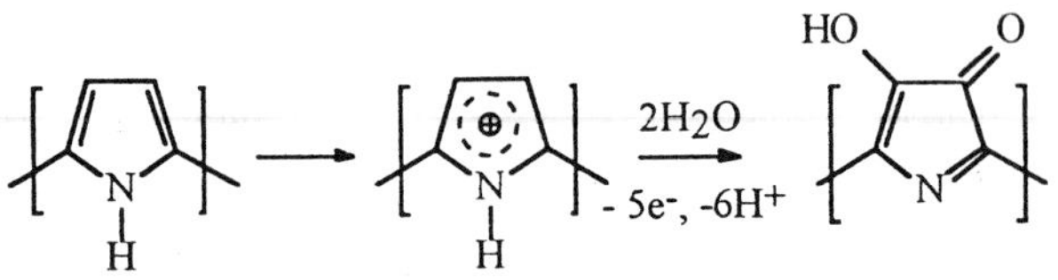

and

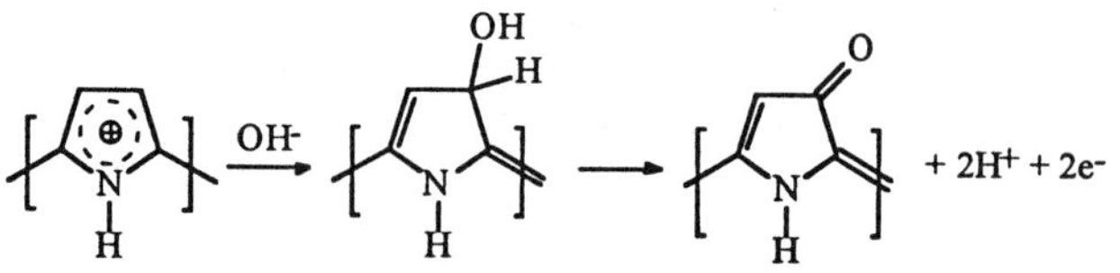

Several reversible electronic structures (oxidation states) are possible for polyaniline (PAni);

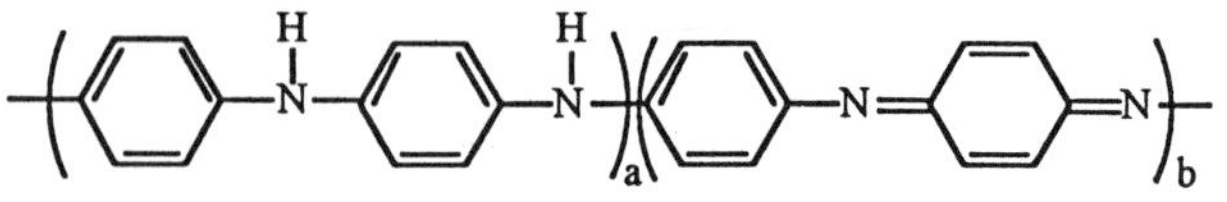

1. PAni° (a=1, b=0) ≡ Leucoemeraldine

2. PAni° (a=b) ≡ Emeraldine

3. PAni° (a=0, b=1) ≡ Pernigraniline

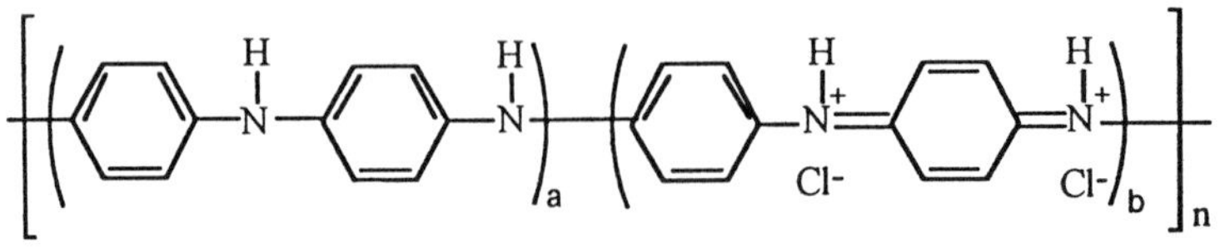

4 PAni$^+$(Cl$^-$)

5

Structures **1**, **2**, and **3** are neutral forms of polyaniline, and structure **4** represents the normal chemical transition after doping by HCl. The more complicated structure **5** includes amine as well as quinoidal and quaternary imine nitrogen subunits. After immersion in electrolytes and the corrosion process begins, the ratio among these three kinds of subunits in structures **4** and **5** will change, and these changes can be detected by XPS.

In aqueous acid solution, two half-reactions of interest involving PAni are

$$(1/4)O_2 + H^+ + e^- = (1/2)H_2O \quad (2b)$$

and

$$PAni^0(a > b) + A^- = PAni^+(a \leq b)(A^-) + e^- \quad (9)$$

The overall reaction can be represented by

$$PAni^0(a > b) + (1/4)O_2 + H^+ + A^-$$
$$= PAni^+(a \leq b)(A^-) + (1/2)H_2O \quad (10)$$

It is important to note that polyaniline containing an excess of the a units (leucopolyaniline subunits) is readily oxidized in air to the emeraldine form of polyaniline.

When the corroding metal is iron, the major electrochemical reaction mechanisms to consider are represented by the following equations:

$$Fe \rightarrow Fe^{2+} + 2e^- \quad (11)$$

$$O_2 + 4H^+ + 4e^- \rightarrow 2H_2O \quad \text{(acid medium)} \quad (2b)$$

$$O_2 + 2H_2O + 4e^- \rightarrow 4OH^- \quad \text{(neutral medium)} \quad (2a)$$

$$2H^+ + 2e^- \rightarrow H_2 \quad (3)$$

$$4PAni^0(a > b) + O_2 + 4H^+ + 4Cl^-$$
$$\rightarrow 4PAni^+(a \leq b)(A^-) + 2H_2O \quad (12)$$

$$Fe^{2+} + O_2 + H_2O \rightarrow Fe_xO_y \cdot H_2O \text{ and/or FeOOH} \quad (5)$$

The mechanism for anodic iron dissolution and passivation in acidic environments that have low concentrations of oxygen and surface-active substances [60] are represented by the following possible reactions:

Active range:

$$Fe + H_2O \rightarrow (FeOH)_{ads} + H^+ + e^- \quad (13)$$

$$Fe + (FeOH)_{ads} \rightleftharpoons [Fe(FeOH)]_{ads} \quad (14)$$

$$[Fe(FeOH)]_{ads} + OH^- \quad (15)$$
$$\rightarrow FeOH^+ + (FeOH)_{ads} + 2e^-$$

$$FeO^+ + 2H^+ \rightleftharpoons Fe^{2+}_{(aq)} + H_2O \quad (16)$$

Transition range:

$$(FeOH)_{ads} + H_2O \rightleftharpoons [Fe(OH)_2]_{ads} + H^+ + e^- \quad (17)$$

Prepassive range:

$$Fe + [Fe(OH)_2]_{ads} \rightarrow FeOH^+ + (FeOH)_{ads} + e^- \quad (18)$$

$$[Fe(OH)_2]_{ads} + H_2O \rightleftharpoons Fe(OH)_3 + H^+ + e^- \quad (19)$$

Passive layer formation range:

$$2Fe(OH)_3 \rightleftharpoons Fe_2O_3 + 3H_2O \quad (20)$$

$$[Fe(OH)_2]_{ads} + Fe_2O_3 \rightleftharpoons Fe_3O_4 + H_2O \quad (21)$$

B. Corrosion Inhibition Mechanisms

1. Molecular Organic Inhibitors

Many of the simple organic compounds found to exhibit corrosion inhibition with steel are heteroaromatic compounds containing nitrogen, such as pyrrole, pyridine, aniline, and their derivatives. It is generally accepted that the mode of inhibition exhibited by these small molecules when they are dissolved in corrosive environments is predominantly absoption and chelation to the iron atoms exposed on the surface of the metal. This absorption process changes the redox properties of the metal surface as well as protecting it from contact with electrolytes and retarding electron transfer reactions [61–63]. With aromatic amine inhibitors, as the electron density at the nitrogen increases, its inhibitor efficiency tends to increase. Antropov [64] suggested that compounds containing the —NH_2 group in appreciable concentrations can be protonated on metal surfaces providing high hydrogen overpotentials. Also, primary and secondary amines have been found to be more effective than tertiary amines owing to steric effects [65]. Solubility of inhibitors also affects inhibition efficiencies. When the structure of the molecules is such that more than one part of it can be absorbed, as with an oligomer or polymer, increased inhibition is observed. It has also been found that the ammonium forms of these molecules can function as good inhibitors in acidic environments. As the corrosion reactions start, the cationic form may become attached to the anodic sites [66], or it is possible that the alkalinity produced at the cathodic sites favors the free base form, which can then be absorbed on the metal surface [53]. The results of Talati and Gandhi [67] show that the efficiency of similar molecular inhibitors such as pyridine, piperidine, and acridine increases with the acid content or with chloride ion concentration of the solution. With pyrrole in acidic environments, the electron-withdrawing power of substituents is a determining factor in inhibition [68]. An increase in electron-withdrawing ability decreases electron density in the pyrrole ring and reinforces its attachment on the negative iron atom sites in cathodic regions. Hackerman [69] demonstrated a correlation between the degree of inhibition for ring-substituted anilines and NMR chemical shift for the amine proton.

2. Conjugated Polymer Inhibitors

Conductive polymers comprising linearly conjugated polymer backbones containing heteroatoms (polypyrrole, polyaniline, polythiophenes, etc.) can function in much the same manner as the molecular inhibitors discussed above for the prevention of corrosion when applied to the surface of a metal. While this normal inhibitor effect arising for surface complexation undoubtedly plays a role in the observed corrosion inhibition of metals with conductive polymer coatings on their surface, there are clearly more powerful corrosion inhibition mechanisms at play. This is especially evident with steel, where corrosion inhibition was observed on bare (uncoated) metal surfaces at significant distances (up to 6 mm) away from the edge of the conductive polymer coating. Further, the nature of the corrosion inhibition on steel with polymers such as polyaniline and polypyrrole arises in large part from the formation of passivating metal oxide films that protect the metal surface from further corrosion and erosion.

The plethora of data presented in this chapter and in the literature leaves no doubt that corrosion processes and mechanisms for inhibition are complex. Nonetheless, it is well established that conductive polymers can impart significant corrosion protection to selected metals such as steel, aluminum, and copper. Also, a picture is beginning to emerge of the corrosion inhibition mechanisms that rationalizes the observations made on corroding cold steel coated with polyaniline. For this system there are two cases to consider. Case I is where a steel surface exposed to a corrosive environment is completely covered by a coating of doped polyaniline. Case II is a case where a steel surface exposed to a corrosive environment is almost completely covered by a coating of doped polyaniline but also has small areas of bare steel exposed.

a. Case I

The overall mechanism for corrosion inhibition for this case is simplistically illustrated in Fig. 31.40. Wessling [20] and Lu et al. [26,29] clearly demonstrated that when doped polyaniline is placed in contact with mild steel, the steel surface undergoes a rapid oxidation process to provide a layer of γ-Fe_2O_3 at the polyaniline/iron interface. This process is shown schematically in Fig. 31.40 by the transformation of (a) to (b) and occurs according to the equation

$$2Fe^0 + PAni^+(a = b)A^- + 3H_2O$$
$$\rightarrow \gamma\text{-}Fe_2O_3 + PAni^+(a + 3 > b - 3)A^- + 6HA$$
(22)

The composition of the iron oxide at the interface has been determined by XPS to be largely γ-Fe_2O_3 next to the polyaniline layer (see Figs. 31.35 and 31.36). There is observed some Fe_3O_4 ($Fe_2O_3 \cdot FeO$) beneath the γ-Fe_2O_3 at the interface with unoxidized iron [27,29]. The γ-Fe_2O_3 formed at the interface is dense and provides a barrier layer that protects the underlying iron from further corrosion. This was nicely demonstrated by

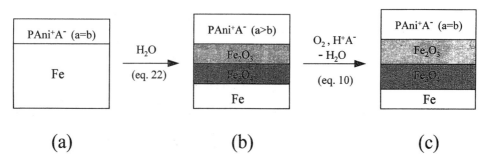

Fig. 31.40 Schematic illustration of the mechanism of corrosion inhibition provided by doped polyaniline on steel with no defects in the coating.

Wessling [19] by removing the polyaniline from the oxidized metal surface and showing that this oxide layer protected the metal surface from further corrosion.

It is important to recognize that in order for a conductive polymer to induce the formation of the passivating oxide layer on steel, it must be of sufficient oxidizing power to directly convert Fe^0 to Fe^{3+}. The oxidation of the iron surface to Fe^{2+} will lead to its dissolution and erosion [Eq. (16)].

It is also important to note that during the formation of the passivating iron oxide layer between the polyaniline and the iron, the metal is oxidized and the polyaniline is reduced (see Fig. 31.40b). The composition of the polyaniline has changed in that more of the backbone has been converted to the leuco form of the polymer (compound 1). This is evident from both XPS [29] and UV-vis spectroscopic analyses [22]. Recognizing that this reduced form of PAni can be readily reoxidized to the emeraldine form by oxygen [Eq. (10) and transformation of (b) to (c) in Fig. 31.40] there is established a mechanism for regenerating the oxidizing power of the polyaniline. Although this air oxidation step is not necessary for formation of the passivating iron oxide layer, it does provide a mechanism that allows the polyaniline to function in a catalytic fashion. Hence, from a practical standpoint, the catalytic nature of polyaniline allows one to apply just small amounts of polymer to the surface of the metal and yet produce a substantial metal oxide layer [20].

The extent to which this metal oxide layer protects the iron surface from corrosion will depend on the nature of the corrosive environment and the stability of the metal oxide to this environment. Hence, quite different extents of protection are expected and observed in going from dilute acid to neutral NaCl corrosive environments. Nonetheless, the presence of the polyaniline and its ability to be oxidatively regenerated by air provide a constant source of high oxidation (chemical) potential on the surface of the metal. Thus, in the event of damage to or dissolution of the metal oxide layer, the oxide layer can be regenerated [15]. This effect is particularly evident in Case II.

b. Case II

The overall mechanism for corrosion inhibition on exposed bare steel surfaces adjacent to the polyaniline coating (overcoated with epoxy) is outlined in Fig. 31.41. It is quite evident that the process of corrosion inhibition and passive oxide layer formation occurs by mechanisms that are more complex that those shown here. However, this figure highlights several important steps that are consistent with the electrochemical, visual, and spectroscopic data.

The first step in this process is shown by the transformation of (a) to (b) in Fig. 31.41. As described for Case I, when the doped polyaniline coating first comes in contact with the iron surface, a quick chemical reaction takes place to form the passivating γ-Fe_2O_3 at the interface. When the polyaniline layer is topcoated with epoxy, a good barrier layer is provided that isolates the polyaniline and metal surface from the corrosive environment. This epoxy layer provides the primary mode of corrosion protection. When this barrier coating is damaged (scratched, drilled, etc.), then areas of bare steel are exposed as shown in (b) of Fig. 31.41. When a sample with this configuration is exposed to a corrosive environment such as dilute HCl, the initial corrosion reaction is one in which the bare steel surface is cathodically polarized and forced to be the cathode while the reduced polyaniline becomes the anode. This is evident from the ratio of the Tafel constants ($\beta c/\beta a$) observed for doped polyaniline samples when they are initially placed in the corrosive environment (Table 31.1). Under these circumstances, reduction reactions occur on the bare steel, thus protecting it from oxidation (cathodic protection). This mechanism prevails as long as sufficient polyaniline is present in its reduced state ($a > b$) relative to its emeraldine form ($a = b$). Eventually, all of the reduced polyaniline will be reoxidized to the emeraldine form as shown in transformation of (b) to (c) in Fig. 31.41. Thus, upon initial exposure to the corrosive environment, the bare steel is cathodically protected and the reduced form of polyaniline acts as a sacrificial anode. Generally, this process lasts only a short time. In principle, coatings containing large amounts of the leuco form of polyaniline galvanically coupled to steel could provide corrosion protection for at least a limited time much in the same fashion as a zinc-rich primer does.

Once the polyaniline layer has been reoxidized back up to its high chemical potential, the mode of corrosion changes on the surface of the bare steel. The region on the bare steel surface adjacent to the polyaniline coating becomes anodically polarized owing to its close proximity to the high potential polymer. This region becomes the anode and supports oxidation reactions while a region further from the conductive polymer edge becomes the cathode, where reduction reactions occur. This localization of anodic and cathodic regions provides for the gradual growth of the passivating iron oxide film on the metal surface as shown in the transformation of (c) to (d) in Fig. 31.41. How far this passive oxide layer extends from the edge of the coating is determined by the oxidizing power of the conductive polymer, the conductance of the electrolyte, oxygen concentration in the solution, pH, and probably other factors. In dilute HCl environments, where the oxidizing power of doped polyaniline remains high, this protective oxide layer can extend as far as 6 mm from the edge of the coating. In dilute neutral pH NaCl solutions, it found to extend only 1–2 mm from the coating. These results indicate that polyaniline coatings on mild steel can provide considerable corrosion protection even to damaged regions in the coating where small areas of bare steel are exposed.

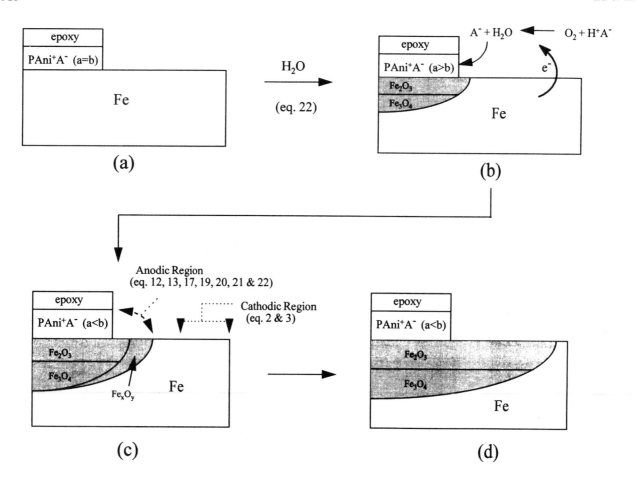

Fig. 31.41 Schematic illustration of the mechanism of corrosion inhibition provided by doped polyaniline on steel overcoated with epoxy and having a small area of exposed bare steel.

SEM analyses on samples corroded for different lengths of time provide evidence that the oxide layer grows out from the edge of the coating [29].

In spite of the understanding that has emerged regarding the mechanisms of corrosion protection provided to steel by polyaniline coatings, much still remains a mystery. For example, it is unclear at this point what the mechanism of corrosion inhibition is for neutral (undoped) polyaniline on steel in 3.5% NaCl. This system even exhibits good protection to bare steel areas in scratches. This may be providing an indication that high electrical conductivity of the polymer may not be necessary for corrosion protection [19]. Even less is understood about mechanisms of corrosion inhibition of other metals, such as aluminum, by conductive polymers. Considerable research effort will be needed to unravel these mysteries and provide a better understanding of the merits and limitations to the use of conductive polymers for the corrosion inhibition of metals.

ACKNOWLEDGMENTS

We thank AlliedSignal, Inc., Americhem, Inc., DSM Research, and Zipperling Kessler Co. for samples of conductive polymers. R.L.E. thanks the University of Texas at Arlington and the Welch Foundation for partial support of this work.

REFERENCES

1. D. E. Fink (ed.), *Aviation Week and Space Technology* 129:29 (1988).
2. M. G. Fontana, *Corrosion Engineering*, McGraw-Hill, New York, 1986.
3. D. A. Jones, *Principles and Prevention of Corrosion*, Prentice Hall, New York, 1996.
4. H. H. Uhlig and R. W. Reive, *Corrosion and Corrosion Control*, Wiley, New York, 1985.

5. R. M. Hudson and C. J. Warning, *Metal Finish. 64*: 63 (1966).
6. R. M. Hudson and C. J. Warning, *Corros. Sci. 10*:121 (1970).
7. E. Stupnisek-Lisac, D. Lencic, and K. Berkovic, *Corrosion 48*:924 (1992).
8. C. C. Nathan, *Corrosion 9*:199 (1953).
9. M. S. Abdel-Aal and F. H. Assaf, *Trans. SAEST 15*: 107 (1980).
10. M. N. Desai and S. M. Desai, *Werkst. Korros. 207* (1966).
11. R. Baskai, A. H. Schroeder, and D. C. Young, *J. Appl. Polym. Sci. 42*:2435 (1991).
12. Z. Szklarska-Smialowska and M. Kaminski, *Corros. Sci. 13*:1 (1973).
13. J. Vosta, J. Eliasek, and P. Knizek, *Corrosion 32*:183 (1976).
14. A. G. MacDiarmid, Short Course on Conductive Polymers, SUNY, New Paltz, NY, 1985.
15. D. W. DeBerry, *J. Electrochem. Soc. 132*:1022 (1985).
16. G. Troch-Nagels, R. Winand, A. Weymeersch, and L. Renard, *J. Appl. Electrochem. 22*:756 (1992).
17. I. Sekine, K. Kohara, T. Sugiyama, and M. Yuasa, *J. Electrochem. Soc. 139*:3090 (1992).
18. S. Sathiyanarayanan, S. K. Dhawan, D. C. Trivedi, and K. Balakrishnan, *Corros. Sci. 33*:1831 (1992).
19. K. G. Thompson, C. J. Bryan, B. C. Benicewicz, and D. A. Wrobleski, Los Alamos Nat. Lab. Rep. LA-UR-92-360, Los Alamos, New Mexico, 1992.
20. B. Wessling, *Advanced Materials 6*:226 (1994).
21. B. Wessling, Personal communication, 1996.
22. B. Wessling, S. Schroder, S. Gleeson, H. Mercle, S. Schroder, and F. Baron, *Mat. Corros. 47*:439 (1996).
23. S. Ren and D. Barkey, *J. Electrochem. Soc. 139*:1021 (1992).
24. G. Singh and G. Kaur, *Bull. Electrochem. 5*:405 (1989).
25. R. C. Ayers and N. Hackerman, *J. Electrochem. Soc. 110*:507 (1962).
26. W. Lu, R. L. Elsenbaumer, and B. Wessling, *Synth. Met. 71*:2163 (1995).
27. S. Jasty and A. J. Epstein, *Polym. Repr.*, ACS Annual Meeting, Anaheim, CA, 1995, p. 565.
28. Y. Wei, J. Wang, X. Jia, J. Yeh, and P. Spellane, *Polymer 36*:4535 (1995).
29. W. K. Lu, Ph.D. Dissertation, The University of Texas at Arlington, 1996.
30. N. Ahmad and A. G. MacDiarmid, *Synth. Met. 78*: 103 (1996).
31. N. V. Krstajic, B. N. Grgur, S. M. Jovanovic, and M. V. Vojnovic, *Electrohim. Acta 42*:1685 (1997).
32. J. Walaszkowski, J. Orlikowski, and R. Juchniewicz, *Corros. Sci. 37*:1143 (1995).
33. ASTM G3, in *Annual Book of ASTM Standards*, Vol. 03.02, American Society for Testing and Materials, Philadelphia, PA, 1991.
34. ASTM G59, in *Annual Book of ASTM Standards*, Vol. 03.02, American Society for Testing and Materials, Philadelphia, PA, 1991.
35. ASTM G61, in *Annual Book of ASTM Standards*, Vol. 03.02, American Society for Testing and Materials, Philadelphia, PA, 1991.
36. Application Note AC-1 in *Electrochemical Corrosion Measurements*, EG&G Princeton Applied Research (1987), P. O. Box 2565, Princeton, NJ 08543.
37. CMS 300 EIS System in *Operators Manual*, Gamry Instruments, Inc., Warminster, PA, p. 2–4 (1993).
38. K. Juettner, W. J. Lorenz, M. W. Kendig, and F. Mansfeld, *J. Electrochem. Soc. 135*:332 (1988).
39. F. Mansfeld, M. W. Kendig, and S. Tsai, *Corrosion 38*:570 (1982).
40. D. C. Silverman and J. E. Carrico, *Corrosion 44*:280 (1988).
41. M. Kendig, F. Mansfeld, and S. Tsai, *Corros. Sci. 23*: 317 (1983).
42. J. R. Scully (ed.), *Wear and Erosion: Metals Corrosion (Metals Handbook*, Vol. 13), ASM Int., Materials Park, Ohio, 1988.
43. A. E. Wiersma, L. M. A. v. Steeg, and T. J. M. Jongeling, *Synth. Met. 71*:2269 (1995).
44. ASTM G102, in *Annual Book of ASTM Standards*, Vol. 03.02, American Society for Testing Materials, Philadelphia, PA, 1991.
45. P. Agarwal and M. E. Orazem, *J. Electrochem. Soc. 139*:1917 (1992).
46. N. Sato, *Passivity of Metals*, The Electrochemical Society, Princeton, NJ, 1978.
47. S. Hino, K. Iwasaki, H. Tatematsu, and K. Matsumoto, *Bull. Chem. Soc. Jpn. 63*:2199 (1990).
48. E. T. Kang, K. G. Neoh, S. H. Khor, K. L. Tan, and B. T. G. Tan, *Surf. Inter. Anal. 20*:833 (1990).
49. J. Yue and A. J. Epstein, *Macromolecules 24*:4441 (1991).
50. M. Aldissi and S. P. Armes, *Macromolecules 25*:2963 (1992).
51. I. D. Welch and P. M. A. Sherwood, *Phys. Rev. 40*: 6386 (1989).
52. K. Asami and K. Hashimoto, *Corros. Sci. 17*:559 (1977).
53. J. D. Talati and J. M. Pandya, *Corros. Sci. 16*:603 (1976).
54. A. Alneami and A. S. Fouda, *Orient. J. Chem. 6*:239 (1990).
55. F. Beck, P. Hulser, and R. Michaelis, *Bull. Electrochem. 8*:35 (1992).
56. R. J. Rabicot, R. Clark, S. Yang, H. Liu, M. Alias, and R. Brown, *Mater. Res. Soc. 413*:529 (K. Jen, ed.), Boston, Masschusetts, 1996.
57. V. Brusic, M. Angelopoulous, and T. Graham, in *Proceedings of 188th Electrochemistry Society Conference*, Chicago, IL, 1995.
58. V. Brusic, M. Angelopoulous, and T. Graham, *J. Electrochem. Soc. 144*:436 (1997).
59. G. Wegner, W. Wernet, and M. Mohammadi, *Synth. Met. 18*:1 (1987).
60. H. Schweikert, W. J. Lorenz, and H. Friedburg, *J. Electrochem. Soc. 127*:1693 (1980).
61. R. G. Pearson, *J. Am. Chem. Soc. 85*:3533 (1963).

62. H. A. Potts and G. F. Smith, *J. Chem. Society*:4018 (1957).
63. G. Singh and O. O. Adeyemi, *J. Surf. Sci. Technol.* 3:1221 (1987).
64. L. I. Antropov, *Corros. Sci.* 7:607 (1967).
65. H. Ohno, H. Nishihara, and K. Aramaki, *Corros. Eng.* 36:335 (1987).
66. F. M. Donahue, A. Akiyama, and K. Nobe, *J. Electrochem. Soc.* 114:1006 (1967).
67. J. D. Talati and D. K. Gandhi, *Corros. Sci.* 23:1315 (1983).
68. A. N. Elneami and A. S. Fouda, *Trans. SAEST 26*: 147 (1991).
69. N. Hackerman, *Corrosion* 18:332t (1962).

32
Conducting Polymers in Microelectronics

Marie Angelopoulos
IBM Research Division, T. J. Watson Research Center, Yorktown Heights, New York

I. INTRODUCTION

Microelectronics, the industry of information processing, has revolutionized our technological society. Electronic products in the form of home entertainment equipment, mobile electronic "gadgets," desktop personal computers, large supercomputers, and so on are pervasive in our everyday world. The electronics revolution began in the 1960s with the fabrication of the first integrated circuits (ICs) [1]. Since then this industry has experienced remarkable growth resulting in significantly more complex ICs that are faster and smaller and whose cost per function has decreased and, in addition, has resulted in a diversity of electronic equipment [2–9].

Integrated circuits consist of devices or single transistor elements that are interconnected by conductors. The number of devices on the ICs has increased dramatically since the 1960s as can be seen in Fig. 32.1, which graphically depicts the number of transistors per chip as a function of dynamic random access memory (DRAM) technology [3]. In today's ICs, millions of devices are interconnected. Figure 32.2 depicts a typical process flow for IC fabrication. IC devices are fabricated directly on a semiconductor wafer, e.g., silicon. They consist of numerous layers, some lying within the body of the silicon and others stacked on top. Each layer has a specific pattern that must be delineated. The various layers must then be precisely positioned one on top of the other and interconnected to form the three-dimensional structure of the IC. This is a complex process involving a succession of deposition, patterning, and etching steps as shown in Fig. 32.2 [4–6]. Numerous chips are made on a single wafer, are subsequently diced, and are then taken through various levels of packaging, i.e., modules, multilayer cards, printed circuit boards, etc., depending on the final sophistication of the electronic product in which they are to be incorporated [4–9]. Packaging basically provides mechanical support and protection for ICs, heat dissipation, and establishes the various interconnections between ICs necessary for information processing [7–9].

Materials representing the entire spectrum of conductivity are at the heart of the microelectronics industry [6,9–12]. Semiconductors are the active device components. Conductors are used extensively for interconnection applications, for electrostatic discharge (ESD) protection of ICs, for electromagnetic interference (EMI) shielding of electronic equipment, and many other purposes. Insulators, most commonly polymers, are widely used as interlevel dielectrics, as encapsulants, as materials for packaging and housing electronic equipment, etc. [6,10–12].

Intrinsically conducting polymers offer a unique combination of properties that make them attractive alternatives for certain materials currently used in microelectronics. The conductivity of these materials can be tuned by chemical manipulation of the polymer backbone, by the nature of the dopant, by the degree of doping, and by blending with other polymers. In addition, they offer light weight, processability, and flexibility. Conducting polymers have potential for an array of microelectronic applications ranging from the device level to the final electronic product, as can be seen in the schematic depicted in Fig. 32.3. This review examines some of these applications. It begins in the area of lithography where conducting polymers have been shown to have numerous applications. This is followed by examining the use of conducting polymers for metallization, as corrosion prevention coatings for metals, and as ESD protective coatings for packages and housings of electronic equipment. The review ends with a look at some of the more futuristic uses of these polymers. In these cases, significant advances in material properties are necessary before they can be realistically considered.

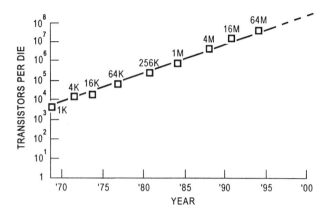

Fig. 32.1 A plot depicting the number of transistors per die in the various dynamic random access memory (DRAM) chips ranging from the 1 Kbit to 64 Mbit chips as a function of the year the chips were first available. (From Ref. 3.)

II. LITHOGRAPHY

A. Background

The various intricate patterns necessary to form, for example, the various doped regions of silicon on a chip or their interconnections or the various interconnections on a package are delineated by lithographic techniques [4,5,12,13]. Lithography relies on radiation-sensitive polymers called resists. When irradiated, these materials undergo chain scissioning, cross-linking, molecular rearrangement, or other process that creates a solubility difference between the irradiated or exposed areas of the polymer and the nonirradiated or unexposed areas [4,5,12,13]. In a subsequent step called developing, the more soluble regions are selectively removed. In positive-acting resists the exposed regions become soluble, whereas in negative-acting resists the exposed regions become insoluble. In either case a pattern has been gen-

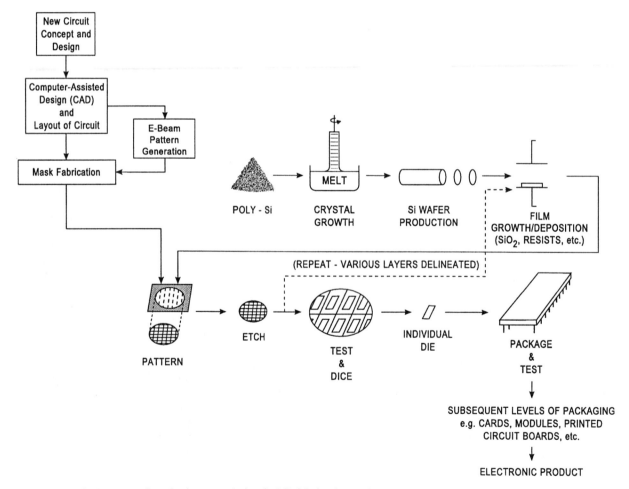

Fig. 32.2 A typical process flow for integrated circuit (IC) fabrication. It is a complex process involving IC design and layout, circuit delineation on a semiconductor wafer, dicing of individual chips, appropriate packaging, and finally incorporating into the electronic product. (From Refs. 4 and 6.)

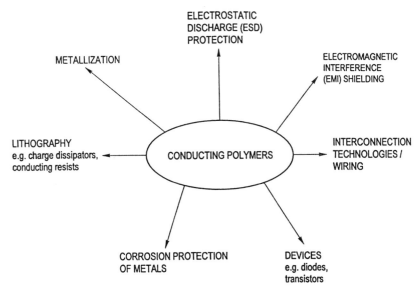

Fig. 32.3 An overview of some of the potential applications of conducting polymers in microelectronics.

erated in the resist as shown in Fig. 32.4. This pattern is subsequently transferred to the underlying substrate (silicon dioxide, metal, etc.) by various etching processes followed by removal of the resist [4,5,14].

Resists are patterned with photons, e.g., ultraviolet radiation, electron beams, X-rays, and ion beams. Photolithography has been the dominant technology in the industry to date. In this case, the resist is exposed to ultraviolet radiation through a quartz/chrome mask containing the pattern to be transferred. Electron beam (e-beam) technology is also currently used but on a much smaller scale. It is used to fabricate high resolution, low volume specialty chips or prototype chips and also to fabricate the masks for photolithography. Ion beam and X-ray as well as e-beam lithography are currently under investigation as possible future alternatives to optical lithography [3–5,13,15,16].

The dimensions that must be delineated are application-dependent. Current DRAM technology features dimensions on the order of <0.5 μm [3,4]. These dimensions will continue to decrease as the industry continues to move towards increased circuit density and microminiaturization [3,4,13]. This trend will require continuous evolution of new materials, processes, and tools that can meet these stringent lithographic requirements.

B. Charge Dissipators for Electron Beam Lithography

Electron beam lithography is a direct-write method in which a focused beam of electrons is directly scanned over the resist [4,5]. No mask is required as the pattern is computer-generated. It is a technology capable of ultimate high resolution as the beam of electrons can be focused to tens of nanometers [17]. It is capable of excellent level-to-level pattern overlay. However, currently it is a low throughput technology and thus, as discussed above, is limited to fabrication of high resolution specialty/prototype chips and to the fabrication of masks [4].

During the e-beam patterning process, charging of the resist is a significant problem [18–20]. Thick insulating resist materials can trap charge and delay bleed-off through the underlying silicon. The trapped charge and any surface charge can deflect the path of the e-beam and result in image distortion as well as level-to-level registration errors. To circumvent this problem, conducting materials that can function as discharge layers are incorporated into resist systems as coatings above or below the imaging resist. Indium tin oxide films [21], amorphous carbon films produced by plasma chemical vapor deposition [22], and, most commonly, thin metal coatings [5,23] have been shown to work quite well in eliminating charging. However, these solutions are not ideal. Evaporative processes are needed to deposit the films. Depending on the actual conditions of evaporation, heat or stray irradiation can be generated that can degrade the lithographic performance of the resist. In addition, the subsequent removal of these layers is difficult if not impossible.

The ionically conducting, water-soluble ammonium poly(p-styrenesulfonate) [24,25] has also been reported as a charge dissipator for e-beam lithography. It has the advantage of ease of processability as it can be spin-

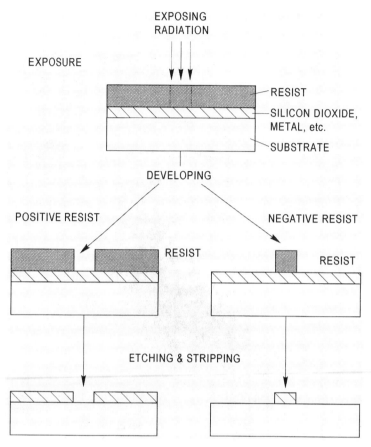

Fig. 32.4 A schematic depicting pattern delineation by lithography.

applied. However, its conductivity is low, and thus its effectiveness at eliminating resist charging is marginal [26].

Intrinsically conducting polymers, in particular the soluble derivatives, are attractive alternative charge dissipators for e-beam lithography. These materials combine high conductivity with ease of processability. The first conducting polymer to be evaluated in this type of application is polyaniline [27]. "Polyaniline" refers to a class of polymers that in the nonconducting or base form have the general composition depicted as follows—[28–33]. These materials are generally doped

(1)

with protonic acids such as aqueous hydrochloric acid (HCl) to give a conductivity on the order of 1 S/cm [28–30]. Polyanilines, from an industrial point of view, are in many applications the preferred conducting polymer system, as they offer a number of advantages over other conducting polymers. They are generally soluble [34–37], environmentally stable polymers that are made by a one-step synthesis involving inexpensive raw materials [38]. They offer extensive chemical versatility, which allows the properties of the polymer to be tuned to more appropriately meet the needs of a given application. Indeed, many polyaniline derivatives exist today as a result of chemical modification of polymer backbone [30,39–44], dopant [45–49], and oxidation state [30,50,51].

In the first report [27] in which polyaniline was evaluated as an e-beam discharge layer, the polymer was incorporated into a multilayer resist system (Fig. 32.5) as a conducting interlayer between the imaging resist and the planarizing underlayer. The multilayer structure was designed to test the efficiency of the conducting polyaniline as charge dissipator. For this reason, relatively thick layers of insulators were used that would tend to enhance charging. The underlayer consisted of a 2.8 μm film of a hard-baked or cross-linked AZ4210 material and 5000 Å of silicon dioxide. Polyaniline in the base form (2000 Å) was spin-applied onto the AZ material.

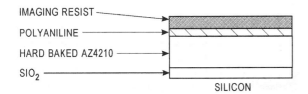

Fig. 32.5 A multilayer resist system incorporating polyaniline as a conducting interlayer.

The base polymer was then doped by dipping the sample into a dilute aqueous hydrochloric acid solution. As this particular conducting salt is insoluble, the resist (a typical diazonaphthoquinone-novolac formulation [4] could be directly coated on top of the polyaniline without any interfacial problems.

The multilayer structure was subjected to a discharge test that evaluates resist charging during the e-beam writing process. Initially, a 20 × 20 matrix of 2 μm squares was written across a 5 mm chip. Because of its low density, charging during the writing of this pattern is negligible. The chip was then completely overwritten except for a 10 × 10 μm area centered on each inner square as illustrated in Fig. 32.6. Following exposure, the resist was developed and the images were inspected to determine the degree of pattern displacement.

In this test, the low density pattern (inner squares) is used as a reference. When no charging occurs, the 2 μm squares are located at the center of the 10 μm squares as shown in Fig. 32.7. However, as charging occurs during the writing of the dense second pattern, the e-beam is offset with respect to the first reference pattern, and as a result the squares are no longer centered. In general, the first square written is the least impacted by charging as only a small area has been written, whereas the last square written is strongly influenced by charging due to the high density of the exposed area. The displacement, which is directly related to charging, is that observed from the first to the last squares written and is measured by taking the difference between the centers of the inner and outer squares as shown in Fig. 32.7.

A pattern displacement greater than 5 μm across a 5 mm chip was observed in the case where a conducting discharge layer was not incorporated into the resist structure (Fig. 32.8) [27]. The use of polyaniline was found to give zero displacement as shown in Fig. 32.9 [27]. These results demonstrate that polyaniline is very effective at eliminating resist charging even though in this particular configuration it is used as a thin interlayer between thick insulating layers. In addition, in this configuration the polyaniline is not grounded. The substrate (i.e., silicon wafer) is grounded, but the connections are made on the top and bottom surfaces of the wafer. The polyaniline functions by bleeding off charges from the resist and preventing charge build-up in the resist layer that would deflect the e-beam and create placement errors.

In this same study [27], the conductivity of polyaniline was varied to determine what level of conductivity is actually required to eliminate resist charging. The conductivity was tuned by changing the pH of the aqueous acid dopant solution [34,52]. It was found that a conductivity greater than 10^{-4} S/cm is required to prevent pattern displacement. It should be noted that this conductivity requirement is for the interlayer structure. The level of conductivity needed for charge dissipation will be dependent on the actual resist structure configuration.

In this first evaluation of polyaniline as a charge dissipator in resist systems, the polymer was processed in two steps. The base form of the polymer, which is generally the more soluble form of polyaniline, was first spin-applied to a surface. The sample was then converted to the conducting form by dipping it into an aqueous acid solution. The doping reaction takes several hours to ensure the diffusion of the dopant into the bulk of the film. This two-step process, in particular the need to soak the sample in aqueous acid solution, is not a desirable process. Not only does it increase the number of processing steps, but the prolonged exposure of substrates to acid solutions also poses contamination, reliability, and corrosion concerns.

In a subsequent study [53], a simplified, one-step process to apply the polyaniline to resist systems was reported. A method of inducing the doping in situ in the

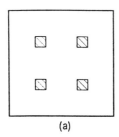

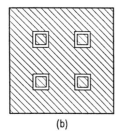

Fig. 32.6 An e-beam discharge test pattern. (a) A 20 × 20 matrix of 2 μm squares is written. (b) The rest of the chip is written except for 10 μm squares around the inner 2 μm squares. (From Ref. 27.)

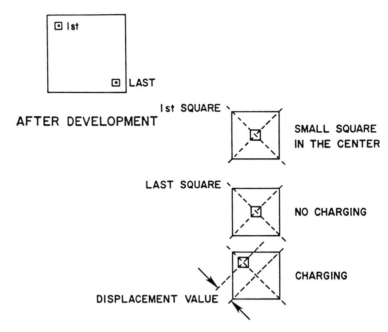

Fig. 32.7 Evaluation of e-beam discharge test results. The pattern displacement is that observed from the first to the last (400th) square written. (From Ref. 27.)

polymer was developed that eliminated the need for external acid solutions. This was accomplished by incorporating salts in the polymer that would decompose upon irradiation or thermal treatment to generate the active dopant species, i.e., a protonic acid, in situ in the material [53–55]. These salts were of two types—onium and amine triflate salts.

Onium salts such as triarylsulfonium and diaryliodonium salts are a class of materials that decompose upon exposure to ultraviolet radiation or to an e-beam to generate a protonic acid [56,57]. These salts are further discussed in Section II.D. Conductivity of 0.1 S/cm was attained when a polyaniline base film containing an onium salt was irradiated [54].

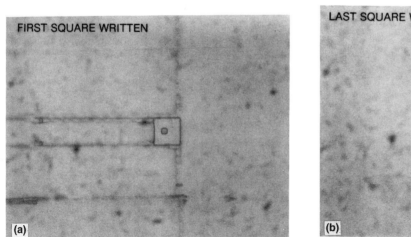
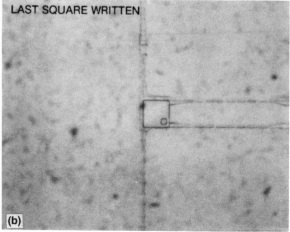

Fig. 32.8 Image placement results for a control resist system with no conducting layer. Greater than a 5 μm displacement is measured across a 5 mm chip. (a) First square written; (b) last square written. (From Ref. 27.)

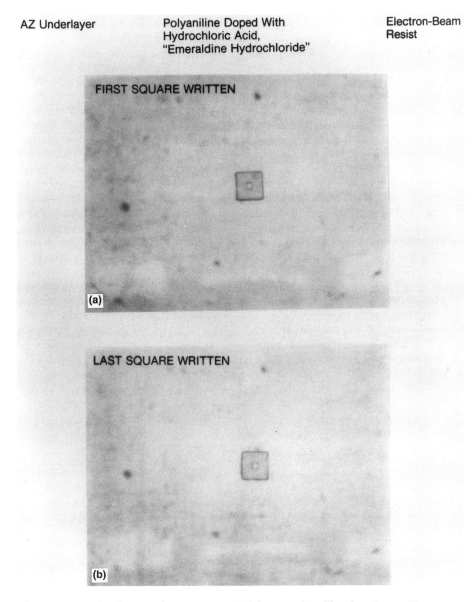

Fig. 32.9 Image placement results for a resist system containing a polyaniline interlayer. Zero pattern displacement is observed. (a) First square written; (b) last square written. (From Ref. 27.)

Amine triflate salts thermally unblock to generate the free triflic acid with the volatilization of the corresponding amine [58,59]:

$$R_3N^+HCF_3SO_3^- \rightarrow CF_3SO_3H + R_3N \uparrow$$

The temperature at which the salts decompose and doping occurs can be tuned by the nature of the amine group. The less volatile or more basic the amine, the greater the temperature required to decompose the salt. Thus, this method offers great latitude in allowing the polyaniline process to be made compatible with the resist process. Triethylammonium triflate, for example, decomposes at 90°C. When a polyaniline film containing this salt was baked for 5 min at 90°C, a conductivity of 1 S/cm was reported, comparable to that measured with aqueous acid solutions [53,55].

The thermally induced doping route is a simpler method than radiation-induced doping as it requires only two steps—application of the film and baking. The baking is normally done on any film to remove solvent. Thus, no additional steps are needed. Polyanilines doped in this fashion were evaluated as charge dissipators in a manner similar to that described above for the aqueous acid–doped materials and were found to be just as effective [53].

In the studies described thus far, the polyaniline is incorporated below the imaging resist. In these systems, once the resist is exposed and developed, the polyaniline remains in the open areas as shown in Fig. 32.10a. To transfer the pattern to the underlying substrate, the exposed polyaniline must be removed. This is done by oxygen reactive ion etching (RIE) [4,14,27] after the imaging resist is made etch-resistant by an appropriate silylation process [60].

Although the method of using polyaniline below the resist works quite well, an easier and more optimum approach is to apply the conducting layer on top of the resist and have the conductor removed simultaneously with the development of the resist. This configuration and process are depicted in Fig. 32.10b. As the conductor in this case is applied directly on the resist, it must meet certain requirements. First, the solvent used to coat the conductor must not dissolve the resist or induce any interfacial problems. Thus, polar solvents such as NMP, which is commonly used to process polyaniline, are not acceptable as they would dissolve most commonly used resists. The conductor must not degrade the lithographic performance of the resist. It should not introduce any contamination. In addition, it should be cleanly removed, if possible during the development of the resist. Most resists currently used in the industry are developed in aqueous solutions, and thus, the conductor must be able to dissolve in these aqueous systems.

In the last few years a number of polyaniline derivatives have been developed that are soluble in the conducting form [44,46,48]. Many of these are based in organic solvents, and although some of these could in principle be used as topcoat discharge layers, water-soluble polyanilines are preferred. First of all, such polymers eliminate environmental concerns of organic solvents. Reduction of solvents is an issue of increasing importance in the electronics industry. In addition, as discussed above, water-based systems would be more compatible with current resists.

A number of water-soluble polyaniline derivatives have been developed in recent years [39–41,49,61–65]. One method used to introduce water solubility has been to incorporate sulfonate groups onto the polymer backbone. This has been accomplished by several routes. One process involved the sulfonation of polyaniline base by treating the polymer with fuming sulfuric acid [39,62]. This results in a sulfonic acid ring-substituted derivative that is alkaline-soluble but only upon conversion to the nonconducting sulfonate salt form. A second method of introducing sulfonate groups was accomplished by de-

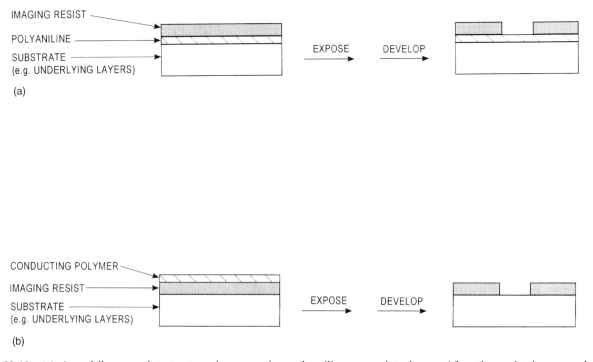

Fig. 32.10 (a) A multilayer resist structure incorporating polyaniline as an interlayer. After the resist is exposed and developed, polyaniline remains in open areas. Etching of polyaniline is required to transfer image. (b) A multilayer resist structure in which the conducting polymer is applied as a topcoat discharge layer. The conducting polymer is removed during resist development.

protonating polyaniline base and reacting with a sultone, i.e., 1,3-propanesultone [63,64]. This gives rise to an N-substituted polyaniline derivative that is water-soluble. Another route involved the polymerization of sulfonated aniline monomers such as the sodium salt of diphenylaminesulfonic acid [40,41]. IBM introduced a family of water-soluble polyanilines [49,65] referred to as PanAquas.* This is a series of polymers that are highly soluble in neutral water in the conducting form. They are made in a one-step straightforward synthesis involving a template-guided polymerization (Fig. 32.11). In this reaction, the aniline monomer is first complexed to a polymeric acid. Once the complex is formed, the aniline is polymerized in a controlled fashion to allow the polyaniline chain as it grows to wrap around the polyacid chain. In this way, the formation of an interpenetrating network is prevented. As a result, the polyaniline/polyacid blend that is isolated directly in the conducting form is water-soluble. A number of different derivatives can be made by this method by variations in the nature of the aniline monomer (variations in R in Fig. 32.11) and in the nature of the polyacid. Conductivity is on the order of 10^{-2}–10^{-3} S/cm for these polymers.

The PanAquas were found to provide a simple and effective discharge solution for e-beam lithography [65]. In particular, the unsubstituted derivative where R = H in Fig. 32.11 was spin-applied onto the surface of a number of common resists used in the industry such as novolacs, acrylates, and chemically amplified systems [4] and was found to be compatible. The performance of the resist was not sacrificed in any way. A 2000 Å layer was found to be quite effective at eliminating charging of the resist, as can be seen in Fig. 32.12, which compares a pattern that was written with no topcoat (a) to one that contained the PanAqua topcoat (b). As can be seen, severe image distortion is observed in Fig. 32.12a, whereas a well-defined image is observed in Fig. 32.12b. The PanAqua can be cleanly removed during the alkaline development of the resist [65].

In a recent study published by Etec Systems [23], a number of conductors were evaluated as charge dissipa-

---

* PanAqua is a trademark of IBM Corporation.

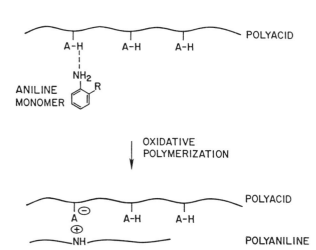

Fig. 32.11 Template-guided polymerization of water-soluble polyanilines (PanAquas). (1) Aniline monomer is complexed to a polyacid template; (2) complex is oxidatively polymerized in a controlled fashion. (From Ref. 49.)

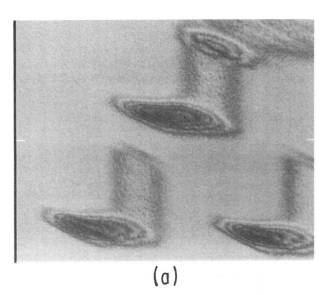

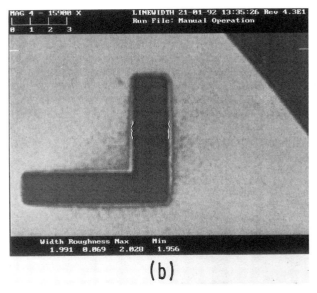

Fig. 32.12 Photomicrographs comparing resist patterns. (a) Resist imaged with no discharge layer; image distortion observed. (b) Resist imaged with a water-soluble polyaniline topcoat; no image distortion is observed. (From Ref. 49.)

tors for phase shift mask (PSM) [4,66,67] e-beam registered writing. Tan [23] looked at a number of materials in parallel and compared their ability to improve the overlay accuracy of the two lithography levels in the PSM process (L0 and L1 in Fig. 32.13). The first level pattern was a reference consisting of butting crosses that were written onto a resist-coated chrome/quartz plate. After the pattern was delineated and etched into the chrome, the second registered level pattern was written using a novolac resist and a charge dissipator including aluminum, chrome, indium tin oxide, and the PanAqua. The overlay accuracy between the two levels was determined from the butting overlay of crosses between the reference level and the registered level as can be seen in Fig. 32.13, which depicts the results for the the PanAqua case. Tan concluded that all the conductors tested provided excellent charge dissipation as an overlay accuracy of less than 0.07 μm (mean + 3σ) was attained. Tan also concluded that the PanAqua provided the simplest solution as it was a spin-apply process that involved no evaporative deposition as was necessary with the other materials. The PanAquas are in the process of becoming commercially available.

Another class of conducting polymers that has been of interest for charge dissipation in e-beam lithography is that of the water-soluble, self-doped polythiophenes such as the sulfonated derivatives, which have the general structure depicted here; [26,68–70]. The one disadvantage with these materials compared to the polyanilines is that they are made by a cumbersome synthetic procedure, as the sulfonic acid–substituted thiophene monomers do not directly polymerize. A typical synthesis for the poly(3-(2-ethanesulfonic acid)thiophene, for example, begins with the monomer, 2-(3-thienyl)ethanol, and is converted in five synthetic steps to the methyl-2-(3-thienyl)-ethanesulfonate [69]. This monomer is then polymerized. The polymethyl ester is in turn converted to the polysodium salt, which is then converted to the self-doped acid version by ion-exchange chromatography [69].

Recently, this polythiophene derivative was evaluated as a topcoat discharge layer that was applied to a typical novolac resist [26]. The polymer was found to be very effective at eliminating resist charging. In another study [71], a water-soluble polythiophene derivative referred to as ESPACER* was applied to a chemically amplified resist, and charging was eliminated during e-beam writing. In this system, the ESPACER was removed by a water wash prior to the development of the resist.

As the microelectronics industry continues to move towards increased circuit density with a concomitant decrease in device geometries, higher precision in e-beam writing will be required in the future. Any surface charge on the resist can jeopardize the precision of the writing. Thus, discharge layers will certainly be necessary in the future. The water-soluble conducting polymers have been shown to provide not only an effective approach to eliminating resist charging but also a much simpler process than currently used materials such as metal coatings. The commercial availability of these materials, in particular the PanAquas and ESPACER, and their ultimate cost will determine whether they are commonly implemented in e-beam lithography in the future.

C. Charge Dissipators for Scanning Electron Microscopic Metrology

Scanning electron microscopy (SEM) offers increased resolution capability compared to optical microscopy. As IC device features continue to decrease into the sub-0.5 μm region, SEM is becoming a commonplace technique for the inspection and dimensional measurements (metrology) of these circuits [72–74]. Measurements are made, for example, on actual device wafers or on high resolution optical and X-ray masks. Charging of the sample during SEM inspection makes accurate metrology difficult. As an electric field builds up on the sample, the electron beam can be deflected. Even a very small beam deflection around a feature can move the beam one or two pixel points. If this occurs, substantial error

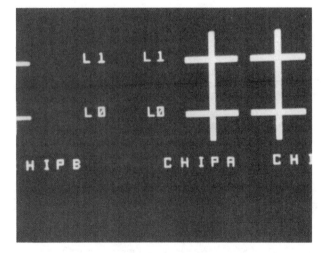

Fig. 32.13 Overlay image of two levels of lithography (L0 and L1) during PSM registered e-beam writing; overlay accuracy of less than 0.07 μm attained. (Courtesy of Z. Tan, Etec Systems; Ref. 23.)

---

* ESPACER is a trademark and product of Showa Denko K.K.

is introduced into the measurement of the critical dimensions.

One method to somewhat alleviate the charging problem is to carry out the microscopy at low accelerating voltages (lower than 2 keV) [72–74]. However, at such voltages the resolution of the measurements is sacrificed. As device geometries continue to shrink, low voltage scanning electron microscopy is not going to be an effective tool for metrology. Another method to prevent charging is to coat the sample with a conducting metal such as gold. Although this permits high resolution metrology, it is a destructive process as the metal cannot be removed from the surface of the substrate. The device or mask cannot be reused.

A nondestructive process that would allow high resolution metrology is needed. Conducting polymers that can be spin-applied onto the sample and subsequently cleanly removed are ideal. Polyaniline has been demonstrated to provide such a solution. Angelopoulos et al. [75] reported that a 1500 Å layer of polyaniline was coated onto the surface of an optical mask. SEM inspection was then done on the coated mask as well as on a control, an uncoated mask. The results are shown in Fig. 32.14. Pronounced charging was observed at 5 kV on the bare mask (Fig. 32.14a) whereas no charging was observed on the polyaniline-coated mask even at 15 kV (Fig. 32.14b). After the measurements, the polyaniline was removed from the surface of the mask by rinsing with a solvent. In this example, the polyaniline was based in NMP, which of course is not the optimum solvent for a practical and useful solution. However, the water-soluble polyanilines are more appropriate for this application and certainly do provide a clean, nondestructive method of doing high resolution metrology. As in the previous section, the implementation of conducting polymers in this area will depend on their future commercial availability and cost.

D. Conducting Resists

In the lithographic applications discussed in the previous sections, the conducting polymers are basically functioning as additive, charge-dissipative layers that are subsequently removed. The resists that are used to delineate the circuitry patterns are insulators. If the resist itself were conducting, then there would not be a need for an additional dissipative layer; only one polymer coating would be necessary. The material could be directly exposed with an e-beam, and the delineated pattern could in turn be inspected with scanning electron microscopy without charging concerns. A sensitive, high resolution, and water-developable conducting resist would be an optimum system that would offer significant process simplification.

There has been a great deal of activity directed toward the development of conducting resists. The first

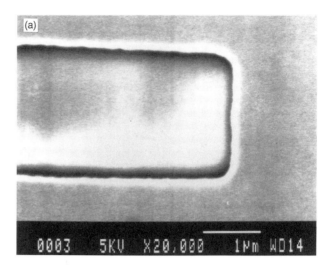

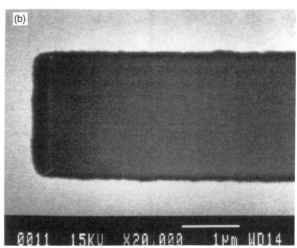

Fig. 32.14 (a) Metrology of an uncoated mask with SEM. Pronounced charging observed. (b) Metrology of a polyaniline-coated mask with SEM. No charging observed. (From Ref. 75.)

report [27] of such a polymer was based on the radiation-induced doping of the unsubstituted polyaniline base with onium salts. These salts have been shown to readily decompose upon irradiation to generate protonic acids [56,57]. They have also been used to photochemically dope polyacetylene [76] and polypyrrole [77]. In these systems, because the polymers are insoluble, the salts were impregnated into the polymer films. In the polyaniline case, the polymer and the onium salt, e.g., triphenylsulfonium hexafluoroantimonate, were dissolved in NMP and processed into a film. Upon exposure of the film to ultraviolet radiation or to an e-beam, the generated acid dopes the polymer. Figure 32.15 depicts the optical changes that are observed in the polyaniline/

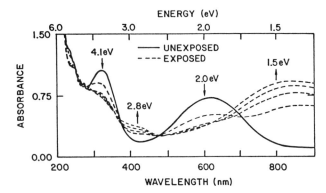

Fig. 32.15 Optical absorption spectra for the polyaniline/onium salt before (———) and after (---) different doses of exposure. The same film was used for all scans. (From Ref. 54.)

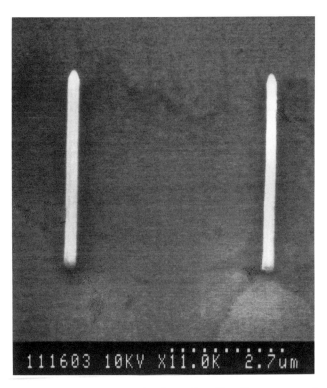

Fig. 32.16 0.25 μm conducting polyaniline lines patterned with e-beam irradiation. (From Ref. 53.)

onium salt film upon ultraviolet irradiation [54]. As can be seen, the polaron absorptions [78,79] characteristic of the conducting form of polyaniline emerge. Conductivity on the order of 0.1 S/cm was reported [54]. As the doped polymer is no longer soluble, a solubility difference is created between exposed and unexposed regions. Thus, a negative conducting resist developed in a mixture of NMP/diglyme was attained [27,53–55]. 0.25 μm conducting lines in a 0.25 μm thick film (Fig. 32.16) were written with this system using an e-beam. The sensitivity of the resist was on the order of 100 μC/cm^2 with an e-beam and 300 mJ/cm^2 for deep UV.

In a later study [80], a similar method was used to pattern the methyl-substituted polyaniline, poly-o-toluidine. This material is a more soluble derivative than the unsubstituted parent polymer used in the previous study and thus allows a broader range of solvents to be used. A nonionic nitrobenzylsulfonate ester was used as the photoacid generator. The polyaniline derivative was mixed in methyl ethyl ketone with the acid generator and exposed to ultraviolet radiation. Upon irradiation, a conductivity on the order of 10^{-7} S/cm was attained. The conductivity was increased to 10^{-3} S/cm by subsequently externally doping the photodoped samples with HCl acid. Although a large increase in conductivity was not observed upon irradiation, the photoinduced doping did create a large enough solubility difference to differentiate doped and undoped regions. Figure 32.17 depicts 2.0 and 1.5 μm features delineated with deep UV. The resist was not very sensitive, as a dose of 1500 mJ/cm^2 was required.

A water-developable negative conducting resist was more recently reported [49,65]. This was based on the PanAquas. Cross-linkable functionality was incorporated into the polyaniline backbone. Upon e-beam irradiation, the polyaniline cross-links and becomes water-

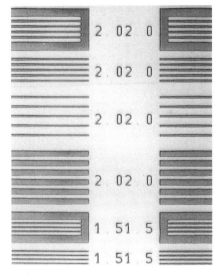

Fig. 32.17 1.5 and 2.0 μm features delineated with the poly-o-toluidine system. (Courtesy of G. Venugopal, Motorola; Ref. 80.)

insoluble. Images are developed with water. 1.0 μm conducting lines in a 0.75 μm thick film were patterned with an e-beam at a dose of 200 μC/cm² (Fig. 32.18). 0.25 μm lines were also delineated (Fig. 32.19); however, this resolution requires further optimization. This resist is quite promising for future use because it is a water-developable conducting resist; however, its lithographic performance needs improvement. Much higher sensitivity is required. Work in this area is ongoing.

Conducting resists have also been reported with polythiophenes. One of the first examples [81,82] was based on poly(3-octylthiophene) (P3OT). The nondoped form of the polymer was combined with a cross-linking reagent, ethylene 1,2-bis(4-azido-2,3,5,6-tetrafluorobenzoate). Upon exposure to deep UV, cross-linking through the octyl side chain was induced as depicted in Fig. 32.20. Cai et al. proposed that this occurred by CH insertion by a triplet nitrene intermediate into the octyl side chain. This was a two-step process involving hydrogen abstraction followed by radical combination. This system was found to cross-link upon e-beam irradiation as well. In addition, the polymer without the cross-linker was also noted to cross-link with an e-beam. Negative images of nondoped P3OT were attained. The developer was xylene. This resist has relatively high sensitivity; an e-beam dose of 30 μC/cm² was reported. 0.2 μm features were delineated. Figure 32.21 is a micrograph depicting a wire pattern structure of cross-linked P3OT written with an e-beam [81,82]. This resist exhibits good lithographic performance. However, the images are attained on the nondoped polymer, and thus it is not a one-step conducting resist. Cai et al. did find that doping could be induced after the images were developed by dipping the patterned film into an FeCl₃ solution. Conductivity of ≈5 S/cm was measured on the doped wire pattern

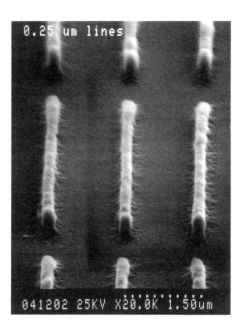

Fig. 32.19 Conducting polyaniline lines 0.25 μm wide in a 0.75 μm thick film patterned with e-beam irradiation. Image developed in water.

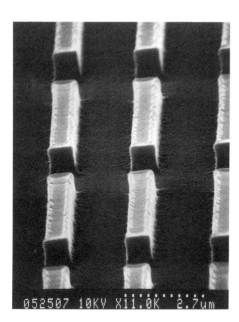

Fig. 32.18 Conducting polyaniline lines 1.0 μm wide in a 0.75 μm thick film patterned with e-beam irradiation. Image developed in water. (From Ref. 65.)

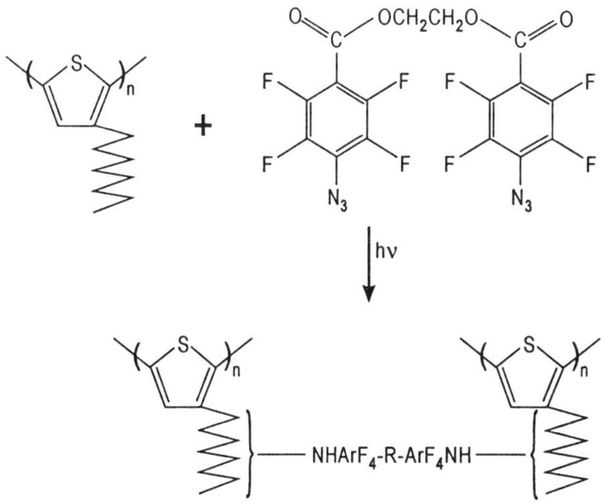

Fig. 32.20 A schematic depicting radiation-induced cross-linking of poly(3-octylthiophene) using the ethylene 1,2-bis(4-azido-2,3,5,6-tetrafluorobenzoate) cross-linker. (From Ref. 81.)

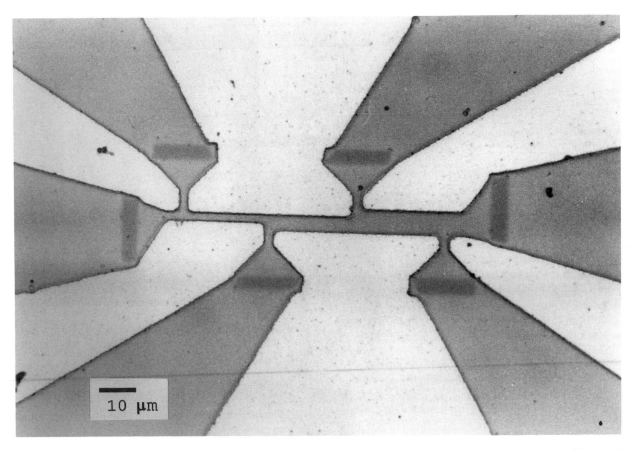

Fig. 32.21 A micrograph depicting a cross-linked poly(3-octylthiophene) wire pattern delineated with e-beam lithography. (Courtesy of J. Keana and M. Wybourne, University of Oregon; Ref. 81.)

depicted in Fig. 32.21. A similar system based on poly(3-hexylthiophene) was later reported by Abdou et al. [83].

An extension of the substituted thiophene work involved the incorporation of methacrylate functionality onto the thiophene backbone [84,85]. Methacrylates are widely known to undergo free radical polymerizations. Poly(3,2-(methacryloyloxy)ethyl)thiophene) as well as

(3)

copolymers with other substituted thiophenes were synthesized. These systems have the basic structure depicted above. The polymers undergo cross-linking through the methacrylate side chains upon irradiation. Negative images developed in organic solvents were obtained. Lines 3 μm wide were written in a 75 nm film. The resist was very sensitive ($\approx$14 mJ/cm^2 at a wavelength of 313 nm) [85]. In this system as in the previous polythiophene examples, the imaged patterns need to be subsequently doped to make the patterns conducting.

A completely different approach to patterning conducting polymers involves the use of photosensitive oxidants [86,87]. In this process, a photosensitive oxidant is mixed with a host polymer such as poly(vinyl chloride), poly(vinyl alcohol), or polycarbonate. The composite is applied to a substrate. Upon irradiation of the film, the oxidant in the exposed regions is made inactive, whereas in the unexposed regions the oxidant can still induce polymerization of appropriate monomers. After exposure, the latent image is exposed to a monomer such as pyrrole either in solution or in the vapor state. Polymerization occurs only in the nonexposed areas where the oxidant is still active. In this fashion, patterns are delineated that consist of conducting composite materials. Some photosensitive oxidants include Fe(III) salts such as iron trichloride and ferrioxalate. Upon exposure, the Fe(III) is converted to Fe(II), which does not induce oxidative polymerization [86,87].

$$Fe(III)(C_2O_4)_3 \longrightarrow Fe(II)(C_2O_4)_2 + 2CO_2$$

A number of different methods of imaging conducting polymers either directly in the conducting form or in the precursor, nondoped form have been developed. These systems may be appropriate in certain applications where the conducting polymer basically needs to be patterned to a certain geometry. However, for these systems to be used in lithography as conducting resists, their lithographic performance needs to be improved. The systems described herein do not compete with conventional resists [4,12] in terms of resolution, sensitivity, contrast, etc., in their current form. They do, of course, offer conductivity, and that is certainly a desirable advantage; however, further work in this area is required to bring the conducting polymer–based resists closer in performance to conventional resists.

III. METALLIZATION

In microelectronics, "metallization" generally refers to the deposition of a patterned film of conducting material on a substrate to form interconnections between electronic components [2,7]. Over the last few years, conducting polymers have been demonstrated to provide a new route to metallization, in particular in printed circuit board (PCB) technology [88–92]. In general, conducting polymers can be used for both electrolytic and electroless metallization. The conducting polymers that have been of interest in this area include polyaniline [88], polypyrrole [89–92], and polythiophene [91].

PCBs [93–95] vary in complexity ranging from single-sided boards where circuitry is found only on one side, to double-sided boards, to boards consisting of multilayers of circuitry. The degree of complexity depends on the specific interconnection needs for a given product. Connections between the two sides of a board and layer-to-layer connections are made with copper-plated through holes (PTHs) [93–95]. PTHs allow greater circuit density because they provide crossover capability. One circuit can cross over another by simply entering a PTH, continuing on the other side, and so on.

The through holes are drilled into the laminate substrate and are then copper plated. Figure 32.22 depicts two common metallization schemes for PCBs [93–96]. In one method (Fig. 32.22a), a conducting strike layer, generally a thin layer of copper, is deposited by electroless plating. This copper layer renders the surface sufficiently conducting to allow a thicker copper layer to be electrolytically deposited in selected regions that are defined by a photoresist process. In an alternative method (Fig. 32.22b), an all electroless process is used. The current processes have certain disadvantages. Elec-

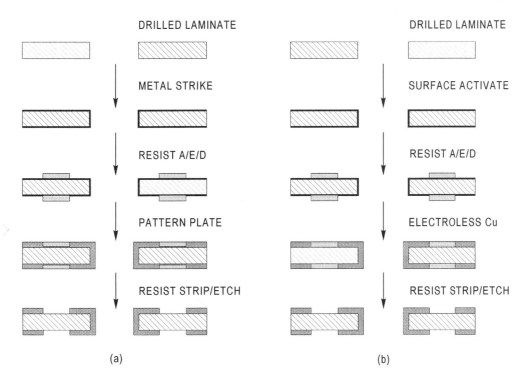

Fig. 32.22 Metallization schemes for PCBs. (a) Initial conducting strike layer deposited followed by electrolytic pattern plate-up. (b) All-electrodeless process. In each scheme, the pattern is defined by a photoresist. (A/E/D = apply, expose, develop.) (Courtesy of A. Viehbeck, IBM T. J. Watson Research Center.)

troless deposition requires the use of noble metal salts, e.g., $PdCl_4^{2-}$, as seeds. The salts are applied to the PCB surface followed by reduction of the noble metal to the zero-valent state. The zero-valent noble metal particles are the active sites for heterogeneous copper reduction in electroless plating. The precious metal seeds are expensive. Electroless baths are generally unstable and require close monitoring. The baths can fluctuate between being too stable, which results in PTH voids, and being too active, which results in homogeneous decomposition of the bath. Formaldehyde, the most commonly used reducing agent in electroless baths, is toxic and poses environmental concerns.

An alternative to the current methods is the use of a conducting polymer as an electrode for direct electrolytic metallization of copper. IBM demonstrated the use of polyaniline in this application [88]. Because of its solubility, the polymer can be directly deposited onto a PCB by a dip-coating process. In one study [88], a PCB was coated with a thick (≈ 1 μm) layer of polyaniline applied from an aqueous acetic acid solution. Today, as a result of the many soluble polyaniline derivatives that have been developed, a variety of solvents including water can be used to apply the polymer. The polyaniline-coated board was electrolytically copper plated. As can be seen in Fig. 32.23b, the copper started to plate on the hole wall from the two contact sides and grew inward until the copper fronts met at the center of the hole wall. As the plating process continued, a thicker, uniform copper coating deposited on the polyaniline surface.

Another method of applying conducting polymers to direct electrolytic metallization of circuit boards was introduced by Blasberg Oberflachentechnik [89]. They reported an in situ polymerization route to deposit polypyrrole [89] and poly-3,4-ethylenedioxythiophene [91] on the surface of a circuit board. The overall process is depicted in Fig. 32.24. The substrate is first selectively coated with an oxidant solution. The monomer (pyrrole or 3,4-ethylenedioxythiophene) is subsequently introduced followed by acid-induced in situ polymerization of the monomer. The conducting polymer–coated board can then be directly electrolytically copper plated. The polypyrrole process referred to as DMS-2 was marketed by Blasberg Oberflachentechnik in 1990 [89]. The polythiophene process, DMS-E, has also since been mar-

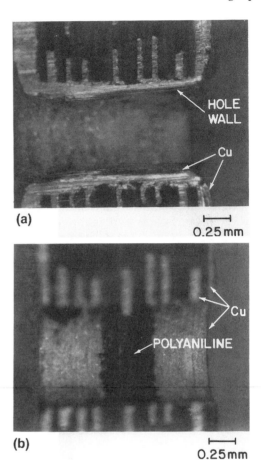

Fig. 32.23 (a) A micrograph depicting a cross-section of a typical circuit board through hole. (b) A micrograph depicting a cross-section of a polyaniline-coated through hole on which electrolytic copper deposition has begun. The copper plates from the two contact sides and continues inward on the polyaniline-coated hole wall. (From Ref. 88.)

keted [91,97]. It has been reported that both of these processes are currently in full-scale production in over 40 companies worldwide [91,97].

Another variation to the use of conducting polymers for metallization was based on spontaneous noble metal

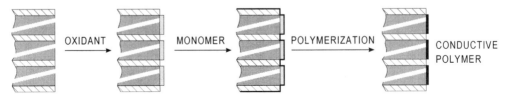

Fig. 32.24 The Blasberg Oberflachentechnik process for the in situ polymerization of a conducting polymer onto a circuit board. (Courtesy of D. Bruce, Uyemura International Corp.)

deposition induced by a conducting polymer [88]. It was found that polyaniline can spontaneously reduce noble metal ions such as Pd^{2+} and Ag^+ to their zero-valent state upon immersion of the polymer into an aqueous solution of the corresponding metal salt. Thin films of Pd^0 and Ag^0 deposit on the polymer surface without an external reducing agent [88]. While this approach does not preclude the use of a precious metal, it does eliminate the need for subsequent activation of the metal. The Pd^0-coated polyaniline can be used for subsequent electrolytic as well as electroless metallization.

It should also be pointed out that the imageable conducting polymers described in the previous section would be quite applicable to metallization. A conducting polymer could be applied to the circuit board surface and directly imaged, thereby eliminating an additional photoresist process. Electroplating can then occur selectively on the patterned conducting polymer.

IV. CORROSION PROTECTION OF METALS

Metals such as copper (Cu) and silver (Ag) are widely used in microelectronics for wiring, EMI shielding, and other purposes. Although both metals are called noble, they readily corrode in a variety of ambients [98–101]. In oxygen-saturated water, Ag and Cu dissolve with a measurable corrosion rate of about 1×10^{-7} and 1×10^{-5} A/cm^2, respectively, or 0.002 and 0.2 nm/min, as can be measured in the potentiodynamic polarization curve shown in Fig. 32.25 [102–104]. With increased anodic potential, metal dissolution rapidly increases to exceed 10^{-1} A/cm^2, corresponding to catastrophic metal removal at a rate of at least 35 nm/s [102–104]. In the presence of an applied potential and humidity, not an uncommon situation for devices in operation, these metals dissolve from the more positive metallic part and plate at the more negative part as dendrites [98–104]. Dendrite formation between two unprotected copper lines is depicted in Fig. 32.26. Dendrites can cause short circuits. In addition, with increasing line density and decreasing dimensions, ion accumulation alone without accompanying dendrite formation can destroy the designed electrical performance of the product. Inhibitors such as benzotriazole (BTA) provide excellent corrosion protection for both metals [98–101]. However, these azole-type inhibitors do not provide protection at high temperature (i.e., in a soldering application) nor do they protect against an applied potential [100,101]. Therefore, new materials are needed to protect Ag and Cu against corrosion and dissolution in particular materials that can provide protection at high temperature and in the presence of an applied potential.

For over a decade, the use of polyaniline for corrosion protection of metals, particularly stainless steel, has been investigated. In the first study [105] by DeBerry, polyaniline was electrochemically deposited on ferritic stainless steels and found to provide anodic protection that significantly reduced corrosion rates in acid solutions. Numerous studies since then have confirmed the corrosion-preventing properties of polyanilines [102–104,106–111]. In addition, nonelectrochemical methods of applying polyaniline have been demonstrated [102–104,106–111]. In one recent study [109], dispersions based on doped polyaniline, Versicon,*

---

* Versicon is a trademark of Allied-Signal Corp.

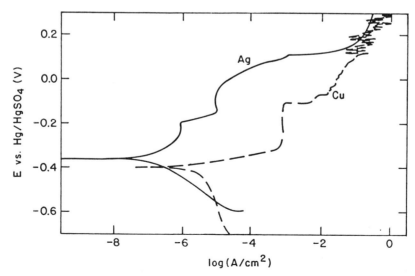

Fig. 32.25 Potentiodynamic polarization curves for unprotected Ag and Cu measured in a droplet of water. (From Ref. 103.)

Fig. 32.26 Dendrite formation between two unprotected copper lines. A water droplet is placed across adjacent metal lines, and a bias of 5 V is applied between the lines.

were found to passivate mild steel, stainless steel, and copper.

The use of polyaniline to protect Cu and Ag, in particular, at high temperature and under an applied potential, which is of interest to the microelectronics industry, was extensively studied by Brusic et al. [102–104]. A number of soluble polyanilines were evaluated in both their nondoped and doped forms. These materials were tested with two methods that were designed to closely resemble conditions to which metals may be exposed during actual use in an electronic product, as for example in a PCB. In most cases, the corrosive environment would vary depending on the relative humidity, i.e., the amount of adsorbed water on the surface. One method uses a three-electrode electrochemical cell with a water droplet as an electrolyte as previously described [112]. This setup basically consists of a sample working electrode, which is the coated metal masked with plating tape to expose a 0.32 cm^2 area, a Pt mesh counter electrode, and a mercurous sulfate reference electrode. A filter paper disk separates the electrodes. A droplet of water is introduced as an electrolyte. The corrosion potential is monitored for ≈15 min, and the polarization resistance is measured by scanning the potential ±20 mV from the corrosion potential. The potentiodynamic polarization curve is measured from 0.25 V cathodic of the corrosion potential. The corrosion rate is evaluated by extrapolation of the cathodic and anodic currents to the corrosion potential. In this study [103], the dissolution rates at high anodic potentials were closely monitored. In a second method, patterned metal lines are tested for ease of dendrite formation by placing a water droplet across adjacent metal lines with an applied potential between the lines.

It was found that the soluble poly-o-phenetidine, particularly in the nondoped form, provided excellent protection for Cu and Ag and was superior to the unsubstituted polyaniline base [102–104]. The nondoped version of this polymer is highly soluble in a variety of organic solvents. Homogeneous and well adherent films of various thicknesses were spin-applied onto Ag and Cu surfaces from an NMP or γ-butyrolactone solution. Films of 184, 339, and 477 nm thickness were evaluated. It was found that even the thinnest film provided a perfect barrier to oxygen, as the electrochemical data indicate no current dependence attributable to oxygen reduction, as can be seen in Fig. 32.27. Oxygen reduction is completely inhibited; generally oxygen reduction is the main cause of Cu and Ag corrosion. The cathodic current of about 2×10^{-7} A/cm^2 is independent of film thickness, type of metal, and ambient atmosphere. The current is diffusion-limited and is most probably caused by reduction of the polymer backbone. The anodic current, metal dissolution, is greatly reduced by a factor that increases with film thickness. The protection is substantial, especially at high potentials where Cu dissolution is about 4 orders of magnitude lower than that measured on bare Cu. Similar results were observed with Ag [102–104].

The corrosion protection offered by the doped version of poly-o-phenetidine is similar to that of the nondoped form except for the measurement of the oxygen reduction rate. As this form is conducting, it allows passage of electrons that are needed for oxygen reduction but prevents passage of metallic ions. At anodic potentials, the protection provided by the film was still excellent.

In terms of dendrite formation, it was found that poly-o-phenetidine provided exceptional protection [102–104]. Bare Cu and Ag were observed to form dendrites within seconds of an applied potential as evidenced by shorting of the lines. BTA was found to be ineffective; BTA-protected metal lines also formed dendrites within seconds. Metal lines that were coated with poly-o-phenetidine base did not form dendrites even after 30 min of applied potential of 5 V. The coated lines were also tested at elevated temperature (220°C for 30 min). No dendrites formed under these conditions. A temperature–humidity study was done in which the poly-o-phenetidine-coated metal was stored for 1000 h at 85°C and 80% relative humidity, and again no dendrites were observed. In addition, a more drastic test was done in which a potential was applied to the coated metal at 85°C and 80% relative humidity for 1000 h. No failure was observed under these conditions either.

The poly-o-phenetidine provides excellent corrosion protection for Cu and Ag at elevated temperatures as well as under an applied potential. The superior protection offered by this polyaniline derivative may stem from its excellent coverage and adhesion to the metal surface. It has good solubility and forms very uniform films. In addition, the ethoxy group can complex with the metal

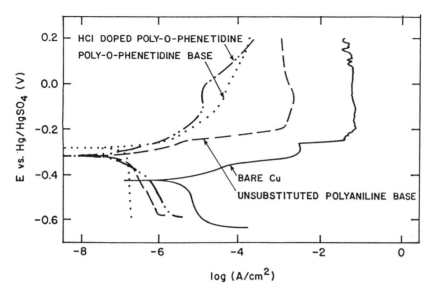

Fig. 32.27 Potentiodynamic polarization curves in a droplet of water for bare Cu and for Cu coated with poly-*o*-phenetidine base, HCl-doped poly-*o*-phenetidine, and unsubstituted polyaniline base. (From Ref. 103.)

and enhance its adhesion characteristics. Indeed, peel test results show adhesion strength in excess of 60 g/mm [104]. Generally, values greater than 50 g/mm for polymer–metal adhesion are recognized as excellent.

V. ELECTROSTATIC DISCHARGE PROTECTION FOR ELECTRONIC COMPONENTS

Electrostatic charge (ESC) and electrostatic discharge (ESD) are a serious and expensive problem for many industries, in particular for microelectronics [113–118]. It has been estimated that $15 billion a year is attributed to ESD damage alone by the U.S. electronics industry [113,114]. Electrostatic charge can accumulate to thousands of volts. The static charge can attract airborne particles, as is often observed on cathode ray tubes (CRTs). This becomes a significant contamination concern if particles are attracted to critical surfaces such as device wafers. The accumulated charge will eventually discharge in the form of a "lightning bolt," which can destroy devices on ICs [113–118]. Figure 32.28 shows an example where a discharge created a "punch-through" effect, exposing lower layers of a circuit. As circuit density continues to increase and the area and thickness of the active device elements continue to shrink, device sensitivity to the destructive effects of ESD will continue to increase in the future.

To protect devices against ESC and ESD, conducting materials are extensively used in clean rooms during their manufacture. In addition, conductors are incorporated into plastic packages that are used to transport sensitive electronic components such as chips, modules, and PCBs. The materials currently used include ionic conductors, carbon- or metal-filled resins, and in certain cases metal coatings [118]. These materials do not offer the ideal solution to ESD protection. The use of ionic conductors, although an inexpensive approach, has significant drawbacks. These materials exhibit very low surface conductivity (10^{-9}–10^{-11} ohms^{-}/square) and

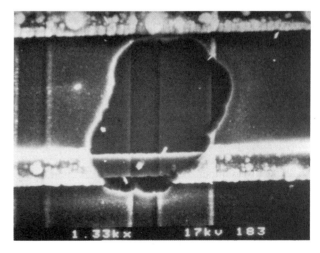

Fig. 32.28 An example of ESD damage to an IC. A "punch-through" effect has occurred that exposes lower layers of a circuit. (From Ref. 113.)

thus are not dissipative. The conductivity is humidity-dependent, as water is needed as an electrolyte for ionic conduction. In addition, ionic conductors can readily be removed by water, which precludes any washing of structural parts containing these materials. The use of ionic conductors is not a reliable method for ESD protection. Electrical conductors, on the other hand, are stable systems. The conductivity is not humidity-dependent. However, these materials are more expensive. Carbon-filled systems pose contamination concerns due to sloughing of the carbon particles. In addition, relatively high loading levels are required to attain a given level of conductivity. The high loading can degrade the mechanical/physical properties of the host polymer. With high loadings, recycling of the plastic carriers becomes more difficult.

Intrinsically conducting polymers offer a new alternative to ESD protection with numerous advantages over current materials. The conductivity can be tuned and can easily meet the high end of the dissipative range. The conductivity is stable compared to that of ionic conductors. By appropriate design of the conducting polymer, contamination concerns can be eliminated. In addition, conducting polymers can offer a high degree of transparency. Polyaniline, polypyrrole, and, more recently, polythiophene have been the predominant conducting polymers of interest for ESD protection. These polymers have been used as fillers in a number of host resins. In addition, coating formulations have been developed that can be applied directly onto plastic surfaces.

Pyrrole has been in situ polymerized onto the surface of textile fabrics [119,120]. A number of polypyrrole-coated fabrics, trade name Contex,* have been produced by Milliken Research Corporation [119,120].

Polyaniline in the form of a dispersible powder (Versicon) has been blended with a number of thermoplastic and thermoset resins to achieve excellent ESD properties [121–125]. Soluble polyanilines have also been blended with appropriate polymers [126,127]. In the latter systems, very low loadings have been reported to be necessary to reach a certain level of conductivity [126,127].

Of particular interest for ESD protection of electronic component packages are coating formulations. The coatings can be applied directly onto already fabricated packages by spray-coating, or they can be applied onto plastic sheets that are subsequently thermoformed into a package. A number of coatings based on conducting polymers have been developed and are currently commercially available. One type of coating is based on dispersions of Versicon [123–125,128,129]. Such coatings have been produced by Americhem [123–125,128]. Coatings based on soluble polyanilines have also been produced. One such system [49,129] is a curable water-based coating reported by IBM. An aqueous coating based on a poly(3,4-ethylenedioxythiophene)/polystyrenesulfonic acid blend [130,131] is reported to exhibit effective antistatic properties. Formulations that are based on soluble conducting polymers have the advantage that they pose no contamination concerns due to particles. This is particularly important for microelectronics where actual devices are in contact with the conducting carriers. Any particle sloughing would contaminate the devices. Some of the conducting polymer coating formulations previously described are currently in commercial use for ESD protection.

VI. FUTURE APPLICATIONS

There are other potential applications of conducting polymers in microelectronics outside of those already discussed. Conducting polymers can in principle be considered candidates for interconnection technology. The use of conducting polymers for wiring has been widely speculated upon since the emergence of these systems in the late 1970s. For such an application, copperlike conductivity would be necessary. Unfortunately, polyacetylene is the only conducting polymer that currently exhibits such conductivity, and its environmental instability and lack of processability hinder its use. Dramatic enhancement in conductivity in some of the more processable and environmentally stable polymers is required before these systems can be realistically considered viable conductors for interconnection technology.

The use of conducting polymers in devices [132–137] is another potential area that may provide new technology in the future for IC fabrication and flat panel displays. It is a very exciting area of research and is the object of considerable academic and industrial interest. However, much progress is required in material properties before these technologies come to fruition. Higher mobilities are required from the semiconducting conjugated polymers before they can compete with silicon-based devices. Enhancement in the long-term stability of the polymers is necessary before polymer-based light-emitting diodes become viable.

Remarkable progress has been made in developing the properties of conducting polymers since the late 1970s. As material properties continue to improve, some of what are currently viewed as "far out" applications may become enabling technologies in the near future.

VII. CONCLUSIONS

Conducting polymers have a broad range of applications in microelectronics. In the area of lithography, they have been shown to provide an effective, simple, spin-apply process for charge dissipation, in particular for e-beam

---

* Contex is a trademark of Milliken Research Corporation.

lithography and for SEM metrology. Implementation of these polymers in this area will depend on their commercial availability and cost, in particular, materials such as the PanAquas and ESPACER. A number of resists based on conducting polymers have been developed. However, the lithographic performance of these systems is not currently at a stage that can compete with conventional resists. They require further improvements for future use. Conducting polymers have been shown to be applicable for metallization of plated through holes for printed circuit board technology. This process is currently in use in a number of companies worldwide. Conducting polymer–based coatings have been developed that offer excellent ESD protection and offer numerous advantages over currently used materials. A number of coating formulations are either already commercially available or in the process of being made so. Corrosion protection of metals such as silver and copper using conducting polymers has been shown to be quite promising.

Conducting polymers have certainly emerged into the marketplace. As the materials become commercially available and their properties continue to evolve, greater potential may be realized with this class of polymers in the future.

REFERENCES

1. J. Kilby, *IEEE Trans. Ed. Devices ED-23*:648 (1976).
2. S. M. Sze (ed.), *VLSI Technology*, McGraw-Hill, New York, 1988.
3. C. Barrett, Semiconductor technology and the growth of the PC industry, Proceedings of 8th Annual Microprocessor Forum, San Jose, CA, 1995.
4. L. F. Thompson, C. G. Wilson, and M. J. Bowden (eds.), *Introduction to Microlithography*, ACS Professional Reference Book, ACS, Washington, DC, 1994, and references therein.
5. W. M. Moreau (ed.), *Semiconductor Lithography*, Plenum, New York, 1988.
6. D. S. Soane and Z. Martynenko (eds.), *Polymers in Microelectronics* Elsevier, Amsterdam, 1989.
7. R. R. Tummala and E. J. Rymaszewski (eds.), *Microelectronics Packaging Handbook*, Van Nostrand Reinhold, New York, 1989, and references therein.
8. M. Pecht (ed.), *Handbook of Electronic Package Design*, Marcel Dekker, New York, 1991, and references therein.
9. C. A. Harper (ed.), *Electronic Packaging and Interconnection Handbook,* McGraw-Hill, New York, 1991, and references therein.
10. R. L. Cadenhead, Materials and electronic phenomena, in *Electronics Materials Handbook* (C. A. Dostal, ed.), ASM International, Materials Park, OH, 1989, pp. 89–120, and references therein.
11. E. D. Feitz and C. W. Wilkins (eds.), *Polymer Materials for Electronic Applications*, ACS Symp. Ser., ACS, Washington, DC, 1982.
12. E. Reichmanis, S. A. MacDonald, and T. Iwayanagi, (eds.), *Polymers in Microlithography*, ACS Symp. Ser., ACS, Washington, DC, 1989.
13. J. M. Shaw, Imaging for microfabrication, in *Imaging Processes and Materials* (J. Sturge, V. Walworh, and A. Shepp, eds.), Van Nostrand Reinhold, New York, 1989, pp. 567–586.
14. B. Chapman (ed.), *Glow Discharge Processes*, Wiley, New York, 1980.
15. H. J. Smith, *J. Vac. Sci. Technol. B 13*(6):2323 (1995).
16. D. Kern (ed.), *Proceedings of 39th International Conference on Electron, Ion, and Photon Beam Technology and Nanofabrication*, American Vacuum Society, American Institute of Physics, New York, 1995.
17. J. M. Ryan, A. C. F. Hoole, and A. N. Broers, *J. Vac. Sci. Technol. B 13*(6):3035 (1995).
18. G. O. Langner, *Proc. Microcirc. Eng. 1979*: 261 (1979).
19. K. D. Cummings and M. Kiersh, *J. Vac. Sci. Technol. B 7*(6):1536 (1989).
20. H. Itoh, K. Nakamura, and H. Hayakawa, *J. Vac. Sci. Technol. B 7*(6):1532 (1989).
21. Y. Todokoro, Y. Takasu, and T. Ohkuma, *Proc. Soc. Photo. Opt. Instrum. Eng. 587*:179 (1985).
22. M. Kakuchi, M. HIkito, A. Sugita, K. Onose, and T. Tamamura, *J. Electrochem. Soc. 133*:1755 (1986).
23. Z. Tan, *Proc. Soc. Photo. Opt. Instrum. Eng. 2322*: 141 (1994).
24. Y. Todokoro, A. Kajiya, and H. Watanabe, *J. Vac. Sci. Technol. B 6*(1):357 (1988).
25. H. Watanabe and Y. Todokoro, *IEEE Trans. Electron Devices 36*(3):474 (1989).
26. W. S. Huang, *Polymer 35*(19):4057 (1994).
27. M. Angelopoulos, J. M. Shaw, R. D. Kaplan, and S. Perreault, *J. Vac. Sci. Technol. B 7*(6):1519 (1989).
28. A. G. MacDiarmid, J. C. Chiang, A. F. Richter, and A. J. Epstein, *Synth. Met. 18*:285 (1987).
29. A. J. Epstein, J. M. Ginder, F. Zuo, H. S. Woo, D. B. Tanner, A. F. Richter, M. Angelopoulos, W. S. Huang, and A. G. MacDiarmid, *Synth. Met. 21*:63 (1987).
30. A. G. MacDiarmid and A. J. Epstein, *Faraday Discuss., Chem. Soc. 88*:317 (1989), and references therein.
31. E. M. Genies, A. Boyle, M. Lapkowski, and C. Tsintavis, *Synth. Met. 36*:139 (1990).
32. S. Stafstrom, W. R. Salaneck, O. Inganas, and T. Hjertberg (eds.), Proceedings of International Conference on Science and Technology of Synthetic Metals, Goteborg, Sweden, 1992, *Synth. Met. 55–57* (1993).
33. S. Stafstrom, W. R. Salaneck, O. Inganas, and T. Hjertberg (eds.), Proceedings of International Conference on Science and Technology of Synthetic Metals, Seoul Korea, 1994 *Synth. Met. 69–71* (1995).
34. M. Angelopoulos, A. Ray, A. G. MacDiarmid, and A. J. Epstein, *Synth. Met. 21*:21 (1987).
35. M. Angelopoulos, G. E. Asturias, S. P. Ermer, A.

Ray, E. M. Scherr, and A. G. MacDiarmid, *Mol. Cryst. Liq. Cryst. 160*:151 (1988).
36. L. W. Shacklette and C. C. Han, *Mater. Res. Soc. Symp. Proc. 328*:157 (1994).
37. K. T. Tzou and R. V. Gregory, *Synth. Met. 69*:109 (1995).
38. A. G. MacDiarmid, J. C. Chiang, A. F. Richter, N. L. D. Somasiri, and A. J. Epstein, in *Conducting Polymers* (L. Alcacer, ed.), Reidel, Dordrecht, 1985, pp. 105–120.
39. J. Yue and A. J. Epstein, *J. Am. Chem. Soc. 112*: 2800 (1990).
40. C. DeArmitt, S. P. Armes, J. Winter, F. A. Uribe, S. Gottesfeld, and C. Mombourquette, *Polymer 34*(1):158 (1993).
41. M. T. Nguyen, P. Kasai, J. L. Miller, and A. F. Diaz, *Macromolecules 27*:3625 (1994).
42. Y. Wei, R. Hariharan, and S. A. Patel, *Macromolecules 23*:758 (1990).
43. M. Leclerc, J. Guay, and L. H. Dao, *Macromolecules 22*:649 (1989).
44. Y. H. Liao, M. Angelopoulos, and K. Levon, *J. Polym. Sci., Polym. Chem. Ed. 33*:2725 (1995).
45. M. Angelopoulos, S. P. Ermer, S. K. Manohar, and A. G. MacDiarmid, *Mol. Cryst. Liq. Cryst. 160*:223 (1988).
46. Y. Cao, P. Smith, and A. J. Heeger, *Synth. Met. 48*: 91 (1992).
47. M. Angelopoulos, N. Patel, and R. Saraf, *Synth. Met. 55*:1552 (1993).
48. K. Tzou and R. V. Gregory, *Synth. Met. 53*:365 (1993).
49. M. Angelopoulos, N. Patel, and J. M. Shaw, *Mater. Res. Soc. Symp. Proc. 328*:173 (1994).
50. E. J. Paul, A. J. Ricco, and M. S. Wrighton, *J. Phys. Chem. 89*:1441 (1985).
51. W. S. Huang, B. D. Humphrey, and A. G. MacDiarmid, *J. Chem. Soc., Faraday Trans. 1, 82*:2385 (1986).
52. J. C. Chiang and A. G. MacDiarmid, *Synth. Met. 13*: 193 (1986).
53. M. Angelopoulos, J. M. Shaw, K. L. Lee, W. S. Huang, M. A. Lecorre, and M. Tissier, *J. Vac. Sci. Technol. B9*(6):3428 (1991).
54. M. Angelopoulos, J. M. Shaw, W. S. Huang, and R. D. Kaplan, *Mol. Cryst. Liq. Cryst. 189*:221 (1990).
55. M. Angelopoulos, J. M. Shaw, and K. L. Leung, *Mater. Res. Soc. Symp. Proc. 214*:137 (1991).
56. J. V. Crivello and J. H. W. Lam, *Macromolecules 10*:1307 (1977).
57. J. V. Crivello and J. H. W. Lam, *J. Polym. Sci., Polym. Chem. Ed. 17*:977 (1979).
58. R. Alm, *Mod. Paint Coatings*, October, 1980.
59. R. Alm, *J. Coating Tech. 53*(683):45 (1981).
60. J. M. Shaw, M. Hatzakis, E. Babich, J. R. Paraszczak, D. Witman, and K. J. Stewart, *J. Vac. Sci. Technol. B7*(6):1709 (1989).
61. J. M. Liu, L. Sun, J. H. Hwang, and S. C. Yang, *Mater. Res. Soc. Symp. Proc. 247*:601 (1992).
62. J. Yue, Z. H. Wang, K. R. Cromack, A. J. Epstein, and A. G. MacDiarmid, *J. Am. Chem. Soc. 113*(7): 2665 (1991).
63. S. A. Chen and G. W. Hwang, *J. Am. Chem. Soc. 116*(17):7939 (1994).
64. S. A. Chen and G. W. Hwang *J. Am. Chem. Soc. 117*(40):10055 (1995).
65. M. Angelopoulos, N. Patel, J. M. Shaw, N. C. Labianca, and S. Rishton, *J. Vac. Sci. Technol. B11*(6): 2794 (1993).
66. M. D. Levenson, N. S. Viswanathan, and R. A. Simpson, *IEEE Trans. Electron Devices ED-59*:1828 (1982).
67. H. Fukuda, A. Imai, and S. Okazaki, *Proc. Soc. Photo. Opt. Instrum. Eng. 14*:1564 (1990).
68. A. O. Patil, Y. Ikenoue, F. Wudl, and A. J. Heeger, *J. Am. Chem. Soc. 109*:1858 (1987).
69. Y. Ikenoue, N. Uotani, A. O. Patil, F. Wudl, and A. J. Heeger, *Synth. Met. 30*:305 (1989).
70. S. A. Chen and M. Y. Hua, *Macromolecules 26*:7108 (1993).
71. F. Mizuno, M. Kato, H. Hayakawa, K. Sato, K. Hasegawa, Y. Sakitani, N. Saitou, F. Murai, H. Siraishi, and S. Uchino, *J. Vac. Sci. Technol. B12*(6): 3440 (1994).
72. M. T. Postek and D. C. Joy, *J. Res. Natl. Bur. Stand. 92*(3):205 (1987).
73. R. D. Larrabee and M. T. Postek, *Solid State Electron. 36*(5):673 (1993).
74. R. I. Scarce, *Solid State Tech. 43* (1994).
75. M. Angelopoulos, J. M. Shaw, M. A. Lecorre, and M. Tissier, *Microelectron. Eng. 13*:515 (1991).
76. T. C. Clarke, M. T. Krounbi, V. Y. Lee, and G. B. Street, *J. Chem. Soc., Chem. Commun. 8*:384 (1981).
77. S. Pitchumani and F. Willig, *J. Chem Soc., Chem. Commun. 13*:809 (1983).
78. A. J. Epstein, J. M. Ginder, F. Zuo, R. W. Bigelow, H. S. Woo, D. B. Tanner, A. F. Richter, W. S. Huang, and A. G. MacDiarmid, *Synth. Met. 18*:303 (1987).
79. S. Stafstrom, J. L. Bredas, A. J. Epstein, H. S. Woo, D. B. Tanner, W. S. Huang, and A. G. MacDiarmid, *Phys. Rev. Lett. 59*:13 (1987).
80. G. Venugopal, X. Quan, G. E. Johnson, F. M. Houlihan, E. Chin, and O. Nalamasu, *Chem. Mater. 7*(2): 271 (1995).
81. S. X. Cai, J. F. W. Keana, J. C. Nabity, and M. N. Wybourne, *J. Mol. Electron. 7*:63 (1991).
82. S. X. Cai, M. Kanskar, J. C. Nabity, J. F. W. Keana, and M. N. Wybourne, *J. Vac. Sci. Technol. B10*(6): 2589 (1992).
83. M. S. A. Abdou, G. A. Diaz-Guijada, M. I. Arroyo, and S. Holdcroft, *Chem. Mater. 3*:1003 (1991).
84. M. S. A. Abdou, Z. W. Xie, J. Lowe, and S. Hold-

croft, *Proc. Soc. Photo. Opt. Instrum. Eng. 2195*: 756 (1994).
85. J. Lowe and S. Holdcroft, *Macromolecules 28*:4608 (1995).
86. J. Bargon, W. Behnck, and T. Weidenbruck, *Synth. Met. 41–43*:1111 (1991).
87. J. Bargon and R. Baumann, *Microelectron. Eng. 20*: 55 (1993) and references therein.
88. W. S. Huang, M. Angelopoulos, J. R. White, and J. M. Park, *Mol. Cryst. Liq. Cryst. 189*:227 (1991).
89. H. Stuckmann and J. Hupe, *Blasberg-Mitteilungen 11*:3–27 (1991).
90. K. Beator, B. Hildesheim, B. Bressel, and H. J. Grapentin, *Metalloberflache 46*:384 (1992).
91. J. Hupe, *Blasberg-Mitteilungen 15*:14–18 (1995).
92. S. Gottesfeld, F. A. Uribe, and S. P. Armes, *J. Electrochem. Soc. 139(1)*:L14 (1992).
93. J. Fjelstad, Printed wiring board technology: current capabilities and limitations, in *Electronics Materials Handbook* (C. A. Dostal, ed.), ASM Int., Materials Park OH, 1989, pp. 507–512 and references therein.
94. L. Lynch and R. A. Nesbitt, Rigid printed wiring board fabrication techniques, in *Electronics Materials Handbook* (C. A. Dostal, ed.), ASM Int., Materials Park, OH, 1989, pp. 538–555 and references therein.
95. D. P. Seraphim, D. E. Barr, W. T. Chen, G. P. Schmitt, and R. R. Tummala, Printed circuit board packaging, in *Microelectronics Packaging Handbook* (R. R. Tammala and E. J. Rymaszewski, eds.), Van Nostrand Reinhold, New York, 1989, pp. 853–922 and references therein.
96. C. A. Deckert, *Plat. Surf. Finish. 82*:48 (1995).
97. DMS-E Technical Information Brochure, available through Uyemura International Corp., Ontario CA.
98. J. J. Steppan, J. A. Roth, L. C. Hall, D. A. Jeannotte, and S. P. Carbone, *J. Electrochem. Soc. 134*:175 (1987) and references therein.
99. R. Walker, *J. Chem. Ed. 57*:789 (1980).
100. V. Brusic, M. A. Frisch, B. N. Eldridge, F. P. Novak, F. B. Kaufman, B. M. Rush, and G. S. Frankel, *J. Electrochem. Soc. 138*:2253 (1991).
101. V. Brusic, G. S. Frankel, J. Roldan, and R. Saraf, *J. Electrochem. Soc. 142*:2591 (1995).
102. V. Brusic, M. Angelopoulos, and T. Graham, *Proc. Soc. Plastics* Eng. 53rd Annual Technical Conference, 1995.
103. V. Brusic, M. Angelopoulos, T. Graham, and P. Buchwalter, Proc. 188th Electrochem. Soc., Chicago, IL, 1995.
104. V. Brusic, M. Angelopoulos, and T. Graham, *J. Electrochem. Soc., 144(2)*:436 (1997).
105. D. W. DeBerry, *J. Electrochem. Soc. 132*:1022 (1985).
106. A. G. MacDiarmid, Personal communication, Int. Conf. Synthetic Metals, Kyoto, Japan, 1986.
107. D. A. Wrobleski, B. C. Benicewicz, K. G. Thompson, and C. J. Bryan, *ACS Polym. Prep. 35*:265 (1994).
108. S. Sathiyanarayanan, S. K. Dhawan, D. C. Trivedi, and K. Balakrishnan, *Corros. Sci. 33*:1831 (1992).
109. B. Wessling, *Adv. Mater. 6(3)*:226 (1994).
110. W. K. Lu, R. L. Elsenbaumer, and B. Wessling, *Synth. Met. 71*:2163 (1995).
111. S. Jasty and A. J. Epstein, *Polym. Mater. Sci. Eng. 72*:565 (1995).
112. V. Brusic, M. Russak, R. Schad, G. Frankel, A. Selius, D. DiMilia, and D. Edmonson, *J. Electrochem. Soc. 136*:42 (1989).
113. S. L. Law, S, Mucha, and S. Banks, *Electron Packag. Prod. 31(5)*:82 (1991).
114. L. Brown and D. Burns, *Electron. Packag. Prod.* May 1990, p. 50.
115. D. Jillie, *Semicond. Int. 16(8)*:120 (1993).
116. J. L. Sproston, *Electrostatic Damage in the Electronics Industry*, Proceedings of Static Electrification Group of Institute of Physics, in IOP, England, 1987.
117. C. Duvvury and A. Amerasekera, *Proc. IEEE 81(5)*: 690 (1993).
118. Proceedings of Electrical Overstress/Electrostatic Discharge Symposium (EOS16), Las Vegas, NV, published by EOS/ESD Assoc., 1994.
119. R. V. Gregory, W. C. Kimbrell, and H. H. Kuhn, *Synth. Met. 28*:823 (1989).
120. H. H. Kuhn, Characterization and application of polypyrrole-coated textiles, in *Intrinsically Conducting Polymers: An Emerging Technology* (M. Aldissi, ed.), Kluwer, Dordrecht, Netherlands, 1993, p. 25.
121. N. F. Colaneri and L. W. Shacklette, *IEEE Trans. Instrum. Meas. 41*:291 (1992).
122. L. W. Shacklette, C. C. Han, and M. H. Luly, *Synth. Met. 57*:3532 (1993).
123. V. G. Kulkarni, W. R. Matthew, J. C. Campbell, C. J. Dinkins, and P. J. Durbin, *Proc. Soc. Plast. Eng.* 665 (1991).
124. V. G. Kulkarni, J. C. Campbell, and W. R. Mathew, *Synth. Met. 57*:3780 (1993).
125. V. G. Kulkarni, Processing of polyanilines, in *Intrinsically Conducting Polymers: An Emerging Technology*: (M. Aldissi, ed.), Kluwer, Dordrecht, 1993, p. 45.
126. C. Y. Yang, Y. Cao, P. Smith, and A. J. Heeger, *Synth. Met. 53*:294 (1993).
127. J. E. Osterholm, J. Laakso, O. Ikkala, H. Ruohonen, K. Vakiparta, E. Virtanen, H. Jarvinen, and T. Taka, *Polym. Prepri 35(1)*:244 (1994).
128. V. G. Kulkarni and M. Angelopoulos, Proceedings of Electrical Overstress/Electrostatic Discharge Symposium, Phoenix, AZ, 1995.
129. M. Angelopoulos, J. D. Gelorme, and J. M. Shaw, Proceedings of Electrical Overstress/Electrostatic Discharge Symposium: Las Vegas, NV, 1994.
130. F. Jonas, W. Kraft, and B. Muys, Proceedings of 5th European Polymer Federation Symposium on Polymeric Materials, 1994, p. 169.
131. G. Heywang and F. Jonas, *Adv. Mater. 4(2)*:116 (1992).

132. J. H. Burroughes, C. A. Jones, and R. H. Friend, *Nature* 335:137 (1988).
133. A. Tsumura, H. Koezuka, and T. Ando, *Synth. Met.* 25:11 (1988).
134. F. Garnier, G. Horowitz, X. Peng, and D. Fichou, Adv. Mater. 2(12):592 (1990).
135. J. H. Burroughes, D. D. C. Bradley, A. R. Brown, R. N. Marks, K. Mackey, R. H. Friend, P. L. Burns, and A. B. Holmes, *Nature* 347:539 (1990).
136. G. Gustafsson, Y. Cao, G. M. Treacy, F. Klavetter, N. Colaneri, and A. J. Heeger, *Nature 357(11)*:477 (1992).
137. D. D. C. Bradley, *Synth. Met. 54*:401 (1993) and references therein.

33
Gas and Liquid Separation Applications of Polyaniline Membranes

Jeanine A. Conklin
PPG Industries, Monroeville, Pennsylvania

Timothy M. Su
City College of San Francisco, San Francisco, California

Shu-Chuan Huang and Richard B. Kaner
University of California at Los Angeles, Los Angeles, California

I. INTRODUCTION

Membrane-based purification systems function as simple, direct methods of purifying commercially important and often difficult to separate mixtures. Membranes are useful in such far-ranging applications as separating oxygen and nitrogen from air, desalinating seawater, removing organics from waste streams, and separating pure ethanol from its azeotropic mixture with water. Membrane purification systems, particularly polymer membranes, have been developed over the past several decades into energy-efficient alternatives to standard industrial purification techniques [1–8].

Polymers have distinct advantages as membranes for purification of many liquids and gases because they are easily synthesized and processed into flexible thin sheets or high surface area hollow fibers that allow large fluxes of gases or liquids through the membranes. Traditionally polymer membranes are tested for their permselectivity. If a given polymer has a low selectivity or insufficient permeability, another polymer is tried. When a promising candidate appears, derivatives with different side chains or copolymers or blends are often synthesized in an attempt to enhance the parent polymer's physical properties. Reviews of polymer membranes for gas separations may be found in Refs. 1–5, and liquid separations are discussed in Refs. 6–8.

Here we explore a new class of polymers for separations, specifically those that are conjugated. These polymers can be modified after synthesis, through doping, to induce structural changes and thus achieve enhanced separating abilities. The doping process, discussed extensively in other chapters of this book with respect to its effect on conductivity, is discussed here in terms of its effects on gas and liquid permeability.

Polyaniline is an excellent example of a conjugated polymer that can be tailored to specific separation applications through the doping process. Since its conducting properties were discovered in the early 1980s [9], polyaniline has been studied for many other potential applications including lightweight battery electrodes [10,11], electromagnetic shielding devices [12], anticorrosion coatings [13,14], and gas and ion sensors [12,15]. The electrically conductive properties of polyaniline in its emeraldine oxidation state stem from the acid doping capability of the imine nitrogens on the polymer backbone [16] (see Fig. 33.1). Dopants can be added in any desired quantity until all imine nitrogens (half of the total nitrogens) are doped, simply by controlling the pH of the dopant acid solution [17]. Polyaniline's conductivity increases with doping from the undoped insulating base form ($\sigma \leq 10^{-10}$ S/cm) to the fully doped, conducting acid form ($\sigma \geq 10$ S/cm). Doping and/or undoping processes are typically done chemically with such common acids and bases as hydrochloric acid and ammonium hydroxide; electrochemical processes can also readily be used [18–22]. In addition to affecting conductivity, doping and undoping processes can have dramatic effects on the morphology of the polymer films [23–25].

Finding large differences in the permeability of gases through doped and undoped polyaniline films led to detailed investigations of polyaniline and other conjugated

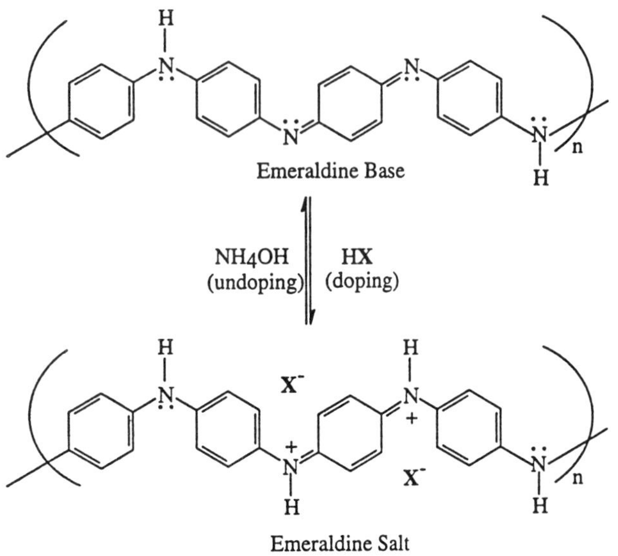

Fig. 33.1 Repeat unit of the emeraldine oxidation state of polyaniline in the undoped, base form (top), and the fully doped, acid form (bottom). HX represents any protonic acid.

polymers for gas separation applications [26–31]. For the full potential of conjugated polymers as separation membranes to be realized, a thorough understanding of fundamental structure–property relationships of the polymers and dopants must be explored. To begin this process, an introduction to polyaniline synthesis and film formation is presented, followed by an overview of membrane separation processes and then a discussion of potential applications of conjugated polymers as gas and liquid separation membranes.

II. POLYANILINE SYNTHESIS AND FILM FORMATION

The chemical synthesis of polyaniline has been widely studied, including types and concentrations of acids for polymerization, effects of different oxidizing agents, and synthesis temperature [32]. For our purposes, a relatively standard synthesis via the chemical oxidation of aniline in an acidic medium was chosen for a consistently pure, high molecular weight polymer following the work by MacDiarmid et al. [33]. A typical synthesis consists of separately dissolving aniline and ammonium peroxydisulfate in 1.0 M hydrochloric acid (HCl), cooling the solution to 0°C in an ice-salt bath, and adding the oxidizing agent dropwise to the monomer solution over a 45 min period. Purification of the resulting green precipitate consists of carrying out alternate washings in 1.0 M HCl and 0.1 M NH_4OH until the filtrate becomes colorless (typically four acid/base washings).

Robust free-standing films of polyaniline are prepared from the emeraldine base polymer powder. The brown powder is dispersed in N-methyl-2-pyrrolidinone (NMP) (5–8% w/v) with continuous grinding. This deep blue viscous polymer solution is uniformly spread onto glass plates and dried in a convection oven at 110°C. The resulting strong free-standing films are removed from the glass plates by soaking in distilled water and dried.

Once the films are made, acids can be used to dope the polymers to achieve the desired properties. Figure 33.1 illustrates the reversible doping properties of polyaniline using any common aqueous protonic acid and ammonium hydroxide base solutions. The dopant counterion (X^-) is electrostatically associated with the polymer backbone to maintain charge neutrality. This close proximity in the preformed film causes local rearrangements in the polymer structure to accommodate the extra space required for the counterions [34]. The size of the counterion will therefore affect the amount of rearrangement and also the amount of "void" space or free volume opened up on its removal by undoping. Increased free volume can be observed after removing dopants by measuring the increase in permeation of gases through the membrane. A correlation between counterion size and permeability has been observed for a range of gases, where the larger the dopant counterion, the higher the permeability of gases through the undoped membrane [31]. A more complete description of membrane separation is helpful before discussing the details of polyaniline's potential utility as a membrane.

III. MEMBRANE SEPARATION FUNDAMENTALS

The selection of a membrane for a particular type of separation is usually determined by its average pore size; e.g., 10–100 μm is useful for conventional filtration, 0.1–10 μm for microfiltration, 50–1000 Å for ultrafiltration, and less than 50 Å for reverse osmosis, gas separation, and pervaporation, as depicted in Fig. 33.2. The latter are also described as nonporous membranes and depend on molecular interactions between the permeant and the membrane itself to affect separation. The basic terminology and theory of membrane-based separation systems are similar for both gases and liquids and are therefore treated together in this section.

The separation process can be divided into three distinct parts, as illustrated in Fig. 33.3: first, sorption of the constituent feed mixture onto the membrane (adsorption); second, movement through the membrane (diffusion); and third, loss out the other side of the membrane (desorption). The term "permeation" is used to describe the overall process of mass transport across the membrane. In liquid separations this process is named

Separation Applications of Polyaniline Membranes

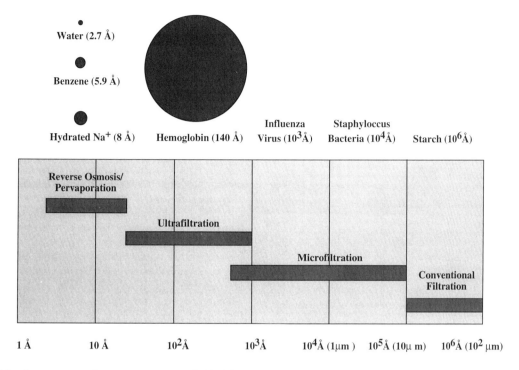

Fig. 33.2 Membrane separation processes are divided into conventional filtration, microfiltration, ultrafiltration, reverse osmosis, gas separation, and pervaporation according to the pore sizes of the membranes. The sizes of selected molecules ranging from water (2.7 Å in diameter) to starch (10^6 Å) are given for comparison. (Modified from Ref. 36.)

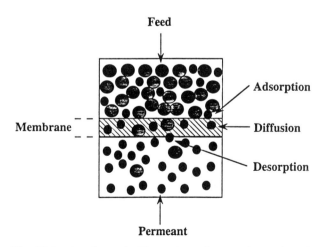

Fig. 33.3 A schematic illustration of a membrane-based separation mechanism indicating adsorption, diffusion, and desorption of feed molecules. The more permeable molecules (small black circles) become the larger component in the permeant.

"pervaporation" to emphasize the fact that the permeant undergoes a phase change from liquid to vapor during transport through the membrane barrier.

The physical process of permeation can be described mathematically in terms of permeability (P), diffusion (D), and solubility (S), which can be calculated using the relationship $P = DS$, assuming the polymer is not plasticized by the penetrant. In a typical experiment analyzing the permeation of a penetrant through a membrane, a plot is made of the amount of penetrant, Q, vs. time t (Fig. 33.4). A time lag before the onset of desorption is observed, allowing time for the permeant to sorb into the membrane, diffuse through it, and finally re-emerge on the other side, where it is measured. The rate of diffusion reaches steady state shortly after the time lag, and the graph attains linearity. The slope of the steady-state flux yields permeability (P). The diffusion coefficient (D) is calculated from the extrapolation of the line at steady-state through the time axis and is described by the equation $\tau = l^2/6D$, where τ is the time lag and l is the thickness of the membrane. Solubility can be determined with the calculated values for P and D from the relationship $P = DS$ [2,4].

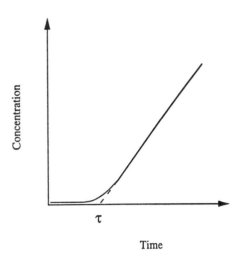

Fig. 33.4 A typical permeation plot of penetrant concentration Q as a function of time t. Extrapolation of the steady-state line through the time axis yields the time lag τ.

Typically, laboratory experiments for gas permeability are performed on one gas at a time to simplify the interpretation of results. The separating ability of the membrane for a particular gas pair is described by the ratio of the individual gases' permeability coefficients and is called the separation factor (α), where $\alpha = P_A/P_B$. This differs from pervaporation experiments, where it is not possible to calculate the composition of the permeant from only a knowledge of the feed composition and the rates of permeation for pure components, because the membranes usually swell in the feed liquid mixture. Therefore, the membrane's condition when it is being permeated by individual compounds can differ from its condition during permeation by a mixture. For liquid separations, mixtures are generally measured to see which components selectively permeate the membrane. The relative composition of the permeant solution is a direct measure of the separating ability of the membrane.

Since the separation of gases and liquids depends on the membrane's permeability to each component, understanding what factors influence permeability, solubility, and diffusion may allow some degree of control over these factors and therefore lead to the development of better separation membranes.

Diffusion is comparable to the molecular sieve method of separation in that it is based on size exclusion; the polymer matrix allows easy passage of smaller molecules and slows larger ones. The diffusivity component is influenced primarily by the physical properties of the polymer, for example, the glass transition temperature, crystallinity, and cross-linking [35].

Solubility is influenced by both chemical and physical factors, such as plasticizing agents and substituent groups, that allow more favorable interactions between the permeant and the polymer. For example, solubility can be improved by adding pendant groups to the backbone of the polymer that interact more favorably with one molecule than another, thus using the "like dissolves like" strategy. However, these modifications can also affect the intersegmental packing of the chains and thus additionally change the diffusivity.

Other factors influencing the efficacy of a polymer membrane for separation are its surface area, the partial pressure differential across it, and its thickness. Generally, the challenge is to manipulate these variables to achieve high flux rates without compromising the membrane's selectivity. Ideally, one wants the thinnest possible separation barrier to achieve the highest flux. This can be achieved using asymmetric membranes that consist of a thin separating layer grown on a porous support. The thinness of the separating layer is limited by the fact that pores or defects cannot be tolerated because unseparated molecules will readily find their way through the defects. If pores are present, Knudsen diffusion is often observed, with the permeation of the gases being inversely proportional to the square root of their masses (Graham's law). Therefore, nitrogen (N_2, mass 28) will permeate slightly faster than oxygen (O_2, mass 32), leading to an O_2/N_2 selectivity of 0.93. Workers at Monsanto introduced the first defect-free self-supporting hollow-fiber membranes in 1979 by combining hollow-fiber spinning techniques with asymmetric membranes, thus achieving very thin, high surface area membranes [35]. These thin hollow fibers had support walls 25–250 μm thick, allowing for relatively high flux rates with a dense skin of only 0.1–1 μm to effect the separation.

Physical features of the separation process, such as surface area and thickness, and physical and chemical characteristics of the membrane itself, such as crystallinity and substituent groups, combine to yield an effective separation membrane. Manipulating these factors and studying the consequences of these changes to flux and separation further the knowledge and understanding of the structure–property relationships of polymer membranes.

IV. GAS SEPARATION

A. Experimental Setup

In a typical laboratory experiment, the polymer membrane, contained in a stainless steel test flange assembly, is subjected to a feed pressure of approximately 3 atm on the upstream side, with static vacuum ($\sim 1 \times 10^{-5}$ torr) maintained downstream. A schematic representation of the permeation apparatus is presented in Fig. 33.5. The permeating gas is measured downstream by a

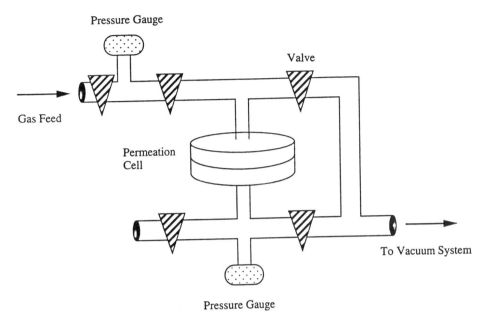

Fig. 33.5 A schematic diagram of a gas permeability testing system. The polymer membrane rests on a porous metal support and is sealed within a stainless steel cell.

pressure transducer until steady state is achieved. Permeability is then calculated from the pressure vs. time curve by applying Fick's first law of diffusion to obtain

$$P = \frac{\theta_p V_d T_0 l}{\Delta p \, A p_0 T_d} \quad (1)$$

where only θ_p, the slope of the curve at steady state; l, the thickness of the membrane; and Δp, the pressure differential, need to be measured. The other values—V_d, the downstream volume of the test cell; T_0 and p_0, reference temperature and pressure at STP (273.15 K and 760 torr, respectively); A, the area of the exposed membrane; and T_d, the temperature of the test—are all known quantities. For a typical experiment,

$$P = \frac{\theta_p (20.24 \text{ cm}^3)(273.15 \text{ K})(0.005 \text{ cm})}{\Delta p (5.171 \text{ cm Hg/psi})(2.607 \text{ cm}^2)(760 \text{ torr})(298.15 \text{ K})} \quad (2)$$

The measured permeability is calculated in barrers, 1 barrer = 10^{-10} [cm³(STP)·cm]/[cm²·s·cm Hg], where cm³(STP) is the volume of permeant at standard temperature and pressure.

B. Membranes for Gas Separations

1. Background

Membrane technology is already a multibillion dollar a year industry and growing at a rate estimated to be over 10% a year [36,37]. Commercial applications for some industrially important gases are listed in Table 33.1. Many common polymers such as polyethylene, polycarbonate, and cellulose acetate have been used as membranes for gas separation with moderate success. Typically, these polymers are structurally modified before they are processed into membranes to test the corresponding effect on P, D, and S.

Perhaps the best example of successful modification of a class of polymers is with polyimides [38] (see Fig. 33.6). When one of the monomers in a polyimide has two substituted trifluoromethyl groups replacing hydrogens, the selectivity remains high while the permeability of the membrane is enhanced, as can be seen in Table 33.2 by comparing PMDA-ODA to 6FDA-ODA. Increasing

Table 33.1 Commercial Applications of Gas Separations

Gas	Source	Use/application
H_2	Refineries	Ammonia synthesis, fuel
CO_2	Gas and combustion by-products	Tertiary oil recover, pollution control
O_2	Air	Medical applications
N_2	Air	Transport of perishable goods, inert atmosphere for air-sensitive materials
CH_4	Landfills, mines	Fuel

Table 33.2 Permeabilities and Selectivities of the Polyimide Membranes BTDA-ODA, PMDA-ODA, and 6FDA-ODA for Selected Gases[a]

Polyimide	P_{CO_2}	P_{O_2}	P_{N_2}	P_{CH_4}	P_{CO_2}/P_{CH_4}	P_{O_2}/P_{N_2}
BTDA-ODA	0.625	0.191	0.0236	0.0109	57.3	8.1
PMDA-ODA	2.71	0.61	0.10	0.059	45.9	6.1
6FDA-ODA	23.0	4.34	0.83	0.38	60.5	5.2

[a] All permeability data are in units of barrers and were performed at 35°C. The structures are shown and abbreviations defined in Fig. 33.6.
Source: Modified from Refs. 38 and 76.

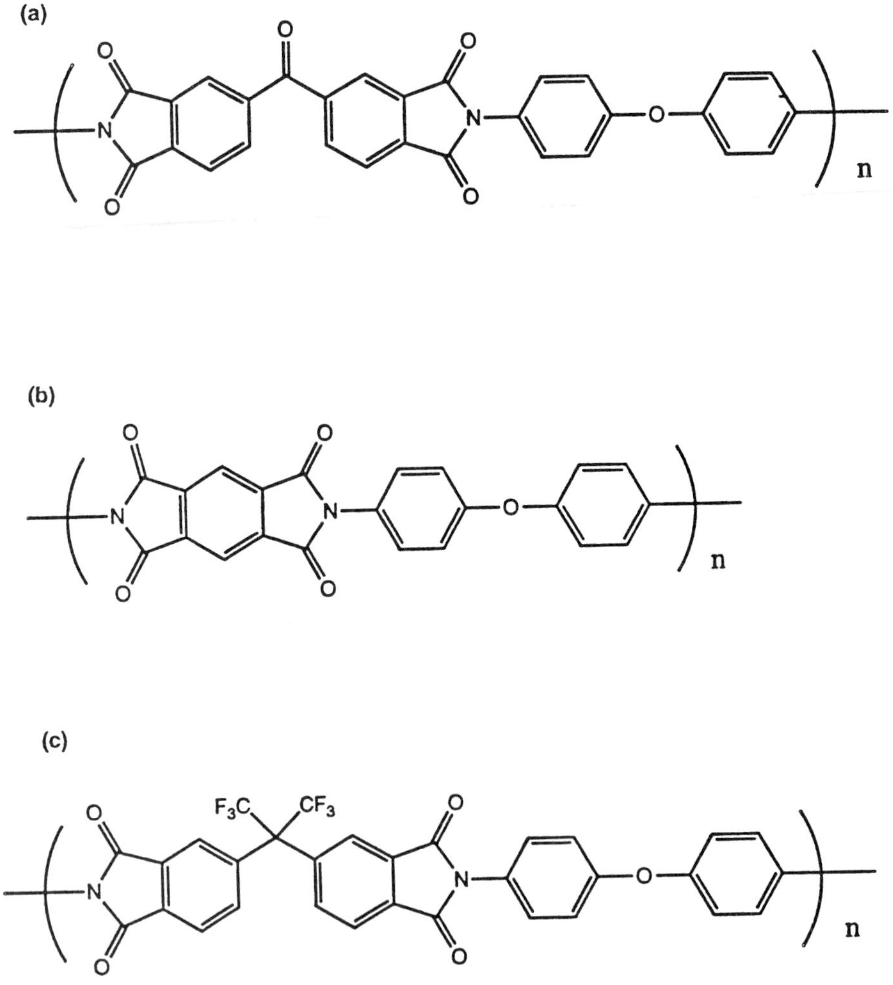

Fig. 33.6 The polyimide structures of (a) BTDA-ODA (3,3′,4,4′-benzophenonetetracarboxylic dianhydride 4,4′-oxydianiline), (b) PMDA-ODA (pyromellitic dianhydride 4,4′-oxydianiline), and (c) 6FDA-ODA [2,2-bis(3,4-dicarboxyphenyl)hexafluoropropane dianhydride 4,4′-oxydianiline].

permeability without adversely affecting selectivity is unusual and very difficult to design. Typically, a counterbalancing effect is observed between permeability and separation factor [39]. By selectively substituting the polar trifluoromethyl groups on the polyimide backbone, Coleman and Koros [40] were able to successfully achieve higher permeabilities without significantly lowering selectivity. However, this case is atypical; other substituted polymers have had only limited success at improving either permeability or selectivity due to the offsetting factors of diffusion and solubility associated with permeability.

Polymer membranes that are effective for the separation of gases must be fully dense, nonporous films. This constraint prevents many conjugated polymers from being useful membranes for separations. For example, polyacetylene synthesized using a Ziegler–Natta type of catalyst (titanium tetrabutoxide and triethylaluminum) produces films of only one-third full density [41]. Although the polyacetylene films appear fully dense, they have a fibrillar morphology with microscopic voids, which can be observed by scanning electron microscopy (SEM; Fig. 33.7A). As a result, this type of polyacetylene cannot be used as a gas separation membrane, since it was found, as expected, to be highly permeable to all gases tested. Many conjugated polymers that are synthesized electrochemically also have a fibrillar morphology with less than full density [42] and are therefore highly permeable and nonselective for gas separations. However, chemically synthesized polyaniline films show no microscopic voids by SEM (Fig. 33.7) and can be used for gas separations. Nitrogen gas adsorption using the BET (Brunauer–Emmett–Teller) method indicates an average pore size for chemically synthesized polyaniline films of less than 20 Å in diameter (the lower limit of BET).

2. Polyaniline

Polyaniline films (as-cast) are selectively permeated by gases, as can be seen in Fig. 33.8. Here, permeabilities of several as-cast polyaniline films are plotted on a log scale in barrers versus the kinetic diameters of seven different gases. Figure 33.9 gives a schematic representation of the kinetic diameters of these gases as determined from sorption in zeolites [43]. In general, it is apparent from Fig. 33.8 that the larger the gas molecule, the lower its permeability through as-cast polyaniline film. The one exception to decreasing permeability with increasing size is oxygen. Oxygen permeability is actually higher than that of argon even though oxygen (kinetic diameter 3.46 Å) is slightly larger than argon (kinetic diameter 3.4 Å). This is most likely due to a relatively high solubility for oxygen in polyaniline [44–47].

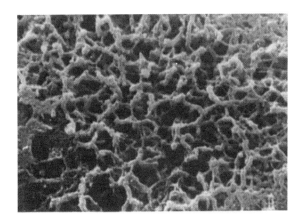

(A) Polyacetylene

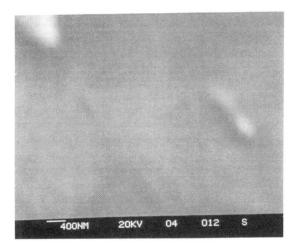

(B) Polyaniline

Fig. 33.7 Scanning electron micrographs of (A) polyacetylene with ~200 Å diameter fibrils (lighter color) and (B) polyaniline illustrating the macroscopic porosities of the films.

Doping polyaniline lowers its permeability to gases, as can be seen in Fig. 33.10 for oxygen and nitrogen. Here the permeability of polyaniline to oxygen and nitrogen is compared in four different states of doping: as-cast (untreated), doped (exposed to acid), undoped (doped film exposed to base), and redoped (undoped film exposed to acid). The as-cast polyaniline shows a higher permeability for oxygen (0.174 barrer) than nitrogen (0.019 barrer) leading to a selectivity α of about 9.1. Fully doping the polyaniline film with aqueous hydrochloric acid leads to lower permeabilities to both oxygen

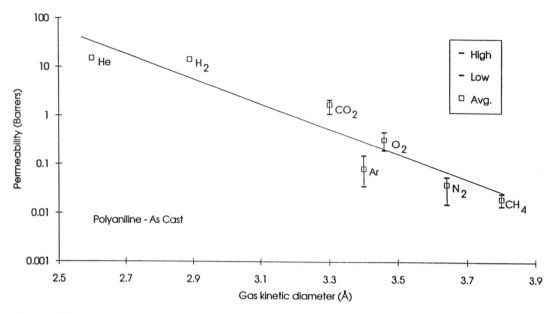

Fig. 33.8 Permeability (in barrers) versus size (in angstroms) for selected gases permeating as-cast polyaniline membranes.

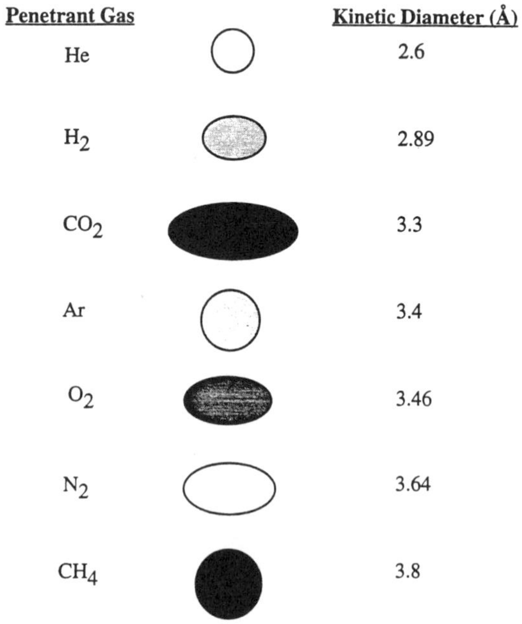

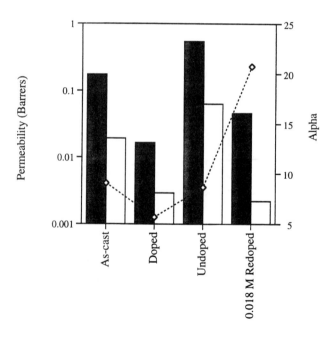

Fig. 33.9 Schematic of kinetic diameters of selected gases. The diameters were determined from sorption in zeolites. (From Ref. 43.)

Fig. 33.10 Effect of redoping, to different levels, on oxygen and nitrogen permeabilities of a polyaniline membrane. (■) O_2; (□) N_2; (◇) α.

and nitrogen as void space in the polymer is filled by dopants that easily penetrate the film to dope each imine nitrogen. Undoping with ammonium hydroxide completely removes the chloride dopants as determined by elemental analysis [31,33]. The undoped polyaniline film is now more permeable to all gases than the as-cast film even though they have chemically equivalent compositions. It should be noted that residual NMP used as the casting solvent is also removed by the doping/undoping process, accounting for a small portion of the increased free volume [48]. The undoped film of Fig. 33.10 has an oxygen permeability of 0.546 barrer, a nitrogen permeability of 0.063 barrers, and a selectivity of 8.7. Redoping can now be used to reduce the permeability of larger nitrogen molecules more so than that of the smaller oxygen, resulting in high selectivity, as illustrated in Fig. 33.10.

A schematic of the processes occurring when an as-cast polyaniline film is doped, undoped, and partially redoped is shown in Fig. 33.11. The hypothetical cross section (Fig. 33.11a) shows as-cast polyaniline possessing some pore connectivity leading to moderate O_2/N_2 selectivity. The pores are representations of microscopic pathways through which gas molecules may penetrate as segments of aniline rings flip due to thermal vibrations [49,50]. On full doping with HCl, the protons find each imine nitrogen, carrying along with them the large solvated chloride counterions to maintain electroneutrality. The dopants and counterions fill void space in the polymer (as shown schematically in Fig. 33.11b) and reduce the permeability of all gases. Undoping completely removes the dopants and counterions. This has the effect of increasing pore connectivity (as shown in Fig. 33.11c), resulting in higher permeabilities for all gases. Note that there is a large driving force on doping for dopants to associate with imine nitrogens, which leads to changes in morphology; undoping, however, serves only to remove dopants as there is no driving force for the polyaniline structure to return to its as-cast form. Redoping will then begin to fill in void spaces left by dopants and affect the diffusion of gases and their solubility by changing gas–polymer interactions. By controlling the amount of dopants put back in polyaniline films, the permselectivity can be tuned.

The mechanism behind high selectivities in partially doped polyaniline likely involves both increased solubility for oxygen and enhanced control over diffusion. The increase in oxygen solubility could be due to interactions of paramagnetic molecular oxygen with unpaired electrons created in the conducting polymer backbone through the doping process [46,47,51]. Interactions of oxygen with free spins have been observed in several other types of materials [52,53].

The oxygen/nitrogen selectivity of as-cast polyaniline (≈ 9) is comparable to that of commercially used poly-

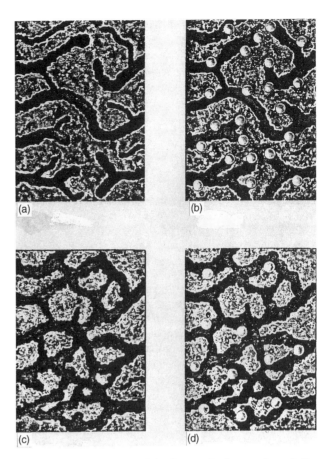

Fig. 33.11 The effects of doping, undoping, and partially redoping a polyaniline membrane are shown schematically in a hypothetical cross section of a polyaniline film: (a) As-cast; (b) doped; (c) undoped; (d) redoped.

mers such as polysulfone ($\alpha = 5.6$ [38]) and other well-studied polymers such as polyimides ($4 < \alpha < 8$ [38]). However, the O_2/N_2 selectivity of redoped polyaniline is the highest reported for any polymer [31]. The commercially applicable polymers generally have higher oxygen permeabilities (e.g., 1.4 barrers for polysulfone) than polyaniline (~ 0.2 barrer). A major challenge for polyaniline is to develop pinhole-free ultrathin membranes (such as asymmetric hollow fibers) that have high enough flux to compensate for their relatively lower permeabilities.

To this end, thin films (1.5–4 μm) of polyaniline grown on alumina (Anopore support membranes) have been studied for gas transport properties by Kuwabata and Martin [54]. When 1.5 μm thick films were tested for permeability, O_2/N_2 selectivities <1 were obtained, indicating Knudsen-type diffusion, likely due to microporous defects. Selectivity coefficients were found to increase with film thickness, to slightly higher than 9 for thicknesses of 3 μm or greater. This selectivity agrees well with earlier work of Anderson et al. [31] on free-standing films and demonstrates that supported films as thin as 3 μm can be made essentially free of defects. When free-standing films were synthesized, those below 10 μm in thickness were found to be microporous, showing Knudsen diffusion [54]. Increasing thicknesses led to increasing selectivity values, reaching >9 for free-standing films thicker than 20 μm.

The effect of doping level was studied for both alumina-supported and free-standing polyaniline films [54]. Doping for thin films was carried out by exposing them to 1.0 M aqueous HCl, corresponding to doping levels of 13, 22, 31, and 38%. Permeability for all gases studied were found to decrease with increasing doping level. Permeability values for oxygen ranged from about 0.5 to 0.15 barrer, while those for nitrogen ranged from about 0.06 to 0.01 barrer. Oxygen/nitrogen selectivities for the supported membranes increased from about 9.5 (undoped) to 15.2 (38% doped) and then fell slightly at full (50%) doping. Free-standing films showed similar trends, with O_2/N_2 selectivities increasing from about 9.5 (undoped) to 14.9 (36% doped) to 15.2 (50% doped). Kuwabata and Martin [54] made the unusual suggestion that repeated doping and undoping of polyaniline films could potentially enhance selectivities. Not surprisingly, they found no discernible effect on selectivities after up to five repeated dopings and undopings.

Other recent work on gas permeability through polyaniline membranes by Rebattet et al. [55] confirm the general effects of doping, undoping, and partial redoping and report an O_2/N_2 selectivity of 14.2. The differences in reported O_2/N_2 selectivities (from Fig. 33.10) could be due to uncertainty in the low nitrogen permeability values, membrane preparation methods, and inhomogeneity in the doping process.

The use of other dopants also leads to relatively high selectivities [54]. For example, a 36% sulfuric acid–doped sample had an O_2/N_2 selectivity of 13.4, while a 36% paratoluenesulfonate-doped sample had an O_2/N_2 selectivity of 14.9. A thin film composite doped with 38% nitric acid had an O_2/N_2 selectivity of 14.8.

In order to look at the effects of dopant sizes in a systematic fashion Anderson et al. [31] studied the halogen acid series HF, HCl, HBr, and HI. Polyaniline films were fully doped with each halogen acid and then completely undoped using ammonium hydroxide. A plot of the resulting permeabilities for each gas from small to large is shown in Fig. 33.12. Looking at the He data, a progression of lower permeabilities is observed on going from HF to HCl to HBr to HI. This is consistent with the sizes of the solvated halogen acids in which solvated HF is actually the largest of the series (in contrast to halides in a crystalline lattice) [56,57]. Figure 33.12 again

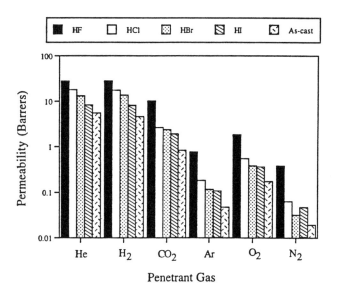

Fig. 33.12 Permeability (in barrers) for gases permeating polyaniline membranes doped with the halogen acids $HX_{(aq)}$ (X = F, Cl, Br, and I) and undoped with NH_4OH.

demonstrates that in general smaller gases will more easily permeate polyaniline than larger gases. One notable deviation is the increased permeation of oxygen through all polyaniline membranes relative to that of argon. This again is most likely due to the high solubility of oxygen in polyaniline [44,45].

3. Polyaniline Blends

In contrast to the small molecule ionically bound protonic acid dopants discussed thus far, large molecule permanently bound dopants are of interest to further elucidate the effects of polyaniline's dopant state on gas permeability. Combining the controlled dopability of polyaniline with the best features of a polymer studied extensively for gas separations, polyaniline–polyimide blends were developed [58–60]. Blending in general consists of physically mixing two components to form a composite. In this case, a polyamic acid (the precursor to polyimide) solution and a polyaniline solution were prepared separately using the same solvent (NMP) and then mixed in a 50/50 weight ratio, cast into films at 100°C, and cured at 200°C for 1 h and at 300°C for 15 min. The idealized structure of a polyamic acid–polyaniline blend is illustrated in Fig. 33.13. The bluish color of the blend, formed from a golden yellow polyamic acid solution and brown polyaniline emeraldine base powder dispersed in NMP, suggests that polyaniline is in its doped acid form, which is typically blue.

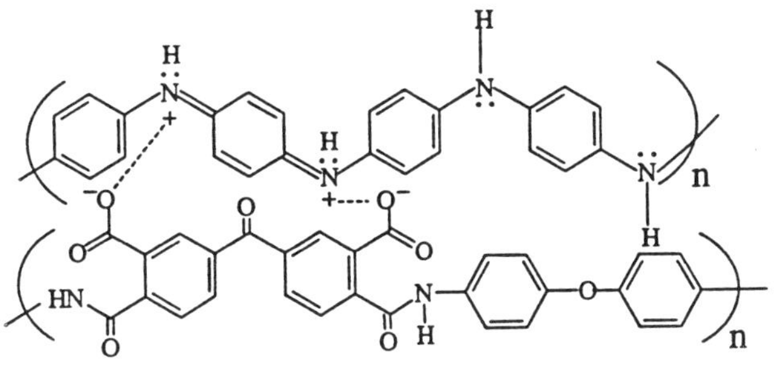

Fig. 33.13 Proposed polyaniline–polyamic acid structure.

The 50/50 mixing of polyaniline and polyamic acid forms a highly miscible blend, with molecular level (acid–base) binding of the two polymers, indicated by FTIR spectroscopy. The blend exhibits all characteristic absorptions (1485, 1504, 1567, 1599, 1660, and 1732 cm^{-1}) of polyamic acid (1660 cm^{-1}), polyimide (1730 cm^{-1}), and both the acid (1486 and 1568 cm^{-1}) and base (1505 and 1595 cm^{-1}) forms of polyaniline. This indicates that chemical doping of polyaniline with the polyamic acid precursor did, in fact, occur. The amide nitrogens from the polyamic acid remain even though full curing temperatures for imidization were used, likely due to the strong interaction with the imine nitrogens of the polyaniline backbone preventing ring closure.

Gas permeability measurements were performed on the blended films as presented in Fig. 33.14. The blend shows a definite increase in permeability for all gases tested, relative to polyimide and the base form of polyaniline [60]. Ordinarily, there is an inverse relationship between permeability of two gases and the separation factor for those gases [61–64]. However, this blend does not hold to this axiom. Figure 33.15 illustrates the separation factors for the blend and the homopolymers. The separation factors for the blend are comparable to those of polyaniline (as-cast) for H_2/N_2 (α = 200) and O_2/N_2 (α = 9) and closer to that of polyimide for CO_2/CH_4 (α = 58). Thus, this blend appears to have achieved an improved combination of properties compared to its parent polymers, with enhanced permeability and good selectivity.

4. Other Conjugated Polymers

The use of polypyrrole and its derivatives was investigated soon after the reported discovery of polyaniline membranes. Liang and Martin [65] demonstrated that thin films of polypyrrole could be grown on alumina (Anopore) support membranes by using interfacial polymerization techniques. Doped polypyrrole films were found to be porous, showing Knudsen diffusion with an O_2/N_2 selection coefficient of 0.94. However, poly(N-methylpyrrole) films were nonporous and showed good gas transport and selectivity. For example, a 1.4 μm poly(N-methylpyrrole) film doped with NO_3^- ions had an oxy-

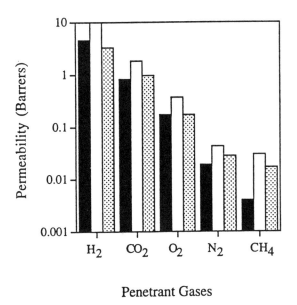

Fig. 33.14 Gas permeabilities (in barrers) for (■) as-cast polyaniline (Pani), (▦) the polyimide BTDA-ODA, and (□) a 50/50 blend of the two polymers.

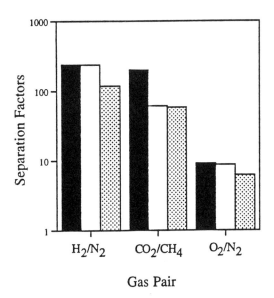

Fig. 33.15 Separation factors for (■) as-cast polyaniline (Pani), (▦) the polyimide BTDA-ODA, and (□) a 50/50 blend of the two polymers.

gen permeability of 1.26 barrers, a nitrogen permeability of 0.15 barrer, and therefore an O_2/N_2 selectivity of 7.9. Undoping of the membrane was carried out using $NaBH_4$ in acetonitrile. The removal of NO_3^- dopants led, as expected, to higher permeabilities. Oxygen permeability increased to 2.04 barrers, while that of nitrogen increased to 0.33 barrer, leading to an O_2/N_2 selectivity of 6.2. Thus the permeability of substituted polypyrrole can also be changed through doping and undoping.

Dimethoxypolyparaphenylenevinylene (DMPPV) is another conjugated polymer that can be used for gas separations [31]. It can be synthesized by slowly reacting dimethylsulfide and 2,5-dimethoxy-p-xylene dichloride in water with a base at 5°C [66]. A 5% aqueous solution of the purified powder was cast onto glass plates, dried for 48 h, and then cured at 200°C for 45 min. Doping with 0.1 M $FeCl_3$ followed by undoping with 1.0 M NH_4OH resulted in increases in gas permeabilities consistent with the results for polyaniline and poly(N-methylpyrrole).

5. Summary

From these studies it appears that polyaniline and other conjugated polymers form an interesting class of membrane materials with tailorable properties. Doping and undoping polyaniline results in increased permeabilities relative to as-cast films owing to the localized morphological changes from the addition and removal of the dopant counterions. Redoping with small amounts of dopants can lead to very high selectivities for important gas pairs such as O_2/N_2, H_2/N_2, and CO_2/CH_4 [31,54]. In fact, three independent research teams (Anderson et al., Kuwabata and Martin, and Rebattet et al.) determined that polyaniline has the highest current O_2/N_2 selectivity of any polymer membrane, falling above a permselectivity threshold for state-of-the-art membrane technology [31,54,55,67]. Changing the properties of polyaniline through the use of other protonic acids and different polymer blends will likely provide fertile ground for future research.

V. PERVAPORATION

A. Background

Pervaporation is a process that can separate liquid mixtures by selectively allowing the passage of one component. Important applications include the separation of azeotropes, the dehydration of organics, and the removal of organic compounds from water [68]. Since most liquids are composed of molecules in the 2–10 Å size regime [69], separations by pervaporation require fully dense pinhole-free membranes of the type used for gas separations. With the ability to separate gases on this size scale, polyaniline films have the potential to separate liquids as well [70]. Earlier work on liquid transport through conjugated polymers was carried out by several groups [70–74].

The transport of water through supported polyaniline membranes grown electrochemically was investigated by Schmidt et al. [71]. A 25–30% increase in water permeation was observed for sulfuric acid–doped polyaniline compared to the undoped polymer. The doped polymer was believed to have a more open structure, accounting for the enhanced water permeation. However, a simpler explanation is the increased hydrophilicity of doped polyaniline. Analogous increases in methanol transport through both polyaniline and polypyrrole were found [72,73].

Ultrafiltration and molecular gating applications using the conjugated polymers poly(N-methylpyrrole) and poly(3-methylthiophene) supported on porous polycarbonate were studied by Feldheim and Elliot [74]. For poly(N-methylpyrrole), acetone was found to have about an order of magnitude higher flux through the undoped polymer (0.2 ppm/min) than through the perchlorate-doped polymer (0.03 ppm/min). Scanning electron microscopy showed that the perchlorate-doped poly(N-methylpyrrole) had a grainy morphology whereas the undoped polymer had a smooth appearance, consistent with their transport properties. This result is also consistent with doped polypyrrole being more dense than its undoped form [75]. Poly(3-methylthiophene), on the other hand, transported acetone much faster through its perchlorate-doped form (~3 ppm/min) than through its undoped (reduced) form (~0.5 ppm/min). Again this was consistent with scanning electron micrographs, which showed that perchlorate-doped poly(3-methylthiophene) had a smooth morphology while the undoped polymer had a rough, grainy appearance. Several dopings and undopings of either polymer gave essentially no change in the transport properties. However, poly(N-methylpyrrole) failed, due to mechanical instability, after four doping/undoping sequences, whereas poly(3-methylthiophene) failed after eight sequences.

The interaction of methanol vapor with doped polypyrrole was examined by Topart and Josowicz [76]. Sorption leading to Langmuir-type isotherms was the driving force for methanol uptake. Adsorption appeared to be the rate-limiting step with diffusion occurring instantaneously.

Pervaporation of methanol and methyl tert-butyl ether (MTBE) using hollow fibers coated with polypyrrole, poly(N-methylpyrrole), and polyaniline was studied by Martin et al. [77]. A microporous hollow-fiber ultrafiltration membrane made of cellulose was coated with a conjugated polymer. An uncoated hollow fiber had only a slight preference for transporting methanol ($\alpha_{methanol/MTBE} = 1.06$), attributed to the more polar methanol being more soluble in the somewhat polar cellulose-based film. The polypyrrole-coated fiber had only a slightly greater selectivity of 1.4, likely due to its porosity as observed in gas selectivity experiments [65]. Poly(N-methylpyrrole)-coated fibers had an improved selectivity of 3.9, while polyaniline-coated fibers showed a selectivity of 4.9.

These studies demonstrate that conjugated polymers can be used selectively to transport liquids. However, many of these studies have been limited by the stability of the membranes and low permeation rates.

B. Free-Standing Polyaniline Membranes

Since chemically synthesized free-standing polyaniline membranes are very robust and quite stable in aqueous solutions, it is interesting to explore their liquid permeation properties to elucidate the transport mechanism and identify potentially useful applications.

A pervaporation setup is shown in Fig. 33.16. A free-standing polyaniline membrane (40–100 μm thick) is placed on a porous stainless steel support and sealed with an O-ring. About 200 mL of feed solution at room temperature is put in the feed reservoir, and the downstream side of the pervaporator is evacuated. The permeant collector is cooled to $-65°C$ to maintain a low vapor pressure on the downstream side. Due to the pressure differential, liquid adsorbs onto, permeates, and desorbs from the membrane in the vapor phase and condenses in a graduated collector. Liquid permeation through the membrane is allowed to proceed for approximately 24 h to ensure the attainment of steady-state flow conditions. The volume of liquid permeating the membrane is monitored with time and recorded from the graduated collection vessel. Flux is then calculated.

Traditionally, either gas chromatography combined with mass spectroscopy (GC-MS), density measurements, or refractometry is used to determine the relative concentration of the permeant mixtures [68]. However, proton nuclear magnetic resonance (^{1}H NMR) is another excellent method to quantify concentrations because of its ease of use and high accuracy. The molar ratio of the mixture is calculated from the ratio of the integration of peaks corresponding to the components in the mixture. Once the ratios of permeants are known, selectivity can be calculated. To demonstrate that ^{1}H NMR is an accurate method for determining the composition of liquid mixtures, a trial run was carried out using measured volumes of water and methanol (Fig. 33.17).

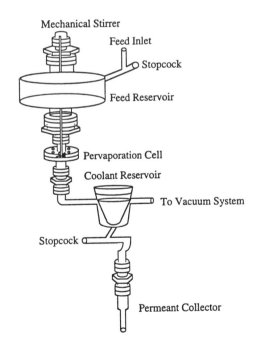

Fig. 33.16 A schematic diagram of the pervaporation setup. The polymer membrane rests on a porous metal support within a stainless steel pervaporation cell. The graduated permeant collector is immersed in an isopropanol bath at $-65°C$ during pervaporation.

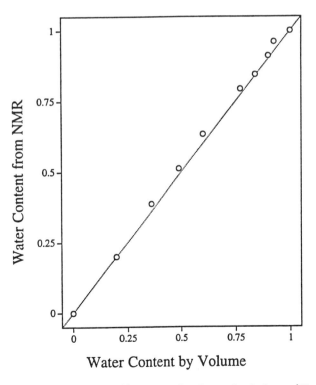

Fig. 33.17 Methanol/water ratio determined from ^{1}H NMR and by volume. The dotted line represents a one-to-one match between the water content measured by volume and the water content determined from NMR.

Pervaporation was next carried out to test the selectivity of polyaniline membranes toward acetic acid–water mixtures. Different feed ratios of acetic acid and water were pervaporated through both undoped and doped polyaniline, as shown in Fig. 33.18, where the feed water content is plotted versus the permeant water content. For comparison the vapor–liquid equilibrium curve for acetic acid–water [78] is plotted just above the line of no separation. From Fig. 33.18 it is clear that undoped polyaniline has a small preference for permeating water over acetic acid at essentially any composition. However, this small preference at an average water permeability of about 0.5 g·mm/(m²·h) is too low to have any utility. More interesting is fully HCl-doped polyaniline, which permeates water over acetic acid in a much more selective fashion. In fact, even with a mixture of 85% acetic acid–15% water, at least 93 wt % of the permeant was water. It should be pointed out that when undoped membranes are used in the presence of acids, they will partially dope the polyaniline, the extent depending on the pH of the acid used. However, when a doped membrane is used, as long as the dopant has a stronger interaction with polyaniline than the feed acid, very little acid–dopant exchange will occur.

The permeability of the fully HCl-doped polyaniline membrane of Fig. 33.18 is shown in Fig. 33.19. Here the feed water content is plotted versus the permeability in gram-millimeters per square meter per hour [g·mm/(m²·h)]. The lower curve indicates the permeability of acetic acid, the middle curve represents the permeability of water, and the upper curve is the overall combined permeability. Clearly the acetic acid permeability of doped polyaniline is exceedingly low except when pure acetic acid is used. Water permeability increases from 0.23 g·mm/(m²·h) at 15 wt % water content in the feed to >1 g·mm/(m²·h) when pure water is the feed. The total permeability essentially parallels the water permeability.

To test the effect of different molecular size permeants on HCl-doped polyaniline, formic, acetic, and propionic acids were used, as shown in Fig. 33.20. Here feed content is again plotted versus permeant content as determined by ^{1}H NMR. The smallest permeant, formic acid (HCOOH), has a molecular size of ~3.5 Å in aqueous solution [69]. Although preferentially held back by doped polyaniline, formic acid is still transported with the water, leading to relatively poor separation. Acetic acid (CH$_3$COOH) with a size of ~4.5 Å [69] is readily

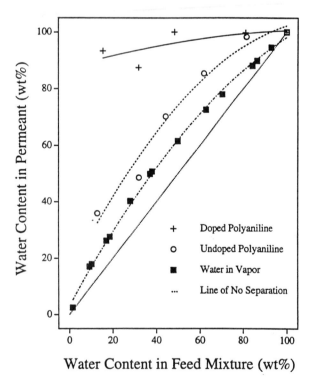

Fig. 33.18 Pervaporation of acetic acid–water mixtures through doped and undoped polyaniline membranes. All polyaniline experiments were carried out at room temperature. The vapor–liquid equilibrium curve is plotted based on data points from Ref. 78 at 110–115.3°C and 760 mm Hg.

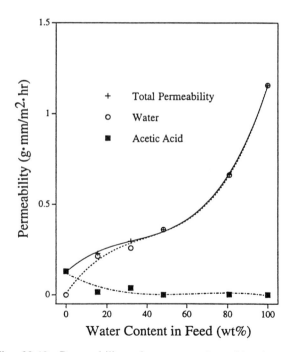

Fig. 33.19 Permeability of water–acetic acid mixtures through fully HCl-doped polyaniline membranes at room temperature.

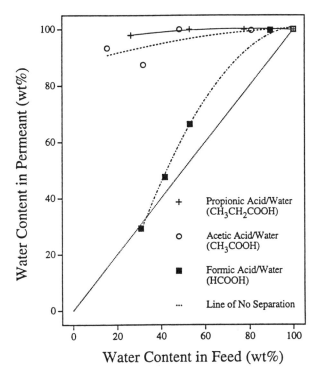

Fig. 33.20 Pervaporation of organic acid–water mixtures through fully HCl-doped polyaniline membranes at room temperature.

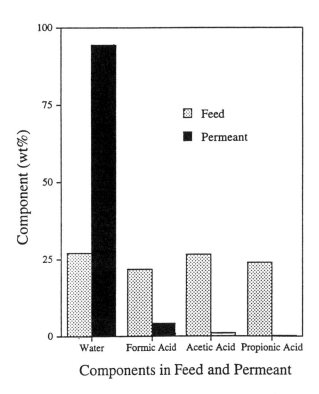

Fig. 33.21 Pervaporation of a mixture of organic acids and water through a doped polyaniline membrane. The feed consists of approximately 25 wt % of each acid in water.

held back by doped polyaniline, while water preferentially permeates. The acid with the largest molecular size (~5.5 Å [69]), propionic acid (CH_3CH_2COOH), is essentially completely rejected by an HCl-doped polyaniline membrane. Water permeates preferentially at all compositions, and at ≥50% water in the feed essentially no propionic acid was detected. Even at the lowest value tested (26 wt %), at lease 98 wt % water was transported through the doped polyaniline membrane.

To test the ability of doped polyaniline to be selectively permeated by mixtures, approximately equal amounts of formic, acetic, and propionic acids along with water were put in the feed reservoir. The actual feed composition was 27.7 wt % water, 21.8% formic acid, 26.6% acetic acid, and 24.0% propionic acid. The results of pervaporation through doped polyaniline are shown in Fig. 33.21. It is clear that water preferentially permeates, going from 27.7% in the feed to 94.5% in the permeant, while formic acid is reduced to 4.3%, acetic acid to 1.1%, and propionic acid to 0.2%. Clearly, HCl-doped polyaniline can separate a mixture, as one might have guessed from the individual permeation experiments shown in Fig. 33.20. While the smaller formic acid molecules permeate to some extent, acetic acid permeation in considerably less, and propionic acid is essentially rejected.

These tests clearly show that HCl-doped polyaniline can be used to prevent the permeation of molecules larger than about 4.5 Å while allowing smaller molecules (especially water) to pass through. This could prove useful for separations ranging from the concentration of fruit juices to the separation of azetotropes. However, limitations due to relatively low flux will have to be compensated for by the formation of asymmetric membranes, which could provide much higher fluxes. As polyaniline is a very thermally stable polymer, increased flux could also be obtained by larger temperature differentials. Since selectively could be adversely affected by increased temperatures, testing is clearly warranted.

C. Summary

The results just presented show that polyaniline, especially in its doped form, can selectively transport liquids by pervaporation. The undoped base form of polyaniline, which is relatively hydrophobic, becomes much more hydrophilic when it is doped. This suggests that both diffusion and solubility will favor water over organic molecules in doped polyaniline, leading to high selectivities.

VI. CONCLUSIONS

Conjugated polymers form a new class of materials with potential applications as separation membranes. The doping process can be used to modify permselectivity after membrane formation, opening up new possibilities for enhancing selectivity. Polyaniline is the most promising candidate among the conjugated polymers for membrane separations because of its high chemical and thermal stability and its simple acid/base doping chemistry. The doping process can be used to control hydrophilicity for liquid separations and gas permselectivities for gas separations. A doping–undoping and partial redoping process leads to the most selective membrane known so far for oxygen–nitrogen separations. A future challenge is to develop pinhole-free ultrathin membranes to achieve high fluxes.

ACKNOWLEDGMENTS

We thank Dr. Mark R. Anderson, Dr. Benjamin R. Mattes, and Prof. Howard Reiss for their contributions to this research. This work was supported by the Office of Naval Research grant N00014-93-1-1307, a Sloan Foundation fellowship, and a Dreyfus Teacher-Scholar award (RBK).

REFERENCES

1. N. Toshima (ed.), *Polymers for Gas Separation*, VCH, New York, 1992.
2. R. E. Kesting and A. K. Fritzsche, *Polymeric Gas Separation Membranes*, Wiley, New York, 1993.
3. S. A. Stern, *J. Membrane Sci. 94*:1 (1994).
4. W. J. Koros and G. K. Fleming, *J. Membrane Sci. 83*:1 (1993).
5. W. J. Koros, M. R. Coleman, and D. R. B. Walker, *Annu. Rev. Mater. Sci. 22*:47 (1992).
6. R. Y. M. Huang (ed.), *Pervaporation Membrane Separation Processes*, Elsevier, New York, (1991).
7. S. M. Zhang and E. Drioli, *Separ. Sci. Technol. 30*:1 (1995).
8. T. M. Aminabhari, R. S. Khinnavar, S. B. Harogoppad, and U. S. Aithal, *J. Macromol. Sci. Rev. Macromol. Chem. Phys. 34*(2):129 (1994).
9. W. S. Huang, B. D. Humphrey, and A. G. MacDiarmid, *J. Chem. Soc. Faraday Trans. 82*:2385 (1986).
10. A. F. Diaz and J. A. Logan, *J. Electroanal. Chem. Interfacial Electrochem. 111*:111 (1980).
11. R. Nonfi and A. J. Nozik, *J. Electrochem. Soc. 129*:2261 (1982).
12. T. J. Skotheim (ed.), *Handbook of Conducting Polymers*, Marcel Dekker, New York, 1986.
13. D. W. DeBerry, *J. Electrochem. Soc. 132*:1022 (1985).
14. D. A. Wrobleski, B. C. Benicewicz, K. G. Thompson, and C. J. Bryan, *ACS Polym. Prepr. 35*:168 (1994).
15. V. I. Krinichniy, O. N. Eremenko, G. G. Rukhman, V. M. Geskin, and Y. A. Letuchy, *Synth. Met. 41*:1137 (1991).
16. A. G. MacDiarmid and A. J. Epstein, *Faraday Discuss. Chem. Soc. 88*:317 (1989).
17. J. C. Chiang and A. G. MacDiarmid, *Synth. Met. 13*:193 (1986).
18. A. G. MacDiarmid, J. C. Chiang, M. Halpern, W. S. Huang, S. L. Mu, N. L. D. Somasiri, W. Wu, and S. I. Yaniger, *Mol. Cryst. Liq. Cryst. 121*:173 (1985).
19. E. M. Genies, A. A. Syed, and C. Tsintavis, *Mol. Cryst. Liq. Cryst. 121*:181 (1985).
20. A. G. MacDiarmid, S. L. Mu, N. L. D. Somasiri, and W. Wu, *Mol. Cryst. Liq. Cryst. 121*:187 (1985).
21. W. R. Salaneck, B. Liedberg, O. Inganas, R. Erlandsson, I. Lundstrom, A. G. MacDiarmid, M. Halpern, and N. L. D. Somasiri, *Mol. Cryst. Liq. Cryst. 121*:191 (1985).
22. J. P. Travers, J. Chroboczek, F. Devreux, F. Genoud, M. Nechtshein, A. Syed, E. M. Genies, and C. Tsintavis, *Mol. Cryst. Liq. Cryst. 121*:195 (1985).
23. L. Xie, L. J. Buckley, and J. Y. Josefowicz, *J. Mater. Sci. 29*:4200 (1994).
24. M. Laridjani, J. P. Pouget, E. M. Scherr, A. G. MacDiarmid, M. E. Jozefowicz, and A. J. Epstein, *Macromolecules 25*:4106 (1992).
25. M. Laridjani, J. P. Pouget, A. G. MacDiarmid, and A. J. Epstein, *J. Phys. I 2*:1003 (1992).
26. R. B. Kaner, M. R. Anderson, B. R. Mattes, and H. Reiss, Gas separation membranes: A novel application of conducting polymers, first presented at the International Conference on Synthetic Metals, Tubingen, Germany, Sept. 4, 1990.
27. M. G. Kanadtzidis, *Chem. Eng. News 68*(49):36 m(1990).
28. B. R. Mattes, M. R. Anderson, H. Reiss, and R. B. Kaner, *ACS Polym. Mater. Sci. Eng. 64*:336 (1991).
29. M. R. Anderson, B. R. Mattes, H. Reiss, and R. B. Kaner., *Synth. Met. 41–43*:1151 (1991).
30. R. B. Kaner, M. R. Anderson, B. R. Mattes, and H. Reiss, U.S. Patents 5,095,586 (Mar. 17, 1992) and 5,358,556 (Oct. 24, 1994).
31. M. R. Anderson, B. R. Mattes, H. Reiss, and R. B. Kaner, *Science 252*:1412 (1991).
32. Y. Cao, A. Andreatta, A. J. Heeger, and P. Smith, *Polymer 30*:2305 (1989).
33. A. G. MacDiarmid, J. C. Chiang, A. F. Richter, and N. L. D. Somasiri, in *Conducting Polymers: Special Applications* (L. Alacer, ed.), Reidel, Dordrecht, Holland, 1987, pp. 105–120.
34. J. Maron, M.J. Winokur, and B. R. Mattes, *Macromolecules 28*:4475 (1995).

35. J. Henis and M. K. Tripoli, *Science 220*:11 (1983).
36. J. Haggin, *Chem. Eng. News* 68(40):22 (1990).
37. U.S. Department of Energy Report No. DE90-01170, Membrane separations systems—A Research needs assessment, U.S. DOE, Washington, DC, April 1990.
38. W. J. Koros, M. R. Coleman, and D. R. B. Walker, *Annu. Rev. Mater. Sci. 22*:47 (1992).
39. T. H. Kim, W. J. Koros, G. R. Husk, and K. C. O'Brien, *J. Membrane Sci. 37*:45 (1988).
40. M. R. Coleman and W. J. Koros, *J. Membrane Sci. 50*:285 (1990).
41. F. E. Karasz, J. C. W. Chien, R. Galkiewicz, and G. E. Wnek, *Nature 282*:15 (1979).
42. G. Tourillion and F. Garnier, *J. Polym. Sci. Polym. Phys. Ed. 22*:33 (1984).
43. D. W. Breck, *Zeolite Molecular Sieves*, Wiley, New York, 1974, pp. 593–724.
44. J. Pellingrino, R. Radebaugh, and B. R. Mattes, *Macromolecules 29*:4985 (1996).
45. M. R. Anderson, An investigation of conducting polymer materials as gas separation membranes, Ph.D. Dissertation, Univ. Calif., Los Angeles, 1992.
46. L. Rebattet, M. Escoubes, E. Genies, and M. Pineri, *J. Appl. Polym. Sci. 58*:923 (1995).
47. L. Rebattet, M. Escoubes, M. Pineri, and E. M. Genies, *Synth. Met. 71*:2133 (1995).
48. J. A. Conklin, Polyaniline films for membrane based gas permeability studies, PhD. Dissertation, Univ. Calif., Los Angeles, 1994.
49. S. Kaplan, E. M. Conwell, A. F. Richter, and A. G. MacDiarmid, *Macromolecules 22*:1669 (1989).
50. S. Kaplan, E. M. Conwell, A. F. Richter, and A. G. MacDiarmid, *Synth. Met. 29*:235 (1989).
51. B. R. Mattes, M. R. Anderson, H. Reiss, and R. B. Kaner, in *Intrinsically Conducting Polymers: An Emerging Technology* (M. Aldissi, ed.), Kluwer, Boston, 1993, p. 61.
52. M. Nechtsein, F. Devreux, F. Genoud, M. Guglielmi, and K. Holczer, *Phys. Rev. B 27*:61 (1983).
53. F. Bensebaa and J. J. Andre, *J. Phys. Chem. 96*:5739 (1992).
54. S. Kuwabata and C. R. Martin, *J. Membrane Sci. 91*:1 (1994).
55. L. Rebattet, M. Escoubes, E. Genies, and M. Pineri, *J. Appl. Polym. Sci. 57*:1595 (1995).
56. E. S. Amis and J. F. Hinton, *Solvent Effects on Chemical Phenoma*, Academic, New York, 1973, pp. 46–181.
57. B. E. Conway, *Ionic Hydration in Chemistry and Biophysics*, Elsevier, Amsterdam, 1981, p. 59.
58. M. Angelopoulos, N. Patel, and R. Saraf, *Synth. Met. 55–57*:1552 (1993).
59. B. R. Mattes, M. R. Anderson, J. A. Conklin, H. Reiss, and R. B. Kaner, *Synth. Met. 55–57*:3655 (1993).
60. S.-C. Huang, J. A. Conklin, T. M. Su, I. J. Ball, S. L. Nguyen, B. M. Lew, and R. B. Kaner, *ACS Polym. Mater. Sci. Eng. 72*:323 (1995).
61. T. H. Kim, W. J. Koros, and G. R. Husk, *Separ. Sci. Technol. 23*:1611 (1988).
62. S. A. Stern, Y. Mi, H. Yamamoto, and A. K. St. Clair, *J. Polym. Sci. Polym. Phys. 27*:1887 (1989).
63. T. A. Barbari, W. J. Koros, and D. R. Paul, *J. Membrane Sci. 42*:69 (1989).
64. L. M. Robeson, W. F. Burgoyne, M. Langsam, A. C. Savoca, and C. F. Tien, *Polymer 35*:4970 (1994).
65. W. Liang and C. R. Martin, *Chem. Mater. 3*:390 (1991).
66. I. Murase, *Polym. Commun. 26* (1985).
67. L. M. Robeson, *J. Membrane Sci. 62*:165 (1991).
68. R. Y. M. Huang (ed.), *Pervaporation Membrane Separation Processes*, Elsevier, New York, 1991.
69. J. Dean (ed.), *Lange's Handbook of Chemistry*, McGraw-Hill, New York, 1992, Table 5-2.
70. S.-C. Huang, J. A. Conklin, T. M. Su, I. J. Ball, S. L. Nguyen, and R. B. Kaner, *ACS Polym. Mater. Sci. Eng. 72*:168 (1995).
71. V. M. Schmidt, D. Tegtmeyer, and J. Heitbaum, *J. Electroanal. Chem. 385*:149 (1995).
72. V. M. Schmidt, D. Tegtmeyer, and J. Heitbaum, *Adv. Mater. 4*:428 (1992).
73. V. M. Schmidt and J. Heitbaum, *Synth. Met. 41*:425 (1991).
74. D. L. Feldheim and C. M. Elliot. *J. Membrane Sci. 70*:9 (1992).
75. P. A. Christensen and A. Hamnet, *Electrochim. Acta 36*:1263 (1991).
76. P. Topart and M. Josowicz, *J. Phys. Chem. 96*:862 (1992).
77. C. R. Martin, N. Liang, V. Menan, R. Parthasarathy, and A. Parthasarathy, *Synth. Met. 55–57*:3766 (1993).
78. J. C. Chu, S. L. Wang, S. L. Levy, and R. Paul, *Ju Chin Vapor–Liquid Equilibrium Data*, J. W. Edwards, Ann Arbor, MI, 1956.
79. K. Tanaka, H. Kita, M. Okano, and K. Okamoto, *Polymer 30*:585 (1992).

34

Chemical and Biological Sensors Based on Electrically Conducting Polymers

Anthony Guiseppi-Elie
ABTECH Scientific, Yardley, Pennsylvania, and Johns Hopkins University School of Medicine, Baltimore, Maryland

Gordon G. Wallace
University of Wollongong, Wollongong, Australia

Tomakazu Matsue
Tohoku University, Sendai, Japan

I. OVERVIEW AND BACKGROUND

The earliest reported application of intrinsically conducting, electroactive polymers has been the use of free-standing polymers as sensor devices. These sensors were designed to detect and measure levels of "doping" within the same material upon exposure to vapor-phase dopants. The early work of Shiwakawa et al. [1], Chiang et al. [2], and Street and coworkers [3,4], and the representative theses of Dury [5], Wnek [6], Guiseppi-Elie [7], and others all show the doping of polyacetylene by vapors of iodine, bromine, and AsF_5 occurring within Schlenk tubes that were outfitted with conducting polymer "sensors." These sensors comprised a strip of the undoped polymer film, of known physical dimensions, pasted, usually using carbon or other inert conducting paste, to a four-wire electrode probe that was suspended within the vapor stream of the "dopant." Placed between the reservoir of polymer to be doped and the source of the vapor-phase dopant, the current within the probe was monitored while permitting exposure of the polymer to the dopant vapor. The level of doping of the polymer within the reservoir could be inferred from a measure of this probe's current and the use of empirically prepared conductivity–composition curves and tables. The doping reaction could then be terminated at exactly the prescribed polymer–dopant composition. This method, while never significant beyond the foregoing description, worked successfully to produce much of the early work on the composition dependence of electrical and optical properties of polyacetylene and other vapor-phase doped conducting polymers. This simple sensor also served to inspire the early pursuit of research into sensor applications of conducting polymers [8–15]. There have since evolved many different schemes for using the transducer-active properties of electrically conducting polymers in chemical and biological sensors, for conferring specificity, and for addressing the technological issues pursuant to the commercialization of electroconductive polymer sensor technology.

Continued growth of interest in intrinsically electrically conducting polymers has been extended to their use as transducer-active polymers in various sensing configurations. This has resulted in a rapidly growing body of research, patent, and product literature on conducting electroactive polymers. Published works appear in diverse forums and are being pursued by researchers in the widely differing fields of electrochemistry, analytical chemistry, materials science, biochemistry, and biotechnology. The type of work pursued spans a broad spectrum that embraces fundamental scientific investigations of redox mediation and electrocatalysis, through sensor device configuration and design, to sensor application and commercialization. Recent reviews provide a useful starting point for an appreciation of trends in this area [16–20]. This chapter provides a status report on this emerging area of research and product development. All of the principles associated with good analytical instrumentation development, such as the use of ratiometric or referencing techniques, the use of analytical controls and blanks, and temperature compensation, are relevant to a discussion of chemical and biological sensors based on electroconductive polymers. However, these are outside the scope of this chapter and are not undertaken here. The chapter seeks first to be prescriptive and instructive in the design, fabrication, and uses of electroconductive polymer sensors; second, to provide a review of the primary literature in this area; third, to provide some assessment of the challenges and research opportunities for the future of these sensors; and finally, to indicate where the commercial and military opportunities for chemical and biological sensors formed from electrically conducting polymers may reside.

II. METHODS OF SENSING EMPLOYING ELECTROCONDUCTIVE POLYMERS

Chemical sensors are analytical devices that convert the chemical potential energy of a targeted analyte into a proportionate measurable signal, usually electrical or optical. Biological sensors are a subset of chemical sensors that employ a biologically active molecule in the recognition function associated with transduction. Defining sensor types has always been a challenge [21]. Sensors are often defined on the basis of the molecular recognition mechanism; thus enzyme sensors use enzymes and immunosensors use antibodies. It is, however, more appropriate to define sensors on the basis of the underlying transduction principle. In fact, a hyphenated recognition-transduction nomenclature would be ideal, e.g., immuno-potentiometric, receptor-impedance, DNA-amperometric sensors. The versatility of conducting electroactive polymers (CEPs) resides in the inherent ability of these materials to undergo a wide range of molecular interactions that have the potential to be "engineered" into the materials during synthesis. The extent of each interaction may be controlled after synthesis, in situ, using simple electrical stimuli. This capacity coupled with the fact that such interactions are intimately involved in determining electrical properties of CEPs provides the basis for a powerful sensor technology we call *electroconductive polymer sensor technology*. Chemical and biological sensors using electroconductive polymers are generally formed from films of the electroconductive polymer fabricated on a pattern of metal or semiconductor electrodes. The goals in electrode modification with conducting electroactive polymers are (1) to improve sensitivity, (2) to impart selectivity, and (3) to suppress the effect of interfering reactions. In addition, the polymer membrane may serve as a support matrix for immobilized indicator molecules. These sensors are based on various transduction principles; typical configurations are illustrated in Fig. 34.1.

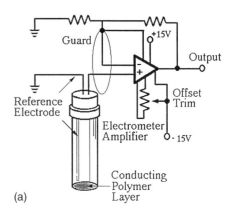

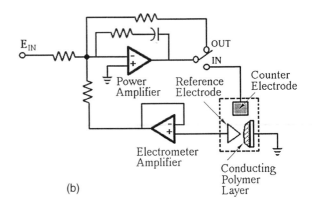

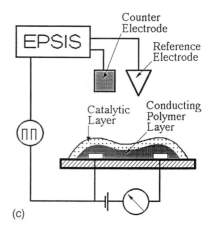

Fig. 34.1 Schematic illustration of chemical and biological sensor configurations involving the use of conducting electroactive polymers. (a) Potentiometric; (b) amperometric; (c) conductimetric; (d) voltammetric; (e) CHEMFET device under constant gate voltage conditions; (f) redox switch. (c, After Refs. 22–24; e, after Ref. 25; f, after Refs. 10, 26.)

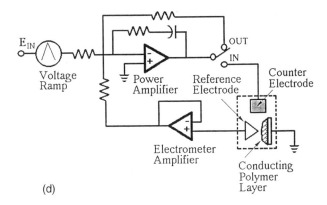

(d)

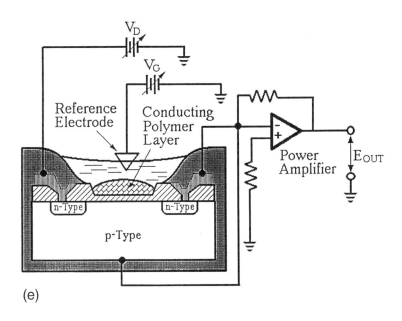

(e)

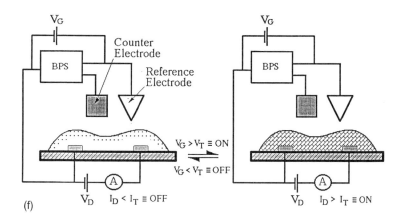

(f)

Potentiometry is the simplest form of sensing that uses electroconductive polymers but may also be the most challenging to control. In potentiometric measurements, the electroconductive polymer serves as a sensing membrane or provides a chemically sensitive surface that is capable of entering into local equilibrium with species of interest in solution. The sensor signal results from changes in the open-circuit potential of a polymer-modified electrode measured versus a suitable reference electrode that is itself in ionic contact with the test solution. The polymer-coated electrode may be gold, platinized platinum, chloridized silver, or other metal. Changes in open-circuit potential result from (1) shifts in the "dopant" anion equilibrium within the polymer film modulated by the concentration of dopant anions within the test solution, (2) ion-exchange processes with other anions within the test solution, (3) redox equilibria at the underlying metal electrode, and (4) further redox reaction of the CEP membrane leading to a change in its redox composition. Potentiometric signals are generally made with near zero current draw through the very high input impedance ($>10^{12}\ \Omega$) of a suitable BIFET operational amplifier that is referenced to the reference electrode. Because electroconductive polymer membranes may span the range from insulator through semiconductor to metal, the potentiometric response to small changes in anion or cation concentration is decidedly influenced by the starting redox composition of the polymer. As perfect insulators, these membranes shield the electrode and produce an open-circuit condition. As pinhole-laden insulators, they may give a potential that corresponds to that of the exposed underlying metal. As semiconductors, the potential results from changes in the Fermi level of the semiconductor relative to that of the reference electrode. As highly conducting metals, they often yield a potential that is measurably invariant with solution anion composition.

Amperometry is the most common approach to sensing using electroconductive polymers. In amperometry the sensor signal is derived from changes in an oxidation/reduction current resulting from an impressed constant voltage maintained on the polymer-modified electrode. Amperometric sensors may have a direct role for the polymer, for example, when the polymer participates in mediated redox chemistry or electrocatalysis. In other cases the polymer may play a passive role, as when the polymer provides a site for covalent, adsorptive, or occluded anchorage for an enzyme or other redox-active indicator molecule. The use of pulsing techniques such as dc pulsed amperometry and the use of ac impedance techniques circumvents some of the limitations of dc amperometry such as ionic polarization. Similarly, the use of current integration to yield a coulometric response allows for the capture of smaller signals but generally does not improve the signal-to-noise ratio.

Conductimetry is a well-known method of sensing to which electroconducitve polymers contribute some unique attributes. In conductimetry the sensor signal is directly traceable to an extent of change ($\Delta G/G$) or rate of change (dG/dt) in the electrical conductance or impedance of the sensor's immediate environment. Electroconductive polymers may serve as transducer-active materials and provide chemical amplification in conductimetric chemical and biological sensors. The transducer generally consists of a thin electroconductive polymer membrane cast on a planar microfabricated interdigitated electrode array of defined cell constant [27]. Conductimetry takes advantage of the large change in electrical conductivity (more than six orders of magnitude in aqueous environment) that accompanies very small changes in redox composition of the electroconductive polymer that is proximal to the device surface [28]. The change in conductivity may result from any phenomenon that changes carrier density or mobility such as gas or vapor sorption that results in polymer swelling that in turn alters interchain transport, carrier or counteranion solvation that affects one-dimensional conductivity, or disruption of conjugation that affects carrier mobility. However, the most dramatic effects are associated with reactions of acceptor or donor species that change both carrier population and mobility. The conductivity change may be followed by using constant current or constant voltage techniques, dc pulsing techniques, ac impedance techniques, or a discontinuous small amplitude voltage pulse technique developed for electroconductive polymers. These sensors offer the advantage of being useful in background environments of high electrolyte concentrations and confer the advantage that the changes ($\Delta G/G$ and dG/dt) may be considerable, leading to chemical signal amplification.

Voltammetry embraces methods in which the sensor signal is derived from a sweep of the electrode potential over a range associated with the redox reaction of the analyte. The response of the sensor derives from a change in the peak current associated with the targeted redox reaction. The conductive polymer may serve as a catalyst, reducing the redox potential at which the analyte of interest is measured and thereby reducing the influence of background and interfering currents, or it may play a passive role such as providing covalent, adsorptive, or occluded anchorage for a redox mediator molecule. Electroconductive polymers, in addition, allow a form of indirect voltammetry of electroinactive but ionic analytes. In these experiments, the analyte ion may induce a redox reaction in the polymer; i.e., the polymer itself may not be efficiently oxidized or reduced under the test conditions except for the presence of the analyte ion.* Thus, the oxidation of polypyrrole in the presence of SO_4^{2-} is less efficient that the oxidation of polypyrrole in the presence of Cl^-.† Voltammetry offers the advantage that reference redox signals associated with standards or controls added to the sample may be simultaneously measured to provide increased accuracy.

CHEMFET devices or chemically sensitive field-effect transistors are potentiometric devices in which the metal layer of a solid-state insulated gate field-effect transistor (IGFET) [29] is replaced with a chemically sensitive electroconductive polymer membrane film. Changes in polymer membrane potential modulate the drain impedance of the space-charge region beneath the insulator. The result is a change in drain current I_D under a fixed drain voltage V_D. The potential of the membrane may be modulated in the standard way (as in potentiometry, above) or may be modulated by an inert electrode placed beneath the film (but isolated from both the source and drain electrodes) and connected to an efficient electrode for the candidate analyte [30].

Redox switches are threshold sensor devices in which the electroconductive polymer is rendered conducting (turned on) under one set of conditions and insulating (turned off) under another. The sharp and rapid change in electrical conductivity through a well-defined resistance transition is essential to the performance of a redox switch. More aptly described as electronic devices, redox switches may be rendered chemically or biologically specific and may be used as simple on/off chemical indicator devices.

III. ROLE OF THE ELECTROCONDUCTIVE POLYMER IN SENSORS

An electroconductive polymer may play a highly varied role as a membrane component of a chemical and/or biological sensor device. It may be active or passive. When active it may serve as a catalytic layer, a redox mediator, an on/off switch, or a resistor with a resistance value that is modulated by a targeted chemical reaction, or it may provide for molecular recognition and/or preconcentration of analyte.

A. Catalysis

Catalysis implies a reduction in the activation barrier for chemical change. For electrochemical reactions, this is manifested in a reduction in the overpotential at which the reaction occurs. An effective polymer catalyst layer, by reducing the overpotential for an electroactive analyte, can reduce interference and greatly increase the signal-to-noise ratio. Conductive polymers are generally poor catalysts for chemical reactions. However, electrodes modified with electroconductive polymers have been reported as catalysts for NADH reduction [31,32]. H_2O_2 oxidation [33], and formic acid oxidation [34]. Conductive polymers are more often used as matrices for the immobilization of truly efficient catalysts such as metalloporphyrins [35–37] for the oxidation of organic compounds, metallophthalocyanins [38–42] for oxygen reduction, metal particles [42–45] for oxygen reduction and hydrogen evolution, and biological catalysts [20,46].

Aizawa and coworkers [47] physically entrapped biological catalysts such as glucose oxidase in polypyrrole by electropolymerization from aqueous solution. Differential pulse voltammetric investigation showed reversible redox peaks that they claim to be indicative of direct electron transfer. In a subsequent study, they immobilized a fructose dehydrogenase monolayer in an electropolymerized pyrrole matrix at a platinum electrode [48]. In both cases, the polymer matrix acted as an interface for direct electron transfer between the Pt electrode and the enzyme molecules. The electrode showed clear catalytic current upon addition of fructose, and the current response was linear with the concentration of fructose from 10 μM to 10 mM. These authors claimed that the activity of the entrapped enzyme depended on the applied potential and could be controlled over a wide potential range. Direct electron transfer between glucose oxidase and polypyrrole was also reported in a track-etched membrane electrode [49,50]. To facilitate direct electron transfer, enzymes should be small so that the conducting polymer backbone can approach the redox site [51]. The enzyme NADH dehydrogenase, a low molecular weight flavine enzyme [52], was a suitable candidate for direct communication. The direct oxidation of diaphorase physically entrapped in a polypyrrole film was evidenced at as low as 0.10 V vs. SCE, where NADH cannot be oxidized directly at polypyrrole [53]. Yamato et al. [54] incorporated the multienzyme system comprising creatininase, creatinase, and sarcosine oxidase into electropolymerized polypyrrole films in their demonstration of a creatinine-sensitive amperometric biosensor. They also suggested a transduction mecha-

---

* For voltammetric analysis of electroinactive ions, see Refs. 29–33.
† The relative ease with which anions partition and diffuse into the CEP membrane therefore becomes the basis for analytical selectivity.

nism of direct electron transfer between the sarcosine oxidase and PPy chains.

The several reports of direct electron transfer from oxidoreductases to polypyrrole and its derivatives have been met with a measure of skepticism. In general, many oxidoreductases, including glucose oxidase, catalyze the oxidation of substrates by ultimate electron transfer to oxygen to form hydrogen peroxide. The H_2O_2 formed may be measured amperometrically [55]. It is therefore difficult to discriminate between the direct electron transfer from the oxidation of hydrogen peroxide at polypyrrole and that at the underlying electrode [56]. Since conductive polymers are infusible and insoluble in aqueous solution, electropolymerization has been frequently used to create a matrix for occluding glucose oxidase at the electrode surface, and in many cases the sensor response was obtained by the oxidation of hydrogen peroxide [15,57–59]. A further complication arises from the observation of Belanger et al. [60], who suggested that the reaction of hydrogen peroxide and polypyrrole decreases the electrical conductivity. Reaction of polypyrrole with H_2O_2 leads to oxidative charge transfer as well as oxidative degradation, which concomitantly increases charge carrier density and reduces carrier mobility. This manifests as a rapid initial rise followed by a slow decay in conductivity [22]. The net result over time is a loss of electrical conductivity relative to that of unreacted polypyrrole films.

An immobilized-enzyme biosensor employing a polymer-modified electrode is illustrated in Fig. 34.2.

B. Redox Mediation

Electroconductive polymer films can potentially provide a three-dimensional reaction layer of surface-confined mediators for high redox efficiency. Charge injection under potential control to polymer species A proximal to the electrode surface ($A + e^- \rightleftharpoons B$) follows the usual Butler–Volmer relationship with defined k_0' and α values. Charge migration through the polymer film to a solution analyte $Y(Y + e^- \rightleftharpoons Z)$ that has partitioned into the film with partition coefficient κ and diffusion coefficient D_p (diffusivity in the polymer) $< D_s$ (diffusivity in solution), react within the reaction layer, there resulting in net $B + Y \rightarrow A + Z$ and $A + e^- \rightarrow B$. Depending on the relative values of the kinetic and transport parameters, the reaction layer may occupy the whole film or be restricted to a defined region along the thickness axis of the film. Two extreme conditions exist. The first is when electron migration is rapid and analyte partitioning and diffusion into the polymer is poor; then reaction occurs at the polymer/solution interface, and the electroconductive polymer is therefore an electrode material. The second is when electron migration is sluggish and analyte partitioning and diffusion are rapid; then reaction occurs at the electrode/polymer interface, and the polymer is only a passive membrane. An example of the former is the conferment of corrosion protection to a reactive metal substrate by electroactive polyaniline [61]. An example of the latter is the amperometric oxidation of hydrogen peroxide at platinized platinum overlaid with a polypyrrole/glucose oxidase film.

In many cases, potentials in excess of the thermodynamically required potentials (overpotentials) are necessary to detect appreciable redox reactions of analytes in electrochemical sensors. The overpotential is mainly derived for slow electron transfer between redox analytes and electrode. The reduction in overpotential brings about improvement in sensitivity and selectivity and is therefore one of the primary goals in the develop-

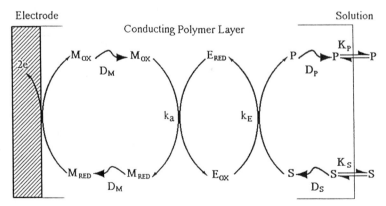

Fig. 34.2 Schematic diagram of an immobilized enzyme biosensor based on conducting electroactive polymer-modified electrode. S and P are the substrate and product species, respectively; E_{OX}/E_{RED} and M_{OX}/M_{RED} are the enzyme and mediator redox couples, respectively; and the constants k_a and k_E are the enzyme–mediator and the enzyme–substrate specific rates. K_i and D_i are the partition and diffusion coefficients of the ith species within the CEP membrane. (After Ref. 20.)

ment of electrochemical sensors; the use of polymer-modified electrodes is one way to approach this challenge. One study [62] using electrochemical and color impedance spectroscopy suggests that the rates of redox processes at polypyrrole surfaces are not particularly fast compared to those at bare metal electrodes. Overpotential can be reduced by using mediators; small molecules that shuttle between electrode and analyte to accelerate electron flows. The mediator accelerates thermodynamically downhill electron transfer. Thus, successful mediators should meet the thermodynamic requirement, which, for oxidation, is that the formal potential of the mediator should be close to or positive to that of the analyte.

The use of redox mediators is particularly effective in the development of biosensors based on redox reactions catalyzed by proteins. The redox-active sites in redox proteins are generally shielded by insulating peptides and sugars. The electron transfer between enzymes and conducting polymers (or electrodes) can be accelerated by using small electron transfer mediators or shuttles, such as quinones and ferrocenes, which enter the clefts of proteins and shuttle between the redox sites of enzymes and electrode surfaces [63]. The use of mediators lowers the overpotential and therefore eliminates the influence of interfering substances in the analyte solutions. This is practically attractive for enzyme sensors that use oxidases, as the mediation reaction can reduce the undesired influence of varying oxygen tension in solution [64–66]. Iwakura et al. [67] simultaneously immobilized glucose oxidase and ferrocene carboxylate, one of the most effective mediators in solution, in polypyrrole films. This electrode showed large and rapid response upon addition of glucose in solution compared with the case where ferrocene carboxylate or oxygen was present in solution. Bartlett and coworkers [68] investigated the redox mediation reaction at glucose oxidase–immobilized poly (N-methylpyrrole) and found that the polymer films were no longer capable of oxidizing mediators. These phenomena indicate the destruction of the conductive polymer due to oxidation by hydrogen peroxide formed by the electron transfer from the mediator to oxygen. Enhancement of the sensor response was also observed for polypyrrole films entrapping dehydrogenase and mediators [63,69,70]. However, the sensor responses decreased slowly with time due to diffusion of the mediator out of the film and into solution [53]. The decrease in the sensor response due to diffusion of the mediator out of the film can be avoided by covalent attachment of the mediator to the conducting polymer backbone [71,72].

C. Chemically and Biologically Sensitive Redox Switches

Due to the enormous conductivity change induced upon changes in redox state, interdigitated array (IDA) and microband electrodes coated with conducting polymers may act as chemically sensitive redox switches responsive to redox species in solution [73,74]. The work of Wrighton et al. [10–14, 75] demonstrated sensors responsive to oxidizing and reducing species in solution. These sensors were based on changes in the electrical conductivity that accompanies oxidation and reduction of polymers such as polypyrrole, poly(3-methylthiophene), and polyaniline. Films were grown by electrochemical polymerization to bridge 1.4 μm gaps between photolithographically defined gold microband electrodes on silicon nitride–coated silicon substrates. Using the analogy of a transistor, the conductivity of the polymer film was then measured by applying a small dc voltage, V_D, between the two microband electrodes and measuring the ensuing drain current, i_D, in the film. A device of this type, based on a poly (3-methylthiophene) film, was shown to detect $[IrCl_6]^{2-}$ and $[Fe(CN)_6]^{4-}$ ions in a flow injection system by measuring the changes in the drain current. In the presence of $[IrCl_6]^{2-}$, the poly(3-methylthiophene) film is readily oxidized (turned on), and this leads, to an increase in conductivity. This in turn is detected as an increase in the drain current under conditions of constant drain voltage. Conversely, on addition of $[Fe(CN)_6]^{4-}$ the polymer is reduced (turned off), the conductivity is decreased, and the drain current falls. In subsequent work, Thackeray et al. [76] extended this approach to the detection of dissolved hydrogen and oxygen by incorporating colloidal platinum particles into electropolymerized films of poly (3-methylthiophene). The metal particles provided catalytic sites for the efficient oxidation of the polymer by molecular oxygen and for its reduction by molecular hydrogen. This therefore led to changes in the conductivity and hence measurable changes in the drain current brought about by gases. Wrighton and coworkers [77] also demonstrated that the IDA electrode coated with polyaniline was turned "on" (conducting state) when $Fe(CN)_6^{3-}$ was added to the solution. Immersion of the device into a $Ru(NH_3)_6^{2+}$ solution turned it "off" (insulating state). Similar responses of the array devices to redox chemicals in solution were found by Nishizawa et al. [74].

It is noteworthy that the redox equilibrium composition and hence the conductivity of these polymers may be controlled by externally impressed potentials. Accordingly, under constant drain voltage conditions, the redox composition and hence the conductivity of the polymer are not entirely under the influence of chemically induced change; a requirement for analytical accuracy. For this reason, devices of this type are limited to on/off configurations and are at best analytically qualitative. The use of non-perturbing dc pulsing techniques, ac techniques, and the discontinuous dc pulsing technique of the electroconductive polymer sensor interrogation system (EPSIS) circumvents this limitation and establishes the basis for an analytically quantitative device based on the kinetics of conductivity change. These chemically sensitive redox switches are reusable with switching speeds on the order of seconds following the addition of a redox species. However, the specificity is poor because many redox species could, in principle, bring about a switching response. Furthermore, the longevity of such a device in terms of the number of repeated redox cycles remains untested.

The threshold potential between insulating and conducting states of electroconductive polymers can be altered by copolymerization [78,79] or chemical modification [80]. Thus an array device coated with polypyrrole can be turned "on" by anthraquinone-2-sulfonate (AQ) as well as by $Fe(CN)_6^{3-}$; on the other hand, the device coated with the 1:1 copolymer of polypyrrole and poly (N-methylpyrrole) responds only to $Fe(CN)_6^{3-}$. Since the threshold potential between insulating and conducting states of polypyrrole is ≈ -0.5 V vs. SCE, both AQ and $Fe(CN)_6^{3-}$ are capable of oxidizing the polypyrrole from the insulating to the conductive state. The 1:1 copolymer (threshold potential, ≈ -0.3 V vs. SCE) can be oxidized only by $Fe(CN)_6^{3-}$ to a sufficiently conductive state. Such engineered variations in the switching potentials of electroconductive polymers may be used as the basis for the development of multielement array sensors, which, when integrated with pattern recognition algorithms, can potentially provide a high degree of specificity [81,82].

When an enzyme-catalyzed reaction produces the threshold conductivity change of a polymer at an IDA or microband electrode, the device acts as an enzyme-based switch. Matsue et al. [53,69] reported an enzyme switch responsive to nicotinamide coenzyme (NADH). Their device was based on a thin film of a copolymer of pyrrole and N-methylpyrrole deposited on an IDA electrode. The film contained physically entrapped diaphorase and anthraquinone-2-sulfonate as an electron mediator. When NADH is present in solution, the diaphorase-catalyzed oxidation of NADH produces the reduced form of the mediator, which subsequently reduces the polymer film. These series of reactions decrease the conductivity of the polymer. Thus, the addition of NADH turned the array device from "on" to "off". The device could be reset to the "on" state by electrochemical reoxidation of the polymer. Bartlett and Birkin [83,84] similarly fabricated an enzyme switch responsive to glucose. The device comprises two microband electrodes coated with anodically polymerized polyaniline film and possessing occluded and immobilized glucose oxidase. Upon addition of glucose and in the presence of tetrathiafulvalenium (TTF) ion, the polyaniline film changed from its insulating state to its conducting state, bringing about a switching response. The devices showed rapid responses for moderate glucose concentrations and were reproducible, being renewed at least 10 times by electrochemical reoxidation.

Immobilization of enzymes that induce local pH changes in the polymer is yet another approach to the fabrication of electroconductive polymer–based enzyme switches and/or sensors. The IDA electrode coated with polypyrrole and penicillinase membranes has served as a penicillin-dependent switch [85]. Penicillinase catalyzes the hydrolysis of penicillin to peniciloic acid, which protonates the polypyrrole membrane to cause an increase in its conductivity. The concentration of penicillin in solution was evaluated from the ohmic current in the PPy film. Flow injection measurements with this device as the detector showed that the response increased linearly with the concentration of penicillin solutions up to 7 mM. A glucose sensor based on a similar principle was reported by Hoa et al. [86]. They immobilized glucose oxidase in a polyaniline film, and the conductivity change of the film upon exposure to glucose solution was investigated by ac impedance measurements. Since many enzymes bring about pH changes through their catalytic reactions, the principle proposed here will be widely applicable to the fabrication of enzyme-based microelectrochemical devices that detect biologically important substances.

D. Chemically and Biologically Modulated Resistor

The early widespread use of conducting polymer "resistance sensors" to monitor levels of "doping" by vapor-phase dopants in Schlenk tubes was closely followed by the description of an aqueous ion-gate membrane. Murray and coworkers [87,88] demonstrated modulation of ion transport properties upon oxidation or reduction of suspended polypyrrole films. Guiseppi-Elie and Wnek [89] were able to distinguish, on the basis of the time rate of decay of the four-probe electrical conductivity of iodine-doped polyacetylene, the nucleophilicity of certain aqueous nucleophiles. However, the first description of a conductimetric sensor device based on chemically modulated resistance with some form of conferred specificity appears ascribable to Malmros [8]. Malmros' 1984 patent describes the use of free-standing polyacetylene films with adsorbed biomolecules to form biosensors. The subsequent and independent work of Wrighton and coworkers [10–14], part of a larger study of microelectrochemical transistors and diodes, demonstrated redox switches that may serve as on/off sensors responsive to oxidizing and reducing species in solution. The electrical conductivity of electroconductive polymer films is greatly influenced by the activity and reactivity of chemical species that partition into the polymer from the surrounding media. Thus, the conductance provides a convenient way to sense chemical species [90] if such influence can be made specific.

The conductimetric device based on chemically modulated resistance (chemoresistance) seeks to relate the chemical potential of an analyte to the rate of change (kinetic response) or extent of change (steady-state or equilibrium response) of conductance. For example, the exposure of polypyrrole films to electron-donating gases such as NH_3 and H_2S compensates free carriers, reduces the carrier density in the polymer, and decreases its conductivity [91]. Conversely, electron-accepting gases such as PCl_3, SO_2, and NO_2 oxidize the reduced form of polypyrrole, generate free carriers, and increase its conductivity [92,93].

Gas sensors based on conducting electroactive polymers have been studied by a number of researchers [91–100], have achieved a high state of engineering development [101,102], and are currently in the early stages of commercialization.* Similar gas-sensing capabilities have also been found in polythiophene [103], polyaniline [104,105], and other conducting films [106,107]. In general, the chemoresistive responses of conducting electroactive polymers to gases and vapors such as alcohols and aromatics that impact carrier mobility without evidence of direct charge transfer compensation or doping tend to be more rapid but display considerably less sensitivity than donor/acceptor gases and vapors. The conductivity change induced by gases has usually been investigated under dry conditions. Since many applications of the polymers to chemical and biological sensors require immersion in electrolyte solution, it is important to clarify the conductivity characteristics under wet conditions.

Immersion measurements under potential control provide a useful format for exploiting the transducer-active characteristics of electroconductive polymers. Under these conditions, the electrolyte medium may be used to regulate the starting composition of the polymer membrane by using an externally impressed potential and an auxiliary current-carrying electrode.† Similar facility is afforded gas-phase sensors when a solid electrolyte is used. Wrighton and coworkers [75,77] used interdigitated microarray (IDA) electrodes consisting of two comb-type microelectrodes for the characterization of conducting polymers under immersion conditions and found that a slight shift in applied potential changes the conductivity of the polymers by several orders of magnitude. These results suggest that the conductivity measurement enables a highly sensitive sensing of chemical species that affect the redox state of the polymer. Likewise, they indicate that the response of the polymer film will itself be sensitive to the interrogation voltage used to measure its conductivity. The quantity of the chemical species in solution can be evaluated from the ohmic current in the film (I_D) under a constant bias potential (V_D). On the basis of this phenomenon, they fabricated various array devices that show gated FET-like behavior [108].

The IDA electrode coated with an electroconductive polymer also serves as a conductimetric pH-sensitive sensor device [75,109]. Measurements of free-standing dry polypyrrole films [110,111] showed that the conductivity of the polymer reversibly changes upon treatment with aqueous bases and acids. The reversible protonation–deprotonation of the pyrrolylium nitrogen atom has been reported as the origin of the pH dependencies [112,113]. The detailed characteristics of the pH dependence of polypyrrole and polyaniline was studied by in situ conductivity measurements using IDA electrodes (A. Guiseppi-Elie, unpublished results). The conductivity of PPy changes by two orders of magnitude upon changing the pH from 3 to 11, with most of the change occurring between pH 10 and 11. The plot of conductivity versus pH showed a sigmoidal shape with an apparent inflection point that was dependent on oxidation level of the polymer and shifted to lower pH values with increasing oxidation level of polypyrrole. This inflection point is also coincident with the $E^{0\prime}$ measured by slow scan rate cyclic voltammetry (Guiseppi-Elie, unpublished). The conductivity of polyaniline similarly changes by several orders of magnitude over the pH range 3–11, with most of the change occurring between pH 3 and 4 [75]. However, polyaniline shows two characteristic inflections. The first corresponds to the $E^{0\prime}$ of the first redox transition at 0.1 V vs. SCE and is associated with an increase in electrical conductivity. The second corresponds to the $E^{0\prime}$ of the second redox transition at 0.7 V vs. SCE and is associated with a decrease in electrical conductivity.

Guiseppi-Elie, in a 1994 U.S. patent [114] described conductimetric sensors using chemically modified polyaniline, polypyrrole, and polythiophene membranes that possess covalently immobilized indicator agents such as biotin, antibodies, amino acid sequences, and DNA probes. Conductimetric responses are derived from the chemical oxidation of the polymer by hydrogen peroxide produced by oxidoreductase enzyme labels upon exposure to substrate [22]. These biosensors are kinetic devices that are interrogated by first electrolyzing the membrane, using a three-electrode arrangement, to a prescribed potential corresponding to a nonconducting (or highly conducting) state. This is followed by the subsequent and instantaneous application of a nonperturbing, small-amplitude, discontinuous dc voltage pulse between the digits of the IDA. The initial rate of change of conductivity following electrolysis produces the analytically significant response result. Similar conductimetric biosensors have since been described by others. Contractor et al. [115] described glucose, urea, lipid, and hemoglobin biosensors based on polyaniline films with occluded enzymes. The response of these sensors is derived form pH changes associated with the production of enzyme transmutation products.

E. Molecular Recognition and Preconcentration

An electrosensing material is functionally enhanced if it has the ability to preconcentrate the analyte using selective molecular interactions. Traditionally, with electrochemical sensing this ability to recognize and collect the

---

* See Sec. VI for a discussion of commercial and military applications of gas sensors.

† This principle is embodied in the Electroactive Polymer Sensor Interrogation System (EPSIS) of ABTECH and is covered by U.S. patent 5,312,762 [23].

target analyte has been somewhat limited. Since these processes must occur at the electrode surface, the limitation can be traced to the availability of only a small number of different electrode substrates. The use of solid metal electrodes (e.g., platinum and gold) and glassy carbon or mercury has been predominamt in the field of electroanalysis. Consequently the ability to collect analyte on the sensor has been limited to species that would preferentially adsorb to these surfaces. Small organic or drug molecules such as diazepam or codeine and, large macromolecules such as proteins can be accumulated via adsorption and determined at low levels. In some cases (e.g., cytochrome c), the fact that the protein is adsorbed actually facilitates the electron transfer process and so aids electroanalysis. Besides those compounds that naturally deposit on the above electrode surfaces, accumulation of others can be induced by either chemical derivatization or in situ electroformation of less soluble compounds. The former approach has given rise to the field of electroanalysis now denoted adsorptive stripping voltammetry (AdSV). For instance, trace metals can be accumulated by derivatizing electrode surfaces with organic complexing agents, as exemplified in Eq. (1).

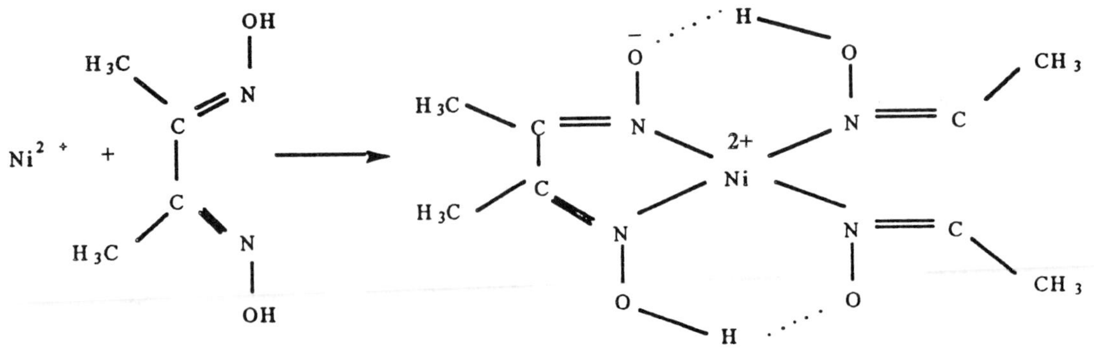

$$Ni^{2+} + \text{dimethylglyoxime} \longrightarrow \text{Ni(dmg)}_2 \text{ complex} \tag{1}$$

This approach is denoted either anodic stripping voltammetry (ASV) if the species is accumulated via a reduction process, e.g.,

$$Pb^{2+} + 2e \underset{}{\overset{Hg}{\rightleftharpoons}} Pb(Hg) \tag{2}$$

or cathodic stripping voltammetry (CSV) if the analyte is accumulated via oxidation,

$$S^{2-} \underset{}{\overset{Hg}{\rightleftharpoons}} Hg(S) + 2e \tag{3}$$

More recently the advent of chemically modified electrodes, particularly polymer-modified electrodes, has greatly increased the scope of recognition and preconcentration modes available. For example, nafion [116] and other ion exchangers such as polyvinylpyridine [117] have been attached to electrode substrates and used to preconcentrate target analytes via anion and cation exchange, respectively.

The use of polymeric modifiers has in general proven extremely popular, and by far the most versatile class of materials in this regard are the conductive electroactive polymers (CEPs). These materials, such as polypyrroles, polythiophenes, and polyaniline (I–III), can be prepared using simple electrochemical or chemical oxidative methods. By incorporating appropriate counterions (A^-) or attaching appropriate functional groups to the primary repeat unit, a range of molecular interactions (H bonding, ion exchange, etc.) can be induced, and these may form the basis of a powerful analyte recognition and preconcentration process [118,119]. Conducting electroactive polymers are capable of recognition/preconcentration using a variety of mechanisms.

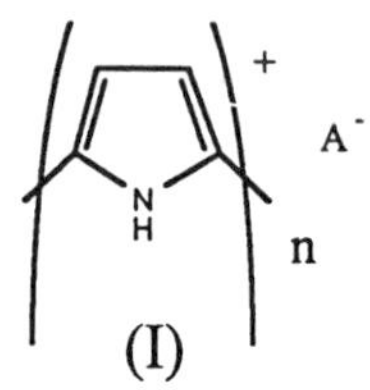

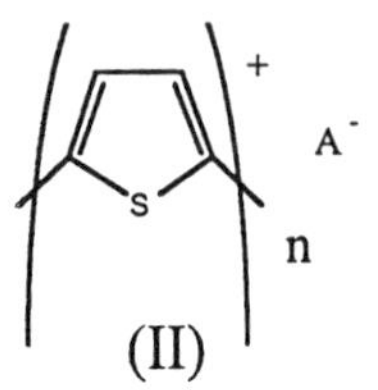

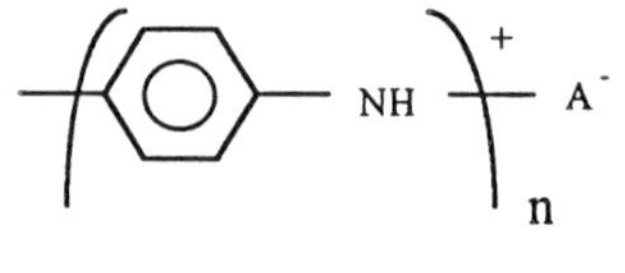

1. Adsorption/Electrodeposition

More conventional approaches such as those discussed above are available with CEPs. Adsorptive accumulation has not been exploited in any detail, presumably due to the fact that it is difficult to get a totally reproducible polymer surface in terms of chemical and physical properties. However, the ability to manipulate the interaction of small molecules with CEP surfaces in flow-through detection cells (subsequent to chromatography) has been used to advantage in the development of a chloramine detection system [120]. The selectivity toward the very similar mono-, di-, and trichloramines was manipulated using different polypyrroles produced by simply incorporating different counteranions (A^-).

Since CEPs are inherently conducting, conventional electrodeposition processes can also be used on the bare polymer [121]. This has not been exploited analytically because at the potentials required to reduce most metals the polymer is less conductive [122]. There is also the fact that oxidation/reduction of the polymer backbone occurs according to reaction (4).

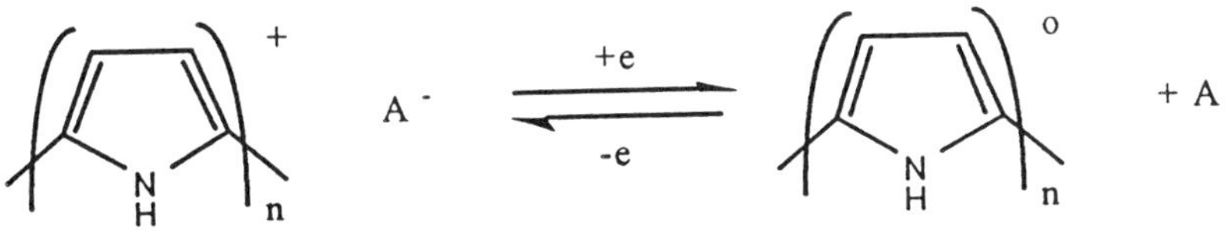

$$\text{[PPy]}^+ \; A^- \; \underset{-e}{\overset{+e}{\rightleftharpoons}} \; \text{[PPy]}^0 \; + \; A^- \tag{4}$$

This results in large background currents. The first problem can be minimized by depositing or incorporating a suitable metal (mercury) on or into the conducting polymer structure. The incorporated metal then provides the conductive pathway when the polymer is reduced to the nonconductive form. This can be achieved by plating the mercury after polymerization [123] or by using a novel procedure that involves formation of mercury "channels" during electropolymerization [124]. Interestingly, the chemical selectivity of the conducting electroactive polymer is retained with such electrode "enhancements" [125].

2. Ion Exchange

As the structure of CEPs suggests, they can function as effective anion exchangers, and this can be used to advantage for the purposes of recognition and preconcentration. As with all ion exchangers, a selectivity series will exist. The series obtained for CEPs is readily modified by changing the anion (A^-) incorporated during polymerization [126–128]. Thus for polypyrrole, both Hailin and Wallace [126] and Dong and coworkers [127,128] found similar ion-exchange series. For example,

Electrode/dopant	Selectivity series
PPy/Cl$^-$	Br$^-$ > SCN$^-$ > SO$_4^{2-}$ > I$^-$ > CrO$_4^{2-}$
PPy/ClO$_4^-$	SCN$^-$ > Br$^-$ > I$^-$ > SO$_4^{2-}$ > CrO$_4^{2-}$

As with all ion-exchange processes, these series vary with pH and ionic strength. The anion-exchange behavior is also observed by using organic acids [129] and amino acids [130]. In those studies the presence of multifunctional interactions, particularly hydrophobic and anion-exchange reactions, was noted. The amino acid studies were interesting in that the strength of the interactions observed for aspartic and glutamic acids (**IV** and **V**) were markedly different, indicating a configurationally sensitive ionic interaction.

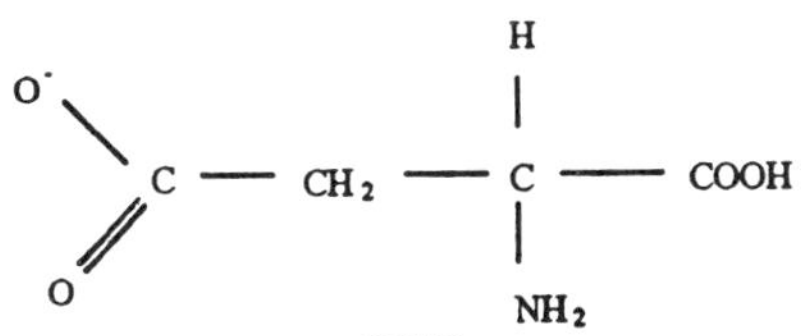

$$\underset{\textbf{(IV)}}{\text{O}=\text{C(O}^-\text{)}-\text{CH}_2-\text{CH(NH}_2\text{)}-\text{COOH}} \qquad \underset{\textbf{(V)}}{\text{O}=\text{C(O}^-\text{)}-\text{CH}_2\text{CH}_2-\text{CH(NH}_2\text{)}-\text{COOH}}$$

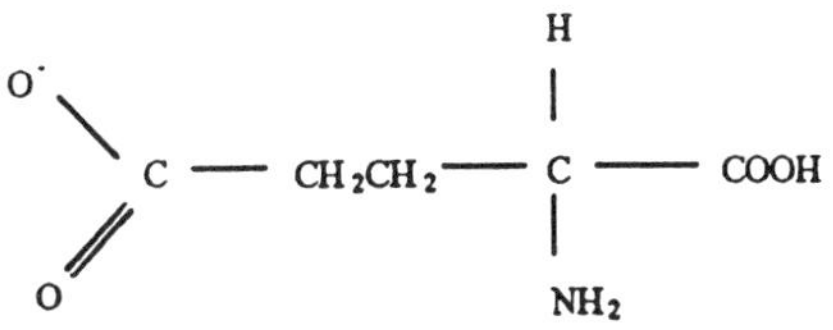

If larger anions such as dodecylsulfate, dodecylbenzenesulfonate, or other anionic polyelectrolytes are incorporated into the CEP during synthesis, then anion-exchange processes will be minimal and cation exchange may become favored [131]. A further, unique aspect of CEPs is that this ion-exchange process can be controlled electrochemically [131]. CEPs such as PPy can be oxidized and/or reduced according to Eq. (4). Polythiophene undergoes similar oxidation/reduction and polyaniline undergoes two redox processes according to Scheme 1. (Note: The exact nature of these processes for polyaniline is pH-dependent.)

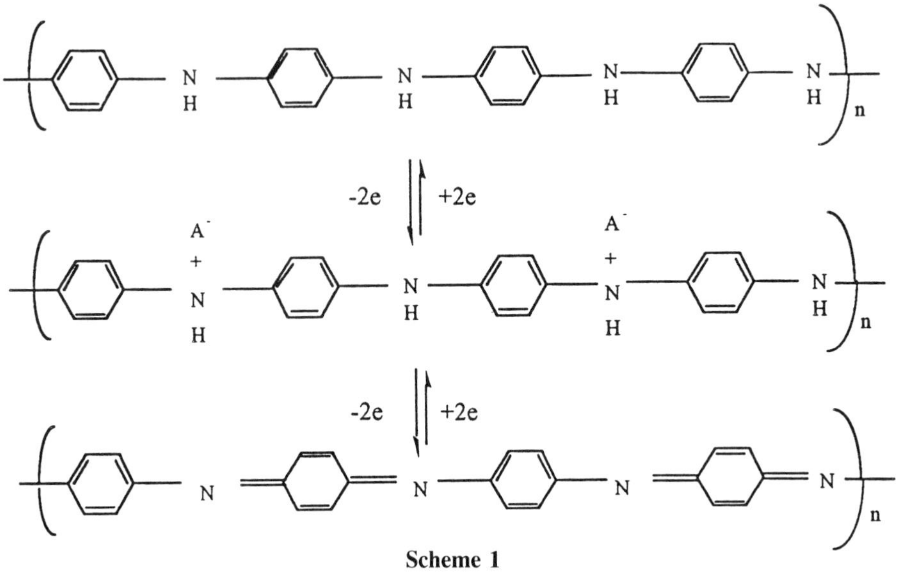

[Scheme 1 structures shown]

Scheme 1

All of these redox reactions are accompanied by ion movement into or out of the CEP membrane. These processes were studied using the electrochemical quartz crystal microbalance (EQCM) technique [132]. EQCM allows small changes in mass of the polymers to be monitored in situ as electrical stimuli are applied. Information regarding the ion-exchange processes occurring can then be deduced [133,134]. These studies revealed that even for relatively simple systems, anion, cation, and solvent processes all occur as the polymer is oxidized/reduced [see Eq. (5)], where

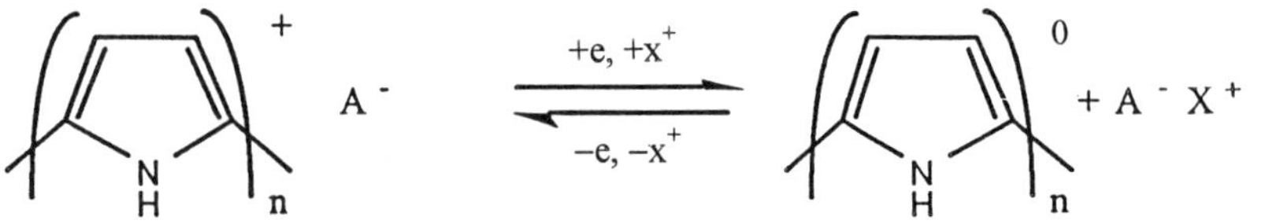

$$\text{[pyrrole]}^+ A^- \underset{-e, -x^+}{\overset{+e, +x^+}{\rightleftarrows}} \text{[pyrrole]}^0 + A^- X^+ \quad (5)$$

[X^+ = cation from the electrolyte solution]. If the anion (A^-) is small and mobile, such as chloride, then anion exchange will predominate; however, if it is large and immobile, such as polystyrenesulfonate, then cation exchange will predominate. The extent to which each process occurs for a given polymer is dependent on both the magnitude and frequency of the electrical stimuli applied [135]. These electrochemically controlled processes have been used to advantage in the development of ion-sensing technologies based on conventional preconcentration/stripping methods [136] or flow injection analysis methods [230–234].

As expected, the selectivity of these detection methods varies with the composition of the eluant [138] and the anion incorporated into the polymer during synthesis [230]. Incorporation of larger molecules, e.g., dodecylsulfate (DS), during synthesis results in a cation detection system [30]. The approach has also been used for the detection of small organic molecules [140] with polydimethylaniline. In addilion to the oxidation/reduction processes discussed above, polypyrrole, polythiophene, and polyaniline can all be "overoxidized" if the potential is too positive. "Overoxidation" is an irreversible oxidative process that results in the deterioration of the mechanical and electrical properties of the polymer. However, interestingly, ion-exchange properties are retained even with overoxidized material [141].

3. Complexation

Conducting electroactive polymers contain functional groups (—NH—, —S—) that make the direct complexation of metals possible. For example, copper ions are readily complexed by polypyrrole to yield the following structure [142]:

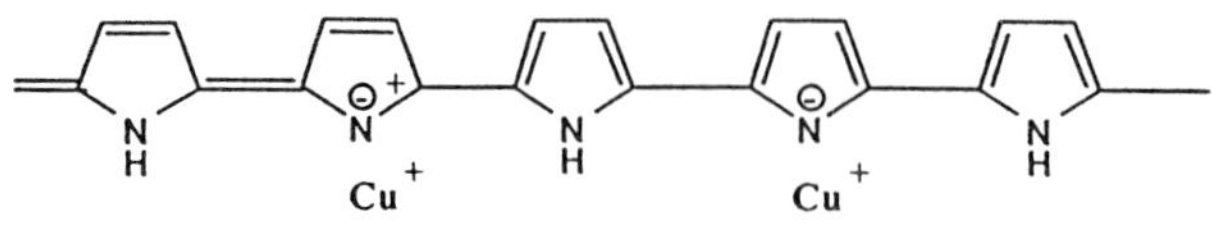

[Cu-polypyrrole complex structure shown]

Although this complexation has been shown to have a stabilizing effect on PPy exposed to aqueous solution, it has not yet proven analytically useful. Metal complexing functionality may also be purposely added to the polymer backbone either prior to or after polymerization. For example, polyether groups capable of selective complexation of alkali earth metals, Li^+ and Na^+, have been attached to the heterocyclic ring prior to polymerization [143,144]. The nitrogen group on polypyrrole has been derivatized after polymerization to form pyrrole-N-carbodithioate (VI) [145] on the surface. These functional

(VI)

groups could then be used to complex metal ions, in particular Hg^{2+} and Cu^{2+}. The problem with derivatizing the polymer backbone is that inevitably the conductivity drops. An alternative approach is to incorporate complexing groups as the counterions (A^-) during polymerization [146–148]. Complexing groups such as PDCA or EDTA can be incorporated using this technique, but this often results in less conductive polymers [147] because there is no high charge density ion present to counterbalance the positive charge on the polymer backbone. The use of sulfonated complexing agents [148,149], as shown in Table 34.1, overcomes this problem. These sulfonated dyes are particularly useful for complexing metal ions such as Al^{3+}, Mn^{2+}, and Ca^{2+}. A truly innovative approach is found in the work of Swager and coworkers wherein host–guest complexation of pseudopolyrotaxanes [150] and crown ethers [151] that are based on polythiophene are linked to conformational changes in the conjugated polymer backbone. Reversible binding of the pseudopolyrotaxanes host with its π-electron-deficient 4,4'-biphenyl guest leads to reversible modulation of the conductivity and a corresponding shift toward higher redox potential ($E^{0'}$). Similar reversible binding of the polythiophene-based crown ether polymers with K^+, Na^+, and Li^+ produced reversible ionochromic shifts that scaled in the order of the binding constants and selectivities.

4. Biorecognition

Conducting electroactive polymers can also support sites for biorecognition via binding and other transmutation reactions. Enzymes provide a high degree of specificity to their substrates. Antibodies (Ab), DNA, and RNA are known to undergo particularly selective binding interactions with their corresponding antigen (Ag), DNA, and RNA complement. Amino acid sequences such as Arg-Gly-Asp (RGD) can provide for binding of whole cells via cell wall adhesion receptor proteins. Techniques that enable direct incorporation of active proteins into CEPs during polymerization have been developed [152]. This may be achieved by using the protein with the monomer in the polymerization bath [227–229]. However, the degree of incorporation is increased by using an electrocatalyst such as Tiron to lower the polymerization potential for the monomer or by using molecular carriers such as dodecylsulfate (DS) or colloidal gold [153]. The latter approach involves attaching the protein to a carrier that increases the overall charge density. Once incorporated, the protein-containing CEP membranes may be used as a conventional voltammetric sensor after exposure to the antigen [154]. Signals can be generated using ac voltammetric methods. However, the technique is not very sensitive nor is the Ab–Ag interaction easily reversed, making this a "single-use" sensor. Much better performance is achieved using the flow injection analysis (FIA) method with pulsed electrical detection. Wallace and coworkers developed methods for determination of p-cresol [155], thaumitin [156], and human serum albumin (HSA) [157] using this approach.

IV. CHEMICAL AND BIOLOGICAL SENSOR SPECIFICITY

A major challenge in the development and use of chemical and biological sensors based on electroconductive polymers has been the conferment of chemical and biological specificity. Specificity implies the response of the sensor to a specific analyte through the actions of an indicator molecule, which may be an inorganic or organic catalyst, enzyme, or a binding molecule such as a macrocycle, antibody, DNA/RNA fragment, peptide sequence, or stabilized receptor. While the polymer may serve as a suitable transducer, its response must be linked to a particular analyte and should display preference for that analyte, i.e., selectivity. The sensor should also display low response to other species, i.e., low interference, and must have an appropriate sensitivity and limit of detection to be useful for a particular measurement task. Several methods have been explored and developed to confer specificity to transducers that are based on electroconductive polymers. Methods of polymer preparation that confer chemical and biological specificity generally fall into one or other of the categories discussed in the following subsections.

A. Physical Adsorption

Adsorption is a useful and simple method for conferring specificity to an electroconductive polymer membrane [8,158–160]. Passive physical adsorption of proteins from a suitable buffered solution onto charged polymer

Table 34.1 Counterions Used for Determination of Al(III)

Counterion	Structure
Eriochrome cyanine RC (EC)	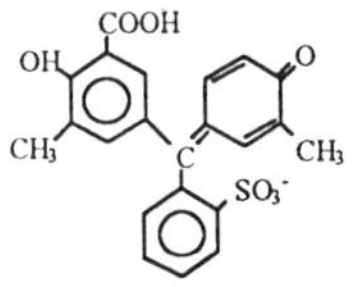
Chrome Azurol-S (CA)	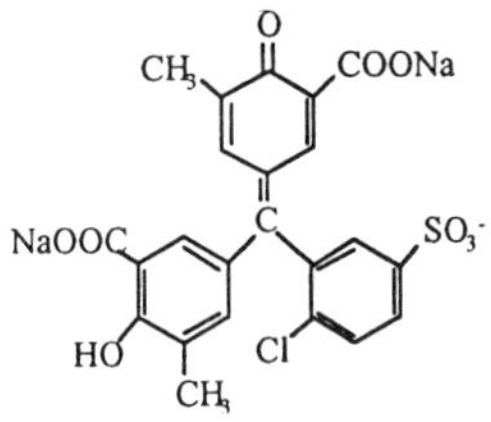
Tiron (Tir)	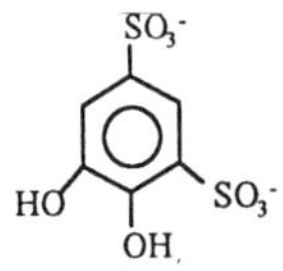
Naphthyl-azoxine-S (NAS)	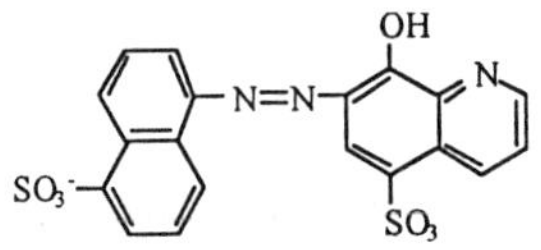
Naphthyl-azoxine (NA)	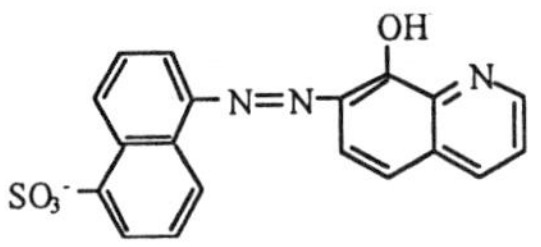
8-Hydroxyquinoline (HQ)	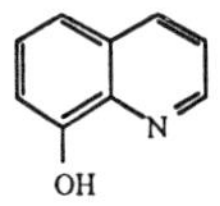
Dodecyl sulfate (DS)	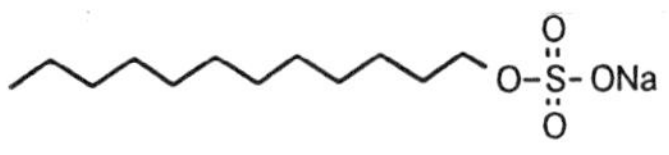

surfaces leads to conferment of the activity of the enzyme to the membrane. While generally poorly regulated and unstable because of desorption and denaturation, it is a convenient means to quickly evaluate the response of the transducer to the specificity of, say, an enzyme or binding of an antibody. This process may be enhanced with the use of cross-linking agents such as glutaraldehyde, benzoquinone, or other cross-linking chemicals that give rise to a protein gel layer. Cross-linking increases the resulting sensor's long-term stability from days to maybe weeks, increases the net activity on the surface as more protein can be retained on the device, but reduces the enzyme's specific activity as enzyme-active sites are often compromised. Additional stability, faster rise times, and more reproducible responses may be possible when the adsorbed enzyme is subsequently subjected to a high electric field [161].

B. Physical Occlusion During Electropolymerization

Electropolymerization under oxidizing conditions that produces a positively charged polymer provides a unique and convenient method to occlude and immobilize anions with indicator properties into the electroconductive polymer membrane of a sensor device [55,57,71,162]. This method has the following important advantages:

1. Control of the location of the polymer film leading to micropatterning of the polymer on the device
2. Direct association of the polymer with metallic or semiconductor electrodes for device interrogation and signal capture
3. Precise control of polymer film thickness by con-

trol of the electropolymerization charge density [163].
4. Control of polymer film morphology by control of the electropolymerization current density
5. Good electrical contact between the polymer and metal or semiconductor
6. Convenient device functionalization by electrophoretic blending or coelectropolymerization with other molecules that may confer chemical and biological specificity to the device.

In addition, because of their considerable in-plane growth compared to their vertical growth [164,165], CEP membranes may be grown by electropolymerization to bridge insulating spaces as wide as 15 μm between conductors. The negative charges required to balance the positive charge of the polymer backbone and maintain charge neutrality may be derived from molecules of biological origin such as polypeptides [166], enzymes [15,53,57,65,67,71,154,167], cofactors (168), antibodies [156,157,169], or oligonucleotides [170] or may be inorganic catalysts such as heteropolyanions [171–173], semiconductor particles [174], and even metallic inclusions [11]. Electrophoretic occlusion of proteins requires that the protein's isoelectric pH result in a net negative charge on the protein under the pH condition for anodic electropolymerization. In this way, both electrophoretic transport to the electrode and the requirement of electroneutrality in the positively charged polymer favor occlusion. Electrochemical doping is also a useful way to immobilize negatively charged catalysts such as metalloporphyrins [35–37], metallophthalocyanines [38–40], and metal chelates [175], at electrode surfaces.

Of the many approaches studied, immobilization of enzymes has been most widely investigated for sensor applications. An enzyme sensor comprises an enzyme membrane that recognizes the substrate to be analyzed and a suitable transducer (in this case, an electroconductive polymer transducer), which, through a physicochemical reaction, converts the chemical signal to an electric signal. A key consideration in the fabrication of reliable enzyme sensors is the method by which the transducer detects the enzyme-catalyzed reactions. Any amperometric, voltammetric, conductimetric, mediated, or direct means may be employed. A highly desirable means to detect oxidoreductase enzyme reactions is to achieve direct redox communication between the entrapped enzyme and the surrounding polymer. Direct biosensors [235, 236] eliminate dependence on the regeneration of consumable reagents such as oxygen or mobile small molecule shuttles.

The electropolymerized polymers formed with large protein molecules or other bulky catalysts often display poorer electrical conductivity than the electroconductive polymers with typical counteranions. Consequently, these polymer membranes do not serve well as the primary sensing layer but more aptly serve as a secondary specificity-conferring layer. The hydrogen peroxide–sensitive transducer of Guiseppi-Elie et al. [22] illustrates this point. The sensor uses an electropolymerized $Mo_7O_{24}^{6-}$ (Mo^{VI})-containing secondary layer of PPy over a primary layer of highly conducting PPy formed with the macroanions poly(styrenesulfonic acid) (PSSA) and dodecylbenzenesulfonate. Exposure to peroxide in the presence of iodide generates molecular iodine, which oxidizes the highly conductive PPy layer. The highly conductive layer is regenerated (reduced) electrochemically. A tertiary layer containing covalently immobilized glucose oxidase makes a glucose biosensor, which, in the presence of glucose, generates peroxide and brings about a conductimetric response.

C. Covalent Methods

Physical occlusion or entrapment of enzymes and other indicator molecules during electropolymerization presents the potential problem of subsequent egress by diffusion from the polymer membrane. Redox cycling, because it is associated with appreciable expansion and collapse of the ionogenic polymer, facilitates this process. Denaturation of occluded proteins also presents a problem and may be more important in the loss of enzyme activity compared to egress. These problems may be addressed by a wide variety of covalent schemes designed to reduce loss of the indicator molecule from the membrane, provide for more intimate and uniform distribution of the indicator molecules within the electroconductive polymer matrix, and impart stability to the polymer–indicator assembly.

1. Chemical Modification and Derivatization of the Preformed Polymer

Chemical modification and derivatization of the preformed polymer backbone has made possible covalent linking methods for the specific attachment of the indicator molecules to the membrane surface or polymer chain in the bulk. Chemical reactions performed on the freestanding or substrate-supported polymer films may lead to the introduction of new functional groups [114,176–178]. These functional groups could then be reacted with appropriate indicator molecules, such as enzymes, as a means of conferring specificity. Heller and coworkers [237, 238] used a chemically modified poly(alkylthiophene) with facile leaving group to achieve covalent attachment of redox enzymes. Martin and coworkers [80] hydroxylated PPy and so altered the redox potential. Epstein and coworkers [176–178] used sulfonation to introduce —SO_3^- groups to polyaniline, and Guiseppi-Elie and Wnek used permanganate, osmium tetroxide, and nitration methods to introduce —OH groups to polyacetylene [179] and —OH, —NO_3, and —NH_2 groups to polyaniline [114]. This approach has the disadvantages of a more costly and complex synthesis, reduction in the overall high electrical conductivity of the polymer, and generally poorer mechanical properties.

2. Polymerization and Copolymerization of Monomers with Reactive Functional Groups

An alternative approach is to synthesize electroactive, electropolymerizable monomers that can retain high electrical conductivity while possessing pendant functional groups suitable for subsequent specific immobilization. By forming an electroconductive polymer as a copolymer of copolymerizable monomers, one of which contains activating functional groups, this may serve as a basis for the subsequent introduction of indicator molecules. The resulting polymer backbone containing such pendant functional groups may be used to covalently attach indicator molecules. Guiseppi-Elie et al. synthesized 3-(1-pyrrolyl)propionic acid [165] and 4-(3-pyrrolyl)butyric acid [180] and made copolymers of varying mole ratios. The pendant carboxylic acid groups of these copolymers were used for the carbodiimide coupling to primary amines of the lysine residues of oxidoreductase enzymes and to 5-(biotinamido)pentylamine. The latter was used for the formation of biotin-avidin complexes. Cooper et al. [181] electropolymerized 3-methyl-4-pyrrole-carboxylic acid, which they showed promotes charge transfer to cytochrome c at both Pt and Au electrodes. Welzel et al. [182] electropolymerized protected 3-substituted thiophene monomer at gold electrodes. The resulting polymer was subsequently deprotected, leading to hydroxyl pendant groups that could be activated with cyanogen bromide and coupled to alcohol oxidase.

3. Copolymerization of Monomers Derivatized with the Indicator Molecule/Moiety

A further solution is found in the copolymerization of the indicator molecule directly with the electroconductive monomer. In this approach, the indicator molecule is first rendered copolymerizable by derivatization with an electroactive monomer—a process we term "monomerization" of the indicator molecule. Lowe and coworkers [183,184] prepared pyrrole films containing covalently immobilized glucose oxidase by coelectropolymerization of the monomerized enzyme and pyrrole. Using carbodiimide cross-linking chemistry, they conjugated 3-(1-pyrrolyl)propionic acid to the free primary amine sites of lysine residues of the enzyme. The modified enzyme was then electropolymerized with free pyrrole monomer, yielding a copolymer of pyrrole and (1-pyrrolyl)-GOX. The covalently immobilized enzyme film showed both higher enzyme activities and larger amperometric responses to glucose compared to physically entrapped enzyme films [41]. Guiseppi-Elie and Wilson [180], in an independent study, used similar cross-linking chemistry to prepare biotinylated pyrrole derivatives. They conjugated 3-(1-pyrrolyl)propionic acid to the free primary amine of 5-(biotinamido)pentylamine. Coelectropolymerization of pyrrole with the biotin derivative produced a copolymer with available biotin sites. These sites then served as the point of attachment of streptavidin or avidin and the subsequent performance of biotin/streptavidin/biotin type immunoassays. A companion approach in this class is the chemical modification of the indicator molecule or moiety to render it directly attachable to the polymer membrane. Marsella et al. [150] and Marsella and Swager [151] used this approach to produce chemically responsive polythiophenes from functionalized bithiophenes. Pseudopolyrotaxanes were produced from bithiophenes that were based on bis(p-phenylene)-34-crown-10, and polythiophenes were produced from bithiophenes functionalized with crown ethers such as 18-crown-6. Binding of the 18-crown-6 derivative with K^+, Ca^{2+} produces conformational changes in the macrocyclic ring that modulate the backbone conjugation and alter polymer conductivity. In this way they were able to build ion-specific sensors based on the principle of conductimetry. Brockmann and Tour [185] likewise synthesized novel zwitterionic N-pyrrolyl-substituted pyrrolinium oxide derivatives. When polymerized these show large solvatochromic and pH-dependent shifts and, for the N-(oligo(ethylene oxide)) monomers, a dramatic ionochromic response. McCullough and Williams [186] achieved similar results with modified polypyrroles.

4. Specific Attachment to Occluded Polymers

Another approach is to occlude into the electroconductive polymer film a second polymer that contains functionalities to which indicator molecules may be conveniently covalently attached. Guiseppi-Elie and Wilson [180] coelectropolymerized pyrrole and 3-(1-pyrrolyl)-propionic acid in the presence of poly(styrenesulfonic acid) and polyvinylamine (PVAm). Incorporation of the PVAm into the polymer film confers free primary amines on the surface of the film. Using aqueous cross-linking chemistry Guiseppi-Elie and Wilson conjugated the succinamide ester of NHS-LC-biotin to the primary amine sites of the polymer surface. A simple variation on this theme can produce a wide variety of chemical and biological sensors. The occlusion of human serum albumin (HSA) [154] or bovine serum albumin (BSA) by electropolymerization in electroconductive polymer can also provide —COOH, —NH_2, and —SH sites for the convenient attachment of indicator molecules. Likewise, macromolecular counteranions may be functionalized prior to or following occlusion into the CEP membrane.

D. Overlayer Membranes

Chemical and biological specificity may also be conferred by the fabrication of a secondary, semipermeable, specificity-conferring membrane over the electroconductive polymer transducer. Thus, a cellulose acetate membrane that is permeable to hydrogen peroxide but not to an enzyme substrate, such as glucose, allows H_2O_2 to be transported through this layer to the transducer without interference from the sample matrix or the substrate. This is a useful approach to distinguish between H_2O_2-induced responses and direct cofactor mediation in oxidoreductase enzyme biosensors. Similar membranes may be formed from biopolymers such as dextran and chitosan or from synthetic polymers such as polyurethanes and polysulfones.

E. Other Methods

Other methods to confer specificity include ion exchange and the deposition of metallic islands and semiconductor particles. Ion exchange occurs spontaneously between counteranions or countercations (in the case of macromolecular anions) and ions that bathe the CEP membrane. Repeated redox cycling encourages the exchange of mobile ions within the film with ions drawn from the solution in which the film is bathed. This becomes a convenient means to introduce catalytic ions into CEPs to confer a measure of specificity [29,126,183]. Another approach used has been to deposit islands of catalytic metal onto the surface or within the electroconductive polymer membrane [187–190], Wrighton et al. [11] deposited platinum particles. Semiconductor particles have also been used [174].

V. ISSUES IN ELECTROCONDUCTIVE POLYMER SENSOR TECHNOLOGY

There are still many unresolved issues on the road toward widespread commercialization of electroconductive polymer sensor technology. A few of the major issues are addressed in this section.

A. Methods to Confer Specificity

Conferring chemical and/or biological specificity remains a wide open area with many opportunities for creative and innovative contributions. In addition to the methods described previously, methods based on reversible sorption or desorption that lead to swelling or collapse of the CEP are receiving some attention [191,192]. These approaches are believed to alter interchain transport and so alter conductivity. By forming blends or composites of conductive electroactive polymers and other well-known selective sorption or diffusion polymers [191,192], this effect can be enhanced and may lead to the development of innovative chemical vapor or gas-phase sensors.

B. Methods for Sensor Interrogation

Electroconductive polymers are a unique class of materials presenting a broad range of material properties. It is clear that analytical applications of electroconductive polymers have become an area of increasingly aggressive research. How these materials are configured into sensing devices and how they are interrogated to obtain analytically significant data will emerge as an area of future technological and engineering innovation. Wrighton et al. [10–14] described a V_G–I_D method analogous to that used in transistor characterization. This method requires a costly bipotentiostat, in which the first potentiostat holds the CEP membrane at one potential (gate voltage) relative to a suitable reversible reference electrode, and the second potentiostat applies a small dc voltage (drain voltage) across the CEP membrane, thereby producing a drain current. Modulation of the gate voltage varies the redox composition of the CEP membrane and accordingly changes its electrical conductivity. In addition, although it is suitable to the characterization of CEP materials under similar test conditions, this method does not yield analytically quantitative results and is thus better suited to interrogation of on/off devices. The CEP, being maintained under continuous drain voltage, has its redox composition, electrode potential, and electrical conductivity influenced by the gate voltage. Sadik et al. [156 and Sadik and Wallace [157] used pulsed amperometry as a means to obtain sensor responses from polypyrrole films. In this method dc pulses are set to bracket the redox transition of the polymer film. Application of the pulse to oxidizing conditions permits anions from the test solution to enter the polymer and support an anodic current. Reversal of the pulse to reducing conditions causes egress of the ions and supports a cathodic current. The magnitude of the current is determined by the selectivity of the polymer toward the particular or dominant anions present. This approach eliminates the ionic polarizations common to the Wrighton method, but analytical reproducibility depends on the reversibility of the polymer redox reaction. Guiseppi-Elie and Wilson [23,24] reported a unique analytical method that embraces a programmed sequence of events. This method includes a diagnostic potentiometric measurement that ensures the integrity of the polymer membrane. A subsequent constant-voltage electrolysis fixes the redox composition and hence electrical conductivity of the polymer membrane at the beginning of each analytical interrogation cycle. This is followed by the application of a burst of discontinuous, small amplitude voltage pulses, typically 25 mV for 50 ms. The conductivity is determined from the current attendant to voltage pulse application. Each pulse is followed by a disconnect or rest period, typically 500 ms (or duty cycle of 10), in which the sensor is completely disconnected from the electronics and is allowed to react, as a reagent would, with the analyte to which it has been rendered specific. At the end of the rest period, the open circuit or rest potential of the sensor is again measured. This technique therefore gives both potentiometric and conductimetric responses. The rest potential, measured at the end of the rest period, forms the reference point for the application of the subsequent pulse. In this way the applied voltage is always a pulse voltage away from the rest potential. This method of conductivity interrogation seeks to be nonperturbing to the spontaneous charge transfer reactions occurring between the analyte and the polymer and is analogous to the polarization resistance technique for corrosion rate determination of Stern and Geary [193,194].

C. Formation of Organized and Adherent Polymer Films at Device Surfaces

The performance of sensor devices using conducting electroactive polymers greatly depends on the ability to prepare uniform, adherent, and fully contiguous membrane films on microfabricated devices or electrode surfaces. The control of polymer growth is essential for developing high performance, molecule-based electronic devices. Of these methods electropolymerization and the casting of soluble polymers are by far the most common. Other methods are available, however. Shimidzu et al. [195] creatively used electropolymerization to produce multilayer thin films with high anisotropic conductivity. Ando et al. [196] used the Langmuir–Blodgett technique to deposit amphiphilic and electropolymerizable monomers into CEP thin films. Yang et al. [197] likewise used the L-B technique to deposit organized thin films of conducting polymers. Cai and Martin [198] and Martin et al. [199] prepared tubular conducting polymers using a track-etch microporous membrane as a template for polymerization. Chemical and electrochemical growth of the conducting polymer proceeds along the inner wall of the micropores. This unique morphological characteristic has been used for fabrication of an amperometric biosensor. Nolte and coworkers [49,50] fabricated a glucose sensor based on a microporous film with polypyrrole tubules and adsorbed glucose oxidase at a Pt electrode. They claimed that the device functioned via direct communication between the adsorbed glucose oxidase and the polymer tubule. However, a subsequent study [200] suggested that the glucose was directly oxidized at the Pt electrode at the back of the film and thus the response was affected by interferents adsorbed to the electrode surface. Innovative application of electropolymerization to yield bilayers with modified ion transport and unique electrorelease properties has been pursued by Pyo and Reynolds [201] and Pyo et al. [202], and potential-programmed electropolymerization of alternating ultrathin multilayers of polypyrrole and poly(3-methylthiophene) from mixed monomer solutions of pyrrole and 3-methylthiophene has been achieved by Iyoda et al. [203] and Fujitsuka et al. [204]

Interdigitated array (IDA) electrodes coated with conducting polymers such as polypyrrole, polythiophene, and polyaniline are a convenient and popular conductimetric format for sensor devices. To fabricate such devices, the formation of uniform, thin polymer films of precisely controlled thickness at the array is essential. To improve adhesion of polypyrrole films to semiconductor devices, Simon et al. [205] used an ω-pyrrolyl functionalized trimethoxysilane. Komori and Nonaka [206,207] also developed CEP adhesion promotion strategies. To improve polymer film uniformity, Nishizawa et al. [208,209] prepared thin polypyrrole and polyaniline films by electropolymerization at hydrophobically pretreated IDA electrodes. Polymer growth rates at hydrophobic surfaces was about 25 times as high for polypyrrole and more than 100 times as high for polyaniline as at hydrophilic surfaces. The presence of a small amount of dodecylsulfate eminently enhances the anisotropic growth [210]. Dodecylsulfate probably forms a bilayer-like structure at the hydrophilic surface that acts as an effective dopant during the electropolymerization. This promotion of lateral growth was used to fabricate micropatterns of conducting polymers at insulating substrates [208]. Using γ-aminopropyltrimethoxysilane, Guiseppi-Elie et al. [165] prepared chemically modified interdigit surfaces of IDA devices. To these primary amines they covalently attached pyrrolyl derivatives formed from 3-(1-pyrrolyl)propionic acid. The result was devices chemically modified with electroactive and electropolymerizable monomer. Electropolymerization of pyrrole from aqueous solution at these devices produced uniform and highly adherent polypyrrole films that were electrochemically indistinguishable from polypyrrole films except for the considerably improved adhesion to the device surface. A vapor deposition technique has also be used for formation of a thin polypyrrole film at IDA electrodes [211]. Thin micropatterns of electropolymerized aniline were produced by photoreaction of a self-assembled monolayer [212]. Formation of micropatterns of conducting polymer at solid substrates can be extended to future fabrication of molecule-based devices.

D. Environmental Stability and Shelf Life

Although little quantitative information is available, some comments on the long-term performance of conducting polymer sensors in terms of their environmental stability and shelf life is warranted. Unlike conventional metallic, semiconductor, and ceramic sensing electrodes, conducting polymer sensors are not inert. In fact, electroconductive polymers are quite reactive and are capable of a host of molecular interactions. It is this reactivity that gives these materials their unique transducer action. To eliminate this as an approach to improve stability would be a retrograde step. Rather, means must be sought to distinguish between desired and undesired molecular interactions and methods adopted to control undesired interactions while promoting the desired analytically important ones. It appears that the use of appropriate potential waveform routines for interrogation is one possible way to achieve this goal.

Several other practical aspects of stability must be considered. The use of appropriate dopant counteranions or macroanions and the control of primary and secondary structure and morphology are avenues that should also be considered.

1. Stability of Electrical Contact

Generally, electrical contact with the electronic control system is achieved by depositing the polymer on metallic, carbon-based, or semiconductor electrodes or substrates. Moreover, the polymer must contact and adhere to vastly different surfaces. Adhesion to these substrates must be maintained to ensure stability. In this regard substrate preparation is important. Surface chemical modification such as the use of functionalized alkane thiols on gold and silane chemistry on glass and oxidized silicon are important steps in this direction. The use of rough versus smooth surfaces will also affect adhesion. The use of large magnitude potential pulses with their attendant electromechanical effects can induce interfacial stresses that result in reduced adhesion. Sorption of some solvents, with attendant swelling of the polymer, may also induce interfacial stresses that result in reduced adhesion.

2. Chemical Stability

To provide reproducible sensor responses, the chemical composition, structure, and morphology of the transducer-active material should remain relatively constant during analysis. Ion-exchange reactions attendant to redox cycling should therefore be considered a potential problem. The ion transport characteristics and pK_a of conjugate bases in the external buffer or analyte matrix and the starting pH are important variables in deciding steady-state behavior [213]. Interrogation voltage pulses should not produce irreversible oxidation or reduction of the polymer. Similarly, the stability of incorporated counteranions and indicator moieties warrants special consideration.

Exposure to extreme pH, temperature, or ionic strengths can result in denaturation of bioactive species and a subsequent loss in bioactivity. As with all sensors, CEP sensors will be subject to chemical fouling and biofouling. The design of polymer surfaces and the use of impressed potentials may be useful in addressing this problem. As mentioned, if exposed to "extreme" positive potentials (>0.80 V), conducting polymers will become overoxidized. This results in a loss of conductivity and electroactivity and a degradation of mechanical properties.

3. Thermal Effects

Conducting electroactive polymers function quite well up to approximately 100°C in aqueous solution. In fact, improved analytical performance can be achieved at higher than ambient temperatures [135]. These sensors must, of course, be calibrated at the temperatures at which they are to be used or temperature coefficients established and made available in look-up tables. In the dry state, conducting polymers are stable up to approximately 200°C with only slow weight loss due to moisture, unreacted monomer, low molecular weight oligomers, or volatilization of small counterions. Breakdown is usually associated with structural decomposition or loss of the counterion incorporated during synthesis.

4. Structural and Morphological Effects

Methods of polymer film fabrication have a profound influence on the structure and morphology of the CEP membrane. Potentiostatically grown polyaniline films display different diffusion coefficients with respect to counteranions compared to potentiodynamically electropolymerized films of similar thickness (Guiseppi-Elie, unpublished). Internally or self-doped polymers provide a source of counteranions and shift the dominant transport from anionic to cationic. The presence of pendant carboxylic groups generally results in lower electrical conductivity in homopolymers relative to other types of reactive functional groups. Although much work has been directed to homopolymers from functionalized monomers and some work has been directed toward copolymers, little has been done regarding the control of the primary repeat unit structure. The preparation of primary repeat unit structures that dictate control over secondary and tertiary levels of structure, such as alternating copolymers, block copolymers, and segmented block copolymers with regions of precisely placed functionality, remains an open area for future research.

VI. CURRENT SENSOR CONFIGURATIONS AND COMMERCIAL APPLICATIONS

A. Potentiometric Sensors

Potentiometric sensors responsive to iodide ion that are based on iodine-doped poly(3-methylthiophene) have been demonstrated [214,215]. Potentiometric sensors, because of the wide range of possible potential-influencing interactions [216], offer no major advantages over conventional functionalized nonconducting polymer membranes.

B. Amperometric Sensors

The goals in electrode modification with conducting electroactive polymers are (1) to improve sensitivity, (2) to impart selectivity, and (3) to suppress the effect of interfering reactions. In addition, the CEP membrane may serve as a support matrix for immobilized indicator molecules in amperometric sensors [55,57,217]. The catalytic action of poly(3-methylthiophene), poly(*N*-methylpyrrole), polyaniline, and polyfuran has been compared for neurotransmitters such as norepinephrine, L-dopa, epinephrine, and dopamine [218,219]. Poly(3-methylthiophene) was found to display the most efficient catalytic activity and could be used as a polymer-modified electrode in amperometric detection by flow injection analysis (FIA). Polypyrrole films have been found to be catalytic for ascorbic acid oxidation [220,221]. Saraceno et al. [220] suggested that the catalytic activity was derived from electrostatic attraction between the positively charged polypyrrole and the negatively charged radical anion intermediate formed in the electrocatalytic oxidation of ascorbic acid. A membrane-entrapped 1,1′-dimethylferrocene outer layer, serving as a redox mediator, was used over a polypyrrole-coated microfabricated electrode with adsorbed GOX in the demonstration of a microglucose sensor. In another configuration, ferrocene mediators were conjugated to *N*-substituted polypyrrole films and used to construct a reagentless amperometric biosensor for glucose [71]. In an alternative configuration [65,67], hydroquinonesulfonate anions were incorporated directly into electropolymerized polypyrrole films and were shown to mediate the regeneration of GOX in a reagentless glucose biosensor. Direct biosensors, involving direct electron transfer between the conducting electroactive polymer and the redox center of enzymes, have also been investigated. Khan et al. [48] immobilized thin layers of fructose dehydrogenase, a PQQ enzyme, onto platinum electrodes via electrophoresis. Onto this they electropolymerized an equally thin layer of polypyrrole. Using constant-potential (0.4 V vs. Ag/AgCl) chronoamperometry, they observed increased anodic current upon addition of aliquots of D-fructose. Along with evidence from differential pulse voltammetry they concluded that direct redox communication between the PQQ prosthetic group and polypyrrole was occurring. Koopal et al. [49,50] also reported on evidence for direct biosensors formed from polypyrrole microcylinders formed in track-etch membranes. While the foregoing illustrates the considerable variety in configurations for amperometric sensors based on CEP membranes, none of those discussed here have yet been commercialized.

C. Conductimetric Sensors

Commercially available sensor systems that are based on the use of conducting electroactive polymers include the Aromascan from Aromascan, plc and the Electronic Nose of Neotronics. Scientific both for gas and vapor phase monitoring. The SmartSense biosensor assay system of Ohmicron Corporation, and the EPSIS and BioSenSys systems of ABTECH Scientificare for liquid-phase monitoring. Each of these systems is based on conductimetry, the measurement of a change in impedance (capacitance, resistance, or both) resulting from interaction with the analyte. Both the Aromascan and the Neotronics Nose are based on nonspecific conductimetric responses of a multiple-element array. In the Aromascan instrument, a plug of analyte vapor is passed over a 32-element conductimetric array. Each array element consists of a polymer bilayer of electroconductive polypyrrole (inner layer) and a sorptive polymer (outer layer). The resulting response histogram when compared to reference air is intended as a signature for vapor-phase analytes. The SmartSense biosensor is based on chemically polymerized polythiophene films cast onto 50 μm spaced banded electrodes. An analyte-specific pair of bands bearing physically adsorbed bioindicator molecule, a reference pair without, and a control pair that is exposed to a known concentration of analyte make up a three-element array. The result is a semiquantitative threshold detector directed at the pesticide testing market. The EPSIS and BioSenSys systems are versions of ABTECH's biospecific, multianalyte, multielement array of microfabricated interdigitated microsensor electrodes. Each array element is rendered biospecific by covalent attachment of bioindicator molecules to the CEP membrane. Both these instruments incorporate novel and patented analytical methodology of sensor interrogation that is unique to conducting electroactive polymer sensors. EPSIS is a research instrument with available polypyrrole sensors possessing surface functional groups such as —COOH, —NH$_2$, and —SH. To these groups the researcher may attach particular indicator molecules. The BioSenSys instrument targets on-site monitoring of specific analytes of environmental importance using immunological methods of recognition.

D. Voltammetric Sensors

Voltammetric measurements are traditionally carried out using a potentiostated three-electrode system. This approach has been used with CEP sensors in both stationary [31,123,124,136,146–148,154,222] and flow-through [30,120,137–140,149,155–157] cell setups. The flow-through cells are used either in conjunction with a chromatographic separation system or in a flow injection analysis (FIA) setup [32,33]. With the flow-through cell, optimal results are obtained using pulsed electrochemical detection. This involves the use of a repetitive pulsed potential waveform. Typically a potential pulse with a range large enough to oxidize/reduce the polymer and so control the polymer–analyte interactions is employed. Using appropriate pulse widths (typically 20–200 ms), the polymer–analyte (even Ab–Ag) interaction can be controlled to the extent that interaction is maintained as reversible. The advantages of the flow-through analysis setup are that background currents are minimized and so better signal-to-noise ratios and, particularly with FIA, quicker sample throughput are obtained.

The use of CEP-coated microelectrodes offers several advantages over the use of conventionally sized electrodes. The use of microelectrodes means that voltammetric measurements can be carried out using much lower concentrations of supporting electrolytes. With the FIA system for detection of ions this enables much lower (sometimes three orders of magnitude lower) detection limits to be obtained [31]. Also, with microelectrodes, faster switching routines can be used without distortion of the voltammetric response [223]. The use of microelectrodes also enables microarrays to be employed [77,224]. This enables multicomponent analyses and pattern recognition to be used for analytical purposes.

E. CHEMFETs and Other Sensor Configurations

Janata and coworkers [225,226] combined CEP membranes of polypyrrole and field effect transistor (FET) devices to fabricate miniaturized solid-state gas sensors. These sensors showed rapid and high sensitivity to aliphatic alcohols at room temperature.

VII. FUTURE CHALLENGES

Major future challenges continue to reside in shelf life or chemical stability of doped electroconductive polymers. One approach to this problem has been to use undoped polymers such as poly(3-methylthiophene) in the SmartSense bioassay of Ohmicron. In the undoped form, the poly(3-alkylthiophenes) are quite stable. Doping occurs only attendant on its analytical application. Reversibility is also a major issue. Poor reversibility, particularly when it is due to the loss of counteranions during redox cycling and to facilitated ion transport resulting in ion exchange, presents a problem in those applications that rely on reuse of the sensor. An approach of some merit is the use of macromolecular anions and internal (self-doped) anions. In this approach, electroneutrality is maintained by the transport of small cations and protons into and out of the polymer during redox cycling. This may serve to improve switching speeds. Structural and morphological changes during cycling leading to dimensional change and internal stress, while highly desirable in actuator applications, is a major limitation in analytical applications of CEPs. Poor reversibility also results from the influences of time, temperature, and redox cycling which induce alterations in the morphology and structure of the CEP membrane. An approach that may address this aspect of reversibility is to be found in the use of polymer composites within which the conducting polymer forms the small volume fraction phase. Hydrogels may serve as a particularly useful matrix in this regard. Analytical applications demand accurate control of film thickness and long-term stability of primary structure and morphology. Sensors depend on the ability to reproducibly fabricate uniform films of precisely controlled thickness. Means are needed to deposit uniformly thin contiguous conducting polymer films at the single molecular layer level. The compability of these polymer films with the device structures with which they will make and must maintain contact will also present engineering challenges. Without doubt, analytical applications of conducting electroactive polymers are currently poised to be a major thrust direction, both commercially and academically, for these novel materials in the immediate future.

VIII. GLOSSARY OF TERMS

Electroactive polymer. A polymer that, when cast on an electrode or in solution, is oxidizable and / or reducible. Examples: Polyvinylferrocence, polyaniline.

Conducting electroactive polymer (CEP). An electroactive polymer that, when oxidized or reduced, displays a significant change in intrinsic electrical conductivity. Examples: Polyaniline, polythiophene. Also called **electroconductive polymer (EP)**.

Electroactive polymer sensor technology. Enabling know-how related to the science, engineering, and commercialization of conducting electroactive polymers in the development and application of chemical and biological sensors.

REFERENCES

1. H. Shirakawa, E. J. Louis, A. G. MacDiarmid, C. K. Chiang, and A. J. Heeger, Synthesis of electrically conducting organic polymers: Halogen derivatives of polyacetylene, $(CH)_x$, *J. Chem. Soc., Chem. Commun.* 578 (1977).
2. C. K. Chiang, C. R. Fincher, Jr., Y. W. Park, A. J. Heeger, H. Shirakawa, E. J. Louis, S. C. Gau, and A. G. MacDiarmid, Electrical conductivity in doped polyacetylene, *Phys. Rev. Lett.* 39(17):1098 (1977).
3. K. Seeger, W. D. Gill, T. C. Clarke, and G. B. Street, Conductivity and Hall effect measurements in doped polyacetylene, *Solid State Commun.* 28(10):873 (1978).
4. J. F. Kwak, T. C. Clarke, R. L. Greene, and G. B. Street, Transport properties of heavily AsF_5 doped polyacetylene, *Solid State Commun.* 31(5):355 (1979).
5. M. E. Dury, Synthesis and characterization of polyacetylene, $(CH)_x$, and its derivatives, Ph.D. Thesis in Chemistry, University of Pennsylvania, 1981.
6. G. E. Wnek, Synthesis and properties of electrically conducting polymers, Ph.D. Thesis in Polymer Science and Engineering, University of Massachusetts, 1980.

7. A. Guiseppi-Elie, Synthesis and characterization of polyacetylene: 1. Stability of doped polyacetylene, 2. Surface chemistry of polyacetylene, Sc.D. Thesis in Materials Science and Engineering, MIT, 1983.
8. M. K. Malmros, Analytical device having semiconductive organic polymeric element associated with analyte binding substance, U.S. Patent 4,444,892 (1984).
9. A. Guiseppi-Elie and G. E. Wnek, Aqueous reactivity of polyacetylene: pH dependence, *J. Phys. Chem.* 97(7):3192 (1990).
10. E. P. Lofton, J. W. Thackeray, and M. S. Wrighton, Amplification of electrical signals with molecule-based transistors: Power amplification up to a kilohertz frequency and factors limiting higher frequency operation, *J. Phys., Chem.* 90:6080 (1986).
11. M. S. Wrighton, H. S. White, Jr., and J. W. Thackeray, Molecule-based microelectronic devices, U.S. Patent 4,717,673 (1988).
12. M. S. Wrighton, H. S. White, Jr., and G. P. Kittlesen, Molecule-based microelectronic devices, U.S. Patent 4,721,601 (1988).
13. M. S. Wrighton, H. S. White, Jr., and G. P. Kittlesen, Molecule-based microelectronic devices, U.S. Patent 4,895,705 (1990).
14. M. S. Wrighton, H. S. White, Jr., and G. P. Kittlesen, Molecule-based microelectronic devices, U.S. Patent 5,034,192 (1991).
15. M. Umana and J. Waller, Protein-modified electrodes. The glucose oxidase/polypyrrole system, *Anal. Chem.* 58:2979 (1986).
16. A. Ivaska, Analytical applications of conducting polymers, *Electroanalysis* 3:247 (1991).
17. C. E. D. Chidsey and R. W. Murray, Electroactive polymers and macromolecular electronics, *Science* 231:25 (1986).
18. R. A. Hillman, Reactions and applications of polymer modified electrodes, in *Electrochemical Science and Technology of Polymers* Vol. 1 (R. G. Linford, ed.), Elsevier, New York, 1987, p. 241.
19. G. K. Chandler and D. Pletcher, The electrochemistry of conducting polymers, in *Electrochemistry*, Vol. 10 (D. Pletcher, ed.), Royal Society of Chemistry, London, 1993, p. 117.
20. P. N. Bartlett and J. M. Cooper, A review of the immobilization of enzymes in electropolymerized films, *J. Electroanal. Chem.* 362:1 (1993).
21. J. Janata and A. Bezegh, Chemical sensors, *Anal. Chem.* 63(7):677 (1988).
22. A. Guiseppi-Elie, A. M. Wilson, C. L. Linden, F. J. Pearce, W. P. Wiesmann, and D. L. Glick, A conductimetric H_2O_2-sensitive electroconductive polymer transducer for development of oxidoreductase enzyme biosensors and oxidoreductase labeled immunosensors, *Proc. ACS Div. Polym. Mater. Sci. Eng.* 71:651 (1994).
23. A. Guiseppi-Elie, Method of measuring an analyte by measuring electrical resistance of a polymer film reacting with the analyte, U.S. Patent 5,312,762 (1994).
24. A. Guiseppi-Elie and A. M. Wilson, Novel analytical method for conductimetric chemical and biosensors formed from electroconductive polymers, *Proc. ACS Div. Polym. Mater. Sci. Eng.* 71:381 (1994).
25. G. F. Blackburn, Chemically sensitive field effect transistors, in *Biosensors: Fundamentals and applications* (A. P. F. Turner, I. Karube, and G. S. Wilson, eds.), Oxford Univ. Press, New York, 1987, p. 481.
26. D. Ofer, R. M. Cooks, and M. S. Wrighton, Potential dependence of the conductivity of highly oxidized polythiophenes, polypyrroles, and polyoniline: Finite windows of high conductivity, *J. Am. Chem. Soc.* 112:7869 (1990).
27. N. F. Sheppard, Jr., R. C. Tucker, and C. Wu, Electrical conductivity measurements using microfabricated interdigitated electrodes, *Anal. Chem.* 65(9):1199 (1993).
28. M. C. Zaretsky, L. Mouyad, and J. R. Melcher, *IEEE Trans. Electr. Insul.* 23:897 (1988).
29. F. Garnier, X. Peng, G. Horowitz, and D. Fichou, Organic-based field-effect thansistors: Critical analysis of the semiconducting characteristics of organic materials, *Mol. Eng.* 1(2): 312 (1991).
30. J. W. Thackeray, H. S. White, and M. S. Wrighton, Poly(3-methylthiophene)-coated electrodes: Optical and electrical properties as a function of redox potential and amplification of electrical and chemical signals using poly(3-methylthiophene)-based microelectrochemical transistors, *J. Phys. Chem.* 89:5133 (1985).
31. N. F. Atta, A. Galal, A. E. Karagozler, H. Zimmer, J. F. Rubinson, and H. B. Mark, Jr., Voltammetric studies of the oxidation of reduced nicotinamide adenine dinucleotide at a conducting polymer electrode, *J. Chem. Soc., Chem. Commun. 1347* (1990).
32. W. J. Albery and P. N. Bartlett, An organic conductor electrode for the oxidation of NADH, *J. Chem. Soc., Chem. Commun. 234* (1984).
33. T. Hepel, Incorporation of redox materials into polyaniline films, *64th ACS Colloid and Surf. Sci. Symp.*, Lehigh, PA, 1990.
34. M. Gholamian, J. Sundaram, and A. Q. Contractor, Oxidation of formic acid at polyaniline-coated and modified-polyaniline-coated electrodes, *Langmuir 3*: 741 (1987).
35. R. Bull, F.-R. Fan, and A. Bard, Polymer films on electrodes. 13. Incorporation of catalysts into electronically conductive polymers: Iron phthalocyanine in polypyrrole, *J. Electrochem. Soc.* 130:1636 (1983).
36. K. Okabayashi, O. Ikeda, and H. Tamura, Electrochemical doping with meso-tetrakis(4-sulphonatophenyl)porphyrincobalt of a polypyrrole film electrode, *J. Chem. Soc., Chem. Commun. 684* (1983).
37. F. Bedioui, C. Bongars, J. Devynck, C. Bied-Charreton, and C. Hinnen, Metalloporphyrin-polypyrrole film electrode: Characterization and catalytic application, *J. Electroanal. Chem.* 207:87 (1986).
38. A. Elzing, A. van der Putten, W. Visscher, and E. Baendrecht, The mechanism of oxygen reduction at iron tetrasulfonato-phthalocyanine incorporated in polypyrrole, *J. Electroanal. Chem.* 233:113 (1987).

39. H. Li and T. F. Guarr, Formation of electrically conductive thin films of metal phthalocyanines *via* electropolymerization, *J. Chem. Soc., Chem. Commun. 832* (1989).
40. H. Li and T. F. Guarr, Electrochemistry at modified electrodes: Electronically conductive metalophthalocyanine coatings, *Synth. Met. 38*(2):243 (1990).
41. M. V. Rosenthal, T. A. Skotheim, and C. A. Linkous, Polypyrrole-phthalocyanine, *Synth. Met. 15*: 219 (1986).
42. C. Choi and H. Tachikawa, Electrochemical behavior and characterization of polypyrrole-copper phthalocyanine tetrasulfonate thin film: Cyclic voltammetry and in situ Raman spectroscopic investigation, *J. Am. Chem. Soc. 112*:1757 (1990).
43. F. T. A. Vork, L. J. J. Janssen, and E. Barendrecht, The reduction of dioxygen at polypyrrole-modified electrodes with incorporated Pt particles, *Electrochim. Acta. 35*:135 (1990).
44. M. E. G. Lyons, P. N. Bartlett, C. H. Lyons, W. Breen, and J. F. Cassidy, Conducting polymer based electrochemical sensors: Theoretical analysis of current response under steady state conditions, *J. Electroanal. Chem. 304*:1 (1991).
45. L. Coche and J.-C. Moutet, Electrocatalytic hydrogenation of organic compounds on carbon electrodes modified by precious metal microparticles in redox active polymer films, *J. Am. Chem. Soc. 109*:6887 (1987).
46. P. C. Pandey, A new conducting polymer-coated glucose sensor, *J. Chem. Soc., Faraday Trans. I 84*(7): 2259 (1988).
47. S. Yabuki, H. Shinohara, and M. Aizawa, Electroconductive enzyme membranes, *J. Chem. Soc., Chem. Commun. 945* (1989).
48. G. F. Khan, E. Kobatake, H. Shinohara, Y. Ikariyama, and M. Aizawa, Molecular interface for an activity controlled enzyme electrode and its application for the determination of fructose, *Anal. Chem. 64*:1254 (1992).
49. C. G. J. Koopal, M. C. Feiters, and R. J. M. Nolte, Amperometric biosensor based on direct communication between glucose and a conducting polymer inside the pores of a filtration membrane, *J. Chem. Soc., Chem. Commun. 1691* (1991).
50. R. Czojka, C. G. J. Koopal, M. C. Feiters, J. W. Gerristsen, R. J. M. Nolte, and H. V. Kempen, Scanning tunneling microscopy study of polypyrrole films and of glucose oxidase as used in a third-generation biosensor, *Bioelectrochem. Bioeng. 29*:47 (1992).
51. F. A. Armstrong, H. A. Hill, and N. J. Walton, Direct electrochemistry of redox proteins, *Acc. Chem. Res. 21*:407 (1988).
52. T. Matsue, H. Yamada, H. Chang, I. Uchida, K. Nagata, and K. Tomita, Electron transferase activity of diaphorase (NADH:acceptor oxidoreductase) from *Bacillus stearothermophilus*, *Biochim. Biophys. Acta 1038*:29 (1990).
53. T. Matsue, N. Narumi, M. Nishizawa, H. Yamada, and I. Uchida, Electron-transfer from NADH dehydrogenase to polypyrrole and its applicability to electrochemical oxidation of NADH, *J. Electroanal. Chem. 300*:111 (1991).
54. H. Yamato, M. Ohwa, and W. Wernet, A polypyrrole/three-enzyme electrode for creatinine detection, *J. Anal. Chem. 67*:2776 (1995).
55. P. N. Bartlett and R. G. Whitaker, Electrochemical immobilization of enzymes. Part II. Glucose oxidase immobilization in poly-*N*-methylpyrrole, *J. Electroanal. Chem. 224*:37 (1987).
56. Z. Sun and H. Tachikawa, Enzyme-based bilayer conducting polymer electrodes consisting of polymetallophthalocyanines and polypyrrole-glucose oxidase thin films, *Anal. Chem. 64*(10):1112 (1992).
57. N. C. Foulds and C. R. Lowe, Enzyme entrapment in electrically conducting polymers, *J. Chem. Soc., Faraday Trans. 1, 82*:1259 (1986).
58. C. Malitesta, F. Palmisano, L. Torsi, and P. G. Zambonin, Glucose fast-response amperometric sensor based on glucose oxidase immobilized in an electropolymerized poly(*o*-phenylenediamine) film, *Anal. Chem. 62*:2735 (1990).
59. N. Bartlett, P. Tebbutt, and C. Tyrrell, Electrochemical immobilization of enzyme. 3. Immobilization of glucose oxidase in thin films of electrochemically polymerized phenols, *Anal. Chem. 64*:138 (1992).
60. D. Belanger, J. Nadreau, and G. Fortier, Electrochemistry of the polypyrrole glucose oxidase electrode, *J. Electroanal. Chem. 274*:143 (1989).
61. S. Jasty and A. J. Epstein, Corrosion prevention capability of polyaniline (emeraldine base and salt): An XPS study, *Proc. ACS Div. Polym. Mater.: Sci. Eng. 72*:565 (1995).
62. T. Amemiya, K. Hashimoto, and A. Fujishima, Dynamics of Faradaic processes in polypyrrole/polystyrenesulfonate composite films in the presence and absence of a redox species in aqueous solutions, *J. Phys. Chem. 97*:4192 (1993).
63. A. E. Cass, G. Davis, G. D. Francis, H. A. O. Hill, W. J. Alton, I. J. Higgins, E. V. Plotkin, L. D. L. Scott, and A. P. F. Turner, Ferrocene-mediated enzyme electrode for amperometric determination of glucose, *Anal. Chem. 56*:667 (1984).
64. Y. Kajiya, R. Tsuda, and H. Yoneyama, Conferment of cholesterol sensitivity on polypyrrole films by immobilization of cholesterol oxidase and ferrocenecarboxylate ions, *J. Electroanal. Chem. 301*:155 (1991).
65. Y. Kajiya, H. Sugai, C. Iwakura, and H. Yoneyama, Glucose sensitivity of polypyrrole films containing immobilized glucose oxidase and hydroquinonesulfonate ions, *Anal. Chem. 63*:49 (1991).
66. P. Janda and J. Weber, Quinone-mediated glucose oxidase electrode with the enzyme immobilized in polypyrrole, *J. Electroanal. Chem. 300*:119 (1991).
67. C. Iwakura, Y. Kajiya, and H. Yoneyama, Simultaneous immobilization of glucose oxidase and a mediator in conducting polymer films, *J. Chem. Soc., Chem. Commun. 15*:1019 (1988).
68. P. N. Bartlett, Z. Ali, and V. Eastwick-Field, Electrochemical immobilization of enzymes. Part 4. Co-immobilization of glucose oxidase and ferro/ferricya-

nide in poly(N-methylpyrrole) films, *J. Chem. Soc., Faraday Trans. 88*:2677 (1992).
69. T. Matsue, M. Nishizawa, T. Sawaguchi, and I. Uchida, An enzyme switch sensitive to NADH, *J. Chem. Soc., Chem. Commun. 1029* (1991).
70. G. F. Khan, E. Kobatake, H. Shinohara, Y. Ikariyama, and M. Aizawa, Molecular interface for an activity controlled enzyme electrode and its application for the determination of fructose, *Anal. Chem. 64*:1254 (1992).
71. N. C. Foulds and C. R. Lowe, Immobilization of glucose oxidase in ferrocene-modified pyrrole polymers, *Anal. Chem. 60*:2473 (1988).
72. C. Horwitz and G. C. Dailey, Ferrocene polymers with polyaniline backbones, *Chem. Mater. 2*(4):333 (1990).
73. T. Matsue, Electrochemical sensors using microarray electrodes, *Trends Anal. Chem. 12*(3):100 (1993).
74. M. Nishizawa, T. Sawaguchi, T. Matsue, and I. Uchida, In-situ characterization of copolymers of pyrrole and N-methylpyrrole at microarray electrodes, *Synth. Met. 45*:241 (1991).
75. E. W. Paul, A. J. Ricco, and M. S. Wrighton, Resistance of polyaniline films as a function of electrochemical potential and the fabrication of polyaniline-based microelectronic devices, *J. Phys. Chem. 89*:1411 (1985).
76. J. W. Thackeray, H. S. White, and M. S. Wrighton, Poly(3-methylthiophene)-coated electrodes: Optical and electrical properties as a function of redox potential and amplification of electrical and chemical signals using poly(3-methylthiophene)-based microelectrochemical transistors, *J. Phys. Chem. 89*:5133 (1985).
77. J. P. Kittlesen, H. S. White, and M. S. Wrighton, Chemical derivatization of electrode arrays by oxidation of pyrrole and N-methylpyrrole: Fabrication of molecule-based electronic devices, *J. Am. Chem. Soc. 106*:7389 (1984).
78. N. S. Sundaresan, S. Basak, M. Pomerantz, and J. R. Reynolds, Electroactive copolymers of pyrrole containing covalently bound dopant ions: Poly{pyrrole-co-[3-(pyrrol-1-yl)propane sulphanate]}, *J. Chem. Soc., Chem. Commun. 8*:621 (1987).
79. J. R. Reynolds, N. S. Sundaresan, M. Pomerantz, S. Basak, and C. K. Baker, Self-doped conducting copolymers: A charge and mass transport study of poly{pyrrole-co-[(pyrrol-1-yl)propanesulfonate]}, *J. Electroanal. Chem. 250*:355 (1988).
80. L. S. Van Dyke, S. Kuwabata, and C. R. Martin, A simple chemical procedure for extending the conductive state of polypyrrole to more negative potentials, *J. Electrochem. Soc. 140*:2754 (1993).
81. M. E. H. Amrani, M. S. Ibrahim, and K. C. Persaud, Synthesis, chemical characterization and multifrequency measurements of poly(N-(2-pyridyl)pyrrole) for sensing volatile chemicals, *Mater. Sci. Eng. C 1*:17(1994).
82. K. C. Persaud, A. A. Qutob, P. Travers, A. M. Pisanelli, and S. Szyszko, Odor evaluation of foods using conducting polymer arrays and neural net pattern recognition, in *Olfaction and Taste*, Vol. II (K. Kurihara, N. Suzuki, and H. Ogawa, eds.), Springer-Verlag, Tokyo, 1994, p. 708.
83. P. N. Bartlett and P. R. Birkin, Enzyme switch responsive to glucose, *Anal. Chem. 65*:1118 (1993).
84. P. N. Bartlett and P. R. Birkin, A microelectrochemical enzyme transistor responsive to glucose, *Anal. Chem. 66*:1552 (1994).
85. M. Nishizawa, T. Matsue, and I. Uchida, Penicillin sensor based on a microarray electrode coated with pH-responsive polypyrrole, *Anal. Chem. 64*(21):2642 (1992).
86. D. T. Hoa, T. N. Suresh Kumar, N. S. Punekar, R. S. Srinvasa, R. Lal, and A. Q. Contractor, Biosensor based on conducting polymers, *Anal. Chem. 64*:2645 (1992).
87. T. Ikeda, R. Schmehl, P. Denisevich, K. William, and R. W. Murray, Permeation of electroactive solutes through ultrathin polymeric films on electrode surfaces, *J. Am. Chem. Soc. 104*(10):2683 (1982).
88. P. Burgmayer and R. W. Murray, An ion gate membrane: Electrochemical control of ion permeability through a membrane with an embedded electrode, *J. Am. Chem. Soc. 104*(10):6139 (1982).
89. A. Guiseppi-Elie and G. E. Wnek, Environmental stability of doped polyacetylene in aqueous solutions, *J. Phys.—Colloq. C3*:193 (1983).
90. G. Bidan, Electroconducting conjugated polymers: New sensitive matrices to build up chemical and electrochemical sensors. A review, *Sens. Actuat. B 6*:45 (1992).
91. J. J. Miasik, A. Hooper, and B. C. Tofield, Conducting polymer gas sensors, *J. Chem. Soc., Faraday Trans. I 82*:1117 (1986).
92. T. Hanawa, S. Kuwabata, and Y. Yoneyama, Gas sensitivity of polypyrrole films to NO_2, *J. Chem. Soc., Faraday Trans. I 84*:1587 (1988).
93. T. Hanawa and H. Yoneyama, Gas sensitivities of polypyrrole films to electron acceptor gases, *Bull. Chem. Soc. Jpn. 62*:1710 (1989).
94. W. Jaikun and M. Hirata, Research into normal temperature gas-sensitive characteristics of polyaniline material, *Sensors Actuators B 12*:11 (1993).
95. P. N. Bartlett, P. B. M. Archer, and S. K. Ling-Chung, Conducting polymer gas sensors. Part I: Fabrication and characterization, *Sensors Actuators B 19*:125 (1989).
96. P. N. Bartlett and S. K. Ling-Chung, Conducting polymer gas sensors. Part II: Response of polypyrrole to methanol vapour, *Sensors Actuators B 19*:141 (1989).
97. J. M. Slater and E. J. Watt, Examination of ammonia–poly(pyrrole) interactions by piezoelectric and conductivity measurements, *Analyst 116*:1125 (1991).
98. J. M. Slater and E. J. Watt, Piezoelectric and conductivity measurements of poly(pyrrole) gas interactions, *Anal. Proc. 29*:52 (1992).
99. J. M. Slater, E. J. Watt, N. J. Freeman, I. P. May,

and D. J. Weir, Gas and vapour detection with poly(pyrrole) gas sensors, *Analyst 117*:1265 (1992).
100. J. M. Slater, J. Paynter, and E. J. Watt, Multi-layer conducting polymer gas sensor arrays for olfactory sensing, *Analyst 118*:379 (1993).
101. P. N. Bartlett and J. W. Gardner, Odour sensors for an electronic nose, in *Sensors and Sensory Systems for an Electronic Nose*, NATO ASI Ser. E: Appl. Sci. Vol. 212 (J. W. Gardner and P. N. Bartlett, eds.), Kluwer, Dordrecht, 1992, Chap. 4.
102. K. C. Persaud, J. Bartlett, and P. Payne, Design strategies for gas and odour sensors, in *Robots and Biological Systems* (P. Dario, ed.), NATO ASI Series, Springer-Verlag, Berlin, 1992, p. 579.
103. T. Hanawa, S. Kuwabata, H. Hashimoto, and H. Yoneyama, Gas sensitivities of polypyrrole films doped chemically in the gas phase, *Synth. Met. 30*: 173 (1989).
104. P. N. Bartlett and S. K. Ling-Chung, Conducting polymer gas sensors. Part III: Results for four different polymers and five dfferent vapours, *Sensors Actuators 20*:287 (1989).
105. S. Dogan, U. Akbulut, T. Yalcin, S. Suzer, and L. Toppare, Conducting polymers of aniline. II. A composite as a gas sensor, *Synth. Met. 60*:27 (1993).
106. H. B. Gu, T. Takiguchi, S. Hayashi, K. Kaneko, and K. Yoshino, Effects of ammonia gas on properties of poly(*o*-phenylene vinylene), *J. Phys. Soc. Jpn. 56*:3997 (1987).
107. B. Chague, J. Germain, C. Maleysson, and H. Robert, Kinetics of iodine doping and dedoping processes in thin layers of poly-*p*-phenylene azomethine, *Sensors Actuators 7*:199 (1985).
108. M. J. Natan and M. S. Wrighton, Chemically modified microelectrode arrays, *Prog. Inorg. Chem. 37*: 391 (1989).
109. M. Nishizawa, T. Matsue, and I. Uchida, Fabrication of a pH-sensitive microarray electrode and applicability to biosensors, *Sensors Actuators B 13–14*: 53 (1993).
110. H. Munstedt, Properties of polypyrroles treated with base and acid, *Polymers 27*:899 (1986).
111. K. L. Tan, B. T. G. Tan, E. T. Kang, and K. G. Neoh, The chemical nature of the nitrogens in polypyrrole and polyaniline. A comparative study by X-ray photoelectron spectroscopy, *J. Chem. Phys. 94*: 5382 (1991).
112. Q. Pei and R. Qian, Electrode potentials of electronically conducting polymer polypyrrole, *Electrochim. Acta 37*:1075 (1992).
113. Q. Pei and R. Qian, Protonation and deprotonation of polypyrrole chain in aqueous solutions, *Synth. Met. 45*:35 (1991).
114. A. Guiseppi-Elie, Electroactive polymers with immobilized active moieties, U.S. Patent 5,352,574 (1994).
115. A. Q. Contractor, T. N. Sureshkumar, R. Narayanan, S. Sukeerthi, R. Lal, and R. S. Srinivasa, Conducting polymer-based biosensors, *Electrochim. Acta 39*(8/9):1321 (1994).
116. M. Szentirmay and C. R. Martin, Ion-exchange selectivity of Nafion films on electrode surfaces, *Anal. Chem. 56*(11):1898, (1984).
117. P. J. Riley and G. G. Wallace, Determination of gold using anion-exchange-based chemically modified electrodes, *Electroanalysis 3*:191 (1984).
118. G. G. Wallace, Dynamic conductors, *Chemistry Britain* Nov. 1993; p. 964.
119. P. R. Teasdale and G. G. Wallace, Molecular recognition using conducting polymers: Basis of an electrochemical sensing technology, *Analyst 118*:329 (1993).
120. Y. Lin and G. G. Wallace, Development of a polymer-based electrode for selective detection of dichloramine, *Anal. Chim. Acta 263*:71 (1992).
121. G. Tourillan, E. Dartyge, H. Dexpert, A. Fontaine, A. Jucha, P. Lagarde, and D. H. Sayers, Electrochemical inclusion of metallic clusters in organic conducting polymers: An in-situ dispersive X-ray absorption study, *J. Electroanal. Chem. 178*:357 (1984).
122. A. Talaie and G. G. Wallace, The effect of the counterion on the electrochemical properties of conducting polymers. A study using resistometry, *Synth. Met. 63*:83 (1994).
123. Z. Wang, A. Galal, H. Zimner, and H. B. Mark, Anodic stripping voltammetry at mercury "films" deposited on conducting poly(3-methylthiophene) electrodes, *Electroanalysis 4*:77 (1992).
124. G. Hailin, H. Zhao, and G. G. Wallace, Development of a polymer dispersed mercury modified electrode, *Anal. Chim. Acta 238*:345 (1990).
125. Z. Qi and P. G. Pickup, *Anal Chem. 65*:696 (1993).
126. G. Hailin and G. G. Wallace, Ion exchange properties of polypyrrole, *React. Polym. 18*:133 (1992).
127. S. Dong, Z. Sun, and Z. Lu, A new kind of chemical sensor based on a conducting polymer film, *J. Chem. Soc, Chem. Commun.* 993 (1988).
128. Z. Lu, Z. Sun, and S. A. Dong, New kind of chemical sensor based on a conducting polymer film, *Electroanalysis 1*:271 (1989).
129. G. Hailin and G. G. Wallace, High performance liquid chromatography on polypyrrole-modified silica, *J. Chromatogr. 588*:25 (1991).
130. P. R. Teasdale and G. G. Wallace, Characterizing the chemical interactions that occur on polyaniline with inverse thin layer chromatography, *Polym. Int. 35*:197 (1994).
131. R. J. Reynolds, P. A. Poropatic, and R. L. Toyooka, Electrochemical copolymerization of pyrrole with N-substituted pyrroles. Effect of composition on electrical conductivity, *Macromolecules 20*:958 (1987).
132. J. R. Reynolds, M. Pyo, and Y.-J. Qiu, Cation and anion dominated ion transport during electrochemical switching of polypyrrole controlled by polymer-ion interactions, *Synth. Met. 55–57*:1388 (1993).
133. W. Lien, W. H. Smyrl, and M. Morita, Cation and anion insertion in separate processes in poly(pyrrole) composite films, *J. Electroanal. Chem. 309*:333 (1991).

134. K. Naoi, M. Lien, and W. H. Smyrl, Quartz crystal microbalance study: Ionic motion across conducting polymers, *J. Electrochem. Soc.* 138:440 (1991).
135. A. Mirmosheni, W. E. Price, and G. G. Wallace, Adaptive membrane systems based on conductive electroactive polymers, *Journal of Intelligent Material Systems and Structures* 4(1):43 (1993).
136. P. R. Teasdale, M. J. Spencer, and G. G. Wallace, Selective determination of Cr(VI) oxyanions using a poly(3-methylthiophene)-modified electrode, *Electroanalysis* 1:541 (1989).
137. Y. Ikariyama, C. Galiatsatos, W. R. Heineman, and S. Yamauchi, Polypyrrole electrode as a detector of anionic substances, *Sensors Actuators* 12:455 (1987).
138. E. Wang and A. Liu, Polyaniline chemically modified electrode for detection of anions in flow injection analysis and ion chromatography, *Anal. Chim. Acta* 252:53 (1991).
139. P. Ward and M. R. Smyth, Development of a polypyrrole-based amperometric detector for the determination of certain anions in water samples, *Talanta* 40:1131 (1993).
140. G. Hailin, J. Zhang, and G. G. Wallace, Use of overoxidized polypyrrole as a chromium (VI) sensor, *Anal. Lett.* 25:429 (1992).
141. T. Ohsaka, K. Taguchi, S. Ikeda, and N. Oyama, *Denki Kagaku* 58:1136 (1990).
142. M. B. Inoue, K. W. Nebesney, Q. Fernando, M. M. Castillo-Ortega, and M. Inoue, Complexation of electroconducting polypyrrole with copper, *Synth. Met.* 38:205 (1990).
143. J. Roncali, L. H. Shi, R. Garreau, F. Garnier, and M. Lemaire, Tuning of the aqueous electroactivity of substituted poly(thiophene)s by ether groups, *Synth. Met.* 36(2):267 (1990).
144. J. Roncali, R. Garreau, D. Delabouglise, F. Garnier, and M. Lemaire, Modification of the structure and electrochemical properties of poly(thiophene) by ether groups, *J. Chem. Soc., Chem. Commun.* 11:679 (1989).
145. M. D. Imisides and G. G. Wallace, Deposition and electrochemical stripping of mercury ions on polypyrrole based modified electrodes, *J. Electroanal. Chem.* 246:181 (1988).
146. H. D. Abruna and A. R. Guadalupe, Electroanalysis with chemically modified electrodes, *Anal. Chem.* 57:142 (1985).
147. G. G. Wallace and Y. Lin, Preparation and application of conducting polymers containing chemically active counteranions for analytical purposes, *J. Electroanal. Chem.* 247:145 (1988).
148. K. K. Shiu, S. K. Pang, and H. K. Cheung, Electroanalysis of metal species at polypyrrole-modified electrodes, *J. Electroanal. Chem.* 367:115 (1994).
149. J. N. Barisci, P. C. Murray, C. S. Small, and G. G. Wallace, *Electroanalysis* 8(4):330 (1996).
150. M. J. Marsella, P. J. Carroll, and T. M. Swager, Conducting pseudopolyrotaxanes: A chemoresistive response via molecular recognition, *J. Am. Chem. Soc.* 116:9347 (1994).
151. M. J. Marsella and T. M. Swager, Designing conducting polymer-based sensors: Selective ionochromic response in crown ether containing polythiophenes, *J. Am. Chem. Soc.* 115:12214 (1993)
152. P. N. Bartlett and P. R. Birkin, The application of conducting polymers in biosensors, *Synth. Met.* 61:15 (1993).
153. A. J. Hodgson, M. J. Spencer, and G. G. Wallace, Incorporation of proteins into conducting electroactive polymers, *React. Polym.* 18:77 (1992).
154. R. John, M. Smyth, M. J. Spencer, and G. G. Wallace, Development of a polypyrrole-based human serum albumin sensor, *Anal. Chim. Acta* 249:381 (1991).
155. D. Barnett, D. G. Laing, S. Skopec, O. A. Sadik, and G. G. Wallace, Determination of *p*-cresol (and other phenolics) using a conducting polymer based electroimmunological sensing system, *Anal. Lett.* 27:2417 (1994).
156. O. A. Sadik, M. J. John, G. G. Wallace, D. Barnett, C. Clarke, and D. G. Laing, Pulsed amperometric detection of taumatin using antibody-containing poly(pyrrole), *Analyst* 119:1997 (1994).
157. O. A. Sadik and G. G. Wallace, Pulsed amperometric detection of proteins using antibody containing polymers, *Anal. Chim. Acta* 279:209 (1993).
158. M. K. Malmros, J. Gulbinski III, and W. B. Gibbs, A semi-conductive polymer film sensor for glucose, *Biosensors* 3:71 (1987/88).
159. T. L. Fare, M. D. Cabelli, C. D. T. Dahlin, S. M. Dalas, V. Narayanswamy, J. L. Schwartz, J. C. Silva, P. H. Thompson, and L. J. Van Houten, Transduction of an immunoassay for pesticides using conducting polymer-based biosensors: Assay development and characterization of electrode response, *Proceedings of the Gordon Conference on Bioanalytical Sensors*, Ventura, CA, March 1993.
160. E. Tamiya, I. Karube, S. Hattori, M. Suzuki, and K. Yokoyama, Micro glucose sensors using electron mediators immobilized on a polypyrrole-modified electrode, *Sensors Actuators* 18:297 (1989).
161. K. W. Johnson and J. J. Mastrototaro, A galvanostatic pretreatment method to reduce the settling time associated with amperometric glucose sensors, *Curr. Separ.* 12:104 (1993).
162. T. Shimidzu, T. Iyoda, H. Segawa, and M. Fujitsuka, Functionalizations of conductive polymers by mesoscopically structural control and by molecular combination of reactive moiety, in *Intrinsically Conducting Polymers: An Emerging Technology* (M. Aldissi, ed.), Kluwer, New York, 1993, p. 13.
163. A. Guiseppi-Elie, S. R. Pradhan, A. M. Wilson, D. L. Allara, P. Zhang, R. W. Collins, and Y.-T. Kim, Growth of electropolymerized polyaniline thin films, *Chem. Mater.* 5(10):1474 (1993).
164. M. Nishizawa, Y. Miwa, T. Matsue, and I. Uchida, Surface pretreatment for electrochemical fabrication of ultrathin patterned conducting polymers, *J. Electrochem. Soc.* 140(6):1650 (1993).
165. A. Guiseppi-Elie, A. M. Wilson, J. M. Tour, T. W.

Brockmann, P. Zhang, and D. I. Allara, Specific immobilization of electropolymerized polypyrrole thin films onto interdigitated microsensor electrode arrays, *Langmuir 11*:1768 (1995).
166. L. A. Prezyna, Y.-J. Qui, J. R. Reynolds, and G. E. Wnek, Interaction of cationic polypeptides with electroactive polypyrrole/poly(styrenesulfonate) and poly(N-methylpyrrole)/poly(styrenesulfonate) films, *Macromolecules 24*:5283 (1991).
167. Z. Sun and H. Tachikawa, Polypyrrole film electrode incorporating glucose oxidase, in *Biosensors and Chemical Sensors: Optimizing Performance Through Polymeric Materials* (P. G. Edelman and J. Wang, eds.), ACS Symp. Ser. 487, Am. Chem. Soc., Washington, DC, 1992, p. 134.
168. M. Hepel, L. Dentrone, and E. Seymour, in *Polymer Solutions, Blends, and Interfaces* (I. Noda and D. N. Rubingh, eds.), Elsevier, New York, 1992, p. 385.
169. D. Barnett, D. G. Laing, S. Skopec, O. Sadik, and G. G. Wallace, Determination of p-cresol (and other phenolics) using a conducting polymer based electroimmunological sensing system, *Anal. Lett. 27*(13):2417 (1994).
170. T. Shimidzu, Functionalized conducting polymers for development of new polymeric reagents, *React. Polym. 6*:221 (1987).
171. G. Bidan, E. M. Genies, and M. Lapkowski, Modification of polyaniline films with heteropolyanions: Electrocatalytic reductions of oxygen and protons, *J. Chem. Soc., Chem. Commun. 533* (1988).
172. G. Bidan, E. M. Genies, and M. Lapkowski, One-step electrochemical immobilization of keggin-type heteropolyanions in poly (3-methylthiophene) film at an electrode surface: Electrochemical and electrocatalytic properties, *Synth. Met. 31*(3):327 (1989).
173. P. N. Bartlett and J. M. Cooper, Modification of polyaniline films with heteropolyanions: Electrocatalytic reduction of oxygen and protons, *J. Chem. Soc., Chem. Commun. 533* (1988).
174. M. Hepel, E. Seymour, D. Yogev, and J. H. Fendler, Electrochemical quartz crystal microbalance monitoring of cadmium sulfide generation in polypyrrole and polypyrrole poly(styrenesulfonate) thin films, *Chem. Mater. 4*:209 (1992).
175. B. R. Saunders, K. S. Murray, R. J. Fleming, D. G. McCulloch, L. J. Brown, and J. D. Cashion, Physical and spectroscopic studies of polypyrrole films containing transition metal EDTA chelates, *Chem. Mater. 6*(5):697 (1994).
176. J. Yue and A. J. Epstein, Synthesis of self-doped conducting polyaniline, *J. Am. Chem. Soc. 112*:2800 (1990).
177. J. Yue, Z. H. Wang, K. R. Cromack, A. J. Epstein and A. G. MacDiarmid, Effect of sulfonic acid group on polyaniline backbone, *J. Am. Chem. Soc. 113*:2665 (1991).
178. A. J. Epstein and J. Yue, Processes for preparation of sulfonated polyaniline compositions and uses thereof, U.S. Patent 5,093,439 (1992).
179. A. Guiseppi-Elie and G. E. Wnek. Introduction of hydrophilicity to polyacetylene surfaces, *J. Polym. Sci.: Polym. Chem. Ed. 23*:2601 (1985).
180. A. Guiseppi-Elie and A. M. Wilson, Electroconductive polymer thin films with internal bioactivity for biosensors, *Proc. ACS Div. Polym. Mater. Sci. Eng. 72*:404 (1995).
181. J. M. Cooper, D. G. Morris, and K. S. Ryder, A bioelectronic interface using functionalized conducting poly(pyrroles), *J. Chem. Soc., Chem. Commun. 697* (1995).
182. P. Welzel, G. Kossmehl, J. Schneider, and W. Plieth, Reactive groups on polymer-covered electrodes. 2. Functionalized thiophene polymers by electrochemical polymerization and their application as polymeric reagents, *Macromolecules 28*:5575 (1995).
183. S. E. Wolowacz, B. F. Y. Yon-Hin, and C. R. Lowe, Covalent immobilization of glucose oxidase in polypyrrole, *Anal. Chem. 64*(14):1541 (1992).
184. B. F. Y. Yon-Hin, M. Smolander, T. Crompton, and C. R. Lowe, Covalent electropolymerization of glucose oxidase in polypyrrole: Evaluation of methods of pyrrole attachment to glucose oxidase on the performance of electropolymerized glucose sensors, *Anal. Chem. 65*:2067 (1993).
185. T. W. Brockmann and J. M. Tour, Planar conjugated pyrrole-derived polymeric sensors. Reversible optical absorption maxima from the UV to the near IR, *J. Am. Chem. Soc. 117*:4437 (1995).
186. R.D. McCullough and S. P. Williams, Toward tuning electrical and optical properties in conjugated polymers using side chains: Highly conductive head-to-tail heteroatom-functionalized polythiophenes, *J. Am. Chem. Soc. 115*:11608 (1993).
187. G. Tourillon and F. Garnier, Inclusion of metallic aggregates in organic conducting polymers: A new catalytic system, (poly(3-methylthiophene)-Ag-Pt), for proton electrochemical reduction, *J. Phys. Chem. 88*:5281 (1984).
188. K. M. Kost, D. E. Bartak, B. Kazee, and T. Kuwana, Electrodeposition of platinum microparticles into polyaniline films with electrocatalytic applications, *Anal. Chem. 60*(1):2379 (1988).
189. G. Arai, K. Matsumoto, T. Murofushi, and I. Yasumori, *Bull. Chem. Soc. Jpn. 63*:121 (1990).
190. N. Takano, M. Nakabayashi, and N. Takeno, Preparation of microparticle palladium incorporating poly[N-(5-hydroxypentyl)pyrrole] film-coated electrode, *Chem. Lett. 219* (1995).
191. F. G. Yamagishi, L. J. Miller, and C. L. van Ast, Conductive polymer-based sensors for application in non-polar media, *Proc. ACS PMSE Div. 71*:656 (1994).
192. L. J. Miller, C. L. van Ast, and F. G. Yamagishi, Reversible sensor for detecting solvent vapors, U.S. Patent 5,417,100 (1995).
193. M. Stern and A. L. Geary, Electrochemical Polarization. I. A theoretical analysis of the shape of polarization curves, *J. Electrochem. Soc. 104*(1):56 (1957).
194. M. Stern, Fundamentals of electrode processes in corrosion, *Corrosion 13*:775.

195. T. Shimidzu, T. Iyoda, M. Ando, T. Kaneko, A. Ohtani, and K. Honda, A novel anisotropic conducting thin film having a conducting and insulating layered structure, *Thin Solid Films 160*:67 (1988).
196. M. Ando, Y. Watanabe, T. Iyoda, T. Honda, and T. Shimidzu, Synthesis of conducting polymer Langmuir–Blodgett multilayers, *Thin Solid Films 179*:225 (1989).
197. X. Q. Yang, J. Chen, P. D. Hale, T. Inagaki, T. A. Skotheim, D. A. Fischer, Y. Okamoto, L. Samuelson, S. Tripathy, K. Hong, I. Watanabe, M. F. Rubner and M. L. DenBoer "Polyheterocycle Langmuir-Blodgett films" *Langmuir 5*(6):1288 (1989).
198. Z. Cai and C. R. Martin, Electronically conductive polymer fibers with mesoscopic diameters show enhanced electronic conductivities, *J. Am. Chem. Soc. 111*:4138 (1988).
199. C. R. Martin, L. S. Van Dyke, Z. Cai, and W. Liang, Template synthesis of organic microtubules, *J. Am. Chem. Soc. 112*: 8976 (1990).
200. S. Kuwabata and C. R. Martin, Mechanism of the amperometric response of a proposed glucose sensor based on polypyrole-tubule-impregnated membrane, *Anal. Chem. 66*(17):2757 (1994).
201. M. Pyo and J. R. Reynolds, Dual ion transport during electrochemical switching of conducting polymer bilayers, *J. Chem. Soc., Chem. Commun. 3*:259 (1993).
202. M. Pyo, G. Maeder, R. T. Kennedy, and J. R. Reynolds, Controlled release of biological molecules from conducting polymer modified electrodes: The potential dependent release of adenosine 5'-phosphate from poly(pyrrole adenosine 5'-phosphate) (PP-ATP) films, *J. Electroanal Chem. 368*(1/2):329 (1994).
203. T. Iyoda, H. Toyoda, M. Fujitsuka, R. Nakahara, H. Tsuchiya, K. Honda, and T. Shimidzu, The 100-Å-order depth profile control of polypyrrole-poly(3-methylthiophene) composite thin film by potential-programmed electropolymerization, *J. Phys. Chem. 95*:5215 (1991).
204. M. Fujitsuka, R. Nakahara, T. Iyoda, T. Shimidzu, and H. Tsuchiya, Optical properties of conjugated polymer superlattices prepared by potential-programmed electropolymerization, *J. Appl. Phys. 74*(2):1283 (1993).
205. R. A. Simon, A. J. Ricco, and M. S. Wrighton, Synthesis and characterization of a new surface derivatizing reagent to promote the adhesion of polypyrrole films to *n*-type photoanode *N*-(3-trimethoxylsilyl)-pyrrole, *J. Am. Chem. Soc. 104*:2031 (1982).
206. T. Komori and T. Nonaka, Electroorganic reactions on organic electrodes. 3. Electrochemical asymmetric oxidation of phenyl cyclohexyl sulfide on poly(2-valine)-coated platinum electrodes, *J. Am. Chem. Soc. 105*:5690 (1983).
207. T. Komori and T. Nonaka, Electroorganic reactions on organic electrodes, 6. Electrochemical symmetric oxidation of unsymmetric sulfides to the corresponding sulfoxides on poly(amino acid)-coated electrodes, *J. Am. Chem. Soc. 106*:2656 (1984).
208. M. Nishizawa, M. Shibuya, T. Sawaguchi, T. Matsue, and I. Uchida, Electrochemical preparation of ultrathin polypyrrole film at microarray electrodes, *J. Phys. Chem. 95*:9042 (1991).
209. M. Nishizawa, Y. Miwa, T. Matsue, and I. Uchida, Surface pretreatment for electrochemical fabrication of ultrathin patterned conducting polymers, *J. Electrochem. Soc. 140*:1650 (1993).
210. M. Nishizawa, Y. Miwa, T. Matsue, and I. Uchida, Ultrathin polypyrrole formed at a twin-microband electrode in the presence of dodecylsulfate, *J. Electroanal. Chem. 371*:L273 (1994).
211. A. S. Fiorillo, C. D. Bartolomeo, A. Nannini, and D. D. Rossi, PPy thin layers grown onto copper salt replica for sensor array fabrication, *Sensors Actuators 7*:399 (1992).
212. L. F. Rozsnyai and M. S. Wrighton, Selective electrochemical deposition of polyaniline via photopatterning of a monolayer modified substrate, *J. Am. Chem. Soc. 116*:5993 (1994).
213. S. Varanasi, S. O. Ogundiran, and E. Ruckenstein, An algebraic equation for the steady state response of enzyme-pH electrodes and field effect transistors, *Biosensors 3*:269 (1988) and references therein.
214. A. E. Karagözler, O. Y. Atama, A. Galal, Z.-L. Xue, H. Zimmer, and H. B. Mark, Jr., Potentiometric iodide ion sensor based on a conducting poly(3-methylthiophene) polymer film electrode, *Anal. Chim. Acta 248*:163 (1991).
215. A. Galal, Z. Wang, A. E. Karagozler, H. Zimmer, H. B. Mark, Jr., and P. B. Bishop, A potentiometric halide ion sensor based on conducting polymer film electrode, II. Effect of electrode conditioning and technical specifications, *Anal. Chim. Acta 299*(2):145 (1994).
216. D. Liu, M. E. Meyerhoff, and H. D. Goldberg, Potentiometric ion- and bioselective electrodes based on asymmetric polyurethane membranes, *Analytica Chimica Acta 274*(1):37 (1993).
217. M. Trojanowicz, W. Matuszewski, and M. Podsiadla, Enzyme entrapped polypyrrole modified electrode for flow-injection determination of glucose, *Biosensors Bioelectron. 5*:149 (1990).
218. N. F. Atta, A. Galal, A. E. Karagozler, G. C. Russell, H. Zimmer, and H. B. Mark, Jr., Electrochemistry and detection of some organic and biological molecules at conducting poly(3-methylthiophene) electrodes, *Biosensors Bioelectron. 6*:333 (1991).
219. A. Galal, N. F. Atta, J. F. Rubinson, H. Zimmer, and H. B. Mark, Jr., Electrochemistry and detection of some organic and biological molecules at conducting polymer electrodes. II. Effect of nature of polymer electrode and substrate on electrochemical behavior and detection of some neurotransmitters, *Anal. Lett. 26*(7):1361 (1993).
220. R. A. Saraceno, J. G. Pack, and A. G. Ewing, Catalysis of slow charge transfer reactions at polypyrrole coated glassy carbon electrodes, *J. Electroanal. Chem. 197*:265 (1986).
221. M. E. G. Lyons, W. Breen, and I. Cassidy, Ascorbic

acid oxidation at polypyrrole-coated electrodes, *J. Chem. Soc., Faraday Trans.* 87(1):115 (1991).
222. J. Wang and R. Li, Highly stable voltammetric measurements of phenolic compounds at poly(3-methylthiophene)-coated glassy carbon electrodes, *Anal. Chem.* 61(24):2809 (1989).
223. L. M. Abrantes, J. C. Mesquita, M. Kalaji, and L. M. Peter, Fast redox switching of polypyrrole films on ultramicroelectrodes, *J. Electroanal. Chem.* 307:275 (1991).
224. J. W. Gardner, T. C. Pearce, S. Friel, P. N. Bartlett, and N. Blair, *Sensors Actuators B* 18:240 (1994).
225. M. Josowicz and J. Janata, Suspended gate field effect transistor modified with polypyrrole as alcohol sensor, *Anal. Chem.* 58:5148 (1986).
226. M. Josowicz, J. Janata, K. Ashley, and S. Pons, Electrochemical and ultraviolet-visible spectroelectrochemical investigation of selectivity of potentiometric gas sensors based on polypyrrole, *Anal. Chem.* 59(2):253 (1987).
227. S. Yabuki, H. Shinohara, Y. Ikariyama, and M. Aizawa, Electrical activity controlling system for a mediator-coexisting alcohol dehydrogenase-NAD conductive membrane, *J. Electroanal. Chem.* 227:179 (1990).
228. T. Tatsuma, M. Gondaira, and T. Watanabe, Peroxidase-incorporated polypyrrole membrane electrodes, *J. Electroanal. Chem.* 301:155 (1991).
229. M. Shaolin, K. Jinquimg, and Z. Jianbing, Bioelectrochemical responses of the polyaniline uricase electrode, *J. Electroanal. Chem.* 334:121 (1992).
230. O. A. Sadik and G. G. Wallace, Effect of polymer composition on the detection of electroinactive species using conductive polymers, *Electroanalysis* 5:555 (1993).
231. R. C. Martinez, F. B. Dominguez, F. M. Gonzalez, and J. H. Mendez, Polypyrrole-dodecyl sulphate electrode as a microsensor for electroinactive cations in flow injection analysis and ion chromatography, *Anal. Chim. Acta* 279:299 (1993).
232. O. A. Sadik and G. G. Wallace, Detection of electroinactive ions using conducting polymer microelectrodes, *Electroanalysis* 6:860 (1994).
233. Y. Ikariyama and W. R. Heineman, Polypyrrole electrode as a detector for electroactive anions by flow injection analysis, *Anal. Chem.* 58:1803 (1986).
234. J. Ye and R. P. Baldwin, Flow injection analysis of electroinactive anions at a polyaniline electrode, *Anal. Chem.* 60:1979 (1988).
235. A. Heller, Electrical connection of enzyme redox centers to electrodes, *J. Phys. Chem.* 96:3579–3587 (1992).
236. W. Schuhmann, C. Kranz., J. Huber, and H. Wohlschläger, Conducting polymer-based amperometric enzyme electrodes. Towards the development of miniaturized reagentless biosensors, *Synthetic Metals* 61:31–35 1993.
237. P. Bauerle, A. Heller, S. Scheib, M. Sokolowski, and E. Umbach, Post-polymerization functionalization of conducting polymers: Novel groups. *Advanced Materials* 8(3):214–218 (1996).
238. A. Heller, C. Kranz, J. Huber, P. Bauerle, and W. Schuhmann, Amperometric biosensors produced by immobilization of redox enzymes at polythiophene-modified electrode surfaces. *Advanced Materials* 8(3):219–222 (1996).

35
Electrically Conducting Textiles

Hans H. Kuhn and Andrew D. Child
Milliken Research Corporation, Spartanburg, South Carolina

I. INTRODUCTION

Textiles are among the oldest materials known to humankind. In addition to their initial use as apparel, they were used in several early structural materials. Composite structures made from cotton fabric and phenolic resins were developed in the early 1900s. Since then, the importance of textile-reinforced materials has grown, and textiles now play a crucial role in many engineering materials including polymers, ceramics, and metals.

With the rapid development of the electrical and particularly the electronics industry, a need arose for flexible conducting and semiconducting materials. Flexible, highly conducting materials have been prepared by weaving thin wires of various metals such as brass and aluminum. Semiconductive textiles, including yarns and woven, nonwoven, and knitted fabrics, have been produced by impregnating textile substrates with conductive carbon or metal powders. The inclusion of appropriate binders allows a sufficient amount of these powders to be incorporated to impart the desired properties. These products have found wide application in the fields of electromagnetic interference (EMI) shielding and static dissipation. The development of filament-size fibers of stainless steel allowed the manufacture of conductive yarns, and subsequently fabrics, by blending various amounts of these fibers with synthetic or natural fibers. Carbon fibers and whiskers have also been incorporated by blending into textiles [1]. In addition, 100% carbon fibers, yarns, and fabrics are widely used in composite structures, particularly in the aerospace industry.

One of the most cost-effective approaches for producing conductive plastics is the incorporation of conductive fillers, particularly carbon [2]. However, the incorporation of a sufficient amount of carbon, up to 40%, to allow percolation causes a significant deterioration of mechanical properties in the polymer/filler blend, which results in considerable processing problems in the production of textile fibers. Commercial products based on nylon and polyester have been developed using highly filled polymers, either in the core or as a sheath of the fiber, that retain at least some of the strength of the unfilled polymer. Products of this nature are available from most commercial fiber producers.

Yet another route to conductive textiles involves the coating of textiles with metals. Silver-, copper-, and nickel-coated textiles using various synthetic fibers are now commercially available. A wide variety of processes can be used, including vapor deposition, sputtering, reduction of complexed copper salts, and electrodeless plating using noble metal catalysts [3,4]. Conductive copper sulfide can be deposited on synthetic fibers to yield moderately conductive fibers that are widely used in static-dissipating carpets due to their relatively light color [5,6].

Conducting polymers offer an interesting alternative to coated or filled plastics and textiles. However, these materials have several severe processing limitations. Although solution-spun fibers of polyaniline and poly(3-alkylthiophene) have been prepared (see Chapter 18) and thin films of many conjugated polymers can be produced electrochemically, these fibers and films are brittle, expensive to produce, and difficult to manufacture on a large scale. Considering these limitations, textiles of various kinds represent a reasonable choice as a substrate for thin coatings of such polymers. As shown in Fig. 35.1, conductive textile composites based on polypyrrole or polyaniline result in structures showing surface resistances of $10-10^3$ ohms/square ($\Omega/\square$). This falls in between metallized fabrics with a surface resistance that is usually below 1 $\Omega/\square$ and carbon-based blends with surface resistances above 10^3 $\Omega/\square$. These textile composites have a considerable advantage over metal-coated fabrics because of their excellent adhesion and

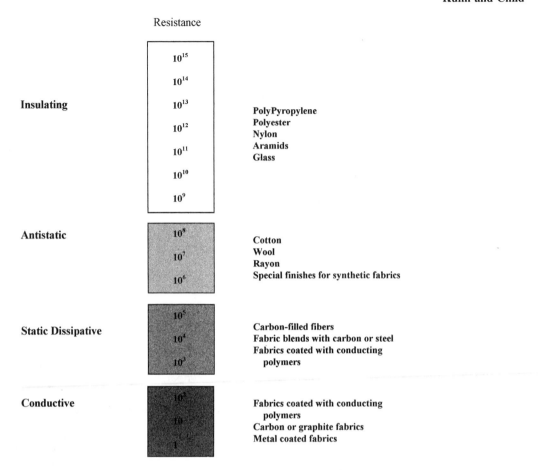

Fig. 35.1 Surface resistance of various textile products. The electrical properties of fabrics are measured by the surface or sheet resistance and expressed in ohms/square ($\Omega/\square$). Although volume resistance or volume conductance of the composite structure may be determined, these values are relatively meaningless as textiles, like foams, may contain large amounts of air.

noncorrosive character. The remainder of this chapter focuses on the synthesis, properties, and applications of these industrially useful products.

II. METHODS OF DEPOSITION

A. In Situ Polymerizations

When pyrrole is chemically polymerized, particularly from aqueous solutions, many researchers have observed the formation of films on either the liquid/air or the liquid/solid interface [7,8]. This spontaneous molecular assembly has been used to polymerize conductive polymers on the surface of numerous materials, including membranes [9], and has been successfully applied to textiles [10–12]. Considering industrial applications, it is fortunate that this in situ adsorption and polymerization proceeds especially well from aqueous solutions. Because of the desirability of aqueous processing, most of the research involving chemical oxidative polymerization on substrates has employed the water-soluble monomers aniline and pyrrole as starting materials.

It is known that the polymerization of pyrrole and aniline proceeds through the formation of radical cations that couple to form oligomers, which are further oxidized to form additional radical cations. These oligomers have been isolated in the electropolymerization of pyrrole [13]. The polymerization on the surface of textiles must proceed through one of these oligomeric intermediates, as neither the monomer nor the oxidizing agent adsorbs to the fabric. In addition, the oxidation of the monomer must be sufficiently slow to obtain uniform adsorption. The polymerization rate can be controlled by the monomer and oxidant concentrations, the reaction temperature, or the addition of an Fe(III) complexing agent. Ideal monomer concentrations for such reactions are 1–2 g/L. Under these conditions the polymerization occurs solely on the surface of the fiber and no polymer is observed in the liquid phase. Industrially, these polymerizations can be performed in textile dyeing equipment.

An interesting modification of the in situ process that results in a fabric with low surface resistance is reported in Refs. 14–16. An aqueous pyrrole solution containing a dopant is combined with a ferric chloride solution in an agitated vessel. The combined liquor is sprayed on the surface of a textile that is supported by a polyethylene film. The textile fabric and the support are rolled up and stored at a reduced temperature in a cylindrical vessel filled with the reaction mixture. After washing and drying, the procedure can be repeated several times to obtain surface resistances as low as 5 $\Omega/\square$.

B. Two-Step Processes

A large body of work describes the polymerization of conducting polymers by a two-step method. In 1983 Bjorklund, at the University of Linköping, showed that paper impregnated with ferric chloride and exposed to a solution of pyrrole yielded a composite structure with ionic and electronic conductance [17]. While in situ polymerizations are normally conducted in aqueous media, a two-step process can be accomplished using a variety of solvents. The choice of solvent may have a profound influence on the properties of the resulting composite structures. Several variations of the process have been reported; for example, the monomer may be applied from the vapor phase [18], or a porous substrate can be impregnated with a polymerizable monomer followed by exposure to an appropriate oxidizing agent [19]. The major advantage of a two-step process is its potential adaptation to a continuous process that is highly desirable for industrial applications. It is, however, more difficult to control and may not produce the ordered structures obtainable by an in situ template polymerization.

C. Other Methods of Deposition

The simplest approach to producing conductive textiles involves the application of conductive polymers from solutions. Highly conductive polymers like polythiophene, polypyrrole, and polyaniline are, for most practical purposes, insoluble in solvents suitable for application to textiles. Although soluble derivatives of some conducting polymers are available, several factors preclude their commercial introduction. First, the organic solvents needed are often not compatible with textile substrates, and many of these derivatives are soluble only in their undoped or insulating form. In addition, for cost reasons, industrial polymerizations must proceed from commercially available monomers. Finally, new environmental regulations have driven industry away from solvent-based coating processes.

Several water-based alternatives have been reported. Numerous papers describe the synthesis of aqueous emulsions of conductive polymers. Such products can be prepared by the emulsion polymerization of monomers such as pyrrole [20] or aniline [21]. A conductive polymer can also be synthesized on the surface of a polymeric emulsified particle [22]. Such aqueous emulsions can be applied to textiles to produce the desired conductive composite structure.

Two fairly recently published approaches allow the combination of a monomer with an oxidizing agent that does not initiate polymerization at room temperature. Pyrrole-2-carboxylic acid can be mixed with ferric chloride or other suitable oxidizing agents in an aqueous or solvent solution. At elevated temperatures decarboxylation of the monomer occurs, causing the instant polymerization of polypyrrole [23]. Similarly, solutions of 3,4-ethylenedioxythiophene can be mixed with ferric chloride or other suitable oxidizing agents. Polymerization occurs only at high concentrations and elevated temperatures, conditions that exist during the drying of textiles [24]. Although the literature does not specifically mention the application to textiles, such stable aqueous solutions are ideal for the continuous industrial application of these monomers to textiles.

III. POLYPYRROLE ON TEXTILES

It is interesting that the bulk of the literature on conductive textiles concerns the use of polypyrrole. This is most likely due to the unique ability of the polymerizing species to adsorb onto hydrophobic surfaces and form uniform coherent films. This reaction is naturally favored in the one-bath process but may also occur to some extent in the two-bath processes. At present time at least three groups of researchers are actively pursuing the in situ polymerization of pyrrole on textiles. Achilles Corporation in Tokyo, Japan, Milliken Research Corporation in Spartanburg, South Carolina, and Centre d'Etudes et de Reserches sur les Materiaux (CEREM) in Grenoble, France have active research groups in this field. Some of the early patent literature is summarized in Tables 35.1 and 35.2. The first reference to an in situ polymerization of pyrrole is described by Bjorklund and Lundström [17], who polymerized pyrrole from a dilute aqueous solution of ferric chloride to the surface of paper pulp. This resulted in a chloride-doped polypyrrole deposit of relatively low conductance and low stability on the cellulose fibrils. The process was later modified by researchers at Milliken Research Corporation and applied to textiles [10]. Since then, numerous investigators have used this process successfully on textiles and other substrates [25–39].

Experiments at Milliken Research Corporation using relatively inert substrates such as polypropylene or polyester fibers have shown that this process is not depen-

Table 35.1 Evolution of Patent Literature on the In Situ Polymerization of Pyrrole on Textiles

Oxidizing agent	Solvent	Substrate	Year	Ref.
Electrochemical	Acetonitrile	Cellulose, silk, linen, acrylic	1985	25
Potassium persulfate; ferric chloride	Water	Polyamide, polyester, cellulose acetate, rayon, poly(vinyl alcohol)	1986	26
Ferric perchlorate	Alcohol/water	Linen	1986	27
Ferric chloride	Tetrahydrofuran	Polyester, polyamide, cotton	1987	28
Ferric chloride	Water	Polyamide, polyester, aramide, aromatic polyester, glass, ceramic fibers	1989	12

Table 35.2 Evolution of Patent Literature on the Polymerization of Pyrrole on Textiles by the Two-Step Process

Step 1 Impregnate	Step 2 React with	Solvent	Substrate	Year	Ref.
Pyrrole	I_2 + KI potassium iodide I	Water	Polyester yarn	1986	29
Pyrrole	Ferric chloride	Water	Silk, rayon, polyamide, wool	1986	30
Pyrrole	Ferric chloride	Acetone/water	Glass	1986	19
Ferric chloride	Pyrrole vapor	Water	Polyamide, polyester, poly(vinyl alcohol)	1986	31
Ferric chloride	Pyrrole vapor	Dry	Glass, Kevlar, Nextel, polyamide, polyester	1986	18
$FeCl_3$ or $K_2S_2O_8$ solution mixed with low MW polymer	Pyrrole or aniline vapor	Dry	Fibers, fabrics	1987	32
Ferric chloride	Pyrrole vapor	Methanol	Polyester	1989	33

dent on an interaction between the chemicals used and the fibers but is controlled by the adsorption of an intermediate. The polymerization in dilute solutions (0.02 mol/L) is quite slow, and in the absence of a surface no precipitation of polymer is initially observed. If textile substrates are added to this mixture, a darkening of the substrate can be immediately observed, and all the polymer is found on the surface of the substrate in the form of a smooth coherent film of rather uniform thickness. In contrast to coatings from solutions or emulsions, no significant fiber-to-fiber bonding is observed, preserving the original strength and flexibility of the substrate (see Fig. 35.2).

Scanning election microscopy has revealed that in the early stages of the polymerization the film starts to grow by an island-type nucleation as shown in Fig. 35.3. This was previously observed in a two-step polymerization of pyrrole [40]. The film formation and the size of the nucleated particles are strikingly similar to those of metal vapor deposition [41] or electrodeless metal plating [42]. In the later stage of the polymerization, these independent nuclei impinge on each other and coalesce to form a continuous film. A typical example of an in situ polymerized polypyrrole film is shown in Fig. 35.4. This in situ polymerization, through adsorption of oligomeric intermediates, leads to ordered molecular assemblies on the surface of the substrate. A clear indication of this increase in order can be found in the X-ray photoelectron spectroscopy (XPS) spectra of polypyrrole polymerized on the surface of a quartz fabric. Compared to a polymer synthesized in the absence of such a surface, the deconvoluted area under the peak assigned to α–α coupling at 284.5 eV increases from about 40% to almost 70% in the presence of quartz fibers [43]. However, attempts to confirm the presence of long-range order by X-ray diffraction have failed to yield positive leads.

The in situ polymerization of pyrrole is relatively independent of the substrate. Fabrics produced by inserting a variety of yarns of equal mass into the fill direction and subsequently coated with polypyrrole show only small variations in conductance. Differences in conductance can be seen, however, if the yarn bundle is significantly different in surface area or hydrophobicity (Milliken Research Corp., unpublished). Surface polarity may also have an effect on the adhesion of the conducting polymer to the surface of the fiber. Porous fibers produced from polyacrylonitrile or nylon may provide better sites for anchoring to the substrates than more dense crystalline structures such as polyester. Fibers that lack polar groups, such as polyethylene or tetrafluoroethylene, make it more difficult to achieve proper adhesion. Relatively little work has been reported in this field, but generally good adhesion has been observed [44,45].

Electrically Conducting Textiles

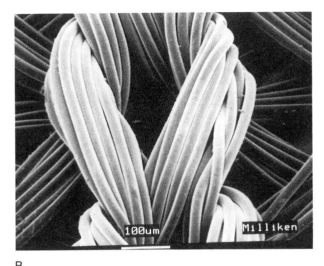

Fig. 35.2 (A) Scanning electron micrograph of polypyrrole-coated knitted nylon fabric. (B) Same fabric at higher magnification.

The situation is quite different for glass fibers. Whereas hydrophilic fabrics made from wool, cotton, or rayon can be successfully coated with polypyrrole, cleaned glass fabrics do not respond to the treatment with pyrrole. Treatment of glass fabrics with commercially available adhesion promoters, such as aminosilanes, followed by treatment with pyrrole and an oxidizing agent leads to conducting materials. Superior results have been obtained by applying an aminosilane-based adhesion promoter that is substituted with a pyrrole ring. The silane group reacts with the glass surface, and the terminal pyrrole ring may copolymerize with the pyrrole, resulting in improved film–fiber adhesion [46].

The kinetics of the polymerization of pyrrole can be followed by measuring the disappearance of the reactants. Pyrrole concentrations can be determined by gas chromatography [10]. The depletion of iron can be followed by the titration of the iron(III) chloride with a complexing agent, and the disappearance of the sulfur-containing doping agents can be determined by inductively coupled plasma (ICP) [47]. Experiments reveal that the reaction follows second-order autocatalytic kinetics [10]. A plot of the concentrations of pyrrole monomer (surface and no surface) versus time at room temperature is shown in Fig. 35.5. The surface resistance of the polypyrrole-coated textiles can be controlled by

Fig. 35.3 Island nucleation in the early stages of the in situ polymerization of pyrrole on S-2 glass fabric.

Fig. 35.4 Scanning election micrograph of polypyrrole on S-2 glass fabric during the later stages of polymerization.

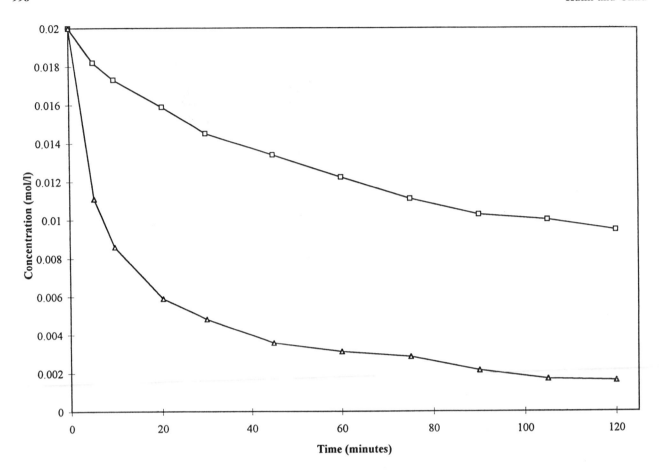

Fig. 35.5 Depletion of pyrrole monomer versus time. (□) Without fibers; (△) with fibers. (Data form Ref. 10.)

altering the concentration of chemicals added to the polymerization bath. However, the resistance of the film does not change linearly with the polymer add-on as expected. Films containing twice the amount of polymer are more than twice as conductive. This effect may be due to the formation of percolating networks, as multiple monolayers are deposited and hopping between conducting islands is facilitated. To oxidize pyrrole to the fully oxidized polypyrrole, at least 2.33 mol of ferric chloride is needed. If the reaction is conducted using a pyrrole concentration of 0.02 mol/L and a lqiuid/fabric ratio of 30:1, about 60% of the pyrrole is converted to the polymer in 3–4 h. Higher molar ratios of ferric chloride may be used to increase the yield of polymer without affecting its quality.

The morphology of the polypyrrole film is highly dependent on its composition. The oxidation of pyrrole in aqueous solutions yields an oxidized polypyrrole with a degree of doping of 0.25–0.33; therefore, every third or fourth repeat unit has a positive charge neutralized with a counterion. If ferric chloride is used as an oxidizing agent, this counterion is most likely Cl^-. Because of the relative mobility of the chlorine ion, the resulting polymer is not very stable, particularly at higher temperatures when thermal dedoping is observed.

Films prepared with the addition of hydrophobic doping agents form denser, more conducting, and more stable films [48]. It is well known that the type of doping agent can have a considerable effect on the conductance and morphology of polypyrrole [49]. For the in situ polymerization, these dopants are traditionally introduced as the sodium salts of arylsulfonates. In the electrochemical oxidation of pyrrole, control over the type of doping agent incorporated can be attained by using the appropriate electrolyte. In the chemical oxidation using ferric chloride, the addition of other dopants leads to a mixture of counterions in the polymer, although doping with the chloride predominates owing to the high concentration of chloride in the reaction mixture. To obtain polymers that are doped with the more desirable aromatic sulfonic acids such as p-toluenesulfonic acid, investigators at Rockwell International have used the iron(III) salts of these acids, which can be prepared in the laboratory with relative ease [50]. It is fortunate, for cost reasons, that some aromatic sulfonic acids show great selectivity in the doping of polypyrrole in the presence of chloride. The hydrophobicity of these molecules may result in a thermodynamic partitioning effect, resulting in preferential incorporation of the organic dopants and an almost chlorine-free product. Preferred dopants for such reactions include anthraquinone-2-sulfonic acid, 2-naphthalenesulfonic acid, and trichlorbenzenesulfonic acid [16,51,52]. The use of these dopants allows the production of conductive textiles with high electrical conductance and respectable environmental stability.

Because of the relatively small amounts of polymer deposited on textiles (1–5%), it is difficult to determine the elemental composition of the polymer produced. There are some indications, however, that the elemental composition of polypyrrole powders produced in the same molar concentrations may be quite similar to that of the deposits on the surface of textiles. The morphology, as well as the degree of order, may be considerably different. A detailed study of the polymerization of pyrrole in aqueous solution was compiled in a doctoral thesis by Marie-France Planche and published in a number of papers [16,53,54]. Many papers discuss the elemental analysis of polypyrrole powders and, in most cases, assign the discrepancies in the carbon, hydrogen, nitrogen, and sulfur determination to oxygen and its possible source as adsorbed water and carbon dioxide. Investigators at Colorado State University [55] showed convincing evidence that the excess oxygen is due to nucleophilic substitution of the polymer backbone with OH groups when the polymerization is conducted in the presence of water. Studies at Milliken Research Corporation indicate that these reports may be correct. A decrease in the excess oxygen attributed to hydroxyl groups is found when certain hydrophobic dopants are used (see Table 35.3).

In an attempt to produce more conducting or more stable products, a variety of additives and cosolvents have been used in these polymerizations with mixed results. Relatively small amounts of certain solvents or surface-active agents may prevent the adsorption of polypyrrole to the surface of textiles. For instance, the presence of 20 g/L of acetone inhibits the deposition, whereas the same amount of methanol allows proper adsorption. Similarly, 3.3 g/L of 1,4-dihydroxybenzene interferes with the adsorption, whereas the same amount of p-nitrophenol allows proper adsorption and may yield benefits similar to those previously published for the chemical synthesis of polypyrrole powders [57]. Cationic and nonionic surface-active agents also interfere with the in situ polymerization on the surface of textiles. Anionic compounds such as dodecylbenzenesulfonic acid interferes slightly with proper adsorption, whereas commercially available surfactants such as alkylnaphthylsulfonate act as hydrophobic dopants [58,59]. Commercially, textiles may be scoured with these types of surfactants prior to treatment with polypyrrole.

In some of the earlier investigations, 2,5-bis-(2-pyrrolyl)pyrrolidine was isolated from the mother liquors of the aqueous polymerization of pyrrole by using ferric chloride [57]. A more recent publication describes the formation of maleimide under similar conditions of polymerization [60]. The presence of such compounds may lead to copolymerization, resulting in undesirable defects in the polymer chain. Although we can confirm the formation of small amounts of maleimide as described in the above literature, we find that its formation is reduced by a factor of 5 if anthraquinone-2-sulfonic acid is used as a dopant for the preparation of polypyrrole powders. Furthermore, the addition of large amounts of maleimide to the in situ polymerization on the surface of textiles (equal weights to pyrrole) does not significantly affect the adsorption or the stability of the resulting coating. Attempts at Milliken Research Corporation to isolate, or observe through gas chromatography, products

Table 35.3 Elemental Analysis of Polypyrrole Powders Polymerized at Room Temperature Using 0.2 mol/L of Pyrrole and a $FeCl_3$/Pyrrole Ratio of 2.4

Dopant	Concentration (g/L)	Atomic ratio			
		Cl/N	S/N	(Cl + S)/N	O/N[a]
Ferric chloride	—	0.22	—	0.22[b]	0.83
p-Toluenesulfonic acid	3.3	0.15	0.12	0.27	0.83
p-Toluenesulfonic acid	10.0	0.11	0.18	0.29	0.57
Anthraquinone-2-sulfonic acid	3.3	0.07	0.27	0.34	0.36

[a] In excess of sulfonate ions.
[b] The lower value obtained for the chlorine ion–doped sample is actually in good accordance with the literature [56] for samples run at the same pyrrole concentrations. This may result from the relative lability of these ions, which may be partially removed by the repeated washings with water.

other than polypyrrole in the mother liquor were unsuccessful. In addition, the weight gain measured on the textile substrate is in good agreement with the sum of the pyrrole and doping agent depleted from the solution.

IV. POLYANILINE ON TEXTILES

A smaller body of work has been reported on polyaniline-coated textiles. This may be due to concerns that the highly toxic benzidine moiety may be formed during the oxidative polymerization of aniline under acidic conditions and may be incorporated in the polymeric structure. The first in situ polymerization of aniline on the surface of textiles was reported by Milliken Research Corporation [10] and has since been used by other investigators [11,61,62]. To obtain polymerization on the surface of the textile substrate only, the polymerization must be carefully controlled. As in the case of polypyrrole, this is best achieved by working in very dilute monomer solutions (about 0.03 mol/L). Usually the polymerization is initiated by using a persulfate catalyst, such as potassium or ammonium persulfate, as has been described for the polymerization of aniline in aqueous media under acidic conditions [63]. This reaction exhibits a considerable induction period that varies from several minutes in concentrations of 0.5 mol/L to rather long periods in the desirable concentration of 0.03 mol/L. If sodium or ammonium vanadate is used for the oxidation, this initiation period is eliminated and the polymerization on the surface of textiles can easily be controlled [64]. In analogy to the pyrrole oxidation with ferric chloride, it is believed that the aniline forms a complex with the vanadium compound to allow a faster reaction to the polymerizing species. It is actually sufficient to use catalytic amounts of vanadium and conduct the bulk of the oxidation with persulfate. No fiber–fiber bonding occurs under these conditions, and the strength and flexibility of the substrate are preserved. No information has been published concerning order in the deposited polymer, but it stands to reason that effects similar to those observed for polypyrrole may be possible. Tzou and Gregory [65] studied the oxidative chemical polymerization kinetics of aniline with ammonium persulfate in aqueous hydrochloric acid solutions with and without textile substrates and noted an increase in reaction rates in the presence of a textile substrate.

In contrast with the reaction with pyrrole, special precautions must be taken with polyaniline-coated textiles during the final washing and rinsing operation. The doping mechanism is somewhat different from the one in polypyrrole, and the counterion, e.g., chlorine or p-toluenesulfonic acid, can easily be partially removed by washing with water. It is therefore necessary to wash with water containing sufficient amounts of the counterion as demonstrated for polyaniline powders [63]. Because of the solubility and processibility of polyaniline doped with camphorsulfonic or dodecylbenzenesulfonic acid, polyaniline can be processed into fibers that may subsequently be incorporated into textiles. These aspects are discussed in Chapter 18.

Several variations on the production of polyaniline-coated textiles have been reported. Researchers at the Huazhong University in China studied the in situ polymerization of aniline on poly(vinyl alcohol) fibers using ammonium persulfate as the oxidizing agent. Conductive fibers with good mechanical properties and stability were obtained [66]. In addition, researchers at the Atomic Energy Commission in Monts, France developed a continuous process to treat textiles with polyaniline. The textile substrate is first impregnated with a relatively concentrated hydrochloric acid solution of aniline (1–20%) and dried. Subsequently, the fabric is exposed to a solution of potassium dichromate in hydrochloric acid. The polyaniline-coated fabric is washed in 2 N hydrochloric acid solution and dried. This process can be repeated up to three times, resulting in a uniform, coherent layer of chloride-doped polyaniline with a thickness of up to 0.5 μm. A single pass takes 10–20 min, and, depending on the number of passes, surface resistances as low as 10 $\Omega/\square$ may be reached. The intrinsic conductivity of the polyaniline deposited by this procedure reaches about 20 S/cm [67,68]. The processing equipment is shown in Fig. 35.6.

A similar process was patented by investigators at the E. I. DuPont de Nemours Co. The fiber to be treated, preferably poly(p-phenylene terephthalamide) is first soaked with aniline hydrochloride and then exposed to the oxidizing agent [69]. Fabrics treated in this fashion exhibit surface resistances of several hundred ohms per square.

V. PROPERTIES OF CONDUCTIVE TEXTILES

A. Mechanical Properties

With few exceptions, the mechanical properties of conductive polymers are rather poor. This is a result of the cross-linked nature and aromatic character of the backbone. One exception is polyaniline, which can be prepared as oriented crystalline fibers of excellent strength (see Chapter 18). The use of a textile substrate represents a convenient method of introducing mechanical strength, flexibility, and processibility to conducting polymers for practical applications.

Fabrics coated with a thin layer of polypyrrole or polyaniline have essentially the same mechanical properties as the textile substrate, with minor variations depending on processing conditions. Even fibers that are highly susceptible to oxidation or hydrolysis under acidic conditions, such as cotton or nylon, do not deteriorate during the in situ polymerization of pyrrole. The resistance of these fibers to degradation in a reaction

Fig. 35.6 Continuous polyaniline deposition range. (Photo courtesy of Atomic Energy Commission, Monts, France.)

bath with a pH of 2 or lower is believed to be due to the rapid formation of a protective coating of the conjugated polymer on the fiber surface [70]. The tactile properties of the textile also remain virtually unchanged for thin coatings of conducting polymers. Thick coatings of polypyrrole have been shown to cause the fabric to become stiff and may also affect the tear strength, but the tensile strength and elongation remain unchanged.

Many applications of conducting textiles involve the formation of laminates or composites with various thermoset resins such as epoxy, imide, and rubber. The adhesion between the various layers of the composite is a critical factor in the utility of these structures. Only a few papers have addressed the adhesion of polypyrrole-coated fabrics with epoxy resins [44,45]. The adhesion at the polypyrrole/textile interface is reasonably strong because of the intermolecular forces between the adsorbed polymer layer and the substrate. The adhesion at the polypyrrole/epoxy interface benefits from the polarity of the polypyrrole layer. Similar results have been observed for polyaniline-coated materials [67].

B. Electrical Properties

The electrical properties of conducting textiles depend on the mass of the substrate, the diameter of the individual textile fibers, the thickness of the adsorbed layer, and the intrinsic volume conductivity of the conducting polymer. Volume conductivities for polyaniline-coated textiles as high as 20 S/cm have been reported, whereas polypyrrole coatings have been measured in the 100 S/cm range. Using multiple coatings, surface resistances as low as 5 $\Omega/\square$ have been reported. By adjusting the concentration of the reaction media, the surface resistance can be tailored to any desired value, with a lower limit somewhat dependent on the substrate weight. Investigators at Milliken Research Corporation reproducibly prepared polypyrrole-coated fabrics with surface resistances ranging from 20 to 2000 $\Omega/\square$ on textile substrates weighing 50–200 g/m^2. The reproducibility within this range has been within 5%. Although textiles with surface resistances of $\approx 10^5$ $\Omega/\square$ can be prepared, the reproducibility is somewhat less reliable due to the small amount of polypyrrole deposited.

The resistance of conducting textiles can be measured using a four-probe technique as in measurements of conducting films or pressed pellets. Since the coating is much thinner than the substrate material, the resistance is normally expressed as a sheet resistance in ohms per square rather than a volume resistivity or conductivity. To obtain reliable measurements of resistance on the uneven surfaces of woven or knitted textiles, significant pressure must be applied at the electrical contacts. For this reason, it may be preferable to use the established two-probe AATCC test method 76-1987, "Electrical Resistivity of Fabrics." We have found that a pressure of 1500 g for a 2 in. wide gold-coated clamp, $\frac{1}{16}$ in. thick, gives accurate and repeatable values for surface resistance on most fabrics. The same instrumentation may be used to measure the resistance per unit length of yarns or fibers. It should be noted that since polypyrrole and polyaniline, like other conducting polymers, are semiconducting materials, their conductance increases with increasing temperature. An increase of 10% in the conductance of polypyrrole-coated polyester fabrics has been observed when the temperature is raised from 25°C to 100°C.

C. Microwave Properties

The range of surface resistance obtainable with conductive textiles based on polypyrrole or polyaniline (10–10,000 $\Omega/\square$) allows these materials to be used for a number of applications. The use of such fabrics as reinforcements to polymeric resins results in composites that have interesting and well-controlled electrical properties. These unusual electrical properties are quite evident in the microwave region of the electromagnetic spectrum. The literature on microwave properties of such composites is not extensive, but some results have been reported by researchers in the United States, the United Kingdom, and France [68,71–74].

The addition of small amounts of a conductive filler such as carbon black or small metal particles to a polymer will do very little to change its dielectric loss until the amount added is sufficient for the particles to be almost in physical contact with one another. Small tunneling currents between adjacent particles render the polymer lossy (able to dissipate energy) and eventually conductive as the amount of filler is increased. This increase in the imaginary part of the dielectric constant (i.e., the conductivity) remains small until the percolation threshold is reached. The addition of the conductive filler does increase the real (capacitive) component of the complex dielectric constant significantly even before any useful dielectric loss is measured. This is the nature of discontinuous, granular, or percolative media.

The continuity of the conductive polymer coatings on textiles leads to a rather different dielectric behavior in polymer composites. In particular, the imaginary part of the complex dielectric constant of such a composite can be increased over a substantial range without a corresponding change in the real part of the dielectric constant. This behavior was reported by Sengupta and Spurgeon [71] for polyester resin–based composites containing polypyrrole-coated fabrics in the 26.5–40 GHz range. Over this frequency range, they report that the measured reflectivity data can be adequately fit by a dielectric function, the real part of which is essentially constant, with the imaginary part given by the bulk conductivity of the fabric divided by 2π times the frequency. In polyester fabric–based composites, the observed increase in the real part of the dielectric constant, ϵ', was comparatively small, although it was somewhat larger in glass fabric–based composites (see Fig. 35.7). The uniformity of the coatings on the particular glass fabrics tested did not appear to be as consistent as that of the coatings on the polyester fabrics when examined with a scanning electron microscope, suggesting that the glass fabrics may have been somewhat granular.

Somewhat similar results were reported by Jousse et al. [68], who found that composites made with polyaniline-coated fabrics show only a small increase in the real part of the dielectric constant above that of the base composite made with untreated fabric. They also report that the imaginary part of the permittivity is frequency-dependent as described above. The conductivity of the polyaniline coating appears to be slightly frequency-dependent whereas the conductivity of the polypyrrole coating is not frequency-dependent in the range of 10–300 GHz. The broad range of permittivity obtained via the conductivity control enables a variety of absorbing structures to be made.

Chambers and coworkers [72–74] also explored the properties of semiconductive sheets of polypyrrole treated paper and cloth, which they characterize as a parallel RC circuit. They too report that the capacitive part of the complex impedance of the sheets depends on the morphology of the polypyrrole coating. Measured reflectivity plots for Salisbury screen (narrowband) and Jaumann (broadband) absorbers fabricated from sheets of polypyrrole-treated material are also presented in their article. The measured results agree well with their model calculations, indicating the potential utility of conductive polymer–treated fabrics in radar-absorbing structures.

D. Stability of Conducting Textiles

In many applications, the stability of conducting materials is an important concern. Metal-coated fabrics, fabrics coated with inorganic conducting particles, and car-

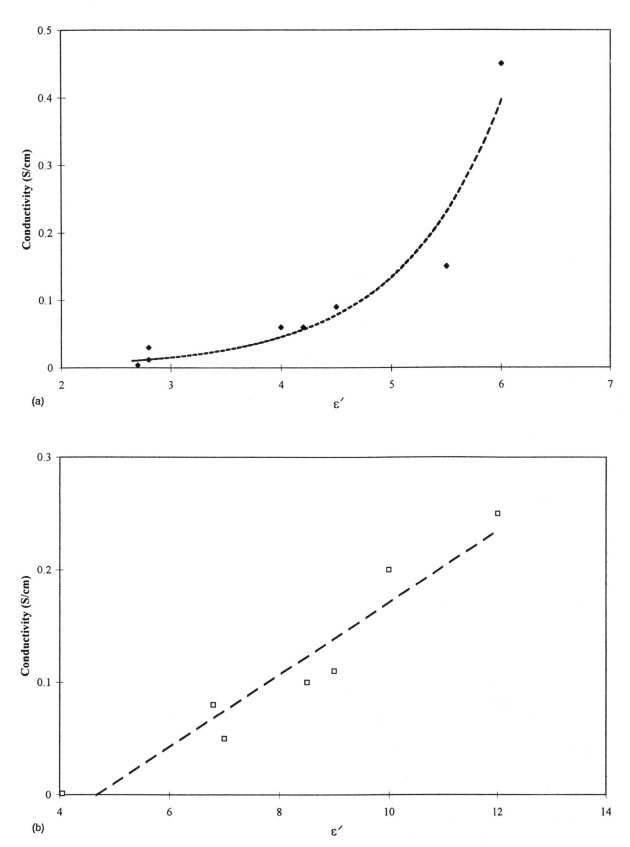

Fig. 35.7 Dielectric constant ϵ' versus conductivity for polypyrrole coatings on (a) polyester fabric and (b) S-2 glass fabric at 35.27 GHz. (Data from Ref. 71, reprinted by permission of the Society for the Advancement of Material and Process Engineering.)

bon-filled materials generally demonstrate stabilities that are equivalent to those of the parent conducting materials when monitored in a noncorrosive environment. Corrosion, or the degradation of the substrate, which causes interruptions in the percolating network, is normally the limiting factor in the ultimate lifetime of these materials. Organic materials, including conducting polymers, are much less stable in air owing to their relatively high reactivity with a variety of atmospheric chemicals, most notably oxygen. The stability of conducting polymers and conducting polymer–coated textiles has been shown to be improved dramatically when the material is isolated from the environment with a protective coating or laminate [44]. Münstedt [75] found no loss of conductivity in a polypyrrole film after 200 days at 80°C in a nitrogen atmosphere. In some applications, the requirement of highly flexible conducting textiles precludes the use of thick coatings, and the intrinsic resistance of the conducting polymer to oxidation becomes a limiting factor.

Investigations into the stability of polypyrrole powders, films, and coatings have been carried out by several groups. The results indicate that the degradation of polypyrrole on textile substrates follows the same rate laws, and presumably the same mechanism, as the degradation of polypyrrole powders and films. In almost all cases, however, the stability of electrochemically grown polypyrrole films is superior to that of chemically synthesized polypyrrole coatings or powders [76]. This can be attributed to two factors: (1) the increased control of oxidation potential in electrochemical polymerizations, and therefore the polymerization rate, leading to films with fewer defects, and (2) the control of the dopant ions by selection of the electrolyte in electrochemical polymerizations. At oxidation potentials achieved during the chemical synthesis of polypyrrole by ferric chloride, overoxidation can occur, leading to nonconjugated moieties and decreased stability within the polymer chain [60]. In addition, significant thermal dedoping has been shown to occur with chloride-doped polypyrrole [77].

The polymerization of pyrrole with ferric chloride leads to materials with a significant amount of chloride doping even with the addition of arylsulfate doping agents. Investigations at Milliken Research Corporation have shown that polypyrrole powders with higher chlorine contents display lower stabilities. In addition, in a masters theseis by James Moreland at Clemson University [78], it was reported that chlorine-free conducting textiles prepared from pyrrole and ferric tosylate show no measurable decrease in conductance after 30 days at 100°C in an argon atmosphere, whereas films produced with ferric chloride are significantly less stable. Although thermal dedoping is observed with some organic sulfonate dopants as well, such as toluenesulfonate, the degradation in the presence of oxygen is significantly faster and dominates the stability kinetics when measured in air.

Samuelson and Druy [79] reported on the stability of an electrochemically grown polypyrrole film in 1985. The data indicate that the degradation proceeds by a first-order process when arylsulfonate dopants are used. The proposed degradation mechanism was said to involve the reaction of the polymer backbone with oxygen or water. These results were confirmed by Moss and Buford in 1991 [80]. XPS data from this study show an increase in carbonyl functionality with increased aging times, indicating oxidation products. The oxidation is thought to involve the reaction of ground-state oxygen with free radical sites on the conducting polymer as described earlier for polyacetylene [81]. Polypyrroles have a large population of free radical sites (polarons) that act as charge carriers, and since the absorption of oxygen is assumed to be faster than its reaction with the polymer backbone, first-order kinetics are observed. Moss and Buford [80] suggest that the degradation may proceed through the formation of hydroperoxy radicals that decompose to form a variety of ketone, aldehyde, and hydroxyl groups. The conductivity loss is due to both the formation of these nonconjugated moieties in the polymer backbone and the loss of the free radical charge carriers. Using XPS, Milliken investigators confirmed a significant increase in carbonyl functionality but observed no mass increase of polypyrrole powders during aging. This suggests that the carbonyl groups may be the oxidation products of pendant hydroxyl groups that were incorporated during the chemical polymerization in water as suggested by Lei and Martin [55].

Several other groups have since published studies involving the stability of polypyrrole and polypyrrole-coated textiles. A plot produced from data obtained at Milliken Research Corporation that displays the resistance as a function of time at several different temperatures for an anthraquinone-2-sulfonic acid doped polypyrrole-coated textile is shown in Fig. 35.8. Similar results were obtained by Truong and coworkers [82,83] and Thieblemont and coworkers [16,84]. Contrary to the earlier reports, these studies concluded that the kinetic of polypyrrole degradation are not first-order but follow a diffusion-controlled $t^{1/2}$ curve similar to a Frickian sorption plot.

A plot of the relative resistance (R_0/R) versus $t^{1/2}$ leads to a straight line for R_0/R values of less than 0.5. The decay can be described by

$$1 - R_0/R = At^{0.5} \tag{1}$$

This implies that the diffusion of the oxidizing species into the polymer film is the rate-limiting step in the degradation of polypyrrole films. From Fick's second law, Thieblemont et al. [16] and Truong et al. [83] calculated diffusion coefficients from the expression

$$D = 16\pi A^2 e^2 \tag{2}$$

where e is the thickness of the film and A is the diffusion constant from Eq. (1). The diffusion coefficient calculated for thin films of polypyrrole (0.2–1.0 μm) range

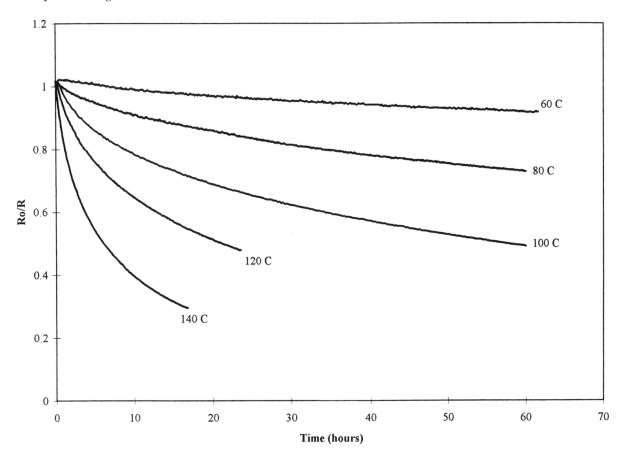

Fig. 35.8 Normalized surface resistance increase versus time for polypyrrole-coated polyester fabrics at various temperatures in a forced air oven.

between 2×10^{-11} cm^2/s [83] and 8×10^{-16} cm^2/s [16] at 90°C. The wide variation in values reported from different groups can be attributed to different polymerization conditions, different aging conditions (i.e., airflow), and the inaccuracies inherent in the measurements of thickness of uneven films. Despite this wide variation in data among research groups, these values are all significantly lower than those of traditional gas diffusion through polymers (10^{-9}–10^{-6} cm^2/s at 25°C) and are more representative of gaseous doping of polyacetylene (10^{-17} cm^2/s) [85]. Such low values may represent the slow diffusion of oxygen through the conjugated chains resulting from interactions between triplet (ground-state) oxygen and polarons [16].

Kinetic studies at Milliken Research Corporation have revealed that although the diffusion-controlled model predicts the conductivity loss during the initial stages of degradation, at longer times the decay follows first-order kinetics [48]. This change in the rate-determining step for degradation implies that the diffusion of oxygen is initially rate-limiting; however, the reaction of oxygen with the polymer free radical becomes rate-limiting after the resistance has approximately doubled.

In addition, some evidence has been offered to explain the variation in stability among various doping agents. SEM studies have revealed a correlation between smooth morphologies and high stabilities. Large planar aromatic dopants, particularly anthraquinone-2-sulfonic acid and naphthalene-2-sulfonic acid, yield polymer coatings with very smooth surfaces [48]. The rate of degradation in the diffusion-controlled regime of these films was found to be significantly lower than in films produced with other arylsulfonate dopants.

The stability of polyaniline-coated textiles has not been extensively studied. Several studies have been published on the thermal degradation of polyaniline films and powders [86–90]. The stability has been found to be highly dependent on the form of the polymer (film, pressed pellet, or powder), its thermal and chemical history, and the dopant employed. The dependence on sample form indicates that the degradation is a diffusion-controlled process [86]. Among conducting polymers, the stability of polyaniline is somewhat better than that of the poly(3-alkylthiophenes) but not as good as that of electrochemically generated arylsulfonate-doped polypyrrole. Polyaniline films that employ methanesulfonic acid as the dopant are more stable in moist air than in nitrogen atmospheres, indicating that oxidation by oxygen is not the primary degradation mechanism in these films [90].

VI. APPLICATIONS

Widespread industrial production of conductive polymers is still lagging considerably behind numerous earlier optimistic forecasts. There are, however, two corporations that are actively marketing conductive textiles. They are the industrial fabrics division of Milliken & Company in Spartanburg, South Carolina, which is marketing polypyrrole-coated textiles under the trademark Contex (see Fig. 35.9) and the Industrial Materials Division of Achilles Corporation in Tokyo, Japan. A summary on applications of conductive polymers including conductive textiles has been published by Miller [91]. Applications of conductive textiles may be divided into static dissipation, EMI shielding, heating elements, composite structures, and military applications. The advantages of using conductive textiles lie in the strength and flexibility of the substrate, the wide range of substrates available, and the wide range of surface conductance available in these fabrics. In addition, anisotropic materials can be constructed with almost any ratio.

A. Static Dissipation

Customers of Milliken report that polypyrrole-coated fabrics show instant dissipation of static electricity with excellent tunneling characteristics. These properties allow a number of industrial applications such as in coated fabrics, abrasive belts [92], conveyor belts, high speed composite rollers, carpets, upholstery fabrics, uniforms, gloves, wrist straps, and filtration. Contex fabrics have been suggested for use as a wrist rest static mats to dissipate static charging near computer keyboards [91].

A special application using a nonwoven Kevlar felt exists in the fabrication of electrical generators. A measure of the quality of generator stator coils is the increase in loss tangent (power factor) with increasing applied voltage. This phenomenon, also known as "tip-up," is attributed to partial discharges occurring in voids in the insulation materials. These discharges degrade the surrounding insulation and could eventually lead to premature coil failure. Moreover, electrical losses in generator coils translate to lost revenue for utilities. One approach for reducing tip-up is to use a partially conducting filler material in the coils. Some manufacturers use carbon-filled pastes, but it is difficult to control the application of the carbon during manufacturing. A polypyrrole-coated felt has the attraction of having a conductivity in the required range and being easy to use in manufacturing. Electrical tests on prototype coil sections have shown

Fig. 35.9 Polypyrrole-coated textile products offered by Milliken Research Corporation (Spartanburg, SC) under the trade name Contex.

reduction in tip-up from 1.37% with an insulating filler to 0.06% with a polypyrrole-coated felt (when measured between 2.8 and 13.8 kV) (E. Schoch, Westinghouse, private communication).

Commercially available polypyrrole-coated products from Achilles are predominantly available as coated films, but a number of products based on nonwoven textiles and used as static dissipating bars, brushes, and rollers for photocopying and fax machines are also being offered (see Fig. 35.10). A similar nonwoven material is used in filtration equipment. They also are offering conductive gloves for applications in clean rooms based on a nonwoven fabric constructed of coated polyurethane fibers.

In static dissipation applications, conductive textiles are meeting or exceeding all physical requirements but may be more expensive than traditional materials. The conductance required for static dissipation is sufficiently low to produce materials with several times the needed conductance. Therefore, even if these materials were used in a hostile environment, their stability would not be a significant concern. Their static dissipative characteristics should not be affected even if the product should lose more than half of its conductance.

B. EMI Shielding

For most effective low frequency EMI shielding applications, surface resistances below 1 $\Omega/\square$ are needed. Po-

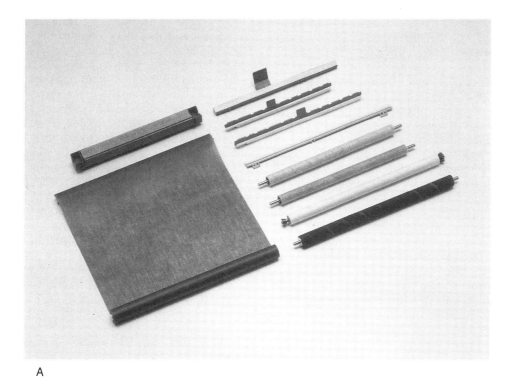

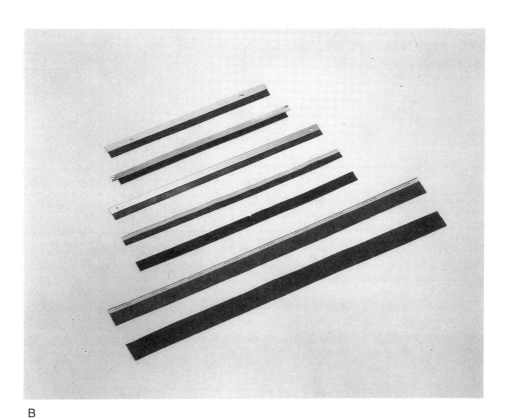

Fig. 35.10 Polypyrrole-coated products for static discharge in office equipment. (A) Rollers; (B) nonwoven brushes. (Photos courtesy Achilles Corporation, Tokyo, Japan.)

lypyrrole-coated textiles have been shown to be fairly transparent to the magnetic component of electromagnetic radiation below 100 MHz. Therefore, metal-coated textiles are preferred in these applications. There are, however, several applications where a higher surface resistance is still useful, such as in gaskets or when corrosion or adhesion prevents the use of metal-coated fabrics. The excellent formability of conductive fabrics and the possibility of using such structured reinforcement in rubber blankets and other polymer composites render these products increasingly useful.

Normal textile production techniques allow the fabrication of material of various thicknesses. Nonwoven felts can easily be produced with a thickness of 1 cm. Carpets with a variety of thicknesses exceeding 1 cm are commercial products. The coating of such products with polypyrrole leads to materials that, because of their thickness, behave differently from thin fabrics in the electromagnetic field. Products of this nature can be successfully used in special EMI shielding applications.

C. Resistive Heaters

Their range of resistance makes conducting textiles suitable for resistive heating applications. Conductance can be provided by incorporating graphite into the fabric or coating it with a metal or a conducting polymer. Regardless of the conducting material, textiles are particularly well suited for large-area radiant or contact heating. A conducting textile offers many potential advantages over traditional heaters consisting of rigid resistive wires. Not the least of these is the diffuse nature of the conducting surface, which gives rise to fast, even heating over the entire surface. It also improves reliability, since a tear in any one area will not disrupt the current. By contrast, traditional heaters are usually wired in series, so that a break anywhere along the wire interrupts the current and renders the heater inoperative. The absence of wires also allows for a thin, flexible construction. If tactile properties are important, for instance in a heated seat, the use of conducting fabric necessitates no additional cushioning.

Considering these advantages, there are many applications where a conducting fabric could be superior. Items that already contain fabric, such as car seats, mattress pads, blankets, and clothing, could easily be made to provide warmth simply by including a piece of conducting fabric in the construction and providing a power source. Other applications could include rapid heating with radiant panels [14], electric heating in floors and

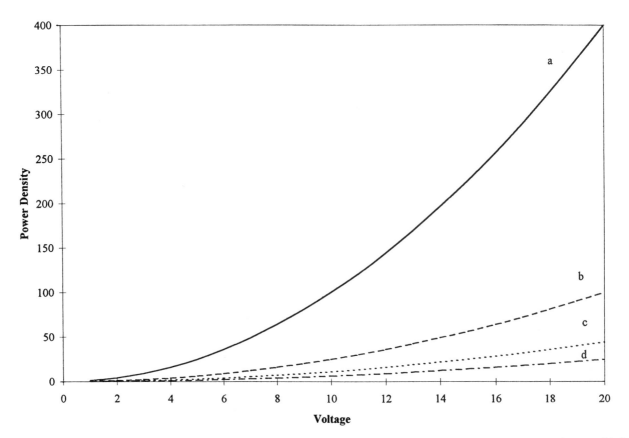

Fig. 35.11 Theoretical power density versus voltage for resistive heaters based on polypyrrole-coated polyester fabrics with various electrode spacings. (a) $l = l_o$; (b) $l = 2l_o$; (c) $l = 3l_o$; (d) $l = 4l_o$.

walls, ice-melting on aircraft [93] or ground surfaces, heating chemical drums, or virtually any need now filled by large-area wire heaters. In some instances, the thinness of the fabric offers a weight and space advantage. In others, the simplicity of the heating component makes construction of the final product easier.

Current, voltage, and power requirements for a heater made of conducting textiles will vary depending on the application. Devices are best defined in terms of the power demands per unit area of heating surface. Contact heaters such as a heated seat or vest will typically require less than 0.1 W/in.^{2}. Radiant heaters may operate in the neighborhood of 0.1–1.0 W/in.^{2}. Other applications could demand up to a few watts per square inch. For simplicity, assume a rectangular heating surface on which two electrodes of length w are separated by a distance l. The total resistance of the surface is given by $R = rl/w$, where r is the surface resistance per unit area (as in $\Omega/\square$). At a given voltage V, the current draw will be $I = Vw/rl$, and the power density (per unit area) is given by $P = V^2/rl^2$. Figure 35.11 relates power density and voltage for various electrode spacings (and hence fabric resistances). The temperature generated by a given power density depends on the configuration and environment of the device. Figure 35.12 shows the radiant temperature measured 3 and 6 in. from the surface of a large sheet of Contex as a function of the power density.

D. Composite Structures

Glass fabrics are widely used for the reinforcement of resins such as polyester, epoxy, imide, and others. The ability to combine the reinforcing properties of such a fabric with the electrical properties of a polypyrrole or polyaniline coating offers intriguing possibilities for a wide variety of industrial applications. The use of textiles based on carbon or graphite fibers offers similar properties, but the range of conductance can be better controlled with coated fabrics. In addition, the wide variety of substrates made from Kevlar, ceramic fibers, glass or quartz, etc. allows applications that cannot be met with carbon fibers. The use of such material has been suggested for airplanes to facilitate deicing [93].

Conductive polymers have been suggested for use in the welding of plastics. A strip of conducting polymer-coated material can be used to convert microwave radiation into thermal energy [94]. Similarly, composite structures can be manufactured using conducting

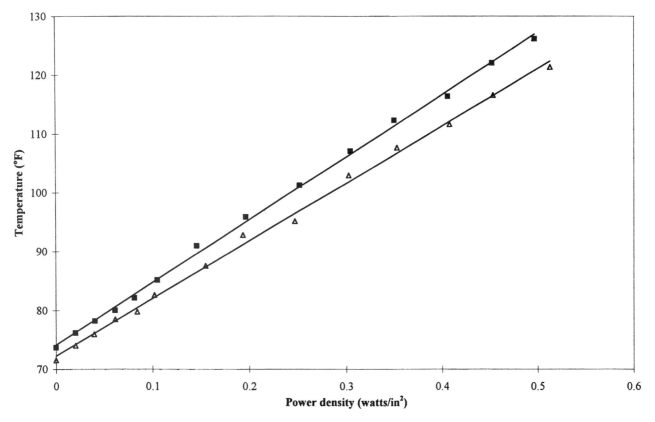

Fig. 35.12 Temperature versus power density for resistive heaters based on polypyrrole-coated polyester fabrics. Temperature probes were placed at a distance of (■) 3 ins. and (△) 6 ins. from the fabric.

polymer–coated fabrics as reinforcement materials for thermoset resins. The advantage lies in the highly efficient microwave absorbance characteristics of conducting polymers.

The use of resins in conjunction with polypyrrole- or polyaniline-coated fabrics yields composite structures that protect the conductive polymer from oxygen and therefore imparts additional stability to these conductive fabrics. Polypyrrole-coated quartz fabrics embedded in epoxy resins have shown sufficient stability in accelerated aging tests, equivalent to 10 years exposure in the field, to be used in the wings of military aircraft [44].

E. Military Applications

The range of conductance attainable with polypyrrole- or polyaniline-coated fabrics make these textiles ideal for use as radar-absorbing materials. These materials show promise for use in a number of military applications. The relatively flat attenuation over a wide frequency range and the overall stability of their microwave absorbance properties allow the effective use of these materials in lightweight camouflage netting applications (see Fig. 35.13). The attenuation can easily be modified by the selection of the base fabric and the thickness of the polypyrrole coating. Thicker fabrics can be employed in applications where high levels of attenuation are desired. Tighter constructions are used when the attenuation at high frequencies is an issue, because the attenuation is affected by the opening size of the textile structure. The microwave absorption characteristics of polypyrrole- and polyaniline-coated fabrics have been further exploited in multilayer products such as Salisbury screens and Jaumann absorbers [93]. The utilization of these fabrics has also been suggested for microwave antennas. Since the conductivity of polypyrrole

Fig. 35.13 Camouflage netting system made from Milliken & Company's Intrigue ultralight, multispecial (visual, infrared, and radar) camouflage materials.

can be adjusted by various methods, materials can be constructed containing patterned areas of conductance [95–98]. This approach has been used in the production of edge-card materials [44] for use in low observable aircraft and in other microwave applications where specific patterned areas of conductivity are required.

VII. CONCLUSIONS

The in situ polymerization of conducting polymers in the presence of a substrate has been shown to offer an alternative to traditional processing techniques to overcome the inherently poor mechanical properties of these polymers. Textiles provide an ideal substrate because of their high surface area, wide range of mechanical properties, high flexibility, and large-scale availability. Fabrics coated with polypyrrole have been shown to retain their mechanical properties and exhibit surface resistivities as low as 10 ohms/square. This technique enables the production of large quantities of conducting polymer composites from relatively inexpensive starting materials in aqueous solution. Large-scale (thousands of linear yards between seams) production of polypyrrole-coated textiles has been accomplished, and further scale-up is technically feasible.

The use of nonreflective, radar-absorbing, lightweight materials for military camouflage remains a high potential application for conductive textiles. At present, either other markets are too small in yardage to support economical manufacturing or competitive materials (carbon fiber, metal foil, sputtered coatings) are better suited for the application. As stability concerns with respect to conducting polymers are addressed further and economies of scale lower the cost of manufacture, conductive textiles based on polypyrrole or polyaniline should become commercially important materials.

REFERENCES

1. W. Löbel, Applications of conductive fibers in textiles, *Mater. Sci. 16*(4):73 (1990).
2. E. K. Sichel (ed.), *Carbon Black Polymer Composites* Marcel Dekker, New York, 1982.
3. H. Ebneth, Metallized fibers and textile fabrics, *Melliand Textilber. 62*:297 (1981).
4. W. C. Smith, Metallized fabrics—techniques and appications, *J. Coated Fabr. 17*:242 (1988).
5. M. Okoniewski, Modifizerung der Synthesefasern zur Erzielung von Leitfahigen, Ferromagnetischen und Anderen Eigenschaften, *Melliand Textilber. 71*:94 (1990).
6. S. Tomibe, R. Gomibuchi, and K. Takahashi, Electrically conductive fiber and method of making same, U.S. Patent 4,410,593 (1983).
7. M. Salmon, K. K. Kanazawa, A. F. Diaz, and M. Krounbi, A chemical route to pyrrole polymer films, *J. Polym. Sci., Polym. Lett. Ed. 20*:187 (1982).
8. X. Chu, V. Chan, L. D. Schmidt, and W. H. Smyrl, Microstructures of chemically polymerized ultrathin polypyrrole films, *Mater. Res. Soc. Symp. Proc. 305*: 177 (1993).
9. C. R. Martin, Nanomaterials: A membrane-based synthetic approach, *Science 266*:1961 (1994).
10. R. V. Gregory, W. C. Kimbrell, and H. H. Kuhn, Conductive textiles, *Synth. Met. 28*:C823 (1989).
11. E. M. Genies, C. Petrescu, and L. Olmedo, Conducting materials from polyaniline on glass textiles, *Synth. Met. 41*:665 (1991).
12. H. H. Kuhn, and W. C. Kimbrell, Electrically conductive textile materials and method for making same, U.S. Patents 4,803,096 (1989) and 4,975,317 (1990).
13. D. Fermin and B. R. Scharifker, Formation and growth of polypyrrole films, *Mem.-Simp. Latinoam. Polym., 3rd*:95 (1992).
14. R. Jolly, C. Petrescu, J. C. Thieblemont, J. C. Marechal, and F. D. Menneteau, Heating panels for accommodation obtained from textiles made electrically conductive by polypyrrole deposit, *J. Coated Fabr. 23*:228 (1994).
15. R. Jolly and C. Petrescu, Process for the production of pourous material web coated with an electronically conductive polymer and product thus obtained, Fr. Patent 2,704,567 (1994).
16. J. C. Thieblemont, M. F. Planche, C. Petrescu, J. M. Bouvier, and G. Bidan, Stability of chemically synthesized polypyrrole films, *Synth. Met. 59*:81 (1993).
17. R. B. Bjorklund and I. Lundström, Some properties of polypyrrole-paper composites, *J. Electron. Mater. 13*:211 (1984).
18. L. Maus, E. F. Witucki, and L. F. Warren, Process for applying an electrically conducting polymer to a substrate, U.S. Patent 4,696,835 (1987).
19. P. R. Newman, Process for producing electrically conductive composites and composites produced therein, U. S. Patent 4,617,228 (1986).
20. S. P. Armes and M. Aldissi, Preparation and characterization of colloidal dispersions of polypyrrole using poly (2-vinylpyridine)-based steric stabilizers, *Polymer 31*:569 (1990).
21. P. Tadros, S. P. Armes, and S. Y. Luk, Preparation and characterization of polyaniline colloids using a monodisperse poly(ethylene oxide)-based steric stabiliser, *J. Mater. Chem. 2*:125 (1992).
22. A. E. Wiersma and L. M. A. Van de Steeg, Dispersion of electrically conductive particles in a dispersing medium, Eur. Patent 0,589,529 A1 (1994).
23. M. J. G. Brouns and R. M. A. M. Schellekens, Process for the preparation of an electrically conducting polymer, World Patent 94/02531 (1994).
24. G. Heywang, Development of stable systems with electric conductivity, *Symp. Materialforsch. 2nd 1*: 390 (1991).
25. H. Naarmann, Electrochemical polymerization of pyrroles, anodes for this process, and products obtained, Eur. Patent 0.133,939 B1 (1985).
26. M. Ito, Y. Katagawa, H. Tajima, Y. Sawachika, M. Kimura, and I. Mizoguchi, Method of making electrically conductive composites, Jpn. Patent 62,275,137 A2 (1987).

27. H. Naarmann and W. Heckmann, Applying films of electroconductive polymers onto other materials, Eur. Patent 206,133 A1 (1986).
28. T. Mitsutake, S. Narisawa, and Y. Yoshii, Manufacture of electrically conductive materials containing polypyrrole, Jpn. Patent 63,213,518 A2 (1988).
29. S. Hishida, Electrically conductive synthetic fibers, Jpn. Patent 61,282,479 A2 (1986).
30. R. Qian, Y. Chen, J. He, M. Wu, and J. Qiu, Preparation and application of composite of conductive polymers, Chinese Patent 86,101,389 A (1987).
31. K. Yasuba and M. Matsunaga, Manufacture of electrically conductive synthetic fiber textiles by coating with polypyrrole, Jpn. Patent 63,042,972 A2 (1988).
32. S. Miyata, and T. Ozio, Process for producing electrically conductive polymer article, U.S. Patent 4,699,804 (1987).
33. T. Mizuki and K. Watanabe, Manufacture of electrically conductive fibers from fibers with surface cavities, Jpn Patent 01,266,280 (1989).
34. G. W. Bartholomew, K. Jongchul, R. A. Volpe, and D. J. Wenzel, Electrically conductive polymeric materials and related method of manufacture, U.S. Patent 5,211,810 (1993).
35. M. Nakata, Y. Shiraishi, M. Taga, and H. Kise, Synthesis of electrically conductive polypyrrole films by interphase oxidative polymerization, *Makromol. Chem. 193*:765 (1992).
36. J. H. Han, T. Motobe, Y. E. Whang, and S. Miyata, Highly electrically conducting polymer blends, *Synth. Met 45*:261 (1991).
37. D. R. Rueda, C. Arribas, F. J. Balta-Calleja, and J. M. Palacios, Polypyrrole grown on the surface of sulfonated polyethylene films. Electrical conductivity and stability, *Synth. Met. 52*:101 (1992).
38. N. V. Bhat and Y. B. Shaikh, Synthesis and structural investigation of conductive composites from cellophane and polypyrrole, *J. Appl. Polym. Sci. 53*:187 (1994).
39. A. Bhattacharya, A. De, and S. N. Bhattacharyya, Preparation of polypyrrole composite with acrylic acid-grafted tetrafluoroethylene-hexafluoropropylene (Teflon-FEP) copolymer, *Synth. Met. 65*:35 (1994).
40. Y. F. Nicolau and M. Nechtschein, Layer-by-layer thin film deposition of polypyrrole and polyaniline, in *Electronic Properties of Conjugated Polymers*, Vol. 3 (H. Kuzmany, M. Mehring, and S. Roth, eds), Springer Ser. Solid-State Sci. 91, Springer, New York, 1989 461.
41. J. Drucker and M. Krishnamurthy, Microstructural evolution of Ag/GaAs (110), *Mater. Res. Soc. Symp. Proc. 355*:59 (1995).
42. G. Stremsdoerfer, Y. Wang, D. Nguyen, P. Clechet, and J. R. Martin, Electroless Ni as a refractory ohmic contact for n-InP, *J. Electrochem. Soc. 140*:2022 (1993).
43. R. V. Gregory, W. C. Kimbrell, and H. H. Kuhn, Conductive textile composites, 3rd Int. SAMPE Electron. Conf., 1989, p. 570.
44. H. H. Kuhn, W. C. Kimbrell, G. Worrell, and C. S. Chen, Properties of polypyrrole treated textiles for advanced applications, *Tech. Pap.-Soc. Plast. Eng. 37*:760 (1991).
45. C. L. Heisey, J. P. Wightman, H. E. Pittman, and H. H. Kuhn, Surface and adhesion properties of polypyrrole-coated textiles, *Textile Res. J. 63*:247 (1993).
46. F. Faverolle, Depot d'un polymere conducteur, le polypyrrole, sur des fibres de verre, Ph.D. Thesis Univ. Paris (1994). See also French Patent 2,697,828 (1994) and 2,667,599 (1992).
47. H. H. Kuhn, W. C. Kimbrell, J. E. Fowler, and C. N. Barry, Properties and applications of conductive textiles, *Synth. Met. 57*:3707 (1993).
48. H. H. Kuhn, A. D. Child, and W. C. Kimbrell, Toward real applications of conductive polymers, *Synth. Met. 71*:2139 (1995).
49. L. F. Warren, J. A. Walker, D. P. Anderson, and C. G. Rhodes, A study of conducting polymer morphology, *J. Electrochem. Soc. 136*:2286 (1989).
50. J. A. Walker, L. F. Warren, and E. F. Witucki, New chemically prepared conducting "pyrrole blacks," *J. Polym. Sci., Part A: Polym. Chem. 26*:1285 (1988).
51. H. H. Kuhn, Polypyrrole coated textiles, properties and applications, *Sen-i Gakkai Symp. Prepr. A*:103 (1991).
52. H. H. Kuhn, Characterization and application of polypyrrole-coated textiles, in *Intrinsically Conducting Polymers* (M. Aldissi, ed.), Kluwer, Dordrecht, 1993, p. 25.
53. M. F. Planche, Etude des mecanismes de viellissement du polymere conducteur polypyrrole et ameliorations possibles, Ph.D. Thesis, Univ. Joseph Fourier, Grenoble, 1994.
54. M. F. Planche, J. C. Thieblemont, N. Mazars, and G. Bidan, Kinetic study of pyrrole polymerization with iron(III) chloride in water, *J. Appl. Polym. Sci. 52*: 1867 (1994).
55. J. Lei and C. R. Martin, Infrared investigations of pristine polypyrrole—is the polymer called polypyrrole really poly(pyrrole-*co*-hydroxypyrrole)? *Synth. Met. 48*:331 (1992).
56. J. Lei, Z. Cai, and C. R. Martin, Effect of reagent concentrations used to synthesize polypyrrole on the chemical characteristics and optical and electronic properties of the resulting polymer, *Synth. Met. 46*: 53 (1992).
57. S. Rapi, V. Bocchi, and G. P. Gardini, Conducting polypyrrole by chemical synthesis in water, *Synth. Met. 24*:217 (1988).
58. Y. Kudoh, M. Fukuyama, T. Kojima, N. Nanai, and S. Yoshimura, A highly thermostable aluminum solid electrolytic capacitor with an electroconducting-polymer electrolyte, in *Intrinsically Conducting Polymers* (M. Aldissi, ed.), Kluwer, Dordrecht, 1993, p. 191.
59. Y. Kudoh, M. Fukuyama, and S. Yoshimura, Stability study of polypyrrole and application to highly thermostable aluminum solid electrolytic capacitor, *Synth. Met. 66*:157 (1994).
60. J. C. Thieblemont, J. L. Gabelle, and M. F. Planche, Polypyrrole overoxidation during its chemical synthesis, *Synth. Met. 66*:243 (1994).
61. D. C. Trivedi and S. K. Dhawan, Shielding of electro-

magnetic interference using polyaniline, *Synth. Met. 59*:267 (1993).
62. M. Ikuo, Manufacture of electrically conductive polymer substrate coated with polyaniline, Jpn, Patent 03,119,612 (1991).
63. A. G. MacDiarmid, J. C. Chiang, A. F. Richter, and N. L. D. Somasiri, Polyaniline: Synthesis and characterization of the emeraldine oxidation state by elemental analysis, in *Conducting Polymers* (L. Alcacer, ed.), Reidel, Dordrecht, 1987, p 105.
64. H. H. Kuhn, and W. C. Kimbrell, Method for making electrically conductive textile materials, U.S. Patent 4,981,718 (1991).
65. K. Tzou and R. V. Gregory, Kinetic study of the chemical polymerization of aniline in aqueous solutions, *Synth. Met. 47*:267 (1992).
66. H. Liu, X. Li, X. Gou, and H. Xie, Preparation and properties of the polyaniline/vinylon conductive composit fibre, *Gaofenzi Cailiao Kexue Yu Gongcheng 10*(6):22 (1994).
67. J. L. Forveille and L. Olmedo, Controlling the quality of deposits of polyaniline synthesized on glass fiber fabric, *Synth. Met. 65*:5 (1994).
68. F. Jousse, L. Delnaud, and L. Olmedo, Processing of conductive fabrics in composite applications, 40th Int. SAMPE Symp. Exhib., 1995, p. 360.
69. C. H. Hsu, Electrically conductive articles, Eur. Patent 0 355 518 A2 (1990).
70. R. V. Gregory, W. C. Kimbrell, and H. H. Kuhn, Electrically conductive non-metallic textile coatings, *J. Coated Fabr. 20*:1 (1991).
71. L. C. Sengupta and W. A. Spurgeon, Dielectric properties of polymer matrix composites prepared with conductive polymer treated fabrics, 6th Int. SAMPE Electron Conf., 1992, p. 146.
72. T. C. P. Wong, B. Chambers, A. P. Anderson, and P. V. Wright, Large area conducting polymer composites and their use in microwave absorbing material, *Electron. Lett. 28*:1651 (1992).
73. T. C. P. Wong, B. Chambers, A. P. Anderson, and P. V. Wright, Fabrication and evaluation of conducting polymer composites as radar absorbers, 8th Int. Conf. Antennas and Propag., 1993, p. 934.
74. P. V. Wright, T. C. P. Wong, B. Chambers, and A. P. Anderson, Electrical characteristics of polypyrrole composites at microwave frequencies, *Adv. Mater. Opt. Electron. 4*:253 (1994).
75. H. Münstedt, Ageing of electrically conducting organic matierals, *Polymer 29*:296 (1988).
76. X. B. Chen, J. Devaux, J.-P. Issi, and D. Billaud, The stability of polypyrrole electrical conductivity, *Eur. Polym. J. 30*:809 (1994).
77. W. A. Gazotti, Jr., V. F. Juliano, and M. A. De Paoli, Thermal and photochemical degradation of dodecylsulfate-doped polypyrrole, *Polym. Degrad. Stab. 42*: 317 (1993).
78. J. H. Moreland, Masters Thesis, Clemson University, 1992.
79. L. A. Samuelson and M. A. Druy, Kinetics of the degradation of electrical conductivity in polypyrrole, *Macromolecules 19*:824 (1986).
80. B. K. Moss and R. P. Burford, Thermal ageing studies of conducting polypyrroles, *Polym. Int. 26*:225 (1991).
81. H. W. Gibson and J. M. Pochan, Chemical modification of polymers, *Macromolecules 15*:242 (1982).
82. V. T. Truong, Thermal degradation of polypyrrole: Effect of temperature and film thickness, *Synth. Met. 52*:33 (1992).
83. V. T. Truong, B. C. Ennis, T. G. Turner, and C. M. Jenden, Thermal stability of polypyrroles, *Polym. Int. 27*:187 (1992).
84. J. C. Thieblemont, M. F. Planche, C. Petrescu, J. M. Bouvier, and G. Bidan, Kinetics of degradation of the electrical conductivity of polypyrrole under thermal aging, *Polym. Degrad. Stab. 43*:293 (1994).
85. F. Beniere, S. Haridoss, J. P. Louboutin, M. Aldissi, and J. M. Fabre, Doping of polyacetylene by diffusion of iodine, *J. Phys. Chem. Solids 42*:649 (1981).
86. T. Haigwara, M. Yamaura, and K. Iwata, Thermal stabiltiy of polyaniline, *Synth. Met. 25*:243 (1988).
87. K. G. Neoh, E. T. Kang, S. H. Khor, and K. L. Tan, Stability studies of polyaniline, *Polym. Degrad. Stab. 27*:107 (1990).
88. Y. Wei and K. F. Hsueh, Thermal analysis of chemically synthesized polyaniline and effects of thermal aging on conductivity, *J. Polym. Sci., Part A: Polym. Chem. 27*:4351 (1989).
89. V. G. Kulkarni, L. D. Campbell, and W. R. Mathew, Thermal stability of polyaniline, *Synth. Met. 30*:321 (1988).
90. Y. Wang and M. F. Rubner, An investigation of conductivity stability of acid-doped polyanilines, *Synth. Met. 47*:255 (1992).
91. J. S. Miller, Conducting polymers—Materials of commerce, *Adv. Mater. 5*:587, 671 (1993).
92. W. L. Harmer, L. Christensen, G. J. Drtina, and H. J. Helmin, Abrasive article with conductive conjugated, polymer coat and method of making same, U.S. Patent 5,061,294 (1991).
93. P. R. Newman and P. H. Cunningham, Electrically heated structural composite and method of its manufacture, U.S. Patent 4,942,078 (1990).
94. A. J. Epstein, J. Joo, C.-Y. Wu, A. Benator, C. F. Faisst, J. Zegarski, and A. G. MacDiarmid, Polyaniline: Recent advances in processing and applications to welding of plastics, in *Intrinsically Conducting Polymers* (M. Aldissi, ed.), Kluwer, Dordrecht, 1993, p.165.
95. E. H. Pittman and H. H. Kuhn, Electrically conductive textile fabric having conductivity gradient, U.S. Patent 5,102,727 (1992).
96. R. V. Gregory, W. C. Kimbrell, and M. E. Cuddihee, Electrically conductive polymer material having conductivity gradient, U.S. Patent 5,162,135 (1992).
97. L. W. Adams, M. W. Gilpatrick, and R. V. Gregory, Method for generating a conductive fabric and associated product, U.S. Patent 5,292,573 (1994).
98. L. W. Adams, M. W. Gilpatrick, and R. V. Gregory, Fabric having non-uniform electrically conductivity, U.S. Patent 5,316,830 (1994).

36
Electrochemomechanical Devices: Artificial Muscles Based on Conducting Polymers

Toribio Fernández Otero and Hans-Jürgen Grande
Universidad del País Vasco, San Sebastián, Spain

I. INTRODUCTION

A. Redox Processes in Conducting Polymers and Concomitant Properties

Most properties of technological interest related to polyconjugated conducting polymers (e.g., polyaniline, polypyrrole, polythiophene, and their derivatives) derive from oxidation–reduction processes that take place in the solid state. When a neutral conducting polymer film is electrochemically oxidized after being submitted to sufficient anodic potential in an electrolytic medium, positive charges are generated along the polymeric backbone and solvated counterions are forced to enter the polymer from the solution in order to maintain the electroneutrality of the solid. This promotes the opening of the polymeric structure and a significant increase in free volume. Opposite processes occur during reduction: electrons are injected into the solid, positive charges are eliminated, and counterions and solvent molecules are expelled to the solution [1–6]. This has two main results: (1) the polymer recovers its neutral state and (2) the volume of the film decreases. In partially degraded or highly reduced polymer films, however, some authors [7–9] have detected an appreciable increase in volume in the first stages of reduction. This was attributed to the initial incorporation of cations from the electrolyte followed by the diffusion of ion pairs from the neutral polymer (salt draining), resulting finally in an overall contraction over time. When polyanion–conducting polymer composites or other immobile counteranions were used, the redox mechanism involved the entry of counterions (and hence an increase in volume) during reduction and their expulsion during oxidation [10–12]. In artificial muscles (described below), composites with immobile counterions are never used and working overpotentials do not reach high cathodic or anodic values. In those conditions, oxidation and reduction processes never attain overall oxidized or neutral states, respectively. Thus, starting from a partially oxidized polymer, oxidation correlates with a continuous increase in volume whereas a continuous decrease in volume occurs during reduction. A single redox process, which involves the interchange of small anions, and related changes in properties are summarized in Fig. 36.1. Most of these properties are reviewed elsewhere in this volume.

In this chapter we focus on the reversible and controllable change in volume linked to electrochemical reactions in the solid state promoted by an electric current. This important feature of conducting polymers was detected by Bourgmayor and Murray [13] in 1982. They described polypyrrole membranes whose permeability could be changed by two orders of magnitude under polarization at various potentials. A direct determination by Okabayashi et al. [14] in polyaniline resulted in a variation in volume by a factor of 2.2 that accompanied oxidation when the experiment was performed in an $LiClO_4$–propylene carbonate solution. In situ bulk measurements by Slama and Tanguy [15] for polypyrrole in the same medium and by Tourillon and Garnier [2] for polythiophene in acetonitrile solutions gave a reversible volume change similar to that observed for polyaniline. Otero and Rodríguez [16] confirmed this result for polypyrrole from density measurements in dry films: oxidized and reduced states showed similar densities (1.51 g/cm^3), with the weight of the oxidized film being 50% higher owing to the incorporation of counterions during oxidation. Relative variations in length of around 50% can be deduced from all these results. Variations in vol-

$$(\text{Polymer})_S + n(\text{ClO}_4^-)_{Aq} \underset{\text{Reduction}}{\overset{\text{Oxidation}}{\rightleftharpoons}} \{(\text{Polymer})^{n+}(\text{ClO}_4^-)_n\}_S + (ne^-)_{Met}$$

REDUCED	- solid state reaction -	OXIDIZED
SEMICONDUCTING	- conductivity -	CONDUCTING
COLOR 1	- electrochromism -	COLOR 2
VOLUME 1	- electrochemomechanics -	VOLUME 2
COMPACT	- permeability -	EXPANDED
SOLUBLE	- electrodissolution -	INSOLUBLE

Fig. 36.1 Redox processes taking place in a conducting polymer in contact with an electrolytic solution. Neutral and oxidized polymer are related to reversible and controllable changes in properties.

ume for polyacetylene films are not so high (around 6.6%, which corresponds to a length variation of 1.8%) as reported by Shacklette and Toth [17] and Murthy et al. [18].

The use and application of such changes in volume is the origin of electrochemomechanical devices [19]—sensors, actuators, electrochemopositioning devices, artificial muscles, and so on—as is explained in the following sections.

B. Artificial Muscles in the Literature

The isothermal conversion of chemical energy into mechanical work underlies the motility of all living systems. These are efficient systems because no intermediate steps that produce heat are present, as was shown early in this century by Van't Hoff. In 1948, Katchalsky and coworkers [20] demonstrated that three-dimensional collagen fibers undergo reversible dimensional changes on transition from cyclic helices to random coils when they are immersed in salt solutions and water. Katchalsky referred to this as a "mechanochemical" system (nowadays the term "chemomechanical" is preferred because it is more precise and avoids confusion with the terminology of chemical reactions induced by mechanical stresses). In general, contraction and expansion of gel fibers provide a means of converting chemical energy into mechanical energy, which can be used to develop artificial muscles and actuators. A great advance in the understanding of chemomechanical systems was made a few years later, when Flory [21] proposed an equation of state for the equilibrium swelling of gels that consists of four terms: a rubberlike elasticity term, a mixing entropy term, a polymer–solvent interaction term, and a term for osmotic pressure due to free counterions [22]. The gel volume is also influenced by temperature, the kind of solvent, the free ion concentration, the degree of cross-linking, and the degree of dissociation of groups on polymer chains. Since the work of Katchalsky and Flory, various polymer gels have been studied as actuators and chemomechanical energy conversion materials [23–25]. Nevertheless, few advances have been made in the design of practical devices. Many recent studies on polymer gels show that chemomechanical properties are related to volume phase transitions [26–30], the volume change at the transition being as much as 1000 times the initial volume of the sample [31]. Those transitions are normally driven by changes in temperature, the response rates being on the order of 10^2–10^3 s [31,32].

Polyelectrolyte gels, which exhibit a bending motion by shrinking in an electric field [33–39], have been called electrochemomechanical devices, electrodriven chemomechanical systems, and muscle-like actuators. They appear to be of particular interest because a natural muscle can be considered the seat of the transformation of chemical energy into mechanical energy triggered by an electric pulse. Mechanical deformation in polyelectrolyte gels can be induced by electric fields via electrodiffusion-induced changes in the intramolecular ionic environment and/or electrokinetically induced pressure gradients. These are very slow processes, so long times (about 10–10^4 s) are needed to complete overall volume variations (if phase transitions are not considered) of up to 250%, that is, fractional length changes of about 50% [40,41]. As the response time depends on both the characteristic length and the diffusion coefficient, the only way to improve response times is to modify the actuator geometry [42,43]. Another disadvantage of such systems is that they require high electric potentials (up to 100 V, which is applied by means of two metal electrodes immersed in aqueous solutions), and no information is available about the consumption of electric energy or about the reversibility of the movement. Some reviews of those systems are available [25,40,44–46].

Another kind of actuator comes from the use of piezoelectric polymers like poly(vinylidene fluoride), as reported by Baughman [47]. Those devices are based on fast and reversible charge polarization processes (no chemical reaction occurs, so they can be called electromechanical actuators) induced by high potentials (around 30 V), so very low response times are obtained (on the order of 10^{-3} s). The main disadvantage of piezoelectric polymers is that the attained dimensional variations are lower than 0.3% in volume, or 0.1% in length, so their commercial applications are limited to high current efficiency microactuators. Photochemomechanical systems based on photoestimulation of conformational levels giving reversible macroscopic changes of volume have also been studied [48–50]. The volume changes reported so far, however, are limited to less than 10% [50]. It is also feasible to generalize about the possibility of constructing molecular machines that function as transductors of intensive variables to mechanical work. Proteins having hydrophilic–hydrophobic temperature transitions have been used with this aim [51]. Such machines can be conceived of as transforming any kind of energy (chemical, electrical, thermal, radiative, etc.) into mechanical energy, giving chemomechanical, thermomechanical, photomechanical, baromechanical, etc. actuators.

Researchers are facing difficulties in attempts to improve properties and response rates of chemomechanical and electrochemomechanical systems based on polymer gels or proteins for practical applications as actuators in robotics. Lack of mechanical toughness and long-term durability are other problems to be solved. The efficiency of energy conversion must also be improved. New polymers that can link reversible chemical reactions to changes in volume are required to produce electrochemomechanical devices of practical interest. From a conceptual point of view, deep discussions are required to clarify and differentiate between chemomechanical, electromechanical, electroosmotic, electrophoretic driven, and electrochemomechanical devices. The main problem is to differentiate the presence and absence of chemical reaction.

C. Artificial Muscles from Conducting Polymers

Polymer gels and proteins do not work as electronic conductors, so electro-osmotic and diffusion processes have to be used as intermediates to convert electric energy to mechanical energy. This results in low actuation rates and high working potentials. The availability of electronically conducting polymers and their redox properties which promote reversible changes in volume, open new possibilities for the development of molecular machines, as reported by Baughman et al. [52]. On the one hand, small changes in potential (of about 0.1–0.5 V, that is, almost two orders of magnitude smaller than previous systems) are sufficient to attain length variations of about 30% with high reversibility. Times required to complete those variations range between 3 and 50 s [53–56], longer than in piezoelectric systems but clearly shorter than in polyelectrolyte gels. On the other hand, conjugated polymers can generate mechanical stresses of around 100 MPa [57] during work as electrochemomechanical devices, whereas polymer gels yield stresses of only 1 MPa [58]. The only negative aspect related to conducting polymers is that the current efficiency in the production of mechanical work is rather low: 0.1–1%, depending on the kind of polymer considered [53,55].

From an overview of all these features, it can be concluded that conducting polymers offer excellent possibilities for application as electrochemomechanical actuators. When such devices are able to describe angular movements of more than 360°, trailing several hundreds of times their own weight, they are called artificial muscles. The use of electrochemomechanical properties related to polyconjugated materials for the fabrication of artificial muscles was reported and patented in 1992 by Otero et al. [56,59]. Since then different devices using polypyrrole [16,54,55,60–64], polyaniline [65–68], and poly(3-alkylthiophene)s [64,69–71] have been described by various researchers. It must be remarked that they will be secondary machines: No direct transformation of electric energy to mechanical work exists. Intermediate electrochemical processes that result in a flow of electrons (involving changes in chemical bonds) and ions through the polymer and conformational changes along the polymeric chains are required. The electrochemical nature of this process will differentiate devices constructed from conducting polymers from those based on polyelectrolyte gels.

II. ARTIFICIAL MUSCLES FORMED BY BILAYERS

As noted above, a physical process occurs during oxidation: volume expansion due to the opening of the polymeric structure allowing counterions to penetrate into the solid. The opposite process occurs during reduction: counterions are expelled from the solid, and conformational changes promote the closing of the polymer network. The construction of an electrochemomechanical device requires the translation of these microscopic conformational changes into macroscopic movements. Among the various possibilities, a bilayer structure was chosen in the first stages of research: a conducting polymer film (3 × 1 cm) was stuck to an adherent, flexible, and nonconducting polymeric film [16]. The bilayer was held at the top with a metallic clamp to allow electrical contact. Once formed the bilayer was checked in aqueous solutions of various salts. An area 1 × 2 cm was introduced into the solution, keeping the clamp out of the liquid. A platinum sheet was used as the counter electrode. The construction of the bilayer is conditioned on the availability of conducting polymer films of high structural homogeneity. Such films can be obtained from electropolymerization techniques that are described elsewhere [72,73].

The controllable volume change taking place only in the conducting polymer film during the current flow is transformed into a stress gradient near the polymer/polymer interface, causing the bilayer to bend. The system is similar to the bilayer thermometers constructed of two metallic sheets having different expansion coefficients. The molecular movement in the conducting film during oxidation and reduction (swelling and shrinking) is transformed into an angular macroscopic movement of the free end of the bilayer around its fixed end (Fig. 36.2). The degree of bending depends on the oxidation level, which can be controlled either through the density of the current flowing across the system or by the electric potential to which the conducting polymer was submitted (in the later case a reference electrode is required).

III. ARTIFICIAL MUSCLES AS ELECTROCHEMOPOSITIONING DEVICES

A typical voltammetric response for a polypyrrole film is shown in Fig. 36.3. A correlation can be established between the electric charge consumed to oxidize the polymer and the potential applied at each moment. If the stress gradient at the polymer/polymer interface is linked to the relative change of volume and this is related to the consumed charge, the angle described by the free end of the bilayer will be related to the electric potential. This fact was experimentally confirmed [16], as can be observed in Table 36.1. The reversibility of the movement is guaranteed by the reversibility of the redox process; each position is recovered when the bilayer is submitted to the corresponding potential.

The position of the free end of the bilayer is independent of the distance between it and the counter electrode. The time required to recover each position depends on the parameters of the electrochemical reaction: concentration of electrolyte, temperature, and kind of electrolyte, mainly.

IV. THE WORKING MUSCLE

Both oxidation and reduction states of the conjugated polymer can be reached at constant potential, at constant current, or through any other variation of potentials or currents such as linear sweeps or pulses. In any case, over a complete oxidation cycle the free end of the bilayer describes an angular movement relative to the vertical position (even greater than 180°). During oxidation, the expansion of the polyconjugated material pushes the flexible nonconducting layer to the concave part of the bend. During reduction, if the conjugated material was in an intermediate oxidation state when the bilayer was constructed, the free end recovers the original vertical position and goes on to describe an angle greater than 180° in the opposite direction. Now the shrinking of the conducting polymer promotes strains of contraction at the interface, giving an opposite bend: the free end of the bilayer moves toward the polypyrrole side.

This movement is able to produce mechanical work. Figure 36.4 shows a bilayer containing 5 mg of polypyrrole with a steel sheet (200 mg) adhering to the bottom of the bilayer. The bilayer was submitted to anodic and cathodic currents of 5 mA. This device took ~30 s to

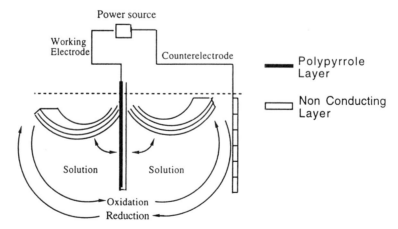

Fig. 36.2 Scheme of the movement of an artificial muscle formed by a polypyrrole film/nonconducting plastic tape bilayer during current flow, starting from the vertical position. The flow of cathodic current through the conducting film promotes its reduction accompanied by the expulsion of counterions and water. The polypyrrole shrinks, and the bilayer bends to the left. The opposite processes and movements occur during oxidation.

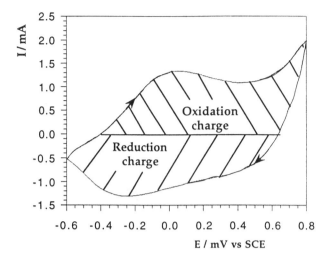

Fig. 36.3 Voltammogram obtained from a polypyrrole film in a 0.1 M LiClO$_4$ aqueous solution at 20 mV/s, using a platinum sheet as counter electrode. This figure shows the reversibility of both anodic and cathodic processes.

cross over 180° from the left side to the right side and vice versa. Movements linked to the reduction process are faster than those related to the oxidation reaction, due to differences in the kinetics of conformational changes, as will be pointed out later. Muscles able to trail a steel sheet (the weight of which was 1000 times the conducting film weight) for more than 200 cycles were built in our laboratory. In spite of this magnitude, the percentage of electric energy transformed to mechanical energy is very low (around 1%) [55]. This seems to be due to the use of very thick films of both conducting and inactive polymer relative to the distance from the polymer/polymer interface where the stress gradient occurs; additional polymer consumes electric charge, does not produce work, and has to be trailed. The use of thinner films, in spite of posing important experimental difficulties, is one of the fields of interest in the search to improve the energy conversion efficiency of these artificial muscles.

V. THE THREE-LAYER MUSCLE

Taking into account that the bilayer requires the presence of a counter electrode to allow current flow through the solution, a new device was developed to produce twice the amount of current: a three-layer device of conducting polymer/flexible layer/conducting polymer, whose movement is depicted in Fig. 36.5. Its construction is easy; once a bilayer is obtained, a second conducting polymer film is adhered to the free side of the nonconducting tape. This inactive tape acts not only as support for stress gradients, but also as an electronic insulator between the conducting films. One of the conducting films acts as the anode, swelling and pushing the device. The second active film acts as the cathode, shrinking by reduction and trailing the device. No direct electronic contact exists between the two films because of the presence of the nonconducting layer in the middle. Therefore the electric current is due to ionic conduction through the solution. The device can be improved by coating it with a film of cellophane or polyacrylamide gel containing an electrolyte, as reported by MacDiarmid and coworkers [66]. A "shell-type" multilayer actuator is obtained in this case that is able to work in air.

VI. CONTROLLING THE RATE OF MOVEMENT

The rate of movement depends on every parameter acting on the electrochemical reaction taking place in the solid film. Both anodic and cathodic overpotentials have a strong influence on the redox kinetics [60,61]. At constant overpotential, any increase in the electrolyte concentration (which is one of the chemical factors in the solid-state reaction) promotes an increase in the rate of movement [62]. The current density has a similar influence: increasing current densities decrease the time required to cross over a constant angle [55] due to the increase in the concentration of oxidation centers in the polymer (the other chemical factor in the process).

On the other hand, movements related to anodic processes are always slower than those related to cathodic processes [55,60]. This fact is related to the extra energy required to open the molecular entanglement and allow the penetration of counterions during oxidation. In contrast, during reduction counterions diffuse along the opened structure toward the solution without any resistance, giving a faster shrinking effect. In other words, oxidation, but not reduction, is controlled by conformational relaxation processes of the polymeric chains. Bending times for a polypyrrole-based bilayer during oxidation (t_{ox}) and reduction (t_{red}) are compared in Table 36.2 at a number of electrolyte concentrations. The difference between t_{ox} and t_{red} becomes an average bending time with a three-layer device.

The internal structure of the polymer plays an important role in the rate of movement; the production of

Table 36.1 Angular Movement of the Free End of a Bilayer Submitted to a Potential Sweep from 400 to −170 mV at 1 mV/s in a 0.1 M LiClO$_4$ Aqueous Solution at Ambient Temperature

E (mV vs. SCE)	400	340	285	210	165	90	20	30	−90	−175
Angle to vertical (deg)	0	10	20	30	40	50	60	70	80	90

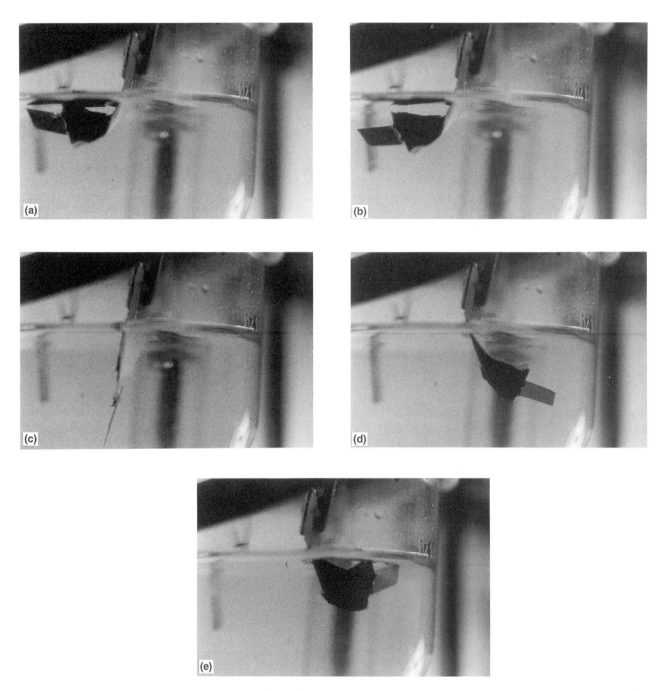

Fig. 36.4 Artificial muscle under work. The weight of the polypyrrole film was 6 mg and the weight of the nonconducting plastic tape was 10 mg. This bilayer trails a steel sheet that weighs 200 mg through 180° under a cathodic current of 5 mA. The time required was 20 s when the electrolyte (LiClO$_4$) concentration was 1 M.

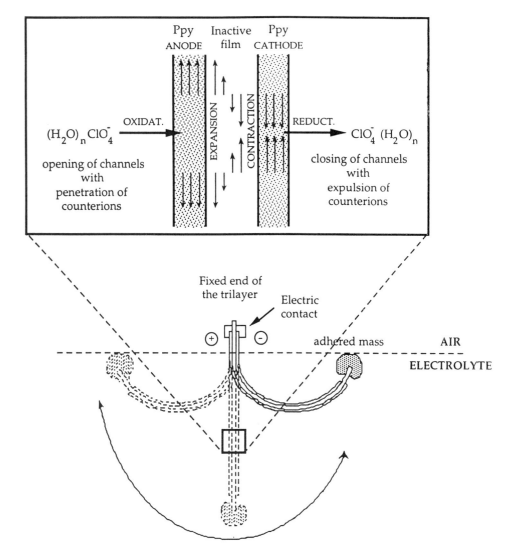

Fig. 36.5 Artificial muscle formed of three layers-polypyrrole/nonconducting tape/polypyrrole. The consumed charge works twice in this device: pushing the free end of the layer when polypyrrole I is oxidized (anodic process) and trailing the layer when polypyrrole II is reduced (cathodic process). Stresses at the polymer/polymer interfaces are summarized in the box.

stress gradients at the polymer/polymer interfaces requires an adequate cross-linking of the polymeric structure. This seems to be the main problem in developing artificial muscles from conducting polymers [74]. When the conditions of synthesis are restrictive, as in the case of the high anodic potential necessary to obtain polythiophene, a dense cross-linking can occur. The resultant muscles move very slowly, describing small angles around the vertical position.

All these effects indicate that a strong interrelation between electrochemistry and polymer science is re-

Table 36.2 Time Required for the Free End of a Bilayer to Cross Over 180° as a Function of Electrolyte (LiClO$_4$) Concentration in Aqueous Solution[a]

	LiClO$_4$ concentration (mol/L)							
Time	0.1	0.2	0.4	0.6	1.0	1.5	2.0	4.0
t_{red}	33	24	17	14	13	12	11	8
t_{ox}	61	52	34	28	25	22	17	15

[a] Anodic steps were from -200 to $+400$ mV, and cathodic steps from $+400$ to -200 mV.

quired to explain redox processes in conducting polymers and related changes in properties (including volume variations). For example, a model for the opening and closing of the polymeric structure during redox switching controlled by conformational relaxation is giving good results in simulating electrochemical responses of polypyrrole-coated electrodes [75,76]. Models including mechanical components do not attain the same fit with experimental results [77].

VII. LIFETIME AND DEGRADATION PROCESSES

Artificial muscles degrade over time as they perform work. In our laboratory, bilayers were checked for over more than 150 cycles (a cycle is considered a movement of the free end of the bilayer from $-90°$ relative to the vertical position to $+90°$ and back) [16]. For a greater number of cycles (150–1000) the movement stops when a fissure appears close to the metallic clamp. The lifetime can be improved by modifying synthesis and control conditions. The key idea is that electropolymerization is a fast though complex mechanism that yields a mixture of materials: linear and cross-linked conducting chains, chemically generated (and therefore nonconducting) polymer, and partially degraded material [72]. By controlling the conditions of synthesis we are able to control the composition of the mixed conducting polymer and obtain the best material for each application. Nevertheless, once obtained, the polymer degrades through two main mechanisms: the Joule effect and overoxidation. As the neutral polymer is a semiconductor, it is difficult to support high current densities when oxidation starts. The flow of current through the high polymeric resistance produces heat (fire points can be observed in the film, mainly at the polymer/electrolyte interface), so fissures appear and the current flow stops. Overoxidation is related to chemical and physical variables. The flow of an anodic current requires an adequate concentration of counterions to allow the oxidation process to take place at low overpotential. Lower concentrations promote thicker diffusion layers and greater overpotentials; new reactions, such as water discharge with the formation of hydroxyl radicals, take place. Those radicals cause nucleophilic attacks on polarons that result in loss of conjugation and slower movements (more dense cross-linking) [78–80].

Reverse oxidation and overoxidation (or degradation) processes can be followed as a function of the applied electric potential by voltammetry. A bilayer (3 × 1 cm) constructed with a 5 mg polypyrrole film was submitted to a potential sweep between -100 mV (vs. SCE) and 3000 mV in a 1 M LiClO$_4$ aqueous solution. To attain an equilibrium state at every potential (the oxidation state of all the solid correlates with each potential), the sweep rate was slow: 3 mV/s. The experimental voltammogram is shown in Fig. 36.6. At -100 mV the poly-

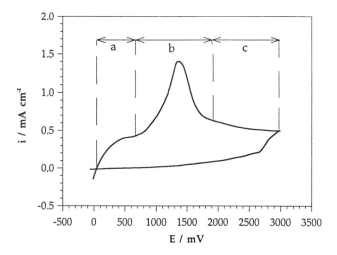

Fig. 36.6 Voltammogram obtained from a bilayer between -100 mV (vs. SCE) and 3 V at 3 mV/s in a 1 M LiClO$_4$ aqueous solution. Region a is related to the reversible oxidation process, region b is linked to an overoxidation/degradation process, and region c is the oxidation process at high overvoltages.

pyrrole film was reduced and the bilayer appeared as in Fig. 36.4a. During the oxidation process, up to 700 mV, a movement of 180° is observed, as in Figs. 36.4a to 36.4d. If the potential sweep is reversed at any potential previous to 700 mV, the movement is reversed. At potentials greater than 700 mV, an overoxidation–degradation process takes place, and a slow reverse movement of the bilayer is observed, from a position like that of Fig. 36.4d to a position like that of Fig. 36.4c, when the potential reaches 2 V. This fact points to a decrease in stress gradient at the polymer/polymer interface, probably due to an increase in cross-linking. Thus all the electrochemical and mechanical processes become irreversible at high anodic potentials.

VIII. THEORETICAL APPROACH

In this section we attempt to develop a theoretical model, based mainly on geometrical considerations, that will enable us to predict experimental results. It is valid for bending angles as high as $\pm 360°$ relative to the vertical position. Other models for small bending angles in actuators are also available in the literature [8,68]. Consider a bilayer consisting of an electroactive and homogeneous polypyrrole film adhered to a nonelectroactive and flexible tape. Anodic or cathodic currents will pass through the polyconjugated film to allow electrochemical oxidation/reduction processes. The attained redox level is a steady state relative to the consumed charge and the applied potential. It is assumed that only one type of dopant ion is present, that the temperature is constant, and that the environment of the conducting polymer (metal clamps, counter electrode, solvent, etc.) is inert.

The oxidation of a microscopic element of volume (of length 1) is shown in Fig. 36.7. The polymer presents a compact structure in the neutral state, being opened during oxidation due to the penetration of counterions. The overall change of volume that occurs during the oxidation of the polymer, expressed in terms of changes in the length of a given polymeric segment (Δl), can be decomposed into two components:

1. During oxidation, electrons are lost from the polymeric chains and solvated counterions penetrate into the polymer to compensate for the generated positive charges. The relative change in the length of a polymeric segment (Δl_1) is proportional to the number of solvated counterions penetrating into the polymer. This number is equal to the number of electrons flowing through the external circuit. Therefore,

$$\frac{\Delta l_1}{l} = h_1 \frac{Q_{\text{segment}}}{l^3} \quad (1)$$

where h_1 is a constant related to the volume of the hydrated anions (which depends on ion–solvent interactions) and Q_{segment} is the anodic charge required to oxidize an unit segment of the polymer.

2. The second component is related to charged polymer–charged polymer, charged polymer–anion, anion–anion, and polymer–solvent interactions. According to the theory of swelling of cross-linked gels [21], the relative increase in the length of the segment can be expressed as

$$\frac{\Delta l_2}{l} = h_2 \frac{Q_{\text{segment}}}{l^3} \quad (2)$$

where h_2 includes all the physical and chemical magnitudes related to the system, such as interaction parameters, dielectric constant of the solvent, screen effects, and electrolyte concentration.

Adding Eqs. (1) and (2), the following expression for Δl is obtained:

$$\Delta l = (h_1 + h_2) \frac{Q_{\text{segment}}}{l^2} = h \frac{Q_{\text{segment}}}{l^2} \quad (3)$$

where $h = h_1 + h_2$ is defined as the electrochemical swelling coefficient. Under steady-state conditions, the relative increase in the length of the polymeric segment is a linear function of the overall charge consumed during the oxidation process (Q), as can be deduced when Eq. (3) is transformed into

$$\Delta l = h \frac{Q_{\text{seg}}}{l^2} = h \left(\frac{Q}{l^2}\right)\left(\frac{V_{\text{seg}}}{V_{\text{pol}}}\right) = h \frac{Ql}{Ae} \quad (4)$$

Here V_{seg} is the volume of the considered segment, V_{pol} is the overall volume of the studied polyconjugated material, A is the area of the polymer/polymer interface, and e is the thickness of the conducting layer.

According to Eq. (4), the relative increase or decrease in length is proportional to the electric charge consumed during the oxidation or reduction processes, respectively. Nevertheless, when the conducting polymer film is adhered to a flexible and nonelectroactive tape, a stress gradient appears across the polymer during the oxidation process (Fig. 36.8): The electron loss induced by the anodic potential tends to cause swelling at the polymer/polymer interface, but fibers are adhered to the nonconducting polymer and the swelling is hindered by a mechanical opposite force. Due to the cross-linking of the polyconjugated material, the stress gradient reaches a distance δ from the polymer/polymer interface. For distances longer than δ, the polyconjugated film expands according to Eq. (4). For this reason, δ is called the distance of influence of the nonelectroactive tape.

Assuming a linear strain gradient from the interface to δ, the elongation of an element of volume present at a distance x from the interface will be

$$h\left(\frac{Ql}{Ae}\right)\left(\frac{x}{\delta}\right) \quad (5)$$

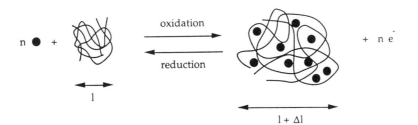

Fig. 36.7 A compact element of volume (of length l) increases in length to $l + \Delta l$ during oxidation. Electrons are lost from the polymer, channels are opened, and hydrated counterions penetrate from the solution to maintain electroneutrality.

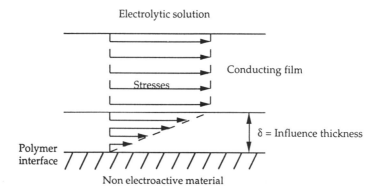

Fig. 36.8 Stress gradient appearing at the polymer/polymer interface as a consequence of the volume increment in the conducting film during oxidation.

The concomitant length variation in our bilayer (ΔL) will be

$$\Delta L = \begin{pmatrix} \Delta P \\ \text{length increment} \\ \text{per segment} \end{pmatrix} \times \begin{pmatrix} L/P \\ \text{number of segments} \\ \text{along } L \end{pmatrix} \quad (6)$$

where L is the overall length of the bilayer. Thus we obtain

$$\Delta L = \begin{cases} h\dfrac{QL}{Ae} & \text{if } x > \delta \\ h\dfrac{QL}{Ae} \cdot \left(\dfrac{x}{\delta}\right) & \text{if } x < \delta \end{cases} \quad (7)$$

Taking into account the change of length as a function of the charge and the distance to the interface, changes in the angle described by the free end of the bilayer can be obtained as a function of the consumed charge (Fig. 36.9). The arc of circumference (L), the bending radius (r), and the thickness of the inactive layer (δ) are related through the bending angle (α) as follows:

$$\left. \begin{array}{l} L = \alpha r \\ L + \Delta L = \alpha (r + e + d) \end{array} \right\} \rightarrow \alpha = \frac{\Delta L}{e + d} \quad (8)$$

Including ΔL in Eq. (8) gives

$$\alpha = \begin{cases} h\dfrac{QL}{Ae\,(e+d)} & \text{if } e > \delta \\ h\dfrac{QL}{A\delta\,(e+d)} & \text{if } e < \delta \end{cases} \quad (9)$$

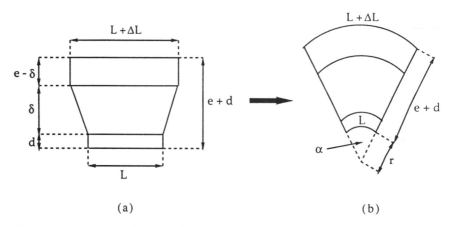

Fig. 36.9 Change of length (a) and concomitant bending (b) in a bilayer muscle. Here d is the thickness of the nonelectroactive film, δ is the thickness influenced by the stress gradient, e is the overall thickness of the polypyrrole layer, and α is the bending angle.

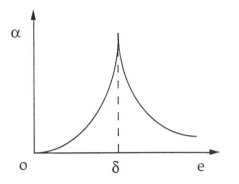

Fig. 36.10 Variation of the bending angle of a muscle, for a constant consumed charge, as a function of the polypyrrole thickness, following Eq. (10). The optimum thickness is δ.

These expressions can be written as a function of the doping level (positive charges per unit volume): $\Psi = Q/V_{pol}$. So the expressions for α become

$$\alpha = \begin{cases} h\dfrac{L}{e+d}\Psi & \text{if } e > \delta \\ h\dfrac{Le}{\delta(e+d)}\Psi & \text{if } e < \delta \end{cases} \quad (10)$$

Therefore, for a constant doping level Ψ, α is a function of the thickness of the polyconjugated film (Fig. 36.10). Moreover, from an electrochemical point of view, δ can be considered the optimum thickness of the conducting film for forming a bilayer. A greater thickness gives an electroactive material that consumes charge but does not participate in the mechanical stress. A lesser thickness gives smaller ΔL and a concomitantly smaller α.

If the doping level is assumed to be proportional to the electric potential to which the conducting polymer is submitted (this approximation sufficiently represents real data), a linear dependence between bending angle and electric potential is obtained, as observed experimentally [16]. On the other hand, the variation of the bending angle α with time as a function of current flow will be given by differentiation of Eq. (9): $d\alpha/dt = kI$, where I is the intensity of the current flowing through the electroactive layer and k is a constant that includes all structural and geometrical parameters of the film. From this result, the developed model can be easily compared with experimental results at constant current [55]. By integration of Eq. (9), the time t required to describe a constant angle is given by

$$t = \frac{\alpha}{k}\left(\frac{1}{I}\right) \quad (11)$$

As can be seen, the bending time decreases when I increases as was proved experimentally (Fig. 36.11). A slope of -1 was obtained from a double logarithmic representation of experimental results, in agreement with theoretical predictions.

IX. SIMILARITIES WITH NATURAL MUSCLES

In a natural muscle an electric pulse is sent from the brain through the nervous system and triggers chemical

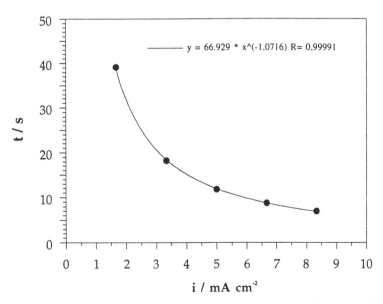

Fig. 36.11 Time required for the free end of a bilayer to cross over 180° during oxidation at different current densities. A 1 M LiClO$_4$ aqueous solution was used as electrolyte, and a platinum sheet as counter electrode. The equation of the solid line is given as $y = 66.929\, x^{-1.0716}$ $R = 0.99991$.

reactions in the muscle cells, promoting changes in volume and concomitant mechanical forces. In artificial muscles constructed from conducting polymers, an electric pulse promotes electrochemical reactions that produce changes in volume that can be transformed into macroscopic movements. Moreover, natural muscles can be artificially activated through an electric pulse (remember Galvani's experiment). Other similarities between natural and artificial muscles can be summarized as follows:

1. Electric currents related to chemical reactions are involved.
2. A change in volume occurs during work.
3. A mechanical movement is produced from chemical energy without mediation of any other transformation or mechanical component.
4. An aqueous solution of ionic salts is required to allow muscles to work.
5. Ionic flows are present during work.

In spite of an important improvement in similarities to artificial muscles based on nonconducting polymer gels, some differences still remain between artificial muscles based on conducting polymers and natural muscles:

1. The driving power in natural muscles is chemical energy, the electric pulse being a trigger. The driving power in artificial muscles is the consumed electric charge, with redox reactions acting as mediators.
2. Natural muscles work only under contraction owing to the irreversibility of chemical reactions. Artificial muscles work under both contraction and expansion because of the reversal of the electrochemical reactions when the sense of the current is changed.

X. FUTURE

From a basic point of view two main causes of action are open: theoretical treatments including all kinds of interactions and experimental approaches to obtain conditions of synthesis that will enable artificial muscles to be developed from any conducting polymer. From a practical point of view, new devices can be constructed as sensors and actuators, muscles for microrobotics and micromachinery, medical instrumentation, mobile electrodes, etc. Difficulties related to conditions of synthesis have to be overcome.

ACKNOWLEDGMENTS

This work has been supported by the Spanish Ministerio de Educación y Ciencia, the Basque government, and the Gipuzkoako Foru Aldundia.

REFERENCES

1. G. Horányi and G. Inzelt, Anion-involvement in electrochemical transformations of polyaniline. A radiotracer study, *Electrochim. Acta 33*:947 (1988).
2. G. Tourillon and F. Garnier, Structural effect on the electrochemical properties of polythiophene and derivatives, *J. Electroanal. Chem. 161*:51 (1984).
3. T. Yeu, K.-M. Yin, J. Carbajal, and R. E. White, Electrochemical characterization of electronically conductive polypyrrole on cyclic voltammograms, *J. Electrochem. Soc. 138*:2869 (1991).
4. Y. Qiu and J. R. Reynolds, Dopant anion controlled ion transport behavior of polypyrrole, *Polym. Eng. Sci. 31*:6 (1991).
5. R. M. Penner, L. S. Van Dyke, and C. R. Martin, Electrochemical evaluation of charge-transport rates in polypyrrole, *J. Phys. Chem. 92*:5274 (1988).
6. A. Talaie and G. G. Wallace, The effect of the counterion on the electrochemical properties of conducting polymers—a study using resistometry, *Synth. Met. 63*:83 (1994).
7. R. John and G. G. Wallace, Doping–dedoping of polypyrrole: A study using current-measuring and resistance-measuring techniques, *J. Electroanal. Chem. 354*:145 (1993).
8. Q. Pei and O. Inganäs, Electrochemical applications of the bending beam method. 1. Mass transport and volume changes in polypyrrole during redox switching, *J. Phys. Chem. 96*:10507 (1992).
9. Q. Pei and O. Inganäs, Electrochemical applications of the bending beam method. 2. Electroshrinking and slow relaxation in polypyrrole, *J. Phys. Chem. 97*:6034 (1993).
10. C. Baker, Y.-J. Qiu, and J. R. Reynolds, Electrochemically induced charge and mass transport in polypyrrole/poly(styrenesulfonate) molecular composites, *J. Phys. Chem. 95*:4446 (1991).
11. D. A. Chesher, P. A. Christensen, and A. Hamnett, Anion movement and carrier type in polypyrrole/dodecyl sulfate, *J. Chem. Soc. Faraday Trans. 89*:303 (1993).
12. M. Elliot, A. Kopetove, and W. J. Albery, Nonaqueous electrochemistry of polypyrrole/(polystyrenesulfonate) composite films: Voltammetric, coulometric, EPR and ac impedance studies, *J. Phys. Chem. 95*:1743 (1991).
13. P. Bourgmayer and R. W. Murray, An ion gate membrane: Electrochemical control of ion permeability through a membrane with an embedded electrode, *J. Am. Chem. Soc. 104*:6139 (1982).
14. K. Okabayashi, F. Goto, K. Abe, and T. Yoshida, Electrochemical studies of polyaniline and its application, *Synth. Met. 18*:365 (1987).
15. M. Slama and J. Tanguy, Influence of the electrolyte on the electrochemical doping process of polypyrrole, *Synth. Met. 28*:C171 (1989).
16. T. F. Otero and J. Rodríguez, in *Intrinsically Conducting Polymers. An Emerging Technology* (M. Aldissi, ed.), Kluwer, Dordrecht, 1993, p. 179.

17. L. W. Shacklette and J. E. Toth, Phase transformations and ordering in polyacetylene, *Phys. Rev. B 32*: 5892 (1985).
18. N. S. Murthy, L. W. Shacklette, and R. H. Baughman, Effect of charge transfer on chain dimensions in *trans*-polyacetylene, *J. Chem. Phys. 87*:2346 (1987).
19. R. H. Baughman and L. W. Shacklette, in *Science and Applications of Conducting Polymers* (W. R. Salaneck, ed.), IOP, 1990, p. 47, Bristol.
20. W. Kuhn, B. Hargitay, A. Katchalsky, and H. Eisenberg, Reversible dilation and contraction by changing the state of ionization of high-polymer acid networks, *Nature 165*:514 (1950).
21. P. J. Flory, *Principles of Polymer Chemistry*, Cornell Univ. Press., Ithaca, NY, 1953, Chap. 13.
22. M. Ilavsky, Effect of electrostatic interactions on phase transition in the swollen polymeric network, *Polymer 22*:1687 (1981).
23. Y. Osada, *Adv. Polym. Sci. 82*:1 (1987).
24. F. Horkay and M. Zrinyi, Mechanochemical energy conversion of neutral polymer gels, *Makromol. Chem., Macromol. Symp. 30*:133 (1989).
25. D. De Rossi, M. Suzuki, Y. Osada, and P. Moraso, *J. Intell. Mater. Syst. Struct. 3*:75 (1992).
26. T. Tanaka, Collapse of gels and the critical endpoint, *Phys. Rev. Lett. 40*(12):820 (1978).
27. Y. Hirokawa and T. Tanaka, Volume phase transition in a non-ionic gel, *J. Chem. Phys. 81*:6379 (1984).
28. S. Fujishige, K. Kubota, and I. Ando, Phase transition of aqueous solutions of poly(*N*-isopropylacrylamide) and poly(*N*-isopropylmetacrylamide), *J. Phys. Chem. 93*:3311 (1988).
29. T. Tanaka, D. Fillmore, S. T. Sun, I. Nishio, G. Swislow, and A. Shah, Phase transitions in ionic gels, *Phys. Rev. Lett. 45*:1636 (1990).
30. S. Katayama, Y. Hirokawa, and T. Tanaka, Reentrant phase transition in acrylamide-derivative copolymer gels, *Macromolecules 17*:2641 (1984).
31. E. S. Matsuo and T. Tanaka, Kinetics of discontinuous volume-phase transitions of gels, *J. Chem. Phys. 89*:1695 (1988).
32. Y. Li and T. Tanaka, Phase transitions of gels, *Annu. Rev. Mater. Sci. 22*:243 (1992).
33. T. Tanaka, I. Nishio, S. T. Sun, and S. Ueno-Nishio, Collapse of gels in an electric field, *Science 218*:467 (1982).
34. Y. Osada and M. Hasebe, Electrically activated mechanochemical devices using polyelectrolyte gels, *Chem. Lett. 9*:1285 (1985).
35. T. Karauchi, T. Shiga, Y. Hirose, and A. Okada, Deformation behaviour of polymer gels in electric field, in *Polymer Gels* (D. De Rossi, K. Kajiwara, Y. Osada, and A. Yamauchi, eds.), Plenum, New York, 1991, p. 237.
36. T. Shiga and T. Kurauchi, Deformation of polyelectrolyte gels under the influence of an electric field, *J. Appl. Polym. Sci. 39*:2305 (1990).
37. M. Suzuki, Proceedings of 12th Annu. Int. Conf. IEEE EMBS, 1990, Vol. 12, p. 1913.
38. D. E. De Rossi, P. Chiarelli, G. Buzzigoli, C. Domenici, and L. Lazzeri, Contractile behaviour of electrically activated mechanochemical polymer actuators, *Trans. Am. Soc. Artif. Intern. Organs 32*:157 (1986).
39. P. Chiarelli, K. Umezawa, and D. De Rossi, A polymer composite showing electrocontractile response, in *Polymer Gels* (D. De Rossi, K. Kajiwara, Y. Osada, and A. Yamauchi, eds.), Plenum, New York, 1991, p. 195.
40. Y. Osada and J. Gong, Stimuli-responsive polymer gels and their application to chemomechanical systems, *Prog. Polym. Sci. 18*:187 (1993).
41. P. E. Grimshaw, J. H. Nussbaum, A. J. Grodzinsky, and M. L. Yarmush, Kinetics of electrically and chemically induced swelling in polyelectrolyte gels, *J. Chem. Phys. 93*:4462 (1990).
42. P. Chiarelli and D. De Rossi, Modelling and mechanical characterization of thin fibers of contractile polymer hydrogels, *J. Intell. Mater. 3*:396 (1992).
43. Y. Osada, Chemical valves and gel actuators, *Adv. Mater. 3*:107 (1991).
44. M. Suzuki and O. Hirasa, An approach to artificial muscle using polymer gels formed by micro-phase separation, in *Polymer Gels* (D. De Rossi, K. Kajiwara, Y. Osada, and A. Yamauchi, eds.), Plenum, New York, 1991.
45. Y. Osada, J. P. Gong, and K. Sawahata, Synthesis, mechanism and application of an electrodriven chemomechanical system using polymer gels, *J. Makromol. Sci. Chem. A 28*:1189 (1991).
46. Y. Osada, H. Okuzaki, and H. Hori, A polymer gel with electrically driven motility, *Nature 355*:242 (1992).
47. R. H. Baughman, Conducting polymers in redox devices and intelligent materials systems, *Makromol. Chem., Macromol. Symp. 51*:193 (1991).
48. G. Smets, New developments in photochromic polymers, *I. Polym. Sci., Polym. Chem. Ed. 13*:2223 (1975).
49. R. Lovrien, The photoviscosity effect, *Proc. Natl. Acad. Sci. U.S. 57*:236 (1967).
50. M. Irie and D. Kunwatchakun, Photoresponsive polymers. Mechanochemistry of polyacrylamide gels having triphenylmethane leuco derivatives, *Macromol. Chem., Rapid Commun. 5*:829 (1985).
51. D. W. Urry, Biophysics of energy converting model proteins, *Angew. Chem. Int. Ed. Engl. 32*:819 (1993).
52. R. H. Baughman, L. W. Shacklette, R. L. Elsenbaumer, E. J. Plichta, and C. Becht, in *Molecular Electronics* (P. I. Lazarev, ed.), Kluwer, Dordrecht, Netherlands, 1991.
53. P. Chiarelli, D. De Rossi, A. Della Santa, and A. Mazzoldi, Doping induced volume change in a π-conjugated conducting polymer, *Polym. Gels Networks 2*: 289 (1994).
54. Q. Pei and O. Inganäs, Electrochemical muscles: Bending strips built from conjugated polymers, *Synth. Met. 55–57*:3718 (1993).
55. T. F. Otero and J. M. Sansiñena, Artificial muscles based on conducting polymers, *Bioelectrochem. Bioenerg. 38*:411 (1995).

56. T. F. Otero et al, Eur. patents 9200095 and 9202628 (1992).
57. R. H. Baughman and L. W. Shacklette, in *Science and Application of Conducting Polymers* (W. R. Salaneck, D. T. Clark, and E. J. Samuelsen, eds.), Adam Hilger, New York, 1990.
58. Y. Itoh, T. Matsamura, S. Umemoto, N. Okui, and T. Sakai, Contraction/elongation mechanism of acrylonitrile gel fibers, *Polym. Prepr. 36*:E184 (1987).
59. T. F. Otero, Conducting polymers (videotape), Instituto de Ciencias de la Educación, University of the Basque Country, BI-1474–1995.
60. T. F. Otero, E. Angulo, J. Rodríguez, and C. Santamaría, Electrochemomechanical properties from a bilayer: Polypyrrole/non-conducting and flexible material artificial muscle, *J. Electroanal. Chem. 341*:369 (1992).
61. T. F. Otero, J. Rodríguez, E. Angulo, and C. Santamaría, Artificial muscles from bilayer structures, *Synth. Met. 57*:3713 (1993).
62. T. F. Otero, J. Rodríguez, and C. Santamaría, Smart muscle under electrochemical control of molecular movement in polypyrrole films, *Mater. Res. Soc. Symp. 330*:333 (1994).
63. T. F. Otero, Electrochemistry and conducting polymers: An emerging and accessible technological revolution, in *New Organic Materials* (C. Seoane and M. Martín, eds.), Univ. Complutense, Madrid, 1994.
64. Q. Pei and O. Inganäs, Conjugated polymers as smart materials, gas sensors and actuators using bending beams, *Synth. Met. 55–57*:3730 (1993).
65. Q. Pei, O. Inganäs, and I. Lundstrom, Bending bilayer strips built from polyaniline for artificial electrochemical muscles, *Smart Mater. Struct. 2*:1 (1993).
66. K. Kaneto, M. Kaneko, Y. Min, and A. G. MacDiarmid, Artificial muscle: Electromechanical actuators using polyaniline films, *Synth. Met. 71*:2211 (1995).
67. W. Takashima, M. Fukui, M. Kaneko, and K. Kaneto, Electrochemomechanical deformation of polyaniline films, *Jpn. J. Appl. Phys. 34*:3786 (1995).
68. W. Takashima, M. Kaneko, K. Kaneto, and A. G. MacDiarmid, The electrochemical actuator using electrochemically-deposited polyaniline film, *Synth. Met. 71*:2265 (1995).
69. S. Morita, S. Shakuda, T. Kawai, and K. Yoshino, Anomalous behaviour of bimorph consisting of conducting polymer gel in electrochemical doping, *Synth. Met. 71*:2231 (1995).
70. X. Chen and O. Inganäs, Doping-induced volume changes in poly(3-octylthiophene) solids and gels, *Synth. Met. 74*:159 (1995).
71. Q. Pei and O. Inganäs, Electroelastomers: Conjugated poly(3-octylthiophene) gels with controlled crosslinking, *Synth. Met. 55–57*:3724 (1993).
72. T. F. Otero and J. Rodríguez, Parallel kinetic studies of the electrogeneration of conducting polymers: Mixed materials, composition and properties control, *Electrochim. Acta 39*:245 (1994).
73. T. F. Otero and E. Angulo, Comparative kinetic studies of polypyrrole electrogeneration from acetonitrile solutions, *J. Appl. Electrochem. 22*:369 (1992).
74. F. P. Bradner, J. S. Shapiro, H. J. Bowley, D. L. Gerrard, and W. F. Maddams, Some insights into the microstructure of polypyrrole, *Polymer 30*:914 (1989).
75. T. F. Otero, H. Grande, and J. Rodríguez, A new model for electrochemical oxidation of polypyrrole under conformational relaxation control, *J. Electroanal. Chem. 394*:211 (1995).
76. T. F. Otero, H. Grande, and J. Rodríguez, Electrochemical oxidation of polypyrrole under conformational relaxation control. Electrochemical relaxation model, *Synth. Met. 76*:285 (1996).
77. T. F. Otero, H. Grande, and J. Rodríguez, An electromechanical model for the electrochemical oxidation of conducting polymers, *Synth. Met. 76*:301 (1996).
78. A. A. Pud, Stability and degradation of conducting polymers in electrochemical systems, *Synth. Met. 66*:1 (1994).
79. A. G. Rangamani, P. T. McTigue, and B. Verity, Slow deactivation of polypyrrole during oxidation-reduction cycles, *Synth. Met. 68*:183 (1995).
80. J. B. Schlenoff and H. Xu, Evolution of physical and electrochemical properties of polypyrrole during extended oxidation, *J. Electrochem. Soc. 139*:2397 (1992).

37
Conductive Polymer/High Temperature Superconductor Assemblies and Devices

John T. McDevitt, Marvin B. Clevenger,* and Steven G. Haupt[†]
The University of Texas at Austin, Austin, Texas

I. INTRODUCTION

A number of researchers have begun to prepare polymer/superconductor composites with the hope of improving the processability and properties of the hybrid materials (see Table 37.1) [1–8]. For example, polymeric matrices loaded with ceramic superconductor components have been used in a plastic extrusion process to prepare superconducting wires and filaments [5,6]. Moreover, hydrophobic polymers have been used as environmentally protective layers [7,8] to slow the parasitic corrosion reactions that occur when cuprate compounds are exposed to water, acids, carbon dioxide, and carbon monoxide.

Other research groups have initiated studies designed to explore conductive polymer systems in the context of high-T_c applications [9]. Here, such polymers offer the prospects for enhanced processability as well as a wide range of electrical conductivities. Polymeric systems of this type can be doped reversibly from neutral, nonconductive forms (10^{-8}–10^{-5} Ω^{-1} cm^{-1}) to oxidized states that display high electrical conductivities (10–10^5 Ω^{-1} cm^{-1}). In their nonconductive form, the electronic interactions between conductive polymers and superconductors should be minimal. Polymers of this type may find utility as passivation layers for the protection of high-T_c structures against corrosion or as insulating dielectric barriers for active devices. However, a doped polymer in intimate contact with a superconductor is expected to display interesting electronic interactions, such as the proximity effect (see Sec. III), in a manner analogous to the well-documented behavior that has been observed for metal/superconductor and semiconductor/superconductor structures. It is this induction of superconductivity into materials that by themselves do not normally support superconductivity that makes this area of research both promising and exciting. These structures provide the opportunity to observe for the first time the superconducting properties of doped organic polymeric conductors. The important challenges in this area are to define conditions that can be exploited to assemble in a chemically compatible manner the hybrid conductive polymer/superconductor structures as well as to develop techniques that can be used to explore the physical and electrical properties of the composite structures.

The study of polymer/superconductor systems may also have important technological implications. With the discovery of a number of cuprate superconductor phases that possess transition temperatures above 130 K, the economic and technological difficulties of cooling superconductor elements for ultrafast electronic devices have been relaxed. Considerable effort has been devoted to the study of Si and GaAs semiconductor devices for operation at 77 K for high speed operation and effective power dissipation. Consequently, it has become important to investigate the prospects for the development of new hybrid semiconductor/superconductor devices. Unfortunately, the initial work completed in the area of high-T_c thin film devices deposited onto Si and GaAs surfaces has shown that these semiconductors are not chemically compatible with the cuprate materials. Although the use of metal oxide buffer layers and low temperature processing has been shown to inhibit somewhat

---

* *Current affiliation*: Westinghouse Electric Corporation, West Mifflin, Pennsylvania
[†] *Current affiliation*: Quantum Magnetics, San Diego, California

Table 37.1 Summary of Polymer/Superconductor Composite Systems

Polymer	Polymer structure	Superconductor used in composite	Application	Ref.
Polythiophene	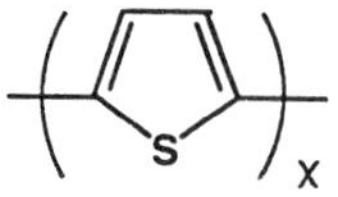 (thiophene)$_x$	$YBa_2Cu_3O_7$	Electrochemical study	1
Poly(3-methylthiophene)	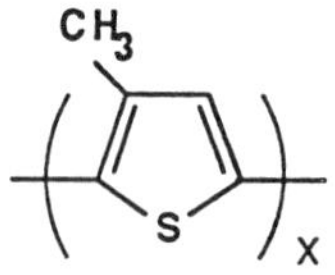 (3-methylthiophene)$_x$	$YBa_2Cu_3O_7$	Electrochemical study	1
Polyaniline	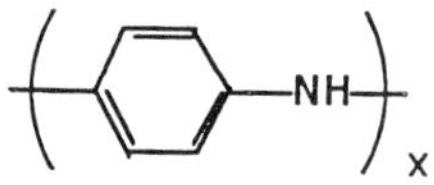 (C$_6$H$_4$–NH)$_x$	$YBa_2Cu_3O_7$	Electrochemical study	2
Polypyrrole	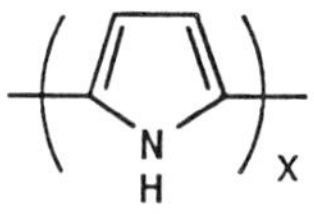 (pyrrole)$_x$	$YBa_2Cu_3O_7$	Electrochemical study; molecular switch	1, 3
Poly(3-hexylthiophene)	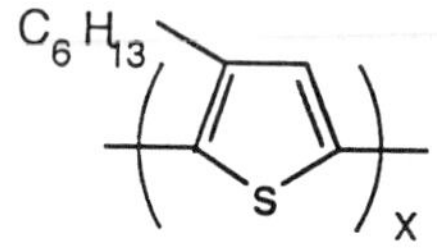 (3-hexylthiophene)$_x$	$YBa_2Cu_3O_7$; $Pb_{0.3}Bi_{1.7}Sr_{1.6}Ca_{2.4}Cu_3O_{10}$	Electrical contacts	4
High density polyethylene	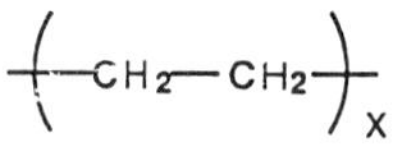 $(-CH_2-CH_2-)_x$	$YBa_2Cu_3O_7$; $Pb_{0.35}Bi_{1.85}Sr_{1.91}Ca_{2.1}Cu_{3.1}O_y$	Processibility	5
Ethyl cellulose	$[C_6H_7O_2(OH)_{3-y}(OC_2H_5)_y]_x$	$YBa_2Cu_3O_7$	Fiber precursor	6
Butadiene/styrene copolymer	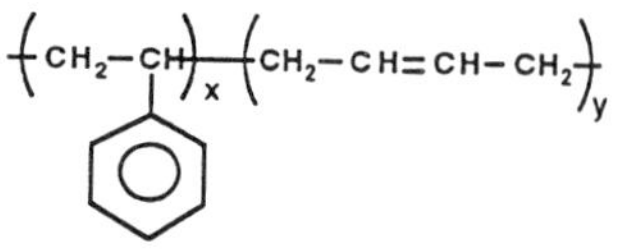 $(-CH_2-CH(C_6H_5)-)_x(-CH_2-CH=CH-CH_2-)_y$	$YBa_2Cu_3O_7$	Surface protection	7
Polyfluorocarbon	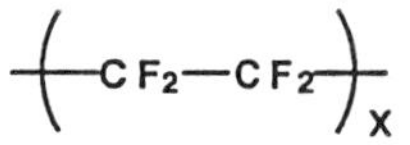 $(-CF_2-CF_2-)_x$	$YBa_2Cu_3O_7$	Surface protection	8

Source: Adapted from Ref. 9.

this adverse reactivity between the semiconductor and cuprate materials [10], the search for new semiconductor materials that do not degrade the cuprate compounds is warranted. In this regard, conductive polymer systems combined with high-T_c superconductor structures may be used for the construction of new types of superconductor circuits, sensors, and devices [3,11].

The purpose of this chapter is to summarize the initial developments related to the preparation and characterization of conductive polymer/high-T_c superconductor assemblies. The study of hybrid conductive polymer/high-T_c superconductor structures is important for the following reasons:

1. These structures provide an excellent platform for the study of novel molecule–superconductor electron and energy transfer phenomena.
2. Chemical compatibility between polymer and su-

perconductor components can be evaluated. Knowledge derived here may aid in the more rapid commercialization of traditional high-T_c superconductor thin-film devices, where, to date, problems with environmental degradation have hampered progress.

3. The assembly of conductive polymer/superconductor structures may lead to the development of new prototype devices and sensors in which the properties of the superconductor are controlled by the influence of the conductive polymer element.

4. Finally, the most exciting new opportunity afforded by research in this area is that under appropriate conditions superconductivity may be induced in conductive polymer structures. Since high-T_c superconductor systems can be used as the source of the superconductivity, it may be possible to drive conductive polymer systems into the superconducting state at temperatures well above 100 K. The search for superconductivity in organic polymeric systems has been an important goal in the field of conductive polymers and has attracted the attention of scientists for more than three decades [12].

This chapter is divided into a number of sections that describe important details related to the conductive polymer/superconductor structures. First, information is provided concerning the preparation and characterization of various polymer/superconductor structures. Chemical and electrochemical deposition methods for localizing the polymers onto a number of cuprate phases are discussed. Section III is devoted to relevant background information related to the induction of superconductivity into metals and semiconductor systems via the proximity effect. More specifically, the four basic methods that have been used to study the occurrence of proximity effects in classical solid-state conductors are described (i.e., contact resistance, modulation of superconductivity in normal/superconductor bilayer structures, passage of supercurrent through superconductor/normal/superconductor systems, and theoretical analyses). Sections IV and V are devoted to experimental studies of conductive polymer/superconductor interface resistances and modulation of superconductivity in the hybrid systems. Finally, there is a discussion of the initial experimental results that explores the possible induction of superconductivity into organic materials.

II. SYNTHESIS AND CHARACTERIZATION OF CONDUCTIVE POLYMER/SUPERCONDUCTOR ASSEMBLIES

Much attention has been paid in recent years to the development and fabrication of electronic sensors and devices from conjugated polymers. Molecular transistors, Schottky diodes, metal-insulator-semiconductor diodes, MIS field-effect transistors, and light-emitting diodes have all been prepared by using these polymeric materials [13–17]. With the discovery of high temperature superconductivity, new opportunities exist for the development of hybrid conductive polymer/superconductor systems in which the unique properties of the conducting polymers complement the properties of high-T_c superconductors for novel applications.

One critical element in the development of conductive polymer/high-T_c superconductor systems is that experimental conditions must be found that can be exploited to combine these very different classes of materials. In order to form well-behaved junctions between high-T_c superconductors and conductive polymers, it is necessary to avoid chemical damage to both the superconductor and conductive polymer components. Unfortunately, both the cuprate compounds and the conductive polymer systems tend to degrade chemically upon exposure to the atmosphere. It is this environmental reactivity that makes fabrication of the hybrid systems so challenging. Consequently, the following section focuses on the preparation and interfacial characteristics of chemically compatible polypyrrole/ $YBa_2Cu_3O_{7-\delta}$ and poly(3-hexylthiophene)/$YBa_2Cu_3O_{7-\delta}$ structures.

A. Superconductor Limitations

In order to prepare useful polymer/superconductor structures, it is necessary to use high-T_c thin films. Of the various available cuprate phases, the compound $YBa_2Cu_3O_{7-\delta}$ is now the preferred material for thin-film applications. This compound, however, degrades rapidly upon exposure to water, CO_2, CO, and acids [18–20]. For example, when $YBa_2Cu_3O_{7-\delta}$ is exposed to the atmosphere it degrades into three insoluble, insulating compounds (Y_2BaCuO_5, CuO, and $BaCO_3$) as shown by the following reaction scheme [1–3].

$$3 H_2O + 2 YBa_2Cu_3O_7(s) \to Y_2BaCuO_5(s) \quad (1)$$
$$+ \frac{1}{2} O_2 + 5 CuO(s) + 3 Ba(OH)_2$$

$$CO_2 + H_2O \to H_2CO_3 \quad (2)$$

$$H_2CO_3 + Ba(OH)_2 \to BaCO_3(s) + 2 H_2O \quad (3)$$

Because of this environmental degradation problem, it is difficult to obtain clean high-T_c superconductor surfaces, a key element for both electrochemical and electronic applications. However, a bulk ceramic superconductor sample can be renewed simply by polishing its surface. Such treatments have been shown to dramatically improve the electrochemical performance of $YBa_2Cu_3O_{7-\delta}$ electrodes [19]. However, thin films of $YBa_2Cu_3O_{7-\delta}$ do not lend themselves to this type of treatment. One technique that has been employed to improve the surface of old or damaged thin films of

$YBa_2Cu_3O_{7-\delta}$ is bromine etching [21]. This is accomplished by immersing the $YBa_2Cu_3O_{7-\delta}$ film in a 1% solution by volume of bromine in dry ethanol. However, the best solution is to minimize delays between deposition of the superconductor film and polymerization of the polymer. Moreover, it has been shown that $YBa_2Cu_3O_{7-\delta}$ has a limited electrochemical potential window of -1.3 to $+1.4$ V vs. SCE [22,23]. Damage to the superconductor surface results when potential values are used beyond these extrema.

B. Polymer Limitations

A number of polymeric compounds and synthetic techniques are available for the growth of conducting polymers. Unfortunately, only a few such candidates are suitable for the preparation of polymer/superconductor composite bilayer structures. Many of the conductive polymer systems degrade chemically when exposed to the atmosphere [24]. This reactivity is most pronounced when the polymers are in their doped form. Furthermore, the conditions required to grow many conducting polymers are detrimental to the superconductor surface. For example, the preparation of polyaniline films is normally accomplished by using an acidic solvent. Although the use of such corrosive fluids is not problematic for the deposition of polyaniline films onto noble metal surfaces such as Pt or Au, growth of polyaniline onto bulk $YBa_2Cu_3O_{7-\delta}$ has been reported to result in the loss of up to 20% of the superconductor pellet used as the working electrode [7].

Although a variety of synthetic techniques have been exploited to prepare composite superconductor systems [9], electrochemical growth of polymers such as polypyrrole onto the surface of $YBa_2Cu_3O_{7-\delta}$ is by far the most common method [22,23]. Electrochemical procedures provide convenient and versatile methods for forming such polymers. Moreover, pyrrole can be electrochemically polymerized at relatively low potentials (1.0 V vs. SCE), and polypyrrole is relatively stable in its conductive form. Thus, pyrrole is an ideal candidate for the direct polymerization onto the surface of cuprate superconductors.

A second approach for the preparation of stable polymer/superconductor systems involves the use of a solution processable polymer such as a poly(3-alkylthiophene). The advantage of this route is that the high oxidation conditions required for polymerization can be completed prior to exposure to the superconductor [25]. After formation, these polymers can then be dissolved in a dry solvent such as tetrahydrofuran, which is compatible with the cuprate superconductor, and thin films of poly(3-alkylthiophene) can be prepared via spin or spray coating [11].

Once a polymer film is deposited onto the surface of the superconductor, electrochemical techniques can be used to cycle the polymer layers between their neutral and oxidized forms. To dope polypyrrole, for example, the polymer/superconductor assembly is immersed in a solution of 0.1 M Et_4NBF_4 in acetonitrile and the electrode potential is raised above 0.5 V vs. SCE [3,11]. While electrochemical polymerization of polythiophene from the monomer cannot be accomplished within the available stable potential window of the superconductor (see below), electrochemical doping of the preformed poly(3-alkylthiophene) material can be achieved readily at low potentials (0.7 V vs. SCE).

C. Polypyrrole Growth onto $YBa_2Cu_3O_{7-\delta}$

Before details related to the growth of conductive polymer materials onto high-T_c surfaces can be understood completely, it is important to examine the growth characteristics of the polymers under similar conditions at noble metal electrodes. Previous voltammetric studies conducted using platinum as a working electrode [26,27] showed that on the initial sweep pyrrole is oxidized at the electrode surface to form solution-dissolved, radical cation species that couple to form dimers. The resulting dimers can be oxidized at lower potentials than the monomers and can combine with other radical ions to form oligomers. These oligomers continue to grow in length to form a polymeric layer that eventually becomes insoluble and coats the electrode surface. More specific chemical or physical interactions may serve to localize polymer strands to the electrode surface prior to the solubility limit for surfaces that are treated with specialized functional groups (see Sec. II.G).

The behavior typical of an unmodified electrode is clearly apparent in the voltammetry acquired at platinum for a neat pyrrole solution. In the initial scan, higher potentials are required to oxidize the pyrrole solution. Subsequent cycles yield oxidative currents at lower potentials. Here, the electrochemical behavior is dominated by the more easily oxidized oligomers. In addition, the initial sweeps on platinum display the characteristic "nucleation loop" [27], in which the current on the reverse sweep is greater than the current on the forward sweep, indicating that nucleation of polymeric structures onto the electrode surface has begun to occur. Subsequent cycles produce successively larger oxidation currents, indicating an increase in the amount of polymer localized at the electrode surface.

Interestingly, the same general behavior is seen when a superconductor thin film is employed as the working electrode. In this regard, nucleation loops are observed for the initial scans (Fig. 37.1A) and are lost in subsequent scans (Fig. 37.1B) where bulk film growth is observed. The similarity of the voltammetry acquired at platinum electrodes [27] and that obtained at $YBa_2Cu_3O_{7-\delta}$ electrodes demonstrates that electric charge flows readily between polypyrrole and $YBa_2Cu_3O_{7-\delta}$ at room temperature and that the growth mechanisms are probably similar. However, careful

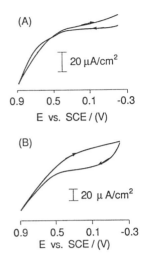

Fig. 37.1 Cyclic voltammetry acquired for the growth of polypyrrole from a solution of 0.25 M Et$_4$NBF$_4$ in neat pyrrole for (A) growth on YBa$_2$Cu$_3$O$_{7-\delta}$ for the first scan and (B) growth on YBa$_2$Cu$_3$O$_{7-\delta}$ for the tenth scan. Data were acquired using a scan rate of 100 mV/s.

analysis of the current densities for the polymer growth reveals that the growth rate is often much slower at the superconductor than that obtained at a platinum electrode. Reasons for the retarded growth of polypyrrole on YBa$_2$Cu$_3$O$_{7-\delta}$ are provided in Sec. II.G.

D. Electrochemical Response of Polypyrrole on YBa$_2$Cu$_3$O$_{7-\delta}$

In addition to providing reliable procedures for depositing polypyrrole layers onto YBa$_2$Cu$_3$O$_{7-\delta}$ films, the electrochemical procedures provide convenient and versatile methods to cycle the polymer between its neutral (insulating) and oxidized (conductive) forms [28]. The fact that polypyrrole films grown on superconductor electrodes display room temperature voltammetry that is similar to that acquired on platinum electrodes (Fig. 37.2) is another indication that electric charge flows readily between polypyrrole and YBa$_2$Cu$_3$O$_{7-\delta}$.

Figure 37.3 displays doping and undoping cycles for a film of polypyrrole on superconductor as studied by chronoamperometry. The close agreement between the magnitudes of the oxidative and reductive currents provides further evidence that the passage of current between the high-T_c and polymer components occurs quite readily, as is the case for noble metal electrodes [24]. Aside from the initial sluggish growth on the superconductor, both chronoamperometric and cyclic voltammetric measurements show that the polymer displays behavior on platinum electrodes almost identical to that on YBa$_2$Cu$_3$O$_{7-\delta}$ thin films. Collectively, these results demonstrate that the techniques used to prepare the polymer–superconductor junctions can be completed without damage to either material.

E. Thin-Film Resistivity

Evaluation of the normal state resistivity values as well as the superconducting transition temperatures of superconductor thin-film specimens before and after modifi-

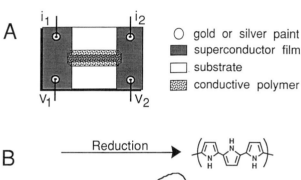

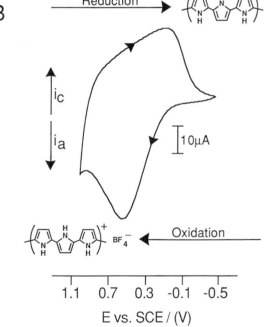

Fig. 37.2 (A) Schematic illustration showing a conductive polymer/high-temperature superconductor sandwich device. To create such a structure, a YBa$_2$Cu$_3$O$_{7-\delta}$ thin film is deposited onto a MgO(100) substrate via laser ablation, a microbridge is patterned on the central portion of the film, and a conductive polymer layer is deposited electrochemically onto the microbridge area. (B) Cyclic voltammetry (5 mV/s) recorded at room temperature in 0.1 M Et$_4$NBF$_4$/acetonitrile for a YBa$_2$Cu$_3$O$_{7-\delta}$ thin-film electrode assembly coated with polypyrrole. Well-behaved voltammetry is observed, indicating that electronic charge flows readily between the superconductor and the polymer layer. (Adapted from Ref. 11.)

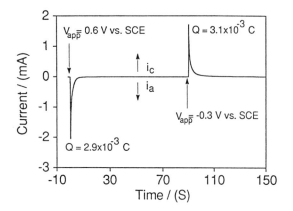

Fig. 37.3 Current vs. time traces acquired for a ~1000 Å thick film of polypyrrole coated onto a $YBa_2Cu_3O_{7-\delta}$ thin-film electrode. Prior to the experiment, the polymer was reduced completely to its neutral form. The polymer layer was then oxidized to its conductive form by stepping the potential to 0.6 V vs. SCE. After the current decayed to the baseline value, the oxidized film was cycled back to the neutral form by applying a voltage of -0.3 V vs. SCE. The experiment was completed in a solution of 0.1 M Et_4NBF_4/acetonitrile. Reversible charging and discharging of the polymer layer is noted here. (Adapted from Ref. 11.)

cation by the polymer layer provides another method for exploring the compatibility of the two conductors. If proper procedures are followed, polypyrrole can be electrochemically polymerized onto the surface of $YBa_2Cu_3O_{7-\delta}$ films as thin as several hundred angstroms with only minor increases in the room temperature resistivity. However, if extreme potentials are used, dramatic increases in room temperature resistivity are noted. Shown in Fig. 37.4 are the resistivity vs. temperature curves for two microbridges modified by different procedures that clearly illustrate these effects. In Fig. 37.4A, the resistivity of a 500 Å microbridge is shown prior to its modification with polypyrrole and after electrodeposition and reduction of the polymer film to its nonconductive form. Here, the polymer is deposited at a modest potential where $YBa_2Cu_3O_{7-\delta}$ is stable (-1.0 V vs. SCE). Only slight increases in normal state resistivity are noted after oxidative polymerization of the monomer. The temperature dependence of the normal state resistivity for the samples before and after polymer modification is metallic. On the other hand, the resistivity vs. temperature curves for a ~900 Å thick microbridge before and after the electrochemical deposition of polythiophene is displayed in Fig. 37.4B, where significant changes in the transport properties are observed. The changes are due to the higher potential (~2.0 V vs. SCE) required for polythiophene polymerization, which is detrimental to the underlying superconductor. The adverse reactivity is clearly evident from the almost two orders of magnitude increase in the room temperature resistivity recorded for the sample after deposition of the polymer. Also due to the deposition of this polymer, there is a change from metallic to semiconducting temperature dependence of the resistivity of $YBa_2Cu_3O_{7-\delta}$. This indicates complete loss of superconductivity in the sample.

F. Conductive Polymer/Superconductor Bilayer Fabrication

While the growth of conductive polymer layers onto bulk high-T_c ceramic pellets can be accomplished readily, the use of thin films of $YBa_2Cu_3O_{7-\delta}$ is preferred for the construction of polymer/superconductor bilayer structures. Consequently, thin films of $YBa_2Cu_3O_{7-\delta}$ (~200–5000 Å in thickness) were deposited onto single-crystal MgO(100) substrates using the pulsed laser ablation method [29], and these films were used to create polymer/superconductor bilayer structures. The result-

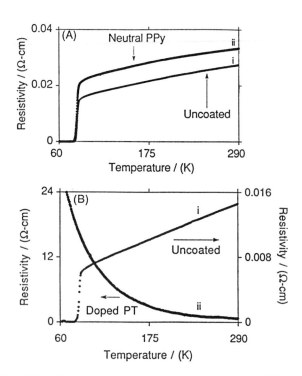

Fig. 37.4 Resistivity vs. temperature curves for a 100 μm wide × 3 mm long microbridge created from films of $YBa_2Cu_3O_{7-\delta}$ (polycrystalline superconductor specimens). (A) Specimen thickness ~500 Å. (i) Uncoated microbridge; (ii) polypyrrole-coated microbridge following its room temperature electrochemical reduction. (B) Specimen thickness ~900 Å. (i) Uncoated microbridge; (ii) microbridge coated with ~2 μm film of doped polythiophene. (Adapted from Refs. 3 and 11.)

ing superconductor films possessed smooth structures (see below). In order to probe the electronic interactions that occur between conductive polymer elements and superconductors, it is necessary to create ultrathin layers of the cuprate materials. Because the polymer–superconductor interactions are strongest at the interface between the two conductors, the use of thin superconductor layers is essential. Moreover, the degree of interaction can be increased if the superconductor element is engineered to possess a number of "weak links." Previous studies have shown that superconductivity is more easily disrupted at such weak links, making these structures more sensitive to the influence of light, magnetism, and heat [3,11]. Thus, rougher films that have disordered structures are well suited for the evaluation of polymer–superconductor interactions.

Two procedures for depositing conductive polymers onto superconductor thin-film structures are considered here. The first involves the use of electrochemical techniques to polymerize the materials directly onto the surface of a $YBa_2Cu_3O_{7-\delta}$ thin-film working electrode. The second approach involves the use of a solution-processible polymer such as poly(3-hexylthiophene). The two different methods used to deposit the conductive polymer materials onto $YBa_2Cu_3O_{7-\delta}$ thin-film assemblies lead to drastically different polymer morphologies. When polypyrrole is grown electrochemically on the surface of the superconductor, the morphology is determined to a large extent by the local conductivity properties of the superconductor electrode surface. The second method produces structures whose morphological details are not influenced by the conductivity of the substrate but rather by the physical method of delivery of the solution-dissolved polymer.

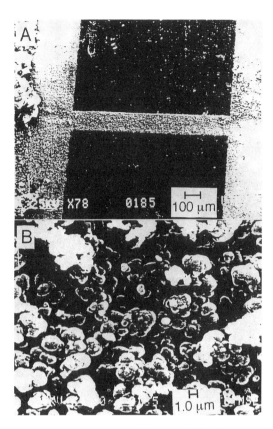

Fig. 37.5 (A) Low and (B) high magnification scanning electron micrographs of a polypyrrole film that was grown electrochemically onto the surface of a 100 μm wide $YBa_2Cu_3O_{7-\delta}$ microbridge supported on a single-crystal MgO(100) substrate. (Adapted from Ref. 11.)

G. Morphology of Polymers on $YBa_2Cu_3O_{7-\delta}$

1. Polypyrrole

The typical polypyrrole film morphology is globular when grown on $YBa_2Cu_3O_{7-\delta}$, and, though similar in some details to films grown on platinum, the superconductor localized polymer layers are normally less uniform than films grown on platinum. Figure 37.5A shows a low-magnification view of a polypyrrole-coated $YBa_2Cu_3O_{7-\delta}$ microbridge. Although the polymer film is relatively smooth, areas of thicker polymer can be seen. Figure 37.5B shows a higher magnification image of the globular morphology of the polypyrrole layer localized on $YBa_2Cu_3O_{7-\delta}$.

The physical structure and electrical properties of $YBa_2Cu_3O_{7-\delta}$ are highly anisotropic. This anisotropy must be considered in both device fabrication using conventional solid-state materials and the construction of polymer/superconductor structures. To explore the extent to which the anisotropy influences the growth of polypyrrole, a number of $YBa_2Cu_3O_{7-\delta}$ thin films were prepared that had different crystallographic orientations. Figure 37.6 shows a series of electron micrographs recorded for $YBa_2Cu_3O_{7-\delta}$ films that possess c-axis, a/c-axis (i.e. mixed alignment), and a-axis orientations, respectively. From the physical appearance of the films, it is clear that the c-axis films are much smoother than their a-axis counterparts. Films with mixed a and c-axis orientations exhibited an intermediate morphology. X-ray diffraction studies (not shown) were used for the assignment of film orientations.

To explore further the deposition of conductive polymer layers onto high-T_c films, a series of experiments were completed using atomic force microscopy (AFM). Measurements of this type can provide nanometer-scale resolution of the polypyrrole surface. Previously, scanning probe measurements were used to image polypyrrole growth on highly oriented pyrolytic graphite (HOPG) [30], polyaniline on platinum [31], polyaniline on gold [32], poly(ethylene oxide) on HOPG [33], and poly-1-aminoanthracene on platinum [34]. Prior measurements in this area showed that polypyrrole grows

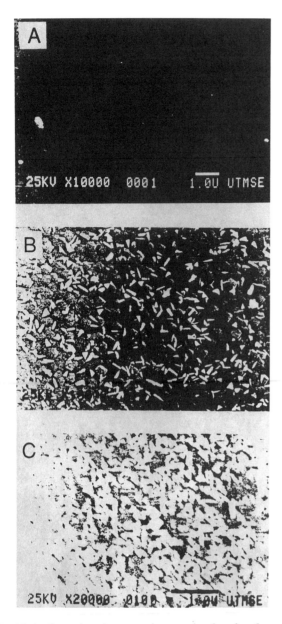

Fig. 37.6 Scanning electron micrographs showing the surface of electrode templates fabricated from (A) *c*-axis-oriented $YBa_2Cu_3O_{7-\delta}$ film, (B) *a/c*-axis-oriented $YBa_2Cu_3O_{7-\delta}$ film, and (C) *a*-axis-oriented $YBa_2Cu_3O_{7-\delta}$ film. All superconductor films were supported on MgO(100) substrates and were ~1500 Å thick.

in strands during the initial stages of growth. These features expand in size as the growth continues to form a semicrystalline film composed of a series of microislands [32]. Longer deposition times, however, result in an abrupt transition to an amorphous structure. These amorphous grains increase in diameter as the film thickens. Similarly, granular morphologies are observed for the other polymers [30,33].

Images obtained from the atomic force microscopy for polypyrrole during the initial stages of deposition reveal that the polymer structures become more dispersed on *c*-axis-oriented $YBa_2Cu_3O_{7-\delta}$ thin films than on platinum, as illustrated in Fig. 37.7A. Both AFM and SEM measurements completed on a large number of *c*-axis films of $YBa_2Cu_3O_{7-\delta}$ supported on MgO(100) and $LaAlO_3$(100) show that the polymer initially collects mostly at localized defect sites. Defects such as laser droplets and film outgrowths are common in films of $YBa_2Cu_3O_{7-\delta}$ [35], especially those prepared by pulsed laser ablation as in these studies. From this work, a consistent trend appears in which the polymer growth is slow initially. At such time intervals, the growth is localized to defect sites. The isolated areas of film growth observed with $YBa_2Cu_3O_{7-\delta}$ during the early stages of formation are probably responsible for the somewhat rougher film morphology (Fig. 37.5B) that is obtained at longer deposition times compared to the polymer morphology on platinum (see below).

Images acquired by atomic force microscopy for the initial stages of polypyrrole growth on Pt film electrodes show that polymer nucleation occurs rapidly and more evenly over the entire exposed surface than on *c*-axis $YBa_2Cu_3O_{7-\delta}$ (Fig. 37.7B). Moreover, chronoamperometric and profilometric measurements indicate that the bulk polymer growth is more rapid on the noble metal electrodes than on *c*-axis $YBa_2Cu_3O_{7-\delta}$. On the other hand, the rougher *a/c*-axis mixed $YBa_2Cu_3O_{7-\delta}$ thin-film electrode fosters more rapid and more uniform growth of the polymer, as shown in Fig. 37.7C. Here, polypyrrole forms a uniform coating over the entire exposed surface. Similarly, polypyrrole growth on *a*-axis $YBa_2Cu_3O_{7-\delta}$ films leads to uniform growth behavior (not shown). Comparison of the different orientations of the $YBa_2Cu_3O_{7-\delta}$ films shows clearly that more favorable electrochemical behavior is obtained at the *a*-axis grains. These data form the basis for a new method that can be used to evaluate the local surface conductivity properties of high-T_c structures. This electrochemical deposition method may find utility in the area of high-T_c device process development.

In an attempt to explore further mechanistic details related to the growth of polymers onto high-T_c surfaces, experiments have been completed to anchor molecular reagents directly to the surfaces of high-T_c structures [36]. The organization of organic molecules onto the surfaces of solid-state materials provides a convenient and rational approach for controlling the interfacial properties of the solid-state structures. Assemblies of this type have been used extensively in the past for the modification of gold surfaces with alkylthiol reagents for the control of chemical and physical phenomena such as adhesion, corrosion reactivity, lubrication, and etching behavior [37–48]. Recently, the McDevitt and Mirkin research groups developed self-assembly methods for the controlled modification of high-T_c structures and devices [49].

An important demonstration of the utility of the high-T_c self-assembly method can be found from the study of the growth of polypyrrole layers on c-axis-oriented films of $YBa_2Cu_3O_{7-\delta}$ that have been modified with amine-substituted pyrrole reagents prior to an electrochemical deposition reactions [50]. Unlike the untreated superconductor surface where sluggish polymer growth is noted (Fig. 37.8A), the amine-tagged surface fosters rapid and uniform polymer growth (Fig. 37.8B). Here, chronoamperometric data following a potential step from 0.0 to 1.0 V vs. SCE are recorded during polymer growth on a c-axis $YBa_2Cu_3O_{7-\delta}$ surface that was modified with a monolayer of N-(3-aminopropyl)pyrrole. A large current of nearly constant magnitude is recorded as the polymer effectively coats the $YBa_2Cu_3O_{7-\delta}$ surface. The behavior is reminiscent of nearly instantaneous nucleation and rapid polymer coupling to the superconductor electrode template.

To explore whether the increase in polymer growth rate is caused by polymeric coupling to the surface-con-

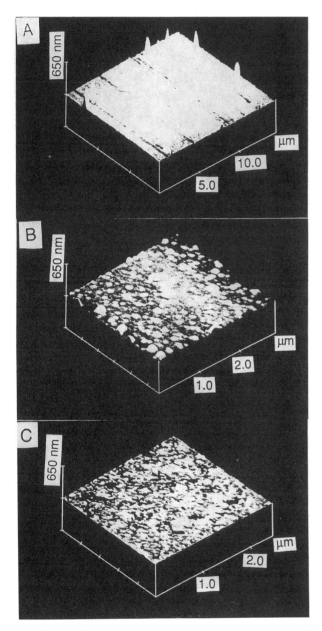

Fig. 37.7 AFM images acquired for a series of high-T_c electrode templates onto which polypyrrole was electrochemically deposited. (A) A c-axis-oriented $YBa_2Cu_3O_{7-\delta}$ thin film: (B) Pt; (C) an a/c mixed axis–oriented $YBa_2Cu_3O_{7-\delta}$ film. Polymer films were grown in all cases by potential steps to 1.2 V vs. SCE for 2 s using a solution composed of 0.10 M Et_4NBF_4 and 0.14 M pyrrole in acetonitrile. (Adapted from Ref. 11.)

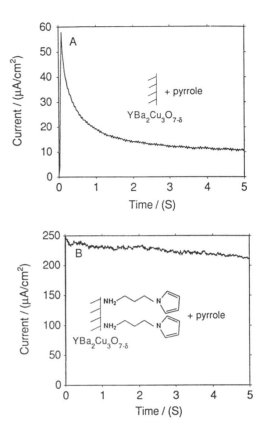

Fig. 37.8 Chronoamperometric data recorded for the growth of polypyrrole onto 1500 Å thick, c-axis-oriented $YBa_2Cu_3O_{7-\delta}$ films on MgO(100). A 0.15 M pyrrole solution with 0.1 M Et_4NBF_4 in acetonitrile was used, and the resulting current transients were recorded following a potential step from 0.0 to 1.0 V vs. SCE. Data are shown for (A) a bare superconductor and (B) the superconductor coated with a monolayer of N-(3-aminopropyl)pyrrole.

fined pyrrole moieties or simply a change in surface wetting and adhesion characteristics, the behavior of the hexylamine-modified surface was evaluated under conditions identical to those detailed above (data not shown). Here, the measured current was found to be much lower in magnitude than for the N-(3-aminopropyl)pyrrole-modified surfaces. It is apparent from these data that the polymerization of pyrrole on bare and hexylamine-modified $YBa_2Cu_3O_{7-\delta}$ thin-film electrodes yields sluggish and nonuniform growth characteristics. The difference in the current–voltage transients for the N-(3-aminopropyl)pyrrole-modified surface compared with the bare and hexylamine-modified surfaces suggests that the presence of the electroactive pyrrole monolayer greatly influences the polymer growth and nucleation behavior.

Images obtained by atomic force and scanning electron microscopy of polypyrrole layers grown under identical conditions onto films of c-axis $YBa_2Cu_3O_{7-\delta}$, with and without the pyrrole self-assembled monolayer, reveal that the N-(3-aminopropyl)pyrrole-modified surface produces thicker, more uniform, and more adherent layers. These differences are remarkable, as illustrated in the AFM images provided in Fig. 37.9A and Fig. 37.9B. The polymer layers grown onto N-(3-aminopropyl)pyrrole/$YBa_2Cu_3O_{7-\delta}$ compare favorably with the smoothest layers of polypyrrole on any substrate reported in the literature [51]. The observation that ultrasmooth polymer layers are obtained with N-(3-aminopropyl)pyrrole-modified $YBa_2Cu_3O_{7-\delta}$ is consistent with the conclusion that polymer growth proceeds through rapid and uniform nucleation on the N-(3-aminopropyl)pyrrole-modified superconductor surface [52].

It should be emphasized that the use of c-axis-oriented films of $YBa_2Cu_3O_{7-\delta}$ serves to expose the poorly conductive crystallographic direction of this anisotropic conductor to the electrolytic solution. While the uniform, highly conductive Au films studied previously foster rapid and uniform polypyrrole growth, untreated c-axis-oriented $YBa_2Cu_3O_{7-\delta}$ surfaces yield only sporadic growth at remotely spaced areas, which we term "electroactive hot spots." These "electroactive hot spots" correspond to the polymer outgrowth features found in Fig. 37.9a. This behavior is consistent with prior studies reporting that electrochemically deposited structures on c-axis $YBa_2Cu_3O_{7-\delta}$ nucleate primarily at defect sites that comprise a small fraction of the overall surface area of the material. Polypyrrole grown directly onto unmodified c-axis $YBa_2Cu_3O_{7-\delta}$ thin films leads to the creation of remotely spaced polymer nodules. Here, polymer nodules 0.4–0.6 μm in width and 0.3–0.5 μm high are observed to cover approximately 4% of the surface (by AFM and SEM measurements). However, polypyrrole grown onto the N-(3-aminopropyl)pyrrole-modified superconductor is observed to have >98% surface coverage with features approximately 0.20–0.25 μm wide and a surface roughness of <0.05 μm. The presence of alter-

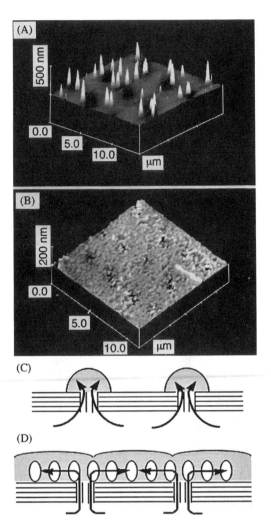

Fig. 37.9 AFM images for c-axis-oriented films of $YBa_2Cu_3O_{7-\delta}$ on MgO(100) (1500 Å superconductor thickness) onto which polypyrrole layers were grown via the potential step procedure. Accordingly, the potential was stepped from a resting value of 0.0 V to 1.0 V vs. Ag wire (~1.1 V vs. SCE) for 5 s. Images are provided for (A) untreated $YBa_2Cu_3O_{7-\delta}$ and (B) N-(3-aminopropyl)pyrrole-modified $YBa_2Cu_3O_{7-\delta}$. (C, D) Schematic drawings showing (C) localized polymer growth that occurs at defect sites for uncoated superconductor and (D) how more uniform nucleation of polymer layers can be accomplished through the influence of a surface-localized electroactive, adsorbed monolayer. The ovals represent the adsorbed pyrrole monolayer.

native crystallographic orientations of the $YBa_2Cu_3O_{7-\delta}$ lattice at localized defect sites in laser-ablated high-T_c films has been noted previously. Accordingly, it is reasonable to conjecture that exposure of such high conductivity sites to the electrolytic solution dominates the polymer growth dynamics during the early stages of polymerization (Fig. 37.9c).

Upon modification of the high-T_c surface with N-(3-aminopropyl)pyrrole and subsequent electrochemically mediated oxidation, it appears that the different surface locations become comparable in their ability to nucleate polymer growth. This behavior is likely due to the presence of the pyrrole monolayer serving to "hardwire" the "electroactive hot spots" on the $YBa_2Cu_3O_{7-\delta}$ surface, thereby forming a modified surface that displays more uniform electrochemical characteristics (Fig. 37.9d). The high surface concentration afforded by the adsorption of the pyrrole monolayer appears to promote this type of behavior. Alternatively, exposure of a-axis features during the adsorption of the amine reagent may be responsible for the observed phenomena.

Comparisons of morphological details of the polypyrrole grown on $YBa_2Cu_3O_{7-\delta}$ (bare and monolayer-modified) reveal a strong correlation between chronoamperometric data and surface morphology. Only those trials that exhibited high constant-current characteristics were found to exhibit the ultrasmooth morphologies. Moreover, evaluation of the early time periods ($t \leq 0.5$ s) of the current–time transients recorded for the polymer growth reveals an important dependence of the nucleation rate on the electrode surface. Here the following trend was noted for the polypyrrole nucleation rates:

N-(3-Aminopropyl)pyrrole-coated c-axis·$YBa_2Cu_3O_{7-\delta}$
> bare gold > bare a-axis $YBa_2Cu_3O_{7-\delta}$
> bare c-axis $YBa_2Cu_3O_{7-\delta}$

Significantly, the N-(3-aminopropyl)pyrrole-modified surface displays more rapid polymer nucleation than that observed at the bare isotropic noble metal conductor, Au. Furthermore, the poorly conductive c-axis surface of $YBa_2Cu_3O_{7-\delta}$ fosters more sluggish growth than the more conductive a-axis orientation.

The fact that ultrasmooth polymer layers are obtained with adsorbed monolayers is consistent with the rapid formation of numerous nuclei during the initial stages of growth. Moreover, the self-assembled monolayer serves also to dramatically increase the adhesion of the polymer to the superconductor surface. Collectively, these results demonstrate that judiciously chosen amine reagents capable of spontaneous adsorption onto cuprate superconductors can be used to localize conductive polymer layers in a controlled fashion onto superconductors, and these systems promote good electrical contact as well as good physical contact between the polymer and superconductor component materials [51,53]. Thus, molecular-scale engineering of conductive polymer/superconductor composite systems is possible with such methodologies.

2. Poly(3-hexylthiophene)

In contrast with the electrochemically derived films, the morphology of poly(3-hexylthiophene) is, for the most part, independent of the quality of the surface of the superconductor sample. Films can be spray-coated onto insulating materials. This property has been exploited to create contacting layers that can electrically bridge superconducting elements over relatively large insulating surfaces. As seen in Fig. 37.10A, the small-scale magnification view of these spray-coated films reveals features that appear to form as a result of the evaporation of the solvent. Interestingly, the higher magnification view shown in Fig. 37.10B demonstrates that the spray-coated film appears to be much denser and more uniform than the electrochemically polymerized pyrrole.

The overriding factor responsible for the quality of the spray-coated films is the size of the microdroplets in the spray. Aspirators with large apertures that produce a relatively coarse mist create the roughest films. The smoothest films have been fabricated by using a commercial artist's airbrush. Another factor that influences the morphology of the spray-coated films is the volatility

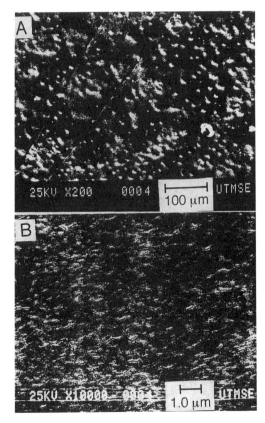

Fig. 37.10 (A) Low and (B) high magnification scanning electron micrographs of a poly(3-hexylthiophene) film that was deposited onto a $YBa_2Cu_3O_{7-\delta}$/MgO(100) thin-film assembly using a spray-coating procedure with tetrahydrofuran as the solvent.

of the solvent. More volatile solvents tend to produce more finely grained films [11]. While spin-coating can create a smoother and more uniform film, the spray-coated films are much thicker and have a higher current-carrying capacity.

H. Crystal Engineering of Organic Superconductor/Inorganic Superconductor Composite Structures

From theoretical considerations that are discussed in further detail in Sec. III.E, it is apparent that conductors that exhibit high carrier mobilities, large carrier concentrations, and low effective masses might be expected to support superconductivity over large dimensions. Although some conductive polymer systems have been shown to exhibit excellent low temperature transport properties, most polymeric conductors suffer from carrier freeze-out, which occurs at cryogenic temperatures. Local inhomogeneities that are normally present in these disordered polymeric systems are to a large extent responsible for this behavior.

On the other hand, crystalline organic metals such as those based on bis(ethylenedithio)tetrathiafulvalene (BEDT-TTF) are less susceptible to these effects. Calculations provided in Table 37.2 (see Sec. III.E) demonstrate the utility of crystalline conductors of this type, as very large superconducting coherence lengths are to be expected for such systems. Based on this expectation, recent efforts have been expended to develop methods that can be used to prepare BEDT-TTF-complex/cuprate superconductor structures.

Although there are now a large number of conductors and superconductors based on BEDT-TTF salts, the majority of these materials have been prepared only as single crystals via electrochemical methods [54]. To make functional systems that can be interfaced readily with high-T_c structures, it is necessary to prepare thin films of these organic conductors. Recently, methods have been developed for the deposition of thin films of $(BEDT-TTF)_2I_3$ via vapor processing steps [55–57].

Because 14 distinct crystallographic phases for iodine complexes of BEDT-TTF [58] have been identified previously, X-ray powder diffraction has proven to be an invaluable technique for analysis of $(BEDT-TTF)_2I_3/YBa_2Cu_3O_{7-\delta}$ hybrid structures. Thin films of α- and β-$(BEDT-TTF)_2I_3$ deposited onto substrates such as glass, Si, MgO, LaAlO$_3$, and Al$_2$O$_3$ were found to display nearly single phase purity and a high degree of crystallinity and c-axis orientation when appropriate precautions were taken to carefully control the source and substrate

Table 37.2 Summary of the Properties of Organic Conductors

Molecular conductor	Property	77 K	4.2 K	Ref.
TTF-TCNQ	μ [m^2/(V·s)]	2.7×10^{-2}	2.7×10^{-2}	76
	N (m^{-3})	4.7×10^{27}	4.7×10^{27}	77
	m^*	3.8	3.8	78
	ξ (nm)	5.9	25.6	
[(CH)I$_{0.16}$]$_x$	μ [m^2/(V·s)]	1×10^{-3}	1×10^{-3}	79
	N (m^{-3})	1.4×10^{27}	1.4×10^{27}	79
	m^*	0.42	0.42	79
	ξ (nm)	2.3		
Poly(3-hexylthiophene) (Undoped)	μ [m^2/(V·s)]	1×10^{-8}	—	80
	N (m^{-3})	2×10^{24}	—	80
	m^*	2	—	80
	ξ (nm)	1×10^{-4}	—	80
Polypyrrole[(4–H$_3$C$_6$H$_4$)SO$_2$]$_{0.3}$	μ [m^2/(V·s)]	4.2×10^{-3}	5.6×10^{-2}	81
	N (m^{-3})	2×10^{27}	1×10^{23}	81
	m^*	0.9	0.9	81
	ξ (nm)	3.6	2.1	
(SN)$_x$	μ [m^2/(V·s)]	6×10^{-3}	1.2×10^{-2}	82
	N (m^{-3})	2×10^{27}	2×10^{27}	82
	m^*	2	2	83
	ξ (nm)	2.7	17.6	
Poly(p-phenylene)/AsF$_5$	μ [m^2/(V·s)]	9×10^{-5}	5×10^{-5}	84
	N (m^{-3})	2×10^{27}	2×10^{27}	84
	m^*	0.4	0.4	84
	ξ (nm)	0.8	2.5	

Source: Adapted from Ref. 85.

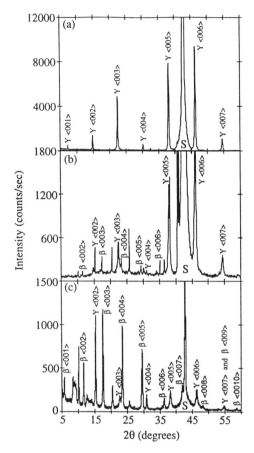

Fig. 37.11 X-ray powder diffraction patterns. (a) (BEDT-TTF)$_2$I$_3$ vapor deposited directly onto bare YBa$_2$Cu$_3$O$_{7-\delta}$, showing only reflections from the high-T_c material. (b) (BEDT-TTF)$_2$I$_3$ deposited onto a dodecylamine monolayer that has been self-assembled to YBa$_2$Cu$_3$O$_{7-\delta}$. Here, reflections from the cation radical salt material are evident along with those of the cuprate superconductor. (c) (BEDT-TTF)$_2$I$_3$ deposited onto an octadecylamine monolayer that has been self-assembled onto YBa$_2$Cu$_3$O$_{7-\delta}$. Reflections from both superconductor materials are evident, with those of (BEDT-TTF)$_2$I$_3$ being, in some cases, as intense as those from YBa$_2$Cu$_3$O$_{7-\delta}$. In all cases, c-axis films of YBa$_2$Cu$_3$O$_{7-\delta}$ (1500 Å) on MgO(100) were used. (Adapted from Ref. 59.)

temperatures [59]. In contrast, (BEDT-TTF)$_2$I$_3$ films deposited onto bare YBa$_2$Cu$_3$O$_{7-\delta}$ show very little, if any, degree of order and crystallinity, as shown in Fig. 37.11A. Here, only peaks for the ⟨001⟩ reflections for the YBa$_2$Cu$_3$O$_{7-\delta}$ substrate are evident. This interesting, highly reproducible result suggests that poorly ordered deposition of (BEDT-TTF)$_2$I$_3$ onto YBa$_2$Cu$_3$O$_{7-\delta}$ is caused by a specific interaction between BEDT-TTF and YBa$_2$Cu$_3$O$_{7-\delta}$ such as sulfur atom lone pairs binding to the ionic copper oxide surface of the high-T_c material.

Sulfur coordination compounds of copper are common in inorganic chemistry, and such an interaction may provide sufficient cause for the (BEDT-TTF)$_2$I$_3$ to deposit in a poorly crystalline, predominantly disordered fashion [60–62].

Using the above-mentioned high-T_c self-assembly method with which monolayers of long-chain alkylamines can be organized on the surface of high temperature superconducting materials, interfacial interactions can be controlled in an impressive manner [63]. In an effort to control the chemical interactions at the interface between (BEDT-TTF)$_2$I$_3$ and YBa$_2$Cu$_3$O$_{7-\delta}$ components, this self-assembly technique was employed to modify the cuprate superconductor surface and eliminate direct contact between the organic and inorganic superconductor materials. Trilayer assemblies incorporating an alkylamine buffer layer between the conductive organic and high-T_c materials were prepared in a three-step process. First, the YBa$_2$Cu$_3$O$_{7-\delta}$ was deposited onto MgO(100) via the laser ablation process. Second, a self-assembled monolayer of dodecylamine or octadecylamine was formed on the surface of the cuprate compound using the previously reported method [49]. Third, the (BEDT-TTF)$_2$I$_3$ was sublimed onto the derivatized superconductor using the vapor deposition method [55–57].

As can be seen in Fig. 37.11B, inclusion of a dodecylamine interlayer reestablishes the crystalline nature of the organic material and results in a substantial degree of c-axis orientation for the deposited (BEDT-TTF)$_2$I$_3$ layer. Also evident in the X-ray powder diffraction pattern are peaks characteristic of a crystalline, c-axis-oriented film of the superconductor β-(BEDT-TTF)$_2$I$_3$, as well as those from the YBa$_2$Cu$_3$O$_{7-\delta}$ substrate. Moreover, the larger octadecylamine molecule, when self-assembled to the cuprate superconductor, results in further improvement in cation radical salt thin-film crystallinity and orientation (Fig. 37.11C). Tenth-order reflections from the now clearly present β-(BEDT-TTF)$_2$I$_3$ can be seen here, along with small reflections from the α-(BEDT-TTF)$_2$I$_3$ phase. Clearly, the elimination of direct contact between the two superconductor materials through incorporation of the self-assembled alkylamine monolayer restores the crystallinity and order that is observed for (BEDT-TTF)$_2$I$_3$ thin films on glass, Si, Al$_2$O$_3$, and other substrates.

Studies of the surface morphology of the (BEDT-TTF)$_2$I$_3$ layers deposited onto the cuprate superconductor substrate provide evidence that supports the conclusions documented in the above-mentioned X-ray diffraction studies [59]. In the case of (BEDT-TTF)$_2$I$_3$ deposited onto glass, MgO, Al$_2$O$_3$, Si, and LaAlO$_3$ substrates, an upright, platelike crystal morphology was obtained. However, only in the case where (BEDT-TTF)$_2$I$_3$ was deposited directly onto the bare high-T_c superconductor surface was a different morphology noted. Here there were platelets that lay flat and parallel to the underlying cuprate superconductor substrate. Interestingly, the self-assembled dodecylamine and octadecylamine monolayer structures served to create large changes in the morphology of the organic charge transfer salt layer deposited on them. On the modified superconductor template, crystallites were found to protrude from the surface in a manner analogous to that obtained for the above-mentioned nonsuperconducting substrates. Thus, the self-assembled monolayers can be used effectively to reestablish the crystallinity of the $(BEDT-TTF)_2I_3$ layer. Studies are now in progress to explore the induction of superconductivity into the $(BEDT-TTF)_2I_3$ systems at elevated temperatures, and the prospects for creating organic superconductor–insulator–cuprate superconductor tunnel junctions are being evaluated.

III. SUPERCONDUCTOR PROXIMITY EFFECTS

Having identified methods to deposit conductive polymer and molecular metal systems onto cuprate superconductor structures without damage to either material, it becomes important now to consider the electronic interactions that occur when the two conductors are in contact with one another. Of particular importance is the interaction that occurs between the polymer-derived charge carriers and the superconducting Cooper pairs. Important background information related to this area can be obtained from the well-documented behavior of the more classical metal/superconductor and semiconductor/superconductor systems. Thus, prior to considering experimental data and theoretical treatments for organic conductor proximity effects, we review previous studies of proximity effects in the more classical systems.

When a normal metal and a superconductor are in intimate contact with each other, there can be a leakage of the superconducting Cooper pairs from the superconductor to the normal metal and quasiparticle (normal electron) leakage from the metal to the superconductor. This effect is known as the superconducting proximity effect and is an important phenomenon that can be used in practical applications such as in electronic devices and as a tool to better understand superconductivity [64]. The proximity effect can occur over distances that are quite large compared to molecular dimensions. Cooper pairs typically extend into normal metals for distances on the order of 100 nm and, in some cases, considerably further.

It should be appreciated that there are a number of prerequisites that must be satisfied before it is possible to observe the induction of superconductivity into materials that by themselves do not exhibit superconducting properties. First, there must be both a source and a sink for the superconducting electrons. Insulating compounds lacking suitable charge carriers cannot support superconductivity. Second, the two conductors must be combined in such a fashion that the formation of insulating degradation layers at the interface is avoided. Third, the composite system must be chilled to a temperature below which one of the two components becomes superconducting. At this point, the superconductor serves as the source of the Cooper pairs and the normal layer serves as the sink for these paired electrons. The relative thickness of the normal and superconductor layers dictates the overall behavior of the composite structure.

Perhaps the most challenging issue related to the construction of systems that exhibit observable proximity effects involves the assembly of the component conductors in a chemically compatible manner. There are two general criteria cited commonly in the literature to avoid this problem: (1) The materials used must be inert or (2) an inert, electrically conductive, buffer layer must be employed at the interface. Chemical compatibility has been a particularly difficult problem in the case of high temperature superconductors, and a noble metal buffer layer has been employed successfully by a number of researchers. For example, M. Gijis et al. [65] used a 40 nm gold buffer layer at the $YBa_2Cu_3O_{7-\delta}$/lead interface and were then able to detect a strong proximity effect in lead. Without the buffer, severe degradation appeared at the high-T_c superconductor/metal interface. Degradation is also minimized when in situ deposition techniques are employed to create the normal metal/superconductor structures. The use of buffer layers effectively prevents degradation layers from forming on either the metal or superconducting surfaces during and between the deposition steps.

Severe complications can also occur from the interdiffusion of the two conductors. To minimize this difficulty, low temperature deposition techniques and the choice of materials that are immiscible with one another can be exploited. One problem unique to some of the high temperature superconductors is their susceptibility to oxygen loss and requirement for high temperature oxygen annealing. Consequently, metal diffusion at the interface can be minimized if the superconductor films are oxygen-annealed prior to the metal deposition [1].

A. Contact Resistance Studies

Manifestations of the proximity effect were observed as early as 1932 when it was noted that the contact resistance between a metal and a superconductor is reduced at the transition temperature of the superconductor [66]. However, significant data on the proximity effect were not collected until Meissner [67] evaluated the contact resistance behavior of over 63 superconductor/metal systems in 1960. Here it was noted that the contact resistance decreases at T_c in the case of many metal–superconductor junctions. Meissner accounted for the observed decrease in contact resistance by deeming it the result of a proximity effect. While the loss of contact resistance at T_c provides useful evidence for the existence of the proximity effect, from this measurement alone it is not possible to determine the depth that the Cooper pairs travel into the normal metal. Measurements of the distance over which the superconducting electrons remain paired in the normal layer provide important information regarding the suitability of the normal layer to support superconductivity. This characteristic distance is called the coherence length, ξ_N. Specific experimental details that influence the magnitude of the contact resistance are provided in Sect. IV.C.

B. Superconductor/Metal/Insulator Sandwiches

In 1962, Hilsch conducted a series of experiments in which the transition temperatures of sandwiches formed from superconductor thin films deposited on top of normal metal thin films were measured [68]. It was noted that the transition temperature of the superconductor/normal metal sandwich was less than that of the bulk superconductor and that the thinner the superconductor layer, the lower the transition temperature of the sandwich. Hilsch also found that if the superconductor layer was sufficiently thin and held at a constant thickness, the observed transition temperature of the sandwich decreased as the normal metal layer thickness increased. Eventually, the metal layer reached a thickness that exceeded the dimensions of the proximity effect, and the transition temperature of the sandwich arrived at a constant value.

Soon after the experimental observations of the modulation of the transition temperature, de Gennes published a significant theoretical treatment of the proximity effect that explained these observations and related the transition temperature of a metal/superconductor sandwich to the electron–electron interaction in the normal metal [69]. In the simplest case, the assumptions made were that the electron–electron interactions in the normal metal (N) are negligible and that the magnetic field is zero ($H = 0$). Then the "clean" and "dirty" cases could be considered. In the clean case, the mean free path (l_N) in the normal metal is large compared to the coherence length (ξ_N) and the penetration depth (k^{-1}). The following expression was derived for this case:

$$k^{-1} = \frac{\hbar v_N}{2\pi k_B T} \quad (4)$$

where v_N is the carrier velocity at the Fermi level and T is temperature. If the normal metal is considered in the dirty case (i.e., the mean free path is less than the coherence length within the normal metal), then the leakage of pairs is controlled by diffusion. A diffusion coefficient (D) is introduced as follows:

$$D = (1/3)v_N l_N \quad (5)$$

In this case, the extent of penetration of the Cooper pairs is equivalent to the coherence length in the normal metal and can then be expressed as

$$\xi_N = \left(\frac{\hbar D}{2\pi k_B}\right)^{1/2} = \left(\frac{\hbar v_N l_N}{6\pi k_B T}\right)^{1/2} \quad (6)$$

From this framework, Werthamer and coworkers [70] developed a theoretical model in which the coherence length in the normal layer can be evaluated by replacing the bulk product $v_N \times l_N$ with the bulk resistivity ρ and the coefficient of normal electronic specific heat γ. Moreover, the theory was rearranged so that all the temperature-dependent terms were combined into a single term. Here, k^{-1} is the temperature-independent coherence length wherein T_c is replaced by T_{cs}, the transition temperature of the bulk superconductor. In the dirty limit, k^{-1} is replaced with ξ_N, and this model yields the expression

$$d_N = \xi_N\left[\frac{1}{2t^{1/2}} \ln\left(\frac{1 + [(1-t)/(1-\theta)](\theta/t)^{1/2}}{1 - [(1-t)/(1-\theta)](\theta/t)^{1/2}}\right)\right] \quad (7)$$

In Eq. (7) the term t is the reduced transition temperature of the sandwich ($T_{c,\text{sandwich}}/T_{c,\text{bulk}}$), which varies with the normal metal thickness for a given thickness of superconductor. The term θ is the reduced transition temperature for the case of a very thick layer of the normal metal.

The temperature-dependent terms in the brackets in Eq. (7) can then be set equal to a single variable, z:

$$z = \left[\frac{1}{2t^{1/2}} \ln\left(\frac{1 + [(1-t)/(1-\theta)](\theta/t)^{1/2}}{1 - [(1-t)/(1-\theta)](\theta/t)^{1/2}}\right)\right] \quad (8)$$

Equation (7) can then be reduced to

$$d_N = \xi_N z \quad (9)$$

By using Eq. (9), it is possible to calculate the coherence length within the normal metal. This is accomplished by plotting the temperature-dependent term z as a function of the thickness of the normal metal, and the slope of the resulting line is the coherence length within the normal metal.

C. Supercurrents in Metals

One of the most dramatic manifestations of the proximity effect is a "supercurrent," or flow of current without resistance, which can be observed to occur over a macroscopically large distance through a normal conductor. The first set of experiments in which a supercurrent in a normal metal was directly observed were conducted in 1969 by Clarke [71]. A three-layered superconductor/metal/superconductor sandwich was created in which copper was employed as the normal metal and lead was used as the superconductor. The samples were prepared by evaporating onto a glass slide a strip of lead followed by a disk of copper and finally a second strip of lead at right angles to the first. Using this configuration, the range of the supercurrent in the copper could be measured. Clarke was able to detect critical currents (i.e., the maximum current supported by the superconductor before it reverts to the normal state) in films as thick as 7000 Å and was able to determine the mean free path of the copper experimentally.

More recent experiments have fostered a number of refinements in the geometry employed by Clarke for detecting proximity effects. One alternative geometry involves the use of step edges to separate the two superconducting structures, which are then bridged with a normal metal layer [72,73]. The advantage of using the step edge geometry is that separation of the two superconducting films is controlled precisely and allows direct access of the metal layer to the *ab* plane of the cuprate superconductor system [74].

D. Supercurrents in Semiconductors

A number of researchers have employed semiconductor structures as the normal layer in superconducting proximity effect experiments using both conventional and high-temperature superconducting materials. The fact that semiconductors have been shown to support superconductivity suggests strongly that similar behavior can be expected from doped conductive polymer systems. Moreover, the use of semiconductors as the normal metal in proximity systems has great potential for application within the electronics industry. In this context, possible new device structures are being evaluated now in which a semiconductor is used as a channel material between a superconducting source and superconducting drain. Such devices serve as superconducting analogs to the three-terminal transistor.

Seto and Van Duzer [64] developed a theory based on the work of de Gennes in which the coherence length within semiconductor systems can be evaluated. In this model, $\xi(T)$ (the coherence length within the semiconductor at temperature T), can be written as

$$\xi(T) = \left(\frac{\hbar^3 \mu}{6\pi k_B T e m^*}\right)^{1/2} (3\pi n)^{1/3} \left(1 + \frac{2}{\ln(T/T_{cn})}\right)^{1/2} \quad (10)$$

where n is the doping concentration, μ is the electron mobility, m^* is the electron effective mass, T is temperature, and T_{cn} is the transition temperature of the normal metal. The final term is used only when a superconducting material is employed as the normal layer at temperatures above its own transition temperature.

Seto and Van Duzer also developed an expression in which the critical current (J_c) through a semiconductor in a superconductor/semiconductor/superconductor sandwich can be predicted as follows:

$$J_c \propto \left[\frac{\Delta_N}{\cosh(L/2\xi)}\right]^2 \left(\frac{1}{\xi_N}\right) \quad (11)$$

where L is the distance separating the two superconductor layers and Δ_N is the pair potential in the semiconductor. Thus, the critical current is expected to display an exponential dependence on the thickness of the semiconducting layer for the case where $L/2 > \xi$, for a given carrier mobility and concentration. Experimental confirmation of this theory was obtained by Seto and Van Duzer through studies of tellurium/lead layers [64].

Hatano and coworkers induced superconductivity in boron-doped silicon by using niobium as the superconductor layer in a Hilsch-type sandwich. Then, by using the Werthamer model [Eq. (7)], they were able to determine the coherence length of silicon as a function of the carrier concentration and found it very close to the predicted theoretical value [75]. Similarly, supercurrents were measured within doped semiconductors by using a planar superconductor/semiconductor/superconductor geometry. The resulting supercurrents measured were found to be dependent on the carrier concentration to the one-third power, and the coherence lengths extrapolated from the temperature dependence of the supercurrent were found to be close to those predicted using Eq. (10). Interestingly, low-dimensional semiconductors have been shown to exhibit a different carrier concentration dependence than that expected for the three-dimensional systems [25].

E. Proximity Effect in Molecular Conductors

An important new direction for conductive polymer research involves the development of techniques whereby molecular conductors are induced to support superconductivity through their intimate contact with high-T_c superconductor systems. There are several reasons why molecular materials are interesting candidates for proximity experiments. First, many molecular materials are low dimensional in nature, making it possible to observe phenomena unique to these systems (see below). Moreover, molecular materials possess the ability to be chemically tailored for specific applications. In addition, many of these materials can be processed at room temperature, thus making the fabrication of novel devices plausible.

The electrical properties of molecular materials differ from those of conventional semiconductors, and these differences will likely influence the properties of induced superconductivity. Whereas molecular materials have carrier mobilities that are lower in value than those of traditional semiconductors, the polymeric conductors possess very high carrier concentrations. It should be noted, however, that the carrier mobilities for many of the molecular conductors display a marked decrease in their magnitudes at lower temperatures compared to the room temperature values. In this regard, high-temperature superconductors may provide a significant advantage over conventional superconductors in molecular conductor/superconductor proximity experiments. The temperature advantage afforded with the cuprate systems allows for the elevated temperature operation of these systems, thereby avoiding carrier freeze-out which occurs within most of the molecular conductors at liquid helium temperatures. Table 37.2 [76–84] provides a summary of carrier mobilities, carrier concentrations, and electron effective mass for a selected number of molecular conductors. In addition, coherence lengths for these materials have been calculated using Eq. (10). Many of the molecular materials in Table 37.2 are low-dimensional and may not conform to the Seto and Van Duzer model. For this reason, actual values for the coherence length may be higher.

It is interesting to note that the predicted coherence lengths for these molecular materials range from 2 to 20 nm and are comparable to the values obtained typically for conventional semiconductors [85]. The values calculated for the molecular materials are for fully doped samples, except for poly(3-hexylthiophene), where data are shown for an as-prepared film that was not doped to its highest conductivity. Here, it is evident that in order to obtain large coherence lengths it is necessary to dope the polymeric conductor. The fact that predicted coherence lengths are larger than molecular dimensions for the doped systems suggests that interesting superconducting phenomena are expected for organic conductor/superconductor structures.

IV. ORGANIC CONDUCTOR/ SUPERCONDUCTOR CONTACT RESISTANCE MEASUREMENTS

A number of researchers have investigated the contact resistance phenomena that occur between high temperature superconductors and conventional materials [63,86–90]. Low contact resistance values are necessary for many practical applications. High contact resistance values can lead to local heating and the loss of superconductivity at the contact interface. In addition, future applications may involve the use of superconductor/normal metal/superconductor (S–N–S) Josephson junctions, and these devices can be fabricated only with materials that exhibit low contact resistance values.

Contact resistance values can also be used as a diagnostic measure of the cleanliness of the normal/superconductor interface. In general, it has been found that in cases where clean normal/superconductor interfaces are achieved, values for the measured contact resistance are small in magnitude and display a metal-like temperature dependence. Moreover, precipitous decreases in the interface resistance occur as the temperature is lowered below the transition temperature, T_c (see Sec. III.A). Such decreases in the interface resistance have been interpreted previously as being the result of the induction of superconductivity into the normal metal contacting layer. Since there are to date no documented examples of conductive organic polymers that by themselves display superconductivity, the study of conductive polymer/superconductor contact resistance phenomena provides a new avenue for research in which conductive polymer proximity effects might be explored.

In the following section, contact resistance experiments are described, based on three- and four-point probe measurements that evaluate the chemical compatibility of organic conductors with a number of p-type cuprate phases. These experiments also explore polymer/superconductor charge transfer phenomena at temperatures above and below T_c and are relevant to organic conductor/superconductor proximity effects.

A. Contributions to Contact Resistance

Low contact resistance values are required for most superconductor applications. Actual values of contact resistance are determined both by intrinsic electronic contributions and by extrinsic barriers. In the former area, problems resulting in higher interface resistance values may result from interface reflections due to carrier density mismatch. For example, the carrier density of $YBa_2Cu_3O_{7-\delta}$ is $\sim 6 \times 10^{21}$ cm^{-3}, while the carrier density of Ag is about an order of magnitude higher. This carrier mismatch can lead to the possibility of interface reflection, which results in an increase in the value of the contact resistance. Lee and Beasley [91] estimated the intrinsic contact resistivity of a $YBa_2Cu_3O_{7-\delta}$/Ag interface based on the carrier mismatch using the expression

$$R_c^{-1} \approx \left(\frac{1e^2}{2\pi\hbar} N_{YBCO}\langle k \rangle\right) D \qquad (12)$$

where N_{YBCO} is the carrier density of $YBa_2Cu_3O_{7-\delta}$ and k is the average momentum of the carriers transverse to the interface (i.e., estimated as the reciprocal of the lattice constant). Here, D is the interface transmission coefficient and is assumed to be much less than 1 due to the carrier mismatch.

In cases where samples were annealed after the Ag deposition, contact resistivities as low as R_c (a axis) $\simeq 5 \times 10^{-11}$ $\Omega \cdot$cm^2 and R_c (c axis) $\simeq 4 \times 10^{-10}$ $\Omega \cdot$cm^2 have been reported, which are close to the theoretical limit [91]. However, most metals aside from silver and gold are unable to form low resistance contacts to cuprate systems due to the formation of extrinsic barriers. Many conventional metals and semiconductors react with the superconductor surface to form insulating oxide layers. These barriers raise the contact resistivity to values that are much higher than the intrinsic contact resistivity. Thus, the measurement of the contact resistance can be used as a direct measurement of the chemical compatibility of the contacting layer with the underlying superconductor.

B. Contact Resistance Measurement Geometries

It is common practice to use linear four-point probe measurements to directly evaluate the sample resistance,

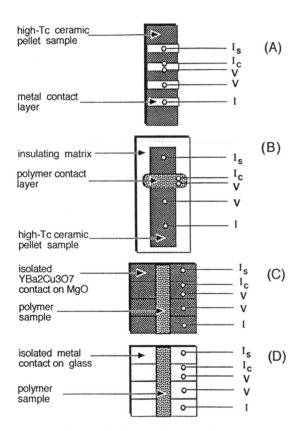

Fig. 37.12 Schematic illustrations showing the electrical attachment geometries that were used for the sample and contact resistance measurements. Terminals labeled I are common current terminals used in both four- and three-point measurements; leads labeled I_c are current leads used in three-point measurements for the contact resistance determinations; those labeled I_s are the current terminals exploited in the four-point measurements to acquire the sample resistance. Terminals labeled V are used to measure voltage in both the three- and four-point modes. (Adapted from Ref. 4.)

thereby eliminating contact resistance contributions to the ascertained values. Accordingly, two outer leads are used to pass current and the potential is measured across two inner leads. However, if the current is forced to pass through one of the voltage leads, the measured resistance will include contributions from both the sample resistance (R_s) and one contact element (R_c). Contact resistance values are acquired simply by subtracting the four-point value from the three-point value. To obtain a geometry-independent value, the magnitude of the contact resistivity is calculated by multiplying the value of the contact resistance by the contact area.

Figure 37.12 displays the different geometries that are used to study the high-T_c superconductor contact resistance phenomenon. In the simplest geometry (Fig. 37.12A), a rectangular segment of a ceramic pellet is coated with metal layers and silver paint is applied to attach the wire electrodes. A slightly different geometry (Fig. 37.12B) is exploited when polymer contacts are made to the ceramic sample. Here the superconductor chip is encapsulated into an insulating epoxy matrix and the polymer is deposited over both the superconductor and the supporting matrix. Consequently, when electrical contact is made with the polymer layer by using gold paste over the insulating segment, the possibility of direct electrical contact between the gold paste and the superconductor is eliminated.

The two geometries described above use superconductor samples and normal contacts. In addition, patterned thin-film contact templates are used to yield the opposite geometry, which consists of a polymer sample with superconductor contacts (Fig. 37.12C). For comparison purposes, the behavior of conductive polymer samples with normal metal contacts is also examined (Fig. 37.12D).

C. Metal/Superconductor Contact Resistance

Prior to studying the organic conductors, the temperature dependence of the contact resistance between a variety of normal metals and a number of cuprate superconductors was examined [4]. The interface resistivity of noble metal contacting layers such as Ag, Au, or Pt was found to have room temperature values of $\sim 1 \times 10^{-5}$ $\Omega \cdot cm^2$, the magnitude of which decreases as the sample is cooled. Typical data for $YBa_2Cu_3O_{7-\delta}$ with Au contacting layers are shown in Fig. 37.13A. It should be noted that contact resistivity values reported here are for samples in which no postdeposition metal annealing was carried out. Further decreases in the contact resistivity by factors of at least 10^3 were noted when the metal/superconductor specimens were heated to temperatures of 500–950°C. The annealing allows for diffusion of the contacting metal past the surface degradation layer that forms during the metal deposition and

Superconductor Assemblies and Devices

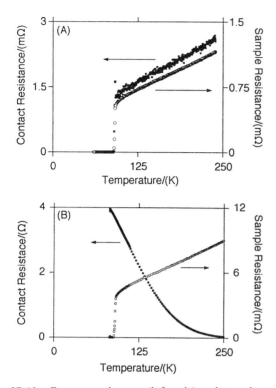

Fig. 37.13 Contact resistance (left axis) and sample resistance (right axis) as a function of temperature recorded for two sample–contact element combinations: (A) $YBa_2Cu_3O_{7-\delta}$ ceramic pellet sample with gold contacts; (B) $YBa_2Cu_3O_{7-\delta}$ ceramic pellet with copper contacts. (Adapted from Ref. 4.)

upon superconductor exposure to the atmosphere, thus improving contact with the underlying superconductor structure. Since the polymer structures are not compatible with such heat treatments, all contact resistivity values reported here are for samples that were not thermally annealed.

On the other hand, when contacting metals such as Cu, Sn, Al, Pb, or In are used, the room temperature contact resistivity is significantly higher [4]. Moreover, these systems display activated temperature dependencies, and no decreases in the interface resistance are observed near T_c. Metals of this type, which form stable metal oxides, are known to react chemically with high-T_c samples to form a degradation layer at the interface between the two conductors [91]. Accordingly, Oka and Iri [92] found a strong correlation between the free energy of the metal oxide of the contact layer and the contact resistance. Metals that are easily oxidized, such as Mg, have the highest contact resistivities. A typical contact resistance versus temperature measurement for the reactive metal case is shown in Fig. 37.13B.

D. Conductive Polymer/Cuprate Superconductor Contact Resistance

The fact that good electrical contact is made between the conductive polymer and the superconductor is noted from decreases in the four-point sample resistance with an onset temperature near 110 K and zero resistance close to 85 K for $Pb_{0.3}Bi_{1.7}Sr_{1.6}Ca_{2.4}Cu_3O_{10}$ samples measured with poly(3-hexylthiophene) contacts, as shown in Fig. 37.14A. Almost identical four-point sample resistance results are acquired with the use of silver contacts on the same specimen. Importantly, these results demonstrate that conductive polymer components can be used to prepare superconductor circuits that operate at temperatures both above and below T_c. Moreover, the values of the contact resistance acquired for the poly(3-hexylthiophene)/$YBa_2Cu_3O_{7-\delta}$ structure are comparable to values acquired for systems in which the superconductor component is replaced with a noble metal material such as platinum (see Table 37.3). Al-

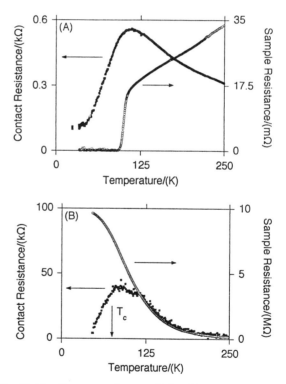

Fig. 37.14 Contact resistance (left axis) and sample resistance (right axis) as a function of temperature recorded for two sample–contact element combinations: (A) $Pb_{0.3}Bi_{1.7}Sr_{1.6}Ca_{2.4}Cu_3O_{10}$ ceramic pellet sample with poly(3-hexylthiophene) contacts; (B) poly(3-hexylthiophene) film sample with $YBa_2Cu_3O_{7-\delta}$ film contacts. (Adapted from Ref. 4.)

Table 37.3 Summary of Contact Resistivity Values

Sample[a]	Contact	ρ_c ($\Omega \cdot cm^2$) 290 K	90 K	45 K	Notes
Highly doped PHT film ρ(RT) = 7.5 $\Omega \cdot$cm T_c (onset) = 73 K; ΔT = 3 K	$YBa_2Cu_3O_{7-\delta}$	250	2100	1200	1,2
Highly doped PHT film ρ(RT) = 51 $\Omega \cdot$cm T_c (onset) = 72 K; ΔT = 45 K	$YBa_2Cu_3O_{7-\delta}$	35	4000	7800	1,2
Lightly doped PHT film ρ(RT) = 1500 $\Omega \cdot$cm T_c (onset) <20 K	$YBa_2Cu_3O_{7-\delta}$	1400	—[b]	—[b]	2,3
$Pb_{0.3}Bi_{1.7}Sr_2Ca_2Cu_3O_{10}$ pellet	Highly doped PHT Film	16	30	7.2	4
Highly doped PHT film ρ(RT) = 12 $\Omega \cdot$cm	Platinum film	30	660	3600	5
$YBa_2Cu_3O_{7-\delta}$ ceramic pellet	Gold film	3.1×10^{-5}	4.9×10^{-6}	1×10^{-7}	6
$YBa_2Cu_3O_{7-\delta}$ ceramic pellet	Copper film	2.0×10^{-3}	3.6×10^{-1}	—[b]	6
$YBa_2Cu_3O_{7-\delta}$ ceramic pellet	Indium film	112.6[c] (300 K)	84[c] (77 K)	—[b]	7

[a] PHT = poly(3-hexylthiophene).
[b] Resistance greater than 10 MΩ.
[c] Measured at specified temperature.

Notes:
1. ΔT = 10–90% transition width.
2. Geometry C, Fig. 37.12.
3. Oxygen-deficient $YBa_2Cu_3O_{7-\delta}$.
4. Geometry B, Fig. 37.12.
5. Geometry D, Fig. 37.12.
6. Geometry A, Fig. 37.12.
7. See Ref. 59.

Source: Adapted from Ref. 4.

though the values obtained for the noble metal/superconductor contact resistance are far superior to those obtained for polymer/superconductor contact resistance, the organic material's performance is comparable to that achieved with indium metal and is much better than that achieved with conventional semiconductors such as silicon. Such results indicate that the oxidized polymer and the cuprate superconductor are chemically compatible with one another.

The temperature dependence of the poly(3-hexylthiophene)/$Pb_{0.3}Bi_{1.7}Sr_{1.6}Ca_{2.4}Cu_3O_{10}$ contact resistance above T_c displays an activated behavior, with the contact resistance increasing as the temperature is lowered from room temperature to 110 K (Fig. 37.14A). However, as T_c(onset) is approached, the contact resistance decreases dramatically. Unlike data acquired with normal metal/high-T_c structures, where the values of the contact resistance become vanishingly small below T_c, the contact resistance for the polymer/superconductor structure possesses a small but finite value at low temperatures. The presence of a short segment of polymer that extends onto the insulating matrix and is not in direct contact with the superconductor material contributes line resistance to the circuit. The presence of this element is likely to be responsible for the majority of the residual resistance associated with the geometry shown in Fig. 37.12b.

Similar decreases in the contact resistance were noted for the opposite geometry, where the polymer resistance is measured and superconducting contacts are employed, as shown in Fig. 37.12C. The contact resistance and sample resistance of a poly(3-hexylthiophene) thin-film sample that was measured using $YBa_2Cu_3O_{7-\delta}$ contacts is shown in Fig. 37.14B. As in the poly(3-hexylthiophene)/$Pb_{0.3}Bi_{1.7}Sr_{1.6}Ca_{2.4}Cu_3O_{10}$ case, the temperature at which the contact resistance decrease occurs correlates well with the transition temperature for the underlying superconductor.

As an important control, contact resistance values for a number of poly(3-hexylthiophene)/normal metal interfaces were investigated (not shown). When a normal metal, such as platinum or gold, was employed as contacts, no decreases in the contact resistance values were observed [4]. When an oxygen-deficient $YBa_2Cu_3O_{7-\delta}$

film with a very broad metal–superconductor transition region ($T_c = 45$ K) was used to make contact with the poly(3-hexylthiophene), only a slight decrease in contact resistance was noted near T_c(onset) (not shown).

The carrier density for most conductive polymers is $\sim 5 \times 10^{21}$ cm^{-3} and is more closely matched to the carrier density of $YBa_2Cu_3O_{7-\delta}$ than is silver. This closer match should, in principle, improve the interface transmission coefficient. However, the low mobility of the carriers in the polymer can lead to an overall higher contact resistance. The fact that the contact resistance of the polymer/platinum interface is comparable to that of the polymer/$YBa_2Cu_3O_{7-\delta}$ interface also suggests strongly that the contact resistance is an electronic phenomenon and not the result of chemical degradation.

From the magnitudes of the polymer/superconductor contact resistivities, it appears unlikely that conductive polymers such as poly(3-hexylthiophene) will be used as contacting materials for high current superconductor applications. The fact that the conductive polymers also display low carrier densities and low mobilities below 100 K is likely to exacerbate the problem. In spite of these limitations, conductive polymers may find utility for the fabrication of hybrid polymer/superconductor devices where low current densities are desirable. It is important in this regard that the conductivity of conductive polymers, such as poly(3-hexylthiophene), can be controlled by varying the polymer oxidation state (i.e., doping level). Moreover, unlike that of silicon and other conventional semiconductors, the doping of these polymers is highly reversible, and these organic materials appear to be reasonably compatible with the cuprate compounds.

The decrease of the poly(3-hexylthiophene)/high-T_c superconductor contact resistance can be attributed possibly to the induction of superconductivity at the interface caused by a proximity effect. This enticing result represents the initial evidence for the induction of superconductivity into doped organic polymeric conductors.

E. (BEDT-TTF)$_2$I$_3$/Cuprate Superconductor Contact Resistance

Further evidence to support the observation of the induction of superconductivity into organic conductors has been derived from studies of hybrid crystalline organic conductor/high-temperature superconductor structures [59]. Bilayer organic superconductor/inorganic superconductor assemblies based on the charge transfer salt (BEDT-TTF)$_2$I$_3$ have been prepared by vapor deposition of the organic material onto the surface of $YBa_2Cu_3O_{7-\delta}$ thin films. Structures of this type have been shown to display behavior similar to that seen for the conductive polymer/high-temperature superconductor assemblies [4,59]. Here, the temperature dependence of the (BEDT-TTF)$_2$I$_3$/$YBa_2Cu_3O_{7-\delta}$ contact resistance above T_c displays an activated behavior with the interfacial resistance increasing as the temperature is lowered from 295 K to ~ 90 K (Fig. 37.15). However, at T_c(onset), the contact resistance decreases in a manner analogous to that seen with the conductive polymer/high-T_c structures. This result provides further evidence to suggest that the decreasing contact resistance phenomenon is related to superconductivity, in that the magnitude of the effect is found to be dependent on the sampling current. In general, the superconducting state is known to be stable under limits of temperature, current, and magnetic field. At lower sampling currents, the decrease in contact resistance at T_c is most pronounced. However, this decrease becomes less prominent with increasing current until the critical current of the interface is reached, at which point the interfacial superconductivity is no longer possible. Similar current dependencies for the detection of the induction of superconductivity into metal contacts have been noted [67].

To further explore the behavior of the (BEDT-TTF)$_2$I$_3$ interfacial resistance phenomena, samples of the organic conductor were deposited onto Pt thin-film structures. The interfacial resistance properties as a function of temperature were also explored for (BEDT-TTF)$_2$I$_3$ samples with Pt contacts. In these structures, an activated temperature dependence was observed for both the four-point resistivity and contact resistance. Here the magnitude of the contact resistance between the two materials is comparable to that between (BEDT-TTF)$_2$I$_3$ and $YBa_2Cu_3O_{7-\delta}$. No decreases in interface resistance were noted at low temperatures as in the (BEDT-TTF)$_2$I$_3$/$YBa_2Cu_3O_{7-\delta}$ bilayer structures. Also, the similarity in the magnitudes of the contact resistance in the (BEDT-TTF)$_2$I$_3$/$YBa_2Cu_3O_{7-\delta}$ and the (BEDT-

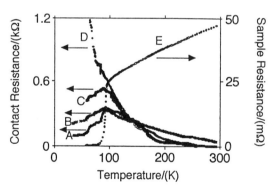

Fig. 37.15 Contact resistance (left axis) and sample resistance (right axis) as a function of temperature recorded for a $YBa_2Cu_3O_{7-\delta}$ thin-film sample (1500 Å) with (BEDT-TTF)$_2$I$_3$ contacts (1 μm) at the following sampling currents: (A) 1×10^{-7} A; (B) 1×10^{-6} A; (C) 1×10^{-5} A; (D) 1×10^{-4} A. Curve E represents the temperature dependence of the $YBa_2Cu_3O_{7-\delta}$ sample resistance. (Adapted from Ref. 59.)

$(TTF)_2I_3$/Pt structures suggests that no interfacial degradation occurs during the preparation or analysis of these hybrid systems.

Although the reduction of contact resistance at T_c provides evidence for the induction of high temperature superconductivity into these organic conductors, the possibility of spreading resistance contributing to the measurement was noted previously [4]. This phenomenon occurs commonly in structures such as these when the resistance of one element, in this case the cuprate superconductor, drops suddenly to zero. Therefore, it is desirable to search for additional evidence for the organic proximity effect in crystalline organic conductors.

V. MODULATION OF SUPERCONDUCTIVITY IN CONDUCTIVE POLYMER/SUPERCONDUCTOR BILAYER STRUCTURES

Following the general methods that were used in the past to prove the existence of proximity effects in normal/superconductor systems, attempts to modulate superconductivity in conductive polymer/superconductor bilayer structures have been completed. Here, the influence of the polymer doping level on the properties of the underlying superconductor were examined [3,11]. Although minor increases in the room temperature bridge resistance values were noted after polymer deposition, the most dramatic changes were observed for the superconducting properties. (Figs. 37.16 and 37.17) display a series of resistance vs. temperature curves for a number of $YBa_2Cu_3O_{7-\delta}$ microbridges of variable superconductor thickness coated in each case with ~2 μm of polypyrrole. When a 1200 Å thick film of $YBa_2Cu_3O_{7-\delta}$ is employed, very little changes in the

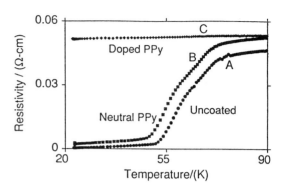

Fig. 37.17 Resistivity vs. temperature curves acquired for a ~100 μm wide × 3 mm long microbridge formed from a $YBa_2Cu_3O_{7-\delta}$ film with thickness of ~200 Å supported on a MgO(100) substrate that possesses a series of step edges. (Film thickness reported here refers to the value in the step-edge region and is estimated from SEM pictures as well as from critical current values.) Data are given for (A) the bare bridge and the bridge coated with ~2 μm polypyrrole in both its (B) undoped and (C) doped states. (Adapted from Ref. 11.)

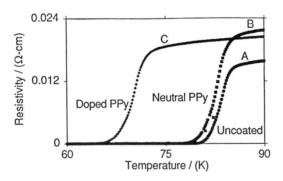

Fig. 37.16 Resistivity vs. temperature curves acquired for a ~100 μm wide × 3 mm long microbridge formed from a $YBa_2Cu_3O_{7-\delta}$ film with a thickness of ~500 Å supported on a polished MgO(100) substrate. Data are provided for (A) the bare bridge and the bridge coated with ~2 μm of polypyrrole, in both its (B) undoped and (C) doped states. (Adapted from Ref. 3.)

properties of the superconductor are noted as the polymer is cycled between its neutral and oxidized forms (Table 37.4). In both cases, the transition temperatures for the polymer-coated films are very close to those obtained for the uncoated structure. For the thinnest superconductor element composed of a ~200 Å film of $YBa_2Cu_3O_{7-\delta}$, superconductivity with an onset temperature close to 75 K is noted for the uncoated bridge (Fig. 37.17A). On the other hand, the same device with the polymer in its oxidized form displays no signs of superconductivity down to 22 K (Fig. 37.17C), the lowest temperature measured. However, after the room temperature reduction of the polymer, the transition temperature of the bridge is found to return to a value very close to that obtained for the bare bridge (Fig. 37.17B). This greater than 50 K modulation of superconductivity is the largest reversible shift in superconductivity reported to date [11]. A superconductor film with an intermediate thickness value of 500 Å (Fig. 37.16) displays a reversible shift of superconductivity on the order of 15 K. In each case, a weakening of superconductivity is noted by a lowering of the transition temperature when the polymer is oxidized. Moreover, the polymer can be cycled several times between its oxidized (conductive) and neutral (insulating) forms, yielding similar behavior (provided that the electrochemical potentials are not taken to extreme values and corrosive reagents such as water are not present in the electrolytic fluid).

In addition to the modulation of T_c, it is noted that the critical current values, J_c, are affected by the oxidation state of the conductive polymer layer. It is evident here that the critical current is suppressed by the pres-

Table 37.4 Summary of Transition Temperatures for $YBa_2Cu_3O_{7-\delta}$ Films of Variable Thickness Coated with Polypyrrole (2 μm)

Superconductor film thickness[a] (Å)	T_c of neutral polymer (K)	T_c of oxidized polymer (K)	ΔT_c	Film axis	Morphology
200	74	≤24	≥50	a/c	Rough
500	84	70	14	a/c	Rough
1500	80	79	1	a/c	Rough
500	87.6	86.8	0.8	c	Smooth
900	87.6	87.0	0.6	c	Smooth
1200	87.0	87.0	0	c	Smooth

[a] Reported thickness represents the estimated value over the step for an 800 Å film of $YBa_2Cu_3O_{7-\delta}$ that was deposited over a 1500 Å step edge.
Source: Adapted from Ref. 11.

ence of the doped polymer relative to the neutral polymer case. (Fig. 37.18) shows a comparison of the temperature dependence of the critical currents for a single device with the polymer in both the doped and neutral forms.

In addition to the polymer doping level and the superconductor film thickness, the morphology of the superconductor, its crystal orientation, and the quality of the superconductor surface appear to be important factors in determining the behavior of the polymer/superconductor structures. High quality superconductor films with high critical currents do not show large modulations of T_c or J_c. Rather, films with low critical currents (10^4–10^5 A/cm² at 77 K), rough surface morphologies, and a large number of weak links tend to yield the largest responses to the polymer layers [11]. In fact, when weak links are purposely introduced into the system by depositing $YBa_2Cu_3O_{7-\delta}$ over the step edges (20 step edges/cm) on cleaved MgO(100) substrates, the modulation of superconductivity is enhanced, as was the case for the

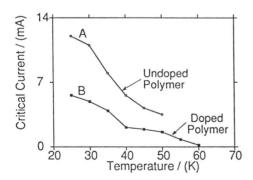

Fig. 37.18 Critical currents obtained at various temperatures for a single 2 μm polypyrrole film coated onto a 100 μm wide × 3 mm long microbridge formed from a 500 Å thick $YBa_2Cu_3O_{7-\delta}$ film. Data are shown for (A) neutral and (B) doped polypyrrole cases. (Adapted from Ref. 11.)

data presented in Fig. 37.17. The use of step edges provides a convenient method to fabricate a variety of interesting macromolecular superconductor devices that display high sensitivities to the properties of molecular compounds. The quality of the superconductor surface also appears to be important in determining the extent of the modulation of T_c. Surfaces with larger amounts of degradation tend to modulate T_c far less than pristine surfaces. This is due to better electrical contact between the polymer and the underlying superconductor.

The reversible changes in the properties of the superconductor (i.e., T_c and J_c) afforded by the conductive polymer layers forms the basis for a "molecular switch for controlling superconductivity" [3,11]. Here the latch for the switch is the polymer oxidation level. In its present form, it is unlikely that these polymer/superconductor structures will be used for the construction of practical devices for active switching applications due to the need to charge and discharge the polymer layers at elevated temperatures. However, it is plausible that the conductive polymer layers could be used to fine-tune the performance characteristics of the superconductor junctions (i.e., for SQUID applications). Exciting possibilities for cryogenic switching using optical techniques are also possible and are now being investigated in our laboratory.

A. Morphology and Orientation of the Superconductor Samples

As mentioned above, the magnitude of the modulation of the superconducting properties in the conductive polymer/high-T_c superconductor bilayer structures is dependent on the superconductor film morphology and sample orientation. Thin films of superconductor with rough morphologies show the greatest change in T_c, whereas the smooth films are influenced by the polymer to only a minor extent. Films with thicknesses greater than 1000 Å, regardless of the film morphology, normally display little or no shift in T_c. Summarized in Table 37.4 for various thickness values and morphologies of the superconductor thin films are transition temperatures for the polymer/superconductor devices in both the neutral and oxidized forms. The first three sets of data are for cases where highly textured, mixed a/c axis films were used, whereas the latter three sets correspond to smooth c-axis films.

There are a number of important differences between the rough and smooth films that are responsible for the differences in the magnitude of the modulation of superconductivity. First, rough films possess regions that are much thinner than the average value. Locally, these weak structures have critical current values that are much less than in the remainder of the film and contribute to the enhanced sensitivity of such regions. Second, the polycrystalline films possess both a-axis and c-axis crystallographic orientations. From a consideration of the structural and electronic properties of the cuprate compounds, it is not surprising that the high-T_c film sample orientation influences the magnitude of the observed polymer-induced superconductivity modulation (see below).

The $YBa_2Cu_3O_7$ compound possesses a layered structure with CuO_2 basal planes and Cu–O chain layers (Fig. 37.19). Motion of hole carriers occurs predominantly

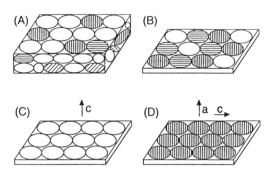

Fig. 37.20 Schematic illustration showing the crystallographic orientations that are found commonly to be associated with various $YBa_2Cu_3O_7$ samples. (A) Polycrystalline bulk ceramic pellets exhibit granular structures with little orientational order between adjacent grains. (B) Thin films deposited under non-ideal conditions or those supported on substrates with poor lattice matching to the cuprate structure can be prepared as polycrystalline samples. These samples, like ceramic pellets, possess largely random orientations. (C) Many substrates and ideal deposition conditions foster growth of c-oriented films. This orientation exposes the weakly superconducting regions of the cuprate lattice. These c-axis-oriented films can be grown with and without additional ordering of the a and b axes. (D) Under certain conditions, films can be prepared having an a-axis orientation perpendicular to the substrate surface. Such films possess the strongly superconducting region exposed at the surface. Samples with a-axis orientations exhibit the strongest interactions with organic conductor layers.

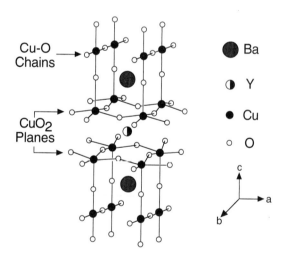

Fig. 37.19 Unit cell diagram for the high-T_c superconductor $YBa_2Cu_3O_7$. The lattice structure is composed of CuO, BaO, CuO_2, Y, CuO_2, BaO, and CuO layers that are stacked along the c axis. It is important to note that there are two distinct types of copper sites: CuO chains and CuO_2 sheets. The former run along the b axis and serve the role of a charge reservoir. The latter run parallel with the ab plane and are responsible for the delocalization of charge carries, which fosters the occurrence of superconductivity in the material.

within the CuO_2 sheets, along the ab planes. Films prepared with the c-axis direction of the $YBa_2Cu_3O_7$ lattice perpendicular to the substrate surface plane are called c-axis or c-oriented films. In a-axis films, the c axis of $YBa_2Cu_3O_7$ is parallel to the substrate's surface plane.

Shown in Fig. 37.20 are the sample orientations that are commonly found for $YBa_2Cu_3O_{7-\delta}$ specimens. Here, the ovals represent individual grains and the lines therein are intended to indicate the edges of the ab planes. Polycrystalline ceramic pellets (Fig. 37.20A) consist of randomly oriented grains. Highly oriented textured thin films (Fig. 37.20B) also exhibit various degrees of order. For example, thin films with both a- and c-axis grains normal to the substrate surface can be prepared readily (Figs. 37.20C, D). The morphological differences between these two crystallographic orientations result in different electrical properties. The a-axis films usually have higher normal state resistivities as well as lower critical currents and transition temperatures than c-axis films. These differences can be attributed to the presence of a large number of junctions or grain boundaries that disrupt the electrical conduction process, which occurs predominantly in the *ab* planes of the $YBa_2Cu_3O_7$ lattice. Additionally, grains having unfavorable orientations or secondary phases at the grain boundaries further frustrate electrical conduction throughout the films.

Since the coherence length along the *c*-axis ($\xi_c \sim 6$ Å) is much shorter than that along the *a* axis ($\xi_{ab} \sim 30$ Å), the inclusion of an *a*-axis component in the hybrid polymer/superconductor assembly serves to increase the sensitivity of the composite structure. This coherence length measures the minimum distance over which the superconducting properties can be altered. Larger values of coherence length along the *ab* plane suggest that superconducting phenomena occur with greater strength along this direction of the lattice relative to that observed along the *c*-axis direction. Thus, it is not surprising that thick $YBa_2Cu_3O_{7-\delta}$ films display little modulation effect when organic conductor layers deposited on them are cycled between conductive and insulating states. Furthermore, it is expected that the greatest modulation will be observed for the thin films that have weak links and *a*-axis components. Clearly, the organic conductor and high-T_c superconductor components can communicate more effectively through interfacial carrier diffusion when the high-T_c component is present with an *a*-axis orientation. Although thicker superconductor samples are expected to exhibit similar interactions with doped organic conductors, the effects for such structures are limited to regions close to the organic conductor/high-T_c superconductor interface. Under such circumstances, little modulation of the superconducting properties is observed experimentally because zero-resistance shunts remotely spaced from the interface can exist. These regions dominate the conduction observed for the composite structure.

Although deposition conditions for the preparation of $YBa_2Cu_3O_{7-\delta}$ films are often optimized to produce *c*-axis-oriented films that exhibit high critical currents, polycrystalline films appear to be better suited for the fabrication of sensitive conductive polymer/superconductor structures. These superconductor thin films are more textured and have lower critical currents than the smooth films, which are prepared with what would normally be considered more optimized deposition conditions. The weak link characteristics of these thin films are also enhanced by depositing the superconductor onto cleaved MgO substrates. These substrates possess natural step edges and can be exploited to further disrupt the connection between selected superconductor grains.

Similarly, Mannhart et al. [93] prepared films rich with weak links for the purpose of making electric field effect devices that exhibit high sensitivities. To prepare such devices, the substrates were polished with 1 μm diamond paste prior to superconductor deposition. The diamond polishing procedure leads to the formation of a network of fine grooves on the substrate and induces the creation of grain boundaries in the film. The electric field devices fabricated from these weak-link-rich films also displayed relatively large shifts in T_c, although not as large as those produced with the conductive polymer layer.

B. Discussion of Polymer/Superconductor Electronic Interactions

There are several possible explanations for the observed modulation of superconducting properties in these bilayer structures. Since $YBa_2Cu_3O_{7-\delta}$ reacts readily with water, CO, CO_2, and acids [18,25,94] chemical degradation processes could, in principle, account for the initial decrease in T_c. However, the subsequent restoration of superconductivity to higher temperatures is not at all consistent with such behavior. All corrosion reactions associated with $YBa_2Cu_3O_{7-\delta}$ known to date involve irreversible transformations that damage the high-T_c lattice. Thus, it is unlikely that reversible corrosion reactions are responsible for the observed effects. Furthermore, uncoated weak-link-rich superconductor films were cycled electrochemically without polymer modifying layers in control studies, and changes of less than 0.5 K were observed in T_c [3,11]. These control studies were completed to determine whether or not the electrochemical treatment might be responsible for a reduction in the oxygen content of the underlying superconductor. Changes in the copper valence values that result from alterations in the oxygen content of $YBa_2Cu_3O_{7-\delta}$ were shown earlier to alter T_c [95]. Likewise, oxygen content changes are known to result in large changes in normal state properties [95], such as room temperature resistivity and the temperature dependence of resistivity. These effects are not noted in the control studies of the polymer/superconductor systems. Thus, changes in the bulk oxygen content do not appear to be responsible for the observed behavior. However, the reversible changes in the chemical structure (i.e., oxygen content) may be limited to the grain boundaries, and this might affect the superconducting properties of the film without causing major changes in the normal state properties. Again, the control studies suggest that such effects do not likely contribute in a significant way to the observed phenomena.

Having largely ruled out the possibility that oxygen content changes and chemical damage are responsible for the reversible modulation in superconductivity, two additional explanations merit consideration. First, if the polymer layer is separated from the superconductor by an insulating barrier formed by superconductor degradation products, the observed modulation of superconducting properties may be the result of an electric field effect [96–98]. Here, the applied field caused by the oxidized polymer may influence the number of mobile charge carriers in the superconductor in a way that is analogous to the field effect in conventional metal oxide semiconducting devices. Electric field effects for the modulation of superconductivity have been discussed by Mannhart et al. [93,98] and Kawabasa et al. [97]. While changes in T_c due to field effects are normally much less than those we have observed, recent studies have shown that the shift can be magnified by using films with weak link characteristics [93]. In the polymer/superconductor structures, the greatest observed shifts are obtained when weak links are introduced into the superconductor component.

The second, and perhaps more likely, possibility is that a superconducting proximity effect [67,68] operates within the composite assembly. In these studies, the doped conductive polymer layer may be acting as the normal metal layer. Collectively, the large dependence on superconductor thickness, lack of modulation of superconductivity in the control studies, and the contact resistance studies present a database consistent with organic conductor proximity effects. Obviously, further research is needed before it will be possible to claim in a definitive manner that elevated temperature superconductivity has been induced to occur in organic conductors. The long-sought observation of elevated temperature organic superconductivity makes the future study of organic conductor/high-T_c superconductor composite assemblies both promising and exciting.

The magnitude of the proximity effect observed for classical studies of metal/superconductor sandwich structures has been shown to be dependent on a number of factors such as the metal thickness, the superconductor thickness, and the coherence length of the superconductor template [70,75]. The shifts in T_c we have observed are inordinately large compared to the very small coherence length ($\xi_c \sim 6$ Å, $\xi_{ab} \sim 30$ Å) of $YBa_2Cu_3O_{7-\delta}$ and the relatively large average thickness values for the superconductor. However, these large shifts in T_c are observed only with films that have a large number of weak links. The presence of these weak links is a key factor in producing the large shifts in the transition temperature that have been observed in conductive polymer/superconductor bilayer structures. Definitive evidence for organic polymer/superconductor proximity effects awaits supercurrent measurement. Experiments of this type are now in progress. Like efforts to commercialize high-T_c digital devices, the study of organic conductor/high-T_c superconductor arrays suitable for supercurrent measurements will require the development and refinement of new processing methods. Here, the creation of devices having ultrasmall features and the minimization of interfacial degradation reactions must be achieved. Despite the formidable materials processing challenges, the prospects for tailoring superconducting devices that are capable of operation at elevated temperature makes the progression of this research particularly important.

VI. CONCLUSIONS

In summary, effective methods have been identified for the preparation of conductive polymer/superconductor and molecular metal/superconductor composite structures. Here both solution-processing strategies and electrochemical deposition techniques for the preparation of the composite structures have been developed. Moreover, a powerful new high-T_c self-assembly method based on the spontaneous adsorption of amine reagents onto cuprate surfaces has been developed that affords precise control of the synthesis of polymer/superconductor composite systems. With these methods, the hybrid structures can be prepared with little chemical or physical damage to either conductor component material. Convincing evidence for the clean combination of the molecular and superconductor components has been obtained from electrochemical, conductivity, contact resistance, and electron microscopic measurements.

Having identified successful strategies for the preparation of well-behaved organic conductor/cuprate superconductor structures, experimental procedures have been developed to explore for the first time the electronic interactions that occur between organic conductors and high-T_c superconductors. Moreover, it has been demonstrated that doped conductive polymers and molecular metals can be used in the context of superconductor measurements at cryogenic temperatures. In fact, these organic conductors have been shown to exhibit performance characteristics superior to those exhibited by conventional semiconductors such as Si and GaAs that suffer from chemical reactivity problems.

Perhaps the most important, most novel, and most significant contribution made by these studies of organic conductor/superconductor systems is the initial evidence for the induction of superconductivity into the organic conductors. Here, several lines of experimental evidence have been acquired following the analogous developments that proved the existence of proximity effects in metal/superconductor and semiconductor/superconductor systems. Contact resistance data, modulation of superconductivity in organic conductor/superconductor bilayer structures, and calculations of coherence length have all been obtained that suggesting that these organic conductors may be suitable to support superconductivity over macroscopically large distances. In the case of the modulation of superconductivity in the bilayer systems, the conductive polymers possess attributes lacking in other conventional conductors that make the analysis of proximity effects at cuprate superconductors more convincing. Accordingly, these polymers can be reversibly cycled between neutral (insulating) and oxidized (conductive) forms, thereby affording an important control for the exploration of the chemical compatibility between the two conductors. Conventional conductors lack this flexibility, making the interpretation of the modulation of superconductivity data less certain.

The preparation of superconductor/organic conductor/superconductor structures suitable for supercurrent measurements is important for the continued growth of this new area of research. Toward this objective, methods have been developed for the controlled growth of conductive polymer and vapor-phase deposition of $(BEDT-TTF)_2I_3$ systems. Molecular engineering of high-T_c structures and devices made possible with the newly discovered high-T_c self-assembly procedure will aid in the further development of these studies.

The strong electronic interactions that occur between molecular conductors and high-T_c superconductors suggest that interesting fundamental interactions as well as the development of new hybrid devices will likely result from future studies of these systems. The combination of molecular conductors and superconductors into functional structures has opened new opportunities in both the fundamental and applied superconductor areas. Using the science base established here, it may be possible in the future to tailor, from the molecular level, organic conductors to support superconductivity through their contact with high-T_c systems. Those systems that exhibit superior properties in the hybrid structures can then be evaluated more closely in monolithic forms to search for superconductivity properties without the need for a high-T_c superconductor template.

ACKNOWLEDGMENTS

The support for our research by the National Science Foundation and the Robert A. Welch Foundation is gratefully acknowledged. We also thank our coworkers, colleagues, and collaborators referenced in joint publications whose insightful comments and enduring efforts have contributed positively to the body of work discussed in this chapter. Finally, we thank Royce W. Murray, Wolfgang J. Lorenz, Allen J. Bard, John T. Markert, Chad A. Mirkin, and Robin McCarley for useful discussions.

REFERENCES

1. K. Kaneto and K. Yoshino, The application of superconducting ceramics as substrates for the electrochemical deposition of conducting polymers and metals, *Jpn. J. Appl. Phys. 26*:L1842 (1987).
2. E. N. Izakovich, V. M. Geskin, and S. V. Stepanov, Composites of polyaniline/superconductor $YBa_2Cu_3O_{7-\delta}$: Production and properties, *Synth. Met. 46*:71 (1992).
3. S. G. Haupt, D. R. Riley, C. T. Jones, J. Zhao, and J. T. McDevitt, Reversible modulation of T_c in conductive polymer/high temperature superconductor assemblies, *J. Am. Chem. Soc. 115*:1196 (1993).
4. S. G. Haupt, D. R. Riley, J. Zhao, and J. T. McDevitt, Contact resistance measurements recorded at conductive polymer/high-temperature superconductor interfaces, *J. Phys. Chem. 97*:7796 (1993).
5. A. K. Sarkar and T. L. Peterson, Fabrication and electrical properties of ceramic superconductor/polymer composites, *Polym. Eng. Sci. 32*:305 (1992).
6. D. Ponnusamy and K. Ravi-Chandar, in *HTS Materials—Bulk Processing and Bulk Application* (C. W. Chu, P. H. Hor, and K. Salama, eds.), World Scientific, Singapore, 1992, p. 389.
7. S.-G. Jin, L.-G. Lio, Z.-Z. Zhu, and Y.-L. Huang, Water reaction of superconducting $YBa_2Cu_3O_{7-x}$ at 0°C and its protection from water corrosion at 100°C, *Solid State Commun. 69*:179 (1989).
8. K. Sato, S. Omae, K. Kojima, T. Hashimoto, and H. Koinuma, Stabilization of $Ba_2YCu_3O_{7-\delta}$ by surface coating with plasma polymerized fluorocarbon film, *Jpn. J. Appl. Phys. 27*:2088 (1988).
9. S. G. Haupt, D. R. Riley, and J. T. McDevitt, Conductive polymers/high-temperature superconductors composite structures, *Adv. Mater. 5*:755 (1993).
10. A. Mogro-Campero, A review of high-temperature superconducting films on silicon, *Supercond. Sci. Technol. 3*:155 (1990).
11. S. G. Haupt, D. R. Riley, J. H. Grassi, R.-K. Lo, J. Zhao, J.-P. Zhou, and J. T. McDevitt, Preparation and characterization of $YBa_2Cu_3O_{7-\delta}$/polypyrrole bilayer structures, *J. Am. Chem. Soc. 116*:9979 (1994).
12. W. A. Little, *Phys. Rev. 134*:A1416 (1964).
13. M. S. Wrighton, Surface functionalization of electrodes with molecular reagents, *Science 231*:32 (1986).
14. J. H. Burroughes, D. D. C. Bradley, A. R. Brown, R. N. Marks, K. Mackay, R. H. Friend, P. L. Burns, and A. B. Holmes, Light-emitting diodes based on conjugated polymers, *Nature 347*:539 (1990).
15. M. J. Sailor, F. L. Klavetter, R. H. Grubbs, and N. S. Lewis, Electronic properties of junctions between silicon and organic conducting polymers, *Nature 346*:155 (1990).
16. S. Chao and M. S. Wrighton, Characterization of a "solid-state" polyanaline-based transistor: Water vapor dependent characteristics of a device employing a poly(vinyl alcohol) phosphoric acid solid-state electrolyte, *J. Am. Chem. Soc. 109*:6627 (1987).
17. F. Garnier, G. Horowitz, X. E. Peng, and N. Fickov, *Adv. Mater. 2*:592 (1990).
18. R. L. Barns and R. A. Laudise, Stability of superconducting $YBa_2Cu_3O_7$ in the presence of water, *Appl. Phys. Lett. 51*:1373 (1987).

19. N. P. Bansal and A. L. Sandkuhl, Chemical durability of high-temperature superconductor $YBa_2Cu_3O_{7-\delta}$, *Appl. Phys. Lett. 52*:323 (1988).
20. M. F. Yan, R. L. Barns, H. M. O'Bryan, P. K. Gallagher, R. C. Sherwood, and S. Jin, Water interaction with the superconducting $YBa_2Cu_3O_7$ phase, *Appl. Phys. Lett. 51*:532 (1987).
21. R. P. Vasquez, B. D. Hunt, and M. C. Foote, Nonaqueous chemical etch for $YBa_2Cu_3O_{7-x}$, *Appl. Phys. Lett. 53*:2692 (1988).
22. J. T. McDevitt, D. R. Riley, and S. G. Haupt, Electrochemistry of high-temperature superconductors. Challenges and opportunities, *Anal. Chem. 65*:535 (1993).
23. J. T. McDevitt, S. G. Haupt, and C. E. Jones, Electrochemistry of high-T_c superconductors, in *Electroanalytical Chemistry: A Series of Advances, Vol. 19* (A. L. Bard and I. Rubinstein, eds.), Marcel Dekker, New York, 1996, p. 337.
24. G. B. Street, in *Handbook of Conducting Polymers* (T. A. Skotheim, ed.), Marcel Dekker, New York, 1986, Chap. 8.
25. R. Sugimoto, S. Takeda, H. Gu, and K. Yoshino, *Chem. Express 1*:635 (1986).
26. A. F. Diaz and J. Bargon, in *Handbook of Conducting Polymers* (T. A. Skotheim, ed.), Marcel Dekker, New York, 1986, Chap. 3.
27. J. R. Reynolds, C. K. Baker, C. A. Jolly, P. A. Poropatic, and J. P. Ruiz, in *Conductive Polymers and Plastics* (J. A. Margolis, ed.), Chapman and Hall, New York, 1989, Chap. 1.
28. B. J. Feldman, P. Burgmayer, and R. W. Marray, The potential dependence of electrical conductivity and chemical charge storage of poly(pyrrole) films on electrodes, *J. Am. Chem. Soc. 107*:872 (1985).
29. D. Dijkkamp, T. Venkatesan, X. D. Wu, S. A. Shaheen, N. Jisrawi, Y. H. Min-Lee, W. L. McLean, and M. Croft, Preparation of Y-Ba-Cu oxide superconductor thin films using pulsed laser evaporation from high T_c bulk material, *Appl. Phys. Lett. 51*:619 (1987).
30. R. Yang, D. F. Evans, L. Christensen, and W. A. Hendrickson, Scanning tunneling microscopy evidence of semicrystalline and helical conducting polymer structures, *J. Phys. Chem. 94*:6117 (1990).
31. Y.-T. Kim, H. Yang, and A. J. Bard, Electrochemical control of polyaniline morphology as studied by scanning tunneling microscopy, *J. Electrochem. Soc. 138*:L71 (1991).
32. D. Jeon, J. Kim, M. C. Gallagher, and R. F. Willis, Scanning tunneling spectroscopic evidence for granular metallic conductivity in conducting polymeric polyaniline, *Science 256*:1662 (1992).
33. R. Yang, X. R. Yang, D. F. Evans, W. A. Hendrickson, and J. Baker, Scanning tunneling microscopy images of poly(ethylene oxide) polymers: Evidence for helical and superhelical structures, *J. Phys. Chem. 94*:6123 (1990).
34. H. Yang, F. R. F. Fan, S. L. Yau, and A. J. Bard, The use of a scanning tunneling microscope to estimate film thickness and conductivity of an electrochemically produced poly-1-aminoanthracene film, *J. Electrochem. Soc. 139*:2182 (1992).
35. M. G. Norton, R. R. Biggers, I. Maartense, E. K. Moser, and J. L. Brown, Surface outgrowths on laser-deposited $YBa_2Cu_3O_7$ thin films, *Physica C 233*:321 (1994).
36. K. Chen, C. A. Mirkin, R.-K. Lo, J. Zhao, J. T. McDevitt, F. Xeng, and C. A. Mirkin, Surveying the surface coordination chemistry of a superconductor: Spontaneous adsorption of monolayer films of redoxactive "ligands" on $YBa_2Cu_3O_{7-\delta}$, *J. Am. Chem. Soc. 117*:6374 (1995).
37. C. D. Bain and G. M. Whitesides, *Angew. Chem. Int. Ed. Engl. 28*:506 (1989).
38. L. H. Dubois and R. G. Nuzzo, Synthesis, structure and properties of model organic surfaces, *Annu. Rev. Phys. Chem. 43*:437 (1992).
39. C. B. Gorman, H. A. Biebuyck, and G. M. Whitesides, Use of a patterned self-assembled monolayer to control the formation of a liquid resist pattern on a gold surface, *Chem. Mater. 7*:202 (1995).
40. M. Itoh, H. Nishihara, and K. Aramaki, A chemical modification of alkanethiol self-assembled monolayer with alkyltrichlorosilanes for the protection of copper against corrosion, *J. Electrochem. Soc. 141*:2018 (1994).
41. E. Kim, A. Kumar, and G. M. Whitesides, Combining patterned self-assembled monolayers of alkanethiolates on gold with anisotropic etching of silicon to generate controlled surface morphologies, *J. Electrochem. Soc. 142*:628 (1995).
42. P. E. Laibinis and G. M. Whitesides, Self-assembled monolayers of *n*-alkanethiolates on copper are barrier films that protect the metal against oxidation by air, *J. Am. Chem. Soc. 114*:9022 (1992).
43. C. A. Mirkin and M. A. Ratner, Molecular electronics, *Annu. Rev. Phys. Chem. 43*:719 (1992).
44. C. A. Mirkin, J. R. Valentine, D. Ofer, J. J. Hickman, and M. S. Wrighton, in *Biosensors and Chemical Sensors* (P. G. Edelman and J. Wang, eds.), American Chemical Society, Washington, DC, 1992, Chap. 17.
45. R. G. Nuzzo and D. L. Allara, Adsorption of bifunctional organic disulfides on gold surfaces, *J. Am. Chem. Soc. 105*:4481 (1983).
46. A. J. Ricco, R. M. Crooks, C. J. Xu, and R. E. Aured, Interfacial design and chemical sensing, ACS Symposium, Washington, DC, 1994, p. 264.
47. A. Ulman, *An Introduction to Ultrathin Organic Films: From Langmuir–Blodgett to Self-Assembly*, Academic, New York, 1991.
48. Y. Yamamoto, H. Nishihara, and K. Aramaki, Self-assembled layers of alkanethiols on copper for protection against corrosion, *J. Electrochem. Soc. 140*:436 (1993).
49. K. Chen, C. A. Mirkin, R.-K. Lo, J. Zhao, and J. T. McDevitt, Surveying the surface coordination chemis-

try of a superconductor: Spontaneous adsorption of monolayer films of redox-active "ligands" on YBa$_2$Cu$_3$O$_{7-\delta}$, *J. Am. Chem. Soc. 117*:6374 (1995).
50. R.-K. Lo, J. E. Ritchie, J.-P. Zhou, J. Zhao, and J. T. McDevitt, Polypyrrole growth on YBa$_2$Cu$_3$O$_{7-\delta}$ modified with a self-assembled monolayer of N-(3-aminopropyl)pyrrole: Hardwiring the "electroactive hot spots" on a superconductor electrode, *J. Am. Chem. Soc. 118*:11295 (1996).
51. R. J. Willicut and R. L. McCarley, *Langmuir 11*:296 (1995).
52. D. D. MacDonald, in *Transient Techniques in Electrochemistry*. Plenum, New York, 1977.
53. R. J. Willicut and R. L. McCarley, Electrochemical polymerization of pyrrole-containing self-assembled alkanethiol monolayers on Au, *J. Am. Chem. Soc. 116*:10823 (1994).
54. D. A. Stephens, A. E. Rehan, S. J. Compton, R. A. Barkhau, and J. M. Williams, *Inorg. Synth. 24*:135 (1986).
55. K. Kawabata, K. Tanaka, and M. Mizutani, Superconducting thin films of (BEDT-TTF) iodide, *Synth. Met. 39*:191 (1990).
56. K. Kawabata, K. Tanaka, and M. Mizutani, Conducting thin films of α-(BEDT-TTF)$_2$I$_3$ by the evaporation method, *Solid State Commun. 74*:83 (1990).
57. K. Kawabata, K. Tanaka, and M. Mizutani, Thin films of (BEDT-TTF) iodide prepared by evaporation method, *Synth. Met. 41–43*:2097 (1991).
58. J. M. Williams, J. R. Ferraro, R. J. Thorn, K. D. Carlson, U. Geiser, H. H. Wang, A. M. Kini, and M.-H. Whangbo, in *Organic Superconductors: Synthesis, Structure, Properties, and Theory* (R. N. Grimes, ed.), Prentice-Hall, Englewood Cliffs, NJ, 1992, p. 44.
59. M. B. Clevenger, J. Zhao, and J. T. McDevitt, Use of a self-assembled monolayer for the preparation of crystalline organic superconductor/high-T_c superconductor structures, *Chem. Mater. 8*:2693 (1996).
60. M. B. Inoue, M. Inoue, Q. Fernando, and K. W. Nebesny, Highly electroconductive tetrathiafulvalenium salts of copper halides, *Inorg. Chem. 25*:3976 (1986).
61. M. B. Inoue, M. Inoue, M. A. Bruck, and Q. Fernando, Structure of bis(ethylenedithio)tetrathiafulvalenium tribromocuprate(1), (BEDT-TTF$^+$)Cu$^1_2$Br$_3$: Coordination of the organic radical cation to the metal ions, *J. Chem. Soc., Chem. Commun. 1992*:515.
62. M. Munakata, T. Kuroda-Sowa, M. Maekawa, A. Hirota, and S. Kitagawa, Building of 2D sheet of tetrakis(methylthio)tetrathiafulvalenes coordinating to copper(I) halides with zigzag and helical frames and the 3D network through the S···S contacts, *Inorg. Chem. 34*:2705 (1995).
63. Y. C. Chen, K. K. Chong, and T. H. Meen, Heat treatment effect on the metal contact of high-T_c (Pb,Bi)SrCaCuO superconductor, *Jpn. J. Appl. Phys. 30*:33 (1991).
64. J. Seto and T. Van Duzer, Theory and measurements on lead-tellurium-lead supercurrent junctions, Proceedings of the 13th International Conference on Low Temperature Physics, 1974, p. 328.
65. M. A. M. Gjis, D. Scholten, T. van Rooy, and R. Isselsteijn, Proximity effect in thin film YBa$_2$Cu$_3$O$_{7-\delta}$-Ag-Pb structures, *Physica C 162*:1615 (1989).
66. R. Holm and H. Meissner, Messungen mit Hilfe von flussigem Helium. XIII. Kontatwidrstand zwischen Supraleitem und Nichtsupraleitem, *Z. Physik 74*:715 (1932).
67. H. Melssner, Superconductivity of contacts with interposed barriers, *Phys. Rev. 117*:672 (1960).
68. P. Hilsch, Zurn Verhalten von Supraleitern im Kontakt mit Normalleitern, *Z. Physik 167*:511 (1962).
69. P. G. de Gennes, Boundary effects in superconductors, *Rev. Mod. Phys. 36*:225 (1964).
70. J. J. Hauser, H. C. Theuerer, and N. R. Werthamer, Superconductivity in Cu and Pt by means of superimposed films with lead, *Phys. Rev. 136*:A637 (1964).
71. J. Clarke, Supercurrents in lead-copper-lead sandwiches, *Proc. Roy. Soc. (Lond.) A 308*:447 (1969).
72. A. de Lozanne, M. Dilorio, and M. Beasley, Fabrication and Josephson behavior of high-T_c superconductors–normal superconductor microbridges, *Appl. Phys. Lett. 42*:541 (1983).
73. R. H. Ono, J. A. Beall, M. W. Cromar, T. E. Harvey, M. E. Johansson, C. D. Reintsema, and D. Rudman, High-T_c superconductor–normal metal–superconductor Josephson microbridges with high resistance normal metal links, *Appl. Phys. Lett. 59*:1126 (1991).
74. A. Fujimaki, T. Tamaoki, T. Hidaka, M. Yanagase, T. Shiota, Y. Takai, and H. Hayakawa, Experimental analysis of YBa$_2$Cu$_3$O$_x$/Ag proximity interfaces, *Jpn. J. Appl. Phys. 29*:L1659 (1990).
75. M. Hatano, T. Nishino, and U. Kawabe, Experiments of the superconducting proximity effect between superconductor and semiconductor, *Appl. Phys. Lett. 50*:52 (1987).
76. J. R. Cooper, M. Miljak, G. Delpanque, D. Jerome, M. Weger, J. M. Fabre, and L. Giral, DC Hall effect measurements on TTF-TCNQ, *J. Phys. 38*:1097 (1977).
77. A. A. Bright, A. F. Garito, and A. J. Heeger, Optical conductivity studies in a one-dimensional organic metal: Tetrathiafulvalene tetracyanoquinodimethane (TTF)(TCNQ), *Phys. Rev. B 10*:1328 (1974).
78. J. E. Eldridge and F. E. Bates, Far-infrared optical properties of semiconducting tetrathiafulvalene tetracyanoquinodimethane (TTF-TCNQ), including the pinned charge-density wave, *Phys. Rev. B 28*:6972 (1983).
79. K. Seeger, W. Markowitsch, and F. Kuchar, Effective mass from magnetoreflection of metallic polyacetylene, *Synth. Met. 17*:527 (1987).
80. J. Paloheimo, H. Stubb, P. Yli-Lahti, and P. Kuivalainen, Field-effect conduction in polyalkylthiophenes, *Synth. Met. 41–43*:563 (1991).
81. K. Kaneto and K. Yoshino, Hall Effect and reflec-

tance in conducting polypyrrole films, *J. Phys. Soc. Jpn.* 55:4568 (1986).
82. K. Kaneto, M. Yamamoto, K. Yoshino, and Y. Inuishi, Electrical conductivity and galvanomagnetic effects in $(SN)_x$ single crystals, *J. Phys. Soc. Jpn.* 47:167 (1979).
83. P. M. Grant, R. L. Greene, and G. B. Street, Optical properties of polymeric sulfur nitride $(SN)_x$, *Phys. Rev. Lett.* 35:1743 (1975).
84. L. W. Shacklette, R. R. Chance, D. M. Ivory, G. G. Miller, and R. H. Baughman, Electrical and optical properties of highly conducting charge-transfer complexes of poly(p-phenylene), *Synth. Met.* 1:307 (1979).
85. S. G. Haupt and J. T. McDevitt, Possible induction of superconductivity in conductive polymer structures, *Synth. Met.* 71:1539 (1995).
86. A. D. Wieck, Superconducting contacts on $YBa_2Cu_3O_{7-x}$ in magnetic fields, *Appl. Phys. Lett.* 53:1216 (1988).
87. R. Caton, R. Selim, A. M. Buoncristiani, and C. E. Byvik, Rugged low-resistance contacts to $YBa_2Cu_3O_x$, *Appl. Phys. Lett.* 52:1014 (1988).
88. J. W. Ekin, T. M. Larson, N. F. Bergren, A. J. Nelson, A. B. Swartzlander, L. L. Kazmerski, A. J. Panson, and B. A. Blankenship, High T_c superconductor/noble-metal contacts with surface resistivities in the 10^{-10} Ω cm^2 range, *Appl. Phys. Lett.* 52:1819 (1988).
89. I. Sugimoto, Y. Tajima, and M. Hikita, Low resistance ohmic contact for the oxide superconductor Eu-Ba$_2$Cu$_3$O$_y$, *Jpn. J. Appl. Phys.* 27:L864 (1988).
90. Y. Suzuki, T. A. Kusaka, T. Aoyama, T. Yotsuya, and S. Ogawa, The contact resistance of the yttrium barium copper oxide ($YBa_2Cu_3O_{7-\delta}$) metal film system, *Jpn. J. Appl. Phys.* 28:2463 (1989).
91. M. Lee and M. R. Beasley, Interface resistance and reduction of I_cR product in $YBa_2Cu_3O_7$-Ag-Pb proximity junctions, *Appl. Phys. Lett.* 59:591 (1991).
92. K. Oka and T. Iri, Contact resistance of several metals on $YBa_2Cu_3O_{7-\delta}$, *Jpn. J. Appl. Phys.* 31:2689 (1992).
93. J. Mannhart, J. Strobel, J. G. Bednorz, and C. Gerber, Large electric field effects in $YBa_2Cu_3O_{7-\delta}$ films containing weak links, *Appl. Phys. Lett.* 62:630 (1993).
94. S. X. Dou, H. K. Liu, A. J. Bourdillon, N. X. Tan, J. P. Zhou, C. C. Sorrell, and K. E. Easterling, Volumetric measurement of labile ions through corrosion of $YBa_2Cu_3O_{7-x}$, *Mod. Phys. Lett. B* 1:363 (1988).
95. M. Schwartz, D. Cahen, M. Rappaport, and G. Hodes, Quantitatively controlled, room temperature reduction of $YBa_2Cu_3O_{7-x}$ by electrochemical methods, *Solid State Ionics* 32/33:1137 (1989).
96. A. T. Fiory, A. F. Hebard, R. H. Eick, P. M. Mankiewich, R. H. Howard, and M. L. O'Malley, Metallic and superconducting surfaces of $YBa_2Cu_3O_7$ probed by electrostatic charge modulation of epitaxial films, *Phys. Rev. Lett.* 65:3441 (1990).
97. U. Kawabasa, K. Asano, and T. Kobayashi, Electric field effect on the Al-MgO-$YBa_2Cu_3O_y$ structure, *Jpn. J. Appl. Phys.* 29:L86 (1990).
98. J. Mannhart, J. G. Bednorz, K. A. Müller, and D. G. Schlom, Electric field effect on superconducting $YBa_2Cu_3O_{7-\delta}$ films, *Z. Phys. B* 83:307 (1991).

38
Transparent Conductive Coatings

Vaman G. Kulkarni
Americhem Inc., Concord, North Carolina

I. INTRODUCTION

The term "conductive coatings" encompasses products derived from a broad range of materials and processes such as

1. Vacuum metallization and metallized films
2. Coatings derived from dispersion of conductive fillers such as carbon black, metal powders, and flakes in polymeric binders
3. Coatings derived from solutions of antistatic materials such as quaternary ammonium compounds, ethoxylated amines, amides, and other antistatic materials that function by absorbing atmospheric humidity
4. Intrinsically conductive polymers (ICPs) such as polyaniline and polythiophene

Conductive coatings derived from ICPs are the theme of this chapter, with special emphasis on transparent coatings.

Since their emergence in the late 1970s, ICPs have held great promise for technological innovation, as these polymers were seen to combine the electronic and magnetic properties of metals with the mechanical properties and processability of conventional polymers [1]. However, by the early 1980s it was discovered that ICPs were infusible and insoluble and did not enjoy the processibility of conventional polymers. Their poor processability became a major roadblock for industrial applications. Over the last decade, significant advances have been made in the processing of ICPs, especially in the area of solution processing. These studies have led to transparent, highly conductive coatings and films. Now, with their ability to be processed into coatings and films coupled with their unique properties, conductive polymers show vast promise for industrial applications ranging from electrostatic dissipation to electrochromic displays. The purpose of this chapter is to discuss the progress in ICP coatings and their applications.

II. RECENT TRENDS IN PROCESSING

Most conducting polymers in their doped form are infusible and insoluble in common organic solvents. In most cases, even the undoped form has limited solubility. Through the vast literature of conductive polymers, ICPs have often been categorized as intractable polymers. Processing into coatings and films on an industrial scale has met with severe limitations. There are two schools of thought on how to process ICPs into coatings and films: (1) Modify or manipulate the chemistry of the conductive polymer to induce solubility in organic solvents, and (2) disperse and/or compatibilize the intractable conductive polymer in conventional film-forming matrices. Most of the work on improving processibility has been focused on the modification or manipulation of the chemistry of the conductive polymer.

Certain polymers, such as polyaniline, in their neutral form are soluble in N-methyl pyrrolidone (NMP) [2], N,N-dimethylpropylene urea (DMPU) [3], and dimethyl sulfoxide (DMSO). Films of neutral polymer can be cast from these solutions. Chemical modification of monomers has proved to be effective in enhancing the modification solubility, especially with polythiophene [4,5]. In other instances, such as poly(p-phenylenevinylene), processibility has been achieved by synthesizing solution-processible precursors, which could be subsequently converted to the conjugated polymer [6]. Nonetheless these techniques have failed to provide a method for processing conductive polymers in their doped form. More recently, the use of functionalized protonic acids that dope polyaniline and simultaneously render the doped polymer soluble in organic solvents has been reported [7,8]. Dispersion of doped polyaniline in nonconductive film-forming matrices has been shown to yield coatings with tailored properties and to offer the most practical route for preparing conductive coatings and films from polyaniline [9–11]. This technique is independent of the chemistry of the ICP and in principle should be applicable to other ICPs.

While the quest for processable conducting polymers continues, polyaniline and more recently polyethylenedioxythiophene have emerged as the most important conducting polymers that can be processed into transparent coatings. Polyaniline in particular has been extensively studied due to its facile chemistry, which lends it to the preparation of coatings using a range of techniques, an in-depth discussion of which is included in the following section. As a prelude to processing, a brief discussion of structure–property relationships and doping is provided. Polypyrroles, polyisothianaphthenes, and certain charge transfer complexes have also shown technological promise and are also discussed in this chapter.

III. POLYANILINES

A. Structure–Property Relationships

Polyaniline represents a family of polymers, rather than a single polymer, that are interconvertible by acid–base and/or oxidation–reduction reactions. It can be prepared in three oxidation states. Leucoemeraldine base is the fully reduced form, emeraldine base is the half-oxidized form, and pernigraniline base is the fully oxidized form. All the base forms are insulators with a conductivity of less than 10^{-10} S/cm. The conductive form of polyaniline, more commonly known as the emeraldine salt, has a conductivity of greater than 1 S/cm, depending on the dopant and processing conditions. A unique feature of polyanilines is that their electronic structure and electrical properties can be reversibly controlled by charge transfer doping and by protonation [12]. The protonic acid doping of emeraldine base is unique to polyaniline and differs from that of other conducting polymers in that it is not accompanied by any change in the number

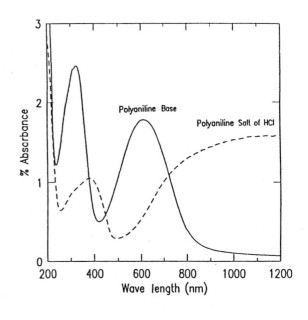

Structure

Emeraldine Base

Emeraldine Salt

Properties

Emeraldine Base (Polyaniline Base)

- Insulator, $< 10^{-10}$ S/cm
- Bluish purple
- Soluble in NMP, DMPU, DMSO and others
- Infusible, plasticized by NMP

Emeraldine Salt (Polyaniline Salt)

- Conductor, > 1 S/cm
- Green
- Insoluble in common organic solvents
- Infusible

Fig. 38.1 Structure–property changes associated with emeraldine base–emeraldine salt interconversion. (Optical spectra from Ref. 19.)

of electrons associated with the polymer backbone. Of particular interest for processing and many applications is the insulator-to-metal transition brought about by simple treatment with protonic acids.

Polyaniline in the form of conductive emeraldine salt can be easily synthesized by chemical oxidation of aniline in aqueous acidic media [12,13]. Both inorganic acids (hydrochloric, sulfuric, etc.) and organic acids (methanesulfonic, toluenesulfonic, etc.,) can be used. Thus synthesized polyaniline salt is a dark blue-green powder with a typical bulk conductivity of 1–10 S/cm. In this form it is infusible and insoluble. The emeraldine salt can be converted to insulating emeraldine base on treatment with an alkaline solution. Ammonium hydroxide is most commonly employed to effect deprotonation. Repeated treatments are often necessary for complete deprotonation. The emeraldine base is a coppery bronze powder with a conductivity of less than 10^{-10} S/cm. Figure 38.1 shows the structure and property changes associated with the emeraldine base–emeraldine salt interconversion that forms the basis of a widely practiced route for processing polyanilines and is the subject of numerous publications.

B. Dopants, Doping, and Solvent Interactions

The emeraldine base can be easily protonated with a variety of protonic acids. A relationship between acid strength and ionization potential of conjugated polymers that will give highly conductive doped complexes is described by Han and Elsenbaumer [14]. Along with the changes in electrical and electronic properties, doping brings about changes in thermal stability, polarity, and solubility parameters, which in turn affect the solubility and dispersibility of doped polyanilines.

Common protonic acids such as hydrochloric and sulfuric acids leave the polymer highly polar. Dopants with long polymer chains such as dodecylbenzenesulfonic acid can lessen the polar character and change the interactions of the polymer [7,15]. The effect is clearly evident from the solubility/dispersibility of polyaniline doped with dodecylbenzenesulfonic acid in nonpolar solvents such as toluene and xylene [7,16]. Certain functionalized protonic acid solutes such as camphorsulfonic acid have been shown to be highly effective in rendering the doped polymer soluble in polar organic solvents such as m-cresol [7].

Sulfonation of emeraldine base produces ring-substituted polyaniline. The ring-substituted polyaniline is "self-doped" and offers solubility in basic aqueous solutions [17,18]. The key feature of this self-doped polymer is its ability to maintain its conductivity up to pH 7, while the unsubstituted polyaniline is converted to an insulator at pH greater than about 4.

Shacklette and Han [15] showed that uncharged emeraldine base has a high propensity for polar and hydrogen bonding interactions and that dopant can influence its solubility behavior. Using empirical measures and group additive calculations, the solubility parameter, d, for emeraldine base ($d = 22.2$ MPa$^{1/2}$) and polyaniline tosylate ($d = 23.6$ MPa$^{1/2}$) have been calculated. Various solvents for emeraldine base and emeraldine salt have been discussed.

C. Processing into Coatings and Films

Due to the ease with which it can be prepared and doped by simple treatment with protonic acids, polyaniline offers versatility in preparation of conductive coatings and films as shown in Fig. 38.2. Much of the work on processing polyanilines has been focused on solutions of emeraldine base [2,3]. Ring-substituted [20,21] and nitrogen-substituted polyanilines [22] have generally failed to yield a soluble doped polyaniline and also reduce the conductivity. Polyanilines can also be dissolved and processed from concentrated sulfuric acid and other strong acids [23]. While this technique has proved useful for preparing fibers, it is not suited for preparation of coatings and films. More recently, counteranion-induced processibility [7,8] has been the subject of intense research. The technique permits preparation of films from solutions of doped polyaniline. Dispersion techniques, in addition to offering processibility in the doped form, offer flexibility in preparing conductive coatings [9–11]. In the following pages, the various techniques for preparing conductive coatings are discussed:

1. Dispersion techniques
2. Processing from emeraldine base solutions
3. Processing from aqueous acids and water-soluble polymers
4. Counterion-induced processibility

1. Dispersion Techniques

In a dispersion technique, as-synthesized conductive emeraldine salt is dispersed in a solution of nonconductive film-forming resin. Thus, the technique offers a one-step postpolymerization process for preparing conductive coatings from doped polyaniline. The film-forming resin can be a polymer or a polymerizable monomer, an oligomer, a polymer precursor, or any other material that is capable of film formation. Depending on the physical form, the neat resin or a solution of the resin in water or organic solvent or a combination thereof can be employed. The formulation components can be varied to adjust the film properties. The dispersion is carried out using any of the variety of commercial dispersion equipment available such as pebble mills or high speed mixers. Since the process relies on dispersion, infusibility and insolubility are not factors in processing nor does it need special dopants. However, it is essential to use the conductive polymer of the right morphology so that high quality dispersions can be produced. The resulting dark green liquid is suitable for preparing transparent films. The technique offers several advantages.

a. Tuned Conductive Properties

The coating is a two-phase system wherein polyaniline is uniformly dispersed in the continuous matrix of the film-forming resin. By varying the concentration of polyaniline in the coating formulation, one can tune the conductive and optical properties of the coating (see Fig. 38.3). Highly transparent coatings with surface resistivities in the 10^3–10^9 ohms/square ($\Omega/\square$) range can be prepared by proper control of the composition and experimental parameters [10,11].

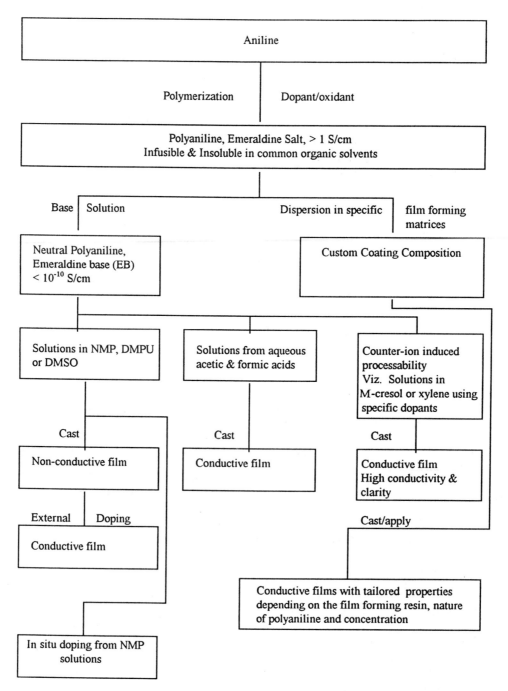

Fig. 38.2 Various routes for preparation of conductive coatings and films from polyaniline.

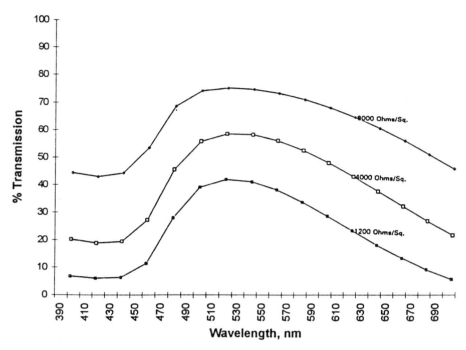

Fig. 38.3 Transmission curves for polyaniline coatings with varying concentrations prepared by dispersing PANI in a solution of a film-forming matrix. (From Ref. 10.)

b. Improved Environmental Stability

Table 38.1 lists the chemical resistance of polyaniline coatings prepared by dispersing polyaniline in a solution of a film-forming matrix. Figure 38.4 shows a plot of the water resistance of these coatings over an extended period of time. By judicious control of the composition and dispersion technique, one can further improve the properties of the coated films. Polyester sheets coated with an approximately 3–4 μm thick film of polyaniline were used for testing. The coated sheets were immersed in the test solution for the prescribed length of time. The sheets were withdrawn at the end of the test period and wiped dry, and their surface resistivity was recorded. The coatings display remarkable resistance to water and dilute alkaline solutions. Polyaniline films that do not contain a film-forming resin are converted to insulator on simple treatment with water [10]. The significant improvement in environmental resistance can be attributed to encapsulation of polyaniline in the film-forming matrix.

c. Direct Technique

The technique uses as-synthesized polyaniline (doped form). All other techniques that are currently available for preparing coatings and films use the emeraldine base form, which must be synthesized from the doped polymer. Thus this technique offers the most direct route for preparing conductive coatings from polyaniline.

It was long ago discovered that conductive polymers are dispersible [24–26]; it is only recently that these techniques have been perfected, and they are now suitable for preparing transparent coatings [9–11]. By careful control of the dispersion technique, extremely fine and

Table 38.1 Properties of Polyaniline Coatings Prepared by Dispersal in a Solution of a Film-Forming Matrix

	Surface resistivity ($\Omega/\square$)	
Property	Coating 1	Coating 2
As prepared	2.3×10^4	1.8×10^6
Resistance to water, 100 h at 23°C	5.4×10^4	5.4×10^5
Resistance to water, 100 h at 60°C	2.2×10^4	2.0×10^6
Resistance to 0.1% NaOH, 15 min at 23°C	1.3×10^7	3.9×10^6
Resistance to 0.1% HCl, 15 min at 23°C	1.4×10^4	5.0×10^5
Resistance to isopropyl alcohol, 15 min at 23°C	1.9×10^4	7.2×10^6

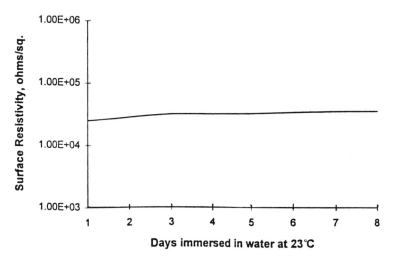

Fig. 38.4 Water resistance of polyaniline coatings prepared by dispersing PANI in a solution of a film-forming matrix.

stable dispersions of conducting polymer can be obtained. These dispersions have the visual appearance of solutions, and particulate matter is evident only under high magnification.

The coating can be applied as a transparent green film on a variety of substrates. The coatings can be applied using conventional coating techniques such as draw bar, spin coating, spray coating, and rotogravure techniques. It is possible to lay down extremely thin films in the range of 0.2–0.4 μm using rotogravure techniques. Coated sheets display the characteristic green color of polyaniline and are highly transparent. Extremely thin coatings (0.2 μm and under) tend to have slightly inferior environmental resistance.

2. Processing from Emeraldine Base Solutions

Emeraldine base readily dissolves in NMP [2]. Processing via NMP solutions has been the most widely practiced route for processing polyaniline [2,3,27,28]. Coppery colored films of emeraldine base cast from NMP solutions can be doped with protonic acids to yield bluish-purple films with conductivity greater than 1 S/cm [2,29]. If one can coat extremely thin layers of the emeraldine base solution, the resulting films will be transparent. Solutions in NMP have been used to make films and fibers [3,30]. One frequently encountered problem with NMP solutions is the extent of solubility of the emeraldine base and the stability of the solution. Concentrated solutions of emeraldine base in NMP tend to gel on relatively short storage [30,31]. Addition of lithium chloride has been shown to prolong the gelation time [31]. N,N-Dimethylpropylene urea (DMPU) offers improved solubility (20%) and longer shelf life [3].

For preparing conductive coatings and films, the use of NMP, DMPU, or other high boiling solvents such as dimethyl sulfoxide and dimethylformamides present two drawbacks. First, because of their high boiling point, it is difficult if not impossible to remove the solvent completely from the cast film without subjecting it to heat and vacuum for prolonged periods. The residual NMP has been shown to plasticize the film as indicated by the lowering of the glass transition temperature [32]. While the plasticized film may offer improved properties, such a high level of solvent retention is undesirable for many applications. Second, the process of preparing a conductive coating requires that a film of the nonconductive form of the polymer, be applied and subsequently doped to render it conductive. This is usually accomplished by treating it with an acid solution. Depending on the film thickness, complete doping may not be possible. The process can be very time-consuming and economically unattractive for many applications. Since most dopants are acidic, external doping poses further complications.

The difficulties associated with external doping can be overcome by the use of in situ dopants such as onium salts [27]. In this technique, the onium salt is dissolved in NMP along with the emeraldine base. After coating, the substrate is exposed to ultraviolet radiation or an electron beam. The onium salt undergoes decomposition, generating protonic acids, which results in in situ doping of polyaniline. In situ doping of polyaniline can also be achieved with amine triflate salts, which thermally dissociate to generate the free acid with volatilization of the corresponding amine [28].

Emeraldine base can also be processed from concentrated protonic acids such as sulfuric acid and methanesulfonic acid [23]. Such solutions are not suitable for coating use due to the corrosiveness of the acids. Furthermore, during the dissolution process, the original dopant is lost and the dissolved polyaniline becomes doped with the counteranion of the acid used for dissolution. In the case of sulfuric acid, it is also possible that the "soluble" product made by this technique is no longer polyaniline but a mixture of ring-sulfonated derivatives [33], Wessling [33] found that polyaniline treated (dissolved) with sulfuric acid and isolated can no longer be completely neutralized.

3. Processing from Aqueous Acids and Water-Soluble Polyaniline

The solubility of polyaniline in aqueous acetic, and formic acids has been known for nearly a century now [34]. It was only recently [35] that this property was used to demonstrate the processability from aqueous solutions. Films with conductivities of 0.5–2 S/cm can be cast from solutions in aqueous acetic acid. The advantage over processing from emeraldine base solutions in NMP, as discussed above, is that as-cast films are conductive and no further doping is necessary. Cast films show very strong adhesion to glass and cannot be removed mechanically without breaking [35]. Similar results could be obtained with 60–88% formic acid solutions. While the technique offers processability in the conductive form, it has limited practical application for preparing conductive coatings due to the high concentration and volatility of the acid.

Water-soluble conducting polyanilines have been synthesized by oxidative polymerization of aniline complexed to a polymeric acid [36] and promise simplified processing to coatings.

"Self-doped" polyanilines provide unique opportunities in processing, since the dopant is an integral part of the polymer backbone and the polymer displays good resistance to water. It offers processability from water and easy conversion to the doped form.

4. Counterion-Induced Processability

Polyaniline can be rendered processible by the use of specific counterions (dopants). In this technique, the emeraldine base is mixed with appropriate dopant and the resulting mixture is dissolved in the appropriate solvent [37]. The solubility of emeraldine base in aqueous acetic acid as discussed above can be viewed as counterion-induced processability. However, it was not until the recent discovery [7,8,37] that polyanilines doped with d,l-camphorsulfonic acid and dodecylbenzenesulfonic acid are soluble in common organic solvents such as m-cresol and toluene that counterion-induced processability sparked interest in the scientific community. The d,l-camphorsulfonic acid/m-cresol system, in particular, has generated considerable interest due to coupled effects of high transparency and conductivity (100–400 S/cm).

The technique offers significant advantages over the use of emeraldine base solutions and aqueous acids. The functionalized protonic acid dopes polyaniline and simultaneously renders the doped polyaniline soluble in common organic solvents. Films can be cast in the conductive form, and hence no external doping is necessary. Due to the high solvency of the organic solvents used, conventional insulating polymers can be dissolved to make coatings with varying concentrations of polyaniline. A special characteristic of transparent coatings prepared from d,l-camphorsulfonic acid/m-cresol is the dramatic change in the electronic spectra, high conductivity, and the optical clarity of the resulting films as shown in Fig. 38.5.

Processability from m-cresol solutions is particularly valuable where optical quality coatings and films are desired. However, the application of coatings is limited to glass and other substrates that are resistant to m-cresol.

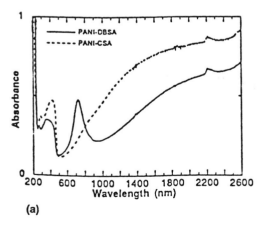
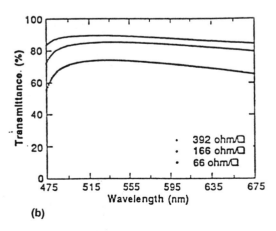

Fig. 38.5 (a) Absorption spectra of spin-cast PANI-CSA and PANI-DBSA films. (b) Transmittance of pure PANI-CSA spin-cast films. (From Ref. 37.)

The technique is very unlikely to have major industrial impact due to the toxicity of the solvent. On the other hand, doping of polyaniline with dodecylbenzenesulfonic acid can be more attractive because the doped polymer can be processed from hydrocarbon solvents such as toluene and xylene.

The ability to induce changes in conductivity and optical properties in thin films of doped polyanilines by exposing them to vapors of *m*-cresol at room temperature [38] offers a practical alternative to solution processing and possibly reduces significant retention of *m*-cresol in the film. Coatings made by dispersing Versicon® in film-forming matrices also show marked improvement in transparency and conductivity on exposure to vapors of *m*-cresol (V. G. Kulkarni, unpublished observations).

D. Secondary Doping and Its Importance in Conductive Coatings

The unusually high conductivity of polyaniline doped with d,l-camphorsulfonic acid has been investigated extensively [38,39]. It has been shown that, d,l-camphorsulfonic acid by itself is a poor dopant for polyaniline and that the combination of d,l-camphorsulfonic acid and *m*-cresol is responsible for high conductivity. The role of *m*-cresol has led to the concept of secondary dopant. A "secondary dopant" has been defined [39] as an apparently inert substance that, when applied to a primary-doped polymer, induces enhanced changes in conductivity, electrical, optical, magnetic, and/or structural properties of the polymer. The enhanced properties may persist even after removal of the secondary dopant. Several primary/secondary dopant systems are being evaluated. The concept can have a major impact in the coatings area in general, especially where high conductivity and optical clarity are desired, provided environmentally friendly primary/secondary dopant systems can be identified.

IV. POLYTHIOPHENES AND OTHER CONDUCTING POLYMERS

The interest in the polythiophene family of polymers stems from their high thermal and conductivity stability. However, like other conducting polymers, unsubstituted polythiophene is infusible and insoluble and hence difficult to process. Solubility and flexibility can be enhanced by choosing the right organic substituents. Perhaps the first major advancement in the preparation of soluble polythiophene was achieved by alkyl substitution on the thiophene ring [4,40]. The solubility was shown to increase with increasing chain length of the

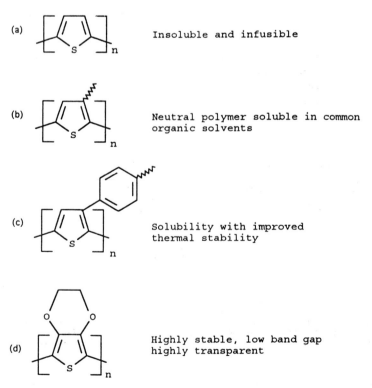

Fig. 38.6 Effect of substitution on the properties of (a) unsubstituted polythiophene; (b) alkyl-substituted polythiophene; (c) poly(3(4-octylphenyl)thiophene; (d) poly(3,4-ethylenedioxythiophene).

substituent, and polymers with substituents larger than the butyl group showed good solubility in common organic solvents. However, alkylthiophenes, especially those with longer side chains such as poly(3-hexylthiophene) and poly(3-octylthiophene), are unstable in the doped form [41,42]. Synthesis of 3-substituted thiophenes containing a benzene ring such as poly(3-(4-octylphenyl)thiophene) or random copolymerization of 3-methylthiophene and 3-octylthiophene have been shown to overcome thermal undoping [43]. Alkoxy-substituted thiophenes, in addition to offering improved processibility show a lower band gap than polyalkylthiophenes [44]. Due to the low band gap, the polymer possesses a higher degree of transparency and hence is better suited for application in electrochromic devices. Processibility into highly transparent coatings suitable for antistatic coating of films has been reported [45]. Polyalkylenedioxythiophenes (PEDTs) are a new class of conducting polymers with high stability and small band gap [46,47]. However, they too are insoluble in both the neutral and doped forms and lack processibility in the neat form. Coatings can be prepared by direct polymerization of the monomer on a substrate. PEDT can also be processed from aqueous solutions by polymerization of ethylenedioxythiophene in aqueous polystyrenesulfonic acid [48]. Such coatings display very low color and excellent hydrolytic stability and are shown to be suitable as antistatic layers. Urethane-substituted polythiophenes offer solubility in tetrahydrofuran, N-methylpyrrolidone, and a host of other solvents, and films can be cast from these solutions [49]. Various substituted polythiophenes and their advantages in processing are summarized in Fig. 38.6.

Unsubstituted polypyrrole is also insoluble and infusible and hence cannot be processed into coatings and films. Plastic films coated with an antistatic layer of polypyrrole have been prepared by polymerizing pyrrole on the surface of the film or by use of polypyrrole grafted onto latex particles [50]. Nonetheless, polypyrroles are less attractive than polyanilines because of their high absorption at practically all wavelengths in the visible spectrum.

Polyisothianaphthenes are interesting because of their low absorption in the visible region. Antistatic layers suitable for photographic materials have been prepared by dispersion of polyisothianaphthene in aqueous λ-carrageenan [51]. Nonetheless, they have not been extensively studied owing to difficulty in synthesis of the polymer.

Transparent conductive films have also been prepared by incorporation of conductive charge transfer complexes such as tetraselenotetracene chloride, $(TseT)_2Cl$, in polymers [52]. Films containing 1% tetraselenotetracene in polycarbonate show surface resistivity as low as 700 $\Omega/\square$ and 65% transparency. The films possess excellent thermo-oxidative stability and high abrasion resistance.

V. MEASUREMENT OF ELECTRICAL PROPERTIES AND TERMINOLOGY

Surface resistivity and static decay measurements have been historically used to characterize conductive coatings. The "static decay" is rather meaningless for coatings with surface resistivities lower than 10^9 $\Omega/\square$, since decay times for such coatings are typically on the order of 1/100 or less. Surface resistivity measurement is by far the most accurate and meaningful technique for measuring conductive polymer coatings and films. ASTM D-257 is the widely used standard for measuring surface resistivity. The values obtained for this property depend on the measurement conditions and vary with the pressure, surface roughness of the coating, etc. Thus, it is essential to use a standard setup for comparison of samples. The surface resistivity of conductive polymer coatings is relatively independent of relative humidity. Nonetheless, the use of a humidity chamber is recommended for comparison of various coatings.

The following are the accepted definitions for surface resistivity and material characterization with varying surface resistivities [53].

Surface resistivity. For electric current flowing across a surface, the ratio of dc voltage drop per unit length to the surface current per unit width. In effect, surface resistivity is the resistance between two opposite sides of a square and is independent of the size of the square or its dimensional units. Surface resistivity is expressed in ohms/square (also written as $\Omega/\square$).

Antistatic. The terms "antistatic" and "static-dissipative" have been used synonymously in the literature. They usually refer to the property of a material that inhibits triboelectric charging [52]. "Antistatic" has also been defined as a property of a material having a surface resistivity of at least 1×10^9 but less than 1×10^{12} $\Omega/\square$.

Static-dissipative. The property of a material having a surface resistivity of at least 1×10^5 $\Omega/\square$ or a volume resistivity of 1×10^4 Ω-cm but less than 1×10^{12} $\Omega/\square$ surface resistivity or 1×10^{11} Ω-cm volume resistivity.

Conductive. The property of a material having a surface resistivity less than 1×10^5 $\Omega/\square$ or a volume resistivity less than 1×10^4 Ω-cm.

VI. APPLICATIONS OF TRANSPARENT CONDUCTIVE COATINGS

At the time of the first edition of this handbook, instability and poor processability were the major obstacles to the commercialization of conducting polymers. Polyacetylene was the most widely studied polymer. Despite the fact that copper-like conductivity has been achieved with polyacetylene, it has remained a laboratory curiosity. It is unstable in ambient atmosphere, and processing problems still remain. The applications touted for intrinsically conductive polymers (ICPs) focused on high conductivity. The possibility of high conductivity-to-weight ratios prompted applications in energy storage systems, electromagnetic interference shielding, and aviation applications such as lightning protection. Electrostatic dissipation was considered a mundane application area for ICPs, and their ability to compete with existing technologies was questioned [54].

With recent advances in solution processing, ICP coatings seem to be poised for use in a broad range of industrial applications [55–58]. Antistatic coatings based on polyaniline are commercially available. These coatings have been shown to provide overall balanced properties for ESD applications [55]. Use of specific dopant–solvent combinations (primary and secondary dopants) has made possible optical quality transparent coatings suitable as electrodes for devices [59,60]. Water-soluble polyanilines have been shown to have applications in lithography [36,61]. Coatings derived from polyethylenedioxythiophene and polyisothianaphthene have been shown to possess clarity and antistatic properties suitable for application in photographic film [51,62]. At the present time, electrostatic discharge protection offers the greatest potential for industrial success.

Various application possibilities for conductive polymer coatings and films are listed below. This list is illustrative of the diverse applications and is by no means complete.

Antistatic, ESD coatings
Absorption of radar frequencies
Corrosion prevention
EMI/RFI shielding
Electrochromic displays
Electrochemical actuators
Lithographic resists
Lightning protection
Microelectronics
Polymer electrolytes
Photovoltaics
Rechargeable batteries
Smart windows
Solar cells
Sensors—pressure, temperature, chemical, and biological

Applications such as electrostatic dissipation and electromagnetic shielding use the conducting polymers in their conductive form. The current level of conductivity is adequate for electrostatic discharge (ESD) applications but falls short of commercial interest for electromagnetic interference shielding (EMI) [63]. Extremely thick parts are needed to achieve industrially acceptable shielding. With recent advances in understanding the effects of secondary doping and the enhancement of conductivity by proper choice of primary and secondary dopants, we might see the use of conductive polymer coatings for EMI.

Applications such as electrochromic windows, displays, sensors, and electromechanical tools are unique to conductive polymers and use the chemical or electrochemical doping and dedoping of conductive polymers. The structure–property changes occurring as a result of the doping/dedoping processes form the basis of operation of these devices [64,65]. For example, electrochromic devices and smart windows use the color change from doped to dedoped polymer. Electromechanical tools, on the other hand, use the dimensional changes.

A. Electrostatic Dissipation

With the rapid advancement in electronic data processing and communication systems, electronic components have become increasingly smaller and are inherently more sensitive to electrostatic discharge. Further, plastics are becoming the materials of choice for packaging due to their clarity, ease of fabrication, and mechanical properties. Since plastics are insulators, they are very receptive to static. The charge accumulated on the surface can discharge suddenly, causing damage to an electronic component that is packaged. Thus there is a need for reliable electrostatic dissipation.

For effective, controlled discharge protection, the ideal situation is one wherein the electrical charges are dissipated nearly as rapidly as they are generated. Thus, at any point there is no accumulation of charge on the surface of the plastic and therefore no danger of hazardous electrostatic discharge. To do this effectively, the surface resistance of the packaging material should be less than about 1×10^{10} $\Omega/\square$, preferably between 10^5 and 10^{10} $\Omega/\square$ even under extreme climatic conditions.

With the ability to offer surface resistivity of less than 10^{10} $\Omega/\square$, independent of humidity, coupled with transparency to allow for identification of electronic parts without opening the package, conductive polymer coatings are ideal materials for electrostatic dissipation. Highly transparent films can be applied on a variety of plastics with good adhesion and surface resistivity ranging from 10^3 to 10^{10} $\Omega/\square$. Depending on the conducting polymer, the coatings display the slight color characteristic of the ICP.

Antistatic coatings for photographic films represent a further challenge in that a very high degree of transparency and resistance to developing fluids is required in addition to humidity independence. Coatings based on polyethylenedioxythiophene and polyisothianaphthene offer low absorption in the visible range due to their low energy band gaps.

B. Computer Manufacturing and Microelectronics

Polyaniline coatings have been shown to be suitable for a number of applications in computer manufacturing and microelectronics. Thin films of conducting polyaniline are shown to be effective discharge layers for e-beam lithography [36,66]. High resolution conducting resists have been developed utilizing in situ doping of polyaniline with onium salts and amine triflate salts [27,67]. It is hoped that the newly developed water-soluble polyaniline [36] and polyethylenedioxythiophene [48] will eliminate the environmental concerns for these applications.

Interest in the use of conducting polymers in electroluminescent devices got an enormous boost with the recent discovery of flexible light-emitting devices [57,68]. For the first time a transparent electrode [68] based on polyaniline was used as the hole-injecting material. Advances in processing into extremely thin and highly transparent coatings are the key to their success. Processible conducting polymers offer many potential advantages [68,69]. These include ease of processing and fabrication of the device. The coatings are flexible and mechanically strong and hence permit twisting and bending of the devices without failure, unlike their inorganic counterparts [68].

C. Electrochromic Devices and Smart Windows

The use of conductive polymers as electrochromic materials was suggested some time ago [70]. Controlled color changes induced as a result of doping and dedoping of conducting polymers form the basis of electrochromic devices. This is accomplished by application of low electric potentials. The lack of adequate processing techniques hampered their development in the past. Transparency and processibility into thin films are the key. With recent progress in the processing of polyanilines leading to optical quality transparent electrodes [59] and synthesis of polymers with high stability and low band gap [46] which offer high transparency in the visible region, renewed interest in this area is certain. Since conducting polymers are now solution-processable, thin films with good adhesion can be applied from solution, which simplifies the preparation of these devices using ICPs. Whether or not conductive polymers will be successful in these applications, their position is certainly advanced.

D. Corrosion Protection

One of the major applications for conductive polymer coatings is corrosion protection. While transparent conductive coatings can be used for the applications, it is not essential for functioning.

Anodic anticorrosion protection of steel with electrochemically deposited polyaniline has been known for nearly a decade [71,72]. Dispersions of polyaniline have also been effective in corrosion protection [73,76]. Wessling [77] reported enhanced passivation of steel and shifts in corrosion potential of 100–300 mV together with decreases in specific corrosion current. This study suggests that interaction of polyaniline with the metal surface (passivation) is responsible for shifts in corrosion potential. Although the passivation proceeds relatively quickly, it is suggested that the polyaniline coating be allowed to remain on the metal surface to allow for regeneration of oxide layer in case of damage. Lu et al. [78] reported on the effect of the form of polyaniline and the corrosive environment on the corrosion protection of mild steel. It seems that corrosion prevention will be a major area of application for polyaniline coatings. Coatings made by dispersion techniques are likely to benefit as they offer easy processing and the possibility of being formulated into environmentally compliant coatings.

E. Other Applications

An interesting example of the use of thin layers of conductive polymers on flexible polymer films is a novel type of loudspeaker [58,79]. A thin conducting membrane is suspended between two grid electrodes. High voltage (1000 V) is applied between the membrane and electrodes, so that the membrane is electrostatically held in an unstable equilibrium position. An unsymmetrical ac field superimposed on the constant high voltage electric field causes the membrane to vibrate.

VII. COMPETING MATERIALS

Conductive polymers are likely to face competition in areas that use only the conductive properties. These applications include electrostatic dissipation and electromagnetic shielding, where conventional materials are already being used. As discussed above, transparent conductive coatings with the ability to offer tuned surface resistivities are ideal candidates for ESD application. Yet, for conductive polymer coatings to be commercially successful, they have to deliver more than just electrical performance. Ionic conductors, despite their shortcomings of humidity dependence and possible corrosion effects on electronic components, are still being used because of their low cost and clarity. For many applications, the surface resistivity of 10^{11}–10^{12} $\Omega/\square$ offered by ionic materials is deemed sufficient. Conductive polymer coatings must provide improved resistance to washing, where the ionics have a drawback. Filled systems such as conductive carbon-loaded polymers are capable of offering lower surface resistivities comparable to those of conductive polymers. They are relatively inexpensive. Their drawbacks include nontransparency and sloughing. Sloughing is the shedding or flaking of carbon particles upon friction. Further, surface resistivities in the range of 10^6–10^8 $\Omega/\square$ are difficult to obtain with filled plastics. Conductive polymers must cash in on these deficiencies of filled plastics. Other competing materials include vacuum metallized films and coatings based on carbon black and other conductors. Conductive polymers need to overcome the drawbacks of competing materials. Their value will be judged by

1. Reusability of coated products
2. Permanence of electrical properties on repeated washing
3. Ease of processing and adaptability to conventional application techniques

Polyaniline coatings have been shown [10] to possess water, heat, and chemical resistance needed to meet industry expectations. A comparison of properties of conductive polymer coatings with those of ionic conductors and carbon blacks is shown in Table 38.2.

It is unlikely that conductive polymer coatings will be used for electromagnetic interference shielding in the near future. Their conductivity falls short, since very thin coatings are often employed. Competing technologies such as copper and nickel-coated graphite provide improved performance and cost effectiveness. With the recent advances in processing leading to higher conductivities, conductive polymer coatings may find a niche, provided environmentally compliant formulations can be designed. It is more likely that injection-moldable ICP blends will benefit from improved conductivity, as other filled plastics present some problems during processing.

Optical quality coatings of polyaniline have been shown to be effective as hole-injecting materials in polymer light-emitting devices (LEDs), where traditionally indium tin oxide (ITO) has been used. Indeed, these coatings offer exceptional flexibility compared to ITOs. Their practical application is still to be assessed.

VIII. ENVIRONMENTAL CONSIDERATIONS IN PROCESSING AND USE OF CONDUCTIVE POLYMER COATINGS

In order for ICP coatings to enjoy commercial success, industrially feasible, economically attractive, and environmentally friendly processing techniques need to be developed. No doubt significant progress has been made in recent years in processing ICPs. However, many of the techniques lack industrial application, especially where bulk handling is required. Use of high boiling and strong solvents, in addition to being environmentally unfriendly, severely limits the application possibilities. Many of the commonly used plastics such as polycarbonate and poly(vinyl chloride) are attacked by strong solvents. At the present time, dispersion techniques offer the most practical route for preparing coatings on an industrial scale. Environmentally compliant formulations need to be developed. These may include 100% solids formulations and water-based systems. While environmental stability is taken for granted with some ICPs, conductive coatings, especially in the form of very thin coatings, demand improved stability because of their reactivity and sensitivity toward water and mild alkaline environments. Conductive coatings based on dispersion in film-forming materials have been shown to improve the stability to water significantly.

Further improvements in processing will come. If the progress in processing in the early 1990s and increased industrial attention are an indication, it appears that conductive polymers are headed for commercial success in the very near future.

IX. SUMMARY

While the potential applications for conductive polymers include exotic areas such as electrochromic windows, optical switches, and light-emitting diodes, it seems that they will find their early commercial success in mundane applications such as electrostatic dissipation. Transparent conductive coatings will be a major area of application for intrinsically conductive polymers. Many of the high-tech applications require very small amounts of polymer. Whether they will be commercially feasible, only time will tell. Recent progress in processing has made the use of conductive polymers possible in several applications. Nonetheless, challenges to the development of industrially sound processes do remain. In the next few years, we will see a push toward environmentally friendly processing techniques, as conductive polymers find more and more applications in industrial processes. One thing is certain: Intrinsically conductive

Table 38.2 Comparison of Selected Antistatic and Electrostatic Dissipative Materials

Property	Ionic conductors	Carbon blacks	ICPs
Transparency	Excellent	No	Good
Surface resistivity ($\Omega/\square$)	>10^{10} (typically 10^{11}–10^{12})	10^3–10^5	10^2–10^9
Resistance to water	Poor	Good	Good
Dependence on humidity	Yes	No	No
Permanency	Poor	Good	Good
Post thermo formability	Yes	Yes	Yes
Reliability of ESD	Questionable	Yes	Yes

polymers offer a variety of novel property combinations that are unattainable with other materials.

REFERENCES

1. T. A. Skotheim (ed.), *Handbook of Conducting Polymers*, Vols. 1 and 2, Marcel Dekker, New York, 1986.
2. M. Angelopoulos, G. E. Asturias, S. P. Ermer, A. Ray, E. M. Scherr, A. G. MacDiarmid, M. Akhtar, Z. Kiss, and A. J. Epstein, Polyaniline: Solutions, films and oxidation state, *Mol. Cryst. Liq. Cryst. 160*: 151–163 (1988).
3. K. T. Tzou and R. V. Gregory, Improved solution stability and spinnability of concentrated polyaniline solutions using N,N'-dimethyl propylene urea as the spin bath solvent, *Synth. Met. 69*:109–112 (1995).
4. K. Y. Jen, G. G. Miller, and R. L. Elsenbaumer, Highly conducting, soluble, and environmentally-stable poly(3-alkylthiophenes), *J. Chem. Soc. Chem. Commun. 1986*:1346.
5. M. R. Bryce, A. Chissel, P. Kathirgamanathan, D. Parker, and N. R. M. Smith, Soluble, conducting polymers from 3-substituted thiophenes and pyrroles, *J. Chem. Soc. Chem. Commun. 1987*:466.
6. D. R. Gannon, J. D. Capistran, F. E. Karasz, and R. W. Lenz, Conductivity anisotropy in oriented poly(p-phenylene vinylene), *Polym. Bull. 12*:293–298 (1984).
7. Y. Cao, P. Smith and A. J. Heeger, Counter-ion induced processability of conducting polyaniline and of conducting polyblends of polyaniline in bulk polymers, *Synth. Met. 48*:91–97 (1992).
8. A. J. Heeger, Polyaniline with surfactant counterions: Conducting polymer materials which are processable in the conducting form, *Synth. Met. 55–57*:3471–3482 (1993).
9. V. G. Kulkarni, Processing of polyanilines, in *Intrinsically Conducting Polymers: An Emerging Technology* (M. Aldissi, ed.), Kluwer, Boston, 1993, p. 45.
10. V. G. Kulkarni, J. C. Campbell, and W. R. Mathew, Transparent conductive coatings, *Synth. Met. 55–57*: 3780–3785 (1993).
11. V. G. Kulkarni, Tuned conductive coatings from polyaniline, *Synth. Met. 71*:2129–2131 (1995).
12. A. G. MacDiarmid, J. C. Chiang, M. Halpern, W. S. Huang, S. L. Mu, N. L. D. Somasiri, W. Wu, and S. E. Yanigar, *Mol. Cryst. Liq. Cryst. 121*:173 (1985).
13. J. P. Travers, J. Chroboczek, F. Devreux, F. Genoud, M. Nechtschein, A. A. Syed, E. M. Genies, and C. Tsintavis, *Mol. Cryst. Liq. Cryst. 121*:195 (1985).
14. C. C. Han and R. L. Elsenbaumer, Protonic acids: Generally applicable dopants for conducting polymers, *Synth. Met. 28*:C437 (1989).
15. L. W. Shacklette and C. C. Han, Solubility and dispersion characteristics of polyaniline, *Mater. Res. Soc. Symp. Proc. 328*:173–178 (1994).
16. S. J. Davies, T. G. Ryan, C. J. Wilde, and G. Beyer, *Synth. Met. 69*:209–210 (1995).
17. J. Yue and A. J. Epstein, Synthesis of self-doping conducting polyaniline, *J. Am. Chem. Soc. 112*:2800 (1990).
18. J. Yue, A. J. Epstein, and A. G. MacDiarmid, Sulfonic acid ring substituted polyaniline, a self-doped conducting polymer, *Mol. Cryst. Liq. Cryst. 189*:255–261 (1990).
19. N. Patel, R. Saraf, and M. Angelopoulos, Polyimide/polyaniline blends, Proceedings of Advances in Polyimide Science and Technology, 1991, p. 165.
20. D. MacInnes, Jr. and L. Funt, Poly-o-methoxyaniline: A new soluble conducting polymer, *Synth. Met. 25*: 235–242 (1988).
21. Y. Wei, W. W. Focke, G. E. Wnek, A. Ray, and A. G. MacDiarmid, Synthesis and electrochemistry of alkyl ring-substituted polyanilines, *J. Phys. Chem. 93*:495 (1989).
22. N. Comisso, S. Daolio, G. Mengoli, R. Salmaso, S. Zecchin, and G. Zotti, Chemical and electrochemical synthesis and characterization of polydiphenyl amine and N-methyl aniline, *J. Electroanal. Chem 255*: 97–110 (1988).
23. A. Andereatta, Y. Cao, J. C. Chaing, P. Smith and A. J. Heeger, Electrically conductive polyaniline spun from solutions in concentrated sulfuric acid, *Synth. Met. 26*:383–389 (1988).
24. B. Wessling, H. Volk, W. R. Mathew, and V. G. Kulkarni, Models for understanding processing properties of intrinsically conducting polymers, *Mol. Cryst. Liq. Cryst. 160*:205–220 (1988).
25. B. Wessling, Electrically conductive polymers, *Kunststoffe/German Plastics 76*(10):930–936 (1986).
26. B. Wessling and H. Volk, Post-polymerization processing of conductive polymers: A way of converting conducting polymers to conducting materials?, *Synth. Met. 15*:183–193 (1986).
27. M. Angelopoulos, J. M. Shaw, W. S. Huang, and R. D. Kaplan, In-situ radiation induced doping, *Mol. Cryst. Liq. Cryst. 189*:221–225 (1990).
28. M. Angelopoulos, J. M. Shaw, and K. L. Lee, Polyanilines: In situ radiation and thermal induced doping, *Mater. Res. Soc. Symp. Proc. 214*:137–142 (1991).
29. A. G. MacDiarmid and A. J. Epstein, Science and technology of conducting polymers, in *Frontiers of Polymer Research* (P. N. Prasad and J. K. Nigam, eds.), Plenum, New York, 1991, p. 259.
30. K. T. Tzou and R. V. Gregory, Mechanically strong, flexible, highly conducting polyaniline structures formed from polyaniline gels, *Synth. Met. 55–57*: 983–988 (1993).
31. A. G. MacDiarmid and A. J. Epstein, *Synth. Met. 55–57* (1993).
32. Y. Wei, G. W. Jang, K. Hsuesh, E. M. Scherr, A. G. MacDiarmid, and A. J. Epstein, Thermal transitions and mechanical properties of films of chemically prepared polyaniline, *Polymer 33*(2):314–319 (1992).
33. B. Wessling, Electrically conducting polymers, *Kunststoffe 80*(3), 323–329 (1990).
34. A. G. Green and A. E. Woodhead, Aniline black and

allied compounds, *J. Chem. Soc. Trans.* 97:2388–2403 (1910).

35. M. Angelopoulos, A. Ray, and A. G. MacDiarmid, Polyaniline: Processability from aqueous solutions and effect of water vapor on conductivity, *Synth. Met. 21*:21–30 (1987).

36. M. Angelopoulos, N. Patel, J. M. Shaw, N. C. Labianca, and S. A. Rishton, Water soluble conducting polyanilines: Applications in lithography, *J. Vac. Sci. Technol. B11*(6):2794–2797 (1993).

37. Y. Cao, P. Smith, and A. J. Heeger, Counter-ion induced processability of conducting polyaniline, *Synth. Met. 55–57*:3514–3519 (1993).

38. A. G. MacDiarmid and A. J. Epstein, The concept of secondary doping as applied to polyaniline, *Synth. Met. 65*:103–116 (1994).

39. A. G. MacDiarmid and A. J. Epstein, Secondary doping in polyaniline, *Synth. Met. 69*:85–91 (1995).

40. R. L. Elsenbaumer, K. Y. Jen, and R. Oboodi, Processable and environmentally stable conducting polymers, *Synth. Met. 15*:169–174 (1986).

41. G. Gustafsson, O. Inganas, and J. O. Nilsson, Thermal stability of doped poly(3-hexyl thiophenes), *Synth. Met. 28*:C427–C434 (1989).

42. M. T. Loponen, T. Taka, J. Laakso, K. Vakiparta, K. Suuronen, P. Valkeinen, and J. E. Osterholm, Doping and dedoping processes in poly(3-alkyl thiophenes), *Synth. Met. 41*:479–484 (1991).

43. Q. Pel, O. Inganas, G. Gustafsson, S. Granstrom, M. Anderson, J. Hjertberg, O. Wennerstrom, J. E. Osterholm, J. Laakso, and H. Jarvinen, The routes towards processable and stable conducting polythiophenes, *Synth. Met. 55–57*:1221–1226 (1993).

44. T. Hagiwara, M. Yamamura, K. Sato, M. Hirasaka, and K. Itawa, *Synth. Met. 32*:367 (1989).

45. K. H. Kochem, H. U. terMeer, and H. Millauer, *Kunststoffe 82*(7):575–579 (1992).

46. Q. Pei, G. Zuccarello, M. Ahlskog, and O. Inganas, Electrochronic and highly stable poly(3,4-ethylenedioxythiophene) switches between opaque blue-black and transparent sky blue, *Polymer 35*(7):1347–1352 (1994).

47. G. Heywang and F. Jonas, Poly(alkylenedioxythiophene)s—New, very stable conducting polymers, *Adv. Mater. 4*(2):116–118 (1992).

48. K. Jonas and W. Krafft, Eur. Patent EP 440957 (1990).

49. M. Liu and R. V. Gregory, Spectral studies and thermal properties of a urethane-substituted polythiophene, *Synth. Met. 72*:45–49 (1995).

50. S. Jasne and C. Chiklis, *Synth. Met. 15*:175–179 (1986).

51. G. Defieuw, R. Samijn, I. Hoogmartens, D. Vanderzande, and J. Gelan, Antistatic polymer layers based on poly(isothianaphthene) applied from aqueous solutions, *Synth. Met. 55–57*:3702–3706 (1993).

52. H. Bleier, J. Finter, B. Hilti, W. Hofherr, C. W. Mayer, E. Minder, H. Hediger, and J. P. Ansermet, Transparent electrically conductive composite material: Methods of preparation and their application, *Synth. Met. 55–57*:3605–3610 (1993).

53. ESD Association, *Electrostatic Discharge Terminology—Glossary*, ESD-ADV1.0-1994, ESD Association, Rome, New York, 1994.

54. J. R. Ellis, in *Handbook of Conducting Polymers* (T. A. Skotheim, ed.), Marcel Dekker, New York, 1986, V2, p. 501.

55. V. G. Kulkarni, Electrostatic dissipative materials. Presented at Conference on Plastics for Portable Electronics, Society of Plastics Engineers, Jan, 5–6, 1995, Las Vegas.

56. M. Angelopoulos, N. Patel, and J. M. Shaw, Application of conducting polyanilines in computer manufacturing, in *Intrinsically Conducting Polymers: An Emerging Technology* (M. Aldissi, ed.), Kluwer, Boston, 1993, p. 147.

57. G. Gustafsson, Y. Cao, G. M. Treacy, F. Klavetter, N. Colaneri, and A. J. Heeger, Flexible light emitting diodes made from soluble conducting polymers, *Nature 357*:477–479 (1992).

58. S. Roth and W. Graupner, Conductive polymers: Evaluation of industrial applications, *Synth. Met. 55–57*:3623–3627 (1993).

59. Y. Cao, G. M. Treacy, P. Smith, and A. J. Heeger, Solution cast films of polyaniline: Optical quality transparent electrodes, *Appl. Phys. Lett. 60*:2711 (1992).

60. Y. Cao, G. M. Treacy, P. Smith, and A. J. Heeger, Optical quality transparent conductive polyaniline films, *Synth. Met. 55–57*:3526–3531 (1993).

61. M. Angelopoulos, N. Patel, and J. M. Shaw, Water soluble polyanilines: Properties and Applications, *Mater. Res. Soc. Symp. Proc. 328*:173–178 (1994).

62. F. Jonas and G. Heywang, *Electrochim. Acta 39*:1345–1347 (1994).

63. L. W. Shacklette, N. F. Coleneri, V. G. Kulkarni, and B. W. Wessling, EMI shielding of intrinsically conductive polymers, *J. Vinyl Technol. 14*(2):118 (1992).

64. R. H. Baugman and L. W. Shacklette, Application of dopant-induced structure–property changes of conducting polymers, in *Science and Applications of Conducting Polymers* (W. R. Salaneck, D. T. Clark, and E. J. Samuelson, eds.), Adam Hilger, New York, 1990, p. 47.

65. R. H. Baughman, L. W. Shacklette, R. L. Elsenbaumer, E. Plichta, and C. Becht, Conducting polymer electromechanical actuators, in *Conjugated Polymeric Materials: Opportunities in Electronics, Optoelectronics and Molecular Electronics* (J. L. Bredas and R. R. Chance, eds.), Kluwer, Boston, 1990, p. 559.

66. M. Angelopoulos, J. M. Shaw, R. Kaplan, and S. Perreault, Conducting polyaniline: Discharge layers for electron beam lithography, *J. Vac. Sci. Technol. B7*(6):1519–1523 (1989).

67. M. Angelopoulos, J. M. Shaw, K. L. Lee, W. S. Huang, M. A. Lecorre, and M. Tisser, Lithographic applications of conducting polymers, *J. Vac. Sci. Technol. B9*(6):3428–3431 (1991).

68. G. Gustafsson, G. M. Treacy, Y. Cao, F. Klavetter,

N. Colaneri, and A. J. Heeger, The "plastic LED": A flexible light-emitting device using a polyaniline transparent electrode, *Synth. Met. 55–57*:4123–4127 (1993).
69. A. B. Holmes, D. D. C. Bradley, A. R. Brown, P. L. Bur, J. H. Burroughes, R. H. Friend, N. C. Greenham, R. W. Gymer, D. A. Halliday, R. W. Jackson, A. Kraft, J. H. F. Martens, K. Pichler, and I. D. W. Samuel, Photoluminescence and electroluminescence in conjugated polymeric systems, *Synth. Met. 55–57*:4031–4040 (1993).
70. R. S. Potember, R. C. Hoffman, H. S. Hu, J. E. Cocchiaro, C. A. Viands, R. A. Murphy, and T. O. Poehler, Conducting organics and polymers for electronic and optical devices, *Polymer 28*:574–580 (1987).
71. A. G. MacDiarmid, Short Course on Electrically Conductive Polymers, SUNY, New Paltz, NY, 1985.
72. D. W. DeBerry, *J. Electrochem. Soc. 132*:1022 (1985).
73. Zipperling Kessler & Co., German patent P33 29 577.7 (1987).
74. Allied-Signal, Americhem, Inc., and Zipperling Kessler & Co. U.S. Patent Appl. 82346 (1992).
75. Allied-Signal, Americhem, Inc., and Zipperling Kessler & Co. U.S. Patent Appl. 823511 (1992).
76. Allied-Signal, Americhem, Inc., and Zipperling Kessler & Co. U.S. Patent Appl. 823512 (1992).
77. B. Wessling, Passivation of metals by coating with polyaniline: Corrosion potential shift and morphological changes, *Adv. Mater. 6(3)*:226–228 (1994).
78. W. K. Lu, R. L. Elsenbaumer, and B. Wessling, Corrosion protection of mild steel by coatings containing polyaniline, *Synth. Met. 69*:2163–2166 (1995).
79. Zipperling Kessler & Co., Germany, unpublished communication.

Index

$2A_g$, 642, 656
mA_g, 644, 645
Absorption, 368, 372-373 727, 729, 731, 736
    spectra, 3, 6
    subgap, 15
Absorption coefficient for polypyrrole, 113, 114
    and evidence of pinned modes at low temperature, 113, 114
    similarity to polyaniline, 113
        formation of polaron lattice in polypyrrole, 113
Acetylene:
    irradiation polymerization of, 205
    polymerization on a surface of metal oxides, 205
    solid-state polymerization of, 205
Acceptor, 727, 729
Acentric alignment, 730
Actuator, 1016
Adhesion, 975, 980, 981
    specific, 970, 978
ADMR, 641-665
    exchange-correlated spin 1/2 pairs, 652
    spin 1/2, 649
    spin 1, 650
Aggregates, 830, 835
    formation of, 368, 370, 376
Alkano loops, cyclic, 376
Alkoxythiophenes:
    chemical stability, 800
    infrared intensity, 800
Amine triflate salts, 927
Ammonium poly(p-styrene sulfonate), 923
Amperometry, 966
Amplitude mode, 166, 171-73
Anderson transition, 30, 32, 87, 91
    comparison with experimental results, 105-108, 117, 118
    evolution of temperature dependent resistivity through, 92, 93
    hopping DC electrical conductivity on insulating side of, 92

    optical frequency dielectric function near, 94
    optical frequency electrical conductivity near, 93
    short scattering time of carriers near, 91, 92
Andrieux-Saveant model, 543
Anisotropy of conductivity:
    in I-$(CH)_x$, 39, 42-44
    in K-$(CH)_x$, 38
Annulenes, 125
Anodic protection, 882, 884
Anthracene, 823
Anthraquinone-2-sulfonic acid (*see* Dopants), 970
Antibodies, 964, 971, 975, 977
Antigen, 975
Applications, 469, 470, 517
    camouflage netting, 1010
    composite structures, 1009
    EMI shielding, 1006
    military, 1010
    resistive heaters, 1008
    static dissipation, 1006
Applications for conducting polymers, 85
Applications of polythiophenes, 226
Applied electric field, 730
Aromatic form, 281-283
Aromatic polyamides, 734
Aromatization, 375-377
    polymer-analogous, 376
Array, interdigitated, 966, 969, 971, 980
Array, micro, 971
Artificial muscle, 1015, 1016, 1017, 1018, 1021, 1026
    able to trail a steel sheet, 1019
    based on polymer gels, 1016
    formed by bilayers, 1017
    formed by three layers, 1019
    similarities with natural muscles, 1016, 1025, 1026
    theoretical approach for, 1022
    working, 1018, 1022

Asymmetric membranes, 948, 963
   multi-element, 970, 982
Arsenic pentafluoride, 205
Arsenic trifluoride, 205
Atomic force microscope, 1035-1038
Austin Model 1 (AM1), 4, 8-9, 13
   AM1/CI (Configuration Interaction) formalism, 12
Azo dye, 731

$1B_u$, 642, 644, 645
Bacteriorhodopsin, vibrational spectra, 815
Band gap (*see* Energy gap)
Band structure:
   of implanted polymers, 628-630
   of polymers with defects, 590-594, 630
Batteries, rechargeable using PPV, 354
BBL (poly benzimidazophenanthroline), 365
(BEDT-TTF) salts 1040, 1041, 1049
   crystalinity, 1040-1042
   morphology, 1041
Bending time, 1019
Benzenoid resonance structure, 226
Benzo[*b*]thiophene (*see* Isothianaphthene), 268
Benzo[*c*]thiophene, 278
5-Benzoylbenzo[*c*]thiophene, 291
Biexciton, 644, 645
Bifurcation, 502
Bilayers, 573
Binding energy, 1, 3
   of the exciton, 10-12
Biosensor, 702, 971, 975
Bipolarons, 1, 226, 643, 659-664, 825-826
   dictions, 748, 757-758
   enhancements of nonlinearity by, 748
   extrinsic (*see* p-Dimers)
   formation of, 750-751
      in bisanthracenylpolyenes, 758-759
      in dithienylpolyenes, 753, 760
      in diphenylpolyenes, 751
   interacting, 15
   isolated, 85, 97, 111
   lattice, 113
   positive, 14
1,4-bis(2-furanylbenzenes), 269
1,2-bis(2-thenoyl)benzene, 295
1,4-bis(2-thienyl)phenylene, copolymer with 1,4-bis(2-thienyl)biphenylene, 270
a-a'-bis(trimethysilyl)thiophenes:
   electrodesilylation of, 316
   tetrathienyl silane, 316
Bithiophene:
   electropolymerization of, 313
   as precursor of small bandgap polymers, 333
   substitution of, 314
Bithiophene, copolymer with:
   aniline, 268
   crown ether linked bithiophene, 263
   isothianaphthene, 265
   terthiophene, 266
Blends, 979
Block copolymers, 262, 273
Bond-length alternation, 4, 8, 16-20, 781
   in bacteriorhodopsin, 817
   and conjugation length, 782
   effect on nonlinearity, 748
   in push-pull polyenes, 790
   and solvent effect, 813-814
Bond order, 166, 172, 180
Breather, 131, 132
Buffer layer, 1041
Butadiene/styrene copolymer, 1030

C_{60}, 838
Carbon black, 471, 478, 489
Carotenoids, 773
   Raman intensities, 777
Carrier:
   concentrations, 1040
   density, 1049
   freeze-out, 1040, 1045
   mismatch, 1045
   mobilities, 737, 1040
Catalysis:
   biological, 967
   definition, 967
      Re catalyst, 204
      Rh catalyst, 204
Cathodic protection, 882, 884
Cavity perturbation technique (microwave transport), 96
$(CD)_X$, 657
Cellophane, 1019
Centrosymmetric media, 729
$(CH)_X$:
   cis-$(CH)_X$, 639, 655, 657-659
   trans-$(CH)_X$, 639, 654-659
$[(CH)I_{0.16}]_X$, 1040
Charge conjugation symmetry, 642, 660
Charge distribution, 19
Charge state:
   incorporation in conjugation sequence, 748
   influence on nonlinearity, 743, 757-760
Charge transfer, 728
Charge transport agent, 737
Charge-trapping, 737
CHEMFET, 967, 983
Chemical corrosion inhibitors:
   acridine, 916
   alkylaniline, 882
   aniline, 882, 900, 910, 915
   n-arylpyrrole, 882
   ethoxyaniline, 900
   piperidine, 916
   pyridine, 915
   pyrrole, 882, 915, 916
   thiophene, 882
Chemical polymerization, 263, 264

Chemoresistance, 970
Chiral polythiophenes:
    conductivity of, 322
    enantioselective properties of, 323
    optical rotation of, 323
Chromism:
    affinitychromic, 703
    electrochromic, 695
    ionochromic, 703
    photochromic, 703
    piezochromic, 700, 703
    solvatochromic, 700, 703
    thermochromic, 696-700, 703
Chronoamperometric data, 1037, 1039
cis-polyacetylene, 176-177
Coating (see also Films), 518, 525, 1059-1070
Coating applications, 1067-1069
    computer manufacturing, 1069
    corrosion protection, 1069
    electrochromic devices, 1069
    electrostatic dissipation, 1068
        comparison, 1070
        materials, 1070
Competing materials, 1069
    dispersion, 1061-1064
    environmental considerations, 1070
    environmental stability, 1063
    films (see also Coatings), 1059-1070
    polyaniline, 1060
        aqueous coatings, 1065
        coatings (films), 1061
        dispersion coatings, 1061-1064
        dispersion coatings, properties, 1063, 1064
        doping, 1061
        emeraldine base coatings, 1064
        secondary doping, 1066
        solvent interactions, 1061
        structure-property relationships, 1060
    polypyrrole, 1067
    polythiophene, 1066
    preparation, 1061-1066
    surface resistivity, 1067
        material classification, 1067
        measurement, 1067
Coherence length, 1040, 1043-1045, 1053-1054
Coil-like morphology, 89, 96
    effect on localization length of, 89
Collisions:
    balistic motion limited by, 154
    of the spin carriers, 147, 157, 162
    rate of, 162
Collagen fiber, 1016
Colloid, definition of, 423, 497
Colloid stability:
    charge stabilization, 423
    steric stabilization, 423
    steric stabilizer, 423
Color conversion, 847, 870

dyes, 867
Color modulation, 10
Commercial products (see Applications)
Complexation, 974
Composites, 571
Composites of polythiophene, 316
Composite system (see also Percolation transition), 87
Computer simulation, 500
Concentration:
    critical (volume), 471, 473, 489, 496, 510
    maximum, 489
Conducting polymers, 1015
    artificial muscles from, 1016, 1026
    film, 1017, 1019
    first-generation, 27
    of high structural homogeneity, 1017
    new generation, 27, 31
    redox processes in, 1015, 1016
Conducting polymer membranes, 945-961
Conductive fibers:
    carbon, 993, 1009
    stainless steel, 993
Conductive polymer corrosion inhibitors:
    corrpassive, 884
    polyaniline, 883, 884, 891, 892, 900, 901, 902, 908, 912, 913, 915, 917
    polyethoxyaniline, 883, 900, 913
    poly(3-hexylthiophene), 883
    poly(3-methylthiophene), 884
    poly(3-octylthiophene), 883
    polypyrrole, 883, 891, 892, 900, 904, 908, 911, 912, 914
    poly(3-thienylacetate), 884
    poly(3-thienylmethylacetate), 883
    versicon, 884
Conductive polymer/superconductor assemblies, 1031
Conductive polymer/superconductor bilayer, 1034
Conductive textiles:
    copper sulfide coated, 993
    from conducting polymer emulsions, 995
    metal coated, 993
    polyaniline coated (see also Polyaniline on textiles), 995, 1000
    polypyrrole coated (see also Polypyrrole on textiles), 995
Conductimetry, 966, 978, 982
Conductivity, 471, 479, 490, 494
    AC conductivity, 611-615, 622-628
    anisotropy of, 159
    DC conductivity, 610-612, 619-622
    intra-, inter-chain, 154
    macro- and microscopic, 159
    mechanism, 480
    microscopic, 161
    parallel to chain axis, 32
    stability of, 616
    temperature dependance of, 508, 514
Conformational changes, 1017, 1019
Conformational level, 1017
Conformational relaxation, 1019, 1022
Conjugated bonds, 727, 737

Conjugated length (CL), 767
    dependence of nonlinearity on, 744
    in doped and photoexcited polyacetylene, 788
    effective, 744, 851, 866
    enhancement of nonlinearity by, 744, 759-760
    and intramolecular potential, 772, 773, 782
    and π electron delocalization, 777
    in polyparaphenylenevinylene, 808
    in polypyrrole, 798
    in polythiophene, 799
    in *trans*-polyacetylene, 775
Conjugated oligomers, 1,3
Conjugated polymers, 1, 737
p-Conjugated systems, 85
Contact resistance, 1031, 1042
    (BEDT-TTF)$_2$I$_3$/cuprate superconductor, 1049
    conductive polymer/cuprate superconductor, 1047
    geometries, 1046
    measurements, 1045
    metal/superconductor, 1046
    values, 1048
Contex™, 940
Cooper pairs, 1042
Coplanarity, 262
Copoly(3-alkoxythiophene)s, 263
Copoly(3-alkylthiophene), 261-263
Copoly(3-aryl thiophene)s, 265
Copolymerization, 970, 978
Copolymers, 572
    bipolaron formation in, 758
    characterization
        microarray electrodes, 260
        X-ray photoelectron spectroscopy, 260
    conductivity, 260
    containing ladder oligomer repeat units, 756
    containing polyene repeat groups, 756-757
    containing PTV oligomer repeat units, 757
    containing redox active groups, 261
    steric interactions, 260, 262, 264, 265, 267
    X-ray structure, 265, 272
Copolythienylpyrroles:
    2,5-di(2-thienyl)-pyrrole, 266
    di-(2-thienyl)silane copolymer with 1,5- and
        1,8-dichlororanthraquinone, 272
    N-phenyl copolythienylpyrrole, 267
Core-shell particles:
    carboxylated styrene-butadiene-methacrylate
        latex-polypyrrole, 431
    ceria-polypyrrole, 431
    hematite-polypyrrole, 431
    polystyrene-polypyrrole, 431
    polyurethane-polypyrrole, 431
    poly(vinyl acetate)-polypyrrole, 431
Corona poling, 730, 734, 738
Correlation function, 141, 146
Corrosion, 881, 885
Corrosion current, 899
Corrosion inhibitors:
    chemical
        acridine, 916
        alkylaniline, 882
        aniline, 882, 900, 910, 915
        n-arylpyrrole, 882
        benzotriazole, 937
        ethoxyaniline, 900
        piperidine, 916
        pyridine, 915
        pyrrole, 882, 915, 916
        thiophene, 882
    polymer, corrpassive, 884
    polyaniline, 883, 884, 891, 892, 900, 901, 902, 908, 912, 913,
        915, 917, 938
    polyethoxyaniline, 883, 900, 913
    poly(3-hexylthiophene), 883
    poly(3-methylthiophene), 884
    poly(3-octylthiophene), 883
    poly-o-phenetidine, 938, 939
    polypyrrole, 883, 891, 892, 900, 904, 908, 911, 912, 914
    poly(3-thienylacetate), 884
    poly(3-thienylmethylacetate), 883
    versicon, 884
Corrosion protection, 470, 518
    aluminum, 910
    copper, 913, 937
    secondary methods, 882
    silver, 913, 937
    steel, 891
Corrosion potential, 886, 894
Corrosion rate, 899
Coulomb gap, 60, 63-65
Counter-ion induced processibility, 440
Counterions, 1015
    concentration of, 1016, 1022
    expulsion of, 1015, 1017, 1019
    incorporation of, 1015, 1017, 1019, 1023
    inmobile, 1015
    osmotic pressure due to, 1016
Critical currents, 1051
Critical regime of the M-I transition, 32, 35, 40
    in K-(CH)$_X$ and I-(CH)$_X$, 38-39
    in PANI-CSA, 36-37
    in PPy-PF$_6$, 40
Cross-coupling:
    Suzuki, 212
    Yamamoto, 211
Cross-linking, 734, 735, 739, 1016, 1021, 1022, 1023
Crown ether substituted polythiophenes:
    cations recognition properties of, 322, 323
    synthesis of, 321
Crystal structure for polyaniline, 96, 101
Crystal structure for polypyrrole, 111, 114
Crystalline coherence lengths for polyaniline, 96
Crystalline coherence lengths for polypyrrole, 111
    comparison with localization lengths estimated from transport
        parameters, 112
Crystallinity, 531

Curie spins, 154
Curie susceptibility, 97, 111
Current efficiency, 1017
Cutoff frequency, 143, 147
Cyanine limit, 17-19
Cyanines, 124-125
Cyano-substituted derivative, 3
Cyclic voltammetry, 1033
Cyclization:
    of suitably substituted single-stranded precursors, 364
    polymer-analogous, 365-366, 370-372
Damping factor, 6
DC electrical conductivity, 87, 89, 90
    behavior near an insulator-metal transition
        in the inhomogeneous disorder model, 89, 90
        in the Anderson transition, 91
        in polyaniline (as a function of temperature), 97, 98
    consistency with inhomogeneous disorder model, 98
    at millikelvin temperatures, 87, 90, 91, 117
        for polyaniline, 97, 98
            as an argument against hopping/Anderson transition, 98, 99, 117
            and relation to presence of free carriers, 103, 104, 117
        for polypyrrole, 112
        for polyacetylene, 117
    non-monotonic variation of through insulator metal transition for polyaniline, 98
    as an argument against Anderson transition, 98, 117
    prediction of ultimate values for, 103, 117
    range for conducting polymers, 85, 86
    d coefficients, 731, 732
    scaling with the free carrier plasma frequency, 103, 104
        in polypyrrole (as a function of temperature), 112
        and comparison with microwave conductivity, 115, 116
Degenerate ground-state polymers, 639-641
Degradation, 968, 981
Density, 490
Density of states:
    for I-$(CH)_x$, 42
    for PANI-CSA, 68
    for $PPy-PF_6$, 69
    vibrational $g(v)$, 773
        two-phonon states of polyacetylene, 774
Density of states at the Fermi level, $N(E_F)$, 85, 88, 117
    in polyaniline, 97
    in polypyrrole, 111, 112
        and relation to structural order, 111, 112, 117
Derivative relations, 19
Derivatization, 972, 977
    by donors and acceptors, 3, 9-10
    effect of, 1, 9-10
Devices, 573, 630
    based on bilayers, 1017
    based on three layers, 1019
    chemomechanical, 1017
    electrochemomechanical, 1015, 1016, 1017
    electrochemopositioning, 1016, 1018
    electroosmotic, 1017
    electrophoretic, 1017
Diaminooctafluorobiphenyl, 858
Diaphorase, 970
Dichroic ratio, 415
Dichroism, infrared, 769
    of polyacetylene, 772
    of polythiophene, 840
Dielectric constants, 72
    of polyaniline, 511
Dielectric function:
    optical frequency for polyacetylene, 117
    optical frequency for polyaniline, 101
        Drude (free carrier) response in, 102, 114, 115
        localized carrier response in, 102
        polarization dependence of, 104, 105
            as an argument against the Anderson model, 104, 117
            comparison with the localization modified Drude model/Anderson transition, 105, 106, 107, 108
            comparison with the inhomogeneous disorder model/percolation transition, 107, 108, 109
            temperature dependence, 103, 104
    for polyacetylene, 117
    for polyaniline, 109
        positive value of for localized carriers, 109
        negative value for free (Drude) carriers, 109, 110
            agreement with optical dielectric function, 110
            differences with optical dielectric function as evidence of distribution in scattering times, 110
            temperature dependence of, 109, 110
    for polypyrrole, 114, 116
        comparison with inhomogeneous disorder model/percolation transition, 114, 115
        localized carrier response in, 114, 115
        negative value for free (Drude) carriers, 116
        positive value for localized carriers, 116, 117
            agreement with optical dielectric function, 116
            estimation of localization length from, 116, 117
            evidence for distribution in scattering times, 116
Dielectric properties (*see* Properties of conductive textiles)
Diels-Alder addition, repetitive, 375
Diels-Alder reaction, retro, 376
Diffusion:
    anomalous, 144
    coefficient, 142, 143, 144, 151, 154
    motion, 143
    one-dimensional, 160
    soliton, 154
    spin diffusion rate, 149, 156, 157, 160
Diffusion coefficient, 947
1,3-Dihydrobenzo[*c*]thiophene, 279, 281
1,3-Dihydrobenzo[*c*]thiophene-S-oxide, 279
1,3-Dihydro-5,6-dioxymethyleneisothianaphthene, 289
E-1,2-(di-2-furyl)ethylene, 306
E-1,2-(di-2-thienyl)ethylene, 306
Dimensionality:
    effective, 159
    random walk and, 142
Dimer, 835

p-Dimers, 15, 644
Dimethoxypolyparaphenylenevinylene (DMPPV), 956
4-(N,N-dimethylamino)-4'-nitrostilbene (DANS), 729, 732
N',N'-dimethylpropylene urea, 444
   solvent for PANI, 443 - 446
   properties of 444-447, 448-454
Diphenylpolyenes, 669, 673
   band progression, 774
   interaction with aluminum, 684
   interaction with calcium, 685
   interaction with sodium, 685
   Raman intensity, 777
Dipole moment, 728-730, 733
Dipolar orientation, 738
   excited-state, 728
   ground-state, 728, 729
      intrinsic, 728
Disorder scattering, 89
   as a cause for Anderson transition, 91
   effect on DC electriucal conductivity, 89
   as scattering mechanism for majority of carriers in polypyrrole, 115
Disperse Orange, 3, 733
Disperse Red 1, 733
Dispersion, 487, 499, 501, 517
   colloidal, 469
   frequency, in Raman spectra, 768, 775, 782-784, 797-799, 809
   hypothesis, 469
   intensity, in Raman spectra, 768, 777, 784
   law, 502
   polymerization, 423
Disordered regions, as medium surrounding ordered (crystalline) regions, 87, 88, 89
   in polyaniline, 96
   in polypyrrole, 111
Distributed Bragg reflector, 837-838
Distribution potential, 558
DMPU (*see* N',N'-dimethylpropylene urea)
DNA, 975
Dodecylamine, 1041
Donnan exclusion, 556
Donnan potential, 555, 559
Donor, 727, 729
Donor-acceptor groups, 728, 729
Dopants, 963, 973, 980, 981
   anthraquinone-2-sulfonic acid, 999
   naphthalene-2-sulfonic acid, 1005
   stability and, 1005
Doped phases, 712
Doping, 1, 13-15, 277
   by ion implantation, 589
   reversibility of, 616
Double layer, electrical, 558, 561
Drude model for metallic carriers, 85, 92
   Drude behavior in optical conductivity for polyaniline, 100
   Drude behavior in optical conductivity for polypyrrole, 114
   Drude behavior in optical dielectric function, 117
      for polyaniline, 102
      at low temperature in samples with finite millikelvin DC conductivity, 103, 104
      for polypyrrole, 115
   Drude behavior in microwave dielectric constant, 117
      for polyaniline, 110
      for polypyrrole, 115, 116
   Drude behavior in microwave conductivity of polypyrrole, 115
Dynamic Nuclear Polarization, 145, 148
Dynamic random access memory, 921
Dynamics, 765-767

Effective Charges and Charge Fluxes (ECCF) model, 767, 772
Effective conjugation coordinate, 167, 170, 174, 178, 769, 782, 794
   damped, 179-180
Effective Conjugation Coordinate (ECC) theory, 773, 782-785
   of polyaromatics and polyetheroaromatics, 794-795
   of polyene systems, 784
   and push-pull polyenes, 789-794
   and vibrational spectra of doped and photoexcited species, 788-789
Effective mass, 91, 92, 103, 1040
   for polyaniline, 101
   for polypyrrole, 114
Electrical conductivity, of PPV and derivatives, 352-354
Electrical Properties (*see* Properties of conductive textiles)
Electric field, 728, 730, 732, 737
Electric field effect devices, 1053-1054
Electric field-induced second harmonic generation, 728
Electroabsorption, 184-187
   Stark shift, 184
Electroactive, 964, 968, 972, 978, 983
Electroactive hot spots, 1038, 1039
Electrochemical cells, 12
Electrochemical copolymerization, 260
Electrochemical impedance spectroscopy, 888
Electrochemical polymerization, 263, 264
Electrochemical synthesis, of PPV, 344, 348
Electrochemistry:
   deposition of polymers, 1034
   doping of polymers, 1033-1034
   nucleation of polymers, 1032
Electrochromism, 268-271
Electrode, 727, 730, 732
   four-wire, 963
   interdigated, 966, 969, 971
   interdigitated microsensor, 980
   micro, 535
   ultramicro, 535
Electrode poling, 730
Electrodeposition, 973
Electroluminescence, 2, 265, 368-370, 372, 823-840, 847
   in PPV and derivatives, 136, 349, 350, 353
   yield, 2, 3
Electromagnetic interference shielding, 921
Electron affinity, 855
Electron correlation energy, 642
   effective, 642, 656, 662, 663
   -electron delocalization, 728

Electron diffusion coefficient, 1043
Electron-electron (e-e) interactions, 41-42, 47-49, 52, 639, 827
Electron energy loss spectroscopy in PPP, 599-601
Electron hopping, redox, 561
Electron injection, 133
Electron paramagnetic resonance, 96
Electron-phonon coupling, 769, 825-827
  and vibrational hyperpolarizabilities, 810, 815
Electron transfer, 1030
Electroneutrality, 554, 1015
  coupling, 532, 563
Electronic:
  correlation effects, 1, 12
  properties, 1
  transitions, 6-7, 14
Electronic specific heat, 1043
Electronic structure of oligoenes, 775
Electronic susceptibility:
  dynamic, NLO, 180 ff, 192
  static, 172-74, 176, 179, 192
Electron-phono interaction, 639
Electron-phonon coupling, 10, 12, 85, 124-125, 169-171, 191
  coupling constants, 166, 171-174
  and NLO, 180-181
  quadratic, 166, 176
  reference state in, 170-171
Electro-optic (EO), 727, 728, 731-733
  applications, 727, 732, 733, 736
  coefficients, 728, 732, 739
  effect, 728
  measurements, 732, 738
    ellipsometric, 732
    Fabry-Perot interferometric, 732
    Mach-Zehnder interferometric, 732
    Michelson interferometric, 732
  modulation, 727, 739
  modulators, 727
Electropolymerization, 976
Electrostatic charge, 939
Electrostatic discharge, 921, 939, 940
Elimination, polymer-analogous, 377
Elliott mechanism, 147, 157, 160, 162
Ellipsoidal cavity, 20-21
EMI shielding, 526
  far-field, 528
  near-field, 528
Energy:
  adhesion, 472
  chemical, 1016
  chemomechanical conversion of, 1016
  cohesion, 472
  electrical, 1016
  interfacial, 473
Energy transfer, 1030
Ennobling, 522, 523
Entropy, 501
Environmental degradation, 1031
Enzyme, 967

EPSIS, 969
Equilibrium:
  ion partitioning, 554, 555
  polymer/electrolyte, 554, 555
Equivalent circuit, 548
ESPACER™, 930
ESR line, 146
  line broadening, 147, 157
  lineshape, 146, 147
  linewidth, 146, 151, 152, 156, 157, 160, 162
Etching of $YBa_2Cu_3O_7$-d, 1032
Ethyl cellulose, 1030
Excimer, 828, 835-836
Exciplex, 828
Excitation, 130
  ionic channel of, 134
  neutral channel of, 134
Excited states, 3, 6-10, 13-19
  correlated, 186 ff
    2A-1B crossover, 187-189
    symmetries, 186
Exciton, 827-828, 830-831, 834-839, 864, 872
  binding energy, 827, 838
  Förster transfer of, 828
  Frenkel, 827
  ionization, 838-839
  Mott-Wannier, 827
  Singlet, 640, 644, 653, 827, 830-831, 834-836
  Triplet, 640, 644, 645, 653, 661, 663-665, 827, 834-835
Extended Hubbard calculations, 13
External electric field, 17-19
Extrinsic barriers, 1045

Fermi energy, 853
Fermi glass, 29, 31, 32
Fiber formation, 441
Fiber spinning (an overview), 441-442
Fibers (see also Conductive fibers), of PPV and derivatives, 351, 354
Fibrils, 531
Field effect transistor, 823, 874
  using PPV, 353
Field-induced crossover:
  in I-$(CH)_X$, 43
  in K-$(CH)_X$, 39
  in PANI-CSA, 36-37
  in PPy-PF_6, 40
Flocculation, 493, 497
Fluctuation-induced tunneling, 33, 43
Fluorescence, 263
Fluorinated polythiophenes:
  fluorophilic interactions in, 326
  3-fluorothiophene, 325
  polyfluorinated 3-alkylthiophenes, 326
    electronic effects in, 326
    synthesis of, 326
  trifluoromethylthiophene, 325
Four probe technique (DC conductivity), 95
  millikelvin temperature measurement, 95

Force field, 766
    of *trans*-polyacetylene, 777
Förster transfer, 828
Fowler Nordheim theory, 831-832
Fractal networks, 70, 75, 78
    superlocalization, 70, 75
Franck-Condon approximation, 4, 10
Free electron (FE-) model, 124
Free volume, 738
Frequency conversion, 727, 739
Frequency doubling, 728, 736, 739
Fresnel reflection coefficients, 96
Fullerene, 838
Fulvene containing polymers, 286
    fulvenoid and quinonoid forms, 286
Functional groups, 727

Galvanic corrosion, 881, 887, 904
Gas separations, 946, 948-956
General corrosion, 881
Geometry relaxation phenomena, 1, 8-9
Glass transition temperature (T_g), 730
Glucose oxidase, 970
Graded copolymers, 266
Guest, 734, 737

Hartree-Fock, 1, 4
    ab initio, 19
    restricted open (ROHF), 13
Heteroaromatics, 259
Heteroaromatic rings, 729
    effects on nonlinearity, 745
Heterojunctions, 832-834
Herzberg-Teller expansion, 170
3-Hexylthiophene copolymer with 3-thiopheneethylacetate, 263
High density polyethylene, 1030
High Temperature (T_c) Superconductor, 1029-1058
    ceramic pellets, 1034
    devices, 1055
    electrode template, 1033
    laser ablation, 1033
Highly oriented pyrolytic graphite, 1035
Hollow fiber membranes, 948, 953
HOMO, 642-645, 660
Homogenous, 31, 59, 61
Hopping frequency, 147
Hopping interchain, 156, 159
Hopping transport, 32, 60, 87, 117
    exponent, 33
    Mott variable range, 92
        for polyaniline, 97, 98
        for polypyrrole, 110, 112
    pairwise, 115, 116
    in the presence of coulombic interaction, 92
Host, 733, 734
Huang-Rhys factor, 4, 6-7
Hückel methods, 826
Hückel Molecular Orbital (HMO-) model, 125-126

Huggins Constant, PANI (in DMPU), 445
Hyperfine coupling, 149
Hyperpolarizabilities, molecular, 809-815
    and two-state model, 810
    of polyenes, 810
    vibrational contributions, 809-815
Hyperpolarizability, 728, 729
Hyper-Rayleigh scattering (HRS), 729
    measurements, 20

IDA, 970
Immunodiagnostic assays, 428, 430, 432
Impedance spectroscopy, 546, 568, 569
Indium-tin oxide, 823, 830, 855
Induced absorption, 827, 836
Induction of superconductivity, 1042
Inelastic scattering length, 41
    for I-$(CH)_x$, 48-50
    for PANI-CSA, 52
    for PPy-PF_6, 57-58
Infrared intensity, 767
    of alkoxythiophenes, 800
    of polythiophene, 799, 800
    of push-pull polyenes, 791
Infrared spectra:
    of doped and photoexcited polythiophene, 801-805
    of doped molecules (DIIR), 768
    of doped polyacetylene, 785
    of doped polyparaphenylenevinylene, 806-808
    of photoexcited molecules (PIIR), 768, 787
    of photoexcited polyacetylene, 788
    of polyconjugated materials, 767
    of polypyrrole, 798
    of polythiophene, 800-801
    of pristine systems, 767-768
Infrared spectroscopy, 416-418
Inhomogeneous disorder model, 87, 88, 117, 118
    comparison with experimental results, 117, 118
        for polyaniline, 98, 99, 100, 103, 107, 108, 109
        for polypyrrole, 113, 114, 115
    differences with a percolation transition, 91
    relation to percolation transition, 89
    relation to random metallic network model, 89
Inorganic materials, 727, 739
In situ polymerization:
    kinetics of, 997, 998
    of polyaniline, 1000
    of polypyrrole, 994
Insulating regime near the M-I transition, 59-61
    in PANI, 65-66
    in PANI-CSA, 65-66
    in polyalkylthiophenes (PATs), 61, 63
    in PPy-PF_6, 54, 61-63
Insulator-metal transitions:
    dependence on structural order, 86, 87
Integrated circuits, 921
    packaging, 921
Interaction length, 41

Interchain interactions, 826, 828
Interface reflections, 1045
Interface transmission coefficient, 1045
Intermediate Neglect of Differential Overlap (INDO), 4, 6-7, 9, 17-19
Interpenetrating networks, 839-840
Intersystem crossing, 3, 8
Intrinsic semiconductor, 861
Intrinsic viscosity, PANI (in DMPU), 445
Ioffe Regel condition for mean free path near an insulator-metal transition, 87, 101, 103, 117
Ion exchange, 972, 973
   equilibria, 532
   ion partitioning, 554, 555
   studies of, 570
Ion implantation, 589
   in $(CH)_X$, 601
   in PPP, 594
   in PPS, 590
   in PPV, 596, 634
Isomerization, thermal, of *cis*-polyacetylene, 769-773
Isothianaphthene, 265, 279

Josephson junctions, 1045
Joule effect, 1022
Junction, p-n, 261, 630-633

Kinetics:
   of degradation, 1004
   of polymerization (*see* In situ polymerization)
Kink:
   collision of, 134
   dynamics of, 129
      strong and weak, 126
Kleinman symmetry, 731
Kohlrausch-Williams-Watts (KWW) equation, 738
Korringa relation, in PANI-CSA, 53
Koutecky-Levich plot, 541
Kraemer Constant, PANI (in DMPU), 445
Kramers-Kronig analysis, 93, 96

Ladder polymers, 287, 288
   containing fused indene rings, 302
Langmuir-Blodgett (LB), 730, 980
   films, 260
   films of PPV derivatives, 351-352
Laser ablation, 1033-1034, 1036, 1038
LEDs (*see also* Light emitting diode), 368, 370
Length change, 1016, 1023
Levich plot, inverse, 541
Light emitting diode, 1, 2, 824-838, 847
   current-voltage characteristics, 830-832
   efficiency, 825, 830-831, 834-838
   electrochemical cell, 853, 858
   emmision colors, 865, 867, 868
   heterostructure devices, 832-834
   lifetimes, 875
   p-n junction, 858
   polymer, 847, 848
   using PPV and derivatives, 353
Light-induced electron spin resonance, 641
Liquid crystal, PPV derivatives, 351-352
Liquid separations, 948, 956-960
Lithography, 922
   electron beam, 923
   photo-, 923
Local field and field factors, 728
Localization, 87, 91
   relation to structural order, 88, 110, 111, 118
Localization-interaction model, 43, 58
Localization length (L_C), 60
   estimated from dc conductivity for polypyrrole, 112
   estimated from microwave dielectric function for polypyrrole, 116, 117
   for PANI-CSA, 40
   for $PPy-PF_6$, 40, 54, 62
   in relation to inhomogeneous disorder model, 88, 89, 104, 118
   in relation to Anderson transition, 91
Localization modified Drude model, 59, 87
   evolution of with changing scattering time (disorder), 94
   optical frequency electrical conductivity in, 93
      comparison with experimental results for polyaniline, 105, 106, 107, 108
      using a distribution of scattering times, 108, 109
   optical frequency dielectric function in, 94
      comparison with experimental results for polyaniline, 105, 106, 107, 108
      as an argument against the Anderson model, 107
      using a distribution of scattering times, 108, 109
Logarithmic derivative of DC conductivity (W), 87
   behavior for insulating and metallic systems
      in the Anderson transition, 91
      in the inhomogeneous disorder model, 90
      in polyaniline, 97
      in polypyrrole, 112
LOMO, 642-645, 660
Low band gap, definition, 278
Low band gap donor-acceptor polymers, 304, 305
Low band gap zwitterionic polymers, 305
LPPP (*see* Poly(*p*-phenylene), methylene-bridged)
Luminescence, 262, 273
Luttinger catalyst, 202

Magnetic length, 36, 40, 60
Magnetic susceptibility in polyaniline, 97
Magnetic susceptibility in polypyrrole, 111
Magnetoconductance (MC), 41-42
   for $I-(CH)_X$, 47-49
   for PANI-CSA, 52
   for $PPy-PF_6$, 57-58
Magnetoresistance:
   for PANI-CSA, 38, 53
   for PANI-CSA networks, 76-77
   for $PPy-PF_6$, 58, 62
   for polyaniline, 99
      in relation to weak localization (due to quantum interference), 99

[Magnetoresistance, for polyaniline]
    in relation to percolating structures, 99
    for polypyrrole, 112
Maker fringe, 731
Materials Science, 468, 470, 517
Mean free path, 91, 92, 101, 105
    long, 108, 115
Mechanical properties (*see* Properties of conductive textiles)
Mediation (*see also* Redox catalysis), 968
    of electron transfer, 531, 564, 566
Mediators, 969
Medium, dependence of the polarizabilities on the nature of, 2, 16-20
MEH-PPV, 824, 831, 838-840
Melt fracture, 505
Membrane, 978, 1015
    anion exchange, 559
    cation exchange, 559
    overlayer, 978
    protective, 978
    semipermeable, 978
    switchable, 559
    track-etch, 989
Metal-insulator-metal structure, 853, 855
Metallic box model, 116, 117
Metallic islands, 31, 34, 59, 66-67, 78
Metallic regime near the M-I transition, 40
    in I-$(CH)_x$, 39, 42-49
    in K-$(CH)_x$, 49
    in PANI-CSA, 51-53
    in PPy-PF_6, 53-58
Metallization, 935
    electrodeless, 935
    electrolytic, 935
Metals:
    mesoscopic, 481, 507
    organic, 516
Metal-semiconductor-metal structure, 853
Metathesis catalyst, 203
Methods:
    electrochemical, 532-554
    Kelvin probe, 553
    quartz crystal microgravimetry, 542, 568, 569
    spectroscopic, 549, 569
3-Methoxythiophene, vibrational spectra, 801
2-Methyl-4-nitroaniline, 733
3-Methylthiophene copolymer with ferrocene esters of 2-(3-thienylethanol), 262
    3-n-octylthiophene, 262;
Microcavities, 836-838
Microelectronics, 921
Microemulsions, 505
Microgravimetry, quartz crystal, 542, 568, 569
Microwave absorption, 508
Microwave properties (*see* Properties of conductive textiles)
Microwave and far-infrared dielectric response, 59
Minimum metallic conductivity, 31
Mobility, 730, 738, 832, 834, 838, 862

Mobility edge (E_C), 29, 31, 87
    in relation to the Anderson transition, 91
Modified Neglect of Differential Overlap (MNDO), 4
Modulation of superconductivity, 1031, 1050, 1053
Molecular machine, 1017
Molecular orbitals:
    effects of derivatization on, 9-10
    frontier, 9-10
    localized and delocalized, 6
    polaronic, 14
Molecular orientation, 730
Molecular switch, 1030, 1051
Monolayer buffers, 1041
Morphology, 483, 484
Morphology of polypyrrole (*see* Polypyrrole on textiles)
Motion:
    anisotropic, 143
    one-dimensional, 142
    spectrum of the, 141, 142, 143, 146
Motional narrowing, 146, 147, 148, 150
Mott variable range hopping model, 418
Multipolar expansion, 20-21
MultiReference Double Configuration Interaction (MRD-CI), 4, 7
Multistep routes:
    "classical", 364
    $\mu\beta$ value(s), 728, 729, 733, 736

NADH, 970
N-(3-aminopropyl)pyrrole, 1037, 1038, 1039
Nanochemistry, 409
Nanomaterials, 409
    possible applications of, 409
Naphtho[2,3-c]thiophene, 292
n-doping, 263, 268
Near infrared spectroscopy, 277
Nearly free electron (NFE-) model, 124
Negative low frequency dielectric constants (*see* Drude model for metallic carriers)
p-Nitroaniline, 733
Nitroaryls, 260
NMR, 957
Noncentrosymmetric alignment, 728, 730, 738
Noncentrosymmetric dipolar orientation, 727, 738
Nondegenerate ground-state (NDGS) polymers, 640, 641
Nonequilibrium, 492, 495
Nonlinear index of refraction, 728, 736
Nonlinear optical (NLO) 727-739
    applications, 727, 736
    chromophores, 727-730, 732-734, 736-738
    dye, 729, 738
    materials, 727, 728, 731, 732, 737-739
    moieties, 730, 733, 734, 736-738
    molecules, 728-730, 733
    process, 728
    properties, 727-731, 736-739
        of PPV derivatives, 354
    responses, 728-730, 732, 733, 735-739
Nonlinear optics, 809-815

Nuclear spin:
    diffusion, 148
    relaxation, 144, 148, 151
    relaxation rate, 144, 155
Nucleation loop, 1032

Octadecylamine, 1041
ODMR, 648, 650, 652-654, 658, 659
Ohms per square (see Surface resistance)
Oligoazomethine, 858
Oligoenes, tert-butyl capped, 773
    hyperpolarizabilities, 810
    infrared spectra, 774
    Raman intensities, 777
    Raman spectra, 784
Oligomer, 265, 564
Oligoparaphenylenes:
    conformation, 805
    vibrational spectra, 805
Oligoparaphenylenevinylenes, vibrational spectra, 807
Oligopyrroles:
    protected, N-BOC, 797-798
    structure, 797
    vibrational spectra, 797-798
Oligothiophenes:
    chain length and electronic properties of, 330
    electropolymerization of, 313
    end-capped oligothiophenes, 331
    functionalization of, 331
    as models of polythiophene, 330
    regiospecific, 801, 804
    spiro-fused, 331
    synthesis of, 328
    vibrational spectra, 798-805
One dimensionality, 85, 87
    effect on electronic localization of, 87, 88
Onium salts, 926
Onsager model, 17, 19-20
Onsager theory, 875
Open circuit potential, 894
Optical band gap, 272
Optical conductivity:
    for polyaniline, 100
        isosbestic behavior in, 100
        Drude dispersion in, 100
        comparison with inhomogeneous disorder model, 101, 107, 108, 109
        comparison with localization modified Drude model, 101, 105, 106, 107, 108
        percolation behavior in, 100
        temperature dependence of, 101
    for polypyrrole, 113
        Drude dispersion in, 113
        relation of structural order to localization, 113
Optical detection of magnetic resonance (see ODMR)
Optical loss, 736, 739
Optical nonlinearities, 727-733, 735, 736, 738, 739
Optical poling, 731

Optical properties, 736, 739
    of implanted polymers, 594
    linear, 1, 6-15
    nonlinear, 1, 16-20
Orbitals, p_Z, 85
Organic conductors, 1040
Organic materials, 727
Organic metals, 1040
Ormecon™, 470
Oscillator strength for conduction band transitions, 106
Overhauser effect, 145, 148, 152, 159
Overoxidation, 974, 1022
Overpotentials, 885, 968
    anodic, 1019, 1022
    cathodic, 1019
    working, 1015, 1019
Oxadiazole, 824, 830
Oxidation (see also Reduction):
    charge consumed during, 1023
    controlled by conformational relaxation, 1019
    incorporation of counterions during, 1015, 1019
    molecular movement during, 1018
    of a microscopic element of volume, 1023
    reaction, 1019, 1022
    state, 1018, 1022
    volume expansion during, 1015, 1017, 1023
Oxidation potential, 259, 264, 272
Oxygen:
    adsorbed, 153
    contamination, 152, 157

PanAquas™, 929
Paracrystallinity, 89, 96
Parahexylphenyl, 847, 871
Particle size of polyaniline, 498, 507
Passivating oxides, 884, 886, 907, 909, 917
    ferric oxide, 884, 911, 916, 917
Passivation, 522, 523
    passivating oxide layer, 520
    reaction, 524
Passivation layers, 1029
Pauli paramagnetic susceptibility, 85, 88
    in polyaniline, 97
    in polypyrrole, 111, 112
$Pb_{0.3}Bi_{1.7}Sr_{1.6}Ca_{2.4}Cu_3O_{10}$, 1030, 1047, 1048
$Pb_{0.35}Bi_{1.85}Sr_{1.91}Ca_{2.1}Cu_{3.1}O_y$, 1030
Pd-catalysts, water-soluble, 217
Peierls distortion, 85
Peierls transition, 32
Penicillinase, 970
Percentage inhibitor efficiency, 899
Percolation, 471, 473, 487
    in PANI-CSA/PMMA blends, 69-78
    theory, 471, 489, 490, 511
Percolation transition, 87, 90
    and behavior of DC electrical conductivity, 90
    and behavior of optical frequency conductivity and dielectric function, 90, 91

[Percolation transition]
  consistency with small fraction of delocalized carriers in polyaniline, 103
  evidence for in optical conductivity of polyaniline, 100
  temperature dependence of in conducting polymers, 103, 104
Permeability, 1015
  diffusion component, 947, 948, 951, 953, 956
  gases, 946-956
  liquids, 948, 956-960
  mechanism, 946-947
  nitrogen, 951-956
  oxygen, 951-956
  solubility component, 947, 948, 953, 956
Perturbational molecular orbital method (PMO), 287
Pervaporation, 946-947, 948, 956-960
Phase-sensitive detector, 548
Phase separation, 492, 495, 506
Phase transition, 495, 497, 500
Phenanthro[9,10-c]thiophene, 292
Phonon-induced delocalization, 87, 118
  effect on localization length of, 89, 104
  relation to temperature dependent transport of, 89, 103, 104, 110
Phonons and dispersion curves, 773
Phonon scattering, 89, 90
  in relation to hopping, 92
  vibrational modes for polyaniline, 100
  vibrational modes for polypyrrole, 112
Phosphorescence, 7
Phosphorus-containing heterocycles, 269
Photoabsorption-detected magnetic resonance (see ADMR, ODMR)
Photoconduction, in PPV and derivatives, 353
Photoconductivity, 641, 737, 838-840
Photodiodes, 838-840
Photoelectron spectroscopy, 669, 832
Photoemission (see Photoelectron spectroscopy)
Photoemission experiments, 12
Photoexcited charges, 737
Photoinduced absorption experiments, 15
Photolithography, of PPV precursor polymer, 348
Photoluminescence, 368, 372, 374, 641, 644, 653, 654, 658, 664, 827-830, 835-838
  decay rate, 836-837, 848
Photomodulation, 641, 646, 654, 659
Photonics, 727
Photorefractive effect, 737
Photorefractive materials, 728, 736, 737
Photorefractive polymers, 737
  diffraction efficiencies, 737
  orientational enhancement, 737
  response time, 737
  two beam-coupling (TBC) gain, 737
Photosensitive oxidants, 934
Photovoltaic devices, 838-840
o-Phthalaldehyde, 295
Phthalic anhydride, 279, 295
Phthalide, 279
Phthalocyanines:
  antracenocyaninatometal-(AncM), 381
  bisisoquinoline complex PcRu(iqnl)$_2$ (iqnl=isoquinoline), 382
  constitutional isomers, separation, 382, 385
  crown ether (CEPcM), 386
    aggregation, 386
  ladder polymers, 401
    dienophilic, 403
    enophilic, 403
    precursor molecules, 401
    repetitive Diels-Alder, 401, 402
    structure, 401
    synthesis statistical, 403
  metallo-, 381
    1,2-naphthalocyaninato-(1,2-NcM), 381, 391
    2,3-naphthalocyaninato-(2,3-NcM), 381, 391
    phenanthrenocyaninato, 381
  Photoconductivity, 398
    bridged metal complexes, 399
      substituted, 399
    metal complexes, 399
  pigment, 381
  polymer, 381, 388
    alkynyl-bridged, 390
    bridged transition metal complexes, 391
      conductivity, 391, 392
      doping, 393
      stability, 393
    bridged trivalent transition metal complexes, 396, 397
    conducting, 389, 390
    fluorine-bridged, 390
    "intrinsic" semiconductive, 394, 395
      substituted, 395
    "shishkebab" [MacM(L)]$_n$, 391
  polyphthalocyaninatometalloxanes [PcMO]$_n$, 388
    doping method, 389
    electrochemical oxidation, 390
    main group, 389, 390
  properties, conductive, 381, 388, 389, 390
  properties, semiconductive, 388
    liquid crystalline behavior, 385
      conductivity, 386
      columnar mesophases, 385
      isotropic liquid (clearing), 385
      liquid discotic, 385
      transition, 385
    photodynamic cancer therapy, 388
    solubilities, 385
      water-soluble, 388
    stability, 381, 382
  structure of PcRu, 382
  substituted
    alkoxy, 384, 386, 393
    alkyl, 385, 386
    octa, 382
    -alkyl, 385
    optically active, 385
    peripherally, 382
    symmetrically, 386
    tetra, 382, 385

-tert-butyl-, 384
  unsubstituted, 381
  unsymmetrically substituted, 381, 388
 sub-, 381, 382
 super- (SPc's), 382
 synthesis, 381, 386, 396
  statistical condensation, 386
Pinned modes, as seen in absorption coefficient of polypyrrole, 113, 114
Pitting corrosion, 881, 902, 913
Pitting potential, 887
Plasma frequency:
 full carrier (W_{p1}), 91, 92, 117
  for polyaniline, 101, 102
   comparison of values from Drude and localization modified Drude models, 106
  for polypyrrole, 114, 115
 percolated (delocalized) carrier (W_p), 91, 117
 percolated (delocalized) carrier for polyaniline, 102
  small fraction of full carrier plasma frequency, 102, 103, 110, 114, 115, 116, 117
   scaling with the DC electrical conductivity, 103, 104, 115
    as an argument against the Anderson transition, 103
    for polyaniline, 103, 104
    for polypyrrole, 115
  presence at low temperature in samples with finite millikelvin conductivity, 103, 104, 110
Plot of W vs. T, 35
 for I-(CH)$_X$, 39, 45
 for K-(CH)$_X$, 38
 for PANI, 65
 for PANI-CSA, 36
 for PANI-CSA network, 74
 for polyalkylthiophenes (PATs), 64
 for PPy-PF$_6$, 40, 62
Pockels effect, 728, 737
Polarizability:
 linear, 4, 16-20
 nonlinear, 16-20
Polarization, 728, 885
 anodic, 885, 895
 bulk, 728
 cathodic, 885, 895
 concentration, 885
 molecular, 728
 nonlinear, 2
Polarization resistance, 889, 890, 906
Polarized infrared spectroscopy, 414, 415
Polarized luminescence, 864
Polaron, 1, 13, 128, 154, 643, 659-665, 825-826
 enhancement of nonlinearity, 748
 formation of in diphenylpolyenes, 750-751
 formation of in dithienylpolyenes, 750, 753
 isolated, 85, 97, 99, 102, 112, 113
 lattice, 97, 99, 113
 pairs, 640, 641, 644, 645, 663-665
 radical-cations, 748, 757-758
Polaron-excitons, 7, 10-12

 singlet, 1, 10-12
  nonradiative decay routes of, 3, 8
 triplet, 1, 8
Poled order, 733, 738, 739
Poled polymer, 731-733, 736-738
Poling, 728, 730, 731, 733, 734, 736-738
 field, 728, 738
Polyacene, 301, 370
 angular, 371-374
 calculations, 287
 containing heteroatoms, 287
 linear, 370, 376
Polyacetylene (see also (CH)$_X$), 278, 423, 467, 695, 669, 675, 709-710, 720, 963
 band gap in, 125
 block copolymers, 432
 bond length alternation in, 125
 bulk powder, film or gel, 423
 chemical repeat unit of, 86
 cis-polyacetylene, 769
 dimerization, 710
 Durham method, 205
 electrical conductivity of, 117, 197
 electrochemical synthesis of, 204
 films, 951
 frequency dependent dielectric function of, 117
 I-(CH)$_X$, 39, 42-49
 isomers, 709
 K-(CH)$_X$, 38-39, 49
 magnetic susceptibility (and density of states at the Fermi level) of, 117
 microwave dielectric function of, 117
 molecular structures, 198-199
 sterically-stabilized particles, 424
 soliton dynamics in, 123-139
 structure of, 772
 trans-polyacetylene, 769
  dynamics of, 773, 782-785
  structure, 772
  vibrational spectra, 775-782
Polyacrylamide, 1019
Polyadducts, Diels-Alder, 375, 376
Polyalkylbithiazoles, 703
Poly(alkylfluorene), 847, 848
Polyalkylthiophenes (PATs), 63-64
Poly 3-alkylthiophenes, 669, 678, 1017, 1032
 branched polyalkylthiophenes, 317
 chemical syntheses of, 318
 electrochemical synthesis of, 318
 interaction with aluminum, 682
 molecular weight of, 318
 solubility of, 318
Poly-1-aminoanthracene, 1035
Polyaniline, 65-68, 423, 467, 832, 855, 924, 925, 935, 938, 940, 969, 1030, 1032, 1035
 bulk powder, 423
 chemical preparations of, 94
 chemical repeat units of different oxidation states of, 86

[Polyaniline]
    comparison of transport models with experiment for, 105
    conductivity, 945
    DC electrical conductivity of, 97
    dielectric constant of, 511
    dopant size, 954
    dopant type, 954
    doping, 945, 946, 953
    doping level, 954
    electrochemical synthesis of particles, 432
    electrochemomechanical devices based on, 1017
    free-standing films, 946
    gas separation membranes, 951-954
    liquid separation membranes, 957-960
    magnetic susceptibility (and density of states at the Fermi level) of, 97
    microwave frequency dielectric function of, 109
    millikelvin electrical conductivity of, 98
    nanocomposites with silica, 428, 429, 430
    optical frequency conductivity of, 100
    optical frequency dielectric function of, 101
    particle size of, 498, 507
    polymer blends of, 493, 499, 500, 506, 510, 514, 515
    properties of, 1015
    reflectance spectrocopy of, 99
    solubility parameter of, 499
    sterically-stabilized particles, 426, 427, 428
    surface-functionalized nanocomposites, 430, 432
    surface tension of, 499
    surfactant-stabilized particles, 432
    volume changes in, 1015
    water soluble, 928
    X-ray diffraction of, 96
Polyaniline-Camphor Sulfonic acid (PANI-CSA):
    in critical regime, 36-38
    in insulating regime, 65-68
    in metallic regime, 51-53, 59
Polyaniline fiber formation, 442, 453-462
    effect of take up speed on conductivity, 454
    effect of take up speed on fiber modulus, 455
    solution properties, 454
        function of shear rate, 452-453
        intrinsic viscosities, 445
        loss (G'') and storage (G') modulus, 460-462
        normalized viscosities, 452
        solubility parameters, 443-444
        solution elasticity (*see* Loss and storage modulus)
    void spaces, effect on conductivity, 457
    void spaces, effect of tenacity, 456
Polyaniline fibers:
    conductivity of, 440-441, 462
    defects in, 456-458, 464
    effect of solution viscoelasticity, 460-463
    electrostatic spinning, 463
    from emeraldine base (EB), 453-459
    from emeraldine salt, 440-442
    from lecoemeraldine base (LEB), 462-463
    hollow fibers, 463
    mechanical properties of, 442-443, 462
    molecular weight, 462
        effect of oxidation state, 462
    rheology of spin solutions 441, 443-446
    X-ray (PANI EB) of, 453
Polyaniline gels, 446
    doping effects on, 450
    effect of shear, 445, 446
    films from, 448, 457
    moisture regain, 450
    properties of, 447-451
    rheology of gel forming solutions, 448
    stress/strain data, 449-450
    thixotrophy, 447
    time of gelation, 448
    yield stresses and elongation of, 449
Polyaniline networks (PANI-CSA/PMMA), 69-78
    transmission electron mocroscopy (TEM), 69-72
Polyaniline on textiles (*see also* Conductive textiles):
    from continuous two-step polymerization, 1000, 1001
    properties of (*see* Properties of conductive textiles)
    stability of (*see* Stability of conductive textiles)
    use of Vanandium pentoxide in, 1000
Poly(anthrothiophene-*alt*-bithiophene) calculations, 285
Poly(anthrothiophene-*alt*-thiophene) calculations, 285
Polyanthrothiophene calculations, 282
Poly(anthrothiophene methine) calculations, 286
Polyarenemethylidenes, 273
Poly(benzo[1,2-*c*:4,5-*c'*]bis(1,2,5-thiadiazole)-4,8-diyl-*alt*-bithiophene), 297
Poly(benzo[1,2-*c*]-1,2,5-thiadiazole-4,8-diyl-*alt*-bithiophene), 298
Poly(benzo[*c*]pyrrole) calculations, 285
Poly(benzo[*c*]thiophene) (*see* Polyisothianaphthene)
Poly(benzo[*c*]thiophene-*alt*-bithiophene), 295
Poly(5-benzoylbenzo[*c*]thiophene), 291
Polybenzvalene, isomerization of, 206
Poly(*trans*-bis(ethylenedioxythienylene vinylene), 303
Polybismaleimides, 734
Poly(2*H*,11*H*-bis[1,4]oxazino[3,2-*b*:3',2'-m]triphenodioxazine-3,12-diyl-2,11-diylidene-11,12-bis(methylidene)), 302
Poly(bis[1,2-*b*:5,6-*b'*]thieno-3,4,7,8-tetrahydrofulvene), 301
Poly[1,4-bis(2-thienyl)phenylene]s, 270
Poly(bithiophenequinodimethane) derivatives, 299
Poly(bithiophenequinodimethane-*alt*-bithiophene) and derivatives, 298
Polycations with ladder structure, 374
Polycondensation:
    Ni-catalyzed, 211
        termination, 211
    Pd-catalyzed, 212
Polycondensation of multifunctional monomers, 365
Polyco(thiophene-aniline)s, 268
Polyco(thiophene-furans), 268
Polyco(thiophene-thiazole)s, 268
Poly(5-cyanobenzo[*c*]thiophene), 291
Poly(4-(cyano(nonafluorobutylsulfonyl)vinylidene)-cyclopenta[1,2-*b*:4,3-*b'*]dithiophene), 301
Poly(cyclobutadiene-1,3-diyl) calculations, 286

Poly(4H,5H-cyclopenta[2,1-b:3,4-b']cyclopenta[2',1'-b":3,4-b"']trithiophene), 301
Poly(cyclopenta[2,1-b:3,4-b']dithiophene-4-one), 300
   calculations, 288
Poly(5-decylisothianaphthene), 290
Poly(4-decylthieno[3,4-b]thiophene), 285, 294
Polydiacetylene, 169, 180, 699-703, 730
   "band mapping", 688
   electroabsorption of, 184-186
   four wave mixing, 183
   NLO spectra of, 182-184
   two-photon spectra, 182
Poly(3,4-dialkoxythienylene vinylene), 303
Poly(5,6-dichlorobenzo[c]thiophene), 291
Poly(5,6-dicyanoisothianaphthene), 282
Poly(4-dicyanovinylidenecyclopenta[2,1-b:3,4-b']dipyrrole)
   calculations, 288
Poly(4-dicyanovinylidenecyclopenta[2,1-b:3,4-b']dithiophene)
   calculations, 288
Poly(5,6-diethylthieno[3,4-b]pyrazine), 293
Poly(E-1,2-(di-2-furyl)ethylene), 306
Polydiheptyl-p-phenylene), 669, 689
Poly(2,3-dihexylthieno[3,4-b]pyrazine), 283, 292, 293
Poly(5,6-dihexylthieno[3,4-b]pyrazine-alt-bithiophene), 296
Polydihydroisothianaphthene, 278, 279
Poly(1,6-dihydropyrazino[2,3-g]quinoxaline-2,3,8-triyl-7(2H)-ylidene-7,8-dimethylidene), 302
   derivatives, 302
Poly(1,3-dihydro-4,5,6,7-tetrafluorobenzo[c]thiophene), 292
Poly(4,4'-dimethoxybithiophene), 303
Poly(5,6-dimethoxyisothianaphthene), 282, 289
Poly(5,6-dimethylisothianaphthene), 282
Poly(5,6-dimethylthieno[3,4-b]pyrazine), 293
Poly(5,6-dimethylthieno[3,4-b]pyrazine-alt-bithiophene), 296
Poly(dioctyloxyphenylene vinylene), 865
Poly(5,6-dioxymethyleneisothianaphthene), 289
Poly(5,6-dioxypropylideneisothianaphthene), 289
Poly(4,6-di(2-pyrryl)thieno[3,4-c]-1,2,5-thiadiazole), 297
Poly(3,4-dioxyethylenethiophene), sterically-stabilized particles, 432
Poly3,4-disubstituted thiophenes:
   cyclopenta[c]thiophene, 315
   3,4-dialkylthiophenes, 315
   3,4-dialkoxythiophenes, 315
   3,4-ethylenedioxythiophene, 315
   3,4-ethylenedithiathiophene, 316
   steric factors in, 316
Poly(dithieno[3,4-b:3',4'-d]thiophene), 288, 294
   calculations, 288
Poly(E-1,2-(di-2-thienyl)ethylene), 301, 306
Poly(5,6-di-2-thienylthieno[3,4-b]pyrazine), 293
Poly(4,6-di(2-thienyl)thieno[3,4-c]-1,2,5-thiadiazole), 295
Poly(4-(dithioethylenevinylidene)cyclopenta[1,2-b:4,3-b']dithiophene) derivatives, 300
Poly(4-(dithioethylenevinylidene)cyclopenta[2,1-b:3,4-b']dithiophene) derivatives, 300
Poly(7-(dithioethylenevinylidene)cyclopenta[1,2-b:4,3-b']dithiophene) derivatives, 300

Poly(5,6-ditridecylthieno[3,4-b]pyrazine), 293
Poly(5,6-ditridecylthieno[3,4-b]pyrazine-alt-bithiophene), 296
Poly(5,6-diundecylthieno[3,4-b]pyrazine), 293
Poly(4-dodecylbenzo[c]thiophene), 290
Poly(5-dodecylbenzo[c]thiophene), 290
Polyelectrolite gel:
   bending motion in, 1016
   electrochemomechanical systems based on, 1017
   fiber, 1016
   swelling of, 1016, 1017
   volume of, 1016
Polyene limit, 17-19
Polyenes, 124-125
   acceptor-acceptor sustituted, 749
   bisanthracenyl substituted, 749, 751, 758-759
   bipolaron formation in, 749-752
   diphenyl substituted, 749-750
   dithienyl substituted, 750-753
      bipolaron formation in, 750-753
      preparation of, 752
      polaron formation in, 749
   donor-acceptor substituted, 749
   donor-donor substituted, 749
   polaron formation in, 749-751
   substituents effects in, 746
Poly(3-(2-ethanesulfonic acid)thiophene), 930
Poly(3-ethoxythienylene vinylene), 289, 303
Poly(4-ethylbenzo[c]thiophene), 290
Poly(4,4-ethylenedioxy-4H-cyclopenta[2,1-b;3,4-b']dithiophene), 301
Poly(3,4-ethylenedioxythiophene), 936, 940
Poly(ethylene oxide), 1035
Poly(4-fluorobenzo[c]thiophene), 291
Poly(5-fluorobenzo[c]thiophene), 291, 293
Polyfluorocarbon, 1030
Poly(furan-co-phenylenes), 272
Polyfurylpyrroles, 267
Poly(heteroarylene-methylene)s, 273
Poly(3-hexylthiophene), 934, 1030, 1035, 1040, 1045, 1048, 1049
   spray-coated film, 1039
Poly(3-n-hexylthiophene-co-3-n-octylthiophene), 262
Polyimides, 733, 734, 737
   blends with polyaniline, 954-955
   gas separation membranes, 949-951, 953
Polyisonaphthothiophene calculations, 282
Polyisothianaphthene (see also Poly(benzo[c]thiophene), 277-279, 281, 282, 289
   antistatic coating, 281
   aromatic form, 280
   band gap, 278
   calculations, 283
   dedoping, 278
   derivatives
      band gaps, 282
      band widths, 282
   ESR spectroscopy, 278, 280
   model compounds
      Raman spectra, 280

[Polyisothianaphthene, model compounds]
    solid-state CP/MAS ^{13}C NMR spectroscopy, 280
    p- and n-electrochemical doping, 280
    photochemical polymerization, 281
    quinonoid form, 280, 283
    resonance Raman spectroscopy, 280
    rotational barriers, 283
    solid-state CP/MAS ^{13}C NMR spectroscopy, 280
    X-ray photoelectron spectra, 280
Poly(isothianaphthene-*alt*-bithiophene), 286
    calculations, 285
Poly(isothianaphthene methine) calculations, 286
Poly(isothianaphthenequinodimethane-*alt*-bipyrrole), 299
Poly(isothianaphthenequinodimethane-*alt*-bithiophene) and derivatives, 299
Poly(isothianaphthene-*alt*-thiophene) calculations, 285
Polyisothionaphthene structure, 801
Polyisothionaphthene vibrational spectra, 801
Polymer blends, 487, 489, 500
    PAni/Polyester, 493, 500
    PAni/PVC, 499, 506
    PAni/PMMA, 514, 515
Polymer laser, 865, 872
Polymer matrix, 730, 732, 733
Polymer networks, 733, 734, 736
Polymer synthesis, precursor concept, 218
Polymers:
    aromatic main chain, 209
    conjugated, 363
    composites, 572
    conjugated, 532
    conjugated ladder (*see* Structures, conjugated ladder)
    copolymers, 572
    crystalline, 531
    decomposition, 1032
    doping, 1034
    growth, 1032
    growth mechanism, 1036-1039
    hydrophilic, 562
    hydrophobic, 562
    ladder (*see* Structures, ladder-type)
    miscellaneous, 573
    morphology, 1035, 1038, 1039
    nucleation, 1036, 1039
    redox, 531
    rigid-rod, 209
Polymer/polymer interface, 1018, 1019, 1021, 1022, 1023
Polymer/superconductor:
    bilayer, 1034
    composites, 1029
    electronic interactions, 1053
    structures, 1031
Poly(3-(2-(methacryloyloxy)ethyl)thiophene), 934
Polymethines, donor-acceptor substituted, 16-19
Poly(3-methoxythienylene vinylene), 289, 303
Poly(5-methylbenzo[*c*]thiophene), 290, 291
Poly(N-methylpyrrole), 970
Poly(methyl methacrylate), 733

Poly(3-methylthiophene), 969, 1030
    composites with latexes, 431
Poly(1-methylthiophenium cation) calculations, 289
Polynaphthopyrrole calculations, 285
Polynaphthothiophene calculations, 282, 283
Poly(naphthothiophene methine) calculations, 286
Poly(naphthothiophene-*alt*-bithiophene), 286
Poly(naphtho[2,3-*c*]thiophene), 292
Poly(naphtho[2,3-*c*]thiophene-*alt*-bithiophene), 295
Poly(5-octylaminobenzo[*c*]thiophene), 291
Poly(5-octylbenzo[*c*]thiophene), 290
Poly(5-octyloxybenzo[*c*]thiophene), 291
Poly(3-octylthiophene), 933
Poly(para-phenylene)s:
    conformation, 805
    derivatives, 213
    solubility, 211
        increase by the attachment of alkyl chains, 211, 213
    synthesis
        direct approach, 211
        indirect approach, 218
        precursor concept, 218
    vibrational spectra, 805
Polyparaphenylenevinylene (PPV), 59, 847
    ESR-ENDOR spectra, 807
    vibrational spectra, 807
Poly(peri-naphthalene), 301
    calculations, 287
Poly(phenanthro[9,10-*c*]thiophene), 292
Poly-o-phenetidine, 938, 939
Polyphenylene:
    conductivity, 619
    devices in, 630
    doping of, 611
    implantation in, 589
    spectroscopy, 594
    thermoelectric power, 622
Poly(*p*-phenylene), 278, 283, 828, 847, 848, 871, 1040
    derivatives, 828, 830, 835
    methylene-bridged, 365-368, 374
Polyphenylenethiophenes, 737
Poly(*p*-phenylene vinylene), 1, 10-13, 278, 824-840
    cyano derivatives, 823-834, 838-830
    oligomers, 6-10
    sulfonium precursor, 824
Polyphenylenevinylene and derivatives, 669, 676
    applications, 353
    conductivity, 349-50
    cross-linking, 350
    defects in, 348
    interaction with calcium, 679
    interaction with aluminum, 680
    interaction with sodium, 680
    mechanism of formation, 346-348
    nomenclature, 343
    solubility, 347
    sulfonium precursor, 346-348
    synthesis, 343-350

Poly(phenylphenylene vinylene), 872
Polypyrrole, 278, 423, 467, 695, 823, 832, 935, 936, 940, 967, 970, 1030, 1032-1036, 1038-1040, 1051
   bulk powder, 423
   chemical preparations for, 95
   chemical repeat unit of, 86
   composites with polystyrene latex, 431
   DC electric conductivity of, 112
   electrochemical synthesis of particles, 432
   electrochemomechanical devices based on, 1017, 1018, 1019, 1022
   gas separation membranes, 955
   liquid separation membranes, 956
   nucleation rates, 1039
   magnetic susceptibility (and density of states at the Fermi level) of, 111
   membranes, 1015
   microwave frequency dielectric function of, 116
   microwave frequency electrical conductivity of, 115
   nanocomposites with silica, 428, 429, 430
   nanocomposites with silica and magnetite, 430
   nanocomposites with tin(IV) oxide, 430
   optical frequency absorption coefficient of, 113
   optical frequency dielectric function of, 114
   optical frequency electrical conductivity of, 113
   properties of, 1015
   protected, N-BOC, 797-798
   reflectance spectroscopy of, 112
   simulation of electrochemical responses of, 1022
   sterically-stabilized particles, 424, 425, 426, 428, 432
   surface-functionalized nanocomposites, 430, 432
   surfactant-stabilized particles, 432
   X-ray diffraction of, 111
   vibrational spectra, 797-798
   volume changes in, 1015
Polypyrrole-PF_6 (PPy-PF_6):
   in critical regime, 40
   in insulating regime, 61-63
   in metallic regime, 53-58
Polypyrrole on textiles (*see also* Conductive textiles):
   adhesion of, 996, 1001
   adsorption to surface, 994, 995
   by decarboxylation of pyrrole-2-carboxylic acid, 995
   diffusion of oxygen into, 1004
   elemental analysis of, 996, 1004
   glass fabrics, 997
   morphology of, 998
   patent literature concerning, 996
   role of oligomers in polymerization, 994
   scanning electron micrographs of, 997
   two-step polymerization, 995
   X-ray photoelectron spectroscopy of, 996, 1004
Poly(pyrrole-benzo[b]thiophene)s, 267
Poly(pyrrole-co-arylene)s, 269
Poly(pyrrole-co-N-(3-bromophenyl)pyrrole), 260
Poly(pyrrole-co-N-methylpyrrole), 260
Poly(pyrrole-co-N-phenylpyrrole), 259
Poly(pyrrole-co-phenylene), 269
Poly(pyrrole methine), derivatives, 284

Poly(pyrrole methine), oxidized form, 284
Poly(*p*-quinodimethane-*alt*-*p*-phenylene) calculations, 287
Polysilanes, 700-701, 703
Poly(silanylene-co-thienylene), 273
Polysulfone gas separation membranes, 953
Poly(a-terthiophene), 268
Poly(terthienyl), 301
Poly(4,5,6,7-tetrafluorobenzo[c]thiophene), 291
Poly(3-thiabicyclo[3.2.0]cyclohepta-1,4,6-trien-2,4-diyl) calculations, 285
Poly(thieno[3,4-*b*]pyrazine), 292, 293
   calculations, 283, 284
   derivatives, quinonoid form, 293
   derivatives, Raman spectra, 293, 294
Poly(thieno[3,4-*b*]pyrazine-*alt*-bipyrrole), 297
Poly(thieno[3,4-*b*]pyrazine-*alt*-bithiophene), 296
Poly(thieno[3,4-*b*]pyrazino-1,2,5-thiadiazole-*alt*-bithiophene), 297
Poly(thieno[3,4-*b*]pyridine), 293
   calculations, 284
Poly(thieno[3,4-*b*]pyridine-*alt*-bithiophene), 296
Poly(thieno[3,4-*b*]quinoxaline) calculations, 283
Poly(thieno[3,4-*b*]thiophene) calculations, 285
Poly(thieno[3,4-*c*]pyridine) calculations, 284
Poly(thieno[3,4-*c*]thiophene) calculations, 282
Poly(thieno[3,4-*d*]pyridazine) calculations, 284
Poly(thienylene vinylene), 289, 303
Polythiophene, 1, 3, 13-15, 278, 282, 283, 639, 641, 662-664, 823, 828, 832, 848, 868, 935, 940, 1030, 1034
   chemical synthesis of, 312
   conductivity of, 312
   conformation, 798
   electrochemical synthesis of, 312
   electronic spectra, 798
   liquid separation membranes, 956
   properties of, 1015
   structure, 798
   vibrational spectra, 798-805
   volume changes in, 1015
Poly(thiophene methine) calculations, 286
Poly(thiophene methine) derivatives, 284
Poly(thiophene-1,1-dioxide) calculations, 289
Poly(thiophene-1-oxide) calculations, 289
Polythiophenes:
   poly(3-alkoxy-4-methylthiophene)s, 695-698, 701
   poly(3-alkylthiophene)s, 695-698
Polythiophenes containing ether groups:
   n-doping with alkali cations, 320
   electronic effects in, 320
   hydrophilicity of, 320
   ionic sensitivity of, 321
   oxidation potential of, 317, 320
   solvatochromism of, 321
Polythiophenes containing mesogenic groups, 326
Polythiophenes containing phenyl groups:
   optical properties of, 325
   steric effects in, 317
   substitution of the phenyl group, 324

Polythiophenes containing redox active groups:
   bipyridine, 323
   ferrocene, 324
   quinones, 324
   viologen, 324
   tetrathiafulvalene, 326
Poly(thiophenequinodimethane) derivatives, 300
Poly o-toluidine, 932
   (POT) fibers, 442
Polyurethane, 734
Poly(N-vinylcarbazole) (PVK), 737
Poly(vinyl cinnamate), 734
Poole-Frenkel effect, 875
Porous alumina membranes, 411
Potential:
   step, 536, 1034
   sweeps, 538
   Volta- , 553
Potentiometry, 966
Power law behavior, 32
   critical exponent (η), 35, 36
PPP-ladder polymers (see Poly(p-phenylene), methylene-ridged)
PPV (see Poly(p-phenylenevinylene)), 639
   oligomers of, 754-755
   oxidative doping of, 755
Pre-concentration, 972
Precursors, poly(n)acene, 375
Pressure dependence of conductivity:
   for I-$(CH)_X$, 39, 43
   for K-$(CH)_X$, 38
   for PPy-PF_6, 40
Prigogine, Y, 501
Process, synchronous, 363
Processing, 469, 484, 489
Properties:
   electrical,. 481, 490
   impact strength, 500
   mechanical, 477, 481
Properties of conductive textiles:
   dielectric, 1002, 1003
   electrical, 1007
   mechanical, 1000
   microwave, 1002
Protein, 969, 975
Protonation level, 155, 156
Proximity effects, 1029, 1031, 1042-1054
2,3-Pyrazinedicarboxylic anhydride, 293
Push-pull polyenes:
   and ECC theory, 789-794
   first-order hyperpolarizability, 811-815
   infrared spectra, 791
   Raman spectra, 790
   structure, 789-794
Pyrrole, 259-261
   copolymer with bithiophene, 266
   copolymer with N-alkyl substitution, 259
   copolymer with N-(2-ferrocenylethyl)pyrrole, 261
   copolymer with N-(4-ferrocenylethyl)pyrrole, 261
   copolymer with N-phenyl maleimide, 272
   copolymer with N-triphenyl-aminophenyl-substituted pyrrole, 261
   copolymer with 3-(pyrrol-1-yl)propanesulfonate, 261
   copolymer with thiophene, 266
Quantum cell models, 186ff
   dimer limit, 187
   Pariser-Parr-Pople, 167
   spin-charge separation, 187, 188
   sum rules, 190
   valence bond basis, 186
Quantum-chemical calculations, 1-22
Quantum efficiency:
   of light-emitting diodes, 825, 830-831, 834-838, 871
   of photodiodes, 838-840
   photoluminescence, 835-836
Quantum mechanical calculations, 281
Quasi-one-dimensionality:
   anisotropic Fermi surface due to, 89
   effect on localization within disordered regions of, 87, 88, 89
   in hopping
      for polyaniline, 97, 98
      for polypyrrole, 112
   long scattering times due to, 89, 91, 117
Quasiparticle, 1042
Quinoid character, 13
Quinonoid form, 281-283
Quioidal resonance structure, 226

Radical cations, 14
Radioisotope labelling, 553, 570
Raman polarizability, 778-782
   two-state model, 780
Raman spectra:
   of doped molecules, 768, 789
   of doped polyacetylene, 786, 788
   of isothionaphtene, 801
   of polyconjugated materials, 767
   of polyenes, 775-782
   of polyparaphenylene, 805
   of polyparaphenylenevinylene, 806-807
   of polypyrrole, 797-798
   of polythiophene, 799-800
   of pristine systems, 768
   of segmented polyacetylene, 787
   of *trans*-polyacetylene, 775-782
Random dimer model, 34
Random metallic network model, 88
   relation to inhomogeneous disorder model of, 89
Random resistor network model, 33, 34
Random walk, 142, 144
r coefficients, 732
Reactions, Diels-Alder, 375
Recognition
   biological, 975
   function, 963
   immunological methods of, 982
   molecular, 963, 971
   pattern, 970, 983

Recombination, 647-665, 834-835
  cross-, 654, 657
  distant pair, 650, 653
  geminate, 649, 653
  kinetics, 647
  nonradiative, 653
  radiative, 653
Redox catalysis (*see also* Mediation):
  electrocatalysis, 561, 564, 571
  by polymer films, 569, 571
Reduced activation energy, 34
Reduction (*see also* Oxidation), 1015, 1019
  and closure of the polymer structure, 1017
  decrease in volume during, 1015, 1023
  expulsion of counterions during, 1015, 1019
  state, 1018
Reference electrode, 533
Reflectance spectroscopy, 95, 96
  temperature dependent spectra for polyaniline, 99, 100
  temperature dependent spectra for polypyrrole, 112, 113
Refractive index, 728, 731, 737
Regiochemical structures, 69
Regioregular, 262
Regioirregular polythiophenes:
  regiochemistry of, 227-228
  synthesis of
    Curtis method, 227
    electrochemical method, 227
    $FeCl_3$ method, 227
    Kumada method, 226-227
Regioregular polythiophenes:
  ab initio calculations, oligothiophenes, 229-230
  chirality in, 253
  chromism
    ionochromism with Pb(II) ions, 252
    ionochromism with ammonium ions, 255
    thermochromism, 236-237
    solvatochromism, 251-252
  conducting polymer sensors, 252, 255
  copolymers of, 248-251
  electrical conductivity of, 247
  electrochemistry of, 247-249
  electroluminescence of, 248
  ion-binding properties (*see* Ionochromism)
  IR characterization of, 235-236
  light scattering studies, 240-241
  McCullough synthetic method, 230-231
  molecular mechanics calculations, oligothiophenes, 229-230
  morphology of, 240-245
  NMR characterization of, 232-235
  photoluminescence of, 248
  polyetheric substituents, 251-253
  Rieke synthetic method, 231
  self-assembly, 229-230. 255
  solid state structure of, 243-246
  solution studies
    melt annealing of, 241
    rod-coil transition of, 239
  thermochromism
  thioalkylpolythiophenes, 253-254
  UV/Vis experiments of, 236-239
  X-ray studies on self-assembled thin films, 243-247
Relaxation energy, 8
Relaxation process, 728, 738, 739
Relaxation times, 738, 739
Resistivity plots, 1034, 1043, 1050
Resists, 923
  conducting, 931, 933
    negative, 932
Resonances, 181-184
  three photon, 181, 182
  two photon, 183
  overlapping, 183
Retinal (*all-trans*), vibrational spectra, 815
Ring closure:
  polymer-analogous (*see* Cyclization, polymer-analogous)
  ring-opening polymerization of, 205
RNA, 975
Rod-like morphology, 89, 96
  effect on localization length of, 89
Rotator phase, 712
Route, Diels-Alder polyaddition, 375, 376
Route, synchronous, 375
Rylenes:
  hyperpolarizabilities, 809
  vibrational spectra, 809

Scanning electrochemical microscopy, 552
Scanning electron micrographs, 1035-1036
Scaling theory of localization, 31
Scattering times (for charge carriers):
  distribution in due to disorder, 108, 109, 116
    comparison of model with experiment for polyaniline, 108, 109
  long, 89, 91, 117
    due to quasi-one-dimensionality (and resulting anisotropic Fermi surface), 89, 103, 110
    for small fraction of carriers
      in polyaniline, 103, 110
      in polypyrrole, 115, 116
    temperature dependence of, in polyaniline, 104
      as an argument against the Anderson transition, 104, 117
      explained in terms of inhomogeneous disorder model/percolation, 104
  short, 87, 117
    form of optical frequency conductivity and dielectric function due to, 94, 105
    for majority of carriers,
      in polyaniline, 101, 106, 107
      in polypyrrole, 114, 115
        and lack of temperature dependence of, 115
        and relation to structural order, 114
    in relation to Anderson model and Ioffe Regel condition, 87, 92
Schottky barrier, 831-832
Second harmonic generation (SHG), 728, 730, 731, 736, 738
  Cerenkov, 736
  measurement, 731

Second hyperpolarizability, 728
Secondary ion mass spectroscopy, 600-605
Second-order nonlinear optical (NLO) polymers, 727, 730, 735, 737, 739
  advantages, 727
  cross-linked network, 735, 736, 739
  cross-linked systems, 733, 734, 739
  electro-optic (EO) coefficient, 728, 732, 739
  full-interpenetrating polymer networks (full-IPN), 736, 739
  guest-host systems, 732, 733, 738, 739
  main-chain polymer, 733
  nonlinear optical (NLO) coefficients, 730, 734
  second harmonic generation (SHG) coefficients, 728, 731, 738
  semi-interpenetrating polymer network (semi-IPN), 736
  side-chain polymers, 733, 735, 738, 739
Selection rules, 14
Self-assembled monolayers, 1037-1039, 1041
Self-assembly, 701, 730
Self-consistency of bond length and p-electron density, 124
Self-consistent reaction field (SCRF) theory, 17, 19-20
Self-doped polythiophenes:
  poly3-w-alkanesulfonatethiophenes, 319
  poly3-w-carboxyalkylthiophenes, 319
Self-doping materials, 260, 264,
  poly(pyrrole-co-pyrrole-N-alkylsulfonate)s, 260
Self-localization, 12
Semiquinoid character, 8
Sensitizers, 737
Sensor, 1016
  amperometric, 981
  baromechanical, 1017
  based on piezoelectric polymers, 1016
  based on polymer gels, 1016
  biological, 964
  chemical, 964
  conductimetric, 982
  definition of, 964
  devices, 980
  electromechanical, 1017
  electrochemomechanical, 1016, 1017
  gas, 971
  interrogation, 979
  ion-specific, 978
  models for small bending angles in, 1022
  muscle-like, 1016
  photomechanical, 1017
  polyacrylamide, 1019
  potentiometric, 981
  in robotics, 1017
  "shell-type" multilayer, 1019
  technology, 963, 979
  theory of swelling of crosslinked, 1023
  thermomechanical, 1017
  voltammetric, 982
Selectivity, 964, 968, 972, 975
  liquids, 956-957
  oxygen/nitrogen, 951-956
  water/carboxylic acids, 958-959

Separation factor, 948
Sexithienylene, 669, 675
Sexithiophene, 640, 641, 659-662
Shear rate, 503, 505
Sheet resistance (see Surface resistance)
Shielding, EMI, 526
Shifts in T_c, 1054
Shrinking
  of conducting polymers, 1018
  of polyelectrolite gels, 1016
  rate of, 1019
Simulation, computer, 500
Single charge rectification layers, 266
Single-component catalyst, 203
Single configuration interaction (SCI), 4, 6-7
Small band gap polythiophenes:
  bithiophenic precursors of, 333
  bridged precursors of, 333
  fused ring systems as precursors of, 332
  polyisothianaphthene, 332
  polythieno[3,4-b]pyrazine, 332
  small band gap oligomers, 335
  trimeric precursors of, 332, 333
$(SN)_x$, 1040
Sol gel, 735, 736
Solid-state effect, 145, 148, 152
Soliton, 1, 7, 85, 639, 825
  charged, 640, 654-659
  diffusion coefficient of, 153
  dynamics of, 123-139
  lattice, 32
  mobility of, 14
  neutral, 147, 148, 153, 154, 162, 640, 654-659
  step potential model of, 123-139
  trapped, 151
  trapping of, 152, 154
Solubility parameter, of polyaniline, 499
Solvatochromic response, 16
Solvent polarity, 16-20
SOMO, 642-645, 660
Space charge fields, 737
Space-charge-limited current, 831
Spectroscopy, 549
  ac impedance, 546
  infrared, 596-598
  in situ, 549
  photon correlation, 497
  Raman, 599
  studies, 570
Specificity, 969, 970
  biological, 977
  definition, 975
  methods to confer, 970, 975
Spin density, 150
Spin dynamics, effect of:
  chain defects on, 144
  chain end on, 143
  chain stretching on, 158, 159

disorder, 144
hydration, 158
polyacetylene, 147, 148, 162
doped polyacetylene, 154
cis (CH)x, 147, 148, 149
trans (CH)x, 147, 149-154
trans (CD)x, 150, 151
polyaniline, 155-159
polypyrrole, 159
polythiophene, polyalkylthiophene, 160, 161
Spin dynamics, temperature dependence of, 154, 157
Spin Hamiltonian, 650, 652
Spin relaxation:
    proton spin-lattice, 149, 158, 159, 161
    electron spin, 145, 146, 150
Spinless species, 15
Spin-orbit coupling, 7-8
Stability, 976, 977, 980
    of alkoxythiophenes, 800
    of polythiophene, 804
Stability of conductive polymers:
    kinetics of degradation (*see* Kinetics)
    polyaniline on textiles, 1005, 1006
    polypyrrole on textiles, 1002-1005
    role of oxidation in, 1004
State dipole moments, 18-19
Step potential model, 123-139
Stokes shift, 13
Stress:
    gradient, 1018, 1019, 1021, 1022, 1023
    mechanical, 1016, 1017, 1025
Structural (morpholological) order:
    homogeneous (*see also* Anderson transition), 87, 91
    inhomogeneous (*see also* Inhomogeneous disorder model, Percolation transition, and Random metallic network), 87, 88, 96, 108
Structure (*see also* Unit cells):
    alkali-metal-doped PA, 713
    alkali-metal-doped PPV, 714
    chain-axis, 715
    channel, 713
    complex forms, 722
    of dispersed phase, 491
    dissipative, 489, 500, 505, 506
    iodine-doped PA, 715
    iodine-doped P3AT's, 718
    ladder-type, 363, 374
    ladder-type, conjugated, 363, 374
    ladder-type, nonconjugated, 374
    ladder-type, with pyracyclene subunits, 376
    layered, 715
    poly(3-alkylthiophene) homopolymers, 717
    regioregular P3AT's, 721
    self-organizing, 500
    side-chain substituted, 717
    stepladder, 369-370
    thermal, 711
    thermotropic, 720

Styrene-acrylonitrile, 733
Substituent effects:
    acceptor-acceptor pairing, 745
    donor-acceptor pairing, 744-745
    donor-donor pairing, 744-745
    electron accepting groups, 744
    electron donating groups, 744
    enhancement of nonlinearity by, 744
Substrate, 535, 542
Sum-over-states (SOS), 4, 17-18
Superconductor, 1029-1058
    anisotropy, 1035, 1037
    coherence lengths, 1040, 1043-1045, 1053-1054
    corrosion, 1031
    electrode, 1035
    electrons pairs, 1042
    limitations, 1031
    /metal/insulator sandwiches, 1043
    microbridge, 1034-1035
    morphology, 1036, 1051
    proximity effects, 1029, 1031, 1042-1054
Supercurrent, 1031, 1043
    metals, 1043
    semiconductors, 1044
Surface morphology, 1039
Surface resistance (*see also* Properties of conductive polymers), 993, 994
Surface tension, 502
    of polyaniline, 499
Susceptibility, 728
    first-order, 728
    second-order, 728, 729, 736
    third-order, 728
Su Schriefer Heeger (SSH-) model, 640
    comparison with nearly free electron (NFE-) model, 125-126
    inconsistency in force constant within, 126
3-Substituted polythiophenes:
    3-cyclohexylthiophene, 317
    electronic and steric effects in, 317
    3-halothiophenes, 317
    3-isoamylthiophene, 317
    3-isobutylthiophene, 317
    3-isopropylthiophene, 317
    3-phenylthiophene, 317
Swelling:
    of conducting polymers, 1018, 1019, 1023
    electrochemical coefficient, 1023
    of polyelectrolyte gels, 1016
    theories of, 1023
Switch enzyme, 970
Switch, redox, 967, 969, 970
System:
    electrochemomechanical, 1017
    electrodriven chemomechanical, 1016
    mechanochemical, 1016
    piezoelectric, 1017

Tafel constants, 885, 897
Tafel plots, 886, 895

Template synthesized nanostructures:
   anatomy of,.415
   conduction mechanism in, 418,419
   enhanced conductivity in, 412-414
   molecular structure of, 416-418
   supermolecular structure of, 414-416
Temporal stability, 733-736, 738, 739
Tensor, 728, 729
   product, 728
Terthiophene, 264-266
   bridging of, 333
   effect of bridging on the electronic properties of, 336
   electrochemical properties of the polymers, 315, 333
   electropolymerization of, 313, 333
   as precursor of self-doped polymer, 319
   reactivity of, 313
   substitution of, 313, 319
Textiles (*see* Conductive textiles)
Thermionic emission, 859
Thermodynamics:
   equilibrium, 500
   irreversible, 501
   nonequilibrium, 500
Thermoelectric power, 605-608, 610
   of implanted polymers, 613-614, 622-628
Thermopower (S), 264, 513
   measurement, 28
   for PANI and PANI-CSA, 67-68
   for PANI-CSA networks, 77-78
   of polymer blends, 515
   for PPy-PF_6, 69
   theory, 67
Thienylene vinylene (PTV), oligomers of, 754-755
Thienylene vinylene (PTV), oxidative doping of, 755
Thin-film resistivity, 1033
Thin-layer cell, 539
Thiophthalic anhydride, 295
Three dimensionality, 87
   effect on electronic localization of, 87
   importance to avoid one-dimensional localization of, 88, 89
   in hopping, for polyaniline, 98
   in hopping, for polypyrrole, 112
Three-dimensional polythiophenes, tetraterthienylsilane as precursor of, 328
Time lag, 947-948
Track-etch membranes, 409
Transistors:
   analogy of, 969
   field effect, 632-634, 967, 983
   interrogation, 979
   microelectrochemical, 970
Transition dipole moments, 18-19
Transition energy, 18-19
Transition temperature, 1043, 1051
Transparent conductive coatings (*see also* Coatings), 1059-1067
   force field, 175-178
   huckel model, 175
   interactions in, 177-178
   processing trends, 1059
   reference force field, 171
   solitons in, 178-180
   vibrations in, 173-80
Transport mechanism, 511, 515
Transport phenomena, 506
Transport properties, 605-611
   of implanted polymers, 611-619
   temperature effect on, 619-628
Traps:
   grain boundaries, 862
   negative polarons, 862
   positive polarons, 862
Triethylammonium triflate, 927
Triphenylsulfonium hexafluoroantimonate, 931
Triplet powder pattern, 650-652
   full-field, 650-652, 658, 661
   half-field, 657, 661
TTF-TCNQ, 1040
Tubules, 411, 412, 415
Tunneling, 87, 516
   in polyaniline, 98
   in polypyrrole, 112
$2k_F$ phonons, 33

Ultraviolet photoelectron spectroscopy, 672
Unit cells, 707-708, 710
   chain-axis, 715
   P3AT's, 719

Valence band, 672
Valence Effective Hamiltonian (VEH) method, 13
Variable range hopping (VRH), 35-37, 59-67, 75
Versicon™, 937, 940
Viscoelasticity
   EB spin solutions, 460
   LEB spin solutions, 461
   theory of, 459-460
Vibrational modes, 4, 6-7
Vibronic spectra, of oligoenes, 775
Vibronic spectra, of oligothiophenes, 799
Vibronic structure, 1, 3, 10
Viscosity, melt, 473, 476, 498
Voltammtery, 966, 967, 971
   adsorptive stripping, 972
   anodic stripping, 972
   cathodic stripping, 972
   cyclic, 538
   differential pulse, 982
   dual-electrode, 551
   hydrodynamic, 540
   linear sweep, 538
   studies, 568
Voltammetric studies, 1032
Volta potential, 553, 560
Volume:
   change in, 1015, 1016, 1017, 1018, 1023, 1026
   decrease in, 1015

increase in, 1015
microscopic element of, 1023
of solvated ions, 1023
of the film, 1023
phase transition, 1016

Warburg impedance, 889, 890, 906
Wave function, 513
Waveguide, 736
Weak links, 1035, 1051, 1053
Weak localization, 41, 47-49
Williams-Landel-Ferry (WLF) expression, 738

Work function, 553, 853

X-ray photoelectron spectroscopy, 671
X-ray powder diffraction patterns, 1041

$YBa_2Cu_3O_7$-d, 1029-1058
   electrodes, 1032
   microbridges, 1035

Zero field splitting parameters, 650-652, 658, 661
Ziegler-Natta catalyst, 199-202
Zwitterionic resonance form, 17-19